HANDBOOK OF SOIL SCIENCES
RESOURCE MANAGEMENT and ENVIRONMENTAL IMPACTS

SECOND EDITION

Handbook of Soil Sciences

Handbook of Soil Sciences: Properties and Processes, Second Edition

Handbook of Soil Sciences: Resource Management and Environmental Impacts, Second Edition

HANDBOOK OF SOIL SCIENCES
RESOURCE MANAGEMENT and ENVIRONMENTAL IMPACTS
SECOND EDITION

Edited by
Pan Ming Huang
Yuncong Li
Malcolm E. Sumner

CRC Press
Taylor & Francis Group
Boca Raton   London   New York

CRC Press is an imprint of the
Taylor & Francis Group, an **informa** business

CRC Press
Taylor & Francis Group
6000 Broken Sound Parkway NW, Suite 300
Boca Raton, FL 33487-2742

© 2012 by Taylor & Francis Group, LLC
CRC Press is an imprint of Taylor & Francis Group, an Informa business

No claim to original U.S. Government works

Printed in the United States of America on acid-free paper
Version Date: 20110722

International Standard Book Number: 978-1-4398-0307-3 (Hardback)

This book contains information obtained from authentic and highly regarded sources. Reasonable efforts have been made to publish reliable data and information, but the author and publisher cannot assume responsibility for the validity of all materials or the consequences of their use. The authors and publishers have attempted to trace the copyright holders of all material reproduced in this publication and apologize to copyright holders if permission to publish in this form has not been obtained. If any copyright material has not been acknowledged please write and let us know so we may rectify in any future reprint.

Except as permitted under U.S. Copyright Law, no part of this book may be reprinted, reproduced, transmitted, or utilized in any form by any electronic, mechanical, or other means, now known or hereafter invented, including photocopying, microfilming, and recording, or in any information storage or retrieval system, without written permission from the publishers.

For permission to photocopy or use material electronically from this work, please access www.copyright.com (http://www.copyright.com/) or contact the Copyright Clearance Center, Inc. (CCC), 222 Rosewood Drive, Danvers, MA 01923, 978-750-8400. CCC is a not-for-profit organization that provides licenses and registration for a variety of users. For organizations that have been granted a photocopy license by the CCC, a separate system of payment has been arranged.

Trademark Notice: Product or corporate names may be trademarks or registered trademarks, and are used only for identification and explanation without intent to infringe.

Library of Congress Cataloging-in-Publication Data

Handbook of soil sciences : resource management and environmental impacts / editors, Pan Ming Huang, Yuncong Li, and Malcolm E. Sumner. -- 2nd ed.
  p. cm.
 Includes bibliographical references and index.
 ISBN 978-1-4398-0307-3 (alk. paper)
 1. Soil science--Handbooks, manuals, etc. 2. Soils--Environmental aspects--Handbooks, manuals, etc. I. Huang, P. M. II. Li, Yuncong. III. Sumner, M. E. (Malcolm E.), 1933- IV. Title.

S591.H232 2011
631.4--dc23   2011028351

Visit the Taylor & Francis Web site at
http://www.taylorandfrancis.com

and the CRC Press Web site at
http://www.crcpress.com

To

Pan Ming Huang (1934–2009)

Editor in Chief

Dr. Pan Ming Huang died on September 13, 2009, after a short illness. He was a towering leader in the field of soil science, and soil chemistry in particular, having served the profession for over 44 years formerly as professor of soil science at the University of Saskatchewan, Saskatoon, Saskatchewan, Canada. He was a dedicated teacher and author of over 300 journal articles published in the most prestigious journals. In the midst of the revision of the *Handbook of Soil Sciences*, having developed and approved the outline, set the standards for the contributions, and solicited the cooperation of all the associate editors, he unexpectedly fell ill. One of his strengths was his ability to involve, organize, and inspire colleagues to cooperate in pursuing scientific endeavors as was evidenced by his initial inputs into this work. Unfortunately, he passed away before being able to see any of the fruits of his labors, the completed manuscripts. Dr. Huang is survived by his wife, Yun Yin, who was gracious in the handover of all the materials required for us to complete the handbook in a timely manner. We dedicate this book to Dr. Huang as a lasting memorial to his many contributions to soil science.

Contents

Preface xi
Editors xiii
Associate Editors xv
Contributors xvii
Introduction xxi

Part I Soil Physical, Chemical, and Biological Interfacial Interactions

Introduction   *Pan Ming Huang (Deceased) and Antonio Violante* I-1

1 **The Role of Synchrotron Radiation in Elucidating the Biogeochemistry of Metal(loids) and Nutrients at Critical Zone Interfaces** 1-1
   Donald L. Sparks and Matthew Ginder-Vogel

2 **Clay–Organic Interactions in Soil Environments** 2-1
   Guodong Yuan and Benny K.G. Theng

3 **Nanoscale Science and Technology in Soil Science** 3-1
   Patricia A. Maurice

4 **Impacts of Environmental Nanoparticles on Chemical, Biological, and Hydrological Processes in Terrestrial Ecosystems** 4-1
   Nikolla P. Qafoku

5 **Enzymatic Activity as Influenced by Soil Mineral and Humic Colloids and Its Impact on Biogeochemical Processes** 5-1
   L. Gianfreda, M.A. Rao, and M. Mora

6 **Biogeochemical, Biophysical, and Biological Processes in the Rhizosphere** 6-1
   Philippe Hinsinger, Davey L. Jones, and Petra Marschner

7 **Mineralogical, Physicochemical, and Microbiological Controls on Soil Organic Matter Stabilization and Turnover** 7-1
   Ingrid Kögel-Knabner and Markus Kleber

8 **Impact of Soil Physical, Chemical, and Biological Interactions on the Transformation of Metals and Metalloids** 8-1
   Antonio Violante, M. Pigna, V. Cozzolino, and Pan Ming Huang (Deceased)

9 **Soil Physicochemical and Biological Interfacial Processes Governing the Fate of Anthropogenic Organic Pollutants** 9-1
   Kun Yang and Baoshan Xing

vii

10 Impact of Soil Physicochemical and Biological Reactions on Transport of Nutrients and Pollutants in the Critical Zone 10-1
Jon Chorover

Part II Soil Fertility and Plant Nutrition

Introduction *James J. Camberato* II-1

11 Bioavailability of N, P, K, Ca, Mg, S, Si, and Micronutrients 11-1
Nanthi Bolan, Ross Brennan, Dedik Budianta, James J. Camberato, Ravi Naidu, William L. Pan, Andrew Sharpley, Donald L. Sparks, and Malcolm E. Sumner

12 Soil Acidity and Liming 12-1
T. Jot Smyth

13 Soil Fertility Evaluation 13-1
J. Thomas Sims and Joshua McGrath

14 Fundamentals of Fertilizer Application 14-1
David B. Mengel and George W. Rehm

15 Nutrient and Water Use Efficiency 15-1
Kefyalew Girma and William R. Raun

16 Nutrient Interactions in Soil Fertility and Plant Nutrition 16-1
William L. Pan

Part III Interdisciplinary Aspects of Soil Science

Introduction *Guy J. Levy* III-1

17 Saline and Boron-Affected Soils 17-1
R. Keren

18 Sodicity 18-1
Guy J. Levy

19 Soil Water Repellency 19-1
Stefan H. Doerr and Richard A. Shakesby

20 Biogeochemistry of Wetlands 20-1
P.W. Inglett, K.R. Reddy, W.G. Harris, and E.M. D'Angelo

21 Acid Sulfate Soils 21-1
L.A. Sullivan, R.T. Bush, E.D. Burton, C.J. Ritsema, and M.E.F. van Mensvoort (Retired)

22 Water Erosion 22-1
Dino Torri and Lorenzo Borselli

23 Wind Erosion 23-1
D.W. Fryrear

24 Land Application of Wastes 24-1
David M. Miller and W.P. Miller

25 Conservation Tillage 25-1
Paul W. Unger and Humberto Blanco-Canqui

26 Soil Quality 26-1
Stephanie A. Ewing and Michael J. Singer

Part IV Soil Databases

Introduction *Marion F. Baumgardner* IV-1

27 Qualitative and Quantitative Aspects of World and Regional Soil Databases and Maps 27-1
    Freddy O. Nachtergaele, Vincent W.P. van Engelen, and Niels H. Batjes

28 United States Soil Survey Databases 28-1
    Jim R. Fortner and Alan B. Price

29 Integrated Digital, Spatial, and Attribute Databases for Soils in Brazil 29-1
    *Carlos Eduardo Pellegrino Cerri, Carlos Gustavo Tornquist, Martial Bernoux, Miguel Cooper, Gerd Sparovek,
    Maria de Lourdes Mendonça-Santos, and Carlos Clemente Cerri*

30 Development and Use of Soil Maps and Databases in China 30-1
    Gan-Lin Zhang and Yun-Jin Wu

31 Soil Geographic Database of Russia 31-1
    *Sergey A. Shoba, Vyacheslav A. Rozhkov, Irina O. Alyabina, Varvara M. Kolesnikova, Inga S. Urusevskaya,
    Erik N. Molchanov, Vladimir S. Stolbovoi, Boris V. Sheremet, and Dmitry E. Konyushkov*

32 Soil Databases in Africa 32-1
    D.G. Paterson and N.M. Mushia

33 Learning about Soil Resources with Digital Soil Maps 33-1
    Darrell G. Schulze, Phillip R. Owens, and George E. Van Scoyoc

Index **Index**-1

Preface

Handbook of Soil Sciences is the second edition of a comprehensive reference work on the discipline of soil science as practiced today. The new edition has been completely revised and rewritten to reflect the current state of knowledge. It contains definitive descriptions of each major area in the discipline, including its fundamental principles, appropriate methods to measure each property, many examples of the variations in properties in different soils throughout the world, and guidelines for the interpretation of the data for various applications (agricultural, engineering, and environmental).

This handbook assembles the core of knowledge from all fields encompassed within the discipline of soil science. It is a resource rich in data, which will provide professional soil scientists, agronomists, engineers, ecologists, biologists, naturalists, and students with their point of first entry into a particular aspect of soil science. The contributions serve those professionals seeking factual reference information on a particular aspect. The handbook provides a thorough understanding of soil science principles and practices based on a rigorous, complete, and up-to-date treatment of the subject matter compiled by the leaders in each field. In general, the following critical elements are present in each part: description of concepts and theories, definitions, approaches, methodologies and procedures, data in tabular and figure forms, and extensive references.

The handbook is organized into two books comprising nine parts, covering the six traditional areas of soil science together with a new part dealing with interfacial interactions among the physical, chemical, and biological regimes within the soil, a part on the interdisciplinary aspects, and a final part on databases. The two books are organized as follows:

Handbook of Soil Sciences: Properties and Processes

Part I: Soil Physics
Part II: Soil Chemistry
Part III: Soil Mineralogy
Part IV: Soil Biology and Biochemistry
Part V: Pedology

Handbook of Soil Sciences: Resource Management and Environmental Interactions

Part I: Soil Physical, Chemical, and Biological Interfacial Interactions
Part II: Soil Fertility and Plant Nutrition
Part III: Interdisciplinary Aspects of Soil Science
Part IV: Soil Databases

The subdivision of each part into a series of chapters was made by the associate editors and, in some cases, may appear to be somewhat arbitrary. The chapters have been arranged in such a way as to produce a thread running through each part. A complete table of contents, which is provided at the beginning of each book, gives a general outline of the scope of the covered subject material. In addition, a comprehensive subject index and an assemblage of units in common usage in soil science are provided at the end.

When the revision of this work was started, Dr. Pan Ming Huang, who was the associate editor for the soil chemistry part in the first edition, took on the role of editor in chief. Tragically and unexpectedly, he passed away in September 2009 in the midst of overseeing this expansive work. At short notice, the publisher approached us to complete this massive task to which we agreed in November 2009. Posthumously, we wish to recognize the vision, hard work, and dedication of our friend and colleague, Dr. Huang, in developing the revised outline of the book and soliciting the participation of all the associate editors and, in turn, the authors.

The chapters of this handbook have been written by many authors, all experts in their own fields, and peer-reviewed by independent reviewers. The nine parts have been carefully edited and integrated by the associate editors, all distinguished soil scientists in their own fields. This handbook is a tribute to the dedication of the authors, associate editors, and the publisher and its editorial associates. We wish to thank all the authors for their valuable contributions, the many nameless reviewers for their useful and helpful comments and criticisms, and the associate editors for all their hard work that they willingly contributed.

We also wish to recognize John Sulzycki who was highly effective in twisting our arms to pick up the threads and continue with the project after Dr. Huang's death, Randy Brehm who continued with the management of this project, and Jill Jurgensen, Vinithan Sethumadhavan, and Suganthi Thirunavukarasu who dealt with the minutiae of the editorial process. In addition, several research associates, graduate and undergraduate students, and visiting scientists, especially Helena Ren, Gaelan Jones, and David Li at the Tropical Research and Education Center, University of Florida, helped us in the editorial process for which we are duly grateful. Finally, we would like to thank our wives, Priscilla and Zhitong, for their patience, sacrifice, understanding, support, and encouragement without which this project would not have been possible.

Editors

Dr. Pan Ming Huang (deceased) was the professor emeritus of soil science at the University of Saskatchewan, Saskatoon, Saskatchewan, Canada, and served for 44 years in that institution. He received his BSc (1957) from the National Chung Hsing University, Taiwan; his MSc (1962) from the University of Manitoba; and his PhD (1966) from the University of Wisconsin, Madison, Wisconsin. His research work significantly advanced the frontiers of knowledge on the formation, chemistry, nature, and surface reactivity of mineral colloids, organic matter, and organomineral complexes in soils and sediments and their role in the dynamics, transformations, and fate of nutrients, toxic metals, and xenobiotics in terrestrial and aquatic environments. His research findings, embodied in well over 300 refereed scientific publications, are fundamental to the development of sound strategies for managing land and water resources in the Earth's critical zone. In addition to developing and teaching courses in soil physical chemistry and mineralogy, soil analytical chemistry, and ecological toxicology, he trained and inspired MSc and PhD students and postdoctoral fellows, and received visiting scientists from all over the world. He served on numerous national and international scientific and academic committees. He also served as a member of editorial boards, including the *Soil Science Society of America Journal*. He served the International Union of Pure and Applied Chemistry and the International Union of Soil Sciences in a number of capacities. He received the Distinguished Researcher Award from the University of Saskatchewan, the Soil Science Research Award from the Soil Science Society of America, the Distinguished Alumnus Award and the Chair Professorship Award from the National Chung Hsing University, and the Y. Q. Tang Chair Professorship Award from Zhejiang University. He was a fellow of the Canadian Society of Soil Science, the Soil Science Society of America, the American Society of Agronomy, the American Association for the Advancement of Science, and the World Innovation Foundation.

Dr. Yuncong Li is the University of Florida Research Foundation (UFRF) professor of soil science in the Department of Soil and Water Science at the Tropical Research and Education Center, Institute of Food and Agricultural Sciences (IFAS), University of Florida in Homestead, Florida. He received his BS (1982) in soil science and agricultural chemistry from the Shandong Agricultural University, China; his MS (1990) in agronomy from the University of Georgia; and his PhD (1993) in environmental science from the University of Maryland. He is also an affiliated professor at the University of Florida's Center for Tropical Agriculture, Hydrologic Sciences Academic Cluster, School of Natural Resources and Environment, and Water Institute, and a courtesy professor at the Shandong Agricultural University. He has received many awards and distinctions including a fellow of both the American Society of Agronomy and the Soil Science Society of America, Food and Agriculture Organization (FAO) Fellow, Wilson Popenoe Award (InterAmerican Society for Tropical Horticulture), Outstanding Paper Award (Florida State Horticultural Society), Jim App Award (University of Florida), Junior Faculty Research Award (Sigma Xi, the international honor society of science and engineering), Senior Faculty Award (Gamma Sigma Delta, the honor society of agriculture), Research Innovation Award (University of Florida), Wachovia Extension Professional Award (Extension Association of Florida), and Research Foundation Professor Award (University of Florida). His research and extension program focuses on water and soil quality monitoring, assessment and remediation, management practices to improve nutrient use efficiency, and nutrient cycling in soils/sediments. He has authored or coauthored over 160 research papers, 70 extension articles, and 15 book chapters. He recently edited a book, *Water Quality Concepts, Sampling, and Analyses* (Taylor & Francis). He serves as an associate editor for *Critical Reviews in Environmental Science and Technology* and *Communications in Soil Science and Plant Analysis*. Additionally, he has chaired or cochaired 20 graduate students, served as a committee member for 24 other graduate students, and supervised 15 postdoctoral fellows and many international visiting scientists.

Dr. Malcolm E. Sumner is the Regents' professor emeritus of environmental soil science in the Department of Crop and Soil Sciences at the University of Georgia, Athens, Georgia. He received his BSc Agriculture (1954) in chemistry and soil science and his MSc Agriculture (cum laude) (1957) in soil physics from the University of Natal, South Africa, and his DPhil (1961) in soil chemistry from the University of Oxford, Oxford, United Kingdom. His alma mater, the University of Natal, awarded him an honorary doctor of science degree in recognition of his career contributions to soil science. Before coming to the University of Georgia, he was professor and head of the Department of Soil Science and Agrometeorology at the University of Natal, South Africa. He has spent sabbatical leaves at the Agricultural University (Wageningen, the Netherlands), University of Missouri (Columbia), University of Wisconsin (Madison), University of Newcastle-upon-Tyne (United Kingdom), Vista University (South Africa), University of Adelaide (Australia), and the Commonwealth Scientific and Industrial Research Organization (CSIRO) (Australia). Over 90 graduate students and postdoctoral associates have studied under his guidance in South Africa and the United States where he taught courses in soil chemistry, soil physics, and soil fertility. In addition to being a fellow of both the American Society of Agronomy and the Soil Science Society of America, he holds the Agronomic Research Award and the Werner L. Nelson Award for Diagnosis of Yield Limiting Factors from the American Society of Agronomy and the Soil Science Research Award, International Soil Science Award, and the Soil Science Distinguished Service Award from the Soil Science Society of America. He is the holder of many other awards and distinctions including the Sir Frederick McMasters Visiting Fellowship (Australia) and the D. W. Brooks Faculty Awards for Excellence in Research and International Agriculture (University of Georgia). He has presented addresses at more than 100 universities and research institutions throughout the world on his research findings. He has also served as associate editor for the *Soil Science Society of America Journal* and is a member of editorial boards of many other journals. His published works cover a wide range of topics, including subsoil acidity, the agricultural uses of gypsum, diagnosis of yield-limiting factors, beneficial use of anthropogenic wastes, and transport of nutrients in soils. A widely respected author, Dr. Sumner's works include *Soil Acidity* (Springer-Verlag, 1991), *Soil Crusting: Chemical and Physical Processes* (Lewis Publishers, 1992), *Suelos de la Agroindustria Cafetalera de Guatemala* (University of Georgia, 1994), *Distribution, Properties and Management of Australian Sodic Soils* (CSIRO Publications, 1995), *Sodic Soils: Distribution, Properties, Management, and Environmental Consequences* (Oxford University Press, 1998), and the *Handbook of Soil Science* (CRC Press, 1999). He has authored or coauthored over 350 scientific papers, including 220 refereed journal articles, and has contributed chapters to over 30 books.

Associate Editors

Dr. Marion F. Baumgardner (Soil Databases) is a fellow of both the American Society of Agronomy and the Soil Science Society of America. He is professor emeritus of agronomy at Purdue University. He has received the International Soil Science Award and the Distinguished Service Award from the Soil Science Society of America. During his 39 years on the faculty at Purdue University, he focused his research interests on the quantitative relationships among the multispectral, physical, and chemical properties of soils and the application of aerospace remote sensing and spatial databases to the inventory and monitoring of soil resources at all scales from local to global. From the mid-1980s until his retirement in December 1997, he was active in research and development of the World Soils and Terrain Digital Database, known as SOTER. During the latter half of his career, his teaching included courses in soil and water conservation; soil, water, and air contamination; and remote sensing of terrestrial ecosystems. In retirement, he continues to be engaged in technical reviews and historical research and writing.

Dr. James J. Camberato (Soil Fertility) received his BS from the University of Massachusetts and his MS and PhD in soil science from North Carolina State University. He is currently associate professor of agronomy at Purdue University. He has conducted research and education programs in soil fertility and plant nutrition for 25 years in the Midwest and Southeast United States for agronomic crops and turfgrasses.

Dr. Pan Ming Huang (see Editors page)

Dr. Guy J. Levy (Interdisciplinary Aspects of Soil Science) is a senior soil scientist; since 1989 with the Institute of Soil, Water and Environmental Sciences, Agricultural Research Organization, Ministry of Agriculture and Rural Development, Israel. He obtained his BSc Agriculture (1981) and MSc Agriculture (1984) from the Hebrew University of Jerusalem, Jerusalem, Israel, and his DSc Agriculture (1988) from the University of Pretoria, Pretoria, South Africa. His main research interests include characterization of soil-structural stability and its improvement with the aid of soil amendments, and sustainable use of treated wastewater and the application of waste material in agriculture. He has authored or coauthored over 100 peer-reviewed scientific papers and book chapters.

Dr. Antonio Violante (Soil Physical, Chemical, and Biological Interfacial Interactions) is professor of soil chemistry at the University of Naples Federico II, Italy, where he received his academic degrees in chemistry. Additionally, he has completed postdoctoral training at the University of Wisconsin, Madison, Wisconsin, and the University of Saskatchewan, Saskatoon, Saskatchewan, Canada. He was head of the Dipartimento di Scienze Chimico-Agrarie and coordinator of the Doctoral School in Agrobiology and Agrochemistry. He taught courses in agricultural chemistry, soil chemistry, environmental biogeochemisty, and soil mineralogy, and has conducted research on the interface between soil chemistry, mineralogy, and soil biology. Dr. Violante's areas of research include the formation mechanisms of Al-hydroxides and oxyhydroxides, the surface chemistry and reactivities of short-range ordered precipitation products of Al and Fe, the influence of biomolecules on the sorption/desorption of nutrients and xenobiotics on/from variable charge minerals and soils, the factors that influence the sorption and residual activity of enzymes on variable charge minerals and organo-mineral complexes, and the chemistry of arsenic in soil environments. He has been the chairman and chief organizer of international and national congresses, and has authored 180 refereed research articles, book chapters, and invited reviews and coedited 7 books. He is a fellow of both the Soil Science Society of America and the American Society of Agronomy. He is also chair of the Commission 2.5 of International Union of Soil Societies (IUSS).

Contributors

Irina O. Alyabina
Institute of Ecological Soil Science
Moscow State University
Moscow, Russia

Niels H. Batjes
World Soil Information
International Soil Reference and
 Information Centre
Wageningen, the Netherlands

Marion F. Baumgardner
Department of Agronomy
Purdue University
West Lafayette, Indiana

Martial Bernoux
Institute of Research for Development
Montpellier, France

Humberto Blanco-Canqui
Western Kansas Agricultural Research
 Center
Kansas State University
Hays, Kansas

Nanthi Bolan
Centre for Environmental Risk
 Assessment Remediation
University of South Australia
Adelaide, South Australia, Australia

Lorenzo Borselli
Instituto de Geologia
Universitad Autonoma de San Luis Potosi
San Luis Potosi, S.L.P., Mexico

Ross Brennan
Department of Agriculture and Food
 Western Australia
Albany Regional Office
Albany, Western Australia

Dedik Budianta
Faculty of Agriculture
Soil Science Department
Sriwijaya University
Indralaya, South Sumatra, Indonesia

E.D. Burton
Southern Cross GeoScience
Southern Cross University
Lismore, New South Wales, Australia

R.T. Bush
Southern Cross GeoScience
Southern Cross University
Lismore, New South Wales, Australia

James J. Camberato
Department of Agronomy
Purdue University
West Lafayette, Indiana

Carlos Clemente Cerri
Center for Nuclear Energy in Agriculture
University of São Paulo
Piracicaba, São Paulo, Brazil

Carlos Eduardo Pellegrino Cerri
Soil Science Department
University of São Paulo
Piracicaba, São Paulo State, Brazil

Jon Chorover
Department of Soil, Water and
 Environmental Science
The University of Arizona
Tucson, Arizona

Miguel Cooper
Soil Science Department
University of São Paulo
Piracicaba, São Paulo, Brazil

V. Cozzolino
Department of Soil Science
Università degli Studi di Napoli
 Federico II
Naples, Italy

E.M. D'Angelo
Plant and Soil Sciences Department
University of Kentucky
Lexington, Kentucky

Stefan H. Doerr
Department of Geography
Swansea University
Swansea, Wales, United Kingdom

Vincent W.P. van Engelen
World Soil Information
International Soil Reference and
 Information Centre
Wageningen, the Netherlands

Stephanie A. Ewing
Department of Land Resources and
 Environmental Sciences
Montana State University
Bozeman, Montana

Jim R. Fortner
Natural Resources Conservation Service
National Soil Survey Center
United States Department of Agriculture
Lincoln, Nebraska

D.W. Fryrear
Custom Products and Consultants LLC
Big Spring, Texas

L. Gianfreda
Department of Soil Science
Università degli Studi di Napoli
 Federico II
Naples, Italy

Matthew Ginder-Vogel
Department of Plant and Soil Sciences
University of Delaware
Newark, Delaware

Kefyalew Girma
Department of Plant and Soil Sciences
Oklahoma State University
Stillwater, Oklahoma

W.G. Harris
Soil and Water Science Department
University of Florida
Gainesville, Florida

Philippe Hinsinger
Institut National de la Recherche
    Agronomique
Ecologie Fonctionnelle et Biogéochimie
    des Sols et Agro-Ecosystèmes
Montpellier, France

Pan Ming Huang (Deceased)
Department of Soil Science
University of Saskatchewan
Saskatoon, Saskatchewan, Canada

P.W. Inglett
Soil and Water Science Department
University of Florida
Gainesville, Florida

Davey L. Jones
Environment Centre Wales
Bangor University
Bangor, United Kingdom

R. Keren
Institute of Soil, Water and
    Environmental Sciences
Agricultural Research Organization
The Volcani Center
Bet-Dagan, Israel

Markus Kleber
Department of Crop and Soil Science
Oregon State University
Corvallis, Oregon

Ingrid Kögel-Knabner
Lehrstuhl für Bodenkunde
Technische Universität München
Freising-Weihenstephan, Germany

Varvara M. Kolesnikova
Institute of Ecological Soil Science
Moscow State University
Moscow, Russia

Dmitry E. Konyushkov
V.V. Dokuchaev Soil Science Institute
Moscow, Russia

Guy J. Levy
Institute of Soil, Water and
    Environmental Sciences
Agricultural Research Organization
The Volcani Center
Bet-Dagan, Israel

Yuncong Li
Department of Soil and Water Science
Tropical Research and Education
    Center
IFAS University of Florida
Homestead, Florida

Petra Marschner
School of Agriculture, Food and Wine
The University of Adelaide
Adelaide, South Australia, Australia

Patricia A. Maurice
Department of Civil Engineering
    and Geological Sciences
University of Notre Dame
Notre Dame, Indiana

Joshua McGrath
Department of Environmental Science &
    Technology
University of Maryland
College Park, Maryland

Maria de Lourdes Mendonça-Santos
EMPRABA-Centro Nacional de Pesquisa
    de Solos
Rio de Janeiro, Brazil

David B. Mengel
Department of Agronomy
Kansas State University
Manhattan, Kansas

M.E.F. van Mensvoort (Retired)
Laboratory of Soil Science and Geology
Wageningen Agricultural University
Wageningen, the Netherlands

David M. Miller
Department of Crop, Soil, and
    Environmental Sciences
University of Arkansas
Fayetteville, Arkansas

W.P. Miller
Crop and Soil Sciences Department
University of Georgia
Athens, Georgia

Erik N. Molchanov
V.V. Dokuchaev Soil Science Institute
Moscow, Russia

M. Mora
Department of Soil Science
Universidad de La Frontera
Temuco, Chile

N.M. Mushia
South African Agricultural Research
    Council
Institute for Soil, Climate and Water
Pretoria, South Africa

Freddy O. Nachtergaele
Land and Water Division
Food and Agriculture Organization
Rome, Italy

Ravi Naidu
Cooperative Research Centre for
    Contaminants Assessment and
    Remediation of the Environment
University of South Australia
Adelaide, South Australia, Australia

Phillip R. Owens
Agronomy Department
Purdue University
West Lafayette, Indiana

William L. Pan
Department of Crop and Soil Sciences
Washington State University
Pullman, Washington

Contributors

D.G. Paterson
South African Agricultural Research Council
Institute for Soil, Climate and Water
Pretoria, South Africa

M. Pigna
Department of Soil Science
Università degli Studi di Napoli Federico II
Naples, Italy

Alan B. Price
Natural Resources Conservation Service
National Soil Survey Center
United States Department of Agriculture
Lincoln, Nebraska

Nikolla P. Qafoku
Fundamental and Computational Sciences Division
Pacific Northwest National Laboratory
Richland, Washington

M.A. Rao
Department of Soil Science
Università degli Studi di Napoli Federico II
Naples, Italy

William R. Raun
Department of Plant and Soil Sciences
Oklahoma State University
Stillwater, Oklahoma

K.R. Reddy
Soil and Water Science Department
University of Florida
Gainesville, Florida

George W. Rehm
Department of Soil, Water and Climate
University of Minnesota
St. Paul, Minnesota

C.J. Ritsema
Winand Staring Centre
Wageningen, the Netherlands

Vyacheslav A. Rozhkov
V.V. Dokuchaev Soil Science Institute
Moscow, Russia

Darrell G. Schulze
Agronomy Department
Purdue University
West Lafayette, Indiana

George E. Van Scoyoc
Agronomy Department
Purdue University
West Lafayette, Indiana

Richard A. Shakesby
Department of Geography
Swansea University
Swansea, Wales, United Kingdom

Andrew Sharpley
Division of Agriculture
Department of Crop, Soil and Environmental Sciences
University of Arkansas
Fayetteville, Arkansas

Boris V. Sheremet
V.V. Dokuchaev Soil Science Institute
Moscow, Russia

Sergey A. Shoba
Faculty of Soil Science
Moscow State University
Moscow, Russia

J. Thomas Sims
Department of Plant and Soil Sciences
University of Delaware
Newark, Delaware

Michael J. Singer
Department of Land, Air and Water Resources
University of California
Davis, California

T. Jot Smyth
Department of Soil Science
North Carolina State University
Raleigh, North Carolina

Donald L. Sparks
Department of Plant and Soil Sciences
College of Agriculture and Natural Resources
University of Delaware
Newark, Delaware

Gerd Sparovek
Soil Science Department
University of São Paulo
Piracicaba, São Paulo, Brazil

Vladimir S. Stolbovoi
V.V. Dokuchaev Soil Science Institute
Moscow, Russia

L.A. Sullivan
Southern Cross GeoScience
Southern Cross University
Lismore, New South Wales, Australia

Malcolm E. Sumner
University of Georgia
Athens, Georgia

Benny K.G. Theng
Landcare Research
Palmerston North, New Zealand

Carlos Gustavo Tornquist
Soil Science Department
The Federal University of Rio Grande do Sul
Porto Alegre, Rio Grande do Sul, Brazil

Dino Torri
National Research Council
Research Institute for Geo-Hydrological Protection
Perugia, Italy

Paul W. Unger
Agricultural Research Service
United States Department of Agriculture
Amarillo, Texas

Inga S. Urusevskaya
Institute of Ecological Soil Science
Moscow State University
Moscow, Russia

Antonio Violante
Department of Soil Science
Università degli Studi di Napoli Federico II
Naples, Italy

Larry P. Wilding (Retired)
Department of Soil and Crop Sciences
Texas A&M University
College Station, Texas

Yun-Jin Wu
State Key Laboratory of Soils and
 Sustainable Agriculture
Institute of Soil Science
Chinese Academy of Sciences
Nanjing, China

Baoshan Xing
Department of Plant, Soil and Insect
 Sciences
University of Massachusetts
Amherst, Massachusetts

Kun Yang
Department of Environmental Science
Zhejiang University
Hangzhou, China

Guodong Yuan
Landcare Research
Palmerston North, New Zealand

Gan-Lin Zhang
State Key Laboratory of Soils and
 Sustainable Agriculture
Institute of Soil Science
Chinese Academy of Sciences
Nanjing, China

Introduction

Malcolm E. Sumner
University of Georgia

Yuncong Li
University of Florida

Larry P. Wilding (Retired)
Texas A&M University

What Is Soil?

Soil, the extremely thin but precious skin covering our planet, supports all terrestrial life forms and contributes nutrients to aquatic and marine environments. This covering of the unweathered and partially weathered geological formations at the Earth's surface is a unique, fragile veneer. No longer rock nor geological sediment, soil has been altered during the process of soil formation by geological, topographical, climatic, physical, chemical, and biological factors to form a living entity, which inextricably links inorganic or mineral particles with organic matter and biota bathed in a milieu of liquid water and gases. Water, the major constituent of all living entities, is the solvent and conveyor of nutrients, which, together with the solid phase, become the fertile substrate from which all planetary life springs. This nurturing and life-supporting pedosphere zone is a biologically active, porous, and structured medium that effectively integrates and dissipates mass fluxes and energy. In its pristine state, soil is a self-regulating, slowly evolving biogeochemical system that weathers with time, simulating sponge-like behavior and regulating and buffering nutrient supply and water quality and quantity for the growth of macro- and microflora and fauna. Furthermore, it determines the partitioning of water into surface and subterranean groundwater reservoirs.

In addition to its primary role in promoting, supporting, and sustaining all forms of life, soil also acts as a living filter for anthropogenic waste. Soil can recycle water through biochemically mediated processes that cleanse, purify, detoxify, and counteract most toxins and pathogens in polluted waters that would irreparably contaminate and degrade our environment. Despite being the repository of human and animal cadavers, including those from epidemics of pestilence and plague, all but a few vectors have been rendered harmless, and it is seldom, if ever, involved in the transmission of diseases. On the contrary, soil has been the source from which great antidotes to disease and infection, the antibiotics, have been developed.

In ancient times, soil and water degradation, resulting in unsustainable crop production often due to the accumulation of salts, led to the downfall of civilizations. Directly or indirectly, the soil resource impacts, undergirds, and transcends all of society's urban, industrial, and agrarian interests. Current local to global policies and issues on conservation and sustainability, land use, energy, environmental quality, taxation, and food, feed, and fiber production are derived largely from the quality and extent of the available soil resource.

Despite the clearly pivotal role that soil plays in supporting life on Earth, a precise definition of soil is elusive. At this stage, one can say that soil is an evolving, living organic/inorganic covering of the Earth's surface, which is in dynamic equilibrium with the atmosphere above, the biosphere within, and the geology below. Soil acts as an anchor for roots, a purveyor of water and nutrients, a residence for a vast and still largely unidentified community of microorganisms and animals, a sanitizer of the environment, and a source of raw materials for construction and manufacturing. Soil is an essential component of terrestrial ecosystems used by nations as long-term capital to grow and develop. Because this foundation is essential to the life forms within and on it, one requires a fundamental understanding of this elastic, porous, three-phase system (solid, liquid, gas), and its components, processes, and reactions in order to be able to effectively manage and exploit this vital resource. In the words of Roy Simonson "Soil resources are the earthen looms that shape the lives of the people. The more completely they are understood, the better can be the fabric of life woven on these earthen looms" (Wilding et al., 1984).

Every soil consists of one to several layers, called horizons, a few to hundreds of centimeters thick that reflect the physical, chemical, and biological processes that have taken place during its formation. Horizons are composed of natural aggregates called peds, which are made up of associations of mineral and organic particles. Peds and particles are often separated from each other by pores that vary widely in size and shape. In addition, individual peds and particles may be coated by materials such as clay, organic matter, sesquioxides, or precipitated salts. Although the internal structure of peds is not readily visible to the naked eye, their spatial arrangement involving particles and pores (soil architecture) greatly influences soil behavior because the organization is frequently systematic and related to macropore distribution. At the microscale, soil architecture governs soil water/solute movement and retention, soil structure/porosity, soil strength/failure, mineral synthesis/weathering, movement of toxic and nontoxic wastes, soil/root environments, root growth/proliferation, nutrient transfers, soil erosion, and oxidation/reduction reactions.

What Are the Soil Sciences?

Soil sciences are a spectrum of Earth sciences that address the importance of soils as very slowly renewable natural resources. They involve the study of soil formation, classification, and mapping; the physical, chemical, biological, and mineralogical properties of soil from microscopic to macroscopic scales of resolution; as well as the processes and behavior of soil systems and their use and management. Soils are integrative links between the atmosphere, biosphere, lithosphere, and hydrosphere. Soil sciences provide tools to integrate the components of Earth science systems, expanding knowledge of the causes and effects of spatial variability, and to take a more holistic approach to the dynamic processes affecting ecosystems.

Soil sciences focus primarily on near-surface processes that govern the quality and distribution of soil resources relative to landform evolution (geomorphology), geochemical environment (geochemistry), and organismal habitat (ecology/biology). Pedogenic processes are interactively conditioned by lithology, climate, and relief through geologic time. Soils are welded together into landscapes like chains; processes that perturb and impact higher topographic surfaces directly affect processes on adjacent surfaces below. Pedologists and other soil scientists study the energy flows and mass fluxes, which are the dynamic driving forces of pedogenesis through and over the three-dimensional soil system. They also quantify renewal and transfer vectors for biomass production, rainfall, and dusts, which counter constituent losses through drainage water, lateral interflow, and downslope migration or erosion products.

Soil sciences have their parentage in geology, chemistry, physics, and biology, but for the past 100 years have evolved as an independent, interdisciplinary body of knowledge with strong underpinnings to agriculture and, more recently, the environment. Because of an unparalleled success that soil sciences have enjoyed in helping to bring ample food, fiber, feed, and fuel to the world, the development of the basic soil sciences has evolved primarily as a by-product of research in agriculture, engineering, and the environment. There is growing evidence that the complexity of these problems requires a much broader approach to the science of the soil than can be stimulated by applied research alone. Soil sciences are taking their place alongside basic research efforts in the biosciences, geosciences, and atmospheric sciences to provide the reservoir of fundamental knowledge needed to develop lasting solutions to the challenges of balanced use and stewardship of the Earth. Despite the knowledge accumulated over the past century, Leonardo da Vinci's statement that "We know more about the movement of the celestial bodies than about the soil underfoot" is as true today as it was then. There is need for much more investigation of our most precious resource to stave off the fate that befell our ancestors in the ancient empires of Babylon, Egypt, China, Europe, and the Americas where soil degradation through nutrient loss, erosion, sodification, and salinization together with siltation of rivers and storage facilities laid waste their land and water resources. The ancients were constrained by their limited knowledge of processes driving the system, but we will be judged by generations to come based on our superior understanding of the soil system and the quality and stewardship of our land resources.

Modern parallels of soil and land degradation continue in both developing and industrialized countries. In addition to the problems facing the ancients that are still prevalent, chemical toxicities, irreversible land use conversion, environmental pollution, and desertification are the modern plagues. Sustainability of today's culture is threatened by loss of biodiversity, population growth, and land degradation, leading to enhanced greenhouse gas emissions culminating in global climate change. Land degradation is a complex technical, socioeconomic, and political issue without simple answers. Although soil scientists can provide technical solutions, without implementation as public policy, they are likely to be futile. Currently, public attention is focusing on soil, air, and water quality in efforts to maintain a clean environment. Soil sciences have a major role to play in this arena.

What Is the Purpose of This Handbook?

Handbook of Soil Sciences is a revision of the first comprehensive reference on the disciplines of soil science as practiced today. It contains definitive descriptions of each major area in the disciplines, including fundamental principles, appropriate methods to measure each property, many examples of the variations in properties in different soils throughout the world, and guidelines for the interpretation of the data for various applications (agricultural, engineering, environmental, biological, regulatory, educational, hydrological, biogeochemical). This handbook assembles the current core of knowledge from all fields encompassed within the soil sciences and provides a resource-rich database that will give professional soil scientists, agronomists, engineers, ecologists, biologists, naturalists, and students their point of first entry into a particular aspect of the soil sciences. The contributions also serve those professionals seeking specific, factual reference information. The handbook provides a thorough understanding of soil science principles and practices based on a rigorous, complete, and cutting-edge treatment of the subject compiled by leading scientists in each field. It is designed as a desk reference book.

This handbook has been extended to two books, *Handbook of Soil Sciences: Properties and Processes* comprises five parts dealing with soil physics, soil chemistry, soil mineralogy, soil biology and biochemistry, and pedology. *Handbook of Soil Sciences: Resource Management and Environmental Impacts* discusses soil physical, chemical, and biological interfacial interactions; soil fertility and plant nutrition; interdisciplinary aspects of soil science; and soil databases.

In the *Handbook of Soil Sciences: Properties and Processes*, the part on soil physics opens with a description of the basic physical properties of soils followed by a treatment of their dynamic properties in relation to tillage and disturbance by machinery. Soil water is discussed, including content/potential relationships, its movement and transfer to the atmosphere both directly and through the plant, and the co-transport of solutes in moving water. The nature of soil structure and its bearing on soil behavior are explored followed by an examination of gas movement under unsaturated conditions. The role of macropores as the superhighway for bypass transport of water and solutes is considered. Thermal regime and heat transfer are discussed. Finally, the heterogeneity of soils is described culminating in an evaluation of spatial variability.

The part on soil chemistry begins with a discussion of the nature and dynamics of soil organic matter. The importance of the soil solution, for transport of nutrients and catalyst for reactions in soil, is then highlighted followed by an evaluation and description of the kinetics of reactions in soils. The reactions taking place when oxygen becomes limiting are then explored followed by a detailed account of soil colloidal phenomena important in predicting the behavior of the finest soil fraction. Ion exchange, sorption, and precipitation reactions are quantitatively treated as a framework to evaluate the behavior of labile constituents in soils. Catalytic reactions promoted by soil and its constituents are examined followed by an account of the effects of acidity and alkalinity on soil reactions and behavior.

Mineralogy of soils is covered in the next part starting with a discussion of the formation and occurrence of minerals in soils followed by three parts in which structure, occurrence, identification, and properties of phyllosilicates, oxide minerals, and poorly crystalline aluminosilicates are discussed. The part on soil biology and biochemistry begins with a discussion of viruses, bacteria, fungi, cyanobacteria and algae, followed by the topic of soil fauna comprising protozoa, nematodes, micro- and macro-arthropods, enchytriads, and vertebrates. The nature, function, and life cycles of each are described. Then the processes mediated by these organisms, including nutrient transformations, are discussed. A major portion of the discussion is devoted to nitrogen transformations because of their importance in soils.

Following a discussion of the geomorphology of soil landscapes and the pedogenic processes and models involved in soil formation, descriptions of the systems used to classify soils are presented in the part on pedology as a framework in which to discuss the soil orders in soil taxonomy and other classification systems. The part continues with a discussion of land evaluation followed by new chapters on hydropedology and subaqueous soils. Thereafter, digital soil mapping and its applications are presented, followed by a discussion of soil change in the anthropocene in which the bridge between pedology, land use, and soil management is built.

In the *Handbook of Soil Sciences: Resource Management and Environmental Impacts*, the new part on soil physical, chemical, and biological interfacial interactions opens with a discussion of the methodology to study metal(oid)s and nutrients at critical zone interfaces followed by insights into the interactions between clay and organic materials in soils. Various aspects of the involvement of nanoparticles in chemical, biological, and hydrological processes are discussed, followed by the effects of mineral and organic colloids on enzymatic activity in soils and its impact on biogeochemical processes. Thereafter, the effects of biophysical, biogeochemical, and biological processes on rhizosphere processes and reactions are discussed. Subsequent chapters deal with mineralogical, chemical, physical, and (micro)biological interactions involved in the turnover and stabilization of organic matter and the transformations of metal(oid)s. The part concludes with a discussion of the physicochemical and biological interfacial processes governing the fate and transport of nutrients and pollutants in soils.

Bioavailability of macro- and micro-nutrients as well as their interactions are discussed in the part on soil fertility and plant nutrition prior to evaluating methods for estimating their potential availability to crops and methods of application as fertilizers. The causes of soil acidity and strategies for its amelioration are also outlined. Finally, the efficiencies of nutrient and water use are discussed. Thereafter, various interdisciplinary aspects of soil science such as salinity, sodicity, water repellency, wetland biogeochemistry, acid sulfate soils, soils and environmental quality, soil erosion by wind and water, land application of wastes, and soil quality and conservation tillage are discussed. Finally, the soil databases available to assess worldwide soil resources are presented and discussed.

Reference

Wilding, L.P., N.E. Smeck, and G.F. Hall (eds.). 1984. Pedogenesis and soil taxonomy. II. The soil orders: Developments in soil science 11B, p. v. Elsevier, Amsterdam, the Netherlands.

Soil Physical, Chemical, and Biological Interfacial Interactions

Pan Ming Huang (Deceased)
University of Saskatchewan

Antonio Violante
Università degli Studi di Napoli Federico II

1. **The Role of Synchrotron Radiation in Elucidating the Biogeochemistry of Metal(loids) and Nutrients at Critical Zone Interfaces** *Donald L. Sparks and Matthew Ginder-Vogel* 1-1
   Introduction • Molecular Environmental Soil Chemistry • Applications of Synchrotron-Based Techniques to Elucidate Soil Chemical Processes and Reactions • Future Needs • References

2. **Clay–Organic Interactions in Soil Environments** *Guodong Yuan and Benny K.G. Theng* 2-1
   Introduction • Structures, Properties, and Reactivity of Soil Clays and Organic Matter • Clay–Organic Interactions • Conclusions and Perspectives • References

3. **Nanoscale Science and Technology in Soil Science** *Patricia A. Maurice* 3-1
   Introduction: Nanoscience and Nanotechnology • Nanoscience Is Changing Our View of the Properties and Behaviors of Nanoparticles • Nanoparticle Mobility in Soils and Sediments • Nanotoxicity • Nanoenabled Sensing • Frontiers in Research in Nanoscale Particles and Processes in Soil Environments • Acknowledgments • References

4. **Impacts of Environmental Nanoparticles on Chemical, Biological, and Hydrological Processes in Terrestrial Ecosystems** *Nikolla P. Qafoku* 4-1
   Introduction • Interactions in the Binary System: Nanoparticles–Soil Solution • Nanoparticle Controls or Impacts on the Extent and Timescale of Soil-/Geo-Processes • Future Research Directions • Acknowledgments • References

5. **Enzymatic Activity as Influenced by Soil Mineral and Humic Colloids and Its Impact on Biogeochemical Processes** *L. Gianfreda, M.A. Rao, and M. Mora* 5-1
   Introduction • Soil Enzymes as Naturally Immobilized Enzymes • Natural and Synthetic Soil Enzymatic Complexes • Ecological Role of Enzyme–Soil Component Interactions: Carbon Turnover and Storage and Transformation of Nutrients • Interaction of Soil-Bound Enzymes with Anthropogenic Organics and Their Potential for Ecosystem Restoration • Conclusions and Future Prospects • References

6   **Biogeochemical, Biophysical, and Biological Processes in the Rhizosphere** *Philippe Hinsinger, Davey L. Jones, and Petra Marschner* 6-1
    Introduction • Rhizosphere Biochemistry • Rhizosphere Biophysics • Rhizosphere Biogeochemistry • Rhizosphere Ecology • Conclusions: A Unique, Multifaceted, and Complex Interface • References

7   **Mineralogical, Physicochemical, and Microbiological Controls on Soil Organic Matter Stabilization and Turnover** *Ingrid Kögel-Knabner and Markus Kleber* 7-1
    Introduction • Modeling of OC Turnover and the Concept of Different SOM Pools in Soils • Recalcitrance as an Index for Intrinsic Organic Matter Quality • Stabilization of SOM • Outlook • References

8   **Impact of Soil Physical, Chemical, and Biological Interactions on the Transformation of Metals and Metalloids** *Antonio Violante, M. Pigna, V. Cozzolino, and Pan Ming Huang (Deceased)* 8-1
    Introduction • Solution Complexation Reactions of Metals in Soils • Sorption/Desorption Processes of Metal(loid)s in Soils • Biomineralization of Metal(loid)s • Chemical Speciation and Bioavailability of Metal(loid)s as Influenced by Physicochemical and Biological Interfacial Interactions • Plant Uptake of Metal(loid)s as Influenced by Physicochemical and Biological Interfacial Interactions • Conclusions and Future Research Prospects • Acknowledgments • References

9   **Soil Physicochemical and Biological Interfacial Processes Governing the Fate of Anthropogenic Organic Pollutants** *Kun Yang and Baoshan Xing* 9-1
    Introduction • Sorption–Desorption Process in Soil–Water Interface • Uptake and Degradation by Plant and Soil Microbes • Effects of Surfactant on Soil Physicochemical and Biological Processes • Conclusions • Acknowledgments • References

10 **Impact of Soil Physicochemical and Biological Reactions on Transport of Nutrients and Pollutants in the Critical Zone** *Jon Chorover* 10-1
    Introduction • Biosynthesis and Humification • Coupled Biogeochemical Weathering Processes • Effects of Organo-Mineral Heteroaggregates on Nutrient and Pollutant Fate and Transport • Microbial Intervention in Metal Cycling • Conclusions and Future Prospects • Acknowledgments • References

In my view, the interfaces are the most underdeveloped areas of the sciences including Soil Sciences.

P.M. Huang

Scientific progress is based ultimately on unification rather than fragmentation of knowledge

F.C. Kafatos and T. Eisner

SOIL IS A DYNAMIC SYSTEM in which minerals constantly interact with organic matter and microorganisms. Minerals, organic components, and microorganisms profoundly affect the physical, chemical, and biological properties and processes of terrestrial systems. However, in soil environments physical, chemical, and biological processes are not independent processes but rather interactive processes. Soils can be defined as complex interactive biogeochemical reactors, reservoirs of organisms (mainly microorganisms), and the major compartment of the terrestrial ecosystems under the influence of anthropogenic activities. To improve our scientific knowledge on soil resources and also its application to remediation and long-term management, it is of major importance and interest to study soil organization and function not only through the subdisciplines of soil sciences but also through interactive approaches. The study of interactions between soil constituents and soil organisms has to be considered at different scales, namely, from the molecular level to field/landscape systems and is, indeed, essential to stimulating further research to uncover the dynamics and mechanisms of soil processes. Therefore, a new Commission 2.5 "Soil Physical/Chemical/Biological Interfacial Reactions" was created within the structure of the International Union of Soil Sciences in 2004. The major goal of Commission 2.5 is to better integrate the four subdisciplines of soil science (soil physics, soil chemistry, soil biology, and soil mineralogy) and to encourage interdisciplinary research. In fact, only by recognizing the complexity of terrestrial systems, it is possible to develop major advances in the understanding of problems related to soil. The fundamental interactive soil processes have enormous impacts on ecosystem productivity, services, integrity, and human welfare. Physical–chemical–biological interfacial interactions govern weathering transformations of minerals; storage and turnover of organic matter; enzymatic activities; the structure, dynamics, and activities of microorganisms; porosity formation and structural stability of aggregates; and biogeochemical transformations and transport of chemical and biological components at different temporal and spatial scales.

This part includes discussions on advances in the employment of synchrotron to elucidate interfacial reactions involving nutrients and pollutants (Chapter 1); in describing the interactions of clay-size minerals in soil with natural organic matter, and the effects of these interactions on some soil properties and functions (Chapter 2); in explaining what makes nanomaterials unique, in providing insight into the special behaviors of nanomaterials, and in highlighting how the soil science community can both benefit from and contribute to nanoscale environmental research (Chapter 3); in describing the impacts of environmental nanoparticles on chemical, biological, and hydrological processes in terrestrial ecosystems (Chapter 4); in explaining the effects of the soil components on the performance and efficiency of enzyme activities, the mechanisms involved in their interactions, the role of soil-bound enzymes from an ecological point of view, as well as their potential for practical purposes (Chapter 5). Furthermore, this part presents issues on the soil physical, chemical, and interfacial processes in the rhizosphere (Chapter 6) and soil mineralogical, physicochemical, and (micro)biological interactions on carbon turnover and sequestration (Chapter 7); on the transformation of metals and metalloids (Chapter 8); on the fate of anthropogenic organic pollutants (Chapter 9); and on the transport of nutrients and pollutants (Chapter 10).

I am grateful to the authors who have contributed chapters to this part. Appreciation is extended to the external referees who have provided invaluable critical inputs to maintain the quality of this part.

This part is dedicated to Dr. P.M. Huang. At the time of his death, he was the editor in chief of the *Handbook of Soil Science* and editor of this part. Dr. Huang leaves to us a great scientific patrimony. We must always remember, particularly the youngest soil scientists, his ideas and suggestions, which were often ahead of their times, in the years to come.

1
The Role of Synchrotron Radiation in Elucidating the Biogeochemistry of Metal(loids) and Nutrients at Critical Zone Interfaces

Donald L. Sparks
University of Delaware

Matthew Ginder-Vogel
University of Delaware

1.1 Introduction 1-1
1.2 Molecular Environmental Soil Chemistry 1-1
   Electromagnetic Spectrum of Light • Synchrotron Radiation • X-Ray Absorption Spectroscopy • Complementary/Alternative Synchrotron-Based Techniques
1.3 Applications of Synchrotron-Based Techniques to Elucidate Soil Chemical Processes and Reactions 1-8
   Adsorption of Metals and Oxyanions on Soil Components • Metal(loid) Surface Precipitation/Dissolution • Redox Reactions on Soil Components • Transition-Metal Oxides • Speciation of Metal(loids) in Soils • Speciation of Nutrients in Soils and Biosolids • Speciation of Metals in Hyperaccumulator Plants
1.4 Future Needs 1-15
References 1-17

1.1 Introduction

It has become increasingly recognized that to accurately predict and model fate/transport, toxicity, speciation (form of), bioavailability, and risk assessment of plant nutrients, toxic metals, oxyanions, radionuclides, and organic chemicals in the Earth's critical zone, we must have fundamental information at multiple spatial and temporal scales, and our research efforts must be multidisciplinary and interdisciplinary (Figure 1.1).

With the advent of state-of-the-art analytical techniques, particularly those that are synchrotron based, one can elucidate reaction mechanisms at small scales. This has been one of the major advances in soil and environmental sciences in the past two decades. The use of small-scale techniques in soil and environmental research has resulted in a new multidisciplinary field of study that soil scientists are actively involved in—molecular environmental science. Molecular environmental science can be defined as the study of the chemical and physical forms and distribution of contaminants in soils, sediments, waste materials, natural waters, and the atmosphere at the molecular level (Sparks, 2002).

There are a number of areas in environmental soil science where the application of molecular environmental science is propelling major frontiers. These include speciation of contaminants, which is essential for understanding release mechanisms, spatial resolution, chemical transformations, toxicity, bioavailability, and ultimate impacts on human health; mechanisms of microbial transformations on mineral surfaces; phytoremediation; humic substance structure and chemistry; air and terrestrial emanated particulate reactivity and composition; soil structure; development of predictive models; effective remediation and waste management strategies; and risk assessment (Sparks, 2002).

The objective of this chapter is to discuss aspects of synchrotron-based radiation, with emphasis on x-ray techniques, and show their application to investigations involving important soil chemical processes and reactions involving soil components and heterogeneous systems such as soils and biosolids. Emphasis is placed on investigations of elements that can be studied in the hard x-ray region of the electromagnetic spectrum.

1.2 Molecular Environmental Soil Chemistry

1.2.1 Electromagnetic Spectrum of Light

The use of synchrotron light to understand mechanisms of soil chemical reactions and processes has revolutionized the field of environmental soil science. The electromagnetic spectrum of light is shown in Figure 1.2. Electromagnetic radiation has both

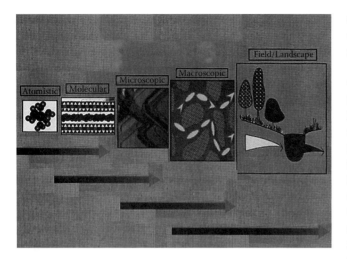

FIGURE 1.1 Illustration of the various spatial scales of interest to environmental scientists. (Reprinted with permission from Bertsch, P.M., and D.B. Hunter. 1998. Elucidating fundamental mechanisms in soil and environmental chemistry: The role of advanced analytical, spectroscopic, and microscopic methods, p. 103–122. *In* P.M. Huang (ed.) Future prospects for soil chemistry. SSSA, Madison, WI.)

particle and wave properties such that light at a particular wavelength corresponds to a particular scale of detection (O'Day, 1999). For example, longer wave radiation detects larger objects while shorter wave radiation detects smaller objects. Light employed to see an object must have a wavelength similar to the object's size. Light has wavelengths longer or shorter than visible light. On the longer side are radio waves, microwaves, and infrared radiation. Shorter wavelength light includes ultraviolet, x-rays, and gamma rays. The shorter the wavelength, the higher the frequency and the more energetic or intense the light. Light generated at shorter wavelengths such as x-rays is not visible to the human eye and must be detected via special means (Sparks, 2002).

Each region of the spectrum is characterized by a range of wavelengths and photon energies that will determine the degree to which light will penetrate and interact with matter. At wavelengths from 10^{-7} to 10^{-10} m, one can explore the atomic structure of solids, molecules, and biological structures. Atoms, molecules, proteins, chemical bond lengths, and minimum distances between atomic planes in crystals fall within this wavelength range and can be detected. The binding energies of many electrons in atoms, molecules, and biological systems fall in the range of photon energies between 10 and 10,000 eV. When absorbed by an atom, a photon causes an electron to separate from the atom or can cause the release or emission of other photons. By detecting and analyzing such electron or photon emissions, scientists can better understand the properties of a sample (Sparks, 2002).

1.2.2 Synchrotron Radiation

Intense light can be produced at a synchrotron facility. Synchrotron radiation is produced over a wide range of energies from the infrared region with energies less than 1 eV to the hard x-ray region with energies of 100 keV or more. There are more than 60 synchrotron light sources in over 30 countries (www.lightsources.org). In the United States, major facilities are found at National Laboratories. These include the National Synchrotron Light Source (NSLS) at Brookhaven National Laboratory, the Advanced Photon Source (APS) at Argonne National Laboratory (Figure 1.3), the Advanced Light Source

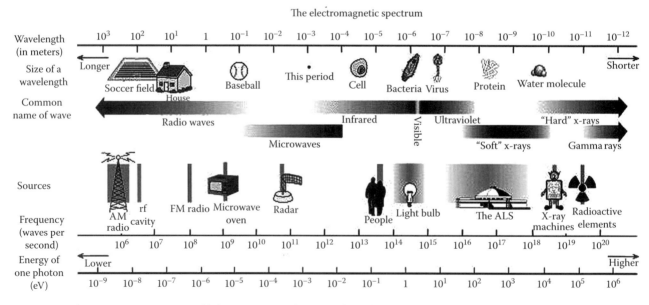

FIGURE 1.2 Electromagnetic spectrum of light covering a wide range of wavelengths and photon energies. Each region of the spectrum is characterized by a range of wavelengths and photon energies that will determine the degree to which light will penetrate and interact with matter. (Courtesy of the Advanced Light Source, Lawrence Berkeley National Laboratory, Berkeley, CA.)

FIGURE 1.3 The Advanced Photon Source (APS), Argonne National Laboratory managed and operated by The University of Chicago for the U.S. Department of Energy under Contract No. W-31-109-ENG-38. (Courtesy of the Advanced Photon Source, Argonne National Laboratory, Argonne, IL.)

(ALS) at Lawrence Berkeley National Laboratory, and the Stanford Synchrotron Radiation Laboratory (SSRL) at Stanford University (Sparks, 2002). For more information on synchrotron user facilities around the globe and their capabilities in the geosciences, the reader should consult Sutton (2006).

Synchrotrons are large machines (Figure 1.4). The APS has a storage ring that is 1104 m in circumference (Figure 1.3). In the synchrotron, charged particles, either electrons or positrons, are injected into a ring-shaped chamber maintained at an ultrahigh vacuum (UHV) (~10^{-9} Torr). The particles enter the ring by way of an injection magnet and then travel around the ring at or near the speed of light, steered by bending magnets. Additional magnets focus and shape the particle beam as it travels around the ring. Synchrotron radiation or light is emitted when the charged particles go through the bending magnets or through insertion devices, which are additional magnetic devices called wigglers or undulators inserted into straight sections of the ring. Beamlines allow the x-rays to enter experimental beamstations, which are shielded rooms that contain instrumentation for conducting scattering, spectroscopy, or imaging experiments (Schulze et al., 1999; Sparks, 2002; Brown et al., 2006).

Recently, several excellent reviews on the use of synchrotron techniques in the environmental sciences have been published. These reviews focused on the use of synchrotron techniques in low-temperature geochemistry and environmental science (Brown and Parks, 2001; Brown and Sturchio, 2002; Fenter et al., 2002), clay and soil science (Schulze et al., 1999; Kelly et al., 2008), and the study of heavy metals in the environment (Sparks, 2005a).

1.2.3 X-Ray Absorption Spectroscopy

One of the most widely used synchrotron-based spectroscopic techniques is x-ray absorption spectroscopy (XAS). XAS was developed in the early 1970s (Sayers et al., 1971). XAS can be used to study most elements in crystalline or noncrystalline solid, liquid, or gaseous states over a concentration range of a few parts per million to the pure element. It is also an in situ technique, which means that one can study reactions in the presence of water. This is a major advantage over many molecular scale techniques,

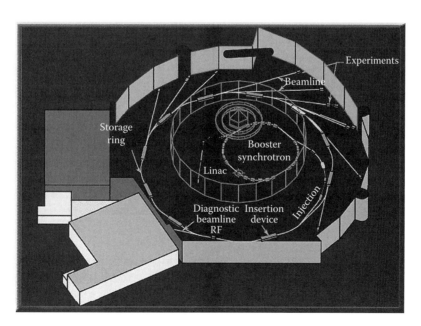

FIGURE 1.4 Schematic diagram of a synchrotron x-ray source. (Courtesy of the Advanced Light Source, Lawrence Berkeley National Laboratory, Berkeley, CA.)

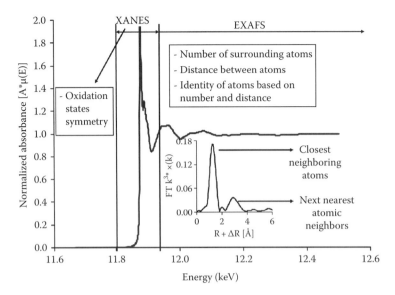

FIGURE 1.5 Energy regions of an XAS spectrum.

which are ex situ, often requiring drying of the sample material, placing it in a UHV, heating the sample, or employing particle bombardment. Such conditions can alter the sample, creating artifacts, and do not simulate most natural soil conditions.

One pitfall in the use of XAS and other synchrotron-based techniques, especially when employed at third- and fourth-generation light sources that have intense flux and brightness, is radiation damage. This can be especially problematic if one is investigating elements that are redox sensitive.

XAS is an element-specific method that yields information about the local structural and compositional environment of an absorbing atom. It "sees" only the two or three closest shells of neighbors around an absorbing atom (<6 Å) due to the short electron mean free path in most substances. Using XAS, one can ascertain important soil chemical information such as the oxidation state, information on next nearest neighbors, bond distances (accurate to ±0.02 Å), and coordination numbers (accurate to ±15%–20%) (Brown et al., 1995; Sparks, 2002).

An XAS experiment, which results in a spectrum (Figure 1.5), consists of exposing a sample to an incident monochromatic beam of synchrotron x-rays, scanned over a range of energies below and above the absorption edge (K, L, M) of the element of interest. When x-rays interact with matter, a number of processes can occur: X-ray scattering production of optical photons, production of photoelectrons and Auger electrons, production of fluorescence x-ray photons, and positron–electron pair production (Sparks, 2002).

The energy region extending from just below to about 50 eV above the absorption edge is the XANES (x-ray absorption near edge structure) portion of the spectrum (Figure 1.5). Fingerprint information (e.g., oxidation states) can be gleaned from this region of the XAS spectrum. The XANES region of the spectrum, although not providing as much quantitative information as the extended x-ray absorption fine structure (EXAFS) region, is often more intense and can provide qualitative or semiquantitative information on the oxidation state of the measured element (Brown et al., 1995). Such information can be obtained by comparing the features of the XANES spectrum of the sample with features of XANES spectra for well-characterized reference compounds (Figure 1.6). Some species, such as Cr,

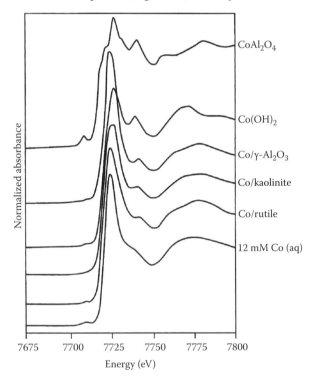

FIGURE 1.6 Co K-edge XANES spectra of $CoAl_2O_4$ spinel, crystalline $Co(OH)_2$, three samples with aqueous Co(II) sorbed on γ-Al_2O_3, kaolinite [$Si_4Al_4O_{10}(OH)_8$], and rutile (TiO_2), and a 12 mM aqueous $Co(NO_3)_2$ solution. (Reprinted by permission from Macmillan Publishers Ltd., Chisholm-Brause, C.J., P.A. O'Day, G.E. Brown, Jr., and G.A. Parks. 1990. Evidence for multinuclear metal-ion complexes at solid/water interfaces from X-ray absorption spectroscopy. Nature 348:528–530.)

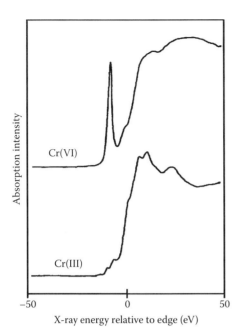

FIGURE 1.7 Comparison of XANES spectra for Cr(III) and Cr(VI). (Reprinted with permission from Fendorf, S.E., G.M. Lamble, M.G. Stapleton, M.J. Kelley, and D.L. Sparks. 1994. Mechanisms of chromium (III) sorption on silica: 1. Cr(III) surface structure derived by extended x-ray absorption fine structure spectroscopy. Environ. Sci. Technol. 28:284–289. Copyright (1994) American Chemical Society).

yield remarkably different, easily recognizable XANES spectra (Figure 1.7). In Figure 1.7, it is easy to differentiate Cr(III) from Cr(VI) as there is a prominent preedge feature for Cr(VI) that is absent for Cr(III) (Sparks, 2002).

The energy region from 50 to 1000 eV above the absorption edge is the EXAFS portion of the spectrum (Figure 1.5). Analysis of an EXAFS spectrum is illustrated in Figure 1.8 and involves extracting structural parameters including interatomic distances (R), coordination numbers (CN), and identity of first, second, and more distant shells of neighbors around an absorber (Brown et al., 1995; Sparks, 2002).

The fine structure refers to the small oscillations in the extended portion of the XAS spectra caused by the interaction between outgoing photoelectrons and those backscattered off atoms neighboring the central absorbing atom. In order to extract information from this region, mathematical procedures must be performed to isolate and amplify these subtle oscillations. The spectrum is first converted from electron volts to photoelectron wave numbers (k in Å^{-1}) based on a specified value for E_0, the threshold energy for x-ray absorption. The oscillatory fine structure data are then isolated by subtracting a background function representing one atomic x-ray absorption event. The resulting function [χ(k)] is often weighted, most commonly by the square or cube of the wave number [χ(k)*k^2 or χ(k)*k^3], to amplify the dampened oscillations occurring at higher k. These data can then be transformed using Fourier filtering methods to separate the contributions from the different atoms surrounding the absorbing atom, resulting in a radial structure function (RSF) with peaks representative of the location and distance of the neighboring atoms (McNear et al., 2005b). To derive accurate structural parameters, it is also necessary to obtain experimental EXAFS data for model or reference compounds that have known structures and contain the absorber and nearest neighbors' backscatters of interest.

XAS experiments can be conducted in several modes that differ in the type of detected particle: transmission (x-rays transmitted through the sample), fluorescence (fluorescent x-rays emitted due to absorption of the incident x-ray beam), or electron yield (emitted photons). In a transmission experiment, the incident (I_0) and transmitted (I_1) x-ray intensities are recorded as a function of increasing incident x-ray energy (E) to yield an absorption spectrum, which is plotted as ln (I_0/I_1) vs. E (in eV) (Figure 1.5). The relationship between these intensities and the linear absorption coefficient μ (in cm^{-1}) of a sample of thickness x (in cm) is ln (I_0/I_1) = μx. In a fluorescence experiment, the x-ray fluorescence from the sample, I_f, can be measured and ratioed with I_0 as I_f/I_0, which is proportional to μ for dilute samples. Fluorescence methods are preferred for elements that may be contained in low concentrations on mineral surfaces (Brown et al., 1995; Sparks, 2002).

Additional detail on XAS principles, methodology, sample preparation, and data analyses can be found in a number of excellent sources (Brown, 1990; Manceau et al., 1992; Fendorf et al., 1994; Schulze and Bertsch, 1995; O'Day, 1999; Bertsch and Hunter, 2001; Brown and Parks, 2001; Brown and Sturchio, 2002; Fenter et al., 2002; Kelly et al., 2008).

1.2.4 Complementary/Alternative Synchrotron-Based Techniques

Soil scientists have recently begun employing variations on traditional bulk XAS including bulk and microfocused synchrotron x-ray diffraction (S-XRD and μ-XRD, respectively) and microfocused-XANES and EXAFS spectroscopies (μ-XANES and μ-EXAFS, respectively). Microfocused diffraction and spectroscopic techniques are often coupled with microfocused x-ray fluorescence (μ-XRF) mapping to determine the areas of interest to be analyzed (Brown and Sturchio, 2002; Ginder-Vogel and Sparks, 2010). Another variation is quick-scanning XAS, which allows the collection of a complete XANES or EXAFS spectrum in as little as 300 ms (Dent, 2002).

1.2.4.1 Synchrotron X-Ray Diffraction

X-ray diffraction (XRD) is one of the most fundamental techniques used in the characterization and identification of crystalline materials in soils and sediments. Conventional (i.e., lab-based) XRD uses monochromatic x-rays generated when high voltage is applied to an anode, generally Al or Cu. Conventional XRD is generally limited to crystalline materials; additionally, conventional XRD is time-consuming and requires a large sample volume. The low flux of lab-based x-ray sources generally limits the geometry of the diffractometer, which requires data collection of reflected x-rays, rather than

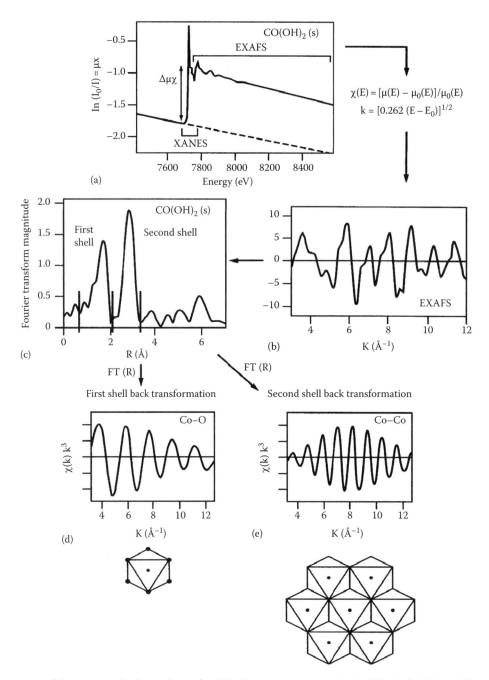

FIGURE 1.8 Illustration of the steps involved in analysis of an EXAFS spectrum, using the Co K-XAS of solid Co(OH)$_2$ as an example. (a,b) Subtraction of background (dashed line in EXAFS spectrum) yields the EXAFS function ($\chi(k) \times k^3$) vs. k (Å^{-1}). (c) Fourier transformation of the background-subtracted EXAFS function yields an RSF (Fourier transform magnitude vs. R (Å)), which contains peaks at different distances associated with various shells of neighboring atoms around Co. (d) Fourier filtering of the different shells and back Fourier transformation yield the frequency contribution of each neighboring shell of atoms around Co. In this example, the first peak (d) in the RSF is caused by backscattering of the photoelectron from six nearest-neighbor O atoms at 2.08 Å from Co and the second peak (e) by backscattering from six second neighbor Co atoms at 3.17 Å from the central Co atom. (Reprinted with kind permission from Springer Science+Business Media: Brown, G.E., Jr., G.A. Parks, and P.A. O'Day. 1995. Sorption at mineral-water interfaces: Macroscopic and microscopic perspectives, p. 129–183. *In* D.J. Vaughan and R.A.D. Pattrick (eds.) Mineral surfaces. Chapman & Hall, London, U.K.)

transmitted x-rays, making analysis of hydrated samples quite difficult. However, S-XRD circumvents many of the limitations of conventional x-ray diffraction. It offers exceptional resolution and sensitivity, even on very small samples. This permits the identification and quantification of trace phases, which are not possible using conventional x-ray sources. Additionally, amorphous materials, thin films, and hydrated samples can be analyzed, due to the high flux of synchrotron x-ray sources and specialized instrumental configuration. With the use of area detectors or CCD cameras, similar to protein crystallography beamlines, it is possible to collect a single XRD pattern in a matter of seconds, allowing for the monitoring of reactions in specially constructed XRD mounts. Finally, S-XRD data can be collected over a very large angular range, which allows analysis of the data using Rietveld refinements. These refinements can then be used to determine unit cell parameters and site occupancy (Ginder-Vogel and Sparks, 2010).

1.2.4.2 Microfocused XAS and XRF

With the advent of third-generation synchrotron light sources that have higher flux and higher brightness x-rays, together with state-of-the-art detectors and better beamline optics, microfocused beams for spectromicroscopy and imaging are available to study heterogeneous natural systems such as soils, biosolids, and plants. With these techniques, one can study areas of square microns and focus the beam to spatial resolutions of less than 5 μm, enabling one to directly speciate contaminants in heterogeneous materials. This has been a major advance in the past 10 years, and coupled with μ-XRF, one can also determine the distribution and association of metals and oxyanions in complex natural systems. Although bulk XAS can and should be used to provide the major species that are found in such systems, one often needs to couple these studies with μ-XAFS and μ-XRF investigations to more precisely quantify major species and determine associations and distributions of elements over small areas. There are several disadvantages in using bulk XAS to speciate contaminants in heterogeneous systems such as soils. Soils are complex and contain an array of inorganic and organic components, including humic substances, clay minerals, metal hydr(oxides), macro- and micropores, and microorganisms, all closely associated with each other. In such systems,

some microenvironments contain isolated phases in higher concentrations relative to the average in the total (bulk) matrix. For example, the oxides, minerals, and microorganisms in the rhizosphere have quite a different chemical environment compared to the bulk soil. These phases may be very reactive and important in the partitioning of metals but may be overlooked using other analytical techniques that measure the average of all phases. Standard bulk XAS techniques probe an area of several square millimeters and provide information on the average local chemical environment. Consequently, where more than one type of surface species is present, bulk XAS will detect only the primary (or average) type of surface product/species in the bulk sample. However, the most reactive sites in soils have particle sizes in the micrometer range, and metal speciation may vary over regions of a few 100 μm². The existence of multiple species in soils results in overlapping atomic shells, and it is difficult to ascertain the precise metal speciation with bulk XAS. Minor metal-bearing phases, even though they may constitute the most reactive or significant species, may not be successfully detected with bulk analyses (McNear et al., 2005b).

Some excellent reviews have appeared on the application of μ-XAS and μ-XRF to speciate contaminants in heterogeneous systems (Bertsch and Hunter, 1998; Hunter and Bertsch, 1998; Bertsch and Seaman, 1999; Manceau et al., 2002). A number of investigations have appeared over the past 10 years on the speciation of metals and oxyanions in soils, plants, biosolids, and fly ash (see references in Brown and Sturchio, 2002; Ginder-Vogel and Sparks, 2010).

A typical μ-XAS experiment (Figure 1.9) begins by collecting x-ray fluorescence elemental maps from which metal correlations can be observed. From these maps, regions of interest

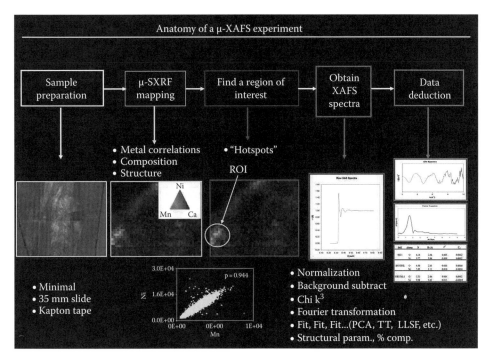

FIGURE 1.9 Anatomy of a μ-XAFS experiment. (Reprinted from McNear, D.H., Jr., R. Tappero, and D.L. Sparks. 2005b. Shining light on metals in the environment. Elements 1:211–216. With permission from Mineralogical Society of America.)

(i.e., hotspots) are selected and further probed using the microfocused x-ray beam. Analysis of XAS spectra collected from multicomponent systems cannot rely on traditional fitting procedures in which the spectra are broken down into individual atomic shells. Thus, to determine the species present within a mixed system, a dataset of spectra from multiple spots throughout a sample are analyzed statistically using principal component analysis (PCA). The PCA technique determines if the dataset can be described as weighted sums of a smaller number of components, which would be the case if each spot in the dataset comprises a smaller number of distinct compounds. Target transformation (TT) is then used to identify the components by taking a spectrum of a known reference compound and mathematically removing from the spectrum anything that does not look like the principal components identified by PCA. If minimal information has to be removed from the known reference spectrum, then one can conclude it is most likely present in the sample. After the contributing standard phases are identified, linear least squares fitting (LLSF) is used to determine the amount (%) of each standard species within the individual sample spectrum making up the dataset. The accuracy of this fitting approach is dependent on the data quality, the completeness of the standard dataset, and the range over which the data were fit. Additional and more thorough discussions on μ-XAS applications and methods of data collection and analysis for contaminants in heterogeneous systems can be found in Bertsch and Hunter (2001), Manceau et al. (2002), and McNear et al. (2005b).

1.2.4.3 Quick XAS

In soil environments, chemical reactions at the mineral/water interface occur over a range of temporal scales, ranging from microseconds to years. Many important processes (e.g., adsorption, oxidation-reduction, precipitation) occurring at mineral surfaces are characterized by a rapid initial reaction on a timescale of milliseconds to minutes (Scheidegger and Sparks, 1996). Knowledge of these initial reaction rates is critical to determining chemical kinetic rate constants and reaction mechanisms, both of which are necessary to understand environmental chemical processes. Kinetic measurements, using traditional techniques, such as batch or stirred-flow techniques, typically yield only a few data points during the initial phases of the reaction and cannot capture important reaction rates on second and slower timescales. Chemical relaxation techniques such as pressure jump (p-jump) and concentration jump (c-jump such as stopped flow) allow rapid data collection on timescales of milliseconds. However, rate "constants" are calculated from linearized rate equations that include parameters that were determined from equilibrium and modeling studies. Consequently, the rate "constants" are not directly determined (Ginder-Vogel et al., 2009; Ginder-Vogel and Sparks, 2010).

Direct, in situ, molecular-scale measurement of rapid reactions has been quite limited until recently. Fendorf et al. (1993) used stop-flow electron paramagnetic resonance spectroscopy to measure Mn(II) sorption to birnessite (δ-MnO_2) on a timescale of milliseconds. More recently, Parikh et al. (2008) used in situ, Fourier Transform infrared (FTIR) spectroscopy to measure As(III) oxidation rates by hydrous manganese(IV) oxide (HMO) at a timescale of ~2.5 s. However, both of these techniques suffer from limitations. EPR can only be used to measure EPR active nuclei, whereas FTIR requires both IR active functional groups and relatively high concentrations of the reactants being examined (Parikh et al., 2008).

Quick-scanning x-ray absorption spectroscopy (QXAS) overcomes both of these limitations. Depending on beamline instrumentation and flux, QXAS can be used to probe most of the atoms in the periodic table and to relatively low concentrations (Khalid et al., 2010). The majority of the quick-scanning beamlines in the world collect a complete EXAFS scan in ~1 min, by slewing the monochromator from low energy to high energy and repeating the process (Mitsunobu et al., 2006). An alternative method for rapidly collecting EXAFS data is to perform energy-dispersive measurements; however, this technique generally suffers from poor sensitivity (Dent, 2002). However, using a unique, cam-operated, continuous-scanning setup at beamline X18B at the NSLS, it is possible to collect XANES and EXAFS spectra as the monochromator travels both up and down in energy (Khalid et al., 2010). Combined with electronics that collect 2000 data points per second, this unique setup allows the collection of a single EXAFS scan in as little as 100 ms. Landrot et al. (2010) used this beamline to examine the kinetics of Cr(III) oxidation by HMO, employing XANES spectroscopy, with a time resolution of 2 s, while Ginder-Vogel et al. (2009) examined the mechanism of As(III) oxidation at the HMO mineral surface, using EXAFS spectroscopy, at similar timescales.

1.3 Applications of Synchrotron-Based Techniques to Elucidate Soil Chemical Processes and Reactions

In situ (under natural conditions where water is present) molecular scale techniques, such as synchrotron-based XAS, XRF, XRD, and microtomography, have been widely used over the past 20 years to study metal(loid) and radionuclide sorption phenomena at the mineral/water interface and to determine the chemical form (speciation) and distribution of environmental contaminants in heterogeneous natural materials, such as minerals, soils, plants, and terrestrial-based airborne particulates. Arguably, they have revolutionized many fields, including soil science, geochemistry, and the environmental sciences.

Synchrotron light sources were first available to general users in 1974. The first published study that employed XAS to directly delineate the type of surface complexes formed at mineral/water interfaces was the seminal research of Hayes et al. (1987). Studying selenite and selenate sorption at the goethite/water interface, selenite was observed to primarily form an inner-sphere surface complex, whereas selenate predominantly formed an outer-sphere complex. This was the first direct confirmation that indeed inner- and outer-sphere surface complexes, first proposed by Werner Stumm and coworkers, form on natural mineral surfaces.

1.3.1 Adsorption of Metals and Oxyanions on Soil Components

Since 1987, hundreds of studies using bulk XAS (Sparks, 1995; Brown and Parks, 2001; Brown and Sturchio, 2002; Sparks, 2005b; Brown et al., 2006) have provided extensive information on metal and oxyanion sorption on metal hydr(oxides), clay minerals, humic substances, and other natural materials, including structure, stoichiometry, attachment geometry (inner- vs. outer-sphere, monodentate vs. bidentate or tridentate), the presence of multinuclear complexes and precipitate phases, and the presence of ternary surface complexes when complexing ligands are present in solution (Brown and Parks, 2001; Brown and Sturchio, 2002; Sparks, 2005b). The type of surface complexes on clay minerals and metal-(oxyhydr)oxides that occur with low atomic number elements, such as Al, B, Ca, Mg, S, and Si, are not easy to ascertain using XAS under in situ conditions. However, major advances are being made in the area of soft x-ray XAS spectroscopy that will enable one to directly determine the types of surface complexes that form with these metal(loids) (Ginder-Vogel and Sparks, 2010).

Based on molecular scale studies, including many using XAS, one can predict that alkaline earth cations, Mg^{2+}, Ca^{2+}, Sr^{2+}, and Ba^{2+} primarily form outer-sphere complexes, whereas the divalent first-row transition metal cations Mn^{2+}, Fe^{2+}, Co^{2+}, Ni^{2+}, Cu^{2+}, and Zn^{2+} and the divalent heavy metal cations such as Cd^{2+}, Hg^{2+}, and Pb^{2+} primarily form inner-sphere complexes. At higher metal loadings and pHs, sorption of metals such as Co, Cr, Ni, and Zn on phyllosilicates and metal-(oxyhydr) oxides can result in the formation of surface precipitates. The formation of these multinuclear and precipitate phases will be discussed in more detail later.

Although there are currently experimental limitations to using in situ molecular scale techniques to directly determine the type of surface complexes that the anions NO_3^-, Cl^-, and ClO_4^- form on mineral surfaces, one can propose that they are sorbed as outer-sphere complexes and sorbed on surfaces that exhibit a positive charge. Some researchers have also concluded that SO_4^{2-} (Zhang and Sparks, 1990) can be sorbed as an outer-sphere complex; however, there is other evidence that SO_4^{2-} can also be sorbed as an inner-sphere complex (Manceau and Charlet, 1994). There is direct spectroscopic evidence to show that selenate can be sorbed as both an outer-sphere and an inner-sphere complex, depending on environmental factors (Hayes et al., 1987; Wijnja and Schulthess, 2000).

Most other anions such as molybdate, arsenate, arsenite, selenite, phosphate, and silicate appear to be strongly sorbed as inner-sphere complexes, and sorption occurs through a ligand exchange mechanism. The sorption maximum is often insensitive to ionic strength changes. Sorption of anions via ligand exchange results in a shift in the point of zero charge of the sorbent to a more acidic value (Sparks, 2002, 2005b).

One must be aware that bulk XAS probes an area of several square millimeters and provides information on the average local chemical environment of a surface. Thus, where more than one type of surface species is present, bulk XAS will detect only the primary (or average) type of surface product/species in the bulk sample (i.e., sums over all geometric configurations of the target atom). Consequently, although it may be concluded that the primary surface complex is inner sphere, this does not mean that outer-sphere complexation is not occurring. Recently, through the use of x-ray scattering measurements to study metal(loid) binding on single crystal surfaces, Catalano et al. (2008) showed that arsenate surface complexation was bimodal, with adsorption occurring simultaneously as inner- and outer-sphere species.

Environmental factors such as pH, surface loading, ionic strength, type of sorbent, and time also affect the type of sorption complex or product. An example of this can be found in the research of Strawn and Sparks (1999) who investigated Pb sorption on montmorillonite over an ionic strength (I) of 0.006–0.1 and a pH range of 4.48–6.77. Employing XAS analysis, at a pH of 4.48 and I of 0.006, it was determined that outer-sphere complexation on basal planes in the interlayer regions of the montmorillonite predominated. At a pH of 6.77 and I of 0.1, inner-sphere complexation on edge sites of montmorillonite was most prominent, and at pH of 6.76, I of 0.006, pH of 6.31, and I of 0.1, both inner- and outer-sphere complexation occurred. These data are consistent with other findings that inner-sphere complexation is favored at higher pH and ionic strength. Clearly, there is a continuum of adsorption complexes that can exist in soils (Sparks, 2002, 2005b, Ginder-Vogel and Sparks, 2010).

1.3.2 Metal(loid) Surface Precipitation/Dissolution

As the amount of metal cation or anion sorbed on a surface (surface coverage or loading, which is affected by the pH at which sorption occurs) increases, sorption can proceed from mononuclear adsorption to surface precipitation (a three-dimensional phase). There are several types of surface precipitates. They can arise via polymeric metal complexes (dimers, trimers, etc.) that form on mineral surfaces and via the sorption of aqueous polymers (Chisholm-Brause et al., 1990). Homogeneous precipitates can form on a surface when the solution becomes saturated and the surface acts as a nucleation site. When adsorption attains monolayer coverage, sorption continues on the newly created sites, causing a precipitate on the surface (Farley et al., 1985; Sposito, 1986; Chisholm-Brause et al., 1990; McBride, 1991; O'Day et al., 1994). When the precipitate consists of chemical species derived from both the aqueous solution and dissolution of the mineral, it is referred to as a coprecipitate. The composition of the coprecipitate varies between that of the original solid and a pure precipitate of the sorbing metal. The ionic radius of the sorbing metal and sorbent ions must be similar for coprecipitates to form. Thus Co(II), Mn(II), Ni(II), and Zn(II) form coprecipitates on sorbents containing Al(III) and Si(IV) but not Pb(II), which is considerably larger (1.20 Å). Coprecipitate formation is most limited by the rate of mineral dissolution rather than the lack of thermodynamic favorability (McBride, 1994;

Scheidegger et al., 1998). If the formation of a precipitate occurs under solution conditions that would, in the absence of a sorbent, be undersaturated with respect to any known solid phase, this is referred to as surface-induced precipitation (Towle et al., 1997; Sparks, 2005a; Borda and Sparks, 2008; Ginder-Vogel and Sparks, 2010).

Thus, there is often a continuum between surface complexation (adsorption) and surface precipitation. At low surface coverages, surface complexation (e.g., outer- and inner-sphere adsorption) tends to dominate. As surface loadings increase, nucleation occurs and results in the formation of distinct entities or aggregates on the surface. As surface loadings increase further, surface precipitation becomes the dominant mechanism.

Using in situ bulk XAS, it has been shown by a number of scientists that multinuclear metal hydroxide complexes and surface precipitates of Co^{2+}, Cr^{3+}, Cu^{2+}, Ni^{2+}, and Pb^{2+} can form on metal oxides, phyllosilicates, soil clays, and soils (Chisholm-Brause et al., 1990, 1994; Roe et al., 1991; Charlet and Manceau, 1992; Fendorf et al., 1994; O'Day et al., 1994; Bargar et al., 1995; Papelis and Hayes, 1996; Scheidegger et al., 1996a, 1996b, 1997, 1998; Towle et al., 1997; Elzinga and Sparks, 1999; Roberts et al., 1999; Thompson et al., 1999a, 1999b; Ford and Sparks, 2000; Scheckel and Sparks, 2001b). These metal hydroxide phases occur at metal loadings below theoretical monolayer coverage and in a pH range well below the pH where the formation of metal hydroxide precipitates would be expected, according to the thermodynamic solubility product (Scheidegger and Sparks, 1996; Sparks, 2002, 2005b; Borda and Sparks, 2008; Ginder-Vogel and Sparks, 2010).

Scheidegger et al. (1997) were the first to show that sorption of metals, such as Ni, on an array of phyllosilicates and Al oxide, could result in the formation of mixed metal-Al hydroxide surface precipitates, which appear to be coprecipitates. The precipitate phase shares structural features common to the hydrotalcite group of minerals and the layered double hydroxides (LDH) observed in catalyst synthesis. The LDH structure is built of stacked sheets of edge-sharing metal octahedra, containing divalent and trivalent metal ions separated by anions between the interlayer spaces (Figure 1.10). The general structural formula can be expressed as $[Me^{2+}Me_x^{3+}(OH)_2]^{x+} \cdot [(x/n)A^n - mH_2O]$, where, for example, Me^{2+} could be Mg(II), Ni(II), Co(II), Zn(II), Mn(II), and Fe(II) and Me^{3+} is Al(III), Fe(III), and Cr(III) (Towle et al., 1997). The LDH structure exhibits a net positive charge x per formula unit, which is balanced by an equal negative charge from interlayer anions A^{n-}, such as Cl^-, Br^-, I^-, NO_3^-, OH^-, ClO_4^-, and CO_3^{2-}; water molecules occupy the remaining interlayer space (Taylor, 1984). The minerals takovite, $Ni_6Al_2(OH)_{16}CO_3 \cdot H_2O$, and hydrotalcite, $Mg_6Al_2(OH)_{16}CO_3 \cdot H_2O$, are among the most common natural mixed-cation hydroxide compounds containing Al (Taylor, 1984; Sparks, 2005b; Borda and Sparks, 2008). Recently, Livi et al. (2009), using an array of microscopic techniques including analytical electron microscopy, high resolution transmission electron microscopy, and powder x-ray diffraction, conducted studies to elucidate the nature of

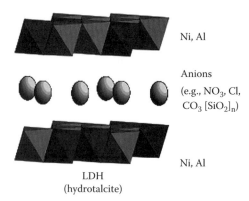

FIGURE 1.10 Structure of Ni–Al LDH showing brucite-like octahedral layers in which Al substitutes for Ni, creating a net positive charge balanced by hydrated anions in the interlayer space. (Reprinted from Scheinost, A.C., R.G. Ford, and D.L. Sparks. 1999. The role of Al in the formation of secondary Ni precipitates on pyrophyllite, gibbsite, talc, and amorphous silica: A DRS study. Geochim. Cosmochim. Acta 63:3193–3203, Copyright 1999. With permission from Elsevier.)

Ni hydroxide precipitates, using the same environmental conditions employed by Scheidegger et al. (1996b, 1997) and reaction times ranging from 1 h to 5 years. Although the precipitate phase had a bonding environment similar to Ni-Al LDH, the precipitate was amorphous.

Mixed Co–Al and Zn–Al hydroxide surface precipitates can also form on aluminum-bearing metal oxides and phyllosilicates (Towle et al., 1997; Thompson et al., 1999a, 1999b; Ford and Sparks, 2000). This is not surprising, as Co^{2+}, Zn^{2+}, and Ni^{2+} all have similar radii to Al^{3+}, enhancing substitution in the mineral structure and formation of a coprecipitate. However, surface precipitates have not been observed with Pb^{2+}, as Pb^{2+} is too large to substitute for Al^{3+} in mineral structures (Sparks, 2002, 2005b; Borda and Sparks, 2008; Ginder-Vogel and Sparks, 2010).

The mechanism for the formation of metal hydroxide surface precipitates is not clearly understood. It is clear that the type of metal ion determines whether metal hydroxide surface precipitates form, and the type of surface precipitate formed, that is, metal hydroxide or mixed metal hydroxide, is dependent on the sorbent type. Additionally, the reaction pH, which controls the degree of metal loading, is also a major factor in determining the formation of metal hydroxide precipitates. Sparks and his group have shown that below pH 6.5, metal hydroxide precipitates do not form on mineral surfaces or in soils in laboratory studies (Elzinga and Sparks, 1999, 2001; Roberts et al., 1999, 2003).

The formation of metal hydroxide surface precipitates appears to be an important way to sequester metals. As the surface precipitates age, metal release is greatly reduced (Figure 1.11). Thus, the metals are less prone to leaching and being taken up by plants and microbes. Peltier et al. (2010) reacted three soils of varying mineralogy and organic matter content with 3 mM Ni at two pHs, 6 and 7.5, and evaluated Ni bioavailability using a biosensor. At pH 6, where surface precipitates did not form, most of the Ni was bioavailable. However, at pH 7.5, where precipitates

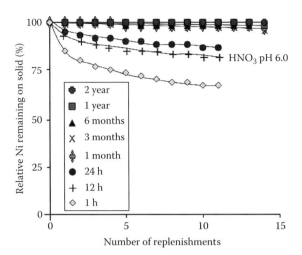

FIGURE 1.11 Dissolution of Ni from surface precipitates formed on pyrophyllite at residence times of 1 h to 2 years. The figure shows the relative amount of Ni^{2+} remaining on the pyrophyllite surface following extraction for a 24 h period (each replenishment represents a 24 h extraction) with HNO_3 at pH 6.0. (Reprinted from Scheckel, K.G., and D.L. Sparks. 2001a. Dissolution kinetics of nickel surface precipitates on clay mineral and oxide surfaces. Soil Sci. Soc. Am. J. 65:685–694. With permission from Soil Science Society of America.)

were observed to form from XAS analyses, Ni bioavailability was markedly reduced.

The decrease in metal release and bioavailability is linked to the increasing silication of the interlayer of the LDH phases with increased residence time, resulting in a mineral transformation from a LDH phase to a precursor phyllosilicate surface precipitate (Ford et al., 1999; Ford and Sparks, 2000). The mechanism for this transformation is thought to be the diffusion of Si, originating from weathering of the sorbent, into the interlayer space of the LDH, replacing the anions such as NO_3^-. Polymerization and condensation of the interlayer Si slowly transforms the LDH into a precursor metal-Al phyllosilicate. The metal stabilization that occurs in surface precipitates on Al-free sorbents (e.g., talc) may be due to Ostwald ripening, resulting in increased crystallization; however, as noted earlier, microscopic analyses have shown that Ni–Al LDH phases are amorphous (Livi et al., 2009). Peltier et al. (2006), using acid-solution calorimetry and results from previous calorimetry studies, showed that the enthalpy of formation of LDH phases is more exothermic, indicating great stability, in the order of $Cl < NO_3 < SO_4 < CO_3 < Si$ of interlayer anionic composition, and that LDH phases were much more stable than a $Ni(OH)_2$ phase.

In short, with time, metal sorption on soil minerals can often result in a continuum of processes, from adsorption to precipitation to solid-phase transformation, particularly in the case of certain metals, such as Co^{2+}, Ni^{2+}, and Zn^{2+}. The formation of metal surface precipitates could be an important mechanism for sequestering metals in soils, such that they are less mobile and bioavailable. Such products must be considered when modeling the fate and mobility of metals such as Co^{2+}, Mn^{2+}, Ni^{2+}, and Zn^{2+} in soil and water environments (Sparks, 2002, 2005b; Borda and Sparks, 2008; Ginder-Vogel and Sparks, 2010).

1.3.3 Redox Reactions on Soil Components

The mobility and environmental threat of many contaminants and nutrients are determined in large part by their redox state and the redox conditions of the soil environments in which they exist. XAS is an ideal tool for determining the oxidation states of elements in complex environmental media including soils and sediments. A change in oxidation state of an element results in a change in the energy required to excite a core electron into the continuum, which manifests itself by a shift in the white-line position for an element and can be determined using XANES spectroscopy. In addition, a change in the element's local coordination environment occurs and is manifested in EXAFS spectra of the element. This is readily exemplified by the reduction of soluble $U(VI)O_2^{2+}$ (and its various complexes) to $U(IV)O_2$, which results in the absorption edge shifting ~3 eV lower in energy (Figure 1.12). Additionally, the transition from the highly linear uranyl ion results in the loss of a large shoulder, due to multiple scattering, that immediately follows the white line (Figure 1.12) (Kelly et al., 2002, 2007, 2008; Ginder-Vogel et al., 2010).

In addition to uranium, XAS has been used to investigate metal and oxyanions cycling in many redox active systems (Fendorf et al., 1992, 2002; Bostick et al., 2002; Manning et al., 2002; Hansel et al., 2003a, 2003b, 2005; Webb et al., 2005; Bank et al., 2007; Ginder-Vogel and Fendorf, 2008). The primary elements of interest include environmentally predominant transition metals, such as iron and manganese, in addition to elements that exist in the environment primarily as oxyanions, such as arsenic, chromium, and selenium. Many of these redox reactions are driven by bacterial metabolism and result in the formation of

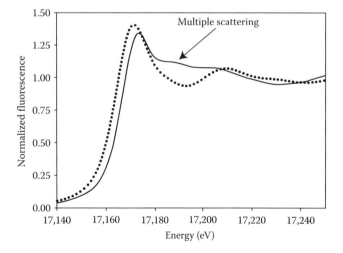

FIGURE 1.12 Uranium L_{III}-edge XANES spectra of uraninite (dotted) and uranyl nitrate hexahydrate (solid). The arrow indicates the contribution of multiple scattering from the uranyl moiety. (Reprinted from Ginder-Vogel, M., and S.E. Fendorf. 2008. Biogeochemical uranium redox transformation: Potential oxidants of uraninite, p. 293–320. *In* M.O. Barnett and D.B. Kent (eds.) Adsorption of metals by geomedia II: Variables, mechanisms, and model applications. Elsevier, Amsterdam, the Netherlands. With permission from Elsevier.)

hydrated mineral phases, making in situ (i.e., wet) analysis critical to determining the operative redox pathways (Ginder-Vogel and Sparks, 2010).

1.3.4 Transition-Metal Oxides

The environmental prevalence of iron and manganese oxides, hydroxides, and oxyhydroxides [(hydr)oxides], coupled with the large quantity of their highly reactive surface area, makes them an important pool of reactivity, with regard to contaminant and nutrient absorption and redox transformation. However, their reactivity, in terms of both sorption and redox chemistry, is radically altered by changes in redox conditions, especially as they impact mineralogy and surface chemistry of iron and manganese oxide phases (Ginder-Vogel and Sparks, 2010).

Ferrihydrite, a poorly crystalline Fe(III) (hydr)oxide, is often found in soils and sediments, oscillating between reducing and oxidizing conditions. Due to its high surface area and intrinsic reactivity, ferrihydrite serves as a critical sink for contaminants (e.g., As and U) and nutrients (e.g., P); however, until recently, the impact of reducing conditions on the evolution of Fe(III) mineral mineralogy and its subsequent ability to retain and interact with contaminants and nutrients has been poorly understood. XAS, coupled with other advanced analytical techniques, has played a key role in furthering our understanding of this important environmental process. Ferrihydrite is considered the most bioavailable Fe(III) (hydr)oxide for dissimilatory metal reducing bacteria, which couple the oxidation of organic matter and H_2 to the reduction of Fe(III) and are ubiquitous in saturated soils and sediments. Upon dissimilatory metal reduction, the production of Fe(II) may result either in Fe(III) mineral dissolution or the rapid transformation of poorly crystalline Fe(III) (hydr)oxides into more crystalline and/or mixed Fe(II)/Fe(III) (hydr)oxides, dependent on the intensity of Fe(III) reduction (Ginder-Vogel and Sparks, 2010).

XAS has played a vital role in the identification and quantification of Fe (hydr)oxide mineralogy. In batch experiments using ferrihydrite, the solid phases formed during microbial iron reduction are dominated by the metabolic byproducts of Fe reduction, including ferrous iron and bicarbonate, and generate minerals rich in Fe(II) and bicarbonate, including siderite, green rusts, and magnetite (Fredrickson et al., 1998; Liu et al., 2001). In general, this type of concentrated batch reaction produced high enough concentrations of the products of biological Fe reduction to be detected using traditional techniques, such as lab-based XRD and Mossbauer spectroscopy. However, in order to investigate the transformation of Fe (hydr)oxide minerals under more natural conditions, Fendorf and coworkers (Hansel et al., 2003a, 2005; Kocar et al., 2006; Nico et al., 2009) transitioned to iron oxide coated sand, which was packed into columns. This allowed Fe biomineralization to be studied under conditions that transported microbial metabolites away from the reaction site. Although a variety of analytical methods were used to analyze the resultant Fe mineralogy, including SEM, HR-TEM, XRD, and Mossbauer spectroscopy, they found that the most accurate and quickest method that allowed the quantification of the various iron mineral phases was EXAFS spectroscopy combined with linear combination analysis (Hansel et al., 2003a, 2005). Using this technique, Hansel et al. (2005) demonstrated that the mineralogical transformation of Fe(III) (hydr)oxides is dependent on the Fe(III) flux resulting from microbial iron reduction. At high Fe(II) concentrations, the dominant mineralogical products include magnetite and goethite, whereas lepidocrocite and goethite are observed at lower Fe(II) concentrations. Interestingly, more reduced phases, such as green rusts and Fe(II)-bearing minerals, were not observed in any experiments where flow conditions were used. This research work has formed the basis for a series of studies that examine the effect of microbial iron reduction on the incorporation of other metals and metalloids, such as uranium and chromium, into iron oxide mineral structures during biological iron reduction and abiotic redox reactions (Ginder-Vogel and Sparks, 2010).

Widespread uranium contamination is present throughout the world as a result of natural uranium deposits, mining activities, and nuclear weapons and fuel production. Environmental uranium speciation is dominated by two oxidation states with markedly different properties. Uranium(VI), as the uranyl cation UO_2^{2+}, is thermodynamically stable under oxidizing conditions and is highly soluble in the presence of dissolved bicarbonate. Although uranyl sorption to solids such as Fe (hydr)oxides may be appreciable, it is largely reversible and is thus subject to changes in aqueous conditions and Fe (hydr)oxide mineralogy. Dissolved uranium can be reduced through the transformation of U(VI) into U(IV), forming solid uraninite (UO_2), which is sparingly soluble under most environmental conditions. However, uraninite may be reoxidized, thereby releasing uranium, by many oxidants including O_2, NO_3^-, and Mn(IV) oxides and Fe(III) (hydr)oxides (Ginder-Vogel and Fendorf, 2008). Given the prevalence of Fe (hydr)oxide minerals in the environment and their critical role in controlling uranium mobility, much research has focused on the interaction of uranium complexes with iron, particularly as (hydr)oxide mineral phases. However, we have only recently begun to understand the role of iron (hydr)oxides in catalyzing uraninite oxidation and incorporation of uranium into the Fe (hydr)oxide mineral structure. Ginder-Vogel et al. (2006) and Ginder-Vogel and Fendorf (2008) used a combination of S-XRD and XAS to identify and quantify iron oxide minerals produced during the oxidation of biologically precipitated uraninite by ferrihydrite. This research revealed that ferrihydrite is capable of oxidizing biogenic uraninite; however, the oxidation reaction is ultimately limited by the conversion of ferrihydrite into more thermodynamically stable forms, such as lepidocrocite and goethite. Interestingly, during Fe(II) catalyzed ferrihydrite remineralization, U(VI) is substituted for between 3% and 6% of iron sites in transformation products (Nico et al., 2009). The substituted uranium is stable even in the presence of strongly complexing solutions (30 mM $KHCO_3$) and during redox cycling (Stewart et al., 2009) and can be environmentally important as a long-term uranium sink (Ginder-Vogel and Sparks, 2010).

Similar to uranium, chromium predominantly exists in the environment in two oxidation states, Cr(III) or Cr(VI). Chromium(VI), chromate, is highly toxic to humans and is quite soluble and hence mobile under most common environmental conditions. Conversely, Cr(III) is less toxic and generally immobile. There are many abiotic and biotic pathways of chromium reduction in the environment; however, under neutrophilic conditions kinetic considerations favor Cr(VI) reduction by either dissolved Fe(II) or Fe(II)-bearing mineral phases (Fendorf et al., 2002). The identification and quantification of chromium oxidation states, using XANES spectroscopy, is quite straightforward, due to a prominent preedge peak at 5993 eV, when Cr is present as Cr(VI) (Peterson et al., 1996, 1997), which is caused by a bound-state 1s to 3d transition. This transition is forbidden for centrosymmetric, octahedral $Cr(II)O_6$ but is allowed for the noncentrosymmetric, tetrahedrally coordinated $Cr(VI)O_4$ molecule, due to mixing of the Cr(3d) and O(2p) orbitals. The intensity of this preedge feature can be used to quantify the proportion of Cr(VI) in a sample, at Cr(VI) concentrations greater than 1%–5%. Using this technique, several studies have examined Cr speciation in reducing Cr-contaminated soils (Fendorf and Zasoski, 1992; Patterson et al., 1997; Kendelewicz et al., 2000; Bank et al., 2007) and determined that less mobile Cr(III) was the dominant species in these systems; however, this may not always be the case in oxic systems. Using a combination of bulk XAS, μ-XRF and μ-XRD to analyze sediments from the arid, oxic, Cr-contaminated Hanford site in Eastern Washington state, in conjunction with flow-through column experiments, Ginder-Vogel et al. (2005) identified several Fe(II)-bearing phyllosilicate minerals as the primary sources of Cr(VI) reducing ability in these sediments. However, depending on geochemical conditions, as much as 50% of the solid-phase chromium remained in the hexavalent form.

In large part, the solubility of Cr(III) that is abiotically reduced by ferrous iron is determined in large part by the ratio of Cr(III) to Fe(III) in the (hydr)oxide precipitate of the general formula $Fe_xCr_{1-x}(OH)_3$ (Sass and Rai, 1987). The solubility of Cr(III)-hydroxide precipitates is proportional to the ratio of Cr(III) to Fe(III), with increased quantities of Fe(III) stabilizing the solid (Sass and Rai, 1987). Generally, the ratio of Cr to Fe in hydroxide solids is determined using extraction methods, which may be compromised by Fe contamination of ex situ environmental samples; however, EXAFS spectroscopy presents an alternative method for determining this molar ratio (Hansel et al., 2003b). The intensity of the corner-sharing peak at 3.48–3.56 Å, normalized to the first Cr–Cr(Fe) shell, has a linear relationship with the mole fraction of Cr(III) within a mixed Cr(III)–Fe(III) solid (Ginder-Vogel and Sparks, 2010).

1.3.5 Speciation of Metal(loids) in Soils

Some examples of studies that have used μ-XAS and μ-XRD to speciate metal(loids) in soils include those by Hunter and Bertsch (1998), Manceau et al. (2000), Isaure et al. (2002), Strawn et al. (2002), Ginder-Vogel et al. (2005), Nachtegaal et al. (2005), Roberts et al. (2005), Arai et al. (2006), McNear et al. (2007), and Gräfe et al. (2008). A few examples of these studies are summarized below.

Nachtegaal et al. (2005) investigated the speciation of Zn in smelter-contaminated soils, employing μ-XAS and μ-XRF, from a large site in Belgium in which part of the site had been remediated by adding beringite, an aluminosilicate material, compost, and planting metal tolerant plants. The other portion of the site was not treated. The objectives of the study were to determine how Zn speciation differed in the remediated (treated) and nonremediated (nontreated) soils. Specifically the study sought to determine if Zn-LDH phases were present in the soils and the stability of the zinc under different environmental conditions.

A large number of μ-XAS spectra for the treated (Figure 1.13b) and nontreated (Figure 1.13d) soils were collected at various regions of interest from the XRF spectra (Figure 1.13a, treated and Figure 1.13c, nontreated soils), as well as spectra for reference mineral, sorbed, and solution phases that were probable species in the soils. The μ-XAS spectra of the soils were analyzed via PCA. TT was then employed to determine spoil values for the known reference materials. The spoil value indicates how well the reference spectrum is matched to the principal components. The lower the spoil value, the better the match, and the more likely the reference material is contained in the soil sample. After the contributing standard phases are identified, LLSF is used to determine the amount (%) of each standard species within the individual sample spectrum making up the dataset. Additional details on μ-XAS applications and data analyses methods for heterogeneous systems can be found in Bertsch and Hunter (2001) and Manceau et al. (2000).

In the study of Nachtegaal et al. (2005), both mineral (e.g., willemite, hemimorphite, sphalerite) and sorbed (Zn-LDH) Zn species predominated in the treated and nontreated soils. The speciation differences in the treated and nontreated soils were slight, with the major difference being the presence of kerolite, a Zn phyllosilicate phase, found in the treated soil (Figure 1.13b). Significant quantities of Zn-LDH phases were formed in the nontreated soil (Figure 1.13d). Desorption studies, using both $CaCl_2$ and HNO_3 at pH 4 and 6, showed that the Zn in both remediated and nonremediated soils was quite stable, reflecting again the role that metal surface precipitates, that is, Zn-LDH phases, play in sequestering metals such that mobility and bioavailability are diminished.

Gräfe et al. (2008) used μ-XAS and μ-XRF to study the speciation of As in a copper-chromated-arsenate-contaminated soil to determine how co-contaminating metal cations (Cu, Zn, Cr) affected the speciation of As. Data analyses revealed that As occurred as a continuum of fully and poorly ordered copper-arsenate precipitates (63%–75% of As species) and surface adsorption complexes on goethite and gibbsite in the presence and absence of Zn. Other non-Cu-based precipitates such as scorodite, adamite, and ojuelaite were also found. These data point out that association of As with Cu suggests that the speciation of As in a contaminated soil is not entirely controlled by surface adsorption reactions but influenced significantly by the co-contaminating metal cation fraction.

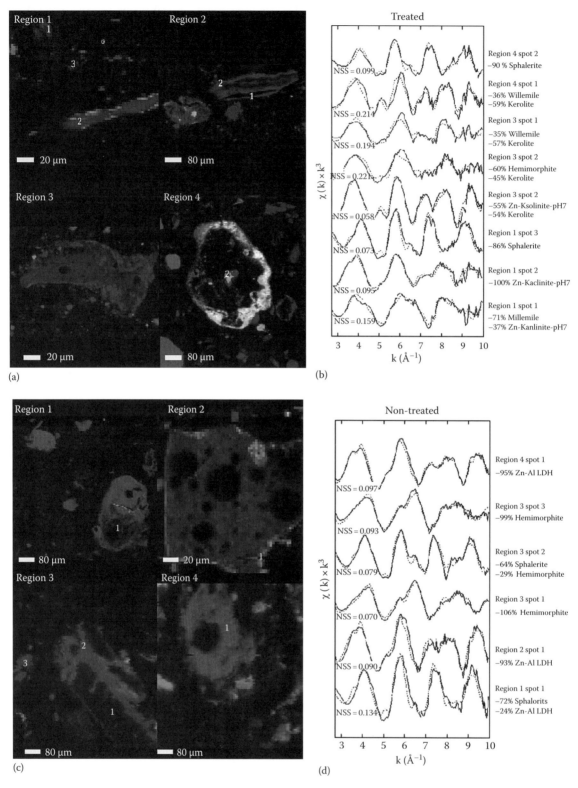

FIGURE 1.13 (**See color insert.**) (a) μ-SXRF tricolor maps for the treated soil samples. The numbers indicate the spots where μ-EXAFS spectra were collected. Red is indicative of the distribution of iron, green of copper, and blue of zinc. (b) μ-EXAFS spectra from selected spots on thin sections from the treated soil. (c) μ-SXRF tricolor maps for the nontreated soil samples. (d) μ-EXAFS spectra from selected spots on thin sections from the nontreated soil. The solid line indicates the raw $k^3c(k)$ data, and the dotted line indicates the best fits obtained with a linear fitting approach. (Reprinted from Nachtegaal, M., M.A. Marcus, J.E. Sonke et al. 2005. Effects of in situ remediation on the speciation and bioavailability of zinc in a smelter contaminated soil. Geochim. Cosmochim. Acta 69:4649–4664. Copyright 2005, with permission from Elsevier.)

1.3.6 Speciation of Nutrients in Soils and Biosolids

A number of studies have used XAS to speciate nutrients such as P in soils and biosolids (see references cited in Toor et al., 2006). Hesterberg et al. (1999) used XANES to investigate adsorbed and mineral phases of phosphate on Al- and Fe-oxides. They also showed that the predominant phase of phosphate in a North Carolina soil sample was dicalcium phosphate. Peak et al. (2002) investigated the speciation of P in alum-amended and nonamended poultry litter samples using XANES. Alum is used in the poultry industry to reduce the level of water-soluble P in poultry litter before it is land applied. First, phosphorus, sulfur, and calcium XANES were collected on the litter samples as well as spectra of reference materials that included aqueous and organic phosphates as well as Ca and Al phosphates and phosphate adsorbed on Al oxides. The experimental XANES spectra of the litter samples were then compared to the reference compounds via fingerprinting to determine the chemical species in the litter samples. In the unamended litter, the predominant P species were a mixture of aqueous phosphate, organic phosphate, and dicalcium phosphate. In the alum-amended litter samples, the major species were P adsorbed on Al oxides, organic phosphate, and aqueous phosphate. These results would indicate that the addition of alum reduces the mobility of P in the litter material. There was no evidence of Al phosphate precipitates in the alum-amended samples. However, such phases could have occurred immediately after alum was added, as the pH is low, ca. 5, but then it increases with time to ca. 7. With the rise in pH, any initial Al phosphate precipitates would undergo dissolution. XANES, along with more detailed speciation analyses, including linear combination fitting, was used by Beauchemin et al. (2003), Toor et al. (2005), Shober et al. (2006), and Seiter et al. (2008) to more quantitatively speciate P in soils, turkey manures, and poultry litter. In order to get a better assessment of organic P species, which are known to be significant in poultry litter, Seiter et al. (2008) used PCA, TT, and LLSF to assess P speciation in alum-amended and nonamended poultry litter samples. They found that alum-amended poultry litter contained higher amounts of Al-bound P and phytic acid, whereas nonamended samples contained Ca–P minerals and organic P compounds.

1.3.7 Speciation of Metals in Hyperaccumulator Plants

Over the past few years, synchrotron-based XAS, XRF, XRD, and microtomography have been employed to enhance understanding of plant/soil interfacial processes and reactivity (Sarret et al., 2002, 2006; Küpper et al., 2004; Scheckel et al., 2004; McNear et al., 2005a, 2005b; Tappero et al., 2007). Of particular interest is determining the speciation and uptake mechanisms of metals into hyperaccumulator plants. Such an enhanced comprehension could prove invaluable in increasing the effectiveness of phytoremediation in decontaminating metal polluted sites around the world. Phytoremediation is a "green" technology that uses plants to remove contaminants from the environment. Phytoextraction depends on unique plants, capable of accumulating higher than normal metal concentrations (e.g., >1000 ppm for Ni and Co and >10,000 ppm for Zn) (Baker, 1981; McNear et al., 2005a, 2005b). To better understand the mechanisms involved in metal hyperaccumulation and tolerance, it is imperative to know whether accumulated metals are bound by strong (specific) ligands or loosely associated with common organic acids (i.e., speciation) as well as where these metals are stored (i.e., localization or compartmentalization). The use of synchrotron-based methods to explore metal speciation and uptake mechanisms in plants is especially attractive because one can use "fresh" hyperaccumulator plant tissues (in vivo) with micrometer resolution and investigate distribution and association of the metals in the plant tissue on two-dimensional (µ-XRF) and three-dimensional (microtomography) scales and determine direct metal speciation (µ-XAS). Several studies have employed these techniques to elucidate Ni and Co speciation in *Alyssum murale*, an important Ni hyperaccumulator plant from the Brassicaceae family that is native to serpentine soils throughout Mediterranean southern Europe (McNear et al., 2005a, 2005b; Tappero et al., 2007).

Tappero et al. (2007) studied metal localization and elemental associations in *A. murale* plants exposed to metal (Ni, Zn, and Co) co-contaminants using synchrotron-based µ-XRF, µ-XAS, and computed microtomography. Two-dimensional XRF images of *A. murale* leaves showed a marked localization pattern for Co compared with Ni and Zn. The Ni distribution was predominately uniform (Figure 1.14), which is consistent with previous findings that Ni is compartmentalized in epidermal tissues, that is, vacuolar sequestration (Kramer et al., 1997; Krämer et al., 2000; Broadhurst et al., 2004a, 2004b). Zinc distribution was similar to Ni and was not hyperaccumulated. Cobalt, however, was preferentially localized at the tips and margins of *A. murale* leaves as precipitate phases (Figure 1.14). Differential absorption computed microtomography images (Figure 1.15) of a hydrated *A. murale* shows leaf Co enrichment in apoplast tissue. Cobalt near leaf tips was localized primarily on the leaf exterior, and the hyperaccumulation mechanism was via exocellular sequestration (Tappero et al., 2007).

1.4 Future Needs

As our knowledge of environmental systems continues to advance, it is critical that synchrotron-based techniques continue advancing with the field. Currently, a primary limitation of the application of these techniques to environmental systems is the availability of adequate amounts of beamtime. This problem can begin to be resolved by advances in both user support and instrumentation. Several improvements in user administration and services, including standardized data collection software, state-of-the-art data analysis software, increased beamline support staff, and improvements in laboratory support facilities, would allow for the more efficient use of limited

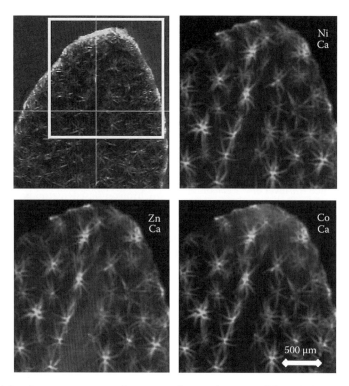

FIGURE 1.14 (See color insert.) Synchrotron x-ray microfluorescence (μ-SXRF) images of the nickel (Ni), cobalt (Co), and zinc (Zn) distributions in a hydrated *A. murale* leaf from the Ni + Co + Zn treatment. Leaf trichomes are depicted in the Ca channel. The camera image shows the leaf region selected for SXRF imaging. (Reprinted from Tappero, R., E. Peltier, M. Grafe et al. 2007. Hyperaccumulator *Alyssum murale* relies on a different metal storage mechanism for cobalt than for nickel. New Phytol. 175:641–654. With permission from Wiley-Blackwell.)

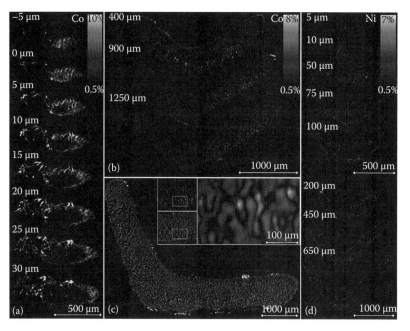

FIGURE 1.15 (See color insert.) Differential absorption (DA-CMT) tomographic projections (5.1 μm slices) of hydrated *A. murale* leaves depicting (a) cobalt (Co) distribution in the leaf-tip region, (b) Co distribution in the bulk-leaf region, (c) Co distribution in relation to the leaf cell structure (gray), and (d) nickel (Ni) distribution in the leaf-tip and bulk-leaf regions. Leaves were collected from a Co-treated plant (a–c) and from a Ni-treated plant (d). Sinograms recorded above and below the Co or Ni K-edge energy (+30 and −100 eV, respectively) were computationally reconstructed, and the resulting projections were subtracted (above − below) to reveal the metal distribution in leaves. Distances are relative to the leaf tissue at the tip as determined from leaf structure images (i.e., below-edge projections). (Reprinted from Tappero, R., E. Peltier, M. Grafe et al. 2007. Hyperaccumulator *Alyssum murale* relies on a different metal storage mechanism for cobalt than for nickel. New Phytol. 175:641–654. With permission from Wiley-Blackwell.)

beamtime. The heterogeneity of environmental samples requires both the application of a wide array of techniques, including x-ray spectroscopic and traditional ones, and more time using each individual technique. The enhanced intensity and flux of third-generation synchrotron light sources allow for the analysis of both lower concentration elements and smaller samples; however, many environmental samples also require a broad energy range (5–50 keV), high energy resolution (0.1 eV), and a range of spatial resolutions (bulk, micro, and nano). Additionally, unique endstation capabilities, including flow-through reaction cells and anaerobic gloveboxes, coupled with fast-scanning, high energy resolution fluorescent detection and simultaneous collection of XRD and XAS data, will take advantage of the unique ability of x-ray absorption to analyze samples in situ, allowing a new generation of complex environmental problems to be solved (Ginder-Vogel and Sparks, 2010).

References

Arai, Y., A. Lanzirotti, S.R. Sutton, M. Newville, J. Dyer, and D.L. Sparks. 2006. Spatial and temporal variability of arsenic solid-state speciation in historically lead arsenate contaminated soils. Environ. Sci. Technol. 40:673–679.

Baker, A.J.M. 1981. Accumulators and excluders—Strategies in the response of plants to heavy metals. J. Plant Nutr. 3:643–654.

Bank, T.L., T.A. Vishnivetskaya, P.M. Jardine, M.A. Ginder-Vogel, S. Fendorf, and M.E. Baldwin. 2007. Elucidating biogeochemical reduction of chromate via carbon amendments and soil sterilization. Geomicrobiol. J. 24:125–132.

Bargar, J.R., G.E. Brown, Jr., and G.A. Parks. 1995. XAFS study of Pb(II) sorption at the α-Al_2O_3-water interface. Physica B 209:455–456.

Beauchemin, S., D. Hesterberg, J. Chou, M. Beauchemin, R.R. Simard, and D.E. Sayers. 2003. Speciation of phosphorus in phosphorus-enriched agricultural soils using X-ray absorption near-edge structure spectroscopy and chemical fractionation. J. Environ. Qual. 32:1809–1819.

Bertsch, P.M., and D.B. Hunter. 1998. Elucidating fundamental mechanisms in soil and environmental chemistry: The role of advanced analytical, spectroscopic, and microscopic methods, p. 103–122. In P.M. Huang (ed.) Future prospects for soil chemistry. SSSA, Madison, WI.

Bertsch, P.M., and D.B. Hunter. 2001. Applications of synchrotron based X-ray microprobes. Chem. Rev. 101:1809–1842.

Bertsch, P.M., and J.C. Seaman. 1999. Characterization of complex mineral assemblages: Implications for contaminant transport and environmental remediation. Proc. Natl. Acad. Sci. USA 96:3350–3357.

Borda, M.J., and D.L. Sparks. 2008. Kinetics and mechanisms of sorption-desorption in soils: A multiscale assessment, p. 94–124. In A. Violante, P.M. Huang, and G.M. Gadd (eds.) Biophysico-chemical processes of heavy metals and metalloids in soil environments. John Wiley & Sons, Inc., New York.

Bostick, B.C., C. Doyle, S. Fendorf, T. Kendelewicz, J. Bargar, and G.E. Brown, Jr. 2002. Role of arsenic in pyrite oxidation. Geochim. Cosmochim. Acta 66:A95–A95.

Broadhurst, C.L., R.L. Chaney, J.S. Angle, E.F. Erbe, and T.K. Maugel. 2004a. Nickel localization and response to increasing Ni soil levels in leaves of the Ni hyperaccumulator *Alyssum murale*. Plant Soil 265:225–242.

Broadhurst, C.L., R.L. Chaney, J.S. Angle, T.K. Maugel, E.F. Erbe, and C.A. Murphy. 2004b. Simultaneous hyperaccumulation of nickel, manganese, and calcium in Alyssum leaf trichomes. Environ. Sci. Technol. 38:5797–5802.

Brown, G.E., Jr. 1990. Spectroscopic studies of chemisorption reaction mechanisms at oxide-water interfaces, p. 309–353. In M.F. Hochella and A.F. White (eds.) Mineral-water interface geochemistry. Mineralogical Society of America, Washington, DC.

Brown, G.E., Jr., G. Calas, and R.J. Hemley. 2006. Scientific advances made possible by user facilities. Elements 2:23–30.

Brown, G.E., Jr., and G.A. Parks. 2001. Sorption of trace elements on mineral surfaces: Modern perspectives from spectroscopic studies, and comments on sorption in the marine environment. Int. Geol. Rev. 43:963–1073.

Brown, G.E., Jr., G.A. Parks, and P.A. O'Day. 1995. Sorption at mineral-water interfaces: Macroscopic and microscopic perspectives, p. 129–183. In D.J. Vaughan and R.A.D. Pattrick (eds.) Mineral surfaces. Chapman & Hall, London, U.K.

Brown, G.E., Jr., and N.C. Sturchio. 2002. An overview of synchrotron radiation applications to low temperature geochemistry and environmental science, p. 1–115. In P.A. Fenter, M.L. Rivers, N.C. Sturchio, and S.R. Sutton (eds.) Applications of synchrotron radiation in low-temperature geochemistry and environmental sciences. Mineralogical Society of America, Washington, DC.

Catalano, J.G., C. Park, P. Fenter, and Z. Zhang. 2008. Simultaneous inner- and outer-sphere arsenate adsorption on corundum and hematite. Geochim. Cosmochim. Acta 72:1986–2004.

Charlet, L., and A. Manceau. 1992. X-ray absorption spectroscopic study of the sorption of Cr(III) at the oxide-water interface. II: Adsorption, coprecipitation and surface precipitation on ferric hydrous oxides. J. Colloid Interface Sci. 148:443–458.

Chisholm-Brause, C.J., S.D. Conradson, C.T. Buscher, P.G. Eller, and D.E. Morris. 1994. Speciation of uranyl sorbed at multiple binding sites on montmorillonite. Geochim. Cosmochim. Acta 58:3625–3631.

Chisholm-Brause, C.J., P.A. O'Day, G.E. Brown, Jr., and G.A. Parks. 1990. Evidence for multinuclear metal-ion complexes at solid/water interfaces from X-ray absorption spectroscopy. Nature 348:528–530.

Dent, A.J. 2002. Development of time-resolved XAFS instrumentation for quick EXAFS and energy-dispersive EXAFS measurements on catalyst systems. Top Catal. 18:27–35.

Elzinga, E.J., and D.L. Sparks. 1999. Nickel sorption mechanisms in a pyrophyllite-montmorillonite mixture. J. Colloid Interface Sci. 213:506–512.

Elzinga, E.J., and D.L. Sparks. 2001. Reaction condition effects on nickel sorption mechanisms in illite-water suspensions. Soil Sci. Soc. Am. J. 65:94–101.

Farley, K.J., D.A. Dzombak, and F.M.M. Morel. 1985. A surface precipitation model for the sorption of cations on metal oxides. J. Colloid Interface Sci. 106:226–242.

Fendorf, S.E., M. Fendorf, D.L. Sparks, and R. Gronsky. 1992. Inhibitory mechanisms of Cr(III) oxidation by δ-MnO_2. J. Colloid Interface Sci. 153:37–54.

Fendorf, S., C.M. Hansel, and B.W. Wielinga. 2002. Operative pathways of chromate and uranyl reduction within soils and sediments, p. 111–130. In P.C. Zhang and P.V. Brady (eds.) Geochemistry of soil radionuclides. SSSA, Madison, WI.

Fendorf, S.E., G.M. Lamble, M.G. Stapleton, M.J. Kelley, and D.L. Sparks. 1994. Mechanisms of chromium (III) sorption on silica: 1. Cr(III) surface structure derived by extended x-ray absorption fine structure spectroscopy. Environ. Sci. Technol. 28:284–289.

Fendorf, S.E., D.L. Sparks, J.A. Franz, and D.M. Camaioni. 1993. Electron paramagnetic resonance stopped-flow kinetic study of manganese (II) sorption-desorption on birnessite. Soil Sci. Soc. Am. J. 57:57–62.

Fendorf, S.E., and R.J. Zasoski. 1992. Chromium(III) oxidation by δ-MnO_2. 1. Characterization. Environ. Sci. Technol. 26:79–85.

Fenter, P.A., M.L. River, N.C. Sturchio, and S.R. Sutton (eds.). 2002. Applications of synchrotron radiation in low-temperature geochemistry and environmental science. Reviews in mineralogy and geochemistry. Vol. 35. The Mineralogical Society of America, Washington DC.

Ford, R.G., A.C. Scheinost, K.G. Scheckel, and D.L. Sparks. 1999. The link between clay mineral weathering and the stabilization of Ni surface precipitates. Environ. Sci. Technol. 33:3140–3144.

Ford, R.G., and D.L. Sparks. 2000. The nature of Zn precipitates formed in the presence of pyrophyllite. Environ. Sci. Technol. 34:2479–2483.

Fredrickson, J.K., J.M. Zachara, D.W. Kennedy et al. 1998. Biogenic iron mineralization accompanying the dissimilatory reduction of hydrous ferric oxide by a groundwater bacterium. Geochim. Cosmochim. Acta 62:3239–3257.

Ginder-Vogel, M., T. Borch, M.A. Mayes, P.M. Jardine, and S. Fendorf. 2005. Chromate reduction and retention processes within arid subsurface environments. Environ. Sci. Technol. 39:7833–7839.

Ginder-Vogel, M., C.S. Criddle, and S. Fendorf. 2006. Thermodynamic constraints on the oxidation of biogenic UO_2 by Fe(III) (hydr) oxides. Environ. Sci. Technol. 40:3544–3550.

Ginder-Vogel, M., and S.E. Fendorf. 2008. Biogeochemical uranium redox transformation: Potential oxidants of uraninite, p. 293–320. In M.O. Barnett and D.B. Kent (eds.) Adsorption of metals by geomedia II: Variables, mechanisms, and model applications. Elsevier, Amsterdam, the Netherlands.

Ginder-Vogel, M., G. Landrot, J.S. Fischel, and D.L. Sparks. 2009. Quantification of rapid environmental redox processes with quick-scanning x-ray absorption spectroscopy (Q-XAS). Proc. Natl. Acad. Sci. USA 106:16124–16128.

Ginder-Vogel, M., and D.L. Sparks. 2010. The impacts of X-ray absorption spectroscopy on understanding soil processes and reaction mechanisms, p. 1–26. In B. Singh and M. Grafe (eds.) Developments in soil science 34: Synchrotron-based techniques in soils and sediments. Elsevier, Burlington, MA.

Ginder-Vogel, M., B. Stewart, and S.E. Fendorf. 2010. Kinetic and mechanistic constraints on the oxidation of biogenic uraninite by ferrihydrite. Environ. Sci. Technol. 44:163–169.

Gräfe, M., R.V. Tappero, M.A. Marcus, and D.L. Sparks. 2008. Arsenic speciation in multiple metal environments—II. Micro-spectroscopic investigation of a CCA contaminated soil. J. Colloid Interface Sci. 321:1–20.

Hansel, C.M., S.G. Benner, and S. Fendorf. 2005. Competing Fe(II)-induced mineralization pathways of ferrihydrite. Environ. Sci. Technol. 39:7147–7153.

Hansel, C.M., S.G. Benner, J. Neiss, A. Dohnalkova, R.K. Kukkadapu, and S. Fendorf. 2003a. Secondary mineralization pathways induced by dissimilatory iron reduction of ferrihydrite under advective flow. Geochim. Cosmochim. Acta 67:2977–2992.

Hansel, C.M., B.W. Wielinga, and S. Fendorf. 2003b. Structural and compositional evolution of Cr/Fe solids after indirect chromate reduction by dissimilatory iron-reducing bacteria. Geochim. Cosmochim. Acta 67:401–412.

Hayes, K.F., A.L. Roe, G.E. Brown, Jr., K.O. Hodgson, J.O. Leckie, and G.A. Parks. 1987. In situ x-ray absorption study of surface complexes: Selenium oxyanions on α-FeOOH. Science 238:783–786.

Hesterberg, D., W.Q. Zhou, K.J. Hutchison, S. Beauchemin, and D.E. Sayers. 1999. XAFS study of adsorbed and mineral forms of phosphate. J. Synchrotron Radiat. 6:636–638.

Hunter, D.B., and P.M. Bertsch. 1998. In situ examination of uranium contaminated soil particles by micro-X-ray absorption and micro-fluorescence spectroscopies. J. Radioanal. Nucl. Chem. 234:237–242.

Isaure, M.P., A. Laboudigue, A. Manceau et al. 2002. Quantitative Zn speciation in a contaminated dredged sediment by μ-PIXE, μ-SXRF, EXAFS spectroscopy and principal component analysis. Geochim. Cosmochim. Acta 66:1549–1567.

Kelly, S.D., D. Hesterberg, and B. Ravel. 2008. Analysis of soils and minerals using X-ray absorption spectroscopy, p. 387–464. In A.L. Ulery and L.R. Drees (eds.) Methods of soil analysis. Part 5—Mineralogical methods. SSSA, Madison, WI.

Kelly, S.D., K.M. Kemner, and S.C. Brooks. 2007. X-ray absorption spectroscopy identifies calcium-oranyl-carbonate complexes at environmental concentrations. Geochim. Cosmochim. Acta 71:821–834.

Kelly, S.D., K.M. Kemner, J.B. Fein et al. 2002. X-ray absorption fine structure determination of pH-dependent U-bacterial cell wall interactions. Geochim. Cosmochim. Acta 66:3855–3871.

Kendelewicz, T., P. Liu, C.S. Doyle, and G.E. Brown, Jr. 2000. Spectroscopic study of the reaction of aqueous Cr(VI) with $Fe_3O_4(111)$ surfaces. Surf. Sci. 469:144–163.

Khalid, S., W.A. Caliebe, P. Siddons, I. So, B. Clay, T. Lenhard, J. Hanson et al. 2010. Quick extended x-ray absorption fine structure instrument with millisecond time scale, optimized for in-situ applications. Rev. Sci. Instrum. 81:015105.

Kocar, B.D., M.J. Herbel, K.J. Tufano, and S. Fendorf. 2006. Contrasting effects of dissimilatory iron(III) and arsenic(V) reduction on arsenic retention and transport. Environ. Sci. Technol. 40:6715–6721.

Krämer, U., I.J. Pickering, R.C. Prince, I. Raskin, and D.E. Salt. 2000. Subcellular localization and speciation of nickel in hyperaccumulator and non-accumulator Thlaspi species. Plant Physiol. 122:1343–1353.

Kramer, U., R.D. Smith, W.W. Wenzel, I. Raskin, and D.E. Salt. 1997. The role of metal transport and tolerance in nickel hyperaccumulation by *Thlaspi goesingense* Halacsy. Plant Physiol. 115:1641–1650.

Küpper, H., A. Mijovilovich, W. Meyer-Klaucke, and P.M.H. Kroneck. 2004. Tissue- and age-dependent differences in the complexation of cadmium and zinc in the cadmium/zinc hyperaccumulator *Thlaspi caerulescens* (Ganges ecotype) revealed by X-ray absorption spectroscopy. Plant Physiol. 134:748–757.

Landrot, G., M. Ginder-Vogel, and D.L. Sparks. 2010. Kinetics of chromium(III) oxidation by manganese(IV) oxides using quick scanning X-ray absorption fine structure spectroscopy (Q-XAFS). Environ. Sci. Technol. 44:143–149.

Liu, C.G., J.M. Zachara, Y.A. Gorby, J.E. Szecsody, and C.F. Brown. 2001. Microbial reduction of Fe(III) and sorption/precipitation of Fe(II) on *Shewanella putrefaciens* strain CN32. Environ. Sci. Technol. 35:1385–1393.

Livi, K.J.T., G.S. Senesi, A.C. Scheinost, and D.L. Sparks. 2009. Microscopic examination of nanosized mixed Ni–Al hydroxide surface precipitates on pyrophyllite. Environ. Sci. Technol. 43:1299–1304.

Manceau, A., and L. Charlet. 1994. The mechanism of selenate adsorption on goethite and hydrous ferric oxide. J. Colloid Interface Sci. 168:87–93.

Manceau, A., L. Charlet, M.C. Boisset, B. Didier, and L. Spadini. 1992. Sorption and speciation of heavy metals on hydrous Fe and Mn oxides. From microscopic to macroscopic. Appl. Clay Sci. 7:201–223.

Manceau, A., B. Lanson, M.L. Schlegel, J.C. Harge, M. Musso, L. Eybert-Berard, J.L. Hazeman, D. Chateigner, and G.M. Lamble. 2000. Quantitative Zn speciation in smelter-contaminated soils by EXAFS spectroscopy. Am. J. Sci. 300:289–343.

Manceau, A., M.A. Marcus, and N. Tamura. 2002. Quantitative speciation of heavy metals in soils and sediments by synchrotron X-ray techniques, p. 341–428. In P. Fenter and N.C. Sturchio (eds.) Applications of synchrotron radiation in low-temperature geochemistry and environmental science. Mineralogical Society of America, Washington, DC.

Manning, B.A., S.E. Fendorf, B. Bostick, and D.L. Suarez. 2002. Arsenic(III) oxidation and arsenic(V) adsorption reactions on synthetic birnessite. Environ. Sci. Technol. 36:976–981.

McBride, M.B. 1991. Processes of heavy and transition metal sorption by soil minerals, p. 149–176. In G.H. Bolt, M.F.D. Boodt, M.H.B. Hayes, and M.B. McBride (eds.) Interactions at the soil colloid-soil solution interface. Kluwer Academic Publishers, Dordrecht, the Netherlands.

McBride, M.B. 1994. Environmental chemistry of soils. Oxford University Press, New York.

McNear, D.H., R.L. Chaney, and D.L. Sparks. 2007. The effects of soil type and chemical treatment on nickel speciation in refinery enriched soils: A multi-technique investigation. Geochim. Cosmochim. Acta 71:2190–2208.

McNear, D.H., E. Peltier, J. Everhart, R.L. Chaney, S. Sutton, M. Newville, M. Rivers, and D.L. Sparks. 2005a. Application of quantitative fluorescence and absorption-edge computed microtomography to image metal compartmentalization in *Alyssum murale*. Environ. Sci. Technol. 39:2210–2218.

McNear, D.H., Jr., R. Tappero, and D.L. Sparks. 2005b. Shining light on metals in the environment. Elements 1:211–216.

Mitsunobu, S., Y. Takahashi, and T. Uruga. 2006. Observation of chemical reactions at the solid-water interface by quick XAFS combined with a column reactor. Anal. Chem. 78:7040–7043.

Nachtegaal, M., M.A. Marcus, J.E. Sonke et al. 2005. Effects of in situ remediation on the speciation and bioavailability of zinc in a smelter contaminated soil. Geochim. Cosmochim. Acta 69:4649–4664.

Nico, P.S., B.D. Stewart, and S.E. Fendorf. 2009. Incorporation of uranium(VI) into Fe (hydr)oxides during Fe(II) catalyzed remineralization. Environ. Sci. Technol. 43:7391–7396.

O'Day, P.A. 1992. Structure, bonding, and site preference of cobalt(II) sorption complexes on kaolinite and quartz from solution and spectroscopic studies. PhD Dissertation. Stanford University, Stanford, CA.

O'Day, P.A. 1999. Molecular environmental geochemistry. Rev. Geophys. 37:249–274.

O'Day, P.A., G.E. Brown, Jr., and G.A. Parks. 1994. X-ray absorption spectroscopy of cobalt(II) multinuclear surface complexes and surface precipitates on kaolinite. J. Colloid Interface Sci. 165:269–289.

Papelis, C., and K.F. Hayes. 1996. Distinguishing between interlayer and external sorption sites of clay minerals using X-ray absorption spectroscopy. Colloid Surf. A 107:89–96.

Parikh, S.J., B.J. Lafferty, and D.L. Sparks. 2008. An ATR-FTIR spectroscopic approach for measuring rapid kinetics at the mineral/water interface. J. Colloid Interface Sci. 320:177–185.

Patterson, R.R., S. Fendorf, and M. Fendorf. 1997. Reduction of hexavalent chromium by amorphous iron sulfide. Environ. Sci. Technol. 31:2039–2044.

Peak, D., S. Hunger, Y. Arai, and D. Sparks. 2002. XANES spectroscopic studies of sulfate and phosphate chemistry in soils. Abstr. Pap. Am. Chem. Soc. 223:U609–U609.

Peltier, E., R. Allada, A. Navrotsky, and D.L. Sparks. 2006. Nickel solubility and precipitation in soils: A thermodynamic study. Clay. Clay Miner. 54:153–164.

Peltier, E., D. van der Lelie, and D.L. Sparks. 2010. Formation and stability of Ni-Al hydroxide phases in Delaware soils. Environ. Sci. Technol. 44:302–308.

Peterson, M.L., G.E. Brown, Jr., and G.A. Parks. 1996. Direct XAFS evidence for heterogeneous redox reaction at the aqueous chromium/magnetite interface. Colloid Surf. A 107:77–88.

Peterson, M.L., G.E. Brown, Jr., G.A. Parks, and C.L. Stein. 1997. Differential redox and sorption of Cr(III/VI) on natural silicates and oxide minerals: EXAFS and XANES results. Geochim. Cosmochim. Acta 61:3399–3412.

Roberts, D.R., R.G. Ford, and D.L. Sparks. 2003. Kinetics and mechanisms of Zn complexation on metal oxides using EXAFS spectroscopy. J. Colloid Interface Sci. 263:364–376.

Roberts, D., M. Nachtegaal, and D.L. Sparks. 2005. Speciation of metals in soils, p. 619–654. *In* M.A. Tabatabai and D.L. Sparks (eds.) Chemical processes in soils. SSSA, Madison, WI.

Roberts, D.R., A.M. Scheidegger, and D.L. Sparks. 1999. Kinetics of mixed Ni-Al precipitate formation on a soil clay fraction. Environ. Sci. Technol. 33:3749–3754.

Roe, A.L., K.F. Hayes, C. Chisholm-Brause, G.E. Brown, G.A. Parks, K.O. Hodgson, and J.O. Leckie. 1991. In situ X-ray absorption study of lead ion surface complexes at the goethite-water interface. Langmuir 7:367–373.

Sarret, G., E. Harada, Y.E. Choi, M.P. Isaure, N. Geoffroy, S. Fakra, M.A. Marcus, M. Birschwilks, S. Clemens, and A. Manceau. 2006. Trichomes of tobacco excrete zinc as zinc-substituted calcium carbonate and other zinc-containing compounds. Plant Physiol. 141:1021–1034.

Sarret, G., P. Saumitou-Laprade, V. Bert, O. Proux, J.L. Hazemann, A. Traverse, M.A. Marcus, and A. Manceau. 2002. Forms of zinc accumulated in the hyperaccumulator *Arabidopsis halleri*. Plant Physiol. 130:1815–1826.

Sass, B.M., and D. Rai. 1987. Solubility of amorphous chromium(III)-iron(III) hydroxide solid solutions. Inorg. Chem. 26:2228–2232.

Sayers, D.E., E.A. Stern, and F.W. Lytle. 1971. New technique for investigating noncrystalline structures: Fourier analysis of extended X-ray—Absorption fine structure. Phys. Rev. Lett. 27:1204–1207.

Scheckel, K.G., E. Lombi, S.A. Rock, and N.J. McLaughlin. 2004. In vivo synchrotron study of thallium speciation and compartmentation in lberis intermedia. Environ. Sci. Technol. 38:5095–5100.

Scheckel, K.G., and D.L. Sparks. 2001a. Dissolution kinetics of nickel surface precipitates on clay mineral and oxide surfaces. Soil Sci. Soc. Am. J. 65:685–694.

Scheckel, K.G., and D.L. Sparks. 2001b. Temperature effects on nickel sorption kinetics at the mineral-water interface. Soil Sci. Soc. Am. J. 65:719–728.

Scheidegger, A.M., M. Fendorf, and D.L. Sparks. 1996a. Mechanisms of nickel sorption on pyrophyllite: Macroscopic and microscopic approaches. Soil Sci. Soc. Am. J. 60:1763–1772.

Scheidegger, A.M., G.M. Lamble, and D.L. Sparks. 1996b. Investigation of Ni sorption on pyrophyllite: An XAFS study. Environ. Sci. Technol. 30:548–554.

Scheidegger, A.M., G.M. Lamble, and D.L. Sparks. 1997. Spectroscopic evidence for the formation of mixed-cation hydroxide phases upon metal sorption on clays and aluminum oxides. J. Colloid Interface Sci. 186:118–128.

Scheidegger, A.M., and D.L. Sparks. 1996. A critical assessment of sorption-desorption mechanisms at the soil mineral/water interface. Soil Sci. 161:813–831.

Scheidegger, A.M., D.G. Strawn, G.M. Lamble, and D.L. Sparks. 1998. The kinetics of mixed Ni-Al hydroxide formation on clay and aluminum oxide minerals: A time-resolved XAFS study. Geochim. Cosmochim. Acta 62:2233–2245.

Scheinost, A.C., R.G. Ford, and D.L. Sparks. 1999. The role of Al in the formation of secondary Ni precipitates on pyrophyllite, gibbsite, talc, and amorphous silica: A DRS study. Geochim. Cosmochim. Acta 63:3193–3203.

Schulze, D.G., and P.M. Bertsch. 1995. Synchrotron x-ray techniques in soil, plant, and environmental research. Adv. Agron. 55:1–66.

Schulze, D.G., J.W. Stucki, and P.M. Bertsch. 1999. Synchrotron x-ray methods in clay science. CMS Workshop Lectures. Vol. 98. Clay Minerals Society, Boulder, CO.

Seiter, J.M., K.E. Staats-Borda, M. Ginder-Vogel, and D.L. Sparks. 2008. XANES spectroscopic analysis of phosphorus speciation in alum-amended poultry litter. J. Environ. Qual. 37:477–485.

Shober, A.L., D.L. Hesterberg, J.T. Sims, and S. Gardner. 2006. Characterization of phosphorus species in biosolids and manures using XANES spectroscopy. J. Environ. Qual. 35:1983–1993.

Sparks, D.L. 1995. Environmental soil chemistry. Academic Press, San Diego, CA.

Sparks, D.L. 2002. Environmental soil chemistry. 2nd edn. Academic Press, San Diego, CA.

Sparks, D.L. 2005a. Toxic metals in the environment: The role of surfaces. Elements 1:193–197.

Sparks, D.L. 2005b. Metal and oxyanion sorption on naturally occurring oxide and clay mineral surfaces, p. 33–36. *In* V.H. Grassian (ed.) Environmental catalyis. Taylor & Francis, Boca Raton, FL.

Sposito, G. 1986. Distinguishing adsorption from surface precipitation, p. 217–229. *In* J.A. Davis and K.F. Hayes (eds.) Geochemical processes at mineral surfaces. American Chemical Society, Washington, DC.

Stewart, B.D., P.S. Nico, and S. Fendorf. 2009. Stability of uranium incorporated into Fe (hydr)oxides under fluctuating redox conditions. Environ. Sci. Technol. 43:4922–4927.

Strawn, D., H. Doner, M. Zavarin, and S. McHugo. 2002. Microscale investigation into the geochemistry of arsenic, selenium, and iron in soil developed in pyritic shale materials. Geoderma 108:237–257.

Strawn, D.G., and D.L. Sparks. 1999. The use of XAFS to distinguish between inner- and outer-sphere lead adsorption complexes on montmorillonite. J. Colloid Interface Sci. 216:257–269.

Sutton, S.R. 2006. User research facilities in the earth sciences. Elements 2:7–8.

Tappero, R., E. Peltier, M. Grafe et al. 2007. Hyperaccumulator *Alyssum murale* relies on a different metal storage mechanism for cobalt than for nickel. New Phytol. 175:641–654.

Taylor, R.M. 1984. The rapid formation of crystalline double hydroxy salts and other compounds by controlled hydrolysis. Clay Miner. 19:591–603.

Thompson, H.A., G.A. Parks, and G.E. Brown, Jr. 1999a. Ambient-temperature synthesis, evolution, and characterization of cobalt-aluminum hydrotalcite-like solids. Clay. Clay Miner. 47:425–438.

Thompson, H.A., G.A. Parks, and G.E. Brown, Jr. 1999b. Dynamic interactions of dissolution, surface adsorption, and precipitation in an aging cobalt(II)-clay-water system. Geochim. Cosmochim. Acta 63:1767–1779.

Toor, G.S., B.J. Cade-Menun, and J.T. Sims. 2005. Establishing a linkage between phosphorus forms in dairy diets, feces, and manures. J. Environ. Qual. 34:1380–1391.

Toor, G.S., S. Hunger, J.D. Peak, J.T. Sims, and D.L. Sparks. 2006. Advances in the characterization of phosphorus in organic wastes: Environmental and agronomic applications. Adv. Agron. 89:1–72.

Towle, S.N., J.R. Bargar, G.E. Brown, and G.A. Parks. 1997. Surface precipitation of Co(II)(aq) on Al_2O_3. J. Colloid Interface Sci. 187:62–82.

Webb, S.M., B.M. Tebo, and J.R. Bargar. 2005. Structural characterization of biogenic Mn oxides produced in seawater by the marine bacillus sp. strain SG-1. Am. Mineral. 90:1342–1357.

Wijnja, H., and C.P. Schulthess. 2000. Vibrational spectroscopic study of selenate and sulfate adsorption mechanisms on Fe and Al (hydr)oxides surfaces. J. Colloid Interface Sci. 229:286–297.

Zhang, P.C., and D.L. Sparks. 1990. Kinetics and mechanisms of sulfate adsorption desorption on goethite using pressure-jump relaxation. Soil Sci. Soc. Am. J. 54:1266–1273.

2
Clay–Organic Interactions in Soil Environments

Guodong Yuan
Landcare Research

Benny K.G. Theng
Landcare Research

2.1 Introduction 2-1
2.2 Structures, Properties, and Reactivity of Soil Clays and Organic Matter 2-2
    Clay-Size Minerals in Soil • Organic Matter
2.3 Clay–Organic Interactions 2-5
    Mechanisms of Clay–Organic Interactions • Methodologies for Investigating Clay–Organic Interactions • Effects of Clay–Organic Interactions
2.4 Conclusions and Perspectives 2-12
References 2-13

2.1 Introduction

Soil may be regarded as a complex bioorganomineral system where minerals, water, air, organic matter (OM) (humus), and microorganisms interact at scales ranging from nanometers to kilometers. Figure 2.1 gives a simplified diagram of the clay–humus–microbe in soil. In this chapter, we focus attention on the interactions of clays with OM, a process that is arguably as important to sustaining life on earth as is photosynthesis (Jacks, 1973). The clay–organic interaction plays a key role in stabilizing soil OM against microbial decomposition, thus making soils suitable for plant growth, while photosynthesis produces biomass for inputs into soil, as well as for human and animal consumption.

It has long been recognized that much of the OM in soil is closely associated with the clay fraction (Schloesing, 1874). The chemistry of clay–organic reactions, however, is a relatively young discipline, dating back to the early 1940s. The publication of Greenland's papers (1965a, 1965b) provided the stimulus for a systematic investigation into the behavior of well-defined small and polymeric organic molecules at specific clay mineral surfaces. The advances in this branch of clay–organic chemistry that have since been made can be traced in the reviews and monographs by Mortland (1970), Theng (1974, 1979, 1982), Theng and Tate (1989), Hayes and Mingelgrin (1991), Clapp et al. (1991), Oades (1989), Violante and Gianfreda (2000), Yariv and Cross (2002), and Lagaly et al. (2006).

Progress in understanding the clay–organic interaction in soil environments has been hampered by the diversity and structural complexity of the reactants, the variety of bonding modes that operate simultaneously, and the microbial participation (Theng, 1980; Huang and Schnitzer, 1986; Oades, 1988; Huang, 2004; Kögel-Knabner et al., 2008). In response, the International Society of Soil Science established Working Group MO in 1990, which became Commission 2.5 (Soil Physical/Chemical/Biological Interfacial Reactions) of the International Union of Soil Sciences in 2004. Working Group MO and Commission 2.5 have since sponsored five international symposia on soil mineral–microbe–organic interactions (ISMOM), the latest one being held in Pucón, Chile, in 2008 (ISMOM, 2008). The titles of the books arising from these symposia—*Environmental Impact of Soil Component Interactions* (Huang et al., 1995), *Effect of Mineral-Organic-Microorganism Interactions on Soil and Freshwater Environments* (Berthelin et al., 1999), *Soil Mineral-Organic Matter-Microorganism Interactions and Ecosystem Health* (Violante et al., 2002), and *Soil Mineral-Microbe-Organic Interactions: Theories and Applications* (Huang et al., 2008)—reflect the issues, concerns, and developments in this field of soil research.

Even after nearly five decades of investigation, some details about the clay–organic interaction await clarification. This is because progress in soil mineralogy, soil OM chemistry, and soil ecology has been slow and incremental, while soil research over the last two decades has focused more on "environmental" issues than basic aspects. The clay–organic interaction has recently attracted renewed interest, however, because of its important role in stabilizing and sequestering organic carbon (OC) in soil (Kaiser and Guggenberger, 2000, 2003; Swift, 2001; Kleber et al., 2005; Chenu and Plante, 2006; Basile-Doelsch et al., 2007; Huang, 2008). This relationship was integral to the development and refinement of models describing the dynamics of carbon in soil and their impact on greenhouse gas emissions (Jastrow and Miller, 1997; Smith et al., 1997; Liski et al., 1999; Jenkinson and Coleman, 2008).

Besides being a principal determinant in the stabilization of OC in soil, the clay–organic interaction is involved in essential ecological services and functions, ranging from regulating water storage and nutrient cycling, through adsorbing and

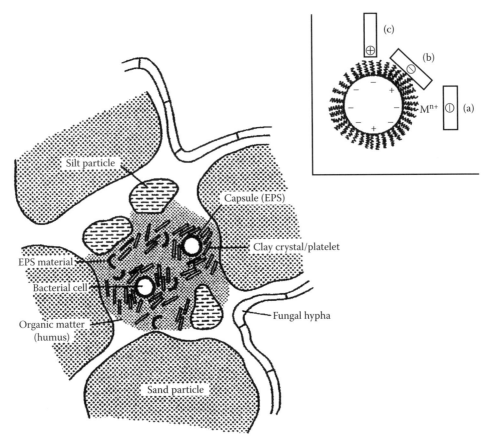

FIGURE 2.1 Diagram showing the association of minerals, OM, and microbes in soil. Bacterial cells with a coat of extracellular polysaccharides (EPS) are enveloped by clay particles. Inset shows an enlarged view of a bacterial cell with its compliment of EPS. Under normal pH conditions, the cell has a net negative charge. Most clay particles adhere to the cell surface by bridging through polyvalent cations, M^{n+} (a) although some may be attached directly by electrostatic interactions, indicated by (b) and (c). (Reprinted from Theng, B.K.G., and V.A. Orchard. 1995. Interactions of clays with microorganisms and bacterial survival in soil: A physicochemical perspective, p. 123–143. In P.M. Huang, J. Berthelin, J.-M. Bollag, W.B. McGill, and A.L. Page (eds.) Environmental impact of soil component interactions. Vol. II. CRC Lewis Publishers, Boca Raton, FL. With permission from Taylor & Francis.)

transporting metal and organic contaminants, to serving as a source or sink of OC and greenhouse gases. Improved understanding of the clay–organic interaction in this context is therefore a key to using soils for food, fiber, and fuel production in a sustainable manner. This chapter describes the interactions of clay-size minerals in soil (phyllosilicates, short-range order minerals, and metal (hydr)oxides) with natural OM and the effects of these interactions on some soil properties and functions.

2.2 Structures, Properties, and Reactivity of Soil Clays and Organic Matter

2.2.1 Clay-Size Minerals in Soil

Being the "living skin of earth," soil contains a wide variety of minerals and organic substances as well as a myriad of microorganisms. Here we focus on the clay fraction, which comprises particles of <2 μm equivalent spherical diameter, including nanoparticles (<100 nm). The clay-size minerals in soil are very reactive not only because of their large surface area, but more so because a high proportion of the structural atoms are exposed on surface sites where chemical bonds are broken and coordination requirements are unsatisfied. The excess energy associated with such surface defects and dislocations lies behind their propensity to interact with water, plant nutrients, metals, and organic substances (Oades, 1988, 1989; Hochella et al., 2008; Theng and Yuan, 2008). By the same token, clay particles are tiny because they have a high density of structural defects, as a result of which nucleation is favored over crystal growth (Meunier, 2006).

The clay fraction of soils consists of various phyllosilicates (clay minerals), short-range order minerals, metal (hydr)oxides, and associated humic substances (HS). The clay fraction may also contain quartz and feldspar. The structures and properties of some common clay-size ("fine-grained") minerals are outlined below. Chapters 20 through 23 of *Handbook of Soil Sciences: Properties and Processes* and the books edited by Dixon and Weed (1989) and Bergaya et al. (2006) should be consulted for more details.

2.2.1.1 Phyllosilicates

The basic structural unit of phyllosilicates (layer silicates) is a layer formed by combining a silica tetrahedral sheet and an alumina octahedral sheet in a given proportion. In 1:1 type phyllosilicates, such as kaolinite, each layer is made up of one tetrahedral and one octahedral sheet, giving a thickness (basal spacing) of about 0.7 nm (Figure 2.2, top). Another species in this group of layer silicates is halloysite whose structure is similar to that of kaolinite but the 1:1 layers are separated by a sheet of water molecules. As a result, the basal spacing of halloysite is 1 nm, which decreases to 0.7 nm on dehydration (Joussein et al., 2005; Brigatti et al., 2006). On the other hand, individual layers in 2:1 type phyllosilcates consist of an octahedral sheet sandwiched between two tetrahedral sheets (Figure 2.2, bottom), giving a layer thickness of about 1 nm. Since the interlayer space may contain inorganic cations with or without water (e.g., smectite, mica), or metal hydroxides (e.g., chlorite), the basal spacing of 2:1 phyllosilicates commonly exceeds 1 nm. Phyllosilicates in soil often occur as mixed-layer or interstratified minerals, such as mica–smectite, mica–chlorite, and kaolinite–smectite rather than as single mineral species (Schulze, 1989).

As a result of isomorphous substitution (e.g., Al^{3+} for Si^{4+} in the tetrahedral sheet and Mg^{2+} for Al^{3+} in the octahedral sheet) within the layers, many clay minerals have a permanent negative charge, which is balanced by inorganic cations (e.g., Na^+, Ca^{2+}) occupying interlayer sites. Since these charge-balancing cations ("counterions") are exchangeable, many phyllosilicates have a permanent cation exchange capacity (CEC) that does not vary with the pH of the ambient solution. Thus, the high CEC values (90–130 $cmol_c$ kg^{-1}) of smectites is a reflection of extensive isomorphous substitution. By comparison, the CEC of kaolinite rarely exceeds 10 $cmol_c$ kg^{-1} because of limited substitution in layer structure, and its CEC can vary with solution pH. Some nanoclays in soil (e.g., allophane) have little, if any, isomorphous substitution. Both the surface charge and CEC of such materials are therefore controlled by solution pH (Table 2.1).

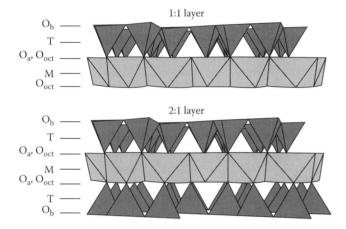

FIGURE 2.2 Models of a 1:1 and 2:1 layer structure. O_a, O_b, and O_{oct} refer to tetrahedral apical, tetrahedral basal, and octahedral anionic position, respectively. M and T indicate the octahedral and tetrahedral cation, respectively. The thickness of a 1:1 layer is about 0.7 nm, while that of a 2:1 layer is about 1 nm. (Reprinted from Brigatti, M.F., E. Galan, and B.K.G. Theng. 2006. Structures and mineralogy of clay minerals, p. 19–86. In F. Bergaya, B.K.G. Theng, and G. Lagaly (eds.) Handbook of clay science. Elsevier, Amsterdam, the Netherlands. With permission.)

2.2.1.2 Short-Range Order Minerals

Imogolite and allophane are short-range order or poorly crystalline nanosize aluminosilicates although imogolite has long-range order along the tubule axis (Wada, 1989; Farmer and Russell, 1990). The unit particle of imogolite is a slender hollow tubule with an outer diameter of about 2 nm. The tubule wall

TABLE 2.1 Composition, Charge Characteristics, and Specific Surface Area of Common Soil Minerals

Minerals	Formula	CEC ($cmol_c$ kg^{-1})	PZC	Specific Area (m^2 g^{-1})
Kaolinite	$Si_2Al_2O_5(OH)_4$	3–15	4–5	5–30
Halloysite (7 Å)	$Si_2Al_2O_5(OH)_4$	40–50		5–10
Halloysite (10 Å)	$Si_2Al_2O_5(OH)_4 \cdot 2H_2O$			10–40
Montmorillonite	$M_{0.33}Si_4(Al_{1.67}Mg_{0.33})O_{10}(OH)_2$	80–150		800 (CPB)
Quartz	SiO_2	0.6–5.3	2–3	1–3.6
Goethite	α-FeOOH	0–6	7–8	14–177
Hematite	α-Fe_2O_3	0–6	8–8.5	30–50
Gibbsite	γ-$Al(OH)_3$	0.2–3	8–9	~50
Imogolite (natural)	$(OH)_3Al_2O_3SiOH$ (Al/Si = 2.0)	Variable pH-dependent	6.8	822–1031 (EGME)
Allophane (synthetic)	Al/Si = 1.15–1.74	Variable pH-dependent	5.5–6.9	638–897 (EG, EGME)
Ferrihydrite (6-line)	$Fe_{1.55}O_{1.66}(OH)_{1.34}$	Variable pH-dependent	5.3–7.5	200–500[b]
Birnessite (synthetic)[a]	δ-MnO_2	Variable pH-dependent	2.7	72 (EGME)

Sources: Data compiled from Feldman, S.B., C. Shang, and L.W. Zelazny. 2008. Soil mineralogy, p. 678–686. In W. Chesworth (ed.) Encyclopedia of soil science. Springer, Dordrecht, the Netherlands; Theng, B.K.G., and V.A. Orchard. 1995. Interactions of clays with microorganisms and bacterial survival in soil: A physicochemical perspective, p. 123–143. In P.M. Huang, J. Berthelin, J.-M. Bollag, W. McGill, and A.L. Page (eds.) Environmental impact of soil component interactions. Vol. II. CRC Lewis Publishers, Boca Raton, FL. Specific surface area values refer to N_2-BET values unless otherwise indicated. EG, ethylene glycol; EGME, ethylene glycol monoethyl ether; CPB, cetylpyridinium bromide.

[a] Taken from Scott and Morgan (1996).
[b] From Schwertmann and Taylor (1989).

consists of an outer curved Al octahedral (gibbsitic) sheet to which O₃SiOH groups are attached on the inside (Cradwick et al., 1972; Farmer and Fraser, 1979). The unit formula of imogolite is usually written as $(OH)_3Al_2O_3SiOH$, showing an Al/Si ratio of 2.0, and indicating the sequence of ions from the periphery to the center of the tubule where the orthosilicate group shares three oxygens with aluminum. Unlike imogolite, allophane has no fixed chemical composition in that its Al/Si ratio can vary between 1 and 2. Irrespective of composition and origin, however, the unit particle of allophane is a hollow spherule with an outer diameter of 3.5–5.0 nm. The 0.7–1.0 nm thick spherule wall is composed of an outer gibbsitic sheet and an inner Si sheet. Defects in the wall structure give rise to perforations of ∼0.3 nm in diameter (Wada, 1989; Parfitt, 1990; Brigatti et al., 2006).

Consisting of nanosize spherules, allophane has a large specific surface area (SSA). A value of about 1000 m² g⁻¹ is calculated for the external (interspherule) and internal (intraspherule) area on the basis of unit particle size and density, while 700–900 m² g⁻¹ is measured by retention of ethylene glycol and ethylene glycol monoethyl ether (Hall et al., 1985; Wada, 1989; Table 2.1). Allophane has a variable (pH-dependent) surface charge attributable to the protonation and deprotonation of $(OH)Al(H_2O)$ groups exposed at perforations along the spherule wall. The point of zero net charge of allophanes ranges from pH 5 to 6 (Theng et al., 1982; Clark and McBride, 1984; Mora et al., 1994; Hashizume and Theng, 2007). The chemical properties of allophane-rich soils reflect the surface charge characteristic of this nanoclay.

Although allophane, imogolite, and halloysite occur in nonvolcanic soils, such as the lower horizons of podzols (Farmer and Russell, 1990), they are especially widespread in volcanic soils (Dahlgren et al., 2004). In humid environments, these minerals form by rapid weathering of volcanic ash ("tephra") where nucleation and precipitation are kinetically favored over crystal growth. Another common nanosize mineral in volcanic soils is ferrihydrite. Like allophane and imogolite, ferrihydrite is a "natural" (soil) nanomineral without microscale analogues. Thus, the size of ferrihydrite particles has never been observed to exceed 20 nm in diameter but is more typically ≤10 nm (Hochella, 2008). The formation and persistence of allophane, imogolite, and ferrihydrite in soil are greatly influenced by OM. Eusterhues et al. (2008), for example, show that even small amounts of water-extractable OM can appreciably reduce the particle size and structural order of ferrihydrite, while the input of large amounts of plant biomass tends to inhibit allophane and imogolite formation (Inoue and Huang, 1986, 1990; Parfitt, 2009).

2.2.1.3 Metal (Hydr)oxides

Most soils also contain a variety of metal "(hydr)oxides" (an umbrella term for oxides, hydroxides, and oxyhydroxides), among which those of Al, Fe, and Mn are very common. Like allophane, Al-, Fe-, and Mn-(hydr)oxides are short-range order, nanosize minerals capable of interacting with OM. These minerals often occur as a "coat" over clay mineral surfaces (Theng and Yuan, 2008), and affect soil properties. Soil color, for example, reflects the presence of Fe (hydr)oxides since these nanosize minerals are strongly colored: hematite ($\alpha\text{-}Fe_2O_3$, red), maghemite ($\gamma\text{-}Fe_2O_3$, reddish-brown), magnetite (Fe_3O_4, black), goethite ($\alpha\text{-}FeOOH$, yellow–brown), lepidocrocite ($\gamma\text{-}FeOOH$, orange), and ferrihydrite ($Fe_{1.5}O_{1.66}(OH)_{1.34}$, red brown). In general, however, the color of surface soils is related to the OC content (Konen et al., 2003). Further, ferrihydrite can transform into hematite through a solid-state reaction, or into goethite by dissolution and precipitation (Schwertmann, 2008). The formation of Al-(hydr)oxides in soil commonly involves the release of Al from primary aluminosilicates, followed by its hydrolysis and precipitation. In terms of soil occurrence, gibbsite [$\gamma\text{-}Al(OH)_3$] is more common than boehmite [$AlO(OH)$]. Like their iron counterparts, the (hydr)oxides of Mn (III), Mn (IV), and Mn (III, IV) in soil form by the oxidation of Mn (II), released through weathering of Mn-containing rocks. The oxidation process may be chemically controlled or mediated by bacteria and fungi. The most common species of Mn-(hydr)oxides in soil are birnessite and vernadite. The surface charge of metal-(hydr)oxides is an important factor affecting their interactions with OM in soil. The point of zero charge of most soil Mn-(hydr)oxides is less than pH 4, while that of their aluminum and iron counterparts is larger than pH 7 (Table 2.1). Thus, in the pH range (4.5–7.5) of most soils, Mn-(hydr)oxides are negatively charged whereas the surface charge of Al- and Fe-(hydr)oxides is positive (Theng and Yuan, 2008).

2.2.2 Organic Matter

A large part of the OC on the earth surface, including the atmosphere, pedosphere, biosphere, and hydrosphere, is to be found in soil. Globally, the top 1 m of soil contains about 1500 Pg (1 Pg = 10¹⁵ g) of carbon, that is, about 2.5 times as much carbon as in the terrestrial vegetation (≈600 Pg), and about twice as much as in the atmosphere (≈750 Pg) (Batjes, 1996; Almendros, 2008). About 60% of the nonliving ("dead") OM in soil occurs in the form of HS. The other 40% consists of particulate OM (i.e., organic fragments with a recognizable cellular structure), dissolved OM (e.g., amino acids, sugars, organic acids, phenolic compounds), and non-humic biopolymers (e.g., enzymes, proteins, polysaccharides, lipids, nucleic acids) (Baldock and Nelson, 2000; Baldock and Nelson, Chapter 11 of *Handbook of Soil Sciences: Properties and Processes*). The content of OM in soil is not directly measurable but may be estimated by multiplying the OC content (obtained by either dry combustion or wet digestion) by 1.724, that is, %OM = 1.724 × %OC.

It is not our intention to summarize the large volume of literature on the separation, characterization, and physicochemical properties of the various components of soil OM (Schnitzer and Khan, 1972, 1978; Flaig et al., 1975; Aiken et al., 1985; Stevenson, 1994; Baldock and Nelson, 2000; Baldock and Nelson, Chapter 11 of *Handbook of Soil Sciences: Properties and Processes*). Rather, the emphasis is on the interactions of OM with clay-size minerals and their impact on soil properties and environmental changes. Further, we exclude plant litter, macroorganic matter (>250 μm in size), charcoal, and the "light fraction" (obtainable by flotation on liquids with densities of 1.6–2.0 Mg m⁻³) from the discussion, but include HS and the microbial biomass.

HS are the most important class of organic materials in the soil environment in terms of quantity, stability, and ubiquity. They are conventionally divided into three fractions based on their solubility in alkali or acid: humic acid (alkali-soluble), fulvic acid (alkali- and acid-soluble), and humin (alkali- and acid-insoluble). Although these definitions are operational rather than chemical, humin represents the fraction that is so strongly bound to clay-size minerals as to resist extraction by alkali and acid. The humin fraction accounts for up to 50% of the total OM in soil, and its radiocarbon (^{14}C) age often exceeds 1000 years (Horwath, 2008).

The microbial biomass accounts for less than 5% of the OC in soil. Nevertheless, microorganisms play a determining role in the turnover and cycling of nutrients as well as in aggregate formation and stabilization (Figure 2.1). The microbial biomass may be estimated by several methods, including chloroform fumigation-incubation/extraction and determination of adenosine triphosphate (Jenkinson and Powlson, 1976; Vance et al., 1987; Oades, 1989).

A long-held view is that HS are a mixture of linear polymers of high molecular weight (>10,000 Da), formed by heteropolycondensation of various biological precursors, such as phenolic acids, peptides, oligosaccharides, and fatty acids, held together by covalent and hydrogen bonding (Kononova, 1966; Flaig et al., 1975; Stevenson, 1994). In basic or low ionic strength solutions, these polymers adopt a random-coil conformation. The coils become denser, more rigid, and more compact in acidic or high ionic strength solutions (Oades, 1989; Swift, 1999). The macromolecular random-coil concept, however, is being displaced by one in which HS consist of aggregates of biologically derived molecules of relatively low molecular weight (200–3000 Da). The various molecules in the supramolecular structure or self-associating aggregate are bound by weak dispersive and hydrophobic interactions, forming micelles in solution in which interior hydrophobic regions are separated from the aqueous surrounding by exterior hydrophilic portions (Piccolo, 2001; Sutton and Sposito, 2005).

We should point out, however, that many low-molecular-weight biomolecules in soil, such as amino acids, sugars, and polyphenols can undergo polymerization and/or polycondensation. Being catalyzed by enzymes and minerals, these processes give rise to the browning reaction and the dark color of HS (Stevenson, 1994; Huang, 2000; Gonzalez and Laird, 2004). Moreover, most biomolecules are crystalline whereas HS are essentially amorphous. Nevertheless, biomolecules may be attached to, or encapsulated by, humic polymers. Thus, HS are not merely aggregates of biologically derived molecules but comprise polymeric materials into which various biomolecules are "trapped" and preserved (Schulten and Schnitzer, 1995; Huang and Hardie, 2009).

2.3 Clay–Organic Interactions

The ability of clays and clay minerals to take up organic substances has been known since antiquity. An example is the use of water slurries of calcium montmorillonite for cleaning raw wool (by removing grease and dirt) that dates back to 5000 BC. This process, known as fulling, gave rise to the term "fuller's earth" (Robertson, 1986). In his *Natural History*, Pliny the Elder (23–79 AD) described the use of clays for curing gastrointestinal ailments. The therapeutic activity of clays is based on their propensity for adsorbing organic toxins and bacteria (Carretero et al., 2006).

2.3.1 Mechanisms of Clay–Organic Interactions

As already remarked, clay-size minerals in soil are very reactive toward external molecules because of their tiny particle size, large surface area, high density of structural defects, and peculiar charge characteristics. We should also add that these minerals seldom exist as discrete, "clean" particles in soil; rather, they are associated with soil OM (humus), including microorganisms (Figure 2.1).

Because of the complexity, diversity, and heterogeneity of clay-size soil minerals and soil OM, the clay–organic interaction has largely been investigated by reacting specific phyllosilicates with well-defined small organic compounds and biomolecules (Mortland, 1970; Theng, 1974; Oades, 1989; Hayes and Mingelgrin, 1991; Yariv and Cross, 2002; Lagaly et al., 2006). The principles and underlying mechanisms that emerge from these studies have been extended to the sorption of organic polymers and HS (Theng, 1979; Clapp et al., 1991; Essington, 2003; Sposito, 2008). Among the phyllosilicates, montmorillonite has been the single, most widely used species because of its large surface area (700–800 $m^2\ g^{-1}$), CEC (~100 $cmol_c\ kg^{-1}$), and propensity for intercalating a wide variety of organic molecules.

2.3.1.1 Interactions with Low-Molecular-Weight Organic Compounds

Nonionic, polar organic molecules must compete with water for the same sites around the exchangeable cation if adsorption is to occur. Thus, little adsorption takes place from dilute aqueous solutions for molecules with less than five carbon atoms (Hoffmann and Brindley, 1960). In order to minimize competition from excess water, the organic solute is often added at a high concentration, in an organic solvent, or as a liquid, vapor, or solid to dry montmorillonite. In conjunction with x-ray diffractometry, infrared spectroscopy has provided strong evidence to show that the intercalated organic is associated with the exchangeable cation through ion–dipole interactions (Mortland, 1970; Theng, 1974). For montmorillonite containing counterions of low hydration energy (e.g., Na^+), the polar group of the organic molecule can associate directly with the cation. This mode of interaction may be regarded as one of ligand exchange for the cation-coordinated water molecules as a result of which the interlayer contracts and the basal spacing decreases. With counterions of high hydration energy (e.g., Ca^{2+}, Mg^{2+}), on the other hand, the organic molecule tends to interact through a water bridge, that is, by hydrogen bonding between its polar group and water in the primary hydration shell of the exchangeable cation. The intercalated organic species (e.g., maleic hydrazide) may simultaneously associate with the interlayer siloxane surface through hydrogen bonding (Cornejo and Hermosin, 1996).

Being essentially hydrophobic, the siloxane surface can also sorb nonionic organic molecules through hydrophobic interactions as Laird et al. (1992) have proposed for atrazine. Further, the entropy change involved in moving fully hydrated atrazine from water to a partially hydrated interlayer space contributes to sorption affinity (Chappell et al., 2005). Sorption is also enhanced when organic intercalation or high solution ionic strength induces the dehydration and collapse of the interlayer space (Li et al., 2006, 2007). Thus, the combination of ion–dipole and hydrophobic interactions with interlayer collapse (by dehydration) enables some smectites to sorb many nonionic organics even from aqueous suspensions. Low-charge smectites containing cations of low hydration energy would have extensive siloxane "patches" between exchangeable cations available for sorption relative to their high-charged counterparts. De Oliveira et al. (2005), for example, found that more carbaryl was sorbed by low- than by high-charged smectites, and sorption decreases in the order Ca > K > Na. Similar results were obtained with p-nitrocyanobenzene (Charles et al., 2006). Molecules containing aromatic π-systems can interact with mineral surfaces (and HS) through the formation of "electron donor–acceptor" complexes, including cation–π and n–π interactions. As the name suggests, the former refers to the interaction between cations and the planar surface of aromatic π-electron donors, while the latter denotes the interaction between the oxygens of the siloxane surface (electron donors) and the ring structures (electron acceptors) of substituted aromatic compounds, notably nitroaromatics (Weissmahr et al., 1996; Keiluweit and Kleber 2009). That some organic compounds (e.g., benzidine, benzene) can donate electrons to the clay surface or interlayer transition metal cations, giving rise to deeply colored complexes, has long been recognized (Doner and Mortland, 1969; Theng, 1971).

Positively charged organic molecules, notably alkylammonium ions, intercalate into montmorillonite by cation exchange, that is, by replacing the inorganic counterions from the interlayer space. Although cation exchange is the principal mechanism involved, van der Waals and hydrophobic interactions between the alkyl chain and the interlayer surface, and entropy effects contribute to the adsorption energy (Theng et al., 1967; Vansant and Uytterhoeven, 1972). With long-chain quaternary ammonium ions, adsorption may exceed the CEC. These ions may intercalate as a monolayer, a bilayer, a pseudotrilayer, or a paraffin-type structure depending on the alkyl chain length, the layer charge density of the clay mineral, and the organic loading (Lagaly et al., 2006).

Organically modified montmorillonites or organoclays have long been used as thickening agents in ink, lubricants, ointments, and paints (Jordan, 1949, 1963). More recently, organoclays have been used for pollution control and the production of clay–polymer nanocomposites (Beall, 2003; Churchman et al., 2006; Theng et al., 2008; Calabi Floody et al., 2009). Because of their superior mechanical, thermal, gas-barrier, and fire-retardant properties, these composite materials have found applications in a wide range of industries, from automobiles and electronics to sports gear and food packaging. A discussion of clay–polymer nanocomposites lies beyond the scope of this chapter. For details about the synthesis, properties, and uses of clay-based composite materials, we refer to the reviews by Alexandre and Dubois (2000), Ray and Okamoto (2003), Zheng et al. (2005), Ruiz-Hitzky and Van Meerbeek (2006), and Chen et al. (2008).

Little, if any, adsorption occurs with negatively charged organic compounds because they tend to be repelled by the negatively charged clay mineral surface. Nevertheless, appreciable uptake can take place in the presence of polyvalent counterions acting as a bridge between the anionic groups of the organic species and the silicate surface. This mode of interaction is therefore referred to as cation bridging. Under acid conditions (pH < 5), the edge surface of clay minerals would become positively charged through protonation of hydroxyl groups coordinated to aluminum and hence can adsorb anionic organic compounds by electrostatic interactions (anion exchange). A more important mode of bonding, however, is ligand exchange between the anionic group of organic species and surface hydroxyl groups of clay minerals and metal (hydr)oxides. The affinity of allophane for 5′-adenosine monophosphate (5′-AMP), for example, is much greater than for either adenine or adenosine. This may be ascribed to ligand exchange between the phosphate of 5′-AMP and the hydroxyl of $(OH)Al(H_2O)$ groups, exposed at surface defect sites (Hashizume and Theng, 2007). As described below, ligand exchange is a major mode of interaction between soil clays and HS.

The principles outlined above would equally apply to the interactions of nonionic pesticides (examples of which have already been given), organic pollutants, and bioorganic compounds with clay mineral surfaces. As would be expected, the sorption of variable-charged biomolecules (e.g., amino acids, purines, nucleosides) is sensitive to medium pH since the charge characteristics of both solute and mineral surface are pH-dependent (Theng, 1974; Hashizume and Theng, 1999, 2007).

2.3.1.2 Interactions with Humic Substances

Whether they exist as macromolecules, adopting random-coil conformations, or as supramolecular structures, forming micelles, HS in solution would behave as polyanions. As such, HS would be repelled by negatively charged clay particles. In the presence of polyvalent cations, however, appreciable adsorption can occur, through cation bridging (inner-sphere bridging complexation), or water bridging (outer-sphere bridging complexation) (Theng, 1976; Theng and Scharpenseel, 1976; Theng and Tate, 1989; Varadachari et al., 1991; Sposito, 2008). In acidic soils, anion exchange involving the positively charged edge surface of clay particles would make an appreciable contribution, while in soils with a high base status, cation bridging (through Ca^{2+}) would be prevalent. Anion exchange is especially important with Al- and Fe-(hydr)oxides since their surfaces are positively charged in the pH range of most soils. Ligand exchange between the carboxylate groups of HS and hydroxyls (or water molecules) coordinated to metal ions exposed on clay particle edges and at the surface of metal (hydr)oxides can occur over a wide range of pH values. Besides being characterized by the release of OH⁻ (or H_2O) into solution (see Equation 2.1), ligand exchange shows a high degree of anion specificity, slow or little desorption, and a shift of the surface charge to more negative values (McBride, 1994). Other (secondary and indirect) modes of interactions include hydrogen bonding, van der Waals, and entropy effects (Table 2.2). Because of the diversity of minerals and HS in soil, and variations in environmental conditions, several bonding modes may operate simultaneously, while the relative importance of each mechanism would change from soil to soil (Jardine et al., 1989; Kögel-Knabner et al., 2008).

Ligand exchange is the dominant mechanism underlying the adsorption of HS by allophane, imogolite, and iron (hydr)oxides because these nanosize minerals have a large concentration of surface hydroxyl groups (Parfitt et al., 1977; Mott, 1981; Gu et al., 1994; Kaiser et al., 1997; Yuan et al., 2000; Farmer and Lumsdon, 2001). Equation 2.1 illustrates the situation for allophane where numerous $(OH)Al(H_2O)$ groups are exposed at defect sites (perforations) along the spherule wall (Theng et al., 1982; Hashizume and Theng, 2007):

$$[\text{allophane}-Al(H_2O)(OH)]^0 + [R.COO]^{n-}$$
$$\rightarrow [\text{allophane}-Al(H_2O)(COO.R)]^{(n-1)-} + OH^- \quad (2.1)$$

where $[R.COO]^{n-}$ denotes a humic molecule containing n carboxylate functional groups.

Thus, volcanic ash soils can accumulate and stabilize OM through interactions with allophane, imogolite, and ferrihydrite together with the formation of Al–humus complexes.

TABLE 2.2 Possible Mechanisms of Bonding between Humic Substances and Clay Mineral Surfaces

	Groups on Humic Substances	Groups and Sites on Mineral	Remarks
A. Primary interactions			
Cation bridging or inner-sphere bridging complexation Water bridging or outer-sphere bridging complexation	Anionic (carboxylate) and uncharged polar groups, such as amino, carbonyl, carboxyl, hydroxyl	Exchangeable cations at external basal surfaces. Also on the interlayer surfaces at low pH (<4.5)	Through weakly hydrated and non-hydrated monovalent cations. Through polyvalent cations in dehydrated system. Desorbable by Na salts Polyvalent cations in hydrated systems. Desorbable by sodium salts, water, and ultra-sonic treatment
Cation exchange	Positively charged nitrogen atoms	Acidic hydroxyl groups	Under acidic pH conditions
Proton transfer	Amine groups	Hydroxyl groups	Under acidic pH conditions
Anion exchange Ligand exchange	Carboxylate groups	Particle edge surfaces. Polyhydroxy compounds of Fe and Al at external, basal surfaces	Under acidic pH conditions and when exchange sites are occupied by Fe and Al ions. Desorbable by other anions in solution, such as chloride and nitrate When exchange sites are occupied by Fe and Al ions. In soils, on hydrous oxides of Fe and Al. Desorbable by other specifically and more strongly adsorbed anions, such as hydroxyl
B. Secondary interactions			
Hydrogen bonding	Amino, carboxyl, hydroxyl	Oxygen of external, basal (siloxane) surfaces	Weak
van der Waals	Uncharged portions of HS	Mainly at external, basal surfaces	Unimportant in hydrated systems but significant at high ionic strengths and in dehydrated systems. Short-range, weak but additive
C. Indirect interactions			
Entropy effects			Arising from entropy gain caused by displacement of many water molecules from the clay surface by a single humic molecule

Sources: Theng, B.K.G. 1979. Formation and properties of clay-polymer complexes. Elsevier, Amsterdam, the Netherlands; Sposito, G. 2008. The chemistry of soils. Oxford University Press, New York.

The OM may further be physically protected from microbial attack by entrapment within nanopores of allophane aggregates (Dahlgren et al., 2004). The stabilization of OM through interaction with clay-size minerals will be further discussed below.

In acidic soils, HS may complex with short-range order minerals (allophane, Fe-(hydr)oxides), or with Al ions in the soil solution (Parfitt et al., 1999; Kleber et al., 2005; Mikutta et al., 2005). A quantitative relationship between pH and adsorption, however, is difficult to obtain. The species and concentration of cations in solution also affect adsorption. Yuan et al. (2000), for example, found that adsorption of humic acid by allophane increased with the concentration of Na^+ or Ca^{2+} in solution, with Ca^{2+} being much more effective than Na^+. Similar results were reported for illite (soil mica), montmorillonite, and kaolinite (Varadachari et al., 1991; Feng et al., 2005).

Assuming a supramolecular structure for soil OM, Kleber et al. (2007) have proposed a conceptual model in which HS interact with clay-size minerals in a discrete zonal sequence (Figure 2.3). The minerals are represented by a low-charge smectite, a 2:1 type phyllosilicate (e.g., illite) coated with Fe-(hydr)oxide, and an uncharged 1:1 type phyllosilicate (e.g., kaolinite) exposing a (hydrophobic) siloxane surface. In the contact zone, amphiphilic fragments of OM may adsorb to charged surfaces through electrostatic interactions, or ligand exchange with hydroxyls of Fe-(hydr)oxide coatings, directing hydrophobic portions outward to the aqueous solution. Proteinaceous materials, adsorbed to low-charge phyllosilicates, can serve as surface "conditioners," adding polar functionality to the hydrophobic siloxane portions on basal surfaces. Hydrophobic organics (e.g., phenanthrene) may also adsorb to the siloxane surface of uncharged minerals. The hydrophobic zone may

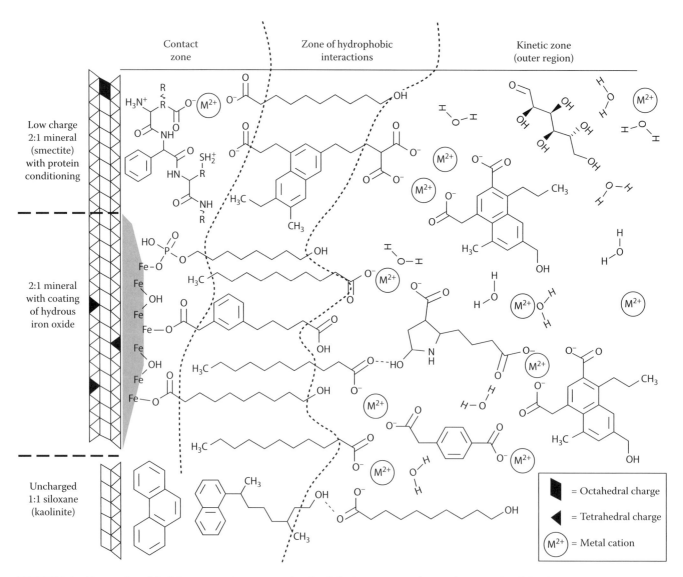

FIGURE 2.3 The zonal model of clay–organic interactions, details of which are given in the text. (Reprinted with kind permission from Springer Science+Business Media: Kleber, M., P. Sollins, and R. Sutton. 2007. A conceptual model of organo-mineral interactions in soils: Self-assembly of organic molecular fragments into zonal structures on mineral surfaces. Biogeochemistry 85:9–24.)

be discontinuous since the components here can exchange with the surrounding solution, while some proteins directly adsorbed on mineral surfaces may well expose ionized or polar functional groups to the solution. In the kinetic zone, the OM is loosely held by cation bridging, hydrogen bonding, and other interactions (Table 2.2). Thus, the thickness of the kinetic zone would depend on OM quality and is essentially controlled by exchange kinetics.

2.3.2 Methodologies for Investigating Clay–Organic Interactions

The clay–humic interaction may be investigated by several approaches (Cheshire et al., 2000). The "addition" approach in which HS are placed in contact with specific clay minerals is well documented (Theng, 1979; Schnitzer, 1986; Oades, 1989; Clapp et al., 1991; Schulthess and Huang, 1991; Stevenson, 1994). In the "separation" approach, the OM is removed from the clay by physical and chemical methods. The separated components are then characterized using various analytical and instrumental techniques. Physical separations include dry and wet sieving, slaking, ultrasonic dispersion, density or magnetic separation, and sedimentation (Elliott and Cambardella, 1991; Christensen, 2001). Chemical fractionation is based on the solubility of soil OM in water or organic solvents (e.g., *n*-hexane), the resistance of soil OM to oxidation, and dissolution of the mineral component by HF, dithionite–citrate–bicarbonate, or oxalate (Baldock and Nelson, 2000; Eusterhues et al., 2003).

The clay–humic interaction may also be probed directly, using solid-state ^{13}C NMR spectroscopy (Preston, 1996; Skjemstad et al., 1997), analytical pyrolysis (Schulten and Leinweber, 1996; Schulten et al., 1996), Fourier transform infrared (FTIR) and near-edge absorption fine structure (NEXAFS) spectroscopy, and scanning transmission x-ray microscopy (STXM) (Kaiser et al., 1997; Solomon et al., 2005; Kinyangi et al., 2006; Lehmann et al., 2007), or indirectly by measuring the SSA of soils by adsorption of N_2 gas at 77 K and applying the Brunauer–Emmett–Teller (BET) equation (Theng et al., 1999; Kaiser and Guggenberger, 2003).

Using FTIR and NEXAFS spectroscopy, Lehmann et al. (2007) found that the OC in 20–250 µm microaggregates (of an Inceptisol and two Oxisols) was unevenly distributed, while the concentration of aliphatic carbon was well correlated with the amount of hydroxyl groups on kaolinite. X-ray microscopy and NEXAFS indicated that the complexity and occurrence of aliphatic carbon forms increased from the exterior to the interior regions of the microaggregate, while particulate OM was occluded in 2–5 µm pores (Kinyangi et al., 2006).

Earlier, Kaiser and Guggenberger (2003) found that the N_2-BET surface area of a wide range of soils (under forest) decreased with increasing surface coverage by OM. This is because OM tends to occupy reactive sites on mineral surfaces (particle edges and dislocations) and "clogs" the openings of micropores, preventing N_2 molecules from adsorbing to micropore surfaces. Such a discontinuous or "patchy" coverage of mineral surfaces by OM has also been observed by several workers (Mayer and Xing, 2001; Chenu and Plante, 2006; Lehmann et al., 2007), and accords with the electron microscopy observation of marine sediments and soil nanoparticles (Ransom et al., 1997, 1998; Tang et al., 2009), and the x-ray photoelectron spectroscopic analysis of soil particles (Yuan et al., 1998).

Besides affecting the surface area of soils, the interaction of HS with soil mineral surfaces through ligand exchange causes the surface charge to become more negative as shown in Equation 2.1 for allophane. Further, OM bound by ligand exchange is more resistant to biodegradation than that held by cation bridging and van der Waals interactions (Mikutta et al., 2007). The effect of mineral surface wettability on OM decomposition will be described below.

2.3.3 Effects of Clay–Organic Interactions

The high reactivity of clay-size minerals toward water and organic compounds in soils (and sediments) is expected to influence biological processes in these environments. Indeed, the work by Müller and Höper (2004) indicates that the stabilizing effect of clay was more in evidence with respect to the microbial biomass than humus. The most important effect would be on the availability of water and substrate to microorganisms. As their metabolism requires water, microorganisms in soil are drawn to sites with a continual supply of water. Being able to retain water against gravity and adsorb organic molecules, clay-size minerals provide a favorable environment for microbial growth and survival (Theng and Orchard, 1995). More recently, Bachmann et al. (2008) have proposed that soil wettability may play an important role in OM stabilization as it controls the microbial accessibility to water, nutrients, and oxygen. The manner in which organic molecules are retained by clays will determine their bioavailability to microorganisms. At the same time, the attachment of clay particles to microbial cells would reduce the effective surface area of microorganisms for transmembrane transfer of nutrients and waste products.

2.3.3.1 Effect on Soil Structure and Aggregate Stability

Soil structure, as indicated by the stability of soil aggregates, is a key factor in the functioning of soil (Bronick and Lal, 2005). OM contributes to aggregate stability through (1) enhancing aggregate cohesion by binding soil particles together and (2) decreasing aggregate wettability by slowing the rate of wetting and extent of slaking (Sullivan, 1990). Aggregate stability is often correlated with soil OM content (Chaney and Swift, 1984; Golchin et al., 1995; Chenu et al., 2000). However, whether there is a defined threshold value below which aggregate stability will decline is still open to question (Loveland and Webb, 2003). Polysaccharides excreted by plant roots and microbial cells can bind small aggregates (<20 µm) into macroaggregates (Theng, 1979; Swift, 1991; Figure 2.1). The effectiveness of polysaccharides, however, tends to be short-lived because these compounds are readily decomposed by microorganisms (Tisdall and Oades, 1982). On the other hand, HS can provide long-term stability to microaggregates (Guckert et al., 1975; Kay and Angers, 2000). Their beneficial effect may be ascribed to the formation of clay–organic complexes by which the anionic groups of HS are oriented toward the interior of the microaggregates, while the hydrophobic components face outward (Piccolo et al., 1997; Figure 2.3). In line with this suggestion, Capriel et al. (1990) observed that the aliphatic fraction of soil OM (extractable with supercritical hexane) was highly correlated with aggregate stability, probably because this hydrophobic material could form a water-repellent "screen" around the aggregates. Similarly, Shepherd et al. (2001) found that the high aggregate stability of an allophanic soil (under long-term cropping) was related to the concentration of polymethylene (alkyl) carbon in the clay fraction. The strong binding of allophane to OM during cropping can confer great stability on HS, thereby arresting further decline in soil OM (Theng and Tate, 1989).

The relationship between OM composition (determined by FTIR spectroscopy), soil wettability (derived from contact angle measurements), and aggregate stability has recently been discussed by Bachmann et al. (2008). Although the coverage of clay-size minerals by OM is patchy, OM can form microaggregates with metal hydr(oxides) and phyllosilicates, and becomes stabilized against microbial decomposition. As the hydrophobicity of the OM increases, the wettability of such microaggregates decreases and their stability is enhanced.

OM as such, or in association with clays, has a positive effect on many other soil physical properties, such as compactability, tillage, friability, tilth, and hard-setting (Ball et al., 2000; Dexter, 2004a, 2004b; Watts et al., 2006). Improvements in such soil physical properties as bulk density, structure, water retention, and content of readily dispersible clay correlates better with clay-associated OM than total carbon content (Dexter et al., 2008).

The clay–organic interaction may occasionally cause soils to become water-repellent. Water repellency occurs when the surface of mineral particles is covered by a film of hydrophobic organic materials (e.g., waxes). This condition can be found in loamy, clayey, peat, and volcanic ash soils (Wallis and Horne, 1992), and even in sandy soils (Franco et al., 2000). Water repellency can reduce water infiltration, enhance surface runoff and erosion, induce the formation of preferential flow patterns in soil, and decrease plant growth. In the absence of a universally applicable solution to water repellency, a number of soil management practices have been proposed to reduce its impact. These include the use of surfactants to reduce surface tension, clay to cover and encourage degradation of hydrophobic surface, water harvesting, and reduced soil drying (Blackwell, 2000).

2.3.3.2 Effect on Organic Matter Dynamics and Turnover

Organic matter in soil consists of a wide range of organic molecules that are stabilized by different mechanisms and have variable turnover rates. Models of soil OM turnover are often based on conceptual pools with turnover rates ranging from less than a year to thousands of years. Terms such as "active," "slow," "passive," and "inert" have been used to describe the relative stabilities of the various functional OM pools. The active pool with turnover rates of less than 10 years is associated with the microbial biomass and "light fraction" (density <1.6–2.0 g cm^{-3}). Most chemical and physical fractionation methods, however, do not yield homogeneous fractions in terms of turnover rates. The conceptual passive pool, in which the majority of soil OM is stabilized, is especially difficult to isolate (von Lützow et al., 2007).

Earlier, Parfitt et al. (1997) compared OM turnover in an Andisol, containing allophane and imogolite, with that in an Inceptisol having mica as the main clay mineral species. Under long-term pasture, the Andisol (0–20 cm) accumulated 144 Mg ha^{-1} of OC, which was twice as much as the Inceptisol. After 20 years of cropping, the Andisol lost 10 T ha^{-1}, representing half the quantity that was lost from the Inceptisol. It would therefore appear that the turnover of carbon was much slower in the Andisol than in the Inceptisol. The most likely mechanism for the slower turnover of soil OM in the allophanic soil could be the stabilization of the microbial biomass and microbial metabolites (Saggar et al., 1994). Modeling (with CENTURY) further indicated that the pool of passive OM in the Andisol was very large. Basile-Doelsch et al. (2007) also found that the minerals of a volcanic soil (in La Réunion) could store large amounts of OM with a very long turnover time. Similar results were reported for some temperate forest soils containing short-range order minerals (Eusterhues et al., 2003; Kögel-Knabner et al., 2008; Rasmussen et al., 2008). In reviewing the literature on stabilization of OM in temperate soils, von Lützow et al. (2006) have concluded that molecular recalcitrance is only important in surface soils during early decomposition. In subsoils and during late decomposition, however, OM is largely stabilized through spatial and physical inaccessibility (occlusion, intercalation, encapsulation) and interactions with mineral surfaces.

The recent extension of single layer models of soil OM turnover to include deeper layers in the soil profile (Jenkinson and Coleman, 2008) provides a reasonable fit between measured and modeled soil C and radiocarbon age for some soils, but not necessarily in soils with clay-rich subsoils. Accounting for the capacity of different clay types and contents to stabilize soil OM pools is still needed to improve such models for use across a range of different soil types.

2.3.3.3 Effect on Nutrient Dynamics

Soil OM is a large reservoir of carbon as well as nitrogen, phosphorus, and other elements that are essential to life. Globally, soils contain 1462–1548 Pg of OC and 133–140 Pg of nitrogen in the upper 100 cm of their profiles, while the mean C:N ratios of soil OM ranges from 9.9 (for arid Yermosols) to 25.8 for Histosols (Batjes, 1996). As most nitrogen in soil is in organic form, soil OM is a huge reservoir of N in comparison with the amount of N that is annually "fixed" by lightning (<10 Tg; 1 Tg = 10^{12} g), by (leguminous) crops (32–53 Tg), and by fertilizer additions (80 Tg in 1992) (Galloway et al., 1995; Vitousek et al., 1997). The amount of N in soil OM is also much greater than that held in (above-ground) vegetation (~10 Pg) (Davidson, 1994). In most terrestrial biomes and many marine ecosystems, net primary production is often limited by nitrogen supply (Vitousek and

Howarth, 1991), and there is a limit of nitrogen fertilizer that can be added to soil without causing adverse impact on water quality. Thus, the importance of soil organic nitrogen to primary production cannot be overstated. Besides being an inexpensive, benign, and reliable source for crop production and forest growth, soil organic nitrogen has long-term availability through interaction with the clay fraction.

The cycling of (organic) nitrogen is closely linked to that of carbon (McGill and Cole, 1981) because most of the nitrogen in OM is bound to carbon. Nitrogen mineralization in soil is primarily driven by the need of microorganisms for carbon. When nitrogen is mineralized, the associated carbon is oxidized to CO_2. In natural systems, the movement of nitrogen from soil to water is mediated by dissolved organic carbon (DOC), most of which originates from plant litter and humus (Kalbitz et al., 2000). A carbon-13 tracer study by Hagedorn et al. (2004) suggests that in forest soils DOC is produced during incomplete decomposition of recalcitrant native carbon. This is because easily degradable carbon components are rapidly consumed by microbes and hence make only a minor contribution to the dissolved carbon fraction.

2.3.3.4 Impact on Sorption and Bioavailability of Heavy Metals and Organic Contaminants

The dynamics and fate of environmental contaminants in soil are greatly altered by their reactions with clay, HS, and clay–humus complexes. Negatively charged phyllosilicates and HS can sorb heavy metal cations through electrostatic interactions (cation exchange), while sorption to short-range order minerals (allophane), Fe-, Al-, and Mn-(hydr)oxides, and the edge surface of phyllosilicates largely occurs through ligand exchange, giving rise to inner-sphere complexes. This process is distinguishable from cation exchange by a high degree of specificity, a tendency for irreversibility, and a shift of the point of zero charge to more positive values (McBride, 1994). Ligand exchange (see Equation 2.1) is also the dominant mechanism underlying the sorption of heavy metal anions (e.g., arsenate) by clay-size soil minerals and HS (Arai et al., 2005; Violante et al., 2008).

Clay–organic interactions affect the mobility and bioavailability of heavy metals in soil by (1) changing the kinetics of metal sorption and desorption and (2) altering the dispersibility of clay particles and OM. As already mentioned, HS (in clay–organic complexes) occupy reactive sites on mineral surfaces and "clog" the openings of micropores. As a result, both the sorption of metal ions to the complex and their diffusion within clay aggregates are diminished. Since desorption is also retarded, the bioavailability of sorbed heavy metals is reduced. Bioavailability also decreases as the contaminant–soil contact time increases. The "aging" effect may be ascribed to slow diffusion of the contaminant molecules into micropores and long-term changes in solute-surface mode of bonding (Chorover et al., 2007). Metal ions in soil may also associate with the soluble fractions of OM (e.g., fulvic acid and DOC). Sorption of these mobile organic colloids to clay-size minerals decreases clay dispersibility as well as the mobility of the associated metal contaminants (Kretzschmar and Schäfer, 2005; Hochella et al., 2008). As a result, leaching of heavy metals from soil to adjacent water bodies is reduced.

The clay–organic interaction also exerts a controlling influence on the sorption of organic contaminants. The correlation of sorption with soil OM content is especially pronounced for nonionic (hydrophobic) organic solutes. Indeed, the sorption of many organic compounds of low polarity to organic-rich topsoils gives rise to linear isotherms and can be rationalized in terms of simple partitioning into OM (Chiou, 1989). This process is akin to solute extraction from water into an organic solvent, such as 1-octanol. In general, however, the sorption of nonionic organic compounds gives rise to curvilinear isotherms, and conforms to the Freundlich equation (e.g., Xing and Pignatello, 1998). The retention of such contaminants (e.g., pesticides, polycyclic aromatic hydrocarbons) by soils with different OC contents may be predicted by the following equation:

$$\log K_{oc} = a + b \log K_{ow} \qquad (2.2)$$

where
  K_{oc} is the organic carbon-normalized sorption coefficient
  K_{ow} is the octanol–water partition coefficient of the organic contaminant
  a and b are empirical constants

In many instances, however, K_{oc} of organic contaminants cannot be derived from the corresponding K_{ow} value because of differences in the quality (aromaticity and polarity) of soil OM (Xing, 1997).

Besides controlling the sorption of organic contaminants in soil, the clay–organic interaction also influences the decomposition and bioavailability of organic contaminants. Thus, sorption and desorption are slow because the organic solute must diffuse through soil OM matrices and intraparticle micropores (Pignatello and Xing, 1996). By promoting aggregate formation and stability, the clay–organic interaction provides a favorable environment, in terms of water and substrate requirements, for microbial decomposition of organic contaminants. Rhodes et al. (2008), for example, reported that the rate and extent of phenanthrene degradation in natural soils was higher than in an artificial soil (comprising 70% fine quartz sand, 20% kaolinite, and 10% fine powdered *Sphagnum* peat), with a comparable sorptive capacity. This might be because the artificial soil lacks stable aggregates where both microorganisms and contaminant reside, allowing for microbial growth and degradation to proceed (Harms and Zehnder, 1995). In other words, the simultaneous enrichment of contaminants and microbes in clay–organic complexes promotes biodegradation. The bioavailability of the contaminants, associated with clay–OM complexes, has been reviewed by Feng and Boyd (2008). As might be expected, the process is influenced by the nature of both contaminant and sorbent. Aging is another important factor in that contaminant bioavailability generally decreases with time.

2.3.3.5 Impact on Greenhouse Gas Emission

Since soils are an important carbon sink, the clay–organic interaction has also attracted the attention of climate change researchers. In temperate, cultivated soils, 50%–75% of the OM exists as clay–organic complexes (Christensen, 2001), while in volcanic soil horizons up to 83% of the OM is associated with mineral surfaces (Basile-Doelsch et al., 2007). Global warming would accelerate the decomposition of mineral-associated OM in soil because the process is temperature-dependent (Lloyd and Taylor, 1994; Davidson and Janssens, 2006). An increase in temperature would therefore act as a positive feedback in the global carbon cycle (Kirschbaum, 1995). On the other hand, Liski et al. (1999) and Jenkinson and Coleman (2008) have suggested that the extent of CO_2 emission from soil in response to climate warming is overestimated, while Giardina and Ryan (2000) have gone so far as to say that the decomposition of OM in forest soils is independent of temperature. The apparent inconsistencies in data reporting and interpretation partly arise from conceptual differences regarding the nature of OM or OC, and the use of single-layer models of soil OC turnover. Further, due caution must be exercised in extrapolating results from short-term laboratory studies (often involving microbes) to long-term impacts at a global scale. For example, the availability of water and organic substrates to microbes during OM decomposition (Gershenson et al., 2009) is an important factor affecting the movement of carbon from soil to the atmosphere. Equally relevant is the finding that the stabilizing effect of clay is apparently much stronger for the microbial biomass than for humus (Saggar et al., 1994; Müller and Höper, 2004). Similarly, Ekschmitt et al. (2005) have suggested that ecological limitations inherent to soil life, such as specialization of organisms with respect to microhabitats and substrates, contribute to lowering decomposition rates and preserving OM in temperate soils.

As already mentioned, there is general agreement that interaction with clays can protect OM in soil from microbial decomposition (e.g., Chenu and Plante, 2006). The protective effect of clays is especially important for soils of tropical regions where climatic conditions favor decomposition (Wattel-Koekkoek et al., 2001). Working with two acid forest soils, Eusterhues et al. (2003) found that the stable fraction of OM with an old radiocarbon age comprised 1%–30% of the total OC in surface horizons, and up to 80% in the subsoil. The stability of the old ("passive") OC was ascribed to interaction with clay mineral and/or iron hydr(oxide) surfaces. Marschner et al. (2008) have come to the same general conclusion saying that "black carbon" and "fossil carbon" may be the only type of OC that can persist in soil without being associated with minerals.

A rather unique example of OM preservation comes from the work by Theng et al. (1986, 1992) who were able to isolate an interlayer clay–organic complex from the surface horizon of a Spodosol. Here the OM was intercalated into a swelling mixed-layer 2:1 type clay mineral and hence was physically protected from microbial attack, giving a radiocarbon age of 6700 years BP. Protection by intercalation into expansible phyllosilicates is uncommon, however (Eusterhues et al., 2003; von Lützow et al., 2007). This is because interlayer complex formation requires a combination of soil factors that is only rarely encountered (Theng et al., 1986).

It is also difficult to establish a quantitative relationship between clay content and OC. Many researchers have found that soil OC content correlates positively with clay content (e.g., Nichols, 1984; Van Veen et al., 1984). On the other hand, Percival et al. (2000) failed to detect such a correlation for New Zealand grassland soils. Instead, they found that pyrophosphate-extractable Al alone or in combination with allophane, Fe-(hydr)oxide, and clay could account for much of the variability in soil carbon contents. Similarly, Müller and Höper (2004) observed a very weak relationship between the concentration of humus carbon and clay content. A recent study of the capacity of soils to store carbon (Kool et al., 2007) shows that there is a hierarchy in C saturation across different soil OM pools exposed to elevated atmospheric CO_2. In this example, there was a shift in the allocation of C from small to large aggregates, leading to the conclusion that soil can become saturated with C, probably because of a limit to physical protection of OC. The effect of clay content and type has been included in some national scale models (Tate et al., 2005). Using a combination of factors such as soil type and climate as model parameters, these models allow changes in soil C with land use change to be simulated.

2.4 Conclusions and Perspectives

The interaction of clay with OM in the soil environment is a ubiquitous natural process and underpins a range of essential ecological services and functions, from regulating water storage and nutrient cycling, through adsorbing and transporting metal and organic contaminants, to serving as a source or sink of OC and greenhouse gases. Until the early 1980s, the focus of research was on the interaction of well-defined small and polymeric organic molecules with specific crystalline clay minerals. Attention has since turned to more complex soil systems where many kinds of minerals, organic substances, and microbes coexist under different environmental conditions. The participation of nanosize short-range order minerals and metal (hydr)oxides, as well as microorganisms, in the interaction process is duly recognized. Also, the concept of HS being a mixture of high-molecular-weight linear polymers, adopting random-coil conformations in solution, has evolved to one of an assembly of low molecular weight, biologically derived molecules held together by weak dispersive forces, forming micelles in solution. As already remarked, however, the two concepts may not be mutually exclusive. Enzymes and clays can catalyze the polymerization and/or polycondensation of low-molecular-weight biomolecules, giving rise to dark-colored humic-like polymers. At the same time, unstable biological molecules and residues may be preserved by incorporation into, and encapsulation by, polymeric humified materials.

Clay-size minerals in soil (phyllosilicates, short-range order alumino-silicates, metal (hydr)oxides) can interact with HS through different modes of bonding, the relative importance of which would vary from soil to soil, and within a given profile, from top to bottom. In temperate soils, for example, the proportion of OC bound to clay-size minerals increases with soil depth as does bonding strength. Microorganisms participate in this process in that the clay–humus complex provides substrates and protection to soil microbes. In excreting polysaccharides and proteins, microorganisms also contribute actively to the clay–organic complex formation (Kögel-Knabner et al., 2008). The clay–organic interaction affects a range of soil properties and functions, including structure (aggregate stability), OM turnover, and nutrient dynamics, as well as the behavior and fate of heavy metals and organic contaminants.

Soil is also an important source and sink of greenhouse gases. By its very nature, the clay–organic interaction would be expected to affect the emissions of greenhouse gases from soil. We have no clear understanding, however, about the operative mechanisms involved. Models of soil carbon dynamics, for example, do not usually account for the capacity of different clay-size mineral species to stabilize OM in various pools. This gap in knowledge needs to be filled in order to store more carbon in soil, reduce greenhouse gas emission from soil, and appreciate the effect of climate change on soil functions.

The quantitative relationship between OM, clays, and environmental conditions, on the one hand, and the extent and rate of OM stabilization in soil, on the other hand, still needs to be clarified. In temperate soils, pedogenic Fe-(hydr)oxides appear to be the principal clay-size minerals involved in the sorption and stabilization of OM, especially in the subsoil. Because of the "patchy" surface coverage by OM, however, total surface area and micropore volume do not generally correlate with the amount of mineral-associated OM (Kögel-Knabner et al., 2008).

No universally applicable relationship has been established between soil OM dynamics and the quantity of various clay-size mineral species. This is because such a relationship is influenced by many variables. Many studies, based on laboratory incubations and statistical analysis, have shown that the content of OM and microbial biomass in soil is positively correlated with that of clay (Nichols, 1984; Kern, 1994; Saggar et al., 1994; Alvarez and Lavado, 1998; Homann et al., 1998). Accordingly, clay content features strongly in a number of soil OM models that have been developed to assess the impact of climate change on soil OM and subsequent feedback effects (e.g., Smith et al., 1997). Other investigators (Percival et al., 2000; Eusterhues et al., 2003; Kleber et al. 2005; Rasmussen et al., 2006), on the other hand, have suggested that short-range order minerals and aluminum are more important than layer silicates in stabilizing OM in soil. Similarly, Spielvogel et al. (2008) have found that OM associated with (oxalate-extractable) Al and Fe is potentially stable for thousands of years.

It would appear that both the quantity and quality of clay-size minerals in soil play a part in OM stabilization and hence should be taken into account in modeling soil OC dynamics. Likewise, the underlying mechanisms need to be considered in quantifying the protective capacity of soil for OM (Baldock and Skjemstad, 2000).

The characterization of the mineral and OM constituents in specific soil fractions, combined with measuring turnover times and soil age would also merit further investigation (von Lützow et al., 2006; Kögel-Knabner et al., 2008). Another topic for future research is the contribution of electron donor–acceptor interactions to the sorption affinity of organic compounds and pollutants containing aromatic moieties and hence to the stability of soil OM against microbial decomposition (Keiluweit and Kleber, 2009).

References

Aiken, G.R., D.M. McKnight, R.L. Wershaw, and P. MacCarthy. 1985. An introduction to humic substances in soil, sediment, and water, p. 1–9. *In* G.R. Aiken, D.M. McKnight, R.L. Wershaw, and P. MacCarthy (eds.) Humic substances in soil, sediment, and water. Wiley-Interscience, New York.

Alexandre, M., and P. Dubois. 2000. Polymer-layered silicate nanocomposites: Preparation, properties and uses of a new class of materials. Mater. Sci. Eng. 28:1–63.

Almendros, A. 2008. Carbon sequestration in soil, p. 97–99. *In* W. Chesworth (ed.) Encyclopedia of soil science. Springer, Dordrecht, the Netherlands.

Alvarez, R., and R.S. Lavado. 1998. Climate, organic matter and clay content relationships in the Pampa and Chaco soils, Argentina. Geoderma 83:127–141.

Arai, Y., D.L. Sparks, and J.A. Davis. 2005. Arsenate adsorption mechanisms at the allophane-water interface. Environ. Sci. Technol. 39:2537–2544.

Bachmann, J., G. Guggenberger, T. Baumgartl, R.H. Ellerbrock, E. Urbanek, M.-O. Goebel, K. Kaiser, R. Horn, and W.R. Fischer. 2008. Physical carbon-sequestration mechanisms under special consideration of soil wettability. J. Plant Nutr. Soil Sci. 171:14–26.

Baldock, J.A., and P.N. Nelson. 2000. Soil organic matter, p. B25–B84. *In* M.E. Sumner (ed.) Handbook of soil science. CRC Press, Boca Raton, FL.

Baldock, J.A., and J.O. Skjemstad. 2000. Role of the soil matrix and minerals in protecting natural organic materials against biological attack. Org. Geochem. 31:697–710.

Ball, B.B., D.J. Campbell, and E.A. Hunter. 2000. Soil compactability in relation to physical and organic properties at 156 sites in U.K. Soil Till. Res. 57:83–91.

Basile-Doelsch, I., R. Amundson, W.E.E. Stone, D. Borschneck, J.Y. Bottero, S. Moustier, F. Masin, and F. Colin. 2007. Mineral control of carbon pools in a volcanic soil horizon. Geoderma 137:477–489.

Batjes, N.H. 1996. Total carbon and nitrogen in soils of the world. Eur. J. Soil Sci. 47:151–163.

Beall, G.W. 2003. The use of organoclays in water treatment. Appl. Clay Sci. 24:11–20.

Bergaya, F., B.K.G. Theng, and G. Lagaly (eds.). 2006. Handbook of clay science. Elsevier, Amsterdam, the Netherlands.

Berthelin, J., P.M. Huang, J.-M. Bollag, and F. Andreux (eds.). 1999. Effect of mineral-organic-microorganism interactions on soil and freshwater environments. Kluwer Academic/Plenum Publishers, New York.

Blackwell, P.S. 2000. Management of water repellency in Australia, and risks associated with preferential flow, pesticide concentration and leaching. J. Hydrol. 231–232:384–395.

Brigatti, M.F., E. Galan, and B.K.G. Theng. 2006. Structures and mineralogy of clay minerals, p. 19–86. In F. Bergaya, B.K.G. Theng, and G. Lagaly (eds.) Handbook of clay science. Elsevier, Amsterdam, the Netherlands.

Bronick, C.J., and R. Lal. 2005. Soil structure and management: A review. Geoderma 124:3–22.

Calabi Floody, M., B.K.G. Theng, P. Reyes, and M.L. Mora. 2009. Natural nanoclays: Applications and future trends—A Chilean perspective. Clay Miner. 44:161–176.

Capriel, P.T., T. Beck, H. Borchert, and P. Härter. 1990. Relationship between soil aliphatic fraction extracted with supercritical hexane, soil microbial biomass, and soil aggregate stability. Soil Sci. Soc. Am. J. 54:415–420.

Carretero, M.I., C.S.F. Gomes, and F. Tateo. 2006. Clays and human health, p. 717–741. In F. Bergaya, B.K.G. Theng, and G. Lagaly (eds.) Handbook of clay science. Elsevier, Amsterdam, the Netherlands.

Chaney, K., and R.S. Swift. 1984. The influence of organic matter on aggregate stability in some British soils. J. Soil Sci. 35:223–230.

Chappell, M.A., D.A. Laird, M.L. Thompson, H. Li, B.J. Teppen, V. Aggarwal, C.T. Johnston, and S.A. Boyd. 2005. Influence of smectite hydration and swelling on atrazine sorption behaviour. Environ. Sci. Technol. 39:3150–3156.

Charles, S., B.J. Teppen, H. Li, D.A. Laird, and S.A. Boyd. 2006. Exchangeable cation hydration properties strongly influence soil sorption of nitroaromatic compounds. Soil Sci. Soc. Am. J. 70:1470–1479.

Chen, B., J.R.G. Evans, H.C. Greenwell, P. Boulet, P.V. Coveney, A.A. Bowden, and A. Whiting. 2008. A critical appraisal of polymer-clay nanocomposites. Chem. Soc. Rev. 37: 568–594.

Chenu, C., Y. Le Bisssonnais, and D. Arrouays. 2000. Organic matter influence on clay wettability and soil aggregate stability. Soil Sci. Soc. Am. J. 64:1479–1486.

Chenu, C., and A.F. Plante. 2006. Clay-sized organo-mineral complexes in a cultivation chronosequence: Revisiting the concept of the 'primary organo-mineral complex'. Eur. J. Soil Sci. 57:596–607.

Cheshire, M.V., C. Dumat, A.R. Fraser, S. Hillier, and S. Staunton. 2000. The interaction between soil organic matter and soil clay minerals by selective removal and controlled addition of organic matter. Eur. J. Soil Sci. 51:497–509.

Chiou, C.T. 1989. Partition and adsorption on soil and mobility of organic pollutants and pesticides, p. 163–175. In Z. Gerstl, Y. Chen, U. Mingelgrin, and B. Yaron (eds.) Toxic organic chemicals in porous media. Springer-Verlag, Berlin, Germany.

Chorover, J., R. Kretzschmar, F. Garcia-Pichel, and D.L. Sparks. 2007. Soil biogeochemical processes within the critical zone. Elements 3:321–326.

Christensen, B.T. 2001. Physical fractionation of soil and structural and functional complexity in organic matter turnover. Eur. J. Soil Sci. 52:345–353.

Churchman, G.J., W.P. Gates, B.K.G. Theng, and G. Yuan. 2006. Clays and clay minerals for pollution control, p. 625–675. In F. Bergaya, B.K.G. Theng, and G. Lagaly (eds.) Handbook of clay science. Elsevier, Amsterdam, the Netherlands.

Clapp, C.E., R. Harrison, and M.H.B. Hayes. 1991. Interactions between organic macromolecules and soil inorganic colloids and soils, p. 409–468. In G.H. Bolt, M.F. DeBoodt, M.H.B. Hayes, and M.B. McBride (ed.). Interactions at the soil colloid-soil solution interface. Kluwer Academic Publishers, Dordrecht, Netherlands.

Clark, C.J., and M.B. McBride. 1984. Cation and anion retention by natural and synthetic allophane and imogolite. Clay. Clay Miner. 32:291–299.

Cornejo, J., and M.C. Hermosin. 1996. Interaction of humic substances and soil clays, p. 595–624. In A. Piccolo (ed.) Humic substances in terrestrial ecosystems. Elsevier, Amsterdam, the Netherlands.

Cradwick, P.D.G., V.C. Farmer, J.D. Russell, C.R. Masson, K. Wada, and N. Yoshinaga. 1972. Imogolite—A hydrated aluminium silicate of tubular structure. Nat. Phys. Sci. 240:187–189.

Dahlgren, R.A., M. Saigusa, and F.C. Ugolini. 2004. The nature, properties and management of volcanic soils. Adv. Agron. 82:113–182.

Davidson, E.A. 1994. Climate change and soil microbial processes: Secondary effects are hypothesised from better known interacting primary effects, p. 156–168. In M.D.A. Rounsevell, and P.J. Loveland (eds.) Soil response to climate change, NATO ASI Series. Vol. 23. Springer-Verlag, Berlin, Germany.

Davidson, E.A., and I.A. Janssens. 2006. Temperature sensitivity of soil carbon decomposition and feedbacks to climate change. Nature 440(7081):165–173.

De Oliveira, M.F., C.T. Johnston, G.S. Premachandra, B.J. Teppen, H. Li, D.A. Laird, D.Q. Zhu, and S.A. Boyd. 2005. Spectroscopic study of carbaryl sorption on smectite from aqueous suspension. Environ. Sci. Technol. 39: 9123–9129.

Dexter, A.R. 2004a. Soil physical quality. Part I: Theory, effect of soil texture, density and organic matter, and the effects on root growth. Geoderma 120:201–214.

Dexter, A.R. 2004b. Soil physical quality. Part II: Friability, tillage, tilth and hard-setting. Geoderma 120:215–225.

Dexter, A.R., G. Richard, D. Arrouays, E.A. Czyz, C. Jolivet, and O. Duval. 2008. Complexed organic matter controls soil physical properties. Geoderma 144:620–627.

Dixon, J.B., and S.B. Weed. (eds.). 1989. Minerals in soil environments. 2nd edn. Soil Science Society of America, Madison, WI.

Doner, H.E., and M.M. Mortland. 1969. Benzene complexes with copper(II) montmorillonite. Science 166:1406–1407.

Ekschmitt, K., M. Liu, S. Vetter, O. Fox, and V. Wolters. 2005. Strategies used by soil biota to overcome soil organic stability—Why is dead organic matter left over in the soil? Geoderma 128:167–176.

Elliott, E.T., and C.A. Cambardella. 1991. Physical separation of soil organic matter. Agric. Ecosyst. Environ. 34:407–419.

Essington, M.E. 2003. Soil and water chemistry. CRC Press, Boca Raton, FL.

Eusterhues, K., C. Rumpel, M. Kleber, and I. Kögel-Knabner. 2003. Stabilisation of soil organic matter by interactions with minerals as revealed by mineral dissolution and oxidative degradation. Org. Geochem. 34:1591–1600.

Eusterhues, K., F.E. Wagner, W. Häusler, M. Hanzlik, H. Knicker, K.U. Totsche, I. Kögel-Knabner, and U. Schwertmann. 2008. Characterization of ferrihydrite-soil organic matter coprecipitates by X-ray diffraction and Mössbauer spectroscopy. Environ. Sci. Technol. 42:7891–7897.

Farmer, V.C., and A.R. Fraser. 1979. Synthetic imogolite, a tubular hydroxy-aluminium silicate, p. 547–553. In M.M. Mortland and V.C. Farmer (eds.) International clay conference 1978. Elsevier, Amsterdam, the Netherlands.

Farmer, V.C., and D.G. Lumsdon. 2001. Interactions of fulvic acid with aluminium and a proto-imogolite sol: The contribution of E-horizon eluates to podzolization. Eur. J. Soil Sci. 52:177–188.

Farmer, V.C., and J.D. Russell. 1990. The structure and genesis of allophanes and imogolite; their distribution in non-volcanic soils, p. 165–178. In M.F. De Boodt, M.H.B. Hayes, and A. Herbillon (eds.) Soil colloids and their associations in soil aggregates, Proc. NATO advanced studies workshop, Ghent, Belgium, 1985. Plenum, New York.

Feldman, S.B., C. Shang, and L.W. Zelazny. 2008. Soil mineralogy, p. 678–686. In W. Chesworth (ed.) Encyclopedia of soil science. Springer, Dordrecht, the Netherlands.

Feng, Y., and S.A. Boyd. 2008. Bioavailability of soil-sorbed pesticides and organic contaminants, p. 259–279. In Q. Huang, P.M. Huang, and A. Violante (eds.) Soil mineral-microbe-organic interactions: Theories and applications. Springer-Verlag, Berlin, Germany.

Feng, X., A.J. Simpson, and M.J. Simpson. 2005. Chemical and mineralogical controls on humic acid sorption to clay mineral surfaces. Org. Geochem. 36:1553–1566.

Flaig, W., H. Beutelspacher, and E. Rietz. 1975. Chemical composition and physical properties of humic substances, p. 1–211. In J.E. Gieseking (ed.) Soil components. Vol. 1. Organic components. Springer-Verlag, Berlin, Germany.

Franco, C.M.M., P.J. Clarke, M.E. Tate, and J.M. Oades. 2000. Hydrophobic properties and chemical characterization of natural water repellent materials in Australian sands. J. Hydrol. 231–232:47–58.

Galloway, J.N., W.H. Schlesinger, H. Levy II, A. Michaels, and J.L. Schnoor. 1995. Nitrogen fixation: Atmospheric enhancement—Environmental response. Global Biogeochem. Cycles 9:235–252.

Gershenson, A., N.E. Bade, and W.X. Cheng. 2009. Effects of substrate availability on the temperature sensitivity of soil organic matter decomposition. Global Change Biol. 15:176–183.

Giardina, C.P., and M.G. Ryan. 2000. Evidence that decomposition rates of organic carbon in mineral soil do not vary with temperature. Nature 404:858–861.

Golchin, A., P. Clarke, J.M. Oades, and J.O. Skjemstad. 1995. The effects of cultivation on the composition of organic matter and the structural stability of soils. Aust. J. Soil Res. 33:975–993.

Gonzalez, J.M., and Laird, D.A. 2004. Role of smectite and Al-substituted goethites in the catalytic condensation of arginine and glucose. Clay. Clay Miner. 52:443–450.

Greenland, D.J. 1965a. Interaction between clays and organic compounds in soils. I. Mechanisms of interaction between clays and defined organic compounds. Soil Fert. 28:415–425.

Greenland, D.J. 1965b. Interaction between clays and organic compounds in soils. II. Adsorption of organic compounds and its effects on soil properties. Soils Fert. 28:521–532.

Gu, B.H., J. Schmitt, Z.H. Chen, L.Y. Liang, and J.F. McCarthy. 1994. Adsorption and desorption of natural organic matter on iron oxide: Mechanisms and models. Environ. Sci. Technol. 28:38–46.

Guckert, A., T. Chone, and F. Jacquin. 1975. Microflore et stabilité structurale du sol. Rev. Ecol. Biol. Sol 12:211–223.

Hagedorn, F., M. Saurer, and P. Blaser. 2004. A C-13 tracer study to identify the origin of dissolved organic carbon in forested mineral soils. Eur. J. Soil Sci. 55:91–100.

Hall, P.L., G.J. Churchman, and B.K.G. Theng. 1985. The size distribution of allophane unit particles in aqueous suspensions. Clay. Clay Miner. 33:345–349.

Harms, H., and A.J.B. Zehnder. 1995. Bioavailability of sorbed 3-chlorodibenzofuran. Appl. Environ. Microbiol. 61:27–33.

Hashizume, H., and B.K.G. Theng. 1999. Adsorption of DL-alanine by allophane: Effect of pH and unit particle aggregation. Clay Miner. 34:233–238.

Hashizume, H., and B.K.G. Theng. 2007. Adenine, adenosine, ribose and 5′-AMP adsorption to allophane. Clay. Clay Miner. 55:599–605.

Hayes, M.H.B., and U. Mingelgrin. 1991. Interactions between small organic chemicals and soil colloidal constituents, p. 323–408. In G.H. Bolt, M.F. De Boodt, M.H.B. Hayes, and M.B. McBride (eds.) Interactions at the soil colloid-soil solution interface. Kluwer Academic Publishers, Dordrecht, the Netherlands.

Hochella, M.F., Jr. 2008. Nanogeoscience: From origins to cutting-edge applications. Elements 4:373–379.

Hochella, M.F., Jr., K.S. Lower, P.A. Maurice, R.L. Penn, N. Sahai, D.L. Sparks, and B.S. Twining. 2008. Nanominerals, mineral nanoparticles, and earth systems. Science 319:1631–1635.

Hoffmann, R.W., and G.W. Brindley. 1960. Adsorption of non-ionic aliphatic molecules from aqueous solutions on montmorillonite. Geochim. Cosmochim. Acta 20:15–29.

Homann, P.S., P. Sollins, M. Fiorella, T. Thorson, and J.S. Kern. 1998. Regional soil organic carbon storage estimates for western Oregon by multiple approaches. Soil Sci. Soc. Am. J. 62:789–796.

Horwath, W.R. 2008. Carbon cycling and formation of soil organic matter, p. 91–97. *In* W. Chesworth (ed.) Encyclopedia of soil science. Springer, Dordrecht, the Netherlands.

Huang, P.M. 2000. Abiotic catalysis, p. B303–B332. *In* M.E. Sumner (ed.) Handbook of soil science. CRC Press, Boca Raton, FL.

Huang, P.M. 2004. Soil mineral-organic matter-microorganism interactions: Fundamentals and impacts. Adv. Agron. 82:391–472.

Huang, P.M. 2008. Soil physical-chemical-biological interfacial interactions: An overview, p. 3–37. *In* Q. Huang, P.M. Huang, and A. Violante (eds.) Soil mineral-microbe-organic interactions: Theories and applications. Springer-Verlag, Berlin, Germany.

Huang, P.M., J. Berthelin, J.-M. Bollag, W.B. McGill, and A.L. Page (eds.). 1995. Environmental impact of soil component interactions. Vols. I and II. CRC Lewis Publishers, Boca Raton, FL.

Huang, P.M., and A. Hardie. 2009. Formation mechanisms of humic substances in the environment, p. 41–109. *In* N. Senesi, B. Xing, and P.M. Huang (eds.) Biophysicochemical processes of nonliving natural organic matter in environmental systems. Vol. 2, Wiley-IUPAC Series. John Wiley & Sons, Hoboken, NJ.

Huang, Q., P.M. Huang, and A. Violante (eds.). 2008. Soil mineral-microbe-organic interactions: Theories and applications. Springer-Verlag, Berlin, Germany.

Huang, P.M., and M. Schnitzer (eds.). 1986. Interactions of soil minerals with natural organics and microbes. Soil Science Society of America, Madison, WI.

Inoue, K., and P.M. Huang. 1986. Influence of selected organic ligands on the formation of allophane and imogolite. Soil Sci. Soc. Am. J. 50:1623–1633.

Inoue, K., and P.M. Huang. 1990. Perturbation of imogolite formation by humic substances. Soil Sci. Soc. Am. J. 54:1490–1497.

ISMOM. 2008. 5th International symposium of Interactions of soil Minerals with Organic Components and Microorganisms. J. Soil Sci. Plant Nutr.—Special issue, 8:1–280.

Jacks, G.V. 1973. The biological nature of soil productivity. Soil Fert. 26:147–150.

Jardine, P.M., N.L. Weber, and J.F. McCarthy. 1989. Mechanisms of dissolved organic carbon adsorption on soil. Soil Sci. Soc. Am. J. 53:1378–1385.

Jastrow, J.D., and R.M. Miller. 1997. Soil aggregate stabilization and carbon sequestration: Feedbacks through organomineral associations, p. 207–223. *In* R. Lal, J.M. Kimble, R.F. Follet, and B.A. Stewart (eds.) Soil processes and the carbon cycle. CRC Press, Boca Raton, FL.

Jenkinson, D.S., and K. Coleman. 2008. The turnover of organic carbon in subsoils. Part 2. Modelling carbon turnover. Eur. J. Soil Sci. 59:400–413.

Jenkinson, D.S., and D.S. Powlson. 1976. The effects of biocidal treatments on metabolism in soil. V. A method of measuring soil biomass. Soil Biol. Biochem. 8:209–213.

Jordan, J.W. 1949. Alteration of properties of bentonite by reactions with amines. Miner. Mag. 28:598–605.

Jordan, J.W. 1963. Organophilic clay-base thickeners. Clay. Clay Miner. 10:299–308.

Joussein, E., S. Petit, J. Churchman, B. Theng, D. Righi, and B. Delvaux. 2005. Halloysite clay minerals—A review. Clay Miner. 40:383–426.

Kaiser, K., and G. Guggenberger. 2000. The role of DOM sorption to mineral surfaces in the preservation of organic matter in soils. Org. Geochem. 31:711–725.

Kaiser, K., and G. Guggenberger. 2003. Mineral surfaces and soil organic matter. Eur. J. Soil Sci. 54:219–236.

Kaiser, K., G. Guggenberger, L. Haumaier, and W. Zech. 1997. Dissolved organic matter sorption on subsoils and minerals studied by ^{13}C-NMR and DRIFT spectroscopy. Eur. J. Soil Sci. 48:301–310.

Kalbitz, K., S. Solinger, J.H. Park, B. Michalzik, and E. Matzner. 2000. Controls on the dynamics of dissolved organic matter in soils: A review. Soil Sci. 165:277–304.

Kay, B.D., and D.A. Angers. 2000. Soil structure, p. A229–A276. *In* M.E. Sumner (ed.) Handbook of soil science. CRC Press, Boca Raton, FL.

Keiluweit, M., and M. Kleber. 2009. Molecular-level interactions in soils and sediments: The role of aromatic π-systems. Environ. Sci. Technol. 43:3421–3429.

Kern, J.S. 1994. Spatial patterns of soil organic carbon in the contiguous United States. Soil Sci. Soc. Am. J. 58: 439–455.

Kinyangi, J., D. Solomon, B. Liang, M. Lerotic, S. Wirick, and J. Lehmann. 2006. Nanoscale biogeocomplexity of the organomineral assemblage in soil: Application of STXM microscopy and C 1s-NEXAFS spectroscopy. Soil Sci. Soc. Am. J. 70:1708–1718.

Kirschbaum, M.U.F. 1995. The temperature-dependence of soil organic-matter decomposition, and the effect of global warming on soil organic-C storage. Soil Biol. Biochem. 27:753–760.

Kleber, M., R. Mikutta, M.S. Torn, and R. Jahn. 2005. Poorly crystalline mineral phases protect organic matter in acid subsoil horizons. Eur. J. Soil Sci. 56:717–725.

Kleber, M., P. Sollins, and R. Sutton. 2007. A conceptual model of organo-mineral interactions in soils: Self-assembly of organic molecular fragments into zonal structures on mineral surfaces. Biogeochemistry 85:9–24.

Kögel-Knabner, I., G. Guggenberger, M. Kleber, E. Kandeler, K. Kalbitz, S. Scheu, K. Eusterhues, and P. Leinweber. 2008. Organo-mineral associations in temperate soils: Integrating biology, mineralogy, and organic matter chemistry. J. Plant Nutr. Soil Sci. 171:61–82.

Konen, M.E., C.L. Burras, and J.A. Sandor. 2003. Organic carbon, texture, and quantitative color measurement relationships for cultivated soils in north central Iowa. Soil Sci. Soc. Am. J. 67:1823–1830.

Kononova, M.M. 1966. Soil organic matter. 2nd edn. Pergamon Press, New York.

Kool, D.M., H.G. Chun, K.R. Tate, D.J. Ross, P.C.D. Newton, and J. Six. 2007. Hierarchical saturation of soil carbon pools near a natural CO_2 spring. Global Change Biol. 13:1282–1293.

Kretzschmar, R., and T. Schäfer. 2005. Metal retention and transport on colloidal particles in the environment. Elements 1:205–210.

Laird, D.A., E. Barriuso, R.H. Dowdy, and W.C. Koskinen. 1992. Adsorption of atrazine on smectites. Soil Sci. Soc. Am. J. 56:62–67.

Lagaly, G., M. Ogawa, and I. Dékány. 2006. Clay mineral organic interactions, p. 309–377. In F. Bergaya, B.K.G. Theng, and G. Lagaly (eds.) Handbook of clay science. Elsevier, Amsterdam, the Netherlands.

Lehmann, J., J. Kinyangi, and D. Solomon. 2007. Organic matter stabilization in soil microaggregates: Implications from spatial heterogeneity of organic carbon contents and carbon forms. Biogeochemistry 85:45–57.

Li, H., T.R. Pereira, B.J. Teppen, D.A. Laird, C.T. Johnston, and S.A. Boyd. 2007. Ionic strength-induced formation of smectite quasicrystals enhances nitroaromatic compound sorption. Environ. Sci. Technol. 41:1251–1256.

Li, H., B.J. Teppen, D.A. Laird, C.T. Johnston, and S.A. Boyd. 2006. Effects of increasing potassium chloride and calcium chloride ionic strength on pesticide sorption by potassium- and calcium-smectite. Soil Sci. Soc. Am. J. 70:1889–1895.

Liski, J., H. Ilvesniemi, A. Makela, and C.J. Westman. 1999. CO_2 emissions from soil in response to climate warming are overestimated—The decomposition of old soil organic matter is tolerant of temperature. Ambio 28: 171–174.

Lloyd, J., and J.A. Taylor. 1994. On the temperature-dependence of soil respiration. Funct. Ecol. 8:315–323.

Loveland, P., and J. Webb. 2003. Is there a critical level of organic matter in the agricultural soils of temperate regions: A review. Plant Soil 76:307–318.

Marschner, B., S. Brodowski, A. Dreves, G. Gleixner, A. Gude, P.M. Grootes, U. Hamer, A. Helm, G. Jandl, R. Ji, K. Kaiser, K. Kalbitz, C. Kramer, P. Leinweber, J. Rethemeyer, A. Schäffer, M.W.I. Schmidt, L. Schwark, and G.L.B. Wiesenberg. How relevant is recalcitrance for the stabilization of organic matter in soils? J. Plant Nutr. Soil Sci. 171:91–110.

Mayer, L.M., and B. Xing. 2001. Organic matter-surface relationships in acid soils. Soil Sci. Soc. Am. J. 65:250–258.

McBride, M.B. 1994. Environmental chemistry of soils. Oxford University Press, New York.

McGill, W.B., and C.V. Cole. 1981. Comparative aspects of cycling of organic C, N, S and P through soil organic matter. Geoderma 26:267–286.

Meunier, A. 2006. Why are clay minerals small? Clay Miner. 41:551–566.

Mikutta, R., M. Kleber, and R. Jahn. 2005. Poorly crystalline minerals protect organic carbon in clay subfractions from acid subsoil horizons. Geoderma 128:106–115.

Mikutta, R., C. Mikutta, K. Kalbitz, T. Scheel, K. Kaiser, and R. Jahn. 2007. Biodegradation of forest floor organic matter bound to minerals via different binding mechanisms. Geochim. Cosmochim. Acta 71:2569–2590.

Mora, M.L., M. Escudey, and G.G. Galindo. 1994. Sintesis y caracterización de suelos alofánicos. Bol. Soc. Chilena Quim. 39:237–243.

Mortland, M.M. 1970. Clay-organic complexes and interactions. Adv. Agron. 22:75–117.

Mott, C.J.B. 1981. Anion and ligand exchange, p. 179–219. In D.J. Greenland, and M.H.B. Hayes (eds.) The chemistry of soil processes. Wiley, Chichester, U.K.

Müller, T., and H. Höper. 2004. Soil organic matter turnover as a function of soil clay content: Consequences for model applications. Soil Biol. Biochem. 36:877–888.

Nichols, J.D. 1984. Relation of organic carbon to soil properties and climate in the southern Great Plains. Soil Sci. Soc. Am. J. 48:1382–1384.

Oades, J.M. 1988. The retention of organic matter in soils. Biogeochemistry 5:35–70.

Oades, J.M. 1989. An introduction to organic matter in mineral soils, p. 89–159. In J.B. Dixon, and S.B. Weed (eds.) Minerals in soil environments. 2nd edn. Soil Science Society of America, Madison, WI.

Parfitt, R.L. 1990. Allophane in New Zealand—A review. Aust. J. Soil Res. 28:343–360.

Parfitt, R.L. 2009. Allophane and imogolite: Role in soil biogeochemical processes. Clay Miner. 44:135–155.

Parfitt, R.L., A.R. Fraser, and V.C. Farmer. 1977. Adsorption on hydrous oxides. 3. Fulvic acid and humic acid on goethite, gibbsite and imogolite. J. Soil Sci. 28:289–296.

Parfitt, R.L., B.K.G. Theng, J.S. Whitton, and T.G. Shepherd. 1997. Effects of clay minerals and land use on organic matter pools. Geoderma 75:1–12.

Parfitt, R.L., G. Yuan, and B.K.G. Theng. 1999. A ^{13}C-NMR study of the interactions of soil organic matter with aluminium and allophane in podzols. Eur. J. Soil Sci. 50:695–700.

Percival, H.J., R.L. Parfitt, and N.A. Scott. 2000. Factors controlling soil carbon levels in New Zealand grasslands: Is clay content important? Soil Sci. Soc. Am. J. 64:1623–1630.

Piccolo, A. 2001. The supramolecular structure of humic substances. Soil Sci. 166:810–832.

Piccolo, A., G. Pietramellara, and J.S.C. Mbagwu. 1997. Use of humic substances as soil conditioners to increase aggregate stability. Geoderma 75:267–277.

Pignatello, J.J., and B.S. Xing. 1996. Mechanisms of slow sorption of organic chemicals to natural particles. Environ. Sci. Technol. 30:1–11.

Preston, C.M. 1996. Applications of NMR to soil organic matter analysis: History and prospects. Soil Sci. 161:144–166.

Ransom, B., R.H. Bennett, R. Baerwald, and K. Shea. 1997. TEM study of in situ organic matter on continental margins: Occurrence and the "monolayer" hypothesis. Mar. Geol. 138:1–9.

Ransom, B., D. Kim, M Kastner, and S. Wainwright. 1998. Organic matter preservation on continental slopes: Importance of mineralogy and surface area. Geochim. Cosmochim. Acta 62:1329–1345.

Rasmussen, C., R.J. Southard, and W.R. Horwath. 2006. Mineral control of organic carbon mineralization in a range of temperate conifer forest soils. Global Change Biol. 12:834–847.

Rasmussen, C., R.J. Southard, and W.R. Horwath. 2008. Litter type and soil minerals control temperate forest soil carbon response to climate change. Global Change Biol. 14:2064–2080.

Ray, S.S., and M. Okamoto. 2003. Polymer/layered silicate nanocomposites: A review from preparation to processing. Prog. Polym. Sci. 28:1539–1641.

Rhodes, A.H., J. Hofman, and K.T. Semple. 2008. Development of phenanthrene catabolism in natural and artificial soils. Environ. Pollut. 152:424–430.

Robertson, R.H.S. 1986. Fuller's earth. A history. Volturna Press, Hythe, Kent, U.K.

Ruiz-Hitzky, E., and A. Van Meerbeek. 2006. Clay mineral- and organoclay-polymer nanocomposite, p. 583–621. In F. Bergaya, B.K.G. Theng, and G. Lagaly (eds.) Handbook of clay science. Elsevier, Amsterdam, the Netherlands.

Saggar, S., K.R. Tate, C.W. Feltham, C.W. Childs, and A. Parshotam. 1994. Carbon turn-over in a range of allophanic soils amended with ^{14}C-labelled glucose. Soil Biol. Biochem. 26:1263–1271.

Schloesing, Th. 1874. Détermination de l'argile dans la terre arable. C.R. Acad. Sci. Paris 78:1276–1279.

Schnitzer, M. 1986. Binding of humic substances by soil mineral colloids, p. 77–101. In P.M. Huang, and M. Schnitzer (eds.) Interactions of soil minerals with natural organics and microbes. Soil Science Society of America, Madison, WI.

Schnitzer, M., and S.U. Khan. 1972. Humic substances in the environment. Marcel Dekker, New York.

Schnitzer, M., and S.U. Khan. 1978. Soil organic matter. Elsevier, Amsterdam, the Netherlands.

Schulten, H.-R., and P. Leinweber. 1996. Characterization of humic and soil particles by analytical pyrolysis and computer modelling. J. Anal. Appl. Pyrolysis 38:1–53.

Schulten, H.-R., P. Leinweber, and B.K.G. Theng. 1996. Characterization of organic matter in an interlayer clay-organic complex from soil by pyrolysis methylation-mass spectrometry. Geoderma 69:105–118.

Schulten, H.-R., and M. Schnitzer. 1995. Three-dimensional models for humic acids and soil organic matter. Naturwissenschaften 82:487–498.

Schulthess, C.P., and C.P. Huang. 1991. Humic and fulvic-acid adsorption by silicon and aluminum-oxide surfaces on clay minerals. Soil Sci. Soc. Am. J. 55:34–42.

Schulze, D.G. 1989. An introduction to soil mineralogy, p. 1–34. In J.B. Dixon, and S.B. Weed (eds.) Minerals in soil environments, 2nd edn. Soil Science Society of America, Madison, WI.

Schwertmann, U. 2008. Iron oxides, p. 363–369. In W. Chesworth (ed.) Encyclopedia of soil science. Springer, Dordrecht, the Netherlands.

Schwertmann, U., and R.M. Taylor. 1989. Iron Oxides. p. 379–438. In J.B. Dixon, and S.B. Weed (eds.) Minerals in soil environments. 2nd edn. Soil Science Society of America, Madison, WI.

Scott, M.J., and J.J. Morgan. 1996. Reactions at oxide surfaces. 2. Oxidation of Se(IV) by synthetic birnessite. Environ. Sci. Technol. 30:1990–1996.

Shepherd, T.G., S. Saggar, R.H. Newman, C.W. Ross, and J. Dando. 2001. Tillage-induced changes to soil structure and organic carbon fractions in New Zealand soils. Aust. J. Soil Res. 39:465–489.

Skjemstad, J.O., P. Clarke, P.A. Golchin, and J.M. Oades. 1997. Characterization of organic matter by solid-state ^{13}C NMR spectroscopy, p. 253–271. In G. Cadisch, and K.E. Giller (eds.) Driven by nature; plant litter quality and decomposition. CAB International, Wallingford, U.K.

Smith, P., J.U. Smith, D.S. Powlson, W.B. McGill, J.R.M. Arah, O.G. Chertov, K. Coleman, U. Franko, S. Frolking, D.S. Jenkinson, L.S. Jensen, R.H. Kelly, H. Klein-Gunnewiek, A.S. Komarov, C. Li, J.A.E. Molina, T. Mueller, W.J. Parton, J.H.M. Thornley, and A.P. Whitmore. 1997. A comparison of the performance of nine soil organic matter models using datasets from seven long-term experiments. Geoderma 81:153–225.

Solomon, D., J. Lehmann, J. Kinyangi, B. Liang, and T. Schäfer. 2005. Carbon K-edge NEXAFS and FTIR-ATR spectroscopic investigation of organic carbon speciation in soils. Soil Sci. Soc. Am. J. 69:107–119.

Spielvogel, S., J. Prietzel, and I. Kögel-Knabner. 2008. Soil organic matter stabilization in acidic forest soils is preferential and soil type-specific. Eur. J. Soil Sci. 59:674–692.

Sposito, G. 2008. The chemistry of soils. Oxford University Press, New York.

Stevenson, F.J. 1994. Humus chemistry: Genesis, composition, reactions. 2nd edn. John Wiley & Sons, New York.

Sullivan, L.A. 1990. Soil organic matter, air encapsulation and water stable aggregates. J. Soil Sci. 41:529–534.

Sutton, R., and G. Sposito. 2005. Molecular structure in soil humic substances: The new view. Environ. Sci. Technol. 39:9009–9015.

Swift, R.S. 1991. Effect of humic substances and polysaccharides on soil aggregation, p. 153–162. In W.S. Wilson (ed.) Advances in soil organic matter research: The impact on agriculture and the environment. The Royal Society of Chemistry, Cambridge, U.K.

Swift, R.S. 1999. Macromolecular properties of soil humic substances: Fact, fiction, and opinion. Soil Sci. 164:790–802.

Swift, R.S. 2001. Sequestration of carbon by soil. Soil Sci. 166:858–871.

Tang, Z.Y., L.Y. Wu, Y.M. Luo, and P. Christie. 2009. Size fractionation and characterization of nanocolloidal particles in soils. Environ. Geochem. Health 31:1–10.

Tate, K.R., R.H. Wilde, D.J. Giltrap, W.T. Baisden, S. Saggar, N.A. Trustrum, N.A. Scott, and J.R. Barton. 2005. Soil organic carbon stocks and flows in New Zealand: System development, measurement and modelling. Can. J. Soil Sci. 85:481–489.

Theng, B.K.G. 1971. Mechanisms of formation of colored clay-organic complexes. A review. Clay. Clay Miner. 19:383–390.

Theng, B.K.G. 1974. The chemistry of clay-organic reactions. Adam Hilger, London, U.K.

Theng, B.K.G. 1976. Interactions between montmorillonite and fulvic acid. Geoderma 15:243–251.

Theng, B.K.G. 1979. Formation and properties of clay-polymer complexes. Elsevier, Amsterdam, the Netherlands.

Theng, B.K.G. (ed.) 1980. Soils with variable charge. New Zealand Society of Soil Science, Lower Hutt, New Zealand.

Theng, B.K.G. 1982. Clay-polymer interactions: Summary and perspectives. Clay. Clay Miner. 30:1–10.

Theng, B.K.G., G.J. Churchman, W.P. Gates, and G. Yuan. 2008. Organically modified clays for pollutant uptake and environmental protection, p. 145–174. In Q. Huang, P.M. Huang, and A. Violante (eds.) Soil mineral-microbe-organic interactions: Theories and applications. Springer-Verlag, Berlin, Germany.

Theng, B.K.G., G.J. Churchman, and R.H. Newman. 1986. The occurrence of interlayer clay-organic complexes in two New Zealand soils. Soil Sci. 142:262–266.

Theng, B.K.G., D.J. Greenland, and J.P. Quirk. 1967. Adsorption of alkylammonium cations by montmorillonite. Clay Miner. 7:1–17.

Theng, B.K.G., and V.A. Orchard. 1995. Interactions of clays with microorganisms and bacterial survival in soil: A physicochemical perspective, p. 123–143. In P.M. Huang, J. Berthelin, J.-M. Bollag, W.B. McGill, and A.L. Page (eds.) Environmental impact of soil component interactions. Vol. II. CRC Lewis Publishers, Boca Raton, FL.

Theng, B.K.G., G.G. Ristori, C.A. Santi, and H.J. Percival. 1999. An improved method for determining the specific surface areas of topsoils with varied organic matter content, texture and clay mineral composition. Eur. J. Soil Sci. 50:309–316.

Theng, B.K.G., M. Russell, G.J. Churchman, and R.L. Parfitt. 1982. Surface properties of allophane, halloysite, and imogolite. Clay. Clay Miner. 30:143–149.

Theng, B.K.G., and H.W. Scharpenseel. 1976. The adsorption of ^{14}C-labelled humic acid by montmorillonite, p. 643–653. In S.W. Bailey (ed.) Proceedings of the International Clay Conference 1975. Applied Publishing, Wilmette, IL.

Theng, B.K.G., and K.R. Tate. 1989. Interactions of clays with soil organic constituents. Clay Res. 8:1–10.

Theng, B.K.G., K.R. Tate, and P. Becker-Heidmann. 1992. Towards establishing the age, location, and identity of the inert soil organic-matter of a spodosol. Z. Pflanzenern. Bodenk. 155:181–184.

Theng, B.K.G., and G. Yuan. 2008. Nanoparticles in the soil environment. Elements 4:395–399.

Tisdall, J.M., and J.M. Oades. 1982. Organic matter and water-stable aggregates. J. Soil Sci. 33:141–163.

Van Veen, J.A., J.N. Ladd, and M.J. Frissel. 1984. Modelling C and N turnover, through the microbial biomass in soil. Plant Soil 76:257–274.

Vance, E.D., P.C. Brookes, and D.S. Jenkinson. 1987. An extraction method for measuring soil microbial biomass C. Soil Biol. Biochem. 19:703–707.

Vansant, E.F., and J.B. Uytterhoeven. 1972. Thermodynamics of the exchange of n-alkylammonium ions on Na-montmorillonite. Clay. Clay Miner. 20:47–54.

Varadachari, C., A.H. Mondal, and K. Ghosh. 1991. Some aspects of clay humus complexation: Effects of exchangeable cations and lattice charge. Soil Sci. 151:220–227.

Violante, A., and Gianfreda, L. 2000 Role of biomolecules in the formation and reactivity towards nutrients and organics of variable charge minerals and organo-mineral complexes in soil environment, p. 207–270. In J.-M. Bollag, and G. Stotsky (eds.) Soil biochemistry. Marcel & Dekker, New York.

Violante, A., A. Del Gaudio, M. Pigna, M. Pucci, and C. Amalfitano. 2008. Sorption and desorption of arsenic by soil minerals and soils in the presence of nutrients and organics, p. 39–69. In Q. Huang, P.M. Huang, and A. Violante (eds.) Soil mineral-microbe-organic interactions: Theories and applications. Springer-Verlag, Berlin, Germany.

Violante, A., P.M. Huang, J.-M. Bollag, and L. Gianfreda (eds.) 2002. Soil mineral-organic matter-microorganism interactions and ecosystem health. Developments in Soils science, 28A and 28B. Elsevier, Amsterdam, the Netherlands.

Vitousek, P.M., J.D. Aber, R.W. Howarth, G.E. Likens, P.A. Matson, D.W. Schindler, W.H. Schlesinger, and D.G. Tilman. 1997. Human alteration of the global nitrogen cycle: Sources and consequences. Ecol. Appl. 7:737–750.

Vitousek, P.M., and R.W. Howarth. 1991. Nitrogen limitation on land and in the sea—How can it occur. Biogeochemistry 13:87–115.

von Lützow, M., I. Kögel-Knabner, K. Ekschmitt, E. Matzner, G. Guggenberger, B. Marschner, and H. Flessa, 2006. Stabilization of organic matter in temperate soils: mechanisms and their relevance under different soil conditions - a review. Eur. J. Soil Sci. 57:426–445.

von Lützow, M., I. Kögel-Knabner, K. Ekschmitt, H. Flessa, G. Guggenberger, E. Matzner, and B. Marschner. 2007. SOM fractionation methods: Relevance to functional pools and to stabilization mechanisms. Soil Biol. Biochem. 39:2183–2207.

Wada, K. 1989. Allophane and imogolite, p. 1051–1087. In J.B. Dixon, and S.B. Weed (eds.) Minerals in soil environments. 2nd edn. Soil Science Society of America, Madison, WI.

Wallis, M.G., and D.J. Horne. 1992. Soil water repellency. Adv. Soil Sci. 20:91–146.

Wattel-Koekkoek, E.J.W., P.P.L. van Genuchten, P. Buurman, and B. van Lagen. 2001. Amount and composition of clay-associated soil organic matter in a range of kaolinite and smectitic soils. Geoderma 99:27–49.

Watts, C.W., L.J. Clark, P.R. Poulton, D.S. Powlson, and A.P. Whitmore. 2006. The role of clay, organic carbon and long-term management on mouldboard plough draft measured on the Broadbalk wheat experiment at Rothamsted. Soil Use Manage. 22:334–341.

Weissmahr, K.W., S.B. Haderlein, R.P. Schwarzenbach, R. Hany, and R. Nüesch. 1996. In situ spectroscopic investigations of adsorption mechanisms of nitroaromatic compounds at clay minerals. Environ. Sci. Technol. 31:240–247.

Xing, B. 1997. The effect of the quality of soil organic matter on sorption of naphthalene. Chemosphere 35:633–642.

Xing, B., and J. Pignatello. 1998. Competitive sorption between 1,3-dichlorobenzene or 2,4-dichlorophenol and natural aromatic acids in soil organic matter. Environ. Sci. Technol. 32:614–619.

Yariv, S., and H. Cross (eds.). 2002. Organo-clay complexes and interactions. Marcel Dekker, New York.

Yuan, G., M. Soma, H. Seyama, B.K.G. Theng, L.M. Lavkulich, and T. Takamatsu. 1998. Assessing the surface composition of soil particles from some Podzolic soils by X-ray photoelectron spectroscopy. Geoderma 86:169–181.

Yuan, G., B.K.G. Theng, R.L. Parfitt, and H.J. Percival. 2000. Interactions of allophane with humic acid and cations. Eur. J. Soil Sci. 51:35–41.

Zheng, Q.H., A.B. Yu, G.Q.M. Lu, and D.R. Paul. 2005. Clay-based polymer nanocomposites: Research and commercial development. J. Nanosci. Nanotechnol. 5: 574–1592.

3
Nanoscale Science and Technology in Soil Science

3.1	Introduction: Nanoscience and Nanotechnology	3-1
	Chapter Goals • Nanoparticles and Nanominerals • Natural and Engineered Nanoparticles	
3.2	Nanoscience Is Changing Our View of the Properties and Behaviors of Nanoparticles	3-4
	Nanoparticle Size and Structure • Nanoparticle Surface Area • The Concept of Surface Free Energy • Surface Free Energy and Classical Mineral Nucleation Theory • Particle Size and Stability: Classical Theory • Nanoparticle Stability: When More Than Size Alone Matters • Stability and the Medium Surrounding a Nanoparticle • A New Concept of Mineral Growth from Nanoparticles: Oriented Attachment • Nanoparticle Sorption, Redox Phenomena, and Catalysis	
3.3	Nanoparticle Mobility in Soils and Sediments	3-10
3.4	Nanotoxicity	3-12
3.5	Nanoenabled Sensing	3-13
3.6	Frontiers in Research in Nanoscale Particles and Processes in Soil Environments	3-13
	Acknowledgments	3-14
	References	3-14

Patricia A. Maurice
University of Notre Dame

3.1 Introduction: Nanoscience and Nanotechnology

Nanoscience and nanotechnology together represent one of the "hottest" research areas in science and engineering at the start of the twenty-first century. With new advances in instrumentation and theory, scientists have come to realize that within the range of 1 to ~100 billionths of a meter, known as the nanoscale, materials may display properties and behaviors that are unique and often cannot be described using our knowledge of either bulk systems or individual molecules. This is not necessarily a surprise to soil scientists who have known for many decades that fine-grained soil minerals such as aluminosilicate clays and ferrihydrite tend to have special properties that are distinct from those of larger minerals. The soil science community also has taken a leadership role in understanding the structures and behaviors of complex nanoscale organic matter components such as humic and fulvic acids.

Many of the natural constituents of soils are nanoparticles—for example, aluminosilicate clays, Fe(hydr)oxides such as ferrihydrite, viruses, and humic substances. Soil nanoparticles may occur as mobile colloids (Kretzschmar and Shäfer, 2005), as coatings on other mineral surfaces, and/or as mixed organic/inorganic aggregates (Nugent et al., 1998; Chorover et al., 2007). The high surface areas of nanoparticles mean that they are particularly influential in controlling the fate and transport of pollutants as well as the bioavailability of key nutrients such as phosphate and Fe (Brown and Parks, 2001; Hochella et al., 2008; Theng and Yuan, 2008). Nanoparticles interact with soil microorganisms at a profound level. Microbial processes are key to both the nucleation and dissolution of Fe and Mn (hydr) oxides (Banfield and Zhang, 2001; Fortin and Langley, 2005) and there have even been reports of biogenic uraninite nanoparticles (Bargar et al., 2008). Synthetic nanoparticles are finding their way into the environment, and how they interact within soils is a key topic of international concern and research (EPA, 2005; Nowcak and Bucheli, 2007).

Nanoparticles are important not only in soils per se but also in the broader near-earth-surface environment known as the critical zone (Hochella et al., 2008; Maurice and Hochella, 2008). Nanoparticles are ubiquitous in aquatic and marine environments, often controlling pollutant transport and nutrient bioavailability. For example, Fe oxide nanoparticles are likely to be a major source of essential Fe to ocean algae, helping to control ocean biological productivity and ultimately the global carbon cycle (Wu et al., 2001). Nanoparticles in surface waters, soils, and aquifers often control the mobilities of pollutants such as heavy metals, radionuclides, and hydrophobic organic compounds (Brown and Parks, 2001). Nanoparticles play a major role in the transport of toxic metals in acid mine drainage environments

(Hochella et al., 2005). Nanoscale precipitates may form on mineral surfaces, helping to control reactivity but often eluding detection and/or characterization by most conventional microscopic and spectroscopic techniques (Nugent et al., 1998). The nanotechnology industry is producing an increasing number of new nanomaterials such as fullerenes and synthetic TiO_2 that are being introduced into thousands of products from cosmetics to targeted drugs to video display screens and finding their way into the environment. Yet, we know very little about the environmental consequences of these materials (NRC, 2008). On the other hand, nanotechnology offers new opportunities for "green" water purification (i.e., with low energy consumption; Shannon et al., 2008) and is leading to an advanced generation of nano-enabled environmental sensors (Potter et al., 2007).

3.1.1 Chapter Goals

This chapter aims to describe our state-of-the art—though rapidly changing—understanding of what makes nanomaterials unique, to provide insight into the special behaviors of nanomaterials, and to highlight how the soil science community can both benefit from and contribute to nanoscale environmental research. A key underlying principle in this chapter is that the high surface areas of nanoparticles mean that their characteristics, including stabilities and reactivities, can be extremely sensitive to surface free energy. Nanoparticles have a relatively high proportion of atoms associated with their surfaces. These surface- and near-surface atoms have different coordination environments than in analogous bulk materials. Nanoparticles may have different surface composition, different types and densities of surface sites, and different reactivities with respect to processes such as sorption, catalysis, and redox reactions (Waychunas et al., 2005). Nanoparticles also may have different overall crystalline order or structure, different stability, and unique thermal, optical, electrical tensile strength, and/or magnetic properties than observed in larger, bulk materials. We are only beginning to understand the controls on the fundamental properties of nanoparticles; how they behave in complex soil environments is even less well understood.

3.1.2 Nanoparticles and Nanominerals

A nanometer is one billionth (10^{-9}) of a meter and the prefix "nano" is used to encompass materials and processes at this general scale. The term "nanoparticle" is used throughout this text, although its definition has been subject to much debate and continues to evolve. One common definition of a nanoparticle is any ultrafine particle that is between 1 and 100 nm in size. This size limit is sometimes applied to just two dimensions. Under this definition, various nanoparticles may or may not have properties that differ from those of larger, bulk materials. Banfield and Zhang (2001) suggested that the upper limit to the size of a nanoparticle might be defined based on the size at which fundamental properties differ from those of a corresponding bulk material. According to this definition, the maximum size range for a nanoparticle is likely to vary for different materials. Hochella et al. (2008) suggested that the size range in which particles of earth materials behave differently is often between one and at most several tens of nanometers.

Hochella et al. (2008) defined nanominerals as minerals such as ferrihydrite that exist only in the nanoparticle size range, or clays that only exist with at least one nanoscale dimension. These authors defined mineral nanoparticles as minerals that are in the nano-size range, but that also exist at larger sizes. Most known minerals probably can and do exist somewhere in nature within the nanoparticle size range, although we are beginning to learn that some minerals may be differently hydrated or undergo other compositional or structural changes at small particle size so that new mineral nomenclature may need to be defined in some instances. In any case, nanominerals challenge the common definition of a mineral as "a naturally occurring homogeneous solid with a definite (but generally not fixed) chemical composition and a highly ordered atomic arrangement" (Klein and Hurlbut, 1985), because composition, structure, and especially physical properties all can change at the nanoscale.

Nanoparticles overlap in size with colloids, which range in diameter from 1 nm to 1 μm (Buffle, 2006; Figure 3.1). Nanoparticles are often included in the dissolved fraction of solution samples, as defined by passage through a 0.45 μm filter, even though nanoparticles are distinct in structure, properties, and reactivity from molecules and ions. Nanoparticles often form as secondary minerals during the weathering of larger particles, for example, kaolinite on feldspar. Such surface adhered nanoparticles can be difficult to detect and characterize by conventional means such as scanning electron microscopy (SEM) and microprobe analysis, but can potentially impact reactivity with respect to adsorption and dissolution (Nugent et al., 1998).

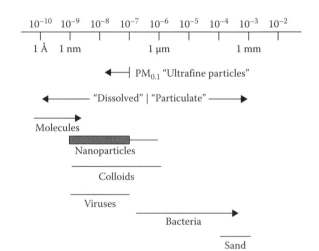

FIGURE 3.1 The size ranges of some environmental particles, including nanoparticles. The box for nanoparticles indicates that nanoparticles are 100 nm or less in at least two dimensions, but can be longer in the third dimension (such as nanotubes and nanowires). (From Maurice, P.A. 2009. Environmental surfaces and interfaces from the nanoscale to the global scale. Wiley, Hoboken, NJ.)

3.1.3 Natural and Engineered Nanoparticles

Soils are in a sense the ultimate nanoenvironments. Nanominerals, mineral nanoparticles, and many nonmineral nanoparticles are commonly found in soils (Hochella et al., 2008; Theng and Yuan, 2008). Viruses generally occur as nanoparticles, and bacteria may contain nanoscale features such as cilia and flagella. Microorganisms may play a role in the formation of a wide variety of nanobiominerals or biomineral nanoparticles, such as Fe and Mn (hydr)oxides and carbonates. According to Tebo et al. (1997) microorganisms appear to be responsible for most Mn(II) oxidation in the environment, and hence they play a dominant role in the formation of Mn oxides, which primarily form as high surface area nanoparticles. The role of microorganisms in the formation of carbonate nanoparticles (Figure 3.2) is receiving a great deal of attention, particularly regarding controls on the carbonate polymorph structure, micromorphology, and microtopography (Warren et al., 2001). Humic substances tend to occur within the nanoscale size range (Figure 3.3; Maurice and Namjesnik-Dejanovic, 1999; Namjesnik-Dejanovic and Maurice, 2001), and their composition, structure, properties and behaviors have long challenged the environmental geochemistry/soil chemistry communities. Black carbon, which forms from the incomplete combustion of fossil fuels, may occur in the nanoscale size range or larger, and is present in many soil environments (Hedges et al., 2000). Given that black carbon tends to be highly unreactive with respect to oxidation, it may represent an important sink of carbon in the global carbon cycle (Kuhlbusch and Crutzen, 1995).

Engineered nanoparticles are becoming increasingly widespread in the environment, and their potential impact on soil and plants is only beginning to be considered (e.g., Nowack and Bucheli, 2007). The U.S. Environmental Protection Agency (EPA, 2005)

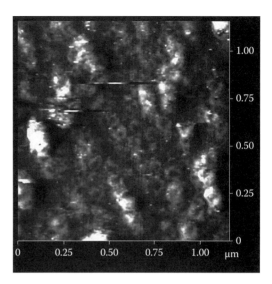

FIGURE 3.3 In-solution AFM image of surface water fulvic acid on muscovite in 0.01 M $CaCl_2$ at pH 5. Some nanoscale structures and porosity are apparent. Sample preparation and imaging parameters as described by Maurice and Namjesnik-Dejanovic (1999). (From Maurice, P., and K. Namjesnik-Dejanovic. 1999. Aggregate structures of sorbed humic substances observed in aqueous solution. Environ. Sci. Technol. 33:1538–1540.)

describes four categories of intentionally manufactured nanoparticles. These are: carbon-based materials such as the fullerenes and carbon nanotubes; metal-based materials such as quantum dots and nano-silver along with reactive metal oxides such as zinc oxide (ZnO) and titanium dioxide (TiO_2); dendrimers, which are nanopolymers constructed of branched units; and nanocomposites. Nanoparticles also may be manufactured unintentionally as a result of grinding, combustion, and other processes.

Gold nanoparticles were used in medieval times in stain glass window production. Nanoparticulate (3.5–100 nm) gold suspensions in water create an intense red color due to plasmon absorption (Brust and Kiely, 2002). Nanoparticulate gold has become the subject of broad industrial interest due to strong catalytic capabilities (Brust and Kiely, 2002). Thiol-stabilized gold nanoparticles may self-assemble into superlattices (Giersig and Mulvaney, 1993) whose change in conductivity in the presence of different organic vapors has led to the development of novel gas sensors (Wohltjen and Snow, 1998). Silver nanoparticles are being widely used as bactericides (Brust and Kiely, 2002). TiO_2 nanoparticles are used in a wide array of applications, including heterogeneous catalysis, photocatalysis, in solar cells for the production of hydrogen and electric energy, gas sensing, pigmentation (paints and cosmetics), sunscreen, protective coatings, and electronics applications (Diebold, 2003).

Discovery, characterization (Kroto et al., 1985), and eventual engineering of fullerenes helped to ignite the nanotechnology revolution. Like many nanomaterials, the definition of the word "fullerene" is subject to some debate and is evolving. Fullerenes are allotropes of carbon, the most famous of which, Buckminsterfullerene, has the formula C_{60} and is about 1 nm in diameter (Figure 3.4). Fullerenes can occur naturally, and have been found in C-rich metamorphic rocks known as shungite

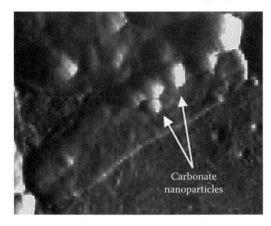

FIGURE 3.2 In-air AFM image of carbonate nanoparticles formed on the surface of *Bacillus pasteurii*. The microbe in the central region of the image is approximately 2.8 μm long and was deposited from solution on polycarbonate membrane filter. Microorganisms play important roles in the formation of mineral nanoparticles and nanominerals. The sawtooth pattern along the edge of the bacteria is an artifact. The experimental conditions are described in Warren et al. (2001). (From Warren, L., P. Maurice, N. Parmar, and G. Ferris. 2001. Microbially mediated calcite precipitation: Implication for solid-phase capture of inorganic contaminants. Geomicrobiol. J. 18:93–115.)

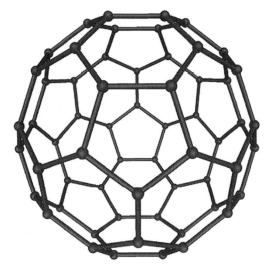

FIGURE 3.4 The structure of C_{60} fullerene, Buckminsterfullerene. The structure of C_{60} resembles a soccer ball consisting of 20 hexagons and 12 pentagons, with a C atom at each vertex and a bond along each polygon edge. (From Wikimedia commons, http://commons.wikimedia.org.)

(Buseck et al., 1992; Parthasarathy et al., 1998); in rocks that have experienced lightning strikes (Daly et al., 1993) or meteorite impacts (Becker et al., 1994), and even in rocks associated with the cretaceous-tertiary (K/T) boundary (Heymann et al., 1994). Fullerenes may occur not only as "buckeyballs" but also as nanotubes, nanosheets, and in other structures, sometimes aggregated. They tend to be insoluble in water, but may be surface functionalized in ways that can greatly affect their solubility, adsorption, aggregation, and other properties.

There are many fundamental questions about the potential impacts of manufactured nanoparticles such as fullerenes on soil and agricultural environments. For example, what are the controls on fullerene mobility in different soil media? What controls fullerene aggregation and how may aggregation affect mobility? What are the potential short- and long-term ecotoxicity effects of fullerenes? How might fullerenes affect soil microbial communities? Are fullerenes taken up by plants? If so, then what are the potential consequences for human health? What are the long-term stabilities of different fullerenes, including various functionalized forms, and how might degradation impact fate and transport, ecotoxicity, and other considerations? In order to begin to answer such questions, it is important to understand that fullerenes and other nanoparticles generally show unique properties and behaviors related to their small size and characteristic high surface area.

3.2 Nanoscience Is Changing Our View of the Properties and Behaviors of Nanoparticles

The large amount of attention and funding focused on nanoscience and technology over the past two decades has led some to question whether it is all just "nano-hype" (Berube, 2005). As will be demonstrated in this chapter, there are real differences in the properties and behaviors of nanoparticles versus ions and molecules or larger, bulk-scale materials, so that the nanoscale represents a true frontier for revolutionary research in science and engineering.

Much of our past understanding of the properties and behaviors of small particles developed over many decades through traditional approaches in (surface) chemistry and physics. Yet, through recent developments in new nanoscale techniques and approaches, we are learning that many nanoparticles and nanoscale processes do not conform to long-held theories and conceptions. In this section, we compare and contrast classical theories and approaches of size, stability, and reactivity with our new and evolving nanoscale understanding.

3.2.1 Nanoparticle Size and Structure

The classical means of determining the structure of a crystalline material is by x-ray diffraction (XRD) as part of x-ray crystallography. Based on a diffraction pattern, the crystallographer can determine detailed information about the mean positions of atoms in a crystal, chemical bonds, and degree of disorder. However, single-crystal XRD, which is the best method for determining structure, is not feasible for nanoparticles because of their small size.

Although it can be difficult to determine nanoparticle structure by classical XRD alone, a number of other techniques can be helpful, particularly when applied in combination. High-resolution transmission electron microscopy (HRTEM) and scanning transmission electron microscopy (STEM) can help in determining whether particles are crystalline or amorphous. The TEM image in Figure 3.5 shows ~10 nm diameter hematite nanoparticles with lattice fringes indicating an ordered crystalline structure.

Other techniques that can be helpful for determining the structure and/or composition of nanoparticles include synchrotron-based x-ray absorption spectroscopy (XAS) and x-ray scattering methods coupled with pair distribution function (PDF) analysis. PDF can allow for determination of changes in interatomic distance as a function of radius, composition, and other

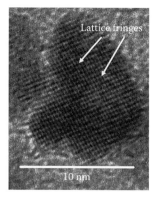

FIGURE 3.5 TEM image of hematite nanoparticles showing lattice fringes. Nanoparticles often are not fully amenable to XRD analysis, and TEM can be essential for determining structure and order. (From Barton, L.E., K.E. Grant, T. Kosel, A.N. Quicksall, and P.A. Maurice. 2011. Size dependent Pb sorption to nanohematite in the presence and absence of a microbial siderophore: Effects of experimental design on sorption edges. Environ. Sci. Technol. 45:3231–3237. Copyright in press American Chemical Society.)

factors, thus providing enhanced analysis of not just short- and long-range structure but also composition and morphology.

For decades, soil scientists and mineralogists have pursued the structure of the Fe hydroxide nanomineral ferrihydrite. Ferrihydrite generally occurs as particles <10 nm, although often aggregated, is poorly ordered, and can have variable structural order. Ferrihydrite is generally designated 2-line or 6-line depending upon the number of reflections that appear in x-ray diffractograms as order increases. Ferrihydrite is not only of interest to the environmental community but also to the biomedical community. This is because ferrihydrite forms the core of the important Fe storage protein, ferritin (Pan et al., 2009). Hence, there is tremendous interest by the broad scientific community in determining the structure of ferrihydrite.

Recently, Michel et al. (2007a, 2007b) suggested that ferrihydrite nanoparticles 6–2 nm in size all have the same composition ($Fe_{10}O_{14}(OH)_2$) and structure described by the hexagonal space group $P6(3)mc$. As particle size decreases from 6 to 2 nm, there is evidence of increasing disorder and distortion of some sites. Michel et al. (2007a, 2007b) showed that for 6 nm particles, the hexagonal unit cell dimensions $a \sim 5.95$ Å and $c \sim 9.06$ Å and the structure contains 20% tetrahedrally and 80% octahedrally coordinated Fe. For 2 nm particles, the hexagonal unit cell dimensions decrease slightly and some of the Fe sites have lesser site occupancy or show evidence of distortion. This model has proven to be controversial and perhaps at least partly incorrect (Rancourt and Meunier, 2008), demonstrating how difficult it can be to determine definitively the structure of nanoparticles of interest to the soil community.

Another common application of XRD has been determination of particle size. Application of the Scherrer equation to observed peak (or line) broadening in XRD patterns been used to determine particle size for particles up to about 0.1 μm. However, the Scherrer equation often cannot be applied reliably to nanoparticles because of issues such as the fact that a large number of atoms in nanoparticles are associated with the surfaces of particles and aggregates, which limits translational symmetry (Hall et al., 2000). New approaches, for example, based on using the Fourier transform of a Debye–Scherrer diffraction pattern, can often provide information on average particle dimensions (Hall et al., 2000). The Fourier method can be used even when only a limited range of scattering angles has been measured. When details of particle size distributions rather than simply average particles size are needed, other techniques such as TEM imaging and small-angle x-ray scattering (Glatter and Krathy, 1982) can prove invaluable. PDF analysis applied to x-ray scattering results can be used to define particle size for nanoparticles of only a few nanometers in size, if shape effects are considered (Kodama et al., 2006). Michel et al. (2007b) found good agreement between results of PDF analysis and TEM-based measurements of particle size for different ferrihydrite samples. Nanoparticle aggregation is common and can complicate nanoparticle size determinations. In such cases, it can be useful to report both apparent individual particle size distributions and aggregate size distributions based on TEM observations.

Detecting and determining the structure and composition of nanoparticles adhered to the surfaces of larger minerals or rocks can be even more challenging. For example, Nugent et al. (1998) observed a patchy coating of nanoscale crystalline and amorphous aluminosilicate material on feldspar that had been naturally weathered in a spodosol. The coating was "largely undetected under scanning electron microscopy" but readily apparent in atomic force microscopy (AFM) images. Composition was determined from results of x-ray photoelectron spectroscopy (XPS), Auger electron spectroscopy (AEM), and secondary ion mass spectrometry (SIMS) analysis. Recently, Livi et al. (2009) characterized nanoscale mixed Ni-Al hydroxide surface precipitates on the clay mineral pyrophyllite using analytical electron microscopy, HRTEM, selected-area electron diffraction (SAED), powder XRD, and x-ray absorption fine structure (EXAFS) spectroscopy. The authors stressed the importance of applying such a combination of state-of-the-art techniques to analyze nanoscale surface precipitates.

3.2.2 Nanoparticle Surface Area

The world of nanoparticles is fundamentally a world of surfaces. Nanoparticles have extremely high specific surface areas for their volume and a significant proportion of atoms associated with nanoparticles occur at surfaces. For example, if we were to take a cube 1 mm on a side and divided it into cubes 1 nm on a side, the total volume would remain the same. But, the surface area would increase a million fold.

For a spherical particle, the ratio of surface area (in nm^2; $A = 4\pi r^2$) to volume (in nm^3; $V = 4/3\pi r^3$) is inversely proportional to the particle radius r (in nm) according to the simple relationship:

$$\frac{A}{V} = \frac{3}{r} \quad (3.1)$$

As a particle becomes smaller, its surface area becomes an increasingly larger component of its overall form or structure.

3.2.3 The Concept of Surface Free Energy

The significance of surface area in environmental systems is readily apparent from the simple experiment of pouring a little oil into water. The oil immediately balls up into spheres (or flattened spheres at the water surface). The hydrophobic oil is taking on the shape that minimizes the surface area in direct contact with water. This and other related commonplace observations suggest that surfaces must have some energy associated with them, known as the surface free energy (or, more correctly in many circumstances, the interfacial free energy). Surface free energy is a driving force for many physical or chemical transformations that act to minimize the surface free energy.

When new surface area is formed, the associated change in surface free energy must be considered in the overall energy change of the system. The (specific) surface or interfacial free energy (γ) is defined as

$$\gamma = \left(\frac{\delta G}{\delta A}\right)_{P,T} \quad \text{(pressure, temp., remain constant)} \quad (3.2)$$

where

G is the Gibbs free energy (in kJ)
A is the surface area (in m²)

An interface will only be stable if the free energy of formation of the interface is positive; otherwise, the interface would eventually disappear. It should be noted that surface free energy refers to the *excess* free energy that molecules or atoms have by virtue of their being associated with a surface.

For a system that contains a surface, the total free energy of the system is the sum of the bulk free energy (G_{bulk}), which in turn is the sum of the molar free energies of formation of its constituents, i, $\Sigma_i \mu_i n_i$, where the chemical potential $\mu = (\delta G / \delta n_i)_{T,p}$, plus the total excess free energy of formation of all surfaces and interfaces j (G_s). The surface (or interfacial) free energy is given by $\Sigma_j \gamma_j A_j$. The total free energy (in kJ) is thus

$$G_{tot} = G_{bulk} + G_{s,j} \quad (3.3)$$

$$= \Sigma_i \mu_i n_i + \Sigma_j \gamma_j A_j \quad (3.4)$$

This concept of surface free energy has important implications for many environmental processes, especially those involving small particles with high specific surface area. A summarizing principle is that for small nanoparticles that have high specific surface area, anything that affects surface free energy can affect overall nanoparticle stability.

3.2.4 Surface Free Energy and Classical Mineral Nucleation Theory

Surface free energy plays an important role in mineral nucleation. Homogeneous nucleation of a new mineral phase from solution necessarily passes through a step in which a nanoparticle forms with very high specific surface area. The free energy of nucleation (G_n) of a crystal that forms from a supersaturated solution can be related to saturation state and surface free energy. The free energy of nucleation of a single crystal (ΔG_n) is the sum of the change in bulk free energy (ΔG_{bulk}) and the change in free energy that occurs upon formation of a new interface (ΔG_{interf}) (Berner, 1980):

$$\Delta G_n = \Delta G_{bulk} + \Delta G_{interf} \quad (3.5)$$

where ΔG values are in kJ mol^{-1}.

For a spherical particle,

$$\Delta G_n = -\left(\frac{4}{V}\right)\pi r^3 k_B T \ln(\Omega) + 4\pi r^2 \gamma \quad (3.6)$$

where

V is the molecular volume of the precipitate (m³)
r is the radius of the nucleus (m)

k_B is Boltzmann's constant (1.3806×10^{-26} kJ K^{-1})
Ω is the saturation ratio
T is the temperature (K)
γ is the surface free energy per unit surface area of the nuclei (kJ m^{-2})

When the solution is supersaturated, $\Omega > 1$. ΔG_n has a maximum at the critical size of the nucleating crystal, which is designated r^*, or the critical nucleus. Once the critical nucleus is formed, any additional growth of the crystal results in a decrease in free energy and the overall process transitions from one of nucleation to growth. At any specific value of Ω, particles with $r > r^*$ will tend to grow and particles with $r < r^*$ will tend to dissolve. Greater Ω corresponds to smaller critical nucleus size, that is, smaller nuclei will be stable in more highly supersaturated solutions.

This element of classical nucleation theory has important implications for the mechanisms of mineral nucleation (Figure 3.6; Lower et al., 1998a, 1998b). At high supersaturation, we can expect that minerals will be likely to nucleate homogeneously in solution. But, at low supersaturation, such homogeneous nucleation is far less likely and heterogeneous nucleation on already existing surfaces (if such are available) is more favorable and thus more likely to occur. Soil environments, which tend not to be highly supersaturated with respect to many mineral phases, should not undergo much if any homogeneous nucleation directly from solution leading to crystal growth, although heterogeneous nucleation on the surfaces of already existing particles can occur more readily. An example of heterogeneous nucleation could potentially be formation of kaolinite as a secondary mineral precipitate on feldspar, although it should be emphasized that we do not fully understand the mechanisms of kaolinite secondary mineral formation in different environments.

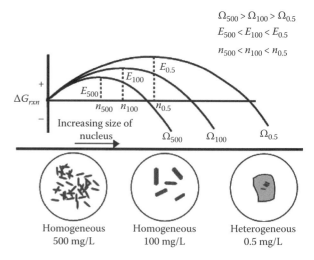

FIGURE 3.6 A schematic showing how saturation state affects nucleation of the Pb-phosphate mineral pyromorphite in the presence of apatite. Ω = saturation state E = activation energy at different saturation states, n = crystal size at different saturation states. From Lower et al. (1998b), after the general construct of Nielsen (1964).

3.2.5 Particle Size and Stability: Classical Theory

Related to this view of particle size and nucleation theory, we can consider how particle size affects stability. The high specific surface areas of small particles has always been taken to indicate that particles become less stable with decreasing size so that nanoparticles should, for the most part, be unstable. To understand the phenomenon of stability relationship to size, we can begin by considering how the pressure within a bubble changes with bubble size (or radius of curvature). According to the Young–Laplace equation, the pressure difference (ΔP) on either side of the surface of a thin film is related to the surface or interfacial tension (γ) and the radii of curvature (R_x, R_y, in m) of the surface:

$$\Delta P = \gamma \left(\frac{1}{R_x} + \frac{1}{R_y} \right) \quad (3.7)$$

For a spherical bubble, $R_x = R_y$, and the Young–Laplace equation reduces to

$$\Delta P = \frac{2\gamma}{R}. \quad (3.8)$$

Thus, the pressure inside a bubble becomes increasingly large as the bubble size decreases. Very small bubbles thus tend to be unstable, and initial bubble formation requires energy to overcome this free energy nucleation barrier.

Expanding this concept from gas bubbles in solution to more general interface phenomena, changes in particle stability with size are traditionally described by the modified Kelvin equation. This equation states that as a hypothetical particle gets smaller in size (r, in m) through the nanoscale range, its solubility (s) increases exponentially:

$$\frac{s(r)}{s_\infty} = \exp\left(\frac{2\gamma V_D}{rRT} \right) \quad (3.9)$$

where
  γ is the surface energy
  V_D is the molar volume of the disperse phase (m³)
  R is the ideal gas constant (8.314 J mol⁻¹ K⁻¹)
  T is the temperature (K)

The term s_∞ refers to the solubility of a hypothetical particle with an infinite planar surface, that is, in the absence of curvature.

The marked decrease in stability with decreasing particle size can lead in some systems to evolution in particle size over time. Consider an aqueous system in which many small crystals have formed from a highly supersaturated solution. The numerous small crystals together have very high combined, total surface area. This system of small, high specific surface area particles is thermodynamically unfavorable relative to a system with fewer large crystals and therefore lower interfacial free energy. Some of the small crystals will dissolve while others grow at their expense to larger dimensions. This spontaneous process of increase in average particle size over time is known as Ostwald ripening, named after Wilhelm Ostwald who first described the process in 1896. Because a system with many small particles has a larger overall surface area-to-volume (or mass) ratio than a system of few large particles, Ostwald ripening minimizes the total free energy of the system by decreasing surface (interfacial) energy.

The above discussion of classical theory suggests that nanoparticles in aqueous environments should be highly unstable and should eventually evolve into larger and more stable crystals. Yet, many nanoparticles can remain stable for long periods of time in aqueous suspensions or in porous media. As we shall see in the section below, particle size and stability can be related in a complex manner at the nanoscale, and classical theories do not always provide good explanations for observed phenomena.

3.2.6 Nanoparticle Stability: When More Than Size Alone Matters

Classical theories of crystal nucleation, growth, and stability as related to particle size withstood decades of inspection, helping to explain many important phenomena in the field and the laboratory. Within the last few years, however, nanoscience is demonstrating that the classical approaches do not always provide a clear understanding of phenomena involving minerals and other natural solids at the nanoscale. Conceptually, soil scientists recognize the general problem immediately: if small particles are inherently unstable, then why are soils, even paleosols, often dominated by mineral or amorphous solid particles less than 1 μm in diameter, including nanoparticles? Is it just that the kinetics of growth are very slow in low-temperature environments? Or, are there thermodynamic controls; that is, can nanoparticles be thermodynamically stable for reasons that we do not yet fully understand? These are questions that we do not yet have answers for in many systems. However, the work of researchers such as Navrotsky and her colleagues (Navrotsky, 2001, 2003; Navrotsky et al., 2008) is starting to suggest that both kinetics and thermodynamics can play a role in nanoparticle longevity in both the lab and the field.

In order to determine the controls on mineral nanoparticle stability as a function of size, one would ideally like to be able to grow nanoparticles of a given mineral phase in which only the size changes, but not the structure or composition. However, for many phases, it is impossible to change size, at least for the smallest nanoparticles, without also changing structure and/or hydration (Navrotsky et al., 2008). Among other factors, very small nanoparticles, in the 1–10 nm size range, have a high proportion of total atoms at or very near the surface. These atoms experience different bonding environments than atoms of the same mineral in the bulk of a larger crystal. Thus, for very small mineral nanoparticles, the high proportion of surface and near-surface atoms can make it virtually impossible to change size independently of structure, and for many minerals, of composition. Moreover, very small nanoparticles of minerals such as hematite often are more hydrated than larger crystals, and this again affects stability (Navrotsky et al., 2008).

Difficulties associated with determining the structures of nanoparticles were discussed in Section 3.2.1. Beyond the overall structure of nanoparticles, we often cannot determine the surface structure and microtopography of mineral nanoparticles; yet, both of these characteristics affect surface free energy and crystal stability, especially within the nanoparticle size range. It can also be difficult to discern the ratios of different crystallographic faces exposed at the surfaces of nanoparticles. Given that different faces may have different structures, reactivities, and surface free energies, this further complicates our understanding of how stability changes as a function of nanoparticle size (A. Navrotsky, University of California at Davis, personal communication). Finally, impurities tend to accumulate at the surfaces of particles, and impurities can thus be expected to play a particularly important role in nanoparticle stability and reactivity (Waychunas et al., 2005).

Recall from above that total free energy of a particle is equal to the sum of bulk free energy plus surface/interfacial free energy,

$$G_{tot} = \Sigma_i \mu_i n_i + \Sigma_j \gamma_j A_j \qquad (3.4)$$

The surface/interfacial free energy may become a major and potentially even dominant component of G_{tot} for nanoparticles, especially in the size range of only a few nanometers in diameter. This means that nanoparticle stability is not just a function of size, alone, but rather of complex structural, compositional, and microtopographic characteristics of the surface/interface.

Navrotsky's group has shown that the different polymorphs of the oxides of Al, Fe, Ti, and Zr can show varying degrees of stability at different particle sizes in the nanoscale versus micron scale (Navrotsky, 2001, 2003; Navrotsky et al., 2008), related to a large extent to differences in surface free energy. Phases that are metastable as micron-scale or "bulk" materials generally have larger surface energies than other polymorphs that are slightly metastable as bulk materials. Because of this competition between bulk metastability and surface energy, nanoparticulate oxides can exist as different stable polymorphs at different size ranges. This process can lead to crossovers in stabilities of different phases as a function of particle size.

Navrotsky et al. (2008) found that stability crossovers can be different for hydrated versus anhydrous surfaces of the Fe(hydr) oxide system. For example, particle size and hydration can play a key role in determining the relative stabilities of hematite and goethite of different nanoparticle sizes (Navrotsky et al., 2008). Surface enthalpy and free energy are generally much higher for anhydrous oxides such as hematite than for oxyhydroxides such as goethite. The lower surface enthalpy/energy for the hydroxide

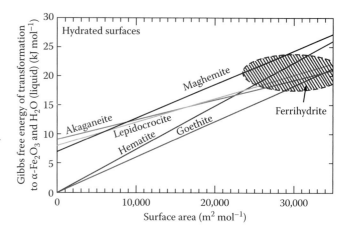

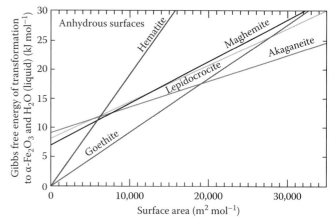

FIGURE 3.7 Enthalpy relative to coarse hematite (α-Fe$_2$O$_3$) plus liquid water as a function of surface area (corrected for compositional/structural differences in terms of Fe atoms). This figure illustrates the crossovers in stability at different surface areas (*note:* smaller particles tend to have higher surface area) and the differences between hydrated and anhydrous systems. (From Navrotsky, A., L. Mazeina, and J. Majzslan. 2008. Size-driven structural and thermodynamic complexity in iron oxides. Science 319:1635. With permission of AAAS.)

phase goethite allows this mineral to be stabilized relative to the oxide hematite at <~60 nm, corresponding to high surface area. Although micron-scale hematite particles in water tend to be more stable than goethite, the nanoparticle size range tends to favor goethite stability (Figure 3.7). The equilibrium relationships between hematite and goethite nanoparticles are sensitive to variations in other factors such as temperature, the activity of water, and Al substitution. Given that impurities tend to associate with surfaces, the high surface areas of nanoparticles may lead to the impurities being a significant proportion of the overall atoms associated with a nanoparticle, so that impurities can be expected to be particularly important to nanoparticle stability.

3.2.7 Stability and the Medium Surrounding a Nanoparticle

Another factor that can strongly impact nanoparticle stability in soils and other porous media is the solution composition, including pH, ionic strength, and concentrations of potential sorbates, in which nanoparticles are found. Because of the small size of nanoparticles and the high specific surface area, materials such as cations, anions, and organic molecules adsorbed to nanoparticles can become a major component of the overall nanoparticle structure and composition. This means that sorbed materials can affect nanoparticle stability, especially for nanoparticles less than a few nanometers in diameter. Because pH affects nanoparticle oxide surface site protonation, pH can affect stability through its effects on surface free energy, alone, beyond considerations of bulk saturation state.

3.2.8 A New Concept of Mineral Growth from Nanoparticles: Oriented Attachment

While Ostwald ripening appears to occur in many systems (Lower et al., 1998a, 1998b), minerals may grow from nanoparticulate nuclei by another mechanism known as oriented attachment. Oriented attachment as a mechanism for coarsening of nanoparticles has been demonstrated through the work of Banfield and her colleagues (Penn and Banfield, 1998a, 1998b). Oriented attachment is a process whereby adjacent nanoparticles spontaneously self-organize so that they share a common crystallographic orientation. The oriented particles are then joined along a planar interface, decreasing the overall surface area and surface free energy. Oriented attachment also can be a mechanism for dislocation generation in otherwise defect-free nanocrystals, when there is a small misorientation at the interface between the joining nanoparticles (Penn and Banfield, 1998b). Oriented attachment may also result in crystals that are separated by twin boundaries or other planar defects (Penn and Banfield, 1998b); the twin boundaries sometimes have atomic arrangements that are distinct from the crystal components on either side of the boundary (Figure 3.8; Penn and Banfield, 1998b). An important question is how prevalent oriented attachment is in actual field environments, including soils. The mechanism can only operate when the particles are free to move in order to align themselves, as in aqueous solutions, and could also potentially occur when nanoparticles nucleate next to each other on a solid substrate, through heterogeneous nucleation.

3.2.9 Nanoparticle Sorption, Redox Phenomena, and Catalysis

Because the basic properties of materials may be different at the nanoscale than at the bulk scale, and given the high surface areas of nanoparticles, an important question concerns the controls on sorption to nanoparticles and whether nanoparticles sorb ions and molecules differently than do larger, bulk materials. There are many different concepts and properties that must be considered in studies of these phenomena.

First, a few comments on "sorption" terminology are needed. The term "sorption" is used when the mechanism by which an adsorbate accumulates in association with a solid the adsorbent is unknown, or to refer to more than one process. Adsorption is the term of choice for a process in which individual species or small clusters accumulate in a two-dimensional fashion at an external surface. Absorption refers to accumulation that goes into a particle, such as into clay interlayers or micropores. Precipitation refers to a three-dimensional accumulation process associated with a surface, known as heterogeneous nucleation, or in the surrounding solution, known as homogeneous nucleation. Sorption phenomena often may play a role in redox reactions and catalysis (Huang, 1999).

Sorption is generally quantified in terms of moles (or grams) of sorbate per meter squared of sorbent surface area, with specific surface area generally measured by gas adsorption (or desorption) using the BET method (Brunauer et al., 1938). However, nanoparticles often aggregate during BET measurements, which can make it difficult to measure specific surface area reliably. Some nanomaterials may recrystallize, dehydrate, or otherwise change in shape and/or structure as a result of the temperature changes that occur during BET sample preparation and measurements, and this process needs to be considered in developing measurement protocols (Maurice et al., 2000).

Other techniques commonly used to characterize particles for use in sorption experiments also require special care when applied to nanoparticles. For example, measurements of electrophoretic mobility, or zeta potential, are often used to determine particle point of zero charge. Most commercial instruments apply the Smoluchowski equation for electrophoretic mobility, but this equation tends to break down for very small particles, and other approaches need to be considered (O'Brien and White, 1978; Maurice, 2009). Particle shape and suspension density can also be important, especially for nanoparticles, and particle aggregation is important because it can affect mobility.

Because the surface properties of nanoparticles change with particle size, we can expect sorption phenomena to be sensitive to particle size, as well. For example, titration experiments showed that the point of zero charge (pHzpc) of titania

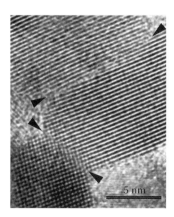

FIGURE 3.8 A high-resolution TEM (HRTEM) image of a particle that formed by oriented attachment of anatase (TiO$_2$). The original crystal boundaries remain as planar defects in the resulting particle. (From Penn, R.L., and J.F. Banfield. 1998b. Imperfect oriented attachment: Dislocation generation in defect-free nanocrystals. Science 281:969–971. With permission of AAAS.)

nanoparticles changes with nanocrystalline size (Finnegan et al., 2007); differences in point of zero charge affect surface potential and can influence sorption behavior as a function of pH. Ha et al. (2009a, 2009b) investigated Zn(II) adsorption to hematite micron- versus nanoscale nanoparticles. They observed differences in adsorption complex structure and in formation of Zn-containing precipitates with hematite particle size.

Many particles in nature may have nanoscale porosity that can be very important for fate and transport of metals, radionuclides, and organic compounds. For example, the d-spacings of swelling clays such as montmorillonite, may swell to the nanometer scale, with thickness of the interlayer region itself approaching 1 nm. Sposito et al. (1999) showed that the structure of water in montmorillonite interlayers appears to be different from the structure of bulk water in solution. Hydrophobic organic compounds and organic cations may absorb in clay interlayers (Lagaly, 1981; Boyd et al., 2001; He et al., 2006) and the conformations of the absorbed organic molecules may be different from their conformations in bulk solution (Haack et al., 2008). Absorption into interlayers and nanopores can be important to contaminant fate and transport, and more research needs to be focused on the hydrology, hydraulics, structure of water, and reactivity of water and solutes in nanoscale spaces.

Because of growing concerns over greenhouse gas emissions and climate change, recent study is focusing on nanoscale sorption phenomena for CO_2 capture. A current method for scrubbing CO_2 from power-plant emissions is the use of liquid amine as a solution absorbent. However, reversible capture by sorption on a solid phase could offer significant benefits. CO_2 may diffuse into the clay mineral pyrophyllite (Wang and Zhang, 2003) and sericite (fine-grained muscovite; Zhang et al., 2005) during dehydroxylation, but the potential extent of uptake and reversibility are not known. Fibrous clays such as palygorskite have channels that can take up water and SO_2 (Zhang et al., 2009); CO_2 uptake seems likely but is yet to be fully explored (R. Cygan, Sandia National Lab, personal communication, 2010). Zeolites and carbon molecular sieves can reversibly sorb significant quantities of CO_2 at room temperature, but they are less efficient at higher temperatures and have poor selectivity with respect to water (Zheng et al., 2005). Considering the success in using liquid amine for CO_2 capture, research is focusing on amine-modified (nano)porous materials such as silica gels and polymers and mesoporous silica (Zheng et al., 2005).

For nanoparticles, size quantization of the electron structure, coupled with changes to surface structure and surface defect site density, may lead to different rates of electron transfer in redox reactions. Madden and Hochella (2005) showed that the rate of heterogeneous manganese oxidation by hematite nanoparticles is up to one and a half orders of magnitude greater for 7.3 nm average diameter particles than for 37 nm particles. This effect was apparent even when surface area was taken into consideration in the rate calculations.

Nanoparticles may have particularly useful catalytic properties because of the high surface areas they expose to the atmosphere or to solution and because of differences in surface structure, electron transfer properties, and other characteristics. For example, TiO_2 nanoparticles are widely used in photocatalysis applications (Aitken et al., 2006), and they have been shown to photocatalyze the degradation of pollutant organics (Armaleo et al., 2007). The heterogeneous Mn oxidation by hematite nanoparticles observed by Madden and Hochella (2005) is an example of a natural catalytic process. Nevertheless, although catalysis by nanoparticles is a topic of significant industrial research, far less is known about the potential catalytic properties of nanoparticles in nature.

3.3 Nanoparticle Mobility in Soils and Sediments

The small size of nanoparticles has important implications for their mobilities in porous media. According to Stokes' Law, particle settling velocity (v_s) (m s^{-1}) increases with the particle size (d_p) (m) and density (ρ_p) (kg m^{-3}):

$$v_s = \frac{g(\rho_p - \rho_w)d_p^2}{18\mu} \quad (3.10)$$

where
  g is the acceleration due to gravity (9.807 m s^{-2})
  ρ_w is the density of water (kg m^{-3})
  ρ_p is the particle density (kg m^{-3})
  μ is the absolute viscosity of the water (Pa s or kg m^{-1} s^{-1})

Thus, when we consider nanoparticle mobility in porous media, it needs to be recognized that individual nanoparticles are subject to translational diffusion and Brownian motion, but that they are too small to be strongly affected by gravity.

DLVO theory, named after work by Derjaguin and Landau (1941) and Verwey and Overbeek (1948), has been widely used to help predict colloid stability as well as the sorption of colloids to sediment (i.e., collector) surfaces in porous media. In classical DLVO theory, the stability of particles in suspension is dependent upon total potential energy V_T:

$$V_T = V_A + V_R + V_S \quad (3.11)$$

which is the sum of
  V_A, potential energy due to attractive van der Waals forces
  V_R, potential energy due to repulsive, electrical double layer, forces
  V_S, potential energy due to the solvent, which is generally small and often neglected in the calculations

This classical DLVO theory has been expanded recently to include other forces of interaction such as Born repulsion and Lewis acid–base interaction (Grasso et al., 2002; Saiers and Ryan, 2005).

One of the challenges in dealing with nanoparticle aggregation and adsorption to surfaces of porous media is that classical and extended DLVO theories were developed for particles whose radius of curvature is greater than the thickness of the double layer, and this is often not the case for nanoparticles. Many interfacial forces, such as electrostatic and van der Waals, can extend over distances that are large in comparison with the size of an individual nanoparticle. Surface and interfacial forces that are not considered in classical DLVO theory, such as hydration/solvation and Born repulsion, are likely to be particularly important for interactions between nanoparticles or between a nanoparticle and a sediment particle (collector) surface (Maurice, 2009).

A great deal of recent research is focused on nanoparticle aggregation and how it affects mobility in porous media. Indeed, the two are linked because aggregated particles have been shown to transport differently than individual (colloidal) particles (Waite et al., 1991). Recently, Gilbert et al. (2007) investigated controls on aggregation of (highly disordered) goethite nanoparticles. Based on extensions to DLVO theory, particles of like charge should be unlikely to form stable aggregates because of repulsion between the like charges. However, aggregation can occur in solutions that have enough thermal energy to overcome this barrier to aggregation. Gilbert et al. (2007) observed the formation of nanoparticle aggregates or nanoclusters between pH 5 and 6.6, with diameters from 25 to ~1000 nm, which remained stable in suspension for at least 10 weeks. This formation of stable nanoclusters of an Fe(III)(hydr)oxide indicates that more attention needs to be paid to the role of nanoclusters in transport in porous media, especially given that many pollutants tend to adsorb to Fe(III)(hydr)oxide minerals.

A good deal of research is also focused on the suspension and aggregation of fullerenes in aqueous solutions. The fullerene C_{60} tends to be highly insoluble in water, but may be induced to form aqueous suspensions by prolonged stirring (Brant et al., 2005). Ma and Bouchard (2009) showed that suspended C_{60} nanoclusters form upon prolonged stirring can remain as stable aggregates in suspension for weeks or longer. The nanoclusters of C_{60} tend to exhibit pH-dependent surface charge, as measured by zeta potential. The mechanism whereby the surface charge is developed on the C_{60} nanoclusters in suspension is not yet understood (Ma and Bouchard, 2009). In any case, there is much that we presently do not know about fullerene surface properties, charge, aggregation behavior, and potential transport through porous media, particularly under hydrologically unsaturated conditions and in long-term environmental exposure.

In hydrologically saturated porous media, it has been shown that colloids may be removed from the solution phase by sorption to the collector surfaces and/or pore straining, which is the trapping of particles in small pores. Many groups are investigating nanoparticle mobility in saturated porous media. Wang et al. (2008) investigated the retention and mobility of C_{60} aggregates in saturated columns packed with Ottawa sand and or glass beads. Retention of the aggregates was considerably greater in the presence of 1.0 mM $CaCl_2$ versus in deionized water without added electrolyte, due at least in part to electrostatic interactions.

In unsaturated conditions, as commonly encountered in soils, colloidal particles may be trapped by a thin film of water on collector surfaces, in a process known as film straining (Chen et al., 2008 and references therein). Colloids may also adhere to the air–water interface, and colloids can be found associated with bubble walls (Wan et al., 1994). Chen et al. (2008) observed that TiO_2 nanoparticle mobility was affected by the saturated/unsaturated ratio in porous media and suggested that the nanoparticles may adsorb to the air–water interface.

Particle aggregation has been shown to play an important role in colloid mobility in porous media, and its effects on nanoparticle mobility are also being explored. Guzman et al. (2006) investigated the effects of pH on TiO_2 nanoparticle aggregation and transport in hydrologically saturated porous media consisting of two-dimensional microfluidic devices. They found that transport was influenced by both surface potential effects and size of nanoparticle aggregates. Nanoparticle aggregates could become clogged in pores (Figure 3.9). An important question raised was whether nanoparticle aggregates would behave similarly to colloidal particles and aggregates of colloids. For example, nanoparticle aggregates might potentially break apart or otherwise be altered by the presence of shear in flowing porous media.

Questions of natural and manufactured nanoparticle aggregation and transport in porous media are reviving interest in humic substances research. Humic substances are well known to affect colloid aggregation, fate, and transport, and are likely to

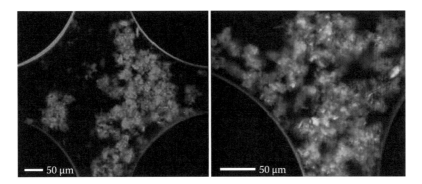

FIGURE 3.9 TiO_2 nanoparticle aggregates in microfluidic pore channels at pH 7. (With permission from Guzman, K.A.D., M.P. Finnegan, and J.F. Banfield. 2006. Influence of surface potential on aggregation and transport of titania nanoparticles. Environ. Sci. Technol. 40:7688–7693. Copyright 2006 American Chemical Society.) Important questions arise regarding the behavior of nanoparticle aggregates in porous media, such as whether new models need to be developed and how aggregates will behave in response to shear forces.

be equally or even more important for nanoparticles. Chen and Elimelech (2008) investigated C_{60} deposition kinetics onto bare silica surfaces or surfaces precoated with natural organic matter (NOM) in the form of humic acid (Figure 3.10) or alginate. They observed that preadsorbed NOM on the silica surface can either retard or enhance the deposition kinetics, depending on the solution ionic composition and the physicochemical properties of the NOM. On the other hand, the presence of NOM in solution tended to decrease deposition, which the authors attributed to steric repulsion upon NOM sorption. This suggests that the effects of NOM, including humic substances, can be complex and different in diverse environments such as soils, groundwaters, or colloid-containing surface waters.

3.4 Nanotoxicity

Many engineered nanoparticles have the potential to be toxic to humans and/or ecosystem components such as microorganisms, plants, aquatic, and other organisms (NRC, 2008; Wiesner et al., 2008). A recent report from the National Research Council (NRC, 2008) stressed the pressing need for a national strategy for nanorisk research to evaluate the environmental health risks of nanotechnologies and engineered nanoparticles. The field of nanotoxicology is one of the new disciplines that has emerged over the last decade as an outgrowth of the explosion of research and development in nanoscience and nanotechnology. As pointed out by Oberdorster et al. (2005), the small size of nanoparticles makes it much easier for them to penetrate pores in skin, to be integrated into the respiratory system, and to penetrate cells. The high surface areas of nanoparticles can make them far more bioactive. Thus, there is much real concern about nanotoxicity, especially given that so many new engineered nanoparticles are being released to the environment without toxicology analysis.

In some instances, the toxicity of nanoparticles is being touted for its potential health benefits, for example, as microbicide agents. Sondi and Salopek-Sondi (2004) showed that Ag nanoparticles have strong antimicrobial effects against *Escherichia coli*. Electron microscopy revealed clear cell damage, including pits in the cell walls and Ag nanoparticle accumulation in the microbial membrane (Figure 3.11). Results such as these are leading to widespread use of Ag nanoparticles as antimicrobial agents in detergents, on clothing, and in other applications.

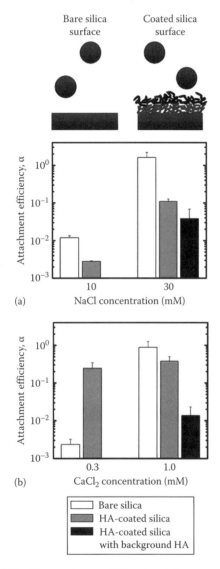

FIGURE 3.10 The attachment efficiencies of C_{60} nanoparticles on silica with and without humic coating in (a) NaCl and (b) $CaCl_2$. (With permission from Chen, K.L., and M. Elimelech. 2008. Interaction of fullerene (C-60) nanoparticles with humic acid and alginate coated silica surfaces: Measurements, mechanisms, and environmental implications. Environ. Sci. Technol. 42:7607–7614. Copyright 2008 American Chemical Society.)

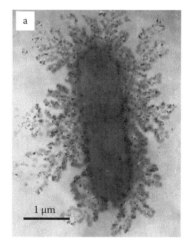

FIGURE 3.11 Silver (Ag) nanoparticles have toxic effects on *E. coli*. A TEM image of *E. coli* following exposure to Ag nanoparticles for 1 h. (With permission from Sondi, I., and B. Salopek-Sondi. 2004. Silver nanoparticles as antimicrobial agent: A case study on E-coli as a model for Gram-negative bacteria. J. Colloid Interface Sci. 275:177–182. Copyright 2004 American Chemical Society.) Some particles have penetrated into the cell. Intracellular materials have leaked into the medium surrounding the cell and Ag nanoparticles are attached to this material. This nanoparticle toxicity is leading to the use of Ag nanoparticles as antimicrobial agents in many products from detergents to dissemination on clothing.

Nanotoxicity to plants, or nanophytotoxicity, is another area of growing research. Lin and Xing (2008) found that ZnO nanoparticles had a (phyto)toxic effect on Ryegrass (*Lolium perenne*). The ZnO particles adsorbed into root tissues and cells and damaged the root tissues, inhibiting seedling growth. The phytotoxicity could not be explained by dissolution of ZnO particles, and dissolved Zn, but rather appeared to result from the actual nanoparticles themselves.

3.5 Nanoenabled Sensing

The revolution in nanotechnology promises to have another major impact on soil science in the form of nanoenabled sensing. There are many nanodevices that can be used for in situ sensing of soil components. Lee et al. (2008) developed colorimetric uranium sensors based on a uranyl (UO_2^{2+})-specific DNAzyme and gold nanoparticles (AuNP). A DNAzyme is essentially a DNA molecule that changes its fluorescence properties upon sorption of a specific component such as uranyl. In the nanosensor approach, uranyl-specific DNAzyme was attached to AuNP, forming functionalized purple aggregates. Uranyl-induced disassembly of the functionalized AuNP aggregates, resulting in individual AuNPs that were red in color. This nanosensor system was shown to have a detection limit of 50 nM uranyl.

Many other groups are using nanotechnology to develop sensors, or sensors that are approaching the nanoscale, specifically for soil applications. Bendikov et al. (2005) developed flexible, miniature, and inexpensive nitrate sensors that are fabricated by electropolymerizing pyrrole onto carbon fiber substrates using nitrate as a dopant. The resulting sensor shown in Figure 3.12 (length 1 cm or less, micron-scale diameter) is ideal for deployment in the pore space of soils and sediments and can quantify nitrate concentrations to just below 10^{-4} M. Although this sensor is not nanoscale per se, it demonstrates the trend in miniaturization that is occurring in environmental sensors for soil applications.

3.6 Frontiers in Research in Nanoscale Particles and Processes in Soil Environments

Soil science is very much at the center of nanoscience and nanotechnology. On the one hand, soils are replete with natural nanoparticles and soil scientists have a long and distinguished history of investigating small particles in nature. On the other hand, nanoengineering is resulting in the release of an enormous number and diversity of nanoparticles to the environment, and they are certain to find their way into soils. Nanotechnology also is contributing to a new generation of sensors that can be used to investigate a host of soil parameters and processes.

Some frontiers for research involving soil science and the nanoscale include the following:

1. What are the unique properties of water in nanopores?
2. What are the controls on nanoparticle aggregation kinetics, and how do soil components such as NOM, mono- and multi-valent ions, and bacteria affect aggregation of natural and engineered nanoparticles?
3. What are the controls on nanoparticle mobility not only in the saturated zone but also in hydrologically unsaturated environments?
4. What are the fundamental controls on nanotoxicity, and how can we prevent potentially toxic engineered nanoparticles from entering the food supply?
5. How well does our new primarily laboratory-based understanding of nanoparticles and nanoprocesses translate to real-world environments, including soils?
6. What implications will our ability to detect and study nanoparticles in soils and water have for various nations' regulatory structures related to pollutants?
7. How can nanotechnology best be used in soil science and broader agricultural applications, from sensing to treatment of crop diseases to controls on soil porosity, structure, and fertility?
8. How can we best harvest the strengths of nanoscience and nanotechnology while at the same time minimizing the risks to environmental and human health?
9. How can the unique abilities and perspectives of soil scientists be harnessed to contribute most successfully to the broader nanoscience and nanotechnology communities?

Given the unique properties and behaviors of nanoparticles, and the burgeoning of nano research and development, it is impossible to predict the potential long-term costs and benefits to soils and agriculture. One certain fact is that the soil community, with its long history of dealing with complex systems of small particles, has a great deal to contribute to understanding complex environmental processes at the nanoscale.

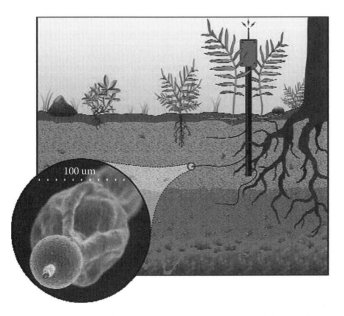

FIGURE 3.12 A soil nitrate microsensor as described by Bendikov et al. (2005). (Courtesy of T. Harmon and U.C. Merced.)

Acknowledgments

P. Maurice thanks the NSF-funded (grant EAR02-21966) Environmental Molecular Science Institute at the University of Notre Dame for funding her group's nanoscale environmental research. This manuscript is dedicated to the memory of Dr. Ksenija Namjesnik-Dejanovic.

References

Aitken, R.J., M.Q. Chaudhry, A.B.A. Boxall, and M. Hull. 2006. Manufacture and use of nanomaterials: Current status in the UK and global trends. Occup. Med. 56:300–306.

Armaleo, L., D. Barreca, G. Bottaro, A. Gasparotto, C. Maccato, C. Maragno, E. Tondello, U.L. Stangar, M. Bergant, and D. Mahne. 2007. Photocatalytic and antibacterial activity of TiO_2 and Au/TiO_2 nanosystems. Nanotechnol. 18: Art. No. 375709.

Banfield, J.F., and H. Zhang. 2001. Nanoparticles in the environment, p. 1–58. In J.F. Banfield, and A. Navrotsky (eds.) Nanoparticles and the environment. Mineralogical Society of America, Washington, DC.

Bargar, R., R. Bernier-Latmani, D.E. Glammar, and B.M. Tebo. 2008. Biogenic uraninite nanoparticles and their importance for uranium remediation. Elements 4:407–412.

Barton, L.E., K.E. Grant, T. Kosel, A.N. Quicksall, and P.A. Maurice. 2011. Size dependent Pb sorption to nanohematite in the presence and absence of a microbial siderophore: Effects of experimental design on sorption edges. Environ. Sci. Technol. 45:3231–3237.

Becker, L., J.L. Bada, R.E. Wians, J.E. Hunt, T.E. Bunch, and B.M. French. 1994. Fullerenes in the 1.85-billion year old Sudbury impact structure. Science 265:642–645.

Bendikov, A., J. Kim, and T.C. Harmon. 2005. Development and environmental application of a nitrate selective microsensor based on doped polypyrrole films. Sens. Actuators B Chem. 106:512–517.

Berner, R.A. 1980. Early diagenesis: A theoretical approach. Princeton series in geochemistry. Princeton University Press, Princeton, NJ.

Berube, D.M. 2005. Nano-hype: The truth behind the nanotechnology buzz. Prometheus books, Amherst, NY.

Boyd, S.A., G. Sheng, B.J. Teppen, and C.T. Johnston. 2001. Mechanisms for the adsorption of substituted nitrobenzenes by smectite clays. Environ. Sci. Technol. 35:4227–4234.

Brant, J., H. Lecoanet, M. Hotze, and M. Wiesner. 2005. Comparison of electrokinetic properties of colloidal fullerenes (n-C60) formed using two procedures. Environ. Sci. Technol. 39:6343–6351.

Brown, G.E., and G.A. Parks. 2001. Sorption of trace elements on mineral surfaces: Modern perspectives from spectroscopic studies, and comments on sorption in the marine environment. Int. Geol. Rev. 43:963–1073.

Brunauer, S., P.H. Emmett, and E. Teller. 1938. Adsorption of gases in multimolecular layers. J. Phys. Chem. 60:309–319.

Brust, M., and C.J. Kiely. 2002. Some recent advances in nanostructure preparation from gold and silver particles: A short topical review. Colloids Surf. A 2–3:175–186.

Buffle, J. 2006. The key role of environmental colloids/nanoparticles for the sustainability of life. Environ. Chem. 3:155–158.

Buseck, P.R., S.J. Tsipursky, and R. Hettich. 1992. Fullerenes from the geological environment. Science 257:215–217.

Chen, K.L., and M. Elimelech. 2008. Interaction of fullerene (C-60) nanoparticles with humic acid and alginate coated silica surfaces: Measurements, mechanisms, and environmental implications. Environ. Sci. Technol. 42:7607–7614.

Chen, L.X., D.A. Sabatini, and T.C.G. Kibbey. 2008. Role of the air-water interface in the retention of TiO_2 nanoparticles in porous media during primary drainage. Environ. Sci. Technol. 42:1916–1921.

Chorover, J., R. Kretzschmar, F. Garcia-Pichel, and D.L. Sparks. 2007. Soil biogeochemical processes within the critical zone. Elements 3:321–326.

Daly, T.K., P.R. Buseck, P. Williams, and C.F. Lewis. 1993. Fullerenes from a fulgurite. Science 259:1599–1601.

Derjaguin, B.V., and L. Landau. 1941. Theory of the stability of strongly charged lyophobic sols and of the adhesion of strongly charged particles in solution of electrolytes. Acta Physicochim. URSS 14:633–662.

Diebold, U. 2003. The science of titanium dioxide. Surface Sci. Rep. 48:53–229.

EPA. 2005. Nanotechnology White Paper. The U.S. Environmental Protection Agency, Science Policy Council, Washington, DC. http://www.epa.gov/osa/pdfs/EPA_nanotechnology_white_paper_external_review_draft_12-02-2005.pdf (downloaded July 2008).

Finnegan, M.P., Z. Hengzhong, and J.F. Banfield. 2007. Phase stability and transformation in titania nanoparticles in aqueous solutions determined by surface energy. J. Phys. Chem. C 111:1962–1968.

Fortin, D., and S. Langley. 2005. Formation and occurrence of biogenic iron-rich minerals. Earth Sci. Rev. 72:1–19.

Giersig, M., and P. Mulvaney. 1993. Preparation of ordered colloid monolayers by electrophoretic deposition. Langmuir 12:3408–3413.

Gilbert, B., G.P. Lu, and C.S. Kim. 2007. Stable cluster formation in aqueous suspensions of iron oxyhydroxide nanoparticles. J. Colloid Interface Sci. 313:152–159.

Glatter, O., and O. Krathy. 1982. Small-angle X-ray scattering. Academic Press, London, U.K.

Grasso, D., K. Subramanian, M. Butkus, K. Strevett, and J. Bergendahl. 2002. A review of non-DLVO interactions in environmental colloidal systems. Rev. Environ. Sci. Bio/Technol. 1:17–38.

Guzman, K.A.D., M.P. Finnegan, and J.F. Banfield. 2006. Influence of surface potential on aggregation and transport of titania nanoparticles. Environ. Sci. Technol. 40:7688–7693.

Ha, J.Y., T.P. Trainor, F. Farges, and G.E. Brown. 2009a. Interaction of aqueous Zn(II) with hematite nanoparticles and microparticles. Part I. EXAFS study of Zn(II) adsorption and precipitation. Langmuir 25:5574–5585.

Ha, J.Y., T.P. Trainor, F. Farges, and G.E. Brown. 2009b. Interaction of aqueous Zn(II) with hematite nanoparticles and microparticles. Part 2. ATR-FTIR and EXAFS study of the aqueous Zn(II)oxalate/hematite ternary system. Langmuir 25:5586–5593.

Haack, E.A., C. Johnston, and P.A. Maurice. 2008. Siderophore sorption to montmorillonite. Geochim. Cosmochim. Acta 72:3381–3397.

Hall, B.D., D. Zanchet, and D. Ugarte. 2000. Estimating nanoparticle size from diffraction measurements. J. Appl. Crystl. 33:1335–1339.

He, H.P., R.L. Frost, T. Bostrom, P. Yuan, L. Duong, D. Yang, X.F. Yunfel, and J.T. Kloprogge. 2006. Changes in the morphology of organoclays with HDTMA(+) surfactant loading. Appl. Clay Sci. 31:262–271.

Hedges, J.I., G. Eglinton, P.G. Hatcher, D.L. Kirchman, C. Arnosti, S. Derenne, R.P. Evershed et al. 2000. The molecularly-uncharacterized component of nonliving organic matter in natural environments. Org. Geochem. 31:945–958.

Heymann, D., L.P.F. Chibante, P.R. Brooks, W.S. Wolbach, and R.E. Smalley. 1994. Fullerenes in the K/T boundary layer. Science 265:645–647.

Hochella, M.F., Jr., S.K. Lower, P.A. Maurice, R.L. Penn, N. Sahai, D.L. Sparks, and B.S. Twining. 2008. Nanominerals, mineral nanoparticles, and Earth chemistry. Science 21:1631–1635.

Hochella, M.F., Jr., J.N. Moore, C. Putnis, A. Putnis, T. Kasama, and D.D. Eberl. 2005. Direct observation of toxic metal-mineral association from a massive acid mine drainage system: Implications for metal transport and bioavailability. Geochim. Cosmochim. Acta 69:1651–1663.

Huang, P.M. 1999. Abiotic catalysis, p. B-303–B-332. In M.E. Sumner (ed.) Handbook of soil science. CRC Press, Boca Raton, FL.

Klein, C., and C.S. Hurlbut. 1985. Manual of mineralogy. Revised. After J.S. Dana. John Wiley & Sons, New York.

Kodama, K., S. Iikubo, T. Taguchi, and S. Shamoto. 2006. Finite size effects of nanoparticles on the atomic pair distribution functions. Acta Crystl. A 6:444–453.

Kretzschmar, R., and T. Shäfer. 2005. Metal retention and transport on colloidal particles in the environment. Elements 1:205–210.

Kroto, H.W., J.R. Heath, S.C. O'Brien, R.F. Curl, and R.E. Smalley. 1985. Buckministerfullerene. Nature 318:162–163.

Kuhlbusch, T.A.J., and P.J. Crutzen. 1995. Toward a global estimate of black carbon in residues of vegetation fires representing a sink of atmospheric CO_2 and a source of O_2. Global Biogeochem. Cycles 9:491–501.

Lagaly, G. 1981. Characterization of clays by organic-compounds. Clay Miner. 16:1–21.

Lee, J.H., Z.D. Wang, J.W. Liu, and Y. Lu. 2008. Highly sensitive and selective colorimetric sensors for Uranyl (UO_2^{2+}): Development and comparison of labeled and label-free DNAzyme-gold nanoparticle systems. J. Am. Chem. Soc. 130:14217–14226.

Lin, D., and B. Xing. 2008. Root uptake and phytotoxicity of ZnO nanoparticles. Environ. Sci. Technol. 42:5580–5585.

Livi, K.J.T., G.S. Senesi, A.C. Scheinost, and D.L. Sparks. 2009. Microscopic examination of nanosized mixed Ni-Al hydroxide surface precipitates on pyrophyllite. Environ. Sci. Technol. 43:1299–1304.

Lower, S.K., P.A. Maurice, and S.J. Traina. 1998a. Simultaneous dissolution of hydroxylapatite and precipitation of hydroxylpyromorphite: Direct evidence of homogeneous nucleation. Geochim. Cosmochim. Acta 62:1773–1780.

Lower, S.K., P.A. Maurice, S.J. Traina, and E.H. Carlson. 1998b. Aqueous Pb sorption by hydroxylapatite: Applications of atomic force microscopy to dissolution, nucleation, and growth studies. Amer. Mineral. 83:147–158.

Ma, X., and D. Bouchard. 2009. Formation of aqueous suspensions of fullerenes. Environ. Sci. Technol. 243:330–336.

Madden, A.S., and M.F. Hochella, Jr. 2005. A test of geochemical reactivity as a function of mineral size: Manganese oxidation promoted by hematite nanoparticles. Geochim. Cosmochim. Acta 69:389–398.

Maurice, P.A. 2009. Environmental surfaces and interfaces from the nanoscale to the global scale. Wiley, Hoboken, NJ.

Maurice, P.A., and M.F. Hochella, Jr. 2008. Nanoscale particles and processes: A new dimension in soil science. 100th Anniversary edition. Adv. Agron. 100:123–153.

Maurice, P.A., Y.-J. Lee, and L. Hersman. 2000. Dissolution of Al-substituted goethites by an aerobic Pseudomonas mendocina var. bacteria. Geochim. Cosmochim. Acta 64:1363–1374.

Maurice, P., and K. Namjesnik-Dejanovic. 1999. Aggregate structures of sorbed humic substances observed in aqueous solution. Environ. Sci. Technol. 33:1538–1540.

Michel, F.M., L. Ehm, S.M. Antao, P.L. Lee, P.J. Chupas, G. Liu, D.R. Strongin, M.A.A. Schoonen, B.L. Phillips, and J.B. Parise. 2007a. The structure of ferrihydrite, a nanocrystalline material. Science 316:1726–1729.

Michel, F.M., L. Ehm, G. Liu, W.Q. Han, S.M. Antao, P.J. Chupas, P.L. Lee et al. 2007b. Similarities in 2- and 6-line ferrihydrite based on pair distribution function analysis of X-ray total scattering. Chem. Mater. 19:1489–1496.

Namjesnik-Dejanovic, K., and P.A. Maurice. 2001. Conformations and aggregate structures of sorbed natural organic matter on muscovite and hematite. Geochim. Cosmochim. Acta 65:1047–1057.

Navrotsky, A. 2001. Thermochemistry of nanomaterials, p. 73–103. In J.F. Banfield, and A. Navrotsky (eds.) Nanoparticles and the environment. Reviews in Mineralogy and Geochemistry 44, Mineralogical Society of America and Geochemical Society, Washington, DC.

Navrotsky, A. 2003. Energetics of nanoparticle oxides: Interplay between surface energy and polymorphism. Geochem. Trans. 4:34–37.

Navrotsky, A., L. Mazeina, and J. Majzslan. 2008. Size-driven structural and thermodynamic complexity in iron oxides. Science 319:1635.

Nielsen, A.E. 1964. Kinetics of precipitation. MacMillan, New York.

Nowack, B., and T. Bucheli. 2007. Occurrence, behavior and effects of nanoparticles in the environment. Environ. Pollut. 150:5–22.

NRC. 2008. Review of federal strategy for nanotechnology-related environmental, health, and safety research. Executive summary. http://books.nap.edu/catalog/12559.html (accessed on May 13, 2011)

Nugent, M., S. Brantley, C. Pantano, and P. Maurice. 1998. The influence of natural mineral coatings on feldspar dissolution. Nature 395:588–591.

O'Brien, R.W., and L.R. White. 1978. Electrophoretic mobility of a spherical colloidal particle. J. Chem. Soc. Faraday Trans. 2:1607.

Oberdorster, G., E. Oberdorster, and J. Oberdorster. 2005. Nanotoxicology: An emerging discipline evolving from studies of ultrafine particles. Environ. Health Persp. 113:823–839.

Pan, Y.-H., K. Sader, J.J. Powell, A. Bleloch, M. Gass, J. Trinick, A. Warley, A. Li, R. Brydson, and A. Brown. 2009. 3D morphology of the human hepatic ferritin mineral core: New evidence for a subunit structure revealed by single particle analysis of HAADF-STEM images. J. Struct. Biol. 166:22–31.

Parthasarathy, G., R. Srninivassan, M. Vairamani, K. Ravikumar, and A.C. Kunwar. 1998. Occurrence of natural fullerenes in low grade metamorphosed Proterozoic shungite from Karelia, Russia. Geochim. Cosmochim. Acta 62:21–22.

Penn, R.L., and J.F. Banfield. 1998a. Oriented attachment and growth, twinning, polytypism, and formation of metastable phases: Insights from nanocrystalline TiO_2. Amer. Mineral. 83:1077–1082.

Penn, R.L., and J.F. Banfield. 1998b. Imperfect oriented attachment: Dislocation generation in defect-free nanocrystals. Science 281:969–971.

Potter, K.W., E.F. Wood, R.C. Bales, L.E. Band, E.A.B. Eltahir, A.W. England, J.S. Famiglietti et al. 2007. Integrating multiscale observations of U.S. Waters. National Academy of Sciences Press, Washington, DC.

Rancourt, D.G., and J.-F. Meunier. 2008. Constraints on structural models of ferrihydrite as a nanocrystalline material. Amer. Mineral. 93:1412–1417.

Saiers, J.E., and J.M. Ryan. 2005. Colloid deposition on non-ideal porous media: The influences of collector shape and roughness on the single-collector efficiency. Geophys. Res. Lett. 32:L21406.

Shannon, M.A., P.W. Bohn, M. Elimelech, J.G. Georgiadis, B.J. Mariñas, and A.M. Mayes. 2008. Science and technology for water purification in the coming decades. Nature 452:301–310.

Sondi, I., and B. Salopek-Sondi. 2004. Silver nanoparticles as antimicrobial agent: A case study on E-coli as a model for Gram-negative bacteria. J. Colloid Interface Sci. 275:177–182.

Sposito, G., S.H. Park, and R. Sutton. 1999. Monte Carlo simulation of the total radial distribution function for interlayer water in sodium and potassium montmorillonites. Clay. Clay Miner. 47:192–200.

Tebo, B.M., W.C. Ghiorse, L.G. van Waasbergen, P.L. Siering, and R. Caspi. 1997. Bacterially-mediated mineral formation: Insights into manganese(II) oxidation from molecular genetic and biochemical studies, p. 225–266. In J.F. Banfield, and K.H. Nealson (eds.) Geomicrobiology: Interactions between microbes and minerals. Reviews in Mineralogy. Vol. 35, Series editor P. H. Ribbe. Mineralogical Society of America, Washington, DC.

Theng, B.K.G., and G. Yuan. 2008. Nanoparticles in the soil environment. Elements 4:395–399.

Verwey, E.J.W., and J.T.G. Overbeek. 1948. Theory of stability of lyophobic colloids. 1st edn. Elsevier, Amsterdam, the Netherlands.

Waite, T.D., A.I. Schafer, A.G. Fane, and A. Heuer. 1991. Colloid fouling of ultrafiltration membranes: Impact of aggregate structure and size. J. Colloid Interface Sci. 212:264–274.

Wan, J., J.L. Wilson, and T.L. Kieft. 1994. Influence of the gas-water interface on transport of microorganisms through unsaturated porous media. Appl. Environ. Microbiol. 60:509–516.

Wang, Y.G., Y.S. Li, J.D. Fortner, J.B. Hughes, L.M. Abriola, and K.D. Pennell. 2008. Transport and retention of nanoscale C-60 aggregates in water-saturated porous media. Environ. Sci. Technol. 42:3588–3594.

Wang, L., and M. Zhang. 2003. Infrared study of CO_2 incorporation into pyrophyllite $[Al_2Si_4O_{10}(OH)_2]$ during dehydroxylation. Clays Clay Min. 51:439–444.

Warren, L., P. Maurice, N. Parmar, and G. Ferris. 2001. Microbially mediated calcite precipitation: Implication for solid-phase capture of inorganic contaminants. Geomicrobiol. J. 18:93–115.

Waychunas, G.A., C.S. Kim, and J.A. Banfield. 2005. Nanoparticulate iron oxide minerals in soils and sediments: Unique properties and contaminant scavenging mechanisms. J. Nanopart. Res. 7:409–433.

Wiesner, M.R., E.M. Hotze, J.A. Brant, and B. Espinasse. 2008. Nanomaterials as possible contaminants: The fullerene example. Water Sci. Technol. 57:305–310.

Wohltjen, H., and A.W. Snow. 1998. Colloid metal-insulator-metal ensemble chemiresistor sensor. Analyt. Chem. 70:2856–2859.

Wu, J., E. Boyle, W. Sunda, and L.-S. Wen. 2001. Soluble and colloidal iron in the oligotrophic North Atlantic and North Pacific. Science 293:847.

Zhang, Q., T. Higuchi, M. Sekine, and T. Imai. 2009. Removal of sulphur dioxide using palygorskite in a fixed bed adsorber. Environ. Technol. 30: 1529–1538.

Zhang, M., L. Wang, S. Hirai, S.A.T. Redfern, and E.K.H. Salje. 2005. Dehydroxylation and CO_2 incorporation in annealed mica (sericite): An infrared spectroscopic study. Am. Mineral. 90:173–180.

Zheng, F., D.N. Tran, B.J. Busche, G.E. Fryxell, R.S. Addleman, T.S. Zemanian, and C.L. Aardahl. 2005. Ethylenediaine-modified SBA-15 as regenerable CO_2 sorbent. Ind. Eng. Chem. Res. 44:3099–3105.

4
Impacts of Environmental Nanoparticles on Chemical, Biological, and Hydrological Processes in Terrestrial Ecosystems

Nikolla P. Qafoku
Pacific Northwest National Laboratory

4.1 Introduction 4-1
  Chapter Objectives, Terminology, and Definitions • Size-Dependent Properties: What Is Unique about NPs • Occurrence and Origin of NPs in Terrestrial Ecosystems • ENP Toxicity
4.2 Interactions in the Binary System: Nanoparticles–Soil Solution 4-5
  Nanoparticle Formation and Stability • Nanoparticle Aggregation • Nanoparticles as Sorbents • Interfacial Electron-Transfer Reactions
4.3 Nanoparticle Controls or Impacts on the Extent and Timescale of Soil-/Geo-Processes 4-9
  Advective Mobility and Diffusive Mass Transfer • Adsorption/Desorption • Dissolution/Precipitation • Electron Transfer Reactions
4.4 Future Research Directions 4-11
Acknowledgments 4-12
References 4-12

4.1 Introduction

4.1.1 Chapter Objectives, Terminology, and Definitions

This chapter seeks to provide some answers to questions about the role of nanoparticles (NPs) in controlling or influencing single and/or coupled chemical, biological, and hydrological processes in terrestrial ecosystems with various physical and mineralogical properties. This is a new and active area of research. For this reason, book chapters or review papers can quickly become outdated. In addition, many topics within this area of research have been explored only superficially, making the writing of critical analyses and comprehensive syntheses on specific or more general topics difficult. Under these circumstances, the aim of this chapter is to present recent results that have contributed to the current understanding of the impact of environmental NPs (ENPs) on these processes. Several important aspects of soil/geo nanoscience as we currently understand it are explained; and then ideas for future efforts and a direction for research in this fascinating, ever-expanding, and quickly advancing area within the soil and geosciences domains are presented. Additional information on NPs' unique properties and behavior and insights on how nanoscience is changing our views in many directions are provided in Chapter 3 (Maurice, 2012).

NPs are defined as particles with at least one of the three dimensions in the range from 1 nm to tens of nanometers (Hochella et al., 2008). These natural NPs (NNPs) are ubiquitous in terrestrial ecosystems and have been present for many years. However, it is very likely that in the near future the NP population (in terms of types and concentration) in soils will grow considerably and even unconventionally, that is, with not easily determined trends, when manufactured NPs (MNPs), which are engineered for different purposes in many areas of human activities, will enter the terrestrial ecosystems and become ENPs. The focus of this chapter is only on ENPs that are present in terrestrial systems and not on those in other environmental media, such as atmosphere or hydrosphere.

Within the context of this chapter, terrestrial ecosystems (e.g., soils, subsoils, deep vadose zone, and unconsolidated aquifer materials) are considered those that are present in the thin layer of the Earth's crust from the surface down to subsurface groundwater. Chemical, biological, and hydrological soil- or geo-processes are those that, directly or indirectly, control or influence the extent and timescale of the mobility and, in some cases, determine the ultimate fate of aqueous or sorbed (adsorbed or precipitated) chemical species of naturally occurring or anthropogenic nutrients and contaminants, in terrestrial ecosystems with different physical and mineralogical properties. These processes may alter the soil mineralogy, change the particle size distribution or the surface and/or internal morphological features of soil minerals (e.g., mineral dissolution followed by neophase precipitation), modify surface chemical properties of the soil/sediment matrix (e.g., adsorption and desorption), or manipulate the chemical speciation (e.g., concentration gradients) and valence state (e.g., oxidation and reduction) of nutrients or contaminants.

4.1.2 Size-Dependent Properties: What Is Unique about NPs

The idea of crystallite size-dependent properties has circulated in the literature for quite some time (Gravie, 1978), and a great deal of time and effort has been spent over the past few decades to understand the properties and behavior of different size fractions that are present in terrestrial systems. These include colloidal fractions, clay fractions, silt fractions, sand fractions, and gravel. Each size fraction has specific properties and roles within the soil/sediment matrix. For this reason, the nanofraction, which is a subset of the soil colloidal fraction, should not be excluded from this rule. However, unlike other soil particles, many aspects of NPs have been little studied (Waychunas and Zhang, 2008), and our understanding of nanomaterials (particles, films, and/or confined fluids) is far from being complete from a fundamental physical chemistry point of view (Hochella, 2007). Recently, research efforts have intensified in this field because of the significant increase in the production of MNPs (Seeger et al., 2009), which almost certainly will find their way to terrestrial systems in substantial amounts and are expected to affect significantly naturally occurring soil- and geo-processes.

The smallest size fraction (fine-grained, submicro, or ultrafine fraction) has always attracted the attention of the soil scientists because of its unique properties, such as the large surface area, the large percentage of surface atoms with unbalanced charge, and the large number of surface functional groups per unit of mass. The smallest size fraction has the ability to dramatically change the soil physical and chemical properties, such as soil water holding capacity, tortuosity, particle aggregation, cation exchange capacity (CEC), anion exchange capacity (AEC), and so on.

It is also true that soil scientists have carefully studied the properties of clay and colloidal fractions in soils for years. However, attention has only recently been drawn toward the nanofraction, and legitimate questions have been raised: Do NP-specific, chemical, physical, and geological properties really exist? How do the properties and behaviors of NPs differ from those of ions and molecules and from those of the larger particles whose properties are understood based on classic physics and chemistry of bulk systems (Maurice and Hochella, 2008)? What is the role of the nanosize fraction in soils, and how does this fraction control or affect the extent and timescales of chemical, biological, and hydrological processes in soils?

The general belief is that NPs can have unique size-related composition, stability, chemical reactivity, molecular and electronic structure, and mechanical behavior (Hochella, 2002; Kahru et al., 2008; Maurice and Hochella, 2008; Waychunas and Zhang, 2008). Especially important are those properties that emerge only after a cluster of atoms has grown beyond some specific size and that disappear once the particle grows beyond the NP range (Banfield and Zhang, 2001). NP properties are often different (sometimes dramatically different) from those of the same material at a larger size and could even show marked departures from their bulk analog materials (Hochella, 2008; Waychunas and Zhang, 2008). In addition, NPs may contain unusual forms of structural disorder that can substantially modify the properties of the materials and, therefore, cannot be considered as small pieces of the bulk material (Gilbert et al., 2004, 2006a). Interestingly, in some cases, for example, ferrihydrite, larger than the nanoscale equivalents, does not even exist (Hochella, 2008).

Considering that for particles on the order of 4 nm in diameter, ca. 50% of the atoms are at the surface, understanding the surface properties and surface chemistry of NPs is essential to fully understand their behavior (Grassian, 2008). Recent investigations have also demonstrated that the electronic and structural properties (Luca, 2009), magnetic properties (Carta et al., 2009), elasticity and compressibility (Gilbert et al., 2006a; Chen et al., 2009a), point of zero charge (PZC) (Vayssieres, 2009), surface free energy (Naicker et al., 2005; Zhang et al., 2009), and melting behavior (Qi et al., 2001) change significantly even within the NP size fraction.

4.1.3 Occurrence and Origin of NPs in Terrestrial Ecosystems

4.1.3.1 Natural NPs

Volcanic dust, most natural waters, soils, and sediments are sources of NNPs, which are generated by a wide variety of soil, geological, and biological processes (Handy et al., 2008a). Natural, accidental, and incidental nanomaterials such as fullerenes have existed on Earth for nearly 2 billion years and are common constituents in combustion exhaust (Posner, 2009). In terrestrial systems, NPs occur as nanominerals [which are defined as minerals that only exist in this size range, for example, certain clays and Fe and Mn (oxyhydr)oxides] (Hochella et al., 2008) or mineral NPs (which are defined as minerals that can also exist in larger sizes, e.g., most known minerals) (Hochella et al., 2008). NPs are widespread components of soils, ranging from the nanomineral ferrihydrite, to a wide variety of mineral NPs, to nanoscale aggregates of natural organic matter (NOM), and to bacterial appendages known as nanowires (Maurice and Hochella, 2008). Soils contain many kinds of inorganic and organic particles, such as clay minerals, metal (hydr)oxides, and humic substances, whereas allophane and imogolite are abundant in volcanic soils (Theng and Yuan, 2008). Nanoparticulate goethite, akaganeite, hematite, ferrihydrite, and schwertmannite are important constituents of soils and

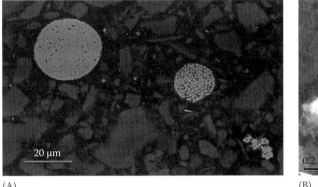
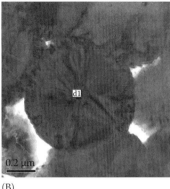

FIGURE 4.1 (A) Framboidal pyrite from a naturally bioreduced alluvial sediment from Rifle, CO. (B) High magnification image of a fibbed section of framboidal microcrystal (nanorim [<100 nm] with a chemical composition that is different from that of the bulk microcrystal). (Taken from Qafoku, N.P., R. Kukkadapu, J.P. McKinley, B.W. Arey, S.D. Kelly, C.T. Resch, and P.E. Long. 2009. Uranium in framboidal pyrite from a naturally bioreduced alluvial sediment. Environ. Sci. Technol. 43:8528–8534. Copyright 2009 American Chemical Society.)

sediments (Waychunas et al., 2005). The soil humic substance consists of a mixture of identifiable nanoscale biopolymers obtained, for example, directly from plant tissues that are added annually by maize plant residues (Adani et al., 2009). Within the soil profile, NPs that are formed during weathering of the primary minerals in the surface horizons where the chemical and biological processes are more intense may be transported downward to deeper horizons or groundwater by advective flow.

Based on a morphological criterion, NPs are classified into three categories: nanofilms (or nanosheets), nanorods, and NPs (Hochella et al, 2008). Nanosheets or thin surface coatings usually cover surfaces of primary minerals partially or fully, or they are products of the weathering processes that accumulate on the soil mineral surfaces, such as, for example, those on the surfaces of the microcrystals of framboidal pyrite found in a naturally bioreduced subsoil from Rifle, CO (Figure 4.1) (Qafoku et al., 2009). Typically, coatings consist of either pure or mixed forms of oxides and oxyhydroxides of Fe or other elements. However, they may also have a remarkably diverse composition (Penn et al., 2001). For example, the yellowish red to strong brown coatings on sediments from an Atlantic coastal plain aquifer in Virginia were predominantly a mixture of Si- and Al-rich nanophases of variable crystallinity with interspersed smectite and agglomerates of goethite NPs (Penn et al., 2001). In addition, clusters of bacterial cells, diatom fossils, sponge spicules, and other trace minerals were also observed during this investigation.

Examples of goethite nanoneedles and sodalite nanorods formed in the Hanford, WA, subsoil are presented in Figures 4.2A and 4.3B and C. NPs of hematite are presented in Figure 4.2B. NPs of ferrihydrite, which is a common soil mineral, are presented in Figures 4.4 and 4.5 (Hochella et al., 2005). In addition, examples of the chemical structure of nanotube imogolite and nanoball allophane, which are two naturally occurring nanoscale aluminosilicate minerals, are presented in Figure 4.6 (Abidin et al., 2007a, 2007b).

A special group of NNPs is formed in terrestrial ecosystems as a result of human activities causing unintentional accelerated weathering of primary soil minerals and formation of secondary minerals. In some terrestrial environments, NPs are formed when soil (or subsurface sediment) is exposed to extreme conditions, which promote accelerated and intensive dissolution of existing minerals and subsequent precipitation of nanoscale neophases (Figures 4.2 and 4.3). One example is the sediment beneath the tank waste farms at the Hanford site, WA, which was exposed to leaking fluids from the waste tanks, creating unusually extreme conditions of high alkalinity (pH ~ 14) and extreme salinity (~1 M $NaNO_3$). As a result, intense dissolution of the soil minerals occurred followed by precipitation of hematite and goethite NPs (Figure 4.2) and feldspathoids in the group of cancrinite and sodalite (Figure 4.3) (Qafoku et al., 2003a, 2003c, 2004a, 2007b; Zhao et al., 2004; Mashal et al., 2005). These chemical and mineralogical transformations significantly affected the mobility of different contaminants present in the waste liquids (Chorover et al., 2003; Qafoku et al., 2003b, 2007a).

Another group of NNPs is formed via biotic processes, for example, biogenic uraninite (UO_2) (Singer et al., 2009) (Figure 4.7). This formation pathway was demonstrated in experiments involving the synthesis of biogenic UO_2. This was accomplished by reduction of aqueous uranyl ions by *Shewanella putrefaciens* CN32 (Singer et al., 2009). The fresh nanoscale UO_2 produced by *Shewanella oneidensis* strain MR-1 was found to exhibit a well-ordered interior core of diameter ca. 1.3 nm and an outer region of thickness ca. ~0.6 nm in which the structure was locally distorted (Schofield et al., 2008). No evidence for lattice strain of the biogenic uraninite NPs was found in this study. The lack of NP strain and structural homology with stoichiometric UO_2 suggests that established thermodynamic parameters for the latter material are an appropriate starting point to model the behavior of nanobiogenic UO_2 (Schofield et al., 2008). Results from another team of researchers showed two distinct types of prokaryotic cells (inferred to be sulfate-reducing bacteria) that precipitated only a U(IV) mineral uraninite (UO_2) or both uraninite and metal sulfides, in a U-contaminated sediment from an inactive U mine anaerobically incubated with organic substrates (Suzuki et al., 2003).

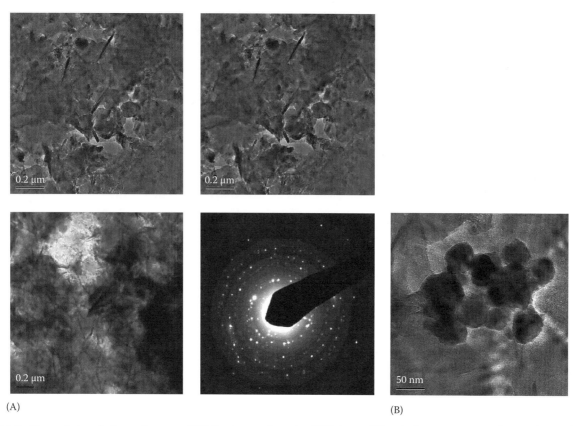

FIGURE 4.2 Transmission electron microscopy (TEM) micrographs and μ-XRD (x-ray diffraction) measurements taken in the Hanford sediments treated with high alkaline and saline solution. (A) Needle-like goethite NPs are shown in these micrographs, while the μ-XRD depicts the crystalline nature of these solids. (B) Hexagonal crystals of hematite NPs. (Taken from Qafoku, N.P., O. Qafoku, C.C. Ainsworth, A. Dohnalkova, and S.G. McKinley. 2007b. Fe-solid phase transformations under highly basic conditions. Appl. Geochem. 22:2054–2064. Copyright 2007, with permission from Elsevier.)

Biological processes may also lead to the formation of minerals via, for example, biologically induced oxidation of Fe(II) by anaerobic Fe-oxidizing bacteria. Fe biomineralization by the anaerobic nitrate-dependent Fe-oxidizing bacterium *Acidovorax* sp. strain BoFeNI in the presence of dissolved Fe(II) was recently studied, and all detected minerals consisted mainly of amorphous Fe phosphates (Miot et al., 2009).

4.1.3.2 Manufactured NPs

The majority of nanotechnologies are, according to one researcher, simply MNP production (Posner, 2009). They are producing and will continue to produce numerous MNPs in the coming years (Johansen et al., 2008). Hundreds of commercial products containing MNPs are used in many areas of human activities. For example, cerium dioxide NPs (CeO_2) are used as a catalyst in the automotive industry (Hoecke et al., 2009); Ag NPs are integrated into consumer and medical products (Bradford et al., 2009); ZnO NPs are used in solar cell device as the photoelectrode (Chen et al., 2009b); TiO_2 NPs are used in cosmetics, sunscreens, paints, and coatings, in addition to photocatalytic degradation of various pollutants in water, air, and soil matrices (Higarashi and Jardim, 2002; Nagaveni et al., 2004; Aarthi et al., 2007; Aarthi and Madras, 2008; Fang et al., 2009). Another group of MNPs (often called functionalized NPs) is used for decontamination purposes, that is, contaminant immobilization (Mattigod et al., 2005). An overview of the classes of NPs relevant to the environment, a comparison of the MNP with NNP and a summary of their formation, emission, occurrence, and fate in the environment is presented in the following reference (Nowack and Bucheli, 2007).

A recent study has confirmed that MNPs are present in the environment (Kaegi et al., 2008). Direct evidence of the release of TiO_2 NPs, with a size of about 20–300 nm (used in large quantities in exterior paints as whitening pigments) from exterior facade paints to the discharge into surface waters, was presented in this study. Analytical electron microscopy and bulk chemical analysis revealed that TiO_2 particles were detached from new and aged facade paints under the effect of natural weather conditions, then were transported by facade runoff and subsequently were discharged into natural, receiving waters. Apparently, this is the first report to present data on the calculated number densities of synthetic TiO_2 particles in the runoff. This and other studies (Kaegi et al., 2008; Mueller and Nowack, 2008; Koelmans et al., 2009) have raised the discussion about ENP to a more quantitative level. Our knowledge on this subject is increasing almost exponentially because of increasing interest and increased funding, although many investigations still have explorative character and raise more new questions (Nowack, 2009b) than provide answers to the "old" ones.

FIGURE 4.3 Sodalite [Na$_4$Al$_3$(SiO$_4$)$_3$NO$_3$] nanorods and cancrinite [Na$_6$Ca$_2$Al$_6$Si$_6$O$_{24}$(CO$_3$)$_2$] NPs formed when the subsoil minerals were exposed to extreme alkaline and saline conditions. (A) A feldspars particle covered with cancrinite balls. (B) Sodalite nanorods. (C) Sodalite nanorods formed on the edge of a mica (biotite) particle. (D) A cancrinite microcrystal composed of nanosheets. (Taken from Qafoku, N.P. 2010. Terrestrial nanoparticles and their controls on soil/geo processes and reactions. Adv. Agron. 107:33–91. Copyright 2010, with permission from Elsevier.)

4.1.4 ENP Toxicity

The field of nanotoxicology, which is a branch of bionanoscience, has become extensive and controversial over the last few years. Although a few years ago data on the potential environmental effects of NPs were scarce (Kahru et al., 2008) and limited information was available on the environmental behavior and associated potential risks of MNPs (Wang et al., 2009), an increasing body of information on this subject is now available in the literature. However, this topic is not discussed in detail in this chapter. For an in-depth coverage of different aspects of nanotoxicology, the reader is referred to the following references: Guzman et al. (2006a), Cañas et al. (2008), Handy et al. (2008a, 2008b), Johansen et al. (2008), Ju-Nam and Lead (2008), Kahru et al. (2008), Lee et al. (2008), Mueller and Nowack (2008), Navarro et al. (2008a, 2008b), Bradford et al. (2009), Drobne et al. (2009), Farre et al. (2009), Hull et al. (2009), Koelmans et al. (2009), Posner (2009), Wang et al. (2009), Zhang and Chen (2009), and Qafoku (2010).

4.2 Interactions in the Binary System: Nanoparticles–Soil Solution

In order to better understand the behavior of ENPs in terrestrial systems, it is essential to initially understand their interactions with different components of natural waters, including dissolved NOM, over a broad range of physicochemical conditions (Ju-Nam and Lead, 2008; Domingos et al., 2009). For this reason, the discussion in this section is focused on the behavior and reactivity of NPs exposed to different conditions of the soil aqueous phase. The binary interaction between NPs and the soil solution is presented and discussed in terms of NP formation, stability and aggregation, and their ability to serve as adsorbents, surface-mediated precipitation catalysts, or electron donors and acceptors. The extent and timescales of these processes depend on geochemical variables such as pH, OM content, and dissolved OM (DOM) concentration and on properties such as NP size, shape, concentration (Akbulut et al., 2007; Hoecke et al., 2009), type, age, and the presence of surface impurities.

4.2.1 Nanoparticle Formation and Stability

4.2.1.1 Formation

Crystals usually grow by attaching ions to inorganic surfaces or organic templates (Banfield et al., 2000). An alternative nucleation and crystal growth mechanism for NPs, namely, the oriented aggregation mechanism, is important coarsening process under a wide range of conditions encountered in the laboratory and in natural environments (Penn and Banfield, 1999). The oriented aggregation is defined as the irreversible and

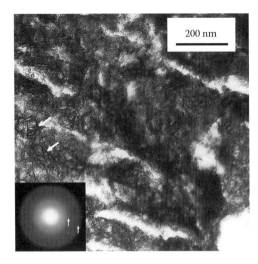

FIGURE 4.4 Transmission electron microscopy (TEM) image of six-line ferrihydrite with a small amount of vernadite-like mineral (dark, fiber-like particles, two of which are indicated by white arrows in the image) from the riverbed at Miles Crossing. The selected-area diffraction pattern of the ferrihydrite shows the broad, very weak lines from this mineral. The inner and outer arrows indicate the positions of the 2.57 and 1.49 Å lines, respectively. The 2.20 and 1.70 Å lines are barely visible between these two and are most easily seen on the left side of center. (Taken from Hochella, M.F., T. Kasama, A. Putnis, C.V. Putnis, and J.N. Moore. 2005. Environmentally important, poorly crystalline Fe/Mn hydrous oxides: Ferrihydrite and a possibly new vernadite-like mineral from the Clark Fork River Superfund Complex. Am. Mineral. 90:718–724. With permission from Mineralogical Society of America.)

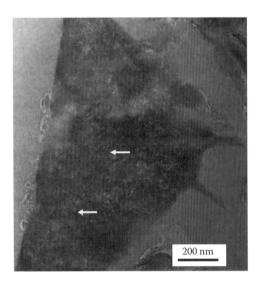

FIGURE 4.5 Transmission electron microscopy (TEM) image of a two-line ferrihydrite with a small amount of a vernadite-like mineral from the upper level of the floodplain of Kohr's Bend, collected at a depth of 15 cm. The arrows point to rare fiber-like features in this image. (Taken from Hochella, M.F., T. Kasama, A. Putnis, C.V. Putnis, and J.N. Moore. 2005. Environmentally important, poorly crystalline Fe/Mn hydrous oxides: Ferrihydrite and a possibly new vernadite-like mineral from the Clark Fork River Superfund Complex. Am. Mineral. 90:718–724. With permission from Mineralogical Society of America)

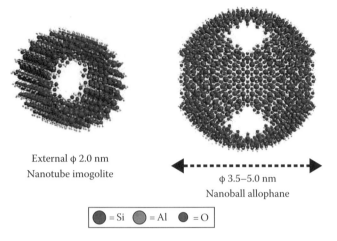

FIGURE 4.6 (See color insert.) Chemical structure of (A) nanotube imogolite and (B) nanoball allophane. (With kind permission from Springer Science+Business Media: Abidin, Z., N. Matsue, and T. Henmi. 2007a. Differential formation of allophane and imogolite: Experimental and molecular orbital study. J. Comput. Aided Mater. Des. 14:5–18.)

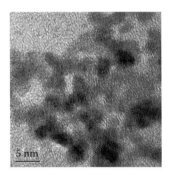 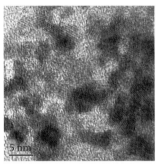

FIGURE 4.7 Images of biogenic uraninite UO$_2$ NPs, a product of microbial U(VI) reduction by a soil bacterium *Shewanella* sp. (Courtesy of Alice Dohnalkova, EMSL, Pacific Northwest National Laboratory, Richland, WA.)

crystallographically specific self-assembly of primary NPs and results in the formation of new single crystals, twins, and intergrowths (Penn et al., 2007). The oriented aggregation has been established as an important nonclassical crystal growth mechanism by which nanocrystals grow, unique and asymmetric morphologies are produced, and defects are incorporated into initially defect-free nanocrystals (Penn et al., 2006). Molecular dynamics simulations demonstrate that lowering of NP energy by changing orientation may drive crystal growth via oriented aggregation (Spagnoli et al., 2008).

A high-resolution transmission electron microscopy and low-temperature magnetometry study of synthetic goethite nanocrystals showed that nanocrystals grow almost exclusively by oriented aggregation of 3–4 nm primary nanocrystals (Guyodo et al., 2003). In another study, however, three coarsening mechanisms were found for anatase depending on solution pH (Finnegan et al., 2008): Ostwald ripening (pH > 11), oriented attachment and Ostwald ripening (pH = 2–11), and oriented attachment growth, followed by anatase dissolution and subsequent rutile formation (pH < 2).

Researchers also believe that to understand how dislocations, which are common defects in solids, are formed during early growth is of fundamental importance to understanding crystal growth (Penn and Banfield, 1998b). Dislocations are created when nanocrystalline materials grow by oriented attachment at crystallographically specific surfaces and there is a small misorientation at the interface (Penn and Banfield, 1998b). Twinning and polytypism in macroscopic crystals can originate at oriented interfaces between primary nanocrystalline particles early in their crystallization history (Penn and Banfield, 1998a). Maghemite (γ-Fe_2O_3)-like surface defects, for example, were found on the hematite (α-Fe_2O_3) NP with sizes of 7, 18, 39, and 120 nm (Chernyshova et al., 2007).

A variety of variables, such as soil solution pH (Myers and Penn, 2007; Finnegan et al., 2008; Isley and Penn, 2008), OM content, and DOM concentration (Eusterhues et al., 2008) in addition to a multitude of NP properties, such as type, size (Penn et al., 2006; Chernyshova et al., 2007), concentration, age, aggregation, size, and formation pathway (biotic or abiotic), may affect (i) the NP crystal structure formation and coarsening and (ii) the kinetics of phase formation and phase transformation. Even surface impurities may inhibit NP growth and phase transformation when they exist as surface clusters (Chen et al., 2007).

NP growth and phase transformation may also be time dependent. Growth by oriented aggregation was found to be consistent with the second-order kinetics with respect to the concentration of the primary NPs (Penn et al., 2007). These researchers also found that the overall rate constant for growth by oriented aggregation increased dramatically with decreasing primary particle size, which explains the common observation that growth by oriented aggregation slows as a function of continued crystal growth. Other investigators have found that the NP aggregation state has a marked influence on the kinetics of phase transformation and particle coarsening of three nanophases of titania (anatase, brookite, and rutile), probably due to the lowering of NP surface energy by interparticle interactions (Zhang and Banfield, 2007).

4.2.1.2 Stability

The stability of NPs is a function of their surface energy, and NPs are more stable when they possess a low surface energy. For example, the surface energies of small rutile particles are higher than those for anatase particles of a similar size, consistent with anatase being the more stable phase of nanocrystalline TiO_2 (Naicker et al., 2005). The surface free energy of NPs in an aqueous solution consists of the electrostatic energy of charged surfaces and interfacial energy (Finnegan et al., 2007). Both these terms can be modified by solution chemistry. They can also be manipulated to control NP phase stability and transformation kinetics (Finnegan et al., 2007). In addition, a better understanding of the stability of nanostructures in different chemical environments can be achieved by the incorporation of a size-, shape-, and temperature-dependent thermodynamic model, into theoretical descriptions of nanomaterials (Barnard and Xu, 2008).

Recently measured thermodynamic data on formation and surface energies of Fe oxides, which occur in almost all terrestrial ecosystems and exist in a rich variety of structures and hydration states, have been presented in a recent publication (Navrotsky et al., 2008). These authors studied the size-driven crossovers in stability, and they claim that this would help to explain patterns of occurrence of different Fe oxides in nature.

A kinetic equation that incorporates the dependence of the rate constant on the particle size for phase transformation via interface nucleation in NPs is proposed (Zhang and Banfield, 2005). These authors found that the temperature dependence of the rate constant of a kinetically controlled phase transformation could be usually described by the Arrhenius equation, which comprises a preexponential factor multiplied by an exponential term involving the activation energy and temperature. They showed that particle size was another factor that was needed in the description of kinetics of NP phase transformations.

Among the geochemical and other variables and/or NP properties that may control the NP stability are pH of solution (Finnegan et al., 2007), NP size (Finnegan et al., 2007; Erbs et al., 2008; Liu et al., 2008), presence of ligands, such as, for example, oxalate, which may cause ligand-promoted NP dissolution (Cwiertny et al., 2009), and the presence of coatings or nanocrystals (Liu et al., 2008; Sun et al., 2009b).

4.2.3 Nanoparticle Aggregation

Aggregation of NPs is expected to be a common phenomenon in soils since NP aggregation is observed on many occasions. For example, CeO_2 NPs, which are used as catalysts in the automotive industry and that are expected to reach the terrestrial system in significant amounts, formed aggregates with a mean size of approximately 400 nm when NPs of three different sizes (14, 20, and 29 nm) were exposed to a solution with a pH of 7.4 (Hoecke et al., 2009).

However, one should be aware of the fact that ENPs may show complex colloid and aggregation behavior because aggregation is likely to be affected by particle shape, size (5 nm particles formed larger aggregates compared with 32 nm particles under the same conditions of pH and solid concentrations; Pettibone et al., 2008), surface area, and surface charge of NPs (Handy et al., 2008a; Pettibone et al., 2008). In addition, other factors such as pH, ionic strength, water hardness, and the presence of DOM or other organic compounds may control NP aggregation (Baalousha et al., 2008; Handy et al., 2008a; French et al., 2009). For example, aggregation of 5 nm TiO_2 NPs increased when the contacting solution's pH values were near the PZC, and at any given pH an increase in ionic strength generally resulted in increased aggregation (Domingos et al., 2009). In addition, NPs of titanium dioxide of 4–5 nm diameter readily formed stable aggregates with an average diameter of 50–60 nm at pH similar to 4.5 in a NaCl suspension adjusted to an ionic strength of 0.0045 M (French et al., 2009).

Contrary effects of surface coatings on aggregation are reported in the literature. On one hand, the surface coatings were found to enhance NP aggregation. In experiments conducted with unpurified manufactured Fe-oxide NP (~7 nm) and standard Suwannee River humic acid (SRHA), results indicated that extensive aggregation of NPs began at approximately pH 5–6 and reached a maximum at approximately pH 8.5, whereas with added SRHA, aggregation, which is mainly due to charge neutralization, was shifted to lower pH values of 4–5 and was affected by SRHA concentration (Baalousha et al., 2008). The same effect was observed in another study, where organic acids were found to destabilize NP suspensions (Pettibone et al., 2008), and, in yet another study, results from experiments involving synthetic zinc sulfide NPs and representative amino acids showed a driving role for cysteine in rapid NP aggregation (Moreau et al., 2007).

On the other hand, a study conducted with TiO_2 NPs of 5 nm and Suwannee River fulvic acid (SRFA) indicated that conditions that favored adsorption of the fulvic acid resulted in less aggregation of the TiO_2 NP presumably due to increased steric repulsion among individual NPs (Domingos et al., 2009).

Another scientific issue that remains unresolved is how aggregation affects the available surface area for sorption (sorption extent) and the timescales of adsorption or desorption of contaminants or nutrients from exposed (surface) and remote sorption sites within the aggregate structures. Results from research have demonstrated that the relative reactivity of 5 and 32 nm particles as determined from Langmuir adsorption parameters did not vary greatly despite differences in NP aggregation for these two different size NPs (5 nm NPs form larger aggregates than the 32 nm NPs under the same conditions of pH and solid concentrations) (Pettibone et al., 2008), suggesting that aggregation did not affect the extent of organic acid sorption by anatase particles. However, these authors believe that challenges remain in assessing the available surface area for sorption in NP aqueous suspensions because of aggregation. This is a very important subject that is also related to the interactions of oppositely charged particles and the double layer overlapping mechanism (Qafoku and Sumner, 2002), which is discussed in Section 4.3.

4.2.4 Nanoparticles as Sorbents

Because the chemical and electrostatic interactions at mineral–water interfaces are of fundamental importance in many geochemical, materials science, and technological processes (Ridley et al., 2006), the topic of contaminant or nutrient sorption on NP surfaces has attracted the attention of many researchers in recent years. NPs have high sorption capacity for metal and anionic contaminants such as As, Cr, Pb, Hg, Se (Waychunas et al., 2005), Cu (Madden et al., 2006), hexavalent uranium [U(VI)] (Zeng et al., 2009), NOM (Domingos et al., 2009), and organic acids (Pettibone et al., 2008; Mendive et al., 2009). Contaminant sequestration is accomplished mainly by surface complexation, but sorbed surface species may be encapsulated within interior interfaces of NP aggregates, a phenomenon with significant consequences for contaminant dispersal or remediation processes (Waychunas et al., 2005).

Although some studies have shown that the sorption behavior of NPs of the same type is not size dependent (Ridley et al., 2006; Pettibone et al., 2008), many researchers now believe that NP sorption behavior is size dependent (Zhang et al., 1999; Waychunas et al., 2005; Madden et al., 2006; Abbas et al., 2008; Auffan et al., 2008; Zeng et al., 2009). Different nanosize polymorphs may also exhibit different sorption behaviors for the same sorbate. For example, results from an experimental and theoretical study of aqueous oxalic acid adsorption on anatase NPs (Mendive et al., 2009) were compared with those from a previous study performed by the same authors with rutile NPs (Mendive et al., 2008). Results indicated that differences between surface complexes on anatase and rutile are mainly in the denticity type: in the case of rutile, the most stable species consist of two bidentate surface complexes followed in the third place by a monodentated form, whereas in the case of anatase the formation of four species in which the stability order is reversed with respect to the denticity type is observed (Mendive et al., 2008).

Studies have also demonstrated that NP crystal structure may be manipulated in some fascinating ways affecting NP sorption properties. For example, ZnS NPs with a mean diameter of 3.2–3.6 nm, which were synthesized and treated with different low-temperature procedures, possessed a dramatic range of interior disorder, and a direct correlation was found between the strength of surface–ligand interactions and interior crystallinity (Gilbert et al., 2006b).

Factors such as soil solution pH and ionic strength, isomorphic (elemental) substitution in the NP crystal structure, and NP aggregation may have an important impact on the sorption behavior of NPs. The pH and ionic strength effect on adsorption was investigated, for example, in experiments with TiO_2 NPs of 5 nm and SRFA (Domingos et al., 2009). The effect of Al substitution on magnetite NP sorption behavior was evaluated in presence of the organic molecules benzoquinone and carbon tetrachloride (Jentzsch et al., 2007). The effect of aggregation on sorption was studied in ZnS NPs (similar to 2–3 nm sphalerite) and the binding energy of water molecules, and, thus, the strength of interaction of water was highest in isolated NPs, lower in NP aggregates, and lowest in bulk crystals (Zhang et al., 2007).

4.2.5 Interfacial Electron-Transfer Reactions

NPs may be involved in interfacial electron-transfer reactions with either redox-sensitive elements that sorb to NP surfaces (heterogeneous electron transfer) or microorganisms that are attached directly to the NP surfaces. In addition to the direct electron transfer mechanisms, it is likely that microorganisms may also employ the indirect route (via intermediate electron carriers) to transfer electrons to NPs. Results from studies conducted with hematite NPs and *S. oneidensis* MR-1 suggested that this bacterium employs both indirect and direct mechanisms of electron transfer to hematite NPs (Bose et al., 2009). The results from this study also suggested that Fe(III) reduction rates depend on many factors, such as the NPs' aggregation state, size, shape, and exposed crystal faces (Bose et al., 2009).

For example, the differences in hematite behavior as a function of size and the rates of Mn^{2+} (aq) oxidation by hematite in the presence of oxygen have been determined in a recent study (Madden and Hochella, 2005). These authors argued that the acceleration of electron transfer rate for the reactions promoted by the smallest particles is rationalized in the framework of electron transfer theory. According to this theory, the rate depends on three factors: the electronic coupling between initial and final electronic states, the substantial reorganization energy for solvent and coordinated ligands between initial and final states, and the free energy of reaction (corrected for work required to bring reactants together) (Madden and Hochella, 2005). In two other studies, the significance and effect of NP size distributions were explored for heterogeneous Mn(II) oxidation on hematite NP surfaces (Madden and Hochella, 2008), and the reactivity of two types of nanosized, zero-valent Fe were found to react more rapidly than microsized zero-valent Fe based on mass-normalized rate constants, although surface area-normalized rate constants did not show a significant nanosize effect (Nurmi et al., 2005).

The aging effect on electron-transfer reactions was investigated in zero-valent nano-Fe (Sarathy et al., 2008), and the changes in (i) the composition and structure of the particles and (ii) the reactivity of the particles (by carbon tetrachloride reaction kinetics, electrochemical corrosion potentials, and H-2 production rates) were studied (Sarathy et al., 2008). Results showed that Fe-H_2 becomes more reactive between 0 and 2 days of exposure to water and then gradually loses reactivity over the next few 100 days.

4.3 Nanoparticle Controls or Impacts on the Extent and Timescale of Soil-/Geo-Processes

The discipline of environmental soil/geo nanoscience is in its infancy, mainly because the study by conventional methods of the complicated behavior of the ENPs that are present in heterogeneous natural media was in the past, and in many respects, it is still nearly impossible. Although review papers on NP behavior and ecotoxicity are now available (Gilbert and Banfield, 2005; Lead and Wilkinson, 2006; Nowack and Bucheli, 2007; Handy et al., 2008a, 2008b; Ju-Nam and Lead, 2008; Klaine et al., 2008; Navarro et al., 2008b; Nowack, 2009a, 2009b), based on the rudimentary knowledge that is currently available (Nowack, 2009b), many challenges remain. The effect of ENPs on the extent and rate of terrestrial chemical, biological, and hydrological processes and reactions is far from being understood.

The complexity of terrestrial ecosystems is remarkable. ENPs may interact in different ways not only with each other but also with a variety of aqueous species and an array of solid particles and soil organic matter (OM) changing significantly or dramatically the nano-, micro-, and macroscale surrounding environment. Soil particles may exhibit permanent surface charge (e.g., micron-size phyllosilicates) or variable-surface charge (e.g., nanosized metal oxides and hydroxides). Interactions between these micron-size and nanosize particles can lead to the formation of surface coatings and colloid assemblages possessing different properties than their individual constituents. However, much of our knowledge of the behavior of soil particles is derived from experiments conducted with single-phase systems (Bertsch and Seaman, 1999). Although these investigations have improved our understanding of the physical and chemical properties of individual soil particles, the results cannot be readily extrapolated and used to understand the behaviors of the complex assemblages that are present within natural soil and sediment systems (Sumner, 1995; Bertsch and Seaman, 1999).

Heterocoagulation reactions between oppositely charged oxide ENPs and larger phyllosilicate colloids may have a considerable impact on colloidal stability, particle size distribution, reactive surface area, and adsorption reactions (Sumner, 1963a, 1963b; Sumner and Davidtz, 1965; Sumner and Reeve, 1966; Qafoku et al., 2000; Qafoku and Sumner, 2002; Qafoku et al., 2004b). For example, positively charged Fe (hydr)oxide ENPs can coat the negatively charged phyllosilicate surfaces blocking or occupying surface adsorption sites and, as a result, affect the extent of solute adsorption. Overlapping electric double layers of the particles may notably induce unique electrostatic conditions, which are a function of the interplay of NP dimensions and the Debye length (controlled by the ionic strength) (Qafoku et al., 2004b). A full understanding of the nature and magnitude of the interactions among soil particles and ENPs of different types, and how these interactions affect the extent and timescales of different soil-/geo-processes, will take some time to achieve. In addition, in order to improve predictive nutrient mobility and contaminant migration in surface and subsurface environments, interactions between ENPs, soil solution, and other soil/sediment matrices require detailed examination.

Nanoscale studies in terrestrial ecosystems have recently intensified for mainly two reasons. First, methods to isolate and characterize ENPs and state-of-the-art techniques and equipment that have been developed only in recent years have increased significantly our ability to probe phenomena, processes, and reactions at the nanoscale, such as those occurring in the nanopores within soil matrices or on the surfaces of the ENP. For example, recently scientists measured the real-time, minute forces that characterize interactions between living cells of *S. oneidensis* (a dissimilatory metal-reducing bacterium) and goethite (α-FeOOH), as a function of the distance (in nanometers) separating the cell from the mineral surface (Lower et al., 2001). In another recent study, the environmental importance of the micrometer-to-submicrometer topography variations of fluid–rock interfaces that govern the fate of chemical elements was demonstrated (Fischer et al., 2009).

Second, as mentioned in the previous sections, the industrial-scale production and the wide variety of applications and use of MNPs in many industrial and household applications will definitely lead to the release of such materials in considerable amounts into environments (Nowack and Bucheli, 2007; Ju-Nam and Lead, 2008; Klaine et al., 2008). Such releases will be controlled, in the form of waste disposal, or uncontrolled. Because MNPs pose potential risks to human health and ecosystems (Wiesner et al., 2006), and because MNPs will affect in different ways chemical, biological, and hydrological processes and physical and mineralogical properties of terrestrial ecosystems (e.g., aggregation and formation of clusters, heteroaggregation, catalysis of precipitation, etc.), the debate on the environmental impact of MNPs has become increasingly important (Hofmann and von der Kammer, 2009), and research efforts have recently been focused in this direction. Assessing these effects and environmental risks requires an understanding of ENP mobility, bioavailability, reactivity, ecotoxicity, and persistency (Nowack and Bucheli, 2007; Kahru et al., 2008).

Discussion in this section is focused on the interactions in the ternary systems composed of ENPs, nutrient/contaminant chemical species, and the soil/sediment matrix. Several aspects of the inhibitory and catalytic effects of ENPs on nutrient/contaminant transport, adsorption and desorption, dissolution and precipitation, and redox reactions are explored and discussed in the subsections of this section.

4.3.1 Advective Mobility and Diffusive Mass Transfer

ENPs may be toxic to different soil living organisms and microorganisms and the chances for them to be exposed to ENPs increase with increasing ENP mobility. Some studies have demonstrated that ENPs may easily move through the soil profiles. For example, results from a study of TiO_2 NP transport behavior through saturated homogeneous soil columns showed that TiO_2 could remain suspended in soil suspensions even after 10 days, and the estimated transport distances of TiO_2 in some soils ranged from 41.3 to 370 cm (Fang et al., 2009).

In addition, topsoils and subsoils are often rich in nutrients or could be contaminated with trace metals, and it is important to understand how different processes govern their transport to fresh and marine waters (Hassellov and von der Kammer, 2008). ENPs have high sorption capacity for aqueous nutrients and contaminants. The sorbed ions neutralize the ENP electrical surface charge and the ENP: sorbed ion complex can potentially travel longer distances through the soil matrix pores than charged ENPs or ions taken separately. ENPs may, therefore, serve as carriers of contaminants and nutrients, increasing their overall mobility in terrestrial ecosystems. Studies have demonstrated that NPs, such as carbon nanotubes (Schierz and Zänker, 2009), can serve as carriers of contaminants, facilitating their transport through soil profiles via advective and diffusive mass transfer. For example, accelerated transport of Pb associated with Fe oxide NNPs was observed in soil and river samples in Germany and Sweden (Hassellov and von der Kammer, 2008). In another study, carbonaceous MNPs were found to enhance the transport of hydrophobic organic contaminants (HOCs) in porous media because HOC was strongly sorbed to them (Hofmann and von der Kammer, 2009).

It is, therefore, essential to understand the fate and transport of ENPs in terrestrial environment in order to evaluate their potential risks to organisms and/or microorganisms and the possibility for increased contaminant and nutrient mobility, human exposure, and damages to the ecosystems.

The movement of ENPs in soils and through the soil profile is controlled by a number of geochemical variables, such as pH (Guzman et al., 2006b; Schierz and Zänker, 2009), ionic strength (Hassellov and von der Kammer, 2008; Fang et al., 2009), the contents of DOM, OM, humic acid and clay (Fang et al., 2009; Johnson et al., 2009; Schierz and Zänker, 2009), sorption kinetics, and residence time (Hofmann and von der Kammer, 2009), in addition to NP-related properties such as surface charge, size (Hofmann and von der Kammer, 2009), shape, aggregation, surface coatings (Moreau et al., 2007; Joo et al., 2009), and impurities. These examples from the literature clearly demonstrate the complexity associated with studies of these types, which involve an array of variables with possible simultaneous or sequential controls or effects on ENP mobility.

4.3.2 Adsorption/Desorption

Within this subject area, there are at least five important topics to be explored. The first topic is the estimation of the extent and timescale of ENP adsorption to and desorption from different soil minerals or soil materials. For example, in a recent study, it was found that the kinetic rate constants of the adsorption reaction of oppositely charged NPs changed with the aqueous NP concentration (Tretiakov et al., 2009).

The second topic covers the estimation of the extent and timescales of adsorption and desorption of the complex ENP: sorbed nutrient/contaminant on different soil minerals and/or soil materials. Such ENP: sorbed ions complexes are formed, for example, in Andosols of the island of Réunion, which have high Ni concentrations due to the natural pedogeochemical background, and high abundances of natural aluminosilicate ENPs (Levard et al., 2009). Ni is chemically bonded to natural short-range ordered aluminosilicate ENPs, and this complex represents about 75% of the total Ni in this soil.

The third topic is assessment of the effects of ENPs on contaminant and nutrient adsorption/desorption in soils. For example, force–volume microscopy and a silicon–nitride probe were used to measure changes in adhesion when a patchy overgrowth of Mn oxide nanostructures formed on the surface of rhodochrosite (Na and Martin, 2009). According to these authors, the quantitative mapping of adhesive force can lead to an improved mechanistic understanding of how nanostructure growth influenced contaminant immobilization and bacterial attachment. Research has also demonstrated that Fe-rich NPs compete efficiently with NOM for Pb binding in soils and rivers (Hassellov and von der Kammer, 2008). Another aspect of this phenomenon is ENP competition with aqueous species of contaminants and/or nutrients for available sorption sites on different sorbents present in soils.

The fourth topic involves the usage of NPs for aqueous contaminant removal and remediation purposes. For example, research work has shown that CuO NPs with a surface area of 85 m² g⁻¹ and 12–18 nm diameter were an effective material for As(III) and As(V) adsorption and may be used to develop a simple and efficient As removal method (Martinson and Reddy, 2009). Other studies have investigated the potential of zerovalent Fe NPs for treatment and remediation of persistent organic pollutants such as hexachlorocyclohexanes (HCHs; $C_6H_6Cl_6$) (Elliott et al., 2008); the removal of radionuclides and contaminants using apatite, a calcium phosphate mineral, with a nanoporous structure and with a strong affinity for contaminants and radionuclides (Moore et al., 2007); application of a new type of alumina-coated magnetite (Fe_3O_4/Al_2O_3) NPs that were modified by the surfactant sodium dodecyl sulfate (SDS) to increase adsorption capacity, for extraction of trimethoprim (TMP) from environmental water samples (Sun et al., 2009b); and the adsorption potential of pure and iron-doped TiO_2 particles with crystal sizes of 108 and 65 nm, which were subsequently immobilized on sand, for removing arsenic from drinking water (Nabi et al., 2009).

The fifth topic is evaluation and quantification of the controls or effects of different variables on all the adsorption and desorption processes that were described in topics 1–4. One can easily realize that a massive research effort is required to address all these scientific issues. The few examples provided above clearly illustrate the fact that the scientific community has just started to work to address these scientific issues.

4.3.3 Dissolution/Precipitation

ENPs that are present in terrestrial systems may influence the extent and timescales of dissolution and precipitation reactions occurring in soils and subsoils, and this might have an effect on nutrient and/or contaminant mobility in these systems. For example, the in situ Fe(III) reduction was monitored in soils, the mineralogical changes of Fe-oxides were initially studied (Fakih et al., 2008), and then the impact of (bio)reduction on both Fe-oxide dissolution and secondary mineral precipitation and the subsequent effects on As mobility were subsequently estimated (Fakih et al., 2009). Again, this is an unexplored area of research, and definitely studies are required to investigate single and coupled reactions and processes that might affect the extent and timescale of dissolution and precipitation and nutrient and contaminant mobility in natural terrestrial systems.

4.3.4 Electron Transfer Reactions

NP catalytic or inhibitory effects on contaminant and/or nutrient reduction or oxidation in soils and sediments is another topic that is not covered well in the literature as there are only few papers published on this topic so far. Two examples of such studies are included below.

In the first study, arsenite [As(III)] adsorption and oxidation were investigated in the presence of TiO_2 NPs, and the influence of TiO_2 NPs on the bioavailability of As(III) was examined in bioaccumulation tests using carp (Cyprinus carpio) (Sun et al., 2009a). Results showed that most of aqueous As(III) was oxidized to As(V) in the presence of sunlight and TiO_2 NPs. In the second study, experiments were conducted with a series of Fe(III) sculptured thin films with varied crystallinity (from hematite to ferrihydrite) and nanowire shapes (slanted columnar, clockwise helical, and counterclockwise helical); the dissimilatory metal-reducing bacterium *S. putrefaciens* CN32 was used to measure the bioreducibility of Fe thin films (Tan et al., 2006). Results indicated that bioreduction was controlled primarily by oxide crystallinity. Postbioreduction characterizations of solid phases demonstrated that the mineralogy of the film materials did not change, but surface roughness generally increased with hematite showing the greatest change. Because Fe(III) phases are excellent sorbents and can get involved in redox reactions with redox-sensitive contaminants or nutrients, a better understanding of the redox reactions involving Fe(III) phases definitely helps in a better understanding of the fate of contaminants and nutrient in soils.

4.4 Future Research Directions

Based on the analysis presented in this chapter, a synthetic presentation of future research needs and directions is provided below:

1. *Studies on NNP occurrence, distribution, and properties are scarce in the literature, and, they are especially rare in the specialized soil literature.* Fundamental questions such as the following remain unanswered: (a) What are the identities and concentrations of soil NNPs in different soil orders and along soil profiles? (b) Are there any differences among temperate climate soils, variable charge soils, subsoils, and unconsolidated sediments in terms of NP occurrence, reactivity, and controls on soil-/geo-processes?
2. *Methods to separate NNP and/or MNP from soil/sediment samples are not developed or described well in the literature.* There is a need for methods or well-established protocols on how to separate NNP and MNP from soil matrices.
3. *The toxic effects of MNP on plants, microbes, and other soil living organisms are only now being assessed.* Additional research work is needed before defensible and scientifically based conclusions can be drawn.
4. *Limited information is available on NNP formation and NNP and MNP stability in the soil systems.* Within this topic: (a) The conditions and geo/soil variables that promote their formation or control their stability are not well studied; (b) There is no database available for MNP with information about the dissolution or phase-transformation processes they might undergo (thermodynamic and kinetic considerations and controlling variables); (c) The effects of type, size, shape, concentration, aggregation, age formation

pathway (biotic or abiotic), and the presence of surface defects and/or impurities, on ENP dissolution and stability, are not well understood and covered in the literature; (d) The effects of soil chemical, biological, and hydrological variables on the NP crystal structure formation, stability, and aggregation and kinetics of phase formation and phase transformation are not extensively covered.

5. *The process of ENP aggregation in terrestrial ecosystems is not well understood and studied.* Even general questions still remain unanswered: (a) What are the driving forces behind this process in terrestrial ecosystems? (b) What are the most important geo/soil variables, in addition to pH and ionic strength, which control self- and/or heteroaggregation of ENPs? (c) What is the aggregation effect on the available surface area for sorption and the timescales of nutrient and/or contaminant adsorption to or desorption from exposed (surface) and remote sorption sites within the aggregate structures? (d) How can one predict aggregation occurrence in different soils?

6. *The molecular-scale mechanisms of contaminants and nutrient adsorption to and desorption from ENPs, and electron transfer reactions involving contaminants and nutrients and ENPs, are not well understood, and this area of research is still developing.* Although researchers are addressing scientific issues in these areas, there is much more to be done considering the wide variety of ENPs and the enormous amounts of contaminants and nutrient chemical species that are present in soils. In addition, the combined effects (synergetic or inhibitory, sequential or simultaneous) that may be significant in a system where more than one type of ENP and more than a single contaminant or nutrient are present have not been explored and understood at this moment.

7. *The information available on ENP advective mobility and diffusive mass transfer and their controls or effects on contaminant and nutrient adsorption/desorption, dissolution/precipitation, and electrons transfer reactions in terrestrial ecosystems is rudimentary and scarce.* The potential number of variables that may affect these processes is large, and there is a need for research efforts in all of the areas mentioned above. For example, studies are needed to assess the mobility of single ENP, aggregated ENPs, or ENP: sorbed nutrient/contaminant complexes in soils as a function of one or more physical, chemical, mineralogical, biological, and hydrological variables and properties, as well as NP properties, such as surface charge, size, shape, aggregation, surface coatings, and impurities.

Acknowledgments

I would like to thank Dr. P. Ming Huang (University of Saskatchewan, Canada) for giving me the opportunity to write this chapter, Dr. Donald R. Baer (Pacific Northwest National Laboratory, PNNL, Richland, WA) for providing guidance and related books and articles, and my manager Dr. Kevin Rosso (PNNL) for supporting me in this effort. PNNL is operated for the DOE by Battelle Memorial Institute under the Contract DE-AC06-76RLO 1830. Some of the figures presented in this chapter were generated in the Environmental Molecular Sciences Laboratory, a national scientific user facility sponsored by the US DOE Office of Biological and Environmental Research and located at the PNNL in Richland, WA.

References

Aarthi, T., and G. Madras. 2008. Photocatalytic reduction of metals in presence of combustion synthesized nano-TiO_2. Catal. Commun. 9:630–634.

Aarthi, T., P. Narahari, and G. Madras. 2007. Photocatalytic degradation of Azure and Sudan dyes using nano TiO_2. J. Hazard. Mater. 149:725–734.

Abbas, Z., C. Labbez, S. Nordholm, and E. Ahlberg. 2008. Size-dependent surface charging of nanoparticles. J. Phys. Chem. C 112:5715–5723.

Abidin, Z., N. Matsue, and T. Henmi. 2007a. Differential formation of allophane and imogolite: Experimental and molecular orbital study. J. Comput. Aided Mater. Des. 14:5–18.

Abidin, Z., N. Matsue, and T. Henmi. 2007b. Nanometer-scale chemical modification of nano-ball allophane. Clay. Clay Miner. 55:443–449.

Adani, F., S. Salati, M. Spagnol, F. Tambone, P. Genevini, R. Pilu, and K.G.J. Nierop. 2009. Nanometer-scale structure of alkali-soluble bio-macromolecules of maize plant residues explains their recalcitrance in soil. Chemosphere 76:523–528.

Akbulut, M., A.R.G. Alig, Y. Min, N. Belman, M. Reynolds, Y. Golan, and J. Israelachvili. 2007. Forces between surfaces across nanoparticle solutions: Role of size, shape, and concentration. Langmuir 23:3961–3969.

Auffan, M., J. Rose, O. Proux, D. Borschneck, A. Masion, P. Chaurand, J.L. Hazemann et al. 2008. Enhanced adsorption of arsenic onto maghemites nanoparticles: As(III) as a probe of the surface structure and heterogeneity. Langmuir 24:3215–3222.

Baalousha, M., A. Manciulea, S. Cumberland, K. Kendall, and J. Lead. 2008. Aggregation and surface properties of iron oxide nanoparticles: Influence of pH and natural organic matter. Environ. Toxicol. Chem. 27:1875–1882.

Banfield, J.F., S.A. Welch, H.Z. Zhang, T.T. Ebert, and R.L. Penn. 2000. Aggregation-based crystal growth and microstructure development in natural iron oxyhydroxide biomineralization products. Science 289:751–754.

Banfield, J.F., and H. Zhang. 2001. Nanoparticles in the environment, p. 1–58, *In* J.F. Banfield and A. Navrotsky (eds.). Reviews in mineralogy and geochemistry: Nanoparticles and the Environment. Vol. 44. Mineralogical Society of America, Chantilly, VA.

Barnard, A.S., and H.F. Xu. 2008. An environmentally sensitive phase map of titania nanocrystals. ACS Nano. 2:2237–2242.

Bertsch, P.M., and J.C. Seaman. 1999. Characterization of complex mineral assemblages: Implications for contaminant transport and environmental remediation. Proc. Natl. Acad. Sci. USA. 96:3350–3357.

Bose, S., M.F. Hochella, Y.A. Gorby, D.W. Kennedy, D.E. McCready, A.S. Madden, and B.H. Lower. 2009. Bioreduction of hematite nanoparticles by the dissimilatory iron reducing bacterium Shewanella oneidensis MR-1. Geochim. Cosmochim. Acta 73:962–976.

Bradford, A., R.D. Handy, J.W. Readman, A. Atfield, and M. Mühling. 2009. Impact of silver nanoparticle contamination on the genetic diversity of natural bacterial assemblages in estuarine sediments. Environ. Sci. Technol. 43:4530–4536.

Cañas, J., M. Long, S. Nations, R. Vadan, L. Dai, M. Luo, R. Ambikapathi, E. Lee, and D. Olszyk. 2008. Effects of functionalized and nonfunctionalized single-walled carbon nanotubes on root elongation of select crop species. Environ. Toxicol. Chem. 27:1922.

Carta, D., M.F. Casula, A. Corrias, A. Falqui, G. Navarra, and G. Pinna. 2009. Structural and magnetic characterization of synthetic ferrihydrite nanoparticles. Mater. Chem. Phys. 113:349–355.

Chen, J., C. Li, J.L. Song, X.W. Sun, W. Lei, and W.Q. Deng. 2009b. Bilayer ZnO nanostructure fabricated by chemical bath and its application in quantum dot sensitized solar cell. Appl. Surf. Sci. 255:7508–7511.

Chen, B., H. Zhang, K.A. Dunphy-Guzman, D. Spagnoli, M.B. Kruger, D.V.S. Muthu, M. Kunz, S. Fakra, J.Z. Hu, Q.Z. Guo, and J.F. Banfield. 2009a. Size-dependent elasticity of nanocrystalline titania. Phys. Rev. B 79:8.

Chen, B., H.Z. Zhang, B. Gilbert, and J.F. Banfield. 2007. Mechanism of inhibition of nanoparticle growth and phase transformation by surface impurities. Phys. Rev. Lett. 98:4.

Chernyshova, I.V., M.F. Hochella, and A.S. Madden. 2007. Size-dependent structural transformations of hematite nanoparticles. 1. Phase transition. Phys. Chem. Chem. Phys. 9:1736–1750.

Chorover, J., S.K. Choi, M.K. Amistadi, K.G. Karthikeyan, G. Crosson, and K.T. Mueller. 2003. Linking cesium and strontium uptake to kaolinite weathering in simulated tank waste leachate. Environ. Sci. Technol. 37: 2200–2208.

Cwiertny, D.M., G.J. Hunter, J.M. Pettibone, M.M. Scherer, and V.H. Grassian. 2009. Surface chem and dissolution of alpha-FeOOH nanorods and microrods: Environmental implications of size-dependent interactions with oxalate. J. Phys. Chem. C 113:2175–2186.

Domingos, R.F., N. Tufenkji, and K.J. Wilkinson. 2009. Aggregation of titanium dioxide nanoparticles: Role of a fulvic acid. Environ. Sci. Technol. 43:1282–1286.

Drobne, D., A. Jemec, and Z. Pipan Tkalec. 2009. In vivo screening to determine hazards of nanoparticles: Nanosized TiO_2. Environ. Pollut. 157:1157–1164.

Elliott, D.W., H.-L. Lien, and W.-X. Zhang. 2008. Zerovalent iron nanoparticles for treatment of ground water contaminated by hexachlorocyclohexanes. J. Environ. Qual. 37:2192–2201.

Erbs, J.J., B. Gilbert, and R.L. Penn. 2008. Influence of size on reductive dissolution of six-line ferrihydrite. J. Phys. Chem. C 112:12127–12133.

Eusterhues, K., F.E. Wagner, W. Hausler, M. Hanzlik, H. Knicker, K.U. Totsche, I. Kogel-Knabner, and U. Schwertmann. 2008. Characterization of ferrihydrite-soil organic matter coprecipitates by X-ray diffraction and Mossbauer spectroscopy. Environ. Sci. Technol. 42:7891–7897.

Fakih, M., M. Davranche, A. Dia, B. Nowack, G. Morin, P. Petitjean, X. Chatellier, and G. Gruau. 2009. Environmental impact of As(V)-Fe oxyhydroxide reductive dissolution: An experimental insight. Chem. Geol. 259:290–303.

Fakih, M., M. Davranche, A. Dia, B. Nowack, P. Petitjean, X. Chatellier, and G. Gruau. 2008. A new tool for in situ monitoring of Fe-mobilization in soils. Appl. Geochem. 23:3372–3383.

Fang, J., X.-Q. Shan, B. Wen, J.-M. Lin, and G. Owens. 2009. Stability of titania nanoparticles in soil suspensions and transport in saturated homogeneous soil columns. Environ. Pollut. 157:1101–1109.

Farre, M., K. Gajda-Schrantz, L. Kantiani, and D. Barcelo. 2009. Ecotoxicity and analysis of nanomaterials in the aquatic environment. Anal. Bioanal. Chem. 393:81–95.

Finnegan, M.P., H.Z. Zhang, and J.F. Banfield. 2007. Phase stability and transformation in titania nanoparticles in aqueous solutions dominated by surface energy. J. Phys. Chem. C 111:1962–1968.

Finnegan, M.P., H.Z. Zhang, and J.F. Banfield. 2008. Anatase coarsening kinetics under hydrothermal conditions as a function of pH and temperature. Chem. Mater. 20:3443–3449.

Fischer, C., V. Karius, and A. Lüttge. 2009. Correlation between sub-micron surface roughness of iron oxide encrustations and trace element concentrations. Sci. Total Environ. 407:4703–4710.

French, R.A., A.R. Jacobson, B. Kim, S.L. Isley, R.L. Penn, and P.C. Baveye. 2009. Influence of ionic strength, pH, and cation valence on aggregation kinetics of titanium dioxide nanoparticles. Environ. Sci. Technol. 43:1354–1359.

Gilbert, B., and J.F. Banfield. 2005. Molecular-scale processes involving nanoparticulate minerals in biogeochemical systems. Rev. Miner. Geochem. 59:109–155.

Gilbert, B., F. Huang, Z. Lin, C. Goodell, H.Z. Zhang, and J.F. Banfield. 2006b. Surface chemistry controls crystallinity of ZnS nanoparticles. Nano Lett. 6:605–610.

Gilbert, B., F. Huang, H.Z. Zhang, G.A. Waychunas, and J.F. Banfield. 2004. Nanoparticles: Strained and stiff. Science 305:651–654.

Gilbert, B., H. Zhang, B. Chen, M. Kunz, F. Huang, and J.F. Banfield. 2006a. Compressibility of zinc sulfide nanoparticles. Phys. Rev. B 74:7.

Grassian, V.H. 2008. When size really matters: Size-dependent properties and surface chemistry of metal and metal oxide nanoparticles in gas and liquid phase environments. J. Phys. Chem. C 112:18303–18313.

Gravie, R.C. 1978. Stabilization of the tetragonal structure in zirconia microcrystals. J. Phys. Chem. 82:218–224.

Guyodo, Y., A. Mostrom, R.L. Penn, and S.K. Banerjee. 2003. From nanodots to nanorods: Oriented aggregation and magnetic evolution of nanocrystalline goethite. Geophys. Res. Lett. 30:4.

Guzman, K.A.D., M.P. Finnegan, and J.F. Banfield. 2006b. Influence of surface potential on aggregation and transport of titania nanoparticles. Environ. Sci. Technol. 40:7688–7693.

Guzman, K.A.D., M.R. Taylor, and J.F. Banfield. 2006a. Environmental risks of nanotechnology: National nanotechnology initiative funding, 2000–2004. Environ. Sci. Technol. 40:1401–1407.

Handy, R.D., R. Owen, and E. Valsami-Jones. 2008a. The ecotoxicology of nanoparticles and nanomaterials: Current status, knowledge gaps, challenges, and future needs. Ecotoxicology 17:315–325.

Handy, R.D., F. von der Kammer, J.R. Lead, M. Hassellov, R. Owen, and M. Crane. 2008b. The ecotoxicology and chemistry of manufactured nanoparticles. Ecotoxicology 17:287–314.

Hassellov, M., and F. von der Kammer. 2008. Iron oxides as geochemical nanovectors for metal transport in soil-river systems. Elements 4:401–406.

Higarashi, M.M., and W.E. Jardim. 2002. Remediation of pesticide contaminated soil using TiO_2 mediated by solar light. Catal. Today 76:201–207.

Hochella, M.F. 2002. Nanoscience and technology the next revolution in the Earth sciences. Earth Planet. Sci. Lett. 203:593–605.

Hochella, M.F. 2007. How nanoscience has changed our understanding of environmental geochemistry. Geochim. Cosmochim. Acta 71:A408.

Hochella, M.F. 2008. Nanogeoscience: From origins to cutting-edge applications. Elements 4:373–379.

Hochella, M.F., T. Kasama, A. Putnis, C.V. Putnis, and J.N. Moore. 2005. Environmentally important, poorly crystalline Fe/Mn hydrous oxides: Ferrihydrite and a possibly new vernadite-like mineral from the Clark Fork River Superfund Complex. Am. Mineral. 90:718–724.

Hochella, M.F., S.K. Lower, P.A. Maurice, R.L. Penn, N. Sahai, D.L. Sparks, and B.S. Twining. 2008. Nanominerals, mineral nanoparticles, and Earth systems. Science 319:1631–1635.

Hoecke, K.V., J.T.K. Quik, J. Mankiewicz-Boczek, K.A.C.D. Schamphelaere, A. Elsaesser, P.V.d. Meeren, C. Barnes et al. 2009. Fate and effects of CeO_2 nanoparticles in aquatic ecotoxicity tests. Environ. Sci. Technol. 43:4537–4546.

Hofmann, T., and F. von der Kammer. 2009. Estimating the relevance of engineered carbonaceous nanoparticle facilitated transport of hydrophobic organic contaminants in porous media. Environ. Pollut. 157:1117–1126.

Hull, M.S., A.J. Kennedy, J.A. Steevens, A.J. Bednar, J. Weiss, A. Charles, and P.J. Vikesland. 2009. Release of metal impurities from carbon nanomaterials influences aquatic toxicity. Environ. Sci. Technol. 43:4169–4174.

Isley, S.L., and R.L. Penn. 2008. Titanium dioxide nanoparticles: Effect of sol-gel pH on phase composition, particle size, and particle growth mechanism. J. Phys. Chem. C 112:4469–4474.

Jentzsch, T.L., C.L. Chun, R.S. Gabor, and R.L. Penn. 2007. Influence of aluminum substitution on the reactivity of magnetite nanoparticles. J. Phys. Chem. C 111:10247–10253.

Johansen, A., A. Pedersen, K. Jensen, U. Karlson, B. Hansen, J. Scott-Fordsmand, and A. Winding. 2008. Effects of C_{60} fullerene nanoparticles on soil bacteria and protozoans. Environ. Toxicol. Chem. 27:1895–1903.

Johnson, R.L., G.O.B. Johnson, J.T. Nurmi, and P.G. Tratnyek. 2009. Natural organic matter enhanced mobility of nano zerovalent iron. Environ. Sci. Technol. 43:5455–5460.

Joo, S.H., S.R. Al-Abed, and T. Luxton. 2009. Influence of carboxymethyl cellulose for the transport of titanium dioxide nanoparticles in clean silica and mineral-coated sands. Environ. Sci. Technol. 43:4954.

Ju-Nam, Y., and J.R. Lead. 2008. Manufactured nanoparticles: An overview of their chemistry, interactions and potential environmental implications. Sci. Total Environ. 400:396–414.

Kaegi, R., A. Ulrich, B. Sinnet, R. Vonbank, A. Wichser, S. Zuleeg, H. Simmler, S. Brunner, H. Vonmont, M. Burkhardt, and M. Boller. 2008. Synthetic TiO_2 nanoparticle emission from exterior facades into the aquatic environment. Environ. Pollut. 156:233–239.

Kahru, A., H.C. Dubourguier, I. Blinova, A. Ivask, and K. Kasemets. 2008. Biotests and biosensors for ecotoxicology of metal oxide nanoparticles: A minireview. Sensors 8:5153–5170.

Klaine, S.J., P.J.J. Alvarez, G.E. Batley, T.F. Fernandes, R.D. Handy, D.Y. Lyon, S. Mahendra, M.J. McLaughlin, and J.R. Lead. 2008. Nanomaterials in the environment: Behavior, fate, bioavailability, and effects. Environ. Toxicol. Chem. 27:1825–1851.

Koelmans, A.A., B. Nowack, and M.R. Wiesner. 2009. Comparison of manufactured and black carbon nanoparticle concentrations in aquatic sediments. Environ. Pollut. 157:1110–1116.

Lead, J.R., and K.J. Wilkinson. 2006. Aquatic colloids and nanoparticles: Current knowledge and future trends. Environ. Chem. 3:159–171.

Lee, W., Y. An, H. Yoon, and H. Kweon. 2008. Toxicity and bioavailability of copper nanoparticles to the terrestrial plants mung bean (Phaseolus radiatus) and wheat (Triticum aestivum): Plant agar test for water-insoluble nanoparticles. Environ. Toxicol. Chem. 27:1915–1921.

Levard, C., E. Doelsch, J. Rose, A. Masion, I. Basile-Doelsch, O. Proux, J.-L. Hazemann, D. Borschneck, and J.-Y. Bottero. 2009. Role of natural nanoparticles on the speciation of Ni in andosols of la reunion. Geochim. Cosmochim. Acta 73:4750–4760.

Liu, J., D.A. Aruguete, J.R. Jinschek, J.D. Rimstidt, and M.F. Hochella. 2008. The non-oxidative dissolution of galena nanocrystals: Insights into mineral dissolution rates as a function of grain size, shape, and aggregation state. Geochim. Cosmochim. Acta 72:5984–5996.

Lower, S.K., M.F. Hochella, and T.J. Beveridge. 2001. Bacterial recognition of mineral surfaces: Nanoscale interactions between Shewanella and alpha-FeOOH. Science 292:1360–1363.

Luca, V. 2009. Comparison of size-dependent structural and electronic properties of anatase and rutile nanoparticles. J. Phys. Chem. C 113:6367–6380.

Madden, A.S., and M.F. Hochella. 2005. A test of geochemical reactivity as a function of mineral size: Manganese oxidation promoted by hematite nanoparticles. Geochim. Cosmochim. Acta 69:389–398.

Madden, A.S., and M.F. Hochella. 2008. Significance of particle size distributions for size-dependent hematite nanomineral reactivity. Goechim. Cosmochim. Acta 72:A583.

Madden, A.S., M.F. Hochella, and T.P. Luxton. 2006. Insights for size-dependent reactivity of hematite nanomineral surfaces through Cu^{2+} sorption. Geochim. Cosmochim. Acta 70:4095–4104.

Martinson, C.A., and K.J. Reddy. 2009. Adsorption of arsenic(III) and arsenic(V) by cupric oxide nanoparticles. J. Colloid Interface Sci. 336:406–411.

Mashal, K., J.B. Harsh, and M. Flury. 2005. Clay mineralogical transformations over time in Hanford sediments reacted with simulated tank waste. Soil Sci. Soc. Am. J. 69:531–538.

Mattigod, S.V., G.E. Fryxell, K. Alford, T. Gilmore, K. Parker, J. Serne, and M. Engelhard. 2005. Functionalized TiO_2 nanoparticles for use for in situ anion immobilization. Environ. Sci. Technol. 39:7306–7310.

Maurice, P.A. 2012. Nanoscale science and technology in soil science, In P.M. Huang, Y.C. Li, and M.E. Summer (eds.). Handbook of soil science: Resource management and environmental impacts. Chapter 3. CRC Press, Boca Raton, Florida.

Maurice, P.A., and M.F. Hochella. 2008. Nanoscale particles and processes: A new dimension in soil science. Adv. Agron. 100:123–153.

Mendive, C.B., T. Bredow, A. Feldhoff, M. Blesa, and D. Bahnemann. 2008. Adsorption of oxalate on rutile particles in aqueous solutions: A spectroscopic, electron-microscopic and theoretical study. Phys. Chem. Chem. Phys. 10:1960–1974.

Mendive, C.B., T. Bredow, A. Feldhoff, M.A. Blesa, and D. Bahnemann. 2009. Adsorption of oxalate on anatase (100) and rutile (110) surfaces in aqueous systems: Experimental results vs. theoretical predictions. Phys. Chem. Chem. Phys. 11:1794–1808.

Miot, J., K. Benzerara, G. Morin, A. Kappler, S. Bernard, M. Obst, C. Ferard et al. 2009. Iron biomineralization by anaerobic neutrophilic iron-oxidizing bacteria. Geochim. Cosmochim. Acta 73:696–711.

Moore, R.C., J.E. Szecsody, M.J. Truex, K.B. Helean, R. Bontchev, and C.C. Ainsworth. 2007. Formation of nanosize apatite crystals in sediment for containment and stabilization of contaminants, p. 89–110. In G.E. Fryxell and C. Gao (eds.) Environmental applications of nanomaterials. Imperial College Press, London, U.K.

Moreau, J.W., P.K. Weber, M.C. Martin, B. Gilbert, I.D. Hutcheon, and J.F. Banfield. 2007. Extracellular proteins limit the dispersal of biogenic nanoparticles. Science 316:1600–1603.

Mueller, N.C., and B. Nowack. 2008. Exposure modeling of engineered nanoparticles in the environment. Environ. Sci. Technol. 42:4447–4453.

Myers, J.C., and R.L. Penn. 2007. Evolving surface reactivity of cobalt oxyhydroxide nanoparticles. J. Phys. Chem. C 111:10597–10602.

Na, C., and S.T. Martin. 2009. Growth of manganese oxide nanostructures alters the layout of adhesion on a carbonate substrate. Environ. Sci. Technol. 43:4967.

Nabi, D., I. Aslam, and I.A. Qazi. 2009. Evaluation of the adsorption potential of titanium dioxide nanoparticles for arsenic removal. J. Environ. Sci. China 21:402–408.

Nagaveni, K., M.S. Hegde, N. Ravishankar, G.N. Subbanna, and G. Madras. 2004. Synthesis and structure of nanocrystalline TiO_2 with lower band gap showing high photocatalytic activity. Langmuir 20:2900–2907.

Naicker, P.K., P.T. Cummings, H.Z. Zhang, and J.F. Banfield. 2005. Characterization of titanium dioxide nanoparticles using molecular dynamics simulations. J. Phys. Chem. B 109:15243–15249.

Navarro, E., A. Baun, R. Behra, N.B. Hartmann, J. Filser, A.J. Miao, A. Quigg, P.H. Santschi, and L. Sigg. 2008b. Environmental behavior and ecotoxicity of engineered nanoparticles to algae, plants, and fungi. Ecotoxicology 17:372–386.

Navarro, E., F. Piccapietra, B. Wagner, F. Marconi, R. Kaegi, N. Odzak, L. Sigg, and R. Behra. 2008a. Toxicity of silver nanoparticles to *Chlamydomonas reinhardtii*. Environ. Sci. Technol. 42:8959–8964.

Navrotsky, A., L. Mazeina, and J. Majzlan. 2008. Size-driven structural and thermodynamic complexity in iron oxides. Science 319:1635–1638.

Nowack, B. 2009a. Is anything out there? What life cycle perspectives of nano-products can tell us about nanoparticles in the environment. Nano Today 4:11–12.

Nowack, B. 2009b. The behavior and effects of nanoparticles in the environment. Environ. Pollut. 157:1063–1064.

Nowack, B., and T.D. Bucheli. 2007. Occurrence, behavior and effects of nanoparticles in the environment. Environ. Pollut. 150:5–22.

Nurmi, J.T., P.G. Tratnyek, V. Sarathy, D.R. Baer, J.E. Amonette, K. Pecher, C.M. Wang, J.C. Linehan, D.W. Matson, R.L. Penn, and M.D. Driessen. 2005. Characterization and properties of metallic iron nanoparticles: Spectroscopy, electrochemistry, and kinetics. Environ. Sci. Technol. 39:1221–1230.

Penn, R.L., and J.F. Banfield. 1998a. Oriented attachment and growth, twinning, polytypism, and formation of metastable phases: Insights from nanocrystalline TiO_2. Am. Mineral. 83:1077–1082.

Penn, R.L., and J.F. Banfield. 1998b. Imperfect oriented attachment: Dislocation generation in defect-free nanocrystals. Science 281:969–971.

Penn, R.L., and J.F. Banfield. 1999. Morphology development and crystal growth in nanocrystalline aggregates under hydrothermal conditions: Insights from titania. Geochim. Cosmochim. Acta 63:1549–1557.

Penn, R.L., J.J. Erbs, and D.M. Gulliver. 2006. Controlled growth of alpha-FeOOH nanorods by exploiting-oriented aggregation. J. Crystal Growth 293:1–4.

Penn, R.L., K. Tanaka, and J. Erbs. 2007. Size dependent kinetics of oriented aggregation. J. Crystal Growth 309:97–102.

Penn, R.L., C. Zhu, H. Xu, and D.R. Veblen. 2001. Iron oxide coatings on sand grains from the Atlantic coastal plain: High-resolution transmission electron microscopy characterization. Geology 29:843–846.

Pettibone, J.M., D.M. Cwiertny, M. Scherer, and V.H. Grassian. 2008. Adsorption of organic acids on TiO_2 nanoparticles: Effects of pH, nanoparticle size, and nanoparticle aggregation. Langmuir 24:6659–6667.

Posner, J.D. 2009. Engineered nanomaterials: Where they go, nobody knows. Nano Today 4:114–115.

Qafoku, N.P. 2010. Terrestrial nanoparticles and their controls on soil/geo processes and reactions. Adv. Agron. 107:33–91.

Qafoku, N.P., C.C. Ainsworth, and S.M. Heald. 2007a. Cr(VI) fate in mineralogically altered sediments by hyperalkaline waste fluids. Soil Sci. 172:598–613.

Qafoku, N.P., C.C. Ainsworth, J.E. Szecsody, D.L. Bish, J.S. Young, D.E. McCready, and O.S. Qafoku. 2003c. Aluminum effect on dissolution and precipitation under hyperalkaline conditions: II. Solid phase transformations. J. Environ. Qual. 32:2364–2372.

Qafoku, N.P., C.C. Ainsworth, J.E. Szecsody, and O.S. Qafoku. 2003a. Aluminum effect on dissolution and precipitation under hyperalkaline conditions: I. Liquid phase transformations. J. Environ. Qual. 32:2354–2363.

Qafoku, N.P., C.C. Ainsworth, J.E. Szecsody, and O.S. Qafoku. 2003b. Effect of coupled dissolution and redox reactions on $Cr(VI)_{aq}$ attenuation during transport in the Hanford sediments under hyperalkaline conditions. Environ. Sci. Technol. 37:3640–3646.

Qafoku, N.P., C.C. Ainsworth, J.E. Szecsody, and O.S. Qafoku. 2004a. Transport-controlled kinetics of dissolution and precipitation in the Hanford sediments under hyperalkaline conditions. Geochim. Cosmochim. Acta 68:2981–2995.

Qafoku, N.P., R. Kukkadapu, J.P. McKinley, B.W. Arey, S.D. Kelly, C.T. Resch, and P.E. Long. 2009. Uranium in framboidal pyrite from a naturally bioreduced alluvial sediment. Environ. Sci. Technol. 43:8528–8534.

Qafoku, N.P., O. Qafoku, C.C. Ainsworth, A. Dohnalkova, and S.G. McKinley. 2007b. Fe-solid phase transformations under highly basic conditions. Appl. Geochem. 22:2054–2064.

Qafoku, N.P., and M.E. Sumner. 2002. Adsorption and desorption of indifferent ions in variable charge subsoils: The possible effect of particle interactions on the counter-ion charge density. Soil Sci. Soc. Am. J. 66:1231–1239.

Qafoku, N.P., M.E. Sumner, and L.T. West. 2000. Mineralogy and chemistry of some variable charge subsoils. Commun. Soil Sci. Plant Anal. 31:1051–1070.

Qafoku, N.P., E. Van Ranst, A. Noble, and G. Baert. 2004b. Variable charge soils: Their mineralogy, chemistry and management. Adv. Agron. 84:159–215.

Qi, Y., T. Cagin, W.L. Johnson, and W.A. Goddard. 2001. Melting and crystallization in Ni nanoclusters: The mesoscale regime. J. Chem. Phys. 115:385–394.

Ridley, M.K., V.A. Hackley, and M.L. Machesky. 2006. Characterization and surface-reactivity of nanocrystalline anatase in aqueous solutions. Langmuir 22:10972–10982.

Sarathy, V., P.G. Tratnyek, J.T. Nurmi, D.R. Baer, J.E. Amonette, C.L. Chun, R.L. Penn, and E.J. Reardon. 2008. Aging of iron nanoparticles in aqueous solution: Effects on structure and reactivity. J. Phys. Chem. C 112:2286–2293.

Schierz, A., and H. Zänker. 2009. Aqueous suspensions of carbon nanotubes: Surface oxidation, colloidal stability and uranium sorption. Environ. Pollut. 157:1088–1094.

Schofield, E.J., H. Veeramani, J.O. Sharp, E. Suvorova, R. Bernier-Latmani, A. Mehta, J. Stahlman et al. 2008. Structure of biogenic uraninite produced by *Shewanella oneidensis* strain MR-1. Environ. Sci. Technol. 42:7898–7904.

Seeger, E., A. Baun, M. Kastner, and S. Trapp. 2009. Insignificant acute toxicity of TiO_2 nanoparticles to willow trees. J. Soils Sediments 9:46–53.

Singer, D.M., F. Farges, and G.E. Brown, Jr. 2009. Biogenic nanoparticulate UO_2: Synthesis, characterization, and factors affecting surface reactivity. Geochim. Cosmochim. Acta 73:3593–3611.

Spagnoli, D., J.F. Banfield, and S.C. Parker. 2008. Free energy change of aggregation of nanoparticles. J. Phys. Chem. C 112:14731–14736.

Sumner, M.E. 1963a. Effect of alcohol washing and pH value of leaching solution on positive and negative charges in ferruginous soils. Nature 198:1018–1019.

Sumner, M.E. 1963b. Effect of iron oxides on positive and negative charges in clays and soils. Clay Min. Bull. 5:218–226.

Sumner, M.E. 1995. Soil chemistry: Past, present, and future, p. 1–38. *In* P.M. Huang (ed.) Future prospects for soil chemistry. SSSA, Madison, WI.

Sumner, M.E., and J.C. Davidtz. 1965. Positive and negative charges in some Natal soils. S. Afr. J. Agric. Sci. 8:1045–1050.

Sumner, M.E., and N.G. Reeve. 1966. The effect of iron oxides impurities on the positive and negative adsorption of chloride by kaolinite. J. Soil Sci. 17:274–278.

Sun, L., C.Z. Zhang, L.G. Chen, J. Liu, H.Y. Jin, H.Y. Xu, and L. Ding. 2009b. Preparation of alumina-coated magnetite nanoparticle for extraction of trimethoprim from environmental water samples based on mixed hemimicelles solid-phase extraction. Anal. Chim. Acta 638:162–168.

Sun, H., X. Zhang, Z. Zhang, Y. Chen, and J.C. Crittenden. 2009a. Influence of titanium dioxide nanoparticles on speciation and bioavailability of arsenite. Environ. Pollut. 157:1165–1170.

Suzuki, Y., S.D. Kelly, K.A. Kemner, and J.F. Banfield. 2003. Microbial populations stimulated for hexavalent uranium reduction in uranium mine sediment. Appl. Environ. Microbiol. 69:1337–1346.

Tan, H., O.K. Ezekoye, J. Van Der Schalie, M.W. Horn, A. Lakhtakia, J. Xu, and W.D. Burgos. 2006. Biological reduction of nanoengineered iron(III) oxide sculptured thin films. Environ. Sci. Technol. 40:5490–5495.

Theng, B.K.G., and G.D. Yuan. 2008. Nanoparticles in the soil environment. Elements 4:395–399.

Tretiakov, K.V., K.J.M. Bishop, B. Kowalczyk, A. Jaiswal, M.A. Poggi, and B.A. Grzybowski. 2009. Mechanism of the cooperative adsorption of oppositely charged nanoparticles. J. Phys. Chem. A 113:3799–3803.

Vayssieres, L. 2009. On the effect of nanoparticle size on water-oxide interfacial. Chem. J. Phys. Chem. C 113:4733–4736.

Wang, H.H., R.L. Wick, and B.S. Xing. 2009. Toxicity of nanoparticulate and bulk ZnO, Al_2O_3 and TiO_2 to the nematode *Caenorhabditis elegans*. Environ. Pollut. 157:1171–1177.

Waychunas, G.A., C.S. Kim, and J.F. Banfield. 2005. Nanoparticulate iron oxide minerals in soils and sediments: Unique properties and contaminant scavenging mechanisms. J. Nanopart. Res. 7:409–433.

Waychunas, G.A., and H.Z. Zhang. 2008. Structure, chemistry, and properties of mineral nanoparticles. Elements 4:381–387.

Wiesner, M.R., G.V. Lowry, P. Alvarez, D. Dionysiou, and P. Biswas. 2006. Assessing the risks of manufactured nanomaterials. Environ. Sci. Technol. 40:4336–4345.

Zeng, H., A. Singh, S. Basak, K.U. Ulrich, M. Sahu, P. Biswas, J.G. Catalano, and D.E. Giammar. 2009. Nanoscale size effects on uranium(VI) adsorption to hematite. Environ. Sci. Technol. 43:1373–1378.

Zhang, H.Z., and J.F. Banfield. 2005. Size dependence of the kinetic rate constant for phase transformation in TiO_2 nanoparticles. Chem. Mater. 17:3421–3425.

Zhang, H.Z., and J.F. Banfield. 2007. Polymorphic transformations and particle coarsening in nanocrystalline titania ceramic powders and membranes. J. Phys. Chem. C 111:6621–6629.

Zhang, H.J., and G.H. Chen. 2009. Potent antibacterial activities of Ag/TiO_2 nanocomposite powders synthesized by a one-pot sol-gel method. Environ. Sci. Technol. 43:2905–2910.

Zhang, H.Z., B. Chen, and J.F. Banfield. 2009. The size dependence of the surface free energy of titania nanocrystals. Phys. Chem. Chem. Phys. 11:2553–2558.

Zhang, H.Z., R.L. Penn, R.J. Hamers, and J.F. Banfield. 1999. Enhanced adsorption of molecules on surfaces of nanocrystalline particles. J. Phys. Chem. B 103:4656–4662.

Zhang, H.Z., J.R. Rustad, and J.F. Banfield. 2007. Interaction between water molecules and zinc sulfide nanoparticles studied by temperature-programmed desorption and molecular dynamics simulations. J. Phys. Chem. A 111:5008–5014.

Zhao, H.T., Y.J. Deng, J.B. Harsh, M. Flury, and J.S. Boyle. 2004. Alteration of kaolinite to cancrinite and sodalite by simulated Hanford tank waste and its impact on cesium retention. Clay. Clay Miner. 52:1–13.

5

Enzymatic Activity as Influenced by Soil Mineral and Humic Colloids and Its Impact on Biogeochemical Processes

L. Gianfreda
Università degli Studi di Napoli Federico II

M.A. Rao
Università degli Studi di Napoli Federico II

M. Mora
Universidad de La Frontera

5.1	Introduction	5-1
5.2	Soil Enzymes as Naturally Immobilized Enzymes	5-1
5.3	Natural and Synthetic Soil Enzymatic Complexes	5-2
	Enzymatic Activities Extracted from Soil • Formation of Synthetic Enzymatic Complexes • Common and Expected Features of Natural and Synthetic Soil Enzyme Complexes	
5.4	Ecological Role of Enzyme–Soil Component Interactions: Carbon Turnover and Storage and Transformation of Nutrients	5-13
5.5	Interaction of Soil-Bound Enzymes with Anthropogenic Organics and Their Potential for Ecosystem Restoration	5-16
5.6	Conclusions and Future Prospects	5-18
References		5-19

5.1 Introduction

Soil enzyme activities are the driving force behind all biochemical transformations occurring in soil. Soil is a complex environment where several inorganic and organic components are simultaneously present and exert their action. These components may affect the synthesis, persistence, stabilization, regulation, and catalytic behavior of enzymatic proteins, present at any moment in the soil environment. In soil, enzyme activities outside the cells exhibit altered catalytic, kinetic, and stability properties. For these peculiar features and properties, they can be regarded as "naturally immobilized enzymes." Several studies have demonstrated that clay minerals, organic matter (OM), and organo–mineral complexes are involved in the interaction with enzyme activities and the immobilization of enzymes in soil. Two methodological approaches have been used for understanding the relationships between immobilized soil enzymes and their clay, humic, or humic–clay supports (Gianfreda and Bollag, 1996). The first approach is based on the isolation, purification, and characterization of active enzymatic fractions from soils. In the second approach, synthetic model systems, in which enzymes are artificially immobilized by their attachment to soil components, have been used for studying the properties of soil-bound enzymes.

This chapter gives a general picture of the interactions possibly occurring between soil inorganic and organic components and soil enzymes. The effects of the soil components on the performance and efficiency of enzyme activities, the mechanisms involved in their interactions, the role of soil-bound enzymes from an ecological point of view, and their potential for practical purposes are addressed.

5.2 Soil Enzymes as Naturally Immobilized Enzymes

Soil is an extremely heterogeneous microenvironment where physical, chemical, and biological processes occur simultaneously among numerous inorganic and organic components and living organisms. In this variety of components, enzymatic proteins, and their activity, exert a predominant role because they (a) catalyze all the biochemical transformations occurring in soil; (b) guarantee the exchange of materials among the biotic and abiotic portions of the soil ecosystem; (c) allow the growth, survival, and activity of soil living organisms. All enzymatic processes occur through coordinated and integrated pathways to ensure the correct energy transfer and nutrient cycling and to preserve soil and environmental quality and crop productivity (Dick, 1994; Tabatabai, 1994).

Although soil is expected to contain a large array of enzymes involved in the cycles of the main nutrients (carbon, nitrogen, phosphorus, and sulfur), only enzymes belonging to the four classes EC 1, EC 2, EC 3, and EC 4, cited in Chapter 3, have been detected and assayed in soil. Moreover, each type of soil presents its own characteristic pattern of specific enzymes and an inherent level of enzymatic activity.

The measurement of enzyme activity in soil is still a bottleneck in soil enzymology. Although many papers, book chapters, and entire books have been dedicated to collect and describe methods for assaying enzyme activities in soil, standardized, universally accepted methodologies for determining the activity of a given soil enzyme are still not available. In addition, the variety of experimental conditions adopted by different authors (i.e., different incubation pH and temperature, assay duration, type and concentration of substrate, treatment of the soil before the assay, etc.) often does not allow enzymatic activity values to be compared. As claimed by Gianfreda and Ruggiero (2006), "the conclusion of all the findings so far reported is that unfortunately methodologies, capable of measuring accurately and unambiguously each of soil enzyme components, are not available to date."

As widely accepted, different intracellular and extracellular enzymatic components contribute quantitatively and qualitatively to the overall enzymatic activity of soil (Burns, 1978, 1982; Nannipieri, 1994; Nannipieri et al., 1990, 2002; Gianfreda and Bollag, 1996; Gianfreda and Ruggiero, 2006). The different components can be from bacterial, fungal, plant, or animal origin. A given enzyme can be produced by all these sources in different quantitative and qualitative amounts. Moreover, the various intracellular and extracellular enzymatic active components may have different locations not only in their producing organism but also in the whole soil (Burns, 1978, 1982). They may exist in different states, present different features and properties, and act under a range of different microenvironmental conditions that affect their catalytic behavior (Gianfreda and Bollag, 1996).

Among enzymatic categories, defined and classified differently by several authors (Skujins, 1976; Kiss et al., 1975; Burns, 1982; Gianfreda and Bollag, 1996; Ruggiero et al., 1996), enzymes associated (e.g., adsorbed, entrapped, polymerized, and/or co-polymerized) with inorganic, organic (including soluble or insoluble substrate), and organo–mineral soil components may contribute a large portion of the overall activity of many enzymes. They are considered soil-bound or "naturally immobilized enzymes" and are recognized as mainly responsible for the enzymatic activity shown by a soil in the absence of living cells (Gianfreda and Bollag, 1996).

As reported in earlier and more recent reviews (Ladd and Butler, 1975; McLaren, 1975; Ladd, 1985; Burns, 1986; Kiss et al., 1986; Boyd and Mortland, 1990; Gianfreda and Bollag, 1996; Naidja et al., 2000; Gianfreda et al., 2002; Ahn et al., 2006; Nielsen et al., 2006; Pietramellara et al., 2009), clays, clay minerals, OM, and organo–mineral complexes have an important role in the immobilization of enzymes in soil. This conclusion was mainly derived from results obtained in model experiments with synthetic clay-, humic-, or clay–humic-enzyme complexes. As discussed recently by Nielsen et al. (2006) and thoroughly studied for the adsorption of extracellular DNA by soil colloids (see review by Pietramellara et al., 2009), proteins can also be released after cell lysis and react with surface reactive particles with other cell constituents.

5.3 Natural and Synthetic Soil Enzymatic Complexes

Two different methodological strategies have been used to investigate soil enzymes, to elucidate their peculiar properties, and to understand the mechanisms involved in the interaction between enzymes in soil and their clay or humic supports. The first approach dealt with the isolation, purification, and characterization of active enzymatic fractions from soils. Various experimental procedures and several biochemical techniques were used for the extraction, separation, and purification of these enzymes. Notwithstanding, very few purified enzymatic proteins were extracted from soils (Tabatabai and Fu, 1992; Nannipieri et al., 1996; Nannipieri, 2006). Indeed, many of the examples reported in literature refer to enzyme-like substances showing enzymatic properties rather than to protein molecules with enzymatic features. Most of these enzyme-like substances contain organic material, very often of carbohydrate nature (Mayaudon, 1986).

Moreover, humus–enzyme complexes are very often extracted from soil by using the experimental procedures usually used in the extraction and purification of humic substances (Nannipieri et al., 1996, Nannipieri, 2006), thus supporting the existence of humus–enzyme associations in soil. No such experimental evidence is available for similar complexes with clay minerals. If short-range ordered mineral colloids are extracted, they are dissolved in the extracting solution. If the enzymes strongly bound to mineral surfaces, they are not extractable.

In the second approach, synthetic model systems prepared under laboratory conditions were used. These consist of placing enzymatic molecules and soil constituents (i.e., pure or "dirty" clays, humic or humic-like substances, organo–mineral complexes) in contact with each other. In this case, adsorption should be the driving mechanism. Although several adsorption studies have been carried out on clay minerals, fewer investigations have been devoted to the adsorption of enzymes to humic substances, probably because of the experimentally established inhibitory effect of humic substances.

The formation of active enzymatic complexes has also been obtained by mixing enzymes and humic precursors and varying the environmental conditions (i.e., nature of the organic constituent [humic acid phenolic component], contact time, ratio between the two components, and presence or absence of an additional biotic or abiotic catalyst).

Particular care must be taken in the preparation of synthetic enzymatic complexes. Indeed, the choice of an enzyme from a source and in a given formulation over the same enzyme from a different source and/or a different formulation may lead to such contrasting results that completely different conclusions may be drawn on the behavior of the enzyme in response to an applied factor. Therefore, although model studies may be of considerable importance to understanding the real soil, care must be taken in the extrapolation of results obtained with synthetic model systems to natural conditions.

5.3.1 Enzymatic Activities Extracted from Soil

Until recently, studies on the extraction of enzymes from soil were important for elucidating the nature, properties, and complexity of enzyme activities in soil and to understanding their role in soil functionality. Nowadays, such studies achieve a further and more important significance as being essential for soil proteomics, which involves protein extraction from soil (Nannipieri, 2006). Soil proteomics, an aspect of the larger field of environmental proteomics, combined with soil genomics can shed light not only on the presence of some genes (and in turn of some peculiar living organisms) in soil but also on protein synthesis and specific soil functions (Ogunseitan, 1993, 2006 and references therein; Nannipieri 2006, and references therein). The aim of soil structural proteomics is the study of proteins and peptides found in water, sediments, and soils. It is devoted to understand the location and stabilization of extracellular proteins in soil. Analysis of proteins extracted from environmental samples may help to characterize the response of microbial communities to stressful conditions such as contamination with toxic chemicals, starvation, heat, or oxygen levels (Ogunseitan, 2006 and references therein). Study of proteins can also be used as a "fingerprint" to type the diversity in the sample and as an index for monitoring the progress of a biocatalytic reaction "in situ" (Ogunseitan, 1993).

Although experimental procedures for extraction of DNA and mRNA from soil samples are quite well standardized and reliable and reproducible data are usually available (Pietramellara et al., 2009), methods affordable for the extraction of proteins and enzymes from soil still have several deficiencies. Although several improvements have been attempted (Ogunseitan, 2006; Benndorf et al., 2007; Solaiman et al., 2007; Masciandaro et al., 2008), these methods have not been standardized. Indeed, several experimental problems are usually encountered, and most of them are still not solved. This is probably a consequence of the existence of proteins and enzymes in soil as immobilized molecules on soil colloids.

Nannipieri (2006) has exhaustively examined the findings about extraction of enzymes from soil. This author showed that studies were mainly devoted to extracellular hydrolases, selected for their importance in nutrient cycling, for their presumed activity in the extracellular environment, and because their activity can be easily determined in soil. One of the most important difficulties encountered in this approach was the choice of the extracting solution. An efficient solvent system should provide high yields of enzymes, without denaturing effects on protein structure, removal of nonenzymatic soil OM, or undesired lysis of soil organisms. Moreover, no artifacts should appear during soil extraction. Indeed, it is possible that the extracted enzymes undergo adsorption or entrapment within the solubilized complexes or the extracting compound (like citrate) may interact with inorganic soil components (such as manganese) and the Mn–citrate complex produce a catalytic activity resembling an enzyme (laccase) (Leonowicz and Bollag, 1987).

Salt solutions such as pyrophosphate, phosphate, acetate, citrate, tris-(hydroxymethyl)-aminomethane (TRIS or THAM), tris-borate, borate, etc. have been used to extract enzymes from soil (Tabatabai and Fu, 1992; Nannipieri et al., 1996; Bonmati et al., 1998; Nannipieri, 2006). The most commonly used extracting solutions were sodium pyrophosphate and phosphate because of their capacity to extract OM under mild conditions. Enzymatic preparations showing urease, β-glucosidase, catalase, protease, and phosphatase activities were extracted from different soil types with pyrophosphate (Nannipieri, 2006 and references therein). Most of these enzymatic preparations were humic–enzyme complexes. Further studies were performed to characterize the enzymatic preparations and different enzymatic fractions showing different catalytic activities, and stability properties were obtained (Nannipieri et al., 1996; Nannipieri, 2006).

In previous studies, Mayaudon (1986), using a different experimental procedure based on phosphate-EDTA as the extractant, concluded that most of the free (i.e., not bound to soil colloids) extracellular enzymes in soil are glycoproteins, as they were strongly associated with carbohydrate residues. Their presence usually confers to the enzyme or protein an increased stability to denaturing agents, although difficulty may arise in the extraction and separation procedures.

The use of 67 mM phosphate buffer at pH 6.0 was very efficient in the extraction of a free cellulase from greenhouse soils (Murase et al., 2003). The efficiency of this extracting solution was evaluated against 0.1 M sodium pyrophosphate (pH 7.0) and 0.5 M potassium sulfate (pH 6.6), in recovering the protein quantity and the β-glucosidase enzyme activity from two natural forest soils (Masciandaro et al., 2008). The pyrophosphate extract showed the highest β-glucosidase activity, thus confirming its capability of extracting humic-bound β-glucosidase enzyme in a stable and active form.

An attempt to increase extraction yield for measuring intracellular soil arylsulfatase was made by Fornasier (2002). Bovine serum albumin (BSA) was added to Tris–HCl buffer used as the extracting solution, and a measurable increase (11%) of extracted enzyme was obtained. An improvement of the method was the simultaneous use of BSA, Triton X-100, and bead beating (Fornasier and Margon, 2007). Acid and alkaline phosphomonoesterase and arylsulfatase were extracted at high yield, with negligible amounts of humic substances, from six soils differing in their properties (Fornasier and Margon, 2007).

5.3.2 Formation of Synthetic Enzymatic Complexes

Most of the information on the behavior, performance, and features of soil enzymes has been gathered by studies with synthetic enzyme systems simulating those possibly occurring in soil.

Many findings are available in the literature on the preparation, characterization, and properties of different enzymes adsorbed, entrapped, and/or complexed with different natural or synthetic soil constituents by means of several mechanisms. Even in this case, standardized methods are, however, not available for the preparation and characterization of immobilized enzymes resembling those naturally occurring in soil. A handbook entitled *Methods of Soil Enzymology* edited by R. Dick for the Soil Science Society of America is in preparation, and a chapter (coauthored by Gianfreda and Rao) is specifically dedicated to the description of standardized methods for the preparation and characterization of synthetic soil enzymatic complexes.

Three main mechanisms may be envisaged in the preparation of synthetic enzymatic model complexes simulating those present in real soils:

- Adsorption on soil components including mineral soil components, organic matter, or organo–mineral complexes.
- Covalent attachment to soil components, as the results of cross-links between reactive groups on the support and external amino or carboxyl groups of the enzyme.
- Entrapment in polymeric organic materials by different types of processes: the enzyme may be embedded into soil OM either as the result of the polymerization of humic or humic-like precursors through the action of biotic or abiotic catalysts and involving enzymatic molecules or by the entrapment of protein molecules in an organic network made by several small or large macromolecules.

Cation-exchange adsorption mechanisms (Harter and Stotzky, 1971), van der Waals type forces (Hamzehi and Pflug, 1981), and ionic or hydrophobic bonds have been suggested as possible mechanisms whereby enzymes are held to clay surfaces (Boyd and Morland, 1990). Ion exchanges, entrapment within organic networks, ionic bonding, or covalent attachment are the main processes accounting for the stable association between enzymes and humic materials (Ladd and Butler, 1975). Electrostatic forces and van der Waals interactions are enthalpic, whereas hydrophobic interactions are mainly entropic (Quiquampoix and Burns, 2007). As claimed by Quiquampoix and Burns (2007) "the affinity of proteins for various types of interfaces originates in the flexibility of the polypeptide chain and in the diversity of the 20 amino acids that can be classified on an electrical scale as positively, neutrally or negatively charged and on a hydrophobic scale from polar to non-polar."

The most abundant findings have probably been obtained by physical adsorption of different enzymes on supports chosen among soil components. The procedure for the preparation of synthetic enzymatic model complexes involving physical adsorption is a very simple process consisting of incubating, very often under stirring, the enzyme molecule and the support and in separating the solid phase and the supernatant after a given incubation time. The amount of adsorbed enzyme can be simply assessed by measuring the amount in the supernatant and washings and by subtracting it from that initially put in contact with the adsorbing support. Adsorbed enzymes can, however, be easily removed from their adsorbing support by changing pH, temperature, ionic strength, substrate concentration, or buffer used in the activity assays.

Natural clays such as montmorillonite, kaolinite, goethite, illite, palygorskite, oxides, "dirty" clays, humic acids, organo–mineral complexes, or even soil fractions and whole soils have been used as supports. Enzymatic proteins from different origins and as different commercial preparations have been employed (see reviews by Harter and Stotzky, 1971; Burns, 1986; Kiss et al., 1986, Boyd and Mortland, 1990; Huang et al., 1995, 1999; Gianfreda and Bollag, 1996; Gianfreda et al., 2002).

5.3.2.1 Clay–Enzyme Complexes

The capacity of proteins, including enzymes, to be sorbed on phyllosilicates is influenced by the surface area of clay minerals (kaolinite, illite, montmorillonite) and their ability to intercalate protein molecules, the charge characteristics of both minerals and proteins, and is influenced by the size, shape, and isoelectric point (IEP i.e., the pH at which the protein carries no net charge) of protein molecules, the cation exchange capacity (CEC) of minerals, the nature of saturating cation of the clays, and the pH and temperature of the system (Mc Laren et al., 1958; Harter and Stotzky, 1971; McLaren, 1978; Theng, 1979; Violante et al., 1995).

The contact and the type of interactions occurring between soil surfaces and protein molecules will dramatically reflect in the conformation and stability of the protein structure and in turn in its functionality. If the protein is an enzyme, the interactions may have severe consequences on its activity and catalytic role. According to Norde (2008), two classes of structural stability for proteins, soft and hard, have been identified. Soft proteins undergo modifications of the structure, whereas hard proteins retain the same structure in the adsorbed state. In the first case, a change in the catalytic activity of the adsorbed enzyme is expected, whereas for hard proteins the observed changes in their catalytic behavior will be the result of other effects such as, for instance, a different orientation of the active site, possibly not accessible to substrate molecules. Both phenomena may be influenced by the pH (Quiquampoix et al., 2002).

When considering the impact of surface area on the adsorption of proteins on clay minerals, not only the surface area of the clay but also the size of the protein molecule and its capability to interstratify into the phyllosilicate have to be taken into account. It may be expected that proteins characterized by smaller size and molecular masses should be more easily stratified than greater protein molecules. However, experimental evidence not always supported a significant correlation between the surface area of the clay mineral and the amount of the enzymes adsorbed (Gianfreda et al., 1991, 1992). For instance, it was not possible to ascribe the greatest amounts of invertase and urease adsorbed onto Na-saturated montmorillonite (Gianfreda et al., 1991, 1992) to its larger specific area (700–800 m^2 g^{-1} when fully dispersed) and capability to adsorb enzyme molecules on both external and internal surfaces (Harter and Stotzky, 1971; Theng, 1979; Fusi et al., 1989), since no interstratification of both enzymes in the interlamellar spaces of montmorillonite was revealed by x-ray diffraction analysis (Gianfreda et al., 1991, 1992). By contrast, a detectable intercalation was observed with acid phosphatase adsorbed on the same clay (Rao et al., 1996) or tyrosinase on Ca-montmorillonite (Naidja et al., 1995). The gradual shift to higher d-spacing values of the complex observed with increasing amounts of tyrosinase indicates that enzyme molecules were interstratified into the interlamellar species of montmorillonite (Naidja et al., 1995) (Figure 5.1a). This may suggest that with some enzymes the surface area may have only a relative influence on their adsorption and that adsorption on the clay minerals is governed by a higher or lower affinity of a given enzyme for the surfaces of a given clay mineral. When the clay interlayers are already partially or totally preoccupied by polymeric species or noncrystalline oxides of aluminum and iron (Figure 5.1b), steric hindrance to the intercalation of enzyme molecules may occur, and a superficial allocation of protein molecules derives (Rao et al., 1996; Naidja et al., 1997; Safari Sinegani et al., 2005). As demonstrated with montmorillonite coated with varying levels of hydroxyaluminum [Al(OH)x-montmorillonite complexes], tyrosinase, and acid phosphatase, the higher the amount of OH-Al species covering the clay, the greater the amount of the adsorbed enzymes (Figure 5.1b). Ligand exchange or secondary interactions such as Van der Waals and hydrophobic or hydrogen bonds were hypothesized for the immobilization of tyrosinase or acid phosphatase on the Al(OH)x-montmorillonite complexes, respectively (Rao et al., 1996; Naidja et al., 1997).

Other important factors to be considered are the electrostatic forces possibly occurring between the charges present on the clay surfaces and the protein. The surface charge of clay can be pH dependent or pH independent, such as that of Al hydroxides and oxyhydroxides, or montmorillonite, respectively. The charge of the protein is strictly influenced by its IEP. Below the IEP, a protein will carry a net positive charge, and above it a net negative charge. Several experimental findings have demonstrated that the adsorption of proteins is often greatest when the pH of the system is approximately equal to the IEP of the protein (Harter and Stotzky, 1971; Theng, 1979), thus ruling out a strong importance of electrostatic forces in the interaction between the protein and the clay.

The results obtained in the adsorption of several enzymes, which have similar or different IEP by clays characterized by different charge properties (different point of zero charge), seem to confirm that non-Columbic forces (hydrophobic forces, hydrogen bonding, van der Waal's forces) may take part in the process, and their sum may be greater than the forces of electrostatic repulsions and/or attractions between the protein molecules and the charged clay minerals (Gianfreda et al., 1991, 1992; Violante et al., 1995; Rao et al., 1996; Quiquampoix, 2000 and references therein). In particular, hydrophobic interactions may occur between the protein molecules and the clay surface. Indeed, the surface of the clay may achieve a hydrophobic character if positively charged amino acid side chains of the protein exchange with the hydrophilic counterions on the clay surface (Staunton and Quiquampoix, 1994). If hydrophobic amino acids come in contact with the hydrophobic siloxane layer, the protein structure may easily reorganize itself on the clay. This could lead to an irreversible modification of the protein conformation driven by thermodynamic factors involved in the process. Dramatic changes in the efficiency of the involved enzyme may arise.

Clay–enzyme complexes have also been obtained by covalent attachment. In this case, chemical binding occurs between

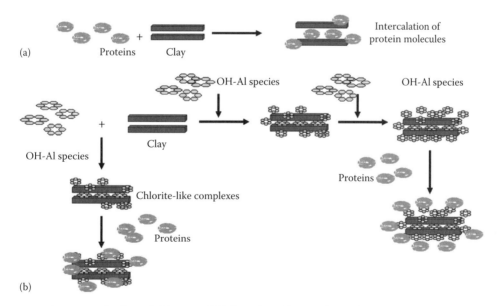

FIGURE 5.1 Interaction of protein molecules with (a) pure and (b) dirty clays. (a) Intercalation of molecules into clay layers occurs with pure clays. (b) Allocation of protein molecules on the external surfaces is favored on dirty clays. The process is much more evident by increasing the amount of OH-Al species on the clay surface. (From Naidja, A., P.M. Huang, and J.-M. Bollag. 2000. Enzyme-clay interactions and their impact on transformation of natural and anthropogenic organic compounds in soil. J. Environ. Qual. 29:678–691; Violante, A., A. De Cristofaro, M.A. Rao, and L. Gianfreda. 1995. Physicochemical properties of protein-smectite and protein-Al(OH)x-smectite complexes. Clay Miner. 30:325–336. With permission.)

enzymes and supports. The enzyme may be directly linked to the support or a spacer or arm is put between the carrier and the enzyme. In the first case, the risk that one or more groups present in the enzymatic active site are involved in the binding is elevated, with evident, negative repercussions on the catalytic activity of the enzyme. Indeed, the active site can be completely masked and therefore not accessible to the substrate. The presence of a spacer, instead, strongly reduces this risk, and the protein can behave as a free enzyme.

In the contest of clay–enzyme complexes, the most used immobilization method has been activation of the support by silanization and grafting through glutaraldehyde (Ruggiero et al., 1989; Sarkar et al., 1989; Gianfreda and Bollag, 1994; Sanjay and Sugunan, 2005; Sanjay and Sugunan, 2007). In particular, the supports were provided by a long arm loading a final chemical group (aldehydic) capable of reacting chemically with reactive amino chain residues of the enzyme (Dick and Tabatabai, 1999). The immobilization process implied a preventive surface modification or an activation step of the immobilizing support. Silanization, that is, the coating of the surface with organic functional groups using an organo-functional silane reagent such as 3-aminopropyltriethoxysilane, was the applied strategy for initial surface modification of inorganic supports, followed by further derivatization to aldehyde groups using glutaraldehyde (Martinek and Mozhaev, 1985).

5.3.2.2 Humic and Humic-Like Enzyme Complexes

Fewer studies have been devoted to the adsorption of enzymes to humic substances probably because of their physicochemical and biochemical complexity. Much evidence is instead available on the entrapment of enzymes in humic fractions or in polymeric aggregates. In both cases, most of the information has been achieved by in vitro studies, although experimental evidence was also provided on humus–enzyme associations extracted from soil (see Section 5.3.1).

Ruggiero and Radogna (1988) studied the interaction of tyrosinase with natural humic acid, extracted from the A1 horizon of an Andosol. Depending on the humic acid concentration, from 32% to 76% of initial tyrosinase was bound to humic acid and retained its catalytic activity with substrate specificity similar to that of the free enzyme (Ruggiero and Radogna, 1988). Infrared spectroscopy suggested that carboxyl and phenolic groups of the humic acid were involved in ionic and hydrogen bond formations with tyrosinase molecules.

To simulate humic substances, a purified humic acid with high anionic properties was put in contact with lysozyme at different pHs (Norde et al., 2008). Lysozyme is characterized by a high IEP (10.5) and thus it is positively charged over a wide pH range, that is, pH < 10.5. The formation of complexes was observed, and it was mainly driven by electrostatic attraction between the oppositely charged components. Dynamic light scattering experiments provided information on the size of the complexes and showed that complexes aggregated to larger structures were found when the IEP was approached. The presence of favorable electrostatic interactions was also supported by calorimetric studies devoted to evaluate enthalpy (≈energy) of complexation (Norde et al., 2008).

One process possibly occurring in soil is the involvement of enzymatic molecules during the formation of humic substances. In the environment, phenolic compounds, possibly present in soil as biomolecules, wastes, pesticide derivatives, or degradation products, may very often act as humic precursors. In the presence of either Fe_2O_3 or MnO_2, soil and clays, behaving as abiotic catalysts, or phenoloxidases and peroxidases (PODs), acting as biotic catalysts (produced by plants, bacteria, and fungi), the phenolic compounds may undergo oxidative polymerization processes (Bollag et al., 1995; Huang, 1995) (Figure 5.2a). Production of polymeric products can proceed by self-coupling or cross-coupling even with other molecules. Extended polymers of different sizes, shapes, and complexity are generally formed. Both processes are affected by pH, temperature, catalyst amount, contact time, and simultaneous presence of different catalysts (Figure 5.2a). The reaction mechanisms, which involve the production of highly reactive free radicals, are different whether catalyzed by oxidative enzymes or by abiotic catalysts (Gianfreda et al., 2006) and strongly depend on the type of phenolic substrate, as the complexity of phenolic compounds may negatively affect the catalyst and decrease its catalytic efficiency. In addition to polyphenols, other molecules present in soil, but extraneous to the oxidative process, may be involved

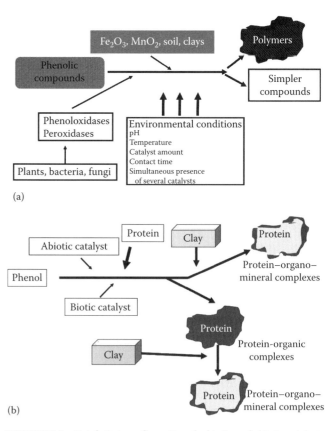

FIGURE 5.2 Catalytic transformations by biotic and abiotic catalysts of (a) phenols and phenolic compounds in natural environments and (b) formation of protein–organo and organo–mineral complexes.

in and/or affect the whole process to a variable extent. In particular, if active proteins are involved, protein–organic or protein–organo–mineral complexes with different structural and functional properties may form (Figure 5.2b).

When the enzyme is involved in the formation of humic substances, several concomitant phenomena may occur and different bonding mechanisms may be envisaged, all affecting the properties of the involved enzyme. As demonstrated by Rao et al. (2007) with a prion protein involved in synthetic humic-like mineral complexes, protein interactions with soluble phenolic polymers, likely involved in the development of humic substances, during or after their formation may occur: (a) through multidentate binding of phenolic polymers to the protein, a process that can become irreversible with time and form insoluble phases (Ladd and Butler, 1975); (b) protein–phenol polymers association through coulombic and/or hydrophobic interactions; (c) protein–phenolic polymeric complexes may transform from flexible soluble structures into insoluble, reticulated structures and microporous micelles in the presence of positive charges (Gianfreda et al., 2002); and (d) quinoid compounds may covalently interact with the protein (Bittner, 2006; Suderman et al., 2006). Multidentate binding is strictly influenced by the structural features of both phenolic polymers and protein and is assisted by conformational flexibility. For instance with protein showing domains with high concentrations of charged residues, like the investigated prion protein (Rao et al., 2007), favorable coulombic interactions with phenol polymers were established. The net result of the above processes is the formation of soluble and insoluble organo–enzymatic complexes that have different catalytic and enzymatic features.

5.3.3 Common and Expected Features of Natural and Synthetic Soil Enzyme Complexes

The main results obtained in the studies with natural and synthetic enzymatic complexes confirm and support what was concluded in two basic studies performed in the late 1980s by Kiss et al. (1986) and Müller-Wegener (1988) and are summarized as follows.

5.3.3.1 Inhibition of the Activity of Enzymes by Both Clays and Humic Substances

In an early but exhaustive review, Kiss et al. (1986) examined the papers published (more than 120 publications) at that time on the interactions between enzymes and clays in soil. The approach followed in the majority of those papers was to investigate the interaction between clays and enzymes under laboratory conditions and to measure the properties of enzymes.

Other studies have since been published in which the same approach or classical inhibition tests are used. For instance, Gianfreda and coworkers (Gianfreda et al., 1991, 1992, 1995a, 1995c; Gianfreda and Bollag, 1994; Rao et al., 2000) investigated the inhibitory effects of pure and "dirty" clays and also of whole soils on the activity of some enzymes (invertase, urease, phosphatase, laccase, etc.). The kinetics of the enzymes were measured in the presence of increasing amounts of various solid phases, and the kinetic parameters K_m and V_{max} and their variations with the presence of clays were evaluated.

In inhibition studies performed with acid phosphatase from various origins (corn roots, wheat germ, and potato) and montmorillonite, the clay exhibited a strong inhibition effect on the activity of the enzyme (Dick and Tabatabai, 1987; Huang et al., 1995; Rao et al., 2000), and the highest the concentration of montmorillonite in the assay mixture, the greatest was the decrease of the enzyme activity. It is interesting to highlight that the presence of Al(OH)x species on the surface of montmorillonite depressed or partially removed the inhibition of the activity of potato phosphatase by montmorillonite. Indeed, when montmorillonite, which was covered by increasing amounts (3, 9, and 18 meq of Al per gram of clay) of OH-Al species, was used as the support, the inhibitory effect on phosphatase activity decreased from the highest value (>70%) measured in the presence of montmorillonite to about a half-value (35%) when the clay was covered by a greater amount of OH-Al species (Rao et al., 1996). All the minerals significantly reduced the catalytic efficiency, as expressed by the V_{max}/K_m ratio of acid phosphatase, and the higher the amount of the clay, the lower was the catalytic efficiency of the enzyme. An inhibition of more than 80% was observed by Huang et al. (1995) when the activity of wheat-germ phosphatase was tested in the presence of 25–100 mg of montmorillonite. The inhibition effect decreased significantly (from 80% in the case of maximum inhibition to about 35%) when Al(OH)x species were held on montmorillonite surfaces. An opposite effect by OH-Al species coated on montmorillonite surfaces was observed with other enzymes (invertase, Gianfreda et al., 1991; urease, Gianfreda et al., 1992; tyrosinase, Naidja et al., 1995).

It is well known that several humic substances such as humic and fulvic acids, tannins, and melanins, their precursors such as phenols and quinones, and related substances may have inhibitory effects on enzyme activity. The pioneers of this research topic were Ladd and Butler (1969a, 1969b, 1970, 1971, 1975) and Butler and Ladd (1969). In their studies on proteolytic enzymes pronase, carboxypeptidase, and trypsin, they showed that tannins, melanins, and humic and fulvic acids had different inhibitory effects on the activity of the investigated enzymes. In order to identify the humic acid groups mainly responsible for the inhibition, studies were carried out with either modified enzymes (after acetylation) or modified humic acids (after methylation) and under different environmental conditions (presence and absence of inorganic cations). The results obtained indicated that carboxyl groups of humic acids were probably the main groups involved in the inhibition mechanism.

Inhibition by humic substances is not limited to proteolytic enzymes. Oxidoreductases such as PODs (Pflug, 1980; Sarkar and Burns, 1984), malate dehydrogenase (Pflug and Ziechmann, 1981), indoleacetic acid oxidase (Mato and Mendez, 1970), tyrosinase and laccase (Sarkar and Bollag, 1987; Filazzola et al., 1999), as well as hydrolases such as phosphatase, invertase, and urease were also affected by the presence of humic or humic-like compounds and humic precursors. For instance, Gianfreda et al. (1993a, 1995a, 1995c) and Rao et al. (1998) demonstrated different effects of tannic acid (a starting humic material) on the activity of invertase, urease, and phosphatase. Invertase, with the high content (>50%) of carbohydrate moieties in the molecule, displayed a great resistance to tannic acid inhibitory effect (Rao et al., 1998), probably because it offered a higher resistance than phosphatase and urease to complexation phenomena by the phenolic compound.

The effects of other organic compounds also including precursors of humic substances (such as catechol, methylcatechol, pyrogallol, resorcinol, and gallic acid) on the activity of a fungal laccase were examined in detail by Filazzola et al. (1999). Classical inhibition tests were carried out by testing laccase activity with two substrates, 2,2-azinobis-(3-ethylbenzthiazoline-6-sulfonate) (ABTS) and syringaldazine, characterized by their structural resemblance and their similar reaction mechanisms with enzyme. All phenols inhibited the oxidation of both the substrates. Humic constituents inhibited laccase activity in the order: pyrogallol > gallic acid > resorcinol, and pyrogallol showed the highest inhibition (>93%) at all concentrations. The number of –OH groups and other substituents and the concentration of phenolic compounds influenced the extent of inhibition of the enzymatic activity.

Activation effects by clays (bentonite and kaolinite) on the activity of soil catalase, α-amylase, and cellulase (Novakova, 1988) as well as stimulation effects such as those shown by low concentrations of humic acids on laccase-mediated transformation of xenobiotic compounds were also detected (Huang et al., 2005 and references therein).

Attempts to derive meaningful correlations between soil enzymatic activities and clay or humus content of soil have been addressed in several papers (Gianfreda and Bollag, 1994 and references therein, Gianfreda and Ruggiero, 2006 and references therein). The data did not, however, indicate any unequivocal evidence on the influence of the presence of clays or humic substances on the formation, existence, and stability of active clay– or humus–enzyme associations on a quantitative basis.

The main conclusion drawn by most of the performed studies is that the observed inhibition is a consequence of enzyme adsorption on clay surfaces and that catalytic and stability properties are different from those of the same enzymes in a purified form (Kiss et al., 1986).

As to humic substances and their humic precursors, their direct action on the active sites of enzymes as well as a modification of the active site by structural conformational changes may be involved in the phenomenon (Müller-Wegener, 1988). As a consequence, an altered, even pronounced, change in performance of the involved enzyme may result. Having high biochemical heterogeneity and complexity, humic substances may behave as possible substrates in the enzymatic reaction and, in turn, interfere with its equilibrium. Moreover, their cation exchange properties may favor their interaction with cations necessary to the enzyme catalysis (cofactors), directly involved in the stabilization of the active enzymatic structure or in the catalytic mechanism. If the formation of polymeric aggregates occurs by the action of catalysts or in the presence of additional ions, different consequences on the entrapped enzyme and its catalytic behavior can result.

5.3.3.2 Activity and Kinetic Parameters as Measured for Enzymes Adsorbed on Clays or Linked to Humic Substances as Compared with Free Enzymes

Immobilization often significantly alters the behavior of catalysts because of the changes in the properties of enzymes upon attachment to a solid support. Lower enzymatic activity and reduced substrate affinity have been generally measured with consequent altered values of the kinetic parameters V_{max} and K_m values (for definition see Chapter 3), thus indicating a lower catalytic efficiency and a reduced substrate affinity.

For instance, adsorption of acid phosphatase on an inorganic, montmorillonite (M), or organic, tannic acid (T), support considerably reduced the activity of the enzyme (by 80% and 67%, respectively) (Table 5.1). The coating of montmorillonite surfaces with increasing amounts of OH-Al species had a beneficial effect on the activity of the adsorbed phosphatase. Phosphatase activity increased to more than 50% on an Al(OH)x–montmorillonite complex containing 18 meq Al g^{-1} clay (AM_{18}) (Table 5.1) (Rao et al., 1996). A similar result was not observed with invertase and urease as the greater the amount of Al species on M surfaces, the lower the activity of urease and invertase (Gianfreda et al., 1991, 1992). The beneficial effect of OH-Al species held on montmorillonite surfaces was also observed when investigating the activity of both acid phosphatase from wheat germ (Huang et al., 1995) and tyrosinase (Naidja et al., 1997).

As shown in Figure 5.1b the allocation of protein molecules on the external surfaces may be favored on "dirty" clays. Indeed, OH-Al species occupying the interlamellar sites of the clay impede the access to protein molecules forced to stay on external surfaces (Rao et al., 2000) or favor the orientation of some protein molecules toward positions that may not affect their catalytic activity (Naidja et al., 1997).

Different results were observed when allophanic clay was used as the support (Rosas et al., 2008). In this case, a consistent increase (>40%) in phosphatase activity was observed, but it was slightly lower when manganese and molybdenum ions were present in the reaction mixture (the presence of the two ions should simulate natural conditions of allophanic soils). Moreover, the response was different when the clay was put in contact with the enzyme before or after its interaction with the ions (Rosas et al., 2008).

Activity levels higher or similar to that of the free enzymes were exhibited by a laccase from *Trametes villosa* when adsorbed on short-range ordered aluminum hydroxide (AL) (Ahn et al., 2007) and by glucoamylase adsorbed on montmorillonite (Sanjay and Sunugan, 2005), respectively. Similarly, higher activities than those of the free enzymes were measured for α-amylase, β-amylase, glucoamylase, acid phosphatase, and β-glucosidase adsorbed on composite supports prepared by equal weights of chitosan and activated clay and cross-linked to each other with glutaraldehyde

TABLE 5.1 Catalytic (V_{max} and K_m) and Stability Properties of Acid Phosphatase Immobilized on Different Supports, by Different Mechanisms, and under Different Environmental Conditions

Support	V_{max} (%)[a]	K_m (mM)	Residual Activity (%) after Exposure at 60°C for 2 h	Residual Activity (%) after Exposure to K Protease for 24 h
Free enzyme (P)	100	0.6 (at 10°C)	21	0
		0.5 (at 30°C)		
Inorganic complexes				
M-P[b]	20	0.4		0
AL-P	44	0.7		4
AM_3-P	21	n.d.	n.d.	n.d.
AM_9-P	32	0.8		1
AM_{18}-P	58	n.d.	n.d.	n.d.
Organic and organo–mineral complexes				
T-P	33	0.6	5	16
AT-P	48	0.3	n.d.	44
ATM-P	55	0.1	n.d.	48
In the presence of soluble species				
T-P + Fe^{3+} (1.0 mM)	33 (41)[c]	1.0 (1.0)[c]	24	29
T-P + Mn^{2+} (1.0 mM)	36 (46)	0.8 (0.9)	3	3
In the presence of insoluble species				
T-P + MnO_2 (10 mg)	33 (29)	0.6 (0.4)	10	35
T-P + Fe_2O_3 (10 mg)	58 (55)	0.2 (0.3)	6	7

M, montmorillonite; AL, aluminum hydrous oxide; AM_3, AM_9, AM_{18}, Al(OH)x–montmorillonite (chlorite) complexes containing 3, 9, or 18 meq of Al g^{-1} clay, respectively; T, tannic acid; AT, OH-Al-tannic acid; ATM, OH-Al-tannic acid-montmorillonite.

n.d., not determined.

[a] Expressed as percentage of the activity of the free enzyme.

[b] Preparation and activity tests of complexes with M, AL, AM_3, AM_9, and AM_{18} were carried out at 10°C. All other preparations and activity tests were performed at 30°C.

[c] Data in parentheses refer to complexes prepared in the presence of also montmorillonite.

(Chang and Juang, 2004, 2005, 2007). A preformed network of Ca-polygalacturonate was used for immobilizing acid phosphatase and urease from both bacterial and plant sources (Ciurli et al., 1996, Marzadori et al., 1998a). The support simulated the mucigel present at the root–soil interface with a similar composition and morphology. The adsorption process was carried out in different organic buffers commonly found in root exudates. The highest enzyme activity for free and adsorbed phosphatase was obtained with Na-maleate buffer, whereas a decrease in activity was observed upon adsorption in the presence of the other investigated buffers. A decrease in activity was also observed with the two ureases although it was lower than that measured for the phosphatase.

The phosphatase activity increased when the enzyme reacted with tannic acid in the presence of OH-Al species (AT-P in Table 5.1), OH-Al species and montmorillonite (ATM-P in Table 5.1), or in the presence of soluble and insoluble species of manganese and iron (Table 5.1). Although the presence of Fe or Mn ions did not significantly change the activities of the complexes compared with tannic acid–phosphatase complex, a higher activity level (increase by 50%) and no effect were measured with Fe_2O_3 and MnO_2, respectively.

The complex T-P formed in the presence of Fe_2O_3 and MnO_2 of Table 5.1 are two examples of humic-like enzyme complexes obtained by contact of phenolic compounds with enzymatic proteins and subsequent polymerization in the presence of biotic or abiotic catalysts (Figure 5.2b). Both the oxides may catalyze the abiotic polymerization of phenolic compounds and Mn oxide is a stronger polymerization catalyst than Fe oxide (Huang, 1990) (Figure 5.2a and b). Evidently, the oxides accelerated the polymerization of tannic acid, but small-size phosphatase–tannic acid–Fe_2O_3 complexes probably formed and phosphatase molecules were adsorbed on the external surfaces of the tannic acid polymers rather than being entrapped within them as probably is the case with Mn oxide. Exposed adsorbed enzymes would give higher activities. The formation of tannic acid–phosphatase polymers and their subsequent adsorption on the oxide surfaces can be also hypothesized (Rao and Gianfreda, 2000). Similar explanations were provided to account for the higher activity shown by urease immobilized on the same phenolic compound and in the presence of the same oxides (Gianfreda et al., 1995c).

On the contrary, the immobilization of acid phosphatase from wheat germ on organo–mineral complexes obtained by intercalation of montmorillonite with representatives of three major classes of biological molecules L-lysine (amino acid), α-D-glucose (carbohydrate), and rhamnolipid (lipid) reduced the activity of the enzyme significantly (Kelleher et al., 2004).

Other precursors of humic substances in soil such as catechol, pyrogallol, resorcinol, etc., have been used for the preparation of organic networks entrapping enzyme molecules (Sarkar and Burns, 1984; Grego et al., 1990; Garzillo et al., 1996; Rao et al., 1999). The phenolic compounds were subjected to polymerization by the addition of the catalyst in the presence of the investigated enzyme, and the properties of the resulting humic-like enzymatic complexes were studied. The catalyst mostly used has been POD. In this case, complexes of only organic nature are obtained. When the catalyst is an oxide, instead of POD, organo–mineral enzymatic complexes are formed.

Experiments performed with phosphatase, when involved in the polymerization of pyrogallol or tannic acid through the action of POD or MnO_2 (Rao et al., 1999), showed that phosphatase interacted much more with tannic acid than with pyrogallol. In the absence of catalysts, pyrogallol was not able to form polymeric aggregates with the enzyme, whereas some polymeric aggregates probably formed with tannic acid. Instead, when the incubation was performed in the presence of POD, the polymerization of pyrogallol was strongly promoted as assessed by the red brown color and the very high E_4/E_6 ratio (equal to 7) of the suspensions, which gives information on the size and molecular mass of humic substances. By contrast, a more efficient effect was demonstrated by the abiotic catalyst on tannic acid. In both the cases, larger enzyme–phenolic copolymers formed, as demonstrated by the browning as well as the decrease in E_4/E_6 ratios of the mixtures (Rao et al., 1999). Stable acid phosphatase–resorcinol complexes showing high levels of activity (~90% of the initial enzymatic activity) were also obtained when the phenolic compound interacted with the enzyme in the presence of POD (Garzillo et al., 1996).

High enzymatic activities were observed when the mechanism used in the immobilization process was the chemical binding between enzymes and supports (Gianfreda and Bollag, 1994; Sanjay and Sunugan, 2005, 2006). The kinetic parameters of three enzymes, acid phosphatase, laccase, and POD immobilized by chemical binding to three supports (a montmorillonite, a kaolinite, and a soil with a high silt content and a moderate amount of OM), after their silanization and grafting through glutaraldehyde, are reported in Table 5.2. Two of the three enzymes, laccase and POD, retained relatively high residual activity levels, and laccase immobilized on montmorillonite showed a residual activity higher than that of the free enzyme (Table 5.2). Only phosphatase showed lower activity on all the supports (Gianfreda and Bollag, 1994). Decreased activities were also measured for α-amylase, glucoamylase, and invertase covalently bound onto montmorillonite using glutaraldehyde as spacer (Sanjay and Sugunan, 2006, 2007).

Different effects were observed on the K_m values. Generally, an increase in the kinetic parameter occurred for most of the enzymes immobilized on the various supports. For instance, acid phosphatase adsorbed on AM_9 and AM_{18} (Table 5.1) showed higher K_m values than the free enzyme, indicating that the affinity between substrate and enzyme was reduced after immobilization. Accordingly, increased K_m values compared with those of the free enzymes were displayed by invertase adsorbed on pure and "dirty" clays (Gianfreda et al., 1991); acid phosphatase and urease adsorbed on the Ca-polygalacturonate network or hydroyapatite (Ciurli et al., 1996; Marzadori et al., 1998a, 1998b); alkaline phosphatase adsorbed on Na-sepiolite (Carrasco et al., 1995); urease immobilized by covalent cross-linking through glutaraldehyde on montmorillonite, kaolinite, and two different soils (Lai and Tabatabai, 1992); and α-amylase, β-amylase, glucoamylase, and acid phosphatase adsorbed on composite chitosan-clay supports (Chang and Juang, 2004, 2005).

Lower K_m values, that is, an increased affinity for the substrate, or values approximating those of the free enzymes were instead observed for acid phosphatase immobilized on some organic and organo–mineral complexes (Table 5.1; Kelleher et al., 2004), urease adsorbed on pure and "dirty" clays (Gianfreda et al., 1992), laccase adsorbed on AL (Ahn et al., 2007), and other enzymes adsorbed on clays and organo–mineral complexes of different nature (see reviews by Burns, 1986; Boyd and Mortland, 1990; Gianfreda et al., 2002).

Different explanations are provided to account for the catalytic and kinetic properties of immobilized enzymes.

An *immobilized enzyme* is a protein physically localized in a certain region of space (Martinek and Mozhaev, 1985). After immobilization, enzymes are converted from a water-soluble mobile state to a water-insoluble immobile state (Klibanov, 1983) and become "static enzymes." As a consequence, catalysis by immobilized enzymes acquires a "heterogeneous nature." In fact, immobilized enzymes represent an individual phase separate from the outer solution. The major components of the immobilized enzyme system, that is, the enzyme, the immobilizing matrix and its properties, and how the enzyme is bound to the matrix, are mainly responsible of the kinetic behavior of the immobilized enzyme. Additional variables, such as pH,

TABLE 5.2 Chemical Binding of Laccase, Peroxidase, and Acid Phosphatase on Two Clays and on a Silt Loam Soil

Enzyme and Support	V_{max} (%)[a]	K_m (mM)	Residual Activity after 4 Months of Storage at 4°C
Laccase	100	0.08	70
Montmorillonite	118	0.07	100
Kaolinite	85	0.08	100
Soil	90	0.90	100
Peroxidase	100	0.22	80
Montmorillonite	105	0.25	100
Kaolinite	80	0.27	100
Soil	95	0.32	100
Acid phosphatase	100	0.52	50
Montmorillonite	31	2.00	100
Kaolinte	35	3.91	100
Soil	30	1.83	100

Source: Adapted from Gianfreda, L., and J.-M. Bollag. 1994. Effect of soils on the behavior of immobilized enzymes. Soil Sci. Soc. Am. J. 56:1672–1681. With permission of Soil Science Society of America.

[a] Expressed as percentage of the activity of the free enzyme.

temperature, ionic strength, pressure, shaking, need for cofactors, substrate delivery, and product removal, may contribute to the environment and thus to the performance of the enzyme.

If the enzyme is placed close to the support surface during the immobilization process, significant surface contributions may influence the performance of the bound enzyme. Indeed, the surface of the support, whether inorganic or organic in nature, may influence the enzyme activity by microenvironmental effects, such as the microenvironmental pH and/or buffering capacity, hydrophobicity and hydrophilicity of the surface, and redox properties.

Moreover, possible conformational changes may occur within the tertiary structure of the protein after immobilization and reflect in a reduced activity of enzymes attached to, or entrapped within, an insoluble support. Denaturation of the enzymatic molecule may arise as a consequence of the high number of linkages between the enzyme and the support and/or the possible interfering interactions between hydrophilic sites and hydrophobic groups of the support and the enzymatic protein, respectively. Stronger negative effects may occur if the immobilization involves groups present in the active site of the enzyme or alter the interactions between the various subunits of the enzymatic molecule, whose integrity is a prerequisite for the catalytic enzymatic function.

The substrate may have restricted mobility and not easily penetrate the active site of the enzyme, possibly because of partial or complete inaccessibility to the substrate as a result of the orientation of the enzyme in relation to the support surface. These effects will be more evident when high-molecular mass substrates such as, for instance, polymeric substrates (e.g., proteins, polysaccharides and nucleic acids) or proteases (which are high-molecular mass proteins) are involved. Immobilized hydrolases, usually, show a marked lower enzymatic activity as well as increased resistance to the attack of proteolytic proteins.

Similar K_m values measured for phosphatase free and immobilized on montmorillonite intercalated with L-lysine, α-D-glucose, and rhamnolipid indicated that the cumulative effect of electrostatic, steric, and conformational changes due to the presence of clay and organic compounds did not affect the access of the enzyme to substrate (Kelleher et al., 2004).

5.3.3.3 Different Activity–pH and Activity–Temperature Profiles Exhibited by Enzymes Adsorbed on Clays or Linked to Humic Substances

Immobilized enzymes usually display different activity–pH profiles, and their optimum pH may move to more alkaline or acidic pH depending on the negative or positive charge of the support, respectively.

For instance, completely different activity–pH profiles were exhibited by phosphatase–clay complexes shown in Table 5.1 (Rao et al., 2000). The activity of the complexes with M, AL, and Al(OH)x–montmorillonite complexes (AM) decreased directly with increasing pH, with maximum activity at pH 4.0 against an optimum pH for the free enzyme in the range 4.5–5.5. However, the decrease in activity was much more acute for phosphatase adsorbed on AM than on AL and M alone, in the order listed. By contrast, the pH of maximum activity as well as the sensitivity to the pH change of the enzyme complexed with Al(OH)x–tannic acid (AT) and Al(OH)x–tannic acid–montmorillonite (ATM) complexes (Table 5.1) were similar to those of the free enzyme. A flatter curve with a shift toward alkaline pH was instead observed for the enzyme complexed with tannic acid.

Activity–pH profiles similar or not substantially different from those of the free enzymes with no differences in pH-optimum values were also observed (Gianfreda et al., 1991, 1992, 1995c; Carrasco et al., 1995; Garzillo et al., 1996; Marzadori et al., 1998a; Chan and Hang, 2004; Ahn et al., 2007; Chan and Hang, 2007). For example, when phosphatase molecules were placed in contact with tannic acid in the presence of Fe ions or OH-Al species, both positively charged, the shift of pH optimum toward the alkaline region observed for the complex with only tannic acid was eliminated, and activity–pH profiles similar to that of the free enzyme were obtained (Rao et al., 2002, unpublished results).

The response to pH of enzymes immobilized on different supports is usually accounted for by microenvironmental effects due to the location of the enzyme in a microenvironment that in the proximity of the support is different from that of the bulk solution. This hypothesis, originating from the double-layer theory, suggests the existence of an interfacial pH effect and presumes that the pH of the thin layer close to the solid surface is different from that of the bulk solution. In other words, the enzyme works at a pH different from that measured. Enzymes immobilized on negative supports usually display a shift in their optimum pH toward the alkaline regions. An opposite shift, that is, versus acidic pH, will probably occur when the enzyme is immobilized in a positively charged support (McLaren and Packer, 1970, Nannipieri and Gianfreda, 1998). In both cases, charged substrates, products, effectors, and/or hydrogen or hydroxyl ions will be differently distributed between the domain of the immobilized enzyme and the outer solution. Electrostatic or hydrophobic interactions between the support and the different components participating in the reaction may occur and reflect in a consequent modification of the microenvironment surrounding the enzyme. Perturbations of the catalytic pathway of the reaction and, consequently, modification of the intrinsic catalytic properties of the bound enzyme may result, and a modification in the catalytic constant or turnover number k_{cat} (see for definition Chapter 3) could occur. This effect can be more pronounced if the kinetic mechanisms involve charged residues of either the substrate or the enzyme, whose charge as well as whose interactions can be enhanced or decreased by the presence of the charged matrix.

A different explanation was provided by Quiquampoix and coauthors in their studies with different proteins and enzymes. Indeed, they demonstrated that a combination of electrostatic and hydrophobic forces determined the pH shift (Leprince and Quiquampoix, 1996; Quiquampoix et al., 1993; Staunton and

Quiquampoix, 1994). The authors provided experimental evidence of pH-dependent modifications of protein conformation, due to attractive electrostatic interactions between positively charged proteins and negatively charged clay surfaces, leading to the unfolding of the enzyme and, in turn, to a decrease in the catalytic activity (Quiquampoix 1987, 2000). Quiquampoix (2008) has recently suggested that the hypothesis of microenvironmental effects has at least three serious drawbacks: (1) "The shift in the optimal pH of activity should give rise to a higher rate of catalytic activity for the adsorbed enzyme than for the enzyme in solution in the alkaline range of pH." For the author, the absence of this experimental evidence is due to "a misleading presentation of the results," which is usually based on "the normalization of enzyme activity values to the maximum value attained for each case, free and bound." (2) "The hypothesis implies that the conformation of the adsorbed enzyme is similar to the conformation of enzyme in solution and can act as a "molecular pH-meter." Experimental evidence to the contrary was provided by the author and coworkers (Leprince and Quiquampoix, 1996; Quiquampoix et al., 1993; Staunton and Quiquampoix, 1994). (3) "The basis of this theory" is not certain. "The tendency of a proton to react with the active site of the enzyme is not given by the proton activity but by its molar free enthalpy…And … the molar free enthalpy of the proton is the same in the bulk of the solution and at the clay surface."

The microenvironmental theory, however, still has its validity, and several authors continue to apply this theory to explain the changes observed in the activity–pH profiles of enzymes after their immobilization on solid supports.

The influence of the support on the catalytic behavior of enzymes is also evident by the activity–temperature curves (Nannipieri and Gianfreda, 1998 and references therein). For instance whereas the activity of free phosphatase increased by increasing the temperature from 0°C up to 60°C and reached its highest value at 50°C–55°C, a less dependence on temperature was, in general, exhibited by phosphatase immobilized on the inorganic supports shown in Table 5.1 (Rao et al., 2000). The values of the thermodynamic parameters E_a and ΔH_a (activation energy and enthalpy of activation) confirmed the existence of only one rate-limiting step at different temperatures, and their increased values, measured for some of the studied complexes, indicated that different reaction mechanisms were probably followed by immobilized phosphatase. Similar results were obtained with other enzymes immobilized on various supports (Carrasco et al., 1995; Chang and Juang, 2004, 2005).

Thermodynamic studies performed on wheat germ phosphatase, when immobilized on montmorillonite intercalated with L-lysine, α-D-glucose and rhamnolipid, suggested that catalytic activity on the lysine and rhamnolipid intercalated surfaces was more spontaneous and favorable than that of pure clay (Kelleher et al., 2004) and was preserved at temperatures as low as 10°C.

The changes in the activity–temperature profiles usually shown by immobilized enzymes are indicative of a different sensitivity (and often of an increased stability) of the immobilized enzyme to temperature exposure (see below).

5.3.3.4 A Stabilization Effect of the Enzyme Molecules against Several Denaturing Agents (Biological Degradation, Proteolysis, Thermal, and Chemical Deactivation) by Immobilization on Clays or Humic Substances

The most remarkable feature of enzymes after their immobilization on solid supports is an achieved resistance and, in turn, an increased stability to various denaturing agents.

The results shown in Table 5.1 indicate the stability of acid phosphatase complexes to two denaturing agents such as temperature and proteolytic enzymes. Increase or decrease in stability to one or both the agents was monitored. In several cases, complex deactivation mechanisms were demonstrated by both free and immobilized enzymes when exposed to temperature (Gianfreda et al., 1991; Gianfreda et al., 1995c; Rao et al., 2000; Rosas et al., 2008). When the residual activity was plotted as log activity vs. deactivation time, a double-slope pattern was observed. Such behavior is usually described by a two-step series mechanism involving the formation of an intermediate less active form from the native enzyme, evolving to a final inactive conformation of the protein. According to fitted mathematical models, it is possible to fully describe and quantify the phenomenon (Gianfreda et al., 1985).

In comparison with the free enzyme, the complexes of phosphatase with AT and ATM showed a gain in activity of about 40% and 48% after thermal deactivation at 60°C and of about 44% and 48% after 24 h exposure to a protease (K protease), respectively (Table 5.1). A similar beneficial effect on the stability of phosphatase–tannate complex was observed when Fe ions or MnO_2 were involved in the complex (Table 5.1). The two complexes with the highest values of activity after 2 h exposure to 60°C showed also the lowest sensitivity to the action of the proteolytic agent (Table 5.1). Higher stability to pH, temperature, and proteolysis was shown by acid phosphatase–polyresorcinol complexes, thus supporting the hypothesis that secondary linkages not affecting the conformation of the enzyme but assuring that protection against denaturing agents occurred (Garzillo et al., 1996). Less sensitivity to temperature was also observed with other enzymes such as tyrosinase (Naidja et al., 1997) or POD (Shen and Tu, 1999).

Experimental evidence of the protective mechanism probably built up by the clay and humus colloids was earlier provided by Lähdesmäki and Piispapen (1992) who studied three exoenzymes (protease, cellulase, and amylase) in untreated humus samples, their alkaline extracts and the protein fractions obtained by gel filtration, subjected to several denaturing factors. Results suggested that the fractionation and the dilution damaged or destabilized the protective mechanism created by the soil colloid (clay and humus, present in the unfractionated samples), which preserve soil enzyme activities against certain environmental and other physicochemical changes.

Reusability (i.e., operational stability) of immobilized enzymes has been thoroughly investigated. For example, acid phosphatase or β-glucosidase immobilized on composite chitosan-activated clay bead maintained high levels (even 90%) of its original activity after 50 times of reuse (Chang and Juang, 2004, 2007). Similar results were also obtained with α-amylase, β-amylase, and glucoamylase (Chang and Juang, 2005).

An increased reusability was often shown by enzymes covalently linked to their support. Continuous reusability for a long period and a greater operational stability were shown by invertase covalently bound or adsorbed on acid activated montmorillonite (Sanjay and Sugunan, 2006). X-ray diffractograms (XRD, nuclear magnetic resonance (NMR), and N_2 adsorption measurements and activity assays performed in a fixed bed reactor gave details on the location of the enzyme onto the clay, on the mechanism and the enzyme groups involved in the immobilization process, and on the catalytic and stability properties of the immobilized enzyme. Similar findings were obtained with α-amylase, glucoamylase, and invertase (Sanjay and Sugunan, 2005, 2007) immobilized onto the same montmorillonite via adsorption or covalent binding (Sanjay and Sugunan, 2005, 2007).

Laccase and POD covalently immobilized on clays and soil showed higher residual activity after 4 month storage at 4°C (Table 5.2) and further peculiar properties, such as reusability in systems simulating real environments, that could be used to solve pollution problems in contaminated soils (Gianfreda and Bollag, 2004) (see below).

An increased stability of enzymatic proteins is the result of an intact three-dimensional structure or conformation. The spatial structure of enzymes has a little plasticity so as to be modified during catalytic reaction as a result of several factors. Dramatic changes in protein conformation may occur and reflect in more or less pronounced reduction in enzyme activity.

Several mechanisms have been envisaged in the stabilization of enzymes upon their immobilization on solid supports (Gianfreda and Scarfì, 1991 and references therein): covalent (mono or multipoint) bindings to an insoluble support may contribute to an increased rigidity and a consequent decreased unfolding of the protein structure; "cage effects" inside a polymeric gel; changes in the enzyme microenvironment; steric hindrances leading to a reduced molecular mobility and capacity of unfolding, and an increased inaccessibility to microbes or proteolytic enzymes.

As also underlined in Chapter 3, immobilization of enzymes on clay minerals and/or humic substances is responsible for the increased stability often shown by soil extracellular enzymes (Burns, 1982; Nannipieri et al., 1994). Whereas the stability of a native enzyme is established by its intrinsic structural features, that of an immobilized enzyme would be affected by several factors dealing with both the properties and nature of the immobilizing support, the type and number of bonds or interactions occurring between the support and the protein molecule, the availability for the protein molecule to preserve some free conformational changes, the environment surrounding the molecule, and the conditions involved in the immobilization process. For instance, the complex formed after interaction of tyrosinase with natural humic acid (Ruggiero and Radogna, 1988) showed a lower stability to both temperature and proteolytic attack probably because the physical state of the organic matrix changed with increasing temperature and a subsequent hindrance of the enzymatic active site occurred.

The increased stability to temperature exposure often shown by immobilized enzymes could also reflect in different activity-temperature profiles. At high temperatures, the immobilized enzyme will show higher activity than that of the corresponding free enzyme, thus confirming that minimization or prevention of conformational changes in the protein's tertiary structure induced by increased temperatures probably occurred.

The stability to protease attack is consistent with the characteristics of an enzyme entrapped in an organic or organo–mineral complex. Proteases are high molecular proteins, and their attack can be prevented by the shield formed by the organic matrix around the enzyme. The shield may hinder the movement of proteases toward the enzymes without hindering the diffusion of substrate to and of product from the active site.

5.4 Ecological Role of Enzyme–Soil Component Interactions: Carbon Turnover and Storage and Transformation of Nutrients

Soil extracellular enzymes have a fundamental role in the cycling of compounds; in the decomposition of OM with production of inorganic forms of carbon, nitrogen, phosphorus, and sulfur; and in the gain of energy from complex substrates. As depicted in Figure 5.3a, microorganisms (and in part plant roots), which are mainly responsible for biotransformation occurring in soil, release enzymes that are active close *to* (ectocellular) or at *a distance from* (extracellular) the cells to degrade insoluble or macromolecular substances such as lignin, carbohydrates, proteins, lipids, and nucleic acids and to allow their use as both nutrient and energy sources (Figure 5.3b). Indeed, although soluble substrates are easily taken up by microbial (and root) cells, insoluble or complex substrates must be broken down into partially degraded or oxidized, simpler compounds that can easily be taken up by cells (Figure 5.3b). Oxidoreductases and hydrolases will act concurrently in the biogeochemical cycles of the main nutrients to assure the biological activity of the soil. Their production is, therefore, controlled by several factors including, at the organism level, the pairing of low-level constitutive synthesis with induction–repression pathways and, at the microbial population and community levels, the quorum-sensing systems that regulate unique and varied cellular responses as well as protection from competing microbial communities (Allison et al., 2007).

In an earlier paper, Burns (1982) described the ecological role of soil enzymes attributing a specific role to the category of extracellular or cell-free enzymes, very often present as soil-bound enzymes. These soil-bound enzymes are persistent in soil, and as such they may be immediately available to transform substrates arriving in the soil. Their activity does not need an induction period to manifest. By contrast, cell-associated enzymes are

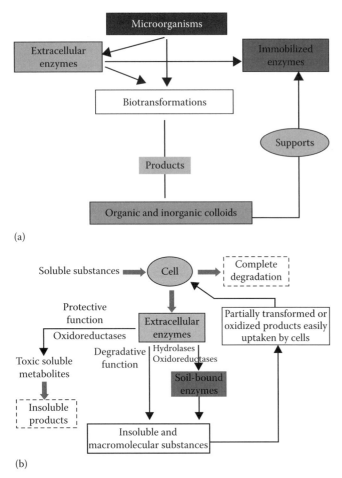

FIGURE 5.3 Role of (a) microorganisms in soil biotransformation and of (b) extracellular enzymes in cell metabolism.

very often produced as inducible enzymes and therefore a short or even long period is necessary before they may develop their activity. Moreover, the rapid transformation of macromolecular substances could lead to simpler compounds possibly acting as inducers of other enzymatic proteins. Since they can easily enter the nearby cells, the synthesis, production, and release of additional extracellular enzymes can be activated, thus enhancing the transformation potential of the soil. Moreover, extracellular enzymes also have the capability of conferring to microorganisms the possibility of reaching and using organic substances embedded in the microporous volume of soil (Quiquampoix and Burns, 2007).

The inorganic and organic derivatives of the more essential macro nutrients C, N, P, and S are involved in complex enzymatic reactions taking place within living (mainly microbial) cells and in the soil solution. A continual interplay of mineralization and immobilization processes brought about by the activities of intracellular and extracellular enzymes affects the availability of nutrients in the soil and thus their importance in crop production. On this basis, a correlation between nutrient availability and enzyme activities in soil do exist, but it is very difficult to identify and quantify the cause–effect relationship between the activity level of a given enzyme and the biochemical fate of a nutrient. This becomes more difficult when complex substrates like different litter substrates are considered. Indeed, the mechanistic link between the production of extracellular enzymes by microorganisms and litter decomposition is still poorly understood (Burns, 1982; Allison and Vitousek, 2004).

As outlined in the previous sections, the interactions of extracellular enzymes with soil components will substantially affect their behavior and consequently their ecological role.

A major contribution to this topic has recently been made by Allison and coworkers (Allison and Vitousek, 2004, 2005; Allison, 2005, 2006a, 2006b; Allison and Jastrow, 2006; Allison et al., 2007) who elaborated most of the concepts already hypothesized by Burns (1982). An interesting approach to the ecological role of extracellular enzymes was suggested by Allison (2005) who introduced the concept of "cheaters," that is, microbes that do not synthesize enzymes but use products. As depicted in Figure 5.3b, the majority of microorganisms rely on extracellular enzymes for their foraging and survival. This observation led the author to pose several questions "When do microbes produce extracellular enzymes? What costs are involved? What is the fate of secreted enzymes and how effective are they?" (Allison, 2005). Reliable answers to these queries can help understanding the importance of the interactions between microbial and chemical substrates in driving "fundamental ecological processes." The products formed by the action of extracellular enzymes produced by some microorganisms can be used by "cheaters," thus competing with the producers and lowering the efficiency of the secreted enzymes for the growth and survival of their originating cells. Results shown in the previous sections, however, demonstrated that several extracellular enzymes might persist in soil as a stable and active (even if reduced) enzymatic pool. Therefore, an integrated model considering all these situations from an economical point of view was proposed by Allison (2005). A simulation model was established in which the influence of competition, nutrient availability, and spatial structure on microbial growth and enzyme synthesis was examined. Simulation was reiterated by changing C, N, or P supply, and the author concluded that cheaters were favored by conditions in which the cost of enzyme is high, whereas lower rates of enzyme diffusion favored producers. Moreover, carbon mineralization, microbial growth, and enzyme production were limited by nitrogen supply because enzymes with a high content of N-units require high levels of N sources (Allison, 2005). In a previous paper, Allison and Vitousek (2005) assumed that resource limitation could constrain the production of enzymes that, according to the economical theory, should increase when simple nutrients are scarce and complex nutrients are abundant. The authors monitored the activity of phosphatase (P cycle), β-glucosidase (C cycle), and glycine aminopeptidase (N cycle) in an infertile tropical soil after supplying it with C, N, and P nutrients under different combinations of complex and simple nutrient additions and evaluated the relationships between resource availability, extracellular enzyme production, and nutrient dynamics. The most relevant results to the topic of the present chapter was that "microorganisms produce enzymes according to 'economic rules,' but a substantial pool of mineral-stabilized or constitutive enzymes mediates this response" (Allison and Vitousek, 2005). A drawback of these studies is that only one enzyme was considered for each nutrient cycle. As clearly established by Nannipieri et al. (1990), a single enzyme activity cannot be representative of complex metabolic processes such as C, P, and N mineralization.

The extracellular production of enzymes and their consequent activity in soil are of pivotal importance in the carbon cycle, that is, in the decomposition and/or storage of carbonaceous substrates, and consequently they may have potential effect on global climate. Indeed, levels of CO_2 in the atmosphere and their effect on global climate are strongly controlled by the cycling of C compounds in the environment, mainly in the soil (soil is a major carbon sink resulting in reduced atmospheric methane and carbon dioxide [two of the main greenhouse gases]). The decomposition of OM and the sequestration of carbon in soil are affected by several phenomena, all acting simultaneously: mineralization by microbial metabolism, protection of OM by its adsorption and entrapment on/in soil minerals, and formation of soil aggregates not easily accessible to microbial attack.

As reported earlier, extracellular enzymes may have a strong influence on all these processes because they convert complex and insoluble substrates into smaller products (Figure 5.3b). On the other hand, organic products or even more complex, undecomposed materials may be sequestered in soil through the action of oxidative enzymes and production of polymeric materials and humus substances (Figure 5.3b). In this complex scenario, not only carbon-degrading enzymes involved in the degradation and/or polymerization of C-substrates but also those involved in the cycling of P- and N-substrates are important, as P and N are essential nutrients for microbial growth and enzyme synthesis (Allison and Jastrow, 2006; Allison et al., 2007). The interaction of these enzymes with soil and their influence and control on C balance were discussed by Allison and Jastrow (2006) by hypothesizing that "minerals and aggregates could enhance C stability by physically impeding the enzymatic degradation of C compounds" and "enzyme activities would be greater in soil fractions with rapid C turnover, such as particulate organic matter (POM), but lower in fractions with slow turnover, such as clay-sized particles." The activities of three enzymes involved in the C-cycle (β-glucosidase, cellobiohydrolase, and polyphenol oxidase), two in the N-cycle (N-acetyl-glucosaminidase and glycine aminopeptidase), and phosphatase for the P-cycle were tested in macroaggregates, microaggregates, and the clay-sized fractions across a prairie restoration chronosequence in Illinois, United States (Allison and Jastrow, 2006). The immobilization of both enzymes and carbon substrates on mineral-dominated fractions reduced the production of enzymes as well as their activity, resulting in slowing down the carbon turnover. By contrast, immobilization of enzymes and carbon substrates occurred rapidly in POM fractions. These results still confirm the main role of the interactions between extracellular enzymes and soil components in the turnover and persistence of carbon in the environment. The presence of the far larger pool of organic C in soils would be a consequence of the constraints to its decomposition by extracellular enzymes as well as by the chemical properties of humic compounds and interactions of C with soil minerals (Allison, 2006a).

To clarify the dynamics of extracellular enzymes as either bound or free within the soil, experiments were performed to determine the turnover rates of some enzymes (phosphatase, β-glucosidase, polyphenol oxidase, N-acetyl-glucosaminidase, and urease) within a well-characterized volcanic soil. The effects of common soil compounds such as ferrihydrite, allophane, and humic acids on enzyme stability and activity were also examined (Allison, 2006b). According to results previously discussed, allophane had a strong positive effect on most enzyme activities, whereas ferrihydrite had a weak positive effect on some enzymes and humic acids strongly inhibited enzyme activity (Allison, 2006b). Probably the most relevant finding from these studies was that stabilization and inactivation of enzymes on soil minerals and humic acids have to be considered and integrated into conceptual models of enzyme function. For instance, the composition of mineral and humic compounds of a soil could justify the ecosystem-specific responses of soil enzymes to environmental perturbations (such as CO_2 or nitrogen fertilization). As Rosas et al. (2008) demonstrated with synthetic enzymatic systems, Andisols, rich in allophanic materials, are expected to have high potential enzyme activity because of their ability to immobilize and stabilize enzymes.

Scheel et al. (2008) studied the effects of precipitation and adsorption of extracellular enzymes on a stable soil OM fraction on the performance of enzyme function for carbon mineralization. The OM fraction was obtained by precipitation of dissolved organic matter (DOM) with aluminum. The activities of eight extracellular enzymes, involved in C degradation (chitinase, cellobiohydrolase, β-glucosidase, glucuronidase, laccase, and xylosidase) and of leucine-aminopeptidase and acid-phosphatase involved in N and P cycles, respectively, were studied in OM fractions that were obtained from two top horizons of two forest soils and precipitated at different Al:C ratios and pH values to achieve a large variation in composition and C mineralization. The investigated enzymes were all functioning in precipitated OM. The results in terms of amounts of precipitated DOM, of strong correlation between enzyme activity levels and C mineralization, and of disappearance of the latter led the authors to conclude that precipitated extracellular enzymes were in a functional state and played a significant role in the turnover of a stable soil organic matter fraction, like precipitated OM, during the initial degradation phase, and that the stability of precipitated OM is mainly governed by chemical properties and not by lack of enzymes.

A confirmation on the effect of soil environment on the performance of enzyme function was provided by George et al. (2007) in comparative studies performed on the mobility, stability, and catalytic activity of phytase from two fungal sources. Phytase is the enzyme responsible for the hydrolysis of inositol phosphates, the largest fraction (>50%) of organic P compounds in soil, and therefore affects the biological availability of these compounds. The enzymes were added to soil suspensions of three soils (a Spodosol, an Oxisol, and an Alfisol), and their activities in the solution and in the solid phase were measured. The two enzymes exhibited different efficiencies in the three soils according to their physicochemical differences; these strongly influenced the adsorption of the phytases, which in turn affected their stabilities and resistance to microbial degradation. Again, the importance of the interactions between enzymes and soil components to the availability and transformation of nutrients in soil was confirmed.

Knight and Dick (2004) suggested that soil-stabilized β-glucosidase activity (e.g., the abiotic fraction of the enzyme stabilized by its association with soil colloids) could act as a soil quality indicator to discriminate between field management effects. Since interaction between enzyme and soil colloids would not show large temporal variability, accumulation or degradation of this stable form of the enzyme will proceed at "a steady pace over time under a given management system."

5.5 Interaction of Soil-Bound Enzymes with Anthropogenic Organics and Their Potential for Ecosystem Restoration

Polluting substances are continuously released in the environment as a consequence of the impact of man's activity on the biosphere. They accumulate in the ecosystem and alter and disturb its natural balance. Among the environmental compartments, soil can be considered a natural sink for anthropogenic contaminants.

The pollution of the environment, and in particular of soil, poses a great concern not only to devise remediation technologies suitable for an effective restoration of polluted sites but also to elucidate the possible influence of pollutants on the many vital functions that soil performs for the environment and society.

In restoration and recovery of soils from pollution as well as in understanding pollution effects on soil functions, free or bound extracellular enzymes have attracted the attention of many researchers (Dick and Tabatabai, 1999; Naidja et al., 2000; Gianfreda and Bollag, 2002; Gianfreda and Rao, 2004, 2008 and references therein).

Any study of the impact of any influencing factor on soil enzymatic activities has to take into account the complexity of these and their existence under several enzymatic categories or "states," each with its "own" catalytic features and characteristics. As a consequence, each enzymatic category may respond in a different way and with a different intensity to the action of a given factor. Moreover, pollutants, and in general xenobiotics, may present irreversible and reversible direct and indirect effects on soil enzyme activities. In the first case, reversible binding may occur between the xenobiotic and/or its degradation products and the intra and extracellular enzymes, with a resulting competitive and/or noncompetitive substrate inhibition or an alteration of the protein conformation. An irreversible inhibition by covalent binding of the xenobiotic (or its derivatives) with functional, catalytic enzymatic groups may also occur with deleterious effects on enzyme functions. Indirect effects are the consequence of possible xenobiotic-induced changes in size, structure, and functionality of the microbial community, in turn of the genesis and synthesis of new intra- and extracellular enzymes, and changes in the ratio between intra- and extracellular enzyme amounts.

These are the main reasons of the often contradictory results obtained on the influence of pesticides on soil enzyme activities (Schaffer, 1993; Gianfreda and Rao, 2008).

To address this aspect and evaluate the possible impact of pesticides on soil enzyme activities, Gianfreda and coworkers (Gianfreda et al., 1993b, 1994, 1995b, 2002; Sannino and Gianfreda, 2001) developed an experimental design that took into account the variability of soil enzyme categories. The effects of four pesticides, three herbicides (atrazine, paraquat, glyphosate), and one insecticide (carbaryl), on the activities of invertase (I), urease (U), and phosphatase (P) under different "states," were studied in the laboratory. The model systems, simulating enzymatic categories possibly occurring in soil, were (1) enzymes free in solution, to simulate enzymes free in soil solution; and (2) synthetic clay-, organo- and organo-clay-enzyme complexes, simulating soil-bound enzymes, and enzyme activities of whole soils, representing natural systems (Table 5.3). The synthetic enzymatic complexes were among those already discussed and presented in Section 5.3.3, that is, montmorillonite-bound enzymes (M-I, M-U, M-P), tannic acid-bound enzymes (T-I, T-U, T-P), and Al(OH)$_x$–tannic acid–montmorillonite–bound enzymes (ATM-I, ATM-U, ATM-P).

TABLE 5.3 Effect of Atrazine, Carbaryl, Glyphosate, and Paraquat (and Methanol Used as Solvent for Carbaryl and Atrazine Preparation) on the Activity of Free and Immobilized Enzymes

Enzymes	Relative Activity[a]				
	Glyphosate	Paraquat	Atrazine	Carbaryl	Methanol
Free enzymes					
Invertase (I)	1.80	1.23	0.71	0.53	0.64
Urease (U)	0.93	0.96	0.58	0.53	0.64
Phosphatase (P)	1.05	—	0.61	0.76	0.74
Immobilized enzymes					
M-I	1.10	1.10	0.62	0.67	0.51
M-U	1.03	1.79	0.54	0.60	0.57
M-P	0.61	—	0.87	0.86	0.87
T-I	0.98	0.85	0.54	0.61	0.43
T-U	0.99	0.95	0.67	0.62	0.73
T-P	0.79	—	0.85	0.80	0.85
ATM-I	0.93	0.92	0.55	0.82	0.43
ATM-U	1.20	1.03	0.58	0.65	0.63
ATM-P	0.59	—	0.70	0.69	0.71

Source: Modified from Sannino, F., and L. Gianfreda. 2001. Pesticide influence on soil enzymatic activities. Chemosphere 45:417–425. Copyright 2001, with permission of Elsevier.

M, montmorillonite; T, tannic acid; ATM, OH-Al–tannic acid–montmorillonite.

[a] The results are expressed as increments or decrements of the activities measured without pesticides (activity in the absence of pesticides = 1).

The results shown in Table 5.3 clearly indicate that the response of each enzyme to the presence of each pesticide was strongly influenced by the type of enzyme and its status, that is, if free or bound to inorganic, organic, or organo–mineral supports, and by the type of pesticide. For example, glyphosate had no effect on free phosphatase but severely depressed its activity (by about 20%–40%, with respect to the activity in the absence of pesticide) when the enzyme was immobilized on synthetic complexes (Table 5.3). By contrast, the same pesticide activated free invertase but had no effect on immobilized invertase. Varied results were also obtained with the other tested pesticides and methanol (the solvent in carbaryl and atrazine solutions) (Table 5.3).

Complex and not easily explainable responses were also obtained when the effects of the four pesticides on the three enzymes were investigated using several whole soils collected at different sites in Italy, which exhibited different physicochemical properties (Gianfreda et al., 1994; Gianfreda et al., 1995b; Sannino and Gianfreda, 2001). For example, in the presence of glyphosate the invertase activity of some soils increased between +4% and +204% and decreased by an average of 30% in other soils. Increases, decreases, or no changes were measured on soil urease activity with paraquat and glyphosate (Gianfreda et al., 1994). By contrast, inhibition by 50% on average was observed on the activity of phosphatase in all soils (Sannino and Gianfreda, 2001). As summarized by Gianfreda et al. (2002), some interpretations can be made by considering the substantial differences of the investigated systems. Indeed, the interaction between a pesticide and a free enzyme occurs in a homogeneous system, and direct interactions at the molecular level may occur. By contrast, with immobilized enzymes a heterogeneous system is present, and different interactions can occur not only between the immobilized enzyme and the pesticide but also between the support and the pesticide. Complex and also opposite effects may occur, all contributing to the measured response.

It is important to emphasize that tests conducted in the laboratory might not be ecologically interpreted. Contrary to model laboratory systems, pesticides in soil may be sorbed onto colloids, degraded rapidly, or exist in widely varying concentrations at different microsites. Therefore, the variability of the results previously mentioned suggests that no generalizable conclusions can be drawn, at least with the experimental systems under laboratory conditions.

The ability of extracellular enzymes in the transformation and degradation of pollutants and xenobiotics is backed by several findings (see reviews by Gianfreda and Bollag, 2002; Gianfreda and Rao, 2004). Oxidoreductases, released by both root or microbial cells, were effective in the transformation of several pollutants including polychlorinated biphenyls (PCBs), polycyclic aromatic hydrocarbons (PAHs), azo dyes, biopolymers (kraft, lignin), bleach plant effluents, chloroderivatives, phenol derivatives (like pentachlorophenol), and trinitrotoluen. Microbial hydrolases, such as esterases and depolymerases, were able to degrade highly recalcitrant and persistent materials such as plastics, including polyurethanes, polyacrylates, polylactides, nylon, starch polymers, and mixed composites of different starting materials. These transformations, however, occurred very often in the presence of enzyme-producing cells, thus relying on the activity and functionality of their metabolism.

By contrast, the use of cell-free enzymes could have several advantages over the use of microbial cells because of their peculiar characteristics, such as higher reactivity and substrate specificity, and ability to function under a wide range of environmental conditions, which are often unfavorable to active microbial cells. They even function in the presence of toxic compounds and are not sensitive to predators or inhibitors of microbial metabolism (Nannipieri and Bollag, 1991; Karam and Nicell, 1997; Dick and Tabatabai, 1999; Gianfreda and Bollag, 2002; Gianfreda and Rao, 2004).

The most prevailing enzymes active as free enzymes in the transformation of pollutants are undoubtedly oxidative enzymes such as phenoloxidases (laccase, tyrosinase) and PODs (lignin and manganese PODs), which are often of fungal origin. These enzymes may be relatively nonspecific and have a very high redox potential. Consequently, they may be able to oxidize recalcitrant pollutants, which are not easily transformed by other enzyme-mediated reactions. For instance, several fungal laccases were tested for the removal of several pollutants from aqueous, solid, and hazardous wastes (Gianfreda et al., 1999 and references therein). Bollag (1992) proposed that laccases were suitable agents for the detoxification of polluted soils. He founded his assumption on the ability of laccase to reduce the toxicity of phenolic substances in the soil by their stabilization through incorporation in humic materials rather than by degradation.

Most of these results were, however, obtained under laboratory conditions, and few results are available on the real effectiveness of enzymes in soil environments. Moreover, the majority of results were obtained with free and unbound enzymes. When extracellular enzymes are applied for the detoxification of soils, their interaction with soil constituents may have two concomitant but contrasting effects on their efficiency in the removal of polluting substances. Their activity and catalytic power may be reduced, but their stability and reusability may be increased. Both processes are the consequence of their immobilization on soil supports.

An attempt to overcome this lack of knowledge was made by Ruggiero et al. (1989) and Gianfreda and Bollag (1994) who used laccase and POD free and immobilized on clays and soils and evaluated their ability to detoxify soils polluted by 2,4-dichlorophenol, a derivative of the pesticide 2,4-D, which is widespread in soils and sediment. Investigations were performed with soils differing in their chemical and physical properties. Besides some interesting properties shown by the immobilized enzymes (e.g., high residual activity and elevated stability to thermal denaturation, storage, and proteolytic attack) (see Section 5.3.3), both free and immobilized enzymes displayed catalytic activity in soil environments. Laccase and POD were able to remove up to 60% of the initial substrate concentration in the presence of soil, having a low content of OM. However, by increasing the OM content, an inhibitory effect on the activity of free and immobilized enzymes was observed (Gianfreda and Bollag, 1994). Another important property of immobilized laccase and POD was that both enzymes retained their activity in the removal of 2,4-DCP for up to six incubation cycles. In the presence of soils, the inhibitory effect decreased after each incubation cycle, indicating that the observed inhibition was probably caused by soil-soluble components, which were easily extracted and removed when the substrate was replaced with a fresh buffered solution (Gianfreda and Bollag, 1994).

As summarized in Gianfreda and Rao (2004) and in Gianfreda et al. (2002), a completely different behavior was observed when laccase was not chemically linked to the support and acted in the reaction assay along with inorganic or organic soil colloids (single or in combination), soils, or sand, (Figure 5.4). In comparison with the control, that is, the enzyme alone, soil components differently affected the 2,4-DCP-laccase activity. When montmorillonite was present in the reaction mixture, less than 50% of activity was measured. A complete inhibition of the enzymatic activity was observed with AM_{18} (Figure 5.4). No effects were evident with humic acids, but the effect of AM_{18} dominated over that of humic acids when the two colloids were simultaneously present. Contrary to what was previously reported for laccase covalently bound to clays (e.g., preservation of high enzymatic activity levels), adsorption of enzymatic molecules on clay surfaces resulted in a concomitant loss of activity of the adsorbed enzyme. These results are in accordance with those previously described and further demonstrate that adsorption of enzymes on clays usually results in a significant decrease in activity of the adsorbed enzyme.

Reusability of enzymes is a necessary prerequisite for their use in practical applications for the restoration of polluted environments (Dick and Tabatabai, 1999). Enzymes covalently immobilized on solid supports may retain their activity much better than adsorbed enzymes as indicated by the investigations of Chang and Juang (2004, 2005, 2007) and Sanjay and Sugunan (2005, 2006, 2007).

5.6 Conclusions and Future Prospects

The findings reported and discussed in this Chapter provide strong experimental evidence not only of the importance of extracellular enzymes in soil but also of the impact of their interactions with soil components on their activity and stability. Both free and soil-bound enzymes may have significant effects on soil biology and biochemistry, microbial diversity, plant growth and nutrient uptake in ecosystems, and environmental management and restoration.

After interaction with inorganic, organic, and organo–mineral soil colloids, enzymes are usually not as catalytically active as enzymes that are free in aqueous solution, but they are frequently more stable to some environmental stresses such as extreme temperatures, digestion by proteases, and other severe conditions. These properties may vary between specific enzymes and confer on them the capability of performing a large variety of chemical reactions vital to soil fertility and soil health.

Several factors may affect the behavior of enzymes immobilized on soil colloids (Nannipieri and Gianfreda, 1998): (1) Changes in the conformation of enzymatic proteins, due to denaturing effects of the supports or of interactions between the enzyme and the support, very often pH dependent; (2) Steric hindrances to the access of substrate to the active center of the enzyme, much more marked with high molecular mass substrates; (3) Microenvironmental effects due to the different local microenvironment, surrounding the enzyme linked to the support, and strongly dependent on its chemical nature; (4) Perturbation of the protein conformation due to combination of electrostatic and hydrophobic forces and pH dependence; (5) Diffusional limitations, since the rate of transformation by immobilized enzymes is influenced by the availability of the substrate and the effectors. Considerable changes in the values of kinetic parameters can be expected, and those observed are only "apparent."

Soil-bound enzymes also have a great potential in maintaining soil biodiversity and assuring a correct and integrated balance between input, assimilation, and utilization of major nutrients and output of inorganic derivatives contributing to greenhouse gases with consequences on global climate changes. They are not sensitive to induction-repression phenomena that usually regulate the production of extracellular enzymes by microorganisms, have a longer turnover time than that of free extracellular enzymes, and therefore represent a stable pool of potential enzyme activity of soil to also function in the absence of living organisms.

Soil-bound enzymes may mediate the transformation and degradation of pollutants, but their reduced catalytic activity can be an obstacle to an effective restoration of polluted soils. On the other hand, their increased stability and the possible involvement of the soil support in the pollutant transformation

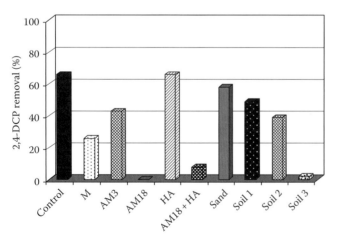

FIGURE 5.4 Removal of 2,4-DCP by laccase in the presence of clays, clay–humic complexes, sand, and soils. M = Na-montmorillonite; AM = montmorillonite covered by different amounts of OH-Al species (AM_3 and AM_{18}, loading 3 and 18 meq of Al g^{-1} clay), HA = humic acid. (From Gianfreda, L., and M.A. Rao. 2004. Potential of extra cellular enzymes in remediation of polluted soils: A review. Enz. Microb. Technol. 35:339–354. Copyright 2004, with permission of Elsevier.)

(i.e., adsorption of either substrates or released products, direct participation to pollutant degradation [Huang, 1990, 1995; Ruggiero et al., 1996; Naidja et al., 2000]) may enhance their performance and provide soil with a persistent, active pollutant degradative capability. Their effective use in soils for environmental cleanup or for modifying biochemical activities still requires, however, an exhaustive understanding of the influence of the soil microenvironment.

Advances in the knowledge of production, regulation, and stabilization of soil enzymes, and in particular of the formation and performance of soil-bound enzymes, of their involvement in the transformation of nutrients, degradation of pollutant, and functioning of soil at a global level, will derive if suitable methodologies are available for (a) measuring and modeling in situ activity of soil enzymes, (b) determining the fraction of stabilized enzymes and their activity, (c) defining the exact mechanisms occurring in vivo in the interactions between enzymes and inorganic and organic soil colloids, (d) determining surface structures of the enzyme–soil colloid complexes, (e) distinguishing the contribution of abiotic reactions to the transformation of nutrients and/or pollutants, and (f) elucidating the concomitant action, if any, of more than one enzyme bound to the same support, in case this occurs.

As reviewed by Wallenstein and Weintraub (2008), emerging tools, including genomic, proteomic, and metabolic tools, are now available for achieving most of the cited needs and for improving "understanding of the drivers of in situ enzyme activities." Genomic studies will assist in elucidating the production of enzymes in soil and the regulating factors. An increased capability for modeling and predicting enzyme pools under different conditions will result. Proteomic studies will provide information on the diversity and source of extracellular enzymes in soil. They will help in discovering and characterizing the production of novel extracellular enzymes independently of assays based on substrate transformation. Imaging methodologies will allow a direct measurement of in situ enzyme activity, thus contributing to a full understanding of the complex interactions between microbial cells, enzymes and clay minerals, or OM. Notwithstanding the fundamental contribution to the visualization of soil microenvironments provided by the studies of Forster and coworkers (Ladd et al., 1996), who developed imaging methodologies to identify the "in situ" activity of several soil enzymes, some limitations are still present in their application. Metabolomics will be useful to quantify the products of enzyme-mediated degradation or transformation of substrates and pollutants and indirectly to evaluate the production and activity of soil enzymes.

Moreover, advanced analytical instrumentation such as atomic force microscopy, scanning tunneling and infrared spectroscopy, and synchrotron-based x-ray spectroscopy will be helpful to clarify the interactions occurring at the molecular level between enzymes, clays, and OM; their catalytic involvement in the transformation of nutrients and pollutants; and their contribution to the whole soil enzyme activity.

The combination and integration of updated information achieved by these methodologies will contribute to reach the goals and objective of the Commission 2.5—Soil, Physical, Chemical and Biological Interfacial Reactions of the International Union of Soil Sciences, that is, "advancing the understanding on physical/chemical/biological interfacial systems at the molecular to field/landscape levels."

References

Ahn, M.-Y., C.E. Martínez, D.D. Archibald, A.R. Zimmerman, J.-M. Bollag, and J. Dec. 2006. Transformation of catechol in the presence of a laccase and birnessite. Soil Biol. Biochem. 38:1015–1020.

Ahn, M.-Y., A.R. Zimmerman, C.E. Martìnez, D.D. Archibald, J.-M. Bollag, and J. Dec. 2007. Characteristics of *Trametes villosa* laccase adsorbed on aluminum hydroxide. Enz. Microb. Technol. 41:141–148.

Allison, S.D. 2005. Cheaters, diffusion, and nutrients constrain decomposition by microbial enzymes in spatially structured environments. Ecol. Lett. 8:626–635.

Allison, S.D. 2006a. Brown ground: A soil carbon analogue for the green world hypothesis? Am. Nat. 167:619–627.

Allison, S.D. 2006b. Soil minerals and humic acids alter enzyme stability: Implications for ecosystem processes. Biogeochemistry 81:361–373.

Allison, S.D., and J.D. Jastrow. 2006. Activities of extracellular enzymes in physically isolated fractions of restored grassland soils. Soil Biol. Biochem. 38:3245–3256.

Allison, S.D., and P.M. Vitousek. 2004. Extracellular enzyme activities and carbon chemistry as drivers of tropical plant litter decomposition. Biotropica 36:285–296.

Allison, S.D., and P.M. Vitousek. 2005. Responses of extracellular enzymes to simple and complex nutrient inputs. Soil Biol. Biochem. 37:937–944.

Allison, S.D., T.B. Gartner, K. Holland, M. Weintraub, and R.L. Sinsabaugh. 2007. Soil enzymes: Linking proteomics and ecological process, p. 704–711. *In* C.J. Hurst, R.L. Crawford, J.L. Garland, D.A. Lipson, A.L. Mills, and L.D. Stetzenbach (eds.) Manual of environmental microbiology. 3rd edn. ASM Press, Washington, DC.

Benndorf, D., G.U. Balcke, H. Harms, and M. von Bergen. 2007. Functional metaproteome analysis of protein extracts from contaminated soil and groundwater. ISME J. 1:224–234.

Bittner, S. 2006. When quinones meet amino acids: Chemical, physical and biological consequences. Amino Acids 30:205–224.

Bollag, J.-M. 1992. Decontaminating soils with enzymes. Environ. Sci. Technol. 26:1876–1881.

Bollag, J.-M., C. Myers, S. Pal, and P.M. Huang. 1995. The role of abiotic and biotic catalysts in the transformation of phenolic compounds, p. 299–315. *In* P.M. Huang, J. Berthelin, J.-M. Bollag, W.B. McGill, and A.L. Page (eds.) Environmental impact of soil component interactions: Natural and anthropogenic organics. CRC Press/Lewis Publishers, Boca Raton, FL.

Bonmati, M., B. Ceccanti, and P. Nannipieri. 1998. Protease extraction from soil by sodium pyrophosphate and chemical characterization of the extracts. Soil Biol. Biochem. 30:2113–2125.

Boyd, S.A., and M.M. Mortland. 1990. Enzyme interactions with clays and clay-organo matter complexes, p. 1–28. In J.-M. Bollag and G. Stozky (eds.) Soil biochemistry. Vol. 6. Marcel Dekker, New York.

Burns, R.G. (ed.) 1978. Soil enzymes. Academic Press, London, U.K.

Burns, R.G. 1982. Enzyme activity in soil: Location and a possible role in microbial ecology. Soil Biol. Biochem. 14:423–427.

Burns, R.G. 1986. Interaction of enzymes with soil mineral and organic colloids, p. 429–451. In P.M. Huang and M. Schnitzer (eds.) Interactions of soil minerals with natural organics and microbes. Soil Science Society of America, Madison, WI.

Butler, J.H.A., and J.N. Ladd. 1969. The effect of methylation of humic acids on their influence on proteolytic enzyme activity. Austr. J. Soil Res. 7:263–268.

Carrasco, M.S., J.C. Rad, and S. Gonzales-Carcedo. 1995. Immobilization of alkaline phosphatase by sorption on Na-sepiolite. Biores. Technol. 51:175–181.

Chang, M.-Y., and R.-S. Juang. 2004. Stability and catalytic kinetics of acid phosphatase immobilized on composite beads of chitosan and activated clay. Process Biochem. 39:1087–1091.

Chang, M.-Y., and R.-S. Juang. 2005. Activities, stabilities, and reaction kinetics of three free and chitosan–clay composite immobilized enzymes. Enz. Microb. Technol. 36:75–82.

Chang, M.-Y., and R.-S. Juang. 2007. Use of chitosan–clay composite as immobilization support for improved activity and stability of β-glucosidase. Biochem. Eng. J. 35:93–98.

Ciurli, S., C. Marzadori, S. Benini, S. Deiana, and C. Gessa. 1996. Urease from the soil bacterium Bacillus pasteurii: Immobilization on Ca-polygalacturonate. Soil Biol. Biochem. 28:811–817.

Dick, R.P. 1994. Soil enzymes activities as indicators of soil quality, p. 107–124. In J.W. Doran, D.C. Coleman, D.F. Bezdicek, and B.A. Stewart (eds.) Defining soil quality for a sustainable environment. Vol. 35. SSSA Special Publication, Madison, WI.

Dick, W.A., and M.A. Tabatabai. 1987. Kinetic parameters of phosphatase-clay complexes. Soil Sci. 143:5–15.

Dick, W.A., and M.A. Tabatabai. 1999. Use of immobilized enzymes for bioremediation, p. 315–338. In D.C. Adriano, J.-M. Bollag, W.T. Frankenberger, Jr., and R.C. Sims (eds.) Bioremediation of contaminated soils. Series no. 37. American Society of Agronomy, Madison, WI.

Filazzola, M.T., F. Sannino, M.A. Rao, and L. Gianfreda. 1999. Effect of various pollutants and soil-like constituents on laccase from Cerrena unicolor. J. Environ. Qual. 28:1929–1938.

Fornasier, F. 2002. Indirect approaches for assessing intracellular arylsulfatase activity in soil, p. 345–352. In A. Violante, P.M. Huang, J.-M. Bollag, and L. Gianfreda (eds.) Soil mineral-organic matter-microorganism interactions and ecosystem health. Vol. 28B. Development in Soil Science. Elsevier, London, U.K.

Fornasier, F., and A. Margon. 2007. Bovine serum albumin and Triton X-100 greatly increase phosphomonoesterases and arylsulphatase extraction yield from soil. Soil Biol. Biochem. 39:2682–2684.

Fusi, P., G.G. Ristori, Calamai, L., and G. Stozky. 1989. Adsorption and binding of protein on "clean" (homoionic) and "dirty" (coated with Fe oxyhydroxides) montmorillonite, illite and kaolinite. Soil Biol. Biochem. 21:911–920.

Garzillo, A.M., L. Badalucco, F. De Cesare, S. Grego, and V. Buonocore. 1996. Synthesis and characterization of an acid phosphatase-polyresorcinol complex. Soil Biol. Biochem. 28:1155–1161.

George, T.S., R.J. Simpson, P.J. Gregory, and A.E. Richardson. 2007. Differential interaction of Aspergillus niger and Peniophora lycii phytases with soil particles affects the hydrolysis of inositol phosphates. Soil Biol. Biochem. 39:793–803.

Gianfreda, L., and J.-M. Bollag. 1994. Effect of soils on the behavior of immobilized enzymes. Soil Sci. Soc. Am. J. 56:1672–1681.

Gianfreda, L., and J.-M. Bollag. 1996. Influence of natural and anthropogenic factors on enzyme activity in soil, p. 123–194. In G. Stotzky and J.-M. Bollag (eds.) Soil biochemistry. Vol. 9. Marcel Dekker, New York.

Gianfreda, L., and J.-M. Bollag. 2002. Isolated enzyme for the transformation and detoxification of organic pollutants, p. 491–538. In R.G. Burns and R. Dick (eds.) Enzyme in the environment. Activity, ecology and applications. Marcel Dekker, New York.

Gianfreda, L., A. De Cristofaro, M.A. Rao, and A. Violante. 1995c. Kinetic behaviour of synthetic organo- and organo-mineral-urease complexes. Soil Sci. Soc. Am. J. 59:811–815.

Gianfreda, L., G. Iamarino, R. Scelza, and M.A. Rao. 2006. Oxidative catalysts for the transformation of phenolic pollutants: A brief review. Biocatal. Biotransfor. 24:177–187.

Gianfreda, L., G. Marrucci, N. Grizzuti, and G. Greco, Jr. 1985. Series mechanism of enzyme deactivation. Characterization of intermediate forms. Biotechnol. Bioengin. 27:872–877.

Gianfreda, L., and M.A. Rao. 2004. Potential of extra cellular enzymes in remediation of polluted soils: A review. Enz. Microb. Technol. 35:339–354.

Gianfreda, L., and M.A. Rao. 2008. Interaction between xenobiotics and microbial and enzymatic soil activity. Crit. Rev. Environ. Sci. Technol. 38:269–310.

Gianfreda, L., M.A. Rao, F. Saccomandi, F. Sannino, and A. Violante. 2002. Enzymes in soil: Properties, behavior and potential applications, p. 301–328. In A. Violante, P.M. Huang, J.-M. Bollag, and L. Gianfreda (eds.) Soil mineral-organic matter-microorganism interactions and ecosystem health. Vol. 28B. Development in Soil Science. Elsevier, London, U.K.

Gianfreda, L., M.A. Rao, and A. Violante. 1991. Invertase (β-fructosidase): Effects of montmorillonite, Al-hydroxide and Al(OH)$_x$-montmorillonite complex on activity and kinetic properties. Soil Biol. Biochem. 23:581–587.

Gianfreda, L., M.A. Rao, and A. Violante. 1992. Adsorption, activity and kinetic properties of urease on montmorillonite, aluminium hydroxide and Al(OH)x-montmorillonite complexes. Soil Biol. Biochem. 24:51–58.

Gianfreda, L., M.A. Rao, and A. Violante. 1993. Interactions of invertase with tannic acid, OH-Al-species and/or montmorillonite. Soil Biol. Biochem. 6:671–677.

Gianfreda, L., M.A. Rao, and A. Violante. 1995a. Formation and activity of urease-tannate complexes as affected by different species of Al, Fe and Mn. Soil Sci. Soc. Am. J. 59:805–810.

Gianfreda, L., and P. Ruggiero. 2006. Enzyme activities in soil, p. 257–311. *In* P. Nannipieri and K. Smalla (eds.) Nucleic acids and proteins in soil. Soil biology. Vol. 8. Springer-Verlag, Berlin, Germany.

Gianfreda, L., F. Sannino, M.T. Filazzola, and A. Violante. 1993b. Influence of pesticides on the activity and kinetics of invertase, urease and acid phosphatase enzymes. Pest. Sci. 39:237–244.

Gianfreda, L., F. Sannino, N. Ortega, and P. Nannipieri. 1994. Activity of free and immobilized urease in soil: Effects of pesticides. Soil Biol. Biochem. 26:777–784.

Gianfreda, L., F. Sannino, and A. Violante. 1995b. Influence of pesticides on the behaviour of free, immobilized and soil invertase. Soil Biol. Biochem. 27:1201–1208.

Gianfreda, L., and M.R. Scarfi. 1991. Enzyme stabilization: State of the art. Molec. Cell. Biochem. 199:7–123.

Gianfreda, L. F. Xu, and J.-M. Bollag. 1999. Laccases: A useful group of oxidoreductive enzymes. Biorem. J. 3:1–25.

Grego, S., A. D'Annibale, M. Luna, L. Badalucco, and P. Nannipieri. 1990. Multiple forms of synthetic pronase-phenolic copolymers. Soil Biol. Biochem. 22:721–724.

Hamzehi, E., and W. Pflug. 1981. Sorption and binding mechanisms of polysaccharide cleaving soil enzymes by clay minerals. Z. Pflanzen. Bodenkunde 144:505–513.

Harter, R.D., and G. Stotzky. 1971. Formation of clay-protein complexes. Soil Sci. Soc. Am. Pro. 35:383–398.

Huang, P.M. 1990. Role of soil minerals in transformations of natural organics and xenobiotics in soil, p. 29–115. *In* J.-M. Bollag and G. Stotzky (eds.) Soil biochemistry. Vol. 6. Marcel Dekker, New York.

Huang, P.M. 1995. The role of short-range ordered mineral colloids in abiotic transformations of organic compounds in the environment, p. 135–167. *In* P.M. Huang, J. Berthelin, J.-M. Bollag, W.B. McGill, and A.L. Page (eds.) Environmental impact of soil component interactions: Natural and anthropogenic organics. Vol. 1. CRC Press/Lewis Publishers, Boca Raton, FL.

Huang, Q., M. Jiang, and X. Li. 1999. Adsorption and properties of urease immobilized on several iron, aluminum oxides and kaolinite, p. 167–174. *In* J. Berthelin, P.M. Huang, J.-M. Bollag, and F. Andreux (eds.) Effect of mineral–organic–microorganism interactions on soil and fresh water environments. Kluwer Academic/Plenum Publishers, New York.

Huang, Q., H. Shindo, and T.B. Goh. 1995. Adsorption, activities and kinetics of acid phosphatase as influenced by montmorillonite with different interlayer materials. Soil Sci. 59:271–278.

Huang, P.M., M.K. Wang, and C.Y. Chiu. 2005. Soil mineral-organic matter-microbe interactions: Impact on biogeochemical processes and biodiversity in soils. Pedobiologia 49:609–635.

Karam, J., and J.A. Nicell. 1997. Potential application of enzymes in waste treatment. J. Chem. Technol. Biotechnol. 69:141–153.

Kelleher, B.P., A.J. Simpson, K.O. Willeford, M.J. Simpson, R. Stout, A. Rafferty, and W.L. Kingery. 2004. Acid phosphatase interactions with organo-mineral complexes: Influence on catalytic activity. Biogeochemistry 71:285–297.

Kiss, S., M. Dragan-Bularda, and D. Radulescu. 1975. Biological significance of enzymes accumulated in soil. Adv. Agron. 27:25–87.

Kiss, S., M. Dragan-Bularda, and D. Pasca. 1986. Activity and stability of enzyme molecules following their contact with clay mineral surfaces. Studia, Univ. Babes-Bolyai (Cluj-Napoca), Biology 2:3–29.

Klibanov, A.M. 1983. Stabilization of enzymes against thermal inactivation. Adv. Appl. Microb. 29:1–28.

Knight, T., and R.P. Dick. 2004. Differentiating microbial and stabilized ß-glucosidase activity in soils. Soil Biol. Biochem. 36:2089–2096.

Ladd, J.N. 1985. Soil enzymes, p. 175–221. *In* D. Vaugham and R.E. Malcom (eds.) Soil organic matter and biological activity. Martinus Nijhoff/Dr. W. Junk Publishers, Dordrecht, the Netherlands.

Ladd, J.N., and J.H.A. Butler. 1969a. Inhibitory effect of soil humic compounds on the proteolytic enzyme pronase. Austr. J. Soil Res. 7:241–251.

Ladd, J.N., and J.H.A. Butler. 1969b. Inhibition and stimulation of proteolytic enzyme activities by soil humic acids. Austr. J. Soil Res. 7:253–261.

Ladd, J.N., and J.H.A. Butler. 1970. The effect of inorganic cations on the inhibition and stimulation of protease activity by soil humic acids. Soil Biol. Biochem. 2:33–40.

Ladd, J.N., and J.H.A. Butler. 1971. Inhibition by soil humic acids of native and acetylated proteolytic enzymes. Soil Biol. Biochem. 3:157–160.

Ladd, J.N., and J.H.A. Butler. 1975. Humus-enzyme systems and synthetic organic polymer-enzyme analogs, p. 143–194. *In* E.A. Paul and A.D. McLaren (eds.) Soil biochemistry. Vol. 3. Marcel Dekker, New York.

Ladd, J.N., R.C. Forster, P. Nannipieri, and J.M. Oades. 1996. Soil structure and biological activity, p. 23–79. *In* G. Stotzky and J.-M. Bollag (eds.) Soil Biochemistry. Vol. 9. Marcel Dekker, New York.

Lähdesmäki, P., and R. Piispapen. 1992. Soil enzymology: Role of protective colloid system in the preservation of exoenzyme activities in soil. Soil Biol. Biochem. 24:1173–1177.

Lai, C.M., and M.A. Tabatabai. 1992. Kinetic parameters of immobilized urease. Soil Biol. Biochem. 24:225–228.

Leonowicz, A., and J.-M. Bollag. 1987. Laccases in soil and the feasibility of their extraction. Soil Biol. Biochem. 19:237–242.

Leprince, F., and H. Quiquampoix. 1996. Extracellular enzyme activity in soil: Effect of pH and ionic strength on the interaction with montmorillonite of two acid phosphatases secreted by the ectomycorrhizal fungus *Hebeloma cylindrosporum*. Eur. J. Soil Sci. 47:511–522.

Martinek, K., and V.V. Mozhaev. 1985. Immobilization of enzymes: An approach to fundamental studies in biochemistry. Adv. Enzymol. 57:179–249.

Marzadori, C., C. Gessa, and S. Ciurli. 1998a. Kinetic properties and stability of potato acid phosphatase immobilized on Ca-polygalacturonate. Biol. Fertil. Soils 27:97–103.

Marzadori, C., S. Miletti, C. Gessa, and S. Ciurli. 1998b. Immobilization of jack bean urease on hydroxyapatite: Urease immobilization in alkaline soils. Soil Biol. Biochem. 30:1485–1490.

Masciandaro, G., C. Macci, S. Doni, B E. Maserti, L.A. Calvo-Bado, B. Ceccanti, and E. Wellington. 2008. Comparison of extraction methods for recovery of extracellular β-glucosidase in two different forest soils. Soil Biol. Biochem. 40:2156–2161.

Mato, M.C., and J. Mendez. 1970. Inhibition of indoleacetic acid-oxidase by sodium humate. Geoderma 3:255–258.

Mayaudon, J. 1986. The role of carbohydrates in the free enzymes in soil, p. 263–206, *In* C.H. Fuchsman (ed.) Peat and water. Springer-Verlag, Berlin, Germany.

McLaren, A.D. 1975. Soil as a system of humus and clay immobilized enzymes. Chem. Scripta 8:97–99.

McLaren, A.D. 1978. Kinetics and consecutive reaction of soil enzymes, p. 97–116. *In* R.G. Burns (ed.) Soil enzymes. Academic Press, London, U.K.

McLaren, A.D., and L. Packer. 1970. Some aspects of enzyme reaction in heterogenous system. Adv. Enzymol. 33:245–308.

McLaren, A.D., G.H. Peterson, and I. Barshad. 1958. The adsorption and reactions of enzymes and proteins on clay minerals: IV Kaolinite and montmorillonite. Soil Sci. Soc. Am. Pro. 22:239–244.

Müller-Wegener, U. 1988. Interaction of humic substances with biota, p. 179–193. *In* F.H. Frimmel and R.F. Christman (eds.) Humic substances and their role in the environment. Wiley-Interscience, New York.

Murase, A., M. Yoneda, R. Ueno, and K. Yonebayashi. 2003. Isolation of extracellular protein from greenhouse soil. Soil Biol. Biochem. 35:733–736.

Naidja, A., P.M. Huang, and J.-M. Bollag. 1997. Activity of tyrosinase immobilized on hydroxylaluminum-montmorillonite complexes. J. Mol. Catal. A: Chem. 115:305–316.

Naidja, A., P.M. Huang, and J.-M. Bollag. 2000. Enzyme-clay interactions and their impact on transformation of natural and anthropogenic organic compounds in soil. J. Environ. Qual. 29:678–691.

Naidja, A., A. Violante, and P.M. Huang. 1995. Adsorption of tyrosinase onto montmorillonite as influenced by hydroxylaluminum coatings. Clay. Clay Min. 43:647–655.

Nannipieri, P. 1994. The potential use of soil enzymes as indicators of productivity, sustainability and pollution, p. 238–244. *In* C.E. Pankhurst, B.M. Doube, V.V.S.R. Gupta, and P.R. Grace (eds.) Soil biota: Management in sustainable farming systems. CSIRO Melbourne, Australia.

Nannipieri, P. 2006. Role of stabilized enzymes in microbial ecology and enzyme extraction from soil with potential applications in soil proteomics, p. 75–94. *In* P. Nannipieri and K. Smalla (eds.) Nucleic acids and proteins in soil. Soil Biology. Vol. 8. Springer-Verlag, Berlin, Germany.

Nannipieri, P., and L. Gianfreda. 1998. Kinetics of enzyme reactions in soil environments, p. 449–479. *In* P.M. Huang, N. Senesi, and J. Buffle (eds.) Structure and surface reactions of soil particles. John Wiley & Sons, New York.

Nannipieri, P., S. Grego, and B. Ceccanti. 1990. Ecological significance of the biological activity in soil, p. 293–355. *In* J.-M. Bollag and G. Stozky (eds.) Soil biochemistry. Vol. 6. Marcel Dekker, New York.

Nannipieri, P., E. Kandeler, and P. Ruggiero. 2002. Enzyme activities and microbiological and biochemical processes in soil, p. 1–33. *In* R.G. Burns and R.P. Dick (eds.) Enzymes in the environment. Activity, ecology and applications. Marcel Dekker, New York.

Nannipieri, P., P. Sequi, and P. Fusi. 1996. Humus and enzyme activity, p. 293–328. *In* A. Piccolo (ed.) Humic substances in terrestrial ecosystems. Elsevier Science, London, U.K.

Nielsen, K.M., L. Calamai, and G. Pietramellara. 2006. Stabilization of extracellular proteins by transient binding to various soil surfaces, p. 141–158. *In* P. Nannipieri and K. Smalla (eds.) Nucleic acids proteins in soil. Series: Soil biology. Vol. 8. Springer-Verlag, Berlin, Germany.

Norde, W. 2008. My voyage of discovery to proteins in flatland.... and beyond. Colloids Surf. B Biointerfaces 61:1–9.

Norde, W., W. Tan, and L. Koopal. 2008. Protein adsorption at solid surfaces and protein complexation with humic acids. J. Soil Sci. Plant Nutr. 8:64–74.

Novakova, J. 1988. Effect of clay minerals on the activity of soil enzymes. Folia Microbiol. 32:504–505.

Ogunseitan, O.A. 1993. Direct extraction of proteins from environmental samples. J. Microbiol. Meth. 17:273–281.

Ogunseitan, O.A. 2006. Soil proteomics: Extraction and analysis of proteins from soils, p. 95–115. *In* P. Nannipieri and K. Smalla (eds.) Nucleic acids and proteins in soil. Series: Soil biology. Vol. 8, Springer-Verlag, Berlin, Germany.

Pflug, W. 1980. Effect of humic acids on the activity of two peroxidases. Z. Pflanz. Bodenkunde 143:432–440.

Pflug, W., and W. Ziechmann. 1981. Inhibition of malate dehydrogenase by humic acids. Soil Biol. Biochem. 13:293–299.

Pietramellara, G., J. Ascher, F. Borgogni, M.T. Ceccherini, G. Guerri, and P. Nannipieri. 2009. Extracellular DNA in soil and sediment: Fate and ecological relevance. Biol. Fertil. Soil 45:219–235.

Quiquampoix, H. 1987. A stepwise approach to the understanding of extracellular enzyme activity in soil. I. Effect of electrostatic interactions on the conformation of a β-D-glucosidase adsorbed on different mineral surfaces. Biochimie 69:753–763.

Quiquampoix, H. 2000. Mechanisms of protein adsorption on surfaces and consequences for extracellular enzyme activity in soil, p. 171–206. In J.-M. Bollag and G. Stotzky (eds.) Soil biochemistry. Vol. 10. Marcel Dekker, New York.

Quiquampoix, H. 2008. Interaction with soil constituents determines the environmental impact of proteins. J. Soil Sci. Plant Nutr. 8:75–83.

Quiquampoix H., and R.G. Burns. 2007. Interactions between proteins and soil mineral surfaces: Environmental and health consequences. Elements 3: 401–406.

Quiquampoix H., S. Servagent-Noinville, and M.H. Baron. 2002. Enzyme adsorption on soil mineral surfaces and consequences for the catalytic activity, p. 285–306. In R.G. Burns and R.P. Dick (eds.) Enzymes in the environment. Activity, ecology and applications. Marcel Dekker, New York.

Quiquampoix, H., S. Staunton, M.H. Baron, and R.G. Ratcliffe. 1993. Interpretation of the pH dependence of protein adsorption on clay mineral surfaces and its relevance to the understanding of extracellular enzyme activity in soil. Colloids Surf. A: Physicochem. Eng. Aspects 75:85–93.

Rao, M.A., and L. Gianfreda. 2000. Properties of acid phosphatase-tannic acid complexes formed in the presence of Fe and Mn. Soil Biol. Biochem. 32:1921–1926.

Rao, M.A., F. Palmiero, L. Gianfreda, and A. Violante. 1996. Interactions of acid phosphatase with clays, organics and organo-mineral complexes. Soil Sci. 161:751–760.

Rao, M.A., F. Russo, V. Granata, R. Berisio, A. Zagari, and L. Gianfreda. 2007. Fate of prions in soil: Entrapment of a recombinant ovine prion protein in humic-like synthetic polymers. Soil Biol. Biochem. 39:493–504.

Rao, M.A., A. Violante, and L. Gianfreda. 1998. Interactions between tannic acid and acid phosphatase. Soil Biol. Biochem. 30:111–112.

Rao, M.A., A. Violante, and L. Gianfreda. 1999. The fate of soil acid phosphatase in the presence of phenolic substances, biotic and abiotic catalysts, p. 175–179. In J. Berthelin, P.M. Huang, J.-M. Bollag, and F. Andreux (eds.) Effect of mineral-organic-microorganism interactions on soils and freshwater environments. Kluwer Academic/Plenum, New York.

Rao, M.A., A. Violante, and L. Gianfreda. 2000. Interaction of acid phosphatase with clays, organic molecules and organo-mineral complexes: Kinetics and stability. Soil Biol. Biochem. 32:1007–1014.

Rosas, A., M.L. Mora, A.A. Jara, R. López, M.A. Rao, and L. Gianfreda. 2008. Catalytic behavior of acid phosphatase immobilized on natural supports in the presence of manganese or molybdenum. Geoderma 145:77–83.

Ruggiero, P., and V.M. Radogna. 1988. Humic acids-tyrosinase interactions as a model of soil humic-enzyme complexes. Soil Biol. Biochem. 20:353–359.

Ruggiero, P., J. Dec, and J.-M. Bollag. 1996. Soil as a catalytic system, p. 79–122. In J.-M. Bollag and G. Stotzky (eds.) Soil biochemistry. Vol. 9. Marcel Dekker, New York.

Ruggiero, P., J.M. Sarkar, and J.-M. Bollag. 1989. Detoxification of 2,4-dichlorophenol by a laccase immobilized on soil or clay. Soil Sci. 147:361–370.

Safari Sinegani, A.A., G. Emtiazi, and H. Shariatmadari. 2005. Sorption and immobilization of cellulase on silicate clay minerals. J. Colloid Interface Sci. 290:39–44.

Sanjay, G., and S. Sugunan. 2005. Glucoamylase immobilized on montmorillonite: Synthesis, characterization and starch hydrolysis activity in a fixed bed reactor. Catal. Commun. 6:525–530.

Sanjay, G., and S. Sugunan. 2006. Fixed bed reactor performance of invertase immobilized on montmorillonite. Catal. Commun. 7:1005–1011.

Sanjay, G., and S. Sugunan. 2007. Enzymes immobilized on montmorillonite K 10: Effect of adsorption and grafting on the surface properties and the enzyme activity. Appl. Clay Sci. 35:67–75.

Sannino, F., and L. Gianfreda. 2001. Pesticide influence on soil enzymatic activities. Chemosphere 45:417–425.

Sarkar, J.M., and J.-M. Bollag. 1987. Inhibitory effect of humic and fulvic acids on oxidoreductases as measured by the coupling of 2,4-dichlorophenol to humic substances. Sci. Total Environ. 62:367–377.

Sarkar, J.M., and R.G. Burns. 1984. Synthesis and properties of β-D-glucosidase-phenolic copolymers as analogues of soil humic-enzyme complexes. Soil Biol. Biochem. 16:619–625.

Sarkar, J.M., A. Leonowicz, and J.-M. Bollag. 1989. Immobilization of enzymes on clays and soils. Soil Biol. Biochem. 21:223–230.

Schäffer, A. 1993. Pesticide effects on enzyme activities in the soil ecosystem, p. 273–340. In J.-M. Bollag and G. Stozky (eds.) Soil biochemistry. Vol. 8. Marcel Dekker, New York.

Scheel, T., K. Pritsch, M. Schloter, and K. Kalbitz. 2008. Precipitation of enzymes and organic matter by aluminum. Impacts on carbon mineralization. J. Plant Nutr. Soil Sci. 171:900–907.

Skujins, J.J. 1976. Extracellular enzymes in soil. CRC Crit. Rev. Microbiol. 4:383–421.

Solaiman, Z., M.A. Kashem, and I. Matsumot. 2007. Environmental proteomics: Extraction and identification of protein in soil, p. 155–166. In A. Varma and R. Oelmüller (eds.) Advanced techniques in soil microbiology. Vol. 11. Springer-Verlag, Berlin, Germany.

Staunton, S., and H. Quiquampoix. 1994. Adsorption and conformation of bovine serum albumin on montmorillonite: Modification of the balance between hydrophobic and electrostatic interactions by protein methylation and pH variation. J. Colloid Interface Sci. 166:89–94.

Suderman, R.J., N.T. Dittmer, M.R. Kanost, and K.J. Kramer. 2006. Model reactions for insect cuticole sclerotization: Cross-linking of recombinant cuticular proteins upon their laccase-catalyzed oxidative conjugation with catechols. Insect Biochem. Molec. Biol. 36:353–365.

Tabatabai, M.A. 1994. Soil enzymes, p. 775–833. In R.W. Weaver, J.S. Angel, and P.S. Bottomley (eds.) Methods of soil analysis. Microbiological and biochemical properties, Part 2, Book Series no. 5. Soil Science Society of America, Madison, WI.

Tabatabai, M.A., and M. Fu. 1992. Extraction of enzymes from soils, p. 197–227. In G. Stozky and J.-M. Bollag (eds.) Soil biochemistry. Vol. 7. Marcel Decker, New York.

Theng, B.K.G. 1979. Formation and properties of clay-polymer complexes. Elsevier, Amsterdam, the Netherlands.

Violante, A., A. De Cristofaro, M.A. Rao, and L. Gianfreda. 1995. Physicochemical properties of protein-smectite and protein-Al(OH)x-smectite complexes. Clay Miner. 30:325–336.

Wallenstein, M.D., and M.N. Weintraub. 2008. Emerging tools for measuring and modeling the in situ activity of soil extracellular enzymes. Soil Biol. Biochem. 40:2098–2106.

6
Biogeochemical, Biophysical, and Biological Processes in the Rhizosphere

Philippe Hinsinger
Institut National de la Recherche Agronomique

Davey L. Jones
Bangor University

Petra Marschner
The University of Adelaide

6.1	Introduction	6-1
6.2	Rhizosphere Biochemistry	6-3
	Exudation and Secretion of Organic Compounds • Organic Nutrient Mobilization: Enzymatic Processes and the Fate of C, N, P, and S	
6.3	Rhizosphere Biophysics	6-7
	Porosity Development, Aggregation, and Compaction • Water and Nutrient Transport Processes	
6.4	Rhizosphere Biogeochemistry	6-10
	Gas Exchange and Redox Processes • Acidification and Alkalization • Inorganic Nutrient Mobilization	
6.5	Rhizosphere Ecology	6-14
	Communication, Signaling, and Sensing • Competition and Facilitation	
6.6	Conclusions: A Unique, Multifaceted, and Complex Interface	6-18
	References	6-19

6.1 Introduction

The rhizosphere is defined as the zone of high biological and chemical activity in the soil that surrounds the living root (e.g., Curl and Truelove, 1986; Pinton et al., 2007; Dessaux et al., 2010). The rhizosphere is chemically, physically, and biologically different from the bulk soil because it has been modified by roots and the associated microorganisms (Table 6.1). Detailed knowledge of the processes in the rhizosphere is fundamental to understanding the interactions of plants with their environment and the development of sustainable agricultural and forestry production systems. Similarly, the rhizosphere is important for many other environmental issues including biosphere C sequestration, greenhouse gas mitigation, human pathogen transfer in the food chain, and the removal, stabilization, and transformation of pollutants in soil (Huang and Gobran, 2004; Cardon and Whitbeck, 2007; Hinsinger et al., 2009; Philippot et al., 2009b). Despite its importance in many ecosystem services, our understanding of the rhizosphere remains relatively poor (Dessaux et al., 2009, 2010). This is due to the intrinsic complexity of the rhizosphere where many thousands of linked processes are operating simultaneously in a highly heterogeneous soil matrix. Unraveling this complexity is a long-term scientific goal that will not be fully realized within our lifetime. Ultimately, however, understanding these processes will allow managing the rhizosphere to harness its natural potential for the sustainable management of ecosystems (Ryan et al., 2009; Wenzel, 2009; Dessaux et al., 2010).

The dimensions of the rhizosphere vary greatly depending upon how it is defined. For example, gaseous chemicals released from roots (e.g., signaling molecules) are known to travel many meters in soil, and theoretically this zone of influence could be termed the *rhizosphere*. In addition, if one considers mycorrhiza, then the mycorrhizosphere may also greatly extend the root's sphere of influence, while the high root density in grasslands means that all the soil can be considered as rhizosphere. To most scientists, however, the rhizosphere encompasses the zone 1–2 mm away from the root surface where maximum microbial activity occurs and where the greatest changes in soil chemistry and physics are apparent (Jensen et al., 2005; Raynaud, 2010). The rhizosphere also extends into the root, the endorhizosphere; numerous studies have shown that symbiotic and nonsymbiotic bacteria and fungi colonize the cortex of the root (Vanpeer et al., 1990; Roesti et al., 2006; Figure 6.1). One of the key features is that every point in the rhizosphere is chemically, physically, and biologically unique because of the simultaneous influx into and efflux of chemicals from the root (e.g., inorganic nutrients, water, H^+, organic compounds), which results in unique diffusion gradient profiles in the soil (Figure 6.2). In addition to these root-induced gradients, rhizosphere processes are also highly

TABLE 6.1 Biophysical, Biochemical, Biogeochemical, and Biological Properties of the Rhizosphere

Root/Rhizosphere Microorganism Biological Function	Soil/Rhizosphere Property	Soil/Rhizosphere Process
Root growth/elongation	Bulk density/porosity Soil strength Soil structure	Fluid (water, gas) transfer (convection, diffusion)
Mucilage/exopolysaccharide production/secretion	Viscosity Surface tension	Adhesion of particles Aggregation Water/solute transfer
Water uptake (by roots)	Water content/tension Tortuosity Aeration	Water flow (along water tension gradient)/gas transfer Mass flow of nutrients
Nutrient/ion uptake (adsorption/absorption)	Nutrient/ion concentration	Ion exchange (adsorption/desorption) Precipitation/dissolution
Proton/hydroxyl release	pH	Acid/base reactions (chemical equilibria) pH-controlled biochemical reactions (e.g., enzymatic)
Respiration (CO_2 emission)	pCO_2	Acidification (see pH effect) except at low pH ($\ll pK_A$ of H_2CO_3)
Respiration (O_2 consumption)	pO_2/redox potential	Reduction of oxidized compounds
Release of reducing compounds (e.g., phenolics)	Reducing compound concentration	
Release of organic ligands (carboxylates, phenolics, (phyto)siderophores)	Organic ligand concentration	Ligand exchange Complexation of metals Chelation of metals
Release of enzymes	Enzyme concentration/activity	Enzymatic hydrolysis of organic compounds
Release of signaling molecules	Signaling molecule concentration	Structure of microbial/plant community

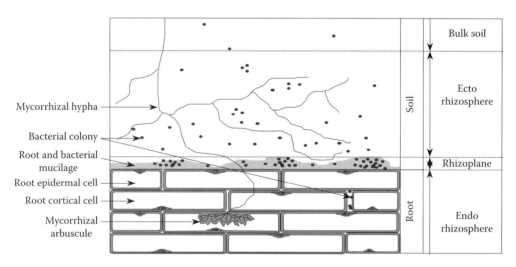

FIGURE 6.1 Schematic representation of an endomycorrhizal root showing the different parts of the rhizosphere.

variable along the root length in response to differences in root morphology (e.g., root hair zone versus lateral root emergence zone), differential exploitation of soil resources (e.g., water, P depletion), changes in the soil atmosphere (e.g., O_2, CO_2), and aging of the rhizosphere microbial community (e.g., in response to a reduction in the quality and quantity of root organic C flow into the soil). Together this spatial heterogeneity makes it extremely difficult to generalize rhizosphere processes occurring on the root system of a single plantlet alone to plants growing in different soil types or regions. This also makes it difficult to devise rhizosphere management techniques that are acceptable to industry and are generally applicable across wide geographical areas. This review aims to summarize current knowledge in the area of biogeochemical, biophysical, and biological processes occurring in the rhizosphere. Due to the number of published works in this area (>15,000), it is impossible to summarize every aspect of the rhizosphere; here we can all but provide an overview of the most important processes (see also Table 6.1).

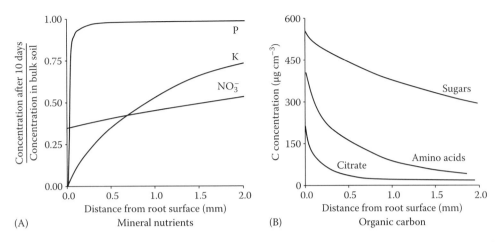

FIGURE 6.2 Schematic representation of the gradients in (A) mineral nutrients and (B) root exuded carbon in the rhizosphere, which makes every point in the rhizosphere chemically unique.

6.2 Rhizosphere Biochemistry

6.2.1 Exudation and Secretion of Organic Compounds

The biological and chemical processes operating in the soil surrounding the root can be dramatically different from those occurring in the bulk soil due to the many interactions occurring between roots, microorganisms, and soil particles. One of the main driving forces in changing the chemical, biological, and physical properties of the soil is the release of organic compounds (Pinton et al., 2007) by the root into the soil (termed root *exudation* or *rhizodeposition* (Nguyen, 2003; Jones et al., 2009). Indeed, this release of C provides the energy for microbial proliferation in the rhizosphere (Figure 6.1). Despite the central importance of C flow in rhizosphere processes, our knowledge of the subject still remains fragmentary and limited to a few well-studied examples, typically young crop plants studied ex situ. To a large extent, this is due to the intrinsic difficulties in studying rhizosphere C flow in situ at sufficient spatial resolution to draw valid conclusions. Although we have a range of isotopic forms of C at our disposal for tracing C in the plant–soil system (i.e., ^{11}C, ^{13}C, ^{14}C), our ability to use them effectively has often been limited by the inability to physically separate roots from soil and the lack of high resolution in situ imaging techniques. As a result of the lack of suitable techniques, the rhizodeposition literature contains many studies that have employed inappropriate methodology and from which erroneous conclusions have subsequently been drawn.

Despite experimental limitations, numerous isotopic tracer studies have quantified the amount of C fixed in photosynthesis that is subsequently partitioned belowground and then lost via exudation. Although there are numerous caveats in the interpretation of the results (Meharg, 1994; Nguyen, 2003; Thornton et al., 2004; Paterson et al., 2005), a meta-analysis of 271 studies undertaken by Jones et al. (2009) has revealed some consensus in C partitioning over a wide range of plant species and experimental conditions. Overall, it appears that when plants are fed with isotopically labeled $^{13}CO_2$ or $^{14}CO_2$, approximately 60% ± 10% of the fixed C (range 20%–98%) remains in the shoot, becoming incorporated into new growth and storage pools. The remaining 40% ± 10% is transferred belowground to the root system. Within the root, the newly fixed C has a number of fates including the transfer to new growing tips and symbionts (e.g., N_2 fixers, mycorrhizal fungi), the repair of older tissues, and loss in respiration or loss into the soil as gaseous, soluble, and insoluble exudates. Once in the soil, the released C can then undergo a number of fates including diffusion away from the root, uptake and respiration by soil microorganisms, uptake and incorporation into new microbial cells, abiotic sorption to soil particles, abiotic mineralization by mineral oxides, leaching through the soil, volatilization from the soil surface, and recapture by the root itself or by neighboring roots (Figure 6.3; Martin, 1970; Jones and Darrah, 1994; Kuzyakov et al., 2003; Paterson et al., 2007; Jones et al., 2009). Separating these numerous fluxes, which occur concurrently and which possess many interrelated feedback loops, has proved impossible, and this has limited a holistic understanding of C flow and the importance of individual processes. On balance, it appears that of the C that is transferred belowground, approximately 19% ± 8% remains in the root (range 5%–55%), 5% ± 1% is lost from the root and incorporated into soil organic matter (SOM) (range 2%–18%), and 12 ± 5 is lost as CO_2 (Jones et al., 2009). This respiratory loss arises from roots and from the microbial turnover of root exudates (range 2%–30%). Distinguishing between root and microbial respiration has been one of the major constraints to gaining an accurate measure of rhizodeposition in soil-grown plants. Using a range of experimental approaches, it is now believed that roots and microorganisms contribute equally to rhizosphere respiration (Kuzyakov, 2002a, 2006), an assumption that must be considered with caution. Carbon loss by rhizodeposition has therefore been estimated to be around 11% of the net fixed C or 27% of C allocated to roots, which corresponds to 0.4–0.6 Mg C ha^{-1} year^{-1} in cereals (Kuzyakov and Domanski, 2000). This can be compared to the average organic C content of a typical agricultural topsoil (0–30 cm) of 50 to 150 Mg C ha^{-1}.

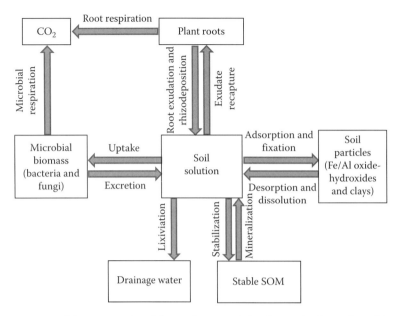

FIGURE 6.3 Schematic representation of the major pools and fluxes of root-exuded carbon in the rhizosphere. SOM indicates soil organic matter.

Typically, it takes minutes for the C fixed in the shoots to start being transferred belowground (Dilkes et al., 2004) and then only a few more minutes before the C is unloaded from the vascular bundle and enters the root metabolism. As root metabolic activity is closely linked to photosynthetic flux (Todorovic et al., 1999; Kuzyakov and Cheng, 2004; Hill et al., 2007), root exudation is very sensitive to C supply although some buffering does exist due to C storage in roots (e.g., as starch). This storage pool has the potential to buffer cytoplasmic metabolism for about 1–3 days if C supply from the shoots is cut off (Saglio and Pradet, 1980). Consequently, exudation into the soil varies dramatically depending upon the physiological state of the plant and the prevailing environmental conditions (Nguyen, 2003).

The composition of root exudates has largely been investigated in sterile hydroponic culture in the absence of soil to prevent loss of exudates by microbial decomposition and to facilitate their collection for analysis (Curl and Truelove, 1986). Virtually all compounds contained in root tissues can be released as root exudates including carbohydrates, organic and amino acids, peptides, phenolics, fatty acids, sterols, and proteins, including enzymes, vitamins, hormones, and nucleosides. Pioneering work in the 1960s and 1970s explored the impact of numerous plant and environmental factors on the amount and composition of root exudates. All studies showed that any change in experimental conditions or ontogenetic stage of the plant resulted in differing net release rates of C and that in each case, the plants produced a unique chemical exudation signature (Vancura, 1967; Rovira, 1969; Vancura and Stotzky, 1976). This has made it very difficult to generalize about the exudation process and its contribution to ecosystem C budgets. Further, it has also been found that exudation is not spatially uniform but varies both along the root length and with root age (McDougal and Rovira, 1970). Generally, rates of exudation are greater at growing root tips in comparison to mature root regions (Hoffland et al., 1989; Ryan et al., 2001; Shane and Lambers, 2005). Reasons for this higher efflux at the tip include preferential site for apoplastic unloading of solutes, higher solute concentrations in the cells, smaller vacuolar volumes, greater surface area of tip cells due to their smaller size, the lack of an endo- or exodermal layer, higher internal C turnover rates, and tips being a major site of plasma membrane vesicle fusion (Jones and Darrah, 1996; Jones et al., 2009). Emergence points for secondary roots can also lead to hotspots of C loss (McDougal and Rovira, 1970). However, some signaling molecules also show a highly targeted release (e.g., flavonoids in the root hair zone released as a chemoattractant for N_2 fixing *Rhizobium* bacteria; Vanrhijn and Vanderleyden, 1995). It should also be noted that while the discussion above is predominantly associated with C loss, root exudation is also synonymous with N loss (e.g., in the form of amino acids, peptides, and proteins).

It is well established that the release of insoluble polysaccharide gel (mucilage) is predominantly associated with root tips and due to secretion by root cap cells (Figure 6.4; Paull and Jones, 1975a, 1975b, 1976a, 1976b). This gel plays a number of roles in enhancing plant growth including (1) reducing frictional resistance at the root tip and thus preventing abrasive damage from soil particles (although this has been debated, e.g., in Read et al., 1999), (2) complexation and detoxification of toxic metals (e.g., Al, Cd, Pb), preventing inhibition of the actively growing root meristem, (3) buffering the pH around the root apex, and (4) promoting aggregate stability and soil–root contact, thereby facilitating water and nutrient transfer toward the root (Horst et al., 1982; Traore et al., 2000; Knee et al., 2001). In recent years, it has become apparent that the release of border cells at the root apex may also play an important role in regulating a range of rhizosphere processes (Hawes et al., 1998). Border cells are cells that are continually sloughed off from the root apex cells in a controlled manner. After their

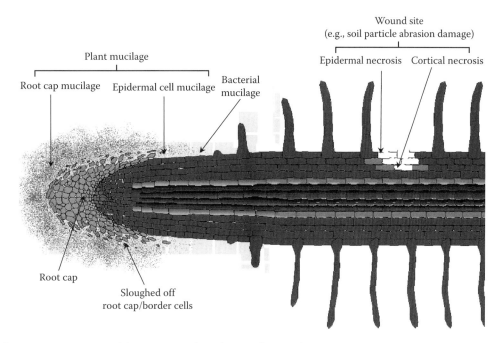

FIGURE 6.4 Schematic representation of the main sites of mucilage production along a root.

release from the root, they can survive for several days in the soil but eventually die, providing a substrate for microbial growth. Although they only represent a small C loss to the plant (0.1%–5% of rhizodeposition depending upon plant species), their role appears to be similar to that of mucilage in that they (1) reduce the frictional force experienced by the root tip (Iijima et al., 2003), (2) bind and keep aluminum and heavy metals away from the root (Horst et al., 1982; Llugany et al., 2003), (3) produce signal compounds that deter plant pathogens (e.g., nematodes and fungi; Hawes et al., 2000), and (4) release signals that attract potential symbionts (Brigham et al., 1995; Hawes et al., 1998; Nagahashi and Douds, 2004).

Until the 1990s it was generally accepted that the release of soluble C exudates from roots was purely a passive process whereby C compounds were simply lost in response to the large concentration gradient that exists between the root (solute concentrations within cells of 1–50 mM) and the soil solution (0.1–50 μM). However, we now know that this is a gross oversimplification of the process, casting doubt on many of the conclusions drawn from earlier studies. This arose as we gained a deeper understanding of the membrane transport processes involved in exudation and specifically that (1) plants can tightly regulate the spatial and temporal release of some compounds (e.g., organic acids, signaling compounds) and (2) that plants can recapture exudates from the soil (Jones et al., 2009). One key question that remains unanswered is why plant roots take up C compounds from soil? Current theories include that it allows (1) capture of C compounds that are delivered apoplastically to root cells (e.g., at growing tips), (2) that direct recapture of exudates is used to enhance the plant's C efficiency and to reduce rhizosphere microbial growth and pathogen attack, (3) direct uptake of C and N compounds allowing the capture of organic nutrients

released from SOM, and (4) that uptake provides a mechanism for inter-root and root–microbial signal exchange.

Of greatest ecological significance is the active root uptake of sugars and organic N compounds (e.g., amino acids and polyamines) from soil. Typically, these compounds are taken up into the root cells by cotransporters, which are constitutively expressed and located throughout the root system (Jones and Darrah, 1994, 1996; Figure 6.3). These cotransporters are powered by the plasma membrane H^+-ATPases, which are predominantly located in the epidermis rather than in the root cortex, although levels of H^+-ATPases are also high in the stellar regions (Samuels et al., 1992; Jahn et al., 1998). The transport proteins simultaneously transport H^+ across the plasma membrane into the cells together with individual organic solutes. The transporters are also relatively solute specific, with transport families for amino acids and sugars being nowadays well characterized at both the physiological and molecular level (Hirner et al., 2006).

6.2.2 Organic Nutrient Mobilization: Enzymatic Processes and the Fate of C, N, P, and S

6.2.2.1 Uptake of Soluble Organic Nutrients

As highlighted above, plant roots have the potential to actively take up low molecular weight (MW) organic compounds from the apoplast and from outside the root (e.g., amino acids, sugars, urea; Phillips et al., 2004; Personeni et al., 2007). From a limited number of studies, it appears that this trait is constitutively expressed and occurs along the entire length of the root and in all root types (Jones and Darrah, 1994, 1996). Although some root uptake can occur purely by passive diffusion down the electrochemical potential gradient across the plasma membrane, these diffusive flux rates appear to be low and largely limited to neutrally charged solutes in a protonated form (e.g., H-acetate^0) rather than in a metal-complexed form (e.g., Al-citrate^0; Jones and Darrah, 1995; Ogiyama et al., 2010). The main proportion of root uptake of amino acids and sugars is mediated by proton cotransporters, which are driven by H$^+$-ATPases. There is evidence that roots may possess multiple transport systems capable of high and low affinity transport to enable the root to respond to a wide range of external concentrations (Xia and Saglio, 1988). In addition, the transporters may be upregulated under extreme anoxia (Gharbi et al., 2009). The ecological significance of this uptake pathway, however, remains controversial. Initially, it was considered that the uptake of low MW solutes from soil was a way for the root to recapture C lost in exudation to the soil or from the apoplast when C is unloaded from vascular tissues (Jones and Darrah, 1994, 1996). However, a different but complementary theory is that transporters are present to enable roots to take up sugars and amino acids released during the microbial breakdown of organic matter in the soil. This could supplement the plant's supply of C and particularly N. In the case of amino acid uptake, it would bypass the need to wait for amino acid mineralization into NH_4^+ and NO_3^- by the soil microbial biomass (i.e., short-circuiting the N cycle). Although there is evidence to support the importance of this N uptake pathway in N limiting ecosystems (e.g., Arctic tundra and boreal forest; Chapin et al., 1993; Lipson and Näsholm, 2001; Näsholm et al., 2009), it appears to be of less importance in temperate and tropical environments (Owen and Jones, 2001; Kuzyakov and Jones, 2006). Further, the methodology for determining the uptake of isotopically labeled (^{13}C/^{15}N) organic solutes has drawn considerable criticism in recent years due to a lack of consideration for isotopic pool dilution, transformation of labeled compounds in soil, use of inappropriate tracer concentrations and experimental times, and difficulties in separating microbial and root uptake and mineralization (Jones et al., 2005; Rasmussen and Kuzyakov, 2009). A special case are phytosiderophores in Poaceae; these plant species have evolved transport systems to take up Fe and Zn, as well as other micronutrients such as Mn and Cu, once chelated by phytosiderophores (i.e., in the metal-complexed form), which are nonproteinogenic amino acid derivatives released by Poaceae roots (Marschner, 1995; Robin et al., 2008).

6.2.2.2 Enzymatic Processes in the Rhizosphere

Typically, key nutrients such as N, P, and S are locked up in insoluble organic matter in most soils. Consequently, for plants to access this nutrient pool SOM is enzymatically broken down into low MW soluble compounds and further into inorganic ions to be transported across the root plasma membrane. In most soils, this breakdown process can occur in the absence of plants by soil microorganisms with the aid of micro-, meso-, and macrofauna. However, the fact that microbial activity is higher in the rhizosphere than in nonrhizosphere soil has led to speculation that enzymatic turnover of SOM might be enhanced around the root (i.e., the rhizosphere priming effect; Ahmad and Baker, 1987; Kuzyakov, 2002b; Vong et al., 2003). This is supported by studies showing that root C inputs stimulate the mineralization of organic pollutants and SOM, which in theory should concomitantly release inorganic nutrients (Gerhardt et al., 2009). In addition, roots are known to release a range of enzymes involved in the breakdown of organic matter. In contrast, more recent studies have indicated a more complex picture where it is difficult to disentangle enzyme-mediated nutrient mobilization of SOM from that associated with enhanced rates of microbial biomass turnover (Weintraub et al., 2007). Irrespective of the source, there is no doubt that enzymatic activity is enhanced around the root. To date most functional enzyme work has focused on P mobilization and to a lesser extent N mobilization with scant attention paid to other nutrients (e.g., Fe, S). In the case of phosphatase, the observation that exudation of acid phosphatase from cluster roots of white lupin coincides with increased exudation of organic acids suggests that plants have coordinated adaptive strategies to maximize their potential synergistic effects under P-deficient conditions (Gilbert et al., 1999). Although efforts have been made to genetically engineer plants to release higher amounts of enzymes into the soil (e.g., phytase), these have so far proved unsuccessful at promoting plant P acquisition in situ (George et al., 2005). However, it is likely that roots will be engineered to release a range of enzymes, particularly in nonmycorrhizal crop plants.

In many instances, the organisms involved in SOM breakdown may also be symbiotically associated with roots, implying a direct connection between the site of breakdown and transfer directly to the root in exchange for C (e.g., mycorrhizas). Indeed, there are many studies showing higher rates of phosphatase activity in soil surrounding mycorrhizal roots and enhanced P and N transfer to the plant (Bolan, 1991; Leake and Read, 1991; Tarafdar and Marschner, 1994; Perez-Moreno and Read, 2000). In these experiments, it is often difficult to distinguish between enzymes released from the root and those from the mycorrhizal fungi. Typically, phosphatases released from fungi are able to mineralize a greater range of organic P forms and at a higher rate than those released from roots (Tarafdar et al., 2001). The magnitude of this response appears to be highly dependent upon both soil type and plant genotype but more importantly, the type of mycorrhizal fungal species (Tadano and Sakai, 1991; Joner and Johansen, 2000; van Aarle and Plassard, 2010). Responses may also be controlled by the composition of the rhizosphere microbial community with whom the mycorrhizal fungi are competing (Baum and Hrynkiewicz, 2006; Phillips and Fahey, 2006).

Mycorrhizas can release a wide range of enzymes involved in SOM breakdown (e.g., proteases, laccases, phenol oxidase, sulfatases, urease, chitinase, acid, and alkaline phosphatase), and many show an upregulated release pattern under nutrient limitation (Vazquez et al., 2000; Raiesi and Ghollarata, 2006). The increase in enzyme activity appears to be associated with the extraradical hyphae away from the root (i.e., the scavenging growth front rather than by hyphae at the root surface; Huang et al., 2009). Ectomycorrhizas may stimulate the dissolution of mineral P and mobilization of soil organic P simultaneously (Liu et al., 2005).

However, it should be noted that it is often difficult to differentiate between cause and effect in many nonfactorial experiments, particularly those involving arbuscular mycorrhizas that cannot be studied in the absence of the host (Kang and Freeman, 2007; Bai et al., 2009). This is particularly true where symbionts enhance plant growth by another mechanism (e.g., making roots less susceptible to H^+, Al^{3+}, or metal toxicity), which in turn affects root–shoot resource partitioning, plant nutrient demand, and microbial community structure in the rhizosphere, all of which can affect enzyme release (Kohler et al., 2007).

6.3 Rhizosphere Biophysics

6.3.1 Porosity Development, Aggregation, and Compaction

The soil matrix is responsible for considerable biophysical constraints on soil living organisms, as related to soil strength, and the distribution/availability of gas and water. These constraints are also faced by roots of higher plants (Bengough et al., 2006); consequently, they have evolved a number of strategies to avoid or minimize these biophysical stresses. Roots of higher plants grow in soil pores, either by (1) using preexisting macropores such as cracks and biopores or (2) building new biopores. In most instances, root growth induces a substantial reorganization of the soil matrix, via aggregation and compaction processes, and thus, higher plants strongly influence the physical structure of soils (Hinsinger et al., 2009).

At the whole plant scale, the spatial extension of the rhizosphere will ultimately depend on (1) the architecture of the root system, which is to a large extent genetically determined (Lichtenegger and Kutschera-Mitter, 1991; Zobel, 1991), and (2) root growth and development, which rely on the allocation of C and environmental factors, either biological, chemical, or physical (Robinson, 1994; Hodge et al., 2009). The plasticity of the root system is indeed quite remarkable, showing the ability of plants to tolerate and respond to adverse soil conditions. A meta-analysis of 475 soil profiles from 209 geographic locations revealed that more than 90% of the soil profiles had at least 50% of their root biomass within the top 30 cm, and that 95% of all roots were in the upper 2 m (Schenk and Jackson, 2002). However, these authors stressed that due to the technical difficulty to acquire quantitative data for large plants grown under field conditions, the extent of the root system may have been underestimated, particularly roots in the subsoil. There are a number of reports showing that many tree species have taproots growing to several tens of meters in depth, particularly in areas where water supply is limited (Canadell et al., 1996). Even fine roots can develop to a considerable depth in fast growing trees such as when growing in nutrient-poor soils (Jourdan et al., 2008). In Congo, Bouillet et al. (2002) showed that the distribution of fine roots of *Eucalyptus grandis* was quite homogeneous throughout the soil profile down to a depth of at least 3 m. Although plant roots growing in the upper horizon encounter rather favorable physical conditions (low bulk density, high organic matter content, and oxygen supply), deeper roots will have to grow through a soil material that is, often, densely packed and waterlogged at least for some time of the year or contains an increasing density of large rock fragments. The use of computer-assisted tomography has recently helped making major steps forward for estimating to what extent plant species can reuse preexisting macropores, particularly to find their way through hard pans at depth (Jassogne et al., 2007; Jassogne, 2008). These studies showed that a large proportion of roots tend to make use of the preexisting macropores where they encounter minimal soil strength, which is in line with previous studies by Stirzaker et al. (1996), Stewart et al. (1999), and Lesturgez et al. (2004). Additionally, some species create their own root macropores more efficiently than others, for example, tap-rooted, perennial legumes such as *Stylosanthes hamata* (Lesturgez et al., 2004), lucerne, and saltbush (*Atriplex nummularia*), a native species of Australia (Jassogne, 2008). However, most plant species will use existing pores, as shown by Williams and Weil (2004) with minirhizotron images of soybean roots growing along channels left by decomposing roots of canola in a compacted soil. In similarly dense soils, Watt et al. (2005) showed that more than half of the entire length of the root system of wheat was in pores formerly occupied by roots. Biopores such as earthworm burrows and root channels, that is, either recent or relic rhizospheres, not only represent areas of lower soil strength and greater aeration but may also favor root growth because they are nutrient-enriched zones with distinct chemical and microbiological properties (Pierret et al., 1999, 2007).

The soil strength around the root apex greatly influences the pressure that a root must exert to penetrate the soil. In the absence of preexisting macropores, a root must exert sufficient pressure to rearrange the soil particles, that is, to push them either aside or ahead of the root apex. The mechanical stress in the soil around the growing root is likely to be greatest in front of the root apex. Kirby and Bengough (2002) modeled that an increase in root diameter decreased the peak axial stress in front of the root tip to facilitate root penetration. Root-induced changes in the rhizosphere bulk density (and thus porosity) will thus be maximal where roots exert most mechanical stress on the surrounding soil, that is, at the root tip. Young (1998) reviewed the variations of soil geometry from rhizosphere to bulk soil and showed a small but significant body of work reporting increased bulk density close to the root–soil interface. Such root-induced changes of soil physical properties can be substantial. Bruand et al. (1996) reported bulk densities as great as $1.8\,g\,cm^{-3}$ next to the surface of maize roots, as compared with $1.5\,g\,cm^{-3}$ in the bulk soil. Soil porosity was shown to decrease almost exponentially away from the root surface (Dexter, 1987). Higher soil density around the root results in increased root–soil contact and thus increased hydraulic contact, which can increase the accessibility to soil resources. However, in front of the root tip any increase in bulk soil density could inhibit root growth, as stressed by Hinsinger et al. (2009).

FIGURE 6.5 Photographs illustrating (A) soil adhering to the root of a *Poaceae* species (i.e., soil rhizosheaths), (B) an SEM image showing rod-shaped bacteria colonizing the root surface of a wheat (*Triticum aestivum*) root, and (C) a TEM image of an endophytic rhizobacteria living between two cells in the root cortex.

Conversely, Martens and Frankenberger (1992) and Alami et al. (2000) showed a significant increase in porosity in the rhizosphere, which was presumably related to exopolysaccharides produced by rhizobacteria altering soil structure in the rhizosphere. Kaci et al. (2005) reported that exopolysaccharides produced by *Rhizobium* had properties that could increase soil aggregation. Besides microbial exopolysaccharides, as stressed above, polysaccharides are also released from roots, the so-called mucilages, that considerably affect rhizosphere biophysics and, thus, structure-dependent processes such as water transport (Czarnes et al., 2000). The combined effect of root hairs and exopolysaccharides produced by root and rhizosphere microorganisms (Watt et al., 1993) can lead to the formation of so-called rhizosheaths (Figure 6.5), which have been shown in a wide range of plant species and particularly grasses (Watt et al., 1994; Young, 1995; North and Nobel, 1997; McCully, 1999; Moreno-Espindola et al., 2007; Hinsinger et al., 2009). Rhizosheaths are remarkably stable and can either facilitate or restrict soil–root water transfer (Watt et al., 1994; Young, 1995). Wetting and drying cycles, which result from the diurnal pattern of water uptake of a growing plant, are likely to play a major role in maintaining the soil stability in the rhizosphere (Hinsinger et al., 2009). However, the factors influencing soil stability in the rhizosphere are largely unknown. This research area has received little attention, compared to rhizosphere biogeochemistry and biology.

6.3.2 Water and Nutrient Transport Processes

Besides anchoring the plants in the soil, the major role of the root system is to provide access to soil resources such as water and nutrients. Water uptake in the soil–plant–atmosphere continuum is a passive water flux driven by water potential gradients, with the root system providing the hydraulic continuity between the soil and the atmosphere, thereby playing a key role in the global water cycle (Hinsinger et al., 2009). The uptake of water by roots results in gradients in soil water content radially around a root if soil hydraulic conductivity becomes limiting for uptake as demonstrated experimentally by Hainsworth and Aylmore (1989) and Hamza and Aylmore (1992) and modeled by Doussan et al. (2003). Such gradients in soil water content or soil matric potential need to be taken into account in modeling plant water uptake. Doussan et al. (2006) combined models describing the architecture of the whole root system with a microscopic approach of describing water uptake based on potential gradients in soil–plant–atmosphere continuum and were able to reproduce the water distribution patterns obtained in experiments with crop species.

Hydraulic resistance during the axial transport of water is a key factor for water uptake by plants. The diurnal fluctuation in root diameter is a well-known phenomenon (e.g., Huck et al., 1970). It may give rise to large changes in root–soil contact (Van Noordwijk et al., 1993) and thus in hydraulic resistance of the root–soil interface. Exopolysaccharides produced by roots and microbes in the rhizosphere are considered to be important for retention and transport of water at the soil–root interface and in maintaining adequate soil–root contact as soils dry. Read et al. (1999) measured that maize mucilage may contain 99.9% water under dry conditions. Under wet conditions, McCully and Boyer (1997) showed that mucilage was able to retain even greater amounts of water than under dry conditions, that is, a typical gel behavior. Read et al. (2003) reported that surfactants in root mucilage, such as lecithin, may change the water-release properties within the rhizosphere, such that more water is released, particularly from relatively wet coarse textured soils. Whalley et al. (2005) showed for barley and maize that the rhizosphere was drier than the bulk soil at the same matric potential, partly as a result of changes in pore-size distribution and angle of wetting in the rhizosphere. On the other hand, hydrophobic compounds can also be produced in the rhizosphere, resulting in increased water repellency relative to bulk soil (Czarnes et al., 2000; Hallett et al., 2003). This further complicates the overall biophysical behavior of the rhizosphere.

As pointed out earlier, root distribution throughout the soil profile varies with season, plant development, and plant species, exhibiting a large plasticity to adapt to environmental conditions and especially to fluctuating availability of water and other soil resources in the various soil horizons (Callaway et al., 2003). When the topsoil dries, deeper roots may take up water from the ground water or soil layers close to it. In addition, roots may redistribute some water from wetter soil regions deeper in the soil profile to drier regions at the top of the soil profile. This so-called hydraulic lift occurs when water is released from roots when transpiration ceases (e.g., at night) and soil water potential in the dry soil becomes more negative than plant water potential. The amount of water transferred by hydraulic lift per night can range from 14% to 30% of daily evapotranspiration. Besides its role in water use efficiency, facilitation, or competition, this process may, more importantly, play a major role in nutrient uptake from dry topsoil and maintenance of root functions (Caldwell et al., 1998; Callaway et al., 2003; Newman et al., 2006).

Plant roots take up inorganic nutrients mainly as dissolved ions in the soil solution. Only a minor fraction of nutrients in the soil solution occur in the immediate vicinity of roots. Thus, nutrient interception is expected to contribute marginally to the overall acquisition of most nutrients by plants. Barber (1995) estimated for maize that interception was amounting to less than 2% of plant's requirement for major nutrients such as N, K, and P. In comparison, the uptake of water by roots, and the resulting gradient of water potential, is the driving force for a convective flow of dissolved nutrients toward the root's surface, often referred to as mass flow, which can supply a large proportion of plant nutrient requirements. For those nutrients that occur at high concentrations in the soil solution, for example, nitrate in many soils, mass flow can contribute a major proportion of the amount of N taken up by plants, about 80% for maize as estimated by Barber (1995). For some cations such as Ca and Mg, mass flow can even exceed plant demand, resulting in accumulation of Ca and Mg in the rhizosphere (Lorenz et al., 1994; Roose and Kirk, 2009). As a consequence of the low concentrations in the soil solution, mass flow fulfills only a minor part of plant's requirement for K, P, and micronutrients. Barber (1995) estimated for maize that mass flow contributed only 18% of K uptake and as little as 5% of P uptake.

When nutrient flux by mass flow is smaller than nutrient uptake, the nutrient concentration in the rhizosphere will decrease. Such depletion is expected to take place for those nutrients that occur at low concentration in the soil solution and/or for which the plant demand is large (Roose and Kirk, 2009). The resulting depletion zones around growing, single roots were first evidenced with autoradiographs of maize roots growing in radiolabeled soils by Lewis and Quirk (1967). The concentration gradient in the rhizosphere is the driving force for diffusion, which is the major transfer mechanism for poorly mobile nutrients at the soil–root interface. The diffusive flux of nutrients will vary considerably with the effective diffusion coefficient in soils, which is much smaller than the diffusion coefficient in water, because of the many physical and chemical interactions that ions encounter within the soil's solid phase (Tinker and Nye, 2000). This effective diffusion coefficient also depends on the soil water content, tortuosity, and buffering capacity, that is, the ability of the soil solid phase to replenish the soil solution. Although the diffusion coefficient in water is fairly similar among major nutrient ions such as NO_3^-, K^+, and $H_2PO_4^-$ (1.9×10^{-9}, 2.0×10^{-9}, and $0.9 \times 10^{-9}\,m^2\,s^{-1}$, respectively), their effective diffusion coefficient in soils spans several orders of magnitude, ranging from 10^{-10}–$10^{-11}\,m^2\,s^{-1}$ for NO_3^- ions, 10^{-11}–$10^{-12}\,m^2\,s^{-1}$ for K^+ ions to only 10^{-12}–$10^{-15}\,m^2\,s^{-1}$ for the least mobile $H_2PO_4^-$ ions. Such differences are mostly due to differences in the buffering capacity of soils. While $H_2PO_4^-$ ions strongly react with soil solid phase, via a whole range of precipitation/dissolution and adsorption/desorption reactions (see below), NO_3^- ions are much less prone to such interactions. The buffering capacity of soils for $H_2PO_4^-$ ions is thus much larger than that for NO_3^- ions, and hence their effective diffusion coefficient is up to 5 orders of magnitude smaller. Consequently, the radial distance from which a nutrient can diffuse toward the root surface, that is, the nutrient depletion zone, varies considerably among nutrients and soils, from centimeters for NO_3^- ions, millimeters for K^+ ions (Figure 6.6), to a fraction of a millimeter for $H_2PO_4^-$ ions. This is in agreement with a number of studies that have provided direct estimates of the extent of K and P depletion zones (e.g., Hendriks et al., 1981; Kuchenbuch and Jungk, 1982; Jungk and Claassen, 1986; Darrah, 1993; Begg et al., 1994; Hinsinger and Gilkes, 1996; Hinsinger, 1998; Jungk, 2002).

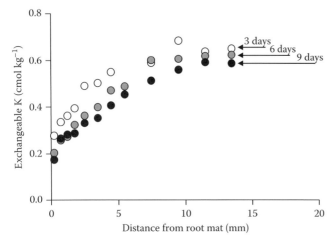

FIGURE 6.6 Depletion of exchangeable K in the rhizosphere of Italian ryegrass (*Lolium multiflorum* cv. Turilo) after 3, 6, and 9 days of growth in a root mat technique in a K-fertilized treatment of the long-term K fertilizer trial in Gembloux, Belgium. (From Hinsinger, P. 1990. Actions des racines sur la libération du potassium et l'altération de minéraux silicatés—Incidences agronomiques. PhD thesis. ENSA Montpellier, France.)

Ge et al. (2000) calculated the extent of the total volume of the rhizosphere based on the extent of the depletion zones and root architecture in young common bean plants. Based on their findings, Hinsinger et al. (2005) estimated that for effective diffusion coefficients ranging from 10^{-11} to $10^{-13}\,\mathrm{m^2\,s^{-1}}$, the volume of the P depletion zones of the whole root system of a young common bean plant is only about 0.4%–3% of the densely rooted topsoil. Such calculations will however underestimate the actual volume of soil supplying $H_2PO_4^-$ ions to plants. Indeed, using root-hairless mutants of *Arabidopsis thaliana* and barley, Bates and Lynch (2001) and Gahoonia et al. (2001) reported that root hairs play a major role in further extending the depletion zone around roots. Based on the SimRoot model, Ma et al. (2001) showed that the density and length of root hairs played a key role in determining P depletion around growing roots. Besides root hairs, arbuscular mycorrhizal (AM) fungi should also be taken into account as their hyphae may provide access to phosphate beyond the rhizosphere depletion zone as stressed by Zhu et al. (2001). Schnepf and Roose (2006) found that the depletion zone around a single hypha was negligibly small when hyphal uptake is small compared with diffusion in soil. More recently, Schnepf et al. (2008) calculated the contribution of AM hyphae to the extent of the P depletion zone. They showed that, provided that hyphae were able to take up P along their entire length, mycorrhizal hyphae can considerably extend the P depletion zone beyond the rhizosphere, which is in agreement with results of Schweiger and Jakobsen (1999). Li et al. (1991) showed that the extent of the depletion zone of P by AM plants is up to several centimeters from the root surface, compared with several millimeters in nonmycorrhizal plants. Schnepf et al. (2008) thus concluded that most P uptake of mycorrhizal plants occurs via the hyphae, which also means that much of the transport of poorly mobile nutrients such as P occurs via the hyphal network. However, this conclusion is based on a model at the scale of a single root and will need to be upscaled to the whole root system for drawing definitive, more ecologically relevant conclusions. Dunbabin et al. (2006) have shown the extra benefit of a whole root system model compared with a single root model for predicting P acquisition in the rhizosphere. Accounting for rapid root growth, as can be done only in a whole root system model, may provide a better indication of the relative contribution of roots and mycorrhizal hyphae to P transport and uptake in the (mycor)rhizosphere.

In contrast with the abundant literature on gradients of major nutrients in the rhizosphere, the transport of micronutrients is much less well documented. Lorenz et al. (1994) estimated that diffusion was a major transport pathway for Zn in the rhizosphere of radish and the depletion profiles of various pools of soil Zn were shown by Kirk and Bajita (1995) in the rhizosphere of lowland rice. Recently, Bravin et al. (2009b) showed that the gradient of Cu concentration (both for soil solution total Cu and Cu^{2+}) in the rhizosphere of durum wheat extended over several millimeters, that is, much beyond the depletion zone expected to occur for a poorly mobile micronutrient such as Cu. However, Bravin et al. (2009b) suggested that the observed decrease in soil solution Cu was the consequence of root-induced increase in rhizosphere pH rather than Cu uptake and diffusion. Hence, mechanisms other than biophysical processes must be taken into account to fully understand nutrient patterns in the rhizosphere.

6.4 Rhizosphere Biogeochemistry

6.4.1 Gas Exchange and Redox Processes

Gas exchange occurs in the rhizosphere primarily as a consequence of respiration of roots and the surrounding microorganisms, which results in a decrease in pO_2 and increase in pCO_2. The decline in pO_2 leads to a decrease in soil redox potential, thereby affecting many biogeochemical cycles. Most plant species are sensitive to hypoxia/anoxia, whereas wetland plant species have evolved specific strategies to cope with hypoxic conditions. The adaptation of wetland plants to hypoxic (submerged) soils is based on the development of a special tissue within their roots, the so-called aerenchyma, which conducts O_2 from the shoots (Armstrong, 1979). Leakage of part of this O_2 can result in an increase in pO_2 in the rhizosphere of wetland plants species (Flessa and Fischer, 1992; Revsbech et al., 1999; Armstrong et al., 2000). Most of the reported direct measurements of redox potential in the rhizosphere of rice plants were conducted only at the surface of the roots (Flessa and Fischer, 1992; Revsbech et al., 1999; Bravin et al., 2008). However, Begg et al. (1994) reported oxidation of Fe^{2+}, which is dominant in the soil solution under reduced conditions of submerged soils, extending several millimeters from the surface of rice roots. In the rhizosphere of wetland plant species, the well-documented formation of the so-called iron plaque is due to (1) the oxidation of Fe^{2+} to Fe^{3+} (Begg et al., 1994; Kirk and Bajita, 1995) and

(2) the subsequent precipitation of ferric iron oxides (e.g., goethite), due to the low solubility of these Fe^{3+} minerals (Bacha and Hossner, 1977). These studies and that of Bravin et al. (2008) have shown that iron plaque occurs in the root apoplasm and on the surface of the roots. Besides O_2 leakage from root aerenchyma, Trolldenier (1988) reported that part of the oxidation in the rhizosphere of lowland rice was due to Fe^{2+}-oxidizing bacteria. Neubauer et al. (2007) showed for *Juncus effusus* L. that rhizosphere oxidation was partly mediated by lithotrophic bacteria. In addition to the formation of iron plaque in wetland plant species, Mn^{2+} can also be oxidized leading to formation of manganese oxides at the surface of roots growing in reduced soil conditions (e.g., Bacha and Hossner, 1977).

In contrast to the wealth of studies in reduced soil conditions, there have been only few published reports of decreased pO_2/redox potential in the rhizosphere of plants growing in oxic conditions. Using arrays of redox microelectrodes, Fischer et al. (1989) found that redox potential decreased from 700 to less than 380 mV in the vicinity of a growing root tip of faba bean. One day later, after the root apex had passed the microelectrode due to root growth, redox potential increased to its initial value of 700 mV. This can be explained by the high respiration rate in the meristematic zone of the root, as shown by Bidel et al. (2000) in roots of peach tree seedlings. The meristematic region of the root is known for its intense metabolic activity and respiration, particularly in growing roots. Bidel et al. (2000) reported that O_2 consumption increased sevenfold in an actively growing root apex compared with that in nongrowing roots, reaching values greater than 35×10^{-14} mol O_2 s^{-1}. These authors also estimated the flux of O_2 due to microbial respiration in the rhizosphere of roots growing in agar gel, which decreased abruptly from 10 to 1 nmol O_2 m^{-3} gel s^{-1} at 300 μm from the root surface near the apex. Cornu et al. (2007) monitored the redox potential in the rhizosphere of tomato with daily resolution, in oxic soil conditions. They showed that the iron status of tomato plants had no significant effect on the redox potential of the rhizosphere, in spite of an increased reductase activity at the root surface as a result of Fe deficiency. Increased reductase activity is part of the strategy of acquisition of Fe of most plant species (all but Poaceae) and is the consequence of the activity of plasmalemma-bound enzymes that are activated under Fe-deficient conditions as frequently occurring in calcareous soils (Marschner, 1995; Robin et al., 2008).

Because of the respiration of belowground organisms and roots, bulk soil pCO_2 is commonly 10- to 100-fold larger than atmospheric pCO_2 (360 μmol mol^{-1}). Given that roots and microbes are major contributors to soil respiration, the rhizosphere should be a hot spot of elevated pCO_2, particularly in the region behind the apex of actively growing roots, in line with the decreased pO_2 reported by Bidel et al. (2000). This is, however, little documented in soils. To our knowledge, the only study is that of Gollany et al. (1993), who measured pCO_2 values as high as about 100,000 μmol mol^{-1} at 1–3 mm from the root surface. As stressed by Hinsinger et al. (2009), it is rather astonishing that in contrast with the current attention on aboveground pCO_2, so little data is available about rhizosphere pCO_2 and its impact on belowground organisms and biogeochemical cycles.

6.4.2 Acidification and Alkalization

In addition to the above-mentioned processes, the respiration of roots and microorganisms can also significantly alter the rhizosphere pH. Respired CO_2 rapidly forms the weak acid H_2CO_3 ($pK = 6.36$) in water. Thus, except for the most acidic soils, where H_2CO_3 remains undissociated, the increase in pCO_2 will result in rhizosphere acidification. Based on the values of pCO_2 in the rhizosphere measured by Gollany et al. (1993) and on the relationship between pH and pCO_2 in calcareous soils based on the calcite dissolution/precipitation equilibrium, Hinsinger et al. (2003) calculated that in situ rhizosphere pH values could be as low as about 6.7–6.8, compared with 8.3 at atmospheric pCO_2. However, such a change in rhizosphere pH by root and microbial respiration is surprisingly little documented and normally not taken into account in studies of rhizosphere biogeochemistry.

In contrast, the control of rhizosphere pH by proton influx/efflux from roots has been extensively studied, as reviewed by Nye (1981) and Hinsinger et al. (2003). When the influx of cations into root cells is greater than that of anions, the excess of positive charges entering the roots is counterbalanced by a net efflux of protons and thus rhizosphere acidification (Hinsinger et al., 2003), whereas alkalization occurs if there is a net influx of excess anions over cations. All charged ionic species that are released from roots, such as carboxylates, which can be exuded in large amounts by roots of some plant species (Jones, 1998; Hinsinger et al., 2003), need to be taken into account in this balance. Among inorganic ions, nitrogen has a major impact on root-mediated pH changes in the rhizosphere because it is taken up in large amounts as either an anion (nitrate) or a cation (ammonium) or even as the uncharged species N_2 in plant species such as legumes, which rely on symbiosis with N_2-fixing bacteria. The uptake of nitrate will result in rhizosphere alkalization, while the uptake of ammonium and, to a lesser extent, N_2 fixation result in rhizosphere acidification (Riley and Barber, 1971; Marschner and Römheld, 1983; Gahoonia et al., 1992; Plassard et al., 1999; Tang et al., 2004). Rhizosphere alkalization confers an adaptative advantage on plants growing in acid soils, as it can alleviate toxicities of aluminum (Pineros et al., 2005) and other metal cations such as copper (Michaud et al., 2007; Bravin et al., 2009a, 2009b). Increased proton efflux from roots, and thus rhizosphere acidification, is well known as an adaptive response of many plant species to both Fe and P deficiencies (Römheld and Marschner, 1989; Hinsinger, 2001; Tang et al., 2004; Robin et al., 2008). Riley and Barber (1971), and many others, have shown that rhizosphere pH changes can dramatically alter the acquisition of P. In most soils, rhizosphere acidification as a result of ammonium supply leads to improved P acquisition by plants, relative to plants fed with nitrate (Riley and Barber, 1971; Gahoonia et al., 1992; Hinsinger and Gilkes, 1996; Hinsinger, 2001). However, as pointed out by Gahoonia et al. (1992) and Hinsinger (2001), rhizosphere alkalization may enhance P acquisition in other soils, for example, acidic oxisols. Using a mechanistic modeling approach, Devau et al. (2009, 2010) suggested that rhizosphere alkalization could also result in a redistribution of P sorbed onto the various positively charged minerals in a neutral soil. In an acidic soil, Pearse et al. (2006a, 2006b) showed that wheat and oilseed rape were more P efficient than white lupin, due to their ability to increase the rhizosphere pH and in spite of their reduced capacity to exude large amounts of carboxylates compared with white lupin cluster roots. Recently, Bravin et al. (2009a) reported in a Cu-contaminated, acidic soil (bulk soil pH = 4.6–4.8) a considerable increase in rhizosphere pH (pH = 7.5) for nitrate-fed durum wheat, compared with ammonium-fed plants (pH = 3.9). As a consequence of the strong alkalization in the rhizosphere of nitrate-fed durum wheat, the soil solution Cu concentration in the rhizosphere and thus Cu uptake by plants were considerably lower than those in ammonium-fed durum wheat. Their study also showed that rhizosphere pH can be more than 2 pH units above bulk soil pH. A number of studies have reported that rhizosphere acidification, which occurs as a consequence of N_2 fixation in legumes or ammonium supply in other plant species, can lead to a pH decrease of up to about 2 pH units compared with the bulk soil (e.g., Gahoonia et al., 1992; Li et al., 2008). Soil pH has a dramatic effect on below-ground organisms as stressed by recent biogeographical studies at various scales, from global to field scale, for example, Fierer and Jackson (2006) and Philippot et al. (2009a). Hence, root-induced pH changes may be a major driver of microbial communities and their function in the rhizosphere.

Beside roots, rhizosphere microorganisms can also produce carboxylates and protons (carboxylic acids), thereby contributing to rhizosphere acidification. This has been well documented for ectomycorrhizal and saprophytic/pathogenic fungi, among which some species have been shown to release massive amounts of oxalic acid (Dutton and Evans, 1996; Wallander, 2000; Casarin et al., 2004). In lowland rice, the coupling of rhizosphere pH changes with redox processes has also been documented for ferrous iron oxidation and the subsequent ferric iron oxide precipitation (e.g., Begg et al., 1994). Begg et al. (1994) calculated that this process accounted for a major proportion of rhizosphere acidification, aside from the uptake of ammonium, which is the major form of nitrogen under reduced conditions.

6.4.3 Inorganic Nutrient Mobilization

6.4.3.1 Desorption and Ligand Exchange

Many nutrient anions (e.g., SO_4^{2-}, HPO_4^{2-}) are adsorbed on positively charged soil constituents with large specific surface areas, mostly metal oxides and clay minerals (Okazaki et al., 1989; Kudeyarova, 1991; Saeki and Matsumoto, 1998; Gustafsson, 2001). This surface sorption reduces anion diffusion to the root; consequently, plants have found ways to accelerate desorption of nutrients held on these surfaces. In most soils, these exchange surfaces are composed of Fe and Al (hydr)oxide minerals (e.g., gibbsite, hematite, ferrihydrite, goethite) as well as clay minerals (e.g., kaolinite, illite, smectite) as recently stressed by the work of Devau et al. (2009, 2010). These minerals present –OH and –OH_2 groups on their surfaces. One or two of the OH^- groups on anions such as $H_2PO_4^-$ can substitute an –OH group on the mineral surface, thereby forming a monodentate or bidentate ligand inner sphere complex. Although pH dependent, ligand exchange takes place even if the net surface charge is the same as that of the adsorbate. These anions are not held irreversibly; however, a higher concentration of other anions in the soil solution may subsequently displace them through ligand exchange. In a rhizosphere context, most interest has focused on the role of organic acid anions (e.g., citrate^{3-}, oxalate^{2-}) in ligand exchange and P release from mineral surfaces (Jungk et al., 1993; De Cristofaro et al., 2000; Hinsinger, 2001; Essington and Anderson, 2008). Although it is quantitatively difficult to separate between mineral dissolution, ligand exchange, and other indirect processes that can occur when organic acids are introduced into soil (e.g., changes in pH and ionic strength), it is clear that ligand exchange plays a role in nutrient mobilization in the rhizosphere. It appears that ligand exchange is greater for di- and tricarboxylic acids than that for monocarboxylic acids and that high concentrations of competing anions are required for the process to be of significance (Jones, 1998). More work is required to quantify the significance of ligand exchange compared with other nutrient mobilization mechanisms. For this purpose, the use of mechanistic modeling will provide new opportunities, as recently exemplified in the work of Devau et al. (2010) who showed that, in addition to P uptake, root-induced alkalization and Ca uptake had to be accounted for in order to properly simulate the observed changes of P availability in the rhizosphere of durum wheat, as a result of complex interactions with surface charges and P adsorption on the various P-sorbing soil minerals.

6.4.3.2 Complexation and Chelation

Complexes and chelates play an important role in nutrient acquisition and therefore in microbe–microbe and plant–microbe interactions. Complexation by microbial and plant carboxylates and phenolic compounds increases the availability of P and trace elements (e.g., Gerke, 1993; Hu et al., 2005). Carboxylates increase P availability by two mechanisms: anion (ligand) exchange and solubilization of Fe and Al (Gerke and Hermann, 1992). Due to the neutral cytoplasmic pH, carboxylates are released as anions (deprotonated); therefore, they do not decrease per se the soil pH (Jones, 1998; Hinsinger et al., 2003). As mentioned above, the carboxylate anions compete with phosphate anions for binding sites, thus releasing P into the soil solution (Gerke, 1993, 1994). A large proportion of P in soils is bound to humic–Fe–complexes and to amorphous Fe-oxides (Gerke and Hermann, 1992). Carboxylates form water-soluble complexes with Fe and Al, thereby decreasing the free Fe and Al ion concentration in the rhizosphere soil solution. This leads to increased solubilization of Fe^{3+} or Al^{3+} and thus release of P bound to Al/Fe oxides or bound to clays and organic matter via Fe/Al bridges (Gerke et al., 2000).

Chelates are small organic molecules that bind metals and change their charge. This changes the behavior of metals in the soil, particularly if the chelate–metal complex has a negative charge because it will be repulsed from negatively charged surfaces, for example, clays and organic matter. Moreover, chelation leads to increased release of metals from soil particles (e.g., metal oxides) to regain the equilibrium concentration of ionic forms in the soil solution.

Chelates that bind Fe have been extensively studied because of their important role in Fe uptake by plants and microbes. Plants can be divided in two groups based on their Fe acquisition mechanism: Strategy I and Strategy II plants (Marschner, 1995). Dicots and nongramineous plants employ Strategy I: under Fe deficiency, they increase the Fe reductase activity of the roots, release protons to decrease the rhizosphere pH, and may also increase the exudation of chelating compounds such as phenolics. Gramineous plants (Poaceae, Strategy II) respond to Fe deficiency by releasing phytosiderophores (nonproteinogenic amino acids such as mugineic acid) (Takagi, 1976; Marschner, 1995; Robin et al., 2008), which preferentially chelate Fe but can also bind Zn, Mn, or Cu. The Fe–PS complexes are taken up intact by root cells, Fe reduction taking place in the cells (Römheld, 1991; Marschner, 1995). Cereal species differ in capacity to release phytosiderophores under Fe deficiency, with barley and wheat releasing more than maize and sorghum (Römheld and Marschner, 1989; Marschner, 1995). Phytosiderophores are also released in response to Zn deficiency (Zhang et al., 1991). However, it is not clear if this is a separate regulation of phytosiderophore release or due to impaired Fe metabolism induced by Zn deficiency.

Microorganisms also release chelating compounds under Fe deficiency: organic acids, but more importantly, siderophores, which can be classified as hydroxamates or catechols (Neilands, 1984; Hoefte, 1993; Robin et al., 2008). A given microbial species may produce one or several siderophores (Bossier et al., 1988; Raaijmakers et al., 1995). The utilization of different siderophores, particularly those released by other microorganisms, appears to be important for the competitive ability of bacteria (Mirleau et al., 2000; Lemanceau et al., 2009). Bacterial siderophores are a poor iron source for plants (Bar-Ness et al., 1991) and can inhibit Fe uptake by plants (Becker et al., 1985). On the other hand, certain fungal siderophores such as rhizoferrin have been shown to increase Fe uptake by plants (Yehuda et al., 1996). The affinity of a given chelator for iron can be expressed as stability constant; the higher the stability constant, the greater the affinity of the chelator to Fe (Table 6.2). Bacterial siderophores such as ferrioxamine have a greater affinity to Fe than phytosiderophore (Table 6.2) (Crowley et al., 1991). Thus, if present at similar concentrations, Fe will bind preferentially to siderophores rather than phytosiderophores. Fe uptake by roots and translocation into shoots of maize is significantly lower if Fe is supplied bound to siderophores compared with Fe–PS (Crowley et al., 1992). On the other hand, phytosiderophores are good Fe sources for microbes (Jurkevitch et al., 1993).

6.4.3.3 Dissolution and Precipitation of Minerals

Soil nutrients occur in a number of minerals that can release their constituents via dissolution processes. Alternatively, nutrients can be immobilized by precipitation processes. These dissolution/precipitation processes can be promoted in the rhizosphere as a consequence of changes in pH and concentrations of cations or anions, including organic ligands. The mobilization of nutrients in the rhizosphere is well documented for K and P (for P, see above). Hinsinger et al. (2001) reported that roots can also increase the dissolution rate of a number of other nutrients (Ca, Mg, Si) when supplied as a basalt rock powder, with the strongest dissolution rate for Fe. Hinsinger and Jaillard (1993) demonstrated in ryegrass that K uptake by roots and the resulting decrease in solution K concentration is the driving force for the selective dissolution of a phlogopite mica, with the rate of release of interlayer K from phlogopite and biotite micas increasing approximately five- to sixfold, compared with unplanted control (Hinsinger et al., 2006). This rhizosphere depletion of K in the soil solution also explains the results of long-term fertilizer trials conducted in Europe showing that the release of nonexchangeable K can amount up to $100\,kg\,K\,ha^{-1}\,year^{-1}$ over several decades (Hinsinger, 2002). The magnitude of the release of nonexchangeable K was expected to be negligible in such temperate, arable soils, because their bulk soil solution K concentration typically ranges from 100 to $1000\,\mu M$. Indeed, Springob and Richter (1998) showed that the rate of release of nonexchangeable K is almost nil at such soil solution concentrations. They also showed that the release of nonexchangeable K is strongly enhanced at soil solution K concentrations below $3\,\mu M$, which is the range of concentration expected to occur close to maize roots, according to the estimations by Claassen and Jungk (1982). Thus, the depletion of soil solution K in the rhizosphere is the driving force of the release of interlayer K and concomitant dissolution of micaceous minerals in soils (Hinsinger and Jaillard, 1993; Hinsinger et al., 2006). Barré et al. (2007) showed that the uptake of K by ryegrass quantitatively matches the increased amount of interstratified illite–smectite minerals in the rhizosphere, forming at the expense of the illite-like clay minerals. Their findings are in agreement with those of Kodama et al. (1994) who reported earlier an increased amount of interstratified illite–vermiculite minerals in the rhizosphere of field-grown maize plants compared with the bulk soil. Accumulation of easily available K in the rhizosphere has also been found in situ for field-grown plants such as forest trees (e.g., Turpault et al., 2005). This is probably due to the combination of uptake-driven depletion and weathering of K-bearing minerals, which, over longer timescales, may contribute a net increase in soil solution and exchangeable K in the rhizosphere.

Changes in rhizosphere pH are also major determinants of dissolution/precipitation equilibria of nutrient-bearing minerals as shown for, for example, Mg (Hinsinger et al., 1993;

TABLE 6.2 Apparent Stability Constants Fe

Chelating Compound	Apparent Stability Constant
Mugineic acid (phytosiderophore)	19.1
Rhizoferrin (fungal siderophore)	19.7
EDTA (ethylene diamine tetra acetic acid; synthetic chelator)	22.2
Ferrioxamine B (bacteria siderophore)	25.1

Source: Taken from Yehuda, Z., M. Shenker, Y. Hadar, and Y. Chen. 2000. Remedy of chlorosis induced by iron deficiency in plants with the fungal siderophore rhizoferrin. J. Plant Nutr. 23:1991–2006. With permission from Taylor & Francis.

Calvaruso et al., 2006) and P (Gahoonia et al., 1992; Begg et al., 1994; Hinsinger and Gilkes, 1996; Bertrand et al., 1999). Rhizosphere acidification induced by oilseed rape induces the formation of hydroxy-interlayered vermiculite and concomitant release of metal cations such as Mg (Hinsinger et al., 1993). The role of rhizosphere pH in the biologically mediated dissolution of silicate minerals has been studied in detail in forest tree seedlings (Wallander and Wickman, 1999; Bakker et al., 2004). Ectomycorrhizal fungi were largely responsible for elevated concentrations of citric and oxalic acids, resulting in mobilization of K and Mg from biotite (Wallander and Wickman, 1999). This mobilization may be due to a direct pH effect or the complexation by carboxylates, which also affects the dissolution of silicate minerals (Huang and Keller, 1970; Jones et al., 1996; Jones, 1998). The direct impact of pH changes on the weathering of silicates such as micas and feldspars in the soil around ectomycorrhizal fungal hyphae was shown by Arocena et al. (1999) and Arocena and Glowa (2000). Ectomycorrhizal fungi supposedly play a major role in these processes, rather than the root, as reviewed by Landeweert et al. (2001) and Hoffland et al. (2004). Fungal hyphae are indeed able to dissolve silicate minerals, forming tunnel-like pores inside mineral grains (Jongmans et al., 1997). Moreover, ectomycorrhizal fungi were shown to select for different bacterial communities that produce acidic or chelating metabolites, which also affect the weathering of minerals (Calvaruso et al., 2007; Uroz et al., 2007). Besides silicate minerals, numerous studies with rock phosphate have further proved that H^+ release by plant roots can considerably increase the dissolution of apatite-like calcium phosphate minerals and hence the availability of P in the rhizosphere (see Hinsinger, 1998, and many references therein). Some species such as buckwheat, oilseed rape, and legumes have been reported to be particularly efficient in using P from calcium phosphates, which is related to their ability to release H^+ (e.g., Aguilar and van Diest, 1981; Amann and Amberger, 1989; Hinsinger and Gilkes, 1995; Zoysa et al., 1998) and/or carboxylic acids (e.g., Hoffland et al., 1989; Ae et al., 1990). The role of P-solubilizing microorganisms (fungi and bacteria) and ectomycorrhizal fungi in enhancing the rates of dissolution of apatite-like calcium phosphate minerals in the (mycor)rhizosphere has been extensively studied (e.g., Wallander, 2000; Reyes et al., 2002).

6.5 Rhizosphere Ecology

6.5.1 Communication, Signaling, and Sensing

In this review, communication is defined as a two-way process between two organisms. For example, one organism releases compounds that are sensed by the other organism, which responds by changing its physiology or by producing other compounds that induce changes in the first organisms' physiology/morphology. Hence, communication requires signaling, sensing, and response. Signaling is the first part of communication. It is a one-way process, that is, a specific compound released by one organism will induce a response in the other. Sensing is a prerequisite for the response. The signal has to be recognized and understood, which then induces a specific or unspecific response in the sensing organism.

6.5.1.1 Communication between Plants and Microbes

The classical example of communication between plants and microorganisms is the complex exchange of signals between legumes and *Rhizobium* or plants and pathogens. In legume–*Rhizobium* interactions, legume roots release flavonoids, which trigger the production of factors that induce nodule formation in the roots, for example, nod factor by *Rhizobium*. In plant–pathogen signaling, certain root exudate components attract species-specific pathogens. A thorough discussion of these classical examples would be beyond the scope of this chapter; the reader is referred to reviews for *Rhizobium* (e.g., Werner et al., 1994; Lugtenberg et al., 1995) and for plant–pathogen signaling (e.g., Bent, 1996; Hahn, 1996; Knogge, 1996; Walton, 1996; Avis et al., 2008). Various facets of communication in the rhizosphere have also been recently reviewed and critically discussed by Bais et al. (2004), Faure et al. (2009), Hartmann et al. (2009), and Sanon et al. (2009).

Another plant–microorganism communication that has attracted considerable interest is the interaction between plants and plant growth–promoting rhizobacteria/fungi (PGPRs). PGPRs, for example, *Azospirillum* sp., enhance plant growth directly via phytohormone production (cytokinins, gibberellins, auxin) and improved nutrient uptake (N_2 fixation, nitrate reduction, P solubilization) or indirectly by inhibiting pathogens or deleterious bacteria (Lugtenberg et al., 1991; Somers et al., 2004; Raaijmakers et al., 2009; Richardson et al., 2009). More recently, they have been shown to alter gene expression in plants (Wang et al., 2005). The relative importance of these factors for plant growth depends on the plant–microbe combination as well as on environmental conditions. Of general importance is the release of phytohormones by PGPRs (Leinhos and Vacek, 1994), which enhance density and length of root hairs and thus root surface area and can stimulate lateral root growth (Martin et al., 1989; Srinivasan et al., 1997; Ryu et al., 2005; Richardson et al., 2009).

Microorganisms can induce changes in quantity and quality of root exudates (Barber and Martin, 1976; Klein et al., 1988). Moreover, microorganisms may induce plants to produce and release compounds that they can use most efficiently, thus giving them a competitive advantage (Somers et al., 2004). Certain root exudates induce expression of a number of genes in microorganisms, for example, nod gene expression induced by flavonoids as mentioned above, and also antibiotic production by some biocontrol bacteria (Somers et al., 2004; Raaijmakers et al., 2009). Root exudates of non–host plants can induce germination of AM spores (Vierheilig and Ocampo, 1990). On the other hand, glucosinolates or other compounds released by roots can inhibit AM colonization (Vierheilig et al., 2000, 2003).

The role of exudates in plant–microbe communication has been studied extensively. However, roots and microbes can also communicate via other mechanisms. For example, root border cells, or sloughed root cap cells, (living plant cells programmed to separate from the periphery of roots into the external environment) are a uniquely differentiated root tissue that is important for plant–microbe interactions (Hawes, 1991). Root border cells have been shown to be crucial for infection by certain pathogens (Hawes et al., 1997; Zhu et al., 1997).

The outer surface of microorganisms consists of exo- and lipopolysaccharides, which are also important for plant–microbe interactions (Dekkers et al., 1998; Somers et al., 2004). These polysaccharides appear to have signaling function, triggering plant development or modulating defense reactions, and they are important for the colonization of roots as well as for biofilm formation where they create a favorable microenvironment (Dow et al., 2000). Interestingly, root exudates can affect the composition of the polysaccharides and thereby their function (Karpati et al., 1999).

For effective interaction between microorganisms and plants, microorganisms have to be able to colonize and persist in the rhizosphere, particularly when foreign microorganisms are introduced to enhance plant growth. Traits that are important for rhizosphere colonization include motility, high growth rate, ability to synthesize amino acids and vitamin B1, ability to utilize organic acids and certain cell surface proteins, and rapid adjustment to changing conditions (Simons et al., 1997; Lugtenberg and Dekkers, 1999; Hartmann et al., 2009).

6.5.1.2 Communication among Microorganisms

The best known compounds involved in communication between microorganisms are autoinducers such as N-acyl homoserine lactones (AHL) produced by Gram-negative bacteria, cyclic dipeptides produced by pseudomonads, butyrolactone-containing autoinducers produced by *Streptomycetes* sp., and oligo peptides in Gram-positive bacteria (Fray, 2002; Hense et al., 2007; Faure et al., 2009). They are called autoinducers because the producer cells respond to their own signals. Bacterial cells produce and secrete these small signaling molecules, which then move away from the cell by diffusion or mass flow.

Of the autoinducers, AHLs have been particularly well studied for their role in quorum sensing (Faure et al., 2009). A number of different AHLs that differ in side chain chemistry are known (Fray, 2002; Hense et al., 2007); different bacterial species may produce the same AHL with each cell producing and releasing a basal level of AHLs. The purpose of quorum sensing is twofold, taking a census (assessing if the minimum density for effective action has been reached) and coordinating or synchronizing behavior among bacteria in a population or among populations (Winans, 2004).

As the cell density increases, AHLs accumulate in the growth medium. Upon reaching a critical threshold concentration, the AHL molecule binds to its receptor, which activates or represses the expression of certain genes (Somers et al., 2004; Hense et al., 2007). Gantner et al. (2006) used a reporter gene approach to evaluate the maximum distance bacteria could communicate to each other in the rhizosphere via AHLs. They found that the "calling distance" could reach up to about 20 μm.

AHLs were initially investigated for their role in density-dependent microbial responses, which occur only if a certain cell density is reached, for example, bioluminescence of *Vibrio fischeri*. They are now known to form a communication network between distinct bacterial populations, for example, inducing of specific genes, gene transfer, or antibiotic production. Other microorganisms may decompose AHLs and thereby intervene in communication (Somers et al., 2004). AHLs also play a role in communication between microbes and plants. They are involved in pathogen attack or infection (or production of cell wall degrading enzymes), which usually occurs only after a critical cell density is reached to ensure that infection is successful. Infection at low density may induce strong and successful defense reactions by the plant, which would reduce infection success and therefore the survival of the pathogen (Fray, 2002). On the other hand, some plants produce compounds that block AHL sensing in bacteria or degrade AHLs, thus preventing the density-dependent response (Hartmann et al., 2009). Moreover, some plants produce AHLs that may induce infection when cell density of pathogen is still too low for successful infection (Fray, 2002).

Quorum sensing as described above assumes that bacteria cooperate and coordinate their activities; that is, that autoinducers have evolved to benefit the population. However, the autoinducer concentration that could function as an estimate of cell density is altered by many factors, including diffusion and mass flow, spatial distribution of the cells, degradation, and the production of the same autoinducer by third parties, whether intentionally or by chance (Hense et al., 2007). This is particularly relevant in the rhizosphere where microbial populations are nonuniformly distributed both in space and time and where, due to the high biodiversity, the same autoinducer may be produced by different microbial species, which could result in chemical manipulation or interference. Moreover, the spatial distribution of cells can have a larger impact on the autoinducer concentration that the cells experience than the density of the cells (Gantner et al., 2006; Hense et al., 2007).

Redfield (2002) proposed that quorum sensing is diffusion sensing, in which the function of secreted autoinducers is to determine whether secreted effectors (e.g., enzymes) would rapidly diffuse away from the cell, thereby allowing bacteria to detect situations in which the disappearance of effectors owing to diffusion is low enough for effector secretion to be efficient. This would mean that autoinducers benefit the individual cell, not the population. Hense et al. (2007) hypothesize that autoinducers are produced for efficiency sensing; that is, the ecologically relevant function of autoinducer sensing is to assess the efficiency of producing diffusible extracellular effectors, as their concentration will be influenced by the same factors that influence the concentration of diffusible extracellular autoinducers. The autoinducer can function as a proxy for the more expensive effectors such as exoenzymes. In the rhizosphere, bacteria are often aggregated in microcolonies. These cell clusters usually arise from the growth and cell division of one cell; they are therefore essentially clonal. The more the cells are clustered together, the sooner the threshold for induction is reached. This, together with the usual positive feedback of autoinducer production, maintains communication within the cell clusters. The fewer and larger these clusters are, the more they will exclude interference from other autoinducer-producing or autoinducer-degrading bacteria, plants, and other eukaryotes.

Induced resistance is an example of interactions between microorganisms occurring via the plant. In induced resistance, one microorganism colonizing the roots of a plant makes the plant resistant to the colonization by certain other microorganisms, for example, AM colonization can induce resistance against root pathogens or even leaf-colonizing fungi but can also reduce colonization by another AM fungus (Demeyer and Hofte, 1997; Vierheilig et al., 2003; Khaosaad et al., 2007; Avis et al., 2008). Compounds involved include salicylic acid, siderophores, and lipopolysaccharides (Somers et al., 2004). Induced resistance is often systemic: the whole plant becomes resistant even if inducing organism and pathogen are spatially separated (Vierheilig et al., 2003; Khaosaad et al., 2007). Therefore, antagonism or direct competition for space or nutrients between microorganisms is unlikely to be involved. Instead, induced resistance is due to changes in host physiology, for example, production of defense compounds. However, the mechanisms involved in induced resistance are in many cases still unclear.

6.5.2 Competition and Facilitation

In competition, two organisms interact for a limited resource, with one organism being more successful (competitive) than the other. Limited resources can be nutrients, water, or space. An overview of strategies to improve competitiveness is given in Table 6.3.

Facilitation means that one organism helps another organism in attaining a certain resource or generally improving growth or fitness.

The two types of interactions are not mutually exclusive: an organism may facilitate uptake of a certain nutrient at a particular point in time but also compete for the same or another nutrient at a different time. The balance between competition and facilitation depends on the partners involved as well as environmental conditions.

An example for the continuum between facilitation and competition between plants and microorganisms are mycorrhiza (Johnson et al., 1997). They can cause growth depression in the early stages of plant growth, when the C drain by the fungus outweighs its benefit to the plant in terms of improved P uptake (Buwalda and Goh, 1982; Koide, 1985; Peng et al., 1993; Graham and Abbott, 2000; Li et al., 2005); but in the later stages of plant growth, mycorrhiza can be beneficial by facilitating P and trace element uptake, reducing heavy metal toxicity, etc. Competition may dominate in soils with high P availability where the benefit from mycorrhiza is low or where photosynthesis is reduced, for example, under low light conditions. Facilitation will dominate in soils with low P availability or when the rate of photosynthesis is high; thus, assimilation supplies sufficient C to meet plant and fungal demand. For a comprehensive review of the costs and benefits of mycorrhiza, the reader is referred to Smith and Read (2008).

Compounds that inhibit microbial growth play an important role in competition between microorganisms. Whereas antibiotics are relatively unspecific, bacteriocins are defined as compounds that are produced by several bacterial species and that act specifically on other strains of the same, or closely related, species. Mazzola et al. (1992) showed that production of antibiotics is important for the survival of introduced strains in soil. Microbes may also inhibit the growth of other microbes by competition for iron, HCN release, chitinases, or glucanases that degrade microbial cell walls (Somers et al., 2004).

6.5.2.1 Competition and Facilitation with Respect to Iron

Competition and facilitation with respect to Fe can occur between microorganisms and between microorganisms and plants (Robin et al., 2008). Interactions between microorganisms

TABLE 6.3 Strategies to Improve Competitiveness

Strategy	Mechanisms	References
Inhibition of competitors	Antibiotics, phenols, hormones	De Boer et al. (2008), Jousset et al. (2008), Kumar et al. (2005), Mazzola et al. (1992), Murphy et al. (1995), Somers et al. (2004)
Stimulation of beneficial organisms	Signals, creation of favorable conditions or release of compounds that stimulate beneficial organisms	Buckling et al. (2007), Duponnois and Garbaye (1991), Poole et al. (2001)
Effective nutrient mobilization	High-affinity chelators, high concentration of chelators/complexes	Langenheder and Prosser (2008), Loper and Buyer (1991), Poll et al. (2008)
Chelation	Utilization of nutrients mobilized by others	Cornelis (2010), Crowley et al. (1991), Raaijmakers et al. (1995)
Growth rate	Rapid response to and high growth rates in favorable conditions; low growth rates under unfavorable conditions to conserve energy	Eilers et al. (2010), Meidute et al. (2008)
Communication	For example, autoinducers to synchronize response of a population or to release enzymes, etc., only when they will be effective	Hense et al. (2007), Somers et al. (2004)

include utilization of own and foreign siderophores (the latter is known as cheating), ligand exchange between siderophores, and decomposition of siderophores (Buyer and Sikora, 1991; Loper and Buyer, 1991; Loper and Henkels, 1999). As mentioned in Section 6.4.3.2, the utilization of different siderophores appears to be important for the competitive ability of bacterial strains (Mirleau et al., 2000). Previously, siderophores were thought to be important for biocontrol of pathogenic fungi in some *Pseudomonas* strains; however, it is now clear that in most cases biocontrol is due to antibiotic production (De Weger et al., 1986; Leong, 1986; Loper, 1988; Thomashow and Weller, 1990; Hamdan et al., 1991; Pal et al., 2000; Raaijmakers et al., 2009).

Competition for Fe between plants and microorganisms appears to be strongly in favor of microorganisms (Figure 6.7). Although plants can utilize Fe bound to bacterial siderophores, they are considered to be poor Fe sources for plants (Wang et al., 1993; Walter et al., 1994). On the other hand, phytosiderophores are good Fe sources for microbes (Jurkevitch et al., 1993) and microorganisms decompose phytosiderophores, thereby decreasing Fe uptake by grasses (Von Wiren et al., 1995). Moreover, Fe uptake by roots and translocation into shoots of maize is much lower for Fe siderophores than that for Fe phytosiderophores (Crowley et al., 1992). This raises the question how plants can take up sufficient Fe. For grasses, Crowley and Gries (1994) hypothesized that due to the high release rates of phytosiderophores at the root tip in the first few hours after onset of light (Zhang et al., 1991), the rhizosphere would be flooded with phytosiderophores. The concentration of bacterial siderophores and decomposition rate of phytosiderophores would be relatively low due to the low microbial density at the root tip.

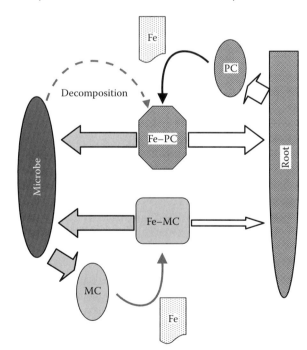

FIGURE 6.7 Interactions between microorganisms and plants for Fe. MC, microbial chelator; PC, plant chelator; Fe–MC, Fe–PC, iron chelated by microbial or plant chelator.

Thus, the concentration of phytosiderophores would be far greater than that of bacterial siderophores, outweighing the higher affinity of Fe for siderophores (Robin et al., 2008). This is in agreement with the model developed by Darrah (1991), which demonstrates that spatially and temporally constrained production of an exudate makes it more efficient. In addition, facilitation occurring between two plant species has been reported in intercropped Strategy I and Strategy II plant species, resulting in improved iron uptake in the legume plants close to rows of the cereal crop (Zuo et al., 2000; Gunes et al., 2007). This suggests that Fe uptake by Strategy I plant species was improved as a consequence of phytosiderophore release by the neighboring Strategy II plant species.

6.5.2.2 Competition and Facilitation with Respect to Phosphorus

Plants and microorganisms in the rhizosphere may also compete for P. Plants and microorganisms have evolved similar mechanisms to increase P availability, such as release of carboxylates or phosphatases or changes in pH (Dinkelaker and Marschner, 1992; Gahoonia and Nielsen, 1992; Jones, 1998; Whitelaw et al., 1999; Hinsinger, 2001; Hinsinger et al., 2009; Richardson et al., 2009). Additionally, plants and fungi can increase the soil volume exploited by growing roots, root hairs, or hyphae (Foehse et al., 1988; Hinsinger et al., 2005). Plants and microorganisms compete for the mobilized P. Additionally, microorganisms can decompose carboxylates released by plant roots (Ström et al., 2001), thereby decreasing the effectiveness of the carboxylates. Microorganisms are highly competitive for P compared with plants, particularly when plant residues are the P source (McLaughlin et al., 1988). McLaughlin et al. (1988) showed that when P was applied as inorganic fertilizer, less than 5% of the applied P was immobilized in the microbial biomass compared with more than 10% of applied P taken up by the plants. However, when P was added as residue, more than 20% of applied P was found in the microbial biomass compared with less than 10% in the plants.

On the other hand, it has been argued that P immobilized in the microbial biomass may facilitate plant P uptake in the long term due to release of P when part of the microbial biomass dies off (Seeling and Zasoski, 1993; Oberson et al., 2001). Hence, the microbial biomass can be regarded as a labile P pool. Indeed, immobilization and release occur simultaneously: part of the microbial population may immobilize P whereas another part dies off and releases P from the cells. This may explain why microbial biomass P in the rhizosphere can be positively correlated with plant P uptake (Marschner et al., 2006). Since microbial growth in soils is predominantly C limited (De Nobili et al., 2001), C availability governs the balance between immobilization and release of P from the microbial biomass. At low C availability, growth rates and thus P immobilization will be low; previously stored P could be released when part of the microbial biomass dies off. On the other hand, when C availability is high, rapid microbial growth and thus P demand will result in strong net immobilization. Turnover of microbial biomass can be very rapid; within a few days, net P uptake can be replaced by net P release (Figure 6.8). Annual biomass P flux has been estimated to range between 7 and 23 kg P ha^{-1} year^{-1} (Brookes et al., 1984). Additionally, microbial P (and N) demand is also affected by growth stage (Tezuka, 1990; Makino et al., 2003). During exponential growth (at high C availability and low cell density), N and P demand is high because they are required for synthesis of DNA, proteins, and cell membranes. As C is depleted, the cells will enter stationary growth; in this phase, N and P demand is low, which can result in net N and P release from microbial biomass.

Moreover, the balance between immobilization and mobilization may also be affected by microbial community structure. A high percentage of strong P mobilizers in a community may result in net mineralization because mobilization exceeds immobilization. On the other hand, a low percentage of strong P mobilizers could lead to net immobilization because mobilization cannot meet P demand. The latter community may utilize P mobilized by plant exudates.

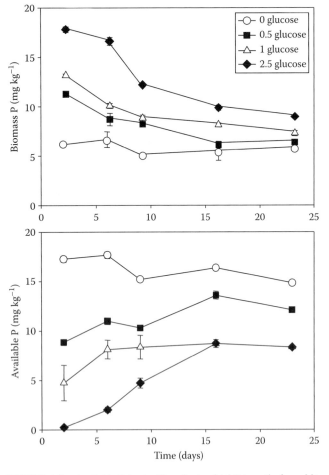

FIGURE 6.8 Available P (resin P) and microbial P in soil after addition of C as glucose at 0, 0.5, 1, and 2.5 g C kg^{-1} soil. (From Springer Science+Business Media: Marschner, P. 2008. The role of rhizosphere microorganisms in relation to P uptake by plants, p. 165–176. *In* P.J. White and J.P. Hammond (eds.) The ecophysiology of plant-phosphorus interactions. Springer, New York.)

Nutrient facilitation among plants with respect to P has been shown in intercropped legume–cereal species (Li et al., 2007). The legume may facilitate P acquisition by the intercropped cereal either via the release of exudates such as carboxylates (Horst and Waschkies, 1987; Ae et al., 1990; Zhang and Li, 2003; Cu et al., 2005) or phosphatase enzymes (Li et al., 2004; Gunes et al., 2007). Another P-mobilizing mechanism may be protons released by N$_2$-fixing legume (Hinsinger et al., 2003; Tang et al., 2004, Li et al., 2008). Part of these complex, positive interactions between two plant species may also be due to their ability to access complementary resources, that is, different pools of soil P, which may also be a consequence of differential rhizosphere microbial communities (Li et al., 2008). Hence, besides rhizosphere-based facilitation, niche complementarity may be a major component of the positive interaction observed between intercropped plant species.

Competition and facilitation between plants and microorganisms, among plants, and among microorganisms play a key role in plant–plant and plant–microbe interactions. However, interactions are rarely solely competitive or beneficial; in most cases, there is a continuum between competition and facilitation, which is governed by signal exchange, nutrient supply, and environmental conditions. Although competition for nutrients between two organisms may be heavily in favor of one of the organisms, the other organism may be able to manipulate the competitive ability of the dominant organism or develop other strategies to acquire sufficient nutrients. The discovery of autoinducers has revealed exciting aspects of the interactions not only among microorganisms but also between plants and microorganisms. The use of microarrays now allows studying gene expression in plants and microorganisms, which will lead to further elucidation of the web of interactions that occur in the interface between plants, soil, and microorganisms.

6.6 Conclusions: A Unique, Multifaceted, and Complex Interface

The rhizosphere is a unique soil compartment in which plant- and microbial-induced processes are most intense. The properties of the soil and environmental factors such as soil water and aeration modulate the processes occurring in the rhizosphere; however, it is the effect of roots and microorganisms that shape this environment. The rhizosphere properties are a function of the activity of roots and microorganisms; compounds released on the one hand and uptake of nutrients and water on the other. However, roots and rhizosphere microorganisms do not act in isolation; they compete for nutrients and exchange signals that may inhibit or promote growth and metabolism. In many cases where nutrients are mobilized, it is currently not possible to distinguish between the plant and microbial effect. Consequently, we need to develop new ways in which to determine the relative contribution of each and their potential interaction (i.e., synergistic, antagonistic effects). Techniques such as stable isotope probing of microbial function, NanoSIMS, high-throughput DNA sequencing, isotopic-enriched metabolomics, or proteomics may allow us to disentangle the complex interactions between plants and microbes. In particular, the rapid development of "omics" platforms is beginning to transform our understanding of rhizosphere ecology (Mochida and Shinozaki, 2010). As never before, these tools have allowed researchers to produce a massive amount of information through in situ measurements and analysis of natural microbial communities, both vital approaches to the goal of unraveling the interactions of microbes with roots and with one another (Dill et al., 2010). While genomics can provide information regarding the genetic potential of microbes and roots, proteomics characterizes the primary end-stage product, proteins, thus conveying functional information concerning microbial and root activity. Advances in mass spectrometry instrumentation and methodologies, along with bioinformatics approaches, have brought this analytical chemistry technique to relevance in the biological realm due to its powerful applications in proteomics (Keller and Hettich, 2009). However, the application of these new techniques also brings new challenges, particularly, the transformation or data mining of these massive datasets (bioinformatics).

Historically, we have been limited by the poor spatial resolution of rhizosphere sampling techniques that fail to differentiate between root and microbial processes. Some of the newly developed techniques such as NanoSIMS provide a partial solution to this, allowing visualization of nutrient flows at the submicron scale via stable isotopes (e.g., ^{13}C, ^{15}N, ^{18}O) and microbes (species/function) via molecular tags. However, as the spatial resolution is so high (nm scale) and the sample preparation time so long, it means that obtaining a complete picture of the whole root system remains logistically impossible.

These examples highlight one of the most important challenges in rhizosphere ecology, which is the scaling up from single root regions, to whole root systems, and then to field and ecosystem level (Standing et al., 2007). While single root scale measurements can be useful for some end users (e.g., plant breeding, bioremediation), typically it is only at the bigger scale that most of our results become meaningful to the majority of end users (e.g., in policymakers, environment regulators, agronomists). Mathematical modeling provides complementary approaches for this purpose (e.g., for predicting nutrient dynamics at the field level to allow better targeting of fertilizers; Roose and Kirk, 2009). The improved understanding of the rhizosphere processes will help plant breeders in selecting genotypes with particular ability to mobilize nutrients themselves or create conditions that maximize the mobilization potential of the rhizosphere microbial community.

References

Ae, N., J. Rihara, K. Okada, T. Yoshihara, and C. Johansen. 1990. Phosphorus uptake by pigeon pea and its role in cropping systems of the Indian Subcontinent. Science 248:477–480.

Aguilar, S.A., and A. van Diest. 1981. Rock phosphate mobilization induced by the alkaline pattern of legumes utilizing symbiotically fixed nitrogen. Plant Soil 61:27–42.

Ahmad, J.S., and R. Baker. 1987. Competitive saprophytic ability and cellulolytic activity of rhizosphere-competent mutants of *Trichoderma harzianum*. Phytopathology 77:358–362.

Alami, Y., W. Achouak, C. Marol, and T. Heulin. 2000. Rhizophere soil aggregation and plant growth promotion of sunflowers by an exopolysaccharide-producing *Rhizobium* sp. isolated from sunflower roots. Appl. Environ. Microbiol. 66:3393–3398.

Amann, C., and A. Amberger. 1989. Phosphorus efficiency of buckwheat (*Fagopyrum esculentum*). Z. Pflanzenernaehr. Bodenkd. 152:181–189.

Armstrong, W. 1979. Aeration in higher plants. Adv. Bot. Res. 7:225–232.

Armstrong, W., D. Cousins, J. Armstrong, D.W. Turner, and P.M. Becket. 2000. Oxygen distribution in wetland plant roots and permeability barriers to gas-exchange with the rhizosphere: A microelectrode and modelling study with *Phragmites australis*. Ann. Bot. 86:687–703.

Arocena, J.M., and K.R. Glowa. 2000. Mineral weathering in ectomycorrhizosphere of subalpine fir (*Abies lasiocarpa* (Hook.) Nutt.) as revealed by soil solution composition. For. Ecol. Manage. 133:61–70.

Arocena, J.M., K.R. Glowa, H.B. Massicotte, and L. Lavkulich. 1999. Chemical and mineral composition of ectomycorrhizosphere soils of subalpine fir (*Abies lasiocarpa* (Hook.) Nutt.) in the Ae horizon of a luvisol. Can. J. Soil Sci. 79:25–35.

Avis, T.J., V. Gravel, H. Antoun, and R.J. Tweddell. 2008. Multifaceted beneficial effects of rhizosphere microorganisms on plant health and productivity. Soil Biol. Biochem. 40:1733–1740.

Bacha, R.E., and L.R. Hossner. 1977. Characteristics of coatings formed on rice roots as affected by iron and manganese additions. Soil Sci. Soc. Am. J. 41:931–935.

Bai, C.M., X.L. He, H.T. Tang, B.Q. Shan, and L.L. Zhao. 2009. Spatial distribution of arbuscular mycorrhizal fungi, glomalin and soil enzymes under the canopy of *Astragalus adsurgens* Pall. in the Mu Us sandland, China. Soil Biol. Biochem. 41:941–947.

Bais, H.P., S.W. Park, T.L. Weir, R.M. Callaway, and J.M. Vivanco. 2004. How plants communicate using the underground information superhighway. Trends Plant Sci. 9:26–32.

Bakker, M.R., E. George, M.P. Turpault, J.L. Zhang, and B. Zeller. 2004. Impact of Douglas-fir and Scots pine seedlings on plagioclase weathering under acidic conditions. Plant Soil 266:247–259.

Barber, S.A. 1995. Soil nutrient bioavailability: A mechanistic approach. 2nd edn. John Wiley, New York.

Barber, D.A., and J.K. Martin. 1976. The release of organic substances by cereal roots into soil. New Phytol. 76:69–80.

Bar-Ness, E., Y. Chen, Y. Hadar, H. Marschner, and V. Roemheld. 1991. Siderophores of *Pseudomonas putida* as an iron source for dicot and monocot plants. Plant Soil 130:231–241.

Barré, P., B. Velde, N. Catel, and L. Abbadie. 2007. Quantification of potassium addition or removal through plant activity on clay minerals by X-ray diffraction. Plant Soil 292:137–146.

Bates, T.R., and J.P. Lynch. 2001. Root hairs confer a competitive advantage under low phosphorus availability. Plant Soil 236:243–250.

Baum, C., and K. Hrynkiewicz. 2006. Clonal and seasonal shifts in communities of saprotrophic microfungi and soil enzyme activities in the mycorrhizosphere of *Salix* spp. J. Soil Sci. Plant Nutr. 169:481–487.

Becker, J.O., R.W. Hedges, and E. Messens. 1985. Inhibitory effect of pseudobactin on the uptake of iron by higher plants. Appl. Environ. Microbiol. 49:1090–1093.

Begg, C.B.M., G.J.D. Kirk, A.F. MacKenzie, and H.-U. Neue. 1994. Root-induced iron oxidation and pH changes in the lowland rice rhizosphere. New Phytol. 128:469–477.

Bengough, A.G., M.F. Bransby, J. Hans, S.J. McKenna, T.J. Roberts, and T.A. Valentine. 2006. Root responses to soil physical conditions; growth dynamics from field to cell. J. Exp. Bot. 57:437–447.

Bent, A.F. 1996. Plant disease resistance genes: Function meets structure. Plant Cell 8:1757–1771.

Bertrand, I., P. Hinsinger, B. Jaillard, and J.C. Arvieu. 1999. Dynamics of phosphorus in the rhizosphere of maize and rape grown on synthetic, phosphated calcite and goethite. Plant Soil 211:111–119.

Bidel, L.P.R., P. Renault, L. Pagès, and L.M. Rivière. 2000. Mapping meristem respiration of *Prunus persica* (L.) Batsch seedlings: Potential respiration of the meristems, O_2 diffusional constraints and combined effects on root growth. J. Exp. Bot. 51:755–768.

Bolan, N.S. 1991. A critical-review on the role of mycorrhizal fungi in the uptake of phosphorus by plants. Plant Soil 134:189–207.

Bossier, P., M. Hoefte, and W. Verstraete. 1988. Ecological significance of siderophores in soil. Adv. Microb. Ecol. 10:385–414.

Bouillet, J.P., J.P. Laclau, M. Arnaud, A.T. M'Bou, L. Saint-André, and C. Jourdan. 2002. Changes with age in the spatial distribution of roots of Eucalyptus clone in Congo—Impact on water and nutrient uptake. Forest Ecol. Manage. 171:43–57.

Bravin, M.N., A. Lara Martí, M. Clairotte, and P. Hinsinger. 2009a. Rhizosphere alkalisation—A major driver of copper bioavailability over a broad pH range in an acidic, copper-contaminated soil. Plant Soil 318:257–268.

Bravin, M.N., P. Tentscher, J. Rose, and P. Hinsinger. 2009b. Rhizosphere pH gradient controls copper availability in a strongly acidic soil. Environ. Sci. Technol. 43:5686–5691.

Bravin, M.N., F. Travassac, M. Le Floch, P. Hinsinger, and J.M. Garnier. 2008. Oxygen input controls the spatial and temporal dynamics of arsenic at the surface of a flooded paddy soil and in the rhizosphere of lowland rice (*Oryza sativa* L.): A microcosm study. Plant Soil 312:207–218.

Brigham, L.A., H. Woo, S.M. Nicoll, and M.C. Hawes. 1995. Differential expression of proteins and mRNAs from border cells and root tips of pea. Plant Physiol. 109:457–463.

Brookes, P.C., D.S. Powlson, and D.S. Jenkinson. 1984. Phosphorus in the soil microbial biomass. Soil Biol. Biochem. 16:169–175.

Bruand, A., I. Cousin, B. Nicoullaud, O. Duval, and J.C. Bégon. 1996. Backscattered electron scanning images of soil porosity for analysing soil compaction around roots. Soil Sci. Soc. Am. J. 60:895–901.

Buckling, A., F. Harrison, M. Vos, M.A. Brockhurst, A. Gardner, S.A. West, and A. Griffin. 2007. Siderophore-mediated cooperation and virulence in *Pseudomonas aeruginosa*. FEMS Microbiol. Ecol. 62:135–141.

Buwalda, J.G., and K.M. Goh. 1982. Host-fungus competition for carbon as a cause of growth depressions in vesicular-arbuscular mycorrhizal ryegrass. Soil Biol. Biochem. 14:103–106.

Buyer, J.S., and L.J. Sikora. 1991. Rhizosphere interactions and siderophores, p. 263–269. *In* D.L. Kleister and P.B. Cregan (eds.) The rhizosphere and plant growth. Kluwer Academic Publishers, Dordrecht, the Netherlands.

Caldwell, M.M., T.E. Dawson, and J.H. Richards. 1998. Hydraulic lift: Consequences of water efflux from the roots of plants. Oecologia 113:151–161.

Callaway, R.M., S.C. Pennings, and C.L. Richards. 2003. Phenotypic plasticity and interactions among plants. Ecology 84:1115–1128.

Calvaruso, C., M.P. Turpault, and P. Frey-Klett. 2006. Root-associated bacteria contribute to mineral weathering and to mineral nutrition in trees: A budgeting analysis. Appl. Environ. Microbiol. 72:1258–1266.

Calvaruso, C., M.P. Turpault, E. Leclerc, and P. Frey-Klett. 2007. Impact of ectomycorrhizosphere on the functional diversity of soil bacterial and fungal communities from a forest stand in relation to nutrient mobilization processes. Microb. Ecol. 54:567–577.

Canadell, J., R.B. Jackson, J.R. Ehleringer, H.A. Mooney, O.E. Sala, and E.D. Schulze. 1996. Maximum rooting depth of vegetation types at the global scale. Oecologia 108:583–595.

Cardon, Z., and J. Whitbeck. 2007. The rhizosphere. An ecological perspective. Elsevier, Amsterdam, the Netherlands.

Casarin, V., C. Plassard, P. Hinsinger, and J.C. Arvieu. 2004. Quantification of ectomycorrhizal fungal effects on the bioavailability and mobilisation of soil P in the rhizosphere of *Pinus pinaster*. New Phytol. 163:177–185.

Chapin, F.S., L. Moilanen, and K. Kielland. 1993. Preferential use of organic nitrogen for growth by a non-mycorrhizal arctic sedge. Nature 361:1550–1553.

Claassen, N., and A. Jungk. 1982. Kaliumdynamik im wurzelnahen Boden in Beziehung zur Kaliumaufnahme von Maispflanzen. Z. Pflanzenernaehr. Bodenkd. 145:513–525.

Cornelis, P. 2010. Iron uptake and metabolism in pseudomonads. Appl. Microbiol. Biotechnol. 86:1637–1645.

Cornu, J.Y., S. Staunton, and P. Hinsinger. 2007. Copper concentration in plants and in the rhizosphere as influenced by the iron status of tomato (*Lycopersicum esculentum* L.). Plant Soil 292:63–77.

Crowley, D.E., and D. Gries. 1994. Modeling of iron availability in the plant rhizosphere, p. 199–224. *In* J.A. Manthey, D.E. Crowley, and D.G. Luster (eds.) Biochemistry of metal micronutrients in the rhizosphere. Lewis Publishers, Boca Raton, FL.

Crowley, D.E., V. Roemheld, H. Marschner, and P.J. Szaniszlo. 1992. Root-microbial effects on plant iron uptake from siderophores and phytosiderophores. Plant Soil 142:1–7.

Crowley, D.E., Y.C. Wang, C.P.P. Reid, and P.J. Szanislo. 1991. Mechanisms of iron acquisition from siderophores by microorganisms and plants, p. 213–232. *In* Y. Chen and Y. Hadar (eds.) Iron nutrition and interactions in plants. Kluwer Academic Publishers, Dordrecht, the Netherlands.

Cu, S.T.T., J. Hutson, and K.A. Schuller. 2005. Mixed culture of wheat (*Triticum aestivum* L.) with white lupin (*Lupinus albus* L.) improves the growth and phosphorus nutrition of the wheat. Plant Soil 272:143–151.

Curl, E.A., and B. Truelove. 1986. The rhizosphere. Advanced series in agricultural science 15. Springer-Verlag, Berlin, Germany.

Czarnes, S., P.D. Hallett, A.G. Bengough, and I.M. Young. 2000. Root- and microbial-derived mucilages affect soil structure and water transport. Eur. J. Soil Sci. 51:435–443.

Darrah, P.R. 1991. Models of the rhizosphere. I. Microbial population dynamics around a root releasing soluble and insoluble carbon. Plant Soil 133:187–199.

Darrah, P.R. 1993. The rhizosphere and plant nutrition: A quantitative approach. Plant Soil 155/156:1–20.

De Boer, W., A.S. De Ridder-Duine, P.J.A.K. Gunnewiek, W. Smant, and J.A. Van Veen. 2008. Rhizosphere bacteria from sites with high fungal densities exhibit greater levels of potential antifungal properties. Soil Biol. Biochem. 40:1542–1544.

De Cristofaro, A., J.Z. He, D.H. Zhou, and A. Violante. 2000. Adsorption of phosphate and tartrate on hydroxy-aluminum-oxalate precipitates. Soil Sci. Soc. Am. J. 64:1347–1355.

De Nobili, M., M. Contin, C. Mondini, and P.C. Brookes. 2001. Soil microbial biomass is triggered into activity by trace amounts of substrate. Soil Biol. Biochem. 33:1163–1170.

De Weger, L.A., R. Van Boxtel, B. Van der Burg, R.A. Gruters, F.P. Geels, and B. Schippers. 1986. Siderophores and outer membrane proteins of antagonistic, plant-growth stimulating, root-colonizing *Pseudomonas* spp. J. Bacteriol. 165:585–594.

Dekkers, L.C., A.J. Van der Bij, I.H.M. Mulders, C. Phoelich, R.A.R. Wentwood, D.C.M. Glandorf, C.A. Wijffelman, and B.J.J. Lugtenberg. 1998. Role of the O-antigen of lipopolysaccharide and possible roles of growth rate and NADH-ubiquinone oxidoreductase (NUO) on competitive tomato root tip colonization by *Pseudomonas fluorescens* WCS 365. Mol. Plant-Microbe Interact. 11:763–771.

Demeyer, G., and M. Hofte. 1997. Salicylic acid produced by the rhizobacterium *Pseudomonas aeruginosa* 7NSK2 induces resistance to leaf colonization by *Botrytis cinerea* on bean. Phytopathology 87:588–593.

Dessaux, Y., P. Hinsinger, and P. Lemanceau. 2009. Rhizosphere: So many achievements and even more challenges. Plant Soil 321:1–3.

Dessaux, Y., P. Hinsinger, and P. Lemanceau. 2010. Rhizosphere: Achievements and challenges. Springer, New York.

Devau, N., E. Le Cadre, P. Hinsinger, and F. Gérard. 2010. A mechanistic model for understanding root-induced chemical changes controlling phosphorus availability. Ann. Bot. 105:1183–1197.

Devau, N., E. Le Cadre, P. Hinsinger, B. Jaillard, and F. Gérard. 2009. Soil pH controls the environmental availability of phosphorus: Experimental and mechanistic modelling approaches. Appl. Geochem. 24:2163–2174.

Dexter, A.R. 1987. Compression of soil around roots. Plant Soil 97:401–406.

Dilkes, N.B., D.L. Jones, and J. Farrar. 2004. Temporal dynamics of carbon partitioning and rhizodeposition in wheat. Plant Physiol. 134:706–715.

Dill, B.D., C. Young Jacque, P.A. Carey, and N.C. VerBerkmoes. 2010. Metaproteomics: Techniques and applications, p. 37–61. *In* W.T. Liu and J.K. Jansson (eds.) Environmental molecular microbiology. Caister Academic Press, Norwich, U.K.

Dinkelaker, B., and H. Marschner. 1992. In vivo demonstration of acid phosphatase activity in the rhizosphere of soil-grown plants. Plant Soil 144:199–205.

Doussan, C., L. Pagès, and A. Pierret. 2003. Soil exploration and resource acquisition by plant roots: An architectural and modelling point of view. Agronomie 23:419–431.

Doussan, C., A. Pierret, E. Garrigues, and L. Pagès. 2006. Water uptake by plant roots: II—Modelling of water transfer in the soil root-system with explicit account of flow within the root system—Comparison with experiments. Plant Soil 283:99–117.

Dow, M., M.A. Newman, and E. von Roepenack. 2000. The induction and modulation of plant defense responses by bacterial lipopolysaccharides. Ann. Rev. Phytopathol. 38:241–261.

Dunbabin, V.M., S. McDermott, and A.G. Bengough. 2006. Upscaling from rhizosphere to whole root system: Modelling the effects of phospholipid surfactants on water and nutrient uptake. Plant Soil 283:57–72.

Duponnois, R., and J. Garbaye. 1991. Mycorrhization helper bacteria associated with the Douglas fir-*Laccaria laccata* symbiosis: Effects in aseptic and in glasshouse conditions. Ann. Sci. For. 48:239–251.

Dutton, M.V., and C.S. Evans. 1996. Oxalate production by fungi: Its role in pathogenicity and ecology in the soil environment. Can. J. Microbiol. 42:881–895.

Eilers, K.G., C.L. Laubner, R. Knight, and N. Fierer. 2010. Shifts in bacterial community structure associated with inputs of low molecular weight carbon compounds to soil. Soil Biol. Biochem. 42:896–903.

Essington, M.E., and R.M. Anderson. 2008. Competitive adsorption of 2-ketogluconate and inorganic ligands onto gibbsite and kaolinite. Soil Sci. Soc. Am. J. 72:595–604.

Faure, D., D. Vereecke, and J.H.J. Leveau. 2009. Molecular communication in the rhizosphere. Plant Soil 321:279–303.

Fierer, N., and R.B. Jackson. 2006. The diversity and biogeography of soil bacterial communities. Proc. Natl. Acad. Sci. 103:626–631.

Fischer, W.R., H. Flessa, and G. Schaller. 1989. pH values and redox potentials in microsites of the rhizosphere. Z. Pflanzenernaehr. Bodenkd. 152:191–195.

Flessa, H., and W.R. Fischer. 1992. Plant-induced changes in the redox potentials of rice rhizospheres. Plant Soil 143:55–60.

Foehse, D., N. Claassen, and A. Jungk. 1988. Phosphorus efficiency in plants. I. External and internal P requirement and P uptake efficiency of different plant species. Plant Soil 110:101–109.

Fray, R.G. 2002. Altering plant-microbe interaction through artificially manipulating bacterial quorum sensing. Ann. Bot. 89:245–253.

Gahoonia, T.S., N. Claassen, and A. Jungk. 1992. Mobilization of phosphate in different soils by ryegrass supplied with ammonium or nitrate. Plant Soil 140:241–248.

Gahoonia, T.S., and N.E. Nielsen. 1992. Control of pH at the soil-root interface. Plant Soil 140:49–54.

Gahoonia, T.S., N.E. Nielsen, P.A. Joshi, and A. Jahoor. 2001. A root hairless barley mutant for elucidating genetic of root hairs and phosphorus uptake. Plant Soil 235:211–219.

Gantner, S., M. Schmid, C. Duerr, R. Schuhegger, A. Steidle, P. Hutzler, C. Langebartels, L. Eberl, A. Hartmann, and F.B. Dazzo. 2006. In situ quantitation of the spatial scale of calling distances and population density-independent N-acylhomoserine lactone-mediated communication by rhizobacteria colonized on plant roots. FEMS Microbiol. Ecol. 56:188–194.

Ge, Z., G. Rubio, and J.P. Lynch. 2000. The importance of root gravitropism for inter-root competition and phosphorus acquisition efficiency: Results from a geometric simulation model. Plant Soil 218:159–171.

George, T.S., A.E. Richardson, J.B. Smith, P.A. Hadobas, and R.J. Simpson. 2005. Limitations to the potential of transgenic *Trifolium subterraneum* L. plants that exude phytase when grown in soils with a range of organic P content. Plant Soil 278:263–274.

Gerhardt, K.E., X.D. Huang, B.R. Glick, and B.M. Greenberg. 2009. Phytoremediation and rhizoremediation of organic soil contaminants: Potential and challenges. Plant Sci. 176:20–30.

Gerke, J. 1993. Solubilization of Fe (III) from humic-Fe complexes, humic/Fe-oxide mixtures and from poorly ordered Fe-oxide by organic acids—Consequences for P adsorption. Z. Pflanzenernaehr. Bodenkd. 156:253–257.

Gerke, J. 1994. Kinetics of soil phosphate desorption as affected by citric acid. Z. Pflanzenernaehr. Bodenkd. 157:17–22.

Gerke, J., L. Beissner, and W. Roemer. 2000. The quantitative effect of chemical phosphate mobilization by carboxylate anions on P uptake by a single root. I. The basic concept and determination of soil parameters. J. Plant Nutr. Soil Sci. 163:207–212.

Gerke, J., and R. Hermann. 1992. Adsorption of orthophosphate to humic-Fe-complexes and to amorphous Fe-oxide. Z. Pflanzenernaehr. Bodenkd. 155:233–236.

Gharbi, I., B. Ricard, S. Smiti, E. Bizid, and R. Brouquisse. 2009. Increased hexose transport in the roots of tomato plants submitted to prolonged hypoxia. Planta 230:441–448.

Gilbert, G.A., J.D. Knight, C.P. Vance, and D.L. Allan. 1999. Acid phosphatase activity in phosphorus-deficient white lupin roots. Plant Cell Environ. 22:801–810.

Gollany, H.T., T.E. Schumacher, R.R. Rue, and S.-Y. Liu. 1993. A carbon dioxide microelectrode for in situ pCO_2 measurement. Microchem. J. 48:42–49.

Graham, J.H., and L.K. Abbott. 2000. Wheat responses to aggressive and non-aggressive arbuscular mycorrhizal fungi. Plant Soil 220:207–218.

Gunes, A., E.G. Bagci, and A. Inal. 2007. Interspecific facilitative root interactions and rhizosphere effects on phosphorus and iron nutrition between mixed grown chickpea and barley. J. Plant Nutr. 30:1455–1469.

Gustafsson, J. 2001. Modelling competitive anion adsorption on oxide minerals and an allophane-containing soil. Eur. J. Soil Sci. 52:639–653.

Hahn, M.G. 1996. Microbial elicitors and their receptors in plants. Annu. Rev. Phytopathol. 34:387–412.

Hainsworth, J.M., and L.A.G. Aylmore. 1989. Non-uniform soil water extraction by plant roots. Plant Soil 113:121–124.

Hallett, P.D., D.C. Gordon, and A.G. Bengough. 2003. Plant influence on rhizosphere hydraulic properties: Direct measurements using a miniaturized infiltrometer. New Phytol. 157:597–603.

Hamdan, H., D.M. Weller, and L.S. Thomashow. 1991. Relative importance of fluorescent siderophores and other factors in biological control of *Gaeumannomyces graminis* var. *tritici* by P*seudomonas fluorescens* 2-79 and M4-80R. Appl. Environ. Microbiol. 57:3270–3277.

Hamza, M.A., and L.A.G. Aylmore. 1992. Soil solute concentration and water uptake by single lupin and radish plant roots. I. Water extraction and solute accumulation. Plant Soil 145:187–196.

Hartmann, A., M. Schmid, D. van Tuinen, and G. Berg. 2009. Plant-driven selection of microbes. Plant Soil 321:235–257.

Hawes, M.C. 1991. Living plant cells released from the root cap: A regulator of microbial populations in the rhizosphere? p. 51–59. In D.L. Kleister and P.B. Cregan (eds.) The rhizosphere and plant growth. Kluwer Academic Publishers, Dordrecht, the Netherlands.

Hawes, M.C., L.A. Brigham, F. Wen, H.H. Woo, and Y. Zhu. 1997. Root border cells: Phenomenology of signal exchange, p. 210–218. In H.E. Flores, J.M. Lynch, and D.M. Eissenstat (eds.) Radical biology: Advances and perspectives on the function of plant roots. American Society of Plant Physiology.

Hawes, M.C., L.A. Brigham, F. Wen, H.H. Woo, and Y. Zhu. 1998. Function of root border cells in plant health: Pioneers in the rhizosphere. Annu. Rev. Phytopathol. 36:311–327.

Hawes, M.C., U. Gunawardena, S. Miyasaka, and X.W. Zhao. 2000. The role of root border cells in plant defence. Trends Plant Sci. 5:128–133.

Hendriks, L., N. Claassen, and A. Jungk. 1981. Phosphatverarmung des wurzelnahen Bodens und Phosphataufnahme von Mais und Raps. Z. Pflanzenernaehr. Bodenkd. 144:486–499.

Hense, B.A., C. Kuttler, J. Müller, M. Rothballer, A. Hartmann, and J.U. Kreft. 2007. Does efficiency sensing unify diffusion and quorum sensing? Nat. Rev. Microbiol. 5:230–239.

Hill, P., Y. Kuzyakov, D.L. Jones, and J. Farrar. 2007. Response of root respiration and root exudation to alterations in root C supply and demand in wheat. Plant Soil 291:131–141.

Hinsinger, P. 1990. Actions des racines sur la libération du potassium et l'altération de minéraux silicatés—Incidences agronomiques. PhD thesis. ENSA Montpellier, France.

Hinsinger, P. 1998. How do plant roots acquire mineral nutrients? Chemical processes involved in the rhizosphere. Adv. Agron. 64:225–265.

Hinsinger, P. 2001. Bioavailability of soil inorganic P in the rhizosphere as affected by root-induced chemical changes: A review. Plant Soil 237:173–195.

Hinsinger, P. 2002. Potassium, p. 1035–1039. *In* R. Lal (ed.) Encyclopedia of soil science. Marcel Dekker, Inc., New York.

Hinsinger, P., A.G. Bengough, D. Vetterlein, and I.M. Young. 2009. Rhizosphere: Biophysics, biogeochemistry and ecological relevance. Plant Soil 321:117–152.

Hinsinger, P., F. Elsass, B. Jaillard, and M. Robert. 1993. Root-induced irreversible transformation of a trioctahedral mica in the rhizosphere of rape. J. Soil Sci. 44:535–545.

Hinsinger, P., O.N. Fernandes Barros, M.F. Benedetti, Y. Noack, and G. Callot. 2001. Plant-induced weathering of a basaltic rock: Experimental evidence. Geochim. Cosmochim. Acta 65:137–152.

Hinsinger, P., and R.J. Gilkes. 1995. Root-induced dissolution of phosphate rock in the rhizosphere of lupins grown in alkaline soil. Aust. J. Soil Res. 33:477–489.

Hinsinger, P., and R.J. Gilkes. 1996. Mobilization of phosphate from phosphate rock and alumina-sorbed phosphate by the roots of ryegrass and clover as related to rhizosphere pH. Eur. J. Soil Sci. 47:533–544.

Hinsinger, P., G.R. Gobran, P.J. Gregory, and W.W. Wenzel. 2005. Rhizosphere geometry and heterogeneity arising from root-mediated physical and chemical processes. New Phytol. 168:293–303.

Hinsinger, P., and B. Jaillard. 1993. Root-induced release of interlayer potassium and vermiculitization of phlogopite as related to potassium depletion in the rhizosphere of ryegrass. J. Soil Sci. 44:525–534.

Hinsinger, P., C. Plassard, and B. Jaillard. 2006. The rhizosphere: A new frontier in soil biogeochemistry. J. Geochem. Explor. 88:210–213.

Hinsinger, P., C. Plassard, C. Tang, and B. Jaillard. 2003. Origins of root-induced pH changes in the rhizosphere and their responses to environmental constraints: A review. Plant Soil 248:43–59.

Hirner, A., F. Ladwig, H. Stransky, S. Okumoto, M. Keinath, A. Harms, W.B. Frommer, and W. Koch. 2006. Arabidopsis LHT1 is a high-affinity transporter for cellular amino acid uptake in both root epidermis and leaf mesophyll. Plant Cell 18:1931–1946.

Hodge, A., G. Berta, C. Doussan, F. Merchan, and M. Crespi. 2009. Plant root growth, architecture and function. Plant Soil 321:153–187.

Hoefte, M. 1993. Classes of microbial siderophores, p. 3–26. *In* L.L. Barton and B.C. Hemming (eds.) Iron chelation in plants and soil microorganisms. Academic Press, San Diego, CA.

Hoffland, E., G.R. Findenegg, and J.A. Nelemans. 1989. Solubilization of rock phosphate by rape. 2. Local root exudation of organic-acids as a response to P-starvation. Plant Soil 113:161–165.

Hoffland, E., T.W. Kuyper, H. Wallander, C. Plassard, A. Gorbushina, K. Haselwandter, S. Holmström et al. 2004. The role of fungi in weathering. Front. Ecol. Environ. 2:258–264.

Horst, W.J., A. Wagner, and H. Marschner, 1982. Mucilage protects root-meristems from aluminum injury. Z. Pflanzenphysiol. 105:435–444.

Horst, W.J., and Ch. Waschkies. 1987. Phosphatversorgung von Sommer-Weizen (*Triticum aestivum* L.) in Mischkultur mit weisser Lupine (*Lupinus albus* L.). Z. Pflanzenernaehr. Bodenkd. 150:1–8.

Hu, H., C. Tang, and Z. Rengel. 2005. Influence of phenolic acids on phosphorus mobilisation in acidic and calcareous soils. Plant Soil 268:173–180.

Huang, P.M., and G.R. Gobran. 2004. Biogeochemistry of trace elements in the rhizosphere. Elsevier, Amsterdam, the Netherlands.

Huang, W.H., and W.D. Keller. 1970. Dissolution of rock-forming silicate minerals in organic acids: Simulated fist-stage weathering of fresh mineral surfaces. Am. Mineral. 57:2076–2094.

Huang, H.L., S.Z. Zhang, N.Y. Wu, L. Luo, and P. Christie. 2009. Influence of *Glomus etunicatum*/*Zea mays* mycorrhiza on atrazine degradation, soil phosphatase and dehydrogenase activities, and soil microbial community structure. Soil Biol. Biochem. 41:726–734.

Huck, M.G., B. Klepper, and H.M. Taylor. 1970. Diurnal variation in root diameter. Plant Physiol. 45:529–530.

Iijima, M., T. Higuchi, P.W. Barlow, and A.G. Bengough. 2003. Root cap removal increases root penetration resistance in maize (*Zea mays* L.). J. Exp. Bot. 54:2105–2109.

Jahn, T., F. Baluska, W. Michalke, J.F. Harper, and D. Volkmann. 1998. Plasma membrane H^+-ATPase in the root apex: Evidence for strong expression in xylem parenchyma and asymmetric localization within cortical and epidermal cells. Physiol. Plant 104:311–316.

Jassogne, L. 2008. Characterisation of porosity and root growth in a sodic texture-contrast soil. PhD Thesis. The University of Western Australia. Perth, Australia.

Jassogne, L., A. McNeill, and D. Chittleborough. 2007. 3D-visualization and analysis of macro- and meso-porosity of the upper horizons of a sodic, texture-contrast soil. Eur. J. Soil Sci. 58:589–598.

Jensen, S.I., M. Kuhl, R.N. Glud, L.B. Jorgensen, and A. Prieme. 2005. Oxic microzones and radial oxygen loss from roots of *Zostera marina*. Mar. Ecol. 293:49–58.

Johnson, N.C., J.H. Graham, and F.A. Smith. 1997. Functioning of mycorrhizal associations along the mutualism-parasitism continuum. New Phytol. 135:575–585.

Joner, E.J., and A. Johansen. 2000. Phosphatase activity of external hyphae of two arbuscular mycorrhizal fungi. Mycol. Res. 104:81–86.

Jones, D.L. 1998. Organic acids in the rhizosphere—A critical review. Plant Soil 205:25–44.

Jones, D.L., and P.R. Darrah. 1994. Amino-acid influx at the soil-root interface of *Zea mays* L. and its implications in the rhizosphere. Plant Soil 163:1–12.

Jones, D.L., and P.R. Darrah. 1995. Influx and efflux of organic acids across the soil-root interface of *Zea mays* L. and its implications in rhizosphere C flow. Plant Soil 173:103–109.

Jones, D.L., and P.R. Darrah. 1996. Re-sorption of organic compounds by roots of *Zea mays* L. and its consequences in the rhizosphere. 3. Characteristics of sugar influx and efflux. Plant Soil 178:153–160.

Jones, D.L., P.R. Darrah, and L.V. Kochian. 1996. Critical evaluation of organic acid mediated dissolution in the rhizosphere and its potential role in root iron uptake. Plant Soil 180:57–66.

Jones, D.L., J.R. Healey, V.B. Willett, J.F. Farrar, and A. Hodge. 2005. Dissolved organic nitrogen uptake by plants—An important N uptake pathway? Soil Biol. Biochem. 37:413–423.

Jones, D.L., C. Nguyen, and R.D. Finlay. 2009. Carbon flow in the rhizosphere: Carbon trading at the soil-root interface. Plant Soil 321:5–33.

Jongmans, A.G., N. Van Breemen, U. Lundström, P.A.W. Van Hees, R.D. Finlay, M. Srinivasan, T. Unestam, R. Giesler, P.-A. Melkerud, and M. Olsson. 1997. Rock-eating fungi. Nature 389:682–683.

Jourdan, C., E.V. Silva, J.L.M. Goncalves, J. Ranger, R.M. Moreira, and J.P. Laclau. 2008. Fine root production and turnover in Brazilian Eucalyptus plantations under contrasting nitrogen fertilization regimes. For. Ecol. Manage. 256:396–404.

Jousset, A., S. Scheu, and M. Bonkovski. 2008. Secondary metabolite production facilitates establishment of rhizobacteria by reducing both protozoan predation and the competitive effects of indigenous bacteria. Funct. Ecol. 22:714–719.

Jungk, A. 2002. Dynamics of nutrient movement at the soil-root interface, p. 587–616. *In* Y. Waisel, A. Eshel, and U. Kafkafi (eds.) Plant roots: The hidden half. 3rd edn. Marcel Dekker, Inc., New York.

Jungk, A., and N. Claassen. 1986. Availability of phosphate and potassium as the result of interactions between root and soil in the rhizosphere. J. Plant Nutr. Soil Sci. 149:411–427.

Jungk, A., B. Seeling, and J. Gerke. 1993. Mobilization of different phosphate fractions in the rhizosphere. Plant Soil 155–156:91–94.

Jurkevitch, E., Y. Hadar, Y. Chen, M. Chino, and S. Mori. 1993. Indirect utilization of the phytosiderophore mugineic acid as an iron source to rhizosphere fluorescent Pseudomonas. Biometals 6:119–123.

Kaci Y., A. Heyraud, M. Barakat, and T. Heulin. 2005. Isolation and identification of an EPS-producing *Rhizobium* strain from arid soil (Algeria): Characterisation of its EPS and the effect of inoculation on wheat rhizosphere soil structure. Research in Microbiology 156:522–531.

Kang, H., and C. Freeman. 2007. Interactions of marsh orchid (*Dactylorhiza* spp.) and soil microorganisms in relation to extracellular enzyme activities in a peat soil. Pedosphere 17:681–687.

Karpati, E., P. Kiss, T. Ponyi, I. Fendrik, M. de Zamaroczy, and L. Orosz. 1999. Interaction of *Azospirillum lipoferum* with wheat germ agglutinin stimulates nitrogen fixation. J. Bacteriol. 181:3949–3955.

Keller, M., and R. Hettich. 2009. Environmental proteomics: A paradigm shift in characterizing microbial activities at the molecular level. Micorbiol. Mol. Biol. Rev. 73:62–70.

Khaosaad, T., J.M. Garcia-Garrido, S. Steinkeller, and H. Vierheilig. 2007. Take-all disease is systematically reduced in roots of mycorrhizal barley plants. Soil Biol. Biochem. 39:727–734.

Kirby, J.M., and A.G. Bengough. 2002. Influence of soil strength on root growth: Experiments and analysis using a critical-state model. Eur. J. Soil Sci. 53:119–127.

Kirk, G.J.D., and J.B. Bajita. 1995. Root-induced iron oxidation, pH changes and zinc solubilization in the rhizosphere of lowland rice. New Phytol. 131:129–137.

Klein, D.A., B.A. Frederick, M. Biondini, and M.J. Trlica. 1988. Rhizosphere microorganisms effects on soluble amino acids, sugars and organic acids in the root zone of *Agropyron cristatum*, *A. smithii* and *Bouteloua gracilis*. Plant Soil 110:19–25.

Knee, E.M., F.C. Gong, M.S. Gao, M. Teplitski, A.R. Jones, A. Foxworthy, A.J. Mort, and W.D. Bauer. 2001. Root mucilage from pea and its utilization by rhizosphere bacteria as a sole carbon source. Mol. Plant–Microbe Interact. 14:775–784.

Knogge, W. 1996. Fungal colonization of plants. Plant Cell 8:1711–1722.

Kodama, H., S. Nelson, F. Yang, and N. Kohyama. 1994. Mineralogy of rhizospheric and non-rhizospheric soils in corn fields. Clay. Clay Miner. 42:755–763.

Kohler, J., F. Caravaca, L. Carrasco, and A. Roldan. 2007. Interactions between a plant growth-promoting rhizobacterium, an AM fungus and a phosphate-solubilising fungus in the rhizosphere of *Lactuca sativa*. Appl. Soil Ecol. 35:480–487.

Koide, R. 1985. The nature of growth depressions in sunflower caused by vesicular-arbuscular mycorrhizal colonization. New Phytol. 99:449–462.

Kuchenbuch, R., and A. Jungk. 1982. A method for determining concentration profiles at the soil-root interface by thin slicing rhizospheric soil. Plant Soil 68:391–394.

Kudeyarova, A.Y. 1991. Ligand exchange as a principal mechanism of phosphate adsorption on the surface of solid soil phase. Agrokhimiya 5:125–137.

Kumar, R.S., N. Ayyadurai, P. Pandiaraja, A.V. Reddy, Y. Venkateswarlu, O. Prakash, and N. Sakthivel. 2005. Characterization of antifungal metabolite produced by a new strain *Pseudomonas aeruginosa* PUPa3 that exhibits broad-spectrum antifungal activity and biofertilizing traits. J. Appl. Microbiol. 98:145–154.

Kuzyakov, Y. 2002a. Separating microbial respiration of exudates from root respiration in non-sterile soils: A comparison of four methods. Soil Biol. Biochem. 34:1621–1631.

Kuzyakov, Y. 2002b. Review: Factors affecting rhizosphere priming effects. J. Plant Nutr. Soil Sci. 165:382–396.

Kuzyakov, Y. 2006. Sources of CO_2 efflux from soil and review of partitioning methods. Soil Biol. Biochem. 38:425–448.

Kuzyakov, Y., and W. Cheng. 2004. Photosynthesis controls of CO_2 efflux from maize rhizosphere. Plant Soil 263:85–99.

Kuzyakov, Y., and G. Domanski. 2000. Carbon input by plants into the soil. Review. J. Plant Nutr. Soil Sci. 163:421–431.

Kuzyakov, Y., and D.L. Jones. 2006. Glucose uptake by maize roots and its transformation in the rhizosphere. Soil Biol. Biochem. 38:851–860.

Kuzyakov, Y., A. Raskatov, and M. Kaupenjohann. 2003. Turnover and distribution of root exudates of *Zea mays*. Plant Soil 254:317–327.

Landeweert, R., E. Hoffland, R.D. Finlay, T.W. Kuyper, and N. Van Breemen. 2001. Linking plants to rocks: Ectomycorrhizal fungi mobilize nutrients from minerals. Trends Ecol. Evol. 16:248–254.

Langenheder, S., and I.J. Prosser. 2008. Resource availability influences the diversity of a functional group of heterotrophic soil bacteria. Environ. Microbiol. 10:2245–2256.

Leake, J.R., and D.J. Read. 1991. Proteinase activity in mycorrhizal fungi. 3. Effects of protein, protein hydrolysate, glucose and ammonium on production of extracellular proteinase by *Hymenoscyphus ericae* (Read) Korf and Kernan. New Phytol. 117:309–317.

Leinhos, V., and O. Vacek. 1994. Biosynthesis of auxins by phosphate-solubilizing rhizobacteria from wheat (*Triticum aestivum*) and rye (*Secale cereale*). Microbiol. Res. 149:31–35.

Lemanceau, P., P. Bauer, S.M. Kraemer, and J.F. Briat. 2009. Iron in the rhizosphere as a case study for analysing interactions between soils, plants and microbes. Plant Soil 321:513–535.

Leong, J. 1986. Siderophores: Their biochemistry and possible role in the biocontrol of plant pathogens. Annu. Rev. Phythopathol. 24:187–209.

Lesturgez, G., R. Poss, C. Hartmann, E. Bourdon, A. Noble, and S. Ratana-Anupap. 2004. Roots of *Stylosanthes hamata* create macropores in the compact layer of a sandy soil. Plant Soil 260:101–109.

Lewis, D.G., and J.P. Quirk. 1967. Phosphate diffusion in soil and uptake by plants. III. P^{31}-movement and uptake by plants as indicated by P^{32}-autoradiography. Plant Soil 27:445–453.

Li, X.-L., E. George, and H. Marschner. 1991. Extension of the phosphorus depletion zone in VA-mycorrhizal white clover in a calcareous soil. Plant Soil 136:41–48.

Li, L., S.M. Li, J.H. Sun, L.L. Zhou, X.G. Bao, H.G. Zhang, and F.S. Zhang. 2007. Diversity enhances agricultural productivity via rhizosphere phosphorus facilitation on phosphorus-deficient soils. Proc. Natl. Acad. Sci. USA 104:11192–11196.

Li, S.M., L. Li, F.S. Zhang, and C. Tang. 2004. Acid phosphatase role in chickpea/maize intercropping. Ann. Bot. 94:297–303.

Li, H., J. Shen, F. Zhang, M. Clairotte, J.J. Drevon, E. Le Cadre, and P. Hinsinger. 2008. Dynamics of phosphorus fractions in the rhizosphere of common bean (*Phaseolus vulgaris* L.) and durum wheat (*Triticum turgidum durum* L.) grown in mono-cropping and intercropping systems. Plant Soil 312:139–150.

Li, H.Y., Y.G. Zhu, P. Marschner, F.A. Smith, and S.E. Smith. 2005. Wheat responses to arbuscular mycorrhizal fungi in a highly calcareous soil differ from those of clover, and change with plant development and P supply. Plant Soil 277:221–232.

Lichtenegger, E., and L. Kutschera-Mitter. 1991. Spatial root types, p. 359–365. *In* B.L. McMichael and H. Persson (eds.) Plant roots and their environment. Elsevier Science, Amsterdam, the Netherlands.

Lipson, D., and T. Nashölm. 2001. The unexpected versatility of plants: Organic nitrogen use and availability in terrestrial ecosystems. Oecologia 128:305–316.

Liu, Q., P. Loganathan, and M.J. Hedley. 2005. Influence of ectomycorrhizal hyphae on phosphate fractions and dissolution of phosphate rock in rhizosphere soils of *Pinus radiata*. J. Plant Nutr. 28:1525–1540.

Llugany, M.E., A. Lombini, C. Poschenrieder, E. Dinelli, and J. Barcelo. 2003. Different mechanisms account for enhanced copper resistance in *Silene armeria* ecotypes from mine spoil and serpentine sites. Plant Soil 251:55–63.

Loper, J.E. 1988. Role of fluorescent siderophore production in biological control of *Pythium ultimum* by a *Pseudomonas fluorescens* strain. Phytopathology 78:166–172.

Loper, J.E., and J.S. Buyer. 1991. Siderophores in microbial interactions on plant surfaces. Mol. Plant–Microbe Interact. 4:5–13.

Loper, J.E., and M.D. Henkels. 1999. Utilization of heterologous siderophores enhances level of iron available to *Pseudomonas putida* in the rhizosphere. Appl. Environ. Microbiol. 65:5357–5363.

Lorenz, S.E., R.E. Hamon, and S.P. McGrath. 1994. Differences between soil solutions obtained from rhizosphere and non-rhizosphere soils by water displacement and soil centrifugation. Eur. J. Soil Sci. 45:431–438.

Lugtenberg, B.J.J., G. Bloemberg, A.A.N. Van Brussel, J.W. Kijne, O.J.E. Thomas, and H.P. Spaink. 1995. Signals involved in nodulation and nitrogen fixation, p. 37–48. *In* I.A. Tikhonovich, N.A. Provorov, V.I. Romanov, and W.E. Newton (eds.) Nitrogen fixation: Fundamentals and applications. Kluwer Academic Publishers, Dordrecht, the Netherlands.

Lugtenberg, B.J.J., L.A. De Weger, and J.W. Bennet. 1991. Microbial stimulation of plant growth and protection from disease. Curr. Opin. Biotechnol. 2:457–464.

Lugtenberg, B.J.J., and L.C. Dekkers. 1999. What makes Pseudomonas bacteria rhizosphere competent? Environ. Microbiol. 1:9–13.

Ma, Z., T.C. Walk, A. Marcus, and J.P. Lynch. 2001. Morphological synergism in root hair length, density, initiation and geometry for phosphorus acquisition in *Arabidopsis thaliana*: A modeling approach. Plant Soil 236:221–235.

Makino, W., J.B. Cottner, R.W. Sterner, and J.J. Elser. 2003. Are bacteria more like plants or animals? Growth rate and resource dependence of bacterial C:N:P stoichiometry. Funct. Ecol. 17:121–130.

Marschner, H. 1995. Mineral nutrition of higher plants. Academic Press, London, U.K.

Marschner, P. 2008. The role of rhizosphere microorganisms in relation to P uptake by plants, p. 165–176. *In* P.J. White and J.P. Hammond (eds.) The ecophysiology of plant-phosphorus interactions. Springer, New York.

Marschner, H., and V. Römheld. 1983. *In vivo* measurement of root-induced pH changes at the soil-root interface: Effect of plant species and nitrogen source. Z. Pflanzenernaehr. Bodenkd. 111:241–251.

Marschner, P., M.Z. Solaiman, and Z. Rengel. 2006. Rhizosphere properties of *Poaceae* genotypes under P-limiting conditions. Plant Soil 283:11–24

Martens, D.A., and W.T. Frankenberger. 1992. Decomposition of bacteria polymers in soil and their influence on soil structure. Biol. Fertil. Soils 13:65–73.

Martin, J.K. 1970. ^{14}C-labelled material leached from rhizosphere of plants supplied with $^{14}CO_2$. Aust. J. Biol. Sci. 24:1131–1142.

Martin, P., A. Glatzle, W. Kolb, H. Omay, and W. Schmidt. 1989. N_2-fixing bacteria in the rhizosphere: Quantification and hormonal effects on root development. Z. Pflanzenernaehr. Bodenkd. 152:237–245.

Mazzola, M., R.J. Cook, L.S. Thomashow, D.M. Weller, and L.S. Pierson. 1992. Contribution of phenazine antibiotic biosynthesis to the ecological competence of fluorescent pseudomonads in soil habitats. Appl. Environ. Microbiol. 58:2616–2624.

McCully, M.E. 1999. Roots in soil: Unearthing the complexities of roots and their rhizospheres. Annu. Rev. Plant Physiol. Plant Mol. Biol. 50:695–718.

McCully, M.E., and J.S. Boyer. 1997. The expansion of maize root-cap mucilage during hydration. 3. Changes in water potential and water content. Physiol. Plant. 99:169–177.

McDougal, B.M., and A.D. Rovira 1970. Sites of exudation of ^{14}C-labelled compounds from wheat roots. New Phytol. 69:999–1002.

McLaughlin, M.J., A.M. Alston, and J.K. Martin. 1988. Phosphorus cycling in wheat-pasture rotations. II. The role of the microbial biomass in phosphorus cycling. Aust. J. Soil Res. 26:333–342.

Meharg, A.A. 1994. A critical-review of labeling techniques used to quantify rhizosphere carbon-flow. Plant Soil 166:55–62.

Meidute, S., F. Demoling, and E. Baath. 2008. Antagonistic and synergistic effects of fungal and bacterial growth in soil after adding different carbon and nitrogen sources. Soil Biol. Biochem. 40:2334–2343.

Michaud, A.M., M.N. Bravin, M. Galleguillos, and P. Hinsinger. 2007. Copper uptake and phytotoxicity as assessed in situ for durum wheat (*Triticum turgidum durum* L.) cultivated in Cu-contaminated, former vineyard soils. Plant Soil 298:99–111.

Mirleau, P., S. Delorme, L. Philippot, J.M. Meyer, S. Mazurier, and P. Lemanceau. 2000. Fitness in soil and rhizosphere of *Pseudomonas fluorescens* C7R12 compared with a C7R12 mutant affected in pyoverdine synthesis and uptake. FEMS Microbiol. Ecol. 34:35–44.

Mochida, K., and K. Shinozaki. 2010. Genomics and bioinformatics resources for crop improvement. Plant Cell Physiol. 51:497–523.

Moreno-Espindola, I.P., F. Rivera-Becerril, M. de Jesus Ferrara-Guerrero, and F. De Leon-Gonzalez. 2007. Role of root-hairs and hyphae in adhesion of sand particles. Soil Biol. Biochem. 39:2520–2526.

Murphy, P.J., W. Wexler, W. Grezemski, J.P. Rao, and D. Gordon. 1995. Rhizopines—Their role in symbiosis and competition. Soil Biol. Biochem. 27:525–529.

Nagahashi, G., and D.D. Douds. 2004. Isolated root caps, border cells, and mucilage from host roots stimulate hyphal branching of the arbuscular mycorrhizal fungus, *Gigaspora gigantean*. Mycol. Res. 108:1079–1088.

Näsholm, T., K. Kielland, and U. Ganeteg. 2009. Uptake of organic nitrogen by plants. New Phytol. 182:31–48.

Neilands, J.B. 1984. Siderophores of bacteria and fungi. Microbiol. Sci. 1:9–14.

Neubauer, S.C., G.E. Toledo-Durán, D. Emerson, and J.P. Megonigal. 2007. Returning to their roots: Iron-oxidizing bacteria enhance short-term plaque formation in the wetland-plant rhizosphere. Geomicrobiol. J. 24:65–73.

Newman, B.D., B.P. Wilcox, S.R. Archer, D.D. Breshears, C.N. Dahm, C.J. Duffy, N.G. McDowell, F.M. Phillips, B.R. Scanlon, and E.R. Vivoni. 2006. Ecohydrology of water-limited environments: A scientific vision. Water Resour. Res. 42:1–15.

Nguyen, C. 2003. Rhizodeposition of organic C by plants: Mechanisms and controls. Agronomie 23:375–396.

North, G.B., and P.S. Nobel. 1997. Drought-induced changes in soil contact and hydraulic conductivity for roots of *Opuntia ficus-indica* with and without rhizosheaths. Plant Soil 191:249–258.

Nye, P.H. 1981. Changes of pH across the rhizosphere induced by roots. Plant Soil 61:7–26.

Oberson, A., D.K. Friesen, I.M. Rao, S. Buehler, and E. Frossard. 2001. Phosphorus transformations in an oxisol under contrasting land-use systems: The role of the soil microbial biomass. Plant Soil 237:197–201.

Ogiyama, S., H. Suzuki, K. Inubushi, H. Takeda, and S. Uchida. 2010. Root-uptake of ^{14}C derived from acetic acid and ^{14}C transfer to rice edible parts. Appl. Radiat. Isot. 68:256–264.

Okazaki, M., K. Sakaidani, T. Saigusa, and N. Sakaida. 1989. Ligand-exchange of oxyanions on synthetic hydrated oxides of iron and aluminium. Soil Sci. Plant Nutr. 35:337–346.

Owen, A.G., and D.L. Jones. 2001. Competition for amino acids between wheat roots and rhizosphere microorganisms and the role of amino acids in plant N acquisition. Soil Biol. Biochem. 33:651–657.

Pal, K.K., K.V.B.R. Tilak, A.K. Saxena, R. Dey, and C.S. Singh. 2000. Antifungal characteristics of a fluorescent Pseudomonas strain involved in the biological control of *Rhizoctonia solani*. Microbiol. Res. 155:233–242.

Paterson, E., T. Gebbing, C. Abel, A. Sim, and G. Telfer. 2007. Rhizodeposition shapes rhizosphere microbial community structure in organic soil. New Phytol. 173:600–610.

Paterson, E., B. Thornton, A.J. Midwood, and A. Sim. 2005. Defoliation alters the relative contributions of recent and non-recent assimilate to root exudation from *Festuca rubra*. Plant Cell Environ. 28:1525–1533.

Paull, R.E., and R.L. Jones. 1975a. Studies on the secretion of maize root cap slime. 2. Localization of slime production. Plant Physiol. 56:307–312.

Paull, R.E., and R.L. Jones. 1975b. Studies on the secretion of maize root-cap slime. 3. Histochemical and autoradiographic localization of incorporated fucose. Planta 127:97–110.

Paull, R.E., and R.L. Jones. 1976a. Studies on the secretion of maize root cap slime. 4. Evidence for the involvement of dictyosomes. Plant Physiol. 57:249–256.

Paull, R.E., and R.L. Jones. 1976b. Studies on the secretion of maize root cap slime. 5. The cell wall as a barrier to secretion. Z. Pflanzenphysiol. 79:154–164.

Pearse, S.J., E.J. Veneklaas, G.R. Cawthray, M.D.A. Bolland, and H. Lambers. 2006a. *Triticum aestivum* shows a greater biomass response to a supply of aluminium phosphate than *Lupinus albus*, despite releasing less carboxylates into the rhizosphere. New Phytol. 169:515–524.

Pearse, S.J., E.J. Veneklaas, G.R. Cawthray, M.D.A. Bolland, and H. Lambers. 2006b. Carboxylate release of wheat, canola and 11 grain legume species as affected by phosphorus status. Plant Soil 288:127–139.

Peng, S., D.M. Eissenstat, J.H. Graham, K. Williams, and N.C. Hodge. 1993. Growth depression in mycorrhizal citrus at high-phosphorus supply. Plant Physiol. 101:1063–1071.

Perez-Moreno, J., and D.J. Read. 2000. Mobilization and transfer of nutrients from litter to tree seedlings via the vegetative mycelium of ectomycorrhizal plants. New Phytol. 145:301–309.

Personeni, E., C. Nguyen, P. Marchal, and L. Pagès. 2007. Experimental evaluation of an efflux-influx model of C exudation by individual apical root segments. J. Exp. Bot. 58:2091–2099.

Philippot, L., J. Cuhel, N.P.A. Saby, D. Chèneby, A. Chronáková, D. Bru, D. Arrouays, F. Martin-Laurent, and M. Šimek. 2009a. Mapping field-scale spatial patterns of size and activity of the denitrifier community. Environ. Microbiol. 11:1518–1526.

Philippot, L., S. Hallin, G. Borjesson, and E.M. Baggs. 2009b. Biochemical cycling in the rhizosphere having an impact on global change. Plant Soil 321:61–81.

Phillips, R.P., and T.J. Fahey. 2006. Tree species and mycorrhizal associations influence the magnitude of rhizosphere effects. Ecology 87:1302–1313.

Phillips, D.A., T.C. Fox, M.D. King, T.V. Bhuvaneswari, and L.R. Teuber. 2004. Microbial products trigger amino acid exudation from plant roots. Plant Physiol. 136:2887–2894.

Pierret, A., C. Doussan, Y. Capowiez, F. Bastardie, and L. Pagès. 2007. Root functional architecture: A framework for modeling the interplay between roots and soil. Vadose Zone J. 6:269–281.

Pierret, A., C.J. Moran, and C.E. Pankhurst. 1999. Differentiation of soil properties related to the spatial association of wheat roots and soil macropores. Plant Soil 211:51–58.

Pineros, M.A., J.E. Shaff, H.S. Manslank, V.M.C. Alves, and L.V. Kochian. 2005. Aluminum resistance in maize cannot be solely explained by root organic acid exudation. A comparative physiological study. Plant Physiol. 137:231–241.

Pinton, R., Z. Varanini, and P. Nannipieri. 2007. The rhizosphere. Biochemistry and organic substances at the soil-plant interface. 2nd edn. CRC Press, Boca Raton, FL.

Plassard, C., M. Meslem, G. Souche, and B. Jaillard. 1999. Localization and quantification of net fluxes of H^+ along roots of maize by combined use of videodensitometry of indicator dye and ion-selective microelectrodes. Plant Soil 211:29–39.

Poll, C., S. Marhan, J. Ingwersen, and E. Kandeler. 2008. Dynamics of litter carbon turnover and microbial abundance in a rye detritusphere. Soil Biol. Biochem. 40:1306–1321.

Poole, E.J., G.D. Bending, J.M. Whipps, and D.J. Read. 2001. Bacteria associated with *Pinus sylvestris-Lactarius rufus* ectomycorrhizas and their effects on mycorrhiza formation in vitro. New Phytol. 151:743–751.

Raaijmakers, J.M., T.C. Paulitz, C. Steinberg, C. Alabouvette, and Y. Moënne-Loccoz. 2009. The rhizosphere: A playground and battlefield for soilborne pathogens and beneficial microorganisms. Plant Soil 321:341–361.

Raaijmakers, J.M., I. Van der Sluis, P.A.H.M. Bakker, P. Weisbeek, and B. Schippers. 1995. Utilization of heterologous siderophores and rhizosphere competence of fluorescent *Pseudomonas* spp. Can. J. Microbiol. 41:126–135.

Raiesi, F., and M. Ghollarata. 2006. Interactions between phosphorus availability and an AM fungus (*Glomus intraradices*) and their effects on soil microbial respiration, biomass and enzyme activities in a calcareous soil. Pedobiologia 50:413–425.

Rasmussen, J., and Y. Kuzyakov. 2009. Carbon isotopes as proof for plant uptake of organic nitrogen: Relevance of inorganic carbon uptake. Soil Biol. Biochem. 41:1586–1587.

Raynaud, X. 2010. Soil properties are key determinants for the development of exudate gradients in a rhizosphere simulation model. Soil Biol. Biochem. 42:210–219.

Read, D.B., A.G. Bengough, P.J. Gregory, J.W. Crawford, D. Robinson, C.M. Scrimgeour, I.M. Young, K. Zhang, and X. Zhang. 2003. Plant roots release phospholipid surfactants that modify the physical and chemical properties of soil. New Phytol. 157:315–326.

Read, D.B., P.J. Gregory, and A.E. Bell. 1999. Physical properties of axenic maize root mucilage. Plant Soil 211:87–91.

Redfield, R.J. 2002. Is quorum sensing a side effect of diffusion sensing? Trends Microbiol. 10:365–370.

Revsbech, N.P., O. Pedersen, W. Reichardt, and A. Briones. 1999. Microsensor analysis of oxygen and pH in the rice rhizosphere under field and laboratory conditions. Biol. Fertil. Soils 29:379–385.

Reyes, I., L. Bernier, and H. Antoun. 2002. Rock phosphate solubilization and colonization of maize rhizosphere by wild and genetically modified strains of *Penicillium rugulosum*. Microbiol. Ecol. 44:39–48.

Richardson, A.E., J.M. Barea, A.M. McNeill, and C. Prigent-Combaret. 2009. Acquisition of phosphorus and nitrogen in the rhizosphere and plant growth promotion by microorganisms. Plant Soil 321:305–339.

Riley, D., and S.A. Barber. 1971. Effect of ammonium and nitrate fertilization on phosphorus uptake as related to root-induced pH changes at the root-soil interface. Soil Sci. Soc. Am. Proc. 35:301–306.

Robin, A., G. Vansuyt, P. Hinsinger, J.M. Meyer, J.F. Briat, and P. Lemanceau. 2008. Iron dynamics in the rhizosphere: Consequences for plant health and nutrition. Adv. Agron. 99:183–225.

Robinson, D. 1994. The responses of plants to non-uniform supplies of nutrients. New Phytol. 127:635–674.

Roesti, D., R. Gaur, B.N. Johri, G. Imfeld, S. Sharma, K. Kawaljeet, and M. Aragno. 2006. Plant growth stage, fertiliser management and bio-inoculation of arbuscular mycorrhizal fungi and plant growth promoting rhizobacteria affect the rhizobacterial community structure in rain-fed wheat fields. Soil Biol. Biochem. 38:1111–1120.

Römheld, V. 1991. The role of phytosiderophores in acquisition of iron and other micronutrients in graminaceous species: An ecological approach. Plant Soil 130:127–134.

Römheld, V., and H. Marschner. 1989. Genotypical differences among graminaceous species in release of phytosiderophores and uptake of iron phytosiderophores. Plant Soil 128:120–126.

Roose, T., and G.J.D. Kirk. 2009. The solution of convection-diffusion equations for solute transport to plant roots. Plant Soil 316:257–264.

Rovira, A.D. 1969. Plant root exudates. Bot. Rev. 35:35–57.

Ryan, P.R., E. Delhaize, and D.L. Jones. 2001. Function and mechanism of organic anion exudation from plant roots. Annu. Rev. Plant Physiol. Plant Mol. Biol. 52:527–560.

Ryan, P.R., Y. Dessaux, L.S. Thomashow, and D.M. Weller. 2009. Rhizosphere engineering and management for sustainable agriculture. Plant Soil 321:363–383.

Ryu, C.M., C.H. Hu, R.D. Locy, and J.W. Kloepper. 2005. Study of mechanisms for plant growth promotion elicited by rhizobacteria in *Arabidopsis thaliana*. Plant Soil 281:285–292.

Saeki, K., and S. Matsumoto. 1998. Mechanisms of ligand exchange reactions involving selenite sorption on goethite labeled with oxygen-stable isotope. Comm. Soil Sci. Plant Anal. 29:3061–3072.

Saglio, P.H., and A. Pradet. 1980. Soluble sugars, respiration, and energy-charge during aging of excised maize root-tips. Plant Physiol. 66:516–519.

Samuels, A.L., M. Fernando, and A.D.M. Glass. 1992. Immunofluorescent localization of plasma-membrane H^+-ATPase in barley roots and effects of K nutrition. Plant Physiol. 99:1509–1514.

Sanon, A., Z.N. Andrianjaka, Y. Prin, R. Bally, J. Thioulouse, G. Comte, and R. Duponnois. 2009. Rhizosphere microbiota interfers with plant-plant interactions. Plant Soil 321:259–278.

Schenk, H.J., and R.B. Jackson. 2002. The global biogeography of roots. Ecol. Monogr. 72:311–328.

Schnepf, A., and T. Roose. 2006. Modelling the contribution of arbuscular mycorrhizal fungi to plant phosphate uptake. New Phytol. 171:669–682.

Schnepf, A., T. Roose, and P. Schweiger. 2008. Impact of growth and uptake patterns of arbuscular mycorrhizal fungi on plant phosphorus uptake—A modelling study. Plant Soil 312:85–99.

Schweiger, P., and I. Jakobsen. 1999. The role of mycorrhizas in plant P nutrition: Fungal uptake kinetics and genotype variation, p. 277–289. *In* G. Gissel-Nielsen and A. Jensen (eds.) Plant nutrition—Molecular biology and genetics. Kluwer Academic Publishers, Dordrecht, the Netherlands.

Seeling, B., and R.J. Zasoski. 1993. Microbial effects in maintaining organic and inorganic solution phosphorus concentrations in a grassland topsoil. Plant Soil 148:277–284.

Shane, M.W., and H. Lambers. 2005. Cluster roots: A curiosity in context. Plant Soil 274:101–125.

Simons, M., H.P. Permentier, L.A. De Weger, C.A. Wijffelman, and B.J.J. Lugtenberg. 1997. Amino acid synthesis is necessary for tomato root colonization by *Pseudomonas fluorescens* strain WCS365. Mol. Plant–Microb. Interact. 10:102–106.

Smith, S.E., and D.J. Read. 2008. Mycorrhizal symbiosis. 3rd edn. Elsevier, Amsterdam, the Netherlands.

Somers, E., J. Vanderleyden, and M. Srinivasan. 2004. Rhizosphere bacterial signalling: A love parade beneath our feet. Crit. Rev. Microbiol. 30:205–240.

Springob, G., and J. Richter. 1998. Measuring interlayer potassium release rates from soil materials. II. A percolation procedure to study the influence of the variable 'solute K' in the $<1…10\,\mu M$ range. Z. Pflanzenernaehr. Bodenkd. 161:323–329.

Srinivasan, M., D.J. Petersen, and F.B. Holl. 1997. Altered root hair morphogenesis in *Phaseolus vulgaris* in response to bacterial coinoculation and the presence of aminoethoxy vinyl glycine (AVG). Microbiol. Res. 152:151–156.

Standing, D., E.M. Baggs, M. Wattenbach, P. Smith, and K. Killham. 2007. Meeting the challenge of scaling up processes in the plant-soil-microbe system. Biol. Fertil. Soils 44:245–257.

Stewart, J.B., C.J. Moran, and J.T. Wood. 1999. Macropore sheath: Quantification of plant root and soil macropore association. Plant Soil 211:59–67.

Stirzaker, R.J., J.B. Passioura, and Y. Wilms. 1996. Soil structure and plant growth: Impact of bulk density and biopores. Plant Soil 185:151–162.

Ström, L., A.G. Owen, D.L. Godbold, and D.L. Jones. 2001. Organic acid behaviour in a calcareous soil: Sorption reactions and biodegradation rates. Soil Biol. Biochem. 33:2125–2133.

Tadano, T., and H. Sakai. 1991. Secretion of acid phosphatase by the roots of several crop species under phosphorus-deficient conditions. Soil Sci. Plant Nutr. 37:129–140.

Takagi, S. 1976. Naturally occurring iron-chelating compounds in oat- and rice-root washings. Soil Sci. Plant Nutr. 22:423–433.

Tang, C., J.J. Drevon, B. Jaillard, G. Souche, and P. Hinsinger. 2004. Proton release of two genotypes of bean (*Phaseolus vulgaris* L.) as affected by N nutrition and P deficiency. Plant Soil 260:59–68.

Tarafdar, J.C., and H. Marschner. 1994. Phosphatase-activity in the rhizosphere and hyphosphere of VA mycorrhizal wheat supplied with inorganic and organic phosphorus. Soil Biol. Biochem. 26:387–395.

Tarafdar, J.C., R.S. Yadav, and S.C. Meena. 2001. Comparative efficiency of acid phosphatase originated from plant and fungal sources. J. Plant Nutr. Soil Sci. 164:279–282.

Tezuka, Y. 1990. Bacterial regeneration of ammonium and phosphate as affected by the carbon:nitrogen:phosporus ratio of organic substrates. Microbiol. Ecol. 19:227–238.

Thomashow, L.S., and D.S. Weller. 1990. Role of antibiotics and siderophores in biocontrol of take-all disease of wheat. Plant Soil 129:93–99.

Thornton, B., E. Paterson, A.J. Midwood, A. Sim, and S.M. Pratt. 2004. Contribution of current carbon assimilation in supplying root exudates of *Lolium perenne* measured using steady-state ^{13}C labelling. Physiol. Plant 120:434–441.

Tinker, B.T., and P.H. Nye. 2000. Solute movement in the rhizosphere. Oxford University Press, Oxford, U.K.

Todorovic, C., C. Nguyen, C. Robin, and A. Guckert. 1999. ^{14}C-assimilate partitioning within white clover plant-soil system: Effects of photoperiod temperature treatments and defoliation. Eur. J. Agron. 11:13–21.

Traore, O., V. Groleau-Renaud, S. Plantureux, A. Tubeileh, and V. Boeuf-Tremblay. 2000. Effect of root mucilage and modelled root exudates on soil structure. Eur. J. Soil Sci. 51:575–581.

Trolldenier, G. 1988. Visualization of oxidizing power of rice roots and of possible participation of bacteria in iron deposition. Z. Pflanzenernaehr. Bodenkd. 151:117–121.

Turpault, M.P., C. Utérano, J.P. Boudot, and J. Ranger. 2005. Influence of mature Douglas fir roots on the solid soil phase of the rhizosphere and its solution chemistry. Plant Soil 275:327–336.

Uroz, S., C. Calvaruso, M.P. Turpault, J.C. Pierrat, C. Mustin, and P. Frey-Klett. 2007. Effect of mycorrhizosphere on the genotypic and metabolic diversity of the bacterial communities involved in mineral weathering in a forest soil. Appl. Environ. Microbiol. 79:3019–3027.

van Aarle, I.M., and C. Plassard. 2010. Spatial distribution of phosphatase activity associated with ectomycorrhizal plants is related to soil type. Soil Biol. Biochem. 42:324–330.

Van Noordwijk, M., D. Schoonderbeek, and M.J. Kooistra. 1993. Root-soil contact of field-grown winter wheat. Geoderma 56:277–286.

Vancura, V. 1967. Root exudates of plants. 3. Effect of temperature and cold shock on exudation of various compounds from seeds and seedlings of maize and cucumber. Plant Soil 27:319–327.

Vancura, V., and G. Stotzky. 1976. Gaseous and volatile exudates from germinating seeds and seedlings. Can. J. Bot. 54:518–532.

Vanpeer, R., H.L.M. Punte, L.A. DeWeger, and B. Schippers. 1990. Characterization of root surface and endorhizosphere pseudomonads in relation to their colonization of roots. Appl. Environ. Microbiol. 56:2462–2470.

Vanrhijn, P., and J. Vanderleyden. 1995. The *Rhizobium*-plant symbiosis. Microbiol. Rev. 59:124–142.

Vazquez, M.M., S. Cesar, R. Azcon, and J.M. Barea. 2000. Interactions between arbuscular mycorrhizal fungi and other microbial inoculants (*Azospirillum, Pseudomonas, Trichoderma*) and their effects on microbial population and enzyme activities in the rhizosphere of maize plants. Appl. Soil Ecol. 15:261–272.

Vierheilig, H., R. Bennett, G. Kiddle, M. Kaldorf, and J. Ludwig-Mueller. 2000. Differences in glucosinolate patterns and arbuscular mycorrhizal status of glucosinolate-containing plant species. New Phytol. 146:343–352.

Vierheilig, H., S. Lerat, and Y. Piche. 2003. Systemic inhibition of arbuscular mycorrhiza development by root exudates of cucumber plants colonized by *Glomus mosseae*. Mycorrhiza 13:167–170.

Vierheilig, H., and J.A. Ocampo. 1990. Role of root extract and volatile substances of non-host plants on vesicular-arbuscular mycorrhizal spore germination. Symbiosis 9:199–202.

Von Wiren, N., V. Roemheld, T. Shiori, and H. Marschner. 1995. Competition between micro-organisms and roots of barley and sorghum for iron accumulated in the root apoplasm. New Phytol. 130:511–521.

Vong, P.C., O. Dedourge, F. Lasserre-Joulin, and A. Guckert. 2003. Immobilized-S, microbial biomass-S and soil arylsulfatase activity in the rhizosphere soil of rape and barley as affected by labile substrate C and N additions. Soil Biol. Biochem. 35:1651–1661.

Wallander, H. 2000. Uptake of P from apatite by *Pinus sylvestris* seedlings colonised by different ectomycorrhizal fungi. Plant Soil 218:249–256.

Wallander, H., and T. Wickman. 1999. Biotite and microcline as potassium sources in ectomycorrhizal and non-mycorrhizal *Pinus sylvestris* seedlings. Mycorrhiza 9:25–32.

Walter, A., V. Roemheld, H. Marschner, and D.E. Crowley. 1994. Iron nutrition of cucumber and maize: Effect of *Pseudomonas putida* YC3 and its siderophore. Soil Biol. Biochem. 26:1023–1031.

Walton, J.D. 1996. Host-selective toxins: Agents of compatibility. Plant Cell 8:1723–1733.

Wang, Y., H.N. Brown, D.E. Crowley, and P.J. Szaniszlo. 1993. Evidence for direct utilization of a siderophore, ferrioxamine B, in axenically grown cucumber. Plant Cell Environ. 16:579–585.

Wang, Y.Q., Y. Ohara, H. Nakayashiki, Y. Tosa, and S. Mayama. 2005. Microarray analysis of the gene expression profile induced by the endophytic plant growth-promoting rhizobacteria, *Pseudomonas fluorescens* FPT9601-T5 in Arabidopsis. Mol. Plant-Microbe Interact. 18:385–396.

Watt, M., J.A. Kirkegaard, and G.J. Rebetzke. 2005. A wheat genotype developed for rapid leaf growth copes well with the physical and biological constraints of unploughed soil. Funct. Plant Biol. 32:695–706.

Watt, M., M.E. McCully, and M.J. Canny. 1994. Formation and stabilisation of maize rhizosheaths: Effect of soil water content. Plant Physiol. 106:179–186.

Watt, M., M.E. McCully, and C.E. Jeffree. 1993. Plant and bacterial mucilages of the maize rhizosphere—Comparison of their soil-binding properties and histochemistry in a model system. Plant Soil 151:151–165.

Weintraub, M.N., L.E. Scott-Denton, S.K. Schmidt, and R.K. Monson. 2007. The effects of tree rhizodeposition on soil exoenzyme activity, dissolved organic carbon, and nutrient availability in a subalpine forest ecosystem. Oecologia 154:327–338.

Wenzel, W.W. 2009. Rhizosphere processes and management in plant-assisted bioremediation (phytoremediation) of soils. Plant Soil 321:385–408.

Werner, D., B. Ahlborn, H. Antoun, S. Bernard, C. Bolanos, J.E. Cooper, E. Goerge et al. 1994. Communication and signal exchange in the *Rhizobium/Bradyrhizobium*-legume system. Endocytobiol. Cell Res. 10:5–15.

Whalley, W.R., B. Riseley, P.B. Leeds-Harrison, N.R.A. Bird, P.K. Leech, and W.P. Adderley. 2005. Structural differences between bulk and rhizosphere soil. Eur. J. Soil Sci. 56:353–360.

Whitelaw, M.A., T.J. Harden, and K.R. Helyar. 1999. Phosphate solubilization in solution culture by the soil fungus *Penicillium radicum*. Soil Biol. Biochem. 31:655–665.

Williams, S.M., and R.R. Weil. 2004. Crop cover root channels may alleviate soil compaction effects on soybean crop. Soil Sci. Soc. Am. J. 68:1403–1409.

Winans, S.C. 2004. Reciprocal regulation of bioluminescence and type III protein secretion in *Vibrio harveyi* and *Vibrio parahaemolyticus* in response to diffusible chemical signals. J. Bacteriol. 186:3674–3676.

Xia, J.H., and P.H. Saglio. 1988. Characterization of the hexose-transport system in maize root-tips. Plant Physiol. 88:1015–1020.

Yehuda, Z., M. Shenker, Y. Hadar, and Y. Chen. 2000. Remedy of chlorosis induced by iron deficiency in plants with the fungal siderophore rhizoferrin. J. Plant Nutr. 23:1991–2006.

Yehuda, Z., M. Shenker, V. Roemheld, H. Marschner, Y. Hadar, and Y. Chen. 1996. The role of ligand exchange in the uptake of iron from microbial siderophores by gramineous plants. Plant Physiol. 112:1273–1280.

Young, I.M. 1995. Variations in moisture contents between bulk soil and the rhizosheath of wheat (*Triticum aestivum* L. cv Wembley). New Phytol. 130:135–139.

Young, I.M. 1998. Biophysical interactions at the root-soil interface: A review. J. Agric. Sci. 130:1–7.

Zhang, F., and L. Li. 2003. Using competitive and facilitative interactions in intercropping systems enhances crop productivity and nutrient-use efficiency. Plant Soil 248:305–312.

Zhang, F., V. Roemheld, and H. Marschner. 1991. Diurnal rhythm of release of phytosiderophores and uptake rate of zinc in iron-deficient wheat. Soil Sci. Plant Nutr. 37:671–678.

Zhu, Y.G., T.R. Cavagnaro, S.E. Smith, and S. Dickson 2001. Backseat driving? Accessing phosphate beyond the rhizosphere depletion zone. Trends Plant Sci. 6:194–195.

Zhu, Y., L.S. Pierson, and M.C. Hawes. 1997. Induction of microbial genes for pathogenesis and symbiosis by chemicals from root border cells. Plant Physiol. 115:1691–1698.

Zobel, R.W. 1991. Genetic control of root systems, p. 27–38. *In* Y. Waisel, A. Eshel, and U. Kafkafi (eds.) Plant roots: The hidden half. Marcel Dekker, New York.

Zoysa, A.K.N., P. Loganathan, and M.J. Hedley. 1998. Phosphate rock dissolution and transformation in the rhizosphere of tea (*Camellia sinensis* L.) compared with other species. Eur. J. Soil Sci. 49:477–486.

Zuo, Y., F. Zhang, X. Li, and Y. Cao. 2000. Studies on the improvement in iron nutrition of peanut by intercropping with maize on a calcareous soil. Plant Soil 220:13–25.

7
Mineralogical, Physicochemical, and Microbiological Controls on Soil Organic Matter Stabilization and Turnover

Ingrid Kögel-Knabner
Technische Universität München

Markus Kleber
Oregon State University

7.1 Introduction 7-1
7.2 Modeling of OC Turnover and the Concept of Different SOM Pools in Soils 7-2
   Modeling of OC Turnover and Stabilization • Mechanisms for the Accumulation of Organic Matter in Labile and Stable Pools • Fractionation Concepts and Isolation of Specific Fractions
7.3 Recalcitrance as an Index for Intrinsic Organic Matter Quality 7-6
   Terms and Mechanisms • Plant and Microbial-Derived Compounds • Fire-Derived Organic Matter (Black Carbon) • Carbonaceous Pollutants and Nanomaterials
7.4 Stabilization of SOM 7-7
   Soil Architecture and Its Effects on C Turnover and Stabilization • Organo-Mineral Interactions
7.5 Outlook 7-15
References 7-15

7.1 Introduction

Soil organic matter (SOM) represents the largest terrestrial carbon (C) pool being almost three times that present in plant and animal biomass. SOM is not a uniform substance, but occurs in a myriad of different molecular forms, whose chemical properties and interactions with the abiotic mineral matrix allow them to be placed in conceptual pools. Knowledge of the size and fluxes of C from these SOM pools is indispensable to predict C turnover in soils as a function of environmental changes. The amount of C stored in a soil is determined by a dynamic equilibrium between C inputs from primary biomass production and C outputs mainly by mineralization, but also from leaching and erosion. Understanding the controls on this equilibrium is an important scientific issue for two reasons:

Feedback: Large quantities of carbon are stored in SOM, and release of this carbon into the atmosphere as CO_2 or methane would have a serious impact on global climate. There are first reports that, as the earth warms, SOM is being oxidized at an increasing rate (Bellamy et al., 2005; Powlson, 2005; Davidson and Janssens, 2006; von Lützow and Kögel-Knabner, 2009). Quantifying these carbon cycle-climate feedbacks is difficult because of the limited understanding of the processes by which carbon and associated nutrients are transformed or recycled within ecosystems, in particular within soils, and exchanged with the overlying atmosphere (Heimann and Reichstein, 2008). Below-ground processes in particular are still poorly understood, yet provide a number of potentially important feedbacks in the carbon-cycle–climate system (Scheffer et al., 2006; Torn and Harte, 2006). It follows that better predictions of the magnitude of soil carbon feedback reactions require an improved mechanistic understanding of processes that protect SOM against decomposition.

Soil carbon sequestration: A large body of literature suggests that terrestrial sequestration of carbon through land management changes should be considered an option for mitigating climate change. Analyses reviewed by the Intergovernmental Panel on Climate Change (IPCC) have estimated that about 40 Gton of carbon could be sequestered in agricultural soil over a 50–100 year span by applying improved agricultural practices (Thomson et al., 2008). In many instances, there are also co-benefits (reduced soil erosion, reduced energy costs, improved soil fertility) that provide incentives for terrestrial sequestration practices (Lal, 2004). It is obvious that sequestration practices must aim to maximize the turnover time of SOM in order to be efficient and useful for mitigation purposes.

Decomposition of organic C (OC) can be retarded by different stabilization processes. The interplay between these processes is complex and requires an understanding of chemical, physical, and biological interactions between organic components and the mineral matrix.

7.2 Modeling of OC Turnover and the Concept of Different SOM Pools in Soils

7.2.1 Modeling of OC Turnover and Stabilization

Turnover models of SOM contribute to our understanding of C dynamics in soils and serve to predict the SOM content under different scenarios. *Turnover time* (τ) is generally defined as the time interval between entering and leaving a reservoir. In carbon biogeochemistry, it is the time interval between assimilation of a C-atom by an ecosystem and the release of the same C-atom as respired CO_2. In many ecosystems (e.g., such as old growth forests), this means that a carbon compound may already be quite old before it enters the decomposition chain in the soil subsystem. Turnover time τ is measured in time units (year). *Carbon stabilization and destabilization* ($\Delta\tau$) is defined as a process that leads to a change in carbon turnover time compared to a reference situation (von Lützow et al., 2006). Prolongation of turnover time, or stabilization, is expressed as

$$\tau_a - \tau_b > 0, \quad (7.1)$$

where

τ_a is the turnover time after a certain stabilization/destabilization process has occurred

τ_b is the turnover time before a certain stabilization/destabilization process has occurred

Reduction of turnover time is defined as destabilization:

$$\tau_a - \tau_b < 0. \quad (7.2)$$

In general, SOM is a complex mixture of substances spanning an extreme range of molecular properties so that there is every reason to expect that the organic fragments and decomposition products in soils would differ widely in their radiocarbon content. Here, the problem arises that while SOM may actually be a continuum best represented by a distribution of residence times (Trumbore, 1997), continuous distributions are difficult to constrain with field measurements (Bruun et al., 2010). One strategy to constrain a heterogeneous population is to divide it into sub-fractions that share enough physical or functional features to allow them to be treated as homogeneous. While most contemporary turnover models share this general approach, predicting the long-term stability of carbon compounds in soil remains difficult because of the wide array of processes that may potentially be involved in SOM turnover, including carbon chemistry and its interaction with microbial enzyme systems, mineral soil surfaces, disturbance, and climate (Grandy and Neff, 2008). Widely used soil carbon models, such as CASA, IBIS-2, CENTURY, and Rothamsted (Jenkinson, 1990; Parton et al., 1994; Randerson et al., 1996; Kucharik et al., 2001), treat SOM as consisting of discrete pools of different turnover times.

The current models subdivide "active" or "labile" pools by assigning different decomposition rate constants to organic input materials (McGill, 1996).

Following Torn et al. (2009), decomposition of a homogeneous reservoir can be treated as a linear, donor-controlled process, that is, the amount of C decomposed is the product of the C stock (C, g C m^{-2}), a decomposition rate constant (k, measured in years), and the time interval (Δt, year). The change in soil C stock between one time point and the next (dC/dt) is the difference between the plant inputs (I) and decomposition outputs (kC) over that period, or $dC/dt = I - kC$, where t = time (year) and I = carbon inputs (g C m^{-2} year^{-1}). We also follow Torn et al. (2009) when we define turnover time (τ) as the reciprocal of the decomposition rate constant ($\tau = 1/k$). At steady state $\tau = I/C$ and the soil C stock is the product of inputs and turnover time. See also Torn et al. (2009) for a detailed discussion on the differences underlying the concept of turnover time and mean residence time (MRT).

The "intermediate" or "slow" organic matter pool is typically assumed to turn over on decadal timescales, while "passive" or "inert" organic matter is defined as such organic materials whose turnover happens on centennial and millennial timescales and is largely unaffected by management or disturbance. The formation of intermediate and passive pools in the models is treated as a selective preservation of recalcitrant substances and basically relies on the assumption that the stable fractions are the leftovers from mineralization of plant residues. While the pool structure is likely adequate for capturing soil C dynamics, research over the past decade has found that the controls of partitioning and turnover are poorly parameterized or missing. An important example is reliance on clay content/soil texture to represent the influence of the mineral matrix on soil C storage and decomposability. In the Century model, for example, soil texture influences the decomposition rate of the active SOM pool and the efficiency of SOM stabilization in the slow pool. Fine textures favor slow decomposition rates and greater stability of SOM. However, a recent sensitivity analysis for the Century model suggests that the effect of clay percentage on the rate of C change is not well understood, leading to overestimations of soil organic C (SOC) by an average of 10% (max of 28% in Bricklemyer et al., 2007) following a management change from tillage to no-tillage.

7.2.2 Mechanisms for the Accumulation of Organic Matter in Labile and Stable Pools

The terms "labile" and "stable" are operational and nearly nondescript terms that are used by the biogeochemistry community to indicate major functional differences between the turnover dynamics of SOM pools. They are used for convenience, but do not correspond to measurable, parametric categories. Table 7.1 gives an overview on the different mechanisms that have been

TABLE 7.1 Controls on Turnover Time and Assignment of Timescales

Mechanisms	Relevance	Timescale
Intrinsic recalcitrance of SOM *Definition*: Molecular characteristics of organic substances that influence their degradation by microbes and enzymes. *Problem*: Characteristics vary independently, that is, molecular size can be small (would facilitate decomposition) while solubility may be low (would retard decomposition)	There is no generally accepted single numerical parameter to express the degree of recalcitrance, but some consensus exists for some controls on "intrinsic recalcitrance": Solubility in H_2O (mass per volume); molecular size Dalton; Stoichiometric O_2 demand (Mol O_2 per mole substrate); Number of functionally different bonds within the molecule; Degree of polymerization (number of monomers within polymer)	Protective effects only of transient importance in environments with a strong redox gradient, oxidation reactions quickly increase solubility, reduce molecular size, and facilitate enzymatic depolymerization
Facultative nonutilization of SOM *Definition*: Specialization of decomposers toward different substrates and microhabitats, which leads to a relatively longer persistence of virtually all kinds of organic substrates in uninhabited soil spaces	Facultative nonutilization is a general mechanism of soil-C preservation that can operate on substrates of any chemical composition. This mechanism can produce a large and stochastically stable C pool with a broad range of components, without relying on an inert status of the C (Ekschmitt et al., 2008)	Mechanism spans whole range from annual to millennial timescale, potential for rapid change with disturbance and introduction of new biota
Spatial accessibility for microorganisms, enzymes and oxygen *Definition*: Physical arrangement of soil particles may restrict or impede access of either decomposers or oxygen or both to organic substrates	Mechanism is ubiquitous, but particularly effective in soil environments with high contents of clay and fine silt-sized particles	Typically in the decadal to lower centennial range. Timescale is related to aggregate size: OM in small aggregates tends to have slower turnover than OM in larger aggregates
Direct interactions with mineral surfaces *Definition*: Protective effect that results from direct attachment to a mineral surface	Mechanism is restricted to such OM that is in direct contact with a mineral surface. A variety of bonding mechanisms are known to operate	Longest residence times are consistently observed in situations where ligand-exchange type bonding mechanisms between protonated Al and Fe surface hydroxyls are favored, however, centennial and millennial turnover has also been observed in kaolinitic and smectitic soils

described to control SOM turnover at different timescales; however, most models do not explicitly include those other than selective preservation. There is consensus in the literature that the active pool is composed of fresh plant residues, root exudates, decomposer feces, and of faunal and microbial residues (Smith and Paul, 1990). The selective preservation of recalcitrant compounds leads to a progressive change of residue composition only during initial stages of decomposition. Recalcitrance alone does not explain longer term stabilization of SOM (von Lützow et al., 2006; Marschner et al., 2008). With further decomposition, spatial inaccessibility due to occlusion within macroaggregates and microaggregates (i.e., biogenic aggregation; >250–20 µm) becomes a relevant process of organic matter (OM) stabilization (Angers and Giroux, 1996; Six et al., 2002). Stability of OM increases with decreasing aggregate size and the stabilized OM is dominated by microbial products (Golchin et al., 1997). Analyses of ^{14}C and ^{13}C revealed that OM bound to mineral surfaces can contribute to the intermediate pool in particular in A horizons with high microbial activity and C turnover (Trumbore, 1993; Balesdent, 1996; Flessa et al., 2008; von Lützow et al., 2008). Also, complexation of OM of different recalcitrance with metal ions and sorption of dissolved organic matter (DOM) to mineral soils leads to stabilization within the intermediate pool (Kalbitz and Kaiser, 2008).

Four process groups have been described to contribute to the formation of the passive pool: Abiotic microaggregation by encrustation of clay particles around OM can lead to the formation of very stable clay–OM microstructures (<20 µm) (Oades, 1993). Another important process that can reduce soil OM accessibility is the formation of hydrophobic properties that reduce surface wettability and thus reduce accessibility for degrading enzymes (Goebel et al., 2005). The formation of hydrophobic surfaces is a major factor in aggregate formation and enhances aggregate stability (Bachmann et al., 2008). Interaction with mineral surfaces is a third process that may increase OM stability in the time frame of the passive pool (Anderson and Paul, 1984; Scharpenseel and Becker-Heidmann, 1994; Torn et al., 1997). The quantification and evaluation of the stability of the different binding mechanisms of organo-mineral interactions (e.g., ligand exchange, cation bridging, van der Waals forces) is difficult because often more than one bonding mechanism or stabilization mechanism are operating simultaneously (Kögel-Knabner et al., 2008). For example, OM within organo-mineral associations may be stabilized simultaneously by ligand exchange at edge sites of phyllosilicates, by polyvalent cation bridges due to their permanent loading (Kahle et al., 2002b) and due to the formation of hydrophobic surfaces (Chenu et al., 2000; Doerr et al., 2000). Finally, passive OM is formed by

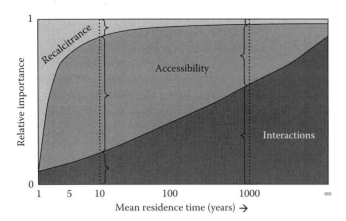

FIGURE 7.1 Current understanding of how the relative contributions of the three major stabilization mechanisms (recalcitrance, accessibility, and interactions) to the protection of organic matter against decomposition change with increasing residence time of organic matter. Only organic matter that is protected by interactions with minerals or physically separated from decomposers (accessibility) achieves long turnover times.

fire (Schmidt et al., 2002). In addition to the selective preservation of these highly condensed aromatic compounds there is evidence that the interaction of black carbon (BC) with minerals, its occlusion particularly in microaggregates and its hydrophobicity are responsible for the high stability of partly degraded charred OM (Brodowski et al., 2007). Figure 7.1 illustrates the notion that the above mechanisms vary in importance for the protection of carbon as a function of residence time.

Some other concepts to explain stabilization within the passive pool have been suggested, but are not convincingly supported by empirical data (von Lützow et al., 2008). The intercalation of DOM into internal phyllosilicate surfaces as proposed by Theng et al. (1986) has not been found to be quantitatively relevant in natural soils. Also, the spontaneous condensation and polymerization of OM as *humic material* was questioned (Knicker and Lüdemann, 1995; Hedges et al., 2000; Burdon, 2001; Sutton and Sposito, 2005). No direct measurement of OM stabilized by encapsulation in larger hydrophobic molecular complexes as often proposed is found (Schaumann, 2006; von Lützow et al., 2006), while speculations about the potential stabilization effects of organo–organo interactions (Wershaw, 1999; Buurman et al., 2002; Ellerbrock et al., 2005; Kelleher and Simpson, 2006; Kleber et al., 2007) have been revived recently. In addition, aromatic π-systems within organic compounds have the capacity to adsorb to minerals and organic soil and sediment components such as natural organic matter (NOM) and fire-derived BC through specific sorptive forces other than hydrophobic interactions (Keiluweit and Kleber, 2009).

7.2.3 Fractionation Concepts and Isolation of Specific Fractions

Our efforts to improve current models of soil carbon turnover would benefit from a greater ability to assign specific stabilization mechanisms to measurable SOM fractions. To achieve this goal, it is necessary to identify and isolate "functionally different but unique (non-composite) SOM pools with a homogeneous decay behaviour" (Smith et al., 2002; Bruun et al., 2004).

7.2.3.1 Aggregate Fractionation

Aggregate fractionation is based on the separation of free SOM and protected SOM that is occluded in organo-mineral assemblages of different sizes. The different aggregate fractions aim to isolate active from intermediate and passive SOM pools (separation of free SOM from SOM occluded by macro- and microaggregates and SOM in clay microstructures, respectively). Aggregate fractions are obtained by dry or wet sieving and slaking (immersion in water of dry aggregates). Aggregates <20 μm are very stable and are only disrupted by ultrasonic treatment (Oades and Waters, 1991). Aggregates <2 μm even survive ultrasonic treatment (Chenu and Plante, 2006). To identify SOM pools that preferentially stabilize SOM in the longer term, Six et al. (2000) suggested a fractionation scheme of wet sieving and slaking to completely break up macroaggregates while minimizing the breakdown of the released microaggregates (53–250 μm). In most temperate surface soils, aggregates generally do not break down into primary particles upon slaking but rather into smaller stable units, indicating that the aggregates are arranged in a hierarchical fashion (Tisdall and Oades, 1982; Oades and Waters, 1991). In these soils, macroaggregates (>250 μm) often contain more OM than microaggregates (<250 μm), because the former include microaggregates plus OM serving as an intra-macroaggregate binding agent. This may not be the case in tropical soils, where other cementing agents stabilize macroaggregates (Oades and Waters, 1991).

Nearly 90% of SOC in surface soils was found to be located within aggregates (Jastrow et al., 1996) with 20%–40% of SOC as intra-microaggregate SOC (Carter, 1996). The degree of decomposition increases with decreasing aggregate size from macro- to microaggregates, as indicated by a decrease of the C/N ratio, plant components or ^{14}C labeled plant residues (Gregorich et al., 2003), an increase of microbial-derived carbohydrates and the ratio of alkyl/O-alkyl C (Steffens et al., 2009). Angers et al. (1997) observed a redistribution of ^{13}C from macroaggregates to microaggregates with time, which clearly indicates that C occlusion in microaggregates only happens after occlusion at the macroaggregate scale. Consequently, microaggregate OM is on average older and in a more decomposed state than macroaggregate OM.

Various studies have shown that turnover times of occluded OM increase with decreasing aggregate size. Turnover times were about 15–50 years for OM stored in macroaggregates (>250 μm) and 100–300 years for OM in microaggregates (<250 μm) (Angers and Giroux, 1996; Besnard et al., 1996; Puget et al., 2000; Six et al., 2002; John et al., 2005). It is considered that SOM is stabilized by different aggregate formation processes, biogenic aggregation in macroaggregates, and abiotic clay flocculation in microaggregates. Biogenic aggregation is a relatively transient process within the active and intermediate OM pools and is sensitive to management practices. Occlusion of SOM within clay microstructures (<20 μm) is operative over long timescales controlled by pedogenic processes (e.g., weathering of clay minerals, formation of iron oxides and hydroxides) (von Lützow et al., 2006).

7.2.3.2 Particle Size Fractions

Particle size fractionation is based on the concept that SOM associated with particles of different size and therefore also of different mineralogical composition differs in structure and function (Christensen, 1992). While quartz particles that dominate the sand fraction exhibit only weak bonding affinities to SOM, the clay-sized particles (e.g., sesquioxides, layer silicates) provide a large surface area and numerous reactive sites where SOM can be associated (Sposito et al., 1999).

A meta-analysis suggests that in temperate arable top soils about 50%–75% of total OC is associated with clay-sized particles (<2 μm), about 20%–40% with silt-sized particles (2–63 μm) and <10% with sand-sized particles (>63 μm) (von Lützow et al., 2007). However, the generally higher allocation of SOC in smaller particles was not always congruent with a longer turnover time. In the sand fraction, ^{13}C turnover times ranged from 0.5 to 374 years, in the silt fraction from 115 to 676 years, and in the clay fractions from 76 to 190 years. The ^{14}C MRT in the silt and clay fraction showed a span of 800–1660 of and 75–4409 years, respectively. Microbial colonization and activity was largely confined to particles coarser than clay (small bacteria such as *Escherichia coli* are about 1 μm in diameter). Fungal activity is strongest in sand, while bacterial activity is strongest in silt. As demonstrated by the hexose-to-pentose ratios, the coarser fractions are dominated by SOM derived from plant material, while the SOM in the clay fraction is dominated by microbially derived metabolites (Guggenberger et al., 1994). The fine clay fraction (<0.2 μm) does not fully follow the general trend of increasing carbon allocation with decreasing particle size. This is also true for the MRT and turnover times, which are shorter and faster, respectively, in this fine fraction than in the coarser clay fraction. The data compiled by von Lützow et al. (2007) show that the proportion of allocated carbon decreased from 30%–50% in coarser clay fractions (0.2–2 μm) to 4%–20% of total OC in the fine clay fraction. This is associated with a lower turnover in fine clay as compared to the coarser clay fraction. The MRT was 130–1741 years (turnover time of 179–277 years) in the coarse clay and MRT 125–352 years (turnover time of 76–190 years) in fine clay. In summary, fractionation by particle sizes provides a rough differentiation between young (active) and older (intermediate and passive) SOM. Slower C turnover rates in clay fractions compared to the sand fraction were explained by a combined action of all three process groups of OM stabilization: the chemical change in OM quality, an increase in spatial inaccessibility (e.g., due to microaggregation) and the adsorption of OM on mineral surfaces. But particle size fractions are not homogeneous in terms of their turnover time and can thus not be used to represent functional pools in turnover models. Several binding mechanisms seem to be responsible for the differentiation of the passive pool, as indicated by the faster turnover of OM associated with the fine clay subfraction compared to OM in the coarse clay subfraction (Laird et al., 2001; Kahle et al., 2003; Wattel-Koekkoek et al., 2003; Kleber et al., 2004; von Lützow et al., 2006).

7.2.3.3 Density Fractionation

Density separation of soils is based on the fact that soil organic materials typically have densities varying from 1.24 to 1.64 kg L^{-1} (Mayer et al., 2004), while minerals have much higher densities (e.g., quartz is at 2.65 kg L^{-1}). Common procedures separate soil materials in (1) a free light fraction with a density smaller than 1.6 kg L^{-1}, (2) an occluded light fraction, which is disaggregated after sonification, and (3) a residual dense fraction representing organo-mineral complexes tightly adsorbed to mineral surfaces (Golchin et al., 1994a). Density fractionation is popular as a technique to isolate proxies for functional SOM pools, with the generally accepted understanding (Sollins et al., 1984) that the light fraction contains predominantly easily decomposable plant debris while the dense fraction represents the mineral-associated and thus protected parts of OM. The light fraction can contain charcoal in addition to particulate plant litters and the dense fraction sometimes shows signs of rapid incorporation of isotopic labels (Strickland et al., 1992; Swanston et al., 2005). These restrictions have led some workers to use density fractionation mainly as a pretreatment with the main objective to remove "light" particulate OM prior to subsequent investigations of particle-size fractions (Eusterhues et al., 2005) or as a pretreatment for chemical extraction procedures (Bird et al., 2002).

Separating soil samples into multiple density fractions has increasingly been used as a technique to investigate the mechanisms behind OM stabilization and the association of organic materials with the mineral matrix. Golchin et al. (1994b) noted that the organic materials in the dense fraction (>2.0 kg L^{-1}) showed predominantly ^{13}C NMR signals of carbohydrate and aliphatic structures, an absence of lignin and tannin structures and a narrow C:N ratio, leading them to conclude that clay minerals are able to adsorb and protect OM rich in nitrogen, suggesting a microbial origin for this fraction (Golchin et al., 1994b). Baisden et al. (2002) used sequential density fractionation to show a progressive decrease of C/N ratios with simultaneous enrichment of ^{13}C and ^{15}N and decreasing $\Delta^{14}C$ with increasing densities in a California soil chronosequence. In line with Golchin et al. (1994b), Baisden et al. (2002) interpreted this phenomenon as suggestive of increasing SOM decomposition with increasing fraction density and increasing age of the mineral-associated OM.

Sollins et al. (2006) hypothesized that sequential density fractionation might remove subsequent layers of organic materials with each fractionation step. Applying sequential density fractionation to a smectitic soil from the Oregon Cascade Range, they observed a "general pattern of an increase in extent of microbial processing with increasing organo-mineral particle density and with increasing soil organic matter turnover time."

Like aggregate and particle size fractionation, the density fractionation allows only for a rough differentiation between active and passive OM. While the free light fraction (<1.6 kg L^{-1}) is a good representation of the active pool (except in soils containing BC), the dense fractions are heterogeneous. Sequential isolation of progressively denser fractions shows trends toward higher nitrogen contents and lower lignin concentrations with increasing density (Sollins et al., 2006). However, the phases separated still contain heterogeneous organic materials and can thus not be equated with functional (active, slow, passive) pools in an unequivocal manner. However, sequential density fractionation has been successfully used to contribute to a mechanistic understanding of carbon turnover dynamics in mineral-organic associations (Golchin et al., 1994b; Baisden et al., 2002; Sollins et al., 2006).

7.3 Recalcitrance as an Index for Intrinsic Organic Matter Quality

7.3.1 Terms and Mechanisms

The term recalcitrance is commonly used when a certain degree of resistance to decomposition is to be conveyed. At present there is a debate about the usefulness of the concept of recalcitrance. Some authors find that the lack of a precise numerical definition for recalcitrance prevents the concept from being useful in support of modeling studies while others consider recalcitrance as a vital control that determines the initial pathways of decomposition. We refer the reader to publications by Kleber (2010) and von Lützow and Kögel-Knabner (2010) for a more detailed discussion of the subject.

Recalcitrance has been defined as the "molecular level characteristics of organic substances that influence their degradation by microbes and enzymes" (Sollins et al., 1996). Throughout this text, we use the term "recalcitrance" as suggested by von Lützow et al. (2006). Several molecular features have been proposed as having the ability to render natural OC compounds refractory, including molecular size (Alexander, 1981), aqueous solubility or polarity (von Lutzow et al., 2006), aromaticity (Campbell et al., 1967), aliphaticity expressed as alkyl/O-alkyl ratio (Baldock et al., 1997), molecular complexity (Bosatta and Angren, 1999), certain N-containing substituents and functional groups (von Lützow et al., 2006), and many others, but these characteristics may vary independently within the same compound. For example, small molecular size is commonly thought to facilitate decomposition, while low aqueous solubility is perceived as an obstacle to decomposition, but molecular size can be small and solubility can be low at the same time, as in benzene.

7.3.2 Plant and Microbial-Derived Compounds

The relative accumulation of recalcitrant molecules leads to a preservation of these components relative to other more easily decomposable components during early stages of litter decomposition. Selective preservation is less important in later stages of decomposition because it was shown that potentially labile organic compounds (polysaccharides and proteins) exhibited longer turnover times than potentially recalcitrant compounds (lignin, lipids) and turnover of recalcitrant compounds was faster or similar as bulk soil OM turnover (see Marschner et al., 2008 for a detailed discussion). Consequently, primary recalcitrance has been defined for plant litter and rhizodeposits as a function of their indigenous molecular characteristic, and secondary recalcitrance for microbial products, humic polymers, and charred materials (von Lützow et al., 2006).

It has been posited that selective preservation due to recalcitrance of OM is the only mechanism by which soil OC can be protected for long periods of time (Krull et al., 2003). It appears, though, that the chemical structure of organic molecules by itself is not sufficient to account for long turnover times. There is ample evidence that the biotic community is able to degrade any OM of natural origin and that we have no indication for the existence of inert OM components. The relatively rapid cometabolic mineralization of charcoal, for example (von Lützow et al., 2006), demonstrates the ability of microorganisms to degrade even complex and "recalcitrant" structures. Relatively fast turnover rates of lipids and lignin and slow turnover rates of polysaccharides and proteins suggest the importance of protective mechanisms (Marschner et al., 2008). Another concept that is increasingly questioned is the notion that humification reactions create recalcitrant humic polymers. Evidence against spontaneous heteropolycondensation theories is mounting (Hedges et al., 2000; Burdon, 2001; Sutton and Sposito, 2005; von Lützow et al., 2006) as state-of-the-art analytical techniques fail to identify chemically distinct humic substances within SOM (Kelleher and Simpson, 2006; Lehmann et al., 2008). Generally, SOM with turnover times equivalent to the passive pools of established soil carbon models (greater then 50–70 years) was only found in association with minerals (Marschner et al., 2008). Thus, recalcitrance alone cannot explain long-term stabilization and is not the major driving force of passive C pool formation. Selective preservation of recalcitrant compounds is only relevant within the active pool (<10 years) and is of particular importance in horizons with large C contents like in topsoil A horizons. This new view of the role of recalcitrance in the hierarchy of protective mechanisms is reflected in the suggestion by von Lützow et al. (2006), who posit that "recalcitrance is only important during early decomposition and in active surface soils." The "facultative nonutilization of substrates" as a result of the specialization of decomposers (Ekschmitt et al., 2005; Fox et al., 2006; Ekschmitt et al., 2008) explains how organic materials can be left undecomposed even when they are neither recalcitrant nor protected by aggregation or adsorption. This implies a reconsideration of the basic concepts underlying existing compartment and cohort models.

7.3.3 Fire-Derived Organic Matter (Black Carbon)

Because exposure of carbon compounds to heat occurs regularly in natural and anthropogenic environments, BC is ubiquitous in terrestrial systems. BC is commonly considered to belong to the carbon pool with millennial turnover time, with no role in short-term carbon cycling (Czimczik and Masiello, 2007), but with an important function as a natural means of long-term carbon storage (Lehmann et al., 2008). BC has received much attention for two major reasons. First, a general lack of knowledge of BC loss processes from soils and sediments has been perceived, which hinders a better understanding of fluxes into and out of the Earth's slow cycling carbon pools (Masiello, 2004). Second, sequestration of anthropogenic BC (biochar) into soils combined with bioenergy production has been suggested as a clean energy technology to mitigate climate change that combines the long turnover time of char with its high CEC to maximize benefits (Lehmann, 2007a, 2007b). In both cases, scientific answers and technological solutions require more information on the dynamics of BC in terrestrial environments. The possibility of the existence of more dynamic forms of BC in soils has been raised repeatedly to explain observations of faster than millennial turnover (Bird et al., 1999; Czimczik and Masiello, 2007; Hammes et al., 2008). Once released into the environment, the original surface properties of BC can be affected by "aging" processes caused by interactions with abiotic and biotic soil constituents. At this point, more evidence points to abiotic, "oxidative weathering-like" alterations than to biotic degradation (Cheng et al., 2006, 2008). However, the extent to which BC is able to persist in the environment is the subject of an ongoing debate.

7.3.4 Carbonaceous Pollutants and Nanomaterials

The intentional and unintentional introduction of synthetic organic chemicals into the environment is a consequence of the rapidly increasing demand for food, materials, and energy of human society. Some of these synthetic chemicals cause problems by affecting nontarget organisms, including humans. A particular concern is associated with the accumulation of persistent chemicals in the food chain. While biouptake of polychlorinated biphenyls (PCB) to toxic levels has been recognized as early as 1968 (Risebrough et al., 1968), the increasing industrial use of organic nanomaterials like fullerenes (e.g., C60) and carbon nanotubes exposes the biosphere to a novel threat of largely unknown dimensions. Fullerenes, which can act as Lewis acids (electron acceptors) despite their inherent hydrophobicity, may exhibit specific interactions with natural soils and sediments due to electron donor–acceptor (EDA) interactions. This has the far-reaching consequence that organic compounds like fullerenes may be able to interact in ways with organic and mineral constituents that deviate from the behavior of other hydrophobic chemicals, creating a large uncertainty about their fate in environmental systems and the timescale for decomposition. There is a clear need for the development of robust methods for the detection and quantification of fullerenes and other carbonaceous nanomaterials in soils and for studies of the controls on their decomposition dynamics in terrestrial environments.

7.4 Stabilization of SOM

Long-term stabilization of labile compounds, especially polysaccharides and proteins shows the importance of active stabilization mechanisms. Some consensus has been reached that the process types that determine long-term stability and lead to long turnover times of SOM (Sollins et al., 1996; von Lützow et al., 2006) are

1. Spatial inaccessibility (due to aggregation)
2. Chemical (sorptive) interactions with mineral particle surfaces and metal ions

Since these mechanisms were first proposed, much effort has gone into elucidating their relative importance for soil carbon content, that is, how much carbon a given soil can protect against decomposition.

7.4.1 Soil Architecture and Its Effects on C Turnover and Stabilization

The soil matrix is separated into variable-sized compartments such that transfer rates of enzymes, substrates, water, oxygen, and microorganisms can be limited. The input pathways as well as the location of OM within these compartments determine accessibility by the decomposer community. Evolutionary selection toward specialized substrate utilization has provided the decomposer community with the chemical and physical tools to degrade any kind of reduced carbon to its fully oxidized state in CO_2. Specialization of decomposers toward preferred substrates and soil spaces (Ekschmitt et al., 2008) convincingly explains longer persistence of C substrates in nonpreferred soil spaces. In A horizons, OM is derived mainly from plant residues that are mixed into surface soils by tillage or by bioturbation, root residues, and exudates (Balesdent and Balabane, 1996). In subsoils, C input occurs mainly through plant roots and through leaching of dissolved OM. Preferential flow paths of dissolved OM may be considered as "hot spots" in soils, because they permit better nutrient and substrate supply compared to the whole soil matrix (Bundt et al., 2001). The importance of preferential flow paths for C input increases with soil depth. With increasing depth there is less probability for any point in a soil to be located near a preferential flow path or hot spot. This is consistent with the generally greater radiocarbon age of subsoil OM compared to surface soils (Chabbi et al., 2009). Physical barriers due to the wetting resistance and chemical heterogeneity of surfaces, hydrophobic interfaces, and instable wetting fronts cause spatial heterogeneity of soil moisture and spatial inaccessibility for decomposer organisms. Important processes that can reduce soil OM accessibility are occlusion of OM by aggregation and hydrophobicity.

7.4.1.1 Accessibility/Aggregation: Physical Protection

Observations of enhanced SOM mineralization following disruption of aggregates showed long ago that occlusion in aggregates has a retarding effect on SOM decomposition. Protection will be greatest where aggregate stability is high and aggregate turnover is low; thus, aggregation is the stabilization mechanism that is potentially most susceptible to disturbance. The fact that soil aggregation is a transient property and that aggregates are continually being formed and destroyed (Baldock and Skjemstad, 2000) suggests further that aggregation/accessibility is the stabilization mechanism controlling the size of the slow or intermediate pool of carbon turnover models, but not the dominant control on centennial or millennial turnover. Measured turnover times for carbon protected by aggregates indicate an inverse relationship between aggregate size and carbon turnover time (Skjemstad et al., 1993; Balesdent, 1996; John et al., 2005; Liao et al., 2006) with the highest reported turnover times being in the lower centennial range (200–320 years) and occurring in the smallest aggregates <20 μm. Occlusion at the clay microstructure level (<20 μm) has been attributed to abiotic mechanisms such as the precipitation of Fe- and Al-oxides or hydroxides (Duiker et al., 2003), but in general soil biota are thought to be strongly involved in the process of occlusion. Thus, microbial cells, secretions, root exudates, and faunal mucus act as cementing agents and are at the same time occluded within microaggregates (von Lützow et al., 2006).

Theories of physical protection (Jastrow et al., 2007) take the presence of old, microbial-type organic materials in very small aggregates as evidence that the OM associated with microaggregates may not only be physically protected but also highly humified and biochemically recalcitrant (Liao et al., 2006). However, if there are no recalcitrant, polymeric humic macromolecules, as suggested by an increasing number of authors (Sutton and Sposito, 2005; Kelleher and Simpson, 2006; Marschner et al., 2008), then we are forced to develop alternative scenarios to explain the mechanisms of long-term stabilization of energy-rich microbial residues or microbial products in small aggregates.

With the assumption that microbiota have an active role in the creation of long-lasting microstructures, we simultaneously infer that operationally defined "passive" soil C pools may at least partially consist of thermodynamically labile C. If confirmed, then this would mean that it is not the thermodynamic properties of OC itself, but rather the association of C with mineral surfaces that provides long-term stabilization. In the model proposed by Lehmann et al. (2007; Figure 7.2), microaggregate formation is initiated by the accumulation of microbial products on clay particle surfaces, not by the occlusion of organic debris between primary particles, a view that is in line with the TEM observations of Chenu and Plante (2006) and the STXM observations by Wan et al. (2007; Figure 7.3). These authors report that many of so called "clay particles" were in fact nanometer- to micrometer-sized microaggregates in which OM was encrusted by minerals. They concluded that these very small microaggregates are major sites of OM stabilization "both by adsorption and by entrapment of organic matter," as demonstrated in Figure 7.4. This concept is supported by measurements of microbial sugars in contact with pedogenic oxides by Spielvogel et al. (2008) and by contemporary theories about the structure of OM coatings in contact with mineral surfaces (Wershaw et al., 1996; Kleber et al., 2007; Bachmann et al., 2008). Further support comes from recent assessments and reports showing that carbon associated with mineral surfaces has a distinct composition related more to microbially processed OM than to plant related compounds (Grandy and Neff, 2008; Spielvogel et al., 2008). Figure 7.5 gives an example for the initiation of aggregation by microbial cells.

7.4.1.2 Hydrophobicity

Hydrophobicity reduces surface wettability and thus the accessibility of OM for microorganisms. This is supposed to result in decreased decomposition rates, as the absence of water directly restricts the living conditions of microorganisms (Goebel et al., 2004; Jandl et al., 2004).

Persistent hydrophobicity has often been attributed to the presence of a coating on soil particles consisting of adsorbed fatty acids (Doerr et al., 2000) that have been identified as mainly plant-derived long-chain compounds (n-C22:0 to n-C34:0) (Jandl et al., 2004). Many other studies have attributed hydrophobicity to the presence of OM, examples are reports by Woche et al. (2005) on hydrophobicity in sandy forest soil, studies by de Jonge et al. (1999) who found the smallest particles (<63 μm) exhibited the highest hydrophobicity, and the work of Chenu et al. (2000) on the impact of OM on the wettability of clays. Hydrophobic properties have been measured for SOM: humic acids, OM bound to clay minerals, plant debris, and microbial residues (Chenu et al., 2000; Doerr et al., 2000). The presence of persistent hydrophobic organic coatings alone, however, does not explain the observed reversibility and *seasonal* changes in water repellency.

Horne and McIntosh (2000) attributed seasonal changes in soil wettability to OM structure, specifically, to the existence of multiple, functionally different layers of OM on sandy grains. In their model, an inner layer of more hydrophobic compounds is covered by a layer

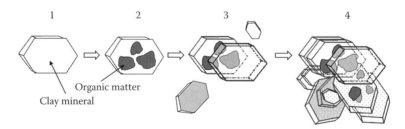

FIGURE 7.2 Model of the formation of free stable microaggregates by, first, the development of an organic coating on a clay minerals and then the physical occlusion of the organic coating (and other entrapped organics by a second mineral). (Reproduced with kind permission from Springer Science+Business Media: Biogeochemistry, Organic matter stabilization in soil microaggregates: implications from spatial heterogeneity of organic carbon contents and carbon forms, 85, 2007, 45–57, Lehmann, J., Kinyangi, J., and Solomon, D., Copyright Springer Verlag GmbH.)

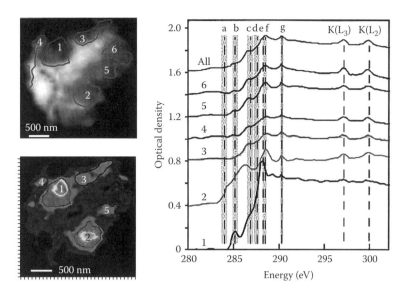

FIGURE 7.3 Variations of organic matter composition in close proximity to mineral surfaces within microaggregate. Carbon NEXAFS spectra obtained within selected areas of the Phaeozem microaggregate (regions numbered 1–6), and for the whole sample. Spectral features identified by the vertical dashed lines correspond to C in (a) quinonic, (b) aromatic, (c) phenolic, (d) aliphatic, (e) peptidic, (f) carboxylic, and (g) carbonate/carbonyl functional groups. The shaded gray bands indicate energy ranges attributed to each functional group. The peaks at the higher energies result from small amounts of K^+ and correspond to its L3 and L2 edges. (Reproduced from Geochim. Cosmochim. Acta, 71, Wan, J., Tyliszczak, T., and Tokunaga, T.K. Organic carbon distribution, speciation, and elemental correlations within soil microaggregates: Applications of STXM and NEXAFS spectroscopy, 5439–5449, Copyright 2007 with permission from Elsevier.)

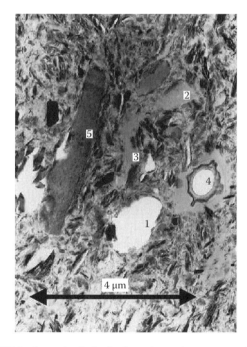

FIGURE 7.4 Examples for both adsorption and entrapment of organic matter. Transmission electron micrograph shows empty voids (1), voids that are filled with an amorphous organic material that appear to be lined with small phyllosilicates (2), spaces that show remnants of membrane-like features (3), features with obvious biological character (4), and a small fragment of charcoal (5). Image magnification ×10,500. (Reproduced from Kleber, M., J. Roessner, C. Chenu, B. Glaser, H. Knicker, and R. Jahn. 2003. Prehistoric alteration of soil properties in a Central German chernozemic soil: In search of pedologic indicators for prehistoric activity. Soil Sci. 168:292–306. With permission from Wolters Kluwer Healt.)

FIGURE 7.5 Aggregation (in the form of "nests" of vermiculite grains caused by *Pseudomanas fluorescens* (sample B1, stain CHA0). (Reproduced from Müller, B., and G. Défago. 2006. Interaction between the bacterium and vermiculite: Effects on chemical, mineralogical, and mechanical properties of vermiculite. J. Geophys. Res. 111: G02017. Copyright 2006 American Geophysical Union. With permission.)

of amphiphilic compounds. When the soil is wettable, the outer surface is well hydrated and the hydrophobic material is effectively screened. Repellence occurs as the soil dries, when the hydrophobic compounds making up the inner layer are more exposed.

The view that soil OM consists largely of relatively small amphiphilic organic molecules has steadily gained ground over the last decade (Wershaw, 1999; Piccolo, 2001; Sutton and Sposito, 2005). Kleber et al. (2007) combined previous evidence in a suggestion of a zonal model of mineral–organic associations and used this basic assumption to derive a multilayered structure for OM accretions on mineral surfaces (Figure 7.6). This model can explain both seasonal (i.e., moisture dependent) as well as permanent hydrophobicity, where the latter would result from a loss of the two outer layers (Figure 7.6) of amphiphilic organic molecules.

In addition to the direct effect of hydrophobic soil OM properties on microbial accessibility, hydrophobicity at the same time enhances aggregate stability and thereby also contributes to the occlusion of OM as a stabilizing mechanism. Goebel et al. (2005) determined CO_2-efflux from aggregates and homogenized soil and found evidence for both mechanisms, that is, stability of hydrophobic OM itself and enhanced stability of OM occluded in aggregates due to hydrophobic aggregate surfaces. More empirical data about the direct effect of hydrophobicity on degradability of OM are necessary.

7.4.2 Organo-Mineral Interactions

7.4.2.1 The Paradigm of Mineral Control

Organo-mineral interactions, or the protection of OM against decomposition through association with mineral surfaces, have received increasing attention over the past two decades and have been identified as the most likely mechanism to achieve long-term, that is, centennial or millennial, protection of OM (Kögel-Knabner et al., 2008).

Among the first to investigate the effects of clay mineralogy on the turnover of SOM were Allison et al. (1949), who studied the decomposition of added corn straw in an incubation experiment. After 1 year, they recovered 23% of added carbon in the unamended soil substrate. With 10% kaolinite, 10% "soil clay," and 10% bentonite added to the "soil substrate," they recovered 26%, 33%, and 41%, respectively, of the added carbon. Early experimental evidence such as the work of Allison et al. (1949) has led a variety of authors (Goh, 1980; Tate and Theng, 1980; Oades, 1989) to suggest that the impact of clay on C stabilization will change as a function of clay mineralogy.

SOM in fine silt and clay fractions has longer turnover times than OM in other soil OM fractions. Kalbitz et al. (2003) showed that sorption of soluble OM to subsoil material (Bw horizon) reduced OM mineralization to less than 30% compared to mineralization in soil solution. A detailed mechanistic understanding of why sorption to soil minerals reduces decomposition rates is lacking and is complicated by artifacts in the experiments. Chenu and Stotzky (2002) suggest that small molecules sorbed to mineral surfaces cannot be utilized by microorganisms unless they are desorbed so that they can be transported into the cell. But they also caution that it is difficult to demonstrate the unavailability of adsorbed molecules because desorption can occur through microbial secretions during the experiments. The adsorption of macromolecules is considered to be associated with conformational changes that render macromolecules unavailable to the action of extracellular enzymes (Theng, 1979; Khanna et al., 1998). But as shown by Demaneche et al. (2001) degradation can also be hindered by adsorption of the relevant enzyme to clay minerals rather than by adsorption of the substrate. There is clearly a need for a better understanding of the effect of adsorption on OM decomposition mechanisms and rates.

Long-term protection of organic molecules by sorptive interactions is limited to those organic materials directly bonded to the protecting mineral surface, which is of a finite size (Kleber et al., 2005). Abundant evidence that substantial parts of mineral surfaces are not covered by OM (Ransom et al., 1998; Arnarson and Keil, 2001; Mayer and Xing, 2001; Kahle et al., 2002a) has led to the insight that organic materials must be stacked or clustered on mineral surfaces (Kaiser and Guggenberger, 2003) in small patches with some vertical extension. Such a multilayer architecture of organic coatings on mineral surfaces suggests that only the inner layer of organic molecules is able to participate in direct, strong mineral–organic interactions (Kleber et al., 2007). The degree of saturation of protective sites should thus affect the preservation potential of newly added carbon to the soil. Consequently, there should be a greater potential for OM preservation in the subsoil where the OM saturation of mineral particles is small.

Clay-sized particles like layer silicates ($<2\,\mu m$), sesquioxides (crystals 5–100 nm), short-range ordered Fe-oxides (3–10 nm) and amorphous Al-oxides (<3 nm) provide the most significant surface area onto which OM can adsorb (Jahn et al., 1992; Schnitzer and Kodama, 1992; Torn et al., 1997; Percival et al., 2000; Basile-Doelsch et al., 2005; Eusterhues et al., 2005; Kleber et al., 2005; Rasmussen et al., 2005; Basile-Doelsch et al., 2007; Kaiser and Guggenberger, 2007; Spielvogel et al., 2008). Mineral reactivity, rather than mineral texture, consistently serves as a better predictor of the residence time and turnover time of stable SOC, and can be associated with two to threefold differences in total soil C storage (Torn et al., 1997; Kahle et al., 2004; Kleber et al., 2004, 2005; Masiello et al., 2004; Mertz et al., 2005; Mikutta et al., 2005; Rasmussen et al., 2005, 2006, 2007). The extent to which microbial metabolites produced from decomposing plant residues are stabilized in different soils is controlled by specific surface area (SSA) provided by clay minerals (Saggar et al., 1996). The OM contents of coarse and fine clay subfractions depend on the mineralogy and, more specifically, on the surface reactivity of the mineral constituents. In the coarse clay fraction, silicate mineral surfaces (montmorillonite > vermiculite > illite > kaolinite) are more important for carbon storage than Fe oxides, which dominate in fine clay fractions and in acid subsoils (Anderson et al., 1981; Kahle et al., 2003; Kleber et al., 2004). Significant debate surrounds the question whether the abundance of certain mineral phases might serve to predict the turnover time of OM in soils. It is noteworthy that no field study has been published to date that indicates a role for unweathered primary minerals in controlling carbon turnover. Mineralogical effects on carbon stabilization have been reported exclusively for secondary minerals, particularly such with abundant hydroxylated surfaces. Examples include Fe oxides (Kaiser and Guggenberger, 2007; Kögel-Knabner et al., 2008),

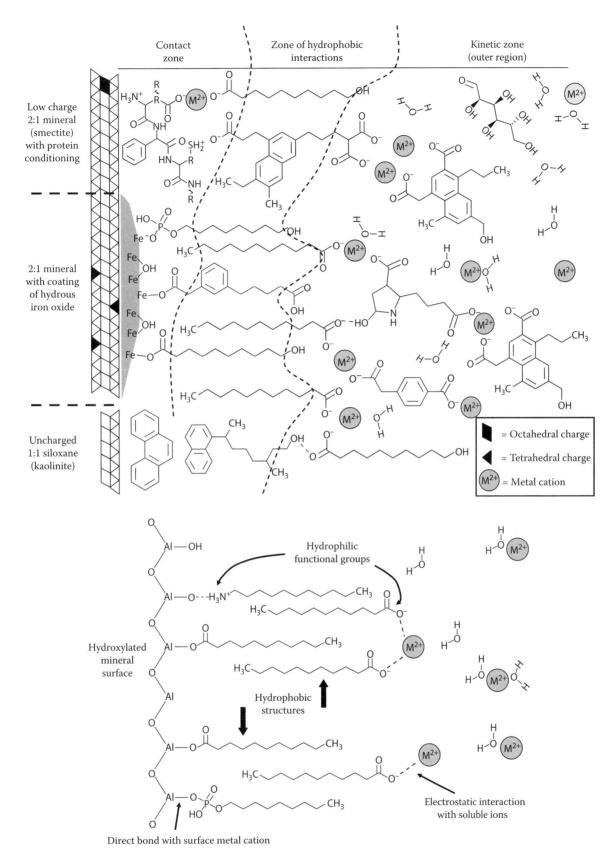

FIGURE 7.6 Multilayer model of amphiphilic organic fragments attached to mineral surfaces. (Reproduced with kind permission from Springer Science+Business Media: Biogeochemistry, A conceptual model of organo-mineral interactions in soils: self-assembly of organic molecular fragments into zonal structures on mineral surfaces, 85, 2007, 9–24, Kleber, M., P. Sollins, and R. Sutton. Copyright Verlag GmbH. With permission.)

Al-rich imogolite type materials (Percival et al., 2000; Basile-Doelsch et al., 2005, 2007), short-range ordered Al hydroxides (Rasmussen et al., 2005; Spielvogel et al., 2008), and poorly crystalline materials in general (Jahn et al., 1992; Schnitzer and Kodama, 1992; Torn et al., 1997; Kleber et al., 2005). Considering the often conflicting reports about the efficacy of individual mineral species for the protection of OM against decomposition, it has become increasingly apparent that the same mineral surface type may function differently in different pedogenic environments, as a result of variations in pH, OM chemistry, cation availability, and numerous other environmental controls.

7.4.2.2 Types of C and N in Organo-Mineral Associations

In situ investigations show that the proportion of the mineral-bound OM and its ^{14}C age generally increase with soil depth. Stabilization by organo-mineral interactions operate at long-term scales and dominate during late decomposition phases and in subsoils (Kögel-Knabner et al., 2008).

Little information is available on the relationship between mineralogy and the chemistry of bound OM. Kögel-Knabner (2000) found that OM in organo-mineral associations of fine fractions and loamy soils has a higher contribution of bacterial polysaccharides whereas mineral-associated OM in acid sandy soils is more aliphatic. Kaolinite-associated OM from grassland and shrub surface soils was enriched in polysaccharide products, whereas smectite-associated OM, supposedly in interlayer spaces, was enriched in aromatic compounds (Wattel-Koekkoek et al., 2001). Laird et al. (2001) attributed differential carbon storage in clay subfractions to a shift in mineral composition from coarse to fine clay. The coarse clay fraction had stronger carboxyl and O-alkyl ^{13}C-NMR peaks and smaller concentrations of extractable amino acids, fatty acids, monosaccharides, and amino sugars than OM associated with the fine clay fraction (Laird et al., 2001; Kahle et al., 2003). Kleber et al. (2004) and Schöning et al. (2005) have found evidence for selective stabilization of O-alkyl C especially by interactions with pedogenic oxides on mineral surfaces within the coarse-clay subfraction. Ligand exchange might be the binding mechanism. In contrast, alkyl C and aromatic C responded to the duration of fertilizer deprivation, but were indifferent to mineral surface reactivity (Kleber et al., 2004). Generally, the OM associated with soil minerals has a low C:N ratio (often around 8–12). This is attributed to the association of mainly proteins and peptides (Knicker, 2004; Kleber et al., 2007; Rillig et al., 2007), but to a smaller extent also of DNA (Pietramellara et al., 2009) with the mineral phase.

7.4.2.3 Phyllosilicate Clay Minerals

Phyllosilicates pose an interesting challenge in this context because it is often unclear whether they function as adsorbents for OM or if they merely represent physical barriers between enzymes and SOM. This is illustrated by the fact that clay content often correlates with SOM content, but not with turnover time (Kleber et al., 2005). Organic anions are repelled from negatively charged surfaces in soils, but binding occurs when polyvalent cations are present on the exchange complex. Unlike Na^+ and K^+, polyvalent cations are able to maintain neutrality at the surface by neutralizing both the charge on the negatively charged surface (e.g., in clay minerals) and the acidic functional group of the OM (e.g., COO^-) and thus act as a bridge between two charged sites. The major polyvalent cations present in soil are Ca^{2+} and Mg^{2+} in neutral and alkaline soils and hydroxy-polycations of Fe^{3+} and Al^{3+} in acid soils. The Ca^{2+} ions do not form strong coordination complexes with organic molecules, relative to Fe^{3+} and Al^{3+}. For a long-chain organic molecule with multiple functional groups, multiple points of attachment to the clay particle (segment-surface contact) on permanent charge sites of layer silicates are possible. Microbially secreted polysaccharides frequently carry a negative charge due to the presence of uronic acids that adsorb strongly to negatively charged clay surfaces through polyvalent cation bridging (Chenu, 1995). The bonding efficiency of OM on phyllosilicates by cation bridges is weaker compared to ligand exchange on Al and Fe hydroxides (Benke et al., 1999; Kaiser and Zech, 2000).

In the case of hydrogen bonds, a hydrogen atom with a positive partial charge interacts with partially negatively charged O or N atoms. Hydrophobic interactions are driven by the exclusion of non-polar residues (e.g., aromatic or alkyl C) from water to force the non-polar groups together. These interactions arise from enthalpy and entropy changes associated with water–water, aromatic compound–water, and sorbent–water interactions, as well as from van der Waals forces between apolar moieties and sorption sites. van der Waals forces are nonspecific attractive forces between uncharged molecules and are superpositions of London dispersive energies, Debye energies, and Keesom energies (Schwarzenbach et al., 2003)

Non-expandable layer silicates (e.g., 1:1 layer silicates like kaolinite) or quartz particles without layer charge and without interlayer spaces exhibit only weak-bonding affinities. The negative charge on the siloxane surface of other clay minerals depends on the type and localization of the excess negative charge created by isomorphic substitution. In the absence of a layer charge, a siloxane surface may be considered uncharged. Uncharged, but polar polysaccharides and extracellular enzymes or other proteins can form linkages via hydrogen bonding or van der Waals forces because of the presence of hydroxyl and other polar groups in the molecule (Quiquampoix et al., 1995). Their typically high molecular weight offers a large number of potential surface-segment contacts and thus strong binding between uncharged polysaccharides and clays can be established (Theng, 1979). Hydrophobic interactions become more favorable at low pH when hydroxyl and carboxyl groups of OM are protonated and the ionization of carboxyl groups is suppressed. Bonding interactions of apolar aromatic ring structures have long been considered as restricted to energetically weak, nonspecific, hydrophobic interactions (Chiou et al., 1979; Karickhoff, 1984; Mader et al., 1997; Schwarzenbach et al., 2003). Over the last decade, an increasing number of authors have suggested specific (i.e., directed) and energetically stronger adsorption mechanisms between aromatic π-systems of organic compounds and sorption sites at mineral surfaces. In particular, noncovalent EDA interactions of aromatic compounds with mineral surfaces have been proposed, including cation–π interactions (Zhu et al., 2004; Mueller et al., 2007), n–π EDA interactions (Haderlein and Schwarzenbach, 1993; Haderlein et al., 1996; Weissmahr et al., 1998), and hydrogen–π interactions (Ringwald and Pemberton, 2000). While these observations suggest that attachment of aromatic SOM to mineral surfaces may involve forces that are stronger than those currently assumed, the quantitative role of EDA interactions for the sorptive preservation of NOM in mineral–organic associations remains to be investigated (Keiluweit and Kleber, 2009).

7.4.2.4 Pedogenic Oxides

It is widely assumed that the energetically strongest associations between OM and mineral surfaces involve the mechanism of ligand exchange between carboxyl groups of OM and hydroxyl groups at the surfaces of mineral phases. Complexation of OM on mineral surfaces via ligand exchange increases with decreasing pH with maximal sorption between pH 4.3 and 4.7, corresponding to pK_a values of the most abundant carboxylic acids in soils. Therefore, ligand exchange between reactive inorganic hydroxyls (OH-groups of Fe- Al- Mn-oxides and edge sites of phyllosilicates) and organic carboxyl and phenolic–OH groups is restricted to acid soils rich in minerals with protonated hydroxyl groups. The sorptive strength of hydroxyl-bearing phases, like Fe-(hydr)oxides and poorly crystalline aluminosilicates increases with decreasing pH (Gu et al., 1994, 1995) and will therefore be particularly relevant in profoundly acidic soils. Kleber et al. (2004) showed that singly coordinated, reactive OH-groups on Fe- Al-oxides and at edge sites of phyllosilicates, which are able to form strong bonds by ligand exchange, are a measure of the amount of OM stabilized in soils in organo-mineral associations. Sorption occurs preferentially at reactive sites such as edges, rough surfaces or micropores (e.g., edges of illite particles where amphoteric Al-OH groups are exposed, crystal surfaces of Fe-oxyhydroxides with singly coordinated OH⁻ groups). Kaiser and Guggenberger (2003) hypothesized that the molecules adsorbed first might be strongly stabilized by multiple ligand attachments. At larger surface loadings, sorption can then take place with fewer ligand attachments involved, which leaves parts of the molecule not attached to the surface and thus renders them more susceptible to degradation.

Investigating correlations between iron oxide contents and OM contents, Wagai and Mayer (2007) discovered that Fe oxides may be critical for OC storage, but likely not via a direct sorption mechanism. A physical role of nano-sized iron oxide particles in carbon stabilization is also suggested by the correlation between crystalline iron oxides and oxidation resistant OM reported by Mikutta et al. (2006), who found that in soils with low contents of poorly crystalline pedogenic oxides, stable OM can be correlated with much less sorptive, crystalline Fe-oxide forms.

7.4.2.5 Allophane and Imogolite

High OM contents are typical for soils derived from volcanic ash (Andosols), containing poorly crystalline aluminosilicates like allophane and imogolite. The striking ability of poorly crystalline aluminosilicate mineral matrices to contribute to soil carbon retention

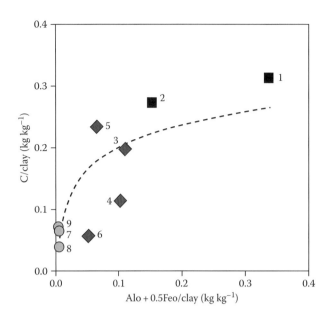

FIGURE 7.7 Soil organic C stored per unit mass of clay as a function of clay composition (Alo + ½Feo = acid oxalate–extractable Al and Fe, index for the abundance of poorly crystalline mineral phases). Values are amounts per square meter and 1 m depth. (Reproduced from Kleber, M., and R. Jahn. 2007. Andosols and soils with andic properties in the German soil taxonomy. J. Plant Nutr. Soil Sci. 170:317–328. Copyright Wiley-VCH Verlag GmbH & Co. KGaA. With permission.) Black squares are andosols (1 = Silandic Andosol, 2 = Aluandic Andosol); grey diamonds are Cambisols and Luvisols (3 = Dystric Cambisol; 4 = Aluandic Luvisol, 5 = Endoskeletic Cambisol, 6 = Orthidystric Cambisol) and circles are Tschernozemic soils (7 and 8 = Haplic Chernozems, 9 = Haplic Kastanozem).

can be illustrated by comparing the OC contents of Andosols with those of other mineral soils (Figure 7.7). On a global scale, Andosols have mean OC contents of 25.4 kg m⁻² in the upper 100 cm (Batjes, 1996), which makes them the most carbon-rich FAO-UNESCO soil unit (organic soils = Histosols excluded). In addition to high OM contents, soils with significant proportions of poorly crystalline minerals also tend to have particularly long OM residence times compared with other soil taxa (Kleber et al., 2005).

Different mechanisms are being discussed to explain high OM storage and long OM residence times in soils containing poorly crystalline aluminosilicate phases:

1. Allophane and imogolite contain both permanent charge and surface hydroxyl groups (Harsh et al., 2002). Their charge characteristics are such that imogolite and Al-rich allophane tend to impart positive charge in acid soils, while Si-rich allophane may be either positively or negatively charged depending on soil pH (Harsh, 2000). The resulting ability to form energetically strong ligand exchange type bonds combined with large SSAs (Harsh et al., 2002) allows OM to form lasting sorptive associations with poorly crystalline aluminosilicates.

2. Unit particles of allophane tend to coagulate in masses of small spheres varying from 3.5 to 5 nm in diameter. Linear assemblages of these particles may combine into secondary floccules, forming microaggregates that render soils highly

porous (Harsh et al., 2002). The preferential formation of aggregates in the fine silt size range (1–10 μm; Buurman and van Doesburg, 2007) creates microcompartments with the potential to physically protect OM trapped in the coagulation process. As metastable, poorly crystalline aluminosilicates can only persist in relatively moist soils, it has also been hypothesized that microaggregation may induce anoxic conditions in water saturated microenvironments, thereby impeding OM decomposition (Buurman et al., 2007).

3. Several other modes of OM preservation other than sorptive protection and physical protection have been suggested to operate in soils containing poorly crystalline aluminosilicates. Among these are aluminum toxicity to microorganisms (Tokashiki and Wada, 1975); sorption of degradation enzymes to active Al and Fe (Tate and Theng, 1980), or phosphorus deficiency of decomposer microorganisms caused by high P-retention (Brahim, 1987). However, Buurman et al. (2007) find that there is no conclusive evidence of the relevance of these processes under field conditions, and Scheel et al. (2008) refuted the notion of Al-toxicity in favor of an Al-OM precipitation mechanism. The latter notion of unrestricted microbial activity is supported by abundant reports that OM in close contact with aluminosilicate surfaces consists of microbially processed material (Rasmussen et al., 2005; Buurman et al., 2007), but exhibits a relatively old radiocarbon age (Rasmussen et al., 2005; Sollins et al., 2006).

7.4.2.6 Interactions with Metal Ions

In comparison with the chemistry of metal binding, less information is available about the effect of metal binding on soil OM stability or about the mechanisms involved. Metal ions that have been considered as potentially stabilizing for soil OM are Ca^{2+}, Al^{3+}, and Fe^{3+} and heavy metals.

The earliest evidence for the stabilizing effect of Ca^{2+} ions on soil OM was given by Sokoloff (1938). Since then, several other studies have shown similar effects of Ca^{2+} ions on the mineralization of soil OM and its solubility (Muneer and Oades, 1989) and the large OM content of calcareous soils is also attributed to the effect of Ca^{2+} ions (Oades, 1988).

There is general consensus that the interaction of soil OM with Al and Fe is the main reason for the stability of soil OM in Podzols (Lundström et al., 2000; Zysset and Berggren, 2001; Nierop et al., 2002). However, this conclusion is mainly based on indirect evidence and experimental data on the stability of OM-Al and -Fe complexes is scarce. The Al/C ratio of DOM seems to be an important parameter for its stability against microbial decomposition. In long-term incubation studies, Schwesig et al. (2003) showed that for natural DOM, Al/C ratios >0.1 increased the half-life of the stable DOM fraction up to fourfold.

For Al (and to a lesser extent for Fe), the inhibitory effect of heavy metals on soil respiration and the subsequent accumulation of soil OM has been shown in numerous field and laboratory studies (Tyler et al., 1989; Giller et al., 1998). The concentrations of heavy metals needed to induce retardation of soil respiration often varied by one to three orders of magnitude, depending on experimental and site conditions and on the metal considered. The accumulation of soil OM in the vicinity of heavy metal emission sources, like smelters, in comparison with unpolluted sites is well documented (McEnroe and Helmesaari, 2001; Johnson and Hale, 2004).

The effects of metals on OM stabilization are still poorly understood and difficult to differentiate. Changes in substrate quality by complexation with metals, direct toxic effect of metals on soil microorganisms, and direct effects of metals on extracellular enzymes have all been described. Changes in molecular size, charge, and steric properties of soil OM induced by metal complexation will likely decrease its accessibility to soil enzymes (McKeague et al., 1986) and thereby reduce their availability as microbial substrates. Evidence for such a mechanism is only available from studies with simple model substances. Aluminum complexes of citrate and oxalate in soils were more stable in comparison with the free compounds, but Al effects on degradation under field conditions were less convincing than in the laboratory (Jones et al., 2001). Similar findings with effects in vitro, but fewer effects under soil conditions were reported for Cu and Zn on citrate degradation (Brynhildsen and Rosswall, 1997).

DOM in soils can be precipitated by metal ions and the precipitated DOM can be more stable than the DOM remaining in solution. Larger DOM molecules are precipitated preferentially, while smaller molecules stay in solution (Römkens and Dolfing, 1998; Blaser et al., 1999; Nierop et al., 2002; Schwesig et al., 2003). Evidence for soil OM stabilization by metal ion interactions is described, but general conclusions on their quantitative relevance in soils are difficult to draw. The mechanisms and their effects depend on concentrations and binding forms, and are specific for each element. Often it is difficult to separate the complexing effect of metal cations (Ca, Mg, Al, Fe) from their ability to form cation bridges. Systematic studies and comparisons of the mechanisms are necessary to resolve this puzzle.

7.4.2.7 Minerals as Catalysts

The role of abiotic catalysis reactions is among the most enigmatic and least understood aspects of OM turnover. The central difficulty in assessing the role of abiotic catalyses is that it can facilitate both synthesis reactions (Shindo and Huang, 1982, 1984; Huang and Hardie, 2009) and degradation processes (Rima et al., 2005, Huang and Hardie, 2009). The situation is complicated by the fact that numerous mineral phases are capable of acting as catalysts. However, these mineral phases routinely occur together at the same time in the same soil, and to date, no systematic study has investigated the simultaneous contribution of abiotically catalyzed syntheses and degradation processes for the persistence and stability of OM in a given soil environment. The situation is further confounded by the fact that traditional concepts viewing humic substances as polymeric, macromolecules are increasingly called into question (Piccolo, 2001; Sutton and Sposito, 2005; Kelleher and Simpson,

2006). Thus, while numerous laboratory experiments (see Jokic et al. 2004 and references therein) have confirmed the general ability of mineral phases to facilitate polymerization reactions of organic monomers, the mechanistic relevance of abiotic catalyses and the quantitative importance of its products for carbon turnover processes in natural soil environments remain to be demonstrated. A detailed description of the concept is given by Huang and Hardie (2009).

7.5 Outlook

Evidence is accumulating that the role of the mineral matrix in carbon stabilization needs to be viewed in a new light. In the past, the effect of mineralogy on SOM turnover has typically been viewed as a matter of variations in surface reactivity, and many are the adsorption studies that illustrate the differential abilities of mineral surfaces to function as sorbents for organic compounds (Kaiser and Zech, 2000; Kahle et al., 2004; Mikutta et al., 2007). Stabilization by organo-mineral interactions operates at long-term timescales and dominates during late decomposition phases and in subsoils (Kögel-Knabner et al., 2008). Pedogenetic processes of mineral formation control the strength of bonding and the amount of SOM associated (Percival et al., 2000; Spielvogel et al., 2008). Sorptive interactions can provide strong and long-lasting mineral-organic bounds, but are limited by the finite size of the reactive mineral surface area available (Hassink, 1997; Kleber et al., 2005; Eusterhues et al., 2005). Additionally, there are further limiting requirements in terms of pH, cation availability, and matching organic functional groups (Gu et al., 1995; Mikutta et al., 2007). The importance of certain mineral phases for OM stabilization is easiest shown for acidic subsoils, where they are strong sorbents (Kleber et al., 2005), but is not as convincing in many topsoils (Eusterhues et al., 2005). Iron oxides, for example, can be strong sorbents for OM in one soil and be negatively correlated with OM content in another (Kahle et al., 2002b).

But differences in mineralogy are not just differences in surface chemistry. As minerals are either weathered into fragments or crystallize from solution, they vary in shape and size, kinetic and thermodynamic stability, abundance and extent of reactive surfaces. Consequently, differently sized specimens of the identical mineral cannot be considered constant in their ability to provide reactive surfaces, their ability to serve as physical barriers or as suppliers of dissolved cations, which in turn may act as complexing and/or precipitating agents (Masiello et al., 2004; Schwertmann et al., 2005). Kaolinites, for example, often have an uncharged basal siloxane plane, but can carry reactive hydroxyls on their edges. It follows that a kaolinite crystal with a large basal extension but of shallow height will function differently than a kaolinite crystal of small basal extension but stacked up high.

In acid subsoils, microporous oxide phases efficiently stabilize OM by ligand exchange while more than one bonding mechanism may operate in neutral soils (Kahle et al., 2003). Over the past 10 years, it became evident that pedogenetic processes of mineral formation control the strength of bonding and the amount of OM sorbed (Spielvogel et al., 2008). Surface coverage of OM on mineral surfaces was found to be discontinuous and SSA was shown to be not always a good predictor for C stabilization (Kaiser and Guggenberger, 2003). Attempts to reconcile these observations include conceptual models that describe the spatial orientation of organo-mineral interactions under different OC contents or a self-assembly of OM into multilayered structures on mineral surfaces (Ellerbrock et al., 2005; Kaiser and Guggenberger, 2007; Kleber et al., 2007).

Although the process groups leading to SOM stabilization—namely, selective preservation of recalcitrant compounds, spatial inaccessibility to decomposer organisms and interactions of OM with minerals and metal ions as well as the factors controlling them—have been identified, a challenge in future research is to describe the process mechanisms quantitatively and to evaluate the relative importance of the different stabilization mechanisms in different soils. Past research has established that the interplay of constructive (OM accumulation) and destructive (mineral weathering and cation release) pedogenetic processes determines the types and strengths of bonding interactions at the mineral–organic interface and ultimately the amount of OM sorbed. Considering that the stabilization potentials of soils for OM are site and horizon specific, we identify the need for better ways to link key stabilization mechanisms to pedogenetic processes.

References

Alexander, M. 1981. Biodegradation of chemicals of environmental concern. Science 211:132–138.

Allison, F.E., M.S. Sherman, and L.A. Pinck. 1949. Maintenance of soil organic matter: I. Inorganic soil colloid as a factor in retention of carbon during formation of humus. Soil Sci. 68:463–478.

Anderson, D.W., and E.A. Paul. 1984. Organo-mineral complexes and their study by radiocarbon dating. Soil Sci. Soc. Am. J. 48:298–301.

Anderson, D.W., S. Saggar, J.R. Bettany, and J.W.B. Stewart. 1981. Particle size fractions and their use in studies of soil organic matter. I. The nature and distribution of form of carbon, nitrogen, and sulfur. Soil Sci. Soc. Am. J. 45:767–772.

Angers, D.A., and M. Giroux. 1996. Recently deposited organic matter in soil water-stable aggregates. Soil Sci. Soc. Am. J. 60:1547–1551.

Angers, D.A., S. Recous, and C. Aita. 1997. Fate of carbon and nitrogen in water-stable aggregates during decomposition of 13C15N-labelled wheat straw in situ. Eur. J. Soil Sci. 48:295–300.

Arnarson, T.S., and R.G. Keil. 2001. Organic-mineral interactions in marine sediments studied using density fractionation and X-ray photoelectron spectroscopy. Org. Geochem. 32:1401–1415.

Bachmann, J., G. Guggenberger, T. Baumgartl, R.H. Ellerbrock, E. Urbanek, M.O. Goebel, K. Kaiser, R. Horn, and W.R. Fischer. 2008. Physical carbon-sequestration mechanisms under special consideration of soil wettability. J. Plant Nutr. Soil Sci. 171:14–26.

Baisden, W.T., R. Amundson, A.C. Cook, and D.L. Brenner. 2002. Turnover and storage of C and N in five density fractions from California annual grassland surface soils. Global Biogeochem. Cycles 16:1117, doi:10.1029/2001GB001822.

Baldock, J.A., J.M. Oades, P.N. Nelson, T.M. Skene, A. Golchin, and P. Clarke. 1997. Assessing the extent of decomposition of natural organic materials using solid-state 13C NMR spectroscopy. Aust. J. Soil Res. 35:1061–1084.

Baldock, J.A., and J.O. Skjemstad. 2000. Role of the soil matrix and minerals in protecting natural organic materials against biological attack. Org. Geochem. 31:697–710.

Balesdent, J. 1996. The significance of organic separates to carbon dynamics and its modelling in some cultivated soils. Eur. J. Soil Sci. 47:485–493.

Balesdent, J., and M. Balabane. 1996. Major contribution of roots to soil carbon storage inferred from maize cultivated soils. Soil Biol. Biochem. 28:1261–1263.

Basile-Doelsch, I., R. Amundson, W.E.E. Stone, D. Borschneck, J.Y. Bottero, S. Moustier, F. Masin, and F. Colin. 2007. Mineral control of carbon pools in a volcanic soil horizon. Geoderma 137:477–489.

Basile-Doelsch, I., R. Amundson, W.E.E. Stone, C.A. Masiello, J.Y. Bottero, F. Colin, F. Masin, D. Borschneck, and J.D. Meunier. 2005. Mineralogical control of organic carbon dynamics in a volcanic ash soil on La Reunion. Eur. J. Soil Sci. 56:689–703.

Batjes, N.H. 1996. Total carbon and nitrogen in the soils of the world. Eur. J. Soil Sci. 47:151–163.

Bellamy, P.H., P.J. Loveland, R.I. Bradley, R.M. Lark, and G.J.D. Kirk. 2005. Carbon losses from all soils across England and Wales 1978–2003. Nature 437:245–248.

Benke, M.B., A.R. Mermut, and H. Shariatmadari. 1999. Retention of dissolved organic carbon from vinasse by a tropical soil, kaolinite, and Fe oxides. Geoderma 91:47–63.

Besnard, E., C. Chenu, J. Balesdent, P. Puget, and D. Arrouays. 1996. Fate of particulate organic matter in soil aggregates during cultivation. Eur. J. Soil Sci. 47:495–503.

Bird, M.I., C. Moyo, E.M. Veenendaal, J. Lloyd, and P. Frost. 1999. Stability of elemental carbon in a savanna soil. Global Biogeochem. Cycles 13:923–932.

Bird, J.A., C. van Kessel, and W.R. Horwath. 2002. Nitrogen dynamics in humic fractions under alternative straw management in temperate rice. Soil Sci. Soc. Am. J. 66:478–488.

Blaser, P., A. Heim, and J. Luster. 1999. Total luminescence spectroscopy of NOM-typing samples and their aluminium complexes. Environ. Inter. 25:285–293.

Bosatta, E., and G.I. Angren. 1999. Soil organic matter quality interpreted thermodynamically. Soil Biol. Biochem. 31:1889–1891.

Brahim, B.H. 1987. Influence des constituants alumineux et ferriques non cristallins sur les cycles du carbone et de l'agote dans les sols montagnards acids. Universite de Nancy I, Nancy, France.

Bricklemyer, R.S., P.R. Miller, P.J. Turk, K. Paustian, T. Keck, and G.A. Nielsen. 2007. Sensitivity of the Century model to scale-related soil texture variability. Soil Sci. Soc. Am. J. 71:784–792.

Brodowski, S., W. Amelung, L. Haumaier, and W. Zech. 2007. Black carbon contribution to stable humus in German arable soils. Geoderma 139:220–228.

Bruun, S., G.I. Agren, B.T. Christensen, and L.S. Jensen. 2010. Measuring and modeling continuous quality distributions of soil organic matter. Biogeosciences 7:27–41.

Bruun, S., J. Six, and L.S. Jensen. 2004. Estimating vital statistics and age distributions of measurable soil organic carbon fractions based on their pathway of formation and radiocarbon content. J. Theor. Biol. 230:241–250.

Brynhildsen, L., and T. Rosswall. 1997. Effects of metals on the microbial mineralization of organic acids. Water Air Soil Pollut. 94:45–57.

Bundt, M., M. Krauss, P. Blaser, and W. Wilcke. 2001. Forest fertilization with wood ash: Effect on the distribution and storage of polycyclic aromatic hydrocarbons (PAHs) and polychlorinated biphenyls (PCBs). J. Environ. Qual. 30:1296–1304.

Burdon, J. 2001. Are the traditional concepts of the structures of humic substances realistic? Soil Sci. 166:752–769.

Buurman, P., and J.D.J. van Doesburg. 2007. Laser-diffraction grain-size analyses of reference profiles, p. 453–468. In O. Arnalds (ed.) Soils of European volcanic regions. Springer, New York.

Buurman, P., B. van Lagen, and A. Piccolo. 2002. Increase in stability against thermal oxidation of soil humic substances as a result of self association. Org. Geochem. 33:367–381.

Buurman P.F. Peterse, and M.G. Almendros. 2007. Soil organic matter chemistry in allophanic soils: A pyrolysis-GC/MS study of a Costa Rican Andosol catena. Eur. J. Soil Sci. 58:1330–1347.

Campbell, C.A., E.A. Paul, D.A. Rennie, and K.J. McCallum. 1967. Applicability of the carbon-dating method of analysis to soil humus studies. Soil Sci. 104:217–224.

Carter, M.R. 1996. Analysis of soil organic matter storage in agroecosystems, p. 3–11. In B.A. Stewart (ed.) Structure of organic matter storage in agricultural soils. CRC Press Inc., Boca Raton, FL.

Chabbi, A.C. Rumpel, and I. Kögel-Knabner. 2009. Stabilised carbon in subsoil horizons is located in spatially distinct parts of the soil profile. Soil Biol. Biochem. 41:256–261.

Cheng, C.H., J. Lehmann, and M.H. Engelhard. 2008. Natural oxidation of black carbon in soils: Changes in molecular form and surface charge along a climosequence. Geochim. Cosmochim. Acta 72:1598–1610.

Cheng, C.-H., J. Lehmann, J.E. Thies, S.D. Burton, and M.H. Engelhard. 2006. Oxidation of black carbon by biotic and abiotic processes. Org. Geochem. 37:1477–1488.

Chenu, C. 1995. Extracellular polysaccharides: An interface between microorganisms and soil constituents, p. 217–233. In P.M. Huang, J. Berthelin, and J.-M. Bollag (eds.) Environmental impact of soil component interactions. Natural and anthropogenic organics. Lewis Publishers, Chelsea, Michigan.

Chenu, C., Y. Le Bissonnais, and D. Arrouays. 2000. Organic matter influence on clay wettability and soil aggregate stability. Soil Sci. Soc. Am. J. 64:1479–1486.

Chenu, C., and A.F. Plante. 2006. Clay-sized organo-mineral complexes in a cultivation chronosequence: Revisiting the concept of the 'primary organo-mineral complex'. Eur. J. Soil Sci. 57:596–607.

Chenu, C., and G. Stotzky. 2002. Interactions between microorganisms and soil particles: An overview. In P.M. Huang et al. (eds.) Interactions between soil particles and microorganisms. John Wiley & Sons, New York.

Chiou, C.T., L.J. Peters, and V.H. Freed. 1979. A physical concept of soil-water equilibria for nonionic organic compounds. Science 206:831–832.

Christensen, B.T. 1992. Physical fractionation of soil and organic matter in primary particle size and density separates. Adv. Soil Sci. 20:1–90.

Czimczik, C.I., and C.A. Masiello. 2007. Controls on black carbon storage in soils. Global Biogeochem. Cycles 21:GB3005.

Davidson, E.A., and I.A. Janssens. 2006. Temperature sensitivity of soil carbon decomposition and feedbacks to climate change. Nature 440:165–173.

de Jonge, L.W., O.H. Jacobsen, and P. Moldrup. 1999. Soil water repellency: Effects of water content, temperature, and particle size. Soil Sci. Soc. Am. J. 63:437–442.

Demaneche, S., L. Jocteur-Monrozier, H. Quiquampoix, and P. Simonet. 2001. Evaluation of biological and physical protection against nuclease degradation of clay-bound plasmid DNA. Appl. Environ. Microbiol. 67:293–299.

Doerr, S.H., R.A. Shakesby, and R.P.D. Walsh. 2000. Soil water repellency: Its causes, characteristics and hydrogeomorphological significance. Earth Sci. Rev. 51:33–65.

Duiker, S.W., F.E. Rhoton, J. Torrent, N.E. Smeck, and R. Lal. 2003. Iron (hydr)oxide crystallinity effects on soil aggregation. Soil Sci. Soc. Am. J. 67:606–611.

Ekschmitt, K., E. Kandeler, C. Poll, A. Brune, F. Buscot, M. Friedrich, G. Gleixner et al. 2008. Soil-carbon preservation through habitat constraints and biological limitations on decomposer activity. J. Plant Nutr. Soil Sci. 171:27–35.

Ekschmitt, K., M. Liu, S. Vetter, O. Fox, and V. Wolters. 2005. Strategies used by soil biota to overcome soil organic matter stability—Why is dead organic matter left over in the soil? Geoderma 128:167–176.

Ellerbrock, R.H., H.H. Gerke, J. Bachmann, and M.-O. Goebel. 2005. Composition of organic matter fractions for explaining wettability of three forest soils. Soil Sci. Soc. Am. J. 69:57–66.

Eusterhues, K., C. Rumpel, and I. Kogel-Knabner. 2005. Organo-mineral associations in sandy acid forest soils: Importance of specific surface area, iron oxides and micropores. Eur. J. Soil Sci. 56:753–763.

Flessa, H., W. Amelung, M. Helfrich, G.L.B. Wiesenberg, G. Gleixner, S. Brodowski, J. Rethemeyer, C. Kramer, and P.M. Grootes. 2008. Storage and stability of organic matter and fossil carbon in a Luvisol and Phaeozem with continuous maize cropping: A synthesis. J. Plant Nutr. Soil Sci. 171:36–51.

Fox, O., S. Vetter, K. Ekschmitt, and V. Wolters. 2006. Soil fauna modifies the recalcitrance-persistence relationship of soil carbon pools. Soil Biol. Biochem. 38:1353–1363.

Giller, K.E., E. Witter, and S.P. McGrath. 1998. Toxicity of heavy metals to microorganisms and microbial processes in agricultural soils: A review. Soil Biol. Biochem. 30:1389–1414.

Goebel, M.-O., J. Bachmann, S.K. Woche, and W.R. Fischer. 2005. Soil wettability, aggregate stability, and the decomposition of soil organic matter. Geoderma 128:80–93.

Goebel, M.-O., J. Bachmann, S.K. Woche, W.R. Fischer, and R. Horton. 2004. Water potential and aggregate size effects on contact angle and surface energy. Soil Sci. Soc. Am. J. 68:383–393.

Goh, K.M. 1980. Dynamics and stability of organic matter, p. 373–393. In B.K.G. Theng (ed.) Soils with variable charge. New Zealand Society of Soil Science, Lower Hutt, New Zealand.

Golchin, A., P. Clarke, J.A. Baldock, T. Higashi, J.O. Skjemstad, and J.M. Oades. 1997. The effects of vegetation and burning on the chemical composition of soil organic matter in a volcanic ash soil as shown by 13C NMR spectroscopy. I. Whole soil and humic acid fraction. Geoderma 76:155–174.

Golchin, A., J.M. Oades, J.O. Skjemstad, and P. Clarke. 1994a. Study of free and occluded particulate organic matter in soils by solid state 13C CP/MAS NMR spectroscopy and scanning electron microscopy. Aust. J. Soil Res. 32:285–309.

Golchin, A., J.M. Oades, J.O. Skjemstad, and P. Clarke. 1994b. Soil structure and carbon cycling. Aust. J. Soil Res. 32:1043–1068.

Grandy, A.S., and J.C. Neff. 2008. Molecular C dynamics downstream: The biochemical decomposition sequence and its impact on soil organic matter structure and function. Sci. Total Environ. 404:297–307.

Gregorich, E.G., M.H. Beare, U. Stoklas, and P. St-Georges. 2003. Biodegradability of soluble organic matter in maize-cropped soils. Geoderma 113:237–252.

Gu, B., J. Schmitt, Z. Chen, L. Liang, and J.F. McCarthy. 1994. Adsorption and desorption of natural organic matter on iron oxide: Mechanisms and models. Environ. Sci. Technol. 28:38–46.

Gu, B., J. Schmitt, Z. Chen, L. Liang, and J.F. McCarthy. 1995. Adsorption and desorption of different organic matter fractions on iron oxide. Geochim. Cosmochim. Acta 59:219–229.

Guggenberger, G., B.T. Christensen, and W. Zech. 1994. Land use effects on the composition of organic matter in particle size separates of soil: I. Lignin and carbohydrate signature. Eur. J. Soil Sci. 46:449–458.

Haderlein, S.B., and R.P. Schwarzenbach. 1993. Adsorption of substituted nitrobenzenes and nitrophenols to mineral surfaces. Environ. Sci. Technol. 27:316–326.

Haderlein, S.B., K.W. Weissmahr, and R.P. Schwarzenbach. 1996. Specific adsorption of nitroaromatic explosives and pesticides to clay minerals. Environ. Sci. Technol. 30:612–622.

Hammes, K., M.S. Torn, A.G. Lapenas, and M.W.I. Schmidt. 2008. Centennial black carbon turnover observed in a Russian steppe soil. Biogeosciences 5:1339–1350.

Harsh, J. 2000. Poorly crystalline aluminosilicate clays, p. F169–F180. In M.E. Sumner (ed.) Handbook of soil science. CRC Press, Boca Raton, FL.

Harsh, J., J. Chorover, and E. Nizeyimana. 2002. Allophane and imogolite. In J.E. Amonette et al. (eds.) Soil mineralogy with environmental applications. Soil Science Society of America Inc., Madison, WI.

Hassink, J. 1997. The capacity of soils to preserve organic C and N by their association with clay and silt particles. Plant Soil 191:77–87.

Hedges, J.I., G. Eglinton, P.G. Hatcher, D.L. Kirchman, C. Arnosti, S. Derenne, R.P. Evershed et al. 2000. The molecularly-uncharacterized component of nonliving organic matter in natural environments. Org. Geochem. 31:945–958.

Heimann, M., and M. Reichstein. 2008. Terrestrial ecosystem carbon dynamics and climate feedbacks. Nature 451:289–292.

Horne, D.J., and J.C. McIntosh. 2000. Hydrophobic compounds in sands in New Zealand—Extraction, characterization and proposed mechanisms for repellency expression. J. Hydrol. 231–232:35–46.

Huang, P.M., and A. Hardie. 2009. Formation mechanisms of humic substances in the environment, p. 41–109. In N. Senesi, B. Xing, and P.M. Huang (eds.) Biophysico-chemical processes of nonliving natural organic matter in environmental systems. Vol. 2. Wiley-IUPAC Series. John Wiley & Sons, Hoboken, NJ.

Jahn, R., M. Zarei, and K. Stahr. 1992. Development of andic soil properties and of clay minerals in the semiarid climate of Lanzarote (Spain). Mineral. Petrogr. Acta XXXV-A:193–201.

Jandl, G., P. Leinweber, H.R. Schulten, and K. Eusterhues. 2004. The concentrations of fatty acids in organo-mineral particle-size fractions of a Chernozem. Eur. J. Soil Sci. 55:459–469.

Jastrow, J.D., J. Amonette, and V. Bailey. 2007. Mechanisms controlling soil carbon turnover and their potential application for enhancing carbon sequestration. Clim. Change 80:5–23.

Jastrow, J.D., T.W. Boutton, and R.M. Miller. 1996. Carbon dynamics of aggregate-associated organic matter estimated by carbon-13 natural abundance. Soil Sci. Soc. Am. J. 60:801–807.

Jenkinson, D.S. 1990. The turnover of organic carbon and nitrogen in soil. Phil. Trans. Roy. Soc. Lond. B 329:361–368.

John, B., T. Yamashita, B. Ludwig, and H. Flessa. 2005. Storage of organic carbon in aggregate and density fractions of silty soils under different types of land use. Geoderma 128:63–79.

Johnson, D. and B. Hale. 2004. White birch (Betula papyrifera Marshall) foliar litter decomposition in relation to trace metal atmospheric inputs at metal contaminated and uncontaminated sites near Sudbury, Ontario and Rouyn-Noranda, Quebec, Canada. Environ. Pollut. 127:65–72.

Jokic, A., M.C. Wang, C. Liu, A.I. Frenkel, and P.M. Huang. 2004. Integration of the polyphenol and Maillard reactions into a unified abiotic pathway for humification in nature: The role of d-MnO_2. Org. Geochem. 35:747–762.

Jones, D.L., T. Eldhuset, H.A. de Wit, and B. Swensen. 2001. Aluminium effects on organic acid mineralization in a Norway spruce forest soil. Soil Biol. Biochem. 33:1259–1267.

Kahle, M., M. Kleber, and R. Jahn. 2002a. Carbon storage in loess derived surface soils from Central Germany: Influence of mineral phase variables. J. Plant Nutr. Soil Sci. 165:141–149.

Kahle, M., M. Kleber, and R. Jahn. 2002b. Predicting carbon content in illitic clay fractions from surface area, cation exchange capacity and dithionite-extractable iron. Eur. J. Soil Sci. 53:639–644.

Kahle, M., M. Kleber, and R. Jahn. 2003. Retention of dissolved organic matter by illitic soils and clay fractions: Influence of mineral phase properties. J. Plant Nutr. Soil Sci. 166:737–741.

Kahle, M., M. Kleber, and R. Jahn. 2004. Retention of dissolved organic matter by phyllosilicate and soil clay fractions in relation to mineral properties. Org. Geochem. 35:269–276.

Kaiser, K., and G. Guggenberger. 2000. The role of DOM sorption to mineral surfaces in the preservation of organic matter in soils. Org. Geochem. 31:711–725.

Kaiser, K., and G. Guggenberger. 2003. Mineral surfaces and soil organic matter. Eur. J. Soil Sci. 54:1–18.

Kaiser, K., and G. Guggenberger. 2007. Sorptive stabilization of organic matter by microporous goethite: Sorption into small pores vs. surface complexation. Eur. J. Soil Sci. 58:45–59.

Kaiser, K., and W. Zech. 2000. Dissolved organic matter sorption by mineral constituents of subsoil clay fractions. J. Plant Nutr. Soil Sci. 163:531–535.

Kalbitz, K., and K. Kaiser. 2008. Contribution of dissolved organic matter to carbon storage in forest mineral soils. J. Plant Nutr. Soil Sci. 171:52–60.

Kalbitz, K., J. Schmerwitz, D. Schwesig, and E. Matzner. 2003. Biodegradation of soil-derived dissolved organic matter as related to its properties. Geoderma 113:273–291.

Karickhoff, S.W. 1984. Organic pollutant sorption in aquatic systems. J. Hydraul. Eng. 110:707–735.

Keiluweit, M., and M. Kleber. 2009. Molecular-level interactions in soils and sediments: The role of aromatic π-systems. Environ. Sci. Technol. 43:3421–3429.

Kelleher, B.P., and A.J. Simpson. 2006. Humic substances in soils: Are they really chemically distinct? Environ. Sci. Technol. 40:4605–4611.

Khanna, M., M. Yoder, L. Calamai, and G. Stotzky. 1998. X-ray diffractometry and electron microscopy of DNA bond to clay minerals. Sci. Soils 3:1–10.

Kleber, M. 2010. What is recalcitrant soil organic matter? Env. Chem. 7:320–322.

Kleber, M., and R. Jahn. 2007. Andosols and soils with andic properties in the German soil taxonomy. J. Plant Nutr. Soil Sci. 170:317–328.

Kleber, M., C. Mertz, S. Zikeli, H. Knicker, and R. Jahn. 2004. Changes in surface reactivity and organic matter composition of clay subfractions with duration of fertilizer deprivation. Eur. J. Soil Sci. 55:381–391.

Kleber, M., R. Mikutta, M.S. Torn, and R. Jahn. 2005. Poorly crystalline mineral phases protect organic matter in acid subsoil horizons. Eur. J. Soil Sci. 56:717–725.

Kleber, M., J. Rossner, C. Chenu, B. Glaser, H. Knicker, and R. Jahn. 2003. Prehistoric alteration of soil properties in a Central German chernozemic soil: In search of pedologic indicators for prehistoric activity. Soil Sci. 168:292–306.

Kleber, M., P. Sollins, and R. Sutton. 2007. A conceptual model of organo-mineral interactions in soils: Self-assembly of organic molecular fragments into multilayered structures on mineral surfaces. Biogeochemistry 85:9–24.

Knicker, H. 2004. Stabilization of N-compounds in soil and organic-matter-rich sediments—What is the difference? Marine Chem. 92:167–195.

Knicker, H., and H.-D. Lüdemann. 1995. N-15 and C-13 CPMAS and solution NMR of N-15 enriched plant material during 600 days of microbial degradation. Org. Geochem. 23:329–341.

Kögel-Knabner, I. 2000. Analytical approaches for characterizing soil organic matter. Org. Geochem. 31:609–625.

Kögel-Knabner, I., G. Guggenberger, M. Kleber, E. Kandeler, K. Kalbitz, S. Scheu, K. Eusterhues, and P. Leinweber. 2008. Organo-mineral associations in temperate soils: Integrating biology, mineralogy and organic matter chemistry. J. Plant Nutr. Soil Sci. 171:61–82.

Krull, E.S., J.A. Baldock, and J.O. Skjemstad. 2003. Importance of mechanisms and processes of the stabilization of soil organic matter for modelling carbon turnover. Funct. Plant Biol. 30:207–222.

Kucharik, C.J., K.R. Brye, J.M. Norman, J.A. Foley, S.T. Gower, and L.G. Bundy. 2001. Measurements and modeling of carbon and nitrogen cycling in agroecosystems of southern Wisconsin: Potential for SOC sequestration during the next 50 years. Ecosystems 4:237–258.

Laird, D.A., D.A. Martens, and W.L. Kingery. 2001. Nature of clay-humic complexes in an agricultural soil. I. Chemical, biochemical, and spectroscopic analyses. Soil Sci. Soc. Am. J. 65:1413–1418.

Lal, R. 2004. Soil carbon sequestration to mitigate climate change. Geoderma 123:1–22.

Lehmann, J. 2007a. A handful of carbon. Nature 447:143–144.

Lehmann, J. 2007b. Bio-energy in the black. Front. Ecol. Environ. 5:381–387.

Lehmann, J., J. Kinyangi, and D. Solomon. 2007. Organic matter stabilization in soil microaggregates: Implications from spatial heterogeneity of organic carbon contents and carbon forms. Biogeochemistry 85:45–57.

Lehmann, J., J.O. Skjemstad, S. Sohi, J. Carter, M. Barson, P. Falloon, K. Coleman, P. Woodbury, and E. Krull. 2008. Australian climate-carbon cycle feedback reduced by soil black carbon. Nature Geosci. 1:832–835.

Liao, J.D., T.W. Boutton, and J.D. Jastrow. 2006. Organic matter turnover in soil physical fractions following woody plant invasion of grassland: Evidence from natural 13C and 15N. Soil Biol. Biochem. 38:3197–3210.

Lundström, U.S., N. van Breemen, and D.C. Bain. 2000. The podzolization process. A review. Geoderma 94:91–107.

Mader, B.T., K. Goss, and S.J. Eisenreich. 1997. Sorption of nonionic, hydrophobic organic chemicals to mineral surfaces. Environ. Sci. Technol. 31:1079–1086.

Marschner, B., S. Brodowsi, A. Dreves, G. Gleixner, A. Gude, P.M. Grotes, U. Hamer et al. 2008. How relevant is recalcitrance for the stabilization of organic matter in soils? J. Plant Nutr. Soil Sci. 171:91–110.

Masiello, C.A. 2004. New directions in black carbon organic geochemistry. Marine Chem. 92:201–213.

Masiello, C.A., O.A. Chadwick, J. Southon, M.S. Torn, and J.W. Harden. 2004. Weathering controls on mechanisms of carbon storage in grassland soils. Global Biogeochem. Cycles 18:GB4023, doi:10.1029/2004GB002219.

Mayer, L.M., L.L. Schick, K.R. Hardy, R. Wagal, and J. McCarthy. 2004. Organic matter in small mesopores in sediments and soils. Geochim. Cosmochim. Acta 68:3863–3872.

Mayer, L.M., and B. Xing. 2001. Organic matter—Surface area relationships in acid soils. Soil Sci. Soc. Am. J. 65:250–258.

McEnroe, N.A., and H.S. Helmesaari. 2001. Decomposition of coniferous forest litter along a heavy metal pollution gradient, south-west Finland. Environ. Pollut. 113:11–18.

McGill, W.B. 1996. Review and classification of ten soil organic matter (SOM) models, p. 111–132. In U.J. Smith (ed.) Evaluation of soil organic matter models. Springer Verlag, Berlin, Germany.

McKeague, J.A., M.V. Cheshire, F. Andreux, and J. Berthelin. 1986. Organo-mineral complexes in relation to pedogenesis, p. 549–592. In P.M. Huang and M. Schnitzer (eds.) Interactions of soil minerals with natural organics and microbes. Soil Science Society of America, Madison, WI.

Mertz, C., M. Kleber, and R. Jahn. 2005. Soil organic matter stabilization pathways in clay sub-fractions from a time series of fertilizer deprivation. Org. Geochem. 36:1311–1322.

Mikutta, R., M. Kleber, and R. Jahn. 2005. Poorly crystalline minerals protect organic carbon in clay subfractions from acid subsoil horizons. Geoderma 128:106–115.

Mikutta, R., M. Kleber, M. Torn, and R. Jahn. 2006. Stabilization of soil organic matter: Association with minerals or chemical recalcitrance? Biogeochemistry 77:25–56.

Mikutta, R., C. Mikutta, K. Kalbitz, T. Scheel, K. Kaiser, and R. Jahn. 2007. Biodegradation of forest floor organic matter bound to minerals via different binding mechanisms. Geochim. Cosmochim. Acta 71:2569–2590.

Mueller, S., K.U. Totsche, and I. Koegel-Knabner. 2007. Sorption of polycyclic aromatic hydrocarbons to mineral surfaces. Eur. J. Soil Sci. 58:918–931.

Müller, B., and G. Défago. 2006. Interaction between the bacterium and vermiculite: Effects on chemical, mineralogical, and mechanical properties of vermiculite. J. Geophys. Res. 111: CiteID G02017.

Muneer, M., and J.M. Oades. 1989. The role of Ca-organic interactions in soil aggregate stability. III. Mechanisms and models. Aust. J. Soil Res. 27:411–423.

Nierop, K.G.J., B. Jansen, and J.A. Verstraten. 2002. Dissolved organic matter, aluminium and iron interactions: Precipitation induced by metal/carbon ratio, pH and competition. Sci. Total Environ. 300:201–211.

Oades, J.M. 1988. The retention of organic matter in soils. Biogeochemistry 5:35–70.

Oades, J.M. 1989. An introduction to organic matter in mineral soils, p. 89–159. *In* J.B. Dixon and S.B. Weed (eds.) Minerals in soil environments. Soil Science Society of America, Madison, WI.

Oades, J.M. 1993. The role of biology in the formation, stabilization and degradation of soil structure. Geoderma 56:377–400.

Oades, J.M., and A.G. Waters. 1991. Aggregate hierarchy in soils. Aust. J. Soil Res. 29:815–828.

Parton, W.J., D. Ojima, C.V. Cole, and D.S. Schimel. 1994. A general model for soil organic matter dynamics: Sensitivity to litter chemistry, texture and management, p. 147–167. *In* B.R. and R. Arnold (eds.) Quantitative modeling of soil forming processes. Soil Science Society of America, Madison, WI.

Percival, H.J., R.L. Parfitt, and N.A. Scott. 2000. Factors controlling soil carbon levels in New Zealand Grasslands: Is clay content important? Soil Sci. Soc. Am. J. 64:1623–1630.

Piccolo, A. 2001. The supramolecular structure of humic substances. Soil Sci. 166:810–832.

Pietramellara, G., J. Ascher, F. Borgogni, M.T. Ceccherini, G. Guerri, and P. Nannipieri. 2009. Extracellular DNA in soil and sediment: Fate and ecological relevance. Biol. Fertil. Soils 45:219–235.

Powlson, D. 2005. Will soil amplify climate change? Nature 433:204–205.

Puget, P., C. Chenu, and J. Balesdent. 2000. Dynamics of soil organic matter associated with particle-size fractions of water-stable aggregates. Eur. J. Soil Sci. 51:595–605.

Quiquampoix, H., J. Abadie, M.H. Baron, F. Leprince, P.T. Matumoto, R.G. Ratcliffe, and S. Staunton. 1995. Mechanisms and consequences of protein adsorption on soil mineral surfaces, p. 321–333. *In* T.A. Horbett and J.L. Brash (eds.) Proteins at interfaces II. Fundamentals and applications, ACS Symposium Series 602. American Chemical Society, Washington, DC.

Randerson, J.T., M.V. Thompson, C.M. Malmstrom, C.B. Field, and I.Y. Fung. 1996. Substrate limitations for heterotrophs: Implications for models that estimate the seasonal cycle of atmospheric CO_2. Global Biogeochem. Cycles 10:585–602.

Ransom, B., K. Dongsom, M. Kastner, and S. Wainwright. 1998. Organic matter preservation on continental slopes: Importance of mineralogy and surface area. Geochim. Cosmochim. Acta 62:1329–1345.

Rasmussen, C., R.J. Southard, and W.R. Horwath. 2006. Mineral control of organic carbon mineralization in a range of temperate conifer forest soils. Global Change Biol. 12:834–847.

Rasmussen, C., R.J. Southard, and W.R. Horwath. 2007. Soil mineralogy affects conifer forest soil carbon source utilization and microbial priming. Soil Sci. Soc. Am. J. 71:1141–1150.

Rasmussen, C., M.S. Torn, and R.J. Southard. 2005. Mineral assemblage and aggregates control carbon dynamics in a California conifer forest. Soil Sci. Soc. Am. J. 69:1711–1721.

Rillig, M.C, B.A. Caldwell, H.A.B. Wösten, and P. Sollins 2007. Role of proteins in soil carbon and nitrogen storage: Controls on persistence. Biogeochemistry 85:25–44.

Rima, J., E. Aoun, K. Hanna, and Q.X. Li. 2005. Degradation of phenol, into mineral compounds, in aqueous solutions using zero-valent iron powder (ZVIP). J. Phys. IV 124:81–89.

Ringwald, S.C., and J.E. Pemberton. 2000. Adsorption interactions of aromatics and heteroaromatics with hydrated and dehydrated silica surfaces by Raman and FTIR spectroscopies. Environ. Sci. Technol. 32:259–265.

Risebrough, R.W., P. Rieche, D.B. Peakall, S.G. Herman, and M.N. Kirven. 1968. Polychlorinated biphenyls in the global ecosystem. Nature 220:1098–1102.

Römkens, P.F.A.M., and J. Dolfing. 1998. Effect of Ca on the solubility and molecular size distribution of DOC and Cu binding in soil solution samples. Environ. Sci. Technol. 38:363–369.

Saggar, S., A. Parshotam, G.P. Sparling, C.W. Feltham, and P.B.S. Hart. 1996. 14C-labelled ryegrass turnover and residence times in soils varying in clay content and mineralogy. Soil Biol. Biochem. 28:1677–1686.

Scharpenseel, H.W., and P. Becker-Heidmann. 1994. 14C-dates and 13C measures of different soil species, p. 72–89. *In* R. Lal, J.M. Kimble, and E. Lavine (eds.) Soil processes and greenhouse effect. USDA, Soil Conservation Service, National Soil Survey Center, Lincoln, NE.

Schaumann, G.E. 2006. Soil organic matter beyond molecular structure Part I: Macromolecular and supramolecular characteristics. J. Plant Nutr. Soil Sci. 169:145–156.

Scheel, T., B. Jansen, A.J. van Wijk, J.M. Verstraten, and K. Kalbitz. 2008. Stabilization of dissolved organic matter by aluminium: A toxic effect or stabilization through precipitation? Eur. J. Soil Sci. 59:1122–1132.

Scheffer, M., V. Brovkin, and P.M. Cox. 2006. Positive feedback between global warming and atmospheric CO_2 concentration inferred from past climate change. Geophys. Res. Lett. 33:L09709.

Schmidt, M.W.I., J.O. Skjemstad, and C. Jäger. 2002. Carbon isotope geochemistry and nanomorphology of soil black carbon: Black chernozemic soils in central Europe originate from ancient biomass burning. Global Biogeochem. Cycles 16:1123, doi:10.1029/2002GB001939.

Schnitzer, M., and H. Kodama. 1992. Interactions between organic and inorganic components in particle-size fractions separated from four soils. Soil Sci. Soc. Am. J. 56:1099–1105.

Schöning, I., H. Knicker, and I. Kögel-Knabenr. 2005. Intimate association between O/N-alkyl carbon and iron oxides in clay fractions of forest soils. Org. Geochem. 36:1378–1390.

Schwarzenbach, R.P., P.M. Gschwend, and D.M. Imboden. 2003. Environmental organic chemistry. Wiley Interscience, New York.

Schwertmann, U., F. Wagner, and H. Knicker. 2005. Ferrihydrite-humic associations: Magnetic hyperfine interactions. Soil Sci. Soc. Am. J. 69:1009–1015.

Schwesig, D., K. Kalbitz, and E. Matzner. 2003. Effects of aluminium on the mineralization of dissolved organic carbon derived from forest floors. Eur. J. Soil Sci. 54:311–322.

Shindo, H., and P.M. Huang. 1982. Role of Mn(IV) oxide in abiotic formation of humic substances in the environment. Nature 298:363–365.

Shindo, H., and P.M. Huang. 1984. Significance of Mn(IV) oxide in abiotic formation of organic nitrogen complexes in natural environments. Nature 308:57–58.

Six, J., R.T. Conant, E.A. Paul, and K. Paustian. 2002. Stabilization mechanisms of soil organic matter: Implications for C-saturation of soils. Plant Soil 241:155–176.

Six, J., K. Paustian, E.T. Elliott, and C. Combrink. 2000. Soil structure and organic matter: I. Distribution of aggregate-size classes and aggregate-associated carbon. Soil Sci. Soc. Am. J. 64:681–689.

Skjemstad, J.O., L.J. Janik, M.J. Head, and S.G. McClure. 1993. High energy ultraviolet photo-oxidation: A novel technique for studying physically protected organic matter in clay- and silt-sized aggregates. Eur. J. Soil Sci. 44:485–499.

Smith, J.L., and E.A. Paul. 1990. The significance of soil microbial biomass estimations, p. 357–397. *In* J.M. Bollag and G. Stotzky (eds.) Soil biochemistry. Marcel Dekker, New York.

Smith, J.U., P. Smith, R. Monaghan, and A.J. McDonald. 2002. When is a measured soil organic matter fraction equivalent to a model pool? Eur. J. Soil Sci. 53:405–416.

Sokoloff, V.P. 1938. Effect of neutral salts of sodium and calcium on carbon and nitrogen of soils. J. Agric. Res. 57:201–216.

Sollins, P., P. Homann, and B.A. Caldwell. 1996. Stabilization and destabilization of soil organic matter: Mechanisms and controls. Geoderma 74:65–105.

Sollins, P., G. Spycher, and C.A. Glassmann. 1984. Net nitrogen mineralization from light-fraction and heavy-fraction forest soil organic matter. Soil Biol. Biochem. 16:31–37.

Sollins, P., C. Swanston, M. Kleber, T. Filley, M. Kramer, S. Crow, B.A. Caldwell, K. Lajtha, and R. Bowden. 2006. Organic C and N stabilization in a forest soil: Evidence from sequential density fractionation. Soil Biol. Biochem. 38:3313–3324.

Spielvogel, S., J. Prietzel, and I. Kögel-Knabner. 2008. Soil organic matter stabilization in acidic forest soils is preferential and soil type-specific. Eur. J. Soil Sci. 59:674–692.

Sposito, G., N.T. Skiper, R. Sutton, S.-H. Park, A.K. Soper, and J.A. Greathouse. 1999. Surface geochemistry of the clay minerals. Proc. Natl. Acad. Sci. USA 96:3358–3364.

Steffens, M., A. Kölbl, and I. Kögel-Knabner. 2009. Alteration of soil organic matter pools and aggregation in semi-arid steppe topsoils as driven by organic matter input. Eur. J. Soil Sci. 60:198–212.

Strickland, T.C., P. Sollins, N. Rudd, and D.S. Schimel. 1992. Rapid stabilization and mobilization of 15N in forest and range soils. Soil Biol. Biochem. 24:849–855.

Sutton, R., and G. Sposito. 2005. Molecular structure in soil humic substances: The new view. Environ. Sci. Technol. 39:9009–9015.

Swanston, C.W., M.S. Torn, P.J. Hanson, J.R. Southon, C.T. Garten, E.M. Hanlon, and L. Ganio. 2005. Initial characterization of processes of soil carbon stabilization using forest stand-level radiocarbon enrichment. Geoderma 128:52–62.

Tate, K.R., and B.K.G. Theng. 1980. Organic matter and its interactions to inorganic soil constituents, p. 225–249. *In* B.K.G. Theng (ed.) Soils with variable charge. New Zealand Society of Soil Science, Lower Hutt, New Zealand.

Theng, B.K.G. 1979. Formation and properties of clay-polymer complexes. Elsevier Scientific Publication Company, Amsterdam, the Netherlands.

Theng, B.K.G., G.J. Churchman, and R.H. Newman. 1986. The occurrence of interlayer clay-organic complexes in two New Zealand soils. Soil Sci. 142:262–266.

Thomson, A.M., R. César Izaurralde, S.J. Smith, and L.E. Clarke. 2008. Integrated estimates of global terrestrial carbon sequestration. Global Environ. Change 18:192–203.

Tisdall, J.M., and J.M. Oades. 1982. Organic matter and water-stable aggregates in soils. J. Soil Sci. 33:141–163.

Tokashiki, T., and K. Wada. 1975. Weathering implications of the mineralogy of clay fractions of two Ando soils, Kyushu. Geoderma 14:47–62.

Torn, M.S., and J. Harte. 2006. Missing feedbacks, asymmetric uncertainties, and the underestimation of future warming. Geophys. Res. Lett. 33.

Torn, M.S., C.W. Swanston, C. Castanha, and S.E. Trumbore. 2009. Storage and turnover of organic matter in soil, p. 219–272, *In* N. Senesi et al. (eds.) Biophysico-chemical processes involving natural nonliving organic matter in environmental systems. John Wiley & Sons, Inc., Hoboken, NJ.

Torn, M.S., S.E. Trumbore, O.A. Chadwick, P.M. Vitousek, and D.M. Hendricks. 1997. Mineral control of soil organic carbon storage and turnover. Nature 389:170–173.

Trumbore, S.E. 1993. Comparison of carbon dynamics in tropical and temperate soils using radiocarbon measurements. Global Biogeochem. Cycles 7:275–290.

Trumbore, S.E. 1997. Potential responses of soil organic carbon to global environmental change. Proc. Natl. Acad. Sci. USA 94:8284–8291.

Tyler, G., A.M. Balsberg-Pahlsson, G. Bengtsson, E. Baath, and L. Tranvik. 1989. Heavy metal ecology of terrestrial plants, microorganisms and invertebrates—A review. Water Air Soil Pollut. 47:189–215.

von Lützow, M., and I. Kögel-Knabner. 2009. Temperature sensitivity of soil organic matter. Biol. Fertil. Soils 46:1–15.

von Lützow, M., and I. Kögel-Knabner. 2010. Opinion essay to the concept article What is recalcitrant soil organic matter. Environ. Chem. 7:333–335.

von Lützow, M., I. Kögel-Knabner, K. Ekschmitt, H. Flessa, G. Guggenberger, E. Matzner, and B. Marschner. 2007. SOM fractionation methods: Relevance to functional pools and to stabilization mechanisms. Soil Biol. Biochem. 39:2183–2207.

von Lützow, M., I. Kögel-Knabner, K. Ekschmitt, E. Matzner, G. Guggenberger, B. Marschner, and H. Flessa. 2006. Stabilization of organic matter in temperate soils: Mechanisms and their relevance under different soil conditions—A review. Eur. J. Soil Sci. 57:426–445.

von Lützow, M., I. Kögel-Knabner, B. Ludwig, E. Matzner, H. Flessa, K. Ekschmitt, G. Guggenberger, B. Marschner, and K. Kalbitz. 2008. Stabilization mechanisms of organic matter in four temperate soils: Development and application of a conceptual model. J. Plant Nutr. Soil Sci. 171:111–124.

Wagai, R., and L.M. Mayer. 2007. Sorptive stabilization of organic matter in soils by hydrous iron oxides. Geochim. Cosmochim. Acta 71:25–35.

Wan, J., T. Tyliszczaka, and T.K. Tokunagaa. 2007. Organic carbon distribution, speciation, and elemental correlations within soil microaggregates: Applications of STXM and NEXAFS spectroscopy. Geochim. Cosmochim. Acta 71:5439–5449.

Wattel-Koekkoek, E.J.W., P. Buurman, J. van der Plicht, E. Wattel, and N. Van Breemen. 2003. Mean residence time of soil organic matter associated with kaolinite and smectite. Eur. J. Soil Sci. 54:269–278.

Wattel-Koekkoek, E.J.W., P.P.L. van Genuchten, P. Buurman, and B. van Lagen. 2001. Amount and composition of clay-associated soil organic matter in a range of kaolinitic and smectitic soils. Geoderma 99:27–49.

Weissmahr, K.W., S.B. Haderlein, and R.P. Schwarzenbach. 1998. Complex formation of soil minerals with nitroaromatic explosives and other pi-acceptors. Soil Sci. Soc. Am. J. 62:369–378.

Wershaw, R.L. 1999. Molecular aggregation of humic substances. Soil Sci. 164:803–813.

Wershaw, R.L., E.C. Llaguno, and J.A. Leenheer. 1996. Mechanism of formation of humus coatings on mineral surfaces. 3. Composition of adsorbed organic acids from compost leachate on alumina by solid-state 13C NMR. Coll. Surf. A: Physicochem. Eng. Asp. 108:213–223.

Woche, S.K., M.O. Goebel, M.B. Kirkham, R. Horton, R.R. Van der Ploeg, and J. Bachmann. 2005. Contact angle of soils as affected by depth, texture, and land management. Eur. J. Soil Sci. 56:239–251.

Zhu, D.Q., B.E. Herbert, M.A. Schlautman, E.R. Carraway, and J. Hur. 2004. Cation-pi bonding: A new perspective on the sorption of polycyclic aromatic hydrocarbons to mineral surfaces. J. Environ. Qual. 33:1322–1330.

Zysset, M., and D. Berggren. 2001. Retention and release of dissolved organic matter in Podzol B horizons. Eur. J. Soil Sci. 52:409–421.

8
Impact of Soil Physical, Chemical, and Biological Interactions on the Transformation of Metals and Metalloids

Antonio Violante
Università degli Studi di Napoli Federico II

M. Pigna
Università degli Studi di Napoli Federico II

V. Cozzolino
Università degli Studi di Napoli Federico II

Pan Ming Huang
(Deceased)
University of Saskatchewan

8.1	Introduction	8-1
8.2	Solution Complexation Reactions of Metals in Soils	8-2
8.3	Sorption/Desorption Processes of Metal(loid)s in Soils	8-2
	Sorption of Heavy Metals in Cationic Form • Sorption of Metal(loid)s in Anionic Form • Kinetics of Sorption • Surface Precipitation and Reduction–Oxidation Reactions • Competition in Adsorption of Metal(loid)s • Influence of Inorganic and Organic Ligands on the Sorption of Metal(loid)s • Desorption of Metal(loid)s	
8.4	Biomineralization of Metal(loid)s	8-14
8.5	Chemical Speciation and Bioavailability of Metal(loid)s as Influenced by Physicochemical and Biological Interfacial Interactions	8-14
	Cations • Anions	
8.6	Plant Uptake of Metal(loid)s as Influenced by Physicochemical and Biological Interfacial Interactions	8-18
	Effects of Plants and Managements • Effect of Arbuscular Mycorrhizal Fungi on the Transfer of Metal(loid)s to Plants	
8.7	Conclusions and Future Research Prospects	8-22
	Acknowledgments	8-22
	References	8-22

8.1 Introduction

Trace elements may be present in solution with positive or negative charges and in different redox states. They occur predominantly in cationic form (Pb, Cu, Zn, Ni, Cd, Hg, Cr(III), Co), but some are present in anionic form (As, Se, Cr(VI), Mo, and B). Heavy metals and metalloids have critically important biological effects, both beneficial and harmful. In trace amounts, many of these elements (e.g., Cu, Zn, Co) are essential to living organisms, but they become poisonous when their concentrations in the environment and in organisms are sufficiently high. Some elements (e.g., Hg and Pb) are not known to perform any essential biochemical functions and then are injurious or lethal to living organisms even at very low ambient concentrations (Jackson, 1998). Due to industrial development and disposal activity, trace elements are considered to be among the most important environmental contaminants that affect all ecosystem components in natural environments. Concerns about these contaminants are justified in view of the evidence that trace elements are continually accumulating in the food chain (Iskandar and Kikham, 2001).

Characterizing the factors that affect the mobility of metals and metalloids in soil is of paramount importance. Because soils are heterogeneous, heavy metals and metalloids can be involved in a series of complex chemical and biological interactions including sorption–desorption reactions, precipitation and dissolution, oxidation–reduction, and solution and surface phase complexation. The heterogeneous nature of the different soil constituents adds further complexity to the interactions of heavy metals and metalloids in the soil (Selim and Sparks, 2001).

In soil environments, sorption–desorption reactions as well as chemical complexation with inorganic and organic ligands and redox reactions, both biotic and abiotic, are of great importance in controlling metal bioavailability, leaching, and toxicity. These reactions are affected by many factors, such as pH, nature of the sorbents, presence, and concentration of organic and inorganic ligands, including humic and fulvic acid, root exudates, microbial metabolites, and nutrients (Kabata-Pendias and Pendias, 2001; Huang and Germida, 2002; Violante et al., 2002; Sparks, 2003; Kabata-Pendias, 2004; Huang and Gobran, 2005; Violante et al., 2008).

In this chapter, we highlight the impact of physical, chemical, and biological interfacial interactions on the transformation, and mobility of metals and metalloids in soil. Special attention is devoted to (1) the sorption/desorption processes of metals and metalloids on/from soil components and soils; (2) their precipitation and reduction–oxidation reactions in solution and onto the surfaces of soil components; (3) their chemical speciation, fractionation, and bioavailability; and (4) their transfer to plants.

8.2 Solution Complexation Reactions of Metals in Soils

The soil solution is the medium through which metals and metalloids are taken up, either actively or passively by plant roots and microbial cells. The soil processes (adsorption–desorption, complexation, precipitation–dissolution, or oxidation–reduction) control solution composition. Microorganisms and organic matter also have an important role in influencing the composition of the soil solution. Microorganisms breakdown organic matter and produce new biochemical reaction products. Soluble organics originate from organisms, including exudates from plant roots and soil microorganisms. The concentrations of these organics in the soil solution are more dynamic and more variable than the inorganic constituents. Enzymes (Bollag et al., 1998) and abiotic catalysts (Huang and Hardie, 2009) are important factors that affect the rate of formation and transformation of organic components. Organic matter, especially low-molecular-mass organic acids (LMMOAs), greatly enhances the solubility and mobility of metals in soil environments. In particular, metal complexation reactions with ligands play a significant role in controlling the chemical reactivity, bioavailability, and toxicity of metals in soil solution. The most important complexing functional groups of soil organic matter can be classified based on their affinity for hard, transition, and soft metals (Buffle, 1988; Sparks, 2003) as will be discussed later (Table 8.1). Furthermore, in view of

TABLE 8.1 Hard and Soft Acids and Bases

Lewis Acids	Lewis Bases
Hard acids: Group I metals	*Hard bases*
H^+, Li^+, Na^+, K^+, Be^{2+}, Mg^{2+}, Ca^{2+}, Sr^{2+}, Fe^{3+}, Al^{3+}, Se^{3+}	H_2O, OH^-, F^-, PO_4^{3-}, SO_4^{2-}, Cl^-, CO_3^{2-}, ClO_4^-, NO_3^-
Transition acids: Group II metals	*Transition bases*
Cr^{2+}, Mn^{2+}, Fe^{2+}, Co^{2+}, Ni^{2+}, Cu^{2+}	Br^-, NO_2^-, N_2
Soft acids: Group III metals	*Soft bases*
Ag^+, Au^+, Tl^+, Cu^+, Zn^{2+}, Cd^{2+}, Hg^{2+}, Pb^{2+}, Sn^{2+}	I^-, CN^-, CO

Source: Modified from Pearson, R.G. 1963. Hard and soft acids and bases. J. Am. Chem. Soc. 85:3533–3539.

the stability constants of these biomolecular ligands with metals (NIST, 1997), a substantial fraction of the soluble metal ions in the soil solution may be complexed with a series of biomolecules commonly present in microbial metabolites and root exudates. The concentrations of metals such as Cu, Mn, Zn, and Co in the rhizosphere vary with seasons and are related to the presence of complexing biomolecules (Nielson, 1976; Linehan et al., 1989). The pH of the rhizosphere soil is generally lower than that of the corresponding bulk soil. The cadmium availability index (CAI) values in the rhizosphere are higher, indicating that more Cd is complexed with biomolecules at the soil–root interface (Krishnamurti et al., 1996). The plant root and prolific microbial activity should result in increased amounts of biomolecules. A larger fraction of the metal contaminant will thus be in a complexed and usually soluble form in the soil–root interface than in the bulk soil.

The research of metal speciation of the soil solution has been encouraged by the free metal ion hypothesis in environmental toxicology (Lund, 1990). This states that the toxicity or bioavailability of a metal is related to the activity of the free aquo ion. This hypothesis is gaining popularity in studies of soil–plant relations (Parker et al., 1995). However, some evidence is now emerging that free metal ion hypothesis may not be valid in all situations (Tessier and Turner, 1995). Plant uptake of metals varies with the kind of chelators present in solution at the same free metal activity. Furthermore, given the same chelate, total metal concentration in solution affects metal uptake by plants. Either kinetic limitations to dissociation of the complex or uptake of the intact complex could explain these observations (Laurie et al., 1991).

8.3 Sorption/Desorption Processes of Metal(loid)s in Soils

The soil components responsible for trace element sorption include soil organic matter, phyllosilicates, carbonates, microorganisms, and variable charge minerals (constituents such as Fe, Al, Mn, and Ti oxides, short-range ordered aluminosilicates as well as phyllosilicates coated by OH–Al or OH–Fe species) whose charge varies with the pH of the soil solution (Parfitt, 1980; Jackson, 1998; Huang and Germida, 2002; Sparks, 2003; Violante et al., 2008). Soil components differ greatly in their sorption capacities, their cation and anion exchange capacities, and the binding energies of their sorption sites. However, in soil environments, organic matter, phyllosilicates, OH–Fe and OH–Al species, and short-range ordered metal oxides are constantly in close association with each other, forming organomineral complexes, which have different shape, size, charge and reactivity toward nutrients and pollutants, including metals and metalloids (Violante and Gianfreda, 2000; Chapter 2). Organomineral complexes have peculiar chemical and physicochemical properties and reactivity toward nutrients and metals and metalloids, which differ from those of each soil component present in these aggregates (Violante and Gianfreda, 2000).

In some soils, particularly those high in humic substances, the clay minerals may be coated with humic substances. It has been estimated that from 52% to 98% of all carbon in soils occurs as clay–soil organic matter complexes (Greedland, 1965; Sparks, 2003; Chapter 2 of Handbook of Soil Sciences: Properties and Processes). It is likely that most of the remaining organic C is linked to metal oxides (Schnitzer and Kodama, 1977).

The sorption/desorption processes of metal(loid)s in cationic form differs greatly from those of elements in anionic form. An excellent review on modeling adsorption of metals and metalloids by soil components has been published by Goldberg and Criscenti (2008).

8.3.1 Sorption of Heavy Metals in Cationic Form

Heavy metals cations show typical ion exchange behavior on layer silicate clays with permanent charge, demonstrating similar affinity for exchange sites on the clays as do alkaline earth metals having the same charge and similar ionic radious. Trace metals are probably not dominantly sorbed on 001 faces of phyllosilicates because they are always vastly outnumbered by other cations with which they compete (Jackson, 1998). Surface bonding is electrostatic, dependent only on the charge and hydration properties of the cation. The high selectivity of some phyllosilicates for trace quantities of heavy metals (relative to metals as Ca or Mg) may indicate the presence of a few sites on the clays capable of chemisorbing these metals. These sites may be –SiOH or –AlOH groups on the edges of the phyllosilicates (McBride, 1991). However, clay minerals have also an important role as carriers of associated oxides and humic substances forming organo-mineral complexes.

Substantial evidence has demonstrated that organic matter and metal oxides are much more effective scavengers of trace elements in cationic form, than even the most efficient sorbent among phyllosilicates, indicating that specific sorption and other complexation processes are the dominant binding mechanisms (Jackson, 1998; Huang and Germida, 2002; Huang, 2008; Violante et al., 2008). Two general surface complexes exist and are described by the configuration geometry of the adsorbate at the adsorbent surface. These include inner- and outer-sphere surface complexes and are defined by the presence, or absence, of the hydration sphere of the adsorbate molecule upon interaction. When at least one water molecule of the hydration sphere is retained upon adsorption the surface complex is referred to as outer-sphere. When the ion is bound directly to the adsorbent without the presence of the hydration sphere, an inner-sphere complex is formed (Sposito, 1986, 2004; Sparks, 2003; Borda and Sparks, 2008).

Besides some noncrystalline minerals that have very high specific surface charge density with highly reactive sites, organic matter appears to have the greatest capacity for sorption of trace elements in cationic form.

Soil organic matter consists of a spectrum of materials, ranging from compounds that mineralize very easily during the first stage of decomposition to more resistant materials that accumulate as microbial by-products as well as humified materials formed by enzymatic and mineral catalyses. Soil organic compounds are of two major types: (1) well-defined biochemical substances produced by microorganisms and plants, such as *simple*, LMMOAs, siderophores, hydroxamate, sugar acids, phenols and phenolic acids, and *complex*, polymeric phenols, lipids, polysaccharides, nucleic acids and proteins; and (2) a series of acidic, yellow- to black-colored substances formed by enzymatic and mineral catalysis reactions, which are referred to as humic and fulvic acids as well as humin materials that are regarded as humic substances (Stevenson, 1976; Sparks, 2003; Huang and Hardie, 2009). Molar mass of organic compounds range from >3000 Da for humic acids, 1000–3000 Da for fulvic acids, and <1000 Da for low-molar-mass organic acids.

In the rhizosphere, root exudates comprise both high and low-molar-mass substances released by the roots and microorganisms. The most important high-molecular-mass compounds are mucilage, polysaccharides, and ectoenzymes, whereas the main constituents of the low-molecular-mass compounds are carbohydrates, organic acids, amino acids, peptides, and phenolics.

Humic substances contain a large number of complexing sites and thus behave as a natural "multiligand" complexing system (Buffle, 1988). The high degree of selectivity of soil organic matter for most of trace metals in cationic form indicates that they form inner-sphere complexes with the functional groups, often forming internal five- or six-member ring on structures (Senesi, 1992; Senesi and Loffredo, 1998; Huang and Germida, 2002; Sparks, 2003). As summarized by Senesi and Loffredo (2008) the most common stoichiometries assessed are humic acid:metal = 1:1 and 2:1, often involving the formation of metal chelates. Aromatic carboxyls and phenolic groups play a prominent role in the 1:1 binding of metal ions by soil humic and fulvic acid, by forming chelates that involve either two COOH groups in a phtalate-type site or one phenolic group and one –COOH group in a salicylate-type site. Nonaromatic carboxyl and hydroxyl sites, including pyruvate-type and glycolate-type sites and carbohydrate moieties, may also be involved in metal ion binding by soil humic and fulvic acid. Other possible combinations involve two phenolic OH, quinone, NH_2, and sulfydryl groups, and conjugated ketonic structures (Stevenson, 1976; Chen and Stevenson, 1986; Senesi and Loffredo, 2005, 2008). Metal ions may also coordinate with ligands belonging to two (or more) humic substances molecules, forming 2:1 complexes and/or chelates, and eventually producing chain structures that may result in the aggregation and precipitation as the chain grows at high metal to humic substances ratios (Stevenson, 1976; Senesi and Loffredo, 2008). The total binding capacity of humic acids for metal ions is about 200–600 mmol kg^{-1}. About 33% of this total is due to retention on cation-complexing sites.

Complexation reactions of metals with ligands in the soil solution are significant in determining the chemical behavior and toxicity of metals in the rhizosphere, where LMMOAs are continuously released as root exudates and microbial metabolites. The hard and soft acid–base rules are useful in predicting the association between cations and ligands. According to the Pearson terminology, Lewis acids and bases can be classified on a scale of "hard" to "soft." Soft bases (ligands) are large molecules that are easily polarized: they readily donate electrons to form covalent bonds. They selectively bond with soft acids, typically metal ions of relatively large radius and low charge. Hard bases tend to be small molecules that are not easily polarized and form less covalent and more ionic bonds, preferentially with hard acids, typically metal ions of small radius and high charge. Therefore, soft acids bond preferentially with soft bases and hard acids bond preferentially with hard bases. The classification of Lewis acids and bases is listed in Table 8.1.

For soft metals (Cu^{2+}, Cd^{2+}, Hg^{2+}), the donor atom affinity follows the order: O < N < S, whereas the reverse order is observed for hard metals. In general, the competitive reactions for a given ligand essentially involve hard cations and transition metals for O sites and transition and soft metals for N and S sites, with competition between hard and soft metals being weak (Sparks, 2003).

Complexation reactions have the following effects: (1) metal ions are prevented from being precipitated; (2) complexing agents can act as carriers for trace elements in soil solution; and (3) their toxicity is often reduced by complexation. Stability constant (K) of trace metal complexes with humic acids increases with increasing pH and decreasing ionic strength. A number of reviews have been published on the chemistry of complexation of metals with organics and the spectroscopic techniques for studying metal–humic complexes in soil, to which the reader can refer for details (Stevenson and Fitch, 1986; Senesi, 1990; Senesi and Loffredo, 2008).

Unlike the phyllosilicates, the ability of variable charge minerals (crystalline and short range ordered Fe-, Al-, Mn-oxides, allophanes, imogolite) to retain trace metals is pH-dependent. On these materials, a hydroxylated or hydrated surface, positive or negative charge is developed by adsorption or desorption of H^+ or OH^- ions (Figure 8.1A). The pH at which the net variable charge on the surfaces of these components is zero is called the point of zero charge (PZC). The reported PZC of Fe-oxides ranges from pH 7.0 to 9.5, whereas that of Al-oxides ranges from pH 8.0 to 9.7. (Kwong and Huang, 1979; Hsu, 1989; Cornell and Schwertmann, 1996; Huang et al., 2002; Violante et al., 2005). Crystalline metal oxides usually have specific surface of 15–70 $m^2\,g^{-1}$, whereas short-range ordered precipitates may have specific surface even greater than 200–300 $m^2\,g^{-1}$. Organic components (fulvic and humic acids) as well as other organic and inorganic ligands have a very important role in the formation of short-range ordered materials. They can interact with Al and Fe perturbing their transformations, retarding or inhibiting the crystallization of Al and Fe precipitates (Huang and Violante, 1986; Violante et al., 2002, 2005). Crystalline and short-range ordered variable charge minerals selectively sorb polyvalent cations even when the net charge of their surfaces are positive (solution pH values lower than the PZC of the sorbent) (Figure 8.1).

Most transition cations (Pb, Cu, Cr, Ni, Co, Zn, Al, Fe, Mn) are often sorbed more strongly than alkaline earth cations. Spectroscopic techniques such as electron spin resonance (ESR) and extended x-ray absorption fine structure spectroscopy (EXAFS) have been used for the identification of metal

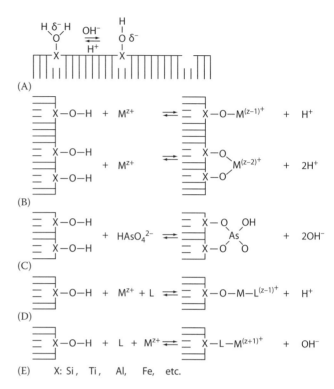

FIGURE 8.1 Coordination phenomena at oxide–water interface: (A) acid–base reactions of surface hydroxyls; (B) coordination of deprotonated groups with dissolved metal ions; (C) exchange of surface hydroxyls by dissolved oxyanions; (D) coordination of dissolved metal ion with both deprotonated surface hydroxyls and dissolved ligands; (E) coordination of dissolved multidentate ligand with both metal ion on the surface (X) and the dissolved metal ion (M). (Modified from Parfitt, R.L. 1980. Chemical properties of variable charge soils, p. 167–194. *In* B.K. Theng (ed.) Soils with variable charge. New Zealand Society of Soil Science, Palmerston North, New Zealand.)

complexes at the surfaces of Al, Fe, or Mn oxides, silicate clays, and soil organic matter (Sparks, 1999). The mechanism of metal ion association with hydrous oxide surfaces involves an ion exchange process in which the sorbed cations replace bound protons (Figure 8.1B). Usually, specifically sorbed cations raise the value of PZC of variable charge minerals. pH affects sorption of trace elements, either by changing the number of sites available for sorption (sorption increases by increasing pH) or by changing the concentration of cation species (Me^{2+}, $MeOH^+$, $Me(OH)_2$).

Sorption of metal cations is pH-dependent (Figure 8.2). Sorption, which increases from 0% to 100% of the amount added over a narrow region of 1–2 pH units, is termed as "sorption edge." The pH at which 50% of the total sorption has occurred is called pH_{50}. The lower the pH_{50} of a trace element for a sorbent the stronger is the element–surface complex (Kinniburgh and Jackson, 1976, 1981). The pH_{50} position of the sorption edge for a given trace element is related to the hydrolysis or acid–base characteristics. In addition to pH, sorption of metals is dependent on sorptive concentration, surface coverage, and the type of the sorbents (Sparks, 2003; Violante et al., 2003, 2008).

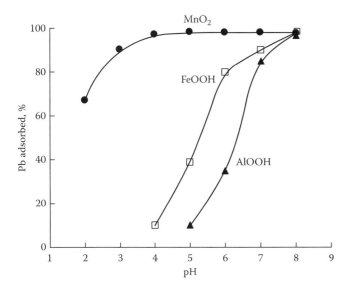

FIGURE 8.2 Effects of pH on the adsorption of Pb by Mn, Fe, and Al oxyhydroxides. (Modified from Gadde, R.R. and H.A. Laitinen. 1974. Studies of heavy metals adsorption by hydrous iron and manganese oxides. Anal. Chem. 46:2022–2026.)

Experiments with various synthetic Fe, Al, and Mn oxides showed that the affinity of trace elements for Mn oxide was usually much greater than that for Fe or Al oxides (Figure 8.2). However, the nature, crystallinity, size of the crystals, surface charge of metal oxides and mixed metal oxides (e.g., Fe–Al oxides) also play an important role in the sorption selectivity of trace elements in cationic form (Kinniburgh and Jackson, 1976, 1981; Pickering, 1979; McKenzie, 1980; McBride, 1982; Sparks, 2003; Violante et al., 2003, 2008).

Pickering (1979) recorded the following affinity series for freshly precipitated Fe, Al, and Mn oxides:

Fe-oxide: Pb > Cu > Zn > Ni > Cd > Co > Sr > Mg

Al-oxide: Cu > Pb > Zn > Ni > Co > Cd > Mg > Sr

Mn-oxide: Cu > Pb > Mn > Zn > Ni

Evidence on the sorption of trace elements on microorganisms has been reported. Free-living bacteria and their extracellular macromolecular products (e.g., fibrils) can accumulate trace elements and may have mineral coatings with bound metals on their surfaces (Beveridge, 1989a, 1989b; Jackson and Leppard, 2002, and references there in). All microorganisms contain biopolymers such as proteins, nucleic acids, and polysaccharides that provide reactive sites for binding metal ions. The cell surfaces of all bacteria are largely negatively charged containing different types of negatively charged functional groups, such as carboxyl, hydroxyl, and phosphoryl that can adsorb metal cations, and retain them by mineral nucleation. *Bacillus subtilis*, for example, has an isoelectric point at pH 2.4, and an average surface charge excess of 1.6 μmol mg^{-1} dry biomass over the pH range of 2.4–9. Accordingly, bacterial cell walls have a strong affinity for metal cations. Yee and Fein (2001) demonstrated that Cd sorption onto various Gram-positive and Gram-negative bacterial species was pH-dependent and the sorption edge behavior was similar to those of trace elements onto oxides. Even dead bacterial cells are also highly efficient in accumulating both soluble and particulate forms of metals. Bacteria therefore play an important role in the speciation, fate and transport of metals, metalloids, and radionuclides in soil and associated environments.

Biosorption comprises a variety of processes including ion exchange, chelation, adsorption, and diffusion through cell walls and membranes, all of which are dependent on the species used, the biomass origin, and solution chemistry (Gavrilescu, 2004). Lopez et al. (2000) reported the following affinity order for the surfaces of microorganisms: Ni ≫ Hg > U ≫ As > Cu > Cd > Co > Cr > Pb. Biosorption is a fast and reversible process for removing toxic metal ions from solution. Many environmental factors influence the chemical reactivity of the binding sites on bacterial cell surfaces and the subsequent biosorption of metals. These factors include pH, ionic strength, temperature, and the presence of other metals and organic compounds. The binding of metals by bacteria is also affected by nutrient and oxygen levels.

The ability of bacteria to accumulate toxic metals also varies with cell age. Huang et al. (2008) reported that 1-day-old cells of *Thiothrix* strain A1 accumulated considerably less Ni or Zn than its 2- to 5-day-old counterparts. However, the biosorptive capacity of dead cells may be greater, equivalent to, or less than that of living cells. For example, biosorption of Cr by the dead cells of *B. circulans* and *B. megaterium* was higher than by living cells (Srinath et al., 2002).

8.3.2 Sorption of Metal(loid)s in Anionic Form

Trace elements that exist in anionic form are sorbed primarily by chemisorption at reactive sites of metal oxides and allophanes (Figure 8.1C) and at the edges of phyllosilicates (Cornell and Schwertmann, 1996; Smith et al., 1998; Kampf et al., 2000; Violante et al., 2002). Usually, they are not sorbed on soil organic matter, but certain elements (e.g., borate, arsenate, arsenite) are found to be bound to humic acids (Thanabalasingan and Pickering, 1986; McBride, 2000). Arsenate and arsenite have maximum adsorption on humic acids at pH 5.5 and 8.0, respectively (Thanabalasingan and Pickering, 1986). Arsenate adsorbs onto solid-phase humic acids more extensively than arsenite, with amine (NH$_2$) groups suspected as the primary functional group responsible for arsenic retention. Some anions may bond indirectly to organic groups through a bridging hydrolytic species of Al, Fe, and Mn. Carbonates are also important sorbing surfaces (Goldberg and Glaubig, 1988a, 1988b).

Sorption of anions onto variable charge minerals and soils varies with pH. With increasing pH, within a certain range, sorption decreases (due to a decrease of positive charge of minerals) or else increases to a maximum close to the pK_a for anions of monoprotic conjugate acids and then decreases. Arsenite and selenite may be sorbed more easily at high pH values because they form weak acids at low pH values and consequently may be dissociated only in alkaline environments (Violante et al., 2008). Anions may be specifically or nonspecifically sorbed, forming inner-sphere or outer-sphere complexes. Ligands that are specifically sorbed replace OH⁻ or OH₂ groups from the surfaces of variable charge minerals, which involve direct coordination to the surface metal atom (Figure 8.1C). Specifically sorbed anions usually lower the PZC of metal oxides; thus, the PZC of a particular oxide may give rise to different values depending on the kind and extent of foreign ion sorption.

Trace elements, which form inner-sphere complexes, are arsenate, arsenite, molybdate, and selenite, which may form different surface complexes on inorganic soil components: monodentate, bidentate-binuclear, and bidentate-mononuclear complex in different proportions depending on pH and surface coverage (Hsia et al., 1994; Sun and Doner, 1996; Fendorf et al., 1997; Manning et al., 1998; O'Reilly et al., 2001; Ona-Nguema et al., 2005). Many studies have demonstrated that arsenite is sorbed on Al-oxides, phyllosilicates, and calcite in a lesser extent than arsenate, whereas the opposite is true for iron oxides (Violante et al., 2008).

Soil selenium content is significantly correlated with total carbonate, free-Fe-oxide, and extractable Al and Fe and clay content. Selenite sorption onto soil components and soils is much greater than selenate (Goldberg and Glaubig, 1988a). Using EXAFS measurements, Hayes et al. (1987) ascertained that selenate formed a weakly bonded, outer-sphere complex and selenite formed a strongly bonded, inner sphere complex when sorbed on goethite. However, Manceau and Charlet (1994) found that selenate may also form inner-sphere complexes onto goethite analogous to sulfate.

The sorption mechanism of chromate is unclear. Zachara et al. (1989) suggested that chromate forms an outer-sphere complex on the surfaces of Fe and Al oxides. However, spectroscopic studies have shown that chromate forms inner-sphere complexes (both bidentate and monodentate) on goethite (Fendorf et al., 1997). Furthermore, in contrast to surface complexation modeling results, competitive ion displacement studies have shown that Cr(VI) is retained much more strongly than anions such as Cl⁻ or SO_4^{2-}, and that its retention strength approaches that of phosphate. There are various sorption mechanisms involving both physical and chemical processes that could occur for heavy metals and metalloids at soil mineral surfaces, as summarized by Sparks (2003, see Table 5.1).

Trace elements in anionic form may be easily sorbed by "anionic clays" (layered double hydroxides [LDHs]), which are present in soils. For example, "green rusts" are green-blue Fe(II) Fe(III) hydroxides, which are formed by a number of abiotic and biotic processes under circumneutral to alkaline conditions in suboxic environments and are postulated to be intermediate phases in the formation of iro(oxyhydro)oxides such as goethite, lepidocrocite, and magnetite (Cornell and Schwertmann, 1996; Violante et al., 2002). They may be important for trace metal mobility because of their high reactive surface area. Many studies have been carried out on the possible use of LDHs, which are also easily synthesized, as powerful filters for removing xenobiotics in anionic forms from polluted waters (Goh et al., 2008; Violante et al., 2009b).

8.3.3 Kinetics of Sorption

Sorption/desorption of trace elements in soils can occur over wide timescales. The mineralogical, chemical, and physicochemical properties of inorganic and organic sorbents and microorganisms of a soil affect the sorption/desorption processes. In particular, trace element sorption kinetics on phyllosilicates, Al, Fe, or Mn oxides and humic substances depend mainly on the nature of the surfaces of the sorbents and trace element being studied. Sorption reactions are often more rapid on kaolinite and smectite than on vermiculitic and micaceous minerals. In fact, kaolinite has readily available planar external sites and smectites have predominantly internal sites that are quite available for retention of sorbates. Thus, sorption reactions on these soil constituents are often quite rapid, even occurring on timescales of seconds and milliseconds (Sparks, 1989, 2003). On the other hand, vermiculite and micas have multiple sites for retention of ions, including planar, edge, and interlayer sites, with some of the latter sites being partially to totally collapsed. Therefore, sorption and desorption reactions on these sites can be slow, tortuous, and mass transfer-controlled. Often, an apparent equilibrium may not be attained even after several days or weeks (Borda and Sparks, 2008).

The use of pressure-jump relaxation and other relaxation techniques have shad light in the study of sorption measurements on soil components (Sparks and Zhang, 1991; Sparks, 1995). An especially attractive approach for ascertaining sorption mechanisms on soils would be to combine relaxation approaches with in situ surface spectroscopic techniques. However, there are a few examples in the literature of studies where sorption reactions on soil components have been hypothesized via kinetic experiments and verified in separate spectroscopic investigations (Fuller et al., 1993; Waychunas et al., 1993; Fendorf et al., 1997; Grossl et al., 1997; Scheidegger et al., 1997; Borda and Sparks, 2008).

Sorption processes of metals and metalloids are initially rapid, with a large amount of the total sorption occurring within a few hours and the remaining sorption often occurring over long periods of time. The rapid step, which occurs over milliseconds to hours, can be ascribed to chemical reaction and film diffusion processes (Sparks, 2003). During this rapid reaction process, a large portion of the sorption may occur. A number of studies have shown that sorption/desorption of metals, oxyanions, radionuclides, and organic chemicals on soils can be slow, and may demonstrate a residence time effect (contact time between trace element and sorbent), as will be discussed later. The mechanisms for these lower reaction rates are not well understood, but have been ascribed to diffusion into micropores of inorganic minerals such as some metal oxides and into humic substances, retention on sites of varying reactivity, and to surface nucleation/precipitation (Borda and Sparks, 2008, and references therein). In some cases, the magnitude of sorption can greatly increase with longer reaction times. For example, Bruemmer et al. (1988) studied Ni^{2+}, Cd^{2+}, and Zn^{2+} sorption on goethite, a porous Fe oxide that has defect structures in which metals can be incorporated to satisfy charge imbalances. It was found at pH 6 that as reaction time increased from 2 h to 42 days, adsorbed Ni^{2+} increased from 12% to 70% of total sorption, and total Cd^{2+} and Zn^{2+} sorption over the same time increased by 21% and 33%, respectively.

An excellent review on kinetics and mechanisms of sorption–desorption in soils includes those of Sparks (1989) and Borda and Sparks (2008).

8.3.4 Surface Precipitation and Reduction–Oxidation Reactions

8.3.4.1 Surface Precipitation

Sorbed trace element species may range from simple cations to complex polynuclear hydroxides formed by hydrolysis and condensation in solution or by nucleation on the mineral surfaces (McBride, 1991; Jackson, 1998). As the amount of metal cation or anion sorbed on a surface (surface coverage or loading, which is affected by the pH at which sorption occurs) increases, sorption can proceed from mononuclear adsorption to surface precipitation (a three-dimensional phase). There is often a continuum between surface complexation (adsorption) and surface precipitation. At low surface coverages, surface complexation (e.g., outer- and inner-sphere adsorption) tends to dominate. As surface loadings increase, nucleation occurs and results in the formation of distinct entities or aggregates on the surface. As surface loadings increase further, surface precipitation becomes the dominant mechanism (Sparks, 2003, 2005). When the precipitate consists of chemical species derived from both the aqueous solution and dissolution of the mineral, it is referred to as a coprecipitate. The composition of the coprecipitate varies between that of the original solid and a pure precipitate of the sorbing metal.

In the last decade, extensive research has been carried out on the heterogeneous precipitation of trace elements on the surfaces of minerals, using modern spectroscopy techniques such as synchrotron-based x-ray absorption spectroscopy (Sparks, 2003). Laboratory studies of metal partitioning identified the incorporation of Cr, Ni, Co, Cu, and Zn into a neoformed surface precipitates. Depending on the reaction conditions, time, and adsorbent phase present, either a metal hydroxide, a mixed LDH, or a phyllosilicate form at the mineral surface. Many studies have shown that sorption of metals, such as Ni, on an array of phyllosilicates and Al oxide could result in formation of mixed metal–Al hydroxide surface precipitates, which appear to be coprecipitates. These neoformed phase shares structural features common to the hydrotalcite group of minerals and the LDHs, also known as anionic clays. The crystal structure of LDHs consists of positively charged octahedral hydroxide layers, which are neutralized by the interlayer anions and water molecules. These minerals have the general formula $[M^{2+}_{1-x} M^{3+}_x (OH)_6]^{x+} [(A^{n-})_{x/n} \, mH_2O]^{x-}$ with x taking values between 0.20 and 0.33 (Miyata, 1975; Cavani et al., 1991) where M^{2+} and M^{3+} are divalent (Ca, Mg, Co, Fe, Cu, Ni, Mn) and trivalent metal cations (Al, Fe, Cr), respectively, and A is an interlayer anion (Cl, NO_3, ClO_4, CO_3, SO_4 and so on) of charge $-n$. The anions in the interlayer spaces can be replaced with any inorganic and organic anion by simple ion exchange methods. These minerals can be considered complementary to the cationic clays whose negative charge of the aluminosilicate layers are counterbalanced by cations.

The ionic radius of the sorbing metal and sorbent ions must be similar for coprecipitates to form. Mixed Co–Al, Ni–Al, and Zn–Al hydroxide surface precipitates form on aluminum-bearing metal oxides and phyllosilicates, having Co^{2+}, Zn^{2+}, and Ni^{2+} similar radii to Al^{3+}, enhancing substitution in the mineral structure and formation of a coprecipitate. However, surface precipitates have not been observed with Pb^{2+}, as Pb^{2+} is too large (0.12 nm) to substitute for Al^{3+} in mineral structures (Sparks, 2003, 2005).

Surface precipitates can start to form on timescale of minutes and then growing in amount for hours or days. The initial precipitates are stabilized upon aging. This stabilization (residence time effect) has been attributed to the transformation of a mixed LDH phase into a more stable phyllosilicate phase. In other words, the metal sorption on soil minerals can often results in a continuum of processes from adsorption to precipitation to solid phase transformation. As a consequence, heavy metals become less mobile and bioavailable with time.

The effect of residence time on the rearrangement of trace elements in anionic form from desorbable into resistant and undesorbable forms has been studied and evidenced (Arai and Sparks, 2002; Pigna et al., 2006).

The effects of humic and fulvic acid on Ni sequestration mechanism in kaolinite and boehmite have been recently investigated (Nachtegaal and Sparks, 2003; Strathmann and Mynemi, 2005). Spectroscopic evidence indicates that the formation of Ni LDH phases can be suppressed by the amount of humic acid coatings on kaolinite surface via direct Ni sorption on humic acid functional groups and/or by the formation of Ni-fulvic acid-boehmite ternary species. In particular, Nachtegaal and Sparks (2003) showed that in the presence of 1 wt% of humic acid a Ni–Al LDH precipitate was formed at the kaolinite surface, but in the presence of 5 wt% of humic acid its formation slowed significantly and the precipitate formed was similar to $Ni(OH)_2$. However, the $Ni(OH)_2$ precipitate was not resistant to proton dissolution while the Ni–Al LDH was. These results augment the findings that the incorporation of first row transition metals into stable surface precipitates is an important sequestration pathway for toxic metals in the environment, despite the presence of ubiquitous coating materials such as humic or fulvic acids.

Yamaguchi et al. (2002) studied the influence of surface area of gibbsite on Ni sorption in the ternary Ni-gibbsite-citrate system over time. At low surface area and corresponding high surface loading, a Ni–Al LDH precipitate formed. At high surface area and lower surface loading, formation of an inner-sphere surface complex prevailed and a small amount of Ni–Al LDH formed only after an extended ageing period. Citrate reduced the amount of Ni sorbed but the effect was more pronounced for the gibbsite with a low surface area than for that with a large surface area. Furthermore, citrate prevented the formation of an LDH phase.

8.3.4.2 Redox Reactions

Changes in the oxidation state of trace metals can occur depending on the redox condition of the environment. Redox reactions are then important in influencing the chemical speciation of a number of metals and metalloids (Sparks, 1995). Trace elements that are directly influenced by electron transfer include Cr [Cr(III), Cr(VI)], Se [Se(IV), Se(VI)], As [As(III), As(V)], as well as Co, Pb, Ni, and Cu.

Clays and oxides often demonstrate the ability to catalyze electron transfer reactions (Huang and Hardie, 2009). Iron and manganese oxides and layer silicates with structural Fe(III) are the most active in this regard. Redox reactions also control the transformation and reactivity of Fe- and Mn-oxides in soils, which are the major sinks for heavy metals and metalloids. Many of the transition metal cations (Ni^{2+}, Zn^{2+}, Pb^{2+}, and Cd^{2+}), which are not always directly impacted by electron transfer, are moderately mobile in an oxidized environment, but they present a relative mobility in a reduced environment very low to immobile. In a reducing environment, sulfur is found in the reduced form (S^{2-}) and metal concentrations in solution are controlled by the solubilities of sulfide minerals (e.g., NiS, ZnS, CdS, and PbS), which are much more stable than the minerals found in oxidized systems (James and Bartlett, 2000; Sparks, 2003).

Redox-reactive metals often do have different degrees of toxicity depending on the specific metal oxidation state. Chromium is one such metal ion that persists in the environment as either Cr(III) and Cr(VI). Chromate is toxic to plants, animals, and humans and is a suspected carcinogen, whereas Cr(III) is not toxic to plants and is necessary in animal nutrition, so that reactions that reduce Cr(VI) to Cr(III) are of great importance. Furthermore, Cr(VI) is mobile in soils and readily available. Organic material, sulfides, and ferrous species appear to be the dominant reductants. Very stable Cr(III)–organic complexes form when Cr(VI) is reduced by soil organic matter (Fendorf, 1995).

The efficiency of the Mn oxides as "electron pump" for a wide range of redox reactions is unique among common soil minerals. Manganese oxides have been proven to be the only naturally occurring oxidant of Cr(III), so that they can enhance the mobility and toxicity of Cr, as Cr(VI) and Pu(VI), which can result from oxidation of Pu(III) by Mn-oxides, is very toxic and mobile in soils and water. Manganese oxides and hydroxides may also catalyze the oxidation of other trace metals such as Co^{2+}, Co^{3+}, Cu^{2+}, Ni^{2+}, Ni^{3+}, and Pb^{2+} by disproportionation to Mn^{2+} and MnO_2. The disproportionation results in vacancies in the Mn oxide structure. Since the Mn^{2+} and Mn^{3+} in the oxides have similar physical sizes as Co^{2+}, Co^{3+}, Cu^{2+}, Ni^{2+}, Ni^{3+}, and Pb^{2+} these metals can occupy the vacancies in the Mn oxide and become part of the structure (Huang and Germida, 2002; Sparks, 2003; Huang and Hardie, 2009).

Manganese oxides can also oxidize As(III) to As(V). This reaction has important implications for the transport, fate, and toxicity of As in natural environments, because As(III) is more toxic to human and more mobile than As(V). The ability of Mn oxides to oxidize As(III) to As(V) varies with their crystallinity and specific surface, but the kinetic of reaction is generally very fast. Under controlled laboratory conditions, the apparent half-lives for oxidation of As(III) are less than 1 h (Oscarson et al., 1981, 1983). In contrast, whereas Sun and Doner (1996) demonstrated that oxidation of sorbed As(III) to As(V) on Fe(III) oxides may occur to a limited (about 20%) extent.

Reduction of some trace elements is promoted by "green rusts." These components are able to reduce Se(VI), Cr(VI), and U(VI), but cannot reduce As(V) to As(III) (Loyaux-Lawniczak et al., 2000; Refait et al., 2000). Transition metal exchange cations on clays can also promote oxidation–reduction reactions. For example, adsorbed Cu^{2+} on smectite can oxidize benzene and benzidine by a single electron reaction, forming radical cations that appear to be stabilized at the surface (McBride, 1989).

Important microbiological processes controlling the oxidation and reduction of metals and metalloids in natural environments have been reviewed by Ehrlich (1997, 2002a, 2002b), Frankenberger (2002), and Gadd (2008).

The changes in soil redox potential and biological activity in the rhizosphere and their effects on the phytoavailability of metal(loid)s will be addressed later.

8.3.5 Competition in Adsorption of Metal(loid)s

Competition in sorption between two or more metals or metalloids is of great importance for understanding their relative affinity for a given sorbent. Lack of competition of different cations may be due to low surface coverage and only partially to the different binding sites having high selectivity for specific metals and metalloids (Kretzschmar and Voegelin, 2001).

Competition between different trace elements is easily ascertained at high surface loading of the sorbents with sorbed cations. Violante et al. (2008) have reviewed the competition of selected trace elements in binary systems onto inorganic and organic soil components and soils. Simine et al. (1998) found that lead competes with Cd for attachment to bacterial surface sites when a solution containing both metals was added.

Few studies have been carried out on the competition of three or more trace elements onto soil components or soils. Recently, the competitive sorption of Cu, Cr(III), and Pb in binary and ternary systems on a ferrihydrite was studied (Zu et al. 2010). Figure 8.3 shows the amounts of each metal sorbed when added alone or in competition with another heavy metal in the systems Cr:Cu, Cu:Pb, and Pb:Cr at pH 4.5 and at cation molar ratio of 1. More Pb (80%) than Cr (67%) or Cu (25%) was sorbed when each cations was added alone (100 mmol added per kg). In binary systems, Pb reduced Cr and Cu sorption by 26% and 39%, Cr reduced Cu and Pb sorption by 35% and 11%, and, finally Cu inhibited Cr and Pb sorption by 11% and 1%, respectively (Figure 8.3). The stronger

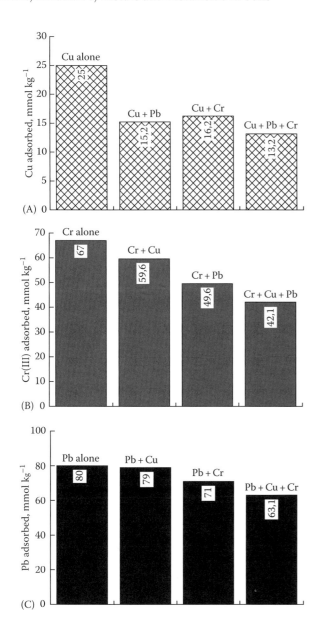

FIGURE 8.3 Total amount of (A) Cu, (B) Cr(III), and (C) Pb sorbed onto ferrihydrite added alone or in binary and ternary systems at equal molar ratio. (Modified from Zhu, J., M. Pigna, V. Cozzolino, A.G. Caporale, and A. Violante. 2010. Competitive sorption of copper(II), chromium(III) and lead(II) on ferrihydrite and two organomineral complexes. Geoderma 159:409–416.).

capacity of Pb in preventing the sorption of the other two cations appeared evident when the three cations were added simultaneously at equimolar amounts (100 mmol kg^{-1}) to the ferrihydrite suspension. The proportion of the cations sorbed on the Fe oxide was 53.3% of Pb, 35.6% of Cr, and 11.1% of Cu. This substantiates the finding that the affinity of the three cations followed the sequence Pb > Cr ≫ Cu. Anyway, by adding the three cations at an initial Pb:Cr:Cu molar ratio of 2:1:1 (not shown), the sorbed cations were for 73.0% Pb, 20.8% Cr, and 6.2% Cu. These findings indicate that even in the presence of high concentrations of Pb, some sites were apparently specific for Cr and Cu.

Elliott et al. (1986) studied the competitive sorption of Cd, Cu, Pb, and Zn onto four soils with different chemical properties. For two mineral soils, sorption under acidic conditions (pH 5.0) followed the sequence Pb > Cu > Zn > Cd, which corresponds to the order of increasing pK for the first hydrolysis product. For two soils with high organic matter content, the order was Pb > Cu > Cd > Zn.

Capasso et al. (2004) demonstrated that the presence of both Zn and Cu did not affect the capacity of Cr(III) to be sorbed onto an humic-like material (polymerin). In contrast, at an equilibrium concentration of 4 meq L^{-1}, the concomitant presence of Zn and Cr(III) decreased Cu sorption by about 50%, with Cr(III) being much more effective than Zn in preventing Cu sorption, whereas Cu and Cr(III) decreased Zn sorption by 79%.

Competition in sorption of trace elements in anionic form has received some attention (Violante et al., 2008). Manning and Goldberg (1996) studied the effects of pH and competing molybdate and arsenate ions onto goethite and gibbsite. Molybdate at 50% surface coverage decreased the sorption of arsenate only at pH < 6.0, whereas arsenate reduced molybdate sorption within a wider pH range (2–9 for goethite and 2–8 for gibbsite).

Goldberg and Glaubig (1988a) studied the competitive effect of selenite on arsenate sorption on a calcareous montmorillonite soil. Arsenate sorption was unaffected by the presence of selenite over a wide range of pH (1–13). These results indicate that these anions sorb on separate sites or, more probably, sorption sites were not limiting at the low concentrations used. However, Zhu et al. (2011) found that the influence of selenate (at about 80% surface coverage for arsenic) in preventing arsenite and arsenate sorption onto ferrihydrite was negligible when the two ligands were added in equimolar amounts at pH 6.0, whereas selenite reduced arsenite and arsenate sorption by 22% and 17%, respectively (Figure 8.4A and B). The efficiency of selenite and selenate increased at lower pH values and at a greater Se(III or V)/As(III or V) molar ratio.

Wu et al. (2001) found that molybdate sorption was insignificantly affected in the presence of equimolar amounts of selenate; however, selenate sorption was significantly reduced in the presence of molybdate at pH < 7.0, where a 30% decrease in sorption was noticed.

8.3.6 Influence of Inorganic and Organic Ligands on the Sorption of Metal(loid)s

Many factors such as pH, surface properties of the sorbents, the number of sites available for sorption, and the nature and charge of species of metals and metalloids in solution influence their sorption onto soil inorganic components (phyllosilicates and variable charge minerals) in the presence of inorganic and biological ligands (Kinniburgh and Jackson 1976, 1981; McBride, 1989; Goldberg et al., 1996; Jackson, 1998; Violante et al., 2003, 2008).

Inorganic and organic ligands (e.g., organic acids), which form strong complexes with trace element cations usually prevent or reverse their association with negatively charged sorbents, as 2:1 expandable phyllosilicates, by forming stable

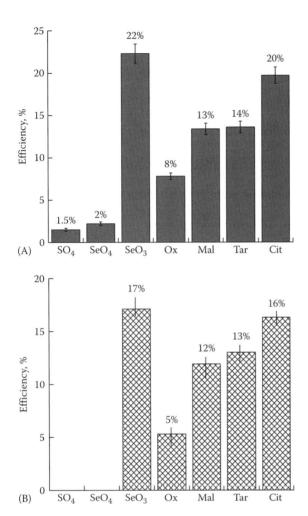

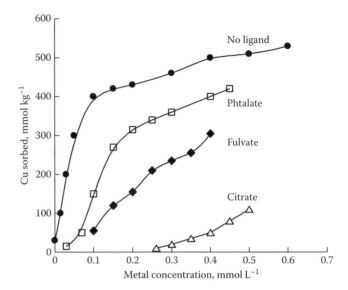

FIGURE 8.4 Influence of organic and inorganic ligands on the sorption of (A) As(III) and (B) As(V) onto ferrihydrite. Arsenite or As(V) and ligands were added together and at an initial molar ratio of As/ligands = 1 at pH 6.0. (Modified from Zhu, J., M. Pigna, V. Cozzolino, A.G. Caporale, and A. Violante. 2011. Sorption of arsenite and arsenate on ferrihydrite: Effect of organic and inorganic ligands. J. Hazard. Mater. 189:564–571.)

FIGURE 8.5 Effects of various organic complexing ligands on the sorption of Cu by montmorillonite. (Redrawn from McBride, M.B. 1991. Processes of heavy and transition metal sorption by soil minerals, p. 149–176. *In* G.H. Bolt, M.F. De Boodt, M.H.B. Haynes, and M.B. McBride (eds.) Interactions at the soil colloid-soil solution interface. Vol. 190. Kluwer Academic Publishers, Dordrecht, the Netherlands.).

dissolved or dispersed negatively charged complexes with the cations (Violante et al., 2008). McBride (1991) showed that various LMMOAs (phthalate, fulvate, and citrate) strongly inhibit the sorption of Cu (Figure 8.5) by montmorillonite.

However, the presence of certain foreign ligands, occurring naturally in the rhizosphere, such as siderophores produced by microorganisms and phytosiderophores exuded by plants, may promote the sorption of trace elements onto phyllosilicates. They may also help to modify the mobility and then the phytoavailability of trace elements at the soil–root interface. Neubauer et al. (2000, 2002) studied the influence of hydroxamate siderophore desferrioxamine-B (DFOB) on the sorption of Zn onto montmorillonite (Figure 8.6A) and ferrihydrite (Figure 8.6B). DFOB showed a strong affinity for montmorillonite at pH < 7.5 (not shown). At low pH and in the absence of DFOB, only small amounts of Zn were sorbed on montmorillonite (Figure 8.6A), but increasing pH, sorption of Zn increased, and especially at pH > 8.0. In the presence of DFOB (DFOB/Zn molar ratio of 4),

Zn sorption edge was shifted to lower pH values, maximum sorption being reached at about pH 7 (Figure 8.6A). Electrostatic interactions are responsible for enhanced Zn sorption onto montmorillonite. The positively charged Zn–DFOB complexes (MeLH$_2^+$) are the dominant species at about pH 7.

The processes, which affect the sorption of trace element cations onto variable charge minerals in the presence of complexing agents, are particularly complex. Complexing ligands may prevent sorption of trace element cations by forming soluble complexes or by being sorbed onto the surfaces of these minerals blocking some sorption sites or facilitating their sorption, under certain conditions, by decreasing the positive charge of the sorbents. Enhanced sorption has been observed for Cd, Cu, and Zn on Al and Fe oxides in the presence of sulfate or of phosphate (Bolland et al., 1977; Kuo, 1986) due to the increased negative charge brought to the surfaces by these inorganic ligands, whereas high levels of phosphate sorption on a noncrystalline metal oxide inhibited Cu sorption, probably due to blocking of surface sites (McBride, 1985). According to many authors (McBride, 1989; Jackson, 1998, and references therein), some ligands enhance trace element sorption on oxides of Al, Fe, and Si by forming stable surface–metal–ligand complexes (ternary complexes).

Certain siderophores (e.g., DFOB) strongly affect the sorption of heavy metals onto variable minerals, but their behavior is different from that of LMMOAs. Neubauer et al. (2002) found that DFOB strongly prevented Zn sorption on ferrihydrite (Figure 8.6B), because positively charged DFOB complexes were dominant up to pH 8.0. Because the complexes are not attracted by the positively charged ferrihydrite, DFOB decreased sorption of Zn, in contrast to montmorillonite suspensions (as discussed before; Figure 8.6A).

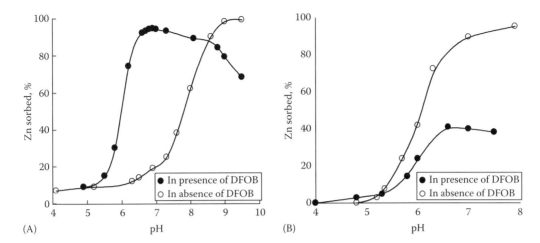

FIGURE 8.6 Sorption of Zn on (A) montmorillonite and (B) ferrihydrite as a function of pH in the absence and presence of DFOB. Dissolved concentration of Zn in montmorillonite suspension was 87.5 μmol L^{-1} without and with 350 mM of hydroxamate siderophore desferrioxamine-B (DFOB), whereas concentration of Zn in ferrihydrite suspension was 10 mmol L^{-1} without and with 300 mM DFOB. (Modified from Neubauer, U., G. Furrer, and R. Schulin. 2002. Heavy metal sorption on soil minerals affected by the siderophore desferrioxamine B: The role of Fe(III) (hydr) oxides and dissolved Fe(III). Eur. J. Soil Sci. 53:45–55.)

Organic and inorganic ligands with a high affinity for Fe, Al, and Mn strongly interact with variable charge minerals and soils and then affect the sorption of trace elements in anionic form by competing for available sorption sites and/or reducing the surface charge of the sorbents (Barrow, 1992). The competition depends on the affinity of the anions for the surfaces of the sorbents as well as the nature and surface properties of the minerals and soils.

Smith et al. (2002) studied the effect of phosphate on the sorption of arsenate and arsenite by an Oxisol, a Vertisol, and two Alfisols. The presence of phosphate (0.16 mmol L^{-1}) greatly decreased arsenate and arsenite sorption by soil containing low amounts of Fe oxides (<100 mmol kg^{-1}: Alfisols. Figure 8.7A and B), indicating competitive sorption between phosphate and arsenate or arsenite for sorption sites. In contrast, the presence of a similar amount of phosphate had a much lower effect on the amount of arsenate or arsenite sorbed by Oxisols with high iron content (>800 mmol kg^{-1}: Figure 8.7C and D).

The influence of different organic and inorganic ligands (sulfate, selenite, selenate, malate, tartrate, and citrate) in preventing the sorption of As(III) and As(V) onto ferrihydrite at pH 6.0 and ligand/As(III) or As(V) molar ratio of 1 has been studied (Figure 8.4). Sulfate and selenate poorly prevented As(III) and As(V) sorption, whereas phosphate (not shown) and selenite showed the highest efficiency. All the ligands showed a stronger capacity in preventing As(III) than As(V) (Figure 8.4). Furthermore, the capacity of the ligands in inhibiting As(III) sorption in comparison to As(V) sorption was much greater when a noncrystalline Al-oxide or a volcanic ash soil containing high amounts of allophane were used as sorbents (not shown).

Balistrieri and Chao (1987) suggested that for a given anion concentration ratio, the competition sequence with selenite on goethite is phosphate > silicate > citrate > molybdate > bicarbonate/carbonate > oxalate > fluoride > sulfate. Dynes and Huang (1997) showed that the ability of 12 LMMOAs to inhibit selenite sorption on poorly crystalline Al-hydroxides was oxalate > malate > citrate > succinate > glycolate > aspartate > salycilate > p-hydroxybenzoate > glycine = formiate = acetate.

There are many reports in published literature showing the effect of competitive anions on oxyanion sorption by LDHs (Goh et al., 2008, and references therein). The effect of competing anions on Cr(VI) and Se(IV) sorption follows the order CO_3^- > Cl^- > water and HPO_4^- > SO_4^{2-} > CO_3^{2-} > NO_3^-, respectively. The influence of different ligands on As(III) and As(V) sorption by LDHs is generally in the order HPO_4^- > SO_4^{2-} > HCO_3^- > F^- > Cl^- > Br^- > NO_3^- and order HCO_3^- > HPO_4^{2-} > SO_4^{2-} > Cl^- (Goh et al., 2008).

As reviewed by Violante et al. (2008), the scientific evidence indicates that the ligand–metal ratio, nature of organic and inorganic ligands and trace elements, and surface properties (surface area, crystallinity, PZC) of the sorbent are apparently critical in determining whether trace element sorption at surfaces is enhanced or inhibited. Future in-depth research on this subject matter at a molecular level is warranted.

8.3.7 Desorption of Metal(loid)s

Sorbed metals and metalloids may be desorbed under favorable conditions, but their desorption is often slow, difficult, and incomplete; these metals and metalloids enclosed in oxide precipitates are probably released only if the precipitates are solubilized. In contrast to sorption studies relatively little information is available on the desorption of metals and metalloids from soils or soil components as affected by organic and inorganic ligands, in spite of the wide occurrence of these reactions in soil environments particularly in the rhizosphere (Hinsinger and Courchesne, 2008; as discussed below).

Many studies have shown biphasic reaction processes for desorption of metals and metalloids (Sparks, 1989, 2003); a fast

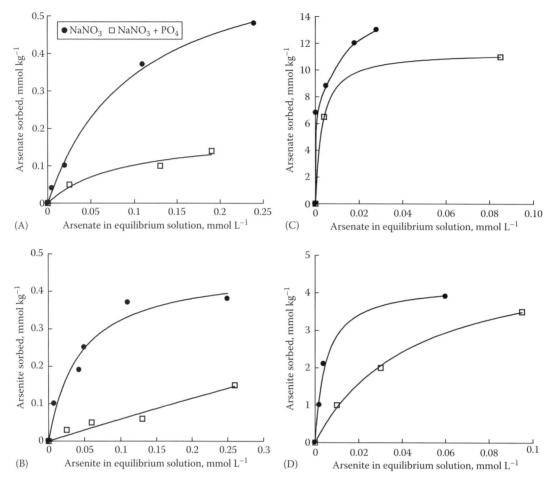

FIGURE 8.7 Arsenate sorption (mmol kg^{-1}) on Alfisol (A) and on Oxisol (C) and arsenite sorption (mmol kg^{-1}) on Alfisol (B) and on Oxisol (D) in the presence of sodium nitrate or sodium nitrate + phosphate (PO$_4$). (Redrawn from Smith, E., R. Naidu, and A.M. Alston. 2002. Chemistry of inorganic arsenic in soils: II. Effect of phosphorus, sodium, and calcium on arsenic sorption. J. Environ. Qual. 31:557–563.)

reaction is followed by a slow reaction. Presence of inorganic ligands, such as phosphate, has a significant impact on the desorption of metals and metalloids. Krishnamurti et al. (1999b) reported the increase in Cd desorption from soils in the presence of phosphate (Table 8.2). The kinetic data on desorption of Cd by phosphate, as related by the amount of Cd released during the initial 30 min reaction period and the overall diffusion coefficients obtained from the desorption kinetics of Cd by 0.1 mol L^{-1} mono ammonium phosphate from the soils reflect well the phytoavailable Cd in the two soils, as shown by the Cd availability index and the grain Cd content of two durum wheat cultivars, Kyle and Arcola, grown on the two soils (Table 8.2).

The amount of contact time between trace elements and soil sorbents (residence time) can dramatically affect the degree of desorption, depending on the metal and sorbent. Pigna et al. (2006) performed kinetic studies on the sorption of arsenate on to crystalline or poorly crystalline metal oxides (noncrystalline Al-hydroxide [Al(OH)$_x$], gibbsite, ferrihydrite, and goethite) and its desorption by phosphate at pH 6.0 as affected by the residence time. By adding phosphate immediately after complete sorption of arsenate onto the oxides (50% surface coverage; PO$_4$ added/AsO$_4$ sorbed molar ratio of 4), a much higher proportion of As(V) was desorbed after 24 h of reaction from Al oxides (48%–56%) than from Fe oxides (18%–23%). The amount of As(V) desorbed decreased with increasing residence time. Figure 8.8 shows the percentage of arsenate desorbed by phosphate from a noncrystalline Al-hydroxide and ferrihydrite as affected by residence time. The kinetics of As(V) desorption by phosphate as a function of residence time can be explained best by Elovich kinetic model (Figure 8.9). The decrease in desorption of arsenate by phosphate with increase in residence time was attributed to the rearrangement of arsenate from desorbable into resistant and undesorbable forms. According to Arai and Sparks (2002), the decrease in desorption of As(V) with increase in residence time could be attributed to different aging mechanisms such as a rearrangement of surface complexes (e.g., from bidentate mononuclear complexes into bidentate binuclear complexes) as well as a conversion of surface complexes into aluminum (or iron) arsenate-like surface precipitates (Jia et al., 2006; Borda and Sparks, 2008). Physicochemical processes may also affect the decrease of As(V) desorption with residence time such as sorption reactions on higher energy binding sites, intraparticle diffusion, and penetration in micropores.

TABLE 8.2 Kinetics of Cd Desorption from Soils by Phosphate

Soil	Cd Desorbed in the Initial 30 min		Overall Diffusion Coefficient (k_d)		CAI[a]	Grain Cd[b]	
	Absence of Phosphate	Presence of Phosphate	Absence of Phosphate	Presence of Phosphate		Kyle	Arcola
	μmol kg^{-1}		μmol kg^{-1} h$^{-0.5}$		μg kg^{-1}		
Jedburgh	5.5×10^{-3}	57.8×10^{-3}	3.2×10^{-3}	7.9×10^{-3}	10	42	28
Luseland	188.0×10^{-3}	1150.0×10^{-3}	46.7×10^{-3}	80.3×10^{-3}	169	384	159

Source: Modified from Krishnamurti, G.S.R., P.M. Huang, and L.M. Kozak. 1999b. Sorption and desorption kinetics of cadmium from soils: Influence of phosphate. Soil Sci. 164:888–898.

Desorption kinetics of Cd on the soils is influenced by 0.1 mol L^{-1} mono ammonium phosphate, described by parabolic diffusion model: $q = a + k_d t^{1/2}$, q is the amount of Cd desorbed in time t (hours) and k_d is overall diffusion coefficient, and a is a constant.

[a] Cadmium availability index: mol L^{-1} NH$_4$Cl-extractable Cd (Krishnamurti et al., 1995a).

[b] Cadmium content in the grain of durum wheat cultivars Kyle and Arcola (Krishnamurti et al., 1999a).

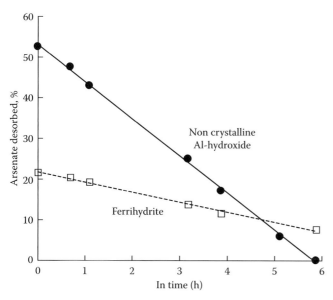

FIGURE 8.8 Arsenate desorbed by phosphate at pH 6.0 from a non-crystalline Al-hydroxide and ferrihydrite. Phosphate was added at initial phosphate/arsenate molar ratio of 4. (Redrawn from Pigna, M., G.S.R. Krishnamurti, and A. Violante. 2006. Kinetics of arsenate sorption-desorption from metal oxides: Effect of residence time. Soil Sci. Soc. Am. J. 70:2017–2027.)

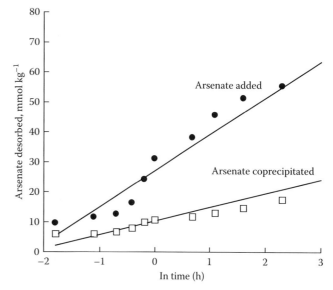

FIGURE 8.9 Kinetics of desorption (Elovich model) of arsenate from 7FeR0.1 (arsenate coprecipitated with iron) and 7FeAR0.1 (arsenate sorbed onto a preformed iron oxide) after addition of phosphate. Phosphate was added at 100% of surface coverage. (Redrawn from Violante, A., M. Pigna, S. Del Gaudio, M. Ricciardella, and D. Banerjee. 2007. Coprecipitation of arsenate with metal oxides: Nature, mineralogy and reactivity of iron(III) precipitates. Environ. Sci. Technol. 41:8275–8280.)

The effect of residence time on Co and Pb desorption from Fe- and Mn-oxides and soil clays has also been studied. Residence time effect is more evident for Co than Pb. In a soil where 2.1% soil organic matter was present, residence time had little effect on the amount of Pb desorbed, but marked hysteresis was observed at all residences times (Sparks, 2003; references therein).

Coprecipitation of metals and metalloids with Al or Fe in soil environments occurs easily. For example, arsenic accumulation in iron plaques (mainly poorly crystalline precipitates; ferrihydrite) of plant roots has been studied (Liu et al., 2006). Unfortunately, few studies have been carried out on the difference in mobility of metals and metalloids sorbed on or precipitated with metal oxides (Violante et al., 2006, 2007, 2009a). Kinetics studies on the sorption/desorption of As(V) from two samples formed at pH 7.0 and As/Fe molar ratio of 0.1 obtained by coprecipitating Fe and As(V) (7FeR0.1) or by adding As(V) immediately after the precipitation of Fe (7FeAR0.1) were performed by Violante et al. (2007). On these two samples, aged 30 days at 50°C, which had similar surface area (147 and 130 m^2 g^{-1}), phosphate was added in order to have a final surface coverage of 100%. The amounts of As(V) desorbed from the two samples were greatly different, being those desorbed from 7FeAR0.1 (As added) greater than those desorbed from 7FeR0.1 (As coprecipitated; Figure 8.9). However, usually low amounts of As(V) coprecipitated with or sorbed onto Al and/or Fe oxides at pH 7.0 were removed by phosphate (5%–25%), attributed to the formation of strong inner-sphere complexes, metal-As(V) precipitates and partial occlusion of As(V) into the coprecipitates.

Finally, desorption of As(V) as well as other oxyanions previously sorbed onto anionic clays (LDH minerals) merits attention. Recently, Violante et al. (2009b) found that more than 80% of As(V) sorbed on a Mg–Al LDH was replaced by phosphate within 15 days of reaction.

8.4 Biomineralization of Metal(loid)s

The formation of minerals by biological processes, that is, biomineralization, is receiving increasing attention (Dove et al., 2003). Microbial biomineralization of metal and metalloids is an important activity of microorganisms. It involves the transformation of metals and metalloids and the subsequent development of fine-grain minerals of tremendous range and kind (McLean et al., 2002; Huang, 2008). Many metalloids bind with various degrees of tenacity to the largely anionic outer surface layers of microbial cells. Some metals are bound to cell walls to a greater extent than by clay minerals, suggesting that bacterial cell walls and membranes may act as foci for accumulation of metals in soils. Important binding sites on the cell walls and membranes apparently include (1) the phosphodiester group of teichoic acids, (2) free carboxyl groups of peptidoglycan, (3) the sugar hydroxyl group of major classes of wall polymers, and (4) the amide groups of peptide chains.

Specific molecular interactions at inorganic–organic interface can result in the controlled nucleation and growth of inorganic crystals (Mann et al., 1993), resulting in biomineral formation and biomimetic synthesis. A central tenet of biomineralization is that the nucleation, growth, morphology, and aggregation (assembly) of the organic crystals are regulated by organized assemblies of organic macromolecules (the organic matrix). Specific processes involving molecular recognition at inorganic–organic interfaces control over the crystallochemical properties of the biominerals. Electrostatic binding or association, geometric matching (epitaxis), and stereochemical correspondence are involved in these recognition processes. Even subtle differences in the kinetics of these recognition processes on different crystal faces may lead to substantial specific changes in crystal morphology.

Although in most environment, soluble metal(loid) ions are present at low concentration, microbial cells, especially bacterial cells that have the highest ratio of surface area to volume have a remarkable ability to concentrate these ions from solutions (McLean et al., 2002). This leads to the precipitation and formation of mineral phases on their cell walls or surface biopolymers. Bacteria surface layers may serve as a nucleation template for metal(loid)s. Bacteria can also initiate mineral formation by producing reactive enzymes, siderophores, and metallothioneins, which bind metals or catalyze their transformations. Moreover, through modification of the chemistry of the microenvironment, bacteria can instigate the precipitation of metal(loid)s (Douglas and Beveridge, 1998).

Microorganisms may produce inorganic ligands such as sulfide and phosphate as cellular metabolic by-products. Sulfide reacts with metals to form metal sulfides, a common reaction in anoxic environments with sulfate-reducing bacteria (Ehrlich, 2002b). Phosphate pumped out of cells can also react with metal ions to form cell-associated and extracellular precipitates (McLean et al., 2002).

Microbial oxidation of Mn(II) is a major process that can produce Mn oxides in soil environments. Microbial formation of Mn oxide coating on soil particles can be 10^5 times faster than can abiotic oxidation (Tebo et al., 1997). Manganese oxides are highly reactive minerals that have the ability to restrict the mobility of metals and influence their bioavailability in soil environment. Biogenic Mn oxides have a significantly larger surface area and higher Pb sorption capacity than those of abiotically formed Mn oxides (Nelson et al., 1999).

Biomineralization should thus have global consequences in the dynamics, transformations, toxicity, and fate of metal(loid)s in soil environments. It should merit close attention in the remediation of metal(loid)-contaminated environments.

8.5 Chemical Speciation and Bioavailability of Metal(loid)s as Influenced by Physicochemical and Biological Interfacial Interactions

The biogeochemical cycle of heavy metals and metalloids has been greatly accelerated by human activities. Accumulation of metal ions and metalloids in different compartments of the biosphere and their possible mobilization under environmentally changing conditions induce a perturbation of both the structure and function of the ecosystem and might cause adverse health effects to biota (Fedotov and Miró, 2008). As a result, heavy metal pollution and the resulting detrimental effects on the environment are widespread throughout the world. A determination of total soil metal content alone is not a good measure of bioavailability and not a very useful tool to quantify contamination and potential environmental and human health risks. Total concentrations of metals in soils are a poor indicator of metal toxicity since metals exist in different solid-phase forms that can vary greatly in terms of their bioavailability (Krishnamurti et al., 1995a, 1995b, 1996; Krishnamurti and Naidu, 2002, 2008; Huang and Gobran, 2005). Most of regulations or guidelines used to protect soil from metal pollution (to set maximum permissible metal concentrations used for definition of contaminated sites, for setting remediation clean-up goals, and for limiting loading of metals to soils in fertilizers and wastes) are still based on assessing the total concentration of metal present in the soil. Recently, some countries have started to introduce the concept of bioavailability in their regulations concerning environmental protection and remediation. The bioavailability of trace metals, their biological uptake, and their ecotoxicological effects on the soil biota can be better understood in terms of their chemical speciation. The mobility and bioavailability and hence potential toxicity of metal in the soil depend on its concentration in soil solution, the nature of its association with other soluble species, and the soil's ability to release the metal from the solid phase to replenish that removed from soil solution by the plants (Christensen and Huang, 1999; Krishnamurti and Naidu, 2002; Huang and Gobran, 2005; Krishnamurti et al., 2007).

A number of chemical equilibrium models are available to predict the solution phase speciation of metals (Loeppert et al., 1995). Solid-phase speciation, however, is particularly important as it determines the rate of replenishment of the metal or metalloid in solution phase and its phytoavailability. The importance of solid-phase speciation has also been stressed in relation to remediation strategies (Krishnamurti, 2000).

8.5.1 Cations

Fractionation of trace metal cations into operationally defined forms under the sequential action of given leachants with increasing aggressiveness is a common approach for distinguishing the operationally defined various species of heavy metals according to their physicochemical mobility and potential bioavailability (Ure and Davison, 2002).

As reported by Fedotov and Miró (2008) there are two well-accepted approaches for the classification of the operationally defined fractions. The first one is based on the specific action of the applied reagents. Hence, fractions can be named as "acid soluble," "easily reducible," "easily oxidizable," and so on. These terms are virtually correct and reflect conditions whereby trace elements associated with certain soil (sediment) components can be released. The second approach is based on the mineralogical phases that are supposed to be dissolved under the action of more or less selective reagents. Therefore, fractions can be termed as "elements co-precipitated with carbonates," "elements bound to Mn oxides," "elements bound to Fe oxides," and so on. Though these names are definitely useful for the interpretation of the fractionation results, in some cases, they might lead to misunderstanding because the reagents applied are not highly selective and might cause the shift or overlapping of fractions.

Although sequential extraction scheme, also termed end-over-end extraction procedure, has been criticized for each harmonization of procedures, insufficient selectivity of a reagent for a given phase, and/or poor reproducibility, sequential extraction procedures (SEP) are an essential tool for distinguishing trace elements of different solubilities related to mineralogical phases.

During the last 30 years, a few dozens of different SEP have been proposed, but the scheme of Tessier et al. (1979) introduced one of the most popular protocols (Table 8.3A). Sequential extraction schemes for metal partitioning in environmental solid samples have been recently summarized by Filgueiras et al. (2002).

A more sophisticated SEP, which significantly improved the specificity and efficiency of extraction, by a carefully designed combination of various extractants in order to identify the specific species contributing to bioavailability was proposed by Krishnamurti et al. (1995a). The differentiation of the metal–organic–complex-bound metal species, as distinct from the other organically bound species, was the innovation in the selective sequential extraction scheme suggested by these authors (Table 8.3B). The particulate-bound metal species in soils were fractionated as exchangeable, carbonate-, metal–organic complex-, easily reducible metal oxide-, organic material-, amorphous mineral colloid-, crystalline Fe oxide-bound and residual. The metal–organic complex-bound was selectively extracted using 0.1 M sodium pyrophosphate (pH 10) as the extractant in the sequential extraction scheme.

Krishnamurti et al. (1995a) determined the distribution of particulate-bound Cd species in a few typical soils of southern Saskatchewan, Canada, following the scheme of Tessier et al. (1979) and their modified scheme (Table 8.3B). They found that following the fractionation scheme of Tessier et al. (1979) (Figure 8.10A), an average of 43.2% of Cd in the soils was observed to be in the form Fe, Mn oxide-bound, whereas following their scheme Cd in these soils was predominantly bound to the metal–organic complexes, accounting for an average of 42.4% of the total Cd present in the soils (Figure 8.10B). Krishnamurti et al. (1995a) also found that the Cd in the surface horizons of temperate soils is predominantly present in the form of metal–organic complex-bound, accounting for 31%–55%, with an average of 40% of the total Cd present in the soils. According to Onyatta and Huang (1999), the metal–organic complex-bound Cd is also generally the highest among the particulate-bound Cd species of the tropical soils, accounting for 25%–46%, with an average of 37% of the total Cd in the soils. However, in certain tropical soils, residual or crystalline Fe oxide-bound Cd is predominant.

TABLE 8.3 Sequential Extraction Scheme

A		B	
Species	Extractant	Species	Extractant
Exchangeable	M MgCl$_2$, pH 7	Exchangeable	M Mg(NO$_3$)$_2$, pH 7
Carbonates	M NaOAc, pH 5	Carbonate-bound	M NaOAc, pH 5
Fe, Mn oxides	NH$_2$OH HCl in 25% HOAc, 95°C	Metal-organic complex-bound	0.1 M Na$_4$P$_2$O$_7$ (pH 10)
Organic	H$_2$O$_2$, pH 2.0, 85°C extract with 3.2 M NH$_4$Oac in 20% HNO$_3$	Easily reducible metal-bound	0.1 M NH$_2$OH HCl in 0.01 M HNO$_3$
Residual	HF-HClO$_4$ digestion	H$_2$O$_2$ extractable organic-bound	30% H$_2$O$_2$, pH 2,0 at 85°C extract with M Mg(NO$_3$)$_2$ in 20% HNO$_3$
		Amorphous oxides	0.2 M NH$_4$Ox/HOX, pH 3 (dark)
		Crystalline Fe oxide-bound	0.2 M NH$_4$Ox/HOX, pH 3 in 0.1 M ascorbic acid; 95°C
		Residual	HF-HClO$_4$ digestion

Sources: (A) Modified from Tessier, A., P.G.C. Campbell, and M. Bissom. 1979. Sequential extraction procedure for the speciation of particulate trace metals. Anal. Chem. 51:844–851; (B) Krishnamurti, G.S.R., P.M. Huang, K.C.J. Van Rees, L.M. Kozak, and H.P.W. Rostad. 1995b. A new soil test method for the determination of plant available cadmium in soils. Commun. Soil Sci. Plant Anal. 26:2857–2867.

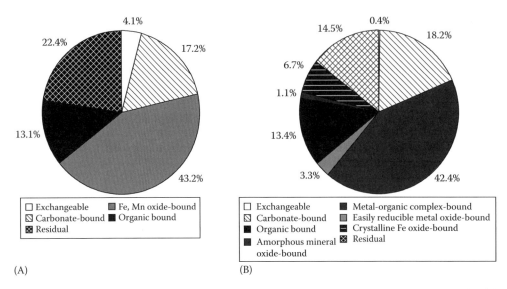

FIGURE 8.10 Average percent distribution of particulate-bound Cd fractions in selected soils of Saskatchewan (Canada), following the method of (A) Tessier et al. (1979) and (B) Krishnamurti et al. (1995a). (Redrawn from Krishnamurti, G.S.R., P.M. Huang, K.C.J. Van Rees, L.M. Kozak, and H.P.W. Rostad. 1995a. Speciation of particulate-bound cadmium of soils and its bioavailability. The Analyst 120:659–665.)

Later, Krishnamurti and Naidu (2000) modified the sequential extraction scheme developed by Krishnamurti et al. (1995a) subfractionating the trace element bound to metal–organic complexes as extracted by 0.1 M sodium pyrophosphate (Table 8.4). This fraction may contain metal associated with both humic and fulvic acid fractions of soil organic matter, which is bound in metal–organic complexes. Following the method suggested by Schnitzer and Schuppli (1989) for separation of metal–fulvate complexes and metal–humate complexes, the metal solubilized by acid (pH 1.0) in the pyrophosphate extracts is termed as metal–fulvate complexes and the metal bound strongly to humic acid and that resist desorption at pH 1.0 is termed as metal–humate complexes. These authors have shown that the modified sequential extraction scheme consisting of eight steps of extractions, subdivides the metal–organic complex-bound Cd into metal–fulvate complex-bound and metal–humate complex-bound Cd.

Therefore, the scheme fractionates the solid components into specific "species" with operationally defined binding mechanisms, that is, exchangeable, specifically adsorbed/carbonate-bound, metal–fulvate complex-bound, metal–humate complex-bound, easily reducible metal oxide-bound, organic site-bound, amorphous metal oxide-bound, crystalline Fe oxide-bound, and residual (aluminosilicate lattice-bound). Their data indicate that the extraction scheme could be used to identify the specific operationally defined "species" contributing to the bioavailability of the trace element. Details of the extraction scheme are presented in Table 8.4.

The distribution of solid-phase fractions of Cu, Zn, and Cd of a few typical surface soils of South Australia, was carried out by Krishnamurti and Naidu (2000) following the fractionation scheme reported in Table 8.4. The trace elements in these soils were dominantly (on an average 40% of Cu, 52.4% of Zn, and

TABLE 8.4 Multistep Selective Sequential Extraction Scheme for Fractionation of Solid Metal Phases

	Target Species	Reagent	Shaking Time and Temperature
Step 1	Exchangeable	10 mL of M NH$_4$NO$_3$ (pH 7)	4 h at 25°C
Step 2	Specifically adsorbed	25 mL of M CH$_3$COONa (pH 5)	6 h at 25°C
Step 3	Metal-organic complex-bound	30 mL of 0.1 M Na$_4$P$_2$O$_7$ (pH 10)	20 h at 25°C
Step 4	Easily reducible metal oxide-bound	20 mL of 0.1 M NH$_2$OH·HCl in 0.01 M HNO$_3$	30 min at 25°C
Step 5	Organic-bound	5 mL of 30% H$_2$O$_2$ (pH 2), 3 mL of 0.02 M HNO$_3$	2 h at 85°C
		3 mL of 30% H$_2$O$_2$ (pH 2), 1 mL of 0.02 M HNO$_3$	2 h at 85°C
		Cool, add 10 mL 2M NH$_4$NO$_3$ in 20% HNO$_3$	30 min at 25°C
Step 6	Amorphous mineral colloid-bound	10 mL of 0.2 M (NH$_4$)$_2$C$_2$O$_4$/0.2 M H$_2$C$_2$O$_4$ (pH 3)	4 h at 25°C (dark)
Step 7	Crystalline Fe oxide-bound	25 mL of 0.2 M (NH$_4$)$_2$C$_2$O$_4$/0.2 M H$_2$C$_2$O$_4$ (pH 3) in 0.1 M ascorbic acid	30 min at 95°C
Step 8	Residual	Digestion with HF-HClO$_4$	

Source: Krishnamurti, G.S.R., and R. Naidu. 2000. Speciation and phytoavailability of cadmium in selected surface soils of south Australia. Aust. J. Soil Res. 38:991–1004. Copyright 2000, with permission from Elsevier.

A measured quantity of 30 mL of 0.1 M Na$_4$P$_2$O$_7$ extract was brought to pH 1.0 with the addition of 6 M HCl and the suspension was left overnight for the coagulation of humic acid. The suspension was centrifuged at 12,000g for 10 min. Metal-fulvate complexes were determined in the supernatant. The residue was solubilized with 0.1 M Na$_4$P$_2$O$_7$ and the metal-humate complexes were determined in the solution.

33.4% of Cd) associated with the alumino-silicate mineral lattices, identified as residual fraction in the scheme, followed by the fraction associated with organic sites (on an average 32.4% of Cu, 28.0% of Zn, and 28.5% of Cd) (see Krishnamurti and Naidu, 2008, Figures 11.3 and 11.4).

The mobility and plant uptake of trace elements proceed through the solution phase. However, the plant uptake of an element depends not only on its activity in the solution, but also on the relation that exists between solution ions and solid-phase ions.

An attempt was made by Krishnamurti and Naidu (2002) to understand the importance of solid-phase fractions in assessing phytoavailability of Cu, Zn, and Cd using multiple regression analysis. Phytoavailable Cu, Zn, and Cd were found to be significantly correlated with the metal–fulvate complex-bound Cu ($r = 0.944$, $p < 0.0001$), exchangeable Zn ($r = 0.832$, $p = 0.002$) and the metal–fulvate complex-bound Cd ($r = 0.824$, $p = 0.002$). It was observed that fulvic complex Cu could explain 89% of variation in phytoavailable Cu, whereas the metal–fulvate complex-bound element together with exchangeable element could explain 79% and 92% of variation in phytoavailable Zn and Cd, respectively. Inclusion of solution element concentration in the regression analysis was not found to improve the predictability of phytoavailable element. The regression analysis indicated that the phytoavailable Cu, Zn, and Cd in these soils are mainly from solid-phase fractions.

A number of extractants have been suggested for estimating phytoavailability indices of metals. One of the most commonly used extractant for Cu is the DTPA-TEA soil test (Lindsay and Norvell, 1978) for near-neutral and calcareous soils. Jackson and Alloway (1991) suggested that neutral salts might be the best means available for predicting the bioavailability of trace elements. Krishnamurti et al. (1995a, b) and Krishnamurti (2000) compared a number of extractants for assessing the phytoavailability of Cd and found that M ammonium chloride (NH_4Cl)–extractable Cd had the best correlation with grain Cd of durum wheat.

Recently, Krishnamurti et al. (2007) have studied the mobility and the phytoavailability of cupper on wheat durum (*Triticum durum*) grown in polluted and unpolluted Italian soils. The study was conducted to determine the solid phase distribution of copper in representative soils of Italy with wide differences in the chemical and physicochemical properties. Sites selected varied in location as well as in current vegetation and land use. Samples with a high load of Cu (132–253 mg kg^{-1}) from a vineyard cultivation area were also included to study the effect of contamination on the distribution of Cu among solid phases. The solid phase fractionation of Cu in the soils was determined using an eight-step selective sequential extraction method (Table 8.4). The results indicated that Cu was dominantly associated with organic binding sites 62.6%–74.8%. The relative importance of the solid-phase fractions in assessing the Cu phytoavailability by durum wheat in a greenhouse setting and the effectiveness of two soil tests, the DTPA-TEA and NH_4Cl extraction method, for predicting the phytoavailable Cu was studied. Most of Cu was retained by roots with very limited translocation to the upper plant parts of wheat. A significant correlation ($r = 0.960$, $p = 0.0001$) was found between the plant Cu content and the Cu associated with the metal–fulvate complexes, indicating that the phytoavailable Cu was mainly from the metal–fulvate complexes. The contaminated soils had a significantly higher proportion (77%) of Cu associated with organic binding sites, in comparison to that of uncontaminated soil (21.3%), resulting in higher proportion of phytoavailable Cu.

8.5.2 Anions

While there are a large number of SEPs available for metal cations, only limited work has been done on the sequential extraction scheme for trace elements in anionic form. Wenzel et al. (2001) have developed an innovative and simple method for arsenic sequential extraction, which provides the following five extraction steps with reagents gradually stronger (Table 8.5):

1. $0.05\,M\ (NH_4)_2SO_4$, 20°C—4 h
2. $0.05\,M\ NH_4H_2PO_4$, 20°C—16 h
3. $0.2\,M\ NH_4^+$-oxalate buffer in the dark. pH 3.25, 20°C—4 h
4. $0.2\,M\ NH_4^+$-oxalate buffer + ascorbic acid pH 3.25, 96°C—0.5 h
5. HNO_3/H_2O_2 microwave digestion

These five as fractions appear to be primarily associated with

1. Nonspecifically sorbed
2. Specifically sorbed
3. Amorphous and poorly crystalline hydrous oxides of Fe and Al
4. Well crystallized hydrous oxides of Fe and Al
5. Residual phases

TABLE 8.5 Sequential Extraction Procedure for Arsenic

Fraction	Extractant	Extraction Conditions	SSR	Wash Step
1	$(NH_4)_2SO_4$ (0.05 M)	4 h shaking, 20°C	1:25	
2	$(NH_4)H_2PO_4$ (0.05 M)	16 h shaking, 20°C	1:25	
3	NH4-oxalate buffer (0.2 M); pH 3.25	4 h shaking in the dark, 20°C	1:25	NH_4-oxalate (0.2 M); pH 3.25 SSR; 1:12.5 10 min shaking in the dark
4	NH4-oxalate buffer (0.2 M); + ascorbic acid (0.1 M) pH 3.25	30 min in a water basin at 96°C ± 3°C in the light	1:25	NH_4-oxalate (0.2 M); pH 3.25 SSR; 1:12.5 10 min shaking in the dark
5	HNO_3/H_2O_2	Microwave digestion	1:50	

Source: Wenzel, W.W., N. Kirchbaumer, T. Prohaska, G. Stingeder, E. Lombi, and D.C. Adriano. 2001. Arsenic fractionation in soils using an improved sequential extraction procedure. Anal. Chim. Acta 436:309–323. Copyright 2001, with permission from Elsevier.

These authors conducted a study using 20 Austrian soils differing in the level of As contamination (from 96 to 218 mg As kg^{-1}) and soil characteristics. The results showed that As was most prevalent in the two oxalate-extractable fractions (30% from step 3 and 27% from step 4), indicating that As is primarily associated with amorphous and crystalline Fe and Al oxides. The fraction of As extracted by $NH_4H_2PO_4$ represented about 10% of total As and may be useful in providing a relative measure of specifically sorbed As in soils that may be potentially mobilized due to changes in pH or P addition. The amount of readily labile As extracted by $(NH_4)_2SO_4$ is generally small (0.3%), but may represent the most important fraction related to environmental risks and has been shown to correlate well with As concentrations in field-collected soil solutions.

Recently, Branco (2008), using the method of Wenzel et al. (2001), have studied the sequential extraction of As from polluted Italian soils collected in Scarlino (Tuscany, Italy), which showed a content of arsenic ranging from 100 to 190 mg As kg^{-1}. Figure 8.11 shows the As distribution in La Botte soil. Arsenic was in the most part recovered in the crystalline oxides (59.8%) and short-range Fe and Al oxides (20%). The As fraction extracted with $NH_4H_2PO_4$ was about 7%, whereas the nonspecifically sorbed (easily exchangeable) fraction that forms outer-sphere complexes onto the mineral surfaces was negligible (<0.2%). The scarce residual As fraction (11%–13%) suggested a low presence of primary minerals rich in this metalloids.

Martin et al. (2007) also used this sequential extraction method for studying the accumulation and potential release of As in a paddy field in Bangladesh irrigated with As-contaminated groundwater. The oxalate-extractable fraction related to amorphous hydrous oxide-bound As represented the dominant As form (47%), and a relatively high amounts of As was removed by phosphate (22%).

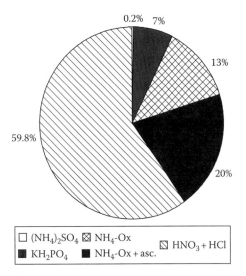

FIGURE 8.11 Arsenate fractionation from one Italian polluted soil (La Botte soil containing 190 mmol kg^{-1} of As), following the method of Wenzel et al. (2001). (Redrawn from Branco, A. 2008. Mobility and bioavailability of arsenic in Italian polluted soils. PhD Thesis, University of Naples (Italy), Federico II, 122pp.)

8.6 Plant Uptake of Metal(loid)s as Influenced by Physicochemical and Biological Interfacial Interactions

8.6.1 Effects of Plants and Managements

Plants require a range of trace elements (e.g., Cu, Mn, Zn, Mo, Ni) as essential micronutrients for normal growth and development (Kabata-Pendias, 2004). These elements have been acquired in the course of evolution because of their chemical properties such as redox activity under physiological conditions (Cu, Fe) or Lewis acid strength (Zn) (Clemens, 2006). When any of these elements are present in short supply, a range of deficiency symptoms can appear and growth is reduced (Marschner, 1995). However, although essential, these elements can also be toxic when present in excess with the production of reactive oxygen species and oxidative injury being particularly important (Schützendübel and Polle, 2002). The physiological range for essential transition metals between deficiency and toxicity is therefore extremely narrow and a tightly controlled metal homeostasis network to adjust to fluctuations in micronutrient availability is a necessity for all organisms (Clemens, 2006, and references therein). Plants can encounter elevated levels of both essential and nonessential metals. Of major concern with respect to plant exposure as well as food chain accumulation are the metalloids (As and Se), and the metals Cd, Cr, Hg, and Pb (McLaughlin et al., 1999).

Excessive levels of many metal(loid)s can result in soil quality degradation, crop yield reduction, and poor quality of agricultural products, consequently posing significant hazards to human, animal, and ecosystem health. Therefore, the bioavailability of essential and nonessential metal(loid)s has been one of the most crucial problems in agricultural and environmental studies. There has been a steady increase in the number of investigations related to both understanding the processes involved in the uptake of an element (nutrient and non-nutrient) by plants, and to finding the most reliable methods for the prediction of availability of a given element to plants, and in particular to crop plants. Physical–chemical soil parameters such as texture, pH, redox status, content, and quality of mineral components and organic matter modify bioavailability of metal(loid)s (Adriano, 2001). Furthermore, metal(loid)s accumulation in plant tissues is influenced by plant-inherent properties and differ between species, ecotypes, and plant parts (Cobb et al., 2000). Plants differ also in specific root-response mechanisms to environmental conditions such as nutrient availability, bulk density, water content, and presence of toxic elements (Fitz and Wenzel, 2006). The composition of root exudates can greatly change under any environmental stress (Kabata-Pendias, 2004). Plants passively and/or actively modify the rhizosphere, the narrow zone at the soil–root interface. This results in altered chemical, physical, and biological characteristics of the rhizosphere relative to bulk soil. The spatial extension of the rhizosphere can be greatly enlarged by mutualistic symbiosis between plants and fungi, known as mycorrhizas (Marschner, 1995). These plant-induced changes in the rhizosphere may influence the bioavailability of metal(loid)s (Hinsinger and Courchesne, 2008, and references therein).

Metal(loid)s can be depleted or accumulated in the rhizosphere along the gradient that increases or decreases from the root toward the bulk soil, depending on whether their rate of uptake is, respectively, greater or lower than the rate of supply by diffusion and mass flow to the soil–root interface (Marschner, 1995). Besides this effect, the concentration gradients measured in various solid-phase metal(loid) fractions at the soil–root interface can also be attributed to the effects of a variety of processes, such as rhizosphere acidification or alkalinization, adsorption or desorption reactions, and precipitation or dissolution phenomena, which are themselves associated with plant uptake and a range of root activities (Hinsinger, 2001).

Changes in rhizosphere pH values have various origins and can severely alter the speciation of most metal(loi)s both in soil solutions and in the solid phase. Factors affecting rhizosphere pH are the source of nitrogen supply (NO_3^- vs. NH_4^+ uptake), nutritional status of plants (e.g., Fe and P deficiency), excretion of organic acids, CO_2 production by roots and microorganisms, and the buffering capacity of the soil (Marschner, 1995).

Changes in the solubility of Cu in the rhizosphere soil solution was shown in tomato plants grown in an acidic-contaminated soil as a consequence of rhizosphere alkalinization (Hinsinger and Courchesne, 2008, and references therein). Acidification of the rhizosphere of tobacco (*Nicotiana tabacum* L. cv. "SR1") increased Zn in the exchangeable fraction of Zn-contaminated soil (Hinsinger and Courchesne, 2008, and references therein). The behavior of oxyanions such as Mo or As is known to be fairly dependent on pH, via its effect on the surface charge of solids and ultimately on the adsorption–desorption of these oxyanions. For instance, rhizosphere acidification that occurred as a consequence of P deficiency in canola (*Brassica napus* L.) grown on As-contaminated kaolinite resulted in enhanced bioavailability of both phosphate and arsenate, due to enhanced desorption of these oxyanions (Quaghebeur and Rengel, 2004).

Increasing soil water content is associated with the decrease of soil redox potential, which induces various chemical changes in the soil solution due to the decreasing availability of O_2. The oxidation of the rhizosphere is a well-known phenomenon that has been extensively described for lowland rice among agricultural plants. In waterlogged or submerged soils, the leakage of O_2 from the roots, as a consequence of its transfer from the shoots through the aerenchyma (Fitz and Wenzel, 2006), provides an adequate supply of O_2 for rice root respiration. It also enables rice plants to alleviate metal toxicity that can occur due to the ambient reducing conditions of the bulk soil and much increased solubility of metal (Fe and Mn, in particular) oxides under such circumstances (Hinsinger, 2001). Besides, these minerals are known to play an essential role in the fate of metal (loid)s in soils as a consequence of adsorption or coprecipitation processes. Several reports showing the role of the iron plaque, consisting mainly of ferrihydrite, in the sequestration of As clearly demonstrate that root-induced changes of redox conditions can have a dramatic effect on the solid-phase speciation of As in the vicinity of roots (Fitz and Wenzel, 2006).

Oxidation of arsenite, the dominant dissolved species in soil of rice plants, to arsenate favors strong adsorption onto the positively charged iron oxides that form the iron-plaque on roots of lowland rice, although there are some evidence that arsenite adsorption is almost similar to arsenate onto the Fe oxides, particularly onto ferrihydrite (Smedley and Kinniburgh, 2002). In contrast to the above, for those plants living under oxic soil conditions, the availability of Fe is a major constraint given the poor solubility of Fe oxides. The reduction of Fe is a major mechanism involved in the acquisition of Fe by most higher plants (Marschner and Römheld, 1996), where the acquisition of Fe is based on a combination of root-induced acidification and enzymatic Fe reduction (Marschner, 1995). A plasma membrane-bound reductase system operates the reduction of Fe^{3+} to Fe^{2+}, which is supposed to be the only available Fe species in all higher plants except for the monocotyledons. The same is likely true for Mn (Hinsinger, 2001). The resulting root-induced changes of reduction of Fe and Mn are expected to shift the dissolution/precipitation equilibria of Fe and Mn oxides and promote their dissolution. This should also lead to the concomitant release of metal(loid)s commonly associated (sorbed or included) with these minerals in soil environments. Furthermore, these redox processes are also expected to play a role in mobilization/immobilization processes involving those metal(loid)s that occur in various oxidation states in soil conditions, such as Cr, Cu, and Se (Hinsinger, 2001). The release by plants of organic C in the rhizosphere, rhizodeposition, represents a significant proportion of the net photosynthetic C. The main components of rhizodeposition are LMMOAs, sugars and polysaccharides, proteins, and enzymes. The LMMOAs are a group of compounds that include carboxylic acids (e.g., citric, malic, oxalic), amino acids (e.g., cysteine and glutamine), and phenolic acids (e.g., caffeic). Rhizodeposition is strongly influenced by exogenous conditions and stresses (Hinsinger and Courchesne, 2008). Nutrient deficiency in the soil, mechanical constraints, and limited water supply all stimulate the release of exudates by plants (Nguyen, 2003). These compounds can play a decisive role in controlling metal(loid)s mobilization and hence their bioavailability. Recent studies have elucidated the role of root exudates in the dissolution of Fe oxides (Hinsinger and Courchesne, 2008). They have shown that Fe-complexing compounds can efficiently dissolve Fe from poorly crystalline Fe oxides and goethite, which is a ubiquitous crystalline oxyhydroxide in soils. In Graminaceous species, the phytosiderophores are released by plants in conditions of Fe deficiency as they are involved in the Fe complexation, but they can form chelates with other metals such as Zn, Cu, and Mn and are therefore able to mobilize other micronutrients in case of deficiency (Treeby et al., 1989). There is evidence that in graminaceous plants Fe^{3+} and Zn^{2+} can be taken up in the form of nondissociated Fe- or Zn-phytosiderophores (Thomas et al., 2005). Other data suggested that uncharged metal–organic complexes can cross cell membranes (Huang and Germida, 2002). Malinowski

TABLE 8.6 Total As Concentration (μg kg^{-1}) and Content in Roots, Shoots, and Grain (μg Pot^{-1}) of Wheat (*T. durum* cv Creso) Exposed to Three As Concentration in Irrigation Water

As Conc. (mg L^{-1})	As Concentrations			As Content			As Root/Shoot
	Roots	Shoots	Grain	Roots	Shoots	Grain	
	mg kg^{-1}			μg Pot^{-1}			
Control P−	0.18 ± 0.05g	0.15 ± 0.04e	0.05 ± 0.03e	0.32 ± 0.02e	1.83 ± 0.06g	0.42 ± 0.02d	1.20d
0.5	0.63 ± 0.02f	0.43 ± 0.01d	0.24 ± 0.02a	0.52 ± 0.04d	5.12 ± 0.10d	1.38 ± 0.05b	1.47d
1	1.10 ± 0.03e	0.44 ± 0.02d	0.19 ± 0.02bc	0.63 ± 0.05cd	3.30 ± 0.09e	0.57 ± 0.04c	2.53c
2	2.06 ± 0.05b	0.62 ± 0.02bc	0.20 ± 0.03ab	0.78 ± 0.05c	2.85 ± 0.06f	0.44 ± 0.02d	3.33b
Control P+	0.22 ± 0.04g	0.15 ± 0.02e	0.05 ± 0.04e	0.52 ± 0.04d	1.95 ± 0.07g	1.06 ± 0.05c	1.46d
0.5	1.40 ± 0.04d	0.47 ± 0.03d	0.11 ± 0.02d	3.22 ± 0.08b	5.83 ± 0.13c	1.94 ± 0.07a	2.98bc
1	1.70 ± 0.05c	0.56 ± 0.04c	0.12 ± 0.03d	3.32 ± 0.10b	6.61 ± 0.12b	2.03 ± 0.12a	3.04bc
2	3.20 ± 0.05a	0.65 ± 0.04a	0.14 ± 0.01cd	5.12 ± 0.13a	7.28 ± 0.15a	2.00 ± 0.14a	4.92a
As level	$p < 0.0001$	$p < 0.0001$	$p < 0.0001$	$p < 0.0001$	$p < 0.0001$	$p < 0.0001$	$p < 0.0001$
P fertilization	$p < 0.0001$	$p < 0.0001$	$p < 0.0001$	$p < 0.0001$	$p < 0.0001$	$p < 0.0001$	$p < 0.0001$
Interaction	$p < 0.0001$	$p < 0.01$	$p < 0.05$	$p < 0.0001$	$p < 0.0001$	$p < 0.0001$	$p < 0.0001$

Source: With kind permission from Springer Science + Business Media: Pigna, M., V. Cozzolino, A. Violante, and A. Meharg. 2009. Influence of phosphate on the arsenic uptake by wheat (*Triticum durum* L.) irrigated with arsenic solutions at three different concentrations. Water Air Soil Pollut. 197:371–380.
P− and P+ stand respectively for without phosphate fertilization and with phosphate fertilization.

et al. (2004) also reported that Cu was bound to extracellular root exudates, as shown by lower dissolved Cu^{2+} activity in the rhizosphere of tall fescue (*Festuca arundinacea* Schreb.) grown in a nutrient solution. Besides their impact on the dissolution of metal-bearing mineral phases and on the complexation of metals, organic substances released by roots can also have an impact on the sorption or desorption of metals at the surface of mineral particles. Organic substances released by roots can compete with sorbed anionic species of metals (e.g., chromate) and metalloids (arsenate) and displace them from the exchange complex of soils. On the other hand, soluble exudates can enhance metal sorption through the formation of metal–organic complexes, especially in the lower pH range (Neubauer et al., 2002). In addition, organic compounds released by roots and involved in the complexation of phytotoxic metals have also been reported to reduce the contamination of the plants, such as in the case of Al detoxification by malate or citrate (Ma et al., 2001). Dessurealt-Rompré et al. (2008) indicated that in *Lupinus albus*, the cluster roots exudation leads on one hand to direct mobilization and complexation of metals like Al, Fe, and Zn by citrate and on the other hand to the mobilization of soil organic matter that complexes and solubilizes Cu and Pb. The complexation of metals by organic substances in the rhizosphere can thus have an entire range of effects and either increase or decrease their bioavailability and phytotoxicity (Hinsinger and Courchesne, 2008). Similar to carboxylic acids released by plant roots, inorganic anions such as phosphate may have a significant impact on the sorption/desorption of metal(loid)s in soil environments. Phosphate additions increase extractable fractions of As in batch experiments (Peryea, 1998) and reduced sorption of arsenate and arsenite onto soils (Smith et al., 2002; Violante and Pigna, 2002). Plant uptake of As has been shown to increase upon P application. Peryea (1998) reported increased As solubility and, thus, its phytoavailability on P-fertilizer application to soils. In wheat plants, Pigna et al. (2009) found that phosphate addition increased As root concentrations but reduced the translocation of As to shoots and grain (Table 8.6).

Rhizosphere microorganisms are key players in the biogeochemistry of metal(loid)s in the rhizosphere. Among them, a prominent role is played by arbuscular mycorrhizal fungi (AMF), which will be discussed in the next section. For future scientific pursuits, it will be necessary to develop a more holistic approach to studying rhizospheric interactions of physical, chemical, and biological processes by integrating advances in microbial and plant ecology and physiology as well as in soil biogeochemistry in order to better predict and eventually manage the mobility and bioavailability of metals and metalloids at soil–root interface.

8.6.2 Effect of Arbuscular Mycorrhizal Fungi on the Transfer of Metal(loid)s to Plants

Soil microorganisms are affected by the presence of high metal concentrations in soil (Gadd, 2005), but the organisms in turn influence the availability of metals in soil either directly, through alterations of pH, Eh, biosorption and uptake, or indirectly in the rhizosphere through their effect on plant growth, root exudation, and the resulting rhizosphere chemistry (Leyval and Joner, 2001).

AMF, phylum Glomeromycota (Schüssler et al., 2001), are one of the most widespread soil microorganisms (Smith and Read, 2008), and constitute an important functional component of the soil–plant system occurring in almost all habitats and climates (Barea et al., 1997), including disturbed soils (McGonigle and Miller, 1996). They form symbiosis with more than 80% of terrestrial plants (Smith and Read, 2008). There has been much interest in the potential use of AMF in agricultural systems by focusing on their ability to improve plant growth through enhancement of nutrient uptake. Many experiments have demonstrated increased plant uptake of macronutrients P (Smith and Read, 2008) and N (Govindarajulu et al., 2005) as well as micronutrients Cu (Li et al., 1991) and Zn (Bürkert and Robson, 1994) by AMF. The mycorrhizal effect can be explained by mechanisms such as enlargement of the absorption area and volume of accessible soil by extraradical hyphal network, which extends from root surface into soil interstices not accessible to roots and the formation of interfaces (arbuscules, coils, and intercellular hyphae) by the intraradical mycelium, which function as nutrient exchange sites between symbionts (Smith and Read, 2008). They form a direct link between the soil and the root, providing a route for the movement of elements in the mycorrhizosphere.

Inoculation with AMF has been repeatedly reported to modify heavy metals uptake by host plants, but the results are contradictory (Göhre and Paszkowski, 2006). In general, AMF plants are more efficient than nonmycorrhizal in the acquisition of micronutrients such as copper, iron, manganese, and zinc when available at low concentrations (Vosatka et al., 2006). However, when grown in excess of micronutrients or in soils contaminated by metalloids or heavy metals with unknown biological function such as Cd, Cr, or Pb, AM plants can show enhanced metals and metalloids uptake and root-to-shoot transfer (phytoextraction, reviewed by Göhre and Paszkowski, 2006; crop biofortification reviewed by Cavagnaro, 2008) while in other cases AMF contribute to heavy metal immobilization in the soil via fungal metal-binding processes, thereby improving phytostabilization (Joner et al., 2000). Audet and Charest (2007) conducted a literature survey and correlated heavy metal uptake and relative plant growth parameters from published data. They concluded that AM symbiosis increases heavy metal phytoextraction via mycorrhizospheric "enhanced uptake" at low soil heavy metal concentrations, and reduces heavy metals bioavailability via AM fungal "metal-binding" processes at high soil heavy metals levels, hence resulting in increased biomass and enhanced plant tolerance through heavy metals stress avoidance. As predicted by "enhanced uptake" hypothesis, the greater volume of the mycorrhizosphere, compared to the rhizosphere alone, provides an increased access to soil resources, including macro, micro, and even nonessential elements. According to the above-cited study, the mycorrhizospheric impact, accounting for nearly 200% greater heavy metals uptake in AM than non-AM plants at the low soil–heavy metal interval, is likely the result of active soil–heavy metals transport to the roots via the extraradical hyphal network (Burleigh et al., 2003; Audet and Charest, 2007), while at high soil–heavy metal interval, the registered nearly 100% lower heavy metals uptake in AM compared with non-AM plants is due to fungal sequestration before transfer to the plant (Joner et al., 2000). The fungus employs strategies similar to those of its host. Among these are immobilization of metals by compounds secreted by the fungus, precipitation in polyphosphate granules in the soil, adsorption to fungal cell walls, and chelation of metals inside the fungus. Glomalin is an example of an insoluble glycoprotein that is produced and released by AMF and binds heavy metals in the soil (Gonzales-Chavez et al., 2004; Göhre and Paszkowski, 2006). Gonzales-Chavez et al. (2004) found that up to 4.3 mg Cu, 0.08 mg Cd, and 1.12 mg Pb g^{-1} glomalin could be extracted from polluted soil that had been inoculated with laboratory cultures of AMF. Passive adsorption of heavy metals to chitin in the fungal cell wall reduces also their local concentrations in the soil. Joner et al. (2000) found 0.5 mg Cd per mg dry biomass. The uptake of Pb and its immobilization were found to be higher in roots of mycorrhizal than nonmycorrhizal plants (Chen et al., 2005). The sequestration of Pb in the roots was found to be correlated with an increase in the number of fungal vesicles in highly colonized species. Similar to plant and fungal vacuoles, fungal vesicles may be involved in storing toxic compounds and thereby could provide an additional detoxification mechanism. In soils heavily contaminated with heavy metals and deficient in nutrients, where nonmycorrhizal plant died, the *Glomus* isolate Br1 obtained from roots of *Viola calaminaria*, a tolerant plant, allowed plant like maize, barley, and alfa–alfa to grow and complete their life cycle. A nonadaptated isolate of *Glomus intraradices* also permitted growth but to a lesser extent (Hildenbrandt et al., 1999). An explanation for this observation is offered by another study in maize where it has been found that heavy metals are selectively retained in the inner parenchyma cells coinciding with fungal structures (Kaldorf et al., 1999). A massive accumulation of heavy metals in inner cortical cells of AM colonized maize roots where arbuscules and intraradical hyphae were located was also confirmed using different microbeam techniques (Dosskey and Adriano, 1993). In addition, Wang et al. (2008) showed that in mycorrhizal maize plants As concentrations in external mycelium of two AMF, *G. mosseae* and *Acaulospora morrowiae*, increased greatly with increasing As addition level in the growth media, highlighting that the deposition in the external mycelium as the possible role of AMF in the detoxification of As in the host plants. Interestingly, in some cases, mycorrhizal fungi may enhance metal(loid)s resistance in the host plants by effluxing the toxic elements to the external medium. This mode of action was clearly demonstrated by Sharples et al. (2000) who reported that the ericoid mycorrhizal fungus *Hymenoscyphus ericae* from an As-contaminated site developed arsenate resistance by extruding arsenite into the medium following arsenate uptake, in a similar way to bacteria and yeasts (Rosen, 1999). Apart from the immobilization of metal(loid)s in fungal structure and products, the colonization by AMF can lead to increased uptake and subsequent accumulation of metal(loid)s in above-ground tissues. For example, it was found that addition of AMF further enhances the uptake and accumulation of As in hyperaccumulator fern *Pteris vittata* (Leung et al., 2006). At the highest As concentration tested (100 mg kg^{-1} soil), noncolonized plants accumulated 60.4 mg As kg^{-1} while colonized by AMF isolated from an As mine accumulated 88.1 mg As kg^{-1}. This was accompanied by enhanced growth, possibly due to improved phosphate nutrition reaching 36.3 mg pot^{-1} in non-colonized and 257 mg pot^{-1} in colonized plants. Both effects combined allow for higher recovery of metal(loid)s.

The AMF are known to change the chemical composition of root exudates and to affect the activities of soil enzymes, and to introduce physical–chemical modifications to the environment surrounding the roots (Smith and Read, 2008) that can influence the status and availability of metal(loid)s in the rhizosphere. These AMF-induced changes in the rhizosphere and the development of AMF mycelia, which can act as a nutrient source for other microorganisms, are likely to influence the structure and functioning of soil microbial communities in the mycorrhizosphere. Specific modification in the environment between the plant roots and the mycorrhizosphere, such as changes in pH and redox potential, may promote the biotransformation processes of metal(loid)s (Jeffries et al., 2003). Ultra et al. (2007) reported the presence of methylated As species in the rhizosphere of sunflower (*Helianthus annuus*) plants, grown on As-polluted soil, when inoculated with the AMF *G. aggregatum*. Dimethylarsinic acid (DMMA) was found at low but detectable concentrations in the rhizosphere soil but not in the bulk soil, nor in the rhizosphere soil of uninoculated plants. Mycorrhizal fungi may be able to mediate As biomethylation and release some of the methylated As in the rhizosphere. However, mycorrhizal effects on plant uptake of metal(loid)s can vary depending on certain factors: the fungal genotype, the metal and concentration and availability of metal(loid)s in soil; the plant species and ecotype; and soil edaphic conditions, including soil fertility (Leyval and Joner, 2001). Reduced metal(loid) concentrations in plant tissues of mycorrhizal plants and translocation from plant roots to shoots can be an important factors for quality and safety of plants that enter food chains such as forage crops, medicinal herbs, and vegetables.

The appropriate management of native fungi and/or application of commercial inocula should be chosen according to the cases. However, it should be considered that the effect of mycorrhizal inoculation may interact with fertilization regime as demonstrated for the plants grown in mine tailing containing high levels of Cr (Vosatka et al., 2006, and references therein) or As (Cozzolino et al. 2010). In both cases, plant growth was best after inoculation combined with nitrogen and phosphorus fertilization, whereas neither mycorrhiza nor fertilization alone had any effect on plant biomass. Therefore, mycorrhizal inoculation cannot be considered as panacea and should be combined with other practices such as appropriate fertilization to maximize plant growth and revegetation successes.

8.7 Conclusions and Future Research Prospects

Soil constituents, organic matter, phyllosilicates, variable charge minerals, and microorganisms govern interactive physical, chemical, and biological processes and affect the mobility, bioavailability, and toxicity of metals and metalloids in soil environments.

The sorption/desorption processes of trace elements on/from soil components is affected by many factors, such as pH, nature of the sorbents, redox reactions, presence, and concentration of organic and inorganic ligands, including humic and fulvic acid, root exudates, and nutrients. The behavior of foreign ligands on the sorption of toxic elements in cationic form is quite different from that toward elements in anionic form. In fact, complexation reaction of trace elements in cationic form with organic and inorganic ligands have an important role on their sorption/desorption processes as well as their toxicity and phytoavailability, whereas competition for available sites and/or reduction of the surface charge of the sorbents between foreign ligands and trace elements in anionic form mainly affect their mobility.

The factors that affect the toxicity or bioavailability of metal(loid)s in soil are still obscure. The dynamics and mechanisms of transformation of metal(loid)s in soil environments and the impact of their mobility, phytoavailability, toxicity, food chain contamination, and ecosystem health should be of increasing concern. Leaching and extraction tests are widely used for assessing trace element mobility and phytoavailability, but the lack of information on speciation and bioavailability in soil arises from the lack of both appropriate techniques for in situ chemical speciation and bioavailability in the field and analytical tools for speciation or fractionation of metal species. Many researchers continue to compare published data without paying much attention to the technique used during speciation. There is therefore a need to standardize speciation and techniques for assessing bioavailability (Krishnamurti and Naidu, 2008). Detailed studies in this area are needed.

Acknowledgments

This study was supported by the Italian Research Program of National Interest (PRIN 2006) and Discovery Grant 2838 of the Natural Sciences and Engineering Research Council of Canada. This paper is dedicated to the memory of Dr. P.M. Huang.

References

Adriano, D.C. 2001. Trace elements in terrestrial environments: Biogeochemistry, bioavailability, and risks of metals. 2nd edn. Springer-Verlag, New York.

Arai, Y., and D.L. Sparks. 2002. Residence time effects on arsenate surface speciation at the aluminum oxide-water interface. Soil Sci. 167:303–314.

Audet, P., and C. Charest. 2007. Dynamics of arbuscular mycorrhizal symbiosis in heavy metal phytoremediation: Meta-analytical and conceptual perspectives. Environ. Pollut. 147:609–614.

Balistrieri, L.S., and T.T. Chao. 1987. Selenium adsorption by goethite. Soil Sci. Soc. Am. J. 51:1145–1151.

Barea, J.M., C. Azcon-Aguilar, and R. Azcon. 1997. Interactions between mycorrhizal fungi and rhizosphere microorganisms within the context of sustainable soil-plant systems, p. 65–78. *In* A.C. Gange and V.K. Brown (eds.) Multitrophic interactions in terrestrial systems. Cambridge University Press, New York.

Barrow, N.J. 1992. The effect of time on the competition between anions for sorption. J. Soil Sci. 43:424–428.

Beveridge, T.J. 1989a. Metal ions and bacteria, p. 1–29. *In* T.J. Beveridge and R.J. Doyle (eds.) Metal ions and bacteria. Wiley, New York.

Beveridge, T.J. 1989b. Role of cellular design in bacterial metal accumulation and mineralization. Ann. Rev. Microbiol. 43:147–171.

Bollag, J.-M., J. Dec, and P.M. Huang. 1998. Formation mechanisms of complex organic structures in soil habitats. Adv. Agron. 63:237–245.

Bolland, M.D.A., A.M. Posner, and J.P. Quirk. 1977. Zinc adsorption by goethite in the absence and presence of phosphate. Austr. J. Soil Res. 15:279–286.

Borda, M.J., and D.L. Sparks. 2008. Mobility of trace elements in soil environments, p. 97–168. *In* A. Violante, P.M. Huang, and G.M. Gadd (eds.) Biophysico-chemical processes of metals and metalloids in soil environments. John Wiley & Sons, Hoboken, NJ.

Branco, A. 2008. Mobility and bioavailability of arsenic in Italian polluted soils. PhD Thesis, University of Naples (Italy), Federico II, 122pp.

Bruemmer, G.W., J. Gerth, and K.G. Tiller. 1988. Reaction kinetics of the adsorption and Desorption of nickel, zinc and cadmium by goethite: I. Adsorption and diffusion of metals. J. Soil Sci. 39:37–52.

Buffle, J. 1988. Complexation reactions in aquatic systems: An analytical approach. Ellis Horwood, Chichester, UK.

Bürkert, B., and A. Robson. 1994. ^{65}Zn uptake in subterranean clover (*Trifolium subterraneum* L.) by three vesicular-arbuscular mycorrhizal fungi in a root-free sandy soil. Soil Biol. Biochem. 26:1117–1124.

Burleigh, S.H., B.K. Kristensen, and I.E. Bechmann. 2003. A plasma membrane zinc transporter from *Medicago truncatula* is up-regulated in roots by Zn fertilization, yet down-regulated by arbuscular mycorrhizal colonization. Plant Mol. Biol. 52:1077–1088.

Capasso, R., M. Pigna, A. De Martino, M. Pucci, F. Sannino, and A. Violante. 2004. Potential remediation of waters contaminated with Cr(III), Cu and Zn by sorption on the organic polymeric fraction of olive mill wastewater (polymerin) and its derivatives. Environ. Sci. Technol. 38:5170–5176.

Cavagnaro, T. 2008. The role of arbuscular mycorrhizas in improving plant zinc nutrition under low soil zinc concentrations: A review. Plant Soil 304:315–325.

Cavani, F., A. Trifirò, and A. Vaccai. 1991. Hydrotalcite-type anionic clays preparation, properties and applications. Catal. Today 11:173–301.

Chen, Y., and F.J. Stevenson. 1986. Soil organic matter interactions with trace elements, p. 73–116. *In* Y. Chen and Y. Avnimelech (eds.) The role of organic matter in modern agriculture. Nijhoff, Dordrecht, the Netherlands.

Chen, X., C. Wu, J. Tang, and S. Hu. 2005. Arbuscular mycorrhizae enhance metal lead uptake and growth of host plants under a sand culture experiment. Chemosphere 60:665–671.

Christensen, T.M., and P.M. Huang. 1999. Solid phase cadmium and the reactions of aqueous cadmium with soil surfaces, p. 65–96. *In* M.J. McLaughlin and B.R. Singh (eds.) Cadmium in soils and plants. Kluwer Academic, Dordrecht, the Netherlands.

Clemens, S. 2006. Toxic metal accumulation, responses to exposure and mechanisms of tolerance in plants. Biochimie 88:1707–1719.

Cobb, P.G., K. Sands, M. Waters, B.G. Wixson, and E. Dorward-King. 2000. Accumulation of heavy metals by vegetables grown in mine wastes. Environ. Toxicol. Chem. 19:600–607.

Cornell, R.M., and U. Schwertmann. 1996. The iron oxides. Structure, properties, reactions and uses. VCH Publishers, New York.

Cozzolino, V., Pigna, M., Di Meo, V., Caporale, A.G., and Violante, A. 2010. Effect of arbuscular mycorrhizal inoculation and phosphorus availability in arsenic polluted soil under non-sterile conditions. Applied Soil Ecol. 45:262–268.

Dessureault-Rompré, J., B. Nowack, R. Schulin, M.L. Tercier-Waeber, and J. Luster. 2008. Metal solubility and speciation in the rhizosphere of *Lupinus albus* cluster roots. Environ. Sci. Technol. 42:7146–7151.

Dosskey, M.G., and D.C. Adriano. 1993. Trace element toxicity in VA mycorrhizal cucumber grown on weathered coal fly ash. Soil Biol. Biochem. 25:1547–1552.

Douglas, S., and T.J. Beveridge. 1998. Mineral formation by bacteria in natural microbial communities. FEMS Microbiol. Ecol. 26:79–88.

Dove, P.M., J.J. De Yoreo, and S. Weiner. 2003. Biomineralization. Reviews in mineralogy and geochemistry 54. The Mineralogical Society of America, Chantilly, VA.

Dynes, J.J., and P.M. Huang. 1997. Influence of organic acids on selenite sorption by poorly ordered aluminum hydroxides. Soil Sci. Soc. Am. J. 61:772–783.

Ehrlich, H.L. 1997. Microbes and metals. Appl. Microbiol. Biotechnol. 48:687–692.

Ehrlich, H.L. 2002a. Geomicrobiology. Marcel Dekker, New York.

Ehrlich, H.L. 2002b. Interactions between microorganisms and minerals under anaerobic conditions, p. 459–494. *In* P.M. Huang, J.-M. Bollag, and S. Senesi (eds.) Interactions between soil particles and microorganisms. Impact on the terrestrial ecosystem. IUPAC series on analytical and physical chemistry of environmental systems. Vol. 8. John Wiley & Sons, Chichester, UK.

Elliott, H.A., M.R. Liberati, and C.P. Huang. 1986 Competitive adsorption of heavy metals by soils. J. Environ. Qual. 15:214–219.

Fedotov, P.S., and M. Mirò. 2008. Fractionation and mobility of trace elements in soils and sediments, p. 467–520. *In* A. Violante, P.M. Huang, and G.M. Gadd (eds.) Biophysico-chemical processes of heavy metals and metalloids in soil environments. Wiley-UPAC series. Vol. 1. John Wiley & Sons, Hoboken, NJ.

Fendorf, S.E. 1995. Surface reactions of chromium in soils and waters. Geoderma 67:55–71.

Fendorf, S.E., M.J. Eick, P.R. Grossl, and D.L. Sparks. 1997. Arsenate and chromate retention mechanisms on goethite. I. Surface structure. Environ. Sci. Technol. 31:315–320.

Filgueiras, A.V., I. Lavilla, and C. Bendicho. 2002. Chemical sequential extraction for metal partitioning in environmental solid samples. J. Environ. Monit. 4:823–857.

Fitz, W.J., and W.W. Wenzel. 2006. Sequestration of arsenic by plants, p. 209–222. In R. Naidu, E. Smith, G. Owens, P. Bhattacharya, and P. Nadebaum (eds.) Managing arsenic in the environment. CSIRO Publishing, Melbourne, Australia.

Frankenberger, W.T., Jr. 2002. Environmental chemistry of arsenic. Marcel Dekker Inc., New York.

Fuller, C.C., J.A. Davis, and G.A. Weychunas. 1993. Surface chemistry of ferrihydrite. II. Kinetics of arsenate adsorption and coprecipitation. Geochim. Cosmochim. Acta 57:2271–2282.

Gadd, G.M. 2005. Microorganisms in toxic metal polluted soils, p. 325–376. In F. Buscot and A. Varma (eds.) Microorganisms in soils: Roles in genesis and functions. Springer-Verlag, Berlin, Germany.

Gadd, J.M. 2008. Transformation and mobilization of metals, metalloids, and radionuclides by microorganisms, p. 53–96. In A. Violante, P.M. Huang, and G.M. Gadd (eds.) Biophysico-chemical processes of metals and metalloids in soil environments. Wiley-IUPAC series. Vol. 1. John Wiley & Sons, Hoboken, NJ.

Gadde, R.R. and H.A. Laitinen. 1974. Studies of heavy metals adsorption by hydrous iron and manganese oxides. Anal. Chem. 46:2022–2026.

Gavrilescu, M. 2004. Removal of heavy metals from the environment by biosorption. Eng. Life Sci. 4:219–232.

Goh, K.H., T.T. Lim, and Z. Dong. 2008. Application of layered double hydroxides for removal of oxyanions: A review. Water Res. 42:1343–1368.

Göhre, V., and U. Paszkowski. 2006. Contribution of the arbuscular mycorrhizal symbiosis to heavy metal phytoremediation. Planta 223:1115–1122.

Goldberg, S., and L.J. Criscenti. 2008. Modelling adsorption of metals and metalloids by soil components, p. 215–264. In A. Violante, P.M. Huang, and G.M. Gadd (eds.) Biophysico-chemical processes of metals and metalloids in soil environments. John Wiley & Sons, Hoboken, NJ.

Goldberg, S., J.A. Davis, and J.D. Hem. 1996. The surface chemistry of aluminum oxides and hydroxides, p. 271–331. In G. Sposito (ed.) The environmental chemistry of aluminum. 2nd edn. Lewis Publishers, Boca Raton, FL.

Goldberg, S., and R.A. Glaubig. 1988a. Anion sorption on a calcareous montmorillonitic soil – Selenium. Soil Sci. Soc. Am. J. 52:954–958.

Goldberg, S., and R.A. Glaubig. 1988b. Anion sorption on a calcareous montmorillonitic soil – Arsenic. Soil Sci. Soc. Am. J. 52:1297–1300.

Gonzales-Chavez, M.C., R. Carrillo-Gonzales, S.F. Whright, and K.A. Nichols. 2004. The role of glomalin, a protein produced by arbuscular mycorrhizal fungi, in sequestering potentially toxic elements. Environ. Pollut. 130:317–323.

Govindarajulu, M., P.E. Pfeffer, H. Jin, J.M. Abubaker, D.D. Douds, J.W. Allen, H. Bücking, P. Lammers, and Y. Shachar-Hill. 2005. Nitrogen transfer in the arbuscular mycorrhizal symbiosis. Nature 435:819–823.

Greedland, D.J. 1965. Interaction between clays and organic compounds in soils. II. Adsorption of soil organic compounds and its effects on soil properties. Soil Fert. Common. Bus Soil Sci. 28:521–532.

Grossl, P.R., M.J. Eick, D.L. Sparks, S. Goldberg, and C.C. Ainsworth. 1997. Arsenate and chromate retention on goethite. II. Kinetic evaluation using a p-jump relaxation technique. Environ. Sci. Technol. 31:321–326.

Hayes, K.F., A.L. Roe, G.E. Brown, K.O. Hodgson, J.O. Lee, and G.A. Parks. 1987. In situ x-ray absorption study of surface complexes: Selenium oxyanions on a-FeOOH. Science 238:783–786.

Hildenbrandt, U., M. Kaldorf, and H. Bothe. 1999. The zinc violet and its colonization by arbuscular mycorrhizal fungi. J. Plant Physiol. 154:709–717.

Hinsinger, P. 2001. Bioavailability of trace elements as related to root-induced chemical changes in the rhizosphere, p. 25–41. In G.R. Gobran, W.W. Wenzel, and E. Lombi (eds.) Trace elements in the rhizosphere. CRC Press, Boca Raton, FL.

Hinsinger, P., and F. Courchesne. 2008. Biogeochemistry of metals and metalloids at the soil-root interface, p. 267–311. In A. Violante, P.M. Huang, and G.M. Gadd (eds.) Biophysico-chemical processes of heavy metals and metalloids in soil environments. John Wiley & Sons, Hoboken, NJ.

Hsia, T.H., S.L. Lo, C.F. Lin, and D.Y. Lee. 1994. Characterization of arsenate adsorption on hydrous iron oxide using chemical and physical methods. Colloids Surf. A: Physicochem. Eng. Aspects 85:1–7.

Hsu, P.H. 1989. Aluminum hydroxides and oxyhydroxides, p. 331–378. In J.B. Dixon and S.B. Weed (eds.) Minerals in soil environments. 2nd edn. ASA, Madison, WI.

Huang, P.M. 2008. Impact of physicochemical biological interaction on metal and metalloids transformation in soils: An overview, p. 3–57. In A. Violante, P.M. Huang, and G.M. Gadd (eds.) Biophysico-chemical processes of heavy metals and metalloids in soil environments. Wiley-IUPAC series. Vol. I. John Wiley & Sons, Hoboken, NJ.

Huang, Q., W. Chen, and K.G. Theng. 2008. Role of bacteria and bacteria-soil composites in metal biosorption and remediating toxic metal-contaminated soil systems, p. 71–98. In Q. Huang, P.M. Huang, and A. Violante (eds.) Soil mineral-microbe-organic interactions: Theories and applications. Springer-Verlag, Berlin, Germany.

Huang, P.M., and J.J. Germida. 2002. Chemical and biochemical processes in the rhizosphere: Metal pollutants, p. 381–438. In P.M. Huang, J.-M. Bollag, and N. Senesi (eds.) Interactions between soil particles and microorganisms: Impact on the terrestrial ecosystem. John Wiley & Sons, New York.

Huang, P.M., and G.R. Gobran. 2005. Biogeochemistry of trace elements in the rhizosphere. Elsevier B.V., Amsterdam, the Netherlands.

Huang, P.M., and A.G. Hardie. 2009. Formation mechanisms of humic substances in the environment. *In* N. Senesi, B. Xing, and P.M. Huang (eds.) Biophysico-chemical processes involving natural nonliving organic matter in environmental systems. Wiley-IUPAC. Vol. 2. John Wiley & Sons, Hoboken, NJ. Chapter 2. p. 41–109.

Huang, P.M., and A. Violante. 1986. Influence of organic acids on crystallization and surface properties of precipitation products of aluminium, p. 159–221. *In* P.M. Huang and M. Schnitzer (eds.) Interactions of soil minerals with natural organics and microbes. Special Publication, No. 17. Soil Science Society of America, Madison, WI.

Huang, P.M., M.K. Wong, N. Kampf, and Q.G. Shulze. 2002. Aluminum hydroxides, p. 261–289. *In* G.B. Dixon and D.G. Shulze (eds.) Soil mineralogy with environmental applications. Soil Science Society of America, Madison, WI.

Iskandar, I.K., and M.B. Kikham. 2001. Trace elements in soil. CRC Press LLC, Boca Raton, FL.

Jackson, T.A. 1998. The biogeochemical and ecological significance of interactions between colloidal minerals and trace elements, p. 93–205. *In* A. Parker and J.E. Rae (eds.) Environmental interactions of clays. Springer-Verlag, Berlin, Germany.

Jackson, A.P., and B.J. Alloway. 1991. The bioavailability of cadmium to lettuce and cabbage in soils previously treated with sawage sludges. Plant Soil 132:179–186.

Jackson, T.A., and G.G. Leppard. 2002. Energy dispersive x-ray microanalysis and its applications in biological research, p. 219–260. *In* A. Violante, P.M. Huang, J.M. Bollag, and L. Gianfreda (eds.) Soil mineral-organic-matter-microorganism interactions and ecosystem health: Dynamics, mobility and transformations of pollutants and nutrients, developments in soil science. Vol. 28. Elsevier, New York.

James, B.R., and R.J. Bartlett. 2000. Redox phenomena, p. B-169–B-194. *In* M.E. Sumner (ed.) The handbook of soil science. CRC Press, Boca Raton, FL.

Jeffries, P., S. Gianinazzi, S. Perotto, K. Turnau, and J.M. Barea. 2003. The contribution of arbuscular mycorrhizal fungi in sustainable maintenance of plant health and soil fertility. Biol. Fertil. Soil. 37:1–16.

Jia, Y.F., L. Xu, Z. Fang, and G.P. Demopoulos. 2006. Observation of surface precipitation of arsenate on ferrihydrite. Environ. Sci. Technol. 40:3248–3253.

Joner, E.J., R. Briones, and C. Leyval. 2000. Metal-binding capacity of arbuscular mycorrhizal mycelium. Plant Soil 226:227–234.

Kabata-Pendias, A. 2004. Soil-plant transfer of trace elements—An environmental issue. Geoderma 122:143–149.

Kabata-Pendias, A., and H. Pendias. 2001. Trace elements in soils and plants. CRC Press, Boca Raton, FL.

Kaldorf, M., A.J. Kuhn, W.H. Schröder, U. Hildenbrandt, and H. Bothe. 1999. Selective element deposits in maize colonized by a heavy metal tolerance conferring arbuscular mycorrhizal fungus. J. Plant Physiol. 154:718–728.

Kampf, N., A.C. Scheinost, and D.G. Schultze. 2000. Oxide minerals, p. F125–F168. *In* M.E. Sumner (ed.) Handbook of soil science. CRC Press, Boca Raton, FL.

Kinniburgh, D.G., and M.L. Jackson. 1976. Adsorption of alkaline earth, transition and heavy metal cations by hydrous oxides gels of iron and aluminum. Soil Sci. Soc. Am. J. 40:796–799.

Kinniburgh, D.G., and M.L. Jackson. 1981. Cation adsorption by hydrous metal oxides and clays, p. 91–160. *In* M.A. Anderson and A.S. Rubin (eds.) Adsorption of inorganics at solid-liquid interfaces. Ann Arbor Science, Ann Arbor, MI.

Kretzschmar, R., and A. Voegelin. 2001. Modeling competitive sorption and release of heavy metals in soils, p. 55–88. *In* H.M. Selim and D.L. Sparks (eds.) Heavy metals release in soils. Lewis Publications, Boca Raton, FL.

Krishnamurti, G.S.R. 2000. Speciation of heavy metals: An approach for remediation of contaminated soils, p. 693–713. *In* D.L. Wise, D.J. Trantolo, E.J. Cichon, H.I. Inyang, and U. Stottmeister (eds.) Remediation engineering of contaminated soils. Marcel Dekker, New York.

Krishnamurti, G.S.R., P.M. Huang, and L.M. Kozak. 1999a. Desorption kinetics of cadmium from soils using M ammonium nitrate and M ammonium chloride. Commun. Soil Sci. Plant Anal. 30:2785–2800.

Krishnamurti, G.S.R., P.M. Huang, and L.M. Kozak. 1999b. Sorption and desorption kinetics of cadmium from soils: Influence of phosphate. Soil Sci. 164:888–898.

Krishnamurti, G.S.R., P.M. Huang, and K.C.J. Van Rees. 1996. Studies on soil rhizosphere: Speciation and availability of Cd. Chem. Speciat. Bioavailab. 8:23–28.

Krishnamurti, G.S.R., P.M. Huang, K.C.J. Van Rees, L.M. Kozak, and H.P.W. Rostad. 1995a. Speciation of particulate-bound cadmium of soils and its bioavailability. The Analyst 120:659–665.

Krishnamurti, G.S.R., P.M. Huang, K.C.J. Van Rees, L.M. Kozak, and H.P.W. Rostad. 1995b. A new soil test method for the determination of plant available cadmium in soils. Commun. Soil Sci. Plant Anal. 26:2857–2867.

Krishnamurti, G.S.R., and R. Naidu. 2000. Speciation and phytoavailability of cadmium in selected surface soils of south Australia. Aust. J. Soil Res. 38:991–1004.

Krishnamurti, G.S.R., and R. Naidu. 2002. Solid-solution speciation and phytoavailability of copper and zinc in soils. Environ. Sci. Technol. 36:2645–2651.

Krishnamurti, G.S.R., and R. Naidu. 2008. Chemical speciation and bioavailability of trace metals, p. 419–466. *In* A. Violante, P.M. Huang, and G.M. Gadd (eds.) Biophysico-chemical processes of heavy metals and metalloids in soil environments. Wiley Interscience, New York.

Krishnamurti, G.S.R., M. Pigna, M. Arienzo, and A. Violante. 2007. Solid-phase speciation and phytoavailability of copper in a few representative soils of Italy. Chem. Speciat. Bioavailab. 19:57–67.

Kuo, S. 1986. Concurrent sorption of phosphate and zinc, cadmium, or calcium by hydrous ferric oxide. Soil Sci. Soc. Am. J. 50:1040–1044.

Kwong, K.F.N.K., and P.M. Huang. 1979. Surface reactivity of aluminum hydroxides precipitated in the presence of low molecular weight organic acids. Soil Sci. Soc. Am. J. 43:1107–1113.

Laurie, S.H., N. Tancock, S.P. McGrath, and J.R. Sanders. 1991. Influence of complexation on the uptake by plants of iron, manganese, copper and zinc: Effect of DTPA in a multi-metal and computer simulation study. J. Exp. Bot. 42:509–513.

Leung, H.M., Z.H. Ye, and M.H. Wong. 2006. Interactions of mycorrhizal fungi with *Pteris vittata* (As hyperaccumulator) in As-contaminated soils. Environ. Pollut. 139:1–8.

Leyval, C., and E.J. Joner. 2001. Bioavailability of heavy metals in the mycorrhizosphere, p. 165–185. *In* G.R. Gobran, W.W. Wenzel, and E. Lombi (eds.) Trace elements in the rhizosphere. CRC Press, Boca Raton, FL.

Li, X.L., H. Marschner, and E. George. 1991. Acquisition of phosphorus and copper by VA-mycorrhizal hyphae and root-to-shoot transport in white clover. Plant Soil 136:49–57.

Lindsay, W.L., and W.A. Norvell. 1978. Development of a DTPA soil test for zinc, iron, manganese, and copper. Soil Sci. Soc. Am. J. 42:421–428.

Linehan, D.J., A.H. Sinclair, and M.C. Mitchell. 1989. Seasonal changes in Cu, Mn, Zn, and Co concentrations in soil in the root zone of barley (*Hordeum vulgare L*). J. Soil Sci. 40:103–115.

Liu, W.J., Y.G. Zhu, Y. Hu, P.N. Williams, A.G. Gault, A.A. Meharg, J.M. Charnock, and F.A. Smith. 2006. Arsenic sequestration in iron plaque, its accumulation and speciation in mature rice plants (*Oryza Sativa L.*). Environ. Sci. Technol. 40:5730–5736.

Loeppert, R.H., A. Schwab, and S. Goldberg. 1995. Chemical equilibrium and reaction models. SSSA, Madison, WI.

Lopez, A., N. Lazaro, J.M. Priego, and A.M. Marques. 2000. Effect of pH on the adsorption of nickel and other heavy metals by *Pseudomonas fluorescens* 4F39. J. Ind. Microbiol. Biotechnol. 24:146–151.

Loyaux-Lawniczak, S., P. Refait, J.J. Ehrhardt, P. Lecomte, and J.M.R. Genin. 2000. Trapping of Cr by formation of ferrihydrite during the reduction of chromate ions by Fe(II)-Fe(III) hydroxysalt green rust. Environ. Sci. Technol. 34:438–443.

Lund, W. 1990. Speciation analysis – Why and how? Fresenius J. Anal. Chem. 337:557–564.

Ma, J.F., P.R. Ryan, and E. Delhaize. 2001. Aluminium tolerance in plant sand the complexing role of organic acids. Trends Plant Sci. 6:273–278.

Malinowski, D.P., H. Zuo, D.P. Beleski, and G.A. Alloush. 2004. Evidence for copper binding by extracellular root exudates of tall fescue but not perennial ryegrass infected with *Neotyphodium* spp. *Endophytes*. Plant Soil 267:1–12.

Manceau, A., and L. Charlet. 1994. The mechanism of selenate adsorption on goethite and hydrous ferric oxide. J. Colloid Interface Sci. 168:87–93.

Mann, S., D.A. Archibald, J.M. Didymus, T. Douglas, R.R. Heywood, F.C. Meldrum, and N.J. Reeves. 1993. Crystallization and inorganic-organic interfaces: Biominerals and biomimetic synthesis. Science 261:1286–1292.

Manning, B.A., S.E. Fendorf, and S. Goldberg. 1998. Surface structures and stability of arsenic (III) on goethite. Spectroscopic evidence for inner-sphere complexes. Environ. Sci. Technol. 32:2383–2388.

Manning, B.A., and S. Goldberg. 1996. Modelling competitive adsorption of arsenate with phosphate and molybdate on oxide minerals. Soil Sci. Soc. Am. J. 60:121–131.

Marschner, H. 1995. Mineral nutrition of higher plants. Academic Press, New York.

Marschner, H., and V. Römheld. 1996. Root-induced changes in the availability of micronutrients in the rhizosphere, p. 557–579. *In* Y. Waisel, A. Eshel, and U. Kafkafi (eds.) Plant roots: The hidden half, 2nd edn. Marcel Dekker, New York.

Martin, M., A. Violante, and E. Barberis. 2007. Fate of arsenite and arsenate in flooded and not flooded soils of southwest Bangladesh irrigated with arsenic contaminated water. J. Environ. Sci. Health A Tox. Hazard. Subst. Environ. Eng. 42:1775–1783.

McBride, M.B. 1982. Cu^{2+}-adsorption characteristics of aluminum hydroxide and oxyhydroxides. Clay. Clay Miner. 30:21–28.

McBride, M.B. 1985. Sorption of copper (II) on aluminum hydroxide as affected by phosphate. Soil Sci. Soc. Am. J. 49:843–846.

McBride, M.B. 1989. Reactions controlling heavy metal solubility in soils. Adv. Soil Sci. 10:1–56.

McBride, M.B. 1991. Processes of heavy and transition metal sorption by soil minerals, p. 149–176. *In* G.H. Bolt, M.F. De Boodt, M.H.B. Haynes, and M.B. McBride (eds.) Interactions at the soil colloid-soil solution interface. Vol. 190. Kluwer Academic Publishers, Dordrecht, the Netherlands.

McBride, M.B. 2000. Chemisorption and precipitation reactions, p. B265–B302. *In* M.E. Sumner (ed.) Handbook of soil science. CRC Press, Boca Raton, FL.

McGonigle, T.P., and M.H. Miller. 1996. Development of fungi below ground in association with plants growing in disturbed and undisturbed soil. Soil Biol. Biochem. 28:263–269.

McKenzie, R.M. 1980. The adsorption of lead and other heavy metals on oxides of manganese and iron. Aust. J. Soil Res. 21:505–513.

McLaughlin, M.J., D.R. Parker, and J.M. Clarke. 1999. Metals and micronutrients—Food safety issues. Field Crops Res. 60:143–163.

McLean, J.S., J.-U. Lee, and T.L. Beveridge. 2002. Interaction of bacteria and environmental metals, fine-grained mineral development and bioremediation strategies, p. 227–261. *In* P.M. Huang, J.-M. Bollag, and N. Senesi (eds.) Interaction between soil particles and microorganism: Impact of the terrestrial ecosystem. IUPAC series on analytical and physical chemistry of environmental systems. Vol. 8. Wiley, Chichester, West Sussex, England.

Miyata, S. 1975. The syntheses of hydrocalcite-like compounds and their structures and physical-chemical properties-1: The systems Mg^{2+}-Al^{3+}-NO_3^-, Mg^{2+}-Al^{3+}-Cl^-, Mg^{2+}-Al^{3+}-ClO_4^-, Ni^{2+}-Al^{3+}-Cl^- and Zn^{2+}-Al^{3+}-Cl^-. Clay. Clay Miner. 23:369–375.

Nachtegaal, M., and D.L. Sparks. 2003. Nickel sequestration in a kaolinite-humic acid complex. Environ. Sci. Technol. 37:529–534.

Nelson, Y.M., L.W. Lion, W.C. Ghiorse, and M.L. Schuler. 1999. Production of biogenic Mn oxides by Leptothrix discophora ss – 1 in a chemically defined growth medium and evaluation of their Pb adsorption characteristics. Appl. Environ. Microbiol. 65:175–180.

Neubauer, U., G. Furrer, and R. Schulin. 2002. Heavy metal sorption on soil minerals affected by the siderophore desferrioxamine B: The role of Fe(III) (hydr)oxides and dissolved Fe(III). Eur. J. Soil Sci. 53:45–55.

Neubauer, U., B. Nowak, G. Furrer, and R. Schulin. 2000. Heavy metal sorption on clay minerals affected by the siderophore desferrioxamine B. Environ. Sci. Technol. 34:2749–2755.

Nguyen, C. 2003. Rhizodeposition of organic C by plant: Mechanisms and controls. Agronomie 23:375–396.

Nielson, N.E. 1976. The effect of plants on the copper concentration in the soil solution. Plant Soil 45:679–687.

NIST. 1997. Standard Reference Database 46. Critically Selected Stability Constants of Metal Complexes, Version 5.0. National Institute of Standards and Technology, U.S. Department of Commerce, Gaithersburg, MD.

Ona-Nguema, G., G. Morin, F. Juillot, G. Calas, and G.E. Brown, Jr. 2005. EXAFS analysis of arsenite adsorption onto two-line ferrihydrite, hematite, goethite, and lepidocrocite. Environ. Sci. Technol. 39:9147–9155.

Onyatta, J.O., and P.M. Huang. 1999. Cadmium speciation and bioavailability index for selected Kenyan soils. Geoderma 91:87–101.

O'Reilly, S.E., D.G. Strawn, and D.L. Sparks. 2001. Residence time effects on arsenate adsorption/desorption mechanisms on goethite. Soil Sci. Soc. Am. J. 65:67–77.

Oscarson, D.W., P.M. Huang, C. Defose, and A. Herbillon. 1981. Oxidative power of Mn(IV) and Fe(III) oxides with respect to As(III) in terrestrial and aquatic environments. Nature 291:50–51.

Oscarson, D.W., P.M. Huang, W.K. Liaw, and U.T. Hammer. 1983. Kinetics of oxidation of arsenite by various manganese dioxides. Soil Sci. Soc. Am. J. 47:644–648.

Parfitt, R.L. 1980. Chemical properties of variable charge soils, p. 167–194. In B.K. Theng (ed.) Soils with variable charge. New Zealand Society of Soil Science, Palmerston North, New Zealand.

Parker, D.R., R.L. Chaney, and W.A. Norvell. 1995. Chemical equilibrium models: Applications to plant nutrition, p. 163–200. In R.H. Loeppert, A.P. Schwab, and S. Goldberg (eds.) Soil chemical equilibrium and reaction models. American Society of Agronomy and Soil Science Society of America, Madison, WI.

Pearson, R.G. 1963. Hard and soft acids and bases. J. Am. Chem. Soc. 85:3533–3539.

Peryea, F.J. 1998. Phosphate starter fertilizer temporarily enhances soil arsenic uptake by apple trees grown under field conditions. Hortic. Sci. 33:826–829.

Pickering, W.F. 1979. Copper retention by sediment/soil components, p. 217–253. In J.O. Nriagu (ed.) Copper in the environment. I. Ecological cycling. John Wiley, New York.

Pigna, M., V. Cozzolino, A. Violante, and A. Meharg. 2009. Influence of phosphate on the arsenic uptake by wheat (*Triticum durum* L.) irrigated with arsenic solutions at three different concentrations. Water Air Soil Pollut. 197:371–380.

Pigna, M., G.S.R. Krishnamurti, and A. Violante. 2006. Kinetics of arsenate sorption-desorption from metal oxides: Effect of residence time. Soil Sci. Soc. Am. J. 70:2017–2027.

Quaghebeur, M., and Z. Rengel. 2004. Phosphate and arsenate interactions in the rhizosphere of canola (*Brassica napus*). Funct. Plant Biol. 31:1085–1094.

Refait, P., L. Simon, and J-M.R. Genin. 2000. Reduction of SeO_4^{2-} anions and anoxic formation of iron(II)-iron(III) hydroxylselenate green rust. Environ. Sci. Technol. 34:819–825.

Rosen, B.P. 1999. Families of arsenic transporters. Trends Microbiol. 7:207–212.

Scheidegger, A.M., G.M. Lamble, and D.L. Sparks. 1997. Spectroscopic evidence for the formation of mixed-cation hydroxide phases upon metal sorption on clays and aluminum oxides. J. Colloid Interface Sci. 62:2233–2245.

Schnitzer, M., and H. Kodama. 1977. Reactions of minerals with soil humic substances, p. 741–770. In J.B. Dixon and S.B. Wood (eds.) Minerals in soil environments. Soil Science Society of America, Madison, WI.

Schnitzer, M., and P. Schuppli. 1989. The extraction of organic matter from selected soils and particle size fractions with 0.5 M NaOH and 0.1 M $Na_4P_2O_7$ solutions. Can. J. Soil Sci. 69:253–262.

Schüssler, A., D. Schwarzott, and C. Walker. 2001. A new fungal phylum, the Glomeromycota: Phylogeny and evolution. Mycol. Res. 105:1413–1421.

Schützendübel, A., and A. Polle. 2002. Plant responses to abiotic stresses: Heavy metal-induced stress and protection by mycorrhization. J. Exp. Bot. 53:1351–1365.

Selim, H.M., and D.L. Sparks. 2001. Heavy metals release in soils. CRC Press, Boca Raton, FL.

Senesi, N. 1990. Molecular and quantitative aspects of the chemistry of fulvic acid and its interaction with metal ions and organic chemicals. Part II. The fluorescence spectroscopy approach. Anal. Chim. Acta 232:77–106.

Senesi, N. 1992. Metal-humic substance complexes in the environment. Molecular and mechanistic aspects by multiple spectroscopic approach, p. 429–451. In D.C. Adriano (ed.) Biogeochemistry of trace elements. Lewis, Boca Raton, FL.

Senesi, N., and E. Loffredo. 1998. The chemistry of soil organic matter, p. 239–370. In D.L. Sparks (ed.) Soil physical chemistry. 2nd edn. CRC Press, Boca Raton, FL.

Senesi, N., and E. Loffredo. 2005. Metal ion complexation by soil humic substances, p. 563–618. *In* D.L. Sparks and M.A. Tabatabai (eds.) Chemical processes in soil. Soil Science Society of America, Madison, WI.

Senesi, N., and E. Loffredo. 2008. Spectroscopic techniques for studying metal-humic complexes in soil, p. 125–168. *In* A. Violante, P.M. Huang, and G.M. Gadd (eds.) Biophysicochemical processes of metals and metalloids in soil environments. John Wiley & Sons, Hoboken, NJ.

Sharples, J.M., A.A. Meharg, S.M. Chambers, and J.W.G. Cairney. 2000. Symbiotic solution to arsenic contamination. Nature 404:951–952.

Simine, D.D., I.C. Finoli, A. Vecchio, and V. Andreoni. 1998. Metal ion accumulation by immobilized cells of *Brevibacterium* sp. J. Ind. Microbiol. Biotechnol. 20:116–120.

Smedley, P.L., and D.G.A. Kinniburgh. 2002. Review of the source, behaviour and distribution of arsenic in natural waters. Appl. Geochem. 17:517–568.

Smith, E., R. Naidu, and A.M. Alston. 1998. Arsenic in the soil environment: A review. Adv. Agron. 64:149–195.

Smith, E., R. Naidu, and A.M. Alston. 2002. Chemistry of inorganic arsenic in soils: II. Effect of phosphorus, sodium, and calcium on arsenic sorption. J. Environ. Qual. 31:557–563.

Smith, S., and D.J. Read. 2008. Mycorrhizal symbiosis. 3rd edn. Elsevier, New York.

Sparks, D.L. 1989. Kinetics of soil chemical processes. Academic Press, San Diego, CA.

Sparks, D.L. 1995. Kinetics of metal sorption reactions, p. 36–58. *In* H.E. Allen, C.P. Huang, G.W. Bailey, and A.R. Bowers (eds.) Metal speciation and contamination of soil. Lewis Publishers, Boca Raton, FL.

Sparks, D.L. 1999. Kinetics and mechanisms of chemical reactions at the soil mineral/water interface, p. 135–191. *In* D.L. Sparks (ed.) Soil physical chemistry. 2nd edn. CRC Press, Boca Raton, FL.

Sparks, D.L. 2003. Environmental soil chemistry. 2nd edn. Academic Press, San Diego, CA.

Sparks, D.L. 2005. Metal and oxyanion sorption on naturally occurring oxide and clay mineral surfaces, p. 3–36. *In* V.H. Grassian (ed.) Environmental catalysis. Taylor & Francis, Boca Raton, FL.

Sparks, D.L., and P.C. Zhang. 1991. Relaxation methods for studying kinetics of soil chemical phenomena, p. 61–94. *In* D.L. Sparks and D.L. Suarez (eds.) Rates of soil chemical processes. Soil Science Society of American Special Publication 27, SSSA, Madison, WI.

Sposito, G. 1986. Sorption of trace metals by humic materials in soils and natural waters. CRC Crit. Rev. Environ. Control 16:193–229.

Sposito, G. 2004. The surface chemistry of natural particles. Oxford University Press, Oxford, UK.

Srinath, T., T. Verna, P.W. Ramteke, and S.K. Garg. 2002. Chromium biosorption and bioaccumulation by chromate resistant bacteria. Chemistry 48:427–435.

Stevenson, F.J. 1976. Binding of metal ions by humic acids, p. 519–540. *In* J.O. Nriagu (ed.) Environmental biogeochemistry. Vol. 2. Metals transfer and ecological mass balances. Ann Arbor Science, Ann Arbor, MI.

Stevenson, F.J., and A. Fitch. 1986. Chemistry of complexation of metal ions with soil solution organics, p. 2–58. *In* P.M. Huang and M. Schnitzer (eds.) Interactions of soil minerals with natural organic microbes. SSSA, Madison, WI.

Strathmann, T.J., and S.C.B. Mynemi. 2005. Effect of soil fulvic acid on nickel (II) sorption and bonding at aqueous-boehmite (γ-AlOOH) interface. Environ. Sci. Technol. 39:4027–4034.

Sun, X., and H.E. Doner. 1996. An investigation of arsenate and arsenite bonding structures on goethite by FTIR. Soil Sci. 161:865–872.

Tebo, B.M., W.C. Ghiorse, L.G. Waasbergen, P.L. Siering, and R. Caspi. 1997. Bacterially mediated mineral formation: Insight into manganese (II) oxidation from molecular genetic and biochemical studies, p. 225–266. *In* J.F. Banfield and K.H. Nealson (eds.) Geomicrobiology: Interactions between microbes and minerals. Reviews in mineralogy and geochemistry 35. Mineralogical Society of America, Washington, DC.

Tessier, A., P.G.C. Campbell, and M. Bissom. 1979. Sequential extraction procedure for the speciation of particulate trace metals. Anal. Chem. 51:844–851.

Tessier, A., and D.R. Turner. 1995. Metal speciation and bioavailability in aquatic systems. IUPAC series on analytical and physical chemistry of environmental systems. Vol. 3. John Wiley & Sons, Chichester, UK.

Thanabalasingan, P., and W.F. Pickering. 1986. Arsenic sorption by humic acids. Environ. Pollut. 12:233–246.

Thomas, S., D. Mahammedi, M. Clairotte, M.F. Benedetti, M. Castrec-Rouelle, F. Persin, P. Peu, J. Martinez, and P. Hinsinger. 2005. Bioavailability and extractability of copper and zinc in a soil amended with pig slurry: Effect of iron deficiency in the rhizosphere of two grasses, p. 337–363. *In* P.M. Huang and G.R. Gobran (eds.) Biogeochemistry of trace elements in the rhizosphere. Elsevier, Amsterdam, the Netherlands.

Treeby, M., H. Marschner, and V. Römheld. 1989. Mobilization of iron and other micronutrient cations from a calcareous soil by plant-borne microbial and synthetic metal chelator. Plant Soil 114:217–226.

Ultra, V.U.Y., S. Tanaka, K. Sakurai, and K. Iwasaki. 2007. Arbuscular mycorrhizal fungus (*Glomus aggregatum*) influences biotransformation of arsenic in the rhizosphere of sunflower (*Helianthus annus* L.). Soil Sci. Plant Nutr. 53:499–508.

Ure, A.M., and C.M. Davison. 2002. Introduction, p. 1–5. *In* A.M. Ure and C.M. Davison (eds.) Chemical speciation in the environment. 2nd edn. Blackwell Science, Oxford, U.K.

Violante, A., V. Cozzolino, S. Del Gaudio, D. Benerjee, and M. Pigna. 2009a. Coprecipitation of arsenate within metal oxides: 3. Nature, mineralogy and reactivity of iron(III)/Al precipitates. Environ. Sci. Technol. 43:1515–1521.

Violante, A., and L. Gianfreda. 2000. Role of biomolecules in the formation of variable-charge minerals and organo-mineral complexes and their reactivity with plant nutrients and organics in soil, p. 207–270. *In* J.-M. Bollag and G. Stotzky (eds.) Soil biochemistry. Vol. 10. Marcel Dekker, New York.

Violante, A., G.S.R. Krishnamurti, and P.M. Huang. 2002. Impact of organic substances on the formation of metal oxides in soil environments, p. 133–188. *In* P.M. Huang, J.-M. Bollag, and N. Senesi (eds.) Interactions between soil particles and microorganism: Impact on the terrestrial ecosystem. John Wiley & Sons, New York.

Violante, A., G.S.R. Krishnamurti, and M. Pigna. 2008. Mobility of trace elements in soil environments, p. 169–213. *In* A. Violante, P.M. Huang, and G.M. Gadd (eds.) Biophysicochemical processes of metals and metalloids in soil environments. John Wiley & Sons, Hoboken, NJ.

Violante, A., and M. Pigna. 2002. Competitive sorption of arsenate and phosphate on different clay minerals and soils. Soil Sci. Soc. Am. J. 66:1788–1796.

Violante, A., M. Pigna, S. Del Gaudio, M. Ricciardella, and D. Banerjee. 2007. Coprecipitation of arsenate with metal oxides: Nature, mineralogy and reactivity of iron(III) precipitates. Environ. Sci. Technol. 41:8275–8280.

Violante, A., M. Pucci, V. Cozzolino, J. Zhu, and M. Pigna. 2009b. Sorption/desorption of arsenate on/from Mg-Al layered double hydroxides: Influence of phosphate. J. Colloid Interface Sci. 333:63–70.

Violante, A., M. Ricciardella, S. Del Gaudio, and M. Pigna. 2006. Coprecipitation of arsenate with metal oxides: Nature, mineralogy and reactivity of aluminum precipitates. Environ. Sci. Technol. 40:4961–4967.

Violante, A., M. Ricciardella, and M. Pigna. 2003. Adsorption of heavy metals on mixed Fe-Al oxides in the absence or presence of organic ligands. Water Air Soil Pollut. 145:289–306.

Violante, A., M. Ricciardella, M. Pigna, and R. Capasso. 2005. Effects of organic ligands on the sorption of trace elements onto metal oxides and organo-mineral complexes, p. 157–182. *In* P.M. Huang and G.R. Gobran (eds.) Biogeochemistry of trace elements in the rhizosphere. Elsevier, New York.

Vosátka, M., J. Rydlová, R. Sudová, and M. Vohník. 2006. Mycorrhizal fungi as helping agents in phytoremediation of degraded and contaminated soils, p. 237–255. *In* M. Mackova, D.N. Dowling, and T. Macek (eds.) Phytoremediation and rhizoremediation. Springer, Berlin, Germany.

Wang, Z., J.L. Zhang, P. Christie, and X.L. Li. 2008. Influence of inoculation with *Glomus mosseae* or *Acaulospora morrowiae* on arsenic uptake and translocation by maize. Plant Soil 311:235–244.

Waychunas, G.A., B.A. Rea, C.C. Fuller, and J.A. Davis. 1993. Surface chemistry of ferrihydrite. I. EXAFS studies of the geometry of coprecipitates and adsorbed arsenate. Geochim. Cosmochim. Acta 57:2251–2269.

Wenzel, W.W., N. Kirchbaumer, T. Prohaska, G. Stingeder, E. Lombi, and D.C. Adriano. 2001. Arsenic fractionation in soils using an improved sequential extraction procedure. Anal. Chim. Acta 436:309–323.

Wu, C.H., L.L. Shang, F.L. Cheng, and Y.K. Chao. 2001. Modeling competitive adsorption of molybdate, sulfate and selenate on γ-Al_2O_3 by the triple-layer model. J. Colloid Interface Sci. 233:259–264.

Yamaguchi, N.U., A.S. Scheinost, and D.L. Sparks. 2002. Influence of gibbsite surface area and citrate on Ni sorption mechanisms at pH 7.5. Clay. Clay Miner. 50:784–790.

Yee, N., and J.B. Fein. 2001. Cd adsorption onto bacterial surfaces: A universal adsorption edge? Geochim. Cosmochim. Acta 65:2037–2042.

Zachara, J.M., C.C. Ainsworth, C.E. Cowan, and C.T. Resch. 1989. Adsorption of chromate by subsurface soil horizons. Soil Sci. Soc. Am. J. 53:418–428.

Zhu, J., M. Pigna, V. Cozzolino, A.G. Caporale, and A. Violante. 2010. Competitive sorption of copper(II), chromium(III) and lead(II) on ferrihydrite and two organomineral complexes. Geoderma 159:409–416.

Zhu, J., M. Pigna, V. Cozzolino, A.G. Caporale, and A. Violante. 2011. Sorption of arsenite and arsenate on ferrihydrite: Effect of organic and inorganic ligands. J. Hazard. Mater. 189:564–571.

9
Soil Physicochemical and Biological Interfacial Processes Governing the Fate of Anthropogenic Organic Pollutants

Kun Yang
Zhejiang University

Baoshan Xing
University of Massachusetts

9.1	Introduction 9-1
	Soil System • Soil Pollution and Organic Pollutants in Soils • Physicochemical and Biological Interfacial Behaviors and Processes in Soil System
9.2	Sorption–Desorption Process in Soil–Water Interface................................. 9-5
	History and Development of AOPs in Soil Sorption Studies • Partition of HOCs to SOM • Adsorption and Isotherm Nonlinearity • Influence of SOM–Mineral Interactions on Sorption • Influence of Sorption and Desorption on Bioavailability of AOPs
9.3	Uptake and Degradation by Plant and Soil Microbes................................. 9-19
	Uptake and Degradation by Soil Microorganisms • Uptake and Degradation by Plant • Rhizosphere Processes of AOPs
9.4	Effects of Surfactant on Soil Physicochemical and Biological Processes...... 9-26
	Surfactant • Solubility Enhancement by Surfactant • Effect of Surfactant on the Sorption by Soil • Effects of Surfactant on Uptake and Degradation by Microorganisms • Effects of Surfactant on Uptake by Plants
9.5	Conclusions... 9-33
	Acknowledgments.. 9-34
	References... 9-34

9.1 Introduction

Anthropogenic organic pollutants (AOPs) may be defined as any organic chemical that is foreign to the natural ecosystem, which is released into the natural ecosystem by human activities and may pose adverse effects, either directly or indirectly, on humans and/or other organisms, and on the natural physical, chemical, and biological equilibrium/processes in the environment. Mostly, AOPs are commonly produced and used for the rapid industrialization in agriculture and expansion in various other industries. Emissions of these harmful AOPs have led to local and global deterioration of the environment when they have accumulated in air, water, sediments, soils, and biota, including man (Schwarzenbach et al., 2003). Most of these compounds are hydrophobic organic compounds (HOCs), having attracted a great deal of research interest for more than three decades because of their persistence in the environment and tendency to bioaccumulate. "Hydrophobic" means low water solubility, which is a major factor of HOCs affecting their fate in the environment (Stokes et al., 2006).

Besides the direct addition and disposal, AOPs from contaminated air and water may also enter the heterogeneous soil environment via means such as precipitation and agricultural runoffs. The fate and behavior of organic pollutants in soil is governed by soil characteristics, compound properties, and environmental factors such as temperature and precipitation. Possible fates of pollutants entering soil environments include leaching into groundwater, biodegradation, volatilization to air, binding to soil solid phases, and transfer to organisms. Soil can act as a "reservoir" for AOPs in environmental systems because soil has the ability to attenuate AOPs with a large capacity. The "saving" and "discharging" of AOPs in soil, governing the fate of AOPs, are controlled by a series of soil physicochemical and biological interfacial processes. These interfacial processes include sorption–desorption in soil–water interfaces, uptake of plants from soil solution in root–soil interface and rhizosphere, and uptake of soil microbe from soil solution in microbe–soil interface. The "saving" of organic pollutants in soil is controlled by sorption and sequestration from soil solution, while the "discharging" is regulated by desorption to soil solution, and uptake and degradation of plant and soil microbes. Among these processes, sorption and desorption, which decide the bioavailability of organic pollutants and arguably responsible for the majority of biogeochemical processes that occur in soil environment, therefore, play the most important role in the fate of organic pollutants. Furthermore, these interfacial processes can be promoted or restrained by the addition of surfactants in soil environment. The techniques to promote or restrain these interfacial processes are expected to be used in different applications such as remediation of contaminated soils.

This chapter will review these physicochemical and biological interfacial processes that govern the fate of AOPs and the effects of surfactants on these processes, with a particular focus on the sorption–desorption behaviors of organic pollutants in soil–water interface. We will primarily discuss the fate of HOCs in the chapter due to their known toxicity, persistence, and ecological risks. We will start out by a brief introduction of soil system, organic pollutants in soils, typical interfacial behaviors and processes in soil system, and environmental significance of these behaviors and processes for the HOC fate. Following these introductory sections, the main text will address sorption and desorption to or from soil minerals and natural organic matter (NOM), and their complexes. The main text includes three parts: sorption–desorption process in soil–water interface, uptake by plant and soil microbes, and effects of surfactants on typical processes at soil–microbe–plant interfaces.

9.1.1 Soil System

Soil is a heterogeneous, but live system. It is a thin surface layer, covering the bedrock of most of the land area of the earth, and is the final product of the weathering action of physical, chemical, geological, hydrological, and biological processes on parent rocks (Brady and Weil, 1996). Though only a tissue-thin layer compared to the earth's total diameter, to humans and most terrestrial organisms, soil is the most important part of the geosphere and a valuable natural resource that provides the basis of human existence. In addition to being the site of most food production, soil is the receptor of large quantities of pollutants and a key component of environmental chemical cycles (Chiou, 2002).

Scientists widely use the system-modeling approach to look at the processes that operate in a soil because in which a series of inputs, outputs, stores, processes, and recycling of materials and energy can be seen. In a soil system, materials are mostly stored in soil solid particles such as minerals and organic matter by sorption, in organisms such as plants and animals by uptake and accumulation, in soil water by dissolution, and in soil air by volatilization. Furthermore, soils are open and dynamic systems that undergo continual exchange of matter and energy with the atmosphere, hydrosphere, and biosphere. In these systems, materials and energy are gained and lost over time. Any change in the inputs or outputs to the system has a profound effect on the way in which the processes within the soil can operate and, consequently, on the characteristics of the resulting soil. AOPs, once released into soil, can be distributed in soil components by the processes that operate in the soil. Due to their toxicity, AOPs such as pesticides in soil may damage and kill soil organisms, cause health and ecological risks to humans and most terrestrial organisms by food supply, and contribute to water contamination by leaching and air pollution by volatilization.

A soil system, generally, includes the following main components: mineral particles, organic matter, water, air, and organisms. Figure 9.1 is a simplified presentation of soil compartments. Their proportions vary among different soils due to processes such as translocation, weathering of parent rock, decay and recycling of dead organisms. Even for a single soil, its components change in the layers and over time as well. Basically, soil is a three-dimensional labyrinth of water- or gas-filled pores and soil solid particles with different sizes, forms, and compositions. The small scales and spatial heterogeneity of soil make the estimations of physical and physicochemical conditions that surround pollutants and soil organisms extremely challenging (Chenu and Stotzky, 2002).

Water is part of the three-phase (solid–liquid–gas) system making up soil, which varies geographically with the soil and with the depth in a vertical soil profile. Soil water supports plants and the soil biota (animal and plant life). Water is probably the most important soil component and responsible for most of the translocation of materials such as nutrients from solid soil particles into plants because phase transfers of a chemical in soil are controlled by its dissolved species. The quantity of water in the soil is of critical importance because plants are sensitive to either a deficiency or excess of moisture. Even if there is water in the soil, however, it may not inevitably be available for plant use because water is located in different soil pores, the spaces between the soil particles. Three types of water (i.e., capillary, hygroscopic, and gravitational water) are commonly observed in the soil pores. Due to the negative charged surface of soil particles and the dipoles of water molecules, electro-molecular forces between the soil and water molecules create a thin layer of water with ordered molecules called *hygroscopic water* surrounding

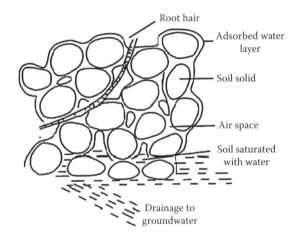

FIGURE 9.1 Fine structure of soil. (Reprinted from Manahan, S.E. 2000. Environmental chemistry. 7th edn. CRC Press, Boca Raton, FL. With permission.)

soil particles (McRae, 1988; Ehrlich, 2002). This water is bound with the inorganic and organic matter to make up the soil, which is held in the soil particles so strongly to prevent its use by plants and does not move as a liquid. Hygroscopic water is surrounded by a layer of *capillary water*, which may move from one soil particle to another by intermolecular attraction, but not by gravity or hydrostatic pressure (Kuznetsov et al., 1963; Ehrlich, 2002). The capillary water that is found in the soil pores and as layers around each grain of soil is most readily available to plants. When a soil is saturated by water, the excess water (i.e., *gravitational water*) fills the pore spaces. Gravitational water is commonly observed in big soil pores (bigger than 0.3 μm) and moves freely in soil pores by gravity or hydrostatic pressure (Standing and Killham, 2007). This water is difficult to be used by plant because it easily drains through the soil. The soil pores, which are not filled by water, are generally occupied by *soil air*. The composition of soil air is regulated by gaseous diffusion and soil respiration because soil pores are not often connected to open atmosphere. Therefore, soil vapor composition varies, but typically soil vapor contains the CO_2 concentration of 0.5%–5.0% and the O_2 concentration of 15%–20% (McRae, 1988).

The principal soil components are minerals and organic matter. The mineral element of the soil is derived from weathered parent rock. Most properties of soil minerals are largely determined by particle size. Particle size distribution determines mineral soil texture (stones >2 mm, sand grains 0.05–2 mm, silt 0.002–0.05 mm, clay particles <0.002 mm) and is often used in the classification of soil (McRae, 1988; Ehrlich, 2002). A decrease of particle size increases soil surface area (Chenu and Stotzky, 2002). Surfaces and edges of soil inorganic particles are often covered with negative charges (McRae, 1988). In most soils, the sand and silt consist largely of grains of resistant minerals, mainly quartz. Clays are composed of aluminosilicates and oxides in various crystalline and amorphous forms.

Soil organic matter (SOM), which originates primarily from biological degradation of plants as well as from animal remains, is composed of plant and animal debris, and intermediates and end-products of the decomposing debris (McRae, 1988). Soil biota is not considered to be a part of the SOM. SOM contains a large amorphous fraction of humus, which has lost most of its original structure and is moderately refractory to biological degradation. The humus fraction, which is soluble in or extractable into aqueous base, is operationally defined *humic substance*, presenting yellow to brown in color (Stevenson, 1994). The water insoluble and organic solvent soluble part of humus is often referred to as humin or kerogen (Schwarzenbach et al., 2003). The humic substances are further divided into humic acids (HAs) that precipitate at acidic pH and fulvic acids that do not (Ehrlich, 2002; Schwarzenbach et al., 2003). Although the molecular structures of humic substances have not been well characterized and defined, they are known to be a mix of low to high molecular weight amorphous materials containing numerous polar and oxygen-containing functional groups including carboxy-, phenoxy-, hydroxyl-, and carbonyl substituents. The number of such polar groups may vary significantly with the type of soils. Highly polar fulvic acids may have oxygen to carbon (O/C) molar ratios of near 0.5, whereas humin/kerogen has O/C ratios around 0.2–0.3 (Schwarzenbach et al., 2003). Except for relatively rare organic-rich soils, ordinary soils are rich in minerals. The organic matter content for most of these mineral-rich soils falls between 0.5% and 3.0% by weight (Chiou, 2002). In ordinary soils, more than 90% of the dry organic matter is made up of carbon, hydrogen, and oxygen, with minor amounts of nitrogen, sulfur, and phosphorus.

Plants and living organisms are also important in soil system due to their roles in forming SOM. Soil is the habitat of millions of living organisms. These living organisms include bacteria, fungi, and animals such as insects and earthworms. They feed on the complex organic compounds and are responsible for the decomposition of organic materials including organic pollutants.

9.1.2 Soil Pollution and Organic Pollutants in Soils

Soil pollution can be defined as entry of materials, mostly chemicals that are out of place or are present at concentrations higher than normal and have the potential to produce adverse effects on humans and/or other organisms. These adverse effects include the danger of acute toxicity, mutagenesis (genetic changes), carcinogenesis, and teratogenesis (birth defects). Soil pollution is often caused by human activities such as pesticide application and industrial waste materials, leading to the contamination of food grown on the soils. The rapid industrialization of agriculture, expansion of the chemical industry, and the need to generate cheap forms of energy have caused the continuous release of manmade chemicals into natural ecosystems. Consequently, the atmosphere, water, and many soil environments have become polluted by a large variety of toxic compounds. Besides the direct addition of xenobiotic chemicals and disposal of industrial waste materials to soil, AOPs from contaminated air and water may also enter the soil systems by precipitation and agricultural runoffs.

Both inorganic (those that do not contain carbon) and organic (those that contain carbon and hydrogen) chemicals are important contaminants in soil. Inorganic contaminants include inorganic acids, nitrates, phosphates, radioactive substances, and heavy metals such as cadmium, chromium, and lead. The most prominent chemical groups of organic contaminants are fuel hydrocarbons, polycyclic aromatic hydrocarbons (PAHs), polychlorinated biphenyls (PCBs), chlorinated aromatic compounds, detergents, and pesticides. Pharmaceuticals and personal care products including antibiotics, endocrine disrupting chemicals, and veterinary pharmaceuticals are emerging organic pollutants with rising concerns in most recent decade due to their adverse effects on wildlife and human (Daughton and Ternes, 1999). Most of these organic compounds are HOCs. They are at high concentrations and resistant to physical, chemical, or biological degradation in the environment, and thus follow prolonged exposure to humans and other organisms and represent an environmental burden of considerable magnitude. Measurements of levels of HOCs such as PCBs in soils that have been archived for several decades provide interesting insight into the contamination of soil and subsequent loss of these substances from soil (Alcock et al., 1993). For PAHs, it was observed that soils can be contaminated with between 1 μg kg^{-1} and 300 g kg^{-1}, depending on the sources of contamination (e.g., old coal gasification sites have the higher levels) (Kanaly and Harayama, 2000).

Soil pollution can lead to the pollution of water (including groundwater and surface water), atmosphere, plant, and soil microorganism through various processes. These processes include leaching into water (groundwater and surface water if contaminated runoff reaches streams, lakes, or oceans), uptake by plant and microorganism that grow on polluted soils, and release into the atmosphere especially by volatilization. Furthermore, these transfer processes can affect the consequent behaviors and fates of organic compounds and lead to serious health risks on human beings and animals through food chains. For example, persistent organic compounds (POPs) that are not water soluble tend to accumulate in plant and microorganism after uptake, and they also tend to accumulate increasingly toward the top of the food chain. Therefore, higher concentrations of POPs are always observed in creatures on the higher level of the food chain. Lake Michigan, as an example, has 2 μg kg^{-1} of DDT in the water, 14 μg kg^{-1} in the bottom mud, 410 μg kg^{-1} in amphipods (tiny water fleas and similar creatures), 3–6 mg kg^{-1} in fish such as coho salmon and lake trout, and as much as 99 mg kg^{-1} in herring gulls at the top of the food chain (www.pollutionissues.com/Re-Sy/Soil-Pollution.html).

9.1.3 Physicochemical and Biological Interfacial Behaviors and Processes in Soil System

An organic compound in the soil can remain as a free phase or distribute among the soil compartments. It can be adsorbed to inorganic or organic particle surfaces, absorbed into organic matter, dissolved to soil water, or vaporized into soil air. In addition, organic molecules dissolved in soil water may move by advection or by diffusion. Diffusion is the only mode of molecular transfer in the hygroscopic and capillary waters (Standing and Killham, 2007). Generally, the transfer of an organic compound from soil particles to other phases such as plants and microorganisms is controlled by the dissolved species and concentration of the compound in soil environments when water is present (Schwarzenbach et al., 2003). This is because soil particles are always surrounded by water films due to their charged surface and the dipoles of water molecules (McRae, 1988; Ehrlich, 2002). Furthermore, soil plants and organisms such as soil bacteria rely on organic and inorganic compounds dissolved in soil water for their nutrition (Chenu and Stotzky, 2002). Therefore, the distribution of an organic compound in soil particles, air, plants, and microorganisms and its transfer among these phases depend strongly on various physicochemical and biological interfacial

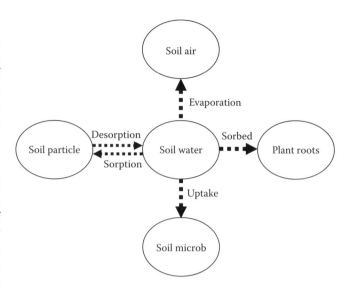

FIGURE 9.2 Physicochemical and biological interfacial behaviors related with soil water.

behaviors related with soil water. These behaviors include sorption–desorption in soil particle–water interface, release to soil air by volatilization from soil water in water–air interface, uptake of plant from soil water in root–water interface, and uptake of soil microorganism from soil water in microbe–soil interface (Figure 9.2).

Distribution of an organic molecule in these interfaces is in a dynamic state. A statistically apparent equilibrium, however, between the concentrations of the organic compound in soil particles and other phases such as soil water can be finally achieved, which helps us to understand the interfacial behaviors of organic compounds and their locations in different soil compartments. Although concentrations of contaminants in soil, water, and other phases in natural systems may deviate from those at equilibrium, the statistical equilibrium data can serve as an essential guide to the direction of contaminant movement at a particular point in time and to the likely consequence of an earlier contamination event. Such information is also valued in elucidating whether a compartment (such as soil particles) functions as a sink (to receive a contaminant) or as a source (to release a contaminant) under a specified condition. Therefore, most studies in this field are often conducted at an equilibrium (or near-equilibrium) condition. The distribution in air–water interface for volatile organic compounds by volatilization, as an example, is well known to be described and estimated by the equilibrium Henry constant. In a soil system, one interface behavior of an organic compound could be affected by another one. For example, uptake of an organic compound by plant root from soil water can be decreased by the sorption of the compound in soil particles from soil water. Among these interfacial behaviors, sorption and desorption, which decide the bioavailability of organic pollutants in contaminated soils and arguably responsible for the majority of biogeochemical processes that occur in soil environment, therefore, play the most critical roles in the fate of organic pollutants.

The understanding of the behaviors of organic compounds at these interfaces can help us to estimate the fate, distribution, transfer, bioavailability, and environmental risks of these compounds in the environment. It can also be of benefit to develop techniques and/or select a technique to alter the distribution and transfer of organic compounds at these interfaces for beneficial purposes. For example, one can select and culture a plant or microbe with high uptake and degradation of organic compounds for the remediation of contaminated soils. Furthermore, we can add an additive to a contaminated soil to enhance the sorption of organic compounds in soil particles to sequestrate them in soil, and consequently decrease its concentrations in soil water and its uptake by plant root for the safety of plant foods that grow on the contaminated soils.

This chapter presents an overview of the physicochemical and biological interfacial processes of nonionic HOCs among soil, water, plant, and microbe because of their environmental significance as described above. The sorption of ionic organic compounds will not be discussed because they constitute a small fraction of all organic contaminants and their most important interfacial process (i.e., sorption by soils) is generally well understood. For example, cationic organic compounds are known to adsorb on soils and clays primarily by a mechanism of cation exchange (Weber et al., 1965; Tahoun and Mortland, 1966), whereas organic anions do not have significant sorption on soils and clays (Harris and Warren, 1964; Weber et al., 1965).

9.2 Sorption–Desorption Process in Soil–Water Interface

9.2.1 History and Development of AOPs in Soil Sorption Studies

Sorption of organic compounds by soil particles is an important distribution process between soil particles and water, controlling the fate, transfer, transport, bioavailability, and eco-environmental risks of these compounds. The distribution of an organic solute between soil particles and water can be described by a soil–water distribution coefficient (K_d),

$$K_d = \frac{Q}{C_e} \quad (9.1)$$

where
Q is the solute concentrations in soil particles at equilibrium
C_e is equilibrium concentration of the solute in water

Studies in this field began in the 1940s for the assessment of the effectiveness and persistence of pesticides in agricultural soils. These studies became more active and intensified in the 1970s for the understanding of the long-term environmental and health risks of organic compounds because wide varieties and large quantities of pesticides and industrial organic wastes were widely found in the environment. It has now come to the characteristics of chemical–soil interactions in terms of soil organic and mineral components and of the properties of organic compounds for the predictions of their sorption behavior under different system conditions.

As early as 1968, it was discovered that the SOM was the principal soil component responsible for the sorption of HOCs, showing isotherm linearity (Lambert, 1968). The insignificant sorption of HOCs by soil minerals was explained by the competitive behavior between organic molecules and water on the mineral surfaces with a strong dipole interaction mechanism because the uptake of organic vapors by relatively dry mineral soils and clays was found to be suppressed strongly by soil moisture (Hanson and Nex, 1953; Chiou and Shoup, 1985). A near constant coefficient, named as soil organic carbon content normalized coefficient ($K_{oc} = K_d/f_{oc}$) or SOM content normalized coefficient ($K_{om} = K_d/f_{om}$), can be achieved by normalizing the soil–water distribution coefficients (K_d, the slopes of linear isotherms) to the organic carbon contents (f_{oc}) or organic matter contents (f_{om}) of soils (Karickhoff et al., 1979; Kenaga and Goring, 1980; Kile et al., 1995), where $f_{om} = 1.72 f_{oc}$ based on the common assumption that SOM contains about 58% of carbon content (Hamaker and Thompson, 1972). The organic carbon content normalized coefficients were largely independent on the nature of soils and SOM for a given HOC. Before 1979, SOM was considered as a high-surface-area adsorbent (Bower and Gschwend, 1952; Bailey and White, 1964) and the sorption of HOCs on SOM was explained by hydrophobic interactions (Weed and Weber, 1974; Browman and Chesters, 1977; Mingelgrin and Gerstl, 1983). A concept of partition was suggested by Chiou et al. (1979) to be the controlling mechanism for the HOC sorption in SOM but not the mechanism of hydrophobic sorption on high-surface-area SOM. The term *partition* refers to a process similar to solution in which an organic molecule penetrates into the entire network of the SOM. Linear relationships between logarithmic K_{oc}, octanol–water partition coefficients (K_{ow}) and water solubility (S_w) of HOCs were observed (Chiou et al., 1982, 1983). These relationships have been applied simply in the estimation of HOC sorption by soils and further applied in the estimation of bioavailability of HOCs and the engineered design of the remediation of HOC contaminated soils. This concept appeared to work at that time likely because most sorption data were obtained at relatively high concentrations of organic compounds and in a limited sorbent–sorbate contact time. This concept, however, was challenged after 1985 because of the findings such as nonlinear isotherms especially at relatively low concentrations of organic compounds (Weber et al., 1992; Pignatello and Xing, 1996; Xing et al., 1996; Huang et al., 1997; Xing and Pignatello, 1997, 1998; Chiou and Kile, 1998; Chiou et al., 2000), variable K_{oc} values (Rutherford et al., 1992; Spurlock and Biggar, 1994), irreversible sorption (Ball and Roberts, 1991; Pignatello and Xing, 1996; Xia and Pignatello, 2001; Lu and Pignatello, 2004a, 2004b; Sander and Pignatello, 2005b; Sander et al., 2006), and competitive sorption (Pignatello, 1991; Xing et al., 1996; Xing and Pignatello, 1997, 1998; Chiou and Kile, 1998; White et al., 1999; White and Pignatello, 1999; Zhao et al., 2001, 2002).

Irreversible sorption suggested the overestimation of bioavailability of soil sorbed HOCs based on the partition concept and limited the bioremediation potential of contaminated soils (Pignatello, 2006). Furthermore, the concentrations of HOCs in natural environment are commonly in the range of relative low concentrations. The above observations as well as multiphasic desorption kinetics (Ball and Roberts, 1991; Pignatello and Xing, 1996) led to the reconsiderations on the HOC sorption by soils. Soil particles are the complex of SOM, minerals, and other materials such as charcoal and soot, showing heterogeneous nature. These soil components, therefore, may play different roles in the sorption of organic compounds with specific mechanisms, that is, the sorption of organic compounds by soil particles is the combination of those on different soil components with a distributed reactivity. A number of conceptual models such as dual-mode model (DMM) and distributed reactivity model, based on the possible sorption by soil multiple domains, therefore, have been postulated to account for the nonlinear sorption on soils in the past decades. In these concept models, in addition to the partition of HOCs in SOM, adsorption on soil multiple domains was proposed to be responsible for the nonlinear sorption. The term *adsorption* refers to the condensation of organic molecule on the surfaces or pores of soil particles. The roles of carbonaceous geo-sorbents such as coal, charcoal, kerogen, coke, and soot in the nonlinear adsorption of HOCs were largely studied and presented in detail in several reviews, for example, Cornelissen et al. (2005). To date, sorption by soil particles is commonly explained by the combination of a linear partition mechanism and a nonlinear adsorption mechanism (Weber et al., 1992; Huang et al., 1997; Xing and Pignatello, 1997; Xia and Ball, 1999; Xia and Pignatello, 2001; Chiou, 2002). The combined isotherm (Figure 9.3) shows apparent linearity at relatively high concentrations of an organic compound and nonlinearity at relatively low concentrations.

9.2.2 Partition of HOCs to SOM

9.2.2.1 Characteristics of Partition in SOM

The partition of organic compounds in a partially miscible solvent–water system is an important behavior in chemistry as the basis for solvent extraction of solutes from water. This partition process has been employed by Chiou and co-workers (Chiou et al., 1979, 1983, 1985; Chiou, 2002) to interpret the sorption of HOCs by soils, where the SOM behaves as a partition medium in addition to the recognized dependence of soil sorption on SOM content. Lambert (1967) proposed that SOM functions in HOC sorption analogous to that of a solvent in partition chromatography, showing a correspondence between sorption coefficient and solvent–water distribution coefficient of an organic compound. It was also noted by Swoboda and Thomas (1968) that the linear sorption of parathion by soil might be partly interpreted by a liquid dissolved process in SOM similar to the partition process of the solvent extraction from water. Based on these two works, Chiou et al. (1979, 1983, 1985) developed and advanced the partition concept of HOC sorption in SOM in a series of studies. A set of unique features in the sorption of HOCs by soils and SOM in aqueous systems were found to associate the sorption behavior of HOCs by SOM with the partition concept in these studies. The first one is that the equilibrium isotherms of HOCs are linear at relatively high concentrations (Chiou et al., 1979; Karickhoff et al., 1979; Schwarzenbach and Westall, 1981; Sun and Boyd, 1991; Rutherford et al., 1992), however, linear isotherm for adsorption can be observed only at very low concentrations of HOCs. The second one is that no apparent sorptive competition was found in binary and multisolutes systems (Schwarzenbach and Westall, 1981; Chiou et al., 1983, 1985). The third is that the equilibrium heats of sorption for solutes are less exothermic than heats of solute condensation from water, showing very slight temperature effect on the sorption (Mills and Biggar, 1969; Spencer and Cliath, 1970; Yaron and Saltzman, 1972; Pierce et al., 1974; Chiou et al., 1979). The fourth is that a solute with higher K_{om} or K_{oc} value exhibits lower

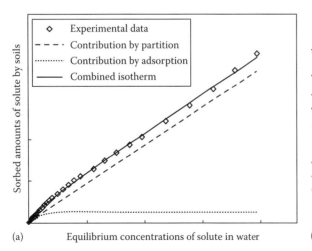
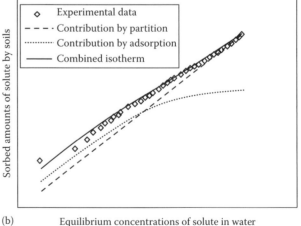

FIGURE 9.3 Schematic plots of contributions by partition and adsorption to the uptake of an organic compound by soils in aqueous system: (a) in normal scale and (b) in log–log scale.

sorption capacity in SOM (Chiou, 1998). For example, the solubility of DDT ($K_{om} \approx 1.5 \times 10^5$) in SOM is about 0.83 g kg^{-1}, which is approximately 40 times lower than that of benzene ($K_{om} \approx 18$) (Chiou, 1998). The fifth one is that negative linear relationships between log K_{om} (or log K_{oc}) and log S_w were observed, indicating that water solubility of organic chemicals is the major determinant of their sorption by SOM (Chiou et al., 1979, 1983; Karickhoff et al., 1979; Kenaga and Goring, 1980; Means et al., 1980; Briggs, 1981; Hassett et al., 1981; Karickhoff, 1984). Moreover, the low actual specific surface area (only about 1 m^2 g^{-1}) of SOM, as measured by the standard Brunauer–Emmett–Teller method with N$_2$ vapor as the adsorbate (Chiou et al., 1990; Pennell et al., 1995), is far too small to account for solute sorption by SOM with the adsorption mechanism established in the earlier views (Bower and Gschwend, 1952; Bailey and White, 1964; Weed and Weber, 1974; Browman and Chesters, 1977; Mingelgrin and Gerstl, 1983) that SOM was a high-surface-area adsorbent capable of adsorbing HOCs. The high surface area (550–800 m^2 g^{-1}) for SOM as measured by the ethylene glycol retention method (Bower and Gschwend, 1952), due to the earlier view that SOM was a high-surface-area adsorbent, was shown to be largely an artifact of the high solubility of polar ethylene glycol in relatively polar SOM (Chiou et al., 1990, 1993).

9.2.2.2 Effect of Polarity and Composition of SOM on Partition

Although the normalized partition coefficients (K_{om} or K_{oc}) of a given HOC was expected to be a constant and largely independent on the nature of SOM for the estimation of its sorption, they were observed varied in many studies (Bailey and White, 1964; Goring, 1967; Kenaga and Goring, 1980; Mingelgrin and Gerstl, 1983; Rutherford et al., 1992; Xing et al., 1994; Kile et al., 1995). Part of this variation may be a result of different experimental procedures using different analytical techniques (Bailey and White, 1964; Goring, 1967). Considering the heterogeneous nature of SOM, which contains various compositions and structures, however, the varied partition coefficients (K_{om} or K_{oc}) can also result actually from the different partitioning abilities of SOM components (Rutherford et al., 1992). Therefore, it is of practical interest to understand how much the SOM medium property affects the partition coefficients (K_{om} or K_{oc}). This information is critical to whether SOM from different geographic locations and with different humification degrees need to be studied individually (if K_{om} or K_{oc} values vary widely) or can be treated rather indiscriminately (if K_{om} or K_{oc} values are relatively invariant).

Based on the selected K_{oc} data from different reports, Kenaga and Goring (1980) observed that the K_{oc} variation between soils is generally less than a factor of 3–4. However, Mingelgrin and Gerstl (1983) noted that the K_{oc} could vary by as large as a factor of 10 or greater for some pesticides. It has also been observed that kerogen material in SOM has the higher K_{oc} values for phenanthrene than the common humic substances in soils and sediments (Ran et al., 2004). Rutherford et al. (1992) observed that the partition coefficients (K_{oc} or K_{om}) of organic chemicals such as carbon tetrachloride (CT) have an inverse relation with the elemental weight ratio of [oxygen + nitrogen] to carbon of SOM (Figure 9.4), that is, the (O + N)/C value, an index describing approximately the polarity of SOM. Four relatively ash-free SOM samples including cellulose, muck, peat, and treated peat (peat washed with 0.1N NaOH) were investigated in this study. This inverse relation shows directly the effect of SOM elemental compositions and polarity on solute partition. Utilizing this correlation, Rutherford et al. (1992) noted that the K_{oc} variation of nonpolar organic chemicals between SOM to be less than a factor of 3 if the range of (O + N)/C values for common SOM is considered. Similar correlations between the sorption coefficients (K_{oc}) of phenanthrene and the (O + N)/C values of sequentially extracted soil-derived HAs and humins from a peat were obtained (Kang and Xing, 2005; Yang and Xing, 2009). These inverse correlations indicate that the partition of a nonpolar solute to SOM is sensitive to SOM polarity (or composition) and the polar SOM is a poor partitioning phase for HOCs. Highly humified organic matter such as humin has low polarity and strong partition of HOCs (Kang and Xing, 2005, 2008). The K_{oc} could vary by as large as a factor of 10 between HAs and humins for phenanthrene (Kang and Xing, 2005). This result suggested that HOC sorption was greatly affected by chemical structure and composition of humic substances, even from the same soil. For a more accurate account of the K_{om} or K_{oc} variation in common natural soils and sediments, Kile et al. (1995) measured the K_{oc} values of two relatively nonpolar solutes, CT and 1,2-dichlorobenzene (DCB), on a large set of soils and sediments with significant SOM contents. Thirty-two "normal" soils and 36 "normal" bed sediments collected from diverse geographic locations in both the United States and China were investigated in this

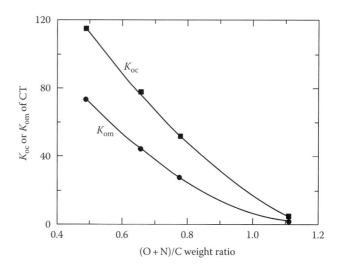

FIGURE 9.4 Plot of the CT K_{om} or K_{oc} value versus the (O + N)/C ratio of natural organic matter. (Reproduced with permission from Rutherford, D.W., C.T. Chiou, and D.E. Kile. 1992. Influence of soil organic matter composition on the partition of organic compounds. Environ. Sci. Technol. 26:336–340. Copyright 1992 American Chemical Society.)

study. The mean K_{oc} value for CT and DCB on 32 normal soils is 60 ± 7 and 290 ± 42, respectively. The mean K_{oc} value for CT and DCB on bed sediments is 102 ± 11 and 502 ± 66, respectively, showing a factor of 1.7 greater than that for CT and DCB on soils. The extreme K_{oc} values for soils (or sediments) differ by less than a factor of 2, implying the high degree of invariance of K_{oc} values between most soils or between most bed sediments. This finding reveals firstly that the sediment NOM has a generally lower polarity than SOM, implying that the eroded process of soils into bed sediments brings a noticeable change in the property of the organic composition. A possible cause for this change is that the sedimentation process fractionates SOM such that the more polar and water-soluble organic components in SOM (e.g., fulvic and HA fractions), or that the soil particles with more polar organic components, are separated out to form dissolved organic matter (DOM) and colloids in water, with the less-polar soil organic constituents preserved in the bed sediment (Chiou, 2002). Second, the (O + N)/C values for normal soils should fall into a relatively narrow range to show the similarity in SOM polarity/composition between soils and between bed sediments, according to relatively constant K_{oc} values of CT and DCB on soils from diverse sources (K_{oc} = 60 ± 7 for CT and 290 ± 42 for DCB). Moreover, the mass ratios of HA to humin for normal soils should vary small based on the large variation of phenanthrene K_{oc} between HAs and humins (Kang and Xing, 2005). The polarity of SOM is often from its polar groups such as –COOH and –OH, as can be obtained from the ^{13}C-NMR data of SOM (Kile et al., 1999). From the ^{13}C-NMR data of SOM, one can also obtain the aromatic and/or aliphatic fraction contents (Chefetz et al., 2000; Kang and Xing, 2005). With pyrene sorption to SOM varying in chemical composition (e.g., high aliphaticity or aromaticity), Chefetz et al. (2000) observed a positive relationship between the K_{oc} and the aliphaticity of sorbents. Moreover, a comparison between phenanthrene K_{oc} values and paraffinic carbon content of various SOM revealed a positive trend between these two variables in general (Salloum et al., 2002). The conclusion reached by these studies was that aliphatic components in SOM contribute substantially to HOC sorption. Although this positive relationship was also observed for sequentially extracted soil HAs, it failed when humin samples were employed in this relationship (Kang and Xing, 2005). Therefore, polarity of SOM seems to mainly regulate the magnitude of HOC partition rather than structure alone (Kang and Xing, 2008).

9.2.2.3 Effect of Water Solubility and the Polarity of Organic Compounds on Partition

Water solubility, for standpoint of organic chemicals, is the major determinant of their partition into an organic partition medium such as organic solvents. Among the studies of the partition behaviors of organic compounds in various solvent–water systems, the partition between octanol and water is the most used behavior because of the success of *n*-octanol as a surrogate for NOM and/or biological organic components such as lipid. Octanol molecule can mimic some extent of NOM and biological organic components through its hydrophilic –OH group and lipophilic alkyl chain. Negative linear relationships between octanol–water partition coefficients (K_{ow}) and water solubility (S_w) of HOCs have been observed for the partition of HOCs between octanol and water (Chiou et al., 1982, 1983). SOM has been accepted to be a partition medium for organic compounds in soil–water system (Chiou et al., 1979, 1983). Negative linear relationships between log K_{om} (or log K_{oc}) and log S_w were also observed in previous works, indicating water solubility is also the major determinant for solute partition into SOM (Chiou et al., 1979, 1983; Karickhoff et al., 1979; Kenaga and Goring, 1980; Means et al., 1980; Briggs, 1981; Hassett et al., 1981; Karickhoff, 1984). Therefore, the solute log K_{om} expresses a linear relationship with solute log K_{ow}.

In addition to the dominant effects of water solubility of solutes on their partition into SOM, polarity of solutes can have significant influence on their partition. This solute polarity influence on partition was reflected by the difference of the linear relationships between solute log K_{om} and solute log K_{ow} for low-polarity solutes and high-polarity solutes (Figure 9.5). Equation 9.2 gives a linear relationship between log K_{om} and log K_{ow} for low-polarity solutes (Chiou et al., 1983), whereas that for high-polarity solutes is given in Equation 9.3 (Briggs, 1981).

For low-polarity solutes

$$\log K_{om} = 0.904 \log K_{ow} - 0.779 \quad (9.2)$$

and for high-polarity solutes

$$\log K_{om} = 0.52 \log K_{ow} + 0.64 \quad (9.3)$$

Compared the slopes and intercepts of Equation 9.3 with Equation 9.2, high-polarity solutes has a much smaller slope but a much larger intercept in the linear relationship between

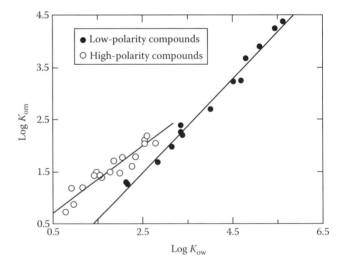

FIGURE 9.5 Plots of log K_{om} versus log K_{ow} for low-polarity solutes (Chiou et al., 1983) and high-polarity solutes (Briggs, 1981). (Chiou, C.T. 2002. Partition and adsorption of organic contaminants in environmental systems. John Wiley & Sons, New York. Copyright Wiley-VCH Verlag GmbH & Co. KGaA. Reprinted with permission.)

log K_{om} and log K_{ow} than low-polarity compounds. For low-polarity compounds with log $K_{ow} < 7$, the K_{om} values are roughly smaller than the corresponding K_{ow} values, suggesting that SOM is not a good solvent for low-polarity solutes as octanol and the SOM polarity is higher than that of octanol. However, SOM is a better solvent for high-polarity solutes than low-polarity solutes, showing the higher K_{om} values of high-polarity solutes than that of low-polarity solutes for a given solute with $K_{ow} < 3.8$ (Figure 9.5). The polar high-polarity solutes exhibiting a considerably higher partition (K_{om}) than the low-polarity solutes in SOM could due presumably to the more powerful molecular interactions between solute and SOM from polar and hydrogen-bond forces over that from London (dispersion) forces (Chiou and Kile, 1994). Moreover, a polar solute could have better compatibility than nonpolar solutes with relatively polar SOM.

9.2.2.4 Effect of Dissolved SOM on Partition

Enhancement of organic solute solubility in water should decrease the solute partition into SOM because partitioning into SOM depends inversely on the water solubility of organic solutes (Chiou et al., 1979, 1983; Karickhoff et al., 1979; Kenaga and Goring, 1980; Means et al., 1980; Briggs, 1981; Hassett et al., 1981; Karickhoff, 1984). It is known that dissolved or suspended SOM in water is able to significantly enhance the water solubility of organic compounds especially for extremely insoluble ones (Wershaw et al., 1969; Carter and Suffet, 1982; Gschwend and Wu, 1985; Chiou et al., 1986). For example, Wershaw et al. (1969) observed that the solubility of DDT in the water solution containing 0.5% SOM is more than 200 times greater than that in pure water (5.5 mg L^{-1}). This solubility enhancement effect expresses a linear correlation between the water solubility of an organic solute in SOM solution (S_w^*) and the dose of SOM in water (X) (Figure 9.6). Solubility of the solute in pure water (S_w) is the intercept of this linear correlation (Figure 9.6). Partition of solutes into the microscopic organic environment of the DOM (Chiou et al., 1986) was suggested to interpret this solubility enhancement. The linear relation between S_w^* and DOM concentration (Figure 9.6), the decrease in solubility enhancement with increasing S_w, and the lack of interference between binary solutes (Figure 9.6), observed by Chiou et al. (1986), are in keeping with the partition interaction between the solute and DOM. Based on the partition theory, one can obtain simply the partition-like coefficient (K_{dom}) of an organic solute distributed between DOM and water to describe the solubility enhancement via the equation established by Chiou et al. (1986), which gives

$$K_{dom} = \frac{C_e^* - C_e}{XC_e} \quad (9.4)$$

that is,

$$C_e^* = C_e + XK_{dom}C_e = C_e(1 + XK_{dom}) \quad (9.5)$$

where C_e^* includes the solute concentrations in pure water (C_e) and in DOM ($C_e^* - C_e$)/X. If the solute saturated concentration in pure water (water solubility, S_w) is considered for its C_e, the

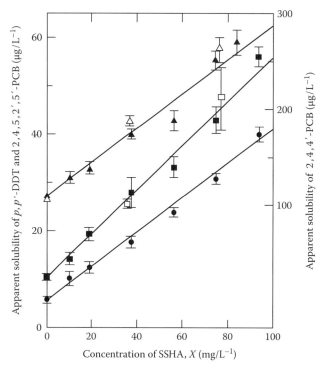

FIGURE 9.6 Apparent water solubility of p,p'-DDT (●), 2,4,5,2'5'-PCB (■,□), and 2,4,4'-PCB (▲,△) as a function of aqueous concentration of Sanhedron soil HA (SSHA) at 24°C–25°C. Solid symbols are for single solutes; open symbols for the two PCBs are for their binary mixtures. (Reprinted with permission from Chiou, C.T., R.L. Malcolm, T.I. Brinton, and D.E. Kile. 1986. Water solubility enhancement of some organic pollutants and pesticides by dissolved humic and fulvic acids. Environ. Sci. Technol. 20:502–508. Copyright 1986 American Chemical Society.)

apparent solute concentration in water with a given dose (X) of DOM should also be saturated. Therefore, a relationship between the solute saturated concentration in water with a given dose (X) of DOM (S_w^*) and solute saturated concentrations in pure water (S_w) can be established based on Equation 9.5, which gives

$$S_w^* = S_w(1 + XK_{dom}) \quad (9.6)$$

Equation 9.6 represents the linear correlation, as shown in Figure 9.6, between the water solubility of an organic solute in DOM solution (S_w^*) and the dose of DOM in water (X) at the same temperature.

Relating Equation 9.5 with Equation 9.1, one can obtain an apparent soil–water distribution coefficient for the solute (K_d^*) at a given dose (X) of DOM in water, which gives

$$K_d^* = \frac{Q}{C_e(1 + XK_{dom})} = \frac{K_d}{1 + XK_{dom}} \quad (9.7)$$

where

Q is the solute concentration in soil particles at equilibrium
K_d is the soil–water distribution coefficient for the solute in pure water

According to Equation 9.7, the more SOM dissolved in water, the lower partition of organic solutes into the solid SOM in soil particles. Moreover, the organic solute with higher K_{dom} should express lower partition into the solid SOM in soil particles if DOM is present in water. With $K_d = f_{om}K_{om}$ or $K_d = f_{oc}K_{oc}$, as defined above for the linear partition of organic solutes into SOM particles, Equation 9.7 turns to

$$K_{om}^* = \frac{K_{om}}{1 + XK_{dom}} \quad (9.8)$$

or

$$K_{oc}^* = \frac{K_{oc}}{1 + XK_{dom}} \quad (9.9)$$

Based on the partition theory, K_{dom} should depend on the type of solute and the composition of DOM. Using a series of solutes with vastly different S_w values and of DOM with varied compositions and structures, Chiou et al. (1986, 1987) examined the effects of solute S_w and DOM composition on K_{dom}. For a given DOM, it was observed that the solute K_{dom} values showed an inverse relation with solute S_w values but a positive relation with respective log K_{ow} values. Solute K_{dom} values become negligible for relatively water-soluble solutes such as lindane and 1,2,3-trichlorobenzene (TCB). An empirical rule, that the concentrations of DOM in water for an enhancement of solute K_{dom} with a factor of 2 must be at least about two orders of magnitude greater than the solute S_w, was suggested by Chiou et al. (1986, 1987) for most nonpolar solutes except for PAHs. Because the DOM concentrations in natural water seldom exceed 100 mg L^{-1}, only those solutes with S_w < 1 mg L^{-1} could possibly exhibit a significant increase in their concentrations by DOM in natural water. The magnitude of K_{dom} was also observed to depend on the polarity as measured by (O + N)/C and molecular size of DOM (Chiou et al., 1986, 1987). Although DOM such as phenylethanoic acid may have low polarity (70.6% C, 5.9% H, and 23.5% O), they are unable to produce a strong solubility enhancement effect on organic solutes as humic and fulvic acids because of the small DOM size (the molecular weight of phenylethanoic acid is 136.15 g mol^{-1}), in agreement with the required DOM size in DOM partition interaction. However, the molecular size of DOM is not the sole factor determining the solubility enhancement of organic solutes. For example, polyacrylic acid with large molecular size (MW = 2000 and 90,000 g mol^{-1}) and comparable elemental composition (50% C, 5.6% H, and 44.4% O) with natural DOM, has not shown solubility enhancement effect on organic solutes (Chiou et al., 1986). The inability of polyacrylic acid in solute solubility enhancement is attributed to the frequent and orderly attachment of hydrophilic carboxyl groups to the carbon chain and to an extended chain structure, which prohibit the formation of a sizable intra-molecular nonpolar environment. Therefore, in order to promote partition interaction with organic solutes, the DOM molecules must possess both a favorable size of nonpolar moiety and a sufficiently large molecular weight.

In addition to the DOM polarity and molecular size, the DOM aromatic content could significantly influence on the solubility enhancement of aromatic chemicals such as PAHs. For example, Chin et al. (1997) found that the magnitude of solubility enhancement of pyrene, by a series of DOM with minor differences in oxygen content, has a good correlation with the DOM aromatic contents and its molecular weight. Furthermore, the K_{dom} value of pyrene by the dissolved Aldrich HA (a commercial HA with a very high aromatic content) is only about a factor of 2 smaller than that of DDT, whereas the K_{ow} value of DDT is about 15 times higher that of pyrene (Chin et al., 1997). The π–π interactions between DOM and PAH aromatic moieties may play an important role in the solubility enhancement of PAHs (Gauthier et al., 1987).

On a unit weight basis, the observed K_{dom} values of organic solutes between DOM and water were about half as large as the K_{om} values of the solutes between solid bulk SOM and water (Chiou et al., 1986), suggesting that the configurations of SOM dissolved in water or adsorbed on soil particle surfaces may also play a critical role in the partition. The dissolved SOM in water, similar to surfactants, could be expected to exist as micelle-like aggregates with relatively small size, whereas adsorbed SOM on soil particle surfaces may exist as semi-micelle-like aggregates with relatively large size. Through the interactions such as the hydrogen-bond most likely between SOM polar fractions and soil particle surfaces, the long chain fractions of adsorbed SOM on soil particle surfaces can come close together through hydrophobic interaction to form the large aggregates. Therefore, adsorbed SOM aggregates could possess a more favorable nonpolar moiety with a sufficiently large size than the dissolved SOM aggregates for solute partition. However, DOM may be different from the bulk SOM in terms of compositions and structures, which can also result in the difference between K_{dom} and K_{om}.

In soil–water systems, the DOM amount (X) by dissolving SOM in water usually increases with increasing soil-to-water (soil/water) ratio. According to Equation 9.7, the enhanced solute solubility by DOM, therefore, offers a logical explanation to the decrease of solute K_d^* with the increasing soil/water ratio in the soil–water systems observed by Gschwend and Wu (1985). The large variation in log K_{om} for highly water-insoluble solutes observed by Means et al. (1982) and Karickhoff (1984) could be interpreted at least in part by the different extents of solute solubility enhancement by DOM because various solid/water ratios were used in the sorption experiment. However, the significant effect of enhanced solute solubility by DOM on the decrease of solute K_d^* could be observed only for highly water-insoluble solutes. For relatively water-soluble solutes such as lindane and TCB, this effect may become negligible because their solubility enhancement by DOM is insignificant and the DOM concentrations in natural water are often less than 100 mg L^{-1} (Chiou et al., 1986, 1987).

9.2.3 Adsorption and Isotherm Nonlinearity

Although the sorption of HOCs by soils commonly shows isotherm linearity especially at relatively high concentrations (C_e/S_w) and is interpreted by the partition-like mechanism of

HOCs into SOM, which is accepted as the partition medium (Chiou et al., 1979; Chiou, 2002), later studies on the HOC sorption by soils and sediments observed the isotherm nonlinearity at relatively low concentrations and the solute concentration-dependent sorption coefficients (Ball and Roberts, 1991; Weber et al., 1992; Spurlock and Biggar, 1994; Xing et al., 1996). This isotherm nonlinearity phenomenon was further proved by the findings such as irreversible sorption (Pignatello and Xing, 1996; Xia and Pignatello, 2001; Lu and Pignatello, 2004a, 2004b; Sander and Pignatello, 2005b; Sander et al., 2006) and competitive sorption (Xing et al., 1996; Xing and Pignatello, 1997, 1998; Chiou and Kile, 1998; White et al., 1999; White and Pignatello, 1999; Zhao et al., 2001, 2002). The chemicals employed in these studies include nonpolar and polar solutes (e.g., phenanthrene, triazines, and trichloroethylene). The isotherm nonlinearity indicates the higher sorption coefficients of solutes at relatively low concentrations compared to the upper linear sorption range. Since the concentrations of organic contaminants in natural environment are often fall into the relatively low concentration range, the linear partition theory may underestimate the adsorbed amount of organic contaminants by soils. Moreover, the irreversible sorption suggested the overestimation of bioavailability of soil sorbed HOCs based on the partition theory and limited the bioremediation potential of contaminated soils (Pignatello, 2006). Therefore, it is of practical interest to examine the cause of such nonlinear sorption for organic solutes at low C_e/S_w.

Considering the heterogeneous nature of soils, a number of conceptual models were postulated to account for the observed isotherm nonlinearity on soils: (1) the unsuppressed adsorption of polar solutes by water on certain soil minerals could result in nonlinear sorption at low C_e/S_w (Laird et al., 1992; Haderlein and Schwarzenbach, 1993; Haderlein et al., 1996; Weissmahr et al., 1997; Chiou and Kile, 1998; Chiou et al., 2000; Chiou, 2002); (2) SOM was considered to be a heterogeneous substance, consisting of two types of domains characterized as "soft carbon, which was envisioned as a highly amorphous and swollen SOM" and "hard carbon, which was envisioned as a condensed and relatively rigid organic matrix" (Huang et al., 1997; Leboeuf and Weber, 1997), where the solute sorption to "soft carbon" SOM is controlled by partition with linear isotherm and that to "hard carbon" portion of SOM is controlled by a surface adsorption with nonlinear isotherm (Weber et al., 1992; Young and Weber, 1995; Huang et al., 1997; Leboeuf and Weber, 1997); (3) SOM was also assumed to be consisted by two structural entities, one in a rubbery state and the other in a glassy state analogous to rubbery and glassy synthetic polymers respectively, where the nonlinear isotherm occurs as a result of the solute adsorption onto the glassy portion of SOM because of the existence of different sets of specific "internal holes" for adsorption, in addition to the linear partition into the rubbery portion of SOM (Pignatello and Xing, 1996; Xing et al., 1996; Xing and Pignatello, 1997); (4) a small amount of carbonaceous material (such as charcoal or soot), or high-surface-area carbonaceous material (HSACM), exists in soils in addition to SOM, the former exhibiting a greater nonlinear adsorption at relatively low concentrations and SOM responsible for linear partition (Chiou, 1995; Gustafson et al., 1997; Chiou and Kile, 1998; Chiou et al., 2000; Cornelissen et al., 2005); (5) the presence of limited active sites in SOM exhibits specific interactions with polar solutes, which is approaching saturation at lower solute concentrations, in addition to the linear partition to SOM (Spurlock and Biggar, 1994). Later, Chiou and Kile (1998) and Chiou et al. (2000) indicated that the combination of the specific-interaction model for polar solutes and the HSACM model for nonpolar solutes may reconcile the outstanding features of the nonlinear and competitive sorption of the solutes with certain soil samples. However, nonlinear isotherms were observed in black carbon (BC)-free NOM samples (Pignatello et al., 2006). Overall, soil is considered to be a dual sorbent for organic compounds. Sorption of organic compounds by soils can be explained by the combination of a linear partition mechanism and a nonlinear adsorption mechanism as shown in Figure 9.3. Therefore, total sorption by NOM is predominated by partitioning at relatively high concentrations showing isotherm linearity, while predominated by adsorption at relatively low concentrations showing commonly isotherm nonlinearity (Figure 9.3).

9.2.3.1 Adsorption of HOCs on Soil Minerals

In soil–water system, SOM was observed to be the principal soil component controlling the sorption of HOCs, while the sorption of HOCs by soil minerals was insignificant (Lambert, 1967; Saltzman et al., 1972; Karickhoff et al., 1979; Kenaga and Goring, 1980; Kile et al., 1995). The lack of adsorption of polar 2,4-dichlorophenol (2,4-DCP) on montmorillonite and amorphous silicon oxide (Yang et al., 2004), phenol on goethite (Yost and Anderson, 1984), and parathion on clay minerals (Saltzman et al., 1972) indicates also the negligible role of soil minerals in sorption of HOCs by soils. In previous studies, it was observed that the vapor uptake of methyl bromide (Chisholm and Koblitsky, 1943), chloropicrin (Stark, 1948), ethylene dibromide (Hanson and Nex, 1953; Wade, 1954), benzene (Chiou and Shoup, 1985), m-dichlorobenzene (Chiou and Shoup, 1985), and 1,2,4-trichlorobenzene (Chiou and Shoup, 1985) was reduced progressively by increasing soil moisture. The suppression of increasing relative humidity on vapor uptake of m-dichlorobenzene by soil (Chiou and Shoup, 1985), as an example, is given in Figure 9.7. Similar suppression of soil moisture was observed for the uptake of parathion and lindane by soils from hexane solution (Yaron and Saltzman, 1972; Chiou et al., 1985). These observations indicate that water can exhibit strong competition with organic compounds on soil mineral surface, thus, the surface of soil minerals is unavailable for adsorption of organic compounds (i.e., the insignificant sorption of HOCs by soil minerals). Even for polar organic solutes such as 2,4-DCP (Yang et al., 2004) and phenol (Yost and Anderson, 1984), the competitive adsorption of water on soil minerals is also so strong to avoid the adsorption of polar solutes.

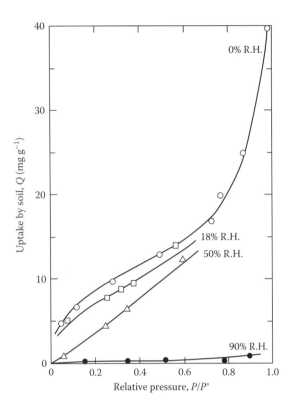

FIGURE 9.7 Vapor uptake of m-dichlorobenzene on dry Woodburn soil as a function of relative humidity (RH) at 20°C. (Reprinted with permission from Chiou, C.T., and T.D. Shoup. 1985. Soil sorption of organic vapors and effects of humidity on sorptive mechanism and capacity. Environ. Sci. Technol. 19:1196–1200. Copyright 1985 American Chemical Society.)

In some cases for polar solutes, adsorption on soil minerals was notably high, even higher than the partition into SOM. For example, Haderlein and Schwarzenbach (1993) and Haderlein et al. (1996) found that a series of polar nitroaromatic compounds (NACs), with electron-withdrawing substituents that are in resonance with the aromatic ring, may adsorb specifically and reversibly to the siloxane surface present in clay minerals such as kaolinite, illite, and montmorillonite. The adsorption of NACs on other natural minerals including aluminum and iron (hydr)oxides, carbonates, and quartz is negligible. Moreover, the affinity of NACs on siloxane surfaces of clay minerals strongly depends on the type of exchangeable cation(s) adsorbed on the siloxane surfaces. For a given clay, the sorption of NACs was relatively high and exhibited notably nonlinear isotherms (Figure 9.8) and competition among NACs (Figure 9.9) when the clay was exchanged with weakly hydrated cations (e.g., K^+, Cs^+, or NH_4^+), while that became negligible when the clay was exchanged with strongly hydrating Na^+, Ca^{2+}, Mg^{2+}, and Al^{3+} ions. The negligible adsorption of NACs on clays exchanged with the latter set of cations is because of the strong hydration interaction between water and strongly hydrated cations on siloxane surfaces, then the clay siloxane surfaces was occupied by water molecules and thereby inaccessible to polar solutes (Weissmahr et al., 1997). The specific adsorption of NACs on siloxane surface of clay minerals

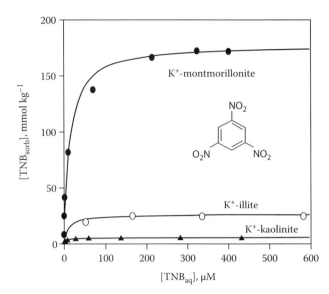

FIGURE 9.8 Adsorption isotherms of the explosive TNB measured in aqueous suspensions of homoionic K^+-clays. (Reprinted with permission from Haderlein, S.B., K.W. Weissmahr, and R.P. Schwarzenbach. 1996. Specific adsorption of nitroaromatic explosives and pesticides to clay minerals. Environ. Sci. Technol. 30:612–622. Copyright 1996 American Chemical Society.)

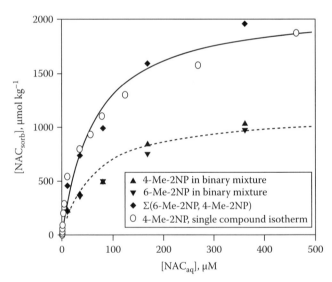

FIGURE 9.9 Example of competitive adsorption among NACs on homoionic K^+-clays. (Reprinted with permission from Haderlein, S.B., K.W. Weissmahr, and R.P. Schwarzenbach. 1996. Specific adsorption of nitroaromatic explosives and pesticides to clay minerals. Environ. Sci. Technol. 30:612–622. Copyright 1996 American Chemical Society.)

was interpreted by the coplanar electron donor–acceptor interactions between clay's siloxane oxygens as electron-donors and NAC's oxygens as electron-acceptor (Haderlein et al., 1996). The role of hydrogen-bond between siloxane oxygens and NAC's hydrogens may be insignificant (Weissmahr et al., 1997). The capacity of clays saturated with K^+, for 1,3,5-trinitrobenzene (TNB) adsorption, follows the order montmorillonite > illite >

kaolinite (Figure 9.8), which could be mainly attributed to the areas of siloxane surfaces present in these clays. However, K$^+$-saturated illite exhibited less effective than K$^+$-saturated montmorillonite for NACs adsorption when normalizing the capacity of TNB (Figure 9.8) to the siloxane surface area of illite (26 m^2 g^{-1}) and montmorillonite (71 m^2 g^{-1}) (Haderlein et al., 1996). The less effective adsorption of illite could probably be because illite has a much higher charge density on some surface sites to make the surface more hydrophilic.

The significant influence of exchangeable cations on the adsorption of NACs indicates that the sites on siloxane surface are only accessible for NACs when these sites are occupied by weakly hydrated cations. On this basis, the high adsorption of atrazine by Ca^{2+}-saturated smectites with low cation exchange capacity, as observed by Laird et al. (1992), may be attributed to the existence of abundance hydration-unaffected sites on siloxane surfaces. The negligible adsorption of 2,4-DCP on montmorillonite and amorphous silicon oxide (Yang et al., 2004), phenol on goethite (Yost and Anderson, 1984) and parathion on clay minerals (Saltzman et al., 1972) could be attributed to the absence of (effective) siloxane surface for minerals, and electron-withdrawing substituents in the aromatic ring for chemicals, particularly for the strongly hydrated cations present on siloxane surfaces. In natural environment, strongly hydrated cations (e.g., Ca^{2+}, Al^{3+}, Mg^{2+}) are usually prevailing. The theory that significant adsorption of polar organic solutes on soil/sediment minerals is based on an assumption that natural soils/sediments are saturated by weak hydrated cations (e.g., NH$_4^+$, C$_s^+$), which is very unusual for ordinary soils. Neglecting the prevailing effect of strongly hydrated cations present in natural environment may lead to the overestimation of adsorption onto natural minerals and misinterpretation of the importance of mineral phases as the controlling factor of nonlinear sorption for natural soils.

Low adsorption of low-polarity nonionic organic solutes (e.g., lindane, chlorobenzenes, and PAHs) and polar nonionic solutes (1,3-dinitrobenzene and 2,4-dinitrotoluene), from single- and binary-solute solutions, on virtually organic-free (<0.01%) hydrophilic minerals (e.g., silica and alumina) and low-organic-carbon clays (e.g., kaolinite and smectite) was also observed (Boucher and Lee, 1972; Huang et al., 1996; Mader et al., 1997; Hundal et al., 2001; Su et al., 2006). Characteristics for this slight adsorption include: (1) isotherm is linear (Boucher and Lee, 1972; Huang et al., 1996; Mader et al., 1997; Hundal et al., 2001; Su et al., 2006) and exhibits practically no competition between solutes (Su et al., 2006); (2) a short time for solution-solid equilibrium (Huang et al., 1996); (3) a dependence of the adsorption coefficient (K_d) on the mineral surface area (Huang and Weber, 1997) and solute water solubility (Mader et al., 1997); (4) no marked effect of solution pH, ionic strength, and solid surface charge and its sign on the adsorption of nonpolar solutes (Mader et al., 1997; Su et al., 2006), while a significant effect of ionic strength and pH on the adsorption of polar solutes (Su et al., 2006); and (5) low exothermic heats compared to the solute condensation heat from water (Huang and Weber, 1997; Su et al., 2006). The adsorption of nonpolar solutes was suggested to happen presumably by London (dispersion) forces onto a water film above the mineral surface rather than the mineral surface directly, while that of polar solutes is also assisted by polar-group interactions in addition to London forces (Su et al., 2006). The reduced adsorptive forces of solutes with hydrophilic minerals, which is responsible for the low solute adsorption, are due to the physical separation by the water film above the mineral surface (Su et al., 2006). The low fractions of the water-film surface covered by solutes offer a theoretical basis for linear solute adsorption, low exothermic heats, and no adsorptive competition (Su et al., 2006). Low adsorption of organic solutes by minerals, as described above, suggests that adsorption on soil minerals may be the major/dominant process if these minerals contains very little or no SOM. However, if soils contain higher SOM, normally higher than 0.1%–0.2%, adsorption on soil mineral/inorganic surfaces may become insignificant compared to the concurrent solute sorption into SOM (Schwarzenbach and Westall, 1981). Therefore, the role of soil minerals in sorption of HOCs by soils could be expected to be negligible because the organic matter content for most soils falls between 0.5% and 3.0% by weight in natural environments. In addition, the linear isotherm and no competition (Boucher and Lee, 1972; Huang et al., 1996; Mader et al., 1997; Hundal et al., 2001; Su et al., 2006), observed for adsorption of solutes on soil mineral surface, indicate that soil minerals in natural environment, when strongly hydrated cations present, cannot be a good candidate accounting for the observed isotherm nonlinearity on soils.

9.2.3.2 Adsorption of HOCs on Carbonaceous Geosorbents

Recently, extensive and nonlinear sorption of HOCs was observed for several inert, rigid, condensed, and aromatic materials with high carbon contents and relatively few polar functional groups (Cornelissen et al., 2005). These materials, named as "carbonaceous geosorbents (CG)" and partly named as "BC" or "HSACM," are distinguishable and identifiable in physically and/or chemically relative to "normal" SOM. They include at least unburned coal, kerogen, coke, cenosphere, soot, char, and charcoal (the remnants of incomplete combustion) (Chiou, 1995; Chiou and Kile, 1998; Chiou et al., 2000; Cornelissen et al., 2005). The observed sorption of HOCs on CG can be highly nonlinear with the Freundlich exponent (n) as low as 0.3–0.7 (Cornelissen et al., 2005). The Freundlich exponent (n) is a common parameter to reflect the isotherm nonlinearity, where linear isotherm is reflected with $n = 1$ and the isotherm nonlinearity commonly reflected with $n < 1$. The highly nonlinear isotherm exhibits a tendency, analogous to the adsorption on activated carbon, that a strong solute adsorption on CG at low C_e/S_w and at moderate to high C_e/S_w values this adsorption is largely saturated. Moreover, sorption of many HOCs on CG is 1–3 orders of magnitude higher than that to SOM especially at relatively low concentrations. These HOCs include PAHs, PCBs, polychlorinated dibenzo-*p*-dioxins/dibenzofurans (PCDD/Fs), polybrominated diphenyl ethers, the pesticide diuron, benzene and chlorobenzenes, and chlorinated short-chain aliphatic compounds (Cornelissen et al., 2005).

Because the ubiquity of CG in soils and sediments has been identified and quantified (Smith et al., 1973; Griffin and Goldberg, 1983; Masiello and Druffel, 1998; Kleineidam et al. 1999; Cornelissen et al., 2005), CG was suggested to be at least one of the causes of the nonlinear sorption and may significantly affect the adsorption of organic compounds in soils and sediments especially for nonpolar compounds and at low C_e/S_w values. The formation, properties, quantification and occurrence of CG as well as the adsorption mechanisms of HOCs on CG has been well reviewed by Cornelissen et al. (2005). In addition, CG could be viewed as an extraneous substance in soils rather than as part of SOM, even though it is often counted as part of the soil organic carbon in SOM analysis by high-temperature combustion methods.

Although CG has higher affinity and capacity than SOM for HOC adsorption at low C_e/S_w values, its importance to exhibit the predominate role in nonlinear sorption of HOCs by soils, relative to that of SOM, should largely depend on the amount of CG in soils. The measured median BC contents as a fraction of total organic carbon (TOC) are 9% for sediments and 4% for soils based on the current quantification methods (Cornelissen et al., 2005). In fire-impacted soils, the content of charcoal BC may be exceedingly high up to 30%–45% of TOC (Schmidt et al., 1999). Most CG quantification methods are based on an isolation procedure firstly to remove the non-CG organic matter. However, the current isolation procedures is extremely difficult for complete isolation of small amounts of CG from ordinary soils and sediments (Chiou, 2002), which could lead to an overestimation of the CG contents in soils. Moreover, the coating of humic materials was observed to decrease the HOC adsorption on CG (Sander and Pignatello, 2005a), which thus may reduce the importance of CG on nonlinear sorption, depending on the accessible surface of CG.

9.2.3.3 Importance and Role of SOM in Isotherm Nonlinearity

Although the unsuppressed adsorption of polar solutes on certain clay fractions of low-organic-content soils (Laird et al., 1992; Haderlein and Schwarzenbach, 1993; Weissmahr et al., 1997) and the highly nonlinear adsorption on CG (Cornelissen et al., 2005) could result in nonlinear sorption at low C_e/S_w, the nonlinear isotherms of polar and nonpolar solutes as noted on soils with relatively high SOM contents (Chiou and Kile, 1998; Chiou et al., 2000) and on CG-free SOM (Pignatello et al., 2006) indicate the importance of SOM not only for solute linear partition but also for nonlinear adsorption. Furthermore, the importance of SOM in nonlinear adsorption for ordinary soils and sediments was also shown by the direct relations existed between nonlinear adsorption capacity (described by Q_0 and $Q_0 \cdot b$ in Figures 9.10 and 9.11) and SOM content for both polar 2,4-DCP and nonpolar phenanthrene (Yang et al., 2005a). Nonlinear adsorption was separated from the total sorption based on the successful fitting and accounted by the following simple DMM (Huang et al., 1997; Xing and Pignatello, 1997; Xia and Ball, 1999; Xia and Pignatello, 2001; Yang et al., 2005a)

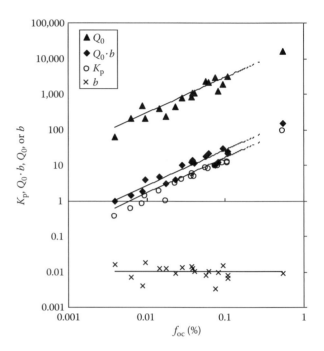

FIGURE 9.10 Values of K_p, Q_0, b, and $Q_0 \cdot b$ for 2,4-DCP as a function of the organic carbon content f_{oc} of 19 soil/sediment sorbents. (Reprinted from Yang, K., L. Zhu, B. Lou, and B. Chen. 2005a. Correlations of nonlinear sorption of organic solutes with soil/sediment physicochemical properties. Chemosphere 61:116–128. Copyright 2005, with permission from Elsevier.)

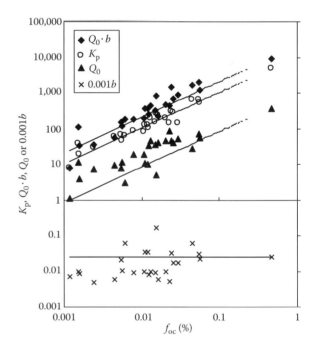

FIGURE 9.11 Values of K_p, Q_0, b, and $Q_0 \cdot b$ for phenanthrene as a function of the organic carbon content f_{oc} of 26 soil/sediment sorbents. (Reprinted from Yang, K., L. Zhu, B. Lou, and B. Chen. 2005a. Correlations of nonlinear sorption of organic solutes with soil/sediment physicochemical properties. Chemosphere 61:116–128. Copyright 2005, with permission from Elsevier.)

$$q_e = q_p + q_{nL} = \frac{K_p C_e + Q_0 \cdot b C_e}{1 + b C_e} \quad (9.10)$$

where
- q_p (µg·g⁻¹) depicts the linear partition uptake
- q_{nL} (µg·g⁻¹) depicts the nonlinear adsorption
- K_p (mL·g⁻¹) is the partition coefficient
- C_e (µg·mL⁻¹) is the equilibrium concentration
- Q_0 (µg·g⁻¹) is the maximum adsorption capacity
- $Q_0 \cdot b$ (mL·g⁻¹) is the Langmuir-type isotherm slope in the low concentration (Henry's law) range
- b (mL·µg⁻¹) is a constant related to the affinity of the surface for the solute

Nonlinear adsorption of HOCs by SOM was interpreted by a pore filling mechanism in a condensed "hard" glassy-like SOM portion with specific "internal holes" (Weber et al., 1992; Young and Weber, 1995; Pignatello and Xing, 1996; Xing et al., 1996; Huang et al., 1997; Leboeuf and Weber, 1997; Xing and Pignatello, 1997) or the specific interactions between the limited active sites in SOM and polar solutes, approaching saturation at low solute concentrations (Spurlock and Biggar, 1994; Chiou and Kile, 1998; Chiou et al., 2000). In addition, the direct relations between nonlinear adsorption capacity and SOM content indicate that SOM, rather than minerals or CG, predominated the nonlinear adsorption for ordinary soils, in addition to the linear partition to SOM. The unsuppressed adsorption of polar solutes on minerals is negligible relative to the nonlinear adsorption by SOM because of the predominance of strongly hydrated cations (e.g., Na^+, Ca^{2+}, Al^{3+}, Mg^{2+}) in natural waters. The reduced influence of CG relative to SOM on nonlinear adsorption may be attributed to the very small amount of CG present in ordinary soils and/or the effect of coated SOM and surface oxidation on the CG.

Because of the predominance of SOM in nonlinear adsorption and linear partition, shown by the linear relationships of K_p, Q_0, and $Q_0 \cdot b$ with f_{oc} for both polar and nonpolar organic solutes (Figures 9.10 and 9.11), not only K_p but also Q_0 and $Q_0 \cdot b$ for a given organic solute can thus be expressed in terms of the f_{oc} of respective sorbents based on a hypothetical natural sorbent of 100% organic carbon.

$$K_{oc} = \frac{K_p}{f_{oc}} \quad (9.11)$$

$$Q_{oc} = \frac{Q_0}{f_{oc}} \quad (9.12)$$

and

$$L_{oc} = \frac{Q_0 \cdot b}{f_{oc}} \quad (9.13)$$

where
- K_{oc} is organic-carbon-content normalized partition coefficient
- Q_{oc} is organic-carbon-content normalized maximum adsorption capacity
- L_{oc} is organic-carbon-content normalized adsorption coefficient

With this type of treatment, the organic-carbon-content normalized K_{oc}, Q_{oc}, and L_{oc} for a given organic solute are largely independent on the properties of soils and expected to be largely constant. Variations in K_{oc}, Q_{oc}, b, and L_{oc} from the difference of organic matter of various soils and sediments were observed for both 2,4-DCP and phenanthrene. However, variations in K_{oc}, Q_{oc}, and L_{oc} are smaller than those in K_p, Q_0, and $Q_0 \cdot b$, which indicates that K_{oc}, Q_{oc}, and L_{oc} are the most important characteristic indices for the nonlinear sorption of a given organic solute with soils/sediments. In addition, variations in K_{oc}, Q_{oc}, and L_{oc} are no greater than variations in K_{oc} reported by different investigators (Briggs, 1981; Schwarzenbach and Westall, 1981; Kile et al., 1995; Yaws et al., 1999).

With the predominance of SOM in nonlinear adsorption, it is also reasonable to observe the influence of SOM composition on isotherm nonlinearity (Chefetz et al., 2000; Xing, 2001a, 2001b; Salloum et al., 2002; Gunasekara and Xing, 2003; Kang and Xing, 2005; Yang and Xing, 2009). For example, it is observed that the isotherm linearity of HOCs by HAs increased with HA aliphaticity (Xing, 2001a; Salloum et al., 2002; Kang and Xing, 2005). Furthermore, Salloum et al. (2002) showed that HA sorption linearity can be increased by reducing the HA aromaticity using chemical oxidation with sodium hypochlorite. These sorption and spectroscopic studies have shown that the condensed (or glassy) SOM portion mainly attributed to SOM aromatic moieties, producing isotherm nonlinearity. The π–π interactions between aromatic moieties of SOM and aromatic compounds such as phenanthrene may play an important role in the nonlinear adsorption of aromatic compounds (Chefetz and Xing, 2009; Yang and Xing, 2009). Therefore, deviation of Q_0 and $Q_0 \cdot b$ from the linear relationships shown in Figures 9.10 and 9.11 could be partly interpreted by the influence of SOM composition and structure on isotherm nonlinearity. The nonlinearity of phenanthrene isotherms on extracted fractions of SOM from a natural soil has an increasing trend with the organic fractions: lipid < fulvic acid < HA < whole soil sample < humin (Liu et al., 2010). SOM configuration may also affect the nonlinear adsorption. For example, Xing (2001b) observed an increasing nonlinearity trend after studying naphthalene and DCB sorption to Florida Pahokee peat and its organic fraction: peat HA < peat < peat humin, which was later explained with the degree of condensed structures expected from organic matter extraction procedures (Cuypers et al., 2002). Gunasekara and Xing (2003) suggested that interaction of SOM with soil mineral surfaces may render the structural configuration of SOM to a more condensed state, which could result in higher nonlinear isotherms of aliphatic-rich humin fractions than HAs (Gunasekara and Xing, 2003; Kang and Xing, 2005). Moreover, the conformation change of SOM may be caused by COOH groups, being oriented in an inaccessible order in a complex with hydrogen-bond acceptors (carbohydrates) contained within SOM (Litvina et al., 2003; Todoruk et al., 2003).

Deviation of Q_0, and $Q_0 \cdot b$ from the linear relationships may also be from the adsorption on CG (Cornelissen et al., 2005) and the unsuppressed adsorption on certain clay fractions for polar solutes (Laird et al., 1992; Haderlein and Schwarzenbach, 1993; Weissmahr et al., 1997) and the highly nonlinear adsorption on CG, though these adsorption effects on nonlinear sorption may be not dominant for ordinary soils relative to that of SOM because of the predominance of strongly hydrated cations (e.g., Na^+, Ca^{2+}, Al^{3+}, Mg^{2+}) in natural waters for clays and the very small amount for CG. pH is also considered to affect the nonlinear sorption especially for ionizable chemicals (Westall et al., 1985; Jones and Tiller, 1999).

9.2.3.4 Effect of Solubility and Polarity of Organic Compounds on Nonlinear Adsorption

Negative linear relationships between log K_{om} (or log K_{oc}) and log S_w were also observed in previous works, indicating water solubility is a major factor for solute partition into SOM (Chiou et al., 1979, 1983; Karickhoff et al., 1979; Kenaga and Goring, 1980; Means et al., 1980; Briggs, 1981; Hassett et al., 1981; Karickhoff, 1984). In a study conducted by Yang et al. (2005b), it was observed that, besides the separated partition coefficients (log K_{oc}), the separated coefficients of nonlinear adsorption fraction (log L_{oc} and log b) have also negative linear relationships with the solute log S_w (Figure 9.12), suggesting water solubility is another determinant for solute nonlinear adsorption by SOM. The deviation of log K_{oc} for low solubility solutes from its negative linear relationship (Figure 9.12) is interpreted by the polarity influence on partition (Briggs, 1981; Chiou et al., 1983; Chiou, 2002). Similarly, the deviation of log L_{oc} for high solubility solutes from its negative linear relationship (Figure 9.12) could also be interpreted by the polarity influence on nonlinear adsorption fraction. The polar solutes with a large water solubility such as 2,4-DCP having a more nonlinear isotherm than the nonpolar and lower solubility solutes, observed by Chiou and Kile (1998), may be an example indicating the effect of water solubility and polarity of organic compounds on nonlinear adsorption.

9.2.3.5 Competitive and Irreversible Sorption

There are many studies (Pignatello, 1991; Xing et al., 1996; Xing and Pignatello, 1997, 1998; Chiou and Kile, 1998; White et al., 1999; White and Pignatello, 1999; Chiou et al., 2000; Zhao et al., 2001, 2002) showing that the nonlinear sorption of a given organic chemical can be reduced in the presence of additional compounds, as shown by the more linear isotherms than that observed in single solute system. The competitive phenomenon has been also observed between organic compounds and naturally occurring aromatic acids (Xing and Pignatello, 1998). Competitive effects are environmentally important because multiple organic pollutants exist far more commonly than single pollutant in natural environments. Competitive phenomenon is one of the important characteristics for nonlinear adsorption against the linear partition (Xing et al., 1996; Xing and Pignatello, 1997). Some important features of competitive effects for nonlinear sorption of solutes, as revealed explicitly by Chiou and Kile (1998), Xing et al. (1996), and Xing and Pignatello (1997), include the following: (1) the suppression of the sorption of a given organic solute at a relatively low concentration by co-solutes is more significant than the solute at a relatively high concentration; (2) the sorption suppression of a given concentration of organic solute by co-solutes is more significant at the relatively high co-solute concentration than that the relatively low concentration co-solute; (3) the sorption suppression of a polar solute (e.g., atrazine) by a nonpolar co-solute (e.g., trichloroethylene) is relatively small, (4) the sorption suppression of a polar solute (atrazine) by other polar co-solutes (e.g., prometon and other triazines) is significant; and (5) the nonlinear sorption of a nonpolar solute could be effectively suppressed by both the polar and nonpolar co-solutes. The results above illustrate that the nonlinear behavior of a given solute in binary- and multi-solute systems is influenced by both the co-solute type and its concentration. Moreover, the polar solutes such as 2,4-DCP are more powerful competitors than nonpolar solutes. A polar co-solute (e.g., 2,4-DCP) of one chemical class may effectively suppress the nonlinear sorption of the given solute of a different class (e.g., diuron) (Chiou and Kile, 1998).

Irreversible sorption phenomenon is also one of the important characteristics for nonlinear adsorption against the linear partition (Xia and Pignatello, 2001; Lu and Pignatello, 2004a, 2004b; Sander and Pignatello, 2005b; Sander et al., 2006).

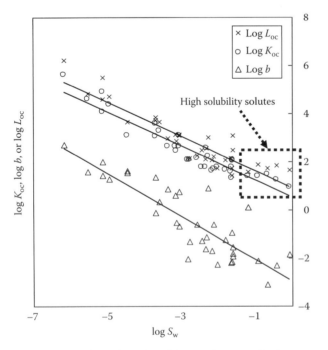

FIGURE 9.12 Values of K_{oc}, b, and L_{oc} as a function of solute water solubility S_w (mol·L^{-1}). (Reprinted from Yang, K., L. Zhu, B. Lou, and B. Chen. 2005a. Correlations of nonlinear sorption of organic solutes with soil/sediment physicochemical properties. Chemosphere 61:116–128. Copyright 2005a, with permission from Elsevier.)

The term "irreversible" means that the pathways for sorption and desorption are different. The irreversible sorption can be verified by the hysteresis in the isotherm (i.e., the desorption branch deviates from the sorption branch for the isotherm) and the conditioning effect (i.e., the enhanced sorption observed in a repeated experiment on the same sorbent after a prior sorption–desorption cycle). Sander and Pignatello (2005b) have verified the true sorption hysteresis using the isotope exchange technique. By comparing the isotherm on the conditioned soils (pre-contaminated) with that on the respective controlled soil (uncontaminated), conditioning effect on irreversible sorption was also observed (Xia and Pignatello, 2001; Lu and Pignatello, 2004a, 2004b; Sander et al., 2006). Pore deformation mechanism has been suggested to interpret the observed irreversible sorption (Xia and Pignatello, 2001; Lu and Pignatello, 2004a, 2004b; Sander and Pignatello, 2005b; Sander et al., 2006). Irreversible sorption in fact is a result of that a fraction of sorbed solute molecules are fixed in soil to resist desorption. However, many studies have shown that only a small fraction of sorbed pollutants are fixed strongly to resist desorption (Braida et al., 2002; Zhao and Pignatello, 2004). For example, Zhao and Pignatello (2004) observed that the fraction of phenanthrene resisted desorption, after 500–600 days of desorption to infinite dilution, ranged only from about 5%–30% of the initial amount added. A significant amount of irreversible sorption could be an apparent result of slow and time-dependent sorption controlled by sorption–desorption kinetics (Ball and Roberts, 1991; Pignatello and Xing, 1996).

9.2.4 Influence of SOM–Mineral Interactions on Sorption

Although soil minerals including clay and oxides alone, at most time, play a minor role in HOC sorption because of the strong suppression of water on their surfaces (Chiou, 2002; Yang et al., 2004), mineral–SOM interactions commonly result in the HOC sorption by mineral-adsorbed SOM being different from that by bulk SOM (Murphy et al., 1990, 1994; Jones and Tiller, 1999; Gunasekara and Xing, 2003; Wang and Xing, 2005; Feng et al., 2006). For example, mineral-adsorbed SOM commonly exhibits a more nonlinear isotherm than the bulk SOM (Murphy et al., 1990, 1994; Jones and Tiller, 1999; Gunasekara and Xing, 2003; Wang and Xing, 2005; Feng et al., 2006). Fractionation and conformation changes of SOM, which occurs when SOM is adsorbed onto mineral surfaces by mineral–SOM interactions, have been suggested to interpret this difference. The assumptions of fractionation and conformation changes of SOM on HOC sorption could be supported by the observed influence of SOM composition and states on HOC sorption as mentioned in the foregoing analysis. For example, SOM characteristics such as polarity, aromaticity and/or aliphaticity have been observed to affect HOC sorption significantly (Rutherford et al., 1992; Xing et al., 1994; Chefetz et al., 2000; Salloum et al., 2002; Gunasekara and Xing, 2003; Kang and Xing, 2005; Yang and Xing, 2009). SOM fractionation is reasonable because SOM is a mixture of complex molecules, and, thus, some fractions of SOM may preferentially adsorb to the inorganic surface, which results in the adsorbed SOM as well as its HOC sorption affinity being different from the original bulk one and the unadsorbed fractions remaining in solution (Jones and Tiller, 1999; Wang and Xing, 2005). SOM conformation changes, that is, the configuration of adsorbed SOM being different from the respective bulk SOM, was used to further explain HOC sorption difference between the adsorbed HA and the respective bulk HA because this HOC sorption difference cannot be explained by HA fractionation alone in some cases (Murphy et al., 1990, 1994; Jones and Tiller, 1999; Gunasekara and Xing, 2003; Wang and Xing, 2005; Feng et al., 2006).

To explain the nonlinearity of SOM, an expanded SOM domain, analogous to rubbery synthetic polymers, is characterized to generate linear partition, while the condensed SOM domain, analogous to glassy synthetic polymers, is characterized to generate nonlinear adsorption. With a study of sorption and desorption for naphthalene by SOM, Gunasekara and Xing (2003) suggested that interaction of SOM with soil mineral surfaces may render the structural configuration of SOM to a more condensed state and result in highly nonlinear isotherms. HA aromatic moieties are suggested to be one important contributor to form the condensed state of HA, while HA aliphatic moieties is a contributor to form the expanded state (Salloum et al., 2002; Gunasekara and Xing, 2003; Kang and Xing, 2005). Hydrophobic interaction is responsible for partitioning and isotherm linearity. Additional interactions such as the possible π-bonds of aromatic molecules with aromatic moieties of SOM may be responsible for adsorption and the isotherm nonlinearity (Yang and Xing, 2009). Functional groups on aromatic rings will enhance the π-polarity/polarizability of aromatic rings and thus π-bonding interaction (Hickey and Passino-Reader, 1991). Therefore, interactions of SOM containing functional groups such as phenolic-OH with minerals could result in the π-polarity/polarizability enhancement of adsorbed SOM, and subsequently the adsorption affinity of adsorbed SOM would increase to show more nonlinear isotherms (Yang and Xing, 2009).

Nonlinear sorption would exhibit concentration-dependent organic carbon-normalized sorption coefficients (K_{oc}), with higher K_{oc} values at low concentrations (Figure 9.13). Since the nonlinear sorption is a combination of linear partition and nonlinear adsorption, the influence of either SOM fractionation or conformation changes on nonlinear sorption may act on the linear partition component or the nonlinear adsorption component or both of them. It is clear that enhancement or reduction in K_{oc} is a constant in all range of the solute concentrations if variation of K_{oc} is originated from the changes of partitioning affinity because of the linear isotherm (Figure 9.13a). If K_{oc} variation comes from adsorption affinity changes, however, K_{oc} variation at low concentrations will be larger than that at high concentrations due to the nonlinear isotherm of adsorption (Figure 9.13b). Increase in adsorption and decrease in partitioning will cause the increase of relative contribution of adsorption to total sorption and isotherm nonlinearity.

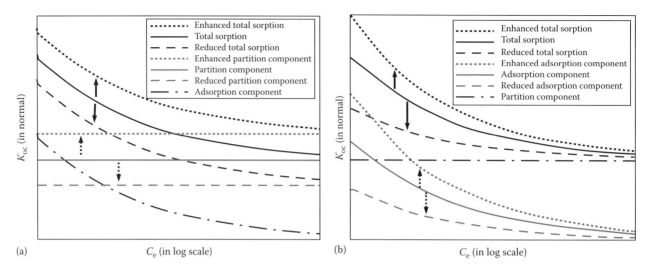

FIGURE 9.13 Enhancement or reduction of the concentration-dependent organic carbon-normalized sorption coefficients (K_{oc}) of total sorption derived from the enhancement or reduction of partition component (a) and adsorption component (b). K_{oc} of partition component was figured out based on the linear equation fitting of the isotherm of partition component, while that of adsorption component was obtained based on the Freundlich equation fitting of the isotherm of adsorption component.

9.2.5 Influence of Sorption and Desorption on Bioavailability of AOPs

The term bioavailability has been and is used in many different scientific fields, which consider interactions between a living cell and a chemical(s). There are many definitions of varying detail and complexity in the scientific literature, which can lead to confusion and inconsistency (Semple et al., 2007). However, the bioavailability should at least represents the "maximum quantity of a contaminant available for uptake by an organism within a given time period" (Semple et al., 2004). Both biological and chemical methods have been developed to assess organic compound bioavailability (Reid et al., 2000; Dean and Scott, 2004). In general, it is accepted that only the organic molecules dissolved in aqueous phase (i.e., water in soil) are bioavailable (Paterson et al., 1990; Scow and Hutson, 1992; White and Alexander, 1996; Cornelissen et al., 1998; Alkorta and Garbisu, 2001; Chiou et al., 2001; Gomez-Lahoz and Ortega-Calvo, 2005), whereas adsorbed molecules by soil solids cannot be bioavailable directly because they do not have direct *contact* with organisms and cannot be taken up. This non-bioavailability is valid whether the sorbent is a natural sorbent such as soil particle (Ehlers and Loibner, 2006) or an artificial sorbent such as activated carbon (Aktas and Cecen, 2007). Therefore, sorption and desorption process of AOPs in the soil–water systems can largely affect the bioavailability of the pollutants in soils.

Sorption of an organic compound by soil can decrease its biodegradation because only dissolved organic molecules in water can be bioavailable. On the one hand, sorption decreases the concentrations of the organic compound in aqueous phase (i.e., maximum quantity of a contaminant available for uptake), and consequently decreases the concentration-dependent uptake and degradation rate. Due to the observed hysteresis (i.e., the difference between sorption and desorption processes) (Pignatello and Xing, 1996; Doucette, 2003), on the other hand, a significant amount of soil adsorbed organic molecules cannot be desorbed and thus is unavailable for microbe uptake and degradation, and also slow kinetics of this desorption process suggests that the desorbed rate of the organic compound from soil to water may fail to match the uptake and degradation rate and thus decrease biodegradation (McLeod and Semple, 2000). The sorption of an organic compound from water to soil solid matter is usually directly proportional to its K_{ow} (Schwarzenbach et al., 2003). Therefore high K_{ow} values indicate limited bioavailability and biodegradation of an organic compound in soil (Cerniglia, 1992; Reid et al., 2000; Semple et al., 2001, 2003).

Among the soil components, SOM is widely accepted to play a predominant role in sorption of AOPs. Consequently, SOM is also the main component that is responsible for the decrease of AOP bioavailability in the soil environment. Furthermore, SOM property that may affect AOP sorption, such as polarity, physical states (e.g., glassy and rubbery state) and aromaticity, could thus be expected to influence the AOP bioavailability. For example, functionalities and surface domain distribution of SOM may have joint effects on the bioavailability of phenanthrene, showing phenanthrene was mainly sequestrated in aromatic domains of lignin (Yang et al., 2010). Adsorbed SOM on minerals, forming a more condensed state than the bulk SOM and having stronger nonlinear sorption for AOPs (Gunasekara and Xing, 2003), may decrease the AOP bioavailability. The small amount of carbonaceous material (such as charcoal or soot) in soils in addition to SOM, exhibits a greater nonlinear adsorption of AOPs (Chiou, 1995; Gustafson et al., 1997; Chiou and Kile, 1998; Cornelissen et al., 2005) and may also decrease the AOP bioavailability (Xia et al., 2010). Therefore, biochars, derived from biomass, can be added as an environmental-friendly engineered sorbents to amend contaminated soils to promote pollutant sorption in soils and thus abate the transfer of AOPs from soil to water, plants, and soil microorganisms (Wang et al., 2010).

Nonlinear sorption of a given organic pollutant by soils and SOM can be reduced by the addition of other organic pollutants because of the competition and displacement, and thus enhance the concentration of the pollutant in water (Xing et al., 1996, Xing and Pignetello, 1998; White and Pignatello, 1999). Therefore, the bioavailability of a given organic pollutant can be enhanced by the addition of competitors (White and Pignatello, 1999). Addition of dissolved organic chemicals (e.g., organic acids, surfactants, and DOM) may not only pose the competition and displacement on the adsorbed pollutants but also enhance the solubility of AOPs in water to give the higher bioavailability (Xing and Pignetello, 1998; Voparil et al., 2003). For example, in field experiments, some surfactants at certain concentrations have been observed to enhance the uptake and accumulation of *p,p′*-DDE by specific *Cucurbita pepo* (zucchini) (White et al., 2006, 2007). Increasing temperature may also enhance the bioavailability of AOPs because the high temperature can promote desorption of sorbed organic chemicals from soil to water. Furthermore, addition of some chelating agents such as citrate, oxalate, or pyrophosphate was found to enhance desorption of hydrophobic compounds such as PAHs from SOM by solubilizing metal ions that are essential for cross-linking humic substances or bridging humic substances to mineral surfaces (Yang et al., 2001; Submmaniam et al., 2004). For example, exudation of chelating agents by plant roots was suggested to be responsible for the observed increased uptake of chlorinated hydrocarbon insecticides (White et al., 2003).

9.3 Uptake and Degradation by Plant and Soil Microbes

When an organic chemical enters into an organism (e.g., soil microorganism and plant) by uptake, it can be accumulated and degraded in the organism. Biodegradation (or to metabolize) is the most important process for the "ultimate" removal of AOPs in the environment, which can be defined as the process by which organic chemicals are decomposed by organisms and/or their extracellular enzymes into simpler substances (United Nations, 1997). In the biodegradation process by microorganism, organic compounds are used as carbon and electron source. Mineralization is the "ultimate" biodegradation and means conversion of organic compounds to inorganics in organism. Biotransformation is another biodegradation that means conversion of organic compounds into other organic compound(s). Unlike in mineralization, the products of biotransformation can be even more harmful than the parent compounds (Alexander, 1999). The products of biotransformation and the un-degraded chemicals may accumulate in organism and then be released into the environment or transferred into other organisms via food chains.

For biodegradation to occur, two prerequisites must be reached: (1) the compound must be bioavailable, that is, accessible to the target organism; and (2) the compound must be inherently biodegradable. The AOP accessibility is responsible for the physical *contact* of organic molecule with organism, which is needed before it is degraded by the organism or its extracellular enzymes (Rosenberg et al., 1992). The entering of organic molecule into organism by the uptake process is to attain this contact, and thus determines the compound *available* for the degrading organism (Reid et al., 2000; Semple et al., 2003). The accessibility of an organic molecule to a biological entity, that is, bioavailability, is of great importance because it is one of the key factors affecting contaminant biodegradation in soils (Mihelcic et al., 1993; Reid et al., 2000; Semple et al., 2003). This bioavailability is also important for the assessment of toxicity and ecological risk of an organic compound in soil. However, bioavailability alone is not enough to account for the possible degradation of AOPs. Other factors affecting the growth of organisms are also required for biodegradation. Moreover, bioavailability is not only organism dependent but also species dependent (i.e., bioavailability can be organism and species specific) (Guerin and Boyd, 1992; Kelsey et al., 1997; White et al., 1997). For example, both White et al. (1997) and Kelsey et al. (1997) noted that the extent of earthworm (*Eisenia foetida*) bioaccumulation and bacterial (*Pseudomonas sp.*) mineralization of soil-associated phenanthrene differed, although both organisms indicated a decline in bioavailability with increasing soil–pollutant contact time.

The organic molecules dissolved in aqueous phase (i.e., water in soil) are bioavailable because they have direct *contact* with organisms (Paterson et al., 1990; Scow and Hutson, 1992; White and Alexander, 1996; Cornelissen et al., 1998; Alkorta and Garbisu, 2001; Chiou et al., 2001; Gomez-Lahoz and Ortega-Calvo, 2005), whereas sorbed molecules by soil solids cannot be bioavailable and for uptake directly because they have not direct *contact* with organisms. The soil water dissolved molecules may also be poorly or not bioavailable if the molecule is dissolved in non-advecting water. Such water is hygroscopic or pellicular or water located in a soil pore with a neck pore diameter smaller than $0.3\,\mu m$ (Standing and Killham, 2007). Organisms such as soil microbes can occupy only a small fraction of the soil surface area.

The uptake of a chemical molecule by an organism occurs after that molecule has passed from the aqueous phase through a biological membrane (e.g., cell membrane) and into the interior of the organism. This uptake for most AOPs is believed to be passive, that is, it occurs by diffusion through the membrane rather than by active transport processes (Pignatello, 2006). Although the uptake and degradation of AOPs by organisms are kinetic processes, an apparent equilibrium between the pollutant concentrations in organisms and water, is commonly considered to reach after a long time contact of pollutants with organisms in the natural systems. Bioconcentration factor, a term, which is given by the ratio of the pollutant concentration in organisms to that in water, has widely been used to describe this apparent equilibrium.

Microorganisms and plants, in soil systems, are the most important organisms for degradation of AOPs. Therefore, the uptake and degradation of AOPs by soil microorganisms and plant do have important effects on the AOP fate in the environment. How and the extent of AOPs to be taken up and degraded by soil microorganisms and plant have been examined in the past.

9.3.1 Uptake and Degradation by Soil Microorganisms

Degrading microorganisms are ubiquitously distributed in the natural environment, such as in soils (bacteria and non-ligninolytic fungi) and woody materials (ligninolytic fungi) (Alexander, 1999; Leung et al., 2007; Roesch et al., 2007). Microbes have developed enzymatic systems for the degradation of compounds emanating from natural processes (Leung et al., 2007). The diversity of soil microbial communities is enormous. It has been proposed that soil may contain 10^9–10^{10} microbial cells cm^{-3} and an estimated number of 10^4–10^6 distinct genomes per gram of soil (Torsvik et al., 2002; Gans et al., 2005; Roesch et al., 2007). However, AOPs have been produced and discharged for only about 100 years. Most of these manmade compounds have a molecular structure never observed in the natural environment, that is, the molecular structures are different from natural chemicals. Therefore, they are foreign and unfamiliar to the microbes and the degrader enzymes, implying the possible low biodegradation. AOPs may be biodegraded when (1) they are compatible with the catabolic enzymatic apparatus of a degrader microbe (Alexander, 1999; Leung et al., 2007), (2) the enzymatic apparatus of a microbe has a wide specificity (Hesselsoe et al., 2005; Baldrian, 2006), and (3) genetic adaptation occurs in a microbe leading to a new catabolic pathway for the pollutants (Janssen et al., 2005). The population of soil microbes, capable of degrading the organic contaminants, is a key for AOP degradation and has received a substantial amount of research (El Fantroussi and Agathos, 2005).

A wide variety of microorganisms including bacteria, fungi, and algae have successfully been cultured to remove the AOPs (especially for the readily biodegraded compounds) in discharged wastewater via anaerobic, aerobic, and sequential anaerobic–aerobic treatment processes called as activated sludge systems. However, conventional activated sludge systems often fail to achieve high efficiency in removal of POPs such as pesticides from wastewater due to the low biodegradability and high toxicity or the inhibition of the compounds to microorganisms. For those un-biodegraded AOPs, microorganisms in conventional activated sludge systems may act as a biosorbent and can remove the AOPs from wastewater by biosorption (Aksu, 2005). The mechanisms of AOPs uptake and degradation by microorganisms in soils could be similar to that as observed in wastewater.

Soil microbes are attached on the surfaces of soil particles in microcolony- or biofilm-like structures (Standing and Killham, 2007; Van Elsas et al., 2007). Dissolved organic molecules to these sites with movement of water fluid promote contacts with degrading microbes. Therefore inherently biodegradable and water soluble AOPs are usually expected to be bioavailable and not persistent in soil if other conditions favor biodegradation (Semple et al., 2003). However, the uptake and degradation of AOPs by microorganisms occurs only after AOP molecule has passed from the aqueous phase through the cell membrane (a phospholipid bilayer) by a passive diffusion and into the interior of the microorganism (Schwarzenbach et al., 2003). This passive diffusion was accepted to be a partitioning process of AOPs between the biomembrane and water (Gobas et al., 1988; Dulfer and Govers, 1995; Escher and Schwarzenbach, 1996; van Wezel et al., 1996; Vaes et al., 1997; Schwarzenbach et al., 2003). This partitioning process is described by the membrane–water partition coefficients (K_{mw}), which is given by the ratio of the pollutant concentration in membrane to that in water. Because chemical partitioning in real biomembranes is experimentally difficult, the artificial bilayer phospholipid vesicles, based on phosphatidylcholine monomers known as *liposome*, are widely accepted in the experiments as models of biomembrane. A linear relationship occurs between the *liposome*–water partition coefficients (K_{mw}) and the octanol–water partition coefficients (K_{ow}) (Gobas et al., 1988; Escher and Schwarzenbach, 1996; van Wezel et al., 1996; Vaes et al., 1997):

$$\log K_{mw} = a \log K_{ow} + b \qquad (9.14)$$

where a and b are regression coefficients. From this equation, an HOC with a high octanol–water coefficient indicates favored partition from water to biological membranes. However, the range where this equation can be applied is limited. First, the slope of this linear relationship for PCB partitioning into liposome seems to decline, and even becomes negative when the chemical log K_{ow} above about 7 (Dulfer and Govers, 1995). Moreover, an HOC to partition into real cellular membranes is most preferred when its log K_{ow} is 1.5–4.0, as reviewed by Sikkema et al. (1995) and Ramos et al. (2002). This is because the multiphase nature of water–membrane lipid bilayers differs from water–octanol two-phase system. The water–lipid bilayer system has a hydrophilic interfacial phase, which creates surface tension between water and lipid "bulk" phases to prohibit the transfer of highly hydrophobic organic chemicals (log $K_{ow} > 4.0$) from water to lipid, whereas the water–octanol system has water and "bulk" (octanol) phases only (De Young and Dill, 1988). In the soil systems, the uptake of AOPs by microorganisms may be decreased by their sorption on soils because the sorption decreases the aqueous concentrations of AOPs. A ratio of K_{mw} to K_{oc}, for a given AOP, can be employed to describe its uptake availability by microorganisms from soils.

AOPs in contaminated soils can be degraded by degrading bacteria or fungi via an aerobic or anaerobic metabolism or both. In the aerobic metabolism process, molecular oxygen is incorporated into the aromatic ring prior to the dehydrogenation and subsequent aromatic ring cleavage. In the anaerobic metabolism process where molecular oxygen is absent, however, alternative electron acceptors such as nitrate, ferrous iron, and sulfate are necessary to oxidize aromatic compounds. The basis of these mechanisms is the oxidation of the aromatic ring, followed by the systematic breakdown of the compound to PAH metabolites and/or carbon dioxide. The mechanisms of anaerobic metabolism of aromatic AOPs are still tentative and thought to occur via the hydrogenation of the aromatic ring (Bamforth and Singleton, 2005).

In the aerobic metabolism of AOPs by bacteria or fungi, there are three fundamentally different mechanisms, described in Figure 9.14 with PAHs as the initial degraded compounds (Bamforth and Singleton, 2005). The principal mechanism for the aerobic bacterial metabolism is the initial oxidation of the aromatic ring by the action of dioxygenase enzymes to form *cis*-dihydrodiols, followed by dehydrogenation to form dihydroxylated intermediates and then be further metabolized via catechols to carbon dioxide and water (Bamforth and Singleton, 2005). There are two main types of fungal metabolism by the non-ligninolytic and ligninolytic fungi, respectively (Bamforth and Singleton, 2005). The first step in the metabolism by non-ligninolytic fungi, the majority of fungi in soils, is to oxidize the aromatic ring in a cytochrome P450 monooxygenase enzyme catalyzed reaction to produce an arene oxide (Sutherland et al., 1995) and subsequently hydrated via an epoxide-hydrolase catalyzed reaction to form a *trans*-dihydrodiol (Jerina, 1983). In comparison to the oxidation in aerobic bacterial metabolism, the monooxygenase enzyme here incorporates only one oxygen atom onto the ring to form an arene oxide and the sequent *trans*-dihydrodiol. Ligninolytic fungi under ligninolytic conditions can produce ligninolytic enzymes, which generates hydroxyl free radicals by the donation of one electron to oxidize AOPs (Sutherland et al., 1995). This oxidation is a nonspecific radical based reaction and generates quinine-like and acid-like structures rather than dihydrodiols (Sutherland et al., 1995). Because the hydroxyl free radical based reaction has low substrate specificity, ligninolytic enzymes are therefore able to degrade even the most recalcitrant compounds. Moreover, ligninolytic enzymes are secreted extracellularly (Kirk and Farrell, 1987), and thus theoretically able to diffuse into the soil matrix and potentially oxidize the soil adsorbed AOPs.

Besides the bioavailability of AOPs and microbial populations, it is apparent that environmental factors, inhibiting the growth of the pollutant-degrading microorganisms or altering the bioavailability of AOPs, could affect biodegradation. These factors mainly include contaminant toxicity, soil moisture, soil temperature, soil pH, and nutrient availability. Toxicity of an organic contaminant can limit biodegradation if the toxic effect of the compound is high enough to limit the microbe's degrading activity (Alexander, 1999). Organic compounds with a log K_{ow} between 1.5 and 4.0 are extremely toxic for microorganisms and other living cells because they partition preferentially in the cytoplasmic membrane, disorganizing its structure and

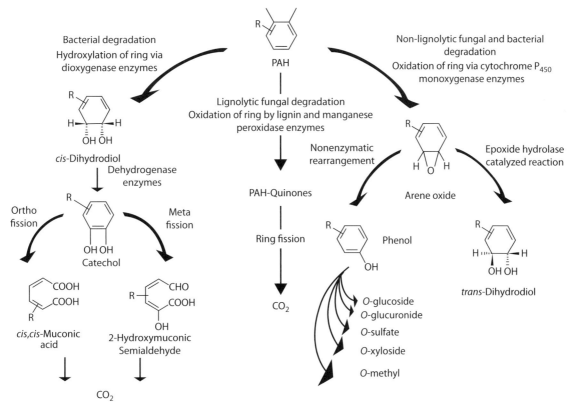

FIGURE 9.14 Three main pathways for PAH degradation by fungi and bacteria. (Bamforth, S.M., and I. Singleton. 2005. Bioremediation of polycyclic aromatic hydrocarbons: Current knowledge and future directions. J. Chem. Technol. Biotechnol. 80:723–736. Copyright Wiley-VCH Verlag GmbH & Co. KGaA. Reprinted with permission.)

impairing vital functions (Sikkema et al., 1995; Ramos et al., 2002). Soil moisture affects not only the supply of water and nutrients for microbes but also the transport of gases such as oxygen and other electron acceptors such as NO_3^- for bio-oxidation. Moreover, total water content of soil affects the bioavailable amount of an organic compound, for example, an increase in the total water content of soil leads to an increased desorption of an organic compound from the soil particles and the increased total amount of the organic compound in water where they are bioavailable (White and Alexander, 1996). Temperature can affect the biodegradation because (1) there are a minimum and a maximum temperature, typically within a range of 20°C–30°C, allowing growth for microbes, outside this range, temperatures make the microbes inactive; (2) Biodegradation of an organic compound in soil is a chemical reaction catalyzed by enzymes in soil, which is affected by temperature; and (3) water solubility and diffusivity of a HOCs are increased with increasing temperature (Nedwell, 1999; Iqbal et al., 2007). Relative to AOPs in temperate soils, therefore, AOPs in tropical soils could be degraded more and faster because tropical soils are under very humid climates with higher temperature allowing desorption and dissolution of AOPs and active growth of microbes, while AOPs in cold soils could be degraded less and slower because cold soils are with lower temperature that makes the desorption and dissolution of AOPs less and the microbes inactive. Inorganic nutrients, as well as organic carbon sources, must be available for the growth of microbes. Large utilization of organic contaminants, as organic carbon sources in the microbial growth process, leads to the depletion of available inorganic nutrients and consequently inhibits the growth of microbes (Morgan and Watkinson, 1989).

9.3.2 Uptake and Degradation by Plant

The pollution of soils by AOPs such as pesticides leads to the subsequent pollution of plants/crops grown in these contaminated soils. Research in contamination of plants by organic compounds began as early as the 1950s, soon after many organochlorine insecticides were introduced for agricultural purposes (Lichtenstein, 1959). The levels of pesticide residues and other organic chemicals in plants accumulated from contaminated soils have since been largely reported (Chiou, 2002). Many soil-incorporated pesticides are now known to translocate into plants following their applications. Contaminated plants especially for the agricultural crops by toxic AOPs may cause the transfer of these AOPs to animals and human beings via food chains, and pose serious health risks. Plant uptake and storage are also a potentially key component in the global cycling of persistent organic pollutants (Dallavalle et al., 2004; Scheringer et al., 2004). Knowledge of contaminant uptake by plants is of vital interest not only because it will improve our understanding and alleviation of the crop-contamination problem but also because it may provide us with a means to effectively remediate contaminated soils by appropriate plantings (Chiou, 2002).

Plants can affect the fates of AOPs in soil through (Figure 9.15): (1) stabilizing and degrading the pollutants in the rhizosphere,

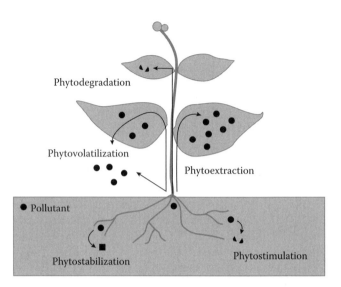

FIGURE 9.15 Possible fates of pollutants during phytoremediation: the pollutant (represented by black circles) can be stabilized or degraded in the rhizosphere, sequestered or degraded inside the plant tissue, or volatilized. (Reprinted from Pilon-Smits, E. 2005. Phytoremediation. Annu. Rev. Plant Biol. 56:15–39. With permission of Annual Reviews.)

(2) taking up the pollutants to plant issues, and (3) subsequent transporting, accumulating, degrading the pollutants in plant issues, or (4) volatilizing the pollutants to the atmosphere. These processes are generally named as phytostabilization, phytodegradation, phytoextraction, phytovolatilization, and phytostimulation (Pilon-Smits, 2005). Phytostabilization is a process that plants stabilize pollutants in soil either simply by preventing erosion, leaching, and/or runoff, or by converting pollutants to less bioavailable forms (Berti and Cunningham, 2000). Phytostimulation or rhizodegradation is a process that plants facilitate microbial degradation of organic pollutants by releasing simple organic molecules to promote the growth of degrading microbes in their rhizosphere (McCutcheon and Schnoor, 2003). Phytodegradation is a process that plants degrade organic pollutants directly through releasing degrading enzymes either inside the plant issues or into the soil (McCutcheon and Schnoor, 2003). Phytovolatilization is a process that certain pollutants taken up in plant tissues leave the plant in volatile form into the atmosphere (Pilon-Smits, 2005). Phytoextraction is a process that pollutants accumulate in plant tissues after uptake in plant tissues (Pilon-Smits, 2005). Phytodegradation and phytoextraction also represent transformation and subsequent accumulation of pollutants inside the plant issues. Uptake of the pollutants from soil and the subsequent transport of the pollutants inside the plant is a first step determining the degradation, transformation, accumulation, and volatilization of AOPs in plants.

9.3.2.1 Plant Uptake Pathways

The uptake of AOPs by plants mainly occurs via two pathways (Collins et al., 2006). One is the soil–water–plant pathway, in this pathway, AOPs translocate into plants by root uptake from soil solution and then transport possibly in the transpiration steam within the xylem (Figure 9.16). The other one is

Soil Physicochemical and Biological Interfacial Processes

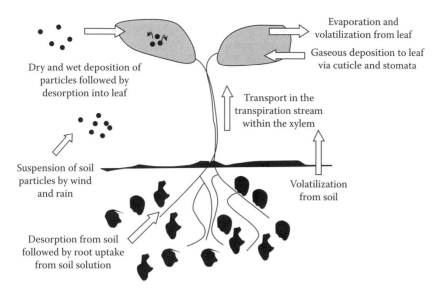

FIGURE 9.16 Principal pathways for the uptake of organic chemicals by plants. (Reprinted with permission from Collins, C., M. Fryer, and A. Grosso. 2006. Plant uptake of non-ionic organic chemicals. Environ. Sci. Technol. 40:45–52. Copyright 2006 American Chemical Society.)

the soil–air–plant pathway, in which AOPs translocate into plants by leaf uptake from the suspended soil particles or the gaseous AOPs volatilized from soil with a deposition process (Figure 9.16). After the uptake process, other plant processes involved in translocation, sequestration, and degradation of AOPs in plants may occur subsequently (Pilon-Smits, 2005). Some key parameters affecting the transport process of contaminants from soil to plant (the soil-to-plant transport process of contaminants) include the levels of contaminants in soil (or water), the contaminant physicochemical properties, the particular plant species, the soil types that sustains the plant, and the time of plant exposure (Chiou, 2002). Although leaf uptake may be important for volatile or semi-volatile organic compounds, only a very small fraction of the gaseous or the suspended soil particle bound-AOPs in atmosphere from contaminated soil can be translocated into plants by leaf uptake because plants can only use a small scale ambient air around their leaves while the AOPs in atmosphere from contaminated soil are mostly dispersed with wind and diluted in the air in a large scale. Therefore, the soil–water pathway through root uptake is considered as the main pathway for most AOPs from soil to plants (Chiou et al., 2001; Chiou, 2002; Pilon-Smits, 2005). In the following section, we also focus on the soil–water pathway through root uptake only.

9.3.2.2 Root Uptake and Translocation

Soil–water–plant pathway consists of two components: the uptake of AOPs from soil solution into plant roots and the transport of AOPs from the root to the shoot. Chemicals, before they can be transported upward to plant shoot via the xylem, must be moved from soil solution to the plant root constituents first and then moved to reach the xylem by penetrating a number of layers including the epidermis, cortex, endodermis, and pericycle (Figure 9.17), where one cell membrane (which is

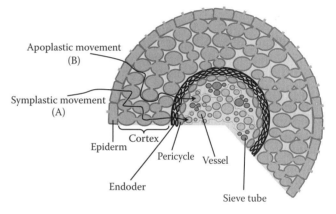

FIGURE 9.17 An illustrative diagram of translocation pathways of organic contaminant molecules with water among plant cells showing both apoplastic (A) and symplastic (B) pathways.

rich in lipids) at least exist to be penetrated at the endodermis (Collins et al., 2006). It is suggested that the solubility of AOPs in water (solution) and in the cell membrane determines their movement into roots and subsequent transport to the plant shoot (McFarlane, 1995).

AOPs can be taken up by plant roots via a passive and/or an active process from soil water, depending on the contaminant and plant types (Shone and Wood, 1974; Briggs et al., 1982). The uptake of most AOPs by plant roots has been shown to be the passive diffusive process, while a few hormone-like chemicals such as the phenoxy acid herbicides may be taken up by the active process (Bromilow and Chamberlain, 1995). The passive plant uptake process of contaminants by plant roots is derived by the water potential gradient created throughout the evapotranspiration within the xylem (Simonich and Hites, 1995; Paraiba, 2007). The passive root uptake process consists of two components (Briggs et al., 1983; Collins et al., 2006): (1) "equilibration" of the aqueous phase in the plant root with the concentration in the surrounding soil solution and (2) "sorption" of the chemical onto lipophilic root solids. Lipophilic root solids include lipids in membranes and cell walls (Paterson et al., 1991). This root uptake of AOPs may be treated as a series of partitions from soil solution to plant water and then to plant organic constituents (Ryan et al., 1988; Chiou et al., 2001), where the partition into lipids is largely responsible for root concentration of relatively water-insoluble compounds.

Movement of water and AOPs across the root epidermis and cortex is the first step for the root uptake. There are two pathways through which AOPs can move across the root epidermis and cortex from soil water (Figure 9.17): one is the apoplastic through which AOPs and water movement involves diffusion between cell walls, not entry into the cells; the other one is the symplastic pathway through which AOPs and water movement is through the cell cytoplasm or vacuoles and to interconnected cells via the plasmodesmata (Wild et al., 2005). Initially, AOPs bound to the epidermis along the zone of elongation, passing through the epidermal cells to reach the cortex within the root hair, and branching zones of the root. The AOPs enter the epidermis radially; however, once within the cortex cells this movement is dominated by slow lateral movement toward the shoot. The lateral movement of AOPs toward the shoot within the cortex cells is dominated by movement within the cell walls, suggesting apoplastic flow through multiple cell walls, but with a low level of symplastic movement to transport compound into the cellular vacuoles (Wild et al., 2005).

Significant accumulation of AOPs including PAHs, chlorobenzenes, PCBs, and PCDD/Fs have been found in plant roots (DuarteDavidson and Jones, 1996). Lipophilic organic chemicals are observed to possess a greater tendency to partition into plant root lipids than hydrophilic chemicals. Briggs et al. (1983) established a linear relationship between the octanol–water partition coefficients (K_{ow}) of nonionized chemicals and the observed root concentration factor (RCF = chemical concentration in the root/concentration in external solution) for the uptake of O-methylcarbamoyloximes and substituted phenylureas by barley plants. For the uptake of organic chemicals by hybrid poplar trees, this relationship was also observed by Burken and Schnoor (1998). It was thus suggested that organic chemicals with log K_{ow} > 3.0 (Briggs et al., 1983) or log K_{ow} > 4 (Wild and Jones, 1992a; Burken and Schnoor, 1998) have a high potential for retention in plant roots. In addition to the hydrophobic nature of AOPs, their root uptake is also suggested to depend on the lipid content of the plant roots because root lipid is the place for storage of AOPs (Bromilow and Chamberlain, 1995; Schwab et al., 1998; Chiou et al., 2001). A linear relationship between the RCF of PAHs and the lipid content of 12 plants was observed by Gao and Zhu (2004). The important influence of root lipid content for the root uptake behavior has been also illustrated by the higher retention of organic compounds in the peels of carrot and potato compared to their pulp (Wild and Jones, 1992b; Fismes et al., 2002).

The AOPs taken into the plant roots can be transported upward as dissolved species with water from the plant roots into other parts through evapotranspiration within the xylem. However, chemicals, before they can be transported upward to plant shoot via the xylem, must dissolve in plant solution from the root membrane/lipids. This process can be treated as a series of partitions from plant organic constituents to plant solution (McFarlane, 1995). Due to this partition behavior, lipophilic organic chemicals are expected to possess a lower tendency to dissolve into plant solution from root lipids than hydrophilic chemicals and subsequent lower transport upward to shoot.

The transpiration stream concentration factor (TSCF = concentration in xylem/concentration in external solution) was developed to describe the total uptake of AOPs in the plant shoot from soil solution by Briggs et al. (1982). A relationship between TSCF and the K_{ow} of AOPs was derived (Figure 9.18), showing a maximum of TSCF for chemicals with log K_{ow} = 1.8 (Briggs et al., 1982). Similar relationships (Figure 9.18) were observed by Burken and Schnoor (1998) and Hsu et al. (1990). A possible explanation for the K_{ow} optima is that hydrophilic chemicals have lower partition into the lipids of cell membrane/walls and thus less get into the plant root, while lipophilic chemicals get stuck in the lipids of cell membranes/walls and cannot enter the cell fluids for the later translocation in the transpiration stream (Bromilow and Chamberlain, 1995). However, organic chemicals with moderate hydrophobicity are hydrophobic enough to move through the lipid bilayer of membranes, and still water soluble enough to travel out into the cell fluids.

It should also be noted that other non-lipid plant components such as waxes, lignin, and suberin may absorb organic chemicals (Rutherford et al., 1992; Mackay and Gschwend, 2000; Chen et al., 2004, 2005; Wang et al., 2007). Therefore, lipid alone may not be enough to interpret the plant uptake of AOPs and thus underestimate their potential accumulation.

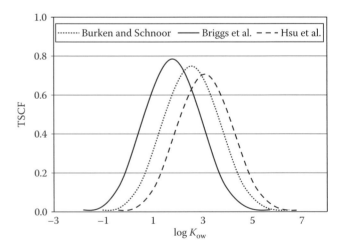

FIGURE 9.18 Variation in the prediction of the TSCF with log K_{ow}. (Reprinted with permission from Collins, C., M. Fryer, and A. Grosso. 2006. Plant uptake of non-ionic organic chemicals. Environ. Sci. Technol. 40:45–52. Copyright 2006 American Chemical Society.)

9.3.2.3 Models for the Plant Uptake by Root

At the present, there are a lot of nonequilibrium models (Riederer, 1990; Trapp et al., 1990; Paterson et al., 1994; Trapp and Matthies, 1995; Tam et al., 1996), formulated on a differential mass-balance due to the rates of contaminant interface transfer, plant growth and transpiration, and contaminant metabolism, along with some estimated transfer coefficients, and equilibrium models (Briggs et al., 1982; Trapp, 1995), assuming an apparent equilibrium between AOP concentrations in plants and water after plant exposure to chemicals over a certain period of time, these models are developed to assess the plant uptake of contaminants from soil and water and to relate the plant contaminant levels to physicochemical properties of the contaminants as well as to the properties and compositions of plants and soils. However, the applications of equilibrium models are limited because the actual state of a contaminant in plants may or may not be at equilibrium. Although nonequilibrium models delineated the uptake rates of contaminants with time, the model calculations are very sensitive to the accuracy of assumed contaminant interface-transfer rates and coefficients. Recently, a quasi-equilibrium partition-limited model, accounting for the passive plant uptake of contaminants from soil or water, has been developed by Chiou et al. (2001) and later demonstrated as an effective model in predicting the plant uptake (Zhu and Gao, 2004; Li et al., 2005; Yang and Zhu, 2007; Zhang and Zhu, 2009). The partition-limited model takes explicit account of the plant contaminant level in relation to the source level and plant composition. Moreover, the model contains both equilibrium and kinetic features and sets the upper (equilibrium) limit for the level of a contaminant in a plant with the respective contaminant level in solution. Therefore, the actual equilibrium between the contaminant in the plant and in the solution at a given time can be estimated (Chiou, 2002). The partition-limited model, for plant uptake of AOPs from a soil-free solution, equates the concentration of a contaminant either in the whole plant or in a specific part of the plant (α_{pt}), expressed as the mass of contaminant per unit wet mass of the plant, with the contaminant concentration in external water (C_w) at the time of sample analyses (Chiou et al., 2001)

$$C_{pt} = \alpha_{pt} C_w [f_{pom} K_{pom} + f_{pw}] \quad (9.15)$$

where
  K_{pom} is the contaminant partition coefficient between plant organic matter and water
  f_{pom} is the total weight fraction of the organic matter in the plant
  f_{pw} is the weight fraction of water in the plant, either for the whole plant or for a specific part of it

The term α_{pt} (≤1), called the *quasi-equilibrium factor*, expresses the extent of approach to equilibrium of any absorbed contaminant in the plant (or in a part of it) with respect to the same contaminant in the external water phase. In this model, α_{pt} is a time-dependent parameter and may be viewed as the ratio of the respective concentrations in plant water and external water at the time of sample analyses. Thus, $\alpha_{pt} = 1$ denotes the attainment of equilibrium. For plants in contaminated soils, the contaminant concentration in external water (C_w) is related to the contaminant concentrations in SOM (C_{som}) and the contaminant partition coefficient between SOM and water (K_{som}), $C_w = C_{som}/K_{som}$, based on the assumption that SOM acts as a partition medium and is responsible for the total soil sorption of the contaminant (Chiou et al., 2001).

Because the organic matter in most root and leaf crops are polar carbohydrates, cellulose, and proteins, and nonpolar lipids, the $f_{pom} K_{pom}$ is simplified as the sum of the contributions by carbohydrates and lipids based on a further assumption that the partition coefficients of the compounds with the relatively polar carbohydrates, cellulose, and proteins are practically the same (Chiou et al., 2001)

$$C_{pt} = \alpha_{pt} C_w [f_{lip} K_{lip} + f_{ch} K_{ch} + f_{pw}] \quad (9.16)$$

where
  f_{lip} and f_{ch} are total weight fraction of the lipids and carbohydrates in the plant, respectively
  K_{lip} and K_{ch} are the contaminant lipid–water partition coefficients and carbohydrate–water partition coefficients, respectively

The application of this partition-limited model requires values of the individual carbohydrate–water partition coefficients (K_{ch}) as well as the individual actual lipid–water partition coefficients (K_{lip}). However, these values are not available. Although octanol–water partition coefficients (K_{ow}) and assumed K_{ch} values (0.1–3 according to the corresponding K_{ow} values) were assumed to represent the actual K_{lip} and K_{ch} values, respectively, this treatment may result in the underestimation of actual K_{lip} and K_{ch} values and thus the deviation of the model predicted results from the experimental data (Zhang and Zhu, 2009). Especially for chemicals with large K_{ow} values, this underestimation on K_{lip} and K_{ch} values and the deviation of the model prediction are very significant (Barbour et al., 2005; Li et al., 2005). Carbohydrates are also important for plant uptake because of its predominant content in plant root though carbohydrates have lower partitioning affinity than lipids (Zhang and Zhu, 2009). The observed relationships of K_{lip}–K_{ow} and K_{ch}–K_{ow} may improve the prediction and help the application of the partition-limited model (Zhang and Zhu, 2009).

9.3.2.4 Degradation and Removal of the AOPs in the Plants

After uptake in plant issues and the subsequent transport, AOPs may accumulate in plant issues. AOPs in the plant may also be degraded or volatilized to atmosphere, thus reducing their concentration in the plant tissues. AOPs can be degraded in the process of plant metabolism via their own enzymatic activities.

This process has been reviewed by Burken (2003). The degradation process includes transformation reactions (e.g., oxidation), conjugation (e.g., with glutathione), and finally sequestration (e.g., into the cell wall). Degradation process and rates are specific to chemicals and plant species (Schnabel et al., 1997; Chekol et al., 2002; Hannink et al., 2002). Plant degradation has been observed for AOPs including herbicides, MTBE, trichloroethylene, benzene, pyrene, and the explosives (Huckelhoven et al., 1997; Ugrekhelidze et al., 1997; Hannink et al., 2002; Shang and Gordon, 2002; Burken, 2003; Winnike-McMillan et al., 2003; Nepovim et al., 2004).

The uptake and degradation of AOPs by plants suggest that plants can be employed in the remediation of contaminated soils by selection of favorable plant species. However, the soil properties, toxicity level, and climate should allow plant growth. Favorable plant properties for phytoremediation in general are to be fast growing, high biomass, competitive, hardy, and tolerant to pollution. Plants, having large and dense root systems and high levels of degrading enzymes would certainly favor the uptake and degradation of soil AOPs (Pilon-Smits, 2005). Moreover, this phytoremediation is limited by root depth because the plants have to be able to reach the pollutant. Root depth is typically 50 cm for herbaceous species or 3 m for trees, although certain phreatophytes that tap into groundwater have been reported to reach depths of 15 m or more, especially in arid climates (Negri et al., 2003).

9.3.3 Rhizosphere Processes of AOPs

The rhizosphere is an area that extends approximately 1 mm around the plant root (Pilon-Smits, 2005). In the rhizosphere zone, the pollutant can be stabilized or degraded by the plant roots, microbes and plant- and microbe-derived chemicals such as bio-surfactants and enzymes. AOPs can be stabilized in the rhizosphere soil by plants either simply through preventing erosion, leaching, and runoff, or through converting pollutants to less bioavailable forms (Berti and Cunningham, 2000). In the rhizosphere soil, microbial densities are 1–4 orders of magnitude higher than in bulk soil from the general rhizosphere effect (Salt et al., 1998), that is, plants release the photosynthesis-derived organic compounds in the rhizosphere for heterotrophic fungi and bacteria as carbon sources and thus promote the growth of microbes (Bowen and Rovira, 1991). In turn, rhizosphere microbes can promote plant health by stimulating root growth (some microorganisms produce plant growth regulators), enhancing water and mineral uptake, and inhibiting growth of other pathogenic soil microbes (Kapulnik, 1996).

The rhizosphere zone is the place for plant root uptake because plant roots can only take up water and the species dissolved in the water that they contact. It is also the important place for the uptake and degradation by soil microbes. Most of the soil physicochemical and biological interfacial processes that governing the fate of AOPs can be found in the rhizosphere zone. Moreover, these soil physicochemical and biological interfacial processes in the rhizosphere may be affected by various plant and/or microbial activities. For example, some bacteria can release biosurfactants (e.g., rhamnolipids) that make hydrophobic pollutants more water soluble and subsequent enhance the bioavailability of these pollutants (Volkering et al., 1998). The observed AOPs that are degraded in the rhizosphere by microbial activity include PAHs, PCBs, and petroleum hydrocarbons (Hutchinson et al., 2003; Olson et al., 2003). Plant exudates or lysates may also enhance the pollutant water solubility and promote biosurfactant-producing microbial growth (Siciliano and Germida, 1998). Furthermore, plant- and microbe-derived enzymes can not only enhance the solubility of organic pollutants but also their degradation through oxidation and modification of side groups (Wolfe and Hoehamer, 2003).

There are two ways by which the degradation of AOPs in the rhizosphere can be promoted by plants: one is the direct oxidation by releasing enzymes and the other is phyto-stimulation of microbial degradation. Plants can stimulate the microbial degradation because (1) plants release carbon compounds into the rhizosphere to facilitate microbial growth and increase the microbial density, and (2) some released plant compounds may also induce specific microbial genes involved in the degradation or act as a cometabolite to facilitate microbial degradation (Fletcher and Hegde, 1995; Leigh et al., 2002; Olson et al., 2003). In general, 20% of carbon fixed by a plant is released into the rhizosphere from its roots (Olson et al., 2003).

Because of the possible enhancement of the uptake and degradation of AOPs by the complex plant–microbe interactions in the rhizosphere, these rhizosphere processes may be optimized in the remediation design to favor the removal of AOPs from the contaminated soil. This remediation taking place in the rhizosphere zone is called as rhizosphere remediation. In this remediation, optimized plant selection based on the rhizosphere processes is the most important step. Expected plants for rhizosphere remediation are those with highly effective uptake and degradation of AOPs, or those that can stimulate microbial degradation, or both of them. A large root surface area may favor phytostimulation, as it promotes microbial growth. Furthermore, production of specific exudate compounds may further promote rhizodegradation via specific plant–microbe interactions (Olson et al., 2003). The biomass, extent and dense of root systems, and the ability to increase the abundance and species degrading microbes should be considered in the selection of plants for rhizosphere remediation. If microbial consortia are responsible for the rhizosphere remediation process, it may be possible to increase the abundance of these species by adding them to the soil (a process called bioaugmentation). It is also possible to regulate the soil conditions such as soil water, nutrients and reducing/oxidizing conditions to provide an optimal environment for rhizosphere processes such as plant and microbial growth.

9.4 Effects of Surfactant on Soil Physicochemical and Biological Processes

Surface-active agents (surfactants), a class of manmade dissolved and/or colloidal organic matter widely used in household and industries as detergent formulations and surface cleaners with huge discharge quantities into the environment, are observed to alter HOC sorption and uptake process in the soil environment. For example, surfactants in water especially with their micelle formation have the ability to enhance the solubility of HOCs and then decrease their sorption by soils (Kile and Chiou, 1989; Edwards et al., 1991). This phenomenon is interpreted by the predominant role of chemical solubility in their physicochemical and biological processes including sorption by soils and uptake by plants and microbes. Therefore, surfactants could be used to remove HOCs from contaminated soils by enhancing chemical solubility (West and Harwell, 1992). However, the adsorbed surfactants on soil mineral surfaces exhibit high sorption of organic chemicals and consequently would attenuate mobility of HOCs (Boyd et al., 1988; Lee et al., 1989; Sheng et al., 1996). It is also observed that cell-associated bio-surfactants facilitate hydrocarbon uptake (Zajic and Seffens, 1984). Humic substances, the most important soil component controlling the physicochemical and biological processes of HOCs, are also surface-active compounds and their dissociated form could be likened to anionic surfactants (West and Harwell, 1992).

9.4.1 Surfactant

Surfactants are surface active because they favor to concentrate at interfacial regions such as air–water, oil–water, and solid–liquid interfaces. Their surface activity derives from their amphiphilic structure, that is, their molecules contain a polar ionic hydrophilic moiety (referred to as the head group) as well as a nonpolar hydrophobic moiety (the tail group). According to the charge nature carried by the hydrophilic portion (head group) of the molecule, surfactants are classified to four types including anionic (negatively charged), cationic (positively charged), nonionic (no charge), and zwitterionic (with both negative and positive charges) surfactants. Bacteria, yeasts, and fungi have the capability to synthesize surfactants, commonly referred to as biosurfactants (Zajic and Seffens, 1984).

A phenomenon unique to surfactants is the self-assembly of molecules into dynamic clusters called micelles above certain concentrations in water (Figure 9.19). The concentration of a surfactant to form micelles is characterized as *critical micelle concentration* (CMC). At concentrations below the CMC, the dissolved surfactant is all in monomeric form; above the CMC, the amount of surfactant in water is in excess of a critical amount of surfactant monomers and to form micelles. CMC is different for every surfactant, typically ranging between 0.1 and 10 mM (West and Harwell, 1992). In a micelle, the individual monomers are oriented with their hydrophilic moieties in contact with the aqueous phase and their hydrophobic moieties tucked into the interior of the aggregate (Figure 9.19). Another phenomenon of surfactants is that the surfactant chemical structure achieves a proper balance of hydrophilic and hydrophobic interactions at the interface to obtain large reductions in the interfacial tension (Bourrel and Schechter, 1988). Surfactants are sensitive to temperature (Rosen, 1989). For ionic surfactants, there is a temperature called Krafft point at which the solubility of the surfactant becomes equal to its CMC. At temperatures below this Krafft point, ionic surfactants exist as monomers without micelles formed. Micelles are formed at temperatures

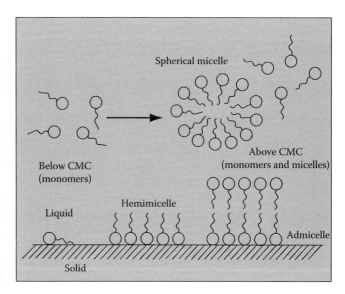

FIGURE 9.19 Examples of surfactant micellization. (Reprinted with permission from West, C.C., and J.H. Harwell. 1992. Surfactants and subsurface remediation. Environ. Sci. Technol. 26:2324–2330. Copyright 1992 American Chemical Society.)

higher than the Krafft point. For nonionic surfactants, there is also a critical temperature called the cloud point, at temperatures above which, the surfactant begins to form a separate surfactant-rich aqueous phase by coacervation and results in a high turbidity of the surfactant solution. Emulsifying organic chemical and oils in water is important for the application of surfactants. Hydrophile–lipophile balance (HLB) scale, which was designed for matching surfactant structure to an organic chemical to be emulsified in water (Rosen, 1989), can be obtained commonly from the manufacturers. Each surfactant has an HLB number indicating the water soluble of the surfactant and the types of organic chemicals it can emulsify. The surfactant with higher HLB number exhibits more water soluble. The surfactant HLB requirement of organic chemicals for their emulsification in the surfactant solution is directly related to their hydrophobicity. Generally, as the K_{ow} of organic chemicals increases (water solubility decreases), the HLB requirement decreases. Concentrating of surfactants at solid–liquid interfaces is commonly exhibited by sorption on solid surface. At very low concentrations, adsorbed surfactant monomers on solid surfaces begin to aggregate and form micelle-like structures called admicelles or hemi-micelles, depending on whether the aggregates have one or two surfactant layers (Figure 9.19). The tendency of surfactant to form admicelles or hemi-micelles is dependent on interactions between the hydrophilic moiety of the surfactant and the solid surface (West and Harwell, 1992).

9.4.2 Solubility Enhancement by Surfactant

Because of the predominant role of chemical solubility in their physicochemical and biological processes including sorption by soils and uptake by plants and microbes, these processes of HOCs could be promoted or attenuated by altering their solubilities. Surfactants could greatly promote the (apparent) water

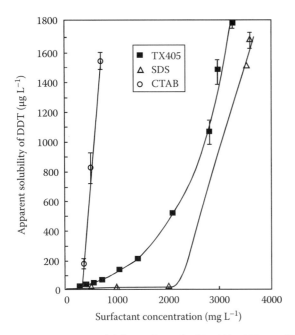

FIGURE 9.20 Water solubility enhanced of DDT by TX-405, SDS, and CTAB. (Reprinted with permission from Kile, D.E., and C.T. Chiou. 1989. Water solubility enhancement of DDT and trichlorobenzene by some surfactants below and above the critical micelle concentration. Environ. Sci. Technol. 23:832–838. Copyright 1989 American Chemical Society.)

solubility of relatively insoluble organic solutes (Figure 9.20) and then decrease their partition into an organic phase such as SOM because the surfactant micelle offers a relatively large microscopic nonpolar environment (i.e., the organic interior of micelles acts as an organic pseudophase) to allow for solute partition (Rosen, 1989). This phenomenon is called solubilization (Rosen, 1989). The increase in the solubility of a contaminant in a surfactant solution can be dramatic as shown in Figure 9.20. According to the dominant role of chemical solubility in their physicochemical and biological processes, solubilization of organic chemicals by surfactants, therefore, provide a potential means to remove contaminants from contaminated soils if the surfactant adsorbed on the soil does not strongly enhance the contaminant sorption (West and Harwell, 1992).

The solubility enhancement is related to surfactant monomer and micelle concentrations and the corresponding solute partition coefficients (Kile and Chiou, 1989), which can be described by a two-phase separation model using the following equation:

$$\frac{S_w^*}{S_w} = 1 + K_{mn}X_{mn} + K_{mc}X_{mc} \quad (9.17)$$

where

S_w^* and S_w are the apparent water solubility of solute in surfactant solution and solubility of solute in pure water, respectively

X_{mn} is the concentration of surfactant monomers in water

X_{mc} is the concentration of surfactant micelles in water

K_{mn} is partition coefficients of a solute between surfactant monomers and water

K_{mc} is solute partition coefficients between the aqueous micellar phase and water

All concentrations are expressed in moles per liter. At a given temperature, S_w, K_{mn}, and K_{mc} are constants. If the surfactant concentration $X < CMC$, then $X_{mn} = X$; if $X > CMC$, $X_{mn} = CMC$ and $X_{mc} = X - CMC$. Kile and Chiou (1989) applied this equation successfully to analyze the apparent water solubilities of extremely water-insoluble DDT (Figure 9.20) and relatively soluble TCB in the presence of nonionic surfactants (Triton series and Brij 35), an anionic surfactant (sodium dodecyl sulfate [SDS]), and a cationic surfactant (cetyltrimethylammonium bromide [CTAB]).

An alternative method to quantify surfactant solubilizing performance in the ranges of surfactant concentrations above CMCs is to measure the molar solubilization ratio (MSR) (Edwards et al., 1991). MSR is defined as the number of moles of compound solubilized per number of moles of micellized surfactant, and can be calculated as follows:

$$MSR = \frac{S_w^* - S_{cmc}}{C_s - CMC} \quad (9.18)$$

where

C_s is the surfactant concentration at which S_w^* is evaluated

S_{cmc} is the compound apparent solubility at CMC

MSR may be obtained from the slope of solubilization curves in the ranges of surfactant concentrations above CMCs. Based on the molar volume of water (1.8×10^{-2} L mol^{-1} at 25°C), K_{mc} can also be calculated from MSR (Zhu and Feng, 2003),

$$K_{mc} = \frac{55.4 \times MSR}{S_{cmc}(1 + MSR)} \quad (9.19)$$

Observed solubility enhancements are related to the properties of organic compounds and surfactants (Kile and Chiou, 1989; Edwards et al., 1991; Diallo et al., 1994; Jafvert et al., 1994; Sun et al., 1995). Theoretical models have also been developed to estimate micellar partitioning coefficients (K_{mc} or MSR) using surfactant and solute specific properties, including surface tension, molecular surface area, solute's structural formula, and Henry's constant (Smith et al., 1987; Liu et al., 2000). In an aqueous system, the extent to which a solute will partition in a surfactant micelle (K_{mc} or MSR) can be related to the octanol–water partition coefficients (K_{ow}) of the solute (Kile and Chiou, 1989; Edwards et al., 1991; Jafvert and Heath, 1991). In general, a solute with the larger K_{ow} exhibits the greater solubilization with higher K_{mc} and MSR. For the nonionic surfactants, their enhancement and the estimated micellar partitioning coefficients of a given solute are closely in proportion to the nonpolar group content of the surfactant, provided that the micelle size is sufficiently large (Kile and Chiou, 1989). For surfactants that have the same nonpolar chain length, the solubilizing power for organic solutes by the inner nonpolar core of the surfactant micelle follows as nonionic > cationic > anionic surfactants (Tokiwa, 1968; Kile and Chiou, 1989). The efficacy of surfactant-enhanced solubility is also limited by the concentration and stability of surfactant micelles. For example, nonionic surfactants can partition into trapped residual phases if their solubilities in an organic solute called as nonaqueous phase liquids (NAPLs) are high (Zhao et al., 2006), whereas anionic surfactants are subject to losses by precipitation with ions (Yang et al., 2007).

Relationships between the surfactant structure and the solubilization extent have been reported for alkanes, monocyclic aromatic hydrocarbons, and substituted monocyclic aromatic hydrocarbons (Diallo et al., 1994; Saitoh et al., 1994). A two-state model described by Mukerjee (1979) considers the hydrophobic core of a micelle and the hydrophilic outer shell as two distinct environments for solubilization, and solutes may be solubilized in either region depending on their molecular properties such as polarity (Mukerjee, 1979; Rosen, 1989). For nonpolar compounds such as alkanes where solubilization takes place only in the core of micelles, relationships between the extent of solubilization and compound properties or the surfactant structure have been reported (Diallo et al., 1994). For monocyclic aromatic hydrocarbons and PAHs, solubilization in nonionic surfactant micelles may occur in the core as well as in the shell region of the micelles because these compounds are slightly polar due to the resonance of π-electrons in the aromatic ring (Ruelle et al. 1992; Graziano and Byungkook, 2001). However, the solubilization and micelle–water equilibrium partition coefficients of the PAHs in various surfactant solutions, due to the resonance of π-electrons in the aromatic rings, increased with the volume and size of the polar shell region of the surfactant micelles rather than the size of the hydrophobic core of the micelle (Bernardez and Goshal, 2004).

Mixed surfactants such as the anionic–nonionic surfactant system exhibit synergistic solubilization for PAHs and NAPLs because the formation of mixed micelles to reduce their CMC and consequently prohibit the partition of nonionic surfactant to NAPLs and the precipitation of the anionic surfactant with ions (Zhu and Feng, 2003; Yang et al., 2005b, 2006; Zhao et al., 2006). When surfactants are combined with oils in certain ratios, they may form stable microscopic mixtures called *microemulsions*, which are stabilized through the interactions with water of the polar head groups of surfactants attached to the oil particles. Microemulsions are especially effective in enhancing the water solubility of nonpolar organic contaminants, due to solute solubilization into the resulting suspended microscopic oil–surfactant phase (Kile et al., 1990). In comparison with normal surfactants, microemulsions exhibit virtually no monomer–micelle transitions (i.e., no CMCs) and hence possess a separate partition phase more effectively than normal surfactants at very low concentrations.

9.4.3 Effect of Surfactant on the Sorption by Soil

Surfactants have a tendency to concentrate (i.e., sorption) on solid surface at the solid–liquid interface and to aggregate and form micelle-like structures such as admicelles or hemimicelles on the surface at very low concentrations (West and Harwell, 1992). Similar to micelles in water, these micelle-like structures can also be expected to exhibit a partitioning ability for organic solutes and consequently result in the uptake of solutes onto or into the adsorbed surfactant phase. In a soil–water system, therefore, surfactant imposes two opposing effects on the sorption of AOPs: the sorbed surfactant increases contaminant sorption, whereas the unsorbed surfactant promotes contaminant dissolution in water (i.e., decreases contaminant sorption by solubilization). This simultaneous effect of soil-sorbed and unsorbed surfactants on contaminant sorption and the consequent change in contaminant apparent sorption coefficient (K_d^*) in a soil–water mixture have been addressed and accounted by the following equation (Sun and Boyd, 1993; Chiou, 1998; Lee et al., 2000), given as

$$K_d^* = \frac{K_d + f_{sf}K_{sf}}{1 + X_{mn}K_{mn} + X_{mc}K_{mc}} \quad (9.20)$$

or, in a more useful alternative form, as

$$\frac{K_d^*}{K_d} = \frac{1 + f_{sf}K_{sf}/K_d}{1 + X_{mn}K_{mn} + X_{mc}K_{mc}} \quad (9.21)$$

where
K_d is as defined before
f_{sf} the mass fraction of the sorbed surfactant in soil
K_{sf} is the solute (contaminant) sorption coefficients between sorbed surfactant and water

The term f_{sf} is for the sum of adsorbed and partitioned surfactant. Here, K_{sf} is not a direct function of f_{sf}, but rather, a function of the aggregation state of the sorbed surfactant molecules. In principle, only the adsorbed surfactant can form a molecular aggregation, the extent being related to the amount adsorbed and mineral surface properties (Gu et al., 1992; Rutland and Senden, 1993; Nayyar et al., 1994).

The magnitude of $(1 + X_{mn}K_{mn} + X_{mc}K_{mc})$ in Equation 9.21, describing the water solubility enhancement of solutes by surfactant, is determined by surfactant type and its concentration in water as well as by contaminant properties. According to the equation, if a surfactant is present, apparent sorption coefficients (K_d^*) of solutes by soil could be higher or smaller than K_d, that is, the $(1 + f_{sf}K_{sf}/K_d)$ is higher or smaller than $(1 + X_{mn}K_{mn} + X_{mc}K_{mc})$, which depends on the amount of surfactant adsorbed on soil (f_{sf}), the partitioning ability of adsorbed surfactant (K_{sf}), the concentrations of surfactant in water (X_{mn} and X_{mc}), and the partition ability of surfactant monomers (K_{mn}) and micelles (K_{mc}) in water. In the presence of a given surfactant,

$K_d^* > K_d$ indicates that the surfactant can be used for the purpose of retarding organic solutes in soil to prohibit their transfers, whereas $K_d^* < K_d$ indicates that the surfactant may be applied in soil remediation to promote desorption of organic contaminants from soil. Surfactant-enhanced desorption and washing process are only effective when the equilibrium surfactant dose in water is greater than its CMC, because (1) most of surfactant molecules are sorbed onto soil below its CMC (Jafvert, 1991; Pennell et al., 1993; Edwards et al., 1994; Kommalapati et al., 1997; Chu and So, 2001; Zhu et al., 2003), (2) the partition ability of surfactant monomers (K_{mn}) is very small for most surfactants (Sun and Boyd, 1993; Chiou, 1998; Lee et al., 2000), and (3) no micelle partition occurs below its CMC.

Cationic surfactants can be expected to adsorb significantly on soil mineral solids (which are generally negatively charged) because of the attractive coulombic interaction between the surfactant head group and the solid surface. The sorption of cationic surfactants on clays and soils that contain significant cation exchange sites is also known to take place primarily by a cation exchange process (Theng et al., 1967; Barrer, 1978; Lee et al., 1989; Xu and Boyd, 1994). Because of the possible incorporation of the significant adsorbed amount of cationic surfactants and the additional microscopic nonpolar phase for solute partition created by the adsorbed surfactant on soils (i.e., to give large $f_{sf}K_{sf}$), cationic surfactants makes $K_d^* \gg K_d$ for nonpolar solutes (Boyd et al., 1988; Smith et al., 1990). Therefore, the addition of cationic surfactants to soil–water systems would normally increase (rather than decrease) the sorption of organic pollutants on soil and hence have been used to reduce pollutant migration through a soil profile and promote pollutant immobilization in soil (Boyd et al., 1988; Lee et al., 1989; Sheng et al., 1996).

In comparison with cationic surfactant, nonionic surfactants adsorb weakly on soil. Anionic surfactants also adsorb weakly on soil due to the repulsive coulombic interaction between its head group and the surfaces of soil minerals (Rouse and Sabatini, 1993). Therefore, both their weak adsorption on soil and high solubilization for AOPs make anionic and nonionic surfactants desirable for soil remediation. However, some nonionic surfactants may adsorb on soils with a significant amount and hence increase AOPs sorption, making them inappropriate for soil remediation (Zhu et al., 2003). In general, nonionic surfactants with higher water solubility show the lower tendency for their adsorption on mineral surfaces. Their adsorption depends on the contents of soil/sediment organic matter as well as the types and contents of clay minerals especially for the clay minerals with 2:1 structure (Zhu et al., 2003). Although anionic surfactants could be expected to adsorb less than nonionic surfactants for most mineral surfaces, they are subject to loss by precipitation (Yang et al., 2007). Clearly, successful surfactant use to enhance remediation goes beyond selection of a surfactant (or surfactant system) that will efficiently solubilize organic solutes and should attempt to minimize surfactant losses by adsorption and precipitation in soil environments. The tendency of anionic surfactants to precipitate and nonionic surfactants to adsorb can be reduced by using mixtures of anionic and nonionic surfactants (Yang et al., 2005b, 2006). Moreover, such mixed anionic–nonionic surfactant system exhibit synergistic solubilization for solutes such as PAHs and NAPLs because the formation of mixed micelles to reduce their CMC and consequently prohibit the partition of nonionic surfactant to NAPLs (Zhu and Feng, 2003; Zhao et al., 2006). Mixed anionic–nonionic surfactant systems, therefore, can be expected to play an important role in the development of commercially viable surfactant systems for enhanced remediation processes.

A key issue of the applications of surfactants in remediation of contaminated soils is the toxicity of synthetic surfactants, which may pose a threat to the environment and ecological systems. Extensive toxicological data on aquatic organisms have been reported for anionic surfactants (Lewis, 1991). Cationic surfactants are potent germicides (Lawrence, 1970). Use of naturally occurring surface-active compounds such as biosurfactants would allay these concerns because they are readily biodegradable (Mulligan et al., 2001; Zhu and Zhang, 2008). Moreover, biosurfactants may be produced in situ and thus cheaper than synthetic surfactants (West and Harwell, 1992).

9.4.4 Effects of Surfactant on Uptake and Degradation by Microorganisms

It is widely accepted that the bioavailability of pollutants is limited to the dissolved portion of the compound in the aqueous phase (Scow and Hutson, 1992). Surfactants, because of their enhanced solubilization of hydrophobic compounds at concentrations above their CMCs (Kile and Chiou, 1989; Edwards et al., 1991; Zhu and Feng, 2003; Zhao et al., 2006), have been successfully used to desorb contaminants from contaminated soils and make contaminants to dissolve in the aqueous phase (West and Harwell, 1992; Tiehm et al., 1997; Yang et al., 2006). Due to the ability of surfactants to increase the apparent aqueous concentration of poorly soluble compounds, various surfactants, therefore, have been expected to increase the bioavailability and the accessibility of HOCs to microorganisms and thus HOC biodegradation (Edwards et al., 1991; Tiehm, 1994; Tsomides et al., 1995; Guha et al., 1998). To better understand the factors involved and the current state of knowledge in this field, a search of the literature concerning the influence of commercial surfactants and biosurfactants on microbial metabolism has been conducted. Due to the direct interaction of surfactants with microorganisms, it appears that steric or conformational compatibility of surfactants with cell membrane lipids and enzymes is an important metabolic factor (Rouse et al., 1994). Although the surfactants increased biodegradation of HOCs in some cases (Aronstein and Alexander, 1993; Rouse et al., 1994; Zhang and Miller, 1994; Guha and Jaffé, 1996a, 1996b; Zhang et al., 1997; Brown et al., 1999), in other cases either no effect or strong inhibition on biodegradation has been observed (Laha and Luthy, 1991, 1992; Falatko and Novak, 1992; Rouse et al., 1994; Deschenes et al., 1996; Bramwell and Laha, 2000; Cort and Bielefeldt, 2002). Because of the toxicity of synthetic surfactants, they may irreversibly damage the cellular envelope and inhibit microbial activity, and consequently reduce the uptake and degradation of the target pollutants (Rouse et al., 1994). Even the bio-surfactants, which is essentially biodegradable and nontoxic relative to synthetic surfactants and can have specific compatibilities with the cell envelope structures of the microorganisms that produce them, they still might produce inhibitory effects on cell structures of other microorganisms (Falatko and Novak, 1992; Rouse et al., 1994). Nontoxic surfactants can also lead to inhibitory effects on biodegradation of HOCs. For readily biodegradable surfactants, their inhibitory effects might be from the lack of the dissolved oxygen concentration in the system because the surfactant degradation lowers the dissolved oxygen concentration (Tiehm et al., 1997; Goudar et al., 1999; Mata-Sandoval et al., 2001). Another possible mechanism, by which surfactant micelles might inhibit biodegradation of contaminants, is the hydrophilic micellar surface, which may form a barrier blocking the interaction of microorganisms with the pollutant partitioned in the micellar core (Rouse et al., 1994), that is, the bioavailability of contaminants is limited to the portion associated with surfactant monomers outside the micellar phase (Volkering et al., 1995; Grimberg et al., 1996). However, other researchers (Guha and Jaffé, 1996a, 1996b; Guha et al., 1998) demonstrated that a fraction of the micellar HOCs is directly bioavailable to bacterial cells (i.e., the micellar HOCs can be directly transferred to the bacterial cell, rather than indirectly bioavailable where it must be released into the aqueous phase prior to being taken up by the bacterial cell). This direct bioavailability of HOCs from the micellar core to the cell can be quantified and depends on the surfactant types, surface characteristics of the biomass, surfactant concentrations, and the mixing conditions (Guha and Jaffé, 1996a, 1996b; Brown, 2007). A recent study by Brown and Al Nuaimi (2005) indicates that sorption of surfactants onto the bacterial cell surface occurs with varying levels of hemi-micelles, which may impact the bioavailable concentration of HOCs at the cell surface and is important governing surfactant-enhanced bioavailability (Brown, 2007).

Three possible pathways were assumed to interpret how and whether HOCs can enter into the bacterial cell (Figure 9.21) and then be available for biodegradation in the presence of surfactants (Guha and Jaffé, 1996a, 1996b; Brown, 2007). The first pathway, depicted as pathways (a) in Figure 9.21, is transfer of the HOCs in aqueous phase into the cell. The second pathway, depicted as pathways (b), is direct transfer of the HOCs in micellar-phase into the cell. The first pathway is considered for the case when biodegradation is controlled by the aqueous-phase HOC transfer into the cell. The second pathway, representing the micellar-phase HOC transport into the bacterial cell, is composed of two mass transfer processes: (1) the transport of the micelles containing HOCs from the bulk fluid to the cell, and followed by dynamic transfer of the micellar HOCs to adsorbed hemi-micelles on the cell surface upon micellar breakdown, and (2) the transport of the HOCs from the hemi-micelles into the bacterial cell. The pathway (b) assumes that all micellar HOCs transferred to the cell upon micellar breakdown is bioavailable and that the HOC gradient across the cell wall is driven by the micellar HOC concentration. The third pathway, depicted as pathway (c), assumes no

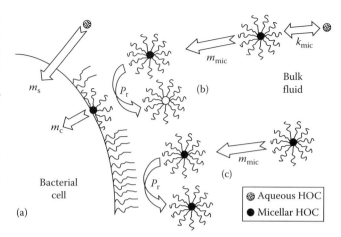

FIGURE 9.21 Pathways for an HOC to enter the bacterial cell. Pathway (a) is transport of aqueous HOC into the cell. Pathway (b) is direct transfer of micellar HOC to adsorbed surfactant hemi-micelles, with subsequent transport into the cell. Pathway (c) depicts a limiting case where no hemi-micelles are formed, resulting in micellar HOC not directly bioavailable. Note that the micellar HOC is always indirectly bioavailable via partitioning into the aqueous phase and subsequent transport into the cell through pathway (a). (Reprinted with permission from Brown, D.G. 2007. Relationship between micellar and hemi-micellar processes and the bioavailability of surfactant-solubilized hydrophobic organic compounds. Environ. Sci. Technol. 41:1194–1199. Copyright 2007 American Chemical Society.)

hemi-micelles formation on the cell surface and thus the micellar HOC cannot be directly transferred to the cell. This assumption, developed from the sorption results of surfactants on cell surface (Brown and Al Nuaimi, 2005; Brown, 2007), is in agreement with the theory that hydrophilic micellar surface may form a barrier that can block the interaction of microorganisms with the pollutant partitioned in the micellar core (Rouse et al., 1994). The pathways (b) and (c), independent of whether hemi-micelles form, describing micellar HOC transport into the cell would vary as a function of the hemi-micelle formation on the cell surface and the ability of those hemi-micelles to partition the HOCs.

The bioavailable HOC concentration (C_{bio}, mg L^{-1}) can be defined as follows (Guha and Jaffé, 1996a, 1996b; Brown, 2007):

$$C_{bio} = C_{aq} + fC_{mic} \qquad (9.22)$$

where C_{mic} (mg L^{-1}) and f (unitless) are the micellar HOC concentration and the fraction of micellar HOC that is directly bioavailable.

The bioavailability factors (f), for the alkyl polyethoxylate surfactants ($C_{12}E_y$), were found to decrease with the increase of surfactant concentrations and the numbers of ethylene oxide groups (Figure 9.22) (Brown, 2007). Therefore, the effect of surfactants on HOC bioavailability enhancement (C_{bio}, in Equation 9.22) is a result of the interplay between C_{mic} and f (Figure 9.22). This interplay results in a range of surfactant concentrations that provides enhanced bioavailability, and surfactant concentrations outside of this range can reduce the overall HOC

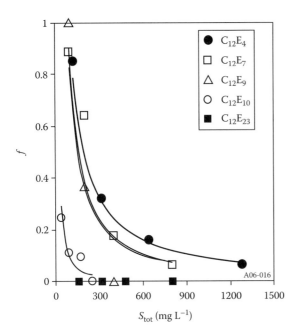

FIGURE 9.22 Bioavailability factor (f) plotted as a function of surfactant structure and concentration (S_{tot}) for the $C_{12}E_y$ surfactants with phenanthrene used as the HOC. (Reprinted with permission from Brown, D.G. 2007. Relationship between micellar and hemi-micellar processes and the bioavailability of surfactant-solubilized hydrophobic organic compounds. Environ. Sci. Technol. 41:1194–1199. Copyright 2007 American Chemical Society.)

bioavailability (Brown et al., 1999). It was also shown by Brown et al. (1999) that both the presence of soil and the total mass of HOCs in the system can strongly affect the HOC bioavailability, which leads to that the addition of surfactants could reduce HOC bioavailability even if f is nonzero. Although the bioavailability is important for HOC degradation, it is also important to note that bioavailability alone is not an indicator of the ability of surfactants to enhance the HOC biodegradation rate. The biodegradation rate is also a function of the ability of the bacterial culture to utilize this enhanced bioavailability. The addition of surfactant may increase the bioavailable concentrations of compounds, but it will have negligible effect on the biodegradation rate when the ability of the bacterial culture is not high enough to utilize the bioavailable HOCs enhanced by the surfactant.

9.4.5 Effects of Surfactant on Uptake by Plants

Traditionally, the major pathway for the plant uptake of HOCs was shown to occur via the aqueous phase of the soil (Paterson et al., 1990; Alkorta and Garbisu, 2001; Chiou et al., 2001). Therefore, the use of surfactants to enhance the apparent aqueous solubility of HOCs has also been expected to enhance their bioavailability for plant uptake. In field experiments, some surfactants at certain concentrations have been observed to enhance the uptake and accumulation of p,p'-DDE by specific C. pepo (zucchini) (White et al., 2006, 2007). In these studies, however, decline of the uptake and accumulation of p,p'-DDE by C. pepo (zucchini) was observed with addition of surfactants, suggesting the enhancement of plant uptake was limited to cultivar specifics, the added surfactants and their concentrations. In a model ecosystem (water–lava–plant–air), Li et al. (2001) and Jiang et al. (2005) also observed reduced uptake of HOCs in plant roots in the presence of a nonionic surfactant (Tween 80) and an anionic surfactant (linear alkylbenzene sulfonate).

Several studies have been conducted in a hydroponic system to examine the uptake rate of HOCs by plant from water in the presence of surfactants at concentrations below or above CMCs (Gao et al., 2006, 2008; Zhu and Zhang, 2008; Sun and Zhu, 2009). As an example shown in Figure 9.23a, it was observed

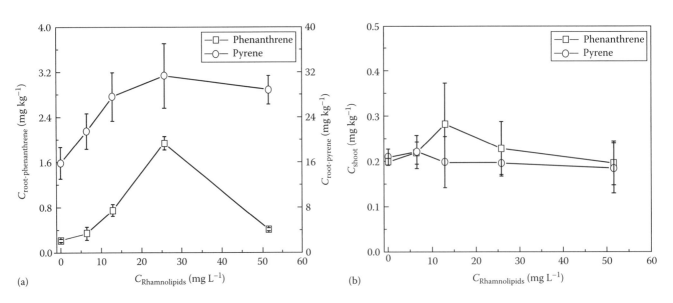

FIGURE 9.23 Concentrations of phenanthrene and pyrene in plant root (a) and shoot (b) after 288 h as a function of rhamnolipid concentration. (Reprinted from Zhu, L., and M. Zhang. 2008. Effect of rhamnolipids on the uptake of PAHs by ryegrass. Environ. Pollut. 156:46–52. Copyright 2008, with permission from Elsevier.)

Soil Physicochemical and Biological Interfacial Processes

in these studies that surfactant at concentrations lower than the respective CMC value could enhance the HOC uptake in plant roots, whereas surfactants at higher concentrations may inhibit the plant uptake. In addition, surfactants have no effects on the HOC levels in plant shoots (Figure 9.23b). The enhancement of HOC uptake in plant roots by relatively low concentration surfactants can be attributed to the permeability increase of ryegrass root cells by surfactants (Zhu and Zhang, 2008). The reduction of HOC uptake in plant roots by relatively high concentration surfactants might result from (1) the phytotoxicity of added surfactants to damage the plant cell and inhibit the plant growth (Gao, 2004; White et al., 2007), (2) the desorption of HOCs from root surface into the culture solution by the surfactants (Li et al., 2001; Jiang et al., 2005; Zhu and Zhang, 2008), and (3) the micellar HOCs are inaccessible and unavailable for the plant uptake (Li et al., 2001; Jiang et al., 2005).

Surfactants in the environment are well documented to have effects on the fate and behavior of organic chemicals by alternating their solubilization and mobilization, sorption/desorption in soil and sediments and microbial bioavailability. Although efforts have been used to explore their effects on the mineralization, metabolism, and uptake of organic chemicals by plants and their transportation in plants (Li et al., 2001; Jiang et al., 2005; Zhu and Zhang, 2008), present experimental data available in the literature are not enough to estimate the rate and the extent of plant uptake for HOCs in the presence of surfactants.

9.5 Conclusions

Overall, the fate of AOPs in the soil environment is a result of the AOP distribution in soil-related phases including soil solid mixtures, air, water, plant, and microorganisms. The aqueous phase, soil water, is the most important one among these phases for the AOP distribution because the aqueous phase acts as a medium that allows AOPs transfer from one phase to another and links with other phases. Without an aqueous phase, no direct transfer of AOPs can take place between soil-related phases. Soil physicochemical and biological interfacial processes describing the direct distribution of AOPs between aqueous phase and other phases, therefore, govern the fate of AOPs in the soil environment. Although these processes for AOPs are dynamic, they are commonly considered to be at an apparent equilibrium because the time for AOP distribution is long enough to reach the equilibrium in the environment. Changes in the equilibrium of AOPs in any of these processes may result in corresponding changes in the others. As such, these processes must be understood both individually and cooperatively in order to successfully predict the fate and risks of AOPs in the soil environment.

Sorption by soil solids determines the immediate biological availability of AOPs by decreasing the aqueous AOP concentration, retarding AOP release from soil to water or sequestrating AOPs in soil matrix, which potentially affects the fate and transport of AOPs in the soil environment. Soil is perceived as a dual sorbent for AOPs, in which the SOM functions not only as a linear partition medium but also a nonlinear adsorption medium.

SOM composition, aromaticity, and polarity have significant influence on the partition and nonlinear adsorption. Other soil components such as carbonaceous materials may also be important for nonlinear adsorption though their amount in normal soils may be low and their adsorption may be suppressed by the coated SOM. Nonlinear adsorption exhibits "competitive" and "irreversible" phenomena. The competitive behavior is expected to result in an increase of bioavailability for soil adsorbed AOPs by other solutes, while the irreversible sorption can decrease AOP bioavailability. In soils, multiple organic chemicals are commonly observed. Therefore, the decrease of AOP bioavailability generated by the irreversible behavior could be partly offset by the competition of the co-existing competitors. In single-solute system, the total nonlinear sorption, including either the linear partition fraction or the nonlinear adsorption fraction, can be estimated approximately from the soil organic carbon/matter content (f_{oc}/f_{om}) and chemical K_{ow}/S_w using the obtained empirical relationships between sorption coefficients and f_{oc}/f_{om} as well as between organic carbon/matter content normalized sorption coefficients and chemical K_{ow}/S_w. This sorption estimation also presents a potential way to predict the bioavailability, transport and fate of AOPs in the environment. However, the estimation of AOP sorption and the consequent bioavailability, transport, and fate of AOPs in natural systems could be more complex than that in the single-solute system because of the competitive sorption of multiple solutes in soils and irreversible sorption.

Plants and microorganisms have the ability to take up, accumulate, and degrade AOPs. The extent of AOPs to be taken up and the consequent accumulation and degradation depends on the bioavailability and toxicity of AOPs as well as the plant and microorganism species and population, which is also controlled by the factors such as nutrient supply, soil moisture, and temperature. A number of researchers have related the uptake of organic chemicals to the lipid content, and employed the lipid content to predict the contaminant uptake and accumulation. However, few data are available for the lipid content of plants and microorganisms and it may not be practicable to determine the lipid content for a wide range of species. Other components of plants and microorganisms such as cellulose and plant lignin may also have the ability for AOP uptake and accumulation, which needs to be addressed in future. Moreover, there is little information on what extent of AOPs in plants could be degraded and on the toxicity of the metabolites and bound AOPs.

Techniques to promote or decrease one interfacial process of AOPs could be employed in the engineered applications of soil and subsurface remediation. For example, cationic surfactants can promote the sorption of AOPs by soils, and thus their addition to soils can inhibit the transfer of AOPs from soil to plant, organisms, and surface/ground water. However, anionic and nonionic surfactants decrease the sorption of AOPs by soils, and thus can be added to water to enhance the washing efficiency and removal of AOPs from contaminated soils. Surfactants may also enhance the bioavailability of AOPs for the uptake and degradation by microorganisms and plants, which can be employed in the phytoremediation and bioremediation of contaminated soils. However, it is important to investigate the limiting steps of AOP bioavailability and degradation rate before surfactants are designed and used for the enhanced remediation purpose. Moreover, minimizing losses from sorption and precipitation, environmental acceptability, balanced biological degradation, and the cost of surfactants must be addressed in designing and selection of surfactants for remediation. More importantly, the addition of surfactants for remediation should not add to the deterioration of soils and pose a new threat to the environment and ecological systems.

Acknowledgments

This work was in part supported by Program for New Century Excellent Talents in University of China (NCET-08-493), NSF of China (20737002 and 20777065), Massachusetts Agricultural Experiment Station (MAS00978), and Agriculture and Food Research Initiative (AFRI) Competitive Grant No. 2009-35201-05819 from the USDA National Institute of Food and Agriculture (NIFA).

References

Aksu, Z. 2005. Application of biosorption for the removal of organic pollutants: A review. Process Biochem. 40:997–1026.

Aktas, Ö., and F. Cecen. 2007. Biogeneration of activated carbon: A review. Int. Biodeterior. Biodegrad. 59:257–272.

Alcock, R.E., A.E. Johnston, S.P. McGrath, M.L. Berrow, and K.C. Jones. 1993. Long-term changes in the polychlorinated biphenyl content of United Kingdom soils. Environ. Sci. Technol. 27:1918–1923.

Alexander, M. 1999. Biodegradation and bioremediation. 2nd edn. Academic Press Inc., San Diego, CA.

Alkorta, I., and C. Garbisu. 2001. Phytoremediation of organic contaminants in soils. Bioresour. Technol. 79:273–276.

Aronstein, B.N., and M. Alexander. 1993. Effect of a nonionic surfactant added to the soil surface on the biodegradation of aromatic hydrocarbons within the soil. Appl. Microbiol. Biotechnol. 39:386–397.

Bailey, G.W., and J.L. White. 1964. Review of adsorption and desorption of organic pesticides by soil colloids, with implications concerning pesticide bioactivity. J. Agric. Food Chem. 12:324–332.

Baldrian, P. 2006. Fungal laccases—Occurrence and properties. FEMS Microbiol. Rev. 30:215–242.

Ball, W.P., and P.V. Roberts. 1991. Long-term sorption of halogenated organic chemicals by aquifer material. 1. Equilibrium. Environ. Sci. Technol. 25:1223–1237.

Bamforth, S.M., and I. Singleton. 2005. Bioremediation of polycyclic aromatic hydrocarbons: Current knowledge and future directions. J. Chem. Technol. Biotechnol. 80:723–736.

Barbour, J.P., J.A. Smith, and C.T. Chiou. 2005. Sorption of aromatic organic pollutants to grasses from water. Environ. Sci. Technol. 39:8369–8373.

Barrer, R.M. 1978. Zeolites and clay minerals as sorbents and molecular sieves. Academic Press, London, U.K.

Bernardez, L.A., and S. Goshal. 2004. Selective solubilization of polycyclic aromatic hydrocarbons from multicomponent nonaqueous-phase liquids into nonionic surfactant micelles. Environ. Sci. Technol. 38:5878–5887.

Berti, W.R., and S.D. Cunningham. 2000. Phytostabilization of metals, p. 71–88. In I. Raskin and B.D. Ensley (eds.) Phytoremediation of toxic metals. Using plants to clean up the environment. John Wiley & Sons, New York.

Boucher, F.R., and G.F. Lee. 1972. Adsorption of lindane and dieldrin pesticides on unconsolidated aquifer sands. Environ. Sci. Technol. 6:538–543.

Bourrel, M., and R.S. Schechter. 1988. Microemulsions and related systems. Marcel Dekker, New York.

Bowen, G.C., and A.D. Rovira. 1991. The rhizosphere—The hidden half of the hidden half, p. 641–669. In Y. Waisel, A. Eshel, and U. Kaffkafi (eds.) Plant roots—The hidden half. Marcel Dekker, New York.

Bower, C.A., and F.B. Gschwend. 1952. Ethylene glycol retention by soils as a measure of surface area and interlayer swelling. Soil Sci. Soc. Am. Proc. 16:342–345.

Boyd, S.A., J.-F. Lee, and M.M. Mortland. 1988. Attenuating organic contaminant mobility by soil modification. Nature 333:345–347.

Brady, N.C., and R.R. Weil. 1996. The nature and properties of soils. Prentice Hall, Upper Saddle Rivers, NJ.

Braida, W., J.C. White, B. Zhao, F.J. Ferrandino, and J.J. Pignatello. 2002. Concentration-dependent kinetics of pollutant desorption from soils. Environ. Toxicol. Chem. 21:2573–2580.

Bramwell, D.A.P., and S. Laha. 2000. Effects of surfactant addition on the biomineralization and microbial toxicity of phenanthrene. Biodegradation 11:263–277.

Briggs, G.G.J. 1981. Theoretical and experimental relationships between soil adsorption, octanol-water partition coefficients, water solubilities, bioconcentration factors, and the parachor. J. Agric. Food Chem. 29:1050–1059.

Briggs, G.G., R.H. Bromilow, and A.A. Evans. 1982. Relationships between lipophilicity and root uptake and translocation of non-ionised chemicals by barley. Pestic. Sci. 13:495–504.

Briggs, G.G., R.H. Bromilow, A.A. Evans, and M. Williams. 1983. Relationships between lipophilicity and the distribution of non-ionized chemicals in barley shoots following uptake by the roots. Pestic. Sci. 14:492–500.

Bromilow, R.H., and K. Chamberlain. 1995. Principles governing uptake and transport of chemicals, p. 38–64. In S. Trapp and J.C. McFarlane (eds.) Plant contamination: Modelling and simulation of organic chemical processes. Lewis Publishers, London, U.K.

Browman, M.G., and G. Chesters. 1977. The solid–water interface: Transfer of organic pollutants across the solid–water interface, p. 49–105. In I.H. Suffet (ed.) Fate of pollutants in the air and water environments, Part I. John Wiley & Sons, New York.

Brown, D.G. 2007. Relationship between micellar and hemimicellar processes and the bioavailability of surfactant-solubilized hydrophobic organic compounds. Environ. Sci. Technol. 41:1194–1199.

Brown, D.G., and K.S. Al Nuaimi. 2005. Nonionic surfactant sorption onto the bacterial cell surface: A multi-interaction isotherm. Langmuir 21:11368–11372.

Brown, D.G., S. Guha, and P.R. Jaffé. 1999. Surfactant-enhanced biodegradation of a PAH in soil slurry reactors. Biorem. J. 3:269–283.

Burken, J.G. 2003. Uptake and metabolism of organic compounds – The green liver model, p. 59–84. *In* S.C. McCutcheon and J.L. Schnoor (eds.) Phytoremediation – Transformation and control of contaminants. John Wiley & Sons, New York.

Burken, J.G., and J.L. Schnoor. 1998. Predictive relationships for uptake of organic contaminants by hybrid poplar trees. Environ. Sci. Technol. 32:3379–3385.

Carter, C.W., and I.H. Suffet. 1982. Binding of DDT to dissolved humic materials. Environ. Sci. Technol. 16:735–740.

Cerniglia, C.E. 1992. Biodegradation of polycyclic aromatic hydrocarbons. Biodegradation 3:351–368.

Chefetz, B., A.P. Deshmukh, P.G. Hatcher, and E.A. Guthrie. 2000. Pyrene sorption by natural organic matter. Environ. Sci. Technol. 34:2925–2930.

Chefetz, B., and B. Xing. 2009. Relative role of aliphatic and aromatic moieties as sorption domains for organic compounds: A review. Environ. Sci. Technol. 43:1680–1688.

Chekol, T., L.R. Vough, and R.L. Chaney. 2002. Plant-soil-contaminant specificity affects phyto-remediation of organic contaminants. Int. J. Phytorem. 4:17–26.

Chen, B., E. Johnson, B. Chefetz, L. Zhu, and B. Xing. 2005. Sorption of polar and nonpolar aromatic organic contaminants by plant cuticular materials: Role of polarity and accessibility. Environ. Sci. Technol. 39:6138–6146.

Chen, Y., D.R.U. Knappe, and M.A. Barlaz. 2004. Effect of cellulose/hemicellulose and lignin on the bioavailability of toluene sorbed to waste paper. Environ. Sci. Technol. 38:3731–3736.

Chenu, C., and G. Stotzky. 2002. Interactions between microorganisms and soil particles: An overview, p. 3–41. *In* P.M. Huang, J.M. Bollag, and N. Senesi (eds.) Interactions between soil particles and microorganisms. John Wiley & Sons, Chichester, U.K.

Chin, Y.-P., G.R. Aiken, and K.M. Danielsen. 1997. Binding of pyrene to aquatic and commercial humic substances: The role of molecular weight and aromaticity. Environ. Sci. Technol. 31:1630–1635.

Chiou, C.T. 1995. Comment on thermodynamics of organic chemical partition in soils. Environ. Sci. Technol. 29:1421–1422.

Chiou, C.T. 1998. Soil sorption of organic pollutants and pesticides, p. 4517–4554. *In* R.A. Meyers (ed.) Environmental analysis and remediation. John Wiley & Sons, New York.

Chiou, C.T. 2002. Partition and adsorption of organic contaminants in environmental systems. John Wiley & Sons, New York.

Chiou, C.T., and D.E. Kile. 1994. Effects of polar and nonpolar groups on the solubility of organic compounds in soil organic matter. Environ. Sci. Technol. 28:1139–1144.

Chiou, C.T., and D.E. Kile. 1998. Deviations from sorption linearity on soils of polar and nonpolar organic compounds at low relative concentrations. Environ. Sci. Technol. 32:338–343.

Chiou, C.T., D.E. Kile, T.I. Brinton, R.L. Malcolm, J.A. Leenheer, and P. MacCarthy. 1987. A comparison of water solubility enhancements of organic solutes by aquatic humic materials and commercial humic acids. Environ. Sci. Technol. 21:1231–1234.

Chiou, C.T., D.E. Kile, D.W. Rutherford, G. Sheng, and S.A. Boyd. 2000. Sorption of selected organic compounds from water to a peat soil and its humic-acid and humin fractions: Potential sources of the sorption nonlinearity. Environ. Sci. Technol. 34:1254–1258.

Chiou, C.T., J.-F. Lee, and S.A. Boyd. 1990. The surface area of soil organic matter. Environ. Sci. Technol. 24:1164–1166.

Chiou, C.T., R.L. Malcolm, T.I. Brinton, and D.E. Kile. 1986. Water solubility enhancement of some organic pollutants and pesticides by dissolved humic and fulvic acids. Environ. Sci. Technol. 20:502–508.

Chiou, C.T., L.J. Peters, and V.H. Freed. 1979. A physical concept of soil-water equilibria for nonionic organic compounds. Science 206:831–832.

Chiou, C.T., P.E. Porter, and D.W. Schmedding. 1983. Partition equilibria of nonionic organic compounds between soil organic matter and water. Environ. Sci. Technol. 17:227–231.

Chiou, C.T., D.W. Rutherford, and M. Manes. 1993. Sorption of N_2 and EGME vapors on some soils, clays, and mineral oxides and determination of sample surface areas by use of sorption data. Environ. Sci. Technol. 27:1587–1594.

Chiou, C.T., D.W. Schmedding, and M. Manes. 1982. Partitioning of organic compounds in octanol–water systems. Environ. Sci. Technol. 18:4–10.

Chiou, C.T., G.Y. Sheng, and M. Manes. 2001. A partition-limited model for the plant uptake of organic contaminants from soil and water. Environ. Sci. Technol. 35:1437–1444.

Chiou, C.T., and T.D. Shoup. 1985. Soil sorption of organic vapors and effects of humidity on sorptive mechanism and capacity. Environ. Sci. Technol. 19:1196–1200.

Chiou, C.T., T.D. Shoup, and P.E. Porter. 1985. Mechanistic roles of soil humus and minerals in the sorption of nonionic organic compounds from aqueous and organic solutions. Org. Geochem. 8:9–14.

Chisholm, R.C., and L. Koblitsky. 1943. Sorption of methyl bromide by soil in a fumigation chamber. J. Econ. Entomol. 36:545–551.

Chu, W., and W.S. So. 2001. Modeling the two stage of surfactant-aided soil washing. Water Res. 35:761–767.

Collins, C., M. Fryer, and A. Grosso. 2006. Plant uptake of nonionic organic chemicals. Environ. Sci. Technol. 40:45–52.

Cornelissen, G., O. Gustafsson, T.D. Bucheli, M.O. Jonker, A.A. Koelmans, and P.M. Van Noort. 2005. Extensive sorption of organic compounds to black carbon, coal, and kerogen in sediments and soils: Mechanisms and consequences for distribution, bioaccumulation, and biodegradation. Environ. Sci. Technol. 39:6881–6895.

Cornelissen, G., H. Rigterink, M.M.A. Ferdinandy, and P.C.M. Van Noort. 1998. Rapidly desorbing fractions of PAHs in contaminated sediments as a predictor of the extent of bioremediation. Environ. Sci. Technol. 32:966–970.

Cort, T.L., and A.R. Bielefeldt. 2002. A kinetic model for surfactant inhibition of pentachlorophenol biodegradation. Biotechnol. Bioeng. 78:606–616.

Cuypers, S., T. Grotenhuis, K.G.J. Nierop, E.M. Franco, A. DeJager, and W. Rulkens. 2002. Amorphous and condensed organic matter domains: The effect of persulfate oxidation on the composition of soil/sediment organic matter. Chemosphere 48:919–931.

Dallavalle, M., J. Dachs, A.J. Sweetman, and K.C. Jones. 2004. Maximum reservoir capacity of vegetation for persistent organic pollutants: Implications for global cycling. Global Biogeochem. Cycle 18:GB4032.

Daughton, C.G., and T.A. Ternes. 1999. Pharmaceuticals and personal care products in the environment: Agents of subtle change? Environ. Health Persp. 107:907–938.

Dean, J.R., and W.C. Scott. 2004. Recent developments in assessing the bioavailability of persistent organic pollutants in the environment. Trends Anal. Chem. 23:609–618.

De Young, L.R., and K.A. Dill. 1988. Solute partitioning into lipid membranes. Biochemistry 27:5281–5289.

Deschenes, L., P. Lafrance, J.P. Villeneuve, and R. Samson. 1996. Adding sodium dodecyl sulfate and *Pseudomonas aeruginosa* UG2 biosurfactants inhibits polycyclic aromatic hydrocarbon degradation in a weathered creosote-contaminated soil. Appl. Microbiol. Biotechnol. 46:638–646.

Diallo, M.S., L.M. Abriola, and W.J. Weber. 1994. Solubilization of nonaqueous phase liquid hydrocarbons in micellar solutions of dodecyl alcohol ethoxylates. Environ. Sci. Technol. 28:1829–1837.

Doucette, W.J. 2003. Quantitative structure–activity relationships for predicting soil—Sediment sorption coefficients for organic chemicals. Environ. Toxicol. Chem. 22:1771–1788.

Duarte-Davidson, R., and K.C. Jones. 1996. Screening the environmental fate of organic contaminants in sewage sludge applied to agricultural soils. 2. The potential for transfers to plants and grazing animals. Sci. Total Environ. 185:59–70.

Dulfer, W.J., and H.A.J. Govers. 1995. Membrane-water partitioning of polychlorinated biphenyls in small unilamellar vesicles of four saturated phosphatidylcholines. Environ. Sci. Technol. 29:2548–2554.

Edwards, D.A., Z. Adeel, and R.G. Luthy. 1994. Distribution of nonionic surfactant and phenanthrene in a sediment/aqueous system. Environ. Sci. Technol. 28:1550–1560.

Edwards, D.A., R.G. Luthy, and Z. Liu. 1991. Solubilization of polycyclic aromatic hydrocarbons in micellar nonionic surfactant solutions. Environ. Sci. Technol. 25:127–133.

Ehlers, G.A.C., and A. Loibner. 2006. Linking organic pollutant (bio)availability with geosorbent properties and biomimetic methodology: A review of geosorbent characterization and (bio)availability prediction. Environ. Pollut. 141:494–512.

Ehrlich, H.L. 2002. Geomicrobiology. 4th edn. Marcel Dekker, New York.

El Fantroussi, S., and S.N. Agathos. 2005. Is bioaugmentation a feasible strategy for pollutant removal and site remediation? Curr. Opin. Microbiol. 8:268–275.

Escher, B.I., and R.P. Schwarzenbach. 1996. Partitioning of substituted phenols in liposome-water, biomembrane-water, and octanol-water systems. Environ. Sci. Technol. 38:266–270.

Falatko, D.M., and J.T. Novak. 1992. Effects of biologically produced surfactants on the mobility and biodegradation of petroleum hydrocarbons. Water Environ. Res. 64:163–169.

Feng, X.J., A.J. Simpson, and M.J. Simpson. 2006. Investigating the role of mineral-bound humic acid in phenanthrene sorption. Environ. Sci. Technol. 40:3260–3266.

Fismes, J., C. Perrin-Ganier, P. Empereur-Bissonnet, and J.L. Morel. 2002. Soil-to-root transfer and translocation of polycyclic aromatic hydrocarbons by vegetables grown on industrial contaminated soils. J. Environ. Qual. 31:1649–1656.

Fletcher, J.S., and R.S. Hegde. 1995. Release of phenols by perennial plant roots and their potential importance in bioremediation. Chemosphere 31:3009–3016.

Gans, J., M. Wolinsky, and J. Dunbar. 2005. Computational improvements reveal great bacterial diversity and high metal toxicity in soil. Science 309:1387–1390.

Gao, Y.Z. 2004. Phytoremediation and its surfactant-enhanced techniques for soil contamination with polycyclic aromatic hydrocarbons (in Chinese). Ph.D. Dissertation. Zhejing University. Hangzhou, China.

Gao, Y.Z., W.T. Ling, and M.H. Wong. 2006. Plant-accelerated dissipation of phenanthrene and pyrene from water in the presence of a nonionic-surfactant. Chemosphere 63:1560–1567.

Gao, Y.Z., Q. Shen, W. Ling, and L.L. Ren. 2008. Uptake of polycyclic aromatic hydrocarbons by *Trifolium pretense* L. from water in the presence of a nonionic surfactant. Chemosphere 72:636–643.

Gao, Y.Z., and L.Z. Zhu. 2004. Plant uptake, accumulation and translocation of phenanthrene and pyrene in soils. Chemosphere 55:1169–1178.

Gauthier, T.D., W.R. Seitz, and C.L. Grant. 1987. Effects of structural and compositional variations of dissolved humic materials on pyrene K_{oc} values. Environ. Sci. Technol. 21:243–247.

Gobas, F.A.P.C., J.M. Lahitme, G. Camfalo, W.Y. Shiu, and D. Mackay. 1988. A novel method for measuring membrane-water partition coefficients of hydrophobic organic chemicals: Comparison with 1-octanol-water partitioning. J. Pharm. Sci. 77:265–272.

Gomez-Lahoz, C., and J.-J. Ortega-Calvo. 2005. Effect of slow desorption on the kinetics of biodegradation of polycyclic aromatic hydrocarbons. Environ. Sci. Technol. 39:8776–8783.

Goring, C.A.I. 1967. Physical aspects of soil in relation to the action of soil fungicides. Annu. Rev. Phytopathol. 5:285–318.

Goudar, C., K. Strevett, and J. Grego. 1999. Competitive substrate biodegradation during surfactant-enhanced remediation. J. Environ. Eng. 125:1142–1148.

Graziano, G., and L. Byungkook. 2001. Hydration of aromatic hydrocarbons. J. Phys. Chem. 105:10367–10372.

Griffin, J.J., and E.D. Goldberg. 1983. Impact of fossil fuel combustion on sediments of Lake Michigan: A reprise. Environ. Sci. Technol. 17:244–245.

Grimberg, S.J., W.T. Stringfellow, and M.D. Aitken. 1996. Quantifying the biodegradation of phenanthrene by *Pseudomonas stutzeri* P16 in the presence of a nonionic surfactant. Appl. Environ. Microbiol. 62:2387–2392.

Gschwend, P.M., and S.-C. Wu. 1985. On the constancy of sediment–water partition coefficients of hydrophobic organic pollutants. Environ. Sci. Technol. 19:90–95.

Gu, T., B.-Y. Zhu, and H. Rupprecht. 1992. Surfactant adsorption and surface micellization. Prog. Colloid Polym. Sci. 88:74–85.

Guerin, W.F., and S.A. Boyd. 1992. Differential bioavailability of soil-sorbed naphthalene to two bacterial species. Appl. Environ. Microbiol. 58:1142–1152.

Guha, S., and P.R. Jaffé. 1996a. Biodegradation kinetics of phenanthrene partitioned into the micellar phase of nonionic surfactants. Environ. Sci. Technol. 30:605–611.

Guha, S., and P.R. Jaffé. 1996b. Bioavailability of hydrophobic compounds partitioned into the micellar phase of nonionic surfactants. Environ. Sci. Technol. 30:1382–1391.

Guha, S., P.R. Jaffé, and C.A. Peters. 1998. Bioavailability of mixtures of PAHs partitioned into the micellar phase of a nonionic surfactant. Environ. Sci. Technol. 32:2317–2324.

Gunasekara, A., and B. Xing. 2003. Sorption and desorption of naphthalene by soil organic matter: Importance of aromatic and aliphatic components. J. Environ. Qual. 32:240–246.

Gustaffson, Ö., F. Haghseta, C. Chan, J. MacFarland, and P.M. Gschwend. 1997. Quantification of the dilute sedimentary soot phase: Implications for PAH speciation and bioavailability. Environ. Sci. Technol. 31:203–209.

Haderlein, S.B., and R.P. Schwarzenbach. 1993. Adsorption of substituted nitrobenzenes and nitrophenols to mineral surfaces. Environ. Sci. Technol. 27:316–326.

Haderlein, S.B., K.W. Weissmahr, and R.P. Schwarzenbach. 1996. Specific adsorption of nitroaromatic explosives and pesticides to clay minerals. Environ. Sci. Technol. 30:612–622.

Hamaker, J.W. and J.M. Thompson. 1972. Adsorption, pp. 49–143. In C.A.I. Goring and J.W. Hamaker (eds.) Organic Chemicals in the Soil Environment. Vol. 1. Marcel Dekker, New York.

Hannink, N.K., S.J. Rosser, and N.C. Bruce. 2002. Phytoremediation of explosives. Crit. Rev. Plant Sci. 21:511–538.

Hanson, W.J., and R.W. Nex. 1953. Diffusion of ethylene dibromide in soils. Soil Sci. 76:209–214.

Harris, C.I., and G.F. Warren. 1964. Adsorption and desorption of herbicides by soil. Weeds 12:120–126.

Hassett, J.J., W.L. Banwart, S.G. Wood, and J.C. Means. 1981. Sorption of a-naphthol: Implications concerning the limits of hydrophobic sorption. Soil Sci. Soc. Am. J. 45:38–42.

Hesselsoe, M., S. Boysen, N. Iversen, L. Jorgensen, J.C. Murrell, I. McDonald, S. Radajewski, H. Thestrup, and P. Roslev. 2005. Degradation of organic pollutants by methane grown microbial consortia. Biodegradation 16:435–444.

Hickey, J.P., and D.R. Passino-Reader. 1991. Linear salvation energy relationships: "Rules of thumb" for estimation of variable values. Environ. Sci. Technol. 25:1753–1760.

Hsu, F.C., R.L. Marxmiller, and A.Y.S. Yang. 1990. Study of root uptake and xylem translocation of cinmethylin and related-compounds in detopped soybean roots using a pressure chamber technique. Plant Physiol. 93:1573–1578.

Huang, W., M.A. Schlautman, and W.J. Weber, Jr. 1996. A distributed reactivity model for sorption by soils and sediments. 5. The influence of near-surface characteristics in mineral domains. Environ. Sci. Technol. 30:2993–3000.

Huang, W., and W.J. Weber, Jr. 1997. Thermodynamic considerations in the sorption of organic contaminants by soils and sediments. 1. The isosteric heat approach and its application to model inorganic sorbents. Environ. Sci. Technol. 31:3238–3243.

Huang, W., T. Young, M.A. Schlautman, H. Yu, and W.J. Weber, Jr. 1997. A distributed reactivity model for sorption by soils and sediments. 9. General isotherm nonlinearity and applicability of the dual reactive domin model. Environ. Sci. Technol. 31:1703–1710.

Huckelhoven, R., I. Schuphan, B. Thiede, and B. Schmidt. 1997. Biotransformation of pyrene by cell cultures of soybean (*Glycine max* L.), wheat (*Triticum aestivum* L.), jimsonweed (*Datura stramonium* L.), and purple foxglove (*Digitalis purpurea* L.). J. Agric. Food Chem. 45:263–269.

Hundal, L.S., M.L. Thompson, D.A. Laird, and A.M. Carmo. 2001. Sorption of phenanthrene by reference smectites. Environ. Sci. Technol. 35:3456–3461.

Hutchinson, S.L., A.P. Schwab, and M.K. Banks. 2003. Biodegradation of petroleum hydrocarbons in the rhizosphere, p. 355–386. In S.C. McCutcheon and J.L. Schnoor (eds.) Phytoremediation: Transformation and control of contaminants. John Wiley & Sons, New York.

Iqbal, J., C. Metosh-Dickey, and R.J. Portier. 2007. Temperature effects on bioremediation of PAHs and PCP contaminated South Louisiana soils: A laboratory mesocosm study. J. Soils Sediments 7:153–158.

Jafvert, C.T. 1991. Sediment- and saturated-soil-associated reactions involving an anionic surfactant (dodecylsulfate). 2. Partition of PAH compounds among phases. Environ. Sci. Technol. 25:1039–1045.

Jafvert, C.T., and J.K. Heath. 1991. Sediment-soil-associated and saturated-soil-associated reactions involving an anionic surfactant (dodecyl-sulfate).1. Precipitation and micelle formation. Environ. Sci. Technol. 25:1031–1038.

Jafvert, C.T., P.L. Van-Hoof, and J.K. Heath. 1994. Solubilization of non-polar compounds by non-ionic surfactant micelles. Water Res. 28:1009–1017.

Janssen, D.B., I.J.T. Dinkla, G.J. Poelarends, and P. Terpstra. 2005. Bacterial degradation of xenobiotic compounds: Evolution and distribution of novel enzyme activities. Environ. Microbiol. 7:1868–1882.

Jerina, D.M. 1983. Metabolism of aromatic hydrocarbons by the cytochrome P450 system and epoxide hydrolase. Drug Metab. Dispos. 11:1–4.

Jiang, X., A. Yediler, and Y.F. Song. 2005. Effect of linear alkylbenzene sulphonate (LAS) on the mineralization, metabolism and uptake of C-14-phenanthrene in a model ecosystem (water-lava-plant-air). Chemosphere 61:741–751.

Jones, K.D., and C.L. Tiller. 1999. Effect of solution chemistry on the extent of binding of phenanthrene by a soil humic acid: A comparison of dissolved and clay bound humic. Environ. Sci. Technol. 33:580–587.

Kanaly, R.A., and S. Harayama. 2000. Biodegradation of high-molecular weight polycyclic aromatic hydrocarbons by bacteria. J. Bacteriol. 182:2059–2067.

Kang, S., and B.S. Xing. 2005. Phenanthrene sorption to sequentially extracted soil humic acids and humin. Environ. Sci. Technol. 39:134–140.

Kang, S.H., and B.S. Xing. 2008. Relationship of polarity and structures of organic matter with sorption capacity of hydrophobic organic compounds, p. 125–143. *In* Q. Huang, P.M. Huang, and A. Violante (eds.) Soil mineral-microbe-organic interactions: Theories and applications. Springer-Verlag, Berlin, Germany.

Kapulnik, Y. 1996. Plant growth promotion by rhizosphere bacteria, p. 769–781. *In* Y. Waisel, A. Eshel, and U. Kaffkafi (eds.) Plant roots—The hidden half. Marcel Dekker, New York.

Karickhoff, S.W. 1984. Organic pollutant sorption in aqueous systems. J. Hydraul. Eng. 110:707–735.

Karickhoff, S.W., D.S. Brown, and T.A. Scott. 1979. Sorption of hydrophobic pollutants on natural sediments. Water Res. 13:241–248.

Kelsey, J.W., B.D. Kottler, and M. Alexander. 1997. Selective chemical extractants to predict bioavailability of soil-aged organic chemicals. Environ. Sci. Technol. 31:214–217.

Kenaga, E.E., and C.A.I. Goring. 1980. Relationship between water solubility, soil sorption, octanol-water partitioning, and concentration of chemicals in biota, p. 78–115. *In* J.C. Eaton, P.R. Parrish, and A.C. Hendricks (eds.) Aquatic toxicology. American Society for Testing and Materials, Philadelphia, PA.

Kile, D.E., and C.T. Chiou. 1989. Water solubility enhancement of DDT and trichlorobenzene by some surfactants below and above the critical micelle concentration. Environ. Sci. Technol. 23:832–838.

Kile, D.E., C.T. Chiou, and R.S. Helburn. 1990. Effect of some petroleum sulfonate surfactants on the apparent water solubility of organic compounds. Environ. Sci. Technol. 24:205–208.

Kile, D.E., C.T. Chiou, H. Zhou, H. Li, and O. Xu. 1995. Partition of nonpolar organic pollutants from water to soil and sediment organic matters. Environ. Sci. Technol. 29:1401–1406.

Kile, D.E., R.L. Wershaw, and C.T. Chiou. 1999. Correlation of soil and sediment organic matter polarity to aqueous sorption of nonionic compounds. Environ. Sci. Technol. 33:2053–2056.

Kirk, T.K., and R.L. Farrell. 1987. Enzymatic 'combustion': The microbial degradation of lignin. Annu. Rev. Microbiol. 41:465–505.

Kleineidam, S., H. Rügner, B. Ligouis, and P. Grathwohl. 1999. Organic matter facies and equilibrium sorption of phenanthrene. Environ. Sci. Technol. 33:1637–1644.

Kommalapati, R.R., K.T. Valsaraj, W.D. Constant, and D. Roy. 1997. Aqueous solubility enhancement and desorption of hexachlorobenzene from soil using a plant-based surfactant. Water Res. 31:2161–2170.

Kuznetsov, S.I., M.V. Ivanov, and N.N. Lyalikova. 1963. Introduction to geological microbiology. McGraw-Hill, New York.

Laha, S., and R.G. Luthy. 1991. Inhibition of phenanthrene mineralization by nonionic surfactants in soil-water systems. Environ. Sci. Technol. 25:1920–1929.

Laha, S., and R.G. Luthy. 1992. Effects of nonionic surfactants on the solubilization and mineralization of phenanthrene in soil-water systems. Biotechnol. Bioeng. 40:1367–1380.

Laird, D.A., E. Barriuso, R.H. Dowdy, and W.C. Koskinen. 1992. Adsorption of atrazine on smectites. Soil Sci. Soc. Am. J. 56:62–67.

Lambert, C.M. 1967. Functional relationship between sorption in soil and chemical structure. J. Agric. Food Chem. 15:572–576.

Lambert, C.M. 1968. Omega (X), a useful index of soil sorption equilibria. J. Agric. Food Chem. 16:340–343.

Lawrence, C.A. 1970. Germicidal properties of cationic surfactants, p. 492–526. *In* E. Jungermann (ed.) Cationic surfactants. Marcel Dekker, Inc., New York.

Leboeuf, E.J., and W.J. Weber, Jr. 1997. A distributed reactivity model for sorption by soils and sediments. 8. Discovery of a humic acid glass transition and argument for a polymer-based model. Environ. Sci. Technol. 31:1697–1702.

Lee, J.-F., J.R. Crum, and S.A. Boyd. 1989. Enhanced retention of organic contaminants by soils exchanged with organic cations. Environ. Sci. Technol. 23:1365–1372.

Lee, J.-F., P.-M. Liao, C.-C. Kao, H.-T. Yang, and C.T. Chiou. 2000. Influence of a nonionic surfactant (Triton X-100) on contaminant distribution between water and several soil solids. J. Colloid Interface Sci. 229:445–452.

Leigh, M.B., J.S. Fletcher, X. Fu, and F.J. Schmitz. 2002. Root turnover: An important source of microbial substances in rhizosphere remediation of recalcitrant contaminants. Environ. Sci. Technol. 36:1579–1583.

Leung, K.T., K. Nandakumar, K. Sreekumari, H. Lee, and J. Trevors. 2007. Biodegradation and bioremediation of organic pollutants in soil, p. 521–552. *In* J.D. Van Elsas, J.K. Jansson, and J.T. Trevors (eds.) Modern soil microbiology. 2nd edn. CRC Press, Boca Raton, FL.

Lewis, M.A. 1991. Chronic and sublethal toxicities of surfactants to aquatic animals: A review and risk assessment. Water Res. 25:101–113.

Li, H., G. Sheng, C.T. Chiou, and O.Y. Xu. 2005. Relation of organic contaminant equilibrium sorption and kinetic uptake in plants. Environ. Sci. Technol. 39:4864–4870.

Li, Y., A. Yediler, Z.Q. Ou, I. Conrd, and A. Kettrup. 2001. Effects of a non-ionic surfactant (Tween-80) on the mineralization, metabolism and uptake of phenanthrene in wheat-solution-lava microcosm. Chemosphere 45:67–75.

Lichtenstein, E.P. 1959. Absorption of some chlorinated hydrocarbon insecticides from soils into various crops. J. Agric. Food Chem. 7:430–433.

Litvina, M., T.R. Todoruk, and C.H. Langford. 2003. Composition and structure of agents responsible for development of water repellency in soils following oil contamination. Environ. Sci. Technol. 37:2883–2888.

Liu, G.G., D. Roy, and M.J. Rosen. 2000. A simple method to estimate the surfactant micelle-water distribution coefficients of aromatic hydrocarbons. Langmuir 16:3595–3605.

Liu, W., S. Xu, B. Xing, B. Pan, and S. Tao. 2010. Nonlinear binding of phenanthrene to the extracted fulvic acid fraction in soil in comparison with other organic matter fractions and to the whole soil sample. Environ. Pollut. 158:566–575.

Lu, Y., and J.J. Pignatello. 2004a. Sorption of apolar aromatic compounds to soil humic acid particles affected by aluminum(III) ion cross-linking. J. Environ. Qual. 33:1314–1321.

Lu, Y., and J.J. Pignatello. 2004b. History-dependent sorption in humic acids and a lignite in the context of a polymer model for natural organic matter. Environ. Sci. Technol. 38:5853–5862.

Mackay, A.A., and P.M. Gschwend. 2000. Sorption of monoaromatic hydrocarbons to wood. Environ. Sci. Technol. 34:839–845.

Mader, B.T., K.U. Goss, and S.J. Eisenreich. 1997. Sorption of nonionic, hydrophobic organic chemicals to mineral surfaces. Environ. Sci. Technol. 31:1079–1086.

Manahan, S.E. 2000. Environmental chemistry. 7th edn. CRC Press, Boca Raton, FL.

Masiello, C.A., and E.R.M. Druffel. 1998. Black carbon in deep-sea sediments. Science 280:1911–1913.

Mata-Sandoval, J.C., J. Karns, and V. Torrents. 2001. Influence of rhamnolipids and Triton X-100 on the biodegradation of three pesticides in aqueous phase and soil slurries. J. Agric. Food Chem. 49:3296–3303.

McCutcheon, S.C., and J.L. Schnoor. 2003. Overview of phytotransformation and control of wastes, p. 3–58. In S.C. McCutcheon and J.L. Schnoor (eds.) Phytoremediation: Transformation and control of contaminants John Wiley & Sons, New York.

McFarlane, J.C. 1995. Anatomy and physiology of plant conductive systems, p. 13–36. In S. Trapp and J.C. McFarlane (eds.) Plant contamination-modelling and simulation of organic chemical processes. Lewis Publishers, Boca Raton, FL.

McLeod, C.J.A., and K.T. Semple. 2000. Influence of contact time on extractability and degradation of pyrene in soils. Environ. Sci. Technol. 34:4952–4957.

McRae, S.G. 1988. Practical pedology. Ellis Horwood limited, Chichester, U.K.

Means, J.C., S.G. Wood, J.J. Hassett, and W.L. Banwart. 1980. Sorption of polynuclear hydrocarbons by sediments and soils. Environ. Sci. Technol. 14:1524–1528.

Means, J.C., S.G. Wood, J.J. Hassett, and W.L. Banwart. 1982. Sorption of amino- and carboxyl-substituted polynuclear aromatic hydrocarbons by sediments and soils. Environ. Sci. Technol. 16:93–98.

Mihelcic, J.R., D.R. Lucking, R.J. Mitzell, and J.M. Stapleton. 1993. Bioavailability of sorbed- and separate-phase chemicals. Biodegradation 4:141–153.

Mills, A.C., and J.W. Biggar. 1969. Solubility-temperature effects on the adsorption of gamma- and beta-BHC from aqueous and hexane solutions by soil materials. Soil Sci. Soc. Am. Proc. 33:210–216.

Mingelgrin, U., and Z. Gerstl. 1983. Reevaluation of partitioning as a mechanism of nonionic chemicals adsorption in soil. J. Environ. Qual. 12:1–11.

Morgan, P., and R.J. Watkinson. 1989. Microbiological methods for the cleanup of soil and ground water contaminated with halogenated organic compounds. FEMS Microbiol. Rev. 63:277–299.

Mukerjee, P. 1979. Solubilization in aqueous micellar systems. In K.L. Mittal (ed.) Solution chemistry of surfactant. Vol. 1. Plenum Press, New York.

Mulligan, C.N., R.N. Yong, and B.F. Gibbs. 2001. Surfactant-enhanced remediation of contaminated soil: A review. Eng. Geol. 60:371–380.

Murphy, E.M., J.M. Zachara, and S.C. Smith. 1990. Influence of mineral-bound humic substances on the sorption of hydrophobic organic compounds. Environ. Sci. Technol. 24:1507–1516.

Murphy, E.M., J.M. Zachara, S.C. Smith, J.L. Phillips, and T.W. Wietsma. 1994. Interaction of hydrophobic organic compounds with mineral-bound humic substances. Environ. Sci. Technol. 28:1291–129.

Nayyar, S.P., D.A. Sabatini, and J.H. Harwell. 1994. Surfactant adsolubilization and modified admicellar sorption of nonpolar, polar, and ionizable organic compounds. Environ. Sci. Technol. 28:1874–1881.

Nedwell, B.D. 1999. Effect of low temperature on microbial growth: Lowered affinity for substrates limits growth at low temperature. FEMS Microbiol. Ecol. 30:101–111.

Negri, M.C., E.G. Gatliff, J.J. Quinn, and R.R. Hinchman. 2003. Root development and rooting at depths, p. 233–262. In S.C. McCutcheon and J.L. Schnoor (eds.) Phytoremediation: Transformation and control of contaminants. John Wiley & Sons, New York.

Nepovim, A., M. Hubalek, R. Podlipna, S. Zeman, and T. Vanek. 2004. In-vitro degradation of 2,4,6-trinitrotoluene using plant tissue cultures of *Solanum aviculare* and *Rheum palmatum*. Eng. Life Sci. 4:46–49.

Olson, P.E., K.F. Reardon, and E.A. Pilon-Smits. 2003. Ecology of rhizosphere bioremediation, p. 317–354. In S.C. McCutcheon and J.L. Schnoor (eds.) Phytoremediation: Transformation and control of contaminants. John Wiley & Sons, New York.

Paraiba, L.C. 2007. Pesticide bioconcentration modelling for fruit trees. Chemosphere 66:1468–1475.

Paterson, S., D. Mackay, E. Bacci, and D. Calamari. 1991. Correlation of the equilibrium and kinetics of leaf air exchange of hydrophobic organic chemicals. Environ. Sci. Technol. 25:866–871.

Paterson, S., D. Mackay, and C. McFarland. 1994. A model of organic chemical uptake by plants from soil and the atmosphere. Environ. Sci. Technol. 28:2259–2266.

Paterson, S., D. Mackay, D. Tam, and W.Y. Shiu. 1990. Uptake of organic chemicals by plants: A review of processes, correlations and models. Chemosphere 21:297–331.

Pennell, K.D., L.M. Abriolar, W.J. Weber, Jr. 1993. Surfactant-enhanced solubilization of residual dodecane in soil columns. 1. Experimental investigation. Environ. Sci. Technol. 27:2332–2338.

Pennell, K.D., S.A. Boyd, and L.M. Abriola. 1995. Surface area of soil organic matter reexamined. Soil Sci. Soc. Am. J. 59:1012–1018.

Pierce, R.H., C.E. Olney, and G.T. Felbeck, Jr. 1974. p,p'-DDT adsorption to suspended particulate matter in sea water. Geochim. Cosmochim. Acta 38:1061–1073.

Pignatello, J.J. 1991. Competitive effects in the sorption of nonpolar organic compounds by soils, p. 293–307. In R.A. Baker (ed.) Organic substances and sediments in water—Vol. 1—Humics and soils. Lewis Publishers, Chelsea, MI.

Pignatello, J.J. 2006. Fundamental issues in sorption related to physical and biological remediation of soils. Nato Science Series 4. Earth Environ. Sci. 69:41–68.

Pignatello, J.J., Y.F. Lu, E.J. LeBoeuf, W.L. Huang, J.Z. Song, and B.S. Xing. 2006. Nonlinear and competitive sorption of apolar compounds in black carbon-free natural organic materials. J. Environ. Qual. 35:1049–1059.

Pignatello, J.J., and B. Xing. 1996. Mechanisms of slow sorption of organic chemicals to natural particles. Environ. Sci. Technol. 30:1–11.

Pilon-Smits, E. 2005. Phytoremediation. Annu. Rev. Plant Biol. 56:15–39.

Ramos, J.L., E. Duque, M.-T. Gallegos, P. Godoy, M.I. Ramos-Gonzalez, A. Rojas, W. Teran, and A. Segura. 2002. Mechanisms of solvent tolerance in Gram negative bacteria. Annu. Rev. Microbiol. 56:743–768.

Ran, Y., B. Xing, P.S.C. Rao, and J. Fu. 2004. Importance of adsorption (hole-filling) mechanism for hydrophobic organic contaminants on an aquifer kerogen isolate. Environ. Sci. Technol. 38:4340–4348.

Reid, B.J., K.C. Jones, and K.T. Semple. 2000. Bioavailability of persistent organic contaminants in soils and sediments—A perspective on mechanisms, consequences and assessment. Environ. Pollut. 108:103–112.

Riederer, M. 1990. Estimating partitioning and transport of organic chemicals in the foliage/atmosphere system: Discussion of a fugacity-based model. Environ. Sci. Technol. 24:829–837.

Roesch, L.F.W., R.R. Fulthorpe, A. Riva, G. Casella, A.K.M. Hadwin, A.D. Kent, S.H. Daroub, F.A.O. Camargo, W.G. Farmerie, and E.W. Triplett. 2007. Pyrosequencing enumerates and contrasts soil microbial diversity. ISME J. 1:283–290.

Rosen, M.J. 1989. Surfactants and interfacial phenomena. 2nd edn. John Wiley & Sons, New York.

Rosenberg, E., R. Legmann, A. Kushmaro, R. Taube, E. Adler, and E.Z. Zon. 1992. Petroleum bioremediation—A multiphase problem. Biodegradation 3:337–350.

Rouse, J.D., and D.A. Sabatini. 1993. Minimizing surfactant losses using twin-head anionic surfactants in subsurface remediation. Environ. Sci. Technol. 27:2072–2078.

Rouse, J.D., D.A. Sabatini, and J.M. Suflita. 1994. Influence of surfactants on microbial degradation of organic compounds. Crit. Rev. Environ. Sci. Technol. 24:325–370.

Ruelle, P., M. Buchmann, H. Nam-Tran, and U.W. Kesselring. 1992. Enhancement of the solubilities of polycyclic aromatic hydrocarbons by weak hydrogen bonds with water. J. Comput. Aided Mol. Des. 6:431–448.

Rutherford, D.W., C.T. Chiou, and D.E. Kile. 1992. Influence of soil organic matter composition on the partition of organic compounds. Environ. Sci. Technol. 26:336–340.

Rutland, M.W., and T.J. Senden. 1993. Adsorption of poly(oxyethylene) nonionic surfactant C12E5 to silica: A study using atomic force microscopy. Langmuir 9:412–418.

Ryan, J.A., R.M. Bell, J.M. Davidson, and G.A. O'Connor. 1988. Plant uptake of non-ionic organic chemicals from soils. Chemosphere 17:2299–2323.

Saitoh, T., H. Hoshino, and T. Yotsuyanagi. 1994. Volume constraint effect on solute partitioning to Triton X-100 micelles in water. J. Chem. Soc. Faraday Trans. 90:479–486.

Salloum, M.J., B. Chefetz, and P.G. Hatcher. 2002. Phenanthrene sorption by aliphatic-rich natural organic matter. Environ. Sci. Technol. 36:1953–1958.

Salt, D.E., R.D. Smith, and I. Raskin. 1998. Phytoremediation. Annu. Rev. Plant Physiol. Plant Mol. Biol. 49:643–668.

Saltzman, S., L. Kliger, and B. Yaron. 1972. Adsorption-desorption of parathion as affected by soil organic matter. J. Agric. Food Chem. 20:1224–1226.

Sander, M., Y. Lu, and J.J. Pignatello. 2006. Conditioning-annealing studies of natural organic matter solids linking irreversible sorption to irreversible structural expansion. Environ. Sci. Technol. 40:170–178.

Sander, M., and J.J. Pignatello. 2005a. Characterization of charcoal adsorption sites for aromatic compounds: Insight drawn from single-solute and bi-solute competitive experiments. Environ. Sci. Technol. 39:1606–1615.

Sander, M., and J.J. Pignatello. 2005b. An isotope exchange technique to assess mechanisms of sorption hysteresis applied to naphthalene in kerogenous organic matter. Environ. Sci. Technol. 39:7476–7484.

Scheringer, M., M. Salzmann, M. Stroebe, F. Wegmann, K. Fenner, and K. Hungerbuhler. 2004. Long-range transport and global fractionation of POPs: Insights from multimedia modeling studies. Environ. Pollut. 128:177–188.

Schmidt, M.W.I., J.O. Skjemstad, E. Gehrt, and I. Ogel-Knabner. 1999. Charred organic carbon in German chernozemic soils. Eur. J. Soil Sci. 50:351–365.

Schnabel, W.E., A.C. Dietz, J.G. Burken, J.L. Schnoor, and P.J. Alvarez. 1997. Uptake and transformation of trichloroethylene by edible garden plants. Water Res. 31:816–824.

Schwab, A.P., A.A. Al-Assi, and M.K. Banks. 1998. Adsorption of naphthalene onto plant roots. J. Environ. Qual. 27:220–224.

Schwarzenbach, R.P., P.M. Gschwend, and D.M. Imboden. 2003. Environmental organic chemistry. John Wiley & Sons, Hoboken, NJ.

Schwarzenbach, R.P., and J. Westall. 1981. Transport of nonpolar organic compounds from surface water to groundwater: Laboratory sorption studies. Environ. Sci. Technol. 15:1360–1367.

Scow, K.M., and J. Hutson. 1992. Effect of diffusion and sorption on the kinetics of biodegradation: Theoretical considerations. J. Am. Soil Sci. Soc. 56:119–127.

Semple, K.T., K.J. Doick, P. Burauel, A. Craven, H. Harms, and K.C. Jones. 2004. Defining bioavailability and bioaccessibility of contaminated soil and sediment is complicated. Environ. Sci. Technol. 38:228A–231A.

Semple, K.T., K.J. Doick, L.Y. Wick, and H. Harms. 2007. Microbial interactions with organic contaminants in soil: Definitions, processes and measurement. Environ. Pollut. 150:166–176.

Semple, K.T., A.W.J. Morriss, and G.I. Paton. 2003. Bioavailability of hydrophobic organic contaminants in soils: Fundamental concepts and techniques for analysis. Eur. J. Soil Sci. 54:809–818.

Semple, K.T., B.J. Reid, and T.R. Fermor. 2001. Impact of composting strategies on the treatment of soils contaminated with organic pollutants. Environ. Pollut. 112:269–283.

Shang, T.Q., and M.P. Gordon. 2002. Transformation of C-14 trichloroethylene by poplar suspension cells. Chemosphere 47:957–962.

Sheng, G., S. Xu, and S.A. Boyd. 1996. Mechanism(s)-controlling sorption of neutral organic contaminants by surfactant-derived and natural organic matter. Environ. Sci. Technol. 30:1553–1557.

Shone, M.G.T., and A.V. Wood. 1974. Uptake and translocation of herbicides. J. Exp. Bot. 25:390–400.

Siciliano, S.D., and J.J. Germida. 1998. Mechanisms of phytoremediation: Biochemical and ecological interactions between plants and bacteria. Environ. Rev. 6:65–79.

Sikkema, J., J.A.M. deBont, and B. Poolman. 1995. Mechanisms of membrane toxicity of hydrocarbons. Microbiol. Rev. 59:201–222.

Simonich, S.L., and R.A. Hites. 1995. Organic pollution accumulation in vegetation. Environ. Sci. Technol. 29:2905–2913.

Smith, G.A., S.D. Christian, E.E. Tucker, and J.F. Scameborn. 1987. Group contribution model for predicting the solubilization of organic solutes by surfactant micelles. Langmuir 3:598–599.

Smith, D.W., J.J. Griffin, and E.D. Goldberg. 1973. Elemental carbon in marine sediments: A baseline for burning. Nature 241:268–270.

Smith, J.A., P.R. Jaffé, and C.T. Chiou. 1990. Effect of ten quaternary ammonium cations on tetrachloromethane sorption to clay from water. Environ. Sci. Technol. 24:1167–1172.

Spencer, W.F., and M.M. Cliath. 1970. Desorption of lindane from soil as related to vapor pressure. Soil Sci. Soc. Am. Proc. 34:574–578.

Spurlock, F.C., and J.W. Biggar. 1994. Thermodynamics of organic chemical partition in soils. 2. Nonlinear partition of substituted phenylureas from aqueous solution. Environ. Sci. Technol. 18:996–1002.

Standing, D., and K. Killham. 2007. The soil environment, p. 1–22. In J.D. Van Elsas, J.K. Jansson, and J.T. Trevors (eds.) Modern soil microbiology. 2nd edn. CRC press, Boca Raton, FL.

Stark, F.L., Jr. 1948. Investigations of chloropicrin as a soil fumigant. New York (Cornell) Agric. Exp. Stn. Mem. 178:1–61.

Stevenson, F.J. 1994. Humus chemistry: Genesis, composition, reactions. 2nd edn. John Wiley & Sons, New York.

Stokes, J.D., G.I. Paton, and K.T. Semple. 2006. Behaviour and assessment of bioavailability of organic contaminants in soil: Relevance for risk assessment and remediation. Soil Use Manage. 21:475–486.

Su, Y.H., Y.G. Zhu, G. Sheng, and C.T. Chiou. 2006. Linear adsorption of nonionic organic compounds from water onto hydrophilic minerals: Silica and alumina. Environ. Sci. Technol. 40:6949–6954.

Submmaniam, K., C. Stepp, J.J. Pignatello, B.F. Smets, and D. Grasso. 2004. Enhancement of polynuclear aromatic hydrocarbon desorption by complexing agents in weathered soil. Environ. Eng. Sci. 21:515–523.

Sun, S., and S.A. Boyd. 1991. Sorption of polychlorobiphenyl (PCB) congeners by residual PCB–oil phases in soils. J. Environ. Qual. 20:557–561.

Sun, S., and S.A. Boyd. 1993. Sorption of nonionic organic compounds in soil–water systems containing a petroleum sulfonate–oil surfactants. Environ. Sci. Technol. 27:1340–1346.

Sun, S., W.P. Inskeep, and S.A. Boyd. 1995. Sorption of nonionic organic compounds on soil-water systems containing a micelle-forming surfactant. Environ. Sci. Technol. 29:903–913.

Sun, L., and L. Zhu. 2009. Effect of anionic-nonionic mixed surfactant on ryegrass uptake of phenanthrene and pyrene from water. Chin. Sci. Bull. 54:387–393.

Sutherland, J.B., F. Rafii, A.A. Khan, and C.E. Cerniglia. 1995. Mechanisms of polycyclic aromatic hydrocarbon degradation, p. 269–306. In L.Y. Young and C.E. Cerniglia (eds.) Microbial transformation and degradation of toxic organic chemicals. Wiley-Liss, New York.

Swoboda, A.R., and G.W. Thomas. 1968. Movement of parathion in soil columns. J. Agric. Food Chem. 16:923–927.

Tahoun, S.A., and M.M. Mortland. 1966. Complexes of montmorillonite with primary, secondary, and tertiary amides-I. Protonation of amides on the surface of montmorillonite. Soil Sci. 102:248–254.

Tam, D.D., W.-Y. Shiu, K. Qiang, and D. Mackay. 1996. Uptake of chlorobenzenes by tissues of the soybean plant: Equilibria and kinetics. Environ. Toxicol. Chem. 15:489–494.

Theng, B.K.G., D.J. Greenland, and J.P. Quirk. 1967. Adsorption of alkylammonium cations by montmorillonite. Clay Miner. 7:1–17.

Tiehm, A. 1994. Degradation of polycyclic aromatic hydrocarbons in the presence of surfactants. Appl. Environ. Microbiol. 60:258–263.

Tiehm, A., M. Stieber, P. Werner, and F.H. Frimmel. 1997. Surfactant-enhanced mobilization and biodegradation of polycyclic aromatic hydrocarbons in manufactured gas plant soil. Environ. Sci. Technol. 31:2570–2576.

Todoruk, T.R., M. Litvina, A. Kantzas, and C.H. Langford. 2003. Low field NMR relaxometry: A study of interactions of water with water-repellent soils. Environ. Sci. Technol. 37:2878–2883.

Tokiwa, F. 1968. Solubilization behavior of sodium dodecylpolyoxyethylene sulfates in relation to their polyoxyethylene chain lengths. J. Phys. Chem. 72:1214–1217.

Torsvik, V., L. Vreas, and T.F. Thingstad. 2002. Procaryotic diversity—Magnitude, dynamics, and controlling factors. Science 296:1064–1066.

Trapp, S. 1995. Model for uptake of xenobiotics into plants. In S. Trapp and J.C. McFarland (eds.) Plant contamination: Modeling and simulation of organic chemical processes. Lewis Publishers, Boca Raton, FL.

Trapp, S., and M. Matthies. 1995. Generic one-compartment model for uptake of organic chemicals by foliar vegetation. Environ. Sci. Technol. 29:2333–2338.

Trapp, S., M. Matthies, I. Scheunert, and E.M. Topp. 1990. Modeling the bioconcentration of organic chemicals in plants. Environ. Sci. Technol. 24:1246–1252.

Tsomides, H.J., J.B. Hughes, J.M. Thomas, and C.H. Ward. 1995. Effect of surfactant addition on phenanthrene biodegradation in sediments. Environ. Toxicol. Chem. 14:953–959.

Ugrekhelidze, D., F. Korte, and G. Kvesitadze. 1997. Uptake and transformation of benzene and toluene by plant leaves. Ecotoxicol. Environ. Saf. 37:24–29.

United Nations. 1997. Glossary of environment statistics, Studies in methods, Series F, No. 67, United Nations, New York.

Vaes, W.J.J., E.U. Ramas, C. Hamwijk, I. van Holsteijn, B.J. Blaauher, W. Seinen, H.J.M. Verhaar, and J.L.M. Hemens. 1997. Solid phase microextraction as a look to determine membrane/water partition coefficients and bioavailable concentrations in in vitro systems. Chem. Res. Toxicol. 10:1067–1072.

Van Elsas, J.D., L. Tam, R.D. Finlay, K. Killham, and J.T. Trevors. 2007. Microbial interactions in soil, p. 177–210. In J.D. Van Elsas, J.K. Jansson, and J.T. Trevors (eds.) Modern soil microbiology. 2nd edn. CRC Press, Boca Raton, FL.

Van Wezel, A.P., G. Cornelissen, J.K. Van Miltenburg, and A. Qpperhuizen. 1996. Membrane burdens of chlorinated benzenes lower the main phase transition temperature in dipalmitoylchloline visicles: Implications for toxicity by narcotic chemicals. Environ. Toxicol. Chem. 15:203–212.

Volkering, F., A.M. Breure, and W.H. Rulkens. 1998. Microbiological aspects of surfactant use for biological soil remediation. Biodegradation 8:401–417.

Volkering, F., A.M. Breure, J.G. Van Andel, and W.H. Rulkens. 1995. Influence of nonionic surfactants on bioavailability of polycyclic aromatic carbons. Appl. Environ. Microbiol. 61:1699–1705.

Voparil, I.M., L.M. Milyer, and A.R. Place. 2003. Interactions among contaminants and nutritional lipids during mobilization by digestive fluids of marine invertebrates. Environ. Sci. Technol. 37:3117–3122.

Wade, P.I. 1954. Soil fumigation. I. The sorption of ethylene dibromide by soils. J. Sci. Food Agric. 5:184–192.

Wang, H., K. Lin, Z. Hou, B. Richardson, and J. Gan. 2010. Sorption of the herbicide terbuthylazine in two New Zealand forest soils amended with biosolids and biochars. J. Soils Sediments 10:283–289.

Wang, K., and B. Xing. 2005. Structural and sorption characteristics of adsorbed humic acid on clay minerals. J. Environ. Qual. 34:342–349.

Wang, X., K. Yang, S. Tao, and B. Xing. 2007. Sorption of aromatic organic contaminants by biopolymers: Effects of pH, copper (II) complexation, and cellulose coating. Environ. Sci. Technol. 41:185–191.

Weber, W.J., Jr., P.M. McGinley, and L.E. Katz. 1992. A distributed reactivity model for sorption by soils and sediments. 1. Conceptual basis and equilibrium assessments. Environ. Sci. Technol. 26:1955–1962.

Weber, J.B., P.W. Perry, and R.P. Upchurch. 1965. Influence of temperature and time on the adsorption of paraquat, diquat, 2,4-D, and prometone by clays, charcoal and an anion-exchange resin. Soil Sci. Soc. Am. Proc. 29:678–688.

Weed, S.B., and J.B. Weber. 1974. Pesticide-organic matter interactions, p. 223–256. In W.D. Guenzi (ed.) Pesticides in soil and water. Soil Science Society of America, Madison, WI.

Weissmahr, K.W., S.B. Haderlein, R.P. Schwarzenbach, R. Handy, and R. Nuesch. 1997. In situ spectroscopic investigations of adsorption mechanisms of nitroaromatic compounds at clay minerals. Environ. Sci. Technol. 31:240–247.

Wershaw, R.L., P.J. Burcar, and M.C. Glodberg. 1969. Interaction of pesticides with natural organic material. Environ. Sci. Technol. 3:271–273.

West, C.C., and J.H. Harwell. 1992. Surfactants and subsurface remediation. Environ. Sci. Technol. 26:2324–2330.

Westall, J.C., C. Leuenberger, and R.P. Schwarzenbach. 1985. Influence of pH andionic strength on the aqueous–nonaqueous distribution of chlorinated phenols. Environ. Sci. Technol. 19:193–198.

White, J.C., and M. Alexander. 1996. Reduced biodegradability of desorption-resistant fractions of polycyclic aromatic hydrocarbons in soil and aquifer solids. Environ. Toxicol. Chem. 15:1973–1978.

White, J.C., M. Hunter, J.J. Pignatello, and M. Alexander. 1999. Increase in the bioavailability of aged phenanthrene in soils by competitive displacement with pyrene. Environ. Toxicol. Chem. 18:1728–1732.

White, J.C., J.W. Kelsey, P.B. Hatzinger, and M. Alexander. 1997. Factors affecting sequestration and bioavailability of phenanthrene in soils. Environ. Toxicol. Chem. 16:2040–2045.

White, J.C., Z.D. Parrish, and M.P.N. Gent, W. Iannucci-Berger, B.D. Eitzer, M. Isleyen, and M.I. Mattina. 2006. Soil amendments, plant age, and intercropping impact p,p′-DDE bioavailability to *Cucurbita pepo*. J. Environ. Qual. 35:992–1000.

White, J.C., R. Peters, and J.W. Kelsey. 2007. Surfactants differentially impact *p,p′*-DDE accumulation by plant and earthworm species. Environ. Sci. Technol. 41:2922–2929.

White, J.C., and J.J. Pignatello. 1999. Influence of biosolute competition on the desorption kinetics of polycyclic aromatic hydrocarbons in soil. Environ. Sci. Technol. 33:4292–4298.

White, J.C., X. Wang, M.P.N. Gent, W. Iannucci-Berpr, B.D. Eitzer, N.P. Schultes, M. Arimzo, and M.I. Mattina. 2003. Subspecies-level variation in the phytoextraction of weathered p,p′-DBE by *Cucurbita pepo*. Environ. Sci. Technol. 37:4368–4373.

Wild, E., J. Dent, G.O. Thomas, and K.C. Jones. 2005. Direct observation of organic contaminant uptake, storage, and metabolism within plant roots. Environ. Sci. Technol. 39:3695–3702.

Wild, S.R., and K.C. Jones. 1992a. Organic chemicals entering agricultural soils in sewage sludges-screening for their potential to transfer to crop plants and livestock. Sci. Total Environ. 119:85–119.

Wild, S.R., and K.C. Jones. 1992b. Polynuclear aromatic hydrocarbon uptake by carrots grown in sludge-amended soil. J. Environ. Qual. 21:217–225.

Winnike-McMillan, S.K., Q. Zhang, L.C. Davis, L.E. Erickson, and J.L. Schnoor. 2003. Phytoremediation of methyl tertiary-butyl ether, p. 805–828. *In* S.C. McCutcheon and J.L. Schnoor (eds.) Phytoremediation: Transformation and control of contaminants. John Wiley & Sons, New York.

Wolfe, N.L., and C.F. Hoehamer. 2003. Enzymes used by plants and microorganisms to detoxify organic compounds, p. 159–187. *In* S.C. McCutcheon and J.L. Schnoor (eds.) Phytoremediation: Transformation and control of contaminants. John Wiley & Sons, New York.

Xia, G., and W.P. Ball. 1999. Adsorption-partitioning uptake of nine low-polarity organic chemicals on a natural sorbent. Environ. Sci. Technol. 33:262–269.

Xia, G., and J.J. Pignatello. 2001. Detailed sorption isotherms of polar and apolar compounds in high-organic soil. Environ. Sci. Technol. 35:84–94.

Xia, X., Y. Li, Z. Zhou, and C. Feng. 2010. Bioavailability of adsorbed phenanthrene by black carbon and multi-walled carbon nanotubes to Agrobacterium. Chemosphere 78:1329–1336.

Xing, B. 2001a. Sorption of naphthalene and phenanthrene by soil humic acids. Environ. Pollut. 111:303–309.

Xing, B. 2001b. Sorption of anthropogenic organic compounds by soil organic matter: A mechanistic consideration. Can. J. Soil Sci. 81:317–323.

Xing, B., W.B. McGill, and M.J. Dudas. 1994. Cross-correlation of polarity curves to predict coefficients of nonionic organic contaminants. Environ. Sci. Technol. 28:1929–1933.

Xing, B., and J.J. Pignatello. 1997. Dual-mode sorption of low-polarity compounds in glassy poly(vinylchloride) and soil organic matter. Environ. Sci. Technol. 31:792–796.

Xing, B., and J.J. Pignatello. 1998. Competitive sorption between 1,3-dichlorobenzene or 2,4-dichlorophenol and natural aromatic acids in soil organic matter. Environ. Sci. Technol. 32:614–619.

Xing, B., J.J. Pignatello, and B. Gigliotti. 1996. Competitive sorption between atrazine and other organic compounds in soils and model sorbents. Environ. Sci. Technol. 30:2432–2440.

Xu, S., and S.A. Boyd. 1994. Cation exchange chemistry of hexadecyltrimethylammonium in a subsoil containing vermiculite. Soil Sci. Soc. Am. J. 58:1382–1391.

Yang, Y., D. Ratte, S.F. Smets, J.J. Pignatello, and D. Grasso. 2001. Mobilization of soil organic matter by complexing agents and implications for polycyclic aromatic hydrocarbon desorption. Chemosphere 43:1013–1021.

Yang, Y., L. Shu, X. Wang, B. Xing, and S. Tao. 2010. Effects of composition and domain arrangement of biopolymer components of soil organic matter on the bioavailability of phenanthrene. Environ. Sci. Technol. 44:3339–3344.

Yang, K., and B. Xing. 2009. Sorption of phenanthrene by humic acid-coated nanosized TiO_2 and ZnO. Environ. Sci. Technol. 43:1845–1851.

Yang, K., L. Zhu, B. Lou, and B. Chen. 2004. Significance of natural organic matter in nonlinear sorption of 2,4-dichlorophenol onto soils/sediments. Water Resour. Res. 40:w01603.

Yang, K., L. Zhu, B. Lou, and B. Chen. 2005a. Correlations of nonlinear sorption of organic solutes with soil/sediment physicochemical properties. Chemosphere 61:116–128.

Yang, K., L. Zhu, and B. Xing. 2006. Enhanced soil washing of phenanthrene by the mixed solutions of TX100 and SDBS. Environ. Sci. Technol. 40:4274–4280.

Yang, K., L. Zhu, and B. Xing. 2007. Sorption of sodium dodecylbenzene sulfonate by montmorillonite. Environ. Pollut. 145:571–576.

Yang, K., L. Zhu, and B. Zhao. 2005b. Minimizing losses of nonionic and anionic surfactants to a montmorillonite saturated with calcium using their mixtures. J. Colloid Interface Sci. 291:59–66.

Yang, Z. and L. Zhu. 2007. Performance of the partition-limited model on predicting ryegrass uptake of polycyclic aromatic hydrocarbons. Chemosphere 67:402–409.

Yaron, B., and S. Saltzman. 1972. Influence of water and temperature on adsorption of parathion by soils. Soil Sci. Soc. Am. Proc. 36:583–586.

Yaws, C.L., S. Nijhawan, L. Bu, and D.R. Balundgi. 1999. Soil sorption coefficient, In C.L. Yaws (eds.) Chemical properties Handbooks. Mcgraw-Hill, New York.

Yost, E.C., and M.A. Anderson. 1984. Absence of phenol adsorption on goethite. Environ. Sci. Technol. 18:101–106.

Young, T.M., and W.J. Weber, Jr. 1995. A distributed reactivity model for sorption by soils and sediments. 3. Effects of diagenetic processes on sorption energetics. Environ. Sci. Technol. 29:92–97.

Zajic, J.E., and W. Seffens. 1984. Biosurfactants. CRC Crit. Rev. Microbiol. 5:87–107.

Zhang, Y., W.J. Maier, and R.M. Miller. 1997. Effect of rhamnolipids on the dissolution, bioavailability, and biodegradation of phenanthrene. Environ. Sci. Technol. 31:2211–2217.

Zhang, Y., and R.M. Miller. 1994. Effect of *Pseudomonas* rhamnolipid biourfactant on cell hydrophobicity and biodegradation of octadecane. Appl. Environ. Microbiol. 60:2101–2106.

Zhang, M., and L. Zhu. 2009. Sorption of polycyclic aromatic hydrocarbons to carbohydrates and lipids of ryegrass root and implications for a sorption prediction model. Environ. Sci. Technol. 43:2740–2745.

Zhao, D., M. Hunter, J.J. Pignatello, and J.C. White. 2002. Application of the dual-mode model for predicting competitive sorption equilibria and rates of polycyclic aromatic hydrocarbons in estuarine sediment suspensions. Environ. Toxicol. Chem. 21:2276–2282.

Zhao, D., and J.J. Pignatello. 2004. Model-aided characterization of Tenax-TA for aromatic compound uptake from water. Environ. Toxicol. Chem. 23:1592–1599.

Zhao, D., J.J. Pignatello, J.C. White, W. Braida, and F. Ferrandino. 2001. Dual-mode modeling of competitive and concentration-dependent sorption and desorption kinetics of polycyclic aromatic hydrocarbons in soils. Water Resour. Res. 37:2205–2212.

Zhao, B., L. Zhu, and K. Yang. 2006. Solubilization of DNAPLs by mixed surfactant: Reduction in partitioning losses of nonionic surfactant. Chemosphere 62:772–779.

Zhu, L., and S. Feng. 2003. Synergistic solubilization of polycyclic aromatic hydrocarbons by mixed anionic-nomonic surfactants. Chemosphere 53:459–467.

Zhu, L., and Y. Gao. 2004. Prediction of phenanthrene uptake by plants with a partition-limited model. Environ. Pollut. 131:505–508.

Zhu, L., K. Yang, B. Lou, and B. Yuan. 2003. A multi-component statistic analysis for the influence of sediment/soil composition on the sorption of a nonionic surfactant (Triton X-100) onto sediments/soils. Water Res. 37:4792–4800.

Zhu, L., and M. Zhang. 2008. Effect of rhamnolipids on the uptake of PAHs by ryegrass. Environ. Pollut. 156:46–52.

10
Impact of Soil Physicochemical and Biological Reactions on Transport of Nutrients and Pollutants in the Critical Zone

Jon Chorover
The University of Arizona

10.1	Introduction	10-1
	Critical Zone Interfaces • Biogeochemical Disequilibria	
10.2	Biosynthesis and Humification	10-3
10.3	Coupled Biogeochemical Weathering Processes	10-4
	Formation of Reactive Interfaces and Fate of Lithogenic Nutrients • Impacts of Organic Matter Infusion on Nutrients and Pollutants	
10.4	Effects of Organo-Mineral Heteroaggregates on Nutrient and Pollutant Fate and Transport	10-16
	Surface Composition of Mineral–Organic Complexes • Sorption of Inorganic Molecules at Mineral–Organic Interfaces • Sorption of Organic Pollutants • Aging Effects	
10.5	Microbial Intervention in Metal Cycling	10-21
	Microbially Mediated Redox Reactions • Biogeochemical Gradients • Biogeophysical Heterogeneities	
10.6	Conclusions and Future Prospects	10-26
	Acknowledgments	10-26
	References	10-26

10.1 Introduction

The fate of nutrients and pollutants in soil and other porous geomedia in the Earth's surface depends on feedbacks and intimate couplings among physical, chemical, and biological processes. At small distances in this realm are the molecular-scale interactions that occur when nutrients and pollutants are introduced to a rich soil geochemistry of lithogenic solutes, organics, and particle surfaces, including those that are formed from organic and inorganic weathering processes. A premise of this chapter is that *the fate and transport of nutrients and pollutants in the environment is highly dependent on the structure, dynamics and interactions of organic and inorganic secondary products of biogeochemical weathering.* If that premise is correct, then once introduced to a soil environment, the fate of nutrients and pollutants must depend on the dynamic behavior of these secondary products. Since aqueous geochemical conditions in soils change on a hydrologic event basis as well as along fluid flow paths, one expects that the metastability and reaction kinetics of these secondary products and co-associated nutrients and pollutants will likewise behave dynamically in time and space.

Physicochemical processes that occur along flow paths in soil include (1) aqueous complexation, (2) redox, (3) sorption and flocculation, (4) nucleation and growth of inorganic and organic weathering products, (5) hetero-aggregation to form micro-aggregates, and (6) pore structure formation and associated limitations on advection and diffusion. These nano- to microscale interactions are then superimposed in the field by consolidation of hetero-aggregates into a macro-structured porous media with emergent properties of pedon- to field-scale heterogeneity and preferential fluid flow.

10.1.1 Critical Zone Interfaces

Soil component interactions lead to the formation of interfaces and aggregates that are central to soil function and therefore affect the fate of pollutants and nutrients (Figure 10.1). Heterogeneous reactions among soil solids, liquids, and gases affect the capacity of this portion of the Earth's surface to contribute to the "membrane" that sustains water and soil quality (Colin et al., 1992). This finite capacity of soil to detoxify pollutants and support ecosystem primary productivity is then modulated by soil

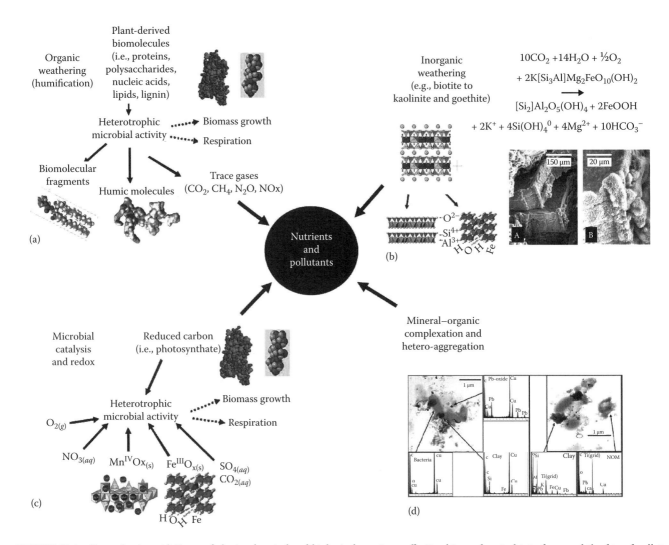

FIGURE 10.1 (See color insert.) Types of physicochemical and biological reactions affecting biogeochemical interfaces and the fate of pollutants and nutrients in soil systems. (a) *Organic weathering* (i.e., humification) processes involve the biotic and abiotic degradation of plant- and microbially derived biomolecules into biomolecular fragments and associated humic substances. (b) *Inorganic weathering* processes include the incongruent dissolution of primary minerals to form high specific-surface-area secondary clays and (oxyhydr)oxides that can coat the surfaces of primary minerals (shown in inset—halloysite coating the surface of biotite mica "books," from Kretzschmar, R., W.P. Robarge, A. Amoozegar, and M.J. Vepraskas. 1997. Biotite alteration to halloysite and kaolinite in soil-saprolite profiles developed from mica schist and granite gneiss. Geoderma 75:155–170. Copyright 1997, with permission from Elsevier.) (c) *Microbial catalysis and redox* reactions drive geochemical transformations as part of the set of alternative TEAPs in microbial respiration, such as the reductive dissolution of manganese (IV) and iron (III) oxides (MnOx and FeOx shown here are birnessite and goethite) with consequences for sorbed contaminants; (d) *Mineral–organic complexation* gives rise to complex hetero-aggregates of secondary mineral particles, humic substances and microbial cells that constitute much of the reactive interface in soils. (Image from Perdrial, N., J. Perdrial, J.-E. Delphin, F. Elsass, and N. Liewig. 2010. Temporal and spatial monitoring of mobile nanoparticles in a vineyard soil: Evidence of nanoaggregate formation. Eur. J. Soil Sci. 61:456–468. In press. Copyright Wiley-VCH Verlag GmbH & Co. KGaA. With permission.).

biogeochemical processes. As discussed in this chapter, finite and dynamic capacity is reflected in the sensitivity of soil biogeochemistry to inputs and changes in environment, including anthropogenic effects. As one open system component of Earth's terrestrial surface, soils change in response to the structure and function of superimposed vegetation and underlying bedrock, as well as the meteoric forcing and interceding fluid fluxes. The entirety of that system, termed the *critical zone* (CZ), exhibits an open system character, strong disequilibrium, and dynamic reaction chemistry (NRC, 2001).

Because of the diversity of surfaces (Figure 10.1b) and the catalytic activity of resident microbial species (Figure 10.1c), geomedia have the capacity for sorption (retention) and chemical transformation (e.g., oxidation–reduction) of a wide range of solutes that affect water quality. Of particular interest to human health is the behavior of nutrients, metals, metalloids, and polar or nonpolar organic pollutants (see Chapter 8). In some cases, pollutants tend to traverse the CZ largely unimpeded (e.g., some polar organics and weakly sorbing oxyanions), whereas in many cases, the CZ appears to have a robust capacity to sequester influent atmospheric, water, or waste pollution (e.g., many transition metals and hydrophobic organic contaminants [HOCs]). In any case, with respect to controlling the transport behavior of nutrients and pollutants in the CZ, soil interfacial reactions are paramount.

10.1.2 Biogeochemical Disequilibria

What drives the reactions is the set of geochemical disequilibria characteristic of open, biologically active, soil systems. Mineral weathering in the presence of biota is fueled by the thermodynamic instability at Earth surface conditions of (i) resident mineral phases and (ii) reduced carbon. These two distinct sets of reactants, one deriving from rock and the other from biomass, interact with water to create the most complex of all Earth surface environments: a multicomponent porous medium with extremely high solid–liquid–gas interfacial area, teeming with microbes and their metabolic products, and hosting large biogeochemical gradients even at micrometer-scale distances (Young and Crawford, 2004; Laverman et al., 2007; Hinsinger et al., 2009; Chapter 6). As a result of these characteristics, soils are capable of sequestering enormous pools of organic carbon, recycling nutrients fixed by the terrestrial biosphere, and remediating anthropogenic toxins. These essential ecosystem services tend to be provided most effectively by those soils whose reaction chemistry promotes the formation of meta-stable, high surface area, biogeochemical weathering products.

Through direct impacts on gas, liquid, and solid phase chemical dynamics, it is now well documented that biota play a prominent role in defining the rate and trajectory of these weathering reactions. What has been less well appreciated until recent years is the extent to which resident microbiota capture the energy liberated from these coupled mineral–organic transformation reactions, while also affecting their rate and trajectory. Biogeochemical reactions among minerals, organics, and microbes in disequilibrium soil systems support the reproduction and growth of diverse microbial communities and their large range of metabolic strategies. If there is energy accessible from an electron transfer reaction (i.e., negative ΔG), there are often microorganisms that have evolved the metabolic capability to promote the reaction and capture the energy liberated (Newman and Banfield, 2002). The intervention of microbiota and their metabolic byproducts in geochemical cycles affects not only reaction rates but also the specific nature (and hence subsequent reactivity) of reaction products. The characteristics affected by these interactions include the chemical composition, crystallinity, size, and surface reactivity of colloidal soil components (Monger et al., 1991; Konhauser, 1997; Toner et al., 2004).

The remainder of this chapter focuses specifically on *coupled* biological, physical, and geochemical processes that affect the transport and fate of nutrients and pollutants in soils. The following topics are discussed in turn: (1) biosynthesis and humification, (2) bio-physicochemical weathering processes, (3) reactivity of organo-mineral heteroaggregates, and (4) microbial intervention in metal cycling. Emphasis is placed on the dynamic nature of the soil system, one where biotic and abiotic components and processes are subjected to tight interplays and feedbacks. A close examination of these interactions indicates that they are a key to better resolving and predicting the fate of nutrients and pollutants in the CZ. Particular emphasis is given to the process-mediating role of organic matter that is mobile in the aqueous phase. Although consisting of both dissolved and colloidal forms, this material is traditionally referred to as "dissolved" organic matter (DOM—that which passes through, for example, 0.45 μm membrane filtration), and this definition is used in the present chapter as well.

10.2 Biosynthesis and Humification

Plants and autotrophic microorganisms that colonize the irradiated surface of soil play a central role in the formation of biogeochemical interfaces by transforming radiant energy to chemical form during photosynthesis. This process harnesses extraterrestrial energy to fuel soil biogeochemical weathering (Kelly et al., 1998; Berner et al., 2003; Taylor et al., 2009). Upon exudation or senescence, biogenic natural organic matter (NOM) is the primary source of carbon and energy for a complex food web of chemoheterotrophic organisms that hydrolyze and oxidize it, either partially to form biomolecular fragments and humic substances, or completely to form CO_2 and H_2O (Figure 10.1a). These products create a geochemical environment that promotes subsurface mineral transformation reactions, alters the chemistry of particle surfaces, and influences the speciation and mobility of redox active nutrient and pollutant elements (e.g., C, N, S, Fe, Mn, Cr, As).

Near the soil surface, NOM is frequently a major component of the porous matrix, particularly in humid environments, because the high influx of plant litter and tissue exudates results in formation of both discrete organic residues and mineral surface coatings. As a result, surface soil exhibits the highest biomass and activity of soil organisms. Organic matter decomposition by soil fauna and microorganisms leads to the release of CO_2 and essential nutrient elements (e.g., N, P, S), and production of dissolved and colloidal organic matter (DOM), including the formation of humic substances by molecular aggregation and extracellular enzymatic or mineral surface-mediated reactions (Piccolo, 2002; Huang, 2004). This region is also influenced by the highest density of plant roots and their mycorrhizal associations that not only take up nutrients from the soil solution, but respire CO_2, and release protons (to maintain charge balance) and organic root exudates (e.g., low-molecular weight organic acids, enzymes, siderophores, polysaccharides, biosurfactants) (Arocena et al., 2004; Taylor et al., 2009). Some of these exudates are strong chelating ligands, including siderophores and low molar mass organic acids, that enhance mineral dissolution and release nutrients to solution (Landeweert et al., 2001; Huang and Germida, 2002; Calvaruso et al., 2006; Kraemer et al., 2006).

Root exudates and root turnover also serve as carbon sources for many microorganisms living in the rhizosphere, which themselves release biomolecules into the local microenvironment via cell exudation and lysis. Extracellular biomolecules may strongly adsorb to nucleating precipitates and mineral surfaces where they influence surface-controlled processes such as crystal growth (Lee and Reeder, 2006), ligand-promoted dissolution (Liermann et al., 2000), ion adsorption, and colloid deposition and release (Kretzschmar et al., 1995). In addition, biological uptake may promote the weathering of certain primary minerals by depleting the pore water of rock-derived nutrient elements. For example, K^+ uptake by plants can increase the rate of mica weathering in soils (Hinsinger et al., 2006). Therefore, the rhizosphere around active plant roots often exhibits steep micrometer to millimeter scale gradients in pH, dissolved ligand and nutrient concentrations, and microbial activity (Marschner et al., 1986; Hinsinger et al., 2006). While plant roots and microorganisms are known to accelerate mineral weathering reactions via release of reactive chemicals, the details of these processes and their distribution in space remain poorly resolved (Jones, 1998; Jones et al., 2003; Akter and Akagi, 2005).

10.3 Coupled Biogeochemical Weathering Processes

10.3.1 Formation of Reactive Interfaces and Fate of Lithogenic Nutrients

The near subsurface environment is composed of highly complex associations among primary and/or secondary minerals, NOM, plant roots, microorganisms, aqueous solutions (pore water), and gases of varying composition (Chorover et al., 2007; Hinsinger et al., 2009). Since many of the relevant reactions are surface-controlled, interfacial characterizations under environmentally relevant conditions are essential. However, accurate characterization of interfacial chemistry in soil and sediment poses unique challenges and it cannot be done by measuring the bulk solid phase composition. In particular, discontinuous thin films of organic and inorganic material often coat underlying mineral surfaces. Although they may be present at trace levels, these "surface modifiers" significantly alter interfacial chemistry relative to that of the underlying solids, and thereby control surface charge, solute partitioning, and colloidal behavior (Davis, 1982; Bertsch and Seaman, 1999; Kretzschmar et al., 1999). Surface sensitive methods (e.g., atomic force microscopy, x-ray photoelectron spectroscopy, x-ray reflectivity, glancing incidence x-ray diffraction, attenuated total reflectance infrared spectroscopy) offer various advantages, and their conjunctive use often provides a more complete "picture" by giving complementary information on parameters such as chemical composition, functional group chemistry, surface topography, and patchiness (Gerin et al., 2003; Giesler et al., 2005; Wogelius et al., 2007; Cheng et al., 2008).

Since many of the reactions controlling nutrient and pollutant fate occur at solid–solution interfaces, the production and prevalence of finely divided mineral–organic particles is paramount. Leaching of water and solutes through the root and vadose zones drives incongruent weathering reactions that affect element mobility and chemical denudation. The secondary minerals formed from rock weathering include mainly layer silicates (clay minerals) and various crystalline and short-range ordered (oxyhydr)oxides of iron, aluminum, and manganese (e.g., Figure 10.1b). Because of their high specific surface area (10–800 m^2 g^{-1}), these phases can occupy a large portion of the solid–water interface even when they are present at a low mass fraction. They also possess high charge and reactive site densities. These characteristics have important implications for wide-ranging environmental issues including metal(loid) and radionuclide fate in the subsurface. For example, (oxyhydr)oxide colloids tend to be "hot spots" for the sequestration of toxic transition metal species into strongly bound, and hence less bioavailable forms (Manceau et al., 2003), and polycrystalline coatings on primary silicate surfaces have been shown to control reactive transport of Pb^{2+} and Zn^{2+} in aquifer sediments (Coston et al., 1995).

10.3.1.1 Organo-Mineral Weathering Incongruency

The incongruent weathering of rock-forming primary silicates to give clay-sized secondary minerals proceeds in the presence of water, irrespective of biological activity. For example, the transformation of K-feldspar to form kaolinite:

$$2KAlSi_3O_{8(s)} + 2H^+ + 9H_2O$$
$$\rightarrow \left[Si_2 \right] Al_2O_5(OH)_{4(s)} + 4Si(OH)_4 + 2K^+ \quad (10.1)$$

occurs abiotically given sufficient fresh water. This incongruent dissolution can lead to surface coating of the feldspar with secondary weathering products. The formation of clay surface coatings on weathering primary minerals alters grain surface chemistry and passivates further mineral transformation. A study of field weathered albite, for example, showed precipitation of kaolinite on the grain surface as detected by a variety of techniques (atomic force microscopy, transmission electron microscopy, scanning ion mass spectrometry, and x-ray photoelectron spectroscopy), whereas the underlying albite surface itself was Al depleted (Nugent et al., 1998). Similarly, mica layers or "books" in the schist saprolite of an Ultisol were observed by SEM to become armored by their colloidal halloysite and kaolinite weathering products (Figure 10.1b) (Kretzschmar et al., 1997).

In the presence of high water flux and biological activity, such reactions can be accelerated or otherwise altered. For example, the production of complexing organic ligands such as dicarboxylic oxalic acid ($C_2H_2O_4$) by plant roots or microorganisms can alter the K-feldspar weathering reaction to give:

$$2KAlSi_3O_{8(s)} + 2C_2H_2O_4 + 4H^+ + 8H_2O$$
$$\rightarrow 2C_2O_4 - Al^+ + 6Si(OH)_4 + 2K^+ \quad (10.2)$$

where $C_2O_4–Al^+$ is the oxalate–Al complex with a net charge of +1. As written, the dissolution reaction in Equation 10.2 is congruent, no secondary solid is formed. However, while experimental studies have shown that feldspar dissolution in the presence of oxalic acid is indeed more rapid than in its absence, it is also non-stoichiometric. Unlike Equation 10.1, however, oxalate reaction resulted in preferential dissolution of Al relative to Si (Stillings et al., 1996). One expects the oxalate–Al complex to be more mobile in the subsurface than monomeric inorganic Al, so its aqueous phase transport should be enhanced. But travel distance of the complex depends on several unknowns such as (1) hydrologic down gradient change in solution chemistry that could cause dissociation of the C_2O_4–Al^+ complex, (2) microbial degradation of the complexing organic at some constant or dynamic rate, (3) adsorption of the metal–organic complex (or some portion thereof) onto soil porous media surfaces, or (4) co-precipitation of the metal–organic complex with other DOM components as observed in B_{hs} horizons of Spodosols.

In the latter case, the resulting soil particles contain variously hydrolyzed Al and Fe bonded into a C backbone in a coagulated organo-metallic solid. Oxidative degradation of oxalate and/or dissociation of the complex down gradient from the weathering front would lead to liberation of Al and potential precipitation as gibbsite:

$$C_2O_4 - Al^+ + 0.5\,O_{2(g)} + 2H_2O$$
$$\rightarrow Al(OH)_{3(s)} + 2CO_{2(g)} + H^+ \qquad (10.3)$$

Thus, coupling between chemical, physical, and biological processes can bear directly on the types of reactive interfaces formed, and their distribution in space.

10.3.1.2 Microbial Role in Primary Mineral Weathering

A microbial role in weathering of primary minerals is now well established. Bacteria play an active role in the weathering of primary silicates by producing metal complexing agents, such as siderophores and other multi-dentate ligands. These ligands enhance the rate of mineral dissolution and increase the bioavailability of lithogenic nutrients. For example, in batch experiments, the rate of release of Fe from hornblende (an amphibole) was increased fivefold in the presence of *Streptomyces sp.*, a rock-surface colonizing bacterial species (Liermann et al., 2000). The enhancement attributed to biosynthesis of a Fe-complexing catechol derivative that was characterized by mass spectrometry was comparable to that observed in the presence of a 24 μM solution of a commercially available Fe-chelating siderophore, desferrioxamine mesylate (DFAM). The role of fungal hyphae and lichen in promoting the physicochemical weathering of primary silicates has also been widely reported. These organisms have the capacity to induce fractures along grain boundaries (Barker and Banfield, 1996), which increases interfacial area for chemical weathering attack. Fungal hyphae have a capacity to "tunnel" into feldspar grains, apparently in pursuit of cationic nutrients (Ca, K, Mg) (Hoffland et al., 2002). Solutions extracted from forest soil in close proximity to mychorrhizae exhibited a significantly greater extent of undersaturation with respect to nutrient-containing primary minerals as compared to non-mycorrhizal soil domains (Arocena and Glowa, 2000). Hence, it is clear that microorganisms contribute significantly to the weathering of rock, whereby lithogenic nutrients are released into bioavailable form.

10.3.1.3 Impact of Vascular Plants on Weathering versus Denudation

In respect to the fate of lithogenic nutrients (Ca, Mg, K, P), the distinction between chemical weathering (mineral transformation) and chemical denudation (element loss), has significant consequences for ecosystem productivity. Leaching losses of these solutes can diminish net primary production, whereas efficient recycling can enhance it. Recent experiments have shown that mass balance ratios of chemical weathering to denudation are higher in the presence versus absence of plants because of effective plant uptake and recycling of lithogenic nutrients (Balogh-Brunstad et al., 2008a). Weathering in the presence of plants results in greater incongruency, a higher proportion of secondary mineral phases and exchangeable nutrient cation concentrations, and diminished nutrient loss to leaching. In a related study of biotite and anorthite mixed into a silica sand (Balogh-Brunstad et al., 2008b), presence of bacteria and fungi were found to increase weathering fluxes compared to abiotic controls. Without a host plant (*Pinus sp.*) present, denudation rates were as large as weathering rates, that is, the weathering to denudation ratio was close to unity. Ectomycorrhizal seedlings produced the greatest weathering to denudation ratios (1.5), followed by non-ectomycorrhizal seedlings (1.3). The authors inferred that lithogenic nutrient loss was minimized in the planted systems because ectomycorrhizal hyphal networks and root hairs of non-ectomycorrhizal trees embedded in biofilm transferred nutrients to the host plant, thereby minimizing drainage losses.

It is now clear that plants enhance chemical weathering through a variety of mechanisms that together with their microbial associations include (1) exudate production (H^+, siderophores, low molar mass organic acids), (2) soil respiration of CO_2, (3) evapotranspiration, and (4) soil physical stabilization (Taylor et al., 2009). However, they also diminish the nutrient losses associated with that weathering process, contributing to the accumulation and retention within soil of biogeochemically active weathering products. To the extent that biota contribute to the retention and increased bioavailability of lithogenic nutrients (e.g., Ca, Mg, K), one expects a positive feedback to net primary production (Berner et al., 2003).

The proliferation and growth of microbial cells and associated biofilms in soils contributes directly to the accretion of reactive surface area for the retention and transformation of nutrients and pollutants (Young et al., 2008). The microbial colonization of mineral surfaces is important for several reasons. Biofilm-derived organic debris impacts hydraulic conductivity by blocking pore throats and decreasing aperture sizes (Baveye et al., 1998). Also, microbial cells and their extracellular matrices increase system surface area, and since most low-temperature mass transfer processes are surface controlled, this may have a large effect on chemical kinetics (Wogelius et al., 2007). More generally, soil biogeochemical weathering involves mineral dissolution–precipitation and oxidation–reduction reactions, and biota play a fundamental role in both.

10.3.2 Impacts of Organic Matter Infusion on Nutrients and Pollutants

Autotrophic photosynthate is the primary source of NOM and, therefore, of protons, electrons, and carbon in soil. Recent work suggests that once introduced to soil and water, plant-derived NOM is transformed into microbial byproducts, such that the majority of NOM resident in mineral soils is microbially derived (Guggenberger et al., 1994; Kogel-Knabner et al., 2008; von Luetzow et al., 2008). The resulting dissolved and colloidal organic matter can, therefore, exhibit similar molecular structures despite deriving from a range of biogeochemical sources, and this has been attributed to partial degradation mediated by heterotrophic microbes. For example, using size exclusion chromatography coupled to electrospray ionization-Fourier-transform ion cyclotron resonance (FTICR)-mass spectrometry, Reemtsma et al. (2006) analyzed fulvic acid (FA) isolates from three diverse sources (river, tarn, peat) and found that they exhibited remarkably similar carboxylated aliphatic molecular structures. This led the authors to conclude that FAs are metastable molecules that characterize a state of transition of more diverse precursor compounds during their oxidation in the environment. They propose that the elemental and structural regularity of FA molecules does not indicate one common precursor material and formation process, but rather the reworking of source materials in the environment and the valency of the three elements (C, H, O) from which most FA molecules are formed.

The infusion of DOM into the subsurface has several important implications for biogeochemical reactivity. First, because of gas diffusion limitations, biologically active porous media can exhibit CO_2 partial pressures that are elevated up to 100 times above that of the atmosphere. As a result, pore water carbonic acid concentrations are likewise elevated (relative to water in equilibrium with the atmosphere), which increases acid attack on primary aluminosilicate minerals and promotes the formation of high-surface-area and surface-reactive, secondary mineral products. Secondly, biological activity is accompanied by aqueous- and solid-phase environments that are enriched in biogenic organic residues and metal-complexing ligands. These organic constituents strongly influence the rate and trajectory of mineral transformation, the speciation and mobility of metal and organic pollutants, and the long-term sequestration of atmosphere-derived carbon into mineral–organic complexes and aggregates. Polar fractions of DOM contain organic ligands that form stable complexes with major and trace metals, thereby increasing the metal mobility, whereas nonpolar moieties can solubilize hydrophobic organic pollutants and hence increase their leaching into groundwater. Thirdly, a large fraction of the DOM leached from the root zone is biologically degradable, which means that it serves as a carbon and energy source for heterotrophic microorganisms living at greater depths. Gas phase O_2 partial pressure may be diminished to trace levels in biologically active subsurface environments and microenvironments (such as within particle aggregates) because of its consumption during microbial and root metabolism as a preferred terminal electron acceptor for respiration.

10.3.2.1 Dissimilatory Reduction Reactions

Microbial catabolism of organic matter (OM) under O_2-limited conditions drives important biogeochemical reactions, including redox transformations directly or indirectly coupled to the oxidation of OM (Figure 10.1c). Heterotrophic oxidation of NOM in O_2-limited systems is coupled to the use of alternative terminal electron acceptors such as NO_3^-, Mn(IV), Fe(III), SO_4^{2-}, CO_2, or organic carbon (Hunter et al., 1998). The use of these electron acceptors depends on their concentration, bioavailability and, importantly, their relative redox potential (with the most electronegative being preferred).

Such "dissimilatory" reduction reactions can also change the valence state and thus chemical speciation of redox-sensitive trace elements (e.g., As, Cr, U), with dramatic effects on their mobility and potential toxicity. These effects are often used as a tool in bioremediation of metal or metalloid contaminants. For example, dissimilatory reduction of U(VI) and Cr(VI) to U(IV) and Cr(III) can promote their removal from solution and incorporation into solid phases, diminishing their potential bioavailability (Valls and de Lorenzo, 2002). In contrast, other contaminant elements, such as arsenic, may become more mobile under anoxic conditions and this can have catastrophic impacts on human and ecosystem health. Mechanisms for As mobilization are multifaceted. Dissimilatory reduction of Fe(III) (oxyhydr)oxide sorbents results in release of sorbed As(V) to solution, and reduction of As(V) to the more mobile As(III) form is also catalyzed microbially (Smedley and Kinniburgh, 2002; Islam et al., 2004). Himalayan fluvial sediments that are naturally enriched in As are contributing to the degradation of anoxic shallow groundwater aquifers through such microbially mediated mechanisms, and this is contributing to a severe mass poisoning of rural populations in south and southeast Asia (Ahmed et al., 2004; Charlet and Polya, 2006; Polizzotto et al., 2008).

10.3.2.2 Reaction and Fate of DOM

The release of organic molecules and colloids from plants and microbes into the aqueous phase results in the transport of dissolved and colloidal organic matter from surface to subsurface soil environments (Herbert and Bertsch, 1995). The source of DOM includes recent photosynthate (throughfall, leaf litter, root exudates, decaying fine roots) and microbial exudates, as well as the leaching or decomposition of older, microbially processed soil organic matter (Guggenberger et al., 1994; Kalbitz et al., 2000; McDowell, 2003). The fact that the measured ^{14}C age of DOM ranges from hours to decades to thousands of years, indicates a correspondingly wide range of sources (Schiff et al., 1990). While it is sometimes assumed that newly abscised leaf litter is a principal source of DOM to subsurface mineral soils, particularly in forests, recent radiocarbon experiments conducted in forested ecosystems suggest that subsurface DOM derives more prevalently from the humified organic horizons or from the mineral soil itself rather than from leaching of the litter layer (Froberg et al., 2007). As a result of diverse sources, the composition of DOM includes plant and microbially derived biomolecules (proteins, polysaccharides, lipids, nucleic acids, lignin), their partial decomposition products (humic substances), and the complexes of these constituents with metals and other DOM components (Figure 10.1a and d).

Soluble DOM components interact via both covalent and noncovalent bonding interaction during the formation of humic substances, and these interactions can be effectively promoted by other soil constituents (Hedges, 1988; Piccolo, 2002; Sutton and Sposito, 2005). Covalent bonding occurs, for example, when amino acids react with sugar molecules in aqueous solution to produce *Maillard* reaction products (Ikan et al., 1996), or when phenolic compounds undergo oxidative coupling. While these reactions tend to be slow in abiotic aqueous solution, the presence of catalytic oxidizing agents, particularly Mn(IV) oxides such as birnessite, or microbially derived exoenzymes such as laccase, can greatly accelerate reaction kinetics (Jokic et al., 2001, 2004).

10.3.2.3 DOM Labile Structure and Reactivity

A thorough understanding of the molecular structure of NOM itself, including the DOM portion, remains a challenge, despite its clear importance to interfacial chemistry in soils and sediments (Hedges et al., 2000; Kelleher and Simpson, 2006). Improving that understanding drives an active area of research and employs state of the art tools in environmental analytical chemistry (Leenheer and Croue, 2003; Sutton and Sposito, 2005). Novel approaches, particularly those that combine computational methods with soft-ionization, mass spectrometry, multidimensional nuclear magnetic resonance (NMR) and other spectroscopic techniques are providing new insights (Diallo et al., 2003; Kelleher and Simpson, 2006). For example, while it has previously been thought that colloidal organic matter (e.g., humic acid [HA]) is dominated by high-molecular-weight polymerized degradation products of biomolecules, studies in the last 10 years suggest that these are instead supra-molecular structures comprising a large number of smaller molecules ranging in size from 100 to 2000 Da (Piccolo, 2001; Ferrer et al., 2002; Piccolo and Spiteller, 2003; Reemtsma and These, 2003; Rostad and Leenheer, 2004; Kelleher and Simpson, 2006). The smaller molecules, which are bound together in clusters via hydrophobic interaction, hydrogen bonding, and polyvalent cation bridging give rise to aggregates with macromolecular characteristics whose *apparent* molecular weight corresponds to 10,000–100,000 Da (Leenheer et al., 2001; Simpson, 2002; Simpson et al., 2002). These clusters undergo aggregation and disaggregation as function of solution chemistry, with clustering being favored in the presence of polyvalent metals such as Ca, Al, and Fe (Romkens et al., 1996; Romkens and Dolfing, 1998) and disfavored by low molar mass acids such as acetate that can form strong H-bonds with DOM constituents (Piccolo, 2001; Simpson, 2002). Studies of such aggregates at the nanometer and larger (colloidal) scale are beginning to reveal details of aggregate structure. Schumacher et al. (2005) used synchrotron scanning transmission electron microscopy (STXM) and C-1s near-edge x-ray absorption fine structure (NEXAFS) spectroscopy to study the chemical heterogeneity and functional group distribution of organic colloidal aggregates isolated from forest soils. Two distinct regions were observed in individual particles confirming the amphiphilic behavior of NOM. A hydrophilic, carboxylic region was concentrated in the outer portion, while a hydrophobic, aromatic region was concentrated in the interior.

Thus, aqueous phase organic matter constitutes a dissolved-to-colloidal continuum that is dominated by a complex mixture of intact and partly degraded biopolymers (proteins, carbohydrates, aliphatic biopolymers, and lignin) and their fragments. The transition from dissolved to colloidal potentially occurs as these molecular fragments and their co-associations are progressively aggregated into labile "supramolecular" structures through hydrogen bonding, cation bridging, and hydrophobic interactions (Piccolo, 2002; Sutton and Sposito, 2005). Biomolecules, constituent metals, and, in particular, the higher order structures that are formed by their intermolecular associations are characteristics of OM. This perspective suggests that the dynamic intermolecular associations of aqueous phase OM can be explored, in part, with tools that probe the nanoscale architecture of its particles. It seems reasonable to expect the self-assembly of NOM colloids in the presence of water affects their reactivity toward solutes and surfaces. For example, the cross-linking of humic molecules by polyvalent aluminum increases the nonlinearity and hysteresis of hydrophobic compound sorption to humic particle surfaces (Lu and Pignatello, 2004).

Mobile DOM participates in a multitude of geochemical reactions during transport that affect not only its own transport and partitioning behavior, but also that of co-associated nutrients and pollutants. These include (1) sorption–desorption, (2) aqueous-phase and surface complexation, (3) mineral dissolution, and (4) oxidation–reduction. These are discussed in turn below.

10.3.2.4 Sorption–Desorption of OM

While the global environmental significance of mineral-stabilized organic matter is apparent (Houghton, 2005), the molecular-scale controls on the formation of stable mineral–organic complexes are not (Guggenberger and Kaiser, 2003; Chorover et al., 2004; Sollins et al., 2006). Unraveling the nature of NOM–mineral bonding, and the incorporation of these complexes into higher-order aggregated structures where carbon is protected from microbial attack, poses a unique challenge to biogeochemists. It requires a multifaceted approach that integrates field,

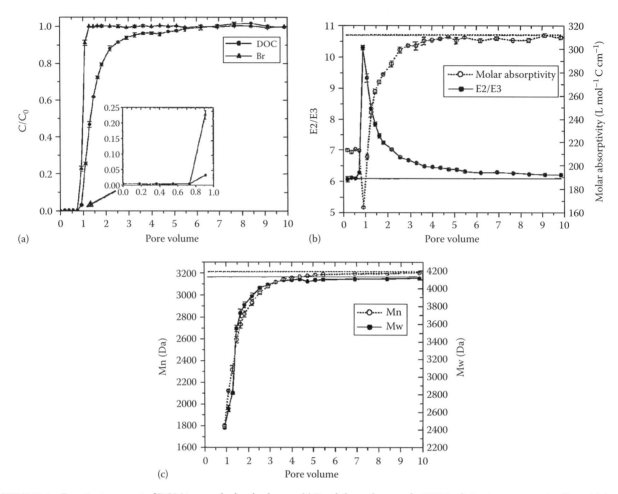

FIGURE 10.2 Reactive transport of DOM in repacked soil columns. (a) Breakthrough curve for DOM relative to an unreactive (bromide) tracer. (b) The time-evolution of UV absorbance properties of the DOM indicate the preferential retention at early times of more aromatic moieties. (c) Molar mass distribution of effluent fraction of DOM increases with time, indicating that higher molar mass compounds are preferentially retained. (From Guo, M.X., and J. Chorover. 2003. Transport and fractionation of dissolved organic matter in soil columns. Soil Sci. 168:108–118. With permission from Wolters Kluwer Health.)

modeling, and laboratory experiments with advanced analytical techniques that are only now becoming readily available.

The reactive transport of DOM can be considered in first approximation as analogous to a chromatographic process where compounds or colloids exhibiting the highest affinity for the solid phase are removed from solution closer to the surface, while more soluble components are transported to depth. The fact that DOM is a multicomponent mixture leads to its *sorptive fractionation* during transport. For example, reactive transport experiments with terrestrial DOM show that the breakthrough of organic carbon in packed Ultisol columns is retarded relative to a bromide tracer (Figure 10.2a). Furthermore, molar absorptivity (Figure 10.2b), apparent molar mass (Figure 10.2c), and DOM hydrophobicity (not shown) were observed to increase toward influent values over the course of the experiment. Hence, those fractions of DOM comprising greater aromaticity, hydrophobicity, and molecular size are preferentially adsorbed during transport, whereas smaller, more aliphatic and hydrophilic molecules remain in aqueous solution and are, therefore, mobilized to greater depth (Guo and Chorover, 2003). Similar DOM fractionation processes have been reported in column-scale and field-scale studies of aquifer sands (Dunnivant et al., 1992; McCarthy et al., 1993), as well as in batch experiments with specimen oxyhydroxides (Zhou et al., 2001).

While retention of DOM on marine sediments is strongly correlated with the specific surface area of mineral sorbents (Mayer, 1994), this parameter alone is not a sufficient predictor of organic carbon in soils (Kaiser and Guggenberger, 2000). Analysis of a large data set on surface and subsurface soils indicates that the prevalence of (oxyhydr)oxides of Al and Fe, in particular, appears to exert a stronger control on organic carbon storage than does mineral specific surface area alone (Figure 10.3). This finding is consistent with an increasing number of reports that highlight the role of Al and Fe oxides, particularly short-range ordered phases, in controlling organic matter stabilization in soils (Torn et al., 1997; Mikutta et al., 2006; Rasmussen et al., 2006; Chapter 2).

Several types of chemical bonding influence DOM uptake during transport. The strong affinity of DOM for the hydroxylated

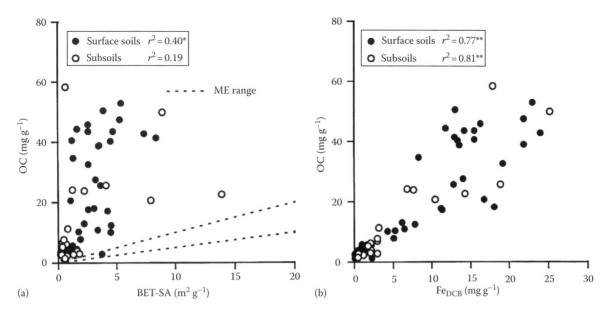

FIGURE 10.3 Organic carbon concentration versus (a) specific surface area (SA) in mineral soils (>10.6 g cm³) and (b) dithionite-citrate-bicarbonate extractable Fe for surface and subsurface forest soils (*, ** significant at $p < 0.05$ and $p < 0.01$ levels, respectively. (From Kaiser, K., and G. Guggenberger. 2000. The role of DOM sorption to mineral surfaces in the preservation of organic matter in soils. Org. Geochem. 31:711–725. Copyright 2000, with permission from Elsevier.)

mineral–water interface of hydrous oxide minerals is often attributed to the formation of inner-sphere complexes between carboxyl groups of DOM and singly coordinated metal centers on mineral surfaces (Filius et al., 2003). However, the extent to which adsorption results from inner-sphere versus outer-sphere mechanisms depends on DOM molecular composition (Yoon et al., 2005; Ha et al., 2008). Evidence for inner-sphere complexation comes from macroscale measurements of solution pH increase during DOM adsorption (Parfitt et al., 1977; Tipping, 1981). This is thought to result from mineral surface hydroxyl release (Gu et al., 1995) through a series of surface reactions such as

$$\equiv Al\text{-}OH_{(s)} + H_{(aq)}^+ \rightarrow \equiv Al\text{-}OH_{2(s)}^+ \quad (10.4)$$

$$x\,(\equiv Al\text{-}OH_{2(s)}^+) + y\,R\text{-}COO_{(aq)}^-$$
$$\rightarrow (\equiv Al\text{-}OOC\text{-}R)_x(R\text{-}COO^-)_{y-x(s)} + x H_2O_{(aq)} \quad (10.5)$$

where

"≡" indicates a surface functional group

x and y are the stoichiometric coefficients of protonated surface hydroxyls and sorptive DOM anionic ligands undergoing reaction

Thus, $y\,R\text{-}COO_{(aq)}^-$ represents y moles of monodentate ligand, one mole of y-dentate ligand, or an intermediate status between those two.

As indicated by Equation 10.5, because of its polyelectrolytic character, adsorption of DOM can result in configurations at the interface where the organic molecules adopt an adsorbed conformation in which a substantial number of their anionic functional groups are not directly interacting with the oxide surface because of steric and electrostatic constraints. This concept has been incorporated into the Ligand and Charge Distribution (LCD) model of Weng et al. (2006). In the LCD model, ion binding to oxides, as described by the Charge Distribution MUlti SIte Complexation (CD-MUSIC) model (Hiemstra and van Riemsdijk, 1999), is integrated with ion binding to humic substances, as described by the nonideal competitive adsorption (NICA) model (Koopal et al., 2005) to effectively predict the sorptive interaction between DOM and iron (oxyhydr)oxides as a function of pH, ionic strength, and DOM concentration. The capability of the LCD model to predict FA adsorption hinges on the formation of inner-sphere carboxylate surface complexes (as depicted in Equation 10.5) as the primary driver. That is, modeling of electrostatic interactions alone is insufficient to fit experimental data (Weng et al., 2006).

A ligand exchange reaction such as that shown above is consistent with several infrared spectroscopy studies of reactions occurring when pristine oxides are reacted with DOM solutions (Gu et al., 1995; Chorover and Amistadi, 2001). Low molar mass organic acids, such as oxalate and lactate, likewise show inner-sphere complexation in attenuated total reflectance Fourier transform infrared (ATR-FTIR) data (Yoon et al., 2005; Ha et al., 2008). However, Yoon et al. (2005) studied adsorption of IHSS reference fulvic and HAs at the boehmite (γ-AlOOH) interface across a wide range of surface coverage using ATR-FTIR spectroscopy. They concluded that ligand exchange, even at low adsorbate concentrations, was a minor contributor to adsorption, and that fulvate/humate was adsorbed primarily in outer-sphere coordination, presumably via electrostatic and H-bonding interaction. Soil-relative humidity likely also plays an important role in complexation, as a shift from predominantly outer-sphere to predominantly inner-sphere coordination was observed for several dicarboxylic organic acids upon drying (Kang et al., 2008). Hence, there remains a need for further investigation of the relative importance of ligand exchange versus outer-sphere complexation as mechanism for DOM adsorption at (oxyhydr)oxide surfaces.

Adsorption of DOM to the surfaces of layer silicate clays may potentially involve inner-sphere complexation at edge aluminol groups (Kubicki et al., 1997), but retention on basal surfaces of expansible clays likely involves weaker cation or water bridging, H-bonding, and/or hydrophobic sorption, especially for low-charged clays (Feng et al., 2005; Sposito, 2008). Spectroscopic studies indicate that organic matter associated with clay minerals either after adsorption experiments or upon sampling from field soils tends to exhibit higher aliphaticity, lower carboxyl group content, and lower overall polarity than that which adsorbs to hydrous oxide surfaces (Parfitt et al., 1997; Feng et al., 2005; Ghosh et al., 2009).

10.3.2.5 Mineral–Organic Solid-Phase Reaction Products

The amphiphilic properties of DOM are postulated to affect not only transport processes, but also the architecture of mineral–organic complexes in the weathering zone (Wershaw, 1993; Kleber et al., 2007). Polar functional groups bond to polar mineral surfaces via metals, oxygen atoms, and intervening water molecules. Exposed nonpolar (aliphatic and aromatic) groups may then interact with similar nonpolar DOM moieties from solution to form bilayer-like or multilayered structures, where the organic coatings are held together by the same hydrophobic, H-bonding, and cation bridging interactions that stabilize DOM aggregates. That is, once an organic coating is established, further adsorption of DOM does not necessarily involve bonding to the mineral surface, but could instead involve complexation with existing adsorbed OM, perhaps via H-bonding, cation bridging, or hydrophobic interactions. As discussed above, such intermolecular associations within organic aggregates have been identified in aqueous suspensions of DOM (Piccolo, 2002; Simpson et al., 2002; Sutton and Sposito, 2005). As a result, models of intra-organic association have been extended to explain adsorption at progressively coated mineral surfaces. For example, Kaiser and Zech (1998) found that progressive adsorption of DOM to the B_s horizon of a forest soil led to displacement of hydrophilic DOM and retention of the hydrophobic fractions, suggesting that organic–organic associations become more important at higher surface loadings.

A conceptual model of this "self-assembly" of organic molecules and fragments into a complex mineral–organic architecture was proposed recently by Kleber et al. (2007). Whereas pristine mineral surfaces are known to vary in their affinity for DOM functional groups and molecular types (e.g., Chorover and Amistadi, 2001), the model of Kleber et al. (2007) suggests that once a mineral surface coating is formed, mineral–organic complexes present a multilayered architecture irrespective of underlying mineral type that includes a zone of hydrophobic interaction derived from the organic sorbate itself. The particle–solution interface in the model then becomes a zone of exchanging DOM fragments that enter and exit the organic realm of the sorbent surface in response to dynamic changes in solution chemistry (Figure 2.3).

Recent studies have implicated, in particular, poorly crystalline Al and Fe oxides in the stabilization of NOM against microbial degradation (Torn et al., 1997; Mikutta et al., 2006). Such stabilization processes affect the capacity of soils to store the largest Earth-surface pool of otherwise labile carbon and, therefore, they exert strong control over atmospheric CO_2 and its radiative forcing effects (Houghton, 2005). The formation of stable NOM–mineral complexes can also have pronounced effects on surface charge and molecular adsorption properties of the solid phase, as well as influence the aggregation behavior of colloidal particles (Kretzschmar et al., 1998; Heidmann et al., 2005). These mineral–organic associations create rough and chemically heterogeneous surfaces in contact with the pore water, thereby affecting the advective–diffusive transport and deposition of solutes and colloidal particles, including sorbed pollutants, in the subsurface (Kretzschmar et al., 1999).

Mineral–organic complexes in soils appear to exist in an array of associations, but most commonly in aggregated form. Constituents of such aggregates include partially humic-coated clay surfaces, and their incorporation into nanometer- to micrometer-sized *micro-aggregates* in which OM is sometimes encrusted by colloidal minerals (Chenu and Plante, 2006). These micro-aggregates assemble into higher-order *macro*-aggregate structures (Six et al., 2000). Both mineral–surface interaction and physical-aggregate occlusion have the potential to impede or diminish enzymatic degradation of OM. Hence, accurate characterization of formation and destruction dynamics of these mineral–organic aggregates is a key to predicting stabilization of organic C in soils. Furthermore, an improved understanding of mineral–organic architecture at the molecular to particle scale will greatly assist in elucidating the important role that OM plays in sequestration of inorganic and organic nutrients and contaminants in soil.

10.3.2.6 Aqueous Phase and Surface Complexation Effects

Upon introduction to soil from plant or microbial tissue abscission, exudation or lysis, photosynthetic biopolymers are attacked, principally by extracellular enzymes that cleave them into smaller (e.g., oligomeric to monomeric) molecular weights, which tend to be more soluble. Once in the aqueous phase, such fragments can adsorb to surfaces, serve as a C source for microbial growth, and undergo reaction with ions or other soluble biotic components. Organic components in soil solution can form complexes with solutes released from rock weathering, nutrient cycling, atmospheric deposition, and anthropogenic sources. By forming such complexes, DOM can play a significant role in the transport of nutrients and pollutants in structured soil. For example, in reactive transport experiments conducted with undisturbed soil columns Camobreco et al. (1996) found that columns treated with

DOM-complexed metals had peak effluent concentrations that were 30%, 26%, 28%, and 27% of influent for Cd, Zn, Cu, and Pb, respectively. In the absence of DOM, peak effluent concentrations were largely unchanged for Cd and Zn, but were substantially reduced to 15% and 12% for Cu and Pb, respectively. This indicates that the relative importance of aqueous organic complexes for mobilizing metal pollutants depends on their stability (i.e., greater for Cu and Pb). However, affinity of the metal–DOM complexes for soil particle surfaces must also be considered. In the same study, homogenized soil columns were found to retain all added metals, whether the metals were added in water or as organic complexes, which highlights the importance of preferential flow paths to the transport of metals in structured soil.

Complexation reactions of DOM can enhance the migration of oxyanion nutrients (e.g., phosphate) or pollutants (e.g., arsenate/arsenite or selenate/selenite), but by different mechanisms than are observed for the cationic transition metals. It is postulated that DOM can compete directly with oxyanions for adsorption sites on hydroxylated minerals. Inner-sphere complexation of phosphate and arsenate has been attributed to ligand exchange at singly coordinated surface hydroxyl groups. Ligand exchange at singly coordinated hydroxyls results in direct bonding of oxyanions to Fe(III) or Al(III) metal centers. Arsenite shows a higher affinity for doubly coordinated hydroxyls, and both As(V) and As(III) are likely adsorbed outer-spherically as well, depending on concentrations (Fuller et al., 1993; Waychunas et al., 1993; Sun and Doner, 1996; Chapter 8). Aluminum metal centers can serve as bonding sites even in poorly crystalline aluminosilicates. Since DOM can compete for adsorption at the same surface hydroxyl groups, its presence in solution can diminish both phosphate and arsenic uptake.

If humics and oxyanions are competing for the same sites, their fate depends on the effects of oxyanion–DOM selectivity. However, sorptive competition can also result from electrostatic effects. Several authors have sought to measure these effects in competitive adsorption experiments and modeling, and the results indicate the importance of both oxyanion speciation and DOM composition. For example, arsenate competes more effectively than arsenite for adsorption to goethite in the presence of peat HA or Suwannee River FA (Grafe et al., 2001). Lower molar mass DOM (e.g., FA) appears to diminish arsenate adsorption to goethite more effectively than does higher molar mass DOM (e.g., HA) as shown in the top two graphs of Figure 10.4 (Weng et al., 2009). Likewise, presence

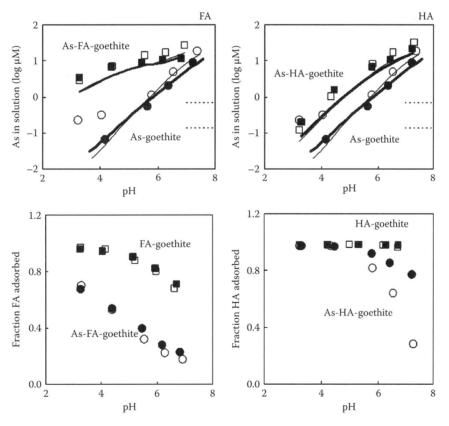

FIGURE 10.4 Competitive adsorption of arsenate (AsO_4^{3-}) and fulvic acid (FA, left side) or HA (right side) to the surface of goethite. (With permission from Weng, L., W.H. van Riemsdijk, and T. Hiemstra. 2009. Effects of fulvic and humic acids on arsenate adsorption to goethite: Experiments and modeling. Environ. Sci. Technol. 43:7198–7204. Copyright 2009 American Chemical Society.) Top graphs show arsenate remaining in solution in the presence and absence of FA or HA, and bottom graphs show fractional adsorption of FA and HA in the presence and absence of arsenate. The FA and HA samples have apparent molar masses of 0.68 and 13.2 kDa, respectively (measured by size exclusion chromatography). Open symbols are for 1.0 mM $NaNO_3$, closed symbols are for 10.0 mM $NaNO_3$. Experimental conditions: 1.0 g L^{-1} goethite, 0.15 mM arsenate, 50 mg L^{-1} FA/HA.

of arsenate more strongly diminishes FA than HA sorption (Figure 10.4, bottom). This may be attributable to the tendency for arsenate to form inner-sphere complexes at goethite surface hydroxyls that are also sites for FA carboxylate complexation (Weng et al., 2006). Alternatively, it has been argued that since FA is smaller in size than HA, it adsorbs more closely to the goethite surface, and thereby more strongly impacts goethite surface charge (Weng et al., 2009). Similarly, Xu et al. (1988) argued that FA that was adsorbed coulombically to aluminum oxide, diminished arsenate adsorption by increasing the net negative charge of the particle surface. Hence, either direct bonding site competition or surface charge repulsion may contribute to diminished arsenate adsorption in the presence of DOM. In addition to these particle surface effects, DOM can also mobilize oxyanions via the formation of ternary aqueous DOM–metal complexes, presumably by metal-bridging mechanisms, which diminishes the tendency of such oxyanions to form surface complexes (Bloom, 1981; Redman et al., 2002; Lin et al., 2004).

10.3.2.7 Bioligand-Mediated Mineral Dissolution

Even in the absence of direct surface interaction, metal-complexing organic ligands can promote mineral transformations by diminishing the relative saturation of the solution phase with respect to mineral constituents. However, organic ligands are also implicated in direct particle surface dissolution to the extent that they can promote the detachment of surface metal centers following ligand-surface-metal complexation. This latter reaction involves first the formation of an *activated precursor complex* wherein a structural metal that is part of the mineral surface becomes coordinated to an adsorbing organic ligand, followed by polarization of the remaining metal–oxygen bonds in the mineral structure, and metal–ligand detachment (Stumm, 1997; Brown et al., 1999).

These two distinct processes—ligand-promoted surface dissolution and aqueous-phase metal–ligand complexation—likely combine in biologically active soils to affect mineral bioweathering. An example of this is provided by the synergistic effect on goethite dissolution of oxalate, a common rhizosphere organic acid, and desferrioxamine B (DFOB), a microbially produced trihydroxamate siderophore (Cheah et al., 2003; Reichard et al., 2007). Batch steady-state and transient experiments indicate that the two types of biomolecules co-operate to enhance the release of Fe(III) from low solubility mineral surfaces as depicted in Figure 10.5. The kinetic model that best fits experimental data is one that includes fast oxalate adsorption to the goethite surface via ligand exchange with goethite surface hydroxyls followed by the slower production of an activated precursor surface complex (labile $\equiv$Fe-L, where L = oxalate). The synergistic effect comes into play because detachment of labile Fe-L into solution is promoted by the presence of "predatory" siderophores that capture Fe(III) from oxalate. This transfer of Fe(III) to siderophore increases the thermodynamic pressure for labile Fe-L desorption, and also recycles the catalyst, oxalate, for re-adsorption at the surface (Figure 10.5). Hence, the pulsed exudation of phytosiderophores that has been observed to occur diurnally likely results in a correlated pulsed bio-availability of nutrient Fe (Reichard et al., 2007).

This preceding example pertains to a relatively simple abiotic experimental system comprising goethite, oxalate, and DFOB in aqueous solution. Nonetheless, it illustrates the emergent complexities and potential synergies that are associated with multicomponent and heterogeneous mixtures. Such complexities are only amplified in biologically active soils as they exist in field settings. This is one reason why translating laboratory-derived understanding to field systems remains a grand challenge to soil scientists and CZ researchers.

10.3.2.8 Oxidation–Reduction

Even at the scale of individual aggregates, soils contain diverse microenvironments comprising minerals, microbes, and organics. Because of diffusion limitations, metabolically catalyzed redox reactions and the generation of acids and complexing agents lead to aqueous phase concentration gradients, mineral nucleation sites, and element speciation that are markedly different from the bulk solution (Templeton et al., 2003; Chan et al., 2004; Hinsinger et al., 2009). The chemical parameters pE and pH govern many reaction trajectories in soils. The fact that photosynthate and its transformation products are a principal source of electrons and protons in soils is readily seen by considering the oxidation half-reaction of glucose (the monomer of the plant polysaccharide cellulose and many other microbial polysaccharides) to form bicarbonate in soil solution:

$$C_6H_{12}O_6 + 12H_2O \rightarrow 6HCO_3^- + 30H^+ + 24e^- \qquad (10.6a)$$

Even when this reaction is coupled with the reduction half-reaction of dissolved O_2 (DO), net proton production is still apparent:

$$C_6H_{12}O_6 + 6O_2 \rightarrow 6HCO_3^- + 6H^+ \qquad (10.6b)$$

Though degassing of $CO_{2(g)}$ can diminish the number of moles of $H_{(aq)}^+$ produced per mole of biomolecule oxidized, the aqueous phase protons produced by microbial oxidation of photosynthate and humics drive numerous reactions in soils, including particle surface charging (in turn affecting solute adsorption and colloidal dynamics) as well as proton-promoted mineral weathering (Equation 10.1).

In addition to serving as the primary source for electrons in soils, DOM plays an important role in soil redox reactions by functioning as an intermediary in microbial metabolic processes (Lovley et al., 1996; Benz et al., 1998) and by facilitating the cycling of metals, for example, Fe, that can exist in multiple oxidation states (Lovley et al., 1998). In particular, quinone–hydroquinone functional groups are thought to serve as reactive moieties in the oxidation–reduction of inorganic pollutants such as mercury, chromium, arsenic, and uranium

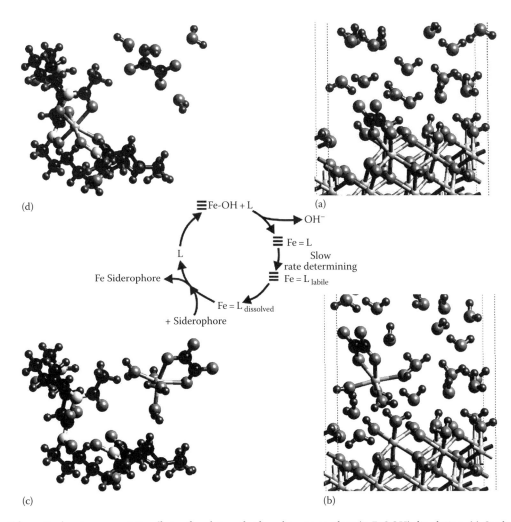

FIGURE 10.5 Schematic showing synergistic effects of oxalate and siderophore in goethite (α-FeOOH) dissolution. (a) Oxalate reacts adsorbs via ligand exchange to form a bidentate-mononuclear surface complex (Duckworth and Martin, 2001); (b) the labile Fe-oxalate surface complex is desorbed; (c) oxalate transfers Fe to DFOB, a high-affinity siderophore ligand; and (d) Fe forms multi-dentate, mononuclear aqueous complex with siderophore, and free oxalate is recycled. (Modified after Reichard, P.U., R. Kretzschmar, and S.M. Kraemer. 2007. Rate laws of steady-state and non-steady-state ligand-controlled dissolution of goethite. Colloids Surf. A Physicochem. Eng. Asp. 306:22–28. Copyright 2007, with permission from Elsevier, with molecular visualization created with Cerius^2 from Accelrys Inc, San Diego, CA; Courtesy of Professor James Kubicki, Pennsylvania State University.)

(Allard and Arsenie, 1991; Redman et al., 2002; Gu and Chen, 2003; Gabriel and Williamson, 2004), as well as organic contaminants such as nitroaromatic and organochlorine compounds (Schwarzenbach et al., 1990; Kappler and Haderlein, 2003). Evidence for the redox activity of quinone-like moieties derives from fluorescence and electron spin resonance (ESR) spectroscopies, as well as electrochemical data (Scott et al., 1998; Struyk and Sposito, 2001; Nurmi and Tratnyek, 2002). Quinone moieties in humic substances form free radical species, such as semiquinone, when they undergo a one-electron reduction. However, electrochemical titrations of humic substances have shown that HAs have electron transfer capacities that are much greater than that which is accounted for based on free radical concentrations alone (from ESR measurements) (Struyk and Sposito, 2001). This suggests the existence of other redox active moieties in DOM, possibly resulting from redox-active heteroatoms (e.g., N or S), or inner-sphere metal-organic charge transfer complexes, such as the strongly bound Fe atoms in humic substances.

Redox character of DOM is not solely a function of its own chemical composition. For example, the electron transfer capacity of humic material has been shown to increase with solution pH (Matthiessen, 1995), which has been attributed to relaxation of humic molecules, which leads to increased availability of redox active functional groups. Aggregate structure of DOM also plays a role. Palmer and von Wandruszka (2009) found that the redox potential of humic substances was strongly dependent on the concentration of bivalent, but not monovalent, cations. Dynamic light scattering measurements of colloidal OM size showed that a redox potential shift was correlated with decreasing radius of the DOM aggregates in response to increasing bivalent cation concentration.

10.3.2.9 Microbial Adhesion

Microbial cells, especially bacteria and protozoa, can be transported over relatively long distances in porous media prior to attaching to mineral or organic surfaces (Hornberger et al., 1992; Schafer et al., 1998; Redman et al., 2004). Bacterial adhesion often involves the coating of substrata and growing cells with extracellular polymeric substances (EPS), prior to an incipient biofilm that evolves into a complex microenvironment at the solid–water interface. Because of the diversity of biomolecular and mineral surface structures and functionalities, adhesion of microbial cells to mineral surfaces has the potential to involve a multitude of biomolecule–surface interactions that can contribute to or diminish cell-surface adhesion. Steric and dispersion forces similar to those that act on bare and polymer-coated surfaces, are superimposed on a wide range of localized molecular bonding possibilities (hydrophobic, van der Waals, ion bridging, covalent).

As a result of the predominance of weakly acidic carboxyl, phosphoryl, and hydroxyl functional groups, microbial surfaces tend to exhibit net negative charge over a wide pH range representative of natural waters. Thus, individual microbial cells may be transported over long distances in porous media. This can be beneficial when contaminant remediation requires microbiological augmentation (Scow and Hicks, 2005). Or, it can be detrimental when it involves the transport of pathogenic microbes (e.g., *Cryptosporidium sp.*, *Giardia sp.*, or *Escherichia coli*). Recent history contains several cases where pathogens have been transported in aqueous suspension through porous geomedia and into community water supplies (Curriero et al., 2001). Hence, an understanding of adhesion processes is essential. Mean field models, based on Derjaguin–Landau–Verwey–Overbeek (DLVO) theory of reactive particle transport and adhesion to surfaces, do not closely match experimental adhesion measurements because key interaction mechanisms of bio-particles and mineral surfaces are not included among the classical DLVO (van der Waals and electrostatic) forces (Ginn et al., 2002). More effective data fits for *E. coli* adhesion to silica, for example, were made when models included a steric term to account for the "polymer brush" surface of microbial

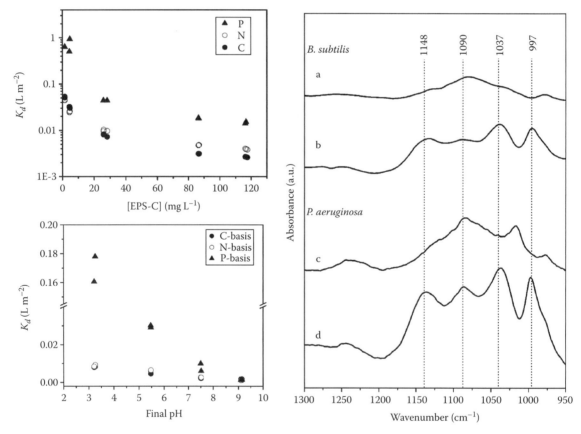

FIGURE 10.6 Left side shows macroscopic data on adsorption of EPS from *Bacillus subtilis* to the surface of α-FeOOH as a function of EPS concentration (carbon basis) (top left) and pH (bottom left). Data indicate preferential adsorption of P-containing moieties. (From Omoike, A., and J. Chorover. 2006. Adsorption to goethite of extracellular polymeric substances from *Bacillus subtilis*. Geochim. Cosmochim. Acta 70:827–838. Copyright 2006, with permission from Elsevier.) Right side shows attenuated total reflectance Fourier transform infrared (ATR-FTIR) spectra of EPS from *B. subtilis* and *Pseudomonas aeruginosa* either free in solution (a, c) or adsorbed to the surface of goethite (b, d). Emergence of new peaks for adsorbed EPS is indicative of formation of inner-sphere PO-Fe bonds as a result of ligand exchange. (With permission from Omoike, A., J. Chorover, K.D. Kwon, and J.D. Kubicki. 2004. Adhesion of bacterial exopolymers to alpha-FeOOH: Inner-sphere complexation of phosphodiester groups. Langmuir 20:11108–11114. Copyright 2004 American Chemical Society.)

cells (Camesano and Logan, 2000). In addition to providing the potential for steric effects and mixing entropy of the polymer brush, surface biomolecules also can enter into direct (e.g., covalent) bonding interactions with hydroxylated mineral surfaces (Parikh and Chorover, 2006). Laboratory and field experiments with microorganisms have shown that ferric oxyhydroxide coatings are important in the removal of microorganisms in saturated porous media, even in the presence of organic matter and phosphate, which can block microorganism attachment (Harvey and Ryan, 2004). As a sorbent for biomolecules, the siloxane surfaces of primary silicates and clay minerals differ significantly from the hydroxylated surfaces of Fe and Al (oxyhydr)oxides. Whereas the basal surfaces of clay minerals tend to effectively sequester aliphatic molecules such as lipids, the hydroxylated surfaces of hydrous oxides exhibit affinity for phosphorylated and carboxylated compounds.

Early adhesion processes likely involve some degree of mineral-surface conditioning film formation by free or cell-bound EPS. Microbial exudates and lysates comprise a mixture of polysaccharides, proteins, nucleic acids, and lipids whose relative affinities for mineral surfaces depend on the structure and functional group chemistry of the substrate surface, as well as solution chemistry conditions. Data on the adsorption of EPS from the bacterial strains *Bacillus subtilis* and *Pseudomonas aeruginosa* to the goethite (α-FeOOH) surface are shown in Figure 10.6. Macroscopic sorption data indicate that reaction of EPS with goethite favors preferential uptake of P-containing biomolecules relative to bulk C or N (left side of Figure 10.6). One explanation for this organic P-selectivity comes from infrared spectroscopic data (right side of Figure 10.6) of goethite-EPS complexes indicating the ligand exchange of biomolecular phosphoryl groups at goethite surface hydroxyls (i.e., $\equiv$Fe-OP-R bonding). These infrared spectroscopic results are consistent with ab initio quantum chemical calculations of complexation of phosphodiester groups, such as those associated with nucleic acids, bonded to Fe metal centers (Omoike et al., 2004; Omoike and Chorover, 2006). ATR-FTIR spectroscopic studies of live bacterial-Fe oxide systems also show the formation of $\equiv$Fe-OP-R$_{(s)}$ bonds, suggesting that cell-bound phosphorylated molecules also play an important role in whole cell adhesion to unconditioned oxide surfaces (Parikh and Chorover, 2006).

The transport of pathogenic microbial contaminants is likewise affected by mineral–microbe bonding interaction. As shown in the left side of Figure 10.7, the progressive coating of quartz sand surfaces with iron oxyhydroxide results in a significant decrease in *Cryptosporidium parvum* oocyst breakthrough from porous media columns, and an increase in adhesion, as reflected in the collision efficiency parameter, α (Abudalo et al., 2005). The right side of Figure 10.7 shows ATR-FTIR studies of oocyst surface functional group chemistry upon adhesion to uncoated (top) and iron-oxide-coated (bottom) surfaces. The data show that at the pH of the transport experiments, oocysts are adhered to Fe oxide surfaces through inner-sphere complexation of carboxylate groups at Fe(III) metal centers (Gao et al., 2009). Hence, in addition to the influence of electrostatic attraction for microbial adhesion to Fe oxide at this pH, direct bonding effects likely also play a role in retarding pathogen transport.

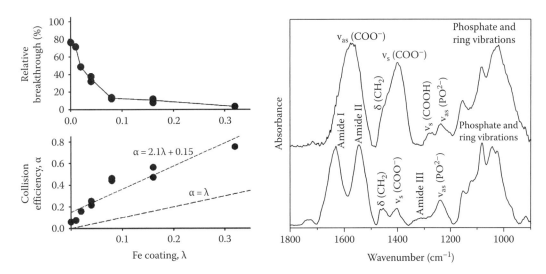

FIGURE 10.7 Macroscopic transport data left side shows that relative breakthrough decreases and collision efficiency increases for *Cryptosporidium parvum* oocysts as a function of the fraction of the Fe(III) oxide surface coating of a quartz sand at pH 5.6–5.8 and 0.1 mM NaCl. (With permission from Abudalo, R.A., Y.G. Bogatsu, J.N. Ryan, R.W. Harvey, D.W. Metge, and M. Elimelech. 2005. Effect of ferric oxyhydroxide grain coatings on the transport of bacteriophage PRD1 and Cryptosporidium parvum oocysts in saturated porous media. Environ. Sci. Technol. 39:6412–6419. Copyright 2005 American Chemical Society.) (Zeta potential of oocysts under these conditions was −10 mV.) Right side uses molecular spectroscopy to depict the mechanism. ATR-FTIR spectra of oocysts adhering to an uncoated (bottom spectrum) and α-Fe$_2$O$_3$-coated (top spectrum) internal reflection element surface indicates the inner-sphere complexation of oocyst carboxylate groups at hematite Fe(III) metal centers. (With permission from Gao, X., D.W. Mettge, C. Ray, R.W. Harvey, and J. Chorover. 2009. Surface complexation of carboxylate adheres *Cryptosporidium parvum* oocysts to the hematite/water interface. Environ. Sci. Technol. 43:7423–7429. Copyright 2009 American Chemical Society.)

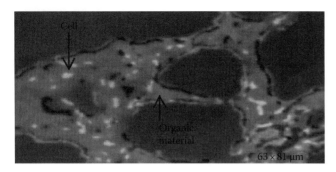

FIGURE 10.8 2D Vertical scanning interferometry image of *Shewanella oneidensis* colonization of a magnesite surface after 9 h. The biofilm consists of taller (white) cells and associated lower (light gray) organic material. (From Davis, K.J., and A. Luttge. 2005. Quantifying the relationship between microbial attachment and mineral surface dynamics using vertical scanning interferometry (VSI). Am. J. Sci. 305:727–751. With permission.)

Following initial adhesion to mineral surfaces, further accumulation of cells and EPS can result in the formation of incipient biofilm, where the nature of biopolymeric exudates are driven partly by cell physiological response to the local interfacial environment. Indeed, Geoghegan et al. (2008) proposed that "cell attachment and related cell growth behaviour is mediated by macromolecular physics and chemistry in the interfacial environment. Ecological success depends on the genetic potential to favourably influence the interface through adaptation of the macromolecular structure." One example of a complex biofilm structure is depicted in Figure 10.8 where colonization of the magnesite ($MgCO_3$) surface by *Shewanella oneidensis* is visualized using vertical scanning interferometry (VSI) (Davis and Luttge, 2005). The image clearly shows local regions of higher cell topography (bright yellow) and lower regions comprised of the organic matrix (darker yellow).

10.4 Effects of Organo-Mineral Heteroaggregates on Nutrient and Pollutant Fate and Transport

The complexation of NOM and microbial cells with inorganic weathering products produces mineral–organic complexes whose architecture and chemistry are distinctly different from the constituent parts in isolation. These complexes and their aggregates control the storage and transport of organic nutrients in soil, and also present heterogeneous surfaces and diverse reactive sites to soluble pollutants. The surface chemistry of organo-mineral hetero-aggregates is sensitively dependent to the biogeochemical environment in which they are formed.

10.4.1 Surface Composition of Mineral–Organic Complexes

The influence of biogeochemical environment on surface chemistry of mineral–organic complexes is illustrated by considering data from a chrono-sequence of basalt-derived tropical forest soils that range in parent material age from 0.3 to 4100 thousand years. The total organic C concentrations for the A horizon (for soil parent material ages of 0.3, 20, 150, 400, 1400, and 4100 ky) are, respectively, 19, 178, 188, 290, 131, and 40 g kg^{-1}. Corresponding organic C concentrations in the B horizon are 23, 114, 121, 81, 20, and 12 g kg^{-1}. Hence, the intermediate-aged soils contain the highest mass fraction of organic matter, consistent with their prevalence of poorly crystalline minerals (Torn et al., 1997). Figure 10.9 shows x-ray photoelectron spectroscopy (XPS) data that give the "surface enrichment" of organic carbon and nitrogen on soil particles from this chrono-sequence as determined from the C_{1S} or N_{1S} peak intensity relative to that measured within the particle bulk, following argon sputter of the surface layers (Mikutta et al., 2009). Surface accumulations of C and N relative to particle bulk are indicated by surface enrichment values greater than unity. Results for A and B horizons are shown. The XPS data show that the youngest soils (0.3 ky) have significant enrichment of organic matter on particle surfaces, consistent with adsorption of DOM to the surfaces of primary minerals (plagioclase) that are predominant at early weathering stage and that are co-associated with smaller quantities of secondary precipitates (allophane and ferrihydrite). Conversely, the intermediate-aged soils (20–400 ky), which are highly enriched in poorly crystalline secondary minerals such as allophane and ferrihydrite, show negligible particle *surface* enrichment in organic matter relative to the particle bulk, despite the fact that these soils contain the highest concentration of organic matter overall. The oldest soils again show particle surface enrichments of organic C, which is consistent with the prevalence in those soils of organic matter adsorption to well crystallized secondary phases kaolinite, gibbsite, and goethite.

These results indicate a change in hetero-aggregate structure over the course of pedogenesis. Surface associated organic matter is prevalent in mineral–organic complexes at early and late weathering stages, when well-crystallized primary and secondary minerals are predominant, respectively. However, in the intermediate weathering stage, when such complexes comprise poorly crystalline minerals and organic matter, the XPS data suggest a solid phase that reflects greater mixing of organic and inorganic components at the nanoscale; that is, the data are consistent with hetero-aggregates wherein colloidal OM is coagulated into floccules with nano-particulate inorganic weathering products (Mikutta et al., 2009).

10.4.1.1 Surface Charge Effects

The impacts of such changes in mineral–organic structure on soil surface chemistry and, therefore, on the transport of nutrients and pollutants, are profound. One way to quantify them is to measure the variation in soil surface charge, which, in turn, affects adsorption–desorption of aqueous phase solutes (Chorover et al., 2004). The point of zero net proton charge (PZNPC) for the chrono-sequence soils is plotted on the right side of Figure 10.9 (Chorover et al., 2004). The PZNPC is defined as the solution pH value where net adsorbed proton surface charge density is equal to zero (Sposito, 2008). Therefore, it represents a measure of Brönsted acidity of soil particle surfaces, and the prevalence of positive or negative variable surface charge at a given pH. The PZNPC data shown here were obtained by discontinuous titrations of the soils

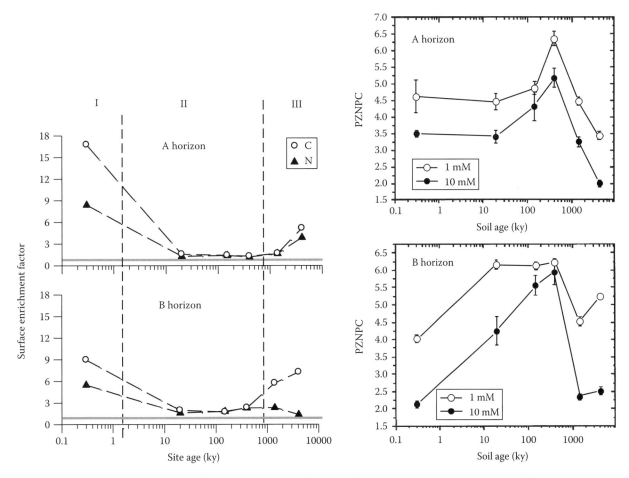

FIGURE 10.9 Left side shows x-ray photoelectron spectroscopy data of surface enrichments in organic C and N on soil particles from Hawaiian chronosequence. (From Mikutta, R., G.E. Schaumann, D. Gildemeister, S. Bonneville, M.G. Kramer, J. Chorover, O.A. Chadwick, and G. Guggenberge. 2009. Biogeochemistry of mineral–organic associations across a long-term mineralogical soil gradient (0.3–4100 kyr), Hawaiian Islands. Geochim. Cosmochim. Acta 73:2034–2060. Copyright 2009, with permission from Elsevier.) Right side shows surface charge (PZNPC) data on particles from the same soils (From Chorover, J., M.K. Amistadi, and O.A. Chadwick. 2004. Surface charge evolution of mineral–organic complexes during pedogenesis in Hawaiian basalt. Geochim. Cosmochim. Acta 68:4859–4876. Copyright 2004, with permission from Elsevier).

suspended in LiCl background electrolyte solutions at 1 and 10 mM ionic strength. The low PZNPC values for early and late weathering stages are consistent with masking of the underlying mineral surfaces by acidic functional groups of NOM (i.e., the pK_a of carboxyl groups is ca. 3–4, whereas the PZNPC of iron or aluminum (oxyhydr)oxides, which predominate at the later weathering stages, is ca. 7–8 (Chorover, 2005). The higher PZNPC values for the intermediate-aged soils suggest that carboxylate and phenolate functionalities in NOM are highly complexed in these soils with monomers, dimers, or higher order polymers of Fe and Al. Such complexes are consistent with recent extendend X-ray absorption fine structure (EXAFS) spectroscopy studies of Fe coordination in organic soils, which show strong binding of hydrolyzed Fe dimers and trimers at organic surface functional groups (Gustafsson et al., 2007). High aqueous phase concentrations of Al and Fe in these soils derive from the solubility of poorly crystalline precipitates and the prevalence of DOM for ligand–metal complexation. The prevalence of Al and Fe on soil particle surfaces shifts the interfacial acidity of the soils to reflect that of OM-adsorbed hydrolyzing cations and their polymerized reaction products (Chorover et al., 2004).

Surface charge properties of mineral–organic complexes are closely linked to their dissolution behavior. For example, the intrinsic surface charge (σ_{in}) of the soil from the 1400 ky site is plotted as a function of pH in Figure 10.11a (open circles). Intrinsic surface charge, which includes the contributions from both proton adsorption–desorption and structural charge resulting from isomorphic substitutions, is negative throughout the pH range studied, but approaches zero (the point of zero net charge) as pH decreases toward 3.5. With increasing pH, σ_{in} increases in magnitude, reflecting increased *negative* surface charge. Correlated with the increase in negative surface charge is the dissolution of DOM (solid circles), indicating that proton dissociation of weakly acidic functional groups (e.g., carboxyl) promotes release of solid phase organic matter into solution (Figure 10.11a). Importantly, the release of DOM is linearly correlated with the dissolution of Fe, Al, and Si (Figure 10.11b), and the regression slopes give the number of moles of metal(loid) complexed with DOM per gram of carbon.

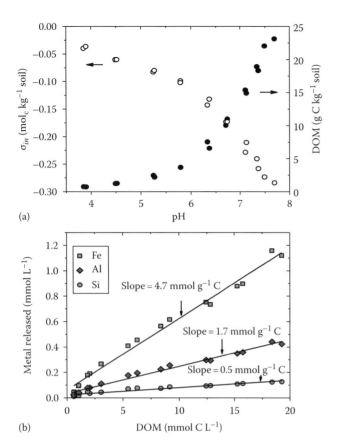

or within solid surfaces. When inorganic ions and molecules are first introduced to a mineral or organic interface, mechanisms of adsorption may include diffuse electrostatic attraction (e.g., Ca^{2+} or Cd^{2+} adsorption to the siloxane surface of montmorillonite or carboxylate groups of DOM), H-bonding (e.g., weakly acidic oxyanions at (oxyhydr)oxide surfaces), inner-sphere or covalent bonding (e.g., phosphate or arsenate ligand exchange at A-type hydroxyls on Fe or Al oxyhydroxides). As shown schematically in Figure 10.11, the adsorption process is a dynamic one, where the adsorbate may desorb back to bulk solution, or diffuse along the surface to higher affinity sites with larger bond number (e.g., ledge and kink sites). Inner-sphere complexes can give rise to a distorted form of sorbent crystal growth, and at higher surface loadings, adsorbate species may polymerize and grow into surface precipitates that alter surface reactivity, masking that of the underlying surface (Sposito, 2004). Hence, in the weathering environment, adsorbate speciation is nonstationary; and it undergoes an evolution subject to biogeochemical forcing. Changes occur on timescales that range from those of rapid water–ligand exchange on solute species ($<10^{-9}$ s) to much slower changes in sorbent structure resulting from weathering reactions ($>10^3$ s) (Chorover and Brusseau, 2008).

Since local coordination strongly affects sorbate stability against chemical perturbations in soil solution (e.g., from hydrologic events), it is likewise expected that the aqueous

FIGURE 10.10 DOM-mediated metal release from a Plinthic Kandiudox Molokai, Hawaii (1400 ky soil shown in Figure 10.9). (a) Top graph shows intrinsic soil surface charge (moles of charge per kilogram of soil, right axis) and DOM release (grams of carbon per kilogram of soil, left axis) in 10 mM LiCl background electrolyte as a function of solution pH (total soil organic carbon = 125 g kg^{-1}. (From Chorover, J., M.K. Amistadi, and O.A. Chadwick. 2004. Surface charge evolution of mineral-organic complexes during pedogenesis in Hawaiian basalt. Geochim. Cosmochim. Acta 68:4859–4876. Copyright 2004, with permission from Elsevier.) (b) Correlated release of lithogenic metal(loid)s Fe, Al, and Si as a function of organic carbon dissolution. Soil PZNPC = 3.3 and PZNC = 2.1. Slopes of lines in (b) give millimoles of metal dissolved per gram of organic C, which for the sum of DOM-Al and DOM-Fe are comparable to the total carboxyl group content of DOM.

The sum of the values for Al and Fe (6.4 mmol g^{-1} C) fall within the range commonly reported for moles of dissociable charge of soil DOM (Tipping, 2002), indicating a high degree of metal ion complexation. These results show that the pH dependence of Fe and Al dissolution from mineral–organic complexes in soil weathering environments can be distinct from that which has been observed for pure mineral phases such as goethite and gibbsite. The latter typically show *decreasing* metal dissolution with increasing pH for the pH range shown in Figure 10.10 (Stumm and Morgan, 1996).

10.4.2 Sorption of Inorganic Molecules at Mineral–Organic Interfaces

The term sorption represents a continuum of interfacial reactions that can lead to accumulation of nutrients or pollutants on

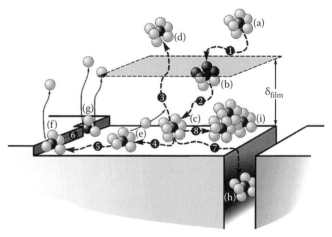

FIGURE 10.11 Schematic of ion sorption at the particle–water interface: (a) hydrated metal ion (water molecules are represented by light colored spheres surrounding the dark colored metal ion) approaching the surface via (1) bulk and (2) film diffusion. At the surface, the hydrated ion may (b) reside in the diffuse ion swarm, (c) form an outer sphere complex, (d) undergo desorption or (e) form an inner-sphere complex in association with loss of hydration water. Adsorbed ions may be desorbed (3) to bulk solution, or undergo (4, 5, 6, 7, 8) surface diffusion to form stronger complexes at higher affinity (f) ledge or (g) kink sites, undergo (h) diffusion into intraparticle pores, and/or to (i) sites of surface polymerization. (With permission from Springer Science + Business Media: Chorover, J., and M.L. Brusseau. 2008. Kinetics of sorption-desorption, p. 109–149. *In* S. Brantley, J. Kubicki, and A.F. White (eds.) Kinetics of water-rock interaction. Springer, New York.)

geochemical environment that accompanies sorbate formation exerts a strong control on the type of product formed. An array of environmental factors—such as sorptive species, pH, ionic strength, surface loading, and contact time—all affect the type of sorption complex or product that results. At short timescales and low sorptive concentrations, diffuse, outer and inner-sphere sorption is common, but as sorption occurs over extended timescales that approach those of sorbent transformation (e.g., mineral weathering), and also higher sorbate loadings, then surface nucleation and co-precipitation reactions are increasingly important to nutrient or pollutant fate (Schindler and Stumm, 1987).

10.4.2.1 Types of Reactive Sites

From the discussion above, it should be apparent that the types and densities of surface functional groups presented to the soil solution will depend on the biogeochemical environment in which the mineral–organic complexes were formed. Given the multicomponent complexity of natural soil particles and aggregates, there is no single analytical method that can provide a complete molecular-scale description of their surface chemistry. However, this is precisely why the conjunctive use of complementary macroscopic, microscopic, and spectroscopic approaches has contributed to significant advances in the past two decades (Brown et al., 1999; Sposito, 2004; Sparks, 2006; Hochella et al., 2008; Maurice and Hochella, 2008). Organic and inorganic moieties contribute to the interface in varying degrees, but not necessarily in proportion to their mass abundance (Bertsch and Seaman, 1999). This has important implications for sorption of metals, oxyanions, and organic pollutants.

For example, organic-rich soils that contain hydrolyzed Al and Fe in carboxylate or phenolate complexes may show much lower affinity for bivalent contaminant transition metals (e.g., Cd, Co, Ni) as a result of competition (Gustafsson et al., 2003, 2007). However, sites of organically complexed Al and Fe in humic-rich soils can serve as effective sorption sites for strongly sorbing oxyanions such as phosphate and arsenate (Bloom, 1981). Combining wet chemical methods with ATR-FTIR spectroscopy, Giesler et al. (2005) found that at high P loadings, phosphate formed precipitates with organically complexed Al and Fe in humus soils, whereas both adsorbed and co-precipitated species were formed when phosphate was added to soils with higher free iron oxide content. While precipitate formation is likely to be less favorable in soils with lower phosphate concentrations, the complexation of oxyanions with organically bound Fe and Al in soil humus may be of general importance to their transport and fate.

10.4.2.2 Diversity of Ion Sorption Mechanisms and Reactive Site Selectivities

Among the bivalent cations, strong differences exist with respect to their affinity for reactive mineral–organic particle surfaces. These differences can be related to their Lewis acid–base properties. Based on numerous molecular-scale spectroscopic studies (using XAFS, FTIR, and NMR), it is generally observed that the non-hydrolyzing Group II cations, Mg^{2+}, Ca^{2+}, Sr^{2+}, and Ba^{2+}, which are hard Lewis acids, are weakly adsorbed in soils as outer-sphere complexes, particularly with hard Lewis base functional groups such as carbonate, sulfate, carboxyl, and phenolic or alcoholic OH. Hydrolyzing metals behave differently. The hydrolyzing first-row transition metal cations, Mn^{2+}, Fe^{2+}, Co^{2+}, Ni^{2+}, Cu^{2+}, and Zn^{2+}, are borderline Lewis acids and they tend to form stronger inner-sphere complexes with borderline and soft base (e.g., sulfite, pyridine) functional groups. Similarly, the soft Lewis acids, which are the hydrolyzing, high-mass metal cations such as Cd^{2+}, Hg^{2+}, and Pb^{2+}, tend to form stronger, inner-sphere surface complexes, particularly with soft base groups such as thiol, sulfhydryl, and lipophilic (e.g., C_2H_4) moieties (Richens, 1997; Sparks, 2005). Although there are currently experimental challenges to using molecular-scale techniques to determine the types of surface complexes that strong acid anions such as NO_3^-, Cl^-, and ClO_4^- form on surfaces, they are generally thought to sorb to positively charged surfaces as outer-sphere complexes. Sulfate and SeO_4^{2-} can be sorbed as both outer-sphere and inner-sphere complexes, depending on environmental conditions, whereas most oxyanions that derive from dissociation of weak acids, such as molybdate, arsenate, arsenite, chromate, selenite, phosphate, and silicate, appear to be sorbed more strongly as inner-sphere complexes on clay edges and oxide surfaces, via a ligand exchange mechanism (Sposito, 2004).

10.4.3 Sorption of Organic Pollutants

Sorption mechanisms for organic pollutants are strongly dependent on the polarity and functional group chemistry of the sorptive compound. For polar molecules, surface interactions are analogous to those for inorganic ions, including ionic and/or covalent (i.e., outer- and inner-sphere complexation with surface metal or oxygen groups) and hydrogen-bonding interactions. For nonpolar compounds, van der Waals interaction and hydrophobic effects are dominant, and for this reason, NOM is the most important sorbent for HOCs. An emerging concern is the ever-increasing number of organic contaminants with widely ranging functional group composition and polarity are introduced into soil and water systems on a daily basis. These derive, in particular, from the widespread use of pharmaceuticals and personal care products, some of which have endocrine disrupting capability even at vanishingly low concentration.

10.4.3.1 Hydrophobic Organic Contaminants

As discussed above, OM's amphiphilic character includes sorption domains that are themselves relatively hydrophobic (e.g., aromatic and aliphatic regions) and these are thought to constitute important sorption sites for HOCs (Semple et al., 2003). HOC sorption–desorption in natural systems is often characterized by a rapid, reversible sorption–desorption process followed by a much slower, nonreversible stage. Observations of rate-limited, nonlinear, and competitive sorption for HOCs have led to the development of the "dual-mode" sorption model, which presumes that solid phase NOM is composed of "soft rubbery" and "hard glassy" domains (Huang et al., 2003). Hydrated carbon components (biomolecules, most humic substances) are considered to comprise the expanded "rubbery" domain, whereas condensed carbon components (black carbon, some humin components, kerogen) are considered to comprise the "glassy" domain. The rubbery domain is represented as flexible and polymeric, wherein organic compound sorption is linear, noncompetitive and controlled by permeation interactions. The glassy domain comprises nanometer-scale pores in which retention—more akin to adsorption—occurs via sorbate interaction with pore walls. Sorption to pores in the glassy NOM is thought to be responsible for the nonlinear and competitive effects that are sometimes observed for HOC sorption. Despite a wealth of macroscopic data highlighting the importance of NOM to HOC sorption, the precise mechanisms responsible for the biphasic behavior and the molecular nature of HOC sorption sites are still poorly understood (Luthy et al., 1997; Ehlers and Loibner, 2006).

10.4.3.2 Trace Organic Contaminants of Emerging Concern

A wide range of emerging contaminants including pharmaceuticals and personal care products (PPCPs) and endocrine disrupting compounds (EDCs) have been measured in water sources throughout the United States (Kolpin et al., 2002). The presence of PPCPs and EDCs in streams has been attributed to widespread domestic and veterinary use of these compounds, their accumulation in wastewater (e.g., sewerage, lagoons), and wastewater treatment that is only partially effective in their removal prior to effluent discharge into the environment. Thousands of different pharmaceutical substances are used daily including painkillers, antibiotics, antidiabetics, β-blockers, contraceptives, lipid regulators, antidepressants, and impotence drugs. Effluent and biosolids from wastewater treatment plants are primary entryways for PPCPs and EDCs into the environment, much occurring with little recognition (Kinney et al., 2006). Only during the last decade has information concerning the concentration and environmental fate of these "emerging contaminants" been reported (Kummerer, 2001). Particular concern is now focused on EDCs because they are capable of eliciting physiological response among continuously exposed organisms at vanishingly low (part-per-trillion to part-per-billion) concentrations (Drewes et al., 2005).

Many EDCs are polar and tend to persist in wastewater effluents. The increased usage of treated wastewater for soil irrigation, as well as its discharge into surface and ground waters allows them to enter soils, the biosphere, and drinking water supplies (Snyder et al., 2003). Biosolids, which are often applied as soil amendments, can contain high concentrations (part-per-million to part-per-thousand on a dry mass basis) of hydrophobic EDCs such as nonylphenol and flame-retarding chemicals. In addition, it has been estimated that thousands of tons of antibiotics are used annually in the United States for livestock farming only (Kemper, 2008). The situation is complicated by the fact that many pharmaceuticals are transformed (oxidized, hydrolyzed, or reduced) or subjected to conjugation processes in vivo before being excreted.

EDC/PPCPs vary considerably in terms of molecular structure, functional group content and physicochemical properties, even within a specific class of compounds. For example, the water solubilities of antibiotics range from the μg L^{-1} level (for polyethers—monensin and salinomycin) to thousands of mg L^{-1} (tetracyclines—chlorotetracycline and oxytetracycline). The log K_{ow} values of these compounds vary commensurately, from 3.5 to −1.2. Thus, it is not surprising that distribution coefficients (K_d values, which indicate the relative affinities of aqueous-phase chemicals for solid-phase sorbents) among antibiotic compounds vary from 70–5600 L kg^{-1} for tetracyclines and quinolones to low values of 0.2–2 L kg^{-1} for sulfonamides (Tolls, 2001). Nonetheless, most EDC/PPCPs tend to be more polar than persistent HOCs and, as a result, their sorption to surfaces is strongly influenced by the interaction of solution chemistry (pH and ionic strength) and sorbent type. For example, protonation of sulfoamines resulted in an increase in K_d (from 1 to 30) when the pH of mineral soil was decreased from 8 to 4 (Boxall et al., 2002). Higher binding of enrofloxacin was reported for montmorillonitic relative to illitic or kaolinitic soils (Tolls, 2001). The accumulation of ofloxacin at particle surfaces exhibits strong pH dependence (maximum adsorption occurring between the pK_{a1} and pK_{a2} of the compound) and binding mechanisms are distinctly different for silica (binding via heterocyclic N) versus alumina (Al chelation by carboxylate and carbonyl group) surfaces (Goyne et al., 2005). These studies underline the importance of considering polar functional groups, and combining macroscopic and spectroscopic studies for elucidating the binding of EDC/PPCPs at environmental surfaces.

Many EDC/PPCPs interact strongly with DOM in preference to hydrophilic mineral surfaces, though the mechanisms are not clear (Zhou et al., 2007). Tolls (2001) and Thiele-Bruhn (2003) suggested that the sorption of antibiotics to organic matter is not related solely to their hydrophobicity because K_d values for complexation with DOM are higher than would be predicted based on simple partitioning theory. Zhou et al. (2007) found that EDC sorption to aquatic colloids was enhanced for those colloids containing a higher mass fraction of organic carbon and, in particular, those colloids that have higher specific UV absorbance (i.e., greater aromaticity). Conversely, it is very likely that sorption to DOM and solid phase OM of more polar EDCs is governed by the same processes that control the "supramolecular" assembly of the humic substances themselves (Piccolo, 2002). Hence, while it is clear that DOM may strongly influence EDC/PPCP transport, the intermolecular interactions are likely to be multifaceted and include not only hydrophobic sorption but also ionic, covalent, and/or hydrogen bonding interactions.

10.4.4 Aging Effects

Prolonged sorbate–sorbent contact in the soil-weathering environment can, in some cases, result in diminished bioavailability of both inorganic and organic contaminants (Hatzinger and

Alexander, 1995; McLaren et al., 1998; Sparks, Chapter 1). The processes that give rise to this time-dependent decrease in availability are known collectively as "the aging effect." Macroscopic observations of the aging effect are attributed to numerous molecular-scale phenomena, including pore and surface diffusion to sorption sites of higher affinity and long-term changes in bonded sorbate–sorbent structure (Figure 10.11). In the case of HOCs, sorption–desorption hysteresis is postulated to result from irreversible structural expansion and the creation of new micropores as a result of HOC molecules penetrating into the NOM sorbent phase. This leads to enhanced solid-phase affinity during subsequent desorption (Sander et al., 2006). Additionally, partial degradation of organic contaminants, as catalyzed by either extracellular enzymes or mineral surfaces, can lead to the formation of reactive intermediates that form covalent bonds with NOM (Dec and Bollag, 1997). These "bound residues" can be detected unambiguously only by using molecular probe methods such as ^{13}C NMR spectroscopy (Hatcher et al., 1993).

Polymeric sorption complexes and surface precipitates of inorganic contaminants can result when high sorbate concentrations are contacted with natural geomedia for extended times. This can lead to slowly decreasing contaminant bioavailability. When such processes involve co-precipitation with major lithogenic elements such as Si, Al, and Fe, they tend to be strongly affected by the rates and trajectories of local mineral weathering. For example, under conditions of high metal loading and pH (>7) sorption of metals such as Co^{2+}, Ni^{2+}, and Zn^{2+} results in formation of mixed (e.g., Al-bearing) hydroxide surface precipitate phases that are of the layered double hydroxide (LDH) form (Scheidegger et al., 1996). Co-contaminant interactions can also be important to this process. For example, using a combination of soil column experiments and EXAFS spectroscopy, Voegelin and Kretzschmar (2005) demonstrated that the presence of Ni can enhance the immobilization of Zn by forming mixed Ni-Zn-LDH phases that are more resistant to dissolution than pure Zn-LDH precipitates. With further weathering time, these LDH phases are shown to transform to stable Si-bearing metal phyllosilicates, where the metal is sequestered in a form that is much less mobile and bioavailable (Scheckel et al., 2000).

Long-term sorption processes may also progressively diminish the bioavailability of cationic radionuclides (^{137}Cs, ^{90}Sr) because of silicate weathering reactions in soil media. This is particularly relevant in cases when radionuclides are released into the subsurface as part of an alkaline liquid waste that promotes native silicate dissolution. This has occurred, for example, at several Department of Energy sites, such as the Hanford site, that employed plutonium purification technologies (Zachara et al., 2007). Laboratory studies indicate that the dissolution of native silicate clays in the presence of caustic wastes gives rise to the precipitation of poorly crystalline aluminosilicate precipitates. These precipitates transform to more stable feldspathoid phases that then undergo "Ostwald ripening" in the liquid waste, which diminishes their subsequent dissolution kinetics when circumneutral pore waters are reintroduced (Qafoku et al., 2003; Chorover et al., 2008).

It should be noted, however, that prolonged residence time in geomedia is not a general solution for removing metals from bioavailable form. For example, although the application of biosolids to soil has the clear benefits of reducing landfill volumes and potentially increasing soil fertility in terms of NOM and nutrient content, there is also concern that sludge-bound metals will be subjected to introduction to the food chain (e.g., via plant uptake) as the organic matter, which serves as the principal metal-complexing component of the applied material, is slowly biodegraded (McBride, 2003). In other cases, metal sequestration is unfavorable thermodynamically, particularly in acidic soils, and contaminant remediation may require proactive intervention. Such approaches can nonetheless take advantage of geochemical phenomena. For example, a common remediation technique for Pb-contaminated soils involves the application of phosphate salts to promote the precipitation of sparingly soluble pyromorphite ($Pb_5(PO_4)_3Cl$) solids.

10.5 Microbial Intervention in Metal Cycling

Recent reviews have highlighted the importance of microbial diversity and its role in direct and indirect microbial mediation of biogeochemical and pollutant processing in the CZ (Newman and Banfield, 2002; Lovley, 2003). Examples of direct microbial mediation include assimilatory or dissimilatory reduction of metals [e.g., Fe(III), Mn(IV), Cu(II), U(VI)] and oxyanions (e.g., NO_3^-, SO_4^{2-} AsO_4^{3-}, CrO_4^{2-}), the oxidative precipitation of metal hydrous oxides, and the biodegradation of organic contaminants. In all of these cases, the organisms derive some benefit from the reaction (energetic, nutritional, detoxification) that increases their chances of survival. These processes are typically very efficient, genetically controlled and thus subject to evolutionary change. Microbes also affect numerous abiotic reactions in an indirect manner. Typically these indirect effects are the consequence of a local change in the geochemical environment due to the presence of metabolic byproducts, and include, for example, the precipitation of carbonates, or dissolution of simple silicates, by proton-consuming metabolic activities such as photosynthesis or dissimilatory sulfate reduction. Biogenic Mn(IV) oxides such as acid birnessite, are another good example of indirect biological mediation in geochemical outcomes: they are believed to be the most abundant and highly reactive Mn oxide phases in the environment and when present can react with a wide range of organic and inorganic compounds and sequester a variety of metals (Tebo et al., 2004).

10.5.1 Microbially Mediated Redox Reactions

Largely because of continuous input of reduced C from autotrophic sources, biologically active soils undergo fluctuations in redox status as a result of hydrologic events or seasonal cycles and these have important implications for biogeochemical reactions. When previously oxic soils are subjected to depletion of dissolved or gaseous O_2 because of soil pore filling with liquid water and associated gas diffusion limitations, heterotrophic microbes utilize alternative terminal electron accepting processes (TEAPs) in the respiration of organic matter. The sequence of TEAPs that occurs in waterlogged soil corresponds to the energetics of the reactions such that use of primary oxidants occurs in the order $O_{2(g)} \rightarrow NO_{3(aq)}^- \rightarrow Mn(IV)_{(s)} \rightarrow Fe(III)_{(s)} \rightarrow SO_{4(aq)} \rightarrow CO_{2(g)}$. While oxygen, nitrate, sulfate, and carbon dioxide are all present in the fluid phases, oxidized forms of manganese and iron are dominantly present in the solid phase because of their low aqueous solubility. In addition to the major elements O, C, N, S, Fe, and Mn, trace elements such as As, Se, Cr, Hg, U, and Mo also undergo speciation change as affected by biologically catalyzed redox reactions. The fluctuating redox conditions characteristic of soil systems have been studied experimentally by numerous investigators (McGeehan et al., 1998; Pett-Ridge et al., 2006; Thompson et al., 2006a,b). Redox fluctuations can be imposed by varying gas (e.g., $N_2/O_2/H_2$) partial pressures over various durations thereby forcing TEAPs to occur.

10.5.1.1 Iron Redox Cycling

The prevalence of iron in the Earth's crust and soils makes iron(III) reduction an especially important alternative TEAP in soils and sediments. Since iron (oxyhydr)oxides are sorbents for a wide range of nutrients and pollutants, this process also has implications for the sequestration and release of sorbed species. Ferric iron may serve as a dominant or exclusive terminal electron acceptor, or it may accompany fermentation as a supplementary terminal electron acceptor. Either case is considered dissimilatory iron reduction because Fe(III) is reduced biotically without being incorporated into specific cellular components (Ehrlich and Newman, 2009).

A wide variety of bacteria and archaea are capable of dissimilatory iron reduction, including both strictly anaerobic and facultative organisms. As a group, these organisms are capable of using dissimilatory Fe(III) reduction as a means to oxidize a wide range of organic molecules (e.g., ranging from acetate and lactate to palmitate and aromatic compounds), but some are unable to degrade the organic electron donor completely, often accumulating acetate (Lovley et al., 1989). The product of Fe(III) reduction is ferrous iron, and when released it can become adsorbed at the surface of the Fe(III) oxide undergoing reduction, interfering with microbial attack. Its removal from the iron oxide surface promotes the reduction of the oxide (Roden and Urrutia, 1999). In addition, Fe(II) is able to catalyze the structural transformation of Fe(III) minerals ferrihydrite and schwertmannite to more crystalline, thermodynamically stable forms, such as lepidocrocite and goethite (Pedersen et al., 2005; Burton et al., 2007). Iron isotope exchange experiments indicate that adsorbed Fe(II) induces electron transfer with mineral surface Fe(III) atoms, resulting in their reductive dissolution. The released Fe(III) then recrystallizes into an Fe(III) mineral species that is generally more crystalline and thermodynamically stable (Yee et al., 2006). In this way, microbial mediation can accelerate Ostwald ripening of mineral solids.

In a soil slurry experiment, Thompson et al. (2006b) subjected triplicate soil reactors of an Inceptisol to alternating one-week oxidation–reduction cycles shifting between E_h conditions either above (oxidizing) and below (reducing) equilibrium with ferrihydrite. Figure 10.12 shows the influence of such fluctuations on aqueous phase (a) pE or E_h, (b) dissolved Fe concentration, and (c) DOM concentration. The pH (not shown) and [Fe(II)] trends are inverse to those of E_h as expected for the reactions whereby dissimilatory Fe reduction (coupled to DOM oxidation) during the reduction cycles is followed by abiotic Fe(II) oxidation (coupled to O_2 reduction) during the oxidizing cycles. The data are consistent with the following reactions (shown using goethite as the Fe(III) solid undergoing dissolution-precipitation):

$$Reduction: FeOOH_{(s)} + \tfrac{1}{4}CH_2O_{(aq)} + 2H_{(aq)}^+$$
$$\rightarrow Fe_{(aq)}^{2+} + \tfrac{1}{4}CO_{2(g)} + 2H_2O_{(l)} \quad (10.6)$$

$$Oxidation: Fe_{(aq)}^{2+} + \tfrac{1}{4}O_{2(aq)} + 1\tfrac{1}{2}H_2O_{(l)}$$
$$\rightarrow FeOOH_{(s)} + 2H_{(aq)}^+ \quad (10.7)$$

The data show that microbially mediated reductive dissolution of Fe(III) minerals can liberate organic matter from mineral–organic complexes. Using Mössbauer spectroscopy and selective chemical extractions, Thompson et al. (2006b) demonstrated that these redox oscillations in soils increase the iron oxide crystallinity, most likely as a result of the formation of aqueous Fe(II) under anoxic conditions and the subsequent onset of Fe(II)-catalyzed crystallization of Fe(III) minerals.

While Fe(III) occurs in a wide variety of secondary mineral types, it has been observed that microbially mediated Fe(III) reduction rates tend to increase with increasing specific surface area (Figure 10.13, Roden, 2008). Hence, ferrihydrite and nano-crystalline goethite tend to be utilized in preference to macro-crystalline goethite and hematite as electron acceptors for microbially mediated reductive dissolution. However, it is clear that Fe oxide mineralogy plays an important role, since different phases with comparable surface areas can show distinctly different Fe(III) reduction rates (Larsen and Postma, 2001). Reactivity can also be strongly impacted by wet-dry cycling and surface "poisoning" of Fe(III) oxyhydroxides. A recent study of Fe(III) reduction rates using ascorbic acid as an electron donor showed that drying of ferrihydrite dramatically diminished its rate of reductive dissolution upon rewetting, whereas formation of ferrihydrite in the presence of dissolved silicon during crystallization significantly increased the reactivity of the dried and rewetted solid (Jones et al., 2009). The influence of Si on ferrihydrite reactivity is particularly intriguing, because it suggests that ferrihydrite formed in an active soil weathering environment is likely to be even more reactive than synthetic specimen minerals that have been most often used in experiments.

10.5.1.2 Contaminant Redox Reactions

Microbial metabolic (redox) processes in colonized soil and aquifer systems can alter the mobility of contaminants with dramatic

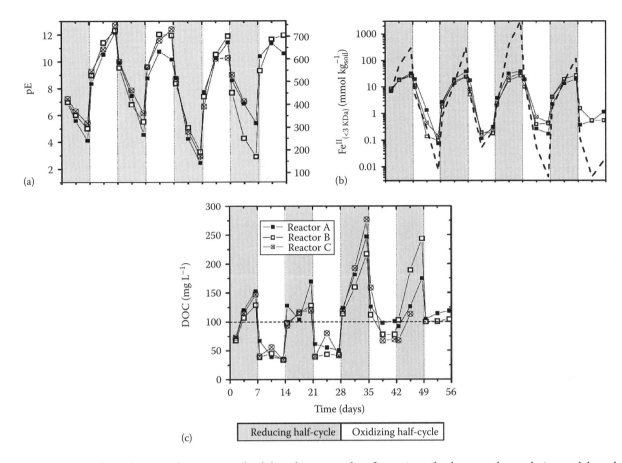

FIGURE 10.12 Iron redox cycling in triplicate Inceptisol soil slurry bioreactors show fluctuating soil redox status that can be imposed through $O_{2(g)}$ depletion during reducing cycles and its reintroduction during oxidizing cycles. Graph (a) shows pE (left axis) or E_h (mV, right axis). These fluctuations result in reductive dissolution of $Fe_{(aq)}^{2+}$ during the reducing half-cycles, and its precipitation as Fe(III) during oxidizing half-cycles (b). Dashed line in (b) shows equilibrium with ferrihydrite. Reductive dissolution of Fe solids results in the (c) dissolution and/or dispersion of organic carbon that was co-associated with aggregates containing Fe oxyhydroxide solids. (From Thompson, A., O.A. Chadwick, D.G. Rancourt, and J. Chorover. 2006b. Iron-oxide crystallinity increases during soil redox oscillations. Geochim. Cosmochim. Acta 70:1710–1727. Copyright 2006, with permission from Elsevier.)

consequences for human and ecosystem health. For example, arsenic release to ground waters in the Bengal delta has been attributed to the microbial oxidation of DOM coupled to the reduction of sorbed As(V) to mobile As(III) (Islam et al., 2004). In this respect, it is noteworthy that, whereas it has been hypothesized that As dissolution is coupled to the reductive dissolution of the Fe(III) sorbents (e.g., iron oxyhydroxides), the study by Islam et al. (2004) shows that Fe and As release was decoupled in their experiments. The principal form of arsenic released to solution was As(III), and this occurred quite some time after Fe(III) bioreduction (Figure 10.14), indicating that As(V) was used as a terminal electron acceptor in respiration of DOM once the labile pools of Fe(III) had been depleted (Islam et al., 2004).

Conversely, microbial reduction of contaminant metals during respiration of organic matter can dramatically *reduce* contaminant bioavailability if the reduced species has a greater tendency to precipitate from solution, as is the case for the contaminant redox couples $Cr^{(VI)}O_{4(aq)}^{2-}/Cr^{(III)}(OH)_{3(s)}$ (Tokunaga et al., 2003a) and $U^{(VI)}O_2^{2+}/U^{(IV)}O_{2(s)}$ (Wilkins et al., 2006). Indeed, the "biomineralization" of contaminant metal species is an active area of research, with clear application to remediation of contaminated soils and sediments (Lovley and Coates, 1997). However, given the dynamic nature of soil biogeochemical systems (e.g., fluctuations in redox status, as discussed above), the long-term stability of solid-phase bioreduction products remains a concern for systems that contain or can generate reactive metal oxidizing agents. For example, Tokunaga et al. (2007) have shown that Cr(III) solids formed by microbial bioreduction in the presence of labile organic carbon can be re-oxidized, most likely by Mn(III) or Mn(IV) containing solids. Manganese oxides are known to be effective oxidants of Cr(III), and they are capable of releasing significant hexavalent chromium into solution upon reaction (Fendorf et al., 1992; Oze et al., 2007). The fact that these solids are rapidly biosynthesized by Mn(II) oxidizing bacteria and fungi under oxic conditions (Tebo et al., 2004), suggests that their highly reactive surfaces may cycle in and out of prevalence in redox fluctuating environments, like biologically active soils, with potential dynamic impacts on Cr redox state.

Similar dynamics may affect the stability of biogenic $U^{(IV)}O_{2(s)}$. In a soil column study of U(VI) bioreduction stimulated by long-term dissolved OM influx, the persistence of Fe(III) solids

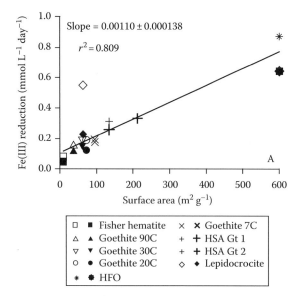

FIGURE 10.13 Initial rates of bacterial (*S. putrefaciens* and *G. sulfurreducens*) reduction of synthetic Fe(III) oxides as a function of oxide specific surface area. (With kind permission from Springer Science+Business Media: Roden, E.E. 2008. Microbial controls on geochemical kinetics 1: Fundamentals and case study on microbial Fe(III) oxide reduction, p. 335–415. *In* S.L. Brantley, J.D. Kubicki, and A.F. White (eds.) Kinetics of water-rock interaction. Springer, New York.)

as an electron acceptor led to re-oxidation and solubilization of biostabilized $UO_{2(s)}$ (Tokunaga et al., 2008). This occurred long after the depletion of both U(VI) and Mn(III and IV). Therefore, while bioreduction can be a very effective means for partitioning contaminants out of mobile solution, it may also require a sustained input of reducing equivalents to maintain thermodynamic stability. In so far as soil biogeochemical processes impose dynamic shifts in environmental conditions, they can also alter the thermodynamic stability, bioavailability, and mobility of redox-active inorganic contaminants.

10.5.2 Biogeochemical Gradients

Biologically active soils have enormous capacity for chemical transformation of element species and organic chemicals. Mixing limitations—and therefore a high degree of spatial and temporal heterogeneity—give rise to steep geochemical gradients that promote reactive transformation of pollutants and nutrients along both biotic and abiotic pathways. Gradients arise temporally in response to input fluctuations such as rainfall events (Silver et al., 1999) and daily changes between photosynthetic and respiratory activity of the biota (Garcia-Pichel and Belnap, 1996). They also arise spatially because of local geochemical character and hydrologic state. A remarkable feature of soil systems is that steep geochemical gradients occur even across short distances (10^{-6}–10^{-3} m) and timescales (Sexstone et al., 1985). Such gradients and their biogeochemical consequences have been particularly well documented by probing aqueous speciation of biological soil crusts in desert environments using micro-electrode techniques (Figure 10.15, Garcia-Pichel et al., 2003; Johnson et al., 2005, 2007). These data show the very large differences in molecular oxygen and nitrogen redox species that occur at the soil–atmosphere interface in desert soils subjected to an influx of water and substrate for microbial growth. Another particularly conspicuous case of such

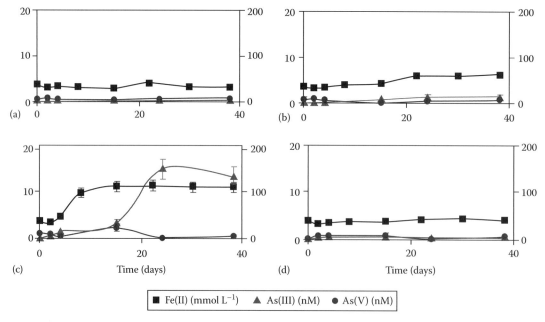

FIGURE 10.14 Reduction of Fe(III) and mobilization of arsenic in microcosms containing Bengali sediments incubated under a range of biogeochemical conditions: (a) oxic, (b) anoxic, (c) anoxic with 4 g L^{-1} Na-acetate as a proxy for DOM, (d) abiotic control sediments with Na-acetate added, but autoclaved before incubation. Black squares, Fe(II); light gray triangles, As(III); gray circles, As(V). Each point and error bar represents the mean and standard deviation of three replicate experiments. (From Islam, F.S., A.G. Gault, C. Boothman, D.A. Polya, J.M. Charnock, D. Chatterjee, and J. R. Lloyd. 2004. Role of metal-reducing bacteria in arsenic release from Bengal delta sediments. Nature 430:68–71. With permission.)

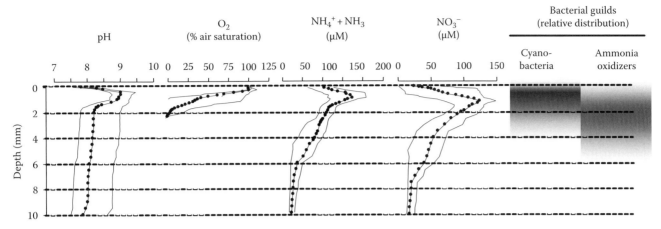

FIGURE 10.15 Vertical chemical gradients developing within the top cm of active (wet) biological soil crust communities as a consequence of microbial metabolic activity, as measured by microelectrodes. Photoautotrophic cyanobacteria close to the surface consume CO_2, driving pH up and creating an internal supply of O_2, which is quickly respired by heterotrophs to the point of anoxia some 2–3 mm below the surface. Leakage of cellular NH_3 produced by cyanobacteria during N_2-fixation, creates a very thin layer where both free NH_3 and O_2 are present, which in turn, constitutes a habitat for chemolithoautotrophic ammonia oxidizing bacteria. These then create a localized source of nitrate as a byproduct of their metabolism. This microscale clockwork constitutes the mechanistic basis for the fertilizing role of soil crust communities in arid lands at the landscape level. (From Chorover, J., R. Kretzschmar, F. Garcia-Pichel, and D.L. Sparks. 2007. Soil biogeochemical processes within the critical zone. Elements 3:321–326. With permission from Mineralogical Society of America. With original data from Garcia-Pichel, F., S.L. Johnson, D. Youngkin, and J. Belnap. 2003. Small-scale vertical distribution of bacterial biomass and diversity in biological soil crusts from arid lands in the Colorado Plateau. Microb. Ecol. 46:312–321; Johnson, S.L., C.R. Budinoff, J. Belnap, and F. Garcia-Pichel. 2005. Relevance of ammonium oxidation within biological soil crust communities. Environ. Microbiol. 7:1–12; Johnson, S.L., S. Neuer, and F. Garcia-Pichel. 2007. Export of nitrogenous compounds due to incomplete cycling within biological soil crusts of arid lands. Environ. Microbiol. 9:680–689.)

gradients is the well known "mottling" of soil color that results from proximal Fe reduction and oxidation processes that occur in adjacent regions of soil subjected to fluctuating redox status. These Fe "redoximorphic features" are indicative of soil systems that can support the metabolic activities of distinct bacterial species that gain energy from either Fe(III) reduction in anoxic microenvironments or Fe(II) oxidation in oxic or anoxic microenvironments that occur in close proximity (Weber et al., 2006).

To the extent that microbial species differ in their metabolic capacity to exploit geochemical gradients and capture energy, the heterogeneous nature of subsurface systems gives rise to a patchwork distribution of distinct microbial associations and geochemical functions from the regional scale (10^1–10^3 m) to that of particle aggregates (10^{-3}–10^{-6} m) (Hunter et al., 1998). For example, the genetic structure of microbial communities has been shown to differ between the exterior and interior of soil aggregates, between various sized aggregates and bulk soil (Ranjard et al., 2000; Pankhurst et al., 2002), and it exhibits vertical patterns of segregation of functional types at the submillimeter scale in biological soil crusts (Figure 10.15) (Garcia-Pichel et al., 2003). Indeed, destruction of aggregate structure through land use change has been shown to diminish microbial diversity and patchiness at the soil aggregate scale (Lupwayi et al., 2001).

This spatial distribution also has implications for pollutant transformations. For example, Tokunaga et al. (2003b) found that the reduction of soluble Cr(VI) to insoluble Cr(III) occurred within the surface layer of soil aggregates where O_2 concentrations were limited by diffusion but where labile NOM concentrations were sufficient to promote high rates of microbial respiration. Spatial distribution of microbially catalyzed redox reactions is also observed up to the field scale in subsurface sediments (Akob et al., 2007). One example is the distribution of redox potential from anoxic to suboxic to oxic as groundwater moves down gradient from an organic contaminant plume. Along this transect, one observes that organic matter decomposition is coupled to a sequence of TEAPs of progressively increasing energy yield as stronger oxidants become available. Common methods of sediment sampling and analysis that average over these geochemical gradients within aggregates and over larger spatial scales can erase important biogeochemical spatial relations necessary for understanding the processing that occurs in these environments (Tokunaga et al., 2003b).

It is now becoming increasingly clear that the processing of matter and energy in soil and aquifers does not only depend on the type of interacting components (biotic or abiotic) but also on their relative positions in space and the transport pathways between them. It is thus imperative that in future studies we strive to understand soils and the larger CZ on the basis of integrative "system architecture" rather than as mere compilations of compositional and distributional components acting independently of position (Hunter et al., 1998; Legout et al., 2005).

10.5.3 Biogeophysical Heterogeneities

In addition to the biogeochemical heterogeneity of interfaces described above, the CZ also exhibits extreme physical heterogeneity. Very wide pore size distributions lead to an associated distribution of water flow velocities and microenvironments. Studies have found that pores are not distributed randomly, but rather that they are networked in ways that depend on the process of their formation. Near the soil surface, large pores can arise from the activities of roots, worms, or other soil organisms, whereas smaller pores within particle aggregates can form from abiotic aggregation processes, as well as root–microbe–soil interactions. Recent work using synchrotron microtomography and computational modeling suggests that microbes and plant roots "microengineer" their soil habitat by changing its porosity and spatial correlation, making it more porous, ordered, and aggregated in structure (Young and Crawford, 2004; Feeney et al., 2006). These studies also suggest that soil porosity and aggregate structure can undergo fluctuations in response to open system dynamics resulting from nutrient flux, biological activity, and microbial growth.

In macroporous systems, mass transfer is often approximated by a "dual porosity model" where solutes are transported advectively in macropores but also undergo diffusive transport into stagnant water associated with the microporous matrix or adsorbed to particle surfaces (Simunek et al., 2003). Hagedorn and Bundt (2002) have shown that such macropores in soils can persist as active preferential flow paths for several decades. In some clay-rich soils, shrinking and swelling can also lead to a pronounced aggregation of the soil and the formation of large cracks. At greater depth penetration into bedrock, large pores can result from rock fractures and physical and chemical weathering. The connectivity and length of these large pores have a major influence on the nature and amount of interface contacted by water, solutes, and colloids as they traverse the CZ (Allaire-Leung et al., 2000). Macropores serve as preferential flow paths where water, nutrients, and pollutants travel much faster than they do in a homogeneous medium of the same total porosity (Reedy et al., 1996). While this diminishes the extent of interfacial contact and reaction with the soil matrix, it also can enrich macropore surfaces with sorbing pollutants (Hagedorn and Bundt, 2002). Similarly, enhanced delivery of nutrients and or metabolic substrates for bacteria may promote their preferential growth on walls of large pores. In extreme cases, such as the disposal of large amounts of organic waste, ensuing enhancement of heterotrophic bacterial growth can lead to pore clogging and decreased hydraulic conductivity (Baveye et al., 1998).

Superimposed on the particle- to pedon-scale distribution of reactive interfaces is the variability in geochemical environment that occurs at the landscape scale because of hydrologic partitioning through geomorphically variable terrain. Landform structure forces migrating water to sample distinct geochemical environments along its transmission through the Earth's surface. One example is the movement of precipitation through oxic weathering environments on hillslope uplands prior to convergence on an organic-rich riparian zone at lower elevation where reducing conditions may predominate. Geomorphic structure then feeds back to affect the nature of interfaces formed at different locations in the landscape. For example, whereas Fe(III) oxyhydroxide formation may be favored in the oxic, upland weathering environment, such phases may become unstable relative to dissolved Fe(II) in the riparian reducing environment. Corresponding "hotspots" of biological activity may also favor local losses of N through denitrification under anoxia (Schade et al., 2002).

10.6 Conclusions and Future Prospects

The transport and fate of nutrients and pollutants in the CZ depend on physical and biogeochemical interactions among living and non-living soil components. In this chapter, we have emphasized the fact that these components do not create a stationary matrix for such interactions. Conversely, soils are a highly dynamic system where feedbacks between biotic and abiotic components create a weathering environment where mineral and organic transformations alter the form and bioavailability of both essential and toxic species.

Significant advances have been made in recent years in the subdisciplines of soil and earth sciences that have enabled much better characterization of these systems from catchment to molecular scales. However, our understanding of key processes remains incomplete, particularly for those phenomena that involve process coupling across the subdisciplines of organic and inorganic soil chemistry, biology, and hydrology. Specifically, research is needed to better resolve (1) the precise mechanisms whereby plants and microbes affect chemical denudation versus mineral transformation rates; (2) the biogeochemical processes leading to the formation of the continuum of mineral–organic complexes and their higher order micro- to macro-aggregate structures; (3) the "architecture" of these complexes, including the nature of bonds that are formed at mineral–organic interfaces, and the types of intermolecular associations that lead to further "layering" above the mineral surface; (4) the role of these complexes in affecting the surface complexation and sorptive retention of nutrients and pollutants, and (5) the kinetic and thermodynamic stability of bio-remediation products, particularly solid phase products of contaminant (e.g., U(VI) and Cr(VI)) bioreduction.

Acknowledgments

Several concepts and ideas in this chapter grew out of an earlier paper that was co-authored by R. Kretzschmar, D. L. Sparks, and F. Garcia-Pichel (Chorover et al., 2007). I am grateful to these individuals in particular for fruitful discussions and interactions over that earlier work. Support for preparation of this manuscript was provided by the National Science Foundation, NSF EAR 0724958.

References

Abudalo, R.A., Y.G. Bogatsu, J.N. Ryan, R.W. Harvey, D.W. Metge, and M. Elimelech. 2005. Effect of ferric oxyhydroxide grain coatings on the transport of bacteriophage PRD1 and Cryptosporidium parvum oocysts in saturated porous media. Environ. Sci. Technol. 39:6412–6419.

Ahmed, K.M., P. Bhattacharya, M.A. Hasan, S.H. Akhter, S.M.M. Alam, M.A.H. Bhuyian, M.B. Imam, A.A. Khan, and O. Sracek. 2004. Arsenic enrichment in groundwater of the alluvial aquifers in Bangladesh: An overview. Appl. Geochem. 19:181–200.

Akob, D.M., H.J. Mills, and J.E. Kostka. 2007. Metabolically active microbial communities in uranium-contaminated subsurface sediments. FEMS Microbiol. Ecol. 59:95–107.

Akter, M., and T. Akagi. 2005. Effect of fine root contact on plant-induced weathering of basalt. Soil Sci. Plant Nutr. 51:861–871.

Allaire-Leung, S.E., S.C. Gupta, and J.F. Moncrief. 2000. Water and solute movement in soil as influenced by macropore characteristics—1. Macropore continuity. J. Contam. Hydrol. 41:283–301.

Allard, B., and I. Arsenie. 1991. Abiotic reduction of mercury by humic substances in aquatic system—An important process for the mercury cycle. Water Air Soil Pollut. 56:457–464.

Arocena, J.M., and K.R. Glowa. 2000. Mineral weathering in ectomycorrhizosphere of subalpine fir (Abies lasiocarpa (Hook.) Nutt.) as revealed by soil solution composition. For. Ecol. Manage. 133:61–70.

Arocena, J.M., A. Gottlein, and S. Raidl. 2004. Spatial changes of soil solution and mineral composition in the rhizosphere of Norway-spruce seedlings colonized by *Piloderma croceum*. Zeit. Pflanzenernah. Bodenk.167:479–486.

Balogh-Brunstad, Z., C.K. Keller, B.T. Bormann, R. O'Brien, D. Wang, and G. Hawley. 2008a. Chemical weathering and chemical denudation dynamics through ecosystem development and disturbance. Global Biogeochem. Cycles 22, GB1007, doi: 10.1029/2007 GB002957.

Balogh-Brunstad, Z., C.K. Keller, R.A. Gill, B.T. Bormann, and C.Y. Li. 2008b. The effect of bacteria and fungi on chemical weathering and chemical denudation fluxes in pine growth experiments. Biogeochemistry 88:153–167.

Barker, W.W., and J.F. Banfield. 1996. Biologically versus inorganically mediated weathering reactions: Relationships between minerals and extracellular microbial polymers in lithobiontic communities. Chem. Geol. 132:55–69.

Baveye, P., P. Vandevivere, B.L. Hoyle, P.C. DeLeo, and D.S. de Lozada. 1998. Environmental impact and mechanisms of the biological clogging of saturated soils and aquifer materials. Crit. Rev. Environ. Sci. Technol. 28:123–191.

Benz, M., B. Schink, and A. Brune. 1998. Humic acid reduction by *Propionibacterium freudenreichii* and other fermenting bacteria. Appl. Environ. Microbiol. 64:4507–4512.

Berner, E.K., R.A. Berner, and K.L. Moulton. 2003. Plants and mineral weathering: Present and past. Treat. Geochem. J. I. Drever. 5:169–188.

Bertsch, P.M., and J.C. Seaman. 1999. Characterization of complex mineral assemblages: Implications for contaminant transport and environmental remediation. Proc. Natl. Acad. Sci. U.S.A. 96:3350–3357.

Bloom, P.R. 1981. Phosphorus adsorption by an aluminum-peat complex. Soil Sci. Soc. Am. J. 45:267–272.

Boxall, A.B.A., P. Blackwell, R. Cavallo, P. Kay, and J. Tolls. 2002. The sorption and transport of a sulphonamide antibiotic in soil systems. Toxicol. Lett. 131:19–28.

Brown, G.E., V.E. Henrich, W.H. Casey, D.L. Clark, C. Eggleston, A. Felmy, D.W. Goodman et al. 1999. Metal oxide surfaces and their interactions with aqueous solutions and microbial organisms. Chem. Rev. 99:77–174.

Burton, E.D., R.T. Bush, L.A. Sullivan, and D.R.G. Mitchell. 2007. Reductive transformation of iron and sulfur in schwertmannite-rich accumulations associated with coastal lowland acid sulfate soils. Geochim. Cosmochim. Acta 71:4456–4473.

Calvaruso, C., M.P. Turpault, and P. Frey-Klett. 2006. Root-associated bacteria contribute to mineral weathering and to mineral nutrition in trees: A budgeting analysis. Appl. Environ. Microbiol. 72:1258–1266.

Camesano, T.A., and B.E. Logan. 2000. Probing bacterial electrosteric interactions using atomic force microscopy. Environ. Sci. Technol. 34:3354–3362.

Camobreco, V.J., B.K. Richards, T.S. Steenhuis, J.H. Peverly, and M.B. McBride. 1996. Movement of heavy metals through undisturbed and homogenized soil columns. Soil Sci. 161:740–750.

Chan, C.S., G. De Stasio, S.A. Welch, M. Girasole, B.H. Frazer, M.V. Nesterova, S. Fakra, and J.F. Banfield. 2004. Microbial polysaccharides template assembly of nanocrystal fibers. Science 303:1656–1658.

Charlet, L., and D.A. Polya. 2006. Arsenic in shallow, reducing groundwaters in southern Asia: An environmental health disaster. Elements 2:91–96.

Cheah, S.F., S.M. Kraemer, J. Cervini-Silva, and G. Sposito. 2003. Steady-state dissolution kinetics of goethite in the presence of desferrioxamine B and oxalate ligands: Implications for the microbial acquisition of iron. Chem. Geol. 198:63–75.

Cheng, S., R. Bryant, S.H. Doerr, P.R. Williams, and C.J. Wright. 2008. Application of atomic force microscopy to the study of natural and model soil particles. J. Microsc. Oxf. 231:384–394.

Chenu, C., and A.F. Plante. 2006. Clay-sized organo-mineral complexes in a cultivation chronosequence: Revisiting the concept of the "primary organo-mineral complex." Eur. J. Soil Sci. 57:596–607.

Chorover, J. 2005. Zero charge points, p. 367–373. *In* D. Hillel and D.L. Sparks (eds.) Encyclopedia of soils in the environment. Academic Press, New York.

Chorover, J., and M.K. Amistadi, 2001. Reaction of forest floor organic matter at goethite, birnessite and smectite surfaces. Geochim. Cosmochim. Acta 65:95–109.

Chorover, J., M.K. Amistadi, and O.A. Chadwick. 2004. Surface charge evolution of mineral-organic complexes during pedogenesis in Hawaiian basalt. Geochim. Cosmochim. Acta 68:4859–4876.

Chorover, J., and M.L. Brusseau. 2008. Kinetics of sorption-desorption, p. 109–149. *In* S. Brantley, J. Kubicki, and A.F. White (eds.) Kinetics of water-rock interaction. Springer, New York.

Chorover, J., S. Choi, P. Rotenberg, R.J. Serne, N. Rivera, C. Strepka, A. Thompson, K.T. Mueller, and P.A. O'Day. 2008. Silicon control of strontium and cesium partitioning in hydroxide-weathered sediments. Geochim. Cosmochim. Acta 72:2024–2047.

Chorover, J., R. Kretzschmar, F. Garcia-Pichel, and D.L. Sparks. 2007. Soil biogeochemical processes within the critical zone. Elements 3:321–326.

Colin, F., G.H. Brimhall, D. Nahon, C.J. Lewis, A. Baronnet, and K. Danti. 1992. Equatorial rain-forest lateritic mantles—A geomembrane filter. Geology 20:523–526.

Coston, J.A., C.C. Fuller, and J.A. Davis. 1995. Pb2+ and Zn2+ Adsorption by a Natural aluminum-bearing and iron-bearing surface coating on an aquifer sand. Geochim. Cosmochim. Acta 59:3535–3547.

Curriero, F.C., J.A. Patz, J.B. Rose, and S. Lele. 2001. The association between extreme precipitation and waterborne disease outbreaks in the United States, 1948–1994. Am. J. Pub. Health 91:1194–1199.

Davis, J.A. 1982. Adsorption of natural dissolved organic-matter at the oxide water interface. Geochim. Cosmochim. Acta 46:2381–2393.

Davis, K.J., and A. Luttge. 2005. Quantifying the relationship between microbial attachment and mineral surface dynamics using vertical scanning interferometry (VSI). Am. J. Sci. 305:727–751.

Dec, J., and J.M. Bollag. 1997. Determination of covalent and noncovalent binding interactions between xenobiotic chemicals and soil. Soil Sci. 162:858–874.

Diallo, M.S., A. Simpson, P. Gassman, J.L. Faulon, J.H. Johnson, W.A. Goddard, and P.G. Hatcher. 2003. 3-D structural modeling of humic acids through experimental characterization, computer assisted structure elucidation and atomistic simulations. 1. Chelsea soil humic acid. Environ. Sci. Technol. 37:1783–1793.

Drewes, J.E., J. Hemming, S.J. Ladenburger, J. Schauer, and W. Sonzogni. 2005. An assessment of endocrine disrupting activity changes during wastewater treatment through the use of bioassays and chemical measurements. Water Environ. Res. 77:12–23.

Duckworth, O.W., and S.T. Martin. 2001. Surface complexation and dissolution of hematite by C-1—C-6 dicarboxylic acids at pH = 5. Geochim. Cosmochim. Acta 65:4289–4301.

Dunnivant, F.M., P.M. Jardine, D.L. Taylor, and J.F. Mccarthy. 1992. Transport of naturally-occurring dissolved organic-carbon in laboratory columns containing aquifer material. Soil Sci. Soc. Am. J. 56:437–444.

Ehlers, G.A.C., and A.P. Loibner. 2006. Linking organic pollutant (bio)availability with geosorbent properties and biomimetic methodology: A review of geosorbent characterisation. and (bio)availability prediction. Environ. Pollut. 141:494–512.

Ehrlich, H.L., and D. Newman. 2009. Geomicrobiology. CRC Press, Boca Raton, FL.

Feeney, D.S., J.W. Crawford, T. Daniell, P.D. Hallett, N. Nunan, K. Ritz, M. Rivers, and I.M. Young. 2006. Three-dimensional microorganization of the soil-root-microbe system. Microb. Ecol. 52:151–158.

Fendorf, S.E., M. Fendorf, D.L. Sparks, and R. Gronsky. 1992. Inhibitory mechanisms of Cr(III) oxidation by Δ-MnO_2. J. Colloid Interface Sci. 153:37–54.

Feng, X.J., A.J. Simpson, and M.J. Simpson. 2005. Chemical and mineralogical controls on humic acid sorption to clay mineral surfaces. Org. Geochem. 36:1553–1566.

Ferrer, I., E.T. Furlong, J.A. Leenheer, C.E. Rostad, and G.R. Aiken. 2002. Analysis of humic and fulvic acids by electrospray/ion trap tandem mass spectrometry. Am. Chem. Soc. Abst. 223:U514.

Filius, J.D., J.C.L. Meeussen, D.G. Lumsdon, T. Hiemstra, and W.H. Van Riemsdijk. 2003. Modeling the binding of fulvic acid by goethite: The speciation of adsorbed FA molecules. Geochim. Cosmochim. Acta 67:1463–1474.

Froberg, M., P.M. Jardine, P.J. Hanson, C.W. Swanston, D.E. Todd, J.R. Tarver, and C.T. Garten. 2007. Low dissolved organic carbon input from fresh litter to deep mineral soils. Soil Sci. Soc. Am. J. 71:347–354.

Fuller, C.C., J.A. Davis, and G.A. Waychunas. 1993. Surface-chemistry of ferrihydrite. 2. Kinetics of arsenate adsorption and coprecipitation. Geochim. Cosmochim. Acta 57:2271–2282.

Gabriel, M.C., and D.G. Williamson. 2004. Principal biogeochemical factors affecting the speciation and transport of mercury through the terrestrial environment. Environ. Geochem. Health 26:421–434.

Gao, X., D.W. Mettge, C. Ray, R.W. Harvey, and J. Chorover. 2009. Surface complexation of carboxylate adheres *Cryptosporidium parvum* oocysts to the hematite/water interface. Environ. Sci. Technol. 43:7423–7429.

Garcia-Pichel, F., and J. Belnap. 1996. Microenvironments and microscale productivity of cyanobacterial desert crusts. J. Phycol. 32:774–782.

Garcia-Pichel, F., S.L. Johnson, D. Youngkin, and J. Belnap. 2003. Small-scale vertical distribution of bacterial biomass and diversity in biological soil crusts from arid lands in the Colorado Plateau. Microb. Ecol. 46:312–321.

Geoghegan, M., J.S. Andrews, C.A. Biggs, K.E. Eboigbodin, D.R. Elliott, S. Rolfe, J. Scholes et al. 2008. The polymer physics and chemistry of microbial cell attachment and adhesion. Farad. Discuss. 139:85–103.

Gerin, P.A., M.J. Genet, A.J. Herbillon, and B. Delvaux. 2003. Surface analysis of soil material by X-ray photoelectron spectroscopy. Eur. J. Soil Sci. 54:589–603.

Ghosh, S., Z.Y. Wang, S. Kang, P.C. Bhowmik, and B.S. Xing. 2009. Sorption and fractionation of a peat derived humic acid by Kaolinite, Montmorillonite, and Goethite. Pedosphere 19:21–30.

Giesler, R., T. Andersson, L. Lovgren, and P. Persson. 2005. Phosphate sorption in aluminum- and iron-rich humus soils. Soil Sci. Soc. Am. J. 69:77–86.

Ginn, T.R., B.D. Wood, K.E. Nelson, T.D. Scheibe, E.M. Murphy, and T.P. Clement. 2002. Processes in microbial transport in the natural subsurface. Adv. Water Resour. 25:1017–1042.

Goyne, K.W., J. Chorover, J.D. Kubicki, A.R. Zimmerman, and S.L. Brantley. 2005. Sorption of the antibiotic ofloxacin to mesoporous and nonporous alumina and silica. J. Colloid Interface Sci. 283:160–170.

Grafe, M., M.J. Eick, and P.R. Grossl. 2001. Adsorption of arsenate (V) and arsenite (III) on goethite in the presence and absence of dissolved organic carbon. Soil Sci. Soc. Am. J. 65:1680–1687.

Gu, B.H., and J. Chen. 2003. Enhanced microbial reduction of Cr(VI) and U(VI) by different natural organic matter fractions. Geochim. Cosmochim. Acta 67:3575–3582.

Gu, B.H., J. Schmitt, Z. Chen, L.Y. Liang, and J.F. Mccarthy. 1995. Adsorption and desorption of different organic-matter fractions on iron-oxide. Geochim. Cosmochim. Acta 59:219–229.

Guggenberger, G., and K. Kaiser. 2003. Dissolved organic matter in soil: Challenging the paradigm of sorptive preservation. Geoderma 113:293–310.

Guggenberger, G., W. Zech, and H.R. Schulten. 1994. Formation and mobilization pathways of dissolved organic-matter—Evidence from chemical structural studies of organic-matter fractions in acid forest floor solutions. Org. Geochem. 21:51–66.

Guo, M.X., and J. Chorover. 2003. Transport and fractionation of dissolved organic matter in soil columns. Soil Sci. 168:108–118.

Gustafsson, J.P., P. Pechova, and D. Berggren. 2003. Modeling metal binding to soils: The role of natural organic matter. Environ. Sci. Technol. 37:2767–2774.

Gustafsson, J.P., I. Persson, D.B. Kleja, and J.W.J. Van Schaik. 2007. Binding of iron(III) to organic soils: EXAFS spectroscopy and chemical equilibrium modeling. Environ. Sci. Technol. 41:1232–1237.

Ha, J.Y., T.H. Yoon, Y.G. Wang, C.B. Musgrave, and G.E. Brown. 2008. Adsorption of organic matter at mineral/water interfaces: 7. ATR-FTIR and quantum chemical study of lactate interactions with hematite nanoparticles. Langmuir 24:6683–6692.

Hagedorn, F., and M. Bundt. 2002. The age of preferential flow paths. Geoderma 108:119–132.

Harvey, R.W., and J.N. Ryan. 2004. Use of PRDI bacteriophage in groundwater viral transport, inactivation, and attachment studies. FEMS Microbiol. Ecol. 49:3–16.

Hatcher, P.G., J.M. Bortiatynski, R.D. Minard, J. Dec, and J.M. Bollag. 1993. Use of high-resolution C-13 NMR to examine the enzymatic covalent binding of C-13-labeled 2,4-dichlorophenol to humic substances. Environ. Sci. Technol. 27:2098–2103.

Hatzinger, P.B., and M. Alexander. 1995. Effect of aging of chemicals in soil on their biodegradability and extractability. Environ. Sci. Technol. 29:537–545.

Hedges, J.I. 1988. Polymerization of humic substances in natural environments, p. 45–58. In F.H. Frimmel and R.F. Christman (eds.) Humic substances and their role in the environment. John Wiley & Sons, Chichester, U.K.

Hedges, J.I., G. Eglinton, P.G. Hatcher, D.L. Kirchman, C. Arnosti, S. Derenne, R.P. Evershed et al. 2000. The molecularly-uncharacterized component of nonliving organic matter in natural environments. Org. Geochem. 31:945–958.

Heidmann, I., I. Christl, and R. Kretzschma., 2005. Sorption of Cu and Pb to kaolinite-fulvic acid colloids: Assessment of sorbent interactions. Geochim. Cosmochim. Acta 69:1675–1686.

Herbert, B.E., and P.M. Bertsch. 1995. Characterization of dissolved and colloidal organic matter in soil, p. 63–88. In W.W. McFee and J.M. Kelly (eds.) Carbon forms and functions in soils. Soil Science Society of America, Madison, WI.

Hiemstra, T., and W.H. van Riemsdijk. 1999. Surface structural ion adsorption modeling of competitive binding of oxyanions by metal (hydr)oxides. J. Colloid Interface Sci. 210:182–193.

Hinsinger, P., A.G. Bengough, D. Vetterlein, and I.M. Young. 2009. Rhizosphere: Biophysics, biogeochemistry and ecological relevance. Plant Soil 321:117–152.

Hinsinger, P., C. Plassard, and B. Jaillard. 2006. Rhizosphere: A new frontier for soil biogeochemistry. J. Geochem. Expl. 88:210–213.

Hochella, M.F., S.K. Lower, P.A. Maurice, R.L. Penn, N. Sahai, D.L. Sparks, and B.S. Twining. 2008. Nanominerals, mineral nanoparticles, and Earth systems. Science 319:1631–1635.

Hoffland, E., R. Giesler, T. Jongmans, and N. van Breemen. 2002. Increasing feldspar tunneling by fungi across a north Sweden podzol chronosequence. Ecosystem 5:11–22.

Hornberger, G.M., A.L. Mills, and J.S. Herman. 1992. Bacterial transport in porous-media—Evaluation of a model using laboratory observations. Water Resour. Res. 28:915–923.

Houghton, R.A. 2005. The contemporary carbon cycle, p. 473–513. In W.H. Schlesinger (ed.) Biogeochemistry. Vol. 8. Elsevier, Dordrecht, the Netherlands.

Huang, P.M. 2004. Soil mineral-organic matter-microorganism interactions: Fundamentals and impacts. Adv. Agron. 82:391–472.

Huang, P.M., and J.J. Germida. 2002. Chemical and biochemical processes in the rhizosphere: Metal pollutants, p. 381–438. In P.M. Huang, J.-M. Bollag, and N. Senesi (eds.). Interactions between soil particles and microorganisms: Impact on the terrestrial ecosystem. John Wiley & Sons, New York.

Huang, W.L., P.A. Ping, Z.Q. Yu, and H.M. Fu. 2003. Effects of organic matter heterogeneity on sorption and desorption of organic contaminants by soils and sediments. Appl. Geochem. 18:955–972.

Hunter, K.S., Y.F. Wang, and P. Van Cappellen. 1998. Kinetic modeling of microbially-driven redox chemistry of subsurface environments: Coupling transport, microbial metabolism and geochemistry. J. Hydrol. 209:53–80.

Ikan, R., Y. Rubinsztain, A. Nissenbaum, and I.R. Kaplan. 1996. Geochemical aspects of the Maillard reaction, p. 1–25. In R. Ikan (ed.) The Maillard reaction: consequences for the chemical and life sciences. John Wiley & Sons, Chichester, U.K.

Islam, F.S., A.G. Gault, C. Boothman, D.A. Polya, J.M. Charnock, D. Chatterjee, and J. R. Lloyd. 2004. Role of metal-reducing bacteria in arsenic release from Bengal delta sediments. Nature 430:68–71.

Johnson, S.L., C.R. Budinoff, J. Belnap, and F. Garcia-Pichel. 2005. Relevance of ammonium oxidation within biological soil crust communities. Environ. Microbiol. 7:1–12.

Johnson, S.L., S. Neuer, and F. Garcia-Pichel. 2007. Export of nitrogenous compounds due to incomplete cycling within biological soil crusts of arid lands. Environ. Microbiol. 9:680–689.

Jokic, A., A.I. Frenkel, M.A. Vairavamurthy, and P.M. Huang. 2001. Birnessite catalysis of the Maillard reaction: Its significance in natural humification. Geophys. Res. Lett. 28:3899–3902.

Jokic, A., M.C. Wang, C. Liu, A.I. Frenkel, and P.M. Huang. 2004. Integration of the polyphenol and Maillard reactions into a unified abiotic pathway for humification in nature: The role of delta-MnO_2. Org. Geochem. 35:747–762.

Jones, D.L. 1998. Organic acids in the rhizosphere—A critical review. Plant Soil 205:25–44.

Jones, A.M., R.N. Collins, J. Rose, and T.D. Waite. 2009. The effect of silica and natural organic matter on the Fe(II)-catalysed transformation and reactivity of Fe(III) minerals. Geochim. Cosmochim. Acta 73:4409–4422.

Jones, D.L., P.G. Dennis, A.G. Owen, and P.A.W. van Hees. 2003. Organic acid behavior in soils—Misconceptions and knowledge gaps. Plant Soil 248:31–41.

Kaiser, K., and G. Guggenberger. 2000. The role of DOM sorption to mineral surfaces in the preservation of organic matter in soils. Org. Geochem. 31:711–725.

Kaiser, K., and W. Zech. 1998. Soil dissolved organic matter sorption as influenced by organic and sesquioxide coatings and sorbed sulfate. Soil Sci. Soc. Am. J. 62:129–136.

Kalbitz, K., S. Solinger, J.H. Park, B. Michalzik, and E. Matzner. 2000. Controls on the dynamics of dissolved organic matter in soils: A review. Soil Sci. 165:277–304.

Kang, S.H., D. Amarasiriwardena, and B.S. Xin., 2008. Effect of dehydration on dicarboxylic acid coordination at goethite/water interface. Coll. Surf. A—Physicochem. Eng. Asp. 318:275–284.

Kappler, A., and S.B. Haderlein. 2003. Natural organic matter as reductant for chlorinated aliphatic pollutants. Environ. Sci. Technol. 37:2714–2719.

Kelleher, B.P., and A.J. Simpson. 2006. Humic substances in soils: Are they really chemically distinct? Environ. Sci. Technol. 40:4605–4611.

Kelly, E.F., O.A. Chadwick, and T.E. Hilinski. 1998. The effect of plants on mineral weathering. Biogeochemistry 42:21–53.

Kemper, N. 2008. Veterinary antibiotics in the aquatic and terrestrial environment. Ecol. Ind. 8:1–13.

Kinney, C.A., E.T. Furlong, S.D. Zaugg, M.R. Burkhardt, S.L. Werner, J. Cahill, and G. R. Jorgensen. 2006. Survey of organic wastewater contaminants in biosolids destined for land application. Environ. Sci. Technol. 40:7207–7215.

Kleber, M., P. Sollins, and R. Sutton. 2007. A conceptual model of organo-mineral interactions in soils: Self-assembly of organic molecular fragments into zonal structures on mineral surfaces. Biogeochemistry 85:9–24.

Kogel-Knabner, I., G. Guggenberger, M. Kleber, E. Kandeler, K. Kalbitz, S. Scheu, K. Eusterhues, and P. Leinweber. 2008. Organo-mineral associations in temperate soils: Integrating biology, mineralogy, and organic matter chemistry. Zeit. Pflanzenernah. Bodenk. 171:61–82.

Kolpin, D.W., E.T. Furlong, M.T. Meyer, E.M. Thurman, S.D. Zaugg, L.B. Barber, and H.T. Buxton. 2002. Pharmaceuticals, hormones, and other organic wastewater contaminants in US streams, 1999–2000: A national reconnaissance. Environ. Sci. Technol. 36:1202–1211.

Konhauser, K.O. 1997. Bacterial iron biomineralisation in nature. FEMS Microbiol. Rev. 20:315–326.

Koopal, L.K., T. Saito, J.P. Pinheiro, and W.H. van Riemsdijk. 2005. Ion binding to natural organic matter: General considerations and the NICA-Donnan model. Colloids Surf. A Physicochem. Eng. Asp. 265:40–54.

Kraemer, S.M., D.E. Crowley, and R. Kretzschmar. 2006. Geochemical aspects of phytosiderophore-promoted iron acquisition by plants. Adv. Agron. 91:1–46.

Kretzschmar, R., M. Borkovec, D. Grolimund, and M. Elimelech. 1999. Mobile subsurface colloids and their role in contaminant transport. Adv. Agron. 66:121–193.

Kretzschmar, R., H. Holthoff, and H. Stiche. 1998. Influence of pH and humic acid on coagulation kinetics of kaolinite: A dynamic light scattering study. J. Colloid Interface Sci. 202:95–103.

Kretzschmar, R., W.P. Robarge, and A. Amoozegar. 1995. Influence of natural organic-matter on colloid transport through saprolite. Water Resour. Res. 31:435–445.

Kretzschmar, R., W.P. Robarge, A. Amoozegar, and M.J. Vepraskas. 1997. Biotite alteration to halloysite and kaolinite in soil-saprolite profiles developed from mica schist and granite gneiss. Geoderma 75:155–170.

Kubicki, J.D., M.J. Itoh, L.M. Schroeter, and S.E. Apitz. 1997. Bonding mechanisms of salicylic acid adsorbed onto illite clay: An ATR-FTIR and molecular orbital study. Environ. Sci. Technol. 31:1151–1156.

Kummerer, K. 2001. Drugs in the environment: Emission of drugs, diagnostic aids and disinfectants into wastewater by hospitals in relation to other sources—A review. Chemosphere 45:957–969.

Landeweert, R., E. Hoffland, R.D. Finlay, T.W. Kuyper, and N. van Breemen. 2001. Linking plants to rocks: Ectomycorrhizal fungi mobilize nutrients from minerals. Trends Ecol. Evol. 16:248–254.

Larsen, O., and D. Postma. 2001. Kinetics of reductive bulk dissolution of lepidocrocite, ferrihydrite, and goethite. Geochim. Cosmochim. Acta 65:1367–1379.

Laverman, A.M., C. Meile, P. Van Cappellen, and E.B.A. Wieringa. 2007. Vertical distribution of denitrification in an estuarine sediment: Integrating sediment flow through reactor experiments and microprofiling via reactive transport modeling. Appl. Environ. Microbiol. 73:40–47.

Lee, Y.J., and R.J. Reeder. 2006. The role of citrate and phthalate during Co(II) coprecipitation with calcite. Geochim. Cosmochim. Acta 70:2253–2263.

Leenheer, J.A., and J.P. Croue. 2003. Characterizing aquatic dissolved organic matter. Environ. Sci. Technol. 37:18A–26A.

Leenheer, J.A., C.E. Rostad, P.M. Gates, E.T. Furlong, and I. Ferre. 2001. Molecular resolution and fragmentation of fulvic acid by electrospray ionization/multistage tandem mass spectrometry. Anal. Chem. 73:1461–1471.

Legout, C., J. Molenat, S. Lefebvre, P. Marmonier, and L. Aquilina. 2005. Investigation of biogeochemical activities in the soil and unsaturated zone of weathered granite. Biogeochemistry 75:329–350.

Liermann, L.J., B.E. Kalinowski, S.L. Brantley, and J.G. Ferry. 2000. Role of bacterial siderophores in dissolution of hornblende. Geochim. Cosmochim. Acta 64:587–602.

Lin, H.T., M.C. Wang, and G.C. Li. 2004. Complexation of arsenate with humic substance in water extract of compost. Chemosphere 56:1105–1112.

Lovley, D.R. 2003. Cleaning up with genomics: Applying molecular biology to bioremediation. Nat. Rev. Microbiol. 1:35–44.

Lovley, D.R., M.J. Baedecker, D.J. Lonergan, I.M. Cozzarelli, E.J.P. Phillips, and D.I. Siegel. 1989. Oxidation of aromatic contaminants coupled to microbial iron reduction. Nature 339:297–300.

Lovley, D.R., and J.D. Coates. 1997. Bioremediation of metal contamination. Cur. Opin. Biotechnol. 8:285–289.

Lovley, D.R., J.D. Coates, E.L. Blunt-Harris, E.J.P. Phillips, and J.C. Woodward. 1996. Humic substances as electron acceptors for microbial respiration. Nature 382:445–448.

Lovley, D.R., J.L. Fraga, E.L. Blunt-Harris, L.A. Hayes, E.J.P. Phillips, and J.D. Coates. 1998. Humic substances as a mediator for microbially catalyzed metal reduction. Acta Hydrochim. Hydrobiol. 26:152–157.

Lu, Y.F., and J.J. Pignatello. 2004. Sorption of apolar aromatic compounds to soil humic acid particles affected by aluminum(III) ion cross-linking. J. Environ. Qual. 33:1314–1321.

Lupwayi, N.Z., M.A. Arshad, W.A. Rice, and G.W. Clayton. 2001. Bacterial diversity in water-stable aggregates of soils under conventional and zero tillage management. Appl. Soil Ecol. 16:251–261.

Luthy, R.G., G.R. Aiken, M.L. Brusseau, S.D. Cunningham, P.M. Gschwend, J.J. Pignatello, M. Reinhard, S.J. Traina, W.J. Weber, and J.C. Westall. 1997. Sequestration of hydrophobic organic contaminants by geosorbents. Environ. Sci. Technol. 31:3341–3347.

Manceau, A., N. Tamura, R.S. Celestre, A.A. MacDowell, N. Geoffroy, G. Sposito, and H.A. Padmore. 2003. Molecular-scale speciation of Zn and Ni in soil ferromanganese nodules from loess soils of the Mississippi Basin. Environ. Sci. Technol. 37:75–80.

Marschner, H., V. Romheld, W.J. Horst, and P. Martin. 1986. Root-induced changes in the Rhizosphere—Importance for the mineral-nutrition of plants. Zeit. Pflanzenernah. Bodenk. 149:441–456.

Matthiessen, A. 1995. Determining the redox capacity of humic substances as a function of pH. Vom Wasser 84:229–235.

Maurice, P.A., and M.F. Hochella. 2008. Nanoscale particles and processes: A new dimension in soil science. Adv. Agron. 100:123–153.

Mayer, L.M. 1994. Surface area control of organic carbon accumulation in continental shelf sediments. Geochim. Cosmochim. Acta 58:1271–1284.

McBride, M.B. 2003. Toxic metals in sewage sludge-amended soils: Has promotion of beneficial use discounted the risks? Adv. Environ. Res. 8:5–19.

McCarthy, J.F., T.M. Williams, L.Y. Liang, P.M. Jardine, L.W. Jolley, D.L. Taylor, A.V. Palumbo, and L.W. Cooper. 1993. Mobility of natural organic-matter in a sandy aquifer. Environ. Sci. Technol. 27:667–676.

McDowell, W.H. 2003. Dissolved organic matter in soils—Future directions and unanswered questions. Geoderma 113:179–186.

McGeehan, S.L., S.E. Fendorf, and D.V. Naylor. 1998. Alteration of arsenic sorption in flooded-dried soils. Soil Sci. Soc. Am. J. 62:828–833.

McLaren, R.G., C.A. Backes, A.W. Rate, and R.S. Swift. 1998. Cadmium and cobalt desorption kinetics from soil clays: Effect of sorption period. Soil Sci. Soc. Am. J. 62:332–337.

Mikutta, R., M. Kleber, M.S. Torn, and R. Jahn. 2006. Stabilization of soil organic matter: Association with minerals or chemical recalcitrance? Biogeochemistry 77:25–56.

Mikutta, R., G.E. Schaumann, D. Gildemeister, S. Bonneville, M.G. Kramer, J. Chorover, O.A. Chadwick, and G. Guggenberge. 2009. Biogeochemistry of mineral-organic associations across a long-term mineralogical soil gradient (0.3–4100 kyr), Hawaiian Islands. Geochim. Cosmochim. Acta 73:2034–2060.

Monger, H.C., L.A. Daugherty, W.C. Lindemann, and C.M. Liddell. 1991. Microbial precipitation of pedogenic calcite. Geology 19:997–1000.

Newman, D.K., and J.F. Banfield. 2002. Geomicrobiology: How molecular-scale interactions underpin biogeochemical systems. Science 296:1071–1077.

NRC. 2001. Basic research opportunities in earth sciences. National Academies Press, National Research Council, Washington, DC.

Nugent, M.A., S.L. Brantley, C.G. Pantano, and P.A. Maurice. 1998. The influence of natural mineral coatings on feldspar weathering. Nature 395:588–591.

Nurmi, J.T., and P.G. Tratnyek. 2002. Electrochemical properties of natural organic matter (NOM), fractions of NOM, and model biogeochemical electron shuttles. Environ. Sci. Technol. 36:617–624.

Omoike, A., and J. Chorover. 2006. Adsorption to goethite of extracellular polymeric substances from *Bacillus subtilis*. Geochim. Cosmochim. Acta 70:827–838.

Omoike, A., J. Chorover, K.D. Kwon, and J.D. Kubicki. 2004. Adhesion of bacterial exopolymers to alpha-FeOOH: Inner-sphere complexation of phosphodiester groups. Langmuir 20:11108–11114.

Oze, C., D.K. Bird, and S. Fendor. 2007. Genesis of hexavalent chromium from natural sources in soil and groundwater. Proc. Natl. Acad. Sci. U.S.A. 104:6544–6549.

Palmer, N.E., and R. von Wandruszka. 2009. The influence of aggregation on the redox chemistry of humic substances. Environ. Chem. 6:178–184.

Pankhurst, C.E., A. Pierret, B.G. Hawke, and J.M. Kirby. 2002. Microbiological and chemical properties of soil associated with macropores at different depths in a red-duplex soil in NSW Australia. Plant Soil 238:11–20.

Parfitt, R.L., A.R. Fraser, and V.C. Farmer. 1977. Adsorption on hydrous oxides. 3. Fulvic-acid and humic-acid on goethite, gibbsite and imogolite. J. Soil Sci. 28:289–296.

Parfitt, R.L., B.K.G. Theng, J.S. Whitton, and T.G. Shepherd. 1997. Effects of clay minerals and land use on organic matter pools. Geoderma 75:1–12.

Parikh, S.J., and J. Chorover. 2006. ATR-FTIR spectroscopy reveals bond formation during bacterial adhesion to iron oxide. Langmuir 22:8492–8500.

Pedersen, H.D., D. Postma, R. Jakobsen. and O. Larsen. 2005. Fast transformation of iron oxyhydroxides by the catalytic action of aqueous Fe(II). Geochim. Cosmochim. Acta 69:3967–3977.

Perdrial, N., J. Perdrial, J.-E. Delphin, F. Elsass, and N. Liewig. 2010. Temporal and spatial monitoring of mobile nanoparticles in a vineyard soil: Evidence of nanoaggregate formation. Eur. J. Soil Sci. 61:456–468.

Pett-Ridge, J., W.L. Silver, and M.K. Firestone. 2006. Redox fluctuations frame microbial community impacts on N-cycling rates in a humid tropical forest soil. Biogeochemistry 81:95–110.

Piccolo, A. 2001. The supramolecular structure of humic substances. Soil Sci. 166:810–832.

Piccolo, A. 2002. The supramolecular structure of humic substances: A novel understanding of humus chemistry and implications in soil science. Adv. Agron. 75:57–134.

Piccolo, A., and M. Spiteller. 2003. Electrospray ionization mass spectrometry of terrestrial humic substances and their size fractions. Anal. Bioanal. Chem. 377:1047–1059.

Polizzotto, M.L., B.D. Kocar, S.G. Benner, M. Sampson, and S. Fendorf. 2008. Near-surface wetland sediments as a source of arsenic release to ground water in Asia. Nature 454:505–508.

Qafoku, N.P., C.C. Ainsworth, J.E. Szecsody, D.L. Bish, J.S. Young, D.E. McCready, and O.S. Qafoku. 2003. Aluminum effect on dissolution and precipitation under hyperalkaline conditions: II. Solid phase transformations. J. Environ. Qual. 32:2364–2372.

Ranjard, L., F. Poly, J. Combrisson, A. Richaume, F. Gourbiere, J. Thioulouse, and S. Nazaret. 2000. Heterogeneous cell density and genetic structure of bacterial pools associated with various soil microenvironments as determined by enumeration and DNA fingerprinting approach (RISA). Microb. Ecol. 39:263–272.

Rasmussen, C., R.J. Southard, and W.R. Horwath. 2006. Mineral control of organic carbon mineralization in a range of temperate conifer forest soils. Global Change Biol. 12:834–847.

Redman, A.D., D.L. Macalady, and D. Ahmann. 2002. Natural organic matter affects arsenic speciation and sorption onto hematite. Environ. Sci. Technol. 36:2889–2896.

Redman, J.A., S.L. Walker, and M. Elimelech. 2004. Bacterial adhesion and transport in porous media: Role of the secondary energy minimum. Environ. Sci. Technol. 38:1777–1785.

Reedy, O.C., P.M. Jardine, G.V. Wilson, and H.M. Selim. 1996. Quantifying the diffusive mass transfer of nonreactive solutes in columns of fractured saprolite using flow interruption. Soil Sci. Soc. Am. J. 60:1376–1384.

Reemtsma, T., and A. These. 2003. On-line coupling of size exclusion chromatography with electrospray ionization-tandem mass spectrometry for the analysis of aquatic fulvic and humic acids. Anal. Chem. 75:1500–1507.

Reemtsma, T., A. These, A. Springer, and M. Linscheid. 2006. Fulvic acids as transition state of organic matter: Indications from high resolution mass spectrometry. Environ. Sci. Technol. 40:5839–5845.

Reichard, P.U., R. Kretzschmar, and S.M. Kraemer. 2007. Rate laws of steady-state and non-steady-state ligand-controlled dissolution of goethite. Colloids Surf. A Physicochem. Eng. Asp. 306:22–28.

Richens, D.T. 1997. The chemistry of aqua ions. John Wiley & Sons, New York.

Roden, E.E. 2008. Microbial controls on geochemical kinetics 1: Fundamentals and case study on microbial Fe(III) oxide reduction, p. 335–415. In S.L. Brantley, J.D. Kubicki, and A.F. White (eds.) Kinetics of water-rock interaction. Springer, New York.

Roden, E.E., and M.M. Urrutia. 1999. Ferrous iron removal promotes microbial reduction of crystalline iron(III) oxides. Environ. Sci. Technol. 33:1847–1853.

Romkens, P.F., J. Bril, and W. Salomons. 1996. Interaction between Ca^{2+} and dissolved organic carbon: Implications for metal mobilization. Appl. Geochem. 11:109–115.

Romkens, P.F.A.M., and J. Dolfing. 1998. Effect of Ca on the solubility and molecular size distribution of DOC and Cu binding in soil solution samples. Environ. Sci. Technol. 32:363–369.

Rostad, C.E., and J.A. Leenheer. 2004. Factors that affect molecular weight distribution of Suwannee river fulvic acid as determined by electrospray ionization/mass spectrometry. Anal. Chim. Acta 523:269–278.

Sander, M., Y.F. Lu, and J.J. Pignatello. 2006. Conditioning-annealing studies of natural organic matter solids linking irreversible sorption to irreversible structural expansion. Environ. Sci. Technol. 40:170–178.

Schade, J.D., E. Marti, J.R. Welter, S.G. Fisher, and N.B. Grimm. 2002. Sources of nitrogen to the riparian zone of a desert stream: Implications for riparian vegetation and nitrogen retention. Ecosystem 5:68–79.

Schafer, A., P. Ustohal, H. Harms, F. Stauffer, T. Dracos, and A.J.B. Zehnder. 1998. Transport of bacteria in unsaturated porous media. J. Contam. Hydrol. 33:149–169.

Scheckel, K.G., A.C. Scheinost, R.G. Ford, and D.L. Sparks. 2000. Stability of layered Ni hydroxide surface precipitates—A dissolution kinetics study. Geochim. Cosmochim. Acta 64:2727–2735.

Scheidegger, A.M., G.M. Lamble, and D.L. Spark., 1996. Investigation of Ni sorption on pyrophyllite: An XAFS study. Environ. Sci. Technol. 30:548–554.

Schiff, S.L., R. Aravena, S.E. Trumbore, and P.J. Dillon. 1990. Dissolved organic-carbon cycling in forested watersheds—A carbon isotope approach. Water Resour. Res. 26:2949–2957.

Schindler, P.W., and W. Stumm. 1987. The surface chemistry of oxides, hydroxides and oxyhydroxide minerals. *In* W. Stumm (ed.) Aquatic surface chemistry. Wiley Interscience, New York.

Schumacher, M., I. Christl, A.C. Scheinost, C. Jacobsen, and R. Kretzschmar. 2005. Chemical heterogeneity of organic soil colloids investigated by scanning transmission X-ray microscopy and C-1s NEXAFS microspectroscopy. Environ. Sci. Technol. 39:9094–9100.

Schwarzenbach, R.P., R. Stierli, K. Lanz, and J. Zeyer. 1990. Quinone and iron porphyrin mediated reduction of nitroaromatic compounds in homogeneous aqueous-solution. Environ. Sci. Technol. 24:1566–1574.

Scott, D.T., D.M. McKnight, E.L. Blunt-Harris, S.E. Kolesar, and D.R. Lovley. 1998. Quinone moieties act as electron acceptors in the reduction of humic substances by humics-reducing microorganisms. Environ. Sci. Technol. 32:2984–2989.

Scow, K.M., and K.A. Hicks. 2005. Natural attenuation and enhanced bioremediation of organic contaminants in groundwater. Curr. Opin. Biotechnol. 16:246–253.

Semple, K.T., A.W.J. Morriss, and G.I. Paton. 2003. Bioavailability of hydrophobic organic contaminants in soils: Fundamental concepts and techniques for analysis. Eur. J. Soil Sci. 54:809–818.

Sexstone, A.J., N.P. Revsbech, T.B. Parkin, and J.M. Tiedje. 1985. Direct measurement of oxygen profiles and denitrification rates in soil aggregates. Soil Sci. Soc. Am. J. 49:645–651.

Silver, W.L., A.E. Lugo, and M. Keller. 1999. Soil oxygen availability and biogeochemistry along rainfall and topographic gradients in upland wet tropical forest soils. Biogeochemistry 44:301–328.

Simpson, A.J. 2002. Determining the molecular weight, aggregation, structures and interactions of natural organic matter using diffusion ordered spectroscopy. Magn. Reson. Chem. 40:S72–S82.

Simpson, A.J., W.L. Kingery, M.H.B. Hayes, M. Spraul, E. Humpfer, P. Dvortsak, R. Kerssebaum, M. Godejohann, and M. Hofmann. 2002. Molecular structures and associations of humic substances in the terrestrial environment. Naturwissenschaften 89:84–88.

Simunek, J., N.J. Jarvis, M.T. van Genuchten, and A. Gardenas. 2003. Review and comparison of models for describing non-equilibrium and preferential flow and transport in the vadose zone. J. Hydrol. 272:14–35.

Six, J., E.T. Elliott, and K. Paustian. 2000. Soil macroaggregate turnover and microaggregate formation: A mechanism for C sequestration under no-tillage agriculture. Soil Biol. Biochem. 32:2099–2103.

Smedley, P.L. and D.G. Kinniburgh, 2002. A review of the source, behaviour and distribution of arsenic in natural waters. Appl. Geochem. 17:517–568.

Snyder, S.A., P. Westerhoff, Y. Yoon, and D.L. Sedlak. 2003. Pharmaceuticals, personal care products, and endocrine disruptors in water: Implications for the water industry. Environ. Eng. Sci. 20:449–469.

Sollins, P., C. Swanston, M. Kleber, T. Filley, M. Kramer, S. Crow, B.A. Caldwell, K. Lajtha, and R. Bowden. 2006. Organic C and N stabilization in a forest soil: Evidence from sequential density fractionation. Soil Biol. Biochem. 38:3313–3324.

Sparks, D.L. 2005. Metal and oxyanion sorption on naturally occurring oxide and clay mineral surfaces, p. 3–36. *In* V.H. Grassian (ed.) Environmental catalysis. Taylor & Francis, Boca Raton, FL.

Sparks, D.L. 2006. Advances in elucidating biogeochemical processes in soils: It is about scale and interfaces. J. Geochem. Explor. 88:243–245.

Sposito, G. 2004. The surface chemistry of natural particles. Oxford University Press, New York.

Sposito, G. 2008. The chemistry of soils. Oxford University Press, New York.

Stillings, L.L., J.I. Drever, S.L. Brantley, Y.T. Sun, and R. Oxburgh. 1996. Rates of feldspar dissolution at pH 3-7 with 0-8 mM oxalic acid. Chem. Geol. 132:79–89.

Struyk, Z., and G. Sposito. 2001. Redox properties of standard humic acids. Geoderma 102:329–346.

Stumm, W. 1997. Reactivity at the mineral-water interface: Dissolution and inhibition. Colloids Surf. A Physicochem. Eng. Asp. 120:143–166.

Stumm, W., and J.J. Morgan. 1996. Aquatic chemistry: Chemical equilibria and rates in natural waters. John Wiley & Sons, New York.

Sun, X.H., and H.E. Doner. 1996. An investigation of arsenate and arsenite bonding structures on goethite by FTIR. Soil Sci. 161:865–872.

Sutton, R., and G. Sposito. 2005. Molecular structure in soil humic substances: The new view. Environ. Sci. Technol. 39:9009–9015.

Taylor, L.L., J.R. Leake, J. Quirk, K. Hardy, S.A. Banwart, and D.J. Beerling. 2009. Biological weathering and the long-term carbon cycle: Integrating mycorrhizal evolution and function into the current paradigm. Geobiology 7:171–191.

Tebo, B.M., J.R. Bargar, B.G. Clement, G.J. Dick, K.J. Murray, D. Parker, R. Verity, and S.M. Webb. 2004. Biogenic manganese oxides: Properties and mechanisms of formation. Annu. Rev. Earth Planet. Sci. 32:287–328.

Templeton, A.S., A.M. Spormann, and G.E. Brown. 2003. Speciation of Pb(II) sorbed by Burkholderia cepacia/goethite composites. Environ. Sci. Technol. 37:2166–2172.

Thiele-Bruhn, S. 2003. Pharmaceutical antibiotic compounds in soils—A review. Zeit. Pflanzenernah. Bodenk. 166:145–167.

Thompson, A., O.A. Chadwick, S. Boman, and J. Chorover. 2006a. Colloid mobilization during soil iron redox oscillations. Environ. Sci. Technol. 40:5743–5749.

Thompson, A., O.A. Chadwick, D.G. Rancourt, and J. Chorover. 2006b. Iron-oxide crystallinity increases during soil redox oscillations. Geochim. Cosmochim. Acta 70:1710–1727.

Tipping, E. 1981. The adsorption of aquatic humic substances by iron-oxides. Geochim. Cosmochim. Acta 45:191–199.

Tipping, E. 2002. Cation binding by humic substances. Cambridge University Press, Cambridge, U.K.

Tokunaga, T.K., J.M. Wan, M.K. Firestone, T.C. Hazen, K.R. Olson, D.J. Herman, S.R. Sutton, and A. Lanzirotti. 2003a. In situ reduction of chromium(VI) in heavily contaminated soils through organic carbon amendment. J. Environ. Qual. 32:1641–1649.

Tokunaga, T.K., J.M. Wan, T.C. Hazen, E. Schwartz, M.K. Firestone, S.R. Sutton, M. Newville, K.R. Olson, A. Lanzirotti, and W. Rao. 2003b. Distribution of chromium contamination and microbial activity in soil aggregates. J. Environ. Qual. 32:541–549.

Tokunaga, T.K., J.M. Wan, Y.M. Kim, S.R. Sutton, M. Newville, A. Lanzirotti, and W. Rao. 2008. Real-time X-ray absorption spectroscopy of uranium, iron, and manganese in contaminated sediments during bioreduction. Environ. Sci. Technol. 42:2839–2844.

Tokunaga, T.K., J.M. Wan, A. Lanzirotti, S.R. Sutton, M. Newville, and W. Rao. 2007. Long-term stability of organic carbon-stimulated chromate reduction in contaminated soils and its relation to manganese redox status. Environ. Sci. Technol. 41:4326–4331.

Tolls, J. 2001. Sorption of veterinary pharmaceuticals in soils: A review. Environ. Sci. Technol. 35:3397–3406.

Toner, B.M., A. Manceau, M. Villalobos, S.C. Fakra, S.M. Webb, M.M. Marcus, J.R. Bargar, B.M. Tebo, and G. Sposito. 2004. Reactivity of a bacterial manganese oxide produced within a biofilm. Am. Chem. Soc. Abst. 227:U1217.

Torn, M.S., S.E. Trumbore, O.A. Chadwick, P.M. Vitousek, and D.M. Hendricks. 1997. Mineral control of soil organic carbon storage and turnover. Nature 389:170–173.

Valls, M., and V. de Lorenzo. 2002. Exploiting the genetic and biochemical capacities of bacteria for the remediation of heavy metal pollution. FEMS Microbiol. Rev. 26:327–338.

Voegelin, A., and R. Kretzschmar. 2005. Formation and dissolution of single and mixed Zn and Ni precipitates in soil: Evidence from column experiments and extended X-ray absorption fine structure spectroscopy. Environ. Sci. Technol. 39:5311–5318.

von Luetzow, M., I. Kogel-Knabner, B. Ludwig, E. Matzner, H. Flessa, K. Ekschmitt, G. Guggenberger, B. Marschner, and K. Kalbitz. 2008. Stabilization mechanisms of organic matter in four temperate soils: Development and application of a conceptual model. Zeit. Pflanzenernah. Bodenk. 171:111–124.

Waychunas, G.A., B.A. Rea, C.C. Fuller, and J.A. Davis. 1993. Surface-chemistry of ferrihydrite. 1. EXAFS studies of the geometry of coprecipitated and adsorbed arsenate. Geochim. Cosmochim. Acta 57:2251–2269.

Weber, K.A., L.A. Achenbach, and J.D. Coates. 2006. Microorganisms pumping iron: Anaerobic microbial iron oxidation and reduction. Nat. Rev. Microbiol. 4:752–764.

Weng, L., W.H. van Riemsdijk, and T. Hiemstra. 2009. Effects of fulvic and humic acids on arsenate adsorption to goethite: Experiments and modeling. Environ. Sci. Technol. 43:7198–7204.

Weng, L., W.H. van Riemsdijk, L.K. Koopal, and T. Hiemstra. 2006. Ligand and charge distribution (LCD) model for the description of fulvic acid adsorption to goethite. J. Colloid Interface Sci. 302:442–457.

Wershaw, R.L. 1993. Model for humus in soils and sediments. Environ. Sci. Technol. 27:814–816.

Wilkins, M.J., F.R. Livens, D.J. Vaughan, and J.R. Lloyd. 2006. The impact of Fe(III)-reducing bacteria on uranium mobility. Biogeochemistry 78:125–150.

Wogelius, R.A., P.M. Morris, M.A. Kertesz, E. Chardon, A.I.R. Stark, M. Warren, and J.R. Brydie. 2007. Mineral surface reactivity and mass transfer in environmental mineralogy. Eur. J. Miner. 19:297–307.

Xu, H., B. Allard, and A. Grimvall. 1988. Influence of Ph and organic-substance on the adsorption of as(V) on geologic materials. Water Air Soil Pollut. 40:293–305.

Yee, N., S. Shaw, L.G. Benning, and T.H. Nguyen. 2006. The rate of ferrihydrite transformation to goethite via the Fe(II) pathway. Am. Mineral. 91:92–96.

Yoon, T.H., S.B. Johnson, and G.E. Brown. 2005. Adsorption of organic matter at mineral/water interfaces. IV. Adsorption of humic substances at boehmite/water interfaces and impact on boehmite dissolution. Langmuir 21:5002–5012.

Young, I.M., and J.W. Crawford. 2004. Interactions and self-organization in the soil-microbe complex. Science 304:1634–1637.

Young, I.M., J.W. Crawford, N. Nunan, W. Otten, and A. Spiers. 2008. Microbial distribution in soils: Physics and scaling. Adv. Agron. 100:81–121.

Zachara, J.M., R.J. Serne, M. Freshley, F. Mann, F. Anderson, M. Wood, T. Jones, and D. Myers. 2007. Geochemical processes controlling migration of high level wastes in Hanford's vadose zone. Vadose Zone J. 4:985–1003.

Zhou, J.L., R. Liu, A. Wilding, and A. Hibberd. 2007. Sorption of selected endocrine disrupting chemicals to different aquatic colloids. Environ. Sci. Technol. 41:206–213.

Zhou, Q.H., P.A. Maurice, and S.E. Cabaniss. 2001. Size fractionation upon adsorption of fulvic acid on goethite: Equilibrium and kinetic studies. Geochim. Cosmochim. Acta 65:803–812.

II

Soil Fertility and Plant Nutrition

James J. Camberato
Purdue University

11 **Bioavailability of N, P, K, Ca, Mg, S, Si, and Micronutrients** *Nanthi Bolan, Ross Brennan, Dedik Budianta, James J. Camberato, Ravi Naidu, William L. Pan, Andrew Sharpley, Donald L. Sparks, and Malcolm E. Sumner*.....**11**-1
Bioavailability of Nitrogen • References • Phosphorus Availability • References • Bioavailability of Soil Potassium • References • Bioavailability of Calcium, Magnesium, Sulfur, and Silicon • References • Bioavailability of Micronutrients • References

12 **Soil Acidity and Liming** *T. Jot Smyth*..**12**-1
Introduction • Nature of Soil Acidity • Acid Soil Constraints to Plant Growth • Lime Requirement Methods • Lime Material Characteristics and Application • References

13 **Soil Fertility Evaluation** *J. Thomas Sims and Joshua McGrath*...**13**-1
Introduction • Soil Testing • Plant Testing • Soil Fertility Evaluation: The Future • References

14 **Fundamentals of Fertilizer Application** *David B. Mengel and George W. Rehm*..**14**-1
Introduction • Application of Mobile Nutrients • Application of Immobile Nutrients • Placement of P and K Fertilizers • Calcium and Magnesium • Micronutrients • References

15 **Nutrient and Water Use Efficiency** *Kefyalew Girma and William R. Raun*..**15**-1
Introduction • Nutrient Use Efficiency • Bray's Nutrient Mobility Concept: What's Nutrient Efficiency Got to Do with It? • Status of Nutrient Use Efficiency • Causes of Low Nutrient Use Efficiency • Strategies to Improve Nutrient Use Efficiency • Water Use Efficiency • Critical Factors Affecting Water Use Efficiency by Crops • Strategies to Improve Water Use Efficiency • Conclusions • References

16 **Nutrient Interactions in Soil Fertility and Plant Nutrition** *William L. Pan*..**16**-1
Historical Perspective • Conceptual and Quantitative Frameworks of Nutrient Interaction Mechanisms • Conclusions • References

THE BIOLOGICAL, CHEMICAL, AND PHYSICAL FACTORS that control the availability of plant essential and beneficial nutrients as well as potentially phytotoxic and zootoxic microelements are reviewed in this part. Nutrient and elemental availability is important not only as it affects crop plant productivity, but as it determines the potential movement of nutrients outside the boundaries of the crop field and their impact on air and water resources and native ecosystems. The complexity of soil acidity and nutrient interactions as they affect plant uptake and utilization of nutrients are also elaborated in this part. The use of fertilizer nutrients by crop plants, as it is affected by fertilizer material and application timing and placement, are discussed with regard to nutrient mobility and reaction with the soil. Ultimately, the continuum of soil–fertilizer–plant–climate interactions determines the efficiency of nutrient and water use. The economically and environmentally sound management of nutrients and toxic elements is facilitated by accurate soil and plant tissue testing methods that account for all these factors impacting availability.

In addition to C, H, and O, at least 14 mineral elements accumulated by root uptake from the soil are considered essential for most plants (N, P, K, Ca, Mg, S, Cu, Mn, Fe, B, Ni, Mo, Cl, and Zn). N, P, and K are often classified as "major" or "macro" nutrients, because they are frequently deficient in agricultural production, and when deficient they can have substantial impact on crop productivity. Although N and P are accumulated in large quantities by plants, the uptake of P is considerably less, on par with the "secondary" nutrients, Ca, Mg, and S. Silicon may not be essential to complete the life cycle of most plants, but its uptake by some plants, like rice and sugarcane, far exceeds that of N or K, and its impact on productivity of these crops can be great. "Minor" or "trace" elements include both biologically essential "micro" nutrients (Cu, Mn, Fe, B, Ni, Mo, Cl, and Zn) and nonessential elements (e.g., Al, Pb, Cd, and Hg). Plant tissue micronutrient sufficiency levels are quite low; however, at high levels of availability, both essential and nonessential elements can become toxic to plants or animals.

The principles of soil fertility and plant nutrition affecting the growth of plants and impacting the environment are discussed in the following six chapters.

11

Bioavailability of N, P, K, Ca, Mg, S, Si, and Micronutrients

Nanthi Bolan
University of South Australia

Ross Brennan
Department of Agriculture and Food Western Australia

Dedik Budianta
Sriwijaya University

James J. Camberato
Purdue University

Ravi Naidu
University of South Australia

William L. Pan
Washington State University

Andrew Sharpley
University of Arkansas

Donald L. Sparks
University of Delaware

Malcolm E. Sumner
University of Georgia

11.1 Bioavailability of Nitrogen 11-1
  Introduction • Nitrogen Cycle • Nitrogen an Essential Plant Nutrient • Nitrogen Transformations Affecting Nitrogen Availability • Factors Reducing N Bioavailability • Ammonium in the Interlayer of 2:1 Clay Minerals • Predicting Nitrogen Bioavailability
  References 11-9
11.2 Phosphorus Availability 11-14
  Introduction • Forms and Amounts in Soil • Principles of Analysis • Cycling in Soil • Optimizing Soil Phosphorus Availability • Water Quality Implications of Soil Phosphorus Availability • Conclusions
  References 11-28
11.3 Bioavailability of Soil Potassium 11-37
  Introduction • Forms of Soil Potassium • Factors Affecting Potassium Availability • Assessing Potassium Extractability and Availability
  References 11-44
11.4 Bioavailability of Calcium, Magnesium, Sulfur, and Silicon 11-47
  Introduction • Calcium and Magnesium • Sulfur • Silicon
  References 11-55
11.5 Bioavailability of Micronutrients 11-61
  Introduction • Sources of Micronutrients • Dynamics of Micronutrients in Soils • Biological Transformation Processes • Bioavailability of Micronutrients in Soils • Factors Affecting Bioavailability of Micronutrients in Soils • Bioavailability Implications for Soil Testing and Risk Assessment
  References 11-75

11.1 Bioavailability of Nitrogen

James J. Camberato

11.1.1 Introduction

The focus of this chapter is to illustrate how the fundamental transformations and transport of nitrogen (N) in the soil alter its bioavailability and impact plant productivity and environmental quality. Nutrient bioavailability or availability can be defined in many ways. Of the 10 definitions described by Blackmer (1999), that of Peck and Soltanpour (1990, p. 4) is most appropriate to this chapter: "By plant-available nutrient, one usually means the chemical form or forms of an essential plant nutrient in the soil whose variation in amount is reflected in variations in plant growth and yield."

The transient nature of N in soil and the large quantity accumulated by plants both contribute to the frequent occurrence of N deficiency in nonlegumes. Maximizing N bioavailability from indigenous soil N and added N (fertilizer, manure, legume residues, etc.) to optimize the yield of crop plants was the initial impetus to understanding N bioavailability. Considerable study of N transformations and transport determined that the processes of mineralization and immobilization, nitrification and denitrification, volatilization, and leaching control N bioavailability. With time, it was realized that N losses from agricultural systems via leaching, denitrification, and ammonia (NH_3) volatilization impacted surface and ground water, atmosphere, and noncultivated ecosystems; and much emphasis was then placed on quantifying and understanding these effects as well.

11.1.2 Nitrogen Cycle

Nitrogen bioavailability to plants is dynamic—the result of several interrelated microbial, chemical, and physical processes in soils, which are contingent upon the nature of the soil as well as temperature, moisture, energy, and other factors. The connectivity of these processes is often displayed as "The Nitrogen Cycle" (Section 27.1 of *Handbook of Soil Sciences: Properties and Processes*).

Nitrogen inputs to soils include N fixation, atmospheric deposition, and purposeful nutrient application (fertilizers, manures, etc.). Although dinitrogen gas (N_2) comprises ≈80% of the atmosphere, it can only be utilized by a few species of free-living bacteria and blue–green algae, and specialized N-fixing bacteria that form symbiotic relationships with leguminous plants (Section 27.1 of *Handbook of Soil Sciences: Properties and Processes*). Historically, the majority of reactive N (N molecules other than N_2) arose from microbial N fixation, until the recent (geologically speaking) development of the Haber–Bosch fertilizer manufacturing process (Galloway and Cowling, 2002; Mosier, 2002). The only other natural processes contributing significantly to the oxidative fixation of N_2 are lightning (Noxon, 1976; Liaw et al., 1990), fire, and volcanic activity (Bandy et al., 1982; Huebert et al., 1999).

With the industrial revolution came an increase in atmospheric deposition of reactive N, mostly due to the burning of N-containing substances. More than 90% of the N in organic residues is lost to the atmosphere with burning (Heard et al., 2006), most as NO_x (nitric oxide [NO] and nitrogen dioxide [NO_2]), nitrous oxide (N_2O), NH_3, hydrogen cyanide (HCN), and higher N-compounds (Lobert et al., 1990). On a global scale, atmospheric N deposition from anthropogenic sources has exceeded that from natural sources since the mid-twentieth century (Galloway, 2001) and was recently estimated to be about fourfold that from natural sources (Fowler et al., 2004). Smil (1999) reported wet deposition of 1 kg N ha^{-1} in the United States, west of the Mississippi, 7 kg N ha^{-1} in the coastal Northeast, and as much as 20 kg N ha^{-1} in the Netherlands, northeastern France, and southern England. Although the deposition of reactive N might be viewed positively in its contribution to plant productivity, atmospheric N deposition also contributes to soil and water acidification (Galloway, 2001; Rodhe et al., 2002), eutrophication of surface waters (Fisher and Oppenheimer, 1991; Bergström and Jansson, 2006), and decreased ecosystem diversity (Bobbink et al., 1998; Lee and Caporn, 1998). Additional beneficial and detrimental effects of reactive N in the environment are overviewed in Section 27.5 of *Handbook of Soil Sciences: Properties and Processes* and Galloway and Cowling (2002).

11.1.3 Nitrogen an Essential Plant Nutrient

"Nitrogen is without doubt the most spectacular of all essential nutrients in its effect on plant growth" (Olson and Kurtz, 1982, p. 568). Plant growth responses to increasing N bioavailability are more frequent and larger than to any other essential nutrient. Viets eloquently stated, "No one can claim that N is more essential than any of the other 15 elements known to be needed for higher plants, but on the basis of relative number of atoms needed, N is at the top of the list of those that come from soil or fertilizers," (Viets, 1965, pp. 503–504). Nitrogen is a component of nucleic acids (RNA and DNA), amino acids (proteins and enzymes), and chlorophyll. Detailed information on the role of N in plant nutrition can be obtained from Lea and Morot-Gaudry (2001) and Amâncio and Stulen (2004).

Plant tissue sufficiency concentrations of N differ among plant species and plant parts, ranging from ≈20 to 50 g N kg^{-1} as reported by Olson and Kurtz (1982). Legumes have higher tissue N concentration than nonlegumes, >40 versus <40 g kg^{-1}. Most plant N is in the organic or reduced form, although nitrate (NO_3^-) will accumulate to some extent in some situations (Grunes and Allaway, 1985, pp. 594–597). Aboveground N accumulation may exceed several hundred kilograms per hectare in productive environments, but in most environments N bioavailability limits the growth of nonleguminous plants.

Inadequate N bioavailability reduces overall plant growth and often results in chlorosis (yellowing) of plant tissues, a result of reduced chlorophyll content. Since N is highly mobile in the plant, chlorosis and necrosis of older plant tissues precedes the chlorosis of younger tissues, as N is translocated from older to younger plant parts. Descriptions and photographs of N deficiency on many important grain, sugar, oilseed, vegetable, and fruit crops as well as turfgrasses can be obtained from Bennett (1993). Excess N can also have detrimental effects on plants; including delayed flowering, reduced fruit set, and increased lodging. Excess N can impair the utilization and quality of crop plants, including poor wear tolerance of turfgrasses, reduced sugar concentration of sugar beets, lower tobacco leaf quality, and excessive NO_3^- content of vegetables and forages.

11.1.3.1 Nitrogen Uptake by Plants and Microorganisms

Nitrogen is accumulated by plants from the soil as the anion NO_3^- and the cation NH_4^+. Most plants readily accumulate both inorganic forms, with perhaps the exception of blueberry (*Vaccinium corymbosum* L.; Claussen and Lenz, 1999) and cranberry (*V. macrocarpon* Ait.; Greidanus et al., 1972), which often demonstrate a strong preference for NH_4^+. Some nondomesticated plants may have a preference for NH_4^+ or NO_3^-, depending on their adaptation to soil pH and the predominant form of N naturally occurring in the soil. Species adapted to acid soil environments may prefer NH_4^+ (Atkinson, 1985); the most prevalent form of N in acid soil, whereas less acid-tolerant species may prefer NO_3^- (Gigon and Rorison, 1972). However, many nondomesticated plants, as do crop plants, grow best on a mixture of NH_4^+ and NO_3^- (Rorison, 1985; Falkengren-Grerup, 1995). Nitrogen accumulation by plants can be rapid. Bowman et al. (1989) calculated a rate of ≈35 kg N ha^{-1} day^{-1} for field-grown, moderately N-deficient Kentucky bluegrass (*Poa pratensis* L.). In Maize (*Zea mays* L.), N accumulation rates peaked between 5 and 15 kg N ha^{-1} day^{-1} (Karlen et al., 1987).

Heterotrophic microorganisms accumulate NH_4^+ over NO_3^- when both are present (Jansson et al., 1955; Wickramasinghe et al., 1985; Rice and Tiedje, 1989), but will readily accumulate NO_3^- if it is the only form present (Davidson et al., 1990; Recous et al., 1990). Microorganisms are more competitive for N than plants in the short term. In 24 h ^{15}N uptake studies conducted in a Mollic Haploxeralf, annual grassland plants accumulated NO_3^- and NH_4^+ at rates of 0.6 and 1.0 kg N ha^{-1} day^{-1}, respectively (average for April and February measures), whereas microbial uptake was two- to fivefold higher at 1.1 and 5.3 kg N ha^{-1} day^{-1} (Jackson et al., 1989).

11.1.3.2 Ammonium and Nitrate Nutritional Effects on Plant Growth and Development

Many greenhouse, growth chamber, and hydroponics experiments demonstrated greater N uptake and/or growth of several crop species when provided both NO_3^- and NH_4^+, compared to a preponderance of one N form over the other (Schrader et al., 1972). Since NO_3^- is the predominant form of N available to plants in most cultivated soils, a nitrification inhibitor is often used to increase the proportion of total N that is in the NH_4^+ form. Crops that respond favorably to mixed N nutrition or enhanced NH_4^+ supply include barley (*Hordeum vulgare* L.; Leyshon et al., 1980); sorghum (*Sorghum bicolor* (L.) Moench; Camberato and Bock, 1989); wheat (*Triticum aestivum* L.; Camberato and Bock, 1990); maize (Below and Gentry, 1992); and Italian ryegrass (*Lolium multiflorum* Lam.; Alderman and Streeter, 1997). Unfortunately, this advantage of enhanced NH_4^+ supply has only infrequently been reported in field grown plants (Smiciklas and Below, 1992). Difficulty inhibiting nitrification to enhance NH_4^+ supply or positional availability of NH_4^+ may minimize the advantage to mixed N nutrition in field situations (Bock, 1986). Plants grown in soils where most of the N is derived from soil organic matter and plant residues and in acid or cold soils where nitrification is slowed may accumulate significant amounts of N in the NH_4^+ form without the aid of an artificial nitrification inhibitor. Otherwise, soils have little propensity to accumulate or store bioavailable NH_4^+-N (Chang et al., 1991; Muñoz et al., 2003).

11.1.4 Nitrogen Transformations Affecting Nitrogen Availability

Nitrogen bioavailability to nonleguminous plants is predicated in the presence of NH_4^+ and NO_3^- for root uptake. Thus, indigenous soil N, applied organic N, uric acid, urea, and other N-containing substances are only available to the plant after they are mineralized to NH_4^+ and NO_3^-. Nitrogen mineralization and immobilization (the microbial transfer of inorganic N to organic N) are discussed in detail in Section 27.2 of *Handbook of Soil Sciences: Properties and Processes*. Although NH_4^+ is subject to conversion to NH_3 in a high pH environment and volatilization to the atmosphere if on the soil surface, the loss of NH_4^+ from the soil is considerably less overall than the loss of NO_3^-. Unfortunately, the nitrification of NH_4^+ to NO_3^- (see Section 27.3 of *Handbook of Soil Sciences: Properties and Processes* for details) occurs readily in most soils and the subsequent loss of NO_3^- by leaching and denitrification substantially reduces N bioavailability.

11.1.4.1 Nitrogen Mineralization

The rate and extent of N mineralization determines the bioavailability of organic N sources, most importantly soil organic matter, crop residues, manures, and by-products of municipal waste treatment and industrial processes.

11.1.4.1.1 Soil Organic Matter

Stevenson (1982) reported that >90% of the N in soils is in the organic form. Organic N in the plowed layer of mineral soils ranges from 0.8 to 4.0 g N kg^{-1} (Bremner, 1965, p. 96) representing 1800–9000 kg N ha^{-1}. Thus, a 2% mineralization rate would provide 36–180 kg inorganic-N ha^{-1}, potentially a significant percentage of crop N accumulation. Cassman et al. (2002) determined indigenous N from maize uptake in 55 field experiments in the North-Central United States to typically range between 80 and 240 kg N ha^{-1}, with most providing between 100 and 140 kg N ha^{-1}. Predicting N bioavailability from soil organic N is important to increase the efficient use of fertilizer and added N and to decrease N losses to the environment (Cassman et al., 2002; Mulvaney et al., 2005). The importance of moisture and temperature (Stanford and Epstein, 1974; Kladivko and Keeney, 1987) and pH (Cornfield, 1952; Dancer et al., 1973; Olness, 1999) on N mineralization are well established. Unfortunately, laboratory tests of N mineralization have largely been unsuccessful (Griffin, 2008) in predicting N mineralization in the field, likely because most do not account for the impact of varying temperature and/or moisture in the field on N mineralization. Honeycutt et al. (1991) and Griffin and Honeycutt (2000) improved the incubation-based prediction of N mineralization from soil organic N, fertilizer, crop residue, sludge, and manure by including cumulative thermal units into equations based on soil properties and moisture.

11.1.4.1.2 Crop Rotation and Residue Effects on Nitrogen Bioavailability

Much effort has been placed on determining the cause of lower yield and higher N requirement in continuous maize compared to maize grown after soybean (*Glycine max* L.). Nitrogen availability is an important factor, although many other factors also contribute. Nitrogen uptake by maize grown after soybean in Wisconsin was 51 kg N ha^{-1}, greater than maize grown after maize on two silt loams, but no different in a sandy soil (Bundy et al., 1993). Nitrogen uptake of a second cereal crop grown 2 years after soybean was reduced to 36 kg N ha^{-1} (Vanotti and Bundy, 1995). The authors proffered that the second year reduction in N bioavailability was due to soybean removing more N from the soil than was fixed symbiotically as described by Heichel and Barnes (1984) and others (David et al., 1997; Jaynes et al., 2001). The yield benefit of maize grown after soybean, compared to maize grown after maize, was the same whether the soybean was nodulating or nonnodulating, and was not eliminated by the application of fertilizer N to a silt loam in Wisconsin (Maloney et al., 1999). These findings suggest that the "rotation effect" was not likely a result of N provided directly by the soybean crop, but an enhancement in the maize plant's ability to accumulate and utilize N. However, Gentry et al. (2001), using a similar experimental approach including nodulating and nonnodulating soybean, had a different conclusion. The "rotation effect" was due both to a decrease in N mineralization in continuous maize and to an increase in soil N arising from symbiotic fixation in maize after soybean.

Alternatively, it has been suggested that the effects of crop residues on N bioavailability may arise from their effect on soil temperature. Normal amounts of corn residue or simulated residue lowered soil temperature 0.7°C–1.7°C and resulted in 32 kg ha^{-1} less N than when residues were removed (Andraski and Bundy, 2008). Since the simulated residue (polypropylene snow fence) had a similar effect on soil temperature and N availability as corn stover, it was concluded that temperature, not microbial immobilization, had altered N bioavailability. This seems plausible since only 3–5 kg ha^{-1} of soybean residue–derived N was taken up by maize grown the following season on two Mollisols in Kansas, even though 144–155 kg N ha^{-1} more N fertilizer had to be supplied to maize grown after maize to achieve the same yield as maize after soybean (Omay et al., 1998). Crookston et al. (1991) pontificated that the "rotation effect" was "a somewhat fickle and elusive phenomenon" and was "easier to exploit … than to explain." Twenty years later this statement is still true.

11.1.4.1.3 Cover Crop Effects on Nitrogen Bioavailability

Legumes grown as cover crops differ in N fixation and residual N accumulation (Heichel and Barnes, 1984). Conditions affecting the growth of a cover crop, such as moisture and temperature, also affect its N fixation and recovery of residual N, thereby affecting the potential N contribution to the subsequent crop. The rate and extent of N mineralization from the cover crops determine the bioavailability to the following crop, which is influenced by the composition of the cover crop, placement on or incorporation in the soil, moisture, and temperature. Plant tissue with high C:N and high concentrations of lignin, cellulose, and hemicelluloses decompose and release N slower than plants with lower levels of these constituents (Wagger et al., 1998).

Legumes can fix considerable amounts of N that are significant when compared to the uptake of the subsequent cereal crop. For example, in Delaware, hairy vetch (*Vicia villosa* Roth.) and crimson clover (*Trifolium incarnatum* L.) grown in combination with spring oats (*Avena sativa* L.) or cereal rye (*Secale cereale* L.) accumulated ≈117 kg N ha^{-1} more N in cover crop roots and shoots than when either cereal was grown alone (≈55 kg N ha^{-1}) (Mitchell and Teel, 1977). An average of 90% of the cover crop N was in the top growth and about 33% of the N was released to the following maize crop. Similarly in North Carolina, crimson clover and hairy vetch, compared to a fallow treatment, increased N recovery by the subsequent maize crop ≈42 kg N ha^{-1}; equivalent to about 33% of the aboveground N content of the cover crops (Wagger, 1989a). Cereal rye, in comparison to legume cover crops, reduced N recovery by about 25 kg N ha^{-1} compared to fallow—likely arising from the high C:N of the rye (≈35:1; Wagger, 1989b) and N immobilization. Hairy vetch, big flower vetch (*Vicia grandiflora* W. Koch), and crimson clover grown as winter cover crops in Kentucky accumulated 173, 24, and 20 kg N ha^{-1} more aboveground plant N than cereal rye (36 kg N ha^{-1}), respectively (Ebelhar et al., 1984). Poor growing conditions, including low rainfall, reduce N accumulation. Four cover crop systems grown in the Brazilian Cerrados contributed 25–64 kg N ha^{-1} to the following maize crop in one season, but had no effect in the second season in which cover crop growth and N uptake were reduced by dry weather (Maltas et al., 2009).

Date of planting (growing season length), killing date, and moisture affect N fixation and N recovery of cover crops. A late August planted wheat-hairy vetch mixture grown 30 km south of Vancouver, British Columbia produced 126 kg N ha^{-1}, whereas the same mixture planted 4 weeks later produced only 70 kg N ha^{-1} (Odhiambo and Bomke, 2001). Many researchers have focused studies on the impact of kill date on cover crop N accumulation (Wagger, 1989a, 1989b; Clark et al., 1994, 1997; Odhiambo and Bomke, 2001; Sainju and Singh, 2001). The longer the cover crop is allowed to grow, the greater the amount of N that is fixed or recovered from the soil. Hairy vetch and a vetch-cereal rye mixture accumulated ≈2 kg N ha^{-1} day^{-1} in the aboveground biomass form late March until early May in Maryland (Clark et al., 1997).

Temperature, moisture, and placement of cover crop residues affect N contribution of the cover crop to the subsequent main crop. For several cover crops, faster N mineralization occurred in the first year of a 2 year study, due to higher rainfall (79 versus 15 mm) and higher mean air temperature (29°C versus 25°C) in the first 4 weeks of residue decomposition in North Carolina (Ranells and Wagger, 1996). Nearly complete N release from the hairy vetch cover crop occurred twice as fast in year 1 than year 2, 8 versus 16 weeks. Incorporating legume cover crop residue into a *Typic Paleudalf* in Kentucky increased the rate and extent of N release to the following crop (Varco et al., 1989). Maize recovered 31 kg ha^{-1} more N when hairy vetch was incorporated into the soil compared to when left on the soil surface, 32% versus 20% of the residue N content, respectively.

Nonlegume cover crops can be effective at recovering residual N from a previous crop or a post-growing season manure application. Nitrogen accumulation and subsequent release to the following crop determine bioavailability and environmental impact. In an irrigated *Xeric Torripsamment* sand in eastern Washington, about 29% of the 92–142 kg N ha^{-1}, captured by a mustard (*Brassica hirta* Moench) cover crop from fertilizer applied to a winter wheat-sweet corn-sweet corn rotation was recovered by the following potato (*Solanum tuberosum* L.) crop (Collins et al., 2007).

Cover crops can increase N bioavailability by reducing percolation losses (Weinert et al., 2002) as well as by recovering residual N. An estimated 50% of the 9 cm of winter rainfall was used by cover crops grown on a *Xeric Torripsamment* in the Columbia Plateau of central Washington. Thus, soil mineral N was reduced by 155 kg N ha^{-1} to a depth of 180 cm by cereal rye, wheat, or rapeseed (*B. napus* L.). However, winter water use by cover crops can sometimes reduce yield and N recovery by the subsequent crop, particularly in semiarid regions (Unger and

Vigil, 1998). Bioavailability of cover crop N can be assessed by aboveground N accumulation, with the PSNT soil test for NO_3^- or NO_3^- and NH_4^+ (Vaughan and Evanylo, 1999), or growing degree day predictive methods (Schomberg and Endale, 2004).

11.1.4.1.4 Nitrogen Bioavailability from Animal Manure and Organic N Sources

The term "animal manure" is used to refer to the array of complex substances arising from animal feces, urine, bedding, unused ration, water, medications, and growth promoters. Organic substances used as N sources for crop plants include by-products from vegetable, meat, oilseed, starch and sugar processing, municipal waste treatment (sewage sludge or biosolids), paper manufacturing, and other processes. A thorough discussion of these by-products can be found in Power and Dick (2000). This section will concentrate on N bioavailability from animal manure, although the same principles apply to other organic N sources. Nitrogen forms in manure include NO_3^-, NH_4^+, NH_3, urea, uric acid, and complex organic N compounds. Managing and predicting N bioavailability from manures is more difficult than that from inorganic fertilizer because of the variable composition of these materials and practical considerations that foster less than optimum timing and placement. The transformations of these compounds as affected by application method, timing, soil cation exchange and buffer capacity, and weather determine N bioavailability, but are difficult to predict. Bioavailability estimates for manures generally include availability factors for N that consider inorganic N loss (primarily NH_3 volatilization and NO_3^- leaching), based on application method and timing and an estimate of organic N mineralization. Nitrogen bioavailability from manure may also be represented in terms of "fertilizer replacement value" or "fertilizer equivalency," which does not take into account N loss that both fertilizer and manure N would be subjected to equally.

Manure N availability algorithms for making manure application rate recommendations for 34 states were reviewed and programmed by Joern et al. (2009). They found that 27 different variables were used to determine plant-available manure N and that for any single state 5 to 12 variables are needed to make this determination. While the number of variables required is indeed large, once the manure source is determined, 12 states use only one factor (e.g., total manure N) and 15 states use only two factors (e.g., total manure N and application method) to determine plant-available manure N. All but one state include method of application to determine plant-available manure N; however, only 13 out of 34 states use month of application to determine manure N availability.

Pratt et al. (1973) developed the "decay series" concept to represent manure N bioavailability over multiple growing seasons. The yearly rates of N mineralization in a decay series are given as the fraction of total N (year 1) or residual N (year 2 and beyond) mineralized. For example, if manure were applied containing 100 kg N ha^{-1}, the decay series 0.35, 0.15, 0.10, and 0.05 would predict mineralized N of 35 kg ha^{-1} in year 1 (0.35 × 100 kg N ha^{-1}), 9.8 kg ha^{-1} in year 2 [0.15 × (100 − 35 kg N ha^{-1})], 5.5 kg ha^{-1} in year 3 [0.10 × (100 − 35 − 9.8 kg N ha^{-1})], and 2.5 kg N ha^{-1} in year 4 [0.05 × (100 − 35 − 9.8 − 5.5 kg N ha^{-1})]. Nitrogen bioavailability from successive annual applications of manure would include N mineralized from previous seasons as well as the current growing season. Thus, if manure containing 100 kg N ha^{-1} were applied annually, N bioavailability in year 4 of the previous example would be 52.8 kg ha^{-1} (35 + 9.8 + 5.5 + 2.5 kg N ha^{-1}). The term "mineralized" was used by Pratt et al. (1973) to represent inorganic N released from urea and uric acid as well as slowly mineralizable organic N. Although this concept was originally put forth to estimate actual mineralization, the author's first experimentally based decay series, 0.40, 0.25, 0.06 from a field trial in the Coachella Valley, California, was based on fertilizer equivalency. Not surprisingly, decay series were found to be soil dependent. Dairy manure N bioavailability determined in a 5 year Vermont study was lower on a somewhat poorly drained clay soil (decay series—0.35, 0.15, 0.10, 0.075, and 0.05) than on a well-drained loam soil (decay series—0.60, 0.15, 0.10, 0.075, and 0.05), perhaps due to denitrification losses in the year of application (Magdoff, 1978).

Plant-available N in poultry litter was estimated in incubation studies to be approximately 80% of the inorganic N and 60% of the organic N contents and this provided a reasonable estimate of N bioavailability in two field experiments with maize, conducted in sandy loam Ultisols in Delaware (Bitzer and Sims, 1988). Nitrogen recovered in the second growing season from the 280 kg organic N ha^{-1} added at the highest rate of poultry manure was equivalent to only 31 and 16 kg N ha^{-1} fertilizer N, 11% and 6% of the organic N applied. Camberato and Frederick (1994) found no difference in residual N from poultry manure and fertilizer N applied at comparable N rates. High N availability shortly after poultry manure application and little residual value suggest that poultry manure application timing should be similar to that recommended for inorganic fertilizer to avoid leaching and denitrification losses.

Swine manure is often handled as a slurry or liquid from a lagoon. In this form 60%–80% of the total N content is in the NH_3/NH_4^+ forms (Mikkelsen, 1997). Thus, NH_3 volatilization from surface applications affects bioavailability. Only 40% of the total N content of surface-applied liquid swine manure slurry is considered available in comparison to 70%, when incorporated. Rapid conversion of NH_4^+ to NO_3^- and subsequent leaching and denitrification losses would be similar to those incurred with NH_4^+ fertilizers. Swine manure lagoon sludges have a higher organic N content (75%), thus predicting that organic N mineralization is a main determinant of bioavailability.

Dairy manure has a higher C:N ratio and more organic N than poultry or swine manure. Net nitrification and immobilization from nine dairy manures was more accurately predicted by considering the impact of manure C or neutral detergent fiber on N transformations (Griffin et al., 2005). The fertilizer N equivalency of dairy manure organic N was determined from maize N accumulation in New York to be 21%, 9%, 3%, 3%, and 2% of the original organic N content in years 1–5 after application, respectively (Klausner et al., 1994).

11.1.5 Factors Reducing N Bioavailability

11.1.5.1 Nitrogen Immobilization

Nitrogen can be added to the soil, remain in the rootzone, yet be unavailable to the crop. Immobilization of inorganic N by microorganisms decomposing high C:N crop residues (>20–30:1), such as maize, oat, or wheat stover or other organic materials can render substantial quantities of inorganic N unavailable for periods of time (Alexander, 1977). Substances such as sawdust or other wood-based by-products can have an even greater effect on N bioavailability than crop residues. For example, a papermaking sludge with a C:N ratio of 480:1 immobilized soil N for at least 250 days when added at 267 g sludge kg^{-1} soil (Zibilske, 1987). Ammonium is preferentially used by heterotrophic microflora, but NO_3^- is also immobilized in the absence of NH_4^+ (Recous et al., 1990).

11.1.5.2 Conversion of Ammonium to Nitrate Increases the Potential for Nitrogen Loss

Losses of NH_4^+-N from soil are generally small but NO_3^--N can be lost in large quantities by leaching and denitrification. Thus, nitrification, the conversion of NH_4^+ to NO_3^- by *Nitrosomonas* and *Nitrobacter* (see Section 27.3 of *Handbook of Soil Sciences: Properties and Processes*), is a critical factor that determines N bioavailability in many situations. Many factors affect nitrification rate, including temperature, pH, and soil type.

In laboratory incubations the nitrification rates of four soils increased with increased temperature from near 0°C to 27°C or 35°C, dependent on soil type (Frederick, 1956). The rate of nitrification is increased by increased temperature, but can proceed at relatively high rates with low temperatures in some soils. For example, equal nitrification rates occurred in a clay loam Ultisol incubated at 32°C or 8°C, but only after a 6 week lag period at the lower temperature (Anderson and Boswell, 1964). Nitrification proceeded in a field study in Missouri as soil temperature approached 0°C, albeit at a reduced rate (Kidwaro and Kephart, 1998).

Low soil pH reduces nitrification rate. A fivefold difference in nitrification during a 15-day incubation at 23°C occurred across a range of pH from 4.5 to 6.5 in a *Typic Argiudoll* (Dancer et al., 1973). Liming an acid soil from 5.0 to 7.4 tripled nitrification during a 3–4 week incubation (Frederick, 1956). In Iowa field experiments, 89% nitrification of fall-applied anhydrous ammonia was achieved in mid-April in soils with pH > 7.5, but only 39% nitrification occurred in soils with pH < 6.0 (Kyveryga et al., 2004).

Nitrapyrin is an effective inhibitor of nitrification (Bundy and Bremner, 1973) that is commonly applied with NH_4^+-containing and NH_4^+-forming fertilizers and manure. The persistence of nitrapyrin ranges from 5 to 50 days and is dependent on soil texture, organic matter, and temperature (Keeney, 1980). The effectiveness of nitrapyrin is greater at lower temperature; therefore, it is often applied with fertilizer after soil temperatures fall below 10°C in autumn for a spring crop (Touchton et al., 1978). Dicyandiamide (DCD) is another commercially available nitrification inhibitor and slow-release N source (Amberger, 1981) that has greater mobility in soil than nitrapyrin. Surface applications of urea and DCD can be moved into the soil with irrigation, shortly after application; but this mobility can also result in separation of DCD from NH_4^+ after urea hydrolysis (Bock et al., 1981).

A nitrification inhibitor can be effective at increasing N bioavailability and reducing NO_3^- leaching and denitrification. Apparent N recovery by maize grown on a sand-textured Ultisol in North Carolina was increased from 17% to 53% by including a nitrification inhibitor with urea versus urea alone (Chancy and Kamprath, 1982). After 22 cm rainfall, substantially more N remained in the upper 55 cm of soil 48 days after application with the nitrification inhibitor, than without. Similarly in a Minnesota study, applying a nitrification inhibitor with urea at 134 kg N ha^{-1} just prior to planting increased maize grain yield 28% on a loamy coarse sand with an available water holding capacity of only 7 cm m^{-1} (Malzer, 1989). Rainfall during the 8 weeks after application averaged 21.5 cm for the 3 years of the study, likely causing substantial NO_3^- leaching when urea was applied without a nitrification inhibitor.

Nitrification inhibitors affect N loss via denitrification as well as via leaching. Bioavailability of fertilizer N applied to a clay loam Ultisol increased when nitrapyrin was added with the fertilizer due to a decrease in denitrification (McElhannon and Mills, 1981). Nitrification of NH_4^+-N in liquid swine manure was complete in 7 weeks without a nitrification inhibitor, but was delayed up to 15 weeks with inclusion of 50 mg nitrapyrin active ingredient per liter of manure (McCormick et al., 1983). Increased N bioavailability was attributed to reduced denitrification as little NO_3^--N was found below the bands of untreated manure.

11.1.5.3 Nitrogen Leaching Reduces Nitrogen Bioavailability

Leaching of NO_3^- from the soil reduces bioavailability to plants and impacts environmental quality. Excess N in surface waters contributes to eutrophication (Keeney and Hatfield, 2008). High NO_3^- and nitrite (NO_2^-) in drinking water and food can cause methemoglobinemia in humans and animals (National Research Council, 1972).

The amount, intensity, and timing of rainfall, evapotranspiration, soil N levels, and many other factors impact the amount of N lost by leaching. In fertilized and manured soils the timing of N uptake by crops in relation to N application and rainfall exceeding evapotranspiration are key factors (Watts and Hanks, 1978). Nitrate movement in sandy soils is thought to occur more or less through the entire soil matrix and, thus, is envisioned as moving downward with the wetting front in proportion to rainfall in excess of evapotranspiration. Nitrogen loss can be substantial and rapid. In structured soil, water moves through an incomplete volume of the soil matrix; thus, NO_3^- is lost from the soil proportional to its concentration and much more slowly and less completely than in sand-textured unstructured soils.

In dry climates and on heavy-textured soils, NO_3^- leaching is limited. Only 7 kg NO_3^--N ha^{-1} year^{-1} of 168 kg N ha^{-1} applied to a field continuously cropped to maize was found in subsurface discharge from loess soils in Iowa (Burwell et al., 1976). Thus, N applied to one crop can "carryover" and be accumulated by the next crop. For example, carryover of fertilizer N from 168 kg N ha^{-1} N applied for maize on a silt loam soil in Wisconsin was about 50% of that applied, ranging from 32 to 106 kg N ha^{-1} using oat to estimate bioavailability (Vanotti and Bundy, 1994). Residual N was highest when N uptake of the maize was reduced by dry weather. Thus, soil NO_3^- testing is often recommended for adjusting N rate recommendations in dry climates (Section 13.2).

In contrast, N leaching occurs readily in warm humid climates on sandy soils, with irrigation, and in heavier-textured soils that are tile drained. A total of 161 kg N ha^{-1} NO_3^--N was leached below 120 cm soil depth in a sandy loam Hapludoll that was irrigated with 50 mm of water each time available soil water decreased to about 50 mm (50% depletion) (Timmons, 1984). Leaching of NO_3^--N from a split application of 185 kg N ha^{-1} beyond 152 cm soil depth in a loamy fine sand Haplustalf was increased by irrigation, 25% above optimum, about 60 mm year^{-1} (Gehl et al., 2005). Only about 10% of the 168 kg N ha^{-1} applied to maize was recovered by wheat (Camberato and Frederick, 1994) or a cereal rye and annual ryegrass (*L. multiflorum* Lam.) cover crop (Shipley et al., 1992).

Nitrate-N leaching occurs in heavier-textured soil if tile drained. Nitrogen leaching ranged from 13 to 61 kg NO_3^--N ha^{-1} year^{-1} in a maize–soybean rotation on Mollisols in Iowa and from 15 to 38 kg N ha^{-1} year^{-1} in silt loam Alfisol (Jaynes et al., 2001). Annual NO_3^--N loss from a tile-drained silt loam *Typic Glossaqualf* in Indiana decreased from 38 to 15 kg N ha^{-1} by including soybean in rotation with corn, reducing N application from ≈300 to ≈200 kg N ha^{-1}, and planting a winter wheat "catch crop" after corn in the rotation (Kladivko et al., 2004). Nitrate leaching from maize grown on Coastal Plain soils in Maryland was reduced to 80% by cereal rye cover crop (Staver and Brinsfield, 1998). Field-scale watershed studies suggested that groundwater NO_3^- was reduced to 60% by the use of cereal rye cover crops.

The environmental benefit of a cover crop is dependent on when the N is mineralized and nitrified and how much is utilized by the subsequent crop. Even though forage and oilseed radish (*Raphanus sativus* L.) were as effective as rapeseed and cereal rye in recovering N from the previous crop, the captured N was released earlier in the spring in Maryland due to winter kill (Dean and Weil, 2009). Thus, high pore water NO_3 in the soil profile under the radishes suggested leaching of NO_3^- from the loamy sand soil was possible unless followed by an early-planted spring crop. Frost-sensitive crops, sudangrass (*S. bicolor* L.) and white mustard, also released N prematurely resulting in soil NO_3^- levels higher than bare fallow (Weinert et al., 2002).

Anion exchange capacity originating at low pH in weathered Alfisols, Ultisols, and Oxisols may slow NO_3^--N loss (Toner et al., 1989; Eick et al., 1999); however, the ability of plants (particularly crop plants) to root into these acid soil layers is questionable due to low Ca availability and/or Al toxicity (Adams and Lund, 1966; Adams and Moore, 1983).

Nitrogen fertilizer, primarily anhydrous ammonia, is often applied in relatively cold and dry climates in late fall for a crop to be planted the following spring. Although nitrification may be limited just after application due to cold soil temperatures, significant nitrification can occur in the spring as the soils warm. Substantial N loss due to leaching and/or denitrification can occur prior to crop N demand. For example, in a corn–soybean rotation, apparent N recovery of anhydrous ammonia applied for corn on a tile-drained Mollisol in southern Minnesota was only 45% for fall-applied N compared to 87% for spring-applied N (Vetsch and Randall, 2004). Nitrate-N loss to tile drainage in this environment was reduced to 14% by spring application in comparison to fall-applied N (Randall and Vetsch, 2005).

11.1.5.4 Denitrification Reduces Nitrogen Bioavailability

Denitrification, conversion of NO_3^- to N_2 or less reduced gaseous N forms (see Section 27.4 of *Handbook of Soil Sciences: Properties and Processes* for details), lowers N bioavailability and contributes to greenhouse gas emissions and global warming. Denitrification losses increase with increasing temperature, but denitrification will occur in winter months with losses to the atmosphere occurring upon thawing (Dusenbury et al., 2008). Anaerobic conditions promote denitrification as do near-neutral soil pH, increasing NO_3^--N (McSwiney and Robertson, 2005; Millar et al., 2010) and soluble C.

Denitrification losses can be agronomically as well as environmentally important on poorly drained soils, with irrigation, application of manures or organic by-products, or reduced tillage. Denitrification losses were substantial on a clay (770 g kg^{-1}) *Typic Humaquept* near Québec City, Canada, on which barley was grown over a 3 year period, ranging from 12 to 45 kg N ha^{-1} during the snow-free season (Rochette et al., 2007). No-tillage, in comparison to moldboard plowing, resulted in a 2.5-fold increase in denitrification due to an increase in water-filled pore space. Denitrification losses in a well-drained loam soil in the same catena as the clay soil were ≤1.5 kg N ha^{-1}. Well-drained soils in dry climates have relatively low denitrification losses. Cumulative denitrification N loss over a 2 year period from four cropping systems in the semiarid northern Great Plains ranged from 0.3 kg N ha^{-1} for alfalfa (*Medicago sativa* L.)-perennial grass to 1.3 kg N ha^{-1} for continuous wheat (Dusenbury et al., 2008).

High rates of N_2O flux from the soil, 0.24–1.06 kg N ha^{-1} day^{-1}, occurred from three vegetable production systems in Santa Barbara County, California (Ryden and Lund, 1980). Peak N_2O emissions followed the first furrow irrigation after N fertilization. Annual emissions were estimated to range from 20 to 42 kg N ha^{-1}. Organic materials high in bioavailable C that stimulate microbial respiration may create an anaerobic microenvironment, creating conditions suitable for denitrification (Rice et al., 1988).

Greater denitrification occurred with alkaline-hydrolyzing fertilizers, anhydrous ammonia, and urea, in comparison to acidic fertilizers, because of increased soil pH and water soluble organic carbon with the alkaline fertilizers (Mulvaney et al., 1997).

Denitrification losses of N from banded urea fertilizer (applied at 218 kg N ha^{-1}) for maize grown on a furrow irrigated, well-drained Aridic Argiustoll in Colorado were 3.2 and 1.7 kg N ha^{-1} in each year of the study (Bronson et al., 1992). Peak N$_2$O emissions were coincident with irrigation and rainfall and were reduced by slowing the conversion of urea to NO$_3^-$ by including the nitrification inhibitor, nitrapyrin, with the banded fertilizer. Cumulative denitrification N losses from maize grown with 210 kg N ha^{-1} as surface-applied urea, no-tillage, and sprinkler irrigation on two Mollisols in Argentina were 9.8 and 2.1 kg N ha^{-1}, for preplant and sidedress N application, respectively (Sainz Rozas et al., 2001). Most of the N loss occurred when water-filled pore space exceeded 80%.

11.1.5.5 Ammonia Volatilization Reduces Nitrogen Bioavailability

Volatilization of NH$_3$ from fertilizer and manure can be substantial, easily >25% of the N content of several fertilizer and manure sources (Lauer et al., 1976; Fillery et al., 1984). Direct loss of anhydrous ammonia and aqua ammonia fertilizers can occur through improper soil application (Parr and Papendick, 1966) or application in irrigation water (Gardner and Roth, 1984). Ammonia losses can also occur from fertilizers that do not initially contain NH$_3$ if they are left on or near the soil surface and conditions are suitable for conversion of NH$_4^+$ to NH$_3$. pH and temperature are the two primary factors affecting NH$_3$ loss. Increasing pH increases NH$_3$ volatilization because it increases the proportion of ammoniacal-N in the NH$_3$ form (du Plessis and Kroontje, 1964).

$$NH_4^+ + OH^- \rightleftharpoons NH_3 + H_2O \quad (11.1)$$

Increasing temperature increases the driving force for NH$_3$ volatilization by decreasing the pK$_a$ for the equilibrium (Koelliker and Kissel, 1988). Wind also increases NH$_3$ loss (Fillery et al., 1984), as long as it does not impede urea hydrolysis by drying the reaction site.

11.1.5.5.1 Ammonia Volatilization from Fertilizers

Ammonium sulfate or (NH$_4$)$_2$HPO$_4$ application to calcareous soils can result in substantial NH$_3$ loss (Fenn and Kissel, 1973). However, when applied to lower pH soils most NH$_4^+$-containing fertilizers result in little NH$_3$ volatilization (Volk, 1959). In contrast, NH$_3$ volatilization is likely with urea-containing fertilizers in many situations because urea hydrolysis increases pH (Overrein and Moe, 1967) as NH$_4^+$ is formed.

$$CO(NH_2)_2 + H^+ + 2H_2O \rightarrow 2NH_4^+ + HCO_3^- \quad (11.2)$$

Low buffer capacity soils result in greater NH$_3$ volatilization due to greater increases in pH and lower retention of exchangeable NH$_4^+$ (Ferguson et al., 1984).

Ammonia volatilization losses from urea are higher when applied to crop residues (e.g., no-tillage cropping systems), pasture, or turfgrass in comparison to the soil surface, partly because residues and plant tissues have high urease activity (Torello and Wehner, 1983; Dick, 1984), resulting in rapid urea hydrolysis and substantial pH changes on low buffered materials (McInnes et al., 1986). Increased temperature promoted urea hydrolysis (Broadbent et al., 1958; Fisher and Parks, 1958) as well and also increased the proportion of ammoniacal-N in the NH$_3$ form (Koelliker and Kissel, 1988). Moisture is required for urea hydrolysis to occur (Ernst and Massey, 1960), but losses are enhanced as drying concentrates the NH$_4^+$ (Hargrove et al., 1977; McInnes et al., 1986). Rainfall of 28 and 33 mm incorporated urea into the soil and halted NH$_3$ volatilization, whereas irrigation of 2.5 mm increased loss (McInnes et al., 1986). Irrigation or rainfall amount must be sufficient to incorporate the N into the soil so as not to enhance losses (Kissel et al., 2004).

Fertilizer application method and form can affect NH$_3$ volatilization. Surface application of urea-containing fertilizers greatly enhanced NH$_3$ volatilization as determined by crop response (Mengel et al., 1982; Touchton and Hargrove, 1982; Fox and Piekielek, 1987). Hargrove (1988) suggested that a depth of 5–10 cm would prevent NH$_3$ loss from urea on most soils. Banding urea versus a broadcast application generally reduces NH$_3$ losses as well. Including a urease inhibitor (Schlegel et al., 1986), coating (Matocha, 1976), or encapsulating urea (Knight et al., 2007) also reduce NH$_3$ volatilization.

11.1.5.5.2 Ammonia Volatilization from Manures

Nitrogen applied to surface-applied manures is subject to NH$_3$ volatilization losses because potentially volatile N compounds are a significant percentage of the total N content. Urea comprised nearly 70% of the N in urine from ruminants (Bristow et al., 1992). Approximately 30%–50% of the total N content of various poultry manures was uric acid- and urea-N (Nicholson et al., 1996; Fujiwara and Murakami, 2007). Manure pH, which fosters NH$_3$ and enhances volatilization, may exceed 8.0 (Nicholson et al., 1996), greatly reducing N availability from poultry manures. Nearly 100% of the initial NH$_4^+$-N content of a poultry litter was lost by NH$_3$ volatilization in ≈8 days after application in summer when conditions were hot, dry, and windy (Sharpe et al., 2004). However, NH$_3$ loss in winter was only 10% of the NH$_4^+$-N content. Rainfall shortly after manure application to the soil surface reduces NH$_3$ volatilization by transporting volatile N forms into the soil (Cabrera and Vervoort, 1998; Sharpe et al., 2004). Ammonia loss from surface-applied cattle slurry, which was not incorporated, exceeded 40% of the initial NH$_4^+$-N content, but immediate incorporation with disk harrow, chisel plow, or moldboard plow substantially reduced loss to 5%, 9%, and 1%, respectively (Thompson and Meisinger, 2002). Peak NH$_3$ loss occurs soon after application (Marshall et al., 1998), so rainfall must also occur soon after application to be effective in reducing NH$_3$ volatilization.

11.1.5.5.3 Ammonia Volatilization from Plant Tissues

Ammonia volatilization directly from plant tissues can be substantial. Harper et al. (1987) estimated that 21% of the 73 kg ha^{-1}

fertilizer N applied in the spring of the year in Georgia was volatilized from wheat tissue. Direct volatilization of NH_3 from maize tissues in Nebraska was equivalent to 10%–20% of the fertilizer N applied, and apparent total N losses from aboveground plant tissues ranged from 45 to 81 kg N ha^{-1} (Francis et al., 1993). This loss was equivalent to 50%–75% of the unaccounted-for N in ^{15}N balance calculations. Although the results of these studies suggest NH_3 volatilization from plant tissue is substantial, rarely is this loss mechanism considered or quantified in N studies.

11.1.6 Ammonium in the Interlayer of 2:1 Clay Minerals

Ammonium ions occur in the interlayer of 2:1 clay minerals, particularly vermiculite and degraded illite (Nommik and Vahtras, 1982). Often, this phenomenon is referred to as "fixation," implying zero bioavailability. In reality, however, interlayer NH_4^+ exchanges with the soil solution to some degree and is, therefore, available for plant uptake. Fixation of nearly 10 mg kg NH_4^+-N (10% of that added as $(NH_4)_2SO_4$) occurred in less than 1 day in two vermiculitic soils (Broadbent, 1965). Green et al. (1994) found up to 23% of the N applied in urea was fixed as NH_4^+ after 20 days of anaerobic incubation. However, nearly half the fixed NH_4^+ was released during nitrification after 15 days of aerobic incubation (Green et al., 1994). Plant (Li et al., 1990) and microbial (Kelley and Stevenson, 1987) utilization of solution NH_4^+ also resulted in the release of fixed NH_4^+.

11.1.7 Predicting Nitrogen Bioavailability

Blackmer (1999) outlined many indices of N bioavailability; including Mitscherlich "b" and "c" values, Fried and Dean "A" values, Dean "a" values, and several biological and chemical tests for estimating N mineralization. Sims (Chapter 13) overviewed several inorganic N bioavailability tests including the residual inorganic N, preplant soil profile nitrate (PPNT), and the presidedress soil nitrate (PSNT) tests. The recently developed Illinois soil nitrogen test (ISNT), a purported indicator of mineralizable N, was also discussed by Sims. Girma and Raun (Chapter 15) discussed the use of optical sensors to make in-season estimates of N bioavailability by assessing crop canopy size and greenness. Tissue testing and hand-held chlorophyll meters were the precursors to optical sensors (Schepers et al., 1992) and achieved much the same goal. *Adapt-N* is a promising web-based tool used to make N recommendations for maize in New York that dynamically simulates the impact of weather on N transformations and transport to estimate crop N accumulation and predict N need (Melkonian et al., 2008). Given the complexity of the N cycle, the importance of temperature and moisture in affecting N bioavailability, and the economic and environmental costs of errors in N management, it is no wonder a sophisticated tool such as *Adapt-N* is needed to predict N bioavailability and improve N management.

References

Adams, F., and Z.F. Lund. 1966. Effect of chemical activity of soil solution aluminum on cotton root penetration of acid subsoils. Soil Sci. 101:193–198.

Adams, F., and B.L. Moore. 1983. Chemical factors affecting root growth in subsoil horizons of Coastal Plain soils. Soil Sci. Soc. Am. J. 47:99–102.

Alderman, S.C., and D.J. Streeter. 1997. Italian ryegrass and nitrogen source fertilization in western Oregon in two contrasting climatic years. I. Growth and seed yield. J. Plant Nutr. 20:19–428.

Alexander, M. 1977. Introduction to soil microbiology. 2nd edn. John Wiley & Sons, New York.

Amâncio, S., and I. Stulen (eds.) 2004. Nitrogen acquisition and assimilation in higher plants. Kluwer Academic Publishers, Dordrecht, the Netherlands.

Amberger, A. 1981. Dicyandiamide as a nitrification inhibitor, p. 3–17. *In* R.D. Hauck and H. Behnke (eds.) Proceedings of the technical workshop on dicyandiamide. Muscle Shoals, AL, December 4–5, 1981. SKW Trostberg, Trostberg, West-Germany.

Anderson, O.E., and F.C. Boswell. 1964. The influence of low temperature and various concentrations of ammonium nitrate on nitrification in acid soil. Soil Sci. Soc. Am. Proc. 28:525–529.

Andraski, T.W., and L.G. Bundy. 2008. Corn residue and nitrogen source effects on nitrogen availability in no-till corn. Agron. J. 100:1274–1279.

Atkinson, C.J. 1985. Nitrogen acquisition in four coexisting species from an upland acidic grassland. Physiol. Plant. 63:375–387.

Bandy, A.R., P.J. Maroulis, L.A. Wilner, and A.L. Torres. 1982. Estimates of the fluxes of NO, SO_2, H_2S, CS_2 and OCS from Mt. St. Helens deduced from in situ plume concentration measurements. Geophys. Res. Lett. 9:1097–1100.

Below, F.E., and L.E. Gentry. 1992. Effects of mixed N nutrition on nutrient accumulation, partitioning, and productivity of corn. J. Fert. Issues 4:79–85.

Bennett, W.F. (ed.). 1993. Nutrient deficiencies & toxicities in crop plants. APS Press, St. Paul, MN.

Bergström, A., and M. Jansson. 2006. Atmospheric nitrogen deposition has caused nitrogen enrichment and eutrophication of lakes in the northern hemisphere. Global Change Biol. 12:635–643.

Bitzer, C.C., and J.T. Sims. 1988. Estimating the availability of nitrogen in poultry manure through laboratory and field studies. J. Environ. Qual. 17:47–54.

Blackmer, A.M. 1999. Bioavailability of nitrogen, p. D3–D18. *In* M.E. Sumner (ed.) Handbook of soil science. CRC Press, Boca Raton, FL.

Bobbink, R., M. Hornung, and J.G.M. Roelofs. 1998. The effects of air-borne nitrogen pollutants on species diversity in natural and semi-natural European vegetation. J. Ecol. 86:717–738.

Bock, B.R. 1986. Increasing cereal yields with higher ammonium/nitrate ratios: Review of potentials and limitations. J. Environ. Sci. Health 21:723–758.

Bock, B.R., J.E. Lawrence, and H.M. Williams. 1981. Relative mobility of dicyandiamide, ammonium and urea by mass flow in soils, p. 25–37. In R.D. Hauck and H. Behnke (eds.) Proceedings of the technical workshop on dicyandiamide. Muscle Shoals, AL, December 4–5, 1981. SKW Trostberg, Trostberg, West-Germany.

Bowman, D.C., J.L. Paul, W.B. Davis, and S.H. Nelson. 1989. Rapid depletion of nitrogen applied to Kentucky bluegrass turf. J. Am. Soc. Hortic. Sci. 114:229–233.

Bremner, J.M. 1965. Organic nitrogen in soils. In W.V. Bartholomew and F.E. Clark (eds.) Soil nitrogen. Agronomy 10:93–149.

Bristow, A.W., D.C. Whitehead, and J.E. Cockburn. 1992. Nitrogenous constituents in the urine of cattle, sheep and goats. J. Sci. Food Agric. 59:387–394.

Broadbent, F.E. 1965. Effect of fertilizer nitrogen on the release of soil nitrogen. Soil Sci. Soc. Am. Proc. 29:692–696.

Broadbent, F.E., G.N. Hill, and K.B. Tyler. 1958. Transformations and movement of urea in soils. Soil Sci. Soc. Am. Proc. 22:303–307.

Bronson, K.F., A.R. Mosier, and S.R. Bishnoi. 1992. Nitrous oxide emissions in irrigated corn as affected by nitrification inhibitors. Soil Sci. Soc. Am. J. 56:161–165.

Bundy, L.G., T.W. Andraski, and R.P. Wolkowski. 1993. Nitrogen credits in soybean-corn crop sequences on three soils. Agron. J. 85:1061–1067.

Bundy, L.G., and J.M. Bremner. 1973. Inhibition of nitrification in soils. Soil Sci. Soc. Am. Proc. 37:396–398.

Burwell, R.E., G.E. Schuman, K.E. Saxton, and H.G. Heinemann. 1976. Nitrogen in subsurface discharge from agricultural watersheds. J. Environ. Qual. 5:325–329.

Cabrera, M.L., and A.A. Vervoort. 1998. Effect of timing of simulated rain on ammonia volatilization from surface-applied broiler litter. Commun. Soil Sci. Plant Anal. 29:575–586.

Camberato, J.J., and B.R. Bock. 1989. Response of grain sorghum to enhanced-ammonium supply. Plant Soil 113:79–83.

Camberato, J.J., and B.R. Bock. 1990. Spring wheat response to enhanced ammonium supply. I. Dry matter and nitrogen content. Agron. J. 82:463–467.

Camberato, J.J., and J.R. Frederick. 1994. Residual maize fertilizer nitrogen availability to wheat on the southeastern Coastal Plain. Agron. J. 86:962–967.

Cassman, K.G., A. Dobermann, and D.T. Walters. 2002. Agroecosystems, nitrogen-use efficiency, and nitrogen management. Ambio 31:132–140.

Chancy, H.F., and E.J. Kamprath. 1982. Effect of nitrapyrin on N response of corn on sandy soils. Agron. J. 74:565–569.

Chang, C., T.G. Sommerfeldt, and T. Entz. 1991. Soil chemistry after eleven annual applications of cattle feedlot manure. J. Environ. Qual. 20:475–480.

Clark, A.J., A.M. Decker, and J.J. Meisinger. 1994. Seeding rate and kill date effects on hairy vetch-cereal rye cover crop mixtures for corn production. Agron. J. 86:1065–1070.

Clark, A.J., A.M. Decker, J.J. Meisinger, and M.S. McIntosh. 1997. Kill date of vetch, rye, and a vetch-rye mixture: I. Cover crop and corn nitrogen. Agron. J. 89:427–434.

Claussen, W., and F. Lenz. 1999. Effect of ammonium or nitrate nutrition on net photosynthesis, growth, and activity of the enzymes nitrate reductase and glutamine synthetase in blueberry, raspberry and strawberry. Plant Soil 208:95–102.

Collins, H.P., J.A. Delgado, A.K. Alva, and R.F. Follett. 2007. Use of nitrogen-15 isotope techniques to estimate nitrogen cycling from a mustard cover crop to potatoes. Agron. J. 99:27–35.

Cornfield, A.H. 1952. The mineralization of the nitrogen of soil during incubation: Influence of pH, total nitrogen, and organic carbon contents. J. Sci. Food Agric. 3:343–349.

Crookston, R.K., J.E. Kurle, P.J. Copeland, J.H. Ford, and W.E. Lueschen. 1991. Rotational cropping sequence affects yield of corn and soybean. Agron. J. 83:108–113.

Dancer, W.S., L.A. Peterson, and G. Chesters. 1973. Ammonification and nitrification of N as influenced by soil pH and previous N treatments. Soil Sci. Soc. Am. Proc. 37:67–69.

David, M.B., L.E. Gentry, D.A. Kovacic, and K.M. Smith. 1997. Nitrogen balance in and export from an agricultural watershed. J. Environ. Qual. 26:1038–1048.

Davidson, E.A., J.M. Stark, and M.K. Firestone. 1990. Microbial production and consumption of nitrate in an annual grassland. Ecology 71:1968–1975.

Dean, J.E., and R.R. Weil. 2009. Brassica cover crops for nitrogen retention in the Mid-Atlantic Coastal Plain. J. Environ. Qual. 38:520–528.

Dick, W.A. 1984. Influence of long-term tillage and crop rotation combinations on soil enzyme activities. Soil Sci. Soc. Am. J. 48:569–574.

du Plessis, M.C.F., and W. Kroontje. 1964. The relationship between pH and ammonia equilibria in soil. Soil Sci. Soc. Am. Proc. 28:751–754.

Dusenbury, M.P., R.E. Engel, P.R. Miller, R.L. Lemke, and R. Wallander. 2008. Nitrous oxide emissions from a northern Great Plains soil as influenced by nitrogen management and cropping systems. J. Environ. Qual. 37:542–550.

Ebelhar, S.A., W.W. Frye, and R.L. Blevins. 1984. Nitrogen from legume cover crops for no-tillage corn. Agron. J. 76:51–55.

Eick, M.J., W.D. Brady, and C.K. Lynch. 1999. Charge properties and nitrate adsorption of some acid southeastern soils. J. Environ. Qual. 28:138–144.

Ernst, J.W., and H.F. Massey. 1960. The effects of several factors on volatilization of ammonia formed from urea in soil. Soil Sci. Soc. Am. Proc. 24:87–90.

Falkengren-Grerup, U. 1995. Interspecies differences in the preference of ammonium and nitrate in vascular plants. Oecologia 102:305–311.

Fenn, L.B., and D.E. Kissel. 1973. Ammonia volatilization from surface applications of ammonium compounds on calcareous soils: I. General theory. Soil Sci. Soc. Am. Proc. 37:855–859.

Ferguson, R.B., D.E. Kissel, J.K. Kolliker, and W. Basel. 1984. Ammonia volatilization from surface-applied urea: Effect of hydrogen ion buffering capacity. Soil Sci. Soc. Am. J. 48:578–582.

Fillery, I.R.P., J.R. Simpson, and S.K. De Datta. 1984. Influence of field environment and fertilizer management on ammonia loss from flooded rice. Soil Sci. Soc. Am. J. 48:914–920.

Fisher, D.C., and M. Oppenheimer. 1991. Atmospheric nitrogen deposition and the Chesapeake Bay estuary. Ambio 20:102–108.

Fisher, W.B., Jr., and W.L. Parks. 1958. Influence of soil temperature on urea hydrolysis and subsequent nitrification. Soil Sci. Soc. Am. Proc. 22:247–248.

Fowler, D., J.B.A. Muller, and L.J. Sheppard. 2004. The GaNE programme in a global perspective. Water Air Soil Pollut. Focus 4:3–8.

Fox, R.H., and W.P. Piekielek. 1987. Comparison of surface application methods of nitrogen solution to no-till corn (*Zea mays* L.). J. Fert. Issues 4:7–12.

Francis, D.D., J.S. Schepers, and M.F. Vigil. 1993. Post-anthesis nitrogen loss from corn. Agron. J. 85:659–663.

Frederick, L.R. 1956. The formation of nitrate from ammonium nitrogen in soils: I. Effect of temperature. Soil Sci. Soc. Am. Proc. 20:496–500.

Fujiwara, T., and K. Murakami. 2007. Application of near infrared spectroscopy for estimating available nitrogen in poultry manure compost. Soil Sci. Plant Nutr. 53:102–107.

Galloway, J.N. 2001. Acidification of the world: Natural and anthropogenic. Water Air Soil Pollut. 130:17–24.

Galloway, J.N., and E.B. Cowling. 2002. Reactive nitrogen and the world: 200 years of change. Ambio 31:64–71.

Gardner, B.R., and R.L. Roth. 1984. Applying nitrogen in irrigation waters, p. 493–506. *In* R.D. Hauck (ed.) Nitrogen in crop production. ASA, Madison, WI.

Gehl, R.J., J.P. Schmidt, L.R. Stone, A.J. Schlegel, and G.A. Clark. 2005. In situ measurements of nitrate leaching implicate poor nitrogen and irrigation management on sandy soils. J. Environ. Qual. 34:2243–2254.

Gentry, L.E., F.E. Below, M.B. David, and J.A. Bergerou. 2001. Source of soybean N credit in maize production. Plant Soil 236:175–184.

Gigon, A., and I.H. Rorison. 1972. The response of some ecologically distinct plant species to nitrate- and to ammonium-nitrogen. J. Ecol. 60:93–102.

Green, C.J., A.M. Blackmer, and N.C. Yang. 1994. Release of fixed ammonium during nitrification in soils. Soil Sci. Soc. Am. J. 58:1411–1425.

Greidanus, T., L.A. Peterson, L.E. Schrader, and M.N. Dana. 1972. Essentiality of ammonium for cranberry nutrition. J. Am. Soc. Hortic. Sci. 97:272–277.

Griffin, T.S. 2008. Nitrogen availability. *In* J.S. Schepers and W.R. Raun (eds.) Nitrogen in agricultural soils. Agronomy 49:616–646.

Griffin, T.S., and C.W. Honeycutt. 2000. Using growing degree days to predict nitrogen availability from livestock manures. Soil Sci. Soc. Am. J. 64:1876–1882.

Griffin, T.S., Z. He, and C.W. Honeycutt. 2005. Manure composition affects net transformation of nitrogen from dairy manures. Plant Soil 273:29–38.

Grunes, D.L., and W.H. Allaway. 1985. Nutritional quality of plants in relation to fertilizer use, p. 589–619. *In* O.P. Engelstad (ed.) Fertilizer technology and use. 3rd edn. SSSA, Madison, WI.

Hargrove, W.L. 1988. Soil, environmental, and management factors influencing ammonia volatilization under field conditions, p. 17–36. *In* B.R. Bock and D.E. Kissel (eds.) Ammonia volatilization from urea fertilizers. Bull. Y-206. National Fertilizer Development Center, Tennessee Valley Authority, Muscle Shoals, AL.

Hargrove, W.L., D.E. Kissel, and L.B. Fenn. 1977. Field measurements of ammonia volatilization from surface applications of ammonium salts to a calcareous soil. Agron. J. 69:473–476.

Harper, L.A., R.R. Sharpe, G.W. Langdale, and J.E. Giddens. 1987. Nitrogen cycling in a wheat crop: Soil, plant, and aerial nitrogen transport. Agron. J. 79:965–973.

Heard, J., C. Cavers, and G. Adrian. 2006. Up in smoke-nutrient loss with straw burning. Better Crops 90:10–11.

Heichel, G.H., and D.K. Barnes. 1984. Opportunities for meeting crop nitrogen needs from symbiotic nitrogen fixation, p. 49–59. *In* D.F. Bezdicek et al. (eds.) Organic farming: Current technology and its role in a sustainable agriculture. ASA Special Publication No. 46. ASA, CSSA, and SSSA, Madison, WI.

Honeycutt, C.W., L.J. Potaro, and W.A. Halteman. 1991. Predicting nitrate formation from soil, fertilizer, crop residue, and sludge with thermal units. J. Environ. Qual. 20:850–856.

Huebert, B., P. Vitousek, J. Sutton, T. Elias, J. Heath, S. Coeppicus, S. Howell, and B. Blomquist. 1999. Volcano fixes nitrogen into plant-available forms. Biogeochemistry 47:111–118.

Jackson, L.E., J.P. Schimel, and M.K. Firestone. 1989. Short-term partitioning of ammonium and nitrate between plants and microbes in an annual grassland. Soil Biol. Biochem. 21:409–415.

Jansson, S.L., M.J. Hallam, and W.V. Bartholomew. 1955. Preferential utilization of ammonium over nitrate by microorganisms in the decomposition of oat straw. Plant Soil 6:382–390.

Jaynes, D.B., T.S. Colvin, D.L. Karlen, C.A. Cambardella, and D.W. Meek. 2001. Nitrate loss in subsurface drainage as affected by nitrogen fertilizer rate. J. Environ. Qual. 30:1305–1314.

Joern, B.C., P.J. Hess, and B. Eisenhauer. 2009. Manure management planner program, version 0.29. Purdue University, West Lafayette, IN.

Karlen, D.L., R.L. Flannery, and E.J. Sadler. 1987. Nutrient and dry matter accumulation rates for high yielding maize. J. Plant Nutr. 10:1409–1417.

Keeney, D.R. 1980. Factors affecting the persistence and bioactivity of nitrification inhibitors, p. 33–46. *In* J.J. Meisinger et al. (eds.) Nitrification inhibitors—Potentials and limitations. ASA and SSSA, Madison, WI.

Keeney, D.R., and J.L. Hatfield. 2008. The nitrogen cycle, historical perspective, and current and potential future concerns, p. 1–18. *In* J.L. Hatfield and R.F. Follet (eds.) Nitrogen in the environment: Sources, problems, and management. 2nd edn. Elsevier Inc., Amsterdam, the Netherlands.

Kelley, K.R., and F.J. Stevenson. 1987. Effects of carbon source on immobilization and chemical distribution of fertilizer nitrogen in soil. Soil Sci. Soc. Am. J. 51:946–951.

Kidwaro, F.M., and K.D. Kephart. 1998. Retention of nitrogen from stabilized anhydrous ammonia in the soil profile during winter wheat production in Missouri. Commun. Soil Sci. Plant Anal. 29:481–499.

Kissel, D.E., M.L. Cabrera, N. Vaio, J.R. Craig, J.A. Rema, and L.A. Morris. 2004. Rainfall timing and ammonia loss from urea in a loblolly pine plantation. Soil Sci. Soc. Am. J. 68:1744–1750.

Kladivko, E.J., J.R. Frankenberger, D.B. Jaynes, D.W. Meek, B.J. Jenkinson, and N.R. Fausey. 2004. Nitrate leaching to subsurface drains as affected by drain spacing and changes in crop production system. J. Environ. Qual. 33:1803–1813.

Kladivko, E.J., and D.R. Keeney. 1987. Soil nitrogen mineralization as affected by water and temperature interactions. Biol. Fertil. Soils 5:248–252.

Klausner, S.D., V.R. Kanneganti, and D.R. Bouldin. 1994. An approach for estimating a decay series for organic nitrogen in animal manure. Agron. J. 86:897–903.

Knight, E.C., E.A. Guertal, and C.W. Wood. 2007. Mowing and nitrogen source effects on ammonia volatilization from turfgrass. Crop Sci. 47:1628–1634.

Koelliker, J.K., and D.E. Kissel. 1988. Chemical equilibria affecting ammonia volatilization, p. 37–52. *In* B.R. Bock and D.E. Kissel (eds.) Ammonia volatilization from urea fertilizers. Bull. Y-206. National Fertilizer Development Center, Tennessee Valley Authority, Muscle Shoals, AL.

Kyveryga, P.M., A.M. Blackmer, J.W. Ellsworth, and R. Isla. 2004. Soil pH effects on nitrification of fall-applied anhydrous ammonia. Soil Sci. Soc. Am. J. 68:545–551.

Lauer, D.A., D.R. Bouldin, and S.D. Klausner. 1976. Ammonia volatilization from dairy manure spread on the soil surface. J. Environ. Qual. 5:134–141.

Lea, P.J., and J.-F. Morot-Gaudry (eds.). 2001. Plant nitrogen. Springer-Verlag, Berlin, Germany.

Lee, J.A., and S.J.M. Caporn. 1998. Ecological effects of atmospheric reactive nitrogen deposition on semi-natural terrestrial ecosystems. New Phytol. 139:127–134.

Leyshon, A.J., C.A. Campbell, and F.G. Warder. 1980. Comparison of the effect of NO_3^- and NH_4-N on growth, yield, and components of Manitou spring wheat and Conquest barley. Can. J. Plant Sci. 60:1063–1070.

Li, C., X. Fan, and K. Mengel. 1990. Turnover of interlayer ammonium in loess-derived soil grown with winter wheat in the Shaanxi province of China. Biol. Fertil. Soils 9:211–214.

Liaw, Y.P., D.L. Sisterson, and N.L. Miller. 1990. Comparison of field, laboratory, and theoretical estimates of global nitrogen fixation by lightning. J. Geophys. Res. 95:22,489–22,494.

Lobert, J.M., D.H. Scharffe, W.M. Hao, and P.J. Crutzen. 1990. Importance of biomass burning in the atmospheric budgets of nitrogen-containing gases. Nature 346:552–554.

Magdoff, F. 1978. Influence of manure application rates and continuous corn on soil-N. Agron. J. 70:629–632.

Maloney, T.S., K.G. Silveira, and E.S. Oplinger. 1999. Rotational vs. nitrogen-fixing influence of soybean on corn grain and silage yield and nitrogen use. J. Prod. Agric. 12:175–187.

Maltas, A., M. Corbeels, E. Scopel, J. Wery, and F.A. Macena da Silva. 2009. Cover crop and nitrogen effects on maize productivity in no-tillage systems of the Brazilian Cerrados. Agron. J. 101:1036–1046.

Malzer, G.L. 1989. Incorporation of nitrification inhibitors with urea and urea-ammonium nitrate for irrigated corn. Fertil. Res. 18:141–151.

Marshall, S.B., C.W. Wood, L.C. Braun, M.L. Cabrera, M.D. Mullen, and E.A. Guertal. 1998. Ammonia volatilization from tall fescue pastures fertilized with broiler litter. J. Environ. Qual. 27:1125–1129.

Matocha, J.E. 1976. Ammonia volatilization and nitrogen utilization from sulfur-coated ureas and conventional nitrogen fertilizers. Soil Sci. Soc. Am. J. 40:597–601.

McCormick, R.A., D.W. Nelson, A.L. Sutton, and D.M. Huber. 1983. Effect of nitrapyrin on nitrogen transformations in soil treated with liquid swine manure. Agron. J. 75:947–950.

McElhannon, W.S., and H.A. Mills. 1981. Inhibition of denitrification by nitrapyrin with field-grown sweet corn. J. Am. Soc. Hortic. Sci. 106:673–677.

McInnes, K.J., R.B. Ferguson, D.E. Kissel, and E.T. Kanemasu. 1986. Ammonia loss from applications of urea-ammonium nitrate solution to straw residue. Soil Sci. Soc. Am. J. 50:969–974.

McSwiney, C.P., and P. Robertson. 2005. Nonlinear response of N_2O flux to incremental fertilizer addition in a continuous maize (*Zea mays* L.) cropping system. Global Change Biol. 11:1712–1719.

Melkonian, J.J., H.M. van Es, A.T. DeGaetano, and L. Joseph. 2008. ADAPT-N: Adaptive nitrogen management for maize using high-resolution climate data and model simulations. *In* R. Kosla (ed.) Proceedings of the 9th International Conference on Precision Agriculture. Denver, CO. July 20–23, 2008 (CD-ROM).

Mengel, D.B., D.W. Nelson, and D.M. Huber. 1982. Placement of nitrogen fertilizers for no-till and conventional till corn. Agron. J. 74:515–518.

Mikkelsen, R.L. 1997. Agricultural and environmental issues in the management of swine waste, p. 110–119. *In* J.E. Rechcigl and H.C. MacKinnon (eds.) Agricultural uses of by-products and wastes. American Chemical Society, Washington, DC.

Millar, N., G.P. Robertson, P.R. Grace, R.J. Gehl, and J.P. Hoben. 2010. Nitrogen fertilizer management for nitrous oxide (N_2O) mitigation in intensive corn (Maize) production: An emission reduction protocol for US Midwest agriculture. Mitig. Adapt. Strat. Global Change 15:185–204.

Mitchell, W.H., and M.R. Teel. 1977. Winter-annual cover crops for no-tillage corn production. Agron. J. 69:569–573.

Mosier, A.R. 2002. Environmental challenges associated with needed increases in global nitrogen fixation. Nutr. Cycl. Agroecosyst. 63:101–116.

Mulvaney, R.L., S.A. Khan, and T.R. Ellsworth. 2005. Need for a soil-based approach in managing nitrogen fertilizers for profitable corn production. Soil Sci. Soc. Am. J. 70:172–182.

Mulvaney, R.L., S.A. Khan, and C.S. Mulvaney. 1997. Nitrogen fertilizers promote denitrification. Biol. Fertil. Soils 24:211–220.

Muñoz, G.R., J.M. Powell, and K.A. Kelling. 2003. Nitrogen budget and soil N dynamics after multiple application of unlabeled or ^{15}nitrogen-enriched dairy manure. Soil Sci. Am. J. 67:817–825.

National Research Council. 1972. Hazards of nitrate, nitrite, and nitrosamines to man and livestock, p. 46–75. In Accumulation of nitrate. National Academy of Science, Washington, DC.

Nicholson, F.A., B.J. Chambers, and K.A. Smith. 1996. Nutrient composition of poultry manures in England and Wales. Bioresour. Technol. 58:279–284.

Nommik, H., and K. Vahtras. 1982. Retention and fixation of ammonium and ammonia in soils. In F.J. Stevenson (ed.) Nitrogen in agricultural soils. Agronomy 22:123–171.

Noxon, J.F. 1976. Atmospheric nitrogen fixation by lightning. Geophys. Res. Lett. 3:463–465.

Odhiambo, J.J.O., and A.A. Bomke. 2001. Grass and legume cover crop effects on dry matter and nitrogen accumulation. Agron. J. 93:299–307.

Olness, A. 1999. A description of the general effect of pH on formation of nitrate in soils. J. Plant Nutr. Soil Sci. 162:549–556.

Olson, R.A., and L.T. Kurtz. 1982. Crop nitrogen requirements, utilization, and fertilization. In F.J. Stevenson (ed.) Nitrogen in agricultural soils. Agronomy 22:567–604.

Omay, A.B., C.W. Rice, L.D. Maddux, and W.B. Gordon. 1998. Corn yield and nitrogen uptake in monoculture and in rotation with soybean. Soil Sci. Soc. Am. J. 62:1596–1603.

Overrein, L.N., and P.G. Moe. 1967. Factors affecting urea hydrolysis and ammonia volatilization in soil. Soil Sci. Soc. Am. Proc. 31:57–61.

Parr, J.F., Jr., and R.I. Papendick. 1966. Retention of ammonia in soils, p. 213–236. In M.H. McVickar et al. (eds.) Agricultural anhydrous ammonia technology and use. Agricultural Ammonia Institute, Memphis, TN and ASA, SSSA, Madison, WI.

Peck, T.R., and P.N. Soltanpour. 1990. The principles of soil testing, p. 1–9. In R.L. Westerman (ed.) Soil testing and plant analysis. 3rd edn. SSSA, Madison, WI.

Power, J.F., and W.A. Dick (eds.). 2000. Land application of agricultural, industrial, and municipal by-products. SSSA Book Series No. 6. SSSA, Madison, WI.

Pratt, P.F., F.E. Broadbent, and J.P. Martin. 1973. Using organic wastes as nitrogen fertilizers. Calif. Agric. 27:10–13.

Randall, G.W., and J.A. Vetsch. 2005. Nitrate losses in subsurface drainage from a corn-soybean rotation as affected by fall and spring application of nitrogen and nitrapyrin. J. Environ. Qual. 34:590–597.

Ranells, N.N., and M.G. Wagger. 1996. Nitrogen release from grass legume cover crop monocultures and bicultures. Agron. J. 88:777–782.

Recous, S., B. Mary, and G. Faurie. 1990. Microbial immobilization of ammonium and nitrate in cultivated soils. Soil Biol. Biochem. 22:913–922.

Rice, C.W., P.E. Sierzega, J.M. Tiedje, and L.W. Jacobs. 1988. Simulated denitrification in the microenvironment of a biodegradable organic waste injected into soil. Soil Sci. Soc. Am. J. 52:102–108.

Rice, C.W., and J.M. Tiedje. 1989. Regulation of nitrate assimilation by ammonium in soils and in isolated soil microorganisms. Soil Biol. Biochem. 21:597–602.

Rochette, P., D.A. Angers, M.H. Chantigny, and N. Bertrand. 2007. Nitrous oxide emissions respond differently to no-till in a loam and a heavy clay soil. Soil Sci. Soc. Am. J. 72:1363–1369.

Rodhe, H., F. Dentener, and M. Schulz. 2002. The global distribution of acidifying wet deposition. Environ. Sci. Technol. 36:4382–4388.

Rorison, I.H. 1985. Nitrogen source and the tolerance of *Deschampsia flexuosa*, *Hocus lanatus* and *Bromus erectus* to aluminium during seedling growth. J. Ecol. 73:83–90.

Ryden, J.C., and L.J. Lund. 1980. Nitrous oxide evolution from irrigated land. J. Environ. Qual. 9:387–393.

Sainju, U.M., and B.P. Singh. 2001. Tillage, cover crop, and kill-planting date effects on corn yield and soil nitrogen. Agron. J. 93:878–886.

Sainz Rozas, H.R., H.E. Echeverría, and L.I. Picone. 2001. Denitrification in maize under no-tillage: Effect of nitrogen rate and application time. Soil Sci. Soc. Am. J. 65:1314–1323.

Schepers, J.S., D.D. Francis, M.F. Vigil, and F.E. Below. 1992. Comparison of corn leaf nitrogen concentration and chlorophyll meter readings. Commun. Soil Sci. Plant Anal. 23:2173–2187.

Schlegel, A.J., D.W. Nelson, and L.E. Sommers. 1986. Field evaluation of urease inhibitors for corn production. Agron. J. 78:1007–1012.

Schomberg, H.H., and D.M. Endale. 2004. Cover crop effects on nitrogen mineralization and availability in conservation tillage cotton. Biol. Fertil. Soils 40:398–405.

Schrader, L.E., D. Domska, P.E. Jung, Jr., and L.A. Peterson. 1972. Uptake and assimilation of ammonium-N and nitrate-N and their influence on the growth of corn (*Zea mays* L.). Agron. J. 64:690–695.

Sharpe, R.R., H.H. Schomberg, L.A. Harper, D.M. Endale, M.B. Jenkins, and A.J. Franzluebbers. 2004. Ammonia volatilization from surface-applied poultry litter under conservation tillage management practices. J. Environ. Qual. 33:183–1188.

Shipley, P.R., J.J. Meisinger, and A.M. Decker. 1992. Conserving corn fertilizer nitrogen with winter cover crops. Agron. J. 84:869–876.

Smiciklas, K.D., and F.E. Below. 1992. Role of nitrogen form in determining yield of field-grown maize. Crop Sci. 32:1220–1225.

Smil, V. 1999. Nitrogen in crop production: An account of global flows. Global Biogeochem. Cycles 13:647–662.

Stanford, G., and E. Epstein. 1974. Nitrogen mineralization-water relations in soils. Soil Sci. Soc. Am. Proc. 38:103–107.

Staver, K.W., and R.B. Brinsfield. 1998. Using cereal grain winter cover crops to reduce groundwater nitrate contamination in the mid-Atlantic coastal plain. J. Soil Water Conserv. 53:230–240.

Stevenson, F.J. 1982. Organic forms of soil nitrogen. In F.J. Stevenson (ed.) Nitrogen in agricultural soils. Agronomy 22:67–122.

Thompson, R.B., and J.J. Meisinger. 2002. Management factors affecting ammonia volatilization from land-applied cattle slurry in Mid-Atlantic USA. J. Environ. Qual. 31:1329–1338.

Timmons, D.R. 1984. Nitrate leaching as influenced by water application level and nitrification inhibition. J. Environ. Qual. 13:305–309.

Toner, C.V., IV, D.L. Sparks, and T.H. Carski. 1989. Anion exchange chemistry of Middle Atlantic soils: Charge properties and nitrate reduction kinetics. Soil Sci. Soc. Am. J. 53:1061–1067.

Torello, W.A., and D.J. Wehner. 1983. Urease activity in a Kentucky bluegrass turf. Agron. J. 75:654–656.

Touchton, J.T., and W.L. Hargrove. 1982. Nitrogen sources and methods of application for no-tillage corn production. Agron. J. 74:823–826.

Touchton, J.T., R.G. Hoeft, and L.F. Welch. 1978. Effect of nitrapyrin on nitrification of fall and spring-applied anhydrous ammonia. Agron. J. 70:805–810.

Unger, P.W., and M.F. Vigil. 1998. Cover crop effects on soil water relationships. J. Soil Water Conserv. 53:200–207.

Vanotti, M.B., and L.G. Bundy. 1994. Frequency of nitrogen fertilizer carryover in the humid Midwest. Agron. J. 86:881–886.

Vanotti, M.B., and L.G. Bundy. 1995. Soybean effects on soil nitrogen availability in crop rotations. Agron. J. 87:676–680.

Varco, J.J., W.W. Frye, M.S. Smith, and C.T. MacKown. 1989. Tillage effects on nitrogen recovery by corn from a nitrogen-15 labeled legume cover crop. Soil Sci. Soc. Am. J. 53:822–827.

Vaughan, J.D., and G.K. Evanylo. 1999. Soil nitrogen dynamics in winter cover crop-corn systems. Commun. Soil Sci. Plant Anal. 30:31–52.

Vetsch, J.A., and G.W. Randall. 2004. Corn production as affected by nitrogen application timing and tillage. Agron. J. 96:502–509.

Viets, F.G., Jr. 1965. The plant's need for and use of nitrogen. In W.V. Bartholomew and F.E. Clark (eds.) Soil nitrogen. Agronomy 10:503–549.

Volk, G.M. 1959. Volatile loss of ammonia following surface application of urea to turf or bare soil. Agron. J. 51:746–749.

Wagger, M.G. 1989a. Cover crop management and nitrogen rate in relation to growth and yield of no-till corn. Agron. J. 81:533–538.

Wagger, M.G. 1989b. Time of desiccation effects on plant composition and subsequent nitrogen release from several winter annual cover crops. Agron. J. 81:236–241.

Wagger, M.G., M.L. Cabrera, and N.N. Ranells. 1998. Nitrogen and carbon cycling in relation to cover crop residue quality. Soil Water Conserv. 53:214–218.

Watts, D.G., and R.J. Hanks. 1978. A soil-water-nitrogen model for irrigated corn on sandy soils. Soil Sci. Soc. Am. J. 42:492–499.

Weinert, T.L., W.L. Pan, M.R. Moneymaker, G.S. Santo, and R.G. Stevens. 2002. Nitrogen recycling by nonleguminous winter cover crops to reduce leaching in potato rotations. Agron. J. 94:365–372.

Wickramasinghe, K.N., G.A. Rodgers, and D.S. Jenkinson. 1985. Transformations of nitrogen fertilizer in soil. Soil Biol. Biochem. 17:625–630.

Zibilske, L.M. 1987. Dynamics of nitrogen and carbon in soil during papermill sludge decomposition. Soil Sci. 143:26–33.

11.2 Phosphorus Availability

Andrew Sharpley

11.2.1 Introduction

Phosphorus (P) is an essential nutrient for plant growth. The low concentration (100–3000 mg P kg^{-1}) and solubility (0.01–0.10 mg P L^{-1}) of P in soils, however, make it a critical nutrient limiting plant growth. In natural ecosystems, P availability in soil is controlled by the sorption, desorption, and precipitation of P released during weathering and dissolution of rocks and minerals of low solubility. Thus, soil P availability is generally inadequate for crop needs in production agriculture. To meet these needs, P is added as fertilizers or animal manures to buildup or maintain soil P availability at predetermined optimum levels. In this section, P availability is defined as that P in soil or water that is available by desorption and dissolution processes for uptake by plants in terrestrial and aquatic ecosystems.

The components, forms, availability, and cycling of P in soil are conceptualized in Figure 11.1. Complex and interrelated processes determine the amounts and availability of several inorganic and organic forms of P in soil. This section will describe these processes occurring in the soil, how they are affected by agricultural management, how we attempt to optimize soil P availability for crop production, and how this can affect water quality.

11.2.2 Forms and Amounts in Soil

Soil P exists in inorganic (P_i) and organic (P_o) forms (Figure 11.1). Inorganic P forms are dominated by hydrous sesquioxides,

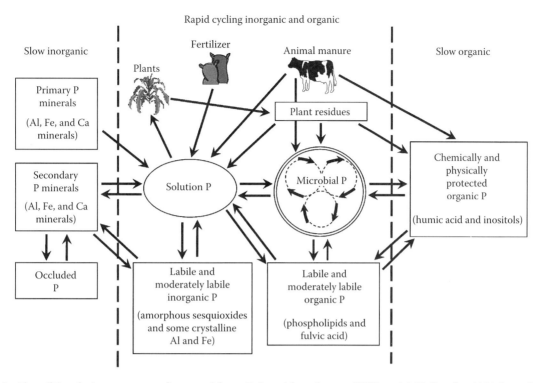

FIGURE 11.1 The soil P cycle: its components, forms, and flows. (Adapted from Stewart, J.W.B., and A.N. Sharpley. 1987. Controls on dynamics of soil and fertilizer phosphorus and sulfur, p. 101–121. *In* R.F. Follett, J.W.B. Stewart, and C.V. Cole (eds.) Soil fertility and organic matter as critical components of production systems. Soil Science Society of America Special Publication No. 19. ASA, CSSA, SSSA, Madison, WI.)

amorphous, and crystalline Al and Fe compounds in acidic, noncalcareous soils, and by Ca compounds in alkaline, calcareous soils (Figure 11.1 and Table 11.1). Organic P forms include relatively labile phospholipids and fulvic acids; more resistant forms comprise inositols and humic acids (Figure 11.1). Forms generalized in Figure 11.1 are not discrete entities, as intergrades and dynamic transformations between forms occur continuously to maintain equilibrium conditions. These approximated forms of P are assigned based on the extent to which sequential extractants of increasing acidity or alkalinity can dissolve soil P (Hedley et al., 1982; Tiessen and Moir, 2007; Zhang and Kovar, 2008).

In most soils, the P content of surface horizons is greater than subsoil, due to the sorption of added P and greater biological activity and accumulation of organic material in surface layers. However, soil P content varies with parent material, extent of pedogenesis, soil texture, and management factors, such as rate and type of P applied and soil cultivation. These factors also influence the relative amounts of P_i and P_o. In most soils, 50%–75% is P_i, although this fraction can vary from 10% to 90% (Table 11.1). Generally, Ca-P_i decreases with weathering, whereas amorphous Al and Fe-P_i and P_o forms tend to increase, due in part to changes in soil clay fraction from basic primary minerals to Al- and Fe-dominated oxides (Table 11.1).

Phosphorus additions are usually needed to maintain adequate available soil P for plant uptake. The level of these additions varies with both soil and plant type (Pierzynski and Logan, 1993). Once applied, P is either taken up by the crop, becomes weakly (physical) or strongly (chemical) adsorbed onto Al, Fe, and Ca surfaces, or incorporated into organic P (McLaughlin et al., 1988; Ottabong et al., 1997; Zhang and MacKenzie, 1997) (Figure 11.2). As P_i generally supplies most of the P taken up by crops in productive agricultural soils, more attention has been focused on the availability of P_i rather than P_o, following P additions.

Overall, soil pH is the main property controlling the nature of P_i forms, although Al, Fe, Mn, and Ca content determine the

TABLE 11.1 Average Amounts of Inorganic P (P_i) and Organic P (P_o) in the A1 Horizon of Virgin Calcareous, Slightly Weathered, and Highly Weathered Soils

P Form	Calcareous (n = 41)	Slightly Weathered (n = 40)	Highly Weathered (n = 39)
		(mg kg^{-1})	
Bioavailable P_i[a]	11	21	22
Amorphous Al and Fe-P_i	37	74	109
Ca-P_i	285	85	16
Labile P_o	8	18	34
Protected P_o	28	60	78
Residual P[a]	152	254	179
Total P	521	512	438

Source: Adapted from Sharpley, A.N., H. Tiessen, and C.V. Cole. 1987. Soil phosphorus forms extracted by soil tests as a function of pedogenesis. Soil Sci. Soc. Am. J. 51:362–365.

[a] Bioavailable P_i is resin P_i and residual P is chemically resistant, mineral, and occluded P_i and P_o as designated on Figure 11.1.

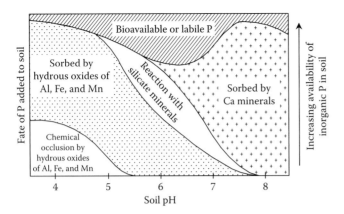

FIGURE 11.2 Approximate representation of the fate of P added to soil by sorption and occlusion in inorganic forms, as a function of soil pH. (Adapted from Buckman, H.O., and N.C. Brady. 1970. The nature and properties of soils. The Macmillan Publishing Co., Collier-Macmillan Limited, London, U.K.)

amounts of these forms (Figure 11.2). In acid soils, amorphous and hydrous oxides of Al, Fe, and Mn dominate P sorption processes, while Ca compounds dominate P sorption and precipitation reactions in alkaline soils. As a result, P availability is greatest at soil pH between 6 and 7 (Figure 11.2). Immobilization of P_i by these processes renders a portion of the added P unavailable for plant uptake (Figure 11.3). Mehlich-3 soil P decreased with time after application of P to a Kingsbury clay (*Aeric Ochraqualfs*; pH of 5.7) and Hagerstown silt loam (*Typic Hapludalfs*; pH of 6.8) incubated at room temperature (25°C) and field moisture (about 30% water). At the same time, P_i becomes more tightly bound with Al and Fe complexes (Figure 11.3). This simple soil-P incubation illustrates why the removal of P_i from soil by crops is generally low. For the United States, an average 29% of P added in fertilizer and manures is removed by harvested crops, ranging from <1% in Hawaii to 71% in Wyoming (National Research Council, 1993). The low recovery reflects the predominance of high P fixing soils in Hawaii.

Even though P_i has generally been considered the major source of plant-available P in soils, the mineralization of labile P_o has also been shown to be important in both low-fertility and high-fertility soils (Stewart and Tiessen, 1987; Oehl et al., 2001; Chen et al., 2002; Condron et al., 2005). Amounts of P_o mineralized in temperate dryland soils range from 5 to 20 kg P ha^{-1} year^{-1} (Stewart and Sharpley, 1987). Mineralization of soil P_o tends to be higher in the tropics (67 to 157 kg P ha^{-1} year^{-1}), where distinct wet and dry seasons and higher soil temperatures enhance microbial activity. In contrast, P_o compounds may also become resistant to hydrolysis by phosphatase through complexation with Al and Fe (Tate, 1984).

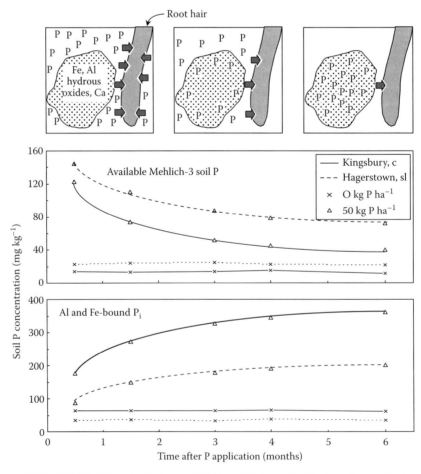

FIGURE 11.3 The change available (Mehlich-3 P) and adsorbed soil P (Al- and Fe-bound) with time after application of P.

11.2.3 Principles of Analysis

The various forms and amounts of P_i and P_o in soil can be estimated by extraction with acids and alkalies that dissolve specific complexes binding P. The most common methods and their background are given by Olsen and Sommers (1982), Turner et al. (2005), Tiessen and Moir (2007), Kovar and Pierzynski (2008). A wide range in methods and principles exists for effective recovery of P based on soil type, environment, and level of detail required. This is particularly true for available soil P_i estimation, which has traditionally been based on acid dissolution (acetic, citric, hydrochloric, lactic, nitric, and sulfuric), anion exchange (acetate, bicarbonate, citrate, lactate, and sulfate), cation complexation (citrate, fluoride, and lactate), or cation hydrolysis (buffered bicarbonate). Several excellent reviews of these methods are available for further reading (Kamprath and Watson, 1980; Fixen and Grove, 1990). In the United States, the most common soil P tests have been Mehlich-1 (Mehlich, 1953), Bray-1 (Bray and Kurtz, 1945), and Olsen P (Olsen et al., 1954); however, most laboratories have converted to the multi-element Mehlich-3 (Mehlich, 1984) or AB-DTPA (Soltanpour and Schwab, 1977) extractants, in the past decade (Carter and Gregorich, 2007; Sims, 2008).

It is unlikely that an extractant would exclusively measure a single pool of soil P_i, although some components of extractants are aimed at specific pools. For example, F in the Bray-1 extractant exchanges with Al-bound P_i, with the assumption that this Al-bound P_i contributes to available P in acid soils, while the inclusion of EDTA in Mehlich-3 means this extractant estimates P associated with Fe and Al amorphous complexes. The success of any extractant to estimate available P depends on the appropriateness of the chemical used relative to dominant soil properties controlling P reactivity. Alternative methods utilize P-sinks, such as anion-exchange resins (AER), ion-exchange membranes, and Fe-oxide impregnated paper, to determine the quantity of soil P available to plants with negligible chemical extraction (Sharpley et al., 1994). These methods more closely mimic rhizosphere conditions and often provide comparable or better correlations with crop response than chemical extractants (Table 11.2).

11.2.3.1 Anion-Exchange Resins

AER are the most common P-sink method for assessing available soil P_i. The procedure typically involves the use of chloride saturated resin at a 1:1 resin-to-soil ratio in 10–100 mL of water or weak electrolyte for 16–24 h (Amer et al., 1955; Olsen and Sommers, 1982). To prevent the diffusion of P from the soil to the resin from being the rate-limiting step, resins should be intimately mixed with the soil, which creates difficulties in separating resin from soil for P analysis. Soil can be ground to a smaller size than the resin, but this probably changes soil P release characteristics. Resin and soil may also be separated by enclosing the resin in a mesh bag, which may limit resin–soil contact or float resin from soil in a sucrose solution (Thien and Myers, 1991). Skogley et al. (1990) encapsulated a mixture of anion- and cation-exchange resins in a mesh sphere. Greenhouse studies indicated that the correlation between P uptake by sorghum-sudangrass and resin-sphere results were as good or better than those with the Olsen P soil test (Table 11.2).

11.2.3.2 Ion-Exchange Membranes

A similar approach using ion-exchange resin impregnated membranes has been investigated by several researchers (Abrams and Jarrell, 1992; Qian et al., 1992; Saggar et al., 1992). Impregnation of the resin onto a plastic membrane facilitates separation of the resin beads from the soil and may eliminate the soil grinding step. Also, an extraction time as short as 15 min can be used, without reducing the accuracy of predicted P availability for a wide range of soils (Qian et al., 1992). In pot studies, the resin membranes have provided a better index of P availability than conventional chemical extraction methods for canola and ryegrass (Table 11.2).

11.2.3.3 Iron-Oxide Impregnated Paper

Another P sink that has received attention is Fe-oxide impregnated filter paper (Fe–O strip), which has successfully estimated available P_i in a wide range of soils and management systems (Menon et al., 1989b, 1990; Sharpley, 1991; Chardon, 2008) (Table 11.2).

Wide-spread adoption of P-sink methods for routine soil testing has not yet occurred in the United States, although parts of Brazil have used the method for the last decade (van Raij et al., 1986). As the P-sink methods operate with limited chemical extraction, they are more suited to a wide range of soils, irrespective of management history (Yang et al., 1991; Qian et al., 1992; Somasiri and Edwards, 1992; Myers et al., 2005). Where fertilizer history is unknown and frequent changes in fertilizer type, including rock phosphate, may have been made, it is difficult to choose the appropriate soil test. For example, Olsen P can underestimate and Bray-1 P overestimate P availability in soils amended with rock P, while P-sink methods have provided accurate estimates when KCl rather than $CaCl_2$ is used as the support medium (van Raij et al., 1986; Menon et al., 1989a; Saggar et al., 1992). Even so, detailed field calibration and improvement in standardized methodology will be essential before any of the P-sink approaches can be used routinely to estimate available soil P and make reliable fertilizer recommendations.

Phosphorus isotopes P^{32} and P^{33} have also been widely used to characterize soil P availability. Using laboratory incubation, greenhouse pot, and limited field plot studies, valuable insights have been gained into soil P availability in terms of exchange kinetics, plant-available forms, and the rate, extent, and direction of P cycling in soils. Readers are directed to excellent reviews by Fardeau et al. (1996), Di et al. (1997), and Frossard and Sinaj (1997) for more information.

11.2.4 Cycling in Soil

Dynamic processes involved in P cycling are chemical and biological (Figure 11.1). Chemical processes include precipitation–dissolution and sorption–desorption. Biological processes involve immobilization–mineralization initiated by P uptake

TABLE 11.2 Relationship between Crop Uptake of P and Bioavailable Soil P Determined by P Sink and Chemical Extraction

Crop	Location	Number of Soils	Soil P Test	Correlation Coefficient	Reference
Canola	Saskatchewan	135	IEM	0.92	Qian et al. (1992)
			Olsen	0.87	
Cotton	Brazil	28	AER	0.85	van Raij et al. (1986)
			0.02 M H_2SO_4	0.68	
Maize	Alabama, USA	32	Fe–O strip	0.87	Menon et al. (1989b)
			AER	0.62	
			Olsen	0.81	
			Bray 1	0.74	
Maize[a]	Australia	2	Fe–O strip	0.91	Kumar et al. (1992)
			IEM	0.91	
			Bray 1	0.87	
Maize	Egypt	10	Fe–O strip	0.90	Monem and Gadalla (1992)
			Olsen	0.91	
Maize	Uganda	2	Fe–O strip	0.85	Butewaga et al. (1996)
			Bray 1	0.77	
			Mehlich 1	0.64	
Rice	Brazil	8	AER	0.98	van Raij et al. (1986)
			AB-DTPA	0.41	
Ryegrass[a]	New Zealand	56	IEM	0.92	Saggar et al. (1992)
			Olsen	0.87	
Ryegrass	Finland	32	Fe–O Strip	0.93	Yli-Halla (1990)
			Olsen	0.87	
Sudangrass	Colorado, USA	23	AER	0.92	Bowman et al. (1978)
Sorghum/barley			Olsen	0.88	
WheatI	Australia	2	Fe–O strip	0.96	Kumar et al. (1992)
			IEM	0.97	
			Bray 1	0.98	
Wheat	China	39	Fe–O Strip	0.84	Lin et al. (1991)
			AER	0.83	
			Olsen	0.83	
			Bray 1	0.56	

Note: AEM, anion–exchange membrane; IEM, ion–exchange resin; AB-DTPA, ammonium bicarbonate diethylene triamine pentaacetic acid; PST, resin phytoavailability soil test.

[a] Relationship between relative crop yield and soil P test.

and decay of plants. Also, soil faunal and floral activities often modify the direction, extent, and rate at which these chemical and biological processes occur.

11.2.4.1 Chemical Processes

Precipitation–dissolution processes differ from sorption–desorption in that the solubility product of the least soluble P compound in the solid phase controls dissolution and, thus solution P concentration, whereas solution P controls the amount of P sorbed (Syers and Curtin, 1989). In reality, retention of P by soil material is a continuum between precipitation and surface reactions.

11.2.4.1.1 Precipitation–Dissolution

In general, Ca controls these reactions in neutral or calcareous environments, while Al and Fe are the dominant controlling cations in acidic environments. Apatite is the most common primary P mineral ($Ca_{10}(PO_4)_6 X_2$, where X is F or OH). Apatite dissolution requires a source of H^+ from soil or biological activity and a sink for Ca and P (Mackay et al., 1986; Smillie et al., 1987):

$$Ca_{10}(PO_4)_2 X_2 + 12H^+ = 10Ca^{2+} + 6H_2PO_4^- + 2X^- \quad (11.3)$$

and occurs during soil development due to weathering (Walker and Syers, 1976). Even though the rate of apatite weathering or dissolution will vary with rainfall and temperature, it is still difficult to predict this first step in the P cycle (Pierzynski, 1991; Pierzynski et al., 2005).

Precipitates in Ca systems occur in the following sequence; monocalcium phosphate [Ca $(H_2PO_4)_2$], dicalcium phosphate dihydrate (Ca HPO_4 $2H_2O$), octacalcium phosphate

[Ca$_8$H$_2$(PO$_4$)$_6$·5H$_2$O], and finally hydroxyapatite [Ca$_{10}$(PO$_4$)$_6$(OH)$_2$] or fluorapatite [Ca$_{10}$(PO$_4$)$_6$F$_2$], which have low solubilities and should, thus, control soil solution P concentration (Lindsay et al., 1989; Syers and Curtin, 1989).

In Al- and Fe-dominated soils, few well-crystallized precipitates have been observed. Generally, P reacts with Al oxides to form amorphous Al P or organized phases such as sterretite [Al(OH$_2$)$_3$ HPO$_4$ H$_2$PO$_4$] and with Fe oxides to such precipitates as tinticite [Fe$_6$ (PO$_4$)$_4$ (OH)$_6$·7H$_2$O] or griphite [Fe$_3$ Mn$_2$ (PO$_4$)$_2$·5(OH)$_2$] (Hsu, 1982; Lindsay et al., 1989). Many other amorphous mixed Al–Fe–Si–P compounds have been observed in soils with high P concentrations from phosphatic parent material or large fertilizer or manure applications (Lindsay et al., 1989; Pierzynski et al., 1990b).

11.2.4.1.2 Sorption/Desorption

In most soils, sorption–desorption processes describe P uptake and release by soils that control P availability better than precipitation–dissolution reactions. In this section, the term "sorption" covers surface "adsorption" and subsequent penetration of P into the retaining component or "absorption." Sorption curves or isotherms have been used extensively to describe the relationship between the amount of P sorbed and that remaining in solution (Figure 11.4). For a given solution P concentration, there is a large difference in the amount of P sorbed by the three soil types given as examples. In general, clay content approximates the reactive surface area of a soil responsible for P sorption (Syers et al., 1971; Juo and Fox, 1977; Sharpley et al., 1984a; Hedley et al., 1995). This surface reactivity is a function of the amount and type of hydrous oxides of Al and Fe and reactive Ca components present, other ions (Ca- or Na-dominated), pH of the system, and reaction kinetics.

Even in calcareous soils, hydrous ferric oxides can influence P sorption reactions (Holford and Mattingly, 1975a, 1975b). The types of bonding associated with hydrous oxides and pH dependency have been described in detail by Sample et al. (1980), Uehara and Gillman (1981), and White (1980) (Figure 11.5). Ligand exchange of P at hydrous OH and Fe oxide surfaces results in the formation of monodentate, bidentate, or binuclear complexes (Figure 11.5). As soil solution pH increases, P sorption is decreased by the greater negative charge at the oxide surface and reduced polarization of the Al or Fe–O bond (Figure 11.5).

Soil P sorption has been characterized by parameters calculated from the Langmuir equation:

$$S = \frac{(kS_{max}C)}{1+kC} \quad (11.4)$$

where
- S is sorbed P
- C is solution P concentration
- S_{max} is the maximum amount of P that can be sorbed
- k is an "affinity" constant describing binding energy

Phosphorus sorption maximum is calculated as the reciprocal of the slope of the plot C/S and C and binding energy as the slope/intercept of the same plot (Olsen and Watanabe, 1957; Syers et al., 1973; Graetz and Nair, 2008) (Table 11.3). Equilibrium P concentration is the solution P concentration at which no net sorption or desorption occurs (Figure 11.4 and Table 11.3). These parameters have been widely used to quantify the extent of specific soil reactions and effects of soil types, counter ion and background electrolyte, P source, and soil management on P sorption (Goldberg and Sposito, 1985; Fixen and Grove, 1990). However, the theoretical assumptions of the Langmuir equation were questioned (White, 1980; Barrow, 1989), as sorbed P carries charges that decrease the surface charge and potential of the sorbing surface, leading to large errors in the estimates of S_{max} (Kuo, 1988). The Freundlich equation often fits data better than the Langmuir equation but does not estimate S_{max} (Barrow, 1978).

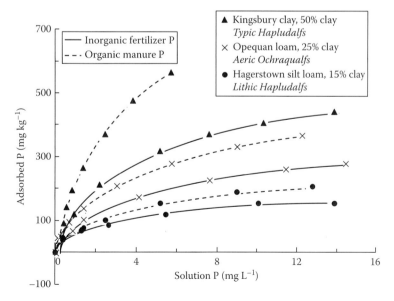

FIGURE 11.4 Phosphorus sorption isotherms as a function of the type of soil and P source.

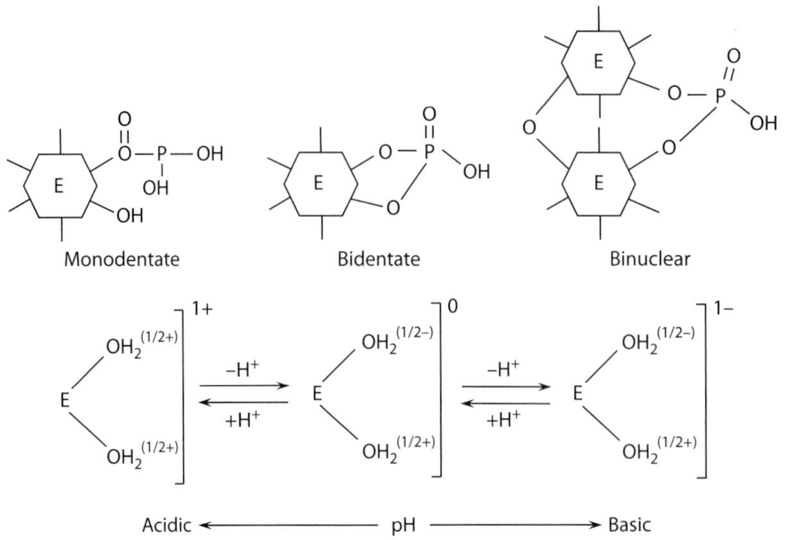

FIGURE 11.5 Types of P bonding associated with hydrous oxides where E is AL or Fe. (Adapted from Fixen, P.E., and J.H. Grove. 1990. Testing soils for phosphorus, p. 141–180. *In* R.L. Westerman (ed.) Soil testing and plant analysis. 3rd edn. SSSA Book Series No. 3. SSSA, Madison, WI.)

TABLE 11.3 Soil P Sorption Properties Calculated from the Langmuir Isotherm as a Function of Soil Type and Source of Added P (Triple Superphosphate and Dairy Manure)

Soil Type	P Sorption Maxima (mg kg^{-1})		Binding Energy (L mg^{-1})		Equilibrium P Concentration (mg L^{-1})	
	Fertilizer	Manure	Fertilizer	Manure	Fertilizer	Manure
Hagerstown, sl *Typic Hapludalfs*	172	245	2.17	2.78	0.019	0.044
Kingsbury, c *Aeric Ochraqualfs*	476	909	2.24	3.82	0.069	0.113
Opequon, l *Lithic Hapludalfs*	303	455	1.97	3.09	0.034	0.114

Note: sl, sandy loam; c, clay; l, loam.

Organic anions can compete with P for similar sorption sites on soil surfaces (Yuan, 1980; Hue, 1991; Ohno and Crannell, 1996). However, P sorption can be increased when the addition of organic compounds to amorphous oxides in soil impedes their crystallization and increases specific surfaces (Table 11.3). Also, humic compounds can complex with Fe, Al, and to a lesser extent Ca and sorb P (Levesque and Schnitzer, 1967; Inskeep and Silvertooth, 1988; Frossard et al., 1995).

The release of sorbed P into solution or desorption is not completely reversible. Thus, desorption curves are displaced to the left of the sorption curves. This hysteresis effect is the result of precipitation (Veith and Sposito, 1977), occlusion (Uehara and Gillman, 1981), solid-state diffusion (Barrow, 1983), and bidentate or binuclear bonding with the colloid surface (Hingston et al., 1974).

Several studies have reported that desorption of soil P during a short period of time is a low-activation energy process (Sharpley et al., 1981). The work of Kuo and Lotse (1974) and Evans and Jurinak (1976) reported that the activation energy for P desorption was between 2 and 3 kcal mol^{-1}. The low-activation energies led these researchers to suggest that the desorption of soil P during short reaction times may be limited more by the diffusion of the desorbed P through the stagnant water films present around the soil particles and within the soil aggregates than by the chemical reaction. Although higher-activation energies (20 kcal mol^{-1}) were reported by Barrow and Shaw (1975), they were determined for a longer reaction period (up to 100 days) and represent the transfer of P between desorbable and fixed forms. It has been suggested that there is a continuous range of activation energies for soil P desorption (Posner and Bowden, 1980). During the initial stages of desorption, P held at low-activation energies is desorbed; during the later stages, P held at higher-activation energies is desorbed.

More detailed analysis and review of these chemical processes controlling soil P availability are given by Barrow (1985), Lindsay et al. (1989), Syers and Curtin (1989), Fixen and Grove (1990), Frossard et al. (1995), Condron et al. (2005), and Pierzynski et al. (2005).

11.2.4.2 Biological Processes

Uptake of soluble P by bacteria and fungi, stimulated by the addition of microbial substrates such as litter and crop residues, and release of P as the result of cell lysis or predation (Coleman et al., 1983; Tiessen et al., 1984; Frossard et al., 2000; Jakobsen et al., 2005) are represented in Figure 11.1 as a revolving wheel. This is done deliberately to emphasize the central role of the microbial population in P cycling and availability. For example, if the wheel is stopped or slowed down by lack of C inputs, the supply of P to plants will be limited to the quantity of labile P_i. If the wheel is operating, then solution P is constantly being replenished from labile P_i and P_o forms. Generally, microbial biomass P ranges from 5 to 100 mg kg^{-1} (Srivastava and Singh, 1991; Joergensen et al., 1995; Oberson et al., 1997). As a result, large amounts of P can turnover through microbial biomass annually (He et al., 1997; Oehl et al., 2001; Jakobsen et al., 2005). For example, in a study of P cycling through soil microbial biomass in England, Brookes et al. (1984) measured annual P fluxes of 5 and 23 kg P ha^{-1} year^{-1} in soils under continuous wheat and permanent grass, respectively. Although biomass P flux under continuous wheat was less than P uptake by the crop (20 kg P ha^{-1} year^{-1}), annual P flux in the grassland soils was greater than P uptake by the grass (12 kg P ha^{-1} year^{-1}).

Within the microbial cell, P exists as a wide variety of compounds, principally RNA (30%–50%), acid soluble P_i and P_o (15%–20%), phospholipids (<10%), and DNA (5%–10%) (Stewart and McKercher, 1982; Condron et al., 2005). If the microbial cell is ruptured or lysed, all these compounds will be released to the soil solution to react with both inorganic and organic soil components to form a host of P_i and P_o compounds of differing solubility or susceptibility to mineralization. The rate of mineralization of P_o forms depends largely on phosphatase activity, which, in turn, can be controlled by solution P concentration (McGill and Cole, 1981; Chen et al., 2002). Stable P_o accumulates in both chemically resistant and aggregate protected forms (Marshall, 1971; Tisdale and Oades, 1982; Condron et al., 2005).

Chemically or physically protected P_o may be slowly mineralized as a by-product of the mineralization of overall soil organic matter or by specific enzyme action in response to the need for P. Therefore, organic matter turnover, as well as solution P_i concentration and the demand for P by microbial and plant components, will be factors controlling the lability of P_o (Richardson, 2000; Turner et al., 2002; Condron et al., 2005). A continuous drain on soil P pools by cultivation and crop removal will rapidly deplete labile P_i and P_o forms and, thereby, reduce available soil P (Hedley et al., 1982; Sharpley and Smith, 1985; Tiessen et al., 1992).

11.2.4.3 Fauna and Flora Processes

The effects of soil fauna (e.g., earthworms and termites) and flora (actinomycetes, bacteria, and fungi) on soil physical, chemical, and biological processes have been extensively studied and reviewed (Reichle, 1977; Lee, 1985; Edwards, 2004). This section outlines the role of the more important of these, earthworms and mycorrhizae, in soil P cycling and their impact on soil P availability.

11.2.4.3.1 Earthworms

Under favorable soil temperature and moisture conditions, earthworms have been found annually to consume 100% of the litter of an evergreen oak forest in Japan (8 Mg ha^{-1} year^{-1} by *A. pheretima*; Sugi and Tanaka, 1978) and mixed deciduous forest in England (3 Mg ha^{-1} year^{-1} by *Lumbricus terrestris*; Satchell, 1967); about 30% of the litter decomposed each year in grass savanna (Lavelle, 1978) and 100% of added cattle manure (17–30 Mg ha^{-1} year^{-1}, Guild, 1955). During this consumption, earthworms commonly ingest 100–500 Mg soil ha^{-1} year^{-1} (equivalent to 0.5–3.0 cm of top soil) (Russell, 1973). Consequently, earthworms can have a major influence on soil physical, chemical, and biological properties through incorporation and assimilation of plant litter and associated ingestion of soil material (Lee, 1985; Lavelle, 1988). Egestion as surface and subsurface cast material can rapidly redistribute P in a soil profile, increasing the availability and potential for uptake. Also, earthworm burrowing can allow a greater soil volume to be exploited by plant roots and decrease susceptibility of plants to water stress. Earthworm burrows open to the soil surface increase infiltration rates, enhancing nutrient movement into the profile via macropore flow (Sharpley et al., 1979; Germann et al., 1984; Edwards et al., 1989; Shuster et al., 2003; Dominguez et al., 2004).

The main source of P affected by earthworms is soil organic matter, which includes plant litter in various stages of decomposition, roots, and organic matter in the soil with its complement of microflora and fauna. Ingestion, maceration, and intimate mixing of organic matter with soil increase the rate of humification and mineralization. Although most organic matter undergoes little chemical change during passage through the earthworms' gut, it is finely ground with the increased surface area exposed to microbial activity, facilitating further decomposition. The 5- to 10-fold increase in P content and availability in earthworm casts results from enhanced mineralization of organic P, enrichment of clay-sized particles in casts, and a reduction in P sorption capacity of soil by organic matter blockage (Sharpley and Syers, 1977; Sharpley et al., 1992). Thus, most of the additional P present in casts is held in more physically sorbed than chemisorbed forms, which are readily available to plants.

Both microbial population and activity are increased during passage through the earthworms' gut as a function of the organic matter content of the initial food source (Tiwari, 1979; Gorbenko et al., 1986; Scheu, 1987). Parle (1963) observed that the numbers of bacteria and actinomycetes increased 1000-fold during passage through the gut, and oxygen consumption remained higher in earthworm casts than in soil for 50 days, indicating an increased microbial activity. This enhanced activity is probably responsible for the increased phosphatase activity found in earthworm casts compared to underlying soil (Sharpley and Syers, 1977).

As a result of the incorporation and decomposition of plant material in soil via earthworm activity, soil fertility and productivity can be improved (Lavelle et al., 1989; Edwards et al., 1995; Katsvairo et al., 2002). For instance, Katsvairo et al. (2002) found that earthworm activity contributed to a 15%–40% increase in corn yields in a soybean (*Glycine max* [L.] Merr.)—wheat (*Triticum aestivum* L.)/clover (*Trifolium pretense* L.)—corn (*Zea mays* L.) rotation on a Kendaia–Lima silt loam in Aurora, New York.

Differences in tillage operation and frequency; amount, type, and quantity of residues left on the soil surface; soil type and climate; and application of manures can influence earthworm populations, species composition, and activity (Pankhurst et al., 1995; Edwards and Bohlen, 1996). For example, several studies have reported an increase in the population and activity of earthworms under reduced compared to conventional tillage practices, concomitant with an increase in food or energy supply (Edwards et al., 1989; Trojan and Linden, 1998; Shipitalo et al., 2000).

The application of manure has been shown to encourage the buildup of earthworm populations in cropped soils to a greater extent than in grass, where there tends to be more decaying organic matter (i.e., food) than in cropped soils (Edwards, 1980; Satchell, 1983). For instance, application of dairy manure (50–400 kg P ha^{-1}) to a rape (*Brassica napus* L.)—sugar beet (*Beta vulgaris* L.)—grass rotation increased the numbers of surface feeding *L. terrestris* by directly increasing their food supply and associated microorganisms (Andersen, 1983). Similarly, the application of dairy manure (20 Mg manure ha^{-1} year^{-1} for 14 year) increased earthworm population, diversity, and activity (mainly *Aporrectodea* and *Lumbricus* genus) in a Le Bras clay loam in Quebec (Estevez et al., 1996).

The potential for earthworm activity to incorporate surface applied manures was demonstrated by Chardon et al. (2007). Disappearance of manure patches deposited by grazing cattle, was partially attributed to earthworm activity, with only 15% of dry matter and 22% of total P in manure patches remaining 234 days after deposition on a permanently grassed sandy soil from Heino, the Netherlands (Chardon et al., 2007). While the manure can remain a long-term source of P to soil leachate, the overall risk of P runoff was reduced as the grassland soil could retain most of the P leached from patches.

11.2.4.3.2 Mycorrhizal Associations

Mycorrhizae are widespread under natural conditions, with vesicular-arbuscular mycorrhizae (VAM) most common in agricultural soils (Smith and Read, 1997; Jakobsen et al., 2005). In fact, VAM are formed by approximately two-thirds of all plants species (Fitter and Moyersoen, 1996) and fungi belonging to the recently established phylum Glomeromycota (Schüssler et al., 2001). Three primary mechanisms by which VAM enhances soil P availability are increased physical exploration of the soil, chemical modification of the rhizosphere, and physiological differences between VAM and plant roots (Read, 2002). For example, extensive hyphal growth of VAM reduced the distance for diffusion of P in soil, thereby increasing uptake. This effect is greater when diffusion limits uptake (Gerdemann, 1968; Facelli and Facelli, 2002). Consequently, a greater response to VAM infection has been exhibited in coarse than in fine rooted plant species (Crush, 1973), in high than in low P sorbing soils (Yost and Fox, 1979), and in soils than in solution culture (Howeler et al., 1982). In addition, the generally smaller diameter of VAM hyphae (2–4 μm) compared to root hairs (710 μm) affords a greater absorptive surface area for hyphae and enables entry of hyphae into soil pores and organic matter that cannot be entered by root hairs (Dodd et al., 2000; Drew et al., 2003).

VAM may chemically modify the rhizosphere through exudation of chelating compounds (Jayachandran et al., 1989; Treeby et al., 1989) or phosphatases (Harley, 1989), which could solubilize poorly soluble soil P. It is clear, however, that VAM utilizes the same sources of P as nonmycorrhizal plants (Mosse et al., 1973; Kucey et al., 1989; Blal et al., 1990), but do so more efficiently.

On a unit weight basis, mycorrhizal plants can rapidly absorb larger amounts of P than nonmycorrhizal plants (Bolan et al., 1987). This difference cannot be accounted for by increased surface area of the hyphae alone; it is, therefore, possible that a greater affinity for P in mycorrhizal plants increased absorption rates. It is also possible that the critical or minimum P concentration, below which there is limited net absorption of P, is lower for mycorrhizal than nonmycorrhizal plants because of an increase in physical contact between hyphae and P, thereby reducing P diffusion distance.

For more detailed information on the role of arbuscular mycorrhizae in influencing soil P availability and plant uptake of P, the reader is directed to the excellent review of Jakobsen et al. (2005).

11.2.5 Optimizing Soil Phosphorus Availability

Critical available soil P concentrations are required to maximize crop yields. These soil concentrations can be optimized by P additions, liming, and cultivation. This section discusses how these factors influence soil P availability; the soil testing and recommendation process that quantifies P additions is discussed in the Chapter 14.

11.2.5.1 Critical Concentrations for Plant Production

Estimates of soil P concentrations, above which little or no crop response to P additions is obtained, vary with the extractant used (Kamprath and Watson, 1980). In many neutral to calcareous soils, 10 mg P kg^{-1} Olsen P was adequate for wheat, alfalfa, and cotton (Olsen et al., 1954). For Bray-1 P, the critical level is about 30 mg P kg^{-1} for Midwestern U.S. soils (Thomas and Peaslee, 1973).

Several additional factors influence the availability of soil P. These include temperature, soil compaction, soil moisture, soil aeration, soil pH, type and amount of clay content, and nutrient (including P) status of soil. When soil temperatures are low during early plant growth, P uptake is reduced. Soil compaction reduces pore space, decreasing the amount of water and oxygen, which, in turn, reduces P uptake. The use of liming materials to increase soil pH, and thereby soil P availability, is an old practice, which has been adequately reviewed by several authors (Adams, 1984; Kamprath, 1984; McLean and Brown, 1984). Reducing subsoil acidity has not been highly successful due to physical limitations of mixing lime into subsoils. However, surface applications of gypsum with sufficient time for transport into the subsoil have been shown to increase crop yields, increase subsoil Ca, and decrease exchangeable Al (Sumner et al., 1986; Farina and Channon, 1988). Soils with high clay content tend to fix more P than sandy soils with low clay content. Thus, more P needs to be added to raise the soil test level of clay soils than loam and sandy soils. In addition, the presence of ammonium enhances P uptake by creating an acid environment around the root when ammonium ions are absorbed. High concentrations of ammonium-N in the soil with fertilizer P may interfere with and delay normal P fixation reactions, prolonging the availability of fertilizer P (Murphy, 1988).

11.2.5.2 Phosphorus Additions

Optimizing soil P availability through P additions should consider application rate, timing, type, placement, and residual availability (Singh and Lal, 2005). Because of the immobility of P in most soils, the timing of fertilizer P application is not as critical as its placement. Even so, small amounts of placed starter fertilizer for vegetable crops have successfully reduced the need for much larger broadcast applications of P (Costigan, 1988), and a similar strategy (e.g., foliar applications) may be appropriate for other crops. In efforts to minimize P inputs in sustainable or low-input management systems, there has been renewed interest in the estimation and utilization of residual P availability from fertilizer or manure amendments (Pierzynski et al., 1990a; Yerokum and Christenson, 1990; McCollum, 1991; Linquist et al., 1997; Oberson et al., 1999; Tiessen, 2005).

11.2.5.2.1 Rate

The application of P either as mineral fertilizer or animal manure increases available soil P (Table 11.4). In many areas of intensive crop and livestock operations, the application of P at rates greater than crop removal has increased available soil P content above critical concentrations for crop production (Kellogg et al., 2000; Kleinman et al., 2005; Lanyon, 2005; Sharpley et al., 2005). The increase in available soil P ranged from 5 to 31 mg kg^{-1} for every 100 kg ha^{-1} of fertilizer P added (an average 18% increase; Table 11.4). Barber (1979), Rehm et al. (1984), and McCollum (1991) also observed 13%–28% increases in available P following mineral fertilizer P application. These values are similar to proportional increases following application of beef (7–23 mg kg^{-1}; an average 14%), poultry (14–28 mg kg^{-1}; an average 20%), and swine manure (5–20 mg kg^{-1}; an average 11%) (Table 11.4).

In general, annual applications of manure in crop production systems generally results in crop yields comparable to those obtained with commercial fertilizer, especially when applied based on its N content (Sutton et al., 1982; Lucero et al., 1995; Stevenson et al. 1998; Eghball and Power, 1999; Randall et al., 1999; Schmitt et al., 1999; Macoon et al., 2002). However, long-term N-based applications of manure can result in the accumulation of available soil P to levels greater than optimum for crop production (King et al., 1990; Sharpley et al., 1993, 2004; Sims et al., 1998; Mullins et al., 2005).

11.2.5.2.2 Type

Traditional soluble P fertilizers are ordinary superphosphate, triple superphosphate, ammonium polyphosphate, monoammonium phosphate, and diammonium phosphate (Young et al., 1985). The use of slow release P fertilizers, such as partially acidulated rock P (RP), on soils other than those with low pH, Ca, and P content have also been evaluated (Hedley et al., 1989; Muchovej et al., 1989). For example, in soils of neutral pH, it may be possible to apply a heavy initial dressing of finely ground RP and include a rotation of fine rooted legumes to generate a low pH rhizosphere with low Ca concentrations and, thus, increase RP dissolution. Other methods designed to increase acidity in the immediate RP-soil environment, and thereby its dissolution, include addition of elemental S (Muchovej et al., 1989), NH_4^+ fertilizers, or organic matter such as animal manure and crop residues (Hedley et al., 1989).

Animal manure itself is a valuable resource of P for crop production. The availability of P in manure in some soils may differ from fertilizer P (Table 11.3; Figure 11.4; Hue, 1991; Frossard et al., 1995). As a major proportion of P in manure can be organic (25%–50%), biological processes in soil will play a greater role in determining P availability than for fertilizer P, when applied at equivalent rates of P. In as much, the slower release of P from manure may make it a longer-term source of P to crops than more readily soluble fertilizer P.

Adding manure can cause an increase in soil pH (Kingery et al., 1994; Iyamuremye et al., 1996; Eghball, 2002), due to input of large amounts of Ca (up to 60 g Ca kg manure^{-1}) and the buffering effects of added bicarbonates and organic acids with carboxyl and phenolic hydroxyl groups (Sharpley and Moyer, 2000; Whalen et al., 2000). This suggests that not only the amounts but also form, solubility, and relative availability of soil P for crop uptake and release to overland flow can change with manure application (Wang et al., 1995). In fact, Sharpley et al. (2004) found that the long-term application (10–25 years) of dairy, poultry, or swine manure to 20 sites in New York, Oklahoma, and Pennsylvania resulted in an average increase in soil pH from 5.9 to 6.6, exchangeable Ca from 0.9 to 6.2 g kg^{-1}, organic C from 15.7 to 32.6 g kg^{-1}, and total P from 407 to 2480 mg kg^{-1}, between untreated and treated sites. What was more important to soil P availability was that as Mehlich-3 P increased (22–662 mg kg^{-1}), the proportion that was water soluble (14%–3%) declined as exchangeable Ca increased ($R^2 = 0.81$). Ion-activity products suggested that addition of manure to soil shifts P from Al– and Fe– to Ca–P reaction products, accounting for the relatively greater Mehlcih-3 P but lower water extractability of soil P (Sharpley et al., 2004). This has important implications for agronomic and environmental soil P testing, where acid-based extractants, such as Mehlich-3 and Bray may dissolve Ca-P complexes in manured soils that may not accurately reflect plant availability or potential for release to runoff. For instance, this may explain

TABLE 11.4 Available Soil P of Soil Treated with Fertilizer or Manure for Several Years and Untreated Soil in the U.S. and U.K. Studies

Soil	Crop	Added P (kg ha^{-1} Year^{-1})	Time (Year)	Method	Available Soil P Untreated (mg kg^{-1})	Treated (mg kg^{-1})	Reference and Location
Fertilizer							
Portsmouth, fsl *Typic Umbraquult*	Mixed veg.	20	9	Mehlich I	18	73	Cox et al. (1981); North Carolina and Rothamsted
Batcombe, cl *Typic Haploborolls*	Mixed veg.	27	19	Olsen	16	44	
Richfield, scl *Aridic Argiustolls*	Mixed veg.	20	14	Bray 1	12	54	Hooker et al. (1983); Kansas
		40	14	Bray 1	12	56	
Pullman, cl *Torrertic Paleustolls*	Sorghum	56	8	Bray 1	15	76	Sharpley et al. (1984b); Texas
Keith, sil *Aridic Argiustolls*	Wheat	11	6	Bray 1	22	31	McCallister et al. (1987); Nebraska
Rosebud, sil *Aridic Argiustolls*	Wheat	33	6	Bray 1	24	47	
		11	6		10	28	
		33	6		10	48	
Beef manure							
Lethbridge, cl *Typic Haploborolls*	Barley	160	11	Bray 1	22	424	Chang et al. (1991); Alberta
		320	11		22	736	
		480	11		22	893	
Pullman, cl *Torrertic Paleustolls*	Sorghum	90	8	Bray 1	15	63	Sharpley et al. (1984b); Texas
		273	8		15	230	
Poultry litter							
Cahaba, vfsl *Typic Hapludults*	Grass	130	12	Bray 1	5	216	Sharpley et al. (1993); Oklahoma
Ruston, fsl *Typic Paleudults*	Grass	100	12	Bray 1	12	342	
Stigler, sl *Aquic Paleudalfs*	Grass	35	35	Bray 1	14	239	
Swine manure							
Norfolk, l *Typic Kandiudults*	Grass	109	11	Mehlich I	80	235	King et al. (1990); North Carolina
		218	11		80	310	
		437	11		80	450	
Captina, sl *Typic Fragiudults*	Grass	101	9	Bray 1	5	121	Sharpley et al. (1991); Oklahoma
Sallisaw, l *Typic Paleudalfs*	Grass	81	15	Bray 1	6	147	
Stigler, sl *Aquic Paleudalfs*	Wheat	37	9	Bray 1	15	82	
Cecil, sl *Typic Kanhapludults*	Grass	160	3	Mehlich I	19	45	Reddy et al. (1980); North Carolina
		320	3		19	100	

Note: vfsl, very fine sandy loam; fsl, fine sandy loam; sl, sandy loam; sil, silt loam; l, loam; scl, silty clay loam; cl, clay loam; c, clay.

the plateau in the relationship between Mehlich-3 P and runoff dissolved P observed by Torbert et al. (2002). Thus, at high soils P levels Mehclih-3 P is likely extracting some P that may not be immediately released from manured surface soil to runoff.

11.2.5.2.3 Placement

Due to the general immobility of P in the soil profile, fertilizer placement is generally more critical for P than N. Depending on soil and environmental factors, band applications of P may or may not be better than broadcast incorporated applications of P. The effect of P application also varies with soil type. For six soils having a 100-fold variation in P sorptivity, Holford (1989) found that fertilizer P effectiveness, as measured by yield response in the first crop (wheat), residual effect in the second crop (clover), or cumulative recovery of applied P, was consistently greater for shallow banding at 5 cm depth compared to banding at 15 cm or broadcast applications. The almost equal effect obtained by mixing P throughout the soil, regardless of P sorptivity, suggested that the positional availability of P in the root zone is important in maximizing fertilizer effectiveness (Holford, 1989) in addition to reducing P sorption.

Positional availability is also influenced by crop type. In order for banding or restricted fertilizer placement to increase potential root extraction of P, the rate of P absorption and growth of roots in fertilized soil must increase to compensate for roots in unfertilized soil. Increased root growth and P uptake in the P-fertilized volume of soil compared to unfertilized soil has been observed for corn (Anghinoni and Barber, 1980), soybeans (Borkert and Barber, 1985), and wheat (Yao and Barber, 1986). In contrast, several studies have shown that flax does not respond to banded fertilizer due to an inability of its root system to expand and proliferate into and efficiently absorb P from high concentrations in the fertilized zone (Soper and Kalra, 1969; Strong and Soper, 1974). In the case of flax, increased P uptake and yield response was obtained when fertilizer P was placed 2–5 cm directly below the seed, ensuring adequate P levels during early growth (Bailey and Grant, 1989).

Field variability in soil properties, crop-yield potential, and topographic differences can affect site-specific requirements of P. For example, mixing a few fertile soil subsamples with any number of subsamples with marginal nutrient availability can produce misleading analytical results. An example from Nebraska, where scientists extensively sampled a 58 ha cornfield, illustrates this situation (Hergert et al., 1994; Peterson et al., 1994). More than 2000 cores collected from a 58 ha cornfield showed that about 75% of the field would likely respond to P fertilization. Yet, composited annual soil samples indicated a need for little, if any, P fertilizer. Further investigation disclosed the existence of an old farmstead that included a swine feeding operation from 20 years earlier and another area where sheep had been fed in confinement more than 70 years ago.

Integration of the spatial variability of soil properties into fertilizer management and placement decisions will receive greater attention as agronomic response models are developed that incorporate the information now readily available through precision agriculture technology. Such technology, coupled with soil and nutrient management information, will facilitate the correct rate and method of N and P application to meet crop needs.

11.2.5.2.4 Residual Availability

Halvorson and Black (1985) showed that soil test P levels were increased above the initial available P level for more than 16 years, by a one-time P application on a Williams loam (*Typic Argiborolls*) in Montana (Figure 11.6). After the initial increase, available P levels declined for about 12 years and then stabilized at a higher available level than was initially present, thus establishing what appears to be a new equilibrium level of available P. Fixen (1986) reported similar changes in available P levels with time. Crop yields, reported by Halvorson and Black (1985), were also improved by the residual P fertilizer for a period of 16 years (Figure 11.6).

The rate of decline in soil P in high P soils when no further P is added varies with soil type and management (Table 11.5). The rate of decline in available soil P ranged from 0.1 to 30 mg kg^{-1} year^{-1}. McCollum (1991) estimated that without further P, 16–18 years of corn (*Z. mays* L.) or soybean (*G. max* (L.) Merr.) production would be needed to deplete soil P (Mehlich-3 P) in a Portsmouth fine sandy loam (*Typic Umbraquults*) from 100 mg P kg^{-1} to the agronomic threshold level of 20 mg P kg^{-1}. Several authors have found the rate of decrease in available soil P with depletion by cropping when no P is added is inversely related to the soil's P buffering capacity (Holford, 1982; Aquino and Hanson, 1984; Dodd and Mallarino, 2005) or P sorption saturation (available soil P/P sorption maximum; Sharpley, 1996).

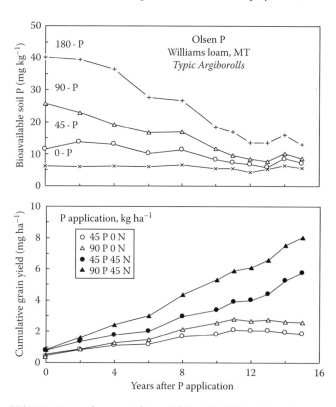

FIGURE 11.6 Changes in bioavailable soil P (Olsen P) and cumulative wheat yields following a single application of fertilizer P. (Data adapted from Halvorson, A.D., and A.L. Black. 1985. Long-term dryland crop responses to residual phosphorus fertilizer. Soil Sci. Soc. Am. J. 49:928–933.)

TABLE 11.5 The Decrease in Available Soil P after P Application Was Stopped in Several North American P Studies

Soil	Crop	Time (Year)	Method	Available Soil P Initial (mg kg⁻¹)	Final (mg kg⁻¹)	Decline (mg kg⁻¹ Year⁻¹)	Reference and Location
Thurlow, l	Small grains	9	Olsen	13	4	1.0	Campbell (1965); Montana
Ustollic Haplargids		9		20	4	1.8	
		9		60	6	6.0	
Georgeville, scl	Small grains	7	Mehlich I	3	1	0.1	Cox et al. (1981); North Carolina and Saskatchewan
Typic Hapludult		7		7	2	0.6	
Haverhill, c	Wheat—fallow	14	Olsen	40	25	1.1	
Typic Epiaquolls l		14		74	33	2.9	
		14		134	69	4.6	
Portsmouth, fsl	Small grains	8	Mehlich I	23	18	0.6	
Typic Umbraquults		9		54	26	3.1	
Sceptre, c	Wheat—fallow	8	Olsen	45	18	3.4	
Typic Haploborolls		8		67	18	6.1	
		8		147	40	13.4	
Williams, l	Wheat—barley	16	Olsen	26	8	1.1	Halvorson and Black (1985); Montana
Typic Argiborolls		16		45	14	1.9	
Richfield, scl	Corn	8	Bray 1	12	8	0.5	Hooker et al. (1983); Kansas
Aridic Argiustolls		8		22	14	1.0	
Carroll, cl	Wheat—flax	8	Olsen	71	10	7.6	Spratt et al. (1980); Manitoba
Typic Haploborolls		8		135	23	14.0	
		8		222	50	21.5	
Waskada, l	Wheat—flax	8	Olsen	48	9	4.9	
Typic Haploborolls		8		88	23	8.1	
		8		200	49	18.9	
Waskada, cl	Wheat—flax	8	Bray 1	140	50	11.3	Wagner et al. (1986); Manitoba
Typic Haploborolls		8		320	80	30.0	

Note: fsl, fine sandy loam; l, loam; scl, silty clay loam; cl, clay loam; c, clay.

11.2.5.3 Land Management

11.2.5.3.1 Cultivation

Cultivation and associated fertilizer applications can influence the amount of soil P_i and P_o and should be considered in optimizing soil P availability. Thompson et al. (1954) measured a decrease in P_i and P_o content of 25 surface (0–15 cm) soils from Iowa, Texas, and Colorado, with cultivation. Similarly, Adepetu and Corey (1976) reported that 25% of the P_o content of the surface of a Nigerian soil was mineralized during the first two cropping periods following cultivation. In fact, P_i changes little with crop removal and no fertilizer P application, but available P declines gradually. The net loss from the system through removal in the harvested crop is primarily accounted for by a decrease in P_o. For example, 60 years (1913–1973) of cotton growth on the Mississippi Delta soil, Dundee silt loam (Typic Endoaqualfs), with no reported P added, resulted in no appreciable effect on P_i (Sharpley and Smith, 1983). However, the P_o content of cultivated (93 mg kg⁻¹) compared to virgin analogue (223 mg kg⁻¹) surface soil (0–15 cm) decreased. Mineralization of P_o slowly replenished the P_i pool, which resulted in a 50% reduction in Bray-1 available P. Where cultivation involved P application, P_i and P_o increased 130% and 227%, respectively, while Bray-1 available P increased 84% and P sorption capacity decreased 33% on an average for 8 U.S. agricultural soils (Sharpley and Smith, 1983).

11.2.5.3.2 Remediation of High P Soils

Large amounts of coal combustion by-products are produced annually by power plants in compliance with clean air legislation (U.S. Environmental Protection Agency, 1988). These by-products can be used safely to increase soil pH and reduce subsoil acidity (Stout and Priddy, 1996; Stout et al., 1999; Callahan et al., 2002). Recent research also suggests that two of these by-products—fluidized bed combustion fly ash and flue gas desulfurization gypsum—can greatly reduce water-soluble P levels in soils and surface runoff without appreciably reducing the plant-available P and plant growth; moreover, heavy metals and arsenic in plants or runoff water are not increased (Stout et al., 1998, 2000).

The use of these amendments on critical areas of a watershed has the potential to make large albeit short-term reductions in P loss (Stout et al., 1999).

11.2.6 Water Quality Implications of Soil Phosphorus Availability

Many studies have reported that the loss of dissolved P in surface runoff is dependent on the available P content of surface soil as measured by soil P test extractants (Sharpley et al., 1996; Pote et al., 1999b; Pierson et al., 2001; Torbert et al., 2002; Andraski and Bundy, 2003; Daverede et al., 2003). For instance, runoff P–soil P relationships were developed within a 40 ha watershed (FD-36) in south-central Pennsylvania (Northumberland Co.), where there was a wide range in soil test P concentration as Mehlich-3 P (20–600 mg L^{-1}) (McDowell and Sharpley, 2001; Sharpley et al., 2001). As Mehlich-3 soil P increased, so did the concentration of dissolved P in surface runoff from 2 m^2 plots subjected to a 30 min rainfall of 7 cm h^{-1} (Figure 11.7). Similarly, the concentration of P in subsurface flow is also related to surface soil P (Figure 11.7). Lysimeters of 30 cm depth were taken from the same Pennsylvania watershed and subjected to simulated rainfall (1 cm h^{-1} for 30 min). The concentration of dissolved P in drainage from the lysimeter increased (0.07–2.02 mg L^{-1}) as the Mehlich-3 P concentration of surface soil increased (Figure 11.7; McDowell and Sharpley, 2001; Sharpley et al., 2001). The dependence of leachate P on surface soil P is evidence of the importance of P transport in preferential flow pathways such as macropores, earthworm holes, and old root channels.

Vadas et al. (2005) conducted a detailed review of the large amount of research published since the late 1990s on the relationship between soil P availability and runoff dissolved P (Table 11.6). Clearly, there was a range in the extraction coefficient representing the slope of the soil P—runoff P relationship or in more simple terms, the extractability of soil P by runoff water. Significant differences among extraction coefficients were not related to any one soil physical, chemical, or management factor (Vadas et al., 2005). The most likely factor influencing soil P extractability was site hydrology, such that a greater proportion of applied rainfall as runoff translated into more P being desorbed from the surface soil and transported in runoff (Vadas et al., 2005). When these studies were combined to include runoff from soil boxes and field plots that were tilled, no-tilled, cropped, and grassed, a single extraction coefficient of 2.0 related Mehlich-3 soil P (mg kg^{-1}) to runoff dissolved P (µg L^{-1}).

This information supports the use of soil test P as a measure of soil P availability in estimating the potential for dissolved P enrichment of runoff, as used in nonpoint source models and more recently in estimating the environmental implications of soil P on runoff P, in relation to impairment of surface waters. Thus, as we move from agronomic to environmental concerns with P, where continued applications of P usually as manure, have led to increased available soil P above optimum levels for crop production, soil P testing is being used to indicate when P enrichment of runoff may become unacceptable. Because of this, a common approach has been to use agronomic soil P standards, following the rationale that soil P in excess of crop requirements is vulnerable to removal by surface runoff or leaching. As agronomic standards already exist for soil test P, this approach required little investment in research and development and can be readily implemented. However, we must be careful how we interpret soil test results for environmental purposes (Figure 11.8).

Interpretations given on soil test reports (i.e., low, medium, optimum, high, etc.) are based on expected crop yield response to P and not on soil P release to surface or subsurface runoff (Sharpley et al., 1994). Some have tried to simply extend crop response levels and say that a soil test that is above the level where a crop response is expected is in excess of crop needs and therefore is potentially polluting (Figure 11.8). However, two factors are of critical importance to the debate on how to use soil test P in environmental risk assessment. First, the gap between crop and environmental soil P thresholds reflects the difference in soil P removed by an acid or base extractant (i.e., Mehlich, Bray, Olsen agronomic tests) and by less invasive water (i.e., simulating extraction of soil P by runoff water), which is soil specific. Secondly, soil P is only one of several factors influencing the

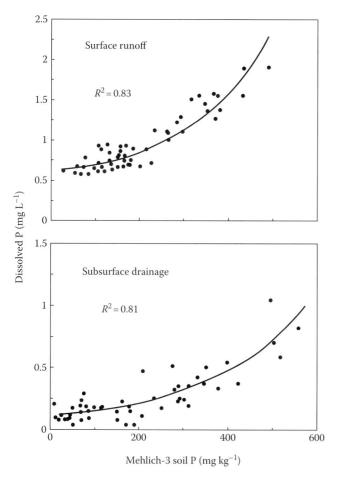

FIGURE 11.7 Relationship between the concentration of dissolved P in surface runoff from 2 m^2 plots and subsurface drainage from 30 cm deep lysimeters and surface soil (0–5 cm depth) Mehlich-3 extractable P concentration from a central PA watershed.

TABLE 11.6 Extraction Coefficients Relating Soil P Availability (mg kg^{-1}) to Runoff Dissolved P (DP; μg L^{-1})

Reference	Location	Land Use	Soil	Number of Observations	Runoff DP	Extraction Coefficient	
						Bray-1	Mehlich-3
Aase et al. (2001)	Indiana	Tilled plots	sl	48	0–0.4	—	—
Andraski and Bundy (2003)	Wisconsin	Cropped plots	sl, cl	126	0–1.0	2.0–13.0	1.3–3.0
Andraski et al. (2003)	Wisconsin	Cropped plots	sl	127	0–0.3	1.9–2.6	1.7
Cox and Hendricks (2000)	North Carolina	Cropped plots	cl, ls	20	0–2.0	—	1.4–3.9
Daverede et al. (2003)	Illinois	Cropped plots	scl	64	0–1.0	0.6–0.8	—
Fang et al. (2002)	Minnesota	Soil boxes	scl, cl, c	10	0–1.0	9.8	5.8
McDowell and Sharpley (2001)	Pennsylvania	Soil boxes, pasture, tilled plots	sl	88	0–1.0	—	1.7–1.9
Pote et al. (1996)	Arkansas	Fescue plots	sl	54	0.2–1.6	2.2	2.6
Pote et al. (1999a)	Arkansas	Fescue plots	sl, sal	36	0–1.2	2.7–4.3	1.6–3.6
Pote et al. (1999b)	Arkansas	Fescue plots	sl	36	0–1.4	—	2.2–2.8
Schroeder et al. (2004)	Georgia	Pasture plots	sal	54	0–1.1	—	1.7–2.0
Sharpley et al. (2001)	Pennsylvania	Pasture tilled plots	sl	—	—	—	2.1
Torbert et al. (2002)	Texas	Bermudagrass plots	sal, cl, c	72	0–1.6	—	0.4–7.0
Turner et al. (2004)	Idaho	Tilled plots	sl, fsl	50	0–0.7	—	1.2–2.8
Weld et al. (2001)	Pennsylvania	Soil boxes	sl	—	—	—	1.5–1.7

Source: Adapted from Vadas, P.A., P.J.A. Kleinman, and A.N. Sharpley. 2005. Relating soil phosphorus to dissolved phosphorus in runoff: A single extraction coefficient for water quality modeling. J. Environ. Qual. 34:572–580.

Note: fsl, fine sandy loam; fas, sandy loam; sl, silt loam; scl, silty clay loam; cl, clay loam; c, clay.

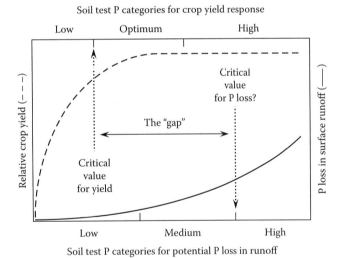

FIGURE 11.8 As soil P increases, so does crop yield and the potential for P loss in surface runoff. The interval between the critical soil P value for yield and runoff P will be important for P management.

potential for P loss; therefore, soil test P should not be used as the sole criteria on which to base P management planning.

11.2.7 Conclusions

The amount and availability of soil P is determined by physical, chemical, and biological processes, which are often managed in attempts to increase or optimize crop uptake of P and yields. In many areas, P cycles have been fragmented by the specialization of agricultural production systems in specific regions. For instance, mineral fertilizer P is imported to areas of crop production from continental United States and overseas deposits, which have been treated to varying degree to increase P solubility. The harvested grain is used to meet human demands, for animal feed, and more recently as biofuel feedstock. In each case, major population areas, confined animal operations, and biorefineries are geographically removed from areas of crop production. For example, most of the corn produced in the Midwest is used as feed in eastern U.S. states. As a result, P is moving from areas where ore deposits are mined, through crop producing regions, and is accumulating in areas of confined animal operations.

Future management of soil P availability must address the impacts that this specialization of agricultural systems is having on regional P requirements and production on a national scale. At the farm scale, however, efforts should continue to find ways of enhancing the efficient use of P, through soil testing, appropriate rates of P application, and utilization of manure sources of P in an increasing number of agricultural production systems. If these goals are not met, soil P availability may become an environmental rather than fertility issue in more areas.

References

Aase, J.K., D.L. Bjorneborg, and D.T. Westermann. 2001. Phosphorus runoff from two water sources on a calcareous soil. J. Environ. Qual. 30:1315–1323.

Abrams, M.M., and W.M. Jarrell. 1992. Bioavailability index for phosphorus using ion exchange resin impregnated membranes. Soil Sci. Soc. Am. J. 56:1532–1537.

Adams, F. 1984. Crop response to lime in the southern U.S. *In* F. Adams (ed.) Soil acidity and liming. 2nd edn. Agronomy 12:211–266.

Adepetu, J.A., and R.B. Corey. 1976. Organic phosphorus as a predictor of plant available phosphorus in soils of Southern Nigeria. Soil Sci. 122:159–164.

Amer, F., D.R. Bouldin, C.A. Black, and F.R. Duke. 1955. Characterization of soil phosphorus by anion exchange resin and adsorption by P-32 equilibration. Plant Soil 6:391–408.

Andersen, N.C. 1983. Nitrogen turnover by earthworms in arable plots treated with farmyard manure and slurry, p. 139–150. In J.E. Satchell (ed.) Earthworm ecology: From Darwin to vermiculture. Chapman & Hall, London, U.K.

Andraski, T.W., and L.G. Bundy. 2003. Relationship between phosphorus levels in soil and in runoff from corn production systems. J. Environ. Qual. 32:310–316.

Andraski, T.W., L.G. Bundy, and K.C. Kilian. 2003. Manure history and long-term tillage effects on soil properties and phosphorus losses in runoff. J. Environ. Qual. 32:1782–1789.

Anghinoni, I., and S.A. Barber. 1980. Predicting the most efficient phosphorus placement for corn. Soil Sci. Soc. Am. J. 44:1016–1020.

Aquino, B.F., and R.G. Hanson. 1984. Soil phosphorus supplying capacity evaluated by plant removal and available phosphorus extraction. Soil Sci. Soc. Am. J. 48:1091–1096.

Bailey, L.D., and C.A. Grant. 1989. Fertilizer phosphorus placement studies on calcareous and noncalcareous chernozemic soils: Growth, P-uptake and yield of flax. Commun. Soil Sci. Plant Anal. 20:635–654.

Barber, S.A. 1979. Soil phosphorus after 25 years of cropping with five rates of phosphorus application. Commun. Soil Sci. Plant Anal. 10:1459–1468.

Barrow, N.J. 1978. The description of phosphate adsorption curves. J. Soil Sci. 29:447–462.

Barrow, N.J. 1983. Understanding phosphate reaction with soil, p. 37–51. Third Int. Cong. on Phosphorus Compounds Proceedings. Institut Mondial du Phosphate, Casablanca, Morocco.

Barrow, N.J. 1983. Reactions of anions and cations with variable charge soils. Adv. Agron. 38:183–230.

Barrow, N.J. 1989. Surface reactions of phosphate in soil. Agric. Sci. 2:33–37.

Barrow, N.J., and T.C. Shaw. 1975. The low reactions between soil and anions: 2. Effect of time and temperature on the decrease in phosphate concentration in the soil solution. Soil Sci. 119:167–177.

Blal, B., C. Morel, V. Gianinazzi-Pearson, J.C. Fardeau, and S. Gianinazzi. 1990. Influence of vesicular-arbuscular mycorrhizae on phosphate fertilizer efficiency in two tropical acid soils planted with micropropagated soil palm (*Elaeis guineensis* jacq.). Biol. Fertil. Soils 9:43–48.

Bolan, N.S., A.D. Robson, and N.J. Barrow. 1983. Plant and soil factors including mycorrhizal infection causing sigmoidal response of plants to applied phosphorus. Plant Soil 73:187–203.

Bolan, N.S., A.D. Robson, and N.J. Barrow. 1987. Effects of phosphorus application and mycorrhizal inoculation on root characteristics of subclover and ryegrass in relation to phosphorus uptake. Plant Soil 104:294–298.

Borkert, C.M., and S.A. Barber. 1985. Soybean shoot and root growth and phosphorus concentration as affected by phosphorus placement. Soil Sci. Soc. Am. J. 49:152–155.

Bowman, R.A., S.R. Olsen, and F.S. Watanabe. 1978. Greenhouse evaluation of residual phosphate by four phosphorus methods in neutral and calcareous soils. Soil Sci. Soc. Am. J. 42:451–454.

Bray, R.H., and L.T. Kurtz. 1945. Determination of total, organic, and available forms of phosphorus in soils. Soil Sci. 59:39–45.

Brookes, P.C., D.S. Powlson, and D.S. Jenkinson. 1984. Phosphorus in the soil microbial biomass. Soil Biol. Biochem. 16:169–175.

Buckman, H.O., and N.C. Brady. 1970. The nature and properties of soils. The Macmillan Publishing Co., Collier-Macmillan Limited, London, U.K.

Butewaga, C.N., G.L. Mullins, and S.H. Chien. 1996. Induced phosphorus fixation and the effectiveness of phosphate fertilizers derived from Sukulu Hills phosphate rock. Fert. Res. 44:213–240.

Callahan, M.P., P.J.A. Kleinman, A.N. Sharpley, and W.L. Stout. 2002. Assessing the efficacy of alternative phosphorus sorbing soil amendments. Soil Sci. 167:539–547.

Campbell, R.E. 1965. Phosphorus fertilizer residual effects on irrigated crops in rotation. Soil Sci. Soc. Am. Proc. 29:67–70.

Carter, M.R., and E.G. Gregorich. (eds.). 2007. Soil sampling and methods of analysis. 2nd edn. Canadian Society of Soil Science, CRC Press, Taylor & Francis Group, Boca Raton, FL.

Chang, C., T.G. Sommerfeldt, and T. Entz. 1991. Soil chemistry after eleven annual applications of cattle feedlot manure. J. Environ. Qual. 20:475–480.

Chardon, W.J. 2008. Phosphorus extraction with iron oxide-impregnated filter paper (P_i test), p. 25–32. In J.L. Kovar and G.M. Pierzynski (eds.) Methods of phosphorus analysis for soils, sediments, residuals, and waters. Southern Cooperative Series Bulletin. Virginia Tech University, Blacksburg, VA.

Chardon, W.J., G.H. Aalderink, and C. van der Salm. 2007. Phosphorus leaching from cow manure patches on soil columns. J. Environ. Qual. 36:17–22.

Chen, C.R., L.M. Condron, M.R. Davis, and R.R. Sherlock. 2002. Phosphorus dynamics in the rhizosphere of perennial ryegrass (*Lolium perenne* L.) and radiata pine (*Pinus radiata* D. Don). Soil Biol. Biochem. 34:487–499.

Coleman, D.C., C.P.P. Reid, and C.V. Cole. 1983. Biological strategies of nutrient cycling in soil systems. Adv. Ecol. Res. 13:1–56.

Condron, L.M., B.L. Turner, and B.J. Cade-Menun. 2005. Chemistry and dynamics of soil organic phosphorus. In J.T. Sims and A.N. Sharpley (eds.) Phosphorus: Agriculture and the environment. Agronomy 46:87–122.

Costigan, P. 1988. The placement of starter fertilizers to improve early growth of drilled and transplanted vegetables. Proc. Fert. Soc. 274:1–24.

Cox, F.R., and S.E. Hendricks. 2000. Soil test phosphorus and clay content effects on runoff water quality. J. Environ. Qual. 29:1582–1586.

Cox, F.R., E.J. Kamprath, and R.E. McCollum. 1981. A descriptive model of soil test nutrient levels following fertilization. Soil Sci. Soc. Am. J. 45:529–532.

Crush, J.R. 1973. The effect of *Rhizophagus tenius* mycorrhizas on rye grass, cocksfoot and sweet vernal. New Phytol. 72:965–973.

Daverede, I.C., A.N. Kravchenko, R.G. Hoeft, E.D. Nafziger, D.G. Bullock, J.J. Warren, and L.C. Gonzini. 2003. Phosphorus runoff: Effect of tillage and soil phosphorus levels. J. Environ. Qual. 32:1436–1444.

Di, H.J., L.M. Condron, and E. Frossard. 1997. Isotope techniques to study phosphorus cycling in agricultural and forest soils: A review. Biol. Fert. Soils 27:1–12.

Dodd, J.C., C.L. Boddington, A. Rodriguez, C. Gonzalez-Chavez, and I. Mansur. 2000. Mycelium of arbuscular mycorrhizal fungi (AMF) from different genera: Form, function and detection. Plant Soil 226:131–151.

Dodd, J.R., and A.P. Mallarino. 2005. Soil-test phosphorus and crop grain yield responses to long-term phosphorus fertilization for corn-soybean rotations. Soil Sci. Soc. Am. J. 69:1118–1128.

Domínguez, J., P.J. Bohlen, and R.W. Parmelee. 2004. Earthworms increase nitrogen leaching to greater soil depths in row crop agroecosystems. Ecosystems 7:672–685.

Drew, E.A., R.S. Murray, S.E. Smith, and I. Jakobsen. 2003. Beyond the rhizosphere: Growth and function of arbuscular mycorrhizal external hyphae in sands of varying pore sizes. Plant Soil 147:105–114.

Edwards, C.A. 1980. Interactions between agricultural practice and earthworms, p. 3–12. *In* D.L. Dindall (ed.) Soil biology as related to land use practices. Proc. 8th Int. Colloquium on Soil Zoology. U.S. Environmental Protection Agency, Washington, DC.

Edwards, C.A. (ed.). 2004. Earthworm ecology. 2nd edn. CRC Press, Boca Raton, FL.

Edwards, C.A., and P.J. Bohlen. 1996. Biology and ecology of earthworms. 3rd edn. Chapman & Hall, London, U.K.

Edwards, C.A., P.J. Bohlen, D.R. Linden, and S. Subler. 1995. Earthworms in agroecosystems, p. 185–213. *In* P.F. Hendrix (ed.) Earthworm ecology and biogeography in North America. Lewis Publishers, Boca Raton, FL.

Edwards, W.M., M.J. Shipitalo, L.B. Owens, and L.D. Norton. 1989. Water and nitrate movement in earthworm burrows within long-term no-till cornfields. J. Soil Water Conserv. 44:240–243.

Eghball, B. 2002. Soil properties as influenced by phosphorus- and nitrogen-based manure and compost applications. Agron. J. 94:128–135.

Eghball, B., and J.F. Power. 1999. Phosphorus- and nitrogen-based manure and compost applications: Corn production and soil phosphorus. Soil Sci. Soc. Am. J. 63:895–901.

Estevez, B., A. N'Dayegamiye, and D. Coderre. 1996. The effect on earthworm abundance and selected soil properties after 14 years of solid cattle manure and N P K Mg fertilizer application. Can. J. Soil Sci. 76:351–355.

Evans, R.L., and J.J. Jurinak. 1976. Kinetics of phosphate release from a desert soil. Soil Sci. 121:205–211.

Facelli, E., and J.M. Facelli. 2002. Soil phosphorus heterogeneity and mycorrhizal symbiosis regulate plant intraspecific competition and size distribution. Oecologia 133:54–61.

Fang, F., P.L. Brezonik, D.J. Mulla, and L.K. Hatch. 2002. Estimating runoff phosphorus losses from calcareous soils in the Minnesota River basin. J. Environ. Qual. 31:1918–1929.

Fardeau, J.C., G. Guiraud, and C. Marol. 1996. The role of isotopic techniques on the evaluation of the agronomic effectiveness of P fertilizers. Fert. Res. 45:101–109.

Farina, M.P.W., and P. Channon. 1988. Acid-subsoil amelioration. II. Gypsum effects on growth and subsoil chemical properties. Soil Sci. Soc. Am. J. 52:175–180.

Fitter, A.H., and B. Moyersoen. 1996. Evolutionary trends in root-microbe symbioses. Philos. Trans. R. Soc. Lond. Ser. B 351:1367–1375.

Fixen, P.E. 1986. Residual effects of P fertilization: Lessons for the '80's, p. 1–8. Proc. of the 16th North Central Extension-Industry Soil Fertility Workshop. St. Louis, MO, October 29–30, 1986. Potash and Phosphate Institute, Atlanta, GA.

Fixen, P.E., and J.H. Grove. 1990. Testing soils for phosphorus, p. 141–180. *In* R.L. Westerman (ed.) Soil testing and plant analysis. 3rd edn. SSSA Book Series No. 3. SSSA, Madison, WI.

Frossard, E., M. Brossard, M.J. Hedley, and A. Metherell. 1995. Reactions controlling the cycling of phosphorus in soils, p. 107–138. *In* H. Tiessen (ed.) Phosphorus in the global environment: Transfers, cycles and management. John Wiley & Sons, New York.

Frossard, E., L.M. Condron, A. Oberson, S. Sinaj, and J.C. Fardeau. 2000. Processes governing phosphorus availability in temperate soils. J. Environ. Qual. 29:15–23.

Frossard, E., and S. Sinaj. 1997. The isotopic exchange kinetic technique: A method to describe the availability of inorganic nutrients. Applications to K, P, S, and Zn. Isot. Environ Health Stud. 34:61–67.

Gerdemann, J.W. 1968. Vesicular-arbuscular mycorrhiza and plant growth. Annu. Rev. Phytopathol. 6:397–418.

Germann, P.F., W.M. Edwards, and L.B. Owens. 1984. Profiles of bromide and increased soil moisture after infiltration into soils with macropores. Soil Sci. Soc. Am. J. 48:237–244.

Goldberg, S., and G. Sposito. 1985. On the mechanism of specific phosphate adsorption by hydroxylated mineral surfaces: A review. Commun. Soil Sci. Plant Anal. 16:801–821.

Gorbenko, A.Y., N.S. Panikov, and D.V. Zvyaginstev. 1986. The effect of invertebrates on the growth of soil microorganisms. Mikrobiologia 55:515–521.

Graetz, D.A., and V.D. Nair. 2008. Phosphorus sorption isotherm determination, p. 33–36. *In* J.L. Kovar and G.M. Pierzynski (eds.) Methods of phosphorus analysis for soils, sediments, residuals, and waters. Southern Cooperative Series Bulletin. Virginia Tech University, Blacksburg, VA.

Guild, W.J. 1955. Earthworms and soil structure, p. 83–98. *In* D.K.McE. Kevan (ed.) Soil zoology. Butterworth, London, U.K.

Halvorson, A.D., and A.L. Black. 1985. Long-term dryland crop responses to residual phosphorus fertilizer. Soil Sci. Soc. Am. J. 49:928–933.

Harley, J.L. 1989. The significance of mycorrhizae. Mycol. Res. 92:129–139.

He, Z.L., J. Wu, A.G. O'Donnell, and J.K. Syers. 1997. Seasonal responses in microbial biomass carbon, phosphorus and sulphur in soils under pasture. Biol. Fertil. Soils 24:421–428.

Hedley, M.J., J.M. Mortvedt, N.S. Bolan, and J.K. Syers. 1995. Phosphorus fertility management in agroecosystems, p. 59–92. *In* H. Tiessen (ed.) Phosphorus in the global environment: Transfers, cycles and management. John Wiley & Sons, New York.

Hedley, M.J., J.W.B. Stewart, and B.S. Chanhan. 1982. Changes in inorganic and organic soil phosphorus fraction induced by cultivation practices and by laboratory incubations. Soil Sci. Soc. Am. J. 46:970–976.

Hedley, M.J., R.W. Tillman, and G. Wallace. 1989. The use of nitrogen fertilizers for increasing the suitability of reactive phosphate rocks for use in intensive agriculture, p. 311–320. *In* R.E. White and L.D. Currie (eds.) Nitrogen fertilizer use in New Zealand agriculture and horticulture. Occasional Report No. 3. Fertilizer and Lime Research Centre, Massey University, New Zealand.

Hergert, G.W., R.B. Ferguson, C.A. Cotway, and T.A. Peterson. 1994. Developing accurate nitrogen rate maps for variable rate application, p. 399. *In* Agronomy abstracts. ASA, Madison, WI.

Hingston, F.J., A.M. Posner, and J.P. Quirk. 1974. Anion adsorption by goethite and gibbsite. II. Desorption of anions from hydrous oxide surfaces. J. Soil Sci. 25:16–26.

Holford, I.C.R. 1982. Effects of phosphate sorptivity on long-term plant recovery and effectiveness of fertilizer phosphate in soils. Plant Soil 64:225–236.

Holford, I.C.R. 1989. Efficacy of different phosphate application methods in relation to phosphate sorptivity in soils. Aust. J. Soil Res. 27:123–133.

Holford, I.C.R., and G.E.G. Mattingly. 1975a. The high- and low-energy phosphate adsorbing surfaces in calcareous soils. J. Soil Sci. 26:407–417.

Holford, I.C.R., and G.E.G. Mattingly. 1975b. Surface areas of calcium carbonate in soils. Geoderma 13:247–255.

Hooker, M.L., R.E. Gwin, G.M. Herron, and P. Gallagher. 1983. Effects of long-term annual applications of N and P on corn grain yields and soil chemical properties. Agron. J. 75:94–99.

Howeler, R.H., C.J. Asher, and D.G. Edwards. 1982. Establishment of an effective mycorrhizal association on cassava in flowing solution culture and its effects on phosphorus nutrition. New Phytol. 90:229–238.

Hsu, P.H. 1982. Crystallization of iron (III) phosphate at room temperature. Soil Sci. Soc. Am. J. 46:928–932.

Hue, N.V. 1991. Effects of organic acids/anions on P sorption and phytoavailability in soils with different mineralogies. Soil Sci. 152:463–471.

Inskeep, W.P., and J.C. Silvertooth. 1988. Inhibition of hydroxyapatite precipitation in the presence of fulvic, humic, and tannic acids. Soil Sci. Soc. Am. J. 52:941–946.

Iyamuremye, F., R.P. Dick, and J. Baham. 1996. Organic amendments and soil phosphorus dynamics: I. Phosphorus chemistry and sorption. Soil Sci. 161:426–435.

Jakobsen, I., M.E. Leggett, and A.E. Richardson. 2005. Rhizosphere microorganisms and plant phosphorus uptake. *In* J.T. Sims and A.N. Sharpley (eds.) Phosphorus: Agriculture and the environment. Agronomy 46:439–494.

Jayachandran, J., A.P. Schwab, and B.A.D. Hetrick. 1989. Mycorrhizal mediation of phosphorus availability: Synthetic iron chelate effects on phosphorus solubilization. Soil Sci. Soc. Am. J. 53:1701–1706.

Joergensen, R.G., H. Kubler, B. Meyer, and V. Wolters. 1995. Microbial biomass phosphorus in soils of beech (*Fagus sylvatica* L.) forests. Biol. Fertil. Soils 19:215–219.

Juo, A.S.R., and R.L. Fox. 1977. Phosphate sorption capacity benchmark soils in West Africa. Soil Sci. 134:370–376.

Kamprath, E.J. 1984. Crop response to lime on soils in the tropics. *In* F. Adams (ed.) Soil acidity and liming. 2nd edn. Agronomy 12:349–368.

Kamprath, E.J., and M.E. Watson. 1980. Conventional soil and tissue tests for assessing the phosphorus status of soils, p. 433–469. *In* F.E. Khawsaneh, E.C. Sample, and E.J. Kamprath (eds.) The role of phosphorus in agriculture. ASA, CSSA, and SSSA, Madison, WI.

Katsvairo, T., W.J. Cox, and H. van Es. 2002. Tillage and rotation effects on soil physical characteristics. Agron. J. 94:299–304.

Kellogg, R.L., C.H. Lander, D.C. Moffitt, and N. Gollehon. 2000. Manure nutrients relative to the capacity of cropland and pastureland to assimilate nutrients: Spatial and temporal trends for the United States. Resource Assessment and Strategic Planning Working Paper 98-1. USDA, Natural Resources Conservation Service and Economic Research Service, General Services Administration, National Forms and Publication Center, Fort Worth, TX. Available online at: http://www.nrcs.usda.gov/technical/NRI/pubs/manntr.pdf (last accessed April 12, 2009).

King, L.D., J.C. Burns, and P.W. Westerman. 1990. Long-term swine lagoon effluent applications on 'Coastal' Bermudagrass: II. Effects on nutrient accumulations in soil. J. Environ. Qual. 19:756–760.

Kingery, W.L., C.W. Wood, D.P. Delaney, J.C. Williams, and G.L. Mullins. 1994. Impact of long-term land application of broiler litter on environmentally related soil properties. J. Environ. Qual. 23:139–147.

Kleinman, P.J.A., A.M. Wolf, A.N. Sharpley, D.B. Beegle, and L.S. Saporito. 2005. Survey of water extractable phosphorus in livestock manures. Soil Sci. Soc. Am. J. 69:701–708.

Kovar, J.L., and G.M. Pierzynski. (eds.). 2008. Methods of phosphorus analysis for soils, sediments, residuals, and waters. Southern Cooperative Series Bulletin. Virginia Tech University, Blacksburg, VA.

Kucey, R.M.N., H.H. Janzen, and M.E. Leggett. 1989. Microbially mediated increases in plant-available phosphorus. Adv. Agron. 42:199–228.

Kumar, V., R.J. Gilkes, and M.D.A. Bolland. 1992. Comparison of seven soil-P tests for plant species with different external-P requirements grown on soils containing rock phosphate and superphosphate residues. Fert. Res. 33:35–45.

Kuo, S. 1988. Application of a modified Langmuir isotherm to phosphate sorption by some acid soils. Soil Sci. Soc. Am. J. 52:97–102.

Kuo, S., and E.G. Lotse. 1974. Kinetics of phosphate adsorption and desorption by hematite and gibbsite. Soil Sci. 116:400–406.

Lanyon, L.E. 2005. Phosphorus, animal nutrition and feeding: Overview. *In* J.T. Sims and A.N. Sharpley (eds.) Phosphorus: Agriculture and the environment. Agronomy 46:561–586.

Lavelle, P. 1978. Les vers de terre de la savane de Lamto (Cote d'Ivoire). Peuplements, populations et fonctions de Pecosysteme. Publ. Lab. Zool. E.N.S. 12:1–301.

Lavelle, P. 1988. Earthworm activities and the soil system. Biol. Fertil. Soils 6:237–251.

Lavelle, P., I. Barois, A. Martin, Z. Zaidi, and R. Schaefer. 1989. Management of earthworm populations in agro-ecosystems: A possible way to maintain soil quality? p. 109–122. *In* M. Charholm and L. Bergström (eds.) Ecology of arable land. Kluwer Academic Publishers, Dordrecht, the Netherlands.

Lee, K.E. 1985. Earthworms: Their ecology and relationships with soils and land use. Academic Press, London, U.K.

Levesque, M., and M. Schnitzer. 1967. Organo-metallic interactions in soils: 6. Preparation and properties of fulvic acid-metal phosphates. Soil Sci. 103:183–190.

Lin, T.H., S.B. Ho, and K.H. Houng. 1991. The use of iron oxide-impregnated filter paper for the extraction of available phosphorus from Taiwan soils. Plant Soil 133:219–226.

Lindsay, W.L., P.L.G. Vlek, and S.H. Chien. 1989. Phosphate minerals, p. 1089–1130. *In* J.B. Dixon and S.B. Weed (eds.) Minerals in soil environment. 2nd edn. SSSA, Madison, WI.

Linquist, B.A., P.W. Singleton, and K.G. Cassman. 1997. Inorganic and organic phosphorus dynamics during a build up and decline of available phosphorus in an ultisol. Soil Sci. 162:254–264.

Lucero, D.W., D.C. Martens, J.R. McKenna, and D.E. Starner. 1995. Poultry litter effects on unmanaged pasture yield, nitrogen and phosphorus uptakes, and botanical composition. Commun. Soil Sci. Plant Anal. 26:861–882.

Mackay, A.D., J.K. Syers, R.W. Tillman, and P.E.H. Gregg. 1986. A simple model to describe the dissolution of phosphate rock in soils. Soil Sci. Soc. Am. J. 50:291–296.

Macoon, B., K.R. Woodard, L.E. Sollengerger, E.C. French III, K.M. Porter, D.A. Graetz, G.M. Prine, and H.H. Van Horn, Jr. 2002. Dairy effluent effects on herbage yield and nutritive value of forage cropping systems. Agron. J. 94:1043–1049.

Marshall, K.C. 1971. Sorptive interactions between soil particles and microorganisms, p. 409–445. *In* A.D. McLaren and J. Skujins (eds.) Soil biochemistry. Vol. 2. Marcel Dekker, New York.

McCallister, D.L., C.A. Shapiro, W.R. Raun, F.N. Anderson, G.W. Rhem, O.P. Engelstadt, M.O. Russelle, and R.A. Olson. 1987. Rate of phosphorus and potassium buildup/decline with fertilization for corn and wheat on Nebraska Mollisols. Soil Sci. Soc. Am. J. 51:1646–1652.

McCollum, R.E. 1991. Buildup and decline in soil phosphorus: 30-year trends on a Typic Umprabuult. Agron. J. 83:77–85.

McDowell, R.W., and A.N. Sharpley. 2001. Approximating phosphorus release from soils to surface runoff and subsurface drainage. J. Environ. Qual. 30:508–520.

McGill, W.B., and C.V. Cole. 1981. Comparative aspects of cycling of organic C, N, S, and P through soil organic matter. Geoderma 26:267–286.

McLaughlin, M.J., A.M. Alston, and J.K. Martin. 1988. Phosphorus cycling in wheat-pasture rotations. III. Organic phosphorus turnover and phosphorus cycling. Aust. J. Soil Res. 26:343–353.

McLean, E.O., and J.R. Brown. 1984. Crop response to lime in the Midwestern U.S. *In* F. Adams (ed.) Soil acidity and liming. 2nd edn. Agronomy 12:267–304.

Mehlich, A. 1953. Determination of P, K, Na, Ca, Mg, and NH_4. Soil Test Division, Mimeo, North Carolina Department of Agriculture, Raleigh, NC.

Mehlich, A. 1984. Mehlich 3 soil test extractant: A modification of Mehlich 2 extractant. Commun. Soil Sci. Plant Anal. 15:1409–1416.

Menon, R.G., S.H. Chien, and L.L. Hammond. 1989a. Comparison of Bray 1 and P_i tests for evaluating plant-available phosphorus from soils treated with different partially acidulated phosphate rocks. Plant Soil 114:211–216.

Menon, R.G., S.H. Chien, L.L. Hammond, and B.R. Arora. 1990. Sorption of phosphorus by the iron oxide-impregnated filter paper (P_i soil test) embedded in soils. Plant Soil 126:287–294.

Menon, R.G., L.L. Hammond, and H.A. Sissingh. 1989b. Determination of plant-available phosphorus by the iron hydroxide-impregnated filter paper (P_i) soil test. Soil Sci. Soc. Am. J. 52:110–115.

Monem, M.A., and A.M. Gadalla. 1992. Evaluation of the Pi soil test in Egyptian soils and its development as a rapid field test, p. 139–144. In J. Ryan and A. Matar (eds.) Fertilizer use efficiency under rain-fed agriculture in West Asia and North Africa. Proc. 4th Regional Workshop, May 1991, Agadir, Morocco.

Mosse, B., D.S. Hayman, and D.J. Arnold. 1973. Plant growth responses to vesicular-arbuscular mycorrhiza. V. phosphate uptake by three plant species from P deficient soils labelled with ^{32}P. New Phytol. 72:809–815.

Muchovej, R.M.C., J.J. Muchovej, and V.H. Alvarez. 1989. Temporal relations of phosphorus fractions in an Oxisol amended with rock phosphate and *Thiobacillus thiooxidans*. Soil Sci. Soc. Am. J. 53:1096–1100.

Mullins, G., B. Joern, and P. Moore. 2005. By-product phosphorus: Sources, characteristics, and management, p. 829–880. *In* J.T. Sims and A.N. Sharpley (eds.) Phosphorus: Agriculture and the environment. Agronomy 46:829–880.

Murphy, L.S. 1988. Phosphorus management strategies for MEY of spring wheat. Proc. Profitable Spring Wheat Production Workshop, January 6–7, 1988, Fargo, ND.

Myers, R.G., A.N. Sharpley, S.J. Thien, and G.M. Pierzynski. 2005. Ion-sink phosphorus extraction methods applied on 24 soils from the continental USA. Soil Sci. Soc. Am. J. 69:511–521.

National Research Council. 1993. Soil and water quality: An agenda for agriculture. National Academy Press, Washington, DC.

Oberson, A., D.K. Friesen, C. Morel, and H. Tiessen. 1997. Determination of phosphorus released by chloroform fumigation from microbial biomass in high P sorbing tropical soils. Soil Biol. Biochem. 29:1579–1583.

Oberson, A., D.K. Friesen, H. Teissen, C. Morell, and W. Stahel. 1999. Phosphorus status and cycling in native savanna and improved pastures on an acid low-P Colombian Oxisol. Nutr. Cycl. Agroecosyst. 55:77–88.

Oehl, F., A. Oberson, S. Sinaj, and E. Frossard. 2001. Organic phosphorus mineralization studies using isotopic dilution techniques. Soil Sci. Soc. Am. J. 65:780–787.

Ohno, T., and B.S. Crannell. 1996. Green and animal manure-derived dissolved organic matter effects on phosphorus sorption. J. Environ. Qual. 25:1137–1143.

Olsen, S.R., C.V. Cole, F.S. Watanabe, and L.A. Dean. 1954. Estimation of available phosphorus in soils by extraction with sodium bicarbonate. USDA Circular No. 939. U.S. Government Printing Office, Washington, DC.

Olsen, S.R., and L.E. Sommers. 1982. Phosphorus. *In* A.L., Page, R.H. Miller, and D.R. Keeney (eds.) Methods of soil analysis, Part 2. 2nd edn. Agronomy 9:403–429.

Olsen, S.R., and F.S. Watanabe. 1957. A method to determine a phosphorus adsorption maximum of soils as measured by the Langmuir isotherm. Soil Sci. Soc. Am. Proc. 21:144–149.

Olsen, S.R., and F.S. Watanabe. 1963. Diffusion of phosphorus as related to soil texture and plant uptake. Soil Sci. Soc. Am. Proc. 27:648–653.

Ottabong, E., J. Persson, O. Iakimenko, and L. Sadovnikova. 1997. The Ultuna long-term soil organic matter experiment. 2. Phosphorus status and distribution in soils. Plant Soil 195:17–23.

Pankhurst, C.E., B.G. Hawke, H.J. McDonald, C.A. Kirkby, J.C. Buckerfield, P. Michelsen, K.A. O'Brien, V.V.S.R. Gupta, and B.M. Doube. 1995. Evaluation of soil biological properties as potential bioindicators of soil health. Aust. J. Exp. Agric. 35:1015–1025.

Parle, J.N. 1963. Microorganisms in the intestines of earthworms. J. Gen. Microbiol. 31:1–11.

Peterson, T.A., J.S. Schepers, C. Chen, C.A. Cotway, R.B. Ferguson, and G.W. Hergert. 1994. Interpreting yield and soil parameter maps in the evaluation of variable rate nitrogen applications, p. 397. *In* Agronomy abstracts. ASA, Madison, WI.

Pierson, S.T., M.L. Cabrera, G.K. Evanylo, H.A. Kuykendall, C.S. Hoveland, M.A. McCann, and L.T. West. 2001. Phosphorus and ammonium concentrations in surface runoff from grasslands fertilized with broiler litter. J. Environ. Qual. 30:1784–1789.

Pierzynski, G.M. 1991. The chemistry and mineralogy of phosphorus in excessively fertilized soils. Crit. Rev. Environ. Control 21:265–295.

Pierzynski, G.M., and T.J. Logan. 1993. Crop, soil, and management effects on phosphorus soil test levels. J. Prod. Agric. 6:513–520.

Pierzynski, G.M., T.J. Logan, and S.J. Traina. 1990a. Phosphorus chemistry and mineralogy in excessively fertilized soils: Solubility equilibria. Soil Sci. Soc. Am. J. 54:1589–1595.

Pierzynski, G.M., T.J. Logan, S.J. Traina, and J. M. Bigham. 1990b. Phosphorus chemistry and mineralogy in excessively fertilized soils: Descriptions of phosphorus-rich particles. Soil Sci. Soc. Am. J. 54:1583–1589.

Pierzynski, G.M., R.W. McDowell, and J.T. Sims. 2005 Chemistry, cycling and potential movement of inorganic phosphorus in soils. *In* J.T. Sims and A.N. Sharpley (eds.) Phosphorus: Agriculture and the environment. Agronomy 46:53–83.

Posner, A.M., and J.W. Bowden. 1980. Adsorption isotherms: Should they split? J. Soil Sci. 31:1–10.

Pote, D.H., T.C. Daniel, D.J. Nichols, P.A. Moore, Jr., D.M. Miller, and D.R. Edwards. 1999a. Seasonal and soil-drying effects on runoff phosphorus relationships to soil phosphorus. Soil Sci. Soc. Am. J. 63:1006–1012.

Pote, D.H., T.C. Daniel, D.J. Nichols, A.N. Sharpley, P.A. Moore, Jr., D.M. Miller, and D.R. Edwards. 1999b. Relationship between phosphorus levels in three ultisols and phosphorus concentrations in runoff. J. Environ. Qual. 28:170–175.

Pote, D.H., T.C. Daniel, A.N. Sharpley, P.A. Moore, Jr., D.R. Edwards, and D.J. Nichols. 1996. Relating extractable soil phosphorus to phosphorus losses in runoff. Soil Sci. Soc. Am. J. 60:855–859.

Qian, P., J.J. Schoenau, and W.Z. Huang. 1992. Use of ion exchange membranes in routine soil testing. Commun. Soil Sci. Plant Anal. 23:1791–1804.

Randall, G.W., M.A. Schmitt, and J.P. Schmidt. 1999. Corn production as affected by time and rate of manure application and nitrapyrin. J. Prod. Agric. 12:317–323.

Read, D.J. 2002. Towards ecological relevance—Progress and pitfalls in the path towards an understanding of mycorrhizal functions in nature, p. 3–32. In M.G.A. van der Heijden and I.R. Sanders (eds.) Mycorrhizal ecology. Springer, Berlin, Germany.

Reddy, K.R., M.R. Overcash, R. Kahled, and P.W. Westerman. 1980. Phosphorus absorption-desorption characteristics of two soils utilized for disposal of animal manures. J. Environ. Qual. 9:86–92.

Rehm, G.W., R.C. Sorensen, and R.A. Wiese. 1984. Soil test values for phosphorus, potassium and zinc as affected by rate applied to corn. Soil Sci. Soc. Am. J. 48:814–818.

Reichle, D.E. 1977. The role of soil invertebrates in nutrient cycling, p. 145–156. In U. Lohn and T. Persson (eds.) Soil organisms as components of ecosystems. Proc. 6th Int. Coll. Soil Zool., Ecol. Bull. 25. Stockholm, Sweden.

Richardson, A.E. 2000. Prospects for using soil microorganisms to improve the acquisition of phosphorus by plants. Aust. J. Plant Physiol. 28:897–906.

Russell, E.W. 1973. Soil conditions and plant growth. Longman Press, London, U.K.

Saggar, S., M.J. Hedley, R.E. White, P.E.H. Gregg, K.W. Perrot, and I.S. Conforth. 1992. Development and evaluation of an improved soil test for phosphorus. 2. Comparison of the Olsen and mixed cation-anion exchange resin tests for predicting the yield of ryegrass growth in pots. Fert. Res. 33:135–144.

Sample, E.C., R.J. Soper, and G.J. Racz. 1980. Reactions of phosphate fertilizers in soils, p. 263–310. In F.E. Khawsaneh, E.C. Sample, and E.J. Kamprath (eds.) The role of phosphorus in agriculture. ASA, CSSA, and SSSA, Madison, WI.

Satchell, J.E. 1967. Lumbricidae, p. 259–322. In A. Burges and F. Raw (eds.) Soil biology. Academic Press, London, U.K.

Satchell, J.E. 1983. Earthworm ecology: From Darwin to vermiculture. Chapman & Hall, London, U.K, 595pp.

Scheu, S. 1987. Microbial activity and nutrient dynamics in earthworm casts (Lumbricidae). Biol. Fertil. Soils 5:230–234.

Schmitt, M.A., M.P. Russelle, G.W. Randall, C.C. Sheaffer, L.J. Greub, and P.D. Clayton. 1999. Effects of rate, timing, and placement of liquid dairy manure on reed canarygrass yield. J. Prod. Agric. 12:239–243.

Schroeder, P.D., D.E. Radcliffe, M.L. Cabrera, and C.D. Below. 2004. Relationship between soil test phosphorus and phosphorus in runoff. J. Environ. Qual. 33:1452–1463.

Schüssler, A., D. Schwarzott, and C. Walker. 2001. A new fungal phylum, the *Glomeromycota*: Phylogeny and evolution. Mycol. Res. 105:1413–1421.

Sharpley, A.N. 1991. Soil phosphorus extracted by iron-aluminum-oxide-impregnated filter paper. Soil Sci. Soc. Am. J. 55:1038–1041.

Sharpley, A.N. 1996. Availability of residual phosphorus in manured soils. Soil Sci. Soc. Am. J. 60:1459–1466.

Sharpley, A.N., L.R. Ahuja, M. Yamamoto, and R.G. Menzel. 1981. The kinetics of phosphorus desorption from soil. Soil Sci. Soc. Am. J. 45:493–496.

Sharpley, A.N., B.J. Carter, B.J. Wagner, S.J. Smith, E.L. Cole, and G.A. Sample. 1991. Impact of long-term swine and poultry manure applications on soil and water resources in eastern Oklahoma. Oklahoma State University, Technical Bulletin No. T169. Oklahoma State University, Stillwater, OK.

Sharpley, A.N., T.C. Daniel, J.T. Sims, and D.H. Pote. 1996. Determining environmentally sound soil phosphorus levels. J. Soil Water Conserv. 51:160–166.

Sharpley, A.N., C.A. Jones, C. Gray, and C.V. Cole. 1984a. A simplified soil and plant phosphorus model. II. Prediction of labile, organic, and sorbed phosphorus. Soil Sci. Soc. Am. J. 48:805–809.

Sharpley, A.N., R.W. McDowell, and P.J.A. Kleinman. 2004. Amounts, forms, and solubility of phosphorus in soils receiving manure. Soil Sci. Soc. Am. J. 68:21049–2057.

Sharpley, A.N., R.W. McDowell, J.L. Weld, and P.J.A. Kleinman. 2001. Assessing site vulnerability to phosphorus loss in an agricultural watershed. J. Environ. Qual. 30:2026–2036.

Sharpley, A.N., J.J. Meisinger, A. Breeuwsma, T. Sims, T.C. Daniel, and J.S. Schepers. 1997. Impacts of animal manure management on ground and surface water quality, p. 173–242. In J. Hatfield (ed.) Effective management of animal waste as a soil resource. Lewis Publishers, Boca Raton, FL.

Sharpley, A.N., J.J. Meisinger, J. Power, and D. Suarez. 1992. Root extraction of nutrients associated with long-term soil management. Adv. Soil Sci. 19:151–217.

Sharpley, A.N., and B. Moyer. 2000. Phosphorus forms in manure and compost and their release during simulated rainfall. J. Environ. Qual. 29:1462–1469.

Sharpley, A.N., J.T. Sims, and G.M. Pierzynski. 1994. Innovative soil phosphorus indices: Assessing inorganic phosphorus, p. 115–142. In J. Havlin, J. Jacobsen, P. Fixen, and G. Hergert (eds.) Soil testing: Prospects for improving nutrient recommendations. SSSA Special Publication No. 40. SSSA, Madison, WI.

Sharpley, A.N., and S.J. Smith. 1983. Distribution of phosphorus forms in virgin and cultivated soil and potential erosion losses. Soil Sci. Soc. Am. J. 47:581–586.

Sharpley, A.N., and S.J. Smith. 1985. Fractionation of inorganic and organic phosphorus in virgin and cultivated soils. Soil Sci. Soc. Am. J. 49:1276–130.

Sharpley, A.N., S.J. Smith, and W.R. Bain. 1993. Nitrogen and phosphorus fate from long-term poultry litter applications to Oklahoma soils. Soil Sci. Soc. Am. J. 57:1131–1137.

Sharpley, A.N., S.J. Smith, B.A. Stewart, and A.C. Mathers. 1984b. Forms of phosphorus in soil receiving cattle feedlot waste. J. Environ. Qual. 13:211–215.

Sharpley, A.N., and J.K. Syers. 1977. Seasonal variation in casting activity and in the amounts and release to solution of phosphorus forms in earthworm casts. Soil Biol. Biochem. 9:227–231.

Sharpley, A.N., J.K. Syers, and J.A. Springett. 1979. Effect of surface-casting earthworms on the transport of phosphorus and nitrogen in surface runoff from pasture. Soil Biol. Biochem. 11:459–462.

Sharpley, A.N., H. Tiessen, and C.V. Cole. 1987. Soil phosphorus forms extracted by soil tests as a function of pedogenesis. Soil Sci. Soc. Am. J. 51:362–365.

Sharpley, A.N., P.J.A. Withers, C. Abdalla, and A. Dodd. 2005. Strategies for the sustainable management of phosphorus. *In* J.T. Sims and A.N. Sharpley (eds.) Phosphorus: Agriculture and the environment. Agronomy 46:1069–1101.

Shipitalo, M.J., W.A. Dick, and W.M. Edwards. 2000. Conservation tillage and macropore factors that affect water movement and the fate of chemicals. Soil Tillage Res. 53:167–183.

Shuster, W.D., M.J. Shipitalo, S. Subler, S. Aref, and E.L. McCoy. 2003. Earthworm additions affect leachate production and nitrogen losses in typical Midwestern agroecosystems. J. Environ. Qual. 32:2132–2139.

Sims, J.T. 2008. Soil test phosphorus: Principles and methods, p. 9–19. *In* J.L. Kovar and G.M. Pierzynski (eds.) Methods of phosphorus analysis for soils, sediments, residuals, and waters. Southern Cooperative Series Bulletin. Virginia Tech University, Blacksburg, VA.

Sims, J.T., R.R. Simard, and B.C. Joern. 1998. Phosphorus loss in agricultural drainage: Historical perspective and current research. J. Environ. Qual. 27:277–293.

Singh, B.L., and R. Lal. 2005. Phosphorus management in low-input agricultural systems. *In* J.T. Sims and A.N. Sharpley (eds.) Phosphorus: Agriculture and the environment. Agronomy 46: 729–759.

Skogley, E.O., S.J. Georgitis, J.E. Yang, and B.F. Schaff. 1990. The phytoavailability soil test—PST. Commun. Soil Sci. Plant Anal. 21:1229–1243.

Smillie, G.W., D. Curtin, and J.K. Syers. 1987. Influence of exchangeable calcium on phosphate retention by weakly acid soils. Soil Sci. Soc. Am. J. 51:1169–1172.

Smith, S.E., and D.J. Read. 1997. Mycorrhizal symbiosis. Academic Press, London, U.K.

Soltanpour, P.N., and A.P. Schwab. 1977. A new soil test for simultaneous extraction of macronutrients and micronutrients in alkaline soils. Commun. Soil Sci. Plant Anal. 8:195–207.

Somasiri, L.L.W., and A.C. Edwards. 1992. An ion exchange resin method for nutrient extraction of agricultural advisory soil samples. Commun. Soil Sci. Plant Anal. 23:645–657.

Soper, R.J., and Y.P. Kalra. 1969. Effect of mode of application and source of fertilizer on phosphorus utilization by buckwheat, rape, oats, and flax. Can. J. Soil Sci. 49:319–326.

Spratt, E.D., F.G. Warder, L.D. Bailey, and D.W.L. Read. 1980. Measurement of fertilizer phosphorus residues and its utilization. Soil Sci. Soc. Am. J. 44:1200–1204.

Srivastava, S.C., and J.S. Singh. 1991. Microbial C, N and P in dry tropical forest soils: Effects of alternative land-uses and nutrient flux. Soil Biol. Biochem. 23:117–127.

Stevenson, F.C., A.M. Johnston, H.J. Beckie, S.A. Brandt, and L. Townley-Smith. 1998. Cattle manure as a nutrient source for barley and oilseed crops in zero and conventional tillage systems. Can. J. Plant Sci. 78:409–416.

Stewart, J.W.B., and R.B. McKercher. 1982. Phosphorus cycle, p. 221–238. *In* R.G. Burns and J.H. Slater (eds.) Experimental microbial ecology. Blackwell Scientific Publications, Oxford, U.K.

Stewart, J.W.B., and A.N. Sharpley. 1987. Controls on dynamics of soil and fertilizer phosphorus and sulfur, p. 101–121. *In* R.F. Follett, J.W.B. Stewart, and C.V. Cole (eds.) Soil fertility and organic matter as critical components of production systems. Soil Science Society of America Special Publication No. 19. ASA, CSSA, SSSA, Madison, WI.

Stewart, J.W.B., and H. Tiessen. 1987. Dynamics of soil organic phosphorus. Biogeochemistry 4:41–60.

Stout, W.L., and W.E. Priddy. 1996. Use of flue gas desulfurization (FGD) by-product gypsum on alfalfa. Commun. Soil Sci. Plant Anal. 27:2419–2432.

Stout, W.L., A.N. Sharpley, W.J. Gburek, and H.B. Pionke. 1999. Reducing phosphorus export from croplands with FBC fly-ash and FGD gypsum. Fuel 78:175–178.

Stout, W.L., A.N. Sharpley, and J. Landa. 2000. Effectiveness of coal combustion by-products in controlling phosphorus export from soils. J. Environ. Qual. 29:1239–1244.

Stout, W.L., A.N. Sharpley, and H.B. Pionke. 1998. Reducing soil phosphorus solubility with coal combustion by-products. J. Environ. Qual. 27:111–118.

Strong, W.M., and R.J. Soper. 1974. Utilization of pelleted phosphorus by flax, wheat, rape and buckwheat from a calcareous soil. Agron. J. 65:18–21.

Sugi, Y., and M. Tanaka. 1978. Number and biomass of earthworm populations, p. 171–178. *In* T. Kira, Y. Ono, and T. Hosokawa (eds.) Biological production in a warm-temperature evergreen oak forest of Japan. J.I.B.P. Synthesis Vol. 18. University of Tokyo Press, Tokyo, Japan.

Sumner, M.E., H. Shahandah, J. Bouton, and J. Hammel. 1986. Amelioration of an acid soil profile through deep liming and surface application of gypsum. Soil Sci. Soc. Am. J. 50:1254–1258.

Sutton, A.L., D.W. Nelson, J.D. Hoff, and V.B. Mayrose. 1982. Effects of injection and surface applications of liquid manure on corn yield and soil composition. J. Environ. Qual. 11:468–472.

Syers, J.K., and D. Curtin. 1989. Inorganic reactions controlling phosphorus cycling, p. 17–29. *In* H. Tiessen (ed.) Phosphorus cycles in terrestrial and aquatic ecosystems. UNDP, Published by Saskatchewan Institute of Pedology, Saskatoon, Canada.

Syers, J.K., T.D. Evans, J.D.H. Williams, and J.T. Murdock. 1971. Phosphate sorption parameters of representative soils from Rio Grande Do Sul, Brazil. Soil Sci. 112:267–275.

Syers, J.K., R.F. Harris, and D.E. Armstrong. 1973. Phosphate chemistry in lake sediments. J. Environ. Qual. 2:1–14.

Tate, K.R. 1984. The biological transformation of P in soil. Plant Soil 76:245–256.

Thien, S.J., and R. Myers. 1991. Separating ion-exchange resin from soil. Soil Sci. Soc. Am. J. 55:890–892.

Thomas, G.W., and D.E. Peaslee. 1973. Testing soils for phosphorus, p. 115–132. In L.M. Walsh and J.D. Beaton (eds.) Soil testing and plant analysis. Revised edition. SSSA, Madison, WI.

Thompson, L.M., C.A. Black, and J.A. Zoellner. 1954. Occurrence and mineralization of organic phosphorus in soils with particular reference to associations with nitrogen, carbon, and pH. Soil Sci. 77:185–196.

Tiessen, H. 2005. Phosphorus dynamics in tropical soils. In J.T. Sims and A.N. Sharpley (eds.) Phosphorus: Agriculture and the environment. Agronomy 46:253–262.

Tiessen, H., and J.O. Moir. 2007. Characterization of available P in sequential extraction, p. 293–306. In M.R. Carter and E.G. Gregorich (eds.) Soil sampling and methods of analysis. 2nd edn. Canadian Society of Soil Science, CRC Press, Taylor & Francis Group, Boca Raton, FL.

Tiessen, H., I.H. Salcedo, and E.V.S.B. Sampaio. 1992. Nutrient and soil organic matter dynamics under shifting cultivation in semi-arid Northeastern Brazil. Agric. Ecosyst. Environ. 39:139–151.

Tiessen, H., J.W.B. Stewart, and C.V. Cole. 1984. Pathways of phosphorus transformations in soils of differing pedogenesis. Soil Sci. Soc. Am. J. 48:853–858.

Tisdale, J.M., and J.M. Oades. 1982. Organic matter and water stable aggregates in soils. J. Soil Sci. 33:141–163.

Tiwari, K.N. 1979. Efficiency of Mussoorie rock phosphate, pyrites, and their mixtures in some soils of UP. Indian Soil Sci. Soc. Bull. 12:519–526.

Torbert, H.A., T.C. Daniel, J.L. Lemunyon, and R.M. Jones. 2002. Relationship of soil test phosphorus and sampling depth to runoff phosphorus in calcareous and noncalcareous soils. J. Environ. Qual. 31:1380–1387.

Treeby, M.T., H. Marschner, and V. Romheld. 1989. Mobilization of iron and other micronutrient cations from a calcareous soil by plant-borne, microbial, and synthetic metal chelators. Plant Soil 114:217–226.

Trojan, M.D., and D.R. Linden. 1998. Macroporosity and hydraulic properties of earthworm-affected soils as influenced by tillage and residue management. Soil Sci. Soc. Am. J. 62:1687–1692.

Turner, B.L., B.J. Cade-Menun, L.M. Condron, and S. Newman. 2005. Extraction of soil organic phosphorus. Talanta 66:294–306.

Turner, B.L., M.A. Kay, and D.T. Westermann. 2004. Phosphorus in surface runoff from calcareous arable soils of the semi-arid western United States. J. Environ. Qual. 33:1814–1821.

Turner, B.L., M.J. Papházy, P.M. Haygarth, and I.D. McKelvie. 2002. Inositol phosphates in the environment. Philos. Trans. R. Soc. Lond. Ser. B 357:449–469.

Uehara, G., and G.P. Gillman. 1981. The mineralogy, chemistry, and physics of tropical soils with variable charge. Westview Press, Boulder, CO.

U.S. Environmental Protection Agency. 1988. Wastes from the combustion of coal by electric utility power plants. EPA/530-SW-88-002. U.S. Environmental Protection Agency, Washington, DC.

Vadas, P.A., P.J.A. Kleinman, and A.N. Sharpley. 2005. Relating soil phosphorus to dissolved phosphorus in runoff: A single extraction coefficient for water quality modeling. J. Environ. Qual. 34:572–580.

van Raij, B., J.A. Quaggio, and N.M. de Silva. 1986. Extraction of phosphorus, potassium, calcium, and magnesium from soils by an ion-exchange resin procedure. Commun. Soil Sci. Plant Anal. 17:547–566.

Veith, J.A., and G. Sposito. 1977. Reactions of aluminosilicates, aluminum hydrous oxides, and aluminum oxide with o-phosphate: The formation of X-ray and amorphous analogs of variscite and montebrasite. Soil Sci. Soc. Am. J. 41:870–882.

Wagner, B.I., J.W.B. Stewart, and J.L. Henry. 1986. Comparison of single large broadcast and small annual seed-placed phosphorus treatments on yield and phosphorus and zinc content of wheat on chernozemic soils. Can. J. Soil Sci. 66:237–248.

Walker, T.W., and J.K. Syers. 1976. The fate of phosphorus during pedogenesis. Geoderma 15:1–19.

Wang, H.D., W.G. Harris, K.R. Reddy, and E.G. Flaig. 1995. Stability of phosphorus forms in dairy impacted soils under column leaching by synthetic rain. Ecol. Eng. 5:209–227.

Weld, J.L., A.N. Sharpley, D.B. Beegle, and W.J. Gburek. 2001. Identifying critical sources of phosphorus export from agricultural watersheds. Nutr. Cycl. Agroecosyst. 59:29–38.

Whalen, J.K., C. Chang, G.W. Clayton, and J.P. Carefoot. 2000. Cattle manure amendments can increase the pH of acid soils. Soil Sci. Soc. Am. J. 64:962–966.

White, R.E. 1980. Retention and release of phosphate by soil and soil constituents, p. 71–114. In P.B. Tinker (ed.) Soils and agriculture. John Wiley & Sons, New York.

Yang, J.E., E.O. Skogley, S.J. Georgitis, B.F. Schaff, and A.H. Ferguson. 1991. Phytoavailability soil test: Development and verification of theory. Soil Sci. Soc. Am. J. 55:1358–1365.

Yao, J., and S.A. Barber. 1986. Effect of one phosphorus rate placed in different soil volumes on P uptake and growth of wheat. Commun. Soil Sci. Plant Anal. 17:819–827.

Yerokum, O.A., and D.R. Christenson. 1990. Relating high soil test phosphorus concentrations to plant phosphorus uptake. Soil Sci. Soc. Am. J. 54:796–799.

Yli-Halla, M. 1990. Comparison of a bioassay and three chemical methods for determination of plant-available P in cultivated soils of Finland. J. Agric. Sci. Finl. 62:213–319.

Yost, R.S., and R.L. Fox. 1979. Contribution of mycorrhizae to P nutrition of crops growing on an Oxisol. Agron. J. 71:903–908.

Young, R.D., D.G. Westfall, and G.W. Colliver. 1985. Production, marketing, and use of phosphorus fertilizers, p. 323–376. In Fertilizer technology and use. 3rd edn. SSSA, Madison, WI.

Yuan, T.L. 1980. Adsorption of phosphate and water-extractable soil organic material by synthetic aluminum silicates and acid soils. Soil Sci. Soc. Am. J. 44:951–955.

Zhang, H., and J.L. Kovar. 2008. Fractionation of soil phosphorus, p. 49–59. *In* J.L. Kovar and G.M. Pierzynski (eds.) Methods of phosphorus analysis for soils, sediments, residuals, and waters. Southern Cooperative Series Bulletin. Virginia Tech University, Blacksburg, VA.

Zhang, T.Q., and A.F. MacKenzie. 1997. Changes in soil phosphorus fractions under long-term corn monoculture. Soil Sci. Soc. Am. J. 61:485–493.

11.3 Bioavailability of Soil Potassium

Donald L. Sparks

11.3.1 Introduction

The role of K in soils is prodigious. Of the many plant nutrient–soil mineral relationships, those involving K are of major, if not prime, significance (Sparks and Huang, 1985; Sparks, 1999).

Since the middle of the seventeenth century, when J.R. Glauker in the Netherlands first proposed that saltpeter (KNO_3) was the principle of vegetation, K has been recognized as being beneficial to plant growth (Russell, 1961). Glauker obtained large increases in plant growth from the addition of saltpeter to the soil that was derived from the leaching of coral soils. The essentiality of K to plant growth has been known since the work of von Liebig.

Of the major nutrient elements, K is usually the most abundant in soils (Reitemeier, 1951). Igneous rocks of the Earth's crust have higher K contents than sedimentary rocks. Of the igneous rocks, granites and syenites contain 46–54, basalts 7, and peridotites 2.0 g K kg^{-1}. Among the sedimentary rocks, clayey shales contain 30, whereas limestones have an average of only 6 g K kg^{-1} (Malavolta, 1985).

Mineral soils generally range between 0.04% and 3% K. Total K contents in soils range between 3000 and 100,000 kg ha^{-1} in the upper 0.2 m of the soil profile. Of this total K content, 98% is bound in the mineral form, whereas 2% is in soil solution and exchangeable phases (Schroeder, 1979; Bertsch and Thomas, 1985).

Potassium, among mineral cations required by plants, is the largest in nonhydrated size ($r = 0.133$ nm) and the number of oxygen atoms surrounding it in mineral structures is high (8 or 12), which suggests that the strength of each K–O bond is relatively weak (Sparks and Huang, 1985). Potassium has a polarizability equal to 0.088 nm^3, which is higher than for Ca^{2+}, Li^+, Mg^{2+}, and Na^+ but lower than for Ba^{2+}, Cs^+, NH_4^+, and Rb^+ ions (Rich, 1968, 1972; Sparks and Huang, 1985). Ions with higher polarizability are preferred in ion-exchange reactions. Potassium has a hydration energy of 142.5 kJ g^{-1} ion^{-1}, which indicates little ability to cause soil swelling (Helfferich, 1962).

11.3.2 Forms of Soil Potassium

Soil K exists in four forms in soils: solution, exchangeable, fixed or nonexchangeable, and structural or mineral (Figure 11.9). Quantities of exchangeable, nonexchangeable, and total K in the surface layer (0–20 cm) of a variety of soils are shown in Table 11.7. Exchangeable K and nonexchangeable K levels comprise a small portion of the total K. The bulk of total soil K is in the mineral fraction (Sparks and Huang, 1985). There are equilibrium and kinetic reactions between the four forms of soil K that affect the level of soil solution K at any particular time and, thus, the amount of readily available K for plants. The forms of soil K in the order of their availability to plants and microbes are solution > exchangeable > fixed (nonexchangeable) > mineral (Sparks and Huang, 1985; Sparks, 1987, 1999).

11.3.2.1 Solution K

Soil solution K is the form of K that is directly taken up by plants and microbes and also is the form most subject to leaching in soils. Levels of soil solution K are generally low, unless recent

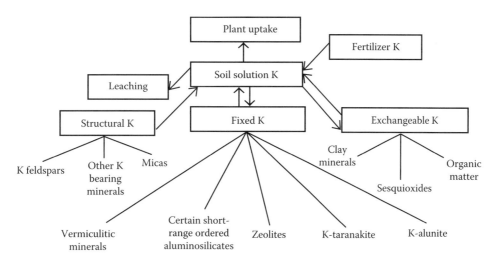

FIGURE 11.9 Interrelationships of various forms of soil K. (From Sparks, D.L., and P.M. Huang. 1985. Physical chemistry of soil potassium, p. 201–276. *In* R.D. Munson (ed.) Potassium in agriculture. ASA, Madison, WI. With permission of American Society of Agronomy.)

TABLE 11.7 Potassium Status of Selected Soils[a]

Origin of Soil	Exchangeable K	Nonexchangeable K	Total K	Source
	(cmol$_c$ kg^{-1})			
Alfisols				
Nebraska, USA	0.40	—	—	Soil Survey Staff (1975)
West Africa	0.46	—	3.07	Juo (1981)
Inceptisols				
California, USA	0.40	—	—	Soil Survey Staff (1975)
Maryland, USA	0.20	—	—	Soil Survey Staff (1975)
Mollisols				
Iowa, USA	0.27	—	—	Survey Staff (1975)
Nebraska, USA	0.40	—	—	Survey Staff (1975)
Ultisols				
Delaware, USA	0.33	0.49	22.5	Parker et al. (1989a)
Florida, USA	0.14	0.25	2.71	Yuan et al. (1976)
Virginia, USA	0.11	0.17	6.5	Sparks et al. (1980)
West Africa	0.24	—	8.06	Juo (1981)

[a] Data are for surface soils (0–20 cm depth).

amendments of K have been made to the soil. The quantity of K in the soil solution varies from 2 to 5 mg K L^{-1} for normal agricultural soils of humid regions and is an order of magnitude higher in arid region soils (Haby et al., 1990). Levels of solution K are affected by the equilibrium and kinetic reactions that occur between the forms of soil K, the soil moisture content, and the concentrations of bivalent cations in solution and on the exchanger phase (Sparks and Huang, 1985; Sparks, 1999).

11.3.2.2 Exchangeable K

Exchangeable K is the portion of the soil K that is electrostatically bound as an outer-sphere complex to the surfaces of clay minerals and humic substances. It is readily exchanged with other cations and also is readily available to plants.

11.3.2.3 Nonexchangeable K

Nonexchangeable or fixed K differs from mineral K in that it is not bonded within the crystal structures of soil mineral particles. It is held between adjacent tetrahedral layers of dioctahedral and trioctahedral micas, vermiculites, and intergrade clay minerals such as chloritized vermiculite (Rich, 1972; Sparks and Huang, 1985; Sparks, 1987). Potassium becomes fixed because the binding forces between K and the clay surfaces are greater than the hydration forces between individual K$^+$ ions. This results in a partial collapse of the crystal structures and the K$^+$ ions are physically trapped to varying degrees, making K release a slow, diffusion controlled process (Sparks, 1987). Nonexchangeable K also can be found in wedge zones of weathered micas and vermiculites (Rich, 1964). Only ions with a size similar to K$^+$, such as NH$_4^+$ and H$_3$O$^+$, can exchange K from wedge zones (Chapter 21 of *Handbook of Soil Sciences: Properties and Processes*). Large hydrated cations, such as Ca^{2+} and Mg^{2+}, cannot fit into the wedge zones. Release of nonexchangeable K to the exchangeable form occurs when levels of exchangeable and soil solution K are decreased by crop removal and/or leaching and perhaps by large increases in microbial activity (Sparks, 1980).

Nonexchangeable K is moderately to sparingly available to plants (Mengel, 1985; Sparks and Huang, 1985; Sparks, 1987). Mortland et al. (1956) showed that biotite could be altered to vermiculite by plant removal of K. Schroeder and Dummler (1966) showed that the nonexchangeable K associated with some German soil illites was an important source of K to crops. The ability of plants to take up nonexchangeable K appears to be related to the plant species. Steffens and Mengel (1979) found that ryegrass (*Lolium perenne*) could take up nonexchangeable K longer without yield reductions, while red clover (*Trifolium pratense*) could not. This was attributed to the ryegrass having longer root length, which would allow it to grow at a relatively low K concentration. A similar concentration would result in a K deficiency in red clover. It may be that the difference in root mass, root length, and root morphology between monocots and dicots explains why monocots feed better from nonexchangeable K than dicots (Mengel, 1985).

11.3.2.4 Mineral K

As noted earlier, most of the total K in soils is in the mineral form, mainly as K-bearing primary minerals such as muscovite, biotite, and feldspars. For example, in some Delaware soils, Sadusky et al. (1987) found that mineral K comprised about 98% of the total K (Table 11.8). Most of the mineral K was present as K feldspars in the sand fractions.

Common soil K-bearing minerals, in the order of availability of their K to plants, are biotite, muscovite, orthoclase, and microcline (Huang et al., 1968; Sparks, 1987). Mineral K is generally assumed to be only slowly available to plants; however, the availability is dependent on the level of K in the other forms, and the degree of weathering of the feldspars and micas constituting the mineral K fraction (Sparks and Huang, 1985; Sparks, 1987).

TABLE 11.8 Potassium Status of Delaware Soils and Sand Fractions

		Soils				Sand Fraction	
Horizon	Depth (cm)	CaCl$_2$ Extractable	HNO$_3$ Extractable	Mineral K[a]	Total K	Total K[b]	K Feldspars[c] Frequency (%)
		(cmol$_c$ kg^{-1})					
Kenansville loamy sand							
Ap	0–23	0.25	0.42	35.02	35.69	30.88	9.5
Bt2	85–118	0.25	0.49	45.30	46.04	33.86	12.0
Rumford loamy sand							
Ap	0–25	0.33	0.49	21.67	22.51	18.62	6.7
BC	89–109	0.21	0.54	23.39	23.96	16.76	8.2
Sassafras fine loamy sand							
Ap	0–20	0.35	0.56	43.54	44.45	28.95	16.0
Cl	84–99	0.13	0.36	45.99	46.68	36.69	24.0

Source: Sadusky, M.C., D.L. Sparks, M.R. Noll, and G.J. Hendricks. 1987. Kinetics and mechanisms of potassium release from sandy soils. Soil Sci. Soc. Am. J. 51:1460–1465. With permission of the Soil Science Society of America.

[a] Mineral K = [(total K) – (CaCl$_2$ ext. K + HNO$_3$ ext. K)].

[b] These data represent the amount of total K in the sand based on a whole soil basis.

[c] Determined through petrographic analyses of the whole sand fractions and represents the percentage of total point counts in a given sample that were K feldspars. The remaining minerals in the sand fraction were quartz, plagioclase, and opaques.

Sadusky et al. (1987) and Parker et al. (1989a, 1989b) have found that a substantial amount of K is released from the sand fractions of Delaware soils that are high in K feldspars. This finding, along with the large quantities of mineral K in these and other Atlantic Coastal Plain soils, could help in explaining the often observed lack of crop response to K amendments on these soils (Liebhardt et al., 1976; Yuan et al., 1976; Sparks et al., 1980; Woodruff and Parks, 1980; Parker et al., 1989a).

11.3.3 Factors Affecting Potassium Availability

11.3.3.1 Solution-Exchangeable K Dynamics

The rate and direction of reactions between solution and exchangeable forms of K determine whether applied K will be leached into lower horizons, taken up by plants, converted into unavailable forms, or released into available forms.

The reaction rate between soil solution and exchangeable phases of K is strongly dependent on the type of clay minerals present (Sivasubramaniam and Talibudeen, 1972; Sparks et al., 1980; Sparks and Jardine, 1981, 1984; Jardine and Sparks, 1984) and the method employed to measure kinetics of K exchange (Sparks, 1989, 1995; Amacher, 1991; Sparks et al., 1996). Vermiculite, montmorillonite, kaolinite, and hydrous mica vary drastically in their ionic preferences, ion binding affinities, and types of ion-exchange reactions. Such fundamental differences in these clay minerals account for the varying kinetics of K exchange.

Kinetics of K exchange on kaolinite and montmorillonite are usually quite rapid (Malcolm and Kennedy, 1969; Sparks and Jardine, 1984; Sparks, 1995, 2002). An illustration of this is shown in Figure 11.10. In the case of kaolin clays, the tetrahedral layers of adjacent clay layers are held tightly by H bonds; thus, only planar external surface and edge sites

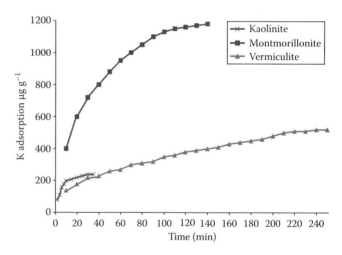

FIGURE 11.10 Potassium adsorption versus time in pure systems. (From Sparks, D.L., and P.M. Jardine. 1984. Soil Sci. 138:115–122. With permission of Wolters Kluwer.)

are available for ionic exchange. With montmorillonite, the inner peripheral space is not held together by H bonds, but instead is able to swell with adequate hydration and thus allow for rapid passage of ions into the interlayer space. Malcolm and Kennedy (1969) found that the rate of Ba exchange on kaolinite and montmorillonite was rapid with 75% of the total exchange occurring in 3 s.

Kinetics of K exchange on vermiculitic and micaceous minerals tends to be extremely slow. Both are 2:1 phyllosilicates with peripheral spaces that impede many ion-exchange reactions. Micaceous minerals typically have a more restrictive interlayer space than vermiculite since the area between layer silicates of the former is selective for certain types of cations (e.g., K$^+$, Cs$^+$). Bolt et al. (1963) theorized the existence of three types of binding sites for K exchange on hydrous mica. The authors hypothesized that slow kinetics were due to internal exchange sites, rapid kinetics to external planar sites, and intermediate kinetics to edge sites.

11.3.3.2 Potassium Fixation

The phenomenon of K fixation or retention significantly affects K availability. The fact that fixation processes are limited to interlayer ions, such as K^+, has been explained in terms of the good fit of K^+ ions (the crystalline radius and coordination number are ideal) in an area created by holes and adjacent oxygen layers (Barshad, 1951). The important forces involved in interlayer reactions in clays are electrostatic attractions between the negatively charged layers and the positive interlayer ions, and expansive forces due to ion hydration (Kittrick, 1966).

The degree of K fixation in clays and soils depends on the type of clay mineral and its charge density, the degree of interlayering, the moisture content, the concentration of K^+ ions as well as the concentration of competing cations, and the pH of the ambient solution bathing the clay or soil (Rich, 1968; Sparks and Huang, 1985).

The major clay minerals responsible for K fixation are montmorillonite, vermiculite, and weathered micas. In acid soils, the principal clay mineral responsible for K fixation is dioctahedral vermiculite. Weathered micas fix K under moist as well as dry conditions, whereas some montmorillonites fix K only under dry conditions (Rich, 1968).

The degree of K fixation is strongly influenced by the charge density on the layer silicate. Those with high charge density fix more K than those with low charge density (Walker, 1957). Weir (1965) noted that K fixation by montmorillonites is limited unless the charge density of the clays is high. Low charge montmorillonite (Wyoming) stays at 1.5 nm when K saturated, unless it is heated (Laffer et al., 1966). Schwertmann (1962a, 1962b) noted that soil montmorillonites have a greater capacity to fix K than do many specimen montmorillonites. Soil montmorillonites have higher charge density and a greater probability of having wedge positions near mica-like zones where the selectivity for K is high (Rich, 1968).

The importance of interlayer hydroxy Al and hydroxy Fe^{3+} material on K fixation was first noted in the classic work of Rich and Obenshain (1955). They theorized that hydroxy Al and hydroxy Fe^{3+} interlayer groups acted as props to decrease K fixation. This theory was later corroborated in the work of Rich and Black (1964) who found that the introduction of hydroxy Al groups into Libby vermiculite increased the Gapon selectivity coefficient (k_G) from 5.7 to 11.1×10^{-2} L $mmol^{-1/2}$.

Wetting and drying as well as freezing and thawing can significantly affect K fixation (Hanway and Scott, 1957; McLean and Simon, 1958; Cook and Hutcheson, 1960). The degree of K fixation or release on wetting or drying is dependent on the type of colloid present and the level of K^+ ions in the soil solution.

Potassium fixation by 2:1 clay minerals may be strongly influenced by the kind of adsorbed cations or the anions within the system. In studies with the silicate ion, Mortland and Gieseking (1951) found that montmorillonite clays dried with K_2SiO_3 were altered in their swelling properties and fixed K in large amounts. Hydrous mica clays also fixed large amounts of K that could not be removed with boiling HNO_3.

Volk (1934) observed a marked increase in K fixation in soils where pH was raised to about 9 or 10 with Na_2CO_3. Martin et al. (1946) showed that at pH values up to 2.5 there was no fixation; between pH 2.5 and 5.5, the amount of K fixation increased very rapidly. Above pH 5.5, fixation increased more slowly. These differences in K fixation with pH were discussed by Thomas and Hipp (1968). At pH values >5.5, Al^{3+} cations precipitate as hydroxy polycations, which increase in the number of OH groups as pH increases, until they have a form like gibbsite (Thomas, 1960). At this pH (~8), Al^{3+} does not neutralize the charge on the clay and cannot prevent K fixation. Below pH 5.5, Al^{3+} and $Al(OH)_x$ species dominate. Below pH 3.5, H_3O^+ predominates (Coleman and Harward, 1953; Thomas and Hipp, 1968).

The increase in K fixation between pH 5.5 and 7.0 can be ascribed to the decreased numbers of $Al(OH)_x$ species, which decrease K fixation (Rich and Obenshain, 1955; Rich, 1960, 1964; Rich and Black, 1964). At low pH, the lack of K fixation is probably due to large numbers of H_3O^+ and their ability to replace K as well (Rich, 1964; Rich and Black, 1964).

11.3.3.3 Potassium Release

The release of K from micas proceeds by two processes: (1) the transformation of K-bearing micas to expansible 2:1 layer silicates by exchanging the K with hydrated cations and (2) the dissolution of the micas followed by the formation of weathering products. The relative importance of these two mechanisms depends on the stability of micas and the nature of soil environments (Sparks and Huang, 1985).

Release of K from feldspars appears to involve a rapid exchange with H, which creates a thin layer of hydrolyzed aluminosilicate. This residual layer ranges in thickness from several to a few tens of nanometers and seems to cause the initial nonstoichiometric release of alkali and alkaline earths, relative to Si and Al. Following this step, there is continued dissolution, which removes hyperfine particles. After these are removed, further dissolution breaks down the outer surface of the residual layer at the same rate that alkalis are replaced by H at the interface between fresh mineral surfaces and the residual layer. This releases all constituents to the solution. Release is now stoichiometric. Thus, the weathering of feldspars appears to be a surface-controlled reaction (Sparks, 1989).

A number of physiochemical and mineralogical factors govern the release of K from micas by both cation exchange reactions and dissolution processes. These include tetrahedral rotation and cell dimensions, degree of tetrahedral tilting, hydroxyl orientation, chemical composition, particle size, structural imperfections, degree of K depletion, layer charge alterations and associated reactions, hydronium ions, biological activity, inorganic cations, wetting and drying, and other factors (Sparks and Huang, 1985). This review will focus on the latter four factors.

Biological activity promotes K release from micas (Mortland et al., 1956; Boyle et al., 1967; Sawhney and Voight, 1969;

Weed et al., 1969). The organisms deplete the K in the soil solution, and their action may be analogous to that of tetraphenylboron (TPB) in artificial weathering of micas. Furthermore, the overall action of organisms is more complex when organic acids are produced (Boyle et al., 1967; Spyridakis et al., 1967; Sawhney and Voight, 1969).

The importance of organic acids in weathering of rock-forming minerals has been recognized for a long time (Sprengel, 1826; Bolton, 1882; Huang and Keller, 1970). All soils contain small but measurable amounts of biochemical compounds such as organic acids. Furthermore, since the time required for soil formation can extend over a period of centuries, the cumulative effect in a soil of even very small quantities of chelating agents will be considerable. The influence of oxalic and citric acids on the dynamics of K release from micas and feldspars was studied by Song and Huang (1988). They found that the sequence of K release from K-bearing minerals by oxalic and citric acids is biotite > microcline orthoclase > muscovite.

The activity of K^+ ions in soil solution around mica particles greatly influences the release of K from micas by cation exchange. When the K level is less than the critical value, K is replaced from the interlayer by other cations from the solution. On the contrary, when the K level is greater than the critical value, the mica expansible 2:1 mineral takes K from the solution. The critical K level is highly mineral dependent, being much higher for the trioctahedral minerals (Smith and Scott 1966; Newman, 1969; von Reichenbach, 1973; Henderson et al., 1976). The critical levels for muscovite are so low that even the K impurities in laboratory chemicals or dissolved from glassware are often sufficient to prevent any K release (Scott and Smith, 1966).

The nature and concentration of the replacing cations also influences the critical K level of the cations tested in Cl^- solutions. Rausell-Colom et al. (1965) found the critical K levels decreased in the order $Ba^{2+} > Mg^{2+} > Ca^{2+} \approx Sr^{2+}$ for the same concentration of these ions and with a constant mica particle size. The activity of all of these replacing ions in the solution phase must be much greater than that of the K for significant K release to occur. The activity of various cations in the soil solution is governed by other minerals in the soil systems, by pedogenic processes, and by anthropogenic activities.

The release of K upon drying a soil is related to the clay fraction (Scott and Hanway, 1960). When a soil is dried, the degree of rotation of weathered soil minerals, such as micas, may be changed. Thus, the K–O bond may be modified. Dehydration of interlayer cations may permit a redistribution of interlayer cations, because Ca could now compete with K for wedge sites. This seems to account for the release of K from soils upon drying. Rich (1972) found that Virginia soils, which contain hydroxy Al interlayers and appreciable amounts of K, did not release K upon drying. The presence of the hydroxy Al interlayers may block or retard the interlayer diffusion of K ions and may change the b-dimension of micas, the degree of tetrahedral rotation, and the length and strength of the K–O bond.

Other factors that can affect K release from soils are leaching, redox potential (Eh), and temperature. Leaching promotes the K release from K-bearing minerals by carrying away the reaction products. Therefore, leaching accelerates the transformation of minerals, for example, micas, to expansible 2:1 layer silicates and other weathering products if the chemistry of leaching water favors the reaction.

Redox potential of soils could influence K release from micas since it has been pointed out that the tenacity with which K is held by biotite is greater after oxidation of its structural Fe. It appears that, other factors being equal, the extent of the K release from biotite should be less in soil environments that oxidize Fe than in soil environments that reduce it. Major elements in K-bearing feldspars do not exist in more than one valence state; thus, the prevailing Eh of a soil may not be of direct concern to chemical weathering of the feldspars. However, the weatherability of feldspars can be affected by complexing organic acids that are vulnerable to oxidation. Therefore, the stability of feldspars may be indirectly related to the prevailing Eh of a soil.

Increasing temperature has been shown to increase the rate of K release from biotite (Rausell-Colom et al., 1965) and K feldspars (Rasmussen, 1972). Under conditions of leaching of biotite with 0.1 mol $NaCl\ L^{-1}$, the rate of K release appears directly proportional to temperature in the range of 293–323 K (Mortland, 1958). Under similar leaching conditions, Mortland and Ellis (1959) observed that the log of the rate constant for K release from fixed K in vermiculite was directly proportional to the inverse of the absolute temperature.

Preheating of micas to high temperatures (1273 K) prior to TPB extraction (Scott et al., 1973) was found to enhance the rate of K extraction from muscovite, to decrease the rate for biotite, and to have little effect on phlogopite, except at very high temperatures. The decrease in K release from biotite by preheating is presumably because of oxidation of Fe at high temperatures. The more rapid rate with muscovite following heating remains unexplained.

11.3.3.4 Leaching of Potassium in Soils

Soil solution K is either leached or sorbed by plants or soils. A number of factors influence the movement of K in soils, including the cation exchange capacity (CEC), soil pH and liming, method and rate of K application, and K absorption by plants (Terry and McCants, 1968; Sparks, 1980).

The ability of a soil to retain applied K is very dependent on the CEC of the soil. Thus, the amount of clay and SOM in the soil strongly influences the degree of K leaching. Soils with higher CEC have a greater ability to retain added K, whereas leaching of K is often a problem on sandy soils (Sparks and Huang, 1985).

Retention of K can often be enhanced in sandy, Atlantic Coastal Plain soils after application of lime, since in such variable charge soils, the CEC is increased as soil pH is increased. Nolan and Pritchett (1960) found that liming a Lakeland fine sand soil (thermic, coated *Typic Quartzipsamment*) to pH 6–6.5 caused maximum retentivity of applied K. Potassium was replaced by Ca on the exchange complex at higher levels of limestone application. Less leaching of K occurred at pH 6.0–6.5 due to enhanced substitution of K for Ca than for Al, which was more abundant at low pH.

Lutrick (1963) found that K leaching occurred on unlimed but not on limed areas when 112–224 kg K ha^{-1} was applied on a Eustis loamy fine sand (sandy, siliceous, thermic Psammentic Paleudult).

Movement of applied K has been related to the method of application. Nolan and Pritchett (1960) compared banded and broadcast placement of KCl applied at several rates to Arredondo fine sand (loamy, siliceous, hyperthermic Grossarenic Paleudult) in lysimeters under winter and summer crops. For the low rate of application, cumulative K removal for both placements was only about 5.0 kg ha^{-1}.

A number of investigations have been conducted to determine the relationship of crop uptake and the rate of K application to leaching of K. Jackson and Thomas (1960) applied up to 524 kg K ha^{-1} prior to planting sweet potatoes (*Ipomoea batatas* L.) on Norfolk sandy loam (fine loamy, siliceous, thermic *Typic Paleudult*). At harvest time, soil and plant K exceeded applied K at the 131 and 262 kg K ha^{-1} rates. However, at the 524 kg K ha^{-1} rate, 38 kg K was unaccounted for by soil and plant K. This deficiency of K was attributed to leaching below sampling depths. During a 2 year study with corn (*Zea mays* L.) on two Dothan (*Typic Paleudult*) soils of Virginia, Sparks et al. (1980) found that 83 and 249 kg of applied K ha^{-1} increased the exchangeable K in the E and B21t horizons of the two soils. These increases were ascribed to leaching of applied K. The magnitude of leaching varied directly with rate of K application. No accumulation of K was found in the top 0.76 m of Leon sand (sandy, siliceous, thermic Aeric Haplaquod) after 40 year of heavy K fertilization (Blue et al., 1955).

11.3.4 Assessing Potassium Extractability and Availability

The extractability and availability of soil and plant K can be assessed by using chemical extractants to quantify the various forms of soil K, soil test extractants, quantity/intensity (*Q/I*) analyses, and plant analysis.

11.3.4.1 Chemical Extractants for Soil K Forms

Methods to determine total K and the other alkali elements in soils use acids or a high temperature fusion to decompose the soil. The most widely employed digestion techniques for total elements in soils and minerals have used combinations of HF and either H_2SO_4 or $HClO_4$ (Helmke and Sparks, 1996).

Exchangeable K is that K that is typically extracted with a neutral normal salt, usually 1.0 M NH_4OAc minus the water soluble K (Knudsen et al., 1982). In soils that are not saline, levels of water soluble K are minimal and can be ignored. However, in saline soils, the levels of water soluble K should be determined from a saturated extract or some similar extract and subtracted from the amount of K determined using NH_4OAc.

It should be noted that in soils that contain weathered vermiculitic and micaceous minerals wedge zones can be present that contain K. This K is not accessible to large index cations such as Ca and Mg, but can be extracted by NH_4, which is of similar size to K. For example, in soils that contain wedge zones, NH_4OAc will extract more K than an extractant like 1 M $CaCl_2$.

It is debatable whether this K is truly exchangeable. Thus, in soils containing wedge zones, exchangeable K could be overestimated with NH_4OAc (Sparks and Huang, 1985; Helmke and Sparks, 1996).

There are a number of chemical methods that can be employed to extract nonexchangeable K. These include boiling HNO_3, H_2SO_4, hot HCl, electroultrafiltration, Na TPB with EDTA, and ion-exchange resins such as H and Ca saturated resins (Hunter and Pratt, 1957; Martin and Sparks, 1985; Helmke and Sparks, 1996).

The most commonly used method for extraction of nonexchangeable K is the boiling HNO_3 technique. Most researchers that use this method boil the soil in M HNO_3 for 10 min over a flame, transfer the slurry to a filter, leach the soil with dilute HNO_3, and then determine the concentration in the filtrate. One of the problems with boiling for only 10 min over a flame is that it is difficult to be precise about the correct boiling time, the time it takes for boiling to occur, and the vigor of boiling (Martin and Sparks, 1985). Some workers have attempted to diminish these problems by using a 386 K oil bath for 25 min, including heating time (Pratt, 1965). This modification is particularly useful when large numbers of samples are being analyzed. Of course, one of the major concerns with using a boiling HNO_3 procedure is the potential to cause partial dissolution of mineral forms of K.

Other researchers have used continuous leaching of the soil with dilute acids such as 0.01 M HCl or with dilute salts such as 0.1 M NaCl, repeated extractions with 3, 0.3, and 0.03 M NaCl, Sr salts, hot $MgCl_2$, and sodium cobaltinitrite (Martin and Sparks, 1985).

Cation-exchange resins saturated with H or Ca also have been widely used to measure nonexchangeable K. These resins have high CECs and when saturated with an appropriate cation and mixed with soil and with a dilute solution, they will adsorb released K. One of the major advantages of using cation-exchange resins to extract K is that they act as a sink for the released K and thus prevent an inhibition of further K release. This is a problem with many batch methods that employ dilute acids and electrolytes. One major disadvantage of cation-exchange resins for extracting K is that the resins are expensive and the procedure is time-consuming and tedious.

In order for electrolyte and acid solutions and cation-exchange resins to be effective in extracting K, the K concentration in the solution phase must be kept very low, or K release must be inhibited (Rausell-Colom et al., 1965; Wells and Norrish, 1968; Feigenbaum et al., 1981; Martin and Sparks, 1983, 1985). The critical concentration above which release is inhibited is 4 mg L^{-1} for soils in general, 2.3–16.8 mg L^{-1} for trioctahedral micas in dilute solution, and as low as <0.1 mg L^{-1} for muscovite and illite (Smith and Scott, 1966; Martin and Sparks, 1985). A low enough concentration of solution K can be maintained by employing a continuous flowing extracting or exchanging solution, cation-exchange resins, or Na TPB (Scott et al., 1960).

One can quantitatively analyze for mineral K (K feldspars and micas) by using a selective dissolution method employing $Na_2S_2O_7$ fusion. The technique and method for calculating the quantities of K feldspars and micas can be found in Helmke and Sparks (1996). A semiquantitative approach for measuring mineral K is to subtract the quantity of nonexchangeable K, using the boiling HNO_3 procedure, from the quantity of total K, using the HF digestion method. One also can quantify K feldspars in the sand fraction of soil, using petrographic analyses (Parker et al., 1989b).

11.3.4.2 Soil Tests for Potassium

Soil test extractants for K were developed to easily and rapidly measure K in soils and to estimate K availability. Based on the amounts of extractable K, recommendations that are based on field test calibrations can then be made on the amount of K that is needed to maximize plant yields. Soil tests for K usually estimate the quantity of solution and exchangeable K, and since acids are usually employed as extractants, some nonexchangeable and mineral K is also extracted (Wolf and Beegle, 1991). The soil tests used to measure extractable K in the United States include Mehlich-1 and Mehlich-3 procedures in the northeastern and southeastern United States, the Morgan and modified Morgan procedures in parts of the northeastern United States, the 1 M NH_4OAc at pH 7 procedure in the north-central United States, and the ammonium bicarbonate-DTPA extraction in the western United States. Procedures for these soil tests are fully described in Helmke and Sparks (1996) and Chapter 14.

11.3.4.3 Q/I Analysis

Schofield (1947) proposed that the ratio of the activity of cations such as K and Ca was defined by the relation $a_K/(a_{Ca})^{1/2}$ where a is the ion activity. He appears to have been the first person to apply the concepts of quantity (Q) and intensity (I) to the mineral nutrient status of soils (Schofield, 1955).

Beckett (1964a), following the consideration of the Ratio law (Schofield, 1947), suggested that the I of K in a soil at equilibrium with its soil solution could best be defined by the ratio $a_K/(a_{Ca} + a_{Mg})^{1/2}$ of the soil solution. This equilibrium activity ratio for K or AR^K (Beckett, 1964a) has often been used as a measure of K availability to plants (Sumner and Marques, 1966; le Roux and Sumner, 1968).

Beckett (1964c) suggested that exchangeable K is held by two distinct mechanisms. The majority is held by general force fields comparable with those that hold exchangeable Na or Ca. A small proportion is held at sites offering a specific binding force for K but not for Ca and Mg. The electrochemical potential of exchangeable K in the diffuse double layer dictates the chemical potential of K in the soil solution. The K activity is also affected by the difference in electrical potential across the diffuse double layer that surrounds the exchange complex. Thus, no simple relationship exists between the activity of K in soil solution and quantity of K on the exchange phase (San Valentin et al., 1973). Moss (1967) and Lee (1973) note that a soil with a given complement of exchangeable K, Ca, and Mg gives rise to an activity ratio for K (AR^K) in the equilibrium soil solution that will be characteristic of that soil and independent of the soil-to-solution ratio and total electrolyte concentration. Moss (1967) noted that the ratio depends only on K saturation and the strength of adsorption of cations.

However, the relation of the amount of exchangeable K to AR^K must be specified to accurately describe the K status of a soil. Beckett (1964b) noted that different soils showing the same value of AR^K may not possess the same capacity for maintaining AR^K while K is removed by plant roots. Therefore, one must include not only the current potential of K in the labile pool but also the form of Q/I or the way in which potential depends on quantity of labile K present. These findings brought about the classic Q/I curves where the ratio of $a_K/(a_{Ca} + a_{Mg})^{1/2}$ is related to the change in exchangeable K to obtain the effect of quantity (exchangeable K) on intensity. The Q/I concept has been widely promulgated in the scientific literature to investigate the K status of soils (Evangelou et al., 1994).

The traditional method for Q/I analyses involves equilibrating a soil with solutions containing a constant amount of $CaCl_2$ and increasing the amounts of KCl (Beckett, 1964a). The soil gains or loses K to achieve the characteristic AR^K of the soil or remains unchanged if its AR^K is the same as the equilibrating solution. The AR^K values are then plotted versus the gain or loss of K to form the characteristic Q/I curve. From the Q/I plot, one can obtain several parameters to characterize the K status of a soil. The AR^K when the Q factor or ΔK equals zero is a measure of the degree of K availability at equilibrium or AR_e^K. The value of ΔK when $AR^K = 0$ is a measure of labile or exchangeable K in soils (ΔK°). The slope of the linear portion of the curve gives the potential buffering capacity of K (PBC^K) and is proportional to the CEC of the soil. The number of specific sites for K (K_x) is the difference between the intercept of the curved and linear portion of the Q/I plot at $AR^K = 0$ (Beckett, 1964b; Sparks and Liebhardt, 1981; Evangelou et al., 1994).

The traditional method described above is too time-consuming for routine analyses. Advances in ion-selective electrode (ISE) technology have allowed for more rapid Q/I analysis (Evangelou et al., 1994). Parra and Torrent (1983) developed an ISE simplified Q/I method whereby a single K-ISE in an electrochemical cell with liquid junction was employed to quantitate the K concentration (C_K) in equilibrated soil suspensions based on a successive addition procedure. The values of AR^K were estimated based on the expression, $AR^K = (11.5 − 0.3b) C_K + 22 \times 10^{−6}$, where b is the CEC (cmol$_c$ kg^{-1}) based on the weight of the soil samples used. The method of Parra and Torrent (1983) is quicker than the traditional Q/I method because equilibration time is reduced to 10 min compared to 24 h for the traditional method. Parra and Torrent (1983) achieved results with their modified procedure that were comparable to the traditional method. Wang et al. (1988) modified the procedure of Parra and Torrent (1983) by making direct measurements of CR_K (concentration ratio: $C_K/(C_{Ca+Mg})^{1/2}$) values with Ca and K-ISEs in an electrochemical cell with or without liquid junction.

11.3.4.4 Plant K Analysis

Plant K analysis will not be discussed in any detail in this review. For extensive discussions on plant analysis, the reviewer

TABLE 11.9 Critical K Concentrations in Agronomic Crops

Crop	Time of Sampling	Plant Part	Critical Concentration[a]
Sugarbeet		Blade	10
Cotton		Leaves	<9–15
Wheat	Jointing (GS6)	Total tops	20–25
	Early boot (GS9)	Total tops	15–20
Rice	Flag leaf to mid-tillering		10–14
Corn	At tassel	Ear leaf	19
	At tassel	Leaves	17–27
	At silk	Sixth leaf from base	13
	At silk	Leaf opposite and just below ear shoot	20
Grain sorghum	Full heading	Second blade below apex	18
Alfalfa		Whole top	8–22
Red clover		Tops	15–22.5
Bermudagrass		Tops	13–15
Orchardgrass		Tops	23–25
Tall fescue		Tops	24–38
Kentucky bluegrass		Tops	16–20

Source: Westerman, R.L. (ed.). 1990. Soil testing and plant analysis. SSSA, Madison, WI. With permission of the Soil Science Society of America.

[a] Critical concentration is that nutrient concentration at which plant growth begins to decrease in comparison with plants above the critical concentration.

is referred to chapters in Westerman (1990). Table 11.9 lists the critical level or concentration of K for various agronomic crops. These levels are usually determined by relating yield (e.g., percentage of the maximum yield or growth rate) to nutrient concentration (g kg^{-1}) for a specific plant part sampled at a given stage of development (Munson and Nelson, 1990). This method is based on the principle that if an element such as K is deficient in a plant, growth rates and yield will be decreased. If one adds increasing amounts of K, the concentration of the element in the plant or plant part increases until an optimum level is attained. Using this approach, growth or yield is expressed as a percentage of the maximum. The zone between the deficient and optimum concentration can be referred to as the transition zone (Ulrich and Hills, 1967). Ulrich and Hills (1967) referred to the transition zone as the zone between nutrient concentrations that produced a 20% reduction in growth or yield and continues to those that cause optimum or 100% in the maximum yield. Dow and Roberts (1982) refer to this latter zone as the critical range where researchers select the yield reduction and nutrient concentrations that are acceptable.

References

Amacher, M.C. 1991. Methods of obtaining and analyzing kinetic data, p. 19–59. *In* D.L. Sparks and D.L. Suarez (eds.) Rates of soil chemical processes. SSSA Special Publication No. 27. SSSA, Madison, WI.

Barshad, I. 1951. Cation exchange in soils: I. Ammonium fixation and its relation to potassium fixation and to determination of ammonium exchange capacity. Soil Sci. 77:463–472.

Beckett, P.H.T. 1964a. Studies on soil potassium: I. Confirmation of the ratio law: Measurement of potassium potential. J. Soil Sci. 15:1–8.

Beckett, P.H.T. 1964b. Studies on soil potassium: II. The 'immediate' Q/I relations of labile potassium in the soil. J. Soil Sci. 15:9–23.

Beckett, P.H.T. 1964c. K-Ca exchange equilibria in soils: Specific sorption sites for K. Soil Sci. 97:376–383.

Bertsch, P.M., and G.W. Thomas. 1985. Potassium status of temperate region soils, p. 131–162. *In* R.E. Munson (ed.) Potassium in agriculture. ASA, Madison, WI.

Blue, W.G., C.F. Eno, and P.J. Westgate. 1955. Influence of soil profile characteristics and nutrient concentrations on fungi and bacteria in Leon fine sands. Soil Sci. 80:303–308.

Bolt, G.H., M.E. Sumner, and A. Kamphorst. 1963. A study between three categories of potassium in an illitic soil. Soil Sci. Soc. Am. Proc. 27:294–299.

Bolton, H.C. 1882. Application of organic acids to the examination of minerals. Proc. Am. Assoc. Adv. Sci. 31:3–7.

Boyle, J.R., G.K. Voight, and B.L. Sawhney. 1967. Biotite flakes: Alteration by chemical and biological treatment. Science 155:193–195.

Coleman, N.T., and M.E. Harward. 1953. The heats of neutralization of acid clays and cation-exchange resins. J. Am. Chem. Soc. 75:6045–6046.

Cook, M.G., and T.B. Hutcheson, Jr. 1960. Soil potassium reactions as related to clay mineralogy of selected Kentucky soils. Soil Sci. Soc. Am. Proc. 24:252–256.

Dow, A.I., and S. Roberts. 1982. Proposal: Critical nutrient ranges for crop diagnosis. Agron. J. 74:401–403.

Evangelou, V.P., J. Wang, and R.E. Phillips. 1994. New developments and perspectives on soil potassium quantity/intensity relationships. Adv. Agron. 52:173–227.

Feigenbaum, S., R. Edelstein, and I. Shainberg. 1981. Release rate of potassium and structural cations from micas to ion exchangers in dilute solutions. Soil Sci. Soc. Am. J. 45:501–506.

Haby, V.A., M.P. Russelle, and E.O. Skogley. 1990. Testing soils for potassium, calcium, and magnesium, p. 181–228. *In* R.L. Westerman (ed.) Soil testing and plant analysis. SSSA, Madison, WI.

Hanway, J.J., and A.D. Scott. 1957. Soil potassium-moisture relations: II. Profile distribution of exchangeable K in Iowa soils as influenced by drying and rewetting. Soil Sci. Soc. Am. Proc. 20:501–504.

Helfferich, F. 1962. Ion exchange. McGraw-Hill Book Co., New York.

Helmke, P.A., and D.L. Sparks. 1996. Lithium, sodium, potassium, rubidium, and cesium, p. 551–574. *In* D.L. Sparks (ed.) Methods of soil analysis: Chemical methods. SSSA, Madison, WI.

Henderson, J.H., H.E. Doner, R.M. Weaver, J.K. Syers, and M.L. Jackson. 1976. Cation and silica relationships of mica weathering to vermiculite in calcareous Harps soils. Clay. Clay Miner. 24:93–100.

Huang, P.M., L.S. Crossan, and D.A. Rennie. 1968. Chemical dynamics of K-release from potassium minerals common in soils. Trans. 9th Int. Congr. Soil Sci. 2:705–712.

Huang, W.H., and W.D. Keller. 1970. Dissolution of rock forming minerals in organic acids. Am. Mineral. 55:2076–2094.

Hunter, A.H., and P.F. Pratt. 1957. Extraction of potassium from soils by sulfuric acid. Soil Sci. Soc. Am. Proc. 21:595–598.

Jackson, W.A., and G.W. Thomas. 1960. Effects of KCl and dolomitic limestones on growth and ion uptake of the sweet potato. Soil Sci. 89:347–352.

Jardine, P.M., and D.L. Sparks. 1984. Potassium-calcium exchange in a multireactive soil system: I. Kinetics. Soil Sci. Soc. Am. J. 47:39–45.

Juo, A.S.R. 1981. Chemical characteristics, p. 51–79. *In* D.J. Greenland (ed.) Characterization of soils in relation to their classification and management for crop production: Examples from some areas of the humid tropics. Clarendon Press, Oxford, U.K.

Kittrick, J.A. 1966. Forces involved in ion fixation by vermiculite. Soil Sci. Soc. Am. Proc. 30:801–803.

Knudsen, D., G.A. Peterson, and P.F. Pratt. 1982. Lithium, sodium, and potassium on a Coastal Plain soil. Commun. Soil Sci. Plant Anal. 7:265–277.

Laffer, B.G., A.M. Posner, and J.P. Quirk. 1966. Hysteresis in the crystal swelling of montmorillonite. Clay Miner. 6:311–321.

le Roux, J., and M.E. Sumner. 1968. Labile potassium in soils: I. Factors affecting the quantity-intensity (Q/I) parameters. Soil Sci. 106:35–41.

Lee, R. 1973. The K/Ca Q/I relation and preferential adsorption sites for potassium. New Zealand Soil Bureau Scientific Report II. Wellington, New Zealand.

Liebhardt, W.C., L.V. Svec, and M.R. Teel. 1976. Yield of corn as affected by potassium on a Coastal Plain soil. Commun. Soil Sci. Plant Anal. 7:265–277.

Lutrick, M. C. 1963. The effect of lime and phosphate on downward movement of potassium in Red Bay fine sandy loam. Proc. Soil Crop Sci. Soc. Fla. 23:90–94.

Malavolta, E. 1985. Potassium status of tropical and subtropical region soils, p. 163–200. *In* R.E. Munson (ed.) Potassium in agriculture. ASA, Madison, WI.

Malcolm, R.L., and V.C. Kennedy. 1969. Rate of cation exchange on clay minerals as determined by specific ion electrode techniques. Soil Sci. Soc. Am. Proc. 33:245–253.

Martin, H.W., and D.L. Sparks. 1983. Kinetics of nonexchangeable potassium release from two Coastal Plain soils. Soil Sci. Soc. Am. J. 47:883–887.

Martin, H.W., and D.L. Sparks. 1985. On the behavior of nonexchangeable potassium in soils. Commun. Soil Sci. Plant Anal. 16:133–162.

Martin, J.C., R. Overstreet, and D.R. Hoagland. 1946. Potassium fixation in soils in replaceable and nonreplaceable forms in relation to chemical reactions in the soil. Soil Sci. Soc. Am. Proc. 10:94–101.

McLean, E.O., and R.H. Simon. 1958. Potassium status of some Ohio soils as revealed by greenhouse and laboratory studies. Soil Sci. 85:324–332.

Mengel, K. 1985. Dynamics and availability of major nutrients in soils. Adv. Soil Sci. 2:65–131.

Mortland, M.M. 1958. Kinetics of potassium release from biotite. Soil Sci. Soc. Am. Proc. 22:503–508.

Mortland, M.M., and B.G. Ellis. 1959. Release of fixed potassium as a diffusion-controlled process. Soil Sci. Soc. Am. Proc. 23:363–364.

Mortland, M.M., and J.E. Gieseking. 1951. Influence of the silicate ion on potassium fixation. Soil Sci. 71:381–385.

Mortland, M.M., K. Lawton, and G. Uehara. 1956. Alteration of biotite to vermiculite by plant growth. Soil Sci. 82:477–481.

Moss, P. 1967. Independence of soil quantity/intensity relationships to changes in exchangeable potassium: Similar potassium exchange constants for soils within a soil type. Soil Sci. 103:196–201.

Munson, R.D., and W.L. Nelson. 1990. Principles and practices in plant analysis, p. 359–387. *In* R.L. Westerman (ed.) Soil testing and plant analysis. SSSA, Madison, WI.

Newman, A.C.D. 1969. Cation exchange properties of micas: I. The relation between mica composition and potassium exchange in solutions of different pH. J. Soil Sci. 20:357–373.

Nolan, C.N., and W.L. Pritchett. 1960. Certain factors affecting the leaching of potassium from sandy soils. Proc. Soil Crop Sci. Soc. Fla. 20:139–145.

Parker, D.R., G.J. Hendricks, and D.L. Sparks. 1989a. Potassium in Atlantic Coastal Plain soils: II. Crop responses and changes in soil potassium under intensive management. Soil Sci. Soc. Am. J. 53:397–401.

Parker, D.R., D.L. Sparks, G.J. Hendricks, and M.C. Sadusky. 1989b. Potassium in Atlantic Coastal Plain soils: I. Soil characterization and distribution of potassium. Soil Sci. Soc. Am. J. 53:392–396.

Parra, M.A., and J. Torrent. 1983. Rapid determination of the potassium quantity-intensity relationships using a potassium-selective ion electrode. Soil Sci. Soc. Am. J. 47:335–337.

Pratt, P.F. 1965. Potassium, p. 1023–1031. *In* C.A. Black et al. (eds.) Methods of soil analysis. Part 2. ASA, Madison, WI.

Rasmussen, K. 1972. Potash in feldspars. Proc. Colloq. Int. Potash Inst. 9:57–60.

Rausell-Colom, J.A., T.R. Sweetman, L.B. Wells, and K. Norrish. 1965. Studies in the artificial weathering of micas, p. 40–70. *In* E.G. Hallsworth and D.V. Crawford (eds.) Experimental pedology. Butterworths, London, U.K.

Reitemeier, R.F. 1951. The chemistry of soil potassium. Adv. Agron. 3:113–164.

Rich, C.I. 1960. Aluminum in interlayers of vermiculite. Soil Sci. Soc. Am. Proc. 24:26–32.

Rich, C.I. 1964. Effect of cation size and pH on potassium exchange in Nason soil. Soil Sci. 98:100–106.

Rich, C.I. 1968. Mineralogy of soil potassium, p. 79–91 *In* V.J. Kilmer et al. (eds.) The role of potassium in agriculture. ASA, Madison, WI.

Rich, C.I. 1972. Potassium in minerals. Proc. Colloq. Int. Potash Inst. 9:15–31.

Rich, C.I., and W.R. Black. 1964. Potassium exchange as affected by cation size, pH, and mineral structure. Soil Sci. 97:384–390.

Rich, C.I., and S.S. Obenshain. 1955. Chemical and clay mineral properties of a red-yellow podzolic soil derived from mica schist. Soil Sci. Soc. Am. Proc. 19:334–339.

Russell, E.W. 1961. Soil conditions and plant growth. Longmans, London, U.K.

Sadusky, M.C., D.L. Sparks, M.R. Noll, and G.J. Hendricks. 1987. Kinetics and mechanisms of potassium release from sandy soils. Soil Sci. Soc. Am. J. 51:1460–1465.

San Valentin, G.O., L.W. Zelazny, and W.K. Robertson. 1973. Potassium exchange characteristics of a Rhodic Paleudult. Proc. Soil Crop Sci. Soc. Fla. 32:128–132.

Sawhney, B.L., and G.K. Voight. 1969. Chemical and biological weathering in vermiculite from Transvaal. Soil Sci. Soc. Am. Proc. 33:625–629.

Schofield, R.K. 1947. A ratio law governing the equilibrium of cations in the soil solution. Proc. Int. Congr. Pure Appl. Chem. 11:257–261.

Schofield, R.K. 1955. Can a precise meaning be given to "available" soil phosphorus. Soils Fertil. 18:373–375.

Schroeder, D. 1979. Structure and weathering of potassium containing minerals. Proc. Congr. Int. Potash Inst. 11:43–63.

Schroeder, D., and H. Dummler. 1966. Kalium-Nachlieferung, Kalium-Festlegung und Tonmineralbestand schleswigholsteinischer Boden. Z. Pflanzenernähr. Dueng. Bodenkd. 113:213–215.

Schwertmann, U. 1962a. Die selective Kationensorption der Tonfraktion einiger Boden aus Sedimenten. Z. Pflanzenernähr. Dueng. Bodenkd. 97:9–25.

Schwertmann, U. 1962b. Eigenschaften und beldung aufwertbarer (quellbarer) Dreischichttonminerale in Böden aus Sedimenten. Beitr. Mineral. Petrogr. 8:199–209.

Scott, A.D., A.P. Edwards, and J.M. Bremner. 1960. Removal of fixed ammonium from clay minerals by cation exchange resins. Nature 185:792.

Scott, A.D., and J.J. Hanway. 1960. Factors influencing the change in exchangeable soil K observed on drying. Trans. 7th Int. Congr. Soil Sci. 4:72–79.

Scott, A.D., F.T. Ismail, and R.R. Locatis. 1973. Changes in interlayer potassium exchangeability induced by heating micas, p. 467–479. *In* J.M. Serratosa (ed.) Proc. 4th Int. Clay Conf. Division de Ciencias, Madrid, Spain.

Scott, A.D., and S.J. Smith. 1967. Visible changes in macro mica particles that occur with potassium depletion. Clays Clay Miner. 15:367–373.

Sivasubramaniam, S., and O. Talibudeen. 1972. Potassium-aluminum exchange in acid soils: I. Kinetics. J. Soil Sci. 23:163–176.

Smith, S.J., and A.D. Scott. 1966. Extractable potassium in Grundite illite: 1. Method of extraction. Soil Sci. 102:115–122.

Soil Survey Staff. 1975. Soil taxonomy. USDA agriculture handbook 436. U.S. Government Printing Office, Washington, DC.

Song, S.K., and P.M. Huang. 1988. Dynamics of potassium release from potassium-bearing minerals as influenced by oxalic and citric acids. Soil Sci. Soc. Am. J. 52:383–390.

Sparks, D.L. 1980. Chemistry of soil potassium in Atlantic Coastal Plain soils: A review. Commun. Soil Sci. Plant Anal. 11:435–449.

Sparks, D.L. 1987. Potassium dynamics in soils. Adv. Soil Sci. 6:1–63.

Sparks, D.L. 1989. Kinetics of soil chemical processes. Academic Press, San Diego, CA.

Sparks, D.L. 1995. Environmental soil chemistry. 1st edn. Academic Press, San Diego, CA.

Sparks, D.L. 1999. Bioavailability of soil potassium, p. D38–D53. *In* M.E. Sumner (ed.) Handbook of soil science. CRC Press, Boca Raton, FL.

Sparks, D.L. 2002. Environmental soil chemistry. 2nd edn. Academic Press, San Diego, CA.

Sparks, D.L., T.H. Carski, S.E. Fendorf, and C.V. Toner, IV. 1996. Kinetic methods and measurements, p. 1275–1307. *In* D.L. Sparks (ed.) Methods of soil analysis: Chemical methods. SSSA, Madison, WI.

Sparks, D.L., and P.M. Huang. 1985. Physical chemistry of soil potassium, p. 201–276. *In* R.D. Munson (ed.) Potassium in agriculture. ASA, Madison, WI.

Sparks, D.L., and P.M. Jardine. 1981. Thermodynamics of potassium exchange in soil using a kinetics approach. Soil Sci. Soc. Am. J. 45:1094–1099.

Sparks, D.L., and P.M. Jardine. 1984. Comparison of kinetic equations to describe K-Ca exchange in pure and in mixed systems. Soil Sci. 138:115–122.

Sparks, D.L., and W.C. Liebhardt. 1981. Effect of long-term lime and potassium applications on quantity-intensity (Q/I) relationships in sandy soil. Soil Sci. Soc. Am. J. 45:786–790.

Sparks, D.L., D.C. Martens, and L.W. Zelazny. 1980. Plant uptake and leaching of applied and indigenous potassium in Dothan soils. Agron. J. 72:551–555.

Sprengel, C. 1826. Uber Pflanzenhumus, Humussaure und Humussaure Salze. Kastners Arch. Naturlehre 8:145–220.

Spyridakis, D.E., G. Chesters, and S.A. Wilde. 1967. Kaolinization of biotite as a result of coniferous and deciduous seedling growth. Soil Sci. Soc. Am. Proc. 31:203–210.

Steffens, D., and K. Mengel. 1979. Das Aneignungsvermögen von *Lolium perenne* im Vergleich zu *Trifolium pratense* für Zwischenschicht-Kalium der Tonminerale. Landwirtschaftliche Forschung Sonderheft 36:120–127.

Sumner, M.E., and J.M. Marques. 1966. Ionic equilibria in a ferrallitic clay: Specific adsorption sites for K. Soil Sci. 102:187–192.

Terry, D.L., and C.B. McCants. 1968. The leaching of ions in soils. North Carolina Agricultural Experiment Station Technical Bulletin No. 184.

Thomas, G.W. 1960. Forms of aluminum in cation exchangers. Trans. 7th Int. Congr. Soil Sci. 2:364–369.

Thomas, G.W., and B.W. Hipp. 1968. Soil factors affecting potassium availability, p. 269–291. In V.J. Kilmer et al. (eds.) The role of potassium in agriculture. ASA, Madison, WI.

Ulrich, A., and F.J. Hills. 1967. Principles and practices of plant analysis, p. 11–24. In G.W. Hardy (ed.) Soil testing and plant analysis. Part II. SSSA Special Publication No. 2. SSSA, Madison, WI.

Volk, N.J. 1934. The fixation of potash in difficultly available forms in soils. Soil Sci. 37:267–287.

von Reichenbach, H.G. 1973. Exchange equilibria of interlayer cations in different particle size fractions of biotite and phlogopite, p. 480–487. In J.M. Serratosa (ed.) Proc. 4th Int. Clay Conf., Division de Ciencias, Madrid, Spain.

Walker, G.F. 1957. On the differentiation of vermiculites and smectites in clays. Clay Miner. Bull. 3:154–163.

Wang, J., R.E. Farrell, and A.D. Scott. 1988. Potentiometric determination of potassium Q/I relationships. Soil Sci. Soc. Am. J. 52:657–662.

Weed, S.B., C.B. Davey, and M.G. Cook. 1969. Weathering of mica by fungi. Soil Sci. Soc. Am. Proc. 33:702–706.

Weir, A.H. 1965. Potassium retention in montmorillonite. Clay Miner. 6:17–22.

Wells, C.B., and K. Norrish. 1968. Accelerated rates of release of interlayer potassium from micas. Trans. 9th Int. Congr. Soil Sci. 2:683–694.

Westerman, R.L. (ed.). 1990. Soil testing and plant analysis. SSSA, Madison, WI.

Wolf, A., and D. Beegle. 1991. Recommended soil tests for macronutrients: Phosphorus, potassium, calcium and magnesium, p. 25–34. In J.T. Sims and A. Wolf (eds.) Recommended soil testing procedures for the Northeastern Region. Delaware Agricultural Experiment Station Bulletin No. 493. Agricultural Experiment Station, University of Delaware, Newark, DE.

Woodruff, J.R., and C.L. Parks. 1980. Topsoil and subsoil potassium calibration with leaf K for fertility rating. Agron. J. 72:392–396.

Yuan, L.L., L.W. Zelazny, and A. Ratanaprasotporn. 1976. Potassium status of selected Paleudults in the lower Coastal Plain. Soil Sci. Soc. Am. Proc. 40:229–233.

11.4 Bioavailability of Calcium, Magnesium, Sulfur, and Silicon

James J. Camberato
William L. Pan

11.4.1 Introduction

Calcium, Mg, and S are essential mineral elements that are classified as macronutrients. While these three elements generally accumulate in plant tissues in higher concentrations (1.5–35 g kg^{-1}) (Table 11.10) than the micronutrients, they are not as commonly limiting in crop production as N, P, and K. Yet, when the soil bioavailability of Ca, Mg, and S is low, crop yield and quality can be suboptimal, depending on crop species and environmental conditions. Although Si is an essential element in the classical sense for diatoms and algal species of the genera Chrysophyceae and Equisetaceae (Epstein, 1999), it is considered a "beneficial" or "quasiessential" element for many other plant species, most notably rice (*Oryza sativa* L.) and sugarcane (*Saccharum officinarum* L.) (Epstein, 1994).

TABLE 11.10 Calcium and Magnesium Concentrations in Plant Tissue Considered Adequate for Plant Growth

Plant	Tissue	Range of Concentration		
		Ca	Mg	S
		g kg^{-1} Tissue		
Maize	Ear leaf	2.1–10.0	2.0–4.0	1.2–2.5
Rice	Leaves (tillering)	1.6–3.9	1.6–3.9	0.9–3.8
Wheat	Leaves (heading)	2.0–5.0	1.5–5.0	1.5–4.0
Soybean	Leaves	3.6–20.0	2.6–10.0	2.0–4.0
Peanut	Leaves and stem	7.5–20.0	3.0–7.5	2.0–3.0
Cotton	Leaves	19.0–35.0	3.0–7.5	2.9–15.0
Fruit trees	Leaves	14.0–21.0	4.1–6.8	1.3–3.0

Sources: Ca and Mg data adapted from Clark, R., Physiological aspects of calcium, magnesium, and molybdenum deficiencies in plants, in F. Adams (ed.), *Soil Acidity and Liming*, SSSA, Madison, WI, 1984, pp. 99–170; S data compiled from Asher, C.J. et al., Sulfur nutrition of tropical annual crops, in G.J. Blair and A.R. Till (eds.), *Sulfur in South East Asian and South Pacific Agriculture*, The University of New England, Armidale, New South Wales, Australia, 1983, pp. 54–64; Fox, R.L. and Blair, G.J., Plant response to sulfur in tropical soils, in M.A. Tabatabai (ed.), *Sulfur in Agriculture*, ASA, Madison, WI, 1986, pp. 405–434; Kamprath, E.J. and Jones, U.S., Plant response to sulfur in the Southeastern United States, in M.A. Tabatabai (ed.), *Sulfur in Agriculture*, ASA, Madison, WI, 1986, pp. 323–343; Jones, M.B., Sulfur availability indexes, in M.A. Tabatabai (ed.), *Sulfur in Agriculture*, ASA, Madison, WI, 1986, pp. 549–566; Tiwari, K.N., *Sulphur Agric.*, 14, 29, 1990.

The bioavailability of Ca, Mg, S, and Si is governed by the following factors: (1) parent material, (2) ion-exchange reactions, (3) biological transformations, (4) losses from the crop root zone by leaching and crop removal from the field, and (5) replenishment via atmospheric deposition, fertilizer, and soil amendments. Factors influencing Ca and Mg bioavailability are collectively discussed due to the similarity in soil chemical and plant nutritional behavior of these two elements. In contrast, S and Si chemistry in soils and plant nutrition are distinctly different, and are discussed separately in the latter half of the section.

11.4.2 Calcium and Magnesium

11.4.2.1 Overview of Calcium and Magnesium Nutrition

General reviews of the function of Ca and Mg in plants have been written by Clark (1984), Hanson (1984), and Poovaiah and Reddy (1996). A large proportion of Ca is found as Ca pectates in cell walls for structural rigidity, or as Ca oxalate and other organic acids in the apoplasm or vacuoles (Kinzel, 1989; Borchert, 1990). Calcium functions in plants include maintenance of membrane integrity (Caldwell and Haug, 1981) and cell wall stability (Konno et al., 1984), root (Poovaiah and Reddy, 1996) and shoot (Slocum and Roux, 1983) gravitropism, callose deposition (Lerchl et al., 1989), and regulation of several enzymes including α-amylase (Mitsui et al., 1984), protein kinases (Raghothama et al., 1987), and ATPases (Clarkson and Hanson, 1980).

Magnesium functions as an enzyme activator, in addition to serving as the coordinating central cation in the chlorophyll molecule. Magnesium activates numerous plant enzymes by bridging enzymes with ligand groups of substrates to optimize geometric conformation of enzyme systems involving the transfer of phosphate or carboxyl groups (Marschner, 1995). Systems activated by Mg include chlorophyll biosynthesis (Walker and Weinstein, 1991), chlorophyll degradation (Langmeier et al., 1993), photosynthesis (Pierce, 1986), protein synthesis (Cammarano et al., 1972), ATP synthesis (Lin and Nobel, 1971), and ATPases (Balke and Hodges, 1975), some of which are involved in phloem loading of sucrose (Williams and Hall, 1987).

11.4.2.2 Occurrence of Soil Calcium and Magnesium

Calcium and Mg occur in soils primarily in mineral forms and as ions on the exchange complex or in the soil solution. Little Ca or Mg occurs in organic complexes in soils (Mokwunye and Melsted, 1972). The relative amounts of Ca derived from parent materials rank as follows: calcareous sedimentary rock > basic igneous rock > acid igneous rock (Jenny, 1941). Parent materials rank with respect to Mg contents: basic igneous rock > acid igneous rock > sedimentary rock (Metson, 1974). Generally, soils with the highest 2:1 lattice clay content have the highest Mg contents (Mokwunye and Melsted, 1972). Contents of Ca and Mg in soils are presented in Table 11.11.

TABLE 11.11 Contents of Calcium and Magnesium in Various Soils

Soil Type	Content Ca	Mg	Source
	(g kg^{-1})		
Temperate soils	0.7–36	1.2–15	Brady (1974)
Humid region soil	4	3	Brady (1974)
Arid region soil	10	6	Brady (1974)
Peat soils	1.1–48.3		Brady (1974)
Muck soils		0.7–5.7	Millar (1955)

11.4.2.3 Additions and Losses of Soil Calcium and Magnesium

Calcium and Mg accession by soils from atmospheric deposition may be significant in comparison to that removed in crop and forest plants (Johnson and Todd, 1987). Calcium and Mg depositions from precipitation in a Tennessee forest were 6.3 and 0.7 kg ha^{-1} year^{-1}, respectively (Johnson and Todd, 1987). Hedin et al. (1994) estimated atmospheric Ca depositions for Sweden, the northeastern United States, and the Netherlands in the 1990s to be 0.5, 1.0, and 4.2 kg Ca ha^{-1} year^{-1}, respectively. Atmospheric deposition of Mg and Ca has garnered more interest lately because of its potential to increase soil inorganic C and thereby reduce atmospheric CO_2 and global warming. Goddard et al. (2007, 2009) reported the highest deposition in the United States to be 1.0–1.4 kg Mg ha^{-1} year^{-1} in the Pacific Northwest and coastal Florida and to be 2.5–3.6 kg Ca ha^{-1} year^{-1} in the central Midwest–Great Plains region.

Considerable amounts of Ca and Mg are added to soils in the form of limestone (21%–32% Ca, 3%–12% Mg), fertilizers, and animal manures. Fertilizers used to specifically supply Ca to soils include gypsum (22% Ca) and calcium nitrate (19% Ca). Calcium may also be added to the soil incidentally while supplying P as superphosphates (14%–20% Ca). Common Mg-containing fertilizers include Epsom salts (10% Mg) and sulfate of potash magnesia (11% Mg).

Loss of Ca and Mg from the soil occurs through leaching and crop removal. Leaching losses of Ca and Mg from an aspen forest (*Populus tremuloides* Michx.) on *Typic Fragiochrept* soils averaged 39 and 9 kg ha^{-1} year^{-1}, respectively (Silkworth and Grigal, 1982). Cropping with perennial species reduces leaching losses in comparison to annuals. Calcium and Mg leaching losses were 193 and 104 kg ha^{-1}, respectively, from a continuous maize (*Zea mays* L.) rotation but only 64 and 48 kg ha^{-1} from a bluegrass (*Poa pratensis*) sod (Bolton et al., 1970). The application of Ca generally increases the leaching of Mg (Pratt and Harding, 1957), partly because the adsorption affinity for Ca is greater than that for Mg in non-vermiculitic soils (Hunsaker and Pratt, 1971). Leaching of Ca and Mg is accelerated by acidification, resulting from the nitrification of ammoniacal N sources (Schwab et al., 1989; Darusman et al., 1991). Under acid conditions, Al^{3+} and H^+ ions displace Ca^{2+} and Mg^{2+} from the exchange complex enabling leaching to occur.

Plant removals of Ca and Mg from the soil are dependent on the plant species and genotype and the harvested fraction that is removed from the field. The Ca concentration of plants varies considerably, but one generalization can be made: the Ca concentration of dicots, legumes, and crucifers (12–18 g kg^{-1}) is far greater than the 4 g kg^{-1} Ca typically found in grasses (Parker and Truog, 1920). In contrast, the Mg concentration of these groups of plants differs only slightly (~2–4 g kg^{-1}). Several authors have provided compilations of plant mineral composition that may be useful in estimating crop removal of Ca and Mg (Truog, 1918; Beeson, 1941). Magnesium removals were 5, 7, 8, and 16 kg Mg ha^{-1} for snap beans (*Phaseolus vulgare* L.), okra (*Hibiscus esculentus* L.), white potatoes (*Solanum tuberosum* L.), and sweet potatoes (*Ipomoea batatas* (L.) Lam.), respectively (Prince, 1951). More recently, representative uptakes and removals of Ca and Mg for several crops were reported by Buol (1995). Crop removal was in the range of 1–30 kg Ca or Mg ha^{-1} year^{-1}. Spring wheat (*Triticum aestivum* L.) accumulated 35 kg Ca ha^{-1} and 13 kg Mg ha^{-1}, whereas sugar beets (*Beta vulgaris* L.) accumulated 104 kg Ca ha^{-1} and 44 kg Mg ha^{-1} (Strebel and Duynisveld, 1989). Barber (1984) reported the total uptake of Mg by maize to be 45 kg ha^{-1}.

Hardwood tree species accumulate substantial amounts of Ca. For example, a mixed oak (*Quercus* sp.) stand grown on *Typic Paleudult* soils contained 1090 kg Ca ha^{-1} (16 kg Ca ha^{-1} year^{-1}) in the aboveground plant parts when harvested (Johnson and Todd, 1987). Loblolly pine (*Pinus taeda* L.) grown on the same soils removed only 196 kg Ca ha^{-1}. Both species removed about 45 kg Mg ha^{-1} from the soil (0.6 and 1.5 kg Mg ha^{-1} year^{-1} for mixed oak and pine, respectively). In a more temperate climate, aspen removed 1034 kg Ca ha^{-1} and 95 kg Mg ha^{-1} from *Typic Fragiochrept* soils in northeastern Minnesota (Silkworth and Grigal, 1982).

11.4.2.4 Factors Affecting Availability of Soil Calcium and Magnesium

The activities of Ca and Mg in the soil solution, which directly influence the rates of uptake into plant roots, are dependent on the level of exchangeable cations (Albrecht, 1941) and the type of colloid. Organic matter and 1:1 clays retain Ca less tightly than 2:1 clay minerals, resulting in greater Ca availability at any given level of Ca saturation (Allaway, 1945; Mehlich and Colwell, 1945; Mehlich, 1946).

The affinity of cations for the exchange complex is dependent on the mineralogy of the colloid. For instance, the affinities of Ca and Mg for montmorillonite are similar, but the affinity of Mg for vermiculite is much greater than the affinity of Ca (Wild and Keay, 1964). Soils with exchange complexes arising from organic matter, peat, allophane, kaolinite, and oxides of Fe and Al have a higher affinity for Ca than for Mg (Hunsaker and Pratt, 1971).

The relative abundance of basic cations on the exchange complex affects the plant availability of these ions. Bear et al. (1945) asserted that the ideal base saturation of the exchange complex was 65% of Ca, 10% of Mg, and 5% of K. This concept was termed the basic cation saturation ratio. Although this ideology of soil test interpretation was widely adopted in the Midwestern United States for making fertilizer and lime recommendations (McLean and Brown, 1984), subsequent research demonstrated that a fairly broad range of basic cation saturations would produce equivalent crop yields (Hunter, 1949; Giddens and Toth, 1951; Key et al., 1962; Simson et al., 1979; Osemwota et al., 2007).

The degree of Ca saturation needed to provide sufficient plant-available Ca to crops is dependent on the colloid. Kamprath (1984) concluded from the results of a number of studies that a Ca saturation of about 25%–30% and an exchangeable Ca level of 1.0 cmol$_c$ kg^{-1} in highly weathered soils, dominated by kaolinitic clays and Fe/Al oxides, were adequate for supplying the Ca requirement of most plants. However, other factors such as Al may still limit plant growth at this level of Ca saturation. Soil solution Ca levels in acid soils ranging from 0.38 to 9.3 mM were reported by Kamprath (1978). The soil solution Ca level required for optimum growth of tropical legumes was 0.125 mM, much lower than the 2.0 mM required by temperate legumes (Kamprath, 1978). The soil solution Ca concentration necessary for maximizing cotton (*Gossypium hirsutum* L.) root growth was between 0.04 and 0.34 mM (Howard and Adams, 1965; Adams and Moore, 1983).

11.4.2.5 Calcium and Magnesium Soil–Plant Relationships

The transport of Ca and Mg ions to the root surface occurs by mass flow and diffusion. The soil solution composition and transpiration have been used to estimate the contribution of mass flow to the movement of these ions to plant roots (Barber, 1962). The difference between total nutrient accumulation in the plant and the estimate of that supplied by mass flow gives the relative contribution of diffusion and root interception. From these estimates, Barber (1962) recognized that mass flow could supply more than the required amount of Ca to plant roots of some soil–plant systems, resulting in a buildup of Ca in the rhizosphere that was observed in subsequent experiments (Barber and Ozanne, 1970). He also surmised that in situations where plant Ca demand is high and soil solution Ca is low, the supply of Ca to plant roots can be diffusion limited (Barber, 1962), resulting in a depletion of soluble Ca around the roots (Barber and Ozanne, 1970). Similarly, the mode of Mg transport is cropping system dependent. While mass flow could account for the Mg accumulated by wheat, a majority of Mg moving to sugar beet roots occurred by diffusion (Strebel and Duynisveld, 1989).

The presence of Ca is required throughout the root profile to maintain the integrity of root cell membranes and nutrient uptake (Haynes and Robbins, 1948). Insufficient Ca limits rooting into the subsoil of many Ultisols (Adams and Moore, 1983) and Oxisols (Ritchey et al., 1982) limiting the volume of soil explored and crop access to soil moisture supplies (Sumner et al., 1986; Sumner, 1995). Subsoil horizons in these soils may also be Al and H toxic (Adams and Moore, 1983). The Ca concentration of the soil solution needed to obtain maximum soybean (*Glycine max* (L.) Merr.) root growth increases as pH decreases and Al activity increases (Lund, 1970). The Ca/Al molar ratio of the soil solution was determined to be an excellent indicator of the potential for Al stress in tree species (Cronan and Grigal, 1995).

The effects of liming on soil Ca levels are initially confined to the zone of incorporation (Ritchey et al., 1980; Pavan et al., 1984, 1987). However, gypsum, phosphogypsum, and soluble Ca applications are effective at providing Ca deep into the soil profile. Forty-four weeks after application, soil Ca was increased in the upper 20 cm of a tropical Inceptisol, but phosphogypsum and $CaCl_2$ increased exchangeable Ca to a depth of 80 cm (Pavan et al., 1987). Growth of alfalfa (*Medicago sativa* L.) was increased in an Appling coarse sandy loam (*Typic Hapludult*) by a surface application of gypsum, which increased subsoil Ca and decreased soluble Al (Sumner et al., 1986; Sumner and Carter, 1988; Sumner, 1995).

Low supplies of Ca inhibit the nodulation, growth, and N fixation of bacteria associated with the roots of legumes (Albrecht, 1931; McCalla, 1937). No N fixation occurs unless the Ca saturation is >40% (Albrecht, 1937). The number of nodules and the amount of N fixed increased to 97% Ca saturation. The number of nodules per plant was positively correlated with the total Ca content of the primary root (Sartain and Kamprath, 1978). The Ca requirement for nodulation is greater at low pH and greater than that for the host plant (Alva et al., 1990). Calcium concentration of 0.5 mM maximized nodule number and weight at pH > 5, but 2.5 mM Ca was required at pH 4.5.

The ratio of Ca and Mg to other cations in the soil solution is important in Ca and Mg sufficiency to the plant. Maximum root length of cotton occurred when the activity of Ca exceeded approximately 15% of the total cation activity in the soil solution (Adams, 1966) and when exchangeable Ca was 13% or more of the total exchangeable cations (Howard and Adams, 1965; Adams and Moore, 1983). These relationships were nearly identical for the two soils examined even though the clay mineralogy of one soil was kaolinitic and that of the other was vermiculitic. Other crops respond similarly to the relative activity of Ca in the soil solution. Root growth of soybean (Lund, 1970) and loblolly pine (Lyle and Adams, 1971) were maximized when the ratio of Ca activity to total cation activity in the soil solution exceeded 10%–20%. Magnesium activity expressed as a function of total cation activity in the soil solution was closely correlated with ryegrass (*Lolium perenne* L.) Mg concentration (Salmon, 1964).

Calcium has low phloem mobility, which results in low Ca redistribution from older plant tissues to growing meristems (Jeschke and Pate, 1991), thereby imposing heavy reliance on concurrent Ca uptake and xylem transport to support new growth and development of vegetative and reproductive tissues (Morard et al., 1996). Plant organs with low rates of transpiration (e.g., new leaves, fruits, tubers) and low rates of xylem flow exhibit Ca deficiency disorders such as pod rot in peanut (*Arachis hypogaea* L.) (Cox et al., 1982; Sumner et al., 1988), internal brown spot in potato (Tzeng et al., 1986), blossom end rot in tomato (*Lycopersicon esculentum* Mill.) (Geraldson, 1957), bitter pit in apple (*Malus* sp.) (Perring, 1986), cork spot in pear (*Pyrus communis* L.) (Raese and Drake, 1993), and blackheart in celery (*Apium graveolens* L.) (Geraldson, 1954). Although low soil Ca levels may increase the frequency and severity of Ca deficiency in these crops, the deficiency is more a function of poor Ca translocation within the plant than of low soil Ca levels.

Calcium uptake and translocation occur at greater rates in actively growing meristematic regions than older root sections (Clarkson, 1984; Marschner, 1995). As a result, soil Ca availability must be optimized in the root zone at the time of active root development (Kratzke and Palta, 1986). Since environmental stresses that curtail root meristematic development can inhibit Ca uptake and translocation to shoot meristems, temperature and moisture stresses can exacerbate low soil Ca availability by imposing transient Ca deficiencies during critical periods of crop development. These environmental stresses by soil–Ca interactions explain why these Ca-related disorders are only displayed in some growing seasons. Foliar and fruit sprays, as well as maintenance of adequate soil Ca, are required to completely alleviate Ca deficiency disorders in sensitive crops (Geraldson, 1954, 1957; Raese and Drake, 1993).

11.4.2.6 Crop Response to Calcium and Magnesium Availability

Optimum soil Ca levels for maximizing yield and quality of peanuts are quite high, 538 kg ha^{-1} for small-seeded cultivars, and 1600 kg ha^{-1} for large-seeded cultivars (Walker et al., 1979; Gaines et al., 1989). Adsorption of Ca for kernel development occurs through the shell. Small-seeded cultivars have relatively more surface area per unit mass for Ca adsorption than large-seeded cultivars, hence the lower soil Ca requirement (Sumner et al., 1988). Reduced peanut yield and quality with low soil Ca are, partly due to destruction of pods by *Pythium* and *Rhizoctonia* fungal pathogens (Hallock and Garren, 1968).

Low Mg uptake by forage grasses can induce grass tetany in ruminants (Grunes et al., 1970), while low Ca in wheat forage leads to wheat pasture poisoning (Bohman et al., 1983). Low rates of Mg and Ca uptake due to suboptimal soil Mg and Ca availability can be exacerbated by high K (Thill and George, 1975; Ohno and Grunes, 1985), low soil temperatures (Miyasaka and Grunes, 1990), low P availability (Reinbott and Blevins, 1994), and wet soil conditions (Karlen et al., 1980). A survey of the incidence of grass tetany in the United States showed greater appearance of the problem in areas where soil parent material is naturally low in Mg and in cooler climates (Kubota et al., 1980). This study concurred with earlier findings of a greater incidence of grass tetany when the Mg concentration in forages is < 2.0 g Mg kg^{-1} and the K/(Ca + Mg) equivalent ratio is > 2.2 (Kemp and 't Hart, 1957).

Exchangeable Mg levels are often poorly related to crop yield response to applied Mg. Incongruities between exchangeable Mg and crop response primarily arise from two sources: a pool of available Mg in the A horizon soil that is not exchangeable, and an accumulation of available Mg below the A horizon. Nonexchangeable Mg may be an important source of plant-available Mg in some soils. Prince and Toth (1937) noted that only about 3%–4% of the total Mg content of Sassafras loam soil was exchangeable, suggesting fixation of Mg in an insoluble form in the colloidal fraction of the soil. Substantial plant uptake of nonexchangeable Mg occurred in five Coastal Plain soils where only 4%–9% of the total Mg content was in the exchangeable form (Rice and Kamprath, 1968). An additional 4%–10% of the total soil Mg content was extractable with dilute acid and was termed nonexchangeable, but plant available.

Significant fixation of Mg occurred in Oxisols and Ultisols when limed to pH above 7 (Sumner et al., 1978; Grove et al., 1981). Magnesium fixation was nearly 80% in a Bradson soil (*Typic Hapludult*) limed to a pH of 7.2 (Grove et al., 1981). Subsurface accumulations of Mg may also satisfy crop requirements for Mg. Adams and Hartzog (1980) noted that Mg accumulated in the subsoil of Ultisols could explain the lack of crop growth response to applied Mg on soils with extremely low exchangeable Mg in the topsoil. Exchangeable Mg levels were 10-fold greater in the subsoil (60–120 cm) of Wagram loamy sand (Arenic Kandiudult) than in the surface 40 cm (Schmidt and Cox, 1992).

Magnesium deficiencies are intensified by high levels of K or Ca (Welte and Werner, 1963). Severe Mg deficiency of potato and tobacco (*Nicotiana tabacum* L.) and mild Mg deficiencies of sugar beet and barley (*Hordeum vulgare* L.) occurred with heavy applications of K, even though exchangeable Mg levels were considered adequate (Walsh and O'Donohoe, 1945). Magnesium deficiency of citrus occurred when the Mg:K ratio was less than 2.5:1 (Pratt et al., 1957). Magnesium deficiency of alfalfa occurred when Mg comprised <6% of the exchangeable cations (Prince et al., 1947). Similarly, Mg deficiency of sudangrass (*Sorghum sudanense* [Piper] Stapf) and clover (*Trifolium repens* L.) occurred when soils had <4% of the CEC occupied by Mg (Adams and Henderson, 1962), and in citrus when Mg was <4%–8% of the CEC (Martin and Page, 1969). Exchangeable Mg in excess of 10% of the CEC generally ensures that Mg availability to crops will not be limiting.

When Mg fertilizer applications are made to the soil surface, as in orchards or pastures, or to crops with high Mg requirements, high-solubility fertilizers are more effective than low-solubility materials (Boynton, 1947; Camp, 1947). For example, $MgSO_4$ and MgO increased grapefruit (*Citrus paradise* Macfad.) leaf Mg and alleviated visual symptoms of Mg deficiency within 9 months, whereas effects of $MgCO_3$ were not detected for more than 24 months (Koo, 1971). McMurtrey (1931) reported that considerably more dolomitic limestone was needed than soluble Mg fertilizer to prevent Mg deficiency in tobacco. Foliar applications of $MgSO_4$ can be more effective than soil applications of Mg in correcting Mg deficiency (Scott and Scott, 1951).

11.4.3 Sulfur

11.4.3.1 Overview of Sulfur Nutrition

Sulfur functions in plants have been reviewed by Duke and Reisenauer (1986), DeKok et al. (1993), and Marschner (1995). Sulfur is a structural constituent of organic compounds, some of which are uniquely synthesized by plants, providing animals with essential amino acids (methionine and cysteine) required to synthesize S-containing proteins. Disulfide bonding plays an important role in regulating the 3D conformation of proteins, affecting enzyme function. Sulfur is also contained in vitamins and coenzymes (thiamin, biotin, coenzyme A, and lipoic acid) (Mengel and Kirkby, 1982; Marschner, 1995). Other S-containing compounds include glutathione, an antioxidant (Bergmann and Rennenberg, 1993), and phytochelatin (Grill et al., 1987) as well as glucosinolates, which can serve as natural plant protectants (Schnug, 1993).

11.4.3.2 Occurrence of Soil Sulfur

Soil S exists in organic compounds, and adsorbed and soil solution SO_4^{2-}. Temperate region soils ordinarily contain between 0.1 and 2.0 g S kg^{-1} (Brady, 1974). Representative soils from arid regions contain more S than those from humid regions (0.8 and 0.4 mg kg^{-1}, respectively) (Brady, 1974). The predominant mineral form of S in arid soils is gypsum. Organic S comprises most of the S in most soils (>90% of the total) (Table 11.12). The C:S and N:S ratios of mineral soils vary. The organic pool is composed of three major sources of S (ester sulfates, amino acid S, and C bonded S). Ester sulfate is considerably more labile than C bonded S and is considered an important source of plant-available S. Ester sulfate and C bonded S accounted for 53% and 14% of total organic S, respectively, in 48 surface soils of Ghana (Acquaye and Kang, 1987). Ester sulfate ranged from 35% to 52% of total S in 54 Canadian soils (Bettany et al., 1973), from 20% to 65% in 6 Brazilian soils, and 43% to 60% in 6 Iowa soils (Neptune et al., 1975). Commonly, more than 30% of the total organic S in soils is not identified as either ester sulfate or C bonded S (Neptune et al., 1975; Acquaye and Kang, 1987). Amino acid S has been reported to account for 21%–31% of the total organic S content of two Australian soils (Freney et al., 1972) and 19%–31% of C bonded S in four mineral soils of Scotland (Scott et al., 1981).

Adsorbed SO_4^{2-} occurs on the positively charged exchange sites at the edges of clay minerals, organic matter, and Fe and Al oxides (Chao et al., 1962). The SO_4^{2-} adsorption capacity of a soil is decreased by increased soil pH and P content (Ensminger, 1954;

TABLE 11.12 Total-S, Organic-S, and C:N:S Ratio of Selected Mineral Soils

Origin	Total-S (mg kg^{-1})	Organic-S (mg kg^{-1})	C:N:S[a]	Reference
Saskatchewan, Canada	296	291	79:7.3:1.0	Bettany et al. (1973)
Iowa, U.S.A	319	305	92:8.3:1.0	Neptune et al. (1975)
Oregon, U.S.A	247	235	144:10:1.0	Harward et al. (1962)
Brazil, South America	166	146	139:7.1:1.0	Neptune et al. (1975)
Ghana, Africa	129	121	66:6.8:1.0	Acquaye and Kang (1987)
New Zealand	717	640	91:8.0:1.0	Perrot and Sarathchandra (1987)

[a] Organic C and S for all entries. Total N for all entries, except organic-N for Bettany et al. (1973).

Kamprath et al., 1956). Ultisols, Alfisols, and Oxisols have a high adsorption capacity due to an abundance of Fe and Al oxides and typically low pH. Binding is due to both electrostatic attraction and ligand exchange mechanisms (Marsh et al., 1987; Marcano-Martinez and McBride, 1989) and is highly dependent on the occurrence of positive charge on the soil colloids. Sulfate adsorption occurs rapidly, with nearly complete adsorption in minutes (Rajan, 1978) to days (Marcano-Martinez and McBride, 1989). However, a slower second stage of SO_4^{2-} adsorption also exists, rendering initially adsorbed SO_4^{2-} less available over time (Barrow and Shaw, 1977). Sulfate retention in acid soils is enhanced by Ca (Barrow, 1972; Ryden and Syers, 1976). Marcano-Martinez and McBride (1989) proposed that the stimulation in SO_4 adsorption arose from the bonding of a SO_4–Ca complex to the oxide surface by an O ligand and attraction of the Ca to a negatively charged hydroxyl.

11.4.3.3 Additions and Losses of Sulfur

Major sources of atmospheric S are oceans (24% of the total), soils (35%), volcanic activity (7%), and industry (35%) (Noggle et al., 1986). Sulfur accretion in soils occurs through direct adsorption of S gases and as rainfall. Distance from the source of S determines the amount of deposition. Sulfur in rainfall in Hawaii was exponentially related to distance from the ocean, averaging from 24 to 1 kg S ha^{-1} year^{-1} from the coast to 24 km inland (Hue et al., 1990). As much as 168 kg S ha^{-1} was deposited on soils near industry, whereas urban and rural locations generally received <15 kg S ha^{-1} (Olson and Rehm, 1986). However, clean air legislation in the United States has substantially decreased S emissions from electric utilities and industrial boilers lowering S deposition >30% in the Northeast and Midwest (USEPA, 2010). Although industrial S emissions continue to be reduced in many industrialized countries, thus reducing S deposition (Schnug, 1991), S deposition in some countries remain quite high (Wang et al., 2004). Areas distant from the sea and devoid of industrial sources may have rainfall concentration of <0.1 mg L^{-1} (<0.1 kg S ha^{-1} per 100 mm rainfall) (Fox and Blair, 1986).

The replacement of ordinary superphosphate (12% S) with concentrated superphosphate (1% S) beginning in the late 1940s as the most common P source resulted in an increase in the occurrence of crop S deficiencies in the southern United States (Mehring and Lundstrom, 1938; Jordan, 1964). The use of gypsum as a Ca source for peanuts is another example of incidental S applications to soils. Nutritive applications of S are most frequently applied with N sources. The primary source of S in both liquid and solid N fertilizers is ammonium sulfate (24% S). Ammonium bisulfite (32% S) and thiosulfate (26% S) are also used to provide S in liquid fertilizers. Sulfur sources used in solid fertilizers are potassium sulfate (18% S), potassium magnesium sulfate (23% S), magnesium sulfate (14% S), and ordinary superphosphate. Elemental S is also used sometimes as a slow release source of nutritive S. The S becomes available to plants when the elemental S is oxidized by bacteria to SO_4^{2-} (Starkey, 1966). Oxidation rate is dependent on the size of the S particles, temperature, and moisture (Burns, 1967).

The retention of fertilizer S in surface soils is dependent on rainfall. Nearly 50% of the SO_4^{2-} added to maize was lost from the soil profile by leaching in the subhumid savannah of West Africa (1500 mm rainfall), whereas 100% of that added to a soil in the semiarid savannah (640 mm rainfall) remained in the soil profile to a depth of 105 cm (Friesen, 1991). Nearly all of 56 kg SO_4^{2-} added to Wagram loamy sand (Arenic Paleudult) was leached from the upper 0.45 m of soil with 445 mm of rainfall in 180 days (Rhue and Kamprath, 1973). Leaching losses (24 kg S ha^{-1} year^{-1}) exceeded crop removal (15 kg S ha^{-1} year^{-1}) even with S inputs from the atmosphere, crop seed, and fertilizer of 37 kg S ha^{-1} year^{-1} and rainfall of 660 mm year^{-1} in clay loam Typic Eutrochrept in central Sweden (Kirchmann et al., 1996). Liming and P fertilization decrease S adsorption and increase SO_4-S leaching (Bolan et al., 1988).

Crop removal of S varies by species. Cruciferous forages, alfalfa, and rapeseed (*Brassica napus* L.) accumulate large amounts of S, about 70 kg S ha^{-1} (Spencer, 1975). Sugarcane, coffee (*Coffea arabica* L.), and coconut (*Cocos nucifera* L.) accumulate moderate amounts of S (50 kg S ha^{-1}) (Spencer, 1975), while field crops and forages accumulate between 20 and 30 kg S ha^{-1} annually (Hoeft and Fox, 1986; Kamprath and Jones, 1986).

11.4.3.4 Factors Affecting the Availability of Soil Sulfur

Sulfate uptake by plants from the rooting solution can be saturated at relatively low concentrations of SO_4^{2-}. Soybean growth was optimized with 0.23 mM SO_4-S in nutrient solution (Elkins and Ensminger, 1971). Wheat achieved maximum S accumulation at 0.01 mM SO_4-S in solution (Reisenauer, 1969), while Fox and Blair (1986) concluded that approximately 0.14 mM SO_4-S was required in the soil solution to optimize growth of some tropical and subtropical crops. Early growth of eight agronomic crops was maximized at 0.06 mM in nutrient solution (Hitsuda et al., 2005). Soil reactions that buffer SO_4^{2-} concentrations above these critical soil solution levels are required to optimize S availability. Replenishment of SO_4^{2-} in the soil solution from organic and adsorbed sources is important in determining S supply to the plant (Ribeiro et al., 2001).

Extractable S fluctuates during the year. Tan et al. (1994b) reported that SO_4^{2-} and C-bonded S were inversely related throughout the year in three New Zealand pasture soils (Inceptisols), with the highest proportions of C-bonded S occurring in the winter and SO_4-S predominating in the spring. Although Watkinson and Kear (1996b) found SO_4-S to be constant throughout the year, in contrast to Tan et al. (1994b), they found extractable organic S concentrations to increase in the fall and winter in agreement with the prior study. Conservation tillage was suspected of limiting organic S mineralization in a silt loam Mollisol, resulting in S deficiency of maize (Rehm, 2005).

Sulfate adsorption is negligible in surface soils because little adsorption occurs at soil pH and phosphate levels conducive to crop plant growth (Kamprath et al., 1956). Therefore, most of the effective SO_4^{2-} adsorption in cultivated soils occurs in acid argillic horizons, which often retain enough S to support plant growth. The range in water soluble SO_4-S concentrations of surface horizons of Ultisols was 0.09–0.16 mM and in the argillic horizon, 0.20–0.36 mM (Camberato and Kamprath, 1986). The depth of the argillic horizon, in part, determines the plant availability of S accumulated in that soil layer. When the argillic horizon was >0.45 m deep, S fertilization increased the yield of tobacco in 2 of 10 site years on four Ultisols, but S deficiency did not occur when the argillic horizon was within 0.30 m of the soil surface (Smith et al., 1987). Linear decreases in maize earleaf S occurred as depth to the argillic horizon in two Ultisols increased from 0.2 to 1.0 m (Cassel et al., 1996).

The depth of crop rooting also influences the amount of S accessible to the crop in soils with subsoil accumulations of SO_4^{2-}. The amount of SO_4-S available to tobacco increased during the growing season and was greater than that of cotton on Durham coarse sandy loam soil (Paleudult) with <18 kg SO_4-S ha^{-1} in the upper 0.45 m of soil, but >72 kg SO_4-S ha^{-1} below 0.45 m (Kamprath et al., 1957). Differences in rooting between these crops explained the difference in S supply. Tobacco was previously shown to have 20% of its root activity below 0.45 m at 7 weeks after transplanting, whereas cotton had a much shallower root system with only 2% of its root activity at this soil depth 11 weeks after planting.

If root growth is prevented by impeding soil layers or high Al^{3+} levels, response to S may occur on soils with high levels of subsoil SO_4^{2-}. Failure to disrupt the tillage pan in a *Typic Kandiudult* prevented rooting into an SO_4^{2-} rich subsoil, and wheat grain yield was increased by S fertilization (Oates and Kamprath, 1985). On an adjacent soil that was subsoiled to allow rooting below the tillage pan, wheat obtained sufficient S for maximum crop productivity. Aluminum saturations in excess of 50% were implicated in reducing rooting in the SO_4^{2-} rich subsoil of Aquic Hapludult, resulting in an increase in maize yield with S fertilization (Kline et al., 1989). Limited crop rooting due to shallow soils causes S deficiency of maize and alfalfa in the Midwestern United States (Hoeft and Fox, 1986).

Sulfur commonly limits crop production in the subhumid and humid regions of the Pacific Northwest. Low atmospheric deposition of S (6 kg ha^{-1} year^{-1}), low S-containing surface irrigation water and basalt, granite, and volcanic ash parent materials have resulted in soils that have low total S content and low S bioavailability (Rasmussen and Kresge, 1986). Sulfur responses in cereals are dependent on having adequate N availability (Koehler, 1965) and high yield potential (Ramig et al., 1975). High rainfall and winter leaching potential in the humid regions of the Western coastal areas coupled with low S deposition create S deficient conditions in cereal, forage, fruit, and vegetable production (Rasmussen and Kresge, 1986). In contrast, arid soils in the western United States with high accumulations of soluble SO_4^{2-}, low leaching potential under nonirrigated conditions, or that are irrigated with high SO_4^{2-}-containing groundwater respond less frequently to S fertilization.

11.4.3.5 Assessing Levels of Soil Sulfur

Although organic S is the predominant form of S in most soils, many studies have demonstrated that total S or organic S are poorly correlated to crop response to applied S. Measures of extractable S, which may include soil solution, exchangeable, and organically bound SO_4^{2-} and S, have been reasonably successful at predicting response to S fertilization. Differences in the chemical extractant, concentration, temperature, and other procedural factors affect the quantity and form of S extracted from the soil (Anderson et al., 1992). The analytical procedure used to determine S in the extract also influences the quantity of S measured. Turbidimetric and anion-exchange chromatography methods measure only solution SO_4^{2-} levels, whereas inductively coupled argon plasma spectrophotometry measures total S in solution, including solution SO_4^{2-}, ester SO_4^{2-}, and soluble organic S (Anderson et al., 1992). The contribution of S in rainfall to the crop (Hoeft et al., 1973) and subsoil SO_4^{2-}, which are not often quantified, may also be reasons why soil S analysis for predicting crop response is not always reliable. An S deficiency risk index based on atmospheric deposition, annual rainfall and soil type, texture, and pH agreed well with the occurrence of reported S deficiency in cereals (McGrath and Zhao, 2007).

Extractable SO_4-S was not a good indicator of S sufficiency on seven silt loam Mollisols in Minnesota (O'Leary and Rehm, 1991). However, the amount of S mineralized during a 4 or 12 week aerobic incubation was related to maize yield response to S fertilization, indicating the importance of organic S to plant-available S in some soils. Approximately 30% of the S extracted from eight field-moist New Zealand soils was in the soluble organic form, while the remainder was SO_4^{2-} (Tan et al., 1994a). Labile organic-S determined S availability in Oxisols with low and medium S adsorption capacity (SAC), but mineral SO_4-S forms determined S availability in high SAC soils (Ribeiro et al., 2001). Extractable organic S is a good predictor of the amount of labile S. Organic S extracted by potassium phosphate was a better predictor of maximum yield in pastoral soils of New Zealand than initial SO_4^{2-}; SO_4^{2-} mineralized during short-term incubation, organic S extracted by calcium phosphate, or organic S extracted by sodium bicarbonate (Watkinson and Kear, 1996a). Potassium phosphate extracted more organic S from high organic matter soils than the commonly used calcium phosphate extractant. Blair et al. (1991) found that extraction with KCl was superior to either calcium phosphate or sodium bicarbonate as an indicator of pasture response to fertilizer S in 18 soils from northern New South Wales, Australia. Critical levels for the three extractants ranged from 6.5 to 8.4 mg kg^{-1}. The bicarbonate extractant in this case was probably inferior to the others because it overestimated the contribution of available S arising from the ester SO_4^{2-} fraction.

11.4.3.6 Crop Response to Sulfur Availability

Sulfur deficiencies occur worldwide, but are most prevalent in areas where S accretions from atmospheric deposition, the S contents of fertilizers and pesticides, and irrigation water are low, and soils are sandy with low soil organic matter content, and rainfall is plentiful (Tisdale et al., 1986).

Numerous examples of crop S deficiencies and responses to S fertilization are presented by Tabatabai (1986). Recently, the incidence of S deficiency appears to be increasing in the United States as a result of less atmospheric S deposition. Sawyer et al. (2009) in Iowa noted only 3 S deficiencies out of 200 research trials with maize and soybean in 40 years prior to 2005. Since then, about two-thirds of more than 50 maize and of 6 alfalfa trials have responded to S fertilization (Sawyer et al., 2009). Many years of cropping without S fertilization may also contribute to the increased occurrence of S deficiency.

Sulfur deficiencies decrease crop yields and in certain instances also reduce crop nutritional value and quality. Sulfur fertilization of subirrigated meadow vegetation (Nichols et al., 1990) and forage maize (O'Leary and Rehm, 1990) increased dry matter accumulation, but did not affect crop quality. In other instances, S fertilization increased both the protein- and S-containing amino acid content of forages, resulting in increased animal performance (Rendig, 1986). In mixed pastures, S application up to 90 kg S ha^{-1} increased total forage production and the proportion of clover to grass (Jones, 1964). Numerous yield responses of coastal bermudagrass (*Cynodon dactylon* (L.) Pers.) to S fertilization have been documented (Kamprath and Jones, 1986). Alfalfa is the most S responsive crop grown in the Midwest and Northeast regions of the United States (Hoeft and Fox, 1986).

Cyst(e)ine and methionine levels of seeds are increased by S fertilization resulting in more nutritious foods (Rendig, 1986). Baking quality of wheat flour is highly dependent on the S concentration of the grain and is increased by S fertilization (Haneklaus et al., 1992). The S status (Zhao et al., 1995) and baking quality (Zhao et al., 1999) of British wheat decreased substantially in the 12 years after 1981 due to reduced atmospheric S deposition. Bread dough made from S-deficient grain resists extension and has lower extensibility (Moss et al., 1981), which is related to low albumin proteins (Wrigley et al., 1984). Sulfur-containing metabolites influence the flavor of several crops including asparagus (*Asparagus officinalis* L.) and various *Allium* sp. (Schnug, 1990).

11.4.4 Silicon

11.4.4.1 Overview of Silicon Nutrition

The roles of Si in plant nutrition and agriculture have been recently reviewed in Datnoff et al. (2001), Korndörfer and Lepsch (2001), Ma et al. (2001), and Snyder et al. (2007). Uptake of Si(OH)$_4$ (Lewin and Reimann, 1969) appears to be active in Si accumulators such as rice, sugarcane, and bamboo [*Sasa veitchii* (Carrière) Rehder], and passive in others such as oats (*Avena sativa* L.) and other "dryland" Gramineae (Jones and Handreck, 1967), while legumes and other dicots appeared to exclude Si(OH)$_4$. Although concentrations of Si in normal plants range from 1 to 100 g kg^{-1} of dry weight (Epstein, 1999), those in Si accumulators greatly exceed the levels of any essential plant nutrients taken up from the soil (e.g., >200 g Si kg^{-1} dry weight in mature bamboo leaves [Motomura et al., 2002]). Silica gel (hydrated amorphous silica, SiO$_2 \cdot n$H$_2$O) is present as specialized "skeletal" structures in the epidermal cells and cell walls of many plants (Lewin and Reimann, 1969). These structures when enhanced by Si fertilization may be responsible for increased resistance to plant diseases and insect pests. Solid silica in plant tissue occurs as opal (Jones et al., 1966) in the form of phytoliths, which persist in soil after plant decay.

11.4.4.2 Occurrence of Soil Silicon

Silicon, next to O$_2$, is the most abundant element in soils, ranging from 2% to 45% and averaging about 31% in the conterminous United States (Shacklette and Boerngen, 1984). Commonly occurring Si minerals include quartz, tridymite, cristobalite, and inorganic and biogenic opal with solubilities ranging from 1 to 70 mg L^{-1} (Wilding et al., 1977). The Si:O ratio is 1:2 in these minerals. Silicon is also a component of other primary and secondary minerals and of the phyllosilicate clay minerals. Silicon occurs in the soil solution as Si(OH)$_4$ at 0.1–0.6 mM levels (Epstein, 1999).

Silicon bioavailability in soil is primarily a function of Si quantity and solubility as determined by weathering and soil pH. Si(OH)$_4$ reacts with Fe, Al, Mn, and other heavy metals to form sparingly soluble silicates (Snyder et al., 2007). Si(OH)$_4$ is sorbed by Fe and Al oxides with the latter sorbing more Si than Fe oxides (Jones and Handreck, 1965). Sorption by both sesquioxides increases with increasing pH (Beckwith and Reeve, 1963; McKeague and Cline, 1963b; Jones and Handreck, 1965). Thus, Si(OH)$_4$ in the soil solution decreases as pH rises (Beckwith and Reeve, 1963, 1964; McKeague and Cline, 1963a, 1963b; Jones and Handreck, 1965). Flooding increases soil solution Si(OH)$_4$, probably due to the reduction of Fe, which releases adsorbed Si(OH)$_4$ (Savant et al., 1997b). Carbonates also sorb Si but not as strongly as sesquioxides (Beckwith and Reeve, 1963). Across 23 paddy soils in China available Si was positively correlated with pH, silt, clay, and CaCO$_3$ content of the soil and negatively correlated with organic matter and amorphous Fe oxide (Liang et al., 1994).

Leaching of Si(OH)$_4$ from the soil profile is a characteristic of soil weathering. Little Si is accrued in soil via atmospheric deposition, but substantial additions may be realized with irrigation. In Hawaii, rainfall and mist Si concentrations were 0.2 and 0.05 mg L^{-1}, respectively, whereas well water contained between 2.5 and 30 mg Si L^{-1} (Fox et al., 1967). From a survey of the literature (Savant et al., 1997a), irrigation water Si concentrations ranged from ~2 to 16 mg Si L^{-1}.

11.4.4.3 Crop Response to Silicon Bioavailability

Silicon application increases crop resistance to disease organisms and insect pests while increasing yield. Rice, sugarcane, and barley show the most consistent response to increased Si bioavailability (Snyder et al., 2007). Beneficial responses occur most often on organic soils (Histisols), highly weathered soils (Ultisols and Oxisols), and sandy Entisols (Datnoff and Rodrigues, 2005).

Silicon increases plant resistance to abiotic stresses including moisture deficit, high temperatures, and salinity (Ma, 2004), while it suppresses several diseases of rice including blast (*Magnaporthe grisea*), brown spot (*Cochliobolus miyabeanus*), and stem rot (*M. salvanii*) (Datnoff and Rodrigues, 2005).

Datnoff (2005) summarized the beneficial effects of Si fertilization in reducing several diseases affecting both warm- and cool-season turfgrasses. Pathogens suppressed include *Bipolaris cynodontis*, *M. grisea*, *Pythium aphanidermatum*, *Rhizoctonia solani*, *Sclerotinia homeocarpa*, and *Sphaerotheca fuliginea*. Reynolds et al. (2009) summarized the indirect and direct impacts of Si augmentation of plants on herbivorous insects. Increased plant Si impacts several insect feeding guilds including lepidopteran borers, folivores, and phloem- and xylem-feeders. Silicon applied to the plant foliage also increased plant resistance to some insect pests (Reynolds et al., 2009).

Yield increases in rice due to Si application ranged from 10% to 30% (Savant et al., 1997b); while in China, wheat was less responsive (4%–9%) than rice (5%–21%) (Liang et al., 1994). These beneficial responses in both crops were partly due to reduced diseases and lodging. Sugarcane responses to Si fertilization in Florida, Hawaii, and Mauritius (Matichenkov and Calvert, 2002) ranged from 17% to 30% in total dry matter and 23% to 58% in sucrose content. Beneficial effects were due to increased disease-, pest-, and frost-resistance. Responses of barley to Si in Russia and England (Snyder et al., 2007) were generally in the range of 10%–20%, but several exceeded 40%.

Slags produced from various metal smelters are often used as sources of Si for crop production. Slag composition can be quite variable in Si concentration ranging from 5%–10% in steel mill to 15%–20% Si in blast furnace slags from pig iron manufacture as well as in other essential elements including Ca, Mg, Mn, and Fe (Savant et al., 1997b). Solubility of Si in slag varies widely (~20% to ~70%) (Savant et al., 1999) with finer particles being more effective than larger pellets as a source of Si for increasing rice plant tissue Si and reducing disease incidence (Datnoff et al., 1992).

Silicon application rates needed to optimize crop yield are substantially higher than those of any other essential plant nutrient. For example, recommended Si fertilization rates (as Ca silicate slag) to optimize rice yield in Florida are 1500 kg Si ha^{-1} at a soil test of <6 mg L^{-1} soil and 1120 kg Si ha^{-1} at 6–24 mg L^{-1} soil (Korndörfer et al., 2001). Suggested rates of slag application for sugarcane production range from 1 to 9 Mg ha^{-1} (approximately 200–1800 kg Si ha^{-1}), dependent on the slag and soil Si concentration (Savant et al., 1999). The high rates of Si application needed to optimize plant bioavailability arise as a result of reduced solubility in the "fertilizer" material, sorption by the soil, and the large quantity needed by the plant. Although significant residual effects of a single Si application have been observed over 5 years of cropping, 50%–75% of that applied was unavailable after 5 years as assessed by chemical and biological techniques (Khalid and Silva, 1978; Khalid et al., 1978), suggesting transformation into unavailable forms. In another study, the residual value of a 3 Mg Si ha^{-1} application for reducing disease and increasing of rice yield in the second growing season was approximately equivalent to 1 Mg Si ha^{-1} applied in that season (Datnoff et al., 1991). Because Si uptake may be as high as 700 kg ha^{-1} for sugarcane, 300 kg ha^{-1} for rice, and 150 kg ha^{-1} for wheat (Snyder et al., 2007), crop removal substantially reduces soil Si (Savant et al., 1997a).

Several soil testing methods to evaluate soil Si sufficiency typically remove a fraction of sorbed Si in addition to water soluble Si. The chemical extractant, solution pH, ionic strength, shaking time, and soil:solution ratio vary among methods and remove different amounts of Si from the soil (Savant et al., 1999). For example, water, phosphate-acetate, acetate, and modified Truog (H_2SO_4) methods were evaluated as potential soil test methods for identifying Si deficient soils for sugarcane production (Fox et al., 1967) with critical values being 9, 50, 20, and 40 mg kg^{-1}, respectively. Although each method extracted different amounts of Si, all methods were equally effective in predicting plant Si accumulation ($r > 0.92$).

References

Acquaye, D.K., and B.T. Kang. 1987. Sulfur status and forms in some surface soils of Ghana. Soil Sci. 144:43–52.

Adams, F. 1966. Calcium deficiency as a causal agent of ammonium phosphate injury to cotton seedlings. Soil Sci. Soc. Am. Proc. 30:485–488.

Adams, F., and D.L. Hartzog. 1980. The nature of yield responses of Florunner peanuts to lime. Peanut Sci. 7:120–123.

Adams, F., and J.B. Henderson. 1962. Magnesium availability as affected by deficient and adequate levels of potassium and lime. Soil Sci. Soc. Am. Proc. 26:65–68.

Adams, F., and B.L. Moore. 1983. Chemical factors affecting root growth in subsoil horizons of Coastal Plain soils. Soil Sci. Soc. Am. J. 47:99–102.

Albrecht, W.A. 1931. The function of calcium in the growth of certain legumes. J. Am. Soc. Agron. 23:1052–1053.

Albrecht, W.A. 1937. Physiology of root nodule bacteria in relation to fertility levels of the soil. Soil Sci. Soc. Am. Proc. 2:315–327.

Albrecht, W.A. 1941. Plants and the exchangeable calcium of the soil. Am. J. Bot. 28:394–402.

Allaway, W.H. 1945. Availability of replaceable calcium from different types of colloids as affected by degree of calcium saturation. Soil Sci. 59:207–217.

Alva, A.K., C.J. Asher, and D.G. Edwards. 1990. Effect of solution pH, external calcium concentration, and aluminum activity on nodulation and early growth of cowpea. Aust. J. Agric. Res. 41:359–365.

Anderson, G., R. Lefroy, N. Chinoim, and G. Blair. 1992. Soil sulphur testing. Sulphur Agric. 16:6–14.

Asher, C.J., E.P.C. Blamey, and C.P. Mamaril. 1983. Sulfur nutrition of tropical annual crops, p. 54–64. *In* G.J. Blair and A.R. Till (eds.) Sulfur in south east Asian and south Pacific agriculture. The University of New England, Armidale, New South Wales, Australia.

Balke, N.E., and T.K. Hodges. 1975. Plasma membrane adenosine triphosphatase of oat roots. Plant Physiol. 55:83–86.

Barber, S.A. 1962. A diffusion and mass-flow concept of soil nutrient availability. Soil Sci. 93:39–49.

Barber, S.A. 1984. Soil nutrient bioavailability. A mechanistic approach. John Wiley & Sons, New York.

Barber, S.A., and P.G. Ozanne. 1970. Autoradiographic evidence for the differential effect of four plant species in altering the calcium content of the rhizosphere soil. Soil Sci. Soc. Am. Proc. 34:635–637.

Barrow, N.J. 1972. Influence of solution concentration of calcium on the adsorption of phosphate, sulfate, and molybdate by soils. Soil Sci. 113:175–180.

Barrow, N.J., and T.E. Shaw. 1977. The slow reactions between soil and anions: 7. Effect of time and temperature of contact between an adsorbing soil and sulfate. Soil Sci. 124:347–354.

Bear, E.E., A.L. Prince, and J.C. Malcom. 1945. Potassium needs of New Jersey soils. New Jersey Agricultural Experimental Station Bulletin No. 721. New Jersey Agricultural Experiment Station, New Brunswick, NJ.

Beckwith, R.S., and R. Reeve. 1963. Studies on soluble silica in soils. I. The sorption of silicic acid by soils and minerals. Aust. J. Soil Res. 1:157–168.

Beckwith, R.S., and R. Reeve. 1964. Studies on soluble silica in soils II. The release of monosilicic acid from soils. Aust. J. Soil Res. 2:33–45.

Beeson, K.C. 1941. The mineral composition of crops with particular reference to the soils in which they were grown: A review and compilation. USDA Miscellaneous Publication No. 369. USDA, Washington, DC.

Bergmann, L., and H. Rennenberg. 1993. Glutathione metabolism in plants, p. 109–123. In L.J. DeKok et al. (eds.) Sulfur nutrition and assimilation in higher plants. SPB Academic Publishing, The Hague, the Netherlands.

Bettany, J.R., J.W.B. Stewart, and E.H. Halstead. 1973. Sulfur fractions and carbon, nitrogen, and sulfur relationships in grassland, forest, and associated transitional soils. Soil Sci. Soc. Am. Proc. 37:915–918.

Blair, G.J., N. Chinoim, R.D.B. Lefroy, G.E. Anderson, and G.J. Crocker. 1991. A soil sulfur test for pastures and crops. Aust. J. Soil Res. 29:619–626.

Bohman, V.R., F.P. Horn, E.T. Littledike, J.G. Hurst, and D. Griffin. 1983. Wheat pasture poisoning. II. Tissue composition of cattle grazing cereal forages and related to tetany. J. Anim. Sci. 57:1364–1373.

Bolan, N.S., J.K. Syers, R.W. Tillman, and D.R. Scotter. 1988. Effect of liming and phosphate additions on sulphate leaching in soils. J. Soil Sci. 39:493–504.

Bolton, E.F., J.W. Aylesworth, and F.R. Hore. 1970. Nutrient losses through tile drains under three cropping systems and two fertility levels on a Brookston clay soil. Can. J. Soil Sci. 50:275–279.

Borchert, R. 1990. Ca^{2+} as developmental signal in the formation of Ca-oxalate crystal spacing patterns during leaf development in *Carya ovata*. Planta 182:339–347.

Boynton, D. 1947. Magnesium nutrition of apple trees. Soil Sci. 63:53–58.

Brady, N.C. 1974. The nature and properties of soils. 8th edn. Macmillan, New York.

Buol, S.W. 1995. Sustainability of soil use. Annu. Rev. Ecol. Syst. 26:25–44.

Burns, G.R. 1967. Oxidation of sulphur in soils. Sulphur Institute Technical Bulletin No. 13.

Caldwell, C.R., and A. Haug. 1981. Temperature dependence of the barley root plasma membrane-bound Ca^{2+} and Mg^{2+}-dependent ATPase. Physiol. Plant 53:117–124.

Camberato, J.J., and E.J. Kamprath. 1986. Solubility of adsorbed sulfate in Coastal Plain soils. Soil Sci. 142:211–213.

Cammarano, P., A. Felsani, M. Gentile, C. Gualerzi, C. Romeo, and G. Wolf. 1972. Formation of active hybrid 80-S particles from subunits of pea seedlings and mammalian liver ribosomes. Biochim. Biophys. Acta 281:625–642.

Camp, A.F. 1947. Magnesium in citrus fertilization in Florida. Soil Sci. 63:43–52.

Cassel, D.K., E.J. Kamprath, and F.W. Simmons. 1996. Nitrogen-sulfur relationships in corn as affected by landscape attributes and tillage. Agron. J. 88:133–140.

Chao, T.T., M.E. Harward, and S.C. Fang. 1962. Soil constituents and properties in the adsorption of sulfate ions. Soil Sci. 94:276–283.

Clark, R. 1984. Physiological aspects of calcium, magnesium, and molybdenum deficiencies in plants, p. 99–170. In F. Adams (ed.) Soil acidity and liming. SSSA, Madison, WI.

Clarkson, D.T. 1984. Calcium transport between tissues and its distribution in the plant. Plant Cell Environ. 7:449–456.

Clarkson, D.T., and J.B. Hanson. 1980. The mineral nutrition of higher plants. Ann. Rev. Plant Physiol. 31:239–298.

Cox, F.R., F. Adams, and B.B. Tucker. 1982. Liming, fertilization, and mineral nutrition. In H.E. Pattee and C.T. Young (eds.) Peanut science and technology. American Peanut Research and Education Society Inc., Yoakum, TX.

Cronan, C.S., and D.F. Grigal. 1995. Use of calcium/aluminum ratios as indicators of stress in forest ecosystems. J. Environ. Qual. 24:209–226.

Darusman, L.R.S., D.A. Whitney, K.A. Janssen, and J.H. Long. 1991. Soil properties after twenty years of fertilization with different nitrogen sources. Soil Sci. Soc. Am. J. 55:1097–1100.

Datnoff, L.E. 2005. Silicon in the life and performance of turfgrass. Online. Appl. Turfgrass Sci. doi: 10.1094/ATS-2005-0914-01-RV.

Datnoff, L.E., R.N. Raid, G.H. Snyder, and D.B. Jones. 1991. Effect of calcium silicate on blast and brown spot intensities and yields of rice. Plant Dis. 75:729–732.

Datnoff, L.E., and F.Á. Rodrigues. 2005. The role of silicon in suppressing rice diseases. Online page 1–28, http://www.apsnet.org/online/feature/silicon. February APS*net* Feature. American Phytopathological Society, St. Paul, MN.

Datnoff, L.E., G.H. Snyder, and C.W. Deren. 1992. Influence of silicon fertilizer grades on blast and brown spot development and on rice yields. Plant Dis. 76:1011–1013.

Datnoff, L.E., G.H. Snyder, and G.H. Korndörfer. (eds.). 2001. Silicon in agriculture. Elsevier Science, New York.

DeKok, L.J., I. Stulen, H. Rennenberg, C. Brunold, and W.E. Rauser. 1993. Sulfur nutrition and sulfur assimilation in higher plants. SPB Academic Publishing, The Hague, the Netherlands.

Duke, S.H., and R.M. Reisenauer. 1986. Roles and requirements of sulfur in plant nutrition, p. 123–168. In M.A. Tabatabai (ed.) Sulfur in agriculture. ASA, Madison, WI.

Elkins, D.M., and L.E. Ensminger. 1971. Effect of soil pH on the availability of adsorbed sulfate. Soil Sci. Soc. Am. Proc. 35:931–934.

Ensminger, L.E. 1954. Some factors affecting the adsorption of sulfate by Alabama soils. Soil Sci. Soc. Am. Proc. 18:259–264.

Epstein, E. 1994. The anomaly of silicon in plant biology. Science 91:11–17.

Epstein, E. 1999. Silicon. Annu. Rev. Plant Physiol. Plant Mol. Biol. 50:641–664.

Fox, R.L., and G.J. Blair. 1986. Plant response to sulfur in tropical soils, p. 405–434. In M.A. Tabatabai (ed.) Sulfur in agriculture. ASA, Madison, WI.

Fox, R.L., J.A. Silva, O.R. Younge, D.L. Plucknett, and G.D. Sherman. 1967. Soil and plant silicon and silicate response by sugar cane. Soil Sci. Soc. Am. Proc. 31:775–779.

Freney, J.R., F.J. Stevenson, and A.H. Beavers. 1972. Sulphur-containing amino acids in soil hydrolysates. Soil Sci. 114:468–476.

Friesen, D.K. 1991. Fate and efficiency of sulfur fertilizer applied to food crops in West Africa. Fert. Res. 29:35–44.

Gaines, T.P., M.B. Parker, and M.E. Walker. 1989. Runner and Virginia type peanut response to gypsum in relation to soil calcium level. Peanut Sci. 16:116–118.

Geraldson, C.M. 1954. The control of black heart of celery. Proc. Am. Soc. Hort. Sci. 63:353–358.

Geraldson, C.M. 1957. Control of blossom-end rot of tomatoes. Proc. Am. Soc. Hort. Sci. 69:309–317.

Giddens, J., and S.J. Toth. 1951. Growth and nutrient uptake of ladino clover grown on red and yellow and grey-brown podzolic soils containing varying ratios of cations. Agron. J. 43:209–214.

Goddard, M.A., E.A. Mikhailova, C.J. Post, and M.A. Schlautman. 2007. Atmospheric Mg^{2+} wet deposition within the continental United States and implications for soil inorganic carbon sequestration. Tellus Ser. B 59:50–56.

Goddard, M.A., E.A. Mikhailova, C.J. Post, M.A. Schlautman, and J.M. Galbraith. 2009. Continental United States atmospheric wet calcium deposition and soil inorganic carbon stocks. Soil Sci. Soc. Am. J. 73:989–994.

Grill, E., E.L. Winnacker, and M.H. Zenk. 1987. Phytochelatins, a class of heavy metal binding peptides from plants are functionally analogous to metallothioneins. Proc. Natl Acad. Sci. USA 84:439–443.

Grove, J.H., M.E. Sumner, and J.K. Syers. 1981. Effect of lime on exchangeable magnesium in variable surface charge soils. Soil Sci. Soc. Am. J. 45:497–500.

Grunes, D.L., P.R. Stout, and J.R. Brownell. 1970. Grass tetany of ruminants. Adv. Agron. 22:331–374.

Hallock, D.L., and K.H. Garren. 1968. Pod breakdown, yield, and grade of Virginia type peanuts as affected by Ca, Mg, and K sulfates. Agron. J. 60:253–257.

Haneklaus, S., E. Evans, and E. Schnug. 1992. Baking quality and sulphur content of wheat. I. Influence of grain sulphur and protein concentrations on loaf volume. Sulphur Agric. 16:31–34.

Hanson, J.B. 1984. The function of calcium in plant nutrition. Adv. Plant Nutr. 1:149–208.

Harward, M.E., T.T. Chao, and S.C. Fang. 1962. The sulfur status and sulfur supplying power of Oregon soils. Agron. J. 54:101–106.

Haynes, J.L., and W.R. Robbins. 1948. Calcium and boron as essential factors in the root environment. J. Am. Soc. Agron. 40:707–715.

Hedin, L.O., L. Granat, G.E. Likens, T.A. Buishand, J.N. Galloway, T.J. Butler, and H. Rodhe. 1994. Steep declines in atmospheric base cations in regions of Europe and North America. Nature 367:351–354.

Hitsuda, K., M. Yamada, and D. Klepker. 2005. Sulfur requirement of eight crops at early stages of growth. Agron. J. 97:155–159.

Hoeft, R.G., and R.H. Fox. 1986. Plant response to sulfur in the Midwest and Northeastern United States, p. 345–356. In M.A. Tabatabai (ed.) Sulfur in agriculture. ASA, Madison, WI.

Hoeft, R.G., L.M. Walsh, and D.R. Keeney. 1973. Evaluation of various extractants for available soil sulfur. Soil Sci. Soc. Am. Proc. 37:401–404.

Howard, D.D., and F. Adams. 1965. Calcium requirement for penetration of subsoils by primary cotton roots. Soil Sci. Soc. Proc. 29:558–562.

Hue, N.V., R.L. Fox, and J.D. Wolt. 1990. Sulfur status of volcanic ash-derived soils in Hawaii. Commun. Soil Sci. Plant Anal. 21:299–310.

Hunsaker, V.E., and P.F. Pratt. 1971. Calcium magnesium exchange equilibria in soils. Soil Sci. Soc. Am. Proc. 35:151–152.

Hunter, A.S. 1949. Yield and composition of alfalfa as affected by variations in the calcium-magnesium ratio in the soil. Soil Sci. 67:53–62.

Jenny, H. 1941. Calcium in the soil: III. Pedologic relations. Soil Sci. Soc. Am. Proc. 6:27–35.

Jeschke, W.D., and J.S. Pate. 1991. Cation and chloride partitioning through xylem and phloem within the whole plant of *Ricinus conununis* L. under conditions of salt stress. J. Exp. Bot. 42:1105–1116.

Johnson, D.W., and D.E. Todd. 1987. Nutrient export by leaching and whole-tree harvesting in a loblolly pine and mixed oak forest. Plant Soil 102:99–109.

Jones, M.B. 1964. Effect of applied sulfur on yield and sulfur uptake of various California dryland pasture species. Agron. J. 56: 235–237.

Jones, M.B. 1986. Sulfur availability indexes, p. 549–566. In M.A. Tabatabai (ed.) Sulfur in agriculture. ASA, Madison, WI.

Jones, L.H.P., and K.A. Handreck. 1965. Studies of silica in the oat plant. III. Uptake of silica from soils by the plant. Plant Soil 23:79–96.

Jones, L.H.P., and K.A. Handreck. 1967. Silica in soils, plants, and animals. Adv. Agron. 19:107–149.

Jones, L.H.P., A.A. Milne, and J.V. Sanders. 1966. Tabashir: An opal of plant origin source. Science 151:464–466.

Jordan, H.V. 1964. Sulfur as a plant nutrient in the southern United States. USDA Technical Bulletin No. 1297.

Kamprath, E.J. 1978. Lime in relation to Al toxicity in tropical soils, p. 233–245. *In* C.S. Andrew and E.J. Kamprath (eds.) Mineral nutrition of legumes in tropical and subtropical soils. The Dominion Press, North Blackburn, Victoria, Australia.

Kamprath, E.J. 1984. Crop response to lime on soils in the tropics, p. 349–368. *In* F. Adams (ed.) Soil acidity and liming. ASA, Madison, WI.

Kamprath, E.J., and U.S. Jones. 1986. Plant response to sulfur in the Southeastern United States, p. 323–343. *In* M.A. Tabatabai (ed.) Sulfur in agriculture. ASA, Madison, WI.

Kamprath, E.J., W.L. Nelson, and J.W. Fitts. 1956. The effect of pH, sulfate and phosphate concentrations on the adsorption of sulfate by soils. Soil Sci. Soc. Am. Proc. 20:463–466.

Kamprath, E.J., W.L. Nelson, and J.W. Fitts. 1957. Sulfur removed from soils by field crops. Agron. J. 49:289–293.

Karlen, D.L., R. Ellis, Jr., D.A. Whitney, and D.L. Grunes. 1980. Influence of soil moisture on soil solution concentrations and the tetany potential of wheat forage. Agron. J. 72:73–78.

Kemp, A., and M.L. 't Hart. 1957. Grass tetany in grazing milk cows. Neth. J. Agric. Sci. 5:4–17.

Key, J.L., L.T. Kurtz, and B.B. Tucker. 1962. Influence of ratio of exchangeable calcium-magnesium on yield and composition of soybeans and corn. Soil Sci. 93:265–270.

Khalid, R.A., and J.A. Silva. 1978. Residual effects of calcium silicate in tropical soils: II. Biological extraction of residual soil silicon. Soil Sci. Soc. Am. J. 42:94–97.

Khalid, R.A., J.A. Silva, and R.L. Fox. 1978. Residual effects of calcium silicate in tropical soils: I. Fate of applied silicon during five years cropping. Soil Sci. Soc. Am. J. 42:89–94.

Kinzel, H. 1989. Calcium in the vacuoles and cell walls of plant tissue. Forms of deposition and their physiological and ecological significance. Flora 182:99–125.

Kirchmann, H., F. Pichlmayer, and M.H. Gerzabek. 1996. Sulfur balances and sulfur-34 abundance in a long-term fertilizer experiment. Soil Sci. Soc. Am. J. 59:174–178.

Kline, J.S., J.T. Sims, and K.L. Schilke-Gartley. 1989. Response of irrigated corn to sulfur fertilization in the Atlantic Coastal Plain. Soil Sci. Soc. Am. J. 53:1101–1108.

Koehler, F.E. 1965. Fertilizer interactions in wheat producing areas of eastern Washington, p. 19–21. Proc. 10th Northwest Ann. Fert. Conf. Tacoma, WA.

Konno, H., T. Yamaya, Y. Yamasaki, and H. Matsumoto. 1984. Pectic polysaccharide break-down of cell walls in cucumber roots grown with calcium starvation. Plant Physiol. 76:633–637.

Koo, R.C.J. 1971. A comparison of magnesium sources for citrus. Soil Crop Sci. Soc. Fla. Proc. 31:137–140.

Korndörfer, G.H., and I. Lepsch. 2001. Effect of silicon on plant growth and crop yield, p. 133–147. *In* Datnoff et al. (eds.) Silicon in agriculture. Elsevier Science, New York.

Korndörfer, G.H., G.H. Snyder, M. Ulloa, G. Powell, and L.E. Datnoff. 2001. Calibration of soil and plant silicon analysis for rice production. J. Plant Nutr. 24:1071–1084.

Kratzke, M.G., and J.P. Palta. 1986. Calcium accumulation in potato tubers: Role of the basal roots. Hort. Sci. 21:1022–1024.

Kubota, J., G.H. Oberly, and E.A. Naphan. 1980. Magnesium in grasses of three selected regions in the United States and its relation to grass tetany. Agron. J. 72:907–914.

Langmeier, M., S. Ginsburg, and P. Matile. 1993. Chlorophyll breakdown in senescent leaves: Demonstration of Mg-chelatase activity. Physiol. Plant. 89:347–353.

Lerchl, D., S. Hillmer, R. Grotha, and D.G. Robinson. 1989. Ultrastructural observations on CTC-induced callose formation in *Riella helicophylla*. Bot. Acta 102:62–72.

Lewin, J., and B.E.F. Reimann. 1969. Silicon and plant growth. Annu. Rev. Plant Physiol. 20:289–304.

Liang, Y.C., T.S. Ma, F.J. Li, and Y.J. Feng. 1994. Silicon availability and response of rice and wheat to silicon in calcareous soils. Commun. Soil Sci. Plant Anal. 25:2285–2297.

Lin, D.C., and P.S. Nobel. 1971. Control of photosynthesis by Mg^{2+}. Arch. Biochem. Biophys. 145:622–632.

Lund, Z.F. 1970. The effect of calcium and its relation to several cations in soybean root growth. Soil Sci. Soc. Am. Proc. 34:456–459.

Lyle, E.S., Jr., and F. Adams. 1971. Effect of available soil calcium on taproot elongation of loblolly pine (*Pinus taeda* L.) seedlings. Soil Sci. Soc. Am. Proc. 35:800–805.

Ma, J.F. 2004. Role of silicon in enhancing the resistance of plants to biotic and abiotic stresses. Soil Sci. Plant Nutr. 50:11–18.

Ma, J.F., Y. Miyake, and E. Takahashi. 2001. Silicon as a beneficial element for crop plants, p. 17–39. *In* Datnoff et al. (eds.) Silicon in agriculture. Elsevier Science, New York.

Marcano-Martinez, E., and M.B. McBride. 1989. Calcium and sulfate retention by two Oxisols of Brazilian Cerrado. Soil Sci. Soc. Am. J. 53:63–69.

Marschner, H. 1995. Mineral nutrition of higher plants. 2nd edn. Academic Press, Inc., San Diego, CA.

Marsh, K.B., R.W. Tillman, and J.K. Syers. 1987. Charge relationships of sulfate sorption by soils. Soil Sci. Soc. Am. J. 51:318–323.

Martin, J.P., and A.L. Page. 1969. Influence of exchangeable Ca and Mg and of percentage base saturation on growth of citrus plants. Soil Sci. 107:39–46.

Matichenkov, V.V., and D.V. Calvert. 2002. Silicon as a beneficial element for sugarcane. J. Am. Soc. Sugarcane Technol. 22:21–30.

McCalla, T.M. 1937. Behavior of legume bacteria (*Rhizobium*) in relation to exchangeable calcium and hydrogen ion concentration of the colloidal fraction of the soil. Missouri Agricultural Experiment Station Research Bulletin No. 256.

McGrath, S.P., and F.J. Zhao. 2007. A risk assessment of sulphur deficiency in cereals using soil and atmospheric deposition data. Soil Use Manage. 11:110–114.

McKeague, J.A., and M.G. Cline. 1963a. Silica in soil solutions I. The form and concentration of dissolved silica in aqueous extracts of some soils. Can. J. Soil Sci. 43:70–82.

McKeague, J.A., and M.G. Cline. 1963b. Silica in soil solutions II. The adsorption of monosilicic acid by soil and by other substances. Can. J. Soil Sci. 43:83–96.

McLean, E.O., and J.R. Brown. 1984. Crop response to lime in the Midwestern United States, p. 267–303. *In* F. Adams (ed.) Soil acidity and liming. ASA, Madison, WI.

McMurtrey, J.E., Jr. 1931. Relation of calcium and magnesium to the growth and quality of tobacco. J. Am. Soc. Agron. 23:1051–1052.

Mehlich, A. 1946. Soil properties affecting the proportionate amounts of calcium, magnesium, and potassium in plants and in HCl extracts. Soil Sci. 62:393–409.

Mehlich, A., and W.E. Colwell. 1945. Absorption of calcium by peanuts from kaolin and bentonite at varying levels of calcium. Soil Sci. 60:369–374.

Mehring, A.L., and E.O. Lundstrom. 1938. The calcium, magnesium, sulphur and chlorine contents of fertilizers. Am. Fert. 88:5–10.

Mengel, K., and E. Kirkby. 1982. Principles of plant nutrition. 3rd edn. International Potash Institute, Bern, Switzerland.

Metson, A.J. 1974. Magnesium in New Zealand soils. I. Some factors governing the availability of soil magnesium: A review. N.Z. J. Exp. Agric. 2:277–319.

Millar, C.E. 1955. Soil fertility. John Wiley & Sons, New York.

Mitsui, T., J.T. Christeller, I. Hara-Nishimura, and T. Akazawa. 1984. Possible roles of calcium and calmodulin in the biosynthesis and secretion of α-amylase in rice seed scutellar epithelium. Plant Physiol. 75:21–25.

Miyasaka, S.C., and D.L. Grunes. 1990. Root temperature and calcium level effects on winter wheat forage. II. Nutrient composition and tetany potential. Agron. J. 82:242–249.

Mokwunye, A.U., and S.W. Melsted. 1972. Magnesium forms in selected temperate and tropical soils. Soil Sci. Soc. Am. Proc. 36:762–764.

Morard, P., A. Pujos, A. Bernadac, and G. Bertoni. 1996. Effect of temporary calcium deficiency on tomato growth and mineral nutrition. J. Plant Nutr. 19:115–127.

Moss, H.J., C.W. Wrigley, R. MacRitchie, and P.J. Randall. 1981. Sulfur and nitrogen fertilizer effects on wheat. II. Influence on grain quality. Aust. J. Agric. Res. 32:213–226.

Motomura, H., N. Mita, and M. Suzuki. 2002. Silica accumulation in long-lived leaves of *Sasa veitchii* (Carrière) Rehder (Poaceae-Bambusoideae). Ann. Bot. 90:149–152.

Neptune, A.M.L., M.A. Tabatabai, and J.J. Hanway. 1975. Sulfur fractions and carbon-nitrogen-phosphorus-sulfur relationships in some Brazilian and Iowa soils. Soil Sci. Soc. Am. Proc. 39:51–55.

Nichols, J.T., P.E. Reece, G.W. Hergert, and L.E. Moser. 1990. Yield and quality response of subirrigated meadow vegetation to nitrogen, phosphorus and sulfur fertilizer. Agron. J. 82:47–52.

Noggle, J.C., J.E. Meagher, and U.S. Jones. 1986. Sulfur in the atmosphere and its effects on plant growth, p. 251–278. *In* M.A. Tabatabai (ed.) Sulfur in agriculture. ASA, Madison, WI.

Oates, K.M., and E.J. Kamprath. 1985. Sulfur fertilization of winter wheat grown on deep sandy soils. Soil Sci. Soc. Am. J. 49:925–927.

Ohno, T., and D.L. Grunes. 1985. Potassium-magnesium interactions affecting nutrient uptake by wheat forage. Soil Sci. Soc. Am. J. 49:685–690.

O'Leary, M.J., and G.W. Rehm. 1990. Nitrogen and sulfur effects on the yield and quality of corn grown for grain and silage. J. Prod. Agric. 3:135–140.

O'Leary, M.J., and G.W. Rehm. 1991. Evaluation of some soil and plant analysis procedures as predictors of the need for sulfur for corn production. Commun. Soil Sci. Plant Anal. 22:87–98.

Olson, R.A., and G.W. Rehm. 1986. Sulfur in precipitation and irrigation waters and its effects on soils and plants, p. 279–294. *In* M.A. Tabatabai (ed.) Sulfur in agriculture. ASA, Madison, WI.

Osemwota, I.O., J.A.I. Omueti, and A.I. Ogboghodo. 2007. Effect of calcium/magnesium ratio in soil on magnesium availability, yield, and yield components of maize. Commun. Soil Sci. Plant Anal. 38:2849–2860.

Parker, E.W., and E. Truog. 1920. The relation between the calcium and the nitrogen content of plants and the function of calcium. Soil Sci. 10:49–56.

Pavan, M.A., E.T. Bingham, and E.J. Peryea. 1987. Influence of calcium and magnesium salts on acid soil chemistry and calcium nutrition of apple. Soil Sci. Soc. Am. J. 51:1526–1530.

Pavan, M.A., E.T. Bingham, and P.F. Pratt. 1984. Redistribution of exchangeable calcium, magnesium, and aluminum following lime or gypsum applications to a Brazilian Oxisol. Soil Sci. Soc. Am. J. 48:33–38.

Perring, M.A. 1986. Incidence of bitter pit in relation to the calcium content of apples: Problems, and paradoxes, a review. J. Sci. Food Agric. 37:591–606.

Perrott, K.W., and S.U. Sarathchandra. 1987. Nutrient and organic matter levels in a range of New Zealand soils under established pasture. NZ. J. Agric. Res. 30:249–259.

Pierce, J. 1986. Determinants of substrate specificity and the role of metal in the reaction of ribulosebiphosphate carboxylase/oxygenase. Plant Physiol. 81:943–945.

Poovaiah, B.W., and A.S.N. Reddy. 1996. Calcium and geotropism, p. 307–321. *In* Y. Waisel et al. (eds.) Plant roots: The hidden half. 2nd edn., revised and expanded. Academic Press, New York.

Pratt, P.F., and R.B. Harding. 1957. Decreases in exchangeable magnesium in an irrigated soil during 28 years of differential fertilization. Agron. J. 49:419–421.

Pratt, P.F., W.W. Jones, and F.T. Bingham. 1957. Magnesium and potassium content of orange leaves in relation to exchangeable magnesium and potassium in the soil at various depths. Proc. Am. Soc. Hort. Sci. 70:245–251.

Prince, A.L. 1951. Magnesium economy in the Coastal Plain soils of New Jersey. Soil Sci. 71:91–98.

Prince, A.L., and S.J. Toth. 1937. Effects of long-continued use of dolomitic limestone on certain chemical and colloidal properties of a Sassafras loam soil. Soil Sci. Soc. Am. Proc. 2:207–214.

Prince, A.L., M. Zimmerman, and F.E. Bear. 1947. The magnesium-supplying powers of 20 New Jersey soils. Soil Sci. 63:69–78.

Raese, J.T., and S.R. Drake. 1993. Effects of preharvest calcium sprays on apple and pear quality. J. Plant Nutr. 16:1807–1819.

Raghothama, K.G., A.S.N. Reddy, M. Friedmann, and B.W. Poovaiah. 1987. Calcium-regulated *in vivo* protein phosphorylation in *Zea mays* L. root tips. Plant Physiol. 83:1008–1013.

Rajan, S.S.S. 1978. Sulfate adsorbed on hydrous alumina, ligands displaced, and changes in surface charge. Soil Sci. Soc. Am. J. 42:39–44.

Ramig, R.E., P.E. Rasmussen, R.R. Allmaras, and C.M. Smith. 1975. Nitrogen-sulfur relations in soft white winter wheat. I. Yield response to fertilizer and residual sulfur. Agron. J. 67:219–224.

Rasmussen, P.E., and P.O. Kresge. 1986. Plant response to sulfur in the Western United States, p. 357–374. *In* M.A. Tabatabai (ed.) Sulfur in agriculture. ASA, Madison, WI.

Rehm, G.W. 2005. Sulfur management for corn growth with conservation tillage. Soil Sci. Soc. Am. J. 69:709–717.

Reinbott, T.M., and D.G. Blevins. 1994. Phosphorus and temperature effects on magnesium, calcium, and potassium in wheat and tall fescue leaves. Agron. J. 86:523–529.

Reisenauer, H.M. 1969. A technique for growing plants at controlled levels of all nutrients. Soil Sci. Soc. Am. Proc. 27:553–555.

Rendig, V.V. 1986. Sulfur and crop quality, p. 635–652. *In* M.A. Tabatabai (ed.) Sulfur in agriculture. ASA, Madison, WI.

Reynolds, O.L., M.G. Keeping, and J.H. Meyer. 2009. Silicon-augmented resistance of plants to herbivorous insects: A review. Ann. Appl. Biol. 155:171–186.

Rhue, R.D., and E.J. Kamprath. 1973. Leaching losses of sulfur during winter months when applied as gypsum, elemental S or prilled S. Agron. J. 65:603–605.

Ribeiro, E.S., Jr., L.E. Dias, V.V.H. Alvarez, J.W.V. Mello, and W.L. Daniels. 2001. Dynamics of sulfur fractions in Brazilian soils submitted to consecutive harvests of sorghum. Soil Sci. Soc. Am. J. 65:787–794.

Rice, H.B., and E.J. Kamprath. 1968. Availability of exchangeable and nonexchangeable Mg in sandy Coastal Plain soils. Soil Sci. Soc. Am. Proc. 32:386–388.

Ritchey, K.D., J.E. Silva, and U.F. Costa. 1982. Calcium deficiency in clayey B horizons of savanna Oxisols. Soil Sci. 133:378–382.

Ritchey, K.D., D.M.G. Souza, E. Lobato, and O. Correa. 1980. Calcium leaching to increase rooting depth in a Brazilian Savannah Oxisol. Agron. J. 72:40–44.

Ryden, J.C., and J.K. Syers. 1976. Calcium retention in response to phosphate sorption by soils. Soil Sci. Soc. Am. J. 40:845–846.

Salmon, R.C. 1964. Cation-activity ratios in equilibrium soil solutions and the availability of magnesium. Soil Sci. 98:213–221.

Sartain, J.B., and E.J. Kamprath. 1978. Aluminum tolerance of soybean cultivars based on root elongation in solution culture compared with growth in acid soil. Agron. J. 70:17–20.

Savant, N.K., L.E. Datnoff, and G.H. Snyder. 1997a. Depletion of plant-available silicon in soils: A possible cause of declining rice yields. Commun. Soil Sci. Plant Anal. 28:1245–1252.

Savant, N.K., G.H. Korndörfer, L.E. Datnoff, and G.H. Snyder. 1999. Silicon nutrition and sugarcane production: A review. J. Plant Nutr. 22:1853–1903.

Savant, N.K., G.H. Snyder, and L.E. Datnoff. 1997b. Silicon management and sustainable rice production. Adv. Agron. 58:151–199.

Sawyer, J., B. Lang, D. Barker, and G. Cummins. 2009. Dealing with sulfur deficiency in Iowa corn production, p. 117–123. Proc. 21st Annual Integrated Crop Manage. Conf. Ames, IA, December 2–3, 2009. Iowa State University, Ames, IA.

Schmidt, J.P., and F.R. Cox. 1992. Evaluation of the magnesium soil test interpretation for peanuts. Peanut Sci. 19:126–131.

Schnug, E. 1990. Sulphur nutrition and quality of vegetables. Sulphur Agric. 14:3–7.

Schnug, E. 1991. Sulphur nutritional status of European crops and consequences for agriculture. Sulphur Agric. 15:7–12.

Schnug, E. 1993. Physiological functions and environmental relevance of sulfur-containing secondary metabolites, p. 179–190. *In* L.J. deKok et al. (eds.) Sulfur nutrition and assimilation in higher plants. SPB Academic Publishing, The Hague, the Netherlands.

Schwab, A.P., M.D. Ransom, and C.E. Owensby. 1989. Exchange properties of an Argiustoll: Effects of long-term ammonium nitrate fertilization. Soil Sci. Soc. Am. J. 53:1412–1417.

Scott, N.M., W. Bick, and H.A. Anderson. 1981. The measurement of sulphur-containing amino acids in some Scottish soils. J. Sci. Food Agric. 32:21–24.

Scott, L.E., and D.H. Scott. 1951. Response of grapevines to soil and spray applications of magnesium sulfate. Proc. Am. Soc. Hort. Sci. 57:53–58.

Shacklette, H.T., and J.G. Boerngen. 1984. Element concentrations in soils and other surficial materials of the conterminous United States. U.S. Geological Survey Professional Paper 1270. U.S. Government Printing Office, Washington, DC.

Silkworth, D.R., and D.F. Grigal. 1982. Determining and evaluating nutrient losses following whole-tree harvesting of aspen. Soil Sci. Soc. Am. J. 46:626–631.

Simson, C.R., R.B. Corey, and M.E. Sumner. 1979. Effect of varying Ca:Mg ratios on yield and composition of corn (*Zea mays*) and alfalfa (*Medicago sativa*). Commun. Soil Sci. Plant Anal. 10:153–162.

Slocum, R.D., and S.J. Roux. 1983. Cellular and subcellular localization of calcium in gravistimulated oat coleoptiles and its possible significance in the establishment of tropic curvature. Planta 157:481–492.

Smith, W.D., G.F. Peedin, W.K. Collins, M.R. Tucker, G.S. Miner, and E.J. Kamprath. 1987. Tobacco response to sulfur on soils differing in depth to the argillic horizon. Tob. Sci. 31:36–39.

Snyder, G.H., V.V. Matichenkov, and L.E. Datnoff. 2007. Silicon, p. 551–568. *In* A.V. Barker and D.J. Pilbeam (eds.) Handbook of plant nutrition. CRC Press, Boca Raton, FL.

Spencer, K. 1975. Sulphur requirements of plants, p. 98–116. *In* K.D. McLachlan (ed.) Sulphur in Australasian agriculture. Sydney University Press, Sydney, Australia.

Starkey, R.L. 1966. Oxidation and reduction of sulfur compounds in soils. Soil Sci. 101:297–306.

Strebel, O., and W.H.M. Duynisveld. 1989. Nitrogen supply to cereals and sugar beet by mass flow and diffusion on a silty loam soil. Z. Pflanzenernahr. Bodenkd. 152:135–141.

Sumner, M.E. 1995. Amelioration of subsoil acidity with minimum disturbance, p. 147–186. *In* N.S. Jayawardane and B.A. Stewart (eds.) Subsoil management techniques. Lewis Publishers, Boca Raton, FL.

Sumner, M.E., and E. Carter. 1988. Amelioration of subsoil acidity. Commun. Soil Sci. Plant Anal. 19:1309–1318.

Sumner, M.E., P.M.W. Farina, and V.J. Hurst. 1978. Magnesium fixation: A possible cause of negative yield responses to lime applications. Commun. Soil Sci. Plant Anal. 9:995–1007.

Sumner, M.E., C.S. Kvien, H. Smal, and A.S. Csinos. 1988. On the Ca nutrition of peanuts (*Arachis hypogaea* L.) I. Conceptual model. J. Fert. Issues 5:97–102.

Sumner, M.E., H. Shahandeh, J. Bouton, and J. Hammel. 1986. Amelioration of an acid soil profile through deep liming and surface application of gypsum. Soil Sci. Soc. Am. J. 50:1254–1258.

Tabatabai, M.A. (ed.). 1986. Sulfur in agriculture. ASA, Madison, WI.

Tan, Z., R.G. McLaren, and K.C. Cameron. 1994a. Forms of sulfur extracted from soils after different methods of sample preparation. Aust. J. Soil Res. 32:823–834.

Tan, Z., R.G. McLaren, and K.C. Cameron. 1994b. Seasonal variations in forms of extractable sulfur in some New Zealand soils. Aust. J. Soil Res. 32:985–993.

Thill, J.L., and J.R. George. 1975. Cation concentration and K to Ca + Mg ratio of nine cool-season grasses and implications with hypomagnesaemia. Agron. J. 67:89–91.

Tisdale, S.L., R.B. Reneau, Jr., and J.S. Platou. 1986. Atlas of sulfur deficiencies, p. 295–322. *In* M.A. Tabatabai (ed.) Sulfur in agriculture. ASA, Madison, WI.

Tiwari, K.N. 1990. Sulphur research and agricultural production in Uttar Pradesh, India. Sulphur Agric. 14:29–34.

Truog, E. 1918. Soil acidity: I. Its relation to the growth of plants. Soil Sci. 5:169–195.

Tzeng, K.C., A. Kelman, K.E. Simmons, and K.A. Kelling. 1986. Relationship of calcium nutrition to internal brown spot of potato tubers and sub-apical necrosis of sprouts. Am. Potato J. 63:87–97.

USEPA. United States Environmental Protection Agency. 2010. Atmospheric deposition, p. 34–36. *In* Our Nation's Air – Status and trends through 2008. EPA-454/R-09-002, February 2010. USEPA, Washington, DC.

Walker, M.E., R.A. Flowers, R.J. Henning, T.E. Keisling, and B.G. Mullinix. 1979. Response of early bunch peanuts to calcium and potassium fertilization. Peanut Sci. 6:119–123.

Walker, C.J., and J.D. Weinstein. 1991. Further characterization of the magnesium chelatase in isolated developing cucumber chloroplasts. Plant Physiol. 95:1189–1196.

Walsh, T., and T.F. O'Donohoe. 1945. Magnesium deficiency in some crop plants in relation to the level of potassium nutrition. J. Agric. Sci. Camb. 35:254–263.

Wang, T.J., Z.Y. Hu, M. Xie, Y. Zhang, C.K. Xu, and Z.H. Chao. 2004. Atmospheric sulfur deposition onto different ecosystems over China. Environ. Geochem. Health 26:169–177.

Watkinson, J.H., and M.J. Kear. 1996a. Sulfate and mineralisable organic sulfur in pastoral soils of New Zealand. II. A soil test for mineralisable organic sulfur. Aust. J. Soil Res. 34:405–412.

Watkinson, J.H., and M.J. Kear. 1996b. Sulfate and mineralisable organic sulfur in pastoral soils of New Zealand. I. A quasi equilibrium between sulfate and mineralisable organic sulfur. Aust. J. Soil Res. 34:385–403.

Welte, E., and W. Werner. 1963. Potassium-magnesium antagonism in soils and crops. J. Sci. Food Agric. 44:180–186.

Wild, A., and J. Keay. 1964. Cation exchange equilibria with vermiculite. J. Soil Sci. 15:135–144.

Wilding, L.P., N.E. Smeck, and L.R. Drees. 1977. Silica in soils: Quartz, cristobalite, tridymite, and opal, p. 471–552. *In* J.B. Dixon and S.B. Weed (eds.) Minerals in soil environments. SSSA, Madison, WI.

Williams, L., and J.L. Hall. 1987. ATPase and proton pumping activities in cotyledons and other phloem-containing tissues of *Ricinus communis*. J. Exp. Bot. 38:185–202.

Wrigley, C.W., D.L. duCros, J.G. Fullington, and D.D. Kasarda. 1984. Changes in polypeptide composition and grain quality due to sulfur deficiency in wheat. J. Cereal Sci. 2:15–24.

Zhao, F.J., S.P. McGrath, A.R. Crosland, and S.E. Salmon. 1995. Changes in the sulphur status of British wheat grain in the last decade, and its geographical distribution. J. Sci. Food Agric. 68:507–514.

Zhao, F.J., S.E. Salmon, P.J.A. Withers, J.M. Monaghan, E.J. Evans, P.R. Shewry, and S.P. McGrath. 1999. Variation in the breadmaking quality and rheological properties of wheat in relation to sulphur nutrition under field conditions. J. Cereal Sci. 30:19–31.

11.5 Bioavailability of Micronutrients

Nanthi Bolan

Ravi Naidu

Ross Brennan

Dedik Budianta

Malcolm E. Sumner

11.5.1 Introduction

Minor or trace elements include both biologically essential (e.g., Cu, Mn, Fe, B, Ni, Mo, Cl, and Zn) and nonessential (e.g., Pb, Cd, and Hg) elements. The essential elements (for plant, animal, or human nutrition) are required in low concentrations and hence are known as "micronutrients" (Table 11.13). Because the nonessential minor elements are phyto- and/or zootoxic even at low concentrations, they are widely known as "toxic elements" or

TABLE 11.13 Sources of Micronutrients in Soils and Their Functions in Plant and Animal

Micronutrient	Ionic Species in Soil Solution	Sources	Functions
Boron (B)	$H_3BO_3^0$	Fertilizers, rocks, fly ash	Plant: required for cell wall strength and development, cell division, fruit and seed development, sugar transport, and hormone development
Chloride (Cl)	Cl^-	Muriate of potash, micas, feldspars	Plant: controls opening and closing of stomata, balances K in cells, required in photosynthesis, for ionic balance and transport, reduces susceptibility to disease
Cobalt (Co)	Co^{3+}, Co^{2+}, $[Co(H_2O)_6]^{2+}$; $[CoCl_4]^{2-}$	Fertilizers, mafic rock, minerals	Plant: required for N fixation, stem and coleoptiles elongation, leaf disc expansion, curvature of slit stems, opening of hypocotyl hooks, and bud development Nonenzymatic: Cobalamine or vitamin B12—erythropoiesis, granulopoiesis, glucose homeostasis
Copper (Cu)	Cu^{2+}, $Cu(OH)_2^0$, $CuOH^+$, $CuSO_4^0$, $CuCO_3^0$, $CuCl^+$, $CuHCO_3^+$	Fertilizers, fungicides, electrical, paints, pigments, timber treatment, mine tailings	Plant: enzyme activator, major function in: photosynthesis, reproductive stage, respiratory enzymes, indirect role in chlorophyll production, increases sugar content, intensifies color, and improves flavor in fruits and vegetables. Animal: Enzymatic: Cytochrome oxidase—Principal terminal oxidase; Lysyl oxidase—Lysine oxidation; Tyrosinase—Skin pigmentation—Cytocuprein: Superoxide (O_2^-) dismutation. Nonenzymatic: Growth promoter in swine and poultry; lameness control in cattle
Iron (Fe)	Fe^{2+}, $FeCl^+$, $Fe(OH)_2^+$, $FeH_2PO_4^+$, $Fe(OH)_3^-$, $Fe(SO_4)_2^-$, $Fe(OH)_4^{2-}$	Fertilizers, iron core, rocks, minerals	Plants: Promotes formation of chlorophyll, enzyme mechanism operates the respiratory system of cells, reactions involving cell division, energy transfer within the plant, a constituent of certain enzymes and proteins, plant respiration, and plant metabolism, and involved in N fixation
Manganese (Mn)	Mn^{2+}, $MnOH^+$, $MnCl^+$, $MnCO_3^0$, $MnHCO_3^+$, $MnSO_4^0$	Fertilizer, rocks	Plant: role in metabolism of organic acids, activates the reduction of nitrite and hydroxylamine to ammonia, role in enzymes involved in respiration and enzyme synthesis, activator of enzyme reactions such as oxidation/reduction, hydrolysis and direct influence on sunlight conversion in the chloroplast, assimilation of carbon dioxide in photosynthesis, synthesis of chlorophyll and in nitrate assimilation, manganese activates fat forming enzymes, formation of riboflavin, ascorbic acid, and carotene, electron transport during photosynthesis, and involved in the Hill Reaction where water is split during photosynthesis Animal: Enzymatic: Arginase—Urea formation; Pyruvate carboxylase—Pyruvate metabolism
Molybdenum (Mo)	MoO_4^{2-}, $HMoO_4^-$, $H_2MoO_4^0$	Fertilizer	Plant: essential for N fixation and nitrate reduction, converting nitrate into amino acids, conversion of inorganic P into organic forms within the plant Enzymatic: Xanthine oxidase—Purine metabolism; Sulfite oxidase—Sulfite oxidation
Nickel (Ni)	Ni^{2+}, $NiSO_4^0$, $NiHCO_3^+$, $NiCO_3^0$	Alloys, batteries, mine tailings	Plant: required for Fe absorption, participate in N metabolism of legume during the reproductive phase of growth, phytoalexin synthesis, and plant disease resistance Nonenzymatic: Increases bone strength in poultry
Selenium (Se)	$HSeO_3^-$, SeO_4^{2-}, SeO_3^{2-}, $MgSeO_4^0$,	Fertilizers, selenide minerals, recycled electronic devices, sulfide ore	Plant: increases the tolerance of plants to UV-induced oxidative stress, delays senescence, promotes the growth of ageing seedlings, regulates the water status of plants under condition of drought Enzymatic: Glutathione peroxidase—Protection against haemoglobin oxidation. Nonenzymatic: Growth promoter
Zinc (Zn)	Zn^{2+}, $ZnSO_4^0$, $ZnCl^+$, $ZnHCO_3^+$, $ZnCO_3^0$	Fertilizers, galvanizing, dyes, paints, timber treatment, mine tailings	Plant: formation of growth hormones (auxin), seed and grain formation, promotes maturity, protein synthesis and transformation, and consumption of carbohydrates Animal: Enzymatic: carbonic anhydrase—CO_2 formation; regulation of acidity; carboxypeptidase—Protein metabolism; alcohol dehydrogenase—alcohol metabolism; cytocuprein—superoxide (O_2^-) dismutation Nonenzymatic: Eczema control in cattle

Bioavailability of N, P, K, Ca, Mg, S, Si, and Micronutrients

"heavy metals" or "heavy metalloids" and include both cationic and anionic forms. Both groups can be toxic to plants, animals, and humans above certain concentrations specific to each element and target biota.

Soil represents the major sink for minor elements released into the biosphere through both geogenic (i.e., weathering or pedogenic) and anthropogenic (i.e., human activities) processes. The mobility and bioavailability of minor elements in soils are affected by adsorption onto mineral surfaces, precipitation as salts, formation of stable complexes with organic compounds, and bioaccumulation and biotransformation by microorganisms (Adriano, 2001). Soil is a biologically active integral component of the terrestrial ecosystem in which higher plants, soil constituents, and soil organisms interact, where the available energy in the form of organic and inorganic compounds promotes microbial activity and microbial weathering processes.

Bioavailability, which refers to how much of an element is available to living biota including plants and soil microorganisms, defines the relationship between its concentration in the terrestrial environment and its level that actually enters the receptor causing either a positive or negative effect on the organism. Bioavailability is species-specific because the dose that reaches an organism's target organs or tissues resulting in a biological response varies among receptors (Stokes et al., 2005). Bioavailability often refers to the extent to which an element can desorb, dissolve, or otherwise dissociate from the solid phase in which it occurs to become accessible (i.e., bioaccessibility) for absorption (Alexander and Alexander, 2000).

Chemical bioavailability is now considered an important environmental consideration because availability may be mitigated once the element comes in contact with the soil or sediment. For this reason, both soil fertility status as well as risk assessment of contaminated sites require quantification of chemical bioavailability as is carried out for other nutrients in fertilizer recommendations and risk calculations (Hrudey et al., 1996). In both fertilizer recommendations and contaminated site assessment, bioavailability, which addresses the fundamental issue of exposure of a receptor (plant) to the element in question, is not dictated by the total element concentration in the soil, but rather by the fraction of the total amount that is biologically available. Consequently for the element to become available, it must be desorbed from the soil particle and transported to the root via the soil solution. The amount of an element in soil that is bioavailable depends on a variety of factors including the properties of both the element and the soil environment (Black, 2002; Adriano et al., 2004).

This chapter covers the following aspects of the essential micronutrients (Cu, Zn, Se, Mo, Fe, Mn, Co, Ni, B, and Cl) in relation to their bioavailability in soils: (1) sources and dynamics, (2) indicators and factors affecting bioavailability, and (3) bioavailability implications to soil testing.

11.5.2 Sources of Micronutrients

Just like the major elements such as P and K, micronutrients reach the soil environment through both pedogenic and anthropogenic processes (Tables 11.14 and 11.15). Most micronutrients

TABLE 11.14 Selected References on Micronutrient Concentrations (g kg^{-1}) in Cattle, Poultry and Swine Manure By-Products, and Municipal Biosolids

Sources	Co	Cu	Mn	Mo	Ni	Fe	Se	Zn	References
Cattle manure	—	0.029	0.372	—	0.009	0.009	—	0.067	de Abreu and Berton (1996)
	—	0.139	—	0.002	0.0008	0.002	0.003	0.191	McBride and Spiers (2001)
	—	0.200	0.700	—	—	—	—	0.800	Eneji et al. (2001)
	0.003	—	0.357	—	0.008	0.005	0.0005	0.164	Raven and Loeppert (1997)
	—	0.016	0.149	—	—	—	—	6.480	Wallingford et al. (1975)
	—	0.037	—	—	0.004	0.003	—	0.153	Nicholson et al. (1999)
	—	0.033	—	—	0.006	0.007	—	0.133	Nicholson et al. (1999)
Poultry manure	—	0.400	1.800	—	—	—	—	2.300	Eneji et al. (2001)
	0.313	—	0.246	—	—	—	—	0.327	Wood et al. (1996)
	0.002	0.743	0.607	0.004	0.010	2.760	0.001	0.501	Jackson et al. (1999)
	0.002	0.031	0.166	0.005	—	—	0.0004	0.158	Capar et al. (1970)
	0.006	0.748	0.956	0.006	0.015	1.095	—	0.718	Moore et al. (1998)
	0.008	0.019	0.271	—	0.014	0.013	—	0.252	Bomke and Lowe (1991)
Swine manure	—	1.338	0.869	—	0.012	0.014	—	1.440	de Abreu and Berton (1996)
		1.000	2.100	—	—	—	—	2.900	Eneji et al. (2001)
	—	1.279	0.197	—	—	—	—	0.231	Mullins et al. (1982)
	—	0.374	—	—	0.007	0.003	—	0.431	Nicholson et al. (1999)
Biosolids	0.017	0.294	6.230	0.011	0.075	48.100	0.001	1.250	Jackson et al. (1999)
	—	0.119	0.214	—	0.015	0.324	—	0.328	de Abreu and Berton (1996)
	—	0.089	0.350	—	4.304	0.085	—	0.354	de Abreu and Berton (1996)
	0.010	1.346	0.194	0.014	—	0.003	0.003	2.132	Capar et al. (1970)
	0.004	—	0.142	—	0.031	0.130	0.001	0.450	Raven and Loeppert (1997)

TABLE 11.15 Total and Water-Soluble Micronutrient Concentration (g kg⁻¹) in Fertilizers, Dairy Cattle, Poultry and Swine Manures, and Biosolids

Element	Dairy Cattle[a] Total	Dairy Cattle[a] Water Soluble	Poultry[b] Total	Poultry[b] Water Soluble	Swine[a] Total	Swine[a] Water Soluble	Sewage Sludge[c] Total	Sewage Sludge[c] Water Soluble	NPK (15-30-15)[d] Fertilizer Total	NPK (15-30-15)[d] Fertilizer Soluble Water
Co			0.0068	0.0031					—	—
Cu	0.356	0.112	0.656	0.314	0.419	0.130	0.148	0.0044	0.700	0.700
Mn	0.345	0.017	0.274	0.006	0.865	0.014			5.000	5.000
Ni	0.009	0.001	0.0087	0.005	0.012	0.003	0.028	0.0001	—	—
Se	—	—	0.0095	0.0038					—	—
Zn	0.765	0.123	0.246	0.018	1.210	0.023	0.330	0.0016	0.600	0.600
Fe									1.500	1.500
Mo									0.005	0.005

[a] Bolan et al. (2003a).
[b] Jackson and Bertsch (2001).
[c] Henry and Harrison (1992).
[d] McCauley (2009).

occur naturally in soil parent materials, chiefly in forms that are not readily bioavailable for plant uptake. Often the micronutrient concentrations released from the soil by natural pedogenic processes are largely related to the origin and nature of the parent material. On the other hand, micronutrients added through anthropogenic activities including fertilization typically have higher bioavailability (Naidu et al., 1996a). Fertilizer and manure additions to agricultural soils are the major sources supplying micronutrients for plant growth (Tables 11.15 and 11.16). Anthropogenic activities, associated with industrial processes, manufacturing and the disposal of domestic and industrial wastes, also contribute to micronutrient enrichment of soils (Adriano, 2001).

11.5.3 Dynamics of Micronutrients in Soils

Micronutrients undergo both chemical and biological transformations including retention, redox and methylation reactions while they are retained in the soil by sorption, precipitation, and complexation, and removed by plant uptake and leaching. Although most micronutrients are not subject to volatilization losses, Se tends to form gaseous compounds through redox and methylation reactions (Frankenberger and Karlson, 1995). When micronutrient solution concentration is low and sorption surfaces large, sorption/desorption processes will govern the soil solution concentration (Tiller, 1989) while the fate of micronutrients in the soil depends on both soil properties and environmental factors.

11.5.3.1 Chemical Transformation Processes

11.5.3.1.1 Sorption and Complexation

Chemical interactions that contribute to micronutrient retention by soil colloids include sorption and complexation with inorganic and organic ligands. Charged ions are attracted to charged soil surfaces by electrostatic and/or stronger covalent bonds (Mott, 1981), which can be specific or nonspecific in nature (Bolan et al., 1999). In nonspecific adsorption, the ion charge balances that on the soil surface by electrostatic attraction while in specific adsorption, chemical bonds form between the ions and the soil surface (Sposito, 1984). Metal ion binding on soil organic matter (SOM) is strongly pH dependent due to competition between metal ions and protons for the available binding sites and to the effect of pH on the electric charge development on SOM. Detailed descriptions of these processes are presented in Chapters 11 and 19 of *Handbook of Soil Sciences: Properties and Processes*.

11.5.3.1.2 Cation-Exchange Reactions

Although most soils have a net negative charge, this charge plays only a minor role in micronutrient reactions in soil. In general, for ions having the same charge, ions with the larger hydrated radius are selectively adsorbed. The many ion exchange selectivity coefficients reported in the literature have been determined in the presence of ligands but the effects of ion-pair formation and complex ions have been generally ignored. Comparisons between reported selectivity coefficients for various systems and exchange reactions can, therefore, not be made with confidence (Sposito, 1984). Very little information is available on the values of the coefficients for ion exchange between dominant and micronutrient cations in soil–water systems.

11.5.3.1.3 Specific Adsorption at Mineral Surfaces

Although most macronutrient cations are weakly retained by soils in exchangeable form, most micronutrient cations are strongly retained as inner sphere complexes with variable charged surfaces by the formation of covalent bonds. Although specific adsorption in the form of inner sphere complexes with variable charged mineral surfaces occurs most readily for metals that hydrolyze in water, such as most of the transition elements (Cu, Fe, Mn, Co, Zn, Ni) and some other divalent

TABLE 11.16 Micronutrient Content of Some Commonly Occurring Fertilizers Used in Agricultural Production

Micronutrient	Fertilizer Compound	Chemical Formula	Concentration (g kg^{-1})
Boron	Boric acid	H_3BO_3	89
	Borax	$NaB_4O_7 \cdot 10H_2O$	110
		NaB_4O_7	214
	Solubor©		175
Cobalt	Cobalt sulfate heptahydrate	$CoSO_4 \cdot 7H_2O$	210
	Cobalt carbonate	$CoCO_3$	460
	Cobalt amino acid chelate	—	20
	Cobalt chelate	CoEDTA	130
Copper	Copper sulfate monohydrate	$CuSO_4 \cdot H_2O$	390
	Copper sulfate pentahydrate	$CuSO_4 \cdot 5H_2O$	250
	Copper oxide	CuO	750
	Copper oxy-chloride	$CuCl_2 \cdot 3Cu(OH)_2$ or $Cu_2Cl(OH)_3$	590–700
	Copper chloride	$CuCl_2$	170
	Copper chelate	CuEDTA	80–130
Chloride	Potassium chloride (muriate of potash)	KCl	500
Iron	Ferrous sulfate	$FeSO_4 \cdot 7H_2O$	400
	Ferric sulfate	$Fe_2(SO_4)_3 \cdot 4H_2O$	230
	Ferrous oxide	FeO	770
	Ferric oxide	Fe_2O_3	690
	Ferrous ammonium phosphate	$Fe(NH_4)PO_4 \cdot H_2O$	290
	Ferrous ammonium sulfate	$(NH_4)_2SO_4 \cdot FeSO_4 \cdot 6H_2O$	140
	Iron ammonium polyphosphate	$Fe(NH_4)HP_2O_7$	220
	Iron chelates	NaFeEDTA	50–140
		NaFeHEDTA	50–90
		NaFeEDDHA	60
		NaFeDTPA	100
	Iron polyflavonoids	—	90–100
	Iron lignosulfonates	—	50–80
	Iron methoxyphenylpropane	—	50
Manganese	Manganese sulfate	$MnSO_4 \cdot 4H_2O$	230–280
	Manganese chelates	MnEDTA	50–120
	Manganous oxide	MnO	410–680
	Manganese methoxyphenylpropane	MnMPP	100–120
	Manganese carbonate	$MnCO_3$	310
	Manganese chloride	$MnCl_2$	170
	Manganese oxide	MnO_2	630
	Manganese frits	—	100–250
Molybdenum	Ammonium molybdate	$(NH_4)_6Mo_7O_{24} \cdot 2H_2O$	540
	Molybdenum frits	Silicates	20–30
	Molybdenum sulfide	MoS_2	600
	Molybdenum trioxide	MoO_3	660
	Sodium molybdate	$Na_2MoO_4 \cdot 2H_2O$	390
Nickel	Nickel plus	Nickel lignosulfonate	54
	Nickel chloride	$NiCl_2 \cdot 6H_2O$	150
Selenium	Sodium selenite	NaSe	450
	Selenium premix	—	2–4
Zinc	Zinc sulfate monohydrate	$ZnSO_4 \cdot H_2O$	360
	Zinc sulfate heptahydrate	$ZnSO_4 \cdot 7H_2O$	224

(continued)

TABLE 11.16 (continued) Micronutrient Content of Some Commonly Occurring Fertilizers Used in Agricultural Production

Micronutrient	Fertilizer Compound	Chemical Formula	Concentration (g kg^{-1})
	Zinc oxy-sulfate	$ZnO-ZnSO_4$	380–500
	Zinc oxide	ZnO	500–800
	Zinc chloride	$ZnCl_2$	500
	Basic zinc sulfate	$ZnSO_4 \cdot 4Zn(OH)_2$	550
	Zinc carbonate	$ZnCO_3$	520
	Zn sulfide	ZnS	670
	Zn frits	—	Varies
	Zinc phosphate	$Zn_3(PO_4)_2$	510
	Zinc chelate	NaZnEDTA	60–140
		NaZnHEDTA	60–100
		NaZnNTA	130
	Zn polyflavonoid	—	100
	Zn ligninsulfonate	—	50

micronutrients, adsorption reactions usually involve the formation of an inner sphere complex between the hydroxo-metal ion and the deprotonated negatively charged surfaces of Al, Mn, and Fe oxides. The relative affinities of the metals tend to follow the *Irving-Williams* order:

$$Cu^{2+} > Zn^{2+} > Ni^{2+} \sim Co^{2+} > Fe^{2+} > Mn^{2+}$$

11.5.3.1.4 Adsorption by Soil Organic Matter

Metals can react with SOM by ion exchange, complexation, and precipitation. Although complexation as inner sphere complexes between the cation and SOM coordinating functional groups is thought to be the most important reaction, the difficulties involved in studying SOM preclude accurate separation of the processes in most experiments. The retention of metals by SOM at pH 5.8 decreases in the order (Schnitzer and Khan, 1978):

$$Fe = Al = Cr = Hg = Pb = Cu > Cd > Zn > Ni > Co > Mn$$

Although multiligand complexes can form with simple organic acids, 1:1 complexes occur mainly with ligands of the structurally complex macromolecules in SOM, but two or more donor atoms in the macromolecule may be involved in the complexation reaction. Because electron donors in SOM are mainly O, N, and S, the important binding groups are as follows: –COOH (carboxyl), –OH (enolic and phenolic), –SH (thiol), –NH$_2$ (amino), =O (carbonyl), and –S (thioether). With increasing pH, carboxyl, phenolic, alcoholic, and carbonyl functional groups dissociate, thereby increasing their affinity for cations. This results in SOM having a high affinity for micronutrients cations.

Micronutrients are known to form organic complexes that affect their sorption onto soil particles (Adriano, 2001; He et al., 2005). For example, Bolan et al. (2003a) demonstrated that the addition of organic manures increased the complexation of Cu in soils. Additionally, they observed that while Cu^{2+} adsorption measured as the change in the total Cu in soil solution was not affected by biosolid addition, Cu complexation measured as the change in free Cu^{2+} concentration, increased with increasing level of biosolid.

Modeling of the binding of metals to humic materials as in MINTEQA2 is in its infancy with most work having been conducted on organic matter dissolved in surface waters rather than on SOM (Susetyo et al., 1991). More research is needed to develop suitable models that quantify metal retention by SOM.

The extent of micronutrients-organic complex formation, however, varies with a number of factors including temperature, steric factors, and concentration. All these interactions are controlled by solution pH and ionic strength, nature of the micronutrients species, dominant cation, and inorganic and organic ligands present in the soil solution.

11.5.3.1.5 Precipitation

At high pH and in the presence of SO$_4^{2-}$, CO$_3^{2-}$, OH$^-$, and HPO$_4^{2-}$, precipitation appears to be the predominant process when micronutrient cation concentrations are high (Naidu et al., 1996b). This occurs when the ionic product in the solution exceeds the solubility product of that phase. In normal soils, precipitation is not very important, but in heavy metal contaminated soils, the precipitation process can play a major role in remediation, especially under alkaline pH. Increasingly, addition of phosphate is being used to precipitate excessive levels of micronutrients such as Zn (He et al., 2005).

11.5.3.1.6 Leaching and Runoff

Micronutrients occur in solution either as free ions or soluble complexes that are prone to leaching. L'Herroux et al. (1997) showed that repeated applications of swine manure slurry increased the concentrations of Mn, Co, and Zn in drainage water manyfold. Similarly, Moore et al. (1998) found increased soluble Cu and Zn in runoff with increasing metal loading through poultry manure application, obtaining a strong relationship between dissolved organic carbon (DOC) and soluble Cu and Zn in the runoff.

However, pretreating the manure with $Al_2(SO_4)_3$ decreased runoff losses of these micronutrients as a result of sorption onto $Al(OH)_3$ formed upon hydrolysis of $Al_2(SO_4)_3$.

Migration of micronutrients in soils after manure slurry applications is greatly enhanced by DOC (Japenga et al., 1992; del Castilho et al., 1993; Li and Shuman, 1997; Hsu and Lo, 2000). Acidification due to nitrification from applied manures releases micronutrients from the solid phase (Japenga et al., 1992; del Castilho et al., 1993). Thus, while organic matter in manure byproducts may provide some buffer against metal bioavailability, it does not prevent the metal from being more mobile.

11.5.4 Biological Transformation Processes

11.5.4.1 Plant Uptake and Bioaccumulation

Micronutrients are applied to overcome deficiencies, thereby increasing crop production. When micronutrients in soil are deficient, yield responses to fertilizer and manure are often observed due to increased uptake. Typical examples of yield responses to micronutrients in several crops are presented by Mortvedt (1999). Although plant uptake is a major sink for micronutrients, only a small fraction of that applied to soil is ever taken up because the greatest proportion is immobilized.

Microorganisms can bioaccumulate micronutrients from substrates containing very low concentrations. Bacteria produce large quantities of extracellular polymers with anionic properties that remove soluble micronutrients ions from solution (Srinath et al., 2002) while many fungal products, such as glucans, mannans, melanins, chitins, and chitosans can act as efficient biosorption agents (Blackwell et al., 1995). Thus, microorganisms are competitors with plant roots for these nutrients.

11.5.4.2 Oxidation/Reduction

Redox reactions, both chemical and biological, affect the bioavailability of micronutrients such as Fe, Mn, and Se. In agricultural soils, the reduced forms of Fe^{2+} and Mn^{2+} are most available while Se is more available in the oxidized [Se(VI)] form. Because of the great heterogeneity in the pore space of most soils, zones of reducing and oxidizing conditions are often in close proximity to one another allowing roots to access the available forms at different points in the soil. Application of manures to soils that are subject to reducing conditions (very wet) can result in losses of micronutrients such as Mn by leaching (L'Herroux et al., 1997), while micronutrients such as Co sorbed on sesquioxides under oxic conditions (McLaren et al., 1984) can be released due to manure-induced reduction of these oxides. Temporary flooding of alfalfa fields often result in substantial yield decreases due to the induced Mn^{2+} toxicity from the reducing conditions.

11.5.4.3 Methylation/Demethylation

While some metals (As, Hg, and Se) can be methylated and lost by volatilization (Cernansky et al., 2009), none of the plant-essential micronutrients are involved. Thus, this process is only of interest in metal contaminated sites.

11.5.5 Bioavailability of Micronutrients in Soils

11.5.5.1 Definition

The bioavailability of an element in the soil environment is defined as the fraction of the total that is present in the soil solution and on the solid phase that is available to the plant (Naidu et al., 2008). Considerable controversy exists as to "what constitutes the bioavailable fraction," including the definition itself and the methods used for its measurement. For instance, microbiologists often regard the bioavailable fraction as the concentration that can induce a change either in morphology or physiology of the organism, whereas plant scientists regard the plant-available pool as the bioavailable fraction, which is the definition that will be used here. Physical accessibility of micronutrients for uptake occurs either through movement of ions to roots by diffusion and mass flow or through root extension to the site of the nutrient. Chemical accessibility requires that nutrients remain in the soil solution in the form in which they are taken up (Marschner, 1995). Because the transformation of micronutrients in soils is a dynamic process, bioavailability changes with time.

11.5.5.2 Indicators of Bioavailability

Micronutrients occur in various fractions in soils that include the following (Shuman, 1991):

1. Structural components of primary and secondary minerals (e.g., Mn in manganous oxide)
2. Precipitated in inorganic forms, including those occluded by Fe, Al, and Mn oxides (e.g., Zn)
3. Complexed by organic matter (e.g., Cu)
4. Incorporated into organic matter including microbial biomass (e.g., B, Se)
5. Specifically adsorbed onto silicate clay minerals and Fe, Al, and Mn oxides (e.g., Zn)
6. On clay and organic exchange sites
7. Water soluble as free and complexed inorganic and organic ions

Despite only small amounts of micronutrients being present in exchangeable (6) and water soluble fractions (7) from which uptake occurs, they are continuously replenished from other less-soluble fractions, ensuring continuity of supply. Micronutrient bioavailability in soils can be assessed using chemical and biological tests.

11.5.5.2.1 Chemical Tests

11.5.5.2.1.1 Single Extraction A range of chemical extractants including mineral acids (e.g., 1 M HCl), salt solutions (e.g., 0.1 M $CaCl_2$), buffer solutions (e.g., 1 M NH_4OAc), and chelating agents (e.g., DTPA) have been used to predict micronutrient bioavailability in soils (Sutton et al., 1984; Payne et al., 1988; Sims and Johnson, 1991) but chelating agents, such as EDTA and DTPA are usually more reliable (Sims and Johnson, 1991), since they are more effective in removing potentially bioavailable fractions. However, they do not actually measure availability (Beckett et al., 1983a, 1983b). For example, although the

DTPA-extractable Cu was linearly related to Cu application rate, no relation with corn grain or leaf tissue Cu existed (Payne et al., 1988). Nevertheless, because DTPA-extractable micronutrients increase with metal level in soil (Wallingford et al., 1975; Haleem et al., 1992; Martinez and Peu, 2000), which is related to crop uptake (Wallingford et al., 1975; Mullins et al., 1982; Bibak, 1994; Duffera et al., 1999), DTPA-extractable micronutrients and plant uptake may in some cases be fortuitously related (Beckett et al., 1983a, 1983b).

11.5.5.2.1.2 Sequential Fractionation Sequential fractionation schemes are often used to examine the redistribution or partitioning of micronutrients in various chemical forms that include soluble, adsorbed (exchangeable), precipitated, organic, and occluded. Although the extraction procedures vary between chemical fractionation schemes, generally the solubility and bioavailability of micronutrients in soils decrease with each successive step in the scheme (Basta and Gradwohl, 2000). Specific chemical pools measured by chemical fractionation have been correlated with plant uptake and have been successful in predicting micronutrient availability in soils (Shuman, 1991). Because chemical extraction is matrix dependent, validation for different micronutrients sources, such as inorganic fertilizer, and organic biosolid and manure by-products must be carried out.

The diversity of reagents used to extract specific metal forms from soils makes comparison of results difficult (Table 11.17). Even when the same reagent is employed, the efficiency of extraction depends on the nature of sample, its particle size distribution, duration of extraction, pH, temperature, strength of extractant, and solid:solution ratio (Miller et al., 1986). Because chemical reagents used for extraction may themselves alter the indigenous speciation of micronutrients, milder extractants are usually more selective albeit less efficient for specific fractions than more aggressive reagents, which may extract other forms as well (Lake et al., 1984; Ross, 1994).

Redistribution of micronutrients as measured by fractionation techniques depends on their source. While fertilizer-applied micronutrients tend to remain in soluble and exchangeable form, those applied in manure are in the organic-bound form. For example, after fractionation of metal-organic components in a manure-amended soil, del Castilho et al. (1993) found that strongly bound organic–metal complexes (Cu and Zn) were mostly associated with hydrophobic acids (phenols) and neutrals (hydrocarbons) while the weakly bound micronutrients were complexed with hydrophilic neutrals (i.e., carbohydrates).

11.5.5.2.1.3 Diffusive Gradient Thin Film Diffusive gradient thin film techniques physically mimic the removal of metals by a plant by having a layer of chelating resin behind a diffusive layer (usually a gel faced with a filter membrane) that is in contact with the soil solution. This chemically and physically well-defined system introduces a sink for ions in the soil, which results in a concentration gradient in the soil solution adjacent to the device and a consequent supply of ions from the solid phase into the locally depleted solution (Smith et al., 2007).

11.5.5.2.1.4 Isotopic Techniques Methods, such as isotopic dilution and the isotopic exchange kinetics (IEK) method have been shown to measure phytoavailable elements in soils (Hamon et al., 2008; Hedley, 2008). Isotopic dilution techniques consist of spiking soil with the metal before cropping after which the specific activity (isotope:nonisotope ratio) of the plant and soil solution are measured; from this, the amount taken up from the isotopically exchangeable pool can be calculated, allowing for a good estimation of the amount of phytoavailable element.

The IEK technique, which is used to study isotopic exchange as a function of time, can be employed in conjunction with other techniques to describe the transfer of Co (Wendling et al., 2009), Cu (Ma et al., 2006), Zn (Sinaj et al., 1999; Zhang et al., 2006), Ni (Echevarria et al., 1998), and Se (Goodson et al., 2003) in soils and is very useful in describing plant availability in terms of quantity/intensity (Q/I) concepts (Hamon et al., 2002).

The labile pool of an element in soil, whether determined by radioactive or stable isotopes, can be considered as a relatively unambiguous assessment of the chemical and biological reactivity of that metal in soil compared with fractions isolated by chemical extractants (Hedley, 2008). However, it should be recognized that isotopic exchange determinations are based on the assumption that the added spike (radioisotope or stable isotope) remains 100% available for exchange. Any fixation ("irreversible" sorption) of some of the spike by reactive soil phases will result in an overestimate of the size of the labile pool (Hamon et al., 2002).

11.5.5.2.2 Biological Tests

11.5.5.2.2.1 Phytoavailability Plant availability of micronutrients largely depends on the source, soil type, and plant species (Martinez and Peu, 2000). While the application of micronutrient fertilizers usually increases the uptake of the elements by plants, application of manure by-products can both increase or decrease micronutrient concentrations in plants. For example, addition of farmyard manure increased the uptake of soil Co and Mn by winter wheat (Bibak, 1994), while acidification of soils by manure increased the B and Zn concentrations in the soil solution and uptake by maize (Jahiruddin et al., 2001). On the other hand, three annual of applications of ~24 kg Cu ha^{-1} as Cu-enriched swine manure increased Cu concentration in the corn ear leaf, but did not affect either corn grain yield or grain Cu concentration (Kornegay et al., 1976). Similarly, the uptake of Se from Se-enriched cattle manure was found to be less than that from an inorganic source (Ajwa et al., 1998) and the addition of cattle manure decreased the concentration of borate anions in soil solution and the subsequent plant uptake of B (Yermiyahu et al., 2001).

11.5.5.2.2.2 Microbial Availability Long-term applications of biosolids to agricultural land can reduce microbial activity in soils as a result of high concentrations of micronutrients (McGrath, 1994). In contrast, only limited studies have examined the direct effect of manure-borne micronutrients on microbial activity. Huysman et al. (1994) showed that in soils receiving

Bioavailability of N, P, K, Ca, Mg, S, Si, and Micronutrients

TABLE 11.17 Selected References on the Array of Reagents Used to Partition the Chemical Forms of Micronutrients Derived from Various Sources in Soils

Source	Element	Soluble	Exchangeable	Adsorbed	Organic	Oxide	Carbonate	Residual	Reference
Fertilizers	Zn	1 mol L^{-1} NH$_4$NO$_3$ (pH 7.0)			0.1 mol L^{-1} Na$_4$P$_2$O$_7$ (pH 10)	0.1 mol L^{-1} NH$_2$OH·HCl in 0.01 mol L^{-1} HNO$_3$	1 mol L^{-1} NaOAc pH 5.0		Gonzalez et al. (2008)
	Cu		0.01 Ca(NO$_3$)$_2$		0.7 M NaOCl (pH 8.5)	0.2 M NH$_4$O, 0.2 Oxalic acid, 0.1 M Ascorbic acid		HNO$_3$, HClO$_4$, HF (3:1:10)	McLaren and Ritchie (1993)
	B	Hot water							Haddad and Kaldor (1982)
	Mn	H$_2$O	NH$_4$OAc						Rayment and Verrall (1980)
	Mn		1 M MgCl$_2$ (pH 7)				1 M NaOAc (pH 5)		Nádaská et al. (2009)
	Co, Mn		0.01 M Ca(NO$_3$)$_2$		3% NaOCl (pH 8.5)	0.1 M NH$_2$OH·HCl (pH 2)		HNO$_3$	Li et al. (2001)
	Ni	H$_2$O	1 M MgCl$_2$		30% H$_2$O$_2$	0.1 M NH$_2$OH·HCl	0.5 M NaOAc + 0.5 M HOAc		Rahmatullah et al. (2001)
						0.175 M (NH$_4$)$_2$C$_2$O$_4$—0.1 M H$_2$C$_2$O$_4$			
	Se	1 M MgCl$_2$		1 M NaH$_2$PO$_4$		10 M HF		Acid	Lim and Goh (2005)
Manures	Cu	0.05 M CaCl$_2$		2% CH$_3$COOH	0.1 M K$_4$P$_2$O$_7$	0.175 M (NH$_4$)$_2$C$_2$O$_4$		HCl, HNO$_3$, HF	Payne et al. (1988)
	Cu	0.5 M CaCl$_2$		2% CH$_3$COOH	0.1 M K$_4$P$_2$O$_7$				Mullins et al. (1982)
	Cd, Cu, Ni, Zn,		1 M MgCl$_2$·6H$_2$O		0.02 M HNO$_3$; 30% H$_2$O$_2$, 3.2 M (NH$_4$)$_2$C$_2$O$_4$	0.04 M NH$_2$OH·HCl in 25% CH$_3$COOH	1 M NaOAc	HCl, HNO$_3$	Narwal and Singh (1998)
	Cd, Pb, Zn	H$_2$O	0.5 M KNO$_3$		0.05 M NaEDTA	0.5 M NaOH		4 M HNO$_3$	Pierzynski and Schwab (1993)
	Cd, Cr, Cu, Ni, Pb, Zn	0.1 M CaCl$_2$			0.5 M NaOH		0.05 M NaEDTA		Canet et al. (1997)
	Cd, Pb, Zn	H$_2$O	1 M Mg(NO$_3$)$_2$		0.7 M NaOCl	0.1 M NH$_2$OH·HCl (Mn oxide); 0.2 M (NH$_4$)$_2$C$_2$O$_4$ – 2 M H$_2$C$_2$O$_4$ – ascorbic		HCl, HNO$_3$, HF	Li and Shuman (1997)
	Cu, Zn	1 M NH$_4$NO$_3$			H$_2$O$_2$	(NH$_4$)$_2$C$_2$O$_4$ – 2 M H$_2$C$_2$O$_4$		4 M HNO$_3$	Han et al. (2001)
	Cu	H$_2$O	0.5 M Ca(NO$_3$)$_2$	0.44 CH$_3$COOH	0.1 M K$_4$P$_2$O$_7$	0.275 (NH$_4$)$_2$C$_2$O$_4$	0.1 M NH$_2$OH·HCl	HCl, HNO$_3$, HF	Miller et al. (1986)

swine manure for 5 years, there was no effect of Cu on anaerobic bacteria or fungal activity, but aerobic Cu-resistant bacteria increased with increasing soil DTPA Cu concentration, indicating that this may provide a sensitive measure of Cu bioavailability in soils. Application of swine and cattle manure slurries to grasslands acidifies soil while increasing total C and N contents, resulting in significant decreases in microbial biomass C and N. This suggests a decrease in the microbial turnover rate of C and N in manure-amended soil, apparently caused by increased solubilization of Cu, Ni, and Zn (Christie and Beattie, 1989). Organic complexation of Cu and other metals in biosolids and manures may be the main reason for the lower level of toxicity to soil biota of this form compared to inorganic salts, indicating that both microbial and plant toxicity is largely controlled by the free metal ion concentration (van Rhee, 1975; Brookes and McGrath, 1984; McGrath, 1994).

11.5.6 Factors Affecting Bioavailability of Micronutrients in Soils

11.5.6.1 Soil Interactions

Properties of the soil and its ambient solution determine the dynamic equilibrium between micronutrients in solution and solid phases, thereby affecting their bioavailability. Micronutrients interact with particulates by metal ion adsorption at surface sites, ion exchange with clay minerals, binding with organic-coated particulate matter or organic colloidal material, and by adsorption of metal–ligand complexes. All of these interactions are controlled by application of various soil amendments, pH, ionic strength, nature of the metal species, dominant cation, and inorganic and organic ligands present in the soil solution.

11.5.6.1.1 pH

Soil pH has a marked effect on metal adsorption increasing from <20% to 100% within the narrow pH range of 3–5 for all metals except those that form oxyanions (Forbes et al., 1976; Kinniburgh and Jackson, 1981; Christensen, 1984; Brümmer et al., 1988; Tiller, 1989; Naidu et al., 1994). This pH range, called the adsorption edge, generally occurs 2–3 pH units below the value of the pK for the hydroxide solubility constant of the hydrated metal ion and is partly a function of the experimental conditions (Barrow, 1986). This pH effect on adsorption of micronutrients depends on initial solution concentration, nature of soil constituents, and metal retention capacity of soils. Increased adsorption of metal ions with increasing pH is due to both increased negative surface charge density (Naidu et al., 1994) and increased concentration of the MOH$^+$ species in the soil system (Hodgson et al., 1964). As discussed above, at pH values less than the value of the point of zero net charge (pznc), an increase in soil pH leads to a marked increase in the negative electrostatic potential of a variable charge system, which promotes cation adsorption. However, the precise nature of this interaction between metal ions and particle surfaces is unclear as different mechanisms have been proposed to explain the increasing metal adsorption with pH. For example, the rapid increase in adsorption of Zn within the adsorption edge was attributed to abrupt changes in the concentration of ZnOH$^+$ species as the pH increases toward the pK$_1$ of the hydrolysis of Zn^{2+} (Barrow, 1986).

11.5.6.1.2 Ionic Strength

The effects of ionic strength on the adsorption of metals by soils are not as well understood as those of pH (Petruzzelli et al., 1985; Naidu et al., 1994). The effect of ionic strength on metal adsorption by soils depends on the surface properties of the particles being more pronounced in variable than permanent charge soils. Metal adsorption decreases with increasing ionic strength for permanent charge minerals (Garcia-Miragaya and Page, 1976; Schindler et al., 1987; Boekhold et al., 1993).

11.5.6.1.3 Index Cation and Solution Composition

Cation adsorption by soils is a competitive process in which the degree to which any cation is adsorbed depends on the concentrations and identity of the other cations present (Bolan et al., 1999). This phenomenon has practical implications on the behavior of micronutrients in soils because the concentration of several metal cations (e.g., Na$^+$, Ca^{2+}, and Mg^{2+}) in soil solution ranges from <0.01 M in Oxisols to >0.1 M in saline sodic soils. Such changes in ionic strength and metal ion concentration can enhance the mobility and, thus, the bioavailability of micronutrients in soils, for example, by liming or gypsum application, both of which increase Ca^{2+} ions that can compete with micronutrients for adsorption sites.

11.5.6.1.4 Inorganic and Organic Ligands

Ligands play important roles in numerous soil chemical processes such as mineral weathering, control of dissolved micronutrient concentrations, and dissolution and plant availability of micronutrients (see Harter and Naidu, 1995). The soil solution also contains inorganic ligand ions such as Cl$^-$, SO$_4^{2-}$, PO$_4^{3-}$, and NO$_3^-$ whose concentrations vary considerably with soil type from >0.1 M in saline to <0.001 M in highly weathered soils. Such variations in ligand ion concentrations modify both the nature and surface interactions of the metal species present, leading to marked variability in metal bioavailability.

Dissolved organic material consists of a range of low molecular weight compounds including water soluble humic and fulvic acids, which are dominant except in the rhizosphere where simple organic compounds may dominate. Despite extensive studies of metal ions with organic matter (Stevenson, 1991), the values of the reaction constants need to be improved because only limited efforts have been directed toward understanding the role of low molecular weight organic acids in the dynamics of metals in the soil solution; this is due partly to the commonly held, but incorrect, belief that such organic material is rapidly degraded, and therefore short-lived, and partly to difficulty assaying the low concentrations of organic acids present. In the vicinity of plant roots, dissolved organic matter concentrations, often reaching millimolar levels, are high due to root exudates and microbial activity (Harter and Naidu, 1995), greatly enhancing the bioavailability and uptake of soil P, Cu, Fe,

Mn, and Zn (Braum and Helmke, 1995). The role of plant and microbial exudates on metal reactions in soils is an area that lacks information and is an active area of research.

11.5.6.2 Soil Amendments

A number of amendments can change micronutrient bioavailability either through mobilization that releases the micronutrient into soil solution for plant uptake or through immobilization where the micronutrient concerned is removed from soil solution either by adsorption, complexation, and precipitation, rendering the micronutrient unavailable for plant uptake and leaching to groundwater.

11.5.6.2.1 Chelating and Complexing Agents

When a micronutrient ion combines with an electron donor containing two or more donor groups, a complex or coordination compound with one or more rings called a micronutrient chelate is formed. A "chelate" (Greek, *chela* meaning claw) is so named because these species can coordinate at several or all positions literally wrapping themselves around a central micronutrient ion. Chelating agents, which have a high affinity for micronutrients ions such as Fe, Cu, and Zn, can be used to enhance the solubilization of micronutrients by the formation of soluble micronutrients chelates such as EDTA, EDHA, DTPA, and EHPG (Table 11.16). The effectiveness of a chelate in mobilizing soil micronutrients depends on several factors, including species, micronutrient:chelate ratio, thermodynamic stability constants, presence of competing cations, soil pH, stability of the micronutrient–chelate complex, the ageing, and the extent of micronutrient on soil constituents (Grčman et al., 2001). Synthetic chelates appear to mobilize micronutrients from the exchangeable, organic, and carbonate-bound fractions, but not from the oxide fraction (Elliott and Shastri, 1999). The use of chelating agents may induce the solubilization of other than the target micronutrients, which may become phytotoxic (e.g., Al and Mn).

11.5.6.2.2 Phosphate Compounds

Phosphate compounds can enhance the immobilization of micronutrients in soils by various processes including direct adsorption by P compounds, phosphate anion-induced adsorption, and precipitation with solution P as phosphates (Bolan et al., 2003b). Because the phosphate anion is very effective in desorbing certain oxyanions, such as selenite and molybdate, it is often used as an extractant to measure the amount of adsorbed oxyanions (James et al., 1995; Aide and Cummings, 1997). Phosphate also competes strongly with molybdate (MoO_4^{2-}) for adsorption sites, thereby resulting in increased desorption of the latter (Barrow, 1973; Xie et al., 1993).

Interaction of P with micronutrients also occurs within the plant. For example, increasing levels of P can induce or accentuate symptoms of Zn deficiency in plants grown in soils or culture media low in available Zn (Loneragan et al., 1979; Cakmak and Marschner, 1987). This disorder is commonly known as "P-induced Zn deficiency," which is the most widely examined P-trace element interaction in soil–plant systems (Loneragan and Webb, 1993). Five plausible mechanisms involving P, which may operate separately or concurrently, can cause decreases in Zn concentrations in the plant, depending upon plant species and environmental conditions (Loneragan et al., 1979; Loneragan and Webb, 1993): (1) dilution of tissue Zn by growth promoted by P addition, (2) inhibition of Zn uptake by roots as a result of competition with the cations accompanying the P, (3) increased P-induced Zn adsorption by Fe and Al oxide-rich variable charge soils, (4) greater Zn requirement in shoots induced by P additions, and (5) inactivation of Zn within the shoot in the presence of high P levels (Loneragan and Webb, 1993).

11.5.6.2.3 Liming Materials

Although liming is primarily aimed at ameliorating soil acidity, bioavailability of Zn, Mn, Cu, and other cationic micronutrients can be reduced as a result of decreased mobility by increased sorption/precipitation and competition between Ca^{2+} and micronutrients ions on the root surface for uptake (Brown et al., 2009). Because liming also increases Mo (molybdate) mobility in soils by decreased sorption on soil surfaces, Mo in soil solution available for uptake is increased.

11.5.6.2.4 Organic Composts

The most important organic fertilizers are biosolids and animal manures. Unlike biosolid application, regulations governing livestock and poultry manure are generally based on total N and P loading so that excessive amounts of some micronutrients are inadvertently applied (Cu and Zn in swine and As in poultry manure) (Bolan et al., 2004), requiring pretreatment to reduce bioavailability. For example, treatment of poultry manure with alum $[Al_2(SO_4)_3]$ decreases the concentrations of water-soluble Zn and Cu. Organic amendments increase the CEC of soils, thereby resulting in increased micronutrients adsorption.

Micronutrients form both soluble and insoluble complexes with organic materials, a process that depends on the nature of the organic matter (Bolan et al., 2003a). With increasing pH, carboxyl, phenolic, alcoholic, and carbonyl functional groups on SOM dissociate, thereby increasing the affinity of ligand ions for micronutrient cations. For example, addition of manure increases the adsorption and complexation of Cu by the soil with DOC being primarily responsible (Bolan et al., 2003a).

11.5.7 Bioavailability Implications for Soil Testing and Risk Assessment

11.5.7.1 Soil Testing and Crop Response

While most soil testing focuses on the bioavailable fraction of micronutrients for fertilizer recommendation purposes, most risk assessment models assume that the target element is 100% available, which is clearly not the case; consequently, a number of soil tests have been developed to predict micronutrient availability. Descriptions of micronutrient deficiency symptoms and the soil conditions in which deficiencies are likely to occur are presented in Table 11.18 while selected references on crop responses to micronutrients are presented in Table 11.19. In this section, we focus on Zn, Mn, Cu, Fe, B, Cl, Mo, and Ni that are essential for crop growth, development, and yield.

TABLE 11.18 Deficiency Symptoms and Crop Response to Micronutrients

Micronutrient	Deficiency Symptoms	Crop Response
Boron	B deficiencies symptoms vary with plant species including impaired meristem activity (dieback of young growth, malformed leaves often manifest as "witches broom"), impaired cell wall development (cracking of stem and petioles), and poor pollen tube development (reduced seed set, fruit fall, and deformed fruit)	B deficiency occurs on a wide range of soils (coarse-textured, highly weathered, shallow, volcanic ash, and thin soils over calcareous material) and crops
Cobalt	Co is essential for N fixation in plants Deficiency causes ill-health in ruminants. Co deficiency is associated with vitamin B12 decreases in animals	Co deficiency occurs in acidic and highly leached soils, soils derived from granites, calcareous, and peat soils Co is usually administered directly to animals
Copper	Catalyst for several enzymes Deficiency effects new growth with wilting, withering of youngest growth Cu deficiency in grain crops is known as reclamation disease, wither-tip, yellow-tip, or blind ear. In woody plants, such as citrus, this is known as dieback or exanthema	Cu deficiency occurs in: peat and muck soil; alkaline and calcareous soils, especially sandy soils or those with high levels of free $CaCO_3$; highly leached soils; or soils heavily fertilized with N, P, and Zn fertilizers Cu deficiency is corrected mostly by soil application rather than foliar application; Cu is also applied as a fungicide spray
Iron	A major element for chlorophyll production. Deficiency turns new leaves pale yellow or white while the veins remain green Difficult for plants to absorb and moves slowly within the plant. High soil pH prevents plants absorption of Fe	Fe deficiency is caused by imbalance of metallic ions (Cu and Mn), excessive P, a combination of high soil pH, high lime application, high soil moisture, cool temperature, and high levels of bicarbonate in the rooting medium Fe deficiency can be corrected by direct application of Fe fertilizer to soils or as foliar application
Manganese	Works with plant enzymes to reduce nitrates before producing proteins Mn deficiency turns young leaves a mottled yellow or brown Mn deficiency is called gray speck in oats, yellow disease in spinach, speckled yellow in sugar beets, marsh spot in peas, crinkle leaf in cotton, stem streak necrosis in potato, streak disease in sugarcane, mouse ear in pecan, and internal bark necrosis in apple	Mn deficiency occurs in: shallow peaty soils overlying calcareous subsoils; alluvial soils and marsh soils derived from calcareous parent materials; poorly drained calcareous soils with high organic matter; acid sandy mineral soils that are low in native Mn or high lime application Mn deficiency is corrected by soil application of Mn fertilizers or by foliar application
Molybdenum	Mo is an essential constituent of enzyme necessary for N_2 fixation. Deficiency causes leaves to turn pale, with scorched-looking edges, and irregular growth. Mo deficiency is known as whiptail disease in brassica plants and yellow spot disease in citrus. Mo deficiency is frequently associated with legume crops as Mo is an essential constituent of N_2 fixation	Mo deficiency occurs in acid sandy soils, highly podsolized soils, highly weathered tropical soils and well drained sandy soils Mo deficiency can be corrected by soil application, foliar spray, and seed treatment with Mo; lime application can overcome Mo deficiency
Nickel	A constituent of four enzymes: urease, methyl coenzyme M reductase, hydrogenase, and carbon monoxide dehydrogenase Without Ni, toxic levels of urea accumulate, leading to the formation of necrotic lesions. In tomato, chlorosis in the youngest leaf leading to necrosis of their meristematic tissues; in cereals (wheat, oats, and barley) growth depression, premature senescence and decreased Fe levels	Occurs mainly in woody perennials such as pecan and coffee. Can be induced as a result of complexation by glyphosate. Foliar sprays are effective in overcoming deficiency
Selenium	Essential for higher animals; Se deficiency causes muscular dystrophy or white muscle disease	Se deficiency occurs in strongly leached soils
Zinc	Zinc is required in a number of enzymes and plays an essential role in DNA transcription Deficiency produces stunting, yellowing, and curling of small leaves A typical symptom of deficiency is the stunted growth of leaves, commonly called "little leaf" and is caused by the oxidative degradation of the growth hormone auxin	Zn deficiency is caused by: low Zn content in soils, high lime application, or unavailability of Zn to the plant, and metabolic disorder within plants (e.g., imbalance between Zn and P) Zn deficiency can be overcome by soil and foliar application and avoiding imbalance between Zn and P

Source: Adriano, D.C., *Trace Elements in Terrestrial Environments: Biogeochemistry, Bioavailability and Risks of Metal(loid)s*, 2nd edn., Springer, New York, 2001.

11.5.7.1.1 Diagnosis and Prognosis of Micronutrient Deficiencies or Toxicities

Dry matter and grain yield of a crop are determined by the ability of the roots of plants to extract micronutrients from the soil at rates that are nonlimiting for crop growth. Therefore, the amount of nutrient taken up by a crop is a measure of availability. Analysis of selected plant parts (e.g., leaves, young growth), or whole shoots (Smith, 1980), is used as a diagnostic tool to identify micronutrients that may be limiting growth in contrast to soil tests that estimate the available pool. Plant analysis as a diagnostic tool is discussed in detail in Chapter 13.

TABLE 11.19 Selected References for Crop Responses to Micronutrients

Micronutrient	Crop Species	Application Rate	Observations	Reference
Cu	Barley (*Hordeum vulgare* L.) and wheat (*Triticum aestivum* L.)	50 kg Cu ha^{-1}	Application levels about 4–5 times the usual for crop application. However, no detrimental effects in yield reduction or phytotoxicity in wheat. For barley, 50 kg Cu ha^{-1} decreased yield by 12% in the first year. The Cu levels in plants were <9 mg kg^{-1}	Gupta and Kalra (2006)
	Wheat	6.7–13.4 kg Cu ha^{-1}	Application of Cu increased grain yield	Rehm (2008)
	Wheat and barley	0.7–2.1 kg Cu ha^{-1}	Application of Cu increased grain yield and concentrations of wheat and barley grown on neutral to acidic soils	Gartrell and Glencross (1968)
Zn	Barley and wheat	50 kg Zn ha^{-1}	Zn applications of 50 kg ha^{-1} for 2 years resulted in crop tissue Zn levels as high as 105 mg kg^{-1} did not cause any phytotoxicity in cereals	Gupta and Kalra (2006)
	Rice	13.5 kg Zn ha^{-1}	Zn at 13.5 kg ha^{-1}, increased dry matter, tissue Zn concentration and grain yield	Slaton et al. (2005)
	Wheat and barley	0.7–2.1 kg Zn ha^{-1}	Application of Zn increased grain yield of wheat and barley grown on neutral to acidic soils	Gartrell and Glencross (1968)
Fe	Soybean and Pigeon pea	20 kg Fe ha^{-1}	Application of Fe at 20 kg Fe ha^{-1} increased grain yield of soybean and pigeon pea by 9% and 414%, respectively	Hodgson et al. (1992)
	Soybean	400 g Fe EDTA (5.5% Fe and 2% EDTA) ha^{-1}	The Fe application with 80 kg N ha^{-1} gave the highest seed yield	Caliskan et al. (2008)
Co	Faba bean (*Vicia faba* L.)	20 mg L^{-1} Co as foliar spray	Co application at 20 mg L^{-1} increased grain yield by 218% and total protein by 48%	Hala (2007)
Mn	*Lupinus angustifolius* L.	1.7 kg Mn ha^{-1} in 200 L foliar application	Foliar Mn prevented deficiency, and increased grain yield	Hannam et al. (1984)
	L. angustifolius L.	3.5–15 kg Mn ha^{-1}	Soil application increased grain yield, decrease amount of split seed at maturity	Brennan (1999)
Ni	Tomato	30 mg Ni kg^{-1} soil	This level, increased quality of fruit and auxin and gibberlin contents	Gad et al. (2007)
	Pecan (*Carya illinoinensis* [Wangenh.] K. Koch)	10–100 mg L^{-1} (plus urea and surfactant)	Decreased "water-stage fruit-split" symptom of pecan reducing crop losses	Wells and Wood (2008); Malavolta and Moraes (2007)
Mo	Tobacco	1.1 kg Mo ha^{-1}	This level of Mo and lime at 7.3 or 14.6 t ha^{-1} increased yield by 16%–22%	Khan et al. (1994)
Se	Wheat	20 g Se ha^{-1}	Se fertilization at 20 g ha^{-1} increased grain Se content by four- to sevenfold	Stroud et al. (2010)

11.5.7.1.2 Soil Analysis to Assess Available Micronutrient Pool

The objectives of a micronutrient soil test are to (1) group soils into classes for fertilizer recommendations, (2) predict the probability of a response to an application, and (3) evaluate soil productivity (Fitts and Nelson, 1956). A good soil test should (1) extract all or a proportionate fraction of the available form or forms of a nutrient, (2) accurately and in a timely manner measure the amount of the nutrient extracted, and (3) be correlated with the growth and response of each crop to each micronutrient under various conditions (Sims and Johnson, 1991). A detailed discussion of soil testing is presented in Chapter 13.

11.5.7.1.2.1 Zinc Zinc deficiency is widespread throughout the world, probably the most common micronutrient deficiency across a range of crop species (Takkar and Walker, 1993). Usually, Zn deficiency in plants occurs in calcareous (high pH) soils because of low Zn availability or in coarse-textured (sandy), highly leached, acid soils because of their low total Zn content. Antagonisms between Zn and several other essential elements (e.g., P, N, and Cu) can also lead to Zn deficiencies (Loneragan and Webb, 1993). Recommended amounts of Zn fertilizer range from about 1 to 15 kg Zn ha^{-1}. Some crops particularly sensitive to inadequate soil Zn are corn, soybean, rice (*Oryza sativa*), field bean (*Vicia faba*), and citrus (*Citrus spp.*).

The sources of Zn used as fertilizers are presented in Table 11.16 with zinc sulfate crystal or granular form being the most common (Mortvedt and Gilkes, 1993). Zinc oxide has low effectiveness in the granular form as it is insoluble in water (Mortvedt, 1991). Because water solubility of the Zn source is important, at least 50% water solubility is required to be effective on calcareous soils (Gangloff et al., 2002; Westfall et al., 2002). However, in acid soils, a wide range of products, including the oxide and sulfate, are equally effective as fine powders mixed thoroughly with the soil (Mortvedt and Gilkes, 1993).

11.5.7.1.2.2 Copper Usually, Cu deficiency occurs on highly leached, acid soils (low total Cu content), on coarse-textured (sandy) soils, and soils rich in organic matter. Recommended amounts of Cu fertilizer range from about 1 to 25 kg Cu ha^{-1}. Because amounts of Cu in excess of crop requirements are often applied particularly in manures and fungicides, Cu has a long residual effect in the soil and remains bioavailable for some considerable time after application; consequently, frequent applications are not required (Gartrell, 1981; Brennan, 2006).

Although soil applications are the commonest method used to correct Cu deficiency (Gartrell, 1981; Martens and Westermann, 1991), foliar sprays are effective in citrus, wheat, corn, and soybean (Martens and Westermann, 1991). Because some fungicides usually supply more Cu than is required by the crop, repeated applications can result in soil accumulation (Graham and Webb, 1991). Inorganic sources applied to the soil such as, Cu oxide, sulfate, carbonate, nitrate, and chloride are effective in correcting deficiencies in crops (Gartrell, 1981).

Bioavailability of Cu fertilizers is affected by (1) the method of placement, (2) ability of the Cu product to produce Cu^{+2} ions in the soil or rhizosphere, (3) reactions of Cu with soil, and (4) different requirements of plant species due to differences in growth rate, efficiency of absorption, or translocation in the plant. Because Cu is immobile in the soil, its availability is greatly influenced by the number of granules per unit volume of soil and the position in which they are placed. Banding Cu is more effective than topdressing, without incorporation into the soil. Copper fertilizer is ineffective in soil profiles that contain few absorbing roots, for example, dry soil. Applied at the same rate, large granules (fewer granules per unit volume) are much less effective than fine material (Gartrell, 1981).

11.5.7.1.2.3 Manganese Usually, Mn deficiency is associated with well-drained neutral to alkaline soils, and those rich in organic matter. In addition, because of their low total Mn content, highly leached, coarse-textured (sandy) acid soils are also frequently deficient for some crop species. Recommended amounts of Mn fertilizer range from about 2 to 20 kg Mn ha^{-1}. Manganese has a short residual effect in most neutral to alkaline soils, because the bioavailable Mn^{+2} is rapidly oxidized to MnO$_2$. On the other hand, for sweet lupin (*Lupinus angustifolius* L) grown on acid sandy soils, the effect of 7.5 kg Mn ha^{-1} has lasted for some considerable time after application, while that of 50 kg Mn ha^{-1} persisted for about 17 years (Brennan et al., 2001). Foliar application is the most efficient and effective method of correcting Mn deficiency in various crops grown on a range of deficient soils. Manganese deficiency is frequently encountered in cereals, cotton, peanuts, soybeans, and a range of fruits and vegetables (Martens and Westermann, 1991).

Band placement of manganese sulfate fertilizer is more effective than broadcasting at the same rate for soybean (*Glycine max*) (Mascagni and Cox, 1985) and sweet lupin (Brennan, 1999) because concentration in the band reduces the rate of oxidation of Mn^{+2} to less available forms. Foliar sprays and band application of Mn sulfate were equally effective for soybeans (Mascagni and Cox, 1985). Soil moisture affects Mn availability with the seed-applied Mn being effective under adequate soil moisture but not drought conditions in soybean (Alley et al., 1978). For some crops (peanut [*Arachis hypogaea*], soybean, and wheat) and conditions, multiple rather than single foliar sprays are required (Cox, 1968; Hallock, 1979; Nayyar et al., 1985). The timing of the foliar spray in relation to crop development stage is also important for soybean (Gettier et al., 1984), corn (Mascagni and Cox, 1984), and sweet lupin (Brennan et al., 2008).

11.5.7.1.2.4 Iron Usually, Fe deficiency is a common problem in numerous crop species grown on alkaline soils containing free CaCO$_3$ in the soil profile (Marschner, 1995). In many cases, the problem stems from various factors that inhibit Fe absorption by plant roots or impair its utilization in metabolic process in Fe-inefficient species. Although rare, acid soils with deficient levels of Fe do occur, for example, in Florida and other similar areas (Welch et al., 1991). Recommended amounts of Fe fertilizer range from about 20 to 100 kg Fe ha^{-1} (Martens and Westermann, 1991). Iron has a short residual effect in most alkaline soils, because Fe^{+2} is rapidly converted to unavailable forms. Foliar application is more effective at correcting Fe deficiency in various crops grown on a range of deficient soils. Iron deficiency is frequently encountered in a range of Fe-sensitive crops and/or Fe-inefficient cultivars (Welch et al., 1991). Growing Fe-tolerant species or varieties is an effective method for controlling Fe deficiency in crops grown on soils where Fe availability is low (Chen and Barak, 1982).

Because soil–fertilizer contact is limited in band placement, iron sulfate fertilizer banded in the soil is more effective than broadcasting at equal rates. However, Fe sulfate is frequently broadcast on pastures (*Trifolium subterannean* L.) where Fe deficiency has been induced by overliming acid soil (Brennan and Highman, 2001). Several other methods such as (1) mixing of Fe sulfate and sulfuric acid (Wallace, 1988; Martens and Westermann, 1991), (2) inclusion of organic residues with Fe sulfate (Mostaghimi and Matocha, 1988), and (3) inclusion of K$_2$SO$_4$ with Fe sulfate that corrected Fe deficiency in peanuts (Shaviv and Hagin, 1987) have been used to reduce the conversion of applied Fe to unavailable forms. Foliar sprays of inorganic (e.g., Fe sulfate) and organic Fe sources (e.g., Fe chelates) are very effective in correcting Fe deficiency (Martens and Westermann, 1991). Foliar applications of Fe sulfate are more effective than soil application of Fe chelate (FeEDDHA). Iron is less available in well-aerated soils due to insolubility of Fe(III) oxides; but availability can be promoted by increasing soil moisture that produces loci where some reduction of Fe(III) to Fe^{2+} can take place (Bjerre and Schierup, 1985). However, excessive soil moisture in calcareous soils may intensify Fe deficiency mainly due to the buildup of HCO$_3$ in the soil solution (Moraghan and Mascagni, 1991). Plant residues, manures, biosolids, charcoal, coal, and a range of by-products from manufacturing (e.g., polyflavoids and lignosulfonates) have been effective in alleviating Fe deficiency (Chen and Barak, 1982).

11.5.7.1.2.5 Boron Boron is an essential micronutrient where the range between deficiency and toxicity is smaller than for most other micronutrients. Plants respond to the activity of B in solution rather than to adsorbed B making the rate of desorption, which follows a first-order rate equation, important in B supply to the root. The bioavailability of B is determined by pH, texture, moisture, temperature, SOM, and clay mineralogy (Goldberg, 1997) while the adsorbing surfaces of importance are sesquioxides, clay minerals, lime, and organic matter. Boron adsorption increases initially with pH on both mineral and organic surfaces up to pH 8–10 and then decreases. Consequently, overliming acid soils can cause a temporary deficiency due to higher adsorption at high pH (Reisenauer et al., 1973). The roles of B in the plant include sugar transport, cell wall synthesis and structure, pollen tube growth, pollen viability, carbohydrate synthesis, respiration, and membrane integrity while excess B can cause impairment in a range of cellular functions, such as phenol and ascorbate metabolism, free radical generation, and detoxification (Cakmak and Romheld, 1997); but its primary role is still unknown. These changes may be direct or indirect functions of B deficiency.

Boron deficiency occurs on a wide range of soils (coarse-textured, highly weathered, shallow, volcanic ash, and shallow soils over calcareous material) and crops (alfalfa [*Medicago sativa*], beets [*Beta vulgaris*], peanut [*A. hypogaea*], brassicas [*Brassica* spp.], coffee [*Coffea* spp.], oil palm [*Elaeis guineensis*], cotton [*Gossypium* spp.], sunflower [*Helianthus annuus*], apple [*Malus domestica*], and grape [*Vitis vinifera*]) (Shorrocks, 1997) as well as many vegetable crops.

Deficiency symptoms vary with plant species including impaired meristem activity (dieback of young growth, malformed leaves often manifest as "witches broom"), impaired cell wall development (cracking of stem and petioles), and poor pollen tube development (reduced seed set, fruit fall, and deformed fruit). The main B sources used to correct B deficiency are borax ($NaB_4O_7 \cdot 10H_2O$ [11% B], NaB_4O_7 [21.4% B]), boric acid ($B(OH)_3$ [17.5% B]), and a range of crushed ores containing variable amount of B (Shorrocks, 1997). Boric acid and Solubor® (17.5% B), products readily soluble in cold water, are used as foliar applications. For annual crops, $1-2\,kg\,B\,ha^{-1}$ is the recommended rate (Shorrocks, 1997).

Boron toxicity commonly occurs in arid and semiarid environments on alkaline and saline soils (Nable et al., 1997). Continued application of irrigation water with high levels of B can lead to toxicity. The typical symptoms of B toxicity are leaf burn, chlorotic, and necrotic patches often on the margins and tips of older leaves (Bennett, 1993).

11.5.7.1.2.6 Chloride Chloride (Cl^-) is essential for many plant functions such as a chemical balancing agent for K^+ during the opening and closing of stomata, photosynthesis, cation balance and transport within the plant, counteracting the effects of fungal infection, and competing with NO_3^-, thus promoting the uptake of NH_4-N (Boyer et al., 1954; Fixen, 1993; Talbott and Zeiger, 1996). Deficiency symptoms vary with plant species including wilting and leaf chlorosis while toxicity manifests itself as leaf margin scorching, leaf drop in excessive situations, and leaf area reduction and thickening, resulting in reduced growth.

Because Cl^- is highly soluble in water, leaching losses occur, where rainfall is high and soil drainage is good. If Cl^--containing fertilizers (e.g., muriate of potash [KCl]) are not regularly used, Cl^- deficiency can occur. In coastal regions with rough surf, atmospheric deposition of Cl^- can be high, decreasing with distance from the coast. Chloride competes with other anions (e.g., NO_3^-, SO_4^{2-}) although little data are available on the specific interactions and competition between each anion.

11.5.7.1.2.7 Nickel Recently Ni, which has been shown to be an essential element for plants, is required in miniscule amounts as a piconutrient. It plays a vital part in the translocation of N in ureide-transporting woody perennials such as pecans (*Carya illinoensis*) and coffee (*Coffea arabica*) where a deficiency of Ni disrupts ureide, amino, and organic acid metabolism but its physiological role is poorly understood (Bai et al., 2006). In addition, it confers disease resistant on plants although much work needs to be done in this field (Wood and Reilly, 2007). Nickel appears to play a key role in the production of secondary plant metabolites that influence resistance to disease. Most soils contain adequate amounts of Ni but under certain specific conditions, for example, where glyphosate that strongly complexes Ni has been used as a herbicide, deficiencies can be induced (Yamada, 2010). Consequently, deficiency symptoms manifest themselves as those of N. Research in this field is still in its infancy.

11.5.7.1.2.8 Molybdenum Because Mo occurs in the soil solution as MoO_4^{2-}, it interacts with other anions, particularly phosphate and SO_4^{2-} to reduce Mo uptake. In addition, it is strongly sorbed on sesquioxide surfaces at pH values below 7, decreasing 100-fold for each unit drop in pH. Thus, Mo deficiencies are mainly found on acid soils where bioavailability is low (Gupta, 1997). Consequently, liming acid soils can often supply sufficient Mo to meet the needs of many crops including legumes, which are most sensitive to deficiency. In legumes, because Mo is required by the Rhizobia in the fixation of N, deficiency of Mo is manifest as those of N. In addition, it is an activator for the nitrate reductase enzyme, crucial to the N nutrition of crops. Where Mo deficiency occurs, correction can be achieved by relatively small applications ($\sim100\,g\,Mo\,ha^{-1}$) as foliar sprays in the form of sodium or ammonium molybdate.

References

Adriano, D.C. 2001. Trace elements in terrestrial environments: Biogeochemistry, bioavailability and risks of metal(loid)s. 2nd edn. Springer, New York.

Adriano, D.C., W.W. Wenzel, J. Vangronsveld, and N.S. Bolan. 2004. Role of assisted natural remediation in environmental cleanup. Geoderma 122:121–142.

Aide, M.T., and M.F. Cummings. 1997. The influence of pH and phosphorus on the adsorption of chromium (VI) on boehmite. Soil Sci. 162:599–603.

Ajwa, H.A., G.S. Banuelos, and H.F. Mayland. 1998. Selenium uptake by plants from soils amended with inorganic and organic materials. J. Environ. Qual. 27:1218–1227.

Alexander, R.R., and M. Alexander. 2000. Bioavailability of genotoxic, compounds in soils. Environ. Sci. Technol. 34:1589–1593.

Alley, M.M., C.I. Rich, G.W. Hawkins, and D.C. Martens. 1978. Corrections of Mn deficiency of soybeans. Agron. J. 70:35–38.

Bai, C., C.C. Reilly, and B.W. Wood. 2006. Nickel deficiency disrupts the metabolism of ureides, amino acids and organic acids in young pecan foliage. Plant Phys. 140:433–443.

Barrow, N.J. 1973. On the displacement of adsorbed anions from soil. I. Displacement of molybdenum by phosphate and by hydroxide. Soil Sci. 116:423–431.

Barrow, N.J. 1986. Testing a mechanistic model. IV. Describing the effects of pH on zinc retention by soils. J. Soil Sci. 37:295–302.

Basta, N.T., and R. Gradwohl. 2000. Estimation of Cd, Pb, and Zn bioavailability in smelter-contaminated soils by a sequential extraction procedure. J. Soil Contam. 9:149–164.

Beckett, P.H.T., E. Warr, and P. Brindley. 1983a. Changes in the extractabilities of the heavy-metals in water-logged sludge-treated soils. Water Pollut. Control 82:107–113.

Beckett, P.H.T., E. Warr, and R.D. Davis. 1983b. Cu and Zn in soils treated with sewage-sludge—Their extractability to reagents compared with their availability to plants. Plant Soil 70:3–14.

Bennett, W.F. 1993. Nutrient deficiencies and toxicities in crop plants. American Phytopathological Society Press, St Paul, MN.

Bibak, A. 1994. Uptake of cobalt and manganese by winter wheat from a sandy loam soil with and without added farmyard manure and fertilizer nitrogen. Commun. Soil Sci. Plant Anal. 25:2675–2684.

Bjerre, G.K., and H.H. Schierup. 1985. Influence of waterlogging on availability and uptake of heavy metals by oat grown in different soils. Plant Soil 88:45–56.

Black, C.A. 2002. Soil fertility evaluation and control. CRC Press, Boca Raton, FL.

Blackwell, K., I. Singleton, and J. Tobin. 1995. Metal cation uptake by yeast: A review. Appl. Microbiol. Biotechnol. 43:579–584.

Boekhold, A.E., E.J.M. Temminghoff, and S. van der Zee. 1993. Influence of electrolyte composition and pH on cadmium adsorption by an acid sandy soil. J. Soil Sci. 44:85–96.

Bolan, N.S., D.C. Adriano, and S. Mahimairaja. 2004. Distribution and bioavailability of trace elements in livestock and poultry manure by-products. Crit. Rev. Environ. Sci. Technol. 34:291–338.

Bolan, N.S., D.C. Adriano, S. Mani, and A.R. Khan. 2003a. Adsorption, complexation and phytoavailability of copper as influenced by organic manure. Environ. Toxicol. Chem. 22:450–456.

Bolan, N.S., D.C. Adriano, and R. Naidu. 2003b. Role of phosphorus in (im)mobilization and bioavailability of heavy metals in the soil-plant system. Rev. Environ. Contam. Toxicol. 177:1–44.

Bolan, N.S., R. Naidu, J.K. Syers, and R.W. Tillman. 1999. Surface charge and solute interactions in soils. Adv. Agron. 67:88–141.

Bomke, A.A., and L.E. Lowe. 1991. Trace element uptake by two British Columbia forages as affected by poultry manure application. Can. J. Soil Sci. 71:305–312.

Boyer, T.C., A.G. Carlton, C.M. Johnson, and P.R. Stout. 1954. Chlorine: A micronutrient element for higher plants. Plant Phys. 29:526–532.

Braum, S.M., and P.A. Helmke. 1995. White lupin utilizes soil-phosphorus that is unavailable to soybean. Plant Soil 176:95–100.

Brennan, R.F. 1999. Lupin grain yields and fertiliser effectiveness are increased by banding manganese below the seed. Aust. J. Exp. Agric. 39:595–603.

Brennan, R.F. 2006. Long-term residual value of copper fertiliser for production of wheat grain. Aust. J. Exp. Agric. 46:77–83.

Brennan, R.F., M.D.A. Bolland, and J.W. Bowden. 2008. Plant nutrition, p. 102–110. In P. White, R. French, and A. McLarty (eds.) Producing lupins. Department of Agriculture and Food, South Perth, Western Australia.

Brennan, R.F., J.W. Gartrell, and K.G. Adcock. 2001. Residual value of manganese fertiliser for lupin grain production. Aust. J. Exp. Agric. 41:1187–1197.

Brennan, R.F., and D.D. Highman. 2001. Residual value of ferrous sulfate for clover production on humic sandy podsols. Aust. J. Exp. Agric. 41:633–639.

Brookes, P.C., and S.P. McGrath. 1984. Effects of metal toxicity on the size of the soil microbial biomass. J. Soil Sci. 35:341–346.

Brown, S.L., A. Svendsen, and C. Henry. 2009. Restoration of high zinc and lead tailings with municipal biosolids and lime: A field study. J. Environ. Qual. 38:2189–2197.

Brümmer, G.W., J. Gerth, and K.G. Tiller. 1988. Reaction kinetics of the adsorption and desorption of Ni, Zn and Cd by goethite. I. Adsorption and diffusion of metals. J. Soil Sci. 39:35–52.

Cakmak, I., and H. Marschner. 1987. Mechanisms of phosphate-induced zinc deficiency in cotton. III. Changes in physiological availability of zinc in plants. Phys. Plant 70:13–20.

Cakmak, I., and V. Romheld. 1997. Boron deficiency-induced impairments of cellular functions in plants. Plant Soil 193:71–83.

Caliskan, S., I. Ozkaya, M.E. Caliskan, and M. Arslan. 2008. The effects of nitrogen and iron fertilization on growth, yield and fertilizer use efficiency of soybean in a Mediterranean-type soil. Field Crops Res. 108:126–132.

Canet, R., F. Pomares, and F. Tarazona. 1997. Chemical extractability and availability of heavy metals after seven years application of organic wastes to a citrus soil. Soil Use Manage. 13:117–121.

Capar, S.G., J.T. Tanner, M.H. Friedman, and K.W. Boyer. 1970. Multielement analysis of animal feed, animal wastes, and sewage sludge. Environ. Sci. Technol. 12:785–790.

Cernansky, S., M. Kolencik, J. Sevc, M. Urik, and E. Hiller. 2009. Fungal volatilization of trivalent and pentavalent arsenic under laboratory conditions. Bioresour. Technol. 100:1037–1040.

Chen, Y., and P. Barak. 1982. Iron nutrition of plants in calcareous soil. Adv. Agron. 35:217–240.

Christensen, T.H. 1984. Cadmium soil sorption at low concentrations: Effect of time, cadmium loading, pH and calcium. Water Air Soil Pollut. 21:105–114.

Christie, P., and J.A.M. Beattie. 1989. Grassland soil microbial biomass and accumulation of potentially toxic metals from long-term slurry application. J. Appl. Ecol. 26:597–612.

Cox, F.R. 1968. Development of a yield response prediction and manganese soil test interpretation for soybeans. Agron. J. 60:521–4.

de Abreu, M.F., and R.S. Berton. 1996. Comparison of methods to evaluate heavy metals in organic wastes. Commun. Soil Sci. Plant Anal. 27:1125–1135.

del Castilho, P., W.J. Chardon, and W. Salomons. 1993. Influence of cattle-manure slurry application on the solubility of cadmium, copper, and zinc in a manure acidic, loamy-sand soil. J. Environ. Qual. 22:689–697.

Duffera, M., W.P. Robarge, and R.L. Mikkelsen. 1999. Estimating the availability of nutrients from processed swine lagoon solids through incubation studies. Bioresour. Technol. 70:261–268.

Echevarria, G., J.L. Morel, J.C. Fardeau, and E. Leclerc-Cessac. 1998. Assessment of phytoavailability of nickel in soils. J. Environ. Qual. 27:1064–1070.

Elliott, H.A., and N.L. Shastri. 1999. Extractive decontamination of metal-polluted soils using oxalate. Water Air Soil Pollut. 110:335–346.

Eneji, E.A., S. Yamamoto, T. Honna, and A. Ishiguro. 2001. Physico-chemical changes in livestock feces during composting. Commun. Soil Sci. Plant Anal. 24:477–489.

Fitts, J.W., and W.L. Nelson. 1956. The determination of lime and fertiliser requirements of soils through chemical tests. Adv. Agron. 8:241–82.

Fixen, P.E. 1993. Crop responses to chloride. Adv. Agron. 50:107–150.

Forbes, E.A., A.M. Posner, and J.P. Quirk. 1976. The specific adsorption of divalent Cd, Co, Cu, Pb and Zn on goethite. J. Soil Sci. 27:154–166.

Frankenberger, W.T., and U. Karlson. 1995. Volatilization of selenium from a dewatered seleniferous sediment—A field study. J. Ind. Microbiol. 14:226–232.

Gad, N., M.H. El-Sherif, and N.H.M. El-Gereedly. 2007. Influence of nickel on some physiological aspects of tomato plants. Aust. J. Basic Appl. Sci. 1:286–293.

Gangloff, W.J., D.G. Westfall, G.A. Peterson, and J.J. Mortvedt. 2002. Relative availability coefficients of organic and inorganic Zn fertilizers. J. Plant Nutr. 25:259–273.

Garcia-Miragaya, J., and A.L. Page. 1976. Influence of ionic strength and inorganic complex formation on sorption of trace amounts of cadmium by montmorillonite. Soil Sci. Soc. Am. J. 40:658–663.

Gartrell, J.W. 1981. Distribution and correction of copper deficiency in crops and pastures, p. 313–349. In J.F. Loneragan et al. (eds.) Copper in soils and plants. Academic Press, Sydney, Australia.

Gartrell, J.W., and R.N. Glencross. 1968. Copper, zinc and molybdenum fertilisers for new land crops and pastures. J. Agric. W. Aust. 9:517–521.

Gettier, S.W., D.C. Martens, D.L. Hallock, and M.J. Stewart. 1984. Residual manganese and associated soybean yield response from $MnSO_4$ application on a sandy loam soil. Plant Soil 81:101–110.

Goldberg, S. 1997. Reactions of boron with soils. Plant Soil 193:35–48.

Goodson, C.C., D.R. Parker, C. Amrhein, and Y. Zhang. 2003. Soil selenium uptake and root system development in plant taxa differing in Se-accumulating capability. New Phytol. 159:391–401.

Gonzalez, A., A. Obrador, L.M. Lopez-Valdivia, and J.M. Alvarez. 2008. Effect of zinc source applied to soils on its availability to navy bean. Soil Sci. Soc. Am. J. 72:641–649.

Graham, R.D., and M.J. Webb. 1991. Micronutrients and disease resistance and tolerance in plants, p. 329–376. In J.J. Mortvedt (ed.) Micronutrients in agriculture. 2nd edn. SSSA, Madison, WI.

Grčman, H., D. Velikonja-Bolta, D. Vodnik, B. Kos, and D. Leštan. 2001. EDTA enhanced heavy metal phytoextraction: Metal accumulation, leaching, and toxicity. Plant Soil 235:105–114.

Gupta, U.C. 1997. Soil and plant factors affecting molybdenum uptake by plants, p. 71–90. In U.C. Gupta (ed.) Molybdenum in agriculture. Cambridge University Press, New York.

Gupta, U.C., and Y.P. Kalra. 2006. Residual effect of copper and zinc from fertilizers on plant concentration, phytotoxicity, and crop yield response. Commun. Soil Sci. Plant Anal. 37:2505–2511.

Haddad, K.S., and C.J. Kaldor. 1982. Effect of parent material, natural available soil boron, and applied boron and lime on the growth and chemical composition of lucerne on some acidic soils of Central Tablelands of New South Wales. Aust. J. Exp. Agric. Anim. Husb. 22:317–323.

Hala, K. 2007. Effect of cobalt fertilizer on growth, yield and nutrients status of Faba bean (*Vicia faba* L.) plants. J. Appl. Sci. Stat. Res. 3:867–872.

Haleem, A.A., W.B. Anderson, M.K. Sadik, and A.A.A. Salam. 1992. Iron availability as influenced by phosphate and ferrous-sulfate and ferric-sulfate fertilizers. J. Plant Nutr. 15:1955–1970.

Hallock, D.L. 1979. Relative effectiveness of several Mn sources on Virginia-type peanuts. Agron. J. 71:685–688.

Hamon, R.E., J.M. McLaughlin, and G. Cozens. 2002. Mechanisms of attenuation of metal availability in in-situ remediation treatments. Environ. Sci. Technol. 36:3991–3996.

Hamon, R.E., D.R. Parker, and E. Lombi. 2008. Advances in isotopic dilution. Adv. Agron. 99:289–343.

Han, F.X., W.L. Kingery, and H.M. Selim. 2001. Accumulation, redistribution and bioavailability of heavy metals in waste-amended soils, p. 145–174. In I.K. Iskandar and M.B. Kirkham (eds.) Trace elements in soils: Bioavailability, flux and transfer. Lewis Publishers, Boca Raton, FL.

Hannam, R.J., W.J. Davies, R.D. Graham, and J.L. Riggs. 1984. The effect of soil- and foliar-applied manganese in preventing the onset of manganese deficiency in *Lupinus angustifolius*. Aust. J. Agric. Res. 35:529–538.

Harter, R.D.R., and R. Naidu. 1995. Role of metal-organic complexation in metal sorption by soils. Adv. Agron. 55:219–264.

He, Z.L.L., X.E. Yang, and P.J. Stoffella. 2005. Trace elements in agroecosystems and impacts on the environment. J. Trace Elem. Med. Biol. 19:125–140.

Hedley, M.J. 2008. Techniques for assessing nutrient bioavailability in soils: Current and future issues, p. 281–325. In R. Naidu et al. (eds.) Chemical bioavailability in terrestrial environments. Elsevier, Amsterdam, the Netherlands.

Henry, C.L., and R.B. Harrison. 1992. Fate of trace metals in sewage sludge compost, p. 195–210. In D. Adriano (ed.) Biogeochemistry of trace metals. Lewis Publishers, Boca Raton, FL.

Hodgson, A.S., J.F. Holland, and E.F. Rogers. 1992. Iron-deficiency depresses growth of furrow irrigated soybean and pigeon pea on Vertisols of northern NSW. Aust. J. Agric. Res. 43:635–644.

Hodgson, J.F., K.G. Tiller, and M. Fellows. 1964. The role of hydrolysis in the reaction of heavy metals in soil forming materials. Soil Sci. Soc. Am. Proc. 28:42–46.

Hrudey, S.E., W. Chen, and C.G. Rousseaux. 1996. Bioavailability and environmental risk assessment. Lewis Publishers, Boca Raton, FL.

Hsu, J.H., and S.L. Lo. 2000. Effect of dissolved organic carbon on leaching of copper and zinc from swine manure compost. Water Sci. Technol. 42:247–252.

Huysman, F., W. Verstraete, and P.C. Brookes. 1994. Effect of manuring practices and increased copper concentrations on soil microbial-populations. Soil Biol. Biochem. 26:103–110.

Jackson, B.P., and P.B. Bertsch. 2001. Determination of arsenic speciation in poultry wastes by IC-ICP-MS. Environ. Sci. Technol. 35:4868–4873.

Jackson, B.P., W.P. Miller, A.W. Schumann, and M.E. Sumner. 1999. Trace element solubility from land application of fly ash/organic waste mixtures. J. Environ. Qual. 28:639–647.

Jackson, B.P., and W.P. Miller. 2000. Soil solution chemistry of a fly ash-, poultry litter-, and sewage sludge-amended soil. J. Environ. Qual. 29:430–436.

Jahiruddin, M., H. Harada, T. Hatanaka, and Y. Sunaga. 2001. Adding boron and zinc to soil for improvement of fodder value of soybean and corn. Commun. Soil Sci. Plant Anal. 32:2943–2957.

James, B.R., J.C. Petura, R.J. Vitale, and G.R. Mussoline. 1995. Hexavalent chromium extraction from soils: A comparison of five methods. Environ. Sci. Technol. 29:2377–2380.

Japenga, J., J.W. Dalenberg, D. Wiersma, S.D. Scheltens, D. Hesterberg, and W. Salomons. 1992. Effect of liquid animal manure application on the solubilization of heavy metals from soil. Int. J. Environ. Anal. Chem. 46:25–39.

Khan, N.A., C.L. Mulchi, and C.G. McKee. 1994. Influence of soil-pH and molybdenum fertilization on the productivity of Maryland tobacco. I. Field investigations. Commun. Soil Sci. Plant Anal. 25:2103–2116.

Kinniburgh, D.G., and M.L. Jackson. 1981. Cation adsorption by hydrous metal oxides and clay, p. 91–160. In M.A. Anderson and A.J. Rubin (eds.) Adsorption of inorganics at solid-liquid interfaces. Ann Arbor Science, Ann Arbor, MI.

Kornegay, E.T., J.D. Hedges, D.C. Martens, and C.Y. Kramer. 1976. Effect of soil and plant mineral levels following application of manures of different copper levels. Plant Soil 45:151–162.

Lake, D.L., P.W.W. Kirk, and J.N. Lester. 1984. Fractionation, characterization, and speciation of heavy metals in sewage sludge and sludge-amended soils. J. Environ. Qual. 13:175–183.

L'Herroux, L., S. Le Roux, P. Appriou, and J. Martinez. 1997. Behaviors of metals following intensive pig slurry applications to natural filed treatment process in Brittany (France). Environ. Pollut. 97:119–130.

Li, Z., J.A. Ryan, J.L. Chen, and S.R. Al-Abed. 2001. Adsorption of cadmium on biosolids-amended soils. J. Environ. Qual. 30:903–911.

Li, Z., and L.M. Shuman. 1997. Mobility of Zn, Cd and Pb in soils as affected by poultry litter extract. I. Leaching in soil columns. Environ. Pollut. 95:219–226.

Lim, T.-T., and K.-H. Goh. 2005. Selenium extractability from a contaminated fine soil fraction: Implication on soil cleanup. Chemosphere 58:91–101.

Lonergan, J.F., T.S. Grove, A.D. Robson, and K. Snowball. 1979. Phosphorus toxicity as a factor in zinc-phosphorus interactions in plants. Soil Sci. Soc. Am. J. 43:966–972.

Lonergan, J.F., and M.J. Webb. 1993. Interactions between zinc and other nutrients affecting the growth of plants, p. 119–133. In A.D. Robson (ed.) Zinc in soils and plants. Kluwer Academic Publishers, Dordrecht, the Netherlands.

Ma, Y.B., E. Lombi, A.L. Nolan, and M.J. McLaughlin. 2006. Determination of labile Cu in soils and isotopic exchangeability of colloidal Cu complexes. Eur. J. Soil Sci. 57:147–153.

Malavolta, E., and M.F. Moraes. 2007. Nickel-from toxic to essential nutrient. Better Crops 91:26–27.

Marschner, H. 1995. Mineral nutrition of higher plants. 2nd edn. Elsevier, New York.

Martens, D.C., and D.T. Westermann. 1991. Fertilizer applications for correcting micronutrient deficiencies, p. 549–592. In J.J. Mortvedt et al. (eds.) Micronutrient in agriculture. 2nd edn. SSSA, Madison, WI.

Martinez, J., and P. Peu. 2000. Nutrient fluxes from a soil treatment process for pig slurry. Soil Use Manage. 16:100–107.

Mascagni, H.J., Jr., and F.R. Cox. 1984. Diagnosis and correction of manganese deficiency in corn. Commun. Soil Sci. Plant Anal. 15:1323–1333.

Mascagni, H.J., Jr., and F.R. Cox. 1985. Effective rates of fertilization for correcting manganese deficiency in soybeans. Agron. J. 77:363–366.

McBride, M.B., and G. Spiers. 2001. Trace element content of selected fertilizers and dairy manures as determined by ICP-MS. Commun. Soil Sci. Plant Anal. 32:139–156.

McCauley, A. 2009. Commercial fertilizers and soil amendments. Nutrient Management Module No 10. Montana State University Extension 4449-10. Montana State University Extension, Bozeman, MT.

McGrath, S.P. 1994. Effects of heavy metals from sewage sludge on soil microbes in agricultural ecosystems, p. 247–260. In S.M. Ross (ed.) Toxic metals in soil-plant systems. John Wiley & Sons, New York.

McLaren, R.G., and G.S.P. Ritchie. 1993. The long-term fate of copper fertilizer applied to a Lateritic sandy soil in Western Australia. Aust. J. Soil Res. 13:39–50.

McLaren, R.G., R.S. Swift, and B.F. Quin. 1984. E.D.T.A.-extractable copper, zinc, and manganese in soils Canterbury Plains. NZ J. Agric. Res. 7:207–217.

Miller, W.P., D.C. Martens, L.W. Zelazny, and E.T. Kornegay. 1986. Forms of solid-phase copper in copper-enriched swine manure. J. Environ. Qual. 15:69–72.

Moore, P.A., Jr., T.C. Daniel, J.T. Gilmour, B.R. Shreve, D.R. Edwards, and B.H. Wood. 1998. Decreasing metal runoff from poultry litter with aluminum sulfate. J. Environ. Qual. 27:92–99.

Moraghan, J.T., and H. Mascagni. 1991. Environmental and soil factors affecting micronutrient deficiencies and toxicities, p. 371–425. In J.J. Mortvedt et al. (eds.) Micronutrient in agriculture. 2nd edn. SSSA, Madison, WI.

Mortvedt, J.J. 1991. Micronutrient fertilizer technology, p. 523–548. In J.J. Mortvedt et al. (eds.) Micronutrient in agriculture. 2nd edn. SSSA, Madison, WI.

Mortvedt, J.J. 1999. Bioavailability of micronutrients, p. D71–D87. In M.E. Sumner (ed.) Handbook of soil science. CRC Press, Boca Raton, FL.

Mortvedt, J.J., and R.J. Gilkes. 1993. Zinc fertilisers, p. 33–44. In A.D. Robson (ed.) Zinc in soil and plants. Kluwer Academic Publishers, Dordrecht, the Netherlands.

Mostaghimi, S., and J.E. Matocha. 1988. Effects of normal and Fe-treated organic matter on iron chlorosis and yields of grain sorghum. Commun. Soil Sci. Plant Anal. 3:593–613.

Mott, C.J.B. 1981. Anion and ligand exchange, p. 179–219. In D.J. Greenland and M.H.B. Hayes (eds.) The chemistry of soil processes. John Wiley & Sons, New York.

Mullins, G.L., D.C. Martens, W.P. Miller, E.T. Kornegay, and D.L. Hallock. 1982. Copper availability, form, and mobility in soils from three annual copper-enriched hog manure applications. J. Environ. Qual. 11:316–320.

Nable, R.O., G.S. Banuelos, and G. Paull. 1997. Boron toxicity. Plant Soil 193:181–198.

Nádaská, G., K. Polčová, and J. Lesný. 2009. Manganese fractionation analysis in specific soil and sediment samples. Nova Biotechnol. 9-3:295–301.

Naidu, R., N.S. Bolan, R.S. Kookana, and K.G. Tiller. 1994. Ionic-strength and pH effects on the adsorption of cadmium and the surface charge of soils. Eur. J. Soil Sci. 45:419–429.

Naidu, R., R.S. Kookana, D.P. Oliver, S. Rogers, and M.J. McLaughlin. 1996a. Contaminants and the soil environment in the Australasia-Pacific region. Kluwer Academic Publishers, London, U.K.

Naidu, R., R.S. Kookana, M.E. Sumner, R.D. Harter, and K.G. Tiller. 1996b. Cadmium adsorption and transport in variable charge soils: A review. J. Environ. Qual. 26:602–617.

Naidu, R., K.T. Semple, M. Megharaj, A.L. Juhasz, N.S. Bolan, S. Gupta, B. Clothier, R. Schulin, and R. Chaney. 2008. Bioavailability, definition, assessment and implications for risk assessment, p. 39–52. In R. Naidu et al. (eds.) Chemical bioavailability in terrestrial environments. Elsevier, Amsterdam, the Netherlands.

Narwal, R.P., and B.R. Singh. 1998. Effect of organic materials on partitioning extractability and plant uptake of metals in an alum shale soil. Water Air Soil Pollut. 103:405–410.

Nayyar, V.K., U.S. Sadana, and T.N. Taskkar. 1985. Methods and rates of application of manganese and its critical levels for wheat following rice on coarse textured soils. Fert. Res. 8:173–178.

Nicholson, F.A., B.J. Chambers, J.R. Williams, and R.J. Unwin. 1999. Heavy metal contents of livestock feeds and animal manures in England and Wales. Bioresour. Technol. 23:23–31.

Payne, G.G., D.C. Martens, E.T. Kornegay, and M.D. Lindemann. 1988. Availability and form of copper in three soils following eight annual applications of copper-enriched swine manure. J. Environ. Qual. 17:740–746.

Petruzzelli, G., G. Guidi, and L. Lubrano. 1985. Ionic strength effects on heavy metal adsorption by soils. Commun. Soil Sci. Plant Anal. 16:971–986.

Pierzynski, G.M., and A.P. Schwab. 1993. Bioavailability of zinc, cadmium and lead in a metal contaminated alluvial soil. J. Environ. Qual. 22:247–254.

Rahmatullah, M., Badr-Uz-Zaman Salim, and Kh. Hussain. 2001. Nickel forms in calcareous soils and influence of Ni supply on growth and N uptake of oats grown in soil fertilized with urea. Int. J. Agric. Biol. 3:230–232.

Raven, K.P., and R.H. Loeppert. 1997. Trace element composition of fertilizers and soil amendments. J. Environ. Qual. 26:551–557.

Rayment, G.E., and K.A. Verrall. 1980. Soil manganese tests and the comparative tolerance of kikuyu and white clover to manganese toxicity. Trop. Grassl. 14:105–114.

Rehm, G.W. 2008. Response of hard red spring wheat to copper fertilization. Commun. Soil Sci. Plant Anal. 39:2411–2420.

Reisenauer, H.M., L.M. Walsh, and R.G. Hoeft. 1973. Testing soils for sulphur, boron, molybdenum, and chlorine, p. 173–200. In Soil testing and plant analysis. SSSA, Madison, WI.

Ross, S.M. 1994. Retention, transformation and mobility of toxic metals in soils, p. 63–152. In S.M. Ross (ed.) Toxic metals in soil-plant systems. John Wiley & Sons, New York.

Schindler, P.W., P. Liechti, and J.C. Westall. 1987. Adsorption of copper, cadmium and lead from aqueous solution to the kaolinite/water interface. Neth. J. Agric. Sci. 35:219–230.

Schnitzer, M., and S.U. Khan. 1978. Soil organic matter. Elsevier Scientific Publications, New York.

Shaviv, A., and J. Hagin. 1987. Correction of lime-induced chlorosis by application of iron and potassium sulphates. Fert. Res. 13:161–167.

Shorrocks, V.M. 1997. The occurrence and correction of boron deficiency. Plant Soil 193:121–148.

Shuman, L.M. 1991. Chemical forms of micronutrients in soils, p. 125–145. *In* J.J. Mortvedt et al. (eds.) Micronutrients in agriculture. SSSA, Madison, WI.

Sims, J.T., and G.V. Johnson. 1991. Micronutrient soil tests, p. 427–476. *In* J.J. Mortvedt et al. (eds.) Micronutrient in agriculture. SSSA, Madison, WI.

Sinaj, S., F. Mächler, and E. Frossard. 1999. Assessment of isotopically exchangeable zinc in polluted and nonpolluted soils. Soil Sci. Soc. Am. J. 63:1618–1625.

Slaton, N.A., E.E. Gbur, Ch.E. Wilson, and R.J. Norman. 2005. Rice response to granular zinc sources varying in water soluble zinc. Soil Sci. Soc. Am. J. 69:443–452.

Smith, F.W. 1980. Interpretation of plant analysis: Concept and principles, p. 1–12. *In* D.J. Reuter and J.B. Robinson (eds.) Plant analysis: An interpretation manual. Inkata Press, Melbourne, Australia.

Smith, I.M., K.J. Hall, L.M. Lavkulich, and H. Schreier. 2007. Trace metal concentrations in an intensive agricultural watershed in British Columbia, Canada. J. Am. Water Resour. Assoc. 43:1455–1467.

Sposito, G. 1984. The surface chemistry of soils. Oxford University Press, New York.

Srinath, T., T. Verma, P.W. Ramteke, and S.K. Garg. 2002. Chromium (VI) biosorption and bioaccumulation by chromate resistant bacteria. Chemosphere 48:427–435.

Stevenson, F.J. 1991. Organic matter-micronutrient reactions in soil, p. 145–186. *In* J.J. Mortvedt et al. (eds.) Micronutrients in agriculture. SSSA, Madison, WI.

Stokes, J.D., A. Wilkinson, B.J. Reid, K.C. Jones, and K.T. Semple. 2005. Validation of the use of an aqueous hydroxypropyl-β-cyclodextrin (HPCD) extraction as an indicator of PAH bioavailability in contaminated soils. Environ. Toxicol. Chem. 24:1325–1330.

Stroud, J.L., H.F. Li, F.J. Lopez-Bellido, M.R. Broadley, I. Foot, S.J. Fairweather-Tait, D.J. Hart et al. 2010. Impact of sulphur fertilisation on crop response to selenium fertilisation. Plant Soil 332:31–40.

Susetyo, W., L.A. Carreira, L.V. Azarraga, and D. Grimm. 1991. Fluorescence techniques for metal-humic interactions. Fresenius J. Anal. Chem. 339:624–635.

Sutton, A.L., D.W. Nelson, V.B. Mayrose, D.T. Kelly, and J.C. Nye. 1984. Effect of copper levels in swine manure on corn and soil. J. Environ. Qual. 13:198–203.

Takkar, P.N., and C.D. Walker. 1993. The distribution and correction of zinc deficiency, p. 151–165. *In* A.D. Robson (ed.) Zinc in soils and plants. Kluwer Academic Publishers, Dordrecht, the Netherlands.

Talbott, L.D., and E. Zeiger. 1996. Central role of potassium and sucrose in guard-cell osmoregulation. Plant Phys. 111:1051–1057.

Tiller, K.G. 1989. Heavy metals in soils and their environmental significance. Adv. Soil Sci. 9:113–142.

van Rhee, J.A. 1975. Copper contamination effects on earthworms by disposal of pig waste in pastures, p. 451–456. *In* J. Vanek (ed.) Progress in soil zoology. Proc. 5th Int. Colloq. Soil Zoology. Prague, Czechoslovakia, September, 1975.

Wallace, A. 1988. Acid and acid–iron fertilizers for iron-deficiency control in plants. J. Plant Nutr. 11:1311–1319.

Wallingford, G.W., V.S. Murphy, W.L. Powers, and H.L. Manges. 1975. Effect of beef-feedlot manure and lagoon water on iron, zinc, manganese and copper content in corn and in DTPA soil extracts. Soil Sci. Soc. Am. J. 39:482–487.

Welch, R.M., W.H. Allaway, W.A. House, and J. Kubota. 1991. Geographic distribution of trace element problems, p. 31–58. *In* J.J. Mortvedt et al. (eds.) Micronutrient in agriculture. 2nd edn. SSSA, Madison, WI.

Wells, M.L., and B.W. Wood. 2008. Foliar boron and nickel applications reduce water-stage fruit-split of pecan. Hortscience 43:1437–1440.

Wendling, L.A., J.K. Kirby, and M.J. Mclaughlin. 2009. Aging effects on cobalt availability in soils. Environ. Toxicol. Chem. 28:1609–1617.

Westfall, D.G., W.J. Gangloff, G.A. Peterson, and J.J. Mortvedt. 2002. Agronomic effectiveness of Zn fertilizers: Solubility and relative availability coefficient relationships. 17th World Cong. Soil Sci. Soil and Fertilizer Society of Thailand, Bangkok, Thailand, August 14–20, 2002.

Wood, B.W., and C.C. Reilly. 2007. Interaction of nickel and plant disease, p. 217–247. *In* L.E. Datnoff et al. (eds.) Mineral nutrition and plant diseases. American Phytopathological Society Press, Minneapolis, MN.

Wood, B.H., C.W. Wood, K.H. Yoo, K.S. Yoon, and D.P. Delaney. 1996. Nutrient accumulation and nitrate leaching under broiler litter amended corn fields. Commun. Soil Sci. Plant Anal. 27:2875–2894.

Xie, R.J., A.F. Mackenzie, and Z.J. Lou. 1993. Causal-modeling pH and phosphate effects on molybdate sorption in three temperate soils. Soil Sci. 155:385–397.

Yamada, T. 2010. Personal communication.

Yermiyahu, U., R. Keren, and Y. Chen. 2001. Effect of composted organic matter on boron uptake by plants. Soil Sci. Soc. Am. J. 65:1436–1441.

Zhang, H., W. Davison, A.M. Tye, N.M.J. Crout, and S.D. Young. 2006. Kinetics of zinc and cadmium release in freshly contaminated soils. Environ. Toxicol. Chem. 25:664–670.

12
Soil Acidity and Liming

T. Jot Smyth
North Carolina State University

12.1	Introduction 12-1
12.2	Nature of Soil Acidity... 12-1
12.3	Acid Soil Constraints to Plant Growth 12-1
	Hydrogen Toxicity • Aluminum Toxicity • Manganese Toxicity • Calcium and Magnesium Deficiencies • Phosphorus and Molybdenum Deficiencies
12.4	Lime Requirement Methods 12-3
	Soil–Lime Incubations • Buffer pH Methods • Exchangeable Aluminum • Comparison of Lime Requirement Methods
12.5	Lime Material Characteristics and Application 12-4
	Neutralizing Value • Fineness of Liming Materials • Effective Calcium Carbonate Rating • Lime Incorporation into the Soil
	References... 12-5

12.1 Introduction

About 26% of the global ice-free land surface is occupied by soils with pH_w (pH determined in water) values of 6.5 or less, of which two-thirds have potential Al toxicity problems. Subsoil Al problems are estimated to encompass 16% of the ice-free land surface (Eswaran et al., 1997). Naturally, acid soils are found in regions where rainfall exceeds evapotranspiration (Jenny, 1941); H^+ and eventually Al^{+3} replace basic cations lost from exchange sites through excess water transport. Most of the acid soils belong to the Alfisols, Histosols, Oxisols, Spodosols, and Ultisols orders of the U.S. Soil Taxonomy (Buol et al., 1997).

Agricultural and industrial processes also contribute to soil acidification. The conversion of ammonium to nitrate produces H^+, which can require up to 3.6 kg $CaCO_3$ to neutralize the acidity produced by each kg of N (Adams, 1984). Plants excrete H^+ from roots when cation uptake exceeds that of anions (Tang and Rengel, 2003). Legumes have excess cation uptake when most of their N is supplied via atmospheric fixation. Neutralization of acidity produced by N-fixing legumes can correspond to 54–550 kg $CaCO_3$ ha^{-1} $year^{-1}$ (Bolan et al., 1991; Coventry and Slattery, 1991). Major sources of acid precipitation are atmospheric emissions of nitrogen and sulfur compounds from burning processes and their eventual transformation into nitric and sulfuric acids (Alewell, 2003).

12.2 Nature of Soil Acidity

Soils with pH_w values <5.5 have appreciable amounts of exchangeable Al^{3+}. The proportion of cation exchange sites occupied with Al^{3+} and the concentration of Al in the soil solution increase with declining soil pH. In mineral soils, the soil solution Al concentration and the %Al saturation of the cation exchange capacity (CEC) approach zero between pH values of 5.5 and 5.8 (Kamprath, 1984; Sumner and Yamada, 2002). Organic soils have significant quantities of both exchangeable Al^{3+} and H^+. Aluminum binds to organic matter in nonexchangeable forms, such that soil solution Al concentrations in organic soils are less than in mineral soils at the same pH value (Evans and Kamprath, 1970). At pH_w 5, organic soils have very little Al in the soil solution and Ca concentrations are high enough to overcome H^+ competitive effects on Ca uptake. Hydrolysis of the Al^{3+} in the equilibrium soil solution impacts the soil solution H^+ concentration and pH (Thomas and Hargrove, 1984).

Exchangeable acidity is extracted from soils with neutral unbuffered salt solutions such as KCl, to minimize changes in soil pH (Thomas and Hargrove, 1984). Soils also contain acidity, which is not extracted with neutral unbuffered salt solutions but is titrated with base. This nonexchangeable acidity represents the acid buffering capacity of soils, and includes Al bound to organic matter, H^+ associated with carboxyl groups of organic matter, hydroxy Al, and hydrated oxides of Fe and Al (Kamprath, 1978).

12.3 Acid Soil Constraints to Plant Growth

Plant growth in acid soils can be limited by a combination of toxicities (H^+, Al^{3+}, and Mn^{2+}) and nutrient deficiencies (namely Ca, Mg, Mo, and P) (Kamprath, 1984; Sumner and Yamada, 2002). Responses to lime can involve amelioration of a combination of factors, wherein experiments in hydroponics often provide insight into the magnitude of individual constraints.

12.3.1 Hydrogen Toxicity

Visual root symptoms of excess H$^+$ usually occur at pH < 4.6 and entail stunted growth, brownish color, and little lateral root development. At low pH and low Ca^{2+} concentrations, root membranes are damaged, leading to loss of organic substrates and absorbed cations. Root injury due to H$^+$ can reduce the uptake of Ca, Mg, and K (Islam et al., 1980; Foy, 1984). Increasing solution Ca^{2+} concentration alleviates root injury due to H$^+$ (Lund, 1970; Moore, 1974; Runge and Rode, 1991; Sanzonowicz et al., 1998). Low pH has detrimental effects on rhizobium survival, legume host plant root infection, and nodule initiation (Andrew, 1978).

12.3.2 Aluminum Toxicity

Aluminum toxicity is a major constraint to plant growth in acid soils. The immediate and visible evidence of Al toxicity is a reduction in root length, which limits plant access to water and nutrients. Aluminum interferes with various root growth processes, namely, disruption of regulatory signals in root cap cells (Bennet and Breen, 1991), interference with cell division, enzyme activities, DNA replication, and P availability at membranes (Foy, 1984; Silva et al., 2002).

Concentrations of soil solution Al^{3+} that reduce crop growth are in the micromolar range (Kamprath, 1984). Soil solution Al^{3+} concentrations are related to the percentage of Al^{3+} saturation of the soil cation exchange sites, with a marked increase in solution concentration when Al^{3+} saturation exceeds 50% (Kamprath, 1978; Kamprath and Smyth, 2005). Determination of Al^{3+} and other cations in soil solutions is tedious and time consuming. It is simpler to measure exchangeable cations and diagnose Al^{3+} toxicity based on the %Al saturation of the cation exchange sites (Kamprath, 1970, 1978). A realistic measure of the soil cation exchange at a given pH value, the effective cation exchange capacity (ECEC), is the sum of exchangeable Ca^{2+}, Mg^{2+}, K$^+$, Na$^+$, and unbuffered salt-extractable acidity (Coleman et al., 1959).

Tolerance to Al differs significantly among plant species and cultivars within species. Reviews of field lime trials suggested that cassava yields were not depressed by %Al saturation values of <75%, whereas best yields for most trials with cotton, mung bean, and wheat were at 0% Al saturation; however, critical Al saturation levels among trials with different varieties ranged from 0% to 40% with corn and soybean and 0% to 60% with upland rice (Osmond et al., 2007). There are various proposed mechanisms of Al tolerance, which can be both internal and external to plants (Taylor 1991; Kochian, 1995). Internal tolerance involves detoxification after Al^{3+} is absorbed, whereas external mechanisms imply the prevention of Al^{3+} uptake and transport. Proposed exclusion mechanisms by various investigators include immobilization of Al^{3+} at the cell wall, low root CEC, selective permeability of the plasma membrane, changes in rhizosphere pH, exudation of Al-chelating compounds, and Al^{3+} efflux. Internal mechanisms may also include chelation of Al^{3+}, compartmentalization in vacuoles and Al-tolerant enzymes (Silva et al., 2002).

12.3.3 Manganese Toxicity

Acid soils with large contents of Fe and Al oxides often contain quantities of Mn that are toxic to plants (Kamprath, 1984). Plant growth is reduced through excess accumulation of Mn in aboveground plant tissues. Plant symptoms of Mn toxicity vary among species, but include marginal chlorosis and necrosis of leaves, leaf puckering and necrotic spots on leaves (Foy, 1984). Soil solution levels of Mn^{+2} are difficult to predict, because they are influenced by total soil Mn content, soil pH and redox potential. Although solubility decreases with increasing pH, Mn toxicity may occur above a soil pH of 5.5 under the appropriate reducing conditions (Sumner and Yamada, 2002). Improved plant growth upon liming soils above pH 5.5, where soil solution and exchangeable Al^{3+} are reduced to essentially zero, is often associated with the alleviation of Mn toxicities (Kamprath, 1984; Sumner and Yamada, 2002).

12.3.4 Calcium and Magnesium Deficiencies

Acid soils with high %Al saturation have limited amounts of Ca^{2+} and Mg^{2+} in either the exchangeable form or in the soil solution. The supply of adequate Ca^{2+} and Mg^{2+} for normal plant growth may also be constrained in soils with low CEC due to their limited capacity to retain these cations (Kamprath, 1984). An external supply of Ca^{2+} is essential for normal root development (Rios and Pearson, 1964; Ferguson and Clarkson, 1976; Clarkson, 1984). The addition of 0.1 cmol$_c$ Ca kg^{-1} (20 mg Ca kg^{-1}) to an Oxisol initially containing 0.02 cmol$_c$ Ca kg^{-1} (4 mg Ca kg^{-1}) increased wheat seedling length by over 50%, regardless of whether the Ca was supplied as lime, calcium chloride or calcium phosphate (Ritchey et al., 1983). In a review of lime experiments in the tropics, Kamprath (1984) suggested that an exchangeable soil Ca level of 1.0 cmol$_c$ kg^{-1} as a minimum value for good plant growth. Plant growth responses to Mg are often linked to conditions of low soil exchangeable Mg^{2+} and/or less than 5% Mg saturation of the CEC and pH values ≤5 (Adams, 1984; Kamprath and Foy, 1985; Kamprath and Smyth, 2005).

The susceptibility of peanut to Ca deficiency is associated with its below ground fruit development. The fruit absorbs most of its Ca from the surrounding soil solution. In the absence of adequate ambient soil Ca, shelling percentage is reduced due to poor fruit development (Cox et al., 1982; Adams, 1984). Soil Ca is often supplemented for peanut by applications of lime and gypsum.

12.3.5 Phosphorus and Molybdenum Deficiencies

Investigations regarding improvement of soil P solubility with liming provide conflicting results, as indicated in several reviews of the subject (Kamprath and Foy, 1985; Sumner and Yamada, 2002). However, there is general agreement that neutralization of exchangeable Al^{3+} increases root growth, exploration of a greater soil volume, and access to more soil P. In acid soils Al^{3+}

at the root surface precipitates P and reduces P transport to plant tops; neutralization of Al^{3+} by liming increases transport of P to plant tops (Kamprath and Smyth, 2005).

Molybdenum is an essential nutrient for N fixation in legumes. The availability of soil Mo is a function of Mo content, pH and MoO_4^{2-} absorption by hydrous oxides of Fe and Al (Adams, 1984; Kamprath and Foy, 1985). In a review of lime and Mo experiments, soil solution Mo increased two- to threefold as soil pH was increased up to values of 6–7; likewise, soybean yields without Mo treatment increased with liming and approached yields of seed Mo treatments at pH values above 6 (Sumner and Yamada, 2002). With Mo added via seed treatments, there was no soybean yield response to lime above pH values of 5.3–5.5 where most of the exchangeable Al^{+3} is neutralized. Increased solubility of Mo with increasing pH is attributed to OH^- displacement of adsorbed MoO_4^{2-} on hydrous oxides of Fe and Al (Barrow, 1978).

12.4 Lime Requirement Methods

For extensive reviews and comparisons between lime requirement methods, the reader is referred to articles by van Lierop (1990), Sims (1996), and Sumner (1997). Some of the most commonly used methods are discussed herein.

12.4.1 Soil–Lime Incubations

Increasing rates of lime are mixed with fixed quantities of soil and equilibrated in a moist state for weeks or months. Lime required to achieve a given soil pH or exchangeable Al^{3+} level can be related to measurements of other soil chemical and physical attributes. Although time and costs of this method are not conducive to lime recommendations via routine soil testing, it is widely used to test and verify most of the rapid lime tests in current use across a number of soils with a range of chemical and physical characteristics within a given region of interest.

12.4.2 Buffer pH Methods

A given volume of a buffer solution is equilibrated with a fixed weight or volume of soil, and pH of the soil–buffer mixture is measured. The molarity of the buffer solution is known and the decrease in its pH, after mixed with soil, is a measure of the soil acidity that must be neutralized by a base in order to achieve a targeted soil pH.

Buffer pH methods used to determine lime requirements in different regions of the United States are described in Table 12.1. Each method was developed for representative soil types in a given region. The composition of the buffer solutions is different and has pH values ranging from 6.6 to 8.0. The targeted soil pH values after liming range from 6.0 to 7.0 among the various methods and the amount of lime required for a pH decrease of 0.1 in the buffer–soil mixture, relative to that of the initial

TABLE 12.1 Characteristics of Buffer pH Methods Used to Determine Lime Requirements in the United States

Method	Buffer pH	Target Soil pH	Intended Soil Use	Lime/0.1 Buffer pH Decrease ton $CaCO_3$ ha^{-1}
Adams and Evans	8.0	6.5	Low CEC Ultisols	0.1
Mehlich	6.6	6.0	Neutralize exch. Al^{+3}, Ultisols	0.16
SMP	6.8	6.8	High exch. Al^{+3} Alfisols	0.6
		6.4		0.5
		6.0		0.4
Woodruff	7.0	6.5–7.0	Mollisols	0.5

Source: Adapted from van Lierop, W. 1990. Soil pH and lime requirement determination, p. 73–126. *In* R.L. Westerman (ed.) Soil testing and plant analysis. 3rd edn. SSSA, Madison, WI.

buffer solution, ranges from 0.1 to 0.6 ton of $CaCO_3$ ha^{-1}. Several investigators have recently proposed modifications to existing methods (Huluka, 2005; Sikora, 2006; Sikora and Moore, 2008) or new buffer methods (Liu et al., 2005).

12.4.3 Exchangeable Aluminum

Lime requirements are based on the amount of exchangeable Al^{3+} extracted from soils with neutral unbuffered salt solutions such as KCl (Kamprath, 1970). Adding lime to mineral soils on an equivalent basis to the amount of exchangeable Al^{3+} raises soil pH to 5.3–5.6 and reduces soil solution Al^{3+} concentration and exchangeable Al to near zero (Kamprath, 1984; Sumner and Yamada, 2002). Lime neutralizes exchangeable acidity, but also reacts with the nonexchangeable acidity that includes Al bound to organic matter and H associated with carboxyl groups of organic matter, hydroxy Al, and hydrated oxides of Fe and Al. Therefore, lime requirements to achieve 0% Al saturation of the ECEC, based on exchangeable soil acidity, include a lime equivalence factor that is greater than 1.0 and usually in the range of 1.5–4.0 (Kamprath, 1984; Sumner, 1997).

There are several proposed modifications of the basic lime requirement determination based on exchangeable Al. Cochrane et al. (1980) proposed the following adjustment that accounts for differential acid tolerance among crops and cultivars:

$$CaCO_3 \text{ equivalent (ton ha}^{-1}) = LF\left[Al - \frac{TAS(Al + Ca + Mg)}{100}\right] \quad (12.1)$$

where

- LF is the product of 1.5 equivalents of $CaCO_3$/equivalent of exchangeable Al plus the conversion to a 20 cm layer for a hectare of soil
- Al, Ca, and Mg are exchangeable cations extracted with a neutral unbuffered salt
- TAS is the %Al saturation tolerated by the crop or cultivar

This approach seeks to only neutralize the quantities of exchangeable Al that exceed the %Al saturation tolerated by the crop. Although useful in regions where lime materials are expensive, data on exchangeable Ca and Mg, and %Al saturation tolerated by the intended crop must also be known in addition to soil exchangeable Al.

In soils with low ECEC, lime required to neutralize the exchangeable acidity may not be sufficient to correct Ca and Mg deficiencies. Therefore, lime requirements may include considerations of minimum targets for soil exchangeable Ca and Mg as illustrated by the following adjustment proposed by Sousa and Lobato (2002) to ensure that soil Ca + Mg achieves a value of 2 cmol$_c$ dm^{-3} after liming:

$$CaCO_3 \text{ equivalent (ton ha}^{-1}) = 2(Al) + (2 - (Ca + Mg)) \quad (12.2)$$

12.4.4 Comparison of Lime Requirement Methods

Given the combination of soil factors associated with a potential acidity constraint, Sumner and Yamada (2002) proposed grouping acid soils into two general categories: soils with pH values <5.2–5.4 where Al and Mn toxicities are the primary constraints to crop growth, and soils with pH values >5.4 where liming impacts crop growth through improved availability of nutrients, namely Ca, Mg, Mo, and P. In the first category, increasing soil pH via liming resolves the toxicity problems, whereas nutrient inputs without liming are an alternative remedial strategy to nutrient availability problems in the latter group. An understanding of the types of acidity constraints in the soils of interest can help in determining the most appropriate remedial strategy.

Differences in soil factors contributing to an acidity constraint and crop species/cultivar tolerances to Al and Mn toxicities reduce the likelihood that a given target pH value can be used to estimate lime requirements under a variety of circumstances. Therefore, it is important that lime requirement methods should be calibrated with crop performance data under field conditions for the representative crops and soil types of interest in a given region (Sumner, 1997). Liming mineral soils to a pH value of 6 ensures neutralization of exchangeable Al (around pH 5.4–5.6) and reduces the frequency for repeated applications of lime as the soil re-acidifies through N fertilizer inputs and plant secretion of H$^+$. Unless Mn toxicity is an issue, liming of acid soils to pH values above 6 will decrease the solubility of Cu, Fe, Mn, and Zn and potentially induce their deficiencies in soils where availabilities are adequate at lower pH values (Kamprath, 1971; Lucas and Knezek, 1972). Modest levels of soil acidity (pH values of 5.0–5.5) can also have economically beneficial impacts by reducing the incidence of pests and diseases such as soybean cyst nematode (*Heterodera glycine*) (Garcia et al., 1999) and scab disease (*Streptomyces scabies*) in potato (Keinath and Loria, 1989).

12.5 Lime Material Characteristics and Application

Various materials derived from natural deposits or industrial by-products can be used for liming. Calcium carbonate (calcitic limestone) and calcium-magnesium carbonate (dolomitic limestone) are frequently used crystalline compounds from natural deposits. Marls are soft, unconsolidated calcium carbonates that may contain some clay impurities. Calcium oxide (burnt or quick lime) and calcium hydroxide (hydrated lime) are white powders that are caustic when wet and difficult to mix uniformly with soil. Basic slag (calcium silicate) is a by-product of the steel industry often containing appreciable amounts of Mg and P. Several factors to consider when selecting and applying lime materials are described in the following sections.

12.5.1 Neutralizing Value

The neutralizing value represents the amount of acid neutralized by a unit weight of lime material. Pure calcium carbonate serves as a standard and has the calcium carbonate equivalent (CCE) value of 100%. The CCE values of several pure liming materials are given in Table 12.2. Most liming materials have impurities, which reduces their CCE value. Lime requirement estimates are often based on 100% CCE and the actual application rate needs to be adjusted for the CCE value of the selected lime material.

12.5.2 Fineness of Liming Materials

Upon contact with acid soils, lime particles dissolve and release Ca^{2+} and HCO$_3^-$, which leads to neutralization of H$^+$ and Al^{3+}. The reaction rate depends on the surface area of the lime particles in contact with the soil. Therefore, it is desirable that small particles are close together so that their zones of neutralization overlap in a short time.

Calcitic and dolomitic limestones are ground to increase the number of particles per unit of soil. Most ground lime materials contain a range of particle sizes from very fine to sand-sizes, which comply with state or province regulations on particle size distribution. A fineness factor for the lime material is the sum of the products of percentages of lime material in each size fraction and its corresponding efficiency factor. Depending on regional regulations, fractions greater than a certain size have

TABLE 12.2 Neutralizing Value (CCE %) of Pure Liming Materials

Material	% CaCO$_3$ Equivalence
Calcium carbonate, CaCO$_3$	100
Calcium-magnesium carbonate, CaMg(CO$_3$)$_2$	109
Calcium oxide, CaO	179
Calcium hydroxide, Ca(OH)$_2$	135
Calcium silicate, CaSiO$_3$	86

Source: Havlin, J.L., J.D. Beaton, S.L. Tisdale, and W.L. Nelson. 2005. Soil fertility and fertilizers: An introduction to nutrient management. 7th edn. Pearson Education, Inc., Upper Saddle River, NJ.

an efficiency factor of zero (e.g., >2 mm) and this value increases to a maximum of 1.0 with decreasing particle fineness (e.g., <0.4 mm) (Havlin et al., 2005). Calcitic limestones are softer than dolomitic limestones; thus, a given size fraction of the former dissolves at a faster rate (Kamprath and Smyth, 2005).

12.5.3 Effective Calcium Carbonate Rating

Effective Ca carbonate rating (ECC) is the product of the CCE and fineness factor for the liming material. The selection among available lime materials often involves comparing costs of sources per unit ECC and whether Mg inputs (via dolomitic limestone) are needed.

12.5.4 Lime Incorporation into the Soil

Lime should be mixed with the volume of soil to which one intends to neutralize acidity. In conventional tillage systems, lime recommendations often assume a soil depth of 15–20 cm. Thorough mixing can be achieved by broadcasting and incorporating half of the lime by disk or plow, followed by inversion and broadcasting and disking the remainder. In no-till systems, however, lime is surface applied and initial acid neutralization will be limited to the 0–5 cm soil depth. Surface applications without incorporation of lime rates to neutralize soil acidity in a 15–20 cm soil depth can lead to excessively high soil pH values and associated nutrient deficiencies in the 0–5 cm soil depth.

The effectiveness of surface-applied lime in ameliorating subsoil acidity below the 0–5 cm depth is an important consideration in no-till systems, especially when root elongation is constrained and crops lose access to soil reserves of water and nutrients. Sumner (1995) has reviewed an extensive collection of investigations regarding amelioration of subsoil acidity through surface application of lime and gypsum. The investigations entail a variety of soil types, climatic regimes, and rates of lime or gypsum. Likewise, the subsoil depth to which lime or gypsum moved was quite variable, although the process of detectable movement was gradual and entailed months and years after surface applications. In some instances, the changes in subsoil acidity consisted of increased exchangeable Ca and Mg without improvements in soil pH values. Although increases in solution Ca^{2+} concentration at constant pH can ameliorate Al^{3+} rhizotoxicity to some extent (Alva et al., 1986; Horst, 1987; Wright and Wright, 1987; Noble et al., 1988; Sanzonowicz et al., 1998), absence of an increased subsoil pH suggested that alkaline anions (HCO_3^- and/or OH^-) were not associated with the downward transport of basic cations from the surface applied lime (Sumner, 1995). In cases where subsoil pH also increased with downward movement of Ca^{2+} and Mg^{2+}, Sumner (1995) proposed two potential processes: (a) downward movement of NO_3^- and plant secretion of OH^- when NO_3^- uptake from subsoils leads to an excess of anion over cation uptake; and (b) complexation of toxic Al^{3+} by downward movement of soluble organic components. The former process supports frequent observations that subsoil Ca^{2+}, Mg^{2+}, and pH increased more rapidly when surface applications of lime were complemented with acidifying fertilizers. The trade-off, however, is a faster acidification of the surface soil when ammoniacal N sources are used.

Several investigators have compared the residual effects of an initial surface application of lime with and without supplementary lime applications in succeeding years (Blevins et al., 1978; Edwards and Beegle, 1988; Godsey et al., 2007). In all cases, lime rate had a greater effect than frequency of lime application in reducing acid characteristics of the subsoil. Nevertheless, frequent surface applications of lime in no-till systems will avoid detrimental effects on crop growth through acidification of the surface soil layers.

References

Adams, F. 1984. Crop response to lime in the Southern United States, p. 211–265. *In* F. Adams (ed.) Soil acidity and liming. ASA, Madison, WI.

Alewell, C. 2003. Acid inputs into soils from acid rain, p. 83–115. *In* Z. Rengel (ed.) Handbook of soil acidity. Marcel Dekker, New York.

Alva, A.K., D.G. Edwards, C.J. Asher, and F.P.C. Blamey. 1986. Effects of phosphorus/aluminum molar ratio and calcium concentration on plant response to aluminum toxicity. Soil Sci. Soc. Am. J. 50:133–137.

Andrew, C.S. 1978. Mineral characterization of tropical forage legumes, p. 93–112. *In* C.S. Andrew and E.J. Kamprath (eds.) Mineral nutrition of legumes in tropical and subtropical soils. Commonwealth Scientific and Industrial Research Organisation, Melbourne, Australia.

Barrow, N.J. 1978. Inorganic reactions of phosphorus, sulphur and molybdenum in soil, p. 189–206. *In* C.S. Andrew and E.J. Kamprath (eds.) Mineral nutrition of legumes in tropical and subtropical soils. Commonwealth Scientific and Industrial Research Organisation, Melbourne, Australia.

Bennet, R.J., and C.M. Breen. 1991. The aluminium signal: New dimensions to mechanisms of aluminium tolerance. Dev. Plant Soil Sci. 45:703–715.

Blevins, R.L., L.W. Murdock, and G.W. Thomas. 1978. Effect of lime application on no-tillage and conventionally tilled corn. Agron. J. 70:322–330.

Bolan, N.S., M.J. Hedley, and R.E. White. 1991. Processes of soil acidification during nitrogen cycling with emphasis on legume based pastures. Plant Soil 134:53–63.

Buol, S.W., F.D. Hole, R.J. McCracken, and R.J. Southard. 1997. Soil genesis and classification. 4th edn. Iowa State University Press, Ames, IA.

Clarkson, D.T. 1984. Calcium transport between tissues and its distribution in plants. Plant Cell Environ. 7:449–456.

Cochrane, T.T., J.G. Salinas, and P.A. Sanchez. 1980. An equation for liming acid mineral soils to compensate for crop aluminum tolerance. Trop. Agric. 57:133–140.

Coleman, N.T., S.B. Weed, and R.J. McCracken. 1959. Cation exchange capacities and exchangeable cations in Piedmont soils of North Carolina. Soil Sci. Soc. Am. Proc. 23:146–149.

Coventry, D.R., and W.J. Slattery. 1991. Acidification of soil associated with lupins grown in a crop rotation in Northeastern Victoria. Aust. J. Agric. Res. 42:391–397.

Cox, F.R., F. Adams, and B.B. Tucker. 1982. Liming, fertilization, and mineral nutrition, p. 139–163. *In* H.E. Pattee and C.T. Young (eds.) Peanut science and technology. American Peanut Research and Education Society, Inc., Yoakum, TX.

Edwards, D.E., and D.B. Beegle. 1988. No-till liming effects on soil-pH, grain yield, and ear leaf nutrient content. Commun. Soil Sci. Plant Anal. 19:543–562.

Eswaran, H., P. Reich, and F. Beinroth. 1997. Global distribution of soils with acidity, p. 159–164. *In* A.C. Moniz et al. (eds.) Plant–soil interactions at low pH: Sustainable agriculture and forestry production. Brazilian Soil Science Society, Viçosa, Brazil.

Evans, C.E., and E.J. Kamprath. 1970. Lime response as related to percent Al saturation, soil solution Al, and organic matter content. Soil Sci. Soc. Am. Proc. 34:893–896.

Ferguson, I.B., and D.T. Clarkson. 1976. Simultaneous uptake and translocation of magnesium and calcium in barley (*Hordeum vulgare* L.) roots. Planta 28:267–269.

Foy, C.D. 1984. Physiological aspects of hydrogen, aluminum, and manganese toxicities in acid soils, p. 57–97. *In* F. Adams (ed.) Soil acidity and liming. ASA, Madison, WI.

Garcia, A., J.F.V. Silva, J.E. Pereira, and P.W. Dias. 1999. Rotação de culturas e manejo do solo para controlo do nematóide de cisto de soja, p. 55–70. *In* Brazilian Nematology Society (ed.) O nematóide de cisto da soja: A experiência Brasileira. Artzinger Editores, Jaboticabal, Brazil.

Godsey, C.B., G.M. Pierzynski, D.B. Mengel, and R.E. Lamond. 2007. Management of soil acidity in no-till production systems through surface application of lime. Agron. J. 99:764–772.

Havlin, J.L., J.D. Beaton, S.L. Tisdale, and W.L. Nelson. 2005. Soil fertility and fertilizers: An introduction to nutrient management. 7th edn. Pearson Education, Inc., Upper Saddle River, NJ.

Horst, W.J. 1987. Aluminium tolerance and calcium efficiency of cowpea genotypes. J. Plant Nutr. 10:1121–1129.

Huluka, G. 2005. A modification of the Adams-Evans soil buffer determination solution. Commun. Soil Sci. Plant Anal. 36:2005–2014.

Islam, A.K., D.G. Edwards, and C.J. Asher. 1980. pH optima for crop growth: Results of a flowing solution culture experiment with six species. Plant Soil 54:339–357.

Jenny, H. 1941. Factors of soil formation. McGraw-Hill Book Company, New York.

Kamprath, E.J. 1970. Exchangeable aluminum as a criterion for liming leached mineral soils. Soil Sci. Soc. Am. Proc. 34:252–254.

Kamprath, E.J. 1971. Potential detrimental effects from liming highly weathered soils to neutrality. Soil Crop Sci. Soc. Fla. Proc. 31:200–203.

Kamprath, E.J. 1978. The role of soil chemistry in the diagnosis of nutrient disorders in tropical situations, p. 313–327. *In* C.S. Andrew and E.J. Kamprath (eds.) Mineral nutrition of legumes in tropical and subtropical soils. Commonwealth Scientific and Industrial Research Organisation, Melbourne, Australia.

Kamprath, E.J. 1984. Crop response to liming on soils in the tropics, p. 349–368. *In* F. Adams (ed.) Soil acidity and liming. ASA, Madison, WI.

Kamprath, E.J., and C.D. Foy. 1985. Lime–fertilizer–plant interactions in acid soils, p. 91–151. *In* O.P. Engelstad (ed.) Fertilizer technology and use. 3rd edn. SSSA, Madison, WI.

Kamprath, E.J., and T.J. Smyth. 2005. Liming, p. 350–358. *In* D. Hillel (ed.) Encyclopedia of soils in the environment, Elsevier Ltd., Oxford, U.K.

Keinath, A.P., and R. Loria. 1989. Management of common scab of potato with plant nutrients, p. 152–167. *In* A.W. Engelhard (ed.) Soilborne plant pathogens: Management of diseases with macro- and microelements. APS Press, St. Paul, MN.

Kochian, L.V. 1995. Cellular mechanisms of aluminum toxicity and resistance in plants. Ann. Rev. Plant Physiol. Plant Mol. Biol. 46:237–260.

Liu, M., D.E. Kissel, M.L. Cabrera, and P.F. Vendrell. 2005. Soil lime requirement by direct titration with a single addition of calcium hydroxide. Soil Sci. Soc. Am. J. 69:522–530.

Lucas, R.E., and B.D. Knezek. 1972. Climatic and soil conditions promoting micronutrient deficiencies in plants, p. 265–288. *In* J.J. Mortvedt et al. (eds.) Micronutrients in agriculture, SSSA, Madison, WI.

Lund, Z.F. 1970. The effect of calcium and its relation to several cations in soybean root growth. Soil Sci. Soc. Am. Proc. 34:456–459.

Moore, D.P. 1974. Physiological effects of pH on roots, p. 135–151. *In* E.W. Carson (ed.) The plant root and its environment. University Press of Virginia, Charlottesville, VA.

Noble, A.D., M.V. Fey, and M.E. Sumner. 1988. Calcium-aluminum balance and growth of soybean roots in nutrient solutions. Soil Sci. Soc. Am. J. 52:1651–1656.

Osmond, D.L., T.J. Smyth, R.S. Yost, D.L. Hoag, W.S. Reid, W. Branch, X. Wang, and H. Li. 2007. Nutrient management support system (NuMaSS), v. 2.2. Soil Management Collaborative Research Support Program, North Carolina State University, Raleigh, NC.

Rios, M.A., and R.W. Pearson. 1964. The effect of some chemical environmental factors on cotton behavior. Soil Sci. Soc. Am. Proc. 28:232–235.

Ritchey, K.D., J.E. Silva, and D.M.G. Sousa. 1983. Relação entre teor de calcio no solo e desenvolvimento de raizes avaliado por um metodo biologico. Revista Brasileira Ciencia Solo 7:269–275.

Runge, M., and M.W. Rode. 1991. Effects of soil acidity on plant associations, p. 183–202. *In* B. Ulrick and M.E. Sumner (eds.) Soil acidity. Springer-Verlag, New York.

Sanzonowicz, C., T.J. Smyth, and D.W. Israel. 1998. Calcium alleviation of hydrogen and aluminum inhibition of soybean root extension from limed soil into acid subsurface solutions. J. Plant Nutr. 21:785–804.

Sikora, F.J. 2006. A buffer that mimics the SMP buffer for determining lime requirement of soil. Soil Sci. Soc. Am. J. 70:474–486.

Sikora, F.J., and K.P. Moore. 2008. The Moore-Sikora buffer for lime requirement determinations. Soil Sci. Soc. Am. J. 72:1163–1173.

Silva, I.R., T.J. Smyth, N.F. Barros, and R.F. Novais. 2002. Physiological aspects of aluminum toxicity and tolerance in plants, p. 277–335. *In* V.H. Alvarez et al. (eds.) Topics in soil science. Vol. 2. Brazilian Soil Science Society, Viçosa, Brazil.

Sims, J.T. 1996. Lime requirement, p. 491–515. *In* D.L. Sparks (ed.) Methods of soil analysis. Part 3. Chemical methods. SSSA and ASA. Madison, WI.

Sousa, D.M.G., and E. Lobato. 2002. Correção da acidez do solo, p. 81–96. *In* D.M.G. Sousa and E. Lobato (eds.) Cerrado: Correção do solo e adubação. Embrapa Cerrados, Planaltina, Brazil.

Sumner, M.E. 1995. Amelioration of subsoil acidity with minimum disturbance, p. 147–185. *In* N.S. Jayawardane and B.A. Stewart (eds.) Subsoil management techniques. Lewis Publishers, Boca Raton, FL.

Sumner, M.E. 1997. Procedures used for diagnosis and correction of soil acidity: A critical review, p. 195–204. *In* A.C. Moniz et al. (eds.) Plant–soil interactions at low pH: Sustainable agriculture and forestry production. Brazilian Soil Science Society, Viçosa, Brazil.

Sumner, M.E., and T. Yamada. 2002. Farming with acidity. Commun. Soil Sci. Plant Anal. 33:2467–2496.

Tang, C., and Z. Rengel. 2003. Role of plant cation/anion uptake ratio in soil acidification, p. 57–81. *In* Z. Rengel (ed.) Handbook of soil acidity. Marcel Dekker, New York.

Taylor, G.J. 1991. Current views of aluminum stress response: The physiological basis of tolerance. Curr. Top. Plant Biochem. Physiol. 10:57–93.

Thomas, G.W., and W.L. Hargrove. 1984. The chemistry of soil acidity, p. 3–56. *In* F. Adams (ed.) Soil acidity and liming. ASA, Madison, WI.

van Lierop, W. 1990. Soil pH and lime requirement determination, p. 73–126. *In* R.L. Westerman (ed.) Soil testing and plant analysis. 3rd edn. SSSA, Madison, WI.

Wright, R.J., and S.F. Wright. 1987. Effects of aluminum and calcium on the growth of subterranean clover in Appalachian soils. Soil Sci. 143:341–348.

13
Soil Fertility Evaluation

13.1 Introduction 13-1
Soil Fertility: A Modern Definition • Soil Fertility Evaluation: Purpose, General Principles, and Practices • Soil Fertility Evaluation for Agricultural and Nonagricultural Systems • Soil Fertility Evaluation: Environmental Issues

13.2 Soil Testing..13-4
Soil Testing: Historical Overview • Soil Testing: Assessing Elemental Availability in Soils • Soil Testing: Overview of the Major Components • Soil Testing Methods for Plant Nutrients • Soil Testing Methods for Soil Chemical, Physical, and Biological Properties • Soil Testing: Interpretation and Recommendations • Environmental Soil Testing

13.3 Plant Testing...13-28
Visual Diagnosis of Deficiency Symptoms • In-Field Evaluation of Plant Nutritional Status • Plant Analysis

13.4 Soil Fertility Evaluation: The Future 13-30
Precision Agriculture and Remote Sensing • New Directions and Uses for Soil Fertility Evaluation Techniques

References..13-31

J. Thomas Sims
University of Delaware

Joshua McGrath
University of Maryland

13.1 Introduction

13.1.1 Soil Fertility: A Modern Definition

Soil fertility is a scientific discipline that integrates the basic principles of soil biology, soil chemistry, and soil physics to develop the practices needed to manage nutrients in a profitable, environmentally sound manner. Historically, the study of soil fertility has focused on managing the nutrient status of soils to create optimal conditions for plant growth. Fertile, productive soils are vital components of stable societies because they ensure that we are able to grow the plants needed for food, fiber, animal feed and forage, medicines, industrial products, energy, and for an aesthetically pleasing environment. Beyond management of soil nutrients, the study of soil fertility rests on two other fundamental principles. First, optimum nutrient status alone will not ensure soil productivity. Other factors such as soil moisture and temperature, soil physical condition, soil acidity and salinity, and biotic stresses (disease, insects, weeds) can reduce the productivity of even the most fertile soils. Second, modern soil fertility practices must stress environmental protection as well as agricultural productivity; that is, we must prevent the pollution of our soils, air, and water as we strive to optimize the nutrient status of soils for plant growth.

13.1.2 Soil Fertility Evaluation: Purpose, General Principles, and Practices

Soil fertility evaluation is a central feature of modern soil fertility management. The fundamental purpose of soil fertility evaluation has always been to quantify the ability of soils to supply the nutrients required for optimum plant growth. Knowing this, we can optimize the nutrient management practices needed to achieve economically optimum plant performance. Related, equally important goals are as follows: (1) to identify other factors that reduce soil productivity (e.g., acidity, salinity, elemental phytotoxicity) and (2) to determine if the intended use of the soil may negatively impact the quality of our environment.

The general principles and practices of soil fertility evaluation are derived from or influenced by many other disciplines (e.g., soil science, plant physiology, plant genetics, crop science, plant pathology, entomology, climatology, hydrology, statistics). This chapter, and other publications (Black, 1993; Tisdale et al., 1993; Foth and Ellis, 1997; Havlin et al., 2004), describes these principles and practices in great detail. At heart, however, we seek to evaluate the fertility of a soil by observations and tests and then to use this information to predict the response of plants, and the larger environment, to our nutrient management efforts. The actual practices that constitute soil fertility evaluation include an impressive array of field and laboratory diagnostic techniques and a series of increasingly sophisticated empirical and/or theoretical models that quantitatively relate these indicators of soil fertility to plant response. Diagnostic techniques include long-standing practices, such as chemical and biological soil tests, visual observations of plant growth for nutrient deficiency or toxicity symptoms, and chemical analyses of plant tissues. New approaches include passive or active optical sensing technologies and geographic information systems (GIS) that facilitate landscape scale, site-specific assessments of soil fertility. Computerized expert systems allow us to rapidly relate these indicators of soil fertility to quantitative or qualitative assessments of plant performance (e.g., yield, composition, quality, color, health) and thus to rapidly adjust soil management practices for the most efficient use of nutrients. These advances in computing, GIS, and sensing technologies have allowed us to better describe and address the temporal and spatial variability of soil fertility.

13.1.3 Soil Fertility Evaluation for Agricultural and Nonagricultural Systems

The study of soil fertility evolved within ecosystems devoted primarily to the production of agricultural crops, particularly plants grown for food, forage, fiber, and industry. Tisdale et al. (1993) traced the relationship of soil fertility management to food production as far back as 2500 BC. The importance of soil fertility to world agriculture continues today as a spiraling world population and a diminishing arable land base create unprecedented pressures on scientists and practicing agriculturalists to produce more food per unit area of land than ever before. Advances in plant genetics and breeding and other agricultural technologies (e.g., irrigation) are increasing agricultural productivity. However, higher crop yields mean greater depletion of soil nutrient supplies, which eventually must be balanced by increased nutrient inputs to maintain the fertile soils needed by our societies. Given this, it is apparent that soil fertility evaluation will play an increasingly important role in the future of global agriculture as we seek to identify new lands that can be brought into production and to maximize the production from existing soils. Land uses other than traditional "production" agriculture also require a thorough, in-depth evaluation of soil fertility for maximum economic and environmental efficiency. Examples are horticultural systems, disturbed lands needing reclamation, and soil conservation and remediation practices.

Horticulture includes an extremely diverse range of situations where nutrients must be managed, often quite intensively. Vegetables for fresh market and processing, greenhouses and nurseries growing ornamental and vegetable plants for urban areas, golf courses, public gardens, parks, athletic facilities, and the managed landscapes around commercial businesses, governmental offices, and cities are only a few examples. The types of plants and growth media ("soilless" growth media are often used for horticultural plants) common to these situations are staggering in breadth and variability, challenging the ability of scientists to provide a systematic process for soil fertility evaluation.

Land reclamation can be equally diverse in terms of the nature of the growth media and the types of plants. Soils at land reclamation sites (e.g., surface mining, highway construction, landfills) are often highly disturbed by human activity and possess extremely unfavorable chemical and physical characteristics, including very low soil fertility. In many cases the soils used to revegetate disturbed lands are in fact subsoils, geologic materials, or low-quality soil materials imported from nearby borrow pits.

In other cases, "synthetic" soils, created from by-products such as biosolids, coal ash, composts, and poor quality soil must be used. Evaluating the fertility status of soil-like materials such as these, with reference to the growth of grasses, shrubs, and trees must often proceed in the absence of a database relating soil nutrient status and/or other soil properties to plant performance.

Related, but slightly different problems are faced when evaluating the fertility of soils used for conservation purposes, such as grassed waterways, terraces, buffer strips, constructed wetlands, and wildlife habitats. In cases such as these, where the goal is not maximum yield but a stable vegetative cover, the objective may often be low to moderate soil fertility, not agronomically optimum nutrient values. Preventing excessive growth of invasive species is also usually a goal in these settings, and many invasive plants become more competitive if soil fertility is high. Finally, the need to ensure optimum soil fertility in soil remediation programs is a new, but increasingly important aspect of soil fertility evaluation. The goals of these programs may be enhancing microbial degradation of an organic contaminant, such as an oil spill, or phyto-remediation (plant-based remediation of a contaminated soil) of an inorganic contaminant, such as cadmium (Cd), lead (Pb), or zinc (Zn) from the soil near an industrial site (Berti and Cunningham, 1994; McCutcheon and Schnoor, 2003).

13.1.4 Soil Fertility Evaluation: Environmental Issues

Environmental quality is inextricably linked with soil fertility. Just as it is essential that we manage soils to optimize plant productivity, it is equally important to manage soil fertility to avoid or minimize the pollution of our waters, atmosphere, and food chain. A wealth of scientific research has clearly documented that some essential plant nutrients can also contribute to environmental problems. Nitrogen (N), arguably the most important plant nutrient, is known to cause human and animal health problems if nitrate-N (NO_3-N) leaches to ground waters used for drinking water supplies. Ammonia-N (NH_3-N) from fertilizers and animal manures can volatilize to the atmosphere as a gas and, upon redeposition in rainfall, cause soil acidification in forest ecosystems. Nitrous oxides (NO_x) produced by the microbial conversion of NO_3-N in soils to gaseous forms of N (NO, N_2O) have been implicated in ozone depletion and global warming. Eutrophication of surface waters, defined as "...enrichment of surface waters by plant nutrients... regarded as a form of pollution as it restricts the potential uses of impacted water bodies" (Foy and Withers, 1995), can be caused by phosphorus (P) and N entering these waters in soil erosion and runoff, or, in the case of NH_3-N by aerial deposition in rainfall. Soil salinity problems can result when the plant nutrients calcium (Ca), magnesium (Mg), potassium (K), sulfate-sulfur (SO_4-S), and chloride (Cl) or nonessential elements such as sodium (Na) accumulate to excess in soils, particularly in arid regions.

Efforts to enhance soil fertility by use of society's wastes or by-products as nutrient sources can also directly or indirectly affect environmental quality. Modern industrialized societies face increasing pressures to land apply wastes to avoid the costs and undesirable environmental impacts of landfilling and incineration. Lesser developed countries often must use wastes and wastewaters as fertilizers and for irrigation because of a lack of resources, equipment, and infrastructure. For example, the use of municipal biosolids (e.g., sewage sludge, compost) as soil amendments is beneficial because it recycles plant nutrients while building soil organic matter. However, this practice is carefully regulated in most countries because biosolids may contain nonessential elements that can be phytotoxic or can accumulate in the food chain, impacting human, animal, or ecosystem health. Additionally, most land application programs for biosolids (and animal manures) base application rates on the amount of N needed for optimum crop yields. However, the unfavorable N:P ratio in most organic wastes, relative to the N:P ratio in harvested crops, usually means that P accumulates in waste-amended soils to concentrations much higher than required for crop production. This creates an environmental dilemma because the beneficial use of organic wastes as N sources causes the build-up of P in soils and the likelihood that significant amounts of P will be lost to surface waters in erosion and runoff (see Section 11.2). Other municipal, agricultural, or industrial by-products that have been shown to have beneficial effects on some aspect of soil fertility and/or productivity often create similar dilemmas. Materials such as paper-mill sludges, municipal composts, wood ashes, coal combustion by-products (e.g., flue gas desulfurization gypsum, coal fly ash) are all sources of plant nutrients. However, if managed improperly, they can also create unfavorable soil pH values, excessive soluble salts, and cause microbial or chemical immobilization of plant nutrients.

Soil fertility evaluation is more complex today because of the need to balance productivity and environmental protection for a wider and more diverse range of land uses, as illustrated conceptually in Figure 13.1. This chapter presents a critical analysis of

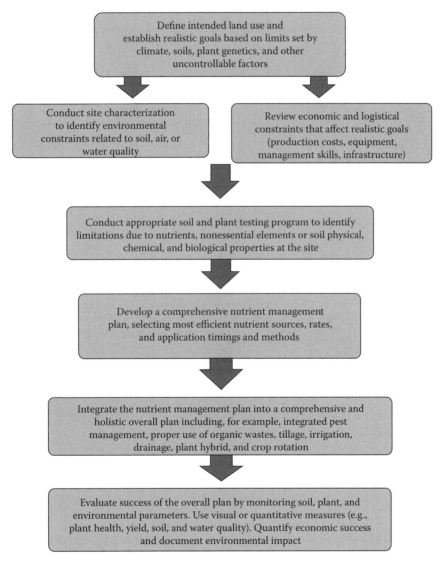

FIGURE 13.1 Conceptual summary of the process of soil fertility evaluation.

the major techniques used in soil fertility evaluation—soil and plant testing, focusing not only on current practices but on the recent advances in science, technology, and interpretive philosophies that will shape soil fertility evaluation in the future.

13.2 Soil Testing

13.2.1 Soil Testing: Historical Overview

Soil testing is defined as "…rapid chemical analyses to assess the plant-available nutrient status, salinity, and elemental toxicity of a soil…a program that includes interpretation, evaluation, fertilizer and amendment recommendations based on results of chemical analyses and other considerations" (Peck and Soltanpour, 1990). The use of soil testing represents perhaps the most significant practical application of our knowledge of soil science to land use management and should be viewed fundamentally as an interpretive process, not simply as a series of laboratory methods. The purpose of soil testing is to provide a quantitative basis to guide soil management decisions, usually, but not always, for agricultural systems (e.g., agronomy, horticulture, silviculture) where plant growth and performance (yield) are the ultimate measures of success. Soil testing has applications to other systems as well; particularly to those where the goal is protection of human health or the environment. Advancing soil testing in nonagricultural systems requires a thorough review of the technical aspects of soil testing and the basis by which we evaluate its success, since plant production is not the issue of greatest importance.

Early efforts at soil testing began in the nineteenth century but systematic studies of the relationship between soil testing and plant growth did not occur until the 1920s. As with soil fertility in general, soil testing evolved within an agricultural setting; hence, its principles and practices have been markedly influenced by the needs and interests of production agriculture. The original purpose of soil testing was rather straightforward—to determine if soils were deficient in nutrients and, based on field studies of crop response to fertilizers and manures, to make recommendations for the nutrient rate, source, and method that would most effectively overcome the limitations caused by the nutrient deficiency. Organized soil testing programs began to be established in the United States in the late 1940s as chemical soil tests were developed that could rapidly assess nutrient availability in soils. The broader infrastructure of soil testing originated first with universities, where research and Extension scientists conducted the laboratory, greenhouse, and field studies needed to verify the accuracy and reliability of soil test methods. Following close upon these studies was the educational aspect of soil testing—publications that described how to test soils and report forms that provided users (primarily farmers) with results and recommendations. Today soil testing is an international activity operating in both the public and private sectors. In the United States, numerous regional and national soil testing organizations, often including public and private soil testing laboratories, are in place to review and revise soil testing methods, update soil test interpretations and recommendations, and ensure the analytical proficiency of soil testing laboratories. Information on the activities and publications of these vitally important advisory and overview groups is available from the web sites of the Soil Science Society of America (SSSA) (https://www.soils.org/), the Soil and Plant Analysis Council (http://www.spcouncil.com/), and the North American Proficiency Testing Program (http://www.naptprogram.org/).

13.2.2 Soil Testing: Assessing Elemental Availability in Soils

One of the fundamental tenets of soil testing is that only a proportion of the total quantity of an element in a soil will be available for assimilation by a biological organism. This means that measuring total elemental concentration in soils is usually of little value. Instead we must have testing methods that can extract (complex, dissolve, desorb, exchange, hydrolyze) a percentage of the total soil nutrient pool that is proportional to the quantity that will become available to the organism of interest during the time period of concern. The term *labile* is often used to describe the chemical and biological forms of an element that are in rapid equilibrium with the soil solution and are thus most likely to be available for biological assimilation. A similar term, *bioaccessible*, is often used when the primary interest is in testing contaminated soils for potentially toxic elements, particularly metals such as arsenic, copper, chromium, and lead, where the main concern is human exposure via ingestion or dermal contact. Much of soil testing research has focused on the development of chemical extracting solutions that can selectively remove labile and bioaccessible elements from the soil in a rapid and reproducible manner.

13.2.2.1 Assessing the Availability of Essential and Nonessential Elements

Insofar as soil fertility is concerned we are most interested in the availability of the 13 elements known to be essential for plant growth that are primarily obtained from soils (N, P, K, Ca, Mg, S, B, Cl, Cu, Fe, Mn, Mo, Zn) as opposed to those obtained from air (C, O) and water (H, O). When we consider human, animal, and environmental health, we have two other concerns. First is the potential environmental impact of essential plant nutrients found in soils that may be transported to other ecosystems (e.g., soil P movement to surface waters). Second is the fate of essential and nonessential elements added to soils in fertilizers or other soil amendments (biosolids, composts, industrial by-products, manures) that may impact human health should they enter the food chain, drinking water supplies or natural ecosystems or become toxic to plants if they accumulate to high concentrations in soils. The nonessential elements of greatest concern are aluminum (Al), arsenic (As), barium (Ba), cadmium (Cd), chromium (Cr), lead (Pb), mercury (Hg), nickel (Ni), and selenium (Se). Potentially toxic essential elements are Cu, Mo, and Zn.

The most common means to assess the availability of elements in soils is with chemical soil test extractants, which are typically dilute solutions or mixtures of acids, bases, salts, and chelates. Biological techniques (e.g., bioassays based on microbial growth) have been used to a limited extent but are generally too expensive and time-consuming for routine use. The most effective concentrations and relative proportions of the reagents in a chemical extracting solution were usually determined empirically by comparison of the amount of an element extracted from the soil with some type of biological response (usually yield) by a target organism (usually a plant). The nature and diversity of chemical solutions used in soil testing is illustrated by comparing the Mehlich 1 soil test, a dilute mixture of two strong acids ($0.05\,M$ HCl + $0.0125\,M$ H_2SO_4), the ammonium bicarbonate-diethylenetriaminepentaacetic acid (AB-DTPA) soil test, a dilute combination of a base ($1\,M$ NH_4HCO_3) and a chelating agent ($0.005\,M$ DTPA), and the Bray P_1 soil test, a dilute mixture of a strong acid ($0.025\,M$ HCl) and a complexing ion ($0.03\,M$ NH_4F).

Soil testing extractants developed for plant nutrients have also been used, with some success, to assess the risk of plant uptake of nonessential elements (O'Connor, 1988, Risser and Baker, 1990; Gaskin et al., 2003; Sukkariyah et al., 2005) or loss of these elements in surface runoff (Zhang et al., 2003). Other soil testing methods are also used for these elements, although not to measure biological availability. Examples include a method to measure total sorbed metals that is often used to monitor the accumulation of elements in the soil up to some defined regulatory limit (EPA-3050; U.S. Environmental Protection Agency, 1986) and the toxicity characteristic leaching procedure (TCLP), sometimes used to determine if a soil is sufficiently polluted with an element or organic compound to be considered a hazardous waste. Recent advances in environmental soil testing for some metals (As, Cd, Pb, Zn) have used extracting solutions that simulate the biological activity within the human digestive system, referred to as the physiologically based extraction test (PBET) (Ruby et al., 1996; Fendorf et al., 2004; Brown et al., 2007).

13.2.2.2 Influence of Soil Properties and Environmental Conditions on Elemental Availability

One of the major challenges in the evaluation of soil fertility is the need to develop tests (chemical or otherwise) that can estimate nutrient availability in soils of widely differing biological, chemical, and physical properties. Decades of soil science research have proved that the availability of elements in soils depends both on soil properties that are essentially permanent (e.g., the percentages of sand, silt, and clay, oxides of Fe and Al, carbonates, and organic matter) and on those that are more sensitive to natural and anthropogenic inputs (pH, cation exchange capacity [CEC], oxidation–reduction status). Elemental availability can also be markedly affected by broader soil properties (drainage class, nature of soil horizons, slope) and by the soil environment (aeration, moisture, temperature). Therefore, if we wish to estimate the present availability of essential and/or nonessential elements in a given soil, and how this availability will change following some management practice, we must also have a good understanding of the basic properties of that soil and of the environmental conditions likely to be present during the time interval of interest.

Two approaches are used to integrate soil properties and environmental conditions with elemental availability provided by soil testing. First, we can directly test soils for some of these properties and use this information to modify our assessment of biological availability. The most common examples of this are soil pH and organic matter content; other parameters (e.g., CEC, texture, oxides, carbonates) are usually too time-consuming to measure routinely. Second, we can obtain information on soil properties that are difficult to measure and on environmental conditions simply by asking a few key questions of the individual submitting the soil sample for analysis. Knowledge of geographic location, soil series, drainage class, slope, and historical information on previous soil management practices (e.g., fertilization, liming, crop rotation) can be invaluable when evaluating soil productivity and thus crop nutrient requirements. Our ability to integrate information such as this into management recommendations has markedly improved due to the now widespread use of computers in soil testing and with the increased ease of using GIS that are capable of "layering" different data bases to provide a more holistic view of the relationship between soil fertility and land use.

13.2.3 Soil Testing: Overview of the Major Components

All modern soil testing programs have four basic components: (1) soil sample collection, handling, and preparation; (2) soil analysis; (3) interpretation of analytical results; and (4) recommendations for action. For soil testing to be successful, each component must be conducted properly, keeping the overall objective (e.g., plant production or environmental protection) firmly in mind, and with an awareness of the potential sources of error that can occur at each step.

13.2.3.1 Soil Sample Collection and Handling

Collection of a sample that is representative of the entire area of interest, whether it is a farm field, a lawn or garden, or a severely disturbed soil at a construction or mining site, is the most important step in any soil testing program. Proper handling of the sample, once collected, is also important to avoid contamination or changes in elemental concentrations due to improper storage and/or the use of incorrect techniques to prepare the soil sample for analysis (drying, grinding, sieving). An effective soil sampling and handling program must be based on an understanding of the natural and anthropogenic sources of soil variation, the proper method of sample collection (depth, time of year, sampling tools), and the sources of error in sample handling and preparation.

13.2.3.1.1 Understanding and Compensating for Variability

A high degree of natural variability in soil chemical and physical properties can exist even within a very small area. The origin of this variability is the soil forming process in which parent material, climate, relief (topography), biota, and time act together to produce soils that are fundamentally different, not only in terms of topsoil properties, but throughout the entire soil profile. Sample collection should reflect this natural variability, to the extent that it is likely to significantly influence the intended use of the soil. In some cases the differences between adjacent soil series in a large field may be pronounced and require not only different soil samples, but entirely different soil and crop management practices. In other situations differences are minor and collecting additional samples and/or altering management practices is not economically justifiable. Information on the spatial distribution of soil series in an area is available in soil surveys such as those published and now available on the internet (http://soils.usda.gov/survey/printed_surveys/) by the U.S. Department of Agriculture's Natural Resources Conservation Service (USDA-NRCS). Internet access to soils information, at least in the United States has increased significantly in the past decade and much useful information is now available at the NRCS "Soil Data Mart" (http://soildatamart.nrcs.usda.gov/). The Soil Data Mart can quickly allow users to determine where soil tabular and spatial data are available and then to download these data and a variety of reports for individual soil survey areas. Consulting soil surveys is a first step in identifying natural sources of variability to consider in a soil sampling program. A simple follow-up to reviewing these soil surveys is to visually inspect the areas of interest, particularly during periods of plant growth or major seasonal changes in climate, to determine how natural soil variability is affecting land use. New technologies are now available to integrate soil survey information and qualitative data on soil spatial variability into GIS-based land use planning including the use of remote sensing and global positioning systems (GPS), to link soil sample location to soil series. For agronomic crops yield, monitoring devices can now be installed on harvesting equipment, providing a spatially based data set that relates soil fertility, soil series, and plant performance (see Section 13.4). Natural variation can often be overshadowed by the variability in soil fertility caused by human activities. Many soil management practices, such as the method of fertilizer or organic waste application, the tillage method, land leveling for irrigation, terracing, and even the year-to-year selection of what plants will be grown and where, can produce marked differences in soil nutrient status and other soil properties. These differences can be even more pronounced in nonagricultural settings, such as land reclamation projects, where soil disturbance can be severe and by-products may be used as soil amendments at unusually high rates. It is also important to remember that anthropogenic as well as natural soil variability can occur in three dimensions. For example, changing from conventional tillage to no-tillage can result in surface accumulation of P and a decrease in pH at the soil surface as well. Therefore, such factors should be accounted for when designing a soil sampling program. Similarly, nutrient management practices that involve subsoil tillage or injection of soil amendments (e.g., lime, fertilizer, manures) can extend spatial variability below the topsoil, a factor that should not be ignored in the soil sampling process.

Many other examples could be cited to make the point that spatial variability in soil properties, natural or anthropogenic, is inevitable. What is more important, however, are the practices that can be used to compensate for any known source of variability. James and Wells (1990) suggested that soil sample collection basically occurs under either uniform or nonuniform conditions and that sampling techniques should reflect this. Uniform fields are those that are reasonably similar in physical properties (e.g., slope, aspect, drainage, soil series,) and management practices (e.g., crop rotation, fertilizer/manure management history, tillage, irrigation). In many agricultural settings this is a rather common scenario. The proper approach to use for uniform fields is to collect a random, composite sample. This is normally done by following a "zig-zag" path across the area to be sampled, collecting enough soil samples to minimize the influence of any localized nonuniformity (Figure 13.2). Typically this involves collecting about 25–30 separate soil cores from a uniform area <8 ha (~20 acres), ensuring that field corners and edges are included in the sample. Individual soil cores are combined into one composite sample that represents the entire area by crushing and mixing and the composite sample is submitted for analysis.

Nonuniform fields were defined by James and Wells (1990) as those with a high degree of either macro-variation (significant variation between sample points separated by >2 m) or meso-variation (significant variation between sample points separated by 0.05–2 m). For areas with high macro-variation a nonrandom sampling process must be followed to characterize the average value for the soil properties of interest and to understand the spatial location of extreme values. This prevents skewing of average soil test results for an area by samples that are extremely high or low in some soil property. It also allows for more site-specific application of fertilizers, organic wastes, lime, etc. at appropriate rates, thus avoiding under- or over-application of these soil amendments. Nonrandom sampling requires a large number of soil samples and is usually done by establishing a field sample grid with a spacing of 15–30 m between grid intersection points. A composite soil sample is collected at each grid point by combining 8–10 soil cores collected from a 1 m diameter circle placed around the sample point. Grid spacing is a function of the intended land use and the anticipated degree of macro-variation. Note that a 15 and 30 m grid spacing will result in about 45 and 12 composite soil samples per ha, respectively, considerably more than the one sample per 8 ha associated with random sampling. Grid sampling has become more common in recent years with the advent of GIS and GPS technology and its use in precision agriculture (see Section 13.4). For areas with significant and identifiable meso-variation, as might occur when fertilizers are consistently placed in bands across a field, a more intensive random sampling pattern can be used to ensure that the average value is not skewed by either the very high value present in the fertilizer band or the lower value for the bulk soil located between the bands. The number of soil cores to be collected in these situations will usually be four or five times as great as those needed for a random, composite sample (i.e., as many as

Soil Fertility Evaluation

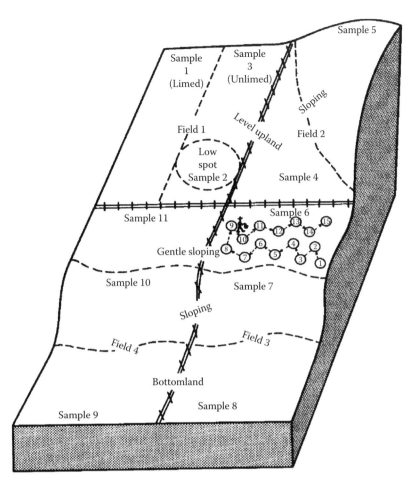

FIGURE 13.2 Illustration of soil sampling practices for uniform fields [Courtesy of Nebraska Agricultural Extension Service].

1–2 soil samples per ha). Finally, it should be noted that random and nonrandom sampling techniques are not appropriate for all aspects of soil fertility evaluation, particularly true for elements that are more mobile in the soil profile (e.g., NO_3-N and SO_4-S).

13.2.3.1.2 Soil Sampling Methods

The most important factors to consider in the soil sampling process are the depth of sampling, the time of year to sample, and the frequency of sampling. The depth to sample depends on the objectives of the soil fertility evaluation, including the crop to be grown, the nutrient of interest, the type of soil test to be performed, and the basis for resulting fertilizer recommendations. Soil samples for routine soil tests of plant nutrient availability and lime requirement are usually obtained from topsoils (0–20 cm depth). There are, however, a number of important exceptions to this general rule such as when soil tests are conducted for mobile soil nutrients (NO_3-N and SO_4-S), or to estimate the effect of pH on herbicide activity in no-till cropping systems, or for soil fertility evaluation for shallow or deep-rooted crops, or to monitor the leaching of potential pollutants downward in the soil profile. Shallow soil sampling is most often recommended for conservation tillage systems where nutrients and lime are not incorporated with the soil by plowing, for permanent pastures and/or turf where root systems rarely extend below a depth of 10 cm, and to estimate the potential for P loss in erosion or runoff in watersheds where eutrophication is an important environmental concern (sample to 0–5 cm). Finally, it is important to review the source of soil test interpretations or fertility recommendations before deciding on a soil sampling depth. In the United States, recommendations often originate from a local or regional Land Grant University and different universities may have calibrated the same tests to different sampling depths. The recommended depth for subsoil sampling varies with the intent of the test. For example, subsoil testing for NO_3-N in arid regions may require that samples be taken to as deep as 1 m (60 cm is often recommended). Testing sandy soils in humid regions for SO_4-S may require a subsoil sample from the B horizon (usually 20–60 cm) where SO_4-S can accumulate by sorption to clays and Fe/Al oxides. Care should be taken to minimize contamination of the subsoil sample by topsoil when collecting subsoil samples, which can seriously influence results and subsequent recommendations.

The proper time of year to collect a soil sample and the frequency of soil sampling should also be considered. Soil samples can be collected at any time of the year when the ground is not frozen, although the ideal time is shortly before making a land management decision because this gives the most current indication of soil properties (fertility, pH). In general, for most agricultural systems routine soil tests for lime and fertilizer recommendations are normally collected 3–6 months prior to planting a crop. This usually provides sufficient time for management decisions to be made and implemented in a timely manner. For example, if a soil test should recommend that limestone is needed to correct a problem with soil acidity, it is important to know this several months in advance of planting because of the time required for the limestone to react in the soil and raise the soil pH to the desired value. Soil testing well in advance of planting also allows time to change the plants to be grown if soil test results indicate that growing conditions are inappropriate for the plant specified. An exception is the pre-sidedress nitrate test for corn (PSNT) (see Section 13.2.4.4), which must be collected when corn plants are about 30 cm in height.

The frequency of soil testing varies somewhat with intended land use, such as the plants to be grown and the nutrient management practices required. Ideally, soils should be sampled at the same time of year (e.g., spring or fall) and at no more than 2–3 year intervals. Sampling at the same time of year minimizes the effect of seasonal variations on soil pH, which decreases during the summer as soluble salts increase in the soil from fertilization and mineralization of organic matter. Some studies have shown similar decreases in P and K during the year, with lower values reported in the fall than spring.

13.2.3.1.3 Soil Sample Handling

Proper handling of soil samples is necessary to prevent contamination and to minimize extreme changes in elemental concentrations or pH between the soil in the field and the sample that is analyzed in the laboratory. Handling includes the actual process of collecting soil cores, the mixing needed to prepare a composite sample, transporting the sample from the field to the laboratory, and the drying, grinding, sieving, and storing that occurs in the laboratory. A variety of tools are available to collect soil cores, including handheld soil probes and augurs, hydraulic soil coring devices, shovels, and hand trowels. The major consideration is to use sampling and mixing tools made of materials that will not contaminate the soil (e.g., stainless steel or plastic). Once collected, individual soil cores from the area of interest are composited by combining them in a clean container, preferably made of plastic to avoid contamination from painted or galvanized metal surfaces, mixing well, and removing a subsample that represents the entire mixture. During the sampling process avoid contamination from dirty or rusty sampling and mixing devices, fertilizer materials, galvanized metals (source of Zn), and paper bags that may contain boron (B). Soil samples should be delivered to the testing laboratory as soon as possible after collection to minimize any changes in elemental concentration that may take place prior to drying. The greatest concern is with NO_3-N, which can increase when soils are kept for extended periods in a warm, moist state (from mineralization of organic N and nitrification of exchangeable NH_4-N), or decrease when soils are maintained in a warm, wet state (from denitrification of NO_3-N). Once received by the laboratory, soil samples are normally dried at low temperatures (ambient to 50°C; avoid higher temperatures), ground and sieved, typically to pass a 2 mm screen. The samples are then ready for analysis.

13.2.3.2 Soil Analyses: Overview of Chemical, Physical, and Biological Methods

Chemical analysis of soils is based on the principle that chemical solutions can rapidly, reproducibly, and inexpensively assess soil nutrient supplying capacity and other soil properties that affect plant growth (pH, soluble salts, organic matter). The most common chemical methods used are extraction and equilibration; others include titration (for acidity) and chemical or thermal oxidation (for organic matter).

Chemical extraction is almost always conducted with dried, ground, and sieved soil samples. For most soil tests the process involves scooping or weighing a small representative portion of the soil sample (from 1 to 10 cm^3 or g) into an extracting vessel (flask, beaker, extraction bottle), adding a known volume of chemical extracting solution (from 10 to 100 mL), shaking rapidly for a short time period (from 5 to 30 min), filtering the sample, and analyzing the filtrate for the elements of interest. For example, extraction by the Mehlich 3 soil test, one of the most widely used extracting solutions in the United States involves scooping 2.5 cm^3 of air-dried, sieved (2 mm) soil into a 100 mL plastic extraction bottle, followed by the addition of 25 mL of the Mehlich 3 solution (0.2 M CH_3COOH + 0.25 M NH_4NO_3 + 0.015 M NH_4F + 0.013 M HNO_3 + 0.001 M EDTA). The soil suspension is shaken for 5 min on a reciprocating shaker, filtered into 25 mL plastic vials, and analyzed by either atomic absorption spectrophotometry (AAS), inductively coupled plasma atomic emission spectroscopy (ICP-AES), or colorimetry to determine the concentration of each element in the soil extract. The principles of these instrumental methods are described in the Soil Science Society of America (SSSA) monograph *Methods of Soil Analysis: Chemical Properties*, Part 2 (Sparks, 1996).

Soil chemical properties are not always assessed by extraction methods. Another common chemical testing method is equilibration, in which a solution is added to the soil, the resulting suspension is shaken (or sits) for a short time period, and some property of the soil suspension is measured. This approach is used to measure soil pH and lime requirement, by use of an electrode that determines the chemical activity of hydrogen ions in a soil–water suspension (pH) or in a suspension of soil and a chemical buffer solution (lime requirement). A similar approach is used to measure soil soluble salts where a conductivity cell or bridge measures the specific conductance of a soil–water suspension. Some soil testing laboratories use titrimetric techniques to measure soil acidity by first extracting soil with a neutral salt solution (e.g., 1 M KCl), followed by titration of the acidic extractant with a dilute base (e.g., 0.1 M NaOH). Soil organic matter tests originally used wet chemical oxidation to estimate organic matter from the amount of carbon in soils that could be oxidized by $K_2Cr_2O_7$. Environmental concerns about the use and disposal of chromium (Cr) have caused most laboratories to now use high temperature oxidation (360°C) to estimate organic matter from weight loss upon ignition of a soil sample.

Most soil testing methods have been standardized by regional and national soil testing organizations and the SSSA. Consequently, numerous publications provide detailed, recommended methods (e.g., soil drying, grinding, sieving; preparation of extracting solutions; proper soil:solution extraction ratio, shaking rate, time and filtration method; and appropriate instrumental method). These publications are both national (Sparks, 1996; Jones, 1999; USDA-NRECS, 2004; Carter and Gregorich, 2007) and regional (SRIEG-18, 1992; NEC-67, 1995; NCR-13, 1998) in scope. Closely following these recommended methods is essential to accurately assess soil fertility.

13.2.3.3 Interpretation of Soil Testing Results

Interpretation of soil testing results may be defined as quantitatively relating the results of a soil analysis to the probability that a soil management activity will have the desired result. For soil fertility evaluation this means using soil test results to accurately predict crop yield without nutrient addition and the probability of a profitable plant response when fertilizers or other soil amendments (e.g., lime, manures, composts) are added. Interpreting soil test results for land uses where the desired result is not economically optimum plant yield (e.g., aesthetic value of ornamental plants in a horticultural setting; biodegradation of an organic pollutant in a bioremediation system; inactivation of a heavy metal in a polluted soil) requires that we are able to identify measures of success that can be quantitatively related to a soil test value and to soil management practices.

Soil test interpretation for agricultural systems begins with research that proves a statistical correlation exists between a soil test value and some aspect of plant response (e.g., nutrient concentration in the plant, crop yield). If the soil test and plant response are correlated the next step in interpretation is soil test calibration. The main goal of soil test calibration is to rate soils in terms of the probability that nutrient additions will be profitable (e.g., to divide the population of soil test values into responsive and nonresponsive categories; Table 13.1 and Section 13.2.6).

13.2.3.4 Recommendations Based on Soil Testing

Soil test recommendations must integrate soil test data with many other factors such as climate, economics, soil and crop management practices, management ability of the soil test user, and any considerations imposed by environmental protection. Soil test values tell us nothing about these other factors, instead we must rely on other sources of information, basic scientific principles, practical experience, and professional judgment to integrate them, with soil test results, into a reasonable recommendation.

The recommendation process starts with knowledge of the intended land use and factors that affect recommendations vary in importance with land use. Production agriculture recommendations usually include the rate, timing, and method of application of fertilizers, liming materials, and other soil amendments such as biosolids and manures. In this situation, where the goals are economically optimum crop yields and minimal environmental impact, recommendations are based on (1) current soil

TABLE 13.1 Generalized Soil Test Categories and Recommendations Based on Crop Response and Environmental Impact

Category Name	Category Definition	Recommendations
Crop response		
Below optimum (very low, low, medium)	The nutrient is considered deficient and will probably limit crop yield. There is a high to moderate probability of an economic yield response to adding the nutrient	Nutrient recommendations are based on crop response and will build soil fertility into the optimum range over time. Starter fertilizer may be recommended for some crops
Optimum (sufficient, adequate)	The nutrient is considered adequate and will probably not limit crop growth. There is a low probability of an economic yield response to adding the nutrient	If soils are tested annually, no nutrient additions are needed for the current crop. For other than annual soil testing, nutrient applications are often recommended to maintain the soil in the optimum range. Starter fertilizer may be recommended for some crops
Above optimum (high, very high, excessive)	The nutrient is considered more than adequate and will not limit crop yield. There is a very low probability of an economic yield response to adding the nutrient. At very high levels, there is the possibility of a negative impact on the crop if nutrients are added	No nutrient additions are recommended. At very high or excessive levels, remedial action may be needed to prevent phytotoxicity or environmental problems
Environmental response		
Potential negative environmental impact (very high, excessive)	Soils testing at this level or above have higher potential to cause environmental degradation and should be monitored closely. The likelihood of environmental problems depends on other site-specific characteristics (e.g., slope, hydrology, rainfall). This soil test level is independent of the crop response categories above and may be above or below the optimum level based on crop response	If other site factors minimize environmental impact, some nutrient additions may be recommended according to crop response guidelines. If other site factors indicate a potential environmental impact is likely, nutrient additions including starter fertilizer are not recommended. Remedial actions may be required to protect the environment

Source: Adapted from Beegle, D. 1995. Interpretation of soil test results, p. 84–91. *In* J.T. Sims, and A.M. Wolf (eds.) Recommended soil testing procedures for the Northeastern United States. Bull. no. 493. University of Delaware, Newark, DE.

test values and any other soil characteristics that affect a recommendation, such as soil drainage class; (2) the crop to be grown and the realistic yield potential for that crop; (3) soil test calibration data that indicate the degree of response expected to any soil amendment; (4) the source of nutrients and/or lime to be used and any restrictions on application method and timing that exist; (5) soil management history, such as the use of animal manures, biosolids, or growth of a leguminous cover crop; and (6) environmental parameters that may require modification of standard recommendations, such as soil leaching potential and depth to ground water and/or soil erosion and runoff potential and proximity to surface waters sensitive to eutrophication.

Not surprisingly, given the rather subjective nature of many components of the soil test recommendation process, several different recommendation philosophies have evolved for agricultural systems. These philosophies, contrasted in detail in Section 13.2.6, vary mainly in the approach used to maintain soil fertility in the range needed for optimum crop production. Some recommend more liberal and more frequent applications of nutrients to ensure that nutrient deficiency does not limit crop production. Others are more conservative, relying heavily on soil tests as the basis for any nutrient addition, often recommending that no nutrients be added when soils are in the optimum or excessive range. The decision on which philosophy to follow is primarily based on the current economics of production. However, for nutrients known to have environmental impacts (N, P), environmental-based recommendations to protect ground and surface water quality, even if the result is suboptimum crop performance, are now mandated or at least being considered in many regions.

13.2.4 Soil Testing Methods for Plant Nutrients

The intent of this chapter is to briefly describe, for each plant nutrient, (1) the major soil processes affecting the nutrient that are relevant to soil testing, (2) current soil test methods and recent advances in research, and (3) any factors that should be taken into consideration when conducting soil tests or interpreting soil test results. Other chapters in the *Handbook of Soil Science* provide reviews of nutrient cycling and should be consulted as necessary.

13.2.4.1 Phosphorus

13.2.4.1.1 Soil Phosphorus

Phosphorus in soils occurs in inorganic and organic forms that primarily originated from the weathering of calcium phosphate minerals, the most common being apatite ($(Ca_{10}(X)_2(PO_4)_6$, where X is either F, Cl, OH, or CO_3). Approximately 30%–50% of the total P in most soils is found in the organic fraction. Chemical weathering of soil minerals and mineralization of organic matter release P into the soil solution where it exists in very low concentrations (0.003–0.3 mg P L^{-1}, average of ~0.05 mg P L^{-1}) almost exclusively as phosphate ions. In acid soils, dihydrogen phosphate ($H_2PO_4^-$) is the prevalent ion while hydrogen phosphate (HPO_4^{2-}) is the main phosphate species found above pH 7.2. Once in solution, phosphate can be assimilated by biological organisms; sorbed (chemically bound) to soil colloids such as clays, oxides of Fe and Al, and $CaCO_3$; precipitate as an insoluble compound by reaction with Al and Fe in acid soils or Ca in calcareous soils; or be lost in surface or subsurface runoff. Phosphorus can be released back into the soil solution by mineralization of organic compounds, desorption of P from soil colloids, or the dissolution of solid phases of P. Plant available forms of soil P are mainly those found in the soil solution, sorbed by soil colloids, or precipitated as relatively soluble minerals. Mineralization of organic P and dissolution of very stable P minerals proceeds too slowly in most soils to provide a large percentage of the available P required for plant growth. Desorption (and dissolution) of P occurs when plant uptake decreases the P concentration in the soil solution, thermodynamically favoring the release of sorbed or solid phase P. Soil tests for P were designed to simulate this process, extracting P from sorbed forms and metastable precipitates by four processes: (1) acid dissolution, (2) anion exchange, (3) cation complexation, and (4) cation hydrolysis (Kamprath and Watson, 1980). For example the Bray P_1 soil test, developed for slightly acid and neutral soils where Al-P and Ca-P are major sources of plant available P, is a mixture of 0.025 M HCl + 0.03 M NH_4F. The F^- ion forms a strong complex with Al^{+3} in solution, causing dissolution of Al-P compounds, and the dilute HCl dissolves a proportion of the Ca-P and lesser amounts of Al-P and Fe-P. Together, NH_4F and the HCl cause labile pools of soil P to release P into solution, similar to what would occur in soils in response to the depletion of soil solution P by plant uptake. In near neutral soils, the F^- also reacts with Ca to form CaF_2, enhancing dissolution of Ca-P. The Olsen soil test (0.5 M $NaHCO_3$, pH 8.5) functions similarly in calcareous soils where HCO_3^- precipitates soluble Ca as $CaCO_3$ causing release of P from $CaHPO_4$. Soil P tests may also extract organic P either by direct acid hydrolysis of organic P esters or by enhancing P release from organo-metallic complexes (Al-OM, Fe-OM). Note that soil tests based on soluble P alone are not an accurate index of a soil's P supplying capacity because they do not reflect the capacity of the solid phase to replenish the solution phase as P uptake occurs. Nor would soil tests using very strong chemical reagents that cause extensive dissolution of mineral phases or oxidation of organic matter because they would overestimate plant available P.

Current soil testing methods for P: the soil testing methods commonly used for P today are shown in Table 13.2 and were reviewed extensively by Fixen and Grove (1990). The Bray P_1, Ca-lactate, Morgan, and Olsen soil tests are only used to extract P while Mehlich 1, Mehlich 3, and AB-DTPA (NH_4HCO_3 + DTPA) are multielement soil tests used to extract P, K, Ca, Mg, and some micronutrients (Cu, Fe, Mn, Zn). As a general rule, the acidic extractants (Bray, Mehlich 1, Mehlich 3, Morgan) are used on acid soils and the alkaline extractants (AB-DTPA, Olsen) on calcareous soils. A number of studies, however, have shown that the Olsen P test may be an accurate test for a broader range of soils.

TABLE 13.2 Summary of Soil Testing Methods Currently Used for Phosphorus (P)

Soil Test	Extractant Composition	Comments, Critical Values,[a] and Sources
AB-DTPA	M NH_4HCO_3 + 0.005Af DTPA—pH 7.5	Multinutrient extractant primarily used with alkaline soils. Critical value: ≥8 mg kg^{-1} (Soltanpour and Schwab, 1977)
Bray P_1	0.03 M NH_4F + 0.025 M HCl	Used only to extract P on acid soils with moderate CEC. Critical value: ≥30 mg kg^{-1} (Bray and Kurtz, 1945)
Mehlich I	0.05 M HCl + 0.0125 M H_2SO_4	Multinutrient extractant used on acidic, low CEC soils. Critical value: ≥25 mg kg^{-1} (Mehlich, 1953)
Mehlich III	0.2 M CH_3COOH + 0.25 M NH_4NO_3 + 0.015 M NH_4F + 0.013 M HNO_3 + 0.001 M EDTA—pH 2.5	Multinutrient extractant suitable for wide range of soils. Well correlated with Bray P_1, Mehlich I, and Olsen P. Critical value: ≥50 mg kg^{-1} (Mehlich, 1984)
Morgan and modified Morgan	Morgan: 0.7 M $NaC_2H_3O_2$ + 0.54 M CH_3COOH + —pH 4.8 modified Morgan: 0.62 M NH_4OH + 1.25 M CH_3COOH—pH 4.8	Multinutrient extractant primarily used in the northeast United States for acid, low CEC soils. Not suitable for calcareous soils. Critical value: ≥4–6 mg kg^{-1} (Morgan, 1941)
Olsen	0.5 M $NaHCO_3$—pH 8.5	Originally developed as P extractant for alkaline soils in the western United States; now also used for acid and neutral soils. Critical value: ≥10 mg kg^{-1} (Olsen et al., 1954)
Egner	P-CAL: 0.01 M Ca lactate + 0.02 M HCl P-AL: 0.10 M NH_4 lactate + HOAc—pH 3.75	Multinutrient extractant used in Europe and Scandinavia but not in the United States. (Egner et al., 1960)

[a] Critical value is defined as the soil test concentration above which the soil test level is considered optimum for plant growth and responses to additions of the nutrient are unlikely to occur. Critical values cited in this table are approximate, can be affected by soil type and crop, and were obtained from several sources.

Two soil testing approaches for P, the P_i soil test and ion exchange resins, do not remove P from soils by chemical extraction. The P_i test was developed and evaluated in the late 1990s and early 2000s and showed considerable promise, but still has not been adopted as a routine soil test (Myers et al., 1995, 1997, 2005; Chardon et al., 1996; Menon et al., 1997). In this method, the soil is equilibrated with 0.01 M $CaCl_2$ in the presence of an Fe-oxide-coated filter paper strip, which acts as an "infinite sink" for soil P. Phosphorus in solution is sorbed to the strip, causing desorption of labile P from the soil. After a specified equilibration period (usually 2 h) the strip is removed, rinsed lightly to remove soil particles, and sorbed P is extracted from the strip by shaking for 2 h with a 1 M H_2SO_4 solution. Ion exchange resins have been used for many years in soil P research to study the desorption of soil P (Amer et al., 1955; Sibbesen, 1978; Wolf et al., 1985; van Raij et al., 1986; Sharpley et al., 1989; Abrams and Jarrell, 1992; Mallarino and Atia, 2005; Saavedra et al., 2007). These studies have shown that anion exchange resins saturated with Cl or HCO_3 and suspended with soils in aqueous suspensions can accurately estimate labile P in soils of widely differing properties. Resins simulate root uptake by removing P from solution by surface sorption processes, with the rate of P sorption controlled by diffusion (Kuo, 1996). Despite this research, ion exchange resins have not been adopted as routine soil tests for P in the United States primarily because of practical difficulties such as separating the resin from the soil after the equilibration process. To overcome these obstacles Skogley et al. (1990) developed an encapsulated ion exchange resin technique for use by routine soil testing laboratories as a multinutrient soil test (commonly referred to as the phytoavailability soil test, or PST; see also Skogley, 1994). Similar efforts have occurred in other countries such as Brazil, where van Raij (1994) successfully adapted ion exchange resin beads for use as multinutrient soil tests by routine soil testing laboratories. Ion exchange resins and chemical extractions can also be used as environmental soil P tests, to identify soils with a higher potential for P loss in soil erosion and surface or subsurface runoff (see Section 13.2.7).

13.2.4.1.2 Considerations in Soil Testing for P

The soil testing process for P is rather straightforward. Soil samples for P can be collected at any time during the year because P is relatively immobile in most soils. The standard sample depth is 0–20 cm, except for permanent pastures and turf (0–10 cm). In fields where the use of banded fertilizers containing P is a common practice it may be necessary to significantly increase the number of soil samples collected per unit area to overcome the high meso-variation that can be present (see Section 13.2.3.1). For situations where the transport of P to surface waters by erosion and runoff is a concern samples should be collected from the 0–5 cm depth, since runoff waters only interact with the upper few centimeters of the soil surface (Sharpley and Smith, 1989; Sharpley et al., 1996). There are no special handling or storage requirements for samples collected for P analysis—air drying, grinding, and sieving with a 2 mm screen is the standard approach. Analysis of P in soil extracts can be done colorimetrically by the molybdenum blue method, which primarily measures orthophosphate (Murphy and Riley, 1962) or by ICP-AES, which measures orthophosphate and some organic P. Note that ICP-AES values for soil test P can often be 5%–10% higher than those measured colorimetrically.

Interpretation of analytical results for P must consider the fact that P moves to roots primarily by diffusion, a process that is highly dependent upon soil moisture and temperature. It is not uncommon to see early season P deficiency in crops grown in soils that are rated as "optimum" or "excessive" in P. Low soil temperatures and dry soil conditions can inhibit diffusion, root growth, and plant uptake of P, creating a temporary P deficiency that often disappears as soils warm and receive rainfall or irrigation. In most cases, if soils have optimum concentrations of plant available P, early season deficiencies have no significant negative effects on crop yield. For the most part, however, soil P tests are interpreted using the sufficiency level approach described in Table 13.1.

13.2.4.2 Soil Testing for Potassium, Calcium, and Magnesium

13.2.4.2.1 Soil K, Ca, and Mg

The cycling and plant availability of K, Ca, and Mg are sufficiently similar that these three nutrients can be considered together for the purposes of soil fertility evaluation. The primary sources of plant available K, Ca, and Mg are soil minerals, which release these ions into solution during the weathering process.

Major K-bearing minerals include the feldspars (orthoclase, microcline, sanidine) and micas (biotite, muscovite, phlogopite). Total K concentrations in soils average 1.9%, but can range from 0.03% in organic soils (peats, mucks), to 0.3% in very sandy soils, to as high as 3.0% in mineral soils derived from feldspars and micas. Once mineral dissolution has occurred, soil K is primarily found in soluble, exchangeable, and nonexchangeable forms. Potassium in the soil solution ranges in concentration from 1 to 80 mg K L^{-1} (average = 2–5 mg K L^{-1}) and is in rapid equilibrium with exchangeable K that is retained by electrostatic attraction to the negatively charged cation exchange sites located on clays and soil organic matter. Usually, less than 5% of the total CEC of a soil is occupied by exchangeable K. Soluble and exchangeable K are the major sources of plant-available K in most soils but together represent less than 1%–2% of total soil K. Nonexchangeable K, also referred to as "fixed" K, is found within the interlayers and on the edges of 2:1 expanding clay minerals and is viewed as a slowly available reserve of K for plant uptake. Soluble K is <1% of exchangeable soil K and can be quickly depleted by plant uptake or leaching. The ability of soils to maintain an adequate concentration of K in solution by releasing K from exchangeable and nonexchangeable forms is referred to as the K buffer capacity and is an important measure of soil fertility. The K buffer capacity of soils depends largely upon the amount and types of clay minerals present and to a lesser extent on soil organic matter content. Soils with higher percentages of clay, especially 2:1 clays, have higher K buffer capacities than sandy or organic soils and more ability to maintain soluble K in an optimum range for plant growth.

Major mineral sources of soil Ca are carbonates, feldspars, and phosphates. The most important Ca-bearing mineral is anorthite ($CaAl_2Si_2O_8$; a plagioclase feldspar), except in calcareous soils of arid regions where calcite ($CaCO_3$) and dolomite ($CaMg(CO_3)_2$) dominate. Total soil Ca varies widely, from <0.1% in highly weathered, tropical soils to as high as 25% in calcareous soils. Typical total soil Ca values in noncalcareous, humid, temperate soils are from 0.7% to 1.5%; total Ca values >3% indicate the presence of $CaCO_3$. In most soils, Ca is the dominant exchangeable cation, occupying 20%–80% of the total CEC. Consequently, soil solution concentrations of Ca are quite high relative to most other plant nutrients, ranging from 30 to 300 mg Ca L^{-1} in noncalcareous soils.

Plant available Mg originates from the weathering of minerals such as biotite, dolomite, hornblende, olivine, and serpentine. Total soil Mg varies from <0.1% in coarse, sandy soils of humid regions to 4% in fine-textured soils formed from Mg-bearing minerals. As with K and Ca, soluble and exchangeable Mg are most important to plant growth. Exchangeable Mg occupies from 4% to 20% of total soil CEC and soluble Mg ranges from 50 to 120 mg Mg L^{-1} in temperate region soils.

13.2.4.2.2 Current Soil Testing Methods for K, Ca, and Mg

Soil chemical extraction is by far the most common approach used today to test soils for plant-available K, Ca, and Mg and most extracting solutions simultaneously extract and analyze soluble and exchangeable K, Ca, and Mg (Table 13.3). Haby et al.

TABLE 13.3 Summary of Soil Testing Methods Currently Used for Potassium (K), Calcium (Ca), and Magnesium (Mg)

Soil Test Extractant	Extractant Composition	Comments, Critical Values,[a] and Sources
Ammonium acetate	M NH_4OAc, pH 7.0	Used for >50 year as soil test for K, Ca, and Mg (Chapman and Kelley, 1930; Schollenberger and Simon, 1945). Suited for wide range of soils, but primarily used in midwestern and western states of the United States. Critical values for K, Ca, and Mg vary widely based on soil type (pH, CEC, clay mineralogy) and crop and reportedly range from K: 110–200 mg kg^{-1}; Ca: 250–500 mg kg^{-1}; and Mg: 30–60 mg kg^{-1} (Haby et al. 1990)
AB-DTPA	M NH_4HCO_3 + 0.005 M DTPA—pH 7.5	Multinutrient extractant primarily used with alkaline soils (Soltanpour and Schwab, 1977)
Mehlich I	0.05 M HCl + 0.0125 M H_2SO_4	Multinutrient extractant used on acidic, low CEC soils (Mehlich, 1953)
Mehlich III	0.2 M CH_3COOH + 0.25 M NH_4NO_3 + 0.015 M NH_4F + 0.013 M HNO_3 + 0.001 M EDTA—pH 2.5	Multinutrient extractant suitable for wide range of soils (Mehlich, 1984)
Morgan and modified Morgan	Morgan: 0.7 M $NaC_2H_3O_2$ + 0.54 M CH_3COOH +—pH 4.8 Modified Morgan: 0.62 M NH_4OH + 1.25 M CH_3COOH—pH 4.8	Multinutrient extractant primarily used in the northeast United States for acid, low CEC soils. Not suitable for calcareous soils (Morgan, 1941)

[a] Critical value is defined as the soil test concentration above which the soil test level is considered optimum for plant growth and responses to additions of the nutrient are unlikely to occur. Critical values cited in this table are approximate, can be affected by soil type and crop, and were obtained from several sources.

(1990) surveyed soil testing laboratories in the United States and Canada and reported that the most common soil test for K, Ca, and Mg was ammonium acetate (1 M NH_4OAc, pH 7.0), which has been used in the United States since the 1930s. Since then, multielement soil testing extractants, such as AB-DTPA and Mehlich 3 have gained wide popularity and are fast becoming the standard soil test method for K, Ca, and Mg. The more widespread use of multielement extractants can be attributed to the fact that many soil testing laboratories formerly used one extracting solution for P (e.g., Bray P_1), another for K, Ca, and Mg (e.g., 1 M NH_4OAc, pH 7.0), and another for micronutrients (e.g., DTPA). Converting to multielement extractants thus reduced the cost and time of analysis. Most soil test extractants for K, Ca, and Mg displace these cations from exchange sites on soil colloids with a replacing cation, usually NH_4^+ (NH_4OAc, AB-DTPA), Na^+ (Morgan), H^+ (Mehlich 1) or a combination of cations (Mehlich 3, Modified Morgan). All extractants also remove any solution K, Ca, and Mg present in the soil. Acidic soil tests (e.g., Mehlich 1, Mehlich 3) may also extract some nonexchangeable K from soils containing 2:1 clays because the H^+ ion is small enough to enter the interlayers of these minerals and displace K from interlayer exchange sites. Acidic extractants may also overestimate exchangeable Ca and Mg in arid region calcareous soils because they may dissolve Ca- and Mg-bearing minerals. One recent advance in soil K testing has been the use of sodium tetraphenylboron (NaTB) to extract plant available K. Cox et al. (1999) showed in greenhouse studies that a 5 min NaTB soil extraction was well correlated with plant K uptake and suggested that NaTB effectiveness was due to its ability to extract both exchangeable and nonexchangeable K. Fernández et al. (2008) in a field study with soybeans found NaTB to work reasonably well but that is was not as effective overall as 1 M NH_4OAc at predicting plant K uptake.

A few alternatives to chemical extraction have been used for routine evaluation of plant available K, Ca, and Mg. Two that have received the most interest have been ion exchange resins, used in a few U.S. states and some South American countries (van Raij et al., 1986; Skogley, 1994), and electro-ultrafiltration (EUF), which has primarily been used in Europe (Nemeth, 1979; Haby et al., 1990). The basic principles of ion exchange resin techniques were discussed earlier in Section 13.2.4.1. The EUF method combines electrodialysis and ultrafiltration and has been used as a multielement extractant for NH_4^+, NO_3^-, P, K, Ca, Mg, Na, S, B, Mn, and Zn. In this approach, a soil suspension is stirred in a central compartment attached to cells on each side containing platinum electrodes. Microfiber filters separate the soil suspension from the electrodes. Voltage is applied to the electrodes and vacuum filtration is used to withdraw water and dissolved ions from the central cell after specified time intervals that characterize different forms of plant nutrients (e.g., the 0–5 min extraction is soluble K, the 5–10 min extraction is exchangeable K, and the 10–30 min extraction reflects K buffering capacity). Solutions collected at the anode and cathode sides of the EUF device are combined and analyzed by standard instrumental techniques. While EUF has been shown to be an effective method to simultaneously extract plant-available nutrients, it is rather slow and expensive and not as widely used as chemical soil tests (Haby et al., 1990; van Lierop and Tran, 1985).

13.2.4.2.3 Considerations in Soil Testing for K, Ca, and Mg

Soil samples for K, Ca, and Mg analysis are collected following the standard approaches described in Section 13.2.3.1. The most important exception is with K where subsoil samples are sometimes recommended for soils with a very sandy surface horizon and a shallow B horizon (zone of clay accumulation). In these situations subsoil K has been shown to be an important source of plant available K and testing the surface horizons alone may underestimate the true K supplying capacity of the soil. This is particularly true if the B horizon has a high percentage of 2:1 clays that can act as a reservoir of slowly available, nonexchangeable K. Subsoil samples are rarely tested for Ca and Mg. Samples should be collected at the same time each year to minimize the effect of natural, seasonal changes in K concentrations caused by processes such as leaching, freezing, and thawing, biological transformations (uptake, mineralization, biocycling of K from subsoils to topsoils), and seasonal differences in soil moisture content.

Soil sample handling, particularly drying method, can markedly and unpredictably alter extractable K. Air-drying will usually cause an increase in exchangeable K except in soils that have very high K values, where drying results in K fixation into nonexchangeable forms, thus decreasing soil test K. Changes caused by drying are greatest in fine-textured soils dominated by 2:1 clays. While it can be argued that soil tests for K would best be conducted on field-moist soils, virtually all laboratories air-dry soils to ease handling, grinding, sieving, mixing, and weighing of a representative subsample. These advantages outweigh the changes in extractable K that occur during drying; however, soils to be analyzed for K should only be air-dried at moderate temperatures (<50°C) because oven-drying, particularly at >60°C greatly enhances the release of K from soils. Analyses of soil extracts for K, Ca, and Mg can be conducted by either AAS or ICP-AES.

Interpretation of soil test results for K, Ca, and Mg follows the sufficiency level approach (Table 13.1) with only a few minor modifications related to soil type, plant to be grown, and soil/crop management. Soil test K interpretations are often modified based on the CEC of the soil. Soils with higher CEC values will often have a higher "critical value" (point above which crop response is not expected and thus no fertilizer is recommended). As an example, the critical value for soil test K in Alabama (Mehlich 1 soil test) increases from 40 to 80 mg K kg^{-1} as CEC increases from 4.5 to 9.0 cmol kg^{-1}. Subsoil K is occasionally considered when interpreting the results of a soil test for K, usually by the use of indirect information on subsoil properties, such as soil survey data on horizonation (e.g., depth to B horizon), texture, and clay mineralogy. Other factors that may alter a soil test-based recommendation for K fertilization include crop and yield goal that affect K removal, tillage practices, and climate, which affects release of K from K-bearing minerals.

Modification of the results of a soil test for Ca or Mg by inclusion of other information is unusual. In general, maintaining soils in an adequate pH range for plant growth by use of the proper type of liming material is adequate to maintain Ca and Mg fertility in most soils.

13.2.4.3 Sulfur

13.2.4.3.1 Soil S

The total S content of temperate zone soils ranges from 0.005% to 0.04%, more than 90% of which is found in organic forms. Total S values can be much higher in arid and semiarid regions where soils can accumulate soluble and mineral forms of sulfate-S (SO_4-S), such as gypsum ($CaSO_4 \cdot 2H_2O$), epsomite ($MgSO_4 \cdot 7H_2O$), and mirabilite ($Na_2SO_4 \cdot 10H_2O$). Important S-bearing minerals in humid regions are pyrite (Fe_2S), sphalerite (ZnS), and chalcopyrite ($CuFeS_2$). Sulfur originating from burning fossil fuels, particularly coal, or S emissions from volcanic activity, wetlands, and oceans can also be added to the soils by wet or dry atmospheric deposition.

The plant available form of S is the sulfate anion (SO_4^{2-}-S), which originates from the dissolution of soluble salts and minerals containing S, oxidation of elemental S, and mineralization of organic S. Solution SO_4-S can be taken up by plants, immobilized in microbial biomass, sorbed by soil colloids, precipitated in an insoluble mineral form by reaction with Ca, Mg, or Na, or leached to subsoils. If plant roots penetrate to subsoils, sorbed SO_4-S can be released into solution and absorbed. Under reducing conditions, SO_4-S can be converted to H_2S gas and lost from the soil by volatilization or precipitated as metal sulfide minerals such as pyrite. Traditionally, S deficiency has been considered uncommon except with high yielding crops grown on deep sandy soils with low organic matter contents or on soils that developed from parent materials low in S. Organic matter mineralization and wet deposition of SO_4-S resulting from the burning of fossil fuels typically have provided enough plant available S for most crops. Concentrations of SO_4-S in the soil solution (A horizon) of most temperate zone soils ranged from 5 to 20 mg SO_4-S L^{-1}, higher than the 3–5 mg SO_4-S L^{-1} required for the optimum growth of most plants. However, recently, S deficiencies have been reported globally due to a combination of factors, the primary factor being a marked decrease in S deposition due to air pollution control measures. Sulfate deposition has clearly decreased over the past 20 years as indicated by the National Atmospheric Deposition Program (NADP, 2008). For example, the average total wet deposition of sulfate at the Huntington Wildlife Station in Essex County, New York, United States, was 22 kg-SO_4 ha^{-1} for the period of 1979–1983 and decreased 43% to 14 kg-SO_4 ha^{-1} for the period of 2003–2007 (NADP, 2009). This decrease in atmospheric deposition combined with increased crop yields and decreased agronomic S inputs are expected to contribute to increasing S deficiencies in crops (McGrath and Zhao, 1995). These predictions have been supported by recent studies showing increased yield responses to S fertilization in previously nonresponsive soils (Chen et al., 2008).

13.2.4.3.2 Current Soil Testing Methods for S

Due to the historic lack of crop response to S fertilization, less effort has been directed toward the development and calibration of soil tests for S than for P and K. Soils tested for S are normally sampled from the A horizon, except in sandy, low organic matter soils with shallow B horizons that can be a significant reserve of plant available SO_4-S. In these situations soil testing laboratories may request a subsoil sample because the subsoil may have enough available S for plant growth.

Soil testing for S relies on chemical extraction. More than 20 extracting solutions have been developed and evaluated as S soil tests, including water and various concentrations of dilute acids (HCl), dilute salts (e.g., $CaCl_2$, LiCl, NaCl, NH_4Cl), acetates (e.g., NH_4OAc, MgOAc, NaOAc), and phosphates (e.g., $Ca(H_2PO_4)_2$, KH_2PO_4) (Johnson and Fixen, 1990). Most extractants remove soluble and sorbed forms of SO_4-S, along with a small percentage of organic S, as these are the soil fractions regarded as plant available. In arid regions where the concentration of SO_4-S is often quite high due to the accumulation of sulfate salts, extraction with deionized or distilled water is used to identify S deficiency. In humid regions, the use of an extractant that contains a replacing anion such as phosphate is often more successful. One of the most widely used extracting solutions for S is a 500 mg P L^{-1} solution of $Ca(H_2PO_4)_2$, sometimes in combination with 2 M HOAc. The phosphate ions displace sorbed SO_4-S, the Ca ions cause flocculation of the soil allowing for ease of analysis of S by either colorimetric or turbidimetric means, and the HOAc extracts some organic S. Recently there has been interest in using Mehlich 3 as a S soil test to eliminate the need for a separate extraction to determine available S. Once extracted, SO_4-S can be analyzed by colorimetry, titrimetry, ion chromatography, and ICP-AES. The most common analytical techniques are turbidimetry (if S is determined alone) and ICP-AES (for S alone and in multielement analyses). A review of the advantages and disadvantages of analytical methods for S is given by Tabatabai (1996).

13.2.4.3.3 Interpretation of S Soil Tests

Sulfur soil testing is a moderately reliable approach to determine the S fertility status of a soil but is best used in conjunction with other information, such as plant analysis, knowledge of soil type and plant yield potential, soil organic matter content, and inputs of S from sources other than fertilizers (e.g., manures, crop residues, rainfall, and irrigation waters). A major concern with soil testing for S is that inputs of S from the atmosphere or irrigation waters, which can be more than the rates of S fertilizer recommended, are not accounted for by soil testing. In situations where S deficiency is probable, based on knowledge of the soils, crops, and management practices it may be more economical to apply a small amount of S fertilizer (10–15 kg S ha^{-1}) than to incur the costs of soil testing for S.

13.2.4.4 Soil Testing for Nitrogen

Nitrogen deficiency is the most common soil fertility problem for nonlegumes. Nitrogen (N) is also well known to negatively impact water and air quality. However, despite the importance of N to agricultural production and environmental quality we have not yet developed a widely accepted method to routinely and rapidly test soils for plant available N, particularly in humid regions. The reasons for this center around the complex transformations undergone by N in soils, referred to as the soil N cycle. A summary of the key aspects of the N cycle most relevant to soil testing follows; C.4 and D.1 in the *Handbook of Soil Sciences*, and elsewhere (Keeney, 1982; Keeney and Nelson, 1982; Power and Schepers, 1989; Tisdale et al., 1993; Pierzynski et al., 1994; Foth and Ellis, 1997; Schepers and Raun, 2008) provide thorough treatments of N cycling, management, and environmental impacts.

Most of the N available for biological assimilation by plants and animals originated from the atmosphere, which is 78% N_2 by volume. Only a small fraction of the global N supply is in soils where total N values typically range from 0.05% to 0.15% and most of the N (>98%) is organic in nature. Atmospheric N is converted to forms of N that can be directly or indirectly used by plants by symbiotic or nonsymbiotic biological N fixation. Bacteria of the genera *Rhizobium* and *Bradyrhizobium* can form symbiotic relationships with plants to assimilate N_2 from the atmosphere while nonsymbiotic N fixation can be conducted by free-living algae, bacteria, and actinomycetes. Electrical discharges and industrial processes that combine N_2 with H from natural gas to produce the ammonia (NH_3) used in fertilizer manufacture also convert atmospheric N to fixed forms that can be used by plants. The burning of fossil fuels and volcanic eruptions are also atmospheric sources of N to soils.

The key components of the soil N cycle are as follows: (1) mineralization, in which soil organic N is converted to inorganic NH_4-N by microbial decomposition; (2) immobilization, in which soil microorganisms assimilate NH_4-N and NO_3-N from the soil solution for population growth and biomass production—essentially the reverse of mineralization; (3) nitrification, in which certain soil bacteria convert NH_4-N to NO_3-N, a rapid process in most well-aerated soils; (4) ion exchange in which NH_4-N is retained by cation exchange sites on soil clays or organic matter (including fixation within the interlayers of 2:1 clays as nonexchangeable NH_4-N) and NO_3-N is retained by any positively charged sites present on soil colloids (rarely of consequence in most soils); (5) denitrification, in which soil bacteria that are more active under reduced conditions convert NO_3-N to gaseous forms of N (N_2O, NO, N_2), which are then lost from soils to the atmosphere; (6) volatilization, in which NH_4-N is converted to gaseous NH_3-N under certain alkaline conditions (high pH) and lost to the atmosphere; and (7) leaching, in which NO_3-N moves downward in the soil profile with percolating waters.

Plants absorb N from the soil solution as NH_4-N and NO_3-N. An accurate soil test for N, therefore, must be able to predict how all components of the soil N cycle that affect the availability of NH_4-N and NO_3-N can be integrated into a quantitative assessment of a soil's N supplying capacity. This is a complex task given the dependence of N cycling on biological activity and environmental conditions (temperature, moisture, rainfall). The fact that the main form of inorganic N in most soils is NO_3-N further complicates the matter given the leachability of NO_3-N and its sensitivity to loss as a gas via denitrification.

13.2.4.4.1 Current Soil Testing Methods for N

Soil testing for N differs markedly between arid and humid regions. In arid (or semiarid) regions evapotranspiration usually exceeds precipitation and inorganic forms of soil N are not as susceptible to leaching or denitrification. For this reason a soil sample collected from a crop's rooting zone shortly before the start of the growing season and analyzed for residual inorganic N (the NH_4-N and NO_3-N remaining from the past year's mineralization, fertilization, and organic waste applications) accurately measures plant available soil N. Nitrogen inputs are then reduced in proportion to the amount of residual inorganic N. In most cases soil samples for residual inorganic N must be collected to deeper depths (as much as 180 cm, minimum of 60 cm) than for standard soil testing (20 cm). Soil samples tested for residual inorganic N are often only analyzed for NO_3-N since this is usually the dominant form of inorganic N in most soils. The best time to take a soil sample for residual inorganic N is usually just before planting or early in the growing season, although in very cold and dry areas, where mineralization and leaching are minimal, samples can be collected the preceding fall or winter.

After sample collection, proper handling is critical to avoid changes during storage that affect the amount of inorganic N present. Moist soils stored under warm conditions can mineralize or immobilize a significant amount of inorganic N; very wet samples can lose NO_3-N by denitrification. To avoid problems soil samples should be rapidly air-dried at ambient temperatures by spreading the soil in a thin layer. Extraction of inorganic N is usually accomplished by shaking a dried, ground soil sample for 30 min to 1 h with a salt solution (e.g., 2 M KCl, 0.01 M $CaSO_4$, 0.04 M $(NH_4)_2(SO_4)$), followed by filtration. The most common methods to determine NH_4-N and NO_3-N in soil extracts are automated colorimetry and ion chromatography. Other, less commonly used methods are steam distillation, ion specific electrodes, and microdiffusion techniques. Details on the various methods used to measure inorganic N in soils are given by Bundy and Meisinger (1996) and an extensive review of chemical extraction methods to assess soil N availability was recently prepared by Griffin (2008).

Soil testing for N in humid regions is a more complex process and a less accurate predictor of soil N fertility, and consequently most soil testing laboratories in humid regions do not offer routine soil N tests. The greater rainfall and warmer temperatures in humid areas can cause rapid seasonal changes in the amount of inorganic N present in the soil profile making direct measures of residual inorganic N estimates of plant available N less reliable.

The general consensus on soil testing for residual inorganic N in humid regions is that this practice has value if conducted at or near planting and if samples are collected to a reasonable depth in the soil profile (not just the topsoil). If residual inorganic N values are high, reductions in N inputs should be made. Bundy et al. (1992) reported on the use of the preplant soil profile nitrate test (PPNT) in the upper Midwestern United States where moderate rainfall and cooler winter temperatures make this approach more likely to be successful than in warmer, higher rainfall humid regions. The economically optimum N fertilizer rate for corn (Zea mays, L.) was shown to decrease in a near-linear manner with increasing PPNT values in soil samples collected to a depth of 1 m. Another situation where residual inorganic N testing has been successful in humid regions has been with short-season crops where there is less likelihood that significant losses of residual inorganic N will occur.

In general, the most promising advance in soil N testing in the humid regions of the United States has been the presidedress soil nitrate test (PSNT) originally developed for corn and later used for a wider range of agronomic and vegetable crops (Magdoff et al., 1984; Bock and Kelley, 1992). The PSNT was conceived and evaluated to address the problem of overfertilization of corn with N in the Northeastern United States, particularly in fields with histories of manure and legume use where residual organic N would likely provide an appreciable percentage of the total N requirement for many nonleguminous crops. The PSNT has four basic tenets: (1) all fertilizer N for corn except for a small amount banded at planting, should be applied by sidedressing when the crop is beginning its period of maximum N uptake, usually early June; (2) soil and climatic conditions prior to sampling integrate the factors influencing the availability of N from the soil, crop residues, and from previous applications of organic wastes; (3) a rapid sample turn-around (<14 d) by a testing laboratory is possible, thus allowing time for farmers to collect the soil sample, submit it to the laboratory, have it an analyzed, receive the results and recommendations, and then apply (or not apply) sidedress N before the corn crop becomes too large for equipment to move through the field; and (4) farmers will normally only sample to a depth of 30 cm, the recommended depth for the PSNT sample. In practice a PSNT sample is collected early during corn growth, when the corn plants are ~30 cm in height. The sample is air-dried rapidly, or oven-dried at <60°C, after spreading the soil in a thin (<1 cm) layer. Extraction and analysis for NO_3-N proceed as described above for residual inorganic N. However, while the PSNT actually measures inorganic soil N, it is not a measure of residual inorganic N, but an indirect, field-based expression of a soil's capacity to provide an adequate supply of inorganic N during the growing season (i.e., of the soil N mineralization potential). The PSNT has been evaluated in over 300 field studies in the Northeastern and Midwestern United States and has been repeatedly shown to be successful in identifying N sufficient soils (Magdoff et al., 1990; Bock and Kelley, 1992; Meisinger et al., 1992; Sims et al., 1995; Jaynes et al., 2004; Muñoz et al., 2008). Some of the logistical difficulties associated with the need for rapid sample analysis have been overcome by the development of "quicktest" kits and specific ion electrodes that can be used in the field (Jemison and Fox, 1988).

Recently, the Illinois soil N test (ISNT) has been proposed as a means to identify fields where corn would not be responsive to N fertilization. Initially, Mulvaney et al. (2001) found that soils that did not respond to fertilizer N mineralized larger quantities of N and this mineralization was correlated to a decrease in soil amino sugar N. However, the fractionation of soil N developed by Mulvaney and Khan (2001) to determine amino sugar N is too cumbersome for routine soil analysis. Therefore, Khan et al. (2001) developed a simpler method for estimating amino sugar N, which has come to be known as the ISNT. The ISNT estimates amino sugar N and has been found by some to reasonably identify soils where no response to fertilizer N would be expected in corn. The procedure as described by Khan et al. (2001) requires incubation of 1 g of air-dried soil with 10 mL of 2 M NaOH in a 472 mL wide mouth jar. The sample is heated for 5 h at 48°C–50°C on a hot plate. Ammonium and amino sugar N is then liberated as NH_3 and collected in an H_3BO_3 indicator solution. After its original publication, the method underwent modest modifications including a recommendation to rotate the sample jars on the hot plate (Mulvaney et al., 2004) or to enclose the hot plate and incubation vessels in a box (Klapwyk and Ketterings, 2005) in order to reduce spatial variability within the hotplate.

Since its release, the effectiveness of the ISNT has received mixed reviews within the soil science literature. Klapwyk and Ketterings (2006) reported that when used in combination with soil organic matter the ISNT appeared to do a good job of identifying corn fields that may be responsive to additional N fertilizer. However, Barker et al. (2006) reported that the ISNT was not able to distinguish between responsive and nonresponsive Iowa soils. A strong correlation was found between the ISNT and total soil N, hydrolysable NH_4-N, and hydrolysable NH_4 + amino sugar-N. They concluded that the ISNT extracted a consistent fraction of soil total N and as a result was not able to predict potentially labile soil N fractions. As a result, the authors recommended against the use of the ISNT to guide N fertilization in Iowa corn production. Similarly, in an on-farm N fertilizer response trial conducted by Spargo et al. (2007), the ISNT was found to extract a consistent percentage of total soil N and was deemed a poor predictor of labile soil N. Osterhaus et al. (2008) conducted 80 corn N response experiments in Wisconsin and reported that ISNT values were not related to the observed economically optimum N rate and that the ISNT had no ability to separate N-responsive from unresponsive sites. They suggested that the ISNT measures a constant fraction of soil organic N instead of measuring the readily available N pool, as would be required for an accurate soil N test designed to predict soil N supplying capability and corn N response. In contrast to these studies, Lawrence et al. (2009) found that adjusting ISNT values based on soil organic matter content (as estimated by loss-on-ignition [LOI]) resulted in accurately identifying 83% of the N responsive sites in a 34-site field study using corn planted after grass-legume sod. They emphasized the importance of the timing of soil sample collection (within 5 weeks of sod plowdown) and the need to include N credits for sod decomposition to the successful use of the ISNT. These mixed results indicate that the ISNT should be used with caution, and perhaps in conjunction with other soil measurements (e.g., organic matter); it is, however currently offered as a routine soil N test by some commercial and state soil testing laboratories.

13.2.4.4.2 Considerations in Interpretation of Nitrogen Soil Tests

Most recommendations for the amount of N required for optimum plant growth are not based on soil N testing, but on field calibration studies quantifying plant performance in response to N inputs (fertilizers, manures, biosolids, etc.). Widespread, commercial-scale use of soil N testing today is confined to certain areas and crops, such as the PSNT and ISNT for corn in humid regions or residual inorganic N testing for grain crops in arid regions. Recommended N rates for the major grain crops, such as corn, wheat (*Triticum aestivum*, L.), and sorghum (*Sorghum bicolor* [L.] Moench) are initially determined from equations using the expected, realistic yield goal and a conversion factor appropriate to that crop. For example, for corn, the fertilizer N rate recommended in many U.S. states is arrived at by directly multiplying the realistic yield goal by an empirically determined factor that ranges from 17 to 18 kg fertilizer N per Mg of expected yield (equivalent to 1.0–1.2 lb fertilizer N per bushel of expected yield). Modifications (reductions) to this recommendation are then made based on PSNT values, residual soil inorganic N, the previous or intended use of other N sources (animal manures), documented N inputs from other sources (high NO_3-N irrigation waters), and credits for N supplied by a previous legume crop in the rotation (e.g., alfalfa, soybeans). Tisdale et al. (1993) summarized the general approach used to make N recommendations:

$$N_{fertilizer} = N_{crop} - N_{soil} - (N_{organic\ matter} + N_{previous\ crop} + N_{organic\ waste}) \quad (13.1)$$

- $N_{fertilizer}$ is the amount of N needed from fertilizers, manures, biosolids, etc.
- N_{crop} is the crop N requirement at realistic yield goal
- N_{soil} is the residual soil inorganic N (NH_4-N + NO_3-N)
- $N_{organic\ matter}$ is the N mineralized from soil organic matter
- $N_{previous\ crop}$ is the residual N available from previous legume crops
- $N_{organic\ waste}$ is the residual N available from previous organic waste use such as animal manures, biosolids, wastewater irrigation, etc.

In some cases, an N availability index, based on soil organic matter content, soil texture (indication of leachability and moisture holding capacity), and climate (indicated by crop being grown) is used to estimate soil N supplying capacity. Fertilizer N recommendations are adjusted accordingly with fine-textured soils with higher organic matter contents presumed to provide more plant available N from mineralization and thus to need less fertilizer N. Mathematical models have also been developed to predict crop N requirements but with only limited success because of the amount of site-specific information required for the models to function with any degree of accuracy (Tanji, 1982).

Interpretations of N soil tests are done routinely and with reasonable success. However, soil N testing may be improved by use with plant analysis or other techniques now available to quantify plant N status in the field, such as remote sensing or leaf chlorophyll meters (LCMs). Plant N testing methods are discussed in Section 13.3; some environmental aspects of soil N testing are discussed in Section 13.2.7.3.

13.2.4.5 Soil Testing for Micronutrients (B, Cl, Cu, Fe, Mn, Mo, Zn)

Micronutrients are essential elements normally present in plants at very low concentrations (<100 mg kg^{-1}) and include boron (B), chlorine (Cl), copper (Cu), iron (Fe), manganese (Mn), molybdenum (Mo), and zinc (Zn). The following overview of micronutrient cycling focuses on soil properties and processes most relevant to soil testing and reflects the fact that four micronutrients exist in soils as cations (Cu^{2+}, Zn^{2+}, $Fe^{2+,3+}$ and Mn^{2+}) while three are found as an uncharged molecule ($H_3BO_3^0$) or anions [$B(OH)_4^-$, Cl^-, MoO_4^{2-}].

Plant available Cu in soils originated from the weathering of igneous and sedimentary rocks. The main Cu bearing primary minerals are chalcopyrite ($CuFeS_2$), chalcocite (Cu_2S), and bornite ($CuFeS_4$). Soil total Cu concentrations typically range from 1 to 40 mg kg^{-1} and concentrations in the soil solution are quite low, from 10^{-8} to 10^{-9} M. More than 99% of the Cu in the soil solution is found as Cu^{2+} complexed with organic matter; above pH 6.9 the dominant inorganic form of Cu is $Cu(OH)_2^0$. Soluble Cu is in equilibrium with Cu complexed by organic matter, exchangeable Cu retained on the CEC sites of soil colloids, and Cu sorbed, occluded, or coprecipitated by soil oxides. Sorption of Cu by organic matter is the primary reaction controlling the plant availability of Cu although Cu solubility is also highly pH dependent, decreasing 100-fold for each unit increase in pH.

Plant available Zn in soils also originates from the weathering of igneous and sedimentary rocks and total soil Zn concentrations usually range from 10 to 300 mg kg^{-1}. The major Zn bearing minerals in soils include franklenite ($ZnFe_2O_4$), smithsonite ($ZnCO_3$), and willemite ($ZnSiO_4$). Zinc concentrations in the soil solution range from 2 to 70 μg L^{-1} (ppb) and Zn^{2+} is the major species below pH 7.7. Approximately 50%–60% of soluble Zn is complexed with organic matter. Plant available Zn also includes exchangeable Zn and Zn sorbed by clays, oxides, and carbonates. As with Cu, the solubility of Zn is highly dependent upon soil pH, decreasing markedly above pH 6.0–6.5 (Wear and Evans, 1968).

Iron is one of the major constituents of the earth's crust (~5%) and total Fe contents in most soils are quite high, ranging from 1,000 to 10,000 mg kg^{-1}. Major mineral forms of Fe include olivene (($Mg,Fe)_2SiO_4$), pyrite (FeS), siderite ($FeCO_3$), hematite (Fe_2O_3), and goethite (FeOOH). Plant available forms of Fe include those that are sorbed by clays and organic matter. Soil solution concentrations of Fe in equilibrium with these minerals are very low and depend greatly upon soil pH, ranging from 10^{-6} M in very acid soils to $<10^{-20}$ M in soils above pH 7.0. The form and solubility of Fe in the soil solution also depends upon soil redox potential; in well-aerated, oxidized soils, ferric Fe (Fe^{3+} or $Fe(OH)_2^+$) predominates, while in reduced, waterlogged soils the major inorganic species is ferrous Fe (Fe^{2+}). Each unit increase in pH decreases the solubility of Fe^{3+} by 1000-fold, but only decreases soluble Fe^{2+} by 100-fold. Of relevance to soil testing for Fe is the fact that the total concentrations of soluble inorganic Fe in most soils are too low to meet the nutritional needs of most plants, even under very acid soil conditions, yet plants are able to obtain adequate Fe for growth. Research has shown that natural organic compounds in soils (chelates) and exuded from plant roots play an important role in preventing the precipitation of Fe as insoluble compounds by forming Fe-chelate complexes that can move to plant roots by mass flow or diffusion. At the root surface, Fe dissociates from the chelate and is taken up by the plant.

Plant available manganese (Mn) in soils originates from the weathering of minerals such as pyrolusite (MnO_2), hausmannite (Mn_3O_4), manganite (MnOOH), rhodochrosite ($MnCO_3$), and rhodonite ($MnSiO_3$). Total soil Mn concentrations are from 20 to 3000 mg kg^{-1} while soluble Mn is usually between 0.01 and 1.0 mg L^{-1}, existing primarily as Mn^{2+} in equilibrium with MnO_2 in oxidized soils and $MnCO_3$ in reduced soils. Exchangeable, sorbed, and organically complexed Mn are the forms important to plant availability. As much as 80% of soluble Mn in some soils is complexed with organic matter. Soil properties and processes affecting Mn solubility include soil pH, complexation/chelation, and redox potential. Manganese solubility decreases about 100-fold as pH increases by one unit and also increases markedly when soils become reduced and Mn oxides (e.g., MnO_2) dissolve.

Plant available boron (B) originates in most soils from the weathering of sedimentary rocks (e.g., shales) and tourmaline (a highly insoluble borosilicate mineral). Total soil B concentrations usually range from 2 to 200 mg kg^{-1} with <5% available to plants. Unlike Cu, Zn, Fe, and Mn, which exist primarily as divalent cations, B in soils is found as undissociated $H_3BO_3^0$ (pH 5–9) or as the $B(OH)_4^-$ anion (dominant species at pH > 9.2). Major sources of plant available B in soils include those that are sorbed by soil clays and oxides or hydroxides of Fe/Al and the B complexed by organic matter. The uncharged nature of the $H_3BO_3^0$ molecule makes it highly mobile in many soils, particularly those low in clays, oxides, and organic matter. However, B availability is well known to decrease at pH > 6.5–7.0 because of an increased affinity of soil clays and oxides for the $B(OH)_4^-$ anion.

Molybdenum is present in soils at very low levels, with total Mo values ranging from 0.2 to 5.0 mg kg^{-1} and soil solution concentrations <0.5 µg L^{-1}. Plant available Mo is anionic in nature, existing in most soil solutions as $HMoO_4^-$ or MoO_4^{2-}, species strongly sorbed by Fe/Al oxides under acidic conditions and/or complexed by organic matter. Unlike all other micronutrients, Mo availability increases with soil pH, due to the greater solubility of several Mo-bearing minerals at pH values >7.0 and to a decreased affinity of most soils for Mo.

Chloride occurs mostly in igneous and metamorphic rocks and, once weathered, is found in soils as precipitates and soluble salts such as NaCl, $MgCl_2$, and $CaCl_2$ and in the soil solution as the Cl^- anion. Dissolution of these salts is the primary process controlling Cl^- availability to plants. Soil solution concentrations of Cl^- vary widely as a function of soil type and geographic location, ranging from <0.5 to >6000 mg L^{-1} in soils from arid regions. Chloride is very mobile in most soils, analogous in many ways to NO_3-N, and significant retention of Cl^- only occurs in highly acid soils that can develop positive, pH-dependent electrostatic charge on 1:1 clay or Al/Fe oxides.

13.2.4.5.1 Current Soil Testing Methods for Micronutrients

Soil tests for micronutrients have historically been conducted as special tests, restricted to situations where soil properties or crop characteristics indicated an economic response to micronutrient fertilization was possible. Consequently, for many years, and even today, separate soil tests were used for each micronutrient, or for groups of micronutrients with similar properties. However, the advent of multielement extractants, such as Mehlich 3, DTPA, and AB-DTPA, and of instruments capable of rapid, simultaneous analysis of soil extracts for several elements (ICP-AES) has increased the use of micronutrient soil testing in many areas of the world.

Soil sampling, handling, and storage for micronutrient soil testing are conducted following the standard techniques outlined in Section 13.2.3. Soil samples are almost always collected from topsoils (0–15 or 0–20 cm). The major consideration for micronutrients is the need to avoid contamination of the soil sample during sample collection, handling, and storage. For example, galvanized sampling tools and mixing buckets and some rubber stoppers contain Zn, metal surfaces on equipment used for grinding and sieving samples may contain Cu, Fe, and Zn, borosilicate glassware should be avoided during the extraction of B from soils as should the use of paper bags and boxes that may contain B in glues, and many common laboratory reagents contain Cl. Sample drying, the length and force of soil grinding, the speed of shaking during the extraction process, and the type of extraction vessel and soil:solution ratio have been shown to affect the quantity of extractable Cu, Fe, Mn, and Zn in some soils (Soltanpour et al., 1976, 1979). Given these somewhat unpredictable potential sources of error, following standardized methods for soil sampling, handling, preparation, and extraction is a vital aspect of micronutrient soil testing. Only a small amount of contamination or a slight alteration in procedure can badly skew a soil analysis, resulting in an erroneous recommendation.

Chemical extraction is the standard approach to assess micronutrient availability in soils. The major types of micronutrient soil tests used for Cu, Fe, Mn, and Zn are dilute acids and extractants that contain chelating agents. Less commonly used tests include neutral salts, which remove only small quantities of most micronutrient cations, and reducing agents, such as hydroquinone, used only for Mn (Table 13.4). In general, soil tests have been designed to remove soluble micronutrient forms

TABLE 13.4 Summary of Major Soil Testing Methods, Interacting Factors, and References for Micronutrient Soil Tests

Micronutrient	Soil Test and Critical Range	Comments and Interacting Factors Used in Soil Test Interpretation
Boron[a]	Hot water: 0.1–2.0 mg kg^{-1} Mehlich III: 0.7–3.0 mg kg^{-1}	Hot water is the most widely used method. Interacting factors include crop yield goal, pH, soil moisture, texture, organic matter, and soil type
Copper	AB-DTPA: 0.5–2.5 mg kg^{-1} DTPA: 0.1–2.5 mg kg^{-1} Mehlich I: 0.1–10 mg kg^{-1} Mehlich III: 0.3–15 mg kg^{-1} 0.1 M HCl: 0.1–2.0 mg kg^{-1}	AB-DTPA and DTPA are used for alkaline soils, Mehlich III for alkaline and acid soils, and Mehlich I and 0.1 M HCl for acid, low CEC soils. Interacting factors include crop type, organic matter, pH, and % $CaCO_3$.
Iron	AB-DTPA: 4.0–5.0 mg kg^{-1} DPTA: 2.5–5.0 mg kg^{-1}	AB-DTPA and DTPA are used for alkaline soils (Fe deficiency is very rare with acid soils). Interacting factors include pH, % $CaCO_3$, CEC, organic matter, and soil moisture

Manganese	AB-DTPA: 0.5–5.0 mg kg^{-1}
	DTPA: 1.0–5.0 mg kg^{-1}
	Mehlich I:
	  5.0 mg kg^{-1} at pH 6
	  10 mg kg^{-1} at pH 7
	Mehlich III:
	  4.0 mg kg^{-1} at pH 6
	  8.0 mg kg^{-1} at pH 7
	0.1 M HCl: 1.0–4.0 mg kg^{-1}
		AB-DTPA and DTPA are used for alkaline soils, Mehlich III for alkaline and acid soils, and Mehlich I for acid, low CEC soils. Interacting factors include pH, texture, organic matter, and % $CaCO_3$

Molybdenum	Ammonium oxalate—pH 3.3: 0.1–0.3 mg kg^{-1}	Soil testing for Mo is rarely done. Interacting factors are pH and crop
Zinc	AB-DTPA: 0.5–1.0 mg kg^{-1} DTPA: 0.2–2.0 mg kg^{-1} Mehlich I: 0.5–3.0 mg kg^{-1} Mehlich IH: 1.0–2.0 mg kg^{-1} 0.1 M HCl: 1.0–5.0 mg kg^{-1}	AB-DTPA and DTPA are used for alkaline soils, Mehlich III for alkaline and acid soils, and Mehlich I and 0.1 M HCl for acid, low CEC soils. Interacting factors are pH, % $CaCO_3$, P, organic matter, % clay, and CEC

Sources: Martens, D.C., and W.L. Lindsay. 1990. Testing soils for copper, iron, manganese, and zinc, p. 229–264. *In* R.L. Westerman (ed.) Soil testing and plant analysis. 3rd edn. SSSA Book Series No. 3. SSSA, Madison, WI; Sims, J.T., and G.V. Johnson. 1991. Micronutrient soil tests, p. 427–476. *In* J.J. Mortvedt et al. (eds.) Micronutrients in agriculture. 2nd edn. SSSA, Madison, WI.

[a] References: Hot water B, Berger and Truog (1940); Mehlich 3 B, Shuman et al. (1992).

(including organically complexed) by solvent action, to displace exchangeable and sorbed forms by ion exchange and desorption reactions with other cations or with hydrogen ions (H_3O^+), to partially dissolve soil minerals or oxides that contain precipitated and occluded forms, and to dissociate or chelate micronutrient cations that are complexed by solid phases of organic matter. In most cases all four cations are extracted and analyzed simultaneously. The most common instrumental techniques used for micronutrient cations are AAS and ICP-AES.

Dilute acids (0.025–0.1 M) have been used for decades as soil tests for micronutrient cations, most successfully on acidic soils because they are not adequately buffered to extract sufficient quantities of micronutrients from calcareous soils. These extractants work primarily by dissociation, displacement, and partial acidic dissolution of cations from soil clays, oxides, and organic matter. The most common dilute acid soil tests today are the Mehlich 1 (dilute double acid: 0.0125 M H_2SO_4 + 0.05 M HCl) and 0.1 M HCl. The Mehlich 1 is a multielement extractant for macro- and micronutrients used in the southeastern and mid-Atlantic United States, and in South America. The 0.1 M HCl soil test has been used for a wider range of soils, primarily as an extractant for Zn, Cu, and Mn.

The other major category of soil tests for micronutrient cations is extractants containing chelating agents, most commonly DTPA and EDTA. Chelating agents reduce the activity of free metal ions in the soil solution by forming metal-chelate complexes, much as occurs in the rhizosphere of a plant. Replenishment of free ion concentrations in the soil solution (during the extraction process) by release from solid phases in the soil (clays, oxides, organic matter) occurs in response to the formation of the soluble metal-chelate complex. The amount of a micronutrient extracted by a chelate-based soil test therefore reflects both the initial quantity present in the soil solution and the ability of the soil to maintain that concentration. Chelate-based extractants thus simulate nutrient removal from the soil by plant uptake and replenishment of the soil solution by labile solid phases. Most chelate-based soil tests were developed for specific physiographic regions and soil types and are buffered at specific pH and ionic strength values to avoid the release of micronutrients from nonlabile solid phases in the soil. Hence, it is important to use them only for the soil type and conditions for which they were originally calibrated. For example, the DTPA soil test, commonly used for calcareous soils, is buffered at pH 7.3 and contains 0.01 M $CaCl_2$ to prevent the dissolution of carbonate minerals that might contain occluded or precipitated Cu, Fe, Mn, and Zn. Clearly, since the DTPA was developed for calcareous soils it would be inappropriate for highly acid soils without careful calibration and perhaps modification of the extractant composition (Lindsay and Norvell, 1978; Norvell, 1984; O'Connor, 1988). EDTA has been successfully used on a wide range of soils either alone or in multielement soil tests (i.e., 0.001 M EDTA is in the Mehlich 3 soil test and the modified Olsen soil test: 0.5 M

NaHCO$_3$ + 0.01 M EDTA, pH 8.6) (Viro, 1955). Similarly, DTPA is included in the AB-DTPA extractant (1 M NH$_4$HCO$_3$ + 0.005 M DTPA, pH 7.6) now widely used in the western United States.

Soil tests for the anionic or uncharged micronutrients (B, Mo, and Cl) have received less attention than those for micronutrient cations because of the relatively rare nature of crop response to fertilization with these elements. Most soil tests have focused on methods that remove soluble, sorbed, or organically complexed forms of these micronutrients.

The most common soil test used for B has been the hot water extraction method of Berger and Truog (1940) in which soil is boiled with water or 0.01 M CaCl$_2$, using a reflux condenser, removing soluble B and organically complexed B. Although shown to be a reasonably good predictor of plant response to B, the cumbersome, time-consuming nature of the Berger and Truog soil test has made its routine use difficult. Research by Mahler et al. (1984) using boiling plastic pouches, by Gestring and Soltanpour (1984) with the ammonium bicarbonate + DPTA soil test, and by Shuman et al. (1992) with the Mehlich 3 extractant has identified some practical alternatives to the original hot water method. Molybdenum is usually extracted with acid ammonium oxalate, primarily via a desorption reaction with the added oxalate, while deionized water or any dilute salt solution (e.g., 0.01 M Ca(NO$_3$)$_2$, 0.5 M K$_2$SO$_4$) can be used as a soil test extractant for Cl because of its high solubility in most soils.

Considerations in interpretation of micronutrient soil tests: Micronutrient deficiencies are, for the most part, associated with specific combinations of soil and plant factors that are reasonably well understood (Tisdale et al., 1993). Copper deficiencies are most common on soils that are extremely high in organic matter (peats, mucks) or with high pH, calcareous soils, while deficiencies with the other micronutrient cations (Fe, Mn, Zn) are almost always confined to calcareous or overlimed soils and sensitive crops. Boron deficiency is most frequently observed on sandy, low organic matter soils or following extremely dry periods that reduce the mineralization of organic matter and thus the release of organically bound B. Molybdenum deficiency rarely occurs except with very acid soils and then only with crops that are highly sensitive to low concentrations of soil Mo (e.g., legumes, crucifers, citrus). Chloride deficiencies are very unusual and confined to certain physiographic regions, such as the Northern Great Plains of the United States.

The critical value approach is the most widely used method to interpret the results of a micronutrient soil test (Tables 13.1 and 13.4). However, when micronutrient deficiencies are a definite concern, it has usually been shown that the predictive value of a micronutrient soil test can be improved by evaluation of more than one soil property or by knowledge of the crop to be grown. For instance, soil tests for Mn and Zn are much more accurate when soil pH is known; other examples of interacting factors that can improve soil test interpretation for micronutrients are given in Table 13.4. In some cases, quantitative "availability indexes," usually multiple regression equations based on the soil test result and another soil property, are calculated and used in place of the critical value approach.

Finally, several micronutrients can be toxic to plants if present in soils at high concentrations. The most common micronutrient phytotoxicities occur with Mn in highly acid soils (pH < 5.2) and B where only a slight over-application of fertilizer B can cause phytotoxicity. Although unusual, Cu and Zn can occasionally become phytotoxic in soils amended with agricultural, municipal, and industrial waste products, such as animal manures (pig and poultry), municipal biosolids, and some by-products of mining industries, if recommended or mandated management practices are not strictly followed (Sukkariyah et al., 2005). Critical phytotoxic levels based on micronutrient soil tests are much more difficult to establish than deficiency values and are usually highly specific to the plant that is grown and soil type.

13.2.5 Soil Testing Methods for Soil Chemical, Physical, and Biological Properties

The availability of essential and nonessential elements to plants, their potential to become phytotoxic, or to cause environmental problems via leaching, erosion, runoff, and/or volatilization depends upon soil physical, chemical, and biological properties. Some of these properties are routinely measured by soil testing laboratories; others are only measured on selected soil samples. Some are rarely measured at all but can be inferred from other soil properties or from information in USDA-NRCS Soil Survey manuals. Since the focus of this chapter is soil fertility evaluation, only the soil properties that are most relevant to plant growth are discussed (soil pH, lime requirement, organic matter, and soluble salts).

13.2.5.1 Soil pH

Soil pH is an index of the hydrogen ion activity in the soil solution that is in equilibrium with H retained by soil colloids (clays, organic matter, oxides) (van Lierop, 1990). It is a measure of the degree of acidity or alkalinity of a soil and is commonly measured electrometrically using a pH meter equipped with glass and reference electrodes (Thomas, 1996). Soil pH is one of the most useful pieces of information for soil fertility evaluation and management because it provides information on (1) the solubility, and thus potential availability or phytotoxicity of some plant nutrients and nonessential elements and (2) the relative biological activity of plants and soil microorganisms. The solubility of most micronutrients, and several nonessential trace elements (e.g., Cd, Ni, Pb) for plant uptake is highly pH dependent because the solubility of the solid phases containing these essential elements changes with soil pH. For most elements, solubility increases as the soil becomes more acidic; exceptions include P, which is most available at pH ranges between 5.5 and 7.5, and Ca and Mo, which are most available at higher pH values (pH > 7.0). Other processes important to nutrient and nonessential element retention in soils (e.g., cation exchange, sorption/desorption) also vary with pH. Acid soil infertility is most severe at pH values < 5.5 and is caused by the increased solubility and toxicity of aluminum (Al) and Mn and by the decreased plant availability of Ca, Mg, Mo, and P. Soil N availability is less under acidic conditions because the bacteria responsible for mineralization of soil organic N are most active at neutral or slightly acid soil conditions. Alkaline soil infertility is most common in calcareous or overlimed soils and results from the reduced availability of several important nutrients including P, B, Cu, Fe, Mn, and Zn due to the decreased elemental solubility (P, Cu, Fe, Mn, Zn) or greater sorption (B) that occurs at higher pH values.

13.2.5.2 Lime Requirement

Soil pH is an index of the soil chemical environment and its general suitability for plant growth. Soil pH, however, provides no information on the amount of soil amendment needed to correct problems with acid or alkaline soil infertility. To determine the rate of lime (or acidulent) needed, it is necessary to measure the buffer capacity of the soil (ability of the soil to resist a change in pH). For acid soils, lime requirement is defined as the amount of agricultural limestone or other basic material needed to increase the soil pH from an unacceptably acidic condition to a value that is considered optimum for the desired use of the soil (Sims, 1996). The acidification requirement is similar to lime requirement and refers to the amount of acidulent (usually elemental sulfur or aluminum sulfate) needed to decrease soil pH to an optimum value. The lime requirement of a soil can be measured by a number of methods, but the most common is the use of soil-buffer equilibrations to determine "buffer pH" (Sims, 1996). A buffer pH measurement is conducted by adding a chemical solution buffered at a high pH (pH 7.5–8.0) to a soil sample, allowing the soil and buffer to equilibrate, and then measuring the pH of the soil-buffer suspension. The decrease in buffer pH that occurs when the acidity in the soil reacts with the buffered chemical solution is an index of the amount of soil acidity that must be neutralized by liming to adjust the soil to the desired pH. Field calibrations between buffer pH measurements and changes in soil pH upon liming are essential to the development of a buffer pH test. The most common buffer pH solutions in the United States are the Shoemaker–McLean–Pratt ([SMP], pH 7.5), the Adams–Evans ([AE] pH 8.0), and the Mehlich (pH 6.6). Recently, environmental and toxicity concerns about chromium and paranitrophenol, constituents of the SMP buffer, led Sikora (2006) to propose a modified SMP buffer. Chemicals chosen to replace chromium and p-nitrophenol were 2-(N-morpholino) ethanesulfonic acid monohydrate (MES) and imidazole. The new buffer is adjusted to pH 7.7. A study comparing soil-buffer pH with the new buffer versus the SMP buffer using 255 Kentucky and 87 soils from throughout the United States showed that the two methods were highly correlated ($r^2 > 0.97$). Similar concerns with the toxicity of paranitrophenol, exist for the AE buffer. Consequently, Sikora and Moore (2009) developed the Moore–Sikora (MS) buffer, which contains no hazardous chemicals and was shown to be highly correlated ($r^2 > 0.98$) with results from the AE buffer. The buffer has been successfully used by Clemson University (South Carolina) for 4 years but the authors suggest further field studies are needed prior to its widespread adoption in other regions of the United States. Another lime requirement technique involves extraction of exchangeable acidity or exchangeable Al^{3+} from a soil with a salt solution (e.g., 2 M KCl), followed by titration of the extract with a standardized base. This is a rapid, inexpensive method that is well adapted to highly acidic, aluminous soils in areas with limited supplies of limestone. Application of sufficient limestone to neutralize 1.5–2.0 times the amount of exchangeable acidity is often adequate to eliminate some of the more serious limitations associated with acid soil fertility (e.g., Al toxicity). Finally, Liu et al. (2005) have shown success with a direct titration method to determine lime requirement based on an initial pH reading in 0.01 M $CaCl_2$ and a second reading following the addition of one dose of $Ca(OH)_2$, followed by extrapolation to the target pH. The more widespread availability of automatic titration systems suggests this approach should receive further investigation.

13.2.5.3 Organic Matter

The organic matter content of a soil is extremely important to soil fertility and many soil testing laboratories now include organic matter as a standard component in the routine soil test (which usually consists of pH, buffer pH, organic matter, and soil test extractable P, K, Ca, Mg). Soil organic matter is the nonliving organic material in the soil and includes both nonhumus (fresh plant, animal, and microbial residues) and humus (amorphous, highly stable, dark colored, organic material). Although organic matter is only a small percentage of most topsoils (1%–5%) and is present at very low levels in subsoils (<0.5%), consideration of organic matter in soil fertility evaluation is critical because it (1) provides plant nutrients, especially N, B, P, and S, as microorganisms decompose organic matter; (2) acts as a chelate, particularly important in maintaining micronutrient cations in a plant available form; (3) effectively complexes Al, reducing its phytotoxic effects in many soils; (4) has a high water holding capacity by weight and thus helps minimize the effects of moisture stress on plants; (5) is the predominant source of pH dependent CEC in most soils and thus contributes to the overall capacity of soils to retain nutrients and nonessential elements; (6) is a significant source of pH buffering, preventing marked and often undesirable changes in soil pH due to anthropogenic inputs; and (7) acts as an aggregating agent, cementing smaller soil particles together and thus improving soil structure, resulting in better aeration and more prolific root growth. Despite the many important roles soil organic matter plays in soil fertility, only recently soil testing laboratories have begun to measure and routinely report this soil property. The long-standing, traditional approach used to estimate soil organic matter content was to measure organic carbon (C) by wet chemical oxidation using dichromate ($Cr_2O_7^{2-}$) as the oxidant. Soil organic matter was then calculated from an empirically derived relationship between organic matter and organic C. This method was too time consuming to use on a routine basis and also generated a significant amount of high Cr waste; hence, it was primarily used as a special test. Most soil testing laboratories now estimate soil organic matter content by the "loss-on-ignition" (LOI) method in which a soil sample is "ignited" in a high temperature oven at ~360°C–400°C for several hours or overnight. The weight loss upon ignition is assumed to be proportional to soil organic matter content. Usually, an empirical relationship between LOI and some direct measure of organic matter (e.g., dichromate oxidation) is used to calculate estimated organic matter content. The LOI method is well suited to modern soil testing laboratories that wish to include an estimate of soil organic matter content in the routine soil test. Advances in electronic weighing and data acquisition and processing have resulted in LOI becoming an efficient, reasonably accurate approach to estimate soil organic matter.

13.2.5.4 Soluble Salts

Soil salinity, defined as the presence of excessive levels of dissolved inorganic solutes in the soil solution, is a global problem and directly affects soil fertility. Soils high in soluble salts, either naturally or due to inputs of salts in fertilizers and irrigation waters, can negatively affect plant growth in several ways. Specific ion toxicities can occur, particularly if high concentrations of Na, Cl, and B are present, causing direct injury to plants, especially young seedlings. Saline soils also disrupt plant water relations by decreasing the osmotic potential of the soil solution, thus making it more difficult for plants to extract water from soils, even to the point of causing plant injury and death. Measuring soluble salts is a fairly easy task but is usually done as a special test even in arid regions where soil salinity is a common problem. Rapid tests for soluble salts are done by measuring the electrical conductivity of a soil:water extract at a 1:2 or 1:5 soil:solution ratio. A more time consuming test, but one that better represents the soluble salts concentration in the soil solution is the saturated paste extract, obtained by mixing soil and deionized water to the point of saturation, followed by filtration and analysis of the extract for electrical conductivity.

13.2.6 Soil Testing: Interpretation and Recommendations

Interpretation is a systematic process, relying primarily on statistical evaluation of research comparing soil test parameters with plant performance, or other indexes of the success of a land use program. Recommendations arising from soil test interpretation must not only be quantitative in nature but also include professional judgment since it is not possible to identify all the factors that control plant performance from analysis of a single soil sample. Individuals responsible for nutrient management recommendations must be thoroughly familiar with the process of soil testing and with all aspects of the intended land use, including soil types, plants to be grown, climate, crop management practices, and any economic or environmental factors that may restrict a recommendation.

13.2.6.1 Correlation and Calibration of Soil Tests

Interpretation of soil analyses begins with soil test correlation, defined as "...the process of determining whether there is a relationship between plant uptake of a nutrient or yield and the amount of nutrient extracted by a particular soil test" (Corey, 1987). To be of value, a soil test must first be shown by laboratory, greenhouse, and field research to be statistically correlated with some measure of plant performance. Greenhouse studies are usually the first step in soil test correlation and can rapidly and inexpensively assess the potential value of a soil test for widely differing soils and plants. The standard approach is to obtain samples of soils representative of the range of soil types where the soil test will be used, measure the amount of soil test extractable nutrient (or nonessential element) in each soil, grow plants in the soils under controlled greenhouse conditions, where moisture, light, and spatial variability are minimized, and then measure plant yield and elemental composition. Statistical correlation and regression methods are then used to assess the relationship between soil test level and plant response. Correlation analysis determines if the change in plant yield or nutrient composition is proportional to the amount of nutrient extracted by a soil test. If a high degree of correlation exists, regression analysis will provide a predictive equation that reliably estimates plant yield or elemental composition at each soil test value. In some cases, multiple correlation and regression analysis are used to develop a predictive equation that quantifies the relationship between plant performance and more than one soil property (e.g., soil test value and pH, organic matter, texture, etc.).

If a soil test is significantly correlated with plant performance in the well-controlled greenhouse environment, field experiments are then conducted to determine how accurate the test will be under normal growing conditions. For greatest reliability, field trials should be conducted at many locations with a range of soil types and soil test values for the nutrient of interest. The trials should be done for several years and should include multiple rates of the nutrient being investigated. Experiments should be replicated at least three times and other variables besides the nutrient being studied that affect plant performance should be controlled to the extent possible (e.g., using irrigation to minimize soil moisture stress, pesticides to prevent insect damage and weed competition, and fertilization or liming to ensure that other nutrients do not limit plant growth and yield). Correlation analysis is again used to determine if there is a statistically significant relationship between the soil test value for a nutrient and plant response. Field experiments usually have poorer correlation coefficients than greenhouse studies because (1) they are conducted at fewer locations due to the time and expense required, and (2) uncontrolled variability is usually present that affects the soil test-plant performance relationship in an unpredictable manner. Recently, the costs and time required to conduct field and greenhouse studies have caused a greater reliance on laboratory correlation studies as a means to evaluate new soil test extractants. For example, several laboratory studies have compared the Mehlich 3 multielement soil test with earlier soil tests (Hanlon and Johnson, 1984; Wolf and Baker, 1985; Sims, 1989). Many of these studies reported highly significant correlations between nutrients extracted by Mehlich 3 and those extracted by the Mehlich 1, Bray P_1, 1 M NH_4OAc, and EDTA, indicating that the Mehlich 3 could be as reliable a means to evaluate soil fertility as the existing soil test. However, while laboratory-based correlation studies may be acceptable for the preliminary evaluation of new soil tests, they should not be used as the sole means to determine soil test reliability. It is fundamental to soil test correlation that the amount of an element extracted by the soil test be correlated with what is of actual interest, usually plant yield or elemental composition (Fixen and Grove, 1990).

The next step in soil test development is calibration, defined as "...ascertaining the degree of limitation to crop growth or the probability of a growth response to applications of a nutrient at a given soil test level" (Dahnke and Olson, 1990). The purpose of calibration studies is to categorize soil test levels in terms of the probability of economic response to applications of a nutrient or for their potential to affect the environment. Traditionally, soil test results have been categorized as very low, medium, high, and very high. More recently, the terms "optimum" and "excessive" have been used to describe soil nutrient status and some regions, such as the northeastern United States have identified two separate classes of soil test category—"crop response" and "environmental impact" (Beegle, 1995). Tisdale et al. (1993) suggested that probability of an economic response to fertilization with P and K for a soil rated as "low," "medium," "high," and "very high" in either nutrient would be 70%–95%, 40%–70%, 10%–40%, and <10%, respectively. Several of the more important approaches used in soil test calibration are briefly described below; for a more thorough review of this subject, refer to Dahnke and Olson (1990) and Black (1993).

The calibration process is essentially an effort to mathematically model the relationship between soil test level and plant response to nutrient additions, which is almost always nonlinear. Consequently, curvilinear regression models (exponential, quadratic, quadratic-plateau) are often used to identify the point where plant performance is optimal (e.g., the plateau yield, usually associated with 93%–95% of maximum attainable yield) and then to determine the soil test value associated with optimum yield, which is referred to as the critical level or critical value (Figure 13.3a). Curvilinear models are often based on relative yield or percent yield, defined as the yield obtained without addition of the nutrient being studied divided by the yield attained at that location when no other factors are limiting. Relative yield data from field trials in the region of interest are combined and plotted against the soil test value from the control treatment for each trial and the critical level is determined mathematically or graphically. Use of relative yields minimizes the influence of uncontrolled variables and allows for more effective interpretation of data collected over many different years, locations, soils, climates, and management settings. Once the critical value has been identified, soil test values below this level are subdivided into categories that are associated with the probability of crop response (e.g., low, medium, optimum) and the nutrient rate required for an optimum yield. It is important to note that critical soil test levels may vary between soils, crops, and climatic regions and will usually differ between soil test extractants. For instance, critical soil test P values by the Bray P_1, Olsen, Mehlich 1, and Mehlich 3 soil tests are about 30, 12, 25, and 50 mg kg^{-1}, respectively (Jones, 1999).

Another calibration approach used to identify critical soil test levels is the Cate–Nelson method (Figure 13.3b; Cate and

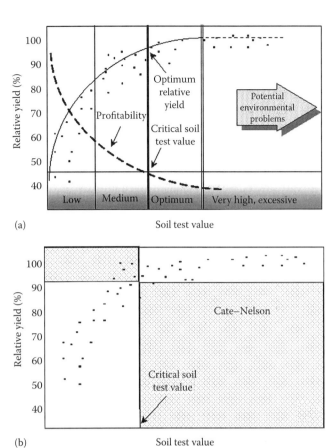

FIGURE 13.3 Generalized illustration of the principles of soil test correlation and calibration using (a) curvilinear models and (b) the Cate–Nelson approach.

Nelson, 1965). In this approach, relative yield is plotted against soil test value and the data are subdivided graphically into four quadrants either visually, using professional judgment, or mathematically (Nelson and Anderson, 1977) by placing a horizontal line at the optimum relative yield (93%–95% of maximum yield) and a vertical line at the soil test value that minimizes the number of points in the upper left and lower right quadrants. If the soil test under consideration is reliable, there will be few points in the upper left or lower right quadrants. Points in the upper left quadrant would mean that a low soil test value was associated with a high relative yield; points in the lower right quadrant are those where a high soil test value was associated with a low yield. Both of these are inconsistent with the basic premise of soil testing, that is, a soil test can accurately and reliably separate responsive from nonresponsive sites. If most points are in the lower left or upper right quadrants, then the soil test accurately predicts plant performance, that is, low soil tests have low relative yields and high soil tests have high relative yields.

As shown in Figure 13.4, the mathematical approach used to model the relationship between plant performance (relative yield) and rate of nutrient added can affect the determination of the economically optimum nutrient rate. Thorough discussions of the mathematical models used to interpret soil test results are provided by Black (1993).

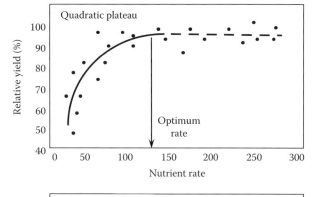
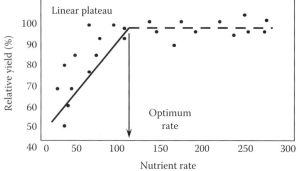
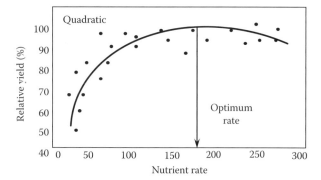

FIGURE 13.4 Illustration of the influence of type of mathematical model selected on the identification of the nutrient rate required for optimum yield.

13.2.6.2 Recommendation Philosophies for Soil Test Interpretation

The final phase in soil test development is the recommendation process in which the actual amount of nutrient to be applied, the application method, and the timing of the application are specified in detail. Individuals making a nutrient recommendation must integrate the quantitative information from soil test calibrations (e.g., the probability and magnitude of response likely to occur at a given soil test value of the nutrient) with other, more subjective aspects of nutrient management. Individuals skilled at making efficient recommendations are able to combine soil testing information with their professional experience, judgment, and scientific understanding of the system of interest in a manner that optimizes the profitability of nutrient use while minimizing any potential impacts on environmental quality. Given the subjective nature of this process, it is not surprising that several different recommendation philosophies have evolved and now receive widespread use. The two most common philosophies, for agricultural crops, are "rapid build-up and maintenance" and "sufficiency level." Both philosophies are more commonly used with "immobile" plant nutrients, those that do not readily leach from topsoils or the root zone (P, K, Ca, Mg, Cu, Fe, Mn, Mo, Zn), than with "mobile" nutrients (B, Cl, NO_3-N, SO_4-S) where more comprehensive soil test approaches, such as subsoil testing, must be relied upon.

The rapid build-up and maintenance approach recommends that soil fertility be built to an optimum level as rapidly as possible, usually within 2 years. Following this initial "build-up" of soil fertility, "maintenance" nutrient applications are made annually at rates equal to the amounts removed in the harvested portion of the crop. This philosophy is sometimes referred to as "fertilization of the soil" and proceeds somewhat independent of soil testing since nutrient applications are made each year, regardless of soil test results. As noted by Dahnke and Olson (1990), "...the rapid build-up and maintenance concept discounts the inherent nutrient delivery capacity of a soil's native mineral reserves which, with most soils other than sands is large for most nutrients" and "Complete adherence to this system... would eliminate the need for further soil testing." While still used, economic and environmental questions about the appropriateness of this recommendation philosophy persist (Olson et al., 1987).

The most widely used recommendation philosophy today is the "sufficiency level" approach, which "...promotes the idea that a measurable soil test level exists below which responses to added fertilizer are probable and above which they are not" (Eckert, 1987). This approach is also sometimes referred to as "fertilization of the crop" and is a more conservative approach than rapid build-up and maintenance because nutrients are only recommended when soil test values are below the critical soil test level and are applied proportionate to the soil test category (i.e., more nutrients are added to soils that are rated low than those that are rated medium). The sufficiency level approach also inherently includes a build-up phase, but once soil fertility is in the optimum range, nutrient additions cease, or are minimal, until subsequent soil testing indicates that soil fertility has declined to the point where an economic response to further nutrient inputs is likely. There is no evidence to support the contention that the sufficiency level approach causes unnecessary depletion of soil nutrient reserves; in fact most studies have shown that adhering to this philosophy results in a slight buildup of soil fertility. Consequently, most soil test calibration research supports the use of the sufficiency level concept for soil test recommendations.

13.2.7 Environmental Soil Testing

13.2.7.1 Overview of the Principles and Purposes of Environmental Soil Tests

Soil testing has traditionally been used to evaluate the soil limitations to agronomic crop performance imposed by nutrient deficiencies (or pH, soluble salts, etc.) and to guide the recommendation process so that these limitations could be eliminated economically and without impacting the quality of our environment. In recent years, interest has grown in environmental soil testing, defined here as "quantitative analysis of soils to determine if environmentally unacceptable levels of nutrients, nonessential elements or organic compounds are present." Environmental soil testing is a much more ambiguous process than agricultural soil testing because it is usually quite difficult to quantify the meaning of the term "environmentally unacceptable." Absent a clear, quantitative measure of success, such as crop yield for agricultural soil testing, the entire process of soil testing, from sample collection to recommendation, becomes more diffuse and more complex. Nevertheless, the rising interest in environmental protection in many areas of the world has prompted an increased effort to use soil testing as an index of the risks posed by soils to other sectors of the environment, particularly ground and surface waters. Given this, it is appropriate to consider the most effective means to restructure soil testing practices to meet environmental goals (Sims et al., 1997).

In the broadest sense, the goals of environmental soil testing are the same as those of routine, agricultural soil testing—rapid, accurate, and reproducible soil analysis by the most appropriate methods, and a reasonable interpretation of results related to environmental risk. Factors to consider when developing an environmental soil testing program are discussed next, using two reasonably common examples, potentially toxic trace elements and plant nutrients that are known to degrade water quality. In both cases, it is critical to consider the conceptual differences in interpretation of an environmental soil test, illustrated in Figure 13.5, compared to an agricultural interpretation (Figure 13.3).

13.2.7.2 Soil Testing for Potentially Toxic Trace Elements

Soil testing for potentially toxic trace elements is an environmental issue because some plant nutrients (Cu, Mo, and Zn) and nonessential elements (As, Cd, Cr, Hg, Ni, Pb, Se) have been shown to be toxic to either plants, animals, or humans. Soils may contain naturally high concentrations of one or more of these elements (very unusual), or concentrations may increase as a result of some anthropogenic activity such as the intentional addition of wastes as beneficial soil amendments (e.g., animal manures, municipal sewage sludges, industrial by-products). Some soils may be highly polluted with toxic elements due to mismanagement of potentially beneficial wastes, by an accidental spill or discharge, or as a result of an industrial activity such as mining or smelting. Note that contamination and pollution are not the same. Contamination occurs when a substance is present at concentrations higher than would occur naturally but no adverse effect on an organism is apparent while pollution implies not only an elevated concentration in the soil, but clearly documented adverse effects on some organism (Pierzynski et al., 1994).

Soil testing for potentially toxic trace elements begins with an understanding of the nature of the risk involved—what organisms may be affected, by what pathway, and by which elements. Primary areas of concern in general are direct soil ingestion,

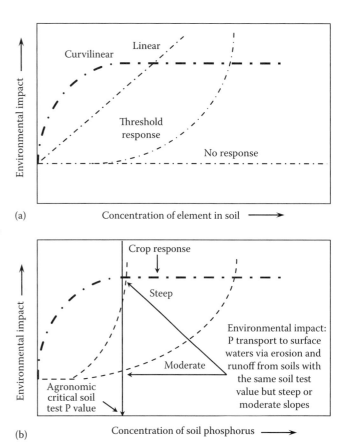

FIGURE 13.5 Illustration of points to consider in interpretation of environmental soil tests. As seen in (a) increasing the concentrations of soil nutrients or nonessential elements can induce no environmental response, cause a response once a threshold is reached, or have a linear or curvilinear effect on the environmental parameter of interest. In (b) crop response to increasing soil concentrations of a nutrient is contrasted with the environmental impact of that element, which can be modified by other soil properties. For example, crop response to inputs of phosphorus (P) is normally curvilinear, while the impacts of P on surface waters due to erosion and runoff can follow a threshold response. Soils with the same soil test P value, but varying slopes (steep or moderate) will have differing environmental impacts. (Figure a adapted from Pierzynski, G.M., Sims, J.T., and Vance, G.F. 1994. Soils and environmental quality. Lewis Publishers, CRC Press, Boca Raton, FL.)

phytotoxicity, plant uptake and food chain contamination, and water pollution from erosion, runoff, or leaching. Direct soil ingestion is normally only a concern for Pb and with young children who are most likely to ingest soil or inhale dust from high Pb soils. Phytotoxicity is rarely an issue with Pb except in highly polluted soils at industrial sites but is a greater concern with Cu, Cr, Ni, and Zn. Food chain contamination and human health effects are most often associated with Cd and Hg and water quality concerns with As, Cu, Hg, and Se. Knowledge of the nature of the risk and the element of concern helps to determine the most effective soil sampling protocol. In the case of Pb in urban soils where human health is the concern, it may be advisable to collect very shallow soil samples (<2 cm) since this is the soil depth most likely to be ingested. A similar depth would be useful if runoff or erosion were the issue since rainfall primarily interacts with only the uppermost few cm of the soils. However, if concerns exist about ground water pollution, sampling into subsoil horizons, perhaps to a depth of 1–2 m would be recommended to determine if elemental leaching is occurring. For elements where the main concern is phytotoxicity and food chain contamination, the normal sampling depths associated with the crops to be grown are usually acceptable (e.g., 0–20 cm). If remediation of the site is the goal, either by soil removal, soil washing, or soil amendment, then systematic deep sampling (e.g., 0–5, 5–10, 10–20, 20–40, 40–60, 60–100 cm) is recommended to determine the depth of soil contamination and thus the extent of remediation required. Careful consideration should also be given to the soil sample handling and preparation steps to avoid contamination from any sampling and mixing tools or grinding and sieving devices (e.g., stainless steel, used in many electric grinding device contains Cr and Ni) and to protect the health and safety of the individual taking the sample. The method of analysis to be used varies with the intent of the test. If the goal is to assess biological availability (e.g., plant uptake, human ingestion) then many dilute acid or chelate-based soil test extractants, or the PBET mentioned earlier (Section 13.2.2.1) may be suitable, providing due consideration is given to the most appropriate test for the intended land use or human health concern (O'Connor, 1988). Interest, and field research, using the PBET has increased in recent years as the need to remediate metal-contaminated sites has grown. Areas of particular interest are simplifying the methodology to adapt the PBET for routine testing and standardizing the interpretation of PBET results. For example, Fendorf et al. (2004) simplified the PBET method by reacting 0.5 g soil and 50 mL of 1 M glycine, adjusted to pH 3 for 1 h in a 35.6°C water bath. They found this method accurately tracked decreases in bioaccessibility of As, Cr, and Pb in a wide range of surface and subsoils from the United States. Other simplifications of the PBET have been evaluated by Yoon et al. (2007) in studies of method to immobilize Pb and reduce leaching, by Brown et al. (2007) in research on the use of soil amendments to reduce Pb bioavailability at a U.S. Environmental Protection Agency (USEPA) Superfund site, by Brown et al. (2003) in studies of factors controlling Pb bioavailability in urban soils, and by Moseley et al. (2008) in investigations of the use of phosphate soil amendments to reduce Pb bioaccessibility in industrial and firing range soils. However, if the goal is to quantify the extent of accumulation of an element, relative to normal soils or natural background levels, or to monitor this accumulation over time, methods that determine or approximate total elemental content are recommended. One example is USEPA Method 3050 that successively digests a soil sample with concentrated HNO_3, H_2O_2, and HCl to measure "total sorbed metals" by acidic dissolution of clays, oxides, and carbonates and oxidation of organic matter; elements associated with silicates are not dissolved. Therefore, for a true measure of total elemental content of a soil sample, complete digestion of the soil with strong acids (e.g., HNO_3-$HClO_4$ for Cd, Hg, and Pb), by carbonate fusion (Cr, Ni), or by alkaline oxidation techniques (As, Se) is required. Given the costs and difficulty of measuring total elemental content of trace elements in soils it is often advisable to use rapid soil testing methods as surrogate monitoring techniques (Sims and Johnson, 1991).

Interpretations and recommendations for soil tests for potentially toxic trace elements are considerably more difficult than for agricultural systems and are often very site and element specific. As mentioned above, the main difficulty lies with soil test calibration—establishing a quantitative relationship between an agreed upon measure of environmental risk and the amount of an element measured by the soil test. In most cases, this has been done by the use of complex risk assessment models that first identify the "target organism" of concern (e.g., human vs. plant) and then evaluate all possible pathways by which the target organism may be exposed to the risk factor (the toxic trace element). If possible, a "most sensitive pathway" is identified, defined as the lowest soil concentration level at which an adverse effect on the organism would be likely to occur (e.g., soil ingestion vs. consumption of contaminated drinking water). Regulatory upper limits may then be established for that pathway, which can be monitored by the soil testing methods described above. This approach was used by USEPA in the formulation of the national rule for the disposal and utilization of municipal sewage sludges, which established regulatory limits for the total amount of several trace elements that could be applied to agricultural soils via land application of these sludges (Ryan and Chaney, 1993). Additional information on the methods used to interpret soil testing results is found in U.S. Environmental Protection Agency (1989), Risser and Baker (1990), and Pierzynski et al. (1994).

13.2.7.3 Soil Testing for Plant Nutrients with Water Quality Impacts

Soil testing for plant nutrients that can degrade water quality is focused primarily on N and P because of their well-documented effects on ground (N) and surface (N, P) water quality. The principles, practices, and problems of soil testing for N were described in Section 13.2.4.4. Other than the methods described in that section (e.g., the PSNT and ISNT), which focus on identifying sites with an adequate N supply and thus avoiding unnecessary applications of fertilizers or organic sources of N (manures, sludges), there are no other approaches currently available for use in an environmental soil testing program for N. Environmental soil testing for P, however, is a considerably different matter. Growing international concerns about the role of P in the eutrophication of surface waters stimulated a large research effort on environmental soil testing for P beginning in the late 1980s and continuing even today. The focus of this effort has been the use of soil testing alone, or as a component of site indexes and nonpoint source pollution computer models, to identify soils that are most likely to be significant nonpoint sources of P pollution of streams, rivers, ponds, lakes, and bays. Some aspects of environmental soil testing for P are described below, illustrating the changes needed in soil sampling, analysis, and interpretation when surface water protection, not agricultural production, becomes the primary goal. Readers are referred to several reviews for more detailed information on this topic (Sims, 1993; Sharpley et al., 1996; Sibbesen and Sharpley, 1997; Sims, 1997; Tunney et al., 1997; Maguire and Sims, 2002).

13.2.7.3.1 Establishing Upper Critical Limits for Soil P Using Soil Testing Methods

One approach proposed for environmental soil testing for P is to simply establish an upper critical limit for soil P using currently available agronomic soil testing methods (e.g., Bray P_1, Mehlich 1, Mehlich 3, Olsen P). Soils that exceeded the upper critical limit for soil test P would no longer receive P inputs from any source (e.g., fertilizers, manures, biosolids) and would be targeted as priority areas for soil and water conservation practices to prevent P loss in erosion, runoff, and leaching. Two reasons are usually given to justify the need for this upper critical limit. First is that an extensive body of research shows that soils that are overfertilized with P relative to crop requirements will create an increased risk of nonpoint source pollution of surface waters (Sharpley, et al. 1994; Sims et al., 1997, 1998, 2002; Tunney et al., 1997). Second is the concern that continuing to apply P to soils well beyond values that are needed for crop production contradicts the principles of sustainable development and sustainable agriculture. The rationale underlying this second concern is the fact that P is obtained from a finite natural resource base, at a cost to society, and that agricultural practices that waste this resource are inconsistent with sustainability. Despite these concerns, until recently there was a reluctance to establish upper critical limits using soil test P measurements primarily because (1) agronomic soil tests were not originally designed or calibrated for environmental purposes and thus there was uncertainty concerning their accuracy at identifying soils with the potential to contribute environmentally significant amounts of P to waters by runoff or leaching and (2) there would be an unjustified reliance upon agronomic soil test P alone by environmental regulatory agencies attempting to control nonpoint source pollution of surface waters, ignoring the complex interaction between soil P and the transport processes and soil/crop management factors that control the movement of P from soil to water.

Because of these questions and concerns, the most appropriate means to use existing agronomic soil P tests, or new soil test methods, to identify soils with a high risk of P transfer to water has received intense research interest in the past decade. Results of many laboratory and field studies conducted in the 1990s and 2000s have clearly shown that agronomic soil P tests are often well correlated with other measures of soil P that suggest increased risks of P loss, such as water soluble P, the equilibrium P concentration in the soil solution (EPC_0), easily desorbable P (e.g., P extracted by Fe-oxide filter paper strips), and algal-available P (Gartley and Sims, 1994; Sibbesen and Sharpley 1997; Sims 1997; McDowell and Sharpley, 2001; Vadas et al., 2005). Additional studies of the relationships between agronomic soil test P and dissolved P concentrations in overland flow or leachate provided direct evidence for increased risks of P loss from overfertilized soils (Heckrath et al., 1995; Westermann et al., 2001; Maguire and Sims, 2002; Little et al., 2007). In some cases, this relationship was near-linear, while in others there appeared to be a "change point" above which there was a more rapid increase in dissolved P concentrations in runoff or leachate with increases in soil test P. Research has also shown that maintaining soil test P in the optimum range for plant growth minimizes dissolved P losses. When combined with information on soil erosion, soil test P can also be used to assess the risk of particulate P loss to water, as is done by risk assessment tools such as the P Site Index, as discussed below.

"Environmental soil P testing," however, is not confined to the use of agronomic soil P tests. It may mean the use of completely different methods for soil sampling, analysis, and interpretation than those done for agricultural purposes. Or, it may mean that an environmental or regulatory interpretation is now being applied to results of an agronomic soil P test, such as establishing an upper limit for soil test P beyond which no further P additions can be made. Changes in soil sample collection may include sampling to a shallower depth (0–5 cm) if overland flow is of greatest concern or to the depth of the mean high water table (e.g., 50–150 cm) if P leaching and subsurface flow is the main pathway for P loss. Different analytical methods may be used such as assessment of the degree of soil P "saturation" (DPS), water-soluble P, easily desorbed P, "algal available" P, and quick tests for P sorption capacity. For example, a DPSox method based on acid ammonium oxalate extraction of P, Al, and Fe was developed in the Netherlands to identify soils where P leaching to shallow groundwaters is a risk. Oxalate extractable P (Pox) represents the amount of P currently sorbed by the soil and oxalate Al and Fe (Alox, Feox) are proportional to the P sorption capacity of noncalcareous soils. The molar ratio of Pox to [α (Alox, + Feox)] is thus an indication of soil P saturation (where α is a parameter used to convert (Alox + Feox) to estimated soil P sorption capacity and typically ranges from 0.4 to 0.6). Upper limits for DPSox of 25%–40% are most commonly used today. Research has found increases in water-soluble and desorbable P in soils above these DPSox values, suggesting a greater risk of P loss to surface waters and shallow groundwaters from P-saturated soils. Research has also shown that DPS measured by the Mehlich 3 (M3) soil test is well correlated with DPSox and also with P concentrations in leachate and overland flow (Sims et al., 2002). The Mehlich 3 soil P test is widely used in the United States and Canada, and some soil testing laboratories now report a soil P fertility assessment (M3-P) and soil P saturation (M3-DPS) using one soil extraction. In addition to using soil P saturation to predict dissolved P losses, research has also shown that using the "safe P storage capacity" (SPSC) of a soil can be accurately estimated using Mehlich 1 or Mehlich 3 DPS values (Nair and Harris, 2004; Chrysostome et al., 2007). The SPSC concept is used to determine how much additional P loading can occur before dissolved P losses become an environmental concern, particularly important in settings where manures or biosolids are regularly applied to soils.

With the possible exception of the M3-DPS test, environmental P tests would probably not be conducted on every sample submitted as part of a nutrient management plan. Instead, they would be part of a more intensive testing procedure used when agronomic soil test P values were considered high enough to warrant further investigation. Even more important, most soil scientists recommend that environmental soil P tests not be used alone, but as a component of a more holistic risk assessment process, such as the P Site Index (Heathwaite et al., 2005) a risk assessment tool that integrates soil P with P transport and management practices to characterize the potential for P loss at a site. For instance, the states of Delaware and Maryland in the United States require that a P Site Index be conducted for fields with an agronomic soil test P (Mehlich 3) value >150 mg P kg^{-1}. Nearly every state in the United States has developed and adopted a P Site Index as the preferred means to assess the risk of P loss from agricultural fields to surface waters, as opposed to the use of a soil P test alone. Many P Site Indices are now available on the internet, such as a very comprehensive index developed by the University of Wisconsin (see http://wpindex.soils.wisc.edu/).

13.3 Plant Testing

Soil fertility evaluation does not rely upon soil analysis alone. Many techniques now exist to assess soil fertility by characterizing the growth and elemental composition of plants. These include visual diagnosis, in-field evaluation techniques, laboratory analysis of plant tissues, and remote sensing. Plant analysis includes both rapid in-field tissue testing and total elemental analysis of plant samples, normally conducted in the laboratory. In both cases, the underlying premise is that the plant concentration of an element is proportional to the availability of the nutrient in the soil and thus is an index of soil fertility. Visual diagnosis and remote sensing do not actually determine the nutrient concentration in a plant, but instead rely upon changes in plant color or growth as indexes of soil fertility.

13.3.1 Visual Diagnosis of Deficiency Symptoms

With experience it is possible to identify visual symptoms that result from nutrient deficiencies. Examples include severely stunted growth and purpling of older leaves (P), chlorosis and necrosis of leaf margins (K), interveinal yellowing of newer leaves by Fe and Mn and of older leaves by Mg, and distorted meristems and blackened internal tissues (B). In general, visual diagnosis should be verified by soil and plant analysis, prior to taking corrective actions. Tisdale et al. (1993) cited four reasons for this: (1) the symptom may be caused by a deficiency of more than one nutrient; (2) the deficiency of one nutrient may be caused by an excess of another; (3) other factors, such as insect and disease injury, can create symptoms similar to nutrient deficiencies; and (4) the symptom may be caused by more than one growth factor (e.g., P deficiency can occur in cold, wet soils that have optimum P fertility levels—the deficiency disappears when soils warm). Other factors to consider are the fact that the same symptom may be caused by different nutrients in differing plants. For example, purpling of older leaves, typically due to P, can be caused in cruciferous plants by S deficiency. The location of a deficiency on the plant can assist in proper identification of the cause. Nitrogen and S can both cause yellowing of plant leaves; however, because N is more mobile in most plants than S, its deficiency symptoms usually occur on older leaves, while S deficiencies are seen as a more overall yellowing. Calcium and most micronutrients (B, Cu, Fe, Mn, Zn) are rather immobile in plants; hence, their deficiencies usually occur on new growth, compared to some of the more mobile macronutrients (N, P, K, Mg) where deficiency symptoms are observed on older plant parts. Finally, for many plants, by the time a nutrient deficiency symptom can be accurately diagnosed, it is either too late for corrective action (unless nutrients can be added via fertigation or sidedressing) or unwise because the deficiency has already damaged the plant beyond the point where nutrient additions can profitably correct the problem.

13.3.2 In-Field Evaluation of Plant Nutritional Status

In addition to visual diagnosis, there are several "in-field" diagnostic techniques that can be used to provide semiquantitative information on plant nutritional status. The most common is the use of rapid tissue testing kits for the colorimetric analysis of plant sap. The color that develops after addition of the appropriate chemical reagents to sap is proportional to the concentration of the nutrient (usually only N, P, and K are tested in these kits). The entire process takes only a few minutes much less than required for total elemental analysis of a plant tissue by an analytical laboratory, which can require a week or more. In general, because of its semiquantitative nature, rapid tissue testing should only be used as a guide, and not as the sole basis for nutrient management recommendations. Tisdale et al. (1993) also cautioned that rapid tissue testing must (1) only be done with the correct plant part, sampled at the proper time of year and, for some nutrients (e.g., NO_3-N), at the proper time of day because nutrient concentrations change during the day; (2) be done in a comprehensive manner, continuing the testing throughout the growing season, not as a one-time activity; (3) focus on periods of maximum vegetative growth and reproductive stages to best assess a fertilizer program; (4) be the average value from the analysis of at least 10–15 plants and be collected from areas of normal and deficient plant growth.

One advance in field-scale plant N testing is the LCM, a small, hand-held spectrometer that directs a beam of light corresponding to the wavelength absorbed by the chlorophyll molecule through a plant leaf, while the leaf is still on the plant. The meter essentially measures leaf "greenness," which has been shown in numerous studies to be correlated with leaf chlorophyll content, plant N, and the likelihood of an economic response to N fertilization. The LCM allows for rapid, ongoing monitoring of plant N nutrition and is particularly well suited to irrigated crops where N can be added via the irrigation system according to LCM results. Since other nutrients affect leaf greenness, especially S, the LCM must be used with caution and with soil N testing where possible to avoid recommending N fertilization when some other factor is responsible for the observed chlorosis (Schepers et al., 1992). Nevertheless, the LCM is an important advance in the field-scale evaluation of plant N nutrition, particularly for agronomic crops.

Remote sensing has long been recognized as a relatively inexpensive method for mapping, monitoring, and classifying vegetative cover. As discussed earlier, the high spatial and temporal variability of soil N has limited development of reliable soil tests for determination of soil N supplying capacity. However, remote sensing shows potential to overcome the shortfalls of traditional plant and soil analysis due to its ability to characterize spatial variability down to resolutions of just a few meters and to identify nutrient deficiencies in real time while applying fertilizers or other nutrient sources. Numerous vegetative indexes have been developed using knowledge of plant spectral reflectance and absorbance characteristics. For example, plants scatter or reflect the majority of near infrared (IR) radiation that they intercept due to internal cellular structures. Conversely, chlorophyll absorbs the majority of red (R) light intercepted to provide energy for photosynthesis. One such index is the normalized differential vegetation index (NDVI):

$$\text{NDVI} = \frac{\text{IR} - \text{R}}{\text{IR} + \text{R}} \qquad (13.2)$$

The NDVI has been widely used to assess land use, monitor crop phenology, and to estimate yields by numerous researchers. More recently, NDVI has been used to relate predicted yield to N requirements of various crops during the growing season (Shanahan et al., 2001; Chang et al., 2003). Nonetheless, remote sensing has often been of limited use in agriculture due to the cost and time associated with collection and processing of remote sensing data generated by satellites or fixed wing aircraft. However, recently portable, active optical sensors have become commercially available. These sensors emit their own light and can be hand-held or mounted on equipment, overcoming many of the shortfalls associated with satellite- and aircraft-based sensors that are reliant on reflected sunlight and vulnerable to atmospheric conditions. These sensors have been shown to reliably predict crop yields and multiple algorithms have been developed to relate various indexes to crop N requirements (Raun et al., 2005; Thomason et al., 2007). Moreover, the sensors can be used to control variable rate N application in real-time, significantly increasing N use efficiency. As sensing technologies continue to develop, they are likely to represent perhaps the most significant advance in the evaluation and adjustment of crop nutrient status.

13.3.3 Plant Analysis

13.3.3.1 Plant Analysis: Basic Principles

Plant analysis is defined as "...the determination of the elemental composition of plants, or a portion of the plant, for elements essential for growth. It can also include determining elements that are detrimental to growth or animals or humans through our food chain" (Munson and Nelson, 1990). The components of a plant analysis program are similar to those used for soil testing: (1) collection of a representative sample from a plant, or the whole plant, at the proper stage of plant development; (2) proper handling of the sample to avoid contamination or damage that could affect interpretation of results, followed by accurate sample analysis using standardized laboratory methods; (3) correlation and calibration studies that establish quantitative relationships between plant analysis and plant performance; and (4) economically and environmentally sound nutrient management recommendations to correct any nutrient deficiencies. Key factors to consider in each step are described below. Many references provide detailed descriptions of the protocols used for plant sampling, handling, and laboratory analysis; hence, these topics will not be discussed (Plank, 1989; Jones and Case, 1990; Westerman, 1990; Mills and Jones, 1997).

13.3.3.2 Plant Analysis: Interpretation of Analytical Results

The two most widely used methods for interpretation of plant analysis are (1) the critical nutrient concentration or range (CNC or CNR) and (2) the diagnosis and recommendation integrated system (DRIS), which relies primarily on nutrient ratios, emphasizing the importance of nutrient balance in the plant to optimum plant performance.

Interpretation by the CNR approach relies, as with soil testing, mainly on correlation and calibration studies that show there is a statistically significant relationship between the nutrient concentration in a plant and plant response to additions of the nutrient (Figure 13.6). In most cases plant analysis results are compared to optimum CNRs, with analytical values expressed in percentages (for major nutrients) and parts per million (mg kg^{-1} or µg g^{-1} for micronutrients), on a dry weight basis. Nutrient concentrations below the CNR are considered deficient, those above it adequate or optimum; the CNR represents a transition zone between these two categories. The CNC is defined as the nutrient concentration where plant performance changes from suboptimum and unsatisfactory to optimum. Because the CNC is a single value that can be difficult to determine experimentally, given the uncertainties and variations associated with field studies, and because it can

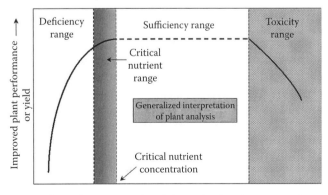

FIGURE 13.6 Illustration of the critical nutrient range approach for plant analysis interpretation.

vary somewhat within a plant species due to genetic differences among hybrids, the CNR approach is more commonly used to interpret plant analysis results. Extensive research conducted to determine CNRs for most plants with economic value clearly show that the CNR for a plant depends upon growth stage and plant part that is sampled (Mills and Jones, 1997). Therefore, comparisons of plant analysis results with CNR values are only valid if the plant sample is taken at the same growth stage and from the same plant part used in the calibration studies conducted to determine CNRs. Interpretation of plant analysis results by this method, therefore, requires careful adherence to recommended protocols for sampling, handling, and analysis.

The balance of nutrients within a plant is often more important than the concentration of any individual nutrient. The DRIS system focuses on nutrient balance as an alternative to the CNR approach to plant analysis interpretation. One advantage of the DRIS system, compared to the CNR, is that nutrient ratios in plants tend to be more constant throughout the growing season than individual nutrient concentrations. Use of DRIS, therefore, allows for greater flexibility in the time of plant sample collection as samples do not necessarily have to be taken at a specific growth stage. The DRIS system first establishes norms for all nutrient ratios (N:P, N:K, P:K, N:S, P:S, etc.) associated with maximum crop yield. These norms can be developed from reviews of the scientific literature or by widespread sampling of a crop in a given physiographic region. As much as possible, all factors that affect crop yield are measured at the time of plant sample collection (soil test values, pest pressure, climate, cultural practices, etc.). In this way, an integrated relationship can be established between the DRIS norms, crop yield, and the other growth-limiting factors. Interpretation of plant analysis via the DRIS system can be done graphically or mathematically by the calculation of DRIS indices. The DRIS system has received intense interest and considerable research in the past decade. Computer programs that can rapidly calculate DRIS indices have been developed from this research, resulting in wider use of DRIS by plant analysis laboratories. Readers are referred to Beaufils (1973), Sumner (1979, 1981), Escano et al. (1981), Amundson and Koehler (1987), Karlen et al. (1988), Walworth et al. (1988), Munson and Nelson (1990), and Beverly (1991) for examples of the use of DRIS.

13.4 Soil Fertility Evaluation: The Future

The principles and practices used to assess soil fertility continue to evolve and it is clear that the process of soil fertility evaluation will take on new dimensions in the next decade. Given this, two important questions should be considered now. First, how should newly emerging technologies for the evaluation of soil fertility be assessed and integrated with current practices? Second, it is clear that in the past decade soil fertility evaluation has expanded well beyond production agriculture to address new challenges, especially those related to environmental protection and human health. Given this, how can we best apply the results of "environmental" soil tests to assess the value of best management practices intended to protect or restore the environment or reduce soil-related human health risk?

13.4.1 Precision Agriculture and Remote Sensing

The past decade has seen the development of several new technologies that have the potential to significantly alter soil fertility evaluation and management. The two most important and closely related examples are "precision agriculture" and remote sensing (defined as "… the science of obtaining information about an area through analysis of data acquired by devices not in contact with the area"; Hergert, 1997). Precision agriculture, also referred to as site-specific management, has become an increasingly accepted and widespread approach to farming that allows us to (1) precisely map soil fertility in a field or on a farm at great detail (typically in 1 ha grids) through the use of hand-held or equipment-mounted GPS that use signals from a network of satellites to instantaneously locate a position on the earth's surface; (2) use variable rate application equipment, equipped with GPS and "on-the-go" sensors of soil nutrient status, to supply nutrients in accordance with these maps; and (3) generate GPS-based maps of crop yields. Factors related to yield that cannot easily be measured by soil and plant testing can be obtained from other computerized data bases (e.g., soil series from soil surveys) or be assessed using remote sensing (e.g., plant health, soil moisture, irrigation and drainage, pest pressure). The most commonly used remote sensing devices today include cameras and other imaging systems mounted on aircraft or satellites. Other, more recent examples are sensors mounted on ground equipment (e.g., tractors, pesticide applicators) or on large permanent structures (irrigation systems, electrical towers). Remote sensing devices acquire electromagnetic energy that is emitted or reflected from plants (or bare soils) and convert this energy into data that can be used in soil fertility evaluation. Each combination of soils, plants, environmental conditions, and management practices has a characteristic spectral "signature"—a specific spectrum of radiation—that can be used to diagnose soil fertility problems. As an example, IR aerial photography has been used for many years to assess plant performance because healthy green plants reflect large amounts of IR radiation while plants damaged by insects, drought, or nutrient deficiencies do not. As with traditional approaches to soil fertility evaluation, the ultimate value of these increasingly sophisticated technologies lies in our ability to interpret the results. Indeed one of the problems with precision agriculture is that vast quantities of spatially located data can be generated very quickly, creating an information overload that can confuse or obscure interpretation. Research is now underway to develop "expert systems" that use computerized GIS to integrate the various "layers" of data in a meaningful way, thus guiding a more holistic approach to crop management, of which soil fertility evaluation is only a part. Some areas in precision agriculture where research is needed include the following: (1) an economic analysis of the value of the additional information obtained by extensive, GPS-based grid soil sampling, relative to traditional, less intensive techniques; (2) a reevaluation of current soil test calibration and correlation models given the enormous, large-scale data bases that are rapidly becoming available; (3) the most effective and rapid means to obtain "ground truth"—verification of the accuracy of remote sensing devices and "on-the-go" sensors as indicators of soil fertility; and (4) the value of site-specific management techniques for application other than crop production, such as minimizing N and P losses via leaching and runoff, irrigation scheduling, erosion control practices, and more efficient use of wastes and by-products. Readers are referred to Robert et al. (1995), Pierce and Sadler (1997), and Bongiovanni and Lowenberg-Deboer (2004), Srinivasan (2006) for reviews of precision agriculture and site-specific management and to the following Web sites from universities and private industry for examples of the latest advances in this field: (1) Australian Centre for Precision Agriculture, http://www.usyd.edu.au/agric/acpa/; (2) Cranfield University (United Kingdom) Centre for Precision Farming, http://www.cranfield.ac.uk/sas/naturalresources/index.jsp; (3) The Ohio State University, http://precisionag.osu.edu/; (4) Purdue University Site Specific Management Center, http://www.agriculture.purdue.edu/ssmc/; and (5) Precision Ag, http://www.precisionag.com/.

13.4.2 New Directions and Uses for Soil Fertility Evaluation Techniques

The formal, institutionalized practice of soil fertility evaluation originated to serve the needs of production agriculture, a function that continues to be of unquestioned importance today. However, some of the needs of the agricultural sector are changing as, for example, in developed countries where the potential environmental impacts of soils that have become over-enriched with nutrients or contaminated by wastes and by-products from industry and urban area has begun to supersede the need to identify and correct nutrient deficiency problems. At the same time, those responsible for nonagricultural land uses are beginning to see the potential value of practices originally developed to optimize soil fertility for crop production. There are several challenges that must be overcome if the process of soil fertility evaluation is to evolve and respond to needs such as these. First is the establishment of better and more direct interfaces between nontraditional users of soil and plant testing and researchers with expertise in these areas. This will help to clarify when, where, and how it is appropriate to use current soil fertility evaluation techniques for purposes other than those for which they were originally designed. It will also provide insight into the advances in soil science research needed to more effectively address the problems faced by those charged with protecting air and water quality, preventing further damage or restoring soil quality to lands disturbed by erosion, salinization, construction, surface mining, and similar activities, and remediation of soils polluted by anthropogenic activities. Second, is the need for the public and private organizations responsible for soil fertility evaluation to recognize the contribution they can make to solving land management problems that are not solely directed at profitable crop production. This will require better interactions with (1) researchers who have developed many new soil testing methods that have value in these areas but that have not been adopted by routine soil testing laboratories; (2) new clientele who often have unusual problems and limited understanding of the principles and practices of soil and plant testing; and (3) a wide range of technical and regulatory agencies and not-for-profit environmental groups who have the responsibility or interest in the protection and restoration of air, water, and soil quality. Strengthening these interfaces will provide the research base essential to support nontraditional uses of soil fertility evaluation techniques and the educational component needed to ensure that only the appropriate analytical methods are used and that proper interpretations of the results are made. This research should then help guide the design and implementation of best management practices that can minimize nutrient loss from land to water, stabilize (or remove) elements in soils that may be toxic to humans or other organisms, and restore ecosystems degraded by nonpoint pollution.

In conclusion soil fertility evaluation is a vital, integral part of global agriculture and plays an increasingly important role in many nonagricultural land uses, especially those related to nonpoint source pollution. Successful application of the principles and practices described in this chapter will increase the profitability and minimize the environmental impacts of nutrient use—the fundamental goal of soil fertility management. Appropriate integration of newly emerging technologies with current practices will further enhance our ability to evaluate soil fertility and make sound nutrient management decisions. Expanding the process of soil fertility evaluation to more fully include nonagricultural settings, and especially to situations where environmental protection or land restoration are the goal, is perhaps the greatest challenge we face today.

References

Abrams, M.M., and W.M. Jarrell. 1992. Bioavailability index for phosphorus using ion exchange resin impregnated membranes. Soil Sci. Soc. Am. J. 56:1532–1537.

Amer, F., D.R. Bouldin, C.A. Black, and F.R. Duke. 1955. Characterization of soil phosphorus by anion exchange resin adsorption and ^{32}P-equilibration. Plant Soil 6:391–408.

Amundson, R.L., and F.E. Koehler. 1987. Utilization of DRIS for diagnosis of nutrient deficiencies in winter wheat. Agron. J. 79:472–476.

Barker, D.W., J.E. Sawyer, M.M. AlKasi, and J.P. Lundvall. 2006. Assessment of the amino sugar-nitrogen test on Iowa soils. II. Field correlation and calibration. Agron. J. 98:1352–1358.

Beaufils, E.R. 1973. Diagnostic and recommendation integrated system (DRIS). A general scheme for experimentation and calibration based on principles developed from research in plant nutrition. Soil Sci. Bull. 1. University of Natal, Pietermaritzburg, South Africa.

Beegle, D. 1995. Interpretation of soil test results, p. 84–91. *In* J.T. Sims, and A.M. Wolf (eds.) Recommended soil testing procedures for the Northeastern United States. Bull. no. 493. University of Delaware, Newark, DE.

Berger, K.C., and E. Truog. 1940. Boron deficiency as revealed by plant and soil tests. J. Am. Soc. Agron. 32:297–301.

Berti, W.R., and S.D. Cunningham. 1994. Remediating soils with green plants, p. 43–51. *In* C.R. Cothern (ed.) Trace substances, environment, and health. Science Reviews, Northwood, U.K.

Beverly, R. 1991. A practical guide to the diagnosis and recommendation integrated system (DRIS). MicroMacro Publishing, Inc. Athens, GA.

Black, C.A. 1993. Soil fertility evaluation and control. Lewis Publishers, Boca Raton, FL.

Bock, B.R., and K.R. Kelley (eds.) 1992. Predicting nitrogen fertilizer needs for corn in humid regions. TVA National Fertilizer Environmental Research Center. Muscle Shoals, AL.

Bongiovanni, R., and J. Lowenberg-Deboer. 2004. Precision agriculture and sustainability. Precis. Agric. 5:359–387.

Bray, R.H., and L.T. Kurtz. 1945. Determination of total organic and available forms of phosphorus in soil. Soil Sci. 59:39–46.

Brown, S.L., R.L. Chaney, J.G. Hallfrisch, and Q. Xue. 2003. Effect of biosolids processing on lead bioavailability in an urban soil. J. Environ. Qual. 32:100–108.

Brown, S.L., H. Compton, and N.T. Basta. 2007. Field test of in situ soil amendments at the tar creek national priorities list superfund site. J. Environ. Qual. 36:1627–1634.

Bundy, L.G., and J.J. Meisinger. 1996. Nitrogen availability indices, p. 951–984. *In* R.W. Weaver et al. (eds.) Methods of soil analysis, Part 2: Microbiological and biochemical properties. SSSA, Madison, WI.

Bundy, L.G., M.A. Schmitt, and G.W. Randall. 1992. Advances in the upper Midwest, p. 73–90. *In* B.R. Bock, and K.R. Kelley (eds.) Predicting nitrogen fertilizer needs for corn in humid regions. TVA National Fertilizer Environmental Research Center, Muscle Shoals, AL.

Carter, M.R., and E.G. Gregorich. 2007. Soil sampling and methods of analysis. 2nd edn. CRC Press, Taylor & Francis Group, Boca Raton, FL.

Cate, R.B., Jr., and L.A. Nelson. 1965. A rapid method for correlation of soil test analyses with plant response data. North Carolina Agric. Exp. Stn. Int. Soil Test. Ser. Tech. Bull. 1, North Carolina Agricultural Experiment Station. Raleigh, NC.

Chang, J., D.E. Clay, K. Dalsted, S. Clay, and M. O'Neill. 2003. Corn yield prediction using multispectral and multidate reflectance. Agron. J. 95:1447–1453.

Chapman, H.D., and W.P. Kelley. 1930. The determination of the replaceable bases and base-exchange capacity of soils. Soil Sci. 30:391–406.

Chardon, W.J., R.G. Menon, and S.H. Chien. 1996. Iron oxide impregnated filter paper (Pi test): A review of its development and methodological research. Nutr. Cycl. Agroecosyst. 46:41–51.

Chen, L., D. Kost, and W.A. Dick. 2008. Flue gas desulfurization products as sulfur sources for corn. Soil Sci. Soc. Am. J. 72:1464–1470.

Chrysostome, M., V.D. Nair, W.G. Harris, and R.D. Rhue. 2007. Laboratory validation of soil phosphorus storage capacity predictions for use in risk assessment. Soil Sci. Soc. Am. J. 71:1564–1569.

Corey, R.B. 1987. Soil test procedures: Correlation, p. 15–22. *In* J.R. Brown (ed.) Soil testing: Sampling, correlation, calibration, and interpretation. SSSA Special Publication 21. SSSA, Madison, WI.

Cox, A.E., B.C. Joern, S.M. Brouder, and D. Gao. 1999. Plant-available potassium assessment with a modified sodium tetraphenylboron method. Soil Sci. Soc. Am. J. 63:902–911.

Dahnke, W.C., and R.A. Olson. 1990. Soil test correlation, calibration, and recommendation, p. 45–71. *In* R.L. Westerman (ed.) Soil testing and plant analysis. 3rd edn. SSSA Book Series No. 3. SSSA, Madison, WI.

Eckert, D.J. 1987. Soil test interpretations: Basic cation saturation ratios and sufficiency levels, p. 53–64. *In* Soil testing: Sampling, correlation, calibration, and interpretation. Special Publication No. 21. SSSA, Madison, WI.

Egner, H., H. Riehm, and W.R. Domingo. 1960. Untersuchungen uber die chemishe bodenanalyse als grundlage fur die beurteilung des nahrstoffzustandes der boden. II. Chemische extraktions-methoden zur phosphor-und kalimbestimmung kungl. Lantbrukshoegsk. Ann. 26:204–209.

Escano, C.R., C.A. Jones, and G. Uehara. 1981. Nutrient diagnosis in corn grown on Hydric Dystrandepts. II. Comparison of two systems of tissue diagnosis. Soil Sci. Soc. Am. J. 45:1140–1144.

Fendorf, S., M.J. La Force, and G. Li. 2004. Temporal changes in soil partitioning and bioaccessibility, of arsenic, chromium, and lead. J. Environ. Qual. 33:2049–2055.

Fernández, F.G., S.M. Brouder, C.A. Beyrouty, J.J. Volenec, and R. Hoyum. 2008. Assessment of plant-available potassium for no-till, rainfed soybean. Soil Sci. Soc. Am. J. 72:1085–1095.

Fixen, P.E., and J.H. Grove. 1990. Testing soils for phosphorus, p. 141–180. *In* R.L. Westerman (ed.) Soil testing and plant analysis. 3rd edn. SSSA Book Series No. 3. SSSA, Madison, WI.

Foth, H.D., and B.G. Ellis. 1997. Soil fertility. 2nd edn. CRC Press, Boca Raton, FL.

Foy, R.H., and P.J.A. Withers. 1995. The contribution of agricultural phosphorus to eutrophication. The Fertilizer Society. Proc. No. 365. Greenhill House, Thorpe Wood, Petersborough, U.K.

Gartley, K.L., and J.T. Sims. 1994. Phosphorus soil testing: Environmental uses and implications. Commun. Soil Sci. Plant Anal. 25:1565–1582.

Gaskin, J.W., R.B. Brobst, W.P. Miller, and E.W. Tollner. 2003. Long-term biosolids application effects on metal concentrations in soil and bermudagrass forage. J. Environ. Qual. 32:146–152.

Gestring, W.D., and P.N. Soltanpour. 1984. Evaluation of the ammonium bicarbonate-DTPA soil test for assessing boron availability to alfalfa. Soil Sci. Soc. Am. J. 48:96–100.

Griffin, T.S. 2008. Nitrogen availability, p. 613–646. *In* J.S. Schepers, and W.R. Raun (eds.) Nitrogen in agricultural systems. Agronomy monograph no. 49. ASA, Madison, WI.

Haby, V.A., M.P. Russelle, and E.O. Skogley. 1990. Testing soils for potassium, calcium, and magnesium, p. 181–227. *In* R.L. Westerman (ed.) Soil testing and plant analysis. 3rd edn. SSSA Book Series No. 3. SSSA, Madison, WI.

Hanlon, E.A., and G.V. Johnson. 1984. Bray/Kurtz, Mehlich III, AB/D and ammonium acetate extractions of P, K, and Mg in four Oklahoma soils. Commun. Soil Sci. Plant Anal. 15:277–294.

Havlin, J.L., S.L. Tisdale, W.L. Nelson, and J.D. Beaton. 2004. Soil fertility and fertilizers: An introduction to nutrient management. 7th edn. Prentice Hall, Upper Saddle Lake, NJ.

Healthwaite, A.L., P.F. Quinn, and C.J.M. Hewett. 2005. Modelling and managing critical source areas of diffuse pollution from agricultural land using flow connectivity simulation. Journal of Hydrology, 304:446–461.

Heckrath, G., P.C. Brookes, P.R. Poulton, and K.W.T. Goulding. 1995. Phosphorus leaching from soils containing different phosphorus concentrations in the Broadbalk experiment. J. Environ. Qual. 24:904–910.

Hergert, G.W. 1997. A futuristic view of soil and plant analysis and nutrient recommendations, p. 24–35. Proc. 5th Intl. Symp. Soil Plant Anal. Minneapolis, MN.

James, D.W., and K.L. Wells. 1990. Soil sample collection and handling: Technique based on source and degree of field variability, p. 25–44. *In* R.L. Westerman (ed.) Soil testing and plant analysis. 3rd edn. SSSA Book Series No. 3. SSSA, Madison, WI.

Jaynes, D.B., D.L. Dinnes, D.W. Meek, D.L. Karlen, C.A. Cambardella, and T.S. Colvin. 2004. Using the late spring nitrate test to reduce nitrate loss within a watershed. J. Environ. Qual. 33:669–677.

Jemison, J.M., Jr., and R.H. Fox. 1988. A quick-test procedure for soil and plant tissue nitrates using test strips and a hand-held reflectometer. Commun. Soil Sci. Plant Anal. 19:1569–1582.

Johnson, G.V., and P.E. Fixen. 1990. Testing soils for sulfur, boron, molybdenum, and chlorine, p. 265–273. *In* R.L. Westerman (ed.) Soil testing and plant analysis. 3rd edn. SSSA Book Series No. 3. SSSA, Madison, WI.

Jones, J.B. (ed.) 1999. Soil analysis. Handbook of reference methods. Soil and Plant Analysis Council, Inc., CRC Press, Boca Raton, FL.

Jones, J.B., Jr., and V.W. Case. 1990. Sampling, handling, and analyzing plant tissue samples, p. 389–427. *In* R.L. Westerman (ed.) Soil testing and plant analysis. 3rd edn. SSSA Book Series No. 3. SSSA, Madison, WI.

Kamprath, E.J., and M.E. Watson. 1980. Conventional soil and tissue tests for assessing the phosphorus status of soils, p. 433–470. *In* F.E. Khawsaneh (ed.) The role of phosphorus in agriculture. SSSA, Madison, WI.

Karlen, D.L., R.L. Flannery, and E.J. Sadler. 1988. Aerial accumulation and partitioning of nutrients by corn. Agron. J. 80:232–242.

Keeney, D.R. 1982. Nitrogen-availability indices, p. 711–733. *In* A.L. Page et al. (eds.) Methods of soil analysis. 2nd edn. Agronomy 9, Part 2. ASA, Madison, WI.

Keeney, D.R., and D.W. Nelson. 1982. Nitrogen—Inorganic forms, p. 643–698. *In* A.L. Page et al. (eds.) Methods of soil analysis. 2nd edn. Agronomy 9, Part 2. ASA, Madison, WI.

Khan, S.A., R.L. Mulvaney, and R.G. Hoeft. 2001. A simple soil test for detecting sites that are nonresponsive to nitrogen fertilization. Soil Sci. Soc. Am. J. 65:1751–1760.

Klapwyk, J.H., and Q.M. Ketterings. 2005. Reducing analysis variability of the Illinois soil nitrogen test with enclosed griddles. Soil Sci. Soc. Am. J. 69:1129–1134.

Klapwyk, J.H., and Q.M. Ketterings. 2006. Soil tests for predicting corn response to nitrogen fertilizer in New York. Agron. J. 98:675–681.

Kuo, S. 1996. Phosphorus, p. 869–919. *In* D.L. Sparks (ed.) Methods of soil analysis, Part 3. Chemical methods. SSSA Book Series No. 5. SSSA and ASA, Madison, WI.

Lawrence, J.R., Q.M. Ketterings, M.G. Goler, J.H. Cherney, W.J. Cox, and K.J. Czymmek. 2009. Illinois soil nitrogen test with organic matter correction for predicting nitrogen responsiveness of corn in rotation. Soil Sci. Soc. Am. J. 73:303–311.

Lindsay, W.L., and W.A. Norvell. 1978. Development of a DTPA soil test for zinc, iron, manganese and copper. Soil Sci. Soc. Am. J. 42:421–428.

Little, J.L., S.C. Nolan, J.P. Casson, and B.M. Olson. 2007. Relationships between soil and runoff phosphorus in small Alberta watersheds. J. Environ. Qual. 36:1289–1300.

Liu, M., D.E. Kissel, M.L. Cabrera, and P.F. Vendrell. 2005. Soil lime requirement by direct titration with a single addition of calcium hydroxide. Soil Sci. Soc. Am. J. 69:522–530.

Magdoff, F.R., W.E. Jokela, R.H. Fox, and G.F. Griffin. 1990. A soil test for nitrogen availability in the Northeastern United States. Commun. Soil Sci. Plant Anal. 21:1103–1115.

Magdoff, F.R., D. Ross, and J. Amadon. 1984. A soil test for nitrogen availability to corn. Soil Sci. Soc. Am. J. 48:1301–1304.

Maguire, R.O., and J.T. Sims. 2002. Measuring agronomic and environmental soil phosphorous saturation and predicting phosphorous leaching with Mehlich 3. Soil Sci. Soc. Am. J. 66:2033–2039.

Mahler, R.L., D.V. Naylor, and M.K. Fredrickson. 1984. Hot water extraction of boron from soils using sealed plastic pouches. Commun. Soil Sci. Plant Anal. 15:479–492.

Mallarino, A.P., and A.M. Atia. 2005. Correlation of a resin membrane soil phosphorus test with corn yield and routine soil tests. Soil Sci. Soc. Am. J. 69:266–272.

Martens, D.C., and W.L. Lindsay. 1990. Testing soils for copper, iron, manganese, and zinc, p. 229–264. *In* R.L. Westerman (ed.) Soil testing and plant analysis. 3rd edn. SSSA Book Series No. 3. SSSA, Madison, WI.

McCutcheon, S.C., and J.L. Schnoor. 2003. Phytoremediation: Transformation and control of contaminants. John Wiley & Sons, Hoboken, NJ.

McDowell, R.W., and A.N. Sharpley. 2001. Approximating phosphorus release from soils to surface runoff and subsurface drainage. J. Environ. Qual. 30:508–520.

McGrath, S.P., and F.J. Zhao. 1995. A risk assessment of sulphur deficiency in cereals using soil and atmospheric deposition data. Soil Use Manage. 11:110–114.

Meisinger, J.J., V.A. Bandel, J.S. Angle, B.E. O'Keefe, and C.M. Reynolds. 1992. Pre-sidedress soil nitrate test evaluation in Maryland. Soil Sci. Soc. Am. J. 56:1527–1532.

Mehlich, A. 1953. Determination of P, Ca, Mg, K, Na and NH_4. North Carolina Soil Test Division. (Mimeo), Raleigh, NC.

Mehlich, A. 1984. Mehlich 3 soil extractant: A modification of Mehlich 2 extractant. Commun. Soil Sci. Plant Anal. 15:1409–1416.

Menon, R.G., S.H. Chien, and W.J. Chardon. 1997. Iron oxide impregnated filter paper (Pi test). II. A review of its application. Nutr. Cycl. Agroecosyst. 47:7–18.

Mills, H.A., and J.B. Jones, Jr. 1997. Plant analysis handbook II. MicroMacro Publishing, Inc. Athens, GA.

Morgan, M.F. 1941. Chemical soil diagnosis by the universal soil testing system. Conn. Agric. Exp. Sta. Bull. 450. University of Connecticut Agriculture Experiment Station, New Haven, CT.

Moseley, R., M.O. Barnett, M.A. Stewart, T.L. Mehlhorn, P.M. Jardine, M. Ginder-Vogel, and S. Fendorf. 2008. Decreasing lead bioaccessibility in industrial and firing range soils with phosphate-based amendments. J. Environ. Qual. 37:2116–2124.

Mulvaney, R.L., and S.A. Khan. 2001. Diffusion methods to determine different forms of nitrogen in soil hydrolysates. Soil Sci. Soc. Am. J. 65:1284–1292.

Mulvaney, R.L., S.A. Khan, R.G. Hoeft, and H.M. Brown. 2001. A soil organic nitrogen fraction that reduces the need for nitrogen fertilization. Soil Sci. Soc. Am. J. 65:1164–1172.

Mulvaney, R.L., S.A. Khan, J.J. Warren, L.C. Gonzini, T.J. Smith, and R.G. Hoeft. 2004. Potential of the Illinois soil nitrogen test to improve nitrogen fertilizer management for corn production, p. 29–37. In R.G. Hoeft (ed.) Illinois fertilizer conference proceedings. Cooperative Extension Service, University of Illinois, Champaign, IL.

Muñoz, G.R., K.A. Kelling, K.E. Rylant, and J. Zhu. 2008. Field evaluation of nitrogen availability from fresh and composted manure. J. Environ. Qual. 37:944–955.

Munson, R.D., and W.L. Nelson. 1990. Principles and practices in plant analysis, p. 359–387. In R.L. Westerman (ed.) Soil testing and plant analysis. 3rd edn. SSSA Book Series No. 3. SSSA, Madison, WI.

Murphy, J., and J.P. Riley. 1962. A modified single solution method for the determination of phosphorus in natural waters. Anal. Chim. Acta 27:31–36.

Myers, R.G., A.N. Sharpley, S.J. Thien, and G.M. Pierzynski. 2005. Ion-sink phosphorus extraction methods applied on 24 soils from the continental USA. Soil Sci. Soc. Am. J. 69:511–521.

Myers, R.G., S.J. Thien, and G.M. Pierzynski. 1995. Improving the iron oxide sink method for extracting soil phosphorus. Soil Sci. Soc. Am. J. 59:853–857.

Myers, R.G., S.J. Thien, and G.M. Pierzynski. 1997. Iron oxide sink method for extracting soil phosphorus: Paper preparation and use. Soil Sci. Soc. Am. J. 61:1400–1407.

NADP. 2008. National Atmospheric Deposition Program 2007 Annual Summary. NADP Data Report 2008-01. Illinois State Water Survey, University of Illinois, Champaign, IL.

NADP. 2009. National Atmospheric Deposition Program: NADP/NTN Monitoring Location NY20. Available online at http://nadp.sws.uiuc.edu/sites/siteinfo.asp?id=NY20&net=NTN (verified June 29, 2009).

Nair, V.D., and W.G. Harris. 2004. A capacity factor as an alternative to soil test phosphorus in phosphorus risk assessment. NZ J. Agric. Res. 47:491–497.

NCR-13. 1998. Recommended chemical soil test procedures for the North Central region. Missouri Agricultural Experiment Station SB 1001, University of Missouri, Columbia, MO.

NEC-67. 1995. Recommended soil testing procedures for the Northeastern United States. Bull. No. 493. 2nd edn. Delaware Agricultural Experiment Station, University of Delaware, Newark, DE.

Nelson, L.A., and R.L. Anderson. 1977. Partitioning of soil test-crop response probability, p. 19–38. In T.R. Peck et al. (eds.) Soil testing: Correlating and interpreting the analytical results. ASA Special Publication 29. ASA, CSSA, and SSSA, Madison, WI.

Nemeth, K. 1979. The availability of nutrients in the soil as determined by electro-ultrafiltration (EUF). Adv. Agron. 31:155–188.

Norvell, W.A. 1984. Comparison of chelating agents as extractants for metals in diverse soil materials. Soil Sci. Soc. Am. J. 48:1285–1292.

O'Connor, G.A. 1988. Use and misuse of the DTPA soil test. J. Environ. Qual. 17:715–718.

Olsen, S.R., C.V. Cole, F.S. Watanabe, and L.A. Dean. 1954. Estimation of available P in soils by extraction with sodium bicarbonate. USDA Circ. 939. U.S. Government Printing Office, Washington, DC.

Olson, R.A., F.N. Anderson, K.D. Frank, P.H. Grabouski, G.W. Rehm, and C.A. Shapiro. 1987. Soil test interpretations: Sufficiency vs. build-up and maintenance, p. 41–52. In Soil testing: Sampling, correlation, calibration, and interpretation. Special Publication No. 21, SSSA, Madison, WI.

Osterhaus, J.T., L.G. Bundy, and T.W. Andraski. 2008. Evaluation of the Illinois soil nitrogen test for predicting corn nitrogen needs. Soil Sci. Soc. Am. J. 72:143–150.

Peck, T.R., and P.N. Soltanpour. 1990. The principles of soil testing, p. 1–10. In R.L. Westerman (ed.) Soil testing and plant analysis. SSSA, Madison, WI.

Pierce, F., and E. J. Sadler (eds.) 1997. The state of site specific management for agriculture. ASA, CSSA, SSSA, Madison, WI.

Pierzynski, G.M., J.T. Sims, and G.F. Vance. 1994. Soils and environmental quality. Lewis Publishers, CRC Press, Boca Raton, FL.

Plank, C.O. 1989. Plant analysis handbook for Georgia. University of Georgia, Athens, GA.

Power, J.F., and J.S. Schepers. 1989. Nitrate concentration of ground water in North America. Agric. Ecosyst. Environ. 26:165–188.

Raun, W.R., J.B. Solie, M.L. Stone, K.L. Martin, K.W. Freeman, R.W. Mullen, H. Zhang, J.S. Schepers, and G.V. Johnson. 2005. Optical sensor-based algorithm for crop nitrogen fertilization. Commun. Soil Sci. Plant Anal. 36:2759–2781.

Risser, J.A., and D.E. Baker. 1990. Testing soils for toxic metals, p. 275–298. In R.L. Westerman (ed.) Soil testing and plant analysis. 3rd edn. SSSA Book Series No. 3. SSSA, Madison, WI.

Robert, P.C., R.H. Rust, and W.E. Larson (eds.) 1995. Site-specific management for agricultural systems. ASA, CSSA, SSSA, Madison, WI.

Ruby, M.V., A. Davis, R. Schoof, S. Eberle, and C.M. Sellstone. 1996. Estimation of lead and arsenic bioavailability using a physiologically based extraction test. Environ. Sci. Technol. 30:422–430.

Ryan, J.A., and R.L. Chaney. 1993. Regulation of municipal sewage sludge under the Clean Water Act Section 503: A model for exposure and risk assessment for MSW-compost, p. 422–506. In H.A. Hoitnik, and H.M. Keener (eds.) Science and engineering of compost: Design, environmental, microbiological, and utilization aspects. Renaissance Publications, Worthington, OH.

Saavedra, C., J. Velasco, P. Pajuelo, F. Perea, and A. Delgado. 2007. Effects of tillage on phosphorus release potential in a Spanish vertisol. Soil Sci. Soc. Am. J. 71:56–63.

Schepers, J.S., T.M. Blackmer, and D.D. Francis. 1992. Using chlorophyll meters, p. 105–114. In B.R. Bock, and K.R. Kelley (eds.) Predicting nitrogen fertilizer needs for corn in humid regions. TVA National Fertilizer Environmental Research Center. Muscle Shoals, AL.

Schepers, J.S., and W.R. Raun (eds.) 2008. Nitrogen in agricultural systems. Agronomy monograph no. 49. ASA, Madison, WI.

Schollenberger, C.J., and R.H. Simon. 1945. Determination of exchange capacity and exchangeable bases in soil-ammonium acetate method. Soil Sci. 59:13–24.

Shanahan, J.F., J.S. Schepers, D.D. Francis, G.E. Varvel, W.W. Wilhelm, J.M. Tringe, M.R. Schlemmer, and D.J. Major. 2001. Use of remote-sensing imagery to estimate corn grain yield. Agron. J. 93:583–589.

Sharpley, A.N., S.C. Chapra, R. Wedepohl, J.T. Sims, T.C. Daniel, and K.R. Reddy. 1994. Managing agricultural phosphorus for protection of surface waters: Issues and options. J. Environ. Qual. 23:437–451.

Sharpley, A.N., T.C. Daniel, J.T. Sims, and D.H. Pote. 1996. Determining environmentally sound soil phosphorus levels. J. Soil Water Conserv. 51:160–166.

Sharpley, A.N., U. Singh, G. Uehara, and J. Kimble. 1989. Modeling soil and plant phosphorus dynamics in calcareous and highly weathered soils. Soil Sci. Soc. Am. J. 53:153–158.

Sharpley, A.N., and S.J. Smith. 1989. Prediction of soluble phosphorus transport in agricultural runoff. J. Environ. Qual. 18:313–316.

Shuman, L.M., V.A. Bandel, S.J. Donahue, R.A. Isaac, R.M. Lippert, J.T. Sims, and M.R. Tucker. 1992. Comparison of Mehlich-1 and Mehlich-3 extractable soil boron with hot-water extractable boron. Comm. Soil Sci. Plant Anal. 23:1–14.

Sibbesen, E. 1978. An investigation of the anion-exchange resin method for soil phosphate extraction. Plant Soil 50:305–321.

Sibbesen, E., and A.N. Sharpley. 1997. Setting and justifying upper critical limits for phosphorus in soils, p. 151–176. In H. Tunney et al. (eds.) Phosphorus loss from soil to water. CAB International, London, U.K.

Sikora, F.J. 2006. A buffer that mimics the SMP buffer for determining lime requirement of soil. Soil Sci. Soc. Am. J. 70:474–486.

Sikora, F.J., and K.P. Moore. 2009. The Moore–Sikora buffer for lime requirement determinations. Soil Sci. Soc. Am. J. 72:1163–1173.

Sims, J.T. 1989. Comparison of Mehlich 1 and Mehlich 3 extractants for P, K, Ca, Mg, Mn, Cu, and Zn in Atlantic coastal plain soils. Commun. Soil Sci. Plant Anal. 20:1707–1726.

Sims, J.T. 1993. Environmental soil testing for phosphorus. J. Prod. Agric. 6:501–507.

Sims, J.T. 1996. Lime requirement, p. 491–515. In D.L. Sparks (ed.) Methods of soil analysis, Part 3. Chemical methods. SSSA Book Series No. 5. SSSA and ASA, Madison, WI.

Sims, J.T. 1997. Phosphorus soil testing: Innovations for water quality protection, p. 47–63. Proc. 5th Intl. Symp. Soil Plant Anal. Minneapolis, MN.

Sims, J.T., and G.V. Johnson. 1991. Micronutrient soil tests, p. 427–476. In J.J. Mortvedt et al. (eds.) Micronutrients in agriculture. 2nd edn. SSSA, Madison, WI.

Sims, J.T., S.D. Cunningham, and M.E. Sumner. 1997. Assessing soil quality for environmental purposes: Roles and challenges for soil scientists. J. Environ. Qual. 26:20–25.

Sims, J.T., R.R. Simard, and B.C. Joern. 1998. Phosphorus losses in agricultural drainage: Historical perspective and current research. J. Environ. Qual. 27:277–293.

Sims, J.T., B.L. Vasilas, K.L. Gartley, B. Milliken, and V. Green. 1995. Evaluation of soil and plant nitrogen tests for maize on manured soils of the Atlantic coastal plain. Agron. J. 87:213–222.

Sims, J.T., R.O. Maguire, A.B. Leytem, K.L. Gartley, and M.C. Pautler. 2002. Evaluation of Mehlich 3 as an agri-environmental soil phosphorus test for the Mid-Atlantic United States of America. Soil Sci. Soc. Am. J. 66:2016–2032.

Skogley, E.O. 1994. Reinventing soil testing for the future, p. 187–201. In J.L. Havlin et al. (eds.) Soil testing: Prospects for improving nutrient recommendations. ASA, Madison, WI.

Skogley, E.O., S.J. Georgitis, J.E. Yang, and B.E. Schaff. 1990. The phytoavailability soil test. Commun. Soil Sci. Plant Anal. 21:1229–1243.

Soltanpour, P.N., and A.P. Schwab. 1977. A new soil test for simultaneous extraction of macro- and micronutrients in alkaline soils. Commun. Soil Sci. Plant Anal. 8:195–207.

Soltanpour, P.N., A. Khan, and W.L. Lindsay. 1976. Factors affecting DTPA-extractable Zn, Fe, Mn and Cu. Commun. Soil Sci. Plant Anal. 7:797–821.

Soltanpour, P.N., A. Khan, and A.P. Schwab. 1979. Effect of grinding variables on the NH_4HCO_3-DTPA soil test values for Fe, Zn, Mn, Cu, P, and K. Commun. Soil Sci. Plant Anal. 10:903–909.

Spargo, J.T., M.M. Alley, W.E. Thomason, and S.M. Nagle. 2007. Illinois soil nitrogen test for prediction of fertilizer nitrogen needs of corn in Virginia. Soil Sci. Soc. Am. J. 73:434–442.

Sparks, D.L. 1996. Methods of soil analysis, Part 2: Chemical properties. SSSA, Madison, WI.

SRIEG-18, 1992. Reference soil and media diagnostic procedures for the Southern region of the United States. Bull. No. 374. Virginia Agricultural Experiment Station. Virginia Polytechnic and State University, Blacksburg, VA.

Srinivasan, A. 2006. Handbook of precision agriculture: Principles and applications. Haworth Press, Binghamton, New York.

Sukkariyah, B.F., G. Evanylo, L. Zelazny, and R.L. Chaney. 2005. Cadmium, copper, nickel, and zinc availability in a biosolids-amended piedmont soil years after application. J. Environ. Qual. 34:2255–2262.

Sumner, M.E. 1979. Interpretation of foliar analysis for diagnostic purposes. Agron. J. 71:343–348.

Sumner, M.E. 1981. Diagnosing the sulfur requirement of corn and wheat using foliar analysis. Soil Sci. Soc. Am. J. 45:87–90.

Tabatabai, M.A. 1996. Sulfur, p. 921–960. *In* D.L. Sparks (ed.) Methods of soil analysis, Part 3. Chemical methods. SSSA Book Series No. 5. SSSA and ASA, Madison, WI.

Tanji, K.K. 1982. Modeling of the soil nitrogen cycle, p. 721–772. *In* F.J. Stevenson (ed.) Nitrogen in agricultural soils. Agronomy monograph no. 22. ASA, CSSA, and SSSA Madison, WI.

Thomas, G.W. 1996. Soil pH and soil acidity, p. 475–490. *In* D.L. Sparks (ed.) Methods of soil analysis, Part 3. Chemical methods. SSSA Book Series No. 5. SSSA and ASA, Madison, WI.

Thomason, W.E., S.B. Phillips, and F.D. Raymond. 2007. Defining useful limits for spectral reflectance measures in corn. J. Plant Nutr. 30:1263–1277.

Tisdale, S.L., W.L. Nelson, J.D. Beaton, and J.L. Havlin. 1993. Soil fertility and fertilizers. 5th edn. MacMillan Publications, New York.

Tunney, H., O.T. Carton, P.C. Brookes, and A.E. Johnston. 1997. Phosphorus loss from soil to water. CAB International. Oxon, U.K.

U.S. Department of Agriculture Natural Resources Conservation Service (USDA-NRCS). 2004. Soil survey laboratory methods manual. Soil survey investigations report no. 42, Version 4.0, November, 2004. USDA, Lincoln, NE.

U.S. Environmental Protection Agency. 1986. Acid digestion of sediment, sludge and soils, *In* Test methods for evaluating solid waste SW-846. USEPA, Cincinnati, OH.

U.S. Environmental Protection Agency. 1989. Development of risk assessment methodology for land application and distribution and marketing of municipal sluges. EPA/600/6-89/001. U.S. Government Printing Office, Washington, DC.

Vadas, P.A., P.J.A. Kleinman, A.N. Sharpley, and B.L. Turner. 2005. Relating soil phosphorus to dissolved phosphorus in runoff: A single extraction coefficient for water quality modeling. J. Environ. Qual. 34:572–580.

van Lierop, W. 1990. Soil pH and lime requirement, p. 73–126. *In*. R.L. Westerman (ed.) Soil testing and plant analysis. 3rd edn. SSSA, Madison, WI.

van Lierop, W., and T.S. Tran. 1985. Comparative potassium levels removed from soils by electro-ultrafiltration and some chemical extractants. Can. J. Soil Sci. 65:25–34.

van Raij, B. 1994. New diagnostic techniques, universal soil extractants. Commun. Soil Sci. Plant Anal. 25:799–816.

van Raij, B., J.A. Quaggio, and N.M. da Silva. 1986. Extraction of phosphorus, potassium, calcium and magnesium from soils by an ion-exchange resin procedure. Commun. Soil Sci. Plant Anal. 17:544–566.

Viro, P.J. 1955. Use of ethylenediaminetetraacetic acid in soil analysis: I. Experimental. Soil Sci. 79:459–465.

Walworth, J.L., H.J. Woodward, and M.E. Sumner. 1988. Generation of corn tissue norms from a small, high-yield data base. Commun. Soil Sci. Plant Anal. 19:563–577.

Wear, J.I., and C.E. Evans. 1968. Relationships of zinc uptake of corn and sorghum to soil zinc measured by three extractants. Soil Sci. Soc. Am. Proc. 32:543–546.

Westermann, D.T., D.L. Bjorneberg, J.K. Aase, and C.W. Robbins. 2001. Phosphorus losses in furrow irrigation runoff. J. Environ. Qual. 30:1009–1015.

Westerman, R.L. (ed.) 1990. Soil testing and plant analysis. 3rd edn. SSSA, Madison, WI.

Wolf, A.M., and D.E. Baker. 1985. Comparison of soil test phosphorus by Olsen, Bray P_1, Mehlich 1 and Mehlich 3 methods. Commun. Soil Sci. Plant Anal. 16:467–484.

Wolf, A.M., D.E. Baker, H.B. Pionke, and H.M. Kunishi. 1985. Soil tests for estimating labile, soluble, and algae-available phosphorus in soils. J. Environ. Qual. 14:341–348.

Yoon, J.K., X. Cao, and L.Q. Ma. 2007. Application methods affect phosphorus-induced lead immobilization from a contaminated soil. J. Environ. Qual. 36:373–378.

Zhang, M., Z. He, D.V. Calvert, P.J. Stoffella, and X. Yang. 2003. Surface runoff losses of copper and zinc in sandy soils. J. Environ. Qual. 32:909–915.

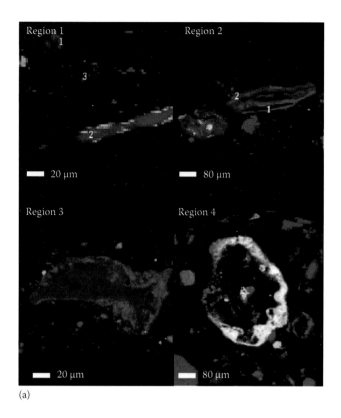

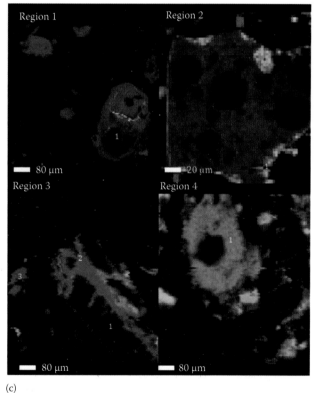

FIGURE 1.13 (a) μ-SXRF tricolor maps for the treated soil samples. The numbers indicate the spots where μ-EXAFS spectra were collected. Red is indicative of the distribution of iron, green of copper, and blue of zinc. (c) μ-SXRF tricolor maps for the nontreated soil samples. (Reprinted from Nachtegaal, M., M.A. Marcus, J.E. Sonke et al. 2005. Effects of in situ remediation on the speciation and bioavailability of zinc in a smelter contaminated soil. Geochim. Cosmochim. Acta 69:4649–4664. Copyright 2005, with permission from Elsevier.)

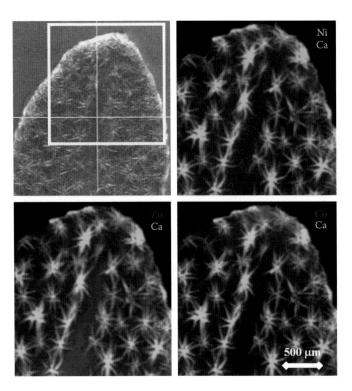

FIGURE 1.14 Synchrotron x-ray microfluorescence (μ-SXRF) images of the nickel (Ni), cobalt (Co), and zinc (Zn) distributions in a hydrated *A. murale* leaf from the Ni + Co + Zn treatment. Leaf trichomes are depicted in the Ca channel. The camera image shows the leaf region selected for SXRF imaging. (Reprinted from Tappero, R., E. Peltier, M. Grafe et al. 2007. Hyperaccumulator *Alyssum murale* relies on a different metal storage mechanism for cobalt than for nickel. New Phytol. 175:641–654. With permission from Wiley-Blackwell.)

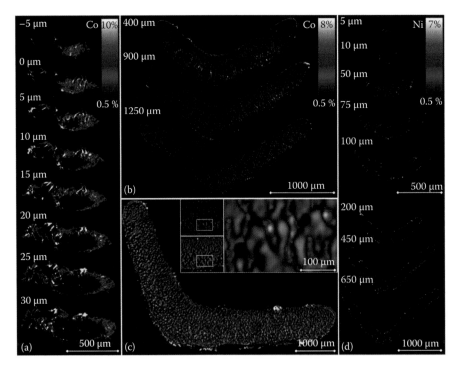

FIGURE 1.15 Differential absorption (DA-CMT) tomographic projections (5.1 μm slices) of hydrated *A. murale* leaves depicting (a) cobalt (Co) distribution in the leaf-tip region, (b) Co distribution in the bulk-leaf region, (c) Co distribution in relation to the leaf cell structure (gray), and (d) nickel (Ni) distribution in the leaf-tip and bulk-leaf regions. Leaves were collected from a Co-treated plant (a–c) and from a Ni-treated plant (d). Sinograms recorded above and below the Co or Ni K-edge energy (+30 and –100 eV, respectively) were computationally reconstructed, and the resulting projections were subtracted (above – below) to reveal the metal distribution in leaves. Distances are relative to the leaf tissue at the tip as determined from leaf structure images (i.e., below-edge projections). (Reprinted from Tappero, R., E. Peltier, M. Grafe et al. 2007. Hyperaccumulator *Alyssum murale* relies on a different metal storage mechanism for cobalt than for nickel. New Phytol. 175:641–654. With permission from Wiley-Blackwell.)

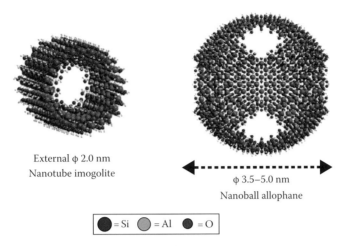

FIGURE 4.6 Chemical structure of (A) nanotube imogolite and (B) nanoball allophane. (With kind permission from Springer Science+Business Media: Abidin, Z., N. Matsue, and T. Henmi. 2007a. Differential formation of allophane and imogolite: Experimental and molecular orbital study. J. Comput. Aided Mater. Des. 14:5–18.)

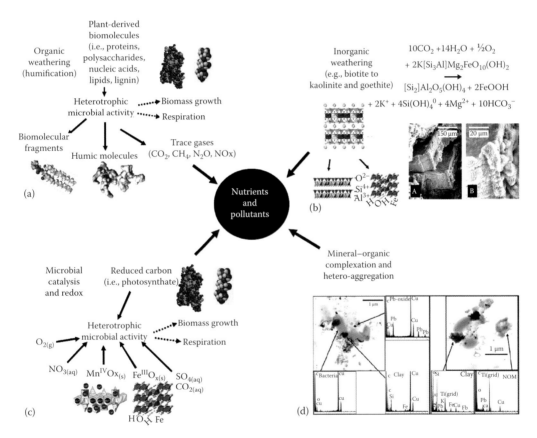

FIGURE 10.1 Types of physicochemical and biological reactions affecting biogeochemical interfaces and the fate of pollutants and nutrients in soil systems. (a) *Organic weathering* (i.e., humification) processes involve the biotic and abiotic degradation of plant- and microbially derived biomolecules into biomolecular fragments and associated humic substances. (b) *Inorganic weathering* processes include the incongruent dissolution of primary minerals to form high specific-surface-area secondary clays and (oxyhydr)oxides that can coat the surfaces of primary minerals (shown in inset—halloysite coating the surface of biotite mica "books," from Kretzschmar, R., W.P. Robarge, A. Amoozegar, and M.J. Vepraskas. 1997. Biotite alteration to halloysite and kaolinite in soil-saprolite profiles developed from mica schist and granite gneiss. Geoderma 75:155–170. Copyright 1997, with permission from Elsevier.) (c) *Microbial catalysis and redox* reactions drive geochemical transformations as part of the set of alternative TEAPs in microbial respiration, such as the reductive dissolution of manganese (IV) and iron (III) oxides (MnOx and FeOx shown here are birnessite and goethite) with consequences for sorbed contaminants; (d) *Mineral–organic complexation* gives rise to complex hetero-aggregates of secondary mineral particles, humic substances and microbial cells that constitute much of the reactive interface in soils. (Image from Perdrial, N., J. Perdrial, J.-E. Delphin, F. Elsass, and N. Liewig. 2010. Temporal and spatial monitoring of mobile nanoparticles in a vineyard soil: Evidence of nanoaggregate formation. Eur. J. Soil Sci. 61:456–468. In press. Copyright Wiley-VCH Verlag GmbH & Co. KGaA. With permission.).

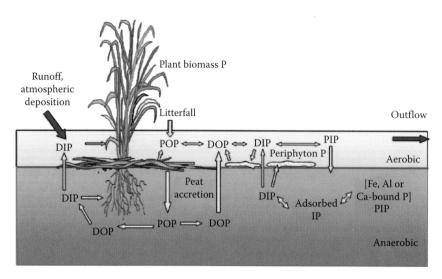

FIGURE 20.18 Schematic diagram of the general phosphorus cycle in wetlands. (From Reddy, K.R., and R.D. DeLaune. 2008. Biogeochemistry of wetlands: Science and applications. CRC Press, Boca Raton, FL.) POP-Particulate organic P, PIP-particulate inorganic P, DOP-dissolved organic P, DIP-dissolved inorganic P, IP-inorganic P.

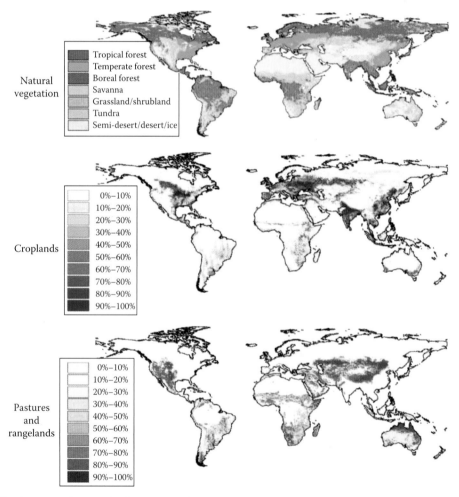

FIGURE 26.1 Worldwide extent of human land-use and land-cover change. (From Foley, J.A., R. DeFries, G.P. Asner, C. Barford, G. Bonan, S.R. Carpenter, F.S. Chapin, M.T. Coe, G.C. Daily, H.K Gibbs, J.H. Helkowski, T. Holloway, E.A. Howard, C.J. Kucharik, C. Monfreda, J.A. Patz, C. Prentice, N. Ramankutty, P.K. Snyder. 2005. Global consequences of land use. Science 309:570–574.) Croplands are 12% and rangelands are 22%. (From Ramankutty, N., A.T. Evan, C. Monfreda, and J.A. Foley. 2008. Farming the planet: 1. Geographic distribution of global agricultural lands in the year 2000. Global Biogeochem. Cycles 22:GB1003, doi: 10.1029/2007GB002952.)

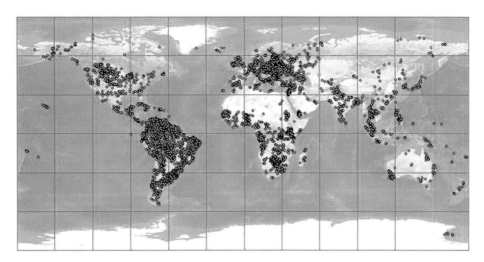

FIGURE 27.5 Global distribution of geo-referenced soil profiles in WISE3. (From Batjes, N.H. 2008. ISRIC-WISE harmonized global soil profile dataset (Version 3.1). Report 2008/02. ISRIC—World Soil Information, Wageningen, the Netherlands. (http://www.isric.org/isric/webdocs/docs/ISRIC_Report_2008_02.pdf).

FIGURE 28.5 Soil interpretive map of weighted average cation-exchange capacity (CEC-7) of the 20–75 cm depth for a portion of Palo Alto County, IA generated from SSURGO data by WSS.

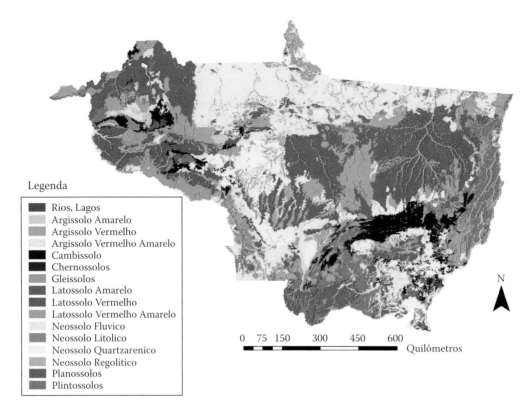

FIGURE 29.5 Soil map of Rondonia and Mato Grosso states. (Modified from Mello, F.F.C. 2007. Estimativas dos estoques de carbono dos solos nos Estados de Rondônia e Mato Grosso anteriores à intervenção antrópica, 88pp. Dissertation (Master Degree). University of São Paulo, Sao Paulo, Brazil.)

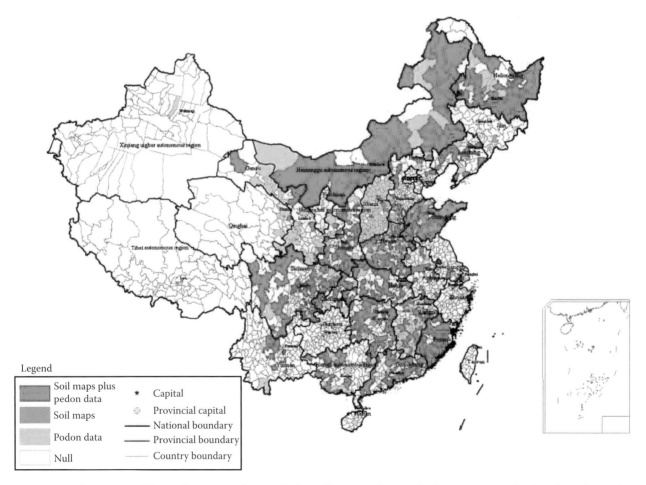

FIGURE 30.2 The coverage of digitized 1:50,000 soil maps of China. (Courtesy of Dr. Weili Zhang, Institute of Soils and Fertilizer, Chinese Academy of Agricultural Science, unpublished.)

FIGURE 31.2 Soil map of the Russian Federation.

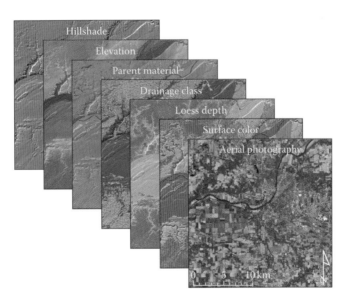

FIGURE 33.2 Examples of some of the different kinds of maps available in our GIS dataset illustrated by a portion of Tippecanoe County, Indiana. Note the obvious correlation between features shown on the different maps.

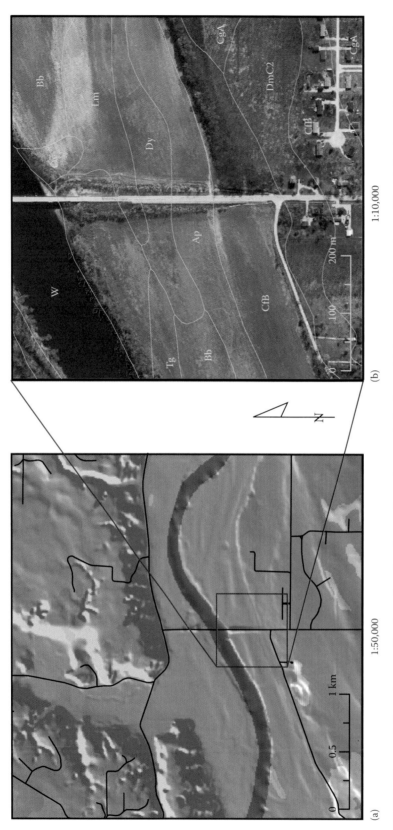

FIGURE 33.3 The most informative map often depends on how fast one is moving over the landscape. (a) A more general thematic map, like this dominant soil parent material map displayed at 1:50,000, is informative and usually most appropriate when one is traveling in a vehicle. (Tan, glacial till; green, glacial outwash; light blue, recent alluvium; yellow, eolian sand and/or other sandy sediments.) (b) An aerial photograph overlain with the detailed soil delineations and displayed at 1:10,000 is often the most informative map when one is walking across the landscape. The polygons are queried individually to obtain information about the soils they represent.

14
Fundamentals of Fertilizer Application

David B. Mengel
Kansas State University

George W. Rehm
University of Minnesota

14.1	Introduction	14-1
14.2	Application of Mobile Nutrients	14-2
	Nitrogen • Sulfur • Boron • Chloride	
14.3	Application of Immobile Nutrients	14-5
	Broadcasting • Banding • Strip Application • Point Fertilization • Application with Irrigation Water • Foliar Application	
14.4	Placement of P and K Fertilizers	14-7
	Field Comparison of Placement Options • Fertilizer Placement in Conservation Tillage Systems	
14.5	Calcium and Magnesium	14-9
14.6	Micronutrients	14-10
	Zinc • Manganese • Iron • Copper • Molybdenum	
References		14-11

14.1 Introduction

Efficiency of nutrient application is one of the major factors affecting the overall effectiveness of a fertilizer program. There are a number of ways that the efficiency of fertilizer use can be measured, such as the percent of the applied nutrient taken up by the plant or the increase in crop yield per unit of applied nutrient. But regardless of the measure used, a number of interacting factors determine the efficiency of a given fertilizer application. The relative mobility of the nutrient of concern in the soil is foremost in importance. Nitrogen (N) is an excellent example of a nutrient that is mobile in the soil and easily lost from the root zone. Devising an application system that can provide N to the crop when it is needed is the challenge in many cropping situations. Sulfur, B, and Cl are other nutrients considered to be mobile in soils. Compared to N usage, uptake of these nutrients is rather small. However, management decisions regarding time and method of application may still be needed to ensure optimum utilization.

Phosphorus, on the other hand, is generally considered to be relatively immobile in soils. While loss of P from the root zone limiting P availability to crops may be unlikely, P is subject to a number of reactions and transformations, which can reduce P availability to the plant. Placement techniques that minimize the effects of these transformations and enhance P availability are potentially valuable to crop managers.

In choosing appropriate application technology, a number of characteristics of the system beyond nutrient mobility must also be considered. These include the nature of the crop being fertilized, weather and climate, soil properties, the form in which the nutrient is applied, the method of application utilized, and the timing of the fertilizer application in relation to nutrient needs of the crop. By understanding how these individual factors interact, one can devise an efficient system of nutrient delivery requiring the least possible amounts of fertilizer, yet achieving optimum growth and subsequent yield.

To arrive at useful and effective decisions with respect to fertilizer application, it is necessary to have some understanding of how plant nutrients get to the root system. There is general agreement that nutrients reach a root by three primary mechanisms: mass flow, diffusion, and root interception.

Mass flow is somewhat self-descriptive. Nutrients dissolved in the soil solution/water move to the surface of the root as water is taken up by the plant. Mobile nutrients such as N, S, B, and Cl, and immobile nutrients present in high concentrations in the soil solution such as Ca and Mg move to the root surface with soil water through mass flow. Diffusion is a process whereby soil nutrients move from an area of high concentration to an area of lower concentration near the root surface created through nutrient uptake. Immobile nutrients such as P, K, and the micronutrient metals move to the surface of the root by diffusion. Diffusion takes place over relatively short distances (<1 mm). Fertilizer application methods, which create zones/areas of high nutrient concentration such as fertilizer banding, can facilitate diffusion. However, only the portions of the root system in, or in direct contact with the high nutrient availability zones are impacted.

Root interception is simply the process where the root intercepts plant nutrients as it grows through the soil. Both mobile and immobile nutrients are made accessible for nutrient uptake by this process.

14.2 Application of Mobile Nutrients

Of the 14 mineral nutrients considered essential for plant growth, N, S, B, and Cl are generally considered to be mobile in soils. Because of the differences in magnitude of uptake and sources available for use in a fertilizer program, the application of each will be discussed separately.

14.2.1 Nitrogen

Application of N is a major concern in the production of most nonlegume crops. When developing an N fertilization program today, the modern grower is faced with decisions about (1) the rate of N to apply, (2) fertilizer source or sources to use, (3) when to apply N to ensure adequate amounts present at key crop growth stages, (4) method of application, and (5) the use of compounds such as nitrification or urease inhibitors, or specialty fertilizer products such as coated urea to prevent N loss.

14.2.1.1 Nitrogen Sources

There are numerous sources of N that have been used for crop production over the years. The three N sources most commonly used for field crop production in the United States today are anhydrous ammonia (82-0-0), urea-ammonium nitrate solutions (28-, 30- or 32-0-0), and granular urea (46-0-0). In addition, significant amount of ammonium nitrate (33-0-0) and ammonium sulfate (21-0-0-24S) are also applied in some areas. Nitrogen is also added as a secondary component in many multiple nutrient fertilizers. These include the dry granular ammonium phosphates diammonium phosphate, DAP (18-46-0), monoammonium phosphate, MAP (11-52-0), and liquid ammonium polyphosphates (10-34-0 and 11-37-0). Many by-product N sources are also marketed as fertilizers, such as dry or liquid ammonium chloride. The choice of the source of fertilizer N is dependent on a number of factors such as the local availability of a specific product, cost of the product per unit of N, safety issues related to using the product, application equipment available and cost of application, time required and availability of custom application, N loss potential of the soils in that particular field, cropping system and residue cover, rooting patterns of the intended crop, and other individual factors.

The relative advantages of one fertilizer N source over another has been widely debated and consequently studied at length. In most studies with field crops, when care was taken to prevent N loss, most N sources have an equal effect on yield when used at rates that supply the same amount of N. Where differences have been observed, the differences can generally be explained by some factor that makes a particular source susceptible to N loss in that particular application. Examples would include (1) a surface application of urea in a high pH or high residue situation leading to N loss from ammonia volatilization; (2) the broadcast surface application of urea ammonium nitrate (UAN) solutions to a soil covered with large amounts of a wide C:N residue such as wheat straw or corn stalks resulting in N immobilization; (3) preplant applications of granular ammonium nitrate for corn on a sandy soil prone to leaching loss; (4) or an early fall preplant application of anhydrous ammonia for corn on a poorly drained soil prone to denitrification in a humid climate.

In each of these examples, N loss could have been reduced and the source successfully used through incorporation of the fertilizer or the addition of a urease inhibitor with the urea in example 1; placement of the UAN below the wide C:N residue through subsurface banding in example 2; use of a high N starter fertilizer followed by sidedressing with the ammonium nitrate in example 3; and delaying the application of the fall preplant ammonia to spring, and/or adding a nitrification inhibitor such as nitrapyrin to reduce denitrification in example 4. As these examples show, N source selection when developing an N management plan is not a decision that can be made independently of many other factors such as N loss potential from leaching, denitrification, immobilization and ammonia volatilization, application equipment available and time available for application, safety, and cost.

14.2.1.2 Time of N Application

The optimum time for N applications to agronomic crops generally revolves around three key factors, (1) the intended crop, architecture of the root system, nutrient uptake pattern, and when physiologically it demands N for key stages in crop growth, (2) the potential for N loss at that particular site, as a function of both climate and soils, and (3) time available for nutrient application in the cropping system. Using time of application as a management tool to avoid loss of N fertilizer will be discussed in detail in the following section.

Growth and development of the crop, and especially of the root system, are important factors to consider in timing decisions. Because of the downward mobility of NO_3-N in soils, more frequent applications of fertilizer N have been most effective for shallow rooted crops such as edible beans and potatoes, especially on sandy soils. Nitrogen uptake by crops is also not constant throughout the growing season. For crops such as wheat, barley, oats, and corn, for example, the rate of N uptake is slow during the early part of the growing season, accelerates before anthesis, and stops or decreases rapidly after pollination. Therefore, N fertilizer should be supplied in such a way as to meet the minimal needs of the young seedling, while providing the majority of the N throughout the later stages of vegetative growth, but be completed before grain fill is started. In the case of corn on sandy soils, where NO_3-N can move quickly below the root zone, a preplant or starter application to support early growth followed by a sidedress application at the four to eight leaf stage is generally adequate on nonirrigated sites. Under high-yielding irrigated environments, additional N applications prior to tasseling can be beneficial, both to ensure adequate N for grain fill, and to reduce the movement of NO_3-N below the root zone facilitated by the application of irrigation water.

With winter wheat, an application of N at seeding, either through broadcasting N or applying N with the planting equipment to enhance fall growth and tillering, followed by a topdress application applied between green-up and "jointing," or initiation of stem elongation, is common. In the high-yielding, intensive management systems used in the Mid-Atlantic and Ohio River Valley regions of the United States, multiple topdress applications are commonly made. In regions where winter wheat is a dual purpose crop used for both forage production and grain, higher rates of N applied prior to or at planting to stimulate fall and winter vegetative growth, followed by a heavy topdressing in early spring to stimulate grain production is common.

On well-drained medium to fine textured soils, timing of the applications of fertilizer N is more flexible for deep rooted crops such as corn and wheat in many environments. In many of these situations, time of application has little effect on yield, as long as adequate N is present to meet the needs of the crop at key physiological growth stages such as ear or head initiation/development. Nitrogen loss and environmental concerns may also be minimal in many environments. In these situations, applying N in a single application prior to planting can be a reasonable system to use for many crops.

Timing of needed fertilizer N is also an important consideration in the production of forage grasses. Because cool season species such as tall fescue and smooth brome grass produce most of their dry matter in late spring and early summer, fertilizer N should be applied before this season of rapid growth. How the forage is utilized, whether harvested for hay or used for grazing, and whether the N is applied alone or in combination with other nutrients can also influence the time of N application. When using cool season grasses for hay, making a single application of N in the early spring before or at green-up is common. When making a combined application of N with P, a late fall application to stimulate tillering and root growth is preferable in many environments. In environments where cool season grasses are used for grazing, especially fall grazing, or a second cutting of hay is made, splitting the N application with the majority of the N and any needed P applied in the fall or early spring followed by an application of the balance of the N in mid- to late summer to stimulate fall growth, can be beneficial.

In contrast, the majority of the growth of warm season species occurs when temperatures are high and therefore application of fertilizer N should be delayed until after the initiation of the warm season grass growth to minimize stimulation of cool season invasive grass and weed species. In the case of high-yielding, intensively managed species such as Bermuda grass, multiple applications of N are generally more efficient than single, early season N applications.

14.2.1.3 Managing Fertilizer N to Prevent Losses

Nitrogen is lost from agricultural soils primarily by (1) leaching, (2) denitrification, and (3) ammonia volatilization. In addition, N can be transformed to unavailable forms through immobilization and ammonium fixation. Detailed reviews of each of these processes can be found in Section 27.1 of *Handbook of Soil Sciences: Properties and Processes*, Hauck (1984), Scharf and Alley (1988), and Schepers and Raun (2008). However, some basic fundamental concepts will be discussed to facilitate the understanding of how application techniques may minimize the impact of these processes.

Leaching is simply the downward transport of NO_3^- (and NH_4^+ in low cation exchange capacity [CEC] soils) with water. Leaching occurs when water reaching the soil surface from precipitation and/or irrigation exceeds evaportranspiration (Nelson and Uhland, 1955). Leaching is a special concern in the more humid climate of the eastern United States, where precipitation exceeds evapotranspiration by 150–500 mm annually, or where irrigation is used. Soil water-holding capacity also plays an important role in determining the importance of leaching losses. Soil with a high water-holding capacity, such as silt loams, can accumulate large quantities of water before NO_3^- is transported below the root zone. Thus, soils with high water-holding capacities are much less prone to leaching losses than coarse textured, low water-holding capacity soils. Tile drainage, such as is commonly used in the midwest, can enhance leaching by providing a shallow outlet for drainage to surface water sources.

Denitrification is a microbial process by which soil bacteria (facultative anaerobes) utilize the O in NO_3^- and NO_x and N_2 gases are released into the soil atmosphere and are subsequently lost. A number of factors or conditions interact to impact the rate of denitrification. These include energy sources such as soil organic matter (SOM), crop residue, and animal manure, moisture/O_2 availability, temperature, and pH (Wijler and Delwhiche, 1954; Bailey and Beauchamp, 1973; Burford and Bremner, 1975; Rolston et al., 1978). Denitrification is the primary N loss process and is a major concern for finer textured poorly drained soils, or soils with high seasonal water tables.

Ammonia volatilization is the gaseous loss of free NH_3 to the atmosphere from soil and fertilizers. There are three general situations where NH_3 volatilization is a significant loss problem. When NH_4^+-based fertilizers such as ammonium sulfate are broadcast on the surface of high pH or alkaline soils and not incorporated, some NH_4 is converted to NH_3, which can volatilize to the atmosphere. This only occurs in significant amounts with pH > 7. When urea-based fertilizers are applied to soils, the naturally present urease enzyme hydrolyzes the urea to NH_3. The hydrolysis reaction generates enough OH^- ions to temporarily raise the pH of the soil around the urea fertilizer, which, in turn, causes NH_3 volatilization. Volatilization of N from urea is enhanced if urea is applied to a soil surface covered with crop residue or vegetation, as in a no-till corn field or a pasture or hayfield. Ammonia volatilization in cases one (volatalization from ammonium sulfate at high pH) and two (volatalization from urea) is exacerbated by the evaporation of water from the soil surface. The third situation deals with the addition of urea- or NH_4-based fertilizers to flooded rice paddies. A diurnal fluctuation in pH due to algal utilization of CO_2 results in a rapid rise in water pH reaching a maximum of 7–10 during midday. The high pH favors the conversion of NH_4 to NH_3, and its volatilization. In all cases, incorporation of urea- or NH_4-based fertilizers into the soil effectively stops volatilization (Ernst and

Massey, 1960). Unfortunately, incorporation is not always possible so alternative management strategies are sometimes utilized.

There are a number of alternative application strategies that can be used to reduce leaching loss of N from agricultural soils. Probably the most commonly used is timing of fertilizer applications to match periods of rapid utilization by crops. The use of sidedressing, delaying applications of N fertilizers until the crop is established and taking up N, or the use of split applications to avoid N leaching losses is common for crops such as corn, wheat, and cotton, particularly on sandy soils. An example of how time of N application has been used to enhance N use efficiency on coarse textured soils by reducing leaching losses can be found in Evanylo (1991).

The use of nitrification inhibitors and slow-release fertilizers is an alternative strategy to application timing as a means of reducing N loss from leaching. While the use of nitrification inhibitors has proven to be a successful means of reducing N loss on fine textured soils in the Eastern Corn Belt (Huber et al., 1982; Stehouwer and Johnson, 1990; Mann, 1995), they have shown mixed results in irrigated areas and the western Corn Belt (Maddux et al., 1985; Cerrato and Blackmer, 1990). A number of slow-release fertilizer products have been studied and have potential for reducing N loss from leaching in coarse textured soils.

Many of the same strategies that are used to avoid leaching losses can be used to avoid losses from denitrification, namely, timing of application, the use of slow-release fertilizers, and nitrifications inhibitors. However, one important difference among soils where these problems occur must be emphasized. Soils prone to high rates of leaching loss tend to be well drained, with low water-holding capacity. These soils dry rapidly after a leaching event and a farmer has a high probability of being able to drive across these fields in a short period of time. Many of these fields may be irrigated, which allows for the application of fertilizer N with irrigation water (fertigation). Soils prone to high rates of N loss through denitrification are not usually irrigated and tend to be poorly drained and slow to dry and trafficability is reduced. Thus, while sidedressing or split N applications are excellent tools to reduce N loss from denitrification, they entail risk to the grower who chooses to use them on poorly drained soils. This is a risk that additional N may not be applied in a timely manner.

Nitrification inhibitors are useful tools for reducing N loss in soils prone to denitrification. A number of studies in fine textured soils of the eastern Corn Belt have shown responses to the use of nitrification inhibitors with fall or spring preplant applied N (Frye et al., 1981; Huber et al., 1982; Stehouwer and Johnson, 1990).

Incorporation of the urea-N into the soil is the most effective strategy for reducing N loss from NH_3 volatilization (Ernst and Massey, 1960). While this is not always possible with no tillage systems, a number of tools have been developed to allow injection or banding of N fertilizers in these production systems. A number of studies have demonstrated that injection of N fertilizers into the soil and below the surface residue in corn and grain sorghum greatly enhances N use efficiency (Mengel et al., 1982; Lamond et al., 1991). Irrigation water can also be used to incorporate urea into the soil to reduce NH_3 volatilization (Mengel and Wilson, 1988).

Choosing a nonvolatile N fertilizer source is another alternative (Bandel et al., 1980; Fox and Hoffman, 1981). Ammonia volatilization losses from ammonium nitrate and ammonium sulfate are negligible compared to those from urea or urea-based products (Keller and Mengel, 1986).

Urease inhibitors are another management tool available to reduce NH_3 production until after the urea has been incorporated into the soil by rain or irrigation water. At present, one urease inhibitor, N-(n-butyl) thiophosphoric triamide (NBPT) is commercially available in the United States. Research has shown that urea impregnated with NBPT is a viable alternative to a nonvolatile N source such as ammonium nitrate. Coated urea products, such as polyurethane coated urea are also useful in reducing NH_3 volatilization.

14.2.2 Sulfur

Sulfur, like N is mobile in soils, yet its transformations and chemistry differ considerably from many of the reactions associated with N. Because S is mobile, many of the best management practices suggested for use of fertilizer N are also appropriate for the management of fertilizer S. Compared to N, S is utilized in relatively small amounts by actively growing crops. Various research projects have evaluated the effect of rate and management of S fertilizers on production and quality of a variety of crops. Specific management practices associated with optimum production vary with the crops that are grown.

The legumes, (alfalfa, clovers, soybeans) and canola remove relatively large amounts of S from the soil system while other crops (corn, small grains, forage grasses) remove smaller amounts. Options for placement of fertilizer S also vary with the intended crop.

Approximately 90% of the total S in soils is found in SOM, which can supply substantial amounts of S for crop production. Traditionally, research has shown that crops grown on fine textured soils with a high SOM do not respond to the application of fertilizer S. On the other hand the use of fertilizer S has produced dramatic increases in yield when applied to crops grown on sandy soils with a low SOM content (Section 11.4). Recent reductions in the atmospheric deposition of S may be altering this relationship however. Reports of S fertilizer responses appear becoming more widespread in recent years.

14.2.2.1 Sulfur Sources and Methods of Application

Application of S in a fertilizer program is most frequently associated with the production of alfalfa and other perennial legumes, corn, and small grains. As reported by Hoeft and Walsh (1975), alfalfa responds favorably to annual topdress applications of fertilizer S. Products containing elemental S (>90% S) or SO_4 form can be used for this method of application. The most common dry sources of SO_4-S are ammonium sulfate (24% S) and the

Fundamentals of Fertilizer Application

double salt of potassium and magnesium sulfate (22% S). If K is needed in a fertilizer program for alfalfa, the double salt is a logical choice for topdress applications. Cogranulated ammonium phosphate products containing both elemental and SO_4 forms of S are also available and would be logical products to use if P were needed. Perennial grasses have also responded to annual topdress applications in situations where S is needed in a fertilizer program (Lamond et al., 1995). In general, annual applications of fertilizer S are suggested for all perennial crops whenever there is a need for S in the fertilizer program. Although legumes such as alfalfa generally remove substantial amounts of S from soils, soybeans have not responded consistently to S fertilization. Yield increases have been noted when fertilizer S is applied on very sandy soils (Matheny and Hunt, 1981) but not on fine textured soils (Brown et al., 1981; Sweeney and Grande, 1993). Sulfur needed for soybeans can be either broadcast before planting or applied in a band at planting. There is no research information to suggest that one method of application is superior.

When corn is grown on sandy soils where responses to S fertilization might be expected, the needed S can either be broadcast and incorporated before planting or applied as a starter fertilizer at planting (Hoeft et al., 1985; Kline et al., 1989). In general, rates of applied S are doubled if S fertilizer is broadcast and incorporated before planting rather than applied in a band near the seed at planting. There is general agreement that the use of split applications of N fertilizers is a best management practice for corn production on irrigated sandy soils. In evaluating the timing of S applications for sandy soils, however, a single application at planting has been as effective as split applications during the first half of the growing season (Rehm, 1993). When wheat and other small grains are considered, responses to S fertilization are not frequently reported. The majority of responses reported have been for the silt loam soils of the Pacific Northwest where broadcast applications have been popular (Ramig et al., 1975; Mahler and Maples, 1987). Responses have also been observed in wheat recently in the sandy, low SOM soils of the central and southern Great Plains.

When fluid fertilizers are used for production of annual crops, S can be supplied as ammonium thiosulfate (26% S) or potassium thiosulfate (17% S). Although these fluid materials are best suited for a band application at planting, caution should be used because germination can be impaired if ammonium thiosulfate is placed in contact with the seed. The safety of potassium thiosulfate when placed in contact with the seed has not been fully documented. Fluid fertilizer containing S can also be injected into irrigation water. This is not a preferred practice and should be limited to correcting identified S deficiencies.

14.2.2.2 Managing Sulfur to Prevent Losses

When present as SO_4-S, downward movement through soils can occur during either heavy rainfall or over-irrigation. The rate of downward movement of SO_4-S is, however, not as rapid as that of NO_3-N. Unless excessive amounts of irrigation water are applied to sandy soils, leaching of SO_4-S should be of minor concern. Because SO_4-S is not associated with problems of water quality, its loss by leaching is of economic concern, which can be minimized by applying fertilizer S, when needed, in a band close to the seed at planting.

14.2.3 Boron

This essential nutrient is also mobile in soils. Boron, which is found in soils as uncharged H_3BO_3, is not strongly sorbed by soil particles and is susceptible to leaching. Boron is classified as a micronutrient because small amounts are required for optimum crop production. The majority of the research conducted with B fertilization on agronomic crops has focused on the production of alfalfa (Brown, 1972), corn (Touchton and Boswell, 1975), small grains (Gupta et al., 1976), cotton, and peanut. Positive responses have not been consistent and are limited to unique soils and/or situations. As with N and S, annual applications of B are suggested often as broadcast preplant applications with annual crops or topdress applications or foliar applications on perennials, such as alfalfa.

Since B applied with the seed can cause germination problems and seedling injuries, especially for oilseeds, broadcast applications incorporated before planting are suggested for all annual crops and placement of B in contact with the seed should be avoided. Foliar application of B is routine for many crops, including cotton (Roberts et al., 2000) and peanut (Gascho and Davis, 1995).

14.2.4 Chloride

Chloride, like B, is mobile in soils behaving like NO_3-N in terms of leaching (Endelman et al., 1974). Although not studied intensively, responses to this nutrient have been documented for corn (Heckman, 1995), sorghum (Mengel et al., 2009), and small grain production in limited situations, and especially where applications of potassium chloride are not routinely made (Fixen et al., 1986a, 1986b; Engel et al., 1994). Because Cl^- is mobile, placement should have little effect on crop response to this nutrient. In contrast to the management of N, there is no indication that split applications would be superior to a single application at or before planting. Common sources for Cl^- include many soluble salts such as KCl, $CaCl_2$, $MgCl_2$, and NaCl, and NH_4Cl.

14.3 Application of Immobile Nutrients

Nutrients such as P, K, Ca, Mg, Zn, Cu, Fe, and Mn that are strongly sorbed by soil components are immobile and do not move easily through soils (Chapter 11). Unless there are substantial losses of soil from the landscape, there are usually no significant losses of immobile nutrients from the soil system other than through crop removal. There are situations however, where nutrients normally considered immobile can move through a soil profile. Two examples are leaching of K^+ and Mg^{2+} in soils with a low CEC and leaching of Mn^{2+} and Fe^{2+} through very sandy soils that are poorly drained.

Placement consideration: Fertilizer placement becomes a major management consideration in the overall management of immobile nutrients with the following options available.

14.3.1 Broadcasting

Broadcasting with or without soil incorporation is the most commonly used method of application for immobile nutrients. Throughout the United States, a very high percentage of P and K fertilizers are applied in this manner. With this placement option, especially when coupled with incorporation through tillage, essentially 100% of the top few cm of topsoil comes in contact with the applied fertilizer. Some of the advantages of broadcast applications are as follows: (1) it is fast and easy; (2) there are several opportunities for application that can reduce the workload at critical times during the year; (3) there is high probability that crop roots will come in contact with fertilized soil especially when incorporated with some form of tillage; and (4) when soils have low CEC, there are more potential sites for adsorption and subsequent retention of K and Mg. Some of the disadvantages of this placement option are as follows: (1) thorough mixing with tillage increases the probability of fixation and reduced availability; and (2) it is difficult to achieve a uniform application of low rates of fertilizers that supply the immobile nutrients.

14.3.2 Banding

Banding fertilizer is a mechanical technique in which some device is used to open a furrow or trench and the fertilizer is applied in a band below the soil surface. With most band applications, <1% of the top soil (15 cm) volume comes in contact with the fertilizers. The common band placement options include (1) placing a low rate of fertilizer in close proximity to the seed to enhance availability of nutrients to young plants (starter); (2) placing a low rate fertilizer in direct contact with the seed at planting (pop-up); and (3) concentrating fertilizer in the root zone to overcome fixation reactions and create zones of high fertility.

In North Carolina, Nelson et al. (1959) compared various band placement options to broadcast applications of P for corn and cotton production. Using low rates of P on P-responsive sites, they found that band placement enhanced early growth and resulted in better utilization of the applied P. There were, however, no differences in yield. Barber (1958) established a long-term study in Indiana and evaluated both the direct and residual responses to broadcast and band placement of P fertilizers. He found that band applications provided the greatest response when the P soil test was low, and that the advantage decreased as the soil test for P increased. This study also showed that the production of optimum corn yields required both broadcast and banded application of P fertilizer at low soil test P levels. Welch et al. (1966a, 1966b) found similar results with the application of both P and K in Illinois. With low fertilizer rates on responsive soils, band application resulted in more efficient utilization of immobile nutrients. However, highest yields were obtained by combining both band and broadcast application at higher rates. For soils with higher soil test values, response to fertilizer was smaller and placement had only limited effects on use efficiency and yield.

The primary advantages of banded application of immobile nutrients are as follows: (1) it creates a zone(s) of enhanced nutrient availability by minimizing contact between soil and fertilizer and (2) it is a simple and efficient method of applying small amounts of fertilizer. Disadvantages include (1) only a limited portion of the root system has high probability of coming in contact with the fertilizer; (2) damage to germination can occur when high rates of some fertilizer products are placed too close to the seed; and (3) the cost of equipment used in banding can be high.

With the continued expansion of no-till production systems, and introduction of strip-till systems, the use of deep banding to place fertilizer 15–20 cm deep below the row area has gained considerable interest. This practice avoids the concentration of nutrients near the soil surface, vertical nutrient stratification, which commonly occurs with broadcast applications when not accompanied with incorporation. In most deep banded situations, a tine or shank device is pulled through the soil doing some tillage with the fertilizer released at some depth below the surface. The depth of placement used varies from 10 to 30 cm. The volume of soil mixed with the fertilizer varies widely, but is still limited in most cases. In most cases, the crop is then planted into the tilled strip, directly over the fertilizer band. Work to date has shown only limited response to deep banding of P and K as compared to traditional starter fertilizer application or broadcasting similar rates of fertilizer in corn, soybean, and wheat production in the United States (Bordoli and Mallarino, 1998; Rehm and Lamb, 2004; Martin, 2009).

14.3.3 Strip Application

In strip application, a compromise alternative to band and broadcast applications, the fertilizer supplying the immobile nutrients is placed in a band on the soil surface that is then incorporated with some tillage. By varying the width of the surface band and the type of tillage used for incorporation, 5%–15% of the volume of the surface soil is mixed with the fertilizer. While the level of nutrient availability in the treated zone may be lower than with other banding options, a much larger portion of the root system can potentially come in contact with the fertilizer. This has enhanced uptake of immobile nutrients. Nelson et al. (1959) were the first to test this concept with corn and cotton and found that a combination of in-row or seed-placed phosphate with stripping resulted in the best utilization of P by young seedlings. In a 5 year study with P and K, Barber (1974) found that strip application produced significantly higher corn yields when compared to either banding or broadcasting equivalent rates.

14.3.4 Point Fertilization

In its simplest form, this consists of opening a hole in the soil with a stick or hoe, and placing a quantity of fertilizer into the soil near the crop to be fertilized. An early study conducted by

Coe (1926) in Iowa compared broadcast application to hill placement, or short bands of fertilizer 7.5–15 cm long and 3–4 cm wide at the same depth as the seed. He concluded that banding the fertilizer in the hills was equal or superior to broadcasting, if contact between fertilizer and seed was avoided, and suggested that the fertilizer should be placed to the side of and below the seed. When used at moderate rates, fertilizer could be placed in the hill with the seed, but at high rates, the fertilizer should be split with part applied in the hill and part broadcast.

The point system is commonly used today in the production of perennial shrub or tree crops. Holes are dug near the outer edge of the canopy (drip line) of trees and fertilizer placed in those holes. Tree stakes, solid blocks of fertilizer shaped to facilitate pushing them into the ground, can also be placed at points along the outer edge of the canopy. In many developing countries, a whole is dug near or in a hill of corn or sorghum and a small quantity of fertilizer is added to maximize nutrient utilization and to minimize competition from weeds. Large individual granules of fertilizer (super granules) have been developed specifically for this purpose. With a point application system, only a very small portion of the soil, generally much <1%, will come in contact with the fertilizer.

14.3.5 Application with Irrigation Water

Fluid fertilizer can be injected into irrigation water but in the case of immobile nutrients, particularly P, this method of application has limited use. Research evaluating this placement has been reviewed by Mikkelsen (1989).

14.3.6 Foliar Application

Although the majority of nutrients needed for growth and development of plants enter through the root system, beneficial effects of foliar fertilization have been reported. For example, Shafer and Reed (1986) studied the foliar absorption of 31 organic and inorganic compounds. Much of the research effort on foliar fertilization has been summarized by Alexander and Schroeder (1987).

14.4 Placement of P and K Fertilizers

The efficiency of use of the immobile nutrients has been the focus of field and greenhouse research projects for many years. In most cases, these studies have focused on the relative advantages or disadvantages of banding versus broadcasting of P and K fertilizers. The impact of the placement of immobile nutrients on uptake by crops and subsequent yield has been studied in both the laboratory and the field. Mechanistic models to explain some of the complex relationships that influence crop response to fertilizer placement were developed by Claassen and Barber (1976), Anghinoni and Barber (1980), and Barber and Cushman (1981).

Using 33 diverse soils to test the Barber–Cushman model, Kovar and Barber (1987, 1988, 1989) found that the concentration of P in the soil solution was the soil parameter that had the greatest effect on P uptake. When P fertilizer was placed in contact with small volumes of soil, banding produced the largest increase in soil solution P for each unit of applied fertilizer. However, the number of roots that can come in contact with the increased solution P is limited by banding. Since roots have a finite capacity to take up nutrients, it is possible that although banding increases the concentration of P in the fertilized zone, restricted root volume in contact with the high P zone could limit uptake. Using the Barber–Cushman model to predict the optimum volume of soil that should be fertilized, they found that the optimum volume varied substantially from a low of about 3% to a maximum of 15%–20%. Borkert and Barber (1985) found a very close relationship between predicted and measured P uptake by soybeans as a function of soil volume fertilized in a pot experiment. Barber (1995) concludes "Placement of P is most important in soils low in available P that sorb or fix large quantities of added P." Phosphorus placement also becomes more important as the rate of applied P decreases.

Although the quantity of P taken up by crops is only a fraction of that of K, fertilizer placement effects are similar for both. The use efficiency of banded K for corn is greater at low soil tests value for K with the difference between banded and broadcast diminishing as soil test K increases (Welch et al. 1966a, 1966b).

14.4.1 Field Comparison of Placement Options

Field comparisons between band and broadcast applications of P and K fertilizers show that banding, especially for P fertilizers, produced greater yield increases when equal rates were used for both. The results summarized by Welch et al. (1966a, 1966b) are typical of many of the results reported.

Working with several forage crops, Sheard (1980) concluded that the banded application of 30 kg P ha^{-1} increased seedling growth as much as fivefold regardless of the species and soil test for P. Working with oats grown on contrasting soils in a greenhouse, Sleight et al. (1984) concluded that the beneficial effects of banding are obtained mainly from placing all the immobile nutrients where contact by active roots is likely, rather than from an increase in availability that may be obtained from the decreased soil/fertilizer contact associated with banding.

Some research utilizing a wide range of application rates has shown that rate can influence the effect of placement. Barber (1958, 1959) has shown that at high fertilization rates, broadcast applications may give higher yields than band applications while Welch et al. (1966a, 1966b) have demonstrated that the optimum combination of banding and broadcasting can be a complex interaction of application rate and soil test level.

With P and K fertilizers, the most frequent placement is to the side of and below the seed at planting (starter fertilizer). In the midwest United States, starter fertilizer has consistently affected early season growth more than yield (Randall and Hoeft, 1988). Starter fertilizer application has consistently been more beneficial in no-till than conventional or clean till corn (Reeves and Touchton, 1986; Mengel, 1992). This benefit, in part, may be attributed to the fact that in no-till the soil is generally cooler in the spring and has a higher bulk density and thus early root growth is inhibited. Starter fertilizer effects are generally found to be greatest when soil temperatures are lowest (Kitcheson, 1968).

Earlier planting dates typically coincide with cooler soil temperature resulting, in slower root growth and metabolism. Although crops typically respond to starter fertilizer under these planting conditions, starter can also increase growth and grain yield on late planted corn, when soil temperatures were very warm (Farber and Fixen, 1986). This was also found to be true on Coastal Plain soils where inherent soil strength inhibits root growth and exploration (Karlen et al., 1984).

Much debate has focused on which nutrients in a starter fertilizer are responsible for the increased early season growth and subsequent increased grain yield, particularly under no-till. In Wisconsin, Motavalli et al. (1993) obtained a response to starters containing N and P in only 1 of the 3 years where manure had been applied. In Illinois, under conventional tillage on a soil with high residual P, an N and P starter fertilizer increased early season growth without increasing final plant dry weight or yield (Bullock et al., 1993). In Indiana, N was the most important nutrient in the starter solution on soils containing adequate levels of P and K (Mengel, 1992). In Alabama, Karim and Touchton (1983) found N primarily responsible for increased growth 14 days after emergence, but P appeared to have the greatest effect at 28–42 days. Wright (1985) found improved yields and earlier maturity in Florida when using a starter fertilizer containing N and P on a high P soil.

Occasionally, fertilizer will be broadcast at planting specifically to enhance early growth. With no-till corn at several locations, Mann (1995) showed that broadcast applications of 44 kg N ha^{-1} as urea-ammonium nitrate solution negated the response to N containing starter fertilizer in no-till corn. This broadcast application of fluid N, many times used as a carrier for burndown herbicides, has become a common practice in some areas of the midwest United States where large corn planters do not routinely have starter fertilizer attachments.

While N and P are the two nutrients most commonly used in starter fertilizers, a number of others are routinely applied including K, S, Mg, and Zn. Starter fertilizer bands provide a convenient delivery system for micronutrients in many crops. For example, Miner et al. (1986) obtained a response to Mn applied in a starter on an Atlantic Coastal Plain soil that was Mn deficient due to excessive liming. The low dissolution pH of common starter fertilizer materials such as MAP can enhance the availability of micronutrient metals such as Zn, Mn, Cu, and Fe. The addition of S compounds in starter fertilizer bands can further lower pH and provide an even greater increase in metal solubility.

In general, the band application of immobile nutrients has proven to be a cost effective method of fertilizer placement, particularly when soil tests values are in the low or very low range. Either band or broadcast applications have been effective when P or K soil test values are in the medium or high range. A combination of broadcast and band applications seem to be most valuable at lower soil test values.

The effect of placement of immobile nutrients has also been evaluated for small grain production. As with corn, band applications have been superior to broadcasting, particularly at lower rates of fertilizer (Sander et al., 1990). The distance between bands is not a concern when immobile nutrients are applied with a drill but could pose a problem when phosphate is applied in a subsurface band before planting small grains. Comparing spacing of 25, 30, and 50 cm, Maxwell et al. (1984) reported that spacing affected P uptake by young plants but had no effect on grain yield.

The residual effects of both broadcast and banded applications of phosphate are also important for small grain production. Following the harvest of spring wheat, Selles (1993) measured greater availability of P in soil samples collected directly over the band. When P was banded in soil having a range of soil test levels established by previous broadcast applications, Alessi and Power (1980) reported that the effects of the combination of band and residual effects of previous broadcast applications are additive.

Starter fertilizer usually involves the application of a small amount of fertilizer close to the seed at planting for the purpose of enhancing the early season growth of crops. In some cases, the starter fertilizer is placed directly in the seed furrow. This pop-up or in-furrow placement should only be used with crops, such as corn and small grains, which are relatively salt tolerant. There are a number of reports of injury to corn seedlings from fertilizer placed in direct contact with the seed (Coe, 1926; Allred and Ohlrogge, 1963; Creamer and Fox, 1980), sometimes resulting in stand reductions (Mengel, 1992; Gordon and Whitney, 2001). The amount of fertilizer that can be placed safely on or in contact with seed varies with crop, row spacing, soil moisture, properties of the fertilizer, and climate. For corn in Indiana, a maximum fertilizer rate of 5.6 kg N plus K ha^{-1} on silt loam or heavier soils is recommended for seed placement (Mengel, 1992). Similar recommendations can be found from many Land Grant Universities. The use of seed-placed fertilizer is a common practice for small grain production and no damage has been reported unless excessive, uneconomical fertilizer rates are used. An exception is urea containing fertilizers. Urea is particularly harmful to seed, due to the formation of free ammonia upon hydrolysis and it is not recommended for placement with the seed. Soybeans are much more sensitive to seed placed, pop-up fertilizer and this practice has severely reduced germination and subsequent yield (Clapp and Small, 1970; Hoeft et al., 1975). Consequently, seed placement of fertilizer is not recommended, regardless of rate, with soybeans.

14.4.2 Fertilizer Placement in Conservation Tillage Systems

The principles for placement of P and K fertilizers (Section 14.4.1) were developed primarily in production systems using a moldboard plow for primary tillage. Because of the emphasis on soil conservation, tillage systems are changing to minimize soil disturbance resulting in less incorporation of broadcast P and K fertilizers with a subsequent reduction in the redistribution of nutrients in crop residues (Larson, 1964).

The stratification of immobile nutrients in the conservation tillage production systems has been documented for a variety of soils (Cruz, 1982; Weil et al., 1988; Robbins and Voss, 1991; Rehm et al., 1995; Hollanda et al., 1998). In no-till or reduced till systems, extremely high levels of immobile nutrients can accumulate in the top 4 or 5 cm of soil because of lack of incorporation of fertilizers and nutrient cycling through the nutrient containing residues remaining on the soil surface. Interestingly, this stratification is not restricted to no-till and ridge-till but is also found with chisel plowing and shallow tillage systems (Cruz, 1982; Hollanda et al., 1998).

In addition to vertical stratification, horizontal stratification can also occur. Work in Indiana clearly shows that the nutrients can accumulate near the old row area when the crop row is consistently placed in the same area, a controlled traffic system. Thus, nutrient stratification can occur in both vertical and horizontal planes. Unless a substantial portion of the root system develops near the soil surface, in the zone of nutrient enrichment, and the root environment is conducive to nutrient uptake, stratification can reduce the amount of nutrients available for uptake by the actively growing crop. A number of studies have shown that roots do concentrate in fertilized zones of soil, particularly in areas of high P (Barber, 1995) and this can be near the soil surface, especially in more humid climates (Cruz, 1982; Kaspar et al., 1991; Hollanda et al., 1998). Thus, the effect of nutrient stratification near the soil surface could be either a negative or positive factor influencing nutrient uptake and is highly dependent on the environment. Schwab et al. (2006) used moldboard plowing to redistribute P in a number of highly stratified soils in Kansas; however, no increased yield was observed.

The advantages of band applications of starter fertilizer for corn in no-till planting systems have been discussed earlier. However, a number of studies have documented the importance of subsurface or deep banding for ridge-till and no-till planting in the northern Corn Belt (Rehm et al., 1995). However, the response to deep banding has not been universal as many studies have obtained no or mixed results in recent years. In one study, placement of P and K had a significant effect on corn yield when soil test levels for P and K were in the low range, but not when they were high. Highest yield responses generally resulted from a combination of subsurface band with a starter fertilizer (Rehm et al., 1988). In contrast, responses observed by Bordoli and Mallarino (1998) appeared more closely related to deficient rainfall in late spring and early summer than with soil test K. Often corn yield advantages to deep band placement of P and K were small (Bordoli and Mallarino, 1998; Mallarino and Murrell, 1998) and would rarely offset the higher application costs (Bordoli and Mallarino, 1998).

A similar range of response to placement of immobile nutrients has been found for other crops grown in conservation tillage production systems. Yields of grain sorghum, for example, were improved when P and K fertilizers were knifed in below the soil surface rather than broadcast on the surface (Sweeney, 1989). Yield increases from this knife placement were larger where soil test values for P and K were in the low rather than the medium range.

Nutrient stratification has not always been associated with reduced yields and response to alternative fertilizer placement techniques, however. In a long-term tillage in Indiana, moldboard plow, chisel plow, ridge-till, and no-till systems were included, in both a continuous corn monoculture and a corn/soybean rotation (Griffith et al., 1988). Extensive stratification of P, K, and acidity, both vertical and horizontal, was found within the surface 30 cm of soil in both rotations (Cruz, 1982; Hollanda et al., 1998). However, differences in corn yields among tillage systems were relatively small in the corn/soybean rotation (<4%), but rather large differences (<17%) were found in the continuous corn yields from 1980 to 1995, indicating that stratification of nutrients alone was not a significant yield limiting factor.

Yibirin et al. (1993) evaluated the effect of mulch on the response of corn to banded K and concluded that its benefits decreased as the amount of mulch on the soil surface increased. The mulch apparently increased soil moisture and reduced soil temperature near the soil surface, thereby increasing root development and subsequent K uptake from that zone.

In contrast to other crops, yield of winter wheat grown in no till planting systems in the Great Plains was not affected by P placement when soil test values for P were in the medium range (Halvorson and Havlin, 1992). The absence of a placement effect may be a consequence of the distribution of roots closer to the soil surface when compared to the corn and grain sorghum crops and a root environment more conducive to nutrient uptake during key periods of growth.

14.5 Calcium and Magnesium

Like P and K, these nutrients are relatively immobile in most agricultural soils. In contrast to K, however, uptake of Ca and Mg by crops is considerably less. As a result, the necessity for these nutrients in fertilizer program is diminished to special or localized situations. Therefore, research that has focused on the application of these two nutrients has not been extensive.

In most agricultural soils, Ca dominates the CEC, which satisfies the Ca requirements of most crops (Mortvedt and Cox, 1985). Under acid conditions, the Ca content of the soil decreases, while the H and Al content increases. The addition of calcitic or dolomitic limestone to correct acidity problems is the primary means by which Ca is supplied to crops. With the exception of the peanut crop (Section 11.4); Ca deficiency is much more common in fruit and vegetable crops such as tomato, apple, and lettuce (Shear, 1975).

Like Ca, availability of Mg to plants decreases under acid soil conditions due to both a pH effect and loss of Mg through leaching from the root zone (Mortvedt and Cox, 1985). Increasing the pH of the soil, even with calcitic lime, can increase Mg availability (Christenson et al., 1973). Dolomitic limestone is the preferred Mg source in areas where it is available.

Applications of $MgSO_4$ or $KMg(SO_4)_2$ fertilizers are made on low Mg soils when dolomitic lime is not available. Soil tests are commonly used to estimate the need for Mg. In the midwest United States, Mg applications for most field crops are recommended when the exchangeable Mg levels are <50 mg kg^{-1} (Vitosh et al., 1995). General recommendations are to apply dolomitic lime if the soil is acid or to apply either 55 kg ha^{-1} soluble Mg, or 0.5 ton dolomitic lime ha^{-1} if the soil pH is adequate.

Since both Ca and Mg are not routinely added to fertilizer programs, the effectiveness of placement of these nutrients for agronomic crops has not been investigated. If lime is routinely used in a crop rotation, these nutrients should not generally be needed in a fertilizer program.

14.6 Micronutrients

The management of the remaining five micronutrients (Zn, Fe, Cu, and Mo), which are taken in small amounts by crop, will be discussed individually.

14.6.1 Zinc

A deficiency of Zn is most likely to occur on sandy soils formed from parent material low in Zn (Krauskopf, 1972), highly weathered tropical soils, highly calcareous soils, and organic soils (Schulte and Walsh, 1982). Overliming (Rehm and Penas, 1982), application of high rates of P (Murphy et al., 1981), and removal of surface soil by erosion, land leveling, and terracing (Frye et al., 1978; Grunes et al., 1961) are management practices that can enhance Zn deficiency.

Crops differ in their sensitivity to Zn deficiency. Corn, edible beans, and sorghum are highly sensitive; barley, sugar beet, soybean, and Sudan grass are moderately sensitive; and wheat, alfalfa, and most forage grasses are not sensitive to Zn deficiency (Laboski et al., 2006). While Zn deficiency can be corrected through either soil or foliar application, soil applications are more common for agronomic crops (Martens and Westerman, 1992) and foliar applications in the fruit and vegetable industry.

Relatively high rates of Zn, broadcast as $ZnSO_4 \cdot 7H_2O$, are commonly used to correct a Zn deficiency in agronomic crops. Other inorganic sources include ZnO and Zn frits, which are not readily available for crop uptake unless finely ground. Application rates vary depending on the demands of the crop being grown, soil properties that could affect Zn availability, and the native supply of Zn in the soil. Higher application rates are commonly needed on calcareous than noncalcareous soils (Wiese and Penas, 1979). Higher rates are also recommended for soils with low levels of extractable Zn (Leikam et al., 2003). Lower rates of Zn are normally applied in band than broadcast applications because of the reduced Zn-soil contact, which slows the reversion of Zn to less available forms and the lowered pH in the band when Zn is applied in combination with N or N–P fertilizers. In addition to inorganic sources, a number of organic Zn sources are used for band applications. These include Zn EDTA, Zn lignosulfate, Zn acetate, and Zn citrate. The chelated materials are commonly used for foliar applications.

14.6.2 Manganese

Manganese deficiencies in crops are found in high organic matter, slightly acidic, and poorly drained sandy soils. Soybeans, wheat, barley, and oats are highly sensitive to Mn deficiency while corn, sugar beets, alfalfa, and forage grasses are moderately sensitive (Vitosh et al., 1994). The availability of Mn is highly influenced by the total Mn content of the soil, drainage, oxidation state, SOM content, and pH. Soil pH is the most important soil property controlling the availability of Mn to plants, which decreases as pH increases. Over liming soil frequently produces Mn deficiencies (Gilbert et al., 1926; Blair and Prince, 1936; Sherman et al., 1942; Snider, 1943; Sanchez and Kamprath, 1959). The fact that Mn availability is influenced by so many factors has led to confusion concerning the effectiveness of various methods of Mn application. Positive yield responses from broadcast application of $MnSO_4$ or MnO, banding of Mn products at planting, banding of Mn in conjunction with an acid forming fertilizer, and foliar application of $MnSO_4 \cdot 7H_2O$ have been used successfully as a means of correcting Mn deficiencies in a number of crops (Gilbert and McLean, 1928; Evans and Purvis, 1948; Anderson and Carstens, 1973; Alley et al., 1978; Gettier et al., 1984; Mascagni and Cox, 1985a, 1985b).

Broadcast applications of Mn were found to be ineffective or less effective than banding as a means of correcting Mn deficiency across a broad range of soils (Harner, 1942; Steckel, 1946; Randall et al., 1975; Gupta, 1986; Eck, 1995). Mascagni and Cox (1985a) found that optimum yield of soybeans were obtained on Atlantic Coastal Plain soils by banding 3 kg Mn ha^{-1} compared to broadcasting 14 kg Mn ha^{-1}. Randall et al. (1975) found similar results in Wisconsin, although an additional foliar application was required for optimum yield when deficiencies were severe. Excellent results have also been obtained when Mn is band applied in conjunction with acid forming fertilizers (Mederski et al., 1960; Petrie and Jackson, 1984) or banding an acid forming fertilizer alone (Steckel et al., 1948).

Foliar applications of Mn have also been used to correct Mn deficiency. Although applications of <1.0–5.0 kg Mn ha^{-1} as $MnSO_4 \cdot 7H_2O$ are commonly recommended (Laboski et al., 2006; Vitosh et al., 1994), the number of applications required has not been unequivocally established. Eck (1995) found a single application of 1.1 kg Mn ha^{-1} adequate for correcting Mn deficiency in soybeans in Indiana while Mascagni and Cox (1985a, 1985b) found that up to three applications were required.

14.6.3 Iron

Like Mn, Fe availability to plants is controlled by a number of soil factors including pH, free $CaCO_3$ content, SOM content, and redox potential. The reader is referred to reviews by Moraghan and Mascagni (1992) and Lindsay (1992) for a detailed discussion of these factors. Iron deficiency (chlorosis) is common in soybeans and grain sorghum on calcareous soils (Martens and Westerman, 1992). Because of large differences in tolerance to Fe chlorosis among and within plant species, development of tolerant cultivars has been possible. Because broadcast applications of inorganic sources of Fe to the soil are not effective in controlling or correcting Fe chlorosis, foliar sprays using chelates such as iron ethylene diamine-N,N′-bis(2-hydroxyphenyl) acetic acid (FeEDDHA) are generally recommended in field crops but multiple applications may be required. Band applications of iron, the application of animal manure and seed treatments of iron have all been used successfully.

14.6.4 Copper

Copper deficiency occurs most commonly on organic soils, but can also occur where sandy soils are highly weathered, on mineral soils with a high SOM content, and on calcareous mineral soils (Martens and Westerman, 1992). Crops differ greatly in their susceptibility to Cu deficiency with wheat, oats, Sudan grass, and alfalfa being highly sensitive; barley, corn, and sugar beet moderately sensitive; and soybeans and most forage grasses tolerant to Cu deficiency (Vitosh et al., 1994).

While a number of fertilizers to supply Cu are available, $CuSO_4 \cdot 5H_2O$ is the most common fertilizer material used because of low cost and high water solubility. Karamanos et al. (1986) evaluated a number of Cu fertilizers and their effect on crop production. Copper oxide was generally ineffective in the year of application but residual effects alleviated Cu deficiency, while $CuSO_4 \cdot 5H_2O$ and chelated products were effective in the year of application. However, there was no residual effect from the application of chelated materials. Applications of animal manure or biosolids containing Cu will also correct deficiencies. Swine manure is an excellent source of Cu in many areas because of the high levels of Cu fed to growing pigs. Soil application is the preferred method of correcting Cu deficiency because of the good residual effects from Cu fertilization. Copper fertilizers are most commonly broadcast at or before planting but can also be banded. Common rates for soil application for field crops are 2.2–3.3 kg Cu ha^{-1} banded or 33–66 kg Cu ha^{-1} broadcast (Schulte and Kelling, 1999). Foliar applications can be used during the growing season to correct Cu deficiencies with chelated materials being favored.

14.6.5 Molybdenum

Molybdenum is required for NO_3 reduction reactions in plants and the symbiotic fixation of N by legumes. Deficiencies of this nutrient have been reported for a number of crops grown on acid soils in the Great Lakes and southern regions and coastal areas of the United States, as well as in New Zealand and Australia. Deficiencies are usually observed with the legume crops.

Molybdenum deficiencies can be corrected by liming, by soil or foliar application of Mo or with seed treatment. While Mo toxicity in plants is rare, high concentrations of Mo in forages may induce Cu deficiency in animals (Miltmore and Mason, 1971). Therefore, care must be exercised in making Mo applications. Application methods that utilize the lowest effective Mo rate, generally seed treatment, are preferred.

References

Alessi, J., and J.F. Power. 1980. Effects of banded and residual fertilizer phosphorus on dryland spring wheat yield in the Northern Plains. Soil Sci. Soc. Am. J. 44:792–796.

Alexander, A., and M. Schroeder. 1987. Modern trends in foliar fertilization. J. Plant Nutr. 10:1391–1399.

Alley, M.M., C.J. Rich, G.W. Hawkins, and D.C. Martens. 1978. Correction of Mn deficiency of soybeans. Agron. J. 70:35–38.

Allred, S.E., and A.J. Ohlrogge. 1963. Principles of nutrient uptake from fertilizer bands. VI. Germination and emergence of corn as affected by ammonia and ammonium phosphate. Agron. J. 56:309–313.

Anderson, W.C., and J.B. Carstens. 1973. Effect of manganese soil and seed treatments on growth and yield of peas. Soc. Hort. Sci. J. 98:581–582.

Anghinoni, I., and S.A. Barber. 1980. Predicting the most efficient phosphorus placement for corn. Soil Sci. Soc. Am. J. 44:1016–1020.

Bailey, L.D., and E.G. Beauchamp. 1973. Effects of temperature on NO_3 and NO_2 reduction, nitrogenous gas production and redox potential in a saturated soil. Can. J. Soil Sci. 53:213–218.

Bandel, V.A., S. Dzienia, and G. Stanford. 1980. Comparison of N fertilizers for no-till corn. Agron. J. 72:337–341.

Barber, S.A. 1958. Relation of fertilizer placement to nutrient uptake and crop yield. I. Interaction of row phosphorus and soil level of phosphorus. Agron. J. 50:535–539.

Barber, S.A. 1959. Relation of fertilizer placement to nutrient uptake and crop yield. II. Effects of row potassium, potassium soil-level and precipitation. Agron. J. 51:97–99.

Barber, S.A. 1974. A program for increasing the efficiency of fertilizers. Fert. Solut. 18:24–25.

Barber, S.A. 1995. Soil nutrient bioavailability: A mechanistic approach. 2nd edn. John Wiley & Sons, New York.

Barber, S.A., and J.H. Cushman. 1981. Nitrogen uptake model for agronomic plants, p. 382–409. *In* I.K. Iskander (ed.) Modeling waste water renovation-land treatment. Wiley-Interscience, New York.

Blair, A.W., and A.L. Prince. 1936. Manganese in New Jersey soils. Soil Sci. 42:327–333.

Bordoli, J.M., and A.P. Mallarino. 1998. Deep and shallow banding of phosphorus and potassium as alternatives to broadcast fertilization for no-till corn. Agron. J. 90:27–33.

Borkert, C.M., and S.A. Barber. 1985. Predicting the most efficient phosphorus placement for soybeans. Soil Sci. Soc. Am. J. 49:901–904.

Brown, J.R. 1972. Micronutrient topdressing of alfalfa (*Medicago sativa* L.) on Udollic Albaqualf. Commun. Soil Sci. Plant Anal. 3:211–221.

Brown, J.R., W.O. Thom, and L.L. Wall, Sr. 1981. Effects of sulfur application on the yield and composition of soybeans and soil sulfur. Commun. Soil Sci. Plant Anal. 12:247–261.

Bullock, D.G., F.W. Simmons, I.M. Chung, and G.I. Johnson. 1993. Growth analysis of corn with and without starter fertilizer. Crop Sci. 33:112–117.

Burford, J.R., and J.M. Bremner. 1975. Relationships between denitrification capacities of soils and total, water-soluble, and readily decomposable soil organic matter. Soil Biol. Biochem. 7:389–394.

Cerrato, M.E., and A.M. Blackmer. 1990. Effects of nitrapyrin on corn yields and recovery of ammonium-N at 18 site-years in Iowa. J. Prod. Agric. 3:513–521.

Christenson, D.R., R.P. White, and E.C. Doll. 1973. Yields and magnesium uptake by plants as affected by soil pH and calcium levels. Agron. J. 65:205–206.

Claassen, N., and S.A. Barber. 1976. Simulation model for nutrient uptake from soil by a growing plant root system. Agron. J. 68:961–964.

Clapp, J.G., Jr., and H.G. Small. 1970. Influence of "pop-up" fertilizers on soybean stands and yield. Agron. J. 62:802–803.

Coe, D.G. 1926. The effects of various methods of applying fertilizers on crop yields. Soil Sci. 21:127–141.

Creamer, F.L., and R.H. Fox. 1980. The toxicity of banded urea or di-ammonium phosphate to corn as influenced by soil temperature, moisture and pH. Soil Sci. Soc. Am. J. 44:298–300.

Cruz, J.C. 1982. The effect of crop rotation and tillage system on some soil properties, root distribution and crop production. Ph.D. Thesis. Purdue University. West Lafayette, IN.

Eck, K.J. 1995. Diagnosing and correcting manganese deficiencies of soybeans in Indiana. M.S. Thesis. Purdue University. West Lafayette, IN.

Endelman, F.J., D.R. Keeney, J.T. Gilmour, and P.G. Saffigna. 1974. Nitrate and chloride movement in the Plainfield loamy sand. J. Environ. Qual. 3:295–298.

Engel, R.E., J. Eckhoff, and R.K. Berg. 1994. Grain yield, kernel weight, and disease responses of winter wheat cultivars to chloride fertilization. Agron. J. 86:391–396.

Ernst, J.W., and H.F. Massey. 1960. The effects of several factors on volatilization of ammonia formed from urea in the soil. Soil Sci. Soc. Am. Proc. 24:87–90.

Evans, H.J., and E.R. Purvis. 1948. An instance of manganese deficiency of alfalfa and red clover in New Jersey. Agron. J. 40:1046–1047.

Evanylo, G.W. 1991. No-till corn response to nitrogen rate and timing in the middle Atlantic Coastal Plain. J. Prod. Agric. 4:180–185.

Farber, B.G., and P.E. Fixen. 1986. Phosphorus response of late planted corn in three tillage systems. J. Fert. Issues 3:46–51.

Fixen, P.E., G.W. Buchenan, R.H. Gelderman, T.E. Shumacher, J.R. Gerwing, F.A. Cholick, and B.G. Farber. 1986b. Influence of soil and applied chloride on several wheat parameters. Agron. J. 78:736–740.

Fixen, P.E., R.H. Gelderman, J. Gerwing, and F.A. Cholick. 1986a. Response of spring wheat, barley, and oats to chloride in potassium chloride fertilizers. Agron. J. 78:664–668.

Fox, R.H., and L.D. Hoffman. 1981. The effect of N fertilizer source on grain yield, N uptake, soil pH and lime requirement in no-till corn. Agron. J. 73:891–895.

Frye, W.W., R.L. Blevins, L.W. Murdock, and K.L. Wells. 1981. Effectiveness of nitrapyrin with surface applied fertilizer nitrogen in no-tillage corn. Agron. J. 73:287–289.

Frye, W.W., H.F. Miller, L.W. Murdock, and D.E. Peaslee. 1978. Zinc fertilization of corn in Kentucky. KY Coop. Ext. Serv. Agron. Notes 11:1–4.

Gascho, G.J., and J.G. Davis. 1995. Soil fertility and plant nutrition, p. 383–418. In H.E. Pattee, and H.T. Stalker (eds.) Advances in peanut science. American Peanut Research and Education Society, Stillwater, OK.

Gettier, S.W., D.C. Martens, D.L. Hallock, and M.J. Stewart. 1984. Residual Mn and associated soybean yield response from $MnSO_4$ application on a sandy loam soil. Plant Soil 81:101–110.

Gilbert, B.E., and F.T. McLean. 1928. A "deficiency disease": The lack of available manganese in a lime-induced chlorosis. Soil Sci. 26:27–31.

Gilbert, B.E., F.T. McLean, and L.J. Hardin. 1926. The relation of manganese and iron to be lime-induced chlorosis. Soil Sci. 22:437–446.

Gordon, W.B., and D.A. Whitney. 2001. Effects of application method and composition of starter fertilizer on irrigated ridge-till corn. KS AES Report of Progress 885. Kansas State University, Manhattan, KS.

Griffith, D.R., E.J. Kladivko, J.V. Mannering, T.D. West, and S.D. Parsons. 1988. Long-term tillage and rotation effects on corn growth and yield on high and low organic matter poorly drained soils. Agron. J. 80:599–605.

Grunes, D.L., L.C. Boawn, C.W. Carlson and R.G. Viets. 1961. Land leveling may cause zinc deficiency, N.D. Farm Res. 21(11):4–7.

Gupta, U.C. 1986. Manganese nutrition of cereals and forages grown in Prince Edward Island. Can. J. Soil Sci. 66:59–65.

Gupta, U.C., J.A. MacLeod, and J.D.E. Sterling. 1976. Effects of boron and nitrogen on grain yield and boron and nitrogen concentration of wheat and barley. Soil. Sci. Soc. Am. J. 40:723–726.

Halvorson, A.D., and J.L. Havlin. 1992. No-till winter wheat response to phosphorus placement and rate. Soil Sci. Soc. Am. J. 56:1635–1639.

Harner, P.M. 1942. The occurrence and correction of unproductive alkaline organic soil. Soil Sci. Soc. Am. Proc. 7:378–386.

Hauck, R.D. (ed.) 1984. Nitrogen in crop production. ASA, Madison, WI.

Heckman, J.R. 1995. Corn responses to chloride in maximum yield research. Agron. J. 84:415–419.

Hoeft, R.G., J.E. Sawyer, R.M. Van Den Heuvel, M.A. Schmitt, and G.S. Brinkman. 1985. Corn response to sulfur on Illinois soils. J. Fert. Issues 2:95–104.

Hoeft, R.G., and L.M. Walsh. 1975. Effect of carrier, rate, and time of application of S on the yield, and S and N content of alfalfa. Agron. J. 67:427–430.

Hoeft, R.G., L.M. Walsh, and E.A. Liegel. 1975. Effects of seed-laced fertilizer on the emergence (germination) of soybeans (*Glycine max* L.) and snapbeans (*Phaseolus vulgaris* L.). Commun. Soil Sci. Plant Anal. 6:655–664.

Hollanda, F.S.R., D.B. Mengel, M.B. Paula, J.G. Caruaha, and J.C. Bertoni. 1998. Influence of crop rotations and tillage systems on phosphorus and potassium stratification and root distribution in the soil profile. Commun. Soil Sci. Plant Anal. 29:2383–2394.

Huber, D.M., H.L. Warren, D.W. Nelson, C.Y. Tsai, M.A. Ross, and D.B. Mengel. 1982. Evaluation of nitrification inhibitors for no-till corn. Soil Sci. 134:388–394.

Karamanos, R.E., G.A. Kruger, and J.W.B. Stewart. 1986. Copper deficiency in cereal and oilseed crops in northern Canadian Prairie soils. Agron. J. 78:317–323.

Karim, F., and J.T. Touchton. 1983. Response of corn seedlings to high concentrations of ammonium phosphates. Commun. Soil Sci. Plant Anal. 14:847–858.

Karlen, D.L., P.G. Hunt, and R.B. Campbell. 1984. Crop residue removal effects on corn yield and fertility of a Norfolk sandy loam. Soil Sci. Soc. Am. J. 48:868–872.

Kaspar, T.C., H.J. Brown, and E.M. Kassmeyer. 1991. Corn root distribution as affected by tillage wheel traffic and fertilizer placement. Soil Sci. Soc. Am. J. 55:1390–1394.

Keller, G.D., and D.B. Mengel. 1986. Ammonia volatilization from nitrogen fertilizers surface applied to no-till corn. Soil Sci. Soc. Am. J. 50:1060–1063.

Kitcheson, J.W. 1968. Effect of controlled air and soil temperature and starter fertilizer on the growth and nutrient composition of corn (*Zea mays* L.). Soil Sci. Soc. Am. J. 32:531–534.

Kline, J.S., J.T. Sims, and K.L. Schilke-Gartley. 1989. Response of irrigated corn to sulfur fertilization in the Atlantic Coastal Plain. Soil Sci. Soc. Am. J. 53:1101–1108.

Kovar, J.L., and S.A. Barber. 1987. Placing phosphorus and potassium for greatest recovery. J. Fert. Issues 4:1–6.

Kovar, J.L., and S.A. Barber. 1988. Phosphorus supply characteristics of 33 soils as influenced by seven rates of P addition. Soil Sci. Soc. Am. J. 52:160–165.

Kovar, J.L., and S.A. Barber. 1989. Reasons for differences among soils in placement of phosphorus for maximum predicted uptake. Soil Sci. Soc. Am. J. 53:1733–1736.

Krauskopf, K.B. 1972. Geochemistry of micronutrients, p. 7–40. *In* J.J. Mortvedt et al. (eds.) Micronutrients in agriculture. SSSA, Madison, WI.

Laboski, C.A.M., J. Peter, and L. Bundy. 2006. Nutrient application guidelines for field, vegetable and fruit crops in Wisconsin. WI Coop. Ext. Serv. Bull. A2809. University of Wisconsin, Madison, WI.

Lamond, R.E., D.A. Whitney, J.S. Hickman, and L.C. Bonezkowski. 1991. Nitrogen rate and placement for grain sorghum production in no-tillage systems. J. Prod. Agric. 4:531–535.

Lamond, R.E., D.A. Whitney, and B.H. Marsh. 1995. Sulfur fertilization of smooth bromegrass in Kansas. Agron. J. 87:13–16.

Larson, W.E. 1964. Soil parameters for evaluating tillage needs and operations. Soil Sci. Soc. Am. Proc. 18:118–122.

Leikam, D.F., R.E. Lamond and D.B. Mengel, 2003. Soil test interpretations and fertilizer recommendations. Kansas State University Ag. Exp. Sta. and Coop. Ext. Ser. Bulletin MF-2586.

Lindsay, W.L. 1992. Inorganic equilibria affecting micronutrients in soils, p. 41–57. *In* J.J. Mortvedt et al. (eds.) Micronutrients in agriculture. 2nd edn. SSSA, Madison, WI.

Maddux, L.D., D.E. Kissel, J.D. Ball, and R.J. Raney. 1985. Nitrification inhibition by nitrapyrin and volatile sulfur compounds. Soil Sci. Soc. Am. J. 49:239–242.

Mahler, R.J., and R.L. Maples. 1987. Effect of sulfur additions on soil and the nutrition of wheat. Commun. Soil Sci. Plant Anal. 18:653–673.

Mallarino, A.P., and T.S. Murrell. 1998. No-till corn grain yield responses to band applications of potassium. Better Crops Plant Food 82:4–6.

Mann, C.L. 1995. Efficiency of nitrogen management systems in no-till corn production. M.S. Thesis. Purdue University. West Lafayette, IN.

Martens, D.C., and D.T. Westerman. 1992. Fertilizer applications for correcting micronutrient deficiencies, p. 549–592. *In* Mortvedt et al. (eds.) Micronutrients in agriculture. 2nd edn. SSSA, Madison, WI.

Martin, K.L. 2009. Effect of phosphorus placement in reduced tillage crop production. Ph.D. Dissertation. Kansas State University. Manhattan, KS.

Mascagni, H.J., Jr., and F.R. Cox. 1985a. Effective rates of fertilization for correcting manganese deficiency in soybeans. Agron. J. 77:363–366.

Mascagni, H.J., Jr., and F.R. Cox. 1985b. Critical levels of manganese in soybean leaves at various growth stages. Agron. J. 7:373–375.

Matheny, T.A., and P.G. Hunt. 1981. Effects of irrigation and sulphur application on soybeans grown on a Norfolk loamy sand. Commun. Soil Sci. Plant Anal. 12:147–159.

Maxwell, T.M., D.E. Kissel, M.G. Wagger, D.A. Whitney, M.L. Cabrera, and H.C. Moser. 1984. Optimum spacing of pre-plant bands of N and P fertilizer for winter wheat. Agron. J. 76:243–247.

Mederski, H.J., D.J. Hoff, and J.H. Wilson. 1960. Manganese oxide and sulfate as fertilizer sources for correcting Mn deficiency in soybeans. Agron. J. 52:667.

Mengel, D.B. 1992. Fertilizing corn grown using conservation tillage. Purdue University Cooperative Extension Service Agronomy Guide AY-268. Purdue University, West Lafayette, IN.

Mengel, D., R. Lamond, V. Martin, S. Duncan, D. Whitney, and B. Gordon. 2009. Chloride fertilization and soil testing—Update for major crops in Kansas. Better Crops 93:20–22.

Mengel, D.B., D.W. Nelson, and D.M. Huber. 1982. Placement of nitrogen fertilizers for no-till and conventional till corn. Agron. J. 74:515–518.

Mengel, D.B., and F.E. Wilson. 1988. Timing of nitrogen for rice in relation to paddy flooding. J. Prod. Agric. 1:90–92.

Mikkelsen, R.L. 1989. Phosphorus fertilization through drip irrigation. J. Prod. Agric. 2:279–286.

Miltmore, J.D., and J.L. Mason. 1971. Copper to molybdenum ratio and molybdenum and copper concentrations in ruminant feeds. Can. J. Anim. Sci. 51:193–200.

Miner, G.S., S. Traoe, and M.R. Tucker. 1986. Corn response to starter fertilizer acidity and manganese materials varying in water solubility. Agron. J. 78:291–295.

Moraghan, J.T., and H.J. Mascagni, Jr. 1992. Environmental and soil factors affecting micronutrient deficiencies and toxicities, p. 371–425. In J.J. Mortvedt et al. (eds.) Micronutrients in agriculture. 2nd edn. SSSA, Madison, WI.

Mortvedt, J.J., and F.R. Cox. 1985. Production, marketing and use of calcium, magnesium and micronutrient fertilizers, p. 455–481. In O.P. Engelstad (ed.) Fertilizer technology and use. 3rd edn. SSSA, Madison, WI.

Motavalli, P.P., K.A. Kelling, T.D. Syverud, and R.P. Wolkowski. 1993. Interaction of manure and nitrogen or starter fertilizer in northern corn production. J. Prod. Agric. 6:191–194.

Murphy, L.S., R. Ellis, Jr., and D.C. Adriano. 1981. Phosphorus-micronutrient interaction effects on crop production. J. Plant Nutr. 3:593–613.

Nelson, W.L., B.A. Krantz, C.D. Welch, and N.S. Hall. 1959. Utilization of phosphorus as affected by placement. II. Cotton and corn in North Carolina. Soil Sci. 68:139–144.

Nelson, L.B., and R.E. Uhland. 1955. Factors that influence loss of fall applied fertilizers and their probable importance in different sections of the United States. Soil Sci. Soc. Am. Proc. 19:492–496.

Petrie, S.E., and T.L. Jackson. 1984. Effect of fertilization on soil solution pH and manganese concentration. Soil Sci. Soc. Am. J. 48:315–318.

Ramig, R.E., P.E. Rasmussen, R.R. Allmaras, and C.M. Smith. 1975. Nitrogen-sulfur relations in soft white winter wheat. I. Yield response to fertilizer and residual sulfur. Agron. J. 67:219–224.

Randall, G.W., and R.G. Hoeft. 1988. Placement methods for improved efficiency of P and K fertilizers: A review. J. Prod. Agric. 1:70–79.

Randall, G.W., E.E. Schulte, and R.B. Corey. 1975. Effect of soil and foliar applied manganese on the micronutrient content and yield of soybeans. Agron. J. 67:502–507.

Reeves, D.W., and J.T. Touchton. 1986. Relative phytotoxicity of dicyandiamide and availability of its nitrogen to cotton, corn and grain sorghum. Soil Sci. Soc. Am. J. 50:1352–1357.

Rehm, G.W. 1993. Timing sulfur applications for corn (Zea mays L.) production on irrigated sandy soil. Commun. Soil Sci. Plant Anal. 24:285–294.

Rehm, G.W., S.D. Evans, W.W. Nelson, and G.W. Randall. 1988. Influence of placement of phosphorus and potassium on yield of corn and soybeans. J. Fert. Issues 5:6–13.

Rehm, G.W., and J.A. Lamb. 2004. Impact of banded potassium on crop yield and soil potassium in ridge-till planting. Soil Sci. Soc. Am. J. 68:629–636.

Rehm, G.W., and E.J. Penas. 1982. Use and management of micronutrient fertilizers in Nebraska. NB Coop. Ext. Serv. Bull. NebGuide G82-596. University of Nebraska, Lincoln, NE.

Rehm, G.W., G.W. Randall, A.J. Scobbie, and J.A. Vetsch. 1995. Impact of fertilizer placement and tillage system on phosphorus distribution in soil. Soil Sci. Soc. Am. J. 59:1661–1665.

Robbins, S.G., and R.D. Voss. 1991. Phosphorus and potassium stratification in conservation tillage systems. J. Soil Water Conserv. 46:298–300.

Roberts, R.K., J.M. Gersman, and D.D. Howard. 2000. Soil- and foliar-applied boron in cotton production: An economic analysis. J. Cotton Sci. 4:171–177.

Rolston, D.E., D.L. Hoffman, and D.W. Toy. 1978. Field measurements of denitrification: I. Flux of N_2 and N_2O. Soil Sci. Soc. Am. J. 42:863–869.

Sanchez, C., and E.J. Kamprath. 1959. Effect of liming and organic matter content on the availability of native and applied manganese. Soil Sci. Soc. Am. Proc. 23:302–304.

Sander, D.H., E.J. Penas, and B. Eghball. 1990. Residual effects of various phosphorus application methods on winter wheat and grain sorghum. Soil Sci. Soc. Am. J. 54:1473–1478.

Scharf, P.C., and M.M. Alley. 1988. Nitrogen loss pathways and nitrogen loss inhibitors: A review. J. Fert. Issues 5:109–125.

Schepers, J.S., and W.R. Raun (eds.) 2008. Nitrogen in agricultural systems. Agronomy monograph no. 49. ASA, Madison, WI.

Schulte, E.E., and L.M. Walsh. 1982. Soil and applied zinc. WI Coop. Ext. Serv. Bull. A2528.

Schwab, G.J., D.A. Whitney, G.L. Kilgore, and D.W. Sweeney. 2006. Tillage and phosphorus management effects on crop production in soils with phosphorus stratification. Agron. J. 98:430–435.

Selles, P. 1993. Residual effect of phosphorus fertilizer when applied with the seed or banded. Commun. Soil Sci. Plant Anal. 24:951–960.

Shafer, W.E., and D.W. Reed. 1986. The foliar application of potassium from organic and inorganic potassium carriers. J. Plant Nutr. 9:143–157.

Shear, C.B. 1975. Calcium related disorders of fruits and vegetables. HortScience 10:361–365.

Sheard, W.W. 1980. Nitrogen in the P band for forage establishment. Agron. J. 72:89–97.

Sherman, G.D., J.S. McHargue, and W.S. Hodgkiss. 1942. Determination of active manganese in soil. Soil Sci. 54:253–257.

Sleight, D.W., D.H. Sander, and G.A. Peterson. 1984. Effect of fertilizer phosphorus placement on the availability of phosphorus. Soil Sci. Soc. Am. J. 48:336–340.

Snider, H.J. 1943. Manganese in some Illinois soils and crops. Soil Sci. 56:186–195.

Steckel, J.E. 1946. Manganese fertilization of soybeans in Indiana. Soil Sci. Soc. Am. Proc. 11:346–348.

Steckel, J.E., B.R. Berstramson, and A.J. Ohlrogge. 1948. Manganese nutrition of plants as related to applied superphosphate. Soil Sci. Soc. Am. Proc. 13:108–111.

Stehouwer, R.C., and J.W. Johnson. 1990. Urea and anhydrous ammonia management for conventional tillage corn production. J. Prod. Agric. 3:507–513.

Sweeney, D.W. 1989. Suspension N-P-K placement methods for grain sorghum in conservation tillage systems. J. Fert. Issues 6:83–88.

Sweeney, D.W., and G.V. Grande. 1993. Yield, nutrient, and soil sulfur response to ammonium sulfate fertilization of soybean cultivars. J. Plant Nutr. 16:1083–1098.

Touchton, J.T., and F.C. Boswell. 1975. Boron application for corn growth on selected southeastern soils. Agron. J. 67:197–200.

Vitosh, M.L., J.W. Johnson, and D.B. Mengel. 1995. Tri-state fertilizer recommendations for corn, soybeans, wheat, and alfalfa. MI State Univ. Ext. Bull. E-2567.

Vitosh, M.L., D.D. Warncke and R.E. Lucas. 1994. Secondary and micronutrients for vegetables and field crops. Michigan State University Extension Bulletin E-486, revised, August 1994.

Weil, R.R., P.W. Bennedetto, L.J. Sikora, and V.A. Bandel. 1988. Influence of tillage practices on phosphorus distribution and forms in three ultisols. Agron. J. 80:503–509.

Welch, L.F., P.E. Johnson, G.E. McKibben, L.V. Boone, and J.W. Pendleton. 1966a. Relative efficiency of broadcast versus banded potassium for corn. Agron. J. 58:618–621.

Welch, L.F., D.L. Mulvaney, L.V. Boone, G.E. McKibben, and J.W. Pendleton. 1966b. Relative efficiency of broadcast versus banded phosphorus for corn. Agron. J. 58:283–287.

Wiese, R.A., and E.J. Penas. 1979. Fertilizer suggestions for corn. NB Coop. Ext. Serv. NebGuide G74-174. University of Nebraska, Lincoln, NE.

Wijler, J., and C.C. Delwhiche. 1954. Investigations on the denitrifying process in soil. Plant Soil 2:155–169.

Wright, D.L. 1985. No-till corn response to starter fertilizer and starter placement, p. 137–140. *In* Proc. South. Region No-Till Conf., Griffin, GA.

Yibirin, H., J.W. Johnson, and D.J. Eckert. 1993. No-till corn production as affected by mulch, potassium placement and soil exchangeable potassium. Agron. J. 85:636–644.

15
Nutrient and Water Use Efficiency

Kefyalew Girma
Oklahoma State University

William R. Raun
Oklahoma State University

15.1	Introduction	15-1
15.2	Nutrient Use Efficiency	15-1
15.3	Bray's Nutrient Mobility Concept: What's Nutrient Efficiency Got to Do with It?	15-1
15.4	Status of Nutrient Use Efficiency	15-2
15.5	Causes of Low Nutrient Use Efficiency	15-3
15.6	Strategies to Improve Nutrient Use Efficiency	15-3
	Adjusting Rate, Timing, Placement and Source of Fertilizer Application to Increase Nutrient Use Efficiency • Urease and Nitrification Inhibitors for N Fertilizers • Foliar Fertilization for Improving Nutrient Use Efficiency • Use of Cover Crops and Crop Rotation • Sensor-Based Variable Rate Nutrient Management • Biotechnology and Variety Selection for Improving NUE	
15.7	Water Use Efficiency	15-8
15.8	Critical Factors Affecting Water Use Efficiency by Crops	15-8
15.9	Strategies to Improve Water Use Efficiency	15-8
	Tillage and Residue Management • Crop Rotations Cover Crops and Fallow System • Crop/Variety/Hybrid Selection • Soil Physical and Chemical Properties • Modeling as a Tool for Improving Water Use Efficiency • Nutrient and Water Interactions	
15.10	Conclusions	15-10
	References	15-10

15.1 Introduction

Nutrient use efficiency (NUE) is an important aspect of plant nutrient management. While discussion of NUE of all essential elements is desirable, because over 10^8 ton of nitrogen (N) fertilizer is produced annually, 60% of which is used in cereals, we will focus on N as an example of a mobile nutrient from both production and environmental perspectives. Phosphorus (P) will represent immobile nutrients. The importance of spatial and temporal variability on improved nutrient and water use efficiency (WUE) will also be discussed. This will highlight the need for better mid-season management practices that can ultimately decrease nutrient loss from different causes. The approaches used to improve NUE of N and P can be applied to other plant nutrients. Recent technological advances that improve WUE will be further highlighted.

15.2 Nutrient Use Efficiency

NUE can be defined in several ways (Moll et al., 1982; Gourley et al., 1994; Baligar et al., 2001; Cassman et al., 2002; Mosier et al., 2004; Fixen, 2006; Roberts, 2008) and at times some definitions are misleading. Roberts (2008) and Cassman et al. (2003) discussed different types of NUE calculations and demonstrated each with data compiled from Asia and the United States. For example, fertilizer use efficiency of 21% or 100% can be calculated using the same input data but different computation methods (Roberts, 2008).

15.3 Bray's Nutrient Mobility Concept: What's Nutrient Efficiency Got to Do with It?

Bray's mobility concept (Bray, 1954) remains a benchmark paper for those truly interested in improving NUE of mobile and immobile nutrients. Further discussion will focus on how this has been applied in precision agriculture, and how Liebig's Law of the Minimum (van der Ploeg et al., 1999) continues to be relevant, especially for those scientists focused on yield prediction as a tool for improved nutrient management.

The term "mobility" in the field of soil fertility and plant relationships refers to the overall process whereby plant nutrient ions reach sorbing root surfaces (Bray, 1954). Bray (1954) proposed that the availability of soil nutrients for plant use might be strongly influenced by their level of mobility in the soil and classified nutrients as mobile or immobile. He also conjectured that the volume of soil from which roots sorbed nutrients also

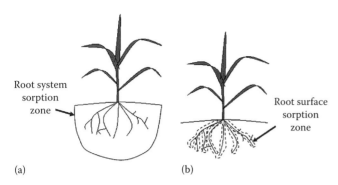

FIGURE 15.1 Root system sorption zone of mobile nutrients (a) and root surface sorption of immobile nutrients (b). (Adapted from Bray, R.H. 1954. A nutrient mobility concept of soil plant relationships. Soil Sci. 78:9–22.)

determined the soil fertility requirements of a plant. Bray's concept of soil nutrient mobility was plausible in the field of soil nutrient–plant relationships.

Bray identified two distinct types of soil sorption zones of plants. The first one is the large volume of soil occupied by the major part of the plant root system called the root system sorption zone (Figure 5.1a) from which mobile nutrients, like nitrate-N (NO_3^-–N), are taken up in large quantities by plants. This results in their "net" requirement being almost equal to the crop content at maturity. Mobile nutrients act as a "limiting nutrient" in the context of the "Law of the Minimum."

The second sorption zone is a relatively thin layer of soil adjacent to each root surface (root surface sorption zone, Figure 5.1b) from which immobile nutrients can be removed by the plant. From this zone, the roots effectively obtain relatively immobile nutrients like P. The sum of these small root surface sorption zones represents only a small part of the soil. Hence, plant roots access only a small fraction of the relatively immobile nutrients present. The amount of nutrient that must be present in the soil to support maximum crop yield is many times larger than the crop nutrient content (Cornforth, 1968). Variation in yield might affect the plant's ability to obtain the relatively immobile nutrients. However, the soil nutrient level needed to support a wide range of plant yield does not vary with yield, because the plant's ability to obtain the nutrient is proportional to its yield.

Bray's mobility concept was a combination of the Mitscherlich percent sufficiency concept (Bray, 1958; Johnson, 1991) and Liebig's Law of the Minimum. Bray showed that Liebig's Law of the Minimum concept applied for mobile nutrients like NO_3^-–N, and that Mitscherlich's percent sufficiency concept worked for immobile nutrients like P and potassium (K). In Liebig's theory of plant response, if all nutrients were adequate except one, then yield would increase in direct proportion to increasing the availability of the deficient nutrient. Bray's concept of how plants responded to soil nutrient availability could be represented as a straight-line response for a nutrient that is 100% mobile in the soil and a curvilinear response for relatively immobile nutrients.

When plants are grown close together, as in intensive agriculture, it becomes clear that the volume of soil from which plants extract mobile nutrients may overlap, while soil volumes supplying immobile nutrients do not. Thus, plants will compete among each other for mobile nutrients if they are closely spaced (Cornforth, 1968; Darrah et al., 2006). As cropping systems increase yield by planting more densely, there will be a direct increase in demand by the crop for the mobile nutrients and it will be necessary to add more of the mobile nutrient to eliminate the competition among plants.

On the other hand, there is no competition among plants for extracting immobile nutrients even when growing close together. This is because the plant root is extracting immobile nutrients from an extremely small volume of soil, often only the soil within a millimeter or two from the root surface (Cornforth, 1968; Eghball and Sander, 1989). As plants grow, they obtain additional supplies of an immobile nutrient by developing more roots that will explore new volumes of soil. If a soil is 100% sufficient in supplying an immobile nutrient for a dry-land crop yield of 3.8 ton ha^{-1} corn (*Zea mays* L.), then it will also be 100% sufficient if the field is irrigated and the yield is 11.3 ton ha^{-1} (Johnson, 1991). Sufficiency of immobile nutrients is independent of yield level. The most limiting of the mobile nutrients will determine the maximum possible yield (as in Liebig's "Law of the Minimum"). Deficiencies of immobile nutrients reduce the potential yield of a site or field (Raun et al., 1998), by a "percent sufficiency" factor, and identify the ultimate potential yield.

In contrast to what Bray documented, Sollins et al. (1988) argued that there are circumstances in which mobile nutrients become immobile. They indicated that variable-charge (v-c) and permanent-charge (p-c) soils differ in the dynamics of nutrient mobility and this dynamic is more complex in v-c than in p-c soils. For example, as the pH of v-c soils decreases, cation exchange capacity (CEC) decreases and anion exchange capacity (AEC) increases. If AEC exceeds CEC, cations such as ammonium (NH_4^+) and K considered as relatively less mobile will be more mobile than anions such as NO_3^-.

Putting into perspective the concept of mobility of nutrients and NUE, it is clear that access of plant roots to nutrients is restricted by the mobility of a nutrient. Nutrient mobility also determines the potential for loss from the nutrient cycle, which consequently affects NUE. The efficiency of nutrient acquisition by roots is dependent on nutrient mobility. Most of the cutting edge nondestructive yield prediction methods such as sensor-based nutrient management depend heavily on this concept.

15.4 Status of Nutrient Use Efficiency

Excessive use of nutrients can lead to environmental pollution and unnecessary fertilizer cost to producers (Cassman et al., 2003). Fertilizer use efficiency (NUE of applied inorganic N, P, and K fertilizers) was reported to be about 50% or lower for N, less than 10% for P, and close to 40% for K in the 1980s (Baligar and Bennett, 1986; Shaviv and Mikkelsen, 1993). Unfortunately, fertilizer use efficiencies have not increased substantially since then. Fertilizer N recovery efficiencies from researcher managed experiments for major grain crops have been reported to range from 46% to 65% (Roberts, 2008) where crops are grown under optimal management practices. For field managed plots, average N recovery did not surpass 30% and 40% under rainfed and irrigated conditions, respectively (Ladha et al., 2005). Globally, cereal N use efficiency is only 33% (Raun and Johnson, 1999; Davis et al., 2003). In corn and sorghum [*Sorghum bicolor* (L.) Moench], Muchow (1998) reported maximum N use efficiency (calculated as grain yield per unit N uptake) of 61 and 48 kg grain^{-1} kg^{-1} N absorbed, respectively.

Phosphorus use efficiency (PUE) is dependent on the method of placement and soil type (Peterson et al., 1981; Sander et al., 1990, 1991). Sander et al. (1990, 1991) reported that PUE averaged 8% when P was broadcast and incorporated and 16% when P was either knifed with anhydrous ammonia or applied with the seed in winter wheat (*Triticum aestivum* L.). Of course, this depends on residual soil P level (Peterson et al., 1981). Several studies (Giskin et al., 1972a, 1972b, 1972c; Bond et al., 2006) showed that P use efficiency of crops decreased with high level of soil residual P level. Residual P can be high due to high soil organic matter or high availability of P from previous applications attributed to low P fixing capacity of a soil. At both soil pH extremes where P could be precipitated, it is likely to see high crop P response, which may not translate to high PUE (Giskin et al., 1972a; Harrison and Adams, 1987).

It is worth mentioning the distinction between NUE and effectiveness. The highest NUE always occurs at the lower parts of the yield response curve, where fertilizer inputs are lowest. There must be a balance between optimal NUE and optimal crop productivity. In fact, in winter wheat, Wuest and Cassman (1992) and Sowers et al. (1994) demonstrated a decrease in N use efficiency with increased N fertilizer rates. Specifically, management systems designed for high protein harvest (Fowler, 2003) in cereals have resulted in low use efficiency since high protein harvest requires higher rates of fertilization than does grain production (Fowler et al., 1990). This suggests that effectiveness of fertilizers in increasing crop yields and optimizing farmer profitability should not be sacrificed for the sake of efficiency alone (Fixen, 2006; Roberts, 2008).

15.5 Causes of Low Nutrient Use Efficiency

One of the major causes of low NUE is the fact that the elemental cycle of most plant nutrients have sinks for loss; meaning one or more forms of a nutrient will leave the cycle through man-made or natural processes. For instance, N can be lost directly from the root system/rhizosphere area by denitrification, volatilization, and leaching or after plant uptake, by ammonia (NH_3) volatilization from plant tissues. Nitrogen losses via denitrification range from 10% to 70% of applied fertilizer N resulting in reduced efficiency (Avalakki et al., 1995; Jambert et al., 1997; Pu et al., 1999). According to Pu et al. (1999), denitrification is a major problem when large amounts of residue are left on the soil. Likewise, Hargrove and Kissel (1979) reported that in a laboratory study using urea, 13%–31% of the applied N was lost as NH_3. Hamid and Mahler (1994) and Mahler and Hamid (1994) reported 4.9%–37.8% loss of surface applied urea fertilizer to the atmosphere through NH_3 volatilization in Northern Idaho soils.

Nitrate leaching occurs more so in well-drained soils than poorly-drained soils and when precipitation/irrigation water exceeds water storage capacity (Stout et al., 2000). Leaching losses increase with increased fertilizer rate especially when the N application rate exceeds the amount required for optimum crop growth (Hauck and Tanji, 1982). Estimates of NO_3^- leaching from different soils and cropping systems ranged from 4 to 80 kg ha^{-1} year^{-1} (Hauck and Tanji, 1982; Dowdell et al., 1984; Jemison and Fox, 1994; Davis et al., 2003).

Ammonia losses to the atmosphere from plant tissues can contribute substantially to decreases in N use efficiency. Francis et al. (1993) reported that 52%–73% of the total unaccounted N was attributed to loss of NH_3 from plant tissue. The authors further showed that N loss from plants increased from 49 to 78 kg N ha^{-1} as N fertilizer rate increased from 50 to 150 kg N ha^{-1} (with 50 kg N ha^{-1} increment). Lees et al. (2000) estimated net plant N loss (determined as the difference of forage N uptake at flowering from total N in the grain and straw at maturity) of 3–42 kg N ha^{-1}. They also found that plant N loss was highest for the highest N rate. Similarly in wheat, Harper et al. (1987) reported that 21% of applied fertilizer N was lost as NH_3, of which 11.4% was from the soil and plant soon after fertilization, and 9.8% from leaves between anthesis and maturity.

Phosphorus fixation by Fe and Al oxides in acid soils and by Ca in alkaline soils and loss of P by erosion in all soils contribute to low PUE. PUE is improved by maintaining soil pH in the optimum range (Blair et al., 1971) to minimize these P fixation reactions. For P, the amount needed to satisfy the precipitation reactions in soils prone to P fixation is a major problem for increasing its use efficiency. It has been reported that only 20%–30% of the P fertilizer applied will be available for crop use and the percentage decreases the longer P is in contact with the soil before plant uptake can occur (Janssen et al., 1987). Soil inorganic P undergoes a very complex system of reactions and compound formation depending on soil pH, type and amount of soil minerals, amount of P in the soil, and several other factors (Sharpley and Sisak, 1997; Slaton et al., 2002). To improve PUE with a reduced input of fertilizer, it is necessary to first develop a strategy to reduce the amount of P fertilizer fixed (Jarvis and Bolland, 1991; Helyar, 1998). Phosphorus fertilizer placement could be critical due to the limited root surface sorption zone as demonstrated by Bray (1954).

15.6 Strategies to Improve Nutrient Use Efficiency

Improving NUE requires the development and adoption of integrated crop and nutrient management methods (Alcoz and Hons, 1993; Arregui and Quemada, 2008) in a sustainable way. Such methods include:

- Appropriate fertilizer application rate (Cochran et al., 1978; Campbell et al., 1993), timing (Bundy, 1986; Shanahan et al., 2004), placement methods (Black and Reitz, 1972; Mahler et al., 1994; Barbieri et al., 2008), and sources (Pan et al., 1984; Salsac et al., 1987; Huffman, 1989)

- Use of inhibitors for N fertilizer (Bremner and Douglas, 1973; Schlegel et al., 1986; Shaviv and Mikkelsen, 1993)
- Foliar applications (Dion et al., 1949; Harder et al., 1982; Mosali et al., 2006; Girma et al., 2007b)
- Cover crops and crop rotations (Raun and Johnson, 1999; Schomberg et al., 2006)
- Tissue and sensor based nutrient management methods (Blackmer et al., 1994; Stone et al., 1996; Raun et al., 2002)
- Use of traditional crop breeding and biotechnology (Kamprath et al., 1982; Bufogle et al., 1997; Kanampiu et al., 1997)

The different methods that must be integrated to maximize NUE are discussed in detail next.

15.6.1 Adjusting Rate, Timing, Placement, and Source of Fertilizer Application to Increase Nutrient Use Efficiency

15.6.1.1 Rate of Nutrient Application

Fertilizer application rate plays an important role in NUE but should be combined with other methods designed to improve NUE. In response to efficiency, environmental, and economic concerns, researchers made significant progress in improving optimum fertilizer rates. Fertilizer application rate evolved from less accurate blanket "recommendation domain," based to site-specific variable rate application (VRA) systems today. Regardless, all fertilizer rates are determined from measurements taken from soil, crop or both.

Worldwide statistical-based nutrient response curves have been widely used for establishing optimum fertilizer rate. Linear or quadratic nutrient response curves and their variants such as liner-plateau, linear-linear, and quadratic-plateau have been fitted to yield data (Neeteson and Wadman, 1987; Cerrato and Blackmer, 1990; Bullock and Bullock, 1994; Girma et al., 2007c). Other nutrient response curve models were also tested but found to be unreliable. For example, for potato (*Solanum tuberosum* L.), Bélanger et al. (2000) compared quadratic, exponential, and square root response models and found that both exponential and square root models were weaker than the quadratic model in determining economically optimum N rate in Atlantic Canada. This approach is affected by temporal and spatial variability (Bullock and Bullock, 1994; Makowski et al., 2001; Liang et al., 2008), varies with change in price of both fertilizer and economic yield (Babcock, 1992), and would not help for in-season fertilization decision. Mechanistic models were also evaluated for predicting N fertilizer requirements (Geist et al., 1970) but were not adopted simply because they were based on too many theoretical assumptions and were not user friendly.

Soil test-based N and P recommendation has been employed in many parts of the United States and elsewhere. For N, total soil N or soil test NO_3^- (Soltanpour et al., 1989) were widely used. Two types of soil test NO_3^-, namely, preplant soil NO_3^-–N test (PPNT) and presidedress soil NO_3^-–N test (PSNT) were employed for drier (Bundy and Meisinger, 1994) and humid areas (Meisinger et al., 1992; Andraski and Bundy, 2002) of corn growing states, respectively. However, PSNT was criticized for lack of correlation with actual crop response and nutrient demand, and resulted in large variability (Andraski and Bundy, 2002). More importantly, the method failed to identify non-N responsive soils when applied to different soils. To overcome these problems, the Illinois soil N test (ISNT) method was developed. This method employs diffusion analytical procedure to measure alkali-hydrolyzable amino sugar, which was highly and inversely correlated with fertilizer N response of corn yield (Khan et al., 2001; Mulvaney et al., 2001). This method has been reported to work in several corn growing states (Khan et al., 2001; Mulvaney et al., 2005; Williams et al., 2007). Yet, in other areas, the method did not improve the prediction of fertilizer N requirements (Barker et al., 2006; Marriott and Wander, 2006; Laboski et al., 2008; Osterhaus et al., 2008) or found to be as good as total soil N (Spargo et al., 2009).

Since the last 20 years, the focus of nutrient rate determination shifted to in-season application rate methods aided by destructive and nondestructive plant tissue analysis (Turner and Jund, 1991; Scharf et al., 2006; Hawkins et al., 2007; Zhang et al., 2008), soil test, and remote sensing (Solie et al., 1996; Raun et al., 2001; Barker and Sawyer, 2010). The latter will be discussed in detail under "sensor-based variable rate nutrient management." Soil test methods such as ISNT discussed above are useful but less convenient for producers as the methods require laboratory testing that might take time. Soil testing remains one of the most accurate methods for determining nutrient requirements, however. At the heart of each in-season N rate decision tool is a level of spatial and temporal accuracy of measurements that aide in increasing NUE of crops. Nondestructive in-season tissue analysis methods used instruments such as chlorophyll meters. Chlorophyll meter readings were correlated with soil NO_3^- or plant tissue N concentration to estimate nutrient needs of crops thereby improving NUE. Both success (Schepers et al., 1992; Hussain et al., 2000) and lack of success (Piekkielek and Fox, 1992; Bavec and Bavec, 2001) with this method were reported. In paddy rice (*Oryza sativa* L.) in the Philippines, greater agronomic efficiency of N fertilizer was obtained with the use of chlorophyll meter sufficiency indices compared with preset timing schemes commonly practiced (Hussain et al., 2000). A study conducted in corn growing areas of north-central United States concluded that chlorophyll meter readings were strongly related ($R^2 = 0.53$–0.79) with economical optimum N rate (EONR) and suggested the method can be applied with reasonable accuracy in wide range of environments (Scharf et al., 2006). Unlike these reports, Piekkielek and Fox (1992) and Bavec and Bavec (2001) showed that the relationship between chlorophyll meter reading and grain yield was weak and growth stage dependent. According to Waskom et al. (1996) portable chlorophyll meter was effective for determining N need but must be accompanied by soil test to determine actual N. Schepers et al. (1992) and Hawkins et al. (2007) used relative chlorophyll meter values to establish N rates in corn.

15.6.1.2 Time of Nutrient Application

The time of application of fertilizer has a significant role in achieving increased NUE for both mobile and immobile nutrients. Providing a mobile nutrient, like N, just prior to a plant's rapid uptake phase increases NUE. The growth stages at which rapid N uptake occurs are well defined for cereals. For winter wheat, Girma et al. (2011) reported that more than 61% of the maximum total N accumulated at later stages of growth could be accounted for by Zadoks 30 (pseudostem) growth stage. Therefore, in wheat, Zadoks 30–32 (2 nodes detectable) was the best time to topdress N to increase use efficiency and optimize final yield (Raun and Johnson, 1999; Girma et al., 2011).

In corn, a steady increase in dry matter (DM) and N accumulation was observed between the V4 (4th leaf collar fully unfolded) and V8 (8th leaf collar fully unfolded) corn growth stages, after which a fast increase in N uptake was measured between V8 and R2 (blister) (Shanahan et al., 2004). Walsh (2006) recommended topdressing/sidedressing N at or before the V10 (10th leaf collar fully unfolded) growth stage to supply the growing corn with adequate N when it is required in the greatest quantities. Similarly, Ma et al. (1999) reported that only a limited amount (20%) of the total plant N was accumulated by V6 (6th leaf collar fully unfolded), and most of the N (60%) was accumulated between V6 and R1 (silking stage). Three studies, Licht and Al-Kaisi (2005), Freeman et al. (2007a), and Girma et al. (2011), reported that greater than half of the total N was accumulated between V8 and VT (tasseling) in corn. These research findings suggest that topdressing or splitting N between V8 and VT growth stages in corn can increase N use efficiency.

Applying a small portion of the N at planting and the remainder of the N just before the rapid growth phase in the spring (a split application) was effective for increasing N use efficiency of wheat (Alcoz and Hons, 1993; Mahler et al., 1994). López-Bellido et al. (2006) reported that in spring wheat, preplant application of 150 kg N ha^{-1} resulted in 13% recovery of N fertilizer, while recovery was as high as 42% when applied at the beginning of stem elongation. For winter cereal crops grown in dry climates, splitting N applications may not increase N use efficiency because N losses are minimal (Arregui and Quemada, 2008). Similarly, in Minnesota, Jokela and Randall (1997) found that nonirrigated corn did not respond to timing of fertilizer N application.

Synchronizing time of N fertilizer with peak crop demand reduces N loss to the environment as well as improves crop yield and N use efficiency (Randall et al., 1997; Karlen et al., 1998). Preplant application of fertilizer especially several months before planting does not seem rational in relation to fertilizer dynamics in the soil. From an environmental health perspective, sidedressing N midseason can significantly reduce NO_3^- leaching into ground water (Bundy, 1986). Also, sidedress N reduces the N that can move by runoff into watersheds impairing quality of surface waters (Allen, 2002). According to Sanchez and Blackmer (1988), 50%–60% of fall-applied N fertilizer is lost from the soil through several of the pathways that lead to N loss from the soil. A winter wheat study conducted to assess the effect of split N applications on N use efficiency using ^{15}N showed that splitting recovered 7%–16% more fertilizer than fall applications (Sowers et al., 1994).

15.6.1.3 Placement of Nutrient

For both mobile and immobile nutrients, the method of application and placement significantly influences the efficiency of the nutrient (Mahler et al., 1994; Barbieri et al., 2008; Kapoor et al., 2008). Generally, subsurface or surface band applications of solid urea and urea ammonium nitrate (UAN) liquid fertilizers in high residue cropping systems can be used to avoid N tie-up in crop residues or N loss by NH_3 volatilization (Mahler et al., 1994). Incorporation of broadcast urea, UAN, and manure into the soil where tillage is practiced can assist in avoiding NH_3 volatilization and run-off losses (Campbell et al., 1993; Freeman et al., 2007b). Banding P is the best strategy instead of broadcasting (Bordoli and Mallarino, 1998).

Tillage and crop residue greatly influence NUE, especially for N. In a no-tillage production system, grain yield improved by 32% when 60 kg N ha^{-1} was banded 8–10 cm below the seed row and 15% when banded between the rows compared to surface broadcast urea (Rao and Dao, 1996). Adaptation of subsurface placement of N fertilizer for no-till winter wheat has the potential to significantly improve N availability to plants and thereby improving N use efficiency and reducing environmental and economic risks (Rao and Dao, 1996). Similarly, crop residue can increase or decrease NUE based on the quantity of residue retained, and type of crop and tillage (Eagle et al., 2000; Kravchenko and Thelen, 2007). In rice, Eagle et al. (2000) reported that straw retention lowered N use efficiency compared with straw removal or burning. Of course, in a no-till setting with total or partial residue retention, fertilizer N use efficiency could be low due to immobilization (Black and Reitz, 1972).

For immobile nutrients the application of all fertilizer at planting is a common practice. NUE can be improved by placing the fertilizer close to the root sorption zone in bands (Sanchez et al., 1991; Jacobsen et al., 1997). PUE is lower when broadcast, compared to banded, because of greater fixation of soluble fertilizer P when broadcast or due to low soil test P (Bell and Black, 1970). For sweet corn (Sanchez et al., 1991) in a low soil test P soils, the relative P fertilizer efficiency was greater than 3:1 for band to broadcast ratio. However, some reports refute the advantage of banding over broadcast P (Borges and Mallarino, 2001, 2003). Borges and Mallarino (2001) found little difference between broadcast and band P on yield and P uptake of corn. In a soybean study, band P had higher P uptake than broadcast for only 3 out of 14 sites (Borges and Mallarino, 2003).

15.6.1.4 Source of Nutrients

Another factor that has a considerable effect on NUE is the nutrient source or form used, which determines the risk of loss and availability of the nutrient. It has been generally agreed that maintaining N in the immobile NH_4^+ form, rather than in the mobile NO_3^- form, will increase N use efficiency. Tsai et al. (1992) suggested that the use of NH_4^+–N fertilizers can reduce leaching and denitrification losses and allow extended availability for late season uptake. Uptake was increased by 35% when N was supplied as NH_4^+ based N sources (Wang and Below, 1992). The N use efficiency of NH_4^+–N based N fertilization has been tied to reduced energy cost for assimilation in an actively growing corn crop compared with NO_3^-–N (Pan et al., 1984; Salsac et al., 1987; Huffman, 1989; Randall and Mulla, 2001). While source of fertilizer plays a crucial role in improving NUE, it should be used as a component of a sustainable nutrient management plan and not as a sole strategy as presented in most reports.

15.6.2 Urease and Nitrification Inhibitors for N Fertilizers

A strategy for slowing down urea hydrolysis and nitrification has been recommended as a method to improve N use efficiency of urea containing fertilizers (Shaviv and Mikkelsen, 1993). Urease inhibitors were reported to delay urea hydrolysis in soils and reduce gaseous loss of urea N as NH_3 (Bremner and Douglas, 1973; Schlegel et al., 1986). Bremner and Douglas (1973) showed that gaseous loss of urea N as NH_3 from urea applied on sandy soil incubated at 20°C for 14 days was reduced from 61% to 0.3% by addition of 2,5-dimethyl-p benzoquinone inhibitor. Likewise, using ammonium lignosulfonate as a urease inhibitor, NH_3 loss from surface-applied urea was reduced up to 85% of the amount lost with plain urea (Al-Kanani et al., 1994).

Nitrification inhibitors (NIs) slow the soil conversion of the less mobile NH_4^+–N form to the leachable NO_3^-–N form. These compounds are especially useful on coarse textured soils where leaching is likely and on fine textured soils where excess water can cause denitrification losses of NO_3^-–N (Shi and Norton, 2000). The use of an NI can be helpful with both preplant and sidedressed N applications. In spring wheat in the northern Great Plains, Goos and Johnson (1999) reported that the apparent N uptake efficiency of grain and straw was 50%–56% with the use of nitrapyrin and ammonium thiosulfate as NI compared to 24% for the control. In corn, Walters and Malzer (1990) reported a significant improvement in N fertilizer use efficiency with NI with the application of 90 kg N ha^{-1} as urea. The use of inhibitors as effective tools for reducing N loss requires certain conditions to be fulfilled. However, the use of inhibitors by themselves may not always translate to improved efficiency. For example, a study in winter wheat and barley (*Hordeum vulgare* L.) revealed that use of fertilizer with an NI (3,4-dimethylpyrazole phosphate, DMPP) did not improve N use efficiency (Arregui and Quemada, 2008).

15.6.3 Foliar Fertilization for Improving Nutrient Use Efficiency

It has been more than six decades since scientists proved the benefits of foliar uptake of nutrients through leaves and other green tissue parts using radioactive and isotopically tagged nutrients (Dion et al., 1949). Studies showed foliar fertilization is more efficient and reduces potential environmental impact than soil fertilization (Harder et al., 1982; Mosali et al., 2006; Girma et al., 2007b). Dixon (2003) reported that foliar-applied nutrients can be 4–30 times more effective than soil applied nutrients. This high efficiency is related to increased enzymatic activities in plant cells with foliar application. Others reported that foliar fertilization was 8–10 times more efficient than root uptake (Wittwer et al., 1963). Foliar applications have provided a highly efficient alternative for supplying nutrients to plants, particularly micronutrients, which are needed in small quantities and are subject to reduced availability when soil applied.

At present, foliar N in corn is not a viable option because of the higher temperatures present throughout the growing season. Liquid N fertilizer sources, such as UAN, adhere to the leaf, when applied directly to the leaves. This coupled with the high temperature of the summer can result in leaf burn (Lohry, 2001). However, many new products are currently available that can be applied foliar, and that have limited leaf burn including urea triazone and urea formaldehyde solutions. In corn, the alternative to foliar application to increase NUE is a traditional sidedress application.

Foliar P fertilization can increase P use efficiency in comparison to soil-applied P because P fixation in the soil is avoided. Averaged over 6 site years, Mosali et al. (2006) reported 80% PUE in winter wheat when applied foliarly at 2 kg P ha^{-1} between stem elongation and heading. Similarly, in rainfed corn, Girma et al. (2006) reported 35% more PUE than soil applications of P when 2 kg P ha^{-1} was applied to the foliage at the V8 growth stage of corn. With new fertilizer formulations and application technologies, the use of foliar fertilization to increase NUE tool may increase in the future.

15.6.4 Use of Cover Crops and Crop Rotation

The use of winter cover crops can help prevent NO_3^-–N leaching in high rainfall areas. Cover crops absorb residual nutrients and return them to the soil for the following crop. Guillard et al. (1995) found that N use efficiency was increased in corn-rye (*Secale cereale* M. Bieb) and oat (*Avena sativa* L.)-tyfon (*Brassica* spp.) cropping systems with the application of 112 kg N ha^{-1}. Pikul et al. (2005) reported N use efficiency of 49, 67, and 70 kg corn grain yield kg^{-1} N for continuous corn, corn-soybean (*Glycine max* L. (Merr.)), and corn-soybean-wheat-alfalfa (*Medicago sativa* L.) rotation systems, respectively, with application of moderate N (N rate predicted to achieve 5.3 Mg ha^{-1} corn yield). In irrigated or high-rainfall production regions, soybean-corn rotations have high N use efficiency and can reduce the amount of residual N available for leaching when compared to continuous corn (Huang et al., 1996). Unfortunately, rotations are not easily adopted by farmers who have become accustomed to monoculture production systems since a new crop often requires purchase of additional equipment and learning to integrate new cultural practices. In irrigated agriculture, the use of high N rates as a substitute for more N use efficient rotation systems (corn-soybean) must be weighed against the increased potential for NO_3^-–N loss (Anderson et al., 1997).

Nitrogen use efficiency for wheat following legumes was 32% and 21% greater than that of wheat following fallow and continuous wheat, respectively (Badaruddin and Meyer, 1994). In another study, Badaruddin and Meyer (1990) found that N use efficiency of wheat was greater following alfalfa or hairy vetch (*Vicia villosa* Roth.) green manure crops than fallow. Average corn N use efficiency derived from soybean and alfalfa residue for the whole plant and grain was 43% and 30%, respectively, (Hesterman et al., 1987). Wheat-corn-fallow production systems are promoted instead of wheat-fallow where only 420 mm precipitation is received per year (Kolberg et al., 1996). The more intensive systems (growing more crops in a given period of time), require greater fertilizer N inputs but are higher in total yield and economically advantageous (Kolberg et al., 1996; Anderson, 2005). More intensive dryland cropping systems lead to increased WUE and better maintained soil quality (Halvorson and Reule, 1994; Anderson, 2005). Alternative dryland systems proposed include spring barley, corn, and winter wheat grown in rotation with adequate N fertilization instead of continuous winter wheat-fallow (Halvorson and Reule, 1994).

15.6.5 Sensor-Based Variable Rate Nutrient Management

Sensor-based variable rate nutrient management employs spectral radiance in red and near-infrared (NIR) regions of the electromagnetic spectrum as a tool to detect the health of an actively growing plant (Lukina et al., 2001). This tool was introduced in nutrient management for cereals particularly in wheat and corn in the mid-1990s. Blackmer et al. (1994) found strong correlation between reflected radiation and relative grain yield at later vegetative growth stages. Stone et al. (1996) demonstrated that an index developed from red and NIR reflectance could be an excellent predictor of total N uptake of a crop at early stages of winter wheat growth. This technique allowed the accurate prediction of fertilizer N requirement, thus improving crop N use efficiency (Raun et al., 2002; Teal et al., 2006).

The technologies that emerged from measurement of normalized difference vegetation index (NDVI), the Greenseeker® (Trimble Navigation Limited, Sunnyvale, CA), and Crop Circle™ (Holland Scientific, Inc., Lincoln, NE) sensors have a functional algorithm developed from measurements based on yield potential and responsiveness of the crop to N fertilization (Johnson and Raun, 2003; Mullen et al., 2003; Solari et al., 2008; Sripada et al., 2008). The specific growth stages at which N uptake and N use efficiency increased have been identified. Accordingly, for winter wheat, the stage between Zadoks 30 and 32 is recommended. In corn, sensor measurements were well correlated with predicted yield when measurements were obtained between V8 and V12 (Teal et al., 2006; Freeman et al., 2007a). Using this technology, N use efficiency in wheat was increased by 10%–20% and savings in fertilizer was about $20 ha^{-1} (Raun et al., 2009). Research in China found that a sensor-based system for determining optimum N rate for winter wheat resulted in 61% N use efficiency, surpassing the soil test-based and farmer methods by about 10% and 48%, respectively (Li et al., 2009).

The variable rate sensor-based system can be used to manage both temporal and spatial variations, which directly influence NUE. It was reported, following extensive soil sampling, optical sensor measurements of plants, and geostatistical analysis, that the spatial scale of N availability in wheat was about 1 m^2 (Raun et al., 1998, 2002; Chung et al., 2008). When N management decisions are made on areas of <1 m^2, the variability that is present beyond that resolution can be detected using optical sensors (Solie et al., 1996; Stone et al., 1996). Simple methods that can manage temporal variability and increase N use efficiency have also been evaluated. The N-rich strip or N reference strip is becoming instrumental since it is simple and can be used with or without a sensor (Girma et al., 2007a; Raun et al., 2008).

Although the use of optical sensors for improving N use efficiency has shown great promise, and there is no doubt about its contribution to improving N use efficiency, the need to develop empirical algorithms for use on different soil types, crop management systems, and climates has limited its adoption (Samborski et al., 2009).

15.6.6 Biotechnology and Variety Selection for Improving NUE

NUE is under genetic and physiological influence of a given crop and is modified by the environment (Baligar et al., 2001). Both biotech and conventional breeding can be used to identify or modify plant traits that can contribute to improvement in NUE. However, there is not much data available in this regard. The early study of N use efficiency was facilitated by identifying individual components that elucidated both uptake and utilization efficiency (Bruetsch and Estes, 1976; Moll et al., 1982). Differences among corn hybrids for N use efficiency are largely due to variation in the utilization of accumulated N before anthesis, especially under low N supply (Moll et al., 1982; Eghball and Maranville, 1991).

Wheat varieties with a high harvest index (grain produced divided by the total dry biomass) and low forage yield have low plant N loss and increased N use efficiency (Kanampiu et al., 1997). Furthermore, N assimilation after anthesis is needed to achieve high wheat yields (Cox et al., 1985) and high N use efficiency. Higher N use efficiency has also been observed in rice varieties with high harvest index (Bufogle et al., 1997). Work by Karrou and Maranville (1993) suggested that wheat varieties that produce more seedling DM with greater N accumulation are not necessarily the ones that use N more efficiently. Genetic selection is often conducted with high fertilizer N input in order to eliminate N as a variable; however, this can mask efficiency differences among genotypes to accumulate and utilize N to produce grain (Kamprath et al., 1982). This is consistent with Earl and Ausubel (1983), noting that high yielding varieties of corn, wheat, and rice released during the Green Revolution were selected to respond to high N inputs. Consequently, continued efforts are needed where plant selection is accomplished under low N, often not considered to be a priority by plant breeders and uncharacteristic of agricultural experiment stations.

15.7 Water Use Efficiency

Water is the single most important regional and global resource management challenge. In rainfed regions, water runoff, surface evaporation, low soil water holding capacity, and random precipitation are major problems constraining WUE among many factors (Kemper, 1993; Hatfield et al., 2001; Condon et al., 2002). Despite its wide use, the word WUE does not reflect losses for which intervention strategies are being developed. Some prefer to use the term "biomass to water ratio (BWR)" to reflect the biomass produced per unit of water used by a crop (Morison et al., 2008). But more importantly, the variation in units used to describe WUE when the term is used in association with different spatial and temporal scales as well as type of yield (biological or economic) calls for careful interpretation of WUE values.

Generally, WUE is the ratio between the amount of water that is used for an intended purpose and the total amount of water input within a spatial domain of interest (Guerra et al., 1998). In this review, we are interested in WUE as it is related to crop biomass production (total DM or economic yield). In view of this, the definition given by Viets (1962) and later modified by Tanner and Jury (1976) is appropriate. Accordingly, WUE is defined as

$$\text{WUE} = \frac{\text{Dry matter or economic crop yield}}{\text{Water used to produce the economic yield}} \quad (15.1)$$

The unit for this computation is usually $kg\,m^{-3}$.

According to World Water Assessment Program (WWAP, 2009), agricultural use accounts for 70% of freshwater withdrawals from different sources. It has been quantified that the process of photosynthesis takes 400–2000 L (average of 1600 L) of evapotranspiration (ET) to produce 1 kg of wheat crop (CAWMA, 2007) depending on climatic conditions and the genetics of crop cultivars. Given this, crop WUE has doubled in the last 40 years (CAWMA, 2007).

It is anticipated that agricultural WUE will improve from 52% in 2003 to 66% in 2050 (FAO, 2008). During the same period, crop water consumption will increase by 64%–83%; almost doubling from its current annual consumption of $7130\,km^3$ (CAWMA, 2007) due to changing weather conditions characterized by unpredictable growing seasons (FAO, 2006). However, population growth, urban expansion, bioenergy, and other economic developments are decreasing the quantity of water available for agricultural use (Müller et al., 2008). Quality of water due to salinization of ground and surface water remains a major problem. In the United States, about 87.5% of crop production is rainfed and this constitutes about 32% of arable land (World Bank, 2007). This clearly demonstrates the need to improve the WUE of this large sector. Current and future water use strategy in crop production must account for these emerging global undertakings (Angus and van Herwaarden, 2001).

15.8 Critical Factors Affecting Water Use Efficiency by Crops

There is not a single factor that controls WUE of crops. It is rather many factors that interactively and simultaneously influence water use. Soil physical and chemical properties (e.g., texture, structure, water holding capacity, salinity, and organic matter), soil management practices (e.g., tillage, residue management, planting date, etc.,), crop and crop genetics, as well as climate are among the factors that determine WUE. However, climate is likely the most influential variable. Seasonal distribution and amount of rainfall, wind, and light intensity are some of the climatic factors that limit WUE by crops. In wheat in Argentina, Abbate et al. (2004) found that WUE was greater when water was limited due to shortage of rainfall than when adequate. They hypothesized that this could be due to stomatal closure when vapor pressure deficit was highest. Genetics (crops and varieties of crops) is another factor that determines the amount of water that can be efficiently used for production of DM. Although crops adapt to specific growing conditions in a given agroecology, some are better users of available water and more efficient in translating it to DM than others (Richards, 2004; Bessembinder et al., 2005; Mueller et al., 2005). For example, Bessembinder et al. (2005) attributed the WUE of crops to their morphology (e.g., leaf angle), anatomy (e.g., leaf cuticle composition), phenology (e.g., growing degree days), and physiology (e.g., protein and fat concentration in different plant parts).

15.9 Strategies to Improve Water Use Efficiency

It is intuitive that it is not possible to manipulate climate but what is possible is to manipulate soil conditions and crop and variety traits to better cope with changes in climate and increased WUE. Therefore, strategies to increase WUE by crops should employ soil and crop management practices and genetic improvements of crops and crop varieties. The intervention needed varies for irrigated and dryland systems. For example, in an irrigated setting, WUE can be achieved by increasing the output per unit of water, reducing losses of water to unusable sinks, reducing water degradation, and reallocating water to higher priority uses (Howell, 2001). In dryland systems, soil and crop management practices can be used to conserve and increase WUE.

15.9.1 Tillage and Residue Management

Conservation tillage, crop residue management, and mulch play a pivotal role in water conservation and improving crop water use. In addition to its benefit for building soil organic matter and reducing soil erosion, no-till has emerged as a useful practice for conserving soil moisture when practiced with appropriate cultural practices (Machado et al., 2008). Such cultural practices include row spacing and seeding rate. For example, WUE was increased by 6% with row spacing of 9 cm compared to 36 cm in no-till wheat (Tompkins et al., 1991). These authors investigated the interaction of seeding rate and row spacing on WUE under no-till and found 9 kg grain yield cm^{-1} higher WUE with seeding rate of 140 kg ha^{-1} and 9 cm row spacing compared to 35 kg ha^{-1} seeding rate and 36 cm row spacing. In corn, no-till increased WUE by 0.96 kg grain yield ha^{-1} mm^{-1} (Norwood, 2000) suggesting reduced irrigation water demand. Reduced tillage and residue retention improved infiltration and enhanced storage of precipitation that in turn increased DM produced per unit of water.

Mulching can improve WUE through its effect on increased water storage of a soil and reduction of evaporation from surface soil. Ji and Unger (2001) found that mulched soils had 10% higher water storage capacity than bare soils when 5 mm simulated rain was applied with ET of 6 mm day^{-1}. They further showed that 2 Mg ha^{-1} mulch is needed to maintain 10% or more water storage with simulated rain water of 10 mm and ET of 12 mm day^{-1}. Similarly, in north China, Zhang et al. (2005) reported that over 12 seasons, mulching improved WUE of corn by 8%–10%. The timing of irrigation also increased WUE for corn and winter wheat in this environment, where 73% of 480 mm precipitation falls during the winter wheat cropping season (Zhang et al., 2005). Beginning irrigation at jointing stage of winter wheat was preferable as prejointing irrigation of all tillers including infertile tillers could reduce crop WUE. However, in a more arid Arizona environment (277 mm seasonal precipitation), irrigation may not be delayed until jointing without yield penalty (Jama and Ottman, 1993). Burning of stubble had no effect on WUE of double cropped wheat and soybean (Daniels and Scott, 1991). In Washington, the interaction of crop residue and tillage showed that soil-stored water level was the same when conventional or no-till were practiced following undisturbed standing wheat residue in the winter (Kennedy and Schillinger, 2006). In a grain sorghum study, Unger and Jones (1981) reported that averaged over 3 years, 8 ton ha^{-1} straw mulch increased WUE by 19% compared with no mulch. These findings suggest that surface residue retention and the amount of soil disturbance are key factors in water conservation.

15.9.2 Crop Rotations Cover Crops and Fallow System

In many parts of the United States, double cropping and other alternatives to continuous corn have been evaluated. Crop-fallow systems have been effective in conserving precipitation water for subsequent crops, in a way, contributing to higher WUE. However, the newer cropping systems can also be as good as or even better than the crop-fallow system in conserving water. In the central Great Plains, Tanaka et al. (2005) reported that wheat WUE was improved 18%–56% with the inclusion of broadleaf crops in a grass-based system. However, in a study designed to identify the most water efficient production system in Texas, Jones and Popham (1997) found that soil water content at planting of the next crop in no-till was increased by 29 mm in a continuous wheat cropping system compared with stubble mulch.

In a study that compared forage and grain systems, Nielsen et al. (2005) found that WUE (kg DM ha^{-1} mm^{-1}) was highest for pea followed by millet and lowest for corn. However, corn grain WUE was second to millet while pea had the lowest grain WUE. In Mediterranean environment, rotation involving barley and wheat with rapeseed (*Brassica napus* L.) and vetch (*Vicia sativa* L.) increased biomass WUE by 7 kg DM mm^{-1} ha^{-1} compared with monoculture barley (Álvaro-Fuentes et al., 2009). Also, precipitation use efficiency is greater for corn grown in rotation when compared to continuous corn (Varvel, 1994).

15.9.3 Crop/Variety/Hybrid Selection

Crop improvement through conventional breeding coupled with appropriate agronomic management has significantly contributed to increased water use by crops and crop varieties under both rainfed and irrigation systems (Richards et al., 2002). However, much work is needed to identify specific physiological and agronomic traits that can significantly control water utilization in crops.

It is a well-established fact that crops vary in their ability to extract soil water and reduce transpiration allowing them to survive in dry environments. In dryland areas of the Great Plains, sorghum and sunflower (*Helianthus annuus* L.) removed 19 mm more water from subsurface soil compared with corn and soybean (Norwood, 1999).

Howell et al. (1998) did not find a difference in grain and DM WUE of two corn hybrids (short and full season) but seasonal ET was reduced with a short season hybrid by 129 mm compared with a full season hybrid. In grain sorghum, Unger (1991) found that hybrids significantly differ in WUE. He concluded that highest yielding cultivar had the highest WUE.

Molecular biology could play a significant role in untangling this in the future. In Australia, isotope discrimination techniques had helped in identifying varieties with high transpiration efficiency (Richards, 2004). Other traits that could be sources of WUE improvement include increased root length (Proffitt et al., 1985; Payne, 2000; Ehdaie et al., 2003) and leaf canopy architecture (Schakel and Hall, 1979; Osborne et al., 1998). Under drought conditions, leaflets of cowpeas [*Vigna unguiculata* (L.) Walp] positioned paraheliotropically (becoming more vertical) in the afternoon to reduce transpiration (Schakel and Hall, 1979). Recently, in rice, researchers identified a mutant that reduced the indole-3-acetic acid (IAA) at the lamina joint, shoot base, and nodes enhancing drought tolerance (Zhang et al., 2009).

15.9.4 Soil Physical and Chemical Properties

Soil physical and chemical properties play a crucial role in soil water conservation and WUE in crops. For example, two decades ago, Stout et al. (1988) studied the effect of N supply and soil physical properties on WUE in tall fescue (*Festuca arundinacea* Schreb.). They found that when rainfall was adequate and evenly distributed over the crop growing season, N supply controlled WUE. In contrast, when rainfall was erratic, water holding capacity controlled WUE.

Soil organic matter is a critical component of soil that "conditions" soil, improving water infiltration rate and water-holding capacity. It serves as a reservoir of water and supplies them to crops when needed, indirectly increasing WUE of crops. With a high level of organic matter soil tilth is improved, and aggregate size tends to be large and bulk density low (Allison, 1973; Six et al., 1998).

Water use by crops is influenced by salt level in irrigation water and the soil itself. In arid climates, irrigation water usually contains a large quantity of salts that over time accumulate in the soil and limit water availability in the root zone. Various strategies have been proposed to address this problem (Ma et al., 2008). Recently variable rate precision irrigation has been investigated. Al-Kufaishi et al. (2006) assessed the feasibility of applying spatially variable irrigation under a center pivot system and found that the loss of water was higher for the uniform application scenarios than that for the VRA scenarios for applications of 20 and 30 mm. Under precision irrigation, water and associated solute movement are designed to vary spatially within the root zone and salt accumulation or leaching of nutrients that influence water balance in the root zone (Mmolawa and Or, 2000; Raine et al., 2007). Many current irrigation systems are open and have relatively low efficiencies of water application (Hagin and Lowengart, 1995, 1999). In addition, water quality relative to management practices has been largely ignored.

15.9.5 Modeling as a Tool for Improving Water Use Efficiency

Since the 1980s, crop production decision support models have been advocated. Models, particularly mechanistic models that strive to simulate water and nutrient transport, and DM production as a function of management and climatic factors are becoming beneficial. Through continued research today there are several models that can be used in decision making regarding water use in specific environments and cropping systems. For example, the AquaCrop model developed by FAO was found to be a good predictor of crop productivity, water requirement, and WUE under water-limiting conditions for maize in different environments (Heng et al., 2009). The SORKAM model predicted an increase in sorghum WUE in the southern High Plains of Texas when a later maturing sorghum cultivar was planted in mid-May at irrigation levels of 3.75 and 5.0 mm day^{-1} (Baumhardt et al., 2007).

15.9.6 Nutrient and Water Interactions

It is well documented that WUE cannot be achieved without addressing fertility constraints. WUE has been constrained by the supply of both N and P. Selles et al. (1992) suggested that soil testing laboratories should modify their fertilizer recommendations according to different levels of available water, rather than the traditional dry, normal, and wet classes used. Halvorson et al. (2004) demonstrated that precipitation use efficiency of winter wheat increased with N rate reaching peak after 56 kg N ha^{-1}. Optimal leaf N concentration promoted higher WUE (Heitholt et al., 1991), and low leaf N led to poor WUE in N limited and drought stressed wheat in the southern Great Plains. NUE and WUE were improved when nutrients were placed beneath the surface of the soil at a depth of 15 cm in turfgrass (Murphy and Zaurov, 1994). Inefficient water use can result if N fertilizer is applied during the vegetative phase since it increases transpiration loss by reducing the soluble carbohydrate reserves available for translocation to grain (Angus and van Herwaarden, 2001).

Likewise, increasing soil P availability resulted in increased water use in both water-stressed and non-water-stressed conditions (Payne et al., 1992; Brück et al., 2000). Another study showed that N and P fertilization was related to WUE (Payne et al., 1995). Generally, in non-water-limiting conditions, WUE was improved with higher rates of both nutrients.

15.10 Conclusions

Nutrient and WUE have a lot to do with energy production and use in agriculture. Improving both NUE and WUE can contribute to efficient and productive agricultural system. Improving NUE requires the use of best management practices in a sustainable way and managing both spatial and temporal variability. There is no single factor that can serve as a silver bullet to improve NUE. Variable rate nutrient management systems, foliar fertilization, and biotechnology are some of the new best management practices (BMPs) that may result in huge improvement in N and other NUE in crops. WUE like NUE will always play a pivotal role in agricultural production. But unlike NUE, WUE, especially in dryland environments is a function of temporal or climatic variability. In the vast majority of rainfed production environments, weather changes dramatically from year to year. As a result, WUE changes from year to year and in these environments it is highly unpredictable. Thus, management systems that can buffer or dampen the effects or impact of water play an increasingly important role.

References

Abbate, P.E., J.L. Dardanelli, M.G. Cantarero, M. Maturano, R.J.M. Melchiori, and E.E. Suero. 2004. Climatic and water availability effects on water-use efficiency in wheat. Crop Sci. 44:474–483.

Alcoz, M.M., and F.M. Hons. 1993. Nitrogen fertilization timing effect on wheat production, nitrogen uptake efficiency, and residual soil nitrogen. Agron. J. 85:1198–1203.

Al-Kanani, T., A.F. MacKenzie, J.W. Fyles, I.P. O'Halloran, and S. Ghazala. 1994. Ammonia volatilization from urea amended with lignosulfonate and phosphoroamide. Soil Sci. Soc. Am. J. 58:244–248.

Al-Kufaishi, S.A., B.S. Blackmore, and H. Sourell. 2006. The feasibility of using variable rate water application under a central pivot irrigation system. Irrig. Drain. Syst. 20:317–327.

Allen, C.J. 2002. Sidedressing nitrogen in corn makes sense. Fluid J. 10:12–13.

Allison, F.E. 1973. Soil organic matter and its role in crop production. American Elsevier Publishing Co. Inc., New York.

Álvaro-Fuentes, J., J. Lampurlanés, and C. Cantero-Martínez. 2009. Alternative crop rotations under Mediterranean no-tillage conditions: Biomass, grain yield, and water-use efficiency. Agron. J. 101:1227–1233.

Anderson, R.L. 2005. Are some crops synergistic to following crops? Agron. J. 97:7–10.

Anderson, I.C., D.R. Buxton, D.L. Karlen, and C. Cambardella. 1997. Cropping system effects on nitrogen removal, soil nitrogen, aggregate stability, and subsequent corn grain yield. Agron. J. 89:881–886.

Andraski, T.W., and L.G. Bundy. 2002. Using the presidedress soil nitrate test and organic nitrogen crediting to improve corn nitrogen recommendations. Agron. J. 94:1411–1418.

Angus, J.F., and A.F. van Herwaarden. 2001. Increasing water use and water use efficiency in dryland wheat. Agron. J. 93:290–298.

Arregui, L.M., and M. Quemada. 2008. Strategies to improve nitrogen use efficiency in winter cereal crops under rainfed conditions. Agron. J. 100:277–284.

Avalakki, U.K., W.M. Strong, and P.G. Saffigna. 1995. Measurements of gaseous emissions from denitrification of applied nitrogen-15. III. Field measurements. Aust. J. Soils Res. 33:101–111.

Babcock, B.A. 1992. The effects of uncertainty on optimal nitrogen applications. Rev. Agric. Econ. 14:271–280.

Badaruddin, M., and D.W. Meyer. 1990. Green-manure legume effects on soil nitrogen, grain yield, and nitrogen nutrition of wheat. Crop Sci. 30:819–825.

Badaruddin, M., and D.W. Meyer. 1994. Grain legume effects on soil nitrogen, grain yield, and nitrogen nutrition of wheat. Crop Sci. 34:1304–1309.

Baligar, V.C., and O.L. Bennett. 1986. Outlook on fertilizer use efficiency in the tropics. Nutr. Cycl. Agroecosyst. 10:83–96.

Baligar, V.C., N.K. Fageria, and Z.L. He. 2001. Nutrient use efficiency in plants. Commun. Soil Sci. Plant Anal. 32:921–950.

Barbieri, P.A., H.E. Echeverría, H.R. Saínz Rozas, and F.H. Andrade. 2008. Nitrogen use efficiency in maize as affected by nitrogen availability and row spacing. Agron. J. 100:1094–1100.

Barker, D.W., and J.E. Sawyer. 2010. Using active canopy sensors to quantify corn nitrogen stress and nitrogen application rate. Agron. J. 102:964–971.

Barker, D.W., J.E. Sawyer, M.M. Al-Kaisi, and J.R. Lundvall. 2006. Assessment of the amino sugar-nitrogen test on Iowa soils: II. Field correlation and calibration. Agron. J. 98:1352–1358.

Baumhardt, R.L., J.A. Tolk, T.A. Howell, and W.D. Rosenthal. 2007. Sorghum management practices suited to varying irrigation strategies: A simulation analysis. Agron. J. 99:665–672.

Bavec, F., and M. Bavec. 2001. Chlorophyll meter readings of winter wheat cultivars and grain yield prediction. Commun. Soil Sci. Plant Anal. 32:2709–2719.

Bélanger, G., J.R. Walsh, J.E. Richards, P.H. Milburn, and N. Ziadi. 2000. Comparison of three statistical models describing potato yield response to nitrogen fertilizer. Agron. J. 92:902–908.

Bell, L.C., and C.A. Black. 1970. Crystalline phosphates produced by interaction of orthophosphate fertilizers with slightly acid and alkaline soils. Soil Sci. Soc. Am. J. 34:735–740.

Bessembinder, J.J.E., P.A. Leffelaar, A.S. Dhindwal, and T.C. Ponsioen. 2005. Which crop and which drop, and the scope for improvement of water productivity. Agric. Water Manage. 73:113–130.

Black, A.L., and L.L. Reitz. 1972. Phosphorus and nitrate-nitrogen immobilization by wheat straw. Agron. J. 64:782–785.

Blackmer, T.M., J.S. Schepers, and G.E. Varvel. 1994. Light reflectance compared with other nitrogen stress measurements in corn leaves. Agron. J. 86:934–938.

Blair, G.J., C.P. Mamaril, and M.H. Miller. 1971. Influence of nitrogen source on phosphorus uptake by corn from soils differing in pH. Agron. J. 63:235–238.

Bond, C.R., R.O. Maguire, and J.L. Havlin. 2006. Change in soluble phosphorus in soils following fertilization is dependent on initial Mehlich-3 phosphorus. J. Environ. Qual. 35:1818–1824.

Bordoli, J.M., and A.P. Mallarino. 1998. Deep and shallow banding of phosphorus and potassium as alternatives to broadcast fertilization for no-till corn. Agron. J. 90:27–33.

Borges, R., and A.P. Mallarino. 2001. Deep banding phosphorus and potassium fertilizers for corn produced under ridge tillage. Soil Sci. Soc. Am. J. 65:376–384.

Borges, R., and A.P. Mallarino. 2003. Broadcast and deep-band placement of phosphorus and potassium for soybean managed with ridge tillage. Soil Sci. Soc. Am. J. 67:1920–1927.

Bray, R.H. 1954. A nutrient mobility concept of soil plant relationships. Soil Sci. 78:9–22.

Bray, R.H. 1958. The correlation of a phosphorus soil test with the response of wheat through a modified Mitscherlich equation. Soil Sci. Soc. Am. J. 22:314–317.

Bremner, J.M., and L.A. Douglas. 1973. Effects of some urease inhibitors on urea hydrolysis in soils. Soil Sci. Soc. Am. J. 37:225–226.

Brück, H., W.A. Payne, and B. Sattelmacher. 2000. Effects of phosphorus and water supply on yield, transpirational water-use efficiency, and carbon isotope discrimination of pearl millet. Crop Sci. 40:120–125.

Bruetsch, T.F., and G.O. Estes. 1976. Genotype variation in nutrient uptake efficiency in corn. Agron. J. 68:521–523.

Bufogle, A., Jr., P.K. Bollich, J.L. Kovar, R.E. Macchiavelli, and C.W. Lindau. 1997. Rice variety differences in dry matter and nitrogen accumulation as related to plant stature and maturity group. J. Plant Nutr. 20:1203–1224.

Bullock, D.G., and D.S. Bullock. 1994. Quadratic and quadratic-plus-plateau models for predicting optimal nitrogen rate of corn: A comparison. Agron. J. 86:191–195.

Bundy, L.G. 1986. Review—Timing nitrogen applications to maximize fertilizer efficiency and crop response in conventional corn production. J. Fert. Issues 3:99–106.

Bundy, L.G., and J.J. Meisinger. 1994. Nitrogen availability indices, p. 951–984. *In* R.W. Weaver, S. Angle, P. Bottomley, D. Bezdicek, S. Smith, A. Tabatabai, and A. Wollum (eds.) Methods of soil analysis. Part 2—Microbiological and biochemical properties. SSSA Book Series 5. SSSA, Madison, WI.

Campbell, C.A., R.P. Zentner, F. Selles, B.G. McConkey, and F.B. Dyck. 1993. Nitrogen management for spring wheat grown annually on zero-tillage: Yields and nitrogen use efficiency. Agron. J. 85:107–114.

Cassman, K.G., A. Dobermann, and D.T. Walters. 2002. Agroecosystems, nitrogen-use efficiency, and nitrogen management. Ambio 31:132–140.

Cassman, K.G., A. Dobermann, D.T. Walters, and H. Yang. 2003. Meeting cereal demand while protecting natural resources and improving environmental quality. Annu. Rev. Environ. Resour. 28:315–358.

CAWMA (Comprehensive Assessment of Water Management in Agriculture). 2007. Water for food, water for life: A comprehensive assessment of water management in agriculture. Earthscan, London, U.K., and IWMI (International Water Management Institute), Colombo, Sri Lanka.

Cerrato, M.E., and A.M. Blackmer. 1990. Comparison of models for describing corn yield response to nitrogen fertilizer. Agron. J. 82:138–143.

Chung, B., K. Girma, K.L. Martin, B.S. Tubaña, D.B. Arnall, and W.R. Raun. 2008. Determination of optimum resolution for predicting corn grain yield using sensor measurements. Arch. Agron. Soil Sci. 54:481–491.

Cochran, V.L., R.L. Warner, and R.I. Papendick. 1978. Effect of N depth and application rate on yield, protein content, and quality of winter wheat. Agron. J. 70:964–968.

Condon, A.G., R.A. Richards, G.J. Rebetzke, and G.D. Farquhar. 2002. Improving intrinsic water-use efficiency and crop yield. Crop Sci. 42:122–131.

Cornforth, I.S. 1968. Relationships between soil volume used by roots and nutrient accessibility. Eur. J. Soil Sci.19:291–301.

Cox, M.C., C.O. Qualset, and D.W. Rains. 1985. Genetic variation for nitrogen assimilation and translocation in wheat. II. Nitrogen assimilation in relation to grain yield and protein. Crop Sci. 25:435–440.

Daniels, M.B., and H.D. Scott. 1991. Water use efficiency of double-cropped wheat and soybean. Agron. J. 83:564–570.

Darrah, P.R., D.L. Jones, G.J.D. Kirk, and T. Roose. 2006. Modelling the rhizosphere: A review of methods for 'upscaling' to the whole-plant scale. Eur. J. Soil Sci. 57:13–25.

Davis, R.L., J.J. Patton, R.K. Teal, Y. Tang, M.T. Humphreys, J. Mosali, K. Girma et al. 2003. Nitrogen balance in the Magruder plots following 109 years in continuous winter wheat. J. Plant Nutr. 26:1561–1580.

Dion, H.G., J.W.T. Spinks, and J. Mitchell. 1949. Experiments with radio phosphorus on the uptake of phosphorus by wheat. Sci. Agric. 29:167–172.

Dixon, R.C. 2003. Foliar fertilization improves nutrient use efficiency. Fluid J. 40:22–23.

Dowdell, R.J., C.P. Webster, D. Hill, and E.R. Mercer. 1984. A lysimeter study of the fate of fertilizer nitrogen in spring barley crops grown as shallow soil overlying chalk: Crop uptake and leaching losses. J. Soil Sci. 35:169–183.

Eagle, A.J., J.A. Bird, W.R. Horwath, B.A. Linquist, S.M. Brouder, J.E. Hill, and C. van Kessel. 2000. Rice yield and nitrogen utilization efficiency under alternative straw management practices. Agron. J. 92:1096–1103.

Earl, C.D., and F.M. Ausubel. 1983. The genetic engineering of nitrogen fixation. Nutr. Rev. 41:1–6.

Eghball, B., and J.W. Maranville. 1991. Interactive effects of water and nitrogen stresses on nitrogen utilization efficiency, leaf water status and yield of corn genotypes. Commun. Soil Sci. Plant Anal. 22:1367–1382.

Eghball, B., and D.H. Sander. 1989. Distance and distribution effects of phosphorus fertilizer on corn. Soil Sci. Soc. Am. J. 53:282–287.

Ehdaie, B., R.W. Whitkus, and J.G. Waines. 2003. Root biomass, water-use efficiency, and performance of wheat–rye translocations of chromosomes 1 and 2 in spring bread wheat 'Pavon'. Crop Sci. 43:710–717.

FAO (Food and Agriculture Organization of the United Nations). 2006. World agriculture towards 2030/2050. Prospects for food, nutrition, agriculture, and major commodity groups. Interim report. FAO, Rome, Italy.

FAO (Food and Agriculture Organization of the United Nations). 2008. Climate change, water and food security. Technical background document from the expert consultation held in Rome, February 26–28, 2008. FAO, Rome, Italy.

Fixen, P.E. 2006. Turning challenges into opportunities. *In* Proceedings of the Fluid Forum, Fluids: Balancing fertility and economics. Fluid Fertilizer Foundation, February 12–14, 2006, Scottsdale, AZ [CD ROM].

Fowler, D.B. 2003. Crop nitrogen demand and grain protein concentration of spring and winter wheat. Agron. J. 95:260–265.

Fowler, D.B., J. Brydon, B.A. Darroch, M.H. Entz, and A.M. Johnston. 1990. Environment and genotype influence on grain protein concentration of wheat and rye. Agron. J. 82:655–664.

Francis, D.D., J.S. Schepers, and M.F. Vigil. 1993. Post-anthesis nitrogen loss from corn. Agron. J. 85:659–663.

Freeman, K.W., K. Girma, D.B. Arnall, R.W. Mullen, K.L. Martin, R.K. Teal, and W.R. Raun. 2007a. By-plant prediction of corn forage biomass and nitrogen accumulation at various growth stages using remote sensing and plant height measures. Agron. J. 99:530–536.

Freeman, K.W, K. Girma, R.K. Teal, D.B. Arnall, B. Tubana, S. Holtz, and W.R. Raun. 2007b. Long-term effects of nitrogen management practices on grain yield, nitrogen uptake and efficiency in irrigated corn. J. Plant Nutr. 30:2021–2036.

Geist, J.M., J.O. Reuss, and D.D. Johnson. 1970. Prediction of nitrogen fertilizer requirements of field crops. II. Application of theoretical models to malting barley. Agron. J. 62:385–389.

Girma, K., C. Mack, R. Taylor, J. Solie, D.B. Arnall, and W. Raun. 2007a. Improving estimation of N top-dressing by addressing temporal variability in winter wheat. J. Agric. Sci. 45:45–53.

Girma, K., S. Holtz, J.B. Solie, and W.R. Raun. 2011. Nitrogen accumulation in shoots as a function of growth stage of corn and winter wheat. J. Plant Nutr. 34:165–182.

Girma, K., K.L. Martin, J. Mosali, K.W. Freeman, R.K. Teal, S.M. Moges, D.B. Arnall, and W.R. Raun. 2007b. Determination of optimum rate and growth stage for foliar applied phosphorus in corn. Commun. Soil Sci. Plant Anal. 38:1137–1154.

Girma, K., W. Raun, H. Zhang, and J. Mosali. 2006. What about foliar P on corn and winter wheat? Fluid J. 14:17–19.

Girma, K., R.K. Teal, K.W. Freeman, R.K. Bowman, and W.R. Raun. 2007c. Cotton lint yield and quality as affected by applications of N, P, and K fertilizers. J. Cotton Sci. 11:12–19.

Giskin, M., J. Hagin, and U. Kafkafi. 1972a. Crop response to phosphate fertilization and to residual phosphate levels: I. Field experiments. Agron. J. 64:588–591.

Giskin, M., J. Hagin, and U. Kafkafi. 1972b. crop response to phosphate fertilization and to residual phosphate levels: II. Evaluation of residual phosphorus "availability" by chemical and plant tests in a greenhouse. Agron. J. 64:591–593.

Giskin, M., J. Hagin, and U. Kafkafi. 1972c. Crop response to phosphate fertilization and to residual phosphate levels: III. Greenhouse Experiment. Agron. J. 64:593–597.

Goos, R.J., and B.E. Johnson. 1999. Performance of two nitrification inhibitors over a winter with exceptionally heavy snowfall. Agron. J. 91:1046–1049.

Gourley, C.J.P., D.L. Allan, and M.P. Russelle. 1994. Plant nutrient efficiency: A comparison of definitions and suggested improvement. Plant Soil 158:29–37.

Guerra, L.C., S.I. Bhuiyan, T.P. Tuong, and R. Barker. 1998. Producing more rice with less water from irrigated systems. SWIM Paper 5. International Water Management Institute. Colombo, Sri Lanka.

Guillard, K., G.F. Griffin, D.W. Allinson, M.M. Rafey, W.R. Yamartino, and S.W. Pietrzyk. 1995. Nitrogen utilization of selected cropping systems in the U.S. Northeast: I. Dry matter yield, N uptake, apparent N recovery, and N use efficiency. Agron. J. 87:193–199.

Hagin, J., and A. Lowengart. 1995. Fertilization for minimizing environmental pollution by fertilizer. Nutr. Cycl. Agroecosyst. 43:5–7.

Hagin, J., and A. Lowengart. 1999. Fertigation—State of the art. Fertiliser Society Proceedings No. 429. The International Fertilizer Society, York, U.K.

Halvorson, A.D., and C.A. Reule. 1994. Nitrogen fertilizer requirements in an annual dryland cropping system. Agron. J. 86:315–318.

Halvorson, A.D., D.C. Nielsen, and C.A. Reule. 2004. Nitrogen fertilization and rotation effects on no-till dryland wheat production. Agron. J. 96:1196–1201.

Hamid, A., and R.L. Mahler. 1994. The potential for volatilization losses of applied nitrogen fertilizers from northern Idaho soils. Commun. Soil. Sci. Plant Anal. 25:361–373.

Harder, H.J., R.E. Carlson, and R.H. Shaw. 1982. Corn grain yield and nutrient response to foliar fertilizer applied during grain fill. Agron. J. 74:106–110.

Hargrove, W.L., and D.E. Kissel. 1979. Ammonia volatilization from surface applications of urea in the field and laboratory. Soil Sci. Soc. Am. J. 43:359–363.

Harper, L.A., R.R. Sharpe, G.W. Langdale, and J.E. Giddens. 1987. Nitrogen cycling in a wheat crop: Soil, plant, and aerial nitrogen transport. Agron. J. 79:965–973.

Harrison, R.B., and F. Adams. 1987. Solubility characteristics of residual phosphate in a fertilized and limed Ultisol. Soil Sci. Soc. Am. J. 51:963–969.

Hatfield, J.L., T.J. Sauer, and J.H. Prueger. 2001. Managing soils to achieve greater water use efficiency: A review. Agron. J. 93:271–280.

Hauck, R.D., and K.K. Tanji. 1982. Nitrogen transfers and mass balance, p. 891–925. In F.J. Stevenson (ed.) Nitrogen in agricultural soils. ASA, CSSA, and SSSA, Madison, WI.

Hawkins, J.A., J.E. Sawyer, D.W. Barker, and J.P. Lundvall. 2007. Using relative chlorophyll meter values to determine nitrogen application rates for corn. Agron. J. 99:1034–1040.

Heitholt J.J., R.C. Johnson, and D.M. Ferris. 1991. Stomatal limitation to carbon dioxide assimilation in nitrogen and drought-stressed wheat. Crop Sci. 31:135–139.

Helyar, K.R. 1998. Sustaining soil fertility with particular reference to phosphorus. Field Crops Res. 56:187–195.

Heng, L.K., T. Hsiao, S. Evett, T. Howell, and P. Steduto. 2009. Validating the FAO AquaCrop model for irrigated and water deficient field maize. Agron. J. 101:488–498.

Hesterman, O.B., M.P. Russelle, C.C. Sheaffer, and G.H. Heichel. 1987. Nitrogen utilization from fertilizer and legume residues in legume-corn rotations. Agron. J. 79:726–731.

Howell, T.A. 2001. Enhancing water use efficiency in irrigated agriculture. Agron. J. 93:281–289.

Howell, T.A., J.A. Tolk, A.D. Schneider, and S.R. Evett. 1998. Evapotranspiration, yield, and water use efficiency of corn hybrids differing in maturity. Agron. J. 90:3–9.

Huang, W., D. Shank, and T.I. Hewitt. 1996. On-farm costs of reducing residual nitrogen on cropland vulnerable to nitrate leaching. Rev. Agric. Econ. 18:325–339.

Huffman, J.R. 1989. Effects of enhanced ammonium nitrogen availability for corn. J. Agron. Educ. 18:93–97.

Hussain, F., K.F. Bronson, Yadvinder-Singh, Bijay-Singh, and S. Peng. 2000. Use of chlorophyll meter sufficiency indices for nitrogen management of irrigated rice in Asia. Agron. J. 92:875–879.

Jacobsen, J.S., S.H. Lorbeer, H.A.R. Houlton, and G.R. Carlson. 1997. Reduced-till spring wheat response to fertilizer sources and placement methods. Commun. Soil Sci. Plant Anal. 28:1237–1244.

Jama, A.O., and M.J. Ottman. 1993. Timing of the first irrigation in corn and water stress conditioning. Agron. J. 85:1159–1164.

Jambert, C., D. Seca, and R. Delmas. 1997. Quantification of N-losses as NH_3, NO, and N_2O and N_2 from fertilized maize fields in southwestern France. Nutr. Cycl. Agroecosyst. 48:91–104.

Janssen, B.H., D.J. Lathwell, and J. Wolf. 1987. Modeling long-term crop response to fertilizer phosphorus. II. Comparison with field results. Agron. J. 79:452–458.

Jarvis, R.J., and M.D.A. Bolland. 1991. Lupin grain yields and fertilizer effectiveness are increased by banding superphosphate below the seed. Aust. J. Exp. Agric. 31:357–366.

Jemison, J.M., Jr., and R.H. Fox. 1994. Nitrate leaching from nitrogen fertilized and manured corn measured with zero tension pan lysimeters. J. Environ. Qual. 23:337–343.

Ji, S., and P.W. Unger. 2001. Soil water accumulation under different precipitation, potential evaporation, and straw mulch conditions. Soil Sci. Soc. Am. J. 65:442–448.

Johnson, G.V. 1991. General model for predicting crop response to fertilizer. Agron. J. 83:367–373.

Johnson, G.V., and W.R. Raun. 2003. Nitrogen response index as guide to fertilizer management. J. Plant Nutr. 26:249–262.

Jokela, W.E., and G.W. Randall. 1997. Fate of fertilizer nitrogen as affected by time and rate of application on corn. Soil Sci. Soc. Am. J. 61:1695–1703.

Jones, O.R., and T.W. Popham. 1997. Cropping and tillage systems for dryland grain production in the southern High Plains. Agron. J. 89:222–232.

Kamprath, E.J., R.H. Moll, and N. Rodriguez. 1982. Effects of nitrogen fertilization and recurrent selection on performance of hybrid populations of corn. Agron. J. 74:955–958.

Kanampiu, F.K., W.R. Raun, and G.V. Johnson. 1997. Effect of nitrogen rate on plant nitrogen loss in winter wheat varieties. J. Plant Nutr. 20:389–404.

Kapoor, V., U. Singh, S.K. Patil, H. Magre, L.K. Shrivastava, V.N. Mishra, R.O. Das, V.K. Samadhiya, J. Sanabria, and R. Diamond. 2008. Rice growth, grain yield, and floodwater nutrient dynamics as affected by nutrient placement method and rate. Agron. J. 100:526–536.

Karlen, D.L., L.A. Kramer, and S.D. Logsdon. 1998. Field-scale nitrogen balances associated with long-term continuous corn production. Agron. J. 90:644–650.

Karrou, M., and J.W. Maranville. 1993. Seedling vigor and nitrogen use efficiency of Moroccan wheat as influenced by level of soil nitrogen. Commun. Soil Sci. Plant Anal. 24:1153–1163.

Kemper, W.D. 1993. Effects of soil properties on precipitation use efficiency. Irrig. Sci. 14:65–73.

Kennedy, A.C., and W.F. Schillinger. 2006. Soil quality and water intake in traditional-till vs. no-till paired farms in Washington's Palouse region. Soil Sci. Soc. Am. J. 70:940–949.

Khan, S.A., R.L. Mulvaney, and R.G. Hoeft. 2001. A simple soil test for detecting sites that are nonresponsive to nitrogen fertilization. Soil Sci. Soc. Am. J. 65:1751–1760.

Kolberg, R.L., N.R. Kitchen, D.G. Westfall, and G.A. Peterson. 1996. Cropping intensity and nitrogen management impact of dryland no-till rotations in the semi-arid western Great Plains. J. Prod. Agric. 9:517–522.

Kravchenko, A.G., and K.D. Thelen. 2007. Effect of winter wheat crop residue on no-till corn growth and development. Agron. J. 99:549–555.

Laboski, C.A.M., J.E. Sawyer, D.T. Walters, L.G. Bundy, R.G. Hoeft, G.W. Randall, and T.W. Andraski. 2008. Evaluation of the Illinois soil nitrogen test in the North Central Region of the United States. Agron. J. 100:1070–1076.

Ladha, J.K., H. Pathak, T.J. Krupnik, J. Six, and C. van Kessel. 2005. Efficiency of fertilizer nitrogen in cereal production: Retrospects and prospects. Adv. Agron. 87:85–156.

Lees, H.L., W.R. Raun, and G.V. Johnson. 2000. Increased plant N loss with increasing nitrogen applied in winter wheat observed with 15N. J. Plant Nutr. 23:219–230.

Li, F., Y. Miao, F. Zhang, Z. Cui, R. Li, X. Chen, H. Zhang, J. Schroder, W.R. Raun, and L. Jia. 2009. In-season optical sensing improves nitrogen-use efficiency for winter wheat. Soil Sci. Soc. Am. J. 73:1566–1574.

Liang, X.Q., H. Li, M.M. He, Y.X. Chen, G.M. Tian, and S.Y. Xu. 2008. The ecologically optimum application of nitrogen in wheat season of rice wheat cropping system. Agron. J. 100:67–72.

Licht, M.A., and M. Al-Kaisi. 2005. Corn response, nitrogen uptake, and water use in strip-tillage compared with no-tillage and chisel plow. Agron. J. 97:705–710.

Lohry, R. 2001. When banding, fluids are the way to go. Fluid J. 9:14–16.

López-Bellido, L., R.J. López-Bellido, and F.J. López-Bellido. 2006. Fertilizer nitrogen efficiency in durum wheat under rainfed Mediterranean conditions: Effect of split application. Agron. J. 98:55–62.

Lukina, E.V., K.W. Freeman, K.J. Wynn, W.E. Thomason, R.W. Mullen, A.R. Klatt, G.V. Johnson, R.L. Elliott, M.L. Stone, J.B. Solie, and W.R. Raun. 2001. Nitrogen fertilization optimization algorithm based on in-season estimates of yield and plant nitrogen uptake. J. Plant Nutr. 24:885–898.

Ma, B.L., L.M. Dwyer, and E.G. Gregorich. 1999. Soil nitrogen amendment effects on nitrogen uptake and grain yield of maize. Agron. J. 91:650–656.

Ma, W., Z. Mao, Z. Yu, M.E.F. van Mensvoort, and P.M. Driessen. 2008. Effects of saline water irrigation on soil salinity and yield of winter wheat–maize in North China Plain. Irrig. Drain. Syst. 22:3–18.

Machado, S., S. Petrie, K. Rhinhart, and R.E. Ramig. 2008. Tillage effects on water use and grain yield of winter wheat and green pea in rotation. Agron. J. 100:154–162.

Mahler, R.L., and A. Hamid. 1994. Evaluation of water potential, fertilizer placement and incubation time on volatilization losses of urea in two northern Idaho soils. Commun. Soil Sci. Plant Anal. 25:1991–2004.

Mahler, R.L., F.E. Koehler, and L.K. Lutcher. 1994. Nitrogen source, timing of application, and placement: Effects on winter wheat production. Agron. J. 86:637–642.

Makowski, D., D. Wallach, and J.M. Meynard. 2001. Statistical methods for predicting responses to applied nitrogen and calculating optimal nitrogen rates. Agron. J. 93:531–539.

Marriott, E.E., and M.M. Wander. 2006. Total and labile soil organic matter in organic and conventional farming systems. Soil Sci. Soc. Am. J. 70:950–959.

Meisinger, J.J., V.A. Bandel, J.S. Angle, B.E. Okeefe, and C.M. Reynolds. 1992. Presidedress soil nitrate test evaluation in Maryland. Soil Sci. Soc. Am. J. 56:1527–1532.

Mmolawa, K., and D. Or. 2000. Root zone dynamics under drip irrigation: A review. Plant Soil 222:163–190.

Moll, R.H., E.J. Kamprath, and W.A. Jackson. 1982. Analysis and interpretation of factors which contribute to efficiency to nitrogen utilization. Agron. J. 74:562–564.

Morison, J.I.L., N.R. Baker, P.M. Mullineaux, and W.J. Davies. 2008. Improving water use in crop production. Phil. Trans. R. Soc. A 363:639–658.

Mosali, J., K. Desta, R.K. Teal, K.W. Freeman, K.L. Martin, J.W. Lawles, and W.R. Raun. 2006. Effect of foliar application of phosphorus on winter wheat grain yield, phosphorus uptake, and use efficiency. J. Plant Nutr. 29:2147–2163.

Mosier, A.R., J.K. Syers, and J.R. Freney. 2004. Agriculture and the nitrogen cycle. Assessing the impacts of fertilizer use on food production and the environment. Scope-65. Island Press, London, U.K.

Muchow, R.C. 1998. Nitrogen utilization efficiency in maize and grain sorghum. Field Crops Res. 56:209–216.

Mueller, L., A. Behrendt, G. Schalitz, and U. Schindler. 2005. Above ground biomass and water use efficiency of crops at shallow water tables in a temperate climate. Agric. Water Manage. 75:117–136.

Mullen, R.W., K.W. Freeman, W.R. Raun, G.V. Johnson, M.L. Stone, and J.B. Solie. 2003. Identifying an in-season response index and midseason fertilizer N. Agron. J. 95:347–351.

Müller, A., J. Schmidhuber, J. Hoogeveen, and P. Steduto. 2008. Some insights in the effect of growing bio-energy demand on global food security and natural resources. Water Policy 10(Suppl.):83–94.

Mulvaney, R.L., S.A. Khan, and T.R. Ellsworth. 2005. Need for a soil-based approach in managing nitrogen fertilizers for profitable corn production. Soil Sci. Soc. Am. J. 70:172–182.

Mulvaney, R.L., S.A. Khan, R.G. Hoeft, and H.M. Brown. 2001. A soil organic nitrogen fraction that reduces the need for nitrogen fertilization. Soil Sci. Soc. Am. J. 65:1164–1172.

Murphy, J.A., and D.E. Zaurov. 1994. Shoot and root growth response of perennial ryegrass to fertilizer placement depth. Agron. J. 86:828–832.

Neeteson, J.J., and W.P. Wadman. 1987. Assessment of economically optimum application rates of fertilizer N on the basis of response curves. Nutr. Cycl. Agroecosyst. 12:37–52.

Nielsen, D.C., P.W. Unger, and P.R. Miller. 2005. Efficient water use in dryland cropping systems in the Great Plains. Agron. J. 97:364–372.

Norwood, C.A. 1999. Water use and yield of dryland row crops as affected by tillage. Agron. J. 91:108–115.

Norwood, C.A. 2000. Water use and yield of limited-irrigated and dryland corn. Soil Sci. Soc. Am. J. 64:365–370.

Osborne, C.P., J. La Roche, R.L. Garcia, B.A. Kimball, G.W. Wall, P.J. Pinter, Jr., R.L. La Morte, G.R. Hendrey, and S.P. Long. 1998. Does leaf position within a canopy affect acclimation of photosynthesis to elevated CO_2? Analysis of a wheat crop under free-air CO_2 enrichment. Plant Physiol. 117:1037–1045.

Osterhaus, J.T., L.G. Bundy, and T.W. Andraski. 2008. Evaluation of the Illinois soil nitrogen test for predicting corn nitrogen needs. Soil Sci. Soc. Am. J. 72:143–150.

Pan, W.L., E.J. Kamprath, R.H. Moll, and W.A. Jackson. 1984. Prolificacy in corn: Its effects on nitrate and ammonium uptake and utilization. Soil Sci. Soc. Am. J. 48:1101–1106.

Payne, W.A. 2000. Optimizing crop water use in sparse stands of pearl millet. Agron. J. 92:808–814.

Payne, W.A., M.C. Drew, L.R. Hossner, R.J. Lascano, A.B. Onken, and C.W. Wendt. 1992. Soil phosphorus availability and pearl millet water-use efficiency. Crop Sci. 32:1010–1015.

Payne, W.A., L.R. Hossner, A.B. Onken, and C.W. Wendt. 1995. Nitrogen and phosphorus uptake in pearl millet and its relation to nutrient and transpiration efficiency. Agron. J. 87:425–431.

Peterson, G.A., D.H. Sander, P.H. Grabouski, and M.L. Hooker. 1981. A New look at row and broadcast phosphate recommendations for winter wheat. Agron. J. 73:13–17.

Piekkielek, W.P., and R.H. Fox. 1992. Use of a chlorophyll meter to predict sidedress nitrogen. Agron. J. 84:59–65.

Pikul, J.L., Jr., L. Hammack, and W.E. Riedell. 2005. Corn yield, nitrogen use, and corn rootworm infestation of rotations in the Northern Corn Belt. Agron. J. 97:854–863.

Proffitt, A.P.B., P.R. Berliner, and D.M. Oosterhuis. 1985. A comparative study of root distribution and water extraction efficiency by wheat grown under high- and low-frequency irrigation. Agron. J. 77:655–662.

Pu, G., P.G. Saffigna, and W.M. Strong. 1999. Potential for denitrification in cereal soils of northern Australia after legume or grass-legume pasture. Soil Biol. Biochem. 31:667–675.

Raine, S.R., W.S. Meyer, D.W. Rassam, J.L. Hutson, and F.J. Cook. 2007. Soil–water and solute movement under precision irrigation: Knowledge gaps for managing sustainable root zones. Irrig. Sci. 26:91–100.

Randall, G.W., T.K. Iragavarapu, and B.R. Bock. 1997. Nitrogen application methods and timing for corn after soybean in a ridge-tillage system. J. Prod. Agric. 10:300–307.

Randall, G.W., and D.J. Mulla. 2001. Nitrate-N in surface waters as influenced by climatic conditions and agricultural practices. J. Environ. Qual. 30:337–344.

Rao, S.C., and T.H. Dao. 1996. Nitrogen placement and tillage effects on dry matter and nitrogen accumulation and redistribution in winter wheat. Agron. J. 88:365–371.

Raun, W.R., and G.V. Johnson. 1999. Improving nitrogen use efficiency for cereal production. Agron. J. 91:357–363.

Raun, W.R., G.V. Johnson, H. Sembiring, E.V. Lukina, J.M. LaRuffa, W.E. Thomason, S.B. Phillips, J.B. Solie, M.L. Stone, and R.W. Whitney. 1998. Indirect measures of plant nutrients. Commun. Soil Sci. Plant Anal. 29:1571–1581.

Raun, W.R., I. Ortiz-Monasterio, and J.B. Solie. 2009. Temporally and spatially dependent nitrogen management for diverse environments. *In* Brett F. Carver (ed.) Wheat: Science and trade. Wiley-Blackwell, Ames, IA.

Raun, W.R., J.B. Solie, G.V. Johnson, M.L. Stone, E.V. Lukina, W.E. Thomason, and J.S. Schepers. 2001. In-season prediction of potential grain yield in winter wheat using canopy reflectance. Agron. J. 93:131–138.

Raun, W.R., J.B. Solie, G.V. Johnson, M.L. Stone, R.W. Mullen, K.W. Freeman, W.E. Thomason, and E.V. Lukina. 2002. Improving nitrogen use efficiency in cereal grain production with optical sensing and variable rate application. Agron. J. 94:815–820.

Raun, W.R., J.B. Solie, R.K. Taylor, D.B. Arnall, C.J. Mack, and D.E. Edmonds. 2008. Ramp calibration strip technology for determining midseason nitrogen rates in corn and wheat. Agron. J. 100:1088–1093.

Richards, R.A. 2004. Physiological traits used in the breeding of new cultivars for water scarce environments: New directions for diverse planet. *In* T. Fischer, N. Turner, J. Angus, L. McIntyre, M. Robertson, A. Borrell, and D. Lloyd (eds.) Proc. 4th Int. Crop Sci. Congr. Brisbane, Australia, Sep. 26–Oct. 1, 2004. The Regional Institute Ltd, Gosford, New South Wales, Australia.

Richards, R.A., G.J. Rebetzke, A.G. Condon, and A.F. van Herwaarden. 2002. Breeding opportunities for increasing the efficiency of water use and crop yield in temperate cereals. Crop Sci. 42:111–121.

Roberts, T.L. 2008. Improving nutrient use efficiency. Turk J. Agric. For. 32:177–182.

Salsac, L., S. Chaillou, J.F. Morot-Gaudry, C. Lesaint, and E. Jolivoe. 1987. Nitrate and ammonium nutrition in plants. Plant Physiol. Biochem. 25:805–812.

Samborski, S.M., N. Tremblay, and E. Fallon. 2009. Strategies to make use of plant sensors-based diagnostic information for nitrogen recommendations. Agron. J. 101:800–816.

Sanchez, C.A., and A.M. Blackmer. 1988. Recovery of anhydrous ammonia derived nitrogen-15 during three years of corn production in Iowa. Agron. J. 80:102–108.

Sanchez, C.A., P.S. Porter, and M.F. Ulloa. 1991. Relative efficiency of broadcast and banded phosphorus for sweet corn produced on Histosols. Soil Sci. Soc. Am. J. 55:871–875.

Sander, D.H., E. J. Penas, and B. Eghball. 1990. Residual effects of various phosphorus application methods on winter wheat and grain sorghum. Soil Sci. Soc. Am. J. 54:1473–1478.

Sander, D.H., E.J. Penas, and D.T. Walters. 1991. Winter wheat phosphorus fertilization as influenced by glacial till and loess soils. Soil Sci. Soc. Am. J. 55:1474–1479.

Schakel, K.A., and A.E. Hall. 1979. Reversible leaflet movements in relation to drought adaptation of cowpeas, *Vigna unguiculata* (L.) Walp. Aust. J. Plant Physiol. 6:265–276.

Scharf, P.C., S.M. Brouder, and R.G. Hoeft. 2006. Chlorophyll meter readings can predict nitrogen need and yield response of corn in the North-Central USA. Agron. J. 98:655–665.

Schepers, J.S., D.D. Francis, M. Vigil, and F.E. Below. 1992. Comparison of corn leaf nitrogen concentration and chlorophyll meter readings. Commun. Soil Sci. Plant Anal. 23:2173–2187.

Schlegel, A.J., D.W. Nelson, and L.E. Sommers. 1986. Field evaluation of urease inhibitors for corn production. Agron. J. 78:1007–1012.

Schomberg, H.H., R.G. McDaniel, E. Mallard, D.M. Endale, D.S. Fisher, and M.L. Cabrera. 2006. Conservation tillage and cover crop influences on cotton production on a Southeastern U.S. Coastal Plain soil. Agron. J. 98:1247–1256.

Selles, F., R.P. Zentner, D.W.L. Read, and C.A. Campbell. 1992. Prediction of fertilizer requirements for spring wheat grown on stubble in southwestern Saskatchewan. Can. J. Soil Sci. 72:229–241.

Shanahan, J.F., J.S. Schepers, D.D. Francis, and R. Caldwell. 2004. Use of crop canopy reflectance sensor for in-season N management of corn. Proc. Great Plains Soil Fertil. Conf. 10:69–74.

Sharpley, A.N., and I. Sisak. 1997. Differential availability of manure and inorganic sources of phosphorus in soil. Soil Sci. Soc. Am. J. 61:1503–1508.

Shaviv, A., and R.L. Mikkelsen. 1993. Controlled-release fertilizers to increase efficiency of nutrient use and minimize environmental degradation—A review. Nutr. Cycl. Agroecosyst. 35:1–12.

Shi, W., and J.M. Norton. 2000. Effect of long-term, biennial, fall-applied anhydrous ammonia and nitrapyrin on soil nitrification. Soil Sci. Soc. Am. J. 64:228–234.

Six, J., E.T. Elliott, K. Paustian, and J.W. Doran. 1998. Aggregation and soil organic matter accumulation in cultivated and native grassland soils. Soil Sci. Soc. Am. J. 62:1367–1377.

Slaton, N.A., C.E. Wilson, Jr., R.J. Norman, S. Ntamatungiro, and D.L. Frizzell. 2002. Rice response to phosphorus fertilizer application rate and timing on alkaline soils in Arkansas. Agron. J. 94:1393–1399.

Solari, F., J. Shanahan, R. Ferguson, J. Schepers, and A. Gitelson. 2008. Active sensor reflectance measurements of corn nitrogen status and yield potential. Agron. J. 100:571–579.

Solie, J.B., W.R. Raun, R.W. Whitney, M.L. Stone, and J.D. Ringer. 1996. Optical sensor based field element size and sensing strategy for nitrogen application. Trans. ASAE 39:1983–1992.

Sollins, P., G.P. Robertson, and G. Uehara. 1988. Nutrient mobility in variable- and permanent-charge soils. Biogeochem. J. 3:181–199.

Soltanpour, P.N., M. El Gharous, A. Azzaoui, and M. Abdelmonem. 1989. A Soil test based N recommendation model for dryland wheat. Commun. Soil Sci. Plant Anal. 20:1053–1068.

Sowers, K.E., W.L. Pan, B.C. Miller, and J.L. Smith. 1994. Nitrogen use efficiency of split nitrogen applications in soft white winter wheat. Agron. J. 86:942–948.

Spargo, J.T., M.M. Alley, W.E. Thomason, and S.M. Nagle. 2009. Illinois soil nitrogen test for prediction of fertilizer nitrogen needs of corn in Virginia. Soil Sci. Soc. Am. J. 73:434–442.

Sripada, R.P., J.P. Schmidt, A.E. Dellinger, and D.B. Beegle. 2008. Evaluating multiple indices from a canopy reflectance sensor to estimate corn N requirements. Agron. J. 100:1553–1561.

Stone, M.L., J.B. Solie, W.R. Raun, R.W. Whitney, S.L. Taylor, and J.D. Ringer. 1996. Use of spectral radiance for correcting in-season fertilizer nitrogen deficiencies in winter wheat. Trans. ASAE 39:1623–1631.

Stout, W.L., G.A. Jung, and J.A. Shaffer. 1988. Effects of soil and nitrogen on water use efficiency of tall fescue and switchgrass under humid conditions. Soil Sci. Soc. Am. J. 52:429–434.

Stout, W.L., S.L. Fales, L.D. Muller, R.R. Schnabel, and S.R. Weaver. 2000. Water quality implications of nitrate leaching from intensively grazed pasture swards in the northeast US. Agric. Ecosyst. Environ. 77:203–210.

Tanaka, D.L., R.L. Anderson, and S.C. Rao. 2005. Crop sequencing to improve use of precipitation and synergize crop growth. Agron. J. 97:385–390.

Tanner, C.B., and W.A. Jury. 1976. Estimating evaporation and transpiration from a row crop during incomplete cover. Agron. J. 68:239–243.

Teal, R.K., B. Tubana, K. Girma, K.W. Freeman, D.B. Arnall, O. Walsh, and W.R. Raun. 2006. In-season prediction of corn grain yield potential using normalized difference vegetation index. Agron. J. 98:1488–1494.

Tompkins, D.K., D.B. Fowler, and A.T. Wright. 1991. Water Use by no-till winter wheat influence of seed rate and row spacing. Agron. J. 83:766–769.

Tsai, C.Y., I. Dweikat, D.M. Huber, and H.L. Warren. 1992. Interrelationship of nitrogen nutrition with maize (*Zea mays*) grain yield, nitrogen use efficiency and grain quality. J. Sci. Food Agric. 58:1–8.

Turner, F.T., and M.F. Jund. 1991. Chlorophyll meter to predict nitrogen topdress requirement for semidwarf rice. Agron. J. 83:926–928.

Unger, P.W. 1991. Ontogeny and water use of no-tillage sorghum cultivars on dryland. Agron. J. 83:961–968.

Unger, P.W., and O.R. Jones. 1981. Effect of soil water content and a growing season straw mulch on grain sorghum. Soil Sci. Soc. Am. J. 45:129–134.

van der Ploeg, R.R., W. Böhm, and M.B. Kirkham. 1999. On the origin of the theory of mineral nutrition of plants and the Law of the Minimum. Soil Sci. Soc. Am. J. 63:1055–1062.

Varvel, G.E. 1994. Monoculture and rotation system effects on precipitation use efficiency in corn. Agron. J. 86:204–208.

Viets, F.G., Jr. 1962. Fertilizers and the efficient use of water. Adv. Agron. 14:223–264.

Walsh, O.S. 2006. Effect of delayed nitrogen fertilization on corn grain yields. M.S. Thesis. Oklahoma State University. Stillwater, OK.

Walters, D.T., and G.L. Malzer. 1990. Nitrogen management and nitrification inhibitor effects on nitrogen-15 urea: I. Yield and fertilizer use efficiency. Soil Sci. Soc. Am. J. 54:115–122.

Wang, X., and F.E. Below. 1992. Root growth, nitrogen uptake, and tillering of wheat induced by mixed-nitrogen source. Crop Sci. 32:997–1002.

Waskom, R.M., D.G. Westfall, D.E. Spellman, and P.N. Soltanpour. 1996. Monitoring nitrogen status of corn with a portable chlorophyll meter. Commun. Soil Sci. Plant. Anal. 27:545–560.

Williams, J.D., C.R. Crozier, J.G. White, R.W. Heiniger, R.P. Sripada, and D.A. Crouse. 2007. Illinois soil nitrogen test predicts southeastern U.S. corn economic optimum nitrogen rates. Soil Sci. Soc. Am. J. 71:735–744.

Wittwer, S.H., M.J. Bukovac, and H.B. Tukey. 1963. Advances in foliar feeding of plant nutrients, p. 429–455. *In* M.H. McVickar (ed.) Fertilizer technology and usage. SSSA, Madison, WI.

World Bank. 2007. World development indictors: Data and statistics, agriculture. Available online at: http://siteresources.worldbank.org/DATASTATISTICS/Resources/wdi3_2.pdf

Wuest, S.B., and K.G. Cassman. 1992. Fertilizer-nitrogen use efficiency of irrigated wheat: II. Partitioning efficiency of pre-plant versus late-season application. Agron. J. 84:689–694.

WWAP (World Water Assessment Program). 2009. The United Nations world water development report 3: Water in a changing world. UNESCO, Paris, France, and Earthscan, London, U.K.

Zhang, J., A.M. Blackmer, J.W. Ellsworth, and K.J. Koehler. 2008. Sensitivity of chlorophyll meters for diagnosing nitrogen deficiencies of corn in production agriculture. Agron. J. 100:543–550.

Zhang, X., S. Chen, M. Liu, D. Pei, and H. Sun. 2005. Improved water use efficiency associated with cultivars and agronomic management in the north china plain. Agron. J. 97:783–790.

Zhang, S.W., C.H. Li, J. Cao, Y.C. Zhang, S.Q. Zhang, Y.F Xia, D.Y. Sun, and Y. Sun. 2009. Altered architecture and enhanced drought tolerance in rice via the down-regulation of indole-3-acetic acid by *TLD1/OsGH3.13* activation. Plant Physiol. 151:1889–1901.

16
Nutrient Interactions in Soil Fertility and Plant Nutrition

William L. Pan
Washington State University

16.1	Historical Perspective 16-1
16.2	Conceptual and Quantitative Frameworks of Nutrient Interaction Mechanisms 16-3
	Nutrient Interactions at the Soil Solid–Solution Interface • Specific Nutrient Interactions at the Solution–Root Interface • Cation–Cation Interactions • Cation–Anion Interactions • Anion–Anion Interactions • Secondary, Nonspecific Nutrient Interactions
16.3	Conclusions 16-9
	References 16-9

Nutrients interact during numerous biological, physical, or chemical processes along the soil–root–shoot continuum such that the level of one nutrient alters the availability, uptake, or plant response to another nutrient (Zhang et al., 2006). Given the very nature of ionic balance in plant nutrition and the ionic interactions in soil chemistry and fertility, virtually all soil–plant processes involve nutrient interactions, making a comprehensive review a boundless task. This review will focus on key concepts and provide examples to illustrate them. The availability and flow of any given nutrient through the continuum are constantly influenced by the activity and form of other nutrients, and the nature and rate of these interactions are sufficiently complex that they are sometimes beyond our current ability to measure and quantify their impacts. The nutrient interactions that are most often recognized and documented have for the most part been observed in soil fertility experiments that characterize crop nutrient uptake and growth response to one or more nutrients added to the system. Thus, this review will focus on the examples of nutrient and other elemental (e.g., heavy metals) interactions documented in the soil fertility literature, with subsequent discussion of plausible chemical and biological mechanisms of those same interactions along the soil–root–shoot continuum as described in the soil chemistry, soil microbiology, and plant nutrition literature.

Nutrient interactions will be delineated as to their specificity, specific or primary interactions being those in which two nutrients directly react in a chemical or biological process. Nonspecific or secondary nutrient interactions occur when the uptake of one nutrient is indirectly affected by the activity of another nutrient through a series of intermediate plant processes.

16.1 Historical Perspective

Our modern concepts of plant nutrient essentiality and their interactions coevolved with the development of modern chemical principles, as described in reviews by Browne (1944), Ihde (1964), Tisdale et al. (1985), Wild (1988), Black (1993), and Epstein and Bloom (2005). The Law of the Minimum (LM), popularized by Justus von Liebig in the mid-nineteenth century but first succinctly stated by Carl Sprengel, recognized a basic fundamental nature of interactions among these essential nutrients. Sprengel conducted experiments on mineral salts extracted from organic matter and soils and published a list of mineral elements essential for plant growth (van der Ploeg et al., 1999). Furthermore, he wrote in 1828 "…when a plant needs 12 substances to develop, it will not grow if any one of these is missing, and it will grow poorly, when one of these is not available in a sufficiently large amount as required by the nature of the plant."

Liebig subsequently stated in 1855 as part of his LM that "by the deficiency or absence of one necessary constituent, all the others being present, the soil is rendered barren for all crops to the life of which that constituent is indispensible" (Wild, 1988). These pioneers of plant nutrition research established a basic concept of nutrient codependence.

The LM and the concept of plant response to a most limiting nutrient was later modeled by Mitscherlich when he published his "Law of Diminishing Returns" (Mitscherlich, 1909). The model depicts a plant yield (y) exponential response to increasing nutrient availability from fertilizer and soil (x + b) that rises to a plateau (the shape of the rise defined by constant [c] as it nears a maximum potential yield [A]). As Black (1993) summarized, while many other empirical polynomial models have been used to represent yield response to nutrient availability, the

Mitscherlich model is unique in that the variables represent biological concepts. The nonfertilizer soil nutrient supply (b), the nutrient use efficiency represented by the rise to the maximum (c), and the maximum yield (A) are all influenced by the status of other growth factors, including other essential nutrients (Figure 16.1).

Comparative analysis of two-way nutrient responses provides insight into the nature of interactions between those nutrients. Historically, there has been considerable debate whether "c" is constant for a given species response to a specific nutrient (N1). Mitscherlich contended that "c" was in fact inherently constant, but van der Pauw (1952) statistically reanalyzed Mitscherlich's data and found a sizable variation in "c" values. Furthermore, subsequent experiments provide compelling evidence and argument for variable "c" values for a given nutrient, notably when a second nutrient (N2) interacts to change the true plant availability of the nutrient N1 or N2 competes with N1 for uptake (see Black, 1993). In retrospect, these two possibilities of having constant or variable "c" need not be mutually exclusive. It is plausible that the alteration of factors "c," "A," or both depends on the nature of the interaction between growth factors and whether they are primary or secondary in the nature.

For the purpose of this review, primary nutrient interactions are defined as specific chemical and biochemical interactions between two nutrients occurring in the bulk soil, soil solution, rhizosphere, or root cell surface that affects the activity, bioavailability, or ion transport of one or both nutrients. In contrast, secondary nutrient interactions are defined as nonspecific nutrient interactions mediated by general physiological responses such as compensatory root growth (Drew and Saker, 1975) to one nutrient that indirectly influence the uptake and assimilation of a second nutrient. Examples of primary and secondary nutrient interactions are listed in Table 16.1.

The constancy or variability of "A" is dependent on whether the second nutrient N2 is supplied in limiting or nonlimiting levels. Obviously, if N2 is nonyield limiting at all levels, then maximum yield "A" will remain unaffected by variation in N2. The constancy or variability of "c" is dependent on the specific or nonspecific nutrient interactions that can alter the nutrient use efficiency of N1. For example, changes in "c" in response to the level of N2 may reflect specific primary chemical interactions affecting nutrient activities that represent their immediate bioavailability as reflected by the ionic activity of the absorbed nutrient form, and how changes would affect nutrient uptake and utilization. Nonspecific secondary interactions, mediated through indirect effects on general plant growth, root, or rhizosphere bacterial activities, might also alter "c" by changing the plant's overall demand for N1 or by changing the N1 nutrient uptake or utilization efficiency. Four hypothetical scenarios, portrayed in Figure 16.1 with modeled yield responses to nutrient 1 (N1) as affected by nutrient 2 (N2) levels, are represented by Mitscherlich equations.

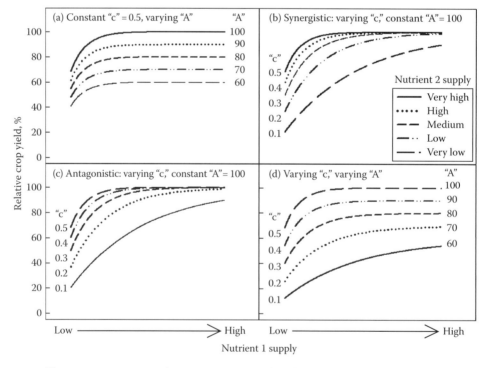

FIGURE 16.1 Depiction of four interaction scenarios between two nutrients (1 and 2) as modeled with the Mitscherlich equation. The legend in Panel (b) indicates level of nutrient 2 supply for all panels. Values for "A" and "c" for each level of nutrient 2 are included in each panel where appropriate. Panel (a): The classic "Mitscherlich" response depicting proportional yield increases in response to incremental levels of nutrients 1 and 2. Panel (b): Increasing levels of nutrient 2 increase the utilization efficiency of nutrient 1 (synergistic). Panel (c): Increasing levels of nutrient 2 decrease the utilization efficiency of nutrient 1 (antagonistic). Panel d: The classic "Law of the Minimum" response, where nutrient 1 is initially most limiting thus nutrient 2 has no effect on crop yield; but with increasing levels of nutrient 1, nutrient 2 has increasing impact on crop yield.

TABLE 16.1 Examples of Primary (Specific) and Secondary (Nonspecific) Nutrient Interactions

Primary	Secondary
Cation–anion exchange	Increased yield potential and nutrient demand
Cation–anion precipitation	Altered nutrient uptake or utilization efficiency
Ion pairing	Modification of rhizosphere chemistry, biology
Ion uptake synergism or antagonism	Modification of soil solution ionic strength

Scenario 1 (Figure 16.1a) depicts Mitscherlich's concept of variable "A" and constant "c." This scenario occurs when N1 and N2 are both initially present at yield-limiting levels, responding simultaneously to incremental increases in both nutrients. Since N2 does not affect N1 bioavailability or uptake and utilization efficiencies, the shape of the rise ("c") to "A" is unaffected. This type of response is referred to as the "Mitscherlich (MTS) response" (Zhang et al., 2007).

Scenario 2 (Figure 16.1b) depicts a specific interaction affecting "c" but not "A" in which N2 levels specifically (e.g., ion uptake synergism) or nonspecifically (e.g., increased root proliferation) increase "c" by improving the N1 uptake or utilization efficiency, while "A" remains unchanged.

Scenario 3 (Figure 16.1c) represents the opposite situation of scenario 2, whereby an increased level of nutrient N2 inhibits the uptake of nutrient N1 due to specific ion antagonism at the N1 influx porter or due to specific or nonspecific reduction in N1 bioavailability. Nutrient 2 is present in nonyield limiting supply, so there is no change in "A," but more N1 supply is required to reach "A" due to the antagonism of N2, which reduces the N1 nutrient use efficiency and shifts "c" to a lower value.

Scenario 4 (Figure 16.1d) depicts situations in which both N1 and N2 are initially limiting, but if N1 is the most limiting nutrient, then the crop will primarily respond to N1 until it reaches a level in which it is no longer the most limiting nutrient and then the crop will respond to increases in N2. The "A" increases with N2 levels, but "c" values decrease. This represents the spirit of Liebig's LM, and has thus been referred to as the LM type response (Zhang et al., 2007) or the Liebig response.

Interactions amongst growth promoting factors were described by Bloom et al. (1985), in the context of a "Multiple Limitation Hypothesis (MLH)," to occur in resource allocation terms, whereby plants invest resources into homeostatic mechanisms for acquiring the most growth limiting resources it needs, at the expense of acquiring less limiting resources. Zhang et al. (2007) applied MLH to nutrient interactions by surmising that a response to a nutrient should increase with increased availability of other nutrients, similar to the MTS response, provided they are not toxic. Rubio et al. (2003) examined 60 responses of *Lemna minor* to combinations of N, P, K, and Mg, and found 23 LM responses, 17 MLH responses, and 20 undefined responses. Type of response was highly dependent on the nutrient pairings. In contrast, Wood et al. (1972) examined 69 experiments and found that 49 matched the LM response and only eight were classified as MTS. These researchers assumed that either of the two theories—MLH or MTS—(Figure 16.1a) or LM (Figure 16.1d) would represent the nature of all nutrient responses and nutrient interactions. However, in both studies there were undefined responses that could not be characterized by either theory. In retrospect, these undefined responses could be rationalized with scenarios 2 (Figure 16.1b) or 3 (Figure 16.1c) in which N1 use efficiencies are altered by specific or nonspecific nutrient interactions.

16.2 Conceptual and Quantitative Frameworks of Nutrient Interaction Mechanisms

Fried and Broeshart (1967) published a conceptual framework depicting nutrient flow through the soil solid–soil solution–root–shoot continuum. They contended root uptake would be the rate-limiting step at equilibrium, a supposition soon after contested (Nye, 1977) and largely disproven with numerous subsequent experiments (see Barber, 1995; Silberbush, 1996) demonstrating that nutrient delivery from and through the soil to the plant root is often rate limiting for sparingly soluble nutrients. A corollary to that concept would be that nutrient interactions affecting availability and mobility are most often important for relatively immobile nutrients such as P and K, whereas nutrient interactions affecting the root uptake and utilization processes are most often important for mobile nutrients such as Ca and NO_3-N.

The essential contribution of the Fried–Broeshart model was that it provided the conceptual basis for subsequent quantitative soil-nutrient uptake models (Baldwin and Nye, 1974; Claassen and Barber, 1976; Rengel, 1993; Barber, 1995; Silberbush, 1996; Claassen and Steingrobe, 1999; Hopmans and Bristow, 2002). These models attempt to simulate soil processes affecting nutrient solubility and movement to plant roots, integrated with simulations of root growth and nutrient uptake. For the sake of simplicity, most models are single nutrient based and have not included modules to account for the multitude of nutrient interactions that can occur along the continuum. Exceptions include a multi-ion uptake model published by Bouldin (1989) that stresses the importance of maintenance of electrical neutrality through balanced cation–anion movement through the system. Relative ratios of cations [K/(Ca + Mg)] on exchange sites and in solution were incorporated into the model. Koenig and Pan (1996b) demonstrated how single ion nutrient uptake models can be used to assess potential effects of soil nutrient interactions on crop nutrient uptake, by modeling cation exchange-driven shifts of quantity/intensity (Q/I) relationships that affect the bioavailability of the absorbed nutrient. Calcium exchange for NH_4 increases solution NH_4 concentrations, a limiting factor in NH_4 diffusion rates to and into plant roots (Table 16.2). This proposed mechanism for Ca-stimulated NH_4 uptake was suggested as an alternative hypothesis to prior theories on membrane level Ca-NH_4 synergies (Fenn et al., 1987; Fenn and Taylor, 1990).

TABLE 16.2 Soil Supply Parameters Used in the Simulation of NH$_4$ Uptake and Predicted NH$_4$ Uptake over a 10 d Period for Rice (*Oryza sativa* L.) and Corn (*Zea mays* L.) as Influenced by Ca

| | | | | | 10 Day N Uptake | |
| | C_s | C_l | B | D_e | Rice | Corn |
Condition	mg N kg^{-1}	mg N L^{-1}	L kg^{-1}	cm^2 s^{-1}	mg N	
−Ca	100	12	5.44	2.10×10^{-7}	415	27
+Ca	100	32	1.79	6.39×10^{-7}	506	29

Source: Koenig, R.T., and W.L. Pan. 1996b. The influence of calcium on ammonium quantity-intensity relationships and ammonium availability in soil. Soil Sci. Soc. Am. J. 60:492–497.

C_s, exchangeable NH$_4$; C_l, solution NH$_4$; b, soil buffer capacity for NH$_4$; D_e, effective diffusion coefficient for NH$_4$.

16.2.1 Nutrient Interactions at the Soil Solid–Solution Interface

Primary nutrient interactions occur at the different stages of the soil–plant continuum (Table 16.1), generally related to cation–cation, anion–anion, or cation–anion chemical interactions.

Chemical and biological interactions affecting nutrient equilibria and flow rates between the labile fraction of the soil solid and solution phases influence solution ionic activities and transport to plant roots (Degryse et al., 2009). Cation availability is reduced by adsorption to negatively charged organic matter, oxides, and clay minerals. Precipitation with reactive anions also reduces soluble cation availability. Sorption into the labile pool occurs within hours following addition, whereas slower "fixation" reactions reduce the labile pool. Fixation of K and NH$_4$ into interlayers of 2:1 layer silicates or "fixation" of metal cations during diffusion into hydroxides or precipitates, ion adsorption, and biological immobilization all remove nutrients from the soil solution, potentially reducing their immediate availability to the plant root. Conversely, depending on the reversibility of these reactions and the solubility of the compounds, those nutrients are held in reserve, preventing nutrient leaching. Upon depletion from the soil solution or with a shift in chemistry or biology affecting the equilibria, these nutrients can be solubilized later in the growing season. For example, the availability of soluble anions promotes cation desorption, solubility, and mobility (Sakuri and Huang, 1996).

The importance of ion composition of the soil solution including cationic micronutrients (McGeorge, 1924), elemental toxicities (Magistad, 1925), and P forms (Pierre and Parker, 1927) as a critical factor in plant response to soil conditions has been recognized for more than a century (Adams, 1974; Sparks, 1984). Solid phase–solution phase solubility, as affected by solubility products of salts and by ion exchange reactions were first characterized with solubility diagrams proposed by Aslyng (1954). Ion activities, as affected by nutrient interactions, were calculated according to the Debye–Huckel equation and related to root growth (Howard and Adams, 1965; Adams, 1966; Adams and Lund, 1966). Subsequently, recognition of nutrient interactions of ion pair formation refined estimates of ion activities (Bennett and Adams, 1970). Tedious iterative calculations of specific ion activities were automated in computer programs such as GEOCHEM (Sposito and Mattigod, 1980) and MINTEQ (Brown and Allison, 1987) despite the fact that these activities are as Sposito (1984) characterized, an immeasurable "thermodynamic illusion." Nevertheless, these models promoted more extensive research that demonstrated correlations between specific ion activities in soil solutions with root growth and ion uptake (Hough et al., 2005). Nutrient interactions in soil solutions that can affect specific ion activities are thereby manifested in this model by either contributing to or interfering with (1) the formation of insoluble precipitates or sorption of ion pairs (e.g., Bolan et al., 1993; Pearce and Sumner, 1997), (2) the formation of soluble ion pairs (e.g., Kinraide and Parker, 1987), or (3) a buildup of ionic strength, which inversely affects ion activities (e.g., Kalis et al., 2008).

16.2.2 Specific Nutrient Interactions at the Solution–Root Interface

Primary and secondary nutrient interactions in the rhizosphere involve both chemical and biological processes. The ion uptake models described above assume ion uptake by plant roots is related to ion concentration or ion activity in the soil solution of the rhizosphere. Thus, primary nutrient interactions in the rhizosphere affecting ion activities would affect plant uptake and plant response. Root metabolism resulting in differential depletion or accumulation of nutrients in the rhizosphere, rhizosphere pH modification, and root release of complexing organic compounds all potentially change the rhizosphere chemistry affecting ion activities (Marschner, 1995; Hinsinger, 1998). In addition, secondary nutrient effects on rhizosphere microbial activities also influence the forms, activities, and uptake of other nutrients.

Much of our knowledge of nutrient interactions at the root surface affecting ion uptake has been derived from solution-grown plants with varied nutrient compositions (Reid, 1999; Epstein and Bloom, 2005). Not coincidentally, ion uptake response to increasing ion activity in the rooting solution resembles crop response to increasing nutrient supply in the field. The diminishing rise to a maximum characterized by the MTS equation is reflective of similarly shaped ion uptake responses to nutrient supply at the root surface, typically characterized by single or multi-phased Michaelis–Menten kinetics. Likewise, some field level nutrient interactions have similar patterns of antagonism or synergism of ion uptake at cellular or root levels (see Epstein and Bloom, 2005). These interactions are expressed in changes in the shape of the rise to maximum, defined by K_m, and/or in changes in V_{max} (Wang et al., 1993). Like-charged ions tend to be antagonistic, while mobile counter-ions can facilitate greater uptake.

Molecular ion transport mechanisms of major nutrients have recently been reviewed (Miller et al., 2009). Families of membrane-bound protein transporters of N, P, and S have been identified, classified, and mapped to gene sequences. Plant regulation and control mechanisms of ion transporters coordinate relative uptake rates among nutrients to maintain internal nutrient balance during shifting nutrient availability regimes (Hesse et al., 2004).

16.2.3 Cation–Cation Interactions

Cation interactions that influence ion uptake rates occur at each stage of the soil–plant continuum. Cation exchange interactions with negatively charged clay and organic matter affect quantities held in reserve, solution cation activities (intensity), and flux rates to plant roots. The Q/I ratio is a measure of this relationship between solid and solution phase, and is a key parameter (buffer power, "b") of soil nutrient uptake models (Barber, 1995; Claassen and Steingrobe, 1999). The Barber and Cushman (1981) model predicted that when the total concentration of a cation in soil is held constant and buffer capacity is reduced in the presence of other exchangeable cations, then the cationic activity in solution increases, which, in turn, can drive an increase in cation uptake rates.

Cation exchange also occurs at the root surface on negative sites residing in the cell wall and plasmalemma. Inflow of cations through the root apoplasm by diffusion or mass flow is influenced by exchange properties of carboxylic groups of cell wall structural components such as polygalacturonic acid (Marschner, 1995). These exchange sites harbor a reserve pool of cations for membrane transport (Epstein and Leggett, 1954). Root cation exchange capacity (CEC) differs among species, and dicotyledonous plants are generally higher in exchange capacity than monocotyledonous plants. As in soils, selective binding of multivalent cations over monovalent cations occurs at the cell wall surface, increasing the concentration of divalent and trivalent cations in the apoplasm of species with high root CEC. The importance of this exchange phenomenon is inferred from correlations between root CEC and cation uptake selectivity differences among plant species (Crooke and Knight, 1962; Haynes, 1980). Conceptually, the cation selectivity on the root exchange sites influence the ratio of cations at the sites of membrane transport. Reid (1999) cautioned that research focused on cell wall electrostatic properties may have overshadowed more important charge properties of the plasmalemma. He argues that the membrane charge is more important due to proximity to the transport sites. Optimal calcium effects of increasing apparent K_m of Rb uptake by yeast cells (Gage et al., 1985) was offered as evidence for this influence of interacting cations on membrane adsorption and absorption. These electrostatic binding interactions in the cell wall and plasma membrane are thought to play an important role in trace element uptake and toxicity as well (Yermiyahu and Kinraide, 2005).

Calcium has long been recognized for its special synergistic role in facilitating ion uptake (Viets, 1944), which is commonly attributed to its positive maintenance of membrane integrity and subsequent membrane transport selectivity of other ions (Marschner, 1995; Epstein and Bloom, 2005). Membrane stabilization is achieved by calcium bridging of carboxylate and phosphate groups of the membrane phospholipids (Caldwell and Haug, 1981). Zinc nutrition has also been linked to membrane integrity and ion retention by roots (Welch et al., 1999). Ca and Mg can reduce the phytotoxicity of Cu, Zn, Al, Na, and H by reducing the electrostatic attraction of the root plasma membrane for these cations (Kinraide et al., 2004).

Commonly, a deficiency in one cationic nutrient will increase the uptake of other cations of similar properties. Direct interference of one cation on the porter-binding of a substrate cation has long been offered as a principle mechanism of cation antagonistic competition between like-charged cations (Epstein, 1972). Aluminum toxicity can be manifested as a primary mechanism of competitive inhibition of divalent cation uptake (Rengel and Robinson, 1989; Rengel, 1990) or secondary effects on root growth and general reduction of nutrient absorptive capacity (Foy, 1988). Primary interference of Al on Mg uptake is rapid (Rengel, 1990) and these types of cation interactions may also be attributed to altered surface charge potential of the cell wall (Gage et al., 1985). Antagonistic interactions among K/Na (Mohammadi et al., 2008; Luan et al., 2009), may relate to both transporter activity and synthesis (Yao et al., 2010). In the field, low K can result in increased Na uptake (Yoshida and Castenada, 1969). Antagonism between NH_4/K is also well recognized during the seedling (Vale et al., 1987; Hoopen et al., 2010) and reproductive (Pan et al., 1986) stages of cereal crops, with this interaction also involving a competition between the two cations for porter sites (Hoopen et al., 2010).

Cation–cation interactions also occur during internal plant transport and mobilization processes (Morgan and Jackson, 1976). For example, excess K is well known to suppress forage Mg accumulation, due to altered Mg translocation from root to shoot (Ohno and Grunes, 1985). Welch et al. (1999) demonstrated that Zn sufficient wheat had lower phloem movement of Cd to the grain compared to Zn deficient plants. Similar observations were made of Zn influences on Mn movement in root phloem (Pearson and Rengel, 1995).

Cation–cation antagonism in micronutrient and heavy metal uptake is also well documented. Calcium-inhibited Zn uptake and translocation was noted in wheat (Hart et al., 1998). Cadmium uptake can be increased (Girling and Peterson, 1981) or decreased (Honma and Hirata, 1978; Cataldo et al., 1983; Keltjens and van Beuschichem, 1998) by the addition of other metals depending on the resulting ratios of Cd and competing metals in solution. Rice accumulated more Fe and Mn under Zn deficient conditions (Sajwan and Lindsay, 1986), and Mn uptake by this crop has also been correlated to Mn/Fe activity ratios, suggesting an antagonism between Mn and Fe (Moore and Patrick, 1989). Excess Cu-induced Fe deficiency in wheat (Aurelia et al., 2008). Zinc deficiency stimulated root exudation and enhanced mobilization of both Zn and Cu that might stimulate cationic micronutrient uptake (Degryse et al., 2008).

16.2.4 Cation–Anion Interactions

Nutrient interactions between cations and anions may either increase or decrease nutrient use efficiency, depending on the dominance of reactions affecting their bioavailability in precipitation/adsorption/dissolution and ion pairing reactions relative to their role in providing ionic balance during uptake and translocation of those nutrients. For example, NH_4/metal interactions depend on opposing reactions at the solid–solution interface versus the solution–root interface. In solution culture experiments, NH_4 antagonizes heavy metal uptake (Chaudhry and Loneragan, 1972) due to cation binding interference in the plasma membrane. In contrast, soil experiments demonstrated that the mobility of the chloride NH_4 salt had the greatest positive influence on increasing Cd absorption by plants compared to less soluble anions, SO_4 and PO_4 (Bingham et al., 1983, 1984; Ohtani et al., 2007), overshadowing possible effects of cation–cation antagonism. The presence of chloride optimized the response to enhanced NH_4 supply in wheat, either by helping to facilitate transport to, or into the root (Koenig and Pan, 1996a). In contrast, cation composition of soils influences phosphate solubility and precipitation, whereby Ca promotes P precipitation while Na increases solution P (Curtin and Syers, 2001).

The formation of soluble ion pairs increases total nutrient concentration in the soil solution, potentially facilitating greater nutrient transport to root surfaces, but at the same time, may affect the activity of charged species that directly interact with roots. For example, the formation of soluble $AlSO_4^+$ was proposed by Kinraide and Parker (1987) to decrease Al toxicity in wheat due to decreased Al^{3+} activity.

At the root surface, the presence of soluble anions can increase cation uptake. For example, Ca uptake by corn roots in solution was greater from NO_3 and Cl salts compared to SO_4 (Maas, 1969).

High P-induced Zn deficiency is often cited in the literature as an important anion–cation antagonism (Jones, 1991), but direct evidence of this type of interaction is sparse (Hernandez and Killorn, 2009) and Zn responses on high P soils have not shown Zn deficiencies (Mallarino and Webb, 1995).

16.2.5 Anion–Anion Interactions

Similar to cation–cation interactions, anions interact at the soil surface as well as the root cell surface to affect anion uptake by plants. Anion concentrations in soil solution influence anion adsorption of variable charged clays (Roy et al., 1986; Manning and Goldberg, 1996) whereby competing anions may increase soil solution anionic concentrations and increase plant uptake. For example, the addition of CO_3 and SO_4 salts to calcareous soil increased PO_4 availability and movement (Olatuyi et al., 2009).

In addition, anions can interact by competing at anion membrane channels that facilitate uptake (White and Broadley, 2001). Leggett and Epstein (1956) demonstrated that Se interferes with SO_4 influx while other anions have no effect. Anion antagonism is best exemplified by mutual antagonisms between Cl and NO_3. High Cl supply inhibits NO_3 uptake and high NO_3 reduces Cl uptake (Mengel and Kirkby, 1987). Interactions vary at high and low affinity porter sites in the root cell membrane (Siddiqi et al., 1990). Inhibition of N uptake by NaCl is often attributed to the role of Cl in antagonizing NO_3 absorption in a scenario 3 type response (Figure 16.2; Massa et al., 2009).

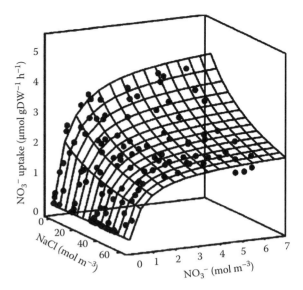

FIGURE 16.2 NaCl and NO_3 effects on NO_3 uptake in *Rosa*. (With kind permission from Springer Science+Business Media: Massa, D., N.S. Mattson, and H.J. Lieth. 2009. Effects of saline root environment (NaCl) on nitrate and potassium uptake kinetics for rose plants: A Michaelis-Menten modeling approach. Plant Soil 318:101–115.)

16.2.6 Secondary, Nonspecific Nutrient Interactions

Nutrients can have secondary effects on the uptake of other nutrients through their alteration of microbial activities in the rhizosphere, plant root activity, and rhizosphere chemical or biochemical modification (Marschner et al., 1987). In contrast to specific nutrient interactions, the uptake of multiple nutrients is affected by these secondary interactions. Furthermore, interactions between any of the macronutrients that have major influence on plant growth and development will include secondary effects on sink demand for other nutrients and/or via increased root proliferation and nutrient absorption capacity (Forde and Lorenzo, 2001; Williamson et al., 2001). Classic nonspecific nutrient interactions include N interactions with other nutrients, when the level of N bioavailability influences growth, yield potential, and corresponding demand for other nutrients.

An example of microbially mediated secondary nutrient interactions involves P-deficiency enhanced wheat responses to arbuscular mycorrhizal fungal inoculation, which increases Zn and Cu uptake and accumulation due to increased sorptive area, but not Fe and Mn (Mohammad et al., 1995). Kothari et al. (1991) suggested that the mycorrhizae decrease the potential for Mn reduction in the rhizosphere by lowering the number of Mn reducing bacteria. Furthermore, Zn binds to polyphosphates, a major P transport form in mycorrhizal hyphae, adding a primary nutrient interaction to this complex system.

Another prominent example of a nonspecific interaction is the rhizosphere acidification of plant roots in response to a nutrient deficiency, which, in turn, affects the solubility and bioavailability of pH-sensitive nutrients such as P and the cationic micronutrients (Marschner et al., 1987). Nitrogen form during N_2 fixation or NH$_4$ and NO$_3$ uptake by plants has the largest impact on cation–anion balance of absorbed charge, which is counterbalanced by net efflux of H$^+$ during excess cation uptake or OH$^-$ equivalents during excess anion uptake (Riley and Barber, 1971; Israel and Jackson, 1982). More recently, rhizosphere acidification has been observed to be a light-mediated reaction, seemingly tightly coupled to photosynthesis and energy transport to roots (Rao et al., 2002). Acidification of the rhizosphere then has secondary effects on nutrient availability and determines the propensity of certain species to utilize sparingly soluble Ca-phosphates (Shen et al., 2004; Zhou et al., 2009) and cationic micronutrients (Tills and Alloway, 1981). Manganese-deficient grass species are predisposed to take-all disease caused by *Gaeumannomyces graminis* var. *tritici* (Graham and Rovira, 1984), and NH$_4$ nutrition-induced acidification and reduction of Mn to the plant-available Mn^{2+} form improves Mn nutrition and reduces disease incidence (Brennan, 1992; Heckman et al., 2003).

Secondary micronutrient interactions can be utilized agronomically to enhance micronutrient uptake. Recent efforts to biofortify food crops with enhanced micronutrients place greater importance in recognizing mechanisms influencing micronutrient uptake, accumulation, and balance in designing agronomic systems (Zhang, 2009). Several researchers have recognized the benefits of intercropping Graminaceae and dicotyledonous crops in improving Zn and Cu accumulation in the dicot plant. This has been demonstrated in wheat/chickpea and lentil (Gunes et al., 2007) and maize/peanut (Zao et al., 2000). It is proposed that the Graminaceae crop, using strategy I in producing phytosiderophores under Fe deficiency to solubilize and acquire Fe and Zn in the comingled rhizosphere of both plant species, thereby enhances the cationic micronutrient uptake of both plants. The supply of FePO$_4$ was shown to increase Ca and Zn uptake in wheat, possibly due to a rhizosphere acidification response (Li et al., 2004).

Nitrogen and S fertility synergistically interact to affect crop grain yield (Wang et al., 1976; Randall et al., 1981), grain protein and grain quality (Randall et al., 1990), N utilization efficiency (Mahli and Gill, 2002), and N use efficiency (Salvagiotti and Miralles, 2008). Optimization of S nutrition increases N use efficiency, mainly by increasing N uptake efficiency in grass species (Brown et al., 2000; Figure 16.3, Salvagiotti et al., 2009). These responses align with scenario 1 (see Figure 16.1a).

In Brassica crops, yield and oil content (Ahmad et al., 1999), glucosinolate (Kim et al., 2002; Schonhof et al., 2007), and isothiocyanate (Gerendas et al., 2008) concentrations are influenced by relative supplies of N and S. Moderate N and S availability optimized production of benzyl-isothiocyanate in cress (Gerendas et al., 2008). In contrast to the previously cited improved nutrient use efficiency responses of grass species, Jackson (2000) observed a Liebig-type N × S interactive response in canola (scenario 4, see Figure 16.1d), in which N was initially more limiting than S, and yield response to S was apparent only at higher N supply (Figure 16.4b). In soil in which S availability was lower (Figure 16.4a), added S increased yield potential, which in turn increased N demand and the N supply required to obtain "A."

The field observations of N and S fertility interactions are well reflected at the cellular and root levels (Reuveny et al., 1980; Hesse et al., 2004). Mineral nutritional studies have revealed the uptake and assimilation of N and S are coregulated by the substrate ions and their assimilatory products (Clarkson et al., 1989; Koprivova et al., 2000; Hesse et al., 2004). Sulfate uptake and assimilation are regulated by O-acetylserine, a cysteine precursor that is in itself regulated by N availability and assimilation (Koprivova et al., 2000). Excess cysteine production when S is high or N is limiting will repress S uptake and assimilation (Zhao et al., 1999). Conversely, N uptake and assimilation is depressed during S starvation (Clarkson et al., 1989; Prosser et al., 2001) as arginine and asparagine accumulate with reduced cysteine and methionine production (Thomas et al., 2000; Prosser et al., 2001).

Nonspecific N–P interactions are commonly observed if for no other reason that they are typically the two most limiting nutrients in crop production and therefore demonstrate additive

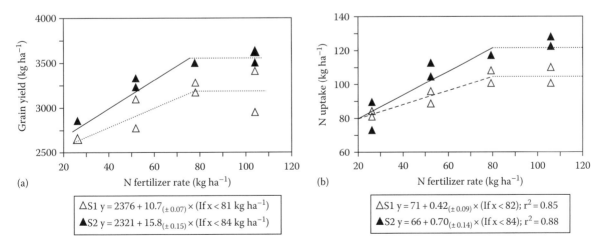

FIGURE 16.3 Wheat (a) grain yield and (b) N uptake responses to N fertilizer rate at two S levels. (From Salvagiotti, F., J.H.M. Castellarín, D.J. Miralles, and H.M. Pedrol. 2009. Sulfur fertilization improves nitrogen use efficiency in wheat by increasing nitrogen uptake. Field Crops Res. 113:170–177. Copyright (2009), with permission from Elsevier.)

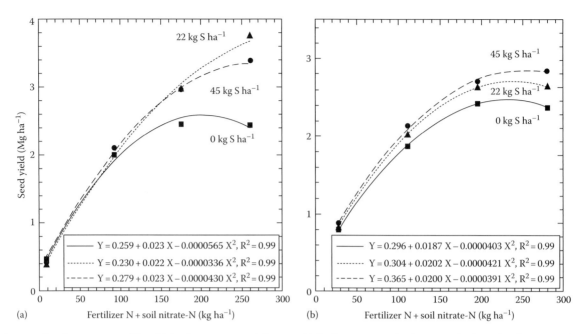

FIGURE 16.4 Canola responses to N and S fertility at two rainfed sites at (a) Sunburst, MT with a lower residual S soil and (b) Conrad, MT with a higher residual S soil. (From Jackson, G.D. 2000. Effects of nitrogen and sulfur on canola yield and nutrient uptake. Agron. J. 92:644–649. With permission from Soil Science Society of America.)

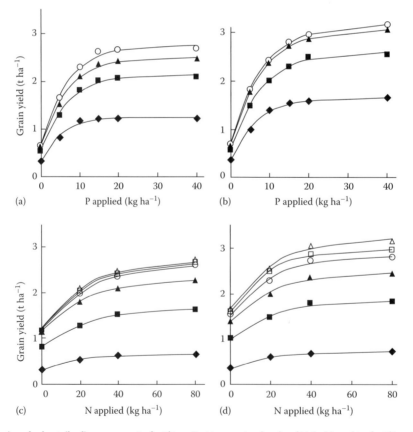

FIGURE 16.5 Canola (a, c) and wheat (b, d) responses to fertilizer P at increasing levels of N (a, b) and to fertilizer N at increasing levels of P (c, d). Nitrogen rates for panels (a) and (b) were 0 (◆), 20 (■), 40 (▲), and 80 (●) kg ha^{-1}. Phosphorus rates for panels (c) and (d) were 0 (◆), 5 (■), 10 (▲), 15 (○), 20 (□), and 40 (△) kg P ha^{-1}. (From Brennan, R.F., and M.D.A. Bolland. 2009. Comparing the nitrogen and phosphorus requirements of canola and wheat for grain yield and quality. Crop Pasture Sci. 60:566–577. With permission from CSIRO Publishing, http://www.publish.csiro.au/nid/40/paper/CP08401.htm)

or synergistic yield effects in bringing both nutrients to sufficient levels of availability. Mechanisms of N–P interactions have been ascribed to all stages of the soil–plant continuum (Miller, 1974). Brennan and Bolland (2009) demonstrated classic Liebig-type responses (see Figure 16.1d) to P at varying levels of N (Figure 16.5a and b) suggesting P was more critically limiting than N in these soils. Nitrogen level did not substantially alter the optimal level of P. The same data plotted differently exhibited an MTS-type response (see Figure 16.1a) to N at varying levels of P in canola and wheat (Figure 16.5c and d), suggesting there were limited specific interactions between the two nutrients that affected their bioavailability and that P availability was at least as critically limiting as N, such that optimization of P increased the N supply required to obtain "A." However, in other instances, a Liebig-type response (see Figure 16.1d) to N and P is observed (Nuttall et al., 1992) in which the level of N increased the optimal level of P supply.

16.3 Conclusions

Nutrient interactions are important factors to consider in accurately predicting nutrient responses in soil fertility management. Both primary and secondary interactions will influence nutrient use, uptake, and utilization efficiencies. Interactions occur at all points along the soil–plant continuum, and multinutrient quantitative models are needed to assess the relative importance and influence of individual interactions that are at times in synchrony and sometimes counter-opposing. Full recognition of the nature of these interactions will better allow us to understand the nature of variable yield responses to nutrient supplies. Shapes of nutrient response curves are influenced by the relative severity of the deficiencies of the interacting nutrients, the specificity with which the nutrients interact in influencing their relative availabilities and ability to react with root membrane surfaces, and their relative ability to influence plant internal processes and overall nutrient demand.

References

Adams, F. 1966. Calcium deficiency as a causal deficiency of ammonium phosphate injury to cotton seedlings. Soil Sci. Soc. Am. Proc. 30:485–488.

Adams, F. 1974. Soil solution, p. 441–481. *In* E.W. Carson (ed.) The plant root and its environment. University of Virginia Press, Charlottesville, VA.

Adams, F., and Z.F. Lund. 1966. Effect of chemical activity of soil solution aluminum on cotton root penetration of acid subsoils. Soil Sci. 101:193–198.

Ahmad, A., G. Abraham, and M.Z. Abdin. 1999. Physiological investigation of the impact of nitrogen and sulphur application on seed and oil yield of rapeseed (*Brassica campestris* L.) and mustard (*Brassica juncea* L.-Czern. and Coss.) genotypes. J. Agron. Crop Sci. 183:19–25.

Aslyng, H.C. 1954. The lime and phosphate potentials of soils: The solubility and availability of phosphates, p. 1–50. *In* Denmark yearbook. Royal Veterinary and Agricultural College, Copenhagen, Denmark.

Aurelia, M., C. Chappellaz, and P. Hinsinger. 2008. Copper phytotoxicity affects root elongation and iron nutrition in durum wheat (*Triticum turgidum durum* L.). Plant Soil 310:151–165.

Baldwin, J.P., and P.H. Nye. 1974. A model to calculate the uptake by a developing root system or root hair system of solutes with concentration variable diffusion coefficients. Plant Soil 40:703–706.

Barber, S.A. 1995. Soil nutrient bioavailability. Wiley, New York.

Barber, S.A., and J.H. Cushman. 1981. Nitrogen uptake model for agronomic crops, p. 382–409. *In* I.K. Iskandar (ed.) Modeling wastewater renovation—Land treatment. Wiley-Interscience, New York.

Bennett, A.C., and F. Adams. 1970. Concentration of NH_3(aq) required for incipient NH_3 toxicity to seedlings. Soil Sci. Soc. Am. Proc. 34:259–263.

Bingham, F.T., J.E. Strong, and G. Sposito. 1983. Influence of chloride salinity on cadmium uptake by Swiss chard. Soil Sci. 135:60–165.

Bingham, F.T., J.E. Strong, and G. Sposito. 1984. The effects of chloride on the availability of cadmium. J. Environ. Qual. 13:71–74.

Black, C.A. 1993. Soil fertility evaluation and control. Lewis Publishers, Boca Raton, FL.

Bloom, A.J., F.S. Chapin, III, and H.A. Mooney. 1985. Resource limitation in plants: An economic analogy. Ann. Rev. Ecol. Syst. 16:363–395.

Bolan, N.S., J.K. Syers, and M.E. Sumner. 1993. Calcium-induced sulfate adsorption by soils. Soil Sci. Soc. Am. J. 57:691–696.

Bouldin, D.R. 1989. A multiple ion uptake model. J. Soil Sci. 40:309–319.

Brennan, R.F. 1992. The role of manganese and nitrogen nutrition in the susceptibility of wheat plants to take-all in Western Australia. Fert. Res. 31:35–41.

Brennan, R.F., and M.D.A. Bolland. 2009. Comparing the nitrogen and phosphorus requirements of canola and wheat for grain yield and quality. Crop Pasture Sci. 60:566–577.

Brown, D.S., and J.D. Allison. 1987. MINTEQA1, an equilibrium metal speciation model: User's manual. US Environmental Protection Agency. Washington, DC.

Brown, L., D. Scholefield, E.C. Jewkes, N. Preedy, K. Wadge, and M. Butler. 2000. The effect of sulphur application on the efficiency of nitrogen use in two contrasting grassland soils. J. Agric. Sci. 135:131–138.

Browne, C.A. 1944. A source book of agricultural chemistry. Vol. 8. *In* F. Vedoorn (ed.) Chronica botanica. The Chronica Botanica Co., Waltham, MA.

Caldwell, C.R., and A. Haug. 1981. Temperature dependence of the barley root plasma membrane-bound Ca^{2+} and Mg^{2+}-dependent ATPase. Physiol. Plant. 53:117–125.

Cataldo, D.A., T.R. Garland, and R.E. Wildung. 1983. Cadmium uptake kinetics in intact soybean plants. Plant Physiol. 73:844–848.

Chaudhry, F.M., and J.F. Loneragan. 1972. Zinc absorption by wheat seedlings. I. Inhibition by macronutrient ions in short-term experiments and its relevance to long-term zinc nutrition. Soil Sci. Soc. Am. Proc. 36:323–327.

Claassen, N., and S.A. Barber. 1976. Simulation model for nutrient uptake from soil by a growing plant root system (maize). Agron. J. 68:961–964.

Claassen, N., and B. Steingrobe. 1999. Mechanistic simulation models for better understanding of nutrient uptake from soil, p. 327–367. *In* Z. Rengel (ed.) Mineral nutrition of crops. The Haworth Press Inc., Binghamton, NY.

Clarkson, D.T., L.R. Sarker, and J.V. Purves. 1989. Depression of nitrate and ammonium transport in barley plants with diminished sulphate status: Evidence of co-regulation of nitrogen and sulphate intake. J. Exp. Bot. 40:953–963.

Crooke, W.M., and A.H. Knight. 1962. An evaluation of published data on the mineral composition of plants in the light of cation exchange capacities of their roots. Soil Sci. 93:365–373.

Curtin, D., and J.K. Syers. 2001. Lime-induced changes in indices of soil phosphate availability. Soil Sci. Soc. Am. J. 65:147–152.

Degryse, F., E. Smoders, and D.R. Parker. 2009. Partitioning of metals (Cd, Co, Cu, Ni, Pb, and Zn) in soils: Concepts, methodologies, prediction and applications—A review. Eur. J. Soil Sci. 60:590–612.

Degryse, F., V.K. Verma, and E. Smolders. 2008. Mobilization of Cu and Zn by root exudates of dicotyledonous plants in resin-buffered solutions and in soil. Plant Soil 306:69–84.

Drew, M.C., and L.R. Saker. 1975. Nutrient supply and the growth of the seminal root system of barley. II. Localized, compensatory increases in lateral root growth and rates of nitrate uptake when nitrate supply is restricted to only part of the root system. J. Exp. Bot. 26:79–90.

Epstein, E. 1972. Mineral nutrition of plants: Principles and perspectives. John Wiley & Sons Inc., New York.

Epstein, E., and A.J. Bloom. 2005. Mineral nutrition of plants: Principles and perspectives. 2nd edn. Sinauer Associates Inc., Sunderland, MA.

Epstein, E., and J.E. Leggett. 1954. The absorption of alkaline earth cations by barley roots: Kinetics and mechanism. Am. J. Bot. 41:785–791.

Fenn, L.B., R.M. Taylor, and G.I. Horst. 1987. *Phaseolus vulgaris* growth in the ammonium-based nutrient solution with variable calcium. Agron. J. 79:89–91.

Fenn, L.B., and R.M. Taylor. 1990. Calcium stimulation of ammonium absorption in radish. Agron. J. 82:81–84.

Forde, B., and H. Lorenzo. 2001. The nutritional control of root development. Plant Soil 232:51–68.

Foy, C.D. 1988. Plant adaptation to acid, aluminum-toxic soils. Commun. Soil Sci. Plant Anal. 19:959–987.

Fried, M., and H. Broeshart. 1967. The soil–plant system in relation to inorganic nutrition. Academic Press, New York.

Gage, R.A., W. Van Wijingaarden, A.P.R. Theunenet, G.W.F.H. Borst-Pauwels, and A.J. Verkleij. 1985. Inhibition of Rb^+ uptake in yeast by Ca^{2+} is caused by a reduction in the surface potential and not in the Donnan potential of the cell wall. Biochim. Biophys. Acta 812:1–8.

Gerendas, J., M. Sailer, M.-L. Fendrich, T. Stahl, V. Mersch-Sundermann, and K.H. Muhling. 2008. Influence of sulfur and nitrogen supply on growth, nutrient status and concentration of benzyl-isothiocyanate in cress (*Lepidium sativum* L.). J. Sci. Food Agric. 88:2576–2580.

Girling, C.A., and P.J. Peterson. 1981. The significance of the cadmium species in uptake and metabolism of cadmium in crop plants. J. Plant Nutr. 3:707–720.

Graham, R.D., and A.D. Rovira. 1984. A role for manganese in the resistance of wheat plants to take-all. Plant Soil 78:441–444.

Gunes, A., A. Inul, M.S. Adak, M. Apaslan, E.G. Bagsci, T. Erol, and P.J. Pilbeam. 2007. Mineral nutrition of wheat, chickpea, and lentil as affected by intercropped cropping and moisture. Nutr. Cycl. Agroecosyst. 73:83–96.

Hart, J.J., W.A. Norvell, R.M. Welch, L.A. Sullivan, and L.V. Kochian. 1998. Characterization of zinc uptake, binding, and translocation in intact seedlings of bread and durum wheat cultivars. Plant Physiol. 118:219–226.

Haynes, R.J. 1980. Ion exchange properties of roots and ionic interactions within the root apoplasm: Their role in ion accumulation of plants. Bot. Rev. 46:75–99.

Heckman, J.R., B.B. Clarke, and J.A. Murphy. 2003. Optimizing manganese fertilization for the suppression of take-all patch disease on creeping bentgrass. Crop Sci. 43:1395–1398.

Hernandez, J.D., and R. Killorn. 2009. Phosphorus fertilizer by-product effect on the interaction of zinc and phosphorus in corn and soybean. Can. J. Soil Sci. 89:189–196.

Hesse, H., V. Nikiforova, B. Gakiere, and R. Hoefgen. 2004. Molecular analysis and control of cysteine biosynthesis: Integration of nitrogen and sulphur metabolism. J. Exp. Bot. 55:1283–1292.

Hinsinger, P. 1998. How do plant roots acquire mineral nutrients? Chemical processes involved in the rhizosphere. Adv. Agron. 64:225–265.

Honma, Y., and H. Hirata. 1978. Noticeable increase in cadmium absorption by zinc deficient rice plants. Soil Sci. Plant Nutr. 24:295–297.

Hoopen, F., T.A. Cuin, P. Pedas, J.N. Hegeland, S. Shabala, J.K. Schjoerring, and T.P. Jahn. 2010. Competition between uptake of ammonium and potassium in barley and *Arabidopsis* roots: Molecular mechanisms and physiological consequences. J. Exp. Bot. 61:2303–2315.

Hopmans, J.W., and K.L. Bristow. 2002. Current capabilities and future needs of root water and nutrient uptake modeling. Adv. Agron. 77:103–183.

Hough, R.L., A.M. Tye, N.M.J. Crout, S.P. McGrath, H. Zhang, and S.D. Young. 2005. Evaluating a 'Free Ion Activity Model' applied to metal uptake by *Lolium perenne* L. grown in contaminated soils. Plant Soil 270:1–12.

Howard, D.D., and F. Adams. 1965. Calcium requirement for penetration of subsoils by primary cotton roots. Soil Sci. Soc. Am. Proc. 29:558–562.

Ihde, A.J. 1964. The development of modern chemistry. Harper and Row, New York.

Israel D.W., and W.A. Jackson. 1982. Ion balance, uptake and transport processes in N_2 fixing and nitrate and urea dependent soybean plants. Plant Physiol. 69:171–178.

Jackson, G.D. 2000. Effects of nitrogen and sulfur on canola yield and nutrient uptake. Agron. J. 92:644–649.

Jones, B.J. 1991. Plant tissue analysis in micronutrients. In J.J. Mortvedt et al. (eds.) Micronutrients in agriculture. 2nd edn. SSSA, Madison, WI.

Kalis, E.J.J., E.J.M. Temminghoff, R.M. Town, E.R. Unsworth, and W.H. van Riemsdijk. 2008. Relationship between metal speciation in soil solution and metal adsorption at the root surface of ryegrass. J. Environ. Qual. 37:2221–2231.

Keltjens, W.G., and M.L. van Beuschichem. 1998. Phytochelatins as biomarkers for heavy metal stress in maize (*Zea mays* L.) and wheat (*Triticum aestivum* L.): Combined effects of copper and cadmium. Plant Soil 203:119–126.

Kim, S.J., T. Matsuo, M. Watanabe, and Y. Watanabe. 2002. Effect of nitrogen and sulphur application on the glucosinolate content in vegetable turnip rape (*Brassica rapa* L.). Soil Sci. Plant Nutr. 48:43–49.

Kinraide, T.B., and D.R. Parker. 1987. Non-phytotoxicity of the aluminum sulfate ion, $AlSO_4^+$. Physiol. Plantarum 71:207–212.

Kinraide, T.B., J.F. Pedler, and D.R. Parker. 2004. Relative effectiveness of calcium and magnesium in the alleviation of rhizotoxicity in wheat induced by copper, zinc, aluminum, sodium, and low pH. Plant Soil 259:201–208.

Koenig, R.T., and W.L. Pan. 1996a. Chloride enhancement of wheat responses to ammonium nutrition. Soil Sci. Soc. Am. J. 60:498–505.

Koenig, R.T., and W.L. Pan. 1996b. The influence of calcium on ammonium quantity-intensity relationships and ammonium availability in soil. Soil Sci. Soc. Am. J. 60:492–497.

Koprivova, A., M. Suter, R. Op den Camp, C. Brunold, and S. Kopriva. 2000. Regulation of sulfate assimilation by nitrogen in Arabidopsis. Plant Physiol. 122:737–746.

Kothari, S.K., H. Marschner, and V. Römheld. 1991. Effect of a vesicular-arbuscular mycorrhizal fungus and rhizosphere micro-organisms on manganese reduction in the rhizosphere and manganese concentrations in maize (*Zea mays* L.). New Phytol. 117:649–655.

Leggett J.E., and E. Epstein. 1956. Kinetics of sulfate absorption by barley roots. Plant Physiol. 31:222–226.

Li, L., C. Tang, Z. Rengel, and F.S. Zhang. 2004. Calcium, magnesium and microelement uptake as affected by phosphorus sources and interspecific root interactions between wheat and chickpea. Plant Soil 261:29–37.

Luan, S., W. Lan, and S.C. Lee. 2009. Potassium nutrition, sodium toxicity, and calcium signaling: Connections through the CBL-network. Curr. Opin. Plant Biol. 12:339–346.

Maas, E.V. 1969. Calcium uptake by excised maize roots and interactions with alkali cations. Plant Physiol. 44:985–989.

Magistad, O.C. 1925. The aluminum content of the soil solution and its relation to plant growth. Soil Sci. 20:181–225.

Mahli, S.S., and K.S. Gill. 2002. Effectiveness of sulphate-S at different growth stages for yield, seed quality and S uptake of canola. Can. J. Plant Sci. 82:665–674.

Mallarino, A.P., and J.R. Webb. 1995. Long-term evaluation of phosphorus and zinc interactions in corn. J. Prod. Agric. 8:52–55.

Manning, B.A., and S. Goldberg. 1996. Modeling competitive adsorption of arsenate with phosphate and molybdate on oxide minerals. Soil Sci. Soc. Am. J. 60:121–131.

Marschner, H. 1995. Mineral nutrition of higher plants. Academic Press, London, U.K.

Marschner, H., V. Römheld, and I. Cakmak. 1987. Root-induced changes of nutrient availability in the rhizosphere. J. Plant Nutr. 10:1175–1184.

Massa, D., N.S. Mattson, and H.J. Lieth. 2009. Effects of saline root environment (NaCl) on nitrate and potassium uptake kinetics for rose plants: A Michaelis-Menten modeling approach. Plant Soil 318:101–115.

McGeorge, W.T. 1924. Iron, aluminum and manganese in the soil solution of Hawaiian soils. Soil Sci. 18:1–12.

Mengel, K., and E.A. Kirkby. 1987. Principles of plant nutrition. 4th edn. IPI, Bern, Switzerland.

Miller, M.H. 1974. Effects of nitrogen on phosphorus absorption by plants, p. 643–668. In E.W. Caron (ed.) The plant root and its environment. University Press of Virginia, Charlottesville, VA.

Miller, A.J., Q. Shen and G. Xu. 2009. Freeways in the plant transporters for N, P and S and their regulation. Curr. Opin. Plant Biol. 12:284–290.

Mitscherlich, E.A. 1909. Das gesetz des minimums und das gesetz des abnehmenden bodentrags. Landwehr Jahrb. 38:537–552.

Mohammad, M., W.L. Pan, and A.C. Kennedy. 1995. Wheat responses to VAM inoculation of soils from an eroded toposequence. Soil Sci. Soc. Am. J. 59:1086–1090.

Mohammadi, H., K. Poustini, and A. Ahmadi. 2008. Root nitrogen remobilization and ion status of two alfalfa (*Medicago sativa* L.) cultivars in response to salinity stress. J. Agron. Crop Sci. 194:126–134.

Moore, P.A., Jr., and W.H. Patrick, Jr. 1989. Manganese availability and uptake by rice in acid sulfate soils. Soil Sci. Soc. Am. J. 53:104–109.

Morgan, M.A., and W.A. Jackson. 1976. Calcium and magnesium in ryegrass. Some differences in accumulation by roots and in translocation to shoots. Plant Soil 44:623–637.

Nuttall, W.F., A.P. Moulin, and L.J. Townley-Smith. 1992. Yield response of canola to nitrogen, phosphorus, precipitation, and temperature. Agron. J. 84:765–768.

Nye, P.H. 1977. The rate-limiting step in plant nutrient absorption from soil. Soil Sci. 123:292–297.

Ohno, T., and D.L. Grunes. 1985. Potassium-magnesium interactions affecting nutrient uptake by wheat forage. Soil Sci. Soc. Am. J. 49:685–690.

Ohtani, T., M. Kawabata, A. Sase, and M. Fukami. 2007. Cadmium and nutrient heavy metals uptake by rice, barley, and spinach as affected by four ammonium salts. J. Plant Nutr. 30:599–610.

Olatuyi, S.O., O.O. Akinremi, D.N. Flaten, and G.H. Crow. 2009. Accompanying cations and anions affect the diffusive transport of phosphate in a model calcareous soil system. Can. J. Soil Sci. 89:179–188.

Pan, W.L., J.J. Camberato, W.A. Jackson, and R.H. Moll. 1986. Utilization of previously accumulated and concurrently absorbed nitrogen during reproductive growth in maize: Influence of prolificacy and nitrogen source. Plant Physiol. 82:247–253.

Pearce, R.C., and M.E. Sumner. 1997. Apparent salt sorption reactions in an unfertilized acid subsoil. Soil Sci. Soc. Am. J. 61:765–772.

Pearson, J.N., and Z. Rengel. 1995. Uptake and distribution of ^{65}Zn and ^{54}Mn in wheat grown in sufficient and deficient levels of Zn and Mn. I. During vegetative growth. J. Exp. Bot. 46:833–839.

Pierre, W.H., and F.W. Parker. 1927. Soil phosphorus studies. II. The concentration of organic and inorganic phosphorus to plants. Soil Sci. 24:119–128.

Prosser, I.M., J.V. Purves, L.R. Saker, and D.T. Clarkson. 2001. Rapid disruption of nitrogen metabolism and nitrate transport in spinach plants deprived of sulphate. J. Exp. Bot. 52:113–121.

Randall, P.J., J.R. Freney, C.J. Smith, H.J. Moss, C.W. Wrigley, and I.E. Galbally. 1990. Effects of additions of nitrogen and sulfur to irrigated wheat at heading on grain yield, composition and milling and baking quality. Aust. J. Exp. Agric. 30:95–101.

Randall, P.J., K. Spencer, and J.R. Freney. 1981. Sulfur and nitrogen fertilizer effects on wheat. I. Concentrations of sulfur and nitrogen and the nitrogen to sulfur ratio in grain, in relation to the yield response. Aust. J. Agric. Res. 32:203–212.

Rao, T.P., K. Yano, M. Iijima, A. Yamauchi, and J. Tatsumi. 2002. Regulation of rhizosphere acidification by photosynthetic activity in cowpea (*Vigna unguiculata* L. Walp.) seedlings. Ann. Bot. 89:213–220.

Reid, R.J. 1999. Kinetics of nutrient uptake by plant cells, p. 41–66. *In* Z. Rengel (ed.) Mineral nutrition of crops. The Haworth Press Inc., Binghamton, NY.

Rengel, Z. 1990. Competitive Al^{3+} inhibition of net Mg^{2+} uptake by intact *Lolium multiflorum* roots. Plant Physiol. 93:1261–1267.

Rengel, Z. 1993. Mechanistic simulation models of nutrient uptake: A review. Plant Soil 152:161–173.

Rengel, Z., and D.L. Robinson. 1989. Competitive Al^{3+} inhibition of net Mg^{2+} uptake by intact *Lolium multiflorum* roots. I. Kinetics. Plant Physiol. 91:1407–1413.

Reuveny, Z., D. Dougall, and P. Trinity. 1980. Regulatory coupling of nitrate and sulfate assimilation pathways in cultured tobacco cells. Proc. Nat. Acad. Sci. 77:6670–6672.

Riley, D., and S.A. Barber. 1971. Effect of ammonium and nitrate fertilization on phosphorus uptake as related to root-induced pH changes at the root–soil interface. Soil Sci. Soc. Am. Proc. 35:301–306.

Roy, W.R., J.J. Hassett, and R.A. Griffin. 1986. Competitive interactions of phosphate and molybdate on arsenate adsorption. Soil Sci. 142:203–210.

Rubio, G., J. Zhu, and J.P. Lynch. 2003. A critical test of the two prevailing theories of plant response to nutrient availability. Am. J. Bot. 90:143–152.

Sajwan, K.S., and W.L. Lindsay. 1986. Effects of redox on zinc deficiency in paddy rice. Soil Sci. Soc. Am. J. 50:1264–1269.

Sakuri, K., and P.M. Huang. 1996. Influence of potassium chloride on desorption of cadmium sorbed on hydroxyaluminosilicate-montmorillonite complex. Soil Sci. Plant Nutr. 42:475–481.

Salvagiotti, F., and D.J. Miralles. 2008. Radiation interception, biomass production and grain yield as affected by the interaction of nitrogen and sulfur fertilization in wheat. Eur. J. Agron. 28:282–290.

Salvagiotti, F., J.H.M. Castellarín, D.J. Miralles, and H.M. Pedrol. 2009. Sulfur fertilization improves nitrogen use efficiency in wheat by increasing nitrogen uptake. Field Crops Res. 113:170–177.

Schonhof, I., D. Blankenburg, S. Muller, and A. Krumbein. 2007. Sulphur and nitrogen supply influence growth, product appearance, and glucosinolate concentration of broccoli. J. Plant Nutr. Soil Sci. 170:65–72.

Shen, J., C. Tang, Z. Rengel, and F. Zhang. 2004. Root-induced acidification and excess cation uptake by N$_2$-fixing *Lupinus albus* grown in phosphorus-deficient soil. Plant Soil 260:69–77.

Siddiqi, M.Y., A.D.M. Glass, T.J. Ruth, and T.W. Rufty, Jr. 1990. Studies of the uptake of nitrate in barley. I. Kinetics of ^{13}NO$_3$-influx. Plant Physiol. 93:1426–1432.

Silberbush, M. 1996. Simulation of ion uptake from the soil, p. 643–658. *In* Y. Waisel et al. (eds.) Plant roots: The hidden half. Marcel Dekker Inc., New York.

Sparks, D.L. 1984. Ion activities: An historical and theoretical overview. Soil Sci. Soc. Am. J. 48:514–518.

Sposito, G. 1984. The future of an illusion. Soil Sci. Soc. Am. J. 48:531–536.

Sposito, G., and S.V. Mattigod. 1980. GEOCHEM: A computer program for the calculation of chemical equilibria of soil solutions and other natural water systems. Kearney Foundation of Soil Science, University of California, Berkeley, CA.

Thomas, S.G., P.E. Bilsborrow, T.J. Hocking, and J. Bennett. 2000. Effect of sulphur deficiency on the growth and metabolism of sugar beet (*Beta vulgaris* cv. Druid). J. Sci. Food Agric. 80:2057–2062.

Tills, A.R., and B.J. Alloway. 1981. The effect of ammonium and nitrate nitrogen sources on copper uptake and amino acid status of cereals. Plant Soil 62:279–290.

Tisdale, S.L., W.L. Nelson, and J.D. Beaton. 1985. Soil fertility and fertilizers. 4th edn. Macmillan Publishing Co., New York.

Vale, F.R., W.A. Jackson, and R.J. Volk. 1987. Potassium influx into maize root systems: Influence of root potassium concentration and ambient ammonium. Plant Physiol. 84:1416–1420.

van der Pauw, F. 1952. Critical remarks concerning the validity of the Mitscherlich effect law. Plant Soil 4:97–106.

van der Ploeg, R.R., W. Bohm, and M.B. Kirkham. 1999. On the origin of the theory of mineral nutrition of plants and the law of the minimum. Soil Sci. Soc. Am. J. 63:1055–1062.

Viets, F., Jr. 1944. Calcium and other polyvalent cations as accelerators of ion accumulation by excised barley roots. Plant Physiol. 19:466–480.

Wang, C.H., T.H. Liem, and D.S. Mikkelsen. 1976. Sulfur deficiency: A limiting factor in rice production in the lower Amazon basin. II. Sulfur requirement for rice production. IRI Res. Inst. 48:9–30.

Wang, M.Y., M.Y. Siddiqi, T.J. Ruth, and A.D.M. Glass. 1993. Ammonium uptake by rice roots. 2. Kinetics of $^{13}NH_4^+$ influx across the plasmalemma. Plant Physiol. 103:1259–1267.

Welch, R.M., J.J. Hart, W.A. Norvell, L.A. Sullivan, and L.V. Kochian. 1999. Effects of nutrient solution zinc activity on net uptake, translocation and root export of cadmium and zinc by separated sections of intact durum wheat (*Triticum turgidium* L. *var durum*) seedling roots. Plant Soil 208:243–250.

White, P.J., and M.R. Broadley. 2001. Chloride in soils and its uptake and movement within the plant: A review. Ann. Bot. 88:967–988.

Wild, A. 1988. Russell's soil conditions and plant growth. 11th edn. Longmna Scientific and Technical, Essex, U.K.

Williamson, L.C., S. Ribrioux, A.H. Fitter, and H.M.O. Leyser. 2001. Phosphate availability regulates root system architecture in Arabidopsis. Plant Physiol. 126:875–882.

Wood, J.T., D.J. Greenwood, and T.J. Cleaver. 1972. Interaction between the beneficial effects of nitrogen, phosphate and potassium on plant growth. J. Agric. Sci. 78:389–391.

Yao, X., T. Morie, S. Xue, H.-Y. Leung, M. Katsuhara, D.E. Brodsky, Y. Wu, and J.I. Schroeder. 2010. Differential sodium and potassium transport selectivities of the rice OsHKT2;1 and OsHKT2;2 transporters in plant cells. Plant Physiol. 152:341–355.

Yermiyahu, U., and T.B. Kinraide. 2005. Binding and electrostatic attraction of trace elements to plant root surfaces, p. 365–389. *In* P.M. Huang and G.R. Gobran (eds.) Biogeochemistry of trace elements in the rhizosphere. Elsevier, Amsterdam, the Netherlands.

Yoshida, S., and L. Castenada. 1969. Partial replacement of potassium by sodium in rice plant under weakly saline conditions. Soil Sci. Plant Nutr. 15:183–186.

Zao, Y.M., F.S. Zhang, X.I. Li, and Y.P. Cao. 2000. Studies on the improvement in iron nutrition of peanut by intercropping maize on a calcareous soil. Plant Soil 220:13–25.

Zhang, Y.Z.F. 2009. Iron and zinc biofortification strategies in dicot plants by intercropping with gramineous species: A review. Agron. Sustainable Dev. 29:63–71.

Zhang, K., D.J. Greenwood, R.J. White, and I.G. Burns. 2007. A dynamic model for the combined effects of N, P and K fertilizers on yield and mineral composition; description and experimental test. Plant Soil 298:81–98.

Zhang, F.S., J. Shen, and Y.-G. Zhu. 2006. Nutrient interactions in soil-plant system. *In* R. Lal (ed.) Encyclopedia of soil science. John Wiley & Sons Inc., New York.

Zhao, F.J., M.J. Hawkesford, and S.P. McGrath. 1999. Sulphur assimilation and effects on yield and quality of wheat. J. Cereal Sci. 30:1–17.

Zhou, L.L., J. Cao, F.S. Zhang, and L. Li. 2009. Rhizosphere acidification of faba bean, soybean and maize. Sci. Total Environ. 407:4356–4362.

III

Interdisciplinary Aspects of Soil Science

Guy J. Levy
Agricultural Research Organization

17 **Saline and Boron-Affected Soils** R. Keren 17-1
   Origin and Distribution of Saline Soils • Water Quality Criteria for Irrigation • Effect of Salinity on Soil Physical
   Properties • Boron–Soil Interaction • Effects of Salinity and Boron on Plants • Reclamation of Saline
   and Boron-Affected Soils • References

18 **Sodicity** Guy J. Levy 18-1
   Introduction • Sodic Soils: Definition and Distribution • Processes Characterizing the Behavior of Sodic
   Soils • Aggregate Stability and Organic Matter in Sodic Soils • Water Flow in Sodic Soils • Crop Responses
   to Sodic Conditions • Amelioration and Management of Sodic Soils • Environmental Aspects of Sodic
   Soils • Conclusions • References

19 **Soil Water Repellency** Stefan H. Doerr and Richard A. Shakesby 19-1
   Introduction • Origin and Classification • Occurrence of Soil Water Repellency • Hydrological and Geomorphological
   Effects of Soil Water Repellency • Conclusions • References

20 **Biogeochemistry of Wetlands** P.W. Inglett, K.R. Reddy, W.G. Harris, and E.M. D'Angelo 20-1
   Introduction • Definitions • Wetland Soils • Redox Gradients • Carbon • Nitrogen • Phosphorus •
   Metals • Toxic Organics • References

21 **Acid Sulfate Soils** L.A. Sullivan, R.T. Bush, E.D. Burton C.J. Ritsema, and M.E.F. van Mensvoort (Retired) 21-1
   Introduction • Characteristics, Occurrence, and General Impacts • Characteristic Minerals and Geochemical
   Processes • Modeling Soil Physical and Chemical Processes in Acid Sulfate Soils • Hazards Arising from
   Acid Sulfate Soil Disturbance • Physical Behavior • Use, Management, and Evaluation of Acid Sulfate Soil
   Landscapes • Identification and Assessment • Conclusions • References

22 **Water Erosion** Dino Torri and Lorenzo Borselli 22-1
   Introduction • Erosion Processes • Soil Erosion Models for Scenario Analysis •
   Conclusions • Acknowledgment • References

23 **Wind Erosion** D.W. Fryrear 23-1
   Introduction • Mechanics of Wind Erosion Processes • Modeling the Wind Erosion Process • Wind Erosion
   Measuring Systems: Field Sampling Instrumentation • Major Control Systems • Conclusions • References

24 **Land Application of Wastes** *David M. Miller and W.P. Miller* 24-1
   A Brief History and Overview of Land Application • Properties of Common Land-Applied By-Products • Soil Fertility Considerations in Land Application • Effects of Land Application on Soil Properties • Environmental Aspects of By-Product Applications • Recommendations • References

25 **Conservation Tillage** *Paul W. Unger and Humberto Blanco-Canqui* 25-1
   Introduction • Definitions • Erosion by Water and Wind • Water Conservation • Residue Management and Decomposition • Biological Activity • Carbon Dynamics and Sequestration in No-Tillage • Biofuel (Cellulosic) Production Implications • Soil Fertility and Nutrition • Surface Sealing, Crusting, and Seedling Emergence • Compaction • Cropping Systems • Cover Crops • Comparison of Conservation Tillage with Conservation Agriculture • Soil by Climate Interactions • Disadvantages • Conclusions • References

26 **Soil Quality** *Stephanie A. Ewing and Michael J. Singer* 26-1
   Introduction: Why Soil Quality? • Defining Soil Quality • Evaluating Soil Quality • Conclusions • References

A nation that destroys its soils destroys itself

Franklin D. Roosevelt

Soil is a dynamic and complex system. Processes in the soil environment are interdependent and interrelated. For a long time, research in soil science has focused mainly on topics along the lines of the traditional disciplines of soil science. This concept led to research efforts that concentrated on well defined, yet limited in number, processes and properties (i.e., studies in soil physics or soil chemistry). However, a comprehensive understanding of the soil environment requires that this approach be accompanied by a more realistic interdisciplinary perspective that considers processes arising from and affected by interactions among properties across these traditional disciplines of soil science.

In light of the above, the current part is devoted to interdisciplinary facets of soil science with special emphasis given to aspects related to soil degradation. Soils are fundamental to the well-being and productivity of both agricultural and natural ecosystems. Hence, the state of soils in various different ecosystems is of concern and has been receiving increased attention globally. The part comprises chapters that discuss, from an interdisciplinary perspective, degradation of soils by salinity (Chapter 17), sodicity (Chapter 18), water repellency (Chapter 19), water logging (Chapter 20), acid sulfate conditions (Chapter 21), water and wind erosion (Chapters 22 and 23), application of wastes (Chapter 24), and tillage (Chapter 25). The final chapter (Chapter 26) discusses the concept of soil quality for agriculture and the environment, and reviews approaches for its evaluation.

I wish to thank all the authors for their valuable contributions and extend my deep appreciation to the reviewers for their invaluable comments and criticism that helped maintain the quality of this part.

17
Saline and Boron-Affected Soils

17.1 Origin and Distribution of Saline Soils... 17-1
  Mineral Weathering • Sea Water Intrusion • Deposition of Salts by Rainfall • Deposit and Secondary Salinization
17.2 Water Quality Criteria for Irrigation.. 17-2
17.3 Effect of Salinity on Soil Physical Properties 17-5
17.4 Boron–Soil Interaction 17-7
  Boron Adsorption by Soil Constituents • Kinetics and Mechanisms of Boron Adsorption • Boron Adsorption Modeling • Boron Transport Modeling
17.5 Effects of Salinity and Boron on Plants.. 17-10
  Crop Tolerance to Salinity • Specific Molecule and Ion Effects on Plants
17.6 Reclamation of Saline and Boron-Affected Soils... 17-13
  Leaching for Salinity Control • Leaching for Boron Control
References.. 17-16

R. Keren
Agricultural Research Organization

17.1 Origin and Distribution of Saline Soils

The history of agriculture has shown that irrigated agriculture cannot survive in perpetuity without adequate salt balance and drainage. The length of time that irrigated agriculture can survive without adequate drainage depends on hydrogeology and water management. Of the world's cultivated lands, about 831 million hectares are saline (Martinez-Beltran and Manzur, 2005). Salt-affected soils occur in more than 100 countries of the world (Szabolcs, 1989). 67% of the agricultural area has a potential for "transient salinity," a type of non-groundwater-associated salinity (Rengasamy, 2006).

Inland saline waters usually contain 500–30,000 mg L^{-1} of dissolved solids (EC 0.7–42 dS m^{-1}), while ocean water has an average dissolved solid concentration of 33,000 mg L^{-1}, and the Dead Sea, Israel, 270,000 mg L^{-1} (hypersaline). Common sources of saline water for agricultural use are ground or surface waters. The main processes by which soluble salts enter the soil and groundwater include weathering of primary and secondary minerals and application of waters containing salts. The importance of each source depends on the type of soil, climate conditions, and agricultural management.

17.1.1 Mineral Weathering

Salt formation takes place during the process of minerals weathering. The accumulation of these soluble salts in groundwater affects water quality, which depends on the natural salinity of the soil and the geologic materials with which the water has been in contact. Many soils from arid and semiarid regions and primary minerals such as olivine, hornblende, oligoclase, and others contribute substantial amounts of salts from weathering (Rhoades et al., 1968). In all cases, the total salt content of the displaced soil solutions was higher than that of the irrigation waters being applied.

Water within sedimentary strata become increasingly saline with increasing depth (Craig, 1970) with the sequence being SO_4^{2-} rich water near the surface, HCO_3^- water at an intermediate level, and more concentrated Cl^- solutions at greater depths. This water becomes saline through weathering of primary and secondary minerals.

Predictions of soil mineral weathering and soil solution compositions can be made using stability diagrams (Kittrick, 1977; Lindsay, 1979). Mineral solubility is calculated from appropriate mineral equilibrium constants. Limitations of this approach are (1) the availability of adequate data on the solubilities or free energies of formation of various minerals as well as dissolved species present in soil solution, and (2) the paucity of information on the kinetics of minerals dissolution. Thermodynamic data generally are more accurate for the primary minerals involved in increasing the salinity of the soil solution (Rhoades et al., 1968) than for the secondary minerals (Kittrick, 1977). The major ions in the dissolved mineral salts are Na, Ca, Mg, K, Cl, SO_4, HCO_3, CO_3, and NO_3.

17.1.2 Sea Water Intrusion

In coastal areas, salt water intrusion and submergence of the low-lying lands by sea water cause salinization of groundwater and soils, while surface waters could become saline through tidal fluctuations. As the high tide moves into a coastal area, sea water moves into streams and drainage canals and travels inland. This upstream migration of sea water alters the quality of water in streams and drainage canals significantly.

17.1.3 Deposition of Salts by Rainfall

The composition of atmospheric salt deposition varies with distance from the source. The salt is predominantly NaCl at the coast, and becomes dominated by Ca^{2+} and SO_4^{2-} inland. Atmospheric contribution to the salt load of arid lands, which is from 10% to 25% of the total yearly contribution from weathering (Bresler et al., 1982), is often overlooked, but is a factor that must be considered in highly weathered landscapes that have poor drainage.

17.1.4 Deposit and Secondary Salinization

Throughout geologic time, sea water has inundated large areas of the continents that have, subsequently, been uplifted and salts deposited when the water evaporated. Salt bodies, which lie beneath the soil surface, formed when inland lakes evaporated (e.g., Searles Lake, CA). These salts were deposited in Pleistocene and Holocene times during major dry episodes. The salts contain varying mixtures of saline minerals such as halite, trona, and nahcolite (Smith, 1979). The term fossil salt has been used to describe the salinity of these deposits, which are substantial sources of salinity.

Irrigation with poor quality water, inadequate leaching, seepage from canals, high water tables, and high evaporation rates all contribute to secondary salinization of irrigated soils. These salts will reduce crop yield if they accumulate in the root zone.

The development of waterlogged soils is mainly associated with low-lying lands having poor physical condition and internal drainage. These soils are mainly found in the flood and deltaic plains of rivers and valleys. Salt accumulation depends on the salinities of the applied water and target soil, and the rate at which salts are leached out of the root zone. If a restricting layer exists close to the soil surface, waterlogging occurs and the accumulation of salt associated with waterlogged soils develops within a comparatively short time. In the absence of such a layer and with large capacity vadose zones, irrigation may be practiced for centuries before problems with surface drainage arise, if ever. Another source of saline water is irrigation drainage effluent, which is often used for irrigation. Although salinity levels may vary, the electrolyte concentrations are usually relatively high.

Under arid or semiarid conditions and in regions of poor natural drainage, salt accumulation is a real hazard. However, salt-affected soils are not confined to semiarid and arid regions. In some other regions, the climate and salt mobility produce saline waters and soil seasonally.

Irrigated agriculture has faced the challenge of sustaining its productivity for generations. Because of natural hydrological and geochemical factors, as well as irrigation-induced activities, soil and water salinity and associated drainage problems continue to plague agriculture.

17.2 Water Quality Criteria for Irrigation

Irrigation water quality is one of the primary considerations ensuring proper water and soil management for crop production. The suitability of water for irrigation is determined by its potential to cause problems to soils and crops and is related to the management practices needed. The quality of water for irrigation is determined mainly by the concentration and composition of solutes present. Most soluble salts in water are composed of the cations Na^+, Ca^{2+}, and Mg^{2+} and anions Cl^-, SO_4^{2-}, and HCO_3^-. Usually, smaller quantities of K^+ and NO_3^- also occur, as do many other ions and molecules (e.g., B). The concentration of HCO_3^- and CO_3^{2-} are pH and CO_2 partial pressure (P_{CO_2}) dependent.

Bicarbonate is an important parameter as it has a tendency to precipitate Ca and Mg as carbonates (Letey et al., 1985; Amrhein and Suarez, 1987). Calcium reacts with the HCO_3^- readily and precipitates as $CaCO_3$. Although Mg is not as easily precipitated as carbonate, it can accelerate the precipitation of $CaCO_3$ by replacing Ca on the exchange complex. As Ca and Mg are precipitated, the relative concentration of Na on the exchange phase increases, resulting in soil dispersion, if the total electrolyte concentration is below the critical flocculation concentration (CFC) of the soil clays.

The relative abundance of various ions such as Ca^{2+}, SO_4^{2-}, and HCO_3^- in irrigation water may affect the concentration of soluble salts in the soil solution due to precipitation or dissolution (common ion effect or changing P_{CO_2}). Expressions proposed to account for these effects include the effective salinity, which represents the total salt concentration of irrigation water minus the fraction expected to precipitate as $CaCO_3$, $MgCO_3$, or $CaSO_4 \cdot 2H_2O$.

In irrigation water, cations must balance anions to maintain electroneutrality. The relative abundance of various anions (such SO_4^{2-}, HCO_3^-) in irrigation water affects the eventual soluble salt concentration and the subsequent cation concentration ratio in the soil solution. Irrigation with saline water containing excess bicarbonate caused soil pH and sodium adsorption ratio (SAR) to increase, due to precipitation of calcium and magnesium carbonates (Gardner, 2004).

Many problems associated with irrigated agriculture arise from dissolved solute concentration and composition in the water applied. Since all waters contain varying concentrations and different species of solutes, considerable effort has been expended to classify the quality of water in terms of its dissolved solutes composition. The pH of irrigation water is usually not an accepted criterion of water quality because of the disparity in buffer capacity between water and soil.

Salinity is usually expressed as a lumped parameter, for example, specific electrical conductance (EC) in dS m^{-1}, or total dissolved solids (TDS) in mg L^{-1}. No exact relationship exists between these two measures, but TDS may be approximated by multiplying EC (dS m⁻¹) by a factor varying from 640 in less saline to 800 in hypersaline water.

Generally, the most important water quality parameter for irrigation is the total salt concentration, most commonly measured as EC. The relation between EC (S m⁻¹) and electrolyte concentration (C) of a given salt (mol$_c$ m⁻³) is given by:

$$\Lambda = \frac{EC}{C} \quad (17.1)$$

where Λ is the equivalent conductance (S mol$_c^{-1}$ m²).

The equivalent conductance can be calculated as

$$\Lambda = \Lambda_0 - bC^{1/2} \quad (17.2)$$

where
Λ_0 is the equivalent conductance at infinite dilution for a given electrolyte
b is a constant whose value depends on the salt

For strong electrolytes, the assumption that b is independent of concentration is valid in the low concentration range. By rearranging Equation 17.1 and introducing Equation 17.2, EC is given by

$$EC = C\Lambda_0 - bC^{3/2} \quad (17.3)$$

The relationships between EC (dS m⁻¹) and electrolyte concentration (mmol$_c$ L⁻¹) for various salt solutions are presented in Figure 17.1. The EC of an aqueous electrolyte solution increases

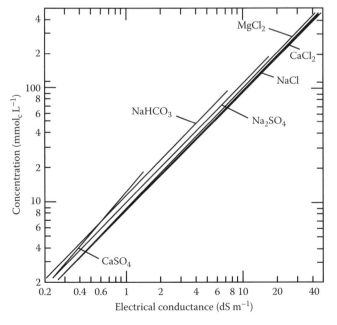

FIGURE 17.1 Relationship between electrolyte concentration and specific EC of various electrolyte solutions at 250°C. (From U.S. Salinity Laboratory Staff. 1954. Diagnosis and improvement of saline and alkali soils. USDA handbook 60. U.S. Government Printing Office, Washington, DC.)

at a rate of approximately 2% C⁻¹. Although the factor for converting EC (dS m⁻¹) to C (mmol$_c$ L⁻¹) depends on the type of electrolyte (e.g., 7.9 for NaCl), a useful approximation is

$$C \,(mmol_c\, L^{-1}) = 10 * EC\,(dS\,m^{-1}) \quad (17.4)$$

when $bC^{3/2} \ll C$.

Usually, classification schemes to define salinity and sodicity hazards are based on broad generalizations regarding crops to be grown, soil properties, irrigation management, and climate. Initial schemes were developed using various expressions for salinity and sodicity criteria (U.S. Salinity Laboratory Staff, 1954) but problems have been encountered, particularly as far as sodicity is concerned (Summer and Naidu, 1998). A classification based on specific use factors that takes salt tolerance of plants into account was proposed by Bernstein (1967) and Carter (1969). The general water classes in relation to their salt concentrations are presented in Table 17.1. This classification was based only on irrigation water quality and plant sensitivity to salinity.

The salinity of irrigation water can be assessed by relating irrigation water salinity (EC$_{iw}$), leaching fraction (LF), average root zone salinity, and crop salt tolerance (Rhoades, 1982). The LF can be estimated using Equation 17.5

$$LF = \frac{EC_{iw}}{5EC_e - EC_{iw}} \quad (17.5)$$

where
EC$_{iw}$ is the electrical conductance of the irrigation water (dS m⁻¹)
EC$_e$ is the average electrical conductance of the soil water in the root zone (dS m⁻¹)

The relationships between the average salinity in the root zone (EC$_e$) and EC$_{iw}$ for each LF are presented in Figure 17.2 in which the horizontal lines indicate crops that can be grown successfully without decreases in yield from salinity. For example, if EC$_{iw}$ is 5 dS m⁻¹ and LF is 0.2, only tolerant crops can be grown without decreasing yield. However, if LF is 0.4, moderately tolerant plants can be grown successfully. Thus, assessing the effects of salinity as a water quality parameter depends on the soil, the crop, available water, irrigation management, and the decrease in yield that can be tolerated.

In addition to the total salinity hazard of irrigation water, the ionic composition must be also considered. A useful index for predicting excessive exchangeable Na is the SAR, defined as

$$SAR = \frac{(Na^+)}{(Ca^{2+})^{1/2}}; \quad 0 \leq SAR \leq \infty \quad (17.6)$$

The activity of Na and Ca ions is given in mmol L⁻¹.

The U.S. Salinity Laboratory (1954) assumed that Mg behaves similar to Ca²⁺ in the adsorbed phase and defined SAR = (Na⁺)/(Ca²⁺ + Mg²⁺). Sposito and Mattigod (1977)

TABLE 17.1 Water Classes in Relation to Their Salt Concentration

Class of Water	Electrical Conductivity dS m^{-1} (25°C)	Comments
Low salinity	0–0.4	These waters can be used for irrigating most crops grown on most soils with a low probability that salt problems will develop. Some leaching is required, but this generally occurs with normal irrigation practices
Moderate salinity	0.4–1.2	These waters can be used if a moderate amount of leaching occurs. Plants with moderate salt tolerance can be grown in most instances without special practices for salinity control. Producing field beans and potatoes with these waters is hazardous and requires special management practices
High salinity	1.2–2.25	These waters should not be used on soils with restricted drainage. Plants tolerant to salinity should be grown. Excess water must be applied for leaching
Very high salinity	2.25–5.0	These waters are suitable for irrigation under special circumstances. Adequate drainage is essential. Only very salt tolerant crops should be grown. Considerable excess water must be applied for leaching

Source: After Carter, D.L. 1969. Managing moderately saline (salty) irrigation waters. University of Idaho Current Information Series 107. University of Idaho, Moscow, ID.

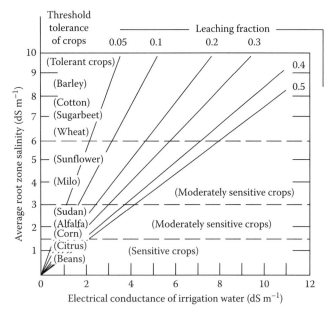

FIGURE 17.2 Relationship between average root zone salinity (saturation extract basis), specific EC of irrigation water, and LF to use for conditions of conventional irrigation management. (From Rhoades, J.D. 1982. Reclamation and management of salt-affected soils after drainage, p. 123–197. *In* Proc. 1st Ann. West. Prov. Conf. Rational. Water Soil Res. Manag. Lethbridge, AB, Canada.)

distinguished between the true SAR (SAR$_t$), which includes free ion activities and practical SAR (SAR$_p$), uncorrected for ion pairs, complexes, or activity coefficients. The relationship between the two parameters is given in Figure 17.3. The SAR$_p$ is generally less than the SAR$_t$ because of the greater effects of ion pair, complex, and activity coefficient corrections on Ca^{2+} and Mg^{2+} than on Na$^+$. The regression equation between the two parameters is

$$(SAR)_t = 0.08 + 1.115 (SAR)_p \qquad (17.7)$$

The ionic strength (I) of a solution is needed in order to calculate the ion activity coefficient. Calculation of I is complicated by the

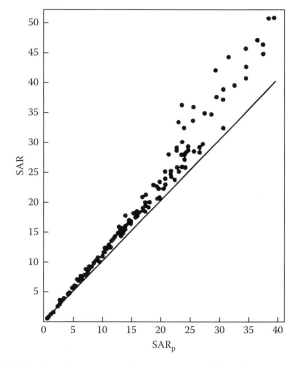

FIGURE 17.3 The relationship between the SAR and the practical SAR for 161 soil solutions and water extracts. The straight line represents a 1:1 functional relationship between SAR and SAR$_p$. (Reprinted from Sposito, G., and S.W. Mattigod. 1977. On the chemical formation of the sodium adsorption ratio. Soil Sci. Soc. Am. J. 41:323–329. With permission of the Soil Science Society of America.)

presence of ion pairs and complexes in soil solutions. However, values of I (mol L^{-1}) can be calculated with sufficient accuracy for most applications from the relation that includes corrections for ion pairs and complexes and is valid for most natural waters (Griffin and Jurinak, 1973):

$$I = 0.013 * EC \qquad (17.8)$$

where EC is the specific electrical conductance (dS m^{-1}).

The relationship between exchangeable sodium percentage (ESP) and SAR is given by the mass action equation:

$$ESP = \frac{100 * K * SAR}{1 + K * SAR} \quad (17.9)$$

where K is the reaction coefficient (mmol L^{-1})$^{-1/2}$.

When K·SAR ≫ 1, 1 can be neglected and

$$ESP \rightarrow 100 \quad (17.10)$$

But when K·SAR ≪ 1, then K·SAR can be neglected in the denominator and

$$ESP \rightarrow 100 * K * SAR \quad (17.11)$$

Therefore, the Na$^+$ concentration of irrigation water alone is not sufficient to estimate the potential sodicity hazard. From Equation 17.9, the effective diagnostic criteria for estimating the sodicity status of a given soil are the SAR of the irrigation water and the reaction coefficient (K) for Na–Ca exchange of the soil. For example, the relationship between ESP and SAR of a soil at equilibrium for various reaction coefficients is given in Figure 17.4. These calculations indicate that the K for Na–Ca exchange in the exchangeable cation complex of the clays is an important criterion in evaluating water quality for irrigation for a given soil. It is important to note that for soils having K > 0.01 (mmol L^{-1})$^{-1/2}$, there is only one SAR value that is equal to ESP for each K (the linear line). Positive and negative deviations are observed when SAR is below or above this value, respectively. However, for many montmorillonitic soils, having a K value ~0.015 (mmol L^{-1})$^{-1/2}$, the deviation of ESP from SAR of the soil solution is small, up to SAR 30.

Since the SAR is usually determined on an irrigation water or saturation extract of a soil (by adding distilled water), there is an inherent problem of relating these values to the condition that exists in the field. By neglecting the effect of salt precipitation or dissolution, mineral weathering, and the uptake of salts by plants, salt concentration in soil solution increases due to root water uptake and water evaporation from bare soil without changing the relative concentrations among the ionic species in soil solution. The SAR value, however, increases proportionally to the square root of the increase in the total concentration. Thus, if the total electrolyte concentration in solution increases by a factor of d, the SAR will increase by the factor of (d)$^{1/2}$. Since the soil ESP increases with the increase in SAR, it is obvious that the resultant ESP would be greater than the ESP expected from the SAR of the irrigation water. This trend has been observed in lysimeters and field experiments (Rhoades, 1968).

It is obvious, therefore, that in attempting to assess the sodicity hazard in soils, texture, clay mineralogy, SAR, and the total electrolyte concentration of the percolating solution should be taken into consideration. Other important parameters needed for water quality assessment for irrigation are soil physical properties, pH, precipitation and dissolution processes of primary and secondary minerals and salts in soil, specific toxic ion concentrations in relation to the affinity of soil constituents to adsorb–desorb, as well as soil and irrigation management (e.g., presence of soil amendments such as gypsum, LF, irrigation methods). A comprehensive irrigation water classification scheme incorporating all these factors has not yet been developed. In contrast to previous studies, Gardner (2004) reported that SAR of the soil extracts was not correlated with ES and soil pH. The SAR values rose with irrigation once pH exceeded 8, suggesting that carbonate formation was incomplete due to insufficient bicarbonate.

17.3 Effect of Salinity on Soil Physical Properties

Considerable progress has been made in managing and controlling salinity and sodicity in irrigated lands. However, because of natural hydrological and geochemical factors, as well as irrigation-induced activities, soil and water salinity and associated drainage problems continue to plague agriculture.

The ionic composition and concentration of the soil solution affects soil physical properties. The accumulation of dispersive cations, such as Na (and sometimes Mg and K), promotes clay swelling and/or dispersion altering the geometry of soil pores which, in turn, affects intrinsic soil permeability, water retention, and crop productivity.

The adverse physical condition imparted to soils by high ESP on the exchange complex can be partially overcome by a sufficiently high concentration of salts. Smectites are the dominant clay mineral in many semiarid and arid regions and they

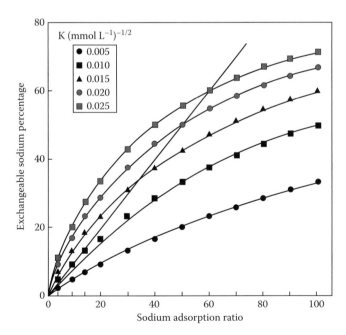

FIGURE 17.4 Relationship between exchangeable sodium percentage (ESP) and sodium adsorption ratio (SAR) for various adsorption coefficients.

determine much of the physical properties of soils, such as swelling and dispersion, due to the large specific surface area and the electrostatic charge. These two processes determine, to a large extent, soil structure, hydraulic properties, and soil water retention.

The Van der Waals attraction energy, V_A, between two semi-infinite plates such as 2:1 clay mineral, is given by

$$V_A = -\frac{A}{12\pi}\left[\frac{1}{h^2} + \frac{1}{(h+2t)^2} + \frac{1}{(h+t)^2}\right] \quad (17.12)$$

where

A is the Hamaker constant (2.2×10^{-20} J, Israelachvili and Adams, 1978)
h is the distance between the surfaces of the plates
t is the thickness of the clay plate

The Van der Waals forces are independent of the charge density on the clay and the composition and concentration (<1 mol dm^{-3}) of the electrolytes in suspension in the distance range 1–15 nm between two adjacent clay plates (Israelachvili and Adams, 1978).

The repulsion energy $V_R(d)$ between two clay platelets can be computed from the diffuse double layer theory (Lyklema, 1982),

$$V_{R(d)} = \frac{64nkT}{\chi}\left[\tanh\left(\frac{Ze\psi_s}{4kT}\right)\right]^2 e^{-2\chi d} \quad (17.13)$$

where

n is the number of cations per unit volume
k is the Bolzmann constant
T is the temperature
Z is the cation valency
ψ_s is the Stern layer potential
d is the distance from clay surface to the midway plane between two parallel platelets
e is the electron charge
χ is the reciprocal Debye length, given by

$$\chi = \left(\frac{2e^2nZ^2}{DkT}\right)^{1/2} \quad (17.14)$$

where D is the dielectric constant of the medium and the other symbols are as defined above.

The repulsion energy depends on charge density of the plate and composition and concentration of the electrolyte in suspension.

Montmorillonite platelets in aqueous suspension may flocculate in three possible modes of particle association (Van Olphen, 1977): (1) association between siloxane planes of two parallel platelets (FF), (2) association between edge surfaces of neighboring particles (EE), and (3) association between an edge surface and a siloxane planar surface (EF). In the presence of electrolyte (e.g., NaCl) at low concentration, the double layers on the Na-montmorillonite particles are well developed and osmotic repulsion prevents particle association. Thus, a stable suspension of individual platelets is obtained (Van Olphen, 1977). As the concentration of NaCl in the suspension increases, double layers at both the planar and the edge surfaces are compressed, and at the CFC, both EF and EE association can occur (Heller and Keren, 2001). As the electrolyte concentration increases further, the FF mode of particle association occurs, and "oriented aggregates" are formed (Keren et al., 1988).

The mechanism of swelling has as its analogy the osmotic characteristics of a semipermeable membrane. The ions between the clay platelets in the exchange phase are constrained there by the electrical attraction of the negatively charged clay surfaces. Thus, under a nonequilibrium condition, the gradient for water movement is from the bulk solution to the inner platelet region since the concentration of ions is greater there. The movement of the water into this region creates a hydrostatic pressure that forces the clay to expand and swell. A Na-soil, with a larger number of ions in the inner platelet region coupled with their weak interaction with the clay platelets, is more susceptible to swelling and dispersion than a Ca soil. Therefore, excess adsorbed Na most adversely affects the permeability of the soil (McNeal et al., 1968; Frenkel et al., 1978; Keren and Singer, 1988), which decreases with the square of the pore radius. Swelling, or the movement of clay, or both, significantly affects permeability. The extent of swelling, or dispersion of clays, or both, depends on the mineralogy of the clays, the composition of the adsorbed ions, the concentration of salt in the solution, and the content of hydroxy-aluminum and hydroxy-iron (Shainberg et al., 1971; Oster et al., 1980; Goldberg and Glaubig, 1987; Keren and Singer, 1988, 1989, 1991). The swelling of clay increases as the electrolyte concentration of the solution decreases. The dispersion of clay occurs only at electrolyte concentrations below the CFC of the clay. The degree to which the clay swells before the solution is replaced with one that has an electrolyte concentration below the clay's FV determines whether clay particles will move into the conducting pores and leave the root zone, or remain trapped in the narrow pores and decrease the hydraulic conductivity (HC) of the soil (Keren and Singer, 1988).

When leached with water of low salt concentrations, the soil's potential for mineral weathering, that is, for releasing electrolytes, influences the HC. Calcareous soils and soils with feldspar minerals release electrolytes into the solution in the presence of exchangeable Na. If the electrolyte concentration of the soil solution is maintained above the CFC, the clay does not disperse and the HC changes minimally. The electrolyte concentration of the applied water controls the HC of soils without weathered minerals. When these soils are leached with low salt waters (below CFC), clay disperses and the HC decreases.

The HC of a soil can be maintained even at high ESP levels provided that the EC of the percolating solution is above a threshold value. Conversely, when irrigation water with very low EC is used, even an ESP = 5 caused a twofold decrease in HC (McIntyre, 1979). Thus, in arid and semiarid regions where irrigation with saline water is practiced and soils are saline and sodic, the deleterious effect of exchangeable Na is more evident during the irrigation with high quality water (or rain) than that during irrigation with saline water.

Infiltration rate (IR) also depends on EC and ESP of the soil (Agassi et al., 1981) but is more sensitive to increasing ESP than the HC, because the energy of impacting raindrops promotes dispersion allowing a seal to form at the surface (Oster and Schroer, 1979). The permeability of the seal decreases with decreasing EC at the soil surface. Thus, when low EC water is applied (rain or snow), salt concentration at the soil surface will be low even in soils containing weatherable minerals such as $CaCO_3$, and clay dispersion takes place. Consequently, sodicity damage to soil will not take place during the irrigation season when EC in the applied water is high enough to counter the dispersive effect of the exchangeable Na, but will occur during the rainy season (or irrigation with high quality water) when EC in the soil decreases to values that allow clay to disperse. Thus, the salinity of the applied water is an important parameter in assessing sodicity hazard.

Successful irrigated agriculture, however, requires permanent control of salinity, sodicity, and B levels in soil and irrigation water. Reclamation of affected soils by means of improved drainage, chemical treatments, and leaching of salts (Keren and Miyamoto, 1990) can reduce natural hazards creating stable conditions for intensive cropping.

17.4 Boron–Soil Interaction

When irrigation water containing boron enters soil, boron can be adsorbed on soil constituents or desorbed into soil solution, depending on the boron concentration in the irrigation water and the level of native adsorbed boron. The time required to reach a steady-state boron concentration in the soil depends on the boron concentration in relation to that before irrigation, the LF, and adsorption capacity of the soil. Even if irrigation water does not contain toxic concentrations of boron, its retention by the soil can lead to accumulation approaching harmful levels in the long term if LF (including rainfall) is insufficient.

The main boron species likely to occur in soil solution are $B(OH)_3$ and $B(OH)_4^-$ and the ratio between them is pH dependent. Boron can be specifically adsorbed by different clay minerals, oxyhydroxides of Al, Fe, and Mg, and organic matter, which vary in their adsorption capacities (Goldberg and Glaubig, 1985; Keren and Bingham, 1985). The adsorption depends on the equilibrium boron concentration in the soil solution and pH (Bingham et al., 1971; Mezuman and Keren, 1981; Goldberg and Glaubig, 1985). Increasing pH enhances boron adsorption by soils, clays, and Al and Fe oxides showing a maximum in the alkaline pH range. Boron adsorption isotherms for a given soil at various pH values are presented in Figure 17.5 (Mezuman and Keren, 1981) as an example. Keren and Bingham (1985) have reviewed the mechanisms and factors that affect the adsorption and desorption of boron by soil constituents.

In assessing boron concentration in irrigation water, the physical–chemical characteristics of the soil must be taken into consideration because of the interaction between boron and soil. Boron sorption and desorption from soil adsorption sites

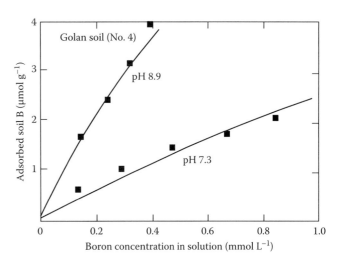

FIGURE 17.5 Adsorbed amount of boron on clay soil as a function of the total boron solution in 0.01 N $BAcl_2$ solution and of pH. (Reprinted from Mezuman, U., and R. Keren. 1981. Boron adsorption by soils using a phenomenological adsorption equation. Soil Sci. Soc. Am. J. 45:722–726. With permission of the Soil Science Society of America.)

regulate the boron concentration in soil solution depending on the changes in solution boron concentration, the affinity of soil for boron species, soil solution pH, and competing ion concentration. Thus, adsorbed boron may buffer fluctuations in solution boron concentration. When irrigation with water high in boron is planned, special attention should be paid to this interaction because of the narrow range between levels causing deficiency and toxicity symptoms in plants.

17.4.1 Boron Adsorption by Soil Constituents

Boron can be specifically adsorbed by different clay minerals that vary in their adsorption capabilities. Boron adsorption isotherms for Ca saturated with montmorillonite, illite, and kaolinite at constant pH and ionic strength were conducted by Keren and Mezuman (1981). It is evident that on a weight basis, illite is the most reactive among these clay minerals, whereas kaolinite is characterized by the lowest level of boron adsorption. Although boron adsorption by illite is much greater than by montmorillonite, the total surface area of montmorillonite is much greater than that of illite. Assuming that (i) most of the boron adsorption by montmorillonite or illite takes places at the broken edges and (ii) the surface area of these broken edges is about $8\,m^2\,g^{-1}$ (Dyal and Hendricks, 1950), the boron concentration at the clay surface (based on the edges alone) should be about 6.6×10^{-5} mol g^{-1} clay. This value is on the same order of magnitude as that found for illite (1.5×10^{-5} mol g^{-1}) by Keren and Mezuman (1981). Moreover, the calculated maximum boron adsorption values based on edge surface areas for montmorillonite and illite are very close: 1.47 and 1.88 μmol m^{-2}, respectively. This suggests that boron is adsorbed on the clay edges rather than on the planar surfaces (Keren and Talpaz, 1984). The boron adsorption process can be explained by the surface complexation approach, in which the surface is considered as a ligand (Keren et al., 1994).

Such specific adsorption, which occurs irrespective of the sign of the net surface charge, can occur theoretically for any species capable of coordination with the surface metal ions. However, because oxygen is the ligand commonly coordinated to the metal ions in clay minerals, the boron species B(OH)$_3$ and B(OH)$_4^-$ are particularly involved in such reactions. Possible surface complex configurations for boron—broken edges of clay minerals—were suggested by Keren et al. (1994).

Hydroxy oxides of Al, Fe, and Mg adsorb large amounts of boron (Keren and Gast, 1983; Goldberg and Glaubig, 1985, 1988). In view of the chemical and structural similarities between clay edges and hydroxy-Al, Keren and Gast (1983) compared boron adsorption by these two surfaces. They showed that the hydroxy-Al exhibited a marked affinity for boron. The amount of boron adsorbed by hydroxy-Al at pH 9.5 was about 7.5 times greater than by Ca-montmorillonite under the same conditions (Keren and Mezuman, 1981; Keren and Gast, 1983). Magnesium hydroxide has an appreciable capacity for boron adsorption, which is not affected by aging (Rhoades et al., 1970). Rhoades et al. (1970) concluded that arid-zone soils can have appreciable boron adsorption capacities in their silt and sand fractions as well as clay fractions. The site of this adsorption appears to be hydroxy-magnesium clusters or coatings that exist on the weathering surfaces of ferro-magnesium minerals. It can be concluded that the presence and amount of hydroxy-Al, -Fe, and -Mg have a substantial role in boron adsorption by soils.

Organic matter is an active portion of a soil. Although most cultivated soils contain only 1%–5% organic matter (mostly in the top 30 cm of soil), this small amount can significantly modify the soil's chemical properties. The humus has a chemical affinity for boron and it appears to have an important role in the retention of boron by soils (Parks and White, 1952). Hydrogen-saturated humus was found to retain boron approximately twice the concentration of the calcium-saturated humus (Parks and White, 1952). The complex formation between diols and boron (Boeseken, 1949) can explain the results observed by Parks and White (1952) for humus. The adsorptivity of organic matter for boron is much higher than that of montmorillonite, kaolinite, or illite (Keren and Bingham, 1985; Yermiyahu et al., 1988). Boron adsorption by dissolved organic matter extracted from effluent of a municipal sewage plan was 294–333 mg kg^{-1} (Communar and Keren, 2008). The interaction of effluent-dissolved organic matter with boron may play an important role in assessing boron solution concentration and boron transport in soil.

17.4.2 Kinetics and Mechanisms of Boron Adsorption

Boron is adsorbed mainly on the edge surfaces of the 2:1 clay minerals (Keren and Talpaz, 1984; Keren and Bingham, 1985; Keren and Sparks, 1994). Since the negative electrical field (e.g., emanating from the particle face of montmorillonite) extends to the edge surface region (Secor and Radke, 1985), it influences edges, making them less accessible to approaching borate anions. The forward adsorption rate constant, k_1, for boron adsorption by pyrophyllite was $10^{4.26}$ L mol^{-1} and the backward rate constant, k_{-1}, for the desorption was $10^{1.11}$ S^{-1} (Keren et al., 1994). The intrinsic equilibrium constant obtained from the kinetic measurements ($\log_{10} K_{kinetic} = 3.15$) agreed relatively well with that calculated from the static studies ($\log_{10} K_{static} = 3.51$). The reaction scheme for the boron–pyrophyllite interaction was suggested as (Keren et al., 1994)

$$S(OH)_2 + B(OH)_4 \leftrightarrow SO_2\text{-}B(OH)_2 + 2H_2O \quad (17.15)$$

or

$$S(OH)_2 + B(OH)_4 \leftrightarrow SO_2H\text{-}B(OH)_3 + H_2O \quad (17.16)$$

17.4.3 Boron Adsorption Modeling

The surface complex approach assumes that the surface consists of discrete sites that interact with the adsorbate specifically. These interactions would include the attractive short-range dispersion forces and hydrogen bonding in the interfacial regions. It is possible that the hydroxyl groups in the coordination sphere of the adsorbate species may act as bridging ligands between the adsorbate and adjacent surface groups. Thus, boron can be specifically adsorbed on mineral surfaces and hydroxy-Al polymers through a mechanism referred to as ligand exchange, whereby the adsorbed species displace OH$^-$ (or H$_2$O) from the surface and form partly covalent bonds with the structural cations (Hingston et al., 1974). Such specific adsorption, which occurs irrespective of the sign of the net surface charge, can occur theoretically for any species capable of coordination with the surface metal ions. However, because oxygen is the ligand commonly coordinated to the metal ions in clay minerals, the boron species, B(OH)$_3$ and B(OH)$_4^-$, are particularly involved in such reactions. Results by Keren and O'Connor (1982) and Keren et al. (1994) support this approach.

The suggested mechanism for the reaction of organic sites with boron is similar to the hydroxyl replacements suggested for clay minerals. Forsyth (1950) studied the composition of the soluble polysaccharide fraction of soil organic matter and found that upon hydrolysis various sugars such as galactose, glucose, mannose, arabinose, xylose, and glucoronic acid were produced. All these compounds meet the requirements for complex formation with boric acid, and most of them have been demonstrated to react with boric acid (Boeseken, 1949).

Boron adsorption by soils (Bigger and Fireman, 1960; Elrashidi and O'Connor, 1982) and clays (Hingston, 1964; Sims and Bingham, 1967) appears to obey the Langmuir equation

$$\frac{x}{m} = \frac{k * b * c}{1 + k * c} \quad (17.17)$$

where
x is the amount of boron adsorbed
m is the amount of adsorbent
k is the affinity coefficient of the adsorbent to boron
b is the maximum capacity of the adsorbents for monolayer coverage
c is the equilibrium boron concentration in solution

This equation was found to be valid for some conditions but deviations occur at higher solution concentrations and at higher pH values (Hingston, 1964). This adsorption model has two disadvantages: (1) it does not account for the existence of two boron species of different affinities for the clay, and their varying proportion in the equilibrium solution with varying pH; and (2) different values of the Langmuir coefficients, namely maximum adsorption and binding energy constants, must be assigned to predict boron adsorption at any given pH value.

Another approach for describing boron adsorption by clays, hydroxy-Al, and soils has been derived by Keren et al. (1981). Their model is based on the assumption that $B(OH)_3$, $B(OH)_4^-$, and OH^- are all competing for the same adsorption sites. The equilibrium constant for adsorption of each of the three species is given by

$$K_i = \frac{(M_i)(1-\theta_i)}{\theta_i} \qquad (17.18)$$

where
  K_i is the equilibrium constant
  M_i is the activity of the i th form
  θ_i is the mole fraction of the i th form adsorbed
  $1 - \theta_i$ is the mole fraction of the adsorber not accounted for

When one combines Equation 17.16 for the three species $B(OH)_3$, $B(OH)_4^-$, and OH^- and introduces $\theta_i \equiv Q_i/T_m$ where Q_i is the adsorption of the species M_i and T_i is the total boron adsorption capacity coefficient, then the adsorption of species M_i can be described by

$$Q_i = \frac{T_m K_i (M_i)}{1 + \sum_{j=1}^{3} K_j (M_j)} \qquad (17.19)$$

where K_i (or K_j) is the equilibrium constant.

The general form of Equation 17.17 written specifically for boron adsorption is given in Equation 17.18 (Keren et al., 1981):

$$Q_{BT} = \frac{T_m(K_{HB}(HB) + K_B(B))}{1 + K_{HB}(HB) + K_B(B) + K_{OH}(OH)} \qquad (17.20)$$

where
  T_m is the maximum boron adsorption
  K_{HB}, K_B, and K_{OH} are affinity coefficients related to the binding energy for the species $B(OH)_3$, $B(OH)_4^-$, and OH^-, respectively
  (HB), (B), and (OH) are the solution activities of the above species, respectively

Equation 17.20 can be rearranged to relate total amount of adsorbed boron, Q_{BT}, to total boron in the suspension, Q_T (adsorbed boron + boron in solution), and to solution-to-adsorbent ratio, R:

$$Q_{BT} = T_m \left\{ 1 + \frac{PR}{F(Q_T - Q_{BT})}[1 + K_{OH}(OH)] \right\}^{-1} \qquad (17.21)$$

Here $P = 1 + K_h 10^{14}(OH)$; K_h is the hydrolysis constant of the reaction

$$B(OH)_3 + 2H_2O \leftrightarrow B(OH)_4^- + H_3O^+ \qquad (17.22)$$

and $F = K_{HB} + K_B(P - 1)$.

Keren et al. (1981) estimated the affinity coefficients as well as the maximum boron adsorption values (Equation 17.20) from experimental results using a best fit procedure (Mezuman and Keren, 1981).

Soil pH is one of the most important factors affecting boron adsorption. Increasing pH enhances boron adsorption, showing a maximum in the alkaline pH range (Mezuman and Keren, 1981). The response of adsorbed boron on soils to variation in pH can be explained as follows. Below pH 7, $B(OH)_3$ predominates and since the affinity of the clay for this species is relatively low, the amount of adsorption is small. Both $B(OH)_4^-$ and OH^- concentrations are low at this pH; thus, their contribution to total boron adsorption is small despite their relatively strong affinity for the soil. As the pH is increased to about 9, the $B(OH)_4^-$ concentration increases rapidly. Since the OH^- concentration is still low relative to the boron concentration, the amount of adsorbed boron increases rapidly. Further increases in pH result in an enhanced OH^- concentration relative to $B(OH)_4^-$, and boron adsorption decreases rapidly due to the competition by OH^- at the adsorption sites.

A constant capacitance model was developed by Stumm, Schindler, and their coworkers (Schindler and Gamsjager, 1972; Hohl and Stumm, 1976; Stumm et al., 1976; Stumm et al., 1980) for hydrous oxides—aqueous solution interface—assuming that the adsorption mechanism is ligand exchange that takes place in the inner sphere and the relationship between surface charge and surface potential is linear. This model can describe boron adsorption by soil over changing conditions of solution pH and boron concentration (Goldberg et al., 2000). The surface complexation reactions for the application of this model to boron are

$$SOH + H^+ \Leftrightarrow SOH_2^+ \qquad (17.23)$$

$$SOH \Leftrightarrow SO^- + H^+ \qquad (17.24)$$

$$SOH + H_3BO_3 \Leftrightarrow SH_2BO_3 + H_2O \qquad (17.25)$$

$$SOH + H_3BO_3 \Leftrightarrow SH_3BO_4^- + H^+ \qquad (17.26)$$

Where SOH represents the surface functional group of surface hydroxyl ion bound to Al or Fe in the oxide mineral or an aluminol group on the clay mineral. The intrinsic conditional equilibrium constant expressions describing these reactions are (Goldberg and Glaubig, 1985)

$$K_+(\text{int}) = \frac{[SOH_2^+]}{[SOH][H^+]}\exp\left[\frac{F\psi}{RT}\right] \quad (17.27)$$

$$K_-(\text{int}) = \frac{[SO^-][H^+]}{[SOH]}\exp\left[\frac{-F\psi}{RT}\right] \quad (17.28)$$

$$K_B(\text{int}) = \frac{[SH_2BO_3]}{[SOH][H_3BO_3]} \quad (17.29)$$

$$K_{B-}(\text{int}) = \frac{[SH_3BO_4^-][H^+]}{[SOH][H_3BO_3]}\exp\left[\frac{-F\psi}{RT}\right] \quad (17.30)$$

where
- square brackets indicate concentrations (mol L^{-1})
- F is the Faraday constant (C mol$_c^{-1}$)
- ψ is the surface potential (V)
- R is the molar gas constant (J mol^{-1} K^{-1})
- T is the absolute temperature (K)

The computer program FITEQL (Westall, 1982) can be used to fit the intrinsic boron surface complexation constant to experimental adsorption data. This model has successfully described boron adsorption on aluminum and iron oxide minerals (Goldberg and Glaubig, 1985, 1988), clay minerals (Goldberg and Glaubig, 1986), and soils (Goldberg et al., 2000).

17.4.4 Boron Transport Modeling

Two-site model (LE-NE) accounting for the existence of equilibrium and nonequilibrium adsorption sites for boron and the two-domain two-rate model (TD-TR) were suggested by Communar and Keren (2005, 2006). The Keren model (Equation 17.20) was introduced into these two models. These models were very successful in the simulation of boron transport in soil.

A transient-state solute transport model TETrans (Corwin et al., 1991) was also used to simulate boron movement through the root zone (Corwin et al., 1999). Because the ability to simulate water flow has been evaluated already for this model, the focus of their study centered around the performance of various functional models of boron adsorption used as subroutines within the TETrans model, including the (1) Freundlich adsorption isotherm equation, (2) Langmuir adsorption isotherm equation (3) kinetic Freundlich equation, (4) temperature-dependent Langmuir equation, and (5) pH-dependent Keren boron adsorption model. They concluded that among these five adsorption models, the pH-dependent Keren boron adsorption model performed the best in simulating boron transport in soil.

17.5 Effects of Salinity and Boron on Plants

17.5.1 Crop Tolerance to Salinity

When soluble salt accumulates in a soil to a level harmful to the growth of particular plants, the soil is termed saline. The composition and concentration of salts in the soil solution adversely influence plant growth through (1) osmotic effects, which limit the ability of plants to absorb water from the soil solution; (2) specific ion effects; and (3) changes in soil physical and chemical properties that can have long-term detrimental effects on crop production.

Threshold concentrations for classifying crop tolerance to salinity are (Maas, 1990) (1) sensitive (<1.5 dS m^{-1}) (bean, clover, carrot, lettuce); (2) moderately sensitive (1.5–3.0 dS m^{-1}) (corn, alfalfa, broccoli, potato); (3) moderately tolerant (3–6 dS m^{-1}) (soybeans, wheat, squash); and (4) tolerant (6–10 dS m^{-1}) (barley, sorghum, sugar beet) (1 dS m^{-1} ~ 350 mg L^{-1} Cl$^-$). Salinity refers to concentrations of soluble salts that are so high as to negatively impact plant growth. In the absence of specific ion effects, crop growth reduction due to salinity is related to the osmotic potential, ψ_o (bar), which is a function of EC$_e$ (dS m^{-1}) at 25°C as follows:

$$\psi_o = -0.39\,EC_e \quad (17.31)$$

This relationship is valid for EC$_e$ in the range 3–30 dS m^{-1}. Because decreasing ψ_o (increasing absolute value) has the net effect of reducing water availability, plants growing on saline soils often appear to be suffering from drought.

Relative crop yield as a function of EC$_e$ for various crops is given in Figure 17.6 (Maas and Hoffman, 1977). Relative yield (Y) was calculated from the relation

$$Y = 100 - B(EC_e - A) \quad (17.32)$$

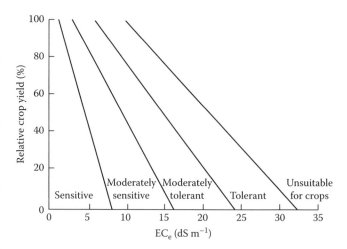

FIGURE 17.6 Divisions for classifying crop tolerance to salinity. (Reprinted from Maas, E.V., and G.J. Hoffman. 1977. Crop salt tolerance-current assessment. J. Irrig. Drain. Div. ASCE 103:115–134. With permission of the Soil Science Society of Civil Engineers.)

where
> EC_e is the electrical conductance of the soil saturation extract solution (dS m^{-1})
> A is the salinity threshold value (dS m^{-1})
> B is the percent yield decrease per unit salinity increase

The threshold salinity value and productivity decrease as a percentage of normal yield for each unit increase in EC_e have been summarized by Maas (1990). Such information is useful in selecting crops for growth under anticipated salinity conditions. However, the sensitivity of crops to soil salinity often changes from one stage of growth to the next.

Any assessment of water suitability for irrigation must be made in relation to crop tolerance to salinity, irrigation and soil management (LF, irrigation methods, soil properties, etc.), and yield decreases that can be tolerated. This can be shown (Figure 17.2) by relating the EC_e in the root zone to EC_{iw} at a LF appropriate for conventional irrigation management (such as furrow or flood), which allows for considerable drying between applications (Rhoades, 1982). Because most plants are relatively salt tolerant during germination, but more sensitive during emergence and early growth, it is imperative to keep salinity in the seedbed low after germination.

Varietal differences in salt tolerance often occur. Rootstocks affect the salt tolerances of tree and vine crops because they regulate the uptake and translocation of potentially toxic ions, such as Na and Cl, to the shoots.

Salt tolerance depends on the method of irrigation and its frequency. As water becomes limiting, plants experience stresses from low matric and osmotic potentials. Available salt tolerance data are most appropriate for crops irrigated by furrow or flood irrigation under conventional management. Because salt concentrations in irrigated profiles change constantly (up to several fold), the plant is most responsive to salinity in that part of the root zone where maximum water uptake occurs.

Sprinkler irrigated crops are potentially subject to additional damage by foliar salt uptake and burn from spray contact on the foliage. Susceptibility to foliar salt injury depends on leaf characteristics and rate of absorption and does not correlate with general salt tolerance. Although injury caused by spray on foliage depends on weather and water stress, relatively little information is available for predicting yield losses. Increase in frequency of sprinkling, temperature, and evaporation leads to increased salt concentration on the leaves and damage (Maas, 1986). Climate is a major factor affecting salt tolerance. Most crops can tolerate greater salt stress under cool humid than hot dry conditions. Yield is reduced more by salinity when humidity is low.

Substantial progress has been made in developing empirical models to relate crop yields and irrigation management under saline conditions. Childs and Hanks (1975) extended the production function model of Bresler and Hanks (1969) and Nimah and Hanks (1973a,b) to consider the effect of salinity and to compute yield. They reported a good correlation between computed and measured transpiration and predictions for crops with different rooting depths, different initial soil salinities, and multiyear cropping. Computed and measured data correlated well when levels of salinity were low but yields were overestimated when salinity was high (Wolf, 1977).

An empirical relationship between yield and water potential (matric and osmotic) after converting salt concentration to osmotic potential is (van Genuchten and Hoffman, 1984)

$$Y = \frac{Y_m}{1+\left(\Pi/\Pi_{50}\right)^p} \qquad (17.33)$$

where
> Y is yield
> Y_m is maximum yield under nonsaline conditions
> Π and Π_{50} are the osmotic potential of the solution and the osmotic potential at which the yield is reduced by 50%, respectively
> p is an empirical constant

This model described salt tolerance data as well or better than the model of Maas and Hoffman (1977). Solomon (1985) and Letey et al. (1985) presented a seasonal water-salinity production function based on the response of crops to water and salts. Measured and computed yields for several crops correlated well (Letey et al., 1985; Letey and Dinar, 1986).

The dynamic models described above assume a unique relationship between yield and evapotranspiration (ET) for a given crop and climate that is independent, regardless of whether the water stress leading to the reduced ET is caused by deficit water supply or excess salinity. Letey and Knapp (1991) discussed the usefulness and limitations of these models. Crop-water production functions can be used to evaluate the losses from increased salinity in soil and water, the potential for reuse of saline drainage waters, the demand for water in irrigated agriculture, and changes in irrigation and drainage policy.

17.5.2 Specific Molecule and Ion Effects on Plants

In addition to the osmotic and soil structure effects, some dissolved molecules and ions (boron, Cl, Na) can cause specific detrimental effects on crops. Toxicity occurs within the crop as the result of uptake and accumulation of these elements within plant tissue. Thus, an excess of these ions in irrigation water may be toxic to various plant physiological processes and may also cause nutritional disorders. In terms of saline soils, B is probably the most important. The effect of salinity on plant physiology and biochemistry is reviewed by Lauchli and Epstein (1990).

Boron has a marked effect on plants, with the optimum range in soil between deficiency and toxicity being very narrow. Boron deficiency is found primarily in humid regions or in sandy soils, while toxicity occurs most frequently in arid and semiarid regions due to high soil boron or boron additions in irrigation water.

The threshold boron concentration for irrigation water (maximum permissible concentration for a given crop that does not reduce yield or lead to injury symptoms) ranged from as low as 0.3 for sensitive to 2.0 mg L^{-1} for tolerant crops (U.S. Salinity Laboratory Staff, 1954). Yield decreases due to boron toxicity can be evaluated from (Bingham et al., 1985) the following equation:

$$Y = 100 - m(X - A) \qquad (17.34)$$

where
Y is the relative yield (for X^{3}A)
m is the decrease in yield per unit increase in boron concentration
X is the boron concentration in the soil solution
A is the maximum boron concentration that does not reduce yield (threshold value)

Boron threshold concentrations and the slopes of yield reduction as a function of boron concentration for a limited number of crops are given in Table 17.2. A more extensive listing is given by Keren and Bingham (1985).

Boron uptake by plants increases with decreasing clay content for a given amount of boron added to soil in irrigation or fertilization (Keren et al., 1985a, 1985b). The reason for this is that

TABLE 17.2 Boron Tolerance Limits for Agricultural Crops

Boron Tolerance	Crop Species Common Name	Threshold Concentration (mg L^{-1})	Slope[a]
Very sensitive	Lemon	<0.5	
Sensitive	Avocado	0.5–0.75	
	Grapefruit	0.5–0.75	
	Orange	0.5–0.75	
	Grape	0.5–0.75	
	Onion	0.5–0.75	
	Wheat	0.75–1.0	3.3
	Sunflower	0.75–1.0	
	Bean, snap	1.0	12
Moderately sensitive	Broccoli	1.0	1.8
	Radish	1.0	1.4
	Potato	1.0–2.0	
	Lettuce	1.3	1.7
Moderately tolerant	Barley	3.4	4.4
	Corn	2.0–4.0	
	Cauliflower	4.0	1.9
Tolerant	Alfalfa	4.0–6.0	
	Sugar beet	4.9	4.1
	Tomato	5.7	3.4
Very tolerant	Cotton	6.0–10.0	
	Celery	9.8	3.2
	Asparagus	10.0–15.0	

[a] % yield reduction resulting from an increase of 1 mg B L^{-1} in the soil solution.

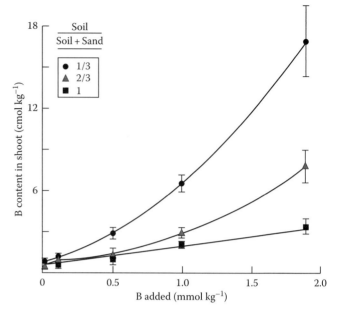

FIGURE 17.7 Relationship between boron content in wheat shoots and the amount of boron added to soil, for three soil–sand mixtures. (Reprinted from Keren, R., F.T. Bingham, and J.D. Rhoades. 1985b. Effect of clay content in soil on boron uptake and yield of wheat. Soil Sci. Soc. Am. J. 49:1466–1470. With permission of the Soil Science Society of America.)

boron absorption increases with increasing clay content as illustrated in Figure 17.7 (Mezuman and Keren, 1981). However, when plant boron content is plotted against boron activity in the soil solution, a linear relationship is obtained (Figure 17.8) indicating that plants obtain their boron solely from that present in the soil solution. Therefore, soil physicochemical characteristics must be taken into consideration when assessing water quality for irrigation in terms of boron. Because adsorption sites may act as a pool from which boron is supplied to solution, adsorbed boron may buffer fluctuations in solution boron concentration, which may change insignificantly by changing the soil water content.

Hanks et al. (1983, 1984) carried out long-term research on the use of saline water (EC$_{iw}$ = 4 dS m^{-1}, boron = 9.4 mg L^{-1}) for irrigation of crops. Over 8 years, no noticeable decrease in yield of forage crops was observed but potato, maize, barley, and wheat yields decreased after the third year mainly due to boron. During the experiment, most of the boron was retained in the soil with levels of 7–10 mg L^{-1} in saturation extracts. These data demonstrate the need to consider also the composition of saline water when judging its suitability for irrigation. Boron concentrations in irrigation water should not exceed the values in Table 17.2 for the crops indicated. However, salinity likely reduces the effects of boron on plant production (Diaz and Grattan, 2009). Interaction may occur that could increase the individual tolerance coefficient for B and salinity when a crop is exposed to both sources of stress at the same time (Ferreyra et al., 1997).

Chloride and sodium are particularly toxic to fruit trees and woody ornamentals. When the leaves of these plants accumulate more than about 0.5% chloride or 0.25% sodium on a dry weight

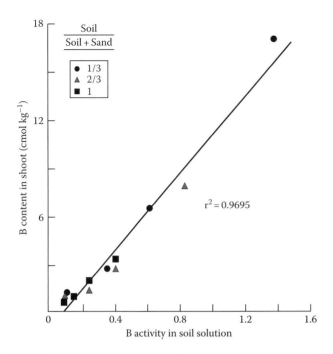

FIGURE 17.8 Relationship between boron content in shoot and the boron activity in soil solution, for three ratios of soil–sand mixtures at a field capacity water content. (Reprinted from Keren, R., F.T. Bingham, and J.D. Rhoades. 1985b. Effect of clay content in soil on boron uptake and yield of wheat. Soil Sci. Soc. Am. J. 49:1466–1470. With permission of the Soil Science Society of America.)

basis, they tend to develop characteristic leaf injury symptoms. Chloride is an essential microelement in very small amounts. Parker et al. (1983) indicated that the decisive attribute appears to be the ability of plants to restrict the transport of Cl⁻ from the root to the shoots. Many woody species are sensitive to chloride ion toxicity, though the degree of sensitivity varies among varieties and rootstocks (Maas, 1990).

High sodium concentrations have been observed to interfere with plant uptake of such essential nutrient element as K or Ca. Tolerance to sodium ion varies widely among species and rootstocks. In the case of avocado, citrus, and stone fruit trees, injury may occur at soil solution concentrations as low as 5 mmol L⁻¹. Sodium is adsorbed and retained for a time in the roots and lower trunk, but after some years the sapwood is converted to heartwood and the accumulated sodium ion is released. It is then transported up the canopy, where it may cause leaf scorch (Bernstein, 1975).

17.6 Reclamation of Saline and Boron-Affected Soils

17.6.1 Leaching for Salinity Control

Reclamation of saline soils is essentially a process where soil solution of high salt concentration is displaced by one less concentrated. Consequently, appropriate natural or installed drainage and disposal systems are essential. For flood and sprinkler irrigation, the flow of water is downward, and leaching is defined with depth. Lateral flow and salt removal from the main root zone typical of drip irrigation can also be considered as temporary leaching. However, rainfall, flood or sprinkler irrigation must be used to occasionally leach out the salt from the entire rooting profile. Leaching efficiency is the amount of salt removed from the root zone in drainage water at a given fraction of the irrigation water. This efficiency depends on salt content and distribution in the soil, solute composition, soil structure, and irrigation method and management.

Soil solution concentration and salt transport mechanisms are the main factors controlling leaching efficiency of noninteracting highly soluble salts. For interacting solutes and less soluble salts, chemical and exchange reactions are also important. For leaching of saline soils, Hoffman (1980) proposed the following empirical equation:

$$\left(\frac{C_i}{C_0}\right)\left(\frac{D_i}{D_s}\right) = K \qquad (17.35)$$

where

C_i and C_0 are final and initial soil salinities, respectively
D_i and D_s are depth of leaching water applied and depth of soil to be leached, respectively
K is an empirical efficiency parameter that ranges from 0.1 (sandy loam) to 0.3 (clay)

This equation is valid for $D_i/D_s > K$. Relationships between the fraction of initial salt concentration remaining in soil (C/C_0) and depth of leaching water applied per unit depth of soil (D_i/D_s) for continuous and intermittent ponding is presented in Figure 17.9 (Hoffman, 1980). Under continuous ponding, more water is required for leaching the clay loam than sandy loam soil. However, under intermittent ponding, the leaching efficiency of both soils was the same (K = 0.1). Intermittent ponding requires less water than continuous ponding to achieve the same degree of leaching (Miller et al., 1965) and sprinkler irrigation is more efficient than other methods at removing salt from small pores in the soil (Nielsen et al., 1966). Many computer programs are widely available to assess salinity hazard and leaching requirements (LRs). For example, the Extract Chem model version 2.0 is based on the chemical routines used in the UNSTCHEM model (Suarez and Simunek, 1997). The model was developed to predict the major ions and boron concentrations, electrolyte conductivity, and osmotic pressure of a soil solution at a desired water content based on the known ion composition at another water content. The WATSUIT model (USDA, Agricultural Research Service, Riverside, CA) is a computer program that predicts the salinity, sodicity, and toxic-solute concentration of the soil water within a simulated crop root zone resulting from the use of a particular irrigation water of given composition and at a specified LF. This model can be used to evaluate the effect of a given salinity level on crop yield and of a given sodicity level on soil permeability.

Effective irrigation and leaching require uniformity in water application, infiltration, and soil-water-holding capacity. Proper management that integrates irrigation methods and scheduling,

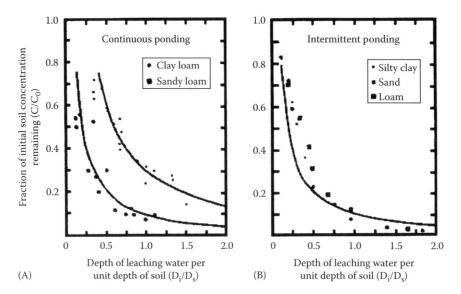

FIGURE 17.9 Relationship between fraction of initial salt concentration remaining in soil and depth of leaching water applied for a given depth of soil. (A) Continuous ponding; and (B) Intermittent ponding. (After Hoffman, G.J. 1980. Guideline for the reclamation of salt-affected soils, p. 49–64. *In* G.A. O'Connor (ed.) 2nd. Inter-Am. Conf. Salin. Water Manag. Tech. NM State University, Las Cruces, NM.)

soil treatments, and special cultivation practices is aimed at improving infiltration uniformity. With flood or furrow irrigation, land leveling, increased watering rates, and soil treatments that improve infiltration are recommended practices. Improvement of infiltration uniformity is obtained by reducing application rates below the soil infiltration capacity, addition of ameliorating chemicals that control crust formation, and use of cultivation practices that increase surface storage and reduce local and long distance runoff, such as check dams in the furrow or rough surface disking.

Furrow or drip irrigation leaches salt below and away from the infiltrating areas but salt accumulates at lateral wetting fronts, most commonly in bed ridges, and at the soil surface between emitters. Long-term use of drip irrigation may result in irregularly distributed salt accumulations, such that may affect crops to be planted in subsequent years. Special cultivation practices that carry salts away from the planting position or adjust planting position to salt distribution in the topsoil are essential (Ayers and Westcot, 1985). With both furrow and drip, occasional leaching by rain, flood or sprinkler irrigation is required to remove salt from the entire root zone.

When fields contain several soil types, leaching is more efficient in sandy than clay-rich portions. If leaching is continued until the clay-rich portions are sufficiently leached, an excessive amount of water would have percolated through the sandy areas. This problem is pronounced under ponded leaching where sandy areas are preferentially leached. In fact, the distribution of salinity in basin, border, or furrow-irrigated fields coincides closely with soil type (Miyamoto and Cruz, 1986, 1987). To increase efficiency, leaching should be based on the distribution of soil types (Miyamoto and Cruz, 1986) or the use of stochastic analysis of solute transport in spatially variable fields to divide a field into small sections that need different amounts of water (Russo, 1984). Dividing highly variable fields into areas of uniform soil type, with irrigation varying according to moisture-holding capacity and infiltration of each area, can significantly increase leaching efficiency.

When salts accumulate in the root zone during the irrigation season, they can be leached by applying water above the amount needed for ET. Over time, salt removal by leaching must equal or exceed salt additions from the applied water. To calculate the LR, both the irrigation water quality and the crop tolerance to soil salinity must be known (Ayers and Westcot, 1985).

Large changes in soil salinity under nonsteady-state conditions arise from rainfall during and out of the irrigation season, changes in irrigation water quality for significant periods, and heavy intermittent leaching rather than a constant LF at every irrigation. If management can synchronize changes in soil salinity with crop tolerance at different growth stages, the use of brackish water may be increased without detriment. When rainfall can accomplish the required leaching without stressing the crop, additional leaching is not required. This may be the case with tolerant crops in humid climates, or during wet years in semiarid zones. With sensitive crops in humid climates and tolerant crops during dry years in semiarid and arid zones, additional leaching by irrigation may be required.

Leaching into tile or open ditch drains rapidly removes salt directly above and near the drains but removes little midway between the drains. Leaching efficiency decreases as soon as the rapid flow reaches the drains. This uneven leaching is more evident in water-saturated (Luthin et al., 1969) than unsaturated fields, where leaching above the water table is a one-dimensional process (Talsma, 1967). To leach more efficiently, it is necessary to avoid ponding water near the drains (Miyamoto and Warrick, 1974) and apply water intermittently to keep the water table as low as possible (Talsma, 1967).

The high salt–water dilution technique (Reeve and Bower, 1960), which involves successive dilutions of a very saline irrigation water containing divalent cations, can be used when (1) the soil physical conditions have deteriorated and HC of the soil is so low that the time required for reclamation or the amount of amendment required is excessive, or (2) if a sodic soil is to be leached with a water so low in salinity that water infiltration decreases adversely. A nomograph is available for predicting reclamation at various stages of leaching (Reeve and Doering, 1966a). In the early phase, the high salinity of the water prevents clay dispersion and promotes flocculation, and the Ca content provides a source of Ca to replace exchangeable Na so that sodicity is decreased. On dilution with high quality water, the SAR of the irrigation water is reduced by the square root of the dilution factor. To ensure reclamation to the desired depth, D_i/D_s should be about 9 (Reeve and Doering, 1966a). This method is particularly effective for soils with swelling clays that have extremely low HC. A theoretical analysis of the use of high salt water in reclaiming sodic soils by the addition of a constant quantity of Ca^{2+} in every step in successive dilutions is presented by Misopolinos (1985).

In the reclamation of a highly sodic clay loam soil, Reeve and Doering (1966b) found that IR for a saturation gypsum solution was only 2×10^{-8} m s^{-1} and the time for reclamation to 0.9 m was 7 years of continuous leaching. However, the profile was reclaimed in 3 days with a series of $CaCl_2$ solutions (300 to 6 mol m^{-3}). The average IR for the $CaCl_2$ solutions was about 150 times that for the gypsum solution. This method has also been shown to reclaim a slowly permeable sodic soil in a humid environment, where HC and IR were increased from 30% to over 100% (Rahman et al., 1974).

The successful application of this method requires a balance between maintaining high EC to reduce reclamation time and low EC to reduce the amount of amendment required. A practical technique is to apply only two-thirds of the solution depth required for exchange equilibria with three dilutions of $CaCl_2$ followed by a gypsum application to satisfy the remaining exchange requirement, and completed by leaching with 0.3 m water m^{-1} soil to be reclaimed. This final step is essential to leach the saline solutions below the root zone.

Leaching reclamation without amendments (Jury et al., 1979, 1987) can be successful when soil drainage is good, adequate leaching water is available, and an internal source of Ca exists (e.g., $CaCO_3$). The rate at which $CaCO_3$ dissolves in water depends on the surface area-solution volume ratio, ionic composition of the solution and solid phases, clay affinity for cations, temperature, and P_{CO_2}. Amrhein et al. (1985) concluded that transfer of atmospheric CO_2 to solution is an important rate-limiting step in the dissolution kinetics of $CaCO_3$ in soils. The ion activity product (IAP) values expected for calcite (Suarez and Rhoades, 1982) are obtained for $CaCO_3$ particles, isolated from the soils, and in water. On the contrary, a supersaturation with respect to calcite that was observed in solution extracts from calcareous soils (Levy, 1981) appears to be due to the presence of silicates more soluble than calcite, and is not the result of unstable $CaCO_3$ phases (Suarez and Rhoades, 1982). Plummer et al. (1979) concluded that there are several uncertainties in modeling the kinetics of carbonate chemistry; at low pH, the rate depends significantly on the thermodynamic transport constant for H$^+$, which is not well defined, while reaction site density and controls on pH at the interface between the crystal and the bulk solution are also not well understood.

Soil $CaCO_3$ may dissolve slowly to contribute Ca, especially in the reclamation of saline sodic soils in which its solubility is enhanced (Oster, 1982). Sodic soils commonly contain $CaCO_3$ that dissolves and maintains the soil solution concentration levels above the flocculation value of the soil clays (Alperovitch et al., 1981). This, in turn, reduces ESP and prevents HC from decreasing when exposed to rain. On the other hand, addition of $CaCO_3$ to nonsaline sodic soils is of doubtful value because its dissolution rate is too slow to provide sufficient Ca for exchange, unless an acid or acid former is applied concurrently. Calcareous soils with moderate ESP levels maintain reasonable physical properties through most of the profile but remain susceptible to dispersion near the surface because EC near the surface may be insufficient to maintain physical structure during raindrop impact (Keren, 1991). Under such conditions, an application of a soil amendment (gypsum) on soil surface is necessary to keep IR sufficiently high (Keren and Shainberg, 1981).

17.6.2 Leaching for Boron Control

Boron can be effectively leached from soil although the rate of removal is much slower for boron than for Cl$^-$ or SO_4^{2-} salts (Bingham et al., 1972). Griffin and Burau (1974) studied the kinetics of boron desorption from soils and observed two pseudo-first-order reactions and a very slow reaction. They speculated that the fast reactions were due to desorption from hydroxy-Fe, -Mg, and -Al materials in the clay size fraction, whereas the slowest reaction rate was due to diffusion of boron from the interior of clay minerals to the solution phase.

The presence of excess soluble boron in many arid soils is usually attributed to the application of boron-containing irrigation waters or the weathering of boron-containing materials. Adsorption/desorption of boron controls its removal from soils. Griffin and Burau (1974) showed that desorption (Figure 17.10)

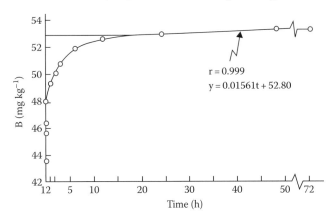

FIGURE 17.10 Solution boron concentration in leachate from soil, naturally high in boron, as a function of time. (Reprinted from Griffin, R.A., and R.G. Burau. 1974. Kinetic and equilibrium studies of boron desorption from soil. Soil Sci. Soc. Am. J. 38:892–897. With permission of the Soil Science Society of America.)

was largely from relatively fast reactions on hydroxy-Fe, Mg, and Al materials in the clay fraction, which are independent of soil texture and initial soil boron content. The slowest reaction rate was due to boron diffusion from the interior of clay minerals to the solution phase. They did not consider the role of clay minerals and primary boron minerals in its desorption and dissolution processes.

If excessive amounts of boron accumulate in soils, reclamation can be accomplished by extensive leaching. Leaching experiments (Reeve et al., 1955; Rhoades et al., 1970; Bingham et al., 1972) show that a large fraction of soil boron can be removed by percolating waters but that the remainder persists even after large amounts of water have been applied. The volume of low boron water needed to reduce it from toxic to nontoxic levels is two- to threefold greater than is needed for a comparable reduction in Cl (Reeve et al., 1955; Bingham et al., 1972).

The relative decrease in soluble boron in field soils during reclamation is given by (Hoffman, 1980)

$$\left(\frac{C}{C_0}\right)\left(\frac{D}{D_s}\right) = 0.6 \qquad (17.36)$$

where

C and C_0 are the final and initial soluble boron concentrations
D/D_s is the depth of leaching water per unit depth of soil

Equation 17.36 is independent of the method of water application (sprinkler or ponding).

Since extensive leaching is required for reclaiming soils with excessive boron, the reclamation should be pursued while tolerant crops are cultivated; and then when boron levels have been reduced, more sensitive crops can be cultivated. Alternatively, high boron soils can be reclaimed by adding amendments such as H_2SO_4. Because boron adsorption is pH dependent (Keren and Bingham, 1985), acidification effectively increases the boron concentration in soil solution thereby making leaching more effective (Prather, 1977). Furthermore, as Si concentration increases with decreasing pH to levels higher than those of boron, Si can compete for adsorption sites (Bingham and Page, 1971; McPhail et al., 1972). During leaching, the presence of Si may decrease readsorption of boron or even cause boron desorption. Applying ~3 Mg H_2SO_4 ha^{-1} reduced the amount of water needed to leach the same amount of boron fivefold (Prather, 1977).

Lime-induced boron deficiency has frequently been observed in acid soils, and is probably due to increased boron adsorption resulting from the increased pH and not a direct effect of Ca per se (Hatcher and Bower, 1967). In acid soils, increasing rates of boron lowered yield and caused increasing typical boron toxicity symptoms (Bartlett and Picarelli, 1973). However, in the presence of lime, boron toxicity was prevented as soil pH increased above 6.

Native soil boron is more difficult to leach than boron accumulated from previous irrigations. The reduced boron concentrations that follow the leaching of native high boron soils may be temporary. Increasing storage times of water in the soil profile results in larger increases in effluent boron concentrations (Rhoades et al., 1970; Bingham et al., 1972). Peryea et al. (1985) showed that this phenomenon, termed boron regeneration, is inversely related to the amount of water used for the initial leaching. Boron regeneration is, therefore, of primary concern during the early stages of soil reclamation, when appreciable residual sources of regenerable boron are present. The potential for reestablishing phytotoxic concentrations in a soil is a function of the relative completeness of the reclamation processes and the rate of boron dissolution after reclamation.

References

Agassi, M., I. Shainberg, and J. Morin. 1981. Effect of electrolyte concentration and soil sodicity on infiltration rate and crust formation. Soil Sci. Soc. Am. J. 45:848–851.

Alperovitch, N., I. Shainberg, and R. Keren. 1981. Specific effect of magnesium on the hydraulic conductivity of sodic soils. J. Soil Sci. 32:543–554.

Amrhein, C., J.J. Jurinak, and W.M. Moore. 1985. Kinetics of calcite dissolution as affected by carbon dioxide partial pressure. Soil Sci. Soc. Am. J. 49:1393–1398.

Amrhein, C., and D.L. Suarez. 1987. Calcite supersaturation in soils as a result of organic matter mineralization. Soil Sci. Am. J. 51:932–937.

Ayers, R.S., and D.W. Westcot. 1985. Water quality for agriculture. FAO Irrig. Drain. Pap. 29.

Bartlett, R.J., and C.J. Picarelli. 1973. Availability of boron and phosphorus as affected by liming an acid potato soil. Soil Sci. 116:77–83.

Bernstein, L. 1967. Quantitative assessment of irrigation water quality, p. 51–64. In Water quality criteria. American Society for Testing Materials Spec. Pub. 416. Philadelphia, PA.

Bernstein, L. 1975. Effects of salinity and sodicity on plant growth. Annu. Rev. Phytopathol. 13:295–312.

Bigger, K.C., and M. Fireman. 1960. Boron adsorption and release by soils. Soil Sci. Soc. Am. J. 24:115–120.

Bingham, F.T., A.W. Marsh, R. Branson, R. Mahler, and G. Ferry. 1972. Reclamation of salt affected high boron soils in western Kern County. Hilgardia 41:195–211.

Bingham, F.T., and A.L. Page. 1971. Specific character of boron adsorption by an amorphous soil. Soil Sci. Soc. Am. Proc. 35:892–893.

Bingham, F.T., A.L. Page, N.T. Coleman, and K. Flach. 1971. Boron adsorption characteristics of selected amorphous soils from Mexico and Hawaii. Soil Sci. Soc. Am. J. 35:546–550.

Bingham, F.T., J.E. Strong, J.D. Rhoades, and R. Keren. 1985. An application of the Maas-Hoffman salinity response model for boron toxicity. Soil Sci. Soc. Am. J. 49:672–674.

Boeseken, J. 1949. The use of boric acid for the determination of the configuration of carbohydrates. Adv. Carbohyd. Chem. 4:189–210.

Bresler, E., and R.J. Hanks. 1969. Numerical method of estimating simultaneous flow of water and salt in unsaturated soils. Soil Sci. Soc. Am. J. 33:827–832.

Bresler, E., B.L. McNeal, and D.L. Carter. 1982. Saline and sodic soils. Springer-Verlag, New York.

Carter, D.L. 1969. Managing moderately saline (salty) irrigation waters. University of Idaho Current Information Series 107. University of Idaho, Moscow, ID.

Childs, E.C., and R.J. Hanks. 1975. Model for soil salinity effects on crop growth. Soil Sci. Soc. Am. Proc. 39: 617–622.

Communar, G., and R. Keren. 2005. Equilibrium and nonequilibrium transport of boron in soil. Soil Sci. Soc. Am. J. 69:311–317.

Communar, G., and R. Keren. 2006. Rate-limited boron transport in soils: The effect of soil texture and solution pH. Soil Sci. Soc. Am. J. 70:882–892.

Communar, G., and R. Keren. 2008. Boron adsorption by soils as affected by dissolved organic matter from treated sewage effluent. Soil Sci. Soc. Am. J. 72:492–499.

Corwin, D.L., S. Goldberg, and A. David. 1999. Evaluation of a functional model for simulating transport in soil. Soil Sci. 164:697–717.

Corwin, D.L., B.L. Waggoner, and J.D. Rhodes. 1991. A functional model of solute transport that accounts for bypass. J. Environ. Qual. 20:647–658.

Craig, J.R. 1970. Saline water: Genesis and relationship to sediments and host rocks. In R.B. Mattox (ed.) Saline water. 46th annual meeting, American Association for the Advancement of Science, Las Vegas, NV.

Diaz, F.J., and S.R. Grattan. 2009. Performance of tall wheatgrass (Thinopyrum cv. 'Jose') irrigated with saline-high boron drainage water: Implications on ruminant mineral nutrition. Agric. Ecosy. Environ. 131:128–136.

Dyal, R.S., and S.B. Hendricks. 1950. Total surface of clays in polar liquids as a characteristic index. Soil Sci. 69: 421–432.

Elrashidi, M.A., and G.A. O'Connor. 1982. Boron sorption and desorption in soils. Soil Sci. Soc. Am. J. 46:27–31.

Ferreyra, R.E., A.U. Aljaro, R.S. Ruiz, L.P. Rojas, and J.D. Oster. 1997. Behavior of 42 crop species grown in saline soils with high boron concentration. Agri. Water Manage. 34:111–124.

Forsyth, W.G.C. 1950. Studies on the more soluble complexes of soil organic matter. 2. The composition of the soluble polysaccharide fraction. Biochem. J. 46:141–146.

Frenkel, H., J.O. Goertzen, and J.D. Rhoades. 1978. Effect of clay type and content, exchangeable sodium percentage, and electrolyte concentration on clay dispersion and soil hydraulic conductivity. Soil Sci. Soc. Am. J. 42:32–39.

Gardner, W.K. 2004. Changes in soils irrigated with saline groundwater containing excess bicarbonate. Aust. J. Soils Res. 42:825–831.

Goldberg, S., and R.A. Glaubig. 1985. Boron adsorption on aluminum and iron oxide minerals. Soil Sci. Soc. Am. J. 49:1374–1379.

Goldberg, S., and R.A. Glaubig. 1986. Boron adsorption on California soils. Soil Sci. Soc. Am. J. 50:1173–1176.

Goldberg, S., and R.A. Glaubig. 1987. Effect of saturating cation, pH and aluminum and iron oxide on the flocculation of kaolinite and montmorillonite. Clay. Clay Miner. 35:220–227.

Goldberg, S., and R.A. Glaubig. 1988. Boron and silicon adsorption on an aluminum oxide. Soil Sci. Soc. Am. J. 52:87–91.

Goldberg, S., S.M. Leschn, and D.L. Suarez. 2000. Predicting boron adsorption by soils using soil chemical parameters in the constant capacitance model. Soil Sci. Soc. Am. J. 64:1356–1363.

Griffin, R.A., and R.G. Burau. 1974. Kinetic and equilibrium studies of boron desorption from soil. Soil Sci. Soc. Am. J. 38:892–897.

Griffin, R.A., and J.J. Jurinak. 1973. Estimation of activity coefficients from the electrical conductivity of natural aquatic systems and soil extracts. Soil Sci. 116:26–30.

Hanks, R.J., R.F. Neilson, R.L. Cartee, and L.S. Willardson. 1984. Use of saline waste water from electric power plants for irrigation, p. 473–492. In R.H. French (ed.) Salinity in watercourses and reservoirs. Butterworth Publishers, Boston, MA.

Hanks, R.J., R.F. Neilson, R.L. Cartee, L.S. Willardson, R.B. Sorenson, A.R. Mitchell, and U. Shani. 1983. Use of saline waste water from electric power plants for irrigation. Utah State University Research Report 103.

Heller, H., and R. Keren. 2001. Rheology of Na-montmorillonite suspension as affected by electrolyte concentration and shear rate. Clay. Clay Miner. 49:286–291.

Hingston, F.J. 1964. Reaction between boron and clays. Aust. J. Soil Res. 2:83–95.

Hingston, F.J., A.M. Posner, and J.P. Quirk. 1974. Anion adsorption by goethite and gibbsite. II. Desorption of anions from hydrous oxide surfaces. J. Soil Sci. 25:16–26.

Hoffman, G.J. 1980. Guideline for the reclamation of salt-affected soils, p. 49–64. In G.A. O'Connor (ed.) 2nd. Inter-Am. Conf. Salin. Water Manag. Tech. NM State University, Las Cruces, NM.

Hohl, H., and W. Stumm. 1976. Interaction of Pb^{2+} with hydrous γ-Al_2O_3. J. Colloid Interface Sci. 55:281–288.

Israelachvili, J.N., and G.E. Adams. 1978. Measurements of forces between two mica surfaces in aqueous electrolyte solutions in the range 0–100 nm. J. Chem. Soc. Faraday Trans. 74:975–1001.

Jury, W.A., W.M. Jarrell, and D. Devitt. 1979. Reclamation of saline-sodic soils by leaching. Soil Sci. Soc. Am. J. 43:1100–1106.

Jury, W.A., W.M. Jarrell, and D. Devitt. 1987. Reclamation of saline-sodic soils by leaching. Soil Sci. Soc. Am. J. 51:1092.

Keren, R. 1991. Specific effect of magnesium on soil erosion and water infiltrations. Soil Sci. Soc. Am. J. 55:783–787.

Keren, R., and F.T. Bingham. 1985. Boron in water, soils and plants. Adv. Soil Sci. 1:229–276.

Keren, R., F.T. Bingham, and J.D. Rhoades. 1985a. Plant uptake of boron as affected by boron distribution between liquid and solid phases in soil. Soil Sci. Soc. Am. J. 49:297–302.

Keren, R., F.T. Bingham, and J.D. Rhoades. 1985b. Effect of clay content in soil on boron uptake and yield of wheat. Soil Sci. Soc. Am. J. 49:1466–1470.

Keren, R., and R.G. Gast. 1983. pH-dependent boron adsorption by montmorillonite hydroxy-aluminum complexes. Soil Sci. Soc. Am. J. 47:1116–1121.

Keren, R., R.G. Gast, and B. Bar-Yosef. 1981. pH-dependent boron adsorption by Na-montmorillonite. Soil Sci. Soc. Am. J. 45:45–48.

Keren, R., P.R. Grossl, and D.L. Sparks. 1994. Equilibrium and kinetics of borate adsorption-desorption on pyrophyllite in aqueous suspensions. Soil Sci. Soc. Am. J. 58:1116–1122.

Keren, R., and U. Mezuman. 1981. Boron adsorption by clay minerals using a phenomenological equation. Clay. Clay Miner. 29:198–204.

Keren, R., and S. Miyamoto. 1990. Reclamation of saline, sodic and boron-affected soils, p. 410–431. *In* K.K. Tanji (ed.) Agricultural salinity assessment and management. American Society of Civil Engineers, New York.

Keren, R., and G.A. O'Connor. 1982. Effect of exchangeable ions and ionic strength on boron adsorption by montmorillonite and illite. Clay. Clay Miner. 30:341–346.

Keren, R., and I. Shainberg. 1981. Effect of dissolution rate on the efficiency of industrial and mined gypsum in improving infiltration of a sodic soil. Soil Sci. Soc. Am. J. 45:103–107.

Keren, R., I. Shainberg, and E. Klein. 1988. Settling and flocculation value of sodium-montmorillonite particles in aqueous media. Soil Sci. Soc. Am. J. 52:76–80.

Keren, R., and M.J. Singer. 1988. Effect of low electrolyte concentration on hydraulic conductivity of Na/Ca-montmorillonite sand systems. Soil Sci. Soc. Am. J. 52:368–373.

Keren, R., and M.J. Singer. 1989. Effect of low electrolyte concentration on hydraulic conductivity of clay-sand hydroxy polymers. Soil Sci. Soc. Am. J. 53:349–355.

Keren, R., and M.J. Singer. 1991. Hydroxy-aluminum's effect on permeability of clay-sand mixtures. Soil Sci. Soc. Am. J. 55:61–65.

Keren, R., and D.L. Sparks. 1994. Effect of pH and ionic strength on boron adsorption by pyrophyllite. Soil Sci. Soc. Am. J. 58:1095–1100.

Keren, R., and H. Talpaz. 1984. Boron adsorption by montmorillonite as affected by particle size. Soil Sci. Soc. Am. J. 48:555–559.

Kittrick, J.A. 1977. Mineral equilibria and the soil system, p. 1–25. *In* J.B. Dixon and S.B. Weed (eds.) Minerals in soil environments. Soil Science Society of America, Madison, WI.

Lauchli, A., and E. Epstein. 1990. Plant responses to saline and sodic conditions, p. 113–137. *In* K.K. Tanji (ed.) Agricultural salinity assessment and management. American Society of Civil Engineers, New York.

Letey, J., and A. Dinar. 1986. Simulated crop-water production functions for several crops when irrigated with saline waters. Hilgardia 54:1–32.

Letey, J., A. Dinar, and K. Knapp. 1985. Crop-water production function model for saline irrigated waters. Soil Sci. Soc. Am. J. 49:1005–1009.

Letey, J., and K. Knapp. 1991. Crop-water production functions under saline conditions. *In* K.K. Tanji (ed.) Agricultural salinity assessment and management. American Society of Civil Engineers, New York.

Levy, R. 1981. Effect of dissolution of aluminosilicates and carbonates on ionic activity products of calcium carbonate in soil extracts. Soil Sci. Soc. Am. J. 45:250–255.

Lindsay, W.L. 1979. Chemical equilibria in soils. John Wiley & Sons, New York.

Luthin, J.N., P. Fernandez, J. Woerner, and F. Robinson. 1969. Displacement front under ponding leaching. J. Irrig. Drain. Div. ASCE 95:117–125.

Lyklema, J. 1982. Fundamentals of electrical double layers in colloidal systems. Royal Society of Chem. Spec. Pub. No. 43. Burlington House, London, U.K.

Maas, E.V. 1986. Salt tolerance of plants. App. Agric. Res. 1:12–26.

Maas, E.V. 1990. Crop salt tolerance, p. 262–304. *In* K.K. Tanji (ed.) Agricultural salinity assessment and management. American Society of Civil Engineers, New York.

Maas, E.V., and G.J. Hoffman. 1977. Crop salt tolerance-current assessment. J. Irrig. Drain. Div. ASCE 103:115–134.

Martinez-Beltran, J., and C.L. Manzur. 2005. Overview of salinity problems in the world and FAO strategies of the international salinity forum. Riverside, CA, April 2005, p. 322–313.

McIntyre, D.S. 1979. Exchangeable sodium, subplasticity and hydraulic conductivity of some Australian soils. Aust. J. Soil Res. 17:115–120.

McNeal, B.L., D.A. Layfield, W.A. Norvell, and J.D. Rhoades. 1968. Factors influencing hydraulic conductivity of soils in the presence of mixed salt solutions. Soil Sci. Soc. Am. J. 32:187–190.

McPhail, M., A.L. Page, and F.T. Bingham. 1972. Adsorption interactions of monosilicic and boric acid on hydrous oxides of iron and aluminum. Soil Sci. Soc. Am. Proc. 36:510–514.

Mezuman, U., and R. Keren. 1981. Boron adsorption by soils using a phenomenological adsorption equation. Soil Sci. Soc. Am. J. 45:722–726.

Miller, R.J., J.W. Biggar, and D.R. Nielsen. 1965. Chloride displacement in Panoche clay loam in relation to water movement and distribution. Water Resour. Res. 1:63–67.

Misopolinos, N.D. 1985. A new concept for reclaiming sodic soils with high-saltwater. Soil Sci. 140:69–74.

Miyamoto, S., and I. Cruz. 1986. Spatial variability and soil sampling for salinity and sodicity appraisal in surface-irrigated orchards. Soil Sci. Soc. Am. J. 50:1020–1025.

Miyamoto, S., and I. Cruz. 1987. Spatial variability of soil salinity in furrow-irrigated torrifluvents. Soil Sci. Soc. Am. J. 51:1019–1025.

Miyamoto, S., and A.W. Warrick. 1974. Salt displacement into drain tiles under ponded leaching. Water Resour. Res. 10:275–278.

Nielsen, D.R., J.W. Biggar, and J.N. Luthin. 1966. Desalinization of soils under controlled unsaturated flow conditions. 6th Congr. Int. Comm. Irrig. Drain. 19:15–24.

Nimah, M.N., and R.J. Hanks. 1973a. Model for estimating soil, water, plant and atmospheric interrelations: I. Description and sensitivity. Soil Sci. Soc. Am. J. 37:522–527.

Nimah, M.N., and R.J. Hanks. 1973b. Model for estimating soil, water, plant and atmospheric interrelations: II. Field test of model. Soil Sci. Soc. Am. J. 37:528–532.

Oster, J.D. 1982. Gypsum usage in irrigated agriculture: A review. Fert. Res. 3:73–89.

Oster, J.D., and F.W. Schroer. 1979. Infiltration as influenced by irrigation water quality. Soil Sci. Soc. Am. J. 43:444–447.

Oster, J.D., I. Shainberg, and J.D. Wood. 1980. Flocculation value and gel structure of Na/Ca montmorillonite and illite suspension. Soil Sci. Soc. Am. J. 44:955–959.

Parker, M.B., G.J. Gascho, and T.P. Gaines. 1983. Chloride toxicity of soybeans grown on Atlantic coast flatwoods soils. Agron. J. 75:439–442.

Parks, W.L., and J.L. White. 1952. Boron retention by clay and humus systems saturated with various cations. Soil Sci. Soc. Am. J. 16:298–300.

Peryea, F.T., F.T. Bingham, and J.D. Rhoades. 1985. Regeneration of soluble boron by reclaiming high boron soils. Soil Sci. Soc. Am. J. 42:782–786.

Plummer, L.N., D.L. Parkhurst, and T.M.L. Wigley. 1979. Critical review of the kinetics of calcite dissolution and precipitation. Am. Chem. Soc. Symp. Ser. 93:537–573.

Prather, R.J. 1977. Sulfuric acid as an amendment for reclaiming soils high in boron. Soil Sci. Soc. Am. J. 41:1098–1101.

Rahman, M.A., E.A. Hiler, and J.R. Runkles. 1974. High electrolyte water for reclaiming slowly permeable soils. Trans. ASAE 17:129–133.

Reeve, R.C., and C.A. Bower. 1960. Use of high-salt waters as a flocculant and source of divalent cations for reclaiming sodic soils. Soil Sci. 90:139–144.

Reeve, R.C., and E.J. Doering. 1966b. The high salt-water dilution method for reclaiming sodic soils. Soil Sci. Soc. Am. Proc. 39:498–504.

Reeve, R.C., and E.J. Doering. 1966a. Field comparison of the high salt-water dilution method and conventional methods for reclaiming sodic soils. 6th Int. Comm. Irrig. Drain. 19:1–14.

Reeve, R.C., A.F. Pillsbury, and L.V. Wilcox. 1955. Reclamation of a saline and high boron soil in the Coachella Valley of California. Hilgardia 24:69–91.

Rengasamy, P. 2006. World salinization with emphasis on Australia. J. Exp. Bot. 57:1017–1023.

Rhoades, J.D. 1968. Leaching requirement for exchangeable sodium control. Soil Sci. Soc. Am. Proc. 32:652–656.

Rhoades, J.D. 1982. Reclamation and management of salt-affected soils after drainage, p. 123–197. *In* Proc. 1st Ann. West. Prov. Conf. Rational. Water Soil Res. Manag. Lethbridge, AB, Canada.

Rhoades, J.D., R.D. Ingvalson, and J.T. Hatcher. 1970. Laboratory determination of leachable soil boron. Soil Sci. Soc. Am. Proc. 34:871–875.

Rhoades, J.D., D.B. Krueger, and M.J. Reed. 1968. The effect of soil-mineral weathering on the sodium hazard of irrigation waters. Soils Sci. Soc. Am. J. 32:643–647.

Russo, D. 1984. Satial variability considerations in salinity management, p. 198–219. *In* I. Shainberg and Y. Shalhevet (eds.) Soil salinity under irrigation. Springer Verlag, Berlin, Germany.

Schindler, P.W., and H. Gamsjager. 1972. Acid-base reactions of the TiO_2 (anatase)-water interface and the point of zero charge of TiO_2 suspensions. Kolloid Z.Z. Polym. 250:759–763.

Secor, R.B., and C.J. Radke. 1985. Spillover of the diffuse double layer on montmorillonite particles. J. Coll. Interf. Sci. 103:237–244.

Shainberg, I., E. Bresler, and Y. Klausner. 1971. Studies on Na/Ca montmorillonite systems. I. The swelling pressure. Soil Sci. 111:214–219.

Sims, J.R., and F.T. Bingham. 1967. Retention of boron by layer silicates, sesquioxides and soil materials: I. Layer silicates. Soil Sci. Soc. Am. Proc. 31:728–732.

Smith, G.J. 1979. Subsurface stratigraphy and geochemistry of late quaternary evaporites, Searles Lake, CA. USGS Prof. Pap. 1043.

Solomon, K.H. 1985. Water-salinity-production functions. Trans. ASAE 28:1975–1980.

Sposito, G., and S.W. Mattigod. 1977. On the chemical formation of the sodium adsorption ratio. Soil Sci. Soc. Am. J. 41:323–329.

Stumm, W., H. Hohl, and F. Dalang. 1976. Interaction of metal ions with hydrous oxide surfaces. Croat. Chem. Acta 48:491–504.

Stumm, W., R. Kummert, and L. Sigg. 1980. A ligand exchange model for the adsorption of inorganic and organic ligands at hydrous oxide interfaces. Croat. Chem. Acta 53:291–312.

Suarez, D.L., and J.D. Rhoades. 1982. The apparent solubility of calcium carbonate in soils. Soil Sci. Soc. Am. J. 46:716–722.

Suarez, D.L., and J. Simunek. 1997. UNSATCHEM: Unsaturated water and solute transport model with equilibrium and kinetic chemistry. Soil Sci. Soc. Am. J. 61:1633–1646.

Sumner, M.E., and R. Naidu. 1998. Sodic soils: Distribution, properties, management and environmental consequences. Oxford University Press, New York.

Szabolcs, I. 1989. Salt affected soils. CRC Press, Boca Raton, FL.

Talsma, T. 1967. Leaching of tile-drained saline soil. Aust. J. Soil Res. 5:37–46.

U.S. Salinity Laboratory Staff. 1954. Diagnosis and improvement of saline and alkali soils. USDA handbook 60. U.S. Government Printing Office, Washington, DC.

van Genuchten, M. Th., and G.J. Hoffman. 1984. Analysis of crop salt tolerance data, p. 258–271. *In* I. Shainberg and J. Shalhevet (eds.) Salinity under irrigation. Springer-Verlag, New York.

Van Olphen, H. 1977. An introduction to clay colloid chemistry. 2nd Edn. John Wiley & Sons, New York.

Westall, J.C. 1982. FITEQL: A computer program for determination of chemical equilibrium constants from experimental data. Rep. 82-01. Department of Chemistry, Oregon State University, Corvallis, OR.

Wolf, J.K. 1977. The evaluation of a computer model to predict the effects of salinity on crop growth. PhD thesis, UT State University. Logan, UT.

Yermiyahu, U., R. Keren, and Y. Chen. 1988. Boron sorption on composted organic matter. Soil Sci. Soc. Am. J. 52:1309–1313.

18
Sodicity

18.1	Introduction	18-1
18.2	Sodic Soils: Definition and Distribution	18-1
	Definition of Sodic Soils • Distribution of Sodic Soils	
18.3	Processes Characterizing the Behavior of Sodic Soils	18-4
	Clay Charge and the Diffuse Double Layer • Swelling and Dispersion	
18.4	Aggregate Stability and Organic Matter in Sodic Soils	18-6
	Aggregation and Its Importance • Sodicity Effects on Aggregate Stability • Organic Matter Effects in Sodic Soils	
18.5	Water Flow in Sodic Soils	18-7
	Hydraulic Conductivity • Infiltration Rate	
18.6	Crop Responses to Sodic Conditions	18-13
	Crop Tolerance to Sodicity • Plant Nutrition in Sodic Soils • Ion Toxicity under Sodic Conditions	
18.7	Amelioration and Management of Sodic Soils	18-15
	Use of Chemical Amendments • High Salt Water Dilution Method • Tillage • Vegetative Remediation	
18.8	Environmental Aspects of Sodic Soils	18-18
	Erosion and Suspended Sediments • Salinity and Pollutants • Carbon Sequestration	
18.9	Conclusions	18-19
	References	18-19

Guy J. Levy
Agricultural Research Organization

18.1 Introduction

Soils in numerous areas of the world are adversely affected by the presence of excess Na in the soil solution and as an exchangeable cation. For example, it is estimated that ~50% of the arable land in Australia suffers from sodicity-related problems (Naidu, 1993). Sodium-affected soils exhibit poor soil water and air relations that adversely affect water movement in the soil, root growth, and plant production, and make the soil difficult to farm when either dry or wet. Consequently, in many cases, such soils have had to be abandoned.

Sodium-affected soils have been referred to in the past as alkali soils. This may have originated from the traditional separation of salt-affected soils into two groups (Szabolcs, 1979): (1) saline soils (soils affected by neutral Na salts, mainly NaCl and Na_2SO_4) and (2) alkali soils (soils affected by Na salts capable of alkaline hydrolysis, including $NaHCO_3$, Na_2CO_3, and Na_2SiO_3). However, at present, Na-affected soils are referred to as sodic soils, with the realization that the level of Na in the soil solution and on the exchange complex, in conjunction with the total electrolyte concentration (TEC), determines whether or not a given soil is sodic, and not the type of the Na salt.

Problems associated with sodic soils have long been recognized (e.g., special issue of Australian Journal of Soil Research [vol. 31 no. 6], 1993; Sumner and Naidu, 1998) and are, unfortunately, expected to increase in the future. The ever increasing need to provide food to an expanding worldwide population, coupled with the increasing demand for good quality water from urban and industrial sectors, results in poorer quality water and soils being used for food production. Wastewater, whether drainage or recycled municipal water, being significantly more saline and sodic than fresh water, is rapidly becoming a common source of water for agriculture in many areas in the world. Consequently, understanding sodic soil behavior and its effects on agriculture and the environment is essential in order to properly manage sodic soils for crop production and sustainable agriculture in many parts of the world. This chapter provides, therefore, a general review of various aspects related specifically to sodic conditions in soils.

18.2 Sodic Soils: Definition and Distribution

18.2.1 Definition of Sodic Soils

Prior to defining sodic soils it is necessary to define the parameters by which sodicity is evaluated. Two important parameters are generally used. The first is the exchangeable sodium percentage

(ESP) that describes the fraction of adsorbed Na from the cation exchange capacity (CEC) of the soil, and is defined as

$$ESP = 100(\text{Exchangeable Na/CEC}) \tag{18.1}$$

where the CEC is normally determined at a reference pH (7.0 or 8.2). The second parameter reflects the sodicity level of the irrigation water or soil solution and is termed the sodium adsorption ratio (SAR) defined as

$$SAR = \frac{[Na]}{([Ca + Mg]/2)^{0.5}} \tag{18.2}$$

where brackets [ ] reflect cation concentrations in $mmol_c\ L^{-1}$. Thus, SAR has units of $(mmol_c\ L^{-1})^{0.5}$. It should be emphasized that the sodicity hazard of solutions is related to the ratio of Na to the divalent cations present in the water and not to Na concentration alone. A more accurate calculation of the SAR is obtained when cation activities are used instead of cation concentrations. Additionally, Sposito and Mattigod (1977) introduced corrections for cation concentrations due to ion pairs or complexes and arrived at the following empirical relation between true (SAR_t) and measured (SAR_p) SARs:

$$SAR_t = 0.08 + 1.15 SAR_p \tag{18.3}$$

Furthermore, in arid and semiarid soils, irrigation may lead to dissolution/precipitation of $CaCO_3$, depending on the pH and the CO_2 partial pressure (pCO_2) of the system. Thus, the SAR of the irrigation water (SAR_{iw}) may differ from the SAR of the drainage water (SAR_{dw}), with the latter being considered as a better indicator for sodicity hazard. Bower et al. (1968) proposed the use of SAR_{iw}, pH, and leaching fraction for calculating SAR_{dw} (also known as SAR_{adj}). Rhoades (1968) added to this relationship the contribution of mineral dissolution. Later, Suarez (1981) suggested that SAR_{dw} be calculated based on the concentration of Na and Mg in the irrigation water, leaching fraction, and pCO_2. The equation proposed by Suarez (1981) is considered more accurate and easier to use than other equations for calculating SAR in the soil and drainage water (Frenkel, 1984).

Exchange reactions take place between the soil solution and the exchange phase. Thus, soil ESP can be estimated from the SAR of saturated paste extracts using the following empirical relationship (USSL Staff, 1954):

$$ESP = 100(-0.0126 + 0.01475 SAR)/(1+[-0.0126 + 0.01475 SAR]) \tag{18.4}$$

or from the nomograph presented in Figure 18.1. When more dilute extracts are used, such as 1:5 soil:water ratio, then a different relationship holds (Rengasamy et al., 1984):

$$ESP = 1.95\ SAR + 1.8 \tag{18.5}$$

The USSL Staff (1954) empirical method has been developed using 59 soil samples from 9 western states of the United States.

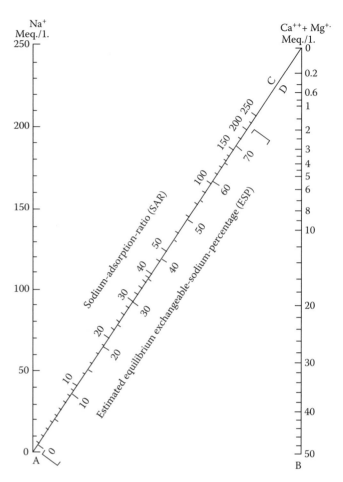

FIGURE 18.1 Nomograph for determining the SAR of a saturation extract and estimating the corresponding ESP value of the soil at equilibrium with the extract. (Courtesy of USSL Staff. 1954. Diagnosis and Improvement of saline and alkali soils. Agricultural Handbook No. 60. USDA, Washington, DC.)

Nevertheless, it has gained worldwide popularity because of its simplicity and has been extensively used to predict the ESPs of Na affected soils, even in recent years (e.g., Condom et al., 1999; Faulkner et al., 2001; Chhabra, 2005).

The use of the USSL Staff (1954) empirical SAR-ESP relation can, at times, lead to erroneous estimation of soil ESP. Ganjegunte and Vance (2006) noted for five sites in Wyoming that the ESP determined by using NH_4OAc was lower than the ESP estimated using the SAR–ESP empirical relationship provided by the USSL Staff (1954). Furthermore, Kopittke et al. (2006) observed that clays in which external exchange sites dominate (kaolinite and pyrophyllite) tend to show an overall preference for Na. In the case of illite (2:1 nonexpanding mineral), the clay displayed an overall preference for Ca. For the expanding 2:1 clay, montmorillonite, Na–Ca exchange varied with ionic strength; at low ionic strength and high ESP, the clay platelets displayed preference for Na. However, as ionic strength increased and ESP decreased, preference for Ca increased. Kopittke et al. (2006) concluded that the relationship between SAR and ESP is not constant and should be determined directly for the soil of interest.

To date there is no widely accepted definition of a sodic soil. The USSL Staff (1954) defined sodic soils as those whose physical properties are adversely affected by the presence of Na and suggested that ESP of 15 should separate sodic from non-sodic soils. The USSL Staff (1954) added the reservation that this limit must be regarded as somewhat arbitrary and tentative. The study of McNeal and Coleman (1966), who used seven soils from the western United States, supported the conclusion that ESP 15 can separate between sodic and nonsodic behavior of the soil; however, it was added that the electrolyte concentration in the percolating solution must exceed 3 mmol$_c$ L^{-1} in order for this to hold true. Greene et al. (1978) suggested that the threshold value separating sodic from nonsodic soils should depend on soil texture; they proposed ESPs of 10 and 20 for fine and coarse textured soils, respectively. In Australia, McIntyre (1979), who used water with TEC of 0.7 mmol$_c$ L^{-1}, proposed that soils with ESP > 5 should be considered as Na-affected (sodic) soils. More recent hydraulic conductivity studies by Mace and Amrhein (2001) and Levy et al. (2005) with calcareous semiarid soils supported the suggestion of McIntyre (1979) that ESP > 5 already imparts a sodic behavior. The differences in the proposed critical ESP levels separating sodic from nonsodic soils could arise, in part, from the fact that in California (USSL Staff, 1954) water with a TEC of >2 mmol$_c$ L^{-1} was being considered, whereas in Australia (McIntyre, 1979), the TEC of the water under consideration was only 0.7 mmol$_c$ L^{-1}. Consequently, a much higher ESP was required before degradation of the soil in the California studies was observed. The above critical ESP levels were determined based on hydraulic conductivity (HC) measurements but studies on soil susceptibility to aggregate slaking (Levy et al., 2003), seal formation, runoff, and soil loss (Agassi et al., 1981; Kazman et al., 1983; Mamedov et al., 2002; Tang et al., 2006) have revealed that soils can exhibit a sodic behavior even at ESPs of 2–4. In view of the continuous effect of Na on soil behavior, from low to high ESP and TEC levels, no satisfactory decision or a critical level of ESP can be made. However, as will be shown later, the interrelationship between the ESP of the soil or the SAR of its equilibrium solution, and the TEC of the soil solution is important in dictating whether a sodic or a nonsodic behavior will be observed. Oster and Jayawardane (1998) present an example for this approach with the assessment of the permeability hazard of infiltrating waters having different TEC values when applied to soils with SAR values in the range of 0–60 (Figure 18.2).

18.2.2 Distribution of Sodic Soils

Before considering the world distribution of sodic soils, it should be pointed out that, for most sodic soils, sodicity is a natural phenomenon related to the nature of the parent material and subsequent pedogenic processes. However, there are also sodic soils where sodicity arises from anthropogenic processes, and is thus termed secondary sodification. Irrigation without proper drainage, forest clearing, and other land management practices

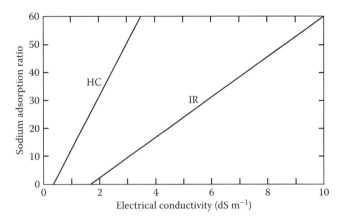

FIGURE 18.2 Threshold values of sodium adsorption ratio and electrical conductivity for hydraulic conductivity (HC) and infiltration rate (IR) levels associated with the likelihood of substantial losses in permeability. (From Oster, J.D., and N.S. Jayawardane. 1998. Agricultural management of sodic soils, p. 143–165. *In* M.E. Sumner and R. Naidu (eds.) Sodic soils. Oxford University Press, New York. With permission.)

that can lead to waterlogging, are the main human activities that yield rapid secondary sodification (Rengasamy and Olssen, 1991; Fitzpatrick et al., 1994). It is expected that secondary sodification of soils will increase considerably in the future, because of the growing use of poorer quality waters and soils for food production to meet the demand of the growing population, and the need to reuse drainage water to compensate for the increasing shortage of good quality water for irrigation.

Sodic soils cover 580 million ha worldwide (Szabolcs, 1989); the areas of sodic soils on each continent and in regions where they form an important proportion of the total area are presented in Table 18.1. The largest areas of sodic soils are concentrated in Australia and north-central Asia. In Australia, it is estimated that 25%–30% of the area is occupied by sodic soils with widespread occurrence (Northcote and Skene, 1972). In North America, many of the sodic soils occur on the Great Plains of western Canada and the northern United States, under cold, semiarid to subhumid climates (FAO,

TABLE 18.1 Global Distribution of Sodic Soils

Continent	Area (Million Hectares)
North America	9.6
Central America	—
South America	59.6
Africa	27.0
South Asia	1.8
North and Central Asia	120.1
Southeast Asia	—
Europe	22.9
Australasia	340.0
Total	581.0

Source: Adapted from Szabolcs, I. 1989. Salt-affected soils. CRC Press, Boca Raton, FL.

1991). In South America, sodic soils are concentrated in the Argentinian and Paraguayan pampas and northeastern Brazil (FAO, 1991) while in Europe, they occur in the Carpathian basin, Hungary (Trans-Tisza region and the Danube-Tisza interfulve), Romania, Serbia, Slovakia, Ukraine (near the Black Sea), and the Transcaucasian plain of Georgia and Azerbaijan (Bui et al., 1998). In northern and central Asia, sodic soils are concentrated mainly (1) west of the Caspian Sea in Kazakhstan, Turkmenia, Uzbekistan, Tadzhikistan, and Kyrghizia (FAO, 1978); (2) south of the Tien Shan Mountains in the basin of the Xinjiang–Gansu–Qinghai provinces, on the Huang and Huari river plains and in northeastern Manchuria, China; (3) in the Russian Republic of Yakutia; and (4) in the forest-meadow-steppe region of western Siberia (Bui et al., 1998). In southern Asia and the Middle East, sodic soils are widespread and can be found in India, Syria, Iran, Pakistan, and Bangladesh (Bui et al., 1998). In Africa, sodic soils occupy only 0.9% of the total land area and are concentrated in the Kalahari Basin, coastal Tunisia, Chad, Nigeria, Somalia, Sudan, Tanzania, and southern Africa (Bui et al., 1998).

The spatial distribution of sodic soils demonstrates that they occur under a range of climates. Among the environmental conditions that promote the formation of sodic soils are the presence of shallow saline groundwater, the occurrence of perched water tables within 1 m of the soil surface, impeded drainage, low slope gradients, and textural discontinuities during deposition of sediments such as eolian, glacial, or alluvial materials (Bui et al., 1998).

The outlined distribution of sodic soils and the conditions promoting their formation are based on a pedological or morphological approach. By contrast, a more useful definition of sodic soils is based on land management considerations (Section 18.2.1). Many agricultural soils that exhibit sodic behavior may not fit into the classical sodic soil group, because typical morphological features including columnar subsurface structure are not present. Consequently, the large areas that these soils cover remain largely undetermined.

18.3 Processes Characterizing the Behavior of Sodic Soils

18.3.1 Clay Charge and the Diffuse Double Layer

Interaction between solid and solution phases plays a dominant role in determining physical behavior of soil. Of the various soil constituents, colloidal clay determines much of the physical behavior of soils because of its large specific surface area and charge, which makes it very reactive in physicochemical processes such as swelling and dispersion. These two processes determine, to a large extent, soil microstructure and thus many of its physical properties. A complete discussion of the processes involved in clay dispersion and swelling is presented in Chapter 15 of *Handbook of Soil Sciences: Properties and Processes*.

18.3.2 Swelling and Dispersion

18.3.2.1 Reference Clay Systems

Upon wetting, swelling takes place in smectites represented most commonly in soils by the mineral montmorillonite. Repulsion forces (swelling pressure) that can be predicted from the diffuse double layer (DDL) theory are responsible for the swelling (van Olphen, 1977; Bresler et al., 1982). When saturated with Na ions and in the presence of a dilute solution, a diffused (wide) DDL forms creating high swelling pressures between the clay platelets, resulting ultimately in dispersion of the platelets (Banin and Lahav, 1968; Shainberg et al., 1971). Conversely, low swelling pressures are observed between Ca-saturated platelets because strong electrical attraction forces between Ca and the clay surfaces (not considered by the DDL theory) prevent complete swelling of Ca montmorillonite, even in distilled water (DW). Instead, Ca platelets aggregate into tactoids or quasicrystals consisting of four to nine platelets each (Aylmore and Quirk, 1959; Blackmore and Miller, 1961) with DDL formation only on the outer and not the internal surfaces of the tactoids (Blackmore and Miller, 1961), which results in a much smaller effective surface area than actually present.

In mixed Na–Ca systems, the ions are not randomly distributed throughout the system as was previously thought (Bresler, 1970), with Ca and Na being located preferentially on inner and outer surfaces, respectively (McAtee, 1961; Shainberg and Otoh, 1968; Bar On et al., 1970; Shainberg et al., 1971). This phenomenon is called ion demixing (Figure 18.3). Shainberg et al. (1971) demonstrated that clay swelling is limited at ESP < 15 because of

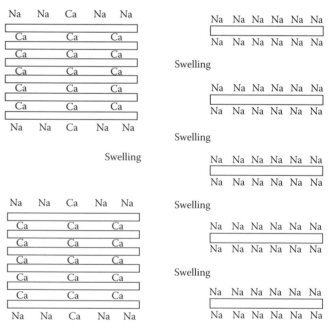

FIGURE 18.3 Comparison of particle arrangements in a homoionic Na montmorillonite (right) with that in a Na–Ca system (left) illustrating the formation of tactoids or quasicrystals with demixing of Na and Ca. (Reprinted from Sumner, M.E. 1993. Sodic soils: New perspectives. Aust. J. Soil Res. 31:683–750.)

demixing; as ESP increases further, a sharp increase in macroscopic swelling occurs as Na enters the tactoids, and at ESP > 50, swelling in mixed Na–Ca and pure Na systems is similar.

In a stable clay suspension, particles frequently collide because of Brownian motion, but separate again because of DDL repulsive forces. When TEC or valency of the counter ions is increased, DDL forces are reduced, allowing particles to remain together after collision to form flocs that subsequently settle. The minimum TEC at which this occurs is called the critical flocculation concentration (CFC), or simply the flocculation value (FV). For Na and Ca montmorillonite, CFCs are 7–20 mmol$_c$ L^{-1} NaCl and 0.25–1.09 mmol$_c$ L^{-1} CaCl$_2$, respectively (Goldberg and Glaubig, 1987). Increasing pH from 4 to 10 has only a small effect on the CFC of Na montmorillonite, but addition of humic substances causes a marked increase over the pH range 4–8 (Tarchitzky et al., 1993).

For Na and Ca illite, CFC values are 40–50 mmol$_c$ L^{-1} NaCl and 0.25 mmol$_c$ L^{-1} CaCl$_2$, respectively (Arora and Coleman, 1979). At low ESP levels, the CFC of illite is higher than that of montmorillonite because of the smaller attractive forces in illite; illite surfaces are terraced, which leads to mismatching and subsequent poor contact so that the particles cannot approach each other as closely (Oster et al., 1980).

Kaolinite saturated with Na remains flocculated in a salt-free solution at pH < 7 because the positively charged edges interact with the negative basal planes (Schofield and Samson, 1954). Elimination of edge positive charge by increasing the pH, addition of small amounts of various salts such as sodium oxalate, sodium pyrophosphate, sodium polymetaphosphate, and sodium alginate (Schofield and Samson, 1954; Durgin and Chaney, 1984; Frenkel et al., 1992), or introducing small impurities of montmorillonite or illite to the kaolinite suspension (Frenkel et al., 1978; Chiang et al., 1987) brings about dispersion (Goldberg and Forster, 1990). Consequently, kaolinite systems cannot be represented by a unique CFC.

A distinction should be made between swelling and dispersion even though both result from the balance between DDL repulsive forces and van der Waals attractive forces. Swelling is a continuous and reversible process, whose magnitude depends on the TEC of the ambient solution and the degree of sodicity. Dispersion is not continuous and may occur even at low ESP levels as long as the TEC < CFC. Dispersion is an irreversible process because flocculation by increasing TEC does not result in the original particle associations and orientations.

18.3.2.2 Soil Clay Systems

At low ESP levels, dispersion rather than swelling is the main mechanism for physical degradation of soils. Only when ESP > 15, is the role of swelling dominant in determining soil physical properties (Shainberg and Letey, 1984).

Flocculation and dispersion behavior of soil clays differs significantly from that of pure clay systems, because they usually occur as mixtures and complexes with other minerals, oxides, and soil organic matter (SOM). Goldberg and Forster (1990) and Frenkel et al. (1992) demonstrated that the CFC values of soil clays are 2- to 10-fold higher than for pure clay systems. Conversely, in the presence of hydrous oxides (McNeal et al., 1968) or sparingly soluble minerals such as CaCO$_3$ (Shainberg et al., 1981a, 1981b), clay dispersivity is much less severe. Hence, extrapolation from pure clay to soil systems is extremely difficult. Goldberg and Forster (1990) demonstrated that the smallest differences in CFC were found between reference and soil illites, indicating that in many dispersive soils that contain illite or weathered mica (Rengasamy et al., 1984; Miller and Baharuddin, 1986; Miller et al., 1990), illite plays an important role in their dispersibility.

Clay dispersion is affected not only by the presence of Na but also by the type of complementary divalent cation. The presence of Mg enhances clay dispersion in soils with mixed mineralogy (Ali et al., 1987; Yousaf et al., 1987), as well as in kaolinitic (Emerson and Smith, 1970) and illitic (Rengasamy et al., 1986) soils, compared to Ca. Furthermore, aggregates saturated with Na and Mg disperse at lower ESPs than those saturated with Na and Ca (Emerson and Bakker, 1973; Ali et al., 1987).

Disintegration of aggregates on wetting and subsequent dispersion of clay from the aggregates cannot be adequately explained by the classical theory of DDL repulsive and van der Waals attractive forces developed for colloidal systems. Rengasamy and Sumner (1998) proposed a schematic model to explain swelling and dispersion of soil clay particles when dry aggregates are wetted (Figure 18.4). Initially, hydration of the adsorbed cations leads to swelling, further hydration of highly sodic soils results in spontaneous dispersion, while mechanical stresses (raindrop impact, tillage, etc.) lead to dispersion even of soils containing predominantly divalent cations. Dehydration of the aggregates and the buildup of an osmotic pressure by increasing the TEC leads to flocculation of the dispersed particles once more (Rengasamy and Sumner, 1998).

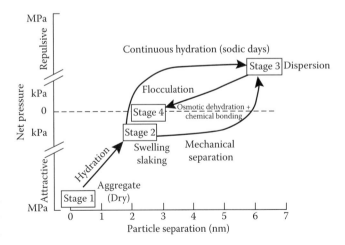

FIGURE 18.4 Schematic illustration of processes that take place and intensity of attractive and repulsive forces involved when a dry aggregate of a soil is wetted. (Reprinted from Rengasamy, P., and M.E. Sumner. 1998. Processes involved in sodic behavior, p. 51–67. *In* M.E. Sumner and R. Naidu (eds.) Sodic soils. Oxford University Press, New York. Copyright Oxford University Press, New York. With permission.)

18.4 Aggregate Stability and Organic Matter in Sodic Soils

18.4.1 Aggregation and Its Importance

Aggregation in soils is the arrangement of primary particles (sand, silt, and clay) into secondary structural units, typically on a millimeter scale. Soil aggregates thus consist of clay particles, in quasicrystals or domains (Quirk and Aylmore, 1971), interspersed with larger quartzitic primary particles and held together by physicochemical forces that are a product of the chemical and microbiological environment of that particular soil.

Recalling Bradfield's early statement that "granulation is flocculation plus," Quirk (1978) emphasized that a necessary condition for the formation of water-stable granules is clay flocculation, followed by stabilization by organic and/or inorganic cementing agents. The principles of flocculation and dispersion are central to the formation of clay quasicrystals and floccules (<20 μm) that become the building blocks of aggregates. Such floccules are further aggregated by inorganic (clay particles or metal oxides) or organic (humified or residual microbial products) materials into microaggregates, ranging from 20 to 250 μm in size. Macroaggregates, typically >250 μm, are agglomerates containing microaggregates plus primary sand and silt particles, held together largely by root hairs, fungal hyphae, and large biomolecules (Tisdall and Oades, 1982). The picture in Figure 18.5, modified from a diagram of Greenland (1979), illustrates these constituents and associations. A further discussion of the role of SOM in structure development is given in Section 18.4.3.

Aggregate breakdown by water may result from a variety of physical and physicochemical mechanisms. Four main mechanisms have been identified: (1) slaking, that is, breakdown caused by compression of entrapped air during fast wetting (Panabokke and Quirk, 1957), (2) breakdown by differential swelling during fast wetting (Kheyrabi and Monnier, 1968), (3) breakdown by impact of raindrops (McIntyre, 1958), and (4) physicochemical dispersion due to osmotic stress upon wetting with low electrolyte water (Emerson, 1967). These mechanisms differ in the type of energy involved in aggregate disruption. For instance, swelling can overcome attractive pressures in the magnitude of mega-Pascals (Rengasamy and Olssen, 1991) while slaking and impact of raindrops can overcome attractive pressures in the range of kilo-Pascals only (Rengasamy and Sumner, 1998). In addition, the various mechanisms may differ in the size distribution of the disrupted products (Farres, 1980; Chan and Mullins, 1994) and in type of soil properties affecting the mechanism (Le Bissonnais, 1996). Regarding the latter, slaking may be affected by porosity and internal cohesion while physicochemical dispersion will be affected by clay mineralogy, ionic composition, and concentration.

18.4.2 Sodicity Effects on Aggregate Stability

The effect of exchangeable Na in inhibiting macroaggregation (i.e., enhancing aggregate breakdown) may occur by weakening the covalent associations between organic materials and soil minerals, and/or by increasing the osmotic/hydration forces that can cause particle repulsion during wetting.

Direct measurements of the effects of sodicity on aggregate stability have yielded inconclusive results. Some studies covering a wide range of soil texture (Coughlan and Loch, 1984; Goldberg et al., 1988; Chappell et al., 1999) observed no correlation between aggregate stability and level of ESP. Other studies, however, have suggested that macroaggregate slaking was affected by fractional Na coverage only in the range of ESP > 5 (Aly and Letey, 1990) or ESP > 10 (Shainberg et al., 1992). Crescimanno et al. (1995) noted for two clay soils that aggregate stability decreased linearly with an increase in ESP in the range of 2–15. Levy and Mamedov (2002) and Levy et al. (2003) noted that ESP ≤ 6 could already have an adverse effect on aggregate stability, especially in fine-textured soils.

Levy et al. (1993) reported that the stability of 105 μm microaggregates was affected by increasing ESP from 1 to 25 and that increasing sodicity had a similar effect on both aggregate breakdown and clay dispersion. They concluded that dispersion is the driving force controlling breakdown of microaggregates at high ESP. Conversely, Abu-Sharar et al. (1987a, 1987b) showed for soils from California that in soils with a high proportion of macropores, sodic conditions (SAR 10) enhance aggregate slaking and that this process was independent of clay dispersion.

Rengasamy et al. (1984) made a distinction between spontaneous dispersion (i.e., slaking) where aggregates in suspension are not physically disturbed, and mechanical dispersion where aggregates are shaken in water. High ESP (>10) and low TEC values are needed for spontaneous dispersion. Under mechanical stress (e.g., impact of raindrops, shear force of flowing water,

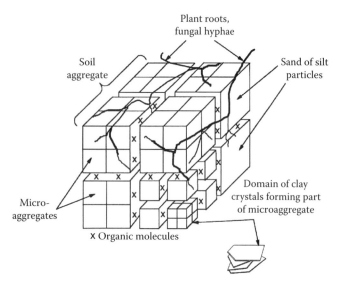

FIGURE 18.5 Heirarchial structure of aggregates. (Adapted from Greenland, D.J. 1979. Structural organization of soils and crop production, p. 47–56. *In* R. Lal and D.J. Greenland (eds.) Soil physical properties and crop production in the tropics. John Wiley & Sons, New York. Copyright John Wiley & Sons, New York. With permission.)

mechanical energy of cultivation machinery), aggregate breakdown and clay dispersion are observed even at low ESP (<10) levels (Shainberg et al., 1992). However, in the presence of Al and Fe oxides, high ESP and low TEC have only a small effect in promoting aggregate breakdown and clay dispersion (McNeal et al., 1968; El-Swaify, 1973).

18.4.3 Organic Matter Effects in Sodic Soils

Organic matter comprises about 1%–10% of a typical soil's mass of which 1%–5% is microbial biomass (Nelson and Oades, 1998). Organic matter acts both as a bonding and dispersing agent in soils. The balance between these opposing effects depends upon the nature of the organic materials and their types of interaction with inorganic colloids. With respect to its effects on soil structure and aggregate stability, SOM can be divided into transient and persistent bonding agents.

Transient components of SOM include plant roots and fungal hyphae. They enmesh macroaggregates (Figure 18.5) and inhibit aggregate breakdown. Transient components are so named because they are present only when plants are growing and fresh organic matter is added, and because they are readily disrupted by some mechanical disturbance (Nelson and Oades, 1998). Soil stability, and especially that of macroaggregates, has often been related to the presence of transient binding agents in soil (Emerson, 1954; Loveland et al., 1987; Celik et al., 2004). However, it has been reported that the stabilizing effect of SOM is only a short-term one and that it diminishes with time (Sort and Alcaniz, 1999).

High levels of SOM seem to promote resistance to sodic conditions. A direct relationship has been noted between soil stability (inferred from modulus of rupture) and SOM content at any given ESP, especially for soils containing 10%–20% clay (Aylmore and Sills, 1982). Addition of pea straw to soils increased macroaggregate stability at ESP values of 3%–36% (Barzegar et al., 1996). The favorable effects of transient organic agents on aggregate stabilization can be explained by nonionic bonding mechanisms by which roots, hyphae, and natural polysaccharides stabilize aggregates, and which are not always directly affected by interactions between water, clay, and exchangeable cations. Nelson and Oades (1998) concluded that the beneficial effects of these transient agents on aggregate stability are most noticeable when ESP is not too high with adequate TEC to offset its effects, clay content is moderate, Mg content is low, mechanical disturbance is limited, and fresh organic materials are added regularly.

Additional sources of transient SOM are mucilages and sugars produced by plant roots and microorganisms. These materials are transient because they decompose readily (Nelson and Oades, 1998) and act as bonding agents in stabilizing smaller structural units (microaggregates) (Tisdall and Oades, 1982).

The persistent components of SOM include polyanionic colloidal materials (plant remains and microbial products), often referred to as humic substances. Their persistence results from their recalcitrance and association with inorganic materials (Stevenson, 1992). These components play an important role in determining the stability of smaller scale aggregates (microaggregates) through a variety of mechanisms, but especially through their dispersing power. Organic anions are known to increase clay dispersivity, in particular when variable charge minerals are present, mainly by increasing the negative charge density of the soil colloid fraction. There is a large body of literature which documents the influence of organic anions on increasing the CFC of various clay minerals, thus giving rise to their susceptibility to dispersion (Frenkel et al., 1992; Tarchitzky et al., 1993; Nelson and Oades, 1998). High ESP, the presence of exchangeable Mg or high pH each increase clay dispersivity in the presence of organic anions (Emerson and Smith, 1970). Because humic substances form chelates with Ca (Sposito et al., 1977), the effective SAR of the soil solution increases, and clay dispersion is promoted.

There are, however, also indications that addition of humic fractions extracted from a variety of sources (permanent pasture soil, cattle manure, and raw oxidized coal, respectively) could be effective in improving aggregate stability (e.g., Piccolo et al., 1997). For instance, Imbufea et al. (2005) observed that addition of potassium-humate to a sodic (ESP = 17) Australian soil significantly improved its aggregate stability.

18.5 Water Flow in Sodic Soils

Sodicity-related degradation of soil structure leads to deterioration of soil water transmission properties, increased susceptibility to crusting, runoff, erosion, and poor aggregate stability. Water transmission properties of soils are commonly characterized by measuring HC and/or infiltration rate (IR), which are distinctly different; HC is measured under conditions where the soil surface in undisturbed, while IR is determined, generally, under conditions where considerable surface disturbance arising from external forces, such as raindrop impact, overhead irrigation, or the shear force of flowing water in furrows, occur. This disturbance often leads to surface seal formation, causing the rate of water movement in the sealed layer to be different from that in the underlying soil.

18.5.1 Hydraulic Conductivity

18.5.1.1 Introduction

The HC of a soil depends on its properties as well as those of the percolating liquid. A detailed discussion of HC is presented in Chapter 5 of *Handbook of Soil Sciences: Properties and Processes*. A number of models have been proposed to describe the relationships between structure of a porous material as reflected by its pore size distribution and permeability (Kozeny, 1927; Childs and Collis-George, 1950; Marshall, 1958). These types of models have not succeeded in adequately predicting soil HC, in part, because they do not take into account the effect of fluid properties on the soil matrix geometry. Clay swelling and dispersion in response to changes in electrolyte composition and concentration of the percolating solution change the size of conducting pores, and hence HC.

18.5.1.2 Sodicity and Electrolyte Concentration

Fireman and Bodman (1939) and Bodman and Fireman (1950) advanced the concept that the higher the SAR and the lower the TEC of the percolating solution, the greater the reduction in soil HC. Subsequently, Quirk and Schofield (1955) developed the concept of threshold concentration, which they defined as the concentration required to prevent a decrease >25% in soil permeability for a given soil ESP or SAR of the percolating solution (Figure 18.6). According to Quirk and Schofield (1955), even a Ca-saturated soil (ESP = 0) may show a reduction in HC, provided that the TEC is below $0.6\,mmol_c\,L^{-1}$ which has been verified in other studies (Emerson and Chi, 1977). When rain or snow water ($<1\,mmol_c\,L^{-1}$) is applied to a soil, even soils with very low ESPs may disperse causing permeability to decrease. The basic approach of Quirk and Schofield (1955) has been extended to a large number of additional soils (e.g., McNeal and Coleman, 1966; McNeal et al., 1966, 1968; Yaron and Thomas, 1968; Rhoades and Ingvalson, 1969; Cass and Sumner, 1982a, 1982b, 1982c; Levy et al., 2005; Dikinya et al., 2007; Bauder et al., 2008).

The importance of clay swelling and dispersion in controlling the HC has been highlighted by numerous studies (Quirk and Schofield, 1955; McNeal et al., 1966; Rowell et al., 1969; Cass and Sumner, 1974; Shainberg et al., 1981a; Radcliffe et al., 1987; Keren and Singer, 1988; Kosmas and Noutakas, 1990; Quirk, 2001). Soils begin to swell as the TEC of the soil solution is reduced, but the effect of swelling on HC becomes apparent only at SAR > 10 in medium and fine textured soils. In coarse textured soils, the effect of swelling on reducing the size of conducting pores, and hence on HC, is hardly noticeable. Because the swelling process is reversible, an increase in HC is observed when TEC is increased again. Clay dispersion, on the other hand, can take place even at low ESPs provided TEC < CFC. When clay dispersion occurs, the dispersed clay particles move through the soil profile and may even cause a complete blockage of conducting pores, and hence irreversible changes in the HC. Consequently, the effects of reducing TEC on the HC are likely to be more severe when clay dispersion is a predominant cause (Shainberg and Letey, 1984). However, the results of the study of Keren and Singer (1988) suggest that interactions between the swelling and the dispersion processes may exist and affect pore size and continuity and thus soil HC.

When Na-affected soils are exposed to rainfall or irrigation water of low TEC, their HC is reduced because the salt concentration in the soil solution is not sufficient to prevent swelling and clay dispersion. For instance, in Australia under rainfed conditions, soils (especially fine textured) often waterlog during the winter rainy season because what little salt is present is leached in the first rains, causing the clay to disperse and seal the soil (Fitzpatrick et al., 1994). In Israel, fields irrigated with water having SAR values as high as 26 with an EC of $4.6\,dS\,m^{-1}$ showed no permeability problems because the TEC in the irrigation water was sufficient to prevent the dispersive effect of Na (Frenkel and Shainberg, 1975). However, upon applying DW to simulate winter rainfall, soil HC dropped to a small fraction of its initial value of $9-12\,mm\,h^{-1}$ (Frenkel and Shainberg, 1975). Similar permeability problems arising from a decrease in TEC of the percolating water have been reported for some areas in the Central Valley of California (Mohammed et al., 1979), and for numerous soils in South Africa (Hensley, 1969; Johnston, 1975). The interaction between SAR, low levels of TEC, and soil HC was studied by Shainberg et al. (1981a) who demonstrated that maintaining a TEC of $3\,mmol_c\,L^{-1}$ can be sufficient to prevent a decrease in HC for a solution having SAR < 12. However, recently Suarez et al. (2008) reported that the HC of undisturbed soil cores decreased when the SAR of the irrigation water was increased from 4 to 6 for solutions with TEC of 10 and $20\,mmol_c\,L^{-1}$.

A number of attempts have been made to relate clay swelling to changes in soil HC. Generally, two types of approaches have been considered, empirical and conceptual. The empirical approach, as represented by the work of Suarez and Simunek (1997), used the experimental results of McNeal et al. (1968), to portray the

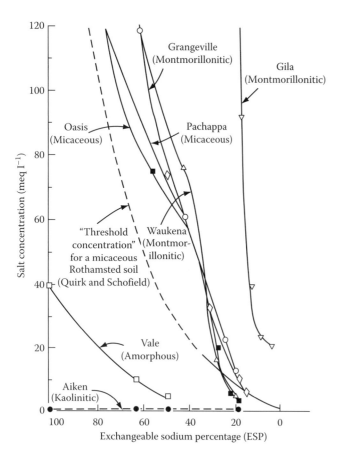

FIGURE 18.6 Combinations of salt concentration and ESP required to produce a 25% reduction in hydraulic conductivity for selected soils. (Reprinted from McNeal, B.L., and N.T. Coleman. 1966. Effect of solution composition on soil hydraulic conductivity. Soil Sci. Soc. Am. Proc. 30:308–312. With permission of the Soil Science Society of America.)

effect of mixed Na/Ca salt solutions on the HC of California soils. The more conceptual, physically based approach was introduced by Russo and Bresler (1977) and is presented in detail by Russo (1988). In this approach, the physicochemical interactions between mixed Na/Ca solutions and the soil matrix are derived from theoretical considerations based on the mixed ion diffuse double-layer theory, the structure of clay particles, the pore size distribution of the soil, and hydrodynamic principles.

18.5.1.3 Soil Properties Affecting Sodic Behavior

18.5.1.3.1 Effect of Mineral Weathering

A major factor causing differences among various sodic soils in their susceptibility to hydraulic failure when leached with solutions having low TEC water is their rate of salt release during mineral dissolution (Rhoades et al., 1968; Felhendler et al., 1974; Shainberg et al., 1981a, 1981b; Shainberg and Gal, 1982; Alperovitch et al., 1985; Naidu et al., 1993b; Keren and Ben-Hur, 2003). Mineral dissolution often determines the TEC of percolating solutions and hence soil HC. Sodic soils containing minerals that readily release soluble electrolytes ($CaCO_3$ and a few primary minerals) will not disperse immediately when leached with DW at moderate ESP values, because a sufficiently high TEC (3 mmol$_c$ L^{-1}) is generally maintained to prevent clay dispersion. In addition, the ESP and SAR values will be reduced because most of the cations released are Ca and Mg. Conversely, soil solution concentrations in soils lacking readily weatherable minerals are likely to be below the CFC, making such soils more susceptible to clay dispersion and a reduction in HC when exposed to DW (Felhendler et al., 1974).

In soils containing $CaCO_3$, the fine $CaCO_3$ particles can, in addition to the above-mentioned dissolution effect, improve the physical condition and permeability of sodic soils by acting as a cementing agent in stabilizing soil aggregates (USSL Staff, 1954; Rimmer and Greenland, 1976; Keren and Ben-Hur, 2003). The relative importance of each of these mechanisms (i.e., dissolution and cementing) can be estimated from their effect on IR and crust formation under rainfed conditions (Agassi et al., 1981; Ben-Hur et al., 1985) (Section 18.5.2.2).

18.5.1.3.2 Effect of Clay Mineralogy

The type of clay mineral also influences the response of soils to sodic conditions. Soils with high contents of expanding 2:1 layer silicates (e.g., montmorillonite) are the most labile while those high in kaolinite and sesquioxides are the least labile (McNeal and Coleman, 1966; Yaron and Thomas, 1968). Acidic kaolinitic soils have been considered insensitive to changes in soil ESP; however, upon addition of smectitic impurities to these soils, their susceptibility to sodic conditions increased markedly (Frenkel et al., 1978). The HC of nonacidic arid and kaolinitic soils has been reported to decrease significantly upon exposure to sodic conditions (Frenkel et al., 1978; Abu-Sharar et al., 1987b). Among the 2:1 layer silicates, montmorillonitic soils have been found to have greater sensitivity to ESP than their vermiculitic counterparts (Rhoades and Ingvalson, 1969).

18.5.1.3.3 Effect of Soil Texture

Mobile clay in the leachate is usually only observed in sandy soils. When clay content is high, the small size of the conducting pores usually ensures that dispersed clay moves only short distances before it clogs the pores, thus leading to low HC. Therefore, in loams and clays, the dispersion mechanism still operates, but no macroscopic movement of the clay particles is observed. Conversely, in sandy soils, the dispersion mechanism and macroscopic clay movement become evident particularly at higher ESP levels, as was shown by Pupisky and Shainberg (1979). The decline and subsequent recovery in HC for a soil of ESP 20 (Figure 18.7) can be explained by a change in flow pattern from that of a solution flowing through a matrix of sand covered with clay which starts to disperse and reduce the HC, to flow of a clay suspension through a pure sandy matrix having large pores resulting in an increase in HC once more (Pupisky and Shainberg, 1979).

Recently, Shainberg et al. (2001) observed a significant interaction between soil clay content (a measure for soil texture) and sodicity with regard to their impact on the HC of semiarid soils. Similarly, Levy et al. (2005) noted that the response of soils' HC to sodic conditions depends on soil texture. However, the combined impact of clay content and soil ESP on the HC is difficult to predict. Clay content acts as a cementing agent binding primary particles into stable aggregates, hence higher clay content supports more stable aggregates (Kemper and Koch, 1966; Kay and Angers, 1999) that are less likely to break down and decrease the HC. Conversely, the greater the clay content the more effective clay swelling is in narrowing the water conducting pores. Changes in soil ESP affect soil swelling and dispersion. The reliance of the HC on both factors depends on the relative weight of each of these factors in determining the mechanism that affects the HC.

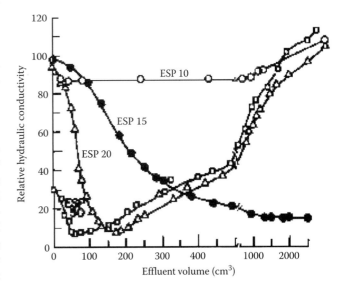

FIGURE 18.7 Relative hydraulic conductivity of soils with various ESPs leached with DW. (Reprinted from Pupisky, H., and I. Shainberg. 1979. Salt effects on the hydraulic conductivity of a sandy soil. Soil Sci. Soc. Am. J. 43:429–433. With permission of the Soil Science Society of America.)

18.5.1.3.4 Effect of Exchangeable Magnesium

Although the exchangeable Ca/Mg molar ratio is high in many soils, the reverse is true for some soils. The effect of adsorbed Mg on soil hydraulic properties is a controversial issue. The USSL Staff (1954) grouped Ca and Mg together with respect to their similarities in promoting and maintaining soil structure, while others have found that a Na–Mg-saturated soil is structurally less stable (van der Merwe and Burger, 1969) and/or has a lower HC (McNeal et al., 1968) than a Na–Ca-saturated soil. Alperovitch et al. (1981, 1985) related the effects of Mg to soil type. In calcareous soils, the presence of Mg enhances the dissolution of $CaCO_3$, thereby producing electrolytes which prevent clay dispersion and HC decay. In noncalcareous soils, Mg causes a decrease in HC of the soils beyond that of a corresponding Na–Ca system (Alperovitch et al., 1981, 1985). Shainberg et al. (1988) concluded that the lower HC values of Na–Mg relative to Na–Ca montmorillonite are related to the effect of Mg on the hydrolysis of montmorillonitic clay. The presence of a high concentrations of Mg at the clay surface slows down the release of octahedral Mg from the mineral lattice (Kreit et al., 1982), thereby lowering the EC of the Na–Mg clay (Shainberg et al., 1988). In kaolinitic soils, this adverse effect of Mg has not been noted (Levy et al., 1989).

A distinction has been made between the direct effect of exchangeable Mg in causing decreases in HC, which has been termed a specific effect, and the inability of Mg in irrigation water to counter the accumulation of exchangeable Na in soils (McNeal et al., 1968; Chi et al., 1977; Emerson and Chi, 1977; Rahman and Rowell, 1979). Curtin et al. (1994) concluded that the specific effect of Mg was dominant, because the observed adverse effects of Mg on clay dispersion and HC were greater than could be explained simply by a slightly higher exchangeable Na level in Mg systems.

18.5.2 Infiltration Rate

18.5.2.1 Introduction

Soil IR is defined as the volume flux of water flowing into the profile per unit surface area under any given set of circumstances. Mechanisms controlling changes in the IR over time depend on the mode of water application to the soil. Under conditions where water is supplied to the soil without appreciable energy input, IR depends on the HC of the soil matrix (Section 18.5.1). The IR is, generally, high during the early stages of infiltration, particularly when the soil is initially quite dry, but decreases monotonically to approach a constant rate asymptotically, due to a decrease in the matric suction gradient that occurs as infiltration proceeds (Baver et al., 1972; Hillel, 1980).

When water is supplied with appreciable energy impact, the IR decreases from its initial high rate due to the formation of a thin layer (<2 mm) at the soil surface, termed surface seal. This seal is characterized by greater density, higher shear strength, finer pores, and lower saturated HC than the underlying soil (McIntyre, 1958; Bradford et al., 1987). A structural seal, usually caused by the impact energy of waterdrops, should be distinguished from a depositional seal formed by translocation and deposition of fine soil particles at a discernible distance from their original location (Chen et al., 1980; Arshad and Mermut, 1988). The discussion to follow will focus on the former type seal.

Structural seal formation is the result of two mechanisms: (1) physical disintegration of surface soil aggregates caused by wetting of the dry aggregates (Loch 1994; Levy et al., 1977) and/or the impact energy of raindrops (McIntyre, 1958; Morin and Benyamini, 1977; Agassi et al., 1981) or sprinkler irrigation (Aarstad and Miller, 1973), and subsequent compaction of the disintegrated aggregates by water drop impact, and (2) physicochemical dispersion of soil clays, which then migrate into the soil with the infiltrating water, clog pores immediately beneath the surface, and form a layer of low permeability (Farres, 1978; Agassi et al., 1981). The two mechanisms are complementary. The physical disintegration of surface soil aggregates enhances clay susceptibility to dispersion under sodic conditions (Shainberg and Letey, 1984), while clay dispersion weakens aggregate stability and renders them greater sensitivity to disintegration by wetting and/or the impact of water drops (Abu-Sharar et al., 1987a; Mamedov et al., 2001).

18.5.2.2 Effect of Sodicity on Infiltration Rate

Kazman et al. (1983) studied under laboratory conditions the sensitivity of seal formation and IR of smectitic soils to low levels of ESP when exposed to DW (simulating rainwater) (Figure 18.8). Even at the lowest sodicity (ESP = 1.0), a seal was formed and the IR dropped from an initial value >100 mm h^{-1} to a final infiltration rate (FIR) of 7.0 mm h^{-1}. An ESP value of 2.2 was sufficient to cause a further drop in FIR of the sandy loam to 2.4 mm h^{-1}. The amount of rain required to approach the FIR was also affected by ESP (Figure 18.8). Because raindrop impact

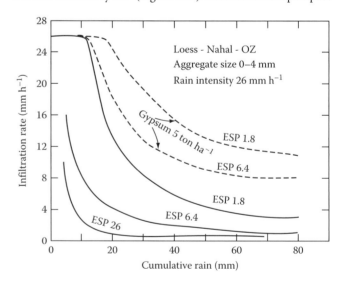

FIGURE 18.8 The effects of the soil ESP and phosphogypsum application on the infiltration rate of the Hamra–Netanya soil as a function of the cumulative rain. (Reprinted from Kazman, Z., I. Shainberg, and M. Gal. 1983. Effect of low levels of exchangeable Na and applied phosphogypsum on the infiltration rate of various soils. Soil Sci. 35:184–192. With permission from Wolters Kluwer Health.)

energy was the same in all experiments, the differences between IR curves for the various treatments reflect the result of chemical dispersion of soil clay caused by sodicity. The high sensitivity of the soil surface to low ESP values is explained by three factors (Oster and Schroer, 1979; Kazman et al., 1983): (1) the mechanical impact of the raindrops, which enhances chemical dispersion; (2) the absence of a surrounding soil matrix (sand particles), which when present slows clay dispersion and movement; and (3) the almost total absence of electrolytes in the applied DW. With respect to the first factor, it is difficult to separate the mechanical and chemical mechanisms that are complementary. Agassi et al. (1985a) noted that in the absence of the former, the chemical mechanism does not come into effect at low ESP levels (<5). Evidently, in the lower range of ESP, the chemical mechanism needs some activation energy before it can begin operating at the soil surface. In the case of rainfall, this energy is provided by the impact of the raindrops.

Clay dispersion is highly sensitive to the TEC and SAR of the applied water. This is particularly true in cases where the soil surface is exposed to the mechanical action of falling water droplets, which enhance clay susceptibility to chemical dispersion. Agassi et al. (1981) noted for two Israeli smectitic soils that, the lower the TEC, the faster the rate at which the IR decreases. Also for the same TEC, increasing soil ESP results in a sharper decrease in IR and a lower FIR. Similarly, Oster and Schroer (1979) found that TEC greatly affected IR even at low SAR values. They observed an increase in final IR from 2 to 28 mm h^{-1} as TEC in the applied water (SAR 2 to 4.6) increased from 5 to 28 mmol$_c$ L^{-1}. These results suggest that IR is far more sensitive than HC to the TEC of the applied water (Shainberg and Letey, 1984). Furthermore, the IR values of calcareous and noncalcareous soils are equally sensitive to low levels of ESP (Agassi et al., 1981; Kazman et al., 1983), whereas the HC of calcareous soils is less sensitive to sodic conditions than that of noncalcareous soils (Section 18.5.1.3). These findings indicate that (1) the rate at which lime dissolves and increases TEC in the soil is insufficient when the soil is exposed to rainfall, and (2) the role of CaCO$_3$ as a cementing material is negligible, because otherwise both types of soil would have had different IR values (Shainberg and Letey, 1984).

Although Ca and Mg are both divalent cations, their effects as the complementary cation to Na on soil stability are often very different. Elevated exchangeable Mg levels can cause deterioration in soil structure resulting in the development of a Mg solonetz (Ellis and Caldwell, 1935). In addition, Mg enhances dispersion of montmorillonitic and illitic clays compared to Ca (Bakker and Emerson, 1973). With respect to IR, the effect of exchangeable Mg has been found to depend on the kinetic energy of the water droplets. When high energy rain (22.9 kJ m^{-3}) was used, exchangeable Mg had an effect similar to that of Ca (Levy et al., 1988), but with low to medium energy rainfall (8.0–12.5 kJ m^{-3}), the IR of Mg–Na-treated soil was lower than that of Ca–Na soil (Keren, 1990). Such findings suggest that the adverse effect of Mg on clay dispersion, and hence on IR, is pronounced only under conditions where chemical dispersion controls IR (i.e., for raindrops with low to medium kinetic energy).

The susceptibility of sodic soils to seal formation and reduction in IR when exposed to rainfall depends on a number of soil properties, and especially on clay content and mineralogy. Soils with 10%–30% clay are considered as the most susceptible to seal formation and have the lowest IR values. With increasing clay content, soil structure is more stable and seal formation diminishes. In soils with lower clay contents (<10%), the amount of clay available to disperse and clog soil pores is limited and, as a result, only a poorly developed seal can be formed (Ben-Hur et al., 1985).

Most of the studies on seal formation and IR have been conducted on soils in which the dominant clay minerals are smectites that are known to be dispersive during both HC and IR determinations. The effect of exchangeable Na on the HC of kaolinitic soils is quite small (Frenkel et al., 1978; Chiang et al., 1987); however, a number of kaolinitic and illitic soils from the southeastern United States (Miller, 1987; Miller and Scifres, 1988) and Africa (Levy and van der Watt, 1988; Stern et al., 1991; Wakindiki and Ben-Hur, 2002) have been found to be susceptible to seal formation. Evidently, upon mechanical agitation by the beating action of raindrops, aggregates from kaolinitic soils may disperse (according to the level of exchangeable Na), and consequently, form a seal with a resultant low IR. Additionally, Stern et al. (1991) observed that the susceptibility of kaolinitic soils to seal formation was positively correlated with their contents of smectitic impurities; kaolinitic soils with small amounts of smectite were dispersive and susceptible to seal formation, whereas kaolinitic soils that contained no smectite impurities were less susceptible.

Seal formation and reduction in IR depend strongly on clay dispersion, which is enhanced by increasing soil ESP and/or decreasing TEC. In addition, a strong relation between dispersible clay, IR, and soil loss has been reported in numerous studies (Miller and Baharuddin, 1986; So and Cook, 1993). Sumner (1993) compiled a table based on published data showing that, for soils from different regions of the world, IR decreases as SAR/ESP increases, making more water available for runoff which, in turn, tends to increase erosion (Table 18.2). A different view of the relation among soil ESP, IR, and erosion was given by Levy et al. (1994), who did not observe a direct relationship between these factors. They argued that with an increase in ESP a lower IR is obtained, but the resultant erosion depends on the balance between the larger volume of runoff and a more dense and structurally more resistant seal; thus the amount of erosion cannot be predicted.

18.5.2.3 Sodicity and Conditions Prevailing in the Soil

The interest in the effects of conditions prevailing in the soil on its susceptibility to seal formation stems from the realization that aggregate stability, which seal formation is closely related to, depends, in addition to its association with various soil properties (e.g., organic matter, clay percentage, and oxides content), on conditions prevailing in the soil such as rate of aggregate wetting, antecedent moisture content (AMC) in the aggregates, and aging duration (Panabokke and Quirk, 1957; Francis and

TABLE 18.2 Effect of Water Quality and ESP on Final Infiltration Rate (IR), Runoff, and Erosion for Soils from a Variety of Locations[a]

Soil Classification and Location	Water Quality (mmol$_c$ L^{-1})	ESP/SAR	Final IR (mm h^{-1})	Runoff (%)	Erosion (kg ha^{-1})
Typic Rhodoxeralf (a), Israel	DW	1.0	7.5	40	
		2.2	2.2	70	
		4.6	0.4	85	
Typic Rhodoxeralf (b), Israel	DW	2.2	5.2		4740
	9.5	2.2	7.3		1070
Calcic Haploxeralf, Israel	DW	1.8	3.0	60	
		6.4	1.0	80	
Typic Chromoxerert, Israel	DW	2.5	2.0	52	
		5.5	1.0	80	
Typic Kandhapludult, Georgia, USA	DW	1.2	8.5	83	157
	5	1.2	23.5	53	130
Typic Halpudult, Georgia, USA	DW	1.2	3.6	92	601
	10	1.2	10.7	77	309
Typic Ochraquult Georgia, USA	DW	1.6	1.6	96	885
	12	1.6	3.1	92	456
Rhodic Paleudult, Georgia, USA	DW	1.37	2		1000
	5	1.37	11		410
Oxic Paleustalf, South Africa	4	1.0	33.0		800
		7.0	10.0		2400

Source: Reprinted from Sumner, M.E. 1993. Sodic soils: New perspectives. Aust. J. Soil Res. 31:683–750. With permission of CSIRO Publications, Melbourne, Australia.
DW, Distilled water.
[a] Reference for data of each individual soil is given in Sumner (1993, table 7).

Cruse, 1983; Kjaergaard et al., 2004). Aggregate disintegration by wetting, termed *slaking*, increases with the increase in their rate of wetting and is ascribed to the explosion of aggregates by entrapped air and differential swelling (Panabokke and Quirk, 1957; Quirk and Panabokke, 1962). Wetting and keeping the soil at a given moisture (i.e., aging) induces an increase in aggregate strength through the development of cohesive bonds between the soil particles (Blake and Gilman, 1970; Kemper and Rosenau, 1984; Attou et al., 1998) and the production of cementing agents by microbial activity (Martin et al., 1995). The effects of aging duration on aggregate stability seem however to depend on the AMC of the soil (Gerard, 1965; Utomo and Dexter, 1981). The effects of these conditions on soil sealing as well as their interactions with one another and with several soil and rain properties has been summarized recently in a review by Ben-Hur and Lado (2008).

Early studies on the effect of wetting rate (WR) on soil vulnerability to sealing clearly indicate that when slow wetting is used (i.e., eliminating aggregate slaking), surface aggregates are considerably less susceptible to disintegration by the impact of raindrops, and thus less developed and more permeable seals are formed (Loch, 1994; Levy et al., 1997). However, results of various studies show that use of slow WR to prevent slaking and maintain high IR occurs mainly in fine textured soils (Mamedov et al., 2001; Shainberg et al., 2003, Lado et al., 2004). Mamedov et al. (2001) studied the combined effects of WR (2, 8, and 64 mm h^{-1}), soil sodicity (ESP levels between 0.9 and 20.4), and texture (clay content from 8.8% to 68.3%) on IR and runoff from six Israeli soils; data for runoff are presented in Figure 18.9. In soils with low clay content (8.8%), effect of WR on seal formation was negligible, whereas effect of ESP was significant. Conversely, in the clay soils (>52.1%), WR had a predominant effect on IR and runoff, while the effect of ESP was notable yet secondary to that of WR. The soils with intermediate clay content (22.5%–40.2% clay) were the soils most susceptible to seal formation, with both WR and ESP having moderate effects on seal formation and runoff (Figure 18.9).

The dependence of aggregate stability on AMC has been linked to aggregate detachment, soil erosion, and seal formation (Farres, 1987; Bradford et al., 1992; Le Bissonnais, 1996; Levy and Mamedov, 2002). Furthermore, moisture content together with aging duration have been found to determine aggregate breakdown mechanism, the resulting particle size distribution, and the evolution of the structure of the seal (Kemper and Rosenau, 1984; Le Bissonnais, 1990). The direct impact of AMC on seal formation and soil loss has received extensive attention (Cruse and Larson, 1977; Al-Durrah and Bradford, 1981; Francis and Cruse, 1983; Luk, 1985; Reichert and Norton, 1994; Le Bissonnais et al., 1995; Froese et al., 1999; Wangemann et al., 2000) but results were inconsistent albeit using nonsodic soils. Conversely, the effect of aging on sealing was studied less. Levy et al. (1997) reported of the existence of an interaction between aging and WR in their effect on the IR. Ageing duration of 18 h was only effective in maintaining higher IR in cases where the

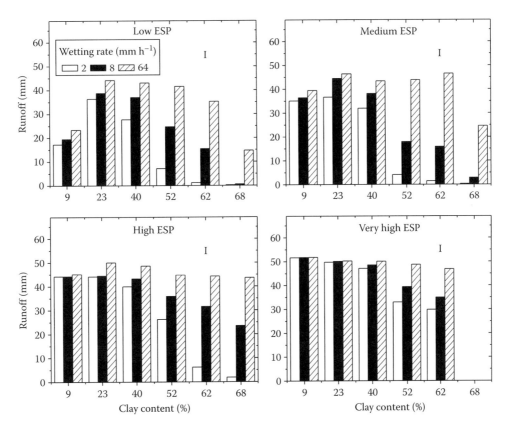

FIGURE 18.9 Effects of wetting rate on the calculated runoff from six soils as a function of clay content and ESP. Bar indicates a single confidence interval value. ESP low, medium, high, and very high refers to 2, 5, 10, and 20, respectively. (Reprinted from Mamedov, A.I., G.J. Levy, I. Shainberg, and J. Letey. 2001. Wetting rate and soil texture effect on infiltration rate and runoff. Aust. J. Soil Res. 30:1293–1305. With permission of CSIRO Publications, Melbourne, Australia.)

WR was not fast (<6 mm h^{-1}). When fast WR was used, similar IR curves were obtained for aged- and non-aged samples (Levy et al., 1997). Ben-Hur et al. (1998) observed for a nonsodic and a sodic soil that 4 days of aging resulted in a more gradual decrease in the IR curve and a higher final IR compared to a nonaged soil.

Recently, Mamedov et al. (2006) studied the combined impact of AMC and aging. These authors reported of the existence of an optimal range of AMC, generally between field capacity and wilting point, at which runoff levels were 30% lower than those obtained below or above this AMC range. They also observed that the level of AMC determined the beneficial effects of aging on improving the resistance of soil to seal formation. Increasing aging duration was most effective in improving soil IR at the aforementioned optimal level of AMC studied. It can be expected, based on current knowledge, that the effects of AMC and aging would be greatly affected by the degree of soil sodicity.

18.6 Crop Responses to Sodic Conditions

Productivity of sodic soils is limited, mainly because the deterioration in their physical properties (Cairns et al., 1962; Minhas, 1996; Oster and Jayawardane, 1998), which causes poor aeration, restricts root development, and enhances root diseases. These not only decrease crop yields but also limit the choice of crops that can be grown on these soils (Ayers and Westcot, 1985; Minhas, 1996; Oster and Jayawardane, 1998). In addition, sodic conditions can also change soil redox potential, pH, dissolved organic C content, and bicarbonate toxicity (Chauhan et al., 2007), all of which could lead to nutritional imbalances (Qadir and Oster, 2004). For a detailed review of crop response to sodicity and sodicity-related fertility and nutritional constraints, readers are referred to Curtin and Naidu (1998).

18.6.1 Crop Tolerance to Sodicity

Research on crop tolerance to sodicity has been conducted under two different types of conditions, in the presence or in the absence of sodicity-related adverse soil physical conditions. In the United States, research has focused mainly on tolerance to sodicity based solely on nutritional factors (Chang and Dregne, 1955; Pearson and Bernstein, 1958), because in the field, sodicity-induced nutritional imbalances may be obscured by unfavorable physical conditions (Bernstein and Pearson, 1956). Data presented by Pearson (1960) indicate that even at relatively low ESP levels (2–10), some crops show signs of Na toxicity. However, most field crops can be classified as moderately tolerant or tolerant to sodicity (Pearson, 1960). Work in India, on the other hand, has been carried out under conditions where soil structure was allowed to be affected by sodicity concurrently

(Abrol and Bhumbla, 1979; Singh et al., 1979). Under such conditions, grasses were tolerant, wheat and barley were moderately tolerant, and pea and beans were sensitive to sodicity (Gupta and Abrol, 1990). Recently, Garg et al. (2005) observed that fennel crop was tolerant to sodicity up to ESP of 25. Use of the concept where the soil is allowed to be affected by sodicity may, however, cause results to be site specific, and hinder their applicability to other locations and soil types. Furthermore, ranking of crops for sodicity tolerance depends on the conditions under which the response was determined. For instance, rice has been found to be highly tolerant under sodic field conditions (Abrol and Bhumbla, 1979), but only moderately tolerant under favorable soil conditions (Pearson, 1960).

18.6.2 Plant Nutrition in Sodic Soils

Field fertility studies on sodic soils often yield inconclusive results, mainly due to high spatial variability (Toogood, 1978). Consequently, most dependable data on sodicity-related nutritional aspects have been obtained from greenhouse and laboratory studies (Curtin and Naidu, 1998; Wright and Rajper, 2000).

Of the three major cations found in the soil in addition to Na (Ca, Mg, K), availability of Mg (Williams and Raupach, 1983) and K (Chhabra, 1985) have not been considered to be of a major concern in sodic soils. Jalali et al. (2008) point, however, to the fact that during irrigation with sodic water, leaching of Mg and K from the root zone can occur, and if uncorrected especially in soils with low CEC, could lead to their deficiencies to plant growth. Unlike Mg and K, impaired uptake of Ca is one of the main adverse nutritional effects of sodic soils, probably due to low Ca concentrations in the soil solution (Wright and Rajper, 2000). Calcium plays a major nutritional and physiological role in plant metabolism. The ratio of Ca to total cation concentration (TCC) is considered to be a better indicator of Ca availability to plants than the actual Ca concentration (Adams, 1974). Carter and Webster (1990) found that Ca/TCC values <0.1 are often found in saturated paste extracts of sodic soils, which for many crops, would result in Ca deficiency (Grieve and Maas, 1988). In addition, high Mg/Ca ratios, which are often found in sodic soils (Williams and Colwell, 1977), can also cause Ca deficiency (Grattan and Grieve, 1992). Extreme ratios of Na:K, Mg:Ca, and Cl:NO_3 (Curtin and Naidu, 1998; Grattan and Grieve, 1999), likely to be found in the soil solution of sodic soils, rather than absolute values, may lead to salt-stressed plants, specific-ion toxicity, and nutritional disorders (Grattan and Grieve, 1999).

Except for the case of weathered sodic soils in Australia, where acute phosphorus (P) deficiency is a frequent occurrence (Naidu and Rengasamy, 1993), P does not pose a nutritional problem in sodic soils; some studies even reported that P is more readily available to plants in sodic than in nonsodic soils (Chhabra, 1985; Gupta et al., 1990). Nitrogen (N), on the other hand, can limit growth in many sodic soils (Cairns et al., 1967; Malhi et al., 1992) because of low rates of N mineralization, denitrification under restricted aeration conditions, volatilization of NH_3 under alkaline conditions, and inhibition of NO_3 uptake by Cl or SO_4 ions, which usually are in abundance (Curtin and Naidu, 1998; Grattan and Grieve, 1999).

Poor productivity of sodic soils can also be ascribed, at least in part, to micronutrient deficiencies in the soil (Naidu and Rengasamy, 1993). Availability of micronutrients in sodic soils is controlled largely by adsorption and precipitation reactions. Neutral pH values generally enhance adsorption of micronutrients. The composition and concentration of the soil solution also determine plant availability of micronutrients. The extent to which these solution parameters affect the availability has not yet been well established (Curtin and Naidu, 1998); the presence of carbonate minerals that provide active sorption sites for the precipitation of many micronutrients should further reduce their availability (Hodgson et al., 1966). Micronutrient deficiencies could also stem from difficulties in their uptake by the plants due to lower dry weight of the various types of tissues of the plant (Wright and Rajper, 2000).

18.6.3 Ion Toxicity under Sodic Conditions

Apart from the osmotic effect on plants due to high TEC generally found in sodic soils, high concentrations of Na have been found to be toxic in some woody species (Maas and Hoffman, 1977) and in avocado, citrus, and stonefruit trees (Lauchli and Epstein, 1990).

High Na/Ca ratios result in high Na uptake and low uptake of Ca by plants and thus cause, in addition to Ca deficiency, some other nutritional imbalances and adverse physiological effects (e.g., membrane permeability and plant stability). High Na uptake can cause deficiencies of elements such as K, Zn, Cu, and Mn, while low uptake of Ca enhances uptake and possible toxicity of elements such as Ni, Pb, Se, Al, and B (Levitt, 1980). Ameliorating sodic soils with gypsum or lime may (1) alleviate heavy metal toxicity, and (2) offset the negative effects created by excessive Na uptake on physiological processes in plants (Rengasamy, 1987). However, in layers where calcium carbonate has accumulated during pedogenesis, sodium bicarbonate and carbonate may accumulate, thereby increasing the soil pH above 9 (Rengasamy et al., 2003). In addition to the toxicity of carbonate and bicarbonate species, high pH also leads to Fe, Mn, Cu, Zn, and P deficiency (Naidu and Rengasamy, 1993). Furthermore, recent observations (Ma et al., 2003) have shown the phytotoxicity of $Al(OH)_4$ in soils with pH above 9.

Boron (B) is fairly soluble in sodic soils with neutral pH; thus B is likely to be toxic rather than deficient in these soils (Naidu and Rengasamy, 1993). Significant B toxicity in cereals was noted in regions of low rainfall of southern Australia; toxic concentrations of B in excess of $10\,mg\,kg^{-1}$ were recorded in subsurface layers of soils from these regions where sodic conditions prevail (Cartwright et al., 1986). Reclamation of sodic soils having pH > 7 with gypsum and leaching reduce B concentrations in the soil solution to levels within the safe limits, and B is no longer toxic to plants (Gupta and Abrol, 1990). For more details on B in soils readers are referred to Chapter 17 in this handbook.

18.7 Amelioration and Management of Sodic Soils

Management of sodic soils, involving the improvement in soil structure (to alleviate problems of water permeability and logging, erosion, inadequate aeration, compaction, etc.), is essential if productivity is to be maintained. Over the past 100 years, several different site-specific approaches involving the use of chemical amendments, tillage, crop diversification, water, and electrical currents have been used to ameliorate sodic soils (Qadir et al., 2001). Of these, chemical amendments have been used most extensively (Oster et al., 1999). A number of tillage options such as deep ploughing and subsoiling have also been used to break up the shallow, dense, sodic clay pans and/or natric horizons that occur within 0–4 m of the soil's surface (Abdelgawad et al., 2004). However, in recent decades, a crop-based approach (phytoremediation, biological reclamation, or vegetative bioremediation) has received considerable attention in many developing countries, as it is much cheaper than chemical amelioration, the costs of which are prohibitively high (Robbins, 1986; Qadir et al., 2001, 2002, 2006, 2007; Ghaly, 2002; Qadir and Oster, 2004).

18.7.1 Use of Chemical Amendments

There are two common practices for ameliorating sodic soils by addition of divalent cations: (1) using chemical amendments that release Ca ions to the soil and (2) application of successive dilutions of a high salt water containing divalent cations. Detailed reviews on the amelioration of sodic soils with those practices have been published by Keren (1995), Oster and Jayawardane (1998), and Qadir et al. (2001).

18.7.1.1 Chemical Amendments

Typical amendments are those containing a source of soluble or potentially soluble Ca when reacted with the soil. Common amendments include gypsum, lime, $CaCl_2$, H_2SO_4, and sulfur.

18.7.1.1.1 Gypsum

Gypsum ($CaSO_4 \cdot 2H_2O$), both from natural and anthropogenic sources (byproducts), is the most commonly used amendment for sodic soil reclamation, primarily because of its low cost, reasonable solubility, and availability. Gypsum added to a sodic soil can increase aggregate stability (Lebron et al., 2002; Mamedov et al., 2007) and soil permeability (Kazman et al., 1983) by means of both EC and cation exchange effects (Loveday, 1976).

The relative importance of each of these effects depends on the purpose for which gypsum is used. If gypsum is used in order to maintain high permeability at the soil surface during application of good quality water (i.e., rainwater or other source of water with low TEC), then the electrolyte effect is important. In this case, surface application of gypsum may be worthwhile, and the dissolution rate of gypsum dictates whether the rate of electrolyte supply to the water is sufficient to prevent clay dispersion and swelling. Thus, because of its higher dissolution rate, gypsum from industrial sources such as by-products of the phosphate (phosphogypsum) and power generation (flue gas desulfurization gypsum) industries is preferable to mined gypsum (Keren and Shainberg, 1981). In this case, the amount of gypsum required depends on the amount of high quality water applied and the rate of gypsum dissolution, and is relatively independent of the amount of exchangeable Na in the soil profile. Field experiments in smectitic Israeli soils in a region of annual precipitation of 350–400 mm indicated that surface application of 5 Mg ha^{-1} of phosphogypsum is effective in maintaining significantly higher IRs and lower runoff and soil loss levels compared with a nontreated soil (Keren et al., 1983; Agassi et al., 1985b). Furthermore, Kazman et al. (1983) reported that addition of phosphogypsum proved effective in maintaining high IR compared to the control, even in soils with ESP < 2 (Figure 18.8). This, once again, demonstrates that some chemical dispersion takes place even at very low ESP values. Use of phosphogypsum was found effective in preventing runoff also in kaolinitic soils (Miller, 1987; Miller and Scifres, 1988). Agassi et al. (1986) suggested that the favorable effect of phosphogypsum on preventing the decline in the IR should be attributed not only to its effect on the EC of the percolating water, but also to (1) physical interference with the continuity of the seal and (2) partial mulching of the surface protecting it from the beating action of raindrops.

Conversely, in soils where the high ESP of the soil profile needs to be reduced, then the cation exchange effect is more important. The amount of gypsum required depends then on the level of exchangeable Na in a selected depth of soil. The amount of exchangeable Na to be replaced per unit land area during reclamation depends on the initial and desired final exchangeable Na levels, CEC, bulk density, and depth of soil to be reclaimed. The amount of gypsum required (GR, Mg ha^{-1}) can thus be calculated by the following equation (Oster and Jayawardane, 1998):

$$GR = 0.0086 \cdot F \cdot D_s \cdot \rho_b \cdot CEC \cdot (ESP_i - ESP_f) \quad (18.6)$$

where
F is the Ca–Na exchange efficiency factor (unitless)
D_s is the soil depth to be reclaimed (m)
ρ_b is the soil bulk density (Mg m^{-3})
CEC is the cation exchange capacity (cmol$_c$ kg^{-1})
ESP_i and ESP_f are the initial and final ESP values, respectively

Efficiency and percentage of Ca exchange for adsorbed Na increase with the increase in soil ESP (Chaudhry and Warkentin, 1968). Thus, the F factor ranges from 1.1 for an ESP_f of ~15 to 1.3 for an ESP_f of 5 (Oster and Frenkel, 1980). In addition, F increases with increasing clay content (Manin et al., 1982).

The cation exchange process, major ion species, complexation reactions, and mineral dissolution have formed the basis of several gypsum requirement models (Dutt and Terkeltoub, 1972; Tanji et al., 1972; Suarez, 2001), which have been field tested and yielded a satisfactory match between calculated and measured data. These models provide a powerful tool for quantitative predictions of water and gypsum required to reclaim a soil to a predetermined level of salinity and sodicity. Recently, Simunek and

Suarez (1997) have proposed a one-dimensional multicomponent transport model, UNSTACHEM, that takes into consideration, in addition to the parameters used by the aforementioned models, soil hydraulic properties and water flow velocities to predict the amendment concentration necessary for significant improvement of soil permeability.

18.7.1.1.2 Calcium Carbonate

Adding $CaCO_3$ to reclaim sodic soils is only of limited value (Gupta and Abrol, 1990), mainly because of its low dissolution rate that fails to provide an adequate amount of Ca for exchange. However, under certain conditions, naturally occurring carbonate minerals [$CaCO_3$, $MgCO_3$, $CaMg(CO_3)_2$] can play a significant role in reclaiming sodic soils (Inskeep and Bloom, 1986; Nadler et al., 1996). Dissolution of $CaCO_3$ can be enhanced by adding an acid or acid former to the soil (Keren, 1995) or by increasing CO_2 levels in the soil (Simunek and Suarez, 1997). The latter can be achieved by enhancing root activity (Bower and Goertzen, 1958; Rao and Ghai, 1985).

Normally, the beneficial effect of naturally occurring $CaCO_3$ on soil reclamation manifests itself mainly by maintaining the TEC of the soil solution at a level high enough to prevent clay dispersion and subsequent HC reductions (Felhendler et al., 1974; Alperovitch et al., 1981). However, $CaCO_3$ is not effective in maintaining high HC at the soil surface during rainfall because it fails to supply electrolytes fast enough at the soil surface, and thus, IR is impeded (Shainberg and Letey, 1984).

18.7.1.1.3 Acids and Sulfur

In calcareous soils, addition of acid amendments will react with $CaCO_3$ to produce a soluble source of Ca ($CaSO_4 \cdot 2H_2O$ or $CaCl_2$). In the case of elemental S additions, the process is somewhat more complicated as it requires an initial phase of microbiological oxidation to produce H_2SO_4 (Keren, 1995). The increasing availability of acid waste products, coupled with the need to dispose of them in a safe manner, increases their potential for use as soil amendments (Miyamoto et al., 1975; Sadiq et al., 2007).

18.7.1.1.4 Calcium Chloride

Calcium chloride (by-product from industry) can at times be an economical amendment for soil reclamation. Under certain conditions, such as soils with high ESP, $CaCl_2$ can be an even more efficient amendment than gypsum because of its higher solubility (Alperovitch and Shainberg, 1973), but it has only a short-term effect. Prather et al. (1978) suggested that for soils with high ESP and low permeability, a combined application of $CaCl_2$ and gypsum might give a more rapid and more effective reclamation than that of either amendment alone.

18.7.2 High Salt Water Dilution Method

In areas where water is not a limiting factor, reclamation of sodic soils can be achieved by leaching the soil with successive dilutions of high salt water containing divalent cations (Reeve and Bower, 1960). In the early stages, the high TEC prevents clay dispersion and induces flocculation of the soil colloids. Simultaneously, the Ca ions in the water decrease sodicity by replacing exchangeable Na. Upon dilution of the saline water with higher quality water, the SAR of the water is reduced. Reeve and Doering (1966) developed a nomograph that predicts reclamation at any stage of leaching. However, to ensure successful soil reclamation, the depth of water added should be ninefold the depth of soil to be reclaimed (Reeve and Doering, 1966).

Among the various methods of water application during reclamation, sprinkler irrigation was found to be the best in removing salts from small pores (Nielsen et al., 1966) followed by intermittent ponding, which requires less water than continuous ponding to obtain a similar degree of leaching (Miller et al., 1965), thereby allowing time for salts to diffuse from soil aggregates to previously well-leached pores prior to the next leaching event.

Qadir et al. (2001) summarized the pros and cons of the high salt water dilution method as follows. The method is effective (1) in smectitic soils with low hydraulic conductivity, (2) when the physical conditions of the soil have been deteriorated to such an extent that use of ameliorants requires excessive amounts, and (3) in the case where the irrigation water to be used after amelioration is of very low electrolyte concentration. Problems associated with this method include (1) use of water containing inadequate concentration of divalent cations, (2) the need to develop facilities that collect, convey, and treat the saline water, and (3) the need to dispose of the highly saline-sodic drainage water in such way that soil, and surface and groundwater contamination is avoided (Qadir et al., 2001).

18.7.3 Tillage

Tilling sodic soils, which often results in improvements in physical and hydraulic properties, reduces bulk density, and increases total porosity (Blackwell et al., 1991), especially macroporosity (Klute, 1982), improves internal drainage and saturated and unsaturated HC. Consequently, aeration is improved and plant growth is enhanced (Oster and Jayawardane, 1998; Sadiq et al., 2007). However, the beneficial effects of tillage on sodic soils are often not long lasting because sodic soils tend to reconsolidate under their own weight when wetted, and their hydraulic properties deteriorate once more (Mead and Chan, 1988).

Deep tillage (1–2 m) or ripping breaks up hardpans and cemented layers, mixes soil layers and causes, in general, a permanent improvement in soil structure and physical properties. In order to obtain optimal results, plowing should be carried out at the correct soil moisture content, 10% for clay, 5% for loam, and 2.5% (w/w) for sand (Wildman, 1981). Sadiq et al. (2007) observed that disc plow was more effective in obtaining high yields and lower soil salinity parameters, in three sites in Pakistan compared to rotavator and cultivator. However, deep tillage can sometime lead to water logging and infiltration problems by bringing sodic subsoil to the surface (McKenzie et al., 1992). Shallow tillage (disking or chiseling) is effective in reducing soil permeability problems associated with surface crusts, hardsetting, and compaction. Shallow tillage requires repetition from time to time (Rawitz et al., 1986).

A combined treatment of tillage and gypsum application, termed gypsum slotting, has recently been developed in Australia, in which gypsum at rates sufficient to reclaim the treated soil is added in tilled slots created by tillage (~150 mm wide, 0.4 m deep and spaced about 1–2 m apart) (Jayawardane and Blackwell, 1986). The advantages of this technique include efficient removal of excess water from the surface layer and its storage in the subsoil for use during dry periods, reclamation of the soil in the slots, and maintenance of adequate macroporosity. In addition, compaction of the soil in the slots is prevented by the use of bridging implements (Blackwell et al., 1989). Results from field studies indicate that gypsum slotting improves aeration and infiltration and results in higher yields (Jayawardane et al., 1987). Gypsum slotting will not be effective where a shallow water table exists, or where subsoil permeability is higher than that of the topsoil, or for crops that require prolonged ponding (Jayawardane and Chan, 1994).

An additional method of combining tillage with gypsum application consists of spreading gypsum on the soil surface at a rate calculated to reclaim the sodic soil to a predetermined depth (Equation 18.6) and tilling to mix the gypsum into the soil, followed by an additional surface application of gypsum to maintain high IR and reduce exchangeable Na at the soil surface (Oster and Jayawardane, 1998).

18.7.4 Vegetative Remediation

Amelioration of sodic soils using vegetation has been extensively reviewed by Qadir et al. (2007). Vegetative remediation treatments are effective largely through their effects on both the physical and the chemical properties of sodic soils. Crop roots can stimulate changes in physical properties of the root zone in several different ways such as removal of entrapped air from larger conducting pores, generation of alternate wetting and drying cycles, and creation of macropores (Qadir and Oster, 2004). Aggregate stability is enhanced because of in situ production of polysaccharides and fungal hyphae in conjunction with differential dewatering at the root–soil interface (Tisdall, 1991). Instead of deep plowing, which at times could aggravate rather than improve soil conditions (Section 18.7.3), vigorous rooted rotation crops can also alleviate subsoil compaction. An alternative to subsoil amelioration is the use of ridge instead of bed farming techniques. With ridges, the plants are located ~0.25 m above the neighboring furrows, and, thus, aeration is improved, especially in fine textured soils suffering from water logging in the subsoil (Hunter et al., 1980; Hearn and Constable, 1984).

Cropping in sodic soils affects the chemistry of the root zone environment, especially in calcareous soils, via a series of processes: (1) enhancing the pCO_2, (2) dissolution of CO_2 in water to form carbonic acid (H_2CO_3) and its immediate dissociation to proton (H^+) and bicarbonate (HCO_3^-), (3) reaction of H^+ with soil $CaCO_3$ to produce Ca, (4) Na–Ca exchange at the soil's cation exchange sites, (5) leaching of the exchanged Na^+ in the percolating water, and (6) subsequent reduction in soil sodicity (Qadir and Oster, 2004).

Removal of aboveground biomass of plant species, used for phytoremediation of sodic and saline-sodic soils, removes salts and Na taken up by the plants and accumulated in their shoots (Qadir et al., 2007). Highly salt-resistant species such as halophytes may accumulate quite high levels of salts and Na in their shoots. Qadir et al. (2003) found, however, that Na removal by shoot harvest of crops such as alfalfa would contribute to only 1%–2% of the total Na removed during phytoremediation of a sodic soil. It was concluded, therefore, that the principal source of sodicity decrease through phytoremediation of calcareous sodic soils is leaching of salts and Na from the root zone to deeper soil depths rather than removal by harvesting the aboveground plant biomass (Qadir et al., 2007).

The efficiency of various plant species for amelioration of sodic soils has been found to vary, possibly due to differences in their tolerance to salinity (Maas and Grattan, 1999), which often accompanies sodicity. In general, the species with greater production of biomass together with the ability to withstand ambient soil salinity and sodicity, and periodic inundation have been found efficient in soil reclamation (Ghaly, 2002; Qadir et al., 2002; Qadir and Oster, 2004). A summary of 17 lysimeter and field experiments showed that vegetative remediation was somewhat less effective than chemical amelioration in reducing soil sodicity (Figure 18.10). The chemical treatment

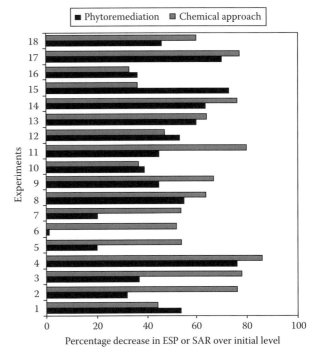

FIGURE 18.10 Summary of 17 experiments where chemical and phytoremediation treatments have been compared for their effects on a decrease in soil sodicity (SAR or ESP). The bars for respective treatments indicate percentage decrease over the respective levels of original soil SAR or ESP values. Number 18 is the mean values of the 17 experiments. (Reprinted from Qadir, M., J.D. Oster, S. Schubert, A.D. Noble, and K.L. Sahrawatk. 2007. Phytoremediation of sodic and saline-sodic soils. Adv. Agron. 96:197–247. Copyright 2007, with permission from Elsevier.)

(application of gypsum in all experiments) caused on average a 60% decrease in original sodicity level (ESP or SAR), whereas a 48% decrease was calculated for the vegetative remediation (Qadir et al., 2007).

Hodgson and Chan (1984) showed that using sunflower in rotation with cotton improved macroporosity and cotton root growth in a clay soil. Jayawardane et al. (1994) showed that growing sunflower in a gypsum slotted soil led to efficient and rapid extraction of water causing the soil to crack; in subsequent irrigation/rainfall events, infiltration thus increased. In other studies, the use of Kaller grass (Akhter et al., 2004), alfalfa (Ilyas et al., 1993), or sordan (Robbins, 1986) was recommended for soil amelioration. Rice is also commonly used, especially in the initial stages of reclamation, because of its tolerance to sodicity as well as the waterlogged conditions that favor reclamation (Abrol and Bhumbla, 1979). Chhabra and Abrol (1977) suggested that in gypsum-treated sodic soils rice should be the first crop. It should, however, be borne in mind that crops used in sodic soils may experience oxygen deficiency especially when excessive irrigations are used to leach the soil. Thus, evaluation of crops for the amelioration of sodic soils should consider also their resistance against hypoxia (Qadir et al., 2001; Qadir and Oster, 2004).

When cropping is not possible, pastures, green manure crops, and agroforestry can also contribute to soil amelioration (Dahiya and Anlauf, 1990; Singh et al., 1994). In their detailed review of the pros and cons of amelioration while cropping with various pasture grasses and trees, Nelson and Oades (1998) suggest that, for maximum benefit, organic matter and crops must be used in conjunction with chemical amendments and irrigation.

Addition of organic ameliorants (farm manure, plant residues), which similar to cropping leads to an increase in the pCO_2 in the soil, is also a widely recognized practice for reclaiming sodic soils. Application of organic matter is most effective in ameliorating alkali (soils with high pH) sodic soils. However, its ameliorative effects are generally restricted to the surface horizons where organic matter inputs tend to concentrate (Nelson and Oades, 1998). Because successful reclamation with organic materials must involve leaching, irrigation must be frequent and in small amounts (Dahiya and Anlauf, 1990). Reclamation under rainfall can be expedited by minimizing runoff and evaporation, and increasing infiltration.

18.8 Environmental Aspects of Sodic Soils

Assessment of Na-related soil degradation has usually been carried out in the context of agriculture and soil productivity. Sodic soils pose numerous environmental hazards in agricultural, engineering, and urban settings (Fitzpatrick et al., 1994; Sumner et al., 1998), but at the same time present some opportunities for enhanced carbon sequestration in the soil (Lal, 2001).

18.8.1 Erosion and Suspended Sediments

Sodic soils are susceptible to clay dispersion and surface sealing, which generate runoff flow and cause erosion; posing a serious problem from crop production point of view. However, despite the fact that ~60% or more of the sediments settle out from the runoff water before it reaches a stream or other water body, there is a consensus that the sediments leaving cultivated fields have substantial negative impacts. Results of a recent study (Ghadiri et al., 2007) indicate that large sediment and salt losses can occur in runoff from saline-sodic soils, even at low slopes and from apparently stable soils, with major downstream water quality consequences. Crosson (1985) estimated that the cost of offsite damage from cropped land in the United States exceeded $1 billion annually. Sediments clog drainage ditches and irrigation canals, and decrease the capacity of water reservoirs, thus increasing maintenance costs. In addition, sediments damage or entirely destroy fish spawning areas. Sedimentation of rivers and streams reduces their capacity to carry water and increases the danger of flooding and consequent damage to property and to human health.

The presence of suspended sediments in water increases its turbidity. Under certain conditions common in sodic soils (high pH and ESP), SOM can dissolve in runoff water and enhance clay dispersion and mobilization (Naidu et al., 1993a) and, hence, increase water turbidity. Turbid waters are less attractive for recreational activities and allow less sunlight to pass through, thus negatively affecting aquatic productivity. In addition, turbid water, which is denser than clean water, is more costly to pump. High turbidity in water bodies used for domestic consumption leads to extra costs in purification. Finally, turbid water increases the wear and tear and maintenance costs of equipment through which it passes.

Enhanced clay dispersivity associated with sodic condition in the soil can lead to surface and subsurface erosion processes such as piping and tunneling. The creation of these subsurface channels may result in the formation of deep gullies that further enhances erosion. Records show that tunnel erosion that occurs in many parts of the world, but mainly in kaolinitic and illitic soils (Sumner et al., 1998), requires readily dispersible clay for its development.

18.8.2 Salinity and Pollutants

Sodic soils commonly contain high levels of salts that are transferred to nearby water bodies at a relatively fast rate when the soil is cultivated (Fitzpatrick et al., 1995). In Australia and America, high saline-sodic watertables contribute significantly to the salt load of rivers (Sharma and Williamson, 1984). In addition, sodic soils require proper drainage systems to remove water, which is much more saline than the irrigation water. When this drainage water enters surface water, its quality deteriorates, rendering it at times unsuitable for further use in irrigation, livestock watering or cities (Sumner et al., 1998).

In addition to the excessive salt load in sodic soils originating from pedogenic processes and irrigation with saline-sodic waters requiring disposal, there is a salt accumulation resulting from soil reclamation. Most reclamation activities require the addition of Ca salts (e.g., gypsum). In the soil, following exchange reactions, the salts become largely Na salts. In general, Na salts dissolve readily and are easily displaced downward into soil strata and groundwater beneath the irrigated field. Extensive reclamation can also cause shallow saline watertables that often require drainage. The saline drainage water ends up at surface water bodies, increases their salinity, and reduces their value. Hence, reclamation of sodic soils may have some negative environmental consequences that should be weighed carefully when planning and execution of reclamation activities are considered.

Transport of various contaminants by soil colloidal materials whether to ground or surface water bodies in runoff is a serious environmental hazard. Metals that usually are fairly immobile in soil and have strong affinity for soil colloids move together with the mobile colloids (Kaplan et al., 1993). Similarly, insoluble pesticides, herbicides, and some nutrients (e.g., P) have a high affinity for soil colloids and can thus be transported with suspended solids to adjacent water bodies and groundwater. The increased concentration of suspended colloids in runoff water due to sodic conditions greatly increases the danger of higher than permitted loadings of these pollutants in ground and surface water bodies.

18.8.3 Carbon Sequestration

Sodic soils have lost over the years a large fraction of their original carbon (C) pool (Lal, 2001). The magnitude of the loss may range between 10 and 30 Mg C ha^{-1}, depending on the antecedent pool and the severity of degradation (Qadir et al., 2006). Amelioration of sodic soils may enhance C accumulation in the soil in both organic and inorganic forms.

With respect to sequestration of C in an organic form, cultivation of sodic soils with suitable crops, shrubs, or trees can lead to the accumulation of organic C in the soil through biomass production (Garg, 1998; Kaur et al., 2002). Qadir et al (2007) report that estimates from various field studies on alkali soils suggest that various land-use systems consisting of a number of grasses and trees can sequester organic C in the range of 0.2–0.8 Mg C ha^{-1} year^{-1}.

Sodic soils are commonly found in arid and semiarid regions. Soils in these regions generally contain large pools of primary (lithogenic) and secondary (pedogenic) inorganic carbonates (Lal, 2002). Dissolution of the carbonates at the root zone could result from the buildup of high pCO$_2$ (root respiration and decomposition of organic matter) or by acidic root exudates. The resultant HCO$_3^-$ can then be leached, especially under irrigation management, down the soil profile, and its subsequent precipitation is a possible pathway for C sequestration in an inorganic form. Wilding (1999) estimated that C sequestration through this mechanism ranges between 0.25 and 1.0 Mg C ha^{-1} year^{-1}.

18.9 Conclusions

Many soils world wide contain sodium at levels that may negatively affect soil physical properties, crop production, and the environment. The area occupied by sodium-affected soils is expected to grow due to the increased use of poor quality soils and water for agricultural activities. However, use of such soils for crop production is also expected to increase with the growth of world population. Hence, careful and detailed assessments of the effects cultivation of sodic soils may have on- and off-site, will become essential.

Sodium-related soil degradation and loss of productivity are, to a certain degree, reversible as sodic soils can be reclaimed through physical interventions (e.g., specific tillage operations), chemical treatments or vegetative remediation. The latter option is effective mainly in moderately sodic conditions and has several economic, environmental, and agronomic advantages; but the remediation process is slow and the presence of a source of Ca in the soil (e.g., calcite) is required. Use of chemical treatment leads to fast soil amelioration and is preferable under extreme sodic conditions. A holistic approach for sustainable amelioration and use of sodic soils that takes into consideration, in addition to agricultural aspects, economic, social, and environmental issues, should be adopted in order to maintain the productivity of these soils for many years.

With a better understanding of sodic soils' behavior coupled with a greater knowledge of the management practices needed to handle such soils, and with the growing demand for food production, it is time to consider sodic soils as a useful resource rather than a problem. Amelioration of these soils will not only increase agricultural productivity but will also carry some environmental benefits such as decrease of desertification and soil erosion, and mitigation of greenhouse effects via C sequestration.

References

Aarstad, J.S., and D.E. Miller. 1973. Soil management practices for reducing runoff under center-pivot sprinkler system. J. Soil Water Conserv. 28:171–173.

Abdelgawad, A., A. Arslan, F. Awad, and F. Kadouri. 2004. Deep plowing management practice for increasing yield and water use efficiency of vetch, cotton, wheat and intensified corn using saline and non-saline irrigation water, p. 67–78. Proc. 55th IEC Meeting Int. Comm. Irrig. Drain., September 9–10, 2004, Moscow, Russia.

Abrol, I.P., and D.R. Bhumbla. 1979. Crop responses to differential gypsum applications in a highly sodic soil and the tolerance of several crops to exchangeable sodium under field conditions. Soil Sci. 127:79–85.

Abu-Sharar, T.M., F.T. Bingham, and J.D. Rhoades. 1987a. Stability of soil aggregates as affected by electrolyte concentration and composition. Soil Sci. Soc. Am. J. 51:309–314.

Abu-Sharar, T.M., F.T. Bingham, and J.D. Rhoades. 1987b. Reduction in hydraulic conductivity in relation to clay dispersion and disaggregation. Soil Sci. Soc. Am. J. 51:342–346.

Adams, F. 1974. The soil solution, p. 441–480. *In* E.W. Carson (ed.) The plant root and its environment. University of Virginia Press, Charlottesville, VA.

Agassi, M., J. Morin, and I. Shainberg. 1985a. Effect of raindrop impact energy and water salinity on infiltration rates of sodic soils. Geoderma 36:263–276.

Agassi, M., J. Morin, and I. Shainberg. 1985b. Infiltration and runoff in wheat fields in the semi-arid region of Israel. Soil Sci. Soc. Am. J. 49:186–190.

Agassi, M., I. Shainberg, and J. Morin. 1981. Effect of electrolyte concentration and soil sodicity on infiltration rate and crust formation. Soil Sci. Soc. Am. J. 45:848–851.

Agassi, M., I. Shainberg, and J. Morin. 1986. Effect of powdered phosphogypsum on the infiltration rate of sodic soils. Irrig. Sci. 7:53–61.

Akhter, J., R. Murray, K. Mahmood, K.A. Malik, and S. Ahmed. 2004. Improvement of degraded physical properties of a saline-sodic soil by reclamation with Kallar grass (*Leptochloa fusca*). Plant Soil 258:207–216.

Al-Durrah, M.M., and J.M. Bradford. 1981. New methods of studying soil detachment due to water drop impact. Soil Sci. Soc. Am. J. 45:949–953.

Ali, O.M., M. Yousaf, and J.D. Rhoades. 1987. Effect of exchangeable cation and electrolyte concentration on mineralogy of clay dispersed from aggregates. Soil Sci. Soc. Am. J. 51:896–900.

Alperovitch, N., and I. Shainberg. 1973. Reclamation of alkali soils with $CaCl_2$ solutions, p. 431–440. *In* A. Hadas et al. (eds.) Physical aspects of soil, water and salts in ecosystems. Springer-Verlag, Berlin, Germany.

Alperovitch, N., I. Shainberg, and R. Keren. 1981. Specific effect of magnesium on the hydraulic conductivity of sodic soils. J. Soil Sci. 32:543–554.

Alperovitch, N., I. Shainberg, and J.D. Rhoades. 1985. Effect of mineral weathering on the response of sodic soils to exchangeable magnesium. Soil Sci. Soc. Am. J. 50:901–904.

Aly, S., and J. Letey. 1990. Physical properties of sodium-treated soil as affected by two polymers. Soil Sci. Soc. Am. J. 54:501–504.

Arora, H.S., and N.T. Coleman. 1979. The influence of electrolyte concentration on flocculation of clay suspensions. Soil Sci. 127:134–139.

Arshad, M.A., and A.R. Mermut. 1988. Micromorphological and physicochemical characteristics of soil crust types in Northwestern Alberta, Canada. Soil Sci. Soc. Am. J. 52:724–729.

Attou, F., A. Bruand, and Y. Le Bissonnais. 1998. Effect of clay content and silt-clay fabric on stability of artificial aggregates. Eur. J. Soil Sci. 49:569–577.

Ayers, R.S., and D.W. Westcot. 1985. Water quality for agriculture. Irrigation and Drainage Paper No. 29, Rev. 1, FAO, Rome, p. 174.

Aylmore, L.A.G., and J.P. Quirk. 1959. Swelling of clay-water systems. Nature 183:1752–1753.

Aylmore, L.A.G., and I.D. Sills. 1982. Characterization of soil structure and stability using modulus of rupture-exchangeable sodium percentage relationships. Aust. J. Soil Res. 20:213–224.

Bakker, A.C., and W.W. Emerson. 1973. The comparative effect of exchangeable calcium, magnesium and sodium on some physical properties of red-brown earth subsoils. III. The permeability of Shepperton soil and comparison methods. Aust. J. Soil Res. 11:159–165.

Banin, A., and N. Lahav. 1968. Particle size and optical properties of montmorillonite in suspension. Isr. J. Chem. 6:235–250.

Bar On, P., I. Shainberg, and I. Mochaeli. 1970. The electrophoretic mobility of Na/Ca montmorillonite particles. J. Colloid Interface Sci. 33:471–472.

Barzegar, A.R., P.N. Nelson, J.M. Oades, and P. Rengasamy. 1996. Organic matter, sodicity and clay type: Influence on aggregation. Soil Sci. Soc. Am. J. 60:583–589.

Bauder, J.W., K.R. Hershberger, and L.S. Browning. 2008. Soil solution and exchange complex response to repeated wetting–drying with modestly saline–sodic water. Irrig. Sci. 26:121–130.

Baver, L.D., W.H. Gardner, and W.R. Gardner. 1972. Soil Physics. 4th Edn. John Wiley & Sons, New York.

Ben-Hur, M., M. Agassi, R. Keren, and J. Zhang. 1998. Compaction, aging and raindrop-impact effects on hydraulic properties of saline and sodic vertisols. Soil Sci. Soc. Am. J. 62:1377–1383.

Ben-Hur, M., and M. Lado. 2008. Effect of soil wetting conditions on seal formation, runoff and soil loss in arid and semiarid soils—A review. Aust. J. Soil Res. 46:191–202.

Ben-Hur, M., I. Shainberg, D. Bakker, and R. Keren. 1985. Effect of soil texture and $CaCO_3$ content on water infiltration in crusted soils as related to water salinity. Irrig. Sci. 6:281–294.

Bernstein, L., and G.A. Pearson. 1956. Influence of exchangeable sodium on the yield and chemical composition of plants: 1. Green beans, garden beets, clover, and alfalfa. Soil Sci. 82:247–285.

Blackmore, A.V., and R.D. Miller. 1961. Tactoid size and osmotic swelling in Ca montmorillonite. Soil Sci. Soc. Am. Proc. 25:169–173.

Blackwell, P.S., N.S. Jayawardane, J. Blackwell, R. White, and R. Horn. 1989. Evaluation of soil recompaction by transverse wheeling of tillage slots. Soil Sci. Soc. Am. J. 53:11–15.

Blackwell, P.S., N.S. Jayawardane, T.W. Greene, J.T. Wood, J. Blackwell, and H.J. Beatty. 1991. Subsoil macropore space of a transitional Red-Brown earth after either deep tillage, gypsum or both. I. Physical effects and short term changes. Aust. J. Soil Res. 29:123–140.

Blake, G.R., and R.D. Gilman. 1970. Thixotropic changes with aging of synthetic soil aggregates. Soil Sci. Soc. Am. Proc. 34:561–564.

Bodman, G.B., and M. Fireman. 1950. Changes in soil permeability and exchangeable cation status during flow of different irrigation waters. Trans. 4th Int. Congr. Soil Sci. 1:397–400.

Bower, C.A., and J.O. Goertzen. 1958. Replacement of adsorbed sodium in soils by hydrolysis of calcium carbonate. Soil Sci. Soc. Am. Proc. 22:33–35.

Bower, C.A., G. Ogata, and J.M. Tucker. 1968. Sodium hazard of irrigation waters as influenced by leaching fraction and by precipitation or solution of calcium carbonate. Soil Sci. 106:29–34.

Bradford, J.M., J.E. Ferris, and P.A. Remley. 1987. Interrill soil erosion processes: I. Effect of surface sealing on infiltration, runoff, and soil splash detachment. Soil Sci. Soc. Am. J. 51:1566–1571.

Bradford, J.M., C.C. Truman, and C. Huang. 1992. Comparison of three measures of resistance of soil surface seals to raindrop splash. Soil Tech. 5:47–56.

Bresler, E. 1970. Numerical solution of the equation for interacting diffuse layers in mixed ionic systems with nonsymmetrical electrolytes. J. Colloid Interface Sci. 33:278–283.

Bresler, E., B.L. McNeal, and D.L. Carter. 1982. Saline and sodic soils: Principles, dynamics, modeling. Springer-Verlag, Berlin, Germany.

Bui, E.N., L. Krogh, R.S. Lavado, F.O. Nachtergaele, T. Toth, and R.W. Fitzpatrick. 1998. Distribution of sodic soils: The world scene, p. 35–50. In M.E. Sumner and R. Naidu (eds.) Sodic soils. Oxford University Press, New York.

Cairns, R.R., W.E. Bowser, R.A. Milne, and P.C. Chang. 1967. Effect of nitrogen fertilization of bromegrass on Solonetzic soils. Can. J. Soil Sci. 47:1–6.

Cairns, R.R., R.A. Milne, and E.W. Bowser. 1962. A nutritional disorder in barley seedlings grown on alkali Solonetz soil. Can. J. Soil Sci. 42:1–6.

Carter, M.R., and G.R. Webster. 1990. Use of calcium-to-total cation ratio as an index of plant-available calcium. Soil Sci. 149:212–217.

Cartwright, B., B.A. Zarcinas, and L.R. Spouncer. 1986. Boron toxicity South Australian barley crops. Aust. J. Agric. Res. 37:351–359.

Cass, A., and M.E. Sumner. 1974. Use of gypsum and high salt water in reclamation of sodic soil. Trans. 10th Int. Congr. Soil Sci. 10:118–127.

Cass, A., and M.E. Sumner. 1982a. Soil pore structural stability and irrigation water quality: I. Empirical sodium stability model. Soil Sci. Soc. Am. J. 46:503–506.

Cass, A., and M.E. Sumner. 1982b. Soil pore structural stability and irrigation water quality: II. Sodium stability data. Soil Sci. Soc. Am. J. 46:507–512.

Cass, A., and M.E. Sumner. 1982c. Soil pore structural stability and irrigation water quality: III. Evaluation of soil stability and crop yield in relation to salinity and sodicity. Soil Sci. Soc. Am. J. 46:513–517.

Celik, I., I. Ortas, and S. Kilic. 2004. Effects of compost, Mycorrhiza, manure and fertilizer on some physical properties of a Chromoxerert soil. Soil Till. Res. 78:59–67.

Chan, K.Y., and C.E. Mullins. 1994. Slaking characteristics of some Australian and British soils. Eur. J. Soil Sci. 45:273–283.

Chang, C.W., and H.E. Dregne. 1955. Effect of exchangeable sodium on soil properties and on growth and cation content on alfalfa and cotton. Soil Sci. Soc. Am. Proc. 19:29–35.

Chappell, N.A., J.L. Ternan, and K. Bidin. 1999. Correlation of physicochemical properties and sub-erosional landforms with aggregate stability variations in a tropical Ultisol disturbed by forestry operations. Soil Till. Res. 50:55–71.

Chaudhry, G.H., and B.P. Warkentin. 1968. Studies on exchange of sodium from soils by leaching with calcium sulfate. Soil Sci. 105:190–197.

Chauhan, S.K., C.P.S. Chauhan, and P.S. Minhas. 2007. Effect of cyclic use and blending of alkali and good quality waters on soil properties, yield and quality of potato, sunflower and Sesbania. Irrig. Sci. 26:81–89.

Chen, J., J. Tarchitzky, J. Morin, and A. Banin. 1980. Scanning electron microscope observations on soil crust and their formation. Soil Sci. 130:49–55.

Chhabra, R. 1985. Crop responses to phosphorus and potassium fertilization of a sodic soil. Soil Sci. Soc. Am. J. 77:699–702.

Chhabra, R. 2005. Classification of salt-affected soils. Arid Land Res. Manage. 19:61–79.

Chhabra, R., and I.P. Abrol. 1977. Reclaiming effect of rice grown in sodic soils. Soil Sci. 124:49–55.

Chi, C.L., W.W. Emerson, and D.G. Lewis. 1977. Exchangeable calcium, magnesium and sodium and the dispersion of illites in water. I. Characterization of illites and exchange reactions. Aust. J. Soil Res. 15:243–253.

Chiang, S.D., D.E. Radcliffe, W.P. Miller, and K.D. Newman. 1987. Hydraulic conductivity of three southeastern soils as affected by sodium, electrolyte concentration and pH. Soil Sci. Soc. Am. J. 51:1293–1299.

Childs, E.C., and N. Collis-George. 1950. The permeability of porous materials. Proc. Roy. Soc. Ser. A 201:392–405.

Condom, N., M. Kuper, S. Marlet, V. Valles, and J. Kijne. 1999. Salinization, alkalinization and sodification in Punjab (Pakistan): Characterization of the geochemical and physical processes of degradation. Land Degrad. Dev. 10:123–140.

Coughlan, K.J., and R.J. Loch. 1984. The relationship between aggregation and other soil properties in cracking clay soils. Aust. J. Soil Res. 22:59–88.

Crescimanno, G., M. Iovino, and G. Provenzano. 1995. Influence of salinity and sodicity on soil structural and hydraulic characteristic. Soil Sci. Soc. Am. J. 59:1701–1708.

Crosson, P. 1985. Impact of erosion on land productivity and water quality in the United States, p. 217–236. In S.A. El-Swaify, W.C. Moldenhauer, and A. Lo (eds.) Soil erosion and conservation. Soil Conservation Society of America, Ankeny, IA.

Cruse, R.M., and W.E. Larson. 1977. Effect of soil shear strength on soil detachment due to raindrop impact. Soil Sci. Soc. Am. J. 41:777–781.

Curtin, D., and R. Naidu. 1998. Fertility constraints to plant production, p. 125–142. In M.E. Sumner and R. Naidu (eds.) Sodic soils. Oxford University Press, New York.

Curtin, D., H. Steppuhn, and F. Selles. 1994. Effects of magnesium on cation selectivity and structural stability of sodic soils. Soil Sci. Soc. Am. J. 58:730–737.

Dahiya, I.S., and R. Anlauf. 1990. Sodic soils in India: Their reclamation and management. Z. Kulturtech. Landentwick. 31:26–34.

Dikinya, O., C. Hinz, and G. Aylmore. 2007. Influence of sodium adsorption ratio on sodium and calcium breakthrough curves and hydraulic conductivity in soil columns. Aust. J. Soil Res. 45:586–597.

Durgin, P.B., and J.G. Chaney. 1984. Dispersion of kaolinite by dissolved organic matter from Douglas-fir roots. Can. J. Soil Sci. 64:445–455.

Dutt, C.R., and R.W. Terkeltoub. 1972. Prediction of gypsum and leaching requirements for sodium-affected soil. Soil Sci. 114:93–103.

Ellis, J.H., and O.G. Caldwell. 1935. Magnesium clay solonetz. Trans. 3rd Int. Congr. Soil Sci. I:348–350.

El-Swaify, S.A. 1973. Structural changes in tropical soils due to anions in irrigation water. Soil Sci. 24:137–144.

Emerson, W.W. 1954. The determination of the stability of soil crumbs. J. Soil Sci. 5:235–250.

Emerson, W.W. 1967. A classification of soil aggregates based on their coherence in water. Aust. J. Soil Res. 5:47–57.

Emerson, W.W., and A.C. Bakker. 1973. The comparative effects of exchangeable calcium, magnesium, and sodium on some physical properties of red-brown earth subsoils. II. The spontaneous dispersion of aggregates in water. Aust. J. Soil Res. 11:151–157.

Emerson, W.W., and C.L. Chi. 1977. Exchangeable Ca, Mg and Na and the dispersion of illites. II. Dispersion of illites in water. Aust. J. Soil Res. 15:255–263.

Emerson, W.W., and B.H. Smith. 1970. Magnesium, organic matter and soil structure. Nature 228:453–454.

FAO. 1978. Soil map of the world. VIII. North and Central Asia. UNESCO, Paris, France.

FAO. 1991. World soil resources. An explanatory note on the FAO world soil resources map at 1:25000000 scale. World Soil Resources Report 66. Food and Agriculture Organization, Rome, Italy.

Farres, P. 1978. The role of time and aggregate size in the crusting process. Earth Surf. Process. 3:243–254.

Farres, P.J. 1980. Some observations on the stability of soil aggregates to raindrop impact. Catena 7:223–231.

Farres, P.J. 1987. The dynamics of rain splash erosion and the role of soil aggregate stability. Soil Sci. Soc. Am. J. 14:119–130.

Faulkner, H., B.R. Wilson, K. Solman, and R. Alexander. 2001. Comparison of three extraction methods and their use in determination of sodium adsorption ratios of some sodic soils. Commun. Soil Sci. Plant Anal. 32:1765–1777.

Felhendler, R., I. Shainberg, and H. Frenkel. 1974. Dispersion and hydraulic conductivity of soils in mixed solution. Trans. 10th Int. Congr. Soil Sci. I:103–112.

Fireman, M., and G.B. Bodman. 1939. The effect of saline irrigation water upon the permeability and base status of soils. Soil Sci. Soc. Am. Proc. 4:71–77.

Fitzpatrick, R.W., S.C. Boucher, R. Naidu, and E. Fritsch. 1994. Environmental consequences of soil sodicity. Aust. J. Soil Res. 32:1069–1093.

Fitzpatrick, R.W., S.C. Boucher, R. Naidu, and E. Fritsch. 1995. Environmental consequences of soil sodicity, p. 163–176. In R. Naidu, M.E. Sumner, and P. Rengasamy (eds.) Australian sodic soils: Distribution, properties and management. CSIRO Publications, Melbourne, Australia.

Francis, P.B., and R.M. Cruse. 1983. Soil water matrix potential effects on aggregate stability. Soil Sci. Soc. Am. J. 47:578–581.

Frenkel, H. 1984. Reassessment of water quality criteria for irrigation, p. 143–172. In I. Shainberg and J. Shalhevet (eds.) Soil salinity under irrigation. Springer-Verlag, Berlin, Germany.

Frenkel, H., M.V. Fey, and G.J. Levy. 1992. Organic and inorganic anion effects on reference and soil clay critical flocculation concentration. Soil Sci. Soc. Am. J. 56:1762–1766.

Frenkel, H., J.O. Goertzen, and J.D. Rhoades. 1978. Effects of clay type and content, exchangeable sodium percentage, and electrolyte concentration on clay dispersion and soil hydraulic conductivity. Soil Sci. Soc. Am. J. 48:32–39.

Frenkel, H., and I. Shainberg. 1975. Chemical and hydraulic changes in soils irrigated with brackish water, p. 175–198. Proc. Intern. Symp. Irrig. with Brackish Water. Beersheva, Israel.

Froese, J.C., R.M. Cruse, and M. Ghaffarzadeh. 1999. Erosion mechanics of soils with an impermeable subsurface layer. Soil Sci. Soc. Am. J. 63:1836–1841.

Ganjegunte, G.K., and G.F. Vance. 2006. Deviations from the empirical sodium adsorption ratio (SAR) and exchangeable sodium percentage (ESP) relationship. Soil Sci. 171:364–373.

Garg, V.K. 1998. Interaction of tree crops with a sodic soil environment: Potential for rehabilitation of degraded environments. Land Degrad. Dev. 9:81–93.

Garg, V.K., P.K. Singh, and P. Pushpangadan. 2005. Exchangeable sodium induced changes in yields, water relation and cation composition of fennel (*Foeniculum vulgare* Mill). J. Environ. Biol. 26:335–340.

Gerard, C.J. 1965. The influence of soil moisture, soil texture, drying conditions and exchangeable cations on soil strength. Soil Sci. Soc. Am. Proc. 29:641–645.

Ghadiri, H., J. Hussein, and C.W. Rose. 2007. A study of the interactions between salinity, soil erosion, and pollutant transport on three Queensland soils. Aust. J. Soil Res. 45:404–413.

Ghaly, F.M. 2002. Role of natural vegetation in improving salt affected soil in northern Egypt. Soil Till. Res. 64:173–178.

Goldberg, S., and H.S. Forster. 1990. Flocculation of reference clays and arid-zone soil clays. Soil Sci. Soc. Am. J. 54:714–718.

Goldberg, S., and R.A. Glaubig. 1987. Effect of saturating cation, pH, and aluminum and iron oxides on the flocculation of kaolinite and montmorillonite. Clay. Clay Miner. 35:220–227.

Goldberg, S., D.L. Suarez, and R.A. Glaubic. 1988. Factors affecting clay dispersion and aggregate stability of arid zone soils. Soil Sci. 146:317–325.

Grattan, S.R., and C.M. Grieve. 1992. Mineral element acquisition and growth response of plants in saline environments. Agric. Ecosyst. Environ. 38:275–300.

Grattan, S.R., and C.M. Grieve. 1999. Salinity–mineral nutrient relations in horticultural crops. Sci. Hort. 78:127–157.

Greene, R.S.B., A.M. Posner, and J.P. Quirk. 1978. A study of the coagulation of montmorillonite and illite suspensions by $CaCl_2$ using the electron microscope, p. 35–40. *In* W.W. Emerson, R.D. Bond, and A.R. Dexter (eds.) Modification of soil structure. John Wiley & Sons, New York.

Greenland, D.J. 1979. Structural organization of soils and crop production, p. 47–56. *In* R. Lal and D.J. Greenland (eds.) Soil physical properties and crop production in the tropics. John Wiley & Sons, New York.

Grieve, C.M., and E.V. Maas. 1988. Differential effects of sodium/calcium ratio on sorghum genotypes. Crop Sci. 28:659–665.

Gupta, R.K., and I.P. Abrol. 1990. Salt-affected soils: Their reclamation and management for crop production. Adv. Soil Sci. 11:223–288.

Gupta, R.K., R.R. Singh, and K.K. Tanji. 1990. Phosphorus release in sodium ion dominated soils. Soil Sci. Soc. Am. J. 54:1245–1260.

Hearn, A.B., and G.A. Constable. 1984. Irrigation for crops in a sub-humid environment. VII. Evaluation of irrigation strategies for cotton. Irrig. Sci. 5:75–94.

Hensley, M. 1969. Selected properties affecting the irrigable value of some Makatini soils. MSc agriculture thesis, University of Natal, Pietermaritzburg, South Africa.

Hillel, D. 1980. Applications of soil physics. Academic Press, New York.

Hodgson, A.S., and K.Y. Chan. 1984. Deep moisture extraction and crack formation by wheat and safflower in a Vertisol following irrigated cotton rotations, p. 299–304. *In* J.W. McGarity, E.H. Hoult, and H.B. So (eds.) The properties and utilization of cracking clay soils. University of New England, Armidale, Australia.

Hodgson, J.F., W.L. Lindsay, and J.F. Trierweiler. 1966. Micronutrient cation complexing in soil solution: II. Complexing of zinc and copper in displaced solution from calcareous soils. Soil Sci. Soc. Am. Proc. 30:723–726.

Hunter, M.N., P.L.M. de Jarbun, and D.E. Byth. 1980. Response of nine soybean lines to soil moisture conditions close to saturation. Aust. J. Exp. Agric. Anim. Husb. 20:339–345.

Ilyas, M., R.W. Miller, and R.H. Qureshi. 1993. Hydraulic conductivity of saline-sodic soil after gypsum application and cropping. Soil Sci. Soc. Am. J. 57:1580–1585.

Imbufea, A.U., A.F. Pattia, D. Burrow, A. Surapaneni, W.R. Jackson, and A.D. Milnerc. 2005. Effects of potassium humate on aggregate stability of two soils from Victoria, Australia. Geoderma 125:321–330.

Inskeep, W.P., and P.R. Bloom. 1986. Calcium carbonate supersaturation in soil solution of calciaquolls. Soil Sci. Soc. Am. J. 50:1431–1437.

Jalali, M., H. Merikhpour, M.J. Kaledhonkar, and S.E.A.T. Van der Zee. 2008. Effects of wastewater irrigation on soil sodicity and nutrient leaching in calcareous soils. Agric. Water Manage. 95:143–153.

Jayawardane, N.S., and J. Blackwell. 1986. Effects of gypsum-enriched slotting on infiltration rates and moisture storage in swelling clay soil. Soil Use Manage. 2:114–118.

Jayawardane, N.S., J. Blackwell, G. Kirchof, and W.A. Muirhead. 1994. Slotting: A deep tillage technique for ameliorating sodic, acid and other degraded soils and for land treatment of waste, p. 109–146. *In* N.S. Jayawardane and B.A. Stewart (eds.) Subsoil management techniques. Lewis Publishers, Boca Raton, FL.

Jayawardane, N.S., J. Blackwell, and M. Stapper. 1987. Effects of changes in moisture profiles of a transitional Red Brown earth due to surface and slotted gypsum applications. Aust. J. Agric. Res. 38:239–251.

Jayawardane, N.S., and K.Y. Chan. 1994. The management of soil physical properties limiting crop production in Australian sodic soils: A review. Aust. J. Soil Res. 32:13–44.

Johnston, M.A. 1975. The effects of different levels of exchangeable sodium on soil hydraulic conductivity. Proc. S. Afr. Sugar Technol. Assoc. 49:142–147.

Kaplan, D.I., P.M. Bertsch, D.C. Adriano, and W.P. Miller. 1993. Soil borne mobile colloids as influenced by water flow and organic carbon. Environ. Sci. Technol. 27:1193–1200.

Kaur, B., S.R. Gupta, and G. Singh. 2002. Bioamelioration of a sodic soil by silvopastoral systems in northwestern India. Agrofor. Syst. 54:13–20.

Kay, B.D., and D.A. Angers. 1999. Soil Structure, p. A229–A276. *In* M.E. Sumner (ed.) Handbook of soil science. CRC Press, Boca Raton, FL.

Kazman, Z., I. Shainberg, and M. Gal. 1983. Effect of low levels of exchangeable Na and applied phosphogypsum on the infiltration rate of various soils. Soil Sci. 35:184–192.

Kemper, W.D., and E.J. Koch. 1966. Aggregate stability of soils from western U.S. and Canada. USDA Technical Bulletin No. 1355.

Kemper, W.D., and R.C. Rosenau. 1984. Soil cohesion as affected by time and water content. Soil Sci. Soc. Am. J. 48:1001–1006.

Keren, R. 1990. Water-drop kinetic energy effect on infiltration in sodium-calcium-magnesium soils. Soil Sci. Soc. Am. J. 54:983–987.

Keren, R. 1995. Reclamation of sodic-affected soils, p. 353–374. *In* M. Agassi (ed.) Soil erosion conservation and rehabilitation. Marcel Dekker Inc., New York.

Keren, R., and M. Ben-Hur. 2003. Interaction effects of clay swelling and dispersion and $CaCO_3$ content on saturated hydraulic conductivity. Aust. J. Soil Res. 41:979–989.

Keren, R., and I. Shainberg. 1981. Effect of dissolution rate on the efficiency of industrial and mined gypsum on surface runoff from loess soil. Soil Sci. Soc. Am. J. 45:103–107.

Keren, R., I. Shainberg, H. Fenkel, and Y. Kalo. 1983. The effect of exchangeable sodium and gypsum on surface runoff from loess soil. Soil Sci. Soc. Am. J. 47:1001–1004.

Keren, R., and M.J. Singer. 1988. Effect of low electrolyte concentration on hydraulic conductivity of Na/Ca montmorillonite-sand system. Soil Sci. Soc. Am. J. 52:368–373.

Kheyrabi, D., and G. Monnier. 1968. Etude experimentale de l'influence de la composition granulometrique des terres leur stabilite structurale. Ann. Agron. 19:129–152.

Kjaergaard, C., L.W. de Jonge, P. Moldrup, and P. Schjonning. 2004. Water dispersible colloids: Effects of measurement method, clay content, initial soil matric potential and wetting rate. Vadose Zone J. 3:403–412.

Klute, A. 1982. Tillage effects on hydraulic properties of a soil: A review, p. 29–43. In P.W. Unger, D.M. van Doren, F.D. Whisle, and E.L. Skidmore (eds.) Predicting tillage effects on soil physical properties and processes. American Society of Agronomy, Madison, WI.

Kopittke, P.M., H.B. So, and N.W. Menzies. 2006. Effect of ionic strength and clay mineralogy on Na-Ca exchange and the SAR-ESP relationship. Eur. J. Soil Sci. 57:626–633.

Kosmas, C., and N. Noutakas. 1990. Hydraulic conductivity and leaching of an organic saline-sodic soil. Geoderma 46:363–370.

Kozeny, J. 1927. Uber kapillare leitung des wesses in boden. Zitzungsber Akad. Wiss. Wein. 136:271–306.

Kreit, J.E., I. Shainberg, and A.J. Herbillon. 1982. Hydrolysis and decomposition of hectorites in dilute solutions. Clay. Clay Miner. 30:223–231.

Lado, M., A. Paz, and M. Ben-Hur. 2004. Organic matter and aggregate size interactions in infiltration, seal formation and soil loss. Soil Sci. Soc. Am. J. 68:935–942.

Lal, R. 2001. Potential of desertification control to sequester carbon and mitigate the greenhouse effect. Clim. Change 51:35–72.

Lal, R. 2002. Carbon sequestration in dryland ecosystems of west Asia and north Africa. Land Degrad. Dev. 13:45–59.

Lauchli, A., and E. Epstein. 1990. Plant responses to saline and sodic conditions, p. 113–137. In K.K. Tanji (ed.) Agricultural salinity assessment and management. American Society of Civil Engineering, New York.

Le Bissonnais, Y. 1990. Experimental study and modeling of soil surface crusting processes. Catena Suppl. 17:13–28.

Le Bissonnais, Y. 1996. Aggregate stability and assessment of soil crustability and erodibility. I. Theory and methodology. Euro. J. Soil Sci. 47:425–437.

Le Bissonnais, Y., B. Renaux, and H. Delouche. 1995. Interactions between soil properties and moisture content in crust formation, runoff and interrill erosion from tilled loess soils. Catena 25:33–46.

Lebron, I., D.L. Suarez, and T. Yoshida. 2002. Gypsum effect on the aggregate size and geometry of sodic soils under reclamation. Soil Sci. Soc. Am. J. 66:92–98.

Levitt, J. 1980. Responses of plants to environmental stress. Vol. II. Academic Press, New York.

Levy, G.J., M. Agassi, H.J.C. Smith, and R. Stern. 1993. Microaggregate stability of kaolinitic and illitic soils determined by ultrasonic energy. Soil Sci. Soc. Am. J. 57:803–808.

Levy, G.J., N. Alperovitch, A.J. van der Merwe, and I. Shainberg. 1989. The hydrolysis of kaolinitic soils as affected by the type of the exchangeable cation. J. Soil Sci. 40:613–620.

Levy, G.J., D. Goldstein, and A.I. Mamedov. 2005. Saturated hydraulic conductivity of semi arid soils: Combined effects of salinity, sodicity and rate of wetting. Soil Sci. Soc. Am. J. 69:653–662.

Levy, G.J., J. Levin, and I. Shainberg. 1994. Seal formation and interrill soil erosion. Soil Sci. Soc. Am. J. 58:203–209.

Levy, G.J., J. Levin, and I. Shainberg. 1997. Prewetting rate and aging effects on seal formation and interrill soil erosion. Soil Sci. 162:131–139.

Levy, G.J., and A.I. Mamedov. 2002. High-energy-moisture-characteristics aggregate stability as a predictor for seal formation. Soil Sci. Soc. Am. J. 66:1603–1609.

Levy, G.J., A.I. Mamedov, and D. Goldstein. 2003. Sodicity and water quality effects on slaking of aggregates from semi-arid soils. Soil Sci. 168:552–562.

Levy, G.J., and H.V.H. van der Watt. 1988. Effects of clay mineralogy and soil sodicity on the infiltration rates of soils. S. Afr. J. Plant Soil 5:92–96.

Levy, G.J., H.V.H. van der Watt, and H.M. du Plessis. 1988. Effect of sodium-magnesium and sodium-calcium systems on soil hydraulic conductivity and infiltration. Soil Sci. 146:303–310.

Loch, R.J. 1994. Structure breakdown on wetting, p. 113–132. In H.B. So, G.D. Smith, S.R. Reine, B.M. Schafer, and R.J. Loch (eds.) Sealing, crusting and hardsetting soils. Australian Society of Soil Science, Queensland, Brisbane, Australia.

Loveday, J. 1976. Relative significance of electrolyte and cation exchange effects when gypsum is applied to a sodic clay soil. Aust. J. Soil Res. 14:361–371.

Loveland, P.J., J. Hazelden, and R.G. Sturdy. 1987. Chemical properties of salt-affected soils in north Kent and their relationship to soil instability. J. Agric. Sci. 109:1–6.

Luk, S.H. 1985. Effect of antecedent soil moisture content on rainwash erosion. Catena 12:129–139.

Ma, G., P. Rengasamy, and A.J. Rathjen. 2003. Phytotoxicity of aluminium to wheat plants in high pH solutions. Aust. J. Exp. Agric. 43:497–501.

Maas, E.V., and S.R. Grattan. 1999. Crop yields as affected by salinity, p. 55–108. In R.W. Skaggs and J. van Schilfgaarde (eds.) Agricultural drainage. ASA-CSSA-SSSA, Madison, WI.

Maas, E.V., and G.J. Hoffman. 1977. Crop salt tolerance–current assessment. J. Irrig. Drain. Div. ASCE 103:115–134.

Mace, J.E., and C. Amrhein. 2001. Leaching and reclamation of a soil irrigated with moderate SAR waters. Soil Sci. Soc. Am. J. 65:199–204.

Malhi, S.S., D.W. McAndrew, and M.R. Carter. 1992. Effect of tillage and N fertilization of a Solonetzic soil on barley production and some soil properties. Soil Till. Res. 2:95–107.

Mamedov, A.I., S. Beckmann, C. Huang, and G.J. Levy. 2007. Aggregate stability as affected by polyacrylamide molecular weight, soil texture and water quality. Soil Sci. Soc. Am. J. 71:1909–1918.

Mamedov, A.I., C. Huang, and G.J. Levy. 2006. Antecedent moisture content and aging duration effects on seal formation and erosion in smectitic soils. Soil Sci. Soc. Am. J. 70:832–843.

Mamedov, A.I., G.J. Levy, I. Shainberg, and J. Letey. 2001. Wetting rate and soil texture effect on infiltration rate and runoff. Aust. J. Soil Res. 30:1293–1305.

Mamedov, A.I., I. Shainberg, and G.J. Levy. 2002. Wetting rate and sodicity effects on interrill erosion from semi arid Israeli soils. Soil Till. Res. 68:121–132.

Manin, M., A. Pissarra, and J.W. van Hoorn. 1982. Drainage and desalinization of heavy clay soil in Portugal. Agric. Water Manage. 5:227–240.

Marshall, T.J. 1958. A relation between permeability and size distribution of pores. J. Soil Sci. 9:1–8.

Martin, J.P., W.P. Martin, J.B. Page, W.A. Raney, and J.D. de Met. 1995. Soil aggregation. Adv. Agron. 7:1–37.

McAtee, J.L. 1961. Heterogeneity in montmorillonites. Clay. Clay Miner. 5:279–288.

McIntyre, D.S. 1958. Permeability measurement of soil crusts formed by raindrop impact. Soil Sci. 85:185–189.

McIntyre, D.S. 1979. Exchangeable sodium, subplasticity and hydraulic conductivity of some Australian soils. Aust. J. Soil Res. 17:115–120.

McKenzie, D.C., D.J.M. Hall, T.S. Abbott, A.M. Kay, and J.D. Sykes. 1992. Soil management for irrigated cotton. NSW Agric. Agfact 6:53.

McNeal, B.L., and N.T. Coleman. 1966. Effect of solution composition on soil hydraulic conductivity. Soil Sci. Soc. Am. Proc. 30:313–317.

McNeal, B.L., D.A. Layfield, W.A. Norvell, and J.D. Rhoades. 1968. Factors influencing hydraulic conductivity of soils in the presence of mixed-salt solutions. Soil Sci. Soc. Am. Proc. 32:187–190.

McNeal, B.L., W.A. Norvell, and N.T. Coleman. 1966. Effect of solution composition on soil hydraulic conductivity and on the swelling of extracted soil clays. Soil Sci. Soc. Am. Proc. 30:313–317.

Mead, J.A., and K.Y. Chan. 1988. Effect of deep tillage and seedbed preparation on the growth and yield of wheat on a hardsetting soil. Aust. J. Exp. Agric. 28:491–498.

Miller, W.P. 1987. Infiltration and soil loss of three gypsum-amended Ultisols under simulated rainfall. Soil Sci. Am. J. 51:1314–1320.

Miller, W.P., and M.K. Baharuddin. 1986. Relationship of soil dispersibility to infiltration and erosion of Southeastern soils. Soil Sci. 142:235–240.

Miller, R.J., J.W. Biggar, and D.R. Nielsen. 1965. Chloride displacement in Panoche clay loam in relation to water movement and distribution. Water Resour. Res. 1:63–73.

Miller, W.P., H. Frenkel, and K.D. Newman. 1990. Flocculation concentration and sodium/calcium exchange of kaolinitic soil clays. Soil Sci. Soc. Am. J. 54:346–351.

Miller, W.P., and J. Scifres. 1988. Effect of sodium nitrate and gypsum on infiltration and erosion of a highly weathered soil. Soil Sci. 148:304–309.

Minhas, P.S. 1996. Saline water management for irrigation in India. Agric. Water Manage. 30:1–24.

Miyamoto, S., J. Ryan, and J.L. Stroelhein. 1975. Potentially beneficial uses of sulfuric acid in southwestern agriculture. J. Environ. Qual. 4:431–437.

Mohammed, E.T.Y., J. Letey, and R. Branson. 1979. Sulphur compounds in water treatment. Sulph. Agric. 3:7–11.

Morin, J., and Y. Benyamini. 1977. Rainfall infiltration into bare soils. Water Resour. Res. 13:813–817.

Nadler, A., G.J. Levy, R. Keren, and H. Eisenberg. 1996. Sodic calcareous soil reclamation as affected by water chemical composition and flow rate. Soil Sci. Soc. Am. J. 60:252–257.

Naidu, R. 1993. Distribution, properties and management of sodic soils: An introduction. Aust. J. Soil Res. 31:681–682.

Naidu, R., R.W. Fitzpatrick, I.O. Hollingsworth, and D.R. Williamson. 1993a. Effect of land use on the composition of through flow water immediately above clayey B horizons in the Warren Catchment, South Australia. Aust. J. Exp. Agric. 33:239–244.

Naidu, R., R.H. Merry, G.J. Churchman, M.J. Wright, R.I. Murray, R.W. Fitzpatrick, and B.A. Zarcinas. 1993b. Sodicity in South Australia: A review. Aust. J. Soil Res. 31:911–929.

Naidu, R., and P. Rengasamy. 1993. Ion interactions and constraints to plant nutrition in Australian sodic soils. Aust. J. Soil Res. 31:801–819.

Nelson, P.N., and J.M. Oades. 1998. Organic matter, sodicity and soil structure, p. 67–91. *In* M.E. Sumner and R. Naidu (eds.) Sodic soils. Oxford University Press, New York.

Nielsen, D.R., J.W. Biggar, and J.N. Luthin. 1966. Desalinization of soils under controlled unsaturated flow conditions, p. 19.15–19.24. Int. Comm. Irrig. Drainage, 6th Congr., New Delhi, India.

Northcote, K.H., and J.K.M. Skene. 1972. Australian soils with saline and sodic properties. CSIRO Soil Publication No. 27.

Oster, J.D., and H. Frenkel. 1980. Chemistry of the reclamation of sodic soils with gypsum and lime. Soil Sci. Soc. Am. J. 44:41–45.

Oster, J.D., and N.S. Jayawardane. 1998. Agricultural management of sodic soils, p. 143–165. *In* M.E. Sumner and R. Naidu (eds.) Sodic soils. Oxford University Press, New York.

Oster, J.D., and F.W. Schroer. 1979. Infiltration as influenced by irrigation water quality. Soil Sci. Soc. Am. J. 43:444–447.

Oster, J.D., I. Shainberg, and I.P. Abrol. 1999. Reclamation of salt affected soils, p. 659–691. *In* R.W. Skaggs and J. van Schilfgaarde (eds.) Agricultural drainage. ASA-CSSA-SSSA, Madison, WI.

Oster, J.D., I. Shainberg, and J.D. Wood. 1980. Flocculation value and gel structure of Na/Ca montmorillonite and illite suspensions. Soil Sci. Soc. Am. J. 44:955–959.

Panabokke, C.R., and J.P. Quirk. 1957. Effect of initial water content on stability of soil aggregates in water. Soil Sci. 83:185–195.

Pearson, G.A. 1960. Tolerance of crops to exchangeable sodium. USDA Information Bulletin No. 216.

Pearson, G.A., and L. Bernstein. 1958. Influence of exchangeable sodium on yield and chemical composition of plants: II. Wheat, barley, oats, rice, tall fescue, and wheatgrass. Soil Sci. 86:254–261.

Piccolo, A., G. Pietramellara, and J.S.C. Mbagwu. 1997. Use of humic substances as soil conditioners to increase aggregate stability. Geoderma 75:267–277.

Prather, R.J., J.O. Goertzen, O. Rhoades, and H. Frenkel. 1978. Efficient amendment use in sodic soil reclamation. Soil Sci. Soc. Am. J. 42:782–786.

Pupisky, H., and I. Shainberg. 1979. Salt effects on the hydraulic conductivity of a sandy soil. Soil Sci. Soc. Am. J. 43:429–433.

Qadir, M., A.D. Noble, S. Schubert, R.J. Thomas, and A. Arslan. 2006. Sodicity-induced land degradation and its sustainable management: Problems and prospects. Land Degrad. Dev. 17:661–676.

Qadir, M., and J.D. Oster. 2004. Crop and irrigation management strategies for saline-sodic soils and waters aimed at environmentally sustainable agriculture. Sci. Total Environ. 323:1–19.

Qadir, M., J.D. Oster, S. Schubert, A.D. Noble, and K.L. Sahrawatk. 2007. Phytoremediation of sodic and saline-sodic soils. Adv. Agron. 96:197–247.

Qadir, M., R.H. Qureshi, and N. Ahmad. 2002. Amelioration of calcareous saline-sodic soils through phytoremediation and chemical strategies. Soil Use Manage. 18:381–385.

Qadir, M., S. Schubert, A. Ghafoor, and G. Murtaza. 2001. Amelioration strategies for sodic soils: A review. Land Degrad. Dev. 12:357–386.

Qadir, M., D. Steffens, F. Yan, and S. Schubert. 2003. Sodium removal from a calcareous saline-sodic soil through leaching and plant uptake during phytoremediation. Land Degrad. Dev. 14:301–307.

Quirk, J.P. 1978. Some physico-chemical aspects of soil structural stability: A review, p. 3–16. In W.W. Emerson, R.D. Bond, and A.R. Dexter (eds.) Modification of soil structure. John Wiley & Sons, New York.

Quirk, J.P. 2001. The significance of the threshold and turbidity concentrations in relation to sodicity and microstructure. Aust. J. Soil Res. 39:1185–1217.

Quirk, J.P., and L.A.G. Aylmore. 1971. Domains and quasi-crystalline regions in clay systems. Soil Sci. Soc. Am. Proc. 35:652–654.

Quirk, J.P., and C.R. Panabokke. 1962. Incipient failure of soil aggregates. J. Soil Sci. 13:60–69.

Quirk, J.P., and R.K. Schofield. 1955. The effect of electrolyte concentration on soil permeability. J. Soil Sci. 6:163–178.

Radcliffe, D.E., W.P. Miller, and S.C. Chiang. 1987. Effect of soil dispersion on surface run-off in Southern Piedmont soils, p. 1–28. Environmental Resource Center, Georgia Institute of Technology, Atlanta, GA.

Rahman, A.W., and D.L. Rowell. 1979. The influence of Mg in saline and sodic soils: A specific effect or a problem of cation exchange. J. Soil Sci. 30:535–546.

Rao, D.L.N., and S.K. Ghai. 1985. Urease and dehydrogenase activity of alkali and reclaimed soils. Aust. J. Soil Res. 23:661–665.

Rawitz, E., W.B. Hoogmoed, and J. Morin. 1986. The effects of tillage practices on crust properties, infiltration and crop response under semi-arid conditions, p. 278–284. *In* F. Callebaut, D. Gabriels, and M. de Boodt (eds.) Assessment of soil surface sealing and crusting. Flanders Research Centre for Soil Erosion and Soil Conservation, Ghent, Belgium.

Reeve, R.C., and C.A. Bower. 1960. Use of high-salt waters as a flocculant and source of divalent cations for reclaiming sodic soils. Soil Sci. 90:139–144.

Reeve, R.C., and E.J. Doering. 1966. The high-salt-water dilution method for reclaiming sodic soils. Soil Sci. Soc. Am. Proc. 30:498–504.

Reichert, J.M., and L.D. Norton. 1994. Aggregate stability and rain-impacted sheet erosion of air-dried and prewetted clayey surface soils under intense rain. Soil Sci. 158:159–169.

Rengasamy, P. 1987. Importance of calcium in irrigation with saline sodic water—A view point. Agric. Water Manage. 12:207–219.

Rengasamy, P., D. Chittleborough, and K. Helyar. 2003. Root-zone constraints and plant-based solutions for dryland salinity. Plant Soil 257:249–260.

Rengasamy, P., R.S.B. Greene, and G.W. Ford. 1986. Influence of magnesium on aggregate stability in sodic red-brown earths. Aust. J. Soil Res. 24:229–237.

Rengasamy, P., R.S.B. Greene, G.W. Ford, and A.H. Mehammi. 1984. Identification of dispersive behaviour and the management of red-brown earths. Aust. J. Soil Res. 22:413–431.

Rengasamy, P., and K.A. Olssen. 1991. Sodicity and soil structure. Aust. J. Soil Res. 29:935–952.

Rengasamy, P., and M.E. Sumner. 1998. Processes involved in sodic behavior, p. 51–67. *In* M.E. Sumner and R. Naidu (eds.) Sodic soils. Oxford University Press, New York.

Rhoades, J.D. 1968. Mineral weathering correction for estimating the sodium hazard of irrigation waters. Soil Sci. Soc. Am. Proc. 32:648–651.

Rhoades, J.D., and R.D. Ingvalson. 1969. Macroscopic swelling and hydraulic conductivity properties of four vermiculite soils. Soil Sci. Soc. Am. Proc. 33:364–369.

Rhoades, J.D., D.B. Kruger, and M.J. Reed. 1968. The effect of soil mineral weathering on the sodium hazard of irrigation waters. Soil Sci. Soc. Am. Proc. 32:643–647.

Rimmer, D.L., and D.J. Greenland. 1976. Effect of $CaCO_3$ on the swelling of a soil clay. J. Soil Sci. 27:129–139.

Robbins, C.W. 1986. Sodic calcareous soil reclamation as affected by different amendments and crops. Agron. J. 78:916–920.

Rowell, D.L., D. Payne, and N. Ahmad. 1969. The effect of the concentration and movement of solutions on the swelling, dispersion and movement of clay in saline and alkali soils. J. Soil Sci. 20:176–188.

Russo, D. 1988. Numerical analysis of nonsteady transport of interacting solutes through unsaturated soil. 1. Homogeneous systems. Water Resour. Res. 24:271–284.

Russo, D., and E. Bresler. 1977. Analysis of the saturated and unsaturated hydraulic conductivity in mixed sodium and calcium soil systems. Soil Sci. Soc. Am. J. 41:706–712.

Sadiq, M., G. Hassan, S.M. Mehdi, N. Hussain, and M. Jamil. 2007. Amelioration of saline-sodic soils with tillage implements and sulfuric acid application. Pedosphere 17:182–190.

Schofield, R.K., and H.R. Samson. 1954. Flocculation of kaolinite due to the attraction of oppositely charged crystal faces. Disc. Farad. Soc. 18:138–145.

Shainberg, I., N. Alperovitch, and R. Keren. 1988. Effect of magnesium on the hydraulic conductivity of sodic smectite-sand mixtures. Clay. Clay Miner. 36:432–438.

Shainberg, I., E. Bresler, and Y. Klausner. 1971. Studies on Na/Ca montmorillonite systems. I. The swelling pressure. Soil Sci. 111:214–219.

Shainberg, I., and M. Gal. 1982. The effect of lime on the response of soils to sodic conditions. J. Soil Sci. 33:489–498.

Shainberg, I., and J. Letey. 1984. Response of soils to sodic and saline conditions. Hilgardia 52:1–57.

Shainberg, I., G.J. Levy, D. Goldstein, A.I. Mamedov, and J. Letey. 2001. Prewetting rate and sodicity effects on the hydraulic conductivity of soils. Aust. J. Soil Res. 39:1279–1291.

Shainberg, I., G.J. Levy, P. Rengasamy, and H. Frenkel. 1992. Aggregate stability and seal formation as affected by drops' impact energy and soil amendments. Soil Sci. 154:113–119.

Shainberg, I., A.I. Mamedov, and G.J. Levy. 2003. Role of wetting rate and rain energy in seal formation and erosion. Soil Sci. 168:54–62.

Shainberg, I., and H. Otoh. 1968. Size and shape of montmorillonite particles saturated with Na/Ca ions. Isr. J. Chem. 6:251–259.

Shainberg, I., J.D. Rhoades, and R.J. Prather. 1981a. Effect of low electrolyte concentration on clay dispersion and hydraulic conductivity of a sodic soil. Soil Sci. Soc. Am. J. 45:273–277.

Shainberg, I., J.D. Rhoades, D.L. Suarez, and R.J. Prather. 1981b. Effect of mineral weathering on clay dispersion and hydraulic conductivity of sodic soils. Soil Sci. Soc. Am. J. 45:287–291.

Sharma, M.L., and D.R. Williamson. 1984. Secondary salinization of water resources in Southern Australia, p. 149–163. In R.H. French (ed.) Salinity in water resources and reservoirs. Butterworth Publishers, Boston, MA.

Simunek, J., and D.L. Suarez. 1997. Sodic soil reclamation using multicomponent transport modeling. J. Irrig. Drain. Eng. 123:367–375.

Singh, S.B., R. Chhabra, and I.P. Abrol. 1979. Effect of exchangeable sodium on the yield and chemical composition of raya (*Brassica juncea* L.). Agron. J. 71:767–770.

Singh, G., N.T. Singh, and I.P. Abrol. 1994. Agroforestry techniques for the rehabilitation of degraded salt-affected lands in India. Land Degrad. Rehabil. 5:223–242.

So, H.B., and G.D. Cook. 1993. The effect of dispersion on the hydraulic conductivity of surface seals in clay soils. Soil Technol. 6:325–330.

Sort, X., and J.M. Alcaniz. 1999. Effects of sewage sludge amendment on soil aggregation. Land Degrad. Dev. 10:3–12.

Sposito, G., K.M. Haltzclaw, and C.S. LeVesque-Madore. 1977. Calcium ion complexation by fulvic acid extracts from sewage sludge-soil mixtures. Soil Sci. Soc. Am. J. 42:600–606.

Sposito, G., and S.V. Mattigod. 1977. On the chemical foundation of the sodium adsorption ratio. Soil Sci. Soc. Am. J. 41:323–329.

Stern, R., M. Ben-Hur, and I. Shainberg. 1991. Clay mineralogy effect on rain infiltration, seal formation and soil losses. Soil Sci. 152:455–462.

Stevenson, F.J. 1992. Humus chemistry: Genesis, composition, reactions. 2nd Edn. John Wiley & Sons, New York.

Suarez, D.L. 1981. Relationship between pH_c and sodium adsorption ratio (SAR) and an alternative method of estimating SAR of soil or drainage water. Soil Sci. Soc. Am. J. 45:469–475.

Suarez, D.L. 2001. Sodic soil reclamation: Model and field study. Aust. J. Soil Res. 39:1225–1246.

Suarez, D.L., and J. Simunek. 1997. UNSATCHEM: Unsaturated water and solute transport model with equilibrium and kinetic chemistry. Soil Sci. Soc. Am. J. 61:1633–1646.

Suarez, D.L., J.D. Wood, and S.M. Lesch. 2008. Infiltration into cropped soils: Effect of rain and sodium adsorption ratio–impacted irrigation water. J. Environ. Qual. 37:169–179.

Sumner, M.E. 1993. Sodic soils: New perspectives. Aust. J. Soil Res. 31:683–750.

Sumner, M.E., W.P. Miller, R.S. Kookana, and P. Hazelton. 1998. Sodicity dispersion and environmental quality, p. 167–231. In M.E. Sumner and R. Naidu (eds.) Sodic soils. Oxford University Press, New York.

Sumner, M.E., and R. Naidu. 1998. Sodic soils. Oxford University Press, New York.

Szabolcs, I. 1979. Review of research on salt affected soils. UNESCO, Paris, France.

Szabolcs, I. 1989. Salt-affected soils. CRC Press, Boca Raton, FL.

Tang, Z., T. Lei, J. Yu, I. Shainberg, A.I. Mamedov, M. Ben-Hur, and G.J. Levy. 2006. Simulated-rain-induced runoff and interrill erosion in sodic soils treated with dry PAM and phosphogypsum. Soil Sci. Soc. Am. J. 70:679–690.

Tanji, K.K., L.D. Doneen, G.V. Ferry, and R.S. Ayers. 1972. Computer simulation analysis on reclamation of salt-affected soil in San Joaquin Valley, California. Soil Sci. Soc. Am. J. 36:127–133.

Tarchitzky, J., Y. Chen, and A. Banin. 1993. Humic substances and pH effects on sodium-and calcium-montmorillonite flocculation and dispersion. Soil Sci. Soc. Am. J. 57:367–372.

Tisdall, J.M. 1991. Fungal hyphae and structural stability of soil. Aust. J. Soil Res. 29:729–743.

Tisdall, J.M., and J.M. Oades. 1982. Organic matter and water-stable aggregates. J. Soil Sci. 33:141–163.

Toogood, J.A. 1978. Fertility status of solonetzic soils, p. 32–50. In J.A Toogood and R.R. Cairns (eds.) Solonetzic soils technology and management in Alberta. University of Alberta Bull., Edmonton, Alberta, Canada.

USSL Staff. 1954. Diagnosis and Improvement of saline and alkali soils. Agricultural Handbook No. 60. USDA, Washington, DC.

Utomo, W.H., and A.R. Dexter. 1981. Age hardening of agricultural top soils. J. Soil Sci. 32:335–350.

van der Merwe, A.J., and R. Burger. 1969. The influence of exchangeable cations on certain physical properties of a saline-alkali soil. Agrochemophysics 1:63–66.

van Olphen, H. 1977. An introduction to clay colloid chemistry. 2nd Edn. John Wiley & Sons, New York.

Wakindiki, I.I.C., and M. Ben-Hur. 2002. Soil mineralogy and texture effects on crust micromorphology, infiltration and erosion. Soil Sci. Soc. Am. J. 66:897–905.

Wangemann, S.G., R.A. Kohl, and P.A. Molumeli. 2000. Infiltration and percolation influenced by antecedent soil water content and air entrapment. Trans. ASAE 43:1517–1523.

Wilding, L.P. 1999. Comments on manuscript by Lal R, Hassan HM, Dumanski J, p. 146–149. In N.J. Rosenberg, R.C. Izauralde, and E.L. Malone (eds.) Carbon sequestration in soils: Science, monitoring and beyond. Battelle Press, Columbus, OH.

Wildman, W.E. 1981. Managing and modifying problem soils. University of California Cooperative Extension Leaflet 2791.

Williams, C.H., and J.D. Colwell. 1977. Inorganic chemical properties, p. 105–126. In J.S. Russell and E.L. Greacen (eds.) Soil factors in crop production in a semi-arid environment. University of Queensland Press, St. Lucia, Australia.

Williams, C.H., and M. Raupach. 1983. Plant nutrients in Australian soils, p. 777–793. In Soils: An Australian viewpoint. Academic Press, London, U.K.

Wright, D., and I. Rajper. 2000. An assessment of the relative effects of adverse physical and chemical properties of sodic soil on the growth and yield of wheat (*Triticum aestivum* L.). Plant Soil 223:277–285.

Yaron, B., and G.W. Thomas. 1968. Soil hydraulic conductivity as affected by sodic water. Water Res. Res. 4:545–552.

Yousaf, M., O.M. Ali, and J.D. Rhoades. 1987. Dispersion of clay from salt-affected arid land soil aggregates. Soil Sci. Soc. Am. J. 51:920–924.

19
Soil Water Repellency

Stefan H. Doerr
Swansea University

Richard A. Shakesby
Swansea University

19.1	Introduction 19-1
19.2	Origin and Classification 19-1
	Fundamental Principles Underlying Water Repellency • Origin of Water Repellency in Soils • Measurement and Classification of Soil Water Repellency
19.3	Occurrence of Soil Water Repellency 19-3
	Distribution of Water Repellency • Factors Affecting Water Repellency Occurrence
19.4	Hydrological and Geomorphological Effects of Soil Water Repellency 19-4
	Hydrological Effects of Water Repellency • Effects of Water Repellency on Erosion
19.5	Conclusions 19-8
References 19-8

19.1 Introduction

Although soils are normally thought of as being readily wettable by rainfall or irrigation, it is not uncommon for soils to behave in a water-repellent (hydrophobic) manner (Figure 19.1). This resistance of soils to wetting when in contact with water can persist from as little as a few seconds to, in extreme cases, months (e.g., King, 1981; Dekker and Ritsema, 1994; Doerr and Thomas, 2000). Water-repellent behavior is typically confined to the organically enriched upper few centimeters or decimeters of the soil and tends to be both spatially and temporally highly variable. It can develop when soil moisture falls below a critical threshold and often disappears after prolonged wet periods (Dekker et al., 2001).

Jamison (1946) was arguably the first to demonstrate that soil water repellency reduced crop productivity, and since the 1980s the use of wetting agents in horticulture and turfgrass industries (Cisar et al., 2000), or clay additions or advanced irrigation techniques in agriculture (Blackwell, 2000), has become a common amelioration measure. Soil water repellency is also viewed as a key factor in enhanced hydrological and erosional hillslope and catchment responses that are often observed following wildfire (Shakesby and Doerr, 2006).

During the last two decades, it has become increasingly apparent that soil water-repellent behavior is relatively common. It has been reported for soils ranging from coarse- to fine-textured, from land uses including plowed cropland, pasture, shrubland, and a wide range of forest types, and under climates ranging from seasonal tropical to subarctic (e.g., Wallis and Horne, 1992; Bauters et al., 2000a; DeBano, 2000; Doerr et al., 2000; Doerr et al., 2006a). Its intensity can vary considerably from extremely high, as observed under many eucalypt species (e.g., Keizer et al., 2005; Doerr et al., 2006b), to being so slight that it is detectable only with a purpose-built micro-infiltrometer (Hallett and Young, 1999). The view that water repellency in soils is the norm rather than the exception, with its intensity being variable (Wallis et al., 1991), has now become widely accepted.

19.2 Origin and Classification

19.2.1 Fundamental Principles Underlying Water Repellency

Over a wettable (i.e., hydrophilic) surface, water will spread in a continuous film, whereas over a hydrophobic surface, it "balls up" into individual droplets. If the hydrophobic surface is a porous medium like sand or soil, water infiltration is inhibited and the surface is considered to exhibit water repellency (Figure 19.1). The affinity between water and solid surfaces originates from mutually attractive forces (adhesion) and the attraction between the water molecules within the liquid (cohesion). In principle, interfacial tensions between the three phases, that is, the liquid (water), a solid, and vapor (air), can affect water repellency (DeGennes et al., 2003). For the sake of simplicity, however, this chapter will deal only with the liquid and solid surface tensions, the underlying principles of which are briefly considered here.

A water molecule comprises an oxygen atom with a partial negative charge and two hydrogen atoms with a partial positive charge. The hydrogen and oxygen atom bonds are positioned 105° apart, giving the water molecule a strongly dipolar structure (Parker, 1987). The attraction of these positive and negative ends causes water molecules to form aggregates, held together by hydrogen bonds. Water adheres to most natural solid surfaces because their positively or negatively charged ions attract the

FIGURE 19.1 Water drops resting on a highly repellent sandy soil. (Photo courtesy of Erik van den Elsen.)

negative or positive ends of a water molecule, respectively. The dipole character of water, however, also results in a comparatively strong force counteracting the attraction to charged surfaces. Within a liquid, the net force acting on an individual molecule is zero as it is surrounded by other molecules and their forces. When a liquid is in contact with air, however, no similar molecules exist beyond the surface of the liquid to oppose the attraction exerted by the molecules within the liquid. Consequently, the surface molecules experience a net attractive force toward the interior, which promotes the reduction of the surface area of water. Thus, if opposing forces are minimal, liquids will assume a spherical shape (i.e., that of a droplet). To enlarge the surface of a liquid, work is necessary. This work is related to the surface tension (γ) or surface free energy of the liquid. Most liquids have surface tensions between 20 and 40 N m × 10^{-3} at 20°C, but that of water is comparatively high at 72.75 N m × 10^{-3} (Parker, 1987). The same principle applies to solid surfaces, although their nature inhibits deformation into a spherical shape. The surface tension of solids therefore leads to lateral forces at the surface. Hard solids such as soil minerals can have surface tensions between 500 and 5000 N m × 10^{-3}, increasing with hardness and melting point (Zisman, 1964). For water to spread on a solid, the adhesive forces between them must exceed the cohesive forces within the body of water. Thus, surfaces with a surface-free energy >72.75 N m × 10^{-3} attract water and are therefore hydrophilic. The higher the surface tension of the solid, the stronger the attraction. All principal soil minerals have a much higher surface tension than water and are therefore hydrophilic (Tschapek, 1984), whereas soft organic solids, such as waxes or organic polymers can exhibit γ values below 72.75 N m × 10^{-3} and are thus hydrophobic (Zisman, 1964).

19.2.2 Origin of Water Repellency in Soils

The wettability of a flat surface is determined by the property of its outermost molecular layer. Thus, in principle, a single layer of hydrophobic molecules can render a hydrophilic mineral surface hydrophobic (Zisman, 1964), although in practise, hydrophobic compounds may be absorbed to soil mineral surfaces as globules rather than uniform monolayers (Cheng et al., 2008). The amount of hydrophobic compound required to induce water repellency in soil is nevertheless very small. For example, Ma'shum et al. (1988) induced severe water repellency in 1000 g of medium-sized sand using only 0.35 g of hydrophobic compound. Soil water repellency can also be caused by the presence of hydrophobic interstitial matter. If hydrophobic particles are present in the pore spaces of a hydrophilic matrix, the wettability of the composite material is reduced. For example, severe water repellency has been induced by intermixing as little as 2%–5% by weight of solid organic matter to wettable sand (McGhie and Posner, 1981).

When considering the water repellency of textured media (such as soils) compared to flat surfaces, it is important to recognize the effect of surface roughness and morphology. When water comes in contact with a rough granular surface, a fraction of the water surface will be in contact with solid material, with the remainder being in contact with the air between granules. Air is very hydrophobic and a granular surface can therefore become substantially more water repellent than a flat surface with identical surface chemistry. This phenomenon has been described as "super-hydrophobicity" when the contact angle between the surface and a water droplet exceeds 150°, and is perhaps better known from other biological surfaces such as the lotus leaf, but its underlying principles are also considered to apply to soils (McHale et al., 2005).

The biological origin of hydrophobic compounds in soils is undisputed. Such compounds are naturally abundant in the biosphere and may be gradually released into the soil, for example, as root exudates (Dekker and Ritsema, 1996; Doerr et al., 1998), from soil fauna fungi or microbes (Hallett and Young, 1999; Schaumann et al., 2007), or directly from decomposing organic matter (McGhie and Posner, 1981). Depending on vegetation cover present and soil temperatures reached, a rapid release or redistribution of hydrophobic compounds often accompanies the burning of vegetation, leading to a distinct hydrophobic layer at the soil surface or at some depth in the soil profile (DeBano et al., 1976; Doerr et al., 2006b).

The specific types of compounds suggested to be a cause of soil water repellency include alkanes, fatty acids and their salts and esters, and other related compounds such as phytanes, phytols, and sterols (Ma'shum et al., 1988; Roy et al., 1999; Franco et al., 2000; Horne and McIntosh, 2000; Atanassova and Doerr, 2010). The presence of these compounds, however, does not necessarily lead to the expression of water repellency in soils. It has been demonstrated that comparable amounts of such compounds occur also in wettable soils (Doerr et al., 2005; Morley et al., 2005). It has been suggested that these compounds only cause hydrophobicity when they form a specific molecular arrangement (Roy and McGill, 2000; Morley et al., 2005), which in turn is influenced by soil physical and chemical properties (Doerr et al., 2000; Graber et al., 2009).

19.2.3 Measurement and Classification of Soil Water Repellency

There are a number of techniques for measuring and classifying soil water repellency (see Wallis and Horne, 1992; Hallett and Young, 1999; Letey et al., 2000; Bachmann et al., 2003). One of the most common methods, the "Water Drop Penetration Time" (WDPT) test (Van't Woudt, 1959), is briefly described here. It involves placing droplets of distilled water onto a sample surface and recording the time for their complete infiltration (Figure 19.1). This test, which can be carried out on undisturbed soil surfaces in the field or on disturbed and typically air-dried bulk samples in the laboratory, broadly determines how long water repellency persists in the contact area of a water droplet. It is particularly relevant to estimating the likely response of a soil surface to rainfall or irrigation. This *persistence* or stability of water repellency is usually somewhat, but not always, well related to the apparent surface tension (i.e., the *severity* of water repellency) of soil (Dekker and Ritsema, 1994; Doerr, 1998; Scott, 2000). Water repellency *severity* will determine how strongly a droplet initially balls up when in contact with the soil and can be determined indirectly by contact angle or critical surface tension measurements (see Letey et al., 2000). Contact angles can, however, change rapidly if the *persistence* of water repellency is low. The apparent surface tension of the soil is particularly relevant when modeling flow processes through the soil matrix. When performing tests in the field or the laboratory, it is important to consider that in situ testing in the field reflects the water repellency measured at a distinct point on a given surface, whereas laboratory testing of bulk samples reflects the apparent water repellency of mixed soil material over the depth and area that has been sampled. Field and laboratory data may therefore not be directly comparable. Furthermore, it has been shown that the physical disturbance to the soil caused by sampling can substantially increase or decrease water repellency (Graber et al., 2006).

Perception of what constitutes high or low level water repellency *persistence* varies widely. To distinguish between wettable and water-repellent soils, an arbitrary WDPT threshold of 5 s (Bisdom et al., 1993; Table 19.1) has been used widely, although considerable effects on water movement at the centimeter-scale have been shown to be caused by lower, so called subcritical, levels of repellency (Hallett et al., 2004). WDPT values found in the literature are generally presented in categories rather than distinct values and are not always directly comparable because, for practical reasons, this test is often terminated well before droplet penetration occurs. In many studies to date, WDPTs exceeding 1 h have been recorded (= classed as extreme persistence; Table 19.1), but *persistence* can in some cases reach levels such that even ponding water may not infiltrate for periods exceeding 1 month (Doerr and Thomas, 2000). Although the aforementioned laboratory tests allow a general classification of soils according to water repellency *persistence* and *severity*, these measurements do not always relate well to the actual wetting behavior of field soils for which infiltrometer measurements or rainfall simulations may provide additional insight (Doerr et al., 2006a).

19.3 Occurrence of Soil Water Repellency

19.3.1 Distribution of Water Repellency

Soil water repellency has been reported in well over 1000 studies, which have focused on soils ranging from coarse- to fine-textured, land uses including plowed cropland, pasture, shrubland, and a wide range of forest types, and from climates ranging from seasonal tropical to subarctic (e.g., Wallis and Horne, 1992; Bauters et al., 2000a; DeBano, 2000; Doerr et al., 2000, 2006a). Although the occurrence of soil water repellency is normally not assessed within general soil surveys and hence little is known about its wider distribution, the view that it is the norm rather than the exception, with its degree being variable (Wallis et al., 1991), is now widely accepted. Regional-scale surveys in The Netherlands (Dekker et al., 2000) and the United Kingdom (Doerr et al., 2006a) indicate that water repellency is particularly common under grass, shrub, and conifer forest land. There are also many reports of high levels of water repellency being naturally common under a range of eucalypt species (e.g., Keizer et al., 2005; Doerr et al., 2006b). Furthermore, it is well established that peat, once dried out, can become highly water repellent (Fuchsmann, 1986). Coarser-scale studies suggest, for example, that in the Netherlands, 75% of the crop- and grassland exhibits water repellency (Dekker and Ritsema, 1994) and that in southern Australia 5 million hectares of land are affected, leading to production losses of up to 80% in Australian agriculture (Blackwell, 2000). In some sectors, such as horticulture and sports turf (e.g., golf greens, playing fields), wetting agents are widely used to increase soil wettability (e.g., Cisar et al., 2000). Where recycled wastewater is used for crop irrigation, it has recently been recognized that this practise can impart water repellency and result in uneven water distribution (Wallach and Graber, 2007).

A substantial body of research has also focused on the presence of water repellency following wildfire where it is thought to play a key role in the often enhanced hydrological and erosional responses of burnt terrain. In such environments, soil water repellency is often considered to be fire-induced although it is clear that it can also be present without the influence of burning (Shakesby et al., 1993; DeBano, 2000; Doerr et al., 2009).

TABLE 19.1 Water Drop Penetration Time (WDPT) Class Intervals in Seconds (Upper Limit) and Associated Repellency *Persistence* Rating

WDPT Interval	≤5	10	30	60	180	300	600	900	1,800	3,600	18,000	>18,000
Persistence rating[a]	—	Slight			Strong			Severe			Extreme	

[a] Based on Bisdom et al. (1993).

The heat generated by burning, in addition to redistributing and concentrating the naturally occurring hydrophobic substances in the soil, is also thought to make these compounds more hydrophobic by pyrolysis and conformational changes in their structural arrangement. Laboratory studies have shown that soil water repellency is intensified when soil temperatures reach 175°C–270°C, but is destroyed at temperatures above 270°C–400°C. The duration of heating can also affect the degree of soil water repellency with longer heating times reducing the temperature at which these changes occur (e.g., DeBano et al., 1976; Doerr et al., 2004). When there is insufficient oxygen, the temperature at which soil water repellency is destroyed may rise to 500°C–600°C (Bryant et al., 2005). Apart from the influence of burning conditions, factors such as the type and quantity of vegetation present or soil properties such as organic matter characteristics, clay content or mineralogy can influence the effects of fire on water repellency (Arcenegui et al., 2007; Mataix-Solera et al., 2008). The effects of burning on soil water repellency can therefore be highly variable, with fire potentially inducing water repellency in soils that were largely nonrepellent, or either enhancing or reducing pre-existing water repellency.

19.3.2 Factors Affecting Water Repellency Occurrence

Apart from the major factors linked to water repellency occurrence such as vegetation type and fire occurrence outlined above, there are a number of other environmental or soil specific factors that can affect water repellency. Coarse-textured soils are generally thought to be more susceptible to the development of soil water repellency than finer-textured ones. This is generally attributed to their relatively smaller particle surface area, which requires less organic material to generate a hydrophobic coating (DeBano, 2000). Hence, clay additions have been used successfully to reduce repellency in sandy soils (Blackwell, 2000; Figure 19.2). The specific mineralogy of the clay applied is of importance here (McKissock et al., 2000; Lichner et al., 2006) and the aforementioned soil surface topography effect (super-hydrophobicity; Section 19.2.2) may also be of relevance. Finer textured soils, however, have also been reported to exhibit high levels of water repellency (Dekker and Ritsema, 1996). This is thought to arise when substantial amounts of hydrophobic organic matter are present (DeBano, 2000; Doerr et al., 2000).

Water repellency appears to be more common in acidic than in calcareous soils (Mataix-Solera et al., 2007). The more widespread occurrence of acidic soils may be a compounding factor here. However, it is also thought that the lower fungal activity and less organic matter content typical of neutral or alkaline soils make them less prone to water repellency development (Hallett and Young, 1999; Mataix-Solera et al., 2007). Furthermore, there is evidence that the pH-related availability and relative abundance of proton active sites in the soil can be another factor affecting water repellency (Diehl et al., 2010). Thus, for example, high pH treatments have been used to reduce water repellency on golf greens (Karnok et al., 1993).

Perhaps the most critical factor affecting the occurrence of water repellency in soils that are prone to water repellency is soil moisture. As outlined earlier, water repellency can be temporally highly variable. Depending on its level of *persistence*, it tends to disappear after prolonged contact with water and typically reappears when soil moisture falls below a critical threshold (Dekker et al., 2001). A fundamental factor thought to be critical in this "switching" behavior is the specific molecular arrangement of the organic coating on soil particle (or soil pore wall) surfaces. It is thought that at low soil moisture contents, the hydrophobic ends of the organic molecules will tend to be oriented away from the surface toward the soil pore space, imparting water repellency. During prolonged water contact, these molecule ends "fold" onto the particle surface, thereby exposing more hydrophilic areas and ultimately making the soil wettable (Tschapek, 1984; Roy and McGill, 2000; Morley et al., 2005). Whether or not this is the main reason for the "switching behavior," several studies have shown that water repellency is not present above a critical soil water threshold. In sandy soils, this threshold can be at a water content of a few volume percent (Dekker et al., 2001; Täumer et al., 2005), but for finer-textured soils, thresholds in the region of 20%–30% have been reported (Doerr and Thomas, 2000; Doerr et al., 2006a). What happens below this threshold is not entirely clear. Although water repellency typically recovers when soils dry out, this effect may be delayed (Crockford et al., 1991; Leighton-Boyce et al., 2005) and some soils may remain wettable after drying (Doerr and Thomas, 2000). This makes the prediction of water repellency occurrence particularly challenging.

19.4 Hydrological and Geomorphological Effects of Soil Water Repellency

The primary hydrologic and erosional effects of soil water repellency include: (a) lower infiltration rates and a corresponding increase in the likelihood and amount of infiltration-excess (Hortonian) overland flow, (b) more spatial variability in infiltration and soil moisture fluxes, causing an uneven distribution of

FIGURE 19.2 Dispersive clay is added to highly water-repellent soil on a farm in southern Australia to improve soil wettability. (Photo courtesy of S. Doerr.)

soil moisture, (c) increased surface erosion aided by the increase in overland flow, and (d) enhanced susceptibility to wind erosion due to drier soil conditions and reduced cohesion of soil particles. The reduction in infiltration can also have secondary effects, such as hindering the germination and growth of vegetation, which can prolong fire impacts on runoff and erosion rates (see reviews by DeBano, 2000; Doerr et al., 2000; Shakesby et al., 2000; Shakesby and Doerr, 2006).

19.4.1 Hydrological Effects of Water Repellency

The most frequently reported effect of soil water repellency is reduced infiltration (e.g., Van Dam et al., 1990; Imeson et al., 1992; Doerr et al., 2003) and thus increased overland flow (e.g., McGhie and Posner, 1980; Crockford et al., 1991; Witter et al., 1991). For example, the infiltration capacity of a water-repellent soil was found to be 25 times lower than for a similar soil rendered hydrophilic by heating (DeBano, 1971). Owing to the reduced surface tension of the soil pore walls, a water-repellent soil matrix will have a *positive* soil water potential. This effectively leads to a capillary depression effect illustrated in Figure 19.3. Infiltration into a relatively dry water-repellent soil matrix will therefore only occur if a ponding depth (hydraulic head; entry pressure) sufficient to exceed the positive soil water potential is reached. The necessary ponding depth may decrease with time if water repellency decays during water contact. These factors acting in combination usually result in infiltration rates *increasing* during rainfall. This contrasts with "normal" soil behavior, where infiltration rates *decrease* over time (Figure 19.4; Letey et al., 1962; Kirkham and Clothier, 2000). The lack of capillary rise in water-repellent topsoil, on the other hand, leads to reduced evaporation from the subsoil, which may be advantageous to plant communities in drier regions (Doerr et al., 2000).

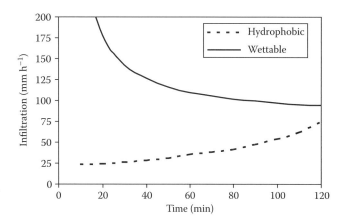

FIGURE 19.4 Infiltration rates into wettable and water-repellent soils. (Adapted from Letey, J., J. Osborn, and R.E. Pelishek. 1962. The influence of the water-solid contact angle on water movement in soil. Bull. Int. Assoc. Sci. Hydrol. 3:75–81.)

On a water-repellent soil surface in the field, rainwater will pond and, if rainfall is sufficient and surface detention exceeded, Hortonian (infiltration-excess) overland flow will occur. The frequency of "gaps" through the water-repellent layer (such as structural or drying cracks, root holes and burrows, and patches of wettable or less repellent soil) will then determine whether overland flow is widespread or only local (Figure 19.5a). Under some conditions, for example, where intense soil heating during a wildfire has destroyed water repellency in the top few centimeters of the soil (see DeBano, 1971), rainfall infiltrating such a topsoil may pond above the water-repellent layer (Figure 19.5b). The commonly observed temporal variability of water repellency also needs to be considered here. Thus reductions in infiltration capacity and increases in overland flow can be expected to be most pronounced following prolonged dry periods, when water repellency tends to be most severe. For example, Burch et al. (1989) recorded infiltration capacities in Australian eucalypt forest of 0.75–1.9 mm h^{-1} when dry, but 7.9–14.0 mm h^{-1} when wet. In many areas, water repellency-linked overland flow may therefore be confined to storm events following dry weather (Sevink et al., 1989; Walsh et al., 1994). The rapid response of Hortonian overland flow in a water-repellent soil following dry weather contrasts sharply with the muted overland flow in moderately wet weather, when soils are generally wettable (Jungerius and De Jong, 1989; Ferreira et al., 2000).

Another common observation in soils prone to water repellency is enhanced preferential flow, which is the concentrated vertical movement of water via preferred pathways through the soil. It may originate for a variety of reasons such as cracks and biopores, textural discontinuities, and unstable wetting fronts, which may result, for example, from soil layering and air entrapment (Ritsema et al., 1993). In stony, water-repellent soils, water movement may be facilitated along sand–stone interfaces and preferential flow may take place where there are contiguous stone-to-stone connections (Urbanek and Shakesby, 2009). Although not restricted to water-repellent soils, repellency can be particularly effective at preventing or

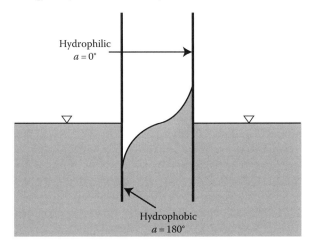

FIGURE 19.3 Shape of the meniscus of water between a hydrophilic and a hydrophobic plate with liquid–solid contact angles of 0° and 180°, respectively. (Adapted from Bauters, T.W.J., T.S. Steenhuis, D.A. DiCarlo, J.L. Nieber, L.W. Dekker, C.J. Ritsema, J.-Y. Parlange, and R. Haverkamp. 2000b. Physics of water repellent soils. J. Hydrol. 231–232:233–243.)

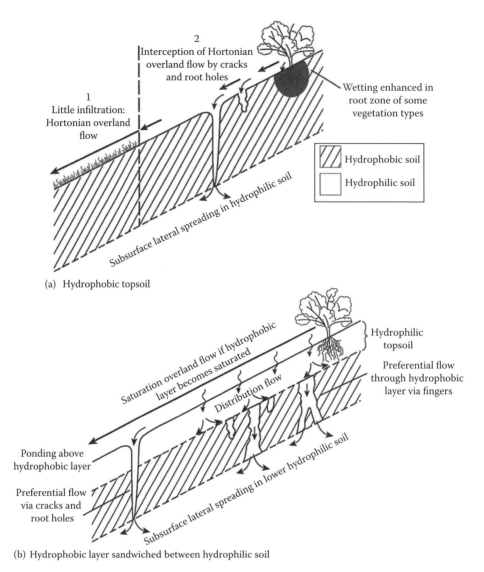

FIGURE 19.5 Schematic illustration of possible hydrological responses of soil with (a) a water-repellent layer located on the surface, and (b) a repellent layer sandwiched between wettable soil. (Adapted from Doerr, S.H., R.A. Shakesby, and R.P.D. Walsh. 2000. Soil water repellency, its characteristics, causes and hydro-geomorphological consequences. Earth Sci. Rev. 51:33–65. Copyright 2000, with permission from Elsevier.)

hindering downward water movement, directing it into structural or textural preferential flow paths (Figure 19.5a and b) and creating an unstable irregular wetting front (Figure 19.6). Consequently, soils may not wet completely with the passage of a wetting front (DeBano, 1971), and water may be channeled via biopores (Shakesby et al., 2007), cracks, and pipes, thereby bypassing the water-repellent soil matrix (e.g., Burch et al., 1989; Ferreira et al., 2000). Walsh et al. (1995) considered that macropores and cracks could explain why even large storms produced little overland flow for extremely water-repellent mature pine and eucalypt forest soils in Portugal. Irrespective of the dominance of enhanced overland or subsurface flow, the reduced soil water storage capacity in catchments affected by soil water repellency is likely to increase flooding particularly in cases where long dry periods are followed by large rainstorm events. This situation can be further exacerbated following the removal of a protective vegetation and litter cover by clearance or fire (Figure 19.7).

On a grass-covered sandy soil in Holland, Ritsema et al. (1993) used tracers to record distribution flow within a thin (<2.5 cm) wettable topsoil, which supplied water via columns of less repellent soil (preferential flow paths) in an otherwise extremely repellent layer to the wettable subsoil, where the water spread laterally. This has been termed "fingered flow" (Ritsema and Dekker, 1994). The fingers formed only after dry weather when soil moisture levels in the sandy, repellency-prone layer were below a "critical" value of 4.75% (vol.). The fingers ranged from 10 to 50 cm in diameter (expanding in wetter weather), and were the sole means of water transport for several hours during sustained rainfalls. Such fingers have been shown to recur at the same places in successive storms following intervening dry weather, possibly aided by preferential leaching of hydrophobic

Soil Water Repellency

FIGURE 19.6 Irregular wetting front and preferential flow pathways in a hydrophobic sandy agricultural soil in Western Australian. (Photo courtesy of R. Shakesby.)

FIGURE 19.7 Enhanced overland flow during intense rain in burnt eucalypt forest terrain in the Victorian Alps, south-east Australia in 2003. (Photo courtesy of Rob Ferguson.)

substances from the finger pathways (Ritsema et al., 1998). In a study in Germany, wet and adjacent dry areas were seasonally persistent in a grass-covered and partially water-repellent soil (Täumer et al., 2005). Preferential flow in general is thought to be reinforced by soil water hysteresis between wetting and drying phases, a feature of wettable soils but more pronounced in water-repellent ones (Ritsema et al., 1998).

Water repellency-induced fingered flow can lead to considerable variations in water content in an initially repellent soil such that zones of very dry soil can abut directly against zones of wet soil. For example, Dekker and Ritsema (1996) found differences in soil moisture of up to 28% (vol.) between closely spaced samples in both clay and sandy soils (see also Figure 19.6). Such differences do not only result in the widely reported poor seed germination and plant growth. Any type of preferential flow path formation can also lead to accelerated leaching of surface-applied agrichemicals and an increased risk of surface- and groundwater contamination (Hendrickx et al., 1993; Ritsema et al., 1993).

19.4.2 Effects of Water Repellency on Erosion

In general, the most important way in which soil water repellency can influence erosion lies in its potential for contributing to an increase in the proportion of rainfall going to overland flow. As the amount of overland flow increases, its depth and velocity also increase and with it the ability of the water to detach and transport particles, initially by sheetwash (Shakesby et al., 2000). Then, as the overland flow becomes concentrated into small rivulets, this can initiate rill erosion (Benavides-Solorio and MacDonald, 2005). At a larger scale, convergence of overland flow in concave parts of hillslopes can lead to the formation of gullies, and to downstream channel bank and bed erosion (e.g., Moody and Martin, 2001).

Soil water repellency can also directly affect erosion rates by altering the erodibility of the soil. Laboratory tests have shown that individual water drops falling on a water-repellent soil produce fewer, slower-moving ejection droplets than those on wettable soils, but the splash ejection droplets developed on the former carried more sediment (Terry and Shakesby, 1993). With successive drops, the surface of the water-repellent soil remained dry and noncohesive, and the soil particles continued to be displaced by rainsplash despite the retention of an overlying film of water. In contrast, drops falling on the surface of the wettable soil sealed and compacted it, which increased its resistance to detachment by rainsplash. Aggregate stability, in contrast, has been shown to be higher in water-repellent soils, which may counter the above effect to some degree in soils that exhibit aggregation (Giovannini and Lucchesi, 1983; Mataix-Solera and Doerr, 2004).

At hillslope and watershed scales, the role of soil water repellency in increasing erosion is less certain. Its overall impact is likely to be reduced where areas with macropores or wettable soil patches promote interception of overland flow generated locally on water-repellent areas.

Soil water repellency can also play a role in other forms of erosion. It can enhance wind erosion as a result of reduced soil moisture, which in turn reduces soil particle cohesion and lowers the threshold wind velocity for particle detachment and entrainment (Whicker et al., 2002). Dry ravel (the rapid, dry particle-to-particle sliding of sediment through gravity) is a form of mass movement that is strongly associated with postfire water-repellent conditions, particularly in chaparral vegetation of the western United States (Wells, 1986). Formation of small-scale debris flows related to failure of a saturated layer of wettable soil only a few millimeters thick overlying a subsurface water-repellent zone can occur (Gabet, 2003). The links between larger debris flows and water-repellent soils seem to be less certain (Cannon, 2001).

19.5 Conclusions

Three decades ago, soil water repellency would have been viewed as a curious aberration by many soil scientists, whereas it has now become a widely acknowledged soil property that is typically (i) highly variable spatially, temporally, and in its degree, (ii) common for many types of soils, vegetation covers, and climates, (iii) confined to the top few centimeters or decimeters of soil, (iv) enhanced, but in some cases unaffected or eliminated following fire, depending on the degree of soil heating, and (v) most pronounced under dry conditions, but reduced or absent after prolonged rainfall.

Water repellency has notable, and in some cases, substantial implications for soils. It can lead to reduced seed germination and plant growth, reduced infiltration and evaporation, enhanced overland flow, accelerated preferential flow and increased soil erosion. In many cases, water repellency is naturally induced and perhaps beneficial in terms of plant competition and reduction of evaporative losses from the soil surface. From a soil management perspective, however, there are a number of detrimental impacts including reduced crop performance, ineffective use of irrigation water, enhanced contamination risk via accelerated overland and preferential flow, and increased risk of flooding, soil erosion, and debris flow. These potential impacts need to be considered carefully when managing water-repellent soils.

References

Arcenegui, A., J. Mataix-Solera, C. Guerrero, R. Zornoza, A.M. Mayoral, and J. Morales. 2007. Factors controlling the water repellency induced by fire in calcareous Mediterranean forest soils. Eur. J. Soil Sci. 58:1254–1259.

Atanassova, A., and S.H. Doerr. 2010. Organic compounds of different extractability in total solvent extracts from soils of contrasting water repellency. Eur. J. Soil Sci. 61:298–313.

Bachmann, J., S.K. Woche, M.-O. Göbel, M.B. Kirkham, and R. Horton. 2003. Extended methodology for determining wetting properties of porous media. Water Resour. Res. 39(12):1353.

Bauters, T.W.J., D.A. DiCarlo, T.S. Steenhuis, and J.-Y. Parlange. 2000a. Soil water content dependent wetting front characteristics in sands. J. Hydrol. 231–232:244–254.

Bauters, T.W.J., T.S. Steenhuis, D.A. DiCarlo, J.L. Nieber, L.W. Dekker, C.J. Ritsema, J.-Y. Parlange, and R. Haverkamp. 2000b. Physics of water repellent soils. J. Hydrol. 231–232:233–243.

Benavides-Solorio, J.D., and L.H. MacDonald. 2005. Measurement and prediction of post-fire erosion at the hillslope scale, Colorado Front Range. Int. J. Wildland Fire 14:457–474.

Bisdom, E.B.A., L.W. Dekker, and J.F.Th. Schoute. 1993. Water repellency of sieve fractions from sandy soils and relationships with organic material and soil structure. Geoderma 56:105–118.

Blackwell, P.S. 2000. Management of water repellency in Australia; and risks associated with preferential flow, pesticide concentration and leaching. J. Hydrol. 231–232:384–395.

Bryant, R., S.H. Doerr, and M. Helbig. 2005. Effect of oxygen deprivation on soil hydrophobicity during heating. Int. J. Wildland Fire 14:449–455.

Burch, G.J., I.D. Moore, and J. Burns. 1989. Soil hydrophobic effects on infiltration and catchment runoff. Hydrol. Proc. 3:211–222.

Cannon, S.H. 2001. Debris-flow generation from recently burned watersheds. Environ. Eng. Geosci. 7:321–341.

Cheng, S., R. Bryant, S.H. Doerr, R.P. Williams, and C.J. Wright. 2008. Application of AFM to the study of natural and model soil particles. J. Microsc. 231(3):384–395.

Cisar, J.L., K.E. Williams, H.E. Vivas, and J.J. Haydu. 2000. The occurrence and alleviation by surfactants of soil-water repellency on sand-based turfgrass systems. J. Hydrol. 231–232:352–358.

Crockford, H., S. Topalidis, and D.P. Richardson. 1991. Water repellency in a dry sclerophyll eucalypt forest—Measurements and processes. Hydrol. Proc. 5:405–420.

DeBano, L.F. 1971. The effect of hydrophobic substances on water movement in soil during infiltration. Soil Sci. Soc. Am. Proc. 35:340–343.

DeBano, L.F. 2000. Water repellency in soils: A historical overview. J. Hydrol. 231–232:4–32.

DeBano, L.F., S.M. Savage, and D.A. Hamilton. 1976. The transfer of heat and hydrophobic substances during burning. Soil Sci. Soc. Am. J. 40:779–782.

DeGennes, P.-G., F. Brochard-Wyart, and D. Quere. 2003. Capillary and wetting phenomena—Drops, bubbles, pearls, waves. Springer, New York.

Dekker, L.W., S.H. Doerr, K. Oostindie, A.K. Ziogas, and C.J. Ritsema. 2001. Water repellency and critical soil water content in a dune sand. Soil Sci. Soc. Am. J. 65:1667–1674.

Dekker, L.W., and C.J. Ritsema. 1994. How water moves in a water repellent sandy soil. 1. Potential and actual water repellency. Water Resour. Res. 30:2507–2517.

Dekker, L.W., and C.J. Ritsema. 1996. Variation in water content and wetting patterns in Dutch water repellent peaty clay and clayey peat soils. Catena 28:89–105.

Dekker, L.W., C.J. Ritsema, and K. Oostindie. 2000. Extent and significance of water repellency in dunes along the Dutch coast. J. Hydrol. 231–232:112–125.

Diehl, D., J.V. Bayer, S.K. Woche, R. Bryant, S.H. Doerr, and G.E. Schaumann. 2010. Reaction of soil water repellency on artificially induced changes in soil pH. Geoderma (available online: doi:10.1016/j.geoderma.2010.06).

Doerr, S.H. 1998. On standardising the 'water drop penetration time' and the 'molarity of an ethanol droplet' techniques to classify soil hydrophobicity: A case study using medium textured soils. Earth Surf. Proc. Landforms 23:663–668.

Doerr, S.H., W.H. Blake, R.A. Shakesby, F. Stagnitti, S.H. Vuurens, G.S. Humphreys, and P. Wallbrink. 2004. Heating effects on water repellency in Australian eucalypt forest soils and their value in estimating wildfire soil temperatures. Int. J. Wildland Fire 13:157–163.

Doerr, S.H., A.J.D. Ferreira, R.P.D. Walsh, R.A. Shakesby, G. Leighton-Boyce, and C.O.A. Coelho. 2003. Soil water repellency as a potential parameter in rainfall-runoff modelling: Experimental evidence at point to catchment scales from Portugal. Hydrol. Proc. 17:363–377.

Doerr, S.H., C.T. Llewellyn, P. Douglas, C.P. Morley, K.A. Mainwaring, C. Haskins, L. Johnsey et al. 2005. Extraction of compounds associated with water repellency in sandy soils of different origin. Aust. J. Soil Res. 43:225–237.

Doerr, S.H., R.A. Shakesby, W.H. Blake, G.S. Humphreys, C.J. Chafer, and P.J. Wallbrink. 2006b. Effects of differing wildfire severity on soil wettability and implications for hydrological response. J. Hydrol. 319:295–311.

Doerr, S.H., R.A. Shakesby, L.W. Dekker, and C.J. Ritsema. 2006a. Occurrence, prediction and hydrological effects of water repellency amongst major soil and land use types in a humid temperate climate. Eur. J. Soil Sci. 57:741–754.

Doerr, S.H., R.A. Shakesby, and R.P.D. Walsh. 1998. Spatial variability of soil hydrophobicity in fire-prone eucalyptus and pine forests, Portugal. Soil Sci. 163:313–324.

Doerr, S.H., R.A. Shakesby, and R.P.D. Walsh. 2000. Soil water repellency, its characteristics, causes and hydro-geomorphological consequences. Earth Sci. Rev. 51:33–65.

Doerr, S.H., and A.D. Thomas. 2000. The role of soil moisture in controlling water repellency: New evidence from forest soils in Portugal. J. Hydrol. 231–232:134–147.

Doerr, S.H., S.W. Woods, D.A. Martin, and M. Casimiro. 2009. 'Natural' soil water repellency in conifer forests of the north-western USA: Its prediction and relationship to wildfire occurrence. J. Hydrol. 371:12–21.

Ferreira, A.J.D., C.O.A. Coelho, R.P.D. Walsh, R.A. Shakesby, A. Ceballos, and S.H. Doerr. 2000. Hydrological implications of soil water-repellency in Eucalyptus globulus forests, north-central Portugal. J. Hydrol. 231–232:165–177.

Franco, C.M.M., P.J. Clarke, M.E. Tate, and J.M. Oades. 2000. Hydrophobic properties and chemical characterisation of natural water repellent materials in Australian sands. J. Hydrol. 231–232:47–58.

Fuchsman, C.H. (ed.). 1986. Peat and water: Aspects of water retention and dewatering in peat. Kluwer, New York.

Gabet, E.J. 2003. Post-fire thin debris flows: Sediment transport and numerical modelling. Earth Surf. Proc. Landforms 28:1341–1348.

Giovannini, G., and S. Lucchesi. 1983. Effect of fire on hydrophobic and cementing substances of soil aggregates. Soil Sci. 136:231–236.

Graber, E.R., O. Ben-Arie, and R. Wallach. 2006. Effect of sample disturbance on soil water repellency determination in sandy soils. Geoderma 136:11–19.

Graber, E.R., S. Tagger, and R. Wallach. 2009. Role of divalent fatty acids in soil water repellency. Soil Sci. Soc. Am. J. 73:541–549.

Hallett, P.D., N. Nunan, J.T. Douglas, and I.M. Young. 2004. Millimeter-scale variability in soil water sorptivity: Scale, surface elevation and subcritical repellency effects. Soil Sci. Soc. Am. J. 68:352–358.

Hallett, P.D., and I.M. Young. 1999. Changes to water repellence of soil aggregates caused by substrate-induced microbial activity. Eur. J. Soil Sci. 50:35–40.

Hendrickx, J.M.H., L.W. Dekker, and O.H. Boersma. 1993. Unstable wetting fronts in water repellent field soils. J. Environ. Qual. 22:109–118.

Horne, D.J., and J.C. McIntosh. 2000. Hydrophobic compounds in sands in New Zealand: Extraction, characterisation and proposed mechanisms for repellency expression. J. Hydrol. 231–232:35–46.

Imeson, A.C., J.M. Verstraten, E.J. van Mulligen, and J. Sevink. 1992. The effects of fire and water repellency on infiltration and runoff under Mediterranean type forest. Catena 19:345–361.

Jamison, V.C. 1946. Resistance to wetting in the surface of sandy soils under citrus trees in central Florida and its effect upon penetration and the efficiency of irrigation. Soil Sci. Soc. Am. Proc. 11:103–109.

Jungerius, P.D., and J.H. De Jong. 1989. Variability of water repellence in the dunes along the Dutch coast. Catena 16:491–497.

Karnok, K.A., E.J. Rowland, and K.H. Tan. 1993. High pH treatments and the alleviation of soil hydrophobicity on golf greens. Agron. J. 85:983–986.

Keizer, J.J., A.J.D. Ferreira, C.O.A. Coelho, S.H. Doerr, M.C. Malvar, C.S.P. Domingues, I.M.B. Perez, C. Ruiz, and K. Ferrari. 2005. The role of tree stem proximity in the spatial variability of soil water repellency in a eucalypt plantation in coastal Portugal. Aust. J. Soil Res. 43:251–259.

King, P.M. 1981. Comparison of methods for measuring severity of water repellence of sandy soils and assessment of some factors that affect its measurement. Aust. J. Soil Res. 19:275–285.

Kirkham, M.B., and B.E. Clothier. 2000. Infiltration into a New Zealand native forest soil, p. 13–26. In A spectrum of achievement in agronomy: Women fellows of the tri-societies. ASA Spec. Pub. 62.

Leighton-Boyce, G., S.H. Doerr, R.A. Shakesby, R.P.D. Walsh, A.J.D. Ferreira, A.K. Boulet, and C.O.A. Coelho. 2005. Temporal dynamics of water repellency and soil moisture in eucalypt plantations, Portugal. Aust. J. Soil Res. 43:269–280.

Letey, J., M.L.K. Carrillo, and X.P. Pang. 2000. Approaches to characterize the degree of water repellency. J. Hydrol. 231–232:61–65.

Letey, J., J. Osborn, and R.E. Pelishek. 1962. The influence of the water-solid contact angle on water movement in soil. Bull. Int. Assoc. Sci. Hydrol. 3:75–81.

Lichner, L., P. Dlapa, S.H. Doerr, and J. Mataix-Solera. 2006. Evaluation of different clay mineralogies as additives for soil water repellency alleviation. Appl. Clay Sci. 31:238–248.

Ma'shum, M., M.E. Tate, G.P. Jones, and J.M. Oades. 1988. Extraction and characterization of water-repellent materials from Australian soils. J. Soil Sci. 39:99–110.

Mataix-Solera, J., V. Arcenegui, C. Guerrero, M.M. Jordán, P. Dlapa, N. Tessler, and L. Wittenberg. 2008. Can terra rossa become water repellent by burning? A laboratory approach. Geoderma 147:178–184.

Mataix-Solera, J., V. Arcenegui, C. Guerrero, A.M. Mayoral, J. Morales, J. González, F. García-Orenes, and I. Gómez. 2007. Water repellency under different plant species in a calcareous forest soil in a semiarid Mediterranean environment. Hydrol. Proc. 21:2300–2309.

Mataix-Solera, J., and S.H. Doerr. 2004. Hydrophobicity and aggregate stability in calcareous topsoils from fire-affected pine forests in southeastern Spain. Geoderma 118:77–88.

McGhie, D.A., and M.A. Posner. 1980. Water repellence of a heavy-textured Western Australian surface soil. Austr. J. Soil Res. 18:309–323.

McGhie, D.A., and A.M. Posner. 1981. The effect of plant top material on the water repellence of fired sands and water repellent soils. Austr. J. Agric. Res. 32:609–620.

McHale, G., M.I. Newton, and N.J. Shirtcliffe. 2005. Water-repellent soil and its relationship to granularity, surface roughness and hydrophobicity: A materials science view. Eur. J. Soil Sci. 56:445–452.

McKissock, I., E.L. Walker, R.J. Gilkes, and D.J. Carter. 2000. The influence of clay type on reduction of water repellency by applied clays: A review of some West Australian work. J. Hydrol. 231–232:323–332.

Moody, J.A., and D.A. Martin 2001. Post-fire, rainfall intensity-peak discharge relations for three mountainous watersheds in the western USA. Hydrol. Proc. 15(15):2981–2993.

Morley, C.P., K.A. Mainwaring, S.H. Doerr, P. Douglas, C.T. Llewellyn, and L.W. Dekker. 2005. Identification of hydrophobic compounds in a sandy soil under permanent grass cover. Aust. J. Soil Res. 43(3):239–249.

Parker, S.D. 1987. Encyclopedia of Science and Technology. McGraw-Hill, New York.

Ritsema, C.J., and L.W. Dekker. 1994. How water moves in a water repellent sandy soil. 2. Dynamics of fingered flow. Water Resour. Res. 30:2519–2531.

Ritsema, C.J., L.W. Dekker, J.M.H. Hendrickx, and W. Hamminga. 1993. Preferential flow mechanism in a water repellent sandy soil. Water Resour. Res. 29:2183–2193.

Ritsema, C.J., L.W. Dekker, J.L. Nieber, and T.S. Steenhuis. 1998. Modeling and field evidence of finger formation and finger recurrence in a water repellent sandy soil. Water Resour. Res. 34:555–567.

Roy, J.L., and W.B. McGill. 2000. Flexible conformation in organic matter coatings: An hypothesis about soil water repellency. Can. J. Soil Sci. 80:143–152.

Roy, J.L., W.B. McGill, and M.D. Rawluk. 1999. Petroleum residues as water-repellent substances in weathered nonwettable oil-contaminated soils. Can. J. Soil Sci. 79:367–380.

Schaumann, G.E., B. Braun, D. Kirchner, W. Rotard, U. Szewzyk, and E. Grohmann. 2007. Influence of biofilms on the water repellency of urban soil samples. Hydrol. Proc. 21(17): 2276–2284.

Scott, D.F. 2000. Soil wettability in forested catchments in South Africa: As measured by different methods and as affected by vegetation cover and soil characteristics. J. Hydrol. 231–232:87–104.

Sevink, J., A.C. Imeson, and J.M. Verstraten. 1989. Humus form development and hillslope runoff and the effects of fire and management, under Mediterranean forest in NE Spain. Catena 16:461–475.

Shakesby, R.A., C. de O.A. Coelho, A.D. Ferreira, J.P. Terry, and R.P.D. Walsh. 1993. Wildfire impacts on soil erosion and hydrology in wet Mediterranean forest, Portugal. Int. J. Wildland Fire 3:95–110.

Shakesby, R.A., and S.H. Doerr. 2006. Wildfire as a hydrological and geomorphological agent. Earth Sci. Rev. 74:269–307.

Shakesby, R.A., S.H. Doerr, and R.P.D. Walsh. 2000. The erosional impact of soil hydrophobicity: Current problems and future research directions. J. Hydrol. 231–232:178–191.

Shakesby, R.A., P.J. Wallbrink, S.H. Doerr, P.M. English, C.J. Chafer, G.S. Humphreys, W.H. Blake, and K.M. Tomkins. 2007. Distinctiveness of wildfire effects on soil erosion in south-east Australian eucalypt forests assessed in a global context. For. Ecol. Manage. 238:347–364.

Täumer, K., H. Stoffregen, and G. Wessolek. 2005. Determination of repellency distribution using soil organic matter and water content. Geoderma 125:10715.

Terry, J.P., and R.A. Shakesby. 1993. Soil hydrophobicity effects on rainsplash: Simulated rainfall and photographic evidence. Earth Surf. Proc. Landforms 18:519–525.

Tschapek, M. 1984. Criteria for determining the hydrophilicity-hydrophobicity of soils. Z. Planz. Bodenk. 147:137–149.

Urbanek, E., and R.A. Shakesby. 2009. Impact of stone content on water movement in water-repellent sand. Eur. J. Soil Sci. 60:412–419.

Van Dam, J.C., J.M.H. Hendrickx, H.C. Van Ommen, M.H. Bannink, M.Th. Van Genuchten, and L.W. Dekker. 1990. Water and solute movement in a coarse-textured water-repellent field soil. J. Hydrol. 120:359–379.

Van't Woudt, B.D. 1959. Particle coatings affecting the wettability of soils. J. Geophys. Res. 64:263–267.

Wallach, R., and E.R. Graber. 2007. Infiltration into effluent irrigation-induced repellent soils and the dependence of repellency on ambient relative humidity. Hydrol. Proc. 21:2346–2355.

Wallis, M.G., and D.J. Horne. 1992. Soil water repellency. Adv. Soil Sci. 20:91–146.

Wallis, M.G., D.R. Scotter, and D.J. Horne. 1991. An evaluation of the intrinsic sorptivity water repellency index on a range of New Zealand soils. Aust. J. Soil Res. 29:353–362.

Walsh, R.P.D., D. Boakes, C. de O.A. Coelho, A.J.B. Goncalves, R.A. Shakesby, and A.D. Thomas. 1994. Impact of fire-induced hydrophobicity and post-fire forest litter on overland flow in northern and central Portugal, p. 1149–1159. In 2nd International Conference on Forest Fire Research. Vol. II. November 21–24, 1994. Domingos Xavier Viegas, Coimbra, Portugal.

Walsh, R.P.D., C. de O.A. Coelho, R.A. Shakesby, A.J.D. Ferreira, and A.D. Thomas. 1995. Post-fire land use and management and runoff responses to rainstorms in northern Portugal, p. 283–308. In D. McGregor, and D. Thompson (eds.) Geomorphology and land management in a changing environment. John Wiley & Sons, Inc., Chichester, U.K.

Wells, W.G. 1986. The influence of fire on erosion rates in California chaparral, p. 57–62. *In* J. DeVries (ed.) Proceedings of chaparral ecosystems conference. Santa Barbara, CA, May 16–17, Water Resources Center Report 62, Davis, CA.

Whicker, J.J., D.D. Breshears, P.T. Wasiolek, T.B. Kirchner, R.A. Tavani, D.A. Schoep, and J.C. Rodgers. 2002. Temporal and spatial variation of episodic wind erosion in unburned and burned semiarid shrubland. J. Environ. Qual. 31:599–612.

Witter, J.V., P.D. Jungerius, and M.J. Ten Harkel. 1991. Modeling water erosion and the impact of water repellency. Catena 18:115–124.

Zisman, W.A. 1964. Relation of equilibrium contact angle to liquid and solid constitution. Adv. Chem. Ser. 43:1–51.

20

Biogeochemistry of Wetlands

P.W. Inglett
University of Florida

K.R. Reddy
University of Florida

W.G. Harris
University of Florida

E.M. D'Angelo
University of Kentucky

20.1 Introduction **20**-1
20.2 Definitions **20**-1
20.3 Wetland Soils **20**-2
  Physical Properties • Chemical and Biological Properties • Organic Soils • Mineral Soils • Soils and Wetland Delineation: Hydric Soils Concept
20.4 Redox Gradients 20-6
20.5 Carbon **20**-7
20.6 Nitrogen **20**-11
20.7 Phosphorus **20**-14
20.8 Metals **20**-17
20.9 Toxic Organics **20**-19
References **20**-21

20.1 Introduction

Wetlands comprise swamps, marshes, fens, peat lands, bogs, potholes, bays, riparian zones, floodplains, and other shallow flooded areas. These areas are located at the interface of terrestrial and aquatic systems and often possess the characteristics of both systems. Wetlands have been estimated to occupy about 2.8×10^6 km^2 or 2.2% of the earth's surface (Post et al., 1982); however, the areal extent may be as high as 6% (Mitsch, 1994). One reason for this discrepancy probably lies in the difficulty of accurately defining wetlands. Within the conterminous United States, wetlands occupy about 5% of the land surface, and many of these wetlands are among the most productive natural systems. Wetlands are critical components of the biosphere because they provide essential ecological functions including habitat for wildlife, groundwater recharge, shoreline stabilization, flood control, and water quality improvement through biogeochemical transformations. Wetland soils specifically function as sinks and as transformers of nutrients, toxic metals, and organics. Recently, the impact of wetlands on the production/consumption of greenhouse gases (CO_2, CH_4, N_2O, and methyl sulfides) has been realized. A major food crop, lowland rice, is also grown under waterlogged soil conditions that are similar to wetland soils in physical, chemical, and biological properties.

20.2 Definitions

Although the definition of soil primarily focuses on upland soils, it includes those that undergo periodic or continuous flooding. Depending on scientific disciplines and ecosystems, soils saturated with water are often called flooded, wetland, waterlogged, marsh, paddy, or hydric soils, but no universally accepted definition for wetland soils exists among various groups.

The term "wetland" implies that the lands are located in wet areas. However, defining a "wetland" is often complex and difficult, because of the diversity in types of wetlands, presence of moisture gradients from clearly defined upland areas to aquatic systems, as well as the observation that many wetlands are not permanently inundated. These difficulties have resulted in the propagation of several legal and nonlegal definitions of wetlands. However, most definitions include reference to three major components of wetlands: (1) hydrology (water), (2) organisms (plants and animals), and (3) hydric soil. (US Army Corps of Engineers, 1987).

The following definitions are probably the most often used to describe and delineate wetlands from other ecosystems:

Webster's Collegiate Dictionary Definition: "Land or areas (tidal flats or swamps) containing much soil moisture."

USEPA Definition: "Those areas that are inundated or saturated by surface or ground water at a frequency and duration sufficient to support, and that under normal circumstances do support, prevalence of vegetation typically adapted for life in saturated soil conditions. Wetlands generally include swamps, marshes, bogs and similar areas." This is a regulatory definition of wetlands used by EPA and U.S. Corps of Engineers for administering Section 404 of the Clean Water Act (EPA, 1988).

U.S. Fish and Wildlife Service Definition: "Wetlands are lands transitional between terrestrial and aquatic systems where the water table is usually at or near the surface or the land is covered by shallow water." For purposes of this classification, wetlands must have one or more of the following three attributes:

(1) at least periodically, the land supports hydrophytes predominately; (2) the substrate is predominately undrained hydric soil; and (3) the substrate is nonsoil and is saturated with water or covered by shallow water at some time during the growing season of each year (Cowardin et al., 1979).

National Research Council Definition: "A wetland is an ecosystem that depends on constant or recurrent, shallow inundation or saturation at or near the surface of the substrate. The minimum essential characteristics of a wetland are recurrent, sustained inundation or saturation at or near the surface and the presence of physical, chemical, and biological features reflective of recurrent, sustained inundation or saturation. Common diagnostic features of wetlands are hydric soils and hydrophytic vegetation. These features will be present except where specific physicochemical, biotic, or anthropogenic factors have removed them or prevented their development" (Lewis, 1995).

Reddy and Patrick (1999) Definition: A much broader definition of a wetland is a biologically active soil or sediment in which the content of water in or the overlapping floodwater is great enough to inhibit oxygen diffusion into the soil/sediment and stimulate anaerobic (oxygen free) biogeochemical processes. This definition does not necessarily contain any reference to plants, but biologically active soil or sediment is necessary, which means the presence of an energy source (usually organic matter). We define wetland soils with or without aquatic vegetation.

As can be seen from this list of potential definitions, wetlands are typically limited to areas with a presence of hydrophytic vegetation. At present this definition is most commonly accepted, although there is considerable disagreement on the boundaries of developing wetlands from upland areas.

20.3 Wetland Soils

Wetland soils form in a variety of climates and parent materials. They represent a broad spectrum of morphological properties and taxonomic classes, and can be dominated by inorganic or organic materials. However, wetland soils have in common the condition of prolonged saturation. The term saturation refers to the condition of zero or positive hydraulic head, during which water would flow into unlined auger holes with a high proportion of pore space filled with water. This condition favors certain physical, chemical, biological, and morphological tendencies that help to distinguish wetland soils from their upland counterparts. A detailed discussion of redox phenomena is presented in Chapter 14 of *Handbook of Soil Sciences: Properties and Processes*.

20.3.1 Physical Properties

Saturation causes a number of physical changes in soils, including (1) reduced soil strength as a result of the weakening effect of water on forces binding soil particles. This physical effect can have several consequences: (i) root penetration by wetland plants is made easier and the soil is much easier to manipulate when wet, advantages that rice farmers have exploited for centuries, and (ii) trafficability of land is much poorer when flooded. (2) increased heat absorption by darkening soil color. Increased water content also increases heat conductivity and stabilizes soil temperate to a more constant value compared to upland soils. Sometimes saturated soils are cooler at the surface as a result of evaporation. In wetlands located in cooler climates, the presence of water may prevent soil temperatures from going below 0°C. (3) decreased soil bulk density (weight of dry soil per unit volume) as a result of flooding. Typical bulk densities of upland and wetland soils range between 0.3–1.5 and 0.1–1.0 g cm^{-3}, respectively, with low values typically observed in organic soils. This is due to high water absorption capacity of organic matter and destruction of soil aggregates.

20.3.2 Chemical and Biological Properties

Wetland soil chemistry is strongly influenced by the chemical reduction normally associated with saturation. Upland soils can be transformed into wetland soils as a result of excessive rainfall, rising water table, and high oxygen demand in the soil. Under these conditions, oxidized chemical species are chemically reduced as a result of abiotic and biotic processes. Similarly, improving the drainage of wetland soils can result in the oxidation of many of the reduced compounds in the soils either by chemical or biochemical reactions (Figure 20.1). The relative abundance of oxidized and reduced chemical species is, therefore, an indicator of the degree of wetness or anaerobic conditions in soils. The elements Fe and Mn are soil components that are particularly influenced by oxidation and reduction. Examples of Fe-bearing minerals likely to be most stable under the reduced conditions of wetlands include pyrite (FeS$_2$), siderite (FeCO$_3$), and vivianite (Fe$_3$(PO$_4$)$_2$ 8H$_2$O). The Fe in these minerals is in reduced (Fe(II), "ferrous") form. The dominant form of Fe under aerobic conditions is Fe(III) or "ferric" form. Common Fe minerals in relatively oxidized soil environments include goethite and lepidocrocite (both FeOOH), and hematite (Fe$_2$O$_3$).

Redox potential is used to measure the intensity of soil anaerobic conditions. Analogous to pH (which measures H$^+$ activity),

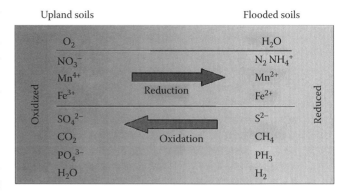

FIGURE 20.1 Predominant forms and changes of key biogeochemical compounds found under aerobic and anaerobic conditions. (From Reddy, K.R., and R.D. DeLaune. 2008. Biogeochemistry of wetlands: Science and applications. CRC Press, Boca Raton, FL.)

redox potential (Eh) measures electron (e⁻) activity in the soil. Redox potential is defined as the tendency of a pair of chemical species to undergo a transfer of electrons, with one species accepting electrons (reduction) and the other donating electrons (oxidation). Redox reactions involve oxidants, reductants, protons, and electrons, as shown below:

$$a(\text{Oxidants}) + b\text{H}^+ + n e^- = c(\text{Reductants}) + d\text{H}_2\text{O} \quad (20.1)$$

The following Nernst equation shows the thermodynamic relationship between redox potential (Eh) and the oxidation–reduction reactions:

$$Eh = E^\circ - \frac{RT}{nF} \ln \frac{[\text{Reductants}]^c}{[\text{Oxidants}]^a [\text{H}^+]^b} \quad (20.2)$$

where
- Eh is redox potential
- E° is the standard electrode potential
- R is the gas constant (8.314 J K⁻¹ mol⁻¹)
- T is the absolute temperature (K)
- F is the Faradays constant (9.65 × 10⁴ C mol⁻¹)
- n is the number of electrons transferred
- Brackets denote the activity of chemical species in mol L⁻¹

Redox reactions are critical because they regulate the fate of many chemical constituents in wetlands. The Eh values of wetland soils range from +700 to −300 mV (Figure 20.2). Negative values represent high electron activity and intense anaerobic conditions typical of permanently waterlogged soils. Under these conditions, there is a low potential for transfer of electrons between oxidized and reduced species, due to lack of oxidized species, such as oxygen. Positive values represent low electron activity and aerobic conditions or moderately anaerobic conditions typical of wetlands in transition zone. Under these conditions, there is a greater potential for electron transfer, due to the presence of oxidized species such as oxygen, nitrate, and oxidized forms of Fe and Mn.

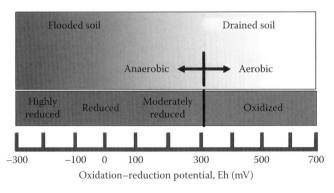

FIGURE 20.2 Association between oxidation–reduction potential and terms used to describe the degree of reduction in a range of flooded and drained soil conditions. (From Reddy, K.R., and R.D. DeLaune. 2008. Biogeochemistry of wetlands: Science and applications. CRC Press, Boca Raton, FL.)

An important redox-related process in wetland soils of marshes involves the formation and potential transformation of the mineral pyrite (FeS₂). Saltwater marsh soils tend to have neutral pH and to support salt tolerant plants (Ponnamperuma, 1972). Pyrite formation occurs as a result of reduction of SO_4^{2-} contained in seawater, high concentrations of Fe(II) in the sediments, and rapid accumulation of organic matter that promotes reduction reactions.

$$\text{CH}_2\text{O}_{(\text{organic matter})} + \text{H}_2\text{O} \rightarrow \text{CO}_2 + 4\text{H}^+ + 4e^- \quad (20.3)$$

$$\text{Fe(OH)}_3 + e^- + \text{H}^+ \rightarrow \text{Fe(OH)}_2 + \text{H}_2\text{O} \quad (20.4)$$

$$\text{SO}_4^{2-} + 6e^- + 8\text{H}^+ \rightarrow \text{S}^\circ + 4\text{H}_2\text{O} \quad (20.5)$$

$$\text{S}^\circ + 2e^- + 2\text{H}^+ \rightarrow \text{H}_2\text{S} \quad (20.6)$$

$$\text{Fe(OH)}_2 + \text{H}_2\text{S} \rightarrow \text{FeS} + 2\text{H}_2\text{O} \quad (20.7)$$

$$\text{FeS} + \text{S}^\circ \rightarrow \text{FeS}_2 \,(\text{pyrite}) \quad (20.8)$$

The drainage of pyritic marsh soils poses an environmental problem, because oxidation of pyrite to ferric hydroxide and sulfuric acid results in severe acidity (to pH less than 2) and dissolution of metal-bearing soil minerals. Bacteria involved in oxidizing FeS₂ are *Thiobacillus ferrooxidans* and *Thiobacillus thiooxidans*.

The chemical and biological processes of wetland soils are strongly mediated by influences and adaptations of the living communities, as exemplified by the redox processes discussed above. The low oxygen environment typical of wetlands inhibits aerobic microbial activity while stimulating activity of facultative and obligate anaerobes, and favors growth of hydrophytic vegetation adapted to living under anaerobic conditions. The presence and types of wetland and aquatic vegetation reflect the degree of soil wetness and intensity of anaerobic conditions.

20.3.3 Organic Soils

Soils vary greatly in their natural organic matter content. However, organic matter tends to accumulate in wetland soils to a greater extent than upland soils because of a high rate of production relative to the rate of decomposition (Mausbach and Richardson, 1994). Thick, dark, organic rich surface layers consisting of slightly to highly decomposed plant remains are therefore common for wetland soils. If these organic rich layers are thick enough such that they essentially comprise the greatest portion of the soil zone, they are referred to as organic soils. The distinction between organic and mineral soils is an arbitrary one, in that depth and organic C content must be specified by a soil taxonomic system. The United States Department of Agriculture (USDA) soil taxonomy (Soil Survey Staff, 1999) specifies an order of organic soils, Histosols, which is defined as meeting specific depth requirements (in most cases, 40 cm or more) of organic soils material. Organic soil material, in turn, must contain at least 12% organic C if no clay is present, and up to 18% if 60% clay is present (i.e., proportional increase in organic C requirement with increasing clay content). If a layer consisting of organic soil material is thick, but not thick enough to qualify as a Histosol, then it may meet the criteria for a Histic Epipedons ("epipedon" refers to a diagnostic surface or near surface horizon as defined within the USDA soil taxonomic system). Both Histosols and Histic Epipedons are, with some exceptions (e.g., organic matter overlying shallow bedrock), mainly restricted to wetlands. Most organic soils are formed from the accumulation of detritus from hydrophytic vegetation, and transformations of this material into stable humic substances. The organic matter in organic soils is generally most highly decomposed near the soil surface.

The three most extensively occurring suborders of Histosols are distinguished based on the amount of identifiable plant material:

Saprists—about two-thirds of the material well decomposed (muck) and less than one-third of the plant material identifiable (muck);

Hemists—about one-half of the material is well decomposed and the other half contains identifiable plant material (mucky peat);

Fibrists—about one-third of the material is well decomposed and greater than two-thirds of the plant material is identifiable (peat).

A fourth suborder, Folists, is less wet than other Histosols. Folists commonly form in decomposing leaves and twigs that overlie rock or fragmented materials (gravel, stones, etc.).

Organic soils are distinguished from mineral soils by processes and conditions affecting the rate and duration of plant detrital accumulations. Long-term dominance by plants that produce abundant below surface biomass under quiescent, shallow water conditions favors the formation of organic soils. Under this scenario, soils rapidly attain intense anaerobic conditions, with redox potential dropping to less than −200 mV within a few days after flooding. Organic matter production rates far exceed the oxidation rates, resulting in organic matter accumulation, which can be substantial. For example, the sawgrass-derived organic deposits of south Florida have reached depths of several meters in some areas over about a 5000 year period.

Organic soils are far less extensive than mineral soils, but where they do occur they are commonly of considerable economic importance. Many have been artificially drained for agricultural purposes. Under drained conditions, aerobic biological oxidation of organic matter results in rapid rates of soil subsidence, with rates of about 2–3 cm year^{-1} in the Everglades (Stephens, 1969; Snyder, 1987). The rate of soil subsidence can be reduced by implementing short-term (1–2 month) or long-term anaerobic conditions, such as through flooding or high water table management strategies.

20.3.4 Mineral Soils

Wetland soils that do not meet the taxonomic requirements for organic soils (e.g., Histosols) are mineral soils. The chemical reduction that accompanies saturated conditions generally results in features that are far more evident in mineral than organic soils, due to the masking effect (darkening) of organic matter for the latter. Reduction promotes dissolution of redox sensitive Fe and Mn oxides in soils, enabling their mobilization in accordance with principles of mass flow and diffusion until they encounter an oxidized zone where they reprecipitate as oxides. This redox-induced redistribution provides visual evidence of saturation and of redox gradients within the soil matrix, since both Fe and Mn are strong coloring agents (shades of yellow, brown, red, and black) in their oxide forms. Gray colored zones, in many cases, are attributable to reduction and depletion of these metals.

The process of gray color formation in mineral soils is sometimes referred to as "gleying" or "gleization." (Glossary of Soil Science Terms, 1997). Gray coloration corresponds to "low chroma" (<2) designations using the conventional Munsell notation for color characterization. A soil horizon dominated by redox-related gray colors is given the subordinate designation "g" (e.g., Btg). High chroma colors of various hues (e.g., browns or reds) in zones intermixed with gray colored zones are usually attributable to the oxidation, and subsequent concentration, of these metals. Over time, these zones of concentration may harden to form nodules and concretions. However, such hardened features, unless they have gradual or diffuse boundaries with the surrounding matrix, are indicative of relict rather than contemporary saturation. Features that reflect current wet–dry cycles tend to remain soft.

Historically, the patterns of gray in zones of periodic saturation were called "gray mottles." Gray coloration was used as a diagnostic criterion for saturation in both formal (e.g., in soil taxonomy) and informal (e.g., soil interpretations) contexts in soil science. However, soil scientists recognized that gray coloration is not always the result of chemical reduction. For example, gray colors are sometimes "inherited" from light colored parent material. Also, carbonates and clean sand grains tend to impart gray colors to soils. Hence, a new terminology was sought that would more explicitly and precisely convey the interpretations of saturation and reduction. Such a terminology was developed by the Committee on Soils with Aquic Moisture Regimes (Bouma, 1991). Interpretive morphological terms recommended by the committee were formally adopted for use in a revision of USDA soil taxonomy (Soil Survey Staff, 1999). Collectively, they are referred to as redoximorphic features. Redoximorphic features are now identified in standard soil descriptions, and include the following (Vepraskas, 1992):

1. Redox concentrations: Bodies interpreted as redox-related concentrations of Fe and Mn.
   a. Nodules and concretions: Partially hardened bodies. Concretions have concentric layers in cross-section, while nodules have a uniform internal fabric.

b. Masses: Soft bodies with reddish or brownish colors.
c. Pore linings: Zones of Fe and/or Mn accumulation along pore surfaces, as inferred from coloration.
2. Redox depletions: Bodies of low chroma (gray colors) corresponding to zones where (i) Fe and Mn oxides have been depleted through reduction, and in some cases, (ii) clay has been depleted by mobilization due to the loss of the oxide cements.
3. Reduced matrices: Soil matrices where low chroma is the result of chemical reduction of Fe, but not total depletion of Fe. A color change resulting from Fe oxidation occurs within 30 min.

Mineral soils, like organic soils, can have thick, dark surface horizons attributable to organic matter accumulation. These horizons may consist entirely of mineral soil material (i.e., below the organic C requirement for organic soil material). For example, many soils of semiarid prairie regions (e.g., Mollisols, USDA taxonomic system) have such surface horizons (called Mollic Epipedons) that are attributable to the climax grass vegetation rather than wetness per se. Alternatively, mineral soils can have organic soil material thick enough to qualify as Histic Epipedons. The distinction between the Mollic Epipedon and a Histic Epipedon, both dark and thick, can be important in wetland delineation (see Section 20.3.5) because the former is not necessarily a wetland indicator while the latter is a very reliable one. However, dark mineral surface horizons can be indicators of wetness in conjunction with immediately underlying redoximorphic features. In humid climates, for instance, surface horizons tend to be thicker and darker in wetlands.

20.3.5 Soils and Wetland Delineation: Hydric Soils Concept

Increased awareness of the important ecological role of wetlands has stimulated policies promoting wetland preservation. It has hence become necessary to delineate wetlands from uplands in a consistent and scientifically sound fashion for jurisdictional purposes. Soils are one of the three key components conventionally used to establish wetland boundaries, the others being vegetation and hydrology (Cowardin et al., 1979). Each component is ideally assessed independently. The soil assessment is particularly critical because soils are the most stable of the three components; for example, vegetation can be quickly altered and hydrology is sensitive to seasonal climatic fluctuations. The influences of prolonged saturation on soil properties, as previously discussed, are the crux of wetland soil delineation. Recent efforts have been made to document and catalog soil properties that are most consistently associated with wetland vegetation and hydrology.

"Wetland soils" could simply be defined as soils that occur in wetlands. However, this would be a circular definition if we intend to use soil characteristics as independent criteria in jurisdictional wetland delineation. Delineating wetland soil boundaries requires the specification of soil criteria that document the types of saturated conditions associated with wetland ecosystems. The term "hydric soils" is commonly used in jurisdictional language to apply to soils that meet such criteria. The USDA-Natural Resources Conservation Service (NRCS), in conjunction with the National Technical Committee for Hydric Soils, has defined hydric soils as "…soils formed under conditions of saturation, flooding, or ponding long enough during the growing season to develop anaerobic conditions in the upper part" (Federal Register, July 13, 1994). The growing season is the period of the year when soil temperature and moisture are most favorable for microbial activity, and hence anaerobiosis if the soil is saturated. Hydric soils, as defined, would tend to be wet even during the season when evapotranspiration loss is at a maximum.

The USDA hydric soils definition serves as a guideline for local technical limits used in jurisdictional wetlands delineation. However, the practical delineation of hydric soils generally requires some indirect assessment and professional judgment based on readily observable indicators associated with near surface saturation. Field indicators of hydric soils have evolved through collective observations and consensus, and have been formally cataloged by the USDA-NRCS (USDA-NRCS, 2006).

The USDA hydric soil indicators are specified as applying to (i) all soils, (ii) sandy soils, and (iii) loamy and clayey soils. Some are applicable to all USDA Land Resource Regions (USDA, 1981), while others are restricted to certain regions. Most of the indicators reflect pronounced organic matter accumulation at the surface, immediately underlain by redoximorphic features or other local subsurface features consistently indicative of saturation and anaerobiosis.

Delineation of hydric soils requires intensive soil investigation at wetland boundaries, since the interior areas are commonly wet enough that the hydric status is indisputable (i.e., the soils in the wettest areas have almost certainly developed "… anaerobic conditions in the upper part"). The uncertainty of hydric soil identification generally increases with proximity to the upland boundary, increasing the challenge of consistent delineation. The indicators that discriminate most effectively at the boundary are, therefore, the most useful. Soils examination generally requires only a shallow excavation (e.g., about 50 cm) using a tiling spade, since evidence for near surface saturation is being sought.

Detailed interpretive "hydric soils criteria" have been developed by USDA-NRCS (Federal Register, February 24, 1995) for the purpose of identifying soil series that could potentially be hydric soils. Series meeting the criteria are placed on a National Hydric Soils List (USDA, 1991). These hydric soils criteria encompass taxonomic and interpretive criteria linked with individual soil series within the USDA-NRCS soils data base. The criteria are not used in field delineation of hydric soils. The occurrence of a soil series on the USDA Hydric Soils List means only that a soil fitting the series in the field is potentially hydric. In effect, hydric and nonhydric soils could both fit the same series, since hydric soil status per se is not a series criterion.

20.4 Redox Gradients

In wetlands, O_2 is introduced into soils by diffusion through the floodwater, from photosynthetic O_2 production, by diffusion and mass flow from the atmosphere through plants into root zone, and by fluctuations in water table depth (Figure 20.3). Oxygen is crucial in regulating plant and microbial respiration rates, speciation, mobility, and bioavailability of chemicals in soils. However, when soils are flooded, O_2 exchange between the air–water–soil phases is severely curtailed, where O_2 diffusion through water is about 10^4 times slower than in air. Dissolved O_2 present in the soil porewater is rapidly consumed during aerobic respiration and approaches zero within a few hours depending on soil O_2 demand. Although not related to saturated soil conditions, application of O_2-demanding wastes (animal wastes, composts, sewage sludge) or ammoniacal fertilizer can also result in O_2 depletion in the soil.

When O_2 is limiting, there is a shift in a number of properties that differentiate wetland and upland soils. Typically, there is an accumulation of organic matter, which may reach several meters of peat in some Histosols. Aerobic microbial activity that uses O_2 as the primary electron acceptor during catabolic activities is replaced by predominantly anaerobic activity, which obtains energy by oxidizing organic and inorganic compounds through several intermediate steps. The electrons released from these compounds pass through an electron transport chain containing several electron carriers that transfer electrons from energy substrates to terminal electron acceptors. In wetlands, O_2 cannot always move into the soil rapidly enough to take care of the biological O_2 demand of organisms because of pore blockage by water. Although O_2 supply to the soil is restricted as a result of flooding, its demand remains high. These conditions result in the development of two distinctly different soil layers: (1) an oxidized or aerobic surface soil layer containing O_2, and (2) an underlying reduced anaerobic soil layer where no free O_2 is present (Figures 20.4 and 20.5). Diffusion of O_2 through floodwater

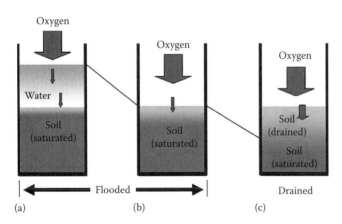

FIGURE 20.3 Patterns of oxygen penetration/movement into soils with varying degrees of saturation ranging from flooded (a) to surface saturated (b) to drained conditions (c). (From Reddy, K.R., and R.D. DeLaune. 2008. Biogeochemistry of wetlands: Science and applications. CRC Press, Boca Raton, FL.)

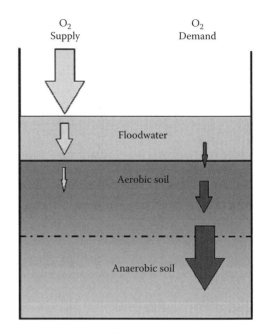

FIGURE 20.4 Comparison of relative differences between oxygen supply and oxygen consumption leading to anaerobic conditions in flooded soils. (From Reddy, K.R., and R.D. DeLaune. 2008. Biogeochemistry of wetlands: Science and applications. CRC Press, Boca Raton, FL.)

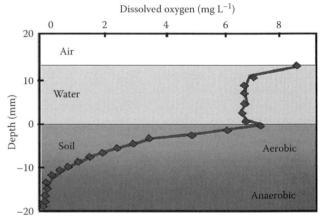

FIGURE 20.5 Vertical profiles of oxygen concentration in flood- and porewaters of a typical wetland soil. (From Reddy, K.R., and R.D. DeLaune. 2008. Biogeochemistry of wetlands: Science and applications. CRC Press, Boca Raton, FL.)

maintains the aerobic interface, which can vary in thickness from a few mm to 2–10 cm. In wetlands with dense layers of periphyton, production of O_2 during photosynthesis may result in a thick aerobic zone during the day, while consumption of O_2 during respiration may convert this layer to an anaerobic zone during the night. Thus, redox gradients at soil/floodwater interface can have diurnal fluctuations. The juxtaposition of aerobic and anaerobic soil layers and the temporal transitions of reducing and oxidizing conditions has major implications for the chemical and biological reactions in wetlands such as nitrification/denitrification, iron oxide-regulated precipitation of P, and oxidation of CH_4, sulfides, and other reduced species.

The depth of the oxidized and anaerobic zones where a given electron acceptor is stable depends on the inflow of electron acceptors, the availability of organic and inorganic substrates, and hydrologic conditions. The sequential reduction of electron acceptors as a function depth follows the order of O_2 reduction at the soil/floodwater interface, followed by other electron acceptors. The soil depths at which these reductions occur are depicted in Figure 20.6. It should be noted that different electron acceptors are utilized in the soil simultaneously at different soil depths. Thus, wetlands can be characterized by measuring the concentration of reduced species (Mn(II), Fe(II), S^{2-}) or oxidized species such as (O_2, NO_3^-, and SO_4^{2-}).

Wetland plants have unique morphological and physiological adaptations for growing in anaerobic soil conditions. One of these adaptations is the development of internal air spaces (aerenchyma) for transporting O_2 into the root zone (Figure 20.7). Depending on plant species, air spaces can occupy up to 60% of the total tissue (Brix, 1996). Oxygen transport in wetland plants occurs through molecular diffusion as a result of partial pressure gradients established within the system. In addition, temperature- and humidity-induced pressurization can move gases through mass flow (Armstrong and Armstrong, 1990; Brix et al., 1992).

Rates of O_2 transport by *Phragmites* range from 0.02 to 12 g m^{-2} day^{-1} (Brix, 1996). Studies have demonstrated the importance of this O_2 release into the root zone for supporting a wide array of processes including heterotrophic respiration and nitrification (Reddy et al., 1989a, 1989b). The wide range of reported O_2 transport rates has been attributed to seasonality and use of different experimental techniques. Low temperatures in colder climates can decrease internal pressurization, thus decreasing overall O_2 transport into the root zone.

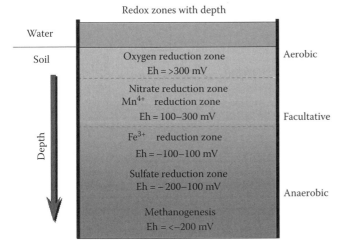

FIGURE 20.6 Typical depth distributions of microbial metabolic pathways in a vertical profile of a flooded soil or sediment.

20.5 Carbon

When compared to upland systems, most wetland soils show an accumulation of organic matter, and therefore, wetlands function as global sinks for C. The C cycle for a typical wetland is shown in Figure 20.8. Accumulation of organic C in wetlands is primarily a result of the balance of two processes: C fixation through photosynthesis and C loss through decomposition. Some natural and constructed wetlands receive significant internal and external loadings of organic C, associated with surface water runoff, carbonaceous wastewater, and diurnal tidal events. Anthropogenic inputs of organic C, including oil spills in coastal wetlands, and toxic organics in agricultural and industrial runoff, have not been well quantified in wetlands. Except for indirect effects on primary production, these inputs probably play a relatively minor role in the overall C budget of most wetlands.

Rates of photosynthesis in wetlands are generally higher than other ecosystems. For example, Houghton and Skole (1990) estimated mean primary productivity of a wetland ecosystem was approximately 1300 g C m^{-2} year^{-1}, which is higher than many terrestrial tropical and subtropical ecosystems. In contrast, rates of decomposition in wetlands are typically lower than other systems due to the presence of anaerobic conditions. This imbalance in inputs and losses results in the net accumulation of organic matter in most wetlands. Net accumulation of C in some peatland ecosystems was found to be in the range 11–105 g C m^{-2} year^{-1} (Schlesinger, 1997). Rates are highly variable between wetlands depending on types of soil and vegetation, hydrology, nutrient inputs, extent of disturbance, presence of toxic substances, solar radiation, length of growing season and other

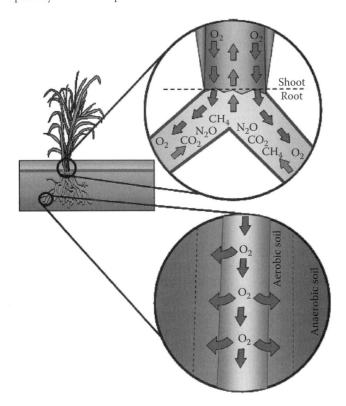

FIGURE 20.7 Schematic view of gas flow pathways in the stems and roots of plants adapted to wetland soil conditions. (From Reddy, K.R., and R.D. DeLaune. 2008. Biogeochemistry of wetlands: Science and applications. CRC Press, Boca Raton, FL.)

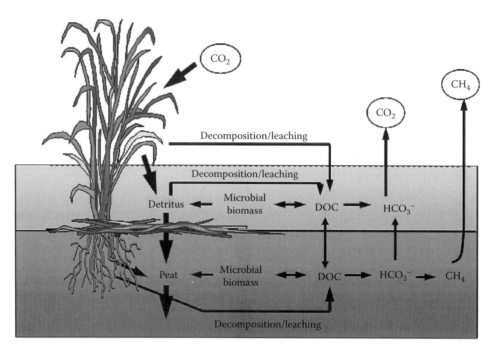

FIGURE 20.8 Schematic diagram of the general carbon cycle in wetlands. (From Reddy, K.R., and R.D. DeLaune. 2008. Biogeochemistry of wetlands: Science and applications. CRC Press, Boca Raton, FL.)

environmental factors. In addition to maintaining proper functioning of wetlands, organic matter storage also plays an important role in protecting other ecosystems and the biosphere. For example, because organic matter immobilizes nutrients such as N, P, and S, its accumulation in wetlands reduces eutrophication of downstream aquatic systems.

Organic C undergoes complex cycling in wetland systems and the fate of C depends on the specific type of molecule and environmental conditions. Easily degradable (labile) fractions are decomposed to inorganic constituents while recalcitrant pools are accreted as new peat layers. Organic matter associated with plant, algal, and microbial biomass in wetlands consists of complex mixtures of substances, including polymers (α-cellulose, hemicellulose, tannins, and lignins, proteins), water soluble monomers (amino acids, sugars, and nucleotide bases), and other extractable components (lipids, waxes, oils). Following frost or seasonal senescence, nonhumic substances are deposited in the water column and surface soil. Root exudates containing ethanol, carbohydrates, and amino acids may also contribute significant amounts of C to subsurface soils (Mendelssohn et al., 1981).

Labile organic C fractions undergo multistep conversions to inorganic constituents including (1) abiotic leaching of water soluble components and fragmentation of tissues into small pieces (<1 mm), (2) extracellular enzyme hydrolysis of biopolymers (nucleic acids, proteins, and carbohydrates) into monomers, and (3) aerobic and anaerobic catabolism of monomers by heterotrophic microorganisms (Figure 20.9). Abiotic leaching of algal and plant biomass is largely complete within days to months after deposition into water, but depends on amount of particulate and types of structural compounds. For example,

Gaur et al. (1992) measured leaching losses of up to 33% of water hyacinth mass after 4 days immersion in water column, while Benner et al. (1985) measured only 10%–20% losses of lignocellulosic components of plants in 1 month. Particulate materials remaining after initial leaching are typically fragmented by meiofauna before undergoing further decomposition. This step is critical because surface area is increased, allowing organisms and enzymes to penetrate tissues. Enzyme hydrolysis is generally considered the rate-limiting step in organic matter decomposition (Sinsabaugh, 1994).

The litter layer is a major support structure being a source of energy and nutrients and containing high numbers (10^9–10^{10} g^{-1} dry matter) of bacteria and fungi (Kjoller et al., 1985). Microbial biomass accounts for a significant amount of C in wetland substrates (DeBusk, 1996), and nutrients released from organic matter decomposition are available for plant uptake and growth. Enzymes such as cellulases and proteases excreted by fungi and bacteria hydrolyze large molecular weight biopolymers into oligomers and monomers that can be assimilated by microorganisms (Figure 20.9). The lack of available O_2 in wetland soils further limits the activity of oxidoreductase enzymes such as peroxidases and phenoloxidases involved in lignin oxidation. For this reason, lignin degradation is most prevalent in aerobic zones such as soil–water and root–soil interfaces. Major fungal genera involved in decomposition in wetland systems include *Alternaria, Cylindrocarpon, Cladosporium, Penicillium, Fusarium, Trichoderma, Alatospora, Tetracladium, Helicodendronm,* and *Helicoon* (Kausik and Hynes, 1971; Given and Dickenson, 1975; Kjoller and Struwe, 1980). Most fungi are aerobes; therefore, bacteria generally dominate decomposition under anaerobic conditions (Alexander, 1977). Important bacterial genera involved in

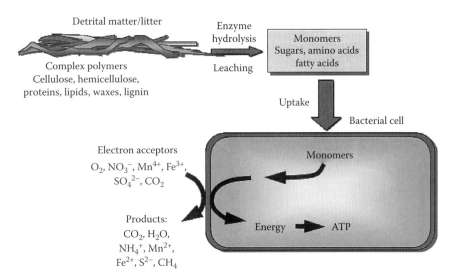

FIGURE 20.9 Diagram illustrating the flow of carbon from organic matter and the conversion of electron acceptors through microbial metabolism. (Adapted From Reddy, K.R., and R.D. DeLaune. 2008. Biogeochemistry of wetlands: Science and applications. CRC Press, Boca Raton, FL.)

extracellular enzyme hydrolysis of organic matter in wetlands include *Cytophaga*, *Vibrio*, *Achromobacter*, *Bacillus*, *Micrococcus*, *Chromobacterium*, *Streptomyces*, *Arthrobacter*, *Actinomyces*, *Clostridia*, *Pseudomonas*, *Flavobacterium*, *Bacteriodes*, *Eubacterium*, and *Peptostreptococcus* (Given and Dickenson, 1975; Molongoski and Klug, 1976; Wheatly et al., 1976).

Production and activity of extracellular enzymes are affected by a number of environmental factors including nutrients, pH, O_2 supply, humic and mineral substances, and inorganic cations. For example, Sinsabaugh et al. (1993) found that production of proteases and phosphatases was enhanced in N- and P-limited wetland systems, while lignocellulase activity and organic C mineralization were promoted under nonlimiting nutrient conditions. Fog (1988) concluded from a study of >60 sites that cellulose degradation was enhanced and lignin degradation was inhibited by inorganic and organic N amendments, probably resulting from N repression of phenoloxidase synthesis. Gordon and Millero (1985) demonstrated that biodegradation rates of low molecular weight organic acids and sugars were decreased due to sorption onto hydroxyapatite mineral surfaces. Goodwin and Zeikus (1987) found that anaerobic decomposition processes in acidic (pH 20.9) bog sediments are decreased, but that C and electron flow pathways are similar to those in neutral sediments. Kim and Wetzel (1993) demonstrated that humic substances complexed and curtailed the activity of extracellular enzymes, which was relieved by the presence of divalent cations such as Ca^{2+} and Mg^{2+}. They speculated that this phenomenon may explain reduced decomposition in soft compared to hard water aquatic systems.

After enzyme hydrolysis, small molecular weight compounds are taken up and utilized as C and energy sources by heterotrophic microorganisms. A multitude of different microorganisms may be involved in this terminal step of decomposition, which depends largely on availability of electron acceptors (Figure 20.10). Diverse types of microorganisms couple oxidation of organic C substrates (electron donors) with the reduction of electron acceptors to generate energy (ATP) required for growth. The most common electron acceptors include O_2, NO_3^-, Mn(IV), Fe(III), SO_4^{2-}, and CO_2, which enter wetlands through both internal and external inputs. For example, internal inputs include oxidation of chemical species (NH_4^+, H_2S, Fe(II), and Mn(II)) that diffuse from anaerobic to aerobic soil zones. External inputs of electron acceptors include atmospheric O_2, seawater (SO_4^{2-}), surface water runoff (NO_3^-), and atmospheric deposition. Most wetlands contain several electron acceptors, so competition exists for electron donors between microbial groups. Due to differences in thermodynamic energy yield, electron acceptors are utilized to oxidize organic compounds in the order $O_2 > NO_3^- > Mn(IV) > Fe(III) > SO_4^{2-} > CO_2$. Organisms that derive the most energy and have the fastest degradation kinetics outcompete other groups. This thermodynamic basis of bacterial competition explains both the temporal pattern after flooding and the vertical stratification of microbial activity and chemicals often observed in wetland soil profiles (Figure 20.11).

For these reasons, aerobic bacteria outcompete anaerobes. For example, DeBusk (1996) and D'Angelo and Reddy (1999) determined that aerobic C mineralization was about three times faster than under anaerobic conditions. Because recently deposited plant and algal detritus occurs typically in the water column and at the soil surface, most (70%–80%) decomposition occurs by aerobic fungi and bacteria (Kjoller and Struwe, 1980). Small zones at the root surface may also promote aerobic microbial activity (Schipper and Reddy, 1996). In contrast, root material and aboveground plant and algal biomass that are buried undergo anaerobic decomposition where microorganisms must utilize alternate electron acceptors to oxidize organic matter. The slower rates of these decomposition processes partially explain why wetlands accumulate organic matter more than upland systems. When wetland soils are drained, however, rapid rates of aerobic decomposition result in soil subsidence (e.g., 2–3 cm year^{-1}

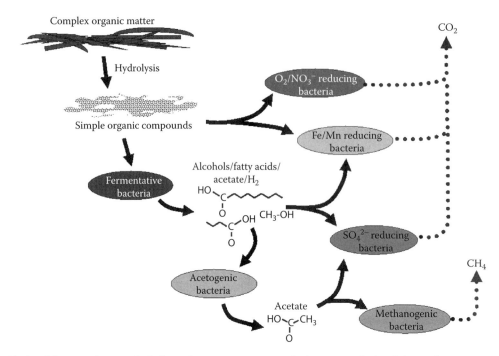

FIGURE 20.10 Idealized diagram showing the linkages between various microbial groups in the breakdown of organic matter to form the end products of respiration and methanogenesis.

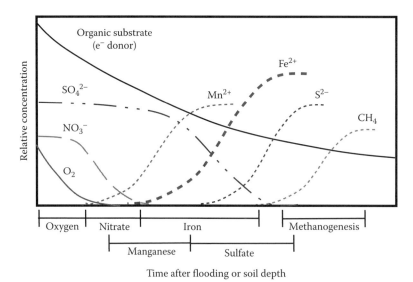

FIGURE 20.11 Hypothetical pattern in concentrations of various oxidized and reduced chemical species following flooding of a soil or with depth into a flooded soil/sediment profile. (From Reddy, K.R., and R.D. DeLaune. 2008. Biogeochemistry of wetlands: Science and applications. CRC Press, Boca Raton, FL.)

in the Florida Everglades) (Snyder, 1987). Under such drained conditions, wetlands act as a source of C to the rest of the biosphere.

Carbon compounds that are recalcitrant to aerobic and anaerobic decomposition tend to accumulate in wetlands either as undecomposed plant tissues (peat) or humic substances (Figure 20.12). Formation of humic substances catalyzed by phenoloxidase enzymes in soils is thought to involve condensation reactions between reactive phenolic groups of tannins and lignins with water soluble nonhumic substances (Francois, 1990).

In the absence of O_2, humic substances are resistant to decomposition, and so represent a significant C and nutrient storage in wetlands. Under drained conditions, humic substances are rapidly degraded, which releases nutrients to the bioavailable pool, thereby affecting downstream water quality. This humification process also accounts for the formation of large molecular weight and heterogenous humic substances that contain significant amounts of N, P, S in their structures. For example, P contained in humic substances may make up between 40% and 90% of the total P in Histosols (DeBusk et al., 1994).

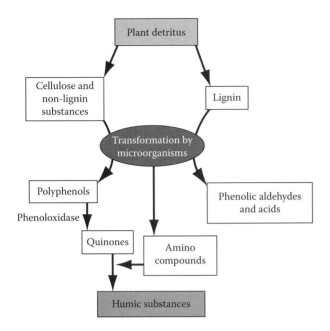

FIGURE 20.12 Diagram of hypothesized reactions and processes resulting in the formation of humic substances from plant detritus. (From Reddy, K.R., and R.D. DeLaune. 2008. Biogeochemistry of wetlands: Science and applications. CRC Press, Boca Raton, FL.)

20.6 Nitrogen

Nitrogen is a major nutrient in wetlands and often controls primary productivity, especially under conditions of excessive P supply. Nitrogen is present in aquatic systems in organic forms and as inorganic forms such as dinitrogen gas (N_2), nitrate (NO_3^-), nitrite (NO_2^-), ammonium (NH_4^+), and trace gases such as N_2O. Organic N is derived primarily from amino-bonded C ($-NH_2$) such as amino acids and proteins, and nucleic acids and amino sugars. Generally, the sum of NO_3^-/NO_2^-, and NH_4^+ measured after filtration is taken to represent dissolved inorganic N (DIN). The difference between the total dissolved N of a system and its DIN is then defined as the dissolved organic N (DON). Particulate forms are removed through settling and burial, while the removal of dissolved forms is regulated by various biogeochemical reactions functioning in soil and water column (Figure 20.13). Traditionally, only NO_3^- and NH_4^+ have been considered biologically available; however, it has also been noted that organic forms of N, such as amino acids and other small molecular weight N compounds (e.g., urea) can be a significant source of N to both microbial/algal communities (e.g., Berman and Bronk, 2003; Linares and Sundback, 2006) and wetland macrophytes (e.g., Nasholm et al., 2009). This is important not only because of its role in nutrition in N limited systems, but also because the presence of organic forms of N (in particular, urea) has been shown to affect algal species composition (e.g., Heil et al., 2007).

Nitrogen is also present at various valence states ranging from −3 (NH_3) to +5 (NO_3^-). The wide range in oxidation numbers provides the possibility for several oxidation–reduction reactions. Relative rates of these reactions are affected by physicochemical and biological characteristics of soils, organic substrates, and the water column. Detailed reviews of N processes functioning in the soil and the overlying water column of wetlands were presented by Reddy and Patrick (1984), Bowden (1987), Johnston (1991), Howard-Williams and Downes (1993), and Reddy and D'Angelo (1994). Several N transformations effectively process inorganic N through nitrification and denitrification, ammonia volatilization, and plant and microbial uptake, and maintain relatively low levels of inorganic N in the water column. The extent of these processes usually increases with N loading.

Nitrogen can enter wetlands through a variety of pathways including atmospheric deposition, biological nitrogen fixation, or through surface- or groundwater inputs. Atmospheric deposition is highly variable and can contribute anywhere from ~0.005 g N m^{-2} year^{-1} in remote areas to >2 g N m^{-2} year^{-1} in areas near urbanized centers (Galloway et al., 2004). By comparison, the process of biological nitrogen fixation, which converts atmospheric N_2 gas into biologically available NH_4-N, can

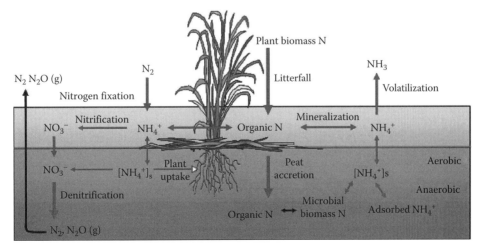

FIGURE 20.13 Schematic diagram of the general nitrogen cycle in wetlands. (From Reddy, K.R., and R.D. DeLaune. 2008. Biogeochemistry of wetlands: Science and applications. CRC Press, Boca Raton, FL.)

add >37 g N m^{-2} year^{-1} to N-limited wetland systems (reviewed by Inglett et al., 2004). In particular, cyanobacterial mats have been shown to exhibit rates of N$_2$ fixation of >74 g N m^{-2} year^{-1} (Howarth et al., 1988). Such high rates of fixation help contribute to the high productivity of these mats and other conspicuous periphytic forms.

Plants derive most of their N from soil porewater with only a small amount of the floodwater N being directly utilized. The ratio of tissue N to tissue P (N:P ratio) can be an effective indicator of N demand/limitation by wetland vegetation (Koerselman and Mueleman, 1996). Efficiency of N utilization (defined as increase in plant N per unit mass of available N) by aquatic vegetation is highly variable, depending on the type of wetland vegetation (woody versus herbaceous), nutrient loading, and temperature. The N-use efficiency of aquatic vegetation and the C/N ratio of plant litter decrease with nutrient loading (Shaver and Melillo, 1984; Reddy and Portier, 1987; Koch and Reddy, 1992). Elevation of temperature from 10°C to 25°C was shown to increase N-use efficiency by *Typha* spp. from 5% to 38% (Reddy and Portier, 1987).

In temperate climates, most N assimilation occurs during the growing season. During winter months, aboveground biomass is killed and accumulates as detrital tissue. At the same time, a significant portion of N is translocated to below ground biomass. Nitrogen assimilation by herbaceous vegetation is usually short term and usually cycled rapidly within the systems. Unlike herbaceous macrophytes, forested wetlands provide long-term storage in the form of woody biomass. However, a significant portion of nutrients stored in the leaves are returned to the forest floor, and eventually incorporated into the soil (Reddy and DeBusk, 1987). Because forest litter decomposition is slow, turnover periods are much longer. Measurement of plant tissue N and biomass can provide an indication of N removal efficiency by wetlands.

Breakdown of detrital tissue results in release of DON to the water column, most of which is resistant to decomposition. Under these conditions, water leaving wetlands may contain elevated levels of N in organic forms. Wetlands usually function as effective sinks for inorganic N, and as sources for organic N. Net release of NH$_4$-N by wetland soils is determined by the balance between ammonification and immobilization, which is controlled by the N requirements of microorganisms involved, the nature of organic N forms, and other soil and environmental factors (Figure 20.14). Because of the low N requirements of anaerobic microorganisms, wetland soils usually accumulate NH$_4$-N, which supports most of the N requirements of wetland plants and microbes. Organic N mineralization can be described as a function of the detritus C/N ratio, activity of extracellular enzymes (e.g., protease), microbial biomass, and soil redox conditions (McLatchey and Reddy, 1998). Stanford et al. (1973) also demonstrated a temperature dependence of organic N mineralization (with Q$_{10}$ values of about 2 between 15°C and 35°C), and that the effect of temperature differed between plant types (greater effect for *Sagittaria* compared to *Typha* tissues). These results suggest that the variety of wetland types and conditions may contribute to a great diversity of organic N mineralization rates and patterns in many systems.

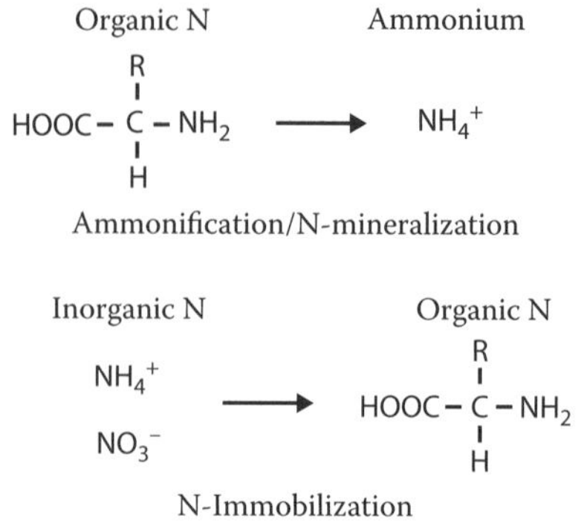

Ammonium released during mineralization has a variety of fates in wetlands (Figure 20.15). The positive charge of the NH$_4^+$ ion results in a significant attraction of this N species with soil particles (in particular, humic substances and clay minerals) having a net negative surface charge. This adsorption process can affect the availability and retention of nitrogen in systems and is comparatively different than that of NO$_3^-$, which is not retained on soil/sediment particles. Free NH$_4^+$ can also undergo a pH-controlled conversion to gaseous NH$_3$, which may rapidly be lost from the watercolumn via ammonia volatilization (Figure 20.16) [Reactions (20.9) and (20.10)].

$$NH_4^+ + OH^- \rightarrow NH_3(aq) + H_2O \qquad (20.9)$$

$$NH_3(aq) \rightarrow NH_3(g) \qquad (20.10)$$

FIGURE 20.14 Processes involved in the conversion between organic nitrogen and ammonium in wetland soils. (From Reddy, K.R., and R.D. DeLaune. 2008. Biogeochemistry of wetlands: Science and applications. CRC Press, Boca Raton, FL.)

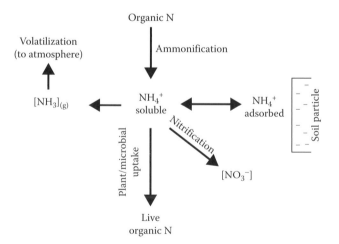

FIGURE 20.15 Diagram depicting the formation and potential fates of ammonium (NH$_4^+$) in wetland soils. (From Reddy, K.R., and R.D. DeLaune. 2008. Biogeochemistry of wetlands: Science and applications. CRC Press, Boca Raton, FL.)

Biogeochemistry of Wetlands

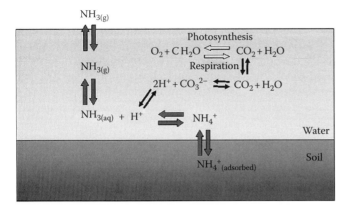

FIGURE 20.16 Schematic view of processes related to the flux of ammonium (NH_4^+) and the pH controlled processes relating to the process of ammonia volatilization. (From Reddy, K.R., and R.D. DeLaune. 2008. Biogeochemistry of wetlands: Science and applications. CRC Press, Boca Raton, FL.)

Ammonia volatilization is an abiotic reaction influenced by the physicochemical characteristics of a wetland water column. Ammonia volatilization is regulated by NH_3 concentration, temperature, vegetation density, air movement above the water surface, mixing in the water column, algal activity, and associated pH fluctuations. In wetlands, this process can play a significant role if the influent water contains high levels of NH_4-N, and photosynthetic activity by algae or submerged aquatic vegetation drives the pH above 8.5.

Ammonium oxidation (nitrification) is a two-step process mediated by aerobic chemoautotrophic bacteria, *Nitrosomonas* (NH_4^+ to NO_2^-) and *Nitrobacter* (NO_2^- to NO_3^-). These bacteria derive energy for biosynthesis from oxidation of NH_4^+ and NO_2^-, while using HCO_3^- as a C source.

$$NH_4^+ + 2O_2 \rightarrow NO_2^- \rightarrow NO_3^- + 2H^+ + H_2O \quad (20.11)$$

Since nitrification is an aerobic process, it occurs in the water column, surface aerobic soil layer, and root zone. The relative importance of these zones in overall nitrification depends on O_2 availability and NH_4-N concentration (Reddy and Patrick, 1984). Nitrification rates are reported to be in the range of 0.01–0.161 g N m^{-1} day^{-1} (mean = 0.048 ± 0.044) (Martin and Reddy, 1997). These values are lower than those reported for mineralization, suggesting that O_2 and NH_4-N availability limit nitrification. Nitrifiers compete with other heterotrophic and lithotrophic organisms for O_2. In wetland environments, O_2 is probably the primary regulator of nitrification.

Nitrate added to or produced in wetlands can be used during assimilatory or dissimilatory NO_3^- reduction. During assimilatory NO_3^- reduction, microorganisms and plants assimilate NO_3^- into their biomass after reduction to NH_4^+ and incorporation into monomers (amino acids) and biopolymers (proteins). Dissimilatory NO_3^- reduction is divided into two main pathways: denitrification (Reaction (20.12)) and dissimilatory NO_3^- reduction to ammonium (DNRA)(Reaction (20.13)). In both cases, NO_3^- is used as an electron acceptor for energy generation by anaerobic bacteria.

$$5C_6H_{12}O_6 + 24NO_3^- + 24H^+ \rightarrow 30CO_2 + 12N_2 + 42H_2O \quad (20.12)$$

$$C_6H_{12}O_6 + 3NO_3^- + 6H^+ \rightarrow 6CO_2 + 3NH_4^+ + 3H_2O \quad (20.13)$$

Denitrification occurs at higher Eh levels (200–300 mV) than dissimilatory reduction (<0 mV). Denitrification is typically the dominant pathway of NO_3^- removal in wetlands. Reported NO_3^- removal rates by wetlands are in the range of 0.003–1.02 g N m^{-2} day^{-1} (mean = 0.097 ± 0.139). In most wetlands, denitrification rates are usually limited by NO_3^- concentrations and diffusion rates of NO_3^- from aerobic to anaerobic sites (Martin and Reddy, 1997). Denitrification rates are usually higher in soils receiving steady loadings of NO_3^- than soils receiving low or negligible levels (Cooper, 1990; Gale et al., 1993). In a system with active denitrification, NO_3^- levels are usually maintained at low levels, thus measurement of NO_3^- in soil and water column may not provide reliable indication of this process. Since denitrification is mediated by heterotrophic microorganisms, its rate may also be regulated by available organic C (electron donor). Significant correlations have been observed between denitrification rates and available organic C (mineralizable organic C) (Burford and Bremner, 1975; Reddy et al., 1982; Gale et al., 1993).

The high energetic potential of using NO_3^- as an electron acceptor is not solely possessed by heterotrophic bacteria, and conversion of NO_3^- can also occur by additional pathways (reviewed by Burgin and Hamilton, 2007) (Figure 20.17). Many of these pathways have yet to be identified in wetlands; however, the potential still exists for these processes to account for significant reduction of NO_3^- in aquatic systems (e.g., systems with high levels of reduced Fe(II), S^0, and S^{2-}) This finding has challenged the classical view of the sole importance of denitrification in aquatic systems. Among the possible pathways, NO_3^- reduction to N_2 can be coupled to the lithotrophic oxidation of either reduced

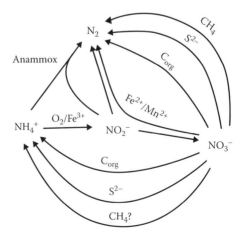

FIGURE 20.17 Diagram depicting the potential pathways of nitrate reduction in aquatic systems.

iron (Fe(II)) or sulfur (S^0, S^{2-}), or even methane. Alternatively, NO_3^- reduction via S^{2-} oxidation can occur through lithotrophic DNRA where the product NH_4^+ is not directly incorporated into biomass. Strongly reducing conditions with high contents of organic matter or high concentrations of reduced Fe and S are required for DNRA. As for lithotrophic denitrification, conditions may also exist for lithotrophic DNRA activity in wetlands with accumulations of reduced iron and sulfur.

Most recently, a novel pathway for N conversion to N_2 has been identified, which adds even more uncertainty to the aquatic N cycle. The process known as anaerobic ammonium oxidation (Anammox) has now been identified in a number of freshwater systems including lakes and wetlands. This process occurs in strictly anaerobic conditions, and couples the reduction of NO_2^- with the oxidation of NH_4^+ to produce N_2 gas. Anammox bacteria appear to be autotrophic, and the presence of organic matter slows the Anammox process. Another important requirement for the Anammox process is the presence of NO_2^-, which is rarely observed in most wetland systems. Nitrite accumulation is possible however, such as in cases of incomplete denitrification (e.g., lack of suitable organic substrates or high concentrations of NO_3^-) (Herbert, 1982; Smith et al., 1997) or partial ammonium oxidation (low O_2 availability or inhibition by ammonia) (Anthonsieu et al., 1976; Smith et al., 1997). Likewise, NO_2^- accumulation is a common trait of DNRA (Cole, 1988) resulting from either inhibitory effects of NO_3^- on NO_2^- reductase (Smith, 1982) or repression of this enzyme (Paul and Beauchamp, 1989). Higher than expected NO_2^- levels (100–200 µg N L^{-1}) have been reported in some rivers, particularly in watersheds receiving large inputs of nitrogenous pollution, and the NO_2^- accumulation corresponded with areas of high DNRA activity (Smith et al., 1995; Kelso et al., 1997). All of these studies suggest NO_2^- may be much more abundant than previously thought, and it is possible that NO_2^- levels may be sufficient to support alternative N cycling pathways in wetlands (such as Anammox).

Whether a wetland acts as a sink or source of N depends on the net effects of the input and loss processes described here. Several studies have used mass balance methods to estimate N retention within a wetland (DeLaune et al., 1989; White and Howes, 1994). These studies showed that N demand of aquatic vegetation, microbial, and periphyton communities is not met by external inputs alone, pointing to the role of internal cycling and turnover. Based on the pathways of N cycling, vegetation can play a significant role in the wetland N cycle by: (1) assimilating N into plant tissue, (2) releasing N through tissue mineralization, and (3) providing an environment in the root zone for nitrification–denitrification to occur. A few studies have quantified the role of remineralization of organic N during plant litter decomposition and belowground biomass turnover. Values of 54%–95% of plant uptake demand being met through this process have been reported (Hopkinson and Schubauer, 1984; DeLaune et al., 1989; White and Howes, 1994). In a greenhouse study, fertilizer inorganic N was the major source during the early part of the cattail growing season, while soil organic N mineralization supplied plant N requirements during the latter part of the growing season (Reddy and Portier, 1987).

The predominance of organic matter accumulation results in the dominant mechanism of N retention in wetlands being through organic N accumulation. Inputs of NO_3^- and NH_4^+ are rapidly taken up or lost via nitrification–denitrification processes. Nitrogen availability to vegetation is thus frequently controlled by the rate of organic N mineralization, and the relative magnitude of N loss mechanisms such as denitrification and NH_3 volatilization. The limiting processes regulating N loss from wetlands by these mechanisms are NH_4^+ diffusion from anaerobic to aerobic zones and nitrification in aerobic zones. Denitrification is limited by NO_3^- availability, since most wetlands have abundant supplies of energy. Nitrification–denitrification pathways are key processes that return N to the atmosphere and complete the N cycle.

20.7 Phosphorus

Unlike C and N, P tends to accumulate in wetlands because there is no significant gaseous loss mechanism. Phosphorus retention by wetlands is regulated by physical (sedimentation and entrainment) and biological mechanisms (uptake and release by vegetation, periphyton and microorganisms) (Figure 20.18). Several reviews have discussed the mechanisms of P retention and cycling in stream and lake sediments, uplands, and wetlands (Khalid et al., 1977; Berkheiser et al., 1980; Bostrom et al., 1982; Logan, 1982; Sonzogni et al., 1982; Howard-Williams, 1985; Froelich, 1988; Reddy and D'Angelo, 1994; Reddy et al., 1999).

Phosphorus entering a wetland system is typically present in both organic and inorganic forms. The relative proportion of each form depends upon the geology, soil, vegetation, and land use characteristics of the watershed. For example, municipal wastewater may contain a large proportion (>75%) as inorganic P in soluble forms, as compared to effluents from agricultural watersheds. To trace the transport and transformations of P, it is convenient to classify forms of P entering these systems as (1) dissolved inorganic P (DIP), (2) dissolved organic P (DOP), (3) particulate inorganic P (PIP), and (4) particulate organic P (POP) (Figure 20.18). The particulate and soluble organic fractions may be further separated into labile and refractory components. DIP is considered bioavailable, whereas organic and particulate P forms generally must undergo transformations to inorganic forms before being considered bioavailable (Figure 20.19).

Phosphorus present in wetland soils can be grouped into: (1) orthophosphate ions sorbed onto the surface of P-retaining components (non-occluded P), (2) P absorbed within the matrices of P-retaining components (occluded P), and (3) phosphate minerals such as apatite or strengite (Williams et al., 1971). These P forms are characterized based on their differential solubilities in various chemical extractants. Several chemical fractionation schemes have been used to identify discrete pools of inorganic P, including exchangeable, Fe/Al bound, and Ca/Mg bound P (Hieltjes and Lijklema, 1980; Olila et al., 1994; Reddy et al., 1995).

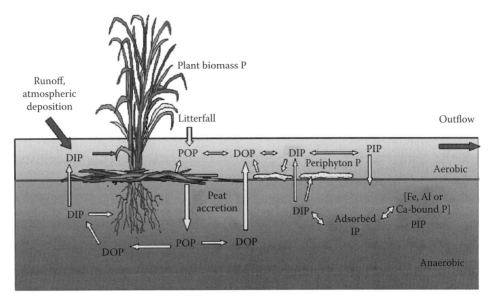

FIGURE 20.18 (See color insert.) Schematic diagram of the general phosphorus cycle in wetlands. (From Reddy, K.R., and R.D. DeLaune. 2008. Biogeochemistry of wetlands: Science and applications. CRC Press, Boca Raton, FL.) POP-Particulate organic P, PIP-particulate inorganic P, DOP-dissolved organic P, DIP-dissolved inorganic P, IP-inorganic P.

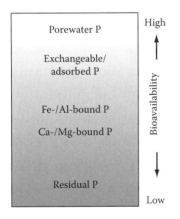

FIGURE 20.19 Schematic showing the relative bioavailability of various P forms in wetland soils. (From Reddy, K.R., and R.D. DeLaune. 2008. Biogeochemistry of wetlands: Science and applications. CRC Press, Boca Raton, FL.)

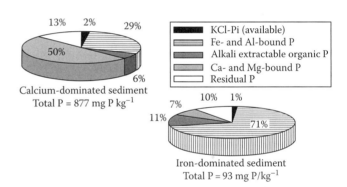

FIGURE 20.20 Comparison of phosphorus distribution in two wetlands on the basis of pools defined by various chemical extractants. (Adapted From Reddy, K.R., and R.D. DeLaune. 2008. Biogeochemistry of wetlands: Science and applications. CRC Press, Boca Raton, FL.)

The example in Figure 20.20 shows labile and nonlabile pools of P in selected stream sediments of south Florida (Reddy et al., 1995). In this example, the inorganic P extracted with neutral salt such as KCl represents loosely absorbed P, which usually accounts for less than 2% of the total P (Figure 20.20). This pool of P is considered bioavailable as it is sufficiently labile to enter porewater for uptake of plants or potential flux into the overlying water column. The NaOH-P_i represents inorganic P associated with Fe and Al, and represents P not readily available. The HCl-P_i represents P associated with Ca/Mg, is also relatively stable and not readily available. Alkali extractable organic P includes both readily available organic P (microbial biomass P) and slowly available organic P (P associated with fulvic and humic acids). Residual P represents highly resistant organic P or unavailable mineral bound P not extracted either with alkali or acid. The Ca-bound P such as apatite has been found to be unavailable (Pettersson, 1986; Gunatilaka, 1988), while the redox-sensitive Fe-bound P may become available under anaerobic conditions (Wildung et al., 1977; Hosomi et al., 1981; Furumai and Ohgaki, 1982).

The organic P fraction in soils and sediments accounts for a high proportion of total P: 20%–60% in mineral soils (Tiessen et al., 1994), 10%–70% in lake sediments, and 40%–90% in Histosols (DeBusk et al., 1994). Most of the organic P in soils is derived from plant detritus and in part synthesized by soil organisms from inorganic sources (Sanyal and DeDatta, 1991). Organic P associated with humic and fulvic acids represents greater than 40% of total soil P (Brannon and Sommers, 1985). The fulvic acid P constitutes a large fraction of the organic P in most soils and it is likely that this pool is derived from plant litter and recently deposited organic matter (Stewart and Tiessen, 1987).

Organic P in biological tissues and cells can be classified into three groups: (1) inositol phosphates, (2) nucleic acids, and (3) phospholipids (Anderson, 1980), with inositol phosphates comprising up to 60% of the soil organic P (Tate, 1984). Phosphorus is bound in most organic compounds as an ester (C-O-P) (Cosgrove, 1977; Golterman, 1984), a form not readily bioavailable until it undergoes enzyme hydrolysis (Harrison, 1983; Hoppe et al., 1988; Chrost, 1991). Extracellular enzyme activity (such as alkaline and acid phosphatases), which hydrolyzes P esters, increases under inorganic P-limiting conditions (McGill and Cole, 1981). In contrast, phosphatase activity is usually lower in areas of high P availability (such as wetlands receiving P inputs); however, eutrophic systems also exhibit high P mineralization rate primarily due to enhanced substrate biodegradability (Newman and Reddy, 1993).

Phosphorus uptake by vegetation is much like that of N, being maximal during the peak growing season, followed by a decrease or cessation in the fall and winter. During winter periods, a significant portion of the P is translocated to belowground biomass (roots and rhizomes) (Davis and van der Valk, 1983). Phosphorus storage in the aboveground biomass of macrophytes is usually short term, with large amounts of P being released during decomposition of detrital tissue. Phosphorus release rates during summer are four to sixfold faster than in winter months. Emergent macrophytes, which have an extensive network of roots, and rhizomes responsible for uptake and storage of P can assimilate little P directly from floodwater (Richardson and Marshall, 1986). Davis (1982) measured only 2%–4% of added floodwater ^{32}P in the living tissue of *Cladium* and *Typha*, however, the active uptake of soil porewater P by these plants can potentially increase the flux of P from the water column into the soil, thus improving overall P retention.

Phosphorus is cycled in soil–water-plant compartments as follows. Herbaceous vegetation rooted in soils obtains most of its P requirements from the soil porewater and translocates it to aboveground vegetation. Upon maturity and senescence, a substantial portion of the P present in the aboveground vegetation can be translocated and stored in belowground biomass (roots and rhizomes). Estimates are that about 45% of the P lost from living *Typha* spp. shoots was translocated to roots and rhizomes (Davis and van der Valk, 1983). Depending on the type of aquatic macrophyte, up to 80% of the remaining P stored in the aboveground detrital tissue in nutrient-rich systems is released into the water column either by initial leaching or as a result of decomposition. The detrital tissue deposited on the soil surface is subjected to decomposition, with residual detrital material becoming an integral part of the soil and providing long-term storage (Caraco et al., 1991; Gachter and Meyer, 1993; D'Angelo and Reddy, 1994a, 1994b; Koch et al., 1994). Over the short term, however, rapid turnover rates and cycling can contribute bioavailable P to the water column and influence water quality.

Apart from plant uptake, inorganic P in the porewater interacts with the soil solid phases through a series of adsorption/desorption processes (Figure 20.21). During these reactions, the PO_4^{3-} ion is held by various positively and neutrally charged sites on

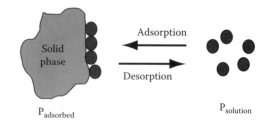

FIGURE 20.21 Schematic representing the process of phosphorus association with the soil solid phase through adsorption/desorption. (From Reddy, K.R., and R.D. DeLaune. 2008. Biogeochemistry of wetlands: Science and applications. CRC Press, Boca Raton, FL.)

soil particles such as sites of permanent charge in clay minerals, pH dependent sites on minerals and organic matter, or through ligand exchange with hydroxyl functional groups of iron and aluminum hydroxides and oxyhydroxides. Precipitation reactions (with calcium, magnesium, iron and aluminum) can also complex P within the soil matrix. The net effect of adsorption/desorption depends on the types of minerals, redox status, pH, organic matter content of each soil/sediment type, and the relative amount of P already stored on these sites. In this manner, adsorption/desorption processes regulate a complex equilibrium between the P in the soil solid and solution phases. The equilibrium P concentration (EPC_0) is often a useful parameter to characterize the P adsorption/desorption behavior of different soils or sediments. The EPC_0 is generally derived using a P isotherm to represent the amount of P lost or retained on the soil solid phase at various P concentrations (Figure 20.22). From this mathematical representation, soils with a higher potential to release P into solution have comparatively higher EPC_0 than soils or sediments that have lower EPC_0 values (Figure 20.23).

Inorganic P added to wetlands at concentrations considerably greater than those present in the soil porewater diffuses to the soil and is retained by oxides and hydroxides of Fe and Al in acid soils and by $CaCO_3$ in alkaline soils. In soils dominated by Fe oxides, P can be readily immobilized through sorption by ferric oxyhydroxide and precipitation as ferric phosphate in the

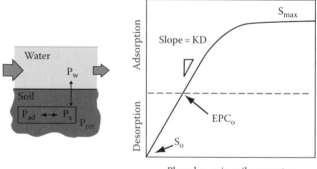

FIGURE 20.22 Illustration of the concept of phosphorus isotherms and the derivation of parameters to characterize phosphorus adsorption and desorption in soils or sediments. (From Reddy, K.R., and R.D. DeLaune. 2008. Biogeochemistry of wetlands: Science and applications. CRC Press, Boca Raton, FL.)

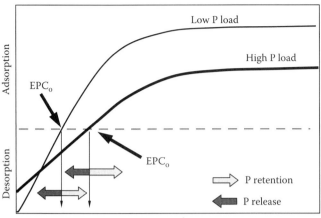

FIGURE 20.23 Comparison of typical isotherms from soils with high and low loading of phosphorus. (From Reddy, K.R., and R.D. DeLaune. 2008. Biogeochemistry of wetlands: Science and applications. CRC Press, Boca Raton, FL.)

The above equation is applicable for [Fe + Al] concentration in the range of 0–100 mmol kg^{-1}. According to this relationship, about 4 mol of Fe + Al are needed to retain 1 mol of P. Other researchers have reported a significant relationship between extractable Fe and Al, and the P-retention capacity of wetland soils (Richardson, 1985; Walbridge and Struthers, 1993; Gale et al., 1994; Lookman et al., 1995; D'Angelo, 2005). These relationships suggest that extractable Fe and Al can be used as a predictor of P retention by wetland mineral soils. In sulfate-dominated wetlands, production of sulfides and removal of Fe(II) through precipitation as ferrous sulfide may preclude P retention by Fe(II) compounds (Caraco et al., 1991).

20.8 Metals

Alterations in Eh and pH in wetland soils affect the reactivity and mobility of metals. Many oxidized stable metallic compounds are reduced into more soluble and bioavailable forms under wetland conditions. Although many metals are present in the wetland environment, two metallic cations (Fe and Mn) play a major role in most wetland mineral soils. Ferric iron (Fe(III)) and Mn(IV) reduction coupled with organic matter turnover is considered as one of the significant mechanisms in anaerobic habitats (Lovley, 1991; Nealson and Myers, 1992). In wetland soil environments, Fe(III) and Mn(IV) reduction is mediated by enzymatic activity, which couples the use of these compounds as electron acceptors during microbial respiration. However, nonenzymatic reduction of Fe(III) and Mn(IV) has also been reported in anaerobic environments (Lovley et al., 1991). Both Fe and Mn are present as solid phases in oxidized form (such as Fe_2O_3, FeOOH, $FePO_4$, and MnO_2), which are reduced to more dissolved forms under anaerobic conditions (Figures 20.25 and 20.26). Typical Fe(III) reduction under wetland soils conditions and its relationship to Eh and pe is shown below:

oxidized zones at the soil–water interface. The rate of adsorption is controlled by soil pH and Eh, adsorptive surface area (active Fe and Al or $CaCO_3$), and temperature. In calcareous systems, P can be precipitated as Ca mineral bound P, especially when the pH of the floodwater is increased by algal photosynthetic activity (pH as high as 10). At night, respiration by algae and bacteria can result in a resolubilization of Ca-precipitated P (Figure 20.24) (Diaz et al., 1994). In calcareous systems, coprecipitation of P with $CaCO_3$ is believed to be a dominant mechanism in the immobilization of soluble P.

The retention of P in noncalcareous systems can be represented by the following empirical relationship that was derived for a number of wetland mineral soils and stream sediments in Florida (Reddy et al., 1998):

$$S_{max} = 0.24 \, [\text{oxalate Fe + Al}] \quad (20.14)$$

where
  S_{max} is maximum P retention capacity
  [oxalate Fe + Al] is ammonium oxalate extractable Fe and Al

$$Fe(OH)_3 + 3H^+ + e^- \rightarrow Fe^{2+} + 3H_2O \quad (20.15)$$

$$Eh = 1.06 - 0.059 \log Fe^{2+} - 0.177 \, pH \quad (20.16)$$

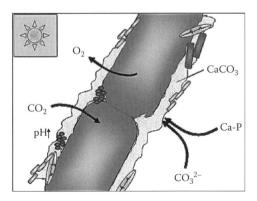
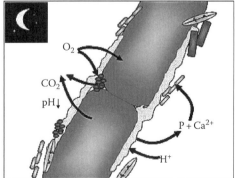

FIGURE 20.24 Depiction of diel processes affecting co-precipitation of phosphorus with calcium in algal dominated systems. (From Reddy, K.R., and R.D. DeLaune. 2008. Biogeochemistry of wetlands: Science and applications. CRC Press, Boca Raton, FL.)

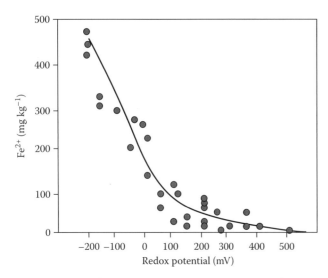

FIGURE 20.25 Influence of redox potential on reduction of oxidized Iron(III) to form Iron(II) (Fe^{2+}). (From Reddy, K.R., and R.D. DeLaune. 2008. Biogeochemistry of wetlands: Science and applications. CRC Press, Boca Raton, FL.)

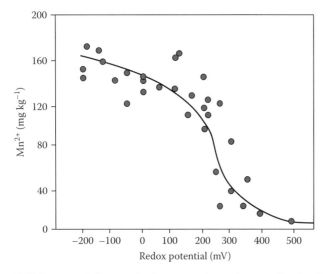

FIGURE 20.26 Influence of redox potential on reduction of oxidized Manganese(IV) to form Manganese(II) (Mn^{2+}). (From Reddy, K.R., and R.D. DeLaune. 2008. Biogeochemistry of wetlands: Science and applications. CRC Press, Boca Raton, FL.)

$$pe = 17.87 + pFe^{2+} - 3\,pH \qquad (20.17)$$

Reduction of metallic compounds is inhibited by the presence of O_2 and to some extent by NO_3^-. The reduction of Fe(III) and Mn(IV) has direct impacts on soil chemistry, such as (1) increased water soluble Fe(II) and Mn(II), (2) increased pH, (3) displacement of other cations from the soil exchange complex into porewater, (4) breakdown of organic matter and nutrient release, (5) increased solubility of P and Si, and (6) formation of new minerals (Ponnamperuma, 1972; Gambrell, 1994).

Reduction of Fe(III) and Mn(IV) depends on the supply of electron donors (primarily availability of labile organic matter), and the concentrations of bioavailable oxidized Fe and Mn. Microbial reduction also depends on the ability of bacteria to solubilize solid phases, attach directly to the substrate and transfer electrons, and/or transport the substrate directly into cells as a solid. Reduction capacity of the oxidized Fe compounds decreases with increasing crystallinity: $FePO_4$ > $Fe(OH)_3$ > $FeOOH$ > Fe_2O_3 (Lovley, 1987). Iron forms more stable organic metal bonds than Mn, which protects it from precipitation reactions. The extent of complexation is greater under wetland soil conditions because of high concentrations of dissolved organic compounds in porewater. Significant amounts of Fe(II) can be present under aerobic conditions for several days due to complexation with organics.

General chemical forms of trace metals in wetland soil are (1) water soluble forms, such as free ions, inorganic and organic complexes, (2) exchangeable metals, (3) metals precipitated as inorganic compounds, (4) metals complexed with large molecular weight humic materials, (5) metals adsorbed or occluded in precipitated hydrous oxides, (6) metals precipitated as insoluble sulfides, and (7) metals bound within the crystalline lattice structure of primary minerals (Gambrell, 1994). These mechanisms often make wetland soils effective sinks for many trace metals (Gambrell, 1994).

One important location of metal binding in wetlands is the rhizosphere where O_2 leaking from roots precipitates Fe (as Fe oxides) in what is known as a root "plaque." Many other metals such as As, Zn, and Ni can also co-precipitate in these root plaques making them effective at metal removal in wastewaters and metal accumulation (e.g., Zn) in wetland plants (reviewed by Jacob and Otte, 2003). In anaerobic soil zones, high levels of S^{2-} result in the effective precipitation of many metals such as Zn, Pb, and Hg. This is particularly important in estuarine wetlands or coastal soils/sediments, but sulfur additions to freshwater systems can also result in the reduction of metal availability. One example is in the Florida Everglades where areas of highest sulfur additions (from agricultural drainages) exhibited the lowest levels of toxic Hg (as methyl-mercury) (Benoit et al., 1999; Krabbenhoft et al., 2000).

Processes which tend to bind metals under anaerobic conditions are not irreversible, however, and are subject to water table fluctuations or wetland drainage which can reverse many of these reactions. For example, in wetland mineral soils, accumulation of Fe(II) displaces many basic cations from the exchange complex, which increases the concentration of basic cations in porewater. Draining these soils removes basic cations from porewater, oxidizes Fe(II) to Fe(III), which results in the release of H^+. In turn, the lower pH releases Al and other associated metals. Similarly, metals bound to sulfides can be directly released under drained conditions when sulfides are oxidized to sulfates, also resulting in the release of H^+. Many other metals such as Ca, Mg, Cu, Zn, Pb, Hg, and Ni are not directly involved in oxidation–reduction reactions, but their solubilities are affected by changes in pH as a result of flooded and drained conditions.

20.9 Toxic Organics

Toxic organics are naturally occurring or synthetic chemicals, which at low concentrations have adverse effects on growth and activity of plants, animals, and/or microorganisms. Figure 20.27 shows the structures of some of the more common toxic organic compounds found in wetlands. Natural and constructed wetlands may receive toxic organics from a number of sources, including (1) direct applications to control aquatic weeds, algae and mosquitoes; (2) accidental spillage, surface runoff, and drainage water from agriculture, urban, and industrial sources; (3) domestic and industrial discharges into natural and constructed wetlands for wastewater improvement; (4) residual toxics remaining in agricultural soils that are converted to wetlands; and (5) in situ production by natural processes, mediated by plants and microorganisms.

These sources are likely to contribute vast amounts and multitudes of different types of toxic organics; however, widespread monitoring and toxicological studies have not been conducted to estimate the extent and consequences of contamination (Catallo, 1993). One estimate indicated that about 5000 wetlands and aquatic systems are impacted by agricultural pesticides in the United States (Cooper, 1993; Taraban et al., 1993). Many soil processes including microbial respiration, nitrification (Sayler et al., 1982), and cellulose and lignin degradation (McKinley et al., 1982; Katayama and Kuwatsuka, 1991; Dusek and Tesarova, 1996) are negatively impacted by toxic organics, and so pose a serious threat to proper functioning of wetland systems.

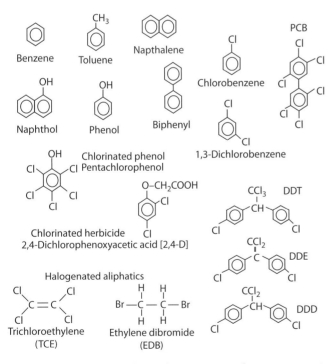

FIGURE 20.27 Basic chemical structures of common toxic organic compounds. (From Reddy, K.R., and R.D. DeLaune. 2008. Biogeochemistry of wetlands: Science and applications. CRC Press, Boca Raton, FL.)

Several processes regulate the fate of toxic organics once they enter wetlands, and most information about these processes has been gained through laboratory experiments that simulate wetland systems. In general, toxic organic compounds undergo similar transformation processes as natural organic matter, including aerobic and anaerobic microbial breakdown, vegetative uptake, volatilization, photolysis, abiotic oxidation and hydrolysis, bioaccumulation, sorption, and burial in the soil (D'Angelo, 2002). Through these processes, many wetlands are capable of attenuating this class of pollutants. However, the extent of these processes depends on the type of compound (water solubility, oxidation state, functional groups, molecular weight), as well as biological and chemical conditions in the soil–water column including pH, Eh, temperature, light intensity, nutrient and electron acceptor availability, type and amount of vegetation, organic matter content, and previous history of contamination. For a review of process models refer to Lyman (1982).

Microbial degradation is generally considered the main pathway for attenuation of toxic organics. This task is usually accomplished by sequential enzymatic reactions mediated by many populations working together, rather than by a single species. In wetlands with characteristically diverse populations of microorganisms ranging from aerobes to extreme obligate anaerobes, there is great potential for degradation of many types of chemicals. In many cases, these transformations are beneficial to microorganisms by reducing toxicity or by acting as C sources, nutrients or electron acceptors. Most enzymatic transformations can be grouped into four main categories: hydrolysis, oxidation, reduction, and synthesis. The type of enzymatic transformation involved depends mainly on the type of chemical and its functional groups, as well as environmental conditions that promote the activity of different enzyme systems. In wetland systems, a combination of reactions are typically required to attain complete mineralization of a chemical.

Extracellular enzymes produced by bacteria and fungi hydrolyze many types of toxic organics including organophosphates, phenoxyacetates, phenylureas, acid anilides, and phenylcarbamates (Burns and Edwards, 1980; Nannipieri and Bollag, 1991). Small molecular weight C, N, and P compounds released from these reactions provide nutrients for degrading populations. Monooxygenase or dioxygenase enzymes incorporate one or two atoms of O_2 into chemicals. Due to the requirement for O_2, this reaction is largely limited to aerobic zones such as soil–water or root–soil interfaces. For alkanes and aromatics without functional groups (benzene and toluene), oxygenase activity is typically required before significant rates of degradation occur, which likely explains the recalcitrance of these chemicals in mostly anaerobic zones. Through oxidative reactions, many toxic organics are eventually utilized as C and energy sources by degrading populations (Figure 20.28). Monooxygenases and dioxygenases are less effective with highly oxidized chemicals, including highly chlorinated pentachlorophenol, 2,4,5 trichlorophenoxyacetic acid, hexachlorobenzene, polychlorinated biphenyls, carbon tetrachloride and tetrachloroethylene which

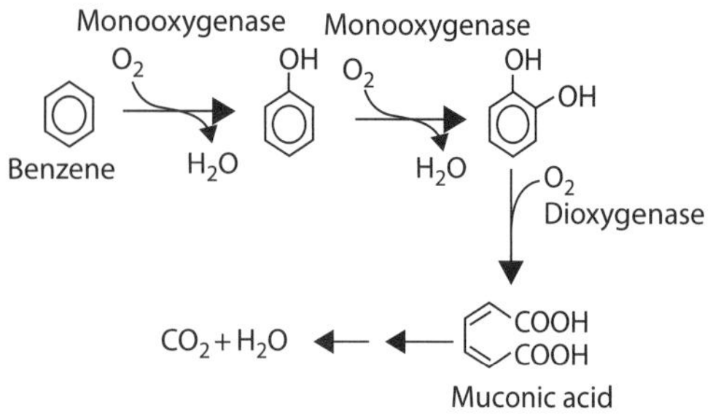

FIGURE 20.28 Depiction of the reactions and pathway in the complete oxidation of benzene to carbon dioxide and water. (From Reddy, K.R., and R.D. DeLaune. 2008. Biogeochemistry of wetlands: Science and applications. CRC Press, Boca Raton, FL.)

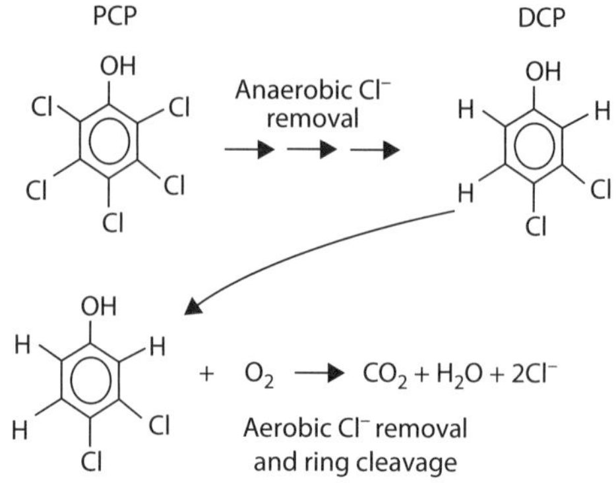

FIGURE 20.30 Pathway of alternating reduction/oxidation steps involved in the complete oxidation of pentachlorophenol to carbon dioxide and water. (From Reddy, K.R., and R.D. DeLaune. 2008. Biogeochemistry of wetlands: Science and applications. CRC Press, Boca Raton, FL.)

explains the recalcitrance of these chemicals under aerobic conditions. For these chemicals, reductive processes may be more effective microbial transformation processes.

Under extremely reduced conditions, electrons from organic matter and intermediate breakdown products (H_2, fatty acids) can be transferred to toxic organics (Gibson and Sewell, 1992; Doong and Wu, 1995). These chemicals, therefore, function as electron acceptors, in which the reaction is coupled to energy generation for degrading populations using similar strategies that are used by aerobes and denitrifiers to gain energy from electron transfer to O_2 and NO_3^-. These redox reactions are important in the process of reductive dehalogenation (Figure 20.29). Mohn and Tiedje (1991) found evidence that reductive dechlorination was beneficial to *Desulfomonile tiedjei*, which used 3-chlorobenzoate as an electron acceptor and was coupled to energy (ATP) production. The presence of alternate electron acceptors (O_2, NO_3^-, SO_4^{2-}) tends to inhibit reduction of toxic organics, since various microbial groups compete with each other for electron equivalents.

One drawback of reduction reactions is that they often result in production of intermediates that are also toxic. For example, D'Angelo and Reddy (2000) showed sequential reductive dechlorination of pentachlorophenol to 2,3,4,5-tetrachlorophenol to 3,4,5-trichlorophenol to 3,5-dichlorophenol to 3-chlorophenol. Complete degradation yielding CO_2 can be achieved if reductive conditions (necessary for dehalogenation) are combined with aerobic reactions (required to break the phenolic ring structure) (Figure 20.30). As toxic organics become reduced, they may more readily undergo oxidation reactions. Therefore, alternately flooding or draining soils, or planting O_2-transporting plants, may promote mineralization compared to one process by itself.

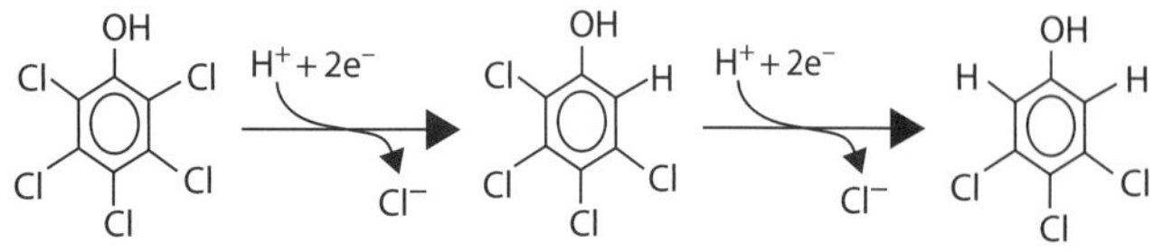

Enhanced degradation by alternate anaerobic-aerobic soil processes has been shown for methoxychlor (Fogel et al., 1982).

In contrast to the above transformations that simplify chemical structures, synthesis reactions result in the addition of functional groups or the polymerization of compounds. Synthesis reactions can be beneficial as they can convert toxic organics into less toxic forms. For example, methylation of chlorophenols results in the production of volatile chloroanisoles, which likely benefits microorganisms by being less toxic (Neilson et al., 1983). Chloroanilines and phenolic compounds can also undergo oxidative coupling, or polymerization with soil humic substances through the enzymatic action of peroxidases, laccases, and tyrosinases (Burns and Edwards, 1980; Bollag and Liu, 1985; Sarkar et al., 1988; Lassen et al., 1994). Shindo and Huang (1984) demonstrated that polymerization of phenols could be abiotically catalyzed by Mn(IV), Fe(III), Al and Si oxides, which result in the formation of humic acids. An example of these types of reactions is shown in Figure 20.31. These processes are thought to be analogous to natural humic substance formation from lignin (polyphenols) and protein molecules (Allard et al., 1994; Pal et al., 1994). Synthesis reactions may also result in production of toxic organics in aquatic environments. For example, Asplund et al. (1993) characterized chloroperoxidases in soils that mediated production of chlorinated humic substances. Natural production of organically

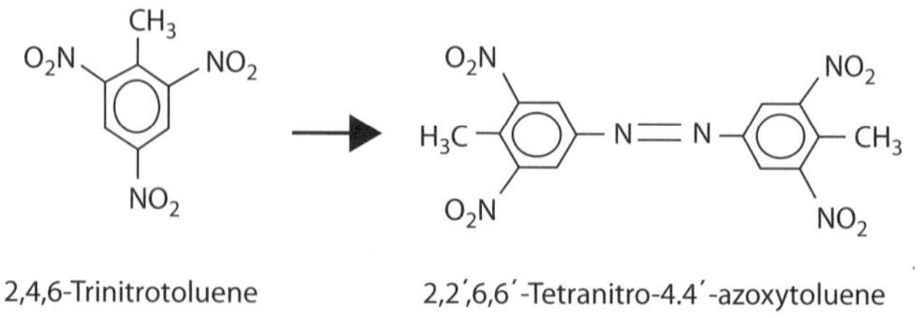

FIGURE 20.29 Steps involved in the reductive dechlorination of pentachlorophenol.

FIGURE 20.31 Example of biotic polymerization of 2,4,6-trinitrotoluene. (From Reddy, K.R., and R.D. DeLaune. 2008. Biogeochemistry of wetlands: Science and applications. CRC Press, Boca Raton, FL.)

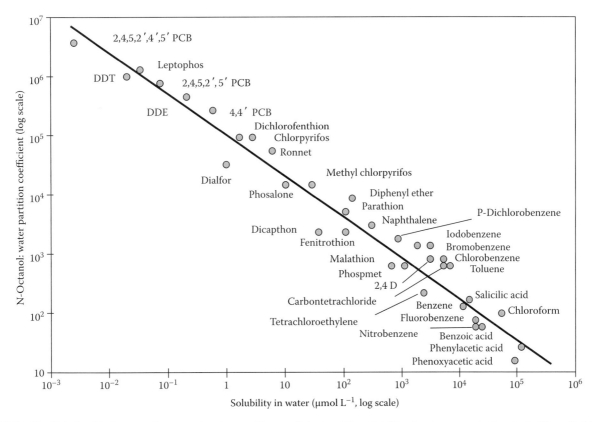

FIGURE 20.32 Relationship between the octanol–water partition coefficient and the solubility of various organic chemicals. (From Reddy, K.R., and R.D. DeLaune. 2008. Biogeochemistry of wetlands: Science and applications. CRC Press, Boca Raton, FL.)

bound chlorine, mediated by white-rot fungi, may partially explain the widespread genetic capacity of microorganisms to degrade man-made chemicals (Oberg et al., 1997).

Sorption to soils and sediments is an important hurdle in transformation processes, since it reduces the amount in the bioavailable (soluble) pool (Ogram et al., 1985; Mihelcic et al., 1993; Bouwer et al., 1994; D'Angelo and Reddy, 2003). Except in the case of extracellular enzymes, which have greater access to soil microsites, desorption and transport of toxic organics to microbial cells are prerequisites for transformation. Sorption to soil particles and organic matter is dependent on the characteristics of each toxic compound and can be described as a partitioning between the soil and water phases. The octanol:water partitioning coefficient is useful to characterize the sorption and solubility of various compounds (Figure 20.32). For strongly sorbed species, such as polychlorinated biphenyls and polyaromatic hydrocarbons with high octanol water partitioning coefficients, desorption may greatly curtail degradation kinetics (Manilal and Alexander, 1991). This process is important in many wetland soils because of accumulation of organic matter, and associated high cation exchange capacity, which promote hydrophobic partitioning and adsorption. Partial removal of toxic organics from the bioavailable pool through sorption may allow degradation to proceed at otherwise toxic levels (Apajalahti and Salkinoja-Salonen, 1984). On the other hand, anaerobic and neutral pH conditions tend to increase solubility of humic substances of wetland soils, which in turn enhances the solubility of toxic organics (Pardue et al., 1993).

References

Alexander, M. 1977. Introduction to soil microbiology. John Wiley & Sons, New York.

Allard, A., P. Hynning, M. Remberger, and A. Neilson. 1994. Bioavailability of chlorocatechols in naturally contaminated sediment and of chloroguaiacols covalently bound to C2-guaiacyl residues. Appl. Environ. Microbiol. 60:777–784.

Anderson, G. 1980. Assessing organic phosphorus in soils, p. 411–413. In F.E. Khasawneh, E.C. Sample, and E.J. Kamprath (eds.) The role of phosphorus in agriculture. Soil Science Society of America, Madison, WI.

Anthonsieu, A.C., Loehr, R.C., Prakasam, T.B.S., and Srinath, E.G. (1976). Inhibition of nitrification by ammonia and nitrous acid. J. Water Pollut. Control Fed. 48, 835–850.

Apajalahti, J.H.A., and M.S. Salkinoja-Salonen. 1984. Absorption of pentachlorophenol (PCP) by bark chips and its role in microbial PCP degradation. Microb. Ecol. 10:359–367.

Armstrong, J., and W. Armstrong. 1990. Light enhanced convective through flow increases oxygenation in rhizomes and rhizosphere of *Phragmites australis* (Cav.) Trin. ex. steud. New Phytol. 114:121–128.

Asplund, G., J.V. Christiansen, and A. Grimvall. 1993. A chloroperoxidase-like catalyst in soil: Detection and characterization of some properties. Soil. Biol. Biochem. 25:41–46.

Benner, R.H., M.A. Moran, and R.E. Hodson. 1985. Effects of pH and plant source on lignocellulose biodegradation rate in two wetland ecosystems, the Okeefenokee Swamp and a Georgia salt marsh. Limnol. Oceanogr. 30:489–499.

Benoit, J.M., C.C. Gilmour, R.P. Mason, and A. Heyes. 1999. Sulfide controls on mercury speciation and bioavailability to methylating bacteria in sediment and pore waters. Environ. Sci. Technol. 33:951–957.

Berkheiser, V.E., J.J. Street, P.S.C. Rao, and T.L. Yuan. 1980. Partitioning of inorganic orthophosphate in soil-water systems. CRC Crit. Rev. Environ. Cont. 10:179–224.

Berman, T., and D.A. Bronk. 2003. Dissolved organic nitrogen: A dynamic participant in aquatic ecosystems. Aquat. Microb. Ecol. 31:279–305.

Bollag, J., and S. Liu. 1985. Copolymerization of halogenated phenols and syringic acid. Pest. Biochem. Physiol. 23:261–272.

Bostrom, B., M. Jansson, and C. Forsberg. 1982. Phosphorus release from lake sediments. Arch. Hydrobiol. Beih. 18:5–59.

Bouma, J. 1991. ICOMAC circular 11, March 15, 1991. Final report of the International Committee for the Classification and Management of Wet Soils. Agricultural University, Wageningen, the Netherlands.

Bouwer, E., N. Durant, L. Wilson, W. Zhang, and A. Cunningham. 1994. Degradation at xenobiotics compounds in situ: Capabilities and limits. FEMS Microb. Rev. 15:307–317.

Bowden, W.B. 1987. The biogeochemistry of nitrogen in freshwater wetlands. Biogeochemistry 4:313–348.

Brannon, C.A., and L.E. Sommers. 1985. Preparation and characterization of model humic polymers containing organic phosphorus. Soil Biol. Biochem. 17:213–219.

Brix, H. 1996. Functions of macrophytes in constructed wetlands. Wat. Sci. Tech. 29:71–78.

Brix, H., B.K. Sorell, and P.T. Orr. 1992. Internal pressurization and convective gas flow in some emergent freshwater macrophytes. Limnol. Oceanogr. 37:1420–1433.

Burford, J.R., and J.M. Bremner. 1975. Relationships between the denitrification capacities of soils and total water soluble and readily decomposable soil organic matter. Soil Biol. Biochem. 7:389–394.

Burgin, A.J., and S.K. Hamilton. 2007. Have we overemphasized the role of denitrification in aquatic ecosystems? A review of nitrate removal pathways. Front. Ecol. Environ. 5:89–96.

Burns, R.G., and J.A. Edwards. 1980. Pesticide breakdown by soil enzymes. Pest. Sci. 11:506–512.

Caraco, N., J.J. Cole, and G.E. Likens. 1991. A cross-system study of phosphorus release from lake sediments, p. 241–258. In J. Cole, G. Lovett, and S. Findley (eds.) Comparative analysis of ecosystems. Springer Verlag, New York.

Catallo, J.M. 1993. Ecotoxicology and wetland ecosystems: Current understanding and future needs. Environ. Toxicol. Chem. 12:2209–2224.

Chrost, R.J. 1991. Environmental control of the synthesis and activity of aquatic microbial ectoenzymes, p. 29–59. In R.J. Chrost (ed.) Microbial enzymes in aquatic environments. Springer-Verlag, New York.

Cole, J.A. 1988. Assimilatory and dissimilatory reduction of nitrate to ammonia, p. 281–329. In J.A. Cole, and S.J. Ferguson (eds.) The nitrogen and sulphur cycles. Cambridge University Press, Cambridge, U.K.

Cooper, A.B. 1990. Nitrate depletion in the riparian zone and stream channel of a small headwater catchment. Hydrobiologia 202:13–26.

Cooper, C.M. 1993. Biological effects of agriculturally derived surface water pollutants on aquatic systems—A review. J. Environ. Qual. 3:402–408.

Cosgrove, D.J. 1977. Microbial transformations in the phosphorus cycle. Adv. Microb. Ecol. 1:95–134.

Cowardin, L.M., V. Cater, F.C. Golet, and E.T. LaRoe. 1979. Classification of wetlands and deepwater habitats of the United States. FWS/OBS-79/31. U.S. Fish and Wildlife Service, Washington, DC.

D'Angelo, E.M. 2002. Wetlands: Biodegradation of organic pollutants, p. 3401–3417. In G. Bitton (ed.) Encyclopedia of environmental microbiology. John Wiley & Sons, New York.

D'Angelo, E.M. 2005. Phosphorus retention and exchange by soils from mitigated and natural bottomland forest wetlands. Wetlands 25:297–305.

D'Angelo, E.M., and K.R. Reddy. 1994a. Diagenesis of organic matter in a wetland receiving hypereutrophic lake water. I. Distribution of dissolved nutrients in the soil and water column. J. Environ. Qual. 23:928–936.

D'Angelo, E.M., and K.R. Reddy. 1994b. Diagenesis of organic matter in a wetland receiving hypereutrophic lake water. II. Role of inorganic electron acceptors in nutrient release. J. Environ. Qual. 23:937–943.

D'Angelo, E.M., and K.R. Reddy. 1999. Regulators of heterotrophic microbial potentials in wetland soils. Soil Biol. Biochem. 31:815–830.

D'Angelo, E.M., and K.R. Reddy. 2000. Aerobic and anaerobic transformations of pentachlorophenol in wetland soils. Soil Sci. Soc. Am. J. 64:933–943.

D'Angelo, E.M., and K.R. Reddy. 2003. Effect of aerobic and anaerobic conditions on chlorophenol sorption in wetland soils. Soil Sci. Soc. Am. J. 67:787–794.

Davis, S.M. 1982. Patterns of radio phosphorus accumulation in the Everglades after its introduction into surface water. Technical Publication 82-2. South Florida Water Management District. West Palm Beach, FL.

Davis, C.B., and A.G. van der Valk. 1983. Uptake and release of nutrients by lining and decomposing *Typha glauca* Godr. tissues at Eagle Lake, Iowa. Aquat. Bot. 16:75–89.

DeBusk, W.F. 1996. Organic matter turnover along a nutrient gradient in the Everglades. PhD Dissertation. University of Florida, Gainesville, FL.

DeBusk, W.F., K.R. Reddy, M.S. Koch, and Y. Wang. 1994. Spatial distribution of soil nutrients in a northern Everglades marsh: Water Conservation Area 2A. Soil Sci. Soc. Am. J. 58:543–552.

DeLaune, R.D., T.C. Feijtel, and W.H. Patrick, Jr. 1989. Nitrogen flows in Louisiana Gulf Coast salt marsh: Spatial considerations. Biogeochemistry 8:25–37.

Diaz, O.A., K.R. Reddy, and P.A. Moore. 1994. Solubility of inorganic P in stream water as influenced by pH and Ca concentration. Water Res. 28:1755–1763.

Doong, R., and S. Wu. 1995. Substrate effects on the enhanced biotransformation of polychlorinated hydrocarbon under anaerobic condition. Chemosphere 30:1499–1511.

Dusek, L., and M. Tesarova. 1996. Influence of polychlorinated biphenyls on microbial biomass and its activity in grassland soil. Biol. Fertil. Soils 22:243–247.

EPA. 1988. EPA Wetland Identification and Delineation Manual (two volumes). In W.S. Sipple (ed.) Office of wetlands protection. U.S. Environmental Protection Agency, Washington, DC.

Federal Register. July 13, 1994. Changes in hydric soils of the United States. Washington, DC.

Federal Register. February 24, 1995. Hydric soils of the United States. Washington, DC.

Fog, K. 1988. The effect of added nitrogen on the value of decomposition of organic matter. Biol. Rev. 63:433–462.

Fogel, S., R.L. Lancione, and A.E. Sewall. 1982. Enhanced biodegradation of methoxychlor in soil under sequential environmental conditions. Appl. Environ. Microbiol. 44:113–120.

Francois, R. 1990. Marine sediment humic substances: Structure, genesis, and properties. Crit. Rev. Aquatic Sci. 3:41–80.

Froelich, P.N. 1988. Kinetic control of dissolved phosphate in natural rivers and estuaries: A primer on the phosphate buffer mechanism. Limnol. Oceanogr. 33:649–668.

Furumai, H. and S. Ohgaki. 1988. Radiochemical analysis of phosphorus exchange kinetics between sediment and water under aerobic conditions. J. Environ. Qual. 17:205-212.

Gachter, R., and J.S. Meyer. 1993. The role of microorganisms in mobilization and fixation of phosphorus in sediments. Hydrobiologia 253:103–121.

Gale, P.M., I. Devai, K.R. Reddy, and D.A. Graetz. 1993. Denitrification potential of soils from constructed and natural wetlands. Ecol. Eng. 2:119–130.

Gale, P.M., K.R. Reddy, and D.A. Graetz. 1994. Phosphorus retention by wetland soils used for treated wastewater disposal. J. Environ. Qual. 23:370–377.

Galloway, J.N., F.J. Dentener, D.G. Capone, E.W. Boyer, R.W. Howarth, S.P. Seitzinger, G.P. Asner et al. 2004. Nitrogen cycles: Past, present, and future. Biogeochemistry 70:153–226.

Gambrell, R.P. 1994. Trace and toxic metals in wetlands: A review. J. Environ. Qual. 23:883–891.

Gaur, S., R.K. Singhal, and S.K. Hasija. 1992. Relative contributions of bacteria and fungi to water hyacinth decomposition. Aquat. Bot. 43:1–15.

Gibson, S.A., and G.W. Sewell. 1992. Stimulation of reductive dechlorination of tetrachloroethylene in anaerobic aquifer microcosms by additions of short chain organic acids or alcohols. Appl. Environ. Microbiol. 58:1392–1393.

Given, P.H., and C.H. Dickenson. 1975. Biochemistry and microbiology of peats, p. 123–211. In E.A. Paul and A.D. McClaren (eds.) Soil biochemistry. Vol. 3. Marcel Dekker, New York.

Glossary of Soil Science Terms. 1997. Soil Science Society of America, p. 138. Madison, WI.

Golterman, H.L. 1984. Sediments, modifying and equilibrating factors in the chemistry of freshwater. Verh. Int. Ver. Theor. Angew. Limnol. 22:23–59.

Goodwin, S., and J.G. Zeikus. 1987. Ecophysiological adaptations of anaerobic bacteria to low pH: Analysis of anaerobic digestion in acidic bog sediments. Appl. Environ. Microbiol. 53:57–64.

Gordon, A.S., and F.J. Millero. 1985. Adsorption mediated decrease in the biodegradation rate of organic compounds. Microb. Ecol. 11:289–298.

Gunatilaka, A. 1982. Phosphate adsorption kinetics of resuspended sediments in a shallow lake. Hydrobiol. 91:293–298.

Harrison, A.F. 1983. Relationship between intensity of phosphatase activity and physicochemical properties of woodland soils. Soil Biol. Biochem. 15:93–99.

Heil, C.A., M. Revilla, P.M. Glibert, and S. Murasko. 2007. Nutrient quality drives differential phytoplankton community composition on the West Florida Shelf. Limnology and Oceanography 52:1067–1078.

Hieltjes, A.H.M., and L. Lijklema. 1980. Fractionation of inorganic phosphates in calcareous sediments. J. Environ. Qual. 9:405–407.

Herbert, R.A. (1982). Nitrate dissimilation in marine and estuarine sediments. In: Nedwell, D. 8., and Brown, C.M. (eds), Sediment Microbiology, pp. 53–71, Academic Press, London.

Hopkinson, C.S., and J.P. Schubauer. 1984. Static and dynamic aspects of nitrogen cycling in the salt marsh graminoid *Spartina alterniflora*. Ecology 65:961–969.

Hoppe, H.G., S.J. Kim, and K. Gocke. 1988. Microbial decomposition in aquatic environments: Combined processes of extracellular activity and substrate uptake. Appl. Env. Microbiol. 54:784–790.

Hosomi, M., Okada, M. and Sudo, R., 1981. Release of phosphorus from sediments. Verh. Int. Ver. Limnol., 21:628–633.

Houghton, R.A., and D.L. Skole. 1990. Carbon, p. 393–408. In B.L. Turner, W.C. Clark, R.W. Kates, J.F. Richards, J.T. Mathews, and W.B. Meyer (eds.) The earth as transformed by human action. Cambridge University Press, Cambridge, U.K.

Howard-Williams, C. 1985. Cycling and retention of nitrogen and phosphorus in wetlands: A theoretical and applied perspective. Freshwater Biol. 15:391–431.

Howard-Williams, C., and M.T. Downes. 1993. Nitrogen cycling in wetlands, p. 141–167. *In* T.P. Burt, A.L. Heathwaite, and S.T. Trudgill (eds.) Nitrate: Processes, patterns and management. John Wiley & Sons, New York.

Howarth, R.W., R. Marino, J. Lane, and J.J. Cole. 1988. Nitrogen fixation in freshwater, estuarine, and marine ecosystems. 1. Rates and importance. Limnol. Oceanogr. 33:669–687.

Inglett, P.W., K.R. Reddy, and P.V. McCormick. 2004. Periphyton chemistry and nitrogenase activity in a nutrient-impacted Everglades ecosystem. Biogeochemistry 67:213–233.

Jacob, D.L., and M.L. Otte. 2003. Conflicting processes in the wetland plant rhizosphere: Metal retention or mobilization? Water Air Soil Pollut. 3:91–104.

Johnston, C.A. 1991. Sediment and nutrient retention by freshwater wetlands: Effects on surface water quality. CRC Crit. Rev. Environ. Control. 21:491–565.

Katayama, A., and S. Kuwatsuka. 1991. Effect of pesticides on cellulose degradation in soil under upland and flooded conditions. Soil Sci. Plant Nutr. 37:1–6.

Kausik, N.K., and H.B.N. Hynes. 1971. The fate of dead leaves that fall into streams. Arch. Hydrobiol. 68:465–515.

Kelso, B.H.L., R.V. Smith, R.J. Laughlin, and S.D. Lennox. 1997. Dissimilatory nitrate reduction in anaerobic sediments leading to river nitrite accumulation. Appl. Environ. Microbiol. 63:4679–4685.

Khalid, R.A., W.H. Patrick, Jr., and R.D. DeLaune. 1977. Phosphorus sorption characteristics of flooded soils. Soil Sci. Soc. Am. J. 41:305–310.

Kim, B., and R.G. Wetzel. 1993. The effect of dissolved humic substances on the alkaline phosphatase and growth of microalgae. Verh. Internat. Verein. Limnol. 25:129–132.

Kjoller, A., and S. Struwe. 1980. Microfungi of decomposing red alder leaves and their substrate utilization. Soil Biol. Biochem. 12:425–431.

Kjoller, A., S. Struwe, and K. Vestberg. 1985. Bacterial dynamics during decomposition of alder litter. Soil Biol. Biochem. 17:463–468.

Koch, M., and K.R. Reddy. 1992. Distribution of soil and plant nutrients along a trophic gradient in the Florida Everglades. Soil Sci. Soc. Am. J. 56:1492–1499.

Koch, M.S., K.R. Reddy, and J.P. Chanton. 1994. Factors controlling seasonal nutrient profiles in a subtropical peatland: Florida Everglades. Soil Sci. Soc. Am. J. 56:1492–1499.

Koerselman, W. and Meuleman, A.F.M. 1996. The Vegetation N:P Ratio: a New Tool to Detect the Nature of Nutrient Limitation. J. Appl. Ecol. 33:1441–1450.

Krabbenhoft, D.P., J.P. Hurley, G. Aiken, C. Gilmour, M. Marvin-DiPasquale, W.H. Orem and R. Harris. 2000. Mercury cycling in the Florida Everglades: A mechanistic field study. Verh. Internat. Verein. Limnol. 27:1–4.

Lassen, P., A. Randall, O. Jorgensen, P. Warwick, and L. Carlson. 1994. Enzymatically mediated incorporation of 2-chlorophenol and 4-chlorophenol into humic acids. Chemosphere 28:703–710.

Lewis, W.M. 1995. Wetlands: Characteristics and boundaries. National Research Council, National Academy Press, Washington, DC.

Linares, F., and K. Sundback. 2006. Uptake of dissolved free amino acids (DFAA) by microphytobenthic communities. Aquat. Microb. Ecol. 42:175–186.

Logan, T.J. 1982. Mechanisms for release of sediment-bound phosphate to water and the effects of agricultural land management on fluvial transport of particulate and dissolved phosphate. Hydrobiologia 92:519–530.

Lookman, R., D. Freese, R. Merck, K. Vlassak, and W.H. van Riemsdijk. 1995. Long-term kinetics of phosphate release from soil. Environ. Sci. Technol. 29:1569–1575.

Lovley, D.R. 1987. Organic matter mineralization with the reduction of ferric iron: A review. Geomicrobiol. J. 5:375–399.

Lovley, D.R. 1991. Dissimilatory Fe(III) and Mn(IV) reduction. Microbiol. Rev. 55:259–287.

Lovley, D.R., E.J.P. Phillips, and D.J. Lonergan. 1991. Enzymatic versus nonenzymatic mechanisms for Fe(III) reduction in aquatic sediments. Environ. Sci. Technol. 25:1062–1067.

Lyman, W.J. 1982. Handbook of chemical property estimation methods. McGraw-Hill, New York.

Manilal, B., and M. Alexander. 1991. Factors affecting the microbial degradation of phenanthrene in soil. Appl. Microbial. Biotechnol. 35:401–405.

Martin, J.F., and K.R. Reddy. 1997. Interaction and spatial distribution of wetland nitrogen processes. Ecol. Model. 105:1–21.

Mausbach, M.J., and J.L. Richardson. 1994. Biogeochemical processes in hydric soil formation, p. 68–127. *In* Current topics in wetland biogeochemistry. Vol. 1. Louisiana State University, Wetland Biogeochemistry Institute, Baton Rouge, LA.

McGill, W.B., and C.V. Cole. 1981. Comparative aspects of organic C, N, S and P cycling through soil organic matter during pedogenesis. Geoderma 26:267–286.

McKinley, V.L., T.W. Federle, and J.R. Vestal. 1982. Effects of petroleum hydrocarbons on plant litter microbiota in an arctic lake. Appl. Environ. Microbiol. 43:129–135.

McLatchey, G.P., and K.R. Reddy. 1998. Regulation of organic matter decomposition and nutrient release in a wetland soil. J. Environ. Qual. 27:1268–1274.

Mendelssohn, I.A., K.L. McKee, and W.H. Patrick. 1981. Oxygen deficiency in *Spartina alterniflora* roots: Metabolic adaptation to anoxia. Science 214:439–441.

Mihelcic, J.R., D.R. Leuking, R.J. Mitzell, and J.M. Stapleton. 1993. Bioavailability of sorbed- and separate phase chemicals. Biodegradation 4:141–153.

Mitsch, W.J. 1994. Global wetlands: Old world and new. Elsevier, New York.

Mohn, W.H., and J.M. Tiedje. 1991. Evidence for chemiosmotic coupling of reductive dechlorination and ATP synthesis in *Desulfomonile tiedjei*. Arch. Microbiol. 157:1–6.

Molongoski, J.J., and M.J. Klug. 1976. Characterization of anaerobic heterotrophic bacteria isolated from freshwater lake sediments. Appl. Environ. Microbiol. 31:83–90.

Nannipieri, P., and J.M. Bollag. 1991. Use of enzymes to detoxify pesticide-contaminated soils and waters. J. Environ. Qual. 20:510–517.

Näsholm, T., Kielland, K. and Ganeteg, U. 2009. Uptake of organic nitrogen by plants. New Phytologist, 182:31–48.

Nealson, K.H., and C.R. Myers. 1992. Microbial reduction of manganese and iron: New approaches to carbon cycling. Appl. Environ. Microbiol. 58:439–443.

Neilson, A.H., A. Allard, P. Hynnig, M. Remberger, and L. Lander. 1983. Bacterial methylation of chlorinated phenols and guaiacols: Formation of veratroles from guaiacols and high-molecular-weight chlorinated lignin. Appl. Environ. Microbiol. 45:774–783.

Newman, S., and K.R. Reddy. 1993. Alkaline phosphatase activity in the sediment-water column of a hypereutrophic lake. J. Environ. Qual. 22:832–838.

Oberg, G., H. Brunberg, and O. Hjelm. 1997. Production of organically bound chlorine during degradation of birch wood by common white-rot fungi. Soil Biol. Biochem. 29:191–197.

Ogram, A.V., R.E. Jessup, L.T. Ou, and P.S.C. Rao. 1985. Effects of sorption on biological degradation rates of (2,4-dichloraphenexyl) bacteria acids in soils. Appl. Environ. Microbiol. 49:582–587.

Olila, O.G., K.R. Reddy, and W.G. Harris, Jr. 1994. Forms and distribution of inorganic phosphorus in sediments of two shallow eutrophic lakes in Florida. Hydrobiologia 129:45–65.

Pal, S., J.M. Bollag, and P.M. Huang. 1994. Role of abiotic and biotic catalysts in the transformation of phenolic compounds through oxidation coupling. Soil Biol. Biochem. 25:813–820.

Pardue, J.H., P.H. Masscheleyn, R.D. Delaune, and W.H. Patrick, Jr. 1993. Assimilation of hydrophobic chlorinated organics in freshwater wetlands: Sorption and sediment-water exchange. Environ. Sci. Technol. 27:875–882.

Paul, J.W., and E.G. Beauchamp. 1989. Denitrification and fermentation in plant-residue-amended soil. Biol. Fertil. Soils 7:303–309.

Pettersson, K. 1986. The fractional composition of phosphorus in lake sediments of different characteristics. In: Sly, P.G. (ed.) Sediments and Water Interactions, Springer, Berlin, 149–155.

Ponnamperuma, F.N. 1972. The chemistry of submerged soils. Adv. Agron. 24:29–96.

Post, W.M., W.R. Emanuel, P.J. Zinke, and A.G. Stangenbegar. 1982. Soil carbon pools and world life zones. Nature 298:156–159.

Reddy, K.R., and E.M. D'Angelo. 1994. Soil processes regulating water quality in wetlands, p. 309–324. In W. Mitsch (ed.) Global wetlands-old world and new. Elsevier Publishers, New York.

Reddy, K.R., E.M. D'Angelo, and T.A. DeBusk. 1989b. Oxygen transport through aquatic macrophytes: The role in wastewater treatment. J. Environ. Qual. 19:261–267.

Reddy, K.R., and W.F. DeBusk. 1987. Plant nutrient storage capabilities, p. 337–357. In K.R. Reddy and W.H. Smith (eds.) Aquatic plants for water treatment and resource recovery. Magnolia Publishing Inc., Orlando, FL.

Reddy, K.R., and R.D. DeLaune. 2008. Biogeochemistry of wetlands: Science and applications. CRC Press, Boca Raton, FL.

Reddy, K.R., O.A. Diaz, L.J. Scinto, and M. Agami. 1995. Phosphorus dynamics in selected wetlands and streams of the Lake Okeechobee Basin. Ecol. Eng. 5:183–207.

Reddy, K.R., R.H. Kadlec, E. Flaig, and P.M. Gale. 1999. Phosphorus assimilation in streams and wetlands. Crit. Rev. Environ. Sci. Tech. 29:1–64.

Reddy, K.R., G.A. O'Connor, and P.M. Gale. 1999. Phosphorus retention capacities of wetland soils and stream sediments impacted by dairy effluent. J. Environ. Qual. 29:1–64.

Reddy, K.R., and W.H. Patrick, Jr. 1984. Nitrogen transformations and loss in flooded soils and sediments. CRC Crit. Rev. Environ. Cont. 13:273–309.

Reddy, K.R., and W.H. Patrick, Jr. 1999. Biogeochemistry of wetlands. CRC/Lewis Publishers, Boca Raton, FL.

Reddy, K.R., W.H. Patrick, Jr., and C.W. Lindau. 1989a. Nitrification-denitrification at the plant-root-sediment interface in wetlands. Limnol. Oceanogr. 34:1004–1013.

Reddy, K.R., and K.M. Portier. 1987. Nutrient storage capabilities of aquatic and wetland plants, p. 337–357. In K.R. Reddy and W.H. Smith (eds.) Aquatic plants for water treatment and resource recovery. Magnolia Publishers, Inc., Orlando, FL.

Reddy, K.R., P.S.C. Rao, and R.E. Jessup. 1982. The effect of carbon mineralization and denitrification kinetics in mineral and organic soils. Soil Sci. Soc. Am. J. 46:62–68.

Reddy, K.R., Wang, Y., Debusk, W.F., Fisher, M.M., and Newman, S. 1998. Forms of soil phosphorus in selected hydrologic units of the Florida everglades. Soil Sci. Soc. Am. J. 62:1134–1147.

Richardson, C.J. 1985. Mechanisms controlling phosphorus retention capacity in freshwater wetlands. Science 228:1424–1426.

Richardson, C.J., and P.E. Marshall. 1986. Processes controlling movement, storage and export of phosphorus in a fen peatland. Ecol. Monogr. 56:279–302.

Sanyal, S.K., and S.K. DeDatta. 1991. Chemistry of phosphorus transformations in soils. Adv. Agron. 16:1–32.

Sarkar, J.M., R.L. Malcolm, and J.M. Bollag. 1988. Enzymatic coupling of 2,4-dichlorophenol to stream fulvic acid in the presence of oxidoreductases. Soil Sci. Soc. Am. J. 52:688–694.

Sayler, G.S., M.P. Shiaris, W. Beck, and S. Held. 1982. Effects of polychlorinated biphenyls and environmental biotransformation products on aquatic nitrification. Appl. Environ. Microbiol. 43:949–952.

Schipper, L.A., and K.R. Reddy. 1996. Methane oxidation in the rhizosphere of *Sagittaria lancifolia*. Soil Sci. Soc. Am. J. 60:611–616.

Schlesinger, W.H. 1997. Biogeochemistry—An analysis of global change. 2nd edn. Academic Press, Orlando, FL.

Shaver, G.R. and J.M. Melillo. 1984. Nutrient budgets of marsh plants: Efficiency concepts and relation to availability. Ecology 65:1491–1510.

Shindo, H., and P.M. Huang. 1984. Catalytic effects of manganese (IV), iron (III), aluminum, and silicon oxides on the formation of phenolic polymers. Soil Sci. Soc. Am. J. 48:927–934.

Sinsabaugh, R.L. 1994. Enzymatic analysis of microbial patterns and processes. Biol. Fertil. Soil 17:69–74.

Sinsabaugh, R.L., R.K. Antibus, A.E. Linkens, C.A. McClaugherty, L. Rayburn, D. Repert, and T. Weiland. 1993. Wood decomposition: Nitrogen and phosphorus dynamics in relation to extracellular enzyme activity. Ecology 74:1586–1593.

Smith, C.J., R.D. DeLaune, and W.H. Patrick, Jr. 1982. Nitrate reduction in *Spartina alterniflora* marsh soil. Soil Sci. Soc. Am. J. 46:748–750.

Smith, R.V., R.H. Foy, S.D. Lennox, C. Jordan, L.C. Burns, J.E. Cooper and R.J. Stevens. 1995. Occurrence of Nitrite in the Lough Neagh River System Published in Journal of Environmental Quality 24:952–959.

Smith, R.V., L.C. Burns, R.M. Doyle, S.D. Lennox, B.H.L. Kelso, R.H. Foy, and R.J. Stevens. 1997. Free ammonia inhibition of nitrification in river sediments leading to nitrite accumulation. J. Environ. Qual. 26:1049–1055.

Snyder, G.H. 1987. Agricultural flooding of organic soils. Univ. FL. Institute of Food and Agricultural Sciences Bull. 870.

Soil Survey Staff. 1999. Soil taxonomy: A basic system of soil classification for making and interpreting soil surveys. 2nd edn. USDA-NRCS, U.S. Govt. Printing Office, Washington, DC.

Sonzogni, W.C., S.C. Chapra, D.E. Armstrong, and T.J. Logan. 1982. Bioavailability of phosphorus inputs to lakes. J. Environ. Qual. 11:555–563.

Stanford, G., M.H. Frere, and D.H. Schwaniger. 1973. Temperature coefficient of soil nitrogen mineralization. Soil Sci. 115:321–323.

Stephens, J.C. 1969. Peat and muck drainage problems. J. Irrig. Drainage Div. Proc. Am. Soc. Civil Eng. 95:285–305.

Stewart, J.W.B., and H. Tiessen. 1987. Dynamics of soil organic phosphorus. Biogeochemistry 4:41–60.

Taraban, R.H., O.F. Berry, D.A. Berry, and H.L. Walker, Jr. 1993. Degradation of dicamba by an anaerobic consortium enriched from wetland soil. Appl. Environ. Microbiol. 7:2332–2334.

Tate, K.R. 1984. The biological transformation of P in soil. Plant Soil 76:245–256.

Tiessen, H., J.W.B. Stewart, and A. Oberson. 1994. Innovative soil phosphorus availability indices: Assessing organic phosphorus, p. 143–162. *In* Soil testing: Prospects for improving nutrient recommendations. Soil Science Society of America Special Publication 40, Madison, WI.

U.S. Army Corps of Engineers. 1987. USACE wetland delineation manual. U.S. Army Engineer Waterway Experiment Station Technical Report Y-87-1, Vicksburg, MS.

U.S. Department of Agriculture (USDA). 1981. Land resource regions and major land resource areas of the United States. USDA-SCS agriculture handbook 286. U.S. Government Printing Office, Washington, DC.

U.S. Department of Agriculture. 1991. Hydric soils of the United States. 3rd edn. SCS Miscellaneous Publication 1491.

U.S. Department of Agriculture, Natural Resource Conservation Service (USDA-NRCS). 2006. Field indicators of hydric soils in the United States. Version 6. G.W. Hurt and L.M. Vasilas (eds.) in cooperation with the National Technical Committee on Hydric Soils.

Vepraskas, M.J. 1992. Redoximorphic features for identifying aquic conditions. Tech. Bull. 301. North Carolina Agricultural Research Service, North Carolina State University, Raleigh, NC.

Walbridge, M.R., and J.P. Struthers. 1993. Phosphorus retention in non-tidal palustrine forested wetlands of the Mid-Atlantic Region. Wetlands 13:84–94.

Wheatly, R.E., M.P. Greaves, and R.H.E. Inkson. 1976. The aerobic bacterial flora of a raised bog. Soil Biol. Biochem. 8:453–460.

White, D.S., and B.L. Howes. 1994. Translocation, remineralization and turnover of nitrogen in the roots and rhizomes of *Spartina alterniflora*. Am. J. Bot. 81:1225–1234.

Wildung, R.E., R.L. Schmidt, and R.C. Rouston. 1977. The phosphorus status of eutrophic lake sediments as related to limnological conditions-phosphorus mineral components. J. Environ. Qual. 6:100–104.

Williams, J.D.H., J.K. Syers, R.F. Harris, and D.E. Armstrong. 1971. Fractionation of inorganic phosphate in calcareous lake sediments. Soil Sci. Soc. Am. Proc. 35:250–255.

21
Acid Sulfate Soils

21.1 Introduction 21-1
21.2 Characteristics, Occurrence, and General Impacts 21-1
Characteristics • Occurrence
21.3 Characteristic Minerals and Geochemical Processes 21-4
Reduction Processes • Oxidation Processes • Pyrite Oxidation Rates
21.4 Modeling Soil Physical and Chemical Processes in Acid Sulfate Soils 21-7
21.5 Hazards Arising from Acid Sulfate Soil Disturbance 21-8
Acidification • Iron Mobilization • Other Metal and Metalloid Mobilization •
Deoxygenation of Water Bodies • Production of Noxious Gases • Scalding
of Acid Sulfate Soil Landscapes
21.6 Physical Behavior 21-12
Consistence and Strength • Permeability
21.7 Use, Management, and Evaluation of Acid Sulfate Soil Landscapes 21-13
Introduction • Land Use on Sulfidic Acid Sulfate Soils • Land Use on Sulfuric Acid
Sulfate Soils • Integrated Soil, Water, and Crop Management • Land Evaluation
21.8 Identification and Assessment 21-16
Identification and Assessment of Sulfuric Material • Identification and Assessment
of Sulfidic Material
21.9 Conclusions 21-20
References 21-20

L.A. Sullivan
Southern Cross University

R.T. Bush
Southern Cross University

E.D. Burton
Southern Cross University

C.J. Ritsema
Winand Staring Centre

M.E.F. van Mensvoort
(Retired)
Wageningen Agricultural University

21.1 Introduction

Acid sulfate soil materials are distinguished from other soil materials by having properties and behavior that have either (1) been affected considerably (mainly by severe acidification) by the oxidation of reduced inorganic sulfur (RIS) or (2) the capacity to be affected considerably (again mainly by severe acidification) by the oxidation of their RIS constituents.

A wide range of environmental hazards can be generated by the oxidation of RIS. These include (1) severe acidification of soil and drainage waters (below pH 4 and often <pH 3), (2) mobilization of metals (e.g., iron, aluminum, copper, cobalt, zinc), metalloids (e.g., arsenic), nutrients (e.g., phosphate), and rare earth elements, (3) deoxygenation of water bodies, (4) production of noxious gases (e.g., H_2S), and (5) scalding (i.e., devegetation) of landscapes. Some of these hazards are caused directly or indirectly by the severe acidification that can occur as a result of the oxidation of RIS, whereas some are not.

The properties and behavior of acid sulfate soils can impact detrimentally on infrastructure such as bridges, drains, pipes, roads—especially when constructed of steel and concrete. Waters draining from acid sulfate soil materials may be enriched in a wide range of potential toxicants, including metals and metalloids, endangering aquatic life and public health. Crops, trees, pastures, and aquaculture may also be severely affected by acid sulfate soil materials. Acid sulfate soils can have detrimental impacts on their surrounding environments as well as on communities who live in landscapes containing these soils.

The assessment and management of acid sulfate soils often require an integration of knowledge from a wide range of disciplines including geochemistry, pedology, hydrology, toxicology, geomorphology, microbiology, geotechnical engineering, geology, agronomy, sedimentology, environmental health, and ecology. The complexities involved with their identification, assessment, classification, and management ensure that acid sulfate soils present a wide range of challenges for soil science.

21.2 Characteristics, Occurrence, and General Impacts

21.2.1 Characteristics

It is useful to distinguish between sulfidic soil materials that, if disturbed sufficiently, will become severely acidified, and sulfuric soil materials that have already become severely acidic as a result of the oxidation of RIS constituents.

FIGURE 21.1 (a) Profile of an acid sulfate soil with sulfuric materials exposed showing jarosite and schwertmannite segregations. (b) Detail of yellow jarosite segregations in root channels (image width = 8 cm). (c) Profile of acid sulfate soil located in acid sulfate soil scald with sulfuric materials exposed. (d) Iron staining (by schwertmannite and goethite) of a drain in an acid sulfate soil landscape. The spade in (a) and (c) is 28 cm in length.

Sulfidic materials may be current or former marine and estuarine sediments, sediments in brackish lakes and lagoons, peats that originally formed in freshwater but have been inundated subsequently by brackish water, or accumulations of sediment in water bodies such as drains or wetlands affected by salinity (especially when sulfate is an appreciable component of that salinity). The required conditions for the formation and accumulation of RIS are (1) a supply of organic matter, (2) reducing conditions sufficient for sulfate reduction brought about by continuous waterlogging, (3) a supply of sulfate from tidewater or other saline ground or surface waters, (the sulfate is reduced to sulfides by bacteria decomposing the organic matter), and (4) a supply of iron from the sediment for the accumulation of iron sulfides, which make up the bulk of the RIS. Detailed discussions of the reactions that arise from these conditions are presented in Section 21.3.1 as well as in Chapter 27 of *Handbook of Soil Sciences: Properties and Processes*.

These conditions are found in tidal swamps and salt marshes where, over the last 10,000 years, thick deposits of sulfidic clay have accumulated in many locations around the globe (Pons and van Breemen, 1982; Dent and Pons, 1995). Figure 21.1a shows an acid sulfate profile highlighting the often characteristic yellow segregations of jarosite ($KFe_3(SO_4)_2(OH)_6$) and red–brown segregations of schwertmannite ($Fe_8O_8(OH)_6(SO_4)$) and goethite ($\alpha FeOOH$) in the uppermost sulfuric layers. Sulfidic layers vary greatly in appearance but often have the gleyed colors typical of soil materials that are dominated by reduced waterlogged conditions.

It is significant that the distribution of RIS within the soil matrix of acid sulfate soils is not uniform; this both affects acid sulfate soil behavior and has implications for their management. Sandy and clayey acid sulfate soils often contain grains of pyrite (FeS_2) transported from elsewhere and deposited within the sediment. Subsequently, especially in slowly deposited clays, masses of framboids or individual crystals of FeS_2 (e.g., Figure 21.2a and b) accumulate in situ in voids such as decomposing roots that provide both organic matter for bacterial decomposition and channels for the diffusion of SO_4^{2-} and dissolved Fe. Although the matrix may contain only relatively few embedded grains of pyrite, the network of neo-formed FeS_2 may raise the overall concentration of pyrite to several percent on a soil dry mass basis. Where freshwater conditions have followed brackish tidal conditions, sulfidic soil materials may have been buried by peat or alluvium devoid of RIS (Diemont et al., 1993).

Disturbance of sulfidic soils by, for example, drainage or excavation often causes dramatic changes in the properties of these soil

Acid Sulfate Soils

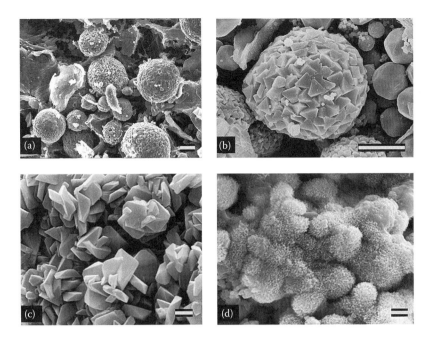

FIGURE 21.2 (a) Pyrite framboids in a root remnant of a sulfidic soil material. (b) Framboids and single crystals of pyrite in a void in a sulfidic material. (c) Jarosite crystals in a sulfuric soil material. (d) Segregations of schwertmannite with characteristic "ball and whisker" morphology. The bar in (a) indicates 10 μm, in (b) 5 μm, and in (c), and (d) 1 μm.

materials and the draining waters. If there are insufficient effective neutralizing materials (such as fine-grained calcium carbonate) in the sediment to neutralize the acidity generated by the oxidation of sulfides, extreme acidity can develop within weeks or months, resulting in sulfuric soil material (Figure 21.3). Sulfuric soil material is characterized by acidic pHs (e.g., pHs < 4), and usually presents yellow segregations of jarosite around pores and on ped faces (e.g., Figure 21.1b and c). Acid sulfate soils of peaty constitution usually do not have visible jarosite segregations, presumably because these soil materials contain only minor amounts of the phyllosilicate clays that act as the main source, upon acid dissolution, of the K^+ that is necessary for jarosite precipitation.

Waters draining from acid sulfate soil landscapes can exhibit a range of colors and clarity. For example, they can be red/brown from precipitation of iron flocs, white from the formation of aluminum flocs (especially if the pH is suddenly increased, for example, by the introduction of drainage waters into tidal waters), blue-green from dissolved ferrous iron compounds, or crystal clear because low pH and Al^{3+} can effectively flocculate clays. In still-water conditions, the surfaces of acid sulfate soil-affected water-bodies can develop a brittle, iridescent film formed of iron-oxidizing bacteria. Acid sulfate soil drainage waters can often have pH < 3.5 and can be the cause of massive fish kills, the death of invertebrates and benthic organisms, the development of chronic fish diseases, and impaired fish recruitment (Sammut et al., 1993).

Acid sulfate soils can also present health hazards to people living in landscapes containing these soils. Ljung et al. (2009) found that acid sulfate soils could impact detrimentally on human health. The human health issues were related not only to the increased mobility of acid and metals from these soils affecting drinking water quality, food production and quality, but also to other issues such as increased dust generation, causing respiratory health issues and acidic pools of surface water in acid sulfate soil landscapes, providing suitable environments for mosquito breeding.

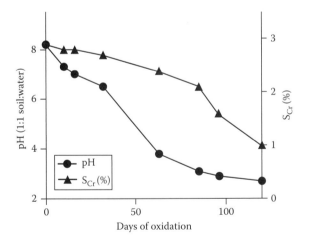

FIGURE 21.3 Typical change in S_{Cr} (mainly pyritic sulfur) and pH (1:1 soil:water) during moist oxidation of 10 mm thick slabs of a clayey gel-like sulfidic soil material from McLeods Creek in eastern Australia. (Modified from Ward, N.J., L.A. Sullivan, and R.T. Bush. 2004. Soil pH, oxygen availability and the rate of sulfide oxidation in acid sulfate soil materials: Implications for environmental hazard assessment. Aust. J. Soil Res. 42:509–514.)

21.2.2 Occurrence

Estimates of the extent and distribution of acid sulfate soils globally suffer from scant field surveys, inadequate laboratory data, and also the lack of uniform, widely accepted definitions of these materials, as will be discussed later. Improvements in these areas have, however, led to better quantification of their extent and, in Australia at least, to better mapping of their distribution. Jennings et al. (1979) estimated the global extent of acid sulfate soil to be 15 million ha; 20 years later, this estimate was revised upward to 50 million ha (Ritsema et al., 2000). The recent Australian Atlas of Acid Sulfate Soils (Fitzpatrick et al., 2008) has greatly improved our understanding of the extent and distribution of acid sulfate soils within Australia.

The location of these soils is even more significant than their extent. Acid sulfate soils are often concentrated in otherwise densely settled coastal regions and floodplains where development pressures are intense and little suitable alternative land exists for the expansion of farming or urban and industrial development. As well as being commonly found in tropical regions (e.g., Vietnam, Thailand, Indonesia, Malaysia, and northern Australia), acid sulfate soils are also commonly found in temperate regions such as (e.g., the United States, Scandinavia, and southern Australia).

Although acid sulfate soils are often thought of as almost exclusively a coastal issue, acid sulfate soils are also widely distributed in inland areas wherever the general conditions for RIS formation—a ready source of sulfate, iron, and organic matter in reducing waterlogged sediments—are met. In Australia, the large areas affected by human-induced salinity, caused by overclearing of trees and suboptimal irrigation practices, have also often been found to be areas affected by the contemporary formation of acid sulfate soil materials (Fitzpatrick et al., 1996; Sullivan et al., 2002; Fitzpatrick et al., 2009).

There is significant spatial variability in the distribution and properties of acid sulfate soils from the microscale (as discussed previously for the distribution of pyrite in relation to soil microstructure) to the regional scale. Dent (1986) gives examples of soil patterns at different scales that have been unearthed by soil surveys. Surveying acid sulfate soils in tropical tidal environments is often arduous and enlivened by mosquitoes and crocodiles: accordingly, remote sensing has been widely used for these activities.

Quantitative methods of analysis are required to support soil survey programs and to provide essential data for modeling the likely response of the land to management options. The required analyses must either be performed in a timely fashion before gross chemical changes take place, or the samples must be preserved quickly by methods such as rapid oven drying or ideally freezing, otherwise, the pH may fall markedly to <4, within days or weeks.

The methods of sampling, sample preparation, and analysis of acid sulfate soil materials vary widely according to the purpose of the study and the corresponding properties required. The methods of analysis vary from standard wet chemical methods [an authoritative, readily available reference for these methods is Ahern et al. (2004)], standard soil physical methods for properties such as texture, hydraulic conductivity, and bulk density, to x-ray diffraction, x-ray fluorescence, analytical electron microscopy, through to advanced synchrotron-based techniques. In terms of management of acid sulfate soil materials, the acid–base accounting approach (see Section 21.7.2.3) has significant advantages over other routine analytical approaches as it allows ready quantification of the acidity hazard, necessary for the rational determination of liming rates and for verification of management practices (Ahern et al., 2004).

21.3 Characteristic Minerals and Geochemical Processes

21.3.1 Reduction Processes

A defining characteristic of sulfidic acid sulfate soil materials is the presence of significant concentrations of RIS. RIS include iron disulfides (most commonly pyrite (FeS_2)) (Bloomfield and Coulter, 1973; Pons, 1973; van Breemen, 1973), and generally lower amounts of other minerals such as monosulfides (Georgala, 1980; Bush et al., 2000), greigite (Fe_3S_4) (Bush and Sullivan, 1997), and elemental sulfur (S_8) (Burton et al., 2006a, 2006b).

The vast majority of RIS in sulfidic acid sulfate soil materials have formed at earth-surface temperatures and pressures under waterlogged, anoxic conditions. Under such conditions, accumulation of RIS species depends on microbially mediated sulfate reduction, which is itself dependent on organic carbon availability, supply of sulfate, and the amount of competing electron acceptors, including reactive Fe^{III}-minerals (Fanning et al., 2002). These variables influence the activity of dissimilatory sulfate-reducing microorganisms, which include phylogenetically diverse anaerobes that oxidize simple organic compounds or hydrogen using sulfate as an electron acceptor. The overall process of dissimilatory sulfate reduction can be shown, for example, by

$$CH_3COO^- + SO_4^{2-} + H^+ \rightarrow H_2S + 2HCO_3^- \quad (21.1)$$

During this process, the sulfur in sulfate is reduced from the S^{6+} oxidation state to S^{2-}. Conditions that are conducive to microbially mediated sulfate reduction include organic-rich coastal and estuarine sediments, such as in tidal marshes and swamps. In such systems, tidal exchange of pore-water supplies sulfate and removes the resultant HCO_3^- produced via the reaction in Equation 21.1. Tidal flushing thereby prevents the accumulation of pore-water alkalinity. In iron-deficient systems, this tidal flushing can also remove pore-water H_2S and lead to its subsequent oxidation to elemental S (and eventually to sulfate).

In contrast, in soils containing appreciable Fe^{2+}, often produced by ferric iron reducing microorganisms, H_2S may react rapidly to form monosulfide (FeS) precipitates as below:

$$H_2S + Fe^{2+} \rightarrow FeS + 2H^+ \quad (21.2)$$

The initial FeS phase to form by reaction between H_2S and Fe^{2+} (Equation 21.2) has proved difficult to characterize, even in well-defined synthetic studies (Rickard and Morse, 2005). Recently, such studies have shown that nanoparticulate mackinawite (tetragonal FeS) is the first condensed phase to

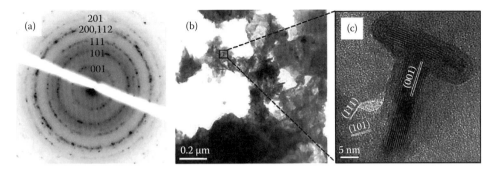

FIGURE 21.4 Transmission electron diffraction pattern and micrographs showing nanoparticulate mackinawite in monosulfidic soil materials. (Image adapted with permission from Burton, E.D., R.T. Bush, L.A. Sullivan, R.K. Hocking, D.R.G. Mitchell, S.G. Johnston, R.W. Fitzpatrick, M. Raven, S. McClure, and L.Y. Jang. 2009. Iron-monosulfide oxidation in natural sediments: Resolving microbially mediated S transformations using XANES, electron microscopy, and selective extractions. Environ. Sci. Technol. 43:3128–3134. Copyright 2009 American Chemical Society.)

FIGURE 21.5 (a) Accumulation of monosulfidic black ooze (MBO) showing the black gel-like MBO after the overlying brown oxidized surface layer has been removed (field of view width is 35 cm). (b) A field trial on an acid sulfate scald showing successful early revegetation in plots treated by a combination of liming, ridging, mulching and stock exclusion.

form through this reaction. In acid sulfate soil materials the occurrence of mackinawite as 5–30 nm nanoparticles has been only recently demonstrated (Figure 21.4) (Burton et al., 2009). The strong black color characteristic of these acid sulfate soil materials (e.g., the monosulfidic black ooze in Figure 21.5a) is largely due to the presence of nanoparticulate mackinawite (Burton et al., 2009).

The H_2S produced by microbial sulfate reduction can also react with Fe^{III} contained in ferric oxide and oxyhydroxide minerals such as goethite, to produce elemental sulfur:

$$H_2S + 2FeOOH + 4H^+ \rightarrow 0.125\, S_8 + 2Fe^{2+} + 4H_2O \quad (21.3)$$

The Fe^{2+} produced via this reaction may then feed into the reaction described by Equation 21.2, thus also resulting in mackinawite formation. This overall process, termed "sulfidization" can be represented as

$$3H_2S + 2FeOOH \rightarrow 0.175\, S_8 + 2FeS + 4H_2O \quad (21.4)$$

In the presence of an oxidant, such as O_2, mackinawite is unstable and can transform readily via a solid-state process to greigite

$$4FeS + 0.5O_2 + 2H^+ \rightarrow Fe_3S_4 + Fe^{2+} + H_2O \quad (21.5)$$

Although frequently mentioned, there are only few studies (e.g., Bush and Sullivan, 1997) that conclusively document the occurrence of greigite in acid sulfate soil materials. On the basis of the limited amount of field data, it appears that greigite occurrence is limited to the oxidation front in mildly acidic soils that are subject to an oscillating groundwater table. Mackinawite and greigite are often described as "iron-monosulfide" minerals because they have Fe:S ratios that are close to 1:1 (Rickard and Morse, 2005). These mineral species are measured analytically by their dissolution in HCl to yield H_2S gas and described as acid volatile sulfide (AVS).

Both mackinawite and greigite have long been implicated as precursors to the formation of iron disulfides such as pyrite and marcasite. For example,

$$Fe_3S_4 + 2H^+ \rightarrow 2FeS_2 + Fe^{2+} + H_2 \quad (21.6)$$

Pyrite can also form without the need for precursory greigite via (1) mackinawite oxidation by polysulfide species (Rickard, 1975; Luther, 1991) and (2) mackinawite oxidation by H_2S (Rickard, 1997; Rickard and Luther, 1997). These two pathways of pyrite formation, which involve an intermediate dissolved FeS cluster complex, can be represented overall as

Polysulfide pathway

$$FeS + S_n^{2-} \rightarrow FeS_2 + S_{n-1}^{2-} \quad (21.7)$$

Hydrogen sulfide pathway

$$FeS + H_2S \rightarrow FeS_2 + H_2 \quad (21.8)$$

Although iron monosulfides are widely believed to be an essential precursor to pyrite formation, this is not necessarily always the case. Pyrite can form quite rapidly in the presence of suitable reactive surfaces such as bacterial surfaces (Canfield et al., 1997) that serve to overcome a significant supersaturation threshold by providing heterogeneous nucleation sites. Other suitable reactive surfaces include preexisting pyrite crystals or organic substrates such as plant material. Accumulation of pyrite in soil can occur rapidly under suitable field conditions (Howarth, 1979; Rosicky et al., 2004a).

Pyrite is by far the most commonly observed RIS species in sulfidic acid sulfate soil materials. In these materials, pyrite presents a range of distinct crystal morphologies (e.g., Figure 21.2a and b). The most remarkable of these morphologies are framboids (from the French term for raspberry—*framboise*). Pyrite framboids consist of spheroidal aggregates of densely packed, individual microcrystals. Earlier research into the origin of pyrite framboids in sediments pointed toward either a bacterial influence or the magnetic aggregation of precursor greigite crystals. However, it now seems that the formation of framboids is more likely a function of the degree of solution supersaturation with regard to pyrite.

While pyrite is normally the most abundant iron disulfide in acid sulfate soil materials, marcasite (orthorhombic FeS_2) may occur in specific situations. Acidic conditions (pH < 6) are required for the initial formation of marcasite instead of pyrite. Such conditions occur in waterlogged soils and sediments that are rich in dissolved organic acids, capable of buffering the low pH. For example, marcasite is a common iron sulfide in some peaty acid sulfate soil materials in eastern Australia (Bush et al., 2004a).

21.3.2 Oxidation Processes

Pyrite and the other iron-sulfide minerals persist in soils only under anoxic, waterlogged conditions. If these conditions become oxic by, for example excavation of the soils, the iron-sulfide components can undergo a series of oxidation reactions. For example, in the presence of oxygen (and water) pyrite oxidizes to ultimately yield sulfuric acid and a poorly soluble Fe^{III} precipitate:

$$FeS_2 + 3.75 O_2 + 3.5 H_2O \rightarrow Fe(OH)_3 + 4H^+ + 2SO_4^{2-} \quad (21.9)$$

While this reaction shows that exposure to oxygen under moist conditions is the driving force for pyrite oxidation, it neglects the great complexity of reaction steps in the overall oxidation process. This complexity includes a number of possible final iron phases as well as the formation of intermediate sulfoxy anions and elemental S. Some of the more important S transformations occurring during pyrite oxidation are shown in Figure 21.6.

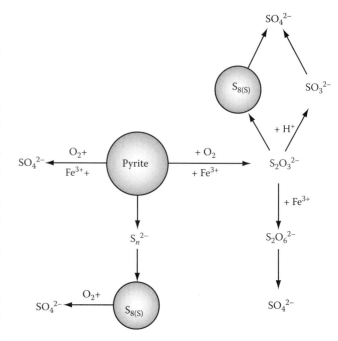

FIGURE 21.6 Conceptual model showing sulfur transformations during pyrite oxidation. (Modified from Druschel, G.K., Baker, B.J., Gihring, T.H., and Banfield, J.F. 2004. Acid mine drainage biogeochemistry at Iron Mountain, California. Geochem. Trans. 5(2):13–32.)

Chemolithotrophic Fe- and S-oxidizing bacteria play an important role in mediating various steps in the overall oxidation process, and in determining the formation and persistence of intermediate S species.

A conceptual model of iron behavior during the pyrite oxidation process is presented in Figure 21.7. This model highlights the wide variety of potential phases that play a role in determining the iron biogeochemistry following pyrite oxidation. Ferrous iron released in the initial stages of pyrite oxidation

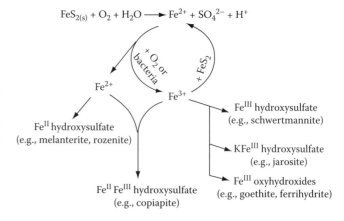

FIGURE 21.7 Conceptual model of iron transformations during and following pyrite oxidation. (Adapted from Rose, A.W., and C.A. Cravotta, III. 1998. Geochemistry of coal mine drainage, p. 1-1, 1–22. *In* K.B.C. Brady, M.W. Smith, and J. Sehueck (eds.) Coal mine drainage prediction and pollution prevention in Pennsylvania. PA Department of Environmental Protection: Commonwealth of Pennsylvania, Harrisburg, PA. 5600-BK-DEP2256.)

may precipitate as Fe^{II} hydroxysulfate minerals (Fanning et al., 2002), most importantly melanterite, rozenite, and szomolnokite. These phases are readily soluble and are rarely observed in acid sulfate soil materials.

Under continuation of oxidizing conditions, the Fe^{2+} released by pyrite oxidation is also subject to oxidation to Fe^{3+}. While the simple oxidation process consumes some acidity, the subsequent hydrolysis of the resulting Fe^{3+} leads to the liberation of acidity. At low pH (e.g., <4), Fe^{3+} is sufficiently soluble that it may serve as a very effective electron acceptor driving further pyrite oxidation (Moses et al., 1987). For this reason, it has been often suggested that the rate of Fe^{2+} oxidation to Fe^{3+} may be the rate-determining step in pyrite oxidation.

Partial oxidation of Fe^{2+} to Fe^{3+} can lead to precipitates of mixed valence Fe salts, such as copiapite. This phase is one of the "soluble salts" that may form in acid-sulfate soils under prolonged dry conditions (Fanning et al., 2002). Dissolution of these minerals during rainfall events may cause a first-flush of stored acidity.

The Fe^{3+} produced via pyrite oxidation also commonly precipitates as a range of Fe^{III}-bearing minerals. In acid sulfate soil conditions at pH < 3, and/or in the presence of abundant K^+, jarosite appears to be the predominant Fe^{III} phase, whereas in the pH range of 3-4, schwertmannite is an important Fe^{III} phase in acid-sulfate soil landscapes (Bigham et al., 1992; Sullivan and Bush, 2004). The widespread occurrence of schwertmannite in acid-sulfate soils has only been confirmed relatively recently (Sullivan and Bush, 2004). Schwertmannite has poor crystallinity and often presents a "ball-and-whisker" micromorphology (Figure 21.2d).

Schwertmannite is metastable and over time transforms, via dissolution-reprecipitation, to form a range of Fe^{III} oxyhydroxides (Bigham et al., 1996). These include ferrihydrite, lepidocrocite, and goethite, with the latter being most stable. The transformation of schwertmannite (an Fe^{III} oxyhydroxysulfate) to these Fe^{III} oxyhydroxides involves the hydrolysis of $Fe^{3+}$and the liberation of acidity. As a consequence, schwertmannite transformation can suppress pH long after the initial source of acidification (i.e., pyrite) has been consumed.

The type of secondary minerals formed from the Fe released during pyrite oxidation determines, to a large extent, the amount of acidity expressed (Dold, 2003; Dold and Fontbote, 2001). For example, if the released Fe precipitates as goethite or ferrihydrite from the Fe^{3+} produced by sulfide oxidation, then 3.0 mol of H^+ are formed for every mole of Fe^{3+} hydrolyzed from pyrite. However, if hydrolysis is incomplete and jarosite is formed, only around 2 mol of H^+ is released for every mole of Fe^{3+} hydrolyzed from pyrite (van Breemen, 1976). If schwertmannite is formed, then approximately 2.575 mol of H^+ is released for every mole of Fe^{3+} hydrolyzed from pyrite (Piene et al., 2000). The "stored" acidity in these two minerals is important as the Fe in both jarosite and schwertmannite can undergo further hydrolysis and result in the release of acidity into the surrounding environment (Dold, 2003; Dold and Fontbote, 2001; Sullivan and Bush, 2004).

21.3.3 Pyrite Oxidation Rates

The oxidation of FeS_2 depends on factors including the supply of O_2, the availability of water, and the physical properties of FeS_2. Pyrite oxidation generates acid and releases heat; consequently, the acidity and temperature of the surrounding solution will affect the overall reaction rates. The oxidation of FeS_2 in the environment is usually ultimately determined by the supply of O_2. Models describing FeS_2 oxidation are often based on the assumption that all other constituents required for the oxidation process are freely available except for O_2, which is supplied through the porous material from the atmosphere (Dent and Raiswell, 1982; Davis and Ritchie, 1986; Pantelis and Ritchie, 1991; Bronswijk et al., 1993). The rate of pyritic oxidation is often assumed to be a linear function of the dissolved O_2 concentration (Bartlett, 1973; Braun et al., 1974) but the Michaelis–Menton equation has also been adopted (Liu et al., 1987; Tan, 1996).

Temperature, which influences both chemical and microbial oxidation, is an important factor in determining the oxidation rate of pyritic materials. Biological oxidation only occurs between 0°C and 55°C (optimum 25°C–45°C) (Lundgren and Silver, 1980) but chemical oxidation can take place above this temperature. Jaynes et al. (1984) modeling acid generation in mine spoil, took account of rates of diffusion of both O_2 and Fe^{3+} and also the activity of the bacteria generating Fe^{3+}, which was estimated from available energy and deviations from ideal temperature, solution pH and O_2 concentration. Pantelis and Ritchie (1992) introduced a ceiling temperature (100°C), above which microorganisms cease to be effective as catalysts in FeS_2 oxidation. The influence of temperature on oxidation rate follows the empirical Arrhenius equation (Ahonen and Tuovinen, 1991). Because the pyritic oxidation reaction is exothermic, temperature rises depending on the rate of reaction and thermal properties of the bulk soil. In acid sulfate soils, temperature profiles might be used to determine pyritic oxidation rates but because pyritic layers are typically shallow (1–2 m below the surface), distinguishing between the calorimetric contribution from pyritic oxidation and solar radiation may prove difficult.

21.4 Modeling Soil Physical and Chemical Processes in Acid Sulfate Soils

Soil and water management are the key factors for sustainable use of acid sulfate soils in agricultural landscapes and natural habitats. Numerous complex physical and chemical processes contribute to the genesis of acid sulfate soils and determine the magnitude and rates of acidification or neutralization and the production/mobilization of (toxic) compounds. Because of this complexity, prediction of the effects of soil and water management and their practical consequences are only feasible by using simulation models. SMASS is one model for acid sulfate soils.

The SMASS model is described in detail in AARD/LAWOO (1993), Bronswijk and Groenenberg (1993), and Ritsema and Groenenberg (1993).

The SMASS model consists of a number of mutually linked submodels in which the various physical and chemical processes occurring in acid sulfate soils are described mathematically. To solve these equations, the soil profile is divided into compartments that may be of variable size. The initial physical and chemical conditions in each compartment and values for the boundary conditions must be given as model inputs for the complete simulation period. The physical and chemical conditions in each compartment, together with the water and solute fluxes at the boundary of the soil system are computed at selected time intervals in the following sequence: (1) The water transport submodel computes vertical water transport, giving the water content profile that complements that of air. (2) In the O_2 transport and FeS_2 oxidation submodel, air contents are used to compute O_2 diffusion coefficients in the air-filled macropores. Oxygen consumption values in the soil are calculated from FeS_2 and organic matter contents, and subsequently, the O_2 content profile in the macropores is computed. (3) Depending on the O_2 concentration at a given depth, the rate of FeS_2 oxidation at that depth is calculated in the O_2 transport and FeS_2 oxidation submodel. For each depth, the oxidized amount of FeS_2 is converted into amounts of H^+, Fe^{3+}, and SO_4^{2-} produced, with the remaining FeS_2 being used for calculations in the next time-step. (4) The solute transport submodel computes solute fluxes between soil compartments in accordance with the calculated water fluxes (from step 1). (5) In the chemical submodel, first the production/consumption terms for the nonequilibrium processes (such as Fe reduction) are calculated. Then the total concentrations of each chemical component are calculated in the soil compartments by summing the production/consumption terms, the inflow/outflow (from step 4), and the total amounts at the end of the previous time-step for each component. From these total concentrations, the equilibrium concentrations in the soil solution, the composition of the exchange complex, and the amount of minerals and precipitates are computed for each compartment.

Time-steps for computations of the water and solute transport submodels are in the order of hours. Pyrite oxidation, O_2 profiles, and chemical equilibria are computed once every day. The output of the SMASS model and its submodels is generally given on a daily basis. Model predictions can be carried out for one or more decades, so that the long-term effects of various water management strategies can be evaluated quantitatively.

21.5 Hazards Arising from Acid Sulfate Soil Disturbance

21.5.1 Acidification

Oxidation of RIS is the primary cause of the extreme acidification that characterizes sulfuric acid sulfate soil materials. By definition, the pH of sulfuric acid sulfate soil is <pH 4 (or <3.5 according to the particular soil taxonomy being employed) but values of pH < 3 in actively oxidizing soils are frequently observed (Dent, 1986). Such extreme acidification significantly alters the soil chemistry, often rendering it hostile to plants and can create a source of contamination to groundwater and surface water run-off. The acid produced can react with clay minerals and oxides to release silica and metal ions, principally, aluminum, iron, potassium, sodium, and magnesium (Nriagu, 1978). Other metal and metalloid ions can also be released (van Breemen, 1973; Sammut et al., 1996b; Åström, 2000).

The impacts of severe acid sulfate soil acidification on agricultural crops have been well documented (Dent, 1986). Many crop plants are highly sensitive to low pH soil conditions and acidification can greatly reduce yields and, in extreme cases, cause complete crop failure. In addition, the formation of acidic secondary iron minerals such as jarosite and schwertmannite can significantly reduce the availability of nutrients such as phosphorus and nitrogen. Farmers have tried many different approaches to ameliorate acidity by techniques, such as the addition of neutralizing agents, other soil amendments, organic mulch, and reconfiguring plant beds to enhance the leaching of acidic products from the soil (Dent, 1986). Success in cropping acid sulfate soil landscapes is mixed and highly dependent on the initial degree of acidification and capacity of the specific crop types to tolerate acidic conditions. Acidity severely constrains farming on acid sulfate soils with some exceptions (White et al., 1997).

Aluminum toxicity is a significant issue linked to acid sulfate soil acidification for terrestrial plants (Dent, 1986) and downstream aquatic flora and fauna (Sammut et al., 1996a, 1996b). The solubility of Al is critically dependent on pH, only becoming soluble at environmentally significant levels at approximately pH < 5. Soluble Al affects plant growth primarily by disrupting root function and is a major concern for food production and agricultural income for rural and regional communities. Severe environmental impacts can occur when acidic Al-rich leachate from acid sulfate soil enters water bodies. The more acute ecological impacts of acid sulfate soil acidification in waterways include fish kills (Sammut et al., 1996a, 1996b; Callinan et al., 2005), loss of native aquatic macrophytes and fauna followed by invasion by acid tolerant species (Sammut et al., 1996a), mass mortality of crustaceans and shell fish (Simpson and Pedini, 1985), and loss of benthic communities (Corfield, 2000). Sublethal exposure of fish to acidity has also been linked to an increased susceptibility to skin diseases (Callinan et al., 2005), whereas depletion of alkalinity has been linked to poor shell development in crustaceans (Dove and Sammut, 2007).

A range of potentially longer-term impacts on aquatic ecosystems arising from acid sulfate soil leachate include disturbance to fish reproduction and recruitment, acidity barriers to fish migration, decline of primary food web, reduction of species diversity, and long-term habitat degradation (Sammut et al., 1996a, 1996b). In assessing the likely impacts of acid sulfate soil acidification on downstream aquatic environments, it is necessary to consider the vulnerability of the aquatic ecosystems, the duration and frequency of acidification episodes, and the potential intensity of acidification based on the properties and quantities of the acidic leachate.

21.5.2 Iron Mobilization

As shown in Figure 21.7, Fe^{2+} is a primary product of pyrite oxidation. At high pH values (pH > 7), Fe^{2+} is chemically rapidly oxidized to Fe^{3+} (Cornell and Schwertmann, 2003). At much lower pHs (i.e., pH < 4.5), the oxidation of Fe^{2+} to Fe^{3+} is catalyzed by acidophilic lithotrophic bacteria such as *Acidithiobacillus ferroxidans* (Pronk and Johnson, 1992), *Thiobacillus ferroxidans*, and *Leptospirillium ferroxidans* (Johnson, 1993). The oxidation of Fe^{2+} has direct environmental consequences arising from the liberation of acidity and the formation of secondary iron minerals that can control soil and water geochemistry.

Accumulations of iron minerals are ubiquitous in acid sulfate soil landscapes. The precipitation and mineralogy of secondary iron minerals has been reviewed elsewhere in this chapter and in detail by Alpers and Nordstrom (1999) and Cornell and Schwertmann (2003).

Understanding the types of iron precipitates that form in acid sulfate soil landscapes during oxidation is important as particular iron mineral phases can exercise a major influence on the environment (e.g., Dold and Fontbote, 2001; Sullivan and Bush, 2004). In a study of surface iron precipitate accumulations associated with waterways in acid sulfate soil landscapes, Sullivan and Bush (2004) found that schwertmannite was the dominant secondary iron mineral. The schwertmannite occurred as coatings on vegetation, accumulations in low depressions, and iron flocs adhering to surfaces in acidified waterways. The potential acidity within the schwertmannite was high, ranging between 1900 and 2580 mol (H^+) t^{-1}, indicating that the schwertmannite was a substantial intermediate store of acidity within these acid sulfate soil landscapes. The retained acidity within both schwertannite and jarosite have recently been included into the quantitative assessment of the net acidity of sulfate soil materials (Ahern et al., 2004).

Iron precipitates in the form of iron flocs within the water column also are known to directly affect gilled organisms, smother benthic communities and aquatic flora (Sammut et al., 1996a, 1996b), diminish the aesthetic values of recreational waterways, and threaten estuarine and marine environments (Powell and Martens, 2005). The accumulation of iron flocs has also been linked to contemporary sulfur cycling and the formation of monosulfidic black ooze (MBO) accumulations in acid sulfate soil affected waterways (Figure 21.8).

21.5.3 Other Metal and Metalloid Mobilization

Mobilization of metals and metalloids to soil porewaters from acid sulfate soil can constitute a major environmental hazard (e.g. Åström et al., 2001; Burton et al., 2006c; Burton et al., 2008). Metals that have been reported at levels exceeding accepted environmental protection thresholds in acid sulfate

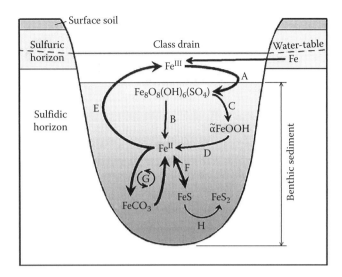

FIGURE 21.8 Conceptual model of in situ Fe transformations in accreting coastal lowland acid sulfate soil (class) waterways. (A) precipitation of schwertmannite ($Fe_8O_8(OH)_6(SO_4)$), (B) reductive dissolution of schwertmannite, (C) transformation of schwertmannite to goethite ($\tilde{\alpha}FeOOH$), (D) reductive dissolution of goethite, (E) upward diffusion and oxidation of Fe^{2+}, (F) precipitation–dissolution of disordered mackinawite (FeS), (G) regulation of porewater Fe^{2+} via precipitation–dissolution of siderite ($FeCO_3$), (H) and formation of pyrite (FeS_2; kinetically retarded due to high pore-water Fe^{2+} concentrations). (After Burton, E.D., R.T. Bush, and L.A. Sullivan. 2006d. Sedimentary iron geochemistry in acidic waterways associated with coastal lowland acid sulfate soils. Geochim. Cosmochim. Acta 70:5455–5468. Copyright (2006d), with permission from Elsevier.)

soil include Al, As, Ba, Cd, Co, Cr, Cu, Fe, Mn, Ni, Pb, Sb, V, Zn (e.g., Åström et al., 2001; Macdonald et al., 2004a; Burton et al., 2006c). Metals in natural soils occur within mineral phases or as ions or ionic complexes sorbed to reactive surfaces (Åström, 1998; Faltmarsch et al., 2008; Claff et al., 2010). Acidification greatly enhances the solubility of many metals, promoting their subsequent release from mineral phases by dissolution or cation exchange. The pH dependence of metal release has received considerable attention (Sammut et al., 1996b; Wilson et al., 1999; Åström, 2001; Preda and Cox, 2001; Macdonald et al., 2004a; Simpson et al., 2010), and there are strong similarities in metal release within acid sulfate soil and acid mine drainage systems (Evangelou and Zhang, 1995).

Numerous studies have documented the impacts from soluble metals on crop production (e.g., Dent, 1986), terrestrial habitats (van Breemen, 1973), and more recently, attention has turned to their impact on aquatic environments (Sammut et al., 1996a, 1996b; Wilson et al., 1999; Johnston et al., 2004; Callinan et al., 2005). Gilled organisms are particularly vulnerable to soluble metals and metal mobilization can lead to rapid mortality rates in these species (Simpson and Pedini, 1985; Sammut et al., 1993; Sammut et al., 1996a, 1996b). Studies of the effects of metals on shellfish (oysters) revealed longer-term, chronic impacts on their growth and survival (Dove and Sammut, 2007). However, the longer-term impacts of metal release from acid sulfate soils to surrounding aquatic environments are poorly understood. Although elevated metal concentrations can be toxic to both aquatic flora and fauna, the consequences of these conditions to algal and phytoplankton production are largely unknown, as is the potential for their bioaccumulation (Macdonald et al., 2004a).

Most reports on the impacts arising from metal release from acid sulfate soil focus on the consequences of metal mobilization under oxic-acidifying conditions. However, metals can also be mobilized when sulfuric acid sulfate soils are subject to prolonged inundation, resulting in the development of anoxic reducing conditions. Acid sulfate soil occurs in low-lying floodplain environments and, therefore, is subject to periodic water logging and oscillating redox conditions. The processes of metal mobilization and behavior of metals are very different under these conditions. The behavior of iron and arsenic are good examples of metal and metalloid mobilization from acid sulfate soil materials following inundation. Accumulations of iron minerals in acid sulfate soils are often concentrated near the ground surface and include goethite, ferrihydrite, jarosite, and schwertmannite. These iron minerals often have a large surface area and are a significant sink for the sorption of metals. Under reducing conditions, these iron oxides are prone to microbial reductive dissolution (van Breemen, 1973; Burton et al., 2007). Microbial iron reduction triggers three major changes that affect metal mobilization. First, it results in the dissolution of Fe^{3+} and transformation to Fe^{2+}, causing the corelease of other metals sorbed to the Fe mineral surfaces. Secondly, the microbial reduction process is proton-consuming and, when accompanied by the formation of bicarbonate as a by-product of microbial respiration, can result in an in situ neutralization (Blodau, 2006). The increase in porewater pH generally reduces the solubility of divalent metals and aluminum. Finally it also facilitates the recently identified Fe^{2+} catalyzed transformation of poorly crystalline iron oxide minerals to more crystalline phases (e.g., rapid transformation of schwertmannite to goethite). Although the overall consequences of these rapid mineral transformations on metal mobility are yet to be quantified (Burton et al., 2010), the mobility of some metals and metalloids can increase under these conditions. For example, arsenic is most soluble at around pH 5 and when associated with iron oxides in acid sulfate soil materials, it is readily mobilized at the onset of microbially mediated iron reduction (Burton et al., 2008). Arsenic contamination of groundwater and surface water is occurring as the result of such processes in acid sulfate soil landscapes (e.g. Johnston et al., 2010). It is important to recognize that metals and metalloids can have a significant impact in acid sulfate soil landscapes not only (1) when acid sulfate soil is allowed to oxidize and acidify, but (2) also following the prolonged inundation of previously oxidized, iron-enriched acid sulfate soil.

21.5.4 Deoxygenation of Water Bodies

Acute deoxygenation of estuaries, lakes, rivers, and drainage channels is a major contributor to catastrophic fish kills (Hamilton et al., 1997; Johnston et al., 2003; Howitt et al., 2007). Many potential factors contribute to deoxygenation events, and they are known to impact a very wide range of environments. Severe deoxygenation of waterways within acid sulfate soil landscapes has been linked directly to the behavior of acid sulfate soil materials (e.g., Sullivan and Bush, 2000).

Deoxygenation results when solids and aqueous compounds with a capacity to react with dissolved oxygen, enter water bodies and consume oxygen more rapidly than it can be replenished. The magnitude of deoxygenation depends on the spatial scale of the event, its persistence and its intensity. Aquatic ecosystems require dissolved oxygen concentrations generally greater than 85% saturation for lowland rivers (e.g., ANZECC/ARMCANZ 2000). Native fish and other large aquatic organisms are known to survive on dissolved oxygen concentration of as little as 2 mg L^{-1}, but may become stressed below 4–5 mg L^{-1} (Hladyz and Watkins, 2009). In recent studies of a major estuarine river system in eastern Australia affected by deoxygenation, Wong et al. (2010) found deoxygenation was confined to downstream of acid sulfate soil affected waterway confluences and occurred during the later phase of the flood recession.

Anaerobic decomposition of floodplain vegetation in backswamps can be a primary process leading to the deoxygenation of large volumes of waters in acid sulfate soil landscapes (Johnston et al., 2003; Wong et al., 2010). Decomposition of flood-intolerant vegetation in drained acid sulfate floodplains can lead to the formation of "blackwater"—a colloquial term used to describe anoxic stagnant floodplain water that develops a distinctive dark color as a result of the accumulation of dissolved organic carbon compounds. Blackwater is typically anoxic, has a high chemical oxygen demand (COD) and high dissolved Fe concentrations, and rapidly consumes dissolved oxygen when it discharges to main water bodies (Johnston et al., 2003). Extensive floodplain drainage networks in acid sulfate soil areas can significantly enhance the transport of hypoxic backswamp blackwater to main river channels, thereby enhancing the magnitude and duration of consequent estuarine deoxygenation.

The propensity for MBO to accumulate and be mobilized by floodwaters in drainage channels has also been identified as a contributing factor to deoxygenation in acid sulfate soil areas (Sullivan et al., 2002; Bush et al., 2004b, 2004c; Burton et al., 2006b, 2006d).

The chemistry of estuarine waters during hypoxic events has indicated elevated concentrations of redox sensitive species associated with acid sulfate soil (e.g., Fe^{2+}, dissolved Mn, and elemental sulfur) (Wong et al., 2010), further implicating acid sulfate soil and MBO materials as a cause of deoxygenation.

The role of MBO in deoxygenation and latter acidification in acid sulfate landscapes has only recently been discovered (Sullivan et al., 2000, 2002). Burton et al. (2006c) have described the oxidation dynamics of MBO when mobilized into oxygenated water. A summary of the changes in sulfur speciation (i.e., pyrite, iron monosulfide (FeS) elemental sulfur and sulfate) and pH during such MBO oxidation is shown in Figure 21.9. The oxidation

Acid Sulfate Soils

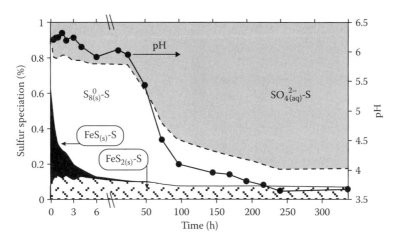

FIGURE 21.9 Changes to the sulfur speciation (pyrite (FeS$_2$), iron monosulfide (FeS) elemental sulfur, and sulfate) and pH during oxidation of monosulfidic black ooze (MBO). (With permission after Burton, E.D., R.T. Bush, and L.A. Sullivan. 2006c. Acid-volatile sulfide oxidation in coastal floodplain drains: Iron-sulfur cycling and effects on water quality. Environ. Sci. Technol. 40:1217–1222. Copyright 2006c American Chemical Society.)

of MBO follows a two-step process with oxygen consumption occurring with each step (after Burton et al., 2006c):

$$\text{Step 1} \quad \text{FeS} \begin{cases} \text{Fe}^{2+} + 0.5\text{O}_2 + 1.5\text{H}_2\text{O} \rightarrow 2\text{H}^+ + \text{FeOOH} \\ \text{S}^{2-} + 0.5\text{O}_2 + 2\text{H}^+ \rightarrow \text{H}_2\text{O} + 0.125\text{S}_8 \end{cases} \quad (21.10)$$

$$\text{Step 2} \quad 0.125\text{S}_8 + 1.5\text{O}_2 + \text{H}_2\text{O} \rightarrow \text{SO}_4^{2-} + 2\text{H}^+ \quad (21.11)$$

The first step is a rapid chemical reaction of iron monosulfide minerals with oxygen, forming iron oxides and elemental sulfur. This initial oxygen-consuming step does not affect pH and is, therefore, nonacidifying. It is probably for this reason that the role of MBO in deoxygenation was overlooked until recently. Acidification associated with MBO oxidation can result from the second step, the microbially mediated oxidation of elemental sulfur, when oxygen is available (Figure 21.9).

Elevated elemental sulfur concentrations in deoxygenated waterways in acid sulfate soil landscapes may be a useful indicator of MBOs as a contributing cause to deoxygenation, although elemental sulfur can also form as a primary product of H$_2$S oxidation, and may be present within MBOs, prior to flood events (Burton et al., 2006a, 2006b).

21.5.5 Production of Noxious Gases

Anthropogenic and biogenic sulfur-containing gases have important impacts on global climate change (Charlson et al., 1987; Lohmann and Feichter, 2005), and atmospheric acid–base chemistry (Berresheim et al., 1995). Coastal estuarine and marine environments are major emitters of biogenic H$_2$S (Aneja, 1990; Bates et al., 1992). Emissions of H$_2$S, and more recently SO$_2$, from floodplains have been linked to acid sulfate soil management (Macdonald et al., 2004b).

H$_2$S is a highly noxious gas that causes distress to humans (Luther et al., 2003; EPA, 2003) and threatens aquatic organisms (Diaz and Rosenberg, 1995; Rabalais, 2002). As described by Equation 21.1, H$_2$S is produced by sulfur-reducing bacteria under anoxic conditions. Even at small concentrations, H$_2$S can be detected by its characteristic rotten-egg odor. In acid sulfate soil landscapes, periodically inundated soil surfaces, shallow waterways, and field drains, where stratified anoxic conditions can develop, are all situations conducive to sulfate reduction and the formation of H$_2$S (Dent, 1986). However, H$_2$S is an unstable phase and its persistence in water and soil and ultimate gaseous emission is highly constrained by a wide range of oxidants in natural sediments and water bodies (Jorgensen et al., 1991). These oxidants include O$_2$, NO$_3$, Mn, and Fe oxyhydroxides (Froelich et al., 1979; Luther et al., 1997). Due to their abundance in acid sulfate soil, iron oxides (Millero et al., 1987) are a particularly effective oxidant of H$_2$S, a process leading to the formation of iron sulfides as described previously. Hydrogen sulfide becomes a problem when the rate of its formation exceeds the catalytic oxidative capacity of the sediments and water bodies to eliminate its gaseous emission. An excess of labile carbon and stagnant water bodies create conditions that favor H$_2$S emissions in acid sulfate soil landscapes (Rozan et al., 2002).

Partially oxidized RIS-containing acid sulfate soil materials are a known source of SO$_2$. Macdonald et al. (2004b) quantified SO$_2$ flux from agricultural acid sulfate soils using both ground chamber and micro metrological methods. In this study, the rates of SO$_2$ emission from the soil was closely linked to soil moisture and evaporative flux, leading the authors to conclude that acidic dissociation of sulfite (SO$_3^{2-}$) occurring within the near-surface soil porewater was probably the major source of SO$_2$. The precise mechanisms for SO$_2$ formation in acid sulfate soil require resolution—bacterial processes that utilize sulfate (Saltzman and Cooper 1989) or organo-sulfur compounds (Freney, 1961) are both possibilities. From relatively few measurements, Macdonald et al. (2004b) estimated global SO$_2$ emissions from acid sulfate soil to be 3.0 Tg S year^{-1}, ~3% of global anthropogenic emissions.

21.5.6 Scalding of Acid Sulfate Soil Landscapes

Scalded (i.e., nonvegetated) land surfaces can be an extreme symptom of land degradation and in low-lying acid sulfate soil landscapes can extend for hundreds of hectares, impacting both the environment and those who live and rely on these areas. Scalded acid sulfate soil land is environmentally damaging, agriculturally unproductive, and difficult to rehabilitate. There are a multitude of causes for the complete and prolonged failure of vegetation to establish. In acid sulfate soil landscapes, extreme acidification and/or salinization are often involved with the initiation of scalds (Rosicky et al., 2004a, 2004b). Peat fires arising from the desiccation of low-lying backswamps can also lead to the formation of scalds, as can the prolonged inundation of low-lying areas with acidic-aluminum-iron rich surface waters.

The size and condition of scalds vary considerably, spatially, and temporally. In a broad study of scalds along the east coast of Australia, Rosicky et al. (2004a) found that even relatively minor changes such as a shift to wetter conditions could instigate the rapid growth of acid tolerant plants such as spike-rush (*Eleocharis acuta*). The establishment of such revegetation typically would advance from the edge of scald, only to die off and recede when drier conditions returned.

Rosicky et al. (2004a, 2004b) found that the surface soil layers of scalds experienced extreme acidification (pH < 3), evaporative accumulation of acidic salts and metals (Al, Fe), high salinities caused by the accumulation of evaporative salts (e.g., gypsum), and accumulations of iron minerals (e.g., schwertmannite, ferrihydrite, goethite and jarosite). Combined with other stresses such as grazing pressure and frosts, such soil conditions generally prevented the long-term establishment of vegetation.

The primary goal for restoring scalds is to establish persistent vegetation. Strategies for revegetating scalds generally revolve around improving the surface soil layers by practical agricultural intervention. Techniques that have been demonstrated to work include the exclusion of stock, the use of ridges and furrows, mulching, liming, addition of fertilizer, pretreating seed with nutrients and neutralizing agents, and, more recently, water management practices that create and maintain wetter conditions. Of particular interest are the simpler interventions such as ridging and furrowing. This remediation involves the forming of ridges and furrows using cultivation, and especially when combined with a mulch layer (e.g., straw), it has proven very effective in facilitating the establishment of vegetation (e.g., Figure 21.5b). Ridges and furrows establish different microhabitats, with the water-tolerant species occupying the wetter furrows (Rosicky et al., 2006). A similar approach for food crop production on acid sulfate soils has been used by farmers in Southeast Asia for decades (Dent 1986).

More recently, landholders have begun experimenting with water table manipulation to provide more persistent wetter conditions to enable plant establishment on scalds. Excessive drainage is generally the most important primary driver of acid sulfate soil scald formation and strategies that reduce evaporation from bare areas and maintain or raise water tables in the near vicinity of scalds can contribute to their restoration and revegetation. The shallow ponding of fresh water can trigger rapid and complete revegetation of scalds (Rosicky et al., 2004b).

21.6 Physical Behavior

21.6.1 Consistence and Strength

Estuarine and marine sediments are usually deposited under saturated conditions. Many of these sediments contain up to 70% water and have a gel-like consistence. These extremely soft sediments pose significant challenges, due to their capacity to shrink on drying, deform under load, and oxidize and acidify when disturbed. The unique physical properties of sulfidic gel-like soil materials have been well described (Dent, 1986). White (2002) has provided a very detailed analysis of the swelling and hydraulic properties of gel-like unoxidized sulfidic clayey acid sulfate soil materials. From their analysis, these soft sulfidic soils are fundamentally different to many rigid soils with both hydraulic conductivities and consolidation coefficients for the gel materials often being generally very low (e.g., hydraulic conductivities <1 m day^{-1}).

Drainage and oxidation of sulfidic soils usually results in irreversible shrinkage and development of fissures and structure. "Soil ripening" is a term used to embrace the physical, chemical, and biological processes that transform unconsolidated sediment to a structured, soil (Dent, 1986). The pedological development of unoxidized sulfidic acid sulfate soil materials to fully oxidized sulfuric materials has been summarized by Dent (1986). Acid sulfate soil profiles often consist of oxidized, sulfuric horizons overlying unoxidized, sulfidic horizons, with a transition zone of variable thickness inbetween. The overlying sulfuric horizon generally has far greater structural development than the underlying sulfidic horizon due to physical ripening processes.

At the time of deposition, clayey marine sediments can have an open "house-of-cards" microporous architecture with water-filled pore space. This type of clay microstructure confers minimal frictional and cohesive strength to the sediment. Even though on a microscale these materials are highly porous, hydraulic conductivities tend to be low due to the lack of connected macropore networks (Johnston et al., 2009). The removal of water by drainage, evaporation, and evapotranspiration causes the microstructure to collapse, increasing cohesive strength and initiating shrinkage-induced fissures. The development of such structure ultimately enhances the potential oxidation of iron sulfides and further microstructural collapse as a result of acidic weathering. Unripe soils are soft, whereas ripe soils are firm and not overly "sticky" (Dent, 1986). The process of ripening is irreversible, and as a consequence, can cause land subsidence of up to 2 m, depending on the depth of previously unripe soil affected by oxidation.

21.6.2 Permeability

Lateral seepage of acidic ground waters is often the major route for the discharge of contaminants from acid sulfate soil to nearby waterways (Wilson et al., 1999; Åström, 2001; Johnston et al., 2003). The hydraulic gradient and hydraulic conductivity (permeability) govern the overall groundwater seepage rates. Factors that affect hydraulic gradient include rainfall, topography, and in coastal low-lying landscapes the water level at the discharge site. In an open estuary, this will be governed by tidal fluctuations. In a controlled drainage system, a designed water level may be set by one-way tidal gates or drainage pumps.

The hydraulic conductivity of acid sulfate soil materials varies greatly, both vertically within an acid sulfate soil profile and laterally across the landscape (White, 2002; Johnston et al., 2009). Ripening processes and spatially variable properties can induce considerable variability in soil structure and consequently, large differences in hydraulic conductivity in sulfuric horizons. The scale of the potential range in the hydraulic conductivity of acid sulfate soil landscapes was demonstrated in a recent study by Johnston et al. (2009) who examined seven coastal floodplains in eastern Australia. The measured saturated hydraulic conductivity (K_{sat}) ranged from <1 to >500 m day^{-1} (Figure 21.10). These large differences in K_{sat} reflect differences in the development, orientation, and density of macropore networks, with the highest K_{sat} occurring in soils with large tubular macropore networks (Johnston et al., 2009).

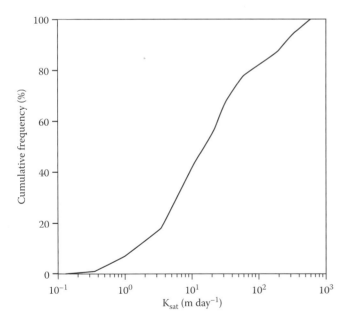

FIGURE 21.10 Cumulative frequency histogram of saturated hydraulic conductivity (K_{sat}) for 148 pits in acid sulfate soil landscapes in eastern Australia. (Modified from Johnston, S.G., P.G. Slavich, P. Hirst, R.T. Bush, and T. Aaso. 2009. Saturated hydraulic conductivity of sulfuric horizons in coastal floodplain acid sulfate soils: Variability and implications. Geoderma 151:387–394. Copyright 2009, with permission from Elsevier.)

21.7 Use, Management, and Evaluation of Acid Sulfate Soil Landscapes

21.7.1 Introduction

Management decisions have been taken, and are still being taken, without adequate knowledge of the unique characteristics of acid sulfate soils. Scientific assessment is often involved only after severe problems have arisen and the development of assessment methods has often focused on improvements in accuracy and precision of specific analytical determinations rather than dealing with soil heterogeneity and data interpretation.

Methods of reclamation and land use/management that have the single purpose of enabling and improving agricultural production may cause adverse environmental effects. In Australia, drained sugarcane fields have released acid into surface waters after heavy rains causing fish kills and ulcerative disease outbreaks in fish due to fungal infections (Callinan et al., 1993). On the Plain of Reeds in the Mekong delta, Minh et al. (1997) found that more Al^{3+} was leached into the surface water early in the rainy season from pineapples and yams (*Dioscorea alata*) in raised beds than rice fields. During a 4-month period, rice, yam, and pineapple fields released 2.2, 41, and 44 kmol Al^{3+} ha^{-1}, respectively.

Existing methodology (Dent, 1992; Xuan, 1993; van Mensvoort, 1996) allows problems to be reliably diagnosed and on-site reclamation and precautionary water table management undertaken to avoid, or at least minimize, acid generation. Despite the severe acidity and associated toxicities of acid sulfate soils, a wide range of land uses is found on these soil types. For generations farmers in tropical coastal wetlands such as the Mekong delta of Vietnam (Xuan et al., 1982; Xuan, 1993; Tri, 1996) have, with great perseverance and skill and often at great cost, developed methods to coexist with the toxicities of acid sulfate soils. Soil and water management is the key to solving the problems of acid sulfate soils (van Breemen, 1976; Tuong, 1993). Surface and water table management, leaching, narrow drain spacing, maintenance of downward water flow, good leveling, timely plowing, and puddling are the most important measures to consider.

21.7.2 Land Use on Sulfidic Acid Sulfate Soils

21.7.2.1 Mangrove Forest Exploitation

Dramatic disappearance of mangrove (*Rhizophora spp.*) forests for charcoal production on sulfidic acid sulfate soils and replacement by shrimp farming has been reported from the Mekong delta (Durang, 1995), and is likely to have happened elsewhere as well. Besides removal, mangrove forests are commonly overfished, resulting in the disappearance of a large number of aquatic species. Assessment of the severity of the acid sulfate hazard on other land uses may provide a powerful argument for retaining and protecting these natural ecosystems.

21.7.2.2 Fish or Shrimp Ponds

Brinkman and Singh (1982) developed a method of reclamation for fishponds on sulfidic acid sulfate soils in the Philippines.

Using the salt water tides, they drained the ponds for forced oxidation of FeS_2, plowed the pond bottom to promote oxidation, allowed salt water to reenter to dissolve the acid formed, redrained the acid water at low tide, and repeated the process. After—three to five treatments, the pond bottom was free of acidity. Dikes were also washed with salt water. Rice and fish yields increased by 2.1 Mg ha^{-1} and 400 kg ha^{-1} year^{-1}, respectively, on reclaimed land (Ligue and Singh, 1988).

21.7.2.3 The Rice and Shrimp System

This system (Xuan et al., 1986) is applied on both sulfidic (where it was developed) and sulfuric acid sulfate soils in the Mekong delta. The monsoon climate with its contrasting dry and wet seasons allows farmers in coastal lands to exploit the fresh and saltwater periods for rice and shrimp production. Land located next to a creek or a canal with large tidal fluctuations is surrounded with a deep ditch and dike of excavated material. At the onset of the rainy season, the water is quickly drained to reduce the salinity so that the rice seedbed is ready early. Rice is sown or transplanted with reported yields of 4 Mg ha^{-1} of paddy. By that time the rains have stopped and surface water becomes saline, the fields are flooded with saltwater and kept submerged or almost submerged throughout the dry season. In the original version of the system, shrimp were caught from natural fry, resulting in yields up to 690 kg ha^{-1} year^{-1}. Fresh sediment was deposited on the field annually from tidal water, raising the land and filling the ditch. This material—composed of good alluvial material without acidity—was used to raise the surrounding dike. Unfortunately, this system was undermined by its own success as too many people applied the system and exhausted the natural waters of fry. Because shrimp farming is economically more attractive, farmers abandoned the system and only cultivate shrimp using purchased fry. Because these shrimp are raised in the early rainy season, the farmers also abandoned rice cultivation. However, declining shrimp yields since 1994–1995 resulted in poverty for many.

21.7.2.4 Drainage

The aim of this technique (Xuan et al., 1982) is to accelerate salt removal during the early rainy season. Each field is provided with a surrounding shallow ditch and dike, plus a number of parallel field drains and a gate. At the start of the rainy season, farmers drain salt quickly from their fields at low tide, and close their gates at high tide. The system enables farmers to have two crops of short duration rice or rice combined with a short-duration upland crop such as watermelons or beans. Banjarese farmers in southern Kalimantan apply a similar shallow drainage system called "handils" (Sarwani et al., 1993).

21.7.3 Land Use on Sulfuric Acid Sulfate Soils

21.7.3.1 Extensive Exploitation

This activity includes of harvesting reeds for matting, and exploitation of indigenous and planted *Melaleuca leucodendron* for timber, firewood, and oil extraction (Brinkman and Xuan, 1991), but few viable options are available on sulfuric acid sulfate soils with severe acidity at shallow depth.

21.7.3.2 Rice

Rice is by far the most important crop cultivated on acid sulfate soils. In the Mekong delta, soils with a sulfuric horizon deeper than 50 cm have been used for many years for traditional or deep water rice cultivation (yielding about 1 Mg ha^{-1}). Until recently, inland freshwater sulfuric acid sulfate soils, where the sulfuric horizon is located near (<50 cm) the soil surface, were scarcely used for cultivation. After the economic reforms of 1986, settlers moved into these uncultivated lands with these soils. Coolegem (1996) estimated that from 1987 to 1995, 53% of the land in a district dominated by sulfuric acid sulfate soils had changed from wasteland or *Melaleuca leucodendron* forest to irrigated rice, during the dry season (28% of the entire district). Husson and Phung (1994) reported that relatively minor changes in elevation (a few centimeters) of fields in relation to the water levels in the growing season were responsible for significant differences in yields. Higher areas suffer from difficulties in maintaining adequate water depth to reduce oxidation and thus acidification, while the lower areas suffer from problems associated with reduction at depth (Fe^{2+} and H_2S toxicities). Many elevation differences may occur within one field, causing great heterogeneity in crop growth. Hanhart et al. (1997) and Hanhart and Ni (1993) developed a set of water management measures for crops, such as rice, recommending good leveling, narrow drain spacing (30–60 m) to create a downward flow of water through the surface soil, and brief drying of the surface soil at flowering to avoid reduction at depth. Yields obtained were 3.6–3.8 Mg ha^{-1} with the economic breakeven point being 2.15 Mg ha^{-1}. Drainage from the rice fields carries an estimated 5 kmol H$^+$ ha^{-1} crop^{-1} into the surface water, so irrigation and drainage waters should be kept separate. In areas with saline surface water in the dry season, traditional rice is cultivated. Moderate fertilizer levels of 100 kg N ha^{-1} and 60 kg P ha^{-1} are recommended, together with a small lime application to improve Ca^{2+} availability (Ren et al., 1993). Poor relationships between uptake and activity of Ca^{2+}, Mg^{2+}, and Fe^{2+} in the soil solution have been found for rice on acid sulfate soils, and Moore and Patrick (1993) recommend instead using the charge fraction of these ions in the soil solution, which show much better relationships with plant uptake.

21.7.3.3 Balanta System

In West Africa (Senegal, Guinea Bissau) farmers use the "Balanta" system for rice cultivation (van Gent and Ukkerman, 1993) in which land is ridged annually to a height in accordance with the expected freshwater flood of the rainy season. The ridging and annual turning of soil speed up removal of soluble toxins from the surface soil, but the system is in decline because of its heavy labor demand and low financial returns.

21.7.3.4 Annual Upland Crops

Yam, sweet potato, cassava, and sugarcane are the most important crops. These crops precede rice cultivation in places where no proper water supply for rice in the dry season is available. Farmers make raised beds by piling up the surface soil before the start of the rainy season. Beds are leached strongly during the wet season. Crops are planted immediately after the recession of the flood. Moisture is supplied by residual water from the soil, water in the ditches between the beds, and from capillary rise. For yam, vegetative growth is luxuriant, but yields are modest (~6 Mg ha^{-1}, compared to yields of 15 Mg ha^{-1} on areas not requiring beds). The disappointing yields are attributed to excess moisture near harvest, when the rainy season starts again (Durang, 1995). After about 3–4 years, the land is leveled and, provided a source of irrigation water is available, rice is cultivated in a double crop system (Tri, 1996).

21.7.3.5 Perennial Upland Crops

Sugarcane, pineapple, and cashew nut, which all require well-drained soil in the flood season, are grown on beds raised well above the level of the flood. Sugarcane on acid sulfate soils suffers from poor tillering, short and thin canes, with total yields averaging ~50 Mg ha^{-1} (Derevier, 1991) while yields on the other soils of the Mekong delta can reach 120 Mg ha^{-1}. Although Al^{3+} toxicity was suspected to be the cause of poor crop yields, no correlation between soil Al^{3+} and plant nutrient content could be established (Nga et al., 1993). For pineapple, fertilization and mulching with grass, a sufficient height above the flood is recommended (Sen et al., 1987; Tri, 1996) leading to yields of 40–50 Mg ha^{-1}, even in areas where irrigation is limited.

21.7.4 Integrated Soil, Water, and Crop Management

Tri (1996) presented an historical account of land use on the Mekong delta showing that the soil–water management measures (raised bed construction, plowing, puddling, leaching, construction of an irrigation and drainage system, field water management during the rice crop) that were taken as changes in land use were being made. This illustrates clearly how soil and water management are of primary importance for the proper management of acid sulfate soil landscapes. The assessments were mainly made by the farmers themselves, but the integration of different soil and water management practices can only be made by the community of land users. Land use has evolved through their learning processes. Significantly, the problems perceived first were those of drainage and water supply, and only later, water table management, in recognition of the acidity problems that subsequently developed.

21.7.5 Land Evaluation

Tuong et al. (1991) used hydrological factors to determine land quality in the Mekong delta for rice cultivation (double or triple cultivation, modern or traditional rice). The hydrological land qualities selected were (1) fresh water flood depth and flood duration during the wet season, (2) rainfall amount and distribution—emphasizing the occurrence of dry spells in the early rainy season, (3) tidal influence, (4) periods during which saline surface water is present in the dry season, and (5) the adequacy of canals for irrigation. For instance, flood depths of <30, <60, and >60 cm allow modern, short-duration, and traditional and deep water rice cultivation, respectively. Tidal influence is judged for its potential in irrigation without pumping; a long period under saline surface water may leave insufficient time for double-cropped rice too.

From a land evaluation study of the acid sulfate soil part of the Mekong delta that, in addition to assessment of hydrology, also included soil conditions and used a wider range of land use types, van Mensvoort et al. (1993) concluded that making fresh water available improves the suitability for rice on all acid sulfate land, and that well-constructed raised beds make moderately acid soils (sulfuric horizon >50 cm) highly suited for pineapple and sugarcane.

Using the same but more detailed methodology, Tri et al. (1993) evaluated two Mekong delta districts, one only partly hampered by acid sulfate soil problems, and the other dominated by sulfuric acid sulfate soils. On moderately acid land, land utilization types such as annual sugarcane in the dry season followed by rice in the wet season, cashew intercropped with pineapple, double rice cultivation and *Melaleuca* forestry, could be recommended. On sulfuric acid sulfate soils with much more difficult hydrological conditions, only single or double rice, or yam could be recommended.

The success or failure of the land utilization types depends, however, on farmers' skills. Present land use and farmer skills form, in fact, the basis for both land evaluation studies. In the studies of Tri (1996) and van Mensvoort (1996), acid sulfate soil knowledge at various levels (farmers, experts, and specialists) is brought together. All knowledge levels are valued equally, but farmer-level knowledge seems to have been underused in the past. Farmer-level knowledge is built on generations of experience in using the land and not only generated interesting land use systems (rice/shrimp, yam, sweet potatoes, pineapples, etc.) but also explained how these systems were historically developed, under what conditions or in what sequence these systems can be used, and what management measures need to be taken (e.g., leaching, mulching, no-till, raised bed construction). Expert-level knowledge such as fertilizer recommendations based on field experiments (Ren et al., 1993), refinement of water management measures (Tuong, 1993; Hanhart and Ni, 1993; Hanhart et al., 1997), and variety selection (Xuan, 1993) can be regarded as complementary, reinforcing the skills of farmers. Finally, true specialist knowledge of a more fundamental nature can also contribute to improved land use, for example, by better survey methods to improve land use planning (Burrough et al., 1988; Bregt et al., 1993), by developing a better grasp of soil–plant relationships through improved methods of chemical characterization (Moore and Patrick, 1993), or by modeling the processes in acid sulfate soils (Bronswijk and Groenenberg, 1993). A coherent and accessible decision support system has been produced by Dent and Dawson (1998) to help assist in this area.

21.8 Identification and Assessment

Accurate identification and assessment of acid sulfate soil materials is critical for acid sulfate soil management if the environmental hazards that these soil materials present are to be avoided or controlled. The basic accepted concept of a sulfuric soil material is one that is strongly acidic as a result of the oxidation of RIS. The concept of what constitutes a sulfidic soil material is, generally, that the soil material has the potential to become a sulfuric material following oxidation of the RIS it contains. However, precise uniform widely accepted definitions of "acid sulfate soil material," and of the related terms such as "sulfidic" and "sulfuric" soil materials, do not exist and in the case of the term "sulfidic," two distinctly different technical uses of this term exist, both within science in general and within soil science in particular, as will be discussed later.

21.8.1 Identification and Assessment of Sulfuric Material

Although essentially unified by a central concept, the definition of sulfuric soil (material or horizon) varies among soil taxonomies.

The definition for sulfuric soil material used in the Australian Soil Classification (Isbell, 1996) is as follows:

Soil material that has a pH less than 4 (1:1 by weight in water, or in a minimum of water to permit measurement) when measured in dry season conditions as a result of the oxidation of sulfidic materials (defined above). Evidence that low pH is caused by oxidation of sulfides is one of the following:

- Yellow mottles and coatings of jarosite (hue of 2.5Y or yellower and chroma of about 6 or more).
- Underlying sulfidic material.

For recognition as a Thionic (i.e., a sulfuric) horizon the World Reference Base (IUSS Working Group WRB, 2006) stipulates that the subsoil horizon that must have

1. pH < 3.5 in a 1:1 water suspension
2. a. either yellow/orange jarosite [$KFe_3(SO_4)_2(OH)_6$] or yellowish-brown schwertmannite [$Fe_{16}O_{16}(SO_4)_3(OH)_{10} \cdot 10H_2O$] mottles
   or—concentrations with a Munsell hue of 2.5Y or more and a chroma of 6 or more
   or
   b. superposition on sulfidic soil materials
   or
   c. 0.05% (by weight) or more water-soluble sulfate
   and
3. Thickness of 15 cm or more

In the USDA Soil Taxonomy (Soil Survey Staff, 2010) the requirements for a sulfuric horizon is as below:

The sulfuric horizon is 15 cm or more thick and is composed of either mineral or organic soil material that has a pH value (1:1 by weight in water or in a minimum of water to permit measurement) of 3.5 or less or less than 4.0 if sulfide or other S-bearing minerals that produce sulfuric acid upon their oxidation are present. The horizon shows evidence that the low pH value is caused by sulfuric acid. The evidence is one or both of the following:

1. The horizon has:
   a. Concentrations of jarosite, schwertmannite, or other iron and/or aluminum sulfates or hydroxysulfate minerals; or
   b. 0.05% or more water-soluble sulfate; or
2. The layer directly underlying the horizon consists of sulfidic materials (defined above).

All three example definitions of sulfuric horizon provided above have much in common, and the differences that do exist—except in the choice of the critical pH—are relatively minor. The critical pH chosen in the Australian Soil Classification (Isbell, 1996) definition of sulfuric is 4 whereas in the WRB (IUSS Working Group WRB, 2006) the critical pH is 3.5.

For assessment and management purposes, both of the following two different acidity-related measures of sulfuric soil materials usually need to be assessed:

1. pH. Determined either in water or dilute solutions, pH measures the "intensity" of the H^+ and is a most useful measure of the geochemical environment.
2. Titratable Actual Activity (TAA). Determined usually by titration of a soil suspension up to a defined pH (usually 6.5), the TAA is a measure of the capacity of the soil material to supply acidity. In acid sulfate soil materials the main components of the TAA are Fe^{3+}, Al^{3+}, as well as H^+ (Ahern et al., 2004).

21.8.2 Identification and Assessment of Sulfidic Material

As recently discussed in Sullivan et al. (2009a) there are several approaches that have been used for the identification and assessment of sulfidic soil materials.

21.8.2.1 Total Potential Acidity/Titratable (or Total) Actual Acidity

The total potential acidity/titratable (or total) actual acidity (TPA/TAA) method of Konsten et al. (1988) assesses the acidity already existing in the soil material (i.e., the TAA) and the acidity after complete oxidation of the sulfides present using hydrogen peroxide (H_2O_2) (i.e., the TPA). The difference between the TPA and the TAA is termed the total sulfidic acidity (TSA) and is assumed to be a measure of the acidity that is generated by the oxidation of the various sulfide mineral fractions. TPA/TSA is now considered as only a semiquantitative and inexpensive method limited by inconsistent relationships between TPA and/or TSA, and pyritic sulfur (Andriesse, 1993; Clark et al., 1996); both the TPA and TSA are no longer effectively considered as suitable stand-alone measures in Australia for the recognition of acid sulfate soil materials (Ahern et al., 2004).

21.8.2.2 Reduced Inorganic Sulfur Determinations

Unlike the incubation methods, RIS determinations are considered to be quantitative (Andriesse, 1993). However, until recently determinations of RIS in soil materials have involved difficult methods that were not compatible with the conduct of soil surveys (Beghein et al., 1978; Andriesse, 1993). However, recent methods for the determination of RIS in acid sulfate soil materials such as the chromium reducible sulfur methods (Sullivan et al., 2000; Burton et al., 2008) have been developed that are simple, rapid, and inexpensive.

Importantly, the concept of sulfidic (as currently widely used in soil science) does not simply imply the presence of appreciable sulfides, but rather that there is a surplus of sulfides in that material (cf. acid neutralizing capacity), such that when oxidized, the soil material is capable of becoming extremely acidic. (Distinctions between the uses of the term *sulfidic* in science will be addressed later.) Thus, soil materials containing even high sulfide mineral contents cannot be classified as sulfidic in many soil taxonomies, if there are sufficient neutralizing materials also present in the soil materials to prevent acidification.

21.8.2.3 Acid–Base Accounting

The long period of time usually required to gain results from the incubation method (i.e., at least several weeks) along with the need for quantitative analyses for calculating liming rates, has resulted in the development of an acid–base accounting (ABA) approach (Ahern et al., 2004). The ABA approach has gained widespread use for acid sulfate soil management. One of the benefits of ABA is that it is quantitative and provides data on the acidification hazard that are suitable for purposes such as acidity hazard prioritization, determination of liming requirements prior to oxidation, verification of liming quantities posttreatment, etc.

While many ABA approaches have been used for acid sulfate soil assessment, they all share a common underlying principle whereby the acidity hazard is the difference between alkalinity and acidity sources as shown below.

$$\text{Acidity Hazard} = \text{Acidity} - \text{Alkalinity} \quad (21.12)$$

There are several sources of acidity and alkalinity in soil materials and in practice the determination of several of these acidity sources is determined and expressed separately in the ABA. A commonly used ABA for acid sulfate soils is the one given by Ahern et al. (2004) as shown below:

$$\text{Net Acidity} = \text{Potential Sulfidic acidity} + \text{Existing Acidity}$$
$$- \text{Acid Neutralizing Capacity} \quad (21.13)$$

In this ABA, the existing acidity is defined as follows:

$$\text{Existing Acidity} = \text{Actual Acidity} + \text{Retained Acidity} \quad (21.14)$$

The Net Acidity in this ABA represents the acidity hazard of the soil material. The Potential Sulfidic Acidity refers to the potential for acidity to develop from oxidation of pyrite. The Potential Sulfidic Acidity is estimated from the RIS determination and assumes both the RIS to be pyritic sulfur and that the overall oxidation reaction occurs in completion according to Equation 21.9 [i.e., 1 mol of pyrite produces 4 mol of (H^+) acidity].

The Existing Acidity comprises both Actual Acidity and Retained Acidity. Actual Acidity is a measure of the readily available soluble and exchangeable acidity in the soil material. Retained acidity is a measure of the more slowly available acidity contained within minerals such as jarosite and schwertmannite—the acidity in these minerals represents incomplete oxidation (cf. the equation above for pyrite oxidation). This acidity can be realized and released when these minerals decompose and this process can be a major source of acidity in drainage. For example, 1 mol of jarosite releases 3 mol of acidity as described by the following reaction:

$$KFe_3(SO_4)_2(OH)_6 \rightarrow 3FeOOH + 2SO_4^{2-} + 3H^+ + K^+ \quad (21.15)$$

The Acid Neutralizing Capacity (ANC) refers to the effective neutralizing sources. In acid sulfate soils acceptable sources of ANC include calcium and magnesium carbonates, exchangeable alkalinity, and organic matter; but sources of buffering that do not act above pH 6.5 are usually considered ineffective (Ahern et al., 2004).

In Australia, the minimum critical Net Acidity levels that initiate the development of detailed management plans, should acid sulfate soil materials be disturbed, vary according to soil texture and the amount of soil disturbed. For sandy soil materials, and/or where large amounts of soil materials are to be disturbed, the critical Net Acidity level is >0.03% S (or alternatively >18 mol (H^+) Mg^{-1} when expressed as acidity) (Ahern et al., 2004).

Limitations of the current ABA methods stem from both our incomplete understanding of the acidifying and neutralizing processes that take place in these soil materials and the lack of reliable methods to quantify Retained Acidity and Acid Neutralizing Capacity. These limitations include the following:

1. It is not clear what proportion of the potential acidity capable of being produced by pyrite oxidation eventually becomes expressed. Similarly, our understanding of the kinetics of acidification and neutralization processes is limited.
2. The currently available methods for quantifying ANC in acid sulfate soil materials require improvement to provide accurate determinations (Ahern et al., 2004). Current methods may either overestimate or underestimate the "real" ANC due to a number of reasons, including the following:
    a. Overestimation may be due to the inclusion of finely ground shell materials in the test sample derived from large shell materials in the field samples. Large shell materials are generally ineffective as a neutralizing agent but if finely ground, such components will be included in the ANC determination inflating the true capacity of the original field samples (Ahern et al., 2004).

b. Overestimation of the ANC may also result from the imposition of extremely low pHs (e.g., pH < 2) when using acid back-titration methods (Ahern et al., 2004) and, hence, the inclusion of acid neutralizing mechanisms such as clay mineral dissolution that may not occur in natural acid sulfate soil environments with less extreme pHs.
3. The current lack of reliable quantification procedures for Retained Acidity can lead to inaccurate estimation of the acidity hazard posed by acid sulfate soil materials.

Despite these limitations, the quantitative capability of the ABA method provides distinct advantages for the purposes of managing acid sulfate soil materials over the other acid sulfate soil identification methods used for soil taxonomic purposes.

The WRB (IUSS Working Group WRB, 2006) also uses an ABA approach to identify sulfidic soil materials. The WRB defines sulfidic materials as having:

1. 0.75% or more S (dry mass) and less than three times as much calcium carbonate equivalent as S; and
2. pH (1:1 in water) of 4.0 or more.

Unfortunately, the widely distributed gypsic soil materials (i.e., soils materials containing appreciable gypsum ($CaSO_4 \cdot 2H_2O$)) are often falsely identified as being sulfidic using the WRB approach, as total sulfur—rather than RIS—is used for identification.

21.8.2.4 Incubation Methods

This approach simulates natural acidification behavior and has been long used for the assessment of these materials. This method is considered to be direct, allowing the soil to "speak for itself" (Dent, 1986) with respect to whether or not the soil material will acidify upon oxidation. However, incubation can also be a protracted method requiring months to give a determination (Andriesse, 1993).

In this method, the incubating soil materials are kept at room temperature and at field wetness by rewetting as required. Critical pH targets and the required duration of the incubation period necessary to identify a soil material as being sulfidic have been widely discussed (Andriesse, 1993) and range from 3.5 to 4.0. The duration of incubation has ranged from a few weeks to over a year.

In the latest soil classification, systems using the incubation method for the recognition of sulfidic materials, both the critical pH target and the duration of incubation have been standardized. For example, the definition used in the Australian Soil Classification (Isbell, 1996) for sulfidic materials is as below:

> A subsoil, waterlogged, mineral or organic material that contains oxidizable sulfur compounds, usually iron disulphide (e.g., pyrite, FeS_2), that has a field pH of 4 or more but which will become extremely acid when drained. Sulfidic material is identified by a drop in pH by at least 0.5 unit to 4 or less (1:1 by weight in water, or in a minimum of water to permit measurement) when a 10 mm thick layer is incubated at field capacity for 8 weeks.

As discussed by Sullivan et al. (2009a), this definition is similar but not identical to that used for sulfidic soil materials currently used in the USDA Soil Taxonomy (Soil Survey Staff, 2003).

Some problems have emerged regarding the duration of incubation being limited to 8 weeks in these definitions. For example, the results of Ward et al. (2004) indicated that 8 weeks of incubation only allowed minor oxidation of the RIS in some gel-like clayey pyritic soil materials. To overcome these issues and to allow "the soil to speak without being interrupted in mid-sentence" Sullivan et al. (2009a) used 2 mm thick incubation slabs to hasten acidification process and proposed changing the recommended duration of incubation (a) until the soil pH changes by at least 0.5 pH unit to below 4, or (b) until a stable pH is reached after at least 8 weeks of incubation. This new definition has been incorporated into recent protocols for recognition of sulfidic soil materials in Australia (e.g., Fitzpatrick et al., 2009; MDBA, 2010; EPHC & NRMMC, 2010).

21.8.2.5 Developments in Classification of Sulfidic and Monosulfidic Materials

As mentioned previously, the distinguishing feature of acid sulfate soil materials has been either the presence of RIS sufficient to cause severe acidification, or severe acidity as a result of the oxidation of those sulfide minerals (Pons, 1973; van Breemen, 1973). More recently, environmental hazards arising from sulfide-containing materials in soils and sediments additional to acidification have been recognized and include deoxygenation, metal and metalloid release, enhanced nutrient release, and release of gases, as described previously. Some of these additional environmental hazards are the result of sulfide-related processes that are redox-driven and not directly associated with acidification. Similarly, deoxygenation of waterways by mobilization of monosulfidic soil materials does not require, nor does it necessarily lead to, acidification (Sullivan et al., 2002; Burton et al., 2006c).

The recognition of the occurrence and importance of monosulfides in soil materials led to the inclusion of monosulfidic materials as a distinguishing property within mapping units of the Australian Atlas of Acid Sulfate Soils (Fitzpatrick et al., 2008). The Australian Atlas of Acid Sulfate Soils is a web-based hazard assessment tool with a nationally consistent legend, which provides information about the distribution and properties of both coastal and inland acid sulfate soil across Australia (Fitzpatrick et al., 2008).

Although severe acidification is a major environmental hazard that can arise from the disturbance or mismanagement of sulfide-affected soil materials, much recent literature on the behavior of sulfide-containing soil materials indicates (1) acidification is not the only important environmental hazard arising from these soil materials, and (2) soil classification systems for acid sulfate soil materials should be modified to accommodate sulfide-containing soil materials that do not have the capacity to acidify, but have the capacity to pose other sulfide-related environmental hazards.

The way that soil classification has used the term "sulfidic" has also proved to be problematic in terms of communicating effectively with the broader scientific community. When describing soil, sediment, rocks and water, scientists in related disciplines (e.g., ecology, geology, geochemistry, and zoology) use the term "sulfidic" differently from the way that soil classification systems have used it. For example, Sullivan et al. (2009b) conducted a literature survey of the 20 scientific papers that appeared first in the literature search results when the term "sulfidic" was associated with each of the terms "soil," "rock," "sediment," and "water." It was clear that when scientists in disciplines related to soil science use the term "sulfidic" to describe their media, they use essentially the plain English meaning of sulfidic (i.e., materials that contain sulfides). In contrast, when soil scientists used the term "sulfidic" in association soil, the soil materials not only contained sulfides, they also had to be capable of becoming strongly acidic (e.g., pH < 4) upon oxidation. As a result, in soil classification, seemingly alone, we cannot describe all sulfide-containing soil materials as being "sulfidic" materials. Identifying soil materials that can severely acidify as a result of sulfide oxidation is clearly a useful concept. However the term "sulfidic" for this concept as currently used in several soil taxonomies (e.g., Isbell, 1996; Soil Survey Staff, 2003; IUSS Working Group WRB, 2006) has been misleading to the broader scientific community who almost uniformly use sulfidic to denote "containing sulfides" (Sullivan et al., 2009b).

A new classification system (Sullivan et al., 2009b, 2010) for sulfide-affected soils has been adopted for use in Australia (e.g., Fitzpatrick et al., 2008, 2009; MDBA, 2010; EPHC & NRMMC, 2010) to accommodate the important group of sulfide-containing soil materials that do not severely acidify upon oxidation and are classified as being "nonsulfidic" by the main soil taxonomies, as well as monosulfidic soil materials.

This new classification system for acid sulfate soil materials (Sullivan et al., 2010) is as below:

Sulfidic
Concept: Soil material containing detectable inorganic sulfides.

Definition: Soil material containing ≥0.01% inorganic sulfidic S.

Hypersulfidic
Concept: Sulfidic soil material that is capable of severe acidification as a result of oxidation of contained sulfides. (Note: hypersulfidic soil materials are conceptually the same soil materials defined previously in soil classifications as "sulfidic")

Definition: (adapted from Isbell (1996) with modifications to inter alia account for recent improvements to the incubation method (Sullivan et al., 2009a):

> Hypersulfidic material is a sulfidic material that has a field pH of 4 or more and is identified by experiencing a substantial* drop in pH to 4 or less (1:1 by weight in water, or in a minimum of water to permit measurement) when a 2–10 mm thick layer is incubated aerobically at field capacity. The duration of the incubation is either: a) until the soil pH changes by at least 0.5 pH unit to below 4, or b) until a stable** pH is reached after at least 8 weeks of incubation.

*A substantial drop in pH arising from incubation is regarded as an overall decrease of at least 0.5 pH unit.

**A stable pH is assumed to have been reached after at least 8 weeks of incubation when either the decrease in pH is < 0.1 pH unit over at least a 14 day period, or the pH begins to increase.

Hyposulfidic
Concept: Sulfidic soil material that is not capable of severe acidification as a result of oxidation of contained sulfides.

Definition: (Adapted from Isbell (1996) with modifications to inter alia account for recent improvements to the incubation method (Sullivan et al., 2009a):

> Hyposulfidic material is a sulfidic material that (i) has a field pH of 4 or more and (ii) does not experience a substantial* drop in pH to 4 or less (1:1 by weight in water, or in a minimum of water to permit measurement) when a 2–10 mm thick layer is incubated aerobically at field capacity. The duration of the incubation is until a stable** pH is reached after at least 8 weeks of incubation.

*A substantial drop in pH arising from incubation is regarded as an overall decrease of at least 0.5 pH unit.

**A stable pH is assumed to have been reached after at least 8 weeks of incubation when either the decrease in pH is <0.1 pH unit over at least a 14 day period, or the pH begins to increase.

Monosulfidic
Concept: Soil material containing detectable monosulfides.

Definition: Soil material containing ≥0.01% acid volatile sulfidic S.

21.8.2.6 Sequential Extraction Methods

The oxidative dissolution of pyrite that occurs when acid sulfate soil materials are disturbed results in the release of metals previously associated with pyrite (van Breemen, 1973). The rate at which these metals are released into the labile fractions is dependent upon the geochemical regime (e.g., pH, redox potential, soil solution composition), as described previously. Sequential extraction procedures are a common method of assessing metal mobility in different operational fractions within soils and sediments. When used within their well-documented limitations (Rao et al., 2008), sequential extraction procedures can provide useful assessments of changes in metal mobility, and in the case of acid sulfate soil, assessing the potential metal hazard that the unoxidized acid sulfate soil materials pose. Sequential extraction methods have been utilized in a comparative fashion to examine metal mobilization in acid sulfate soils. Both Åström (1998) and Sohlenius and Öborn (2004) compared metal mobility in different fractionations in the oxidized and unoxidized zones of in situ acid sulfate soil profiles. These studies showed that a range of metals were mobilized from the oxidized zones of acid sulfate soils, especially from the pyritic and residual soil metal fractions. More recently, a sequential extraction method featuring the separate partitioning of metals in the pyritic fraction has been specifically designed for acid sulfate soils (Claff et al., 2010).

21.9 Conclusions

Acid sulfate soils can greatly affect the environment of and societies within landscapes that contain these materials. A wide range of environmental hazards are presented by these soil materials. These include (1) severe acidification of soil and drainage waters (below pH 4 and often <pH 3); (2) mobilization of many metals (e.g., iron, aluminum, copper, cobalt, zinc), metalloids (e.g., arsenic), nutrients (e.g., phosphate), and rare earth elements; (3) deoxygenation of water bodies; (4) production of noxious gases (e.g., H_2S); and (5) scalding (i.e., denudation) of landscapes. Some of these hazards are caused directly or indirectly by the severe acidification, whereas others are not.

These environmental hazards can impact on infrastructure such as bridges, drains, pipes, and roadways. Waters draining from acid sulfate soil materials may be enriched in a wide range of potential toxicants, including metals and metalloids, endangering aquatic life and public health. Crops, trees, and pastures may all be severely affected by acid sulfate soil materials, as can aquaculture. The extent, location, properties, and the often intense demands made on acid sulfate soils for water and food supply dictate that the development of our understanding of these soils will need to continue apace. Certainly, the identification, assessment, classification, and management of acid sulfate soils will continue to present a wide range of challenges for soil science.

References

AARD/LAWOO. 1993. Acid sulfate soils in the humid tropics: Simulation model of physical and chemical processes to evaluate water management strategies. Acid sulfate soils in the humid tropics. Vol. 3. International Institute for Land Reclamation and Improvement, Wageningen, the Netherlands.

Ahern, C.R., A.E. McElnea, and L.A. Sullivan. 2004. Acid sulfate soils laboratory methods guidelines. Queensland Department of Natural Resources, Mines and Energy, Indooroopilly, Queensland, Australia.

Ahonen, L., and O. Tuovinen. 1991. Temperature effects on bacterial leaching of sulfide minerals in shake flask experiments. Appl. Environ. Microbiol. 57:138–145.

Alpers, C.N., and D.K. Nordstrom. 1999. Geochemical modelling of water-rock interactions in mining environments, p. 289–323. In G.S. Plumee and M.J. Logsdon (eds.) The environmental geochemistry of mineral deposits. Rev. Econ. Geo. 6.

Andriesse, W. 1993. Acid sulphate soils: Diagnosing the ills, p. 11–29. In D.L. Dent and M.E.F. van Mensvoort (eds.) Selected papers of the Ho Chi Minh City symposium on acid sulphate soils. International Institute for Land Reclamation and Improvement, Wageningen, the Netherlands.

Aneja, V.P. 1990. Natural sulfur emissions into the atmosphere. JAPCA J. Air Waste Manag. 40:469–476.

ANZECC/ARMCANZ. 2000. Australian and New Zealand guidelines for fresh and marine water quality; Australian and New Zealand Environment and Conservation Council, Agricultural and Resource Management Council of Australia and New Zealand.

Åström, M. 1998. Partitioning of transition metals in oxidised and reduced zones of sulphide-bearing fine-grained sediments. Appl. Geochem. 13:607–617.

Åström, M. 2001. Effect of widespread severely acidic soils on spatial features and abundance of trace elements in steams. J. Geochem. Explor. 73:181–191.

Åström, M., and N. Corin. 2000. Abundance, source and speciation of trace elements in humus-rich streams affected by acid sulfate soils. Aquat. Geochem. 6:367–383.

Bartlett, R.W. 1973. A combined pore diffusion and chalcopyrite dissolution kinetics model for in situ leaching of a fragmented copper porphyry, p. 331–374. In D.J.I. Evans and R.S. Shoemaker (eds.) International symposium on hydrometallurgy. Chicago, IL.

Bates, T.S., B.K. Lamb, A. Guether, J. Dignon, and R.E. Stoiber. 1992. Sulfur emissions to the atmosphere from natural sources. J. Atmos. Chem. 14:315–337.

Beghein, L.T., N. van Breemen, and E.J. Velthorst. 1978. Analysis of sulphur compounds in acid sulphate soils and other recent marine sediments. Soil Sci. Plant Anal. 9:203–211.

Berresheim, H., P.H. Wine, and D.D. Davis. 1995. Sulfur in the atmosphere, p. 251–307. In H.B. Singh (ed.) Composition, chemistry, and climate of the atmosphere. Van Nostrand Reinhold, New York.

Bigham, J.M., U. Schwertmann, and L. Carlson. 1992. Mineralogy of precipitates formed by the biogeochemical oxidation of Fe(II) in mine drainage, p. 219–232. In H.C.W. Skinner and R.W. Fitzpatrick (eds.) Biomineralization processes of iron and manganese—Modern and ancient environments. Catena Supplement 21.

Bigham, J.M., U. Schwertmann, S.J. Traina, R.L. Winland, and M. Wolf. 1996. Schwertmannite and the chemical modeling of iron in acid sulfate waters. Geochim. Cosmochim. Acta 60:2111–2121.

Blodau, C. 2006. A review of acidity generation and consumption in acidic coal mine lakes and their watersheds. Sci. Total Environ. 369:307–332.

Bloomfield, C., and J.K. Coulter. 1973. Genesis and management of acid sulfate soils. Adv. Agron. 25:265–326.

Braun, R.L., A.E. Lewis, and M.E. Wadsworth. 1974. In-place leaching of primary sulfide ores: Laboratory leaching data and kinetics model. Metall. Trans. 5:1717–1726.

Bregt, A.K., J.A.M. Janssen, and P. Alkasuma. 1993. Survey strategies for acid sulfate soils, p. 43–50. In D.L. Dent and M.E.F. van Mensvoort (eds.) Selected papers of the Ho Chi Minh City symposium on acid sulfate soils. ILRI Pub. 53. International Institute for Land Reclamation and Improvement, Wageningen, the Netherlands.

Brinkman, R., and V.P. Singh. 1982. Rapid reclamation of fishponds on acid sulfate soils, p. 318–330. *In* H. Dost and N. van Breemen (eds.) Proceedings of the Bangkok symposium on acid sulfate soils. ILRI Pub. 31. International Institute for Land Reclamation and Improvement, Wageningen, the Netherlands.

Brinkman, W.J., and V. Xuan. 1991. Melaleuca leucodendron s.l., a useful and versatile tree for acid sulfate soils and some other poor environments. Int. Tree Crop J. 6:61–274.

Bronswijk, J.J.B., and J.E. Groenenberg. 1993. A simulation model for acid sulfate soils I: Basic principles, p. 341–357. *In* D.L. Dent and M.E.F. van Mensvoort (eds.) Selected papers of the Ho Chi Minh City symposium on acid sulfate soils. ILRI Pub. 53. International Institute for Land Reclamation and Improvement, Wageningen, the Netherlands.

Bronswijk, J.J.B., K. Nugroho, I.B. Aribawa, J.E. Groenenberg, and C.J. Ritsema. 1993. Modeling of oxygen transport and pyrite oxidation in acid sulfate soils. J. Environ. Qual. 22:544–554.

Burrough, P.A., M.E.F. van Mensvoort, and J. Bos. 1988. Spatial analysis as a reconnaissance survey technique: An example from acid sulfate regions in the Mekong delta, Vietnam, p. 68–91. *In* H. Dost (ed.) Selected papers of the Dakar symposium on acid sulfate soils. ILRI Pub. 44. International Institute for Land Reclamation and Improvement, Wageningen, the Netherlands.

Burton, E.D., R.T. Bush, and L.A. Sullivan. 2006a. Elemental sulfur in drain sediments associated with acid sulfate soils. Appl. Geochem. 21:1240–1247.

Burton, E.D., R.T. Bush, and L.A. Sullivan. 2006b. Reduced inorganic sulfur speciation in drain sediments from acid-sulfate soil landscapes. Environ. Sci. Technol. 40:888–893.

Burton, E.D., R.T. Bush, and L.A. Sullivan. 2006c. Acid-volatile sulfide oxidation in coastal floodplain drains: Iron-sulfur cycling and effects on water quality. Environ. Sci. Technol. 40:1217–1222.

Burton, E.D., R.T. Bush, and L.A. Sullivan. 2006d. Sedimentary iron geochemistry in acidic waterways associated with coastal lowland acid sulfate soils. Geochim. Cosmochim. Acta 70:5455–5468.

Burton, E.D., R.T. Bush, L.A. Sullivan, R.K. Hocking, D.R.G. Mitchell, S.G. Johnston, R.W. Fitzpatrick, M. Raven, S. McClure, and L.Y. Jang. 2009. Iron-monosulfide oxidation in natural sediments: Resolving microbially mediated S transformations using XANES, electron microscopy, and selective extractions. Environ. Sci. Technol. 43:3128–3134.

Burton, E.D., R.T. Bush, L.A. Sullivan, S.G. Johnston, and D.R.G. Mitchell. 2008. Mobility of arsenic and selected metals during re-flooding of iron- and organic-rich acid-sulfate soil. Chem. Geol. 253:64–73.

Burton, E.D., R.T. Bush, L.A. Sullivan, and D.R.G. Mitchell. 2007. Reductive transformation of iron and sulfur in schwertmannite-rich accumulations associated with acidified coastal lowlands. Geochim. Cosmochim. Acta 71:4456–4473.

Burton, E.D., S.G. Johnston, K. Watling, R.T. Bush, A.F. Keene, and L.A. Sullivan. 2010. Arsenic effects and behavior in association with the Fe(II)-catalysed transformation of schwertmannite. Environ. Sci. Technol. 44:2016–2021.

Bush, R.T., D. Fyfe, and L.A. Sullivan. 2004b. Occurrence and abundance of monosulfidic black ooze in coastal acid sulfate soil landscapes. Aust. J. Soil Res. 42:609–616.

Bush, R.T., R. McGrath, and L.A. Sullivan. 2004a. Occurrence of marcasite in an organic-rich Holocene mud. Aust. J. Soil Res. 42:617–622.

Bush, R.T., and L.A. Sullivan. 1997. Morphology and behaviour of greigite from a Holocene sediment in eastern Australia. Aust. J. Soil Res. 35:853–861.

Bush, R.T., L.A. Sullivan, D. Fyfe, and S.J. Johnston. 2004c. Redistribution of monosulfidic black oozes by floodwaters in a coastal acid sulfate soil floodplain. Aust. J. Soil Res. 42:603–607.

Bush, R.T., L.A. Sullivan, and C. Lin. 2000. Iron monosulfide distribution in three coastal floodplain acid sulfate soils, eastern Australia. Pedosphere 10:237–246.

Callinan, R.B., G.C. Frazer, and M. Melville. 1993. Seasonally recurrent fish mortalities and ulcerative outbreaks associated with acid sulfate soils in Australian estuaries, p. 403–410. *In* D.L. Dent and M.E.F. van Mensvoort (eds.) Selected papers of the Ho Chi Minh City symposium on acid sulfate soils. ILRI Pub. 53. International Institute for Land Reclamation and Improvement, Wageningen, the Netherlands.

Callinan, R.B., J. Sammut, and G.C. Fraser. 2005. Dermatitis, bronchitis and mortality in empire gudgeon *Hypseleotris compressa* exposed naturally to run-off from acid sulfate soil. Dis. Aquati. Organ. 63:247–253.

Canfield, D.E., B. Thamdrup, and S. Fleischer. 1997. Isotope fractionation and sulfur metabolism by pure and enrichment cultures of elemental sulfur-disproportionating bacteria. Limnol. Oceanogr. 43:253–264.

Charlson, R.J., J.E. Lovelock, M.O. Andreae, and S.G. Warren. 1987. Oceanic phytoplankton, atmospheric sulfur, cloud albedo and climate. Nature 326:655–661.

Claff, S.R., L.A. Sullivan, E.D. Burton, and R.T. Bush. 2010. A sequential extraction for acid sulfate soils: Partitioning of iron. Geoderma 155:224–230.

Clark, M.W., G. Lancaster, and D. McConchie. 1996. Total sulphide acidity for the definition and quantitative assessment of the acid sulphate hazard: Simple solution or a new suite of problems. Sci. Total Env. 183:249–254.

Coolegem, L. 1996. Land use changes in the Plain of Reeds, Vietnam (1987–1995). MSc. Thesis. Department of Soil Science and Geology, Wageningen Agricultural University. Wageningen, the Netherlands.

Corfield, J. 2000. The effects of acid sulfate soil run-off on a subtidal estuarine macrobenthic community in the Richmond river, NSW, Australia. ICES J. Mar. Sci. 57:1517–1523.

Cornell, R.M., and U. Schwertmann. 2003. The iron oxides. Wiley-VCH, Weinheim, Germany.

Davis, G.B., and A.I.M. Ritchie. 1986. A model of oxidation in pyritic mine wastes: Part 1. Equations and approximate solution. Appl. Math. Model. 10:314–322.

Dent, D.L. 1986. Acid sulfate soils: A baseline for research and development. ILRI Pub. 39. International Institute for Land Reclamation and Improvement, Wageningen, the Netherlands.

Dent, D.L. 1992. Reclamation of acid sulfate soils. Adv. Soil Sci. 17:79–121.

Dent, D.L., and B. Dawson. 1998. The acid test. Expert system for the identification, assessment and management of acid sulphate soils. CD:ROM Users Manual. EWMAN Ltd., Norwich, U.K.

Dent, D.L., and L.J. Pons. 1995. Acid sulphate soils: A world view. Geoderma 67:263–276.

Dent, D.L., and R.W. Raiswell. 1982. Quantitative models to predict the rate and severity of acid sulphate development, p. 73–95. In H. Dost and N. van Breemen. Proceedings of the Bangkok symposium on acid sulfate soils. ILRI Pub. 31. International Institute for Land Reclamation and Improvement, Wageningen, the Netherlands.

Derevier, A. 1991. Mission d'evaluation des possibilites de culture de la canne a sucre au Vietnam. Institut de Recherches Agronomiques Tropicales er des Cultures Vivrieres, Montpellier, France.

Diaz, R.J., and Rosenberg, R. 1995. Marine benthic hypoxia: A review of its ecological effects and the behavioural responses of benthic macrofauna. Oceanogr. Mar. Biol. Ann. Rev. 33:245–303.

Diemont, W.H., L.J. Pons, and D.L. Dent. 1993. Standard profiles of acid sulfate soils, p. 51–60. In D.L. Dent and M.E.F. van Mensvoort (eds.) Selected papers of the Ho Chi Minh City symposium on acid sulphate soils. International Institute for Land Reclamation and Improvement, Wageningen, the Netherlands.

Dold, B. 2003. Dissolution kinetics of schwertmannite and ferrihydrite in oxidized mine samples and their detection by differential X-ray diffraction (DXRD). Appl. Geochem. 18:1531–1540.

Dold, B., and L. Fontbote. 2001. Element cycling and secondary mineralogy in porphyry copper tailings as a function of climate, primary mineralogy, and mineral processing. J. Geochem. Explor. 74:3–55.

Dove, M.C., and J. Sammut. 2007. Impacts of estuarine acidification on survival and growth of Sydney rock oysters Saccostrea glomerata (Gould, 1850). J. Shellfish Res. 26:519–527.

Druschel, G.K., Baker, B.J., Gihring, T.H., and Banfield, J.F. 2004. Acid mine drainage biogeochemistry at Iron Mountain, California. Geochem. Trans. 5(2):13–32.

Durang, T. 1995. Yam cultivation on acid sulfate soils. MSc. Thesis. Department of Soil Science and Geology, Wageningen Agricultural University. Wageningen, the Netherlands.

EPA. 2003. Toxicological review of hydrogen sulfide. http://www.epa.gov/iris/toxreviews/0061-tr.pdf

EPHC & NRMMC. 2010. National guidance for the management of acid sulfate soils in inland aquatic ecosystems. Environment Protection and Heritage Council and the Natural Resource Management Ministerial Council, Canberra, Australia.

Evangelou, V.P., and Y.L. Zhang. 1995. A review: Pyrite oxidation mechanism and acid mine drainage prevention. Crit. Rev. Environ. Sci. Technol. 25:141–199.

Fältmarsch, R.M., M.E. Åström, and K.-M. Vuori. 2008. Environmental risks of metals mobilized from acid sulphate soils in Finland: A literature review. Boreal Env. Res. 13:444–456.

Fanning, D.S., M.C. Rabenhorst, S.N. Burch, K.R. Islam, and S.A. Tangren. 2002. Sulfides and sulfates. In J.B. Dixon, D.G. Schulze, and W.L. Daniels (eds.) Soil mineralogy with environmental applications. Soil Science Society of America, Madison, WI.

Fitzpatrick, R.W., E. Fritsch, and P.G. Self. 1996. Interpretation of soil features produced by ancient and modern processes in degraded landscapes. V. Development of saline sulfidic features in non-tidal seepage areas. Geoderma 69:1–29.

Fitzpatrick, R.W., B. Powell, and S. Marvanek. 2008. Atlas of Australian acid sulfate soils, p. 90–97. In R.W. Fitzpatrick and P. Shand (eds.) Inland acid sulfate soil systems across Australia. CRC LEME Open File Report 249. (Thematic Volume), CRC LEME, Perth, Australia.

Fitzpatrick, R.W., P. Shand, and R.H. Merry. 2009. Acid sulfate soils, p. 65–111. In J.T. Jennings (ed.) Natural history of the Riverlands and Murraylands. Royal Society of South Australia (Inc.), Adelaide, South Australia.

Freney, J.R. 1961. Some observations on the nature of organic sulfur compounds in soils. Aust. J. Agric. Res. 12:424–432.

Froelich, P.N., G.P. Klinkhammer, M.L. Bender, N.A. Luedtke, G.R. Heath, D. Cullen, P. Dauphin, D. Hammond, B. Hartman, and V. Maynard. 1979. Early oxidation of organic matter in pelagic sediments of the eastern equatorial Atlantic: Suboxic diagenesis. Geochim. Cosmochim. Acta 43:1075–1090.

Georgala, D. 1980. Paleoenvironmental studies of post-glacial black clays in northern Sweden. PhD. Thesis. University of Stockholm. Stockholm, Sweden.

Hamilton, S.K., S.J. Sippel, D.F. Calheiros, and J.M. Melack. 1997. An anoxic event and other biogeochemical effects of the Pantanal wetland on the Paraguay river. Limn. Oceanog. 42:257–272.

Hanhart, K., and D.V. Ni. 1993. Water management of rice fields at Hoa An, Mekong delta, Vietnam, p. 161–176. In D.L. Dent and M.E.F. van Mensvoort (eds.) Selected papers of the Ho Chi Minh City symposium on acid sulfate soils. ILRI Pub. 53. International Institute for Land Reclamation and Improvement, Wageningen, the Netherlands.

Hanhart, K., D.V. Ni, N. Bakker, F. Bil, I. Postma, and M.E.F. van Mensvoort. 1997. Surface water management under varying drainage conditions for rice on an acid sulfate soil in the Mekong delta, Vietnam. Agric. Water Manag. 33:99–116.

Hladyz, S., and S. Watkins. 2009. Understanding blackwater events and managed flows in the Edward-Wakool river system. Fact Sheet p. 2. Murray-Darling Freshwater Research Centre, Wodonga, Vic.

Howarth, R.W. 1979. Pyrite: Its rapid formation in a salt marsh and its importance in ecosystem metabolism. Science 203:49–51.

Howitt, J.A., D.S. Baldwin, G.N. Rees, and J.L. Williams. 2007. Modelling blackwater: Predicting water quality during flooding of lowland river forests. Ecol. Model. 203: 229–242.

Husson, O., and M.T. Phung. 1994. Synthetic report, Research methodology and main technical results. Farming systems research project, plain of Reeds. IAS/FOS, Fund for Development Cooperation, Brussels, Belgium.

Isbell, R.F. 1996. The Australian soil classification. CSIRO Publishing, Melbourne, Australia.

IUSS Working Group WRB. 2006. World reference base for soil resources 2006. World Soil Resources Report 103. FAO, Rome.

Jaynes, D.B., A.S. Rogowski, and H.B. Pionke. 1984. Acid drainage from reclaimed coal strip mines. I. Model description. Water Resour. Res. 20:233–242.

Jennings, P.R., W.R. Coffman, and H.E. Kauffman. 1979. Rice improvement, p. 186. International Rice Research Institute, Los Baños, Laguna.

Johnson, D.B. 1993. Biogeochemical cycling of iron and sulfur in leaching environments. FEMS Microbiol. Rev. 11: 63–70.

Johnston, S.G., A.F. Keene, E.D. Burton, R.T. Bush, L.A. Sullivan, A.E. McElnea, C.R. Ahern, C.D. Smith, and B. Powell. 2010. Arsenic mobilisation in a seawater inundated acid sulfate soil. Environ. Sci. Technol. 44:1968–1973.

Johnston, S.G., P.G. Slavich, and P. Hirst. 2004. The acid flux dynamics of two artificial drains in acid sulfate soil backswamps on the Clarence river floodplain, Australia. Aust. J. Soil Res. 42:623–637.

Johnston, S.G., P.G. Slavich, P. Hirst, R.T. Bush, and T. Aaso. 2009. Saturated hydraulic conductivity of sulfuric horizons in coastal floodplain acid sulfate soils: Variability and implications. Geoderma 151:387–394.

Johnston, S.G., P.G. Slavich, L.A. Sullivan, and P. Hirst. 2003. Artificial drainage of floodwaters from sulfidic backswamps: Effects on deoxygenation in an Australian estuary. Marine Freshwater Res. 54:781–795.

Jorgensen, B.B., H. Fossing, C.O. Wirsen, and H.W. Jannasch. 1991. Sulfide oxidation in the anoxic Black Sea chemocline. Deep-Sea Res. 38:1083–1103.

Konsten, C.J.M., R. Brinkman, and W. Andriesse. 1988. A field laboratory method to determine total potential andactual acidity in acid sulphate soils, p. 106–134. In H Dost (ed.) Selected papers of the Dakar symposium on acid sulphate soils. Dakar, Senegal, January 1986. ILRI Pub. 44. International Institute for Land Reclamation and Improvement, Wageningen, the Netherlands.

Ligue, D.L., and V.P. Singh. 1988. Social and economic status of farmers in acid sulfate soil areas in the Philippines, p. 238–249. In H. Dost (ed.) Selected papers of the Dakar symposium on acid sulfate soils. ILRI Pub. 44. International Institute for Land Reclamation and Improvement, Wageningen, the Netherlands.

Liu, M.S., R.M.R. Branion, and D.W. Duncan. 1987. Oxygen transfer to Thiobacillus cultures, p. 375–384. In P.R. Norris and D.P. Kelly (eds.) Biohydrometallurgy: Proceedings of the international symposium. Warwick, U.K.

Ljung, K., F. Maley, A. Cook, and P. Weinstein. 2009. Acid sulfate soils and human health—A millennium ecosystem assessment. Env. Int. 35:1234–1242.

Lohmann, U., and J. Feichter. 2005. Global indirect aerosol effects: A review. Atmos. Chem. Phys. 5:715–737.

Lundgren, D.G., and M. Silver. 1980. Ore leaching by bacteria. Ann. Rev. Microbiol. 34:263–283.

Luther, III, G.W. 1991. Pyrite synthesis via polysulfide compounds. Geochim. Cosmochim. Acta 55:2839–2849.

Luther, III, G.W., B. Glazer, S. Ma, R. Trouwborst, B.R. Shultz, G. Drushcel, and C. Kraiya. 2003. Iron and sulfur chemistry in a stratified lake: Evidence for iron-rich sulfide complexes. Aquat. Geochem. 9:87–110.

Luther, III, G.W., B. Sundby, B.L. Lewis, P.J. Brendel, and N. Silverberg. 1997. Interactions of manganese with the nitrogen cycle: Alternative pathways to dinitrogen. Geochim. Cosmochim. Acta 61:4043–4052.

Macdonald, B.C.T., O.T. Denmead, I. White, and M.D. Melville. 2004b. Natural sulfur dioxide emissions from sulfuric soils. Atmos. Env. 38:1473–1480.

Macdonald, B.C.T., J. Smith, A.K. Keene, M. Tunks, A. Kinsela, and I. White. 2004a. Impacts from runoff from sulphuric soils on sediment chemistry in an estuary lake. Sci. Total Environ. 329:115–130.

MDBA. 2010. Detailed assessment of acid sulfate soils in the Murray–Darling basin: Protocols for sampling, field characterisation, laboratory analysis and data presentation. Murray Darling Basin Authority, Publication 57/10. Canberra, Australia.

Millero, F.J., S. Hubinger, M. Fernandez, and S. Garnett. 1987. Oxidation of H_2S in seawater as a function of temperature, pH and ionic strength. Environ. Sci. Technol. 21:439–443.

Minh, L.Q., T.P. Tuong, M.E.F. van Mensvoort, and J. Bouma. 1997. Contamination of surface water as affected by land use in acid sulfate soils in the Mekong river delta, Vietnam. Agric. Ecosyst. Eviron. 61:19–27.

Moore, P.A., and W.H. Patrick, Jr. 1993. Metal availability and uptake by rice in acid sulfate soils, p. 205–224. In D.L. Dent and M.E.F. van Mensvoort (eds.) Selected papers of the Ho Chi Minh City symposium on acid sulfate soils. Pub. 53. International Institute for Land Reclamation and Improvement, Wageningen, the Netherlands.

Moses, C.O., D.K. Nordstrom, J.S. Hermann, and A.L. Mills. 1987. Aqueaous pyrite oxidations by dissolved oxygen and by ferric iron. Geochim. Cosmochim. Acta 51:1561–1571.

Nga, T.T., D.V. Ni, and V.T. Xuan. 1993. Cultivation of sugar cane on acid sulfate soils in the Mekong delta, p. 123–128. *In* D.L. Dent and M.E.F. van Mensvoort (eds.) Selected papers of the Ho Chi Minh City symposium on acid sulfate soils. ILRI Pub. 53. International Institute for Land Reclamation and Improvement, Wageningen, the Netherlands.

Nriagu, J.O. 1978. Biogeochemistry of lead in the environment, ecological cycles. Elsevier, Amsterdam, the Netherlands.

Pantelis, G., and A.I.M. Ritchie. 1991. Macroscopic transport mechanisms as a rate-limiting factor in dump leaching of pyrite ores. Appl. Math. Model. 15:136–143.

Pantelis, G., and A.I.M. Ritchie. 1992. Rate limiting factors in dumps leaching of pyritic ores. Appl. Math. Model. 16:553–560.

Piene, A., A. Tritschler, K. Kusel, and S. Peiffer. 2000. Electron flow in an iron-rich acidic sediment-evidence for an acidity-driven iron cycle. Limnol. Oceanogr. 45:1077–1087.

Pons, L.J. 1973. Outline of the genesis, characteristics, classification and improvement of acid sulphate soils, p. 3–27. *In* H Dost (ed.) Acid sulphate soils. Proc. Int. Symp. Acid Sulphate Soils August 13–20, 1972. I. Introductory Papers and Bibliography. Pub. 18. International Institute for Land Reclamation and Improvement, Wageningen, the Netherlands.

Pons, L.J., and N. van Breemen. 1982. Factors influencing the formation of potential acidity in tidal swamps, p. 37–51. *In* H. Dost and N. van Breemen (eds.) Proceedings of the Bangkok symposium on acid sulfate soils. ILRI Pub. 31. International Institute for Land Reclamation and Improvement, Wageningen, the Netherlands.

Powell, B., and M. Martens. 2005. A review of acid sulfate soil impacts, actions and policies that impact on water quality in the Great Barrier Reef catchments, including a case study on remediation at East Trinity. Marine Poll. Bull. 51:149–164.

Preda, M., and M.E. Cox. 2001. Trace metals in acid sediments and waters, Pimpama catchment, southeast Queensland, Australia. Environ. Geol. 40:755–768.

Pronk, J., and D.B. Johnson. 1992. Oxidation and reduction of iron by acidophilic bacteria. Geomicrobiol. J. 10:153–171.

Rabalais, N.N. 2002. Nitrogen in aquatic ecosystems. Royal Swedish Academy of Sciences. Ambio 31:102–112.

Rao, C.R.M., A. Sahuquillo, and J.E. Sanchez. 2008. A review of different methods applied in environmental geochemistry for single and sequential extraction of trace elements in soils and related materials. Water Air Soil Poll. 189:291–333.

Ren, D.T.T., V.T. Guong, N.M. Hoa, V.Q. Minh, and T.T. Lap. 1993. Fertilisation of nitrogen, phosphorus, potassium and lime for rice on acid sulfate soils, p. 147–154. *In* D.L. Dent and M.E.F. van Mensvoort (eds.) Selected papers of the Ho Chi Minh City symposium on acid sulfate soils. ILRI Pub. 53. International Institute for Land Reclamation and Improvement, Wageningen, the Netherlands.

Rickard, D.T. 1975. Kinetics and mechanisms of pyrite formation at low temperatures. Am. J. Sci. 275:636–652.

Rickard, D. 1997. Kinetics of pyrite formation by the H_2S oxidation of iron (II) monosulfide in aqueous solutions between 25 and 125°C: The rate equation. Geochim. Cosmochim. Acta 61:115–134.

Rickard, D., and G.W. Luther. 1997. Kinetics of pyrite formation by the H_2S oxidation of iron (II) monosulfide in aqueous solutions between 25 and 125°C: The mechanism. Geochim. Cosmochim. Acta 61:135–147.

Rickard, D., and J.W. Morse. 2005. Acid volatile sulfide (AVS). Mar. Chem. 97:141–197.

Ritsema, C.J., and J.E. Groenenberg. 1993. Pyrite oxidation, carbonate weathering, and gypsum formation in a drained potential acid sulfate soil. Soil Sci. Soc. Am. J. 57:968–976.

Ritsema, C.J., M.E.F. Van Mensvoort, D.L. Dent, Y. Tan, H. Van den Bosch, and A.L.M. Van Wijk. 2000. Acid sulfate soils, p. G121–G154. *In* M.E. Sumner (ed.) Handbook of soil science. CRC Press, Boca Raton, FL.

Rose, A.W., and C.A. Cravotta, III. 1998. Geochemistry of coal mine drainage, p. 1–1, 1–22. *In* K.B.C. Brady, M.W. Smith, and J. Sehueck (eds.) Coal mine drainage prediction and pollution prevention in Pennsylvania. PA Department of Environmental Protection: Commonwealth of Pennsylvania, Harrisburg, PA. 5600-BK-DEP2256.

Rosicky, M.A., L.A. Sullivan, P.G. Slavich, and M. Hughes. 2004a. Soil properties in and around acid sulfate soil scalds in the coastal floodplains of New South Wales, Australia. Aust. J. Soil Res. 42:587–594.

Rosicky, M.A., L.A. Sullivan, P.G. Slavich, and M. Hughes. 2004b. Factors contributing to the acid scalding process in the coastal floodplains of New South Wales. Aust. J. Soil Res. 42:587–594.

Rosicky, M.A., L.A. Sullivan, P.G. Slavich, and M. Hughes. 2006. Techniques for the revegetation of acid sulfate soil scalds in the coastal floodplains of New South Wales, Australia: Ridging, mulching, and liming in the absence of stock grazing. Aust. J. Exp. Agr. 46:1589–1600.

Rozan, T.F., M. Taillefert, R.E. Trouwborst, B.T. Glazer, S. Ma, J. Herszage, L.M. Valdes, K.S. Price, and G.W. Luther, III. 2002. Iron, sulfur and phosphorus cycling in the sediments of a shallow coastal bay: Implications for sediment nutrient release and benthic macroalgal blooms. Limn. Oceanogr. 47:1346–1354.

Saltzman, E.S., and W.J. Cooper. 1989. Biogenic sulfur in the environment. American Chemical Society, Washington, DC.

Sammut, J., R.B. Callinan, and G.C. Fraser. 1993. The impact of acidified water on freshwater and estuarine fish populations in acid sulphate soil environments, p. 26–40. *In* R.T. Bush (ed.) Proc. Natl. Conf. Acid Sulphate Soils. Coolangatta, Qld. June 24–25, 1993. CSIRO, NSW Agriculture, Tweed Shire Council.

Sammut, J., R.B. Callinan, and G.C. Fraser. 1996a. An overview of the ecological impacts of acid sulfate soils in Australia, p. 140–143. *In* R.J. Smith (ed.) Proc. 2nd Natl. Conf. Acid Sulfate Soils. R.J. Smith & Associates and ASSMAC, Alstonville, Australia.

Sammut, J., I. White, and M.D. Melville. 1996b. Acidification of an estuarine tributary in eastern Australia due to drainage of acid sulphate soils. Marine Freshwater Res. 47:669–684.

Sarwani, M., M. Lande, and W. Andriesse. 1993. Farmers' experiences in using acid sulfate soils: Some examples from tidal swampland of southern Kalimantan, Indonesia, p. 113–122. In D.L. Dent and M.E.F. van Mensvoort (eds.) Selected papers of the Ho Chi Minh City symposium on acid sulfate soils. ILRI Pub. 53. International Institute for Land Reclamation and Improvement, Wageningen, the Netherlands.

Sen, L.N., D.A. Duong, V.T. Guong, N.N. Hung, and T.K. Tinh. 1987. Influence of different water management and agronomic practices on pineapple yield in an acid sulfate soil at Binh Son 3 State Farm, Kien Giang. Fin. Rep. Proj. STD 302-NL. Department of Soil Science and Geology, Wageningen Agricultural University, Wageningen, the Netherlands.

Simpson, S.L., R.W. Fitzpatrick, P. Shand, B.M. Angel, D.A. Spadaro, and L. Mosley. 2010. Climate-driven mobilisation of acid and metals from acid sulfate soils. Marine Freshwater Res. 61:129–138.

Simpson, H.J., and M. Pedini. 1985. Brackishwater aquiculture in the tropics: The problem of acid sulfate soils. Food and Agriculture Organisation of the United Nations, Rome, Italy.

Sohlenius, G., and I. Öborn. 2004. Geochemistry and partitioning of trace metals in acid sulphate soils in Sweden and Finland before and after sulphide oxidation. Geoderma 122:167–175.

Soil Survey Staff. 2010. Keys to soil taxonomy. 11th edn. U.S. Department Agriculture Natural Resources Conservative Service, Washington, DC.

Sullivan, L.A., and R.T. Bush. 2000. The behaviour of drain sludge in acid sulfate soil areas: Some implications for acidification of waterways and drain maintenance, p. 43–48. In P. Slavich (ed.) Proceedings of workshop on remediation and assessment of broadacre acid sulfate soils. Acid Sulfate Soil Management Advisory Committee (ASSMAC). Southern Cross University, Lismore, NSW, Australia.

Sullivan, L.A., and R.T. Bush. 2004. Iron precipitate accumulations associated with waterways in drained coastal acid sulfate landscapes of eastern Australia. Marine Freshwater Res. 55:727–736.

Sullivan, L.A., R.T. Bush, and D. Fyfe. 2002. Acid sulfate soil drain ooze: Distribution, behaviour and implications for acidification and deoxygenation of waterways, p. 91–99. In C. Lin, M. Melville, and L.A. Sullivan (eds.) Acid sulfate soils in Australia and China. Science Press, Beijing, China.

Sullivan, L.A., R.T. Bush, and D.M. McConchie. 2000. A modified chromium-reducible sulfur method for reduced inorganic sulfur: Optimum reaction time for acid sulfate soil. Aust. J. Soil Res. 38:729–734.

Sullivan, L.A., R.T. Bush, D. McConchie, G. Lancaster, P.G. Haskins, and M.W. Clark. 1999. Comparison of peroxide oxidisable sulfur and chromium reducible sulfur methods for determination of reduced inorganic sulfur in soil. Aust. J. Soil Res. 37:255–265.

Sullivan, L.A., R.W. Fitzpatrick, R.T. Bush, E.D. Burton, P. Shand, and N.J. Ward. 2009b. Modifications to the classification of acid sulfate soil materials. Southern Cross GeoScience Technical Report 309. Southern Cross University, Lismore, NSW, Australia.

Sullivan, L.A., R.W. Fitzpatrick, R.T. Bush, E.D. Burton, P. Shand, and N.J. Ward. 2010. The classification of acid sulfate soil materials: Further modifications. Southern Cross GeoScience Technical Report 310. Southern Cross University, Lismore, NSW, Australia.

Sullivan, L.A., N.J. Ward, R.T. Bush, and E.D. Burton. 2009a. Improved identification of sulfidic soil materials by a modified incubation method. Geoderma 149:33–38.

Tan, Y. 1996. Comments on modeling of oxygen transport and pyrite oxidation in acid sulfate soils by Bronswijk et al. J. Environ. Qual. 25:928–930.

Tri, L.Q. 1996. Developing management packages for acid sulfate soils based on farmer and expert knowledge. Field study in the Mekong delta, Vietnam. PhD. Thesis. Department of Soil Science and Geology, Wageningen Agricultural University. Wageningen, the Netherlands.

Tri, L.Q., N.V. Nhan, H.G.J. Huizing, and M.E.F. van Mensvoort. 1993. Present land use as basis for land evaluation in two Mekong delta districts, p. 299–320. In D.L. Dent and M.E.F. van Mensvoort (eds.) Selected papers of the Ho Chi Minh City symposium on acid sulfate soils. ILRI Pub. 53. International Institute for Land Reclamation and Improvement, Wageningen, the Netherlands.

Tuong, T.P. 1993. An overview of water management on acid sulfate soils, p. 265–281. In D.L. Dent and M.E.F. van Mensvoort (eds.) Selected papers of the Ho Chi Minh City symposium on acid sulfate soils. ILRI Pub. 53. International Institute for Land Reclamation and Improvement, Wageningen, the Netherlands.

Tuong, T.P., C.T. Hoanh, and N.T. Khiem. 1991. Agro-hydrological factors as land qualities in land evaluation for rice cropping patterns in the Mekong Delta, Vietnam. In P. Deturck and F.N. Ponnamperuma (eds.) Rice production on acid soils of the tropics. Institute for Fundamental Studies, Kandy, Sri Lanka.

van Breemen, N. 1973. Soil forming processes in acid sulphate soils, p. 66–129. In H. Dost (ed.) Proc. Int. Symp. Acid Sulphate Soils. August 13–20, 1972. International Institute for Land Reclamation and Improvement, Wageningen, the Netherlands.

van Breemen, N. 1976. Genesis and solution chemistry of acid sulphate soils in Thailand. Agricultural Research Report 848. PUDOC, Wageningen, the Netherlands.

van Gent, P.A.M., and R. Ukkerman. 1993. The Balanta rice farming system in Guinea Bissau, p. 103–112. In D.L. Dent and M.E.F. van Mensvoort (eds.) Selected papers of the Ho Chi Minh City symposium on acid sulfate soils. ILRI Pub. 53. International Institute for Land Reclamation and Improvement, Wageningen, the Netherlands.

van Mensvoort, M.E.F. 1996. Soil knowledge for farmers, farmer knowledge for soil scientists: The case of the acid sulfate soils in the Mekong delta, Vietnam. PhD. Thesis. Department of Soil Science and Geology, Wageningen Agricultural University. Wageningen, the Netherlands.

van Mensvoort, M.E.F., N.V. Nhan, T.K. Tinh, and L.Q. Tri. 1993. Coarse land evaluation of the acid sulfate soil areas in the Mekong delta based on farmers' experience, p. 321–330. *In* D.L. Dent and M.E.F. van Mensvoort (eds.) Selected papers of the Ho Chi Minh City symposium on acid sulfate soils. ILRI Pub. 53. International Institute for Land Reclamation and Improvement, Wageningen, the Netherlands.

Ward, N.J., L.A. Sullivan, and R.T. Bush. 2004. Soil pH, oxygen availability and the rate of sulfide oxidation in acid sulfate soil materials: Implications for environmental hazard assessment. Aust. J. Soil Res. 42:509–514.

White, I. 2002. Swelling and hydraulic properties and management of acid sulfate soils, p. 24–48. *In* C. Lin, M.D. Melville, and L.A. Sullivan (eds.) Acid sulfate soils in Australia and China. Science Press, Beijing, China.

White, I., M.D. Melville, B.P. Wilson, and J. Sammut. 1997. Reducing acidic discharges from coastal wetlands in eastern Australia. Wetland Ecol. 5:55–72.

Wilson, B.P., I. White, and M.D. Melville. 1999. Floodplain hydrology, acid discharge and changing water quality associated with a drained acid sulfate soil. Marine Freshwater Res. 50:149–157.

Wong, V.N.L., S.G. Johnston, R.T. Bush, L.A. Sullivan, C. Clay, E.D. Burton, and P.G. Slavich. 2010. Spatial and temporal changes in estuarine water quality during a post-flood hypoxic event. Estuarine, Coastal Shelf Sci. 87:73–82.

Xuan, V.T. 1993. Recent advances in integrated land uses on acid sulfate soils, p. 129–136. *In* D.L. Dent and M.E.F. van Mensvoort (eds.) Selected papers of the Ho Chi Minh City symposium on acid sulfate soils. ILRI Pub. 53. International Institute for Land Reclamation and Improvement, Wageningen, the Netherlands.

Xuan, V.T., N.K. Quang, and L.Q. Tri. 1982. Rice cultivation on acid sulfate soils in the Vietnamese Mekong delta, p. 251–259. *In* H. Dost and N. van Breemen (eds.) Proceedings of the Bangkok symposium on acid sulfate soils. ILRI Pub. 31. International Institute for Land Reclamation and Improvement, Wageningen, the Netherlands.

Xuan, V.T., N.V. Sanh, D.V. Ni, N.T. Ut, and H.M. Hoang. 1986. The rice-shrimp cropping system on potential and actual acid sulfate soils in the Mekong delta. Paper presented at the Dakar conference on acid sulfate soils. Farming Systems Research and Development Center, Can Tho University, Can Tho, Vietnam.

22
Water Erosion

Dino Torri
Research Institute for Geo-Hydrological Protection

Lorenzo Borselli
Universidad Autonoma de San Luis Potosi

22.1	Introduction 22-1
22.2	Erosion Processes 22-1
	Wetting of the Soil • Drop Impact and Splash Erosion • Overland Flow • Concentrated Flow Erosion (Rills, Pipes, Gullies) • Pipe Erosion and Dispersive Soil Behavior • Shallow Mass Movement • Other Factors and Processes Affecting Soil Erosion by Water
22.3	Soil Erosion Models for Scenario Analysis 22-11
22.4	Conclusions 22-13
	Acknowledgment 22-14
	References 22-14

22.1 Introduction

Soil erosion refers to a series of processes leading to soil depletion *in situ* and the export of sediment toward downstream areas. This chapter will try to identify the main processes and their interrelationship with soil and soil properties. Some extremely important factors in the definition of soil behavior, such as vegetation, land use and wildfire, land leveling, and tillage erosion will be briefly discussed. The focus of this chapter is on processes of soil erosion rather than soil erosion models, each of which would require a chapter to itself. Our objective here is to review current scientific understanding of the processes that are important from the standpoint of soil science and soil observation, rather than the sum of their parts as described by a model.

22.2 Erosion Processes

Soil erosion by water takes place through three main processes: (1) wetting and detachment of soil particles, aggregates, clods, and large soil volumes from the soil mass; (2) movement of detached material (e.g., by gravity or by overland flow); and (3) deposition. Processes are also subdivided based on other criteria into (1) diffuse surface erosion (e.g., splash and interrill erosion), (2) linear erosion (rill and gully), (3) subsurface erosion (piping), and (4) shallow mass movements. Furthermore, the translocation of soil with machinery (tillage, land leveling, crop harvesting) or export of soil when harvesting root or tuber crops locally produces net soil loss or soil accumulation and, generally, a diffuse degradation of soil properties with increase of soil susceptibility to water erosion. In the following discussion, primary soil particles and aggregates (usually <5 mm) will be often grouped together under the term "grains."

22.2.1 Wetting of the Soil

As drops impact, detach, and compact the soil surface, the water content of soil changes. Soil wetting diminishes matrix forces because the capillary component disappears as water content increases. The van der Waals forces binding soil particles (especially clay particles, Frydman and Baker, 2009) decrease in intensity as interparticle distance increases due to water uptake. At the same time, chemical processes take place due to the interplay between soil particles, water, and ions. This can cause soil particle dispersion. Dispersion, which can vary in intensity, is always active (Borselli et al., 2001). It takes place at the soil surface mainly because infiltration and runoff water are usually low in ions (high ion concentration tends to enhance flocculation). Dispersion favors processes of pore clogging, which decrease water infiltration and enhance overland flow production. Ignoring dispersion has led to incorrect conclusions, usually in the nature of lower runoff intensities, lower erodibilities, longer time to ponding and time to runoff (Shainberg and Levy, 1996; Borselli et al., 2001). Values of these characteristics obtained using rainfall simulation made with low-quality water (e.g., tap water) should not be used for field application without correction.

Slaking of aggregates occurs during the first stages of soil wetting. This is due to the air trapped in the aggregates and compressed by the water driven into the soil by the matrix forces. The velocity of air diffusion through water (particularly in micropores) is lower than that at which water penetrates into the aggregates. This causes localized high pressure that can break interparticle bonds. The process is extremely rapid; a submerged dry clod can disintegrate in several minutes. Obviously, it ends as the soil water content approaches saturation. Usually, slaking diminishes and becomes negligible within 15–30 min after rainfall starts. The presence of slaking explains why erosion is often more intense when the soil is dry before rainfall. The processes involved in slaking have been discussed in detail by Rengasamy and Sumner (1997).

22.2.2 Drop Impact and Splash Erosion

Drop impact is responsible for much of the detached sediment, changes in surface morphology, and sealing. Drop impact forces depend on the drop mass and its impacting velocity. Falling drops accelerate until a maximum velocity value (V_F) is reached (Laws, 1941). Final fall velocity (m s^{-1}) varies with drop diameter (Φ, mm) approximately as follows:

$$V_F = 9.38 \tanh (0.44\Phi) \qquad (22.1)$$

When the drop hits the soil surface, it compresses the soil and removes particles and aggregates entrained by a crown of lateral droplets that are ejected laterally and upward (Engel, 1955), excavating a micro-crater. Grains are ejected around with an ejection angle (measured between the ejection trajectory and the soil surface) increasing with the depth of the water film that covers the soil surface (Mutchler, 1967). In the absence of any water film the angle seems to vary inversely to soil resistance (Al Durrah and Bradford, 1982). The presence of a water layer at the surface protects the soil as the layer absorbs part of the drop impact force (Palmer, 1963; Torri et al., 1987). When a drop hits soil grains on a sloping surface, the ejected grains are propelled further in the downslope than in the upslope direction. This causes a net downslope movement of grains. Locally, this causes depression to be filled by splashed sediments while protruding spots are eroded. Substantially, this process gives rise to two types of crust: those due to soil compaction by raindrop impact in the protruding areas (i.e., structural crusts), and the sedimentary crust where deposition prevails (Valentin and Bresson, 1992). Besides these crusts, drop impact is recognizable by the typical erosion forms it leaves on the soil surface, that is, splash pedestals. Splash pedestals are generated where an un-erodible element (rock fragment, plant residue) protects the underlying soil grains so that drop impact forces are dispersed. Hence, the spot is not eroded and slowly protrudes above the surrounding eroded zone.

In order to estimate the amount of splashed material, the forces produced by the impinging drops must be estimated. Many researchers studied drop impact on rigid surfaces (e.g., Engel, 1955) or modeled it (e.g., Harlow and Shannon, 1967). Nearing et al. (1987) measured impact forces on soil surfaces produced by impacting drops, 5–6 mm in diameter, falling from 14 m. Using 1 mm piezoelectric transducers, they measured pressure peaks of about 300–400 kPa located on a ring at about 2–2.3 mm from the impact centre. Elsewhere in the impact crater, pressure values were close to 40 kPa.

Since Ellison's studies (Ellison, 1947), raindrop energy has been the winning surrogate of the force of impact for describing drop detachment. Kinnel (1990) in his studies showed that drop impact increases overland flow transport capacity, enhancing soil erosion.

22.2.2.1 Vegetation Interception

Splash erosion is greatly modified by vegetation. Vegetation intercepts raindrops and subdivides water into three compartments: (1) stored on leaves and branches (which will finally evaporate), (2) re-dripped drops from leaves and branches (water in excess of storage), and (3) concentrated toward the "centre" of the plant (trunk, stem) where it reaches the ground as a flow along the trunk-stem (stem flow). This partitioning causes a decrease in rain impact energy because storage and stem flow water are subtracted from impacting raindrops. Re-dripping is characterized by drops that are generally larger than usual (4 mm in diameter approx., smaller in the presence of intense wind; Brandt, 1989). The average falling height of a drop is generally lower than 4–7 m (which is typical of medium-size trees) and is often close to zero (under small bushes or grass). Hence, drop kinetic energy is usually lower than that of raindrops (final velocity is reached in less than 20 m of falling height even by the largest drops).

This aspect becomes particularly important when natural and seminatural vegetation stands or agroforestry fields are considered because they are usually composed of a mixture of many plant species, living one under the other, where re-dripped drops from trees are intercepted and then partly re-dripped by bushes, to be caught by moss or litter or hit the bare soil. This is true also for many new and traditional crops (e.g., Lott et al., 2000a, 2000b; Diemont et al., 2006). The speed at which new varieties are adopted (and created), and new techniques invented makes it challenging to determine empirically the ability of vegetation to reduce splash erosion and protect the soil. Hence, models that make use of vegetation structure, partitioning raindrops and recalculating drop impact energy have a big advantage over models where vegetation is parameterized in a substantially empirical way such as the Universal Soil Loss Equation (USLE) C-factor (Renards et al., 1997; see Gabriels et al., 2003, as an example of C-factor adaptation).

Models describing in detail the behavior of plants in partitioning rain water and in protecting the soil are few. That developed by De Ploey (1982, 1984) and Van Elewijck (1989), was partly implemented in the runoff-erosion models EUROSEM (Morgan et al., 1998) and LISEM (De Roo et al., 1996; Jetten et al., 2006). Other similar models were developed independently by Gash et al. (1995, 1999), Navar et al. (1999), Crockford and Richardson (2000) and van Dijk and Bruijnzeel (2001). Plants are described through a series of parameters. They include crop cover and mean height of fall of re-dripped drops. Plant water storage represents the amount of water that is trapped at the plant surface: it is proportional to leaf area per unit of soil surface, which is often expressed through the leaf area index (or LAI). The water that is intercepted can be directed toward the inner part of the plant: this depends on the angle at which branches or stems help to concentrate intercepted rain water toward the central part of the plant (plant basal area). This central part can be either permeable (e.g., *Stipa tenacissima*) or impermeable (any tree's plant base) to water: in the former case infiltration and water storage in the soil increase, in the latter case some erosion may be encouraged by the concentration of clear water at the plant intersection with the soil surface (as is the case with maize).

Roots increase soil shear resistance (*T*) substantially, acting probably through their exudates, binding soil grains among themselves and to the roots, and through mechanical action. A series of papers have recently tackled the effect of roots on concentrated flow erosion rates (Gyssels et al., 2005; De Baets et al., 2007, 2008) proposing relationships between root density and erosion reduction.

22.2.3 Overland Flow

Water flow detaches and transports soil particles and aggregates. Hence, the main characteristics of overland flow (diffuse and concentrated) must be understood and correctly calculated if we want to understand erosion and which characteristics of soil and vegetation are relevant.

22.2.3.1 Flow Velocity

Flow is characterized by its unit discharge rate, q (m² s⁻¹); velocity, u (m s⁻¹); and depth, h (m). Darcy–Weisbach and Manning equations provide links between these variables:

$$q = hu \qquad (22.2)$$

$$u = \frac{h^\beta s^{0.5}}{m} \qquad (22.3)$$

where for the Manning equation m is equal to the Manning friction coefficient, n, and $\beta = 2/3$, and for the Darcy–Weisbach equation $1/m = \sqrt{8g/f}$ (where f is the Darcy–Weisbach friction factor) and $\beta = 0.5$. The Darcy–Weisbach equation has the advantage that the friction factor is dimensionless, which is not true of the Manning equation where n is measured in m^{1/3} s⁻¹.

If channels have a large depth-to-width ratio then depth must be substituted by the hydraulic radius (cross section occupied by the discharge divided by the wetted perimeter).

Friction factors have been studied in detail by many researchers, but the main problem is that they depend on water depth or some derived parameter such as the inundation ratio (h/D) where D is a characteristic of the bed grain size (e.g. the diameter corresponding to the median or some other percentile of the grain-size distribution). Lawrence (1997) applied this concept to the Darcy–Weisbach friction factor f, from sheet to river flow.

Smart et al. (2002) and Aberle and Smart (2003) found the following formula, which is based on data collected in fixed non-erodible beds

$$\sqrt{\frac{8}{f}} = 0.91 \frac{h}{RR} \qquad (22.4)$$

where *RR* is the standard deviation of channel bed elevations.

A more complex approach to f can be found in Gilley and Weltz (1995) where vegetation, litter, and moss play different roles in rill vs. interrill flows, and in rangeland vs. cropland.

A simple way of estimating Manning's n is the one proposed by Gilley and Finkner (1991):

$$n = \frac{0.172 RR^{0.74}}{Re^{0.282}} \qquad (22.5)$$

where

Re is the Reynolds number ($Re = hu/\eta$)
η is the kinematic viscosity of water (m² s⁻¹)
RR (mm) is a measure of soil surface random roughness (i.e., surface irregularity such as clods, which cannot be perceived as oriented in any preferential direction).

In this case, *RR* is obtained by calculating the standard deviation of soil elevations at the surface measured with sticks or laser-equipped profilemeters (Zobeck and Onstad, 1987) once form trends have been removed.

Equation 22.3 is valid for clear water flow when the interaction between the fluid in motion and the channel are negligible, that is, no erosion. When runoff interacts with the soil surface it becomes sediment laden. Hence, overland flow uses part of its energy to detach and transport grains. It uses another part to modify the soil surface (and change roughness). Moreover, fluid density and viscosity are affected by the presence of sediment and ions (Vanoni and Nomicos, 1960). The following are some of the empirical equations found by different authors to describe local experimental data better than Equation 22.3:

$$u = 0.097 Q^{0.137} \qquad (22.6.1)$$

calculated using data from Abrahams and Parsons (1990), for interrill runoff;

$$u = 3.52 Q^{0.294} \qquad (22.6.2)$$

for rill channels in erosion (Govers, 1992) in the absence of constraints such as rock fragments, or important bed forms;

$$u = 16 Q^{0.28} S^{0.48} \qquad (22.6.3)$$

for an unerodible fiber-glass replica of a rill channel (our calculation from data by Foster et al., 1984);

$$u = 2.88 \left(\frac{Q}{w}\right)^{0.68} \qquad (22.6.4)$$

for steep step-pool channels (Comiti et al., 2007); where u is the mean velocity in m s⁻¹, Q is flow discharge in m³ s⁻¹, and S is bed gradient.

In a field study Hessel et al. (2003) found that Manning's n increases with slope, which is a clear effect of the erosion interaction with the eroding overland flow, in agreement with laboratory observations made by Gimenez and Govers (2001) who found that surface roughness increases with erosion. The data of Hessel et al. (2003), collected in eroding rills in the presence of some physical constraints limiting erosion potential, plot below but parallel to Equation 22.6.2.

As the four Equations 22.6 show, they are quite different depending on the conditions under which they were calculated. The only relationship where slope gradient is present is Equation 22.6.3, which was calculated on a fixed non-erodible channel bed. The other equations were determined in field conditions while erosion was taking place, demonstrating that slope and hydraulic roughness are not present (or scarcely important) when erosion occurs. Obviously there must be some other parameter playing a similar role if we want all the equations to merge into a more general one.

22.2.3.2 Flow Detachment, Transport, and Sedimentation

The forces resisting detachment are the grain weight, its angle of friction with surrounding grains, and the cohesive forces with which each grain sticks to the soil mass. Detachment operates by drag and lift forces, which are roughly proportional to the flow velocity squared.

Flow detachment occurs when forces produced by runoff (drag and lift) are able to remove grains (detachment phase). As soon as the grain is entrained runoff accelerates it toward a velocity approximating the local flow velocity. At the same time, gravity tends to bring the grain back to the soil surface until the grain is eventually deposited.

Detachment is due to raindrop impact (which is dominant in interrill erosion, with diffuse and shallow overland flow), and to the forces produced by runoff. The latter case is often modeled using equations that estimate runoff forces using fluid stream power (p) or fluid shear stress (τ_F) applied to the channel bed surface:

$$p = \rho g q \sin \gamma \quad (22.7)$$

and

$$\tau_F = \rho g h \sin \gamma \quad (22.8)$$

where

p is in W m^{-2}
γ is the gradient (rad) of the channel bed (often equal to the local slope gradient)
$q = Q/w$ is unit discharge
τ_F is fluid shear stress (Pa)

These equations are rough proxies of the real situation, which is characterized by a local time distribution of shear stresses, wildly oscillating around a mean value (Naden, 1987). Consequently, particle detachment occurs even if the average value of the detaching forces is smaller than that of the bonding forces (Naden, 1987; Nearing, 1991). Soil grain bonding forces are similarly (but spatially) distributed around a mean value. Hence, a partial overlap of the two distributions is sufficient to guarantee that soil erosion will occur. Moreover, removing a grain has the effect of making other grains less protected and more easily erodible. As a consequence, while mean soil resistance is usually measured in kPa, the flow shear stress eroding it may be close to a few pascals (Torri et al., 1987).

The trajectories followed by entrained grains during transport and sedimentation can be quite complex; some grains are transported in contact with the soil surface, by rolling or sliding, others in suspension or by saltation. Consequently, sedimentation equations, such as Stokes', are poor approximations of reality.

The concept of flow transport capacity is a means for overcoming this problem. Transport capacity is defined as the amount of sediment that a given flow can transport over an infinite length when detachment and deposition rates are equal. Many (>10) equations have been proposed mainly for rivers and channels. Among these, we can refer to those proposed by Govers (1990), which were specifically developed for rill conditions. These equations are used in EUROSEM and LISEM (Morgan et al., 1998; De Roo and Jetten, 1999).

Once transport capacity (T_C) is known, net erosion rate due to overland flow scouring can be assessed as a transport deficit. In this case, the amount of soil that will be detached (D_f) in excess of what is deposited is given by the following:

$$D_f = \beta u_S (T_C - T_A) \quad (22.9)$$

where

T_A is the present sediment load
u_S is sedimentation velocity
β is a coefficient of proportionality, which takes into account soil shear strength (τ_S, measured with a pocket torvane at saturation) as follows (Morgan et al., 1998):

$$\beta = 0.79 e^{-0.85 \tau_S} \quad (22.10)$$

When Equation 22.9 gives negative values, then sedimentation is dominant and $\beta = 1$ by definition.

Other, nevertheless similar, detachment equations were developed starting from different concepts (Foster et al., 1989; Woolhiser et al., 1990). Some detachment equations were compared by Zhu et al. (2001) who found that power relationships seem to explain measured data better.

The general picture is further complicated by the fact that the raindrop-detached soil is easily entrained by the flow because these particles are already detached. As a result, splashed particles compete with overland flow detachment because both add to T_A (of Equation 22.9) modifying D_f.

As it has already been said, part of the drop kinetic energy is spent on impact with the overland flow. This causes an increase in turbulence, which is important in shallow flows, that is, in interrill erosion. Kinnel (1990) developed a set of equations to take into account this increase in turbulence to which there corresponds an increase in transported material. He developed a set of equations containing instantaneous rain intensity, flow mean velocity, and a complex function depending on raindrop size distribution. Interrill is further affected by raindrop impact as shown by the Australian school (Yu and Rose, 1999; Yu, 2003; Asadi et al., 2007) During splash erosion the soil surface becomes armored by the coarser grain fraction, which slowly caps the soil surface.

Obviously, interrill flow is influenced by the presence of small obstacles because overland flow is usually thin. This causes the flow to divert, turn around obstacles and follow paths that do not go along the steepest mean interrill gradient. Findeling et al. (2003) proposed an algorithm describing how the hydraulics of sheet flow is influenced by mulch cover. They included a flow tortuosity and a gradient along the tortuous interrill flow path as a function of the fraction of the surface covered by mulch.

22.2.4 Concentrated Flow Erosion (Rills, Pipes, Gullies)

Overland flow tends to rapidly coalesce into rivulets whose paths are unstable and can quickly change due to instances of local deposition or erosion. They wind around small protuberances until they reach local micro-depressions where they merge with other rivulets forming well-defined flow patterns that can eventually develop into rills. This aspect has been dynamically modeled by Favis-Mortlock (1998) who produced a model (RILLGROW) able to generate natural rill patterns very similar to those observed (Favis Mortlock et al., 2000).

Rill erosion is often treated by using Equation 22.9, and balancing erosion and deposition within the rill bed with the erosion inflow from interrill areas.

Rills are cut into the soil when the flow is sufficiently erosive. Smerdon and Beasley (1961) found that critical flow shear stress (computed using Equation 22.8), at which incision begins, increases linearly with clay content. This trend was not confirmed by data collected by Torri et al. (1987), who found that critical shear velocity for rill excavation increases with soil shear strength measured with a torvane. Rauws and Govers (1988) found a linear relationship (passing through the measured data) between flow shear velocity and soil shear strength (at saturation), which defined the threshold shear velocity to be overcome for the rill to be excavated. A more recent review (Leonard and Richard, 2004), mainly basing its deductions on the above-mentioned studies and data-sets, suggests a linear relationship between soil shear strength and critical flow shear stress.

When a linear incision becomes too deep or too wide to be easily removed by normal tillage operations, the channel changes from a rill to a gully. If the gully is removed shortly after its formation (which is common in cultivated fields), then it is called ephemeral. For a closer definition of gully formation, independent of tillage equipment, any channel with ephemeral runoff and maximum cross-sectional area equal to or above $0.0929\,m^2$ may be considered to be a gully (Hauge, 1977; Poesen et al., 2003). These definitions do not give differences between rills and gullies apart from the difference in size. While rills form on slopes, gullies usually form in thalwegs and the wall dynamics is important (mass movements and rills). Given their larger size, gullies usually drain larger catchments than rills. Hence, runoff is greater and lasts longer. Gully channels include both the incisions disappearing gradually upslope and those ending abruptly with a step. A classification of channel heads was proposed by Dietrich and Dunne (1993) on the basis of the morphology of the head: (1) gradual through small steps (<0.1 m), (2) large steps (0.1–1 m), (3) small headcuts (1–10 m), and (4) large headcuts (>10 m). Gullies have been classified in several ways, following genetic or morphologic criteria. They can be described as digitated if they repeatedly bifurcate while retreating until no dominant channel is clearly visible; axial when there is a main channel; or frontal, if the width of the retreating head is evidently larger than the length of the gully (Morgan, 1995).

Gullies are generated by concentrated overland flow sufficiently erosive to excavate deep or large channels that generally occur where several concentrated flow lines merge or where a concentrated flow of relatively clear water (e.g., an outflow from a forest, pasture or paved road) intersects a cultivated field. Furthermore, gullies may form because of local base-level lowering due to human activities or river dynamics. Even sea level lowering can trigger gully formation (Leyland and Darby, 2009). A retreating gully head can intercept un-eroded flow paths and cause tributary gullies to form. More generally, gullies may develop where the flow path intersects an earth bank, such as road or terrace banks, lynchets, etc. (Poesen, 1993). In addition, gullies form easily in highly dispersible soils or when pipes and tunnels collapse after reaching a critical size. This frequently occurs in sodic soils (Sumner et al., 1997).

Gully erosion has been documented as being responsible for 10%–94% of soil erosion (Poesen et al., 2003). Hence, it is of paramount importance to understand the process.

Gully erosion is characterized by an upslope retreat of the gully head. Gully sidewalls are usually reshaped by rills, slumps, and mud flows. The large quantities of sediment produced at the heads, walls, and over the bed tend to rapidly fill the transport capacity of the gully, reducing further bed scouring downstream. Hence, gullies are often characterized by an upstream segment where erosion is active and a downstream portion where sedimentation prevails. Where gully walls are made of easily weathered materials, early pedogenetic processes allow the development of alterites and entisols. Contact between the weathered and relatively unweathered layers makes for a discontinuous surface where permeability decreases abruptly, favoring soil saturation by water. As soon as soil formation reaches a certain depth that depends on local slope and soil shear strength values, new mass movements rejuvenate the gully walls. This process is typical of *calanchi*, ravines, arroyos. The deep incision of permanent gullies can trigger seepage erosion anywhere along its length. This, in turn, can cause the formation of new tributary gullies. If tributaries of different orders are numerous, they can intercept one another (with "valley" capture).

In this case, residual surfaces of the original slope are confined between interconnected gullies (Bryan and Jones, 1997).

Topics of major interest for gully erosion are whether and where gullies form in a given landscape, how fast they will erode, and how far upslope they can retreat. Their location in the landscape is determined by local precipitation (drier areas require larger catchments than wetter ones), runoff contributing area, mean valley gradient, and local gradient at gully head. Montgomery and Dietrich (1994) showed that a threshold exists for channel head location, which depends on the area (A) of the drainage basin and the gradient (S) at gully head. Several sets of observations conducted in different geographic areas are discussed by Poesen et al. (1998, 2003). The general relationship expressing the threshold is as follows:

$$S \geq aA^{-b} \qquad (22.11)$$

where
  S is the local slope (m m^{-1})
  A (ha) is the catchment area at the gully head
  $a \in [0.07; 0.11]$ in cropland, $a \approx 0.2$ in rangeland (Poesen et al., 1998)
  $b \in [0.1; 0.8]$ often being lower than 0.3

Recent data by Parkner at al. (2006) suggest smaller a and larger b values for pastures in the Waiapu catchment, New Zealand. Poesen et al. (2002) suggested a similar topographic threshold for rill initiation with and $a = 0.0027$ and $b = 0.59$.

Land use and soil erodibility play an important role in defining the threshold area. Retreat rates of gully heads depend on the amount of water that the gully head drains, that is to say, that gully heads retreat at a rate proportional to the catchment area (A). Seginer (1966) found the following relationship for gully retreat rate (G_R) in Southern Israel:

$$G_R = cA^\xi \qquad (22.12)$$

The empirical constant c depends on local characteristics (soil, land use, topography, precipitation) while the value of the exponent ξ is generally close to 0.5. Stocking (1981) reported more complex relationships for Southern Africa.

A second threshold can be expected to exist in relation to where gullies end, due to the great amount of sedimentation that counteracts gully development (Poesen et al., 1998, 2002). Generally, whatever can cause a sudden increase in sedimentation can trigger an end to gully entrenchment. Hence, a change from arable to pasture or rangeland may cause a gully to end. More generally, this process can occur with any passage from a lesser to a more protective land use (Beuselinck et al., 2000). A decrease of slope gradient often corresponds to a decrease of runoff detachment and transport, which causes a decrease in gully channel depth and, eventually, the end of the gully. Attempts to define relationships like (22.11) have produced a set of data usually plotting somewhat below the corresponding data for gully initiation (Poesen et al., 2002). In cropland areas of Northern Europe, ephemeral gully processes are usually dominated by sedimentation at a local slope gradient equal to or smaller than 0.02–0.04 in cropland. If rock fragment content is relatively large, then intense deposition may occur at gradients as large as 0.25–0.30 m m^{-1} (Poesen et al., 2002).

More complex relationships have been proposed by others: for example, De Santisteban et al. (2005) developed two more elaborated topographic indices to characterize the influence of watershed topography on gully erosion, obtaining reasonable agreement between observation and prediction.

Usually, local equilibrium is established between the erosion processes that make sidewalls retreat and those that further scour the gully bed. Hence, local relationships between the width, depth and cross section of gullies can be established. For example, in the United States gullies are characterized by a width to depth ratio close to 3 on cohesive material and 1.75 in non-cohesive sediments (Reid and Dunne, 1996). Poesen et al. (1998) showed that the depth of an ephemeral gully increases rapidly beyond a critical slope (>3%–4%) unless an erosion-resistant horizon (e.g., Bt, plough pan, fragipan, etc.) is present at shallow depth in the profile.

Other processes, typical of linear incision, have been tackled in studies presented in a still on-going series of conferences on gully erosion, which was started in 2000 (Poesen and Valentin, 2003). Some of the major processes for gully erosion are still active fields of research (e.g., Prasad and Römkens, 2003; Torri and Borselli, 2003). Among gully models that stand out are those developed by Sidorchuk (1999), Alonso et al. (2002), and Casalì et al. (2003). Processes include knick point evolution and head-cut retreat (Robinson et al., 2000). Øygarden (2003) observed all the aforementioned processes at work in various rills and gullies that developed during snowmelt in agricultural fields in Norway.

As already pointed out, there are discrepancies between theory and field observations. Govers et al. (2007) summarized some of them (we have already reported those regarding flow velocity, Equations 22.6) in the form of a set of equations, which constitute the rill-equivalent of the hydraulic geometry equations used in rivers (i.e., power equations). Nachtergaele et al. (2002) showed that rills and gullies may be characterized by an equation relating channel width (w) with the channel formative runoff discharge (Q). Torri et al. (2006) and Salvador et al. (2009) developed a channel width—discharge relationship and evaluated it in over more than 1500 cross sections of rills and gullies in badlands, rangeland, cropland and mountain forests located in Italy, Spain, Belgium, and Canada. Based on these studies, they expanded the relationship into the following:

$$w = \chi \left(\frac{Q}{Q_0}\right)^\alpha \qquad (22.13)$$

where the parameters Q_0 has the dimension of a discharge and χ of a length, while the exponent α varies with channel width

$$\alpha = 0.534(1 - 0.354 e^{-0.0255\bar{w}}) \qquad (22.14)$$

χ can be calculated using the junction method proposed by Torri et al. (2006) (see below) or deduced from tables (Salvador et al., 2009). Reliable methods or equations for the denominator Q_0 have not yet been developed (Salvador et al., 2009).

22.2.5 Pipe Erosion and Dispersive Soil Behavior

Pipes develop in particular situations. Some are linked to karstic processes, involving soil chemical dissolution and removal by subsurface flow, while others are linked to excess subsurface water that drains through cracks or macropores in which the water can flow fast enough to transport sediment. Subsoil sodicity further promotes clay dispersion and transport. These processes result in a pipe or a karst-like depression. The formation of pipes is facilitated by the presence of cracks or of impervious layers. If a soil horizon overlaying a less permeable layer has adequate horizontal hydraulic conductivity, water flows rapidly and becomes erosive. In other words, everything that promotes rapid accumulation of water beneath the soil surface (e.g., tectonic joints, micro-faults, cracks, biological macropores, etc.) or rapid water exfiltration (e.g., artificial and natural trenches such as gullies, landslides, abrupt increase of hill-slope gradient, etc.) promotes tunnel scouring and seepage erosion. Tunnel scouring takes place when subsurface water flows sufficiently fast to entrain soil particles.

Seepage erosion takes place when the water head can be dissipated at a scarp or at any other discontinuity, detaching particles and causing the discontinuity to retreat. An overview of pipe erosion was presented by Jones and Bryan (1997) depicting the various situations in which pipes can develop and their significance.

When soil dispersion occurs, seepage and piping can significantly modify processes and rates (Stocking, 1981) because dispersivity increases soil erodibility (Fitzpatrick et al., 1995). Faulkner et al. (2004) observed that peaks of sodium absorption ratio (SAR) correspond to zones of maximum incision (in Mocatan, Almeria, Spain). They also found that SAR increases with clay content. A well swelled clay can reduce, even suppress, infiltration. In this case water cannot penetrate any further and must move laterally, causing the formation of pipes subparallel to soil surface. Sometimes pipes are located under calcium-rich crusts. This local enrichment in calcium is due to the replacement of sodium during leaching. The description by Faulkner et al. (2004) is consistent with the observation that bed incision in pipes and rills in *biancana* badlands (Italy) is limited by the water front penetration velocity in the soil: only when saturated or near to saturation the material loses its high shear resistance and reaches soil shear strength values often below 0.2 kPa (Torri et al., 1994). Under these conditions, the presence of vegetation (pioneering species, such as *Artemisia cretacea*), the roots of which cause the formation of large connected porosity, accelerates sodium leaching.

These dispersion processes strongly undermine soil resistance to erosion. Faulkner et al. (2008) used these observations to explain the coupling of pipes to a rejuvenating channel in Mocatan; a series of incisions reconnects slopes and channels causing an increase in the hydraulic gradient along the slopes; this, in turn, favors piping and pipe enlargement with the subsequent collapse of pipe roofs. Hence, lateral gullies can develop easily. The new (lateral) gullies have walls along which similar processes can occur, first piping, followed by gully formation again. This constitutes a feedback mechanism that ends a period of slope-base stability. Examples of this type of process can be found in many places where badlands exist, such as the *biancana* badlands of Tuscany (Torri and Bryan, 1997). Wilson (2009) discussed some of the processes involved in this pipe-gully interaction using laboratory experiments.

A basic classification for evaluating the risk of pipe erosion on the basis of the dispersion potential of a soil was given by Fitzpatrick et al. (1995) for soils in Southeastern Australia. They suggested that a soil is at risk of pipe erosion when the following condition is fulfilled:

$$EC < 0.062 ESP + 0.33 \qquad (22.15)$$

where
    EC is the electrical conductivity at 25°C in a 1:5 soil:water suspension
    ESP is the exchangeable Na percentage

Other criteria, including those to use in the field, were discussed by Bell and Walker (2000).

22.2.6 Shallow Mass Movement

Shallow mass movements are important in water erosion because (1) they are responsible for the evolution of rill and gully slopes during and between rainstorms; (2) mass movements may trigger gullies and pipes and *vice versa*.

Local mass movements of soil result from local severe degradation processes and are usually associated with other forms of erosion such as rills, pipes and gullies. In agriculture, mass movements inhibit farm operations by loss of accessibility, exposure of infertile subsoil and unprotected soil surface layers to splash and rill erosion, and net downslope movement of the soil mass. Gravity is the principal force producing slides, slips, slumps, flows, and landslides. At a particular water content soil becomes unstable and can slide downslope. Landslides and mass movements are usually classified by type and velocity of movement (Varnes, 1978). Rapid movements of soil mass over a distinct sliding surface are termed "landslides." Slower long-term deformations with a series of sliding surfaces and exhibiting viscous movement are termed "creep."

Generally mass movement occurs when the weight (shear stress) of the surface material on the slope exceeds the retaining ability (shear strength) of the material. Risk of mass movement is increased by erosion or excavation undermining the foot of a slope, loss of stabilizing roots through removal of vegetation, and increase in pore water pressure within the soil profile. Increased pore water pressure or greater water absorption may weaken intergranular bonds, reducing internal friction, therefore lessening the cohesive strength of the soil and, ultimately, slope stability.

Usually slope stability is determined quantitatively by the ratio between available shear strength of soil mass and imposed shear stress computed along the assumed sliding surface:

$$F_s = \frac{T}{S} \quad (22.16)$$

where

F_s is the stability factor and depends on the shape of the sliding surface and on the type of movement
T is the shear strength (the sum of resisting forces)
S is the shear stress (the sum of driving forces)

Values of $F_s < 1$, $F_s > 1$, and $F_s = 1$ correspond to unstable, stable and metastable slopes. Considering the spatial and temporal variability of soil properties, values of $F_s > 1.2$ are considered satisfactory for stable conditions in a conservative analysis. Usually, sliding surfaces are supposed to be circular, planar or generic. Two-dimensional analysis is commonly used to evaluate overall slope stability (Huang, 1983). Basic inputs are slope transect, interface profile of the discontinuities present in the soil mass, mechanical properties of each stratum on the slope portion being considered, and the profile of the final phreatic surface separating saturated from unsaturated soil.

Mass movements may be classified using the following relationship (Janbu, 1973):

$$d_1 = \frac{d}{L} \quad (22.17)$$

where

d is the maximum depth of the sliding surface
L the length of the sliding mass (Figure 22.1)

Landslides characterized by $d_l < 0.05$ are considered planar movements with sliding surfaces subparallel to the slope. Higher values are typical of circular or pseudo-circular sliding surfaces.

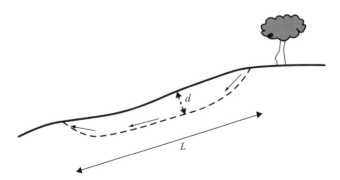

FIGURE 22.1 Shallow mass movement: definition of the d and L parameters.

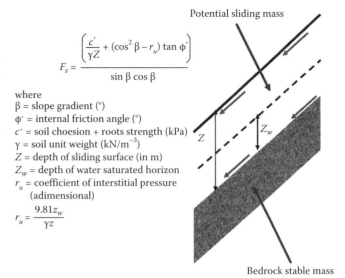

$$F_s = \frac{\left(\dfrac{c'}{\gamma Z} + (\cos^2 \beta - r_u)\tan\varphi'\right)}{\sin\beta\cos\beta}$$

where
β = slope gradient (°)
φ' = internal friction angle (°)
c' = soil choesion + roots strength (kPa)
γ = soil unit weight (kN/m^{-3})
Z = depth of sliding surface (in m)
Z_w = depth of water saturated horizon
r_u = coefficient of interstitial pressure (adimensional)

$$r_u = \frac{9.81 z_w}{\gamma z}$$

FIGURE 22.2 Planar shallow mass movement—infinite slope stability model.

For shallow mass movements, the infinite slope approach assuming planar movements (Figure 22.2) is commonly used. The stability factor may be evaluated from

$$F_s = \frac{\dfrac{c'}{\gamma z}(\cos^2\beta - r_u)\tan\varphi'}{\sin\beta\cos\beta} \quad (22.18)$$

and

$$r_u = \frac{9.81 z_w}{\gamma z} \quad (22.19)$$

where

z is the depth of the soil layer or the slip surface in the topsoil used to evaluate stability
z_w is the height of the water table above the slip surface
β is the local slope angle
φ' is the angle of soil internal friction
c' is the internal cohesion of the soil
γ is the unit weights of soil

Equation 22.18 applies to planar sliding of the soil mass where sliding surface and seepage flow are approximately parallel to the land surface. Such conditions only occur in some instances, whereas the infinite slope criterion is used for evaluating stability over large areas or watersheds. Some computer models use the infinite slope method under probabilistic and deterministic approaches (LISA) (Hammond et al., 1992).

The soil strength parameters φ' and c' are commonly measured in the laboratory on undisturbed soil cores, based on the Mohr–Coulomb soil strength criterion (Whitlow, 1990) or estimated indirectly from empirical correlations with penetrometer or vane test values measured in the field (Huang, 1983; Whitlow, 1990).

Particular situations such as concave–convex slopes, gully walls, tension cracks, thawing slopes, and vegetation effects necessitate modification of Equation 22.18. Alternatively, limit equilibrium methods should be employed for nonplanar slopes with tension cracks (Janbu, 1973). Several features, such as evidence of previous mass movement, poor vegetation cover, and linear features across slopes related to cracks or minor terracing in the landscape can indicate whether a certain area deserves study in order to assess the risk of mass movement. Old or dormant landslides characterized by long, uneven hummocky slopes, cracks, and fissures and bent tree trunks on steep slopes indicate a history of mass movement. Vegetation has the effect of reducing shallow mass movements; roots add resistance to the soil if they pass through the sliding surface, fixing the potentially sliding portion to the layers underneath (Hammond et al., 1992). Sliding potential may increase if trees are rooted only above the sliding surface as previously discussed.

22.2.7 Other Factors and Processes Affecting Soil Erosion by Water

22.2.7.1 Vegetation

We have already pointed out that vegetation has a strong influence on soil erosion, intercepting and redistributing rain water. Vegetation has an effect on linear erosion through an increase in the hydraulic friction factor (Manning n and Darcy–Weisbach f) and in the soil shear strength. The detachment rate, as expressed in Equations 22.9 and 22.10, shows that the factor β decreases exponentially with increasing soil shear strength. As root density increases soil resistance, then erosion decreases with it (De Baets et al., 2008). Roots also affect mass movements when they pass through the surface of failure. Roots increase macroporosity and, hence, enhance water infiltration and decrease overland flow. Protecting soil surface against direct drop impact (e.g., litter, low grasses, moss), vegetation reduces soil crusting. As the environment protected by vegetation is more suitable to life, vegetation also enhances the positive effects of microorganisms, fungi, etc., on soil structure and soil resistance to erosion (e.g., Salvador Sanchis et al., 2008).

Due to its dynamics, natural vegetation is also a candidate for driving changes in soil characteristics relevant to erosion. Particularly on abandoned land, but also on overgrazed or degraded land (Trimble and Mendel, 1995), species composition changes with time as local conditions stabilize or improve. An example of integrated approach to soil erosion and vegetation dynamics is the PATTERN model (Mulligan, 1998). It uses high temporal resolution meteorological data to calculate water budget and vegetation growth. It simulates seed dispersal, which is usually absent in most models that deal only with cultivated crops. For medium- to long-term studies and predictions it is also necessary to include the effects of animals and of the dynamics of animal species on plants and plant succession (Smith et al., 2003).

A long-term effect of vegetation is the enrichment of the organic matter component of the soil, which, in a healthy soil, becomes a ligand, which increases soil aggregate stability. Frequent tillage operations or other sources of disturbance (such as overgrazing) interfere with the positive effect of vegetation and make the soil more erodible.

Soil is dynamic, with changes that accumulate over years or with seasonal variations. This is apparent in rangeland soil (e.g., aggregate stability seasonal variation—see Salvador et al., 2008, for a short review) but it is much more evident in cropland. Many system parameters, such as the soil erodibility factor of USLE, show seasonal dynamics, which can be inferred through mean monthly air temperature oscillation during the year (Salvador et al., 2008). Bulk density of cropland soils changes because soils after tillage resettle and most of the tillage macropores are reduced to smaller sizes, aided by rainfall and cycles of wetting and drying (Knapen et al., 2008). with modifications of erodibility, shear strength and aggregates stability.

22.2.7.2 Rock Fragments

When non-erodible (or scarcely erodible) grains, such as rock fragments, are present, the soil surface is carved by erosion into pedestals surmounted by grains. Erosion can continue until a pavement of rock fragments protects the soil from any further loss. The first direct effect of rock fragments is that of intercepting raindrops. Consequently, drop detachment decreases proportionally to the fraction of soil surface covered by rock fragments. A further effect is that of concentrating runoff between stones, locally increasing the flow capacity for detachment and transport. More specifically, water flowing around isolated rock fragments excavates a horseshoe-shaped incision on the upstream side of the stone due to whirls (Poesen et al., 1994). The rock fragments can be considered isolated from other fragments until they cover less than 20%–25% of the surface. After such a threshold, eddies interfere and tend to disperse their energy onto nearby non-erodible rock fragments rather than erodible grains. As a consequence, erosion may initially increase and then decrease while the rock fragment cover increases. The initial possible increase in erosion occurs following the balance between reduced drop detachment and increased overland flow detachment. Reduction occurs when rock fragments are not embedded into the soil so that water can infiltrate under it, hence maintaining a high infiltration rate and a low overland flow. On the contrary, erosion may increase when the rock fragments are embedded into the soil surface so that overland flow per unit of erodible soil surface is increased.

Rock fragments also show a protective effect in reducing rill erosion. This occurs because rock fragments are often too large to be entrained by the flow (e.g., Rieke-Zapp et al., 2007). Selective erosion leaves more and more rock fragments on the soil surface, armoring it. This effect is clearly shown on soil erodibility (Salvador et al., 2008).

22.2.7.3 Erosion due to Tillage and Land Leveling and Its Interaction with Water

Customary farming activities are an important source of change in soil characteristics and in local slope morphology. For this reason it is important that a person surveying an area should be aware of the modification caused by such practices and of their effects on soil erosion. This is also important in scenario analysis, for example, when a change of land use or 10 years of tillage are hypothesized to occur.

Whenever soil is cultivated, a displacement of the tilled layer takes place (Govers et al., 1994; Van Oost et al., 2006). The term "tillage erosion" is used when the soil translocation produces a net soil loss or a net soil accumulation locally (Figure 22.3). Tillage erosion is influenced by slope gradient, depth and speed of tillage, type of tillage tool used, and direction with respect to the local slope gradient (Torri and Borselli, 2002; De Alba et al., 2006). Tillage is responsible for the creation of false terraces at field borders and of macro-pedestals under isolated poles or trees.

Tillage erosion rates are comparable to or higher than soil erosion by water, even if the net erosion due to tillage is concentrated mainly in convex morphology and deposition in concavities. So, where soil erosion by tillage is maximal, soil erosion by water is minimal (Van Oost and Govers, 2006). Annual erosion rates due to tillage often range between 10 and 50 Mg ha^{-1} year^{-1} (Van Oost and Govers, 2006). In some cases, localized in convexities, detailed surveys indicate a lowering rate of soil surface ranging approximately between 2 and 4 cm year^{-1} due to deep ploughing (Borselli et al., 2002). For the purpose of comparison with the customary unit of measures used in soil erosion, these values correspond to losses ranging between 200 and 600 Mg ha^{-1} year^{-1}. These values indicate that soils in cultivated fields can change their properties (including soil resistance to erosion and infiltration capacity) in a few years when the tillage tool starts cutting into a B or C horizon and mixes it with the original Ap material.

Let us now examine land leveling. Since the beginning of agriculture, farmers have acted to adapt plants and slope morphology to their needs. Today, reshaping the slope morphology has become a sort of *condicio sine qua non* for growing new crops in many areas of the world. The earth movements (land leveling) change the slope gradients, the concentration lines of surface and subsurface water fluxes, and the shapes and sizes of local catchments. The depth affected by earth movement often reaches the C and R soil horizons. Generally, upslope soils and soils on convexities are cut and the material moved to concavities or

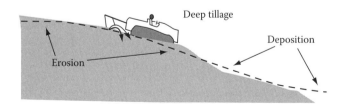

FIGURE 22.3 Soil erosion by tillage due to deep ploughing.

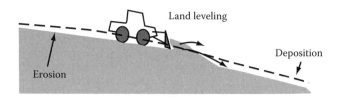

FIGURE 22.4 Land leveling: soil redistribution, erosion, and deposition.

down the slope (Figures 22.3 and 22.4). These processes produce a truncation of original soil profiles in the convex surfaces and a degradation of soil properties and qualities (e.g., water conductivity, organic matter content, fertility, erodibility).

Tillage erosion and land leveling cause soil depth and soil characteristics to be modified within the slope. This modifies the spatial variability of runoff generation and often promotes larger runoff concentrations. Rill and ephemeral gully formation is thus generally enhanced. The concavities that have been filled with additional material are increasingly unstable; subsurface water fluxes concentrate water in the same concavity, but in this case, the weight of the soil column above the potential sliding surface is larger, enhancing the risk of shallow mass movements (Borselli et al., 2006). Increased frequencies of ephemeral gully generation and of shallow mass movement trigger further land leveling, establishing positive feedback mechanisms. Row crops can intensify the mechanism because each row acts as a preferential flow path, further concentrating runoff.

We should consider that 1 m of land leveling corresponds to 13,000 Mg ha^{-1} of net erosion or accumulation, which is at least two orders of magnitude more intense than any concentrated erosion or sedimentation rate due to overland flow. Land leveling often has on and off-site effects produced by water erosion on fresh leveled surfaces: the extremely high erodibility of the leveled surface, coupled with an intense rainfall can produce instances of extreme erosion with total soil losses up to 400–600 Mg ha^{-1} (Borselli et al., 2006).

A further implication linked to tillage and leveling is that an intensively tilled or recently leveled slope hides gully and mass movement occurrence, presenting a landscape that appears to be stable and well managed. To understand the real processes occurring in such a slope, a surveyor must use sets of historical photos (air photos, now also very high definition satellite images) to identify processes and hazards.

22.2.7.4 Wildfires

Wildfires damage vegetation and litter and can severely affect the soil (e.g., Giovannini and Lucchesi, 1997; Rengasamy and Sumner, 1997). Soil biota and soil organic matter can be destroyed or damaged, clay minerals modified, and water repellency generated or enhanced (e.g., De Bano et al., 1976; Giovannini and Lucchesi, 1983).

The heat developed by the fire causes textural changes through the aggregation of clay and silt particles into sand sized grains resistant to the chemical and physical actions of water (Giovannini and Lucchesi, 1997). This leads to changes in soil structure with disintegration of macro-aggregates and migration of micro-aggregates through the profile (Giovannini, 1994; Ubeda and Sala, 1998), which, in turn, modifies the water retention properties of the soil profile. If temperatures reach sufficiently high levels, clays are denatured and lose their charge rendering them less dispersible (Rengasamy and Sumner, 1997). On the other hand, high temperatures can cause the distillation of organic matter. The nonvolatile compounds, in particular, move downward into the soil profile until they condense in a cooler portion of the soil. Under certain conditions of combustion and soil moisture content, a subsurface hydrophobic layer develops (De Bano et al., 1976; Giovannini and Lucchesi, 1983). Usually, these transformations give rise to changes in the behavior of burnt soils with loss in water storage capacity, and changes in the rate of infiltration and hydrophobic behavior (DeBano et al., 1970; Le Bissonnais, 1996). The intensity and type of deterioration caused by fire are closely related to the physical and chemical properties of the soil that is burnt, above all texture, size, and composition of the structural aggregates (e.g., Mataix-Solera and Doerr, 2004). The net effect of these mechanisms is a decrease in organic matter content in the soil surface layer and an increase in water permeability, which is contrasted by an increased in hydrophobicity. If a hydrophobic layer forms, water permeability decreases dramatically and overland flow production is enhanced.

These modifications are accompanied by a decreased stability of soil aggregates. Usually this favors the decrease in infiltration capacity but mainly causes soil erodibility and erosion to increase substantially. Measurements of soil erosion and runoff in equipped plots showed that maxima are reached where the fire had been most intense (Rubio et al., 1997). Branson et al. (1981) and, more recently, Pierson et al. (2008) reported values indicating that the predominant effect of fire is to reduce infiltration and increase erosion. In particular, soil loss (on an annual basis) after a moderate to intense fire may increase by something between 2 and 10 Mg ha^{-1}, with exceptional values of more than 100 up to 500 Mg ha^{-1} in a fine sandy loam over pumice.

The powerful effect of fire on aggregate stability is used by Varela et al. (2010) who based their estimates of erodibility modification caused by fire on the modification of soil aggregate stability in water. Following the intensity of a fire and of its immediate consequences, the local ecosystem requires time to recover, sometimes several decades even in the absence of other wildfires (Andreu et al., 2001). Hence, it is important to be able to recognize where erosion can be more intense in order to identify sites where to protect the soil and help the system to recover (Varela et al., 2010). It is an important goal to study the soil properties discussed so far (particularly induced hydro-repellence, new soil-aggregate characteristics, and colloidal fraction modification) and describe them through appropriate algorithms in order to improve runoff-erosion models.

22.3 Soil Erosion Models for Scenario Analysis

Reliable scenario analyses can be conducted with the use of models. Models are needed because erosion is determined by many processes and factors interwoven in a non-linear fashion. Practically all models can be used for scenario analysis. Some are more reliable and easy to adapt to newly designed situations than others.

The first model able to estimate soil loss over a wide range of situations was the USLE (Wischmeier and Smith, 1965) initially limited to the United States, East of the Rocky Mountains, and later extended to the Pacific coast (Wischmeier and Smith, 1978). The model, which was recently revised (RUSLE), is expressed as (Renard et al., 1997):

$$A = RKLSCP \qquad (22.20)$$

where
   A is the average annual soil loss over a 20–22 year period
   K is soil erodibility
   R is rainfall erosivity
   L is the field length factor
   S is the gradient factor
   C is the crop and management factor
   P is the effect of the soil protective practices

Soil erodibility is determined experimentally as the soil loss per unit R measured on a 22.1 m long plot with a 9% uniform slope gradient that must be kept under continuous fallow with frequent up and down cultivations to remove weeds and erosion features. Erosivity is defined as

$$R = 0.01 \sum EI_{30} \qquad (22.21)$$

where
   E is raindrop kinetic energy over the storm (MJ ha^{-1}) before hitting either the soil surface or vegetation
   I_{30} (mm h^{-1}) is the maximum rain intensity over 30 min during the storm

In other parts of the world, maximum intensities over shorter periods have been found to be better suited for predictions.

Given that RUSLE relies on generally available or easily found input data, it is often used outside its proper application scale (field scale). Terranova et al. (2009) exemplify this type of application of RUSLE at a territorial level, where the local erosion value becomes a sort of local index of potential erosion.

Recent reexamination of a worldwide erodibility dataset has shown the marked effect of climate, with soils developed in cooler climates being more erodible than soil developed in warmer climates (Salvador et al., 2008).

Many models have been developed since USLE was first published, many of which are much more physically based. Merritt et al. (2003), Singh (1995), and Torri et al. (2007) described more than 30 different models. Reviews are always relatively inaccurate and rarely up to date unless the models they are describing are old models on which nobody is working anymore. Moreover, a model cannot be said to be properly known until its physics (i.e., the approximations contained in each model's components) and its mathematics (especially the routine of numerical approximation) are known together with the code that translates them into the computational reality of the model. Generally, the manuals that are distributed with the models are not sufficiently informative. Models are a synthesis of the knowledge on a given subject. Their submodels (algorithms) represent processes that feed into each other well or need bridges, types of heuristic solutions to missing links between processes. Obviously not all we know on a subject can be put into a model. Hence, models are approximations of our knowledge at a given moment.

Distributed models usually bring in computational problems. This is an aspect that generally is not described in manuals but must be known because a robust numerical approximation is a prerequisite for any decent prediction. The problems are usually due to the numerical approximations needed to solve the equations that guarantee conservation of mass, momentum, and energy in the fluxes from cell to cell. All this imposes limits on the time-step length, based on cell spatial size and local discharge: roughly, all that comes out of a cell during one time-step cannot travel beyond the downslope cell during the same time-step. Information on these math problems are usually scanty or completely missing. The LISEM manual mentions Courant's condition only to state that it is the user's responsibility to understand if the calculations have actually converged at all stages, in every cell (http://www.itc.nl/lisem/ > lisem basic > basic theory > kinematic wave > routing procedure in LISEM). Other models, such as KINEROS and EUROSEM, in their earlier versions, advised users to change time-step lengths when convergence was at risk. As the transfer of energy and matter often requires applications of numerical methods, approximations become significant and weak tests on convergence allow major errors in estimates. A review of methods for modeling kinematic wave and water transfer can be found in Singh (1996).

The scale at which a catchment is represented affects the way in which processes can be mathematically described because the model scale is usually inadequate to surrogate the process scale (e.g., cellslope gradient generally decreases with increasing cell size). Once the range of scales at which a given model is set to work is decided, inputs and outputs are defined. Many relevant input parameters of a hydrological-erosive model depend on the interaction between different types of information. For example soil parameters have a daily, weekly or seasonal variation and are modified by the cumulative effects of erosion, land use and soil management. If the model runs over several months, vegetation must be examined in relation to temporal dynamics. If land use or management changes are included, then soil parameter values also need adjustment. As already discussed, land leveling and tillage erosion may also require morphology to be adjusted. If climatic changes are involved, then the resilience of natural vegetation to withstand the new situation must also be considered.

Landscapes must be properly described in order to model flows (water and sediment) over the terrain. This requires (1) the division of landscape (terrain surface and subsurface) into a set of connected elements; (2) the use of conservation of mass, energy and momentum within neighboring elements, of the fluid as such and of the different components of the fluid (sediment and water); and (3) the identification of flow paths.

Cascading planes, raster and triangulated irregular network (TIN) are three different representations of the landscape. They can produce different outcomes and may need different degrees of data aggregation. A "cascading planes" representation requires the user to simplify the landscape into series of planes and channel segments. A raster representation is obtained by automatically subdividing the landscape into a series of equal square cells. The TIN approach allows the description of the landscape using a variable planar density of coordinates.

Linear forms such as cliffs, vertical gully walls or terrace borders, are either poorly or not represented in the above-mentioned systems. They have to be taken into account in the model software. Simulations, including local landscape modification processes (e.g., gully or rill generation, land use change, years of tillage) are difficult to represent with cascading planes without the need to manipulate them during the simulation period. This type of simulation is better achieved if the landscape is represented as a raster or a TIN (Hofierka et al., 2002; Shary et al., 2002; Jetten et al., 2006).

Major models (continuous, distributed, and physically based) are those derived from the Système Hydrologique Europeenne (SHE, Abbot et al., 1986; MIKE SHE, Refsgaard and Storm, 1995; SHESED, Bathurst et al., 1995). In a raster landscape the soil is described with its horizons, through which water can flow vertically (Richards equation). When saturation is reached, or a water table is met, then the flow runs parallel to the saturated surface. Both saturated and hortonian flow can be described. It is not clear how well they perform in the presence of cracks and other macropores that cannot be described by Richards equation.

Cascading planes landscape representation is used in several models (e.g., OPUS, Smith, 1995; WEPP, Flanagan and Nearing, 1995; EUROSEM, Morgan et al., 1998) OPUS is a continuous simulation model with time steps varying to match process dynamics and numerical approximations. WEPP too, is a continuous and event-based model, containing a noticeable series of innovations derived from WEPP specific research. A raster representation is used in models such as ANSWER (Beasley et al., 1980), which, initially developed for describing single rainfall event, is now a continuous physically based simulation model.

A single-event model KINEROS (Woolhiser et al., 1990) was developed to test different research hypotheses such as transport equations. KINEROS was also used as a basis on which to develop EUROSEM (Morgan et al., 1998). Its hydraulics is limited to hortonian overland flows (rainfall exceeding infiltration intensity). Its numerical approximations converge fairly well.

KINEROS evolved into a KINEROS2 model (Smith et al., 1999). Still based on cascading planes, it has a better infiltration routine based on a 2-layer infiltration equation (Smith, 1990).

LISEM (De Roo et al., 1996,) uses most of the equations developed in EUROSEM for erosion. LISEM describes rills and ephemeral gullies, (Jetten et al., 2006) while flow direction can follow either the slope gradient downslope direction or the direction of tillage rows (Takken et al., 1999, 2001).

The model SIMWE (Mitasova et al., 2001) presents original solutions of the equations responsible for water and sediment fluxes, treated with the stochastic methods described by Gardiner (1985).

The European project MEDALUS produced two models (MEDALUS and MEDRUSH), which include long-term modification of soil and vegetation (Kirkby et al., 1998; Mulligan, 2004). Both models are targeted to the Mediterranean environment in natural or seminatural (i.e., abandoned) conditions. Recent European funded research has produced PESERA, which works on cells 1 km² in size, disconnected, in which the erosion processes take place without export or import of sediment from adjacent cells (http://pesera.jrc.it; Kirkby et al., 2008). The TEST model (van Dijk and Bruijnzeel, 2003) deals with soil erosion in terraced lands (bench terraces).

Common GIS software (GRASS, IDRISI, and ESRI) can calculate parameters linked to runoff, such as the catchment area of each pixel. This allows the use of simple algorithms for sediment production in a distributed way (Moore and Burch, 1986a, 1986b, 1986c; Mitasova et al., 1995; Desmet and Govers, 1996). Examples are RUSLE3D and USPED.

Other models (e.g., GEOWEPP) are equipped with a special interface for using GIS-based data-sets, GRASS can produce data for many models that are integrated in it: For example, ANSWERS (Rewerts and Engel, 1991), KINEROS and SIMWE. Siepel et al. (2002) proposed a simplified erosion model running in a GIS environment in which rills are included as a separate contributing process to erosion. In Table 22.1 some links to GIS based models are given.

TABLE 22.1 Web Links for GIS-Based Soil Erosion Models

Model Acronym	Web Link
RUSLE3D/USPED	http://skagit.meas.ncsu.edu/~helena/gmslab/reports/CerlErosionTutorial/denix/denixstart.html
GEOWEPP	http://www.geog.buffalo.edu/~rensch/geowepp/index.html
GRASS/ANSWERS	at http://grass.itc.it/gdp/html_grass5/html/r.answers.html
CASC2D/GRASS	http://www.engr.uconn.edu/~ogden/casc2d/casc2_desc.htm
	http://www.grass-kr.org/gdp/html_grass4/html/r.hydro.CASC2D.htm
KINEROS/GRASS	http://grass.itc.it/gdp/html_grass4/html/r.kineros.html
SIMWE	http://grass.itc.it/pipermail/grass-commit/2003-January/007314.html

The hydraulics of the quoted models usually does not deal properly with cracks or fluxes through macropores. Hence, tunneling and piping cannot be described despite their importance.

Models are often afflicted by problems of this kind, that is, missing processes. Most ignore feedback on soil characteristics and flow connectivity. As missed modifications tend to accumulate, divergences between simulation and reality are bound to increase with simulation time length. Hence, care must be taken in choosing a model and it is certainly better not to make a decision without having first read carefully what the model contains and how it works.

Generally, models do not describe the effects of other causes of degradation and erosion that can severely affect the way that the soil and landscape respond to intense rainfall. Examples are overgrazing, which can rapidly worsen soil characteristics, tillage, and land leveling, both causing soil degradation and morphology modification. When the crop is of a type that grows underground (e.g., beetroot, potatoes), soil is exported during harvesting (Poesen et al., 2001; Ruysschaert et al., 2004). This also causes a soil profile truncation in a short time span. In all these cases, soil characteristics must be changed accordingly.

Tillage erosion is already the core process of some models among which one of the most evolved is WATEM (van Oost and Govers, 2000). It is spatially distributed and simulates erosion and deposition by water (with USLE-based algorithms) and tillage translocation in a two-dimensional landscape (raster type).

Land leveling, on the other hand, is absolutely ignored. The only attempt has been made by Borselli et al. (2003), but no full description of the model (BULLRUN) is available.

22.4 Conclusions

Water erosion is characterized by many processes induced by raindrops, overland flow, pipe flow, and mass movements. The main equations describing those processes highlight the main soil characteristics involved directly: grain size, colloidal content, chemical composition of soil solution, structural stability, infiltration capacity, rock fragment content, soil surface roughness, and soil profile characteristics. In addition, factors such as slope morphology, catchment size, flow concentration lines, vegetation, land use, and land use techniques are the cofactors that can control or speed up erosion. Additional processes such as soil erosion by tillage and land leveling have a severe impact on soil erosion by water.

In order to use soil in a sustainable way and to conserve it for future generations, soil erosion processes must be better understood together with all their links to the biophysical system of which the soil is an important part. A major tool for soil conservation and land-use planning is scenario analysis, Current erosion models are tools for helping experts to devise possible conservation measures and scenario analyses although they must be used with caution, bearing in mind that they are just approximations.

Acknowledgment

The authors are grateful to Mark Nearing and Jean Poesen for their suggestions that helped in improving the original manuscript.

References

Abbot, M.B., J.C. Bathurst, J.A. Cunge, P.E. O'Connell, and J. Rasmussen. 1986. An introduction to the European hydrological system—Systeme hydrologique Europeen, "SHE", 2: Structure of a physically based, distributed modelling system. J. Hydrol. 87:61–77.

Aberle, J., and G.M. Smart. 2003. The influence of roughness structure on flow resistance on steep slopes. J. Hydraul. Res. 41:259–269.

Abrahams, A.D., and A.J. Parsons. 1990. Determining the mean depth of overland flow in field studies of flow hydraulics. Water Resour. Res. 26:501–503.

Al Durrah, M.M., and J.M. Bradford. 1982. Parameters for describing soil detachment due to single waterdrop impact. Soil Sci. Soc. Am. J. 46:836–840.

Alonso, C.V., S.J. Bennett, and O.R. Stein. 2002. Predicting headcut erosion and migration in concentrated flows typical of upland areas. Water Resour. Res. 38(12):39–1, 39–15.

Andreu, V., A.C. Imeson, J.L. Rubio. 2001. Temporal changes in soil aggregates and water erosion after a wildfire in a Mediterranean pine forest. CATENA, 44:69–84.

Asadi, H., H. Ghadiri, C.W. Rose, and H. Rouhipour. 2007. Interrill soil erosion processes and their interaction on low scope. Earth Surf. Proc. Landforms 32:711–724.

Bathurst, J.C., J.M. Wicks, and P.E. O'Connel. 1995. The SHE/SHESED basin scale water flow and sediment transport modelling system, p. 595–626. In V.P. Singh (ed.) Computer models of watershed hydrology. Water Resources Publications, Highlands Ranch, CO.

Beasley, D.B., L.F. Huggins, and E.J. Monke. 1980. ANSWERS—A model for watershed planning. Trans. Am. Soc. Agric. Eng. 23:938–944.

Bell, F.G., and D.J.H. Walker. 2000. A further examination of the nature of dispersive solid in Natal, South Africa. Q. J. Eng. Geol. Hydroge. 33:187–199.

Beuselinck, L., A. Steegen, G. Govers, J. Nachtergaele, I. Takken, and J. Poesen. 2000. Characteristics of sediment deposits formed by intense rainfall events in small catchments in the Belgian Loam Belt. Geomorphology, 32, 1–2, 69–82.

Borselli, L., S. Pellegrini, D. Torri, and P. Bazzoffi, 2002. Tillage erosion and land levelling evidences in Tuscany (Italy). In Rubio, J.L., Morgan, R.P.C., Asins, S. and Andreu, V. (eds.) Man and Soil at the Third Millennium; Vol. II, Geoforma Ediciones, S.L., Logroño, Spain, 1341–1350.

Borselli, L., P. Bazzoffi, D. Torri, and S. Pellegrini. 2003. Modelling DTM modification by land levelling: Simulation using BULLRUN model. Abstract, in Book of abstract and field guide, COST 623 "Soil Erosion and Global Change", p. 32. Final Meeting Conf., Budapest, Hungary, July 5–7, 2003.

Borselli, L., D. Torri, L. Øygarden, S. De Alba, J.A. Martìnez-Casasnuevas, P. Bazzoffi, and G. Jakab. 2006. Land levelling, p. 643–658. In J. Boardman and J. Poesen (eds.) Soil erosion in Europe. Wiley & Sons, Chichester, U.K.

Borselli, L., D. Torri, J. Poesen, and M.P. Salvador Sanchis. 2001. Effect of water quality on infiltration, runoff and interrill erosion processes during simulated rainfall. Earth Surf. Proc. Landforms 26:339–342.

Brakensiek, D.L., and W.J. Rawls. 1983. Agricultural management effects on soil water processes. Part II: Green and Ampt parameters for crusting soils. Trans. ASAE 26:1753–1757.

Brandt, C.J. 1989. The size distribution of throughfall drops under vegetation canopies. Catena 16:507–524.

Branson, F.A., G.F. Gifford, K.G. Rendard, and R.F. Hadley. 1981. Rangeland hydrology, p. 645. Society for range management, range science series 1. 2nd Edn. Kendal Hunt Publishing Company, Dubuque, IA.

Bryan, R., and J.A.A. Jones. 1997. The significance of soil piping processes: Inventory and prospect. Geomorphology 20:209–218.

Casalì, J., J.J. Lopez, and J.V. Giraldez. 2003. A process-based model for channel degradation: Application to ephemeral gully erosion. Catena 50:435–448.

Comiti, F., L. Mao, A. Wilcox, E.E. Wohl, and M.A. Lenzi. 2007. Field-derived relationships for flow velocity and resistance in high-gradient streams. J. Hydrol. 340:48–62.

Crockford, R.H., and D.P. Richardson. 2000. Partitioning of rainfall into throughfall, stemflow and interception: Effect of forest type, ground cover and climate. Hydrol. Process 14:2903–2920.

De Alba, S., L. Borselli, D. Torri, S. Pellegrini, and P. Bazzoffi. 2006. Assessment of tillage erosion by mouldboard plough in Tuscany (Italy). Soil Till. Res. 85:123–142.

De Baets, S., J. Poesen, P. Galindo-Morales, and A. Knapen. 2007. Impact of root architecture on the erosion-reducing potential of roots during concentrated flow. Earth Surf. Proc. Landforms 23:1323–1345.

De Baets, S., D. Torri, J. Poesen, M.P. Salvador, J. Meersmans, 2008. Modelling increased soil cohesion due to roots with EUROSEM. Earth Surface Processes and Landforms 33(13): 1948–1963.

De Bano, L.F., S.M. Savage, and D.A. Hamilton. 1976. The transfer of heat and hydrophobic substances during burning. Soil Sci. Soc. Am. J. 40:779–782.

De Ploey, J. 1982. A stemflow equation for grasses and similar vegetation. Catena 9:139–152.

De Ploey, J. 1984. Stemflow and colluviation: Modeling and implications. Pedologie XXXIV-2:135–146.

De Roo, A.P.J., and V.G. Jetten. 1999. Calibrating and validating the LISEM model for two data sets from the Netherlands and South Africa. Catena 37:477–493.

De Roo, A.P.J., C.G. Wesseling, and C.J. Ritsema. 1996. LISEM: A single event physically-based hydrologic and soil erosion model for drainage basins: I. Theory, input and output. Hydrol. Proc. 10:1107–1117.

De Santisteban, L.M., J. Casalí, J.J. López, J.V. Giráldez, J. Poesen, and J. Nachtergaele. 2005. Exploring the role of topography in small channel erosion. Earth Surf. Proc. Landforms 30:591–599.

Desmet, P.J.J., and G. Govers. 1996. A GIS procedure for automatically calculating the USLE LS factor on topographically complex landscape units. J. Soil Water Cons. 51:427–433.

Diemont, S.A.W., J.F. Martin, and S.I. Levy-Tacher. 2006. Emergy evaluation of Lacandon Maya indigenous Swidden agroforestry in Chiapas, Mexico. Agroforest. Syst. 66:23–42.

Dietrich, W.E., and T. Dunne. 1993. The channel head, p. 177–219. In K. Beven and M.J. Kirkby (eds.) Channel network hydrology. John Wiley & Sons, Chichester, U.K.

Ellison, W.D. 1947. Soil erosion studies. Agric. Eng. 28:145–146.

Engel, O.G. 1955. Waterdrop collisions with solid surfaces. J. Res. Nat. Bur. Stand. 54:281–298.

Faulkner, H., R. Alexander, R. Teeuw, and P. Zukovskyj. 2004. Variations in soil dispersivity across a gully head displaying shallow subsurface pipes. Earth Surf. Proc. Landforms 29(9):1143–1160.

Faulkner, H., R. Alexander, and P. Zukowskyj. 2008. Slope–channel coupling between pipes, gullies and tributary channels in the Mocatán catchment badlands, Southeast Spain. Earth Surf. Proc. Landforms 33:1242–1260.

Favis-Mortlock, D.T. 1998. A self-organising dynamic systems approach to the simulation of rill initiation and development on hillslopes. Comput. Geosci. 24:353–372.

Favis-Mortlock, D.T., J. Boardman, A.J. Parsons, and B. Lascelles. 2000. Emergence and erosion: A model for rill initiation and development. Hydrol. Proc. 14:2173–2205.

Findeling, A., S. Ruy, and E. Scopel. 2003. Modeling the effects of a partial residue mulch on runoff using a physically based approach. J. Hydrol. 275:49–66.

Fitzpatrick, R.W., S.C. Boucher, R. Naidu, and E. Fritsch. 1995. Environmental consequence of soil sodicity, p. 163–176. In R. Naidu, M.E. Sumner, and P. Rengasamy (eds.) Australian sodic soils: Distribution, properties and management. CSIRO Publications, Melbourne, Australia.

Flanagan, D.C., and M.A. Nearing. 1995. USDA—Water Erosion Prediction Project: Hillslope profile and watershed model documentation. NSERL Report No. 10, West Lafayette, IN.

Foster, G.R., L.F. Huggins, and L.D. Meyer. 1984. A laboratory study of rill hydraulics: I. Velocity relationships. Trans. ASAE 27:790–796.

Foster, G.R., L.J. Lane, M.A. Nearing, S.C. Finkner, and D.C. Flanagan. 1989. Erosion component. In L.J. Lane and M.A. Nearing (eds.) USDA—Water Erosion Prediction Project: Hillslope profile model documentation. NERSL Report 2. USDA-ARS National Soil Erosion Research Laboratory, West Lafayette, IN.

Frydman, S., and R. Baker. 2009. Theoretical soil-water characteristic curves based on adsorption, cavitation, and a double porosity model. Int. J. Geomech. 9:250–257.

Gabriels, D., G. Ghekiereì, W. Schiettecatte, and I. Rottiers. 2003. Assessment of USLE cover-management C-factors for 40 crop rotation systems on arable farms in the Kemmelbeek watershed, Belgium. Soil Till. Res. 74:47–53.

Gardiner, C.V. 1985. Handbook of stochastic methods for physics, chemistry and natural sciences. Springer, Berlin, Germany.

Gash, J.H.C., C.R. Lloyd, and G. Lachaud. 1995. Estimating sparse forest rainfall interception with an analytical model. J. Hydrol. 170:79–86.

Gash, J.H.C., F. Valente, and J.S. David. 1999. Estimates and measurements of evaporation from wet, sparse pine forest in Portugal. Agri. Forest Meteorol. 94:149–158.

Gilley, J.E., and S.C. Finkner. 1991. Hydraulic roughness coefficient as affected by random roughness. Trans. ASAE 34:897–903.

Gilley, J.E., and M.A. Weltz. 1995. Hydraulics of overland flow. WEPP handbook, Chapter 10, p. 10.1–10.7. www.ars.usda.gov/SP2UserFiles/ad_hoc/36021500WEPP/chap10.pdf; last accessed: 23 June 2011.

Giménez, R., and G. Govers. 2001. Interaction between bed roughness and flow hydraulics in eroding rills. Water Resour. Res. 37:791–799.

Giovannini, G. 1994. The effect of fire on soil quality. In Sala, M., Rubio, J.F. (Eds.), Soil Erosion and Degradation as a Consequence of Forest Fires. Selection of Papers from the International Conference on Soil Erosion and Degradation as Consequence of Forest Fires. Barcelona, Spain, 1991. Geoforma Ediciones, Logrono, Spain, pp. 15–27 (275p).

Giovannini, G., and S. Lucchesi. 1983. Effect of fire on hydrophobic and cementing substances of soil aggregates. Soil Sci. 136:231–236.

Giovannini, G., and S. Lucchesi. 1997. Modifications induced in soil physico-chemical parameters by experimental fires at different intensities. Soil Sci. 162:479–486.

Govers, G. 1990. Empirical relationships on the transporting capacity of overland flow: A laboratory study. IAHS Spec. Pub. 189:45–63.

Govers, G. 1992. Relationship between discharge, velocity and flow area for rills eroding loose, non-layered materials. Earth Surf. Proc. Landforms 17:515–528.

Govers, G., K. Vandaele, P. Desmet, J. Poesen, and K. Bunte. 1994. The role of tillage in soil redistribution on hillslopes. Eur. J. Soil Sci. 45, 469–478. www.ars.usda.gov/SP2UserFiles/ad_hoc/36021500WEPP/chap10.pdf; last access: 23 June 2011

Govers, G., R. Giménez, and K. Van Oost. 2007. Rill erosion: Exploring the relationship between experiments, modelling and field observations. Earth Sci. Rev. 84:87–102.

Govers, G., K. Vandaele, P. Desmet, J. Poesen, and K. Bunte, 1994. The role of tillage in soil redistribution on hillslopes. Eur. J. Soil Sci. 45, 469–478.

Gyssels, G., J. Poesen, E. Bochet, and Y. Li. 2005. Impact of plant roots on the resistance of soils to erosion by water: A review. Prog. Phys. Geog. 29:189–217.

Haan, C.T., B.J. Barfield, and J.C. Hayes. 1994. Design hydrology and sedimentology for small catchments. Academic Press, San Diego, CA.

Hammond, C., D. Hall, S. Miller, and P. Swetik. 1992. Level I stability analysis (LISA): Documentation for version 2.0. USDA Forest Service General Technical Report INT-285.

Harlow, F.H. and J.P. Shannon. 1967. The splash of a liquid drop. J. Appl. Phys. 38:3855–3866.

Hauge, C. 1977. Soil erosion definitions. Calif. Geol, 30: 202–203.

Hessel, R. 2002. Modelling soil erosion in a small catchment on the Chinese loess plateau—Applying LISEM to extreme conditions. PhD thesis. Koninklijk Nederlands Aardrijkskundig Genootschap, Fakulteit Ruimtelijke Wetenschappen Universiteit Utrecht, Labor Grafimedia, Utrecht, the Netherlands.

Hessel, R., V. Jetten, and Z. Guanghui. 2003. Estimating manning's n for steep slopes. Catena 54:77–91.

Hofierka, J., J. Parajka, H. Mitasova, and L. Mitas. 2002. Modeling impact of terrain on precipitation using 3-D spatial interpolation. Trans. GIS 6:135–150.

Huang, Y.H. 1983. Stability analysis of earth slopes. Van Nonstrand Reinhold Company, New York.

Janbu, N. 1973. Slope stability computation, p. 47–86. In H. Hirshfield and H.G. Poulos (eds.) Embankment dam engineering: Casagrande memorial volume. John Wiley & Sons, New York.

Jetten, V., J. Poesen, J. Nachtergaele, and D. van de Vlag. 2006. Spatial modelling of ephemeral gully incision: A combined empirical and physical approach, p. 328. In P.N. Owens and A.J. Collins (eds.) Soil erosion and sediment redistribution in river catchments. CABI Publishing Series, Wallingford, Oxfordshire, U.K., ISBN0851990509.

Jones, J.A.A., and R.B. Bryan (eds.). 1997. Piping erosion. Geomorphology 20:114.

Kinnel, P.I.A. 1981. Rainfall intensity-kinetic energy relationships for soil loss prediction. Soil Sci. Soc. Am. J. 45:153–155.

Kinnel, P.I.A. 1990. The mechanics of raindrop induced flow transport. Aust. J. Soil Res. 28:497–516.

Kirkby, M.J., R. Abrahart, M.D. McMahon, J. Shao, and J.B. Thornes. 1998. MEDALUS soil erosion models for global change. Geomorphology 24:35–49.

Kirkby, M.J., B.J. Irvine, R.J.A. Jones, G. Govers, and PESERA team. 2008. The PESERA coarse scale erosion model for Europe I.—Model rationale and implementation. Eur. J. Soil Sci. 59(6):1293–1306.

Knapen, A., J. Poesen, and S. De Baets. 2008. Rainfall-induced consolidation and sealing effects on soil erodibility during concentrated runoff for loess-derived topsoils. Earth Surf. Proc. Landforms 33:444–458. DOI: 10.1002/esp.1566.

Lawrence, D.S.L. 1997. Macroscale surface roughness and frictional resistance in overland flow. Earth Surface Processes and Landforms 22:365–382.

Laws, J.O. 1941. Measurements of fall velocity of water-drops and rain-drops. Am. Geoph. Un. Trans. 22:709–721.

Le Bissonnais, Y., 1996. Aggregate stability and assessment of soil crustability and erodibility: I. Theory and methodology. European Journal of Soil Science, 47:425–437.

Leonard, J., and G. Richard. 2004. Estimation of runoff critical shear stress for soil erosion from soil shear strength. CATENA, n. 57, 233–249.

Leyland, J., and S.E. Darby. 2009. Effects of Holocene climate and sea-level changes on coastal gully evolution: Insights from numerical modelling. Earth Surf. Proc. Landforms 34:1878–1893. DOI: 10.1002/esp.1872.

Lott, J.E., S.B. Howard, C.K. Ong, and C.R. Black. 2000a. Long-term productivity of a Grevillea robusta-based overstorey agroforestry system in semi-arid Kenya. 1. Tree growth. Forest Ecol. Manage. 139:175–186.

Lott, J.E., S.B. Howard, C.K. Ong, and C.R. Black. 2000b. Long-term productivity of a Grevillea robusta-based overstorey agroforestry system in semi-arid Kenya. 2. Crop growth and system performance. Forest Ecol. Manage. 139:187–201.

Mataix-Solera, J., and S.H. Doerr. 2004. Hydrophobicity and aggregate stability in calcareous topsoils from fire-affected pine forest in southeastern Spain. Geoderma 118:77–88.

Merritt, W.S., R.A. Letcher, and A.J. Jakeman. 2003. A review of erosion and sediment transport models. Environ. Model. Software 18:761–799.

Mita, C., N. Catsaros, and N. Gounaris. 2001. Runoff cascades, channel network and computation hierarchy determination on a structured semi-irregular triangular grid. J. Hydrol. 244:105–118.

Mitasova, H., W.M. Brown, M. Hohmann, and S. Warren. 2001. Using soil erosion modeling for improved conservation planning: A GIS-Based tutorial. (http://skagit.meas.ncsu.edu/~helena/gmslab/reports/CerlErosionTutorial/denix/default.htm; last access 23 June 2011).

Mitasova, H., L. Mitas, and W.M. Brown. 2001. Multiscale simulation of land use impact on soil erosion and deposition patterns, p. 1163–1169. In D.E. Stott, R.H. Mohtar, and G.C. Stainhardt (eds.) Sustaining the global farm. Selected papers from the international soil conservation organization meeting, May 24–29, 1999. Purdue University and USDA-ARS National Soil Erosion Research Laboratory, West Lafayette, IN.

Mitasova, H., L. Mitas, B.M. Brown, D.P. Gerdes, and I. Kosinovsky. 1995. Modeling spatially and temporally distributed phenomena: New methods and tools for GRASS GIS. Int. J. GIS 9:443–446.

Montgomery, D.R., and W.E. Dietrich. 1994. Landscape dissection and drainage area-slope thresholds, p. 221–246. In M.J. Kirkby (ed.) Process models and theoretical geomorphology. John Wiley & Sons, Chichester, U.K.

Moore, I.D., and G.J. Burch. 1986a. Physical basis of the length-slope factor in the universal soil loss equation. Soil Sci. Soc. Am. J. 50:1294–1298.

Moore, I.D., and G.J. Burch. 1986b. Sediment transport capacity of sheet and rill flow: Application of unit stream power theory. Water Resour. Res. 22:1350–1360.

Moore, I.D., and G.J. Burch. 1986c. Modeling erosion and deposition: Topographic effects. Trans. ASAE 29:1624–1640.

Morgan, R.P.C. 1995. Soil erosion and conservation. 2nd Edn. Longman, London, U.K.

Morgan, R.P.C., J.N. Quinton, R.E. Smith, G. Govers, J.W.A. Poesen, K. Auerswald, G. Chisci, D. Torri, and M.E. Styczen. 1998. The European soil erosion model (EUROSEM): A process-based approach for predicting soil loss from fields and small catchments. Earth Surf. Proc. Landforms 24:527–544.

Mulligan, M. 1998. Modelling the geomorphological impact of climatic variability and extreme events in a semi-arid environment. Geomorphology 24:59–78.

Mulligan, M. 2004. A review of European Union funded research into modelling Mediterranean desertification. Adv. Environ. Monit. Model. 1:1–78.

Mutchler, C.K. 1967. Parameters for describing raindrop splash. J. Soil Water Conser. 22:91–94.

Nachtergaele, J., J. Poesen, A. Sidorchuk, and D. Torri. 2002. Prediction of concentrated flow width in ephemeral gully channels. Hydrol. Proc. 16:1935–1953.

Naden, P. 1987. An erosion criterion for gravel bed rivers. Earth Surf. Proc. Landforms 12:83–93.

Navar, J., D.E. Carlyle-Moses, and M.A. Martinez. 1999. Interception loss from the Tamaulipan matorral thornscrub of north-eastern Mexico: An application of the Gash analytical interception loss model. J. Arid Environ. 41:1–10.

Nearing, M.A. 1991. A probabilistic model of soil detachment by shallow turbulent flow. Trans. ASAE 34:81–85.

Nearing, M.A., J.M. Bradford, and R.D. Holtz. 1987. Measurement of waterdrop impact pressures on soil surfaces. Soil Sci. Soc. Am. J. 51:1302–1306.

Øygarden, L. 2003. Rill and gully development during an extreme winter runoff event in Norway. Catena 50:217–242.

Palmer, R. 1963. The influence of a thin water layer on waterdrop impact forces. IAHS Pub. 65:141–148.

Parkner, T., M.J. Page, T. Marutani, and N.A. Trustrum. 2006. Development and controlling factors of gullies and gully complexes, East Coast, New Zealand. Earth Surf. Proc. Landforms 31:187–199. DOI: 10.1002/esp1321.

Pierson, F.B., P.R. Robichaud, C.A. Moffet, K.A. Spaeth, P. Stuart, S.P. Hardegree, P.E. Clark, and C.J. Williams. 2008. Fire effects on rangeland hydrology and erosion in a steep sagebrush-dominated landscape. Hydrol. Process 22:2916–2929.

Poesen, J. 1993. Gully topology and gully control measures in the European loess belt, p. 221–239. In S. Wicherek (ed.) Farm land erosion in temperate plains environment and hills. Elsevier Science Publishers B.V., Amsterdam, the Netherlands.

Poesen, J., J. Nachtergaele, G. Verstraeten, and Ch. Valentin. 2003. Gully erosion and environmental change: Importance and research needs. Catena 50:91–133.

Poesen, J.W., D. Torri, and K. Bunte. 1994. Effects of rock fragments on soil erosion by water at different spatial scales: A review. Catena 23:141–166.

Poesen, J., and Ch. Valentin. 2003. Preface. Catena 50:87–89.

Poesen, J., K. Vandaele, and B. Van Wesemarl. 1998. Gully erosion: Importance and model implication. In J. Boardman and D. Favis-Mortlock (eds.) Modeling soil erosion by water. NATTO ASI Series Vol. 55. Springer Verlag, Berlin, Germany.

Poesen, J., L. Vandekerckhove, J. Nachtergaele, D. Oostwoud Wijdenes, G. Verstraeten, and B. van Wesemael. 2002. Gully erosion in dryland environments, p. 229–262. In L.J. Bull and M.J. Kirkby (eds.) Dryland rivers: Hydrology and geomorphology of semi-arid channels. Wiley, Chichester, U.K.

Poesen, J.W.A., G. Verstraeten, R. Soenens, and L. Seynave. 2001. Soil losses due to harvesting of chicory roots and sugar beet: An underrated geomorphic process? Catena 43:35–47.

Prasad, S.N., and M.J.M. Romkens. 2003. Energy formulation of headcut dynamics. Catena 50:469–488.

Rauws, G., and G. Govers. 1988. Hydraulics and soil mechanical aspects of rill generation in agricultural soils. J. Soil Sci. 39:111–124.

Refsgaard, J.C., and J.R. Storm. 1995. MIKE SHE, p. 809–846. In V.P. Singh (ed.) Computer models of watershed hydrology. Water Resources Publications, Highlands Ranch, CO.

Reid, L.M., and T. Dunne. 1996. Rapid evaluation of sediment budgets. Catena Verlag, Reiskirchen, Germany.

Renard, K.G., G.R. Foster, G.A. Weesies, D.K. McCool, and D.C. Yoder. 1997. Predicting soil erosion by water—A guide to conservation planning with the revised universal soil loss equation (RUSLE). Agricultural handbook no. 703, US Government Printing Office, Washington, DC.

Rengasamy, P., and M.E. Sumner. 1997. Processes involved in sodic behavior. In M.E. Sumner and R. Naidu (eds.) Sodic soils. Oxford University Press, New York.

Rewerts, C.C., and B.A. Engel. 1991. ANSWERS on GRASS: Integrating a watershed simulation with a GIS. ASAE Paper No. 91-2621, ASAE, St. Joseph, MI.

Rieke-Zapp, D., J. Poesen, and M.A. Nearing. 2007. Effects of rock fragments incorporated in the soil matrix on concentrated flow hydraulics and erosion. Earth Surface Processes and Landforms, 32:1063–1076.

Robinson, K.M., J. Sean, S.-J. Bennet, J. Casali, and G.J. Hanson. 2000. Processes of headcut growth and migration in rills and gullies. Int. J. Sediment Res. 15:69–82.

Rose, C.W., P.B. Hairsine, A.P.B. Proffitt, and R.K. Misra. 1990. Interpreting the role of soil strength in erosion processes. Catena Suppl. 17:153–165.

Rubio, J.L., J. Forteza, V. Andreu, and R. Cerni. 1997. Soil profile characteristics influencing runoff and soil erosion after forest fire: a case of study (Valencia, Spain). Soil Technology 11, 67–78. https://www.kuleuven.be/geography/frg/modelling/erosion/watemhome/index.htm; last access 23 June 2011.

Rousseva, S. 1989. A laboratory index for soil erodibility assessment. Soil Tech. 2:287–299.

Ruysschaert, G., J. Poesen, G. Verstraeten, and G. Govers. 2004. Soil loss due to crop harvesting: Significance and determining factors. Prog. Phys. Geog. 28:467–501.

Salvador Sanchis, M.P., D. Torri, L. Borselli, R. Bryan, J. Poesen, M.S. Yañez, and C. Cremer. 2009. Rill and gully junctions: Estimating the parameters of the width-discharge relation in the field. Earth Surf. Proc. Landforms 34:2023–2030. DOI: 10.1002/esp.

Salvador Sanchis, M.P., D. Torri, L. Borselli, and J. Poesen. 2008. Climate effects on soil erodibility. Earth Surf. Proc. Landforms 33(7):1082–1097. DOI: 10.1002/esp.1604.

Seginer, I. 1966. Gully development and sediment yield. J. Hydrol. 4:236–252.

Shainberg, I., and G.J. Levy. 1996. Infiltration and seal formation processes, p. 1–24. *In* M. Agassi (ed.) Soil erosion, conservation and rehabilitation. Marcel Dekker, New York.

Shary, P.A., L.S. Sharaya, and A.V. Mitusov. 2002. Fundamental quantitative methods of land surface analysis. Geoderma 107:1–32.

Shuguang, L. 1997. A new model for the prediction of rainfall interception in forest canopies. Ecol. Model. 99:151–159.

Sidorchuk, A. 1999. Dynamic and static models of gully erosion. Catena 37:401–414.

Siepel, A.C., T.S. Steenhuis, C.W. Rose, J.Y. Parlange, and G.F. McIsaac. 2002. A simplified hillslope erosion model with vegetation elements for practical applications. J. Hydrol. 258:111–121.

Singh, V.P. 1995. Watershed modeling, p. 1–22. *In* V.P. Singh (ed.) Computer models of watershed hydrology. Water Resources Publications, Highlands Ranch, CO.

Singh, V.P. 1996. Kinematic wave modelling in water resources: Surface water hydrology. Wiley, New York.

Smart, G.M., M.J. Duncan, and J.M. Walsh. 2002. Relatively rough flow resistance equations. J. Hydraul. Eng. 128:568–578.

Smerdon, E.T., and R.P. Beasley. 1961. Critical tractive forces in cohesive soils. Agric. Eng. 42:26–29.

Smith, R.E. 1990. Analysis of infiltration through a two-layer soil profile. Soil Sci. Soc. Am. J. 54:1219–1227.

Smith, R.E. 1995. Opus simulation of a wheat/sugarbeet plot near Neuenkirchen, Germany. Ecol. Model. 81:121–132.

Smith, R.E., D.C. Goodrich, and C.L. Unkrich. 1999. Simulation of selected events on the Catsop catchment by KINEROS2: A report for the GCTE conference on catchment scale erosion models. Catena 37:457–475.

Smith, D.W., R.O. Peterson, and D.B. Houstin. 2003. Yellowstone after wolves. BioScience 53:330–340.

Stocking, M.A. 1981. Causes and prediction of the advance of gullies, p. 30–41. Proc. South-East Asian Reg. Symp. Probl. Soil Eros. Sedimen. Bangkok, Thailand.

Sumner, M.E., W.P. Miller, R.S. Kookana, and P. Hazelton. 1997. Sodicity, dispersion, and environmental quality, p. 167–190. *In* M.E. Sumner and R. Naidu (eds.) Sodic soils. Oxford University Press, New York.

Takken, I., L. Beuselinck, J. Nachtergaele, G. Govers, J. Poesen, and G. Degraer. 1999. Spatial evaluation of a physically-based distributed erosion model (LISEM). Catena 37:431–447.

Takken, I., G. Govers, A. Steegen, J. Nachtergaele, and J. Guérif. 2001. The prediction of runoff flow directions on tilled fields. J. Hydrol. 248:1–13.

Terranova, O., L. Antronico, R. Coscarelli, and P. Laquinta. 2009. Soil erosion risk scenarios in the Mediterranean environment using RUSLE and GIS: An application model for Calabria (southern Italy). Geomorphology 112:228–245.

Torri, D., D. Bartolini, and M.P. Salvador Sanchis. 2007. Hydrological and erosion models—Chapter 2.1, p. 19–29. *In* Conditions for restoration and mitigation of desertified areas using vegetation (RECONDES)—Review of literature and present knowledge. EC-DG Res, Environment.

Torri, D., and L. Borselli. 2002. Clod movement and tillage tool characteristics for modelling tillage erosion. J. Soil Water Conser. 57:24–28.

Torri, D., and L. Borselli. 2003. Equation for high-rate gully erosion. Catena 50:449–468.

Torri, D., and R.B. Bryan. 1997. Micropiping processes and biancana evolution in southeast Tuscany, Italy. Geomorphology 20:219–235.

Torri, D., A. Colica, and D. Rockwell. 1994. Preliminary study of the erosion mechanisms in a biancane badland (Tuscany, Italy). Catena 23:281–294.

Torri, D., J. Poesen, L. Borselli, and A. Knapen. 2006. Channel width—Flow discharge relationships for rills and gullies. Geomorphology 76:273–279.

Torri, D., M. Sfalanga, and M. Del Sette. 1987. Splash detachment: Runoff depth and soil cohesion. Catena 14:149–155.

Trimble, S.W., and A.C. Mendel. 1995. The cow as a geomorphic agent—A critical review. Geomorphology 13:233–253.

Tucker, G.E., S.T. Lancaster, N.M. Gasparini, R.L. Bras, and S.M. Rybarczyk. 2001. An object-oriented framework for distributed hydrologic and geomorphic modeling using triangulated irregular networks. Comput. Geosci. 27:959–973.

Ubeda, X., and M. Sala. 1998. Variations in runoff and erosion in three areas with different fire intensities. Geoö dynamik 19:179–188.

Valentin, Ch., and L.M. Bresson. 1992. Erosional crusts: Morphology, genesis and classification of surface crusts in loamy and sandy soils. Geoderma 55:225–245.

van Dijk, A.I.J.M., and L.A. Bruijnzeel. 2001. Modelling rainfall interception by vegetation of variable density using an adapted analytical model. Prat 1. Model description. J. Hydrol. 247:230–238.

van Dijk, A.I.J.M., and L.A. Bruijnzeel. 2003. Terrace erosion and sediment transport model: A new tool for soil conservation planning in bench-terraced steeplands. Environ. Model. Software 18:839–850.

Van Elewijck, L. 1989. Influence of leaf and branch slope on stemflow amount. Catena 16:525–533.

van Oost, K., and G. Govers. 2000. WATEM. https://www.kuleuven.be/geography/frg/modelling/erosion/watemhome/index.htm; last accessed: 23 June 2011.

Van Oost, K., G. Govers, S. deAlba and T.A. Quine. 2006. Tillage erosion: a review of controlling factors and implications for soil quality. Prog. Phys. Geog. 30(4): 443–466.

Vanoni, V.A., and G.N. Nomicos. 1960. Resistance properties of sediment-laden streams. Trans. Am. Soc. Civ. Eng. 125/I:1140–1175.

Varela, M.E., E. Benito, and J.J. Keizer. 2010. Wildfire effects on soil erodibility of woodlands in NW Spain. Land Degradation & Development 21:75–82.

Varnes, D.J. 1978. Slope movement: Type and processes. *In* R.L. Shuster and R.J. Kriziek (eds.) Landslides: Analysis and control. Transportation research board special report. 176:11–33.

Whitlow, R. 1990. Basic soil mechanics. Longman, Hong Kong.

Williams, J.R. 1995. The EPIC model, p. 909–1000. *In* V.P. Singh (ed.) Computer models of watershed hydrology. Water Resources Publications, Highlands Ranch, CO.

Wilson, G.V. 2009. Mechanisms of ephemeral gully erosion caused by constant flow through a continuous soil-pipe. Earth Surf. Proc. Landforms 34:1858–1866. DOI: 10.1002/esp.1869.

Wischmeier, W.H., and D.D. Smith. 1965. Predicting rainfall-erosion losses from cropland east of the Rocky Mountains. Agr. Handbook No. 282, U.S. Dept. Agr., Washington, DC.

Wischmeier, W.H., and D.D. Smith. 1978. Predicting rainfall erosion losses. Agriculture handbook no. 537, USDA-ARS, Washington, DC.

Woolhiser, D.A., R.E. Smith, and D.C. Goodrich. 1990. KINEROS, A kinematic runoff and erosion model: Documentation and user manual. U.S. Department of Agriculture, Agricultural Research Service, ARS-77.

Yu, B. 2003. A unified framework for water erosion and deposition equations. Soil Sci. Soc. Am. J. 67:251–257.

Yu, B., and C.W. Rose. 1999. Application of a physically based soil erosion model, GUEST, in the absence of data on runoff rates, I. Theory and methodology. Aust. J. Soil Res. 37:1–11.

Zhu, J.C., C.J. Gantzer, S.H. Anderson, R.L. Peyton, and E.E. Alberts. 2001. Comparison of concentrated-flow detachment equations for low shear stress. Soil Till. Res. 61:203–212.

Zobeck, T.M., and C.A. Onstad. 1987. Tillage and rainfall effects on random roughness. Soil Till. Res. 9:1–20.

23
Wind Erosion

23.1 Introduction 23-1
23.2 Mechanics of Wind Erosion Processes..23-2
 Wind Speed • Soil • Plants • Humankind
23.3 Modeling the Wind Erosion Process... 23-6
 Wind Erosion Equation • The Pasak Model • Texas Erosion Analysis Model •
 Wind Erosion Assessment Model • Revised Wind Erosion Equation • The Wind
 Erosion Prediction System • Wind Erosion on European Light Soils
23.4 Wind Erosion Measuring Systems: Field Sampling Instrumentation........23-12
 Air Quality Monitoring • Threshold Velocities
23.5 Major Control Systems 23-14
 Crop Residues • Soil Roughness • Wind Barriers/Shelterbelts • Soil Amendments
23.6 Conclusions..23-15
References..23-15

D.W. Fryrear
Custom Products and Consultants

23.1 Introduction

Wind erosion is a basic geomorphologic process that erodes the land surface and shapes yardangs, ventifacts, pedestal rocks, blowouts, ergs, dunes, lag deposits, desert pavement, and desert armor (Thornbury, 1957). On every continent, aeolian features document the persistence and magnitude of wind forces. While the transport capacity of the wind is much less than that of water, dust concentrations may exceed 6000 μg m^{-3} in severe events (Song et al., 2007). Dust accumulations over a 16 year period in the Mojave Desert varied from 2 to 20 g m^{-2} year^{-1}, but were affected by amount and seasonal distribution of rainfall (Reheis, 2006).

The recorded history of wind erosion extends over a few hundred years; however, peat bog cores in southern Chile indicate that dust accumulations were effectively constant over a period of six millennia (Sapkota et al., 2007). A chronology of dust storm activity in Northern China over the last 1000 years showed two marked dust peaks between AD 1430–1570 and between AD 1770 and 1990 in the arid west. In China the dust storm maxima occurred at AD 1140–1220, 1430–1500, and 1680–1780 (Yang et al., 2007). All of these were during periods of cold–dry climatic conditions.

Wendelin reported a purple or dark red rain in Brussels in 1646 (Wendelin, 1646). The red rain was the result of dust deposition on October 6, 1646, and caused considerable concern. Based on current knowledge, red rains in Europe may result from dust produced by storms in Africa (Stout et al., 2008).

In the wind erosion process, fine particles and nutrients are removed, leaving coarse particles on the eroded surface. As the surface continues to erode, the remaining soil becomes less fertile. The loss of fertile soil material from agricultural lands degrades the soils in the source area, reduces crop production, and further intensifies wind erosion. Dust in the atmosphere may affect air temperatures, cloud formation, and convectional activity (Goudie, 2009). Hurricane peak intensity in the North Atlantic Region has been linked to dust outbreaks in the Sahara (Wu, 2007; Wong et al., 2008). Dust clouds contaminate the atmosphere in the transport area and impact citizenry throughout the depositional area. It is not uncommon for the source and depositional areas to be thousands of kilometers apart. Dust-emission models (Zhang et al., 2003; Tegen et al., 2004) suggest that the global dust load is primarily related to climate change and not in response to land use changes.

Erosion control practices that minimize erosion have attracted farmers long before the basic processes were understood. In 1790 Deane used hedges and locust trees and in 1824 Drown used the addition of 50–80 mm of clay to blowy sand soil surfaces in efforts to reduce wind erosion in the United States (McDonald, 1971). Deep plowing to increase clay content of the surface soil controls wind erosion on thousands of hectares of sandy soil in the Southern Great Plains (Chepil and Woodruff, 1963). The benefits of deep plowing persist for several years, depending on management practices (Fryrear, 1984). Straw barriers in a checkerboard pattern (Xu et al., 1982), raised fences (Liu et al., 1983), or forest belts, also called shelterbelts (Dong et al., 1983), have been used in China to protect railroads and highways from blowing sand. Shelterbelts are used in India to reduce wind speeds, erosion, and evaporation losses (Gupta et al., 1981). The success of

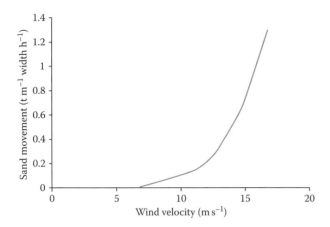

FIGURE 23.1 Transport flux of sand as a function of wind velocity. (Reprinted from Bagnold, R.A. 1941. The physics of blown sand and desert dunes. Methuen, London, U.K. Fig. 22. With kind permission of Kluwer Academic Publishers.)

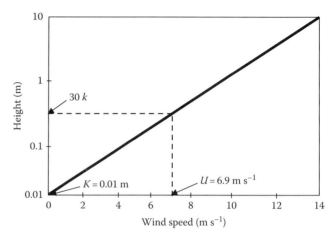

FIGURE 23.2 Determining aerodynamic roughness (k) and drag speed (U_*) graphically from plot of wind speed and height of wind speed measurement ($U_* = 6.9/8.493 = 0.8124$ m s^{-1}).

shelterbelts depends on dominant erosive wind directions, sufficient annual rainfall to sustain the trees, and other potential uses of the trees (lumber, pulpwood, firewood, etc.). Any system that reduces the speed of the wind at the soil surface can significantly reduce soil erosion (Figure 23.1).

Wind erosion concerns can be grouped into three categories: environmental, agricultural, and transportation (Greeley and Iverson, 1985). Environmental concerns include effects of dust on human, animal, and plant health and on visibility and climate. Agricultural concerns include the impact of soil erosion by wind on the sustainability of agricultural production. Transportation concerns include reduced visibility from blowing soil on highways, railroads, and airports.

23.2 Mechanics of Wind Erosion Processes

Wind is defined as the motion of the atmosphere relative to the surface (Greeley and Iverson, 1985). As wind flows over the land surface, the surface exerts a drag such that an atmospheric surface boundary layer approximately 1 m thick is formed. This atmospheric boundary layer impacts the detachment (also called deflation), movement, and deposition of soil particles. When soils erode, the wind is turbulent and turbulence facilitates the transfer of momentum from the wind to the soil surface and exerts a shear effect on the surface. This shear or friction can be computed from wind speeds as follows:

$$\tau = \rho(U_*)^2 \quad (23.1)$$

where
  τ is the friction, dynes m^{-2}
  ρ is the air density, g m^{-3}
  U_* is the friction velocity, m s^{-1}

Friction velocity can be computed by

$$U_* = \frac{U_z}{5.75 \log(z/k)} \quad (23.2)$$

where
  U_* is the friction velocity, m s^{-1}
  U_z is the wind speed at height z, m s^{-1}
  z is the height where U_z is measured, m
  k is the aerodynamic surface roughness, m

Shear velocity can also be determined graphically using the relationships in Figure 23.2. The wind speed (U_z) is plotted as a function of the $\log_{10}$ of the height (z). From this plot the aerodynamic surface roughness (k) and friction velocity (U_*) can be determined at the intercept and slope, respectively.

The relationship between speed and height (Equation 23.2 and Figure 23.2) is valid for clean air and neutral atmospheres. As the concentration of transported material increases, the relationships are no longer valid (Anderson et al., 1991).

Weather systems that erode soils vary in their time frames, space scale, and predictability. The author's estimates are listed in Table 23.1 and may not be valid for all regions of the world.

Dust devils are very common on Earth and on Mars, and are an abbreviated wind–sand transport system that may be numerically modeled (Hess, 1973; Fernandez, 1997). Within the dust devil, the moving particles may become electrically charged to the point of causing electromagnetic interferences (Huang et al., 2008b). The sign of the electrical charge is negative when the diameter of sand particles is smaller than 250 microns and positive when diameter is larger than 250 microns (Zheng et al., 2003). The dust devil is a common sight in semiarid regions, but the time is short and spatial scales are very small.

On Mars dust storms are considered the seeds of global dust storms that may even engulf the entire planet (Hess, 1973). The haboob is a very dramatic cloud of dust that appears to roll over the landscape. While a haboob can blot out the sun for short periods of time they usually do not cover as large an area or last

TABLE 23.1 Hierarchy of Weather-Duststorm Systems

Duststorm	Weather System	Duration Time (h)	Aerial Scale (km)	Predictability
1. Dust devil	Microtemperature differences	0.1–0.5	0.01–0.50	Observe only in real time
2. Haboob	Thunderstorm downdraft into dry air, that is, gravity flow	0.05–6	25–75	1–12 h
3. Severe mountain downslope windstorm	Complex terrain enhancement of downward transport of midtropospheric momentum	0.5–18	25–250	12–36 h
4. Frontal	Gravity flow, pressure gradient with dynamic assist	1–8	500–1000	24–48 h
5. Cyclogenic				
a. Low level jet	Boundary layer thermal differences, momentum transfer within shallow adiabatic planetary boundary layer	6–12	500–1000	24–48 h
b. Upper level jet	Deep adiabatic heating of troposphere through dynamic subsidence	8–24	500–1000	24–72 h
c. Surface storm circulation	Deeper-gradient winds	8–18	50–150	12–36 h
d. Severe mountain downslope windstorm	(Mechanism No. 3 above)[a]			

[a] The mechanism for the cyclogenic severe mountain downslope windstorm is the same as for the severe mountain downslope windstorm, but the duration is usually longer and the aerial scale is generally larger.

as long as dust storms produced by down slope, frontal, or cyclogenic weather systems (Table 23.1). Thirty years of dust storm data at Big Spring, TX, were analyzed. The average number of storm days was 26 per year with an average of 192 h with dust per year. Severe storms with visibilities of less than 300 m produced an estimated 54.9% of the total dust load with 11% of the total dust hours. Most of the dust storms included in these analyses resulted from frontal weather systems (Fryrear, 1981).

23.2.1 Wind Speed

Wind erosion is possible when the velocity of the wind at the soil surface exceeds the threshold velocity required to move the least stable soil particle. The detached particle may move a few millimeters before finding a more protected site on the landscape. An exception would be the direct suspension of dust without saltation or creep, which may occur on some loessial soils (Kjelgaard et al., 2002). The wind speed required to move this least stable particle is called the static threshold (Bagnold, 1941). If the wind speed increases, soil movement begins. When soil movement is sustained, the dynamic threshold speed has been attained and the quantity of soil that can be transported by the wind varies as the cube of the velocity (Bagnold, 1941; Chepil and Woodruff, 1963). Many transport equations have been reported that include a term for wind speed (Greeley and Iverson, 1985; Sorensen, 1991).

Soil roughness, erodibility, wetness, surface disturbance, and crop residue quantity and orientation are parameters that impact the vulnerability of soils to wind erosion. If the erodible soil surface is covered with vegetation or residues from a previous crop, the force of the wind is transferred to the nonerodible cover and erosion is controlled. Unfortunately, rainfall in arid and semiarid regions may be insufficient to grow enough vegetation to protect the soil. Soils in these regions must be protected with other methods or the soil may erode until a desert pavement is formed. With deep sands, the surface may never stabilize and active sand dunes will dominate the landscape.

Computation of friction velocity from wind profiles during severe wind erosion events is difficult because of the influence of soil material in the wind profiles. Soil particles may roll over the surface in a transport mode called creep, may bounce over the surface in saltation, or may float over the surface in suspension (Chepil and Woodruff, 1963). Chepil (1941) reported that saltation movement reduces momentum and speed of the wind. The greater the concentration of moving sand grains, the greater the apparent wind speed reduction near the ground (Chepil and Milne, 1941). Anderson et al. (1991) also reported that the presence of sediment transport in the windstream influences the wind speed in the boundary layer. Sand flux frequencies respond to fluctuations in wind frequencies less than 0.03 Hz (Shi et al., 2008).

For every soil condition there is a minimum wind speed called dynamic threshold that initiates sustained soil movement. Plant residues and soil roughness increase the threshold wind speed, but soil wetness temporarily stabilizes the soil surface (Saleh and Fryrear, 1995). Any non-erodible material such as stones or soil clods on the soil surface resists the force of the wind protecting a portion of the surface.

23.2.2 Soil

For mineral soils, the larger the soil particle, the greater the wind speed required to dislodge and transport the particle. Most field erosion occurs from a crusted surface with an abundance of loose sand. Individual sand particles are more easily eroded than particles with a median grain diameter of 40–160 microns. When median diameter of soil particles was <40 microns the coarsest particles were more easily eroded (Goossens and Gross, 2002). For the same wind, the transport rate may be twice as large for particles of mixed size as for particles of uniform size. The speed and turbulence of the wind, density ratio of particles to air, and surface roughness dictate the largest soil particle that can be eroded (Chepil and Woodruff, 1963; Greeley and Iverson, 1985; Batt and Peabody, 1994). The erosion of soil aggregates by impacting sand particles increases with the square of the particle velocity. Increasing the size of the abrading particles from 125 to 715 μm gave a small increase in erosion of the soil aggregates (Hagen, 1984). Electromagnetic and static forces influence detachment and transport of small particles. When coupled with cohesive forces, micron size particles may coalesce into a larger aggregate less susceptible to transport by wind. Cyanobacterial biomass on California desert soils significantly increased surface stability, increased the threshold velocity, and reduced sediment yield (Belnap et al., 2006).

23.2.2.1 Soil Crust

After tillage the soil surface is loose and unconsolidated, but subsequent rainfall tends to create a consolidated zone in the soil surface called a crust. Crusts are more compact and mechanically stable than the soil below (Chepil and Woodruff, 1963). Crust may be more resistant to wind erosion than loose particles, but can erode when blowing soil particles are present in the wind stream. Crust properties important to estimate wind erosion include the thickness and stability of the crust layer, fraction of surface crust covered by nonerodible stones or clods, and the amount of loose erodible material present (Zobeck, 1991). Soil properties important in crust development include clay content (Chepil, 1958), $CaCO_3$ and organic-matter content (Chepil, 1954a), and soluble-salt concentration (Nickling and Ecclestone, 1981). Not all crusts protect the soil from wind erosion. For soils with less than 3% silt and clay the crust formed is not a significant factor in reducing wind erosion (Rajot et al., 2003).

23.2.2.2 Dust Production and Transport

Wind-eroded particles are normally transported in one of three modes: creep, saltation, or suspension (Chepil and Woodruff, 1963). The specific mode of transport depends on the speed and turbulence of the wind and soil surface conditions. Creep particles 1000–2000 μm in diameter roll along the surface. The creep component of the total soil flux can be a significant portion or almost zero. The texture and condition of the soil surface determines if large soil aggregates can roll over the soil or become trapped in microdepressions.

Saltating particles 100–1000 μm in diameter move in a series of short hops depending on surface roughness, particle size, and wind speed. For speeds slightly above threshold, a saltation size particle may roll on the surface in a creep mode. For high wind speeds, saltation size particles may be transported at a considerable height in a mode normally called suspension. The saltating particles are largely responsible for the abrasion of soil aggregates and accelerate the erosion process until the wind has attained its maximum transport capacity.

Suspended particles up to 100 μm in diameter represent the visible material in dust clouds that move into the upper atmosphere, and are subject to long-range transport. The size and binding energies between particles influence their release to the atmosphere, but are not dependent on the soil texture or its mineralogical composition (Alfaro et al., 1998). As particles rise from the soil surface they may be subjected to numerous midair collisions in the saltation zone and the effect on wind speed is appreciable (Huang et al., 2007). Data suggest that relatively large particles may be transported considerable distances (Betzer et al., 1988). For most wind-eroded surfaces, the majority of the mass being transported is close to the soil surface. That portion in suspension may represent a small percentage of the total mass being transported but may have a major impact on air quality (Saxton, 1995, 1996; Stetler and Saxton, 1995). For soils the source of the material may be determined by the magnetic properties of the dust (Reynolds et al., 2006).

The vertical flux of dust divided by the horizontal flux is defined as the sandblasting efficiency (Gomes et al., 2003). The fine-textured soils may emit more vertical flux in relation the horizontal flux than coarse-textured soil, but the total quantity of material eroded by wind is much lower for the fine-textured soil. In NW Germany over a 2 year period dust emission due to tillage was four times the dust emission due to wind erosion. The total dust transport was 17.1 t $km^{-1} a^{-1}$ in the first 10 m of the atmosphere. The horizontal dust flux increased from 50 μg $m^{-2} s^{-1}$ in January to 100 g $m^{-2} s^{-1}$ in April and then dropped to 20 g $m^{-2} s^{-1}$ in August (Goossens, 2004).

In the twenty-first century there is increased concern about air quality and how it is impacted by airborne particles. A Dust Production Model (DPM) (Alfaro and Gomes, 2001) was developed by combining a sandblasting model (Alfaro et al., 1998) with a saltation model (Marticorena and Bergametti, 1995). The DPM estimates the size and amount of aerosols from a given soil during a wind event. The model identifies the importance of aerodynamic surface roughness, dry aggregate size, and wind friction velocity.

23.2.2.3 Soil Erodible Fraction

For most agricultural fields, the soil is composed of sand, silt, and clay particles. Organic matter and $CaCO_3$ are constituents of soil that are important in describing the soil's susceptibility to wind erosion (Fryrear et al., 1994). Using the standard compact rotary sieve (Chepil, 1962) and a sample of the surface 20 mm of dry topsoil, the erodible fraction (*EF*) of the soil can be determined. The erodible fraction is that portion smaller than 0.84 mm in diameter easily eroded by wind. Several thousand samples were analyzed to develop the following equation:

$$EF = \frac{29.09 + 0.31Sa + 0.17Si + 0.33Sa/Cl - 2.59OM - 0.95CaCO_3}{100}$$

(23.3)

where
  EF is the erodible fraction (0.077–0.822)
  Sa is the sand % (5.5–93.6)
  Si is the silt % (0.5–69.5)
  Cl is the clay % (1.2–53.0)
  OM is the organic matter % (0.18–4.79)
  $CaCO_3$ is the calcium carbonate % (0.0–25.2)

The range of values in the dataset is in parentheses. The percentage of sand, silt, and clay are based on a dispersed mechanical

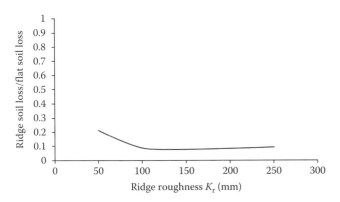

FIGURE 23.3 Soil loss ratio for various soil ridge roughness (k_r) values when wind is perpendicular to soil ridge. (From Fryrear, D.W. 1984. Soil ridges—Clods and wind erosion. Trans. ASAE 27(2):445–448.)

TABLE 23.2 Exposure Time to Windblown Sand for 25%, 50%, or 75% Survival

	Survival (%)		
	25	50	75
Crop	Exposure Time (mi)		
Carrots	11.0	7.5	3.9
Onions	31.0	21.4	11.0
Peppers	12.0	7.9	3.8
Cabbage	14.8	10.1	5.4
Cucumbers	21.2	14.1	7.1
Southern peas	49.3	32.7	16.0
Cotton	18.3	124	6.6

Source: Fryrear, D.W., and J.D. Downes. 1975. Consider the plant in planning wind erosion control systems. Trans. ASAE 18:1070–1072.

analysis procedure where clay particles are smaller than 2 μm, silt particles are 2–50 μm, and sand particles are 50–1000 μm (Taylor, 1960).

With this equation, the potential erodibility of a mineral soil can be estimated from the physical and chemical properties. Equation 23.3 allows scientists to evaluate wind erosion control practices on small plots. If the soil surface is covered with erodible material, theoretically, erosion will continue as long as the wind speed is above the threshold. Large sand dunes can result, but even in desert dune regions, the interdunal area is normally composed of more resistant clay or rock material.

23.2.2.4 Soil Roughness

In most agricultural systems, the farmer tills the soil to prepare a seedbed, control weeds, or improve rainwater infiltration. Tilling normally increases soil surface roughness. Soil roughness may consist of uniformly spaced ridges or furrows and randomly distributed soil aggregates. The protection from soil ridges persists longer than random roughness, but soil ridge roughness is sensitive to wind direction. When the wind is perpendicular to ridges 0.06–0.25 m high, erosion may be controlled (Figure 23.3). The roughness of the soil surface other than that caused by clods or vegetation has been described as the Soil Ridge Roughness, K_r. When the wind is parallel to the ridge, the ridge is not effective in controlling wind erosion (Saleh, 1994). Aerodynamic roughness with ridges or soil aggregates controls or, at least, reduces soil erosion until rainfall smoothes the surface (Fryrear, 1984; Saleh et al., 1997). Roughness also reduces dust emissions from agricultural fields compared to dust from a flat, dry, loose soil (Batt and Peabody, 1994). A computational fluid dynamics model (CFD) has been used to describe particle stability depending on its position on a ridged surface (Huang et al., 2008a).

23.2.3 Plants

23.2.3.1 Seedlings

Emerging crop seedlings are sensitive to wind damage. In fact, each year thousands of hectares of crops are replanted because of windblown sand damaging or destroying the young crop (Woodruff, 1956b; Skidmore, 1966; Armbrust et al., 1969; Fryrear and Armbrust, 1969; Fryrear, 1971; Armbrust, 1972, 1984). The first 4 weeks after planting are critical to many crops. Some crops are so sensitive that a 20 min exposure to windblown sand may destroy 75% of the seedlings (Table 23.2). As the plants mature, they are less sensitive to wind damage and the developing plant canopy affords some protection against wind erosion (Fryrear and Downes, 1975; Armbrust, 1977). Windblown sand injury delays maturity, increases susceptibility to diseases, and impacts the physiology of plants (Claflin et al., 1973; Armbrust et al., 1974; Armbrust, 1982). Completely defoliated cotton plants may recover due to increased shoot growth efficiency because the relative growth rate was related to net assimilation rate and not leaf area ratio (Baker, 2007).

23.2.3.2 Growing Plants

As crops mature, they can tolerate more sand injury and can reduce the wind speed at the soil surface (Fryrear et al., 1975; Bilbro, 1989, 1991, 1992). However, some crops such as cabbage or tobacco may be of inferior quality if sand is contained within the plant leaves (Fryrear and Downes, 1975; Armbrust, 1979).

23.2.3.3 Flat Residues

When crops are harvested, some plant material remains in the field. This residue or stubble has value as a source of nutrients for the soil, as animal feed, as building material or fuel, and as protection against wind erosion. Studies were conducted to determine the relationship between the proportion of soil surface covered with plant material and soil erosion (Fryrear, 1985; Bilbro et al., 1994). The available published data were combined into a single value that expresses soil loss ratio (SLR_f) as a function of flat soil cover. The SLR allows one to test many different sizes and types of material and combine the results into one expression (Findlater et al., 1990). The SLR_f is the mass of soil erosion with residue cover divided by the mass of soil erosion without residue cover. Figure 23.4 illustrates that covering a 30% of the soil surface can significantly reduce wind erosion. These studies assume that the flat residues are uniformly distributed over the soil surface.

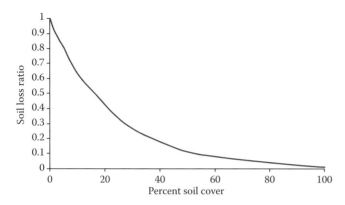

FIGURE 23.4 Soil loss ratio (soil loss with partial cover divided by soil loss for same soil without any cover) as a function of percent soil cover with nonerodible elements such as flat residues or rocks.

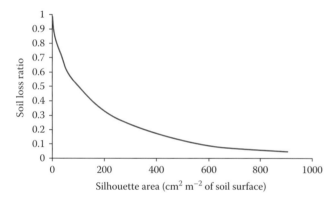

FIGURE 23.5 Soil loss ratio (soil loss with partial cover divided by soil loss for same soil without any cover) as a function of silhouette area; silhouette area is the profile area of all standing residues in a 1 m² area. (From Bilbro, J.D., and D.W. Fryrear. 1994. Wind erosion losses as related to plant silhouette and soil cover. Agron. J. 86:550–553.)

23.2.3.4 Standing Residues

When small grains, corn, sunflowers, soybeans, or sorghum are harvested, a portion of the crop residue remains standing in the field. Standing residues are at least six times more effective than the same mass of material flat on the soil surface (Englehorn et al., 1952; van de Ven et al., 1989; Bilbro and Fryrear, 1994) (Figures 23.5 and 23.6). These results emphasize that management practices that maintain residues in an erect position are more effective in controlling wind erosion than practices that flatten the residues or incorporate residues into the soil. The silhouette of a standing crop residue is the vertical area of the plants in 1 m² of ground area. The effect of silhouette area on soil erosion is expressed as a soil loss ratio of standing residues (SLR_s). The SLR_s is the mass of soil loss with standing residues divided by the mass of soil loss when the surface is bare.

23.2.4 Humankind

Although wind erosion is a normal process, it can be enhanced by the irresponsible activities of humans (Chepil and Woodruff, 1955). When fragile soil or plant resources in arid regions are mismanaged by man's activities, restoration requires many years or may be impossible. The mismanagement may be the result of surface disturbance from overgrazing, over cultivation, vegetative burning, woodcutting, salinization, or the use of animal dung as fuel instead of fertilizer (Middleton, 1987). When effective management practices are used, the soil, crop, and climatic resources can be utilized indefinitely. The most effective management practice must be flexible enough to respond to variable climatic conditions.

The clearing of forests, woodlands, or grasslands reduced soil fauna and organic matter and impacted soil temperature and soil permeability Strzelecki (1845). In addition, man-induced changes in surface and atmospheric conditions may impact local and regional climatic conditions (Saxton, 1995, 1996; Stetler and Saxton, 1995; Williams and Balling, 1996; Zobeck et al., 1996). Drift sand areas were preserved when undesirable vegetation was removed to expose the original sandy surface restoring the ecosystem (Riksen and Goossens, 2005). The impact of humans may be overshadowed by natural fluctuations in the global climatic system.

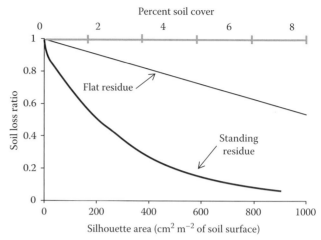

FIGURE 23.6 Comparison of soil loss ratio with the same quantity of residue standing as silhouette and flat on the soil surface. (From Bilbro, J.D., and D.W. Fryrear. 1994. Wind erosion losses as related to plant silhouette and soil cover. Agron. J. 86:550–553.)

23.3 Modeling the Wind Erosion Process

The wind erosion process is a complex interaction of weather–soil–plants–man. To evaluate the effect of any single factor, the interactions between factors must be understood. Wind erosion models can be empirically, physically (also called process) or theoretically based. The application of empirically based models is limited to the range of conditions under which the model was developed. Also, empirically based models require carefully controlled experiments to develop the essential coefficients. Physically or theoretically based models may be more widely adapted but require that the theory and assumptions imbedded within the model are correct. Because of the complex interactions involved in wind erosion, the physical processes may not be well understood. For example, an analytical model developed by Sorensen (1991) utilizes processes for trajectory of the particles, wind modification, and grain bed collisions. Formulae were derived for transport rate and flux of the sand grains from the surface into the air. In developing a conceptual transport model, McEwan and Willetts (1993) concluded that inaccurate wind speeds and sand trap inefficiencies may have been responsible for scatter in wind tunnel data. They also cautioned wind erosion scientists that extrapolation of sand transport rate formulae from the uniform laboratory wind tunnel conditions to field conditions may not be justified.

An empirically derived differential equation of dust concentrations in a thin layer near the ground to predict dust concentrations at the top of the surface layer was developed by Berkofsky and McEwan (1994). Unfortunately, the equation neglects horizontal transport, which is a major component of the wind erosion process; however, they recognized that the detachment plus a source term was inversely proportional to surface roughness.

While the above studies have been valuable sources of new approaches and have expanded our understanding of the wind erosion process, they do not have essential input parameters to develop a complete wind erosion model. Essentially all the models developed to date are a combination of physical, empirical, or theoretical approaches. Depending on the intended use the operator must determine if the model produces the desired output with the available input data.

23.3.1 Wind Erosion Equation

Wind erosion equation (WEQ; Woodruff and Siddoway, 1965) has been widely used to plan conservation systems and estimate annual soil erosion. WEQ was one of the first models to estimate soil erosion in the field using annual input values for weather, soil erodibility, soil roughness, field length, and crop residues. WEQ, which was developed from wind tunnel studies with limited field verification, assumes that the wind erosion process is similar to an avalanche of snow and debris rolling down a mountain (Chepil, 1957). As an avalanche rolls down the side of a mountain, the mass of material in the avalanche may increase as long as there is a source of snow and the slope is long and uniform. The quantity of material being transported by the wind can increase until the capacity of the wind to detach and transport soil has been reached (Bagnold, 1941; Chepil, 1957).

The WEQ consists of a functional relationship among five factors:

$$E = f(IKCLV) \quad (23.4)$$

where
- E is the annual soil erosion, Mg ha^{-1}
- f is the functional relationship
- I is the soil erodibility, Mg ha^{-1} year^{-1}
- K is the soil roughness factor
- C is the climatic factor
- L is the unprotected field length, m
- V is the vegetation factor expressed as small grain equivalent

WEQ estimates average soil loss (note units of E and I) for an unprotected field length (L). If soil loss is divided by field length, the units are mass per unit width or an expression of transport mass.

Concerns that the climatic factor did not describe wind erosion in high or low rainfall regions were addressed and a new climatic factor was developed (Woodruff and Armbrust, 1968; Skidmore, 1986). The WEQ was modified for shorter time periods, but as stated by Bondy et al. (1980, p. 176), "no experimental data base exists for using the wind erosion equation for periods of less than one year."

WEQ was coupled with the Erosion Productivity Impact Calculator (EPIC) (Williams et al., 1984) to estimate daily wind erosion soil losses (Cole et al., 1983). The I, K, C, and V factors in WEQ are annual values, but in the EPIC wind erosion model they are considered constant for a single day. Estimates of wind erosion with EPIC were compared with WEQ estimates. When compared with observed erosion, WEQ under-predicted erosion 11 of 14 periods by as much as nine times (Van Pelt and Zobeck, 2004). While a model may perform satisfactorily under one set of conditions, it may not be suitable for all field conditions.

Sometimes models are used for conditions beyond the limits of the original research. For example, a Soil Quality Index (SQI) was developed using the WEQ model (Zobeck et al., 2008) even though the WEQ model has been shown to be inaccurate for many regions of the country and for many management systems (Fryrear et al., 2001).

23.3.2 The Pasak Model

A soil loss model reported by Pasak in 1973 was intended as a single event model (Holy, 1980):

$$E = 22.02 - 0.72P - 1.69V + 2.64Rr \quad (23.5)$$

where
- E is the erodibility, kg ha^{-1}
- P is the nonerodible particles, %
- V is the relative soil moisture
- Rr is the wind speed, km h^{-1}

The lack of inputs for crop residues and soil roughness limits the use of this model. Also wind speed and soil moisture conditions are not constant during an erosion event.

23.3.3 Texas Erosion Analysis Model

The Texas Erosion Analysis Model (TEAM) developed by Gregory et al. (1988) is a computer program developed to simulate wind profile development and soil movement over multiple field lengths. The basic equation is

$$X = C(SU_*^2 - U_{*t}^2)U_*(1 - e^{-0.000169 A_a IL_f}) \quad (23.6)$$

where

    X is the rate of soil movement at length L_f, M L^{-1} T^{-1}
    $C(SU_*^2 - U_{*t}^2)U_*$ is the maximum rate of soil movement, which occurs when the surface is covered with fine noncohesive material
    C is a constant, which depends on width sampled and units used for U_*
    L_f is the length of unprotected field in the wind direction
    A_a is the abrasion adjustment factor
    I is the soil erodibility factor involving soil shear strength
    S is the surface cover factor, which expresses the amount of detachment energy at the top of the cover, which is transferred to the soil
    U_* is the friction velocity
    U_{*t} is the threshold friction velocity

The abrasion adjustment factor is computed using the equation:

$$A_a = (1 - A_l)(1 - e^{-0.0072 e^{0.00079 I L_f}}) \quad (23.7)$$

where A_l is the lower limit of abrasion effect (assumed 0.23).

23.3.4 Wind Erosion Assessment Model

This process-based model developed in Australia is a combination of established and theoretical results (Shao et al., 1996). It synthesized research results on sand drift and dust entrainment, approximated quantitative assessment of wind erosion, and evaluated the current limits to our knowledge of the wind erosion process. The Wind Erosion Assessment Model (WEAM) model does not predict the evolution of soil surface properties due to wind erosion, natural weathering processes, or management interventions. WEAM treats erosion of a dry, bare soil in a given wind. To model daily or annual erosion, it is essential that changes in soil and crop conditions be described by the model.

When coupled with large computers, WEAM does have the capability of estimating the movement of large dust clouds across complex landscapes. Input requirements are such that GIS or other information databases are essential to utilize this model effectively. The lack of management intervention inputs limits WEAM's usefulness for field or short-period erosion estimates. It does express the regional or continental impacts of wind erosion.

23.3.5 Revised Wind Erosion Equation

To expedite the availability of new technology, the revised wind erosion equation (RWEQ) effort was requested by Natural Resources Conservation Service national staff. The RWEQ model requires simple inputs to estimate daily, weekly, monthly, seasonal, or long-term soil loss from agricultural lands. RWEQ has been field tested against measured transport data for a broad range of soil, field, management and climatic conditions. A physically based mass transport equation was derived by Stout (1990) for RWEQ. This equation is based on the concept that a self-balancing mechanism controls wind erosion and is the major relationship in RWEQ. RWEQ utilizes the principle that wind erosion is the result of the erosive force of the wind exceeding resisting forces from the surface. The resisting forces include the soil inherent wind erodibility, the surface roughness, the reduction in wind velocity with barriers, and the protection from flat or standing plant residues.

With the self-balancing concept, momentum is transferred from the wind to the soil surface (Owens, 1964). The equation derived by Stout (1990) is

$$\frac{Q_x}{Q_{max}} = 1 - e^{-(x/b)} \quad (23.8)$$

where

    Q_x is the mass being transported by wind at field length x, kg m^{-1}
    Q_{max} is the maximum transport capacity of wind over that field surface, kg m^{-1}
    x is the field length, m
    b is the field length where wind attains 63.2% of Q_{max}, m

Equation 23.8 assumes if the wind speed is above threshold, the erosion process detaches and picks up soil particles at an increasing rate across a field. Equation 23.8 is a negative exponential decay curve that gives increasing transport mass as field length increases. The highest soil loss per unit area occurs immediately downwind of the nonerodible upwind boundary, assuming $Q_x = 0$ at $x = 0$ (Owens, 1964). While developed for transport layers close to the soil surface, Stout's (1990) basic equation was used to describe the total mass being transported between the soil surface and a height of 2 m. Equation 23.8 was modified to the form:

$$Q_x = Q_{max}(1 - e^{-(x/s)^2}) \quad (23.9)$$

where s is the inflection point where the slope of the line (Figure 23.7) changes from positive to negative.

Equation 23.9 gives a sigmoid curve where the mass being transported is zero at the upwind boundary, and then increases until the wind attains 63.2% of its capacity at field length "s" (Figure 23.7) (Fryrear and Saleh, 1996). Beyond field length "s" the rate of increase in transport mass is limited by the capacity of the wind to detach and transport additional soil. Therefore, the rate of increase in mass being transported diminishes until the maximum transport capacity has been achieved. The basic relationship between field length and transport mass in RWEQ reflects that as the quantity of material being

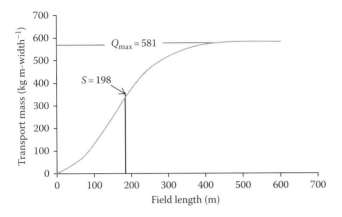

FIGURE 23.7 Measured mass flux (20 min periods, sampler 12.5 mm wide × 50.8 mm high) from western Canada. (From Chepil, W.S., and R.A. Milne. 1941. Wind erosion of soil in relation to roughness of surface. Soil Sci. 52:417–433.)

transported approaches the maximum transport capacity of the wind, any additional material picked up must result in the deposition of a portion of the original load. The field length required for the wind to become saturated with sediment varies with the soil surface conditions and the velocity of the wind. Rarely does the atmosphere become saturated with sediment before surface conditions or wind velocity changes. With measurements of transport mass and field lengths, the coefficients Q_{max} and s (Figure 23.7) can be determined using nonlinear regression analysis.

The input data required to use the RWEQ model on noninstrumented fields include factors for weather (WF), soil (EF, SCF), crop (COG), and tillage (K). Measured values for the input factors were correlated with computed values for Q_{max} and s from single erosion events. These events were for fields of the same size and shape. The soil surface was flat and dry ($SW = 1$) with no snow cover ($SD = 1$).

From this analysis, the relationship between the input parameters and Q_{max} is described as follows:

$$Q_{max} = 109.8(WF \times EF \times SCF \times K \times COG) \qquad (23.10)$$

where
Q_{max} is the maximum transport capacity, kg m-width^{-1}
WF is the weather factor, kg m-width^{-1} (contains SW and SD values)
EF is the soil erodible fraction
SCF is the soil crust factor
K is the soil roughness
COG is the vegetation factor including flat residues, standing residues, and canopy
SW is the soil wetness
SD is the snow depth

The slope s of the empirical relationship in (8.9) is computed from the input factor using the equation:

$$s = 150.71(WF \times EF \times SCF \times K \times COG)^{-0.3711} \qquad (23.11)$$

where s is the critical field length where Q_x is 63.2% of Q_{max}, m.

When input data are available, the estimated Q_{max} and s values can be computed for single storms (Table 23.3). With the coefficients from Equations 23.10 and 23.11 as input for Equation 23.9, soil erosion can be computed for circular fields up to 60 ha and for square or rectangular fields up to 145 ha. The WF is time based, so erosion can be computed for any time interval when the input parameters are constant. Within the RWEQ computer program, the WF is adjusted for rainfall and snow cover and, crop residues are decomposed as a function of soil moisture and soil temperature

TABLE 23.3 Wind Factor (WF), Soil Erodible Fraction (EF), Soil Crust Factor (SCF), Soil Roughness (K), Flat and Standing Residues (COG), Maximum Transport Capacity (Q_{max}), and Critical Field Length (s) for Selected Erosion Events

Site	Date	WF	EF	SCF	K	COG	Q_{max} kg m^{-1}	S M
BS	01/27/90[a]	2.3	0.64	0.77	0.95	0.90	112	123
BS	01/29/90	2.8	0.64	0.77	0.95	0.90	133	88
BS	02/08/90	0.6	0.64	0.77	0.95	0.90	96	289
BS	03/06/90	2.8	0.64	0.77	0.95	0.90	226	149
BS	03/29/93[b]	3.6	0.77	0.77	1.00	0.96	402	84
MW	04/02/91	8.4	0.79	0.91	0.82	0.43	168	43
EK	03/09/92	41.9	0.70	0.65	0.91	0.65	1430	98
KM	03/13/93	15.3	0.85	0.90	0.85	1.00	751	109
EC	03/12/91	179.9	0.26	0.21	0.80	0.48	648	179

Sites are coded Big Spring, Texas (BS); Mabton, Washington (MW); Elkhart, Kansas (EK); Kennett, Missouri (KM); and Eads, Colorado (EC).
[a] Includes January 27 and 28, 1990 wind data.
[b] Includes March 28 and 29, 1993 wind data.

TABLE 23.4 Comparison of Total Soil Loss Measured on Instrumented Sites and Soil Loss Estimated with RWEQ for Various Time Periods

	Time Period		Soil Loss	
			Measured	Estimated
Site	Beginning	End	kg m^2	
Akron, CO	10/27/88	05/26/89	0.83	6.70
Akron, CO	10/20/89	04/25/90	1.10	0.38
Eads, CO	10/30/90	05/07/91	2.43	1.57
Eads, CO	09/15/91	04/15/92	0.67	0.81
Crown Point, IN	01/10/90	12/31/90	31.21	23.42
Crown Point, IN	01/01/91	12/31/91	23.95	11.23
Crown Point, IN	01/01/92	06/04/92	0.42	0.67
Elkhart, KS	02/27/90	12/30/90	0.29	0.63
Elkhart, KS	01/01/91	12/30/91	1.32	3.41
Elkhart, KS	01/01/92	10/15/92	15.50	20.73
Elkhart, KS	01/01/93	05/25/93	2.06	13.72
Crookston, MN	11/07/88	11/01/89	0.21	0.00
Crookston, MN	11/02/89	05/06/90	0.32	0.00
Swan Lake, MN	04/11/91	12/31/91	1.14	2.51
Swan Lake, MN	01/01/92	12/31/92	0.13	0.04
Swan Lake, MN	01/01/93	12/31/93	0.00	0.00
Kennett, MO	12/02/92	06/17/93	13.73	5.42
Kennett, MO	11/08/93	05/05/94	0.64	1.75
Havre, MT	10/28/92	05/05/93	0.01	4.44
Havre, MT	10/19/93	03/30/94	0.01	0.52
Lindsey, MT	10/18/90	05/21/91	0.03	1.97
Lindsey, MT	10/08/91	04/08/92	0.09	2.02
Scobey, MT	10/03/88	05/10/89	4.68	7.37
Scobey MT Fallow	10/04/89	04/21/90	1.34	0.07
Scobey, MT Stubble	10/04/89	04/21/90	0.39	0.00
Sidney, NE	10/25/88	05/24/89	0.52	0.52
Sidney, NE	10/24/89	04/24/90	0.38	0.04
Sidney, NE	10/31/90	05/07/91	2.29	4.44
Portales, NM	11/24/94	04/06/95	0.01	0.09
Fargo, ND	12/06/94	05/07/95	0.00	0.00
Fargo, ND	10/24/95	05/07/96	0.00	0.00
Fargo, ND	10/24/96	05/15/97	0.00	0.00
Big Spring, TX	01/12/89	05/03/89	21.54	25.26
Big Spring, TX	01/05/90	05/04/90	20.96	17.06
Big Spring, TX	01/28/91	05/15/91	0.10	0.11
Big Spring, TX	03/16/93	06/01/93	28.78	29.85
Big Spring, TX	01/06/94	05/18/94	17.16	11.23
Big Spring, TX	01/11/95	05/15/95	26.29	39.15
Big Spring, TX	01/12/96	05/16/96	3.99	10.51
Big Spring, TX	01/23/97	05/23/97	13.63	11.99
Martin-C, TX #2	01/24/95	06/06/95	0.30	0.20
Martin-C, TX #3	01/11/95	05/23/95	0.80	0.52
Martin-C, TX #4	02/11/95	05/19/95	0.30	0.43
Plains E, TX	11/15/94	06/03/95	2.20	1.82
Plains B, TX	12/13/94	05/24/95	1.60	0.36
Plains, TX E	12/12/95	06/04/96	3.83	0.61
Plains, TX B (800 m)	12/13/95	05/29/96	2.02	0.22
Plains, TX B (1600 m)	12/13/95	05/29/96	1.55	0.54
Mabton, WA	12/13/90	04/28/91	3.68	2.98
Prosser, WA #1	12/03/91	03/25/92	0.17	0.13
Prosser, WA #2	06/10/92	06/15/93	0.32	1.05

(Steiner et al., 1994). Soil roughness is decayed by rainfall amount and intensity (Saleh et al., 1997).

The same equation used to estimate single event soil erosion has been used with routines to adjust for rainfall, to degrade soil roughness, and to decompose crop residues to estimate erosion for the entire erosion season (Table 23.4). With the RWEQ model, erosion is computed for a field length along the dominant wind direction. The transport mass from Equation 23.9 is important in evaluating potential plant injury from blowing sand. The critical field length from Equation 23.9 can be used to plan field barrier systems or determine field widths that minimize soil erosion.

RWEQ was tested on cropland where irrigation water was removed at Reno, Nevada, and in the Northwest Coastal Zone (NWCZ) of Egypt. The presence of stones (>50 mm diameter) modifies the force of the wind. An additional subroutine is required to describe the silhouette of surface stones in the NWCZ. (The impact of surface M.M. Massif). The estimates with RWEQ were tested against field measurements for 52 erosion events at Fuka and Abu Lahu (Fryrear et al., 2008). The soil at Fuka had an erodible fraction of 0.4. The maximum measured transport mass was 738 kg m-width^{-1} and the estimated maximum transport was 650 kg m-width^{-1}. The erodible fraction of the soil at Abu Lahu was 0.2, the maximum measured transport mass 262 kg m-width^{-1}, and the maximum estimated transport mass was 286 kg m-width^{-1}. When stones are present, the RWEQ with the stone silhouette coefficient can be used to describe wind erosion in arid regions when the small grain was seeded and harvested by hand.

23.3.6 The Wind Erosion Prediction System

As the WEQ was used limitations of its empiricism were recognized (Skidmore, 2007). The climatic factor (Woodruff and Armbrust, 1968; Skidmore, 1986) and management periods (Bondy et al., 1980) for WEQ were modified but the results were never validated with field erosion data. As erosion measuring equipment (Fryrear, 1986) and procedures (Fryrear et al., 1991) were developed, it became apparent that there were significant limitations to the empiricisms in WEQ. A concentrated effort was made to assemble new technology into a model called the Wind Erosion Prediction System (WEPS). WEPS uses a daily time step that simulates weather and field conditions to estimate wind erosion. The WEPS program includes an erosion submodel, hydrology submodel to simulate water movement within the soil and evaporation from the soil surface, a plant growth submodel to grow vegetation, a decomposition submodel to decompose residues, and a management submodel to represent tillage and other operational decisions by the land manager. WEPS is modular in design so the model can be upgraded with improved technology (Hagen, 1991). WEPS is a very complex model requiring numerous input data (Hagen et al., 1999). To illustrate some of the input requirements, the coefficients for the WEPS erosion submodel are as follows:

C_{ani}	Coefficient of abrasion
C_{bk}	Coefficient of breakage of saltation/creep
C_{dp}	Coefficient of deposition of suspension size
C_{en}	Coefficient of emission
C_i	Coefficient of plant interception
C_m	Coefficient of mixing at soil surface
C_s	Saltation transport parameter
C_t	Coefficient of surface trapping
F_{ani}	Mass fraction of saltation impacting clods and crust
G_{an}	Vertical flux from abrasion of surface clods and crust
G_{en}	Vertical flux from emission of loose aggregates
G_{tp}	Vertical flux from trapping of saltation/creep aggregates
Gss_{an}	Vertical flux from suspension-size aggregates created by abrasion of clods and crust
Gss_{bk}	Vertical flux of suspension-size aggregates created by breakage of saltation/creep
Gss_{dp}	Vertical flux (deposition) of suspension-size aggregates above a non-eroding surface
GSS_{en}	Vertical emission flux of loose suspension-size aggregates
I_1, I_2	Least and greatest input values, respectively used in the sensitivity analysis
I_{12}	The average of I_1 and I_2
O_1, O_2	Output values for I_1 and I_2 respectively in sensitivity analysis
O_{12}	The average of O_1 and O_2
Q	Horizontal saltation/creep discharge
q_{cp}	Transport capacity of the surface, when 40% or more is removed
q_{en}	Transport capacity of saltation/creep
q_s	Horizontal discharge of primary (nonbreakable) sand particles
qss	Horizontal discharge of suspension component
qss_o	Maximum value of qss entering deposition region
$Sfer$	Soil mass fraction of loose, erodible size, less than about 2.0 mm diameter at soil surface
$Sfss$	Soil mass fraction of loose, suspension size less than about 0.1 mm diameter at soil surface
$SFss_{an}$	Mass fraction of suspension size (<0.10 mm) of total (<2.0 mm diameter) created by abrasion
$SFss_{en}$	Mass fraction of suspension size (<0.10 mm) emitted from loose aggregates (<2.0 mm diameter)
U_*	Friction velocity
U_{*t}	Dynamic threshold friction velocity
X	Downwind distance from non-erodible boundary

The initial soil input conditions when using the WEPS erosion submodel (Hagen, 2008) for bare aggregated or crusted soils void of residue cover include the following:

Intrinsic soil properties

    Sand
    Silt
    Clay
    Rock volume
    1.5 mPa water content
    Top soil layer depth

Temporal soil clod parameters

    Geometric mean diameter
    Geometric standard deviation
    Maximum diameter
    Minimum diameter
    Dry stability
    Density

Temporal crust properties

    Crust depth
    Crust density
    Crust cover
    Crust loose cover
    Crust loose cover mass
    Crust dry stability

Temporal surface roughness

    Random roughness
    Ridge height
    Ridge spacing
    Ridge orientation

Biomass cover

    Flat residues

Field length

    Strip
    Small size
    Medium size
    Large size

When WEPS (Hagen, 2008) estimates changes in soil surface conditions during wind erosion events the following surface conditions are updated:

SM_{los}	Soil mobile mass on crusted surface
SF_{cr}	Soil fraction of surface crust cover
SF_{los}	Soil fraction loose cover on crust
SZ_{cr}	Soil depth of crust (consolidated zone)
SM_{aglos}	Soil mobile mass on aggregated surface
$SF84$	Soil mass fraction <0.84 mm on aggregated surface and mobile cover fraction on aggregated surface
$SF200$	Soil mass fraction <2.00 mm on aggregated surface
$SF10$	Soil mass fraction <0.1 mm on aggregated surface
SV_{roc}	Soil volume rock >2.0 mm diameter
SZ_{rg}	Ridge height
SL_{rr}	Random roughness height (standard deviation)

Additional databases are required for climate, management, and crop and deposition to supply information for the hydrology, management, crop, decomposition and weather submodels. Because of WEPS modular design, new features can be incorporated.

The WEPS erosion submodel was tested against 46 single event storms from six states (Hagen, 2004). The WEPS model under-predicted soil loss, particularly for some of the largest events with daily weather data inputs of wind speed, wind direction, solar radiation, relative humidity, air temperature, and rainfall. Temporal field data includes surface soil roughness, plant/residue cover, and dry aggregate size distribution. The possible causes for the under prediction were essential WEPS model inputs were not measured and weathering and prior erosion may have increased soil erodibility.

The WEPS model simulation may begin at any time and may cover any period of time. Heterogeneous regions are simulated by dividing into homogenous subregions where soil type, biomass, and management are similar (Hagen et al., 1995). WEPS permits sub-hourly estimates of soil movement and describes changes in soil surface conditions during erosion events.

Wind erosion scientists must also be aware that using a model that has not been verified with field measurements can give misleading results. Estimates of wind erosion using the WEPS erosion submodel (Wind Erosion Prediction System) in the Pacific Northwest (Feng and Sharratt, 2007) did not correspond with WEG (Wind Erodibility Group) values assigned by NRCS. The general grouping of soil erodibility in the USDA SSURGO database may not hold under all management and climatic conditions.

23.3.7 Wind Erosion on European Light Soils

The Wind Erosion on European Light Soils (WEELS) model consists of modular components of wind, wind erosivity, soil moisture, soil erodibility, soil roughness, and land use. The first two components describe the temporal variations of climatic erosivity. The last four components predict the temporal soil and vegetation cover. The first version was limited to sandy soil because of the simplified parameterizations of the soil water budget. Future improvements in WEELS will include estimation of dust emission and expansion to other soils (Bohner et al., 2003).

23.4 Wind Erosion Measuring Systems: Field Sampling Instrumentation

Humans have lived with the hazards of wind erosion for millennia. Efforts to accurately measure wind erosion from natural erosion events span almost 70 years. To determine the effectiveness of control systems, methods of measuring wind erosion in the field must be available. Laboratory wind tunnel erosion samplers have been developed, but they were not suitable for field studies because some required external power, some were difficult to construct and maintain, and some lacked the capacity or efficiency for field use.

Permanent elevation pins have been used to provide a rough estimate of accumulated erosion over many years (Gibbons et al., 1983). For extremely severe events, the depth of crop roots exposed is an indicator of the depth of soil removed. To provide more detailed information on the wind erosion process, more precise field sampling equipment is required.

A field sampler was used but not described in 1939 (Chepil and Milne, 1939). A vertical slot sampler designed by Bagnold (1941) was the first described instrument for collecting wind-eroded sand in the field, but it did not rotate as wind direction changed. Bagnold's sampler was modified by Chepil (1957) to rotate, but Chepil's sampler was not vented nor suited for highly erodible soils. A single pipe with a vertical slot was described and used because it was simple and cheap to construct and suitable for short-term monitoring of aeolian transport (Leatherman, 1978). Later samplers were vented and rotated into the wind (Fryberger et al., 1979; Merva and Peterson, 1983). These samplers had a metal edge that slid over the eroding soil surface presenting a problem in collecting surface creep.

A glass or plastic bottle with an inlet and outlet tube was designed by Wilson and Cooke (1980). These samplers are now called the Modified Wilson and Cooke (MWAC) and are easy to construct and many can be mounted on a single pole. The MWAC was slightly modified so the bottle was in a horizontal position. When tested with 132–287 micron material the efficiency was between 90% and 120% for wind velocities of 6.6–14.4 m s^{-1} (Goossens et al., 2000).

The Suspended Sediment Trap (SUSTRA) was designed to trap airborne dust, but can be used to trap sand and soil as well (Janssen and Tetzlaff, 1991). The sand laden air enters the instrument via a horizontal metal tube 50 mm in diameter and rebounds onto a metal plate inside a central vertical pipe. The sand grains settle onto an electronic balance under the pipe. The balance is in a metal box buried in the ground and the inlet tube is 23 cm above the soil surface. The efficiencies were somewhat lower for the finest-grain sand and for winds above 11 m s^{-1} (Goossens et al., 2000; Kuntze et al., 2000).

A wedge-shape metal sedimentation chamber (POLCA) was developed in 1995 (Pollet, 1995). The sampler chamber is 100 mm long and 100 mm (front side) and 210 mm on the back with a rectangular inlet. The sides diverge at 28.2° at both sides of the collector. Air discharges from the chamber through a 10 mm-high and 120 mm-wide screened outlet at the rear. The efficiency was between 80% and 120% (Goossens et al., 2000).

Fryrear (1986) developed the BSNE (Big Spring Number Eight) sampler that collects airborne material above the soil surface. The BSNE does not contact the soil surface and is the first field wind-aspirated sampler to accurately sample eroded material. Extensive field wind erosion data have been collected from these samplers (Fryrear et al., 1991). Wind-eroded material can be collected from a height of 0.05 m to any accessible height. The BSNE's sampling efficiency averaged over 89% for washed sand or sieved soil (sieved to remove all aggregates larger than 250 mm) and wind velocities of 10.4–15.7 m s^{-1} (Fryrear, 1986). The same sampling efficiencies were obtained by Goossen in 2000. For particles less than 10 μm, the efficiency was 40% (Shao et al., 1993). Samples from field erosion equipment can be used to identify nutrient losses, to validate wind erosion models, and to test assumptions on the wind erosion process (Fryrear et al., 1991; Fryrear and Saleh, 1993).

Stout and Fryrear (1989) developed a surface creep and saltation sampler for collecting eroded material from the soil surface to a height of 0.2 m. Stout's sampler has an efficiency of 98% in wind tunnel tests with fine sand and the rotating portion does not contact the soil surface. The sampler is suited for field use and has operated successfully in extremely erodible soil; however, the sampler must be lowered as the soil surface erodes. The combination of the BSNE and Stout's sampler permits the measurement of eroding material from the soil surface to any accessible height. For most studies, the BSNE samplers are located from 0.05 to 1.0 m above the soil surface.

With the development of good field erosion measurement (Fryrear, 1986; Fryrear et al., 1991), significant improvements have been made in describing wind erosion mechanics (Vories and Fryrear, 1991). Procedures have been developed to analyze measured erosion in numerous fields (ranging from 2.5 to 145 ha) to test a working wind erosion model on a variety of weather, crop, soil, and farming systems. With the measurements of field erosion, it is possible to establish baseline conditions for future comparisons. Field measurements with the BSNE (Table 23.4) and 5 min data on wind velocities verified that soil mass transport varies as the cube of the wind velocity. Mass transport increases with field length until the wind stream becomes saturated. Mass transport as high as 1231 kg m-width^{-1} has been measured from a single erosion event (Table 23.3 Elkhart, Kansas March 9, 1992). This level of transport is sufficient to destroy most crop seedlings within 15 min. Measured soil losses from an entire erosion season varied from 0.00 to 28.9 kg m^{-2} (Table 23.4).

23.4.1 Air Quality Monitoring

Most of the dust and sand sampling instrumentation described above was designed to collect a sample of the eroded material within the saltation/creep zone. While some suspended material may be collected, the efficiency of all the samplers decreases with grain sizes below 100 microns. A PM_{10} sampler must be tested according to procedures and performance specifications set by the U.S. Environmental Protection Agency. The sampler must be tested in a wind tunnel to show a 10 ± 0.5 micron cut point in aerodynamic diameter for each wind speed of 2, 8, and 24 km h^{-1}. The Kimota 180 and the Wedding ambient beta gauge PM_{10} samplers were tested against an Anderson SA 1200 high-volume sampler (Tsai et al., 1996). The Kimoto 180 significantly underestimated PM_{10} because the cut point of its inlet is 3.5 microns. Deviations ranging from 0% to 50% were measured and increased with water vapor content. The daily average PM_{10} concentrations of the Wedding were 7.2%–9.3% lower than the SA 1200. To collect samples of material smaller than 50 microns requires instrumentation with a calibrated and controlled vacuum pump. The Partisol Dichotomous Aerosol Air Sampler (Model 2025) may be used in conjunction with wind speed and direction sensors, data logger, and a source of power to collect filters with 2.5 and 10 micron particles. To collect a sample of 10 micron material the TSI Dust Trac aerosol monitor may be used if power is available. These instruments are expensive and collect data at only one height unless multiple units are used.

23.4.2 Threshold Velocities

Visual observations can be used to identify when soils begin to erode, but it is not always possible to be in the field when erosion begins. Gillette and Stockton (1986) developed a piezoelectric quartz crystal sensor to detect the exact minute soil movement begins. With this device, the number of particles and the momentum of the particles being transported by the wind can be measured. This sensor, called SENSIT, also permits the researcher to accurately determine the start and duration of wind erosion events and the intensity of blowing sand during the erosion event. Coupled with wind direction sensors, the direction of erosive winds can be determined. A similar device called the Saltiphone was developed to determine when soil movement begins and to obtain a relative measure of sand flux (Spaan and van den Abele, 1991). Depending on soil texture these sensors can be very useful in measuring threshold velocities and mass transport.

23.5 Major Control Systems

The resources available influence the type of practices used to control wind erosion. The major wind erosion control practices include manipulation of surface residues, soil roughening, planting, wind barrier/shelterbelts, and the use of cover crops and soil amendments. No single wind erosion control practice is equally suited for all regions of the country because the climate and management options are so variable. The recommendations are to understand the principles of erosion control practices and use combinations of practices. For example, surface residues can be very effective, except in the more arid climates where normal residue levels may not be sufficient to provide complete protection. Also, in some regions termites may consume all dead surface residues. Soil roughening may be effective until rainfall or irrigation has destroyed surface aggregates. Wind barriers or shelterbelts can provide excellent protection, but like soil ridges they are not effective when the wind is parallel to the barrier. Trees require many years before they are tall or dense enough to provide complete protection from perpendicular winds. In the Fuka area of the Northwest Coastal Zone of Egypt, tillage perpendicular to the erosive wind constitutes an effective wind erosion control practice. It is highly recommended because the local Bedouins favor soil conservation methods that are low cost, simple to implement, and rely on local resources and skills (Wassif et al., 1999).

23.5.1 Crop Residues

The effectiveness of crop residues in controlling soil erosion can be expressed as a percentage of the soil surface that is covered. After small grains, corn, or sorghum are harvested, there is usually sufficient residue in the field to protect the soil. When these same residues have value as fodder for animals, fuel (Williams and Balling, 1996), or building materials, there are economic pressures to remove more and more of the residue.

The relationship between soil erosion and soil covered with residues flat on the soil surface illustrate that covering even 30% of the surface reduces soil losses by 70% (Figure 23.4). The challenge is to leave sufficient residue in the field to maintain the soil resource.

The impact of standing stubble has been extensively researched (Englehorn et al., 1952; Siddoway et al., 1965; Skidmore et al., 1966). van de Ven et al. (1989) developed a silhouette index to describe the relationship between standing plant material (stubble) and wind erosion (Figure 23.5). This way of describing the effect of standing residue is currently used in the RWEQ wind erosion model. Standing residues present a silhouette to the wind that reduces the wind stress at the soil surface. Standing residues are about six times more effective in reducing wind erosion than the same mass lying flat on the soil surface (Figure 23.6). The combined effect of crop residues and soil roughness can be expressed as the product of the coefficients (Horning et al., 1996).

As seedlings develop and mature, the taller plants provide some protection to the soil surface. This protection has been related to the portion of the soil surface covered with plant canopy (Bilbro et al., 1994; Armbrust and Bilbro, 1997). As relationships that include rigidity, leafiness, and canopy height are improved, these will be incorporated into the new models.

23.5.2 Soil Roughness

Managing crop residues is an important method of controlling wind erosion but if crops other than small grains, corn, or sorghum are grown, residue production may not be sufficient to protect fragile soils. During drought periods or when residue levels are not adequate to protect erodible soils, other erosion control methods must be employed.

Roughening the soil with tillage implements can be an effective alternative depending on soil texture of the soil surface (Chepil et al., 1963a; Fryrear, 1984) (Figure 23.3). The impact of effective tillage operations was described by Chepil and Woodruff (1954). They reported soil losses of 25 Mg ha^{-1} were reduced to 0.25 Mg ha^{-1} with a single listing operation. The same magnitude was reported by Fryrear (1969). Soil roughening is most effective when performed perpendicular to the erosive wind and before the soils begin to erode (Woodruff et al., 1957). Timing is critical, and may be more important than the implement used to roughen the soil (Chepil et al., 1952; Woodruff and Chepil, 1956).

Since tillage can be effective in controlling wind erosion, methods of describing soil roughness (Zobeck and Onstad, 1987; Potter and Zobeck, 1990), tillage effects (Zobeck and Popham, 1992; Saleh et al., 1997) and methods of measuring surface roughness (Zobeck and Onstad, 1987; Saleh, 1993) have been developed.

23.5.3 Wind Barriers/Shelterbelts

Hedges (McDonald, 1971), trees (Woodruff and Zingg, 1955; Woodruff, 1956a), and annual crops (Fryrear, 1963) have been used as barriers to reduce wind speeds to the lee of the barrier. In general wind barriers protect the soil surface 10 times the height of the barrier. However, the precise length of protection offered by a barrier depends on wind speed, barrier density, and surface condition. Field research is underway to develop a wind barrier model that relates leeward wind reduction and the shape and extent of the protected zone to wind velocity and barrier density (J.D. Bilbro, Personal communication). The efficiency of a shelterbelt to control shifting sand is governed by the height, width, and porosity of the trees (Draz and Zaghioul, 2007). Bilbro and Stout (1999) studied the effect of uniform plastic pipe barriers on wind patterns in the field, and found that a barrier with 12.5% optical density would reduce downwind velocities over a 30 m distance by 4.3%. Increasing the number of pipes to achieve 75% optical density reduced wind velocities over the same field by 32.5%. Multiple rows of "pipe barriers" were also tested and equations developed to establish relationships among wind barrier conformation, barrier density, number of rows, and wind velocities. Wind tunnel studies of models of multi-row wind barriers showed that the optical density of the first row on the windward side of the barrier system was most important in defining the effectiveness of the barrier system (Fryrear et al., 2000), but normally the protected zone for the barrier is measured from the leeward row. Wind barriers can be very effective in controlling wind erosion in higher rainfall regions, but in limited rainfall areas or areas without winter snowfall, perennial barriers extract essentially all the soil water for a distance of three times the barrier height on both sides of the barrier (Bilbro and Stout, 1999).

23.5.4 Soil Amendments

Economical control of wind erosion usually requires the use of the available soil and plant resources. In many regions wind erosion is closely associated with drought conditions. During prolonged droughts plant residue levels decline and soil roughness techniques are ineffective. If the drought continues over multiple years, annual crop barriers may not grow and trees may die. The options available for reducing wind erosion become very limited. To provide one additional option, research has been and is being conducted on soil amendments (Chepil, 1954b, 1955; Chepil et al., 1963b; Lyles et al., 1969; Armbrust and Dickerson, 1971). The problems include cost of the amendment; excessive quantities needed, and limited effectiveness.

23.6 Conclusions

Wind erosion is one of the basic geological processes that shape and mold the landscape. Deep deposits of rich loess soil are evidence that wind erosion and deposition have played a major role in modifying the earth's surface. These loess soils are vital agricultural regions in modern agriculture.

Research on wind erosion began in the 1700s and continues today. This research is aimed at understanding the mechanics of wind erosion, measuring the products of erosion, modeling the processes, and understanding interactions. Field measuring methods have been developed to describe plant residues and soil roughness conditions.

Modeling wind erosion in the United States began with WEQ. New models (WEAM, TEAM, RWEQ, WEPS, and WEELS) are much more flexible and reflect new science not available in WEQ. These models represent the current knowledge of wind erosion mechanics and provide the cornerstone for evaluating wind erosion control systems. These systems take advantage of the available soil, plant, management, and weather resources. When used to plan management systems, the wind erosion model must make accurate estimates for extended periods of time and not just for single events. Therefore, it is imperative that models be field tested and validated over many months or multiple years. Also, when a model is used beyond its original scope the user must be aware of the mechanics and limitations of the model.

Field measuring techniques have progressed to the point where it is possible to detect the moment soil erosion begins and mass is being transported. With these field samples the vertical and horizontal distribution of eroded material within an eroding field can be described. These field data permit the verification of estimated erosion.

The new models permit designing systems that effectively utilize the available resources and efficiently control wind erosion. This does not mean that erosion will no longer be a problem, because wind erosion is a basic geomorphologic process.

References

Alfaro, S.C., A. Gaudichet, L. Gomes, and M. Maille. 1998. Mineral aerosol production by wind erosion: Aerosol particle sizes and binding energies. Geophys. Res. Lett. 25(7):991–994.

Alfaro, S.C., and L. Gomes. 2001. Modeling aerosol production by wind erosion: Emission intensities and aerosol size distributions in source areas. J. Geophys. Res. 106:18075–18084.

Anderson, R.S., M. Sorensen, and B.B. Willetts. 1991. A review of recent progress in our understanding of the aeolian sediment transport. Acta Mechanica [Suppl] 1:1–19.

Armbrust, D.V. 1972. Recovery and nutrient content of sandblasted soybean seedlings. Agron. J. 64:707–708.

Armbrust, D.V. 1977. Chemical tests to evaluate plant sandblast damage. Agron. J. 69:523–525.

Armbrust, D.V. 1979. Wind- and sandblast-damage to tobacco plants at various growth stages. Tobacco Sci. XXIII:117–119; Tobacco Int. 181:63–65.

Armbrust, D.V. 1982. Physiological responses to wind and sandblast damage by grain sorghum plants. Agron. J. 74: 133–135.

Armbrust, D.V. 1984. Wind and sandblast injury to field crops: Effect of plant age. Agron. J. 76:991–993.

Armbrust, D.V., and J.D. Bilbro. 1997. Relating plant canopy characteristics to soil transport capacity by wind. Agron. J. 89:157–162.

Armbrust, D.V., and J.D. Dickerson. 1971. Temporary wind erosion control: Cost and effectiveness of 34 commercial materials. J. Soil Water Cons. 26:154–157.

Armbrust, D.V., J.D. Dickerson, and J.K. Greig. 1969. Effect of soil moisture on the recovery of sandblasted tomato seedlings. J. Am. Soc. Hort. Sci. 94:214–217.

Armbrust, D.V., G.M. Pawlsen, and R. Ellis. 1974. Physiological responses to wind- and sand blast-damaged winter wheat plants. Agron. J. 66:421–423.

Bagnold, R.A. 1941. The physics of blown sand and desert dunes. Methuen, London, U.K.

Baker, J.T. 2007. Cotton seedling abrasion and recovery from wind blown sand. Agron. J. 99:556–561.

Batt, R.G., and S.A. Peabody. 1994. Dust sweep-up experiments. Defense Nuclear Agency, Alexandria, VA, Technical Report No. 94.

Belnap, J., S.L. Phillips, J.E. Herrick, and J.R. Johnansen. 2006. Wind erodibility of soils at Fort Irwin, California (Mojave Desert), USA, before and after trampling disturbance: Implications for land management. Earth Surf. Process. Landforms 32:75–84.

Berkofsky, L., and I. McEwan. 1994. The prediction of dust erosion by wind: An interactive model. Boundary Layer Meteorol. 67:385–406.

Betzer, P.R., K.L. Carder, R.A. Duce, J.T. Merrill, N.W. Tindale, M. Uematsu, D.K. Costello, R.W. Young, R.A. Feely, J.A. Breland, R.E. Bernstein, and A.M. Greco. 1988. Long-range transport of giant mineral aerosol particles. Nature 336(6199):568–571.

Bilbro, J.D. 1989. Evaluation of sixteen fall-seeded cultivars for controlling wind erosion. J. Soil Water Conserv. 44:228–231.

Bilbro, J.D. 1991. Relationships of cotton dry matter production and plant structural characteristics for wind erosion modeling. J. Soil Water Conserv. 46:381–384.

Bilbro, J.D. 1992. Sunflower dry matter production and plant structural relationships for wind erosion modeling. J. Soil Water Conserv. 47:194–197.

Bilbro, J.D., and D.W. Fryrear. 1994. Wind erosion losses as related to plant silhouette and soil cover. Agron. J. 86:550–553.

Bilbro, J.D., E.L. Harris, and O.R. Jones. 1994. Erosion control with sparse residue, p. 30–32. USDA-ARS, Conservation Research Report No. 37.

Bilbro, J.D., and J.E. Stout. 1999. Wind velocity patterns as modified by plastic pipe windbarriers. J. Soil Water Conserv. 54:551–556.

Bohner, J., W. Scafer, O. Conrad, J. Gross, and A. Ringeler. 2003. The WEELS model: Methods, results and limitations. Catena 52:289–308.

Bondy, E., L. Lyles, and W.A. Hayes. 1980. Computing soil erosion by periods using wind-energy distribution. J. Soil Water Conserv. 35:173–176.

Chepil, W.S. 1941. Relation of wind erosion to the dry aggregate structure of a soil. Sci. Agric. 21:488–507.

Chepil, W.S. 1954a. Factors that influence clod structure and erodibility of soil by wind III. Calcium carbonate and decomposed organic matter. Soil Sci. 77:473–480.

Chepil, W.S. 1954b. The effect of synthetic conditioners on some phases of soil structure and erodibility by wind. Soil Sci. Soc. Am. Proc. 18:386–391.

Chepil, W.S. 1955. Effects of asphalt on some phases of soil structure and erodibility by wind. Soil Sci. Soc. Am. Proc. 19:125–128.

Chepil, W.S. 1957. Width of field strips to control wind erosion. Kansas State Agricultural Experiment Station Technical Bulletin No. 92.

Chepil, W.S. 1958. Soil conditions that influence wind erosion. USDA-ARS Technical Bulletin No. 1185. U.S. Government Printing Office, Washington, DC.

Chepil, W.S. 1962. A compact rotary sieve and the importance of dry sieving in physical soil analysis. Soil Sci. Soc. Am. Proc. 26:4–6.

Chepil, W.S., E. Burnett, and F.L. Duley. 1963a. Management of sandy soils in the Central United States. USDA Farmers' Bulletin No. 2195, November 1963.

Chepil, W.S., C.L. Englehorn, and A.W. Zingg. 1952. The effects of cultivation on erodibility of soils by wind. Soil Sci. Soc. Am. Proc. 16:19–21.

Chepil, W.S., and R.A. Milne. 1939. Comparative study of soil drifting in the field and in a wind tunnel. Sci. Agric. 19:249–257.

Chepil, W.S., and R.A. Milne. 1941. Wind erosion of soil in relation to roughness of surface. Soil Sci. 52:417–433.

Chepil, W.S., and N.P. Woodruff. 1954. Estimations of wind erodibility of field surfaces. J. Soil Water Conserv. 9:257–265, 285.

Chepil, W.S., and N.P. Woodruff. 1955. How to reduce dust storms, p. 11. Kansas State Agricultural Experiment Station Circular No. 318.

Chepil, W.S., and N.P. Woodruff. 1963. The physics of wind erosion and its control. Adv. Agron. 15:211–301.

Chepil, W.S., N.P. Woodruff, F.H. Siddoway, D.W. Fryrear, and D.V. Armbrust. 1963b. Vegetative and non-vegetative materials to control wind and water erosion. Soil Sci. Soc. Am. Proc. 27:86–89.

Claflin, L.E., D.L. Stuteville, and D.V. Armbrust. 1973. Windblown soil in the epidemiology of bacterial leaf spot of alfalfa and common light of beans. Phytopathology 63:1417–1419.

Cole, G.W., L. Lyles, and L.J. Hagen. 1983. A simulation model of daily wind erosion soil loss. Trans. ASAE 26:1758–1765.

Dong, G., G. Zou, C. Li, and F. Cheng. 1983. Preliminary observation of the efficiency of the wind-preventing and sand-resisting forest belt in the western part of the Great Bend of the Huanghe River. J. Desert Res. 3:9–19.

Draz, M.Y., and A.K. Zaghioul. 2007. Impact of different designs of shelterbelt on the control of shifting sand at Toshka, South Egypt. J. Agric. Sci. Mansoura Univ. 32(2):845–856.

Englehorn, C.L., A.W. Zingg, and N.P. Woodruff. 1952. The effects of plant residue cover and clod structure on soil loss by wind. Soil Sci. Soc. Am. Proc. 16:19–21.

Feng, G., and B. Sharratt. 2007. Scaling from field to region for wind erosion prediction using the Wind Erosion Prediction System and geographical information systems. J. Soil Water Conserv. 62:321–328.

Fernandez, W. 1997. Martian dust storms: A review. Earth Moon Planets 77:19–46. doi:10.1023/A:1006134805135.

Findlater, P.A., D.J. Carter, and W.D. Scott. 1990. A model to predict the effects of prostrate ground cover on wind erosion. Aust. J. Soil Res. 28:609–622.

Fryberger, S.G., T.S. Ahlbrandt, and S. Andrews. 1979. Origin, sedimentary features, and significance of low-angle eolian sand sheet deposits, Great Sand Dunes National Monument and Vicinity, Colorado. J. Sediment. Petrol. 49:733–746.

Fryrear, D.W. 1963. Annual crops as wind barriers. Trans. ASAE 6:340–342, 352.

Fryrear, D.W. 1969. Reducing wind erosion in the Southern Great Plains. Texas Agricultural Experimental Station Miscellaneous Publication No. 929.

Fryrear, D.W. 1971. Survival and growth of cotton plants damaged by windblown sand. Agron. J. 63:638–642.

Fryrear, D.W. 1981. Dust storms in the Southern Great Plains. Trans. ASAE 24:991–994.

Fryrear, D.W. 1984. Soil ridges—Clods and wind erosion. Trans. ASAE 27(2):445–448.

Fryrear, D.W. 1985. Soil cover and wind erosion. Trans. ASAE 28:781–784.

Fryrear, D.W. 1986. A field dust sampler. J. Soil Water Conserv. 41:117–120.

Fryrear, D.W., and D.V. Armbrust. 1969. Cotton gin trash for wind erosion control. Texas Agricultural Experimental Station Miscellaneous Publication No. 928.

Fryrear, D.W., D.V. Armbrust, and J.D. Downes. 1975. Plant response to wind erosion damage. Proc. 30th Annual Meeting SCSA, August 10–13, 1974. San Antonio, TX.

Fryrear, D.W., J.D. Bilbro, C.E. Yates, and E.G. Berry. 2000. Modeling multirow wind barrier density. J. Soil Water Conserv. 55:385–392.

Fryrear, D.W., and J.D. Downes. 1975. Consider the plant in planning wind erosion control systems. Trans. ASAE 18:1070–1072.

Fryrear, D.W., C.A. Krammes, D.L. Williamson, and T.M. Zobeck. 1994. Computing the wind erodible fraction of soils. J. Soil Water Conserv. 49:183–188.

Fryrear, D.W., and A. Saleh. 1993. Field wind erosion: Vertical distribution. Soil Sci. 155:294–300.

Fryrear, D.W., and A. Saleh. 1996. Wind erosion: Field length. Soil Sci. 161:398–404.

Fryrear, D.W., J.E. Stout, L.J. Hagen, and E.D. Vories. 1991. Wind erosion: Field measurement and analysis. Trans. ASAE 34:155–160.

Fryrear, D.W., P.L. Sutherland, G. Davis, G. Hardee, and M. Dollar. 2001. Wind erosion estimates with RWEQ and WEQ, p. 760–765. In D.E. Stott, R.H. Mohtar, and G.C. Steinhardt (eds.) Sustaining the global farm. Selected papers from the 10th International Soil Conservation Organization Meeting held May 24–29, 1999 at Purdue University and the USDA–ARS National Soil Erosion Research Laboratory May 24–29, 1999, Purdue University.

Fryrear, D.W., M.M. Wassif, S.F. Tadrus, and A.A. Ali. 2008. Dust measurements in the Egyptian Northwest Coastal Zone. Trans. Am. Soc. Agric. Biol. Eng. 51:1–8.

Gibbons, R.P., J.M. Tromble, J.T. Hennessy, and M. Gardenas. 1983. Soil movement in mesquite dunelands and former grasslands of Southern New Mexico from 1939 to 1980. J. Range Manage. 36:145–148.

Gillette, D.A., and P.H. Stockton. 1986. Mass, momentum, and kinetic energy fluxes of saltating particles, p. 88–103. In W.G. Nickling (ed.) Aeolian geomorphology. Binghamton Symposium. Allen and Unwin, Boston, MA.

Gomes, L., J.L. Arrue, M.V. Lopez, G. Sterk, D. Richard, R. Garcia, M. Sabre, A. Gaudichet, and J.P. Frangi. 2003. Wind erosion in a semiarid agricultural area of Spain: The WELSONS project. Catena 52:235–256.

Goossens, D. 2004. Wind erosion and tillage as a dust production mechanism in North European farmland, p. 15–40. In D. Goosens and M. Riksen (eds.) Wind erosion and dust dynamics: Observations, simulations, modeling. ESW Publications, Wageningen, the Netherlands.

Goossens, D., and J. Gross. 2002. Similarities and dissimilarities between the dynamics of sand and dust during wind erosion of loamy sandy soil. Catena 47:269–289.

Goossens, D., Z. Offer, and G. London. 2000. Wind tunnel and field calibration of five aeolian sand traps. Geomorphology 35:233–252.

Goudie, A.S. 2009. Dust storms: Recent developments. J. Environ. Manage. 90:89–94.

Greeley, R., and J.D. Iverson. 1985. Wind as a geological process on Earth, Mars, Venus, and Titan. Cambridge University Press, New York.

Gregory, J.M., J. Borrelli, and C.B. Fedler. 1988. TEAM: Texas erosion analysis model, p. 88–103. In Proc. 1988 Wind Erosion Conference. Texas Tech University, Lubbock, TX.

Gupta, J.P., R.K. Aggarwal, and N.P. Raikhy. 1981. Soil erosion by wind from bare sandy plains in western Rajasthan, India. Arid Environ. 4:15–20.

Hagen, L.J. 1984. Soil aggregate abrasion by impacting sand and soil particles. Trans. ASAE 27:805–808, 816.

Hagen, L.J. 1991. A Wind Erosion Prediction System to meet user needs. J. Soil Water Conserv. 46:106–111.

Hagen, L. 2004. Evaluation of the Wind Erosion Prediction System (WEPS) erosion submodel on cropland fields. Environ. Modell. Softw. 19:171–176.

Hagen, L.J. 2008. Updating soil surface conditions during wind erosion events using the Wind Erosion Prediction System (WEPS). Trans. ASABE 51(1):129–137.

Hagen, L.J., L.E. Wagner, and E.L. Skidmore. 1999. Analytical solutions and sensitivity analysis for sediment transport in WEPS. Trans. ASAE 42(6):1715–1721.

Hagen, L.J., L.E. Wagner, and J. Tatarko. 1995. Wind Erosion Prediction System (WEPS). BETA Release 95-08. USDA-ARS, Manhattan, KS.

Hess, S.L. 1973. Martian winds and dust clouds. Planet. Space Sci. 21:1549–1557. doi:10.1016/0032-0633(73)90061-5.

Holy, M. 1980. Environmental sciences and applications. Vol. 9: Erosion and environment. Pergamon Press, New York.

Horning, L.B., L.D. Stetler, and K.E. Saxton. 1996. Surface residue and roughness for wind erosion protection. ASAE Paper 96-2116. American Society of Agricultural Engineering, Phoenix, AZ.

Huang, N., F. Shi, and R.S. Van Pelt. 2008a. The effect of slope and slope position on local and upstream fluid threshold friction velocities. Earth Surf. Process. Landforms 33:1814–1823.

Huang, N., G. Yue, and X. Zheng. 2008b. Numerical simulation of a dust devil and the electric field in it. J. Geophys. Res. 113:D20203. doi:10.1029/2008JD10182.

Huang, N., Y. Zhang, and R. D'Adamo. 2007. A model of the trajectories and midair collision probabilities of sand particles in a steady state saltation cloud. J. Geophys. Res. 112:1–11.

Janssen, W., and G. Tetzlaff. 1991. Entwicklung und Eichung einer registrierenden suspensionsfalle. Z. Kulturtech. Landesentwickl. 32:167–180.

Kjelgaard, J., D. Chandler, and K. Saxton. 2002. Evidence for direct suspension of loessial soils, p. 38. In T.M. Zobeck (ed.) Proceedings of ICARS5/GCTE-SEN Joint Conference. USA Publication 02-2. International Center for Arid and semi-arid Lands Studies, Texas Tech University, Lubbock, TX.

Kuntze, H., R. Beinhouer, and G. Tetlaff. 2000. Wind tunnel and field calibration of five aeolian sand traps. Geomorphology 35:233–252.

Leatherman, S.P. 1978. A new aeolian sand trap design. Sedimentology 25:303–306.

Liu, X., Y. Lin, D. He, and F. Chen. 1983. A study on the experiment of the under-leading fence in wind tunnel. J. Desert Res. 3:25–34.

Lyles, L., D.V. Armbrust, J.D. Dickerson, and N.P. Woodruff. 1969. Spray-on adhesives for temporary wind erosion control. J. Soil Water Conserv. 24:190–193.

Marticorena, B., and G. Bergametti. 1995. Modeling the atmospheric dust cycle. J. Geophys. Res. 100:16415–16430.

McDonald, A. 1971. Early American soil conservationists. USDA Miscellaneous Publication No. 449.

McEwan, I.K., and B.B. Willetts. 1993. Sand transport by wind: A review of the current conceptual model, p. 7–16. In K. Pye (ed.) The dynamics and environmental context of aeolian sedimentary systems. Geological Society of London, Special Publication No. 72.

Merva, K.R., and G. Peterson. 1983. Wind erosion sampling in the North Central Region. ASAE Paper 83-2133. American Society of Agricultural Engineering, Phoenix, AZ.

Middleton, N.J. 1987. Desertification and wind erosion in the western Sahel: The example of Mauritania. School of Geography, University of Oxford, Research Paper 40.

Nickling, W.G., and M. Ecclestone. 1981. The effects of soluble salts on the threshold shear velocity of fine sand. Sedimentology 28:505–510.

Owens, P.R. 1964. Saltation of uniform grains in air. J. Fluid Mech. 20:225–242.

Pollet, I. 1995. Meten van windsnelheden en zandtransport in een wind tunnel. MSc thesis. Universiteit Gent. Ghent, Belgium.

Potter, K.N., and T.M. Zobeck. 1990. Estimation of soil microrelief. Trans. ASAE 33:156–161.

Rajot, J.L., S.C. Alfaro, L. Gomes, and A. Gaudichet. 2003. Soil crusting on sandy soils and its influence on wind erosion. Catena 53:1–16.

Reheis, M.C. 2006. A 16-year record of eolian dust in Southern Nevada and California, USA: Controls on dust generation and accumulation. J. Arid Environ. 67:487–520.

Reynolds, R.S., M. Reheis, J. Yount, and P. Lamothe. 2006. Composition of aeolian dust in natural traps on isolated surfaces of the central Mohave Desert–Insights to mixing, sources, and nutrient inputs. J. Arid Environ. 66:42–61.

Riksen, M.J.P.M., and D. Goossens. 2005. Tillage techniques to reactivate aeolian erosion on inland drift-sand. Soil Till. Res. 83:218–236.

Saleh, A. 1993. Soil roughness measurement: Chain method. J. Soil Water Conserv. 48:527–529.

Saleh, A. 1994. Measuring and predicting ridge-orientation effect on soil surface roughness. Soil Sci. Soc. Am. J. 58:1228–1230.

Saleh, A., and D.W. Fryrear. 1995. Threshold wind velocities of wet soils as affected by windblown sand. Soil Sci. 160:304–309.

Saleh, A., D.W. Fryrear, and J.D. Bilbro. 1997. Aerodynamic roughness prediction from soil surface roughness measurements. Soil Sci. 162:205–210.

Sapkota, A., A.K. Cheburkin, G. Bonani, and W. Shotyk. 2007. Six millennia of atmospheric dust deposition in southern South America (Isla Navarino, Chile). The Holocene 17:561–572.

Saxton, K.E. 1995. Wind erosion and its impact on offsite air quality in the Columbia Plateau: An integrated research plan. Trans. ASAE 38:1031–1038.

Saxton, K.E. 1996. Agricultural wind erosion and air quality impacts: A comprehensive research program. J. Altern. Agric. 11:64–70.

Shao, Y., G.H. McTainsh, J.F. Leys, and M.R. Raupach. 1993. Efficiencies of sediment samplers for wind erosion measurement. Aust. J. Soil Res. 31:519–532.

Shao, Y., M.R. Raupach, and J.F. Leys. 1996. A model for predicting aeolian sand drift and dust entrainment on scales from paddock to regions. Aust. J. Soil Res. 34:309–342.

Shi, F., N. Huang, and X. Zheng. 2008. Some investigations into near surface wind and saltation intensity in Minqin Area. Proc. of the 12th Asian Congress of Fluid Mechanics, August 18–21. Daejeon, Korea.

Siddoway, F.H., W.S. Chepil, and D.V. Armbrust. 1965. Effect of kind, amount, and placement of residues on wind erosion control. Trans. ASAE 8:327–331.

Skidmore, E.L. 1966. Wind and sandblast injury to seedling green beans. Agron. J. 58:311–315.

Skidmore, E.L. 1986. Wind erosion climatic erosivity. Clim. Change 9:195–208.

Skidmore, E.L. 2007. History of wind erosion prediction technologies [abstract]. ASA-CSSA-SSSA Annual Meeting Abstracts. 2007 CDROM.

Skidmore, E.L., L.L. Nossaman, and N.P. Woodruff. 1966. Wind erosion as influenced by row spacing, row direction, and grain sorghum population. Soil Sci. Soc. Am. Proc. 30:505–509.

Song, Z., J. Wang, and S. Wang. 2007. Quantitative classification of northeast Asian dust events. J. Geophys. Res. 112:D04211. doi:1029/2006JD007048.

Sorensen, M. 1991. An analytic model of wind-blown sand transport. Acta Mech. [Suppl] 1:67–81.

Spaan, W.P., and G.D. van den Abele. 1991. Wind borne particle measurements with acoustic sensors. Soil Technol. 4:51–63.

Steiner, J.L., H.M. Schomberg, C.L. Douglas, and A.L. Black. 1994. Standing stem persistence in no-tillage small-grain fields. Agron. J. 86:76–81.

Stetler, L.D., and K.E. Saxton. 1995. Fugitive dust (PM_{10}) emissions and eroded soil relations from agricultural fields on the Columbia Plateau. Particulate matter: Health and regulatory issues. Proc. AWMA International Conference on Particulate Matter. April 4–6, 1995. Pittsburgh, PA. Air Waste Management Association, Pittsburgh, PA.

Stout, J.E. 1990. Wind erosion within a simple field. Trans. ASAE 33:1597–1600.

Stout, J.E., and D.W. Fryrear. 1989. Performance of a windblown particle sampler. Trans. ASAE 32:2041–2045.

Stout, J.E., A. Warren, and T.E. Gill. 2008. Publications trends in aeolian research: An analysis of the bibliography of Aeolian research. Geomorphology 1–12. doi:10.1016/j.geomorph.2008.02.015

Strzelecki, P.E. De. 1845. Physical description of New South Wales and Van Diemen's Land. Longman, Brown, Green, and Longmans, London, U.K.

Taylor, D.W. 1960. Fundamentals of soil mechanics. 11th edn. John Wiley & Sons, New York.

Tegen, I., M. Werner, S.P. Harrison, and K.E. Kohfeld. 2004. Relative importance of climate and land use in determining present and future global soil dust emission. Geophys. Res. Lett. 31, L 05105, doi:10.1029/2003GL019216, 2004.

Thornbury, W.D. 1957. Principles of geomorphology. John Wiley & Sons, New York.

Tsai, Chuen-Jinn, and Yu-Hisang Cheng. 1996. Comparison of two ambient beta gauge PM_{10} samplers. J. Air Waste Manage. Assoc. 46:142–147.

van de Ven, T.A.M., D.W. Fryrear, and W.P. Spaan. 1989. Vegetation characteristics and soil loss by wind. J. Soil Water Conserv. 44:347–349.

Van Pelt, R.S., and T.M. Zobeck. 2004. Validation of the Wind Erosion Equation (WEQ) for discrete periods. Environ. Modell. Soft. 19:199–203.

Vories, E.D., and D.W. Fryrear. 1991. Vertical distribution of wind-eroded soil over a smooth, bare field. Trans. ASAE 34:1763–1768.

Wassif, M.M., S.F.T. Sharkawy, M.R. Bayoumi, A.Y. Genead, and S.K. Atta. 1999. Wind erosion as related to some soil conservation practices in the Northwest Coastal Zone, Egypt. Proceedings Sixth International Conference on the Development of dry lands, ICARDA.

Wendelin, G., 1646. Pluvia purpurea Bruxellenmsis. Apud Lodovicum de Heuqueville, Parislis.

Williams, M.A.J., and R.C. Balling. 1996. Interactions of desertification and climate. Arnold Publishers, London, U.K.

Williams, J.R., C.A. Jones, and P.T. Dyke. 1984. A modeling approach to determining the relationship between erosion and soil productivity. Trans. ASAE 27:129–144.

Wilson, S.J., and R.U. Cooke. 1980. Wind erosion, p. 217–252. In M.J. Kirkby and R.P.S. Morgan (eds.) Soil erosion. John Wiley & Sons, New York.

Wong, S., A.E. Dessler, N. Mahowald, P.R. Colarco, and A. da Silva. 2008. Long-term variability in Saharan dust transport and its link to North Atlantic sea temperatures. Geophys. Res. Lett. 35:L07812. doi:10.1029/2007GL32297.

Woodruff, N.P. 1956a. The spacing interval for supplemental shelterbelts. J. For. 54:115–122.

Woodruff, N.P. 1956b. Wind-blown soil abrasive injuries to winter wheat plants. Agron. J. 48:499–504.

Woodruff, N.P., and D.V. Armbrust. 1968. A monthly climatic factor for the wind erosion equation. J. Soil Water Conserv. 23:103–104.

Woodruff, N.P., and W.S. Chepil. 1956. Implements for wind erosion control. Agric. Eng. 37:751–754, 758.

Woodruff, N.P., W.S. Chepil, and R.D. Lynch. 1957. Emergency chiseling to control wind erosion. Kansas State Agricultural Experiment Station Technical Bulletin No. 90.

Woodruff, N.P., and F.H. Siddoway. 1965. A wind erosion equation. Soil Sci. Soc. Am. Proc. 29:602–608.

Woodruff, N.P., and A.W. Zingg. 1955. A comparative analysis of wind tunnel and atmospheric air flow patterns about single and successive barriers. Trans. Am. Geophys. Union 36:203–208.

Wu, L. 2007. Impact of Saharan air layer on hurricane peak intensity. Geophys. Res. Lett. 34:L09802. doi:10.1029/2007GL029564.

Xu, J., Z. Pei, and R. Wang. 1982. A research on the width of the protection belt of half-hidden straw checkerboard barriers. J. Desert Res. 2:16–22.

Yang, B., A. Brauning, Z. Zhang, Z. Dong, and J. Epser. 2007. Dust storm frequency and its relation to climatic changes in Northern China during the past 1000 years. Atmos. Environ. 41:9288–9299.

Zhang, X.Y., S.L. Gong, T.L. Zhao, R. Arimoto, Y.Q. Wang, and Z.J. Zhou. 2003. Sources of Asian dust and role of climate change versus desertification in Asian dust emission. Geophys. Res. Lett. 30.

Zheng, X., N. Huang, and Y. Zhou. 2003. Laboratory measurements of electrification of wind-blown sands and simulation on its effect on sand saltation movement. J. Geophys. Res. 108(D10):4322. doi:10.1029/2002JD002572,2003.

Zobeck, T.M. 1991. Abrasion of crusted soils: Influence of abrader flux and soil properties. Soil Sci. Soc. Am. J. 55:1091–1097.

Zobeck, T.M., A.D. Halverson, B. Wienhold, V. Acosta-Martinez, and D.L. Karlen. 2008. Comparison of two soil quality indexes to evaluate cropping systems in Northern Colorado. J. Soil Water Conserv. 63:329–338.

Zobeck, T.M., and C.A. Onstad. 1987. Tillage and rainfall effects on random roughness: A review. Soil Till. Res. 9:1–20.

Zobeck, T.M., and T.W. Popham. 1992. Influence of microrelief, aggregate size and precipitation on soil crust properties. Trans. ASAE 35:487–492.

Zobeck, T.M., J.E. Stout, T.E. Gill, and D.W. Fryrear. 1996. Airborne dust and sediment measurements in agricultural fields. Int. Conf. on Air Pollution from Agric. Operations, Kansas City, MO, February 7–9, 1996. Midwest Plan Service, Iowa State University, Ames, IA.

24
Land Application of Wastes

David M. Miller
University of Arkansas

W.P. Miller
University of Georgia

24.1	A Brief History and Overview of Land Application 24-1
	Historical Perspective • Types of By-Products Considered • Rationale for Land Application
24.2	Properties of Common Land-Applied By-Products 24-3
	Agricultural Materials • Forestry Products • Industrial By-Products • Biosolids and Municipal Solid Waste • Composting and Mixing of By-Products
24.3	Soil Fertility Considerations in Land Application 24-10
	Nutrient Content and Bioavailability
24.4	Effects of Land Application on Soil Properties 24-11
	Physical Properties • Biological Properties • Chemical Properties
24.5	Environmental Aspects of By-Product Applications 24-14
	Governmental Regulations Concerning Land Application • Soil and Food-Chain Contamination • Water Quality Considerations
24.6	Recommendations 24-19
	References 24-19

24.1 A Brief History and Overview of Land Application

Before the introduction and widespread use of commercial fertilizers in the early twentieth century, application of manures and other by-products to soils was an important way of maintaining soil fertility and productivity. These applications were made based on the empirical observation that plant growth was stimulated by such applications. The importance of these materials in agricultural production declined during the twentieth century in response to the lower cost of commercial (inorganic) fertilizer per mass of nutrient applied and the concentration of animals in confined animal feeding operations (CAFOs) removed from the farm. The term "waste" was applied to manures and other materials, since these materials were considered to have a negative (i.e., disposal) value in lieu of their use as soil amendments. However, increasing energy costs have recently caused dramatic increases in the cost of commercial fertilizers, especially N fertilizers. In addition, global reserves of phosphate rock are dwindling. At current rates of extraction, known United States reserves will be depleted in 40 years and known global reserves in about 90 years (Vaccari, 2009). This, coupled with current environmental regulations regarding release of nutrients into the environment and the rising costs associated with construction of landfills, has revived interest in land application of waste materials to crop and forest lands as a way of avoiding disposal costs while recycling nutrients to soils. The organic material present in many of these by-products also has the potential to build humus levels in soils and thereby improve long-term productivity.

A wide range of materials derived from both agricultural and industrial activities have been proposed for land application. From the point of view of the producer of these materials, they are wastes if they have no or negative value. If they can be shown to have positive growth effects on cultivated plants that justify their land application, and do not have adverse environmental impacts with such a use, a better term may be by-products rather than wastes (Powers, 1998). The focus of current research is to provide guidelines for the environmentally safe use of such materials in agriculture as substitutes for commercial fertilizers and to avoid the cost of landfilling of a potential resource.

24.1.1 Historical Perspective

The oldest stable agricultural systems known, in both the West and East, employed additions of organic matter, primarily animal manures and nightsoil (human excrement), to soil to maintain productivity. Asian agriculture dating from 2500 BCE employed these materials extensively. At about the same time, Greek and later Roman agriculture used manures as well, but did not find human waste as suitable, perhaps due to its tendency to cause disease (Tisdale et al., 1993).

During the Middle Ages, agriculture appeared to follow the Roman model, and continued use of organic materials (manure, food waste, and perhaps human excrement) over time created distinctive topsoil materials high in carbon (C) and extractable nutrients (plaggens). Similar soil horizons have been found in Amazonia at about the same time associated with village sites of long habitation, presumably enriched with food processing wastes and charcoal added intentionally (or not) to garden plots near habitations.

Industrialization in the West during the eighteenth and nineteenth centuries brought many changes to agriculture, which had been operating in Europe and the United States with only modest change since Rome. The emphasis on tillage in order to allow plant uptake of humus continued to encourage manure additions, but Liebig's clarification of the role of soil-derived nutrients in the 1860s opened the way to widespread use of fertilizers for their nutrient, rather than their humus content.

Prior to the development of industrial processes to manufacture inorganic fertilizers in the later nineteenth century, imported guano (sea bird manure) from Chile and the Pacific islands was a major fertilizer used in row crop production in Europe, and particularly in the southern United States for cotton production as an N source and a P source (Wines, 1985). Once the Haber process was developed to manufacture ammoniacal salts on an industrial scale, and phosphatic mineral deposits were discovered and mined along the eastern seaboard and Florida, the rapidly depleting guano reserves were abandoned.

During the period 1920–1940, inexpensive supplies of inorganic nutrients coupled with other changes greatly increased yields of crops and encouraged specialization in farming, which often meant separation of row crop and animal production. In confined production facilities, manures became wastes, and an increasing disposal problem arose. While much of the focus in the 1970s was on point source pollution (i.e., on industrial and municipal discharges), by the 1990s it had become clear that stockpiled or overapplied manures were a major contributor to nonpoint source N and P pollution of surface and groundwater.

The environmental movement of the 1970s initiated the modern emphasis on environmental quality and recycling, which, coupled with economic constraints on producers of a variety of by-product materials, has led to a new focus on utilization of a wide range of these materials on agricultural lands.

24.1.2 Types of By-Products Considered

Animal manures, largely from cattle, poultry, and swine production, still make up the bulk of by-product materials applied to soils. The increasing centralization and industrialization of animal production has led to problems with distribution of such manures back to cropland, primarily due to high transport cost relative to cost of equivalent inorganic nutrients. Despite the long history of use, rates of application are currently being studied to determine water quality impacts and optimal application rates for crop production.

Domestic waste, the nightsoil of preindustrial days, is now handled in wastewater treatment plants. The treated solids are separated from the wastewater. When municipal sewage solids (sewage sludge) are treated and meet standards so that they are suitable for land application, they are known as biosolids. This material is increasingly land applied, either in dedicated land-treatment facilities or on farmers' fields. Governmental regulation is encouraging application to land through publication of guidelines for acceptable nutrient and heavy metal loadings, thereby avoiding landfilling of this material. Other domestic wastes include yard trimmings (grass clippings, chipped limbs, etc.), now commonly composted or mixed with other by-products instead of consuming landfill space.

Industrial by-products are increasingly being viewed as potential soil amendments due to their negative value to industries and their capacity to substitute for costly inorganic fertilizers. Some industrial by-products such as blast furnace slags, wood ash, and kiln dusts have been used locally as lime substitutes due to their alkaline nature; metal refining baghouse dusts are commonly used to formulate micronutrient fertilizers. Other materials such as coal fly ash, by-product gypsums, food processing, and pulp and paper manufacturing by-products are being considered for land application based on either alkalinity, nutrient content, and/or organic matter content. With these new materials, there is a greater uncertainty concerning possible environmental hazards from trace metal and organic contaminants, and this has prompted ongoing research.

The Resource Conservation and Recovery Act (RCRA) and the Clean Water Act (CWA) and subsequent USEPA rulings contain regulations governing the definitions of waste types, discharge or loading limits for pollutants, and management practices for land application of biosolids (Walker et al., 1997). In particular, hazardous waste is defined as materials that are corrosive, ignitable or explosive, or toxic, the latter based on the leaching potential of trace metals. While under certain circumstances materials defined as hazardous may be land applied, in general, these materials cannot be used as soil amendments.

24.1.3 Rationale for Land Application

Land application in the sense used here refers to addition of by-products to soil for the purposes of improving crop growth, either through addition of nutrients or improvement of soil physical or chemical properties, either in the short or long term. Materials used for land application should be studied and documented to be effective for use in agriculture while posing no threat to the environment. This is distinctly different from land treatment, which suggests that materials are being applied to land primarily in order to dispose of (treat) them. There is a long history of land treatment of waste materials such as biosolids, secondary municipal waste water, and industrial solids and waste waters on dedicated treatment facilities, which are managed and monitored for such uses. The design and operation of such facilities are typically under permit from governmental regulatory agencies, and are not discussed in this review.

The key to land application is the idea of beneficial reuse, which is the major criterion used by the USEPA and many state environmental agencies to consider materials for land application. Animal manures typically have not been closely scrutinized due to their long history of use and accepted value in agriculture, although increasing use of hormones and high metal levels as feed additives has raised some questions regarding environmental safety. Other materials, however, require testing to document the absence of significant levels of contaminants, and to demonstrate some positive effect on crop yield or potential soil productivity. This latter criterion may be addressed by demonstrating improvement in long-term productive potential by increasing soil water-holding capacity, improving soil structure or tilth, or by overall increase in nutrient levels that may not affect yield immediately.

The United States and the European Union (EU) have taken different regulatory approaches to determining if by-products can be land applied. The USEPA uses risk assessment to determine acceptable concentrations of potential contaminants. In 1993, USEPA promulgated 40 CFR Part 503 as the standards for land application of municipal biosolids. Part 503 specifies the degree of treatment for organic material stabilization and pathogen reduction as well as acceptable concentrations and annual and cumulative loadings for nine contaminant metals applied to crop land (USEPA, 1993a, 1993b). While this ruling has been applied to other types of by-products, it is not clear how the specifications apply to other types of wastes. Regarding animal manures, a growing emphasis on comprehensive nutrient management plans (CNMP) in the United States has had a significant impact on addressing issues of nutrient export from fields due to over- or mis-application.

Despite these successes, questions have increasingly been raised regarding the overall safety of land application programs, particularly the risk assessment approach used by EPA, both in its general outline and specifics of the data used in the models (Harrison et al., 1999; National Research Council, 2002). Public opposition to land application of by-products, particularly biosolids, has escalated over the past 10 years, evidenced by a citizen petition to ban biosolids application nationally in the United States (Center for Food Safety, 2003; the petition was denied by EPA) and numerous critical, even vitriolic web sites on the topic (displayed by any Internet search of "sewage sludge"). The range of criticisms includes regulatory incompetence, collusion of regulators with industry, and the failure of research to prove the safety of land-applied products. Biosolids are increasingly portrayed in the media as a "hazardous" or "toxic" waste, lumped with industrial by-products that have occasionally been misapplied to soils and resulted in serious contamination (e.g., Wilson, 1997). Using Beecher et al.'s (2005) terminology, the public's "perceived risk" with respect to biosolids has grown, regardless of the "actual risk." It is likely that EPA will need to account for such perceptions if land application programs are to be "sustainable" in the United States, a process that will involve engaging stakeholder groups in project planning, management, and in research (Bastian, 2005).

The EU has adopted much more stringent standards regarding land application of wastes than those of the United States. This may be due to Europeans' longer view of history and hence resource conservation (Harrison et al., 1999), or to an adherence to some form of the "precautionary principle," which states that scientific uncertainty shall not be used to justify action (or inaction) that might result in severe or irreversible environmental harm (Goldstein, 2004). This idea can be applied to many environmental questions where lack of "complete" scientific understanding limits our confidence in regulating the environment. The European ban on genetically modified crops and use of hormone treatment of food animals are examples, and illustrate that in the absence of "certainty" of safety, "public perception of the potential for harm is sufficient to ban a food or chemical with little if any need for scientific data" (Goldstein, 2004, p. 490). Such an approach is the antithesis of risk assessment, dose–response relationships, and modern environmental and toxicological science. The evolution of environmental regulation of land application in the United States will depend on the interplay of these different approaches.

24.2 Properties of Common Land-Applied By-Products

24.2.1 Agricultural Materials

Strictly speaking, the single most important type of agricultural by-product applied to soils is crop residues. However, crop residues are not often classified as a waste material because they are naturally returned to the soil during the course of tillage operations, and because it is universally acknowledged that returning residues to the soil enhances soil productivity. Therefore, crop residues will not be discussed further in this chapter.

By virtually any standard the most important agricultural by-products applied to soils are animal manures. Because of the amounts applied and their relatively high nutrient content, it could even be argued that agricultural animal manure is the single most important land-applied by-product of agricultural or nonagricultural origin. Huge quantities of agricultural animal manures are produced each year in the United States and throughout the developed countries of the world. In 1978, a USDA task force estimated that just over 0.5 billion Mg (fresh wt.) of manure, or 39% of all the agricultural animal manure generated in the United States, was produced in confined areas and was thus available for land application. Brady and Weil (1996) estimate that over 1 billion Mg (fresh wt.) of animal manure is now produced annually in confined livestock and poultry operations in the United States.

The geographic distribution of animal manure production in the United States, as well as total quantities produced, has changed markedly since the middle of this century. Prior to the 1940s, animals and crops were for the most part raised together on millions of small farms scattered throughout the world. In this precommercial fertilizer era, manure was a valuable resource. It was faithfully collected, stored, and applied to crop land at rates that seldom supplied nutrients in excess of crop needs, primarily because there was an appropriate balance between the number of animals produced and the amount of land available. With the advent of chemical fertilizers, manure was no longer seen as an economical source of plant nutrients, so small farmers divested themselves of their livestock. Corn, wheat, and sorghum that were produced inexpensively using the new chemical fertilizers became an economical source of feed grains for a new form of animal agriculture that was characterized by economy of scale, high density, confined production and centralized processing and distributing facilities. Higher land prices, increased cost of labor and other factors have prompted animal production facilities to locate in areas not well suited to row crop agriculture. The end result is a concentration of animals in specific regions of the country where limited areas of crop land cannot accommodate the quantities of manure produced. The situation is exemplified by the poultry industry in the southeastern United States (AR, AL, GA, NC, DE, MD) where intensive poultry production generates more manure than the local region's crop lands and pastures can accommodate.

The chemical and physical properties of the manures produced by today's animal agriculture vary widely. A property of particular importance as far as storage and handling of manure is concerned is water content (or, conversely, solids content), which is sometimes referred to as consistency. Manures containing less than 5% solids are considered liquids, while those containing between 5% and 12% solids are referred to as slurries; those containing between 12% and 20% solids are semisolid; and those containing more than 20% solids are solids. When they are first excreted, many agricultural animal manures are classed as semisolids. Liquid manures and manure slurries typically are produced when water is used to wash semisolid manures out of the animal rearing areas into pits or lagoons for storage. Left undisturbed in a lagoon, solids will settle out, creating a situation in which manure slurry is overlain by liquid manure. Solid manures are produced either by drying of slurries or semisolids, or by mixing of manures with a carbonaceous bedding material such as straw or wood chips. An example is chicken broiler litter, which is a physical mixture of poultry (broiler) manure and either rice hulls, wood chips or sawdust.

The chemical composition of animal manure is highly variable and depends on many factors including (1) type of animal, (2) nature and composition of feed, (3) nature and composition of added bedding material, (4) manure/bedding ratio, and (5) the manner in which the manure was collected, stored, and handled prior to land application (Sutton, 1994; Wright et al., 1998). Representative chemical composition data for a number of different types of animal manures are presented in Table 24.1. Poultry and swine manures generally have higher N and P contents than either beef or dairy manure. In addition, poultry litters tend to have lower N and P contents than poultry manures because of the diluting effect of the carbonaceous bedding materials present in litters. These data also illustrate the fact that when animal manures are applied to soil, significant quantities of nutrients other than N and P are added to the soil. For example, application of 4 Mg ha^{-1} (dry wt.) of swine manure having a composition similar to that shown in Table 24.1 would provide, in addition (kg ha^{-1}) to 252 N and 80 P, 100 K, 180 Ca, 40 Mg, 4.4 Fe, 1.56 Zn, and 0.6 Cu. It should also be noted that because of the high C content of manures, one of the major benefits of adding animal manures to soil is the resulting increase in soil organic matter (SOM), which exerts a beneficial influence on so many soil physical properties, such as water infiltration rate, aggregate stability, aeration and so on.

It should be emphasized that the data in Table 24.1, while representative of actual analyses, are for comparative purposes and should not be used to calculate actual manure application rates. The variability in the chemical makeup of manures is so high that the use of average data to calculate manure application rates can result in significant over or under application of nutrients (Mullins et al., 2005; Kleinman et al., 2005). In actual practice, a sample of the manure to be land applied should be analyzed to determine its exact chemical composition, which should be used in application rate calculations. Sims (1995) has discussed both the need for and problems associated with obtaining accurate chemical analyses of agricultural animal manures.

TABLE 24.1 Representative Chemical Composition of Animal Manures

Manure	Total N	P	K	Ca	Mg	Fe	Zn	Cu
	g kg^{-1}					mg kg^{-1}		
Poultry[a]	52	22	20	81	10	890	1032	155
Poultry[b]	40	16	23	23	5	1000	315	473
Swine[c]	63	20	25	45	10	1100	390	150
Beef[d]	19	7	20	13	7	5000	8	2
Dairy[e]	24	7	21	14	8	1800	165	30

[a] Layer manure (no bedding); after Edwards and Daniel (1992) and Sims and Wolf (1994).
[b] Broiler litter; after Edwards and Daniel (1992) and Brady and Weil (2008).
[c] After Choudhary et al. (1996) and Brady and Weil (2008).
[d] After Eghball and Power (1994).
[e] After Brady and Weil (2008).

24.2.2 Forestry Products

Processing of forest products generates large amounts of by-products, most of which are currently incinerated, landfilled or stockpiled near production facilities. The production of pulp for papermaking is the major regional source of by-products, with industries concentrated in the Southeast (AL, GA, SC, TN), New England (MA, NY, PA), and Pacific Northwest (WA, OR). Total by-product production in the United States was approximately 13 million Mg (dry wt.) in 1989, but has likely declined with the reduction in domestic pulp production since 2000. About 50% of total by-products is composed of organic-based sludges and the remainder of inorganic materials such as by-product lime and boiler ash (NCASI, 1997; Unwin, 1997).

During the pulping process, alkaline dissolution of lignin is employed to separate the cellulose for further processing. The solid by-products produced consist of inorganic residues from the wood and processing chemicals (grits, dregs), reject cellulose and wood knots, spent alkali used in regenerating the pulping chemicals (lime mud), and ash from boilers fired with wood by-product and/or fossil fuel. The reject cellulose taken directly from the pulping digestor is termed primary sludge; often this material is sluiced to a lagoon and mixed with other by-product streams (ash and alkali by-product), and may undergo some forced aerobic decomposition. This latter material is referred to as secondary or activated sludge. Pulp and paper mills use steam generated in boilers that typically burn wood by-product, bark, reject pulp and accessory fossil fuels, generating ash that is variable depending on the fuel source. A variety of proprietary processes are used in pulp and paper making, and individual plants differ substantially in their handling of by-product streams; some mills combine all by-product streams prior to disposal. Therefore, by-products are extremely heterogeneous and typically vary in composition within the same mill over time as operating conditions change.

Some general properties of the major classes of pulp and paper by-products are given in Table 24.2. Ranges are shown rather than average values, given the extreme variation in products between different mills. Both primary and secondary pulp sludges are largely composed of organic materials (reject wood and pulp), while the other classes of by-products are primarily inorganic. Nutrient contents are in general low, particularly for N, P, and K. Some ashes and the lime muds are quite alkaline ($pH_{H_2O} > 12$), and have good potential as agricultural lime substitutes. Calcium carbonate equivalences (CCE) for lime muds are often >90%, and in the range 10%–65% for primary sludges, which often contain free $CaCO_3$ (Morris et al., 1997).

Trace element contents, including regulated metals such as Hg, Cd, and As, are quite variable but typically quite low in most of these materials. Ash from boilers using nonwood fuel (coal, shredded tires, oil) are not reported here, but are often higher in trace elements such as As and Zn. Dioxin content of ash and pulp by-product from mills where Cl_2 process bleaching is used has been a subject of concern. Studies have shown significant levels of dioxin or furan compounds in some bleached kraft mills where Cl_2 bleaching is used, but this method is rapidly being replaced by other technologies (Camberato et al., 1997; Miner, 1997). Older stockpiled sludges in lagoons may locally contain higher levels of dioxins and furans. Except where salt laden (i.e., salt water-saturated) wood was used as a feedstock, boiler ash from pulp and paper mills is quite low in these compounds (Someshwar, 1996).

Land-applied forestry by-products are used principally as liming agents, fertilizers, soil conditioners, and mulches.

24.2.3 Industrial By-Products

There is obviously a large range of by-product materials produced by industries, some of which may have potential for use as soil amendments. Many have no appreciable nutrient, organic or alkali content that might benefit crop production, or have levels of toxic contaminants that rule out land application. Several classes of by-products described in this section have been used, or have been suggested for use on soils. Several types have been tested for crop response or agronomic efficacy or for potential environmental impacts, but certainly much work remains to be done in this area. The use of any of these materials is not recommended in the absence of careful chemical characterization and consultation with agricultural and environmental experts or state officials.

24.2.3.1 Slags

Iron and steel making slags are perhaps the best known industrial by-products historically used in agriculture. These materials are produced by addition of lime to furnaces to remove Si impurities, and are composed primarily of $CaSiO_3$. Blast furnace (Fe) slags have high CCE values (80%–100%), and have been used for many years locally as lime substitutes in the northeastern United States and Europe. Basic (steel) slags are produced from high P ores, and have a high P content (4%–8%) as well as considerable CCE (40%–60%; Tisdale et al., 1993). Electric furnace slags have intermediate CCE and P values.

In 1995, 21 million Mg of slag were produced in the United States, with about 60% being Fe slags and 40% steel or electric furnace slag (USGS, 1997). In the past, much of the basic slag was land applied as a P source (10 million Mg in 1974), but current use is low, with only 0.25 million Mg land applied worldwide in 1995. In the United States, production is centralized in the industrial Midwestern states (IL, IN, OH, PA), and competing

TABLE 24.2 Chemical Composition of Pulp and Paper Mill Waste Products

Element	Primary Sludge	Secon./Mixed Sludge	Wood Ash
	g kg^{-1} by wt.		
N	0.7–20 (3)[a]	6–90 (23)	0.3–3
P	0.1–4 (0.2)	0.4–17 (4)	0.6–16
K	0.1–10 (2)		23–162
Ca	0.3–210 (14)		211–466
Mg	0.2–19 (1.6)		25–117
	mg kg^{-1} dry wt.		
Cu	3.9–1590		3–210 (75)
Zn	13–3780 (188)		63–2200 (443)
Co	<0.1–9.7		0.5–20 (8.7)
Mo	<2–14		<0.1–114 (15)
As	<0.05–8.3 (1.2)		3–63 (23)
Pb	<0.05–880 (28)		22–220 (66)
Hg	0.001–3.5 (0.35)		<0.1–2.8 (0.4)
Cr	3–2250 (42)		3–130 (39)
Ni	1.3–133		<1–97 (24)
Cd	0.1–56 (1.2)		<0.01–21 (5)

Sources: Sludge data from NCASI. 1984. Land application and related utilization of pulp and paper mill sludges. NCASI Bulletin 439. NCASI, Research Triangle Park, NC; Morris, L.A., W.P. Miller, M.E. Sumner, and W.L. Nutter. 1997. Characterization and utilization of pulp mill wastes for forest land application. Proc. NCASI 1997 Southern Regional Mtg., Asheville, NC, June 1997. NCASI, Research Triangle Park, NC; Camberato, J.J., E.D. Vance, and A.V. Someshwar. 1997. Composition and land application of paper manufacturing residuals, p. 185–203. *In* J.E. Rechcigl and H. MacKinnon (eds.) Agricultural uses of byproducts and wastes. American Chemical Society, Washington, DC; Someshwar, A.V. 1996. Wood and combination wood-fired boiler ash characterization. J. Environ. Qual. 25:962–972.

[a] Mean values in parenthesis where available.

uses (cement production, lightweight aggregate) have increased prices to $30–$100 Mg^{-1} (USGS, 1997), largely pricing agricultural uses out of the available market.

No environmental studies dealing with contaminant (i.e., heavy metal) contents of slags or their fate in soils appear in the literature; it is likely that certain metals alloyed in steel (Ni, Cr, Mo) may be present in the slag and should be carefully monitored in any land application.

24.2.3.2 Fly Ash

Coal-burning electric-generating plants produce fly ash in all industrial countries. In the United States in 2001, utilities produced 57 million Mg of fly ash, as well as 18 million Mg of bottom ash and boiler slag materials (Kalyoncu, 2001). Worldwide, 480 million Mg are produced, with China, India, North America, and Europe responsible for about 100 million Mg year^{-1} each (Feuerborn, 2005). Fly ash is a fine (typically <0.05 mm diam) largely silicious material trapped in electrostatic precipitators prior to discharge of flue gases from smoke stacks. Nationally, about 50% of this material is reused as a cement additive or used in structural fills, roadbases, mineshaft backfilling and other engineering applications; the remainder is stored on site in lagoons or landfills. While fly ash has traditionally been exempt from regulation, new landfill construction is likely to require liners and monitoring of leachates to prevent groundwater contamination with metals, which has been observed at a number of existing facilities. These higher disposal costs have spurred an interest in use of fly ash as a soil amendment.

The composition of coal fly ash differs significantly from that of wood ash (Table 24.2) due to the nature of the fuel; coal has a greater ash content due to inclusions of minerals in the coal as it was deposited and then modified over geologic time. Clay minerals deposited within the coal bearing strata give rise to high Si and Al contents, and Fe precipitated as pyrite (FeS) within the coal adds Fe and a range of associated trace metals. In the United States, coals mined in KY, IN, AL, WV, IL, and PA tend to be higher in pyrite (high S, bituminous coals), with a resulting ash that is acidic to neutral in pH, high in Fe, and relatively low in Ca and Mg, classified as Class F in the ASTM system (Mattigod et al., 1990). Ash from sub-bituminous and lignite western United States coals (low S) has lower Fe and associated metals, and higher Ca, Mg and pH values (Class C), and is more pozzolanic (hardens irreversibly, similar to portland cement), and therefore is more desirable as a cement additive. Representative chemical analyses are given in Table 24.3.

The Class F ashes, constituting about one-half of the annual production in the United States, have limited nutrient value; major nutrient concentrations are low, and studies of availability indicate that K in particular is not readily water soluble, limiting its effect as a fertilizer. These ashes typically have low CCE (1%–10%) and, therefore, have limited potential as lime substitutes. A wide range of microelements are present that have a potential benefit for agricultural use. However, some Class F ashes have As levels that are a cause for concern environmentally. Using USEPA metal limits for biosolids (Section 24.5.1)

TABLE 24.3 Chemical Analysis (Total Digestion) of 9 Class C and 49 Class F Fly Ashes

	Class C Ashes (9)			Class F Ashes (49)		
	Mean	Min.	Max.	Mean	Min.	Max.
Major Elements	g kg^{-1}			g kg^{-1}		
Al	68	51	104	110	63	144
Fe	51	29	95	80	17	174
Si	137	93	177	231	198	282
P	3.6	1.0	6.3	1.4	0.8	2.6
K	6.3	2.7	11.6	15.4	3.1	25.3
Ca	172	101	225	32	2	115
Mg	21.9	4.4	41.8	8.9	3	27
Trace elements	mg kg^{-1}			mg kg^{-1}		
As	39.9	8.0	96.0	93	8.0	391
B	nd	nd	nd	425	55.0	1108
Cd	nd	nd	nd	1.0	0.1	3.2
Co	nd	nd	nd	47	14	120
Cr	66.9	40.0	123.0	171	25	651
Cu	116.4	45.0	223.0	133	48	242
Pb	45.6	19.0	92.0	63	13.0	273
Mo	14.6	5.3	32.0	29.8	3.7	139
Ni	54.0	34.0	78.0	121	14.0	309
Se	11.4	6.0	14.2	11.4	1.0	47
V	nd	nd	nd	259	58	470
Zn	266.0	25.0	658.0	270	27	2050

Sources: Data combined from Ainsworth, C.C., and D. Rai. 1987. Chemical characterization of fossil fuel combustion wastes. EPRI Report EA-5321. Electric Power Research Institute, Palo Alto, CA; Miller, W.P., and M.E. Sumner. 1997. Agricultural and industrial uses of byproduct gypsums, p. 226–239. *In* J.E. Rechcigl and H.C. MacKinnon (eds.) Agricultural uses of byproducts and wastes. American Chemical Society, Washington, DC.
nd = no data available.

as a comparison, As concentrations above the 75 mg kg^{-1} ceiling occurred in about one-half of ashes in this group. In comparison, Class C ashes are much higher in Ca, Mg, and K, and have a significant liming potential due to the higher CCE values (20%–60%). Trace metals are overall lower, with only 1 sample in 9 having >75 mg kg^{-1} As. Boron is elevated in all ashes relative to levels in soils, particularly in Class C ashes. Boron phytotoxicity may injure plants temporarily, although this effect often disappears as B is leached out of the root zone by rainfall. There are no human health effects of B in food crops or water supplies.

Finally, it should be noted that the solubility of P in high P soils (e.g., resulting from over-manuring,) has been shown to decrease following application of fly ash (Stout et al., 1998, 1999, 2000). Thus, fly ash shows promise in the remediation of high P soils.

24.2.3.3 Gypsum (CaSO$_4$·2H$_2$O)

Gypsum is produced as a by-product by a number of industries, most notably the phosphate fertilizer industry and electric generating utilities as well as by pigment processing and electroplating industries. Nearly 1 billion Mg of phosphogypsum (PG) from P fertilizer processing are stockpiled in Florida and the Mississippi Valley, with an annual production of 40 million Mg, mostly in central Florida (Miller and Sumner, 1997). In addition, electric utilities are producing increasing amounts of flue gas desulfurization gypsum (FDG) in scrubbers designed to remove SO_2 from stack gases in response to amendments to the Clean Air Act. About 30 million Mg were produced in 2006 in the United States from forced air scrubbers that fully oxidize SO_2 to gypsum, which is a large increase over the past 10 years (American Coal Ash Association, 2006). Other scrubbing technologies produce materials containing calcium sulfite ($CaSO_3$) mixed with fly ash (Korcak, 1995). Some of the gypsum produced is used in wallboard manufacture, but continued emphasis in the United States and Europe on SO_2 removal from coal-fired power plants suggests increased production in the future (Kalyoncu, 2001).

Some typical analyses for various gypsiferous by-products are given in Table 24.4. Fluidized bed combustion (FBC) material is produced from a dry scrubber that injects lime directly into the furnace, one common technology that produces a mixed by-product composed of calcium sulfite/sulfate material containing unreacted lime. It has a variable composition depending on the boiler design often containing considerable residual CaO or $Ca(OH)_2$ and some fly ash (evidenced by the Si content). No calcium carbonate equivalent (CCE) data were located, but it would likely be a reasonable lime substitute. PG and FDG are relatively pure materials, approaching the theoretical 23% Ca and 19% S of $CaSO_4 \cdot 2H_2O$. Impurities in PG are largely Si sand and a minor amount of fly ash in FDG. Trace elements are quite low in all materials, with B being the exception in the case of FBC and FDG materials.

Both PG and FDG have a history of use in agriculture; in the southeastern United States, by-product $CaSO_4 \cdot 2H_2O$ materials are applied to peanuts as a Ca supplement, and various types of $CaSO_4 \cdot 2H_2O$ have been used in the western United States, Australia, and South Africa on high Na soils to improve physical properties. Higher application rates (up to 10 Mg ha^{-1}) have been shown to improve root penetration on acid subsoils with no enhanced metal uptake on a range of crops (Shainberg et al., 1989; Sumner, 1993; Ritchey et al., 1995). Field studies using FDG have demonstrated limited uptake of metals over 3 years on a range of crops from 10 Mg ha^{-1} application rates on three soils in Georgia (Sumner et al., 1996; Miller and Sumner, 1997). The USEPA, however, has restricted the use of PG containing >10.0 pCi ^{226}Ra g^{-1} due to trace contamination with radionuclides of the ^{238}U decay series, principally ^{226}Ra and ^{222}Rn gas, which carry certain health risks (Alcordo and Rechcigl, 1995; Miller and Sumner, 1997). The risk assessments on which this decision were based involve inhalation hazards to persons living on highly amended soils, and are being challenged by fertilizer and phosphate industry groups. Currently, PG containing >10 pCi g^{-1} of total radiation is restricted and cannot be land applied.

Finally, it should be mentioned that, like fly ash, FDG application has been shown to sequester P in high P soils (Stout et al., 1999; Callahan et al., 2002).

24.2.3.4 Flue and Kiln Dusts

These are fine airborne particulates collected from manufacturing facilities, particularly metal processing/refining and cement manufacturing. Flue dusts often refer to fines from metal processing (especially electric furnaces), and are often ferrous with admixtures of heavy metals; something in the range of 5 million Mg year^{-1} are produced in the United States, most of which is either recycled within the facility or landfilled. If such dusts do not contain contaminant metals (Cd, Pb, etc.), they are a valuable source of trace nutrients, and significant quantities are used by the micronutrient industry as a source of Cu and Zn fertilizer (Wyatt, 1997). However, many flue dusts and similar products contain contaminant metals (up to 20% heavy metals such as Pb, Cd, and Zn) and are classified as either hazardous wastes or are derived from hazardous wastes, thus restricting their use. It should be noted that several instances of ill-informed use of flue dusts as liming materials and/or micronutrient sources have resulted in Zn and Pb poisoning of crop land to the point of near sterilization (Davis et al., 1995b), and the use of such materials should be carefully monitored for metal loading rates applied to soils.

TABLE 24.4 Chemical Composition of Various Gypsum Containing By-Products

	Fluidized Bed Combustion By-Product	Phosphogypsum	Flue Gas Gypsum
Macroelements		g kg^{-1} dry wt.	
P	0.1–0.5	8–16	<1–2
K	0.5–8	0.5–2	1
Ca	240–460	180–235	210–235
Mg	1–16	0.4–2	0.2–2
S	72–140	128–181	166–186
Si	Nd	15–50	0.7
Al	3–20	1–20	0.3–2
Microelements		mg kg^{-1} dry wt.	
B	95–170	<3	75
Mo	0.1–0.3	2–11	1.3
Cu	12–20	9	8.0
Zn	12–20	Nd	36
Ni	13–20	10	9.7
Pb	1.5–7.5	15	<0.1
Cd	0.5	4	0.01
Cr	9–23	20	10
Se	0.2–0.6	1–2	2.9
Hg	Nd	0.3–0.4	0.01–0.4
As	Nd	1–4	3.1

Sources: Summarized from Korcak, R.F. 1995. Utilization of coal combustion byproducts in agriculture and horticulture, p. 107–130. *In* D.L. Karlen, R.J. Wright, and W.O. Kemper (eds.) Agricultural utilization of urban and industrial byproducts. American Society of Agronomy, Madison, WI; Miller, W.P. 1995. Environmental considerations in land application of byproduct gypsums, p. 183–208. *In* D.L. Karlen, R.J. Wright, and W.O. Kemper (eds.) Agricultural utilization of urban and industrial byproducts. American Society of Agronomy, Madison, WI; Alcordo, I.S., and J.E. Rechcigl. 1995. Phosphogypsum and other byproduct gypsums, p. 365–426. *In* J.E. Rechcigl (ed.) Soil amendments and environmental quality. Lewis Publishers, Boca Raton, FL.

Cement kiln dust is a fine-grained mixed calcium silicate/oxide/carbonate material produced during kilning of portland cement. Roughly 6 million Mg year^{-1} are collected in baghouses at cement kilns, with most being recycled to the process or disposed of onsite. About 1 million Mg are shipped off-site as a by-product additive/stabilizer (for hazardous waste or biosolid stabilization) or as a lime substitute for use in agriculture. No specific data on the latter use appears to be available, nor any agronomic studies on use of the material. In 1992 USEPA ordered a study of kiln dusts prompted by the increasing use of by-product or waste fuels, including some hazardous petroleum and solvent by-products, as an accessory fuel in heating the kilns, which must reach very high temperatures in order to calcine the cement. In a report to Congress under RCRA, data indicated significant levels of dioxins, furans, metals (particularly Pb and Cd), and radionuclides in a range of kiln dust samples (USEPA, 1993a, 1993b). In 2002, regulation of this material was still under consideration by USEPA. Until further analytical data on contaminant levels and their fates in soils are available, caution is clearly indicated in use of kiln dusts as lime substitutes.

24.2.4 Biosolids and Municipal Solid Waste

No other by-product material has been as extensively studied for its agricultural value and environmental impact as has sewage sludge (biosolids), the solid residual material resulting from sewage digestion at municipal wastewater treatment facilities. Biosolids were used in Europe in the 1950s and 1960s on sewage farms to grow a variety of crops, and excessive uptake of heavy metals was eventually observed. Many studies conducted during the 1970s showed that high rates of sludge from industrial cities containing high metal levels resulted in enhanced plant uptake of Cd and other metals. Regulations followed that limited industrial metal inputs into water treatment plants, which considerably reduced metal levels in sludges (Chang et al., 1987). USEPA has encouraged land application of biosolids in order to recycle nutrients and avoid landfill costs, and has provided regulations to define high-quality biosolids for unrestricted use on crop land (Section 24.5.1). About 6.5 million Mg of biosolids (dry wt.) are generated annually in the United States, and currently 50% is land applied on agricultural, forestry, or reclaimed mined land; of the remainder, two-thirds is landfilled and the rest incinerated (NEBRA, 2007).

Trace elements in biosolids have been extensively documented in the United States and Europe, given the historical interest in land application of this material (Table 24.5). Mercury and Cd are the most toxic of the metals typically monitored in biosolids, and are elevated in some materials compared to background levels in soils (<0.05–1 mg kg^{-1}). Other metals listed in Table 24.5 can be hazardous when sludges of higher concentration are added at high rates to soils over a period of years. Potential hazards are discussed in Section 24.5.2. Regulatory limits have been established in the United States and the EU, as discussed in Section 24.5.1.

Organic contaminants in biosolids have not been regulated in the United States, on the rationale that insufficient data exists on concentrations and that risk assessments have not been conducted for the wide range of potential compounds. The recent Targeted National Sewage Sludge Survey (USEPA, 2009) presents data for selected polyaromatic compounds, organobromine flame retardants, pharmaceuticals, steroids, and hormones; while concentrations vary over a wide range, mean levels are in the low ppm (mg kg^{-1}) range for most compounds, but reach 10–100 ppm for some common steroids and pharmaceuticals. Kester et al. (2005) report values for some volatile organics, polychlorinated biphenyl (PCB)'s, and dioxin-class compounds in biosolids, and suggest some new approaches to assessing risk of these compounds in biosolids-amended soils that may be implemented in the future.

The presence of potentially harmful microorganisms (bacteria, viruses, parasites) in sewage sludge was long thought to be the most significant public health risk associated with land application, particularly when biosolids were surface applied without incorporation. Airborne dispersal during application was thought to be the most common vector, since disease organisms were not expected to survive very long when mixed with

TABLE 24.5 Major and Trace Elements in Sewage Sludge and MSW Compost

	United Kingdom	Biosolids the United States	Australia	MSW
		g kg^{-1}		
N		1–200 (37)	21–82 (39)	
P		1–150 (23)	11–89 (30)	
K		0.2–65 (4.5)	0.9–4 (2)	
Ca		1–250 (45)	10–43 (19)	
Mg		0.3–25 (6)	2–7 (4)	
		mg kg^{-1}		
As	<2–123 (6)	0.3–316 (6)	3–21 (9)	1–5 (2.6)
Cd	<2–152 (9)	0.7–8,220 (7)	3–21 (9)	1–13 (2.9)
Co	<2–620 (10)			
Cr	4–23,000 (197)	2–3,750 (40)	81–1,620 (805)	8–130 (35)
Cu	69–6,100 (590)	7–3,120 (460)	571–1,427 (915)	31–620 (154)
Hg	<2–140 (4)	0.2–47 (4)	1–5 (3)	0.5–3.7 (1.3)
Mo	<2–154 (5)	2–68 (11)		
Ni	9–930 (61)	2–980 (29)	25–179 (94)	7–101 (25)
Pb	43–2,600 (398)	9–1,670 (106)	140–579 (320)	22–913 (215)
Sβ	<2–15 (3)	0.5–70 (5)	0.5–11(3)	
Zn	280–27,600 (1,140)	38–68,000 (725)	834–1,826 (1,300)	152–1,360 (503)

Sources: U.K. data from Smith, S.R. 1996. Agricultural recycling of sewage sludge and the environment. CAB International, Wallingford, U.K; Hue, N.V. 1995. Sewage sludge, p. 199–247. *In* J.E. Rechcigl (ed.) Soil amendments and environmental quality. Lewis Publishers, Boca Raton, FL; McLaren, R.G., and C.J. Smith. 1996. Issues in the disposal of industrial and urban wastes, p. 183–212. *In* R. Naidu et al. (eds.) Contaminants and the soil environment in the Australasia-Pacific region. Kluwer Academic Publishers, Dordrecht, the Netherlands; Epstein, R., R.L. Chaney, C. Henry, and T.J. Logan. 1992. Trace elements in municipal solid waste compost. Biomass Bioenerg. 3:227–238.

soil. Recent research confirms that pathogens from biosolids do not survive long in the soil (Rusin et al., 2003; Zaleski et al., 2005). Furthermore, the risk of bioaerosol infection during application may be significantly lower than previously thought (Brooks et al., 2005). USEPA defines Class A biosolids as having undergone advanced treatment (via heat, composting, or lime treatment) to eliminate pathogens, using fecal coliforms as the indicator organism. Class B biosolids have undergone treatment to reduce the number of human pathogens, but are likely to still contain them. According to the NRC (2002) about 60% of all biosolids generated in the United States are land-applied, with most of these being Class B solids. Pepper et al. (2008) have recently argued that risks to human health posed by human pathogens in land-applied Class B biosolids are low. In the United States, regulations regarding incorporation of land-applied biosolids vary from state to state. In Pennsylvania, for example, biosolids must be incorporated within 24 h of application, while in Arkansas Class B biosolids must be incorporated within 6 h of application, but Class A biosolids do not need to be incorporated. Regulations regarding incorporation are generally more strict in Europe than in the United States (NRC, 2002).

Biosolids are a good source of N and P for agronomic crops (Table 24.5), and many examples of their use on agronomic crops, forest trees, horticultural crops, and plants grown on reclaimed lands have been reported in the literature. Typically, application rates are based on either N or P content depending on relative concentrations and crop requirements. Some inorganic N is present, but because most must be mineralized by soil organisms, release occurs over several years. Potassium is low in most sewage sludges, and must be added as supplemental K fertilizer. Sludges have a variable effect on soil pH, depending on how they are processed (some are lime stabilized, and thus have considerable alkalinity); most sludges not containing free lime will acidify soil over time due to acid generated by nitrification, and pH must be monitored and adjusted.

Nearly 250 million Mg of municipal solid waste (MSW), including household garbage and some municipal and inert industrial wastes, are produced annually in the United States, and finding ways to reduce landfilling costs of this material is a growing industry (USEPA, 2007). Technological improvements in separation of plastic, glass, and metal from the MSW stream have suggested that the remaining 40%–60% of the waste, which is largely paper and food waste, might be composted and land applied. There are a number of firms with operational separation and composting units in the United States, subsidized by municipalities hoping to avoid capital expenses of building new Subtitle D (lined and monitored) landfills. Current production levels of compost (derived from MSW and/or manures) were estimated at 22 million Mg year^{-1} in 2007 (USEPA, 2007), but are likely to increase dramatically in the future (Stratton et al., 1995; Barker, 1997). Data are scarce, but the N and P contents of MSW composts are likely to be only 50% of the levels in biosolids, and mineralization rates are appreciably slower (Barker, 1997).

24.2.5 Composting and Mixing of By-Products

Composting refers to the process of stockpiling by-product materials under conditions that favor decomposition of organic materials to form a more stable product for use on soils. Carbon is lost as CO_2 during composting, resulting in a loss of bulk and concentration of nutrients compared to the original material (Barker, 1997). Nitrogen is also lost (as ammonia) during composting, but because more of the original C is lost than of the original N, composts generally have narrower C:N ratios than the corresponding original materials. The heat release accompanying composting kills many pathogens, and is an approved process for producing Class A biosolids for land application or for reducing pathogens in animal manures. Composting also eliminates odor from some materials, increases the density of the material (i.e., reduces bulk) and improves handling characteristics for land spreading. On the negative side, it may also narrow the N:P ratio of the finished product to such an extent that if the compost is used as the principal N source for a crop, over-fertilization with P could result. Composting is common for processing of MSW and biosolids prior to land application.

A wide range of literature is available on formulating and processing compost, as well as describing the microbiology involved; these specifics are beyond the scope of this review. It should be noted, however, that composting can nearly always benefit materials destined for land application, due to the changes noted above (a notable exception to this is the loss of N that occurs during composting). For some materials with active pathogens (particularly biosolids, MSW and food waste), wide C:N ratios, and/or objectionable odors, the advantages are significant, and if the material is to be used by home gardeners or near urban areas, proper composting is crucial to the successful marketing of the material (Stratton et al., 1995; Barker, 1997).

The concept of mixing by-product materials to obtain more optimum physical, chemical, or nutritional characteristics, in combination or apart from composting, has been used in formulating biosolid mixes with ash and cement kiln dust (Logan et al., 1997). Blending of organic (sludge) and inorganic (ash) materials has advantages in balancing nutrient content, adjusting pH, optimizing water content, and immobilizing contaminant metals. Currently, the additional costs of handling, blending, and storage of mixtures and composts are likely greater than the market value; costs of a manure/pulp mill sludge composting operation were estimated at $50–$120 Mg^{-1} (NCASI, 1984). Subsidies by contributing industries, as well as increases in demand and cost of landfilling, may make such operations feasible in the future. Hyatt (1995) has summarized information on potential marketing of composts in agriculture, and concludes that improving compost quality (particularly N content and availability) should increase farmer demand and establish pricing at a reasonable level for compost producers.

24.3 Soil Fertility Considerations in Land Application

The primary beneficial effect of land application of many by-products is the supply of essential plant nutrients to crops, particularly macronutrients (N, P, K). There are instances where the supply of other elements (Ca, S, or micronutrients) may be considered as the primary effect (such as in the case of gypsum), or the liming value may be appreciable. Additionally, some applications (particularly of organic materials) may increase SOM, improve soil structure, or increase water-holding capacity. Assessing the type and extent of beneficial effect to be achieved with a given type and rate of application is largely at the discretion of state officials in departments of agriculture or environmental regulation. It should be reiterated that a beneficial effect must be demonstrated or inferred for a given type of application, or such application may be construed as simply disposal.

24.3.1 Nutrient Content and Bioavailability

As mentioned in the previous section, most by-products contain significant quantities of several macro- and micronutrients, but it is the N content that is used most often as the basis for calculating application rates. This has historically been true because of the high N requirements of most crops coupled with the low N-supplying power of most soils. In order to understand the reasoning behind N-based application rates, it is necessary to know something about the chemistry of N in organic by-products.

Most organic by-products (i.e., manures and biosolids) contain two forms of N: organic and inorganic N. Organic N is chemically combined with C, H, and O while inorganic N occurs as nitrate (NO_3^-) and ammonium (NH_4^+). Both NO_3^- and NH_4^+ can be taken up and used by plants, but organic N cannot. Ammonium and NO_3^- can be lost in large quantities from the soil, the former by denitrification (conversion to gaseous N_2O and N_2) and the latter by NH_3 volatilization. Denitrification is a microbially mediated process, which occurs under reducing (wet) conditions, while NH_3 volatilization is an abiotic process, which occurs at high pH. The organic N fraction is not readily lost from the soil or by-product through these processes, but many microorganisms convert it into inorganic N through a process called mineralization. Finally, inorganic N can be converted into organic N by microorganisms that take up NH_4^+ and NO_3^- ions and convert these into N-containing organic compounds. This process, called immobilization, temporarily reduces the amount of plant-available N (PAN) in the soil. The losses resulting from NH_3 volatilization and denitrification are permanent. A detailed discussion of the N cycle in soils and by-products is presented by Paul and Clark (1989), Sims and Wolf (1994), Brady and Weil (2008), and in Chapter 13.

Most of the N in organic by-products is in the organic form. The relative proportion of inorganic N in beef, swine, and poultry manure samples is approximately 8.1%, 5.8%, and 23%, respectively (Cabrera and Gordillo, 1995). When manures are mixed with water and stored as a liquid or slurry, mineralization occurs to such an extent that in aged slurries and liquids there is more inorganic N than organic N (Cabrera and Gordillo, 1995). In virtually all manures the inorganic N pool is dominated by NH_4^+, although instances of high NO_3^- levels have also been reported.

Sims (1986) has proposed the following model for estimating PAN from soil applied manure:

$$PAN = A_i N_i + P_m N_o \quad (24.1)$$

where
- N_i and N_o are the quantities of inorganic and organic N in the manure
- A_i is the fraction of inorganic N available for plant uptake
- P_m is the fraction of organic N that is mineralizable

The ability of this model to accurately predict the amount of PAN that will result from application of an organic by-product material to soil is dependent on the accuracy of the values assigned to the coefficients A_i and P_m. The values assigned to N_i and N_o also have a major impact on the calculated PAN, but it is possible, at least in theory, to accurately measure these quantities. Much of the N in manures may be lost prior to land application during storage of the manure, which will decrease the values of both N_i and N_o. The USDA Agricultural Waste Management Field Handbook (USDA, 1992) contains estimates of N, P, and K losses for several different types of animal manure stored under a variety of conditions. Organic by-products should be sampled and analyzed immediately prior to land application in order to account for storage losses not only of N, but of other nutrients as well.

There is a great deal of uncertainty involved in assigning values to the coefficients A_i and P_m. The coefficient A_i would have a value of 1 if there were no losses of inorganic N following application of the by-product, but this is seldom the case. Nitrogen losses can be significant depending on a host of environmental and management variables such as soil type and moisture, temperature, and method of by-product application (e.g., incorporated versus unincorporated). The combined losses of NO_3^- by leaching (Davis et al., 1995a) and denitrification (Johnson and Wolf, 1995) and losses of NH_4^+ by NH_3 volatilization (Thompson et al., 1990) and immobilization can cause significant losses (>50%) of the manure's inorganic N pool.

The proportion of the organic N (P_m) that is mineralizable will likewise depend on a variety of factors, including temperature and moisture conditions, the C:N ratio of the by-product, and the amount of N in the urea or uric acid forms, which are particularly susceptible to mineralization (Cabrera et al., 2005; Cusick et al., 2006; Diaz et al., 2008). In practice, P_m is not used per se in the calculation of application rates. Rather, the mineralization rate of organic N is represented by a decay series that gives the percentage of the organic N that is mineralized as a function of time after initial application of the by-product. The data in Table 24.6 are representative of typical decay series for organic N in several by-product materials. These data illustrate the following general concepts: (1) in fresh manures, a large

TABLE 24.6 Cumulative Mineralization of Organic N in a Variety of By-Products at Several Times after Initial Application

Material	Cumulative N Mineralized after		
	Year 1	Year 2	Year 3
	%		
Poultry manure (layer)	80	82	83
Swine manure (fresh)	75	79	81
Swine manure (aged)	60	66	68
Lagoon effluent	40	46	49
Paper mill sludge (second.)	30	45	50
Biosolids (anerobic)	20	30	35
Biosolids (aerobic)	15	25	30
Compost	10	15	20

Sources: Eghball, B., B.J. Weinbold, J.E. Gilley, and R.A. Eigenberg. 2002. Mineralization of manure nutrients. J. Soil Water Conserv. 57:470–473; USDA. 1992. Agricultural waste management field handbook. National Engineering Manual, Part 651. United States Department of Agriculture, Washington, DC; NCASI. 1984. Land application and related utilization of pulp and paper mill sludges. NCASI Bulletin 439. NCASI, Research Triangle Park, NC; King, L.D. (ed.). 1986. Agricultural use of municipal and industrial sludges in the southern U.S. NC State University Southern Cooperative Series Bulletin 314. North Carolina State University, Raleigh, NC.

percentage (>75%) of the organic N is mineralized during the first year after application; (2) rates of organic-N mineralization are generally higher for poultry manures than for swine or cattle manures; and (3) storage of manures causes decreased rates of mineralization following application because most of the easily mineralizable fractions of the organic-N pool have already been mineralized during storage. While data such as those in Table 24.6 can provide general guidance in nutrient management planning, in the United States, local Cooperative Extension Service or Natural Resource Conservation offices should be contacted for additional information that is tailored to specific, local conditions.

For biosolids, N mineralization rates vary considerably with treatment process; nitrogen in anaerobic sludges tends to mineralize more rapidly than in aerobic sludges (King, 1986; Table 24.6), but both release N less rapidly than manures. This is related to the partial decomposition that takes place during wastewater treatment, during which N is incorporated into more recalcitrant compounds. In general, composting of manures and other organic by-products leads to lower N mineralization rates and decreased N availability (Cabrera and Gordillo, 1995; Preusch et al., 2002; Gale et al., 2006; Munoz et al., 2008). Gilmour's mechanistic computer simulation model, which uses site-specific weather data and biosolids analytical data, has been used successfully to predict N mineralization from a variety of organic by-products under a variety of environmental conditions (Gilmour and Skinner, 1999; Gilmour, 2009).

In computing application rates based on N content, mineralization rates such as those in Table 24.6 should be used to predict N availability. Application rates should also account for any inorganic N present and N mineralized from previous years' applications to the same field. Losses of N from volatilization of stored manures can be appreciable, so reanalysis for N should be done if the material is stored for appreciable time periods.

By-products such as pulp and paper sludges, waste paper, and composts of various materials that contain low amounts of N (<1%) with C:N ratios of >50 probably have little potential to mineralize sufficient N for crop production. These materials will, in fact, immobilize N from the soil and may result in serious N deficiencies in crops planted soon after application. Additional (inorganic fertilizer) N should be applied to these soils to prevent N deficiency, or a period of several months should be allowed before planting to allow for decomposition to recycle N back into available forms.

By-product application rates are usually based on the N requirement of the crop that is being fertilized. However, many by-products are also good sources of other macro- and micronutrients, particularly P, Ca, Zn, and Cu. Perhaps because they contain so many different plant essential elements in available forms, manures have been found to be much more effective than synthetic fertilizers at restoring agronomic productivity to eroded and otherwise disturbed soils (Whitney et al., 1950; Aina and Egolum, 1980; Frye et al., 1985; Dormaar et al., 1988; Larney and Janzen, 1996). While some of the restorative ability of manures and other organic by-products may be due to improvements in the physical and biological properties of degraded soils, most researchers cite improved availability of N and P as the likely mechanism. Freeze et al. (1993) have pointed out that the restorative value of manures should make it economically feasible to transport manure farther than if it were to be applied strictly as an NPK source to nondegraded soils.

Gypsum is an example of a by-product used as a source of Ca, which can be a limiting nutrient on sandy and/or very acidic soils. Shainberg et al. (1989), Sumner (1993) and Ritchey et al. (1995) have reviewed the literature on the use of gypsum on acidic soils to increase available Ca in the subsoil in order to promote deeper rooting and water extraction by a range of crops. Yield increases are routinely obtained on a range of soils and crops with single applications of up to 10 Mg ha^{-1}, and the increases are maintained over >10 year. Such applications give a good economic return over time, and show no adverse environmental effects.

Wood ash has been used as a liming material (Lundström et al., 2003), a source of phosphorus and potassium (Erich, 1991), and as a micronutrient source (Nieminen et al., 2005).

24.4 Effects of Land Application on Soil Properties

Improvements in soil properties such as structure (aggregation), permeability, water-holding capacity, and CEC often do not have an immediate or consistent effect on crop yields, although soil scientists probably would agree that they are highly desirable. By-product additions, particularly of organic materials, have the potential to improve many soil properties over a period of years, and these effects are consistent with the idea of beneficial reuse of these materials.

24.4.1 Physical Properties

Land application of by-products generally has been shown to improve soil physical properties (Wallingford et al., 1985), although there are several reports describing studies in which animal manures had either no effect or deleterious effects on some soil physical properties. Hileman (1967), for example, found no changes in either soil organic C content or bulk density as a result of 3 year of poultry litter applications to a silt loam soil at rates as high as 18 Mg ha^{-1} annually. Similarly, Sharpley et al. (1992) found no consistent effect of long-term (up to 15 year) poultry or swine manure applications on soil physical properties. Weil and Kroontje (1979), on the other hand, reported that after 5 year of incorporating poultry litter at an annual rate of 110 Mg ha^{-1} the bulk density of the plow layer of a clay loam soil decreased from 1.1 to 0.8 g cm^{-1}. However, these authors also reported that a layer of partially decomposed manure had accumulated at the bottom of the plow layer, which greatly reduced the water percolation rate. Cattle manure has been shown to improve soil aggregation (Hafez, 1974; Adeli et al., 2007), to reduce soil bulk density (Tiarks et al., 1974), and to improve water-holding capacity of soils (Unger and Stewart, 1974).

A comprehensive study of the effects of long-term manure applications on soil physical properties was conducted by Schjønning et al. (1994) in Denmark. Farm yard manure (FYM) or inorganic fertilizer was applied for 90 year to plots on a sandy loam soil. The FYM treatments had significantly higher organic C contents, lower bulk densities and decreased compactibilities compared to the inorganic fertilizer treatments. The results of this study provide particularly compelling evidence for the beneficial effects that long-term manure applications have on soil physical properties. It is expected that other types of organic by-products might have similar effects.

In a review of literature on effects of biosolid application on physical properties, increases in soil organic C of 0.5%–2.5% were noted over 2–5 year at higher rates (20–100 Mg ha^{-1}) of sludge additions (Hill and James, 1995). Increases in C are typically correlated with lower bulk density, greater aggregation and aggregate stability, and improved water retention. Preliminary results on a sandy soil in Georgia showed improved aggregate stability and water-holding capacity with primary and secondary pulp mill sludge added at rate of 40 Mg ha^{-1} (Cochran et al., 1997).

Infiltration and hydraulic conductivity of sludge-amended soils is highly variable. The time and rate of application can have large effects. Freshly applied sludges may lower saturated hydraulic conductivity at the surface due to pore clogging, but incorporated sludge additions over time tend to improve infiltration due to better aggregation and macroporosity, leading to less surface crusting under rainfall (Hill and James, 1995). Price and Voroney (2007) found high rates (100 Mg ha^{-1}) of a paper mill biosolid increased infiltration rate and aggregate size and decreased bulk density on four different soils over 3 year in a field experiment in Canada.

Shiralipour et al. (1992) reviewed the literature on the impacts of MSW compost on soil physical properties. Typically, high (>100 Mg ha^{-1}) rates of application have been added to degraded or very sandy soils. In most cases, SOM levels and water retention increased and bulk density decreased with compost additions.

Several studies have suggested that fly ash applications may be used to increase water-holding capacities of very sandy soils at high rates of application (100–300 Mg ha^{-1}) due to addition of fine particles (Ghodrati et al., 1995). Gypsum is widely used in semi arid regions on sodic soils to improve aggregation and permeability, and by-product gypsum should perform equally well as the mined type mostly in use (Shainberg et al., 1989). Surface-applied gypsum also improves infiltration rates of weathered acidic soils over several rainfall events due to its flocculating effect on soil clays (Miller, 1987). Radcliffe et al. (1986) demonstrated that given sufficient time to move into the subsoil, surface applications of gypsum reduced the penetration resistance of subsoil hardpan layers.

24.4.2 Biological Properties

There do not appear to be a large number of published reports on the effects of manure applications on soil biological properties. Because of the great diversity of soil microorganisms and the complexity of the interactions within the soil microbial community, studies tend to focus on how manure applications affect either a single group of soil organisms or the soil microbial biomass as a whole. For example, Christie and Beattie (1989) found that soil microbial biomass C and N decreased with increasing rates of applied swine manure slurry, but increased with increasing rates of applied cattle manure slurry. Adeli et al. (2007) reported that 3 year of broiler litter application at a rate of 6.7 Mg ha^{-1} increased soil microbial biomass C 28% relative to the unlittered control. Eight years of pig slurry or solid cattle manure applications were reported to increase numbers of bacteria, actinomycetes and fungi in soils (N'Dayegamiye and Côté, 1989), while 19 year of pig and cattle slurry additions to grassland significantly decreased vesicular arbuscular (VA) mycorrhiza infection of plant roots (Christie and Kilpatrick, 1992). The reduction in VA mycorrhiza infection apparently caused by application of manure may not be a serious cause for concern, as the impact of mycorrhizal infection on plant nutrition is greatest in low fertility soils, and declining mycorrhizal populations in nutrient-rich, manured soils may have little or no impact on plant growth. Mårtensson and Witter (1990) concluded that the changes that they observed in populations of N-fixing soil organisms brought about by applications of manure were, in fact, caused by manure-induced changes in soil pH.

A potential concern in by-product application is the possibility of chronic toxicity to soil microbial populations caused by low levels of added contaminants. Such concerns have been voiced particularly in the European scientific community. Although no systematic body of information is currently available to evaluate this possibility, data from Europe suggests there are levels of some trace metals that do inhibit overall microbial activity and specific groups of organisms (Smith, 1996). Typically, these levels are greater than regulatory limits on metal loadings of soils (Section 24.5.1), but in certain cases (e.g., Zn; see Smith, 1996), lower metal levels may inhibit some soil microbes. Cela and Sumner (2002) showed that nitrification in three soils was inhibited in the laboratory by Cu and Ni added as inorganic salts at rates of about 100 ppm, but solubility of metals was very high in these experiments. Current risk assessment models are not specifically designed to consider these kinds of effects.

A recurring question is the safety of land-applied biosolids and manures in terms of disease vectors. Treatment methods to eliminate pathogenic organisms from biosolids to create Class A (pathogen-free) biosolids are used on about one-half of the land-applied biosolids in the United States, and are mandatory in Britain and other EU countries. The available data support the contention that these are effective treatments when executed properly, although the study of microbial and virus agents in biosolids is perhaps the most difficult aspect of assessing risk to the public, due to the difficulty of quantifying these organisms (Angle, 1994). Recent applications of risk assessment to pathogens in land-applied wastes has been reviewed by Gerba and Smith (2005), along with documentation of types of disease organisms present and past disease outbreaks associated with manure applications. Anecdotal evidence of human illness related to Class B biosolids applications have been reported in the literature (Harrison and Oakes, 2002; Khuder et al., 2007), and the National Research Council's (2002) report on biosolids emphasized the need for better information on pathogen survival, dispersal, and relationship to human disease, as well as improved enforcement and monitoring of current regulations. As mentioned previously in Section 24.2.4, recent research indicates that the public health risks associated with land application of pathogen-containing biosolids may be lower than originally thought (Zaleski et al., 2005; Brooks et al., 2005; Pepper et al., 2008).

According to Guan and Holley (2003), the major pathogens present in swine manure (*Escherichia coli* O157:H7, *Salmonella*, *Campylobacter*, *Yersinia*, *Cryptosporidium*, and *Giardia*) can be eliminated by holding the manure at 25°C for 90 days. Holley et al. (2006) reported that under optimal conditions *Salmonella* can survive for up to 180 days in soil following application of swine manure. Soupir et al. (2006) reported finding large numbers of *E. coli* in runoff from fields fertilized with "cowpies," liquid dairy manure, or turkey litter. *Campylobacter* spp. and *E. coli* have also been found in runoff from poultry litter amended fields (Duggan et al., 2001).

24.4.3 Chemical Properties

The impact of by-product amendments on soil chemical properties depends strongly on the types and rates of by-products applied. Materials might be separated into those that are largely organic, those that are inorganic and alkaline, and gypsiferous by-products.

Organic materials (manures, biosolids, pulp mill sludge, MSW compost) typically increase soil organic C, and therefore increase CEC due to the high negative charge on organic matter. The amount of charge varies with the level of C increase, as well as the pH, since the CEC is of the variable charge type. Increases in CEC have been measured in amended soils in proportion to the C increases, but have also been noted to decline as decomposition processes reduce the organic C levels in years following a single application (Hill and James, 1995). The capacity of soils to specifically (strongly) adsorb metal cations is similarly a function of organic matter, and partially explains the lower plant availability of metals added in biosolids compared to inorganic salts (Smith, 1996).

Organic by-products also cause short-term shifts in soil pH and Eh (redox potential) due to the highly reducing nature of many organic by-products and microbial activity (and hence O_2 depletion) resulting from their addition. Anaerobically digested biosolids and uncomposted animal manures added at high rates (>20 Mg ha^{-1}) may result in temporary pH increases of 1–2 units, and decreases in Eh to 0 V or below within a week of application. In a study of swine manure added at 100 Mg ha^{-1}, considerable solubilization of Fe and Mn oxides was observed during this period, although trace metal solubility did not increase appreciably due to adsorption onto organic matter (Miller et al., 1985). Within 2–3 weeks, Eh and pH returned to near initial levels. Although biosolids may be applied at high rate (>20 Mg ha^{-1}) to supply adequate N for crops, animal manures are usually applied in the 4–10 Mg ha^{-1} range. These lower rates will help prevent the effects discussed above.

Inorganic alkaline by-products such as fly and wood ash, kiln dusts, and lime obviously impact soil pH, and rates should be adjusted based on the CCE of the material to reach a desired pH level. Some of these materials (such as alkaline fly ashes) also contain considerable soluble salts, which initially increase electrical conductivity (EC) of soil pore water when land applied that can injure salt-sensitive plants or seedlings. These salts are rapidly leached out of the root zone within a few rainfall events, and typically have no residual effect on plant growth.

Gypsum is soluble in water (2.5 g L^{-1}) and this solubility has been used to great advantage on soils with acidic subsoils to encourage deeper rooting of a range of crop plants (Shainberg et al., 1989). High application rates (10–20 Mg ha^{-1}) of by-product gypsum result in appreciable increases in exchangeable Ca, with a decrease in exchangeable Al in subsoil horizons of Ultisols and Oxisols over a several-year period. Overall, CEC also increased possibly by the ligand exchange of SO_4^{2-} for OH^- (Souza and Ritchey, 1986); pH in the subsoil is not appreciably changed. The increase in base saturation and Ca availability increased root distribution down to 1 m, resulting in greater water extraction and yield improvements for alfalfa of up to 100% compared to unamended control plots (Shainberg et al., 1989; Sumner, 1993). Both PG and FDG are similarly soluble, and have identical effects on subsoil chemistry and crop yields.

24.5 Environmental Aspects of By-Product Applications

The benefits of recycling nutrients and organic material to crop production via land application must be balanced against the potential for environmental degradation and threats to human health of such applications. The risks of such effects can be serious, even for seemingly innocuous materials such as animal manures that have been used in agriculture for centuries. Despite volumes of research findings compiled over the past 50 years, the assessment of these risks is a major topic of research today.

The degree of both scientific debate and public contention over the public health effects of land application practices has increased significantly over the past decade. Some of this controversy is due to reevaluation of the risk of well-studied factors (trace metals, nutrient export, pathogens), as well as to the consideration of relatively "new" risk factors in land-applied materials (hormones, pharmaceuticals, trace organics). Debates within the scientific community (McBride, 2003) and critical reviews of policies by independent agencies (National Research Council, 2002) have fueled public skepticism of land application regulations, which some perceive as industrial/governmental attempts to "dump toxic waste" on rural farmland.

Response to such criticism in the United States has been continual reevaluation of risk assessment models designed to quantify human health response to loadings of contaminants resulting from land application (Kester et al., 2005). These models used "risk pathways" (e.g., transfer from soil → plant → human) to quantify contaminant transfer from amended soil to humans, using the idea of a "most exposed individual" as one who (for example) obtains most (or all) of their food from waste-amended soil. Maximum allowable contaminant loading limits are set based on an acceptable increase in human health impacts (e.g., increase in cancer occurrence of 1 per 10^5 or 10^6). Such an analysis requires large amounts of basic data on contaminant behavior and toxicological response, and involves assumptions and judgment decisions that introduce varying degrees of uncertainty into the final result.

European countries have set (in many cases) more stringent rules for soil contaminant loadings than has EPA in the United States, based on other considerations including relation of contaminant increases to native or background levels in soils (Smith, 1996). This may in part be due to application of the "precautionary principle," as discussed previously (Section 24.1.3).

In the United States, state environmental agencies are largely responsible for implementation of federal regulations regarding land application. Where no such regulations exist, state agencies develop their own guidelines or consider materials on a case by case basis. State departments of agriculture are involved in registering by-products as fertilizers or lime substitutes, and may not effectively interact with state environmental agencies to ensure safe use of these materials. There are examples of quite hazardous materials being land applied where state agencies have failed to coordinate (Wilson, 1997).

24.5.1 Governmental Regulations Concerning Land Application

The USEPA regulates the handling, disposal, and land application of wastes under Congressional authority, and through rule making associated with legislative acts. Nearly all such regulation deals with industrial and municipal by-products, with agricultural by-products being expressly exempt (this exemption may change as manure applications are linked with water quality degradation). State departments of agriculture and/or environmental affairs implement federal regulations, and may modify them in some cases. The EU has also adopted resolutions, implemented by member countries, regulating handling and use of wastes.

The primary legislation affecting industrial wastes in the United States is the RCRA of 1976, which defines waste and, in particular, the concept of hazardous waste, which must be specially handled and typically cannot be land applied (40 CFR 264-5). Hazardous waste is liquid or solid material derived from a manufacturing process that has one of the following characteristics: (1) corrosivity, typically due to pH > 12 or pH < 2 of liquid or liquid derived from a solid; (2) ignitability or explosivity largely associated with organic solvents and petroleum wastes; or (3) toxicity, as determined by trace contaminant levels of leachates as measured in an acetate buffer at pH 5.2 (Toxicity Characteristic Leaching Procedure, TCLP). The presence of certain toxic organics and radionuclides automatically classifies a waste as hazardous. Hazardous wastes fall under Subtitle C of RCRA, and must be transported and disposed of in a proscribed manner (e.g., hazardous waste landfills); nonhazardous waste falls under Subtitle D, and can be disposed of in less secure (but still lined and monitored) landfills, and may be potentially land applied.

MSWs, FGD, most kiln dusts (not containing restricted trace contaminants), most pulp and paper by-products (except highly alkaline residues or high dioxin sludges), and many other more or less inert by-products are classified nonhazardous. Biosolids are exempt, being covered under special rules of 40 CFR 503. Fly ash is also exempt, but may be restricted in use if it fails the TCLP. Flue dusts typically fail the TCLP, but may be used as micronutrient fertilizers if Pb and Cd are low. PG has been regulated by USEPA due to its radionuclide content. Animal manures and most agricultural by-products are currently exempt and unrestricted.

Biosolid use on land is regulated separately under authority of the CWA in 40 CFR 503 (USEPA, 1993a, 1993b). This ruling specifies treatment for pathogen reduction as well as sets upper limits for metal concentrations, as well as application methods for pathogen vector control and use of materials at agronomically appropriate rates (i.e., based on nutrient contents). "Exceptional quality" (Class A) biosolids are defined as having been treated (via heat, lime stabilization, etc.) to stabilize organic matter and reduce total pathogen load, and also must have metal concentrations below certain limits (Table 24.7); these materials can be used without restriction or record-keeping. Use of materials that do not meet "exceptional" criteria (referred to as Class B) is more

TABLE 24.7 USEPA and EU Limits for Metals in Biosolids and Loading Rates for Land Application of Biosolids (Annual and Lifetime)

	USEPA			EU		
	Metal[a] Concentrations	Annual Loading	Cumulative Loading	Metal[d] Concentrations	Annual Loading	Cumulative Loading
Element	mg kg⁻¹	kg ha⁻¹		mg kg⁻¹	kg ha⁻¹	
As	41–75	2	41	—	(0.7)[e]	(100)[e]
Cd	39–85	1.9	39 {21}[f]	20–40	0.15	2–6
Cu	1500–4300	75	1500	1000–1750	12	100–280
Cr	[3000][b]	150	3000	—	(15)	(800)
Pb	300–840	15	300	750–1200	15	100–600
Hg	17–57	0.85	17	16–25	0.1	2–3
Mo	[18–75][c]	[0.9][c]	[18][c]	—	(0.2)	(8)
Ni	[420][b]	21	420	300–400	3	60–150
Se	[100][b]	5	100	—	(0.15)	(6)
Zn	2800–7500	140	2800	2500–4000	30	300–600

Sources: USEPA. 1993a. Cement kiln dust-Report to Congress. Available at http://www.epa.gov/osw/nonhaz/industrial/special/ckd/cement2.htm (accessed May 2009); Smith, S.R. 1996. Agricultural recycling of sewage sludge and the environment. CAB International, Wallingford, U.K.

[a] Lower value is for high quality unrestricted use; higher value is ceiling for application.
[b] Limits for these metals are tentative.
[c] Limits on Mo are being removed by USEPA.
[d] Lower value for soils at pH < 6, higher value for pH > 7.
[e] Values in parenthesis are U.K. limits; no EU values specified for these metals.
[f] USEPA is considering reducing Cd loading to 21 kg ha⁻¹ at USDA's request.

restricted in terms of application conditions in order to limit pathogen vectors, and metal additions must be recorded to limit annual and lifetime loading rates (kg metals ha⁻¹) for a given site.

Current USEPA regulations allow metal additions that are considerably higher than previous regulations, and are, in general, greater by a factor of 5–15 for most metals than EU soil-loading rates (EU Directive 86/278/EEC; Smith, 1996; Table 24.7). Allowable metal concentrations do not differ significantly between the two, but maximum rates of application in Mg ha⁻¹ allowed (computed as [loading in kg ha⁻¹]/[concentration in mg kg⁻¹] × 1000) are quite different. Using the lower metal concentration limits for EU sludges and corresponding annual metal loading rates, only about 5–10 Mg ha⁻¹ year⁻¹ of sludge can be applied, compared to 50 Mg ha⁻¹ year⁻¹ for the USEPA limits. Depending on N and P needs of the crop on a given site and the available nutrients in the biosolids, more than 10 Mg ha⁻¹ may be needed to meet annual nutrient needs, therefore requiring supplemental inorganic fertilizer (Section 24.3.1). Rates based on N or P required would seldom exceed the 50 Mg ha⁻¹ limit for metals based on USEPA values.

Metal limits in 40 CFR 503 are often applied to the utilization of other by-products for land application. This is probably justified for other organic by-products such as MSW compost or pulp and paper sludges, which have a similar matrix and impact on soil chemistry. However, for inorganic materials such as gypsum, fly ash, and by-product dusts with a largely inorganic matrix, this could lead to increases in leachable or plant-available metals above safe levels, due to the fact that metals in an organic matrix are generally less soluble (Section 24.5.2).

24.5.2 Soil and Food-Chain Contamination

Contamination of soil and subsequent uptake by plants and transfer to animals are major risk pathways for metals and organic contaminants in by-products applied to land. A large body of information is available concerning such risks for biosolid applications, summarized recently in a number of books and review articles (Chang et al., 1987; Chaney, 1994; Hue, 1995; Smith, 1996). For other types of by-products, there is no systematic evaluation of contaminant effects, but there are data that allow some interpretation of potential risks, particularly using the experience with biosolids as a guiding example.

24.5.2.1 Trace Inorganic Contaminants

Trace metals and metalloids have a long history of research and regulation in land application. Risk assessment analysis of trace inorganics with respect to leaching, food chain impact, and direct soil ingestion form the basis of USEPA biosolids regulation. These elements are typically insoluble in soils and generally considered to have limited plant availability and leaching potential due to their strong affinity to organic materials when applied in manures and biosolids. Basta et al. (2005) note that this strong adsorption, coupled with inherent limits to plant uptake, allow relatively high loadings of trace elements to soils without environmental consequences. However, there are many "exceptions" for specific element/plant/soil combinations that may short-circuit these processes (McBride, 2003).

Zinc, copper, and nickel tend to be toxic to plants before they accumulate sufficient tissue concentrations to affect animals or humans. As a result, over application tends to kill or stunt plants, preventing poisoning of animals or humans consuming them. Zinc is readily translocated to plant shoots, while Cu and Ni are toxic to plant roots without much movement to leaves. None of the elements moves appreciably into plant seeds. All are more soluble at pH < 6, and therefore, careful monitoring of pH can limit plant availability on amended soils. The 40 CFR 503 limits for these metals are quite high, and many by-products contain appreciable concentrations. In organic materials, solubility is limited by complexation by organic matter; in inorganic materials such as flue dusts or fly ash, solubilities may be significantly higher (Ainsworth and Rai, 1987; Hue, 1995). Solubility decreases with time after application due to reactions with soil mineral surfaces.

Because swine and poultry manures may contain elevated Cu and Zn due to feed supplementation, repeated applications can result in elevated levels of total and extractable Cu (Anderson et al., 1991; del Castilho et al., 1993; Kingery et al., 1994) and Zn (Christie and Beattie, 1989; Kingery et al., 1994) in soils, but it is still not clear whether this is a significant problem in terms of plant uptake and subsequent transfer through the food chain.

Similarly, lead, chromium, and mercury are not taken up by plants to any degree, and thus are not limited by risk pathways that involve plants. These elements are strongly adsorbed by soil and are quite insoluble. Rather, direct soil ingestion by children living on amended soils is the limiting pathway for these metals. This scenario assumes 1–6 year old children ingest 200 mg day^{-1} of contaminated soil, increasing risks of Pb-related neurological damage (Smith, 1996). Similarly, animals in grazing pastures may ingest considerable amounts of by-product directly as soil and material coating leaves, and thus are exposed to these metals even without direct plant uptake. For these metals, 40 CFR 503 limits should be fairly readily transferred to other by-products, given that no soil or plant chemistry is involved in the risk assessment.

Cadmium is perhaps the most important potential contaminant of food supplies on amended lands. Being relatively soluble in soils (particularly at low soil pH) and chemically similar to Zn, it is readily taken up by crop plants and is quite toxic to humans. Cereal grains in the EU must contain less than 0.1 mg kg^{-1} Cd to be marketed (Smith, 1996). The 40 CFR 503 limit of 39 mg kg^{-1} Cd in biosolids has been criticized as overly liberal, and is based on a risk pathway involving direct ingestion of sludge or soil–sludge mixtures by children, rather than plant uptake. Chaney (1994) cites studies that indicate that Cd uptake by plants on biosolids-amended soil is seldom excessive due to the high Zn levels usually present, but also notes that some grains in the United States exceed the EU Cd limit, and that lower Cd loadings might be desirable. He indicates that 21 mg kg^{-1} Cd in biosolids (and a 21 kg ha^{-1} loading limit) may be necessary to limit plant uptake to acceptable levels.

Arsenic, selenium, and molybdenum typically are present in lower concentrations in biosolids, but may be elevated in wood and fly ash and in various by-product dusts. Fly ash from high S coals in particular has elevated As contents, as well as appreciable Se and Mo (Table 24.3). These elements are anionic in soils, which means they are more soluble at higher pH (opposite of the metallic trace elements) due to strong adsorption on Fe oxides at low pH. Both Se and Mo are required trace elements for animals, and Se has been cited as deficient in human diets in the eastern United States (Adriano, 1986). Both are toxic at higher concentrations. Levels of As or Se in food or feed crops above 0.25–1 mg kg^{-1} are considered a potential hazard. Arsenic is similar in chemistry to P and is therefore taken up by plants, although a wide range of factors influence uptake. While poultry litter contains 15–30 mg kg^{-1} As due to additions to feed (Toor et al., 2007), Staed et al. (2009) found no accumulation of As in the top 8 cm of soil after repeated applications of poultry litter. Molybdenum is a concern for grazing ruminants, which suffer Mo-induced Cu deficiency if the Cu/Mo ratio of forage falls below a value in the range of 2 (Miltimore and Mason, 1971; Chang et al., 1987) to 5 (Suttle, 1991). It is taken up in large amounts by legumes at soil pH > 6.5, however, and O'Connor et al. (2001) have suggested a limit of 40 kg ha^{-1} be established as the maximum Mo loading on biosolids-amended soils using a risk assessment based on livestock grazing.

24.5.2.2 Trace Organic Contaminants

These are present in many types of by-products, and may classify an individual material as hazardous under RCRA if present in sufficient quantities. Safe levels in by-products for land application have not been determined, although future revisions of the 40 CFR 503 rule for biosolids may address limits for specific compounds. Rule making on these contaminants has been limited by difficulty and cost of analysis, and limited studies in the literature on occurrence, reactions, and transformations in the environment (Hue, 1995).

Smith (1996) has summarized types of toxic organics likely to be found in biosolids. The most important classes, each of which contains an array of individual compounds, are polychlorinated dibenzo-p-dioxins and furans (PCDD/Fs), polynuclear aromatic hydrocarbons (PAHs), and PCBs. All are moderately high molecular weight, nonvolatile and insoluble compounds that adsorb strongly to soil, show little tendency to be taken up by plants, and degrade slowly in the soil environment. Biosolids typically contain <1 mg kg^{-1} of PCBs and PAHs, and considerably less PCDD/Fs.

Pulp and paper mill sludges, as well as wood and fly ash have been suspected of containing dioxins and furans (PCDD/Fs) originating from Cl_2 treatment during pulp bleaching and from combustion, particularly in the presence of Cl salts. A 1988 survey of 104 kraft mill sludges derived from Cl_2 bleaching processes gave TCDD (dioxin) values averaging 70 ng kg^{-1} and 600 ng kg^{-1} TCDF (furans) (Camberato et al., 1997; Miner, 1997). Provisional standards for materials to be land applied have been set at 10–20 ng kg^{-1} based on toxic equivalent (TEQ) of 2,3,7,8-TCCD, the most toxic of the dioxin compounds. Non-Cl_2 bleaching methods are being widely adopted by the paper industry, and result in TEQs uniformly <10 ng kg^{-1}. Interestingly, six processed municipal biosolids contained from 13 to 75 ng kg^{-1} TEQs, significantly higher than the clean technology pulp sludges (Miner, 1997).

Fly ash from boilers burning wood and fossil fuels at pulp mills may contain significant dioxins depending on boiler configuration and fuel type (Someshwar, 1996). Computed TCDD TEQs mean values ranged from 20 to 400 ng kg^{-1}, with one data set averaging 1800 ng kg^{-1} when salt-saturated wood was used as fuel.

Overall, considerably more study on organic contaminants is required, and USEPA is likely to consider organics in revisions of the 40 CFR 503 rule, despite difficulties in obtaining data and assessing risks for these compounds (Smith, 1996).

24.5.2.3 Antibiotics, Hormones, and Pesticides

Some land-applied by-products, notably animal manures and biosolids, contain variable levels of antibiotics, hormones, and pesticides. They are discussed separately here because an extensive amount of recent research has revealed that these compounds, particularly antibiotics and hormones, may pose unique, largely unanticipated and very serious threats to the environment, including potential threats to human health.

Antibiotics are used in livestock and poultry production in two ways. In therapeutic use, antibiotics are administered at relatively high dosages to animals that are exhibiting symptoms of a particular disease. Subtherapeutic use, on the other hand, involves adding antibiotics at lower rates to the animals' feed, regardless of whether they are exhibiting symptoms, to promote weight gain and growth rate in the animals. In their early review of this subject, Sims and Wolf (1994) list 13 different antibiotics (along with 5 coccidiostats and 7 larvacides) that have been added to poultry diets. It is estimated that 75% of these antibiotics are not absorbed by the animals and are excreted in the feces and/or urine (Chee-Sanford et al., 2009). When these wastes are land applied, the antibiotics they contain may enter the soil, move through the vadose zone to groundwater, or enter surface waters through runoff and erosion (Halling-Sørensen et al., 1998). Among the many factors that control their ultimate fate in the environment are water solubility, tendency to adsorb to soil colloids, and resistance to both biotic and abiotic degradation (Gavalchin and Katz, 1994; Campagnolo et al., 2002; Burkhardt et al., 2005; Hamscher et al., 2005; Kester et al., 2005; Sarmah et al., 2006a, 2006b; Dolliver and Gupta, 2008a, 2008b; Sanders et al., 2008; Chee-Sanford et al., 2009). Antibiotic uptake by plants from soil fertilized with animal manure has also been reported (Tietjen, 1975; Kumar et al., 2005; Dolliver et al., 2007).

Bacteria inhabiting soil repeatedly treated with animal manures containing antibiotics (e.g., tetracycline) develop resistance to these antibiotics (Khachatourians, 1998; Sengelov et al., 2003; Heuer and Smalla, 2006). This is of particular concern when the resistant organisms are also human pathogens (e.g., salmonella, streptococcus, etc.). According to Gilchrist et al. (2007), the reduced effectiveness of several classes of antibiotics for treating infections in humans may be a result of the increasing land application of animal manures containing antibiotics. Along with the World Health Organization, Gilchrist et al. (2007) has called for a gradual phasing-out of the subtherapeutic use of antibiotics in animal agriculture. Until that is achieved, others have suggested that all manure that is to be land applied should be composted first. Composting and in some cases just stockpiling reduces the number of antibiotic resistant bacteria in manure (Dolliver et al., 2008; Sharma et al., 2009).

Much less information is available in the literature regarding the presence of antibiotics in biosolids. This is presumably due to the fact that the quantity of antibiotics used therapeutically to treat human infectious diseases is small compared to the quantity used in animal agriculture. As a result, biosolids contain much lower levels of antibiotics than do animal manures. However, biosolids do contain what are referred to as pharmaceuticals and personal care products (PPCPs). Xia et al. (2005) list the most common PPCPs found in the effluent of wastewater treatment plants. Information on the relationship between concentrations of PPCPs in soil and land application of biosolids containing PPCPs is scarce, as is information on the fate and transport of PPCPs in biosolids-treated soils. Additional research is needed in this area.

Biosolids and animal manures contain natural steroidal hormones (e.g., 17β-estradiol, testosterone) and synthetic compounds with estrogen-like activity (e.g., 17α-ethynylestradiol, 4-nonylphenol) (Bolan, 2004; Lorenzen et al., 2006; Sarmah et al., 2006a, 2006b). Such chemical compounds are collectively referred to as endocrine disrupting chemicals (EDCs) because of their effects on endocrine system function in both humans and other animals (Colburn et al., 1993; Sonnenschein and Soto, 1998). Land application of biosolids and manures may be an avenue by which these EDCs are introduced into the environment. Shore et al. (1995) and Nichols et al. (1997) were among the first to report runoff losses of 17β-estradiol from fields fertilized with poultry litter. Several recent studies indicate that decomposition of the EDCs present in manures and biosolids is relatively rapid under conditions representative of a well-aerated, temperate region soil (Colucci et al., 2001; Jacobsen et al., 2005; Lorenzen et al., 2006). On the other hand, Sangsupan et al. (2006) and Lægdsmand et al. (2009) reported significant leaching of EDCs through intact soil columns, suggesting that EDC contamination of groundwater could occur under conditions of preferential flow. What effects, if any, these land applied EDCs have on aquatic organisms and humans is unknown at this time.

The most common class of pesticides found in beef cattle manures are parasiticides such as organophosphates, carbamates, and synthetic pyrethroids (Khan et al., 2008). Sims and Wolf (1994) report that a variety of coccidiostats and larvacides are added to poultry diets. Very little is known about the environmental fate and transport of these substances once they are introduced into the environment through the land application of animal manures. Additional research is urgently needed.

24.5.2.4 Nitrate and Ammonium

These are included here as they can cause toxicities to plant and animals on by-product-amended soils. Manures (especially poultry) or biosolids may contain appreciable inorganic NH_4, or rapidly mineralize N as NH_4 under optimal conditions. This NH_4 can be converted to NH_3 gas at high pH > 7. Ammonia is quite toxic to plants, and may cause appreciable damage to crops, especially seedlings (USDA, 1992). In addition, high levels of NH_3 can cause NO_2^-, which is very toxic to plants, to accumulate in soil (Schumann and Sumner, 1998).

Nitrate may accumulate in plant tissues during certain times of the year in response to high N levels in sludge- or manure-amended soils. Forage crops on soils heavily amended with biosolids have been found to contain high levels of nitrate (400 mg kg^{-1} N as NO_3), particularly when the forage is drought stressed. Ruminants are quite susceptible to NO_3 poisoning, and significant mortality of cows fed with hay at such NO_3 levels has been observed. The level at which NO_3 is toxic depends on the animals' condition and whether they have been acclimated to nitrates in the diet.

24.5.3 Water Quality Considerations

The adverse impacts of improper land application of by-products on both surface and groundwater quality can be severe. However, by far the greatest incidence of such impacts involves N and P in animal manures, rather than trace contaminants in industrial by-product, and such problems are commonly traced to poor management or over-application of manures. On the other hand, many instances of water contamination can be traced to disposal of wastes in landfills or stockpiling of by-products at the point of production. The reasonable reuse of these materials on crop land, where appropriate, would therefore reduce the likelihood of water contamination.

Applications of animal manure can cause leaching of high levels of NO_3^- to groundwater and enrichment of surface waters with P contained in runoff. Instances of groundwater pollution with NO_3^- in England are now so common that legislation is being considered to limit N applications to crop land to <200 kg ha^{-1} (Smith, 1996). Both the United States and United Kingdom limits are 10 mg NO_3-N L^{-1} in groundwater, and soils receiving high rates of manures or high N sludges often release NO_3^- rapidly enough to pollute shallow aquifers, especially on sandy soils (Bergström and Kirchmann, 2006).

In recent years, as point sources of water pollution have been identified and eliminated, the importance of nonpoint source water pollution has become more apparent. Agricultural activities are believed to cause a considerable amount of the nonpoint source pollution of both surface and groundwater in the United States and other developed countries. Furthermore, land application of animal manures is now suspected of being a major P source, which is frequently the nutrient limiting biological productivity in freshwater lakes (Sharpley and Rekolainen, 1997). The recent outbreak of *Pfiesteria*-like organisms in Chesapeake Bay, whose toxins are believed to have caused memory loss in humans and the death of thousands of fish (Martin and Cooke, 1994), may be linked to elevated P levels in the Bay, which is surrounded by an area of intensive animal production (MacIlwain, 1997).

The nutrient that is of primary concern with regard to water pollution resulting from land application of animal manures is P (Daniel et al., 1994; Sharpley et al., 1994). The relative amounts of N and P contained in many animal manures are such that when application rates of manure are calculated based on the N requirement of the crop, the amount of P added often exceeds the P requirement of the crop. This results in accumulation of P in the soil (Sharpley et al., 1993; Slaton et al., 2004). While most soils have the capacity to fix, or immobilize, large amounts of P, this capacity is finite and can be exceeded if N-based manure applications are made repeatedly over a period of years. As the soil's ability to fix P declines, a greater proportion of the P in the soil exists in a water-soluble form (as the orthophosphate anion), which is highly susceptible to edge of field loss in runoff. Except in very sandy soils, which have low P-fixation capacities, leaching of P to groundwater is generally not considered to be a major loss mechanism of P from manure-amended soils. Runoff of soluble P from surface-applied manure can be a significant contributor to soluble P in surface waters.

Loss of N from fields receiving animal manures is also a concern. While most of the N in animal manures is in the form of organic N compounds or NH_4^+, mineralization and nitrification can result in the formation of large amounts of NO_3^- following manure application if the soil is warm and well-aerated (Johnson and Wolf, 1995). Unlike the $H_2PO_4^-$ anion, NO_3^- is not fixed in appreciable quantities by most soils and is, therefore, more likely to be leached through the soil to groundwater (Unwin, 1986; Adams et al., 1994; Bergström and Kirchmann, 2006).

In addition to N and P, it has been shown that runoff from manure-amended soils can also contain pesticides and hormones. Pote et al. (1994) found concentrations of cyromazine, a feed-through larvacide, as high as 101 g L^{-1} in runoff from plots that had been treated with 17.7 Mg ha^{-1} of caged layer manure. Cyromazine was not detected in water samples collected from lysimeters placed at a depth of 60 cm in the soil for up to 1 year after the manure was applied.

Nichols et al. (1997) found elevated levels of the hormone 17β-estradiol in runoff from poultry litter-amended pastures. The effects that compounds such as these may have on aquatic ecosystems are not currently known.

The potential for pollution of surface waters resulting from land application of animal manures can be minimized if certain management practices are followed (Daniel et al., 1994). The primary factors affecting quality of runoff from animal manure-amended fields are application rate, method of application, and timing of application (Sharpley, 1997). In the case of P, to this list should be added the soil P status (or the degree to which the P-fixation capacity is already saturated) because it is well documented that the amount of P in the surface soil directly influences the amount of P in runoff (Daniel et al., 1997). Research has shown that concentrations of N and P in runoff increase as the manure application rate increases (Edwards and Daniel, 1993a, 1993b) and that runoff N and P concentrations decrease as the interval of time between manure application and the rainfall event increases (Westerman and Overcash, 1980; Edwards and

Daniel, 1993a, 1993b). Incorporation of manure into the soil generally decreases concentrations of N and P in runoff (Mueller et al., 1984), although Nichols et al. (1994) reported that shallow (2–3 cm) incorporation of poultry litter applied to established fescue pasture had no effect on runoff N and P concentrations. Furthermore, incorporation may increase erosion and edge of field losses of particulate P.

With regard to trace metals and organic contaminants in land-applied products, leaching of contaminants to groundwater or transport to surface waters is unlikely with good management of amended soils (Smith, 1996). This assumes application rates have been chosen based on limited metal loadings (less than 40 CFR 503 limits), that trace organics are limited to mg kg^{-1} levels or less, and that soils are properly limed and runoff and erosion controlled. The solubility of contaminants under these conditions is so low that movement by leaching is very slow, and with control of surface runoff, little export to surface water is expected.

24.6 Recommendations

1. Before considering a land-applying by-product material, check with state departments of agriculture and/or environmental divisions to see if the product has been registered or approved for use on land, or is likely to be unsuitable (classified as hazardous waste based on testing or known levels of contaminants).
2. By-products have a wide range of properties that affect potential agronomic value and environmental limitations, even within one class of materials. A current, accurate analysis of important chemical properties of the material (nutrient content, pH, liming value, contaminant levels) is the most important piece of information needed to make decisions on use and application rates. Ask the producer to provide this information, or be ready to assume the cost of such an analysis.
3. By-products are applied based on one of the following rationales
   a. as a nutrient source, typically either for N or P
   b. as a liming agent, determined by the CCE
   c. as a soil conditioner, in order to increase SOM, water-holding capacity, etc.

   Application rates should be calculated for nutrient sources and liming material based on (available) nutrients supplied or lime required to reach a target pH. For soil conditioners, specific rates are difficult to determine agronomically, but should be environmentally conservative (see point 4).
4. Compute annual loading rates of metals or other contaminants resulting from the application; for biosolids, use the 40 CFR 503 regulations to insure loadings are safe. For other by-products, use 40 CFR 503 limits as guidelines. Be aware of conditions that may make these limits not applicable for low C by-products. Keep accurate records of loadings of metals and nutrients for individual fields over time, and do not exceed 40 CFR 503 lifetime cumulative limits.
5. Use good management practices on fields to prevent chemical (metals, organics) or microbial (pathogen) losses through runoff or erosion. Follow soil and water conservation practices.
6. Soil test for pH and nutrients annually at cooperative extension service or private laboratories. Maintain pH > 6 with lime additions, especially with organic by-products, to minimize metal solubility. Monitor soil P and metal levels, and reduce applications if soil test levels exceed high ratings over time. Check soluble (NO_3^- and NH_4^+) N levels in soils periodically to ensure that excess soluble N is not accumulating in soil.

References

Adams, P.L., T.C. Daniel, D.R. Edwards, D.J. Nichols, D.H. Pote, and H.D. Scott. 1994. Poultry litter and manure contributions to nitrate leaching through the vadose zone. Soil Sci. Soc. Am. J. 58:1206–1211.

Adeli, A., K.R. Sistani, D.E. Rowe, and H. Tewolde. 2007. Effects of broiler litter applied to no-till and tillage cotton on selected soil properties. Soil Sci. Soc. Am. J. 71:974–983.

Adriano, D. 1986. Trace elements in the terrestrial environment. Springer-Verlag, New York.

Aina, P.O., and E. Egolum. 1980. The effect of cattle feedlot manure and inorganic fertilizer on the improvement of subsoil productivity of two soils. Soil Sci. 129:212–217.

Ainsworth, C.C., and D. Rai. 1987. Chemical characterization of fossil fuel combustion wastes. EPRI Report EA-5321. Electric Power Research Institute, Palo Alto, CA.

Alcordo, I.S., and J.E. Rechcigl. 1995. Phosphogypsum and other byproduct gypsums, p. 365–426. In J.E. Rechcigl (ed.) Soil amendments and environmental quality. Lewis Publishers, Boca Raton, FL.

American Coal Ash Association. 2006. Coal combustion product survey: Production and use. Available at http://acaa.affiniscape.com/index.cfm (accessed May 2009).

Anderson, M.A., J.R. McKenna, D.C. Martens, S.J. Donohue, E.T. Kornegay, and M.D. Lindemann. 1991. Long-term effects of copper rich swine manure application on continuous corn production. Commun. Soil Sci. Plant Anal. 22:993–999.

Angle, J.S. 1994. Sewage sludge: Pathogenic considerations, p. 35–39. In C.E. Clapp, W.E. Larson, and R.H. Dowdy (eds.) Sewage sludge: Land utilization and the environment. Soil Science Society of America, Madison, WI.

Barker, A.V. 1997. Composition and uses of compost, p. 140–163. In J.E. Rechcigl and H.C. MacKinnon (eds.) Agricultural uses of byproducts and wastes. American Chemical Society, Washington, DC.

Basta, N.T., J.A. Ryan, and R.L. Chaney. 2005. Trace element chemistry in residual-treated soil: Key concepts and metal bioavailability. J. Environ. Qual. 34:49–63.

Bastian, R.K. 2005. Interpreting science in the real world for sustainable land application. J. Environ. Qual. 34:174–183.

Beecher, N., E. Harrison, N. Goldstein, M. McDaniel, P. Field, and L. Susskind. 2005. Risk perception, risk communication, and stakeholder involvement for biosolids management and research. J. Environ. Qual. 34:122–128.

Bergström, L., and H. Kirchmann. 2006. Leaching and crop uptake of nitrogen and phosphorus from pig slurry as affected by different application rates. J. Environ. Qual. 35:1803–1811.

Bolan, N.S. 2004. An overview of the environmental effects of land application of farm effluents. N. Z. J. Agric. Res. 47:389–403.

Brady, N.C., and R. Weil. 1996. The nature and properties of soils (11th Ed.). Prentice Hall, Upper Saddle River, NJ.

Brady, N.C., and R. Weil. 2008. The nature and properties of soils, 14th Edn. Prentice Hall, Upper Saddle River, NJ.

Brooks, J.P., B.D. Tanner, C.P. Gerba, C.N. Haas, and I.L. Pepper. 2005. Estimation of bioaerosol risk of infection to residents adjacent to a land applied biosolids site using an empirically derived transport model. J. Appl. Microbiol. 98:397–405.

Burkhardt, M., C. Stamm, C. Waul, H. Singer, and S. Müller. 2005. Surface runoff and transport of sulfonamide antibiotics and tracers on manured grassland. J. Environ. Qual. 34:1363–1371.

Cabrera, M.L., and R.M. Gordillo. 1995. Nitrogen release from land-applied manures, p. 393–403. In K. Steele (ed.) Animal waste and the land-water interface. CRC Press, Boca Raton, FL.

Cabrera, M.L., D.E. Kissel, and M.F. Virgil. 2005. Nitrogen mineralization from organic residues: Research opportunities. J. Environ. Qual. 34:75–79.

Callahan, M.P., P.J.A. Kleinman, A.N. Sharpley, and W.L. Stout. 2002. Assessing the efficacy of alternative phosphorus sorbing soil amendments. Soil Sci. 167:539–547.

Camberato, J.J., E.D. Vance, and A.V. Someshwar. 1997. Composition and land application of paper manufacturing residuals, p. 185–203. In J.E. Rechcigl and H. MacKinnon (eds.) Agricultural uses of byproducts and wastes. American Chemical Society, Washington, DC.

Campagnolo, E.R., K.R. Johnson, A. Karpati, C.S. Rubin, D.W. Kolpin, M.T. Meyer, J.E. Esteban, R.W. Currier, K. Smith, K.M. Thu, and M. McGeehin. 2002. Antimicrobial residues in animal waste and water resources proximal to large-scale swine and poultry feeding operations. Sci. Total Environ. 299:89–95.

Cela, S., and M.E. Sumner. 2002. Critical concentrations of copper, nickel, lead and cadmium in soils based on nitrification. Comm. Soil Sci. Plant Anal. 33:19–30.

Center for Food Safety. 2003. Petition seeking an emergency moratorium on the land application of sewage sludge. Submitted to U.S. Environmental Protection Agency. Available online at http://www.centerforfoodsafety.org/pubs/FinalPetitionSludge.pdf (accessed May 2009).

Chaney, R.L. 1994. Trace metal movement: Soil-plant systems and bioavailability of biosolids-applied metals, p. 27–32. In E.E. Clapp, W.E. Larson, and R.H. Dowdy (eds.) Sewage sludge: Land utilization and the environment. Soil Science Society of America, Madison, WI.

Chang, A.C., T.D. Hinesly, T.E. Bates, H.E. Doner, R.H. Dowdy, and J.A. Ryan. 1987. Effects of long-term sludge application on accumulation of trace elements by crops, p. 53–66. In Land application of sludge. Lewis Publishers, Chelsea, MI.

Chee-Sanford, J.C., R.I. Mackie, S. Koike, I.G. Krapac, Y.-F. Lin, A.C. Yannarell, S. Maxwell, and R.I. Aminov. 2009. Fate and transport of antibiotic residues and antibiotic resistance genes following land application of manure waste. J. Environ. Qual. 38:1086–1108.

Choudhary, M., L.D. Bailey, and C.A. Grant. 1996. Review of the use of swine manure in crop production: Effects on yield and composition and on soil and water quality. Waste Manage. Res. 14:581–595.

Christie, P., and J.A.M. Beattie. 1989. Grassland soil microbial biomass and accumulation of potentially toxic metals from long-term slurry application. J. Appl. Ecol. 26:597–607.

Christie, P., and D.J. Kilpatrick. 1992. Vesicular-arbuscular mycorrhiza infection in cut grassland following long-term slurry application. Soil Biol. Biochem. 24:325–331.

Cochran, R., W. Miller, and L. Morris. 1997. Decomposition of pulp mill biosolids and effects on soil properties. Agron. Abstr., p. 275.

Colburn, T., F.S. vom Saal, and A.M. Soto. 1993. Developmental effects of endocrine-disrupting chemicals in wildlife and humans. Environ. Health Prospect. 101:378–389.

Colucci, M.S., H. Bork, and E. Topp. 2001. Persistence of estrogenic hormones in agricultural soils. I. 17β-estradiol and estrone. J. Environ. Qual. 30:2070–2076.

Cusick, P.R., K.A. Kelling, J.M. Powell, and G.R. Munoz. 2006. Estimates of residual dairy manure nitrogen availability using various techniques. J. Environ. Qual. 35:2170–2177.

Daniel, T.C., O.T. Carton, and W.L. Magette. 1997. Nutrient management planning, p. 297–309. In H. Tunney et al. (eds.) Phosphorus loss from soil to water. CAB International, Wallingford, U.K.

Daniel, T.C., A.N. Sharpley, D.R. Edwards, R. Wedepohl, and J.L. Lemunyon. 1994. Minimizing surface water eutrophication from agriculture by phosphorus management. J. Soil Water Conserv. 49:30–38.

Davis, J.G., G. Vellidis, R.K. Hubbard, J.C. Johnson, G.L. Newton, and R.R. Lowrance. 1995a. Nitrogen uptake and leaching in a no-till forage rotation irrigated with liquid dairy manure, p. 405–410. In K. Steele (ed.) Animal waste and the land-water interface. CRC Press, Boca Raton, FL.

Davis, J.G., G. Weeks, and M.B. Parker. 1995b. Use of deep tillage and liming to reduce zinc toxicity in peanuts grown on flue dust contaminated land. Soil Technol. 8:85–95.

del Castilho, P., W.J. Chardon, and W. Salomons. 1993. Influence of cattle-manure slurry application on the solubility of cadmium, copper, and zinc in a manured, acidic, loamy sand soil. J. Environ. Qual. 22:689–696.

Diaz, D.A.R., J.E. Sawyer, and A.P. Mallarino. 2008. Poultry manure supply of potentially available nitrogen with soil incubation. Agron. J. 100:1310–1317.

Dolliver, H., and S. Gupta. 2008a. Antibiotic losses from unprotected manure stockpiles. J. Environ. Qual. 37: 1238–1244.

Dolliver, H., and S. Gupta. 2008b. Antibiotic losses in leaching and surface runoff from manure-amended agricultural land. J. Environ. Qual. 37:1227–1237.

Dolliver, H., S. Gupta, and S. Noll. 2008. Antibiotic degradation during manure composting. J. Environ. Qual. 37:1245–1253.

Dolliver, H., K. Kumar, and S. Gupta. 2007. Sulfamethazine uptake by plants from manure-amended soil. J. Environ. Qual. 36:1224–1230.

Dormaar, J.F., C.W. Lindwall, and G.C. Kozub. 1988. Effectiveness of manure and commercial fertilizer in restoring productivity of an artificially eroded Dark Brown Chernozemic soil under dryland conditions. Can. J. Soil Sci. 68:669–679.

Duggan, J., M.P. Bates, and C.A. Phillips. 2001. The efficacy of subsurface flow reed bed treatment in the removal of *Campylobacter* spp., faecal coliforms and *Escherichia coli* from poultry litter. Inter. J. Environ. Health Res. 11:168–180.

Edwards, D.R., and T.C. Daniel. 1992. Environmental impacts of on-farm poultry waste disposal: A review. Biores. Technol. 41:9–33.

Edwards, D.R., and T.C. Daniel. 1993a. Drying interval effects on runoff from fescue plots receiving swine manure. Trans. ASAE 36:1673–1678.

Edwards, D.R., and T.C. Daniel. 1993b. Effects of poultry litter application rate and rainfall intensity on quality of runoff from fescuegrass plots. J. Environ. Qual. 22:361–365.

Eghball, B., and J.F. Power. 1994. Beef cattle feedlot manure management. J. Soil Water Conserv. 49:113–122.

Eghball, B., B.J. Weinbold, J.E. Gilley, and R.A. Eigenberg. 2002. Mineralization of manure nutrients. J. Soil Water Conserv. 57:470–473.

Epstein, R., R.L. Chaney, C. Henry, and T.J. Logan. 1992. Trace elements in municipal solid waste compost. Biomass Bioenerg. 3:227–238.

Erich, M. 1991. Agronomic effectiveness of wood ash as a source of phosphorus and potassium. J. Environ. Qual. 20:576–581.

Feuerborn, H.J. 2005. Coal ash utilization over the world and in Europe. Workshop on environmental and health aspects of coal ash utilization. Tel-Aviv, Israel, November 23–24, 2005. Available at http://www.coal-ash.co.il/sadna/Abstract_Feuerborn.pdf (accessed May 2009).

Freeze, B.S., C. Webber, C.W. Lindwall, and J.F. Dormaar. 1993. Risk simulation of the economics of manure application to restore eroded wheat cropland. Can. J. Soil Sci. 73:267–274.

Frye, W.W., O.L. Bennett, and G.J. Buntley. 1985. Restoration of crop productivity on eroded or degraded soils, p. 335–356. *In* R.F. Follett and B.A. Stewart (eds.) Soil erosion and crop productivity. American Society of Agronomy, Madison, WI.

Gale, E.S., D.M. Sullivan, C.G. Cogger, A.I. Bary, D.D. Hemphill, and E.A. Myhre. 2006. Estimating plant available nitrogen release from manures, composts, and specialty products. J. Environ. Qual. 35:2321–2332.

Gavalchin, J., and S.E. Katz. 1994. The persistence of fecal-borne antibiotics in soil. J. Assoc. Off. Anal. Chem. 77:481–485.

Gerba, C.P., and J.E. Smith, Jr. 2005. Sources of pathogenic microorganisms and their fate during land application of wastes. J. Environ. Qual. 34:42–48.

Ghodrati, M., J.T. Sims, and B.L. Vasilas. 1995. Evaluation of fly ash as a soil amendment for the Atlantic Coastal Plain: 1. Soil hydraulic properties and elemental leaching. Water Air Soil Pollut. 81:349–361.

Gilchrist, M.J., C. Greko, D.B. Wallinga, G.W. Beran, D.G. Riley, and P.S. Thorne. 2007. The potential role of concentrated animal feeding operations in infectious disease epidemics and antibiotic resistance. Environ. Health Perspect. 115:313–316.

Gilmour, J.T. 2009. Estimating yield and yield response using computer simulation of plant available nitrogen from soil organic matter and manure. Soil Sci. Soc. Am. J. 73:328–330.

Gilmour, J.T., and V. Skinner. 1999. Predicting plant available nitrogen in land-applied biosolids. J. Environ. Qual. 28:1122–1126.

Goldstein, B. 2004. The precautionary principle, toxicological science, and European-U.S. scientific cooperation. Drug Metab. Rev. 36:487–495.

Guan, T.Y., and R.A. Holley. 2003. Pathogen survival in swine manure environments and transmission of human enteric illness—A review. J. Environ. Qual. 32:383–392.

Hafez, A.A.R. 1974. Comparative changes in soil physical properties induced by admixtures of manures from various domestic animals. Soil Sci. 118:53–59.

Halling-Sørensen, B., S.N. Nielsen, P.F. Lanzky, F. Ingerslev, H.C.H. Lützhøft, and S.E. Jørgensen. 1998. Occurrence, fate and effects of pharmaceutical substances in the environment—A review. Chemosphere 36:357–393.

Hamscher, G., H.T. Pawelzick, H. Höper, and H. Nau. 2005. Different behavior of tetracyclines and sulfonamides in sandy soils after repeated fertilization with liquid manure. Environ. Toxicol. Chem. 24:861–868.

Harrison, E.Z., and S.R. Oakes. 2002. Investigation of alleged health incidents associated with land application of sewage sludges. New Solutions 12:387–408.

Harrison, E.Z., M.B. McBride, and D.R. Bouldin. 1999. Land application of sewage sludges: An appraisal of the U.S. regulations. Int. J. Environ. Pollut. 11:1–36.

Heuer, H., and K. Smalla. 2006. Manure and sulfadiazine synergistically increased bacterial antibiotic resistance in soil over at least two months. Environ. Microbiol. 9:657–666.

Hileman, L. 1967. The fertilizer value of broiler litter. AR Agricultural Experiment Station Report Series 158. University of Arkansas, Fayeteville, AR.

Hill, R.L., and B.R. James. 1995. Influence of waste amendments on soil properties, p. 311–326. *In* J.E. Rechcigl (ed.) Soil amendments and environmental quality. Lewis Publishers, Boca Raton, FL.

Holley, R.A., K.M. Arrus, K.H. Ominski, M. Tenuta, and G. Blank. 2006. Salmonella survival in manure-treated soils during simulated seasonal temperature exposure. J. Environ. Qual. 35:1170–1180.

Hue, N.V. 1995. Sewage sludge, p. 199–247. *In* J.E. Rechcigl (ed.) Soil amendments and environmental quality. Lewis Publishers, Boca Raton, FL.

Hyatt, G.W. 1995. Economic, scientific, and infrastructure basis for using municipal composts in agriculture, p. 19–72. *In* D.L. Karlen, R.J. Wright, and W.O. Kemper (eds.) Agricultural utilization of urban and industrial byproducts. American Society of Agronomy, Madison, WI.

Jacobsen, A.-M., A. Lorenzen, R. Chapman, and E. Topp. 2005. Persistence of testosterone and 17β-estradiol in soils receiving swine manure or municipal biosolids. J. Environ. Qual. 34:861–871.

Johnson, W.F., Jr., and D.C. Wolf. 1995. Nitrogen transformations in soil amended with poultry litter under aerobic conditions followed by anaerobic periods, p. 27–34. *In* K. Steele (ed.) Animal waste and the land-water interface. CRC Press, Boca Raton, FL.

Kalyoncu, R. 2001. Coal combustion products. *In* USGS minerals handbook, 2001. Available at http://minerals.usgs.gov/minerals/pubs/commodity/coal/874400.pdf (accessed May 2009).

Kester, G.B., R.B. Brobst, A. Carpenter, R.L. Chaney, A.B. Rubin, R.A. Schoof, and D.S. Taylor. 2005. Risk characterization, assessment, and management of organic pollutants in beneficially used residual products. J. Environ. Qual. 34:80–90.

Khachatourians, G.G. 1998. Agricultural use of antibiotics and the evolution and transfer of antibiotic-resistant bacteria. Can. Med. Assoc. J. 159:1129–1136.

Khan, S.J., D.J. Roser, C.M. Davies, G.M. Peters, R.M. Stuetz, R. Tucker, and N.J. Ashbolt. 2008. Chemical contaminants in feedlot wastes: Concentrations, effects and attenuation. Environ. Intl. 34:839–859.

Khuder, S., S. Milz, M. Bisesi, R. Vincent, W. McNulty, and K. Czajkowski. 2007. Health survey of residents living near farm fields permitted to receive biosolids. Arch. Environ. Occup. Health 62:5–11.

King, L.D. (ed.). 1986. Agricultural use of municipal and industrial sludges in the southern U.S. NC State University Southern Cooperative Series Bulletin 314. North Carolina State University, Raleigh, NC.

Kingery, W.L., C.W. Wood, D.P. Delaney, J.C. Williams, and G.L. Mullins. 1994. Impact of long-term land application of broiler litter on environmentally related soil properties. J. Environ. Qual. 23:139–148.

Kleinman, P.J.A., A.M. Wolf, A.N. Sharpley, D.B. Beegle, and L.S. Saporito. 2005. Survey of water extractable phosphorus in livestock manures. Soil Sci. Soc. Am. J. 69:701–708.

Korcak, R.F. 1995. Utilization of coal combustion byproducts in agriculture and horticulture, p. 107–130. *In* D.L. Karlen, R.J. Wright, and W.O. Kemper (eds.) Agricultural utilization of urban and industrial byproducts. American Society of Agronomy, Madison, WI.

Kumar, K., S.C. Gupta, S.K. Baidoo, Y. Chander, and C.J. Rosen. 2005. Antibiotic uptake by plants from soil fertilized with animal manure. J. Environ. Qual. 34:2082–2085.

Lægdsmand, M., H. Andersen, O.H. Jacobsen, and B. Halling-Sørensen. 2009. Transport and fate of estrogenic hormones in slurry-treated soil monoliths. J. Environ. Qual. 38:955–964.

Larney, F.J., and H.H. Janzen. 1996. Restoration of productivity to a desurfaced soil with livestock manure, crop residue, and fertilizer amendments. Agron. J. 88:921–927.

Logan, T.J., B.J. Lindsay, and S. Titko. 1997. Characteristics and standards for processed biosolids in the manufacture and marketing of horticultural fertilizers and soil blends, p. 63–71. *In* J.E. Rechcigl and H. MacKinnon (eds.) Agricultural uses of byproducts and wastes. American Chemical Society, Washington, DC.

Lorenzen, A., K. Burnison, M. Servos, and E. Topp. 2006. Persistence of endocrine-disrupting chemicals in agricultural soils. J. Environ. Eng. Sci. 5:211–219.

Lundström, U.S., D.C. Bain, A.F.S. Taylor, and P.A.W. van Hees. 2003. Effects of acidification and its mitigation with lime and wood ash on forest soil processes: A review. Water Air Soil Pollut. Focus 3:5–28.

MacIlwain, C. 1997. Scientists close in on cell from hell lurking in Chesapeake Bay. Nature 389:317–318.

Mårtensson, A.M., and E. Witter. 1990. Influence of various soil amendments on nitrogen-fixing soil microorganisms in a long-term field experiment, with special reference to sewage sludge. Soil Biol. Biochem. 22:977–985.

Martin, A., and G.D. Cooke. 1994. Health risks in eutrophic water supplies. Lake Line 14:24–26.

Mattigod, S.V., D. Rai, L.E. Eary, and C.C. Ainsworth. 1990. Geochemical factors controlling the mobilization of inorganic constituents from fossil fuel combustion residues: I. Review of major elements. J. Environ. Qual. 19:188–201.

McBride, M.B. 2003. Toxic metals in sewage sludge-amended soils: Has promotion of beneficial use discounted the risks? Adv. Environ. Res. 8:5–19.

McLaren, R.G., and C.J. Smith. 1996. Issues in the disposal of industrial and urban wastes, p. 183–212. *In* R. Naidu et al. (eds.) Contaminants and the soil environment in the Australasia-Pacific region. Kluwer Academic Publishers, Dordrecht, the Netherlands.

Miller, W.P. 1987. Infiltration and soil loss of three gypsum-amended Ultisols under simulated rainfall. Soil Sci. Soc. Am. J. 51:1314–1320.

Miller, W.P. 1995. Environmental considerations in land application of byproduct gypsums, p. 183–208. In D.L. Karlen, R.J. Wright, and W.O. Kemper (eds.) Agricultural utilization of urban and industrial byproducts. American Society of Agronomy, Madison, WI.

Miller, W.P., D.C. Martens, and L.W. Zelazny. 1985. Effects of manure amendment on soil chemical properties and hydrous oxides. Soil Sci. Soc. Am. J. 49:856–861.

Miller, W.P., and M.E. Sumner. 1997. Agricultural and industrial uses of byproduct gypsums, p. 226–239. In J.E. Rechcigl and H.C. MacKinnon (eds.) Agricultural uses of byproducts and wastes. American Chemical Society, Washington, DC.

Miltimore, J.E., and J.L. Mason. 1971. Copper to molybdenum ratio and molybdenum and copper concentrations in ruminant feeds. Can. J. Anim. Sci. 51:193–200.

Miner, R.A. 1997. National solid and hazardous waste regulatory initiatives of interest of the forest products industry. Proc. NCASI 1997 Southern Regional Mtg., Asheville, NC, June 1997. NCASI, Research Triangle Park, NC.

Morris, L.A., W.P. Miller, M.E. Sumner, and W.L. Nutter. 1997. Characterization and utilization of pulp mill wastes for forest land application. Proc. NCASI 1997 Southern Regional Mtg., Asheville, NC, June 1997. NCASI, Research Triangle Park, NC.

Mueller, D.H., R.C. Wendt, and T.C. Daniel. 1984. Phosphorus losses as affected by tillage and manure application. Soil Sci. Soc. Am. J. 48:901–905.

Mullins, G., B. Joern, and P. Moore. 2005. By-product phosphorus: Sources, characteristics and management, p. 829–880. In J.T. Sims and A.N. Sharpley (eds.) Phosphorus: Agriculture and the environment. Agronomy monograph 46. American Society of Agronomy, Madison, WI.

Munoz, G.R., K.A. Kelling, K.E. Rylant, and J. Zhu. 2008. Field evaluation of nitrogen availability from fresh and composted manure. J. Environ. Qual. 37:944–955.

N'Dayegamiye, A., and D. Côté. 1989. Effect of long-term pig slurry and solid cattle manure application on soil chemical and biological properties. Can. J. Soil Sci. 69:39–45.

National Research Council. 2002. Biosolids applied to land: Advancing standards and practices, 282pp. National Academy Press, Washington, DC.

NCASI. 1984. Land application and related utilization of pulp and paper mill sludges. NCASI Bulletin 439. NCASI, Research Triangle Park, NC.

NCASI. 1997. Solid waste management and disposal practices in the U.S. paper industry. NCASI Tech. Bulletin 641. NCASI, Research Triangle Park, NC.

NEBRA (North East Biosolids and Residuals Association). 2007. A national biosolids regulation, quality, end use and disposal survey. NEBRA, Tamworth, NH. Available at http://www.nebiosolids.org (accessed May 2009).

Nichols, D.J., T.C. Daniel, and D.R. Edwards. 1994. Nutrient runoff from fescue pasture after incorporation of poultry litter and inorganic fertilizer. Soil Sci. Soc. Am. J. 58:1224–1228.

Nichols, D.J., T.C. Daniel, P.A. Moore, Jr., D.R. Edwards, and D.H. Pote. 1997. Runoff of estrogen hormone 17ß-estradiol from poultry litter applied to pasture. J. Environ. Qual. 26:1002–1006.

Nieminen, M., S. Piirainen, and M. Moilanen. 2005. Release of mineral nutrients and heavy metals from wood and peat ash fertilizers: Field studies in Finnish soils. Scand. J. Forest Res. 20:146–153.

NRC. 2002. Biosolids applied to land: Advancing standards and practices. National Academy Press, Washington, DC.

O'Connor, G.A., R.B. Brobst, R.L. Chaney, R.L. Kincaid, L.R. McDowell, G.M. Pierzynski, A. Rubin, and G.G. van Ripper. 2001. A modified risk assessment to establish molybdenum standards for land application of biosolids. J. Environ. Qual. 30:1490–1507.

Paul, E.A., and F.E. Clark. 1989. Soil microbiology and biochemistry. Academic Press, San Diego, CA.

Pepper, I.L., H. Zerzghi, J.P. Brooks, and C.P. Gerba. 2008. Sustainability of land application of Class B biosolids. J. Environ. Qual. 37:S58–S67.

Pote, D.H., T.C. Daniel, D.R. Edwards, J.D. Mattice, and D.B. Wickliff. 1994. Effect of drying and rainfall intensity on cyromazine loss from surface-applied caged-layer manure. J. Environ. Qual. 23:101–104.

Powers, J.E. 1998. Land application of byproducts. American Society of Agronomy, Madison, WI.

Preusch, P.L., P.R. Adler, L.J. Sikora, and T.J. Tworkoski. 2002. Nitrogen and phosphorus availability in composted and uncomposted poultry litter. J. Environ. Qual. 31:2051–2057.

Price, G.W., and R.P. Voroney. 2007. Papermill biosolids effect on soil physical and chemical properties. J. Environ. Qual. 36:1704–1714.

Radcliffe, D.E., R.L. Clark, and M.E. Sumner. 1986. Effect of gypsum and deep-rooting perennials on subsoil mechanical impedance. Soil Sci. Soc. Am. J. 50:1566–1570.

Ritchey, K.D., D. de Sousa, C.M. Feldhake, and R.B. Clarke. 1995. Improved water and nutrient uptake from subsurface layers of gypsum-amended soils, p. 157–182. In D.L. Karlen, R.J. Wright, and W.O. Kemper (eds.) Agricultural utilization of urban and industrial byproducts. American Society of Agronomy, Madison, WI.

Rusin, P.A., S.L. Maxwell, J.P. Brooks, C.P. Gerba, and I.L. Pepper. 2003. Evidence for the absence of *Staphylococcus aureus* in land applied biosolids. Environ. Sci. Technol. 37:4027–4030.

Sanders, S.M., P. Srivastava, Y. Feng, J.H. Dane, J. Basile, and M.O. Barnett. 2008. Sorption of the veterinary antimicrobials sulfadimethoxine and ormetoprim in soil. J. Environ. Qual. 37:1510–1518.

Sangsupan, H.A., D.E. Radcliffe, P.G. Hartel, M.B. Jenkins, W.K. Vencill, and M.L. Cabrera. 2006. Sorption and transport of 17β-estradiol and testosterone in undisturbed soil columns. J. Environ. Qual. 35:2261–2272.

Sarmah, A.K., G.L. Northcott, F.D.L. Leusch, and L.A. Tremblay. 2006a. A survey of endocrine disrupting chemicals (EDCs) in municipal sewage and animal waste effluents in the Waikato region of New Zealand. Sci. Total Environ. 355:135–144.

Sarmah, A.K., M.T. Meyer, and A.B.A. Boxall. 2006b. A global perspective on the use, sales, exposure pathways, occurrence, fate and effects of veterinary antibiotics (VAs) in the environment. Chemosphere 65:725–759.

Schjønning, P., B.T. Christensen, and B. Cartensen. 1994. Physical and chemical properties of a sandy loam receiving animal manure, mineral fertilizer or no fertilizer for 90 years. Eur. J. Soil Sci. 45:257–266.

Schumann, A., and M.E. Sumner. 1998. Potential ammonia and nitrate toxicity from flyash-sewage-poultry manure mixtures. Proc. World Congress of Soil Science No. 1289. Montpellier, France.

Sengelov, G., Y. Agerso, B. Halling-Sørensen, S.B. Baloda, J.S. Andersen, and L.B. Jensen. 2003. Bacterial antibiotic resistance levels in Danish farmland as a result of treatment with pig manure slurry. Environ. Int. 28:587–595.

Shainberg, I., M.E. Sumner, W.P. Miller, M.P. Farina, M.A. Pavan, and M.V. Fey. 1989. Use of gypsum on soils: A review. Adv. Soil Sci. 9:1–111.

Sharma, R., F.J. Larney, J. Chen, L.J. Yanke, M. Morrison, E. Topp, T.A. McCallister, and Z. Yu. 2009. Selected antimicrobial resistance during composting of manure from cattle administered sub-therapeutic antimicrobials. J. Environ. Qual. 38:567–575.

Sharpley, A.N. 1997. Rainfall frequency and nitrogen and phosphorus runoff from soil amended with poultry litter. J. Environ. Qual. 26:1127–1132.

Sharpley, A.N., B.J. Carter, B.J. Wagner, S.J. Smith, E.L. Cole, and G.A. Sample. 1992. Impact of long-term swine and poultry manure application on soil and water resources in eastern Oklahoma. OK Agricultural Experiment Station Technical Bulletin T-169. Oklahoma State University, Stillwater, OK.

Sharpley, A.N., S.C. Chapra, R. Wedepohl, J.T. Sims, T.C. Daniel, and K.R. Reddy. 1994. Managing agricultural phosphorus for protection of surface waters: Issues and options. J. Environ. Qual. 23:437–451.

Sharpley, A.N., and S. Rekolainen. 1997. Phosphorus in agriculture and its environmental implications, p. 1–53. *In* H. Tunney et al. (eds.) Phosphorus loss from soil to water. CAB International, Wallingford, U.K.

Sharpley, A.N., S.J. Smith, and W.R. Bain. 1993. Nitrogen and phosphorus fate from long-term poultry litter applications to Oklahoma soils. Soil Sci. Soc. Am. J. 57:1131–1137.

Shiralipour, A., D.B. McConnell, and W.H. Smith. 1992. Physical and chemical properties of soils as affected by municipal solid waste compost application. Biomass Bioenerg. 3:261–266.

Shore, L.S., D.L. Correll, and P.K. Chakraborty. 1995. Relationship of fertilization with chicken manure and concentrations of estrogens in small streams, p. 155–162. *In* K. Steele (ed.) Animal waste and the land-water interface. CRC Press, Boca Raton, FL.

Sims, J.T. 1986. Nitrogen transformations in a poultry manure amended soil: Temperature and moisture effects. J. Environ. Qual. 15:59–67.

Sims, J.T. 1995. Characteristics of animal wastes and waste-amended soils: An overview of the agricultural and environmental issues, p. 1–13. *In* K. Steele (ed.) Animal waste and the land-water interface. CRC Press, Boca Raton, FL.

Sims, J.T., and D.C. Wolf. 1994. Poultry waste management: Agricultural and environmental issues. Adv. Agron. 52:1–83.

Slaton, N.A., K.R. Brye, M.B. Daniels, T.C. Daniel, R.J. Norman, and D.M. Miller. 2004. Nutrient input and removal trends for agricultural soils in nine geographic regions in Arkansas. J. Environ. Qual. 33:1606–1615.

Smith, S.R. 1996. Agricultural recycling of sewage sludge and the environment. CAB International, Wallingford, U.K.

Someshwar, A.V. 1996. Wood and combination wood-fired boiler ash characterization. J. Environ. Qual. 25:962–972.

Sonnenschein, C., and A.M. Soto. 1998. An updated review of environmental estrogen and androgen mimics and antagonists. J. Steroid Biochem. Mol. Biol. 65:143–150.

Soupir, M.L., S. Mostaghimi, E.R. Yagow, C. Hagedorn, and D.H. Vaughan. 2006. Transport of fecal bacteria from poultry litter and cattle manures applied to pastureland. Water Air Soil Poll. 169:125–136.

Souza, D.M.G., and K.D. Ritchey. 1986. Uso do gesso no solo de Cerrado. Ann. Sem. Uso Fosfogesso Agricultura, p. 119–144. IBRAFOS, São Paulo, Brazil.

Staed, J.B., D.M. Miller, K.R. Brye, T.C. Daniel, C. Rom, and E.E. Gbur. 2009. Land use effects on near-surface soil arsenic in the Ozark Highlands. Soil Sci. 174:121–129.

Stout, W.L., A.N. Sharpley, W.J. Gburek, and H.B. Pionke. 1999. Reducing phosphorus export from croplands with FBC fly-ash and FGD gypsum. Fuel 78:175–178.

Stout, W.L., A.N. Sharpley, and J. Landa. 2000. Effectiveness of coal combustion by-products in controlling phosphorus export from soils. J. Environ. Qual. 29:1239–1244.

Stout, W.L., A.N. Sharpley, and H.B. Pionke. 1998. Reducing soil phosphorus solubility with coal combustion by-products. J. Environ. Qual. 27:111–118.

Stratton, M.L., A.V. Barker, and J.E. Rechcigl. 1995. Compost, p. 249–310. *In* J.E. Rechcigl (ed.) Soil amendments and environmental quality. Lewis Publishers, Boca Raton, FL.

Sumner, M.E. 1993. Gypsum and acid soils: The world scene. Adv. Agron. 51:1–32.

Sumner, M.E., W.P. Miller, D.E. Radcliffe, C.S. Hoveland, U. Kukier, C. Fauzia, and R. Arnold. 1996. Use of byproduct flue gas desulfurization gypsum as an ameliorant for Southeastern soils. Final Report. DOE Clean Coal Technology Project. Department of Energy, Washington, DC.

Suttle, N.F. 1991. The interactions between copper, molybdenum and sulphur in ruminant nutrition. Annu. Rev. Nutr. 11:121–141.

Sutton, A.L. 1994. Proper animal manure utilization. *In* Nutrient management, supplement to J. Soil Water Conserv. 49:65–70.

Thompson, R.B., B.F. Pain, and D.R. Lockyer. 1990. Ammonia volatilization from cattle slurry following surface application to grassland. Plant Soil 125:109–115.

Tiarks, A.E., A.P Mazurak, and L. Chesnin. 1974. Physical and chemical properties of soil associated with heavy applications of manure from cattle feedlots. Soil Sci. Soc. Am. Proc. 38:826–830.

Tietjen, C. 1975. Influence of antibiotics and growth promoting feed additives on the manuring effect of animal excrements in pot experiments with oats, p. 328–330. *In* Managing livestock wastes. Proc. Third International Symposium, Champaign-Urbana, IL.

Tisdale, S.L., W.L. Nelson, J.D. Beaton, and J.L. Havlin. 1993. Soil fertility and fertilizers. 5th Edn. Macmillan Publishers, New York.

Toor, G.S., B.E. Haggard, and A.M. Donoghue. 2007. Water extractable trace elements in poultry litters and granulated products. J. Appl. Poult. Res. 16:351–360.

Unger, P.W., and B.A. Stewart. 1974. Feedlot waste effects on soil conditions and water evaporation. Soil Sci. Soc. Am. Proc. 38:954–957.

Unwin, R.J. 1986. Leaching of nitrate after application of organic manures. Lysimeter studies, p. 158–167. *In* A.D. Kofoed et al. (eds.) Efficient land use of sludge and manure. Elsevier Scientific Publishers, London, U.K.

Unwin, J.P. 1997. Results of the NCASI 1995 wastewater and solid waste survey. Proc. NCASI 1997 Southern Regional Mtg., Asheville, NC, June 1997. NCASI, Research Triangle Park, NC.

USDA. 1992. Agricultural waste management field handbook. National Engineering Manual, Part 651. United States Department of Agriculture, Washington, DC.

USEPA. 1993a. Cement kiln dust-Report to Congress. Available at http://www.epa.gov/osw/nonhaz/industrial/special/ckd/cement2.htm (accessed May 2009).

USEPA. 1993b. Part 503 Standards for the use or disposal of sewage sludge. Fed. Reg. 58:9387–9404.

USEPA. 2007. Municipal solid waste generation, recycling and disposal in the United States: Facts and figures for 2007. Available at http://www.epa.gov/osw/nonhaz/municipal/msw99.htm (accessed May 2009).

USEPA. 2009. Targeted national sewage sludge survey, Vol. 2. Statistical analysis report. Report EPA-822-R-08-018. USEPA, Washington, DC.

USGS. 1997. Commodity minerals information: Iron and steel slags. Available at http://minerals.er.usgs.gov/minerals/0pubs/commodity/iron_&steel_slag (accessed May 15, 2010).

Vaccari, D.A. 2009. Phosphorus: A looming crisis. Sci. Am. 300(6):54–59.

Walker, J.M., R.M. Southworth, and A.B. Rubin. 1997. U.S. Environmental Protection Agency regulations and other stakeholder activities affecting agricultural use of byproducts and wastes, p. 28–49. *In* J.E. Rechcigl and H.C. MacKinnon (eds.) Agricultural uses of byproducts and wastes. American Chemical Society, Washington, DC.

Wallingford, G.W., W.L. Powers, and L.S. Murphy. 1985. Present knowledge on the effects of land application of animal waste, p. 580–586. *In* Managing livestock wastes. Proc. Third International Symposium, Champaign-Urbana, IL.

Weil, R.R., and W. Kroontje. 1979. Physical condition of a Davidson clay loam after five years of heavy poultry manure applications. J. Environ. Qual. 8:387–392.

Westerman, P.W., and M.R. Overcash. 1980. Short-term attenuation of runoff pollution potential for land-applied swine and poultry manure, p. 289–292. *In* Livestock waste—A renewable resource. American Society of Agricultural Engineering, St. Joseph, MI.

Whitney, R.S., R. Gardener, and D.W. Robertson. 1950. The effectiveness of manure and commercial fertilizer in restoring productivity of subsoils exposed to leveling. Agron. J. 42:239–245.

Wilson, D. 1997. Fear in the fields, Part 1 and 2. Seattle Times, July 3, 1997.

Wines, R.A. 1985. Fertilizer in America. Temple University Press, Philadelphia, PA.

Wright, R.J., W.D. Kemper, P.D. Millner, J.F. Power, and R.F. Korcak (eds.). 1998. Agricultural uses of municipal, animal, and industrial byproducts. USDA, ARS, Conservation Research Report No. 44. Beltsville, MD.

Wyatt, J.M. 1997. Byproduct usage in fertilizer micronutrients, p. 255–267. *In* J.E. Rechcigl and H.C. MacKinnon (eds.) Agricultural uses of byproducts and wastes. American Chemical Society, Washington, DC.

Xia, K., A. Bhandari, K. Das, and G. Pillar. 2005. Occurrence and fate of pharmaceuticals and personal care products (PPCPs) in biosolids. J. Environ. Qual. 34:91–104.

Zaleski, K.J., K.L. Josephsen, C.P. Gerba, and I.L. Pepper. 2005. Potential regrowth and recolonization of *Salmonellae* and indicators in biosolids and biosolid-amended soil. Appl. Environ. Microbiol. 71:3701–3708.

25
Conservation Tillage

25.1 Introduction	25-1
25.2 Definitions	25-2
25.3 Erosion by Water and Wind	25-3
Erosion by Water • Erosion by Wind	
25.4 Water Conservation	25-5
25.5 Residue Management and Decomposition	25-7
25.6 Biological Activity	25-8
25.7 Carbon Dynamics and Sequestration in No-Tillage	25-9
25.8 Biofuel (Cellulosic) Production Implications	25-11
Negative Aspects of Crop Residue Removal • Positive Aspects of Crop Residue Removal • Threshold Levels of Residue Removal	
25.9 Soil Fertility and Nutrition	25-13
Nitrogen • Phosphorus and Potassium	
25.10 Surface Sealing, Crusting, and Seedling Emergence	25-16
25.11 Compaction	25-17
25.12 Cropping Systems	25-18
25.13 Cover Crops	25-20
25.14 Comparison of Conservation Tillage with Conservation Agriculture	25-20
25.15 Soil by Climate Interactions	25-21
Problems Related to Cold Soils and Excessive Soil Water • Weed Control Problems Related to Soil, Climate, and Tillage Choice • Surface Residue and Other Conservation Measures	
25.16 Disadvantages	25-22
25.17 Conclusions	25-23
References	25-23

Paul W. Unger
United States Department of Agriculture

Humberto Blanco-Canqui
Kansas State University

25.1 Introduction

"For thousands of years, agriculture and tillage were considered synonymous. It was simply not thought possible to grow crops without first tilling the soil before planting and for weed control" (Triplett and Dick, 2008). Indeed, soil tillage as developed over the centuries permitted farmers to grow more and better crops by loosening and mixing soil and controlling weeds. However, with the introduction of herbicide 2,4-D[(2,4-dichlorophenoxy) acetic acid] in 1942 and continued development of new herbicides since then (Unger et al., 2009), the need to mechanically control weeds has slowly, but greatly, declined. Furthermore, the need to prepare seedbeds by plowing and disking has decreased with the increased adoption of no-tillage (NT) since the 1970s. Originally, the use of the moldboard plow was deemed necessary to produce high crop yields, but such plowing (clean tillage) in many cases accelerated erosion by water and wind, organic matter (OM) oxidation, and land degradation. Because of severe erosion, government conservation compliance regulations were implemented in an effort to reduce erosion on highly erodible lands. To meet these regulations and in an attempt to increase profit, many farmers have switched to using conservation tillage (CS) systems, including NT, during the last half century or so. Conservation tillage now is widely recognized as a best management practice that improves soil productivity, reduces runoff and erosion, protects water quality, and improves environmental quality (Sullivan et al., 2008).

Conservation tillage is a broad term used to define any tillage system that has as its primary objective reduction of soil and water losses. The primary reason, however, why farmers in many U.S. regions change from conventional tillage (CT) to some form of CS is to reduce farm input costs and to increase profits (Smart and Bradford, 1996). Secondary reasons include soil and water conservation and insulation of the soil surface from temperature extremes. The use of CS reduces traffic operations, thereby decreasing soil compaction and costs for labor, fuel, tractors, and other equipment. In some cases, however, costs are increased because herbicides cost more than the tillage they replace. Even so, farmers often adopt the new technology because yields increase at a faster rate than the costs (Bradford and Peterson, 1999).

Timeliness of cultural operations is improved when using CS in many situations. With the development of effective herbicides and improvements in sprayer design (e.g., the hooded sprayer), weed control without tillage can be accomplished, even in subtropical areas of the United States where freezing temperatures seldom occur (Smart and Bradford, 1997). Possibly the most important benefit associated with CS is soil restoration (Langdale et al., 1992; Lal et al., 2004) or in the words of Crovetto (1996), "The great economic advantage of no-till is that low value soils can be brought into higher production without erosion risks."

In this chapter, we discuss the soil science principles of CS, but believe the ultimate goal of CS, especially NT, should be to conserve soil and water while improving food security and environmental quality. If herbicides fail to control weeds, alternative methods such as biological control, crop rotations, or other techniques should be pursued. Where soil compaction becomes a major crop limiting factor, deep-chiseling or zone tillage may be needed to loosen the soil. When surface residue is limited, emergency tillage that can quickly roughen the surface may be needed to control ongoing erosion by wind. Where cold soil temperatures and excess water at spring planting time are problems, it may be necessary to remove most residue from the seeding row in the fall (e.g., while band-applying fertilizer). With time, as soil resilience improves and as farmers become familiar with the total system, soil management becomes easier.

A large number of reports in the literature pertain to the performance of NT relative to that with CT. Most studies pertained to short-term (2–5 years) evaluations of soil and crop parameters. Only few of the studies pertained to long-term trends. Long-term evaluations provide a better overall assessment because many of the benefits or disadvantages develop after several years of using NT. The major benefits of NT agriculture can be more accurately assessed only after NT has been in place for a decade or more. Adoption of NT is an investment in the soils of our future. Adopting CS, particularly NT, has significant agronomic and environmental implications. NT agriculture reduces soil erosion, enhances soil and water quality, sustains crop production, reduces off-site transport of pollutants (e.g., sediment and sediment-bound chemicals), and sequesters soil organic carbon (C), thereby reducing net emissions of greenhouse gases. The organic C enhanced by CS is vital to soil and agricultural sustainability. It improves soil resilience, soil structural stability, microbial processes, water retention, and crop productivity.

25.2 Definitions

The term CS often is used interchangeably with minimum tillage, mulch tillage, reduced tillage, NT planting, and others, thereby creating confusion. We, therefore, begin by defining the various terms used in this chapter. The definitions are from the Glossary of Soil Science Terms (SSSA, 2001) and the Conservation Technology Information Center (Schertz and Becherer, 1994; Hill, 1996). These definitions were also discussed by Bradford and Peterson (1999).

Conservation tillage: Any tillage sequence, the object of which is to minimize or reduce loss of soil and water; operationally, a tillage or tillage and planting combination that leaves a 30% or greater cover of crop residue on the surface.

Conventional tillage: Primary and secondary tillage operations normally performed in preparing a seedbed and/or cultivating for a given crop grown in a given geographical area, usually resulting in <30% cover of crop residue remaining on the surface after completion of the tillage sequence.

No-tillage (or zero tillage; also know as slot planting): A procedure whereby a crop is planted directly into the soil with no primary or secondary tillage since harvest of the previous crop; usually, a special planter is necessary to prepare a narrow, shallow seedbed immediately surrounding the seed being planted. NT is sometimes practiced in combination with subsoiling or in-row chiseling to facilitate seeding and early root growth, whereby the surface residue is left virtually undisturbed, except for a small slot in the path of the subsoiler shank. The soil is left undisturbed, except for nutrient injection and the planting operation. Planters or drills may be equipped with coulters, row cleaners, disk openers, in-row chisels, or rototillers. Weeds are controlled primarily with herbicides. Cultivation may be used for emergency weed control. In essence, NT is the extreme CS method because it results in the least soil disturbance and greatest retention of crop residue on the soil surface.

Ridge tillage: A tillage system in which ridges are reformed atop the planted row by cultivation, and the ensuing row crop is planted into ridges formed in the previous growing season. The soil is left undisturbed from harvest to planting, except for nutrient injection. Planting on ridges is accomplished with planters having sweeps, disk openers, coulters, or row cleaners. Residue is left on the surface between ridges. Weeds are controlled with herbicides and/or by cultivation.

Strip tillage: Tillage operations performed in isolated bands separated by bands of soil remaining essentially undisturbed by the particular tillage equipment.

Mulch tillage: Tillage or preparation of the soil so that plant residue or other materials are left to cover the surface, also known as mulch farming, trash farming, stubble mulch tillage, or plowless farming. Operationally, mulch tillage is a full-width tillage or tillage and planting combination that leaves >30% of the surface covered with crop residue. The soil is disturbed prior to planting. Tillage implements such as chisels, field cultivators, disks, sweeps, or blades are used. Weeds are controlled with herbicides and/or cultivation.

Crop residue management: Disposition of stubble, stalks, and other crop residue (aboveground biomass) by tillage operations to remove them from the soil surface (burying), to partially anchor them in the surface soil while leaving them partially exposed at the surface (mulch tillage), or to leave them at the surface intact or cut into smaller pieces. With regard to crop residue management, CS pertains to using any tillage and planting system that retains all or a portion of the previous crop's residue on the soil surface. The goal is to retain adequate crop residue on the surface to control erosion, increase water infiltration and its storage in soil, and decrease soil water evaporation. On some soils, particularly in semiarid regions, crop residue production is limited; therefore, most residue should be retained on the soil surface to provide protection against erosion by water and wind. The portion required depends on other conservation practices that may be included in the farmer's total conservation plan.

25.3 Erosion by Water and Wind

25.3.1 Erosion by Water

Soil erosion by water is a process of particle detachment, transport, and deposition. Sediment must first be detached from the soil mass before it can be transported. In interrill areas, soil is detached by impacting raindrops and transported by sheet flow. Sheet flow concentrates in low areas, thereby forming small channels or rills and detaching and transporting more particles. Reduced erosion by water is achieved, therefore, by limiting either the detachment or transport processes. Complete erosion control seemingly would be achieved by eliminating runoff or reducing it to a flow rate that cannot transport detached particles. Splash erosion that involves detachment and aerial movement of small soil particles due to raindrop impact (SSSA, 2001), however, may still occur when runoff is prevented or reduced to a low rate.

Tillage and surface residue influence soil erosion because they create soil surface conditions and affect surface processes that determine infiltration and runoff rates. Erosion is reduced under CS conditions due to greater crop residue cover that protects the soil surface from raindrop impact, increased consolidation and greater strength of the surface soil layer, increased tortuosity and hydraulic resistance of runoff flow paths, decreased capacity of runoff to detach and transport soil particles, less runoff due to greater infiltration, and increased deposition of sediment in small ponded areas created by surface residue (McGregor et al., 1990; Meyer et al., 1999; Dabney et al., 2004; Wilson et al., 2004; Knapen et al., 2008).

Results given for many studies have shown remarkable reductions in runoff and soil loss when using CS rather than CT. Norton and Brown (1992) reported significantly lower interrill and rill erosion rates in old consolidated ridges in a ridge tillage system as compared with freshly formed ridges. Interrill erosion rates were 40% and 59% lower with older than with freshly formed ridges for Hoytville silty clay (Mollic Ochraqualf) and Rossmoyne silt loam (Aquic Fragiudalf), respectively. For both soils, the rill erosion rate was 72% lower for older ridges than for freshly formed ridges. Interrill soil loss on a moderately well-drained Saybrook silt loam (Typic Argiudoll) under NT for 15 years was reduced to 0.01 kg m^{-2} h^{-1} compared with 2.45 kg m^{-2} h^{-1} for CT when water was applied at an intensity of 100 mm h^{-1} (Bradford and Huang, 1994). Runoff was 12 mm h^{-1} for the NT condition and 68 mm h^{-1} for the tilled soil condition with residue removed. In a study by Andraski et al. (1985), soil loss was consistently 80%–90% less with NT than with CT (moldboard plowing) on Griswold silt loam (Typic Argiudoll) near Arlington, WI. On some soils, however, the smooth consolidated surface with NT may encourage runoff, but the runoff leaving NT fields contains less sediment and sediment-bound chemicals than that leaving plowed fields. Runoff with NT on Pullman clay loam (Torrertic Paleustoll) was 56% greater than with CT in the semiarid environment of northwest Texas, but sediment loss was 54% less with NT than for the tilled treatment (Jones et al., 1994a).

In some cases, sediment loss from NT soils can be practically eliminated as compared with that from plowed soils. Harrold and Edwards (1972) reported negligible soil losses (0.07 Mg ha^{-1}) from a NT watershed having a slope three to four times greater than on nearby plowed watersheds. Soil losses were 7.2 and 50.7 Mg ha^{-1} from plowed contour- and sloping-row watersheds, respectively, in Ohio. In Nigeria, soil losses were negligible (<0.005 Mg ha^{-1}) from NT watersheds having slopes up to 15% (Rockwood and Lal, 1974). In contrast, soil losses were 16.0 and 3.9 Mg ha^{-1} from bare fallow and plowed watersheds on 15% slope. Raczkowski et al. (2009) reported a 2.6 Mg ha^{-1} year^{-1} soil loss rate with NT for a soil having a 7.0 Mg ha^{-1} year^{-1} tolerable loss rate. With CT (chisel plow/disk), soil losses totaled 448 Mg ha^{-1} for the 6 year study with 397 Mg ha^{-1} (89%) of the loss occurring during 3 months having multiple storms of high-intensity rainfall. The average loss rate with CT was almost 11 times greater than the tolerable rate for the soil.

NT management often reduces runoff and soil loss due to the accumulation of OM on or near the soil surface, which promotes soil aggregation and improves water infiltration. West et al. (1992) found that NT with residue removed resulted in a 60%–70% decrease in rill erodibility compared with CT systems. They attributed the reduction to increases of organic C from 9.3 to 12.9 g kg^{-1} and of water-stable aggregates from 50% to 76%, and to greater soil strength due to consolidation. Another study for which soil erosion decreased significantly with NT because of greater aggregation and infiltration was reported by Rhoton et al. (2002). In their study, average annual soil loss was 3.9 Mg ha^{-1} with CT while no soil loss was observed with the NT treatment. Greater aggregate stability and size for NT than for CT soils were reported also by Bruce et al. (1990), Cambardella and Elliott (1993), Drees et al. (1994), Beare et al. (1994a, 1994b), and Lal et al. (1994). The greater structural stability with NT is attributed to increased OM content, surface residue mulch, and biological activity (Doran, 1980; Rhoton et al., 2002). Surface cover provided by crop residue has a major effect on soil erosion by water (Figure 25.1). It is possible to obtain nearly complete surface cover by residue on most soils when using NT and, thereby, achieving maximum control of erosion.

The soil detachment rate due to surface flow in rills is a linear function of hydraulic shear stress. Foster et al. (1980) estimated that the critical shear stress required to initiate erosion in rills in an untilled soil is about 10–15 times greater than that in a freshly tilled soil. Rill erodibility was significantly greater for freshly tilled NT than for undisturbed NT Saybrook silt loam (Typic Argiudoll) in north-central Illinois. The critical shear stress of freshly tilled

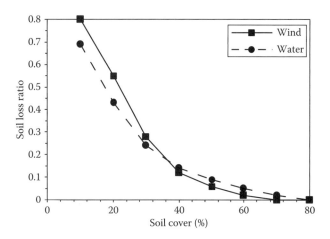

FIGURE 25.1 Relationship between soil loss ratio (loss with cover divided by loss from bare soil) and percentage of surface covered with residues. (Redrawn from Papendick, R.I., J.F. Parr, and R.E. Meyer. 1990. Managing crop residues to optimize crop/livestock production systems for dryland agriculture. Adv. Soil Sci. 13:253–272.)

NT soils is much less than for undisturbed NT soils. The critical shear stress increases with residue removal in NT soils, thus indicating the importance of surface residue for slowing runoff, reducing sediment entrainment by surface flow, and reducing raindrop impact forces due to greater ponding and greater flow depth. If a critical flow rate, slope steepness, or slope length is exceeded, massive rilling may occur in an NT soil, but these critical values are much greater for an NT than for a tilled soil.

Transport of detached particles occurs via surface runoff under both interrill and rill erosion processes. Hence, soil erosion can be reduced by creating soil surface conditions that reduce surface runoff. Runoff from a 0.5 ha watershed on Rayne silt loam (Typic Hapludault) at Coshocton, Ohio, farmed 20 years for continual NT corn (Zea mays L.), averaged <2 mm year^{-1} when average rainfall was >1000 mm year^{-1} (Edwards et al., 1988). Storms with 1 min rainfall intensities >76 mm h^{-1} were required to generate surface runoff with NT (Dick et al., 1989). The high infiltration rates were attributed to large macropores resulting from earthworm (Lumbricus spp.) activity and old root channels. These macropores are destroyed to the depth of tillage under CT conditions each year. Earthworm populations are greater under high residue (CS) conditions than in CT soil (Kladivko, 1994; Buman et al., 2004).

The effectiveness of crop residue for controlling erosion with any tillage method ultimately depends on the amount of residue retained on the soil surface. Residue retained on the surface obviously is much more effective than incorporated residue for controlling erosion (Wischmeier and Smith, 1978). As indicated earlier, however, in some climates such as in semiarid regions (e.g., southwestern United States) and with some crops [e.g., cotton (Gossypium hirsutum L.)], it is difficult to achieve adequate residue cover to protect the soil from erosion by water (Bradford and Peterson, 1999). Also, when a crop such as corn is harvested for silage, too little residue may remain to provide protection against erosion by water.

Residue cover plays an important role in reducing runoff and sediment loss when using NT. On a soil with a loamy sand surface (Plinthic Paledudult-Typic Hapludult complex) in Alabama, Truman et al. (2005) evaluated CT and NT with and without surface residue and with and without soil loosening by paratillage under simulated rainfall conditions. Runoff was greatest from residue-removed, non-paratilled CT plots (58% of rainfall); it was 4% of rainfall from residue-retained, paratilled NT plots. Sediment loss was four times greater with CT than with NT. The use of NT with residue retained along with paratillage to loosen the soil improved soil properties, reduced runoff and soil loss, and resulted in the potential for greater plant available water for row crops. Also in Alabama, runoff from Decatur silt loam (Rhodic Paleudult) varied between times of water application, but it was 34% to 10-fold less from residue-retained, paratilled NT plots than from CT plots (without paratillage and without surface residue). Soil loss was 1.5–5.4 times greater with CT than with other tillage systems (Truman et al., 2003). For that location, NT along with a rye (Secale cereale L.) cover and paratillage was best for reducing runoff and soil losses and for increasing water infiltration and plant available water.

Soils that freeze and thaw may be highly susceptible to erosion in late winter and early spring, especially when the subsurface layer remains frozen and, thereby, limits water flow into or through the soil profile. For a study under simulated rainfall conditions, soil losses were 2.58 and 0.86 kg m^{-2} with 0% and 80% surface residue cover, respectively, from boxes inclined at 13% and having a frozen subsurface layer. Losses were 1.24 and 0.58 kg m^{-2} with 0% and 80% residue cover from boxes inclined at 5% and soil not frozen (Cruse et al., 2001). Residue cover was considered to be highly important for reducing erosion during the thawing period, especially on steeply sloping soils, and the residue cover was of greater importance when the subsurface was frozen (Cruse et al., 2001). Under field conditions in Minnesota, residue cover was shown to reduce runoff and soil erosion during the spring thaw period (Sharratt et al., 2008). Under many conditions, the use of CS, especially NT, should result in retaining adequate surface residue to reduce soil losses under frozen soil conditions.

25.3.2 Erosion by Wind

The effectiveness of surface residue cover for protecting soil against erosion by wind is similar to its effectiveness for protecting soil against erosion by water (Figure 25.1). In wind tunnel experiments, erosion by wind was reduced 57% with a 20% surface cover and 95% with a 50% cover (Fryrear, 1985). For a sandy Sahelian soil (Psammentic Paleustalk) covered with 1.5 Mg ha^{-1} of flat millet (Pennisetum spp.) stalks, sediment transport by wind was reduced 64%; a cover of 1.0 Mg ha^{-1} reduced sediment transport 42% (Sterk and Spaan, 1997).

Effectiveness of residue cover for controlling erosion by wind depends on the amount, kind, and orientation of the residue in relation to the soil surface (Siddoway et al., 1965; Lyles and Allison, 1976, 1981). Among seven crops and orientations tested in a wind tunnel, standing winter wheat (*Triticum aestivum* L.) residue was most effective and flat, randomly positioned sunflower (*Helianthus annuus* L.) stalks were least effective for protection against erosion by wind (Lyles and Allison, 1981). On a weight basis, 5.5 and 8.7 times more standing grain sorghum [*Sorghum bicolor* (L.) Moench] and corn stubble, respectively, was needed to provide the same protection as standing wheat (*Triticum* spp.) stubble (Lyles and Allison, 1976).

Controlling erosion by wind may be unsuccessful with NT where the amount of plant residue produced is low or only a small amount is retained. Continued use of NT for several years, however, usually allows residue to accumulate because of positive feedbacks in the system. For example, residue accumulates over time with continuous NT in a semiarid environment because of its water conservation effect (see next section).

Small grain production (e.g., wheat) in semiarid regions sometimes results in short plants and low plant populations and, therefore, low amounts of plant residue remaining after harvest. When harvesting such crops with combines having cutter-type headers, plants are cut low to the ground, thus leaving short standing residue having low effectiveness for controlling erosion. Harvesting such crops with stripper-header combines has potential to leave taller residue and, therefore, maximize soil conservation while minimizing harvest losses of grain (McMaster et al., 2000; Henry et al., 2008).

Plant material production by crops can be extremely low during dry years. In a study in North Dakota by Merrill et al. (1999), residue yield by spring wheat (*Triticum* spp.) was 0.93 Mg ha^{-1} in 1988 compared with an average of 3.64 Mg ha^{-1} in other years (1989–1994). The low amount of residue produced was inadequate to control erosion by wind, even with NT. The spring wheat-fallow (SW-F) system was considered non-sustainable with regard to controlling soil erosion.

Controlling erosion by wind is highly important at the time of crop establishment and early crop growth. On sandy soils of the U.S. southern Great Plains and the Rio Grande Valley of Texas, sandblasting and burying of young cotton and grain sorghum seedlings requires that those crops be reseeded about 2 in 5 years. In the Lower Rio Grande Valley, however, fewer air-borne soil particles are observed under NT than under mulch-tillage conditions (Bradford and Peterson, 1999). In northern Mexico along the Rio Grande River, erosion by wind was much greater under CT than under NT conditions (de Quattro, 1997).

Soil erosion by wind can be greatly reduced by managing residue cover through the use of CS, but data on comparisons of the susceptibility of soils to erosion by wind under specific residue coverings are not available. Modeling of erosion by water and wind under different scenarios of tillage, cropping, and residue management systems can promote an understanding and estimation of soil erosion dynamics on a local and regional basis (van Donk et al., 2008).

Severe erosion by wind often occurs during fallow after wheat when CT is used for the wheat-fallow system in semiarid regions. Schillinger (2001) in the U.S. Pacific Northwest found that using minimum tillage and delayed minimum tillage systems significantly increased surface residue and clod retention, which improved control of erosion by wind.

The severe erosion in the Pacific Northwest (in eastern Washington and Oregon) results in air-borne soil particles that cause major concern regarding air quality (Saxton, 1995; Janosky et al., 2002). Erosion during the dust mulch (CT) fallow phase after winter wheat is the major source of the air-borne particles (Thorne et al., 2003). By maintaining a soil surface cover in spring and fall when the potential for erosion was high, cropping of spring cereals under NT conditions reduced the susceptibility to erosion as compared with the winter wheat-fallow system. Cover with NT results from both plant canopy and residue maintained on the soil surface; cover is the dominant factor regarding the reduction in soil erodibility. Adoption of spring cropping under NT conditions would significantly reduce erosion in crop-fallow areas (Thorne et al., 2003).

25.4 Water Conservation

Residue management and CS effects on soil water storage and use are highly dependent on climate and soil type. Soil water content generally is greater under NT than under CT conditions (Smika, 1976; Dickey et al., 1983; Unger, 1984; Díaz-Zorita et al., 2004). Greater soil water contents with NT result from increased infiltration and reduced evaporation with adequate residue cover. In contrast, greater water loss occurs from tilled soils because tillage exposes moist soil to the atmosphere, thereby increasing evaporative water losses.

In some situations, using NT has resulted in lower infiltration rates due to the development of a highly consolidated soil surface. Such was the case on a sandy loam Alfisol in Nigeria. Differences were not significant under CT and minimum tillage conditions for infiltration rate or accumulated infiltration, but were greater than with NT (Akinyemi and Adedeji, 2004). Apparently, benefits of surface residue remaining with NT regarding accumulative infiltration were negated on the consolidated sandy loam.

In the United States, the relative benefits of residue management for conserving water become greater when going from the wetter, more humid regions into the drier, semiarid regions. In humid or subhumid regions, additional soil water resulting from CS has increased crop yields relative to those obtained with CT, particularly when soil water becomes limited during the growing season. The water conservation benefits of CS normally decrease the adverse effects of short midseason droughts. In mid-summer, soil surface temperatures often are several degrees cooler under plant residue than those of bare surfaces because the surface residue insulates the soil surface against solar radiant energy. As a result, soil water evaporation is less from the residue-covered soil (Lascano and Baumhardt, 1996).

The water conservation benefits of CS have resulted in significant crop growth and yield increases relative to those obtained when using CT. In a study conducted in the Upper Piedmont region of North Carolina, corn and soybean (*Glycine max* L.) yields were 32% and 43% greater, respectively, with NT than with CT on a Pacolet sandy clay loam (Typic Hapludult) (Wagger and

Denton, 1989). The crusted soil surface with CT limited infiltration, thus resulting in lower soil water contents than with NT. In a study in the Lower Piedmont of North Carolina, total water-use efficiency (units of crop produced per unit of water used) for corn grain production increased from 0.18 Mg ha^{-1} cm^{-1} with CT to 0.22 Mg ha^{-1} cm^{-1} with NT in the fourth study year on a Hiwassee clay loam (Rhodic Kanhapludult) (Wagger and Cassel, 1993). For corn silage, water-use efficiency was 0.29 Mg ha^{-1} cm^{-1} with CT and 0.34 Mg ha^{-1} cm^{-1} with NT. Yields of both corn grain and silage were about 20% greater with NT than with CT.

In some cases, cooler and wetter soils can lead to slower crop development and lower yields. On a very poorly drained Hoytville silty clay loam (Mollic Ochraqualf) in Ohio, 10 year average yields of continual corn were 13% lower with NT than with plowed treatments (van Doren et al., 1976). For a study in Michigan where no residue, root only residue, and total wheat residue treatments were compared for NT corn production, soil water content was greater and soil temperature was lower with the total residue treatment (Kravchenko and Thelen, 2007). Residue mulch has mixed effects on crop yields, depending on soil nutrient availability and soil temperature fluctuations. In Colorado, lower grain yield of irrigated corn with NT than with CT probably results from slow early growth and delayed tasseling due to cooler springtime soil temperatures with NT (Halvorson et al., 2006).

In semiarid regions where water availability is a greater limiting factor for plant growth, increased water conservation with NT normally produces greater crop yields. Water content in a silty clay loam (Udic Boroll) in North Central Alberta, Canada, was greater with NT than with CT (Nyborg and Malhi, 1989). In Australia, greater and deeper water movement occurred with NT than with CT in Hermitage clay (Udic Pellustert) (Dalal, 1989). Infiltration on seven soils of the Pacific Northwest was 30% greater under NT or grass conditions than under CT conditions (Wuest et al., 2006), thereby providing for greater water availability for crop use. Over winter water infiltration and soil water storage in eastern Washington, however, are similar with CS and CT where stubble from the previous wheat crop is undisturbed and left standing (Kennedy and Schillinger, 2006).

The increased amount of water stored in soil by retaining crop residue on the surface is highly effective for increasing grain yields of subsequent crops. At Bushland, Texas, soil water contents were determined at grain sorghum planting after fallow following wheat harvest. The water contents and sorghum grain yields are shown in Figure 25.2a and b, respectively. The grain yield increases were 7 and 10 kg mm^{-1} of additional water stored with the sweep tillage and NT treatments, respectively, as compared with the moldboard treatment (Unger, 1984). The greater response with NT than with sweep tillage indicates an additional water conservation benefit due to the greater amounts of residue present with NT during the growing season. For an irrigated wheat-dryland grain sorghum study at Bushland, soil water storage efficiencies during fallow after wheat were 35% with NT and 15% with disk tillage (Unger and Wiese, 1979). Average grain yields of dryland grain sorghum planted after fallow were 3.14 and 1.93 Mg ha^{-1} with the respective treatments.

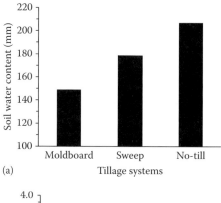

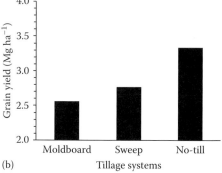

FIGURE 25.2 Soil water contents at planting (a) and grain yields of sorghum (b) as affected by different tillage systems at Bushland, Texas. All differences due to tillage systems were significant. (Adapted from Unger, P.W. 1984. Tillage and residue effects on wheat, sorghum, and sunflower grown in rotation. Soil Sci. Soc. Am. J. 48:885–891.)

Surface residue maintained by using stubble mulching in U.S. Great Plains wheat-fallow systems increased precipitation use efficiency from 0.09 to 0.14 Mg ha^{-1} cm^{-1} (Greb, 1979). Conversion to NT, however, did not lead to further increases in the productivity of wheat-fallow systems. Other studies have shown that water is stored more rapidly with NT than with tilled systems during the early part of the fallow period, and that NT fallow systems usually result in as much stored water by early spring as when fallowing is done for an additional 3 months with tilled systems (Peterson et al., 1996). It has been clearly demonstrated from North Dakota to Texas that adding spring crops to the cropping system is the only way to more efficiently use the extra precipitation stored with NT. Conversion from wheat-fallow to systems such as wheat-corn (or sorghum)-fallow increased water use efficiency from 0.06 Mg ha^{-1} cm^{-1} with wheat-fallow to 0.08 Mg ha^{-1} cm^{-1} for wheat-corn (or sorghum)-fallow (Peterson et al., 1996). The more intensive cropping (IC) systems not only resulted in a 43% increase in water use efficiency, but also greatly reduced the potential for soil erosion by water and wind. Tanaka and Anderson (1997) observed that NT and minimum tillage stored 12% more water and increased precipitation capture by 16% compared with mulch tillage in wheat production systems on a Williams loam (Typic Argiboroll) near Sidney, Montana.

Precipitation amount is not the only issue involved in water conservation with NT. Its distribution is equally important when comparing effectiveness of NT in semiarid regions. In the

Pacific Northwest where average annual precipitation is 290 mm, greater water loss occurs during the fallow period with NT than with CT (Schillinger and Bolton, 1993). The major reason for different results with NT fallow in the Great Plains versus results in the Pacific Northwest is the long dry period (no summer rainfall) in the Pacific Northwest compared with mostly summer rainfall in the Great Plains. Hammel (1995) reported that depth and amount of water extracted by winter wheat grown under NT conditions were considerably less than with CT (moldboard plow) and minimum tillage (chisel plow) systems in the Pacific Northwest. The decrease may have resulted from higher surface layer impedance and possibly due to root diseases (Hammel, 1995).

In Oklahoma, Dao (1993) found water storage to be greater with NT than with tilled systems for continual wheat. In the northern Great Plains, Tanaka (1989) found that the response of spring wheat to tillage treatments varies yearly, depending on soil water storage during fallow and early spring temperatures. He pointed out that during years with little soil water storage after fallow and above average early spring temperatures, using chemical fallow (essentially NT) results in greater spring wheat yields than using stubble-mulch fallow. However, during years with abundant soil water storage after fallow and below-average spring temperatures, use of stubble-mulch fallow produces more spring wheat than use of chemical fallow. Based on a 4 year average, water use efficiency was greater with stubble-mulch and reduced tillage than with chemical fallow.

Soil desiccation that results from dry and strong winter winds that occur from south Texas to western Canada is reduced by the presence of surface residue. In south Texas and northern Mexico, greater soil water contents at planting under NT conditions often result in adequate seedling emergence as compared with no germination under other conditions (Bradford and Peterson, 1999). Conservation tillage practices that increase water storage have been developed for cotton in the central Texas High Plains (Harman et al., 1989) and for dryland grain sorghum at Bushland, Texas (Unger, 1991; Jones et al., 1994b).

During prolonged dry periods in the U.S. semiarid regions, cultivated soils and NT soils will dry out with time and reach the same water content. In some instances, the water content of upper soil layers even may become lower in NT than in CT soils because of capillary drying with NT. Surface cultivation under CT conditions decreases such drying in some soils (Bradford and Peterson, 1999).

A possible negative effect of greater water infiltration and higher soil water contents with CS is the potential for increasing saline seep problems on some soils. Saline seeps develop where relatively large amounts of water enter soils that have a relatively impermeable subsoil. As a result, the water table rises and water flows and seeps to the surface at down-slope sites, carrying with it dissolved salts. According to Halvorson (A. Halvorson, personal communication, 2009), the solution to the saline seep problem is more intensive dryland cropping to utilize the precipitation more efficiently.

Fresh water availability is becoming an increasingly important worldwide issue (Unger and Howell, 1999). It is a major agricultural concern where water is naturally limited and where the increasing needs of other users often clash with agriculture for available supplies (Rothfeder, 2001; Levy, 2003; Kuhn et al., 2007). Improved water conservation, therefore, is becoming increasingly important for agriculture and for society as a whole so that adequate water will be available for all users. Improved water conservation is possible by increased adoption of CS, which is important for rainfed and irrigated crop production. Greater soil water storage with CS can reduce irrigation requirements (Sullivan et al., 2007).

25.5 Residue Management and Decomposition

Unlike with CT, the main goal with CS is to minimize residue incorporation into the soil. For CS systems, farm implements are specifically designed to leave the maximum amount of residue on the soil surface, whereas for plow systems, the intent often is to incorporate the residue as quickly as possible to minimize problems associated with subsequent tillage operations. Residue amounts remaining on the surface after one pass with various implements are presented in Table 25.1. Implements such

TABLE 25.1 Percentages of Surface Residue Cover Remaining after One Pass with Various Implements

Tillage and Planting Implements	Percentages
Moldboard plow	0–10
Machines that fracture soil	70–90
Chisel plows	
Straight points	60–80
Twisted points	50–70
Sweeps and field cultivators	60–90
Disk machines	
One-way disk	55–80
Tandem or offset disk	
25 cm or greater blade spacing	25–50
18–25 cm blade spacing	30–70
Harrows and packers	60–95
Row cultivators	
Sweeps	75–90
Finger wheel cultivator	65–75
Rolling disk cultivator	45–55
Ridge-till cultivator	20–40
Drills and planters	60–95
Natural weathering	
Over-winter following summer harvest of small grain	70–90
Over-winter following fall harvest of summer crop	80–95

Sources: Adapted from Fenster, C.R., N.P. Woodruff, W.S. Chepil, and F.H. Siddoway. 1965. Performance of tillage implements in a stubble mulch system: III. Effects of tillage sequences on residues, soil cloddiness, weed control, and wheat yield. Agron. J. 57:52–55; Hill, P.R., K.J. Eck, and J.R. Wilcox. 1994. Managing crop residue with farm equipment. Agronomy Guide AY-280. Purdue University, West Lafayette, IN; Bradford, J.M., and G.A. Peterson. 1999. Conservation tillage, p. G-247–G-270. *In* M.E. Sumner (ed.-in-chief) Handbook of soil science. CRC Press, Boca Raton, FL.

as the moldboard plow and the tandem disk that invert and/or vigorously stir the soil leave less than 5% and 60% of the residue on the surface, respectively. Repeated passes with a tandem disk result in essentially no residue on the surface. In contrast, slicing-action implements such as chisels and sweep plows leave 50%–90% and 85%–90% of the residue, respectively. Managers thus have some control over residue cover, depending on the implement they use. NT relies on herbicidal weed control and no primary tillage; therefore, it provides for maximum possible surface residue cover.

Residue decomposition rates are affected directly by the amount of residue incorporation that occurs (Schomberg et al., 1994). Both surface and buried residue immobilized N, but surface residue immobilized N three times longer than buried residue, which reflects the residue decomposition status. Maximum decomposition rates occurred when residue was completely incorporated into the soil; slower decomposition occurred with NT. In the study by Schomberg et al. (1994), grain sorghum and wheat residue decomposed at about the same rate, and decomposition rates were highest under the wettest regimes. Residue decomposition rates with NT and other CS systems were highly dependent on surface soil water content and temperature. Decomposition rates were highest when soil temperatures were between 20°C and 30°C and the water content was near field capacity just under the residue cover.

Decomposition of surface residue is maximized in climatic regions where atmospheric water demand is low, relative humidity is high, warm temperatures prevail, and growing seasons are long. Peterson et al. (1995), working across a potential evapotranspiration (ET) gradient in Colorado under NT conditions, showed that surface residue loss from corn planting to harvest varied from 32% in the lowest to 62% in the highest ET zone. Higher temperatures in the high ET zone and the longer warm season promoted faster decomposition despite the drier conditions. Obviously, temperature controlled decomposition more than the relative humidity of the air at the soil surface.

The objective of CS farming is to maintain surface residue cover throughout the growing season, but especially during the preplant and seedling stages of a crop cycle because the soil is most vulnerable to erosion by either wind or water at those times. When the crop canopy provides adequate cover, residue cover becomes less important until after crop harvest. The importance of a residue cover after harvest depends on the particular climate. With high erosion potential after harvest, obviously residue cover is needed to help control erosion. On an average, a crop such as soybean leaves less cover than a crop such as corn. Furthermore, soybean residue that has a favorable C:N ratio degrades 20%–30% over winter whereas corn stalk residue that has a less favorable C:N ratio degrades 5%–20% over winter (Hill et al., 1994).

Planting equipment being used often dictates how much tillage is needed for seedbed preparation. Prior to the advent of CS, seedbeds had to be free of residue to ensure proper planter operation. Now, NT planters for both small grain and row crops are commercially available. Those planters minimally disturb the soil and leave most residue on the surface (Table 25.1).

25.6 Biological Activity

Biological activity generally is greater in CS than in CT soils. First, with CS, especially NT, there is a continuous and uniform supply of C as an energy source for organisms because the residue is not introduced all at once as is the case with CT. Secondly, the soil water content remains higher longer with CS, thereby allowing for more organism activity. Blevins et al. (1984) summarized the NT influence on soil biology.

High residue levels favor the activity of macroorganisms such as earthworms. Results of the following studies illustrate the large differences between NT and CT soils as well as the effects of crop residue management on earthworm populations. Kladivko et al. (1997) observed more earthworms in NT than in CT soils in paired sites under corn-soybean rotations across Indiana and Illinois. Similarly, Johnson-Maynard et al. (2007) found that when averaged across spring and summer seasons, NT soils had more earthworms (104 m^{-2}) than CT soils (27 m^{-2}) under spring pea, wheat, and barley in Palouse silt loam (Pachic Ultic Haploxeroll) in Idaho. In Ohio, Blanco-Canqui and Lal (2007a) reported that the number of earthworms decreased with a systematic increase in corn stover removal from Rayne silt loam (Typic Hapludult), Celina silt loam (Aquic Hapludalfs), and Hoytville clay loam (Mollic Epiaqualfs). In silt loams, earthworms averaged 78 m^{-2} in plots with <25% of stover removed and 32 m^{-2} in plots with >25% of stover removed, whereas 100% stover removal eliminated earthworms from a clayey soil.

Earthworm activity creates soil physical conditions that favor water infiltration. Runoff was <2 and 700 mm $year^{-1}$ from NT and CT watersheds, respectively, used for continual corn production for 20 years in an area that receives over 1000 mm of precipitation annually (Edwards et al., 1988). The difference was attributed to earthworm burrows in the NT watershed. Use of NT increases earthworm populations and allows their burrows to remain intact from season to season, thus providing continuous pathways for water infiltration. Furthermore, earthworm burrow diameters are much larger than the usual size of soil pores and, therefore, promote rapid intake of large volumes of water. This water is immediately transmitted past the surface soil and is stored at a depth from which evaporation is slight. Rainstorms causing the most flow in burrows were high intensity, short duration events (Edwards et al., 1989), which indicates that burrows would enhance seasonal water infiltration substantially in regions where high intensity thunderstorms are prevalent. Trojan and Linden (1994) showed that earthworm activity could more than double water penetration into NT soil as compared with CT soil under high intensity rainfall conditions.

Greater water infiltration due to earthworm activity decreases the potential for erosion and rapidly transfers water to greater depths, thereby enhancing water storage in the profile. Possible negative aspects of this rapid transfer are soluble fertilizer and pesticide movement to depths where they are less effective for their intended purposes (Hawkins et al., 2008) and groundwater contamination.

Biological activity at the microscale is also affected greatly by adoption of reduced tillage and NT systems, and these effects can be positive or negative. On the positive side, Dick (1984) observed increased activity of many enzymes in the top 7.5 cm of soil after 18 years of NT in Ohio, which was related to soil organic C contents. Furthermore, the herbicides and pesticides used for NT management did not appear to adversely affect enzyme activity. Other evidence for increased biological activity under NT was provided by Staley et al. (1988), who reported that soil biomass-C reached a maximum in surface soil after just 1 year under NT. Biomass-C approached levels found in pasture environments, and equilibrated in 10 years at a level 30% greater than with CT.

Processes that involve N are excellent indicators of microbiological activity and several authors have made interesting observations regarding contrasts between CS and CT systems. For example, Lamb et al. (1987) showed that N_2 fixation potentials in wetter NT soils were twofold greater than in drier plowed soils. Although the quantities of N_2 fixed were small (0.33 kg ha^{-1} year^{-1}), the data show the distinct difference that persists in microbial environments with NT as compared with those with CT. The increased microbial biomass present under NT conditions reduced potential N losses during periods of low crop demand because the N was incorporated into organic compounds that were less subject to loss through leaching and denitrification (Bremer and van Kessel, 1992). Wood et al. (1990) reported that adopting NT after long-term use of CT quickly increased the amount of surface residue and of potentially mineralizable N. When they used more intense cropping systems in conjunction with NT, they obtained the highest N mineralization.

Possible negative effects of microbial activity on soil-plant systems include greater loss of N from urea fertilizers, greater denitrification, and production of phytotoxic compounds. Dick et al. (1991) showed that soils with a 25 year history of NT had increased urease activity at the surface, making urea-based fertilizers more susceptible to N volatilization losses. Rice and Smith (1982) showed that the denitrification potential increased under NT because of the more conducive environment provided by the large amounts of C at the soil surface, which resulted in the high biological activity, and the wetter soil conditions under the residue cover.

Phytotoxic compounds synthesized by microorganisms during residue decomposition were identified many years ago. NT environments have regenerated interest in these compounds and in the processes involved in their production. The slower residue decomposition, the prolonged production period, and the colder temperatures in the soil or at the soil surface and very near the emerging plant seedlings create potentially enhanced phytotoxicity problems under NT conditions. Buchanan and Kin (1993) clearly demonstrated that the decomposition period with NT was longer than with tillage systems. Working in a warm North Carolina environment, they found that 45% and 20% of the C remained under NT and CT systems, respectively, after 50 weeks. The results were similar for wheat and soybean stalk residue. Cochran et al. (1977) demonstrated that surface residue of lentil (*Lens culinaris*), pea (*Pisum* spp.), wheat, barley, and bluegrass (*Poa* spp.) all produced phytotoxic materials at some point during the spring season. Toxicities were worst when wet weather and low temperatures prevailed. Lentil and pea materials were more toxic to wheat seedlings than the other materials, but toxin production fortunately was short-lived. Martin et al. (1990) found that microbially synthesized phytotoxic materials from corn residue were more damaging to corn seedlings than were the byproducts of soybean and oat (*Avena sativa* L.) residue decomposition. Corn planted back into corn residue, therefore, would be more prone to phytotoxin damage than corn planted into soybean or oat residue. These findings illustrate that to avoid potential seedling damage, it may be more important to use crop rotations when using NT than when using CT systems.

25.7 Carbon Dynamics and Sequestration in No-Tillage

In contrast to agriculture with tillage, NT soil management greatly alters the C cycle because plant residue C is added to the soil in a different manner. The traditional moldboard plow system is the antithesis of NT. Moldboard plowing inverts and incorporates most or all residue of the previous crop into the soil in one operation. In such case, microbes responsible for the residue decomposition process are stimulated by the instantaneous addition of large amounts of organic C. No matter what system is used, however, decomposition rates are controlled by soil water conditions, soil temperature, and inorganic N supply. If the water content is above the wilting point and the temperature is above 4°C, inorganic N supply limits decomposition. In tilled systems, if appropriate amounts of N fertilizer are added, soil organisms can complete decomposition of the incorporated crop residue and roots in the surface soil layer within a few months after plowing. Often, the surface soil is disked before plowing and/or the crop residue is shredded to enhance farm implement function, which creates more residue surface area for soil organisms and hastens the decomposition rate. Large quantities of CO_2 evolve from soils shortly after plowing. Soils usually are tilled several more times after plowing to prepare a fine, firm seedbed for planting. These operations stir the soil and residue mixture, and further stimulate decomposition of the organic materials if the soil contains adequate water. Tillage, however, also hastens soil drying, which may retard decomposition if no rainfall or irrigation occurs.

Continued use of any type of inversion tillage for a long time decreases soil aggregate size, which results in organisms having access to C that once was physically inaccessible within the larger aggregates. Therefore, organic C that literally required centuries to be stored is released as CO_2 within a few years. For example, when temperate-zone soils of North America were first cultivated, as much as 10^9 Mg of C were released as CO_2 (Paustian et al., 1997). Intensive tillage accelerates release of CO_2 and thereby increases organic C losses (Blanco and Lal, 2008). On Barnes loam (Udic Haploboroll) in west central Minnesota, Reicosky and Archer (2007) observed that the CO_2 release from CT soils was consistently greater than from NT soils under spring wheat, particularly immediately following tillage. Deep and frequent tillage increases the release of CO_2 relative to shallow tillage. Retaining crop residue after harvest and reducing tillage depth are strategies to enhance organic C sequestration and reduce emissions of greenhouse gases.

When soils that have been managed with maximum tillage such as by moldboard plowing are converted to NT, a dramatic change occurs in the way organic C is introduced into the soil system, both in terms of timing and position. Furthermore, soil under NT management remains wetter for longer periods. The new system (NT) results in little or no mixing of residue with the soil. Crop residue remains on the surface and decomposition occurs much more slowly than under tillage conditions because much less surface area of residue is exposed to soil organisms, even if the producer shreds the residue. Roots at shallow soil depths also remain basically intact, again reducing their area exposed to decomposition. Soil surfaces with NT are generally moist enough, especially in the spring and early summer, so that water availability does not limit decomposition. The residue cover, however, keeps the soil cooler longer than for a bare surface soil, especially in northern latitudes, thus slowing decomposition.

Under NT conditions, N supply is more limiting for decomposition because the organisms have access only to the inorganic N in a much smaller soil volume than in a tilled system. Obviously, root mass decomposers have access to more N than surface residue decomposers, but overall residue decomposition is delayed with NT relative to that in a tilled soil.

After 5–10 years, the amount of surface residue in an NT system appears to reach equilibrium for a given soil and climate combination (Sherrod et al., 1996). Resident residue levels increase more in cooler than in warmer climates due to slower decomposition rates in the former. Within climatic zones, soil types also affect residue levels. Excessively drained soils have lower residue levels relative to well-drained soils primarily because less biomass is produced. At the other extreme, wetter soils have greater residue levels because biomass production generally is high and because the cooler soil environment in the wet soil delays decomposition, which, in turn, is accentuated by the heavy residue cover.

Generally speaking, the total C sequestered in surface residue plus the organic C in the uppermost soil layer (0–5 cm depth) is higher with NT than with the same soil under tilled conditions. Furthermore, substantial amounts of C still reside as crop residue in the upper 2.5 cm of soil. Ortega (1995) found residue C amounts within the upper 2.5 cm of NT soil equal to that present on the soil surface and Dell et al. (2008) found about 50% more C in particulate and mineral-associated pools in the upper 5 cm of NT soils than of CT soils.

Soil organic C tends to increase with NT relative to that with tilled systems, but the increase rate is slow and usually confined to the upper soil layers with little effect on deeper layers (Dick, 1983; Potter et al., 1998). Changes in soil C after 18 years were 0% to −11% with NT, −12% to −23% for minimum tillage, and −14% to −23% for CT (Dick, 1983). Organic P also was greater in surface soil layers with NT than with tillage. Wood and Edwards (1992) found that NT relative to CT resulted in C and N concentrations after 10 years that were 67% and 66% greater, respectively, than those with plow tillage at the 0–10 cm depth. Macroaggregates (>250 mm) under NT were more numerous and more stable than those under CT, probably due to greater biological activity under NT conditions. Microbiological activity was more episodic in the tilled system, which had a drier surface soil than NT. CT also caused more physical disruption coupled with less production of aggregate-stabilizing materials. For a drier climate (eastern Colorado), Wood et al. (1991) concluded that higher equilibrium levels of soil C and N could be maintained by increasing cropping intensity, which increased C input to the system. These changes were measurable even after only 4 years of NT. Less inorganic NO_3–N was present in the intensive system, indicating that the N loss due to leaching or denitrification was reduced.

Data from Colorado show that with some NT systems, OM contents have increased in the upper 5 cm of soil, but have declined in the 5–10 cm depth. The net result for the 0–10 cm depth generally was a decline after 7 years of NT, especially in systems that include fallow. Under climate and soil conditions that permit continual cropping, adoption of NT is likely to increase soil OM much more. This means that the more humid the climate (high precipitation coupled with low potential ET), the greater the probability that NT will promote soil OM accumulation. Conversely, under climates with low precipitation and high ET that usually require fallow periods in the cropping systems, there would be less chance of increasing soil OM; and net losses of C may occur, even under NT management, because of less plant material production under such conditions.

NT adoption promotes restoration of some of the C lost when soils were broken from their native condition. Potential C sequestration, however, can be small and slow relative to the losses from the original soil OM level, depending on soil, management, and climate. It will take years of NT farming to increase soil C because the amounts of available N and P required to provide the correct C:N and C:P ratio for C sequestration are limited. For example, producers usually apply adequate N for the crop, but plan to leave no excess in the soil system. Therefore, little N is available to balance the high C input from crop residue. The same situation can occur with P, especially in soils with low P availability. Paustian et al. (1997) provided an excellent summary of tillage effects on soil OM. They concluded that reduction in tillage, and especially using NT, provided an environment conducive to C storage in soils. Soil organic C storage is generally greater with NT than with CT at the 0–5 cm depth (West and Post, 2002). In a few cases, where NT reduces biomass production compared with CT, there may be no differences between NT and CT.

What is the net effect of NT on the C cycle? It is the amount sequestered in the soil. Carbon sequestration in soil has received much attention in recent years because of the possible effect that increasing atmospheric CO_2 levels may have on global climate change. More soil organic C can be sequestered by using reduced tillage and NT in combination with diversified cropping systems than by using CT. It is estimated that about 60% of organic C has been lost due to intensive plowing, monocropping, simplified crop rotations, and short-sighted management systems (IPCC, 1996; Benjamin et al., 2007). These losses make up nearly 23% of the total greenhouse gas emissions in the atmosphere (Lal, 2004). Reducing or eliminating tillage is a strategy to restore the organic C pool. For example, NT, in addition to providing many other benefits, can enhance soil organic C sequestration. Across five soils in Kansas, McVay et al. (2006) showed that NT management resulted in more soil C at the 0–5 cm depth than CT management. The increase of residue-derived C in a silt loam in Minnesota was 2.3 times higher with NT than with CT or chisel-plowing (Allmaras et al., 2004).

NT is superior to CT for storing C in soil. In most cases, however, organic C in NT soils is concentrated near the surface (West and Post, 2002). The marked stratification of organic C in NT soils is partly due to surface residue mulching. The soil organic C with NT can be higher at the 0–5 cm depth and lower below this depth compared with plowed soils. Furthermore, substantial amounts of C still reside as crop residue in the upper 2.5 cm of soil. Ortega (1995) found residue C amounts within the upper 2.5 cm of NT soil equal to that present on the soil surface and Dell et al. (2008) found about 50% more C in particulate and mineral associated pools in the upper 5 cm of NT soils than in CT soils.

The difference in total organic C between NT and plowed soil profiles may not be significant if there is considerable stratification. It is important to note that organic C stored in deeper layers is important to long-term C sequestration (Lorenz and Lal, 2005). The organic C near the surface layers is subject to abrupt fluctuations in soil temperature and soil water content, which causes rapid mineralization of soil organic materials.

The NT impacts on C sequestration must be characterized for the entire soil profile and not only for shallow surface soil (<20 cm depth) (West and Post, 2002). Plant roots of most crops often extend to a 50 cm depth (Qin et al., 2006), which is much deeper than the typical soil sampling depth for organic C characterization (West and Post, 2002). Plant roots in plowed soils may penetrate to greater depths than in NT soils because of relatively loose soil following tillage. Recently, some researchers have argued that the greater C sequestration in NT systems reported in many studies may be simply due to shallow sampling protocol that biases the results (Baker et al., 2007). At this point, the limited information on organic C distribution for the whole soil profile hinders our ability to conclusively state the benefits of NT farming to sequester C at lower profile depths.

A study conducted under on-farm conditions in the eastern United states by Christopher et al. (2009) showed that organic C with 5–35 years NT systems was not greater than with CT systems in most soils at the 0–5 cm depth. Total organic C in the profile with NT and CT did not differ in 8 of the 12 soils. It was actually greater with CT than with NT in three soils after 8, 15, and 30 years of NT management. For a similar on-farm study for the same temperate region, Blanco-Canqui and Lal (2008) observed that NT management increased organic C concentration over CT in 5 out of 11 soils, but the increases were significant only at the 0–5 cm depth. Beneficial effects of NT for increasing organic C pools tend to disappear with increases in soil depth.

Impacts of NT farming on organic C can thus be variable and soil-specific (VandenBygaart et al., 2003). Length of NT management, degree of soil erosion, crop grown, and residue incorporation with CT are among the factors that cause differences between NT and CT. Increases in organic C with NT often are more pronounced with long-term (>10 years) management and for erosion prone soils. Indeed, NT farming may favor greater organic C storage in sloping terrains and soils with low clay content. NT benefits for increasing organic C concentrations, particularly at lower profile depths, are complex and should not be generalized for all soils.

25.8 Biofuel (Cellulosic) Production Implications

Crop residue retained on the soil surface greatly reduces the potential for soil erosion by wind and water (Figure 25.1) and enhances water conservation under many conditions. Such residue retention, therefore, is highly important for conserving the soil resource base to maintain its sustainability for crop production and to enhance production by effectively using the conserved water for greater crop production for an ever-increasing world population. These benefits generally increase with increases in the amounts of residue retained on the surface, as discussed in previous sections. Removing even some residue, at least from some soils, therefore, would negate the conditions essential for sustained or improved soil and water conservation and crop production.

One of the main differences between CS and plow tillage (CT) is that most crop residue remains on the soil surface when using CS, thereby enhancing soil and water conservation. Today, however, there are many competing uses for crop residue including soil and water conservation, animal feed, industrial raw material, and as feedstock for cellulosic ethanol production. Particularly, producing biofuel from crop residue is generating a lot of interest (Perlack et al., 2005). A number of commercial-scale cellulosic ethanol plants are being built in the United States to produce the second generation of biofuel (USDOE, 2007). Production of ethanol from cellulosic plant materials is projected to increase exponentially in the near future as concerns over increased food prices may slow ethanol production from grain. At present, corn stover is the main candidate as cellulosic feedstock for biofuel production (Graham et al., 2007). Residue of crops such as wheat and sorghum is also being considered as feedstock for biofuel as technologies for cellulosic conversion progress (Sarath et al., 2008). Production of cellulosic ethanol from renewable energy sources is important and should be pursued. But removing crop residue as biofuel feedstock must be objectively examined and contrasted against its positive and negative impacts on soil conditions, crop production, and environmental quality. Dedicated energy crops (e.g., warm season grasses and fast-growing woody plants) are potential alternatives to crop residue removal.

25.8.1 Negative Aspects of Crop Residue Removal

Removal of crop residue can have more negative than positive impacts in the long term for conserving soil and water resources (Wilhelm et al., 2007). Although sometimes dubbed as "low-cost waste," crop residue is a valuable commodity. Indiscriminate removal of crop residue is not advisable particularly in semiarid regions with low and variable precipitation and biomass production. Crop residue mulch is important for precipitation capture and storage, and water use efficiency. By increasing soil erosion, residue removal can also accelerate losses of nutrients in runoff, thereby further increasing risks of non-point source pollution of downstream water sources such as ponds, lakes, rivers, and streams. Sediment and sediment-borne chemicals (e.g., fertilizers and pesticides) are the major pollutants in runoff. Total N and total P transport in runoff often increases with increases in crop residue removal (Lindstrom, 1986; Blanco-Canqui et al., 2009). Residue removal can also increase soil susceptibility to erosion by wind by leaving the soil partially or completely unprotected. Soils under NT conditions can be as susceptible to erosion by wind as CT soils unless adequate residue cover is maintained on the surface.

Crop residue mulch reduces the susceptibility of the surface soil to compaction by buffering impacting raindrops and traffic. Residue mulched soils are more resilient than unmulched soils (Wilhelm et al., 2004). Residue cover also reduces fluctuations in soil temperature and wetting and drying cycles. Soil aggregate stability, water infiltration, water retention, saturated hydraulic conductivity, macroporosity, and biological activity decrease, whereas soil cone index and bulk density increase with increases in crop residue removal (Karlen et al., 1994; Blanco-Canqui and Lal, 2007a).

Crop residue removal reduces soil organic C because residue contains about 45% C. Removal of crop residue for alternative uses reduces not only organic C, but also plant nutrient pools. Nutrients also may be lost via increased erosion (Blanco-Canqui et al., 2009). The magnitude of losses depends on soil and climatic conditions. For example, losses of organic C pools in some soils increase linearly with increasing removal rates of corn stover (Karlen et al., 1994; Clapp et al., 2000). Blanco-Canqui and Lal (2007a) observed that removal of corn stover from long-term NT soils decreased organic C pools in sloping and erosion-prone soils, but had little or no effects on nearly level, clayey, and cold soils 3 years after residue removal.

Crop residue removal influences crop production because it reduces plant available water and causes abrupt fluctuations in soil surface temperature. Crop residue is a source of soil OM and essential plant nutrients and thus its removal can reduce the supply of vital nutrients. On a silty clay loam in Nebraska, complete removal of stover from a 4 year NT system reduced corn grain and biomass yields by about 23% (Wilhelm et al., 1986). Grain yield is often higher from residue mulched soils than from unmulched soils due to increased soil water content and favorable soil temperature in the growing season. Rates of crop residue mulch can explain as much as 95% of the variability in grain and biomass yield (Wilhelm et al., 1986; Blanco-Canqui and Lal, 2007a). Corn stover removal reduced corn yield by 1 Mg ha^{-1} during 3 years of a 12 year NT continual corn study on a silt loam in Minnesota (Linden et al., 2000). Residue removal impacts on corn production are site-specific and depend on soil, topography, duration of stover management, tillage, and climate. Crop production in sloping, erosion-prone, and well drained soils can be more adversely affected by residue removal than in flat and clayey soils.

25.8.2 Positive Aspects of Crop Residue Removal

In some soils or ecosystems, it may be feasible to remove a portion of the crop residue for alternative uses in the short term without increasing risks of erosion by water and wind, reducing the soil OM pools, or reducing crop production. Indeed, removal of some of the total residue produced may be beneficial to improve seed germination, facilitate planting, increase N mineralization, and reduce pest infestations. Dam et al. (2005) observed that corn emergence in mulched plots was reduced by as much as 30% relative to unmulched plots. In a similar study, Blanco-Canqui et al. (2006) observed that corn stover removal from long-term NT soils enhanced seed germination on three soils in Ohio. Without stover removal, emergence was delayed by up to 3 days as compared with that on soils where all stover was removed. These studies showed, however, that the delayed emergence did not reduce crop yields. Plants in plots with low amount of residue cover often grow taller during the first few weeks, but the height differences often diminish rapidly with time.

Excessively wet and cold soils during the germination period, proliferation of weeds and pests, and nutrient immobilization under mulched soils may lower crop yields. On two silt loams in southwestern Wisconsin, corn yield decreased when stover cover was doubled (Swan et al., 1994). On a silty clay loam in Iowa, corn yield decreased during the last 4 years of a 13 year continual corn system with the addition of 2, 4, 8, and 16 Mg ha^{-1} of stover mulch (Morachan et al., 1972).

25.8.3 Threshold Levels of Residue Removal

The magnitude of residue removal impacts on soils and crop production is site specific and information on maximum permissible rates of residue removal is limited. The positive and negative impacts of crop residue removal must, therefore, be characterized for different scenarios of tillage and cropping systems, soil characteristics, and climatic zones. Some computer-simulation model estimates in the United States, mainly for the Corn Belt region, show that 20%–50% of the residue produced may be available for removal (Graham et al., 2007). These estimates, however, are mainly based on the residue requirements to control erosion. Permissible levels of residue removal must account for the residue requirements not only to control soil erosion, but also to maintain soil OM and nutrient pools, sustain corn yields, and maintain or enhance soil quality. Depending on the soil and climate, between 5.3 and 12.5 Mg ha^{-1} of crop residue is needed to maintain soil organic C levels (Johnson et al., 2006; Varvel and Wilhelm, 2008). This amount exceeds that needed for controlling erosion (Wilhelm et al., 2007). Thus, the amount that can be harvested may be limited by the amount needed to maintain soil organic C pools.

The amount of stover that can be harvested for biofuel production must be established for different soils and ecoregions based on the multiple roles of residue. Process-based models are valuable tools for predicting the impacts of different scenarios of crop residue management across regional and national scales on the potential for erosion by water and wind. These models allow the scaling up of information on residue management to larger geographic areas. Use of current models combined with advanced tools such as remote sensing and GIS is a promising approach to enhance our understanding of residue removal impacts on soil and water resources (Green et al., 2003).

25.9 Soil Fertility and Nutrition

25.9.1 Nitrogen

All soil biological processes are greatly affected by the organic and inorganic portions of the N cycle. The N cycle is inextricably linked to the C cycle. The presence of mineral N in soil for plant uptake depends on what is happening within the C cycle. Because N is the most plant-growth-limiting nutrient and because it has been widely demonstrated that conversion to NT and other reduced tillage systems alters the C cycle, it is important to understand the specific effects of tillage systems on N availability to plants. As a result, soil testing and plant analyses are highly important components of effective nutrient management for successful crop production under CS conditions (Schlegel and Grant, 2006).

25.9.1.1 Immobilization of Fertilizer N

All CS and CT systems tend to immobilize N, but CS systems do not immobilize more N than CT systems; they only keep the N immobilized longer. Net N mineralization is eventually equal in all systems, but the timing of N release is delayed when residue is either slowly incorporated into the soil or left on the soil surface. Rice and Smith (1984) reported that within the first 35 days after application, N immobilization was 19% and 11% with NT and CT, respectively. Use of subsurface N-fertilization techniques may minimize immobilization in NT systems. Separating N and crop residue by placing fertilizer in soil below the residue was considered more important under CS than under CT systems (Schlegel and Grant, 2006). Because of immobilization differences, crop yields could be affected. Kelley and Sweeney (2007) showed that wheat grain yields averaged 3.68, 3.40, and 3.19 Mg ha^{-1} with subsurface knife, surface band, and surface broadcast applications, respectively, of N and P fertilizer in Kansas on Parsons silt loam (Mollic Albaqualf). Subsurface band application of N at greater rates was considered especially important when wheat followed grain sorghum in a rotation. Because of reduced immobilization, band applications of N fertilizer were more effective for production of durum wheat (*Triticum* spp.) (Grant et al., 2001) and canola (*Brassica* spp.) (Grant et al., 2002) in Canada. In Kansas, winter wheat yields were 8% greater with springtime point-injected than with broadcast N fertilizer application (Schlegel et al., 2003).

According to Schomberg et al. (1994), surface and buried residue both immobilized N, but net immobilization was longer than 1 year for surface residue and about 0.33 year for buried residue. Residue type, either grain sorghum or wheat, made little difference in terms of decomposition rate. Smith and Sharpley (1990, 1993) studied effects of residue type and placement on N immobilization and mineralization. Surface residue caused temporary N immobilization and the C:N ratio of the materials governed the duration of immobilization. Alfalfa (*Medicago sativa* L.) residue had the maximum mineralization followed by peanut (*Arachis* spp.), soybean, oat, sorghum, wheat, and corn in decreasing order. Although surface-placed residue had less N mineralized than buried residue, the difference between placements was minor compared with differences among residue types.

Soil OM contents are often very low under dryland conditions. In New Mexico, NT soils had less inorganic N and more organic C and N than a stubble mulch tillage soil after 5 years (Christensen et al., 1994). Unfertilized grain sorghum was very N deficient with the NT compared with the stubble mulch treatment. Soil organic C increased with time with NT and stubble mulch systems, but a point was not reached in the 5 year study where N mineralization with NT equaled that with stubble mulch tillage.

Immobilization delays N availability to plants, but the impact of N fertilizer management appears not to be as significant as once thought. Crop yields where no N is applied can be lower with NT than for other tillage systems due to immobilization with NT (Bundy et al., 1992). Nitrogen immobilization, however, may be affected by residue amounts on the soil surface. In Nebraska, increased residue levels resulted in greater grain yields in a tillage experiment and the additional residue increased stover production and N uptake by subsequent crops (Maskina et al., 1993). Nitrogen immobilization seemed not to be a factor. Tillage per se, that is, NT versus disked, did not affect grain yields. Vigil and Kissel (1991) developed equations to predict net N mineralization for a season based on N content and/or N + lignin/N ratio of the residue and concluded that net N mineralization would not occur if the C:N ratio of the residue was above 40.

25.9.1.2 Soil N and C Mineralization with Time with No-Tillage

Converting crop production to reduced tillage and NT systems affects the formation, size distribution, and strength of soil aggregates. Changes in aggregate properties alter the rates at which both physical and biological processes proceed. When soils are not tilled, macroaggregates that form from natural processes tend to remain longer and have an opportunity to become strengthened. The more stable the soil aggregates, the greater the organic C stabilization in soil.

Soil organic C mineralization is a function of the interactive physical, chemical, and physicochemical processes that occur inside soil aggregates. Soil organic C is physically protected inside macro- and microaggregates through entrapment, and chemically through adsorption and recalcitrance (Blanco-Canqui and Lal, 2004). The confinement of C inside of aggregates prevents OM from rapid mineralization by microbial processes while enhancing long-term C sequestration. There is a mutual interrelationship between soil C and aggregates. The protected C pool enhances macro- and microaggregate formation and stabilization, while the microaggregates, in turn, protect the organic C from rapid mineralization.

The soil organic C residence times (mineralization rates) range from a few minutes to hundreds of years. Particulate OM or relatively labile OM can be mineralized in 5 or 10 years, while stable C can be mineralized in 500 or 1000 years (Blanco-Canqui and Lal, 2004). The residence times of C in soil aggregates is influenced by the type of organic binding agents, type of soil, size of aggregates, tillage, and climate. Mineralization of C normally increases with increases in aggregate size because of the differences in physical and chemical protection (Jastrow et al., 1996). In some cases, the C inside microaggregates can be protected physically, but be highly susceptible to rapid mineralization. The mechanisms of C mineralization or residence times inside aggregates are complex.

Macroaggregate formation often is greater in NT than in CT soils. These large aggregates with NT store more C because they are made up of a large number of C-enriched microaggregates. Beare et al. (1994a, 1994b) found that macroaggregates from NT soil protected soil OM from decomposition. In their study, 19% of the total mineralizable C and N in NT soils came from these larger aggregates, while only 10% and 5% of the C and N, respectively, came from them in CT soils. This aggregate-soil OM association under NT conditions may explain why conversion to NT results in a need for higher N fertilizer rates at the outset (Bradford and Peterson, 1999). Adjustments in N application rates during early years of adoption of NT were considered important also by Schlegel and Grant (2006). Apparently, a new steady state is reached and the immobilization capacity due to macroaggregate formation is fulfilled. In fact, Follett and Schimel (1989) inferred from data collected from a 15-year NT experiment that less tillage helped conserve mineral N. In other words, the greater, but temporary, immobilization of N under NT conditions reduced the opportunity for both leaching and denitrification losses of mineral N.

How long does the net immobilization period of NT last before mineralization processes equilibrate? Rice et al. (1984) observed that after 10 years, N mineralization with NT was equal to that with CT. The authors suggested that the net immobilization phase is transitory when NT is adopted. For a wheat-fallow system in western Nebraska, Lamb et al. (1985) reported that for the first 5–7 years after adoption of reduced tillage and/or NT, less NO_3–N accumulated during the fallow period than in plowed soils. Approximately 7 years after adoption, NT and stubble mulch tillage soils had NO_3–N accumulations during fallow equal to those in plowed soils and continued to have equal contents in subsequent years.

25.9.1.3 Fertilizer N Conservation and Efficiency

Perhaps of more concern than immobilization of N fertilizer is the potential loss of N from urea fertilizers applied directly to a residue-covered NT soil. In Missouri, substantial N losses occurred from urea-based N fertilizers under NT conditions when the fertilizer was not injected into the soil (Stecker et al., 1993a, 1993b). When N was injected below the residue, time and source of application had no effect on corn yield. In Wisconsin, ammonia losses reduced the effectiveness of urea-containing fertilizers, but such loss did not fully explain the observed N source differences. The N source effects were similar with all residue levels, which suggested that applying extra N can overcome yield reductions with high residue levels (Andraski and Bundy, 2008).

Under drier climatic condition, the effect of placing N below the residue depends on the season when crops are grown. For fall-planted wheat, hot and dry weather conditions at planting time in fall apparently stimulate N losses from urea compounds that are broadcast over the residue. In contrast, cooler and wetter weather at corn planting time diminishes the loss mechanisms, thereby causing minimal N losses (Kolberg et al., 1996).

Cold springtime soil temperatures are the rule in northern regions such as, for example, the U.S. northern Great Plains and, therefore, influence crop responses to N fertilizer. Spring wheat in a wheat-fallow system in the northern Great Plains responded most to N fertilizer in years when springtime NO_3–N levels were lowest, apparently because of greater N immobilization due to lower soil temperatures with CS systems (Halvorson et al., 2000). The NT and minimum tillage systems are considered suitable for spring wheat production in the northern Great Plains, but yields are slightly reduced when compared with CT systems in some years.

25.9.1.4 Crop Responses to Starter Fertilizer

Cool soil temperatures under NT conditions often slow early growth of spring-planted crops and, therefore, may affect their yield. To improve early growth and crop yields under such conditions, starter fertilizers are sometimes applied, but there appears to be no clear evidence indicating where the use of starter fertilizer is effective for improving crop growth and yields. Studies under varying conditions at different locations showed various responses to the use of starter fertilizers.

Wortmann et al. (2006b) in southeastern Nebraska applied NP and NP + S starter fertilizers as in-furrow, over-the-row, and 5 cm deep and 5 cm to the side of the seed treatments for grain sorghum planted under NT conditions. The soil orders were Mollisols, Alfisols, and an Entisol. The use of starter fertilizer increased early growth by 48% in some trials where soil P was low, but grain yield responses to starter fertilizer were not related to the soil P content. Including S did not affect yield. Fertilizer placement effects on yields varied, but the frequency and magnitude of NT sorghum responses to starter fertilizer were not adequate to be profitable at the southeastern Nebraska sites. In contrast, in a similar study in eastern Nebraska where soil orders were Eutrudepts, Ustorthents, and Mollisols, application of starter fertilizer increased early corn growth by 30% in irrigated trials and by 10% in some rainfed trials. The use of starter fertilizer increased grain yields by 0.86 Mg ha^{-1} with irrigation, but only slightly without irrigation. Yields were similar with the different placement treatments. The yield response to starter fertilizer was greatest and most profitable under irrigated conditions where soil P was low. The conversion of early growth response to starter fertilizer into a grain yield response appeared to depend on soil water availability and on the soil P level (Wortmann et al., 2006a).

Under precision agriculture conditions for NT corn in Iowa, application of liquid NPK starter fertilizer in the seed furrow and beside and below the seed was compared with a no starter fertilizer treatment. The use of starter fertilizer often increased early growth and increased yield in some fields having a low soil P value, but decreased yield in one field. With high P values, yield responses were 0.08–0.19 Mg ha^{-1} with starter fertilizer applied in the furrow and 0.16–0.46 Mg ha^{-1} when applied beside and below the seed. Across fields, the early growth response was linearly but poorly correlated with the yield response (2.4%). With NT, corn yield responses to starter fertilizer are apt to be greatest when the soil P value is below optimum or when preplant or side-dress N rates are deficient (Bermudez and Mallarino, 2002).

To evaluate corn responses to starter fertilizer and tillage treatments, Bermudez and Mallarino (2004) in Iowa had starter or no starter fertilizer treatments in addition to farmers' normal broadcast application rates of NPK. Tillage treatments were spring disking, strip tillage on previously managed NT fields, and non-tilled NT fields. The use of starter fertilizer increased corn grain yield in some fields (up to 0.52 Mg ha^{-1}) and tillage increased yield up to 0.50 Mg ha^{-1} in some fields, but the overall average yield increases were small. Dry matter and nutrient uptake responses to both treatments did not result in large or consistent yield responses. Soil test results or soil series could not always be used to identify fields where an application of starter fertilizer would increase corn yield.

In Kansas, Niehues et al. (2004) applied starter fertilizer containing N, P, K, and S (with subsurface placement) as direct seed contact, dribble over-the-row, and subsurface band 5 cm below and 5 cm to the side of the seed row treatments for NT corn production. For all treatments, N was balanced at 0.17 Mg ha^{-1} by applying ammonium nitrate (AN) at planting, even for the nonstarter check treatment. Regardless of placement, the use of starter fertilizer often increased early dry matter production and significantly increased corn grain yields. Including S sometimes increased early dry matter production, grain yield, and nutrient uptake. For this study, the use of a starter fertilizer was an effective and efficient way to stimulate early growth and improve grain yields of NT corn grown continually in Kansas.

The aforementioned studies showed that there is no universal response to the use of starter fertilizer for improving crop growth and yields. The different studies under varying conditions at different locations clearly showed different responses to the use of starter fertilizers. Early growth often was greater where such fertilizer was applied and yields increased in some cases, but the yield increases were slight in most cases. Consequently, the question regarding whether starter fertilizer should be applied remains unanswered.

25.9.1.5 Crop Responses to Fertilizer (Nonstarter)

As for any other tillage method, crops grown under CS conditions respond to applied fertilizer. The degree of response depends on the inherent soil nutrient level. A response difference with CS, especially NT, relative to that with CT may occur due to the method of application. With CT, the fertilizer generally is placed into the soil before or when planting occurs. With NT, the fertilizer may be broadcast on the surface. Broadcast application, however, is satisfactory under NT conditions as indicated for a study with cotton in Tennessee (Howard et al., 2001). Relative to no N application, broadcast application of AN at 67 kg N ha^{-1} increased average lint yields from 0.74 to 1.28 Mg ha^{-1} on Loring silt loam (Oxyaquic Fradiudalfs) and from 1.09 to 1.54 Mg ha^{-1} on Lexington silt loam (Utlic Hapludalfs). A rate of 101 kg N ha^{-1} was needed to increase average yields on Memphis silt loam (Typic Hapludalfs) from 0.82 to 1.17 Mg ha^{-1}. Broadcasting AN was satisfactory for producing lint yields equal to or higher than those obtained by injecting urea-AN or splitting AN applications on the loess soils having different surface covers or residue.

25.9.2 Phosphorus and Potassium

Reduced and NT systems tend to keep soils cooler than clean tillage systems (see explanation in previous sections). For spring-planted crops in northern latitudes of the northern hemisphere, this results in slower plant root growth with reduced tillage systems. Because P is an immobile nutrient and uptake depends greatly on root interception, cold soils can create temporary P deficiencies. Starter fertilizers placed near the seed at planting may be beneficial in such cases. Failure to correct the temporary P deficiency stunts plant growth and delays development so that the crop may fail to mature before frost occurs in the fall. For a wheat-fallow system, placing P fertilizer with the seed was more beneficial under NT than for plowed and/or reduced tillage (stubble mulch) systems (Stecker et al., 1988). Apparently, cooler soil temperatures under high residue conditions in the spring were responsible for this difference.

In contrast to P, K is much more mobile in soils. Uptake of K and its effect on crops, however, may be influenced by placement as a fertilizer under NT conditions. In Iowa, P and K at various rates were broadcast, banded with the planter, or deep banded (at 15–20 cm depth) to determine their effect on early growth, grain yield, and nutrient uptake by NT soybean. The study involved nine soil series on Argiudolls, Endoaqualfs, Endoaquolls, and Hapludolls. Yields increased with P fertilization when soil-test P was <9 and <12 mg kg^{-1} at the 0–15 and 0–7.5 cm depths, respectively. Band placement of K resulted in slightly higher yield than broadcast placement. Responses to K were not related to soil-test K values or stratification in the soil. Placement of P and K had little effect on early soybean growth, but affected early P and K uptake. Banding with the planter was more effective than broadcast application for P uptake, but both depths of band placement were more effective than broadcast application for K uptake (Borges and Mallarino, 2000).

No clear-cut recommendation can be made regarding when and how to apply K fertilizer. Therefore, producers should rely on soil test results and associated recommendations for their particular soil, cropping system used, and crop to be grown. Xinhua and Vyn (2002a) in Indiana found that NT soybean responded to K applications on some soils, but there was no yield difference whether the K was banded or broadcast. In another study involving different row spacing for soybean, band placement was superior to broadcast application of K fertilizer (Xinhua and Vyn, 2003). The response was greatest when soybean was seeded over the K bands. Also in Indiana, Vyn and Janovicek (2001) found that NT or zone-tilled corn responded to starter K fertilizer applied at a high rate, even when no K fertilizer was applied the previous fall. Fall or spring application of K fertilizer for corn was equally beneficial for the subsequent soybean crop under continuous or intermittent NT conditions (Xinhua and Vyn, 2002b).

When soils are sampled to determine P fertilizer requirements, the test results may be influenced by the previous P application method. The P fertilizer application method under NT conditions may affect P distribution in soil for months or even for years because P is highly immobile in soils. To quantify the spatial distribution and temporal dynamics of band-applied P and to assess its availability to crops, Stecker et al. (2001) obtained soil samples in a grid pattern around bands where 10 and 20 kg ha^{-1} of starter fertilizer P was applied about 6, 12, and 24 months previously. Soils were Mexico silty clay loam (Mollic Epiaqualf), Putman silty clay loam (Vertic Albaqualf), and Dockery silt loam (Aquic Udifluvent). The P concentrations decreased outward from the band center, were highest at either 6 or 12 months, and varied substantially along the direction of band application. The results indicated that including bands in soil tests would increase the risk of overestimating the amount of P that is available to crops.

Application of P and K fertilizers under continuous NT conditions may result in their stratification similar to that which occurs with band application. As a result, special sampling techniques may be needed to adequately determine the nutrient status of NT soils. For a study on three silt loams (Memphis, Lexington, and Loring) in Tennessee to which K had been broadcast at various rates for NT cotton production, Howard et al. (1999) determined P and K distribution with depth at in-row and between-row positions. Sampling depths were 0–8, 8–15, and 15–30 cm. The soil-test P level varied with soil, sampling position, and soil depth, but differences due to sampling positions (in-row or between-row) did not affect soil test ratings. Additional years under NT cotton production, however, could increase position influences and, therefore, affect P fertilizer recommendations. The soil-test K level was greater at the in-row position at the 0–8 cm depth of all soils. As a result, sampling only between rows could give a lower test value in some cases. The differences varied with fertilizer K rates and could increase with time of NT cotton production.

25.10 Surface Sealing, Crusting, and Seedling Emergence

Conservation tillage, particularly NT, reduces surface sealing and crusting compared with CT by maintaining permanent residue cover. Seal formation on a freshly cultivated soil exposed to the beating action of falling raindrops is due to two mechanisms, namely, the breakdown of soil aggregates by the impact force of raindrops and a physiochemical dispersion of soil clays. The initial water condition of aggregates at the beginning of a rainstorm affects their resistance to breakdown or dispersion (le Bissonnais, 1990). For initially dry aggregates, the breakdown is mainly due to slaking. The dry surface slakes rapidly under wetting and forms a seal of lower permeability. If the water content of the soil surface is high before rainfall, the degree of aggregate breakdown and surface sealing is low and results primarily from the mechanical impact of raindrops (Bradford and Huang, 1992). Wetting and drying of newly formed crusts (dry seals) normally act to weaken the crust and it may largely disappear as the soil surface granulates. Even though drying forms new aggregates, their size and stability are less than those of the original surface, and the final infiltration rates at the end of subsequent storms are lower than rates at the end of the first storm (Ben-Hur et al., 1985).

Soils crusts reduce seedling emergence, but their final effect on yield is crop dependent. For example, in the coarser-textured soils in the Lower Rio Grande Valley of Texas, cotton under CT conditions often must be replanted due to crusting following an intense rainstorm that prevents or greatly limits seedling emergence (Bradford and Peterson, 1999). The same is true for many soils in the southeast United States (Miller and Radcliffe, 1992).

The extent of soil crusting and its effect on seedling emergence are highly dependent on the tillage system used and the amount of residue cover. Maximum crusting usually occurs with moldboard plowing followed by disking several times. While surface sealing and crust formation are minimized in CS systems by maintaining high crop residue cover percentages, the high levels may have some negative effects, for example, reduced and uneven seedling emergence, particularly in regions with cooler and wetter climates. In regions where spring planting is delayed by snow melt and warming of the soil, seedling emergence is less with CS than with CT unless crop residue is partially removed from the row. Uneven crop emergence and reduced stand density were reported in the northern Corn Belt in Wisconsin (Carter and Barnett, 1987), Iowa (Mock and Erbach, 1977), Minnesota (Ford and Hicks, 1992), and Illinois (Nafziger et al., 1991). By using injection knives to apply fertilizer and till the soil ahead of planting, Greiner (2008) reported that the tilled zone at planting time usually is warmer, the seeding zone is somewhat drier, and seedling emergence occurs sooner than where residue is not removed. Row-zone tillage in which narrow bands are tilled to satisfy the seedbed requirements allows most of the area between rows to be managed to improve infiltration and control runoff and erosion (Burwell et al., 1968).

25.11 Compaction

Soil compaction is defined as the volume change produced by momentary load application caused by rolling, tamping, or vibration. It is the process by which soil grains are rearranged into closer contact to decrease void space. Compaction removes air from the soil without significantly changing the amount of water in the soil. Compaction can result from external forces such as traffic by tractors, implements, and other farm machinery or from internal forces (sometimes termed consolidation) resulting from increased drying, overburden, or changes in the groundwater table. The usual effects of compaction are increased soil bulk density, shear strength, resistance to probe penetration, and decreased soil compressibility and permeability. Soil drying or desiccation not only results in increased soil bulk density, but also soil cracking and development of macropores if expanding lattice clays are present (Bradford and Peterson, 1999).

One of farmers' primary concerns in switching to NT systems is the problem of soil compaction. Bulk density and soil strength can be less for plowed than for NT soil (Mielke et al., 1986; Bruce et al., 1990; Hill, 1990; Cassel et al., 1995). Bulk densities in the surface 75 mm were higher in four Midwestern U.S. NT soils, no differences in two, and less in one (Mielke et al., 1986). At the 75–150 mm depth, bulk density was greater with NT for only two soils. Clay contents were similar, ranging from 23% to 36%, and time in NT ranged from 6 to 13 years. After 28 years, bulk density of Wooster silt loam (Typic Fragiudalf) at Wooster, Ohio, was lower for continual corn with NT than with other tillage (moldboard plow and chisel plow) and crop rotation treatments (Lal et al., 1994). For a regional study across Ohio, Indiana, and Pennsylvania, Blanco-Canqui and Lal (2007b) reported that NT farming had moderate impacts on soil compaction. They observed that cone index, a parameter of soil compaction, was greater with NT than with CT in 9 out of 13 soils, while bulk density was greater with NT only in 2 soils. These increases in soil compaction parameters with NT were, however, small and well below the threshold levels of excessive compaction.

In poorly structured soils with low OM, long-term NT can lead to higher soil strengths that may limit root growth and crop yields. In most soils, however, even though strength and bulk density under NT conditions are often greater than under CT conditions, crop yields are seldom reduced (Radcliffe et al., 1988). Detrimental effects of greater compaction are often negated by more positive effects of increased water infiltration, reduced soil temperature, reduced soil water evaporation, and greater water storage. Also, roots under NT conditions grow into existing root channels, worm holes, or fractures between soil structural units. As a result, root growth may actually be greater under NT conditions (Merrill et al., 1996).

Although use of NT provides various benefits on soils subject to compaction as mentioned earlier, cotton yield reductions have jeopardized NT adoption in the Tennessee Valley region of northern Alabama. Subsoiling reduced the compaction index 12% and 15% compared with non-subsoiled CT and NT soils, respectively. Fall deep non-inversion tillage by either paratilling or in-row subsoiling with a narrow-shank subsoiler resulted in cotton yields 16% and 10% greater than with CT and NT, respectively, for which the soil was not loosened. Non-inversion in-row deep tillage in fall coupled with a rye cover crop that produces adequate residue to provide for erosion control and water conservation is a highly competitive and practical CS system for the region (Schwab et al., 2002).

In addition to soil compaction that may result from performing cultural operations for crop production, compaction may result from other traffic in fields such as spreading fertilizer or manure, pesticide applications, and harvesting operations (harvester and grain hauling). In many cases, the weight of such equipment plus their loads is greater than that of equipment used for cultural operations and, therefore, may lead to severe soil compaction. To some extent, the degree of compaction may be controlled by the tire pressure or type of tires on the equipment. For a study on Hublersburg silt loam (Typic Hapludult) in Pennsylvania, NT soil was not compacted or compacted annually with a truck having road tires (700 kPa inflation pressure) or flotation tires (250 kPa pressure) that provided a 10 Mg axle load. In-row tilling the compacted soil 40 cm deep resulted in corn yields similar to those with NT without compaction in 3 years and greater yields in 1 year. The results showed little need for in-row tillage to manage compaction under long-term NT conditions when axle loads do not exceed 10 Mg and when using flotation tires that have inflation pressures <250 kPa (Sidhu and Duiker, 2006). Also, practicing controlled-traffic farming minimizes the potential for the development of severe compaction problems (Tullberg et al., 2007). Clearly, careful management regarding equipment used for cultural operations and other equipment traversing fields is important for minimizing soil compaction.

The degree of soil compaction is gradually reduced with time under NT conditions. Macroporosity and earthworm channels generally are greater in NT soils (Drees et al., 1994). In the subtropical, semiarid regions of Texas, earthworms are abundant in NT soils, but absent in CT soils (Bradford and Peterson, 1999). Even in colder semiarid regions such as North Dakota, changing to an NT system greatly enhances earthworm populations (Deibert and Utter, 1994).

NT soils develop an intrinsic ability to resist excessive soil compaction with time, depending on OM input, soil type, management, and climate. Accumulation of soil organic C is one of the main causes for the reduced susceptibility of NT soils to compaction in the long term. Soil OM confers elasticity and resilience to soil because of its low density and high specific surface area. Accumulation of organic C near the soil surface in combination with residue mulch cover can reduce compaction and compression. In turn, CT accelerates decomposition of soil OM, reducing the soil's ability to resist compaction and compression. Thomas et al. (1996) and Assis and Lancas (2003) reported that soil compactibility as determined by the Proctor test was much lower with NT than with CT in temperate and tropical soils, due to the greater organic C accumulation in NT soils. Soil compactibility is negatively correlated with organic C content.

The potential for plow pans or other mechanically compacted layers to develop is sharply reduced when using smaller tractors, performing less tillage, and making fewer trips across the field, which occurs for crop production under CS, especially NT, conditions. Furthermore, compaction of soil surface layers by raindrop forces or surface sealing is less under CS due to protection of surface soil layers by greater residue cover. Consolidation of surface layers due to desiccation is reduced with greater residue cover under CS because of less soil water evaporation.

The degree of reconsolidation and compaction that occurs when plowing is eliminated depends on the crop rotation, amount of surface residue, control of wheel-traffic, climatic factors (rainfall and temperature), and soil variables. Soil variables include soil texture and structure, clay mineralogy, OM content, and others. Compaction due to tractor and machinery traffic and consolidation from desiccation are more of a problem in coarser-textured soils with massive structure and low OM levels, especially with monocultures of low residue crops such as cotton and soybeans, than in fine-textured soils. For example, reduced yields due to soil compaction have been reported for the highly weathered soils of the southeastern United States (Touchton and Johnson, 1982), whereas compaction is less of a problem for deep loessial soils in Illinois and Indiana (Bradford and Peterson, 1999).

Soil compaction problems must be corrected by mechanical means before starting an NT system (Radcliffe et al., 1988), or crop yields on some soils may be reduced for several years until biological forces loosen the soil. To reduce soil reconsolidation or if traffic compaction occurs, a paratill or similar tool that loosens the soil from beneath, leaves the surface relatively smooth, and maintains residue of the previous crop on the surface can be used (Unger, 1993). Soil compaction in the upper horizons can also be reduced with strip-tillage in the row (Radcliffe et al., 1989; Raper et al., 1994) or while applying fertilizer with fertilizer knives.

By controlling traffic patterns, compaction is generally confined to the wheel track areas (Larney and Kladivko, 1989; Liebig et al., 1993; Sweeney et al., 2006). About 70%–90% of traffic compaction occurs during the first pass of a tractor or machinery. By not controlling traffic patterns, therefore, compaction in the entire field is possible. With time, soil compaction is gradually alleviated by biological means when using a CS system. After 3 years on Willacy fine sandy loam (Udic Argiustolls) in south Texas, soil structure was improved, yields were equal to those with CT, and the compaction problem was gradually reduced by using NT (Bradford and Peterson, 1999). Wheel track compaction can also be reduced by performing chisel tillage (Sweeney et al., 2006).

25.12 Cropping Systems

Production of any crop involves some type of cropping system. The system may involve growing the same crop each year (continual cropping), the same crop grown alternately with fallow periods, or two or more different crops grown in a rotation. Some results were given and some comments were made regarding cropping systems in previous sections. In this section, we stress, among other factors, including the use of CS, the importance of cropping systems for storing and making better use of soil water, crop production by using shorter fallow periods, recognizing concern about herbicide carryover to succeeding crops in a sequence, identifying possible substitute crops if one crop is destroyed (hail damage, delayed planting, etc.), and cropping system effects on soil properties.

In the 1930s, the combination of a major drought and clean tillage used for crops at that time resulted in catastrophic erosion by wind in the semiarid U.S. central and southern Great Plains. Stubble mulch tillage was developed during that period to help control erosion and it also improved soil water conservation (Greb, 1979). When stubble mulch tillage is used repeatedly to control weeds, it can result in less than 30% surface residue cover when the next crop is planted. This is especially the case for a wheat-fallow cropping system that results in one crop in 2 years with a fallow period of about 15 months between crops. The low amount of residue remaining increases the potential for erosion and has generally resulted in low water storage efficiencies, often <25% (Haas et al., 1974; Johnson et al., 1974; Farahani et al., 1998). As a result, fallowing that has long been a controversial practice because of the low water storage efficiencies associated with it has, therefore, been replaced by more IC systems in some cases.

With the development of herbicides in the 1940s and continued development of even more effective herbicides through the years, reduced tillage and even NT crop production became possible. This resulted in further improvements in soil water conservation because of less water use by weeds, improved water infiltration, and decreased soil water evaporation. As a result, the controversial practice of using long fallow periods has been replaced by more IC systems in some cases.

With adequate surface residue, which is possible by using CS, especially NT, water conservation is improved and conditions become favorable for reducing the length of fallow periods and increasing cropping intensity. For the southern Great Plains, a winter wheat-grain sorghum-fallow (WSF) system that results in two crops in 3 years is well adapted. This system involves fallow periods of about 11 months between the crops and increases annualized grain yields relative to those with a wheat-fallow system (Baumhardt and Anderson, 2006). For a WSF study at Bushland, Texas, wheat straw was placed on Pullman clay loam at rates ranging from 0 to 12 Mg ha⁻¹ at the time of wheat harvest. Plots were not tilled during the ensuing fallow period. When grain sorghum was planted about 11 months later, fallow period precipitation storage as soil water averaged 23% and 46% with the 0 and 12 Mg ha⁻¹ residue treatments, respectively. Dryland sorghum grain yields averaged 1.78 and 3.99 Mg ha⁻¹ with the respective treatments (Unger, 1978).

The use of some type of CS has been shown to be suitable for various cropping systems in the central and northern Great Plains. A winter wheat-corn (or grain sorghum)-millet-fallow cropping system avoids long fallow periods and results in three crops in 4 years in the central Great Plains (Wood et al., 1991). Cropping systems that reduce fallow length in the northern Great Plains include spring wheat-winter wheat-fallow (two crops in 3 years); safflower (*Carthamus tinctorius* L.)-barley-winter wheat; spring wheat-corn-peas; spring wheat-winter wheat-sunflower; and spring wheat in rotation with soybean, peas, safflower, sunflower, buckwheat (*Fagopyrum esculentum* Moench), or canola (Black, 1986; Black and Tanaka, 1996; Unger and Vigil, 1998).

By increasing cropping frequency, the proportion of fallow that occurs in fall, winter, and spring months is increased, thereby resulting in the highest precipitation storage efficiency. Also, it greatly decreases or avoids the fallow extending in the second summer when no additional precipitation is stored as soil water (Peterson and Westfall, 2004).

The foregoing examples stressed the importance of improving soil water storage in an effort to increase cropping intensity by reducing the length of fallow in the cropping system. Even with fallow, however, water availability for dryland crop production is strongly influenced by precipitation amount and time of occurrence, which are highly unpredictable. Rigid cropping systems rely on growing certain crops at predetermined times. Occasionally, however, substantial amounts of precipitation occur late in the growing season or soon after harvesting a crop, thereby possibly negating the opportunity for storing much additional water during the ensuing fallow period. When such condition develops, opportunity cropping can be used to grow an adaptable crop to use the water that has become available and to eliminate or shorten the originally planned fallow period. In the southern Great Plains, for example, short-season grain sorghum can be grown after harvesting winter wheat and winter wheat can be grown after harvesting grain sorghum. Other crops evaluated for opportunity cropping in a study at Bushland, Texas, were forage sorghum (*Sorghum* spp.), pearl millet, oat, triticale (X *Triticosecale* Wittmack), kenaf (*Hibiscus cannabinus* L.), pinto bean (*Phaseolus vulgaris* L.), and fall and spring canola (Unger, 2001). Crops suitable for opportunity cropping were winter wheat, grain sorghum, triticale, forage sorghum, pearl millet, and oat. Opportunity cropping is a more intensive system than fixed cropping systems. As a result, precipitation use efficiency is greater than with systems involving long fallow periods. For a 5 year study in the central Great Plains, economic precipitation use efficiency (i.e., value of crops produced) was 45% greater with opportunity cropping than for set rotations that included fallow (Nielsen et al., 2006).

The producer's goal for cropping systems generally is maximum crop yields, but the economics of production are important also with tillage system often influencing the economic returns. For a dryland study in North Dakota, DeVuyst and Halvorson (2004) evaluated the economics of SW-F and spring wheat-winter wheat-sunflower IC systems under CT, minimum tillage, and NT conditions and various N levels. For the IC system, minimum tillage resulted in greater profits than CT and NT under the same N treatments, but profits were more variable with minimum tillage. The highest N rates resulted in the greatest profits. The IC system with minimum tillage or NT was more profitable than the best SW-F system with CT at that location. Analyses revealed that the SW-F system and the IC system with CT were economically inefficient as compared with the IC system using minimum tillage or NT.

Under more humid conditions than in the Great Plains, crops are grown without fallow, but crop rotations are still widely used. For a corn-soybean rotation study in Minnesota, NT after soybean reduced corn yields in 2 of 4 years compared with full-width fall chisel tillage and spring field cultivation. Also, full-width tillage for soybean in rotation with zone tillage or fall strip tillage for corn resulted in greater corn yields and economic returns than annual full-width tillage. Soybean yields were greater with rotational tillage, but the difference among tillage systems was not large enough to provide an advantage to full-width tillage. Considering both crops, rotational tillage provided for greater yields, but not greater economic returns (Vetsch et al., 2007).

Simplified crop rotations lead to deterioration of soil physical, chemical, and biological attributes and adversely impact crop production. Diverse and consequently more IC systems in combination with reduced tillage and NT that improve soil attributes and sequester C are potential alternatives to crop-fallow systems (Pikul et al., 2006; Benjamin et al., 2007). Impacts of cropping systems on soil properties differ from those of tillage treatments. Cropping system impacts on soil conditions and C sequestration are often detectable for longer periods than with tillage treatment impacts (Benjamin et al., 2007).

Cropping systems alter soil functions because of differences in the amount of above- and below-ground biomass input and rooting patterns. Differential surface cover among cropping systems influences changes in surface soil processes (e.g., surface crusting) and properties (e.g., water infiltration, aggregate stability). High-biomass producing crops in combination with reduced tillage and NT systems improve soil surface conditions. Plant roots through deep extension and proliferation can alter pore-size distribution by creating fine and coarse root channels within the soil profile (Benjamin et al., 2007).

Impacts of cropping systems on soil properties are not always consistent, depending on soil, length of management, and climate. After 15 years of management, Benjamin et al. (2007) observed no differences in soil bulk density, pore-size distribution, water retention capacity, and saturated hydraulic conductivity due to using wheat-fallow, wheat-corn-fallow, wheat-corn-sunflower-fallow, and wheat-corn-millet rotations under NT on a silt loam in eastern Colorado. Also in eastern Colorado, Shaver et al. (2003) found that the use of NT for 12 years in IC systems including continual cropping, wheat-corn-fallow, and wheat-sorghum-fallow reduced soil bulk density and increased soil porosity, but had no affect on sorptivity when compared with a wheat-fallow system. On a regional study across eight locations in the Great Plains, Pikul et al. (2006) observed that cropping system effects on soil structural properties were highly soil dependent.

25.13 Cover Crops

Cover crops may play an important role within a CS system by protecting the soil against erosion by wind and water, increasing soil OM, improving soil structure and other soil physical properties, altering soil temperature and soil water content, and increasing fertility by recycling nutrients and providing biologically fixed nitrogen (in the case of leguminous cover crops). Cover crops also can improve weed control through competition (Vasilakoglou et al., 2006) or allelopathic effects, and may improve environmental quality through the protection of surface and ground waters. Efficient use of cover crops in management systems, however, depends on the purpose of the cover crop and on factors such as prevailing temperatures, length of growing season, and rainfall amount and distribution (Bradford and Peterson, 1999).

Cover crops generally are more adaptable to farming systems in the southeastern United States because of greater rainfall than in drier climatic regions (Bradford and Peterson, 1999). The goal for a study by Schomberg et al. (2006) in Georgia was to achieve a better understanding of cover crop and tillage system interactions that can lead to greater productivity and economic returns for cotton production. That study on Bonifay fine sand (Grossarenic Plinthic Paleudults) involved seven different cover crops and strip tillage or NT. Cover crop biomass was greatest with rye and hairy vetch (*Vicia villosa* Roth) and the most N (80 kg ha^{-1}) resulted from Austrian winter peas (*Pisum sativum* ssp. *arvense*). Other cover crops were balansa clover (*Trifolium michelianum* ssp. *balansae*), crimson clover (*Trifolium incarnatum* L.), oil seed radish (*Raphanus* spp.), and black oat (*Avena strigosa* Schreb.). Strip tillage increased yields and annual returns over those with NT, probably because of improved availability of water. Strip tillage with black oat as a cover crop was the best combination for maximizing profit under the conditions studied. It increased profits for cotton by $50–$75 ha^{-1} over those for systems where rye was the cover crop.

Although cover crops seemingly are more adaptable to the more humid regions, they may have advantages in other regions also. In Missouri, maximum dry matter yield in spring resulted from a hairy vetch cover crop that provided 44 and 57 kg N ha^{-1} in corn and grain sorghum experiments, respectively. Greatest corn and grain sorghum yields, however, were obtained with either hairy vetch or Austrian winter pea cover crops (Reinbott et al., 2004). In California, the presence of cover crops increased the steady infiltration rate by 37%–41% and cumulative infiltration by 20%–101% as compared with conditions without cover crops (Folorunso et al., 1992). Timely killing of cover crops is essential to achieve maximum benefits from their use (Wagner-Riddle et al., 1994; Duiker and Curran, 2005).

In northern climates, cover crops often are not used because they generally result in soils being cooler in the spring, reduce the water supply for subsequent crops, and tend to shorten the growing season for subsequent crops. In U.S. semiarid regions, greater water stress and, therefore, potentially lower yields of subsequent crops are major constraints to using cover crops (Unger and Vigil, 1998; Baumhardt and Lascano, 1999). Also, farmers are reluctant to use cover crops because of the additional cost for their establishment and termination.

Intense management is required to achieve success with cover crops (Unger and Vigil, 1998). Time to kill the cover crop in the spring is critical to yield of the ensuing crop due to factors such as soil drying and warming, insect and weed pressure, and planter performance. Late April or early May kill dates consistently resulted in greater corn yield than earlier kill dates in Maryland (Clark et al., 1997). Soil water conservation during summer resulting from the presence of cover crop residue was more important than spring water depletion by growing cover crops in determining final corn yield.

25.14 Comparison of Conservation Tillage with Conservation Agriculture

Conservation agriculture (CA) "is a concept for resource-saving agricultural crop production that strives to achieve acceptable profits together with high and sustained production levels while concurrently conserving the environment" (FAO, 2008). Based on several reports, CA is a system of integrated soil, water, and agricultural resources management. The main objective is economical, ecological, and socially sustainable agricultural production while the soil is regenerated or soil degradation is reversed. The system involves maintaining a permanent soil cover, practicing minimum soil disturbance, and using crop rotations. Benefits include fuel, implement maintenance and replacement, labor (time) savings, greater and more stable crop yields, and a lower requirement for heavy work. Other benefits are crop diversification; food security and diet improvements; improved water availability, amount, and quality; improved soil fertility and regeneration; less erosion; improved air quality; and greater agricultural biodiversity, namely, more diverse crop rotations, enhanced soil biodiversity, and less pressure on marginal lands, forests, and natural reserves (Hobbs, 2006; Unger, 2006). In essence, many characteristics of CA are also embodied in CS, especially the NT type of CS.

The key principles of CA are minimum mechanical soil disturbance, permanent soil surface cover with organic materials, and use of crop rotations involving more than two crop species (FAO, 2008). In its maximum condition, NT could meet the principles of CA, but, according to its definition, requirements for NT do not meet the requirements for CA. For example, soil disturbance with a subsoiler is possible with NT whereas no such disturbance is appropriate with CA. Also, complete cover of the surface is required for CA, but not for NT. It is, however, important to recognize that a high level of residue cover is also essential for NT if soil and water conservation and soil C sequestration are priorities. Finally, CA requires use of crop rotations with more than two crop species. In contrast, NT can be practiced with one crop grown continually. Both systems, however, can provide for excellent control of erosion by wind and water and improve water conservation, crop production, and environmental conditions as well as many of the benefits mentioned earlier. As a result, either system may have its place in a given locale.

Although CA is a highly beneficial system, as previously mentioned, there are problems associated with its use (FAO, 2006, 2008). First, there may be financial constraints for producers to adopt CA because of the need to purchase new equipment and the initially lower crop yields with the system (CA is based on establishing an organic layer and producing its own fertilizer, which may take time and thereby cause yields to be lower than those previously obtained). Second, there seemingly is not enough pressure to adopt CA in some parts of the world because the need for an intensive conservation system as CA is not fully recognized and, therefore, it is not adopted. Finally, with the initial reduction in crop yields when changing to CA, it may be difficult to produce adequate food for the ever-increasing world population.

25.15 Soil by Climate Interactions

Surface residue has major benefits for agriculture, among the foremost being improved soil and water conservation. The level of benefits is controlled by the particular soil and climatic conditions at a given location. Surface residue also may have effects on agroecosystems that are detrimental to plant growth and require management adjustments to ameliorate. For example, soil temperatures are reduced by residue cover, which can negatively affect spring-planted crops in northern latitudes.

The interactive effects of surface residue with soil and climatic factors determine whether the residue effects are positive, neutral, or even negative. Primary issues are problems with cold soil temperatures, excessively wet soil conditions, and high runoff situations; interactions affecting crop rotation choices and weed control; and using surface residue management in conjunction with other conservation measures.

25.15.1 Problems Related to Cold Soils and Excessive Soil Water

In cold and humid climates, lower crop yields obtained with NT are often attributed to lower soil temperatures and excessive soil water contents during the early growth of spring crops, especially when crop residue is not removed from the row (Fausey and Lal, 1989). McCalla and Duley (1946) showed that corn residue mulches decreased soil temperatures in midsummer (July–August) at Lincoln, NE, by 3°C–7°C as mulch rates varied from 5 to 20 Mg ha^{-1}. This exemplifies an ameliorating effect of residue on a negative process, namely, summertime evaporation. In contrast, Al-Darby and Lowery (1987) showed that a cover of 55%–87% under NT conditions lowered soil temperatures by as much as 4°C compared with those with CT under spring conditions in Wisconsin. Corn seedling emergence (100%) was delayed by 2–8 days with NT compared with those with CT systems. Although emergence was delayed and plant growth was slower early in the season, corn grain yield differences were not significant. Al-Darby and Lowery (1987) concluded that the additional water saved with NT compensated for the early delays in plant growth. Also in Wisconsin, Carter and Barnett (1987) reported that for continual corn with NT, soil temperatures were cooler, emergence percentage and stand establishment were lower, and phenological development was delayed compared with corn produced using a moldboard plow system. Soil temperature at the 2-cm depth in barley plots in the semiarid region of southern Alberta, Canada, averaged 13.9°C with NT and 14.7°C with CT in spring (May), but these differences disappeared later in the growing season and grain yield was not affected (Carefoot et al., 1990).

Residue cover does delay plant development. For example, corn grown in residue-covered plots in Michigan required 3.5 more days to reach the VT (designated as the transitional stage between the vegetative and reproductive stages, namely, tasseling) stage in 1987 and 8 more days in 1988 (Fortin and Pierce, 1990). Producers must be prepared to deal with delayed growth of spring crops in reduced and NT systems because of low soil temperatures. Delayed emergence, for example, may reduce crop competition with weeds and allow them to flourish; such weeds use water and nutrients (e.g., N) and could ultimately decrease crop yields. Therefore, proper attention to weed control, planting dates, cultivar selection, and possibly starter fertilizer application is needed when it is known that cold stress will occur. In general, the need for intense management increases whenever additional stresses are imposed.

Soil properties that affect water infiltration, permeability, and drainage must always be properly assessed when making residue management decisions. Factors that cause soils to remain wetter longer generally cause them to be colder. Again, this is only a problem with spring-planted crops such as spring cereals, soybean, corn, and grain sorghum at northern latitudes. The latter two species are much more sensitive than small grains under these conditions. Research in the Corn Belt of the United States, showed that use of NT on poorly drained soils results in lower yields than use of CT. Long-term research has shown, however, that continued use of NT and disease-resistant cultivars has overcome the negative response. Also, grain yields after 18 years of NT are now equal to or greater than those with CT. Grain yields on well-drained soil increased with time under the NT system (Dick et al., 1991). Producers with poorly drained soils should expect lower productivity in the initial years after switching to high residue systems. It even may be best to avoid complete NT in these situations. Better choices may be ridging and bedding that reduce adverse effects of high residue systems on poorly drained soils.

Soil texture, slope, and drainage classification are the key factors in making decisions about tillage system choice (Peterson, 1994). Poorly drained soils are not easily adapted to high residue systems. Coarse-textured soils, even when poorly drained, are less of a problem than fine-textured soils in the same drainage category. Well drained soils are good candidates for NT or reduced tillage if they have a silt loam or coarser texture. Silty clay loam and silty clay soils may present problems for planting equipment if they are wet and, therefore, would be best managed with a more disturbed-soil system; perhaps ridge tillage. All steeply sloping soils regardless of textural class are best managed with reduced or NT systems because their runoff and erosion potentials are high.

25.15.2 Weed Control Problems Related to Soil, Climate, and Tillage Choice

Weed control problems usually increase, at least in initial years, when converting to high residue (CS) systems. With less burial of weed seeds, greater weed emergence seems to be the norm. Grassy weeds like downy brome (*Bromus tectorum*) are particularly favored when tillage is reduced. Furthermore, substitution of sweep tillage for mixing and inverting operations (disks and plows) decreases ease of grassy weed control. The problem is greatest when weeds and crops have the same or similar growth cycles. This is the case with downy brome, which germinates at about the same time that winter wheat is planted. Such weed is especially difficult to control in monoculture systems like continual wheat or wheat-fallow.

Crop rotation is one of the best ways to combat weed problems in any system and this is especially true with reduced tillage and NT. By changing crop type, weeds can be controlled at a different phase of their growth cycle. For example, following wheat (a cool season crop) with a warm season crop (e.g., grain sorghum, sunflower, millet, or corn) provides an opportunity to control weeds that are a problem for wheat.

Herbicides are used to replace part or all of the tillage operations in high residue (CS) systems. Close adherence to specifications for these products is essential to achieve maximum weed control benefits. Also, application methods may need refinement to minimize interception of the herbicide by the residue. Required rates of material can vary with soil texture and OM content. In general, finer-textured soils and those of higher OM content require higher rates of soil-applied herbicides to satisfactorily control weeds than coarse-textured soils. Soil clays and OM inactivate these herbicides and, therefore, rates must be adjusted as prescribed by the manufacturer. Also, caution must be exercised regarding herbicide carryover to the next crop. Coarse-textured, low OM content, and high pH soils are particularly susceptible to carryover problems. Again, applying herbicides at rates no greater than label specifications minimizes these potential problems (Bradford and Peterson, 1999).

When converting to high residue systems, producers must have or must acquire herbicide management skills. Correct sprayer operation from calibration to nozzle type selection is a key requirement to success with residue management systems. New technology, namely, site-specific application for which herbicides are applied only where weeds are present in a field, will greatly aid in judicious use of herbicides (Wilkerson et al., 2004; Goudy et al., 2008).

25.15.3 Surface Residue and Other Conservation Measures

Maximizing snow catch is a vital water conservation measure in dry regions where snow constitutes 20%–50% of the annual precipitation and, therefore, represents a valuable resource for agriculture. Stubble height strongly influences snow catch. Taller stubble retains more snow, but also can increase runoff from snowmelt in spring. Soils that freeze deeply prior to snowfall may be non-conductive to water and, therefore, potentially provide the worst scenario for runoff from snowmelt. With warmer climates like in the central and southern Great Plains, snowmelt is more easily captured because soils do not freeze as deeply and often thaw under the snow; thus, substantial snow and its melt water may be captured after spring thawing occurs. Greb (1979) reported that the efficiency of storing water from snowmelt often is double that of water received as rain.

Contour farming in conjunction with residue maintenance (CS) further maximizes water capture from either rain or snowmelt. Terraces also effectively capture water, but are less necessary on land with good residue cover and contour farming. Obviously, land on slopes >7%–8% may still be best managed with terraces in addition to residue maintenance. Use of contour farming, terracing, and possibly strip cropping in conjunction with CS may be especially important in regions where major high-intensity rainstorms commonly occur and, therefore, may lead to severe soil erosion where only CS, even the NT system, is used.

Maintaining residue on the soil surface by using CS is an invaluable management practice. It enhances water capture and retention, but the cost of storing the additional water with improved residue management can decrease profits if producers are not prepared to effectively and efficiently use the water. Finally, producers should be aware that residue may have negative effects in terms of cold soil temperatures at the time of spring planting. Appropriate system choices including planting method, crop choice, possibly starter fertilizer use, etc. can help ameliorate the potentially negative cold temperature effects.

25.16 Disadvantages

By definition, at least 30% of the soil surface must be covered with residue when the next crop is planted in order to qualify as CS. With NT, a CS method, much greater amounts of residue often are present. Surface residue is highly important for conserving water, controlling erosion, and minimizing environmental pollution under many conditions, but the residue along with the NT method itself also may cause disadvantages under some conditions. Through proper management, most disadvantages usually can be avoided or overcome, thus generally leading to favorable results when using CS. Potential disadvantages of CS include:

1. Conservation tillage involves an increased use of herbicides, which may increase production expenses that are not offset by decreased tillage costs.
2. Increased use of herbicides may increase the risk of non-point-source pollution, for example, of runoff water and groundwater under some climatic conditions and on some more conductive soils.
3. Some weeds are difficult to control with herbicides or have become tolerant of herbicides. As a result, tillage may be required to control such weeds.
4. Greater amounts of N fertilizer may be needed when first switching to CS (especially NT) systems; also, reduced N mineralization due to heavy residue mulch can require the use of increased amounts of fertilizers. Changes in fertilizer application methods may be needed when using CS systems and it may be necessary to apply a starter fertilizer under some conditions.
5. High earthworm populations and preferential water flow through the deep burrows may cause leaching of agricultural chemicals to depths beyond the reach of plant roots and into groundwater.
6. Soil compaction may occur under some conditions, particularly in nearly level and clayey soils, thus requiring occasional tillage or subsoiling to overcome the problem.
7. The NT method of CS may not be suitable on hard-setting soils because water infiltration often is lower than where such soils are loosened by tillage.
8. The NT method may not always be better than CT on clayey, cold, and wet soils. Cold soil (delayed warming) due to large amounts of surface residue may delay crop planting, reduce the seed germination rate, cause low plant populations, and possibly result in freeze damage if maturity is delayed. Also, reduced runoff and evaporation aggravate excessive water problems on poorly drained soils.
9. Leachates from residue of some crops may cause phytotoxicity problems for subsequent crops.
10. New crop-planting and herbicide-application equipment may be required, thus initially resulting in additional expenses for the producer.
11. Crop planting may be difficult where large amounts of residue are on the surface. For example, NT planting under irrigated conditions at Bushland, Texas, was difficult because the large amount of surface residue kept the clay loam moist, thus causing the soil to severely adhere to the planting equipment (S. Evett, Personal communication, 2008).
12. Surface residue provides habitat for some rodents, insects, etc., which may damage some crops.
13. Conservation tillage, especially NT, is a major departure from CT and, therefore, requires good management for it to be successful. As a result, special training may be required, which may not be readily available or may entail additional expenses for the producer.

25.17 Conclusions

Conservation tillage, especially the NT method, represents the most dramatic change in soil management for crop production in the history of agriculture. Historically, farmers prepared their whole field as a seedbed, even though seeds were placed only in a small area. Now, major soil disturbance is no longer necessary because only a small area needs to be disturbed to prepare a seedbed or plant seeds. Quite obviously, these changes decrease labor and power requirements for the farmer. By not disturbing most of the soil, fields are kept in a more "natural setting." When disturbance is minimized, soil aggregate integrity is maintained and soil physical, chemical, and biological components respond positively. As a result, water conservation is improved, erosion is minimized, plant nutrients are retained, and soil biological processes proceed with fewer interruptions. The greatest benefits of CS, especially NT, agriculture will be achieved when farmers, technology-transfer persons, and scientists learn how to fully exploit this ecologically beneficial system. NT is a pioneering method of agriculture. The large benefits of NT for minimizing soil erosion, reducing tillage costs, improving soil quality, and many others outweigh the disadvantages associated with this tillage method. Conservation tillage combined with residue mulching, cover crops, crop rotations, and other best management practices is a win-win strategy for soil and water conservation and for sustained agronomic production (Blanco and Lal, 2008).

References

Akinyemi, J.O., and A.O. Adedeji. 2004. Water infiltration under no-tillage, minimum tillage and conventional tillage systems on a sandy loam Alfisols. Paper No. 042111, *2004 ASAE Annual Meeting*. ASAE, St. Joseph, MI.

Al-Darby, A.M., and B. Lowery. 1987. Seed zone soil temperature and early growth with three conservation tillage systems. Soil Sci. Soc. Am. J. 51:768–774.

Allmaras, R.R., D.R. Linden, and C.E. Clapp. 2004. Corn-residue transformations into root and soil carbon as related to nitrogen, tillage, and stover management. Soil Sci. Soc. Am. J. 68:1366–1375.

Andraski, T.W., and L.G. Bundy. 2008. Corn residue and nitrogen source effects on nitrogen availability in no-till corn. Agron. J. 100:1274–1279.

Andraski, B.J., D.H. Mueller, and T.C. Daniel. 1985. Effects of tillage and rainfall simulation date on water and soil losses. Soil Sci. Soc. Am. J. 49:1512–1517.

Assis, R.L. de, and K.P. Lancas. 2003. Effect of the adoption time of the no till system in the soil maximum bulk density and in the optimum moisture content for soil compaction in a red dystroferric Nitosol. Energia na Agricultura 18:22–33.

Baker, J.M., T.E. Ochsner, R.T. Venterea, and T.J. Griffis. 2007. Tillage and soil carbon sequestration—What do we really know? Agric. Ecosyst. Environ. 118:1–5.

Baumhardt, R.L., and R.L. Anderson. 2006. Crop choices and rotation principles, p. 113–139. *In* G.A. Peterson, P.W. Unger, and W.A. Payne (eds.) Dryland agriculture. 2nd Edn. Agron. Mono. 23. ASA, CSSA, and SSSA, Madison, WI.

Baumhardt, R.L., and R.J. Lascano. 1999. Water budget and yield of dryland cotton intercropped with terminated winter wheat. Agron. J. 91:922–927.

Beare, M.H., M.L. Cabrera, P.F. Hendrix, and D.C. Coleman. 1994a. Aggregate-protected and unprotected OM pools in conventional- and no-tillage soils. Soil Sci. Soc. Am. J. 58:787–795.

Beare, M.H., P.F. Hendrix, and D.C. Coleman. 1994b. Water-stable aggregates and organic matter fractions in conventional- and no-tillage soils. Soil Sci. Soc. Am. J. 58:777–786.

Ben-Hur, M., I. Shainberg, D. Bakker, and R. Keren. 1985. Effect of soil texture and $CaCO_3$ content on water infiltration in crusted soil as related to water salinity. Irrig. Sci. 6:281–294.

Benjamin, J.G., M. Mikha, D.C. Nielsen, M.F. Vigil, F. Calderon, and W.B. Henry. 2007. Cropping intensity effects on physical properties of a no-till silt loam. Soil Sci. Soc. Am. J. 71:1160–1165.

Bermudez, M., and A.P. Mallarino. 2002. Yield and early growth responses to starter fertilizer in no-till corn assessed with precision agriculture technologies. Agron. J. 94:1024–1033.

Bermudez, M., and A.P. Mallarino. 2004. Corn response to starter fertilizer and tillage across and within fields having no-till management histories. Agron. J. 96:776–785.

Black, A.L. 1986. Resources and problems in the northern Great Plains area, p. 25–38. *In* Proc. Workshop on Planning and management of water conservation systems, October 1985. USDA-SCS Midwest National Technical Center, Lincoln, NE.

Black, A.L., and D.L. Tanaka. 1996. A conservation tillage-cropping systems study in the northern Great Plains of the USA, p. 335–342. *In* E.A. Paul, K.A. Paustian, E.T. Elliott, and C.V. Cole (eds.) Soil organic matter in temperate agroecosystems. Lewis Publishers, Boca Raton, FL.

Blanco, H., and R. Lal. 2008. Principles of soil management and conservation, p. 620. Springer Publishing Co., New York.

Blanco-Canqui, H., and R. Lal. 2004. Mechanisms of carbon sequestration in soil aggregates. Crit. Rev. Plant Sci. 23:481–504.

Blanco-Canqui, H., and R. Lal. 2007a. Soil and crop response to harvesting corn residues for biofuel production. Geoderma 141:355–362.

Blanco-Canqui, H., and R. Lal. 2007b. Regional assessment of soil compaction and structural properties under no-till farming. Soil Sci. Soc. Am. J. 71:1770–1778.

Blanco-Canqui, H., and R. Lal. 2008. No-tillage and carbon sequestration: An on-farm assessment. Soil Sci. Soc. Am. J. 72:693–701.

Blanco-Canqui, H., R. Lal, W.M. Post, R.C. Izaurralde, and L.B. Owens. 2006. Changes in long-term no-till corn growth and yield under different rates of stover mulch. Agron. J. 98:1128–1136.

Blanco-Canqui, H., R.J. Stephenson, N.O. Nelson, and D.R. Presley. 2009. Wheat and sorghum residue removal for expanded uses increases sediment and nutrient loss in runoff. J. Environ. Qual. 38:2365–2372.

Blevins, R.L., M.S. Smith, and G.W. Thomas. 1984. Changes in soil properties under no-tillage, p. 190–230. *In* R.E. Phillips and S.H. Phillips (eds.) No-tillage agriculture: Principles and practices. Van Nostrand Reinhold Co, New York.

Borges, R., and A.P. Mallarino. 2000. Grain yield, early growth, and nutrient uptake of no-till soybean as affected by phosphorus and potassium placement. Agron. J. 92:380–388.

Bradford, J.M., and C. Huang. 1992. Mechanisms of crust formation: Physical components, p. 55–72. *In* M.E. Sumner and B.A. Stewart (eds.) Soil crusting: Physical and chemical processes. Lewis Publishers, Boca Raton, FL.

Bradford, J.M., and C. Huang. 1994. Interrill soil erosion as affected by tillage and residue cover. Soil Till. Res. 31:353–361.

Bradford, J.M., and G.A. Peterson. 1999. Conservation tillage, p. G-247–G-270. *In* M.E. Sumner (ed.-in-chief) Handbook of soil science. CRC Press, Boca Raton, FL.

Bremer, E., and C. van Kessel. 1992. Seasonal microbial biomass dynamics after addition of lentil and wheat residues. Soil Sci. Soc. Am. J. 56:1141–1146.

Bruce, R.R., G.W. Langdale, and A.L. Dillard. 1990. Tillage and crop rotation effect on characteristics of a sandy surface soil. Soil Sci. Soc. Am. J. 54:1744–1747.

Buchanan, M., and L.D. Kin. 1993. Carbon and phosphorus losses from decomposing crop residues in no-till and conventional till agroecosystems. Agron. J. 85:631–638.

Buman, R.A., B.A. Alesii, J.L. Hatfield, and D.L. Karlen. 2004. Profit, yield, and soil quality effects of tillage systems in corn-soybean rotations. J. Soil Water Conserv. 59:260–270.

Bundy, L.G., T.W. Andraski, and T.C. Daniel. 1992. Placement and timing of nitrogen fertilizers for conventional and conservation tillage corn production. J. Prod. Agric. 5:214–221.

Burwell, R.E., L.L. Sloneker, and W.W. Nelson. 1968. Tillage influences water intake. J. Soil Water Conserv. 23:185–187.

Cambardella, C.A., and E.T. Elliott. 1993. Carbon and nitrogen distribution in aggregates from cultivated and native grassland soils. Soil Sci. Soc. Am. J. 57:1071–1076.

Carefoot, J.M., M. Nyborg, and C.W. Lindwall. 1990. Tillage-induced soil changes and related grain yield in a semi-arid region. Can. J. Soil Sci. 7:203–214.

Carter, P.R., and K.H. Barnett. 1987. Corn-hybrid performance under conventional and no-tillage systems after thinning. Agron. J. 79:919–926.

Cassel, D.K., C.W. Raczkowski, and H.P. Denton. 1995. Tillage effects on corn production and soil physical conditions. Soil Sci. Soc. Am. J. 59:1436–1443.

Christensen, N.B., W.C. Lindemann, E. Salazar-Sosa, and L.R. Gill. 1994. Nitrogen and carbon dynamics in no-till and stubble mulch tillage systems. Agron. J. 86:298–303.

Christopher, S.F., R. Lal, and M. Umakant. 2009. Regional study of no-till effects on carbon sequestration in the midwestern United States. Soil Sci. Soc. Am. J. 73:207–216.

Clapp, C.E., R.R. Allmaras, M.F. Layese, D.R. Linden, and R.H. Dowdy. 2000. Soil organic carbon and C-13 abundance as related to tillage, crop residue, and nitrogen fertilization under continuous corn management in Minnesota. Soil Till. Res. 55:127–142.

Clark, A.J., A.M. Decker, J.J. Meisinger, and M.S. McIntosh. 1997. Kill date of vetch, rye, and a vetch-rye mixture. 2. Soil moisture and corn yield. Agron. J. 89:434–441.

Cochran, V.L., L.F. Elliott, and R.I. Papendick. 1977. The production of phytotoxins from surface crop residues. Soil Sci. Soc. Am. J. 41:903–908.

Crovetto, C. 1996. Stubble over the soil. American Society of Agronomy, Madison, WI.

Cruse, R.M., R. Mier, and C.W. Mize. 2001. Surface residue effects on erosion of thawing soils. Soil Sci. Soc. Am. J. 65:178–184.

Dabney, S.M., G.V. Wilson, K.C. McGregor, and G.R. Foster. 2004. History, residue, and tillage effects on erosion of loessial soil. Trans. ASAE 47:767–775.

Dalal, R.C. 1989. Long-term effects of no-tillage, crop residue, and nitrogen application on properties of a Vertisol. Soil Sci. Soc. Am. J. 53:1511–1515.

Dam, R.F., B.B. Mehdi, M.S.E. Burgess, C.A. Madramootoo, G.R. Mehuys, and I.R. Callum. 2005. Soil bulk density and crop yield under eleven consecutive years of corn with different tillage and residue practices in a sandy loam soil in central Canada. Soil Till. Res. 84:41–53.

Dao, T.H. 1993. Tillage and winter wheat residue management effects on water infiltration and storage. Soil Sci. Soc. Am. J. 57:1586–1595.

Deibert, E.J., and R.A. Utter. 1994. Earthworm populations related to soil and fertilizer management practices. Better Crops 78:9–11.

Dell, C.J., P.R. Salon, C.D. Franks, E.C. Benjamin, and Y. Plowden. 2008. No-till and cover crop impacts on soil carbon and associated properties on Pennsylvania dairy farms. J. Soil Water Conserv. 63:136–142.

de Quattro, J. 1997. Sustaining agriculture in drought years. Agric. Res. 45:4–9.

DeVuyst, E.A., and A.D. Halvorson. 2004. Economics of annual cropping versus crop-fallow in the northern Great Plains as influenced by tillage and nitrogen. Agron. J. 96:148–153.

Díaz-Zorita, M., J.H. Grove, L. Murdock, J. Herbeck, and E. Perfect. 2004. Soil structural disturbance effects on crop yields and soil properties in a no-till production system. Agron. J. 96:1651–1659.

Dick, W.A. 1983. Organic carbon, nitrogen, and phosphorus concentrations and pH in soil profiles as affected by tillage intensity. Soil Sci. Soc. Am. J. 47:102–107.

Dick, W.A. 1984. Influence of long-term tillage and crop rotation combinations on soil enzyme activities. Soil Sci. Soc. Am. J. 48:569–574.

Dick, W.A., E.L. McCoy, W.M. Edwards, and R. Lal. 1991. Continuous application of no-tillage to Ohio soils. Agron. J. 83:65–73.

Dick, W.A., R.J. Roseberg, E.L. McCoy, W.M. Edwards, and F. Haghiri. 1989. Surface hydrologic response of soils to no-tillage. Soil Sci. Soc. Am. J. 53:1520–1526.

Dickey, E.C., C.R. Fenster, J.M. Laflen, and R.H. Mickelson. 1983. Effects of tillage on soil erosion in a wheat-fallow rotation. Trans. ASAE 26:814–820.

Doran, J.W. 1980. Soil microbial and biochemical changes associated with reduced tillage. Soil Sci. Soc. Am. J. 44:765–771.

Drees, L.R., A.D. Karathanasis, L.P. Wilding, and R.L. Blevins. 1994. Micromorphological characteristics of long-term no-till and conventionally tilled soils. Soil Sci. Soc. Am. J. 58:508–517.

Duiker, S.W., and W.S. Curran. 2005. Rye cover crop management for corn production in the Northern Mid-Atlantic Region. Agron. J. 97:1413–1418.

Edwards, W.M., L.D. Norton, and C.E. Redmond. 1988. Characterizing macropores that affect infiltration into no-tilled soil. Soil Sci. Soc. Am. J. 52:483–487.

Edwards, W.M., M.J. Shipitalo, L.B. Owens, and L.D. Norton. 1989. Water and nitrate movement in earthworm burrows within long-term no-till cornfields. J. Soil Water Conserv. 44:240–242.

FAO (Food and Agriculture Organization of the United Nations). 2006. Conservation agriculture. http://www.fao.org/ag/magazine/0110sp.htm (accessed October 10, 2008).

FAO (Food and Agriculture Organization of the United Nations). 2008. Conservation agriculture. http://www.fao.org/ag/ca/la.html (accessed July 24, 2009).

Farahani, H.J., G.A. Peterson, and D.G. Westfall. 1998. Dryland cropping intensification: A fundamental solution to efficient use of precipitation. Adv. Agron. 64:197–223.

Fausey, N.R., and R. Lal. 1989. Drainage-tillage effects on Crosby-Kokomo soil association in Ohio. Soil Technol. 2:371–383.

Fenster, C.R., N.P. Woodruff, W.S. Chepil, and F.H. Siddoway. 1965. Performance of tillage implements in a stubble mulch system: III. Effects of tillage sequences on residues, soil cloddiness, weed control, and wheat yield. Agron. J. 57:52–55.

Follett, R.F., and D.S. Schimel. 1989. Effect of tillage practices on microbial biomass dynamics. Soil Sci. Soc. Am. J. 53:1091–1096.

Folorunso, O.A., D.E. Rolston, T. Prichard, and D.T. Louie. 1992. Soil surface strength and infiltration rate as affected by winter cover crops. Soil Technol. 5:189–197.

Ford, J.H., and D.R. Hicks. 1992. Corn growth and yield in uneven emerging stands. J. Prod. Agric. 5:185–188.

Fortin, M.C., and F.J. Pierce. 1990. Developmental and growth effects of crop residues on corn. Agron. J. 82:710–715.

Foster, G.R., C.B. Johnson, and J.D. Nowlin. 1980. A model to estimate sediment yield from field sized areas: Application to planning and management for control of nonpoint source pollution, p. 193–281. In Creams: A field scale model for chemicals, runoff, and erosion from agricultural management systems. USDA Conservative Research Report 26. U.S. Government Printing Office, Washington, DC.

Fryrear, D.W. 1985. Soil cover and wind erosion. Trans. ASAE 28:781–784.

Goudy, H., F. Tardiff, R. Brown, and K. Bennett. 2008. Site-specific herbicide applications. http://www.ontariocorn.org/magazine/Issues/pre%20Nov%202005/ocpmag/feb99art5.html (accessed November 21, 2008).

Graham, R.L., R. Nelson, J. Sheehan, R.D. Perlack, and L.L. Wright. 2007. Current and potential U.S. corn stover supplies. Agron. J. 99:1–11.

Grant, C.A., K.R. Brown, G.J. Racz, and L.D. Bailey. 2001. Influence of source, timing and placement of nitrogen on grain yield and nitrogen removal of Sceptre durum wheat under reduced- and conventional-tillage management. Can. J. Plant Sci. 81:17–27.

Grant, C.A., K.R. Brown, G.J. Racz, and L.D. Bailey. 2002. Influence of source, timing and placement of nitrogen on seed yield and nitrogen accumulation in the seed of canola under reduced- and conventional-tillage management. Can. J. Plant Sci. 82:629–638.

Greb, B.W. 1979. Reducing drought effects on croplands in the west-central Great Plains, p. 420. USDA Information Bulletin. U.S. Government Printing Office, Washington, DC.

Green, T.R., L.R. Ahuja, and J.G. Benjamin. 2003. Advances and challenges in predicting agricultural management effects on soil hydraulic properties. Geoderma 116:3–27.

Greiner, L. 2008. Manage residue using strip till. http://magissues.farmprogress.com/WAL/WF02Feb08/wal034.pdf (assessed August 24, 2009).

Haas, H.J., W.O. Willis, and J.J. Bond. 1974. Summer fallow in the northern Great Plains (spring wheat), p. 12–35. In Summer fallow in the Western United States. USDA-ARS Conservative Research Report 17. U.S. Government Printing Office, Washington, DC.

Halvorson, A.D., A.L. Black, J.M. Krupinsky, S.D. Merrill, B.J. Wienhold, and D.L. Tanaka. 2000. Spring wheat response to tillage system and nitrogen fertilization within a crop-fallow system. Agron. J. 92:288–294.

Halvorson, A.D., A.R. Mosier, C.A. Reule, and W.C. Bausch. 2006. Nitrogen and tillage effects on irrigated continuous corn yields. Agron. J. 98:63–71.

Hammel, J.E. 1995. Long-term tillage and crop rotation effects on winter wheat production in northern Idaho. Agron. J. 87:16–22.

Harman, W.L., G.J. Michels, and A.F. Wiese. 1989. A conservation tillage system for profitable cotton production in the central Texas High Plains. Agron. J. 81:615–618.

Harrold, L.L., and W.M. Edwards. 1972. A severe rainstorm test of no-till corn. J. Soil Water Conserv. 27:30.

Hawkins, C.L., M.J. Shipitalo, E.M. Rutledge, M.C. Savin, and K.R. Brye. 2008. Earthworm populations in septic system filter fields and potential effects on wastewater renovation. Appl. Soil Ecol. 40:195–200.

Henry, W.B., D.C. Nielsen, M.F. Vigil, F.J. Calderon, and M.S. West. 2008. Proso millet yield and residue mass following direct harvest with a stripper-header. Agron. J. 100:580–584.

Hill, R.L. 1990. Long-term conventional and no-tillage effects on selected soil physical properties. Soil Sci. Soc. Am. J. 54:161–166.

Hill, P.R. 1996. Conservation tillage: A checklist for U.S. farmers. Conservation Technology Information Center, West Lafayette, IN.

Hill, P.R., K.J. Eck, and J.R. Wilcox. 1994. Managing crop residue with farm equipment. Agronomy Guide AY-280. Purdue University, West Lafayette, IN.

Hobbs, P.R. 2006. Conservation agriculture: What is it and why is it important for future sustainable food production? In M.P. Reynolds, D. Poland, and H.J. Braun (eds.) Challenges to international wheat improvement. CIMMYT, Mexico.

Howard, D.D., M.E. Essington, and D.D. Tyler. 1999. Vertical phosphorus and potassium stratification in no-till cotton soils. Agron. J. 91:266–269.

Howard, D.D., C.O. Gwathmey, M.E. Essington, R.K. Roberts, and M.D. Mullen. 2001. Nitrogen fertilization of no-till cotton on loess-derived soils. Agron. J. 93:157–163.

IPCC (Intergovernmental Panel on Climate Change). 1996. Climate change 1995: The science of climate change. Cambridge University Press, Cambridge, U.K.

Janosky, J.S., D.L. Young, and W.F. Schillinger. 2002. Economics of conservation tillage in a wheat-fallow rotation. Agron. J. 94:527–531.

Jastrow, J.D., T.W. Boutton, and R.M. Miller. 1996. Carbon dynamics of aggregate-associated organic matter estimated by carbon-13 natural abundance. Soil Sci. Soc. Am. J. 60:801–807.

Johnson, J.M.-F., R.R. Allmaras, and D.C. Reicosky. 2006. Estimating source carbon from crop residues, roots and rhizodeposits using the national grain-yield database. Agron. J. 98:622–636.

Johnson, W.C., C.E. Van Doren, and E. Burnett. 1974. Summer fallow in the southern Great Plains, p. 86–109. In Summer fallow in the western United States. USDA-ARS Conservative Research Report 17. U.S. Government Printing Office, Washington, DC.

Johnson-Maynard, J.L., K.J. Umiker, and S.O. Guy. 2007. Earthworm dynamics and soil physical properties in the first three years of no-till management. Soil Till. Res. 94:338–345.

Jones, O.R., V.L. Hauser, and T.W. Popham. 1994a. No-tillage effects on infiltration, runoff, and water conservation on dryland. Trans. ASAE 37:473–479.

Jones, O.R., S.J. Smith, and L.M. Southwick. 1994b. Tillage system effects on water conservation and runoff water quality—Southern High Plains drylands, p. 67–76. Proc. Great Plains Residue Mgmt. Conf., August 15–17, 1994. Amarillo, TX. Great Plains Agricultural Council.

Karlen, D.L., N.C. Wollenhaupt, D.C. Erbach, E.C. Berry, J.B. Swan, N.S. Eash, and J.L. Jordahl. 1994. Crop residue effects on soil quality following 10 years of no-till corn. Soil Till. Res. 31:149–167.

Kelley, K.W., and D.W. Sweeney. 2007. Placement of preplant liquid nitrogen and phosphorus fertilizer and nitrogen rate affects no-till wheat following different summer crops. Agron. J. 99:1009–1017.

Kennedy, A.C., and W.F. Schillinger. 2006. Soil quality and water intake in traditional-till vs. no-till paired farms in Washington's Palouse Region. Soil Sci. Soc. Am. J. 70:940–949.

Kladivko, E.J. 1994. Residue effects on soil physical properties, p. 123–141. *In* P.W. Unger (ed.) Managing agricultural residues. Lewis Publishers, Boca Raton, FL.

Kladivko, E.J., N.M. Akhouri, and G. Weesies. 1997. Earthworm populations and species distributions under no-till and conventional tillage in Indiana and Illinois. Soil Biol. Biochem. 29:613–615.

Knapen, A., J. Poesen, G. Govers, and S. De Baets. 2008. The effect of conservation tillage on runoff erosivity and soil erodibility during concentrated flow. Hydrol. Process. 22:1497–1508.

Kolberg, R.L., N.R Kitchen, D.G. Westfall, and G.A. Peterson. 1996. Cropping intensity and nitrogen management impact of dryland no-till rotations in the semiarid western Great Plains. J. Prod. Agric. 9:517–522.

Kravchenko, A.G., and K.D. Thelen. 2007. Effect of winter wheat residue on no-till corn growth and development. Agron. J. 99:549–555.

Kuhn, T.J., K.W. Tate, D. Cao, and M.R. George. 2007. Juniper removal may not increase overall Klamath River Basin water yields. Calif. Agric. 61:166–171.

Lal, R. 2004. Soil carbon sequestration impacts on global climate change and food security. Science 304:1623–1627.

Lal, R., M. Griffin, J. Apt, L. Lave, and G. Morgan. 2004. Managing soil carbon. Science 305:5690.

Lal, R., A.A. Mahboubi, and N.R. Fausey. 1994. Long-term tillage and rotation effects on properties of a central Ohio soil. Soil Sci. Soc. Am. J. 58:517–522.

Lamb, J.A., J.W. Doran, and G.A. Peterson. 1987. Nonsymbiotic dinitrogen fixation in no-till and conventional wheat-fallow systems. Soil Sci. Soc. Am. J. 51:356–361.

Lamb, J.A., G.A. Peterson, and C.R. Fenster. 1985. Fallow nitrate accumulation in a wheat-fallow rotation as affected by tillage system. Soil Sci. Soc. Am. J. 49:1441–1446.

Langdale, G.W., L.T. West, R.R. Bruce, W.P. Miller, and A.W. Thomas. 1992. Restoration of eroded soil with conservation tillage. Soil Technol. 5:81–90.

Larney, F.J., and E.J. Kladivko. 1989. Soil strength properties under four tillage systems at three long-term study sites in Indiana. Soil Sci. Soc. Am. J. 53:1539–1545.

Lascano, R.J., and R.L. Baumhardt. 1996. Effects of crop residue on soil and plant water evaporation in a dryland cotton system. Theor. Appl. Climatol. 54:69–84.

le Bissonnais, Y. 1990. Experimental study and modeling of soil surface crusting processes. Catena Supp. 17:13–28.

Levy, S. 2003. Turbulence in the Klamath River basin. BioScience 53:315–320.

Liebig, M.A., A.J. Jones, L.N. Mielke, and J.W. Doran. 1993. Controlled wheel traffic effects on soil properties in ridge tillage. Soil Sci. Soc. Am. J. 57:1061–1066.

Linden, D.R., C.E. Clapp, and R.H. Dowdy. 2000. Long-term corn grain and stover yields as a function of tillage and residue removal in east central Minnesota. Soil Till. Res. 56:167–174.

Lindstrom, M.J. 1986. Effects of residue harvesting on water runoff, soil-erosion and nutrient loss. Agric. Ecosyst. Environ. 16:103–112.

Lorenz, K., and R. Lal. 2005. The depth distribution of soil organic carbon in relation to land use and management and the potential of carbon sequestration in subsoil horizons. Adv. Agron. 88:35–66.

Lyles, L., and B.E. Allison. 1976. Wind erosion: The protective role of simulated standing stubble. Trans. ASAE 19:61–64.

Lyles, L., and B.E. Allison. 1981. Equivalent wind-erosion protection from selected crop residues. Trans. ASAE 24:405–408.

Martin, V.L., E.L. McCoy, and W.A. Dick. 1990. Allelopathy of crop residues influences corn seed germination and early growth. Agron. J. 82:555–560.

Maskina, M.S., J.F. Power, J.W. Doran, and W.W. Wilhelm. 1993. Residual effects of no-till crop residues on corn yield and nitrogen uptake. Soil Sci. Soc. Am. J. 57:1555–1560.

McCalla, T.M., and F.L. Duley. 1946. Effect of crop residues on soil temperature. Agron. J. 38:75–89.

McGregor, K.C., C.K. Mutchler, and M.J.M. Romkens. 1990. Effects of tillage with different crop residues on runoff and soil loss. Trans. ASAE 33:1551–1556.

McMaster, G.S., R.M. Aiken, and D.C. Nielsen. 2000. Optimizing wheat harvest cutting height for harvest efficiency and soil and water conservation. Agron. J. 92:1104–1108.

McVay, K.A., J.A. Budde, K. Fabrizzi, M.M. Mikha, C.W. Rice, A.J. Schlegel, D.E. Peterson, D.W. Sweeney, and C. Thompson. 2006. Management effects on soil physical properties in long-term tillage studies in Kansas. Soil Sci. Soc. Am. J. 70:434–438.

Merrill, S.D., A.L. Black, and A. Bauer. 1996. Conservation tillage affects root growth of dryland spring wheat under drought. Soil Sci. Soc. Am. J. 60:575–583.

Merrill, S.D., A.L. Black, D.W. Fryrear, A. Saleh, T.M. Zobeck, A.D. Halvorson, and D.L. Tanaka. 1999. Soil wind erosion hazard of spring wheat-fallow as affected by long-term climate and tillage. Soil Sci. Soc. Am. J. 63:1768–1777.

Meyer, L.D., S.M. Dabney, C.E. Murphree, W.C. Harmon, and E.H. Grissinger. 1999. Crop production systems to control erosion and reduce runoff from upland silty soils. Trans. ASAE 42:1645–1652.

Mielke, L.N., J.W. Doran, and K.A. Richard. 1986. Physical environment near the surface of plowed and no-tilled soils. Soil Till. Res. 7:355–366.

Miller, W.P., and D.E. Radcliffe. 1992. Soil crusting in the southeastern United States, p. 233–266. *In* M.E. Sumner and B.A. Stewart (eds.) Soil crusting: Chemical and physical processes. Lewis Publishers, Boca Raton, FL.

Mock, J.J., and D.C. Erbach. 1977. Influence of conservation-tillage environments on growth and productivity of corn. Agron. J. 69:337–340.

Morachan, Y.B., W.C. Moldenhauer, and W.E. Larson. 1972. Effects of increasing amounts of organic residues on continuous corn: I. Yields and soil physical properties. Agron. J. 64:199–203.

Nafziger, D., P.R. Carter, and E.E. Graham. 1991. Response of corn to uneven emergence. Crop Sci. 31:811–815.

Niehues, B.J., R.E. Lamond, C.B. Godsey, and C.J. Olsen. 2004. Starter nitrogen fertilizer management for continuous no-till corn production. Agron. J. 96:1412–1418.

Nielsen, D.C., M.F. Vigil, and J.G. Benjamin. 2006. Evaluating decision rule for dryland crop selection. In 2006 Agronomy abstracts [CD-ROM]. ASA, Madison, WI.

Norton, L.D., and L.C. Brown. 1992. Time-effect of water erosion for ridge tillage. Trans. ASAE 35:473–478.

Nyborg, M., and S.S. Malhi. 1989. Effect of zero and conventional tillage on barley yield and nitrate nitrogen content, moisture and temperature of soil in north-central Alberta. Soil Till. Res. 15:1–9.

Ortega, R.A. 1995. Residue distribution and potential C and N mineralization in no-till dryland agroecosystems. MS Thesis. Colorado State University. Fort Collins, CO.

Papendick, R.I., J.F. Parr, and R.E. Meyer. 1990. Managing crop residues to optimize crop/livestock production systems for dryland agriculture. Adv. Soil Sci. 13:253–272.

Paustian, K., H.P. Collins, and E.A. Paul. 1997. Management controls on soil carbon, p. 15–49. In E.A. Paul, K. Paustian, E.T. Elliott, and C.V. Cole (eds.) Soil organic matter in temperate agroecosystems. Lewis Publishers, Boca Raton, FL.

Perlack, R.D., L.L. Wright, A.F. Turhollow, R.L. Graham, B.J. Stokes, and D.C. Erbach. 2005. Biomass as feedstock for a bioenergy and bioproducts industry: The technical feasibility of a billion-ton annual supply. DOE/GO-102005-2135. U.S. DOE, Oak Ridge, TN.

Peterson, G.A. 1994. Interactions of surface residues with soil and climate, p. 9–12. In W.C. Moldenhauer and A.L. Black (eds.) Crop residue management to reduce erosion and improve soil quality. USDA Conservative Research Report 38. U.S. Government Printing Office, Washington, DC.

Peterson, G.A., A.J. Schlegel, D.L. Tanaka, and O.R. Jones. 1996. Precipitation use efficiency as affected by cropping and tillage systems. J. Prod. Agric. 9:180–186.

Peterson, G.A., and D.G. Westfall. 2004. Managing precipitation use in sustainable dryland agroecosystems. Ann. Appl. Biol. 144:127–138.

Peterson, G.A., D.G. Westfall, L. Sherrod, R. Kolberg, and D. Poss. 1995. Sustainable dryland agroecosystem management. Technical Bulletin TB 95-1. Colorado State University and Agricultural Experiment Station. Fort Collins, CO.

Pikul, J.L., R.C. Schwartz, J.G. Benjamin, R.L. Baumhardt, and S. Merrill. 2006. Cropping system influences on soil physical properties in the Great Plains. Renew. Agric. Food Syst. 21:15–25.

Potter, K.N., H.A. Torbert, O.R. Jones, J.E. Matocha, J.E. Morrison, Jr., and P.W. Unger. 1998. Distribution and amount of soil organic C in long-term management systems in Texas. Soil Till. Res. 47:309–321.

Qin, R.J., P. Stamp, and W. Richner. 2006. Impact of tillage on maize rooting in a Cambisol and Luvisol in Switzerland. Soil Till. Res. 85:50–61.

Raczkowski, C.W., M.R. Reyes, G.B. Reddy, W. Buscher, and P. Bauer. 2009. Comparison of conventional and no-tillage corn and soybean production on runoff and erosion in the southeastern US Piedmont. J. Soil Water Conserv. 64:53–60.

Radcliffe, D.E., G. Manor, R.L. Clark, L.T. West, G.W. Langdale, and R.R. Bruce. 1989. Effect of traffic and in-row chiseling on mechanical impedance. Soil Sci. Soc. Am. J. 53:1197–1201.

Radcliffe, D.E., E.W. Tollner, W.L. Hargrove, R.L. Clark, and M.H. Golabi. 1988. Effect of tillage practices on infiltration and soil strength of a Typic Hapludult soil after ten years. Soil Sci. Soc. Am. J. 52:798–804.

Raper, R.L., D.W. Reeves, E.C. Burt, and H.A. Torbert. 1994. Conservation tillage and traffic effects on soil condition. Trans. ASAE 37:763–768.

Reicosky, D.C., and D.W. Archer. 2007. Moldboard plow tillage depth and short-term carbon dioxide release. Soil Till. Res. 94:109–121.

Reinbott, T.M., S.P. Conley, and D.G. Blevins. 2004. No-tillage corn and grain sorghum response to cover crop and nitrogen fertilization. Agron. J. 96:1158–1163.

Rhoton, F.E., M.J. Shipitalo, and D.L. Lindbo. 2002. Runoff and soil loss from midwestern and southeastern US silt loam soils as affected by tillage practice and soil organic matter content. Soil Till. Res. 66:1–11.

Rice, C.W., and M.S. Smith. 1982. Denitrification in no-till and plowed soils. Soil Sci. Soc. Am. J. 46:1168–1173.

Rice, C.W., and M.S. Smith. 1984. Short-term immobilization of fertilizer nitrogen at the surface of no-till and plowed soils. Soil Sci. Soc. Am. J. 48:295–297.

Rice, C.W., M.S. Smith, and R.L. Blevins. 1984. Soil nitrogen availability after long-term continuous no-tillage and conventional tillage corn production. Soil Sci. Soc. Am. J. 50:1206–1210.

Rockwood, W.G., and R. Lal. 1974. Mulch tillage: A technique for soil and water conservation in the tropics. Span 17:77–79.

Rothfeder, J. 2001. Every drop for sale. Penguin Putnam, Inc., New York.

Sarath, G., R.B. Mitchell, S.E. Sattler, D. Funnell, J.F. Pedersen, R.A. Graybosch, and K.P. Vogel. 2008. Opportunities and roadblocks in utilizing forages and small grains for liquid fuels. J. Ind. Microbiol. Biotechnol. 35:343–354.

Saxton, K.E. 1995. Wind erosion and its impact on off-site air quality in the Columbia Plateau—An integrated research plan. Trans. ASAE 38:623–630.

Schertz, D.L., and J. Becherer. 1994. Terminology. In B.A. Stewart and W.C. Moldenhauer (eds.) Crop residue management to reduce erosion and improve soil quality 3.

USDA, Agricultural Research Service, Conservative Research Report 37. U.S. Government Printing Office, Washington, DC.

Schillinger, W.F. 2001. Minimum and delayed conservation tillage for wheat-fallow farming. Soil Sci. Soc. Am. J. 65:1203–1209.

Schillinger, W.F., and F.E. Bolton. 1993. Fallow water storage in tilled vs. untilled soils in the Pacific Northwest. J. Prod. Agric. 6:267–269.

Schlegel, A.J., K.C. Dhuyvetter, and J.L. Havlin. 2003. Placement of UAN for dryland winter wheat in the central High Plains. Agron. J. 95:1532–1541.

Schlegel, A.J., and C.A. Grant. 2006. Soil fertility, p. 141–194. In G.A. Peterson, P.W. Unger, and W.A. Payne, (eds.) Dryland agriculture. 2nd Edn. Agronomy monograph 23. ASA, CSSA, and SSSA, Madison, WI.

Schomberg, H.H., R.G. McDaniel, E. Mallard, D.M. Endale, D.S. Fisher, and M.L. Cabrera. 2006. Conservation tillage and cover crop influences on cotton production on a southeastern U.S. Coastal Plain soil. Agron. J. 98:1247–1256.

Schomberg, H.H., J.L. Steiner, and P.W. Unger. 1994. Decomposition and nitrogen dynamics of crop residues: Residue quality and water effects. Soil Sci. Soc. Am. J. 58:372–381.

Schwab, E.B., D.W. Reeves, C.H. Burmester, and R.L. Raper. 2002. Conservation tillage systems for cotton in the Tennessee Valley. Soil Sci. Soc. Am. J. 66:569–577.

Sharratt, B.S., M.J. Lindstrom, G.R. Benoit, R.A. Young, and A. Wilts. 2008. Runoff and soil erosion during spring thaw in the northern U.S. Corn Belt. J. Soil Water Conserv. 55:487–494.

Shaver, T.M., G.A. Peterson, and L.A. Sherrod. 2003. Cropping intensification in dryland systems improves soil physical properties: Regression relations. Geoderma 116:149–164.

Sherrod, L., G.A. Peterson, and D.G. Westfall. 1996. No-till rotational residue dynamics across an ET gradient, p. 282. In 1996 Agronomy abstracts. ASA, Madison, WI.

Siddoway, F.H., W.S. Chepil, and D.V. Armbrust. 1965. Effect of kind, amount, and placement of residue on wind erosion control. Trans. ASAE 8:327–331.

Sidhu, D., and S.W. Duiker. 2006. Soil compaction in conservation tillage: Crop impacts. Agron. J. 98:1257–1264.

Smart, J.R., and J.M. Bradford. 1996. No-tillage and reduced tillage cotton production in south Texas. Proc. Beltwide Cotton Conf. 2:1397–1400.

Smart, J.R., and J.M. Bradford. 1997. Cotton weed management in no-till maize and sorghum stubble. Proc. Beltwide Cotton Conf. 1:801–802.

Smika, D.E. 1976. Seed zone soil water conditions with reduced tillage in the semi-arid central Great Plains. Proc. 7th Conf. Int. Soil Till. Res. Org. Uppsala, Sweden.

Smith, S.J., and A.N. Sharpley. 1990. Soil nitrogen mineralization in the presence of surface and incorporated crop residues. Agron. J. 82:112–116.

Smith, S.J., and A.N. Sharpley. 1993. Nitrogen availability from surface-applied and soil-incorporated crop residues. Agron. J. 85:776–778.

SSSA (Soil Science Society of America). 2001. Glossary of soil science terms. SSSA, Madison, WI.

Staley, T.E., W.M. Edwards, C.L. Scott, and L.B. Owens. 1988. Soil microbial biomass and organic component alterations in a no-tillage chronosequence. Soil Sci. Soc. Am. J. 52:998–1005.

Stecker, J.A., J.R. Brown, and N.R. Kitchen. 2001. Residual phosphorus distribution and sorption in starter fertilizer bands applied in no-till culture. Soil Sci. Soc. Am. J. 65:1173–1183.

Stecker, J.A., D.D. Buchholz, R.G. Hanson, N.C. Wollenhaupt, and K.A. McVay. 1993a. Application placement and timing of nitrogen solution for no-till corn. Agron. J. 85:645–650.

Stecker, J.A., D.D. Buchholz, R.G. Hanson, N.C. Wollenhaupt, and K.A. McVay. 1993b. Broadcast nitrogen sources for no-till continuous corn and corn following soybean. Agron. J. 85:893–897.

Stecker, J.A., D.H. Sander, F.N. Anderson, and G.A. Peterson. 1988. Phosphorus fertilizer placement and tillage in a wheat-fallow cropping sequence. Soil Sci. Soc. Am. J. 52:1063–1068.

Sterk, G., and W.P. Spaan. 1997. Wind erosion control with crop residues in the Sahel. Soil Sci. Soc. Am. J. 61:911–917.

Sullivan, D.G., D. Lee, J. Beasley, S. Brown, and E.J. Williams. 2008. Evaluating a crop residue cover index for determining tillage regime in a cotton-corn-peanut rotation. J. Soil Water Conserv. 63:28–36.

Sullivan, D.G., C.C. Truman, H.H. Schomberg, D.M. Endale, and D.H. Franklin. 2007. Potential impact of conservation tillage on conserving water resources in Georgia. J. Soil Water Conserv. 62:145–152.

Swan, J.B., R.L. Higgs, T.B. Bailey, N.C. Wollenhaupt, W.H. Paulson, and A.E. Peterson. 1994. Surface residue and in-row treatment effects on long-term no-tillage continuous corn. Agron. J. 86:711–718.

Sweeney, D.W., M.B. Kirkham, and J.B. Sisson. 2006. Crop and soil response to wheel-track compaction of a claypan soil. Agron. J. 98:637–643.

Tanaka, D.L. 1989. Spring wheat plant parameters as affected by fallow methods in the northern Great Plains. Soil Sci. Soc. Am. J. 53:1506–1511.

Tanaka, D.L., and R.L. Anderson. 1997. Soil water storage and precipitation storage efficiency of conservation tillage systems. J. Soil Water Conserv. 52:363–367.

Thomas, G.W., G.R. Hazler, and R.L. Blevins. 1996. The effects of organic matter and tillage on maximum compactibility of soils using the Proctor test. Soil Sci. 161:502–508.

Thorne, M.E., F.L. Young, W.L. Pan, R. Bafus, and J.R. Alldredge. 2003. No-till spring cereal cropping systems reduce wind erosion susceptibility in the wheat/fallow region of the Pacific Northwest. J. Soil Water Conserv. 58:250–257.

Touchton, J.T., and J.W. Johnson. 1982. Soybean tillage and planting method effects on yield of double-cropped wheat and soybean. Agron. J. 74:57–59.

Triplett, G.B., Jr., and W.A. Dick. 2008. No-tillage crop production: A revolution in agriculture. Agron. J. 100:153–165.

Trojan, M.D., and D.R. Linden. 1994. Tillage, residue, and rainfall effects on movement of an organic tracer in earthworm-affected soils. Soil Sci. Soc. Am. J. 58:1489–1494.

Truman, C.C., D.W. Reeves, J.N. Shaw, A.C. Motta, C.H. Burmester, R.L. Raper, and E.B. Schwab. 2003. Tillage impacts on soil property, runoff, and soil loss variations from a Rhodic Paleudult under simulated rainfall. J. Soil Water Conserv. 58:258–267.

Truman, C.C., J.N. Shaw, and D.W. Reeves. 2005. Tillage effects on rainfall partitioning and sediment yield from an Ultisol in central Alabama. J. Soil Water Conserv. 60:89–98.

Tullberg, J.N., D.F. Yule, and D. McGarry. 2007. Controlled traffic farming—From research to adoption in Australia. Soil Till. Res. 97:272–281.

Unger, P.W. 1978. Straw-mulch rate effect on soil water storage and sorghum yield. Soil Sci. Soc. Am. J. 42:486–491.

Unger, P.W. 1984. Tillage and residue effects on wheat, sorghum, and sunflower grown in rotation. Soil Sci. Soc. Am. J. 48:885–891.

Unger, P.W. 1991. Ontogeny and water use of no-tillage sorghum cultivars on dryland. Agron J. 83:961–968.

Unger, P.W. 1993. Reconsolidation of a Torretic Paleustoll after tillage with a Paratill. Soil Sci. Soc. Am. J. 57:195–199.

Unger, P.W. 2001. Alternative and opportunity dryland crops and related soil conditions in the southern Great Plains. Agron. J. 93:216–226.

Unger, P.W. 2006. Soil and water conservation handbook: Policies, practices, conditions, and terms. The Haworth Press, Inc., New York.

Unger, P.W., and T.A. Howell. 1999. Agricultural water conservation—A global perspective, p. 1–36. In M.B. Kirkham, (ed.) Water use in crop production. The Haworth Press, Inc., New York.

Unger, P.W., M.B. Kirkham, and D.C. Nielsen. 2009. Water conservation for agriculture, p. 1–45. In T.M. Zobeck and W.F. Schillinger (eds.) Soil and water conservation in the USA: Past lessons—Future directions. SSSA, Madison, WI.

Unger, P.W., and M.F. Vigil. 1998. Cover crop effects on soil water relationships. J. Soil Water Conserv. 53:200–207.

Unger, P.W., and A.F. Wiese. 1979. Managing irrigated winter wheat residues for water storage and subsequent dryland grain sorghum production. Soil Sci. Soc. Am. J. 43:582–588.

USDOE (U.S. Department of Energy). 2007. DOE selects six cellulosic ethanol plants for up to $385 million in federal funding. DOE, Washington, DC. http://www.energy.gov/news/4827.htm (accessed January 31, 2009).

VandenBygaart, A.J., E.G. Gregorich, and D.A. Angers. 2003. Influence of agricultural management on soil organic carbon: A compendium and assessment of Canadian studies. Can. J. Soil Sci. 83:363–380.

van Donk, S.J., S.D. Merrill, D.L. Tanaka, and J.A. Krupinsky. 2008. Crop residue in North Dakota: Measured and simulated by the wind erosion prediction system. Trans. ASAE 51:1623–1632.

van Doren, D.M., G.B. Triplett, and J.E. Henry. 1976. Influence of long term tillage, crop rotation, and soil type combinations on corn yield. Soil Sci. Soc. Am. J. 40:100–105.

Varvel, G.E., and W.W. Wilhelm. 2008. Soil carbon levels in irrigated western Corn Belt rotations. Agron. J. 100:1180–1184.

Vasilakoglou, I., K. Dhima, I. Eleftherohorinos, and A. Lithourgidis. 2006. Winter cereal cover crop mulches and inter-row cultivation effects on cotton development and grass weed suppression. Agron. J. 98:1290–1297.

Vetsch, J.A., G.W. Randall, and J.A. Lamb. 2007. Corn and soybean production as affected by tillage systems. Agron. J. 99:952–959.

Vigil, M.F., and D.E. Kissel. 1991. Equations for estimating the amount of nitrogen mineralized from crop residues. Soil Sci. Soc. Am. J. 55:757–767.

Vyn, T.J., and K.J. Janovicek. 2001. Potassium placement and tillage system effects on corn response following long-term no-till. Agron. J. 93:487–495.

Wagger, M.G., and D.K. Cassel. 1993. Corn yield and water-use efficiency as affected by tillage and irrigation. Soil Sci. Soc. Am. J. 57:229–234.

Wagger, M.G., and H.P. Denton. 1989. Tillage effects on grain yields in a wheat, double-crop soybean, and corn rotation. Agron. J. 81:493–498.

Wagner-Riddle, C., T.J. Gillespie, and C.J. Swanton. 1994. Rye cover crop management impact on soil water content, soil temperature and soybean growth. Can. J. Plant Sci. 74:485–495.

West, L.T., W.P. Miller, R.R. Bruce, G.W. Langdale, J.M. Laflen, and A.W. Thomas. 1992. Cropping system and consolidation effects on rill erosion in the Georgia Piedmont. Soil Sci. Soc. Am. J. 56:1238–1243.

West, T.O., and W.M. Post. 2002. Soil organic carbon sequestration rates by tillage and crop rotation: A global data analysis. Soil Sci. Soc. Am. J. 66:1930–1946.

Wilhelm, W.W., J.W. Doran, and J.F. Power. 1986. Corn and soybean yield response to crop residue management under no-tillage production systems. Agron. J. 78:184–189.

Wilhelm, W.W., J.M.F. Johnson, K.L. Douglas, and D.T. Lightle. 2007. Corn stover to sustain soil organic carbon further constrains biomass supply. Agron. J. 99:1665–1667.

Wilhelm, W.W., J.M.F. Johnson, J.L. Hatfield, W.B. Voorhees, and D.R. Linden. 2004. Crop and soil productivity response to corn residue removal: A literature review. Agron. J. 96:1–17.

Wilkerson, G.G., A.J. Price, A.C. Bennett, D.W. Krueger, G.T. Roberson, and B.L. Robinson. 2004. Evaluating the potential for site-specific herbicide application in soybean. Weed Technol. 18:1101–1110.

Wilson, G.V., S.M. Dabney, K.C. McGregor, and B.D. Barkoll. 2004. Tillage and residue effects on runoff and erosion dynamics. Trans. ASAE 47:119–128.

Wischmeier, W.H., and D.D. Smith. 1978. Predicting rainfall erosion losses—A guide to conservation planning, p. 537. In Agriculture Handbook. U.S. Government Printing Office, Washington, DC.

Wood, C.W., and J.H. Edwards. 1992. Agroecosystem management effects on soil carbon and nitrogen. Agric. Ecosystems Environ. 39:123–138.

Wood, C.W., D.G. Westfall, and G.A. Peterson. 1991. Soil carbon and nitrogen changes on initiation of no-till cropping systems. Soil Sci. Soc. Am. J. 55:470–476.

Wood, C.W., D.G. Westfall, G.A. Peterson, and I.C. Burke. 1990. Impacts of cropping intensity on carbon and nitrogen mineralization under no-till dryland agroecosystems. Agron. J. 82:1115–1120.

Wortmann, C.S., S.A. Xerinda, and M. Mamo. 2006a. No-till row crop response to starter fertilizer in eastern Nebraska. II. Rainfed grain sorghum. Agron. J. 98:187–193.

Wortmann, C.S., S.A. Xerinda, M. Mamo, and C.A. Shapiro. 2006b. No-till row crop response to starter fertilizer in eastern Nebraska. I. Irrigated and rainfed corn. Agron. J. 98:156–162.

Wuest, S.B., J.D. Williams, and T.R. Johlke. 2006. Effects of tillage on water infiltration. Spec. Rep. 1068. Corvallis, OR: Oregon Agric. Exp. Stn.

Xinhua, Y., and T.J. Vyn. 2002a. Soybean responses to potassium placement and tillage alternatives following no-till. Agron. J. 94:1367–1374.

Xinhua, Y., and T.J. Vyn. 2002b. Residual effects of potassium placement and tillage systems for corn on subsequent no-till soybean. Agron. J. 94:1112–1119.

Xinhua, Y., and T.J. Vyn. 2003. Potassium placement effects on yield and seed composition of no-till soybean seeded in alternative row widths. Agron. J. 95:126–132.

26
Soil Quality

Stephanie A. Ewing
Montana State University

Michael J. Singer
University of California

26.1 Introduction: Why Soil Quality?..26-1
26.2 Defining Soil Quality..26-1
  Soil Quality as Soil Function • Soil Quality Is Not a New Idea • Agricultural Productivity and Sustainability • Rangeland Health • Soil Degradation and Resilience • Reference States for Soil Quality • Global Climate and Energy Production
26.3 Evaluating Soil Quality...26-12
  Selecting Soil Quality Indicators • Examples of Soil Quality Indicators • Approaches for Evaluating Soil Quality
26.4 Conclusions..26-20
References...26-21

26.1 Introduction: Why Soil Quality?

Soils are fundamental to the well-being and productivity of agricultural and natural ecosystems. The concept of soil quality was developed to characterize the usefulness and health of soils as a means of evaluating sustainable soil management practices. The concept has been applied throughout the world for a multitude of purposes (e.g., Lamarca, 1996; Ouedraogo et al., 2001; Shepherd et al., 2001; Tian and Feng, 2008). Recent reviews have emphasized the importance and success of the concept for sustainable agriculture and ecological management (Karlen et al., 2001; Wienhold et al., 2004; Kibblewhite et al., 2008), while others have critiqued the concept as counterproductive to the goals of soil science and the challenges of modern agriculture (Sojka and Upchurch, 1999; Singer and Sojka, 2001; Letey et al., 2003; Sanchez et al., 2003; Sojka et al., 2003). This chapter summarizes recent literature on the soil quality concept and touches upon some of the larger issues raised in the course of its discussion.

Since the 1970s, there has been heightened attention paid to the sustainability of human uses of soil based on concerns that soil quality may be declining as global population and pressure on environmental resources increase (Fournier, 1989; Parr et al., 1990; Doran et al., 1996; Boehn and Anderson, 1997). Lal (1995) described the land resources of the world (of which soil is one component) as "finite, fragile, and nonrenewable," and reported that only about 22% (3.26 billion ha) of the total land area on the globe was suitable for cultivation, and only about 3% (450 million ha) had a high agricultural production capacity. Soil as a natural, undisturbed body is in large but finite supply, and the condition of soils in agriculture and the environment is an issue of global concern given the extensive human reliance on soil resources (Howard, 1993; FAO, 1997; Foley et al., 2005; Figure 26.1). Well-known and more recently recognized issues include soil losses from erosion, maintaining agricultural productivity, protecting natural areas, protecting water quality, adverse effects of soil degradation on human health, and the role and management of soils in the global climate system (Crosson and Stout, 1983; Siegrist, 1989; Denneman and Robberse, 1990; Haberern, 1992; Howard, 1993; Sims et al., 1997; Melillo et al., 2002; Tilman et al., 2002; CAST, 2004; Robertson and Swinton, 2005; Palm et al., 2007; IPCC, 2007). In a general sense, soil quality has effectively become a central issue in the global climate debate because soils regulate greenhouse gases, store carbon, and will require careful management if they are to produce biofuels or effectively sequester carbon (CAST, 2004; Hill et al., 2009). However, despite the importance of understanding and managing the condition of soil resources globally, the goal of evaluating "soil quality" as a measurable attribute at the global scale remains elusive.

Even at a local scale, priorities for assessing soil quality in an agricultural context may be somewhat different from priorities for assessing soil quality in a natural ecosystem. For agriculture, soil quality may be managed to optimize production of a crop without adverse environmental effect, while in a natural ecosystem, soil quality may be managed relative to a baseline condition or potential against which future changes in the system may be compared, and with management decisions structured to maintain the natural potential or function of soils supporting plant communities, animal habitats, etc. This chapter discusses the concept of soil quality for agriculture and the environment and reviews approaches for evaluating soil quality.

26.2 Defining Soil Quality

Useful evaluation of soil quality requires agreement about why soils are important, how soil quality is defined and how it should be measured, and how to respond to soil quality assessments

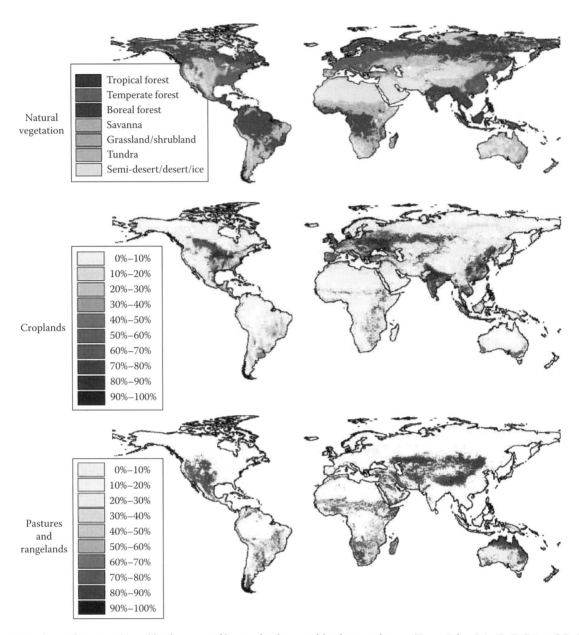

FIGURE 26.1 (See color insert.) Worldwide extent of human land-use and land-cover change. (From Foley, J.A., R. DeFries, G.P. Asner, C. Barford, G. Bonan, S.R. Carpenter, F.S. Chapin, M.T. Coe, G.C. Daily, H.K Gibbs, J.H. Helkowski, T. Holloway, E.A. Howard, C.J. Kucharik, C. Monfreda, J.A. Patz, C. Prentice, N. Ramankutty, P.K. Snyder. 2005. Global consequences of land use. Science 309:570–574.) Croplands are 12% and rangelands are 22%. (From Ramankutty, N., A.T. Evan, C. Monfreda, and J.A. Foley. 2008. Farming the planet: 1. Geographic distribution of global agricultural lands in the year 2000. Global Biogeochem. Cycles 22:GB1003, doi: 10.1029/2007GB002952.)

with management, restoration, or conservation practices. Because evaluating soil quality requires one or more value judgments, and because we have both much to learn about soil and much to lose as a result of poor soil management, these issues are not easily addressed (Sojka et al., 2003; Palm et al., 2007). Gregorich et al. (1994) stated, "Soil quality is a composite measure of both a soil's ability to function and how well it functions, relative to a specific use." This emphasis on use is consistent with the older articulation of "land quality" (FAO, 1976, 1995, 1997; Rossiter, 1996; Sombroek, 1997). However, the relationship between broadly defined soil functions and diverse human uses of soils can be difficult to comprehensively address (Sanchez et al., 2003). Soil scientists do not all agree that soil quality is a useful concept. For opposing views of the topic, refer to thorough discussions and expansive literature reviews in Karlen et al. (2001) and Sojka et al. (2003).

The soil quality concept was developed as a way to integrate existing and developing ideas about sustainable agriculture and environmental health in the last few decades (Arshad and Coen, 1992; Sheppard et al., 1992; Karlen et al., 1992, 2001; NRC, 1993; Doran and Parkin, 1994; Herrick et al., 2002). Discussion of soil quality invokes issues critical to human survival: the

> # Box 26.1 Three Functions of Soil
>
> Soils are living systems that are vital for producing the food and fiber humans need and for maintaining the ecosystems on which all life ultimately depends. Soil directly and indirectly affects agricultural productivity, water quality, and the global climate through its function as a medium for plant growth, a regulator, and partitioner of water flow, and an environmental buffer.
>
> *Soils make it possible for plants to grow.* Soils mediate the biological, chemical, and physical processes that supply nutrients, water, and other elements to growing plants. The microorganisms in soils transform nutrients into forms that can be used by growing plants. Soils are the water and nutrient storehouses on which plants draw when they need nutrients to produce roots, stems, and leaves. Eventually, these become food and fiber for human consumption. Soils—and the biological, chemical, and physical processes that they make possible—are a fundamental resource on which the productivities of agricultural and natural ecosystems depend.
>
> *Soils regulate and partition water flow through the environment.* Rainfall in terrestrial ecosystems falls on the soil surface where it either infiltrates the soil or moves across the soil surface into streams or lakes. The condition of the soil surface determines whether rainfall infiltrates or runs off. If it infiltrates the soil, it may be stored and later taken up by plants, move into groundwaters, or move laterally through the earth, appearing later in springs or seeps. This partitioning of rainfall between infiltration and runoff determines whether a storm results in a replenishing rain or a damaging flood. The movement of water through soils to streams, lakes, and groundwater is an essential component of recharge and base flow in the hydrological cycle.
>
> *Soils buffer environmental change.* The biological, chemical, and physical processes that occur in soils buffer environmental changes in air quality, water quality, and global climate. The soil matrix is the major incubation chamber for the decomposition of organic wastes including pesticides, sewage, solid wastes, and a variety of other wastes. The accumulation of pesticide residues, heavy metals, pathogens, or other potentially toxic materials in the soil may affect the safety and quality of food produced on those soils. Depending on how they are managed, soils can be important sources or sinks for carbon dioxide and other gases that contribute to the greenhouse effect (greenhouse gases). Soils store, degrade, or immobilize nitrates, phosphorus, pesticides, and other substances that can become pollutants in air or water.
>
> *Source:* NRC (National Research Council), *Soil and Water Quality: An Agenda for Agriculture*, Committee on Long-Range Soil and Water Conservation, Board on Agriculture, National Academy Press, Washington, DC, 1993.)

environmental cost of agricultural production and grazing, the potential for reclamation of degraded soils, and the sustainability of human land use practices in a larger environmental context. In general, definitions of soil quality have attempted to comprehensively inventory soil functions and relate them to diverse human benefits from soil resources (Box 26.1). In practice, however, the soil quality concept has mainly been applied to agricultural and rangeland management from a local to regional scale.

The definition of soil quality currently used by the Soil Science Society of America (1996; https://www.soils.org/publications/soils-glossary#) is that of Doran and Parkin (1994): "the capacity of a soil to function within ecosystem boundaries to sustain biological productivity, maintain environmental quality, and promote plant and animal health." Other function-based definitions of soil quality have been reviewed and evaluated in Karlen et al. (2001) and Sojka et al. (2003).

Multifunctional soil quality definitions are challenging to evaluate because agricultural and environmental functions (e.g., nutrient status and water quality) may conflict (Sojka et al., 2003), and because our understanding of soil functions and sustainability in the Earth system continues to expand. Since the first writing of this chapter (Singer and Ewing, 2000), the management of soils for carbon sequestration and greenhouse gas regulation has become a major topic of research and discussion in the global climate debate (e.g., CAST, 2004; IPCC, 2007). We also know a great deal more about atmospheric mobilization and deposition processes affecting soils (reactive N, dust, global transport of pollutants, etc.; e.g., Derry and Chadwick, 2007; Bowman et al., 2008). Development of a biofuels industry will put enormous pressure on soil resources (e.g., CAST, 2008; FAPRI, 2008; Fargione et al., 2008; Searchinger et al., 2008; Hill et al., 2009; Tilman et al., 2009). Agricultural and land management practices are coming under increased scrutiny as these connections are drawn (Tilman et al., 2002; Foley et al., 2005; Robertson and Swinton, 2005; Rockstrom et al., 2009a, 2009b). The challenge for soil resource management lies in connecting what we know and value about soils with how we realistically manage the demands of carbon regulation and food and energy production that we now face with agricultural and other land management choices, and the pressures of a growing population on both agricultural and natural systems (Tilman et al., 2002; Foley et al., 2005; Gomiero et al., 2008; Goulding et al., 2008; Rockstrom et al.,

2009a). A multifunctional view of soils is critical for evaluating global ecological management goals that incorporate human and societal well-being (MEA, 2005; Palm et al., 2007).

26.2.1 Soil Quality as Soil Function

The 1993 National Research Council report "Soil and water quality: An agenda for agriculture" (NRC, 1993) articulated the soil quality concept in the United States, basing it firmly in agriculture (Box 26.1; Figure 26.2). Yet the report included natural systems in its justification, finding that "…national policies to protect soil resources should be based on the fundamental functions that soils perform in natural and agroecosystems." It is worth noting that this evaluation of soil functions is fundamentally different from the philosophy of soil taxonomical classification in the U.S. system, which seeks to characterize the diversity of soils by properties observable in a field setting or with limited laboratory analysis, without the interpretation required to assess soil functions. A limitation of the U.S. taxonomy—or any descriptive system—is that it does not address soil degradation, which can only be successfully evaluated in terms of ecological functions broadly defined (MEA, 2005; Foley et al., 2005; Palm et al., 2007). Recent work has sought to unify soil taxonomic mapping with the evaluation of soil degradation (Tugel et al., 2005; Palm et al., 2007; Sanchez et al., 2009).

The functions of soils within agricultural and natural ecosystems can be characterized as ecosystem services (Cassman, 1999; Table 26.1) that are intimately tied to human and societal well-being (Figure 26.3) (Tilman et al., 2002; MEA, 2005; Robertson and Swinton, 2005; Palm et al., 2007; Kibblewhite et al., 2008; Carpenter et al., 2009). Increases over the last 40 years in agricultural production—in terms of both yield and cultivated land area—have come with trade-offs that include soil degradation, that is, reduction of many of the regulatory and supporting services in which soils play a central role (Figure 26.4) (Foley et al., 2005; Palm et al., 2007; Carpenter et al., 2009). Issues of sustainable

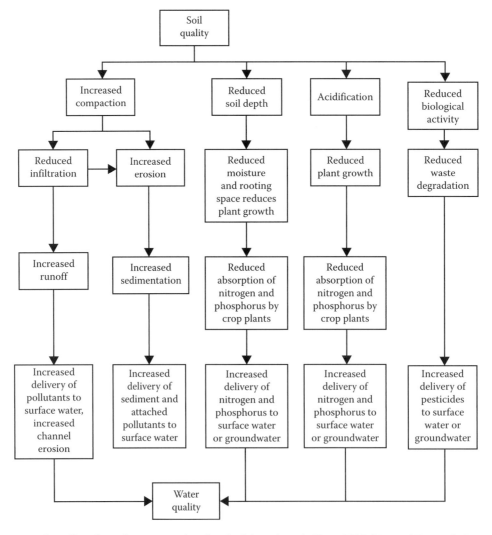

FIGURE 26.2 Water quality effects from changes in soil quality (soil degradation). (From NRC (National Research Council). 1993. Soil and water quality: An agenda for agriculture. Committee on Long-Range Soil and Water conservation, Board on Agriculture, National Academy Press, Washington, DC.)

Soil Quality

TABLE 26.1 Relationships among Provisioning Ecosystem Services, Soil Processes and Properties, and Core Soil Determinants

Provisioning Ecosystem Service	Ecosystem/Soil Process	Soil Property	Core Soil Determinants
Physical support for plants	Soil formation	Depth, texture, structure	State factors of soil formation, clay mineralogy
Provision of nutrients/ biogeochemical cycles	Mineral weathering	Mineral distribution	Parent material
	Soil organic matter mineralization	Soil organic matter quantity and quality	Texture, water balance
	Decomposition of organic additions	Soil biota	Texture, water balance
	Ion retention and exchange	Effective cation and anion exchange capacity	Texture, mineralogy, soil organic matter
	Toxicities	Percent Al saturation, electrical conductivity, percent exchangeable Na, toxic levels of Fe, Mn, B	Clay mineralogy, pH
Provision of water	Infiltration	Surface macroporosity, hydraulic conductivity	Aggregation, texture, bulk density, soil organic matter, soil biota
	Storage in soil	Aggregation, bulk density, depth	Texture, mineralogy, soil organic matter
	Drainage	Macroporosity, hydraulic conductivity	Texture, mineralogy, soil organic matter

Source: Adapted from Palm, C., P. Sanchez, S. Ahamed, and A. Awiti. 2007. Soils: A contemporary perspective. Annu. Rev. Environ. Resour. 32:99–129.

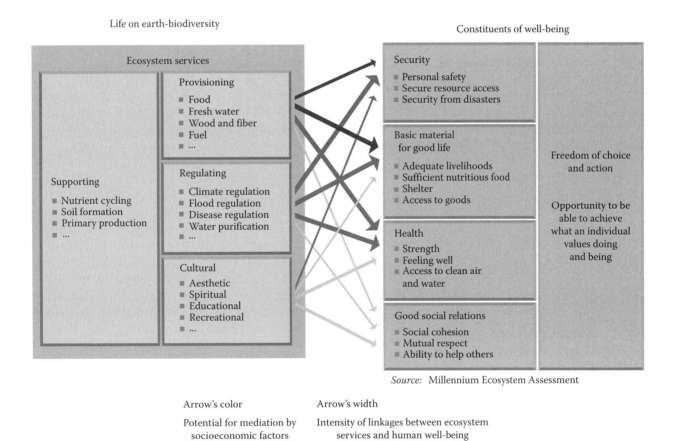

FIGURE 26.3 Ecosystem services and human well-being. (From Millenium Ecosystem Assessment (MEA). 2005. Ecosystems and human well-being: Synthesis. Island Press, Washington, DC.)

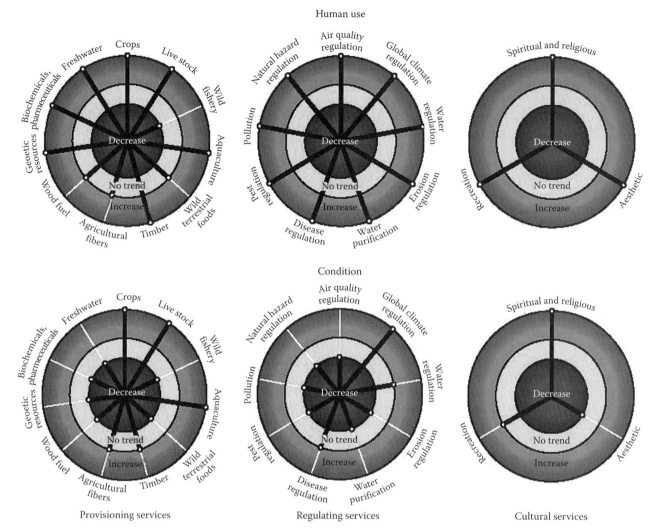

FIGURE 26.4 Trends in human use (upper) and condition (lower) of ecosystem services. (From Carpenter, S.R., H.A. Mooney, J. Agard, D. Capistrano, R.S. DeFries, S. Diaz, T. Dietz, A.K. Druaiappah, A. Oteng-Yeboah, H.M. Pereira, C. Perrings, W.V. Reid, J. Sarukhan, R.J. Scholes, and A. Whyte. 2009. Science for managing ecosystem services: Beyond the millennium ecosystem assessment. Proc. Natl. Acad. Sci. 106:1305–1312.)

soil management, soil degradation, and soil resilience are central to the soil quality concept, and are further discussed below. Our review of the literature suggests that these issues have been more forcefully articulated independently of the formalized concept of "soil quality." Yet the need for the soil expertise in discussion of these issues and in global change science has never been greater.

26.2.2 Soil Quality Is Not a New Idea

Soil quality has historically been equated with agricultural productivity in the United States, whereas in Europe, the concept was developed to address soil pollution. Soil conservation practices to maintain soil productivity are as old as agriculture itself, with documentation dating to the Roman Empire (Jenny, 1961). Beginning in the 1930s, soil productivity ratings were developed in the United States and elsewhere to help farmers select crops and management practices that would maximize production and minimize erosion or other adverse environmental effects (Huddleston, 1984). The Storie Index (Storie, 1932, 1964) and the USDA Land Capability Classification (Klingebiel and Montgomery, 1973) were developed to separate soils into different quality classes based on inherent and "static" soil properties such as depth to a restrictive layer. These rating systems are important predecessors of more recent attempts to quantitatively assess soil quality. In the 1970s, attempts were made to identify and protect soils of the highest productive capacity by defining "prime agricultural lands" (Miller, 1979; Reganold and Singer, 1979; Singer et al., 1979). Alexander (1971) suggested developing soil quality criteria. Warkentin and Fletcher (1977) followed by introducing a soil quality concept based on dynamic soil properties relating to organic matter and biology, which sought to improve upon evaluations based solely on productivity and erosion control. Carter et al. (1997) articulated the distinction between "static" and "dynamic" aspects of soil quality. Farmers typically consider both in a working evaluation of soil health—a term often used interchangeably with soil quality (Romig et al., 1995; Table 26.2).

TABLE 26.2 Soil Health Properties Derived from Farmer Interviews

Rank	Soil Health Property	Description Healthy	Description Unhealthy
1	Organic matter	As high as possible, at soil's potential, manure, compost >3%, 2%, 7%–8%, putting more back	Rough, lack of organic matter, less, low
2	Crop appearance	Green, healthy, uniform, lush, dense stand, tall, larger, sturdy, stout, proper color, darker, good crop	Yellow, stunted corn, small, poor color, poorer, lack of green, light green, streaks in field
3	Erosion	Would not erode, water, and wind not taking soil, prevented, stays in place, less, slowed down, delayed	Blows sooner, washes, topsoil is lost, erodes more, clouds of dust, ravines, runs bad, any, easier
4	Earthworms	Fishing and red worms present, see after rain, a lot, angle worms, see holes and castings, see during plowing	Not there, do not work, cannot find, no holes, lack of, killed by insecticides or anhydrous, void
5	Drainage	Water goes away, fast, better, no ponding, moves through, takes a lot of rain, drains properly, dries out	Tight, waterlogged, drains too fast, ponding, no outlet for water, will not drain, slop, poor, saturated
6	Tillage ease	One pass and ready, breaks up, mellow, easier, smooth, crumbles, flows, plow a gear faster, minimum	Never works down, needs more disking, lumps, slabs, shiney, pulls hard, worked wet, overworked
7	Soil structure	Will not roll out of hand, crumbly, loose, holds together, granular	Hard, does not hold together, lumpy, falls apart, massive, cloddy, lumpy, clumpy, tight, compacted, powder
8	pH	7.0, 6.7–6.8, 6.2–6.7, balanced, neutralize	<6.0, high, nothing works, wrong, too low, high acidity
9	Soil test	Up to recommendations, high, elevated, complete, where it belongs, every year or two, stay up with soil test	Law of minimum at work
10	Yield	9.5–11.5 Mg ha^{-1} corn, 4 Mg ha^{-1} soybeans, 30%–40% higher, +0.7 Mg ha^{-1}, better 5 years average, significantly higher	7 Mg ha^{-1} corn, 9.5 Mg ha^{-1} corn, 2.5 Mg ha^{-1} soybeans, 20%–50% less, do not get much off, down, reduced, low
11	Compaction	Does not pack down, not compacted, stays loose, not out there when wet	Compacted, plow layer, packs down, hardpan, plowsoil, tight, cannot get into it, packed
12	Infiltration	Water does not stand, absorbs, water moves into soil, soaks, rapid, no ponding, fast, spongy	Water runs off soil, sits on top, water stands, does not absorb, puddles, nonporous
13	Soil color	Dark, black, dark brown, gray, holds dark color	Orange, brown, light, white, red, blue-gray, subsoil color, bleached, sandy colored, light brown, pale, anemic, gray
14	Nitrogen	Put on less, manure, as required, compost, slurry, more available, organic N, organic matter	Too much N, chemical N, commercial fertilizers burn ground, anhydrous, sludge
15	Water retention	Holds moisture, get by with less, retains more, moisture travels, gives and takes water freely, conserving	Too much water, does not hold water, dries out, too wet or dry, droughty, stays wet, runs out of moisture, poor
16	Phosphorus	As required	—
17	Nutrient deficiency	Has what it needs, no shortage of elements, no spots on leaves	Yellow, purple, discoloration in leaves, lodging, crop falls off, stripping, brown streaks, firings on bottom, blight
18	Decomposition	Breaks down, decays, rots in 4–5 months, manure part of soil in 1 year, disappears, not too fast, 2/3 gone in year	See stalks from last year, does not break down, manure plows up next year
19	Potassium	As required	—
20	Roots	Larger, spread out, grown down, white, deep, numerous, good penetration, full, lots of feeders, branched out	Do not penetrate, undeveloped, balled up, grow crossways, discolored, diseased, at hard angles, shallow, short

Source: Romig, D.E., M.J. Garlynd, R.F. Harris, and K. McSweeney. 1995. How farmers assess soil health and quality. J. Soil Water Conserv. 50:229–236. With permission of the Soil and Water Conservation Society of America.

Many attempts to define soil quality and to develop indices to measure it have been similar to earlier soil productivity ratings (Doran and Parkin, 1994; Doran and Jones, 1996; Snakin et al., 1996; Seybold, et al., 1997; Karlen et al., 2001; Sojka et al., 2003). As the soil quality concept has evolved, researchers have moved from qualitative definitions to more quantitative models that combine diverse variables and spatial characteristics of soils. They have also moved from assessing static properties to those that are dynamic (Sanchez et al., 2003; Tugel et al., 2005; Tugel et al., 2008). By the 1990s, soil quality had become a national policy in several countries. In 1990, the Canadian Soil Quality Evaluation Program was initiated, and benchmark sites established for long-term monitoring (Walker et al., 1997). In New Zealand, the 1991 Resource Management Act required sustainable land management practices (Lilburne et al., 2002). In 1994, the U.S. Soil Conservation Service was reorganized as the Natural Resources Conservation Service, and the Soil Quality Institute (http://soils.usda.gov/sqi/) was created within NRCS "to cooperate with partners in the development, acquisition and dissemination of soil quality information and technology to help people conserve and sustain our natural resources and the environment."

26.2.3 Agricultural Productivity and Sustainability

To many in agriculture and agricultural research, productivity is still the primary indicator of "soil quality" because it is the net result of soil degradation processes and soil conservation practices (Parr et al., 1990). Worldwide, agriculture is the most extensive human land use, amounting to ~35% of the land area on Earth (Foley et al., 2005; Ramankutty et al., 2008; Figure 26.1) and including irrigated and rain fed cultivated cropland, permanent crops, such as orchards and vineyards, irrigated pasture, range, and managed forests. Each cropping system has distinct soil and soil management conditions for optimal production.

Since the early 1960s, the world population has doubled, and global food production has increased by 145% (Pretty, 2008). Of that increased food production, ~70% has been through intensification of agricultural yields rather than through expansion of production area (MEA, 2005; Pretty, 2008). At the same time, increasing meat consumption has led to increasing intensity of livestock production through use of inexpensive grains for feed in concentrated animal feeding operations (CAFOs) (Tilman et al., 2002; Thorne, 2007; Pretty, 2008). In industrialized countries, ~70% of cereal products are now used for animal feed (Pretty, 2008). Intensification of livestock production at CAFOs has raised concerns about sustainability given resources used for feed, increased disease incidence related to these operations, and pollution due to concentrated animal waste (Thorne, 2007). Moreover, long-term yield declines in high-input cereal cropping systems suggest slow degradation of highly productive soils (Cassman, 1999; Cassman et al., 2003).

Sustaining food production for the world's growing population will require careful evaluation of the relationship between soil quality (degradation) and yield (productivity) (Cassman, 1999; Cassman et al., 2003) (Figure 26.5a). It has recently been argued that at the global scale, loss of high-productivity land to degradation, biofuel production, or urbanization could lead to widespread agricultural expansion into marginal lands with lower yields and higher risk of degradation (Rockstrom et al., 2009b). Evaluating soil quality vs. yield is part of optimizing management practices to sustain increases in food production while optimizing efficient use of inputs (N, water) and C sequestration, and minimizing greenhouse gas emissions (Cassman et al., 2003) (Figure 26.5b).

The resources available to farmers dictate strategies for increasing agricultural productivity and the ability to manage soil quality for agriculture (Cassman, 1999). In the United States, resources may be readily available for management of dynamic soil properties such as nutrient or water status. In other countries, farmers may be resource poor, and agricultural systems are generally low input, meaning that large-scale irrigation is absent, use of fertilizers, pesticides, and herbicides is minimal, and high energy, mechanized equipment is not available (Eswaran et al., 1997). Under these circumstances, soil quality for agriculture will depend more on climate than if the same soils were part of a highly managed, irrigated system (Eswaran et al., 1997). Increases in agricultural production must come primarily through increases in production area rather than yield. In sub-Saharan Africa, for example, two thirds of increased production in the last 40 years has come from agricultural expansion (MEA, 2005). In low-input agriculture, sustainability is more dependent on maintenance of inherent soil fertility and other attributes, because resources may not exist to remedy soil degradation (Sanchez et al., 2003).

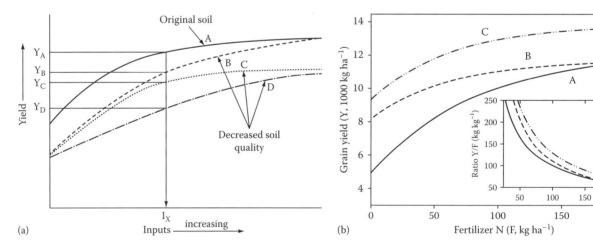

FIGURE 26.5 (a) Conceptual model illustrating the relationship between crop yields and input requirements as influenced by soil quality. (From Cassman, K.G. 1999. Ecological intensification of cereal production systems: Yield potential, soil quality, and precision agriculture. Proc. Natl. Acad. Sci. 96:5952–5959.) A decrease in soil quality from an initial state (curve A) can result in the need for greater inputs of energy, nutrients, water, seed, and pest control measures to achieve the same yield. The slope and asymptote of the shifted response (shown by curves B, C, and D) depend on the type of soil degradation and can result in a reduction in input use efficiency, yield potential, or both. (b) Hypothetical relationship between maize yield (Y) and the N application rate (F) for average soil quality and average yield (curve A), average yield and increased soil organic matter content and associated indigenous N supply (curve B), and increased soil organic matter content and indigenous N supply with improved crop management to achieve greater N fertilizer efficiency at all rates of applied N (curve C). Scenarios B and C assume an increase of 50 kg N ha^{-1} in indigenous N supply from the increase in soil organic matter. Insert shows the overall N use efficiency (Y/F) for each scenario. (From Cassman, K.G., A. Dobermann, D.T. Walters, and H. Yang. 2003. Meeting cereal demand while protecting natural resources and improving environmental quality. Annu. Rev. Environ. Resour. 28:315–358.)

The ability to irrigate also has a profound impact on management of soil resources for food production. One-third of the total global harvest of food comes from the 17% (255 million ha) of the world's cropland that is irrigated (Hoffman et al., 1990; Gleick, 1993; Doll and Siebert, 2002), three quarters of which is in developing countries (Tribe, 1994; Doll and Siebert, 2002; FAO, 2008). The area of cropland that is irrigated doubled between 1961 and 2002 (Pretty, 2008), with the greatest area of irrigated land in India, China, the United States and Pakistan (Vorosmarty et al., 2000; Doll and Siebert, 2002; FAO, 2008; Wisser et al., 2008). Sojka (1996) suggested that the arid and semiarid soils that support most irrigated agriculture have thin erodible surfaces, characteristics that could classify such soils as having poor quality. Yet under irrigation, they feed much of the world. Without irrigation, for example, in many African soils, moisture stress becomes a significant factor limiting production, and the water holding capacity of a soil becomes crucial (Eswaran et al., 1997; Cassman et al., 2003; Sanchez et al., 2003).

These practices associated with cultivation dictate regionally specific soil management strategies. Yet soils that are not cultivated are a much larger component of agriculture, broadly defined, than those that are cultivated. About 65% of the land in the United States is forest (284 million ha) or rangeland (312 million ha), with only about 284 million ha cultivated (NRC, 1994). Herrick and Whitford (1995) suggested that rangeland soils, which often serve multiple uses, present unique challenges and opportunities for assessing soil quality because spatial and temporal variability are higher than in cropped systems. On rangelands and forestlands, food, fiber, timber production, biomass for fuel, wildlife, biodiversity, recreation, and water supply are all potential uses that may have diverse criteria for quality soils.

In the past, many of the arguments about the sustainability of agricultural systems have related to the form in which nutrients are most sustainably returned to soils. Increasing pressure on soil resources globally has made it clear that no agricultural system will be sustainable in the long run without management that considers larger scale nutrient cycling, water use, and energy budgets (Tilman et al., 2002; Cassman et al., 2003). Increasing yields typically require more energy and resources, although increased efficiencies of water, fertilizer, and other input use are possible (Tilman et al., 2002; Pretty, 2008). Yet the sustainability of tradeoffs inherent in all decisions about soil and land management must ultimately be evaluated with respect to larger human benefits including opportunity, security, and health (McMichael et al., 2003). Soil fertility and resilience are among ecosystem services that profoundly benefit all of human society. Excessive human exploitation of ecosystem services, including soil quality undermines the capacity of the biosphere to both sustain food production and provide a continued flow of other goods and services essential to long-term human survival, including maintenance of freshwater resources, regulation of climate and air quality, and amelioration of infectious diseases (McMichael et al., 2003; Foley et al., 2005). It has been argued that in the interest of balancing the regulatory and provisioning functions of soils globally, a limit on cultivated land area must be set at 15% (cultivated area is now 12%; Rockstrom et al., 2009b). As we consider global limits on food production, the management of finite soil resources for adequate yields and sustained soil quality is clearly critical (Cassman et al., 2003).

Thus, while our need for sustainable soil management only increases with time, in some respects, the lesson of the soil quality debate has been that an increasingly urban population (Pretty, 2008) is increasingly ignorant of soil conservation practices known since the beginning of agriculture. Our collective ability to practice sustainable soil management diminishes in the face of exploitative agricultural capabilities and the culinary and energy demands of urban society. The "soil quality" concept emerged in wealthy, industrialized countries following the rise of organic farming and the sustainable agriculture movement, which has largely catered to urban populations looking to remedy the real and perceived environmental ills of modern agriculture through purchasing choices. The underlying issues of true sustainability and widespread soil degradation are profound and of global importance (Rockstrom et al., 2009b, Sanchez et al., 2009). The application of "soil quality" evaluation to rangeland management in the western United States (summarized below) is evidence that all land and ecosystem management choices rely on sustainable soil management. Yet the specifics of formalized "soil quality" evaluation have remained parochial (Sanchez et al., 2003; Palm et al., 2007). There is a real and pressing need to apply the strongest aspects of the soil quality discussion to the goal of sustainable soil, land, and ecosystem management (Cassman et al., 2003; Foley et al., 2005; Palm et al., 2007).

26.2.4 Rangeland Health

Soil quality is a component of rangeland health (NRC, 1994; Herrick et al., 2002; Pyke et al., 2002; Herrick et al., 2005a, 2005b; Pellant et al., 2005). In 1994, the NRC defined rangeland health as "…the degree to which the integrity of the soil and the ecological processes of rangeland ecosystems are sustained," and recommended a coordinated national monitoring effort by the USDA, DOI, and EPA. Soil integrity was defined primarily in terms of susceptibility to erosion, as "the capacity of the site to limit redistribution and loss of soil resources (including nutrients and organic matter) by wind and water" (USGS, 2002). Herrick et al. (2002) recommended an integrated approach for soil quality and rangeland monitoring. They proposed site-specific indicators closely tied to resource concerns, generally soil and site stability, hydrologic function, and biologic integrity. They emphasized an understanding of spatial variation in relating indicators to ecological functions as well as nonlinearity of indicator response and threshold effects in monitoring and managing ecological systems. This approach is developed in subsequent interagency guidelines for assessing rangeland health in the United States (Herrick et al., 2005a, 2005b; Pellant et al., 2005). The same fundamental priorities of preventing erosion and maintaining site productivity are evident in the assessment of soil quality indicators in Iranian rangelands (Rezaei et al., 2006).

26.2.5 Soil Degradation and Resilience

Soil degradation has been defined as "a reduction in soil quality as a result of human activities" (Cassman, 1999) and as "the adverse changes in soil properties and processes leading to a reduction in ecosystem services" (Palm et al., 2007). The increased provisioning of food, fuel, and fiber realized over the past 40 years has resulted in the degradation of soils and the ecosystem services they provide (Palm et al., 2007). Salinization causes the loss of ~1.5 million ha of arable land per year, and up to 40% of global croplands—mostly in developing countries—may also be experiencing some degree of degradation due to erosion, reduced fertility, or overgrazing (Cassman, 1999; Foley et al., 2005). Erosion continues to be a primary adverse human impact on soils (Pimentel et al., 1995; Sojka et al., 2003; Montgomery, 2007), affecting over 80% of degraded soils (Cassman, 1999). Moreover, downstream effects of soil erosion and increased fertilizer use can be profound (Pimentel et al., 1995; EEA, 1999, 2005; Foley et al., 2005).

Willis and Evans (1977) had the problem of soil erosion in mind when they argued that soil is not renewable over the short term based on studies that suggest that 30 to more than 1000 years are required to develop 25 mm of surface soil from parent material by natural processes. Jenny (1980) also argued that soil is not renewable over the time scale to which humans relate. Yet under less extreme circumstances of soil degradation such as contamination or loss of fertility, some soil functions can be restored even if the original, natural state of the soil cannot be entirely recreated. For agriculture, some important (dynamic) soil characteristics are slowly renewable through management practices. Organic matter, most nutrients, and some physical properties may be renewed through careful long-term management. Certain chemical properties (pH; salinity; N, P, K content) may be altered to a more satisfactory range for agriculture within a growing season or two, while immobilization or bioremediation of unwanted chemicals may depend strongly on the interaction of management techniques and intrinsic soil properties.

However, the consequences of soil degradation may be most dire in the tropics, where resources for soil rehabilitation are limited and primary concerns include food insecurity and rural poverty tied to inadequate nutrient inputs to maintain soil fertility (Sanchez, 2002; Sanchez et al., 2003; Virtuosic et al., 2009). The soil quality concept was developed in temperate regions to address environmental issues arising from agricultural systems with large nutrient and energy inputs (Sanchez et al., 2003). While the holistic approach implicit in debate about the soil quality concept is relevant to all regions, work in the tropics has shown that useful assessment of sustainable soil management and rehabilitation practices requires informed simplification of goals and prioritization of measured attributes (Sanchez et al., 2003). Arguably, what is needed for better global management of soil resources is improved global mapping of soil taxa, function, and degradation (Tugel et al., 2005; Sanchez et al., 2009).

The capacity of soil to recover its structural and functional integrity after a disturbance has been called soil "resilience"

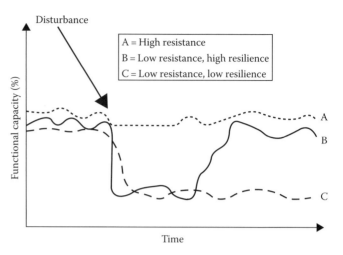

FIGURE 26.6 Soil function, disturbance, resistance, and resilience. (From Tugel, A., J.E. Herrick, J.R. Brown, M.J. Mausbach, W. Puckett, and K. Hipple. 2005. Soil change, soil survey, and natural resources decision making: A blueprint for action. Soil Sci. Soc. Am. J. 69:738–747.)

(Blum, 1994; Seybold et al., 1999; Figure 26.6). It has been argued that the stability and resilience of soils upon disturbance is not well understood, and, hence, that soil degradation is poorly characterized from regional to global scales (Palm et al., 2007). In ecology, the concept of ecosystem resilience was developed starting in the 1970s, leading to the idea of alternative stability states and "catastrophic shifts" in major ecosystems as a result of climate change or other perturbation (Scheffer et al., 2001). An example is an irreversible change in a plant community resulting from soil degradation due to grazing (van de Koppel et al., 1997). Key points in this thinking are that disturbance is a natural component of ecosystems that promotes diversity and renewal, and that soil properties are a primary factor in ecosystem resilience against disturbance (Scheffer et al., 2001; Folke et al., 2004; Rockstrom et al., 2009b). Hence, soil functions and land use are a major component of successful ecosystem management to maintain ecosystem services critical to human survival (Folke et al., 2004; Foley et al., 2005; Ellis and Ramankutty, 2008; Carpenter et al., 2009). The limits of agricultural productivity and the regulatory capacity of the Earth system must somehow be considered at a global scale in evaluating sustainable soil management choices (Rockstrom et al., 2009a, 2009b).

Buffering is a function related to soil resilience. The term "buffer" refers collectively to processes that constrain shifts in the dissolved concentration of any ion when it is added to or removed from the soil system (Singer and Munns, 2006). Soils "buffer" nutrients as well as contaminants and other solutes, via sorption to or incorporation into clay and organic materials. The extent to which a soil immobilizes or chemically alters substances that are toxic, thus effectively detoxifying them, reflects "quality" in the sense that humans or other biological components of the system are protected from harm. This is the basis for the concept of soil quality developed in Europe (Moen, 1988; Siegrist, 1989; Denneman and Robberse, 1990). Lack of soil buffering results in direct toxicity or contamination of air or water.

Identifying substances that qualify as "contaminants" can be challenging because some, such as nitrates and phosphates, are important plant nutrients as well as potential water pollutants. An example is agricultural runoff containing nitrate or soluble phosphorus (Yli-Halla et al., 1995).

Holden and Firestone (1997) defined soil quality in this context as "the degree to which the physical, chemical, and biological characteristics of the soil serve to attenuate environmental pollution." Howard (1993) defined the ecological risk of a chemical in the environment as "the probability that a random species in a large community is exposed to a concentration of the chemical greater than its no-effect level." The extent to which a soil is capable of reducing the probability of exposure is a measure of its quality and hence in some sense a component of its "resilience."

A well-studied example of a common soil contaminant is lead (Pb) (McBride et al., 1997). Although legislated limits may be on a concentration basis in soil (e.g., $500 \mu g\ kg^{-1}$), risk assessment techniques have attempted to account for the chemical form of Pb present, as well as the observed relative relationship between the amount of Pb present in soil and blood levels in local residents (Bowers and Gauthier, 1994). Critics have questioned analytical techniques used to determine bioavailable levels of Pb in soil as well as the degree to which toxicity data account for its chemical fate and ecologically damaging properties (Cook and Hendershot, 1996). Natural variability of soils and variation within a soil series make average values or average background values inadequate for soil quality assessments. In addition, bioaccumulation and toxicity need to be considered when establishing levels of toxicants that may not be exceeded in a "high-quality" soil for a given use (Jongbloed et al. 1996; Traas, et al. 1996). Speir et al. (1996) proposed an "ecological dose value" that represents the inhibitory effects of a heavy metal (in this case Cr(VI)), based on the kinetics of soil biological properties, and serves as a generic index for the determination of permissible concentration levels for heavy metals in soils. Lair et al. (2007a, 2007b) showed that long-term management of soil organic matter in Sweden led to differential sorption of metals among treatments (e.g., manure vs. chemical fertilizer).

26.2.6 Reference States for Soil Quality

Ultimately, any assessment of soil quality must be made relative to desired functional outcomes or management goals. Granatstein and Bezdicek (1992) considered whether the standard reference state of a soil in agricultural systems should be "native" soil conditions or conditions given maximum agronomic, environmental, and economic performance. Seybold et al. (1997) considered three potential baseline values (reference points) for soil quality: the native, the intensively managed, and the altered (reclaimed) states. Given that native conditions may be unknown or difficult to determine, standard conditions for agricultural systems could be based on high-yielding, low environmental impact systems using progressive management practices that do not degrade soils over the longer term. This designation requires careful evaluation of yield potential vs. soil quality (Cassman et al., 2003; Figure 26.3). Natural systems may be compared to observed baseline or potential states, considering previous human influences or other disturbance (Habich, 2001).

Historical analyses of land use and soil property trends are a means of assessing soil degradation relative to some initial state (Lindert et al., 1996a, 1996b). Studies comparing uncultivated forest and range soils with cultivated soils also provide a measure of the effects of cultivation on soil quality (e.g., Velasquez et al., 2007). A number of studies have found lower organic C and N in cultivated vs. uncultivated soils (e.g., Amundson, 2001; Miller et al., 2004; Golchin and Asgari, 2008) depending on management practices (Olson, 2007). Others have observed increases in soil organic C and N following conversion from agricultural management back to uncultivated land (Poulton et al., 2003; McLauchlan, 2006). These studies have important implications for climate regulation and land management strategies for global sustainability (Rockstrom et al., 2009b).

Baseline values of soil properties have been determined using long-term research plots in many parts of the world (Rasmussen et al., 1998; Richter et al., 2007), including Canada (Beke et al., 1994; Bremer et al., 1994; Biederbeck et al., 1996; Zentner et al., 2001), China (Wu et al., 2004; Ding et al., 2007), Denmark (Schjønning et al., 1994, 2005; Munkholm et al., 2002), England (Johnston, et al., 1986; Jenkinson et al., 1994), Germany (Merker, 1956), India (Masto et al., 2008), New Zealand (Murata et al., 1995; Lilburne et al., 2002), Nigeria (Oluwatosin et al., 2008), Sweden (Gerzabek et al., 1997; Gerzabek et al., 2001; Kirchmann et al., 2004; Gerzabek et al., 2006), Switzerland (Fließbach et al., 2007; Birkhofer et al., 2008), and the United States (Odell et al., 1984; Varvel et al., 2006; Khan et al., 2007). These studies have shown how management of inputs, crop rotation, and tillage practices can change the physical, chemical, and biological properties of soil. In considering soil management strategies for agriculture globally, these studies and others provide important constraints on climate regulation and potential productivity vs. degradation of soil resources (Cassman et al., 2003; Sanchez et al., 2009; Rockstrom et al., 2009b).

26.2.7 Global Climate and Energy Production

As the potential climate effects of rising atmospheric CO_2 gained public attention in the 1990s, the potential for carbon sequestration and hence climate mediation by soils through agricultural management practices was recognized (Schlesinger, 1999; Amundson, 2001; Lal, 2004). This concept dovetailed with ideas presented here (Table 26.3) of soil quality based on organic matter content and fertility, facilitated by management practices favored for sustainable agriculture, such as manuring, crop rotation, and reduced tillage (Tilman et al., 2002; Cassman et al., 2003; West and Post, 2002). The potential for carbon sequestration in soils has motivated the renewed examination of the effects of agricultural management practices on soil organic carbon (e.g., West and Post, 2002; Liu et al., 2006),

TABLE 26.3 Soil Quality, Management, and Climate

Soil Functions/Ecosystem Services[a]	Agricultural and Human Health Management Goals[b]	Influence of Climate Change[c]
Food and fiber production	Increasing yields	Shifting yields: increase in currently colder regions, decrease in mid-latitudes
Nutrient cycling Atmospheric composition and climate control	Increasing nutrient use efficiency (precision agriculture) Maintaining and restoring soil fertility	Positive feedback from GHG production by agricultural practices; mitigation by changing land management practices to sequester C, reduce N_2O and CH_4 emissions
Water relations Water quality and supply	Increasing water use efficiency	Drought effects in currently productive regions, e.g., Central Valley of California
Physical stability and support	Increasing yields Erosion control	Damage to crops and soil erosion due to increased intensity of storm events, rising sea level, continuation and/or expansion of conventional tillage
Filtering and buffering Pollutant attenuation and degradation Resistance and resilience Biodiversity and habitat	Reduce exposure to pollutants Disease and pest control	Change external variables influencing soil resilience

[a] Andrews et al. (2004); Kibblewhite et al. (2008).
[b] Tilman et al. (2002); Montgomery (2007); Cassman (1999); Cassman et al. (2003).
[c] IPCC (2007).

with additional awareness of global N dynamics (e.g., Gerzabek et al., 2001; Neff et al., 2002). However, efforts to quantify agricultural C sequestration potential face methodological issues associated with the use of average yields and average management practices in field experiments, as well as underestimates of biomass turnover that have kept models from accurately predicting the C sequestration effects resulting from changes in crop management and climate (Cassman et al., 2003). It has been argued that the finite and uncertain potential for carbon sequestration by agriculture is unlikely to significantly offset fossil fuel consumption in the developed world (Lal, 2008), and carbon sequestration potential may itself be offset by the associated production of greenhouse gases, such as N_2O (Cassman et al., 2003). Substantial uncertainties remain about the design of optimal management practices to sustain increases in food production while optimizing N use efficiency and C sequestration, and minimizing greenhouse gas emissions; arguably, resolution of these uncertainties will require improved field studies at production scale, using progressive management practices (Cassman et al., 2003).

Nonetheless, soil management practices for both increased carbon storage and reduced greenhouse gas production (CO_2, CH_4, N_2O)—with increased yields through nutrient and water use efficiency—have been proposed for the mitigation of global climate change effects, along with the development of energy crops to replace fossil fuels (CAST, 2004; Hutchinson et al., 2007; IPCC, 2007). Debate about the relative merits of various candidate biofuel crops has led to comprehensive analysis of the environmental and health costs of biofuel production and use (e.g., Hill et al., 2006; Donner and Kucharik, 2008; Landis et al., 2008; Moebius-Clune et al., 2008; Schmer et al., 2008; Hill et al., 2009; Melillo et al., 2009; Tilman et al., 2009). Arguably, an approach focusing on carbon and energy balances provides a more complete evaluation of the environmental costs and benefits of agriculture (Cassman et al., 2003), yet agricultural policies to address concerns raised by this approach are still lacking in the United States and elsewhere (Robertson et al., 2004; Robertson and Swinton, 2005; Lant et al., 2008), and true sustainability remains elusive (Tilman et al., 2002; Cassman et al., 2003; McMichael et al., 2003; Pretty, 2008: Rockstrom et al., 2009a).

Development of biofuels is just one aspect of soil- and land-based energy production receiving increased attention in the current energy debate (e.g., Tilman et al., 2009), along with other sources such as wind and solar power. These efforts to move energy production from subsurface resource extraction to highly visible and spatially extensive solar panels, turbines, and energy crops at the earth's surface will increase pressure on soil resources and ecosystems both directly and indirectly through changing land use patterns (Melillo et al., 2009). These pressures are likely to further test our ability to sustainably manage soil resources.

26.3 Evaluating Soil Quality

The definition of soil quality and the selection of soil characteristics needed to evaluate soil quality continue to evolve (Idowu et al., 2008; Kibblewhite et al., 2008). Bouma (1989a) recognized that an essential problem with definitions that produce carefully limited suitability classes is that empirical decisions must be made to separate the classes along what is essentially a continuum. Nonetheless, evaluation of soil conditions remains a critical exercise globally (Sanchez et al., 2009).

26.3.1 Selecting Soil Quality Indicators

Carter et al. (1997) suggested a framework for evaluating soil quality that includes (1) describing each soil function on which quality is to be based, (2) selecting soil characteristics or properties that influence the capacity of the soil to provide each function, (3) choosing indicators of characteristics that can be measured, and (4) using methods that provide accurate measurement of those indicators.

No soil property is permanent, but rates and frequency of change vary widely among properties. Soil properties also vary with ecosystems, arguably depending mostly on climate. In rangelands, for example, temporal variability is high and relatively unpredictable due to the strong dependence of soil properties on soil moisture (Herrick and Whitford, 1995).

Arnold et al. (1990) suggested that changes in soil properties can be nonsystematic, periodic, or trend. Carter et al. (1997) distinguished between dynamic soil properties that are most subject to change through human use and are strongly influenced by agronomic practices, and intrinsic or static properties that are not subject to rapid change or management.

Examples of dynamic soil characteristics are the size, membership, distribution, and activity of a soil's microbiological community; the soil solution composition, pH, and nutrient ion concentrations, and the exchangeable cation population. Soils respond quickly to changes in conditions such as water content. As a result, the optimal frequency and distribution of soil measurements vary with the property being measured.

Soil mineralogy, particle size distribution, and soil depth are static soil quality indicators. Although changes occur continuously, they are slow under natural conditions. Other soil physical properties such as aggregate stability and hydraulic conductivity are spatially and temporally highly variable (Moebius et al., 2007). Organic matter content may be a dynamic variable, but the chemical properties of organic matter may change only over periods on the order of 100–1000 years depending on texture.

Soil properties that change quickly present a problem because many measurements are needed to determine an average value and to evaluate whether changes in the average indicate improvement or degradation of soil quality. Conversely, properties that change very slowly are insensitive measures of short-term changes in soil quality. Papendick et al. (1995) argued that the minimum data set (MDS) required for soil quality analysis includes a mix of "dynamic" and relatively "static" properties.

A soil quality assessment must specify area. One could use the pedon (the three-dimensional soil individual) as the unit of measure, or a soil map unit component, a landscape, a field, or an entire watershed. The choice will depend to some degree on what property is of interest and spatial variability of the property. Karlen et al. (1997) proposed that soil quality can be evaluated at a wide range of scales, while Pennock et al. (1994) discussed scaling up data from discrete sampling points to landscape and regional scales.

One example of investigating spatial variability of soil quality is the study by Cambardella et al. (2004). They investigated the effect of topographic position on soil quality changes produced by different tillage practices in three small Iowa watersheds using terrain analysis methods. They found differences in soil properties used to assess soil quality among the different landform elements and found differences in soil quality among the three experimental watersheds. In their analysis, soil erosion played a major role in reducing soil quality on steeper slopes in the three watersheds. Karlen et al. (2008)—also working at the watershed scale in Iowa—used a number of tools to examine soil quality and to consider differences in soil quality at different landscape positions. They found that a soil quality index (SQI) was helpful in assessing the potential for soil degradation and for making management decisions.

Sparling and Schipper (2002) evaluated soil quality as part of an environmental assessment on a national scale for New Zealand using 222 sites in five regions. As in many studies, only 10 cm of the surface were sampled and analyzed. The chemical, physical, and biological properties of the soil over 200 sites were measured and then statistically analyzed to determine the relationships between land use and soil quality indicators at this location. Variability was high, as might be expected for the range of soil types, landscapes, and land uses in the survey. This demonstrates some of the difficulty of extrapolating soil quality concepts from a pedon to larger spatial scales. It also indicates the value of evaluating the soil quality for a "whole soil," which means sampling throughout the profile and not just the top 10 cm.

26.3.2 Examples of Soil Quality Indicators

Soil quality is an attribute of a soil that is inferred from soil characteristics or indirect observations. To proceed from a definition to a measure of soil quality, an MDS of soil characteristics that represents soil quality must be selected and quantified (Larson and Pierce, 1991, 1994; Gregorich et al, 1994; Papendick et al., 1995; Karlen et al., 2001; Idowu et al., 2008; Rodrigues de Lima et al., 2008). The MDS may include biological, chemical, or physical soil characteristics (Table 26.4). For agriculture, the measurement of properties should lead to a relatively simple and accurate way to rank soils based on potential plant production without soil degradation or excessively adverse environmental effects. Unfortunately, commonly identified soil quality parameters may not correlate well with yield (Reganold, 1988).

Four points are relevant to the selection and quantification of soil characteristics: (1) soil characteristics may be desirable or undesirable, (2) soil renewability involves judgment of the extent to which soil characteristics can be controlled or managed, (3) rates of change in soil characteristics vary, and (4) there may be significant temporal or spatial variation in soil characteristics.

26.3.2.1 Physical Indicators

The physical characteristics of soil (Table 26.4) are a necessary part of soil quality assessment because they often cannot be easily improved (Karlen and Cambardella, 1996; Wagenet and Hutson, 1997; Idowu et al., 2008). Larson and Pierce (1991) summarized the physical indicators of soil quality as those properties that influence crop production by determining: (1) whether a soil can accommodate unobstructed root growth and provide

TABLE 26.4 Commonly Used Indicators of Soil Quality

Physical Indicators	Chemical Indicators	Biological Indicators
Aeration	Percent base saturation	Organic carbon content
Aggregates	Cation-exchange capacity and exchangeable acidity	Microbial biomass
Type (structure)		Total
Stability		Bacterial
Size distribution		Fungal
Bulk density	Contaminants	Total C and N content[a]
	Types (Zn, Cu, Pb)	
	Availability	
	Concentrations	
	Mobility	
Clay mineralogy	Salinity (electrical conductivity)	Biomass C/total organic carbon
Color	Sodicity (ESP or SAR)	Potentially mineralizable N
Consistence (dry, moist, wet)	pH	Earthworm population
Thickness	Nutrients	Enzymes
Depth-to-root limiting layer	Content	Dehydrogenase
Profile thickness	Availability	Phosphatase
Surface layer thickness	Cycling rates	Arlysulfatase
Hydraulic conductivity	Total C and N content[a]	Nematode population
		Beneficial
		Parasitic
Oxygen diffusion rate		Substrate utilization
Particle size distribution		Microbial community
		Composition
		Size (biomass)
		Distribution
		Respiration
Soil strength (penetration resistance)		Fatty acid composition
Pores		Nucleic acid composition
Total porosity		
Pore size distribution		
Macro porosity		
Meso porosity		
Micro porosity		
Connectivity		
Soil tilth		Weed seed bank
Temperature		Glomalin content
Available water-holding capacity and water retention at wilting point		

[a] Carbon is sometimes considered a biological indicator, and a chemical indicator of soil quality.

pore space of sufficient size and continuity for root penetration and expansion, (2) the extent to which the soil matrix will resist deformation, and (3) the capacity of soil for water supply and aeration. Factors such as effective rooting depth, porosity or pore size distribution, bulk density, hydraulic conductivity, soil strength, and particle size distribution capture these soil functions. However, many of the processes that contribute to soil structure, aggregate stability, bulk density, and porosity are not well understood, making them difficult to manage. Proposed suites of physical parameters for soil quality evaluation are given in Letey (1985), Karlen et al. (1992), and Reganold and Palmer (1995). An example of regional and crop-dependent variation is evident in the physical parameters proposed by Hulugalle and Scott (2008) for sustainable cotton growing in Australia.

Soil depth is a good example of an easily measured and independent property that provides direct information about a soil's ability to support plants. *Effective* soil depth is the depth available for roots to explore for water and nutrients. Effective soil depth is a problem for agricultural use of over 50% of soils in Africa (Eswaran et al., 1997). Soil depth requirements vary with crop or species. Many vegetable crops, for example, are notably shallow rooted while grain crops and some legumes, like alfalfa, are deep rooted. Variation will be even greater in unmanaged, natural systems. Wheat yield in Colorado was shown to decrease from 2700 to 1150 kg ha^{-1} over a 60 year period of cultivation primarily due to decrease in soil depth (Bowman et al., 1990).

Letey (1985) suggested that identification of a range of water content that is nonlimiting to plant productivity might be a good way of assessing the collective effect of soil's physical characteristics that contribute to crop productivity. For soils of decreasing quality, the width of the nonlimiting water range decreases. This and many of the physical parameters are highly variable over time and space (Moebius et al., 2007), thus making them difficult properties to assess when defining soil quality.

26.3.2.2 Chemical Indicators

Components of soil quality definitions may include desirable and undesirable chemical characteristics (Table 26.4). Desirable soil characteristics may either be the presence of a property that benefits soil productivity and/or other important soil functions, or the absence of a property that is detrimental to these functions. A soil characteristic may include a range of values that contributes positively to quality and a range that contributes negatively. Soil pH, for example, may be a positive or negative characteristic depending on its value. Larson and Pierce (1991) suggested that ranges of property values can be defined as optimal, suboptimal, or superoptimal. A pH range of 6–7.5 is optimal for the production of most crops, but the choice of optimal pH range can be crop dependent.

Undesirable soil characteristics may be either the presence of contaminants or a range of values of soil characteristics that contribute negatively to soil function. The presence of chemicals that inhibit plant root growth or the absence of nutrients that result in low yields or poor crop quality are examples of undesirable soil characteristics that lower soil quality.

Assessment of soil quality based on soil chemistry, whether the property is a contaminant or part of a healthy system, requires a sampling protocol, a method of chemical analysis, an understanding of how its chemistry affects biological systems and interacts with mineral forms, methods for location of possible contamination, and standards for soil characterization.

Nutrient availability depends on the physical and chemical processes of soil, such as weathering and buffering, and properties such as organic matter content, CEC, and pH. At low and high pH, for example, some nutrients become unavailable to plants and some toxic elements become more available. Larson and Pierce (1991) chose those chemical properties that either inhibit root growth or that affect nutrient supply due to the quantity present or the availability. Reganold and Palmer (1995) used chemical parameters related to nutrient availability as measures of soil quality, while Karlen et al. (1992) suggested that total and available plant nutrients and nutrient cycling rates should be included in soil quality assessments.

Soil properties may be severely compromised by intended or unintended human additions of chemical compounds, and soil productivity reduces if unwanted chemicals exceed safe thresholds. Data are required to determine whether a site is significantly polluted and if it requires cleanup (Beck et al., 1995; Sims et al., 1997). International standard methods have been created to maintain the quality of measurements (Hortensius and Welling, 1996). A difficult determination is the level of each chemical that is considered an ecological risk. Beck et al. (1995) provided a list of levels for organic chemicals adopted by the Netherlands and Canada. Sims et al. (1997) argued that clean and unclean are two extremes of a continuum and that it is more appropriate to define the physical, chemical, and biological states of the soil as acceptable or unacceptable.

In the Netherlands, soil quality reference values have been created for heavy metals and organic chemicals based on a linear relationship with soil clay and organic matter content. The Dutch Ministry of Housing, Physical Planning, and Environment has used the maximum of a range of reference values for a given substance as a provisional reference value for good soil quality (de Bruijn and de Walle, 1988; Howard, 1993).

26.3.2.3 Biological Indicators

The focus of many soil quality definitions is soil biology, including microbial and biochemical soil properties (Table 26.4) (Filip, 2002; Schloter et al., 2003; Bending et al., 2004; Tian and Feng, 2008). Early work by Waksman (1927), who studied measurements of soil microorganisms that could indicate soil fertility, demonstrated that physical and chemical factors as well as soil biology were needed to predict soil fertility.

Soil microbiological properties are dynamic, highly responsive to management practices, and often difficult to measure or effectively evaluate. Soil quality evaluations have commonly included bulk measures of microbial activity such as microbial biomass, microbial respiration, and organic matter mineralization and denitrification (Karlen et al., 1992; Visser and Parkinson, 1992; Reganold et al., 1993; Franzluebbers et al., 1995; Yakovchenko et al., 1996; Boehn and Anderson, 1997; Franzluebbers and Arshad, 1997); or soil microbial C, phospholipid analyses, and soil enzymes (Franzluebbers et al., 1995; Jordan, et al., 1995; Gregorich et al., 1997).

Microbial communities are highly sensitive to changes in soil conditions (Kennedy and Papendick, 1995). Visser and Parkinson (1992) suggested evaluating microbiological criteria for soil quality at three levels; population, community, and ecosystem. Stork and Eggleton (1992) discussed species richness as a powerful but elusive indicator of invertebrate community and soil quality. Meso- and macrofauna populations have also been considered as part of soil quality evaluations (Berry, 1994; Reganold and Palmer, 1995; Hulugalle and Scott, 2008; van Eekeren et al., 2008).

More recently, Mele and Crowley (2008) investigated the use of statistical tools to make use of complex soil biology data sets. They noted that the biological properties of soil can range from single variables to complex multiparametric data sets, and reviewed an assortment of biological indicators and biological, structural, and functional indicators that have been used as soil quality parameters. They proposed the development of "ecologically derived algorithms that measure microbial community stability, richness, diversity, and functional redundancy resilience," acknowledging that this is an "outstanding challenge."

Total organic C and N are common, easily measured biochemical parameters in soil quality evaluations (Karlen et al., 1992; Franzluebbers et al., 1995; Franco-Vizcaíno, 1997; Gregorich et al., 1997; Sanchez et al., 2003). Measurement of one or more components of the N cycle including ammonification, nitrification, and nitrogen fixation may be used to assess soil fertility (Visser and Parkinson, 1992). Bending et al. (2004) argued that a wide variety of biochemical and microbial analyses should be used when comparing impacts of management on soil quality.

26.3.3 Approaches for Evaluating Soil Quality

Methods for evaluating soil quality are both qualitative and semiquantitative. Using more than one quantitative variable requires some system for combining the measurements into a useful index (Halvorson et al., 1996). This is one of the unmet challenges of creating a useful SQI. The region, crop, or general soil use for which an index was created will likely limit its effectiveness outside the scope of its intended application. Even an index designed only to rate productivity is not likely to be useful for all crops and soils, leading Gersmehl and Brown (1990) to advocate regionally targeted systems.

Rice is a good example of a crop requiring significantly different soil properties than other crops. It is a food staple for a large proportion of the world population. Approximately 146 million ha were in rice production in 1989 (FAO, 1989) mainly (90%) in Asia. High-quality soils for rice production may be poor quality for most other irrigated and dryland crops because they may be saline or sodic and high in clay with slow infiltration and permeability. These physical and chemical properties often constrain the production of other crops. Although they are not reviewed here, various land suitability classifications, specifically for rice, have been developed since the turn of the century (Dent, 1978).

Examples of several soil quality indexing systems are presented in the following sections. To some extent, recent attempts to enumerate the factors of soil quality resemble Jenny's (1941) introduction of the interrelated factors of soil formation.

26.3.3.1 Nonquantitative Evaluations

An index is categorized here as "nonquantitative" if it does not combine the evaluated parameters into a numerical index that rates soils along a continuous scale. Examples are the USDA Land Capability Classification (LCC) and the U.S. Bureau of Reclamation (USBR) Irrigation Suitability Classification.

The purpose of the Land Capability Classification is to place arable soils into groups based primarily on physical attributes affecting their ability to sustain common cultivated crops that do not require specialized site conditioning or treatment. Nonarable soils, unsuitable for long term, sustained cultivation are grouped according to their ability to support permanent vegetation and according to the risk of soil damage if mismanaged. Inherent fertility, organic matter, and other biological attributes are not evaluated. Several studies showed that lands of higher LCC have higher productivity than lands of lower LCC (Patterson and MacIntosh, 1976; van Vliet et al., 1979; Reganold and Singer, 1984).

The USBR Irrigation Suitability was a frequently used system of land evaluation for irrigation in the western United States during the period of rapid expansion of water delivery systems (USBR, 1953; Maletic and Hutchings, 1967; FAO, 1979; McRae and Burnham, 1981). As the area of irrigated agriculture continues to rise (Tilman et al., 2002; Pretty, 2008), and with potentially increasing water scarcity due to rising global temperatures (IPCC, 2007), the fundamental assessment of soil irrigability and management outcomes for water resources will continue to be relevant for soil quality assessment.

An emerging critique of older soil rating systems is that the true costs were not comprehensively evaluated (Robertson and Swinton, 2005). For example, the high energy and carbon inputs required to industrially produce inorganic N fertilizers using fossil fuels and the pollution effects of inefficient and excessive fertilizer application were not included in an economic analysis of management choices.

26.3.3.2 Quantitative Evaluations

We use "quantitative" to refer to systems that result in a numerical index or a descriptive category indicating quantitative assessment of key attributes. Indexing systems may be additive, multiplicative or have more complex functions. They have two important advantages over nonquantitative systems: (1) they are easier to use with GIS and other automated data retrieval and display systems, and (2) they typically provide a continuous scale of assessment. No single national system is presently in use, but several state or regional systems exist. Not all quantitative systems have functional relationships between indices and productivity. Some simply result in a numerical score that is easier to manipulate than nonnumerical rating systems.

Soil quality indices attempt to connect the multiple soil functions identified by soil quality definitions with intrinsic or quantifiable properties of soils as a means of evaluating the sustainability of management practices over time (Karlen et al., 2001; Table 26.5). Two criticisms of this approach have been that broad definitions of soil quality lead to simplistic soil indexing approaches, and that a better approach would simply evaluate individual functions using established metrics for, for example, productivity or susceptibility to erosion (Sojka et al., 2003). A second criticism is that the cost of collecting data to evaluate the complex multiple functions of soil makes such soil quality definitions impractical (Sanchez et al., 2003).

Compilation of an SQI requires an MDS of critical soil properties that are generally specific to a region and a particular land use. Examples of MDS can be found in Larson and Pierce (1991), Lilburne et al. (2002), Sparling and Schipper (2002), Rodrigues de Lima et al. (2008), and Villamil et al. (2008). Pedotransfer functions (Bouma, 1989a, 1989b) can be used to evaluate soil properties that are difficult to measure (Palm et al., 2007). Research on soil quality under different land uses and in different parts of the world has resulted in specific MDSs useful in defining and interpreting soil quality. Target values identify the magnitude of change constituting a significant effect on soil quality (e.g., Lilburne et al., 2002).

TABLE 26.5 Soil Characteristics and Soil Quality

Soil Attributes	Elements of Soil Quality—Ability to				
	Accept, Hold and Release Nutrient	Accept, Hold and Release Water	Promote Growth	Provide a Suitable Soil Biotic Habitat	Resist Degradation
Surface horizons					
Organic matter					
Total	×	×	×	×	×
Labile	×		×	×	×
Nutrient supply	×		×	×	×
Texture	×	×	×	×	
Soil depth	×	×	×		
Structure		×	×		×
pH	×	×	×	×	×
Electrical conductivity	×		×	×	×
Limiting subsoil horizon					
Texture	×	×	×	×	×
Depth	×	×	×		×
Structure		×	×		×
pH	×		×		×
Electrical conductivity	×		×	×	×

Source: Larson, W.E., and F.J. Pierce. 1991. Conservation and enhancement of soil quality, p. 175–203. *In* J. Dumanski, E. Pushparajah, M. Latham, and R.J.L. Myers (eds.) Evaluation for sustainable land management in the developing world. International Board for Soil Research and Management, Bangkok, Thailand. With Permission from the International Board for Soil Research and Management.

The *Storie Index* rating (SIR) is an early example of a numerical index of potential soil productivity. Although Storie considered the productivity of the land to be dependent on 32 soil, climate, and vegetative properties (Table 26.6), only nine properties were used in the SIR, because incorporating a greater number of factors made the system unwieldy. The nine factors, rated from 1% to 100%, are soil morphology (A), surface texture (B), slope (C), and six variables (Xi) that rate drainage class, sodicity, acidity, erosion, micro-relief, and fertility. The percentages are converted to their decimal value and multiplied together (Storie, 1932, 1964; Wier and Storie, 1936) (Equation 26.1):

$$\mathrm{SIR} = \left(A \times B \times C \times \sum_{i=1}^{6} X_i \right) \times 100 \quad (26.1)$$

Values for each factor were derived from Storie's experience of mapping and evaluating soils in California, and in soil productivity studies in cooperation with California Agricultural Experiment Station cost-efficiency projects relating to orchard crops, grapes, and cotton. In describing the SIR, Storie (1932, 1964) explicitly mentioned "soil quality." Soils that were deep had no restricting subsoil horizons, and held water well had the greatest potential for the widest range of crops. O'Geen et al. (2008) modified the Storie Index using modern mathematical techniques such as fuzzy logic to make the system less "operator dependent" and more useful with digital soils information.

The *productivity index model* (PI) was developed to evaluate soil productivity in the top 100 cm, especially with reference to potential productivity loss due to soil erosion (Neill, 1979; Kiniry, et al., 1983; Pierce et al., 1983). The PI model rates soils on the sufficiency for root growth based on potential available water storage capacity, bulk density, aeration, pH, and electrical conductivity. A value from zero to one is assigned to each property describing the importance of that parameter for root development. The product of these five index values is used to describe the fractional sufficiency of any soil layer for root development.

Pierce et al. (1983) modified the PI to include the assumption that nutrients were not limiting and that climate, management, and plant differences are constant. A number of authors found that it is useful to various degrees (Pierce et al., 1984; Gantzer and McCarty, 1987; Lindstrom et al., 1992).

Parr et al. (1992) suggested that an SQI could take the following form:

$$\mathrm{SQI} = f(\mathrm{SP, P, E, H, ER, BD, FQ, MI}) \quad (26.2)$$

where SQI is a function of soil properties (SPs), potential productivity (P), environmental factors (E), human and animal health (H), erodibility (ER), biological diversity (BD), food quality and safety (FQ), and management inputs (MI). Determination of the specific measurable indicators of each variable and the interactions among these diverse variables is a daunting task. Moreover, the mathematical method of combining these factors, as well as the resulting value that would indicate a high-quality

TABLE 26.6 Soil and Land Properties for Land Evaluation

Property Group	Individual Property
Surface conditions	Physiographic position
	Slope
	Microrelief
	Erosion
	Deposition
	External drainage, runoff
Soil physical conditions	Soil color
	Soil depth
	Soil density and porosity
	Soil permeability
	Soil texture
	Stoniness
	Soil structure
	Soil workability-consistence
	Internal drainage
	Water-holding capacity
	Plant-available water
Soil chemical conditions	Organic matter
	Nitrogen
	Reaction
	Calcium carbonate, bases
	Exchange capacity
	Salts: Cl, SO_4, Na
	Toxicities, e.g., B
	Available P
	Available K
	Minor elements, e.g., Zn, Fe
	Fertility
Mineralogical conditions	Mineralogy
Climate	Precipitation
	Temperature
	Growing season
	Winds
Vegetative cover	Natural vegetation

Source: Storie, R.E. 1964. Handbook of soil evaluation. Associated Students Store, University of California, Berkeley, CA.

soil, is not specified. The inclusion of variables BD, FQ, and MI makes this a land quality index as suggested by FAO (1997), and also raises issues of the scale of analysis for soil quality assessment. However, this holistic approach is similar to that of the Millennium Ecosystem Assessment (MEA, 2005), and it has been argued that this kind of approach is critical for the useful evaluation of soil management practices globally (Palm et al., 2007).

Larson and Pierce (1991) defined soil quality (Q) as the state of existence of soil relative to a standard or in terms of a degree of excellence. Q is defined as the sum of individual soil qualities q_i and is expressed as

$$Q = f(q_i \ldots q_n) \quad (26.3)$$

These authors did not identify the best subset of properties or their functional and quantitative relationship, but suggested that an MDS should be selected from those soil characteristics in which changes are measurable and relatively rapid (i.e., "dynamic" properties), arguing that it is more important to know about changes in soil quality (dQ) than the magnitude of Q (Larson and Pierce, 1991). Changes in soil quality are a function of changes in soil characteristics (q_i) over time (t):

$$dQ = f\left[(q_{i,t} - q_{i,t0}) \ldots (q_{n,t} - q_{n,t0})\right] \quad (26.4)$$

If dQ/dt is ≥ 0, the soil or ecosystem is improving relative to the standard at time t_0. If dQ/dt < 0, soil degradation is occurring. Time zero can be selected to meet management needs or goals. If there is a drastic change in management, time zero can be defined as prior to the change. If a longer time period of comparison is considered more appropriate, properties of an uncultivated or pristine soil could be used.

Karlen et al. (1994) developed their SQI based on a 10 year crop residue management study. This SQI is based on four soil functions: (1) accommodating water entry, (2) retaining and supplying water to plants, (3) resisting degradation, and (4) supporting plant growth. Numerous properties were measured and values normalized based on standard scoring functions. Priorities were then assigned to each value. For example, aggregate stability was given the highest weight among factors important in water entry. After normalizing, each value is then multiplied by its weighting factor (wt) and products are summed

$$SQI = q_{we}(wt) + q_{wt}(wt) + q_{rd}(wt) + q_{spg}(wt) \quad (26.5)$$

Subscripts refer to the four main functions described earlier. It should also be noted that resisting degradation (rd) and sustaining plant growth (spg) are assigned secondary and tertiary levels of properties that are themselves normalized and weighted before a final value is calculated and incorporated into Equation 26.5.

The resulting index resulted in values between 0 and 1. Of the three systems in the study, the one with the highest rate of organic matter return to the soil had the highest index value, and the soil with the lowest had the lowest value. The authors suggested that this demonstrates the usefulness of the index for monitoring the status and change in status of a soil as a function of management. They also suggested that the index and the soil characteristics that go into the index might change as the index is refined (Karlen et al., 1994).

Andrews et al. (2004) took the Karlen model a step further by creating a framework for incorporating multiple variables into an index. They recognized explicitly that soil quality varied depending on how a soil is to be used. They defined the indicators of soil quality as "those properties and processes that have greatest sensitivity to changes in soil function." They created a "Soil Management Assessment Framework (SMAF)" that utilizes a three-step framework of indicator selection, indicator interpretation, and integration into an SQI value. The database includes 81 indicators and 169 selection rules that allow the user to combine a soil management goal with supporting soil function into an SQI. Indicators have scoring curves that relate to concepts such as more of a property is better or less of a property is better. An example of a more is better curve includes total organic carbon and water stable aggregation (Andrews et al., 2004). An example of a less is better curve is for bulk density. A single SQI score is then calculated as an average of the indicator scores.

As an example of the use of the SQI, Sharma et al. (2008) computed an SQI for sorghum (Sorghum bicolor (L.) Moench)-mung bean (Vigna radiata L.) using key soil quality indicators. Twenty-one physical, chemical, and biological soil properties were measured and statistically treated to determine the MDS necessary to calculate the SQI. For the Alfisols being studied, they found that the available N, Zn, and Cu, microbial biomass carbon, aggregate mean weight diameter, and hydraulic conductivity were the variables most closely related to yield. Weighting factors were developed for each indicator, and these were used to calculate the SQI for different treatments on the soil. One finding was that tillage alone did not have a significant effect on the calculated SQI. Other researchers have also tested the SQI with varying results (e.g., Erkossa et al., 2007; Zobeck et al., 2008).

Smith et al. (1993) and Halvorson et al. (1996) proposed a multiple-variable indicator transform procedure to combine values or ranges of values that represent the best estimate of soil quality. Their system converts measured data values into a single value according to specified criteria. They did not attempt to define soil quality or specify what soil characteristics should be used. They combined this procedure with kriging to develop maps that indicated the probabilities of meeting a soil quality criterion on a landscape level. Critical threshold values must be known, assumed, or determined in order to separate different soil qualities.

Numerous additive productivity rating systems have been developed for specific states, as reviewed by Huddleston (1984). In these systems, soil properties are assigned numerical values according to their expected impact on plant growth. The index is usually calculated as the sum of the values assigned to each property with 100 the maximum value. Huddleston (1984) notes advantages and disadvantages to such a system, which are similar to those for many of the soil quality indices previously discussed. Additive systems become complex as the number of factors, cropping systems, and soil and climatic conditions increases (e.g., Andrews et al., 2004; Velasquez et al., 2007). Other numerical systems have been proposed (Zornoza et al., 2007; Masto et al., 2008), using sophisticated statistics to develop relationships between the land use of interest and the MDS.

The evaluation of land quality for forestry is a well-known practice. Indices range from quantitative through semiquantitative to qualitative. Quantitative evaluations, such as the *site index*, use regression equations to predict tree height at a predetermined tree age based on soil and climate data. Qualitative evaluations assign land to classes based on soil and climate properties.

The *Fertility Capability Classification* (*FCC*) was developed in the early 1970s to interpret soil taxonomy and additional soil attributes in terms of their relevance to plant growth (Sanchez et al., 2003). Sanchez et al. (2003) argue for the use of an updated version of the FCC for a quantitative assessment of soil quality in the tropics. Table 26.7

TABLE 26.7 Distribution of Soil Attributes by Latitudinal Belt[a,b]

Attributes and Constraints[c]	Tropical[d]		Temperate[d]		Boreal[d]		Total Area under Each Attribute[e]	
	%	10^6 ha	%	10^6 ha	%	10^6 ha	%	10^6 ha
Soil moisture stress d	43.6	2154.9	20.9	1381.4	0.3	4.8	27.1	3541.1
Aridic d+	25.8	1272.7	38.4	2535.6	0.2	3.6	29.2	3811.9
High erosion risk y	15.4	758.1	18.1	1194.0	17.3	264.0	17.0	2216.1
Low nutrient capital reserves k	36.5	1803.5	9.9	652.6	7.7	116.9	19.7	2573.0
Calcareous b	6.9	342.4	22.2	1464.1	2.7	40.6	14.1	1847.0
Permafrost t+	0.3	12.7	12.2	805.1	84.7	1289.4	16.1	2107.2
Aluminum toxic a	27.8	1374.6	7.7	507.4	5.9	90.4	15.1	1972.3
Cold t	0.4	20.9	17.7	1168.4	11.2	170.4	10.4	1359.7
Waterlogged g	8.5	419.0	9.6	634.7	34.3	522.2	12.1	1576.0
High P fixation i	10.5	519.8	1.7	112.6	0.0	0.0	4.8	632.4
High leaching potential e	9.0	443.2	1.8	117.7	0.0	0.1	4.3	561.0
High organic content o	0.9	44.6	2.4	160.9	12.7	194.2	3.1	399.7
Cracking clays v	4.5	220.2	1.6	104.2	0.0	0.7	2.5	325.1
Sodic n	1.6	81.3	3.0	197.0	0.3	4.4	2.2	282.7
Saline s	0.7	35.3	2.4	157.5	0.0	0.0	1.5	192.8
Volcanic x	0.5	24.2	0.5	33.1	0.0	0.3	0.4	57.6
Sulfidic c	0.3	12.5	0.0	0.5	0.0	0.0	0.1	13.1

Source: Palm, C., P. Sanchez, S. Ahamed, and A. Awiti. 2007. Soils: A contemporary perspective. Annu. Rev. Environ. Resour. 32:99–129.

[a] Arranged in descending order of world total.
[b] Estimates exclude areas not covered by soils (e.g., rocks, water bodies, shifting sands, ice).
[c] Letters are the FCC modifier symbols (Sanchez et al., 2003).
[d] Definitions: tropical, <23.5°; temperate, 23.6°–60°; boreal, >60°
[e] The sum of percentages exceeds 100 because a single soil usually has more than one FCC attribute.

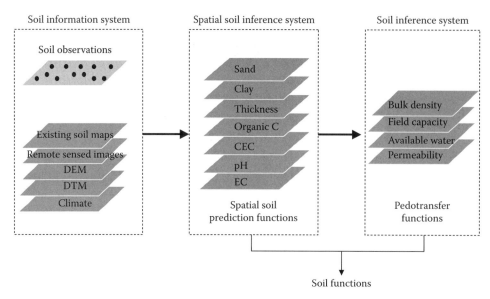

FIGURE 26.7 Schematic of digital mapping approach. (From Palm, C., P. Sanchez, S. Ahamed, and A. Awiti. 2007. Soils: A contemporary perspective. Annu. Rev. Environ. Resour. 32:99–129.)

(Palm et al., 2007) quantifies FCC attributes and constraints globally by latitudinal belts.

Palm et al. (2007) offer a re-evaluation of the soil quality debate that incorporates both an inventory of global soils and an evaluation of their current condition. They argue that "the natural capital of soils that underlies ecosystem services is primarily determined by three core soil properties: texture, mineralogy, and soil organic matter." They emphasize connections between soil types and biomes globally, and attempt to link specific soil processes, properties, constraints, and attributes with ecosystem services (Table 26.1) and quantify attributes and constraints globally (Table 26.7). They propose global mapping of soil properties in conjunction with the assessment of soil functions (Figure 26.7) (Palm et al., 2007). Subsequently, Sanchez et al. (2009) have argued that an ongoing effort to map soil properties globally and evaluate the condition of soils in conjunction with management recommendations for preventing soil degradation is critical for the protection of ecosystem services that rely on soil quality (Palm et al., 2007; Sanchez et al., 2009).

26.4 Conclusions

Soil quality is a concept developed to characterize the usefulness and health of soils because soils are fundamental to the well-being and productivity of agricultural and natural ecosystems. It is a compound characteristic that cannot be directly measured. Many definitions of soil quality can be found in the literature. Regional variation in soil properties and management ensures that no single set of soil characteristics can be universally adopted to quantify these definitions. Soil quality is often equated with agricultural productivity and sustainability, and is most commonly defined as a soil's potential to perform ecological functions in a system, for example, to maintain biological productivity, partition and regulate water and solute flow through an ecosystem, serve as an environmental buffer, sequester carbon and cycle nutrients, water, and energy through the biosphere. Air and water quality standards are usually based on maximum allowable concentrations of materials hazardous to human health. A definition of soil quality based on this concept would encompass only a fraction of the important roles soils play in agriculture and the environment.

To proceed from a definition to a measure of soil quality, an MDS of soil characteristics that relate to soil functions is commonly selected and quantified. Many physical, chemical, and biological properties of soil have been suggested to distinguish soils of different qualities. These include desirable and undesirable properties. In selecting characteristics, it is necessary to recognize that some soil properties are static—in the sense that they change slowly over time—and others are dynamic. In addition, the spatial and temporal variability of soil properties must be considered when selecting the properties used to assess soil quality. Numerous approaches to soil quality assessment have been proposed. Recent efforts to inventory the global state and function of soils provide a useful reassessment of soil quality in the broader context of the intimate connection between global ecosystem health and human and societal well-being.

We face a profound challenge in evaluating and sustaining the broader ecological services that soils can provide with proper management, while still maintaining sufficient food and energy production for a population projected to reach 9 billion by 2050 (Tilman et al., 2002; Robertson and Swinton, 2005; Rockstrom et al., 2009a, 2009b), in the context of a changing climate likely to affect food production and ecosystem health in the near future (IPCC, 2007). Soil science research has helped to demonstrate that soils regulate global climate, food production, and ecological services critical to human survival. Our future now depends on our ability to effectively reform agricultural and other land use policies for vastly more effective management of soil resources.

References

Alexander, M. 1971. Agriculture's responsibility in establishing soil quality criteria, p. 66–71. *In* Environmental improvement—Agriculture's challenge in the seventies. National Academy of Sciences, Washington, DC.

Amundson, R. 2001. The carbon budget of soils. Annu. Rev. Earth Planet. Sci. 29:535–562.

Andrews, S.S., D.L. Karlen, and C.A. Cambardella. 2004. The soil management assessment framework: A quantitative soil quality evaluation method. Soil Sci. Soc. Am. J. 68:1945–1962.

Arnold, R.W., I. Zaboles, and V.C. Targulian (eds.). 1990. Global soil change. Report of an IIASA-ISSS-UNEP Task Force on the Role of Soil in Global Change. International Institute for Applied Systems Analysis, Laxenburg, Austria.

Arshad, M.A., and G.M. Coen. 1992. Characterization of soil quality: Physical and chemical criteria. Am. J. Altern. Agr. 7:25–31.

Beck, A.J., S.C. Wilson, R.E. Alcock, and K.C. Jones. 1995. Kinetic constraints on the loss of organic chemicals from contaminated soils: Implications for soil-quality limits. Crit. Rev. Environ. Sci. Technol. 25:1–43.

Beke, G.J., H.H. Janzen, and T. Entz. 1994. Salinity and nutrient distribution in soil profiles of long-term crop rotations. Can. J. Soil Sci. 74:229–234.

Bending, G.D., M.K. Turner, F. Rayns, M.-C. Marx, and M. Wood. 2004. Microbial and biochemical soil quality indicators and their potential for differentiating areas under contrasting agricultural management regimes. Soil Bio. Biochem. 36:1785–1792.

Berry, E.C. 1994. Earthworms and other fauna in the soil, p. 61–90. *In* J.L. Hatfield and B.A. Stewart (eds.) Soil biology: Effects on soil quality. Advanced Soil Science Lewis Publisher, Boca Raton, FL.

Biederbeck, V.O., C.A. Campbell, H.U. Krainetz, D. Curtain, and O.T. Bouman. 1996. Soil microbial and biochemical properties after ten years of fertilization with urea and anhydrous ammonia. Can. J. Soil Sci. 76:7–14.

Birkhofer, K., T.M. Bezemer, J. Bloem, M. Bonkowski, S. Christensen, D. Dubois, F. Ekelund, A. Fließbach, L. Gunst, K. Hedlund, P. Mäder, J. Mikola, C. Robin, H. Setälä, F. Tatin-Froux, W.H. Van der Putten, and S. Scheu. 2008. Long-term organic farming fosters below and aboveground biota: Implications for soil quality, biological control and productivity. Soil Bio. Biochem. 40:2297–2308.

Blum, W.E.H. 1994. Soil resilience—General approaches and definition, p. 233–237. Proc. 15th World Cong. of Soil Sci. Acapulco, Mexico.

Boehn, M.M., and D.W. Anderson. 1997. A landscape-scale study of soil quality in three prairie farming systems. Soil Sci. Soc. Am. J. 61:1147–1159.

Bouma, J. 1989a. Using soil survey data for quantitative land evaluation. Adv. Soil Sci. 9:177–213.

Bouma, J. 1989b. Land qualities in space and time, p. 3–13. *In* J. Bouma and A.K. Bregt (eds.) Land qualities in space and time. Pudoc, Wageningen, the Netherlands.

Bowers, T.S., and T.D. Gauthier. 1994. Use of the output of a lead risk assessment model to establish soil lead cleanup levels. Environ. Geochem. Health 16:191–196.

Bowman, W.D., C.C. Cleveland, L. Halada, J. Heresko, and J.S. Baron. 2008. Negative impact of soil nitrogen deposition on soil buffering capacity. Nature Geosci. 1:767–770.

Bowman, R.A., J.D. Reeder, and G.E. Schuman. 1990. Evaluation of selected soil physical, chemical and biological parameters as indicators of soil productivity, Vol. 2, p. 64–70. Proc. Int. Conf. on Soil Quality in Semi-arid Ag. University of Saskatchewan, Saskatoon, Saskatchewan, Canada.

Bremer, E., H.H. Janzen, and A.M. Johnston. 1994. Sensitivity of total, light fraction and mineralizable organic matter to management practices in a Lethbridge soil. Can. J. Soil Sci. 74:131–138.

Cambardella, C.A., T.B. Moorman, S.S. Andrews, and D.L. Karlen. 2004. Watershed-scale assessment of soil quality in the loess hills of southwest Iowa. Soil Till. Res. 78:237–247.

Carpenter, S.R., H.A. Mooney, J. Agard, D. Capistrano, R.S. DeFries, S. Diaz, T. Dietz, A.K. Druaiappah, A. Oteng-Yeboah, H.M. Pereira, C. Perrings, W.V. Reid, J. Sarukhan, R.J. Scholes, and A. Whyte. 2009. Science for managing ecosystem services: Beyond the millennium ecosystem assessment. Proc. Natl. Acad. Sci. 106:1305–1312.

Carter, M.R., E.G. Gregorich, D.W. Anderson, J.W. Doran, H.H. Janzen, and F.J. Pierce. 1997. Concepts of soil quality and their significance. *In* E.G. Gregorich and M. Carter (eds.) Soil quality for crop production and ecosystem health. Elsevier Science Publishers, Amsterdam, the Netherlands.

Cassman, K.G. 1999. Ecological intensification of cereal production systems: Yield potential, soil quality, and precision agriculture. Proc. Natl. Acad. Sci. 96:5952–5959.

Cassman, K.G., A. Dobermann, D.T. Walters, and H. Yang. 2003. Meeting cereal demand while protecting natural resources and improving environmental quality. Annu. Rev. Environ. Resour. 28:315–358.

CAST (Council for Agricultural Science and Technology). 2004. Climate change and greenhouse gas mitigation: Challenges and opportunities for agriculture. CAST, Ames, IA.

CAST (Council for Agricultural Science and Technology). 2008. CAST commentary QTA2008-2 convergence of agriculture and energy: III. Considerations in biodiesel production. http://www.cast-science.org/publications/index.cfm/

Cook, N., and W.H. Hendershot. 1996. The problem of establishing ecologically based soil quality criteria: The case of lead. Can. J. Soil Sci. 76:335–342.

Crosson, P.R., and A.J. Stout. 1983. Productivity effects of cropland erosion in the United States. Resources for the future. Washington, DC.

de Bruijn, P.J., and F.B. de Walle. 1988. Soil standards for soil protection and remedial action in the Netherlands. *In* K. Wolf, W.J. van den Brink, and F.J. Colon (eds.) Contaminated soil '88, Vol. I. Kluwer Academic Publishers, Dordrecht, the Netherlands.

Denneman, C.A.J., and J.G. Robberse. 1990. Ecotoxicological risk assessment as a base for development of soil quality criteria, p. 157–164. *In* F. Arendt, M. Hinsenveld, and W.J. van den Brink (eds.) Contaminated soil '90. Proc. Intl. KfK/TNO Conf. on contaminated soil, Karlsruhe, Germany. Kluwer Academic Publishers, Dordrecht, the Netherlands.

Dent, F.J. 1978. Land suitability classification, p. 273–293. *In* Soils and rice. International Rice Research Institute, Manila, Philippines.

Derry, L.A., and O.A. Chadwick. 2007. Contributions from Earth's atmosphere to soil. Elements 3:333–338.

Ding, W., L. Meng, Y. Yin, Z. Cai, and X. Zheng. 2007. CO_2 emission in an intensively cultivated loam as affected by long-term application of organic manure and nitrogen fertilizer. Soil Bio. Biochem. 39:669–679.

Doll, P., and S. Siebert. 2002. Global modeling of irrigation water requirements. Water Resour. Res. 38:1037, doi: 10.1029/2001WR000355.

Donner, S.D., and C.J. Kucharik. 2008. Corn based ethanol compromises goal of reducing nitrogen export by the Mississippi River. Proc. Natl. Acad. Sci. 105:4513–4518.

Doran, J.W., and A.J. Jones (eds.). 1996. Methods for assessing soil quality. Soil Sci. Soc. Am. Spec. Pub. 49. SSSA, Inc., Madison, WI.

Doran, J.W., and T.B. Parkin. 1994. Defining soil quality for a sustainable environment. *In* J.W. Doran, D.C. Coleman, D.F. Bezdicek, and B.A. Stewart (eds.) Soil Sci. Soc. Am. Spec. Pub. 35. SSSA, Madison, WI.

Doran, J.W., M. Sarrantonio, and M.A. Liebig. 1996. Soil health and sustainability. Adv. Agron. 56:1–54.

EEA (European Environment Agency). 1999. Nutrients in European ecosystems. Environmental Assessment Report No. 4. EEA, Copenhagen, Denmark.

EEA (European Environment Agency). 2005. The European environment—State and outlook 2005. EEA, Copenhagen, Denmark.

Ellis, E.C., and N. Ramankutty. 2008. Putting people in the map: Anthropogenic biomes of the world. Front. Ecol. Environ. 6:439–447.

Erkossa, T., F. Itanna, and K. Stahr. 2007. Indexing soil quality: A new paradigm in soil science research. Aust. J. Soil Res. 45:129–137.

Eswaran, H., R. Almaraz, E. van den Berg, and P. Reich. 1997. An assessment of the soil resources of Africa in relation to productivity. Geoderma 77:1–18.

FAO (Food and Agricultural Organization). 1976. A framework for land evaluation. FAO Soils Bulletin 32. Rome, Italy.

FAO (Food and Agricultural Organization). 1979. Land evaluation criteria for irrigation. Report of an expert consultation. World Soil Resour. Report 50. Rome, Italy.

FAO (Food and Agricultural Organization). 1989. FAO yearbook. Rome, Italy.

FAO (Food and Agricultural Organization). 1995. Planning for sustainable use of land resources: Towards a new approach. *In* W.G. Sombroek and D. Sims (eds.) Land and Water Bulletin 2. FAO, Rome, Italy.

FAO (Food and Agricultural Organization). 1997. Land quality indicators and their use in sustainable agriculture and rural development. Land and Water Bulletin 5. FAO, Rome, Italy.

FAO (Food and Agriculture Organization of the United Nations). 2008. AQUASTAT: FAO's information system of water and agriculture, http://www.fao.org/nr/water/aquastat/main/index.stm, Food and Agric. Organ. of the U.N., Rome, Italy.

FAPRI (Food and Agricultural Policy Research Institute). 2008. U.S. and World Agricultural Outlook, Staff Report 08-FSR 1, Ames, IA.

Fargione, J., J. Hill, D. Tilman, S. Polasky, and P. Hawthorne. 2008. Land clearing and the biofuel carbon debt. Science 319(5867):1235–1238, doi: 10.1126/science.1152747.

Filip, Z. 2002. International approach to assessing soil quality by ecologically-related biological parameters. Ag. Ecosyst. Environ. 88:169–174.

Fließbach, A., H.-R. Oberholzer, L. Gunst, and P. Mäder. 2007. Soil organic matter and biological soil quality indicators after 21 years of organic and conventional farming. Ag. Ecosyst. Environ. 118:273–284.

Foley, J.A., R. DeFries, G.P. Asner, C. Barford, G. Bonan, S.R. Carpenter, F.S. Chapin, M.T. Coe, G.C. Daily, H.K Gibbs, J.H. Helkowski, T. Holloway, E.A. Howard, C.J. Kucharik, C. Monfreda, J.A. Patz, C. Prentice, N. Ramankutty, P.K. Snyder. 2005. Global consequences of land use. Science 309:570–574.

Folke, C., S. Carpenter, B. Walker, M. Scheffer, T. Elmqvist, L. Gunderson, and C.S. Holling. 2004. Regime shifts, resilience, and biodiversity in ecosystem management. Annu. Rev. Ecol. Evol. Syst. 35:557–581.

Fournier, F. 1989. The effect of human activity on soil quality, p. 25–32. *In* J. Bouma and A.K. Bregt (eds.). Land qualities in space and time. Pudoc, Wageningen, the Netherlands.

Franco-Vizcaíno, E. 1997. Comparative soil quality in maize rotations with high or low residue diversity. Biol. Fertil. Soils 24:32–38.

Franzluebbers, A.J., and M.A. Arshad. 1997. Soil microbial biomass and mineralizable carbon of water stable aggregates. Soil Sci. Soc. Am. J. 61:1090–1097.

Franzluebbers, A.J., D.A. Zuberer, and F.M. Hons. 1995. Comparison of microbiological methods for evaluating quality and fertility of soil. Biol. Fertil. Soils 19:135–140.

Gantzer, C.J., and T.R. McCarty. 1987. Predicting corn yields on a claypan soil using a soil productivity index. Trans. ASAE 30:1347–1352.

Gersmehl, P.J., and D.A. Brown. 1990. Geographic differences in the validity of a linear scale of innate soil productivity. J. Soil Water Conserv. 45:379–382.

Gerzabek, M.H., R.S. Antil, I. Kögel-Knabner, H. Knicker, H. Kirchmann, and G. Haberhauer. 2006. How are soil use and management reflected by soil organic matter characteristics: A spectroscopic approach. Eur. J. Soil Sci. 57:485–494.

Gerzabek, M.H., G. Haberhauer, and H. Kirchmann. 2001. Nitrogen distribution and 15N natural abundances in particle size fractions of a long-term agricultural field experiment. J. Plant Nutr. Soil Sci. 164:475–481.

Gerzabek, M.H., F. Pichlmayer, H. Kirchman, and G. Haberhauer. 1997. The response of soil organic matter to manure amendments in a long-term experiment at Ultuna, Sweden. Eur. J. Soil Sci. 48:273–282.

Gleick, P.H. 1993. An introduction to global fresh water issues. In P.H. Gleick, (ed.) Water in crisis. Oxford University Press, New York.

Golchin, A., and H. Asgari. 2008. Land use effects on soil quality indicators in north-eastern Iran. Aust. J. Soil Res. 46:27–36.

Gomiero, T., M.G. Paoletti, and D. Pimentel. 2008. Energy and environmental issues in organic and conventional agriculture. Crit. Rev. Plant Sci. 27:239–254.

Goulding, K., S. Jarvis, and A. Whitmore. 2008. Optimizing nutrient management for farm systems. Phil. Trans. Roy. Soc. B: Biol. Sci. 363:667–680.

Granatstein, D., and D.F. Bezdicek. 1992. The need for a soil quality index: Local and regional perspectives. Am. J. Alt. Ag. 7:12–16.

Gregorich, E.G., M.R. Carter, D.A. Angers, C.M. Monreal, and B.H. Ellert. 1994. Towards a minimum data set to assess soil organic matter quality in agricultural soils. Can. J. Soil Sci. 74:367–386.

Gregorich, E.G., M.R. Carter, J.W. Doran, C.E. Pankhurst, and L.M. Dwyer. 1997. Biological attributes of soil quality. C. 4. In E.G. Gregorich and M. Carter (eds.) Soil quality for crop production and ecosystem health. Elsevier Science Publishers, Amsterdam, the Netherlands.

Haberern, J. 1992. A soil health index. J. Soil Water Conserv. 47:6.

Habich, E.F. 2001. Ecological site inventory. U.S. Bureau of Land Management, Inventory and Monitoring Technical Reference, pp. 1734–1737.

Halvorson, J.J., J.L. Smith, and R.I. Papendick. 1996. Integration of multiple soil parameters to evaluate soil quality: A field example. Biol. Fertil. Soils 21:207–214.

Herrick, J.E., J.R. Brown, A.J. Tugel, P.L. Shaver, and K.M. Havstad, 2002. Application of soil quality to monitoring and management: Paradigms from rangeland ecology. Agron. J. 94:3–10.

Herrick, J.E., J.W. Van Zee, K.M. Havstad, L.M. Burkett, and W.G. Whitford. 2005a. Monitoring manual for grassland, shrubland and savanna ecosystems, volume I: Quick start. USDA—ARS Jornada Experimental Range, Las Cruces, NM.

Herrick, J.E., J.W. Van Zee, K.M. Havstad, L.M. Burkett, and W.G. Whitford. 2005b. Monitoring manual for grassland, shrubland and savanna ecosystems, volume II: Design, supplementary methods and interpretation. USDA—ARS Jornada Experimental Range, Las Cruces, NM.

Herrick, J.E., and W.G. Whitford. 1995. Assessing the quality of rangeland soils: Challenges and opportunities. J. Soil Water Conserv. 50:237–242.

Hill, J., E. Nelson, D. Tilman, S. Polasky, and D. Tiffany. 2006. Environmental, economic and energetic costs and benefits of biodiesel and ethanol biofuels. Proc. Natl. Acad. Sci. 103:11206–11210.

Hill, J., S. Polasky, E. Nelson, D. Tilman, H. Huo, L. Ludwig, J. Neumann, H. Zheng, and D. Bonta. 2009. Climate change and health costs of air emissions from biofuels and gasoline. Proc. Natl. Acad. Sci. 106(6):2077–2082.

Hoffman, G.J., T.A. Howell, and K.H. Solomon. 1990. Introduction. In G.J. Hoffman, T.A. Howell, and K.H. Solomon (eds.) Management of farm irrigation systems. ASAE, St. Joseph, MI.

Holden, P.A., and M.K. Firestone. 1997. Soil microorganisms in soil cleanup: How can we improve our understanding? J. Environ. Qual. 26:32–40.

Hortensius, D., and R. Welling. 1996. International standardization of soil quality measurements. Commun. Soil Sci. Plant Anal. 27:387–402.

Howard, P.J.A. 1993. Soil protection and soil quality assessment in the EC. Sci. Total Environ. 129:219–239.

Huddleston, J.H. 1984. Development and use of soil productivity ratings in the United States. Geoderma 32:297–317.

Hulugalle, N.R., and F. Scott. 2008. A review of the changes in soil quality and profitability accomplished by sowing rotation crops after cotton in Australian Vertosols from 1970 to 2006. Aust. J. Soil Res. 46:173–190.

Hutchinson, J.J., C.A. Campbell, and R.L. Desjardins. 2007. Some perspectives on carbon sequestration in agriculture. Ag. Forest Met. 142:288–302.

Idowu, O.J., H.M. Van Es, G.S. Abawi, D.W. Wolfe, J.I. Ball, B.K. Gugino, B.N. Moebius, R.R. Schindelbeck, and A.V. Bilgili. 2008. Farmer-oriented assessment of soil quality using field, laboratory, and VNIR spectroscopy methods. Plant Soil 307:243–253.

IPCC. 2007. Summary for policymakers. In S. Solomon, D. Qin, M. Manning, Z. Chen, M. Marquis, K.B. Averyt, M. Tignor, and H.L. Miller (eds.) Climate change 2007: The physical science basis. Contribution of working group I to the fourth assessment report of the intergovernmental panel on climate change. Cambridge University Press, Cambridge, U.K.

Jenkinson, D.S., N.J. Bradbury, and K. Coleman. 1994. How the Rothamsted classical experiments have been used to develop and test models for the turnover of carbon and nitrogen in soil, p. 117–138. In R.A. Leigh and A.E. Johnston (eds.) Long-term experiments in agricultural and ecological sciences. CAB International, Wallingford, U.K.

Jenny, H. 1941. Factors of soil formation. McGraw-Hill Book Company, Inc., New York.

Jenny, H. 1961. E.W. Hilgard and the birth of modern soil science. Industrie Grafiche V. Lischi & Figli, Pisa, Italy.

Jenny, H. 1980. The soil resource. Springer-Verlag, New York.

Johnston, A.E., K.W.T. Goulding, and P.R. Poulton. 1986. Soil acidification during more than 100 years under permanent grassland and woodland at Rothamsted. Soil Use Manage. 2:3–10.

Jongbloed, R.H., T.P. Traas, and R. Luttik. 1996. A probabilistic model for deriving soil quality criteria based on secondary poisoning of top predators. II. Calculations for dichlorodiphenyltrichloroethane (DDT) and cadmium. Ecotox. Environ. Safety 34:279–306.

Jordan, D., R.J. Kremer, W.A. Bergfield, K.Y. Kim, and V.N. Cacino. 1995. Evaluation of microbial methods as potential indicators of soil quality in historical agricultural fields. Biol. Fertil. Soils 19:297–302.

Karlen, D.L., S.S. Andrews, and J.W. Doran. 2001. Soil quality: Current concepts and applications. Adv. Agron. 74:1–40.

Karlen, D.L., and C.A. Cambardella. 1996. Conservation strategies for improving soil quality and organic matter storage. Adv. Soil Sci. CRC Press, Inc., Boca Raton, FL.

Karlen, D.L., N.S. Eash, and P.W. Unger. 1992. Soil and crop management effects on soil quality indicators. Am. J. Altern. Agr. 7:48–55.

Karlen, D.L., M.J. Mausbach, J.W. Doran, R.G. Cline, R.F. Harris, and G.E. Schuman. 1997. Soil quality: A concept, definition and framework for evaluation. Soil Sci. Soc. Am. J. 61:4–10.

Karlen, D.L., M.D. Tomer, J. Neppel, and C.A. Cambardella. 2008. A preliminary watershed scale soil quality assessment in north central Iowa, USA. Soil Till. Res. 99:291–299.

Karlen, D.L., N.C. Wollenhaupt, D.C. Erbach, E.C. Berry, J.B. Swan, N.S. Eash, and J.L. Jordahl. 1994. Crop residue effects on soil quality following 10-years of no-till corn. Soil Till. Res. 31:149–167.

Kennedy, A.C., and R.I. Papendick. 1995. Microbial characteristics of soil quality. J. Soil Water Conserv. 50:243–248.

Khan, S.A., R.L. Mulvaney, T.R. Ellsworth, and C.W. Boast. 2007. The myth of nitrogen fertilization for soil carbon sequestration. J. Environ. Qual. 36:1821–1832.

Kibblewhite, M.G., K. Ritz, and M.J. Swift. 2008. Soil health in agricultural systems. Phil. Trans. R. Soc. B Biol. Sci. 363:685–701.

Kiniry, L.N., C.L. Scrivner, and M.E. Keener. 1983. A soil productivity index based upon predicted water depletion and root growth. University of Missouri, College of Agriculture, Agriculture Experiment Station Bulletin 1051.

Kirchmann, H., G. Haberhauer, E. Kandeler, A. Sessitsch, and M.H. Gerzabek. 2004. Effects of level and quality of organic matter input on carbon storage and biological activity in soil: Synthesis of a long-term experiment. Global Biogeochem. Cycles 18:1–9.

Klingebiel, A.A., and P.H. Montgomery. 1973. Land-capability classification. Agriculture Handbook No. 210. Soil Conservation Service USDA, Washington, DC.

Lair, G.J., M.H. Gerzabek, and G. Haberhauer. 2007a. Sorption of heavy metals on organic and inorganic soil constituents. Environ. Chem. Lett. 5:23–27.

Lair, G.J., M.H. Gerzabek, and G. Haberhauer. 2007b. Retention of copper, cadmium and zinc in soil and its textural fractions influenced by long-term field management. Eur. J. Soil Sci. 58:1145–1154.

Lal, R. 1995. Trends in world agricultural use: Potential and constraints, p. 521–536. In R. Lal and B.A. Stewart (eds.) Soil management, experimental basis for sustainability and environmental quality. Advanced Soil Science 1. CRC Press, Boca Raton, FL.

Lal, R. 2004. Soil carbon sequestration impacts on global climate change and food security. Science 304:1623–1627.

Lal, R. 2008. Carbon sequestration. Phil. Trans. R. Soc. B 363:815–830.

Lamarca, C.C. 1996. Stubble over the soil: The vital role of plant residue in soil management to improve soil quality. American Society of Agronomy, Madison, WI.

Landis, D.A., M.M. Gardinera, W. van der Werf, and S.M. Swinton. 2008. Increasing corn for biofuel production reduces biocontrol services in agricultural landscapes. Proc. Natl. Acad. Sci. 105:20555–20557.

Lant, C.L., J.B. Ruhl, and S.E. Kraft. 2008. The tragedy of ecosystem services. BioScience 58:969–974.

Larson, W.E., and F.J. Pierce. 1991. Conservation and enhancement of soil quality, p. 175–203. In J. Dumanski, E. Pushparajah, M. Latham, and R.J.L. Myers (eds.) Evaluation for sustainable land management in the developing world. International Board for Soil Research and Management, Bangkok, Thailand.

Larson, W.E., and F.J. Pierce. 1994. The dynamics of soil quality as a measure of sustainable management, p. 37–51. In J.W. Doran, D.C. Coleman, D.F. Bezdicek, and B.A. Stewart (eds.) Defining soil quality for a sustainable environment. SSSA Spec. Pub. No. 35. ASA and SSSA, Madison, WI.

Letey, J. 1985. Relationship between soil physical properties and crop production. Adv. Soil Sci. 1:227–294.

Letey, J., R.E. Sojka, D.R. Upchurch, D.K. Cassel, K.R. Olson, W.A. Payne, S.E. Petrie, G.H. Price, R.J. Reginato, H.D. Scott, P.J. Smethurst, and G.B. Triplett. 2003. Deficiencies in the soil quality concept and its application. J. Soil Water Conserv. 58:180–187.

Lilburne, L.R., A.E. Hewitt, G.P. Sparling, and N. Selvarajah. 2002. Soil quality in New Zealand: Policy and the science response. J. Environ. Qual. 31:1768–1773.

Lindert, P.H., J. Lu, and W. Wu. 1996a. Trends in the soil chemistry of North China since the 1930s. J. Environ. Qual. 25:1168–1178.

Lindert, P.H., J. Lu, and W. Wu. 1996b. Trends in the soil chemistry of South China since the 1930s. Soil Sci. 161:329–342.

Lindstrom, M.J., T.E. Schumacher, A.J. Jones, and C. Gantzer. 1992. Productivity index model comparison for selected soils in North Central United States. J. Soil Water Conserv. 47:491–494.

Liu, X., S.J. Herbert, A.M. Hashemi, X. Zhang, and G. Ding. 2006. Effects of agricultural management on soil organic matter and carbon transformation—A review. Plant, Soil Environ. 52:531–543.

Maletic, J.T., and T.B. Hutchings. 1967. Selection and classification of irrigable lands, p. 125–173. In R.M. Hagan et al. (eds.) Irrigation of agricultural lands. Agronomy 10. ASA, CSSA, SSSA, Madison, WI.

Masto, R.E., P.K. Chhonkar, D. Singh, and A.K. Patra. 2008. Alternative soil quality indices for evaluating the effect of intensive cropping, fertilization and manuring for 31 years in the semi-arid soils of India. Environ. Monit. Assess. 136:419–435.

McBride, M., S. Sauvé, and W. Hendershot. 1997. Solubility control of Cu, Zn, Cd, and Pb in contaminated soils. Eur. J. Soil Sci. 48:337–346.

McLauchlan, K. 2006. The nature and longevity of agricultural impacts on soil carbon and nutrients: A review. Ecosystems 9:1364–1382.

McMichael, A.J., C.D. Butler, and C. Folke. 2003. New visions for addressing sustainability. Science 302:1919–1920.

McRae, S.G., and C.P. Burnham. 1981. Land Evaluation. Clarendon Press, Oxford, U.K.

Mele, P.M., and D.E. Crowley. 2008. Application of self-organizing maps for assessing soil biological quality. Ag. Ecosystems Environ. 126:139–152.

Melillo, J.M., D.W. Kicklighter, T.W. Cronin, B.S. Felzer, J.M. Reilly, A.C. Gurgel, T.W. Cronin, S. Paltsev, A.P. Sokolov, C.A. Schlosser. 2009. Indirect emissions from biofuels: How important? Science 23 October 2009.

Melillo, J.M., P.A. Steudler, J.D. Aber, K. Newkirk, H. Lux, F.P. Bowles, C. Catriciala, A. Magill, T. Ahrens, and S. Morrisseau. 2002. Soil warming and carbon-cycle feedbacks to the climate system. Science 298:2173–2176.

Merker, J. 1956. Untersuchungen an den Ernten und Böden des versuches 'Ewiger Roggebau' in Halle (Saale). Kühn-Archiv 70:154–215.

Millenium Ecosystem Assessment (MEA). 2005. Ecosystems and human well-being: Synthesis. Island Press, Washington, DC.

Miller, F.P. 1979. Defining, delineating and designating uses for prime and unique agricultural lands, p. 291–318. *In* M.T. Beatty, G.W. Petersen, and L.D. Swindale (eds.) Planning the uses and management of land. Agronomy 21. ASA, CSSA, SSSA, Madison, WI.

Miller, A.J., R. Amundson, I.C. Burke, and C. Yonker. 2004. The effect of climate and cultivation on soil organic C and N. Biogeochemistry 67:57–72.

Moebius, B.N., H.M. van Es, R.R. Schindelbeck, O.J. Idowu, D.J. Clune, and J.E. Thies. 2007. Evaluation of laboratory-measured soil properties as indicators of soil physical quality. Soil Sci. 172:895–912.

Moebius-Clune, B.N., H.M. van Es, O.J. Idowu, R.R. Schindelbeck, D.J. Moebius-Clune, D.W. Wolfe, G.S. Abawi, J.E. Thies, B.K. Gugino, and R. Lucey. 2008. Long-term effects of harvesting maize stover and tillage on soil quality. Soil Sci. Soc. Am. J. 72:960–969.

Moen, J.E.T. 1988. Soil protection in the Netherlands, p.1495–1503. *In* K. Wolf, W.J. van den Brink, and F.J. Colon (eds.) Contaminated soil '88. Kluwer Academic Publishers, Dordrecht, the Netherlands.

Montgomery, D.R. 2007. Soil erosion and agricultural sustainability. Proc. Natl. Acad. Sci. 104:13268–13272.

Munkholm, L.J., P. Schjønning, K. Debosz, H.E. Jensen, and B.T. Christensen. 2002. Aggregate strength and mechanical behaviour of a sandy loam soil under long-term fertilization. Eur. J. Soil Sci. 53:129–137.

Murata, T., M.L. Nguyen, and K.M. Goh. 1995. The effects of long-term superphosphate application on soil organic matter content and composition from an intensively managed New Zealand pasture. Eur. J. Soil Sci. 46:257–264.

Neff, J.C., A.R. Townsend, G. Gleixner, S.J. Lehman, J. Turnbull, and W.D. Bowman. 2002. Variable effects of nitrogen additions on the stability and turnover of soil carbon. Nature 419:915–917.

Neill, L.L. 1979. An evaluation of soil productivity based on root growth and water depletion. MS thesis. University of Missouri.

NRC (National Research Council). 1993. Soil and water quality: An agenda for agriculture. Committee on Long-Range Soil and Water conservation, Board on Agriculture, National Academy Press, Washington, DC.

NRC (National Research Council). 1994. Rangeland health. New methods to classify, inventory, and monitor rangelands. National Academy Press, Washington, DC.

Odell, R.T., S.W. Melsted, and W.M. Walker. 1984. Changes in organic carbon and nitrogen of Morrow plot soils under different treatments, 1904–1973. Soil Sci. 137:160–171.

O'Geen, T., S. Southard, and R.J. Southard. 2008. A revised Storie index for use with digital soils information. UC div. Ag. Natural Res. Pub. ANR 8355. Berkeley, CA.

Olson, K.R. 2007. Soil organic carbon storage in southern Illinois woodland and cropland. Soil Sci. 172:623–630.

Oluwatosin, G.A., D.O. Adeyolanu, O.J. Idowu, and L.B. Taiwo. 2008. Changes in soil quality as influenced by cropping in the forest ecology of Southwestern Nigeria. J. Food Agric. Environ. 6:529–533.

Ouedraogo, E., A. Mando, and N.P. Zombre. 2001. Use of compost to improve soil properties and crop productivity under low input agricultural system in West Africa. Ag. Ecosyst. Environ. 84:259–266.

Palm, C., P. Sanchez, S. Ahamed, and A. Awiti. 2007. Soils: A contemporary perspective. Annu. Rev. Environ. Resour. 32:99–129.

Papendick, R.I., J.F. Parr, and J. van Schilfgaarde. 1995. Soil quality: New perspective for a sustainable agriculture. *In* Proc. International Soil Conservation Organization, New Delhi, India, 1994.

Parr, J.F., R.I. Papendick, S.B. Hornick, and R.E. Meyer. 1992. Soil quality: Attributes and relationship to alternative and sustainable agriculture. Am. J. Altern. Ag. 7:5–10.

Parr, J.F., B.A. Stewart, S.B. Hornick, and R.P. Singh. 1990. Improving the sustainability of dryland farming systems: A global perspective. Adv. Soil Sci. 13:1–8.

Patterson, G.T., and E.E. MackIntosh. 1976. Relationship between soil capability class and economic returns from grain corn production in Southwestern Ontario. Can. J. Soil Sci. 56:167–174.

Pellant, M., P. Shaver, D.A. Pyke, and J.E. Herrick. 2005. Interpreting indicators of rangeland health, version 5. Interagency technical reference 1734-6, USDI, Bureau of Land Management, National Science and Technology Center, Denver, CO.

Pennock, D.J., D.W. Anderson, and E. De Jong. 1994. Landscape-scale changes in indicators of soil quality due to cultivation in Saskatchewan, Canada. Geoderma 64:1–19.

Pierce, F.J., W.E. Larson, R.H. Dowdy, and W.A.P. Graham. 1983. Productivity of soils: Assessing long-term changes due to erosion. J. Soil Water Conserv. 38:39–44.

Pierce, F.J., W.E. Larson, R.H. Dowdy, and W.A.P. Graham. 1984. Soil productivity in the corn belt: An assessment of erosion's long term effects. J. Soil Water Conserv. 39:131–136.

Pimentel, D., C. Harvey, P. Resosudarmo, K. Sinclair, D. Kurz, M. McNair, S. Crist, L. Shpritz, L. Fitton, R. Saffouri, and R. Air. 1995. Environmental and economic costs of soil erosion and conservation benefits. Science 267:1117–1123.

Poulton, P.R., E. Pye, P.R. Hargreaves, and D.S. Jenkinson. 2003. Accumulation of carbon and nitrogen by old arable land reverting to woodland. Global Change Biol. 9:942–955.

Pretty, J. 2008. Agricultural sustainability: Concepts, principles and evidence. Phil. Trans. Roy. Soc. 363:447–465.

Pyke, D.A., J.E. Herrick, P. Shaver, and M. Pellant. 2002. Rangeland health attributes and indicators for qualitative management. J. Range Manage. 55:584–597.

Ramankutty, N., A.T. Evan, C. Monfreda, and J.A. Foley. 2008. Farming the planet: 1. Geographic distribution of global agricultural lands in the year 2000. Global Biogeochem. Cycles 22:GB1003, doi: 10.1029/2007GB002952.

Rasmussen, P.E., K.W.T. Goulding, J.R. Brown, P.R. Grace, H.H. Janzen, and M. Korschens. 1998. Long-term agroecosystem experiments: Assessing agricultural sustainability and global change. Science 282:893–896.

Reganold, J.P. 1988. Comparison of soil properties as influenced by organic and conventional farming systems. Am. J. Altern. Agr. 3:144–155.

Reganold, J.P., and A.S. Palmer. 1995. Significance of gravimetric versus volumetric measurements of soil quality under biodynamic, conventional, and continuous grass management. J. Soil Water Conserv. 50:298–305.

Reganold, J.P., A.S. Palmer, J.C. Lockhart, and A.N. Macgregor. 1993. Soil quality and financial performance of biodynamic and conventional farms in New Zealand. Science 260:344–349.

Reganold, J.P., and M.J. Singer. 1979. Defining prime farmland by three land classification systems. J. Soil Water Conserv. 34:172–176.

Reganold, J.P., and M.J. Singer. 1984. Comparison of farm production input/output ratios of two land classification systems. J. Soil Water Conserv. 39:47–53.

Rezaei, S.A., R.J. Gilkes, and S.S. Andrews. 2006. A minimum data set for assessing soil quality in rangelands. Geoderma 136:229–234.

Richter, D.D. Jr., M. Hofmockel, M.A. Callaham, D.S. Powlson, and P. Smith. 2007. Long-term soil experiments: Keys to managing Earth's rapidly chancing ecosystems. Soil Sci. Soc. Am. J. 71:266–279.

Robertson, G.P., J.C. Broome, E.A. Chornesky, J.R. Frankenberger, P. Johnson, M. Lipson, J.A. Miranowski, E.D. Owens, D. Pimentel, and L.A. Thrupp. 2004. Rethinking the vision for environmental research in US agriculture. BioScience 54:61–65.

Robertson, G.P., and S.M. Swinton. 2005. Reconciling agricultural productivity and environmental integrity: A grand challenge for agriculture. Front. Ecol. Environ. 3 (1 SPEC. ISS.):38–46.

Rockstrom, J., W. Steffen, K. Noone, A. Persson, F.S. Chapin, E.F. Lambin, T.M. Lenton, M. Scheffer, C. Folke, H. Joachim Schellnhuber, B. Nykvist, C.A. deWit, T. Hughes, S. van der Leeuw, H. Rodhe, S. Sorlin, P.K. Nsyder, R. Costanza, U. Svedin, M. Falkenmark, L. Karlberg, R.W. Corell, V.J. Fabry, J. Hansen, B. Walker, D. Liverman, I. Richardson, P. Crutzen, J.A. Foley. 2009a. A safe operating space for humanity. Nature 461:472–475.

Rockstrom, J., W. Steffen, K. Noone, A. Persson, F.S. Chapin, E.F. Lambin, T.M. Lenton, M. Scheffer, C. Folke, H. Joachim Schellnhuber, B. Nykvist, C.A. deWit, T. Hughes, S. van der Leeuw, H. Rodhe, S. Sorlin, P.K. Nsyder, R. Costanza, U. Svedin, M. Falkenmark, L. Karlberg, R.W. Corell, V.J. Fabry, J. Hansen, B. Walker, D. Liverman, I. Richardson, P. Crutzen, J.A. Foley. 2009b. Planetary boundaries: Exploring the safe operating space for Humanity. Ecol. Soc. 14:1–31.

Rodrigues de Lima, A.C., W. Hoogmoed, and L. Brussard. 2008. Soil quality assessment in rice production systems: Establishing a minimum data set. J. Environ. Qual. 37:623–630.

Romig, D.E., M.J. Garlynd, R.F. Harris, and K. McSweeney. 1995. How farmers assess soil health and quality. J. Soil Water Conserv. 50:229–236.

Rossiter, D.G. 1996. A theoretical framework for land evaluation. Geoderma 72:165–190.

Sanchez, P.A. 2002. Soil fertility and hunger in Africa. Science 295:2019–2020.

Sanchez, P.A., S. Ahamed, F. Carre, A.E. Hartemink, J. Hempel, J. Huising, P. Lagacherie, A.B. McBratney, M.J. McKenzie, M. de Lourdes Mendoncap-Santos, B. Minasny, L. Montararella, P. Okoth, C.A. Palm, J.D. Sachs, K.D. Shepherd, T.-G. Vagen, B. Vanlauwe, M.G. Walsh, L.A. Winowiecki, and G.-L. Zhang. 2009. Digital soil map of the world. Science 325:680–681.

Sanchez, P.A., C.A. Palm, and S.W. Buol. 2003. Fertility capability soils classification: A tool to help assess soil quality in the tropics. Geoderma 114:157–185.

Scheffer, M., S. Carpenter, J.A. Foley, C. Folke, and B. Walker. 2001. Catastrophic shifts in ecosystems. Nature 413:591–596.

Schjønning, P., B.T. Christensen, and B. Carstensen. 1994. Physical and chemical properties of a sandy loam receiving animal manure, mineral fertilizer or no fertilizer for 90 years. Eur. J. Soil Sci. 45:257–268.

Schjønning, P., B.V. Iversen, L.J. Munkholm, R. Labouriau, and O.H. Jacobsen. 2005. Pore characteristics and hydraulic properties of a sandy loam supplied for a century with either animal manure or mineral fertilizers. Soil Use Manage. 21:265–275.

Schlesinger, W.H. 1999. Carbon sequestration in soils. Science 284:2095.

Schloter, M., O. Dilly, and J.C. Munch. 2003. Indicators for evaluating soil quality. Agric. Ecosyst. Environ. 98:255–262.

Schmer, M.R., K.P. Vogel, R.B. Mitchell, and R.K. Perrin. 2008. Net energy of cellulosic energy from switchgrass. Proc. Natl. Acad. Sci. 105:464–469.

Searchinger, T., R. Heimlich, R.A. Houghton, F. Dong, A. Elobeid, J. Fabiosa, S. Tokgoz, D. Hayes, and T.-H. Yu. 2008. Use of U.S. croplands for biofuels increases greenhouse gases through emissions from land-use change. Science 319:1238–1240.

Seybold, C.A., J.E. Herrick, and J.J. Brejda. 1999. Soil resilience: A fundamental component of soil quality. Soil Sci. 164:224–234.

Seybold, C.A., M.J. Mausbach, D.L. Karlen, and H.H. Rogers. 1997. Quantification of soil quality, p. 387–404. *In* R. Lal, J.M. Kimble, R.F. Follett, and B.A. Stewart (eds.) Soil processes and the carbon cycle. CRC Press, Boca Raton, FL.

Sharma, K.L., J. Kusuma Grace, U.K. Mandal, P.N. Gajbhiye, K. Srinivas, G.R. Korwar, V.H. Bindu, V. Ramesh, K. Ramachandran, and S.K. Yadav. 2008. Evaluation of long-term soil management practices using key indicators and soil quality indices in a semi-arid tropical Alfisol. Aust. J. Soil Res. 46:368–377.

Shepherd, T.G., L.J. Bird, M.R. Jessen, D.J. Bloomer, D.J. Camaeron, S.C. Park, and P.R. Stephens. 2001. Visual soil assessment of soil quality—Trial by workshops, p. 119–126. *In* L.D. Currie and P. Loganathan (eds.) Precision tools for improving land management. Occasional report number 14. Fertilizer and Lime Research Centre, Massey University, Palmerston North, New Zealand.

Sheppard, S.C., C. Gaudet, M.I. Sheppard, P.M. Cureton, and M.P. Wong. 1992. The development of assessment and remediation guidelines for contaminated soils, a review of the science. Can. J. Soil Sci. 72:359–394.

Siegrist, R.L. 1989. International review of approaches for establishing cleanup goals for hazardous waste contaminated land. The Agricultural Research Council of Norway, Institute of Georesources and Pollution Research, Oslo, Norway N-1432 Aas-NHL.

Sims, J.T., S.D. Cunningham, and M.E. Sumner. 1997. Assessing soil quality for environmental purposes: Roles and challenges for soil scientists. J. Environ. Qual. 26:20–25.

Singer, M.J., and S. Ewing. 2000. Soil quality, p. G-271–G-298. *In* M.E. Sumner (editor-in-chief) Handbook of soil science. CRC Press, Boca Raton, FL.

Singer, M.J., and D.N. Munns. 2006. Soils: An introduction. 6th Edn. Prentice-Hall, Inc., Upper Saddle River, NJ.

Singer, M.J., and R.E. Sojka. 2001. Soil quality, p. 312–314. *In* M. Licker (ed.) Yearbook of Science and Technology 2002. McGraw-Hill, New York.

Singer, M.J., K.K. Tanji, and J.H. Snyder. 1979. Planning uses of cultivated cropland and pastureland, p. 225–271. *In* M.T. Beatty, G.W. Petersen, and L.D. Swindale (eds.) Planning the uses and management of land. Agronomy 21. ASA, CSSA, SSSA, Madison, WI.

Smith, J.L., J.J. Halvorson, and R.I. Papendick. 1993. Using multiple-variable indicator kriging for evaluating soil quality. Soil Sci. Soc. Am. J. 57:743–749.

Snakin, V.V., P.P. Krchetov, T.A. Kuzovnikova, I.O. Alyabina, A.F. Gurov, and A.V. Stepichev. 1996. The system of assessment of soil degradation. Soil Technol. 8:331–343.

Soil Science Society of America. 1996. Glossary of soil science terms. SSSA. Madison, WI.

Sojka, R.E. 1996. A PAM Primer: A brief history of PAM and PAM-issues related to irrigation, p. 11–20. *In* R.E. Sojka and R.D. Lentz (eds.) Managing irrigation-induced erosion and infiltration with polyacrylamide. University of Idaho Miscellaneous Publication 101-96, Idaho, Moscow.

Sojka, R.E., and D.R. Upchurch. 1999. Reservations regarding the soil quality concept. Soil Sci. Soc. Am. J. 63:1039–1054.

Sojka, R.E., D.R. Upchurch, and N.E. Borlaug. 2003. Quality soil management or soil quality management: Performance versus semantics. Adv. Agron. 79:1–68.

Sombroek, W.G. 1997. Land resource evaluation and the role of land-related indicators, p. 9–17. *In* Land quality indicators and their use in sustainable agriculture and rural development. FAO Land and Water Bulletin 5. Rome, Italy.

Sparling, G.P., and L.A. Schipper. 2002. Soil quality at a national scale in New Zealand. J. Environ. Qual. 31:1848–1857.

Speir, T.W., H.A. Kettles, A. Parshotam, P.L. Searle, and L.N.C. Vlaar. 1996. A simple kinetic approach to derive the ecological dose value for the assessment of Cr(VI) toxicity to soil biological properties. Soil Biol. Biochem. 27:801–810.

Storie, R.E. 1932. An index for rating the agricultural value of soils. Cal. Agr. Exp. Sta. Bulletin 556.

Storie, R.E. 1964. Handbook of soil evaluation. Associated Students Store, University of California, Berkeley, CA.

Stork, N.E., and P. Eggleton. 1992. Invertebrates as determinants and indicators of soil quality. Am. J. Altern. Agr. 7:38–47.

Thorne, P.S. 2007. Environmental health aspects of concentrated animal feeding operations. Environ. Health Perspect. 115:296–297.

Tian, Y., and Y. Feng. 2008. Application of microbial research in evaluation of soil quality. Chinese J. App. Environ. Bio. 14:132–137.

Tilman, D., K.G. Cassman, P.A. Matson, R. Naylor, and S. Polasky. 2002. Agricultural sustainability and intensive production practices. Nature 418:671–677.

Tilman, D., R. Socolow, J.A. Foley, J. Hill, E. Larson, L. Lynd, S. Pacala, J. Reilly, T. Searchinger, C. Somervillle, and R. Williams. 2009. Beneficial biofuels—The food, energy and environment trilemma. Science 325:270–271.

Traas, T.P., R. Luttik, and R.H. Jongbloed. 1996. A probabilistic model for deriving soil quality criteria based on secondary poisoning of top predators I. Model description and uncertainty analysis. Ecotoxicol. Environ. Safety 34:264–278.

Tribe, D. 1994. Feeding and greening the world, the role of agricultural research. CAB International, Wallingford, U.K.

Tugel, A., J.E. Herrick, J.R. Brown, M.J. Mausbach, W. Puckett, and K. Hipple. 2005. Soil change, soil survey, and natural resources decision making: A blueprint for action. Soil Sci. Soc. Am. J. 69:738–747.

Tugel, A.J., S.A. Wills, and J.E. Herrick. 2008. Soil change guide: Procedures for soil survey and resource inventory, Version 1.1. USDA, Natural Resources Conservation Service, National Soil Survey Center, Lincoln, NE.

USBR. 1953. Land classification handbook. USDI, Bureau of Reclamation Publication V, Part 2 US Government Printing Office.

USGS. 2002. Assessing Rangelands Fact Sheet (USGS FS-125-02).

van de Koppel, J., M. Reiekirk, F.J. Weissing. 1997. Catastrophic vegetation shifts and soil degradation in terrestrial grazing systems. Trends Ecol. Evol. 12:352–356.

van Eekeren, N., L. Bommelé, J. Bloem, T. Schouten, M. Rutgers, Ron de Goede, D. Reheul, and L. Brussaard. 2008. Soil biological quality after 36 years of ley-arable cropping, permanent grassland and permanent arable cropping. Appl. Soil Ecol. 40:432–446.

van Vliet, L.J.P., E.E. Mackintosh, and D.W. Hoffman. 1979. Effects of land capability on apple production in Southern Ontario. Can. J. Soil Sci. 59:163–175.

Varvel, G., W. Riedell, E. Deibert, B. McConkey, D. Tanaka, M. Vigil, and R. Schwartz. 2006. Great Plains cropping system studies for soil quality assessment. Renew. Agric. Food Syst. 21:3–14.

Velasquez, E., P. Lavelle, and M. Andrade. 2007. GISQ, a multifunctional indicator of soil quality. Soil Bio. Biochem. 39:3066–3080.

Villamil, M.B., F.E. Miguez, and G.A. Bollero. 2008. Multivariate analysis and visualization of soil quality data for no-till systems. J. Environ. Qual. 37:2063–2069.

Visser, S., and D. Parkinson. 1992. Soil biological criteria as indicators of soil quality: Soil microorganisms. Am. J. Altern. Agr. 7:33–37.

Vitousek, P.M., R. Naylor, T. Crews, M.B. David, L.E. Drinkwater, E. Holland, P.J. Johnes, J. Katzenberger, L.A. Martinelli, P.A. Matson, G. Nziguheba, D. Ojima, C.A. Palm, G.P. Robertson, P.A. Sanchez, A.R. Townsend, F.S. Zhang. 2009. Nutrient imbalances in agricultural development. Science 324:1519–1520.

Vorosmarty, C.J., P. Green, J. Salisbury, and R.B. Lammers. 2000. Global water resources: Vulnerability from climate change and population growth. Science 289:284–288.

Wagenet, R.J., and J.L. Hutson. 1997. Soil quality and its dependence on dynamic physical processes. J. Environ. Qual. 26:41–48.

Waksman, S.A. 1927. Principles of soil microbiology. The Williams and Wilkins Co., Baltimore, MD.

Walker, B.D., C. Wang, and H.W. Rees. 1997. Benchmark sites for monitoring agricultural soil quality. Soil Quality Program Research Factsheet CSQ01, Agriculture and Agri-Food Canada.

Warkentin, B.P., and H.F. Fletcher. 1977. Soil quality for intensive agriculture, p. 594–598. *In* Proceedings of the international seminar on soil environment and fertilizer management in intensive agriculture. Society for Science of Soil and Manure, National Institute of Agricultural Science, Tokyo, Japan.

West, T.O., and W.M. Post. 2002. Soil organic carbon sequestration rates by tillage and crop rotation: A global data analysis. Soil Sci. Soc. Am. J. 66:1930–1946.

Wienhold, B.J., S.S. Andrews, and D.L. Karlen. 2004. Soil quality: A review of the science and experiences in the USA. Environ. Geochem. Health 26:89–95.

Wier, W.W., and R.E. Storie. 1936. A rating of California soils. California Agricultural Experiment Station Bulletin 599.

Willis, W.O., and C.E. Evans. 1977. Our soil is valuable. J. Soil Water Conserv. 32:258–259.

Wisser, D., S. Frolking, E.M. Douglas, B.M. Fekete, C.J. Vorosmarty, and A.H. Schumann. 2008. Geophys. Res. Lett. 35, L24408, doi: 10.1029/2008GL035296.

Wu, T., J.J. Schoenau, F. Li, P. Qian, S.S. Malhi, Y. Shi, and F. Xu. 2004. Influence of cultivation and fertilization on total organic carbon and carbon fractions in soils from the Loess Plateau of China. Soil Till. Res. 77:59–68.

Yakovchenko, V., L.J. Sikora, and D.D. Kaufman. 1996. A biologically based indicator of soil quality. Biol. Fertil. Soils 21:245–251.

Yli-Halla, M., H. Hartikainen, P. Ekholm, E. Turtola, M. Puustinen, and K. Kallio. 1995. Assessment of soluble phosphorus load in surface runoff by soil analyses. Agric. Eco. Environ. 56:53–62.

Zentner, R.P., C.A. Campbell, V.O. Biederbeck, P.R. Miller, F. Selles, and M.R. Fernandez. 2001. In search of a sustainable cropping system for the semiarid Canadian prairies, J. Sustainable Agric. 18:117–136.

Zobeck, T.M., A.D. Halvorson, B. Wienhold, V. Acosta-Martinez, and D.L. Karlen. 2008. Comparison of two soil quality indexes to evaluate cropping systems in northern Colorado. J. Soil Water Conserv. 63:329–338.

Zornoza, R., J. Mataix-Solera, C. Guerrero, V. Arcenegui, F. García-Orenes, J. Mataix-Beneyto, and A. Morugán. 2007. Evaluation of soil quality using multiple lineal regression based on physical, chemical and biochemical properties. Sci. Total Environ. 378:233–237.

IV

Soil Databases

Marion F. Baumgardner
Purdue University

27 **Qualitative and Quantitative Aspects of World and Regional Soil Databases and Maps** *Freddy O. Nachtergaele, Vincent W.P. van Engelen, and Niels H. Batjes* 27-1
Introduction • Global Soil Maps and Databases • Regional Soil and Terrain Databases (SOTER) • ISRIC–WISE Harmonized Global Soil Profile Database • Digital Soil Mapping • Conclusions and Recommendations • References

28 **United States Soil Survey Databases** *Jim R. Fortner and Alan B. Price* 28-1
Introduction • Description of Soil Survey Databases • The Evolution of Soil Survey Map Unit Databases • Standards for Soil Survey Data • Future Directions for Soil Information Systems • Acknowledgments • References

29 **Integrated Digital, Spatial, and Attribute Databases for Soils in Brazil** *Carlos Eduardo Pellegrino Cerri, Carlos Gustavo Tornquist, Martial Bernoux, Miguel Cooper, Gerd Sparovek, Maria de Lourdes Mendonça-Santos, and Carlos Clemente Cerri* 29-1
Introduction • Brief Biophysical Setting of Brazil • Soil Maps of Brazil and Description of the Soil Attribute Database • Integration and Application of Soil Digital Databases in Brazil • Future Directions and Opportunities • References

30 **Development and Use of Soil Maps and Databases in China** *Gan-Lin Zhang and Yun-Jin Wu* 30-1
Development of Digitized Soil and Land Information Systems in China • Use of Soil and Land Information Systems in China • Future Development and Opportunities • Acknowledgments • References

31 **Soil Geographic Database of Russia** *Sergey A. Shoba, Vyacheslav A. Rozhkov, Irina O. Alyabina, Varvara M. Kolesnikova, Inga S. Urusevskaya, Erik N. Molchanov, Vladimir S. Stolbovoi, Boris V. Sheremet, and Dmitry E. Konyushkov* 31-1
Introduction • Soil and Land Resources of Russia • History of Collection of Soil Information in Russia • Soil Geographic Database of Russia • Conclusions • Acknowledgments • References

32 **Soil Databases in Africa** *D.G. Paterson and N.M. Mushia* 32-1
Introduction • Background • Soil Characterization • Soil Databases • Soil Profile Databases • Database Interpretation and Utilization • Conclusion • References

33 **Learning about Soil Resources with Digital Soil Maps** *Darrell G. Schulze, Phillip R. Owens, and George E. Van Scoyoc* 33-1
Introduction • Teaching and Learning with Web Soil Survey • Teaching and Learning Using GIS and Soil Geodatabases in the Field • Teaching and Learning Using a Web-Based Geobrowser • Acknowledgments • References

Terrestrial Ecosystem of Planet Earth

For many years, soil scientists have addressed the question of the carrying capacity of the Earth. During the past 150 years crop and soil scientists have made extensive advances in increasing yields of crops and animal products to feed rapidly increasing populations. During more recent decades, environmental scientists have given increasing attention to the issue of sustainability of the Earth's terrestrial ecosystems, especially those ecosystems that are occupied and managed by human populations.

Within the last half century, remarkable advances have been made in several strategic technologies of importance to soil and other terrestrial ecosystem scientists. These include the computer, electronic digital information systems, and polar-orbiting Earth observation satellites with a wide array of digital sensors, including the digital camera. In addition to these new measuring devices, advanced digital analysis systems, and storage media for almost limitless quantities of data, two other remarkable developments have occurred. One is the U.S. military declassification in the 1990s of Global Positioning Systems (GPS) data for use by the civilian community. This means that any soil scientist (or anyone else) with rather inexpensive equipment can locate a site anywhere on the Earth surface within an accuracy of a fraction of a meter. A second development is the capacity to transmit and share large volumes of digital data instantaneously from one address on the Earth surface to potentially any other address on the Earth surface.

Role of Soil Databases in Understanding and Managing Terrestrial Ecosystems

Selected soil science colleagues from around the world were invited to participate in the development of Part IV of this handbook. The authors were invited to prepare manuscripts that provide a current view of the soil databases of their country, region, or the world. This description of their soil databases was to be followed by a discussion of how the new information technologies are being used by the soil science community to make soil data and information available to the User Community. This User Community is defined to include agricultural producers, foresters, private industries, government agencies, and any others who have responsibilities in the use and management of terrestrial ecosystems anywhere.

Part IV includes seven chapters on soil databases. In Chapter 27, Nachtergaele, FAO, Rome, and colleagues present "Qualitative and Quantitative Aspects of World and Regional Soil Databases and Maps." Chapters 28 through 31 describe the development and use of national and state soil databases in four of the world's largest agricultural and timber producers—Brazil, China, Russia, and the United States. A study of these four large and complex soil databases from three continents suggests the need for information systems which may assist in bridging the gaps among the inconsistencies and incompatibilities of different soil classification systems. Chapter 32 by Paterson and Mushia from South Africa presents an overview of the database situation for the continent of Africa while in Chapter 33, D.G. Schulze and colleagues, Purdue University, West Lafayette, Indiana, complete this part with their presentation on "Learning about soil resources with digital soil maps." Their illustrations provide an idea of the almost unlimited possibilities for distributing useful and understandable soils information to anyone in the world involved in the management and conservation of terrestrial ecosystems.

Many other countries, regions, and organizations have activities related to the development of soil databases. Two other areas of the world where significant progress is being made in the development of soil databases are Europe and Africa. The effort to produce an integrated soil database for Europe is well along in its complex task. The creation and development of a useful and compatible soil database to meet the needs of the newly developing Alliance for Green Revolution in Africa (AGRA) are in the initial stages.

27

Qualitative and Quantitative Aspects of World and Regional Soil Databases and Maps

Freddy O. Nachtergaele
Food and Agriculture Organization

Vincent W.P. van Engelen
International Soil Reference and Information Centre

Niels H. Batjes
International Soil Reference and Information Centre

27.1 Introduction 27-1
27.2 Global Soil Maps and Databases 27-2
 The FAO-UNESCO *Soil Map of the World* • The Harmonized World Soil Database
27.3 Regional Soil and Terrain Databases (SOTER) 27-4
27.4 ISRIC–WISE Harmonized Global Soil Profile Database 27-6
27.5 Digital Soil Mapping 27-6
27.6 Conclusions and Recommendations 27-8
 Conclusions • Recommendations
References..27-8

27.1 Introduction

Soil databases take various forms and have evolved quite rapidly recently. About 20 years ago, most soil data were produced as paper soil maps accompanied by a soil survey report, which usually contained thick annexes with soil profile descriptions and laboratory data. Nowadays, however, soil information is most likely to come stored in a Geographical Information System (GIS) accompanied by a digital database containing the soil profile information linked as point data to the corresponding soil mapping unit polygon. The data sets are stored in various media ranging from stand-alone media like CD-ROMs to globally accessible Internet GIS servers. Queries can vary as well: from locally operated GIS software for the fully trained GIS operators toward full internet GIS server functionality for nonspecialists.

While this technological revolution is in no way a guarantee for an enhanced quality of the data itself, it has assured that presentation of soil data has become more accessible to potential users and decision makers. This is true regardless of the geographic location, as computers and GIS software are nowadays to be found in all but the most remote places. Fast Internet access, on the other hand, still remains largely limited to the developed world, although developing countries are fast catching up.

Soil survey is at a crossroad. Over the last several years, pessimistic views have been expressed about its future. Some of the reasons put forward are external to soil survey and are strongly influenced by the general economic situation. Inappropriate presentation and the poor accuracy of soil information, together with high survey costs, are often to blame. The worldwide crisis in collecting primary field data in general may also partly be blamed by the overreliance on the use of satellite imagery as the ideal tool to carry out natural resources inventories. Although satellite imagery is an extremely useful tool to assist surveys, even at the highest resolution it limits observation mainly to surface features like land cover. The actual land use and underlying soil layers are largely undetectable to even the high resolution sensors of recent satellites, although a number of soil properties can now be detected with a certain degree of confidence (Hartemink et al., 2008).

Another factor responsible for this field data crisis is political and is particularly true for the collection of soil data in Western Europe where most national soil survey institutes have closed down or were privatized and have all abandoned systematic soil mapping. This is particularly critical for a region that with few exceptions (Belgium, Denmark, the Netherlands, United Kingdom) had a scant and insufficient soil information system before privatization, compared to say, Eastern Europe and the United States.

The situation in tropical countries, or in the developing world in general, is more varied in this respect, as overall there are still relatively strong soil institutes in most countries backed by, albeit insufficient, government funds. Systematic surveys are limited but still continue in most countries and have resulted in national maps and databases that with regard to detail of scale and number of available soil profiles can easily compete with those of countries in the industrialized world. Curiously, the so-called recession in carrying out soil surveys thus appears to have affected developing countries less than developed ones

(Hodgson, 1991). However, when data concerning the European Community (Lee, 1991) are compared to those of the less advanced countries (Zinck, 1990), the differences in soil survey status are less than might be initially expected. This is because there is renewed interest at the EC level as a whole (the EU Soil Thematic Strategy, the Soil Protection law in Germany, new national soil mapping program in Italy). Nowadays, inequalities between developing and developed countries appear to be deepening rather than shrinking.

A last introductory remark concerns the problem of information access, which is not unique for soil data but applies for many other natural resources data (climate, geology, etc.) as well. The problem appears most acute in Europe, but recent developments point to an increase of data access problem in Africa and Asia as well.

In Europe, several phenomena have accelerated data access problems. The first event was the privatization process that started in the late 1970s and resulted in national soil institutes being no longer subsidized as was the case in the past. The only valuable money source present in these institutes, apart from human resources, appeared to be the data and maps that had been compiled over the years. These were consequently marketed and commercialized and seen as an extra source of income. Strict copyright rules were put in place to avoid misuse and illegal reproduction of the data.

Another, less well-documented phenomenon causing data access problems appears to be the accelerated regionalization within the EU in which the power of some regions within countries (particularly in Germany, but also in Italy, France, and Belgium) have become much more expressed. These countries have legally limited the data access that national institutes are allowed to extend to interested researchers or commercial companies. As international organizations can only discuss data availability with national entities, the problem becomes more difficult to solve.

The approach of UN agencies such as the Food and Agriculture Organization (FAO) is that data should be made available at their production and maintenance costs only.

27.2 Global Soil Maps and Databases

27.2.1 The FAO-UNESCO *Soil Map of the World*

At the global level, the 1:5 M scale *Soil Map of the World* (FAO-UNESCO, 1971–1981) was until recently, nearly 30 years after its finalization, the only worldwide, consistent, harmonized soil inventory that was readily available in digital format and came with a set of estimated soil properties for each mapping unit.

The *Soil Map of the World* project originated by a motion of the International Society of Soil Science (ISSS) at the Wisconsin congress in 1960, started in 1961 and was completed over a span of 20 years. The first draft of the *Soil Map of the World* was presented to the Ninth Congress of the ISSS, in Adelaide, Australia, in 1968. The first sheets, those covering South America, were issued in 1971. The last and final map sheet for Europe appeared in 1981 (FAO-UNESCO, 1971–1981).

With the rapidly advancing computer technology and the expansion of GISs during the 1980s, the *Soil Map of the World* was first digitized by ESRI (1984) in vector format. Subsequently, a first rasterized version of the soil map was prepared by Zobler (1986) using the ESRI map as a base and using 1 × 1 grid cells. Only the dominant FAO soil unit in each cell was indicated; this digital product gained popularity because of its simplicity and ease of use, particularly in the United States, but should no longer be used.

FAO (1995a) produced its own raster version that had the finest resolution with a 5′ × 5′ cell size (9 km × 9 km at the equator) and contained a full database corresponding with the information in the paper map in terms of composition of the soil units, topsoil texture, slope class, and soil phase in each of the more than 5000 mapping units. In addition to the vector and raster maps discussed above, the CD-ROM contains a large number of databases and digital maps based on statistically derived soil properties (pH, organic carbon, C/N, soil moisture storage capacity, soil depth, etc.) for predefined depth zones (0–30 and 30–100 cm). The CD-ROM also contains interpretation by country on the extent of specific problem soils, the fertility capability classification results by country, and corresponding maps (FAO and IIASA, 2002).

27.2.2 The Harmonized World Soil Database

Recognizing the urgent need for improved soil information worldwide, particularly in the context of the Climate Change Convention and the Kyoto Protocol for soil carbon measurements and the immediate requirement for the update of the FAO/International Institute for Applied Systems Analysis (IIASA) Global Agro-ecological Assessment study (FAO-IIASA, 2002), the FAO of the United Nations and the IIASA took the initiative of combining the recently collected vast volumes of regional and national updates of soil information with the information already contained within the 1:5,000,000 scale FAO-UNESCO Digital *Soil Map of the World* into a new comprehensive Harmonized World Soil Database (HWSD; FAO/IIASA/ISRIC/ISSCAS/JRC, 2008).

This state-of-the-art database was achieved in partnership with

- The International Soil Reference and Information Centre (ISRIC-World Soil Information) together with FAO, responsible for the development of regional soil and terrain databases (SOTER, see Section 27.3) and the ISRIC-WISE soil profile database (see Section 27.4)
- The European Soil Bureau Network (ESBN), which had recently completed a major update of soil information for Europe and northern Eurasia
- The Institute of Soil Science, Chinese Academy of Sciences that provided the recent 1:1,000,000 scale soil map of China

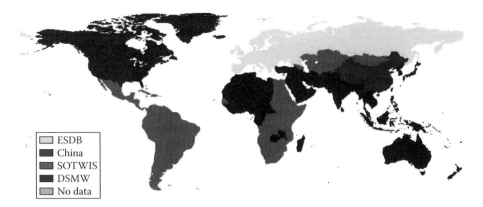

FIGURE 27.1 Source material for the Harmonized World Soil Database. (ESDB: European Soil Database; China: Soil Map of China 1:1 M; SOTWIS: secondary databases derived from SOTER and WISE; DSMW: Digital *Soil Map of the World*.)

In order to estimate soil properties in a harmonized way, the use of actual soil profile data and the development of pedo-transfer rules were undertaken in cooperation with ISRIC and ESBN drawing on the ISRIC-WISE soil profile database and the earlier work of Van Ranst et al. (1995), Batjes et al. (1997, 2007), and Batjes (2002).

The harmonization and data entry in a GIS was assured at IIASA and verification of the database was undertaken by all partners. As the main aim of the product is to be of practical use to modelers and to serve perspective studies in agro-ecological zoning, food security, and climate change impacts (among others), a resolution of about 1 km (30 as × 30 as) was selected. Over 15,000 different soil mapping units are recognized in the HWSD; source materials are shown in Figure 27.1.

The resulting raster database consists of 21,600 rows and 43,200 columns, which are linked to harmonized attribute data.

The use of a standardized structure allows linkage of the attribute data with GIS to display or query the map unit composition in terms of soil units and the characterization of selected soil parameters (organic carbon, pH, soil moisture storage capacity, soil depth, cation exchange capacity of the soil and the clay fraction, total exchangeable nutrients, lime and gypsum contents, sodium exchange percentage, salinity, textural class, and granulometry) that can be used in a wide range of environmental assessments at global level. Soil reaction (pH), for example (Figure 27.2), provides a measure for the solubility of various compounds in a soil, the relative bonding of nutrient ions to exchange sites, and the activity of various microorganisms.

The HWSD map, database, and a viewer are available on CD-ROM and freely downloadable from FAO/IIASA/ISRIC/ISSCAS/JRC (2008).

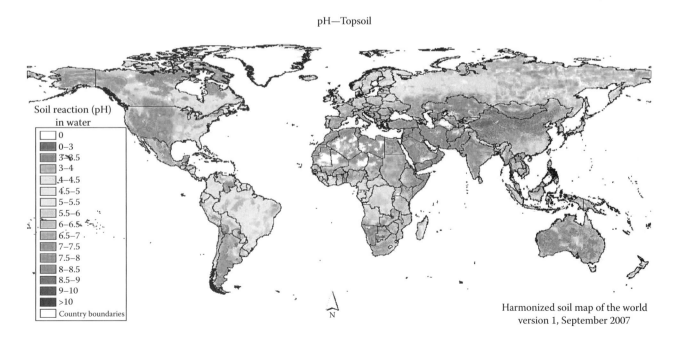

FIGURE 27.2 Global distribution of soil reaction (pH; 0–30 cm) according to HWSD.

Reliability of the information presented on the HWSD is variable: the parts of the database that still make use of the *Soil Map of the World* such as North America, Australia, West Africa, and South Asia are considered less reliable, while most of the areas covered by SOTER databases are considered to have the highest reliability (Southern Africa, Latin America and the Caribbean, Central and Eastern Europe).

Further expansion and update of the HWSD is foreseen for the near future, notably with the information held in the U.S. Natural Resources Conservation Service's U.S. General Soils Map (STATSGO, 2009), the Canadian Soil Database (Agriculture and Agrifood Canada, 2009), and the soil database for Australia (CSIRO/ACLEP, 2009).

27.3 Regional Soil and Terrain Databases (SOTER)

The discussion paper "Towards a Global Soil Resources Inventory at Scale 1:1 Million" (Sombroek, 1984) was an initiator for an ISSS (now International Union of Soil Sciences—IUSS) program to update and expand the global spatial soil information. In 1987, the idea was further elaborated and resulted in the development of the SOTER methodology (FAO, 1995b; Van Engelen and Wen, 1995).

FAO recognized that a rapid update of the *Soil Map of the World* would be a feasible option if the original map scale of 1:5 M was retained, and started, together with the United Nations Environment Programme (UNEP) and ISRIC, to fund national updates at a 1:5 M scale of soil maps in Latin America and Northern Asia following the standardized SOTER approach. These parallel programs of ISRIC, UNEP, and FAO merged together in mid-1995, when the three major partners agreed to join the concerned resources and work toward a common world SOTER approach covering the globe at a 1:1 M scale. Since then, other international organizations have shown support and collaborated to develop SOTER databases for specific regions. This is, for instance, the case for Northern and Central Eurasia, where IIASA joined FAO and the national institutes involved (FAO and IIASA, 1999).

Basic in the SOTER approach is the mapping of tracts of land (SOTER units) with a distinctive, often repetitive, pattern of landform, soil parent material, surface form, slope, and soil (Van Engelen and Wen, 1995). Each SOTER unit represents a unique combination of terrain and soil characteristics; Figure 27.3 shows how the terrain and soils are represented by a SOTER unit in the database.

It should be noted that although the information is collected according to the same SOTER methodology, the specific level of information in each region results in a variable scale of the end products presented. The soils and terrain database for northeastern Africa (FAO et al., 1998a), for instance, contains information at equivalent scales between 1:1 and 1:2 M, but the soil profile information is not geo-referenced. For north and central Eurasia, profile information is very limited (FAO and IIASA, 1999). Fully comprehensive information is available for the SOTER database of South and Central America and the Caribbean (FAO et al., 1998b; Dijkshoorn et al., 2005) and includes more than 1800

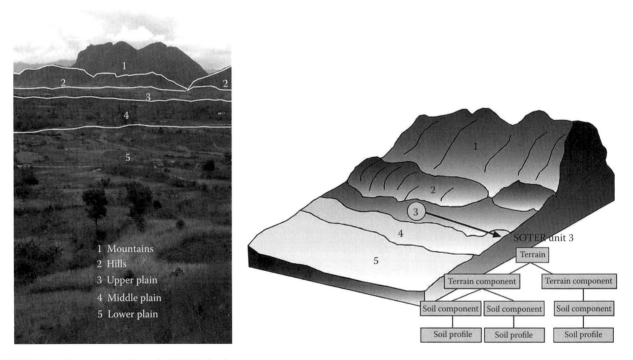

FIGURE 27.3 Terrain and soils in the SOTER database.

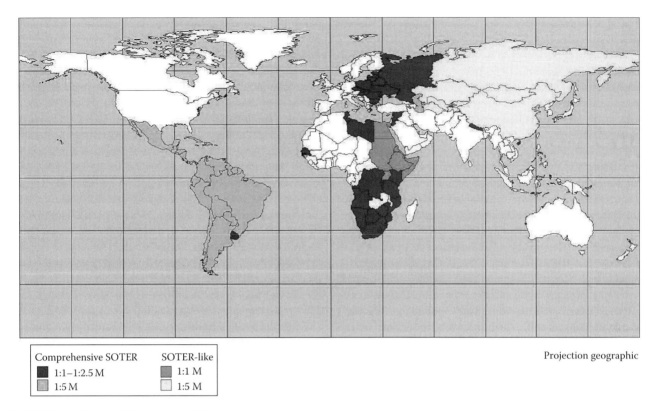

FIGURE 27.4 World SOTER status in 2009.

geo-referenced soil profiles. The SOTER database for Central and Eastern Europe (1:2.5 M scale) contains over 600 geo-referenced soil profiles (FAO and ISRIC, 2000). The scales of the SOTER database of Southern Africa (FAO and ISRIC, 2003) and of Central Africa (FAO et al., 2007) range between 1:1 M for most countries and 1:2 M for Angola and the Democratic Republic of the Congo.

The current world coverage by SOTER is given in Figure 27.4; comprehensive SOTER databases are developed according to the methodology described by Van Engelen and Wen (1995). Conversely, SOTER-like databases characterize the soil and terrain conditions but they do not contain any measured soil profile data.

Soil analytical data held in comprehensive SOTER have been analyzed according to a range of methods and in various laboratories, and these are documented in the various data sets. Generally, however, there are many gaps in the measured soil analytical data stored in primary SOTER databases—typically, not all the attributes that can be handled in SOTER have been collected during the ground surveys that provided the legacy profile data. Such gaps in the attribute data will often preclude the direct use of primary SOTER databases for modeling.

For practical reasons, a distinction is therefore made between (1) primary SOTER databases that provide full information on terrain unit composition in terms of landform, parent materials, as well as a suite of geo-referenced soil profile data with inherent gaps in the soil analytical data, and (2) *secondary* or *SOTWIS* databases in which gaps in the measured profile data have been filled using consistent taxotransfer procedures for selected soil attributes thus facilitating modeling; for details see Section 27.4. Both types of databases may be downloaded through the ISRIC web site (www.isric.org) using the "soil data on-line" button.

An update of the SOTER methodology started in 2008 within e-SOTER, an EU Seventh Framework Program financed research project with 14 partners from Europe, Morocco, and China. The project addresses four major barriers to a comprehensive soil observing system:

- Morphometric descriptions—enabling quantitative mapping of landforms (Dobos et al., 2005) as opposed to crude slope categories as well as newly developed DEM analysis using natural breaks
- Soil parent material characterization and pattern recognition by remote sensing and using legacy data will enable a separation of soil processes within the landscape and will generate a parent material classification relevant for soil development
- Soil pattern recognition by remote sensing will generate additional predictors of soil properties
- Standardization of methods and measures of soil attributes to convert legacy data already held in the European Geographical Soil Database and various national databases to a common standard so that they may be applied, for example, in predictive and descriptive models of soil behavior

27.4 ISRIC–WISE Harmonized Global Soil Profile Database

In the early 1990s, staff at ISRIC developed a uniform methodology for a global soil database in the framework of a project entitled "World Inventory of Soil Emission Potentials" (ISRIC–WISE). WISE was especially conceived for a geographical quantification of main soil factors that control processes of global change on a broad scale (Batjes and Bridges, 1994; Batjes et al., 1995).

Profiles in WISE have been classified according to the original and the revised Legend of the *Soil Map of the World* (FAO-Unesco, 1974; FAO, 1988). Thereby, interpretations derived from primary (measured) data held in WISE can be linked to the spatial units of both digitized and digital soil maps that use either of these Legends (e.g., Batjes, 2002; Batjes et al., 2007; FAO/IIASA/ISRIC/ISSCAS/JRC, 2008).

In the process of collating materials for compilation of the ISRIC-WISE profile database, the quality and validity of the original data had to be evaluated carefully, while at the same time recognizing that these are the only materials available. The description status of the various profiles has been documented in WISE to provide a coarse indicator for the inferred reliability of the source data (Table 27.1); typically, status four (low) profiles are purged from WISE when better data become available for the corresponding soil units.

The latest public domain release of WISE, version 3.1, comprises data for some 10,250 profiles with some 47,800 horizons, from 149 countries (Batjes, 2008, 2009), as opposed to some 4,350 profiles in an earlier version (Batjes, 1999). Most profiles are from Africa (41%), followed by Asia (18%), South America (18%), and Europe (13%) (Table 27.1); their approximate location is shown in Figure 27.5.

TABLE 27.1 Number of Profiles in WISE3 by Continent and Their Description Status

Continent	Profile Description Status[a]				Total
	1	2	3	4	
Africa	421	1337	2392	23	4173
Asia	441	970	426	10	1847
Antarctica	4	6	0	0	10
Europe	225	712	359	20	1316
North America	495	222	127	11	855
Oceania	50	49	106	4	209
South America	149	1380	313	1	1843
Total	1785	4676	3723	69	10253

Source: Batjes, N.H. 2009. Harmonized soil profile data for applications at global and continental scales: Updates to the WISE database. Soil Use Manage. 25:124–127.

[a] The number code under *profile description status* refers to the completeness and apparent reliability of the soil profile descriptions and accompanying analytical data for the specified profile in the original source; the status is the highest for 1 and the lowest for 4 (see FAO, 2006). Continents are defined according to Times Atlas (2003).

Profile data in WISE3 originate from over 260 different sources, both analog and digital. Some 40% of the profiles were extracted from auxiliary datasets, including the ISRIC Soil Information System (ISIS), the NSSC Soil Survey Laboratory Soil Characterization Database (USDA-NRCS, 2008), various SOTER databases, and the FAO Soil Database (FAO-SDB).

Overall, chemical and physical analyses have taken place in at least 150 laboratories worldwide, using a range of methods; these are described in broad terms in WISE. Analytical methods used in ISIS and the NSSC collection may be considered to be similar (van Reeuwijk, 1983; Vogel, 1994). Conversely, for the other sources, methods typically vary from one laboratory to the next, even within one country, and may change over time within a single laboratory. These methodological differences complicate the worldwide comparison of soil analytical data—no single solution for addressing this issue has been found as yet (Wilson, 1979; van Reeuwijk, 1983; Vogel, 1994; FAO, 1995a, 1995b; Batjes et al., 1997; Breuning-Madsen and Jones, 1998; Hollis et al., 2006; FAO/IIASA/ISRIC/ISSCAS/JRC, 2008). As a result, the amount of measured data available for modeling is sometimes much less than expected. Conversely, the issue of comparability of soil analytical data should always be considered in relation to the scale at which the derived data are to be used. Adroit use of the data stored in WISE will permit a wide range of agricultural and environmental applications at global and continental scales (1:500,000 and broader). Published examples of such applications include, the development of harmonized sets of derived soil properties for the main soil types of the world, gap-filling in primary SOTER databases using consistent pedo-transfer procedures (resulting in so-called secondary SOTWIS databases), global modeling of environmental change, analyses of global ecosystems, upscaling and downscaling of greenhouse gas emissions, and crop simulation and agro-ecological zoning (see ISRIC-WISE, 2008).

An important application of large, harmonized profile databases such as WISE remains making other databases, such as primary SOTER sets, widely usable. Without supporting auxiliary soil profile data that permit thorough statistical analyses and the development of pedo-transfer procedures, such databases would remain fraught with numerous gaps in the measured data thereby precluding modeling applications. Further, by implication, this would preclude the expansion of derived soil products, such as the HWSD, that draw heavily on WISE-derived data. Large profile databases such as WISE are also needed for studying relationships in soil distribution and for providing soil legacy data for the Digital Soil Mapping project discussed in Section 27.5.

27.5 Digital Soil Mapping

In 2006, a consortium of scientific institutes, centers, and universities launched an appeal to use the latest satellite technology and new information layers, such as the recently

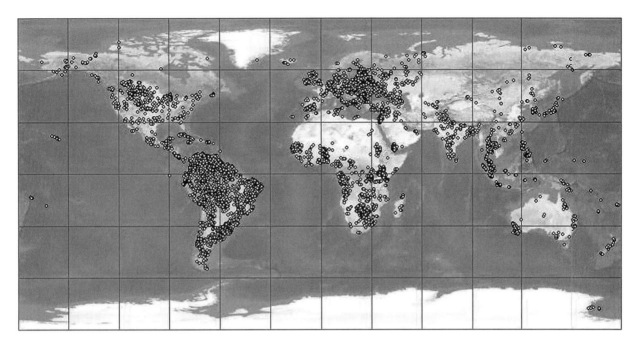

FIGURE 27.5 (See color insert.) Global distribution of geo-referenced soil profiles in WISE3. (From Batjes, N.H. 2008. ISRIC-WISE harmonized global soil profile dataset (Version 3.1). Report 2008/02. ISRIC—World Soil Information, Wageningen, the Netherlands. (http://www.isric.org/isric/webdocs/docs/ISRIC_Report_2008_02.pdf).

released topographic information at 90 m resolution, to achieve a spatial database of soil properties, based on a statistical sample of landscapes (see GlobalSoilMap.net, 2009). Within the sample or satellite sites, field sampling is used to determine the spatial distribution of soil properties in order to develop reflectance spectral libraries for characterization of soil properties (Shepherd and Walsh, 2002). These are then used to predict soil properties in areas not sampled, also using covariates reflecting state factors of soil formation; details may be found elsewhere (see McBratney et al., 2003; Lagacherie et al., 2006; Hartemink et al., 2008). The resulting digital soil maps describe the uncertainties associated with such predictions and, when based on time-series data, can also provide information on dynamic soil properties. Maps derived from digital soil mapping differ from conventional, polygon-based maps in that they are pixel-based and thus can be more easily displayed at a higher resolution than those currently used by other earth and social sciences (Sanchez et al., 2009).

The GlobalSoilMap.net (2009) project, funded by the Bill & Melinda Gates Foundation in 2008, started with a pilot covering sub-Saharan Africa (AfSIS). Regional nodes in other continents were also established.

As summarized by Sanchez et al. (2009), GlobalSoilMap.net proceeds in five steps with the ultimate aim to develop evidence-based soil management recommendations at very high resolution. The first three steps are relevant here as they concern the establishment of a digital soil map. The first step, Data Input, starts with the production of base maps that are compiled from topography, climatic information, land cover, and geological variables relating to soil parent materials (using airborne gamma radiometric spectroscopy). Pre-existing geo-referenced field and laboratory measurements, also known as conventional or soil legacy data, are needed at this stage (Carre et al., 2007). Secondly, soil properties are estimated using soil probability functions that express the probabilities of occurrence of a certain predictor value. During the initial stages, six soil properties will be considered: clay content, organic carbon content, pH, estimated cation-exchange capacity, electrical conductivity, and bulk density. These may be used to generate a range of maps as discussed by Sanchez et al. (2009). Finally, during the third step, spatially inferred soil properties are used to predict more difficult-to-measure soil functions such as available soil water storage, carbon density, and phosphorus fixation. This is achieved using pedo-transfer functions.

An innovative element of the approach is that the overall uncertainty of the prediction will be determined by combining uncertainties of the input data, a spatial inference model, and soil functions used.

Several pilot projects are underway to support the theories that underlie GlobalSoilMap.net of which work in Australia is probably the most advanced. Experimentation with various approaches is gradually leading toward consensus. Conversely, as with any new approach, a number of scientific and operational challenges still need to be resolved; these have been discussed in detail by various authors (see Lagacherie et al., 2006; Hartemink et al., 2008).

27.6 Conclusions and Recommendations

27.6.1 Conclusions

1. Conventional soil survey and mapping have declined considerably in many areas of the globe, while at the same time the development of new technologies have favored the enhanced development of soil property databases and models to predict, at high resolution, certain soil characteristics of interest for agriculture or the environment.
2. The most significant global and regional soil maps and databases are nowadays freely available and downloadable online. Conversely, national and subnational soil data access remains a problem or comes at a cost in many countries.
3. The main purpose of the FAO-UNESCO *Soil Map of the World* was to illustrate the geographical distribution of different major soil types globally. More recent regional efforts have focused on quantifying a selected number of soil properties for defined soil units. Soil classification remains an important element in these mapping approaches. Generally, the spatial and attribute uncertainty is still poorly quantified in these products and the spatial resolution remains coarse.
4. SOTER and HWSD are close to achieve satisfactory developing world coverage, except for South East Asia and West Africa, but are lacking recent information on the industrial world with the exception of Europe.
5. Soil profile databases such as WISE have had to overcome numerous problems of data access and the lack of harmonized soil analytical methods, but have reached a stage where impressive world coverage is achieved. Significant geographical areas (Central and Northern Eurasia, Deserts) still lack adequate soil data however.
6. The digital soil mapping project has just started with a pilot focusing on Africa. Once tested and validated, the resulting techniques will allow quantifying the overall uncertainty of the predictions by combining uncertainties of the input data, spatial inference model, and soil functions. Conversely, in these approaches, one may move away from understanding the soil distribution in a landscape to a statistical representation of soil properties for soil management purposes.

27.6.2 Recommendations

1. Policy makers should be made aware of the importance of good quality soil data to address major challenges facing the world be it climatic change mitigation, securing food security, or remedying water scarcity. The soil science community has a key role to play in this process—cross-correlations between uncertain variables will determine how uncertainties will propagate in a modeling study, and accuracy levels considered to be acceptable will vary with the scale and type of questions being asked (Finke, 2006; Heuvelink and Brown, 2006).
2. Sufficient resources should be made available to complete the SOTER/HWSD approach with data from the industrial world and some of the other geographical areas still missing (notably West Africa and Southeast Asia). Ultimately, this information, complemented with WISE profile data for so far underrepresented regions, should be used to compile a world covering SOTER product.
3. In conjunction with the above, new digital soil mapping approaches, as currently being developed, tested, and validated for sub-Saharan Africa in the framework of GlobalSoilMap.net, should be expanded to the other continents.

References

Agriculture and Agrifood Canada. 2009. Canada national soil database (http://sis.agr.gc.ca/cansis/nsdb/intro.html). Accessed on 6 July 2011.

Batjes, N.H. 1999. Development of a 0.5 by 0.5 resolution global soil database, p. H29–H40. *In* M.E. Sumner (ed.) Handbook of soil science. CRC Press, Boca Raton, FL.

Batjes, N.H. 2002. Revised soil parameter estimates for the soil types of the world. Soil Use Manage. 18:232–235.

Batjes, N.H. 2008. ISRIC-WISE harmonized global soil profile dataset (Version 3.1). Report 2008/02. ISRIC—World Soil Information, Wageningen, the Netherlands. (http://www.isric.org/isric/webdocs/docs/ISRIC_Report_2008_02.pdf). Accessed on 6 July 2011.

Batjes, N.H. 2009. Harmonized soil profile data for applications at global and continental scales: Updates to the WISE database. Soil Use Manage. 25:124–127.

Batjes, N.H., R. Al-Adamat, T. Bhattacharyya, M. Bernoux, C.E.P. Cerri, P. Gicheru, P. Kamoni, E. Milne, D.K. Pal, and Z. Rawajfih. 2007. Preparation of consistent soil data sets for SOC modelling purposes: Secondary SOTER data sets for four case study areas. Agric. Ecosyst. Environ. 122:26–34.

Batjes, N.H., and E.M. Bridges. 1994. Potential emissions of radiatively active gases from soil to atmosphere with special reference to methane: Development of a global database (WISE). J. Geophys. Res. 99:16479–16489.

Batjes, N.H., E.M. Bridges, and F.O. Nachtergaele. 1995. World inventory of soil emission potentials: Development of a global soil data base of process-controlling factors, p. 102–115. *In* S. Peng, K.T. Ingram, H.U. Neue, and L.H. Ziska (eds.) Climate change and rice. Springer-Verlag, Heidelberg, Germany.

Batjes, N.H., G. Fischer, F.O. Nachtergaele, V.S. Stolbovoy, and H.T. van Velthuizen. 1997. Soil data derived from WISE for use in global and regional AEZ studies (Version 1.0). Interim Report IR-97-025. FAO/IIASA/ISRIC, Laxenburg, Austria. http://www.iiasa.ac.at/Admin/PUB/Documents/IR-97-025.pdf. Accessed on 6 July 2011.

Breuning-Madsen, B.H., and R.J.A. Jones. 1998. Towards a European soil profile analytical database, p. 43–49. *In* H.J. Heineke et al. (eds.) Land information systems: Developments for planning the sustainable use of land resources. Office for Official Publications of the European Community, Luxembourg.

Carre, F., A.B. McBratney, and B. Minasny. 2007. Estimation and potential improvement of the quality of legacy soil samples for digital soil mapping. Geoderma 141:1–14.

CSIRO/ACLEP. 2009. ACLEP Natural Heritage Trust and National Land and Water Audit. ASRIS at http://www.asris.csiro.au/index_other.html. Accessed on 6 July 2010.

Dijkshoorn, J.A., J.R.M. Huting, and P. Tempel. 2005. Update of the 1:5 million soil and terrain database for Latin America and the Caribbean (SOTERLAC; Version 2.0). ISRIC Report 2005/01. ISRIC—World Soil Information, Wageningen, the Netherlands.

Dobos, E., L. Daroussin, and L. Montanarella. 2005. An SRTM-based procedure to delineate SOTER Terrain Units on 1:1 and 1:5 million scales. EUR 21571 EN, Institute for Environment and Sustainability, Joint Research Centre, Ispra, Italy.

ESRI. 1984. Final Report UNEP/FAO World and Africa GIS Database. Environmental Systems Research Institute, Redlands, CA.

FAO-UNESCO. 1974. Soil map of the world, 1:5,000,000. Vol. 1. Legend, United Nations Educational, Scientific, and Cultural Organization, Paris, France.

FAO-UNESCO. 1971–1981. Soil map of the world, 1:5,000,000. Vol. 1–10. United Nations Educational, Scientific, and Cultural Organization, Paris, France.

FAO. 1988. FAO-UNESCO soil map of the world, revised legend, with corrections and updates. World Soil Resources Report 60. FAO, Rome; reprinted with updates as Technical Paper 20 by ISRIC, Wageningen, the Netherlands, 1997.

FAO. 1995a. Digital soil map of the world and derived soil properties. Land and water digital media series 1. Food and Agriculture Organization of the United Nations, Rome, Italy.

FAO. 1995b. Global and national soil and terrain databases (SOTER). Procedures manual (revised edition). World Soil Resources Report 74. FAO, Rome, Italy.

FAO. 2006. Guidelines for soil description. 4th Edn. FAO, Rome, Italy.

FAO, Cooperazione-Italiana, and IGAD. 1998a. The soil and terrain database for northeastern Africa. Crop production system zones of the IGAD subregion. Land and water digital media series 2. FAO, Rome, Italy.

FAO and IIASA. 1999. Soil and physiographic database for north and central Eurasia at 1:5 Million scale. Land and water digital media series 7. FAO, Rome, Italy.

FAO and IIASA. 2002. Global agro-ecological assessment for Agriculture in the 21st Century. Research Report RR-02-02. IIASA, Laxenburg, Austria.

FAO and ISRIC. 2000. Soil and terrain database, land degradation status and soil vulnerability assessment for central and eastern Europe (Version 1.0), 1:2.5 million scale. Land and water digital media series 10. FAO, Rome, Italy.

FAO and ISRIC. 2003. Soil and terrain database for southern Africa (1:2 million scale). Land and water digital media series 25. FAO, Rome, Italy.

FAO, ISRIC, and Ghent University. 2007. Soil and terrain database for central Africa. Land and water digital media series 33. FAO, Rome, Italy.

FAO, ISRIC, UNEP, and CIP. 1998b. Soil and terrain database for Latin America and the Caribbean—SOTERLAC (Version1.0). Land and water digital media series 5. FAO, Rome, Italy.

FAO/IIASA/ISRIC/ISSCAS/JRC. 2008. Harmonized world soil database (Version 1.0) Food and Agriculture Organization of the United Nations, International Institute for Applied Systems Analysis, ISRIC—World Soil Information, Institute of Soil Science—Chinese Academy of Sciences, Joint Research Centre of the European Commission, Laxenburg, Austria. (http://www.iiasa.ac.at/Research/LUC/External-World-soil-database/HTML/index.html?sb=1) Accessed on 6 July 2011; also available as: Land and Water Digital Media Series 35. FAO, Rome, Italy.

Finke, P. 2006. Quality assessment of digital soil maps: Producers and users perspectives, p. 523–541. *In* P. Lagacherie, A. McBratney, and M. Voltz (eds.) Digital soil mapping: An introductory perspective. Elsevier, Amsterdam, the Netherlands.

GlobalSoilMap.net. 2009. Project overview available at: http://www.globalsoilmap.net/

Hartemink, A.E., A. Mc Bratney, and M. Mendonca-Santos. 2008. Digital soil mapping with limited data. Springer-Verlag, New York.

Heuvelink, G.B.M., and J.D. Brown. 2006. Towards a soil information system for uncertain soil data, p. 97–106. *In* P. Lagacherie, A. McBratney, and M. Voltz (eds.) Digital soil mapping: An introductory perspective. Elsevier, Amsterdam, the Netherlands.

Hodgson, J.M. (ed.). 1991. Soil survey, a basis for European soil protection. CEC, Brussels, Belgium.

Hollis, J.M. et al. 2006. SPADE-2: The soil profile analytical database for Europe (Version 1.0). European Soil Bureau Research Report No. 19 (EUR 22127 EN). Office for Official Publications of the European Communities, Luxembourg.

ISRIC-WISE. 2008. References and examples of WISE database applications. http://www.isric.org/content/wise-publications. Accessed on 6 July 2011.

Lagacherie, P., A. McBratney, and M. Voltz. 2006. Digital soil mapping: An introductory perspective. Elsevier, Amsterdam, the Netherlands.

Lee, J. 1991. The European soil map and beyond, p. 197–209. *In* J.M. Hodgson (ed.) Soil survey, a basis for European soil protection. CEC, Brussels, Belgium.

McBratney, A.B., M.L. Mendonça-Santos, and B. Minasny. 2003. On digital soil mapping. Geoderma 117:3–52.

Sanchez, P.A. et al. 2009. Digital soil map of the world. Science 325:680–681.

Shepherd, K., and M. Walsh. 2002. Development of reflectance spectral libraries for characterization of soil properties. Soil Sci. Soc. Am. J. 66:988–998.

Sombroek, W.G. 1984. Towards a global soil resource inventory at scale 1:1 million. Working Paper and Preprint 84/04, ISRIC, Wageningen, the Netherlands.

STATSGO. 2009. US General Soil Map. http://soils.usda.gov/survey/geography/statsgo/. Accessed on 6 July 2011.

Times Atlas. 2003. The comprehensive times atlas of the world. Times Books, Harper Collins Publishers, London, U.K.

USDA-NRCS. 2008. NSSC soil survey laboratory soil characterization database. United States Department of Agriculture, Natural Resources Conservation Service, Lincoln, U.K. http://ssldata.nrcs.usda.gov/. Accessed on 6 July 2011.

van Engelen, V.W.P., and T.T. Wen. 1995. Global and national soils and terrain digital databases (SOTER), Procedures Manual (revised edition). FAO, ISSS, ISRIC, Wageningen, the Netherlands.

Van Ranst, E., L. Vanmechelen, A.J. Thomasson, J. Daroussin, J.M. Hollis, R.J.A. Jones, M. Jamagne, and D. King. 1995. Elaboration of an extended knowledge base to interpret the 1:1 000 000 EU Soil Map for environmental purposes, p. 71–84. *In* D. King, R.J.A. Jones, and A.J. Thomasson (eds.) European land information systems for agro-environmental monitoring. EUR 16232 EN, Office for Official Publications of the European Communities, Luxemburg.

van Reeuwijk, L.P. 1983. On the way to improve international soil classification and correlation: The variability of soil analytical data. Annual Report 1983. ISRIC, Wageningen, the Netherlands. http://www.isric.org/isric/webdocs/docs/ISRIC_TechPap09_2002.pdf. Accessed on 6 July 2011.

Vogel, A.W. 1994. Comparability of soil analytical data: Determinations of cation exchange capacity, organic carbon, soil reaction, bulk density, and volume percentage of water at selected pF values by different methods. Working Paper 94/07. ISRIC, Wageningen, the Netherlands http://www.isric.org/isric/webdocs/docs/ISRIC_Report_1994-07.pdf. Accessed on 6 July 2011.

Wilson, A.L. 1979. Approach for achieving comparable analytical results from a number of laboratories. Analyst 104:1273–1288.

Zinck, J.A., 1990. Soil survey: Epistemology of a vital discipline. ITC J. 1990:335–351.

Zobler, L. 1986. A world soil file for global climate modeling. Technical Memorandum 87802, NASA Goddard Institute for Space Studies (GISS), New York.

28
United States Soil Survey Databases

Jim R. Fortner
United States Department of Agriculture

Alan B. Price
United States Department of Agriculture

28.1 Introduction 28-1
28.2 Description of Soil Survey Databases 28-1
  Soil Characterization Database • National Soil Information System • Official Series Descriptions and Soil Classification Database • Soil Survey Geographic Database • U.S. General Soil Map • Data Access
28.3 The Evolution of Soil Survey Map Unit Databases 28-6
  The First Generation • The Second Generation • The Third Generation
28.4 Standards for Soil Survey Data 28-9
28.5 Future Directions for Soil Information Systems 28-11
Acknowledgments 28-11
References 28-11

28.1 Introduction

Soil surveys have been conducted in the United States since about 1886. An organized soil survey program began in 1899. In 1953, the Secretary of Agriculture established the National Cooperative Soil Survey (NCSS), a group of federal and state agencies and universities loosely knit through cooperative agreements, in part to accelerate the completion of soil surveys. The USDA Natural Resources Conservation Service (NRCS), formerly the Soil Conservation Service (SCS), provides federal leadership to the NCSS. Soil surveys and supporting pedagogical studies conducted through the NCSS are complete for about 91% of all lands in the United States and its territories. About 96% of the privately owned land has a detailed soil survey, generally at a scale of 1:12,000–1:24,000. Information about soils is available in approximately 3000 county-level soil survey reports and databases. A digital soil survey at a scale of 1:250,000 for the United States (1:1 million for Alaska) is also available. Each year, between 50 and 100 new soil surveys with digital spatial and attribute data and corresponding soil survey reports are developed and published. Most completed soil surveys have been digitized and are available online.

Beginning in the late 1960s and early 1970s, NCSS soil scientists recognized the need for and the potential utility of automating the large quantity of soil survey information that was being generated. In the more than 30 years since, the NCSS has been an innovative leader in the adaptation and use of information technology. As a result, the demand for soil survey information is greater than it has ever been. A National Soil Information System (NASIS) was introduced in 1994 to improve the acquisition and management of these data. To improve the distribution of digital data, the Soil Data Mart (SDM) was implemented in 2003 and the Web Soil Survey (WSS) was implemented in 2005.

28.2 Description of Soil Survey Databases

The NRCS and NCSS develop and maintain six databases. These are the (1) Soil Characterization Database (SCD), (2) NASIS, (3) Official Series Descriptions (OSDs), (4) Soil Classification Database (SC), (5) Soil Survey Geographic Database (SSURGO), and (6) the U.S. General Soil Map (STATSGO). Each set of data was developed for a specific need or purpose. A simplified schema of how these data are related is presented in Figure 28.1. Information contained within these databases has been reviewed and certified in accordance with NCSS data-quality standards (Soil Survey Staff, 1995a, 2009).

28.2.1 Soil Characterization Database

This database stores the laboratory characterization of about 20,000 pedons from the United States and about 1,100 pedons from other countries. Soil profile descriptions are available for about 15,000 of these pedons. These data were produced primarily by the Soil Survey Laboratory (SSL) at the National Soil Survey Center (NSSC), Lincoln, Nebraska, and at previous SCS laboratories located in Beltsville, Maryland, and Riverside, California.

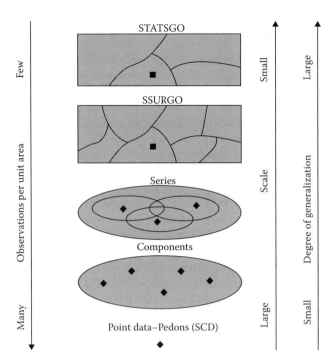

FIGURE 28.1 Spatial data relationships. (Modified from Waltman, 1997, personal communication.)

Sample preparation and analytical procedures are carried out according to standard methods (Soil Survey Staff, 1996). The NSSC has produced a guide for interpreting the information (Soil Survey Staff, 1995a). These pedons are sampled by NRCS scientists and NCSS cooperators at universities and at state and other federal agencies. Sites are usually selected to represent the concept of a soil series. Over time, as they are sampled and characterized, these pedons bracket a range of soil properties for a soil series or landscape. These data are used to populate the values for physical and chemical soil properties in the soil map unit data and OSDs and for research purposes, such as the development of environmental models.

Programming for the development of this database began at the SSL as early as 1966, when the calculation and reporting of bulk density, particle size, chemical, and mineralogical analysis were automated. These early efforts evolved into the development of a Laboratory Information Management System (LIMS) that collects, controls quality, calculates, and reports laboratory analysis results. As analytical data from sampling projects are completed, they are reviewed and verified. Data are then published in the Soil Characterization Database. These data are available via a public Web site at http://ssldata.nrcs.usda.gov/.

As part of the NCSS, many universities have developed SSL characterization data for the soil surveys in their respective states. Institutions with significant databases include the University of California at Davis, Clemson University, Colorado State University, University of Florida, Kansas State University, University of Idaho, University of Illinois, Louisiana State University, University of Maine, Michigan Technological University, Mississippi State University, University of Missouri, Montana State University, Ohio State University, Oregon State University, Pennsylvania State University, University of Tennessee, Texas A&M University, Virginia Polytechnic Institute & State University, and West Virginia University. These data are being accumulated into a national database and will be merged with data from SSL and made available on the Internet. They are also used to document physical and chemical soil property ranges of soil map unit components and OSDs.

28.2.2 National Soil Information System

NASIS is the transactional database and application used to collect, develop, store, and manage the soil survey data for the NCSS. Information about soil map units and map unit components is stored in NASIS. The data are developed from observations made during field mapping, including reconnaissance observations, transects, grid samples, and field measurements of soil properties. Each soil map unit is composed of one or more map unit components (phases of soil series, other taxa of higher categories, or nonsoil miscellaneous areas). The NASIS database does not limit the number of components within a soil map unit. Each component can have data stored about component percentages, slope, climate, elevation, erosion, flooding, ponding, and performance. Each component can have an unlimited number of soil horizons. Each horizon of a component is characterized by data about physical, chemical, and morphological properties and qualities. Analytical data from the SCD are used to assist in the population of horizon data. Soil interpretation ratings can be generated dynamically for components based on these data. The NASIS design accommodates the relatively easy addition of new data elements to the data model as new needs are identified.

In the NASIS database, soil survey area legends are linked to soil map units. Soil map units are linked to data map units that contain component and horizon level data. This capability is useful in correlation across political boundaries and for the retention of symbols found on earlier versions of maps. This structure also provides the capability of having a set of data available for each delineation (polygon), area, point, or line feature on the soil survey map. The relationship between horizon properties, horizons, components, and map unit polygons and point and line features is shown in Figure 28.2. Component objects and other attributes are assigned to polygons, points, or lines. Table 28.1 lists the physical, chemical, and morphological properties and other data stored either for horizons, components, or map units. To date, there are over 300 data elements.

The NASIS database also stores information about individual detailed soil profile descriptions (e.g., pedons), the site information associated with these descriptions, the relationship or grouping of associated sites, and how soil descriptions relate to transects and traverses. The collection and analysis of these data are essential in the development of soil map unit concepts and their associated components. Soil profile descriptions are linked to associated analytical data in the SCD and can be linked to soil map unit component records.

The soil map unit and corresponding component data are developed in conjunction with soil survey maps, generally at a

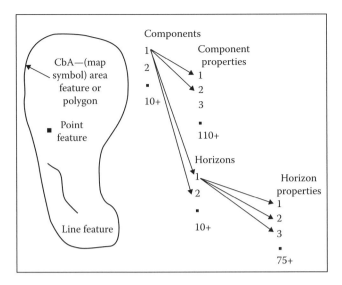

FIGURE 28.2 Data relationships.

scale of 1:12,000–1:24,000. Periodically, datasets are exported from NASIS and certified as official soil survey datasets. These datasets are the source from which most soil survey publications are produced and are provided to NRCS field office computers for use in natural resource planning activities. The data are also used by landowners, private and public land managers, federal, state, and local land use planners, environmental modelers, regulators, and others. The datasets are available through the SDM and WSS.

28.2.3 Official Series Descriptions and Soil Classification Database

The OSDs and associated SC contain the taxonomic classification of each soil series identified in the United States together with other information about the soil series, such as office of responsibility, series status, dates of origin and establishment, and geographic areas of usage. These databases are maintained by the soils staff of the NRCS Major Land Resources Area (MLRA) Soil Survey Regional Offices. Additions and changes are continually being made, resulting from the ongoing soil survey work and refinement of the soil classification system. The series is the lowest level of the soil taxonomic system used in the United States. (Soil Survey Staff, 2006). As the databases are updated, the changes are immediately available to the user, so the data retrieved are always the most current. The data are available through the Internet, which provides the capability of viewing the contents of individual series records, querying the database, and producing a report with the selected soils or national reports with all soils in the database. The standard reports available allow the user to display the soils by series name or by taxonomic classification. Maps showing the geographic distribution of each series can also be displayed. Future plans include the integration of these two databases into NASIS. OSDs are available for more than 23,000 soil series recognized in the United States and are maintained and made available over the Internet.

TABLE 28.1 Soil Data Elements and Interpretive Ratings Included in SSURGO and STATSGO Data

Component data

Component: name, percent, kind
Albedo, dry
Climatic data (MAAT, MAP, FFD)
Drainage class
Elevation
Flooding: frequency, duration, months of the year
Geomorphic setting description
Hydric soil rating and reasons
Hydrologic soil group
Land use: earth cover kind
Parent material
Ponding: depth, duration, frequency, months of the year
Restrictive layers: depth, hardness, kind
Slope: gradient, length, aspect
Soil moisture: monthly depth and status
Soil temperature: monthly depth
Surface rock fragments: percent, size, kind, shape, hardness
USDA Soil Taxonomy classification

Horizon data

Horizon designation and depths
AASHTO class and group
Available water capacity
Bulk density: 1/3, 1/10, 15 bar and oven dry
Calcium carbonate equivalent
Cation-exchange capacity: CEC, ECEC
Electrical conductivity
Erosion factors: Kw, Kf
Excavation difficulty class
Extractable Al and acidity
Free iron oxides
Gypsum
Linear extensibility percent
Liquid limit and plasticity index
Organic matter percent
Particle density
Percent of soil passing #4, 10, 40, 200 sieves
Percent sand, silt, and clay and fractions
Rock fragments: percent, size, kind, shape, hardness
Saturated hydraulic conductivity
Sodium absorption ratio (SAR)
Soil reaction (pH): water, $CaCl_2$
Unified soil classification
USDA texture
Water content: 1/3, 1/10, 15 bar, and satiated

Interpretive and performance data

Crop yields, common crops
Erosion class
Land capability class and subclass: irrigated, nonirrigated
Potential frost action

(continued)

TABLE 28.1 (continued) Soil Data Elements and Interpretive Ratings Included in SSURGO and STATSGO Data

Prime farmland classification
Soil loss tolerance (T factor)
Soil slippage potential
Subsidence depth: total and initial
Wind erodibility index and group
Building site development:
    Shallow excavations
    Dwellings w/o basements
    Dwellings with basements
    Small commercial buildings
    Local roads and streets
    Lawns
    Landscaping
    Golf fairways
    Paths and trail
Construction material:
    Road fill source
    Sand source
    Gravel source
    Topsoil source
Forestland management limitations and ratings:
    Hand planting suitability
    Harvest equipment operability
    Haul road and landing construction
    Log landing suitability
    Mechanical planting suitability
    Mechanical site preparation (deep)
    Mechanical site preparation (surface)
    Natural surface
    Potential erosion hazard (off road/trail)
    Potential erosion hazard (roads/trails)
    Road suitability
    Soil rutting hazard
Range and woodland vegetation:
    Conservation tree shrub group classification
    Ecological site correlation
    Plant species composition percent
    Plant species, cover percent
    Potential production
    Site index
    Tree species: existing/potential
Recreation development:
    Camp areas
    Picnic areas
    Playgrounds
Sanitary facilities:
    Septic tank absorption fields
    Sewage lagoon areas
    Sanitary landfills
    Daily cover for landfills

TABLE 28.1 (continued) Soil Data Elements and Interpretive Ratings Included in SSURGO and STATSGO Data

Waste management:
    Land application of sewage sludge
    Manure and food processing waste
    Waste water disposal by irrigation
    Waste water treatment (overland flow)
    Waste water treatment (rapid infiltration)
    Waste water treatment (slow)
Water management:
    Drainage
    Embankments, dikes, levees
    Excavated ponds, aquifers
    Grassed waterways
    Irrigation
    Pond and reservoir areas
    Terraces and diversions
Water quality:
    Leaching
    Runoff adsorbed
    Runoff solution
Wildlife habitat suitability:
    Burrowing mammals and reptiles
    Crawfish aquiculture
    Desertic herbaceous plants
    Domestic grasses and legumes
    Freshwater wetland plants
    Grain and seed crops
    Herbaceous tundra
    Irrigated domestic grasses and legumes
    Irrigated freshwater wetland plants
    Irrigated grain and seed crops
    Irrigated saline water wetland plants
    Riparian herbaceous plants
    Riparian shrubs, vines, and plants
    Saline water plants
    Sedge-grass tundra
    Tussock tundra
    Upland conifer trees
    Upland deciduous trees
    Upland desertic herbaceous
    Upland mixed conifer/deciduous trees
    Upland shrub tundra
    Upland shrubs and vines
    Upland wild herbaceous plants

OSD is a term applied to the soil description that defines a specific soil series. The OSDs are in a text format and describe general and detailed information about each recognized soil series in the United States, including location, author's initials, introductory paragraph, taxonomic classification, a detailed soil profile description, location of the typical soil profile, range in characteristics, competing series, geographic setting, geographically associated soils, drainage and permeability, use and vegetation, distribution and extent, where and when the series was proposed and/or established, remarks, and additional data. While doing survey work, field soil scientists use official soil series descriptions that are applicable to their soil survey areas or projects. Other official soil series descriptions that include soils in adjacent or similar survey areas or projects are also commonly used. Thus, OSDs serve as a national standard for identifying and classifying soils and are the basis for the placement of a soil series in the soil taxonomic system. Soil series provide a central concept for soil map unit components, and their names are commonly used in the names of soil map units and map unit components. Scientists in other disciplines, such as agronomists, horticulturists, engineers, planners, and extension specialists, also use the OSDs to learn about the properties of soils in a particular area.

28.2.4 Soil Survey Geographic Database

SSURGO is the digitized version of detailed soil surveys in the United States. The SSURGO spatial data are generally compiled, digitized, and archived at a scale of 1:12,000 or 1:24,000. The minimum-sized area delineated on a map ranges from about 0.6 to 4.0 ha at these scales. The base maps used meet national map accuracy standards and are generally either digital orthophotoquads, quarter-quads, or U.S. Geological Survey 7.5 min quadrangles. The SSURGO database contains digital soil maps, an extensive attribute dataset of physical and chemical soil properties and interpretations that describe each soil map unit and map unit component, and a standard Federal Geographic Data Committee (FGDC) compliant metadata file. SSURGO attribute data are obtained by an export from the NASIS database (Soil Survey Staff, 1995b). SSURGO data are structured for use in a geographic information system (GIS) and represent the most up-to-date, official data of the NCSS for each soil survey area.

Efforts to digitize soil surveys in the United States started in 1971 with the grid cell Map Information Assembly and Display System (MIADS). Soil surveys were digitized in 4.05- and 10.12 ha grids, often along with land use information for the same grid cell. Beginning in 1976, Computervision® software was used to vector digitize soil surveys by either scanning or manual methods. Currently, soil surveys are vector digitized by scanning or heads-up digitizing to meet specified SSURGO standards, using a variety of commercial and public domain software. Some field soil scientists use tablet computers and GIS software to capture soil map unit boundaries in digital format while in the field. This process eliminates the need to later recompile the lines to stable base maps and digitize them.

Starting in 1997, the NRCS began an accelerated effort to digitize all detailed soil surveys and a Presidential initiative provided funding to complete the work. As of 2009, essentially all areas of the United States that have a completed soil survey have been digitized and are included in the SSURGO database. About 90% of the United States is included. These data are available for download from the SDM and can be viewed online via the WSS. The SSURGO data are continually being updated as additional information is gathered by ongoing field operations. An example of an interpretative map based on SSURGO data generated by WSS is shown in Figure 28.3.

28.2.5 U.S. General Soil Map

The U.S. General Soil Map (STATSGO) was designed to be used for broad planning and management in a GIS. An example of an interpretative map generated from STATSGO data is shown in Figure 28.4. Typically, STATSGO is used for state- or region-wide

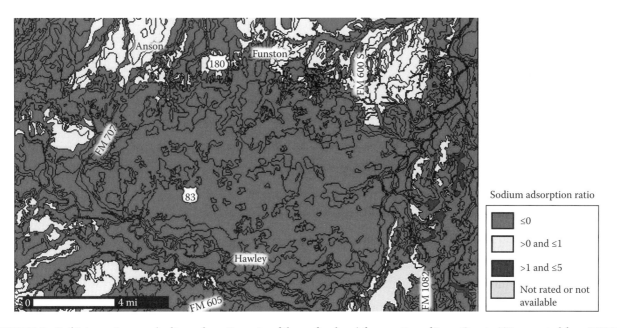

FIGURE 28.3 Soil interpretive map (sodium adsorption ratio of the surface layer) for a portion of Jones County, TX, generated from SSURGO data by WSS.

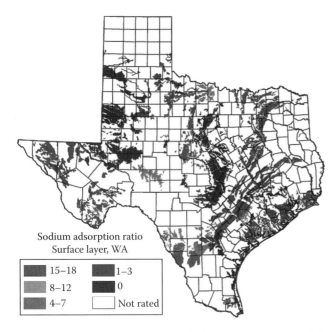

FIGURE 28.4 Soil interpretive map (sodium adsorption ratio of the surface layer) from STATSGO data for the State of Texas.

analysis to identify natural resource and environmental problem areas, prime farmland, potential productivity, global change modeling (carbon maps, etc.), and other uses. STATSGO has also been used to develop resource (staffing and funding) allocation models for NRCS. Several federal agencies, such as the U.S. Environmental Protection Agency, U.S. Geological Survey, and Agriculture Research Service, and many university researchers use STATSGO extensively for small-scale modeling and assessment projects.

STATSGO was developed as the result of NCSS cooperator efforts to create regional general soil maps. The idea for STATSGO was generated at an NCSS work planning conference in 1978. Work on STATSGO began in the mid 1980s and was completed in the mid 1990s. In areas of the United States where detailed soil maps were available, STATSGO maps were made by generalizing detailed county soil survey area maps, which were then compiled and digitized from U.S. Geological Survey 1:250,000 base maps. In unmapped areas, the STATSGO maps were developed using reconnaissance-scale soil survey knowledge and information.

Like the SSURGO data, STATSGO consists of digital soil maps, associated attribute data, and standard format metadata file. STATSGO attribute data are obtained from NASIS exports. The STATSGO map units consist of up to 21 map unit components. NRCS staff assigned a percent composition to each of these components from information in the detailed county-level soil surveys. The minimum-size area delineated on an STATSGO map is about 607 ha (Soil Survey Staff, 1994).

In 2006, the soil property attribute data associated with the STATSGO digital map were updated and converted to the same data structure as that of the SSURGO data. The STATSGO map data were updated somewhat, and the combined dataset was posted to the SDM for distribution. Coverage for the nation is available as are clipped datasets for individual states.

28.2.6 Data Access

SSURGO and STATSGO datasets and associated metadata can be accessed and downloaded over the Internet at http://soildatamart.nrcs.usda.gov/. SSURGO data are also displayed online via the WSS at http://websoilsurvey.nrcs.usda.gov/. Official Soil Series data can be accessed at http://soils.usda.gov/technical/classification/osd/index.html. Soil Characterization data are available at http://soils.usda.gov/survey/nscd/. Linkages to other soil information and to information about the NCSS and the NRCS Soil Survey Program are available at http://soils.usda.gov/.

28.3 The Evolution of Soil Survey Map Unit Databases

28.3.1 The First Generation

The NRCS first established a national soil database in the early 1970s through a cooperative agreement with the Statistical Laboratory at Iowa State University (ISU). ISU was chosen because of its long history of cooperative work with the SCS, dating back to the 1940s. Programming work for a soil database began in 1972 with automation of the soil interpretations records or Form SOI-5, which was used primarily as an input form to generate tables on engineering uses of soils for soil survey reports. Computer programs were developed to store, check, and print the data. The record for the Cecil soil series (NC0018) was the first one stored on the ISU mainframe in 1973. In 1974, the generation of manuscript tables of soil properties for inclusion in soil survey reports was introduced. Initially, all data processing was done at ISU, and a printed copy of the tables was sent through the mail. The SOI-5 forms, along with the SOI-6 forms, which were used to enter map unit information for the soil surveys, were mailed from SCS offices to ISU for processing. Printed copies of revised records and generated tables were mailed back to the SCS office requesting the tables. This automated table generation system replaced a very tedious, time-consuming manual process of creating tables for the published reports.

With the availability of this useful product came a much greater interest in getting data stored in the computer system. In 1977, the capability was added to automatically generate soil interpretations for selected uses from the soil data stored in the database using programmed criteria. After 1977, other enhancements were developed, including the addition of the OSD and Soil Series Classification (SC) databases.

Computerization in SCS offices for processing soil survey data began in 1977 with Linolex word-processing equipment in SCS regional offices. This equipment was used to prepare manuscript tables received on magnetic tape from ISU for final publication. Remote access to ISU from SCS began in the early 1980s with Harris Remote Job Entry equipment in both state and regional offices. Communication was through 4800-baud dial-up commercial ports. This was a time of significant change as batch software had to be redesigned for remote usage and data entry. Processing and printing of manuscripts shifted from ISU to SCS offices.

28.3.2 The Second Generation

Work on the second generation of the national soil database began in 1978, when SCS developed a computer program to rate soils for prime farmland and other important farmland classes and create maps for the Colorado Important Farmlands project. This project required the rating of about 4500 soil map units in Colorado. It used national criteria for prime farmland and state criteria for farmland of state importance and for unique farmland. The most difficult problem was making consistent ratings across soil survey areas. The program evaluated 10 soil characteristics and was fairly accurate in its ratings, but a large database was required to make the ratings, and the effort required to develop the database made the project unfeasible. The need for a large database, which also had to be readily accessible and easy to manipulate, resulted in the development of concepts for the second generation of soil information management.

These concepts were first documented in 1980 in the first technical report for the Colorado Soil Resource Information System (SRIS). SRIS demonstrated the feasibility of integrating several natural resource databases into a common, easy-to-use data environment. SRIS included (1) a soil map unit component database, (2) a soil interpretation database, (3) a pedon characterization database, (4) a climatology database, (5) a plant database, (6) a soil management component, and (7) a schema for the data and description of the system. SRIS was the first effort to manage soil data using a new technology called Database Management Systems (DBMS). This new information system allowed questions relating to more than one natural resource to be answered. It facilitated easy access to soil information and allowed the data to be managed independently of the application software that accessed it, in contrast to the SOI-5 database, which required a computer program to be written for each unique request. SRIS used the System 2000 DBMS, which provided a query language that could be understood and used by anyone with a basic understanding of soil information. In 1982, the soil database was implemented using MicroRim®, later renamed Rbase®, relational database management software (RDBMS).

As an outgrowth of the SRIS effort, SCS established software development staff at Fort Collins, Colorado, in 1985. The mission of this staff was to develop computer software to assist the SCS field office. In 1987, this effort produced the Computer Assisted Management and Planning System (CAMPS) field office software and the State Soil Survey Database (SSSD). SSSD, which used Prelude® RDBMS software, was the culmination of the SRIS effort.

With the release of SSSD, state SCS offices were equipped with UNIX computers. The SSSD software allowed the SCS state offices to manage their portion of the soil survey databases, which were downloaded from ISU via telecommunications. The primary function of the first release of SSSD was to clean up soil data and provide a download of the Map Unit Interpretation Record (MUIR) database to CAMPS. The first release of SSSD provided the ability to develop reports through standard database queries and manage nontechnical soil descriptions. With this software release also came the recognition that a soil scientist (soil dataset manager) position was needed at each SCS state office to manage the soil information system.

With SSSD, the SCS state offices could edit the soil map unit property and interpretation (MUIR) data at ISU and thus more accurately represent local conditions. The offices then returned a copy of the edited data to ISU. This editing capability provided for a national collection of MUIR data in 1993. SSSD releases in 1988–1993 added additional capabilities. In 1988, the Pedon Description Program, version 1.0, and the Official Soil Series management and soil reports modules were released. In 1989, the interface between the SSURGO database and the Geographic Resource Analysis Support System (GRASS) GIS was released. In 1989, a UNIX mail system called SoilNet and an automated version of the SOILS-6 form, which was used to record map unit data and facilitate the downloading and managing of MUIR data from ISU, were released. In 1991, the Soil Survey Schedule module was released. This module provided management, scheduling, and record-keeping software for SCS state and national offices to use in production soil survey efforts. In 1993, the Hydric Soils, Range Site, and MUIR incremental update modules were released.

Although table generation remained the primary purpose, the MUIR database soon began to be used for more than developing soil interpretation tables for reports. SCS began to use the database to answer questions on a wide range of soil-related issues across the United States, for example, the extent of salt-affected soils, soil loss tolerance, and erosion potential for the determination of highly erodible land, and, more recently, identification of hydric soils (wetlands). The uses of the soil database continued to expand and change until it became apparent in 1988 that SSSD and MUIR could not meet the changing needs. New information systems technology was available that could advance the use of soil survey information.

The MUIR soil database system was remarkable in that it was able to evolve in many ways over time but still kept its basic system design for about 25 years, until it was retired in 1996. At that time, MUIR contained data from about 2,900 soil survey areas and included approximately 250,000 soil map units. Implementation of the replacement system, the NASIS, began in 1994. Before the SSSD system was retired in 1996, the soil information in the MUIR database was converted to the NASIS database.

28.3.3 The Third Generation

Development of NASIS began with the analysis and documentation of the business of soil survey from beginning to end. Teams from various levels in the NCSS were established to complete the requirements analysis. Using structured systems analysis, these teams documented requirements that were passed on to contract software programmers. This analysis documented the important shift of the NCSS from producing static, printed soil survey reports to providing a dynamic database of soil information that could meet a wide range of needs.

A field data collection system was needed to ensure the integrity and completeness of the data collected, including the geographic coordinates. The system was designed to provide users accurate and complete soil survey information based on what was observed during the soil survey process. Implicit in this idea was the ability to describe accurately the variability of soils and their properties as they occur on the landscape. This new system had to provide for a continuous update of the database as new information is gathered, so that one version of these data was available to users at the field, state, and national levels.

NASIS had to provide a means for a variety of scientists to develop interpretations criteria, and generate soil interpretations based on local, state, or national requirements. For example, at the local level, there might be a need for an interpretation of suitability for animal waste disposal, and at the national level there may be a need for a soil productivity index. To ensure consistency, these interpretations must be applied to only one nationally consistent version of the data. The system had to provide for effective and efficient data delivery, including easy access by both internal and external (non-NCSS) users. This information needed to be delivered with a common data structure, data dictionary definitions, and appropriate metadata so that users could understand the information and use it appropriately.

28.3.3.1 NASIS System Objectives

Many weeks of analysis (discussion) early on and numerous follow-up meetings identified the following specific system objectives (Soil Survey Staff, 1991):

- The placement of automated tools in the hands of front-line field office staff
- One-time data entry, so that data could be retrieved by multiple software modules in various computer programs
- A simple means of entering data in same format as that used during data collection
- Validations to ensure proper entry of data and algorithms to provide default values
- Automated procedures for correlation and quality assurance
- Flexibility of the system to adapt to changes in procedure and standards and to new data needs and policies
- Capability to aggregate large-scale digital soil maps to smaller scales based on user-defined criteria
- Data manipulation and retrieval options for all databases and software modules that include modeling capability
- The ability to use single property values or representative values in addition to ranges to be used in models
- Capability to indicate confidence limits and the reliability of map unit data
- Continuous update of national, state, and field office soil survey databases
- Field access to state and national databases at the discretion of the users
- Permanent storage of all soil survey documentation
- Capability to transfer data files between various kinds of equipment
- Two-way linkages to other natural resource databases
- Software modules that are interactive, menu driven, and user-friendly
- Training on how to use the new system

28.3.3.2 NASIS Software Design

NASIS was developed according to an isolated design philosophy that separates the user interface, the NASIS engine, the INFORMIX® engine, and the concepts driven by the data dictionary so that changes can be easily made without reprogramming the entire system. This approach has allowed the easy addition of new capabilities and has permitted the programmers to reuse previously developed codes. NASIS was developed in the C+ programming language using the X Window system, a UNIX-based graphics window system. Similar to the Microsoft Windows® application that comes on many personal computers, the X Window system is a graphical user interface (GUI). This GUI allows users to manipulate the application by means of graphical objects, such as buttons, scroll bars, and icons. A pointing device, typically a mouse, is used. Because NASIS operates in this windows environment, several applications or processes can be run at the same time, each appearing in a separate pane of the window on the screen. NASIS users see a cascading or overlapping that mirrors the hierarchical structure of the data in the soil database.

28.3.3.3 NASIS Security

INFORMIX was selected as the NASIS database management software (DBMS) largely because of its security features. This proprietary design enabled the construction of a system that prevents most accidental or intentional corruption of data. NASIS allows different data records in the database to be owned by different individual users or groups of users, so that only qualified scientists can edit or create data. The owner of an object has the authority to change data as needed. Individual or group ownership can be established as needed.

28.3.3.4 NASIS Software, Hardware, and Network Requirements

The NASIS software and database are located on a central server, and access to it by users requires the same configuration regardless of the user. It requires only a personal computer, an Internet connection, and Citrix® client software for Internet access to NASIS. NASIS is operational in all NRCS state, regional, national, and soil survey project offices, and in the federal, state, and university offices of other NCSS cooperators.

28.3.3.5 NASIS Software Development and Implementation Status

As with the SSSD software, the initial releases of the NASIS software were in successive yearly versions. Version 1.0 was released in 1994. It provided validation and conversion of MUIR to the NASIS database structure, a security system and controls, an operational data dictionary, editors for areas, legends, and data map units, and online help. Version 2.0, released in 1995, provided cut, copy, and paste functions for data objects, a query editor and manager, global assign functions, report generation, and an enhanced online help system. Version 3.0, released in 1996, provided calculation and validation routines for data and the ability to create interpretations criteria and generate interpretations. The version released in 1997 provided for the replacement of the national MUIR and consolidated NASIS databases from individual state offices to the 18 MLRA Soil Survey Regional Offices. It provided downloads to the NRCS field office computing system (FOCS) and downloads of SSURGO-format datasets. These releases dealt primarily with the development and management of map unit data.

The NASIS version, released in 1998, incorporated capabilities for input of and access to pedon descriptions and soil site information and replicated storage of national map unit data via the Internet at ISU and data sharing via the Internet. In 2001, the NASIS database was consolidated further to a central server environment at the NRCS Information Technology Center in Ft. Collins, Colorado, and data storage at ISU was discontinued. In 2003, the SSURGO data model was adopted and implemented. Periodic releases have continued to add new functionality to the system and to refine the data model as needs change.

Beginning in 2004, development of a new generation of NASIS began. This system is scheduled for release in 2009. It will be a Microsoft Windows system using a .NET® operating system and SQL Server® DBMS. This version will begin managing data by projects rather than the traditional soil survey areas (usually county-based legends). This design will promote designing map units on the basis of their natural occurrence rather than limiting the spatial extent to geopolitical boundaries. As a result, properties, qualities, and interpretations of map units will extend across geopolitical boundaries to their full natural extent.

28.3.3.6 Soil Data Warehouse and Soil Data Mart

Early business analysis for a NASIS identified the need for a single point of delivery of official soil survey data and information and the ability to archive versions of official data. The NASIS database and application are intended primarily for internal use in developing and managing soil survey data, not for public access or delivery of data. To meet this need, the Soil Data Warehouse (SDW) and SDM were deployed in 2003. By that year, significant progress was being made in digitizing soil survey maps. The SDW is designed to hold all versions of official soil survey data (both SSURGO and STATSGO) produced since 2003, including not only tabular attribute data and digital spatial data but also a metadata file that complies with the standards of the FGDC. The SDM contains only the current version of official data and serves as the data-distribution site. It provides a public access point for the data and allows the user either to download digital SSURGO datasets in a standard format for use in a local GIS or to run standard soil survey reports on selected datasets. The SDM can be accessed at http://soildatamart.nrcs.usda.gov/.

28.3.3.7 SSURGO Access Template Database

When data are downloaded from the SDM, the attribute data tables are in a series of unrelated text files. For the data to be used, they first must be loaded into a relational database format of the user's choosing. NRCS has developed a database template in Microsoft Access format for this purpose. The template includes macros for loading the data as well as standard queries and reports for viewing the data. It is included with each download from the SDM.

28.3.3.8 Soil Data Viewer

Soil Data Viewer (SDV) is an application developed as a plug-in extension of ESRI ArcMap for viewing digital soil maps downloaded from the SDM. It requires the SSURGO Access Template Database (described above) for accessing the attribute data. It was developed to help shield the user from some of the complexity of the attribute data structure. SDV includes a series of rules for aggregating soil properties and interpretations for individual map unit components to a single value for the respective map units for display in the GIS-generated thematic maps. This tool is available to the public and can be downloaded from http://soildataviewer.nrcs.usda.gov/.

28.3.3.9 Web Soil Survey

As the digitizing of soil survey maps progressed and the SDM became more fully populated with data, users began to ask why they could not view the soil maps from the SDM online and why they needed to download the data. In response to this need, the WSS was developed. The WSS provides a publicly accessible online interface to the national collection of SSURGO datasets in the SDM. The WSS users can specify their area of interest (AOI), display the soil map for the AOI, generate interpretive maps for a wide variety of uses or selected soil properties, print individual maps or accumulate them into a composite report, or download the SSURGO data for the AOI. The AOI is not limited to part of a single soil survey area but can span multiple survey areas. WSS merges the datasets and displays data and maps in a single layer. WSS uses the same rule set as does SDV for aggregating data for display at the map unit level. A sample interpretive map generated by WSS is shown in Figure 28.5. WSS also provides the capability to download the underlying SSURGO dataset clipped to the AOI boundary for use in a local GIS. Download options are the same as those from the SDM. The WSS can be accessed at http://websoilsurvey.nrcs.usda.gov/.

28.4 Standards for Soil Survey Data

Under the Office of Management and Budget Circular A-16, revised in 1990, the lead agency responsible for coordinating soil data-related activities is the NRCS of USDA. This government-wide leadership for soil data coordination is carried out under the policy guidance and oversight of the FGDC.

NRCS is responsible for developing and implementing a plan to coordinate soil data-related activities among federal

FIGURE 28.5 (See color insert.) Soil interpretive map of weighted average cation-exchange capacity (CEC-7) of the 20–75 cm depth for a portion of Palo Alto County, IA generated from SSURGO data by WSS.

and nonfederal agencies. A soil data subcommittee was established to assist the NRCS in coordinating federal and nonfederal interests in soil data. The subcommittee consists of representatives designated by member agencies of the NCSS that collect or finance the collection of soil data as part of their mission or are responsible for the direct application of these data through legislated mandate. The subcommittee is chaired by the member from the NRCS, Soil Survey Division. It coordinates (1) the exchange of information and transfer of data, (2) the establishment and implementation of standards for quality, content, and transferability, and (3) the identification of the requirements for the collection of spatial data in a manner that minimizes the duplication of effort where practical and economical. Specific responsibilities of the subcommittee include the following:

- To facilitate the coordination of the activities of the agencies related to soil data and the exchange of soil data, by formal and informal means.
- To facilitate the collection and compilation of information for soil data activities.
- To participate in the development and evaluation of data definitions and standards used by the United Nations and other international organizations and make recommendations for their inclusion in soil datasets, as appropriate.
- To identify ways in which soil data from any source may be included in the National Spatial Data Infrastructure.
- To recommend any changes in the definition of soil data that more accurately describe the scope of the data.
- To assist the NRCS in establishing and publishing standards and specifications for the data (e.g., incorporating conversion to metric units) and to assist in establishing priorities for soil data production.
- To assist in the development and adoption of common standards for the content, format, and accuracy of soil data for use by all federal agencies and for encouragement of use by nonfederal organizations. These common standards can increase the interchangeability of the data and enhance the potential for multiple uses.
- To promote government-wide use of defined and published standards for the transfer of spatial soil data.
- To facilitate the economical and efficient application of soil data through the sharing of experiences involving applications.
- To support higher order or cross-cutting activities established or recognized by the FGDC.
- To encourage agencies to schedule the disposition of soil data in coordination with the National Archives and Records Administration in order to facilitate the permanent preservation of historically valuable data.

The subcommittee may create work groups, task groups, or further subdivisions as appropriate to carry out its activities and meet its responsibilities. A work group of this subcommittee has been assigned the responsibility of developing and maintaining the definitions, terminology, and content portion of the Soil Geographic Data Standard. The subcommittee chairperson coordinates activities with other FGDC subcommittees, working groups, and other components by participating in the FGDC's Coordination Group. The FGDC subcommittees and working groups are listed in Table 28.2. To date, standards have been created for the soils data dictionary, soil survey digitizing, and metadata. The *Keys to Soil Taxonomy* (Soil Survey Staff, 2006), the *National Soil Survey Handbook* (Soil Survey Staff, 2009), and the *Soil Survey Manual* (Soil Survey Division Staff, 1993) have

TABLE 28.2 FGDC Subcommittees and Working Groups

Subcommittees	Working Groups
Base cartographic data	Biological data
Bathymetric data	Data clearinghouse
Cadastral data	Earth cover
Base cartographic data	Geospatial enterprise architecture
Bathymetric data	Facilities
Cadastral data	Historical data
Cultural and demographic data	Homeland security
Federal geodetic control data	Metadata
Geologic data	Marine boundary
Ground transportation	Standards
International boundaries and sovereignty	Sample inventory and monitoring of natural resources and the environment
Marine and coastal spatial data	
Soils data	Sustainable forest data
Spatial climate	Tribal
Transportation	
Vegetation data	
Water data	
Wetlands data	

Source: http://www.fgdc.gov/participation/working-groups-subcommittees.

also been designated as standards. These standards are accessible at http://soils.usda.gov/technical/.

28.5 Future Directions for Soil Information Systems

Research, development, and testing are being conducted in several areas that hold a great deal of promise for further use of information system technology in soil survey. The first area is in field data collection. Tablet computers and global positioning system (GPS) technology have been deployed to soil survey offices for use in data collection for both point and map unit data. Pilot projects are underway in which digitized soil survey information is used in conjunction with a grid-point sampling scheme and GPSs in order to provide information on the variability of soils and soil properties on the landscape. The tablet computers will provide for on-screen mapping (digitizing) and the collection of soil data. GIS is being used in the field for premapping and quality assurance of mapping. GIS tools and procedures are being used to generalize large-scale mapping to smaller scales. For example, an updated STATSGO data layer is being proposed to be developed by reclassification or generalization of SSURGO data.

The modern soil survey effort began in the early 1950s in the United States. From then to the present, work by many different people on individual soil survey areas has resulted in a somewhat inconsistent product. Because of a nearly complete SSURGO data layer for the nation and the ability to view the data in GISs, inconsistencies within the data, primarily along artificial soil survey area boundaries, are becoming more readily apparent. The NCSS has started a significant effort of evaluating existing soil survey data across the nation and updating the data as needed to remove as many of the inconsistencies as possible. New tools are being used by field soil scientists to investigate and remove the inconsistencies.

The NCSS is working to merge the spatial soil data with the attribute database into a common database for further development, enhancement, and management. Pilot studies are underway to identify the workflow processes needed. The goal is to have a seamless SSURGO data layer for the nation. This product will enable more uniform application of various programs and provide better soil data and information to land use planners and decision makers at all levels.

To date, NCSS digital soil maps have been provided in vector format. There is a growing user community who are asking for the data in raster format, which allows much faster analysis operations. Efforts are ongoing with NCSS cooperators, academia, and international colleagues to investigate the feasibility of providing the data in this format.

The biggest challenge for the future will be keeping up with the rapid changes in technology and making the necessary cultural changes in the NRCS and NCSS and to train employees in the use of new technologies. The United States has an estimated $5 billion replacement cost investment in the soil survey. Information technology will be increasingly used to manage and capitalize on this investment. Finally, information technology and soil survey information will increasingly help to provide solutions as the U.S. population grows and the demand on natural resources increases.

Acknowledgments

The authors gratefully acknowledge the contribution of the following individuals to this chapter and to the development of the U.S. Soil Information System: Harvey P. Terpstra, Statistical Laboratory, Ames, IA; David L. Anderson, Dennis J. Lytle, Gale W. TeSelle, Russell J. Kelsea, Terry Aho, Dorn Egley, Ricky J. Bigler, Sharon W. Waltman, Adrian Smith, Jon D. Vrana, Fred E. Minzenmayer, Ken E. Harward, Ellis C. Benham, Thomas Reinsch, and Dennis Darling, NRCS; and many very dedicated contract programmers.

References

Soil Survey Division Staff. 1993. Soil survey manual. USDA agriculture handbook 18. USDA, Soil Conservation Service. U.S. Government Printing Office, Washington, DC. http://soils.usda.gov/technical/manual/

Soil Survey Staff. 1991. National soil information system—Soil interpretation and information dissemination sub-system. Draft requirements statement. USDA, Natural Resources Conservation Service. National Soil Survey Center, Lincoln, NE.

Soil Survey Staff. 1994. State soil geographic (STATSGO) database data use information. USDA, Natural Resources Conservation Service. USDA Misc. Pub. 1492.

Soil Survey Staff. 1995a. Soil survey laboratory information manual. Soil Survey Investigations Report 45, Ver. 1.0. USDA, Natural Resources Conservation Service. U.S. Government Printing Office, Washington, DC. http://soils.usda.gov/survey/nscd/lim/index.html

Soil Survey Staff. 1995b. Soil survey geographic (SSURGO) database data use information. USDA Misc. Pub. 1527.

Soil Survey Staff. 1996. Soil survey laboratory methods manual. Soil Survey Investigations Report 42, Ver. 3.0. USDA, Natural Resources Conservation Service. U.S. Government Printing Office, Washington, DC. http://soils.usda.gov/technical/lmm/

Soil Survey Staff. 2006. Keys to soil taxonomy. 10th Edn. USDA, Natural Resources Conservation Service. U.S. Government Printing Office, Washington, DC. http://soils.usda.gov/technical/classification/tax_keys/

Soil Survey Staff. 2009. National soil survey handbook. USDA, Natural Resources Conservation Service. http://soils.usda.gov/technical/handbook/

29
Integrated Digital, Spatial, and Attribute Databases for Soils in Brazil

Carlos Eduardo
Pellegrino Cerri
University of São Paulo

Carlos Gustavo Tornquist
The Federal University of Rio Grande do Sul

Martial Bernoux
Institute of Research for Development

Miguel Cooper
University of São Paulo

Gerd Sparovek
University of São Paulo

Maria de Lourdes
Mendonça-Santos
EMPRABA-Centro Nacional de Pesquisa de Solos

Carlos Clemente Cerri
University of São Paulo

29.1	Introduction	29-1
29.2	Brief Biophysical Setting of Brazil	29-2
29.3	Soil Maps of Brazil and Description of the Soil Attribute Database	29-2
	The National Soil Profile Database • Digital Soil Properties Database of Brazil • Digital Spatial and Attribute Databases for Soils in Rondonia and Mato Grosso States • Digital Mapping of Soil Classes in Rio de Janeiro State	
29.4	Integration and Application of Soil Digital Databases in Brazil	29-10
	Organization and Structure of the Brazilian Soil Information System • Regional Organic Carbon Storage Maps of Western Brazilian Amazon Based on Prior Soil Maps and Geostatistical Interpolation • Soil Organic Carbon Stocks of Rio Grande do Sul State • SOTER Database as Basis for Modeling Soil Organic Carbon Stocks and Changes in the Brazilian Amazon	
29.5	Future Directions and Opportunities	29-25
References		29-25

29.1 Introduction

Large-scale phenomena and impacts, discussed on a global scale under a multiperspective analysis, are increasingly on society's agenda (Pearce and Warford, 1993). Examples directly related to soil science are (1) the expansion of agriculture to areas of tropical forest; (2) the contamination of water resources with residues of pesticides, phosphate, and nitrate used in agriculture; and (3) the degradation of soils through soil erosion (Habern, 1992; Doran et al., 1996; Doran and Safley, 1997).

Additionally, soil scientists are increasingly concerned about global climate changes, where soil organic carbon (SOC) storage is a key issue. To analyze global phenomena related to soil science, we need comprehensive, consistent, georeferenced, and quantitative databases on national or continental scales.

The emergence of integrated digital, spatial, and attribute databases as a credible alternative to fulfill the increasing worldwide demand in spatial soil data is conditioned not only to enhance its ability to increase spatial resolutions and enlarge extents but also to deliver relevant information. Integrated digital databases are an essential part of a soil assessment framework that supports soil-related decision- and policy-making, and therefore, it is of crucial importance that the mentioned databases are of known quality.

Therefore, the objective of the present work is to gather available information on integrated digital, spatial, and attribute databases for soils in Brazil. In addition, soil maps at different scales at national, regional, and state levels are presented. Some applications of soil digital databases are discussed in order to illustrate potential uses of soil databases in Brazil. Finally, future directions and opportunities are suggested.

29.2 Brief Biophysical Setting of Brazil

The Federative Republic of Brazil lies mainly below the equator, in South America, and is broadly bounded by 5°16′N and 33°45′S latitudes and 35°W and 74°W longitudes. It covers 8,514,879 km^2 or about 48% of the continent, and has over 165 million inhabitants.

A detailed account of the geology is given in Schobbenhaus et al. (1984). Altitude generally does not exceed 500 m as Brazil occurs on an old geologic base without recent tectonics. The highest peak, the Pico da Neblina (3014 m), occurs on the plateau of Guyana near the border with Venezuela.

Brazil can be characterized into five main landscapes: the plateau of Guyana, the Brazilian plateau, the Amazon plain and low lands, the plain of the Pantanal, and plains and coastal lowlands (Ab'Saber, 1996). Knowledge on landforms, soil conditions, and the biota of the Amazon region had remained sketchy until the 1960s (Ab'Saber, 1967; FAO-Unesco, 1971; Sombroek, 1996). A wealth of new information has been gained in the context of the RADAM Project (1973–1982), which covered the entire Brazilian Amazon. It included surveys of the geology, geomorphology, vegetation, soils, and land use. The resulting reports contain results of analyses for over 2500 soil profiles (see Bernoux et al., 2003). While detailed information is available about organic carbon, nitrogen, and phosphorus concentration, soil texture, major cations, and other chemical soil properties, there are few determinations of bulk density (BD) (Bernoux et al., 1998b).

RADAM and other materials were used by EMBRAPA (1981) to compile a 1:5M scale soil map of the country, showing about 2700 map units representing about 70 soil types. Subsequently, EMBRAPA compiled a 1:5M scale soil and terrain database for the country in the framework of the world-encompassing SOTER project (Peters and Van Engelen, 1993).

The 1:5M scale digital soil map of EMBRAPA (1981) uses the Brazilian classification system. On the other hand, SOTER follows the Revised Legend and WRB terminology, facilitating international exchange and correlation. Adoption of a uniform methodology allows using similar procedures for plugging gaps in the measured soil data (Batjes, 2005) and for linking soil parameter estimates to models. The SOTER methodology, albeit with some regional modifications, has also been used to compile a 1:1M database for the state of Rondonia (Cochrane and Cochrane, 1998).

Ab'Saber (1996) and Sombroek (1996) studied the diversity of Amazon landforms and their soils, in relation to biological diversity. Sombroek (2001) also discussed spatial and temporal patterns of Amazon rainfall and possible consequences for the planning of agricultural occupation, and the protection of primary forests. Land use change, the main driving factor that has been described by Lambin et al. (2003), in combination with climate change may have a varying impact on net primary production and, thus, soil carbon stocks in the various natural regions of Brazil (e.g., Moraes et al., 1996; Fearnside, 1998, 1999; Houghton et al., 2000; Cardille and Foley, 2003; Sisti et al., 2004). Many of these aspects are being studied in the context of LBA, the Large Scale Biosphere-Atmosphere experiment (Kabat et al., 1999). Several authors have reviewed perspectives on sustainable development of the Brazilian Amazon (Clustener-Godt and Sachs, 1995; Cattaneo, 2002; Vosti et al., 2002).

29.3 Soil Maps of Brazil and Description of the Soil Attribute Database

29.3.1 The National Soil Profile Database

Many countries already have soil electronic databases available such as the United States (USDA-National Resources Conservation Service, 2004), Canada (Agriculture and Agri-Food Canada, 2000), and FAO (Food and Agricultural Organization, 1995), but in most countries these data are not available. This is especially true in countries of the tropics.

A digital soil profile database of Brazil was compiled and made available through the internet. Most of the soil data were obtained from the Radambrasil project and other regional surveys. The database contains information from 5,086 profiles distributed over the whole Brazilian territory corresponding to data from 10,034 horizons, each with 31 variables (Figure 29.1).

The variables were chosen to represent different areas of soil science, embracing soil morphological, chemical, mineralogical, and physical attributes. The distribution uniformity of the data was low with sampling densities varying from one profile per 10,000 km^2 to one profile per 1,370 km^2. The access to the database is free and its design allows its use not only by soil scientists but also by those working with agricultural, environmental, and land use issues.

FIGURE 29.1 Brazil's database soil profile location. (Modified from Cooper, M., L.M.S. Mendes, W.L.C. Silva, and G. Sparovek. 2005. A national soil profile database for Brazil available to international scientists. Soil Sci. Soc. Am. J. 69:649–652.)

Due to the large amount of data that were collected and organized, and the objective of developing the database as a multiuser tool, the basic information was compiled in a single text (ASCII format) table to allow a more flexible and user-friendly environment. In this way, any piece of data can be partially or totally added, deleted, edited, or extracted using the database system and structure of the final user's preference.

The sequence used to show the data follows the Radam volumes or that of the institution responsible for the regional soil surveys. This main table contains data from all the soil profiles extracted from the above-mentioned surveys. The profiles are numbered and georeferenced (latitude and longitude in geographic coordinates using decimal degrees with unspecified datum). Each profile contains data of the surface and the diagnostic subsurface horizons. The variables or soil attributes (Table 29.1) are listed according to a logical order that depends on the type of soil attribute (physical, chemical, or morphological).

Auditing of the database was performed to control the quality of the information. The main errors found in the original publications were typing mistakes, wrong locations, and analytical or calculation errors. The auditing routines were based on those attributes that can be calculated using other analytically determined soil attributes such as cation exchange capacity (CEC) and particle size distribution, or those that can be geographically represented (e.g., latitude and longitude). No quality control was done on analytically determined data. In all the auditing routines, when possible, corrections were made. In the case of various or difficult to correct and missing data, the profiles were eliminated from the final database.

TABLE 29.1 List of Variables Used in the Database with Their Description and Analytical Method of Determination

Database Code	Variables and Soil Attributes	Description	Analytical Method
ClassifSolo	Soil classification	Soil name according to the 1970 and 1980s Brazilian soil classification system	Field and analytical observations
DecVal	Slope	Inclination of the surface in relation to a flat reference	Field observation
Drenagem	Drainage	Velocity at which water is removed from the soil profile	Field observation
ProfSolo	Soil depth	Maximum soil depth indicated in the descriptions	Field observation
HzSimb	Horizon type	According to pedological convention	Field observation
HzProf	Horizon depth	Depth of appearance and disappearance of the horizon	Field observation
CorMunsell	Color	Described according to Munsell Soil Color Chart	Field observation
CG	Coarse gravel	Material >20 mm in diameter	Sieving
FG	Fine gravel	Material between 20 and 2 mm in diameter	Sieving
Areia	Sand	Particles between 2 and 0.05 mm in diameter	Sieving
Silte	Silt	Particles between 0.05 and 0.002 mm in diameter	(Sand – Clay)
Argila	Total clay	Particles <0.002 mm in diameter	Dispersion with NaOH and Sodium Hexametaphosphate, Bouyoucos or Pipette method for determination
ArgilaNat	Natural clay	Particles <0.002 mm in diameter	Dispersion with H_2O, Bouyoucos or Pipette method for determination
Floc	Degree of flocculation	Proportion of clay that is flocculated	100 (total clay – natural clay)/total clay
BD	Bulk density	Mass of dry soil per unit volume	Volumetric ring method
SiO_2	SiO_2	Amount of silicon dioxide in soils	Sulfuric attack
Al_2O_3	Al_2O_3	Amount of aluminum oxide in soils	Sulfuric attack
Fe_2O_3	Fe_2O_3	Amount of ferric oxide in soils	Sulfuric attack
pH H_2O	pH H_2O	pH at a specified soil-water ratio	Soil-water suspension, potentiometer
pHKCl	pH KCl	pH at a specified soil KCl ratio	Soil-KCl suspension, potentiometer
C	C	SOC content	Oxidation with $K_2Cr_2O_7$ 0.4 N, Tiurim method
N	N	Soil total nitrogen content	Kjeldahl digestion
Ca	Ca^{++}	Exchangeable calcium	Extracted with KCl N
Mg	Mg^{++}	Exchangeable magnesium	Extracted with KCl N
K	K^+	Exchangeable potassium	Extracted with HCl 0.05 N
Na	Na^+	Exchangeable sodium	Extracted with HCl 0.05 N
Al	Al^{+++}	Exchangeable aluminum	Extracted with KCl N
H	H^+	Exchangeable hydrogen	(Total acidity-Al^{+++})
Tac	Total acidity	Total acidity including residual and exchangeable acidity	Extracted with KCl N
CEC	CEC	Cation exchange capacity	Bases extracted with NH_4OAc at pH 7
EC	EC	Electrical conductivity	Conductivity meter in a saturated paste of soil and water

Source: Adapted from Cooper, M., L.M.S. Mendes, W.L.C. Silva, and G. Sparovek. 2005. A national soil profile database for Brazil available to international scientists. Soil Sci. Soc. Am. J. 69:649–652.

One objective for assembling this database was to make available the most comprehensive soil profiles information as possible while using only published data as sources. In this way, information that was restricted to personal or public libraries is being made available through the Internet for noncommercial use. Several variables reflecting chemical, physical, mineralogical, morphological, and pedogenetic features of soil were included, useful for a wide range of topics related to soil science. Of the 31 chosen variables (Table 29.1), 6 correspond to soil morphological attributes, 8 correspond to soil physical attributes, and 16 correspond to soil chemical attributes. To these soil attributes, soil classification of each profile is also included, using the original terminology, completing the 31 variables contained in the database.

The database is useful for soil scientists and other professionals working with agronomy, land-use planning, environmental management, soil process modeling, or any other area in which basic soil data are necessary.

The database is available for free download at http://www.esalq.usp.br/gerd/. The soil taxonomy was updated based on the latest version of the Brazilian Soil Classification System. The coverage and potential for quantitative analyses of the updated database was evaluated by Benedetti et al. (2008). For this purpose, the data were arranged in an electronic database and soil classification updated up to the fourth categorical level of the current taxonomic system. Mainly in the first three levels, the name updating procedure was reliable. Restrictions were observed at the fourth level (subgroup), as measured by the degree of confidence used for the assessment of taxonomic classification accuracy.

By updating the soil profiles to the current and standard taxonomic criteria, the databases were internally standardized and became comparable to external data, for example, spatial distribution, altitude, and climate types. Profiles of the different taxonomic groups nationwide were well represented as well as most federal States (Benedetti et al., 2008).

Furthermore, soil profiles from several taxonomic groups were related to the Köppen climate zones and types (Table 29.2). A relation between the first soil taxonomic level (soil type) and climate was not always evident, with some exceptions, for example, a substantial portion of Vertisols and Alfisols occur in semi-arid climate (BS). The altitude of soil type occurrence was also variable, with some indications of quartile and median preferences for some soils. Inceptisols and Oxisols were more frequent at higher altitudes, whereas 75% of the Spodosols and Ultisols were found below 200 masl. The analysis of the database proposed by Benedetti et al. (2008) can be further complemented for specific study needs, including adaptations to ongoing modifications of the taxonomy of the Brazilian Soil Classification System.

29.3.2 Digital Soil Properties Database of Brazil

The soil and terrain database for Latin America and the Caribbean provided the basis for preparing the present 1:5M SOTER database for the country (BRASOTER). The original geographical and attribute data have been collated into SOTER format by staff of the Empresa Brasileira de Pesquisa Agropecuária—Centro Nacional de Pesquisa de Solos and ISRIC between 1993 and 1997 (Peters and Van Engelen, 1993).

TABLE 29.2 Distribution of the First Soil Taxonomic Level in Function of the Zones and Climatic Types According to the Köppen Classification

Köppen Classification	Soil Class[a]												
	P	C	M	E	G	L	T	R	N	O	S	F	V	Total
	%													
Zone														
A	85.6	71.0	64.7	97.7	93.5	8.3	63.2	82.0	69.4	72.7	69.0	96.5	64.9	81.4
B	4.5	8.2	8.6	0.0	0.0	3.8	26.0	10.5	1.5	0.0	20.1	0.9	28.8	6.3
C	9.9	20.8	26.7	2.3	6.5	14.0	10.8	7.5	29.2	27.3	10.8	2.6	6.3	12.3
Type														
Af	12.8	3.4	1.1	43.0	18.0	6.8	1.7	7.1	6.2	9.1	5.5	11.5	7.2	9.0
Am	32.3	16.2	1.1	25.6	32.9	19.0	16.0	20.3	24.8	13.6	4.6	28.3	3.6	23.1
As	1.2	0.0	2.1	0.0	0.0	0.5	5.2	0.9	1.1	0.0	1.8	0.0	3.6	0.9
As'	1.5	1.4	4.8	2.3	2.0	0.7	5.2	2.0	0.0	13.6	9.1	1.3	8.1	1.8
Aw	25.8	44.7	39.6	23.3	30.7	43.1	11.3	38.7	31.7	36.4	36.3	33.8	35.1	34.7
Aw'	12.0	5.2	16.0	3.5	10.0	12.3	23.8	13.0	5.5	0.0	11.7	21.6	7.2	11.9
BS	4.3	7.7	8.6	0.0	0.0	3.7	25.1	10.0	1.5	0.0	17.5	0.9	28.8	6.0
BW	0.2	0.5	0.0	0.0	0.0	0.1	0.9	0.5	0.0	0.0	2.7	0.0	0.0	0.3
Cf	8.5	15.2	25.7	0.0	6.0	10.0	10.0	5.9	26.1	27.3	10.8	2.6	6.3	10.0
Cs	0.0	0.0	0.0	0.0	0.0	0.1	0.0	0.4	0.4	0.0	0.0	0.0	0.0	0.1
Cw	1.5	5.6	1.1	2.3	0.5	3.8	0.9	1.2	2.8	0.0	0.0	0.0	0.0	2.2

Source: Modified from Benedetti, M.M., G. Sparovek, M. Cooper, N. Curi, and A. Carvalho Filho. 2008. Representatividade e potencial de utilizacao de um banco de dados do Brasil. R. Bras. Ci. Solo 32:2591–2600.

[a] P, Argissolos; C, Cambissolos; m, Chernossolos; E, Espodossolos; G, Gleissolos; L, Latossolos; T, Luvissolos; R, Neossolos; N, Nitossolos; O, Organossolos; S, Planossolos; F, Plintossolos; V, Vertissolos.

The SOTER methodology allows mapping and characterization of areas of land with a distinctive, often repetitive, pattern of landform, lithology, surface form, slope, parent material, and soils. The approach resembles physiographic or land systems mapping. The collated materials are stored in a SOTER database linked to geographical information system (GIS), permitting a wide range of environmental applications (see Falloon et al., 1998; Batjes and Dijkshoorn, 1999). The SOTER methodology is mainly applied at scales ranging from 1:250,000 to 1:5M.

According to Batjes et al. (2004) each SOTER database comprises two main elements, a geographical component and an attribute data component. The geographical database holds information on the location, extent, and topology of each SOTER unit. The attribute database describes the characteristics of the spatial unit and includes both area data and point data. A GIS is used to manage the geographic data, while the attribute data are handled in a relational database management system (RDBMS).

Each SOTER unit in the geographic database has a unique identifier, called SOTER unit-ID (SUID). This primary key provides a link to the attribute data for its constituent terrain, terrain component(s) (TCID), and soil component(s) (SCID). Each soil component within a SOTER unit is described by a profile (PRID), identified by the national soil experts as being regionally representative. Representative profiles are selected from available soil survey reports as the SOTER program does not involve new ground surveys. A comprehensive description of the methodology and coding conventions is given by Van Engelen and Wen (1995).

Brazil has been described using 298 unique SOTER units, corresponding with 520 mapped polygons. At the small scale under consideration, most SOTER units will be compound units. Some of the spatially minor soil units, however, may be of particular relevance. For example, soils in the Amazon basin (Figure 29.2) may be of great importance for national inventories of carbon stocks and change. It is therefore recommended that end-users consider all component soil units of a SOTER unit in their assessments or model runs. Ultimately, the type of research purpose will determine which parameter estimates or single value maps are of importance in a special case. Therefore, the full map unit composition can best be addressed with tailor-made programs depending on the scope of the application.

Gaps in the measured soil profile data have been filled using a step-wise procedure that includes three main stages (Batjes et al., 2004): (1) collating additional measured soil analytical data where available, (2) filling gaps using expert knowledge and common sense, and (3) filling the remaining gaps using a scheme of taxotransfer rules.

Parameter estimates are presented by soil unit for fixed depth intervals of 0.2–1 m depth for organic carbon, total nitrogen, pH(H$_2$O), CEC soil, CEC clay, base saturation, effective CEC, aluminum saturation, CaCO$_3$ content, gypsum content, exchangeable sodium percentage (ESP), electrical conductivity of saturated paste (ECe), BD, content of sand, silt, and clay, content of coarse fragments, and available water capacity. These attributes have been identified as being useful for agro-ecological zoning, land evaluation, crop growth simulation, modeling of soil carbon stocks and change, and analyses of global environmental change (Batjes et al., 2004).

The current parameter estimates should be seen as best estimates based on the current selection of soil profiles and data clustering procedure. Taxo-transfer rules have been flagged to provide an indication of the possible confidence in the derived data.

Results are presented as summary files and can be linked to the 1:5M scale SOTER map for Brazil in a GIS, through the unique SOTER-unit code. Most SOTER units in Brazil comprise at least two soil components. In the primary database, the associated information is stored in a range of relational databases to enhance data storage and management efficiency.

All geographic data in SOTER are presented in vector format. However, should grid-based soil layers be required, these can be generated using the convert-to-grid module of the spatial analyst extension to ArcView (ESRI, 1996). The secondary data set is appropriate for studies at the national scale. Correlation of soil analytical data, however, should be done more rigorously when more detailed scientific work is considered.

29.3.3 Digital Spatial and Attribute Databases for Soils in Rondonia and Mato Grosso States

Spatial data were extracted from SEPLAN-RO (1998) and SEPLAN-MT (2002) as sets of electronic files on hypsometry, geology, geomorphology, hydrography, vegetation and soil maps at 1:500,000 and 1:250,000 scale. Hypsometry, geology, geomorphology and vegetation maps were used without any changes. The legend of the soil maps was slightly adapted for consistency.

A single general soil map was made for Mato Grosso using the soil maps at the 1:250,000 scale. All the delineations were

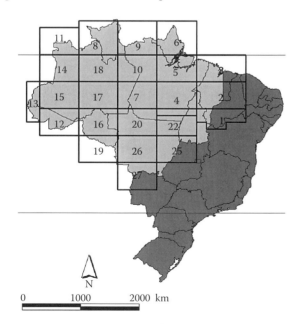

FIGURE 29.2 RADAMBRASIL volume numbers and locations for the Brazilian Amazon.

conserved. As most map units were similarly coded but their soil content was different, a general legend was established assembling all the original legends of the elementary maps. In the Mato Grosso soil map, the number of soil components of the map units usually ranges from one to two or three, seldom exceeding four. When various components are present, their relative proportions are given in attached tables. Soil components are characterized by their taxonomic type (generally the equivalent of the "soil order" in the Brazilian soil taxonomy), plus texture, vegetation, and slope class attributes.

For Rondônia, we used the unique SIGTERON (Sistema de Informação Geográfico de Terrenos e Solos para o estado de Rondônia) map, which provides spatial units called SOTRO (Solos e Terrenos do Estado de Rondônia). SOTRO are physiographic units delineated on Landsat-5 TM satellite images at the 1:250,000 scale. They are differentiated by a landform type composed of various topographic ("terrain") elements characterized by one soil association. A taxonomic classification (Brazilian and WRB "great soil groups") of the soil components and their relative proportions in the associations were available from attached files. Neither texture and slope nor vegetation is directly referenced. However, information is found in additional SIGTERON files. Slope and vegetation are "terrain" attributes, while texture can be assessed through a soil profile data base. It is noteworthy that in the Rondônia soil map, map units may consist of quite a large number of soil components.

With the complete legend, that is, with all the taxonomic levels and all the classes (with texture and vegetation phases), the number of the soil units is fairly high (425). Without vegetation, the number drops to 147. Only 10 soil units (soil units without vegetation) have a significant extension, more than 2% of the total Rondônia–Mato Grosso area. The spatial extent of the top 8 soil units represents nearly 70% of the total area (Figure 29.3).

Table 29.3 shows the two top "great soil groups" for each landscape region (Figure 29.4) and the number of soil units that belong to each great soil group. The North Rondônia Peneplain (region 1) can be considered as composed basically of two great soil groups—Oxisols ("Latossolo") and Ultisols ("Argissolos"), that represent 77% of the total area (Table 29.3). All are dystrophic, but they are of various color types and various textures. They are represented by 13 and 16 soils units, respectively (Table 29.3).

TABLE 29.3 The Top Great Soil Types of the Landscape Regions

LR	% Area	NB Soil Units[a]	Great Soil Groups
1.	52	13	Latossolo
	26	77	16	Solos Podzólicos
2.	89	8	Solos Podzólicos
	4	93	7	Latossolo
3.	56	3	Solos Podzólicos
	17	73	5	Latossolo
4.	43	23	Latossolo
	23	66	16	Solos Podzólicos
5.	69	13	Latossolo
	19	88	1	Areias Quartzosas
6.	40	18	Latossolo
	20	60	14	Solos Podzólicos
7.	33	6	Latossolo
	32	66	1	Plintossolo
8.	30	3	Cambissolo
	23	53	8	Latossolo
9.	37	6	Latossolo
	29	66	1	Areias Quartzosas
10.	46	2	Planossolo
	33	79	2	Plintossolo

LR, Landscape region.
[a] Number of soil units belonging to the great soil group.

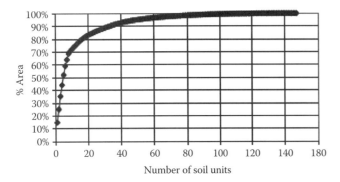

FIGURE 29.3 Proportion of the Rondonia–Mato Grosso area covered by number of soil units.

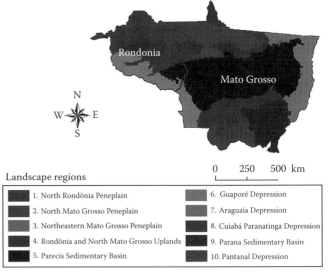

FIGURE 29.4 Rondonia and Mato Grosso states main landscape regions. (Adapted from Volkoff, B., F.F.C. Mello, S.M.F. Maia, and C.E.P. Cerri. 2009. Landscape and soil regionalization in southern Brazilian Amazon and contiguous areas: Spatial pattern for environmental modelling. Scientia Agricola, 2011.)

Many other soil types like Ultisols, gravely soil with ferruginous pisoliths, Inceptisols, and Entisols, are observed (Figure 29.5).

Overall, 69 soil units were identified. The occurrence of Latossolo Amarelo (Xanthic Ferralsol) is one of the characteristics of the region. Oxisol ("Latossolo") soil units have frequently

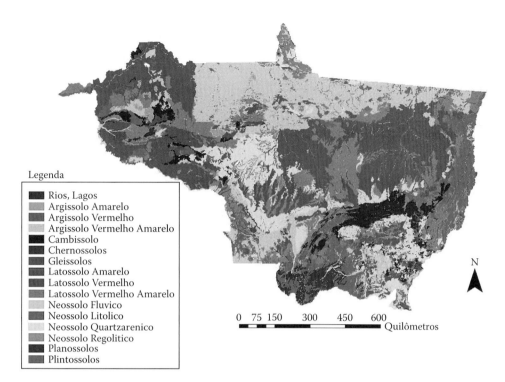

FIGURE 29.5 (See color insert.) Soil map of Rondonia and Mato Grosso states. (Modified from Mello, F.F.C. 2007. Estimativas dos estoques de carbono dos solos nos Estados de Rondônia e Mato Grosso anteriores à intervenção antrópica, 88pp. Dissertation (Master Degree). University of São Paulo, Sao Paulo, Brazil.)

clayey or heavy clayey texture, which is another regional soil characteristic (Figures 29.4 and 29.5).

Regarding its soil cover, the North Mato Grosso Peneplain (region 2) is very homogeneous. It has only one dominant soil unit, the Ultisol "*Podzólico Vermelho-Amarelo distrófico textura media*" (Table 29.3). For this reason, the region can be clearly differentiated from the North Rondônia Peneplain, where Oxisols represent twice the area occupied by Ultisol soils.

The most widespread soil unit in the Northeastern Mato Grosso Peneplain (region 3) is also "Ultisol" *Podzólico Vermelho-Amarelo distrófico textura media*. This soil is associated with various types of Oxisols, "Solos Concrecionários" and Entisols for a total of 23 soil units. Due to differences in parent rocks, slope and incipient pediment development the soils of the Rondônia and North Mato Grosso Uplands (region 4) show great variability (Figure 29.5). Although the Great Soil Groups are identical, 66% of the area is covered by Oxisols and Ultisols, and the number of soil units is very large, with 100 types listed. There are a great range of Ultisols, and a large variety of Oxisols and Inceptisols and 15 soil units cover 70% of the area. The most frequent soil unit, Ultisol "*Podzólico Vermelho-Amarelo distrófico textura média*," represents only 13% of the area. This complex spatial soil pattern is one of the characteristics of the region, along with the occurrence of eutrophic soils (mainly, Eutrophic Oxisol and Eutrophic Inceptisol). In addition, it is noteworthy that Oxisol and Ultisols are not always related to leveled topographic surfaces as it is the case in most of the other northern Landscape regions, but they are found here as well as on plane surfaces and on the slopes of rounded hills. They are then of probably differentiated origins. If the topographic parameter were taken into account in the definition of the soil units, the number of soil units should be significantly increased in this region.

The Parecis Sedimentary Basin (region 5) is principally covered by Oxisols (69%) and Entisols. Within the Great Soil Group of Oxisols, several (13) soil units are found. The most represented unit is *Latossolo Vermelho-Escuro distrófico textura media* (Table 29.3). The region is characterized by the predominant occurrence of the Oxisol types, which are all dystrophic. *Latossolo Vermelho-Escuro* and *Latossolo Vermelho-Amarelo* are found indistinctly (Figure 29.5). The texture ranges from clayey to sandy and the heavy clayey texture is atypical. The whole region is almost totally covered by 5 rather important soil units.

The Guaporé Depression (region 6, Figure 29.4) is a very heterogeneous region. The great soil groups Oxisol and Ultisol represent 60% of the total surface (Table 29.3). The variety of soil units is very high, as 86 soil units were identified. For Oxisol and Ultisol, the dominant texture is medium. The large soil diversity and the medium texture of the top soil units are the main characteristics of this region.

The Araguaia Depression (region 7) is another heterogeneous region. The most widespread soils are soils of the great group—Oxisol, with 6 soil units and the most represented soil unit is the great group Ultisol (Table 29.3). The *Plintossolo distrófico textura media* soil unit covers 32% of the area. The great group Alfisol is almost absent. For both the Oxisol and the Ultisol, the dominant texture is medium.

Inceptisol is the main great soil group in the Cuiabá Paranatinga Depression region (region 8). Together with Oxisol, they cover no more than 50% of the total area. Inceptisol is the soil of the central and eastern hilly and mountainous parts. Oxisol occupies the leveled areas of the west (Figure 29.5). Inceptisol and Oxisol units are normally dystrophic and have a medium texture. They are associated with a large range of other soil units, the number of which exceeds 50. The occurrence of Alfisols/Ultisols, "Solos Concrecionários" in the plains and Entisols in the mountain areas must also be pointed out.

In the Paraná Sedimentary Basin region (region 9, Figure 29.4), like in the Parecis Sedimentary Basin region, the dominant great soil groups are Oxisols and Entisols. Altogether they account here for 66% of the total region area while the total was 88% in the Parecis region. This difference is explained by the occurrence of soils (mainly Inceptisols and Alfisols/Ultisols) developed on exposed underlying Paleozoic layers of the sedimentary basin on eroded parts of the plateau. All Oxisols are dystrophic. *Latossolo Vermelho-Escuro* and *Latossolo Vermelho-Amarelo* are found indistinctly. The texture ranges from clayey to medium. The main soil unit is *Areias Quartzosas distróficas* and covers 29% of the total area (Figure 29.5).

In the Pantanal Depression (region 10), Alfisol (essentially 2 soil units) and Ultisol (1 soil unit) are the main represented soils. They account for 80% of the surface (Table 29.3). Excluding the singular Pantanal region, the other landscape regions appear not to be exactly similar with regard to their great soil groups. Although some have only one predominant group (the region 2), they generally have two, that is, Oxisol and Ultisol, that can each occupy the first position depending on the region considered. A better differentiation between the landscape regions can be obtained by considering the soil units. Rarely a single soil unit, sometimes a soil unit association characterizes the soil cover of a landscape region. The soil unit association is generally combined with the occurrence, even if not in predominant position, of one or several specific other soil units. The soil cover can then hardly be summarized by a single soil criterion.

Finally, it is possible to say that for each subdivision, a detailed inventory of the soil units, with their area distribution, can be obtained from soil survey maps at scales of 1:250,000–1:500,000. But this inventory can hardly be used for ecological purposes because of the great number of the soil units, which are generally sparsely sampled. However, this inventory might provide the framework for a stratification of the areal data extracted from soil maps in a new hierarchical classification, the pertinence of which might be tested with related profile data bases. The ensuing increase in the number of observations can then be used in a statistical description of each stratum, making possible the selection of the most relevant stratum to be used for the studied properties.

29.3.4 Digital Mapping of Soil Classes in Rio de Janeiro State

The Rio de Janeiro State is located in the eastern portion of southeast of Brazil, between 41° and 45° of West longitude and 20°30′ and 23°30′ of South latitude (Figure 29.6). With an area of 43.696,054 km², its economy is principally based on services, also with a significant part of industry and little influence in the agriculture sector. In general, the state is divided into two great geomorphologic features; the coastal plain and mountains area, of which eight large landscape types can be identified: coastal plains, north-northwest fluminense, Rio Paraíba do Sul middle valley, mountainous areas, upper Itabapoana river plateau, Serra dos Órgãos, Bocaina, and Mantiqueira (Rio de Janeiro, 2001; Lumbreras et al., 2003).

The Rio de Janeiro State was selected for this work due to the availability of massive georeferenced profile data organized in a data bank, which is unusual in Brazil. For this work, a database was built in Access and Delphi, containing 431 soil profiles derived from the soil survey accomplished by Embrapa Solos in different times. The soils were classified according to Brazilian Soil Classification System at the order level. Of the 13 orders of the Brazilian Classification, 9 orders were identified within the 431 soil profile datasets. Figure 29.7 shows the soil profile distribution in the Rio de Janeiro State. Information about vegetation and landform were also obtained from different works performed in the area over time, which were previously compiled by Carvalho Filho et al. (2003).

In this application, the following covariates were used as predictor variables to build the models: Geocover mosaic (bands 7, 4, and 2 in RGB), freely distributed by NASA (https://zulu.ssc.nasa.gov/mrsid/); the NDVI index (using band 2 instead of 3); Land Use/Land Cover (LULC) map of Rio de Janeiro State, performed by Mendonça-Santos et al. (2003); Lithology class map (Rio de Janeiro, 2001), and srtm DEM 90 m, obtained from the *cgiar* database at http://srtm.csi.cgiar.org (Jarvis et al., 2006). Detailed information on the digital processing and statistical data analysis are given in Mendonça-Santos and Santos (2008).

Soil classes were predicted by a regression tree (Rulequest Research, 2003). The dataset was partitioned into prediction and validation datasets, being ¾ and ¼, respectively. The program gives the error assessment for both datasets. A tree structure was generated by partitioning the data recursively into a number of groups, each division being chosen to maximize some measure of difference in the response variable in the resulting groups.

Table 29.4 illustrates the main data and information used to build every prediction model (M1, M2, M3, M4 and M5) of soil classes (at the order level), according to *scorpan* factors. All the prediction models were built by regression tree, using the See5 software (RuleQuest Research, 2003), varying the predictor variables. Information was obtained in almost all the factors of *scorpan model* as shown in Table 29.4, including a preexisting polygon-based soil class map (Carvalho Filho et al., 2003) and a Lithology class map.

In order to compare the performance of the models accounting for the number of parameters used, Akaike's Information Criterion (AIC) (Akaike, 1973) was used as a measure index:

$$AIC = -2\log like + 2m \quad (29.1)$$

where
  loglike is the log-likelihood of the prediction
  m is the number of parameters used in the model

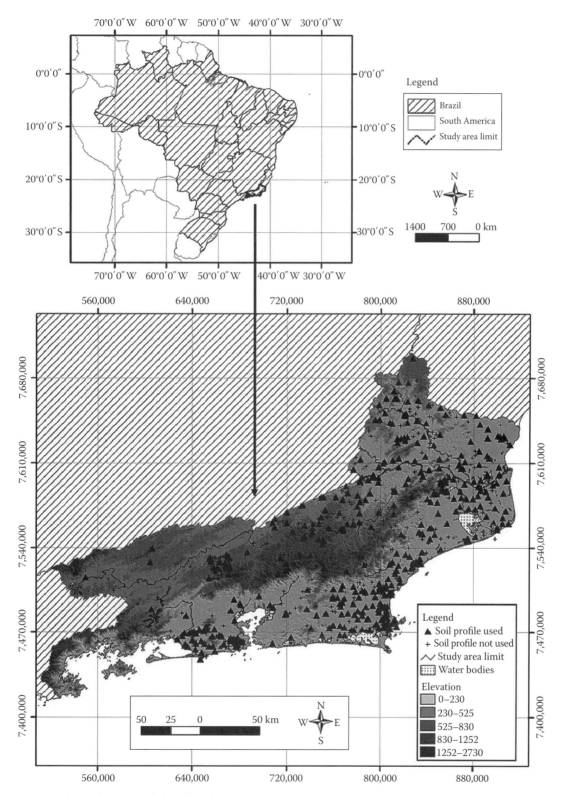

FIGURE 29.6 The study area location and the soil profile distribution on the elevation map, extracted from the SRTM DEM (Jarvis et al., 2006) at 90 m pixel resolution. (Adapted from Mendonça-Santos, M.L., Santos, H.G. dos. 2008. Digital Mapping of Soil Classes in Rio de Janeiro State, Brazil: Data, Modelling and Prediction pp. 381–396. *In* A.E. Hartemink, A.B. McBratney, and M.L. Mendonça Santos (eds.) Digital Soil Mapping With Limited date. Springer, New York.)

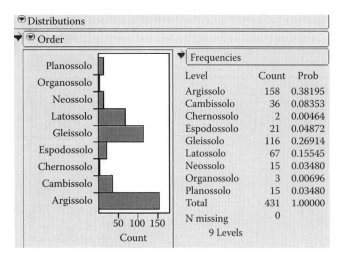

FIGURE 29.7 Soil order and frequency of the 431 soil profiles used in the modeling process. (Adapted from Mendonça-Santos, M.L., Santos, H.G. dos. 2008. Digital Mapping of Soil Classes in Rio de Janeiro State, Brazil: Data, Modelling and Prediction pp. 381–396. *In* A.E. Hartemink, A.B. McBratney, and M.L. Mendonça Santos (eds.) Digital Soil Mapping With Limited date. Springer, New York.)

This index is a compromise between the goodness of fit and the parsimony of the model. The best model is the one that has the smallest AIC.

The log-likelihood for class prediction (k = 1, ..., K) is calculated as follows (Hastie et al., 2001):

$$\text{loglike} = \sum_{k=1}^{K} \log \hat{p}_k \quad (29.2)$$

where $\hat{p}_k$ is the proportion correctly classified as class k:

$$\hat{p}_k = \frac{1}{N_k} \sum I(y_i = k) \quad (29.3)$$

N is the total number of data (soil profiles).

In a general way, all models were able to give a good prediction of soil classes at higher categorical levels (order) of the Brazilian Classification showing reasonable error values, which means that soil classes in the study area could be predicted from environmental covariates, that can be easily, cheaply, or even freely acquired, like Geocover images and the srtm DEM 90 m (Models M4 and M5). However, the best model to estimate soil class was the M1, basically due to the use of the existing soil information (polygon soil map). The use of the **S** factor of the *scorpan* model was first proposed by McBratney et al. (2003) but has not been tested as a source to predict soil classes.

Figure 29.8 illustrates the resulting maps for each model. The M1 model was able to predict all the nine soil classes included in the model, with a very good performance (the smallest AIC), as well as their spatial distribution all over the study area (soil classes appear where expected), in a fairly generalized soil map (order level of the Brazilian Classification System).

This work demonstrates that the soil–landscape relationship can be predicted in a quantitative and efficient way, using available information on data sources and existing methodologies in the *scorpan* procedure, which can be useful in areas with lack of or little soil information, as is the case in much of Brazil.

29.4 Integration and Application of Soil Digital Databases in Brazil

29.4.1 Organization and Structure of the Brazilian Soil Information System

The Brazilian Georeferenced Soil Information System (SIGSOLOS) has been developed at the National Center for Soil Research of Embrapa, the Brazilian Enterprise for Agriculture Research, with the support of modern methodologies, techniques, and tools for information management (Chagas et al., 2004). It arose as a recognized necessity from soil scientists, research and development agencies, and agribusiness and environment organizations. The old system for storing information on soil surveys, which had been developed in the 1970s with data

TABLE 29.4 Predictor Variables, Predictive Model, Predicted Variable, and Soil Dataset Information for Each Model Built (M1, M2, and M3)

Models	Predictive Model	Predictor Variables for SCORPAN Model	Predicted Variables	Soil Profiles	Pred_Dataset (3/4)	Valid_Dataset (1/4)
M1	Regression tree	**S** (Soil classes polygon map) **O**; (Landsat ETM + 7 Images, bands 7, 4, 3, and NDVI); **R** (ELEV, ASPECT, PLAN, PROF, QWETI, SLOPE); **P** (Lithology classes polygon map)	Soil class at order level (13 soil class according to Embrapa 2006)	445	334	111
M2		**O** (Landsat ETM + 7 Images, bands 7, 4, 3, NDVI and LULC map); **R** (ELEV, ASPECT, PLAN, PROF, QWETI, SLOPE), **P** (Lithology classes polygon map)		445	334	111
M3		**O** (Landsat ETM + 7 Images, bands 7, 4, 3 and NDVI); **R** (ELEV, ASPECT, PLAN, PROF, QWETI, SLOPE), **P** (Lithology classes polygon map)		445	334	111

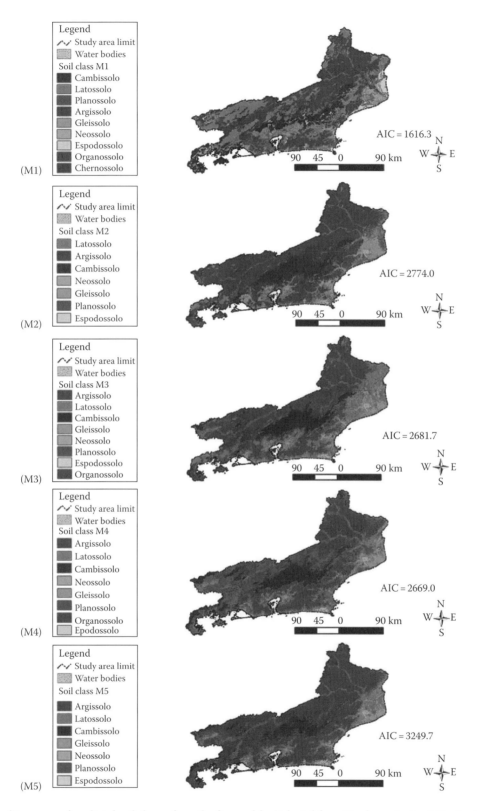

FIGURE 29.8 Resulting maps of predicted soil classes from the five models. (Adapted from Mendonça-Santos, M.L., Santos, H.G. dos. 2008. Digital Mapping of Soil Classes in Rio de Janeiro State, Brazil: Data, Modelling and Prediction pp. 381–396. *In* A.E. Hartemink, A.B. McBratney, and M.L. Mendonça Santos (eds.) Digital Soil Mapping With Limited date. Springer, New York.)

processing techniques oriented to text and numbers, did not satisfy the new requirements from soil science and its applications (Tanaka et al., 1997).

According to Tanaka et al. (1997) and Chagas et al. (2004), the main objectives of SIGSOLOS are (1) the construction of a database of soil surveys and other published works on soil science, as a support for the Brazilian Soil Classification System; (2) the development of new soil mapping models; and (3) the generation of interpretative maps.

The development of SIGSOLOS is the result of the evolution of information technology, based on database management systems (DBMSs) and GISs, in conjunction with the availability of data and metadata through the Internet World Wide Web. Intended, at first, for soil science researchers and students, but designed also for other users interested in a variety of applications, the system provides a visual presentation interface through maps, making a natural interaction with the users possible (Tanaka et al., 1997).

29.4.1.1 The Database

The solution for this kind of problem, through an integrated system with both DBMS and GIS features, is still a research issue in Information Technology. There is no single commercial software product that provides the functionality required by a system like SIGSOLOS (Tanaka et al., 1997). On one hand, DBMSs do not have data types and operators appropriate for all kinds of geographic objects and images. On the other hand, GISs do not have the capability for efficiently managing large amounts of data (Chagas et al., 2004). In order to make the solution adequate to the reality of Embrapa, a strategic design decision was taken to use a dual system architecture that consists in the interaction of an object-relational DBMS—Open Ingres (CA, 1995)—and a proprietary GIS—Spring (INPE, 1995).

Users can access the database directly using DBMS interfaces, through a data entry/query application or through the geographical interface provided by the GIS. The data entry/query is performed with a form-/report-based visual subsystem. Although it has targeted the above-mentioned software products, due to its modularity, this approach allows the use of different DBMSs and/or GISs (Tanaka et al., 1997).

According to Chagas et al. (2004), the database conceptual schema was developed using the well-known entity-relationship model in a database design software tool, making it easy to map the model to any DBMS. The data model was defined during Joint Application Design (JAD) sessions with Embrapa Soil Science researchers and users. The resulting schema is similar to the ones found in the international literature (DeGloria and Wagenet, 1993; Lytle, 1993), with the peculiarities of the Brazilian methodology for soil classification and mapping (Bhering, 1995).

Figure 29.9, from Tanaka et al. (1997), shows the relevant entity and relationship types in the database conceptual schema. The central entity type is WORK that represents three kinds of published works on soil. The most important works are the soil surveys in various levels of detail that describe soil sample profiles,

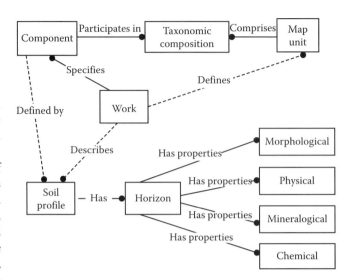

FIGURE 29.9 SIGSOLOS database (simplified) conceptual schema. (Adapted from Tanaka, A.K., S.B. Bhering, and C.S. Chagas. 1997. SIGSOLOS—Conceptual model and future perspectives (in Portuguese), Brazilian Symp. Inf. Agric. Belo Horizonte, Brazil.)

their horizons with morphological, physical, mineralogical, and chemical properties, as well as define map units that comprise taxonomic components based on described or derived profiles. The other kinds of work that SIGSOLOS addresses are technical–scientific ones such as articles and theses that usually deal only with profile information, and reports on derived map generation such as agricultural aptitude maps that use soil profile data described in soil surveys.

In addition to the entity types shown in Figure 29.9, the detailed schema contains diverse data on institutions, authors, municipal districts, and states, as well as dozens of auxiliary tables with attribute domain values. For the sake of simplicity, the diagram in Figure 29.9 is presented at entity type level; some entity types such as WORK, SOIL PROFILE, and COMPONENT may have dozens of attributes. The entity types MAP UNIT and SOIL PROFILE have geographical object identifications that are used as links between the database and its spatial expression—areas for map units and points for soil profiles—in the GIS (Tanaka et al., 1997).

29.4.1.2 Final Considerations

The initially designed application in SIGSOLOS was the support for the parameterization of the Brazilian Soil Classification System. It serves as a tool for classification methodology, at the time of the production soil survey bulletins (CNPS, 1995). Generation of soil maps and derived maps is largely facilitated by SIGSOLOS. Every database-centered information system presupposes, at design time, a set of useful applications. In general, after implementation, data availability and manipulation capability provided by the system raise new, unforeseen demands. With SIGSOLOS, because of the combination of database with GIS features, such new demands have been voluminous (Chagas et al., 2004).

Various versions of the system have been developed, both for client–server workstations and personal computers, and for portable computers to be used in the field. An early prototype of the database was presented at the Latin–American Soil Congress, in August 1996, in Brazil. The system and its conceptual model were presented and discussed during the XXVI Brazilian Congress on Soil Science, in July 1997 (Tanaka et al., 1997).

29.4.2 Regional Organic Carbon Storage Maps of Western Brazilian Amazon Based on Prior Soil Maps and Geostatistical Interpolation

Bernoux et al. (1998a, 2006) and Cerri et al. (1994) studied an area of 334,000 km^2 of the western Brazilian Amazon basin (Figure 29.10). This area was defined as a 6°-square part of the Brazilian Amazon, from 60° to 66° West longitude and 8° to 14° South latitude, and corresponding to 6.7% of the Brazilian Amazon basin. Administratively, this region corresponds to nearly all the Rondônia state, and small portions of the Amazonas and Mato Grosso states.

The authors organized a specific database from soil profile information reported in published and unpublished soil surveys. The spatial components were derived from two digitized soil maps: 1:5,000,000 soil map and 1:1,000,000 soil map of the region. At this regional scale, the main soil types and their relative extents are similar to that at continental scale. The Ultisols and the Oxisols are the dominant types covering respectively 40% and 34% of the area. The group of Inceptisols, Oxisols, and Alfisols, and the group of Entisols still represent a noticeable extent, accounting for 5% and 3% of the area. The Entisols and Molisols and the *Areais quartzosas* (Entisols) show extents representing 7.5% and 5% of the total area. In total, these six soil types cover 95% of the area. The main difference between the continental and the regional levels concerns the extent of the Spodosols that represent less than 0.5% in Rondônia and around 2% at the continental scale.

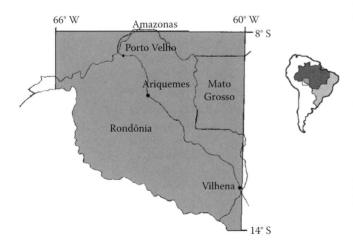

FIGURE 29.10 Location of the western Brazilian Amazon basin.

29.4.2.1 Database and Carbon Stock Calculation for Each Soil Profile

Regarding the soil database and carbon stock calculation, these authors applied a first correction assuming that the soil fraction >2 mm is carbon free. Soil bulk densities were often lacking, and soil carbon content was not always determined for several horizons. Several methods were used to estimate this lacking information. A stepwise multiple regression procedure was developed to predict soil bulk density (SBD) from other properties using the data of 323 soil horizons spread over the whole Amazon basin (Bernoux et al., 1998a). Clay and carbon contents were the best predictors to estimate SBD. For instance, in the case of Oxisols, the use of carbon and clay contents as predictors resulted in a percentage of explained variation near 70%. The following equations were used to estimate soil bulk densities for Oxisols and Ultisols/Alfisols:

1. Oxisols:

$$BD = 1.419 - 0.0037\,Clay\% - 0.061\,OC\% \text{ for A horizons} \quad (29.4)$$

$$BD = 1.392 - 0.0044\,Clay\% \text{ for B horizons} \quad (29.5)$$

2. Ultisols/Alfisols:

$$BD = 1.133 - 0.041\,OC\% + 0.0026\,S \text{ and } \% \text{ for A horizons} \quad (29.6)$$

$$BD = 1.718 - 0.0056\,Clay\% - 0.068\,pH_{water} \text{ for B horizons} \quad (29.7)$$

The problem of missing soil carbon data was resolved using "power" (Bennema, 1974) and "exponential" (Arrouays and Pelissier, 1994) type equations to model soil carbon distribution with depth. Bernoux et al. (1998b) determined that both models were well adapted for tropical forested Oxisols, but that the "exponential" model is more precise and produces better data. Bernoux et al. (1998a) extend the use of the exponential model to the other Amazonian soil types, mainly the Ultisols/Alfisols. The model equation is

$$\frac{(C(X) - C_2)}{(C_1 - C_2)} = \frac{(e^{-bX} - e^{-bX_2})}{(e^{-bX_1} - e^{-bX_2})} \quad (29.8)$$

where
 X and C(X) are the depth and the C content
 X_1 and C_1 the depth and the C content of a fixed upper position
 X_2 and C_2 for a fixed deeper position

X_1 was set to 0 cm, that is, the theoretical contact litter/soil, and X_2 was set to 100 cm; therefore, C(X) could be written as

$$\frac{(CX - C_2)}{(C_1 - C_2)} = \frac{(e^{-bX} - e^{-100b})}{(1 - e^{-100b})} \quad (29.9)$$

TABLE 29.5 Statistics on C Stock Values (P100 and M100 in kg C m^{-2})

Variable	n	Average	Median	Minimum	Maximum	SD
P100	324	7.30	6.30	0.61	41.62	4.51
M100	424	7.32	6.55	1.50	25.19	3.52

Sources: Modified from Bernoux, M., D. Arrouays, C.C. Cerri, and H. Bourennane. 1998a. Modeling vertical distribution of carbon in Oxisols of the Western Brazilian Amazon (Rondônia). Soil Sci. 163:941–951; Bernoux, M., D. Arrouays, C.C. Cerri, P.M. Graça, B. Volkoff, and J. Trichet. 1998b. Estimation des stocks de carbone des sols du Rondônia (Amazonie brésilienne). Etudes Gestion Sols 5:31–42.

For each soil profile of the database, the parameters of the fitted equation were determined using the nonlinear regression module. Corresponding C stock (M100 in kg m^{-2}) was obtained integrating CX:

$$M100 = 10\left(C_2 - (K_1 \cdot K_2) + \left(\frac{K_2}{b}\right) \cdot (1 - K_1)\right) \quad (29.10)$$

where

$$K_1 = e^{-100b} \quad \text{and} \quad K_2 = \frac{(C_1 - C_2)}{(1 - K_1)} \quad (29.11)$$

Soil Carbon stocks (C mass per area) for a given depth were also calculated in a "classical way" for each profile, when data were all available. This classical way of calculating C stock for a given depth consists in summing C stock by horizon, determined as a product of BD, C concentration, and horizon thickness. For each profile, when possible, the corresponding C stock (P100 in kg m^{-2}) was calculated not only for the first top meter (Table 29.5) but also for the top 30 cm depth (P0.3 in kg m^{-2}), as shown in Table 29.6. Statistics for M100 and P100 gave similar values. The coefficient of correlation calculated among them was 0.972, for 304 profiles with both P100 and M100.

29.4.2.2 Regional Stocks

Cerri et al. (2000) evaluated the regional carbon stocks for the 0–30 and 0–100 cm layers using two different approaches. The first

TABLE 29.6 Carbon Stored in the First 0.3 m (P0.3) and in the First Meter (P1.0) by Soil Taxonomic Unit

	P0.3						P1.0					
	N	Means	Min	Max	SD	Median	N	Means	Min	Max	SD	Median
Unit[a]		kg C m^{-2}						kg C m^{-2}				
AQ	18	4.12	2.19	9.2	1.88	4.05	17	8.96	4.68	19.13	4.29	6.88
AQH	7	3.01	0.32	5.83	2.3	1.88	4	6.02	0.61	13.88	5.78	4.79
BA	8	7.53	4.7	13.31	2.83	7.12	6	11.14	6.61	19.59	4.78	10.37
CTd	36	3.64	1.04	6.23	1.35	3.75	14	6.5	2.3	13.21	2.71	5.51
CTe	6	4.91	2.87	11.88	3.45	3.57	5	10.61	5.19	23.04	7.63	6.23
GH	6	16.33	7.59	24.31	7.08	15.59	4	23.32	9.57	33.79	10.25	24.96
GPH	12	4.67	2.84	7.16	1.35	4.36	9	7.99	3.71	10.86	2.54	7.77
LA	27	3.65	1.97	6.38	1.03	3.53	26	7.3	4.94	10.71	1.48	7.32
LR	4	6.46	3.49	12.49	4.1	4.93	3	19.29	7.12	41.62	19.36	9.14
LVA	53	3.78	1.93	9.57	1.35	3.61	34	7.49	4.4	19.46	3.03	6.97
LVE	13	3.73	2.31	5.56	1.13	3.49	10	6.9	3.78	11.52	2.37	6.65
PH	6	4.96	0.76	12.54	5.52	2.02	4	3.18	1.34	7.62	2.98	1.88
P	19	3.41	1.01	9.57	2.08	3.16	12	7.06	1.8	27.69	6.77	5.78
PVA	202	3.41	0.61	13.51	1.95	2.96	154	6.4	1.44	23.12	3.14	5.84
PVE	9	3.09	1.11	6.23	1.42	2.75	4	5.9	4.16	9.96	2.73	4.74
SA	10	2.7	1.07	5.26	1.08	2.54	9	5.57	3.45	11.66	2.45	5.13
SL	22	3.95	0.97	7.45	1.87	4.08	1	6.73				6.73
TR	16	4.89	1.99	8.96	2.07	4.34	8	9.52	4.57	15.76	3.8	8.44

Sources: Adapted from Bernoux, M., D. Arrouays, C.C. Cerri, and H. Bourennane. 1998a. Modeling vertical distribution of carbon in Oxisols of the Western Brazilian Amazon (Rondônia). Soil Sci. 163:941–951; Bernoux, M., D. Arrouays, C.C. Cerri, P.M. Graça, B. Volkoff, and J. Trichet. 1998b. Estimation des stocks de carbone des sols du Rondônia (Amazonie brésilienne). Etudes Gestion Sols 5:31–42.

PVA, Podzolico Vermelho Amarelo distrofico (Ultisols); LVA, Latossolo Vermelho Amarelo (Oxisols); LA, Latossolo Amarelo (Oxisols); AQ, Areia Quartzosas (Psamments); Ca, Cambissolo alico (Inceptisols); Pa, Planossolo alico (Alfisols, Mollisols); LVE, Latossolo Vermelho Escuro (Oxisols); GPH, Gley Pouco Humico (Aquic suborder); SA, Solo Aluvial (Entisols-Fluvents); TRE, Terra Roxa Estruturada (Ultisols, Alfisols); BA, Brunizem Avermelhado (Chernozems); Ce, Cambissolo eutrofico (Inceptisols); PVE, Podzolico Vermelho Escuro (Alfisols); PH, Podzol Hidromorfico (Spodosols); AQH, Areia Quartzosas Hidromorfica (Psamments); LR, Latossolo Roxo (Oxisols); Cd, Cambissolo distrofico (Inceptisols); SL, Solo Litolico (Lithic subgroup); Pe, Planossolo eutrofico (Ultisols); Pd, Planossolo distrofico (Alfisols, Mollisols).

[a] Soil unit.

approach is based on carbon stocks averaged by soil type multiplied by the extent of these soil types. It corresponds to the methodology used by Moraes et al. (1995) and Bernoux et al. (2002).

Using the soil maps, the regional stocks were obtained multiplying the carbon content of each taxonomic unit by its area. This calculation leads to similar results using either the soil map at a 1:5,000,000 scale or the soil map at 1:1,000,000. Regional stocks ranged between 2100 Tg, using median values, and 2400 Tg, using means, and exhibit an associated error (based on standard deviation [SD]) of 900 Tg. Using the map from Embrapa (1:5,000,000) instead of the Radam (1:1,000,000) leads to an increase of only 25 Tg. It is striking that the results are very similar using (1) the median value of all P100 or M100 values or (2) the medians of M100 and P100 segregated by soil type. But in the second calculation, the associated errors based on SD are reduced to less than 50% amount, whereas it reached 71% in the first calculation.

The second approach is based on interpolation of carbon stocks using geostatistics. Geostatistical treatments were carried out using the theory first presented by Matheron (1965) and applied to pedometrics by Burgess and Webster (1980a, 1980b).

Experimental semivariogram for the variables P0.3 and P1.0 were all best (i.e., more satisfactory results of the validations) modeled using a nugget effect plus a spherical model according to the function:

$$\gamma(h) = c_0 + c_1 \left\{ \frac{3h}{2a} - \frac{1}{2}\left(\frac{h}{a}\right)^3 \right\} \quad \text{for } 0 < h \leq a \quad (29.12)$$

$$\gamma(h) = c_0 + c_1 \quad \text{for } h > a \quad (29.13)$$

$$\gamma(0) = 0 \quad (29.14)$$

where
 a is the range of the model
 c_0 is the nugget
 c_1 is the sill values

The experimental semivariogram for P0.3 (Figure 29.11) was fitted with a nugget effect of 1.84 (kg m^{-2})^2 plus a spherical model using a sill value of 1.93 (kg m^{-2})^2 and range of 772 km. This result shows that carbon densities in the first 0.3 m exhibit a spatial structure at the regional scale. However, the high nugget effect reveals that about half of the spatial variability may appear within distances inferior to a few tens of kilometers. Results of the fitting of the experimental semivariogram for P1.0 show the same pattern with a nugget effect of 4.70 (kg C m^{-2})^2 and a sill value of 4.26 (kg C m^{-2})^2, but the range is reduced to 513 km.

Resulting estimates are very similar, ranging from 2100 to 2400 Tg C in the top meter of soil with the first approach. The geostatistics furnish an estimate of 2220 Tg C, with an associated error of only 13%, compared to a 40% error in the case of soil map-based methods.

Maps of estimated carbon densities and their associated kriged standard deviation (KSD) were calculated by block kriging (Figure 29.12). Each map is made up of 334,107 one km^2-square blocks. The P100 kriged values ranged from 4.27 to 10.40 kg m^{-2}, and KSD from 0.54 to 1.31 kg C m^{-2}, the mean values being 6.665 kg m^{-2}. The mean of the kriged values is close to the median of the population of the carbon densities without segregation of soil type. This results in a regional soil carbon content of 2220 Tg C, indicating an intermediary level between the previous estimates, but now the associated error is considerably reduced to 295 Tg, that is, only 13.5% of the amount.

For the carbon stocks in the upper 0.3 m, the kriged values ranged from 2.09 to 6.00 kg m^{-2}, and KSD from 0.36 to 0.76 kg C m^{-2}, their average values being respectively 3.58 kg m^{-2} (mean) and 0.51 kg m^{-2} (KSD). The mean of the kriged values is close to the median of the population of the carbon densities for the first 0.3 m without segregation of soil type. This results in a regional soil carbon content of 1195 Tg, very close to the one previously calculated with the median, but now the associated error is considerably reduced to 170 Tg, that is, only 14.2% of the amount. The geostatistical approach gives regional estimates very similar to those derived from the classical approach based on a soil map, but the advantage of the geostatistical method is to lead to a much lower global associated error and to provide a map of the punctual kriged error.

Bernoux et al. (2006) proposed a third methodology running geostatistical interpolation separately on the parameters of the exponential model fitted for each carbon profile. The three parameters of

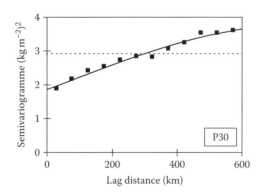
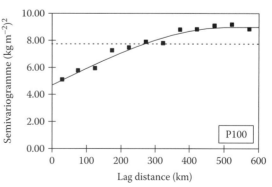

FIGURE 29.11 Experimental semivariograms and fitted models for variables P0.3 and P1.0. (Adapted from Bernoux, M., D. Arrouays, B. Volkoff, C.C. Cerri, and C. Jolivet. 1998c. Bulk densities of Brazilian Amazon soils related to other soil properties. Soil Sci. Soc. Am. J. 62:743–749.)

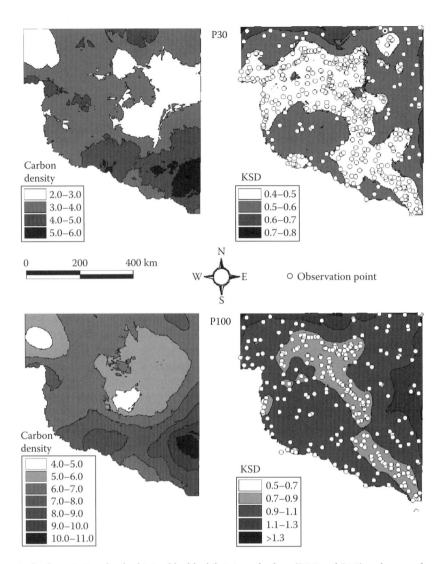

FIGURE 29.12 Carbon stocks down to 1 m depth obtained by block kriging of values (P0.3 and P1.0) and maps of associated errors. (Adapted from Bernoux, M., D. Arrouays, C.C. Cerri, P.M. Graça, B. Volkoff, and J. Trichet. 1998b. Estimation des stocks de carbone des sols du Rondônia (Amazonie brésilienne). Etudes Gestion Sols 5:31–42.)

the vertical model were estimated for 424 soil profiles. Variographic analysis was done independently for each parameters. Results of the variographic analyses showed that the parameters C1 and C2 were spatially structured, but with a high nugget effect. On the contrary, the parameter b relative to the profile curvature showed no spatial dependency. Experimental variograms, corresponding to parameters C_1 and C_2, were modeled with a linear model with respective nuggets effects of 0.85 and 0.0198 (kg m^{-2})^2 and slopes of 6.83 10^{-4} and 1.65 10^{-5} (kg m^{-2})^2. Even with such high nugget effects, the internal and external validations presented good results.

After validation, the grid of the parameters were obtained by block kriging. Because of the nonlinearity in the profile depth model, conditional simulation might be better than kriging, but kriging procedures were chosen because they can be considered as a more generic tool usable by nonspecialists. Values varied from 1.61 to 3.19 for parameter C1, with a mean of 2.19. Parameters C2 varied from 0.23 to 0.50, with a mean value of 0.31. Associated kriging error represented in both cases from 9% to 23% with a mean error of about 15%.

From the grid parameters it was then possible to calculate the C stocks corresponding to the first top meter. As parameter b showed no structure, two solutions were possible: the use of a fixed value (mean or median of the population) or a moving window average. Population of parameter b is log-normal; therefore, the use of the median was judged more adequate.

This approach was validated by the authors in two different ways. First, a classical external validation using reserved data was employed. The second validation was based on comparing the grid obtained by calculating C stock with the grids for the three parameters, with the grid corresponding to kriging P100 values: the mean deviation was an overestimation of 8.2%.

Finally, it is possible to stress that classical geostatistical approach furnished similar estimates, but with a reduced associated error compared to base-map approach. The new geostatistical method appeared to be satisfactory, giving an estimate of 2400 Tg C (0–100 cm layer), but with the advantage of estimating soil C stored at any given depth.

29.4.3 Soil Organic Carbon Stocks of Rio Grande do Sul State

The International Panel of Climate Change (IPCC) recommended the implementation of national inventories of greenhouse gas emissions to establish an empirical basis for climate change monitoring (IPCC, 2006). These inventories also provide an invaluable tool to establish mitigation and adaptation strategies in face of climate change. An important usage of inventories is the assessment of temporal changes in SOC stocks by comparison of current with baseline data, usually spatially explicit SOC stocks data prior to human-induced changes. Information on SOC stocks changes derived from different land uses and soil management practices is also critical for the assessment of country-level greenhouse gas budgets (Bernoux et al., 2001; Ministério da Ciência e Tecnologia, 2006).

A protocol suggested by the IPCC for the implementation of national inventories that assigns default SOC stocks to generic soil classes and vegetation types is defined as a Tier 1 method (IPCC, 2006). Tier 2 and 3 methods are increasingly refined approaches using soil pedon datasets from existing soil surveys at finer scales or produced by new field sampling campaigns.

Recent national estimates of baseline and temporal changes of SOC stocks using soil survey databases include Bernoux et al., 2002; Krogh et al., 2003; Bellamy et al., 2005; Lettens et al., 2005. In these studies, regional or national SOC stock maps were derived from soil data of pedon databases with geoprocessing techniques. However, this approach has recognized limitations: soil survey sampling densities might not be adequate for the intended scale of the SOC stock estimate, and sometimes might not cover the diversity and extent of soils; soil sampling protocols and analytical techniques are seldom consistent within databases (IPCC, 2006).

In fact, the most accurate available C analytical techniques, such as dry combustion, were rarely applied in older surveys, and SBD, a critical soil attribute for C stocks calculation was quite often missing. These issues become more critical in less developed countries where systematic soil surveys and soil mapping are generally scarce. For example, in Brazil, a countrywide soil map derived from several regional surveys is available only at a small scale (1:5,000,000) with sparse soil profile coverage. State-level soil surveys and maps are mostly in 1:750,000–1:500,000 scale (Giasson et al., 2006).

An important aspect of soil mapping and database management in recent years is the great expansion of personal computers and GIS, which has provided incentive for the conversion of soil maps and databases previously available only in paper format into digital formats. Once converted and validated, these electronic maps and associated databases provide an excellent means for assessment of SOC stocks at spatial and temporal scales. The end-products provide valuable insights for policymakers to assess compliance with international climate treaties and to establish greenhouse gas mitigation strategies linked to land use and soil management changes.

29.4.3.1 Previous Estimates of SOC Stocks of Rio Grande do Sul State

Rio Grande do Sul (RS), the southernmost state in Brazil, has land area of about 268,897,387 km² (Figure 29.13). The climate is predominantly subtropical with mean annual temperatures ranging from 14°C at higher elevations (>1000 m) in the northeast to 20°C in the western border with Argentina (IBGE, 1986).

The presettlement land covered comprised mostly subtropical and temperate forests, savannas, and wetlands. Current land uses reported include approximately 80,000 km² under agriculture and perennial crops (mainly grain crops, such as soybeans, corn, rice, and small grains) and about 110,000 km² under managed grasslands interspersed with agriculture and plantation forestry. The remaining land area includes native woodlands, savannas, and wetlands (Secretaria Estadual do Meio Ambiente, 2001).

The first estimate SOC stocks in RS is part of the Brazilian National Greenhouse Inventory using the soil map of Brazil (1:5,000,000) and a soil pedon database built from a national natural resources survey (RADAMBRASIL Project) to derive representative stocks for all Brazilian federal states. Tornquist et al. (2004) obtained a new provisional estimate of 1167 Tg original SOC stocks of RA using the Reconnaissance Soil Survey map of 1973 (1:750,000 scale) and the associated soil pedon properties for the 75 mapping units defined in the survey (Ministério da Agricultura, 1973).

FIGURE 29.13 Location of Rio Grande do Sul in South America. (Adapted from Tornquist, C.G., J. Mielniczuk, and E. Giasson. 2004. Estimativa de estoques de carbono orgânico em solos do RS. Anais da XV Reunião Brasileira de Manejo e Conservação do Solo e Águas. Departamento de Solos. UFSM, Santa Maria, Brazil. [CD-ROM] (in Portuguese).)

29.4.3.2 Refined Assessment of SOC Stocks of Rio Grande do Sul

A new estimation of soil C stocks was conducted as a part of a soil C dynamics project in RS, using the most comprehensive soil data available (Tornquist et al., 2009a, 2009b). The Natural Resources Survey of Brazil (Levantamento de Recursos Naturais, better known as the RADAMBRASIL Project) was a wide-ranging national natural resources survey executed in the 1980s. Within the scope of the project, a soil survey was conducted that originated a written compendium—the Natural Resources Report—that included a pedology chapter, which listed soil sampling site descriptions, analytical data of pedons, soil classification, and a state soil map published at 1:1,000,000 scale. The soil map was compiled from field worksheets produced at 1:750,000, preserving the mapping unit delineation. These paper products, including the soil map, have only recently been digitized to GIS-compatible format (IBGE, 2002). However, this digitization process did not include the analytical data of soil classes or soil profiles.

29.4.3.2.1 Data Preparation and GIS Operations

To produce a soil SOC map to a depth of 30 cm, a digital soil pedon database was constructed based on the published RADAMBRASIL survey report. The digital soil maps provided by IBGE consisted of a GIS polygon file containing 1478 records with the attributed soil mapping units, water bodies, urban areas, and sand dunes. For SOC stock estimates, only the 1200 soil polygons representing the spatial distribution of the 44 mapping units were selected.

The calculation of statewide C stocks with soil profile data required compilation of a georeferenced table in GIS-compatible format, in this case ArcGIS (ESRI, 2003). This table included the location of soil pedon (in latitude-longitude format), soil classification, horizon depths, SBD, organic carbon content, and texture of soil profiles listed in the Natural Resources Report. SBD data, which is required for the calculation of soil C stocks, was not included in the RADAMBRASIL soil survey. SBD data was obtained with a pedotransfer function developed for Brazilian soils (Benites et al., 2007):

$$SBD = 1.568 - 0.0005 \text{ clay content (g kg}^{-1}) - 0.009 \text{ C concentration (g kg}^{-1}) \quad (29.15)$$

$$[R^2 = 0.63]$$

Although strict validation was not conducted for the function, estimated SBD were generally adequate based on SBD from previous field studies conducted in the Soil's Department at Federal University of Rio Grande do Sul. However, estimated SBD of the "Organosolos Tiomórficos" (Hemists) was greatly overestimated, probably because these Hemists lie outside the range of C content and densities of the pedons originally used to develop the function (Benites et al., 2007). A SBD of 0.19 Mg m^{-3} reported for Histosols in RS by Kämpf and Schneider (1989) was used.

The georeferenced table with pedon properties was converted to a GIS point layer and overlaid on the IBGE soil map polygon layer. Soil pedon location and description were visually compared to the underlying soil map polygon soil classification. Pedons with soil classification that did not spatially match with the soil map unit classification were rejected for this project. Pedon data were then transferred to the soil map polygons. The result of this GIS operation was a vector layer (the soil map) with an attribute table that included soil analytical data for each mapping unit. One hundred and twenty-four soil pedons listed in the RADAMBRASIL soil survey report whose geographical location and soil classification matched the mapping units in the soil map were utilized. The soil classes Spodosols and Psamments were not represented in the RADAMBRASIL pedon database and were obtained from other larger-scale surveys in RS.

SOC stocks were calculated by soil horizon as mass of C per area (kg C m^{-2}), following the usual procedure: multiplying C concentration, BD, and horizon depth. To attain the depth of interest of 30 cm, in most cases, calculations required either a portion of the surface horizon or the entire surface horizon, and a portion of the next underlying horizon. As a final step, soil classes were aggregated in most cases to the second categorical level of the Brazilian soil classification system. Oxisols and Inceptisols that occur in a very large area of the state, were reported in lower categorical levels. The Spodosol class, which was not sampled and classified in lower categorical levels in the RADAMBRASIL survey, was only reported in the first level. SOC stocks were then averaged by aggregated soil class. SOC stocks for each soil map polygon were calculated by multiplying area (ha) and assigned average SOC stock. Total state C stock in soils was estimated by summing each polygon stock. Mean and median stocks and area weighted mean and median stocks were calculated for each soil class.

29.4.3.2.2 Statewide Estimate of SOC Stocks from Soil Profiles

One of the advantages of a GIS approach to regional SOC stocks estimation is the quick synoptic visualization of the spatial distribution of C stocks (Figure 29.14). In this case, it is clear that SOC stocks were greater in the northeastern part of the state, where a combination of higher elevations (>1000 m), higher rainfall (mean annual precipitation of approximately 220 cm), and temperate climate (mean daily temperatures <22°C) occur. Other areas with high SOC stocks occur near the Atlantic coast and river floodplains in the south central region of RS, were soils are periodically affected by waterlogging.

The geospatial database with SOC information allows the quick calculation of total C in soils of the state, by adding a calculated field with areas under the soil mapping units multiplied by the above C stock estimates. Other calculated fields of interest, including basic statistics, are also possible (Table 29.7). Taking into account the area covered by the soil

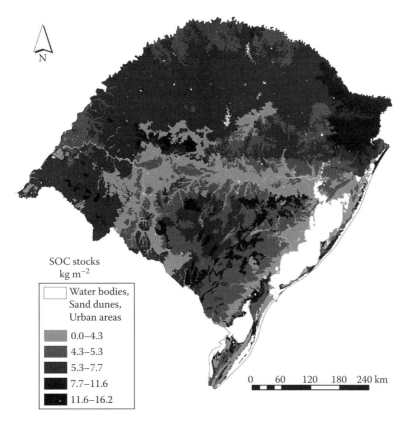

FIGURE 29.14 Soil organic C map of Rio Grande do Sul (0–30 cm depth). (Modified from Tornquist, C.G., E. Giasson, J. Mielniczuk, C.E.P. Cerri, and M. Bernoux. 2009a. Soil organic carbon stocks of Rio Grande do Sul, Brazil. Soil Sci. Soc. Am. J. 73:975–982.)

classes of RS, more SOC was originally present in "Latossolos Vermelhos" (Udox) (294.1 Tg C), "Neossolos Regolíticos" (Orthents) (249.5 Tg C), and "Argissolos Vermelhos" (Udults) (163.7 Tg C), which corresponds to 18%, 15%, and 10% of the total stock. Total terrestrial SOC stock of RS was estimated at 1610 ± 458 Tg C (1.6 ± 0.3 Pg C) calculated from soil pedons from the RADAMBRASIL dataset. Mean C density for soils of the state in this project was 7.7 ± 1.9 kg C m^{-2}. The total from mean pedon SOC was approximately 1.9% lower and the median was 5.7% than the total stock of 1641 ± 232 Tg C reported for RS state in the Brazilian National Greenhouse Gases Inventory (Bernoux et al., 2002).

On one hand, this result could be expected, as the dataset used was essentially the same in both studies. The differences were in few soil profiles discarded in this project—those that did not match spatially with the soil mapping units. The result is somewhat surprising, given the finer delineation of soil classes in a larger-scale soil map (1:250,000, compiled to 1:1,000,000) as compared to the Brazilian National Inventory map (1:5,000,000). However, total C stock of RS is mostly determined by a few soil classes, such as Oxisols and Entisols. These two classes are represented by similar areas and SOC stocks in both soil maps.

This pedon-based total SOC stock is 28% larger than a similar earlier attempt to estimate RS state's total SOC stock (1167 Tg C) with a smaller pedon database (Tornquist et al., 2004). The small number of available pedons (75) and inaccuracies of the Reconnaissance Survey soil map used in that project are possible causes of this underestimation. Mean C density of all soil classes was 7.7 ± 1.9 kg C m^{-2} and area-weighted mean was 6.1 kg C m^{-2}. The Brazilian National Inventory reported the same area-weighted median SOC stock of 6.1 ± 0.9 kg C m^{-2} for RS.

The SOC digital map and SOC database developed in this research project provide crucial baseline information for conducting state-level contemporary assessment of C stocks and soil carbon sequestration programs and initiatives.

29.4.3.2.3 Uncertainty

Soil carbon accounting initiatives have inherent uncertainties that will determine the final accuracy of the estimation of carbon stocks. A general concern with existing databases is the ubiquitous utilization of the Walkley–Black wet combustion analytical method for determining C concentration of soils. It has been shown that a single correction factor, as normally applied in soil C analyses, might not be appropriate for all soil classes, and will lead to the underestimation of C stocks (Skjemstad et al., 2000; Pereira et al., 2006). Soil classes with few complete profiles or missing analytical data also compromise the accuracy of the estimate. Pedotransfer functions are helpful to fill in data gaps, but indiscriminate application is unwarranted and local validation highly recommended.

TABLE 29.7 Presettlement Surface (0–30 cm) Soil Organic Carbon Stocks in Soils of Rio Grande do Sul

Soil Classes				SOC Stocks				
		Pedon	Area	Mean	SD	Total	Median	Total
Brazilian Soil Classification	Soil Taxonomy[a]	n	km²	kg m⁻²		Tg	kg m⁻²	Tg
Argissolo Amarelo	Udults	7	5,181	4.9	1.2	25.4 ± 6.2	4.7	24.3
Argissolo Bruno-Acinzentado	Udults	6	10,670	5.0	2.7	50.3 ± 27.2	4.6	49.1
Argissolo Vermelho	Udults	21	32,106	5.1	1.7	163.7 ± 54.6	4.6	147.7
Argissolo Vermelho-Amarelo	Udults	11	20,967	3.8	1.9	79.7 ± 39.8	3.1	65.0
Cambissolo Háplico	Udepts	1	10,414	5.0	—	52.1	5.0	52.1
Cambissolo Hístico	Udepts	3	9,930	11.6	2.7	115.2 ± 26.8	12.2	121.2
Chernossolo Ebânico	Udolls	8	7,735	7.6	1.1	58.8 ± 8.5	7.7	59.6
Espodossolos	Aquods	—	742	11.3	—	8.4	11.3	8.4
Gleissolo Háplico	Fluvents	4	4,083	5.3	2.7	21.6 ± 11.0	4.7	19.2
Gleissolo Melânico	Fluvents	1	1,357	9.1	—	12.3	9.1	12.3
Latossolo Bruno	Udox	4	1,559	13.4	4.0	20.9 ± 6.2	13.4	20.9
Latossolo Vermelho	Udox	6	47,440	6.2	1.6	294.1 ± 7.6	6.3	298.9
Luvissolo Crômico	Udalfs	3	3,928	7.7	0.3	30.2 ± 1.2	7.8	30.6
Neossolo Flúvico	Fluvents	—	1,188	4.1	—	4.9	4.1	4.9
Neossolo Litólico	Orthents	2	11,462	9.3	0.3	106.6 ± 3.4	9.3	106.6
Neossolo Quartzarênico hidromórfico	Psamments	1	829	16.2	—	13.4	16.2	13.4
Neossolo Quartzarênico órtico	Psamments	1	454	3.6	—	1.6	3.6	1.6
Neossolo Regolítico	Orthents	8	39,601	6.3	2.5	249.5 ± 99.0	6.1	241.6
Nitossolo Bruno	Udox	4	3,169	7.0	0.8	22.2 ± 2.5	6.9	21.9
Nitossolo Vermelho	Udox	9	15,821	6.5	1.9	102.8 ± 30.0	5.7	90.2
Organossolo Tiomórfico	Hemists	2	1,105	14.1	2.3	15.6 ± 2,5	14.1	15.6
Planossolo Háplico	Aquults	15	25,897	4.3	2.0	111.4 ± 51.8	3.6	93.2
Plintossolo Argilúvico	Udults	6	4,421	5.1	2.6	22.5 ± 11.5	4.9	21.7
Vertissolo Ebânico	Uderts	1	1,980	13.5	—	26.7	13.5	26.7
Mean				7.7	1.9[b]		7.6	
Total		124	261,436			1,610.1 ± 458		1546.7

Source: Modified by Tornquist, C.G., E. Giasson, J. Mielniczuk, C.E.P. Cerri, and M. Bernoux. 2009a. Soil organic carbon stocks of Rio Grande do Sul, Brazil. Soil Sci. Soc. Am. J. 73:975–982.

[a] Correlation between the Brazilian Classification System and Soil Taxonomy is only possible at higher levels.

[b] Pooled standard deviation.

29.4.4 SOTER Database as Basis for Modeling Soil Organic Carbon Stocks and Changes in the Brazilian Amazon

Land use and land cover changes in the Brazilian Amazon have major implications for regional and even global carbon cycling. Cattle pasture represents the largest single use (about 70%) of this once-forested land in most of the region. The aim of this study was to predict soil carbon stocks and changes in the Brazilian Amazon during the period between 2000 and 2030, using the GEFSOC Modeling System.

The GEFSOC Modeling System (Easter et al., 2007) was assembled under a cofinanced project supported by the United Nations Environment Programme (UNEP) Global Environmental Facility (GEF). The tool was built to provide scientists, natural resource managers, policy analysts, and others with the tools necessary to conduct regional- and country-scale soil carbon inventories. It is intended to allow users to assess the effects of land use change on soil carbon stocks, soil fertility, and the potential for soil carbon sequestration.

The GEFSOC Modeling System was used to integrate the data presented below and to conduct analysis with three well-recognized models and methods: (1) the Century© general ecosystem model, (2) the RothC© soil carbon decomposition model, and (3) the Intergovernmental Panel on Climate Change (IPCC) method for assessing soil carbon at regional scales. The modeling system requires six basic data classes to build the datasets necessary for a regional simulation: native vegetation, historic and current land use management, climate, soils, and latitude/longitude. The Century and RothC models generate large amounts of output. To reduce the size of the dataset while still generating meaningful results for users, the modeling system generates a series of regression statistics for user-defined breakpoints in the management sequence.

The user specifies the years for which soil organic C data are needed in the output dataset, and the modeling system calculates the intercept value, slope, and r^2 statistic for each time interval between the year breakpoints for the following output data (Easter et al., 2007): (1) Century sum of C in the active, slow, and passive soil organic matter pools (*somsc*), (2) Century total soil C (*somtc*), including belowground structural and metabolic fractions, (3) Century minimum annual total nonliving C (*totc*), and (4) RothC *soil organic C*. These data are followed by the mean values for each of the following output variables for the time period between breakpoints: (1) Century growing season accumulator for aboveground C production (*agcacc*), (2) Century growing season accumulator for belowground C production (*bgcacc*), (3) Century annual C inputs (*cinputs*), and (4) Century *grain yields*. Another output is the results from the IPCC method simulation runs. The soil C stocks and change rates calculated from the IPCC method are obtained using information (land management, climate and soils data) entered into the Graphical User Interface (GUI).

In order to achieve the mentioned goal, Cerri et al. (2007a, 2007b) devised current and future land use scenarios for the Brazilian Amazon, taking into account: (1) deforestation rates from the past three decades, (2) census data on land use from 1940 to 2000, including the expansion and intensification of agriculture in the region, (3) available information on management practices, primarily related to well-managed versus degraded pasture and conventional versus no-tillage systems for soybean, and (4) FAO predictions on agricultural land use and land use changes for the years 2015 and 2030. The land use scenarios were integrated with spatially explicit soils data, climate, potential natural vegetation, and land management units using the recently developed GEFSOC Modeling System.

For soil data, both the soil map of the Brazilian Amazon and the dataset that we describe below were used. The soil map was digitized from EMBRAPA data (1981) at the scale of 1:5,000,000 containing 2698 map units representing 69 soil types for the whole of Brazil. The Soil and Terrain (SOTER) dataset presents a harmonized set of soil parameter estimates for the Brazilian Amazon, developed to permit the modeling of soil carbon stocks and changes (Batjes et al., 2007). The SOTER database for Brazil (of which the Brazilian Amazon data is part), compiled in the framework of the 1:5M SOTER work for Latin America and the Caribbean, provided the basis for the current study. Results are presented as summary files and can be linked to the 1:5M scale SOTER map for Brazil and Brazilian Amazon in a GIS through the unique SOTER-unit code (Batjes et al., 2004).

29.4.4.1 Regional Simulations

The Brazilian Amazon plays an important role in the global cycle due to its large extent, its relatively high carbon density, and high deforestation rates. As it is well known, there are multiple variables that determine SOC distribution in the Brazilian Amazon. The GEFSOC Modeling System was used to provide estimates of current and future SOC for the entire Brazilian Amazon as shown in Figures 29.15 through 29.17.

Figure 29.15 shows the current SOC stock distribution in the Brazilian Amazon for the 0–20 cm layer, obtained by the Century model. SOC stocks vary from 20 to 150 Mg C ha^{-1} in the top 20 cm of soil. The majority of the area is under the third class of SOC stocks, that is, from 60 to 80 Mg C ha^{-1}. Figure 29.15 shows agreement with information from other maps. For instance, soils adjacent to the Amazon River have the highest SOC stocks (100–150 Mg C ha^{-1}) due to their hydric conditions. The most northern and southeastern parts of the map show the lowest SOC stocks, which is as expected as the native vegetation is savanna, rather than evergreen forest.

The GEFSOC Modeling System allows estimating SOC stocks, not only for different land uses but also for different soil hydrological conditions. According to SOTER and other data sets for Amazonian soils, the majority of Amazonian soils are nonhydric

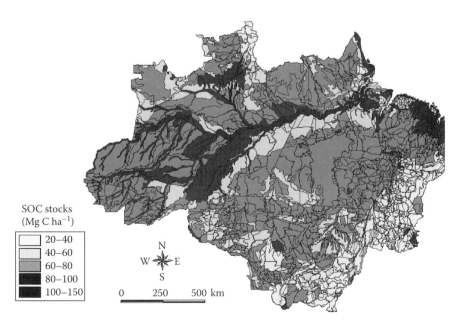

FIGURE 29.15 Map of year 2000 SOC stocks from the Century model for the 0–20 cm layer in the Brazilian Amazon.

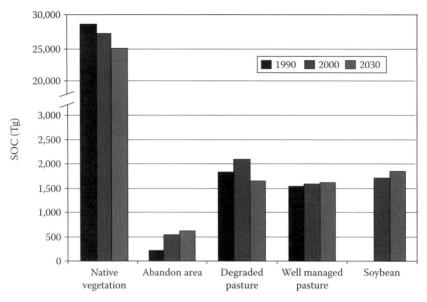

FIGURE 29.16 Modeled SOC stocks in the 0–20 cm layer for the Brazilian Amazon derived from Century (1990–2030).

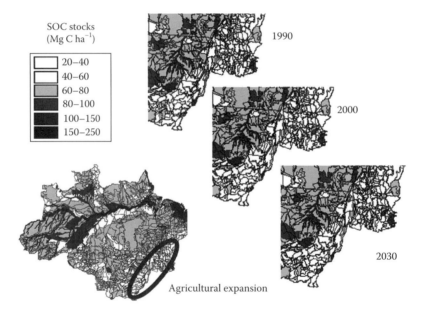

FIGURE 29.17 Estimated SOC stocks in the agricultural expansion area within the Brazilian Amazon (1990–2030).

or freely drained. Among the different land uses under hydric conditions, the largest contribution to SOC stocks comes from soils under native vegetation (mainly evergreen forests, classes colored with black and dark gray, Figure 29.15).

Estimated SOC stocks (0–20 cm) for the years 1990, 2000, and 2030 from Century are presented in Figure 29.16. Stocks of SOC under native vegetation are projected to decrease through time due to the area reduction according to the land use scenarios described earlier. If the projected deforestation in 2030 is realized, approximately 4200 Tg C will be lost from the soil vis-à-vis the estimate for 1990 (Figure 29.16). It should be stressed that following deforestation, large amounts of C are also lost from biomass, much of which is emitted to the atmosphere as CO_2.

According to Dias-Filho et al. (2001), forest-to-pasture conversion releases 100–200 Mg C ha^{-1} from aboveground forest biomass to the atmosphere.

In the Brazilian Amazon, areas that have been converted from pasture to agriculture are mainly cropped with soybean, associated with a cover crop (usually millet), and cultivated under conventional tillage during the first 1–2 years before moving to a no-tillage system. Conversion of native vegetation to cultivated cropland under a conventional tillage system has resulted in a significant decline in soil organic matter content (Paustian et al., 2000; Lal, 2002). Farming methods that use mechanical tillage, such as the moldboard plough for seedbed preparation or disking for weed control, can promote soil C loss by several mechanisms—they disrupt soil aggregates that protect soil organic matter from decomposition (Karlen and Cambardella, 1996; Six et al., 1999), they stimulate short-term microbial activity through enhanced aeration, resulting in increased emissions of CO_2 and other radiatively active gases to the atmosphere (Bayer et al., 2000a, 2000b; Kladivko, 2001), and they mix fresh residues into the soil where conditions for decomposition are often more favorable than on the surface (Karlen and Cambardella, 1996). Furthermore, tillage can make soils more prone to erosion, resulting in further on-site loss of soil C (Lal, 2002). Conversely, no-tillage practices cause less soil disturbance, often resulting in significant accumulation of soil C (Sa et al., 2001, Schuman et al., 2002) and consequent reduction of gas emissions (especially CO_2) to the atmosphere (Lal, 1998; Paustian et al., 2000). There is considerable evidence that the main impact occurs in the topsoil with little overall effect on C storage in deeper layers (Six et al., 2002). In the humid tropics, minimum tillage methods may need to be combined with herbicide use for weed control, possibly leading to environmental pollution and a reduction in biodiversity (Batjes and Sombroek, 1997).

Since the adoption of no-tillage systems, SOC stocks are projected to have increased under soybean cultivation (Figure 29.16). The effects of agricultural expansion are evident in the modeled SOC stock change rates (data not shown), where soybean shows the highest change rate among the different land uses in the Brazilian Amazon. There are no estimates for SOC stock change rate or SOC stocks for soybean in the year 1990, as soybean production was not widespread in the region at the time. The projected increase in areas cultivated with soybean that adopt no-tillage systems would lead to an increase in SOC stocks. This is illustrated (Figure 29.17) for the years 2000 and 2030. In 1990, the area of agricultural expansion was dominated by SOC stock class of 20–40 Mg C ha^{-1}. In the year 2000, some areas began to be cultivated with soybean and finally in 2030, most of this region is projected to move to a higher class of SOC stock (40–60 Mg C ha^{-1}; Figure 29.17).

The estimates presented here are comparable with those obtained from other studies considering no-till systems in the midwestern part of Brazil (Riezebos and Loerts, 1998; Vasconcellos et al., 1998; Peixoto et al., 1999; Resck et al., 2000). These studies reported SOC stock accumulation rates of 0–1.2 Mg C ha^{-1} year^{-1} for the surface soil layer.

Increased soil C concentrations in surface horizons are a common consequence of pasture formation after forest clearance in the Amazon basin (Bonde et al., 1992; Moraes et al., 1995; Trumbore et al., 1995; Moraes et al., 1996; Neill et al., 1997).

The estimated SOC stocks derived from the GEFSOC Modeling System show SOC stocks slowly increase for well-managed pasture from 1990 to 2030 (Figure 29.16). The area under degraded pasture, that is, areas dominated by herbaceous and woody invaders (Uhl et al., 1988, Serrão and Toledo, 1990) should diminish according to Cerri et al. (2007b) scenarios, mainly due to land prices. This is reflected in the reductions in SOC stocks in the year 2030.

29.4.4.2 Comparison of Current Stocks with Outcome from Existing Methods

There are few studies on estimates of SOC stocks for the Brazilian Amazon (Batjes and Dijkshoorn, 1990; Moraes et al., 1995; Bernoux et al., 2002; Batjes, 2005). Those that are there have focused on soil carbon stocks under native vegetation. According to Cerri et al. (2007a), no prior study has considered stocks of SOC under actual land use (e.g., including areas under pasture and agriculture). Therefore, comparisons between estimates of current SOC stocks obtained using the GEFSOC Modeling System and existing studies do not truly compare like with like. However, given the fact that the outputs from the GEFSOC system for the Brazilian Amazon are unique in terms of the land uses encompassed, it was deemed appropriate to compare them with other available studies, as these represent the state-of-the-art.

29.4.4.2.1 Estimates by Moraes et al. (1995)

Moraes et al. (1995) determined stocks of carbon and nitrogen for soils under undisturbed vegetation across the Brazilian Amazon basin using 1162 soil profiles from the Radam Brasil survey and a digitized Brazilian soil survey map. Mean basin soil carbon density was 10.3 kg C m^{-2} to 1 m depth. About 47 Pg C and 4.4 Pg N were contained in the top 1 m of soil. According to the authors, 45% of the total basin soil C (21 Pg C) and 41% of total soil N (1.8 Pg N) were contained in the top 20 cm over a 500 Mha area of the Brazilian Amazon. As these data represent sites with forest vegetation in the absence of significant disturbances, they represent a valuable baseline for evaluating the effects of land-use changes on soil carbon stocks in the Amazon.

29.4.4.2.2 Estimates by Batjes and Dijkshoorn (1999)

Soil nitrogen and organic carbon stocks, to a depth of 0.3 and 1 m respectively, were determined for the entire Amazon Region (about 700 Mha) using the soil and terrain database for Latin America and the Caribbean (SOTERLAC, FAO et al., 1995). Mean carbon densities, to a depth of 1 m, range from 4.0 kg m^{-2} for coarse-textured Arenosols to 72.4 kg m^{-2} for the poorly drained Histosols. Mean carbon density for the mineral soils, excluding Entisols and Andosols (30.5 kg C m^{-2}), is 9.8 kg m^{-2}. In total, the top 1 m holds 66.9 Pg C and 6.9 Pg N.

Approximately 52% of this carbon pool (about 34.8 Pg) is held in the top 0.3 m of the soil, the layer that is most prone to changes upon land use conversion and deforestation. For the Brazilian Amazon (about 500 Mha), the organic carbon stock calculated is 25 Pg C (0–30 cm) and 46.5 Pg C (0–100 cm). For total nitrogen this is 2.3 Pg N (0–30 cm) and 4.9 Pg N (0–100 cm). The figures found for the top 1 m of soil are similar to the ones reported by Moraes et al. (1995).

29.4.4.2.3 Estimates by Bernoux et al. (2002)

The objective of the study by Bernoux et al. (2002) was to give reliable values and a distribution map, for the 0–30 cm reference layer used by IPCC/UNEP/OECD/IEA (1997), for the Brazilian Amazon (an area that was "clipped" from the map of Brazil's total SOC stock under undisturbed vegetation). In this study, the following steps were performed: (1) elaboration of a map of representative Soil-Vegetation Associations (SVA) for Brazil; (2) organization of a soil profile data base gathering information, such as C concentration, BD, soil type, and native vegetation; (3) calculation of representative C stock (RCS) values for each SVA category and subsequent production of a map of their distribution, and (4) extraction of part of the Brazilian Amazon from the stock produced for the whole of Brazil (Bernoux et al., 2002).

1. The Soil-Vegetation Association (SVA) *map* was derived by a combination of the Brazilian soil map (EMBRAPA, 1981) which comprises 2698 map units, split into 69 soil types and vegetation map of Brazil (IBGE, 1988) at the 1:5,000,000 scale. The latter contains 2020 map units divided into 94 vegetation classes by both major groups and subgroups. This map represents the "potential vegetation," that is, the vegetation that could be achieved under current (at the time of the map elaboration) climatic and soil conditions without human disturbance. The strategy of simplification was to use the 12 original major groups as a starting point and to divide them according to vegetation specificity and/or geographical localization. These maps are distributed in digital form by FAO.
2. *Soil profiles database elaboration:* A digitized soil profile database was elaborated by using information published in soil inventories at regional or national scale. Most of the soil profile information came from descriptions of soil pits surveyed by the RADAM-BRASIL project and by EMBRAPA projects at the state level. The soil survey of the RADAMBRASIL project initially focused on the Brazilian Amazon basin and was later extended to the whole of Brazil. The profiles for Amazon Brazil are now available in digital format (Bernoux et al., 2003).
3. Calculation of SOC stocks (0–30 cm, kg m^{-2}) by profile by summing the C stock of successive horizons in a given profile. C stock in a soil horizon was determined as a product of BD, C concentration, and horizon thickness (for those horizons located in the 0–30 cm depth). Finally, an area was extracted from the Brazil soil carbon stock map (Bernoux et al., 2002) that was consistent with the Brazilian Amazon.

The potential total carbon stock of Brazilian Amazon soils under native vegetation for the 0–30 cm layer, was calculated as 22.7 Pg C stored in approximately 500 Mha of the SVA. Using the standard error (SE) calculated for each SVA category (using an arbitrary level of 25% of the representative value when it is derived from only one carbon stock value) as a mean to estimate the accuracy of the total carbon stock, the entire error would be 2.3 Pg C.

29.4.4.2.4 Estimates by Batjes (2005)

SOC stocks were calculated using an updated 1:5,000,000 scale SOTER database for Brazil (Batjes et al., 2004). Each SOTER unit is composed of 1–5 soil components, depending on the complexity of the mapping unit. Individual soil components were characterized by a typical profile selected as being regionally representative. Use of a single representative profile, however, ignores the variability within a soil component.

Therefore, the procedure included the simulation of phenoforms, using the typical profile as the genoform. The information resulting from the simulations was linked to the soil geographical information through the unique profile identifier for each soil component, to arrive at n (300) realizations of the regional organic carbon stocks. Results are presented as 95% confidence limits for the population median. For the Brazilian Amazon, median SOC stocks to 1 m were estimated to be 42.3–43.8 Pg C of which 23.9–24.2 Pg C occur in the top 30 cm (Batjes, 2005).

Estimates of SOC stocks for the Brazilian Amazon obtained from the different mapping methods are presented in Table 29.2; they are considered to be comparable to, yet lower than those obtained using the GEFSOC Modeling System. The discrepancies may be related to the fact that our modeling results consider possible changes in land use (pasture and agriculture) and management practices (well-managed versus degraded pasture and conventional versus no-tillage soybean cultivation) as opposed to native vegetation only.

Conversely, the mapping approaches implicitly account for possible effects of soil properties other than soil wetness and clay content on SOC stocks, for example, activity of the clay minerals (Wattel-Koekkoek et al., 2001), aluminum toxicity, and content of coarse fragments that are not yet considered in the dynamic models. So, according to Kogel-Knabner et al. (2005), it remains difficult to make unambiguous projections concerning the development of the different carbon pools in soils under changing environmental conditions and land use/management.

A clear advantage of the GEFSOC Modeling System is that it permits the estimation of SOC stocks not only for current conditions, but also for past (historic) conditions and future scenarios. The mapping methods discussed earlier, however, can only calculate SOC stocks for the period for which data is available. They are not dynamic and, hence, do not allow presenting "windows-of-opportunity" for future land use/management scenarios, as required for policy making and in the context of the international conventions.

Estimates of mean SOC density in the top 20 cm layer from RothC and Century models using the GEFSOC Modeling System (4 and 6 kg C m^{-2}, respectively) are within the range of values presented by previous studies in the Brazilian Amazon. For instance, Bernoux et al. (2002) reported that three quarters of all areas in Brazil had C density varying between 3 and 6 kg C m^{-2} for the first 30 cm of soil. Batjes (2005) found mean C densities varying from 2.4 (Entisols) to 9.3 (Oxisols) kg C m^{-2} for the 0–30 cm layer for the Brazilian territory.

29.4.4.3 Final Remarks

Pasture productivity and longevity, and agricultural expansion in the Amazon basin are closely related to soil organic matter dynamics. Thus, understanding the major biogeochemical cycles that influence soil organic matter under cleared lands is vital for predicting the consequences of continued conversion of tropical forest to cattle pastures and agriculture. This understanding is also important for devising management technologies that enhance the sustainability of these areas and thus slows further deforestation.

Therefore, more field data is needed on the effect of human-induced land use changes on tropical SOC dynamics and their relationship to soil type and climate condition. Such data can be used together with remotely obtained imagery and geographic information systems and base maps of land use conversions to improve overall estimates of current and future stocks of SOC in the Brazilian Amazon. The GEFSOC system worked well for the conditions in the case study area providing a means of efficiently handling complex interactions among biotic-edapho-climatic conditions (>363,000 combinations) in a very large area (~500 Mha).

Modeling studies, using, for example, the GEFSOC Modeling System, in these pasture and agricultural systems can help us to evaluate the magnitude of these impacts and to determine the effect of possible new management strategies on pasture and cropland sustainability. Our findings for land conversion from forest to pasture/agriculture have important implications, for example, for calculating CO_2 emissions from land-use change in national greenhouse gas inventories.

29.5 Future Directions and Opportunities

In Brazil, research and development has been carried out in several related areas to generate information and technology in integrated digital, spatial, and attribute databases for soils in the country. The first area is in field data collection, in which researchers are exploring new sampling schemes and environmental covariates in digital soil mapping. The second area is related to the use of integrated sensors or some other new technologies for inferring soil properties or status. GIS are being used for pre-mapping and quality assurance of mapping in specific areas of the country. Other areas of interest can be mentioned, such as (1) innovative inference systems (new methodologies for predicting soil classes and properties, (2) GIS tools and procedures are being used to generalize large-scale to smaller scale digital, spatial, and attributes of soils in Brazil, and (3) use of integrated digital, spatial, and attribute database products and their uncertainties for soil assessment and environmental applications.

References

Ab'Saber, A.N. 1967. Problemas geomorfologicos da Amazônia Brasileira. Atos do Simposio sobre biota amazônica, Rio de Janeiro. Geomorfologia I:35–67.

Ab'Saber, A.N. 1996. Dominios morfoclimaticos e solos de Brasil. In V.H. Alvarez, L.E.F. Fontes, and M.P.F. Fontes (eds.) O solo nos grandes dominios morfoclimaticos do Brasil e o desenvolvimento sustentado, Sociedade Brasileira de Ciencia do Solo, Viçosa, Brazil.

Agriculture and Agri-Food Canada. 2000. Canadian Soil Information System. Available at http://sis.agr.gc.ca/cansis/. CanSIS, Ottawa, Ontario, Canada. (accessed May 23, 2002)

Akaike, H. 1973. Information theory and an extension of maximum likelihood principle, p. 267–281. In B.N. Petrov, and F. Csaki (eds.) Second Int. Symp. Inf. Theory. Akademia Kiado, Budapest, Hungary.

Arrouays, D., and P. Pélissier. 1994. Modeling carbon storage profiles in temperate forest humic loamy soils of France. Soil Sci. 157:185–192.

Batjes, N.H. 2005. Organic carbon stocks in the soils of Brazil. Soil Use Manage. 21:22–24.

Batjes, N.H., R. Al-Adamat, T. Bhattacharyya, M. Bernoux, C.E.P. Cerri, P. Gicheru, P. Kamoni, E. Milne, D.K. Pal, and Z. Rawajfih. 2007. Preparation of consistent soil data sets for SOC modelling purposes: Secondary SOTER data sets for four case study areas. Agr. Ecosyst. Environ. 122:26–34.

Batjes, N.H., M. Bernoux, and C.E.P. Cerri. 2004. Soil data derived from SOTER for studies of carbon stocks and change in Brazil (Version 1.0; GEFSOC Project). Report 2004/03, ISRIC—World Soil Information. Wageningen, the Netherlands.

Batjes, N.H., and J.A. Dijkshoorn. 1999. Carbon and nitrogen stocks in the soils of the Amazon region. Geoderma 89:273–286.

Batjes, N.H., and W.G. Sombroek. 1997. Possibilities for carbon sequestration in tropical and subtropical soils. Global Change Biol. 3:161–173.

Bayer, C., L. Martin-Neto, J. Mielniczuk, and C.A. Ceretta. 2000a. Effect of no-till cropping systems on soil organic matter in a sandy clay loam Acrisol from southern Brazil monitored by electron spin resonance and nuclear magnetic resonance. Soil Till. Res. 53:95–104.

Bayer, C., J. Mielniczuk, T.J.C. Amado, L. Martin-Neto, and S.V. Fernandes. 2000b. Organic matter storage in a sandy clay loam Acrisol affected by tillage and cropping systems in southern Brazil. Soil Till. Res. 54:101–109.

Bellamy, P.H., P.J. Loveland, I. Bradley, R.M. Lark, and G.J.D. Kirk. 2005. Carbon losses from all soils across England and Wales 1978–2003. Nature 437:245–247.

Benedetti, M.M., G. Sparovek, M. Cooper, N. Curi, and A. Carvalho Filho. 2008. Representatividade e potencial de utilizacao de um banco de dados do Brasil. R. Bras. Ci. Solo 32:2591–2600.

Benites, V.M., P.L.O.A. Machado, E.C.C. Fidalgo, M.R. Coelho, and B.E. Madari. 2007. Pedotransfer functions for estimating soil bulk density from existing soil survey reports in Brazil. Geoderma 139:90–97.

Bennema, J. 1974. Organic carbon profiles in Oxisols. Pédologie 2:119–146.

Bernoux, M., D. Arrouays, C.C. Cerri, and H. Bourennane. 1998a. Modeling vertical distribution of carbon in Oxisols of the Western Brazilian Amazon (Rondônia). Soil Sci. 163:941–951.

Bernoux, M., D. Arrouays, C.E.P. Cerri, and C.C. Cerri. 2006. Regional organic carbon storage maps of western Brazilian Amazon based on prior soil maps and geostatistical interpolation. pp. 337–346. In A.E. Hartemink, A.B. McBratney, and M.L. Mendonça Santos (eds.) Digital Soil Mapping With Limited Data. Springer. 2006.

Bernoux, M., D. Arrouays, C.C. Cerri, P.M. Graça, B. Volkoff, and J. Trichet. 1998b. Estimation des stocks de carbone des sols du Rondônia (Amazonie brésilienne). Etudes Gestion Sols 5:31–42.

Bernoux, M., D. Arrouays, B. Volkoff, C.C. Cerri, and C. Jolivet. 1998c. Bulk densities of Brazilian Amazon soils related to other soil properties. Soil Sci. Soc. Am. J. 62:743–749.

Bernoux M., M.C.S. Carvalho, B. Volkoff, and C.C. Cerri. 2001. CO_2 emission from mineral soils following land-cover change in Brazil. Global Change Biol. 7:779–787.

Bernoux, M., M.C.S. Carvalho, B. Volkoff, and C.C. Cerri. 2002. Brazil's soil carbon stocks. Soil Sci. Soc. of Am. J. 66:888–896.

Bernoux, M., C.E.P. Cerri, and C.C. Cerri. 2003. Digital soil properties database of the Amazon part from the RADAMBRASIL project (Version 1, September 25, 2005, ftp://lba.cptec.inpe.br/lba_archives/CD/CD-208/Cerri/). Centro de Energia Nuclear Na Agricultura (CENA), Universidade de Sao Paulo, Sao Paulo, Brazil.

Bhering, S.B. 1995. Soil information systems—Directions for development (in Portuguese). Master Thesis. University of São Paulo. Sao Paulo, Brazil.

Bonde, T.A., B.T. Christensen, and C.C. Cerri. 1992. Dynamics of soil organic matter as reflected by natural ^{13}C abundance in particle size fractions of forested and cultivated Oxisols. Soil Biol. Biochem. 24:275–277.

Burgess, T.M., and R. Webster. 1980a. Optimal interpolation and isarithmic mapping of soil properties, I. The semi-variogram and punctual kriging. J. Soil Sci. 31:315–331.

Burgess, T.M., and R. Webster. 1980b. Optimal interpolation and isarithmic mapping of soil properties, II. Block kriging. J. Soil Sci. 31:333–341.

CA. 1995. Computer Associates. CA-OpenIngres. SQL Language Reference, 1995.

Cardille, J.A., and J.A. Foley. 2003. Agricultural land-use change in Brazilian Amazonia between 1980 and 1995: Evidence from integrated satellite and census data. Remote Sens. Environ. 87:551–562.

Carvalho Filho, A., J.F. Lumbreras, K.P. Wittern, A.L. Lemos, R.D. dos Santos, B. Calderano Filho, S.B. Calderano, R.P. Oliveira, M.L.D. Aglio, J.S. Souza, and C.E. Chaffin. 2003. Mapa de reconhecimento de baixa intensidade dos solos do estado do Rio de Janeiro. Rio de Janeiro: Embrapa Solos, 2003. 1 mapa, color. Escala 1:250.000. Available at http://www.cnps.embrapa.br/solosbr/sigweb.html (accessed January 18, 2006)

Cattaneo, A. 2002. Balancing agricultural development and deforestation in the Brazilian Amazon. Research Report 129 (Available on-line via: http://www.ifpri.org/pubs/pubs.htm#rreport). International Food Policy Institute (IFPRI), Washington, DC. (accessed December 14, 2002)

Cerri, C.C., M. Bernoux, D. Arrouays, B. Feigl, and M.C. Piccolo. 2000. Carbon stocks in soils of the Brazilian Amazon, p. 33–50. In R. Lal, J.M. Kimble, and B.A. Stewart (eds.) Global climate change and tropical ecosystems. CRC Press, Boca Raton, FL. Advances in Soil Science, 456 p. Matheron, G. 1965. Les Variables Régionalisées et Leur Estimation, 306 p. Masson, Paris, France.

Cerri, C.C., M. Bernoux, and G.J. Blair. 1994. Carbon pools and fluxes in Brazilian natural and agricultural systems and the implication for the global CO_2 balance, p. 399–406. Trans. 15th Int. Cong. of Soil Sci. Vol. 5a. Acapulco, Mexico.

Cerri, C.E.P., M. Easter, K. Paustian, K. Killian, K. Coleman, M. Bernoux, P. Falloon, D. Powlson, N. Batjes, E. Milne, and C.C. Cerri. 2007b. Predicted soil organic carbon stocks and changes in the Brazilian Amazon between 2000–2030. Agric. Ecosyst. Environ. 122:58–72.

Cerri, C.E.P., M. Easter, K. Paustian, K. Killian, K. Coleman, M. Bernoux, D. Powlson, E. Milne, and C.C. Cerri. 2007a. Evaluation and modification of two soil carbon models, using 11 land use change chronosequences from the Brazilian Amazon. Agric., Ecosyst. Environ. 122:45–57.

Chagas, C.S., W. Carvalho Junior, S.B. Bhering, A.K. Tanaka, and J.F.M. Baça. 2004. Estrutura e organização do sistema de informações georreferenciadas de solos do Brasil (sigsolos—versão 1.0). R. Bras. Ci. Solo 28:865–876.

Clustener-Godt, M., and I. Sachs. 1995. Brazilian perspectives on sustainable development of the Amazon region, p. 311 Man and biosphere series, 15. UNESCO and The Parthenon Publishing Group, Casterton Hall, U.K.

CNPS. 1995. National Center for Soil Research of Embrapa, the Brazilian Enterprise for Agricultural Research. Normative Procedures for Pedological Surveys (in Portuguese).

Cochrane, T.T., and Cochrane, T.A. 1998. SIGTERON: Geographic information system of land and soils for Rondônia state, Brazil. Consultancy consortium TECNOSOLO-DHV report. Porto Velho, RO, Brazil.

Cooper, M., L.M.S. Mendes, W.L.C. Silva, and G. Sparovek. 2005. A national soil profile database for Brazil available to international scientists. Soil Sci. Soc. Am. J. 69:649–652.

DeGloria, S.D., and R.J. Wagenet. 1993. Soil information requirements for modeling and visualizing soil behavior at multiple scales. In Proc. Workshop Integr. Geogr. Inf. Syst. Environ. Model. Breckenridge, CO.

Dias-Filho, M.B., E.A. Davidson, and C.J.R. Carvalho. 2001. Linking biogeochemical cycles to cattle pasture management and sustainability in the Amazon basin, p. 84–105. In M.E. McClain, R.L. Victoria, and J.E. Richey (eds.) The biogeochemistry of the Amazon basin. Oxford University Press, New York.

Doran, J.W., and M. Safley. 1997. Defining and assessing soil health and sustainable productivity, p. 1–28. In C.E. Pankhurst (ed.) Biological indicators of soil health, CAB International, Wallingford, U.K.

Doran, J.W., M. Sarrantonio, and M. Liebig. 1996. Soil health and sustainability. Adv. Agron. 56:1–54.

Easter, M., K. Paustian, K. Kilian, S. Williams, E. Milne, R. Adamat, T. Bhattacharyya et al. 2007. The GEFSOC soil carbon modeling system©: A tool for conducting regional-scale soil carbon inventories and assessing the impacts of land use change on soil carbon. Agric., Ecosyst. Environ. 122:13–25.

EMBRAPA. 1981. Mapa de Solos do Brasil, escala 1:5,000,000. Serviço Nacional de Levantamento e Conservação de Solos. Rio de Janeiro, Brazil.

ESRI. 1996. Geoprocessing in ArcGIS. Redlands, CA. Environmental Systems Research Institute, Inc.

ESRI. 2003. ArcGIS 9. Environmental Systems Research Institute, Redland, CA.

Falloon, P.D., P. Smith, J.U. Smith, J. Szabó, K. Coleman, and S. Marshall. 1998. Regional estimates of carbon sequestration potential: Linking the Rothamsted carbon model to GIS databases. Bio. Fertil. Soils 27:236–241.

FAO, ISRIC, UNEP, CIP, 1995. Soil and terrain digital database for Latin America and the Caribbean at 1:5 million scale. Land and water digital media series no. 5. Food and Agriculture Organization of the United Nations, Rome, Italy.

FAO-Unesco. 1971. Soil map of the world, 1:5,000,000. Vol. 4—South America. United Nations Educational, Scientific, and Cultural Organization, Paris, France.

Fearnside, P.M. 1998. Forests and global warming mitigation in Brazil: Opportunities in the Brazilian forest sector for responses to global warming under the "clean development mechanism". Biomass Bioenergy 16:171–189.

Fearnside, P.M. 1999. Plantation forestry in Brazil: The potential impacts of climatic change. Biomass Bioenergy 16:91–102.

Food and Agricultural Organization. 1995. Digital soil map of the world and derived soil properties (Version 3.5). FAO, Rome, Italy.

Giasson, E., R.T. Clarke, A.V. Inda, Jr., G.H. Merten, and C.G. Tornquist. 2006. Digital soil mapping using logistic regression on terrain parameters: A case study in Southern Brazil. Scientia Agricola 63:263–268.

Habern, J. 1992. Viewpoint: A soil health index. J. Soil Water Conserv. 47:6.

Hastie, T., R. Tibshirani, and J. Friedman. 2001. The elements of statistical learning: Data mining, inference, and prediction, p. 533 Springer, New York.

Houghton, R.A., D.L. Skole, C.A. Nobre, J.L. Hackler, K.T. Lawrence, and W.H. Chometowski. 2000. Annual fluxes of carbon from deforestation and regrowth in the Brazilian Amazon. Nature 403:301–304.

IBGE. 1986. Folha SH 22-Porto Alegre e parte das folhas SH 21-Uruguaiana E SI 22-Lagoa Mirim: Geologia, Geomorfologia, Vegetação, Uso Potencial da Terra. Levantamento De Recursos Naturais 33. Instituto Brasileiro de Geografia e Estatística. Rio de Janeiro, Brazil.

IBGE. 1988. Mapa de vegetação do Brasil, escala 1:5,000,000. Fundação Instituto Brasileiro de Geografia e Estatística. Rio de Janeiro, Brazil.

IBGE. 2002. Mapa exploratório de solos do Rio Grande Do Sul, escala 1:1.000.000. Instituto Brasileiro de Geografia e Estatística. Rio de Janeiro, Brazil.

INPE. 1995. National Institute of Spatial Researches. SPRING 2.0 User Manual (in Portuguese).

IPCC. 2006. Guidelines for national greenhouse gas inventories. IV. Agriculture, forestry and other land uses. National Greenhouse Gas Inventories Programme. Institute for Global Environmental Strategies, Hayama, Japan. (Available on-line at www.ipcc-nggip.iges.or.jp/public/2006gl/vol4.htm) (Verified May 4, 2009).

IPCC/UNEP/OECD/IEA. 1997. Revised 1996 IPCC guidelines for national greenhouse gas inventories: Reporting instructions (Vol. 1); workbook (Vol. 2); reference manual (vol. 3). Intergovernmental Panel on Climate Change, United Nations Environment Programme, Organization for Economic Co-Operation and Development, International Energy Agency. Paris, France.

Jarvis, A., H.I. Reuter, A. Nelson, and E. Guevara. 2006. Hole-filled SRTM for the globe (Version 3). Available from the CGIAR-CSI SRTM 90m database: http://srtm.csi.cgiar.org (accessed May 6, 2007)

Kabat, P., R.W.A. Hutjes, A.J. Dolman, J. Adis, M.O. Andreae, J.H.C. Gash, J. Grace et al. 1999. The effects of changes in land use and climate on the sustainability of natural and man made ecosystems in Amazonia—The European contribution to the Large Scale Biosphere Atmosphere Experiment in Amazonia (LBA). International Activities Report 74.1. DLO Winand Staring Centre. Wageningen, the Netherlands.

Kämpf, N., and P. Schneider. 1989. Caracterização de solos orgânicos do Rio Grande do Sul: Propriedades morfológicas e físicas como subsídio à classificação. R. Bras. Ci. Solo 13:227–236.

Karlen, D.L., and C.A. Cambardella. 1996. Conservation strategies for improving soil quality and organic matter storage, p. 395–420. In M.R. Carter, and B.A. Stewart (eds.) Structure and organic matter storage in agricultural soils. Advances in Soil Science. CRC Press, Boca Raton, FL.

Kladivko, E. 2001. Tillage systems and soil ecology. Soil Till. Res. 61:61–76.

Kogel-Knabner, I., M.V. Lutzow, G. Guggenberger, H. Flessa, B. Marschner, E. Matzner, and K. Ekschmitt. 2005. Mechanisms and regulation of organic matter stabilisation in soils. Geoderma 128:1–2.

Krogh, L., A. Norgaard, M. Hermansen, M.H. Greve, T. Thomas Balstroem, and H. Breuning-Madsen. 2003. Preliminary estimates of contemporary soil organic carbon stocks in Denmark using multiple datasets and four scaling-up methods. Agric. Ecosyst. Environ. 96:19–28.

Lal, R. 1998. Soil processes and the greenhouse effect, p.199–212. In R. Lal, W.H. Blum, C. Valentine, and B.A. Stewart (eds.) Methods for assessment of soil degradation. CRC Press, Boca Raton, FL.

Lal, R. 2002. Soil carbon dynamic in cropland and rangeland. Environ. Pollut. 116:353–362.

Lambin, E.F., H.J. Geist, and E. Lepers. 2003. Dynamics of land-use and land-cover change in tropical regions. Ann. Rev. Environ. Resour. 28:205–2041.

Lettens, S., J.V. Orshoven, B.V. Wesemael, B. Vos, and B. Muys. 2004. Stocks and fluxes of soil organic carbon for landscape units in Belgium derived from heterogeneous datasets for 1990 and 2000. Geoderma 127:11–23.

Lumbreras, J. F., U.J. Naime, A. Carvalho Filho, K.P. Wittern, E. Shinzato, M.E. Dantas, F. Palmieir et al. 2003. Zoneamento Agroecológico do Estado do Rio de Janeiro, 113 p. Embrapa Solos. Boletim de Pesquisa e Desenvolvimento n. 33, Rio de Janeiro, Brazil.

Lytle, D.J. 1993. Digital soils databases for the United States. *In* Environmental modeling with GIS. Oxford University Press, New York.

Matheron, G. 1965. Les Variables Régionalisées et Leur Estimation, 306pp. Masson, Paris, France.

McBratney, A.B., M.L. Mendonca-Santos, and B. Minasny. 2003. On digital soil mapping. Geoderma 117:3–52.

Mello, F.F.C. 2007. Estimativas dos estoques de carbono dos solos nos Estados de Rondônia e Mato Grosso anteriores à intervenção antrópica, 88pp. Dissertation (Master Degree). University of São Paulo, Sao Paulo, Brazil.

Mendonça-Santos, M.L., P.D. Jacques, M.R. Coelho, J. Pimentel, H.G. dos Santos, P. Almeida, E.L. de A. Barbosa, T.C. Da, A'vila, E. Shinzato, and P.C.M.P. Branco. 2003. Mapeamento do Uso Atual e Cobertura Vegetal dos Solos do Estado do Rio de Janeiro, 44pp. Boletim de Pesquisa e Desenvolvimento n. 22. Embrapa Solos, Rio de Janeiro, Brazil.

Mendonça-Santos, M.L., and Santos, H.G. dos. 2008. Digital Mapping of Soil Classes in Rio de Janeiro State, Brazil: Data, Modelling and Prediction pp. 381–396. *In* A.E. Hartemink, A.B. McBratney, and M.L. Mendonça Santos (eds.) Digital Soil Mapping With Limited Data. Springer. 2008.

Ministério da Agricultura. 1973. Levantamento de Reconhecimento de Solos do Rio Grande Do Sul. Boletim Técnico 30. Serviço Nacional de Levantamento de Solos, Recife, Brazil.

Ministério da Ciência e Tecnologia. 2006. Primeiro Inventário Brasileiro de Emissões Antrópicas de Gases de Efeito Estufa. Emissões e remoções de dióxido de carbono pelos solos por mudanças de uso da terra e calagem (Relatório de Referência). Ministério da Ciência e Tecnologia, Brasília.

Moraes, J.L., C.C. Cerri, J.M. Melillo, D. Kicklighter, C. Neill, D.L. Skole, and P.A. Steudler. 1995. Soil carbon stocks of the Brazilian Amazon basin. Soil Sci. Soc. Am. J. 59:244–247.

Moraes, J.F.L., B. Volkoff, C.C. Cerri, and M. Bernoux. 1996. Soil properties under Amazon forest change due to pasture installation in Rondônia, Brazil. Geoderma 70:63–81.

Neill, C., C.C. Cerri, J.M. Melillo, B.J. Feigl, P.A. Steudler, J.F.L. Moraes, and M.C. Piccolo. 1997. Stocks and dynamics of soil carbon following deforestation for pasture in Rondonia, p. 9–28. *In* R. Lal, J.M. Kimble, R.F. Follett, and B.A. Stewart (eds.) Soil processes and the carbon cycle. CRC Press, Boca Raton, FL.

Paustian, K., J. Six, E.T. Elliott, and H.W. Hunt. 2000. Management options for reducing CO_2 emissions from agricultural soils. Biogeochemistry 48:147–163.

Pearce, D., and J. Warford. 1993. World without end. World Bank, Washington, DC.

Peixoto, R.T., L.M. Stella, A. Machulek Junior, H.U. Mehl, and E.A. Batista. 1999. Distibução das frações granulométricas da matéria orgânica em função do manejo do solo, p. 346–348. Anais 3° Encontro brasileirosobre substâncias húmicas. Santa Maria, Brazil.

Pereira, M.G., G.S. Valladares, L.H.C. Anjos, V.M. Benites, A. Espíndula, Jr., and A.G. Ebeling. 2006. Organic carbon determination in Histosols and soil horizons with high organic matter content from Brazil. Scientia Agricola 63:187–193.

Peters, W.L., and V.W.P. Van Engelen. 1993. Guia para la Recomplacion de la Base de Datos SOTER a una escala de 1:5 M. Papel de Trabajo 93/7, ISRIC, Wageningen, the Netherlands.

Resck, D.V.S., C.A. Vasconcellos, L. Vilela, and M.C.M. Macedo. 2000. Impact of conversion of Brazilian Cerrados to cropland and pastureland on soil carbon pool and dynamics, p. 169–196. *In* R. Lal, J.M. Kimble, and B.A. Stewart (eds.) Global climate change and tropical ecosystems. CRC Press, Boca Raton, FL.

Riezebos, H.T.H., and A.C. Loerts. 1998. Influence of land use change and tillage practice on soil organic matter in southern Brazil and eastern Paraguay. Soil Till. Res. 49:271–275.

Rio de Janeiro. 2001. Geologia, geomorfologia, geoquímica, geofísica, recursos minerais, economia mineral, hidrogeologia, estudos de chuvas intensas, aptidão agrícola, uso e cobertura do solo, inventário de escorregamentos, diagnóstico geoambiental. Companhia de Pesquisa de Recursos Minerais—CPRM/EMBRAPA SOLOS/DRM-RJ. 1CD-ROM.

RuleQuest Research. 2003. See5/C5.0 version 1.20. RuleQuest Research Pty Ltd., Sydney, Australia. http://www.rulequest.com/see5-info.html (accessed November 29, 2009)

Sa, J.C.M., C.C. Cerri, R. Lal, W.A. Dick, S.P. Venzke Filho, M.C. Piccolo, and B.J. Feigl. 2001. Organic matter dynamics and carbon sequestration rates for a tillage chronosequence in a Brazilian Oxisol. Soil Sci. Soc. Am. J. 65:1486–1499.

Schobbenhaus, C., D. Almeud Campos, G.R. Derze, and H.E. Asmus. 1984. Geologia do Brasil (escale 1:2 5000 000), 500pp. Departemento Nacional da Producao Mineral, Ministerio das Minas e Energia, Republica Federativa do Brasil, Barsilia.

Schuman, G.E., H.H. Janzen, and J.E. Herrick. 2002. Soil carbon dynamics and potential carbon sequestration by rangelands. Environ. Pollut. 116:391–396.

Secretaria Estadual do Meio Ambiente. 2001. Relatório Final do Inventário Florestal do Rio Grande Do Sul. Departamento de Florestas e Áreas Protegidas. Porto Alegre, Brazil.

SEPLAN-RO 1998. Socio economic and ecological zoning of Rondonia state. Secretaria de Estado de Planejamento. Porto Velho, RO, Brazil.

SEPLAN-MT 2002. Socio economic and ecological zoning of MatoGrosso state. Secretaria de Estado de Planejamento. Cuiabá, MT, Brazil.

Serrão, E.A.S., and J.M. Toledo. 1990. The search for sustainability in Amazonian pastures, p. 195–214. *In* A.B. Anderson (ed.) Alternatives to deforestation: Steps towards sustainable utilization of Amazon forests. Columbia University Press, New York.

Sisti, C.P.J., H.P. dos Santos, R. Kohhann, B.J.R. Alves, S. Urquiaga, and R.M. Boddey. 2004. Change in carbon and nitrogen stocks in soil under 13 years of conventional or zero tillage in southern Brazil. Soil Till. Res. 76:39–58.

Six, J., E.T. Elliott, and K. Paustian. 1999. Aggregate and soil organic matter dynamics under conventional and no-tillage systems. Soil Sci. So. Am. J. 63:1350–1358.

Six, J., C. Feller, K. Denef, S.M. Ogle, J.C.M. Sa, and A. Albrecht. 2002. Soil organic matter, biota and aggregation in temperate and tropical soils—Effects of no-tillage. Agronomie 22:755–775.

Skjemstad, J.O., L.R. Spouncer, and A. Beech. 2000. Carbon conversion factors for historical soil carbon data. Australia Greenhouse Office. National Carbon Accounting System Technical Report 15. Australia Greenhouse Office, Canberra, Australia.

Sombroek, W.G. 1996. Amazon landforms and soils in relation to biological diversity. Acta Amazonica 30:81–100.

Sombroek, W.G. 2001. Spatial and temporal patterns of Amazon rainfall: Consequences for the planning of agricultural occupation and the protection of primary forests. Ambio 30:388–396.

Tanaka, A.K., S.B. Bhering, and C.S. Chagas. 1997. SIGSOLOS—Conceptual model and future perspectives (in Portuguese), Brazilian Symp. Inf. Agric. Belo Horizonte, Brazil.

Tornquist, C.G., P.W. Gassman, J. Mielniczuk, E. Giasson, and T. Campbell. 2009b. Spatially explicit simulations of soil C dynamics in Southern Brazil: Integrating century and GIS with i_Century. Geoderma 150:404–414.

Tornquist, C.G., E. Giasson, J. Mielniczuk, C.E.P. Cerri, and M. Bernoux. 2009a. Soil organic carbon stocks of Rio Grande do Sul, Brazil. Soil Sci. Soc. Am. J. 73:975–982.

Tornquist, C.G., J. Mielniczuk, and E. Giasson. 2004. Estimativa de estoques de carbono orgânico em solos do RS. Anais da XV Reunião Brasileira de Manejo e Conservação do Solo e Águas. Departamento de Solos. UFSM, Santa Maria, Brazil. [CD-ROM] (in Portuguese).

Trumbore, S.E., E.A. Davidson, P.B. Camargo, D.C. Nepstad, and L.A. Martinelli. 1995. Below-ground cycling of carbon in forest and pastures of Eastern Amazonia. Global Biogeochem. Cycles 9:515–528.

Uhl, C., R. Buschbacher, and E.A.S. Serrão. 1988. Abandoned pastures in eastern Amazonia. I. Patterns of plant succession. J. Ecol. 76:663–681.

USDA-National Resources Conservation Service. 2004. Online soil surveys. Available at http://soils.usda.gov/. USDA-NRCS, Washington, DC. (accessed September 24, 2006)

Van Engelen, V.W.P., and T.T. Wen. 1995. Global and National Soils and Terrain Digital Databases (SOTER): Procedures Manual. (Published also as FAO World Soil Resources Report No. 74), UNEP, IUSS, ISRIC and FAO, Wageningen, the Netherlands.

Vasconcellos, C.A., A.P.M. Figueiredo, G.E. França, A.M. Coelho, and W. Bressan. 1998. Manejo do sols e a atividade microbiana em latossolo vermelho-escuro da região de Sete Lagoas, MG. Pesq. Agropec. Bras. 33:1897–1905.

Volkoff, B., F.F.C. Mello, S.M.F. Maia, and C.E.P. Cerri. 2011. Landscape and soil regionalization in southern Brazilian Amazon and contiguous areas: Spatial pattern for environmental modelling. Scientia Agricola, in press.

Vosti, S.A., J. Witcover, and C.L. Carpetnier. 2002. Agricultural intensification by smallholders in the western Brazilian Amazon: From deforestation to sustainable land use. Research Report 130 (Available on-line http://www.ifpri.org/pubs/pubs.htm#rreport). International Food Policy Institute (IFPRI). Washington, DC. (accessed July, 2009)

Wattel-Koekkoek, E.J.W., P.P.L. van Genuchten, P. Buurman, and B. van Lagen. 2001. Amount and composition of clay-associated soil organic matter in a range of kaolinitic and smectitic soils. Geoderma 99:27–49.

30
Development and Use of Soil Maps and Databases in China

30.1	Development of Digitized Soil and Land Information Systems in China	30-1
	The Reconnaissance Soil Survey and Marbut Soil Classification Stage • The Russian School Stage • The National Soil Survey Stage • Soil Database and Information System Stage	
30.2	Use of Soil and Land Information Systems in China	30-4
	Soil Status Assessment • Land Suitability Assessment • Assessment of Land Degradation and Improvement in China	
30.3	Future Development and Opportunities	30-13
	Near-Future Tasks in Soil Correlation • Digital Soil Mapping and Its Practice in China	
	Acknowledgements	30-14
	References	30-14

Gan-Lin Zhang
Chinese Academy of Sciences

Yun-Jin Wu
Chinese Academy of Sciences

30.1 Development of Digitized Soil and Land Information Systems in China

Soil mapping is a technique showing the spatial relation of soil types and soil properties visibly, on a sheet of paper or on computer screen. Soil maps, as the product of soil mapping, are important tools in soil resource management and regional planning.

Diversified natural conditions and intensive human activities have endowed China with abundant soil types and land use patterns. Mapping the soil resources in as great a detail as possible has long been a task of the country, striving for a better management of this valuable basis of existence and development, no matter how primitive or advanced the technologies involved. There are several important stages regarding the development of mapping soils of China or inventory of soil data. In general, the generation of soil maps is closely related to the soil knowledge obtained by corresponding activities in a certain time period in meeting the societal demand. Several stages of soil survey and soil mapping can be identified in China up to now.

30.1.1 The Reconnaissance Soil Survey and Marbut Soil Classification Stage

Modern soil science came to China only in the early twentieth century. With the establishment of the Soil Survey Division in the China Geological Survey in 1930, modern soil survey and soil research began. Since then, there have been several major stages of soil survey and soil mapping in China (Li, 1997; Zhang et al., 2008).

The first phase is represented by the early regional soil survey and soil transect studies in the 1930s and 1940s. Since early 1930s, soil survey and soil mapping have been an important part of soil research, among which some regional small and medium map scale soil mapping was conducted. Meanwhile, the national reconnaissance soil maps were planned and tested. At that time, the major guideline followed the (agro)geological point of view for soil regionalization and soil mapping. With the introduction of modern soil classification concepts and principles by American pedologists, such as J. Thorp, who helped conduct soil survey in several regions in China, the Marbut soil classification was adopted. In the county and smaller areas, soil series were used as mapping legends, with soil group becoming the most accepted. For small-scale soils maps, "soil region" and "subregion" were adopted. In 1935, J. Thorp published his Soil Geography of China, which is the first monograph of its kind documenting the major soil types and their general distribution pattern, according to Marbut soil classification system. In order to show the abundant and complicated soil types and soil resources of China, Ma Y.T. and his colleagues started with many studies about soil cover and soil cover structure, soil complexes, and soil association. In 1934, Ma Y.T. proposed "soil area" concept and illustrated it in his early mapping practices. The 1:10 M Soil Map of China classified soils into 11 great groups and created 2 distinct categories: pure soil or complex soil. That map covered the whole territory of China, except for a small portion of the high and frigid regions in the Qinghai-Tibetan Plateau, which for the first time showed more or less the complete picture of the general soil distribution pattern of China and put forward a classification of alpine soils. Moreover, special delineations were also made to incorporate paddy soils in the lower Yangtze and Huai River valleys. The naming of paddy soils was first proposed by Hou and Ma, (1941). In 1949, another small-scale soil map was created by Ma and others. The 1:8 M soil map categorized 60 soil areas and subareas for which major bioclimate zones and soil types were used to classify soil areas and major soil associations for subareas. In these soil maps, soil classification and soil regionalization were combined and the mapping legends were mainly related to regions. Between 1946 and 1949, in the soil regionalization of the Yellow River Valley and that of China, bioclimatic zone, soil zone, soil complex, and soil catena were used. All these reflected the main thoughts of small-scale soil mapping in that stage (Zhang et al., 2008).

Other than the above mentioned maps (see Figure 30.1), many other mapping activities and products were developed during this stage. In 1947, Zhu compiled The Generalized Soil Map of Jiangxi Province, with a scale of 1:6 M (Zhu, 1948). Zhu and Hesung also compiled the 1:6 M Generalized Soil Map of China, using soil region and region complex.

30.1.2 The Russian School Stage

The second phase started with the foundation of the People's Republic of China in 1949 and spanned from the 1950s to the 1960s. This phase is characterized by the use of zonal soil distribution theory and deepening soil mapping studies. Russian genetic soil zonal distribution theory fully embraced in China as well as soil mapping. Legends used in small map scale soil mapping such as "soil region" and "subregion" were shifted to pure soil classification units. On the other hand, there was an accumulation of knowledge about soils of China, especially the classification of soils going into more detail by which more and more soils were identified and classified that greatly improved the understanding of soil types and their distribution. In 1955, the newly compiled 1:4 M "Generalized Soil Map of China" by the Russian soil scientist Gerasimov (co-authored by Ma) clearly reflected the pedogenetic point of view and the zonal distribution of soils in China and the shift of mapping legend from complex and association to pure soil types (Li, 2007). This map classified 42 mapping units at the great group levels, among which 32 were located on plain area, the remainder being found in mountainous regions. The map pointed out that the pedogenetic features of mountain soils are different from that of plain soils. This facilitated soil classification and mapping basis. Therefore, in the first published soil map of China, mountain soils and plain soils were divided into two major groups. Besides the general soil map showing soil distribution, the idea of providing service

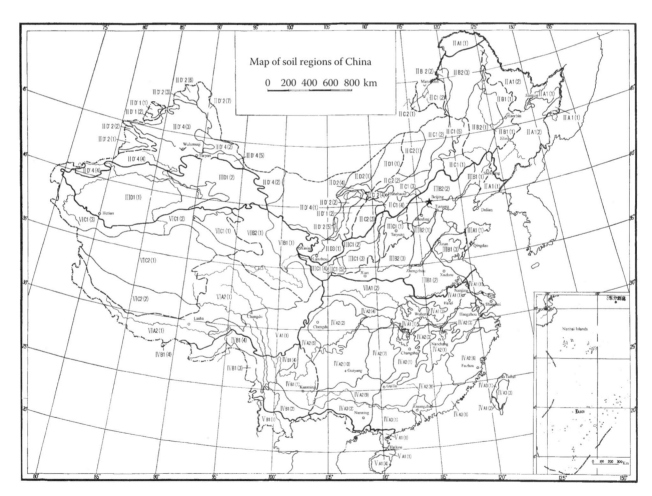

FIGURE 30.1 The soil region maps of China, generated in 1956, by Ma Y.T., Institute of Soil Science, Chinese Academy of Sciences.

to agriculture was adopted into the project and medium to large map scales maps of soil texture, soil nutrient content, soil reaction, carbonate content, etc., were compiled. Those thematic maps provided important soil information for regional soil management and agricultural development (Ma, 1960; Zhang et al., 2008).

With the increasing demand for soil information, potential forestry and agricultural land exploitation, soil and water conservation, and soil survey and mapping programs were implemented rigorously, which enabled the rapid accumulation of regional soil geographical data. This progress was apparent in the 1959 version 1:4 M Soil Map of China, which clearly showed the geographical distribution with a great deal more data. Up to 1965, soil maps were formally collected in the National Atlas of China in which 129 mapping units were established. Many upland soils, in addition to several paddy soils, were emphasized. However, due to the overemphasis of the zonal distribution principle, the reality that soil distribution is complicated and spatially variable was not well reflected, especially on cultivated soil types.

Many regional soil maps were produced during this stage, in order to meet the demand of regional natural resource exploitation and economic development. For example, a series of soil maps and soil erosion maps, ranging from a map scale of 1:5000, 1:10,000 1:200,000, to 1:1,000,000, were produced for soil and water conservation in the loess plateau of the middle Yellow River valley (Ma, 1960). Similarly, in the middle and lower reaches of the Yangtze River, soil survey programs generated medium-scale soil maps to meet the demand of catchments planning and irrigation programs. In the Huai River valley, the major problem in land reclamation was the flooding hazard in those low lying areas. In order to develop irrigation program, several soil survey programs were conducted during several different time periods, which generated 1:200,000 soil maps and soil regionalization maps as well as soil geology maps of the major irrigation regions. These maps provided a very important service to water resource planning and irrigation programming. In northeastern China, due to the need of potential arable land exploration and wind erosion control as well as that of forest development and watershed planning, regional soil reconnaissance survey and detailed survey in key areas were conducted. The detailed and medium-scale soil maps and other thematic maps, including those indicating soil parent materials, land use, groundwater mineralization degree and quality, soil salinization level, and soil reclamation regionalization, were produced based on the surveys. The detailed soil maps were mainly made for state farms and forestry farms in northeast China, on which maps covering the whole area were compiled. In Xinjiang, comprehensive soil survey programs were conducted, which completed the provincial small- and medium-scale soil maps and soil regionalization maps of several key areas, with emphasis on the genesis, classification, properties, distribution, and amelioration of saline soils. The area of potentially arable land was estimated and land management measures were proposed (Ma, 1960).

South and southwest China was mapped with medium-scale soil maps. For the key areas, the detailed maps were again provided. The small-scale Central and South China soil maps and soil regionalization maps were compiled based on surveys in those areas. The main applications of these maps were to meet the requirement of tropical and subtropical crop suitability regionalization and to provide resource statistics and planning of land resources.

Based on the regional soil survey and soil mapping practices, the national soil distribution principle was studied and the 1:4,000,000 soil maps and soil regionalization maps of China were compiled (Ma, 1957; Gerasimov et al., 1958). Soil regionalization classification system was established, which included Soil Climate Zones (First Class), Soil Bioclimate Regions (Second Class), Soil Regions or Subregions (Third Class), Soil Areas or Provinces (Fourth Class), Soil Groups (Fifth Class), etc., facilitating national, provincial, and even regional soil regionalization.

30.1.3 The National Soil Survey Stage

The third phase refers to a rapid development period that occurred between the late 1970s and 1980s, during which soil mapping theory and methodology were emphasized, revolutionizing the soil mapping system with the development of remote sensing application in soil resource survey and soil mapping. Begun in the middle and late 1970s, with the recovery of basic and applied research in China, soil mapping research, including both mapping theory and methodology, was closely followed. This allowed new/modern concepts such as the soil mapping system, its division, polygenetic mapping principle, as well as the comprehensive legend system to be incorporated into national soil research techniques (Li, 1988). Until the mid-1980s, a number of national, regional, and multiple-scale soil maps were compiled, among which the *Soil Atlas of China* was compiled using the serial mapping method. This atlas reflected the up-to-date knowledge of all disciplinary results of soil science in China (Hseung and Li, 1986).

In this period, experimental remote sensing based soil mapping was conducted, especially using visual interpretation of remotely sensed images. Automated mapping according to videotaped remote sensing data was also used commonly. Regional soil maps and land use maps were compiled using satellite imagery data (Lin, 1981). The most updated national 1:1 M *Soil Map of China*, which was based on nationwide soil survey as well as extensive remotely sensed data, has become the most detailed national soil map.

The second national soil survey program mobilized all levels of soil science in China, covering the whole mainland territory and all land uses. Township level, county level, and province level soil maps, typically in the scale of 1:10,000, 1:50,000, and 1:250,000, were compiled frequently during this period. These maps have become the most important sources of many current soil information systems. In the meantime, very small-scale maps, such as 1:4,000,000, 1:10,000,000, and 1:18,000,000 maps, were made for the whole country. The *Soil Atlas of China* by ISSAS (1986) collected a series of soil type and soil attribute maps.

30.1.4 Soil Database and Information System Stage

The fourth phase is the soil information system and digital soil mapping stage that started in the late 1980s and extends to the present. From the late 1980s to the early 1990s, technological advancement led to the integration of geographical information and soil information as a geo-referenced soil information system. This has dramatically changed the methods of traditional soil mapping. Thanks to the ever-increasing processing power of computers, soil maps have become soil information systems by which large amounts of attribute data become readily available for widespread use. Mapping technology, objects to be included, attribute data, as well as the scope of map interpretation had undergone a revolutionary change. Many important soil databases have been built since then.

In the 1980s the Remote Sensing Center of Peking University tried to establish a soil erosion information system. In the early 1990s, the creation of such a soil database began a new era of soil research. In 1991, a Regional Soil Information System (RSIS) was built and applied by the Institute of Applied Ecology, Chinese Academy of Sciences (CAS) (He and Hou, 1991). In 1992, the Institute of Soil Science, CAS (ISSCAS) established 1:500,000 Hainan soil and terrain database (SOTER) (Zhou, 1994). Actually, with the introduction of SOTER methodology into China, several SOTER databases, including 1:4,000,000 national SOTER and 1:500,000 provincial SOTER of Jiangsu and Shandong were constructed (Zhang et al., 2001a). A more detailed scale, the 1:200,000 SOTER of Hainan Island–HaiSOTER, was started in 1998, under the collaborative efforts of ISSCAS and ISRIC (Gong et al., 2001; Zhang et al., 2001d). As a more complete land information system, SOTER integrates soil and terrain information and is regarded as a useful tool for land suitability evaluation (van Engelen and Wen, 1995; Dijkshoorn et al., 2008). SOTER methodology was introduced into China in the mid-1990s, and by 2002, the 1:200,000 provincial Hai-SOTER was built and applied for agricultural land management (Mantel et al., 2003; Zhao et al., 2003), which was almost the first medium-scale SOTER in the world.

The 1:1 M SOTER of China was recently constructed and has been used for land degradation study in combination with other data sources. The topography of the 1:1 M SOTER was interpreted according to the 90 m resolution SRTM data (CGIAR-CSI, 2004) that was available at the time. Based on the images, parameters like altitude and slope and relief intensity were extracted according to standard SOTER description (Dobos et al., 2005). The overlay of the above three parameters can delineate the terrain units for the SOTER database. Soil data comes from the 1:1 M soil information system of China (Shi et al., 2004b) and with reference to FAO soil mapping legend (FAO-Unesco, 1988). Point data of more than 3000 pedons, including morphological data and physical, chemical, and mineralogical data, were collected and linked with the SOTER units by referring to the parent materials and landform as described in the soil database. By doing so, every SOTER unit has at least one typical soil profile linked, which allows further thematic soil property mapping and other sophisticated land evaluation.

The early twenty-first century has witnessed the rapid development of soil information systems in China. The digitized 1:4,000,000 soil map was not adequate to meet the ever-increasing demand of digital soil information. A more detailed soil information system, the Soil Information System of China (SISChina) was initiated and completed by ISSCAS, covering 1:1,000,000 spatial data, pedon attribute data, as well as the soil reference system of soil classification (Shi et al., 2007b). Up to this point, the serial soil databases of China covered 1:14,000,000, 1:4,000,000, 1:2,500,000, and 1:1,000,000 scale soil maps and their corresponding pedon data. For provincial and local regions, 1000 soil geochemical maps and trace element distribution maps were digitized and constructed as usable databases. SISChina incorporated more than 7000 pedons and more than 1 million bits of attribute data, enabling the creation of secondary soil property maps measuring attributes such as soil pH and soil texture (Shi et al., 2004a). The 1:1 M *Soil Map of China* is now a part of the Harmonized World Soil Database (HWSD) (FAO/IIASA/ISRIC/ISSCAS/JRC, 2009).

Based on the second national soil survey of China conducted in the 1980s, a much more detailed soil information system, using 1:50,000 map scales, is being built. As many of those detailed paper maps are stored in counties across the country, they are being lost or damaged physically due to inappropriate storage, which implied the overlooking of soil information in the past years. The rescuing work of collecting and digitizing the valuable maps (more than 2000 sheets of maps) is a huge project. Currently, more than 60% of county-scale maps in the 1:50,000 scale have been merged into an integrated soil database, covering about half of the country territory geographically (Figure 30.2).

In short, the development of soil mapping and soil information systems have undergone drastic changes since the 1930s and 1940s and is outlined as Figure 30.3, which shows how soil mapping has evolved and what uses the soil data has facilitated.

Entering the new millennium, the development of the ever-recognized digital soil mapping (DSM) technology has become one of the most important advances in basic soil geography. The core of DSM is the quantitative soil–environmental relationships and their application in soil mapping. Through the integration of digitized soil-forming covariates (McBratney et al., 2003), it is possible to automate soil identification and soil mapping in various resolutions, according to expert rules and/or quantitative algorithms. Similarly in China, DSM is being tested in various pilot regions (Yang et al., 2007; Zhao et al., 2007), to provide a methodological and technical basis for future digital soil mapping.

30.2 Use of Soil and Land Information Systems in China

30.2.1 Soil Status Assessment

Soil databases are important bridges, linking basic soil studies and their applications to other agriculture fields. Since the late 1980s, a series of regional and national soil databases have

Development and Use of Soil Maps and Databases in China

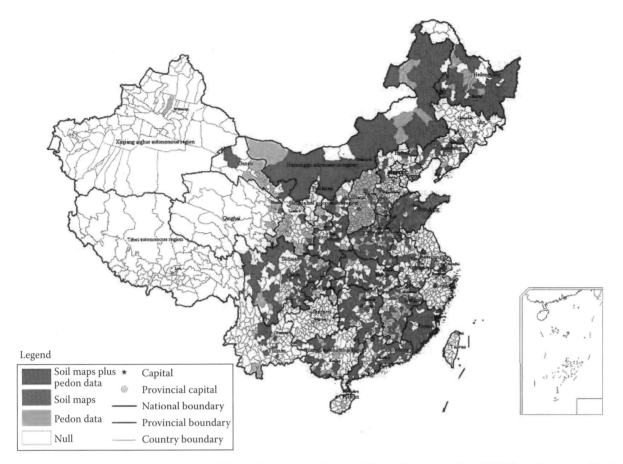

FIGURE 30.2 (See color insert.) The coverage of digitized 1:50,000 soil maps of China. (Courtesy of Dr. Weili Zhang, Institute of Soils and Fertilizer, Chinese Academy of Agricultural Science, unpublished.)

been established, which are used extensively in natural resource exploitation, land use and agricultural planning, environmental protection, and ecological restoration, on regional and national scales. Given their flexibility in manipulating data using various application models, applications of soil databases were employed extensively. Basically, the applications covered the single factor analysis, synthesized analysis, and model-based applications.

30.2.1.1 Single Factor Analysis

Single factor analysis mainly involves the spatial variability of soil properties, such as pH and soil texture as well as nutrients and their stocks. These applications primarily involve the soil database itself and require nearly no other related data.

The soil pH value varies across the country, clearly reflecting the general decrease of soil pH from the north and the northwest to the south and the southeast (Figure 30.4). A series of important soil management measures can be drawn from such maps, for example, for the allocation of chemical fertilizer resources or for the use of suitable fertilizer types. For example, because low soil pH is often related to high soil iron and aluminum oxides, phosphorus fertilizer efficiency is low in south China. Therefore, calcium magnesium phosphate was recommended as early as the 1950s.

This kind of single factor analysis also applies to many other soil properties, such as soil organic matter variation, and soil phosphorus content and availability. However, it should be pointed out that the areal result of rasterized soil pH values, and other similar soil properties, is obtained by the interpolation of point data of more than 3000 pedons. The uncertainty of the extrapolation is to be assessed.

30.2.1.2 Multiple Factor Analysis

Other than a single soil property manipulation, multiple soil factor analysis involves the calculations of more than one soil property. Taking soil organic carbon stock assessment as another example, soil database application provides an important role in systematically calculating the stock and the spatial distribution of soil organic carbon. The basic formula for calculating SOC stock is

$$S = \sum C_i D_i \rho_i \tag{30.1}$$

where
S (kg m^{-2}) represents total SOC stock within one square meter of a given soil down to a certain depth
C_i is the SOC content (kg kg^{-1})
D_i is the depth of horizons (m)
ρ_i is the soil bulk density (kg m^{-3})

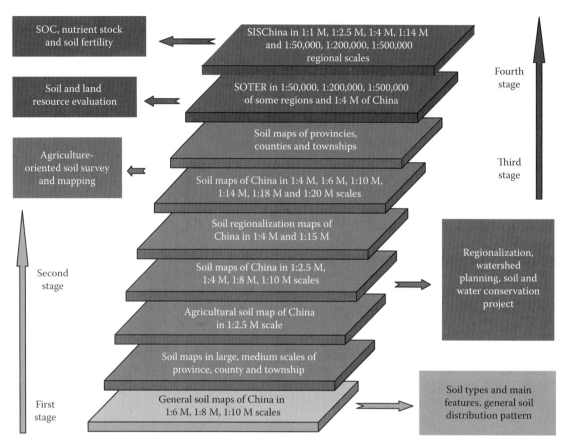

FIGURE 30.3 The major stages of soil mapping and soil database construction in China and their main uses.

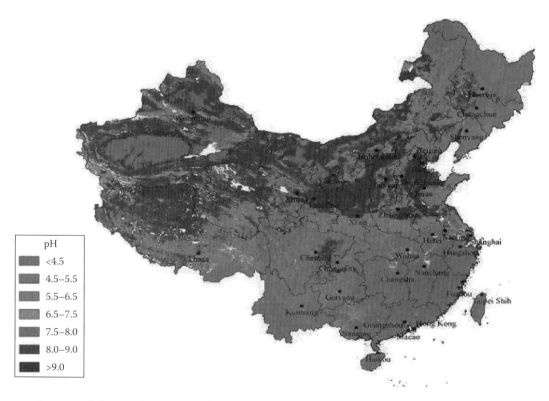

FIGURE 30.4 Soil pH map of China as obtained by database operation. The data source comes from the digitized 1:1 M Soil Map of China.

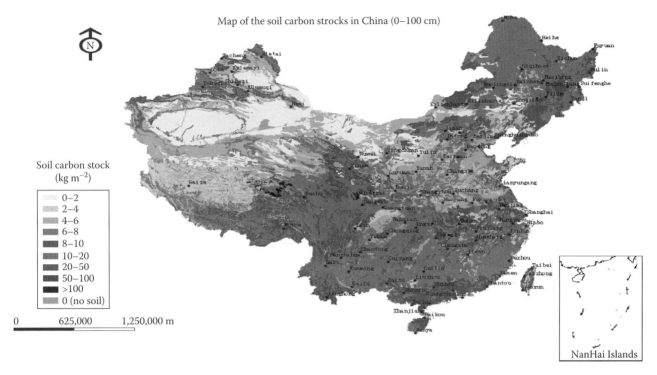

FIGURE 30.5 Soil carbon density at 100 cm depths in China. (Courtesy of Dr. XIE Xianli, Institute of Soil Science, Chinese Academy of Sciences.)

All necessary variables of such a calculation, including soil organic carbon content of soil layers, bulk density, gravel content, as well as the depth of layers, are stored in a typical soil database. The basic manipulation of the soil database allows the calculation of the overall density (kg m^{-2}) of organic carbon at a given depth. Multiplying the density by the area the soil represented in a GIS context, the total SOC reserve of a given area is easily obtained. Moreover, the spatial distribution of such a result can be visibly illustrated via a map.

Figure 30.5 shows the calculated results of SOC density (Kg m^{-2}, 1 m depth) of China, based on the digitized 1:4 M soil map and the supporting pedon data (Xie et al., 2004). The averaged SOC density is 9.14 kg·m^{-2}, varied from less than 2 kg·m^{-2}, to more than 100 kg·m^{-2}. The total stock of SOC amounted to 84.4 Pg. Another similar study, facilitated by the 1:1 M soil map of China, showed that the average density is about 9.60 kg·m^{-2} and the total stock about 89.14 Pg (Yu et al., 2005). These calculations gave a fairly good estimation of SOC storage in China and clearly showed the spatial distribution of the density, which provided policy makers with an idea of how much carbon was sequestered in soils while also showing potential regions for further carbon sequestration by possible agricultural and engineering measures. However, both of those results were not able to show any uncertainty estimation during the calculations, because normally, there were only a very limited number of pedon data available for a given map unit or delineation, which prevent further intra-unit analysis.

30.2.2 Land Suitability Assessment

Land quality evaluation is important for soil and land management. The qualitative land evaluation methods, such as ALES (automated land evaluation system) (Rossiter, 1990), were used quite often. Land quality assessment involves multiple factors and criteria, such as soil fertility assessment, soil quality assessment, land suitability assessment, and soil erosion risk assessment. The basic procedure for such analysis is to first determine the assessment entity, for example, a map unit, followed by evaluation factors and selection and assessment models (Figure 30.6). The core data source of such assessment relies normally on soil information, which is an often limiting factor.

The basic mapping unit for such an evaluation varies according to the objectives. Besides the normal soil mapping units, which are either soil types or association of complex soil types, SOTER units have become new assessment objectives. Given the characteristics of combining terrain classes, geology (lithology), and soil types, SOTER units can best describe the physically homogenous areas where similar soils developed on same geological background in similar terrain segments (van Engelen and Wen, 1995). The advantage of SOTER units in land evaluation is that they are more operational than a single soil area in terms of being more complete in reflecting actual land conditions.

A series of land evaluations were conducted in China and applied to various regions and map scales, including soil fertility, soil quality, soil erosion risk, and the suitability of crops in different scenarios. In order to facilitate a sustainable agricultural land management policy, comprehensive land evaluation was recently conducted in the tropical Hainan Island.

Based on the integrated SOTER databases at the map scales of 1:500,000, 1:200,000, and 1:50,000 (county level only), the suitability of tropical crops, including banana, coffee, rubber,

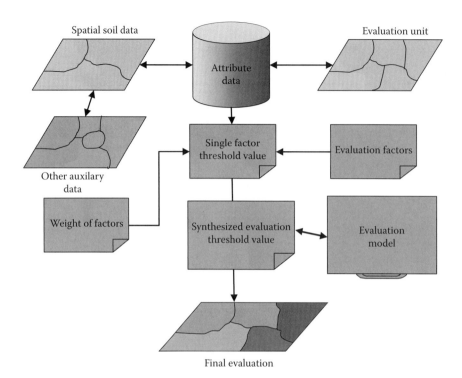

FIGURE 30.6 Land suitability assessment based on multiple factor analysis of soil databases.

mango, and several other tropical fruits, was assessed, by using ALES as the basic land evaluation model (Zhang et al., 2001c). Moreover, other evaluation tests, including soil quality as defined mainly by physical and chemical fertility (Zhang and Zhang, 2005), the current soil erosion risk and the potential erosion under different land use scenarios (Zhao et al., 2003), and soil diversity (Zhang et al., 2003), were also conducted.

30.2.2.1 Soil Quality Assessment

Soil quality has been viewed in two ways: first, as an intrinsic or static property of a soil governed by natural soil-forming processes, and second, as a dynamic state or health of a soil that reflects its condition under a specific management system (Karlen et al., 1997). Numerical soil quality assessment is an effective decision-making tool for land users to identify the best management practices (Karlen et al., 1997). During the last 10 years, a systematic method of soil quality assessment has been developed. This method includes three main steps: (1) identifying appropriate soil quality indicators for a minimum data set, (2) transforming those indicators into dimensionless standard scores, and (3) incorporating the indicator scores into soil quality indices (Andrews et al., 2002).

Inappropriate use and management have caused severe degradation of soil quality in the tropics (Lal, 1990). Therefore, the assessment of soil quality is urgently needed to reverse degradation (Stocking, 2003). Here we take Hainan Island, the second largest island and the largest tropical region of China, as an example, to show how regional soil quality variation can be differentiated based on the existing SOTER database.

Standard scoring functions (SSFs) suggested by Karlen and Stott (1994) were used to transform soil properties into dimensionless scores (Zhang and Zhang, 2005). For each soil quality indicator, the types and parameters for the SSFs were determined by expert opinion as recommended in the literature (Wang and Gong, 1998; Hussain et al., 1999; Glover et al., 2000).

In this study, soil quality was evaluated in terms of four soil functions: water availability, nutrient availability, rooting suitability, and erosion resistance with reference to a combination of the methods of Karlen and Stott (1994) and Harris et al. (1996). Indicators in minimum data sets were grouped to determine the four functions. The weightings of each indicator and of each soil function to the overall soil quality index were assigned based on expert opinion (Table 30.1).

According to above soil functions and indicators as well as their weights in the calculation, the soil quality of Hainan Island was estimated (Figure 30.7). From the result it became apparent that the outer part of the island has a low soil quality, which is explained by their very low nutrient score (Zhang and Zhang, 2005).

30.2.2.2 Soil Erosion Risk Assessment

The assessment of soil erosion risk is also a multiple factor analysis. Currently, soil water erosion models, such as SWEAP (Soil Water Erosion Assessment Program), USLE (Universal Soil Loss Equation), and Revised USLE are being used, although they are not tailored for regional use. SWEAP was developed specifically for SOTER, from which soil erosion factors, such as rainfall, soil organic matter content and texture (erodibility), slope and slope length, etc., can be extracted from the database. Figure 30.8 shows

TABLE 30.1 Table Additive Weightings Used to Calculate Indices of Soil Functions and Overall Soil Quality

Function	Weight	Indicator	Weight
Water availability	0.25	Clay content	0.3
		Bulk density	0.3
		Total organic C	0.2
		Water stable aggregates	0.2
Nutrient availability	0.25	Total organic C	0.3
		CEC	0.2
		pH	0.2
		Extractable P	0.1
		Exchangeable K	0.1
		Microbial biomass C	0.1
Rooting suitability	0.25	Soil depth	0.3
		Bulk density	0.3
		% Clay	0.2
		pH	0.2
Erosion resistance	0.25	Water stable aggregates	0.35
		Total organic C	0.35
		% Clay	0.3

the general procedure of soil erosion risk assessment using USLE. The result is a rough estimation as USLE is based on a single slope. Nevertheless, such a result is useful in knowing the relative risk categories (van den Berg, 1992).

For the actual operation, the spatial database is composed of SOTER unit, climate zone, and land use maps. Three kinds of base maps (1:100,000 landform map, 1:200,000 lithology map, and 1:200,000 soil map) were selected to compile the SOTER unit map. Two routes for field check were conducted to confirm the reasonability of the SOTER unit map. In total, 185 SOTER units were identified on Hainan Island. Relating to 19 county climate stations, 19 climate zones were delineated based on average temperature, accumulated temperature, and average 30 year typhoon rainfall data. Climate data, monthly precipitation, minimum and maximum temperatures, relative humidity, radiation, and evaporation, from the 19 stations, were included in the database. Climate zone and SOTER unit maps were integrated to get the agricultural ecological unit (AEU) map. Land use data were obtained from SPOT satellite images where land-cover classes were converted to land-use classes according to the standard SOTER methodology.

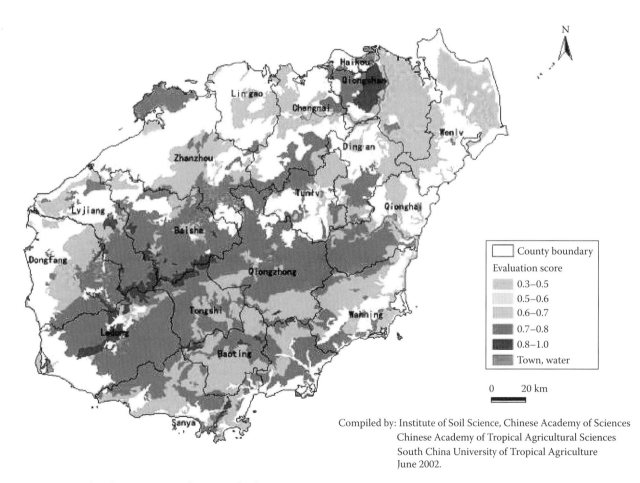

FIGURE 30.7 Soil quality assessment of Hainan Island.

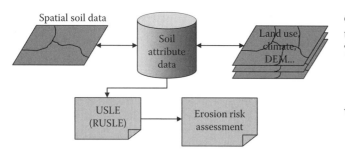

FIGURE 30.8 Scheme of soil water erosion risk assessment using SOTER and USLE.

Many bits of attribute data were needed for soil erosion evaluation. The evaluation model obtains climate data from a climate database, land use information from the base map as well as basic data (such as slope length, slope gradient, organic carbon content in the fine earth fraction, weight of silt, weight of very fine sand, soil structure and profile-permeability) from the SOTER database. Classification names and definitions are fully illustrated in van Engelen and Wen's SOTER manual (van Engelen and Wen, 1995).

The USLE was specially designed for erosion assessment at landscape level. For larger scales, like SOTER units, uniform slopes hardly occur and some places will be eroded whereas others will receive sediments. However, it can be used generally to compare regional erosion hazard classes (van den Berg, 1992). The basic equation of USLE is

$$A = R \times K \times LS \times C \times P \tag{30.2}$$

where
    A is the soil erosion modulus
    R is the rain erosivity factor
    K is the soil erodibility factor
    LS is the slope length and slope gradient factor
    C is the vegetation factor
    P is the engineering control factor

Figure 30.9 shows the result of the assessment. Water erosion risk in Hainan Island is low, and has a very low c value due to dense vegetation cover in the tropics. Slightly higher water erosion happens in the central mountainous region due to steep region slopes. However, given a little or no vegetation cover scenario, the erosion in this area would be extremely high (Zhao et al., 2003).

30.2.2.3 Land Suitability for Tropical Crops

The suitability of a crop depends on several major factors, namely, soil, site, and climate factors. Therefore, physical suitability

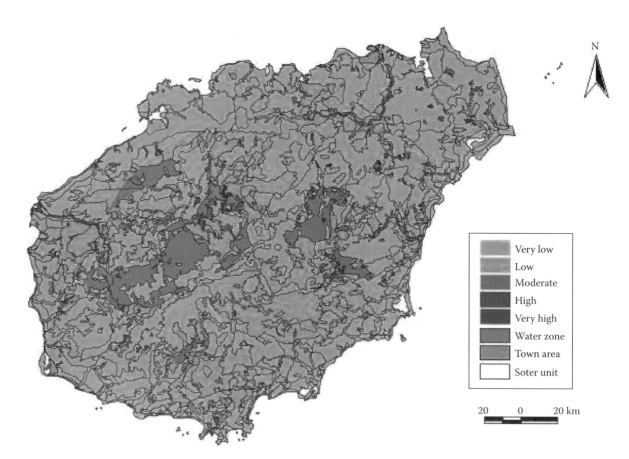

FIGURE 30.9 Soil water erosion risk of Hainan Island under current land use condition.

Development and Use of Soil Maps and Databases in China

TABLE 30.2 Land Qualities and Land Characteristics Used in the HAILES Model

Land Quality	Land Characteristics
Availability of moisture	Length of growing period (LGP), defined as P > 0.5 ET0 + SM; the period that rainfall (P) exceeds half the potential evapotranspiration (ET0) + soil moisture (SM) reserve, and temperatures are conducive to crop growth
Availability of nutrients	Soil reaction (pH), organic carbon, cation exchange capacity, base saturation percentage
Resistance to erosion	Slope, slope length, sol structure grade, sensitivity to capping/sealing
Flooding hazard	Frequency of flooding, flooding duration
Potential for mechanization	Slope, soil gravel content, surface rockiness
Availability of oxygen to roots	Soil drainage class
Soil toxicity	Aluminum saturation of the subsoil

assessment mainly considers these aspects. Crop suitability is also related to socioeconomic conditions; therefore, management at the input and technology level does matter (Mantel et al., 2003).

The framework for crop suitability assessment is set according to ALES (Rossiter, 1990). For land suitability assessment of tropical crops, such as banana, coffee, mango, and rubber, more specific grading systems were incorporated into HAILES (Mantel et al., 2003) Soil moisture condition, nutrient status, physical status, site condition, and management levels were all considered as variables. Table 30.2 shows the major land characteristics used in the evaluation. Besides these factors, climate suitability (rainfall and accumulative temperature) was also divided according to the requirement of the crops.

Figure 30.10 is an example of such an assessment, which shows the classes of land suitability for rubber plantation in Hainan Island. A series of such maps were compiled, according

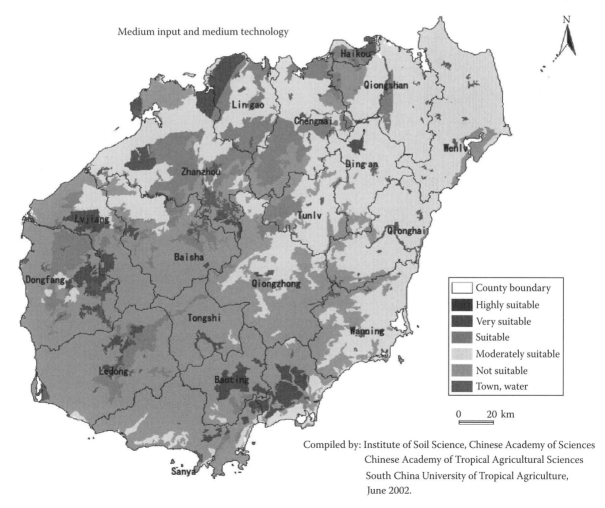

FIGURE 30.10 Rubber suitability in Hainan Island at medium input and technology level.

to different scenarios of management level, considering mainly chemical fertilizer input possibility and machine use potential (technology level).

The establishment of comprehensive soil databases is the base of modern land quality assessment. Up to now, several major databases are available for country and province level. At the local level, there are many more regional soil datasets. By comparing the newly sampled data with the original data, changes in soil properties and qualities can be monitored. Recently, the typical areas of major agricultural land were studied to determine the soil quality change in the lower Yangtze delta region, the low hill red soil region, the Huang-Huai-Hai plain region, and the northeast Mollisols region (Cao and Zhou, 2008).

30.2.3 Assessment of Land Degradation and Improvement in China

Land degradation is a global problem and a national issue in China. The Global Assessment of Land Degradation and Improvement (GLADA) under the LADA program indicates that over the last 25 years, 24% of the land surface has been degrading, in addition to the historical legacy of degradation since human settlement (Bai et al., 2008). In China, degradation of dry lands has received much attention and reclamation programs have achieved some success, but over the last 25 years, 23% of the country has suffered a decline of net primary productivity, 24% of the croplands and 40% of the forests, not mainly in dry lands but in the high-rainfall areas of South China (Bai and Dent, 2008) (Figure 30.11).

GLADA analyses long-term trends in biomass productivity using remotely sensed NDVI data. The assumption is that these may indicate land degradation or improvement if false alarms due to climatic variability and land use change can be accounted for. In the preliminary analysis, climatic variability was taken into account by an analysis of rain-use efficiency and energy-use efficiency. Land use change, by itself, leads to changes in biomass that are not necessarily land degradation as normally understood. Recently, a study compiling NDVI changes, taking advantage of soil and terrain (SOTER) database of China, was conducted (Wu et al., 2010).

By overlaying land degradation results on a SOTER map, factors of land degradation can be analyzed. Comparison of degrading areas with land cover indicates that about half of the burnt, deciduous needle-leaved, evergreen broadleaved and mixed forests are all suffering degradation. Seventeen percent of cultivated and managed areas and about 34% of mosaic cropland are degrading. The most degraded land cover is grassland, occupying 31.1%; the second is needle-leaved forest, amounting to 19.3%; the third is cultivated land with 14.9% of the degrading land; the fourth is broadleaved forest, accounting for 14.2%; and just 4.8% of the degrading area is bare.

Of the improving area, 30% is cultivated and managed, 16.7% is mosaic cropland, 21% is grassland, and 16.8% is closed deciduous broadleaved forest. Cultivated and managed areas and herbaceous

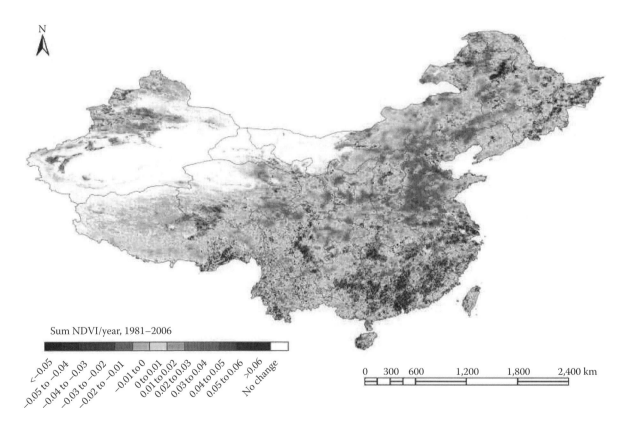

FIGURE 30.11 NDVI change of China over 1981–2006.

Development and Use of Soil Maps and Databases in China

cover account for 63% of the whole improving area, which occupy 32.1% and 30.9% of the improving land respectively.

Relationships between soil properties and land degradation trends can be established too. Taking soil organic matter as an example, classes defined by 0%–5%, 5%–15%, 15%–20%, and >20%, occupy 72%, 25%, 2%, and 0% of the degrading area, respectively (and almost the same percentage of the degrading area at the 95% confidence level [CL]), similar to the national extent of each class (Figure 30.12a). There is almost no relationship between total soil organic matter and land degradation. The percentages of each pH class of degrading area are quite different from those of the whole country. With the decrease of pH, the percentage of each class of the degrading area or at 95% confidence level becomes bigger than that of the whole country, which shows the influence of lower pH on degrading land (Figure 30.12b). Clay classes are defined by 0%–15%, 15%–25%, 25%–45%, 45%–65%, and >65% based on international system of texture. The higher percentage of clay class 25%–45% of degrading area or at 95% confidence level compared with the whole country may show some tendency to degrade on this kind of texture (Figure 30.12c). In short, some soil attributes may influence land degradation.

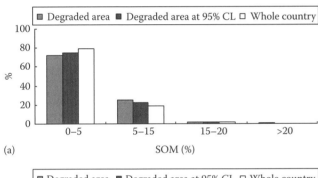
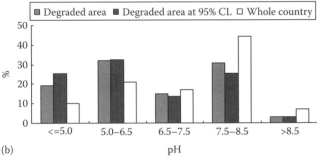
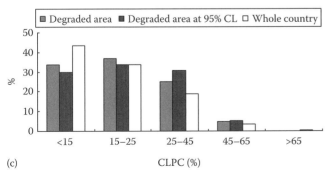

FIGURE 30.12 Area percentage of each class of soil attributes. (a, Soil organic matter [SOM]; b, pH; c, clay percentage [CLPC])

30.3 Future Development and Opportunities

30.3.1 Near-Future Tasks in Soil Correlation

Looking back at the history of soil mapping and soil database achievement, it is not difficult to draw the conclusion that for the vast agriculturally based country like China, soil information has been a prerequisite for any proper land management. However, the level of soil data acquired and manipulated are far from satisfactory.

First, the soil classification system has changed several times due to the influence of different academic schools in the world. It changed from the Marbut system, to Russian genetic system, and recently to the diagnostic system. It should be pointed out that soil classification has generally followed international trend, and the newly established taxonomic system does represent the current development of our knowledge. Influenced by soil classification, mapping legends changed as well. The legend systems and mapping scales for soil maps up to now have varied at different stages, which has led to problems in soil database construction, due to inconsistency and standardization problems with soil names, analytical methods, description standards, and so on. The new attempt to map the country at a small scale follows the new Chinese Soil Taxonomy and the corresponding legend system. For consistent soil database building, more correlation work must be done.

Second, the national soil maps usually adopt high soil classification categories as mapping legends, so the delineation of soil area is fairly heterogeneous. This is of course because of the size of the country, but it calls for a more detailed soil record. Similarly, the corresponding pedon data serving to explain soil delineations is not enough in terms of its adequacy to show the soil variability. A recent soil series inventory project, led by the Institute of Soil Science, Chinese Academy of Sciences, will hopefully provide many new and standardized datasets to update the original soil databases for various map scales. The new data will be precisely located with correct taxonomic names making it easy to pinpoint on a map. Such a soil correlation and inventory work will serve a better record of soil information and long-term application.

Third, the quality, standard, and interpretation of soil data need to be strengthened (Zhang et al., 2001b; Shi et al., 2007a). This is a huge task facing soil scientists. For a long time, there have been changes in soil survey terminology, analytical methods, soil classification, and so on. The data we have are not always comparable. This is, of course, an issue not only in China but in the international community as well. There is a pressing demand for soil scientists across the world to establish a real international standard. On the other hand, study of the models and expert systems applicable to soil data interpretation is fragmented in China. We may borrow some more or less universally used models elsewhere and modify them for use in China, but we also need to develop models that fit regional conditions in China.

Last but not the least, there is a room for the improvement of methods linking spatial data and attribute data. The currently used matching method normally takes the so-called central concept soil profile to represent the whole area and links the pedon data with the mapping unit. There is a high variation of soil properties within a certain area, as we know. The linking method of SOTER, which considers more than soil types, improves the precision of soil attribute data and spatial data matching, but remains with the similar problem of one pedon. How the spatial uncertainty can be reduced by linking and using multiple profile data is a work to be done.

30.3.2 Digital Soil Mapping and Its Practice in China

The need for more accurate, easy to use, and systematic soil information is increasing. Digital soil mapping (DSM) is an effort to provide high resolution rasterized soil data. It can be defined as "the creation and population of spatial soil information systems by numerical models inferring the spatial and temporal variations of soil types and soil properties from soil observation and knowledge and from related environmental variables" (Lagacherie and McBratney, 2007). The core of DSM is the soil–landscape relationship and its visualization through various models in the context of a geographical information system. Recently, many modeling methods have been tested in different locations. Scull et al. (2005) classified them into geostatistical, statistical, decision-tree, and expert system. Beside those, fuzzy logic method is also widely used (Zhu and Band, 1994; Zhu et al., 1996, 1997; Zhu, 1997a, 1997b).

Attempts to map soils digitally in China have also made some progress. By using geostatistical methods, Xu and Webster (1983) mapped soil nitrogen distribution in a county. Similarly, Wang et al. (2001) and Li et al. (2001) mapped soil thickness and soil water content by kriging, combining this method with remote sensing data. By establishing a linear regression relationship between soil properties and landscape parameters, Sun et al. (2008b) predicted soil organic carbon content in the low hill region of subtropical China and compared the impact of DEM resolution (Sun et al., 2008a). The decision-tree method was tested by Zhou et al. (2003, 2004) for soil property and type mapping in Zhejiang province. Sun et al. (personal communication) recently tried to map soils of Hong Kong by using decision-tree and discriminate analysis of various environmental covariates combinations, and found that the predicting models, rather than the covariate combinations, determined the preciseness of the final map. As fuzzy logic method is suggested to be very useful in predicting soil variation (McBratney et al., 1992; Moore et al., 1993; Zhu, 1997a, 1997b), much effort has been put into its use in soil mapping in China. Yang et al. (2007) and Zhao et al. (2007) used the fuzzy set of topographical data and conducted predictive soil property mapping based on soil–landscape relationships. Sun et al. (2008b) discussed the optimization of fuzzy clustering parameters.

There is an increasing belief that DSM will play a major role in the future. With the recent establishment of GlobalSoilMap.net, a global consortium has formed and pledges to contribute to the full coverage of high resolution rasterized soil data for the globe. It is believed that DSM will be soon put into the implementation stage (Lagacherie, 2008). The readily available technologies, including that of remote sensing, spatial database, spatial modeling, and so on, will promote the efficient acquisition of soil information. As an important participant of GlobalSoilMap.net, there are reasons to believe that China will develop in this direction too, following the international trend.

Acknowledgements

The authors appreciate the great help of Prof. Yuncong Li, University of Florida, and acronym reviewer(s) for the improvement of this paper. Supporting materials of the authors are from projects funded by Natural Science Foundation of China (40625001) and Jiangsu Natural Science Foundation (BK2008058).

References

Andrews, S.S., D.L. Karlen, and J.P. Mitchell. 2002. A comparison of soil quality indexing methods for vegetable production systems in Northern California. Agric. Ecosyst. Environ. 90:25–45.

Bai, Z.G., and D.L. Dent. 2008. Land degradation and improvement in China 1. Identification by remote sensing GLADA Rept 1d/ISRIC Rept 2007/6. ISRIC—World Soil Information, Wageningen, the Netherlands.

Bai, Z.G., D.L. Dent, L. Olsson, and M.E. Schaepman. 2008. Proxy global assessment of land degradation. Soil Use Manage. 24:223–234.

Cao, Z.H., and J.M. Zhou. 2008. Soil quality of China (In Chinese). Science Press, Beijing, China.

CGIAR-CSI. 2004. SRTM 90 m digital elevation data. Consortium for Spatial Information.

Dijkshoorn, J.A., V.W.P. van Engelen, and J.R.M. Huting. 2008. Soil and landform properties for LADA partner countries (Argentina, China, Cuba, Senegal and The Gambia, South Africa and Tunisia) SRIC report 2008/06 and GLADA report 2008/03. ISRIC—World Soil Information, Wageningen, the Netherlands.

Dobos, E., J. Daroussin, and L. Montanarella. 2005. An SRTM-based procedure to delineate SOTER Terrain Units on 1:1 and 1:5 million scales EUR 21571 EN. Office for Official Publications of the European Communities, Luxembourg.

FAO/IIASA/ISRIC/ISSCAS/JRC. 2009. Harmonized world soil database (version 1.1). FAO, Rome, Italy and IIASA, Laxenburg, Austria.

FAO-Unesco. 1988. FAO/Unesco soil map of the world, revised legend world resources report 60. FAO, Rome, Italy.

Gerasimov, K., Y.C. Ma, and Y.T. Ma. 1958. The genetic types of soils of China and their geographical distribution (In Chinese) Soil Bull. 32:1–52.

Glover, J.D., J.P. Reganold, and P.K. Andrews. 2000. Systematic method for rating soil quality of conventional, organic, and integrated apple orchards in Washington State. Agric. Ecosyst. Environ. 80:29–45.

Gong, Z.T., X.L. Zhang, G.B. Luo et al. 2001. The creation of SOTER and its expansion to the world (In Chinese). Geogr. Sci. 21:217–223.

Harris, R.F., D.L. Karlen, and M.J. David. 1996. A conceptual framework for assessment and management of soil quality and health, p. 61–82. In J.W. Doran and A.J. Jones (eds.) Methods for assessing soil quality. Soil Science Society of America, Madison, WI.

He, H.S., and Y.L. Hou. 1991. The establishment and applications of regional micro-computer based soil information system (In Chinese). Acta Pedol. Sin. 28:345–354.

Hou, K.C., and Y.T. Ma. 1941. On the morphological aspects of the podzolic rice paddy soils in Nanchang region, Kiangsi, China. Spec. Soil Publ. Ser. A 3:1–17.

Hseung, Y., and J. Li. 1986. Soil atlas of China (in Chinese). Cartographic Publishing House, Beijing, China.

Hussain, I., K.R. Olson, M.M. Wander, and D.L. Karlen. 1999. Adaptation of soil quality indices and application to three tillage systems in southern Illinois. Soil Till. Res. 50:237–249.

ISSAS. 1986. The soil atlas of China (in Chinese). Cartographic Publishing House, Beijing, China.

Karlen, D.L., M.J. Mausbach, J.W. Doran, R.G. Cline, R.F. Harris, and G.E. Schuman. 1997. Soil quality: A concept, definition, and framework for evaluation. Soil Sci. Soc. Am. 61:4–10.

Karlen, D.L., and D.E. Stott. 1994. A framework for evaluating physical and chemical indicators of soil quality, p. 53–72. In J.W. Doran et al. (eds.) Defining soil quality for a sustainable environment. Soil Science Society of America, Madison WI.

Lagacherie, P. 2008. Digital soil mapping: A state of the art. In A.E. Hartemink et al. (eds.) Digital soil mapping with limited data. Springer, the Netherlands.

Lagacherie, P., and A.B. McBratney. 2007. Spatial soil information systems and spatial soil inference systems: Perspectives for Digital soil mapping, p. 3–24. In P. Lagacherie et al. (eds.) Digital soil mapping: An introductory perspective. Developments in soil science. Vol. 31. Elsevier, Amsterdam, the Netherlands.

Lal, R. 1990. Soil erosion in the tropics. Principles and Management Inc., New York.

Li, J. 1988. Theory and methodology of small scale soil mapping (in Chinese). Acta Pedol. Sin. 25:336–348.

Li, J. 1997. Soil map (in Chinese). Fujian Cartographical Publishing House, Fuzhou, China.

Li, J. 2007. Contribution of Ma Yongting to soil mapping of China (in Chinese). In Institute of Soil Science Chinese Academy of Sciences (ed.) Ma Yongting and soil Science of China. Jiangsu Science and Technology Press, Nanjing, China.

Li, H.B., Z.H. Lin, and S.X. Liu. 2001. Application of krigging technique in estimating soil moisture in China (in Chinese). Geogr. Res. 20:446–452.

Lin, P. 1981. Visual interpretation of Landsat images for medium scale soil mapping (a case study in Handan, Hebei). In China Remote Sensing Center (ed.) Selected papers on remote sensing (in Chinese). Science Press, Beijing, China.

Ma, Y.T. 1957. General principles of geographical soil distribution (in Chinese). Acta Pedol. Sin. 5.

Ma, Y.T. 1960. Retrospect of Chinese sciences in the last ten years: Soil science (1949–1959) (in Chinese). Science Press, Beijing, China.

Mantel, S., X.L. Zhang, and G.L. Zhang. 2003. Identification of potential for banana in Hainan Island, China. Pedosphere 13:147–155.

McBratney, A.B., J.J. Gruijter De, and D.J. Brus. 1992. Spacial prediction and mapping of continuous soil classes. Geoderma 54:39–64.

McBratney, A.B., M.L.M. Santos, and B. Minasny. 2003. On digital soil mapping. Geoderma 117:3–52.

Moore, I.D., P.E. Gessler, G.A. Nielsen, and G.A. Peterson. 1993. Soil attribute prediction using terrain analysis. Soil Sci. Soc. Am. 57:443–452.

Rossiter, D.G. 1990. Ales—A framework for land evaluation using a microcomputer. Soil Use Manage. 6:7–20.

Scull, P., J. Franklin, and O.A. Chadwick. 2005. The application of classification tree analysis to soil type prediction in a desert landscape. Ecol. Model. 181:1–15.

Shi, X.Z., G.X. Yang, D.S. Yu, H.J. Wang, W.X. Sun, Y.C. Zhao, and P. Gao. 2007a. Future of soil information Science. In Soil Science Society of China (ed.) Actual and prospect of soil science of China. Hehai University Press, Nanjing, China.

Shi, X.Z., D.S. Yu, P. Gao, H.J. Wang, W.X. Sun, Y.C. Zhao, and Z.T. Gong. 2007b. Soil information system of China (SIS China) and its application (Chinese). Soil 39:329–333.

Shi, X.Z., D.S. Yu, X. Pan, Z.W. Sun, X.H.J. Wang, and Z.T. Gong. 2004a. 1:1 million soil database of China and its application. In Soil Science Society of China (ed.) Soil science for agricultural and environment (in Chinese). Science Press, Beijing, China.

Shi, X.Z., D.S. Yu, E.D. Warner, X.Z. Pan, G.W. Petersen, Z.G. Gong, and D.C. Weindorf. 2004b. Soil database of 1:1,000,000 digital soil survey and reference system of the Chinese genetic soil classification system. Soil Surv. Horiz. 45:129–136.

Stocking, M.A. 2003. Tropical soils and food security: The next 50 years. Science 302:1356–1359.

Sun, X.L., Y.G. Zhao, C.Z. Qin, D.C. Li, L. Zhao, and G.L. Zhang. 2008a. Effects of DEM resolution on multi-factor linear soil-landscape models and their application in predictive soil mapping. Acta Pedol. Sin. 45:971–977.

Sun, X.L., Y.G. Zhao, G.L. Zhang, and D.C. Li. 2008b. Optimization of clustering parameters in predictive mapping of soil organic matter. Trans. CSAE 24:31–37.

van den Berg, M. 1992. SWEAP. A computer program for water erosion assessment applied to SOTER: Documentation Version 1.4. SOTER Report No. 7. International Society of Soil Science, Wageningen, the Netherlands.

van Engelen, V.W.P., and T.T. Wen. 1995. Global and national soil and terrain digital database (SOTER) procedures manual. ISRIC—World Soil Information, Wageningen, the Netherlands.

Wang, X.J., and Z.T. Gong. 1998. Assessment and analysis of soil quality changes after eleven years of reclamation in subtropical China. Geoderma 81:339–355.

Wang, S.Q., S.L. Zhu, and C.H. Zhou. 2001. Characteristics of spatial variability of soil thickness in China. Geogr. Res. 20:161–169.

Wu, Y.J., G.L. Zhang, Z.G. Bai, D.L. Dent, V. Van Engelen, and Y.G. Zhao. 2011. Land degradation in China during 1981 and 2006: An NDVI SOTER approach. Unpublished.

Xie, X.L., B. Sun, H.Z. Zhou, and Z.P. Li. 2004. Organic carbon density and storage in soils of China and spatial analysis (in Chinese). Acta Pedol. Sin. 41:35–43.

Xu, J.Y., and R. Webster. 1983. Optimal estimation of soil survey data by geostatistical method—Semi-variogram and block krigging estimation of top soil N of Zhangwu county. Acta Pedol. Sin. 20:419–430.

Yang, L., A.X. Zhu, B.L. Li, C.Z. Qin, T. Pei, B.Y. Liu, R.K. Li, and Q.G. Cao. 2007. Extraction of knowledge about soil-environment relationship for soil mapping using fuzzyc-means(FCM) clustering. Acta Pedol. Sin. 44:784–791.

Yu, D.S., X.Z. Shi, W.X. Sun, H.J. Wang, Q.H. Liu, and Y.C. Zhao. 2005. Estimation of China soil organic carbon storage and density based on 161 000 000 soil database. Chin. Appl. Ecol. 16:2279–2283.

Zhang, X.L., J. Chen, G.L. Zhang, M.Z. Tan, and J.J. Ibáñez. 2003. Pedodiversity analysis in Hainan Island. Geogr. Sci. 13:181–186.

Zhang, G.L., Z.T. Gong, G.B. Luo, and X.L. Zhang. 2001b. Structure, contents and possible applications of China National soil information system. Sci. Geogr. Sin. 21:401–406.

Zhang, G.L., X.Z. Shi, and Z.T. Gong. 2008. Retrospect and prospect of soil geography in China. Acta Pedol. Sin. 45(5):792–801.

Zhang, D.X., D.S. Yu, and X.Z. Shi. 2001a. SOTER construction and its application in the assessment of paddy soil productivity in southern Jiangsu Province (in Chinese). Anhui Agri. Univ. 28:119–124.

Zhang, H., and G.L. Zhang. 2005. Landscape-scale soil quality change under different farming systems of a tropical farm in Hainan, China. Soil Use Manage. 21:58–64.

Zhang, X.L., G.L. Zhang, and Z.T. Gong. 2001c. Evaluation for tropical crops in Hainan province by using ALES based upon HaiSOTER (in Chinese). Sci. Geogr. Sin. 21:343–349.

Zhang, X.L., G.L. Zhang, and Z.T. Gong. 2001d. Soil quality assessment supported by SOTER—A case study in Hainan (in Chinese). Mountain Sci. 19:377–380.

Zhao, Y.G., G.L. Zhang, and Z.T. Gong. 2003. SOTER-based soil water erosion simulation in Hainan Island. Pedosphere 13:139–146.

Zhao, L., Y.G. Zhao, D.C. Li, X.L. Sun, and G.L. Zhang. 2007. Digital soil mapping by extracting quantitative relationships between soil properties and terrain factors based on fuzzy set theory (in Chinese). Acta Pedol. Sin. 44:961–967.

Zhou, H.Z. 1994. Soil and terrain digital database of Hainan Island and its mapping (in Chinese). Science Press, Beijing, China.

Zhou, B., F. Wang, and R.C. Wang. 2004. Automated soil mapping by using classification tree algorithm (in Chinese). Soil Water Conserv. 18:140–147.

Zhou, B., H.W. Xu, and R.C. Wang. 2003. Soil organic matter mapping based on classification tree modeling (in Chinese). Acta Pedol. Sin. 40:801–808.

Zhu, X.M. 1948. Soil map at 1:2 000 000 scales of Jiangxi, general soil of Jiangxi (in Chinese). Quart. Soil 7:1–18.

Zhu, A.X. 1997a. Measuring uncertainty in class assignment for natural resource maps under fuzzy logic. Photogramm. Eng. Remote Sens. 63:1195–1202.

Zhu, A.X. 1997b. A similarity model for representing soil spatial information. Geoderma 77:217–242.

Zhu, A.X., and L.E. Band. 1994. A knowledge-based approach to data integration for soil mapping. Can. J. Remote Sens. 20:408–418.

Zhu, A.X., L.E. Band, B. Dutton, and T.J. Nimlos. 1996. Automated soil inference under fuzzy logic. Ecol. Model. 90:123–145.

Zhu, A.X., L. Band, R. Vertessy, and B. Dutton. 1997. Derivation of soil properties using a soil land inference model (SoLIM). Soil Sci. Soc. Am. 61:523–533.

31
Soil Geographic Database of Russia

Sergey A. Shoba
Moscow State University

Vyacheslav A. Rozhkov
V.V. Dokuchaev Soil Science Institute

Irina O. Alyabina
Moscow State University

Varvara M. Kolesnikova
Moscow State University

Inga S. Urusevskaya
Moscow State University

Erik N. Molchanov
V.V. Dokuchaev Soil Science Institute

Vladimir S. Stolbovoi
V.V. Dokuchaev Soil Science Institute

Boris V. Sheremet
V.V. Dokuchaev Soil Science Institute

Dmitry E. Konyushkov
V.V. Dokuchaev Soil Science Institute

31.1 Introduction 31-1
31.2 Soil and Land Resources of Russia 31-2
  Soil Resources • Land Resources
31.3 History of Collection of Soil Information in Russia 31-5
  History of Soil Mapping • History of Soil Information Science and the Development of Digital Soil Mapping in Russia
31.4 Soil Geographic Database of Russia 31-8
  Cartographic Block: Digital Maps • Attribute Block: Soil Profile Database • Software Support of the SGDB
31.5 Conclusions 31-17
Acknowledgments 31-17
References 31-17

31.1 Introduction

Russia, the largest country in the world, encompasses nearly 30% of the world area of permafrost-affected soils, more than 60% of the world area of soils of the Boreal zone, and about 40% of the world area of Chernozems. More than a century since the shaping of modern pedology by V.V. Dokuchaev, Russian soil scientists have gained a vast amount of information about the soils of Russia. Some of the data are integrated in thematic information systems and databases, although a larger portion of it is preserved in paper form and cannot be readily applied in modern information technologies.

In the last decade, specialists from several organizations have digitized a great number of various thematic maps. In particular, the digital Soil Map of the Russian Federation on a scale of 1:2.5 M (1988) was developed, as well as a new digital map of the soil-ecological zoning (SEZ) of Russia on the same scale.

One of the major challenges facing Russian soil scientists is the creation of the national Soil Geographic Database (SGDB). The SGDB of Russia is considered a crucial element of soil inventory at the federal level. The SGDB should facilitate the regulation of land management and soil conservation policies in the country. The SGDB should be applied toward the development of legislative acts for soil protection and federal standards of soil quality for the purposes of soil-ecological monitoring. It should also serve as the basis for cadaster assessments of land resources.

Work on the SGDB project was initiated by the Dokuchaev Soil Science Society in 2008 at the All-Russia Conference in Rostov-on-Don. Since then, the concept and ideology of the project have been developed rigorously, and several major principles and goals of the SGDB have been formulated. The SGDB has been discussed in several papers and reports. The V.V. Dokuchaev Soil Science Society has presented soil scientists throughout the country with a request to contribute their time and research to the development of the SGDB. Special software aimed to aid in the completion of the SGDB has been developed; work on filling out informational fields in the SGDB is still in progress.

31.2 Soil and Land Resources of Russia

31.2.1 Soil Resources

The total area of Russia is 17,098,242 km², extending nearly 3,000 km from north to south and nearly 7,000 km from west to east. These large dimensions allow for a diversity of environmental conditions throughout the country and many differences in land management practices. The environmental differences manifest themselves in different ways: from gradual geographic transitions to sharply pronounced contrasting changes.

The diversity of climatic and biotic conditions in Russia is the first group of factors determining the diversity of soils and soilscapes throughout the country. The second group of differentiating factors is related to the geological history of particular territories and their lithological composition; the chemical, mineralogical, and physical properties of surface deposits serving as parent materials; the hydrogeological conditions; and the modern dynamic processes shaping the surface topography.

The superposition of bioclimatic, litho-geomorphic, and historical–geological factors is responsible for the diversity and complexity of the soil cover in Russia.

General regularities of the soil cover of Russia and its regional differentiation are reflected in the most explicit form on the maps of SEZ.

The digital *Map of Soil-Ecological Zoning of Russia* (Figure 31.1, Table 31.1) was developed at the Faculty of Soil Science of Moscow State University on the basis of the *Soil Map of the Russian Federation* developed by the V.V. Dokuchaev Soil Science Institute under the supervision of V.M. Fridland (1988). SEZ is based on the taxonomic system with several hierarchical levels: (1) geographic belts and (2) soil-bioclimatic areas. These levels are further subdivided into mountainous and plain areas. Plain areas comprise (1) plain soil zones (subzones), (2) plain soil provinces, (3) plain soil regions, and (4) plain soil districts; and mountainous areas comprise (1) mountain soil provinces, (2) mountain soil regions, and (3) mountain soil subregions or mountain soil districts (not obligatory level).

The types of Russian soils are numerous, and extend from Arctic soils located on islands in the Arctic Ocean to Brown Semidesert soils of the Caspian Lowland, as well as Subtropical Yellow Ferrallitic soils from the foothills of the Caucasus in the south. Soils of the taiga zone occupy more than 50% of Russia, including Podzols and Gleyed Podzols (10.2%),

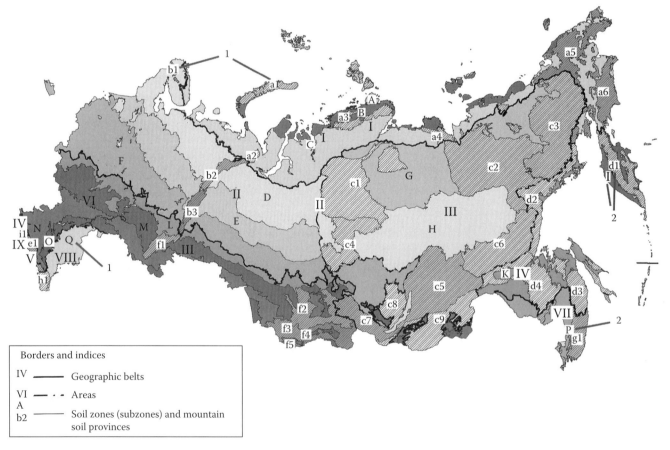

FIGURE 31.1 Scheme of the SEZ of Russia.

TABLE 31.1 SEZ of Russia

I—Polar Geographic Belt

I—Eurasian Soil-Bioclimatic Area

Plain Soil Zones (Subzones)	Mountain Soil Provinces
A—Arctic soils of Arctic zone	a1—Arctic islands
B—Arctic-tundra soils of subarctic subzone	a2—Polar–Urals
	a3—Byrranga
C—Gley tundra soils and podburs of subarctic subzone	a4—East-Siberian
	a5—Chukotka
	a6—Koryak–Taigonos

II—Boreal Geographic Belt

II—European–West-Siberian Taiga-Forest Soil-Bioclimatic Area

Plain Soil Zones (Subzones)	Mountain Soil Provinces
D—Gley podzolic soil, gleyzems, and podzols of northern-taiga	b1—Khibiny
	b2—North-Ural
E—Podzolic soils of middle taiga	b3—Central Ural
F—Soddy-podzolic soils of southern taiga	

III—East-Siberian Permafrost Taiga Soil-Bioclimatic Area

Plain Soil Zones (Subzones)	Mountain Soil Provinces
G—Peaty-muck cryogenic taiga gleyzems of northern taiga	c1—Anabar–Putoran
	c2—Verkhoyansk
H—High-humus peaty-muck nongleyed and pale cryogenic taiga soils of middle taiga	c3—Kolyma
	c4—Yenisei
	c5—Cisbaikal
	c6—Aldan
	c7—East-Sayan
	c8—Lena–Angara
	c9—Transbaikal

IV—Far-Eastern Taiga-Forest Soil-Bioclimatic Area

Plain Soil Zones (Subzones)	Mountain Soil Provinces
I—Forest volcanic soils	d1—Kamchatka
K—Raw-humus burozems and podzols	d2—Okhotsk
	d3—Sikhote-Alin–Sakhalin
	d4—Bureya

III—Subboreal Geographic Belt

V—Western Forest Soil-Bioclimatic Area of Brown Forest Soils

Plain Soil Zones (Subzones)	Mountain Soil Provinces
—	e1—North-Caucasian

VI—Central Broadleaved-Forest, Forest-Steppe, and Steppe Soil-Bioclimatic Area

Plain Soil Zones (Subzones)	Mountain Soil Provinces
L—Gray forest soils of deciduous forests	f1—South-Ural
	f2—Salair–Kuznetsk–Sayan
M—Podzolized, leached, and typical chernozems and gray forest soils of forest-steppe	f3—Altai
	f4—South-Sayan
	f5—South-Altai
N—Ordinary and southern chernozems of steppe	
O—Dark chestnut and chestnut soils of dry steppe	

TABLE 31.1 (continued) SEZ of Russia

VII—Eastern Forest Soil-Bioclimatic Area of Brown Forest Soils

Plain Soil Zones (Subzones)	Mountain Soil Provinces
P—Burozems and podzolic-burozemic soils of coniferous-broadleaved and broadleaved forests	g1—South-Sikhote-Alin

VIII—Semidesert Soil-Bioclimatic Area

Plain Soil Zones (Subzones)	Mountain Soil Provinces
Q—Light chestnut and brown soils of semidesert	h1—East-Caucasian

IV—Subtropical Geographic Belt

IX—Subtropical Humid-Forest Soil-Bioclimatic Area

Plain Soil Zones (Subzones)	Mountain Soil Provinces
—	i1—West-Caucasian

Source: Dobrovol'skii, G.V., and I.S. Urusevskaya (eds.). 2007. The map of soil-ecological zoning of Russia (digital version), scale 1:2.5 M. Moscow State University, Moscow, Russia [in Russian].

permafrost-affected Taiga and Gleyed Taiga soils (9%), Podburs (6.9%), Brown Taiga soils (2.8%), permafrost-affected Pale (Palevye) soils (3.1%), Gley-Podzolic and Podzolic soils (2.2%), Soddy-Podzolic soils (7.9%), and Bog-Podzolic and Peat Bog soils (11.8%). Smaller areas are occupied by Gray Forest soils (3%), Brown Forest soils (Burozems) (1.4%), Chernozems and Meadow-Chernozemic soils (7%), Chestnut (Kastanozemic), and Brown Semidesert Soils and their complexes with Solonetzes (3%). Mountain soil provinces occupy about one-third of the territory of Russia.

31.2.2 Land Resources

Land resources of Russia are classified with respect to land categories and land use types.

According to Article 7 of the Land Code of the Russian Federation (2001), seven land categories are specified in regard to land designation:

1. Agricultural lands
2. Settlement lands
3. Lands of industry, power generation, transport, communications, radio broadcasting, television, and information media; lands for ensuring spacecraft launches; lands of defense and security, and lands of other special designation
4. Lands of specially protected territories
5. Forest lands
6. Water-protection lands
7. Reserved lands

Data on the areas of different land categories in Russia for 2007 (Rosnedvizhimost', 2008) are given in Table 31.2. A majority of the country (about 65%) is classified as forest land. Agricultural

TABLE 31.2 Distribution of Russian Lands by Land Use Categories (as of January 1, 2008)

Land Category	Thousand Hectares	Percentage of the Total Area
Agricultural lands	403177.0	23.58
Settlement lands	19258.5	1.13
Lands of industry, electric power, transport, communications, radio broadcasting, television, and information and lands for ensuring outer space activity, lands of defense and security, and lands of other special designation	16687.4	0.98
Lands of specially protected territories and objects	34393.8	2.01
Forest lands	1104975.9	64.62
Water-protective lands	27942.3	1.63
Reserved lands	103389.3	6.05
Total	1709824.2	100

Source: Data from Rosnedvizhimost'. 2008. National report on the state and use of land in the Russian Federation in 2007. Federal Cadaster Center, Moscow, Russia [in Russian].

TABLE 31.3 Distribution of Russian Lands by Land Use Types (as of January 1, 2008)

Land Type	Thousand Hectares	Percentage of the Total Area
Agricultural lands, including	220567.9	12.90
Plowlands	121573.5	
Fallow	5105.7	
Perennial plantations	1794.2	
Hayfields	24004.1	
Pastures	68090.4	
Forest lands	870761.2	50.93
Forests beyond the category of forest lands	26594.5	1.56
Lands under water	72049.8	4.21
Lands under construction	5604.9	0.33
Lands under roads	7937.8	0.46
Bogs	152936.9	8.94
Disturbed lands	1107.5	0.07
Other lands	352263.7	20.60
Total	1709824.2	100

Source: Data from Rosnedvizhimost'. 2008. National report on the state and use of land in the Russian Federation in 2007. Federal Cadaster Center, Moscow, Russia [in Russian].

TABLE 31.4 Soil Cover of Agricultural Lands in the Russian Federation

Soil Types	Percentage in the Total Area of	
	Agricultural Land	Plowland
Podzolic and soddy-podzolic soils	12.3	14.7
Soddy and soddy-calcareous soils	2.4	1.4
Gray and brown forest soils	11.8	14.9
Chernozems, including	42.9	52.6
Leached chernozems	10.5	14.7
Ordinary chernozems	11.4	15.1
Chestnut soils	12.8	10.6
Solonetzes, solonchaks, solods	7.0	3.4
Floodplain soils (alluvial)	4.9	0.6
Other soil types	5.9	1.8
Total	100	100

Source: Romanenko, G.A., N.V. Komov, and A.I. Tyutyunnikov. 1996. Land resources of Russia and the efficiency of their use. Rossel'khozakademiya, Moscow, Russia [in Russian].

lands occupy more than 23% of Russia's territory, while specially protected territories occupy 2% of the area of Russia.

Land use type is defined as an area regularly used for specific economic purposes and having specific natural, historical, and other properties. In accordance with current economic use, all the lands are subdivided into two major groups: (1) agricultural lands, that is, lands actually or potentially used for agricultural production, and (2) nonagricultural lands, that is, other lands not used in agriculture.

Agricultural lands include the following land use types:

1. Plowland
2. Idle land (fallow, temporarily abandoned land)
3. Perennial plantation
4. Hayfield
5. Rangeland (pasture)

Nonagricultural lands are subdivided into the following land use types:

1. Land subjected to reclamation measures
2. Land under forest
3. Land under woody-shrub vegetation
4. Land under water bodies
5. Land under bogs
6. Land under construction
7. Land under roads
8. Disturbed land
9. Other lands

Data on land use in Russia (as on January 1, 2008) are given in Table 31.3.

Only about 13% of the country is actually used for agriculture (land use types 1–5). More than 50% is considered forest land. Other lands not used for agricultural purposes compose one-fifth of the total land reserve. These are mainly glaciers; lands occupied by landslides, slumps, and ravines; tundras; windblown sands; landfills; and dumps.

Table 31.4 contains data on the percentage of major soil types in the total agricultural, as well as the plow land of Russia.

Nearly 75% of Russia lies in the cold zone and in the mountainous regions with an insufficient heat supply. Only focal crop farming is possible in these areas. Rangelands are also limited in these regions.

A combination of sufficient heat supply (with the accumulated mean daily temperatures above 10°C of more than 2500 degree days) and sufficient moistening (with the precipitation to the potential evapotranspiration ratio of more than 0.75) is only observed on about 1% of Russian land.

Territories with optimum combinations of high heat supply and sufficient moisture are virtually absent in Russia.

31.3 History of Collection of Soil Information in Russia

31.3.1 History of Soil Mapping

Soil mapping in Russia has a long history. The first estimates of soil resources were based on the written descriptions of real estate prepared by local officials in the time of Ivan the Terrible in the sixteenth century. The first soil map of European Russia on a scale of 1:8 400,000 was compiled and edited by academician N.S. Veselovsky on the basis of information sent by local land committees; it was published in the *Economic-Statistical Atlas of European Russia* in 1851. Despite the schematic character of this map, its publication was a noteworthy milestone in the study of soil resources of the country.

In 1879, a new soil map of European Russia on a scale of 1:2,520,000 was published. This map was developed by a statistician V.I. Chaslavsky on the basis of completed questionnaires received from local land committees. After the death of Chaslavsky, V.V. Dokuchaev was entrusted with the preparation of explanatory notes to this map. They were published in 1879 under the title *The Cartography of Russian Soils*. In this work, the basic principles of pedology and soil cartography were outlined by Dokuchaev for the first time. Field studies on land assessment for a "better justified taxation" were performed by Dokuchaev and his team in the 1880s and 1890s. With other special soil survey works, they served as the basis for the first soil map of European Russia based on factual results of soil surveys and compiled on a scale of 1:2,520,000 by N.M. Sibirtsev, G.I. Tanfil'ev, and A.R. Ferkhmin under the supervision of V.V. Dokuchaev in 1900.

A new period in the study of soil resources began in the 1920s. Large-scale soil surveys of agricultural territories were organized with the aim of providing the cartographic basis for rational land use planning, crop rotation, agrochemical service, and soil reclamation.

The scales of Russian soil maps are subdivided into detailed (1:5000 and larger), large-scale (1:10,000–1:50,000), medium-scale (1:100,000–1:300,000), small-scale (1:500,000–1:1,000,000), and general (smaller than 1:1 M) maps.

Detailed soil maps have been developed for experimental fields and strain-testing stations. Large-scale soil maps are compiled for land use planning and land reclamation purposes at the level of particular farms, while medium-scale (sometimes, large-scale) soil maps are used for agricultural and forestry planning at the levels of administrative districts, oblasts, and regions. Small-scale and general soil maps are used for strategic economic planning at the federal level as well as for educational purposes. Soil maps of different scales are used to assess soil resources and land quality, and to perform soil zoning for various applied and scientific purposes.

Since the 1960s, the data provided by aerial surveys and satellite imagery have been applied to large-scale soil mapping. Their application has made it possible to improve the accuracy and objectivity of mapping large areas of soil. Large-scale soil maps, developed by local research institutes for land surveys and land management, have been supplied with explanatory notes containing the descriptions of representative soil pits, analytical data on the soil properties, and recommendations on the optimum use of soil resources. By the 1990s, three to four rounds of large-scale soil surveys were performed for agricultural lands of Russia.

In the 1970s and 1980s, zonal institutes for land management and land planning initiated the work of compiling medium-scale soil maps for the administrative regions and oblasts of Russia. These maps were developed based on large-scale soil maps. In some cases, new remotely sensed data were used for their compilation. The legends to these maps are rather detailed (normally about 150 mapping units are shown on them). Inset maps show the agricultural zoning of the territory as well as soil erosion and other hazardous processes. In general, the degree of detail on such maps is much higher than that on medium-scale soil maps developed in the first half of the twentieth century. Recently, a new methodology for compiling medium-scale soil maps with the use of GIS technologies has been prepared by the V.V. Dokuchaev Soil Science Institute (Simakova, 2008). Regional soil maps on the scales of 1:50,000–1:100,000 have also been developed by various research institutes for about a half of the administrative regions of Russia (Simakova and Andronikov, 1999). Unfortunately, most of them have not been published and are stored in various archives.

The accumulated data have also been generalized in a series of general soil maps. These maps include the Soil Map of the Soviet Union (1:4 M) developed by N.N. Rozov and E.V. Lobova with I.P. Gerasimov as editor in 1954, soil maps included in the Physico-Geographic Atlas of the World (Ed. I.P. Gerasimov, 1964), Soil Map of the World (1:15 M) developed by Glazovskaya and Fridland (1982), and others.

In the 1930s, L.I. Prasolov initiated the development of the State Soil Map of the Soviet Union on a scale of 1:1 M. This work was resumed in 1949. Presently, this is the most detailed soil map for the entire territory occupied by Russia and the FSU countries, as well as the main reference for assessing soil resources at the federal level. This map developed by the V.V. Dokuchaev Soil Science Institute synthesizes knowledge on the genesis and geography of soils gained by Russian pedologists in the second half of the twentieth century.

In the past 60 years, 111 separate pages of this map have been published. They encompass the territory of all the republics of the former Soviet Union. For Russia, 83 pages of this map have been published, mainly for the agricultural regions. The total list of soils and soil complexes shown on separate map pages exceeds 1000 mapping units. The work on unification of the legend to the map and harmonization of separate pages is in progress.

In 1988, the Soil Map of the Russian Soviet Federal Socialist Republic (*Soil Map of the Russian Federation*) on a scale of 1:2.5 M,

FIGURE 31.2 (See color insert.) Soil map of the Russian Federation.

was published. This map was created in the V.V. Dokuchaev Soil Science Institute on the basis of data provided by a large team of scientists from various research and educational institutes in Russia. This project was carried out under the supervision of V.M. Fridland. It consists of 16 pages (Figure 31.2) and presents the most detailed information on the soil cover of the entire Russia in published form.

In the 1960s and 1970s, important work on the generalization of soil maps and the development of soil geographic and soil-agrochemical zoning maps was performed. The *Soil Geographic Zoning of the Soviet Union* (1962), *Soil-Agricultural Zoning of the Soviet Union* (1975), the *Soil-Agrochemical Zoning of the Soviet Union* (1976), the *Map of Soil Geographic Zoning of the Soviet Union* (1983), and the recently developed digital *Map of Soil-Ecological Zoning of Russia* (edited by Dobrovol'skii and Urusevskaya (2007)) are the major accomplishments of this effort.

31.3.2 History of Soil Information Science and the Development of Digital Soil Mapping in Russia

The application of modern informational tools to soil studies began in the 1970s, with the appearance of the first data banks and their structural and functional components (databases) (Rozhkov, 2002). By that time, mathematical methods had actively been applied by soil scientists and agrochemists. Data archives on notched cards became prototypes of data banks (Fridland and Il'ina, 1972). The first steps in the development of information technology in soil science (or soil informatics as referred to by B.G. Rozanov) were made at the V.V. Dokuchaev Soil Science Institute.

In 1980, a complete system of soil classification and coding (SCC) was created. The final description contained five volumes and included all known soil characteristics and methods of their determination, a classification table, and the forms of data input. Being a formalized language of soil description, this system was designed for the development of a national bank of soil data with the application of a commercial system of database management INES for ES-1020 computers. This language was used in the creation of the Automated Information Search System (AISS) SOIL.

In order to improve the classification system, its basic principles with abbreviated forms for soil description and with a system of data coding were published (Rozhkov et al., 1982). Later on, the concept of the information base of soil classification was suggested (Shishov et al., 1985), and a special *Guide to Soil Description in the Soil Classification Database System* was published (Shishov, 1986). In accordance with the suggested descriptive structure, the information system MERON with formalized descriptions of individual soil types was created (Rozhkov and Stolbovoi, 1988).

By the early 1990s, several information systems and separate databases were developed by the V.V. Dokuchaev Soil Science Institute. With these as a base, the bank of soil fertility models was created (Frid, 1999).

In the 1990s, work on applying geographic information systems to soil science was initiated by the V.V. Dokuchaev Soil Science Institute using EPPL7 (U.S.) software. An original GIS envelope called LESSA was developed by V.B. Vagner. Based on LESSA, and with the support of the Russian Foundation for Basic Research, the first bank of maps and their related attribute databases was shaped. In 4 years, an ecological soil atlas was developed in the GIS environment.

In 1995, the Dokuchaev Soil Science Institute purchased the world-standard program ARC/INFO and the Digital Chart of the World (DCW) (1:1 M) showing relief, hydrographic network,

TABLE 31.5 Digital Cartographic Materials on the Territory of Russia

Maps and Schematic Maps
Administrative division of Russia
Soil map
Soil-geographic zoning
Accumulated daily air temperatures above 10°
Number of days with the mean daily air temperature above 10°
Duration of frostless period
Climatic continentality
Accumulated daily soil temperatures above 10° at the depth of 20 cm
Annual amplitude of soil temperatures at the depth of 20 cm
Mean annual precipitation
Precipitation during the warm period (April–October)
Annual potential evaporation from the soil surface
Annual difference between precipitation and potential evaporation
Bulk moistening of the territory
Soil-forming potential of the climate (as estimated for reference zonal soils)
Soil-forming potential of the climate for humus accumulation and mineral transformation
Vegetation
Forests of the Soviet Union
Net phytomass production for actual and restored (natural) phytocenoses
Phytomass reserves for actual and restored (natural) vegetation
Dead organic mass (mortmass) of ecosystems
Capacity of the biological turnover of ash elements in natural phytocenoses
Ecological functions of the vegetation cover of Russia
Geomorphologic zoning
Age, genesis, and composition of surface deposits
Texture of surface deposits
Bulk calcium content in surface deposits
Parent materials
Texture and petrographic-mineralogical and chemical compositions of soils
Age of soils and types of soil rejuvenation
Soil-geological countries
Soil cover patterns
Geochemical landscapes
Pedodiversity
Soil temperature regimes
Pools of organic matter in terrestrial ecosystems
Types of humus in soil humus horizons
Contents of organic carbon in soil humus horizons
Soil quality estimates for soil regions
Pools of organic carbon in the 1-m-deep soil layer
CO_2 emission from soils
Sustainability of soil functioning
Soil capacity for self-purification from technogenic substances
Soil capacity for oxidation
Soil sorption capacity
Budget of organic carbon in biomes
Budget of organic compounds of nitrogen in biomes
Soil erosion
Erosion control measures
Zoning of the territory with respect to the types of soil cover changes upon oil extraction and transportation
Natural-agricultural zoning of Russia
Boundaries of specially protected natural territories of the federal level
Soils of specially protected natural territories
Major soil types in the reserves and national parks of Russia

and infrastructures. The Soil Map of the Russian Federation and the Landscape Map of the Soviet Union (1:2.5 M scale) were some of the first to be digitized. Thus, the first serious soil GIS at the federal level appeared. The attribute databases of these maps are being updated constantly. The cartographic base of the Dokuchaev Soil Science Institute was registered in the international network GRID UNEP. Information on some of the maps developed by the institute can be found on the Web site http://agro.geonet.ru.

Representatives of the V.V. Dokuchaev Soil Science Institute (V.S. Stolbovoi and E.N. Rudneva) participated in the international project on the Global Assessment of Human-Induced Soil Degradation (GLASOD). This work was continued from 1997 to 2000 within the framework of the Soil and Terrain Vulnerability in Central and East Europe (SOVEUR) project aimed at the creation of GIS for the assessment and prognostic modeling of different types of soil degradation in this vast region.

An important issue was the integration of the Russian soil classification system into the soil database of the European Community (Stolbovoi et al., 2001). This step was intended to facilitate the creation of a unified policy regarding land use and soil monitoring. Russian classification was linked to the World Reference Base on Soil Resources. The soil map of Russia was also interpreted in terms of the FAO legend to the Soil Map of the World. The Dokuchaev Soil Science Institute suggested that the World Reference Database for Soil Resources should be organized as an information system in 1990.

Active work on the creation of new digital maps, as well as the use of cartographic and statistical databases for the assessment of soil resources, is being performed by the Faculty of Soil Science and by the Institute of Ecological Soil Science of Moscow State University. Soil maps of Russia on the scales of 1:25 M, 1:22 M, and 1:15 M and a digital version of the Map of Soil Ecological Zoning of Russia on a scale of 1:2.5 M have been developed.

The geographic information system has been created using the MapInfo format; it includes differently scaled maps for the entire territory as well as for separate regions of Russia. Some of the initial sources of cartographic information used in this work were published materials, unpublished original materials, and specially purchased digitized maps. Some of the digitized cartographic materials available on different scales for the entire territory of Russia are listed in Table 31.5.

In recent years, geographic information systems have been extensively applied in Russian agriculture. Since 2006, the Computing Center of the Ministry of Agriculture has been involved in the development of a remote-sensing system for the monitoring of agricultural lands. This system's objective is to integrate previously accumulated archive data and recently obtained data to refine the inventory of agricultural lands, to evaluate the efficiency of their use, and to predict crop yields. This system includes the federal and regional geographic information systems arranged according to the hierarchical principle: from the federal level to the level of local farms (specially selected representative farms) and fields. This process is based on digitized cartographic data in the ArcGIS format and satellite imagery subjected to continuous renewal (http://www.gisa.ru/pd_1_09.html).

31.4 Soil Geographic Database of Russia

The development of the SGDB of Russia is aimed at providing the scientific basis for the federal strategy of sustainable land use, soil monitoring, and soil conservation (Shoba et al., 2008).

Thus, the major objectives of the SGDB are as follows:

- The inventory and formalization of data on Russian soils. Most information on soil profiles is presently kept as hard copies—published articles, monographs, collections of papers, dissertations, and field records. These data should be unified and prepared for collective use.
- Information support of research projects and educational programs.
- The integration of Russia into a unified soil-information space of the European Community and the world and in various global and regional soil programs.

The main principles of SGDB development and functioning are as follows:

- The integration of soil information is based on digital maps using modern GIS technologies.
- The base map scale is 1:2.5 M. This scale is selected because this is the largest scale of digitized soil map for the entire country developed on the basis of a unified legend and common methodological principles.
- Openness of the database and the possibility for its further augmentation.
- Availability of the database.
- Involvement of a wide range of specialists in soil science and related disciplines possessing original information on the morphological and physicochemical properties of representative georeferenced soil profiles to be included in the SGDB.
- Online data transfer via the Internet.
- Offline administration and edition of input materials (for ensuring data security and quality).
- Involvement of experts in the work with input materials from separate contributors.

The main blocks of the SGDB are the Geographic Database and the Specialized Attribute Database. A relational database management system is used for data storage and processing.

31.4.1 Cartographic Block: Digital Maps

The cartographic base of the SGDB—the Geographic Information Soil Database—consists of two digitized maps in the MapInfo format. COVERAGE 1 is a digital map uniting the Soil Map of the Russian Federation on a scale of 1:2.5 M edited by V.M. Fridland (1988) (corrected digital version, 2007) and the Map of Soil-Ecological Zoning of Russia on the same scale edited by Dobrovol'skii and Urusevskaya (2007). COVERAGE 2 is the digital map of the administrative division of Russia on a scale of 1:1 M (2007).

31.4.1.1 Soil Map of the Russian Federation, 1:2.5 M Scale

The digital version of this map consists of 25,711 polygons (soil delineations). Each polygon contains information on the soil cover (color) and parent materials (hatching). The legend to the digitized map includes 200 names of individual soil units, 70 names of different soil complexes, 5 names for nonsoil formations, and 30 legend units for textural classes and petrographic composition (for hard rocks) of soil parent materials. Correlation of the Legend to the Soil Map of the Russian Federation with international and foreign soil classification systems can be found in Fridland (1982), Shishov et al. (1985), Rozhkov et al. (1990), Stolbovoi and Sheremet (1995), Stolbovoi (2000), Stolbovoi and Sheremet (2000), and CD-ROM Land Resources of Russia (2002).

A given polygon may include up to four soil units (or soil complexes). The main soil (soil complex) unit occupies 100% of the polygon area in the absence of accompanying soils (complexes), 75% in the presence of one accompanying soil (complex), 65% in the presence of two accompanying soils (complexes), and 55% in the presence of three accompanying soils (complexes). The accompanying soils are considered to occupy equal areas. In soil complexes, the percentages of component soils are not calculated. Parent materials in soil polygons may be represented by one (major) or two (major and accompanying) units.

31.4.1.2 Digital Map of Soil-Ecological Zoning of Russia, 1:2.5 M Scale

In the digital map of SEZ, 1377 polygons are separated (Figure 31.3).

The digital version of this map contains information about soil cover and the factors of its differentiation at different levels. Thus, at the soil district level, information about soils and their texture is given; at the soil region level, information about the genetic types of relief, parent materials, land use pattern, and soil quality classes is provided. For larger typological units—soil provinces—parameters of atmospheric and soil regimes are given for plain territories and the patterns of the vertical soil zonality are indicated for mountainous territories.

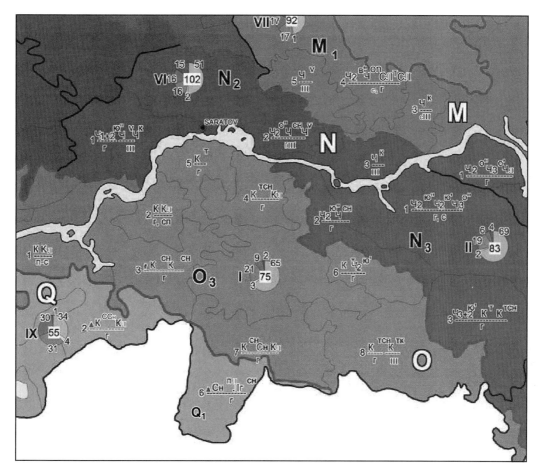

FIGURE 31.3 Fragment of the map of SEZ of Russia.

31.4.1.3 Soil Map Combined with the Map of Soil-Ecological Zoning of Russia, 1:2.5 M Scale: COVERAGE 1

Digitized versions of the soil map and the map of SEZ on a scale of 1:2.5 M are combined into COVERAGE 1. The number of polygons on this integral map reaches 32,605. Each polygon contains information from both maps (Table 31.6).

31.4.1.4 Digital Scheme of the Administrative Division of the Russian Federation, 1:1 M Scale: COVERAGE 2

At present, there are 83 subjects of the Russian Federation, including oblasts, krais (regions), republics, autonomous areas, autonomous oblast, and two cities. They are combined into eight federal okrugs: Tsentral'nyi (Central), Severo-zapadnyi (Northwestern), Yuzhnyi (Southern), Severo-Kavkazskii (North Caucasian), Privolzhskii (Volga), Ural'skii (Ural), Sibirskii (Siberian), and Dal'nevostochnyi (Far Eastern) (Figure 31.4).

COVERAGE 2 includes 2394 polygons representing the administrative division of Russia. It contains data of the Federal Statistical Survey of land resources performed by the Federal Agency for Realty Cadaster for January 1, 2006 (Table 31.7).

31.4.1.5 Additional Digital Cartographic Materials

Currently, work on inclusion of additional materials in the geographic database is being performed. In particular, these are characteristics of relief and parameters of soil temperature regimes.

Digital elevation model (DEM) with a horizontal grid spacing of 30 as (approximately 1 km) is taken from the Land Resources of Russia CD-ROM (2002).

Long-term soil temperature data (Table 31.8) recorded at 370 weather stations in 1960–2000 were provided for the database by D.A. Gilichinskii (the Laboratory of Cryogenic Soils, Institute of Physicochemical and Biological Problems of Soil Science, Russian Academy of Sciences, Pushchino).

The large volume of information concerning soils, their properties, functions, soil factors of soil formation, soil cover patterns, and land use is also available from digitized maps on different scales included in the GIS of Russia (Table 31.5).

TABLE 31.6 Information Structure of the SGDB. COVERAGE 1

No.	Information Description	Source of Information	Level of Information
1	Identification field	COVERAGE 1	COVERAGE 1 polygons
2	Major soil (soil complex)	Soil map	Soil map polygons
3	First accompanying soil (soil complex)	Soil map	Soil map polygons
4	Second accompanying soil (soil complex)	Soil map	Soil map polygons
5	Third accompanying soil (soil complex)	Soil map	Soil map polygons
6	Soil-forming rocks	Soil map	Soil map polygons
7	Soil-forming rocks	Soil map	Soil map polygons
8	Soil cover pattern	SEZ map	SEZ map polygons
9	Soil-forming rocks	SEZ map	SEZ map polygons
10	Genetic relief types	SEZ map	Districts
11	Litho-geomorphic characteristics	SEZ map	Districts
12	Land use pattern	SEZ map	Districts
13	Soil cover quality class	SEZ map	Districts
14	Vertical zonality pattern	SEZ map	Mountain provinces
15	Average July temperature (°C)	SEZ map	Plain provinces
16	Accumulated daily air temperatures >10°C		
17	Duration of period with temperatures >10°C (days)		
18	Duration of frostless period (days)		
19	Mean January temperature (°C)		
20	Mean annual precipitation (mm)		
21	Annual humidity factor (Vysotskii–Ivanov)		
22	Accumulated daily soil temperatures >10°C at the depth of 20 cm		
23	Penetration depth of temperatures >10°C into the soil (cm)		
24	Period with soil temperatures >10°C at the depth of 20 cm (months)		
25	Accumulated daily soil temperatures <0°C at the depth of 20 cm		
26	Penetration depth of temperatures <0°C into the soil (cm)		
27	Predominant water regime		

31.4.2 Attribute Block: Soil Profile Database

31.4.2.1 Contents and Structure of Soil Profile Database

The database of representative soil profiles is an attribute part of the SGDB. Representative soil profiles should be provided with information on their geographical location and with a complete set of data on soil morphology and analytic properties to be included in the database.

The main component of the SGDB database is a soil profile with a set of soil horizons characterized by attribute data. Necessary information on the properties and composition of soils is obtained via selecting of representative soil profiles characterizing the main soil types shown on the Soil Map of the Russian Federation (1:2.5 M scale). A series of representative soil profiles for a given soil type makes it possible to calculate averaged characteristics for this soil type. Information on the regional features of soil cover patterns, vegetation and relief conditions, soil-forming rocks, and climatic parameters is taken from the Geographical Database.

The systematization of data on the selected reference soil profiles implies a unified form of data presentation. For this purpose, classifiers (systematized lists of parameters) were developed based on existing concepts of soil morphology and basic scales used in the current descriptions of soil profiles. The form and format of data presentation are described in detail in the attachment to the database. The necessity for including both the results of recent studies and the vast archival materials into the database is taken into consideration. Most soil attributes are ranked using specially developed scales (the database envisages the description of each attribute and its range). The developed systematized list of parameters includes two large sections:

1. Parameters characterizing the location of soil profiles and the morphological characteristics of soils
2. Parameters characterizing the physical, chemical, physicochemical, and other soil properties

The database has a hierarchical structure ensuring soil description at several levels: SOIL–PIT–PROFILE–HORIZON–SAMPLE (Table 31.9).

Soil Geographic Database of Russia

FIGURE 31.4 Scheme of the administrative division of Russia.

TABLE 31.7 Information Structure of the SGDB. COVERAGE 2

No.	Information Description
1	Identification field
2	Total area
3	Total agricultural land
4	Including plowlands
5	Including fallow
6	Including perennial plantations
7	Including hayfields
8	Including pastures
9	Land of reclamation construction and fertility restoration
10	Total forest lands
11	Including forest lands
12	Including forest-free lands
13	Tree–shrub vegetation beyond forest lands
14	Including shelterbelts
15	Land under water
16	Total land under construction
17	Including lands under industrial structures
18	Total lands under roads
19	Including lands under ground roads
20	Bogs
21	Disturbed lands
22	Total other lands
23	Including landfills and dumps
24	Including sands
25	Including ravines
26	Including lands with tundra vegetation not entered in other lands
27	Including other lands
28	Reindeer pastures in all lands

TABLE 31.8 Meteorological Archive Data

No.	Information Description
1	Penetration depth of 0°C into the soil for months and years (thawing)
2	Penetration depth of 0°C into the soil for months and years (freezing)
3	Soil freezing depth for months and years
4	The number of days per month with soil temperature ≤0°C for standard depths
5	The number of days per month with soil temperature ≤0°C for non-standard depths
6	Dates of the first and last soil frosts for standard depths
7	Dates of the first and last soil frosts for nonstandard depths
8	Dates of the first and last soil surface frosts
9	Average maximum temperatures of soil surface for months and year
10	Average minimum temperatures of soil surface for months and year
11	Average monthly and annual soil surface temperatures
12	Average monthly soil temperature for standard depths (bent thermometers)
13	Average monthly soil temperature for nonstandard depths (bent thermometers)
14	Average monthly soil temperature for standard depths (drawing thermometers)
15	Average monthly soil temperature for nonstandard depths (drawing thermometers)
16	Average annual soil temperature for standard depths
17	Average annual soil temperature for nonstandard depths

TABLE 31.9 Hierarchical Model of Soil Description

Soil	Classification Position	
PIT	Environmental conditions	Date
	Relief	Information source
	Vegetation	Georeference
	Groundwater table	Coordinates
	Soil-forming rocks	Height above sea level
		Economic use
		Erosion signs
Profile	Profile type	
	Disturbance	
	Number of horizons	
Horizon	Horizon description	
	Index	
	Small additional index	
	Water content	
	Color	
	Texture	
	Structure	
	Density	
	Consistence	
	Effervescence with HCl	
	Boundaries and transition, thickness	
	Description of soil morphological elements	
	Roots	
	Mycelium	
	Alga film	
	Plant residues	
	Zoogenic elements	
	Pedons	
	Hard rock fragments	
	Inclusions	
	Neoformations (pedofeatures)	
	Pores	
Sample	Chemical properties	
	Physical properties	

TABLE 31.10 Structure of Soil Attribute Database

Soil—Pit

Description of environmental conditions

Relief	Macrorelief
	Mesorelief
	Microrelief
	Slope aspect
	Slope gradient (degrees) of the pit site
	Slope shape
Vegetation	Association name
	Stratification and species composition, projective cover
Groundwater	Groundwater level
	Groundwater salinity
Soil-forming rocks	Genetic type of rocks
	Weathering level of soil-forming rocks
	Hard rock outcrops
General information	Date
	Information source, author
	Administrative-territorial location
	Coordinates
	Height above sea level
Economic use	
Erosion signs	Erosion type
	Erosion intensity

Profile—Horizon

Profile description

Profile disturbance	Disturbance source
	Disturbance degree
Profile type	
Total number of genetic horizons	

Description of genetic horizons

Horizon index
Small additional index
Water content
Color

Texture	Additional characterization of texture
	Stone content

Mineral matrix composition
Structure
Density
Consistence

Effervescence with HCl	Effervescence depth
	Effervescence intensity
Boundaries and thickness	Boundary shapes
	Transition type
	Horizon and sampling depths
	Horizon thickness

Description of soil morphological elements

Roots
Mycelium
Alga film on soil surface

The set of semantic soil data is given in Table 31.10.

At the SOIL level, the classification position of the soil is given. For the purposes of correlation between the existing classifications, the most complete descriptions of soils are given within the frameworks of current national and world soil classification systems.

The description of a pit includes the characterization of the pit site, the main factors of soil formation (including relief, parent materials, and vegetation), and the source of information. The number of genetic horizons, the degree of disturbance of the profile, and its genetic type are also indicated. The HORIZON level includes a description of the most common morphological properties of the horizons, as well as the description of soil morphological elements at the next hierarchical level.

At the SAMPLE level, data on the physicochemical properties of soils are systematized. In this section, the values of the

TABLE 31.10 (continued) Structure of Soil Attribute Database

Plant residues
Zoogenic elements
Pedons
Hard rock fragments
Inclusions
Neoformations
Pores

Sample—Chemical Properties

Chemical composition of soils

Elemental composition of mineral soil component (bulk composition)	Ignition loss
	Silicon
	Sesquioxides
	Aluminum
	Iron
	Calcium
	Magnesium
	Titanium
	Manganese
	Phosphorus
	Sulfur
	Potassium
	Sodium
Elemental composition of organic soil component	Organic carbon
	Humus
	Nitrogen (total)
	C:N ratio
Material soil composition	Alkaline earth carbonates
	Gypsum
	Ion composition of readily soluble salts soil salinization parameters
Group (fractional) composition of chemical element compounds in soils	Group composition of iron compounds
	Group composition of aluminum compounds
	Group composition of phosphorus compounds
	Group composition of potassium compounds
	Group composition of microelement and heavy metal compounds
	Group and fractional humus composition

Mobility of chemical elements in soils

Mobility of phosphorus compounds (intensity factor)	Degree of phosphate mobility
	Reserve of mobile phosphorus compounds (capacity factor)
Mobile compounds of chemical elements	Mobile phosphorus compounds
	Mobile potassium compounds
	Mobile nitrogen compounds
	Mobile compounds of microelements and heavy metals

TABLE 31.10 (continued) Structure of Soil Attribute Database

Cation exchange capacity of soils

Cation exchange capacity
Total exchangeable bases
Composition (content) of exchangeable bases
Degree of base saturation

Acid–base properties of soils

Acidity	Actual acidity (pH_{water})
	Actual acidity (pH_{KCl})
	Exchangeable acidity
	Total acidity
Alkalinity	Water suspension pH
	Water-saturated paste pH
	Total alkalinity (Alk_{tot})
	Carbonate alkalinity
	Organic alkalinity
	Borate alkalinity
	Difference between the total alkalinity and the sum of calcium and magnesium (water extraction method)

Sample—Physical Properties

Particle-size composition
Aggregate composition
Microaggregate composition
Specific surface area
Soil density
Solid phase density
Density of pedons and aggregates
Porosity

Soil-hydrological constants	Hygroscopic moisture (HM)
	Maximum hygroscopic moisture (MHM)
	Field capacity (FC)
	Wilting point (WP)
	Total moisture capacity (TMC)
	Water permeability

attributes measured, the methods of their determination, and the measurement units (with due account for the currently and formerly used units) are presented.

Each position in the systematized list of parameters characterizing soil morphology, environmental conditions, and soil physicochemical characteristics have several levels of parameter detailing.

31.4.2.2 Soil-DB: Software for Filling in the Attribute Database on Soil Profiles

A special software, Soil-DB (version 1.0) has been developed for database management (Alyabina et al., 2009). This software is intended for the input of initial soil information by contributors to the SGDB and transmission to the central server of the system, where the collective SGDB is being created. Information on the SGDB, User's manual, and the Soil-DB software are available from the Web site http://db.soil.msu.ru.

Development of the system includes several stages. The first stage is registration of potential contributors. Registered contributors may fill in the cards of soil description using the suggested software tools and send these cards to the central server of the system via the internet. At the next stage, moderators of the system verify information received from the particular contributors and decide (with the help of experts) on the feasibility of its inclusion in the SGDB. The contributors are informed of this decision. In certain cases, not all the fields of the card concerning the classification position of the soil and its exact geographic coordinates are filled in by the contributor. These fields may be filled in by experts who link information about soil profiles to the geographic database.

The Soil-DB program, together with tens of built-in information lists, is rather compact and occupies less than 500 Kb of disc space. The files created with the use of this program are also small (2–3 Kb). This allows one to keep and launch the program from a flash card.

Filling in of Personal Data and Registration in the System

To register in the system, the registration form should be filled in (Figure 31.5a). Three fields of this form are obligatory for registration: (1) personal data (name/family name), (2) address, and (3) email. This information is necessary for contact between moderators of the database and the registered contributors, and is confidential. Registered persons acquire the status of official participants to the project and may contribute to the collective database. A participant may fill in the cards for soil description and send them to the central server of the system to supplement the database.

Information Units

The major information unit of the Soil-DB program is a card with the description of a given soil horizon (sample). Numerous separate cards can be organized by a user of the system into a card file. Any card can be supplemented and/or edited at any time by saving the results on the system. Because a considerable volume of information is in separate cards, it is subdivided into separate pages (thematic fragments) on the monitor. A transition from any given card to another card is accomplished with the help of a menu bar, from which necessary page names can be selected. Each page contains several rubricated fields represented either by a menu from which necessary items can be selected, or a window, in which necessary textual or numerical information can be filled in. Many ways of completing the fields are available in the program: keyboard typing, copy–paste procedure, choice from the list of variants (including multistep lists), and the input of nonformalized data that do not fit the developed data structure.

Pull-down menus are the main means of data input; these menus contain necessary lists of classifiers of soil properties, so that a user can select required information from the menu window without typing it. This approach makes the completion of the cards more efficient and allows one to avoid typing errors. There are simple and complex (hierarchical) pull-down menus. In a simple menu, only one window for the choice of appropriate information is opened. In a complex menu, two or three windows with hierarchically organized information lists are opened. Complex menus differ from simple menus in the font color and background color. The left panel (red font) is used to select the appropriate term for a general description of the macrorelief (e.g., plain territories), and the right panel (white font against blue background color) is used to specify the description (alluvial and ancient alluvial plains).

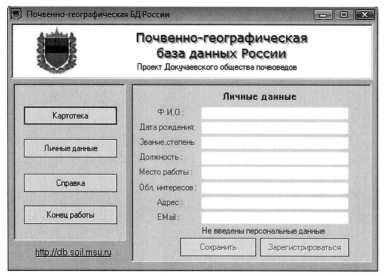

(a)

FIGURE 31.5 (a) Registration form

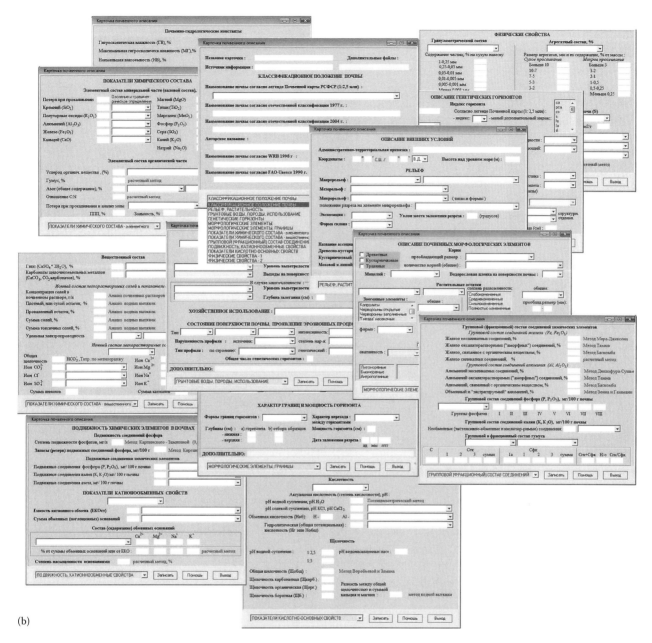

FIGURE 31.5 (continued) (b) Pages of the card file.

In some cases, several entries from the menu can be selected. For this purpose, multiple-selection menus are suggested. They consist of three–four visible information lines against the pink-colored background. Most of these menus have scroll bars allowing the selection of appropriate information.

In a blank (new) card, all the pull-down menus are initially empty (nothing is selected). If the required information for a given menu is absent, its window should be left empty. Upon the submission of the completed card to the central database, such empty windows are considered as a lack of data (no data). Upon the secondary opening of the earlier filled card, previously selected information is shown in all the menu fields. It can be changed at any time, and the last choice will be saved on the disk and opened during the next opening of the card.

Textual Fields

Some of the windows opened on card pages are designed for the entry of textual information (names, descriptions, etc.). Where necessary, the maximum length of the text is limited, though such limits are absent in most cases.

Numerical Fields

The windows designed for the entry of *integers* have certain limits for the number of digits. These limits are set up in order to avoid unfortunate misprints and errors because of the inadvertency of the users. With the same aim, only digits can be inserted into such windows; letters and symbols are not allowed. The windows designed for the entry of *fractional numbers* (e.g., for the values of some physical and chemical soil properties) do not have special limitations for the number of digits and for their format. However, only digits and the decimal sign (period) can be inserted into such windows.

Card Maintenance and Transmitting to the Server

After the launch of the program, a panel of the most frequently used (default) regime of card filling is opened. To begin the work, the blank card has to be created and saved as a file. Each card is a separate file stored in the same directory as the Soil-DB program.

The program ensures the following options for the work with the cards:

Card editing (menu bar "Edit"). After pressing this bar, a panel for data entry and editing of information is opened.

Card copying (menu bar "Copy"). A copy of the selected card is saved under another name. This option is useful upon the availability of analytical results in several duplicate cards, for the description of different horizons in the same soil profile, or for the description of several soil profiles within the same area, (for which a larger part of the information is identical, and only several information fields are different). In such cases, a copy of the first card is made, and then information in necessary fields is changed.

Menu bar "Delete" is used to delete unnecessary cards.

Menu bar "Send to the Database" is used for sending the card to the central server and to the moderator of this work. After the card is verified by the moderator, its information becomes available from the collective database.

Each soil profile's geographic location should be indicated. Coordinates of the pit (longitude and latitude) are given, and the radius of uncertainty (in meters or kilometers) should also be indicated. In case such information is unavailable (for archive data, literature sources, dissertations, etc.), approximate location in relation to a nearby settlement or other notable geographic feature is acceptable (e.g., 2 km to the northwest from Peskovka railway station).

The multipage format of the card assumes that not all the fields are obligatory. Some fields may remain empty in the case of missing corresponding information. Overall, the card consists of 13 pages (Figure 31.5b). The first six pages are designed for the description of the classification position of a soil, the environmental conditions in the area of described soil profile, and the soil morphology.

Pages 7–13 are designed for the description of soil chemical and physicochemical properties. For each characteristic, special fields for the numerical values of particular properties, measurement units, and the methods of their determination (to be selected from the pull-down menu) are provided.

The program also suggests fields for the input of nonformalized data. Thus, the "Additional Files" field on the first page of the soil profile description is used to indicate the names of files sent to the moderator of the system with the descriptions of soil attributes that are not included in the standard list. Also, each page of the standard soil descriptions ends with the line "Additional Information" against the yellow background color. This line should be filled in with any relevant information about the particular soil and not included in standard menus.

31.4.2.3 Development of the Attribute Database on Soil Profiles

The development of an attribute database on soil profiles was initiated by specialists from the Dokuchaev Soil Science Institute. Appropriate data are derived from published and archived sources, including the archive of dissertations at the Higher Attestation Commission of the Russian Federation. Representative soil profiles characterizing the soil cover of Russia are typically selected. An obligatory requirement to the data sources is the exact knowledge of the geographic position of a given soil profile, or its allocation to a given delineation on the Soil Map of the Russian Federation (1:2.5 M scale). The selected profile should also have a full morphological description and be supplied with analytical data on the main physical, physico-chemical, and chemical properties. The quality and completeness of these data are assessed by the experts.

Currently, the National Soil-Geographic Database of Russia includes data on 584 representative soil profiles with nearly 3000 samples from the genetic soil horizons.

Additionally, data on 688 representative soil profiles on the European part of Russia have been selected. At present, an expert evaluation of these data is performed, and the classification position of corresponding soils is verified. After completion of this work, this data will also be included in the Soil-Geographic Database. The search of representative soil profiles for the Asian part of Russia is still in progress.

31.4.3 Software Support of the SGDB

Work on filling in the SGDB is performed using cartographic data presented in the MapInfo format (MapInfo Professional, version 9.5 and earlier versions) and with attribute soil data in the Microsoft Office 2007 format; the database management on the server is ensured by the PHP, MySQL, and MapServer software (Windows platform).

The SGDB management will be ensured by several programs.

- Software for the local input of soil data (Soil-DB)
- Modulus of information input–output on the server
- Relational database management system on the server
- Program for administrating, editing, and conversion of the input information (TXT-format) into the format designed for storage and transfer of structured data (XML-format)
- Modulus of remote database query (search for necessary data from metadata) and data output via internet

At present, the first two software tools are ready. The third and fourth software tools are being developed. The work on database query software is at the stage of selection of the particular indices to be displayed on key fields.

Technical documentation to the program with a detailed description of the work of the system and with an instruction for filling in soil description cards can be obtained from the website devoted to the SGDB of Russia (http://db.soil.msu.ru).

31.5 Conclusions

At present, one of the challenges for Russian soil scientists is to ensure the legal status of soil as a natural resource via the preparation of the corresponding legislative act for its adoption by federal authorities. In this context, the inventory of soils of the country and the creation of the All-Russia SGDB as the scientific and information basis for the policy of sustainable land management are necessary. A computer-based inventory of extensive soil information linked with digitized soil maps and the use of modern GIS technologies will make it possible to create this database and to apply it for soil monitoring purposes and for the development of a unified system of soil management, certification, and conservation.

The development of the collective SGDB as an available information center for collecting, storage, and processing of soil information is in progress. This work is performed by specialists in soil science and related sciences. The SGDB of Russia should be compatible with the European Soil Database and the World Reference Base for Soil Resources. This will make it possible for Russian soil scientists to be integrated into the common European soil information system and to participate in the development of global soil programs.

Acknowledgments

This chapter summarizes the work of many people. The authors are particularly grateful to academician G.V. Dobrovol'skii for his moral support and valuable consultations; to Professor of the Department of Soil Chemistry of Moscow State University L.A. Vorob'eva, who developed the conceptual scheme of the database on the chemical properties of soils; and to program engineer V.V. Artyukhov, who contributed to the development of the software support of the SGDB of Russia.

References

Alyabina, I.O., V.V. Artyukhov, V.M. Kolesnikova, and S.A. Shoba. 2009. Soil-geographic database of Russia: Database management system soil-DB. Moscow State University, Moscow, Russia (deposited in VINITI August 3, 2009, No. 512-B2009) [in Russian].

Dobrovol'skii, G.V., and I.S. Urusevskaya (eds.). 2007. The map of soil-ecological zoning of Russia (digital version), scale 1:2.5 M. Moscow State University, Moscow, Russia [in Russian].

Frid, A.S. 1999. Annotated list of fertility models in the automated database PLOMOD. Dokuchaeva Soil Science Institute, Moscow, Russia [in Russian].

Fridland, V.M. (ed.). 1982. Major principles and elements of the basic soil classification and the program for its development. Dokuchaeva Soil Science Institute, Moscow, Russia [in Russian].

Fridland, V.M. (ed.). 1988. Soil map of the RSFSR. Scale 1:2.5 M. Gos. Uprav. Geodes. i Kartograf., Moscow, Russia [in Russian].

Fridland, V.M., and L.P. Il'ina. 1972. The punchcard model for the description of soil combinations and soil cover patterns, p. 195–201. In Soil combinations and their genesis. Nauka, Moscow, Russia [in Russian].

Gerasimov, I.P. (Ed.). 1964. Physico-Geographic Atlas of the World. Akad. Nauk SSSR–Gos. Uprav. Geodes. i Kartograf., Moscow, Russia [in Russian].

Glazovskaya, M.A. and V.M. Fridland. 1982. Soil Map of the World for the Institutes of Higher Education, 1 : 15 M Scale. Gos. Uprav. Geodes. i Kartograf., Moscow, Russia [in Russian].

Romanenko, G.A., N.V. Komov, and A.I. Tyutyunnikov. 1996. Land resources of Russia and the efficiency of their use. Rossel'khozakademiya, Moscow, Russia [in Russian].

Rosnedvizhimost'. 2008. National report on the state and use of land in the Russian Federation in 2007. Federal Cadaster Center, Moscow, Russia [in Russian].

Rozhkov, V.A. 2002. The development of soil informatics. Eur. Soil Sci. 35:761–769.

Rozhkov, V.A., V.M. Fridland, and S.V. Ovechkin. 1982. A system for the description and encoding of soil profiles. In V.M. Fridland (ed.) Major principles and elements of the basic soil classification and the program for its development. Dokuchaeva Soil Science Institute, Moscow, Russia [in Russian].

Rozhkov, V.A., and V.S. Stolbovoi. 1988. The development of classification of soils of the USSR using an automated system. Dokuchaeva Soil Science Institute, Moscow, Russia [in Russian].

Rozhkov, V.A., V.S. Stolbovoi, and L.L. Shishov. 1990. Expert system of soil classification, p. 123–128. In Problems of soil science. Nauka, Moscow, Russia [in Russian].

Rozov, N.N. and E.V. Lobova. 1954. Soil Map of the USSR, 1 : 4 M Scale (ed. by I.P. Gerasimov), Gos. Uprav. Geodes. i Kartograf., Moscow, Russia [in Russian].

Shishov, L.L. 1986. Methodological guidelines on soil description in the information database of soil classification. Dokuchaeva Soil Science Institute, Moscow, Russia [in Russian].

Shishov, L.L., V.A. Rozhkov, and V.S. Stolbovoi. 1985. Information database of soil classification. Pochvovedenie 9:9–20 [in Russian].

Shoba, S.A., V.S. Stolbovoi, I.O. Alyabina, and E.N. Molchanov. 2008. Soil-geographic database of Russia. Eur. Soil Sci. 41:1029–1036.

Simakova, M.S. (ed.). 2008. Manual on medium-scale soil mapping with GIS application. Dokuchaev Soil Science Institute, Moscow, Russia [in Russian].

Simakova, M.S., and V.L. Andronikov. 1999. Soil maps. In A.A. Lyutyi and N.N. Komedchikov (eds.) Cartographic materials in Russia (topographic and thematic maps). Institute Geografii RAN, Moscow, Russia [in Russian].

Stolbovoi, V. 2000. Soils of Russia: Correlated with the revised legend of the FAO soil map of the world and world reference base for soil resources. Research Report, RR-00-13. p. 112. International Institute of Applied Systems Analysis (IIASA), Laxenburg, Austria.

Stolbovoi, V., L. Montanarela, V. Medvedev et al. 2001. Integration of data on the soils of Russia, Belarus, Moldova, and Ukraine into the soil geographic database of the European community. Eur. Soil Sci. 34:687–703.

Stolbovoi, V., and I. McCallum. 2002. CD-ROM Land Resources of Russia, International Institute for Applied Systems Analysis and the Russian Academy of Sciences, Laxenburg, Austria.

Stolbovoi, V.S., and B.V. Sheremet. 1995. A new soil map of Russia compiled in the FAO system. Pochvovedenie 2:149–158 [in Russian].

Stolbovoi, V.S., and B.V. Sheremet. 2000. Correlation between the legends of the 1:2.5 M soil map of the Soviet Union and the FAO soil map of the world. Eur. Soil Sci. 33:239–248.

32
Soil Databases in Africa

D.G. Paterson
South African Agricultural Research Council

N.M. Mushia
South African Agricultural Research Council

32.1 Introduction 32-1
32.2 Background..32-1
32.3 Soil Characterization 32-1
32.4 Soil Databases..32-5
   Reconnaissance Scale • Detail Scale • Ultra-Detailed Scale
32.5 Soil Profile Databases 32-6
32.6 Database Interpretation and Utilization..32-6
32.7 Conclusion 32-6
References..32-7

32.1 Introduction

Africa has long been known as the "Dark Continent," due to such factors as the difficulties posed by travel across its unforgiving terrain. Even today, in the twenty-first century, because this situation persists in places, it has been said that more is known about the soils on the Moon or Mars than many parts of Africa. The colonization factor in Africa is well-known, but one of its consequences was that when a need for increased characterization and cataloguing of the natural resources of the various countries on the continent arose in the mid-twentieth century, these tasks were carried out largely by colonial powers or their representatives.

The main result of this was that, with few exceptions, soil maps and other data concerning the natural resources of Africa were collected and stored outside the continent and became difficult to trace and inventory. However, because there is now a much greater spirit of cooperation between both African countries and their former colonial (mainly European) partners, much of the information has been, or is currently being, incorporated into robust and readily available source databases.

32.2 Background

Africa covers an area of just over 30 million km² and has a population of 877 million, second only to Asia. The continent comprises 53 countries, with the largest in area being Sudan (2.505 million km²) and in population being Nigeria (149 million). The level of development varies greatly, being limited by topographical features such as the Sahara desert and the Congo River basin but being driven by the occurrence of mineral or other resources, which provided the incentive for early development, such as the gold and diamonds in South Africa, copper in Zambia, and, at a later stage, oil and gas in Nigeria.

32.3 Soil Characterization

About half of all the world's "tropical soils" lie in Africa and are associated with unique features such as high rainfall, warm to hot temperatures, intense weathering, and rapid leaching, to name but a few. These soils thus present significant challenges to those wishing to utilize them for food and fiber production.

Although the first soil survey was carried out in Madagascar in the late 1890s (van Ranst and Verdoodt, 2010) and several surveys of smaller areas, such as irrigation schemes in South Africa, were carried out in the 1920s and 1930s (MacVicar, 1980), the first significant studies of African soils originated from Russia (Schokalskaya, 1944) and America (Shantz and Marbut, 1923), both at scales of approximately 1:30 million. The first real impetus to improve on these maps came from the Inter-African Pedological Service, under the leadership of J.L. d'Hoore, which recommended production of a map at 1:5 million. The first edition of this map was produced in 1964 (d'Hoore, 1964), reflecting the dominance of French and Belgian soil scientists, who had been most active up until then in most parts of Central Africa. The level of information varied, but it was estimated that around half of the area was covered by high quality maps, while the remainder was covered by less precise maps or extrapolations (d'Hoore, 1964).

It is not easy to obtain information on the progress of soil mapping within Africa or information as to how much has been incorporated into one or other type of database, but information that could be obtained is presented in Table 32.1.

The existing soil information was incorporated into the first global soil map (FAO-UNESCO, 1981). However, due to its global coverage, as well as the need to correlate the many existing soil classification systems, the resolution of the map was low (1:5 million scale) and lacked much of the qualitative information on soil properties that would allow a soil map to be further interpreted.

TABLE 32.1 African Soil Information at Small (Reconnaissance) Scale

Country	Scale	Map Type	Area (km²)	Profile Classification	Soil Analyses	Application	Information Content	Reference
West Africa								
Cameroon	1:1 million	Surveyed areas					West Cameroon	Hawkins and Brunt (1965)
Cape Verde	1:200,000	Crusting soils				Erosion risk mapping	Volcanic soils, erosion	Tavares and Amiotte-Suchet (2007)
Ghana	1:250,000	Soil information system					Soil information system and land suitability mapping	Boateng et al. (1998)
	1:250,000–1:1.5 million	Ghana regions					Soil suitability, fertility, degradation, land use	Asiamah and Dwomo (1995)
	1:1.5 million	Great soil groups					Great soil groups	Brammer (1958)
Guinea	1:500,000	Reconnaissance soil map				Reconnaissance soil survey	Pedology	Diallo and Condé (1982)
Ivory coast	1:3.3 million	Pedology map				Soil survey		ORSTOM (1960)
Liberia	1:500,000	Land suitability				Land suitability	Land suitability	van Mourik (1976)
Mali	1:1 million	Soil map						Stroosnijder (1979)
Niger	1:500,000	Reconnaissance soil map	1,267,000			Reconnaissance soil survey	Soil survey	Bocquier and Gavaud (1960)
Nigeria	1:5.5 million	Provisional soil map						Vine (1952)
	1:5 million	Soil map				Calcium deficient and sufficient soils		FAO World Soil Resources Office (1954)
	1:5 million	Soil potentialities and productivity				Soil potentialities and productivity		FAO World Soil Resources Office/ORSTOM (1964)
Sao Tome e Principe	1:250,000	Soil map	1,001					Cardoso (1958)
Senegal Gambia	1:1 million	SENSOTER v 1.0 (149 SOTER units)	208,000	FAO-UNESCO (1988)	90 representative profiles			Batjes (2008a)
Sierra Leone	1:1 million	Soil map	71,740					Allbrook (1973)
Togo	1:500,000	Pedology map	56,785					Lamouroux (1962)

Soil Databases in Africa

Country	Scale	Map name	Number	Classification	Purpose	Description	Reference
Central Africa							
Burundi and Rwanda	1:1 million	SOTERCAF v 1.0			Mapping soil carbon stock	Soil carbon stock, SOTER database	Batjes (2008b)
Congo	1:5 million	Kwango soils					INEAC (1959)
Democratic Republic of Congo	1:2 million	SOTERCAF v 1.0			Mapping soil carbon stock	Soil carbon stock, SOTER database	Batjes (2008b)
Gabon	1:2 million	Soil survey map					Martin (1981)
North Africa							
Algeria	1:500,000		2,381,741				
Egypt	1:5 million	National soil association map	1 million	U.S. soil taxonomy	Agriculture	National soil association map	Ghaith and Tanios (1965)
Ethiopia	1:2 million	Soil regions of Ethiopia					Dainelli (1941)
Libya	1:250,000	SOTER map (21 SOTER units)	43,000				Greiff and Ritter, 2002, USOM (1957).
	1:2 million	Provisional soil map					Cavallar (1950)
Morocco	1:1.5 million						
Somalia	1:1 million	Dominant soils		FAO-UNESCO (1974)		Soil classification, dominant soil types	FAO Africover (2002)
Sudan	1:2 million	Soil map					Asamoa (1982)
Tunisia	1:500,000	Soil map	164,000	FAO-UNESCO (1974)		Soil organic carbon stocks	Tunisian Ministry of Agriculture (1973), Brahim et al., 2010.
East Africa							
Kenya	1:1 million	KENSOTER (397 SOTER units)	582,646	FAO-UNESCO (1988)	Soil carbon stock	495 representative profiles	Batjes and Gicheru (2004)
Rwanda	1:1 million	Exploratory soil map	582,646	FAO-UNESCO (1988)		Soil carbon stock, SOTER database	Sombroek et al. (1980)
	1:50,000	Soil map				1834 profiles	Imerzekene and van Ranst (2001)
Tanzania	1:2 million and 1:2.5 million	National soil resources inventory	883,749	FAO-UNESCO (1974)		Soil survey	Hathout (1983)
Uganda	1:1.5 million	Soil map					Uganda Department of Lands and Survey (1962)

(*continued*)

TABLE 32.1 (continued) African Soil Information at Small (Reconnaissance) Scale

Country	Scale	Map Type	Area (km²)	Profile Classification	Soil Analyses	Application	Information Content	Reference
Southern Africa								
Angola	1:3 million	General soil map	1,247,700					Ministerio do Ultramar (1965)
Botswana	1:2 million	Soil map	1,247,700	FAO-UNESCO (1988)		Food security		SADCC (1991)
	1:250,000	Inventory	582,000		3500 soil profiles	Drought mitigation	Soil resources	De Witte and Nachtergaele (1990); Remmelzwaal and van Waveren (1988)
Lesotho	1:250,000	Reconnaissance soil map	30,000	FAO and MacVicar et al, 1977			Soil inventory	
Madagascar	1:1 million	Reconnaissance soil map	309,000	CPCS (1967)	500 soil samples collected	Agricultural and forestry development, engineering purposes, regional planning	Soil survey and classification	Riquier (1968)
Malawi	1:1 million	Soil resources		Agro-ecological surveys			Natural resource inventory	Lowole (1983)
Mozambique	1:1 million	National soil map		(FAO-UNESCO-1988)	800 soil profiles	Drought mitigation	Soil survey, Limpopo Basin	INIA (1995)
Namibia	1:1 million	Soil map	825,418					Coetzee (1999)
South Africa	1:250,000	Land type survey	122,000	MacVicar et al. (1977)	2,500 modal profiles	Agricultural potential, district and regional planning	Soil types, terrain units, macroclimate information	Land type survey staff (1972–2002)
Swaziland	1:250,000	Soil map	17,364					Murdoch (1970)
Zambia	1:1 million	Reconnaissance soil map						FAO and UNSF (1964)
Zimbabwe	1:500,000	Physical resource inventory	163,500	Zimbabwe soil classification system			Soil survey	Anderson et al. (1993); Thompson and Purves (1978)
	1:3 and 1:1 million	Provisional soil maps					Soil map	Ellis (1951); Thomas and Ellis (1955); Thompson (1960)

32.4 Soil Databases

The aim in creating a soil database is to move from the previous system of paper maps and reports to one where a geographic information system (GIS) can be used to produce a polygon-based digital map with geo-referenced point data, with the digital report and soil profile database hyperlinked, so that all the information is available from a CD-ROM or on the web (van Ranst and Verdoodt, 2010).

32.4.1 Reconnaissance Scale

In the early 1990s, there was a movement, started at ISRIC in the Netherlands, for a global inventory of soil and terrain resources, at an approximate scale of 1:1 million, so that the SOTER (SOil and TERrain database) approach came into being. Applications arising from the SOTER databases have been developed to evaluate a range of district and regional development issues. Early examples of these applications were an erosion hazard map under current land use for the Kajiado District of Kenya, and the physical suitability map for rain-fed sorghum in western Kenya (Van Engelen and Mantel, 1995). In each case, the FAO Framework for Land Evaluation (FAO, 1976) was used together with digital SOTER data to focus on development issues by adopting mixed quantitative and qualitative land evaluation approaches (van Lanen et al., 1991). The applications for Kenya (1995, updated in 2007) were followed by similar projects in northeastern Africa (1998), southern Africa (2003), central Africa (2007), and a host of other country databases (Senegal/Gambia, South Africa, Tunisia) in 2008.

Two projects concerning the improvement of digital soils coverage in Africa have recently commenced. They are the "Soil Atlas of Africa" project and the "African Soil Information Service" (AfSIS).

The "Soil Atlas of Africa" project (www.eusoils.jrc.ec.europa.eu/library/maps/africa_atlas/index.html) is a joint initiative between the Joint Research Centre (JRC) of the European Union, ISRIC, the FAO, and the African Soil Science Society, where the aim is to collect as much as possible of the significant soil information for Africa that does exist and to digitize it to create a soil atlas, similar to the "Soil Atlas of Europe" (European Soil Bureau Network of the European Commission, 2005). The atlas is planned to be published during 2011.

The AfSIS project (www.africasoils.net) is based on the concept of supplying information on "soil health," which can then be related to soil productivity. The project, which is due to run from 2008 to 2012, has obtained funding from the Bill and Melinda Gates Foundation and the "Alliance for a Green Revolution in Africa" (AGRA). One of the project aims is to establish a total of 60 soil fertility field trials in five countries containing significant agro-ecological zones called "sentinel landscapes" and then to link the field data to remotely sensed information so that the findings are applicable to the widest possible area.

Remote sensing techniques are possibly one of the most valuable techniques for improving the quality of soil-based databases in Africa, due to the size of the areas involved. By using such tools as the 250 × 250 m resolution MODIS imagery, zones can be identified, which can be further investigated by focused ground-truthing programs. One area where this has been applied is in Egypt, where changing salinity levels in the vicinity of the Nile delta have been studied (Hamdi and Abdelhafez, 1998).

The AGRA project (www.agra-alliance.org), whose board is chaired by former UN Secretary General Kofi Annan, has as a goal to support smallholder farmers in food production, while protecting the environment and adapting to climate change. One of the programs to achieve this is concerned with enhancing soil health by improving the levels of soil testing and mapping, and it is surely here where the implementation of digital soil data, especially at more detailed scales in specific focus areas, might make a significant contribution to food security on the continent.

32.4.2 Detail Scale

Instances where more detailed soil information (1:5,000–1:50,000 scale) has been collected and digitized are scarcer, but those that could be sourced are listed in Table 32.2. The priority

TABLE 32.2 Selected Soil Databases at Large (Detailed) Scale

Region	Country	Scale	Map Type	Area (km²)	Profile Classification	Soil Analyses	Information Content	Reference
West Africa	Cameroon	1:50,000	Mbonge–Boa (Soil suitability)				Soil associations and land suitability	Anderson (1982)
North Africa	Egypt	1:50,000	Soil physiographic maps, upper and lower Egypt				Soil map and land use map; reclaimed areas	Hamdi and Abdelhafez (1998)
	Ethiopia	1:50,000	Jelo catchment, Chercher highlands	3,250	FAO (1988)	50 profiles	Land suitability information	Ahmed (2003)
Southern Africa	Botswana	1:20,000	Semi-detailed soil survey				Soil suitability for irrigation	Mitchell (1968)
	South Africa	1:50,000	PWV Peri-urban survey, Gauteng province	14,800	MacVicar et al. (1977)	40 soil profiles	Soil types with depth and texture class for land use planning	Yager et al. (1990)

in many countries was to have any existing reconnaissance scale data converted into digital format, so that large-scale digital soil information is not common.

32.4.3 Ultra-Detailed Scale

In areas of the continent where high-value, intensive agriculture is practiced (such as several areas in South Africa), soil information (descriptive and analytical), which is spatially defined using a GPS, is collected at a very detailed scale. These data, which may be at an interval as small as 2×2 m, may then be used in precision farming applications for accurate and cost-effective application of fertilizers and other ameliorants. However, most of these datasets are commercially based and are privately operated so that they are not currently available for public use.

Precision agriculture embraces many different permutations and varieties of philosophies. The theoretical considerations should be balanced with practically implementable operations within the established farming system. The production potential of rain-fed arable cropping can often only be achieved when the physical and chemical limitations have been reduced or eliminated. This implies access to comprehensive and reliable soil and climate information, enhanced by digital information technologies, and supported by sound scientific research.

Precision agriculture technologies are gaining prominence in the central and western grain-producing regions of South Africa with a sub-humid climate. Water extraction by plant roots and specific nutrient deficiencies are the primary soil limitations to crop growth (Muller, 2008). Soils comprise fine to medium aeolian sandy profiles, often underlain by slowly permeable plinthic or gleyed horizons. The latter are responsible for improved delivery of plant available water, particularly in low rainfall seasons, and hence, significantly reduce crop yield risk. The digital capture of detailed farm scale soil survey information identifies the land areas most suitable for sustained grain crop production. Soil fertility sampling at increasing levels of detail is used to overcome soil fertility and acidity imbalances, while GPS technology at harvest provides a visual record of grain yields across agricultural fields. These database technologies coupled with research and agricultural advice systems are contributing towards economically sustainable production systems.

32.5 Soil Profile Databases

In addition to soil mapping information at its various scales, another type of information contained in soils databases is that of profiles, which is usually linked to the surrounding soil pedon. Nachtergaele and van Ranst (2003) state that of the 4353 soil profiles contained within the Global Pedon Database created by ISRIC, a total of 1799, or 41%, lie in Africa. In addition, several countries have their own database of soil profile information, such as South Africa, where approximately 2500 geo-referenced soil profiles, each with a full profile description and analytical data for each horizon, have been collated as part of their national land type survey program.

32.6 Database Interpretation and Utilization

A soil database is only as good as its ability to be accessed and utilized, so that the basic soil data contained therein can be interpreted for a variety of uses. Two examples of national soil databases (at varying scales) where such use has been made are those of Rwanda and South Africa. Rwanda is a small country, with an area of just under 25,000 km^2, so that it was possible to carry out soil mapping between 1981 and 1994 of the whole country at 1:50,000 scale (Imerzekene and van Ranst, 2001), with over 1800 profiles described and analyzed and all of the map polygons digitized using ArcView. The final maps are used not only for traditional agricultural purposes, but also for such purposes as settlement planning, disaster-preparedness planning and high-input agriculture. The system allows easy updating, modification or reorientation depending on the requirements of the different Ministerial departments who are going to use it. As such, the land resources information can be used more efficiently and accurately allowing the Government of Rwanda to investigate specific agricultural and environmental issues in a timely manner.

In contrast, South Africa is a much larger country, with an area of approximately 1.2 million km^2, so that the soil mapping program, which was carried out between 1972 and 2002, was done at a scale of 1:250,000 (Land Type Survey Staff, 1972–2002). The project, known as the Land Type Survey, has over 7000 unique mapping units, or *land types*, each one of which is a combination of soil pattern, terrain type and macroclimate, and around 2500 fully described and analyzed soil profiles. One of the benefits of the database is that algorithms can be written to assess each map unit for specific crop suitability, and the algorithm inputs can be altered very readily for ease of comparison of results. The information is widely used for municipal and regional planning, identification of potentially problematic areas (such as erodible soils), and even multi-resource research projects, such as possible links with the occurrence of soil-borne pathogens or weeds. More and more applications are being discovered as a wider range of potential users (e.g., engineers, planners, and agronomists) become aware of the content of the database.

In addition, South Africa has a more detailed (1:50,000 scale) soil database for the densely populated area around Johannesburg and Pretoria in Gauteng Province (Yager et al., 1990). Here, the soil information is used to identify valuable areas of productive agricultural land in order to protect them from peri-urban expansion against the background of ever-increasing urbanization and consequent competition for land resources.

32.7 Conclusion

Despite the inconsistent history of soil database compilation in Africa, coupled with communication and other limitations of the past, it seems that the supply of digital information is improving and that there are a number of initiatives underway to collect, check, and distribute the information to land users who require it. For the benefit of the population of Africa as well as the need to ensure their future food security using a fragile soil resource, it is to be hoped that this situation continues.

References

Ahmed, M.A. 2003. Land suitability evaluation in the Jelo Catchment, Chercher Highlands, Ethiopia. Unpublished PhD thesis, University of the Free State, Bloemfontein, South Africa.

Allbrook, R. 1973. Soil map of Sierra Leone. Department of Agriculture, Sierra Leone, West Africa.

Anderson, I.P. 1982. Cameroun (Mbonge-Boa). Map 1. Soil associations and land suitability. Land Resources Development Centre, Surrey, England, U.K.

Anderson, I.P., P.J. Bruin, M. Moyo, and B. Nyamwanza. 1993. Physical resources inventory of the communal areas of Zimbabwe: An overview. NRI Bulletin 60. Natural Resources Institute, Chatham, U.K.

Asamoa, G.K. 1982. Soil map of the East African sub-region (Tanzania). ISRIC, Wageningen, the Netherlands.

Asiamah, R.D., and O. Dwomo. 1995. National soil reference collection and exposition, Ghana. In Batjies, N.H., J.H. Kauffman, and O.C. Spaargaren (eds.) National soil reference collections and databases (NASREC—Vol. 3). ISRIC, Wageningen, the Netherlands.

Batjes, N.H. 2008a. Soil parameter estimates for Senegal and The Gambia derived from SOTER and WISE (SOTWIS-Senegal, version 1.0). Report—ISRIC World Soil Information, Wageningen, the Netherlands.

Batjes, N.H. 2008b. SOTER-based soil parameter estimates for Central Africa—DR of Congo, Burundi and Rwanda (SOTWIS-Caf, version 1.0). Report—ISRIC World Soil Information, Wageningen, the Netherlands.

Batjes, N.H., and P. Gicheru. 2004. Soil data derived from SOTER for studies of carbon stocks and change in Kenya (version 1.0). Report—ISRIC World Soil Information, Wageningen, the Netherlands.

Boateng, E., T.R.E. Chidley, D.J. Savory, and J. Elgy. 1998. Soil information system development and land suitability mapping at the Ghana Soil Research Institute. Poster present. 16th World Congr. Soil Sci. August 20–26, 1998, Montpelier, France.

Bocquier, G., and M. Gavaud. 1960. Carte pédologique de reconnaissance de la Republique du Niger (Feuille Zinder). ORSTOM, Centre de Rec. Péd. de Hann. Dakar, Senegal.

Brahim, N., M. Bernoux, D. Blavet, and T. Gallali. 2010. Tunisian soil organic carbon stocks. Int. J. Soil Sci. 5:34–40.

Brammer, H. 1958. Great soil groups (Provisional) of Ghana. Soil and Land Use Survey Division, Department of Agriculture, Ghana (Text GHA/4).

Cardoso, C. 1958. Dos sols e Sao Tome e Principe. Junta e Export. do Café (PTE/1).

Cavallar, W. 1950. Esquisse préliminaire de la carte de sols du Maroc. Centre de Recherches Agronomiques du Maroc, Rabat, Morocco.

Coetzee, M.E. 1999. National soil survey of Namibia. Proc. 1999 Ann. Res. Rep. Conf. Directorate Agric. Res. Train. Ministry of Agriculture, Water and Rural Development, September 1999, Swakopmund, Namibia.

C.P.C.S. (Commission de Pédologie et de Cartographie des Sols), 1967. Classification des sols. Orléans. Multicopie.

Dainelli, G. 1941. Carta geologica dell'Africa Orientale. 1:2 million scale (coloured). Accompanying Geologica dell'Africa Orientale 4. Reale Accademia d'Italia (1943), Roma.

De Witte, O., and F. Nachtergaele. 1990. General soil map of Botswana: Drought impact mitigation and prevention in the Limpopo River Basin. http://www.fao.org/docrep/008/y5744e/y5744e06.html (Accessed November 2010)

d'Hoore, J.L. 1964. Soil map of Africa—Scale 1:5 million. Commission for technical co-operation in Africa. Joint Project No. 11. Lagos, Nigeria.

Diallo, C., and S. Condé. 1982. Carte de reconnaissance pedologique au 1:500,000 de la R.P.R. de Guinee. Feuille Tougué. Project PNUD/FAO GUI 72/004. Army Map Service USA NC.29.1. La Direction du Service Nationale des Sols et la Direction Technique de J.H. Molfino.

Ellis, B.S. 1951. Provisional soil map of southern Rhodesia. Office of the Chief Chemist, Federal Ministry of Agriculture, Salisbury, U.K.

European Soil Bureau Network of the European Commission. 2005. A. Jones, L. Montanarella, and R. Jones (Principal eds.). Office for Official Publications of the European Community, Luxembourg.

FAO. 1976. A framework for land evaluation. Soils Bulletin No. 32. FAO, Rome, Italy.

FAO Africover. 2002. Soil map of Somalia at scale of 1:1,000,000. http://www.faoswalim.org/downloads/DominantSoilTypesSomalia.pdf (Accessed November 2010)

FAO-UNESCO. 1974. Unesco soil map of the world (1:5 M scale). Unesco, Paris.

FAO-UNESCO. 1981. Soil map of the world: Revised legend (with corrections and updates). World Soil Resources Report No. 60. FAO, Rome, Italy.

FAO-UNESCO. 1988. FAO/UNESCO soil map of the world. Revised legend with corrections. World Resources Report No. 60, FAO, Rome.

FAO and UNSF. 1964. Reconnaissance soil survey of Zambia (sheets: Mesh-Teshi Namwala, Mwebheshi Kwembe). FAO, Rome, Italy.

FAO World Soil Resources Office. 1954. Soil map of Nigeria (based on CCTA soil map of Africa) showing calcium deficient and calcium sufficient soils. FAO World soil Resources Office, Rome, Italy.

FAO World Soil Resources Office. 1964. Soil resources of Nigeria (based on CCTA soil map of Africa) map of present soil productivity and potentialities. FAO World Soil Resources Office, Rome, Italy.

Ghaith, A., and M. Tanios. 1965. Preliminary soil association map of the United Arab Republic. General Organisation for Government Printing Office, Cairo, Egypt.

Greiff, P., and M. Ritter. 2002. Soil classes (FAO): Libya, Egypt, Chad and Sudan. SFB 389 Project E1.

Hamdi, H., and S. Abdelhafez. 1998. Agriculture and soil survey in Egypt. Options Méditerranéennes Serie B, No. 34, Mediterranean Agronomy Institute, Bari, Italy.

Hathout, S.A. 1983. Soil atlas of Tanzania. Tanzania Publishing House, Dar es Salaam, Tanzania.

Hawkins, P., and M. Brunt. 1965. The soils and ecology of West Cameroon, p. 285. Vol. 1, Part 2. FAO, Rome, Italy.

Imerzekene, S., and E. van Ranst. 2001. Une banquet de données pédologiques et son S.I.G. pour une nouvelle politique agricole au Rwanda. Bull. Séanc. Acad. R. Sci. Outre-Mer. 47:299–329.

INEAC. 1959. Carte des sols du Congo Belge et du Ruanda-Urundi. (Text RCO/23). ISRIC, Wageningen, Netherlands.

INIA. 1995. Carta Nacional de solos, 1:1 m. Série Terra e Água do INIA, Communicação no 73. Maputo, Mozambique.

Lamouroux, M. 1962. Carte pédologique du Togo. Inst. de Rec. du Togo (TOG/1). ORSTOM, Paris, France.

Land Type Survey Staff. 1972–2002. Land type survey of South Africa, 1:250,000 scale. ARC-Institute for Soil, Climate and Water, Pretoria, South Africa.

Lowole, M.W. 1983. Soil map of Malawi. Department of Agricultural Research, Lilongwe, Malawi.

MacVicar, C.N., de Villiers, J.M., Loxton, R.F., Verster, E., Lambrechts, J.J.N., Merryweather, F.R., le Roux, J., van Rooyen, T.H. & Harmse, H.J. von M., 1977. Soil classification. A binomial system for South Africa. ARC-Institute for Soil, Climate & Water, Pretoria.

MacVicar, C.N. 1980. Advances in soil classification and genesis in South Africa. Proc. 8th Congr. SSSSA, Pietermaritzburg. Tech. Comm. 165:22–40. Department of Agriculture and Technical Services, Pretoria, South Africa.

Martin, D. 1981. Carte pédologique du Gabon. ORSTOM, Paris, France.

Ministerio do Ultramar. 1965. Carta generalizada dos solos de Angola (Aproximacao Escala 1:3 m (Generalized soil map of Angola, scale 1:3 m). Junta das Missoes Geograficas e des Investigacoes do Ultramar, Missao de Pedologia de Angola e Mocambique and Centro de Estudos de Pedologia Tropical.

Mitchell, A.J.B. 1968. Botswana semi-detailed soil survey (+ Irrigation Potential of Soils). Sheet A: Masia-Dieme, Kolobeng, Mmaikowane; Sheet B: Moshupa, Mahalapye, Tonota; Sheet C: Manyana, Tobane; Sheet D: Lotlakane, Bobonong. Land Resources Division, Overseas Development Administration, Tolworth Tower, Surbition, Surrey, England, U.K.

Muller, B. 2008. Measurement of soil profile water: A handy management tool. (Afr). Senwes Koöperasie, Klerksdorp, South Africa.

Murdoch, G. 1970. Soil and land classification in Swaziland. Ministry of Agriculture, Mbabane, Swaziland.

Nachtergaele, F., and E. van Ranst. 2003. Qualitative and quantitative aspects of soil databases in tropical countries. Workshop, "Evolution of tropical soil science: Past and future". March 6, 2002, Brussels, Belgium.

ORSTOM. 1960. Carte pédologique de la Republique de Côte d'Ivoire. ORSTOM, Inst. d'Enseignement et de Rec. Trop., Adiopodoume, Paris, France.

Remmelzwaal, A., and E. van Waveren. 1988. Botswana soil database. FAO, Gaborone, Botswana.

Riquier, J. 1968. République de Madagascar. Carte Pédologique de Madagascar. Feuille Centrale. ORSTOM, Paris, France.

SADCC (Southern African Development Coordination Conference). 1991. Soil map of Angola, 1:2 m. Draft (revised legend FAO-UNESCO 1988). Food Security Programme, regional inventory of agricultural resource base. Harare, Zimbabwe.

Schokalskaya, Z.J. 1944. A new soil map of Africa. Pochvovedenie 9:424–475.

Shantz, H.L., and C.F. Marbut. 1923. Vegetation and soils of Africa. American Geographical Society, New York.

Sombroek, W.G., H.M.H. Brown, and B.J.A. Van Der Pouw. 1980. Exploratory soil map and agro-climatic zone map of Kenya. Exploratory Soil Survey Report No. E, (Including Map). Kenya Soil Survey (KSS), Nairobi, Kenya.

Stroosnijder, L. 1979. Carte des sols du Mali. Bassée sur l'interprétation des Images Satelite 1972–1975 avec Observations du Terrain en 1976–1979.

Tavares, J.D., and P. Amiotte-Suchet. 2007. Rainfall erosion risk mapping in volcanic soils of Santiago Island, Cape Verde Archipelago. Africa Geosci. Rev. 14:399–414.

Thomas, R.G., and B.S. Ellis. 1955. Provisional soil map of southern Rhodesia, Office of the Chief Chemist, Federal Ministry of Agriculture, Salisbury, U.K.

Thompson, J.G. 1960. Provisional soil map of southern Rhodesia. Soil Survey Lab, Dept. of Resources and Special Services, Federal Ministry of Agriculture, Salisbury.

Thompson, J.G., and W.D. Purves. 1978. A guide to the soils of Rhodesia. Rhodesian Agricultural Journal, Technical Handbook No. 3. Salisbury, U.K.

Tunisian Ministry of Agriculture. 1973. Carte pédologique 1:500,000 de la Tunisie. Ministère de l'Agriculture de Tunisie, Tunis, Tunisia.

Uganda Department of Land and Surveys. 1962. Soil and land use map, Atlas of Uganda, Kampala, Uganda.

USOM. 1957. Provisional soil map of Libya. FAO World Soil Resource Office, Rome, Italy.

van Engelen, W.V.P., and S. Mantel. 1995. World soils and terrain digital database (SOTER). In J.H. Kaufmann (ed.) National soil reference collections and databases (NASREC) Vol. 2. Use of ISRIC's databases for the characterization of soils of major agro-ecological zones. NASREC Workshop, November 6–17, 1995. Wageningen, the Netherlands.

van Lanen, H.A.J., J. Hack-ten Broeke, J. Bouma, and W.J.M. de Groot. 1991. A mixed qualitative/quantitative physical land evaluation methodology. Geoderma 55:37–54.

van Mourik, D. 1976. Land suitability of the Mano River Union project area in Liberia (Map 3). Cartograph, Intermap, Enschede, the Netherlands.

van Ranst, E., and A. Verdoodt. 2010. Soil mapping in Africa at the crossroads: Work to make up lost ground. Soil Society of South Africa Combined Congress, January 19–21, 2010, Bloemfontein, South Africa.

Vine, H. 1952. Provisional map of Nigeria (simplified). Survey Department, Lagos.

Yager, T.U., L.C. Kaempffer, and C.J. Coetzee. 1990. 1:50,000 scale peri-urban soil database of the Pretoria-Witwatersrand-Vereeniging area. ARC-Institute for Soil, Climate and Water, Pretoria, South Africa.

33
Learning about Soil Resources with Digital Soil Maps

Darrell G. Schulze
Purdue University

Phillip R. Owens
Purdue University

George E. Van Scoyoc
Purdue University

33.1 Introduction 33-1
33.2 Teaching and Learning with Web Soil Survey..33-2
33.3 Teaching and Learning Using GIS and Soil Geodatabases in the Field......33-2
  Hardware Considerations • GIS Software • Development of the GIS Dataset • Pedagogic Aspects
33.4 Teaching and Learning Using a Web-Based Geobrowser..........................33-7
Acknowledgments..33-7
References..33-7

33.1 Introduction

Many of the concepts that students learn in soil, crop, and environmental science courses are inherently spatial. Soils vary across landscapes in predictable, repeating patterns. Certain soils and landscapes are better for particular crops than others. An environmental problem may impact a whole watershed, not just the point at which a contaminant is introduced. Patterns of land uses, whether for crop production, forestry, wildlife habitat, or urban development, vary spatially in response to soils, topography, geology, human infrastructure, and many other factors. Although we *implicitly* acknowledge the existence of spatial patterns in our soil, crop, and environmental science courses, our ability to *explicitly* observe these spatial patterns visually and make them clear to our students has, in the past, been limited. Geographic Information Systems (GISs) and the recent availability of detailed digital soil survey data, particularly in the United States, along with increasingly detailed Digital Elevation Models (DEMs) worldwide, allow us to visualize and analyze such complex geospatial information in new ways, and to use them to enhance teaching and learning. Research has repeatedly shown that a dynamic, active, visually rich learning environment significantly increases comprehension and retention relative to a more passive, auditory environment (Bransford et al., 2000).

Aside from the pedagogic reasons outlined earlier, two recently published reports, *Learning to Think Spatially: GIS as a Support System in the K-12 Curriculum*, and *Beyond Mapping: Meeting National Needs Through Enhanced Geographic Information Science* (National Research Council, 2006a, 2006b), make compelling arguments for the need to incorporate spatial thinking and spatial technologies throughout our curricula. Teaching and learning soil science using a geospatial approach is one way to accomplish this.

Advances in computer hardware now allow us to take powerful, reliable computers to the field that can run state-of-the-art GIS software and have the capacity to store large GIS datasets. This makes it possible to have at one's fingertips a wide variety of different maps that can be zoomed in to show detail or zoomed out to show the overview at any time. The addition of a Global Positioning System (GPS) receiver allows for the display of one's exact location on the map at all times. The result is an engaging and informative set of maps through which one can see and understand the soil landscape in a new geospatial context. We have only begun to take advantage of the powerful, learner-centered instructional tools that GIS and soil geodatabases have to offer. The purpose of this chapter is twofold: (1) to describe how we use, and continue to evolve our use of, these new tools and databases in our undergraduate and graduate soil, crop, and environmental science courses; and (2) to provide practical suggestions for those wishing to implement a similar approach in their own teaching program.

This chapter is divided into three parts. First, we begin with a description of how we have incorporated USDA's Web Soil Survey into several of our soils courses in addition to or in lieu of paper maps. This is the easiest approach to incorporate soil geodatabases into a teaching program, although Web Soil Survey covers only the United States and its territories. Second, we describe how we have incorporated a "teaching with GIS" approach into our field-based pedology course by taking GIS to the field on rugged tablet PCs equipped with GIS software and GPS receivers. This approach requires considerable initial work on the part of the instructor, but with the result that the maps can be customized and optimized to teach particular concepts.

Third, we conclude with a brief description of how geospatial soils information can be delivered using Web 2.0 technology such as Google Earth.

33.2 Teaching and Learning with Web Soil Survey

We began using Web Soil Survey in our introductory Soil Science course as soon as Web Soil Survey became available in August 2005. Students are taught in lecture how to use both printed copies of County Soil Surveys and the Web Soil Survey to access valuable soils information. Following an introduction in lecture, students come to our Soils Resource Center where they spend approximately 2 h using both County Soil Surveys and Web Soil Survey to answer questions about their home communities or about Tippecanoe County where Purdue University is located. Students develop skills in finding information about soil types, land use, soil classification, engineering properties, and crop, forest, and wildlife recommendations for each soil in a county. Since Soil Science is a prerequisite for all other crops and soils courses in the department, all instructors in the department are able to now use the Web Soil Survey database as a primary tool in any class project utilizing soil and landscape information. Students who are not from Indiana can access information about their home county, or instructors can assign projects related to land use anywhere in the United States This provides students with a direct understanding of how soils and their use vary not only in Indiana but across the country. Examples of projects assigned to students include golf course construction and fertility planning, nutrient management plans for farms and livestock operations, and urban and suburban building projects. Soil and suitability maps and data tables can be directly printed from the web and included in student project reports without retyping them. Since Web Soil Survey became available, students have been more likely to utilize soil survey data because they are accessible from anywhere at any time, making it easy to include soils information as part of their project reports in multiple courses. Students rate the use of the Web Soil Survey for these projects as an extremely valuable part of their soils educational experience and many of them indicate in their senior exit interviews that they use the Web Soil Survey in courses outside of the Agronomy Department as a tool for class projects.

Web Soil Survey and the printed County Soil Surveys that preceded it are designed to deliver detailed soil survey information for relatively small "areas of interest," generally an individual field, farm, or other area of less than a few thousand hectares, where the goal is to obtain information for a particular use or management of the area. Both Web Soil Survey and County Soil Surveys are excellent for this purpose, but they are not designed to illustrate spatial aspects of soils and geomorphology that may occur over tens or hundreds of kilometers. The vast amount of publicly available soil and other geospatial data now available digitally provides an opportunity for new approaches to teaching and learning about soils and the landscapes in which they occur.

33.3 Teaching and Learning Using GIS and Soil Geodatabases in the Field

In the summer of 2005, we began experimenting with how we could utilize the newly available soil and geomorphology data in our teaching program, and in the fall of that year, we initiated a "teaching with GIS" approach in our *Soil Classification, Genesis, and Survey* course using three tablet PCs and a minimal GIS dataset. We now routinely take our class on weekly 3 h field labs and on two, all-day field trips along with 14 tablet PCs and a GIS dataset that has grown to encompass the entire state of Indiana (92,895 km^2) (Figure 33.1). Our focus is to teach students how soils and landscapes vary spatially over many different scales, and how the soil geomorphic concepts illustrated with diagrams in the classroom correspond to actual features observed in the field. Students can zoom out for an overview and zoom in to see details on the maps displayed on the computer they have in their hands. They can quickly access maps of topography, soil parent material, soil drainage class, presettlement vegetation, and other maps, thereby learning how the soil that they just examined in a pit is distributed over larger areas than they can see from their current location. They can query the map to obtain more information on individual soil delineations, and, soon, to view a diagram of the dominant soil within the map unit. This approach has significantly altered how we teach field soil science and what we teach (we have learned many new things ourselves), and has significantly impacted student learning.

We continue to incorporate the tablet PCs and "teaching with GIS" into an increasing number of our soil, crop, and environmental science courses. For those wishing to incorporate a similar approach into their own teaching program, the following sections provide information on computer hardware considerations, development of the GIS dataset, and how we teach with GIS.

FIGURE 33.1 A student using a tablet PC and GIS software on a bus during an all-day soils field trip. The small, gray device mounted on the upper right corner of the tablet PC is the GPS receiver. The pen in the student's right hand is used to manipulate the map.

33.3.1 Hardware Considerations

Almost any laptop computer can be used in the field, but whether or not it will be a useful tool for teaching and learning depends on a number of factors.

Reliability is obviously an important issue. If the computers will only be used in a vehicle and can always be kept clean and dry, then normal laptops will probably work. If, on the other hand, the computers are to be used while walking in fields, where dust, sand, or seeds are encountered, or if the computers will be handled a lot, as when people get into and out of a vehicle frequently, then more robust "rugged" laptops or tablet PCs are more appropriate.

We use the PCs almost exclusively for displaying maps. For this, the major user interaction with the PC consists of manipulating check boxes to turn data layers on and off, zooming and panning the map, and choosing items from dropdown boxes. Since there is almost no need to enter text, tablet PCs are considerably more practical than laptop computers with a keyboard. A tablet PC consists of one rectangular unit the size of a large book (Figure 33.1). There is no physical keyboard (or the keyboard is folded under and not used) and the user interacts with the tablet PC exclusively through the use of a pen. A laptop computer with its flip up screen is less practical, particularly in the confined space of a van or bus.

The display screen is an important consideration. If the screen cannot be seen under the conditions under which the computers are to be used, very little learning can occur. If the computers are to be used frequently in bright sunlight, then a bright sunlight-readable screen is essential. If the computers are used mainly inside of a van or bus, a screen that does not perform well in bright sunlight may still work well in the lower light conditions inside the vehicle. If one computer is to be shared by two or more students, it is important that all viewers can see the image on the screen. The image on some screens is difficult to see when viewed at approximately 45° from the normal, while on other screens it is not. Testing a computer under the conditions under which it will be used prior to purchase can prevent disappointment and frustration later.

Knowing where one is on the map at all times is essential to the educational experience of using GIS in the field. There are many options in choosing a GPS receiver. All that is needed, however, is an inexpensive, consumer-grade GPS sensor that can be connected to the computer running the GIS software. There is no need for the maps or additional software that is often sold with the receiver. Also, positional accuracy of a few meters is sufficient. It is important, however, to use a "high sensitivity" GPS receiver as it will typically maintain a GPS "fix" in wooded areas and when the receiver is inside of a vehicle. Older "standard sensitivity" receivers will not. The receiver can be connected to the computer via a wired Universal Serial Bus (USB) or serial port connection, or via Bluetooth® wireless technology. The wired receiver may be easier to set up initially, but has the disadvantage of requiring a connector plugged into the side of the computer, which may be susceptible to damage as the computer is handled frequently in the field. A wireless connection eliminates the cord, but requires Bluetooth® connectivity on the host computer and may be a bit more difficult to setup in some situations. The battery in the wireless GPS receiver will need to be charged separately from the host computer, whereas a wired GPS receiver may or may not have a separate battery.

Learning stops when the computer battery dies, so power management to assure that the computers remain functional for the duration of the time in the field is essential. High-capacity batteries, if available, should be considered. Some computers offer hot-swappable batteries. If the computers will be used in a vehicle, 12 V power adapters are more convenient and less expensive than additional batteries. Setting the power management so that the screen goes blank and the hard disk is shut off after a few minutes of inactivity will greatly extend battery life.

There is no substitute for experience when taking GIS to the field. When just beginning, use of an existing laptop and an inexpensive GPS receiver is a good way to obtain experience without much additional expense. When considering purchase of new equipment, it makes sense to obtain a demo unit and use it under field conditions prior to placing an order. The complexity of getting the computers, GIS software, and GPS units working together correctly may require specialized assistance from computer professionals. It is important, however, that the instructor or a teaching assistant be familiar enough with the hardware and software so that minor problems that pop up can be addressed in the field. An inoperative computer or a computer that is not operated effectively by the instructor is not useful for learning. In addition to planning for specific teaching goals, one must plan ahead to be sure that the computer hardware will be ready to go when students arrive for class, and that the hardware will function for the duration of the class time. We typically have the computers ready to go when the class begins so that we do not waste valuable class time setting up computer hardware.

Computer hardware and software continues to evolve rapidly. When appropriate software becomes available, new devices such as Apple's iPad (http://www.apple.com/), for example, may make it even easier to teach and learn with GIS in the field.

33.3.2 GIS Software

We use ArcGIS 9 software (ESRI, Redlands, California) running under the Windows XP operating system. ArcGIS 9 includes support for GPS input. Other GIS software might be suitable as well, so long as it supports GPS input and the large datasets which one is likely to develop, but we have worked only with the various versions of ArcGIS 9 to date.

33.3.3 Development of the GIS Dataset

Over time, we have developed an approach for structuring our GIS dataset so that it functions as an effective teaching tool in our courses. Two distinctly different datasets are essential to making a GIS-based approach effective in teaching. One is the soil survey data that cover the area of interest, the other is a DEM of the same area. Each dataset alone is informative by itself, but in combination, the information conveyed is vastly greater than the sum of the individual parts. Additional data layers are important as well, but the soils–DEM combination is central to learning about soils using GIS in the field.

33.3.3.1 The Digital Elevation Model

A DEM consists of a raster dataset in which the value for each cell represents elevation. DEM data are available for all parts of the world. Elevation data at 90 m resolution is available from the Shuttle Radar Topography Mission (USGS, 2008) and a 30 m worldwide DEM has just become available from the Advanced Spaceborne Thermal Emission and Reflection Radiometer (ASTER) program (NASA, 2009). Data at 30 and 10 m resolution are available for the entire United States (USGS, 2009) and data at even higher resolution may be available for certain areas. Elevation data for the state of Indiana, for example, are available at 1.5 m resolution (ISDP, 2009). New data are constantly becoming available, and in many cases this data can be downloaded from the Internet.

In general, one should use the highest quality, highest resolution data available for the area of interest. The available data should be evaluated carefully, however, because the highest spatial resolution data may not necessary provide the highest elevational quality. If very high resolution data are available, it is advantageous to determine whether or not it can be resampled to a lower resolution without loss of significant information. We have found, for example, that the 1.5 m DEM dataset for Indiana can be resampled to 5 m resolution without compromising our ability to see subtle geomorphic features, but with a substantial reduction in the size of the dataset.

A DEM hillshade should be calculated from the original DEM data. A hillshade is a black and white image that shows the hypothetical shadows and highlights that would occur on the Earth's surface for a given sun position (usually 45° above the horizon and 45° west of the viewing direction). The hillshade, therefore, shows slope but no longer contains information on elevation. The hillshade is one of the most useful layers in the dataset because it graphically illustrates the topography. The hillshade is used as a base layer for other data, particularly soils data.

The hillshade should be viewable by itself because some topographic features are more easily seen on the opaque black and white hillshade. To best illustrate elevation, however, a semitransparent (50% transparency) sequential color scheme should be applied to the DEM data and then overlain on the opaque hillshade. These two layers are illustrated in Figure 33.2.

33.3.3.2 Soil Data

Soil survey data at this time are typically available in a polygon format, and most, if not all, of the soil survey data currently available have been digitized from existing paper maps. The data consist of two parts: (1) spatial data, which consists of the polygons that delineate the map units within the survey area, and (2) tabular data, which consists of a table (the attribute table) that tabulates the various attributes associated with each map unit. A map unit

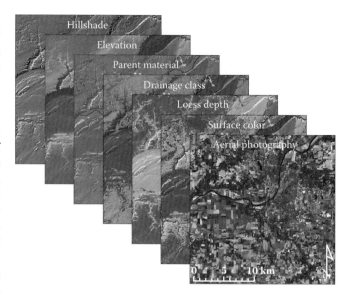

FIGURE 33.2 (See color insert.) Examples of some of the different kinds of maps available in our GIS dataset illustrated by a portion of Tippecanoe County, Indiana. Note the obvious correlation between features shown on the different maps.

may consist of thousands of individual polygons, but each polygon of the same map unit is associated with only one unique set of values in the attribute table. The attribute table typically contains information on many different properties of the map unit, such as map unit name, soil classification, soil drainage class, moisture holding capacity, etc. Thus, by selecting different columns in the attribute table, one can easily prepare maps for any property represented in the attribute table. If a soil property of interest is not represented in the attribute table, one can add additional columns to the attribute table with one's own interpretation, although this can be quite time consuming if the dataset is large, and it requires an intimate knowledge of the soils of the area.

Soil survey data are available from a variety of sources and an Internet search may be all that is needed to locate useful data. Detailed soil survey data for the entire United States can be downloaded from the U.S. Department of Agriculture, Natural Resources Conservation Service's Soil Data Mart (Soil Survey Staff, 2009). Two types of data are available: (1) the U.S. General Soil Map (STATSGO2), and (2) the Soil Survey Geographic (SSURGO) Database. The SSURGO database consists of the detailed order 2 soil survey data covering almost the entire United States. We have found the SSURGO data to be the most useful for our purposes.

Most of Indiana was glaciated during the Pleistocene, and the northern two-thirds was glaciated during the Wisconsin Glaciation, which reached its maximum extent in Indiana only ~21,000 years ago. Thus, there is a very close correspondence between the various glacial deposits and the properties of the soils that formed in and on them. As a result, we rely heavily on a soil geomorphological approach in teaching soil science in Indiana. The single most useful map in our GIS dataset is a *dominant soil parent material* map that we developed from our own interpretation of the SSURGO data. Since many soils in

Indiana developed in two and sometimes even three different parent materials, we define the dominant soil parent material as that material with the greatest influence on other soil properties, or a material that highlights a particular distinguishing feature. Usually, the dominant soil parent material is the lowest one in the profile, but it may be at the surface, for example, surface layers composed of organic soil material. By carefully grouping soil map units with similar soil parent materials, we are able to visualize the soil landscape in new ways. By making the dominant soil parent material layer transparent and overlaying it on the DEM, outwash plains, flood plains, dune fields, and other geomorphic features stand out in stark contrast to the surrounding till plains (Figure 33.2).

A separate loess depth map (Figure 33.2) provides information about the loess (wind-blown silt) that covers large areas of the state. Maps of soil drainage class and surface color (which generally correlates with presettlement vegetation) (Figure 33.2) provide information on other soil properties or soil forming factors that we discuss in class.

33.3.3.3 Other Maps and Data Layers

A number of other maps and data layers round out our GIS dataset. Color aerial photography (Figure 33.2) is particularly useful when one is walking over the landscape and the focus is on one's immediate surroundings. Topographic maps, referred to as Digital Raster Graphics (DRGs), are scanned versions of traditional paper topographic maps. DRGs can often provide information on the names of physical features not found in other databases. Since many were last updated 20, 30, or more years ago, traditional topographic maps provide a useful historical perspective. Topographic lines can be used to show topography, although their use is largely supplanted by the DEM. Hydrography, the centerlines or boundaries of rivers, streams, and lakes, shows where water bodies are located. Roads and towns are important for orienting oneself relative to observable anthropogenic features. Historic maps (we have, e.g., a 1852 map of the state of Indiana) provide a historical perspective and show, in some cases, how humans have influenced the landscape. Maps of bedrock geology and topography of the bedrock surface provide insights on what is below the upper 2 m of the Earth's surface that soil scientists usually focus on. In short, any map or data layer can and should be included in the GIS dataset, so long as it serves to further one's teaching and learning objectives.

33.3.3.4 Hierarchy of Data Layers

Table 33.1 lists the hierarchy of data layers that we use most frequently. Layers at the top of the list, if they are turned on, overlie layers below them. The user typically turns on only a few layers at any one time, essentially customizing the map as needed. We have found that soil layers are most informative when they are semitransparent and paired with an underlying opaque hillshade. The two are turned off and on as a unit. Although it is tempting to overlay several semitransparent layers, this quickly becomes visually confusing. It is far more informative to turn an overlying layer off and on to compare it with an underlying layer.

TABLE 33.1 A Typical Hierarchy of Data Layers

Layer Name	Type	Comments
Cities and towns	Points	Important for locating oneself on a map and for describing locations to others
Reference		Political boundaries
State outline	Lines	
County outlines	Lines	
County names	Labels only	Turn on as needed
Land survey		
Townships	Lines	Used only occasionally
Sections	Lines	Used only occasionally
Roads and railroads		Important reference features, particularly when driving
Interstate highways	Lines	
Primary roads	Lines	
Secondary roads	Lines	
Railroads	Lines	
Rivers and streams	Lines	Hydrography, center lines of rivers and streams, outlines of lakes
Contours	Lines	Usually derived from traditional topographic maps, used occasionally but largely replaced by DEMs
SSURGO soils	Lines	Outlines of soil polygons from the detailed soil survey
Aerial photography	Raster, opaque	Color photography, if available, is generally more interesting than black and white photography
Historic maps	Raster, opaque	Selected maps that provide important or interesting historical context
Topographic maps	Raster, opaque	Scanned versions of traditional paper topographic maps
Other maps		Additional maps such as geology, natural regions, etc., as needed
Soil drainage class[a]		
Soil drainage class	Raster, 50% transparent	Soil drainage class from detailed soil survey (SSURGO) data
Hillshade	Raster, opaque	
Surface soil color[a]		
Surface soil color	Raster, 50% transparent	A surrogate for presettlement vegetation over much of the state, derived from soil survey (SSURGO) data
Hillshade	Raster, opaque	
Loess depth[a]		
Loess depth	Raster, 50% transparent	Derived from detailed soil survey (SSURGO) data
Hillshade	Raster, opaque	
Dominant soil parent material[a]		
Dominant soil parent material	Raster, 50% transparent	Derived from detailed soil survey (SSURGO) data
Hillshade	Raster, opaque	

(continued)

TABLE 33.1 (continued) A Typical Hierarchy of Data Layers

Layer Name	Type	Comments
Elevation[a]		
Elevation	Raster, 50% transparent	DEM data rendered as a sequential color scheme
Hillshade	Raster, opaque	
Hillshade	Raster, opaque	The DEM hillshade alone is often useful for seeing subtle geomorphic features

[a] These layers function as a unit and are turned off and on together rather than individually.

33.3.3.5 Other Considerations

Most maps and data layers "work" best within a particular range of scales or "zoom" levels. In general, there should be enough information on the screen to make the map interesting and informative, yet not so much as to be confusing. The speed at which one is moving is an additional consideration. Ideally, most people wish to see noticeable movement of the GPS position marker across the map as they move over the landscape so that they can see how the physical features of the landscape correspond with the information depicted on the map. Thus, when driving across the landscape at 80–100 km h^{-1}, a more general map, like the dominant soil parent material map that we described earlier, viewed at a scale of around 1:50,000 is more informative and appropriate (Figure 33.3a). On the other hand, when walking over a field at a speed of 4–5 km h^{-1}, an aerial photograph overlain by detailed soil delineations at a scale of 1:10,000 may be the most informative map (Figure 33.3b). With experience, one quickly discovers which map and scale are most appropriate for a particular situation. The instructor, however, is in control of what can be displayed by the GIS and careful design of the various map layers is critical to making GIS an effective teaching tool in the field. Considerable thought, as well as trial-and-error, is generally necessary to create effective maps for a particular landscape. Students should be able to turn most map layers on and off as needed. The location of roads and towns may not be necessary to understand soil and landscape features, but they are essential to making the viewer feel correctly oriented on the map even though the GPS shows the position at all times. The types of soil features highlighted by the GIS will likely be different in different landscapes. Every landscape has a story to tell. It is the role of the instructor to determine what that story is and how best to tell it.

The number of classes displayed on any one layer should be as small as possible or it becomes very difficult to select distinctly different colors to symbolize the map. Five to seven classes is generally considered optimal (Brewer, 2002), but more classes may be necessary to prevent the map from becoming too general. We have used up to 16 classes, but at the risk of occasional ambiguity in a map. One should keep in mind that the screens on different computers may not display colors with the same intensity or fidelity, so colors that look distinctly different on one computer may appear almost the same on another. ColorBrewer (Brewer, 2002) is useful in providing guidance on selecting colors to symbolize a map. Poorly selected colors can make an otherwise informative data layer incomprehensible. Colors should be selected with an eye toward assuring that the map is clear and meaningful at a glance. If possible, colors should evoke the particular soil property being displayed. Our color scheme for soil drainage class, for example, assigns a deep brown color to the "well drained" class to evoke the homogeneous brown colors typical of our well drained soils, while increasingly grayer colors are assigned to wetter drainage classes to evoke the gray colors of soils with wetness characteristics. About 8% of men and 0.4% of

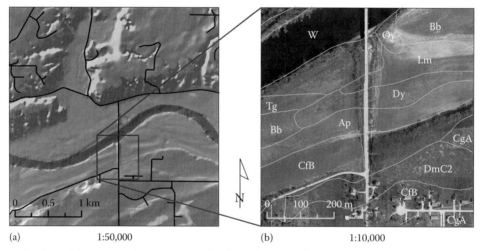

FIGURE 33.3 (See color insert.) The most informative map often depends on how fast one is moving over the landscape. (a) A more general thematic map, like this dominant soil parent material map displayed at 1:50,000, is informative and usually most appropriate when one is traveling in a vehicle. (Tan, glacial till; green, glacial outwash; light blue, recent alluvium; yellow, eolian sand and/or other sandy sediments.) (b) An aerial photograph overlain with the detailed soil delineations and displayed at 1:10,000 is often the most informative map when one is walking across the landscape. The polygons are queried individually to obtain information about the soils they represent.

women are affected by red-green colorblindness (Brewer, 2002). Ideally, the color scheme should be designed with this in mind so that all students will be able to interpret the maps correctly.

To provide the best user experience, the map should refresh as quickly as possible when it is zoomed and panned and when layers are turned on and off. Raster data are usually displayed very quickly, but polygon data can take a considerable time to display, particularly when more than a few thousand polygons are within the field of view. To increase the speed at which the map refreshes, polygon data should be rasterized. Raster data display more quickly at various resolutions when successive lower resolution copies of the data are stored along with the original data, a process called pyramiding. GIS software typically contains different algorithms for pyramiding raster data. It is worthwhile to try different algorithms for a particular dataset because some produce more pleasing small scale maps than others.

33.3.4 Pedagogic Aspects

Even though we now regularly use GIS while we are in the field, like most pedology classes, we still have our students study soil profiles and write morphological descriptions of them. We have not appreciably reduced the number of soil profiles examined or the number of auger holes bored. What we have done, however, is to make the time that we spend driving to field sites more productive by using the tablet PCs. To structure the learning process, we provide worksheets that ask our students to view particular layers at given scales at specific locations along our route so that we can be sure that they can locate and identify the features that we wish them to see. When we reach our field destination, we discuss the worksheets and go over what we expected the class to see along the route. Thus, over the course of the semester we are able to guide our students to formulate their own generalizations of how soils occur spatially and how the few soils that they saw in pits or auger borings are distributed more widely over the landscape. We also encourage our students to explore the maps on their own by zooming in and out and examining different layers as they see fit, thereby becoming active participants in their own learning.

The response of our students has been very enthusiastic and all have reported significant increases in comprehension and understanding of the course material, which our assessment has shown as well. The power of GIS to teach soil science became clear to us as instructors because we have learned many new and interesting things ourselves. For example, although we already understood the general glacial geomorphology of the area near campus, the full extent of the rich variety of subtle glacial features was not evident until we developed the GIS dataset. The soils vary from one geomorphic feature to the next, sometimes only slightly, sometimes greatly. We are now able to explain differences in soil properties in a geomorphic context rather than just describing and interpreting each pedon individually. The result is a rich new dimension in how we describe, and how our students understand, the soil landscape around them.

33.4 Teaching and Learning Using a Web-Based Geobrowser

Although taking GIS to the field on dedicated rugged tablet PCs has significantly enhanced teaching and learning, it has the disadvantage of reaching only a relatively small number of students in a few classes. To integrate a geospatial component into more of our undergraduate and graduate soil, crop, and environmental science courses, we have just launched the Integrating Spatial Educational Experiences (Isee) web site (http://isee.purdue.edu/), which will allow our students to access geospatial soils information via the Internet at any time or place. In addition to use in our own courses, we anticipate that the Isee web site will be used by students, teachers, and the general public throughout the state.

Acknowledgments

This work has been partially supported by grants from the Purdue University College of Agriculture Instructional Equipment Grants program, the Information Technology at Purdue (ITaP) Digital Content Development Grants program, and by U.S. Department of Agriculture Higher Education Challenge Grant no. 2008-01955.

References

Bransford, J.D., A.L. Brown, and R.R. Cocking (eds.). 2000. How people learn: Brain, mind, experience, and school. National Academy Press, Washington, DC.

Brewer, C. 2002. (Copyright 2002–2009). ColorBrewer 2.0. http://colorbrewer.org/ (accessed June 28, 2010).

ISDP. 2009. Indiana spatial data portal. http://gis.iu.edu (accessed June 28, 2010).

NASA. 2009. Advanced spaceborne thermal emission and reflection radiometer (ASTER) global digital elevation model. https://wist.echo.nasa.gov/ (accessed June 28, 2010).

National Research Council. 2006a. Learning to think spatially: GIS as a support system in the K-12 curriculum. The National Academies Press, Washington, DC. Available online at: http://www.nap.edu/catalog.php?record_id=11019 (accessed June 28, 2010).

National Research Council. 2006b. Beyond mapping: Meeting national needs through enhanced geographic information science. The National Academies Press, Washington, DC. Available online at: http://www.nap.edu/catalog.php?record_id=11687 (accessed June 28, 2010).

Soil Survey Staff. 2009. Soil data mart, Natural Resources Conservation Service, United States Department of Agriculture. http://soildatamart.nrcs.usda.gov (accessed June 28, 2010).

USGS. 2008. Shuttle radar topography mission (SRTM). http://srtm.usgs.gov/ (accessed June 28, 2010).

USGS. 2009. U.S. Geological Survey, Elevation products. http://eros.usgs.gov/#/Find_Data/Products_and_Data_Available/Elevation_Products (accessed June 28, 2010).

Index

A

Acaulospora morrowiae, 8-21
Acidification/alkalization, 6-11–6-12
Acid sulfate soils
    characteristic minerals and geochemical processes
        oxidation processes, 21-6–21-7
        pyrite oxidation rates, 21-7
        reduction processes, 21-4–21-6
    characteristics impact
        acid sulfate profile, 21-2
        pyrite, 21-2
        RIS, 21-2
        sulfuric soil material, 21-3
        tidal swamps and salt marshes, 21-2
    consistence and strength, 21-13
    detrimental impacts, 21-1
    hazards
        acidification, 21-8–21-9
        deoxygenation, water bodies, 21-10–21-11
        iron mobilization, 21-9
        metal and metalloid mobilization, 21-9–21-10
        noxious gases production, 21-11
        scalding, 21-12
    landscape (*see* Soil landscape)
    occurrence impact, 21-3–21-4
    permeability, 21-13
    RIS, 21-1
    soil physical and chemical processes, 21-7–21-8
    sulfidic material
        acid–base accounting, 21-17–21-18
        incubation methods, 21-18
        monosulfidic materials, 21-18–21-19
        reduced inorganic sulfur determinations, 21-17
        sequential extraction methods, 21-19
        TPA/TAA method, 21-16
    sulfuric material, 21-16
Advanced Photon Source (APS), 1-3
African soil information, 32-1–32-4
Anthropogenic organic pollutants (AOPs), 9-1
    degradation/removal, 9-25–9-26
    rhizosphere processes, 9-26
    soil sorption, soil-water interface, 9-5–9-6

Arabidopsis thaliana, 6-10
Areias Quartzosas distróficas, 29-7–29-8
Atriplex nummularia, 6-7
Automated land evaluation system (ALES), 30-7

B

Bacillus subtilis, 8-5, 10-14
Big Spring Number Eight (BSNE) sampler, 23-13
Biogeochemical weathering processes
    hot spots, 10-4
    organic matter infusion impacts
        aqueous phase and surface complexation effects, 10-10–10-12
        autotrophic photosynthate, 10-6
        biogeochemical reactivity, 10-6
        bioligand-mediated mineral dissolution, 10-12
        dissimilatory reduction reactions, 10-6
        DOM reaction and fate, 10-6–10-7
        fulvic acid, 10-6
        labile structure and reactivity, DOM, 10-7
        microbial adhesion, 10-14–10-16
        mineral-organic solid-phase reaction products, 10-10
        natural organic matter, 10-6
        oxidation–reduction, 10-12–10-13
        sorption–desorption, 10-7–10-10
    organo-mineral weathering incongruency, 10-4–10-5
    primary mineral weathering, 10-5
    solid-solution interfaces, 10-4
    solid-water interface, 10-4
    surface sensitive methods, 10-4
    vascular plants, weathering *vs.* denudation, 10-5–10-6
Biomass to water ratio (BWR), 15-8
Boron–soil interaction
    adsorption capacity, 17-7
    boron adsorption modeling, 17-8–17-10
    kinetics and mechanisms, 17-7–17-8
    soil constituents, 17-7–17-8
    steady-state boron concentration, 17-7
    transport modeling, 17-10

Brazil soil databases, *see* Soil database
Brunauer–Emmett–Teller (BET) equation, 2-9

C

Calcium and magnesium bioavailability
    atmospheric deposition, 11-48
    crop respons, 11-50–11-51
    enzyme activator, 11-48
    exchangeable cations, 11-49
    fertilizers, 11-48
    hardwood tree species, 11-49
    leaching and crop removal, 11-48
    loblolly pine, 11-49
    macronutrients, 11-47
    occurrence, 11-48
    plant mineral composition, 11-49
    plant removals, 11-49
    plant tissues, 11-47
    soil–plant relationships, 11-49–11-50
    soil solution, 11-49
    Tennessee forest, 11-48
Canadian Soil Quality Evaluation Program, 26-7
Carbon sequestration, 18-19
Cation exchange capacity (CEC), 2-3
Clay-organic interaction
    1:1 and 2:1 layer structure, 2-3
    bioorganomineral system, 2-1
    clay-size minerals
        bacterial cells, aggregate formation/stabilization, 2-2, 2-5
        metal (hydr)oxides, 2-4
        phyllosilicates, 2-2–2-3
        short-range order minerals, 2-3–2-4
    effects
        greenhouse gas emission impact, 2-12
        nutrient dynamics, 2-10–2-11
        organic matter dynamics/turnover, 2-10
        soil structure/aggregate stability, 2-9–2-10
        sorption/bioavailability impact, 2-11
    humic substances-clay mineral surfaces bonding, 2-7
    mechanisms
        humic substances, 2-6–2-9

Index-1

low molecular-weight organic compounds, 2-5–2-6
methodology, 2-9
organic matter, 2-4–2-5
soil environment/stabilization, 2-1
SSA, composition/charge characteristics, 2-3
zonal model, 2-8
Clean Water Act (CWA), 24-2
Co K-edge XANES spectra, 1-4
Comprehensive nutrient management plans (CNMP), 24-3
Concentrated animal feeding operations (CAFOs), 26-8
Confined animal feeding operations, 24-1
Conservation tillage
   agriculture and tillage, 25-1
   agronomic production, 25-23
   biofuel (cellulosic) production implications
      cellulosic ethanol production, 25-11
      corn stover, 25-11
      crop residue removal, 25-12
      threshold levels, 25-12–25-13
      wind and water soil erosion, 25-11
   biological activity, 25-8–25-9
   carbon dynamics and sequestration, no-tillage
      agriculture, 25-9
      macroaggregates, 25-10
      management, 25-11
      moldboard plowing, 25-9–25-10
      root mass decomposers, 25-10
      soil organic carbon and matter, 25-10
      soil organisms, 25-9
      surface and decomposition, 25-10
      surface residue mulching, 25-11
   climate interactions, 25-21–25-22
   compaction, 25-17–25-18
   conservation agriculture, 25-20–25-21
   cover crops, 25-20
   cropping systems, 25-18–25-20
   crop residue management, 25-2–25-3
   CT, 25-1–25-2
   disadvantages, 25-22–25-23
   herbicide, 25-1–25-2
   moldboard plow, 25-1
   mulch tillage, 25-2
   natural setting, 25-23
   no-tillage (NT), 25-1–25-2
   residue management and decomposition, 25-7–25-8
   ridge tillage, 25-2
   soil fertility and nutrition (see Soil fertility and nutrition)
   soil science principles, 25-2
   strip tillage, 25-2
   surface sealing, crusting, and seedling emergence, 25-16–25-17
   water conservation
      chemical fallow, 25-7
      corn grain production, 25-6
      crusted soil surface, 25-6
      fresh water availability, 25-7
      Hoytville silty clay loam, 25-6
      humid/subhumid regions, 25-5
      intensive cropping systems, 25-6
      precipitation, 25-6
      sandy loam Alfisol, 25-5
      silty clay loam, 25-6
      soil desiccation, 25-7
      soil water contents, 25-6
      soil water storage and use, 25-5
      stubble mulching, 25-6
      water infiltration and storage, 25-7
   water erosion, 25-3–25-4
   wind erosion, 25-4–25-5
Conventional tillage (CT), 25-1–25-2
Critical flocculation concentration (CFC), 17-2
Critical micelle concentration (CMC), 9-27

D

Derjaguin, Landau, Verwey, and Overbeek (DLVO) theory, 3-10–3-11
Digital Elevation Models (DEMs), 33-4
Digital spatial and attribute databases
   Areias Quartzosas distróficas, 29-7–29-8
   great soil groups, 29-6
   inceptisol, 29-8
   Mato Grosso soil map, 29-5–29-6
   oxisol, 29-8
   Pantanal depression, 29-8
   Paraná Sedimentary Basin region, 29-8
   Rondônia, 29-6
   SEPLAN-RO and SEPLAN-MT, 29-5
   SIGTERON, 29-6
   soil components, 29-6
   soil survey maps, 29-8
   Solos Concrecionários, 29-8
   SOTRO, 29-6
Dissimilatory NO_3^- reduction to ammonium (DNRA), 20-14
Dissolved organic carbon (DOC), 2-11
Dissolved organic matter (DOM), 5-15

E

Endocrine disrupting compounds (EDCs), 10-20, 24-17
Environmental nanoparticles (ENPs) impacts
   aggregation, 4-12
   occurrence distribution/properties, 4-11
   molecular-scale mechanism, 4-12
   occurrence/origin, terrestrial ecosystem
      manufactured, 4-4–4-5
      natural, 4-2–4-4
      toxicity, 4-5
   size-dependent properties, 4-2, 4-6
   soil-/geo-processes
      adsorption, 4-10–4-11
      advective mobility, 4-10
      desorption, 4-10–4-11
      diffusive mass transfer, 4-10
      dissolution, 4-11
      electron transfer reactions, 4-11
      heterocoagulation reaction, 4-9
      industrial-scale production, 4-9–4-10
      isolate/characterization, 4-9
      organic matter/particles, 4-9
      precipitation, 4-11
   soil solution interaction
      crystal growth mechanism, 4-5–4-6
      dislocations, 4-7
      interfacial electron-transfer reactions, 4-8–4-9
      kinetic equation, 4-7
      nanoparticle aggregation, 4-7–4-8
      nanoparticles, sorbents, 4-8
      pH, 4-7
      phase transformation, 4-7
      stability, 4-7
      TEM/magnetometry, 4-6
   terrestrial ecosystems, 4-1–4-2
   toxic effect, 4-11
Extended x-ray absorption fine structure (EXAFS), 1-4–1-5
Extracellular polysaccharides (EPS), 2-2

F

Federal Geographic Data Committee (FGDC), 28-10–28-11
Fertility Capability Classification (FCC), 26-19
Fertilizer application
   calcium and magnesium, 14-9–14-10
   immobile nutrients
      banding, 14-6
      broadcasting, 14-6
      CEC and leaching, 14-5
      fertilizer placement, 14-6
      foliar application, 14-7
      irrigation water, 14-7
      point fertilization, 14-6–14-7
      soil components, 14-5
      strip application, 14-6
   mass flow and diffusion, 14-1
   micronutrients
      copper, 14-11
      iron, 14-10–14-11
      manganese, 14-10
      molybdenum, 14-11
      zinc, 14-10
   mobile nutrients (*see* Mobile nutrients)
   nitrogen and phosphorus, 14-1
   nutrient application, 14-1
   phosphorus and potassium placement
      band and broadcast applications, 14-7–14-8
      Barber–Cushman model, 14-7
      conservation tillage system, 14-8–14-9
      forage crops, 14-7
      no-till corn, 14-8
      pop-up/in-furrow placement, 14-8
      seed-placed fertilizer, 14-8
      starter fertilizer, 14-7–14-8
   root interception, 14-2
Food and Agriculture Organization (FAO), 27-2
Fourier transform infrared (FTIR), 1-8, 2-9

G

Geographical Information System (GIS), 27-1
Global Assessment of Land Degradation and Improvement (GLADA), 30-12
Global positioning system (GPS) technology, 28-11
Glomus intraradices, 8-21

H

Harmonized World Soil Database (HWSD), 27-2–27-4
Helianthus annuus, 8-22
Hydraulic conductivity (HC)
    clay swelling and dispersion, 18-7
    porous material structure, 18-7
    sodicity and electrolyte concentration, 18-8–18-9
    soil properties
        clay mineralogy, 18-9
        exchangeable magnesium, 18-10
        mineral weathering, 18-9
        soil texture, 18-9
Hydrophile–lipophile balance (HLB) scale, 9-27
Hydrophobic organic compounds (HOCs), 9-1
Hymenoscyphus ericae, 8-21

I

Illinois soil nitrogen test (ISNT), 11-9, 13-16–13-17
Industrial by-products
    flue and kiln dusts, 24-7–24-8
    fly ash, 24-6
    gypsum, 24-6–24-7
    slags, 24-5–24-6
    toxic contaminants, 24-5
Infiltration rate (IR)
    sodicity and conditions prevailing, 18-11–18-13
    sodicity effect, 18-10–18-11
    structural seal formation, 18-10
    volume flux, water flowing, 18-10
International Soil Reference and Information Centre (ISRIC), 27-6

K

Kriged standard deviation (KSD), 29-15–29-16

L

Land application
    agricultural systems, 24-1–24-2
    by-products types, 24-2
    CAFOs, 24-1
    CNMP, 24-3
    environmental aspects, by-product applications
        dump toxic waste, 24-14
        governmental regulations concerning, 24-14–24-15
        public health effects, 24-14
        recycling nutrients and organic material, 24-14
        soil and food-chain contamination (*see* Soil and food-chain contamination)
        water quality considerations, 24-18–24-19
    EPA, 24-3
    Haber process, 24-2
    hazardous/toxic waste, 24-3
    industrial and agricultural activities, 24-1
    land-applied by-products
        agricultural materials, 24-3–24-4
        biosolids and municipal solid waste, 24-8–24-9
        composting and mixing, 24-9
        forestry products, 24-4–24-5
        industrial by-products (*see* Industrial by-products)
    soil fertility
        agriculture/environmental regulation, 24-10
        nutrient content and bioavailability, 24-10–24-11
        plant nutrients, 24-10
        primary effect, 24-10
        productivity, 24-1
    soil horizons, 24-1
    soil properties
        biological properties, 24-12–24-13
        chemical properties, 24-13
        physical properties, 24-12
    USEPA, 24-2–24-3
    waste materials, 24-2
Land capability classification (LCC), 26-16
Layered double hydroxides (LDH), 1-10
Linear least squares fitting (LLSF), 1-8

M

Major land resources area (MLRA), 28-3
Mato Grosso soil map, 29-5–29-6
Metals and metalloids
    arbuscular mycorrhizal fungi effect
        Acaulospora morrowiae, 8-21
        biotransformation processes, 8-22
        Glomus intraradices, 8-21
        Helianthus annuus, 8-22
        Hymenoscyphus ericae, 8-21
        inoculation, 8-21
        metal-binding processes, 8-21
        physical–chemical modification, 8-21–8-22
        soil concentration/contamination, 8-20–8-21
        soil–plant system, 8-20–8-21
        structure and product, 8-21
        Viola calaminaria, 8-21
    biomineralization, 8-14
    hard/soft acids and base, 8-2
    mobility, 8-1
    physicochemical-biological interfacial interactions
        anions, 8-17–8-18
        biogeochemical cycle, 8-14
        chemical equilibrium models, 8-14–8-15
        DTPA-TEA soil test, 8-17
        end-over-end extraction procedure, 8-15
        environment/human health risks, 8-14
        fractionation, 8-15
        metal–organic complex, 8-15–8-16
        mineralogical phases, 8-15
        mobility/plant uptake, 8-17
        multistep selective sequential extraction scheme, fractionation, 8-16
        phytoavailability/mobility, 8-17
        protection and remediation, environment, 8-14
        solid-phase fraction, 8-16–8-17
    plants and managements effects
        acidification, 8-19
        affecting factors, 8-19
        deficiency and toxicity, 8-18
        dissolution/precipitation, 8-19
        hazards, 8-18
        mobilization/complexation, 8-20
        oxidation, 8-19
        physical–chemical soil parameters, 8-18
        rhizosphere depletion/accumulation, 8-19
        rhizosphere microorganisms, 8-20
        root-response mechanisms, 8-18–8-19
        soil nutrient deficiency, 8-19–8-20
        soil water content, 8-19
    solution complexation reaction, 8-2
    sorption–desorption reactions, 8-1–8-2
    sorption processes (*see* Sorption processes)
    trace elements, 8-1
Micronutrients bioavailability
    animal and plant sources, 11-61–11-62
    bioaccumulation and plant uptake, 11-67
    biologically essential and nonessential elements, 11-61
    biological tests
        microbial availability, 11-68, 11-70
        phytoavailability, 11-68
    chelating and complexing agents, 11-65–11-66, 11-71
    chemical bioavailability, 11-63
    chemical tests
        diffusive gradient thin film, 11-68
        isotopic techniques, 11-68
        sequential fractionation, 11-68–11-69
        single extraction, 11-67–11-68
    chemical transformation processes
        cation-exchange reactions, 11-64
        leaching and runoff, 11-66–11-67

mineral surfaces adsorption, 11-64, 11-66
    precipitation, 11-66
    soil organic matter adsorption, 11-66
    sorption and complexation, 11-64–11-65
definition, 11-67
geogenic and anthropogenic processes, 11-63
heavy metals/heavy metalloids, 11-63
index cation and solution composition, 11-70
inorganic and organic ligands, 11-70–11-71
ionic strength, 11-70
liming materials, 11-71
methylation/demethylation, 11-67
organic composts, 11-71
oxidation/reduction, 11-67
pH, 11-70
phosphate compounds, 11-71
plants and soil microorganisms, 11-63
primary and secondary minerals, 11-67
soil testing and crop response, 11-71, 11-73
    boron, 11-74
    copper, 11-74
    deficiency symptoms, 11-71–11-72
    diagnosis and prognosis, deficiencies/toxicities, 11-72
    iron, 11-74
    manganese, 11-74
    molybdenum, 11-74
    nickel, 11-74
    zinc, 11-65–11-66, 11-73–11-74
sources, 11-63–11-64
trace elements, 11-61

Mineral weathering, 17-1

Mobile nutrients
    boron, 14-5
    chloride, 14-5
    nitrogen
        application time, 14-2–14-3
        fertilization program, 14-2
        managing fertilizer nitrogen, 14-3–14-4
        nonlegume crops, 14-2
        sources, 14-2
    sulfur
        legumes, 14-4
        rainfall/over-irrigation, 14-5
        SOM, 14-4
        sources and methods, 14-4–14-5
        transformations and chemistry, 14-4

Modified Wilson and Cooke (MWAC), 23-13
Multiple limitation hypothesis (MLH), 16-3

N

Nanoscale science and technology
    microorganisms interaction, 3-1
    nanoenabled sensing, 3-13
    nanomaterials behaviors, 3-2
    nanoparticles and nanominerals, 3-2
    nanotoxicity, 3-12–3-13
    natural and engineered nanoparticles, 3-3–3-4
    natural constituents, 3-1
    near-earth-surface environment, 3-1–3-2
    particle size and stability
        classical theory, 3-7
        nanoparticle stability, 3-7–3-8
    properties/behaviors nanoparticles
        classical mineral nucleation theory, 3-6
        mineral growth, 3-9
        nanoparticle sorption, 3-9
        nanoparticle surface area, 3-5
        nanoscale porosity, 3-10
        particle size/stability, 3-7
        redox phenomena, 3-9–3-10
        size and structure, 3-4–3-5
        stability, 3-8–3-9
        surface free energy, 3-5–3-6
    soil environment, 3-13
    soils and sediments
        air–water interface, 3-11
        DLVO theory, 3-10–3-11
        nanoparticle aggregation, 3-11
        natural/manufactured aggregation, 3-11–3-12
        particle aggregation, 3-11
        sorption, 3-11
        Stokes' Law, 3-10
        suspension/aggregation, 3-11
        TiO_2 nanoparticle aggregates, 3-11

NASIS, see National Soil Information System (NASIS)
National Cooperative Soil Survey (NCSS), 28-1
National Soil Information System (NASIS)
    field data collection system, 28-8
    SDW, 28-9
    security and software design, 28-8
    software development and implementation status, 28-8–28-9
    software, hardware, and network requirements, 28-8
    soil data warehouse and soil data mart, 28-9
    soil map units and map unit components, 28-2
    soil profile descriptions and interpretation, 28-2
    system objectives, 28-8
    transactional database and application, 28-2

National soil profile database
    auditing routines, 29-3
    CEC, 29-3
    Köppen climate zones and types, 29-4
    Radambrasil project, 29-2
    Radam volumes, 29-3
    soil electronic databases, 29-2
    soil taxonomy, 29-4
    variables/soil attributes, 29-3

National Synchrotron Light Source (NSLS), 1-2
Natural Resources Conservation Service (NRCS), 28-1
Near-edge absorption fine structure (NEXAFS), 2-9

Nitrogen bioavailability
    Adapt-N, 11-9
    ammonia volatilization
        anhydrous ammonia and aqua ammonia fertilizers, 11-8
        manures, 11-8
        pH and temperature, 11-8
        plant tissues, 11-8–11-9
    ammonium-nitrate conversion, 11-6
    denitrification, 11-7–11-8
    Fried and Dean "A" values, 11-9
    interlayer clay minerals, ammonium, 11-9
    ISNT, 11-9
    Mitscherlich "b" and "c" values, 11-9
    nitrogen cycle, 11-1–11-2
    nitrogen immobilization, 11-6
    nitrogen leaching reduces, 11-6–11-7
    nitrogen mineralization
        animal manure and organic nitrogen sources, 11-5
        cover crop effects, 11-4–11-5
        crop rotation and residue effects, 11-3–11-4
        soil organic matter, 11-3
    plant nutrient
        ammonium and nitrate nutritional effects, 11-3
        plant growth, 11-2
        plants and microorganisms, 11-2–11-3
        plant tissue sufficiency concentrations, 11-2
    PPNT, 11-9
    PSNT, 11-9
    soil phosphorus cycling, 11-14–11-15
    tissue testing and hand-held chlorophyll meters, 11-9
    transformations and transport, 11-1

Northwest Coastal Zone (NWCZ), 23-11
NUE, see Nutrient use efficiency (NUE)

Nutrients and pollutants
    biogeochemical disequilibria, 10-3
    biogeochemical weathering processes (see Biogeochemical weathering processes)
    biosynthesis and humification, 10-3–10-4
    critical zone interfaces, 10-1–10-3
    microbial intervention, metal cycling
        biogeochemical gradients, 10-24–10-25
        biogeophysical heterogeneities, 10-25–10-26
        contaminant redox reactions, 10-22–10-24
        iron redox cycling, 10-22
        major and trace elements, 10-22
        microbial mediation, 10-21
        oxic soils, 10-21

photosynthesis/dissimilatory sulfate
reduction, 10-21
redox fluctuations, 10-22
TEAPs, 10-22
molecular-scale interactions, 10-1
organo-mineral heteroaggregates (*see*
Organo-mineral heteroaggregates)
NOM and microbial cells, 10-16
surface chemistry, 10-16
Nutrient use efficiency (NUE)
adjusting rate, timing, placement, and
source, fertilizer application
nutrient application rate, 15-5
nutrient application time, 15-5
nutrient source, 15-5–15-6
placement, nutrient, 15-5
biotechnology and variety selection, 15-7
Bray's nutrient mobility, 15-1–15-2
causes, 15-3
cereal, 15-2
cover crops and crop rotation use,
15-6–15-7
fertilizer, 15-2
foliar fertilization, 15-6
integrated crop and nutrient management
methods, 15-3
nitrogen and phosphorus, 15-1
plant nutrient management., 15-1
PUE, 15-3
sensor-based variable rate nutrient
management, 15-7
urease and nitrification inhibitors, N
fertilizers, 15-6

O

Official Series Descriptions (OSDs), 28-3–28-5
Organo-mineral heteroaggregates
aging effects, 10-20–10-21
inorganic molecules sorption, 10-18–10-19
mineral–organic complexes surface
composition
biogeochemical environment, 10-16
organic carbon concentrations, 10-16
surface charge effects, 10-16–10-18
XPS, 10-16
NOM and microbial cells, 10-16
organic pollutants sorption
hydrophobic organic contaminants,
10-19–10-20
polarity and functional group
chemistry, 10-19
trace organic contaminants, 10-20
van der Waals interaction and
hydrophobic effects, 10-19
Organo-mineral interaction
allophane and imogolite, 7-13–7-14
carbon/nitrogen, 7-12
catalysts minerals, 7-14–7-15
metal ions interaction, 7-14
mineral control paradigm, 7-10–7-12
pedogenic oxides, 7-13
phyllosilicate clay minerals, 7-12–7-13

P

Paraná Sedimentary Basin region, 29-8
Particulate organic matter (POM), 5-15
Pharmaceuticals and personal care products
(PPCPs), 10-20, 24-17
Phosphorus availability
anion-exchange resins, 11-17
biological processes, 11-15, 11-21
chemical processes
precipitation–dissolution, 11-18–11-19
sorption/desorption, 11-19–11-20
complex and interrelated processes, 11-14
components, forms, availability, and
cycling, 11-14
crop uptake, 11-17–11-18
fauna and flora processes
earthworms, 11-21–11-22
mycorrhizal associations, 11-22
ion-exchange membranes, 11-17
iron-oxide impregnated paper, 11-17
land management, 11-26
plant growth, 11-14
soil forms and amounts
Al, Fe, Mn, and Ca content,
11-15–11-16
inorganic (P_i) and organic (P_o),
11-14–11-15
inositols and humic acids, 11-15
Kingsbury clay and Hagerstown silt
loam, 11-16
mineralization, 11-16
pH, 11-15
phospholipids and fulvic acids, 11-15
phosphorus content, 11-15
plant uptake, 11-16
soil phosphorus availability
critical concentrations, plant
production, 11-22–11-23
placement, 11-25
rate, 11-23
residual availability, 11-25
type, 11-23–11-25
sorption, desorption, and precipitation,
11-14
water quality implications, 11-27–11-28
Phosphorus use efficiency (PUE), 15-3
Plant and soil microbes, uptake/degradation
AOP degradation/removal, 9-25–9-26
aqueous phase, dissolved organic
molecules, 9-17
bioavailability, 9-17
biodegradation, 9-19
biotransformation, 9-17
chemical molecules, 9-17
crop-contamination problem, 9-22
microorganisms and plants, 9-18
models, plant uptake, 9-25
phytostabilization, 9-22
plant uptake pathways, 9-22–9-23
pollutants, 9-22
rhizosphere processes, AOPs, 9-26
root uptake/translocation, 9-23–9-24

soil microorganisms
aerobic/anaerobic metabolism,
9-20–9-21
bioavailability, 9-21–9-22
biodegradation, 9-22
degradation, 9-20
fungal metabolis, 9-21
microcolony-/biofilm-like structure,
9-20
octanol–water partition coefficient,
9-20
variety, 9-20
soil pollution, 9-22
Plant nutrients
nitrogen
agricultural production and
environmental quality, 13-15
arid and humid regions, 13-15
automated colorimetry and ion
chromatography, 13-15
biological assimilation, 13-15
chemical extraction methods, 13-15
ISNT, 13-16–13-17
plants absorb nitrogen, 13-15
PSNT, 13-16
residual inorganic nitrogen, 13-15
Rhizobium and *Bradyrhizobium,* 13-15
soil nitrogen cycle, 13-15
soil profile nitrate test, 13-16
soil tests interpretation, 13-17
phosphorus, 13-10–13-12
potassium, calcium, and magnesium soil
testing
anorthite and dolomite, 13-12
buffer capacity, 13-12
cycling and plant availability, 13-12
feldspars and micas, 13-12
interpretation, 13-13
sandy surface horizons, 13-13
soil sample handling, 13-13
soil samples, 13-13
soil testing methods, 13-12–13-13
soluble, exchangeable, and
nonexchangeable forms, 13-12
weathering minerals, 13-12
sulfur, 13-14
Preplant soil profile nitrate (PPNT), 11-9
Presidedress soil nitrate test (PSNT), 11-9,
13-16
Principal component analysis (PCA), 1-8
Pseudomonas aeruginosa, 10-14

Q

Quick-scanning X-ray absorption
spectroscopy (QXAS), 1-8

R

Redox reactions, 1-11–1-12
Reduced inorganic sulfur (RIS), 21-1
Regional organic carbon storage maps
Areais quartzosas, 29-13

Brazilian Amazon basin, 29-13
carbon stocks
　classical geostatistical approach, 29-16–29-17
　experimental semivariogram, 29-15
　geostatistical interpolation, 29-15
　geostatistical treatments, 29-15
　KSD, 29-15–29-16
　soil maps, 29-15
　variographic analysis, 29-16
database and carbon stock calculation, 29-13–29-14
spatial components, 29-13
Regional soil and terrain databases (SOTER), 27-4–27-5
Resource Conservation and Recovery Act (RCRA), 24-2
Revised wind erosion equation (RWEQ)
　critical field length (s), 23-9
　cropland, 23-11
　erosion season, 23-10–23-11
　field length and transport mass, 23-8–23-9
　mass flux, 23-8–23-9
　mass transport equation, 23-8
　maximum transport capacity (Q_{max}), 23-9
　noninstrumented fields, 23-9
　nonlinear regression analysis, 23-9
　NWCZ, 23-11
　soil loss, agricultural lands, 23-8
　soil roughness, 23-11
　wind factor, soil erodible fraction and soil crust factor, 23-9
　wind speed, 23-8
Rhizosphere
　biochemistry
　　abiotic mineralization, 6-3
　　cytoplasmic metabolism, 6-4
　　ecology significance, 6-5
　　environmental factors, 6-4
　　enzymatic processes, 6-6–6-7
　　exudation, 6-4
　　membrane transport processes, 6-5
　　microbial decomposition, 6-4
　　mucilage production, 6-4–6-5
　　N_2 fixing, 6-4
　　plant growth enrichment, 6-4–6-5
　　plant–soil system, 6-3
　　plant species role, 6-5
　　root metabolic activity, 6-4
　　soil microorganisms, 6-3
　　soil physical properties, 6-3
　　soluble organic nutrients uptake, 6-5–6-6
　　SOM, 6-3
　biogeochemistry
　　acidification/alkalization, 6-11–6-12
　　complexation and chelation, 6-16–6-18
　　desorption and ligand exchange, 6-12
　　gas exchange and redox processes, 6-10–6-11
　　minerals, dissolution and precipitation, 6-13–6-14
　biophysics
　　Arabidopsis thaliana, 6-10
　　buffering capacity, 6-9
　　computer-assisted tomography, 6-7
　　effective diffusion coefficient, 6-9
　　Eucalyptus grandis, 6-7
　　hydraulic resistance, 6-9
　　inorganic nutrients, 6-9
　　nutrient flux, 6-9
　　plasticity, 6-7
　　porosity, 6-8
　　root distribution, 6-9
　　soil adhering, 6-8
　　soil matrix/strength, 6-7
　　soil–root interface, 6-9
　　spatial extension, 6-7
　　water/nutrients, 6-8
　biosphere carbon sequestration, 6-1
　ecology
　　competition and facilitation, 6-16
　　microorganisms communication, 6-15–6-16
　　plants-microbes communication, 6-14–6-15
　endomycorrhizal root, schematic representation, 6-1–6-2
　microorganisms, 6-1
　mineral nutrients/root, 6-1, 6-3
Rondonia–Mato Grosso, 29-6
RWEQ, *see* Revised wind erosion equation (RWEQ)

S

Saline and boron-affected soils
　boron control, leaching, 17-15–17-16
　boron–soil interaction (*see* Boron–soil interaction)
　irrigation water quality
　　bicarbonate, 17-2
　　CFC, 17-2
　　dissolved solute concentration and composition, 17-2
　　electrical conductance and total dissolved solids, 17-2–17-3
　　electrolyte concentration, 17-3
　　exchangeable sodium percentage (ESP), 17-4
　　ions, 17-2
　　irrigated agriculture, 17-2
　　root zone salinity, 17-3–17-4
　　salinity and sodicity hazards, 17-3
　　SAR, 17-3–17-5
　　water classes, 17-3–17-4
　origin and distribution, 17-1–17-2
　plants
　　crop tolerance, 17-10–17-11
　　specific molecule and ion effects, 17-11–17-13
　salinity control, leaching
　　flood and sprinkler irrigation, 17-13
　　furrow/drip irrigation leaches, 17-14
　　lateral flow and salt removal, 17-13
　　leaching efficiency, 17-13
　　leaching reclamation, 17-15
　　rainfall, flood/sprinkler irrigation, 17-13
　　salt–water dilution technique, 17-14–17-15
　　sodic clay loam soil, 17-14–17-15
　　soil $CaCO_3$, 17-15
　　soil solution concentration and salt transport mechanisms, 17-13
　　tile/open ditch drains, 17-14
　　UNSTCHEM model, 17-13
　　WATSUIT model, 17-13
　soil physical properties, 17-5–17-7
　water quality parameter, 17-3
Scanning transmission x-ray microscopy (STXM), 2-9
Sea water intrusion, 17-1–17-2
Sequential extraction procedure (SEP), 8-15
SGDB, *see* Soil geographic database (SGDB), Russia
Silicon bioavailability
　crop response, 11-54–11-55
　nutrition, 11-54
　occurrence, 11-54
Sistema de Informação Geográfico de Terrenos e Solos para o estado de Rondônia (SIGTERON), 29-6
SOC, *see* Soil organic carbon (SOC) stocks
Sodicity
　aggregate stability and organic matter
　　aggregation, 18-6
　　organic matter effects, 18-7
　　sodicity effects, 18-6–18-7
　amelioration and management
　　acids and sulfur, 18-16
　　calcium carbonate, 18-16
　　calcium chloride, 18-16
　　gypsum, 18-15–18-16
　　salt water dilution method, 18-17
　　tillage, 18-16–18-17
　　vegetative remediation, 18-17–18-18
　clay charge and diffuse double layer, 18-4
　crop production and sustainable agriculture, 18-1
　crop responses
　　ion toxicity, 18-14
　　plant nutrition, 18-14
　　tolerance, crop, 18-13–18-14
　definition, 18-1–18-3
　distribution, 18-3–18-4
　environmental aspects
　　carbon sequestration, 18-19
　　erosion and suspended sediments, 18-18
　　salinity and pollutants, 18-18–18-19
　saline and alkali soils, 18-1
　sodium-affected soils, 18-1
　sodium-related soil degradation, 18-19
　swelling and dispersion
　　reference clay systems, 18-4–18-5
　　soil clay systems, 18-5
　water flow
　　HC (*see* Hydraulic conductivity (HC))
　　IR (*see* Infiltration rate (IR))

Sodium adsorption ratio (SAR), 17-2
Soil acidity and liming
    agricultural and industrial processes, 12-1
    Al concentrations, 12-1
    exchangeable acidity, 12-1
    ice-free land surface, 12-1
    lime material characteristics and application
        ameliorating subsoil acidity, 12-5
        calcium carbonate rating, 12-5
        calcium-magnesium carbonate, 12-4
        conventional tillage systems, 12-5
        liming materials fineness, 12-4–12-5
        natural deposits/industrial by-products, 12-4
        neutralize acidity, 12-5
        neutralizing value, 12-4
        no-till systems, 12-5
    lime requirement methods
        buffer pH methods, 12-3
        crop growth, 12-4
        exchangeable aluminum, 12-3–12-4
        liming mineral soils, 12-4
        potential acidity constraint, 12-4
        scab disease, 12-4
        soil–lime incubations, 12-3
        soybean cyst nematode, 12-4
    plant growth, acid soil constraints
        aluminum toxicity, 12-2
        calcium and magnesium deficiencies, 12-2
        hydrogen toxicity, 12-2
        manganese toxicity, 12-2
        phosphorus and molybdenum deficiencies, 12-2–12-3
Soil and food-chain contamination
    antibiotics, hormones, and pesticides, 24-17
    metals and organic contaminants, 24-15
    nitrate and ammonium, 24-17–24-18
    trace inorganic contaminants, 24-15–24-16
    trace organic contaminants, 24-16–24-17
Soil and land information systems, China
    land degradation and improvement, 30-12–30-13
    land suitability assessment
        ALES, 30-7
        basic mapping unit, 30-7
        land evaluations, 30-7
        physical and chemical fertility, 30-8
        selection and assessment models, 30-6–30-7
        soil erosion risk assessment, 30-8–30-10
        soil quality assessment, 30-8–30-9
        SOTER units, 30-7
        tropical crops, 30-10–30-12
    national soil survey stage, 30-3
    resources, 30-1
    Russian school stage, 30-1–30-2
    soil database and information system stage, 30-4–30-6
    soil status assessment
        multiple factor analysis, 30-5, 30-7
        regional and national soil databases, 30-4
        single factor analysis, 30-5–30-6
    soil survey and Marbut soil classification stage, 30-1–30-2
    soil types and soil properties, 30-1
Soil characterization database (SCD), 28-1–28-2
Soil conservation service (SCS), 28-1
Soil database
    Africa
        detail scale, 32-5–32-6
        interpretation and utilization, 32-6
        reconnaissance scale, 32-5
        Sahara desert and the Congo River basin, 32-1
        soil characterization, 32-1–32-4
        soil profile databases, 32-6
        ultra-detailed scale, 32-6
    biophysical setting, 29-2
    integration and application (*see* Soil integration and application)
    maps
        accelerated data access, 27-2
        digital soil mapping, 27-6–27-7
        FAO-UNESCO, 27-2
        GIS software, 27-1
        global and regional soil maps, 27-8
        HWSD, 27-2–27-3
        ISRIC–WISE, 27-6
        policy makers, 27-8
        soil survey, 27-1–27-2
        SOTER, 27-4–27-5
        WISE, 27-8
    soil attribute
        digital soil properties, 29-4–29-5
        digital spatial and attribute databases (*see* Digital spatial and attribute databases)
        national soil profile database (*see* National soil profile database)
        Rio de Janeiro, digital mapping, 29-8–29-10
    soil organic carbon storage, 29-1
Soil Data Mart (SDM), 28-9
Soil Data Viewer (SDV), 28-9
Soil Data Warehouse (SDW), 28-9
Soil-ecological zoning (SEZ), 31-1
Soil enzymatic activity
    anthropogenic organics, soil-bound interaction
        catalytic features and characteristics, 5-16
        cell-free enzymes usage, 5-17
        2,4-DCP by laccase, removal, 5-17–5-18
        effects/ atrazine, carbaryl, glyphosate, and paraquat, 5-16–5-17
        environment pollution, 5-16
        oxidative enzymes, 5-17
        pesticides impacts, 5-16
        physicochemical properties, 5-17
        pollutants/xenobiotics, production and degradation, 5-17
    clays-humic substances, 5-7–5-8
    components, 5-1
    denaturing agents effects, clays/humic substances, 5-12–5-13
    ecological role
        carbon mineralization, 5-14
        DOM, 5-15
        enzymes and soil components interaction, 5-16
        enzymes, extracellular production, 5-15
        extracellular/cell-free enzymes, 5-13–5-14
        extracellular enzymes, microorganisms, 5-13, 5-14
        inorganic and organic derivatives, 5-14
        microbial and chemical substrate interaction, 5-14
        N&P-cycle, 5-15
        oxidoreductases and hydrolases, 5-13
        POM, 5-15
        precipitation and adsorption effects, 5-15
        soil biotransformation, microorganisms, 5-13, 5-14
    features and properties, 5-1
    kinetic parameters
        catalytic/stability properties, 5-8–5-9
        chemical binding, 5-10
        clay molecules interaction, 5-5, 5-8
        free enzymes measurement, 5-8–5-9
        humic-like enzyme, 5-9
        immobilization, 5-9–5-10
        phosphatase activity, 5-8
        polymerization, 5-10
        proteins tertiary structure, 5-11
        static enzymes, 5-10–5-11
    natural and synthetic
        cation-exchange adsorption mechanisms, 5-4
        clay–enzyme complexes, 5-4–5-6
        humic-humic like enzyme complex, 5-6–5-7
        humus–enzyme complex, 5-2
        mechanisms, synthetic enzymatic model, 5-4
        natural clays, 5-4
        physical adsorption, 5-4
        preparation/characterization, 5-4
        soil extraction, 5-3
        synthetic model systems, 5-2
    natural immobilized enzymes, 5-1–5-2
    pH/temperature ctivity, 5-11–5-12
Soil erosion processes
    concentrated flow erosion (rills, pipes, gullies), 22-5–22-7
    deposition, 22-1
    drop impact and splash erosion, 22-2–22-3
    overland flow

flow detachment, transport, and sedimentation, 22-4–22-5
flow velocity, 22-3–22-4
pipe erosion and dispersive soil behavior, 22-7
rock fragments, 22-9
shallow mass movements
   definition, 22-8
   degradation processes, 22-7
   gravity and landslides, 22-7
   Mohr-Coulomb soil strength criterion, 22-8
   rill and gully slopes, 22-7
   sliding potential, 22-9
   slope stability, 22-8
   soil strength parameters, 22-8
   two-dimensional analysis, 22-8
   vegetation, 22-9
susceptibility, 22-1
tillage and land leveling, 22-10
wetting and detachment, 22-1
wildfires, 22-10–22-11
Soil fertility and nutrition
nitrogen
   crop responses, 25-14
   fertilizer nitrogen conservation and efficiency, 25-14
   fertilizer nitrogen immobilization, 25-13
   nitrogen cycle, 25-13
   soil biological processes, 25-13
   soil nitrogen and carbon mineralization, 25-13–25-14
   soil testing and plant analyses, 25-13
phosphorus and potassium, 25-15–25-16
Soil fertility and plant nutrition
crop nutrient uptake and growth response, 16-1
Law of the Minimum, 16-1
Mitscherlich equation, 16-2–16-3
Mitscherlich response, 16-3
MLH, 16-3
multinutrient quantitative models, 16-9
N1 uptake/utilization efficiency, 16-3
nutrient interaction mechanisms
   anion–anion interactions, 16-6
   arbuscular mycorrhizal fungal inoculation, 16-6
   Brassica crops, 16-7
   canola, 16-7–16-8
   cation–anion interactions, 16-5–16-6
   cation–cation interactions, 16-5
   cysteine production, 16-7
   Fried–Broeshart model, 16-3
   intercropping Graminaceae and dicotyledonous crops, 16-7
   ion nutrient uptake models, 16-3
   Liebig type responses, 16-2, 16-9
   manganese-deficient grass species, 16-7
   multi-ion uptake model, 16-3
   N2 fixation/ NH4 and NO3 uptake, 16-6–16-7
   nitrogen and sulfur fertility, 16-7
   nitrogen and sulfur fertility interactions, 16-7
   nonspecific nitrogen–phosphorus interactions, 16-7
   pH-sensitive nutrients, 16-6
   rhizosphere acidification, 16-6–16-7
   root proliferation and nutrient absorption capacity, 16-6
   secondary micronutrient interactions, 16-7
   soil solid–solution interface, 16-4
   soil supply parameters, 16-3–16-4
   solution–root interface, 16-4
   wheat, 16-7–16-8
nutrients 1 and 2, 16-2
primary and secondary nutrient interactions, 16-2–16-3
soil–root–shoot continuum, 16-1
specific/nonspecific nutrient interactions, 16-2
Soil fertility evaluation
agricultural and nonagricultural systems, 13-2
definition, 13-1
environmental issues, 13-2–13-4
plant testing
   nutritional status, 13-28–13-29
   plant analysis, 13-29–13-30
   visual symptoms, 13-28
precision agriculture and remote sensing, 13-30–13-31
purpose, principles, and practices, 13-1–13-2
soil testing (*see* Soil testing)
uses, 13-31
Soil geographic database (SGDB), Russia
cartographic block, digital maps
   digital cartographic materials, 31-9–31-10
   Russian Federation, 31-8–31-9
   soil-ecological zoning, 31-8–31-9
HP, MySQL, and MapServer software, 31-16
land resources, 31-2–31-3
MapInfo format, 31-16
Microsoft Office 2007 format, 31-16
objectives and principles, 31-8
principles, 31-8
SEZ, 31-1
soil conservation, 31-8
soil information
   science and development, digital soil mapping, 31-6–31-8
   soil mapping history, 31-5–31-6
soil monitoring, 31-8
soil profile database (*see* Soil profile database)
soil resources, 31-2–31-3
sustainable land use, 31-8
Soil integration and application
Brazilian soil information system
   database-centered information system, 29-12
   Latin–American Soil Congress, 29-13
   SIGSOLOS, 29-10, 29-12
   regional organic carbon storage maps (*see* Regional organic carbon storage maps)
   SOC (*see* Soil organic carbon (SOC) stocks)
Soil landscape
integrated soil, water, and crop management, 21-15
land evaluation, 21-15
land use
   sulfidic acid sulfate soils, 21-14
   sulfuric acid sulfate soils, 21-14–21-15
Soil Management Assessment Framework (SMAF), 26-18
Soil maps and databases, China
digital soil mapping, 30-14
soil and land information systems (*see* Soil and land information systems, China)
soil correlation, 30-13–30-14
Soil organic carbon (SOC) stocks
Brazilian Amazon, SOTER database
   Batjes and Dijkshoorn estimation, 29-23
   Bernoux estimation, 29-23–29-24
   biogeochemical cycle, 29-24
   EMBRAPA data, 29-21
   GEFSOC carbon modeling system, 29-20
   land use and land cover changes, 29-20
   land use scenarios, 29-21
   Moraes estimation, 29-23
   regional simulations, 29-21–29-23
   soil organic carbon data, 29-20
Rio Grande do Sul
   Brazilian National Greenhouse Inventory, 29-17
   carbon analytical techniques, 29-17
   data preparation and GIS operations, 29-18
   greenhouse gas emissions, 29-17
   IPCC, 29-17
   pedotransfer functions, 29-19
   protocol, 29-17
   Reconnaissance Soil Survey map, 29-17
   soil mapping and database management, 29-17
   statewide estimate, 29-18–29-19
   Tier 1, 2 and 3 method, 29-17
   Walkley–Black wet combustion analytical method, 29-19
Soil organic matter (SOM)
characteristics, partition, 9-6–9-7
decomposition organic carbon, 7-1–7-2
definition, 7-1
dissolved effect, 9-9–9-10
feedback, 7-1
importance/role, 9-14–9-16

labile and stable pools, accumalation
mechansims
adsorption and entrapment, 7-8–7-9
aggregate fractionation, 7-4–7-5
clay–organic matter microstructures,
7-3
composition variation, 7-8–7-9
definition, 7-2–7-3
density fractionation, 7-5–7-6
humic material, 7-4
macroaggregates/microaggregates, 7-3
multilayer model, amphiphilic organic
fragments, 7-10–7-11
organo-mineral interactions, 7-3–7-4
particle size fractions, 7-5
recalcitrance, 7-3
time/assignment, timescales, 7-2–7-3
mineral interactions, sorption, 9-17
organo-mineral interactions
allophane and imogolite, 7-13–7-14
carbon/nitrogen organo-mineral
associations, 7-12
catalysts minerals, 7-14–7-15
metal ions interaction, 7-14
mineral control paradigm, 7-10–7-12
pedogenic oxides, 7-13
phyllosilicate clay minerals, 7-12–7-13
polarity and composition, 9-7–9-8
recalcitrance
black carbon, 7-6–7-7
carbonaceous pollutants/
nanomaterials, 7-7
plant and microbial-derived
compounds, 7-6
terms and mechanisms, 7-6
soil architecture/effects
accessibility/aggregation, physical
protection, 7-7–7-9
hydrophobicity, 7-8–7-10
soil matrix, 7-7
soil carbon sequestration, 7-1
turnover/stabilization, modeling, 7-2
Soil physicochemical-biological interfacial
processes
air–water interface, 9-4
AOP, 9-1
diffusion, 9-4
HOCs adsorption
carbonaceous geosorbents, 9-13–9-14
soil minerals, 9-11–9-13
human activity, 9-1
hydrophobic, 9-1
leaching, 9-1
organic molecule interface, 9-4
organic pollutants/pollution, 9-3–9-4
plant and soil microbes, uptake/
degradation (see Plant and soil
microbes, uptake/degradation)
soil system, 9-2–9-3
soil–water interface
adsorption/isotherm nonlinearity,
9-10–9-11
AOPs, soil sorption, 9-5–9-6

competitive/irreversible sorption,
9-16–9-17
dissolved SOM effect, 9-9–9-10
HOCs adsorption, 9-11–9-13
importance/role, SOM, 9-14–9-16
organic compounds, solubility/polarity
effect, 9-16
polarity and composition, SOM,
9-7–9-8
SOM characteristics, partition,
9-6–9-7
water solubility/polarity effects,
9-8–9-9
SOM–mineral interactions, sorption, 9-17
sorption/desorption, bioavailability,
9-18–9-19
surfactant effects
cloud point, 9-27
CMC, 9-27
HLB number, 9-27
micelles, 9-27
plants uptake, 9-32–9-33
solubility enhancement, 9-27–9-29
sorption effects, 9-29–9-30
uptake/degradation microorganism,
9-30–9-32
Soil potassium bioavailability
equilibrium and kinetic reactions, 11-37
exchangeable and nonexchangeable
potassium, 11-38
leaching, 11-41–11-42
mineral soils, 11-37
nutrient– soil mineral relationships, 11-37
plant growth, 11-37
potassium extractability and availability
chemical extractants, soil potassium
forms, 11-42–11-43
plant potassium analysis, 11-43–11-44
Q/I analysis, 11-43
soil tests, 11-43
potassium fixation, 11-40
potassium mineral, 11-38–11-39
potassium release, 11-40–11-41
soil variety, 11-37–11-38
solution-exchangeable K dynamics,
11-39–11-40
solution K, 11-37–11-38
Soil profile database
attribute database development, 31-16
contents and structure
hierarchical model, soil description,
31-10, 31-12
HORIZON level, 31-12
SAMPLE level, 31-12
soil attribute database, 31-12–31-13
soil morphology and analytic
properties, 31-10
soil types, 31-10
soil-DB software
card maintenance and transmitting,
31-14–31-16
information units, personal data and
registration, 31-14

numerical and textual fields, 31-15
Soil quality
agricultural productivity and
sustainability
CAFOs, 26-8
crop yields and input requirements,
26-8
food production, 26-8
irrigation, 26-9
soil degradation processes and soil
conservation, 26-7
soil fertility and resilience, 26-9
soil properties, 26-8
sustainable soil management, 26-9
definitions, 26-2
finite, fragile, and nonrenewable, 26-1
global climate and energy production,
26-11–26-12
indicators selection
biological indicators, 26-15–26-16
chemical indicators, 26-15
dynamic soil characteristics, 26-13
environmental assessment, 26-13
mineralogy, particle size distribution,
and soil depth, 26-13
minimum data set (MDS), 26-13
physical indicators, 26-13–26-15
soil property, 26-13
terrain analysis methods, 26-13
land quality, 26-2
land use and soil property, 26-11
native soil conditions, 26-11
nonquantitative evaluations, 26-16
plant growth, 26-3
prime agricultural lands, 26-6
qualitative and semiquantitative, 26-16
quantitative evaluations
digital mapping approach, 26-20
FCC, 26-19
indexing systems, 26-16
MDS, 26-16, 26-19
productivity index model, 26-17
SIR, 26-17
site index, 26-19
SMAF, 26-18
soil and land properties, 26-17–26-18
soil characteristics and soil quality,
26-16–26-17
SQI, 26-17–26-19
rangeland health, 26-9
rice, 26-16
soil and water quality, 26-3
soil degradation and resilience,
26-10–26-11
soil function, 26-4–26-6
soil pollution, 26-6
soil quality assessments, 26-1
soils buffer environmental change, 26-3
soils regulate and partition water flow,
26-3
Storie Index, 26-6
sustainable agriculture and ecological
management, 26-1

USDA Land Capability Classification, 26-6
Soil quality index (SQI), 26-17–26-19
Soil Survey Geographic Database (SSURGO), 28-5
Soil testing
    agricultural setting, 13-2
    chemical, physical, and biological properties
        lime requirement, 13-21
        organic matter, 13-21–13-22
        pH, 13-20–13-21
        soluble salts, 13-22
    definition, 13-2
    elemental availability
        bioaccessible, 13-4
        biological organism, 13-4
        elemental concentration, 13-4
        essential and nonessential elements, 13-4
        soil properties and environmental conditions, 13-5
    environmental
        plant nutrients, water quality impacts, 13-26–13-28
        potentially toxic trace elements, 13-25–13-26
        principles and purposes, 13-24–13-25
    interpretation and recommendations
        calibration process, 13-22
        Cate-Nelson method, 13-22
        correlation and regression methods, 13-22
        curvilinear models, 13-22
        greenhouse, 13-22
        intended land use and factors, 13-9
        laboratory-based correlation, 13-22
        mathematical approach, 13-22–13-23
        nutrient recommendation, 13-22–13-23
        plant yield/elemental composition, 13-23
        production agriculture, 13-9
        rapid buildup and maintenance approach, 13-24
        soil test correlation, 13-22
        sufficiency level approach, 13-24
    nutrient deficiency, 13-2
    plant nutrients (*see* Plant nutrients)
    public and private sectors, 13-2
    soil analysis, 13-6
    soil sample collection, handling, and preparation
        agronomic crops yield, 13-6
        chemical and physical properties, 13-5
        collecting soil cores, 13-8
        composite soil sample, 13-6
        contamination, 13-8
        natural and anthropogenic sources, 13-5
        natural variability, 13-6
        nonuniform fields, 13-6
        SDM, 13-6
        soil sampling methods, 13-7–13-8
        soil sampling practices, 13-6–13-7
Soil water repellency
    erosion effect, 19-7
    fundamental principles, 19-1–19-2
    hydrological and erosional hillslope and catchment responses, 19-1
    hydrological effect
        burnt eucalypt forest terrain, 19-6–19-7
        capillary depression, 19-5
        clay and sandy soils, 19-7
        fingered flow, 19-6–19-7
        grass-covered sandy soil, 19-6
        infiltration capacity, 19-5
        irregular wetting front, 19-6–19-7
        overland flow, 19-5
        rainfall infiltrating, 19-5
        sand-stone interfaces and preferential flow, 19-5
        structural/textural preferential flow paths, 19-5
    hydrophobic manner, 19-1
    measurement and classification, 19-3
    occurrence
        calcareous soils, 19-4
        distribution, 19-3–19-4
        golf greens, 19-4
        soil moisture contents, 19-4
        soil surface topography effect, 19-4
        vegetation type and fire, 19-4
    origin, 19-2
SOM, *see* Soil organic matter (SOM)
Sorption processes
    adsorption, competition, 8-8–8-9
    desorption
        biphasic reaction processes, 8-11–8-12
        coprecipitation, 8-13
        kinetics, 8-12–8-13
        non-crystalline Al-hydroxide and ferrihydrite, 8-12–8-13
        oxyanions, 8-13
        residence time effects, 8-13
        soil components, 8-11
        trace elements-soil sorbents contact time, 8-12
    heavy metals, cationic form
        Bacillus subtilis, 8-5
        biopolymers, 8-5
        biosorption, 8-5
        cation transition, 8-4
        complexation reactions, 8-3–8-4
        humic substances, 8-3
        hydration sphere, 8-3
        Lewis acid classification, 8-4
        noncrystalline minerals, 8-3
        oxide–water interface, 8-4
        pH-dependent effects, 8-4–8-5
        rhizosphere, 8-3
        surface bonding, 8-3
    inorganic-organic ligands
        ferrihydrite/arsenite, 8-10–8-11
        ligand–metal ratio, 8-11
        minerals-soils interaction, 8-11
        pH, 8-9–8-10
        siderophores, 8-10
        soil–root interface, 8-10
        volcanic ash soil, 8-11
        zinc montmorillonite/ferrihydrite, 8-10–8-11
    kinetics, 8-6–8-7
    metalloids anionic form, 8-5–8-6
    redox reactions, 8-8
    surface precipitation, 8-7–8-8
Specific surface area (SSA), 2-4
Storie Index rating (SIR), 26-17
Stylosanthes hamata, 6-7
Sulfur bioavailability
    accretion, 11-52
    crop removal, 11-52
    crop response, 11-53–11-54
    crop rooting depth, 11-53
    elemental sulfur, 11-52
    extractable sulfur, 11-52
    gypsum, 11-52
    industrial sulfur emissions, 11-52
    nutrition, 11-51
    occurrence, 11-51–11-52
    root growth, 11-53
    sulfate adsorption, 11-52
    superphosphate, 11-52
    turbidimetric and anion-exchange chromatography methods, 11-53
Suspended sediment trap (SUSTRA), 23-13
Synchrotron radiation
    applications
        hyperaccumulator plants, 1-15
        metal(loids), 1-13–1-14
        metals/oxyanions adsorption, 1-9
        metal(loid) surface precipitation/dissolution, 1-9–1-11
        redox reactions, 1-11–1-12
        soil nutrients/biosolids, 1-15
        transition-metal oxides, 1-12–1-13
    Co K-edge XANES spectra, 1-4
    complementary/alternative techniques
        microfocused XAS and XRF, 1-7–1-8
        quick XAS, 1-8
        synchrotron X-ray diffraction, 1-5–1-7
        XANES/EXAFS spectrum, 1-5
    soil chemistry, molecular environment
        electromagnetic spectrum, 1-1–1-2
        synchrotron radiation, 1-2–1-3
        X-ray absorption spectroscopy, 1-3–1-5
Synchrotron x-ray microfluorescence (μ-SXRF), 1-15
Synchrotron x-ray source, 1-3

T

Teaching and learning
    field-based pedology, 33-1
    geospatial approach, 33-1
    GIS and soil geodatabases, 33-1
        data layers hierarchy, 33-5–33-6
        DEM, 33-4

GIS-based approach, 33-3
hardware considerations, 33-2–33-3
informative map, 33-6
maps and data layers, 33-4–33-5
pedagogic aspects, 33-7
raster data, 33-7
software, 33-3
soil and geomorphology data, 33-2
soil parent material map, 33-6
tablet PC and GIS software, 33-2
landscapes, 33-1
soil, crop, and environmental science, 33-1
USDA's Web Soil Survey, 33-1
web-based geobrowser, 33-7
web soil survey, 33-2
Web 2.0 technology, Google Earth, 33-2
Terminal electron accepting processes (TEAPs), 10-22
Texas Erosion Analysis Model (TEAM), 23-7–23-8
Total potential acidity/titratable actual acidity (TPA/TAA), 21-16
Transmission electron microscopy (TEM), 4-6

U

United States Soil Survey Databases
data access, 28-6
FGDC, 28-10–28-11
first generation, 28-7
information system technology, 28-11
NCSS, 28-1
NRCS, 28-1, 28-9–28-10
OSDs and soil classification database, 28-3–28-5
SCD, 28-1–28-2
second generation, 28-7
soil data subcommittee, 28-10
spatial data relationships, 28-1–28-2
SSURGO, 28-5
STATSGO, 28-5–28-6
tablet computers and GPS technology, 28-11
third generation
NASIS (see National Soil Information System (NASIS))
SDV, 28-9
SSURGO access template database, 28-9
WSS, 28-9–28-10
U.S. General Soil Map (STATSGO), 28-5–28-6

V

Viola calaminaria, 8-21

W

Water erosion
erosion processes (*see* Soil erosion processes)
soil conservation and land-use planning, 22–13
soil erosion models
BULLRUN, 22-13
cascading planes landscape, 22-12
distributed models, 22-12
erosivity, 22-11
EUROSEM, 22-12–22-13
GIS-based soil erosion models, 22-13
GIS software, 22-13
GRASS, 22-13
KINEROS, 22-12–22-13
LISEM, 22-12–22-13
MEDALUS, 22-13
raster landscape, 22-12
RUSLE, 22-11
soil erodibility, 22-11
triangulated irregular network (TIN), 22-12
USLE, 22-12
WATEM, 22-18
Water use efficiency (WUE)
agricultural, 15-8
AquaCrop model, 15-10
BWR, 15-8
crop and crop genetics, 15-8
crop biomass production, 15-8
crop rotations cover crops and fallow system, 15-9
crop/variety/hybrid selection, 15-9
decision support models, 15-10
nutrient and water interactions, 15-10
soil physical and chemical properties, 15-9–15-10
SORKAM model, 15-10
tillage and residue management, 15-8–15-9
Web Soil Survey (WSS), 28-9–28-10
Wetlands biogeochemistry
carbon
aerobic and anaerobic decomposition, 20-10
aerobic carbon mineralization, 20-9
anthropogenic inputs, 20-7
carbon cycle, 20-7–20-8
decomposition, carbon loss, 20-7
enzyme hydrolysis, 20-9
extracellular enzymes, 20-9
fungal genera, 20-8
humification process, 20-10
labile organic carbon fractions, 20-8
microbial biomass, 20-8
microorganisms, 20-8–20-9
organic matter storage, 20-8
oxidoreductase enzymes, 20-8
photosynthesis, carbon fixation, 20-7
soil profiles, 20-9–20-10
terrestrial tropical and subtropical ecosystems, 20-7
definitions, 20-1–20-2
earth and land surface, 20-1
ecological functions, 20-1
metals, 20-17–20-18
nitrogen
adsorption/desorption processes, 20-16
algal species composition, 20-11
ammonium, 20-12
Anammox process, 20-14
aquatic vegetation, 20-12
Ca-precipitated phosphorus, 20-16
cyanobacterial mats, 20-12
denitrification, 20-13
dissolved inorganic N (DIN), 20-11
DNRA, 20-14
EPC_0 values, 20-12
nitrification-denitrification pathways, 20-14
Nitrobacter, 20-13
Nitrosomonas, 20-13
noncalcareous systems, 20-16
organic and inorganic forms, 20-11
phosphorus isotherms, 20-12
potential pathways, 20-13
soil and environmental factors, 20-12
soil and water column, 20-11
soil-water-plant compartments, 20-16
watercolumn via ammonia volatilization, 20-12–20-13
phosphorus
apatite, 20-16
extracellular enzyme activity, 20-16
labile and nonlabile pools, 20-16
macrophytes biomass, 20-16
organic and inorganic forms, 20-14
organic phosphorus fraction, 20-16
phosphorus cycle, 20-14–20-15
physical and biological mechanisms, 20-14
relative bioavailability, 20-15–20-16
wetland soils, 20-15
redox gradients, 20-6–20-7
terrestrial and aquatic systems, 20-1
toxic organics, 20-19–20-21
wetland soils
chemical and biological properties, 20-2
hydric soils, 20-5
mineral soils, 20-4–20-5
organic soils, 20-3–20-4
physical properties, 20-2–20-3
Wind erosion
atmospheric surface boundary layer, 23-2
China dust storm, 23-1
conceptual transport model, 23-7
crop residues, 23-14
dust devils, 23-2
dust-emission models, 23-1
environmental, agricultural, and transportation, 23-2
field measuring techniques, 23-15
field sampling instrumentation
air quality monitoring, 23-13
Bagnold's sampler, 23-13
BSNE sampler, 23-13
glass/plastic bottle, 23-13

laboratory wind tunnel erosion samplers, 23-12
MWAC, 23-13
surface creep and saltation sampler, 23-13
SUSTRA, 23-13
threshold velocity, 23-14
geomorphologic process, 23-1
humankind, 23-6
Mars dust storm, 23-2
Pasak model, 23-7
persistence and magnitude wind forces, 23-1
plants, 22-5–22-6
red rain, 23-1
RWEQ (*see* Revised wind erosion equation (RWEQ))
sand transport flux, 23-2
shear/friction, 23-2
shear velocity, 23-2
soil
  aggregate, 23-4
  amendments, 23-15
  crust, 23-4
  cyanobacterial biomass, California desert, 23-4
  dust production and transport, 23-4
  electromagnetic and static forces, 23-4
  erodible fraction, 23-4–23-5
  roughness, 23-5, 23-14
  sand particles, 23-3–23-4
TEAM, 23-7–23-8
WEAM, 23-8
weather-duststorm systems, 23-2–23-3
weather- soil-plants-man, 23-7
WEELS, 23-12
WEPS, 23-11–23-12
WEQ, 23-7
wind barriers/shelterbelts, 23-14–23-15
wind speed, 23-2–23-3
Wind Erosion Assessment Model (WEAM), 23-8
Wind erosion equation (WEQ), 23-7
Wind Erosion on European Light Soils (WEELS), 23-12
Wind Erosion Prediction System (WEPS), 23-11–23-12
Wind speed, 23-2–23-3
World Inventory of Soil Emission Potentials (WISE), 27-6
WUE, *see* Water use efficiency (WUE)

X

X-ray absorption near edge structure (XANES), 1-4–1-5
X-ray absorption spectroscopy, 1-3–1-5
X-ray diffraction (XRD), 1-5–1-7
X-ray photoelectron spectroscopy (XPS), 10-16